# 28th International Conference on Low Temperature Physics (LT28)

AA006117

Journal of Physics: Conference Series Volume 969

Gothenburg, Sweden
9 - 16 August 2017

**Editors:**

**Jonas Bylander**
**Tomas Lofwander**
**Maciej Misiorny**

ISBN: 978-1-5108-6236-4
ISSN: 1742-6588

**Printed from e-media with permission by:**

Curran Associates, Inc.
57 Morehouse Lane
Red Hook, NY 12571

CURRAN ASSOCIATES INC. proceedings .com

**Some format issues inherent in the e-media version may also appear in this print version.**



Printed by Curran Associates, Inc. (2018)

For permission requests, please contact the Institute of Physics
at the address below.

Institute of Physics
Dirac House, Temple Back
Bristol BS1 6BE UK

Phone: 44 1 17 929 7481
Fax: 44 1 17 920 0979

techtracking@iop.org

**Additional copies of this publication are available from:**

Curran Associates, Inc.
57 Morehouse Lane
Red Hook, NY 12571 USA
Phone: 845-758-0400
Fax: 845-758-2633
Email: curran@proceedings.com
Web: www.proceedings.com

# Table of contents

## Volume 969

**28th International Conference on Low Temperature Physics (LT28)**

**9–16 August 2017, Gothenburg, Sweden**

**Accepted papers received: 13 February 2018**
**Published online: 19 April 2018**

### Preface



### Papers

#### Quantum fluids and solids

**Superconductivity**

**Cryogenic techniques and applications**

**Magnetism and quantum phase transitions**

**Quantum transport and quantum information in condensed matter**

IOP Conf. Series: Journal of Physics: Conf. Series **969** (2018) 011001 doi:10.1088/1742-6596/969/1/011001

**Preface to the Proceedings of the LT28**

LT28 was held in Gothenburg, Sweden, on August 9-16, 2017, and was a great success. A total of about 900 abstracts were submitted for oral and poster presentations, including invited talks. The summary by Professors William Halperin and John Saunders, representing the International Union of Pure and Applied Physics (IUPAP), highlights some of the scientific achievements presented at the conference.

The present Proceedings of LT28 comprise 155 papers, describing some of the currently most interesting works in the field of low temperature physics, subdivided into five categories:

1. Quantum fluids and solids
2. Superconductivity
3. Cryogenic techniques and applications
4. Magnetism and quantum phase transitions
5. Quantum transport and quantum information in condensed matter

All papers published in this volume of *Journal of Physics: Conference Series* have been peer reviewed through processes administered by the Editors. Reviews were conducted by expert referees to the professional and scientific standards expected of a proceedings journal published by IOP Publishing. Most reviews resulted in one or more rounds of revision. The reviewers were specifically asked to call out any self-plagiarism in the submitted manuscripts. The ultimate acceptance rate was 90%. Parenthetically, the present editors were themselves co-authors of submitted papers: in these cases the submissions in question were exclusively handled by one of the other Editors.

We would like to thank all authors and referees who participated in the compilation of these Proceedings, and to congratulate all authors on their work.

Jonas Bylander, Tomas Löfwander, and Maciej Misiorny
*Chalmers University of Technology*
*Editors of the Proceedings of LT28*

## Conference summary

The International Conference on Low Temperature Physics is an international meeting held every three years, under the auspices of the International Union of Pure and Applied Physics (IUPAP) through its Commission C5 on Low Temperature Physics. The aim of these conferences is to exchange information and views among the members of the international scientific community in the general field of low temperature physics. It is a tradition that the breadth of scientific research that takes place at low temperatures and makes use of, or contributes to, low temperature physics technologies, finds a commonality in this international conference series. The LT28 conference was organized in this tradition to provide updates on various topics by distinguished researchers in their fields, including plenary and half-plenary talks, invited and contributed oral presentations, and posters. The conference covered five subtopics which have evolved since the last few conferences in this series. These are Quantum fluids and solids, Superconductivity, Cryogenic techniques and applications, Magnetism and quantum phase transitions, and Quantum transport and quantum information.

Gothenburg is situated in the center of Scandinavia, on the Swedish west coast, and is easily accessed by air. The city's two universities, Chalmers University of Technology and the University of Gothenburg, were the official conference hosts. Both have a long tradition in low temperature physics research, particularly superconductivity, quantum transport, and quantum information.

In this report we summarize some of the scientific achievements that were presented at LT28. Although this is not a comprehensive account of the meeting, it should be considered as representative of its program which is detailed in the schedule and abstracts available on-line.

*Quantum fluids and solids.* The significance of topology and topological defects in phase transitions in two-dimensional systems was recognized through the award of the 2016 Nobel Prize to Michael Kosterlitz, David Thouless, and Duncan Haldane. Its experimental manifestation in thin films of superfluid $^4$He was precisely described by the theory of the Kosterlitz-Thouless transition which has also found application in a wide range of quantum condensed states including superconductivity, two dimensional melting and two dimensional magnetism. Topological structure and topological defects was a recurring theme of emphasis in this conference including superfluid $^3$He and unconventional superconductors. A physical object within a superfluid medium, in relative motion with respect to the fluid, is expected to exhibit a threshold for onset of dissipation called the Landau critical velocity. Novel results with rather different interpretations in the case of superfluid $^3$He-B, both theory and experiment, were compared and discussed. The properties of the superfluid near the surface of the object play a crucial role in this case, specifically the surface-Andreev bound states of the superfluid that can be identified with Majorana fermionic excitations. Such exotic quasiparticles are of great current interest, and are of potential importance in topological quantum information processing. They were discussed in a wide variety of quantum materials: superfluid $^3$He-B (meaurements of ion mobility, heat and mass transport, specific heat); ad-atom chains; Cooper pair transmission at an interface between a normal metal and a nodal superconductor; spin liquids; topological insulators; magnetic atom chains on elemental superconductors; half-quantum vortices in the recently discovered polar state of superfluid $^3$He-B.

Superfluid $^3$He was discovered in 1972 and was first reported at the LT13 conference held later that year. Its importance is underscored by Nobel prizes in 1996 and 2003. This superfluid is a paradigm for a wide range of physics of unconventionally paired fermionic systems. Recent work on topological superfluid $^3$He was reported at LT28 in quasi two-dimensions for slabs; quasi one-dimension for pores, and in three dimensions for superfluid $^3$He imbibed in various kinds of aerogel. In the latter case the effect of quenched disorder on superfluid phase stability and symmetry is remarkable including the first observation of the polar phase of superfluid $^3He$ and, in this phase, the existence of half-quantum vortices. The discovery and understanding of superfluid $^3He$ in aerogel was recognized at LT28 by the award to William Halperin, Jeevak Parpia, and James Sauls of the Fritz London Memorial prize sponsored by the C5 commission of IUPAP and Oxford Instruments and administered by Duke University, "For pioneering work on the influence of disorder on the superfluidity of Helium-3". Two dimensional films on atomically precise surfaces of graphite show intertwined superfluid and density order for $^4He$. In contrast, there is evidence that a $^3He$ film on graphite at ultra-low temperatures is a spin liquid at certain coverages. Superfluid $^4He$ counterflow in square channels evolves from Poiseuille flow and is related to the problem of onset of turbulence that has been investigated using linear displacement of a grid. In bulk $^4He$, phonon and vortex ring emission have been modeled; however, from recent work vortex reconnections are found to be primarily identified with roton emission. The full development of the excitation spectrum of $^3He$ has been determined from inelastic neutron scattering, now well-understood theoretically and consistent with experiment. The community of researchers focused on cold atomic gases has strong intellectual links with that working on liquid helium, while using very different experimental techniques. Progress was reported on using cold atoms in optical lattices to engineer many-body systems as analogue quantum simulators. Studies of mesoscopic transport, including the observation of quantized conductance for neutral particles and quantum interference in transport, vortex dynamics and turbulence, and the observation of rotons in cold atom systems with both short and long-range (dipolar) interactions, illustrate the strong link to condensed matter physics.

*Superconductivity*. An important goal is topological superconductivity, where topological insulators (TI) are a promising host as indicated in the superconductor $Cu_xBi_2Se_3$ which has been shown to have nematic order. The behavior of TI-based materials is one of the most active areas today in condensed matter physics, largely motivated by the expectation that the topological character of superconductivity that can be stabilized will be a major step toward quantum information processing. Recent work in this area was recognized with the award at LT28 to Vlad Pribiag of the Young Investigator Prize from the C5 commission of IUPAP on "Electronic Transport in Semiconductor Nanowires and 2D Topological Insulators". Improvement in cuprate superconducting materials, both electron and hole doped, are giving better resolution and understanding of transition temperatures such as the role of the apical oxygen site probed by ARPES and also by NMR. New theoretical work indicates the existence of vortex-antivortex pairs at the edge of a *d*-wave superconductor which should be verifiable experimentally. Anti-perovskite oxides like $Sr_{3-x}SnO$ open the possibility of topological odd-parity superconductivity. The superconducting ruthenates $Sr_2RuO_4$ and $Ca_2RuO_4$ have special properties. For the former, a transition from odd-parity to even parity under strain has been reported. The measurements of strain dependence were acknowledged at LT28 by the award to Clifford Hicks of the Young Investigator Prize from the C5 commission of IUPAP, "Correlated Electron Materials Under Uniaxial Stress". In the latter case of the calcium compound, closing of the Mott gap leads to emergence of a semimetal state with giant diamagnetic behavior. A classic example of quantum criticality is manifest in hole-doped cuprates. Heavy fermion compounds provided the first paradigms of quantum critical behavior, where in many cases unconventional superconductivity appears at the quantum critical point. Examples that have been discussed at LT28 include $CeIrSi_3$, $CeCu_2Si_2$, $CeCoIn_5$ and uranium based compounds, $URu_2Si_2$ URhGe, and $UPt_3$. The use of tuning parameters such as magnetic field or pressure is important for demonstrating quantum criticality. Ferromagnetic uranium-based superconductors such as $UGe_2$, URhGe, or UCoGe, are thought to be odd-parity; ferromagnetic fluctuations are evident in NMR and inelastic neutron scattering. Coexistence of superconductivity and magnetic order is well documented in these materials. Similarly, coexistent spin density wave order is evident in both hole and electron doped pnictide superconducting compounds. These results from reports at LT28 are consistent with what is now the very well-accepted understanding in many different classes of compounds that suppression of magnetic order, at a quantum critical point, by doping or pressure is concomitant with the emergence of superconductivity.

*Cryogenic techniques and applications.* The development of commercially available cryogen-free cryogenics, notably dilution refrigerators, has had an important impact on low temperature physics research, especially in the past three years. This technology has expanded to include cryogenic platforms for nuclear demagnetization cooling to ultra-low temperatures, superconducting magnets, and development of various protocols for reduction of vibrations from the dry coolers that replace liquid helium as a temperature reservoir. Applications of dry refrigerators provide a means to mitigate the increasing cost and limited availability of liquid helium, which will become more severe with the closing of the helium reserve in the United States in 2021 that supplies a major fraction of the world supply of helium (30% in 2013) used by the international low temperature research community. For example, dry refrigerators operating near 0.02 K have been used extensively in the rapidly developing field of research on quantum information processing that has the overarching goal of quantum computation and the development of quantum computers. Basic research in cryogenic techniques has the dual impact of supporting programs in fundamental physics as well as for its applications, both of which were significantly represented at LT28. The development of micro-calorimetric methods using modern clean room fabrication techniques to create two dimensional arrays with high resolution readout have potential metrological application in nuclear physics and particle physics. Also for these purposes, microwave detectors with unprecedented energy resolution have been demonstrated using sub-gap superconducting resonators. High sensitivity SQUIDs for various types of magnetic measurements have been developed for measurements of systems in the extreme environments of ultra-low temperatures and high magnetic fields in excess of 10 Tesla. Among many types of measurement based on the SQUID, are NMR, magnetization, torque, and thermometry methods having many orders of magnitude greater sensitivity than in the earliest work (a SQUID application at ultra-low temperatures was first announced at LT12, 48 years ago). Cryogenic detectors now take advantage of nanofabrication techniques, including on-chip demagnetization coolers and ultra-low noise amplifiers discussed at LT28. Opto-mechanical devices operating near the mechanical quantum ground state have important technological application for frequency transduction, an essential component for quantum communication. These fabrication tools have led to development of high sensitivity mechanical devices at very low temperatures, with strong interest in nano-electromechanical devices and nano-superfluidics, for example.

*Magnetism and quantum phase transitions.* Research on quantum materials with exotic ground states requires, in many cases, low temperatures for their investigation. Frustrated magnetism in correlated oxides with strong spin-orbit interactions is an example. New compounds and those with higher quality crystal structure or purity have revealed dimensional effects that have been determined using neutron scattering. Iron dichalcogenide ladder compounds with competing magnetic and superconducting interactions have been synthesized, with $BaFe_2Se_3$ being the first in its family found to be superconducting. Their magnetic dynamics have been reported from neutron and muon scattering experiments at low temperatures. Intriguing recent results on the Kondo insulator $SmB_6$, which appears to show a bulk Fermi surface, were reported. The spin liquid is a highly quantum entangled state of matter in search of a material which unambiguously realizes it: the current status of this search was reviewed. For example, magnetic frustration induced by dimensional or geometrical constraints, as with the topological honeycomb materials, are candidates for spin-liquid behavior. Similarly spin-stripe materials have been of great interest in both experimental and theoretical research, an example being the elementary excitations in the vanadate compound $TeVO_4$ that were investigated by muon relaxation and dielectric loss spectroscopy. The counterpoint between spin-ice and spin-liquid ground states has been of great interest at LT28. The magnetism in the high field Q-phase of the heavy fermion compound $CeCoIn_5$ and a possible connection to orbital order was reported from scanning tunneling experiments with potential significance for the mixed spin and orbital orders observed in the superconducting cuprates. Inelastic neutron scattering is a measure of magnetic fluctuations. A good example from the family of uranium based heavy fermion ferromagnetic superconductors is URhGe where small angle neutron measurements give a complementary, and possibly inconsistent, picture as compared with recent NMR results on the same material. Improvements in crystal quality will be very helpful for better understanding the interplay between magnetism and superconductivity in this special class of quantum materials.

*Quantum transport and quantum information in condensed matter.* Quantum transport and mesoscopic physics include important and traditional topics presented and discussed at most LT conferences and LT28 was no exception. What is new is the significant growth in research presented concerning quantum information in condensed matter systems. The latter ranges from the invention and investigation of new qubit systems including spin-qubits from semiconducting nanodots and superconducting qubits based on coupling Josephson junction to microwaves in superconducting resonators. Multi-terminal Josephson junctions are predicted to have topological properties with different phases of what can be considered as a form of quantum matter. Majorana fermionic excitations have been actively sought in condensed matter materials associated with nanowires. They are also expected at the surface of an unconventional superconductor, such as the B-phase of superfluid $^3$He, otherwise known as surface-Andreev bound states. These particle-like excitations, or anyons, have non-abelian statistics and are expected for the $\nu = 5/2$ state in the fractional quantum Hall effect. Although this statistical behavior has not yet been demonstrated, a possible manifestation of the latter by an entropic test has been reported at LT28 in the context of thermodynamic measurements on a high mobility heterostructure. Superconducting qubits based on microwave cavities coupled to Josephson junctions have been improved, increasing the coherence time through significant improvements in device and materials physics and engineering. It has also been shown that super-adiabatic protocols can be a very robust route to efficient quantum state manipulation of quantum information. For a flux qubit, the role of universal conductance fluctuations limiting the coherence time have been studied theoretically. Circuit-QED experiments were presented demonstrating photon entanglement achieved using a two-mode resonant circuit. It was also found that quantum dots in a hybrid light-matter state coupled to cavity photons can be sufficiently well-coupled that they protect entanglement over ultra-long distances. Mesoscopic devices that control heat transfer using superconducting qubits have been developed that promise the experimental realization of a quantum refrigerator. Quantum transport theory suggests that quantum criticality and super ballistic transport can be achieved in a tunable circuit. Electrical transport in a graphene bilayer with superconducting $NbSe_2$ contacts have been shown to exhibit specular Andreev reflection. There is broad interest in quantum matter consisting of semiconducting two-dimensional $Bi_2O_2Se$. Notable for understanding the electronic behavior in nano-plates of this material are the reported observations of quantum oscillations, paving the way for future device fabrication. Recent work on quantum oscillations in cuprate superconductors, using high magnetic fields to suppress superconductivity was presented, with evidence for a quantum critical point. This and prior work was celebrated at LT28 by the award to Louis Taillefer of the Simon Memorial Prize in Low Temperature Physics administered by the Institute of Physics, "For pioneering transport measurements at high magnetic fields and low temperature in heavy-fermion and cuprate superconductors."

LT28 conference summary respectfully submitted by,

William Halperin, Secretary
John Saunders, Chair
*IUPAP Commission C5*

IOP Conf. Series: Journal of Physics: Conf. Series **969** (2018) 011003 doi:10.1088/1742-6596/969/1/011003

**Local organizing committee**

| | | |
|---|---|---|
| Conference Chair | Per Delsing | Chalmers University of Technology |
| Co-Chair | Jonas Bylander | Chalmers University of Technology |
| Co-Chair | Mikael Fogelström | Chalmers University of Technology |
| Co-Chair | Floriana Lombardi | Chalmers University of Technology |
| | Thilo Bauch | Chalmers University of Technology |
| | Susannah Carlsson | Chalmers University of Technology |
| | Tord Claeson | Chalmers University of Technology |
| | Henrik Johannesson | University of Gothenburg |
| | Göran Johansson | Chalmers University of Technology |
| | Sergey Kubatkin | Chalmers University of Technology |
| | Tomas Lofwander | Chalmers University of Technology |
| | Vitaly Shumeiko | Chalmers University of Technology |
| | Janine Splettstoesser | Chalmers University of Technology |
| | Dag Winkler | Chalmers University of Technology |
| | Avgust Yurgens | Chalmers University of Technology |
| | Stellan Östlund | University of Gothenburg |

**Program Committee**

**1. Quantum Fluids and Solids**

| | | |
|---|---|---|
| Mikael Fogelström (Chair) | Chalmers University of Technology, Gothenburg | Sweden |
| James Sauls | Northwestern University, Evanston Il | USA |
| Kimitoshi Kono | Riken, Wako | Japan |
| Päivi Törmä | Aalto University, Helsinki | Finland |
| Vladimir Dmitriev | Kapitsa Institute, Moscow | Russia |
| Jeevak Parpia | Cornell University, Ithaca, NY | USA |

**2. Superconductivity**

| | | |
|---|---|---|
| Floriana Lombardi (Chair) | Chalmers University of Technology, Gothenburg | Sweden |
| Hermann Suredow | Universidad Autonoma de Madrid | Spain |
| Manfred Sigrist | ETH, Zürich | Switzerland |
| Yoshiteru Maeno | Kyoto University | Japan |
| Yoichi Ando | University of Cologne, Köln | Germany |
| Laura Greene | University of Illinois, Urbana Champaign | USA |
| Aharon Kapitulnik | Stanford University | USA |

**3. Cryogenic techniques and applications**

| | | |
|---|---|---|
| Thilo Bauch (Chair) | Chalmers University of Technology, Gothenburg | Sweden |
| Osamu Ishikawa | Osaka City University | Japan |
| Hervé Courtois | CNRS and Institute Néel, Grenoble | France |
| Cathy Foley | CSIRO Manufacturing, Lindfield, NSW | Australia |
| Carmine Senatore | University of Geneva | Switzerland |
| Christian Enss | University of Heidelberg | Germany |

**4. Magnetism and quantum phase transitions**

| | | |
|---|---|---|
| Stellan Östlund (Chair) | Gothenburg University, Gothenburg | Sweden |
| Tim Ziman | CNRS and Universite Joseph Fourier, Grenoble | France |
| Dragana Popovic | National High Field Magnet Lab, Talahassee | USA |
| Christian Ruegg | Paul Scherrer Institute, Villigen | Switzerland |
| Piers Coleman | Rutgers University, Piscataway | USA |
| Nan-Lin Wang | Peking University | China |

**5. Quantum transport and quantum information in condensed matter**

| | | |
|---|---|---|
| Göran Johansson (Chair) | Chalmers University of Technology, Gothenburg | Sweden |
| Dale van Harlingen | University of Illinois, Urbana Champaign | USA |
| Liliana Arrachea | University of Buenos Aires | Argentina |
| Franco Nori | Riken, Wako | Japan |
| Hélène Bouchiat | Université Paris Sud | France |
| Hongqi Xu | Peking University | China |
| Janine Splettstoesser | Chalmers University of Technology, Gothenburg | Sweden |

**Editors of the Proceedings of LT28**

| | |
|---|---|
| Jonas Bylander | Chalmers University of Technology |
| Tomas Löfwander | Chalmers University of Technology |
| Maciej Misiorny | Chalmers University of Technology |

**International Advisory Committee**

| | | |
|---|---|---|
| Hernan Pastoriza | Centro Atómico Bariloche | Argentina |
| Liliana Arrachea | University of Buenos Aires | Argentina |
| Susana Hernandez | University of Buenos Aires | Argentina |
| Tim Duty | UNSW | Australia |
| Ernst Bauer | TU-Wien | Austria |
| Francois M. Peeters | University of Antwerpen | Belgium |
| Victor V. Moshchalkov | INPAC Katholieke Universiteit Leuven | Belgium |
| Albino Aguiar | Universidade Federal de Pernambuco, Recife | Brazil |
| Carlos Egues | Universidade de Sao Paulo | Brazil |
| Elisa Baggio-Saitovich | CBPF, Rio de Janeiro | Brazil |
| Enrique Anda | PUC- Rio de Janeiro | Brazil |
| Silvio Vitiello | UNICAM | Brazil |
| Vanderlei Bagnato | University of San Carlos, Sao Paulo | Brazil |
| John Beamish | University of Alberta | Canada |
| Christopher Wilson | University of Waterloo | Canada |
| Louis Taillefer | Sherbrooke | Canada |
| HongQi Xu | Department of Electronics, Peking Univeersity | China |
| Jianlin Luo | Chinese Academy of Sciences | China |
| Li Lu | Chinese Academy of Sciences | China |
| Nan-Lin Wang | Peking University | China |
| Tao Xiang | Chinese Academy of Sciences | China |
| Wu-Ming Liu | Chinese Academy of Sciences | China |
| Z.X. Zhao | Chinese Academy of Sciences | China |
| Jeroen Custers | Charles University, Prague | Czech Republic |
| Ladislav Skrbek | Charles University in Prague | Czech Republic |
| Antti-pekka Jauho | TDU | Denmark |
| Karsten Flensberg | Niels Bohr Copenhagen | Denmark |
| Matti Krusius | Helsinki University of Technology | Finland |

IOP Conf. Series: Journal of Physics: Conf. Series **969** (2018) 011003 doi:10.1088/1742-6596/969/1/011003

| | | |
|---|---|---|
| Erkki Thuneberg | University of Oulu | Finland |
| Grigory Volovik | Aalto University, Helsinki | Finland |
| Jukka Pekola | Aalto University, Helsinki | Finland |
| Pertti Hakonen | Aalto University, Helsinki | Finland |
| Sergey A Vasiliev | University of Turku | Finland |
| Antoine Georges | Ecole Politechnique | France |
| Cristian Urbina | CEA, Saclay | France |
| Daniel Esteve | CEA, Sacaly | France |
| Didier Poilblanc | Université Paul Sabatier | France |
| Georg Knebel | CEA | France |
| Gilles Montambaux | Universite Paris-Sud | France |
| Helene Bouchiat | Université Paris Sud | France |
| Henri Alloul | Université Paris Sud | France |
| Jean Dalibard | LKB and Collège de France, Paris | France |
| Julia Meyer | CEA Grenoble | France |
| Karyn Le Hur | CNRS and Ecole Polytechnique, Paris | France |
| Lara Faoro | CNRS Jussieu | France |
| Manuel Nuñez Regueiro | Institut Néel – CNRS-UJF | France |
| Pierre-Etienne Wolf | CNRS/UJF – Institut Néel – Grenoble | France |
| Rodolfo Jalabert | Université de Strasbourg | France |
| Henri Godfrin | Institue Néel, Grenoble | France |
| Yury Bunkov | Institue Néel, Grenoble | France |
| Christian Enss | University of Heidelberg | Germany |
| Dagmar Gerthsen | Karlsruhe Institute of Technology (KIT) | Germany |
| Dieter Vollhardt | University of Augsburg | Germany |
| Frank Steglich | Max Planck Institute for Chemical Physics of Solids | Germany |
| Hilbert von Löhneysen | KIT, Karlsruhe | Germany |
| Jochen Mannhart | U Augsburg | Germany |
| Milena Grifoni | Universität Regensburg | Germany |
| Paul Leiderer | University of Konstanz | Germany |
| Peter Fulde | Max Planck Institut, Dresden | Germany |
| E.V. Sampathkumaran | Tata Institute | India |
| Srinivasan Ramakrishnan | Tata Institute of Fundamental Research | India |
| Ady Stern | Weizman Institute | Israel |
| Ora Entin-Wohlman | Ben Gurion University | Israel |
| Yuval Gefen | Weizman Institute | Israel |
| Elisabetta Paladino | Catania University | Italy |
| José Lorenzana | University of Rome | Italy |
| Lara Benfatto | Sapienza Rome | Italy |
| Maura Sasetti | Universita' di Genova | Italy |
| Atsushi Fujimori | University of Tokyo | Japan |
| Franco Nori | Riken | Japan |
| Hideo Hosono | Tokyo Institute of Technology | Japan |
| Hiroshi Fukuyama | University of Tokyo | Japan |
| Yung Woo Park | Seoul National University | Korea |
| Pier Mello | UNAM | Mexico |
| Anne de Visser | University of Amsterdam | Netherlands |
| Leo Kouwenhoven | Delft Univ. of Technology | Netherlands |
| Lieven m.k. Vandersypen | Delft University of Technology | Netherlands |
| Miriam Blaauboer | TU Delft | Netherlands |
| Teun Klapwijk | TUD | Netherlands |

| | | |
|---|---|---|
| Dariusz Kaczorowski | Polish Academy of Sciences | Poland |
| Tomasz Dietl | IOP PAS | Poland |
| Alexaner Smirnov | Kapitza Institute for Physical | Russia |
| Gregory Goltsman | Moscow State Pedagogical University | Russia |
| Igor Fomin | Kapitza, Moscow | Russia |
| Oleg Snigirev | Moscow State University | Russia |
| Olga Volkova | Moscow State University, Moscow | Russia |
| Vladimir Pudalov | Lebedev Physical Institute | Russia |
| Vladimir V. Dmitriev | Kapitza Institute for Physical | Russia |
| Jordi Boronat | UPC, Barcelona | Spain |
| Alfredo Levy-Yeyati | Universidad Autonoma de Madrid | Spain |
| Gloria Platero Coello | ICMM-CSIC Madrid | Spain |
| Rosa López | Universidad de Balears | Spain |
| Xavier Obradors | ICMAB | Spain |
| David Haviland | KTH, Stockholm | Sweden |
| Gianni W. Blatter | ETH, Zürich | Switzerland |
| Didier Jaccard | University Geneva | Switzerland |
| Dirk Van der Marel | University Geneva | Switzerland |
| Janusz Karpinski | ETH, Zurich | Switzerland |
| Jean Marc Triscone | Université de Genève | Switzerland |
| Manfred Sigrist | ETH, Zürich | Switzerland |
| Thierry Giamarchi | University of Ginebra | Switzerland |
| Thomas Ihn | ETH Zurich | Switzerland |
| Tilman Eslinger | ETH, Zurich | Switzerland |
| Juhn-jong Lin | National Chiao Tung University | Taiwan |
| Ali Gencer | Ankara University | Turkey |
| Andie Mackenzie | University of St Andrews | UK |
| Andrei Golov | University of Manchester | UK |
| Carlo Barenghi | Newcastle University | UK |
| George Pickett | Lancaster University | UK |
| Jan Nyeki | Royal Holloway | UK |
| John Saunders | Royal Holloway University of London, Surrey | UK |
| Malte Grosche | Cavendish Laboratory | UK |
| Natalia Berloff | University of Cambridge | UK |
| Peter B. Littlewood | University of Cambridge | UK |
| Stephen Hayden | University of Bristol | UK |
| Allen.M. Goldman | University of Minnesota | USA |
| Collin Broholm | Johns Hopkins University | USA |
| David Ceperley | University of Illinois Urbana-Champaign | USA |
| Eduardo Fradkin | University of Illinois at Urbana-Champaign | USA |
| George Seidel | Brown University | USA |
| Joe D. Thompson | Los Alamos National Laboratory | USA |
| Laura H. Greene | University of Illinois, Urbana-Champaign | USA |
| Michel Devoret | Yale University | USA |
| Aharon Kapitulnik | Stanford University | USA |
| Allan H. Macdonald | University of Texas at Austin | USA |
| Andrew Cleland | University of Chicago | USA |
| Eva Andrei | Rutgers | USA |
| Gabriel Kotliar | Rutgers University | USA |
| Gary Ihas | University Florida | USA |
| Humphrey Maris | Brown University | USA |

| | | |
|---|---|---|
| Ivan Schuller | UCSD | USA |
| J.C. Seamus Davis | Cornell University | USA |
| John Clarke | Univ. California Berkeley | USA |
| John Martinis | UC Santa Barbara | USA |
| Katja Nowack | Cornell | USA |
| Leon Balents | UCSB | USA |
| Leonardo Civale | Los Alamos National Laboratory | USA |
| Leonid Glazman | Yale University | USA |
| Milton Cole | Penn State University | USA |
| Moses Chan | Penn State University | USA |
| Paul Chu | University of Houston | USA |
| Wai-Kwong Kwok | Argonne National Laboratory | USA |
| Valerii Vinokur | Argonne National Laboratory | USA |
| Warren Pickett | University California Davis | USA |
| William P. Halperin | Northwestern University, Evanston | USA |
| Wolfgang Ketterle | MIT | USA |

# Peer review statement

All papers published in this volume of *Journal of Physics: Conference Series* have been peer reviewed through processes administered by the proceedings Editors. Reviews were conducted by expert referees to the professional and scientific standards expected of a proceedings journal published by IOP Publishing.

# Probing the dynamics of $^3$He atoms adsorbed on MCM-41 with pulsed NMR

**C. Huan, N. Masuhara, J. Adams, M. Lewkowitz and N. S. Sullivan**

Department of Physics, University of Florida, Gainesville, Fl 32611, USA

E-mail: huan@phys.ufl.edu

**Abstract.** We report measurements of the nuclear spin-spin and spin-lattice relaxation times for $^3$He adsorbed on MCM-41 for temperatures $0.08 < T < 1.2$ K. Deviations from Curie behavior are observed at low temperatures. The relaxation times exhibit a two-component behavior representing the differing dynamics of the mobile quasi-free molecules in the center of the tubes compared to the adsorbed layer on the walls. The amplitudes of the two components provide an accurate measure of the number of fluid-like molecules traveling in the center of the nanotubes.

## 1. Introduction

Interesting new quantum states have been predicted for quantum fluids ($^3$He, $^4$He, $H_2$, HD) constrained to nanoscale dimensions where the de Broglie wavelength and/or the Fermi length become comparable to or larger than the available pore or channel size[1–6]. 1D $^4$He superfluidity has been reported for low density $^4$He in nanotubes[4, 7] and the onset of degeneracy has been seen for $^3$He in the hexagonal channels of FSM16[8, 9]. The mesoporous structure of MCM-41 is similar to that of FSM-16 but with small changes in dimensions. The typical pore size for MCM-41 is 2-4 nm in diameter and 300 nm in length[5, 10]. Previous research[11] has observed the crossover from 2D motion to 1D motion for adsorbed $^3$He atoms. However, to observe the predicted 1D Luttinger liquid behavior, much lower temperatures are necessary[12]. Studies of $^3$He in these mesoporous structures to temperatures below 50mK provide a framework that allows the exploration of the unique properties of quantum liquids or solids in confined states.

We have used pulsed NMR techniques to measure the dynamics of 1.08 monolayers of $^3$He on MCM-41 for temperatures $0.08 < T < 1.2$ K. The nuclear spin relaxations rates are very sensitive to the motion of the $^3$He atoms and it is possible to distinguish the different components of the NMR signal according to slow and fast motions.

## 2. Experimental considerations

The MCM-41 sample was loaded into a polycarbonate NMR cell with modest pressure ($\sim$ 10 N/m$^2$ against a silver end cap that formed the end of an extension from a nuclear demagnetization refrigerator. The NMR probe consisted of a single coil tuned and matched to 50$\Omega$ at the cell site and connected to a room temperature hybrid tee bridge[13, 14]. Fine tuning was carried out using small variable capacitances at room temperature to correct for small changes when the NMR probe is cooled. The sample was pumped to a high vacuum and flushed with $^4$He gas several times to purge adsorbed gases. Fig. 1 shows a typical $^3$He isotherm measured at 2.5 K. The volume of the NMR cell is small and with the long (~ 2 m.) input capillary the accuracy of the isotherm is limited. Nevertheless, the compressibility deduced from the derivative given in the inset to Fig. 1 does show the minima attributed in Ref. [11] to the wall monolayer coverage and pore completion. The first minimum corresponds to 9.5 ± 1.5 atoms per $(nm)^2$ comparable to the values reported by Taniguchi *et al.*[8].

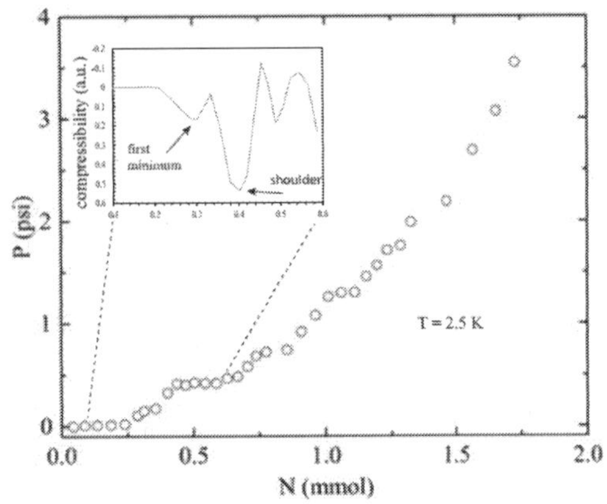


**Figure 1.** Isotherm of $^3$He on MCM-41 in NMR cell at 2.5 K. The inset shows the compressibility derived from the isotherm with the characteristic two minima near wall completion.

For the NMR studies gas was admitted to cover the wall plus an additional 9.0% that would be expected to be free to travel inside the nanotubes. The NMR measurements provide a very reliable measure of the fraction of the $^3$He atoms that are mobile as the nuclear spin-spin relaxation times $T_2$ of the atoms bound to the wall are very different to those for mobile atoms. The motion of the atoms modulates the nuclear spin-spin interactions and the spin-spin relaxation times are proportional to the frequency of the atomic motion, leading to a long $T_2$ for the mobile atoms compared to those bound to the wall. This difference is shown in Fig. 2 from which we deduce that the mobile $^3$He atoms form 0.77 ± 0.05 % of the total sample, much smaller than expected from the amount of $^3$He added. A similar observation was made by Taniguchi *et al.*[15] in studies of $^3$He on FSM-16. They attributed the difference to the formation of an amorphous solid $^3$He layer on top of the the wall coating similar to the behavior observed by Golub and Pobell[16] for $^3$He in vycor, leaving a very small gas component in the center.

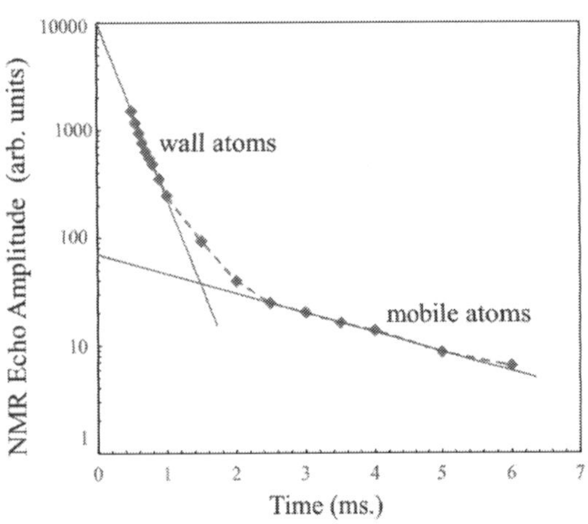


**Figure 2.** Time dependence of NMR echo decays for $^3$He in MCM-41 showing the difference in relaxation for the mobile $^3$He atoms (slow decay) compared to the relatively immobile (fast decay) atoms bound to the walls of the MCM-41.

The nuclear spin magnetization $M$ was measured as a function of temperature using $(\pi/2 - \tau - \pi)$ echos with $\tau = 250\mu s$. A small deviation from Curie's law was observed but is believed to follow the observation of Taniguchi *et al.*[15] who showed that the amorphous layer can have a broad distribution of very large exchange interactions (up to $J \sim 200$mK) for which the antiferromagnetic susceptibility has pronounced peaks for $T \sim J$[17].

## 3. Results

The nuclear spin-spin relaxation times, $T_2$, were measured by following the echo decay of $\pi/2 - \tau - \pi$ RF pulse sequences with variable $\tau$ and long waiting times between sequences. A distinctive two-component relaxation was observed for almost all temperatures (see inset of Fig. 4), and the echo decay could not be fit with an ordinary power law decay. The behavior is attributed to a slow relaxation expected for the quasi-fixed atoms adsorbed at the wall and the long-time relaxation component ($\sim 0.8\%$) to atoms free to travel axially along the MCM-41 channels. The mobile fraction is much smaller than expected from the amount of $^3$He adsorbed. As discussed above this is attributed to the observation of Taniguchi *et al.* that a layer of amorphous solid $^3$He is formed on top of the layer of wall atoms. The solid black line represents the best fit to an expected $T^{3/2}$ temperature dependence for the long time relaxation for $T > T_F$, where $T_F$ is the Fermi temperature. At low temperatures, $T < T_F$, a linear dependence is expected and this leads to a small departure from the $T^{3/2}$ dependence observed for $0.2 < T < 0.8$K. One does not expect any significant temperature dependence for the short-time relaxation but there is small increase with temperature.

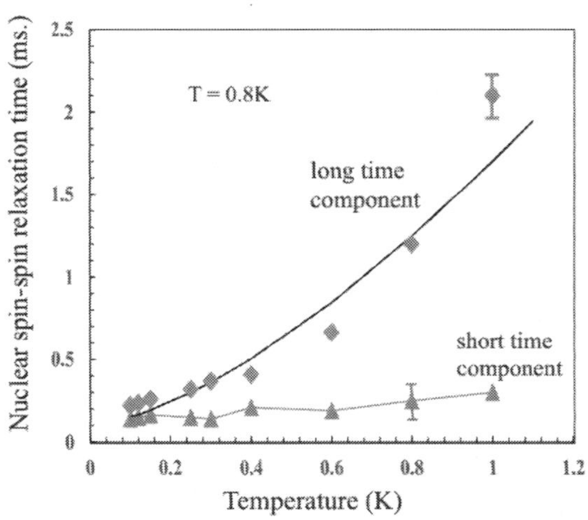


**Figure 3.** Temperature dependence of the nuclear spin-spin relaxation times for $^3$He on MCM-41; orange diamonds long-time component, blue triangles short-time component. Typical error bars are shown.

The nuclear spin lattice relaxation rates were determined from repetitive $\pi/2 - \tau - \pi$ RF pulse sequences and observing the recoveries to equilibrium. Figure 5 shows the observed temperature dependence for the same sample used for the $T_2$ measurements. The solid black line corresponds to the expected $T^{3/4}$ dependence deduced from the $T_2$ behavior with no adjustable parameters.

## 4. Discussion

In this experiment the nuclear spin relaxation is determined by modulation of the nuclear dipole-dipole interactions between the mobile atoms and the atoms coating the wall, rather than by collisions in the nanochannel. The correlation time $\tau_c$ can be estimated from the diffusion constant $D$ using $\tau_c^{-1} = D/a_0^2$ where $a_0$ is the lattice spacing of the wall atoms. The temperature dependence is given by that of $D(T) = \frac{1}{2}\langle V^2\rangle\tau_{coll}$ where $\tau_{coll}$ is the time during the hard core collisions of the mobile atoms. We therefore need to evaluate the mean energy $\langle E\rangle = \int E g_{1D}(E)F(1-F)dE$ where $g_{1D}(E) = 4n/(EE_F)^{1/2}$ is the 1D density of states and $F$

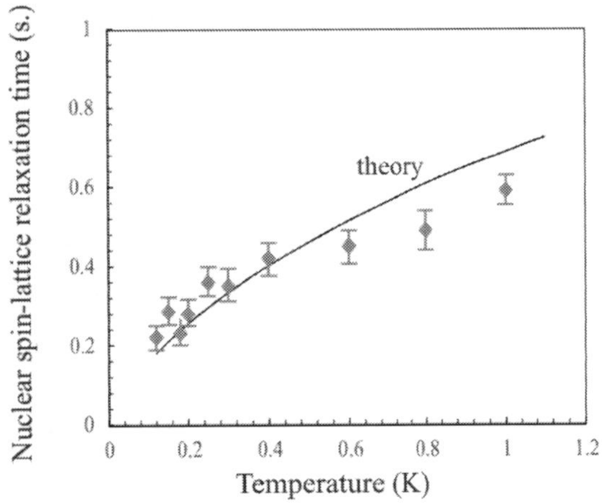


**Figure 4.** Temperature dependence of the nuclear spin-lattice relaxation time for $^3$He on MCM-41. A typical error bar is shown. The solid line is the dependence calculated from the $T_2$ temperature dependence of Fig. 3 without any adjustable parameters.

is the Fermi function. We find $\tau_c^{-1} = Ana_\perp \frac{k_b}{\hbar} \frac{T^{3/2}}{T_F^{1/2}}$ for $T > T_F$. $n$ is the linear density, $a_\perp$ is the transverse dimension and A is a numerical constant. The nuclear spin-spin relaxation time is given by $T_2 = M_2^{-1}\tau_c^{-1}$ where $M_2$ is the NMR second moment for the $^3$He-$^3$He wall interactions. We find $T_2 = BT^{3/2}$ where $B$ is a constant $\sim 10^{-3}$ in excellent agreement with Fig. 4.

As noted by Yager *et al.*[11] the correlation function $g(t) = (t/\tau_c)^{1/2}$ in the hydrodynamic limit and the nuclear spin lattice relaxation time $T_1$ is therefore given by $T_1^{-1} = 10.5M_2(\frac{\tau_c}{\omega_L})^{1/2}$. This relation predicts a $T^{3/4}$ temperature dependence in agreement with the observations shown in Fig. 5. Furthermore, we have $T_2/T_1 = 10.5/(\omega_L\tau_c)^{1/2}$ independent of $M_2$.The solid line in Fig. 5 was determined from the best fit of Fig. 4, using $\tau_c = 8.3\,10^{-2}T^{-3/2}$ s. The agreement with the predictions for a 1D Fermi gas is excellent. For higher dimensions the temperature dependencies for $T_2$ and $T_1$ would have opposite trends, with one increasing while the other decreases with temperature.

It should be noted that the relaxation we observe here is very different from the spin drag relaxation for Fermi gases[18–20] for which $\tau_D \propto T$ at very low temperatures ($T < T_F$) and $\propto T^{-(1/2)}$ at high temperatures ($T > T_F$) with a peak near $T = T_F$.

## 5. Conclusion

NMR studies of $^3$He constrained to the interior of nanochannels of MCM-41 have shown the onset of degeneracy below 220 mK. The nuclear spin-spin relaxation has two components corresponding to atoms in the wall and atoms in the center of the tube. The temperature dependence above 220 mK is consistent with that expected for 1D dynamics in the classical regime. Future experiments are planned using $^4$He and HD plating and measurements at lower temperatures to test for the dynamics predicted for a 1D degenerate Fermi system.[12, 21–23]

## Acknowledgments

The research was carried out at the NHMFL High B/T Facility which is supported by NSF DMR 1157490 and by the State of Florida. We thank the referee for bringing reference [15] to our attention. We also thank Bill Halperin, Brian Cowan and Aldo Migone for many useful discussions. Support from the National Science foundation through the award DMR-1303599 is also gratefully acknowledged.

## References

[1] Boninsegni M and Moroni S 2000 *J. Low Temp. Phys.* **118** 1–6 ISSN 1573-7357
[2] Krotscheck E and Miller M D 1999 *Phys. Rev. B* **60**(18) 13038–13050

[3] Wada N, Hieda M, Toda R and Matsushita T 2013 *Low Temp. Phys.* **39**(9) 786–792
[4] Wada N and Cole M W 2008 *J. Phys. Soc. Jp* **77**(11) 111012–1 – 111012–11
[5] Wada N, Matsushita T, Hieda M and Toda R 2009 *J. Low Temp. Phys.* **157** 324–351 ISSN 1573-7357
[6] Inagaki S, Fukushima Y and Kuroda K 1993 *J. Chem. Soc., Chem. Commun.* **8** 680–682
[7] Nakashima Y, Minato Y, Matsushita T, Hieda M and Wada N 2012 *Journal of Physics: Conference Series* **400** 012055
[8] Taniguchi J, Yamaguchi A, Ishimoto H, Ikegami H, Matsushita T, Wada N, Gatica S M, Cole M W, Ancilotto F, Inagaki S and Fukushima Y 2005 *Phys. Rev. Lett.* **94**(6) 065301
[9] Matsushita T, Kurebayashi K, Shibatsuji R, Hieda M and Wada N 2016 *J. Low Temp. Phys.* **183** 251–257 ISSN 1573-7357
[10] Kresge C T, Leonowicz M E, Roth W J, Vartuli J C and Beck J S 1992 *Nature* **359** 710–712
[11] Yager B, Nyéki J, Casey A, Cowan B P, Lusher C P and Saunders J 2013 *Phys. Rev. Lett.* **111**(21) 215303
[12] Astrakharchik G E and Boronat J 2014 *Phys. Rev. B* **90**(23) 235439
[13] Sullivan N and Pound R V 1973 *Cryogenics* **13** 28–29
[14] Kim K, Bodart J R and Sullivan N S 1996 *J. Mag. Res A* **118**(1) 28–32
[15] Taniguchi J, Tanaka D and Suzuki M 2014 *Journal of Physics: Conference Series* **568** 012022
[16] Golov A and Pobell F 1996 *Phys. Rev. B* **53**(19) 12647–12650
[17] Bonner J C and Fisher M E 1964 *Phys. Rev.* **135**(3A) A640–A658
[18] Polini M and Vignale G 2007 *Phys. Rev. Lett.* **98** 266403
[19] Rainis D, Polini M, Tosi M P and Vignole G 2008 *Phys. Rev B* **77** 035113
[20] Ebling U, Krauser J S, Fläschner N, Sengstock K, Becker C, Lewenstein M and Eckardt A 2014 *Phys. Rev. X* **4**(2) 021011
[21] Girardeau M D and Astrakharchik G E 2010 *Phys. Rev. A* **81**(6) 061601
[22] Imambekov A, Schmidt T L and Glazman L I 2012 *Rev. Mod. Phys.* **84**(3) 1253–1306
[23] Chitra R and Giamarchi T 1997 *Phys. Rev. B* **55**(9) 5816–5826

IOP Conf. Series: Journal of Physics: Conf. Series **969** (2018) 012002 doi:10.1088/1742-6596/969/1/012002

# Observation of the spiral flow and vortex induced by a suction pump in superfluid $^4$He

**H Yano*, K Ohyama, K Obara and O Ishikawa**

Graduate School of Science, Osaka City University, Osaka 558-8585, Japan

E-mail: hideo@sci.osaka-cu.ac.jp

**Abstract.** A suction flow generates a whirlpool, namely a bathtub vortex, in a classical fluid; in contrast, rotating containers, which are usually used for studies of superfluid helium, can produce only simple solid rotation. In the present work, the superfluid flow and concentrated quantized vortices induced by a cryogenic motor immersed in superfluid $^4$He were investigated. Using a motor with six blades in a cylinder caused the free surface of the superfluid $^4$He to take on a parabolic shape, indicating that the motor produces a rotating superfluid flow. To drive a suction flow in superfluid helium, the motor was mounted in a cylindrical container with a small hole at the center of the top and a slit at the side, acting as a superfluid pump. This pump was successfully used to generate a spiral flow and a vortex with a funnel-shaped core in superfluid $^4$He, suggesting that the resulting suction flow transports and centralizes quantized vortices to the suction hole, increasing the vortex circulation and sucking the free surface of the superfluid down.

## 1. Introduction

Superfluid vortices have been investigated extensively because they have a simpler structure than classical vortices in viscous fluids [1]. Hall and Vinen have suggested that in uniformly rotated superfluid helium, the arrangement of vortex lines can be found by minimizing the free energy of the rotating system [2]. Vinen also performed an experiment to observe quantized circulation under rotation [3]. When a container containing liquid helium is rotated, a regular arrangement of aligned quantized vortices forms [4]. Even under very high-speed rotations on the order of $10^6$ revolutions per second (rps), quantized vortices remain aligned [5], indicating that a simple rotating flow cannot centralize quantized vortices in the fluid.

In a viscous fluid under rotation, the vorticity is, or virtual element vortices are, also distributed uniformly throughout the fluid. However, an additive axial flow may produce a whirlpool, namely a bathtub vortex, in a rotating fluid [6]. A flow drained through a pipe drives a concentration gradient in the fluid, transporting and centralizing each element vortex in the fluid during this process. Consequently, the vorticity increases near the drain, producing a large vortex in the fluid.

Superflow is also expected to transport and centralize quantized vortices and produce a large vortex, though such a large vortex has not yet been studied for superfluid helium. The centralization may produce a vortex bundle, or even might produce a quantized vortex with a winding number larger than one. A new state of quantized vortices enables to extend the study of the superfluidity of helium. In this study, a large vortex with a funnel-shaped core was successfully generated in superfluid helium by producing a suction superflow. This paper reports an experimental setup of a cryogenic motor and a superfluid pump and discuss the vortex and superfluid flow induced by the pump in superfluid helium.

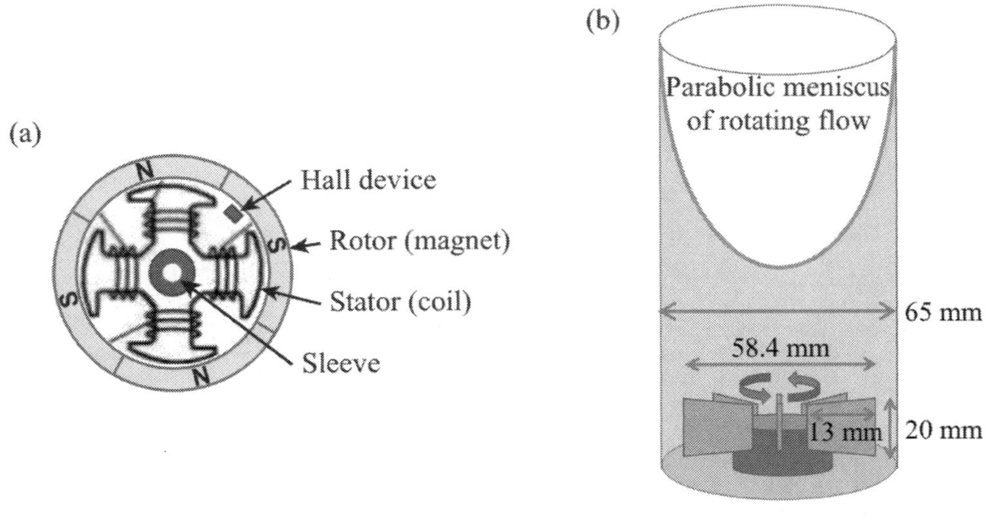


**Figure 1.** Schematics of (a) the cryogenic motor and (b) the setup of the rotating flow experiment. The outer part of the cryogenic motor is a permanent magnet rotor. The rotor is connected to the stator via a sliding bearing. In the rotating flow experiment, the cryogenic motor with six attached blades is mounted in the cylinder. The rotating superfluid flow produces a centrifugal force, resulting in the formation of a parabolic meniscus (see text).

## 2. Cryogenic motor

The cryogenic motor used in this study was a brushless motor with a stator coil, as shown in Fig. 1(a). An electrode brush is not acceptable for cryogenic use, because it produces friction heat and electric sparks. The stator coil is composed of a wound copper wire or a superconducting NbTi wire. A rotor composed of a permanent magnet is connected to the stator by a revolving ceramic shaft. The shaft is supported by a sleeve with no grease and spins smoothly even at low temperatures. To control the rotation of the brushless motor, a position sensor is required. A Hall device (SHARP, LT135A) that operates at low temperatures was used as a position sensor. The magnetic properties of the stator were also measured to ensure that the stator ferrite does not show magnetic hysteresis at low temperatures.

Through these adjustments, the motor was successfully rotated at low temperatures. The rotational velocity ranged from 2 to 40 rps at a temperature of 2 K for the copper coil. The rotor of the cryogenic motor spins smoothly at high velocities, though the friction between the shaft and the sleeve causes the rather stepping motion of the rotor at 2 rps. The dissipation power of the motor in saturated vapor at 2 K increases linearly with the rotational velocity at a rate of 0.25 W per 40 rps, allowing the motor to operate in superfluid helium.

## 3. Rotating superfluid flow driven by cryogenic motor

To produce a rotating flow in superfluid helium, the cryogenic motor was mounted in a cylinder with an inner diameter of 65 mm, as shown in Fig. 1(b). Six blades were attached to the rotor of the motor to agitate the liquid helium in the cylinder. The cylinder was mounted in a conventional glass dewar and filled with liquid helium. A thermometer was immersed in the liquid helium at the outside wall of the cylinder near the bottom to measure the temperature. By pumping helium vapor in the container, liquid helium can be cooled to a superfluid state. The temperature can reach a minimum of 1.6 K in this cryostat system, corresponding to a superfluid ratio of $\rho_s/\rho \approx 0.8$, where $\rho_s$ and $\rho$ are the densities of the superfluid helium and the liquid helium, respectively.

After the helium was cooled to a superfluid state, the rotor was rotated with the blades immersed

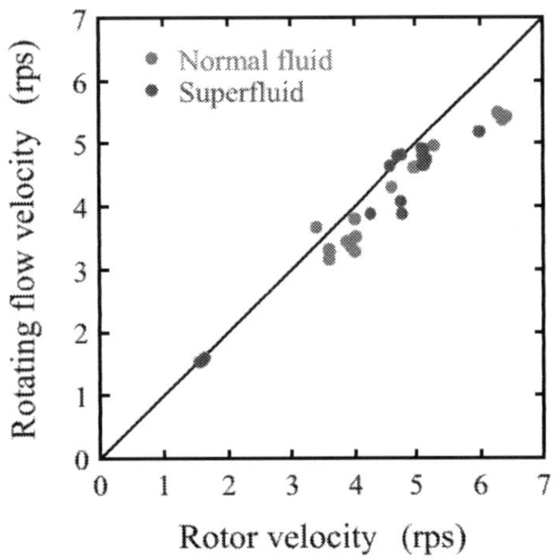


**Figure 2.** Rotational velocity of normal liquid and superfluid helium flow. The data were taken at temperatures above 1.6 K, which corresponds to superfluid components ranged from $\rho_s/\rho = 0$ to 0.8 for the superfluid state. The solid line indicates a rotational velocity of the flow equal to the rotor velocity.

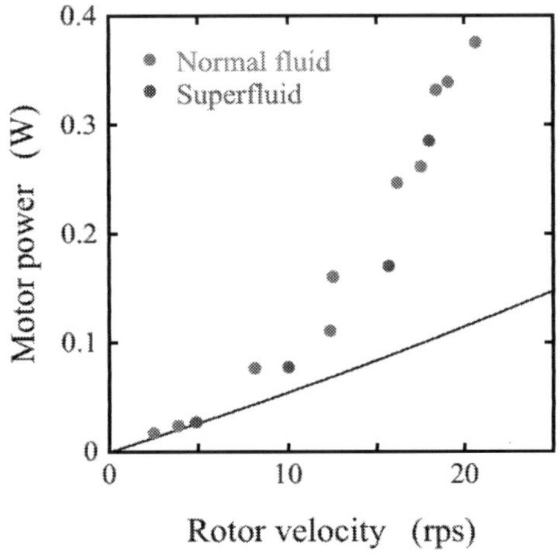


**Figure 3.** Dissipation power of the cryogenic motor rotating in normal liquid and superfluid helium. The dissipation powers for the two states are very similar. The solid line indicates the intrinsic dissipation of the motor measured in saturated vapor at 2 K.

in the superfluid helium, and it was found that the surface of the fluid took on a parabolic shape in the cylinder. The formation of a parabolic meniscus has been also observed in superfluid helium in a rotating container [7], indicating that the rotating blades produce a solid rotating flow even in a superfluid state. During the rotation, remnant vortices attached to the blades are expected to be stretched by boundary superflows along the surfaces of the blades, producing a macroscopic rotating flow in the cylinder. The parabolic meniscus forms because of the centrifugal force produced by the rotation, and its height $h$ from the bottom is given by

$$h = \frac{R^2\omega^2}{2g}, \tag{1}$$

where $R$ is the distance from the axis of rotation, $\omega$ is the angular velocity of the rotation, and $g$ is the acceleration due to gravity.

The parabolic meniscus was photographed to estimate the velocity of the rotating flow from Eq. (1), and the flow velocity was found to be nearly equal to the rotor velocity, as shown in Fig. 2. Moreover, although a viscous laminar drag decreases in a superfluid state, the rotation velocities for the superfluid state in the range from $\rho_s/\rho = 0$ to 0.8 were almost equal to those for the normal fluid. Interestingly, the dissipation powers of the motor for the two states were also very similar, as shown in Fig. 3. The dissipation powers were estimated from the applied currents and the induced voltages of the motor to increase in the fluid whatever its state. The rotational flow velocity appears to be influenced little by the viscosity of the normal fluid. However, the rotating flows are slightly slower than the rotor velocity for both states, suggesting that a stationary cylinder wall may decrease the flow velocity. Quantized vortices attached to the wall in the superfluid may contribute to the drag braking force observed in studies of oscillating objects, such as a tuning fork or a vibrating wire, in superfluid helium [8, 9].

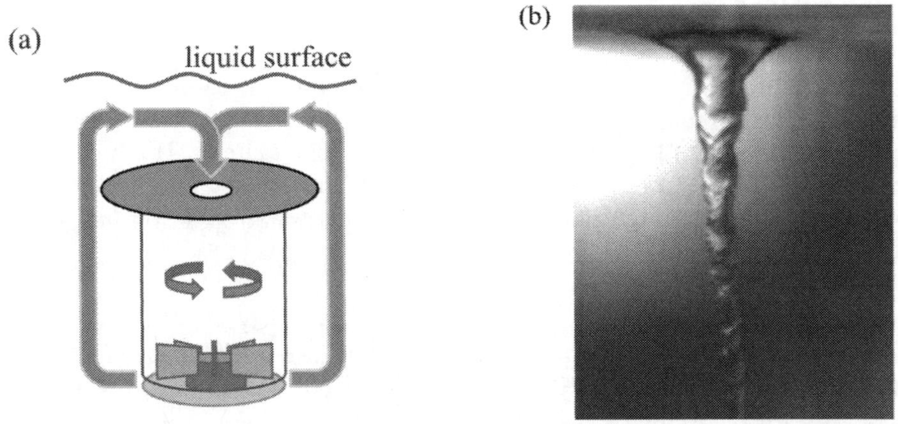


**Figure 4.** Suction vortex experiment. (a) Schematic of the suction pump and superfluid flows (arrows). The cylinder length of the pump is 90 mm, and the hole diameter and the side slit gap are 5 mm and 2 mm, respectively. (b) Photograph of a suction vortex produced by the pump in superfluid helium at a temperature of 1.6 K. As shown in this photograph, superfluid flows from the side slit to the top hole of the cylinder may enhance vortex circulation above the hole, sucking the free surface down to the hole.

## 4. Superfluid flow and vortex

A solid rotating fluid with angular velocity $\omega$ produces a centrifugal force that causes a pressure difference $\Delta p$ given by

$$\Delta p = \frac{1}{8}\rho\omega^2 R^2, \tag{2}$$

where $\rho$ is the fluid density and $R$ is the distance from the axis of rotation. The pressure difference in a rotating superfluid may provide the feature of a superfluid pump, namely, a centrifugal pump. A cryogenic motor was mounted in a cylinder with a small hole at the center of the top and a slit along the side, as shown in Fig. 4(a). The cylinder containing the motor was immersed in superfluid helium. The rotating flow in the cylinder ejects the superfluid out through the side slit and forces the fluid in through the top hole, which acts as a drain, as a result of the pressure difference given by Eq. (2), generating the superfluid circulation loop shown in Fig 4(a).

Interestingly, a tornado-like vortex emerged in the superfluid helium above the top hole of the cylinder during the rotation of the motor, as shown in Fig. 4(b). The fluid ejected from the side slit has an angular momentum as a result of the vorticity generated by the rotating flow in the cylinder, causing it to flow to the upper area above the cylinder. Because the flow transports the vorticity of the fluid, the flow drained into the hole appears to centralize the vorticity, enhancing the vortex circulation [6]. The enhanced circulation just above the drain hole produces a large vortex that sucks the free surface of the helium fluid. Therefore, the tornado-like vortex is visualized as a pipe, as shown in Fig. 4(b). Thus, a large vortex with a funnel-shaped core was successfully generated in superfluid helium. The motion of the surface of the vortex core reveals the existence of a spiral flow around the core. The vortex core extends to the drain hole with a diameter of approximately half that of the hole. The descent height $h - h_0$ of the core surface of a suction vortex is given at a distance $R$ from the axis by

$$h - h_0 \approx -\frac{\Gamma^2}{8\pi^2 g R^2}, \tag{3}$$

where $h_0$ and $\Gamma$ are the normal level of the fluid and the circulation, respectively. The circulation of the vortex was estimated from Eq. (3) to be on the order of $10^3$ mm$^2$/s or $10^4\kappa$, where $\kappa$ is the circulation of a single quantized vortex of superfluid $^4$He. Thus, a suction flow enables the generation of a vortex with a high circulation even in a superfluid.

## 5. Summary

The superfluid flow and vortex induced by a newly proposed suction pump in superfluid $^4$He were investigated in this study. A cryogenic motor immersed in superfluid helium in a cylindrical container produces a rotating flow in the cylinder. The pressure difference in the rotating superfluid generates a superflow circulation outside of the cylinder, acting as a superfluid pump. The suction superflow into the pump was successfully used to generate a vortex with a high circulation much larger than that of the single quantized vortex in superfluid $^4$He. It is expected that this new method will find wide applications in superfluid studies.

## Acknowledgments

The authors are very grateful to K. Fujimoto, M. Tsubota, and H. Takeuchi for their stimulating discussions. The research was supported by JSPS KAKENHI Grant Numbers 15H03694 and 17K18761.

## References

[1] Donnelly R J 1991 *Quantized Vortices in Helium II* (Cambridge, England: Cambridge University Press)
[2] Hall H E and Vinen W F 1956 *Proc. Roy. Soc. A* **238** 215
[3] Vinen W F 1961 *Proc. Roy. Soc. A* **260** 218
[4] Yarmchuk E J, Gordon M J V and Packard R E 1979 *Phys. Rev. Lett.* **43** 214
[5] Gomez L F *et al.* 2014 *Science* **345** 906
[6] Andersen A, Bohr T, Stenum B, Rasmussen J J and Lautrup B 2003 *Phys. Rev. Lett.* **91** 104502
[7] Andronikashvili E L and Mamaladze Y G 1966 *Rev. Mod. Phys.* **38** 567
[8] Blažková M, Schmoranzer D, Skrbek L and Vinen W F 2009 *Phys. Rev. B* **79** 054522
[9] Hashimoto N, Goto R, Yano H, Obara K, Ishikawa O and Hata T 2007 *Phys. Rev. B* **76** 020504(R)

IOP Conf. Series: Journal of Physics: Conf. Series **969** (2018) 012003 doi:10.1088/1742-6596/969/1/012003

# Strongly Interacting Multi-component Fermions: From Ultracold Atomic Fermi Gas to Asymmetric Nuclear Matter in Neutron Stars

**Hiroyuki Tajima[1], Tetsuo Hatsuda[1,2], and Yoji Ohashi[3]**

[1] RIKEN, Nishina Center, 2-1, Hirosawa, Wako, Saitama, 351-0198, Japan
[2] RIKEN iTHES and iTHEMS, 2-1, Hirosawa, Wako, Saitama, 351-0198, Japan
[3] Department of Physics, Keio University, 3-14-1 Hiyoshi, Kohoku-ku, Yokohama, Kanagawa 223-8522, Japan

E-mail: hiroyuki.tajima@riken.jp

**Abstract.** We investigate an asymmetric nuclear matter consisting of protons and neutrons with spin degrees of freedom ($\sigma = \uparrow, \downarrow$). By generalizing the Nozières and Schmitt-Rink theory for two-component Fermi gases to the four-component case, we analyze the critical temperature $T_c$ of the superfluid phase transition. Although the pure neutron matter exhibits the dineutron condensation in the low-density region, the superfluid instability toward the deuteron condensation is found to take place as the proton fraction increases. We clarify the mechanism of the competition between the deuteron condensation and dineutron condensation. Our results would serve for understanding the properties of asymmetric nuclear matter realized in the interior of neutron stars.

## 1. Introduction

Two-component ultracold Fermi gases (such as $^6$Li and $^{40}$K) are known to be the good testing grounds to study various strong coupling phenomena, because the strength of the pairing interaction between atoms can be controlled by a Feshbach resonance [1, 2, 3]. Indeed, the crossover from the weak-coupling BCS Fermi superfluid to the Bose-Einstein condensate (BEC) of tightly bound molecules, which was originally proposed theoretically [4, 5, 6, 7], has been realized by this novel technique [8, 9]. Such crossover behavior can be confirmed with high precision in various thermodynamic quantities such as the pressure, internal energy, chemical potential, as well as single-particle and collective excitations [10, 11, 12].

The BCS-BEC crossover of the ultracold Fermi gas is essentially dictated by the $s$-wave interaction with large scattering length in the atomic scale. This situation is analogous to the low-density pure neutron matter where the $^1$S$_0$ channel (spin-singlet and $s$-wave) has a scattering length $a_{\rm nn}^{\rm T} = -18.5$ fm in the nuclear scale [13]. Indeed, the current experimental results on the equation of state of the $^6$Li superfluid Fermi gas near the $s$-wave Feshbach resonance and the numerical simulations of the pure neutron matter in the low density region have quantitative agreement after appropriate rescaling [12]. Although there are some differences between these two systems such as the effect of the effective range and the three-body force, the ultracold Fermi gas is expected to be a useful reference system for the study of pure neutron matter [14].

The above-mentioned agreement makes us expect that a two-component Fermi gas in the BCS-BEC crossover region may also be used as a good starting point for the study of an asymmetric nuclear matter, where protons (with spin $\sigma =\uparrow,\downarrow$) mix with neutrons (with spin $\sigma =\uparrow,\downarrow$). Recently, thermodynamic properties of an asymmetric nuclear matter have attracted much attention, such as astrophysical simulations for supernova explosion [15, 16, 17, 18, 19], the problem of massive neutron stars [20, 21], as well as the so-called symmetry energy [22, 23, 24] in nuclear physics.

In cold atomic Fermi gases, strong-coupling theories that explain various experimental results have been developed in the BCS-BEC crossover region [25, 26, 27]. Therefore, it is an interesting challenge to extend those theories to the four-component system, which may be applied to the physics of the proton-neutron mixture. Even in ultracold atomic Fermi gas, exploration toward the multi-component Fermi gases beyond the two-component would be one of the interesting future directions.

In this paper, we investigate strong-coupling properties of the low-density asymmetric nuclear matter, consisting of protons and neutrons. Extending the strong-coupling BCS-BEC crossover theory developed by Nozières and Schmitt-Rink (NSR) [5] for two-component Fermi gases, we examine the superfluid phase transition temperature $T_{\rm c}$. We clarify how $T_{\rm c}$ varies with the density increased, as well as how the change of proton fraction affects the pairing-type (neutron-neutron pair vs. proton-neutron pair) responsible for the superfluid instability. Throughout this paper, we take the system volume to be unity for simplicity, and set $\hbar = k_{\rm B} = 1$.

## 2. Formulation

We consider a mixture of interacting protons (I = p) and neutrons (I = n), described by the Hamiltonian

$$
\begin{aligned}
H \;=\; & \sum_{\boldsymbol{k},\sigma=\uparrow,\downarrow,{\rm I=n,p}} \xi_{\boldsymbol{k},{\rm I}} c^\dagger_{\boldsymbol{k},\sigma,{\rm I}} c_{\boldsymbol{k},\sigma,{\rm I}} - \sum_{\boldsymbol{k},\boldsymbol{k}',\boldsymbol{q},{\rm I=n,p}} U^{\rm T}_{\rm II} c^\dagger_{\boldsymbol{k}+\boldsymbol{q}/2,\uparrow,{\rm I}} c^\dagger_{-\boldsymbol{k}+\boldsymbol{q}/2,\downarrow,{\rm I}} c_{-\boldsymbol{k}'+\boldsymbol{q}/2,\downarrow,{\rm I}} c_{\boldsymbol{k}'+\boldsymbol{q}/2,\uparrow,{\rm I}} \\
& -\frac{U^{\rm T}_{\rm np}}{2} \sum_{\boldsymbol{k},\boldsymbol{k}',\boldsymbol{q}} \left[ c^\dagger_{\boldsymbol{k}+\boldsymbol{q}/2,\uparrow,{\rm n}} c^\dagger_{-\boldsymbol{k}+\boldsymbol{q}/2,\downarrow,{\rm p}} + c^\dagger_{\boldsymbol{k}+\boldsymbol{q}/2,\uparrow,{\rm p}} c^\dagger_{-\boldsymbol{k}+\boldsymbol{q}/2,\downarrow,{\rm n}} \right] \\
& \qquad \times \left[ c_{-\boldsymbol{k}'+\boldsymbol{q}/2,\downarrow,{\rm p}} c_{\boldsymbol{k}'+\boldsymbol{q}/2,\uparrow,{\rm n}} + c_{-\boldsymbol{k}'+\boldsymbol{q}/2,\downarrow,{\rm n}} c_{\boldsymbol{k}'+\boldsymbol{q}/2,\uparrow,{\rm p}} \right] \\
& -\frac{U^{\rm S}_{\rm np}}{2} \sum_{\boldsymbol{k},\boldsymbol{k}',\boldsymbol{q}} \left[ c^\dagger_{\boldsymbol{k}+\boldsymbol{q}/2,\uparrow,{\rm n}} c^\dagger_{-\boldsymbol{k}+\boldsymbol{q}/2,\downarrow,{\rm p}} - c^\dagger_{\boldsymbol{k}+\boldsymbol{q}/2,\uparrow,{\rm p}} c^\dagger_{-\boldsymbol{k}+\boldsymbol{q}/2,\downarrow,{\rm n}} \right] \\
& \qquad \times \left[ c_{-\boldsymbol{k}'+\boldsymbol{q}/2,\downarrow,{\rm p}} c_{\boldsymbol{k}'+\boldsymbol{q}/2,\uparrow,{\rm n}} - c_{-\boldsymbol{k}'+\boldsymbol{q}/2,\downarrow,{\rm n}} c_{\boldsymbol{k}'+\boldsymbol{q}/2,\uparrow,{\rm p}} \right] \\
& -U^{\rm S}_{\rm np} \sum_{\boldsymbol{k},\boldsymbol{k}',\boldsymbol{q},\sigma=\uparrow,\downarrow} c^\dagger_{\boldsymbol{k}+\boldsymbol{q}/2,\sigma,{\rm n}} c^\dagger_{-\boldsymbol{k}+\boldsymbol{q}/2,\sigma,{\rm p}} c_{-\boldsymbol{k}'+\boldsymbol{q}/2,\sigma,{\rm p}} c_{\boldsymbol{k}'+\boldsymbol{q}/2,\sigma,{\rm n}}. 
\end{aligned}
\tag{1}
$$

where $\xi_{\boldsymbol{k},{\rm I}} = k^2/(2m) - \mu_{\rm I}$ is the kinetic energy of a nucleon measured from the nucleon chemical potential $\mu_{\rm I}$. In this paper, we ignore difference between the proton mass $m_{\rm p}$ and neutron mass $m_{\rm n}$, for simplicity ($m_{\rm p} = m_{\rm n} = m$). Also, we assume that the system is not spin-polarized. $c_{\boldsymbol{k},\sigma,{\rm I}}$ is the annihilation operator of a nucleon, where $\sigma =\uparrow,\downarrow$ represents the spin. $-U^{\rm T}_{\rm II'}$ and $-U^{\rm S}_{\rm np}$ originate from the strong nuclear forces [28] and correspond to the attractive interactions in the isospin-triplet (spin-singlet) and the isospin-singlet (spin-triplet) channels, respectively. In this paper, we focus on the low-density and low-energy region, so that the nuclear forces are simply treated as the contact-type $s$-wave pairing interactions. We use the experimental values of the $s$-wave scattering lengths, $a^{\rm T}_{\rm nn} = -18.5$ fm, $a^{\rm T}_{\rm pp} = -7.8$ fm, $a^{\rm T}_{\rm np} = -23.75$ fm, and $a^{\rm S}_{\rm np} = 5.42$ fm [13], that are related to the"bare" interactions $-U^{\rm Z}_{\rm II'}$ (Z = S, T) as

$$
\frac{4\pi a^{\rm Z}_{\rm II'}}{m} = \frac{-U^{\rm Z}_{\rm II'}}{1 - U^{\rm Z}_{\rm II'} \sum_{\boldsymbol{k}}^{k_{\rm c}} \frac{m}{k^2}}, \tag{2}
$$

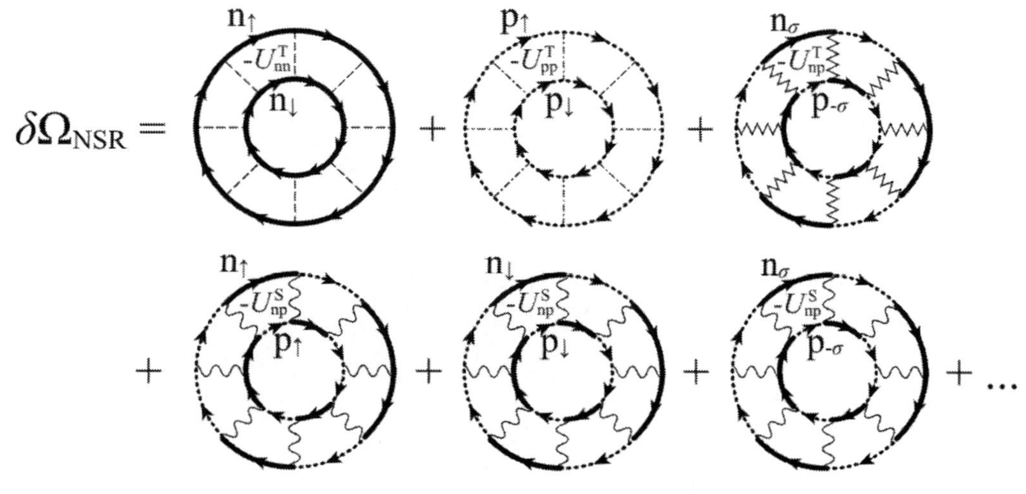


**Figure 1.** Feynman diagrams describing strong-coupling corrections ($\delta\Omega_{\rm NSR}$) to the thermodynamic potential $\Omega$ of an interacting proton-neutron mixture. The solid (dotted) line shows the bare propagator of a neutron (proton). The neutron-neutron interaction $-U^{\rm T}_{\rm nn}$, proton-proton interaction $-U^{\rm T}_{\rm pp}$, isospin-triplet neutron-proton interaction $-U^{\rm T}_{\rm np}$, as well as isospin-singlet neutron-proton interaction $-U^{\rm S}_{\rm np}$ are all included in these diagrams on an equal footing, within the framework of NSR theory.

with $k_{\rm c}$ being an ultra-violet cutoff momentum. Note that the difference between $a^{\rm T}_{\rm nn}$ and $a^{\rm T}_{\rm pp}$ comes mainly from the Coulomb interaction. Our setup is justified in the low density region where the effective range or higher angular-momentum pairs are negligible. Our theory involves only nucleon masses and scattering lengths as input parameters, in contrast to the case with more realistic finite-range nuclear potential [22]. If we neglect the existence of protons, our model given by Eq. (1) is reduced to the ordinary BCS Hamiltonian in a two-component ultracold Fermi gas.

To examine many-body effects associated with the interactions $-U^{\rm Z}_{\rm II'}$ in Eq. (1), we extend the strong-coupling theory developed by Nozières and Schmitt-Rink (NSR) to the present proton-neutron mixture. The resulting strong-coupling corrections ($\delta\Omega_{\rm NSR}$) to the thermodynamic potential ($\Omega = \Omega_0 + \delta\Omega_{\rm NSR}$ with $\Omega_0$ being the non-interacting part) are diagrammatically described in Fig. 1 . The total $\Omega$ is then written as

$$\begin{aligned}\Omega &= -2T\sum_{\rm I=n,p}\sum_{\boldsymbol{k}}\ln\left[1+e^{-\frac{\xi_{\boldsymbol{k},\rm I}}{T}}\right]\\ &\quad +T\sum_{\boldsymbol{q},i\nu_l}\Big[\ln\left[1-U^{\rm T}_{\rm nn}\Pi_{\rm nn}(\boldsymbol{q},i\nu_l)\right]+\ln\left[1-U^{\rm T}_{\rm pp}\Pi_{\rm pp}(\boldsymbol{q},i\nu_l)\right]\\ &\quad +\ln\left[1-U^{\rm T}_{\rm np}\Pi_{\rm np}(\boldsymbol{q},i\nu_l)\right]+3\ln\left[1-U^{\rm S}_{\rm np}\Pi_{\rm np}(\boldsymbol{q},i\nu_l)\right]\Big],\end{aligned}\tag{3}$$

where $\nu_l$ is the boson Matsubara frequency. In Eq. (3),

$$\Pi_{\rm II'}(\boldsymbol{q},i\nu_l) = -\sum_{\boldsymbol{k}}\frac{1-n_{\rm F}(\xi_{\boldsymbol{k}+\boldsymbol{q}/2,\rm I})-n_{\rm F}(\xi_{-\boldsymbol{k}+\boldsymbol{q}/2,\rm I'})}{i\nu_l-\xi_{\boldsymbol{k}+\boldsymbol{q}/2,\rm I}-\xi_{-\boldsymbol{k}+\boldsymbol{q}/2,\rm I'}},\tag{4}$$

is the two-particle pair propagator. In the NSR scheme, the proton ($\mu_{\rm p}$) and neutron ($\mu_{\rm n}$) chemical potential are determined from the equations for the number density $\rho_{\rm I=p,n}$ of protons and neutrons, that are derived from the NSR thermodynamic potential in Eq. (3) as,

$$\rho_{\rm I} = -\frac{\partial \Omega}{\partial \mu_{\rm I}}. \quad (5)$$

The superfluid phase transition temperature in the dineutron channel is determined by the Thouless criterion,

$$1 - U^{\rm T}_{\rm nn}\Pi_{\rm nn}(\boldsymbol{q}=0, i\nu_l=0) = 0. \quad (6)$$

On the other hand, the superfluid instability being accompanied by deuteron (the isospin-singlet proton-neutron pair) condensation occurs when the following condition is satisfied,

$$1 - U^{\rm S}_{\rm np}\Pi_{\rm np}(\boldsymbol{q}=0, i\nu_l=0) = 0. \quad (7)$$

For later convenience, we define the proton fraction $Y_{\rm p}$ by

$$Y_{\rm p} = \frac{\rho_{\rm p}}{\rho_{\rm n}+\rho_{\rm p}}. \quad (8)$$

In principle, other superfluid instabilities such as the proton superfluid are possible. However, we have numerically confirmed that only the dineutron condensation or the deuteron condensation occur.

## 3. Results

Figure 2(a) shows the superfluid phase transition temperature $T_{\rm c}$ in a proton-neutron asymmetric nuclear matter as a function of $(k_{\rm F,n}a^{\rm T}_{\rm nn})^{-1}$, where $k_{\rm F,n}$ is the Fermi momentum of neutrons. In cold Fermi gas physics, similar interaction strength $(k_{\rm F}a_s)^{-1}$ is frequently used, where $k_{\rm F}$ is the Fermi momentum of a Fermi atomic gas and the scattering length $a_s$ describes a pairing interaction between Fermi atoms. In our model in Eq. (1), the case of pure neutron matter ($Y_{\rm p}=0$) is formally the same as the two-component Fermi gas, where $a_s$ corresponds to $a^{\rm T}_{\rm nn}$. However, $a^{\rm T}_{\rm nn} = -18.5$ fm is fixed in a nuclear matter, unlike the tunable $a_s$ in ultracold Fermi gases. Thus, the scaled interaction $(k_{\rm F,n}a^{\rm T}_{\rm nn})^{-1}$ changes via the change of the neutron density $\rho_{\rm n} = k^3_{\rm F,n}/(3\pi^2)$. In this sense, the $x$-axis in Fig. 2 (a) may be regarded as indicating the density dependence, that is, the neutron density increase as one moves from the left toward the unitary regime, $(k_{\rm F,n}a^{\rm T}_{\rm nn})^{-1} \sim 0$ in Fig. 2 (a).

Around the unitarity limit, or the region of high neutron density ($(k_{\rm F,n}a^{\rm T}_{\rm nn})^{-1} \simeq 0$), Fig. 2 (a) shows that $T_{\rm c}$ decreases with increasing proton fraction $Y_{\rm p}$, indicating that the existence of protons disfavors the dineutron condensation and promotes the deuteron (proton-neutron pair) condensation, especially in the low-density regime $(k_{\rm F,n}a^{\rm T}_{\rm nn})^{-1} \lesssim -0.5$.

To understand the competition between the dineutron condensation and deuteron condensation in the asymmetric nuclear matter shown in Fig. 2 (a), it is important to recall that the isospin-singlet proton-neutron (isospin-triplet neutron-neutron) scattering length is positive (negative). Because of this difference, the "weak-coupling regime" $(k_{\rm F,n}a^{\rm T}_{\rm nn})^{-1} \lesssim -1$ in the neutron-neutron Cooper channel corresponds to the "strong-coupling regime" $(k_{\rm F,n}a^{\rm S}_{\rm np})^{-1} \gtrsim +1$ in the proton-neutron Cooper channel, giving high superfluid phase transition temperature of deuteron condensation. In "the strong-coupling limit" of the deuteron channel, $(k_{\rm F,n}a^{\rm S}_{\rm np})^{-1} \gg +1$, or $(k_{\rm F,n}a^{\rm T}_{\rm nn})^{-1} \ll -1$, $T_{\rm c}$ is conventionally given by the Bose-Einstein condensation temperature $T^{\rm deuteron}_{\rm BEC}$ in an ideal Bose gas with $\rho_{\rm n}Y_{\rm p}/(1-Y_{\rm p})$ spin-1 deuterons

$$T^{\rm deuteron}_{\rm BEC} = \frac{\pi}{m}\left[\frac{\rho_{\rm n}}{3\zeta(3/2)}\frac{Y_{\rm p}}{1-Y_{\rm p}}\right]^{\frac{2}{3}}, \quad (9)$$

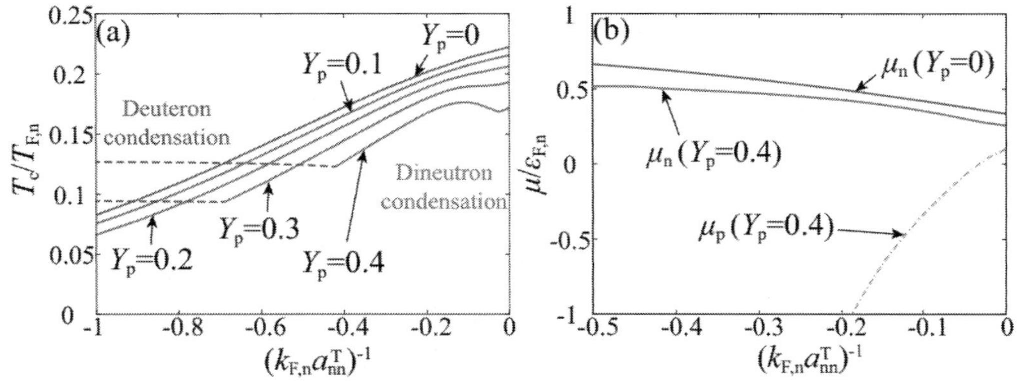


**Figure 2.** (a) Calculated superfluid phase transition temperature $T_c$, as a function of $(k_{F,n}a_{nn}^T)^{-1}$. The solid and dashed lines show $T_c$ for the dineutron and the deuteron condensation, respectively. (b) Proton ($\mu_p$) and neutron ($\mu_n$) chemical potential at $T_c$. $k_{F,n}$ and $\varepsilon_{F,n} = k_{F,n}^2/(2m)$ are the Fermi momentum and the Fermi energy of the neutron component. In these figures, the neutron density $\rho_n$ is fixed instead of the total density $\rho = \rho_n + \rho_p$.

where $\zeta(s)$ is the zeta function. Eq. (9) gives $T_{BEC}^{deuteron} = 0.126T_{F,n}$ when $Y_p = 0.4$.

Although the proton-neutron pairing interaction ($a_{np}^S = 5.42$ fm) is stronger than the neutron-neutron interaction ($a_{nn}^T = -18.5$ fm), one sees in Fig. 2 (a) that the superfluid instability is still dominated by dineutrons around the unitarity limit ($(k_{F,n}a_{nn}^T)^{-1} \simeq 0$), even when the system is close to the symmetric matter ($Y_p = 0.4$). This is because the neutron chemical potential $\mu_n$ is different from the proton chemical potential $\mu_p$ as shown in Fig. 2 (b), as a result of the nucleon density difference as well as the charge dependence, that is, the difference between $a_{nn}^T$ and $a_{pp}^T$. We note that in the low density region, since the system is dominated by the deuteron formation, the proton chemical potential is given by $\mu_p \simeq -\mu_n + E_b$ where $E_b(< 0)$ is the binding energy of a deuteron. Then, the deuteron pairing is influenced by the "effective magnetic field" $h_{eff} = \mu_n - \mu_p$, which suppresses the superfluid instability as well known for metallic superconductor under an external magnetic field. Thus, to realize the deuteron condensation, the scaled interaction $(k_{F,n}a_{np}^S)^{-1}$ must be strong enough to overwhelm this depairing effect. On the other hand, $h_{eff}$ does not suppress the neutron-neutron pairing, leading to the dineutron condensation near the unitarity limit even when $Y_p = 0.4$.

We note that $T_c$ would be affected by the effective range which becomes non-negligible in the high density region ($(k_{F,n}a_{nn}^T)^{-1} \gtrsim -0.1$) [29]. Thus, to clarify the superfluid phase transition, as well as the competition between the dineutron condensation and deuteron condensation, we need to extend our present work to include such effect. In addition, the so-called Gorkov-Melik-Barkhudarov (GMB) correction [30] needs to be included to make quantitative prediction of $T_c$ in the weak-coupling regime.

## 4. Summary

To summarize, we have theoretically investigated strong-coupling properties of a mixture of protons and neutrons with spin degrees of freedom ($\sigma =\uparrow, \downarrow$). Extending the theory of Nozières and Schmitt-Rink for a two-component Fermi gas to the four-component case, we have calculated the superfluid phase transition temperature. We showed that the deuteron condensation becomes more favorable than the dineutron condensation as the proton fraction is increased.

In this paper, we take only two-body pairing interactions and the associated two-particle pairs in the Cooper-channels. Although this treatment is reasonable for small $Y_p$, we would need to consider triton and alpha particles when the system is close to the symmetric matter ($Y_\mathrm{p} \simeq 0.5$) [31, 32]. Also, the three-body interaction is known to be crucial [22, 33] when the nucleon density approaches to the nuclear saturation density $\rho_0 \simeq 0.17$ fm$^{-3}$. These are the important future problems to be examined. Besides the importance in nuclear physics, studying strong-coupling properties of multi-component Fermi gases beyond the two-component systems is an interesting future direction in the physics of cold Fermi gas.

## Acknowledgments

We thank G. Baym, P. Naidon, P. van Wyk, S. Furusawa, and H. Togashi for useful discussions. H. T. was supported by a Grant-in-Aid for JSPS fellows (No. 17J03975). T. H. was supported by RIKEN iTHES Project and RIKEN iTHEMS Program. Y. O. was supported by the KiPAS project at Keio University. This work was supported by Grant-in-Aid for Scientific Research from MEXT and JSPS in Japan (Nos. 15K00178, 15H00840, 15H03663, 16K05503).

## References

[1] Giorgini S, Pitaevskii S, and Stringari S, 2008 *Rev. Mod. Phys.* **80** 1215
[2] Bloch I, Dalibard J and Zwerger W, 2008 *Rev. Mod. Phys.* **80** 885
[3] Chin C, Grimm R, Julienne P, Tiesinga, 2010, *Rev. Mod. Phys.* **82** 1225
[4] Eagles D M 1969 *Phys. Rev.* **186** 456
[5] Nozières P and Schmitt-Rink S, 1985 *J. Low Temp. Phys.* **59** 195
[6] Sa de Melo C A R, Randeria M, and Engelbrecht J R, 1993 *Phys. Rev. Lett.* **71** 3202
[7] Ohashi Y and Griffin A, 2002 *Phys. Rev. Lett.* **89** 130402
[8] Regal C, Greiner M, Jin D S, 2004 *Phys. Rev. Lett.* **92** 040403
[9] Zwierlein M W, Stan C A, Shunck C H, Raupach S M F, Kerman A J, and Ketterle W, 2004 *Phys. Rev. Lett.* **92** 120403
[10] Navon N, Nascimbène S, Chevy F, Salomon C, 2010 *Science* **328** 729
[11] Sagi Y, Drake T E, Paudel R, Chapurin R, and Jin D S, 2015 Phys. Rev. Lett. **114** 075301
[12] Horikoshi M, Koashi M, Tajima H, Ohashi Y, and Kuwata-Gonokami M, 2016 arXiv:1612.04026v1
[13] Wiringa R B, Stoks V G J, and Schiavilla, 1995 *Phys. Rev.* C **51** 38
[14] Gezerlis A and J. Carlson, 2008 *Phys. Rev* C **77** 032801(R)
[15] Shen H, Toki H, Oyamatsu K, and Sumiyoshi K, 2011 *Astrophys. J* **197** 20
[16] Togashi H and Takano M, 2013 *Nucl. Phys.* A, **902** 53
[17] Furusawa S, Nagakura H, Sumiyoshi K, and Yamada S, 2013 *Astrophys. J.* **774** 78
[18] Furusawa S, Sumiyoshi K, Yamada S, and Suzuki H, 2017 *Nucl. Phys.* A **957** 188
[19] Oertel M, Hempel M, Klähn T, and Typel S, 2017 *Rev. Mod. Phys.* **89** 015007
[20] Demorest P, Pennucci T, Ransom S, Roberts E, and Hessels J, 2010 *Nature* **467** 1081
[21] Antoniadis J, Freire P C C, Wex N, *et al.*, 2013 *Science* **340** 6131
[22] Akmal A, Pandharipande V R, and Ravenhall D G, 1998 *Phys. Rev.* C **58** 1804
[23] Lattimer J M and Prakash M, 2000 *Astrophys. J.* **550** 426
[24] Hebeler K, Lattimer J, Pethick C, and Schwenk, 2013 *Astrophys. J.* **773**, 11
[25] Hu H, Liu X-J, and Drummond P D, 2006 *Europhys. Lett.* **74** 574
[26] Haussmann R, Rantner W, Cerrito S, and Zwerger W, 2007 *Phys. Rev.* A **75**, 023610
[27] Tajima H, van Wyk P, Hanai R, Kagamihara D, Inotani D, Horikoshi M, and Ohashi Y, 2017 *Phys. Rev.* A **95** 043625
[28] Ishii N, Aoki S, and Hatsuda T, 2007 *Phys. Rev. Lett.* **99** 022001
[29] Ramanan S and Urban M, 2013 *Phys. Rev.* C **88** 054315
[30] Tanizaki Y, Fejős G, and Hatsuda T, 2014 *Prog. Theor. Exp. Phys.* 043I01
[31] Röpke G, Schnell A, and Schuck P, 1998 *Phys. Rev. Lett.* **80** 3177
[32] Schuck P, 2013 *J. Phys.: Conf. Ser.* **436** 012065
[33] Carbone A, Cipollone, Barbieri C, Rios A, and Polls A, 2013 *Phys. Rev.* C **88** 054326

IOP Conf. Series: Journal of Physics: Conf. Series **969** (2018) 012004 doi:10.1088/1742-6596/969/1/012004

# Decay of grid turbulence in superfluid helium-4: Mesh dependence

**J Yang** [1] **and G G Ihas**
Department of Physics, University of Florida, Gainesville, FL 32611, USA
E-mail: jhyang@ufl.edu

**Abstract.** Temporal decay of grid turbulence is experimentally studied in superfluid $^4$He in a large square channel. The second sound attenuation method is used to measure the turbulent vortex line density ($L$) with a phase locked tracking technique to minimize frequency shift effects induced by temperature fluctuations. Two different grids (0.8 mm and 3.0 mm mesh) are pulled to generate turbulence. Different power laws for decaying behavior are predicted by a theory. According to this theory, $L$ should decay as $t^{-11/10}$ when the length scale of energy containing eddies grows from the grid mesh size to the size of the channel. At later time, after the energy containing eddy size becomes comparable to the channel, $L$ should follow $t^{-3/2}$. Our recent experimental data exhibit evidence for $t^{-11/10}$ during the early time and $t^{-2}$ instead of $t^{-3/2}$ for later time. Moreover, a consistent bump/plateau feature is prominent between the two decay regimes for smaller (0.8 mm) grid mesh holes but absent with a grid mesh hole of 3.0 mm. This implies that in the large channel different types of turbulence are generated, depending on mesh hole size (mesh Reynolds number) compared to channel Reynolds number.

## 1. Introduction

There has been intense interest in grid- and thermal counterflow-generated quantum turbulence (QT) in liquid $^4$He contained in channels of late [1–5]. There is an accumulating set of data which seems to hold the promise of understanding some of the similarities and differences between classical and quantum turbulence in such situations. Some theories and simulations have been advanced to begin this understanding [5–9].

A quasi-classical theory predicts the decay of vortex line density of grid turbulence [9, 10]. From the classical energy spectrum, the line density of grid turbulence in a confined geometry is expected to exhibit power law behaviors which evolve with time (with development of length scale of the turbulent eddies). When the energy containing eddies are smaller than channel size (early time), the line density should decay as $t^{-11/10}$. After the eddy size becomes comparable to the size of the confinement (later time), the decay should follow $t^{-3/2}$ behavior which developed from the classical Komogorov-like energy spectrum [9, 11].

We contribute pulled-grid data from by far the largest square channel ever used (4.5 cm x 4.5 cm inner dimension of the channel) for QT over the temperature range 1.5 K to 2.1 K using 2 different grid mesh sizes. Second sound attenuation is used to probe the turbulent intensity. This paper focuses on mesh size and mesh surface finish. The decay of turbulence generated by two square hole mesh grids (Fig. 1) are found to be qualitatively different. However, the decay

[1] Present address: Department of Physics, University of Washington, Seattle, WA 98195, USA

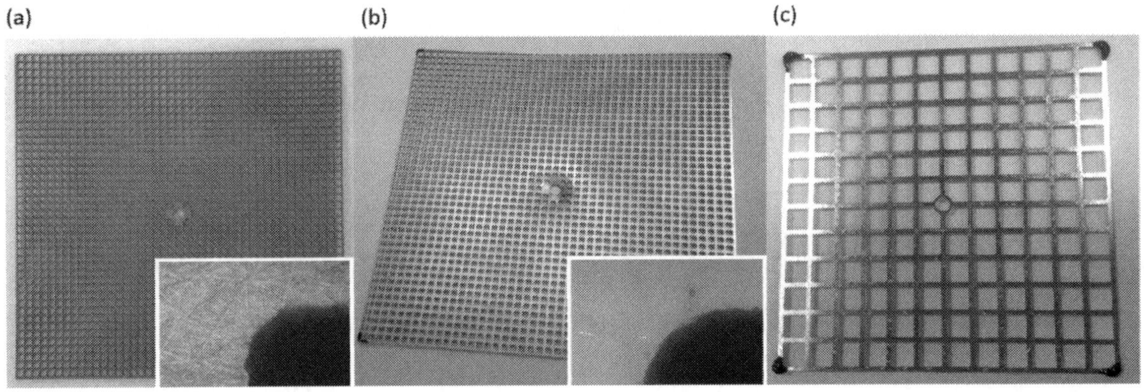


**Figure 1.** Photographs of the three grids in this study. Material is 0.4 mm thick stainless steel. Holes are made by Electric Discharge Machining (EDM) method. (a) 0.8 mm square hole grid with surface finished with #220 emery paper; (b) 0.8 mm hole grid electro-polished; and (c) 3.0 mm hole grid with as-supplied finish. Insets in (a) and (b) show microscopic pictures (x400) of grids. Pull rod is attached at the hole in center of each grid. Black bumpers on corners are Stycast 2850FT to allow smooth transit of the grid through the channel without interfering with sound transducers, which are mounted flush with the inside channel wall.

of turbulence from the grid which is first (a) finished with #220 grit (50 micron) abrasive paper and then later (b) electro-polished to an atomically smooth surface is found to be unchanged.

One interesting aspect of many data is an apparent increase in QT intensity which occurs during the general decay of the QT after the heat has been terminated (in the case of counterflow) or after the pulled grid has stopped (in our work). This appears as a bump in the otherwise monotonic decay of the turbulent intensity under these conditions. We find that the size of this bump is dependent on mesh size in our so-far limited studies. This is surprising since it is not commonly expected that mesh size plays a critical role in the nature of the QT generated. This discovery suggests that, at least in the range of mesh size (0.8 mm and 3.0 mm) and mesh Reynolds numbers (3 x $10^4$ and $10^5$ respectively), fundamentally different types of turbulent structures are being generated in the two situations.

## 2. Experimental setup and procedure

A versatile apparatus has been constructed to investigate quantum turbulence decay inside a 4.5 cm square-cross-section channel. In this paper, we present results with two different square-grid meshes (Fig. 1) pulled through superfluid $^4$He to generate turbulence. The whole apparatus assembly, including the channel, is submerged in a stationary helium bath (Fig. 2). The relatively long length (40 cm) of this channel allows high velocities of grid motion and minimizes possible interfering effects from the top and bottom channel ends on the second sound signal. The size of second sound transducers (diameter= 3.8 cm), mounted on opposite sides of the channel facing each other, are comparable to the channel width. For sensitive temperature regulation, a Ruthenium oxide thermometer and a resistive heater are installed on the channel. They are attached to the outside, on opposite walls, 10 cm from the bottom of channel. All the decay data is taken when the channel is completely submerged in liquid helium II. The bath temperature is regulated by a resistance bridge AVS-46 [12] and a homemade PID controller. Typical temperature regulation range is $\pm 0.06$ mK at 1.8 K except near the instant of grid motion, where the temperature may vary up to $\pm 0.2$ mK.

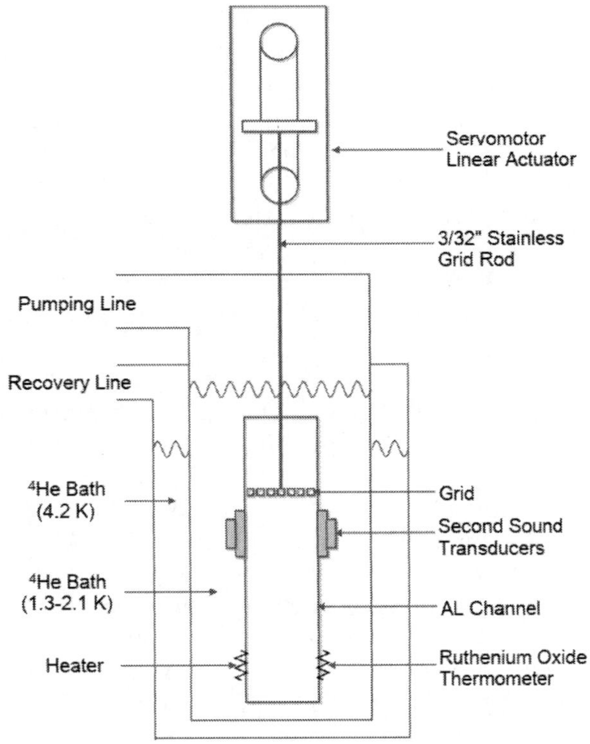


**Figure 2.** Schematic of experimental apparatus including linear servo motor and two dewar system.

The channel and support structure is made of commercial extruded aluminum square pipe to lower its mass. The main support structure immediately above the channel is made of the same commercial pipe material, cut into a triangular lattice shape maintaining robust support through both rotational and translational strains with less mass and little heat leak. Plates with partial sections of finger stocks are placed in between apparatus structures to help center and stabilize them in the dewar.

To extend the run time by reducing heat leak from radiation and conduction, the experimental setup is placed in a double dewar system. The inner dewar contains the main experimental apparatus, including the channel and the regulation heater, etc. The bath temperature of the inner dewar may be varied for the turbulence study between 1.5 - 2.1 K. The outer dewar contains a helium bath at 4.2 K that behaves as a radiation shield (lowering heat leak) and greatly reduces the heat leak coming down the support structure for the inner dewar by thermally connecting the neck of the inner dewar with the outer dewar bath.

A motor system and actuator rod for pulling the grid are mounted on top of the cryostat. A commercial motor controller (Roboclaw 2x15A) [13] and batteries for the system are also mounted right next to the controller. All cables and wires are kept as short as possible to minimize electrical noise. A linear servo motor is used along with an encoder and a tachometer for position and velocity measurements of the actuator rod. The encoder is an optical quadrature encoder that is used for position feedback to establish constant velocities for the grid. A PID control feedback mechanism is used to reach and maintain a given velocity. A Labview code was written to set PID constants, desired stroke distance, velocity profile, and maximum accelerations so the grid could be pulled at constant speed over the longest possible distance, thus promoting the creation of isotropic and homogeneous turbulence in the neighborhood of the sound transducers. A linear power supply was used initially for the motor controller, but it introduced intolerable cross talk noise into other signals, such as temperature and position measurements, so they were replaced with batteries. However, the lower capacity of the batteries limited the maximum velocity of the grid to 60 cm/s.

Two different grids are used for turbulence generation. Both grids have a square cross sectional outline (Fig. 1) with 4.4 cm side length, and the same transparency, 67 %. The gap between grid and the channel wall is about 0.4 mm on each side and the grid has corner bumpers made of Stycast 2850FT [14]. Those bumpers are designed to keep the grid in the center of the channel and minimize rotation of the grid and therefore interference with the transducers when passing them.

The amplitude of a second sound resonance produced spanning the turbulence is measured and converted into turbulence intensity [15, 16]. During the experiment, resonance frequencies could be shifted by temperature fluctuations, since the speed of second sound is strongly temperature dependent ($v_{ss} = f\lambda$, where $v_{ss}$: second sound velocity, $f$: frequency, and $\lambda$: wavelength). This leads to apparent but fictitious vortex line density measurements. To overcome this issue, a second sound tracking system was developed [17] and used in the system. With this technique, the transmitting signal frequency adjusts (tracks) so it always stays on the peak of a desired resonance. A similar tracking scheme was used in other second sound experiments [18, 19]. Our tracking system is a positive feedback circuit. Instead of providing a fixed frequency signal input to drive the second sound resonance, a stable AC oscillation is self-generated inside a closed loop circuit which includes the second sound cavity. When the circuit condition satisfies the Barkhausen criterion [20], the output signal can occur without an external signal generator. It can be designed to oscillate only at a desired resonant frequency and maintaining a constant transmitter signal amplitude.

One directional grid stroke motion is used with constant velocity to produce turbulence. The grid starts from the bottom of the channel and is pulled up at a desired speed by the linear servo motor. The stroke length is 37 cm, which gives a final grid location $\sim$ 0.8 cm below the top of the channel. Once a grid motion is completed, it takes about 5 - 10 minutes to collect the data and return the grid to the original bottom position with speed of 5 cm/s. The SS signal data is taken by the tracking system for the whole duration of the procedure above. Before another motion, the system is allowed to settle for $\sim$ 5 minutes to ensure the turbulence generated in previous motions is completely decayed to the same background level (remnant vorticity). The line density is calculated from individual second sound attenuation data from each grid motion.

## 3. Results and conclusions

We report electro-polishing the grid surfaces made no significant difference in the measured turbulence decay, while changing mesh size showed rather striking and interesting effects on decay behaviors. Figure 3 shows typical vortex line density measurements after grid pulls of 30 cm/s at 1.65 K (the peak in the second sound velocity vs. temperature curve) for two different grid hole sizes (0.8 mm and 3.0 mm). The tracked resonance is 21.6 kHz with Q-value of 1600. The t=0 for vortex line density calculation is set when the grid is passing through the center of transducers assuming that is the average time when turbulence is generated initially throughout the space between the large transducers. The plots are averages of 12 independent grid pulls as described above. Measurements show high reproducible tendency between each grid pull and look very similar at 1.5 K, 1.8 K, and 2.1 K [5]. The straight lines on the log-log plots represent slopes that both are predicted by theory and fit different parts of the data.

The early decay (t<0.3 s) is not shown because of the grid position effect on the sound signal [5] and re-locking time of the lock-in amplifier used in the tracking circuit [17]. The $t^{-11/10}$ dashed line for early time and $t^{-3/2}$ for later decay are for comparison with the theory [9]. The

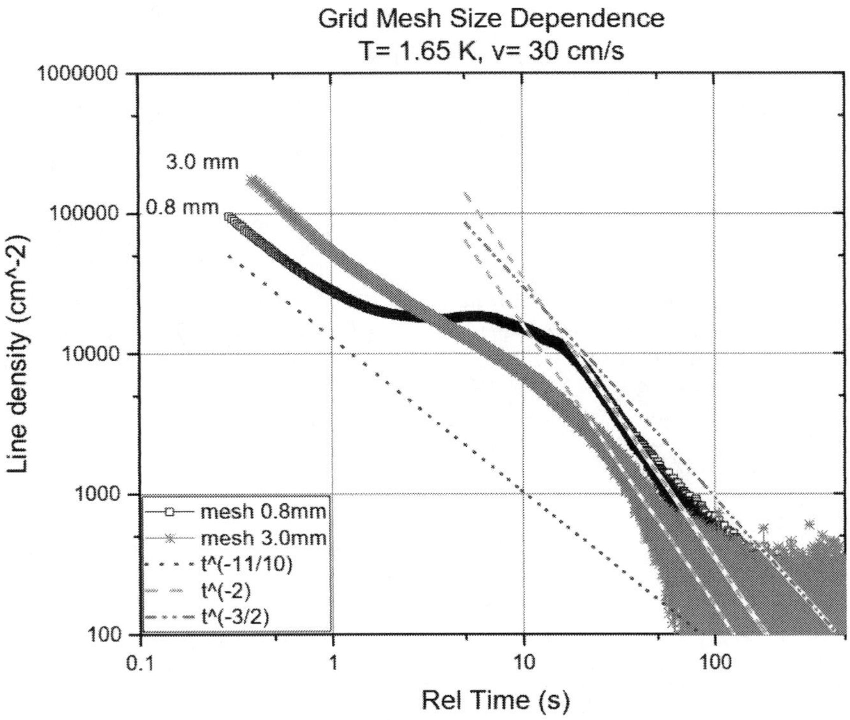


**Figure 3.** Vortex line densities vs. time for turbulence generated by two different mesh sizes (as labeled) at 1.65 K with grid velocity of 30 cm/s. Lines of 3 different slopes are shown for comparison.

later time slope tends toward $t^{-3/2}$ for counterflow turbulence decay (1 cm square channel) [21,22] and our 1 cm square channel measurements [1,5]. These 1 cm channel data display the bump as in our smaller grid hole data in the 4.5 cm channel shown in Figure 3. Another straight line $t^{-2}$ is a better fit to both of our data sets in the large channel at later time. It had been suggested that the $t^{-2}$ later time behavior was due to the bump pushing up the turbulent intensity before the later time decay was measured. By comparing with smaller channel data, we believe the later time behavior depends on the channel size (bigger channel means faster decay), since the smaller grid mesh has a prominent bump and the larger grid mesh does not, but both decay faster than $t^{-3/2}$. This is unexpected. It could be interpreted that the turbulence in the larger channel contains a non-Kolmogorov-like energy spectrum. This is quite different from classical grid turbulence. Clearly more grid mesh sizes need to be studied.

## Acknowledgments

We want to thank Joe Vinen for many insightful discussions. Also we thank Russell Donnelly, Mike Smith, and Kris Johnson for sending pieces of their apparatus. We would like to acknowledge contributions of Pete Axson and Art Greenberg from electronics shop, Greg Labbe and John Graham from cryogenics shop, and Marc Link, Bill Malphurs, Ed Storch, and John Van Leer from machine shop at the University of Florida. Partial support from the US NSF (DMR #1007937) is acknowledged in the early stages of this work.

IOP Conf. Series: Journal of Physics: Conf. Series **969** (2018) 012004 doi:10.1088/1742-6596/969/1/012004

## References

[1] Munday L L 2014 *Quantum Turbulence Generated by Moving Grids* Ph.D. thesis Lancaster University
[2] Marakov A, Gao J, Guo W, Van Sciver S W, Ihas G G, McKinsey D N and Vinen W F 2015 *Phys. Rev. B* **91**(9) 094503
[3] Babuin S, Varga E, Skrbek L, Lvque E and Roche P E 2014 *EPL (Europhysics Letters)* **106** 24006
[4] Varga E, Babuin S and Skrbek L 2015 *Phys. Fluids* **27** 065101
[5] Yang J 2016 *Growth and Decay of Quantum Turbulence in Superfluid Liquid Helium above 1 K* Ph.D. thesis University of Florida (unpublished thesis)
[6] Barenghi C F, Gordeev A V and Skrbek L 2006 *Phys. Rev. E* **74**(2) 026309
[7] Barenghi C F and Skrbek L 2007 *J. Low Temp. Phys.* **146** 5–30
[8] Ikawa S and Tsubota M 2016 *Phys. Rev. B* **93** 184508
[9] Stalp S R, Skrbek L and Donnelly R J 1999 *Phys. Rev. Lett.* **82**(24) 4831–4834
[10] Smith M R, Donnelly R J, Goldenfeld N and Vinen W F 1993 *Phys. Rev. Lett.* **71**(16) 2583–2586
[11] Kolmogorov A N 1962 *J. Fluid Mech.* **13** 82–85
[12] 1991 *Model AVS-46 Resistance Bridge Instruction Manual* (RV-Elektroniikka Oy)
[13] 2014 *RoboClaw Series User Manual* 1st ed (Ion Motion Control)
[14] Loctite stycast 2850 ft description `http://www.henkel-adhesives.com/product-search-1554.htm?nodeid=8802585018369` accessed: 2017-06-18
[15] Hall H E and Vinen W F 1956 *Proc. R. Soc. Lond. A* **238** 204–214
[16] Hall H E and Vinen W F 1956 *Proc. R. Soc. Lond. A* **238** 215–234
[17] Yang J, Ihas G G and Ekdahl D 2017 *Rev. Sci. Instrum.* **88** 104705 (*Preprint* `http://dx.doi.org/10.1063/1.5005948`) URL `http://dx.doi.org/10.1063/1.5005948`
[18] Barenghi C F, Swanson C E and Donnelly R J 1982 *Phys. Rev. Lett.* **48**(17) 1187–1189
[19] Barenghi C 1982 *Experiments on quantum turbulence* Ph.D. thesis Univ. of Oregon,Eugene, OR
[20] Brophy J J 1983 *Basic Electronics for Scientists* 4th ed (McGraw-Hill Book Company) ISBN 0070081336
[21] Skrbek L, Gordeev A V and Soukup F 2003 *Phys. Rev. E* **67**(4) 047302
[22] Gordeev A, Chagovets T, Soukup F and Skrbek L 2005 *J. Low Temp. Phys.* **138** 549–554

IOP Conf. Series: Journal of Physics: Conf. Series **969** (2018) 012005 doi:10.1088/1742-6596/969/1/012005

# Torsional Oscillator Measurements of Liquid $^4$He Confined in 2.5-nm Channel of FSM

**Kento Taniguchi, Junko Taniguchi, Masaru Suzuki**

Department of Engineering Science, University of Electro-Communications, Chofu, Tokyo 182-8585, Japan

E-mail: tany@phys.uec.ac.jp

**Abstract.** We measured the superfluid response of liquid $^4$He confined in 2.5-nm channel under low pressure by means of the torsional oscillator to clarify how the superfluid response varies with decreasing channel diameter. In the previous measurements, in the 2.8-nm channel under 0.13 MPa, the superfluid fraction appears at 1.8 K, and then shows a sharp rise at 0.9 K. On the other hand, in the 2.2-nm channel, no sharp rise is observed although the superfluid fraction appears. In the present measurements for 2.5-nm channel, the superfluid appears at around 1.5 K, and then shows a sharp rise at 0.13 K. By comparison, with decreasing channel diameter, the temperature of superfluid appearance is suppressed with gentle acceleration, which is attributable to the suppression of falling into a BEC-like state. In contrast, the temperature of the rise in the superfluid is suppressed drastically compared with that of superfluid appearance, by decreasing the channel diameter from 2.8 to 2.5 nm.

## 1. Introduction

One-dimensional (1D) quantum many-body systems have attracted many researchers' interest due to large quantum fluctuations. In particular, Liquid $^4$He confined in 1D nanometer-size channel is a promising system to study the superfluidity specific to 1D Bosonic Tomonaga–Luttinger (TL) liquid. [1]

We have experimentally studied the superfluid response of pressurized liquid $^4$He confined in a nanoporous medium called FSM, using the torsional oscillator technique. FSM has a uniform straight channel whose diameter ranges from 2 to 5 nm and can be precisely controlled.[2] In the channel 2.8 nm in diameter, we found that a sharp rise in superfluid at 0.9 K under 0.13 MPa is accompanied by a broad dissipation peak at a resonance (observation) frequency of 2 kHz.[3] When the observation frequency is lowered from 2 to 0.5 kHz, both of the sharp rise and the dissipation peak were shifted to the low temperature by 40 mK.[4] The observed frequency dependence is qualitatively explained by the theory based on the bosonic TL liquid model.[5]

Another important property of superfluid response in the 2.8-nm channel is a two-stage growth of the superfluid; under 0.13 MPa the superfluid fraction already appears at 1.8 K, much higher than the temperature of sharp rise, 0.9 K, and keeps gradual growth between theses two temperatures.[6] Such a two-stage growth is not observed in the system of either 3D bulk liquid or 2D film, suggesting a property specific to $^4$He in 1D channel.

Regarding the two-stage growth, there have been two theoretical studies by means of a quantum Monte Carlo method. Kelchytskyy et al. demonstrated that the superfluid response to the rotational motion of walls of the channel survives to the high temperature, compared with

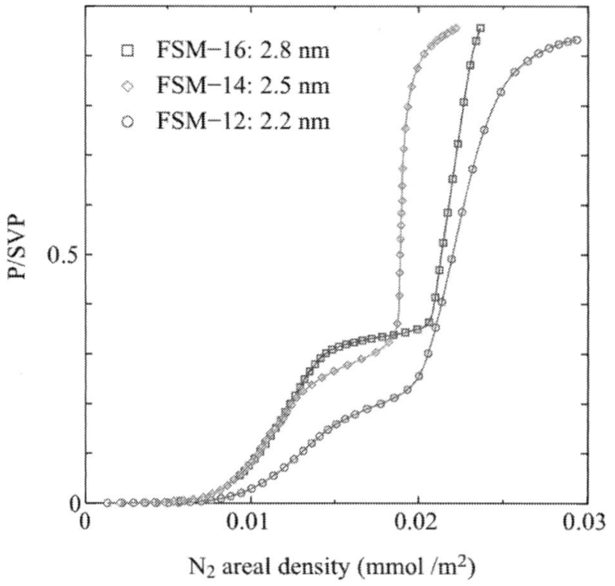


**Figure 1.** $N_2$ isotherms at 77 K.□, ◇, ○ symbols correspond to FSM-16 (2.8 nm), FSM-14 (2.5 nm), and FSM-12 (2.2 nm), respectively. SVP is the saturation vapor pressure.

that to the longitudinal one.[7] Kiriyama et al. also reported the same tendency and showed that the superfluid response to the rotational motion appears at the same temperature as BEC takes place.[8] Both groups pointed out that the superfluid response is strongly suppressed by decreasing channel diameter.

Up to the present, we have studied the superfluid response in the channels whose diameters are 4.7, 2.8, and 2.2 nm.[3, 9] Between 2.8 and 2.2 nm, the superfluid response changes drastically. In the 2.2-nm channel, no sharp rise was not observed and only gradual growth was observed below about 1.2 K under 0.1 MPa. To clarify how the two-stage growth of superfluid varies between 2.8- and 2.2-nm channels, we started the torsional oscillator measurements of $^4$He confined in FSM with 2.5-nm channel. In this paper, we report the superfluid response in the 2.5-nm channel under low pressure, in comparison with the responses in the 2.8- and the 2.2-nm channel, and then discuss the channel diameter dependence of the superfluid response.

## 2. Experimental details

The used substrate is one of FSM-N series. Here, N is the number of alkyl chain in the alkyltrimethylammonium used as a template in synthesis. The channel diameter of FSM is controlled by this number. In the present measurements, we used FSM-14, whose nominal channel diameter is 2.5 nm.[10] The 2.8- and the 2.2-nm channels correspond to FSM-16 and -12, respectively.

When certain kinds of adatoms are introduced into such a narrow channel, a capillary condensation occurs, causing a plateau in the adsorption isotherm. The pressure of the plateau lowers monotonously with decreasing channel diameter. In order to examine how clearly the channel diameter of FSM-14 is different from FSM-12 and -16, we measured the $N_2$ adsorption isotherm. The measurements were performed using a pellet which was sintered after mixing FSM-14 powder with silver one for thermal contact.

Figure 1 shows the adsorption isotherm of FSM-14 at 77 K as a function of $N_2$ areal density, accompanied by those of FSM-12 and FSM-16. Every isotherm has a plateau, whose center

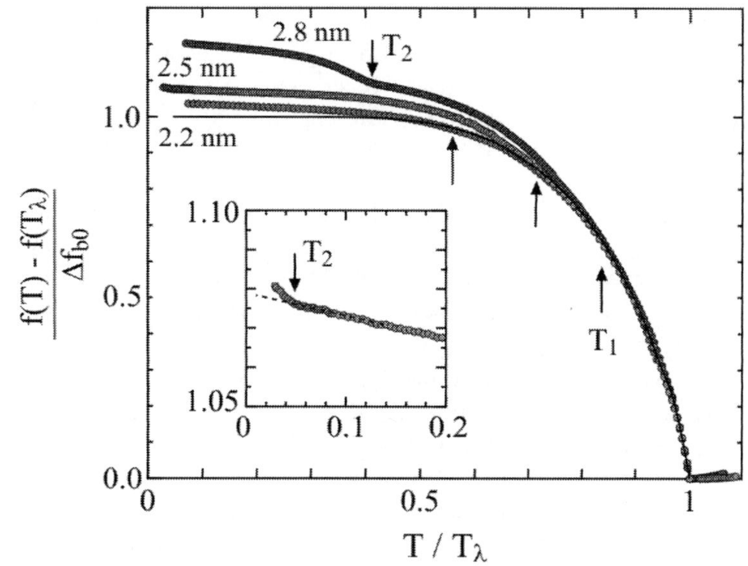


**Figure 2.** Change in the resonance frequency from $T_\lambda$ for 2.5-nm channel filled with liquid $^4$He under 0.05 MPa is plotted against the temperature divided by $T_\lambda$. The data for 2.8-nm (0.01 MPa) and 2.2-nm (0.1 MPa) channels are from Ref. 3. The frequency change is divided by the contribution of bulk superfluid outside the FSM powder at absolute zero. The solid curve is the contribution of bulk superfluid. Inset is the enlarged figure of the low-temperature range for 2.5-nm channel. The dotted line is the extrapolation from the high-temperature side. Upward and downward arrows point $T_1$ and $T_2$ in the text, respectively.

pressure is clearly separated. The values of center pressure indicate that the difference of channel diameter between the neighboring samples is 0.3–0.4 nm. The plateau of FSM-16 ranges the widest areal density region, corresponding to the largest channel volume. As the channel diameter decreases, the areal density region of the plateau gets narrow, indicating that the ratio of the channel volume to the surface area becomes small.

We put this pellet into a torsional oscillator to measure the superfluid response. The total surface area of the pellet is evaluated as 77 m$^2$ by fitting the $N_2$ adsorption isotherm at 77 K to the Brunauer–Emmett–Teller equation in the pressure range of $0.05 \leq$ P/SVP $\leq 0.15$. Here SVP is the saturation vapor pressure. The torsional oscillator is made of Be-Cu, and oscillates at a resonance frequency $f$ of 1774.977 Hz, with a quality factor $Q$ of $7.3\times10^5$, at 0.1 K under the condition that the channel is filled with liquid $^4$He under 0.05 MPa.

## 3. Results and Discussion

Figure 2 shows the change in resonance frequency from $T_\lambda$ for the samples of 2.8-, 2.5-, and 2.2-nm channel filled with liquid $^4$He at low pressure. The abscissa is the normalized temperature divided by $T_\lambda$, and the frequency change in the ordinate is divided by the contribution of superfluid outside powders at absolute zero, $\Delta f_{b0}$. Here, as the contribution of superfluid outside powders, we used the superfluid density calculated from the Landau expression by Brooks and Donnelly [11] and fitted the experimental data at the high-temperature side.

As the temperature is decreased, the resonance frequency for 2.8-nm starts to deviate upward from the bulk contribution at 0.83 $T_\lambda$ (1.8 K), due to the appearance of superfluid in the channel. This deviation temperature ($T_1$) is suppressed with decreasing channel diameter, as 0.71 $T_\lambda$

(1.5 K) for 2.5-nm and 0.56 $T_\lambda$ (1.2 K) for 2.2-nm. Further decreasing temperature, a rise of frequency appears at 0.41 $T_\lambda$ for 2.8-nm, corresponding to the second rapid growth of superfluid. In the case of 2.5-nm channel, as shown in the inset of Fig. 2, a small rise appears at around $0.06T_\lambda$ (0.13 K). For 2.2-nm, no rise was observed at least down to 0.16 K, the lowest temperature we measured.

First, we discuss the appearance of superfluid at $T_1$. In the case of 2.8-nm channel, it was made clear that liquid $^4$He in the channel falls into a "BEC-like" state[12] at almost the same temperature as $T_1$, from the heat capacity measurements.[6] On the other hand, in the quantum Monte Carlo calculation by Kiriyama et al., the BEC temperature is about 1.7 K for 2.8-nm channel, and suppressed with decreasing channel diameter. [8] In the present experiments, $T_1$ is suppressed with gentle acceleration, and the obtained values of $T_1$ are close to the calculated BEC temperature. This coincidence supports that in the channel 2~3 nm in diameter the superfluid appears at the same time as the "BEC-like" state appears.

Next, we focus on the sharp rise in superfluid at $T_2$. $T_2$ is suppressed drastically by decreasing the channel diameter from 2.8 to 2.5 nm. It may be related to a change in the number of liquid layers in the channel. In the channel, inside the two-atomic inert solid layer tubes, liquid layers are formed. According to the calculation of Kulchytskyy et al., there exist the third and the fourth liquid layer tubes in the 2.8-nm channel. As the channel diameter is decreased from 2.6 to 2.4 nm, the fourth layer tube shrinks into a line, and then disappears at 2.2 nm. In 2.0-nm channel, only the third liquid layer tube exists. In order to detect the superfluid along the channel, the superfluid coherence has to be established throughout the channel length. The decrease in the number of liquid layers can make this superfluid coherence fragile drastically, since it drops the number of candidates for particle exchange discretely. In fact, in the calculation, the superfluid response also varies discretely between 2.6 and 2.4 nm.

Another point to be noted is that the observed $T_2$ (0.13 K) is significantly lower than the temperature of the superfluid rise, ~0.4 K, for a thick film observed by Matsushita et al. [13] In the case of 2.8 nm, $T_2$ is suppressed by about 0.2 K by increasing $^4$He amount from 20.4 atoms/nm$^2$ to the liquid under 0.13 MPa.[14] This suppression is attributable to the increase of density in the channel. Although the reason why the suppression is stronger for 2.5-nm channel than for 2.8-nm, this tendency implies that the second rise of liquid $^4$He in the 2.2-nm channel is further suppressed.

In 2.8-nm channel, the second rise has a strong observation frequency dependence, and is thought as a signature of 1D TL liquid behavior.[4] It is a future issue to study whether the $^4$He in the 2.5-nm channel also has a strong frequency dependence.

## 4. Summary

We measured the superfluid response of liquid $^4$He confined in 2.5 nm channel under low pressure and compared with those for 2.8- and 2.2-nm channels. We found that the temperature where the superfluid appears in the channel, is suppressed with gentle acceleration, which is attributable to the suppression of BEC. On the other hand, the temperature of the sharp rise in the superfluid is suppressed much faster than that of superfluid appearance, by decreasing the channel diameter from 2.8 to 2.5 nm.

*Acknowledgments*
We would like to thank S. Inagaki for supply of the sample of FSM. This work was partly supported by KAKENHI 26400352.

## References

[1] Cazalilla M A, Citro R, Giamarchi T, Orignac E and Rigol M 2011 *Rev. Mod. Phys.* **83** 1405
[2] Inagaki S, Fukushima Y and Kuroda K 1993 *J. Chem. Soc. Chem. Commun.* **22** 680

[3] Taniguchi J, Aoki Y and Suzuki M 2010 *Phys. Rev.* B **82** 104509
[4] Taniguchi J, Demura K and Suzuki M 2013 *Phys. Rev.* B **88** 014502
[5] Eggel T, Cazalilla M A and Oshikawa M 2011 *Phys. Rev. Lett.* **107** 275302
[6] Taniguchi J, Fujii R and Suzuki M 2011 *Phys. Rev.* B **84** 134511
[7] Kulchytskyy B, Gervais G and Maestro A D 2013 *Phys. Rev.* B **88** 064512
[8] Kiriyama H, Taniguchi J, Suzuki M and Takagi T 2014 *J. Phys. Soc. Jpn.* **83** 044601
[9] Taniguchi J and Suzuki M 2008 *J. Low Temp. Phys.* **150** 347
[10] When the sample was supplied to us, the channel diameter was evaluated as 2.5 nm. However this value seems to have an ambiguity of about 0.2 nm, becuase the precise estimation method has not been established yet. The used sample is the same batch of that used in the work by Matsushita et al.[13], where the channel diameter is displayed as 2.4 nm.
[11] Brooks J S and Donnelly R J 1977 *J. Phys. & Chem.* Reference Data **6** 51
[12] At present, it is not clear whether the BEC state and a bosonic TL liquid state can coexist. From the heat capacity measurements, the existence of a low-entropy state is confirmed, which we call a "BEC-like" state.
[13] Matsushita T, Shinohara A, Hieda M and Wada N 2016 *J. Low Temp. Phys.* **183** 273
[14] Submitted for publication.

IOP Conf. Series: Journal of Physics: Conf. Series **969** (2018) 012006 doi:10.1088/1742-6596/969/1/012006

# Recombination of radiation defects in solid methane: neutron sources and cryo-volcanism on celestial bodies

**O. Kirichek[1], E.V. Savchenko[2], C.R. Lawson[1], I.V. Khyzhniy[2], D.M. Jenkins[1], S.A. Uyutnov[2], M.A. Bludov[2], D.J. Haynes[1]**

[1]ISIS Neutron and Muon Source, Science and Technology Facilities Council, Rutherford Appleton Laboratory, Didcot, OX11 0QX, UK
[2]Institute for Low Temperature Physics & Engineering NASU, Kharkiv 61103, Ukraine

oleg.kirichek@stfc.ac.uk

**Abstract**. Physicochemical properties of solid methane exposed to ionizing radiation have attracted significant interest in recent years. Here we present new trends in the study of radiation effects in solid methane. We particularly focus on relaxation phenomena in solid methane pre-irradiated by energetic neutrons and electron beam. We compare experimental results obtained in the temperature range from 10K to 100K with a model based on the assumption that radiolysis defect recombinations happen in two stages, at two different temperatures. In the case of slow heating up of the solid methane sample, irradiated at 10K, the first wave of recombination occurs around 20K with a further second wave taking place between 50 and 60K. We also discuss the role of the recombination mechanisms in "burp" phenomenon discovered by J. Carpenter in the late 1980s. An understanding of these mechanisms is vital for the designing and operation of solid methane moderators used in advanced neutron sources and could also be a possible explanation for the driving forces behind cryo-volcanism on celestial bodies.

## 1. Introduction

Solid methane is proven to be the most promising material for moderating hot neutrons produced by the target of a spallation neutron source, or in the core of nuclear reactor based neutron source. The conversion of short wavelength to long wavelength neutrons using solid methane is ~ 3.5 times more efficient than that achieved in liquid hydrogen based moderators [1]. However the history of using solid methane in cryogenic moderators knows numerous failures caused by large, seemingly spontaneous, releases of energy. This energy release is due to recombination of frozen-in products of radiolysis and the expansion of hydrogen, which builds up in the solid methane during irradiation by neutrons [2].

The physiochemical properties of solid methane and the recombination events which occur due to radiation may also prove interesting to the planetary physics community, particularly when describing comets.

According to R. Miles [3, 4] there are large cavities below a considerable fraction of the comet's surface containing methane rich solid mixtures under high pressure. This complex physicochemical system can sustain exothermic reactions at cryogenic temperatures from 40 to 200K which can provide the energy necessary for cryo-volcano eruption [3]. Energy released by the recombination of radicals generated by cosmic radiation could play significant role in this mechanism.

Here we present and discuss the results of commissioning tests of solid methane moderators at the ISIS neutron source, along with recent results obtained by experiments with solid methane samples irradiated by fast electrons. The temperature range of the moderator commissioning tests was 37 – 100K and in fast electron irradiation experiments 10 – 100K. The evidence presented here allow us to suggest that the process of radiolysis defect recombination happens in two stages, at two different temperatures. The first "low temperature" recombinations occur at around 20K, with the second "high temperature" process taking place between 50K and 60K. Consequences of this behavior for designs of solid methane moderators and cryo-volcanism on celestial bodies are also discussed.

## 2. Neutron moderators based on solid methane

The first neutron moderator based on solid methane was developed, commissioned and successfully operated for more than a decade by the IPNS neutron source at Argon National Laboratory [1, 2]. After a few initial tests it became clear that solid methane possesses superior neutronic properties to other materials, but at the same time it has reviled itself as a very challenging material from an operational point of view. As discovered by J. Carpenter in the late 1980's the accumulation of radiolysis products in low temperature (below 20K) solid methane, when exposed to a flux of neutrons, can lead to a spontaneous self-accelerated heating process, also known as the 'burp' phenomenon. It was also suggested that the 'burp' effect, together with the expansion of hydrogen (also one of the products of radiolysis), was responsible for the large number of failures of solid methane moderators. Later, solid methane exposed to neutron radiation was thoroughly studied at the IBR neutron source in Dubna [5]. The results of this research have confirmed the relevance of Carpenter's model, and give some quantitative estimation of the scale of the process.

The design of the Target Station 2 (TS2) solid methane moderator, commissioned by ISIS neutron source in 2009, was based on the conclusion that all recombination processes are completed at temperatures below 40K [1, 2, 5]. However the very first test of the moderator, operated at 38K, ended due to a burp-like effect which damaged the moderator. This observation suggested that there might be other species of radiation defects, recombining at different temperatures.

As shown in [1, 2, 5 - 7] the most substantial final products of sequential reactions of methane radiolysis are molecular hydrogen $H_2$ and ethane $C_2H_6$. Fast neutron collision with methane molecules results in a damage cascade in the irradiated methane, producing atomic hydrogen, $CH_3$ radicals and other species of radiolysis that remain dormant or 'frozen' into the cryogenic matrix of the solid methane. With rising temperature the defects start to move and (once they get close to each other) recombine, releasing significant energy in the exothermic reactions: $H + H \rightarrow H_2 + 218$ kJ/mole and $CH_3 + CH_3 \rightarrow C_2H_6 + 368$ kJ/mole [2]. These are the most probable reactions, however there are others [7] some of which lead to the production of $H_2$ and $C_2H_6$. The release of energy in the recombination events initiate heating of the cryogenic matrix and, as a result, to the acceleration of thermally activated diffusion which in turn drives the 'burp' effect. The key parameter of this process is the mobility of radiolysis products H and $CH_3$ in the cryogenic matrix of solid methane.

Another mechanism which might activate the 'burp' effect occurs when the density of frozen-in defects (at low temperatures) exceeds a critical value, when defects become close enough to each other to react. In this situation the 'burp' effect starts spontaneously, even at the lowest temperatures.

As it was suggested in [2] all recombination reactions are governed by the same activation energy for defect diffusion; $\Delta E$ in equation (3) of that paper, which leads to self-accelerated heating at around 20K [2]. However during commissioning tests of the ISIS TS2 solid methane moderator we have observed a similar 'burp' effect which starts above 40K and reaches a maximum in heat release at around 60K. This allows us to suggest that the radiolysis defect recombination process happens in two stages, at different temperatures, and is therefore governed by different activation energies. In order to check this supposition we have modified the model used by Carpenter (equations (1 – 3) in [2]) based on the theory of thermal explosion [8]. We have introduced a second reaction with the rate coefficient having Arrhenius-form temperature dependence and an activation energy $\Delta E_2 \approx 200$K. Our model uses the solid methane thermal conductivity data from [9] and heat capacity from [10]. The amount of solid methane in the first prototype of the ISIS solid methane moderator was 12.5 moles. During tests the moderator was irradiated by a neutron flux ~ $10^{16}$ neutron $cm^{-2}s^{-1}$ at a temperature of 38K for periods of between 8 to 24 hours. In three instances the self-accelerating heating was triggered at ~ 50K due to warming of the moderator with the intention to anneal the solid methane. In all three cases the moderators were damaged at temperatures around 65K.

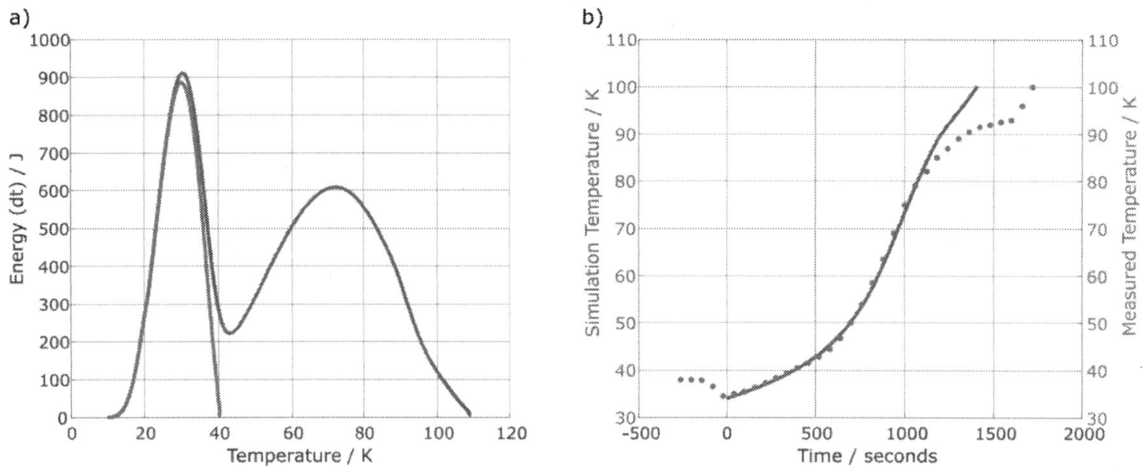


Fig 1 (a) Energy release during slow heating of the solid methane irradiated at 10K. The blue line represents our model, red line is the Carpenter model; (b) TS2 solid methane moderator test data (red dots) and time dependence of solid methane temperature obtained as an outcome of our model (blue line).

The temperatures corresponding to two maxima of energy release during slow heating of the solid methane irradiated at 10K can be determined from the Fig. 1. The blue curved line (our model) shows two distinctive maxima of energy release at 30K and 70K, contrary to the red curved line obtained from the Carpenter model showing only one maximum at 30K. One of the interesting observations is that the energy release at 30K happens much faster (hence the sharper feature) than that which happens at higher temperatures. This different dynamic could be explained by the interplay between temperature dependent parameters of the model, such as heat capacity, thermal conductivity or mobility of radiation defects.

In Fig. 1 (b) we compare experimentally obtained self-accelerating heating of the ISIS TS2 solid methane moderator (dots) with our model (solid curved line). During testing the solid methane moderator was exposed to neutron flux for 8 hours at a temperature of 38K. The proton beam was then switched of and the circulation of coolant within the moderator reduced. This causes a rise in temperature with the intention to anneal the solid methane and release hydrogen accumulated in the cryogenic matrix. This process is necessary as the buildup of defects causes a deterioration of the methane's thermal conductivity, and therefore significantly reduces the performance of the moderator over time. During this procedure initial slow heating lead to an accelerated event at around 50K and achieved the highest heating rate at around 65K. The model agrees well with experimental results and starts to deviate only above 80K where solid methane approaches its melting point at 90.7K. This deviation might be explained by not including the latent heat of solid methane melting into the model.

## 3. Relaxation phenomena in solid methane pre-irradiated with fast electrons

In the second series of experiments we study the radiation defect relaxation mechanisms in solid methane films pre-irradiated with an electron beam.

The methane solids were grown by deposition of $CH_4$ gas (of purity 99.97%) onto a cooled Cu substrate mounted in a high-vacuum chamber with a base pressure of $10^{-8}$ mbar. The technique was similar to that described in [11]. Untrapped electrons (promoted to the conduction band by heating) escape from the sample yielding thermally stimulated exo-electron emission. Stimulated currents were detected with an electrode kept at a small positive potential $V_F = + 9$V and connected to the current amplifier. The sample was irradiated with a 1keV electron beam of 1.2 mAcm$^{-2}$ current density in dc regime. The sample heating under electron beam did not exceed 0.5K. The cathodoluminescence (CL) spectra were detected concurrently in the visible range and in the vacuum ultraviolet (VUV) using two spectrometers. The CL of solid methane steeply increased upon switching on the beam, reached a maximum and then decayed exponentially, indicating radiation-induced chemical modification of the film. Upon completion of the electron beam exposure an after-emission of electrons was detected, pointing to an electrostatic charging of the $CH_4$ film. When the after-emission current had decayed to essentially zero, measurements of thermally stimulated relaxation emissions were performed in the temperature range 5 – 80K. During a controlled warm-up of the solid methane sample (with a heating rate of 5K min$^{-1}$) three thermally stimulated effects were monitored: VUV and visible photon emission, an exo-electron emission and the ice's particle ejection. In view of the absence of thermally stimulated luminescence (TSL) from pure solid $CH_4$ only the emissions of electrons and particles have been measured in a correlated fashion.

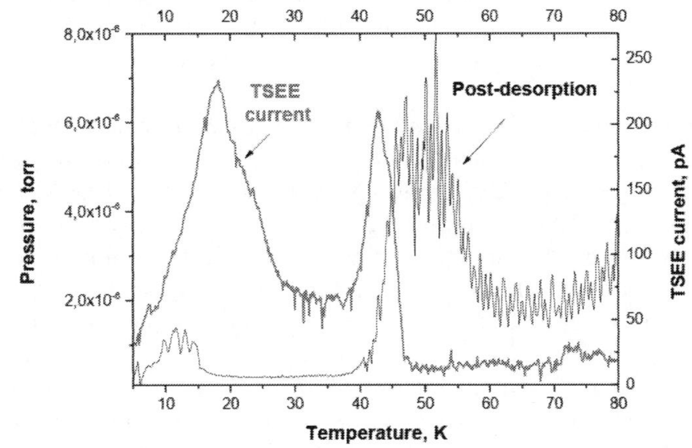


Fig 2 Temperature dependences of thermally stimulated exo-electron emission TSEE (red line) and post-desorption of solid methane or products of radiolysis (black line).

In Fig.2 we present temperature dependences of thermally stimulated exo-electron emission (TSEE) (red line) and post-desorption of solid methane or products of radiolysis (black line). The TSEE signal (two peaks at about 18 and 43K) was detected from solid methane for the first time. The first peak related to defects of structure represents in fact a multi-peak curve with unresolved structure. A rough estimation of the averaged activation energy $E_{act}$ by the half-width method [12] gives $E_{act}\sim$ 10meV. $E_{act}$ for the second peak at 43K is about 70 meV. A check on the presence of TSL made in the visible and VUV range verified its absence in pure solid methane. First peak of particle ejection (post-desorption) was found in solid methane at temperatures much lower than the characteristic sublimation temperatures. Another double peak in post-desorption signal was observed at higher temperatures – between 45 and 60K.

## 4. Discussion and conclusions

The radiation defect recombination process observed around 20K is well known and has been a subject of thorough investigation since the 1980s [1, 2, 5 - 7]. However the second exothermal recombination around 50K has manifested itself quite unexpectedly and has given rise to a number of

IOP Conf. Series: Journal of Physics: Conf. Series **969** (2018) 012006 doi:10.1088/1742-6596/969/1/012006

questions about the physical mechanisms behind this process. It is worth mentioning that anomalies in the same temperature range have also been observed in other parameters of solid methane. As shown in [13, 14] the transport and mechanical properties of solid methane undergo significant changes at temperatures around 50K [14]: in the 50K – 90K temperature range solid methane is soft and sticky, but below 50K solid methane loses its stickiness and ductility and behaves more or less like an ordinary window glass. In [13] it was recently suggested that this change might be caused by quantum effects of rotational degrees of freedom and their collectivization. However, irrespective to the microscopic mechanism which governs this process, it is clear that dynamics of crystal defects such as dislocations, vacancies or impurities experience dramatic change in this temperature range.

According to the scenario suggested here, in the low temperature limit only 'light' defects (such as atomic hydrogen, protons or electrons) are capable of thermal diffusion, and therefore recombination. All massive defects, such as $CH_3$, remain motionless. If the temperature starts to increase above 20K the recombination rate of 'light' mobile defects rises, following an Arrhenius thermal activation dependence. In the case where the accumulated number of defects is significant, this process could self-accelerate: Triggering the 'burp' phenomenon. After all mobile defects have recombined the heating driven by 'light' defect recombination stops. If the temperature increases further and reaches 50K, where the 'heavy' defects begin to move, the self-accelerated heating process repeats itself again, with a different thermal activation energy in the Arrhenius temperature dependence. We accept that the suggested scenario is quite speculative and more information, particularly related to microscopic dynamics of radiation defects is required. Nevertheless, the suggested model describes test results reasonably well and allows us to propose some modifications to the design of solid methane moderators.

The first, and most important, issue is that the higher temperature self-accelerated heating happens in the temperature range 40K to 70K where solid methane undergoes dramatic thermal expansion [15]. The expansion (if it happens quickly) might generate significant pressure on the walls of the moderator vessel, which could lead to structural damage. In order to prevent this buildup of pressure it may be possible to rapidly warm the coolant circulated through the cooling loop, prior to beginning of the annealing process. This quick heating could allow warming up of the solid methane layer close to the surface of the heat exchanger. The rest of the bulk solid methane is expected to remain at low temperature due to the poor thermal conductivity of solid methane [9]. At temperatures above 50K solid methane becomes soft [14] and can be squeezed out of the warm layer by the rising pressure, releasing the pressure on the moderator walls. The second option would be to never fill the moderator with methane up to 100% and always leave some empty space, where soft methane from the warm layer can be squeezed out. The last approach has been already incorporated into the ISIS TS2 solid methane moderator operational procedure.

In addition to this, in the case of the ISIS TS2 moderator, it was decided to increase the thickness of the aluminum walls of the vessel from 3 mm to 10 mm. Such radical design change has completely eliminated the possibility of moderator damage, but at the same time significantly reduced the cold neutron flux.

In addition to the application of defect recombination models in solid methane, Carpenter has previously suggested that the 'burp' phenomenon might play some role in generating jets, such as those observed from comets [2]. Comets spend a significant amount of time in the low temperature environment of the Kuiper Belt and the Oort Cloud, exposed to the continuous radiation of outer space. This allows bulk solid methane inclusions (whose existence is suggested in [3]) to accumulate a significant amount of radiation defects, which in turn could lead to spontaneous self-accelerated heating. However the role of this phenomenon in cryo-volcanism on comets was downplayed later [3] mainly on the basis of differences in temperature ranges: the 'burp' effect takes place around 20K, but the temperature of a comet is never thought to drop below 30K, with cryo-eruptions usually happening above 50K. According to R. Miles [3, 4] this cryo-volcanism is mostly driven by the release of the enthalpy of solution, also known as 'heat of solution' [3]. However, to start the process the system would still require an initial heat release in combination with the generation of pressure, in many cases around 1 kPa. The higher temperature self-accelerated heating effect observed here at 50K provides the necessary heating and pressure to trigger the eruption of a cryo-volcano. Furthermore, we would suggest that radiation defect recombination in water ice [5] might also play noticeable role in cryo-volcanism on outer celestial bodies.

We conclude that the process of radiolysis defect recombination happens in two stages at different temperatures. First the "lower temperature" recombination (also known as the 'burp' phenomenon) occur at around 20K and the second "higher temperature" self-accelerated heating process takes place between 50 and 60K. We also discussed the consequences of this observation in the design and operation of solid methane moderators, as well as its role as a possible trigger of cryo-volcanism on celestial bodies such as comets.

**Acknowledgments**
We thank our colleagues J. M. Carpenter, E. P. Shabalin and S. A. Kulikov for their active interest and useful discussions and M. I. Bagatskii for sharing the solid methane thermodynamic data with us. The work done at ISIS was supported by the project TUMOCS. This project has received funding from the European Union's Horizon 2020 research and innovation program under the Marie Skłodowska-Curie Grant No. 645660.

**References**
[1] Scott T L, Carpenter J M and Miller M E 1999 The development of solid methane neutron moderators at the intense pulsed neutron source facility of Argonne National Laboratory *Preprint* ANL/IPNS/CP-98533
[2] Carpenter J M 1987 *Nature* **330** 358
[3] Miles R 2016 *Icarus* **272** 356-86
[4] Milesl R 2016 *Icarus* **272** 387-413
[5] Kulagin E, Kulikov S, Melikhov V and Shabalin E 2004 *Nuclear Instruments and Methods in Physics Research B* **215** 181-6
[6] Shabalin E P 1995 *JINR Commun.* E17-95-142
[7] Foti G, Calcagno L, Sheng K and Strazzulla G 1984 *Nature* **310** 126
[8] Bowden F P and Yoffe A D 1958 *Fast reactions in Solids* (London: Butterworth)
[9] Manzhelii V G and Krupskii I N 1968 *Fiz. Tverd. Tela* **10** 281
[10] Clusius K *Z.* 1929 *Phys. Chem.* **B3** 41
[11] Savchenko E V, Khyzhniy I V, Uyutnov S A, Ponomaryov A N, Gumenchuk G B and Bondybey V E 2013 *Low Temp. Phys.* **39** 446-50
[12] Vij D R 1998 *Thermoluminescence in: Luminescence of Solids* ed D R Vij (New York: Plenum Press) 271
[13] Zakharov A Yu, Leont'eva A V, Prokhorov A Yu and Erenburg A I 2017 *J. of Low Temp. Phys.* **187** 140-7
[14] Kirichek O, Church A J, Thomas M G, Cowdery D, Higgins S D, Dudman M P and Bowden Z A 2012 *Cryogenics* **52** 325–30
[15] Bol'shutkin D N, Gasan V M and Prokhvatilov A I 1971 *Zhurnal Strukturnoi Khimii* **12** 734-6

# Luminescence of molecular nitrogen nanoclusters containing stabilized nitrogen, oxygen, hydrogen, and deuterium atoms.

**McColgan P T, Meraki A, Boltnev R E*, Sheludiakov S, Lee D M and Khmelenko V V**

Department of Physics and Astronomy and Institute for Quantum Science & Engineering, Texas A&M University, College Station. TX 77843, USA
* Branch of Talrose Institute for Energy Problems of Chemical Physics, Russian Academy of Sciences, Chernogolovka, Moscow Region, 142432, Russia

E-mail: PatrickMcColgan@physics.tamu.edu

**Abstract.** We observed the appearance of a broad feature at 360 nm in spectra obtained during the destruction of ensembles of molecular nitrogen nanoclusters containing stabilized nitrogen, oxygen, hydrogen and deuterium atoms. We attribute this broad feature with a maximum at 360 nm to the $N_4$ polynitrogen molecules which are the product of the neutralization reaction of $N_4^+$ ions with electrons in solid $N_2$.

## 1. Introduction

Relaxation processes in solid nitrogen containing stabilized atoms are of high interest in diverse fields of sciences. Nitrogen and nitrogen containing ices are present in large abundance in cosmic space [1], being exposed to solar wind and galactic cosmic rays. Another aspect of interest is related to polynitrogen compounds which are high-energy density materials (HEDM)[2, 3]. HEDM materials with local concentrations of stabilized nitrogen atoms as high as $2\times10^{21}$ cm$^{-3}$ have been obtained in porous collections of nitrogen nanoclusters immersed in superfluid helium [4, 5]. Most of the spectroscopic results obtained from studies of solid nitrogen containing stabilized atoms were explained by diffusion and chemical reactions of atoms and molecules in the process of annealing of solid matrices [6–9]. Only recently was the important role of radiation-induced charged particles (electrons and ions) suggested for these processes [10–15]. The explanation for some features in the observed luminescence spectra are still controversial. As an example, observed emission band at 360 nm had been assigned to the emission from the $N_4^*$ polymer resulting from an $N_4^+$ neutralization process in solid $N_2$ irradiated by electrons [14]. However, earlier the band at 360 nm had been assigned to the $A^3\Pi, v' = 0 \rightarrow X^3\Sigma^-, v'' = 1$ transition of NH or ND radicals [16]. The NH radical has been observed by several methods. Also detected in the interstellar medium in 1991 [17], this radical has sparked the interest of astronomers and astrophysicists.

In this work we performed experimental studies of luminescence of nitrogen nanoclusters containing stablilized nitrogen atoms with small admixtures of hydrogen or deuterium atoms. The sample preparation method we used was the injection of the discharge products of nitrogen-helium or nitrogen-rare gas-helium mixtures into bulk superfluid (HeII) [18, 19]. This method creates nanoclusters of characteristic size $\sim$ 5 nm with an overall density of impurity atoms and molecules in HeII of order $10^{20}$ cm$^{-3}$ [20–22]. Nitrogen, oxygen, hydrogen, and deuterium atoms were formed in the discharge zone and were captured mostly on the surfaces of the molecular nitrogen nanoclusters [5, 23]. We found that a weak band at $\lambda = 360$ nm was present in the spectra observed during destruction of the nitrogen nanoclusters containing stabilized N and small admixtures of O atoms [9, 24]. This band is enhanced if hydrogen or deuterium atoms are present in the system. The addition of rare gases in the system did not significantly change the position of this band. On the basis of experimental observations described in this work, we conclude that the band at $\lambda = 360$ nm in the spectra obtained during the destruction of molecular nitrogen nanoclusters containing stabilized atoms, results from emission by $N_4$ polynitrogen.

## 2. Experimental Apparatus and Technique

The cryogenic part of the experimental setup consists of two silvered glass Dewars. Each Dewar has two slits for optical access. The inner Dewar is filled with liquid helium. By pumping on the helium bath with an Edwards E2M80 rotary vacuum pump, temperatures $\sim$ 1.1 K are achievable.

The samples are created by injecting gas mixtures through a radio frequency (RF) discharge into a quartz beaker filled with HeII. The hydrogen-nitrogen-helium gas mixtures are prepared at room temperature in a gas handling system using Linde Electronics&Specialty research grade helium gas (99.9999% purity). We also used research purity nitrogen and hydrogen gases (99.9999%) from Matheson and deuterium gas (99.6% $D_2$ and 0.4% HD) from Cambridge Isotope Laboratory Inc. The oxygen content in the gas mixtures $\sim$ 1 ppm results from contamination in this helium gas. A steady flux of $5 \times 10^{19}$ atoms and molecules per second is maintained by a Brooks Instrument 5850E flow controller. The gas mixture enters the cryogenic system through a quartz capillary cooled by liquid nitrogen ($LN_2$). The RF discharge ($f \sim 50$ MHz, power $\sim$ 75 W) is provided by a HP 8556B signal generator amplified by an E&I 3100L amplifier connected to electrodes surrounding the bottom of the quartz capillary. After passing through the discharge zone, the dissociated impurity atoms and excited molecules along with helium gas exit the quartz capillary through an orifice diameter $\sim$ 0.75 mm and enter the helium Dewar. The pressure gradient $\sim$ 2 Torr creates a well-formed jet which is injected into a beaker of superfluid HeII placed 2 cm below the orifice. The helium level in the beaker is maintained by a thermo-mechanical fountain pump which pumps HeII from the bottom of the main helium bath. During sample preparation a temperature of 1.5 K was maintained. The temperature inside the beaker was measured using a germanium thermometer.

After the sample is created, the fountain pump is turned off, ceasing the flow of HeII into the sample collection beaker. Over the course of $\sim$ 15 - 20 minutes the liquid helium exits the beaker via film creep and evaporation, leaving a "dry" sample. At this moment, the pumping line to the Dewar is closed off, allowing the temperature of the sample to increase from 1.2 to $\sim$15 K in $\sim$ 50 s, and to initiate destruction of the sample, which is accompanied by thermoluminescence with a series of bright flashes.

The registration system for the luminescence spectra in the experimental setup consists of a bifurcated optical fiber which transfers the light emitted from the sample to the Andor Shamrock SR-500i and Ocean Optics HR2000+ spectrometers [24]. The Ocean Optics spectrometer can record spectra over a large wavelength range from 200 to 1100 nm, with spectral resolution 1.3 nm (generally used with a registration time from 100 to 300 ms). The Andor Shamrock SR-500i spectrometer with Newton EMCCD camera was used to obtain high resolution (0.53 nm, 1st grating) spectra. The higher sensitivity of the Andor spectrometer allows for a very short registration time (3 ms).

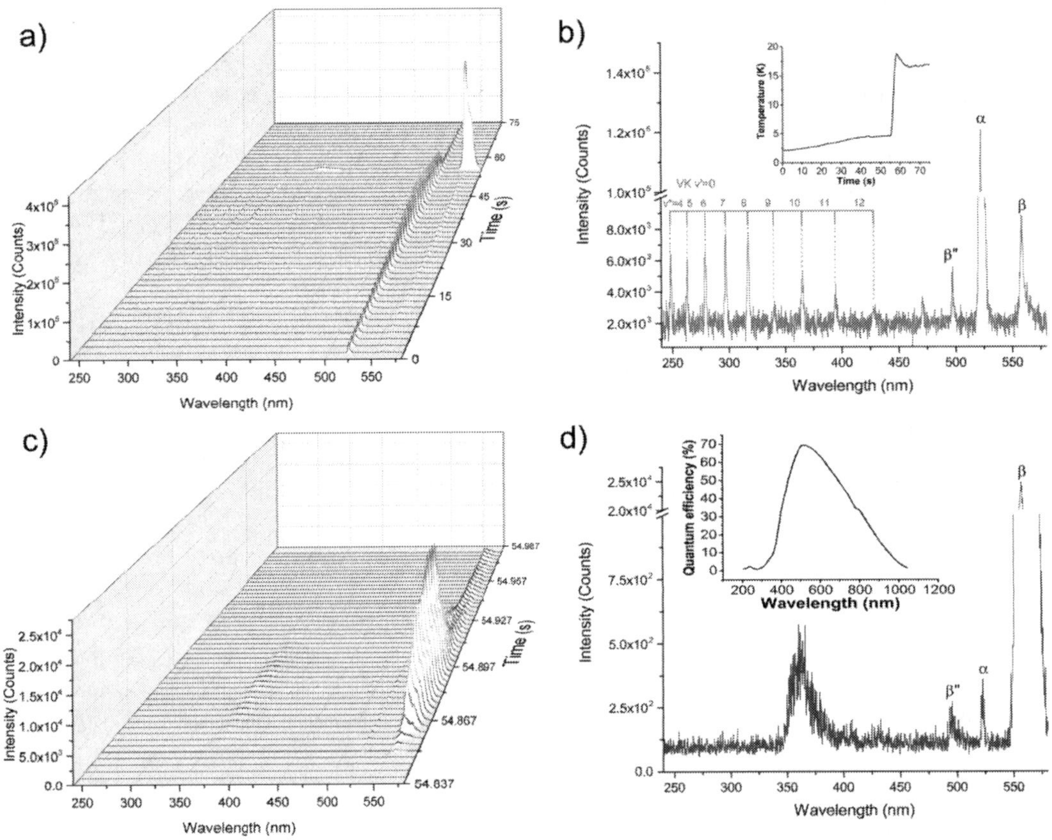


**Figure 1.** Spectra observed by the Andor Spectrometer during destruction of the sample prepared from gas mixture $[H_2]$:$[N_2]$:[He] = 1:330:33000. a) Dynamics of luminescence spectra for the entire destruction process which were accumulated during period 1.5 s. b) Spectrum taken during destruction at t = 37.5 s. Temperature dependence on time during sample destruction is shown in the inset. c) Dynamics of emission taken during the final destruction of the sample with exposure time 3 ms. d) Spectrum taken at the end of sample destruction corresponds to t = 54.891 s with the most intense 360 nm band emission. Quantum efficiency of the Andor spectrometer is shown in the inset.

## 3. Experimental Results

First we studied the destruction process of the sample prepared from the gas mixture $H_2$:$N_2$:He=1:330:33000. Figure 1a shows the time dependence of the emission spectra of this sample in the spectral range from 240 to 580 nm. Spectra were obtained by the Andor spectrometer with exposure time 3 ms, but each spectrum shown in the Fig.1a is the sum of 500 spectra. Figure 1b shows the spectrum of sample luminescence corresponding to the 37th second after the beginning of destruction. The intense $\alpha$-group of N atoms (transition $^2$D $\rightarrow$ $^4$S), $\beta$-group of O atoms (transition $^1$D $\rightarrow$ $^1$S) and Vegard-Kaplan (VK) bands of $N_2$ molecules (transitions $A^3\Sigma_u \rightarrow X^1\Sigma_g^+$) are present in the spectrum. Increasing the temperature of the sample initiates recombination of stabilized nitrogen atoms accompanied by luminescence. The warm-up thermogram during sample destruction is shown in the inset of Fig.1b. At the beginning of sample destruction, only the luminescence of the $\alpha$-group of atomic nitrogen and VK bands of $N_2$ can be seen in the spectra, and their intensities grow with increasing temperature. The intensity of the $\beta$-group emission of atomic oxygen is relatively small at the beginning of destruction but is increasing very strongly at the final stage of destruction. We also observed the appearance of the band with maximum at $\lambda$=360 nm at the end of destruction. All these features can be clearly seen in Figs. 1a and 1c. The spectrum taken at the end of destruction is shown in Fig. 1d. The main new effect observed in these experiments is the appearance of the strong band at $\lambda = 360$ nm.

As a next step we performed studies of the influence of the presence of molecular hydrogen in the initial gas mixtures on the appearance of the band at $\lambda = 360$ nm. We studied hydrogen-nitrogen-helium samples prepared from gas mixtures with different $H_2/N_2$ ratios: 1:200, 1:330, and 1:500. Figure 2 compares spectra of the largest flash observed during the destruction of samples prepared from different hydrogen-nitrogen-helium gas mixtures. All spectra were taken using the Ocean Optics spectrometer with the integration time 250 ms. In all samples the broad feature at $\sim$ 360 nm, the $\alpha$-group of N atoms, $\beta$-group of O atoms, and $\gamma$-line emission of $N^-$ anions (transition $^1D\rightarrow^3P$) [25] were observed. The largest intensity of the band at $\lambda = 360$ nm was observed for the sample prepared from gas mixture $[H_2]$:$[N_2]$:[He] = 1:330:33,000 (see Fig 2b), where it was found that reducing or increasing the hydrogen content present in this nitrogen-helium gas mixture decreased the intensity of all observed lines in the spectra of the largest flashes.

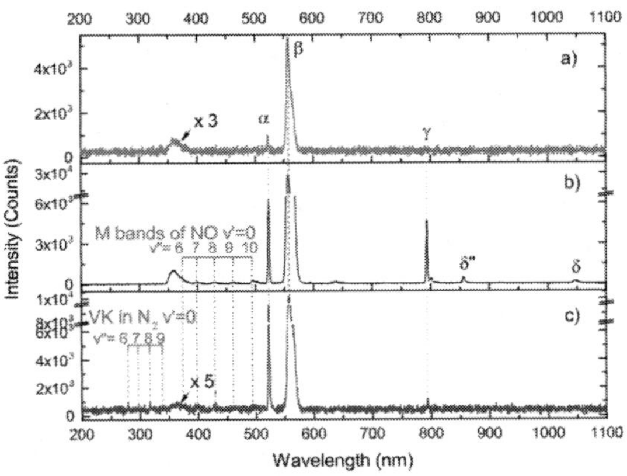


**Figure 2.** Comparison of spectra taken with the Ocean Optics spectrometer of the largest flash during the destruction of samples formed by different hydrogen-nitrogen-helium gas mixtures. a) $[H_2]$:$[N_2]$:[He] = 1:200:20,000 b) $[H_2]$:$[N_2]$:[He] = 1:330:33,000 c) $[H_2]$:$[N_2]$:[He] = 1:500:50,000.

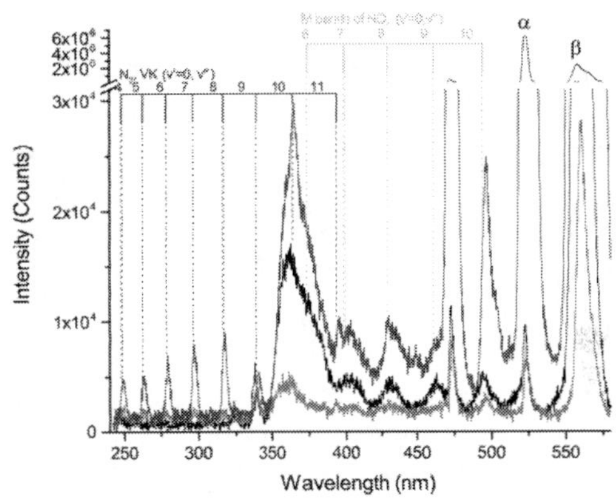


**Figure 3.** Graph comparing spectra observed by the Andor Spectrometer with resolution 0.5 nm during the destruction of samples prepared from different gas mixtures: $[D_2]$:$[N_2]$:[Ne]:[He] = 1:500:5000:100,000 (blue), $[D_2]$:$[N_2]$:[Ar]:[He] = 1:500:4500:225,000 (red), and $[D_2]$:$[N_2]$:[He] = 1:2,000:100,000 (black).

We also studied the influence of addition of molecular deuterium to the nitrogen-helium make-up gas mixture on the luminescence spectra during sample destruction. It was found that the addition of $D_2$ molecules also leads to the appearance of the band at $\lambda = 360$ nm in the spectra. We studied spectra of thermoluminescence during destruction of samples prepared from deuterium-nitrogen-helium gas mixtures as well as mixtures also containing rare gases Ne and Ar. Figure 3 shows a comparison of spectra observed during the destruction of samples prepared from three different gas mixtures: deuterium-nitrogen-helium, deuterium-nitrogen-argon-helium, and deuterium-nitrogen-neon-helium. All spectra were taken using the Andor Spectrometer with the integration time 300 ms. All spectra contain the 360 nm band, as well as the $\alpha$, $\delta$, and $\delta''$ -groups of N atoms, the $\beta$-group of O atoms, the VK bands of $N_2$ molecules and the M bands of NO molecules.

## 4. Discussion

We have studied the optical luminescence spectra during the destruction of molecular nitrogen nanoclusters containing stabilized nitrogen, oxygen, hydrogen and deuterium atoms. As in earlier work [9, 24], the $\alpha$-group of N atoms, $\beta$ and $\beta''$ groups of O atoms and VK bands of $N_2$ molecules were present in the spectra of emitted light. The strong broad band with maximum at $\sim$ 360 nm was observed in the spectra of samples which were prepared from gas mixtures containing hydrogen or deuterium. After analysis of our previous results we found that we observed only weak bands at 360 nm during the destruction of nitrogen nanoclusters containing only stabilized nitrogen atoms and small ($10^{-3} - 10^{-4}$) admixtures of oxygen atoms [9, 24]. The addition of $H_2$ or $D_2$ molecules in the gas mixture used during sample preparation resulted in enhancement of the intensity of the band appearing at $\lambda = 360$ nm in the luminescence spectra at the end of sample destruction (see Fig. 1c and Fig. 2). We found that the maximum intensity of the 360 nm band corresponds to some optimal quantity of $H_2$ or $D_2$ present in the make-up gas mixture. An important observation is that the addition of rare gases in the make-up gas mixture does NOT influence significantly the position of this band (see Fig. 3).

In the literature there are two explanations for the emission bands at $\lambda = 360$ nm. The first observation of this broad band was obtained when solid nitrogen was irradiated by 400 eV electrons, and the band was assigned to an unidentified impurity [26]. Later this luminescence band was studied during the excitation of solid nitrogen films by 500 eV electrons [14]. The band was assigned to the emission of polynitrogen $N_4^*$ which was formed as a result of a neutralization reaction of $N_4^+$ with electrons. The scenario of "hole self-trapping" for $N_2^+$ ions with formation of $N_4^+$ in a nitrogen matrix was quite recently proposed [12, 13] because of the localized character of positive charge carriers in solid $N_2$[27]. This suggestion is in good accordance with both the study of gas-phase equilibria of solvation reactions of $N_2^+$ with $N_2$ molecules which revealed electrostatic bonding in the core clusters $N_4^+$ [28] and the laser-induced dissociation experiments showing $N_4^+$ as the ionic core for the even ion clusters $N_2^+ - (N_2)_n$ [29].

Another interpretation of the 360 nm band corresponding to the transition ND ($A^3\Pi\ v' = 0 \rightarrow X^3\Sigma^- v'' = 1$) was suggested from the analysis of the luminescence spectra of the 0.1% $N_2$ doped solid deuterium irradiated by electrons [16]. Emission of NH (ND) radicals was also studied in different solid rare gases. It was found that the position of the (0-1) NH($A^3\Pi \rightarrow X^3\Sigma^-$) transition is shifted by tens of nanometers for different solid rare gases and that the band had a resolved structure [30].

In our experiments the spectra of the 360 band were not significantly influenced by the addition of rare gases or by the replacement of hydrogen isotopes in the nitrogen nanoclusters. This would give a preference for the interpretation of the observed band at 360 nm in our experiments as an emission of $N_4$ polynitrogen. However the question remains open as to why this emission was enhanced when the impurities of $H_2$ or $D_2$ were present in the molecular nitrogen nanoclusters. The exact identification of the transition of $N_4$ polynitrogen responsible for emission at $\lambda = 360$ nm requires additional experimental and theoretical work.

## 5. Conclusions

(i) The process of stabilizing ground state atoms in molecular nitrogen nanoclusters provides a unique opportunity to study low temperature reactions and to produce a variety of unusual molecules in excited states.

(ii) A weak broad band at 360 nm is observed during the destruction of ensembles of molecular nitrogen nanoclusters containing stabilized nitrogen and small admixtures of oxygen atoms.

This band was enhanced in the spectra of samples prepared from gas mixtures that contained hydrogen or deuterium.

(iii) Since this band is not changed significantly in rare gas matrices, we suggest that $N_4^*$ polynitrogen molecule is responsible for the emission of the band at 360 nm. The exact shape and location of the 360 nm band is not altered by hydrogen isotope substitution These $N_4$ polynitrogen compounds are formed during the process of sample destruction accompanied by fast chemical reactions of nitrogen atoms and molecules.

## 6. Acknowledgments
This work was supported by NSF grant No DMR 1707565.

## 7. References
[1] Clark R N, Carlson R, Grundy W and Noll K 2013 Observed ices in the solar system *The Science of Solar System Ices* (*Astrophysics and Space Science Library* vol 356) ed Gudipati M S and Castillo-Rogez J (New York: Springer) pp 3–46 1st ed
[2] Nguyen M T 2003 *Coordination Chemistry Reviews* **244** 93 – 113
[3] Zarko V E 2010 *Combustion, Explosion, and Shock Waves* **46** 121–131
[4] Boltnev R E, Krushinskaya I N, Pelmenev A A, Popov E A, Stolyarov D Y and Khmelenko V V 2005 *Low Temp. Phys.* **31** 547–555
[5] Mao S, Boltnev R E, Khmelenko V V and Lee D M 2012 *Low Temp. Phys.* **38** 1313–1319
[6] Bass A M and Broida H P 1956 *Phys. Rev.* **101**(6) 1740–1747
[7] Oehler O, Smith D A and Dressler K 1977 *J. Chem. Phys.* **66** 2097–2107
[8] Khmelenko V V, Kunttu H and Lee D M 2007 *J. Low Temp. Phys.* **148** 1–31
[9] Khmelenko V V, Lee D M, Krushinskaya I N, Boltnev R E, Bykhalo I B and Pelmenev A A 2012 *Low Temp. Phys.* **38** 688–699
[10] Ponomaryov A N, Savchenko E V, Gumenchuk G B, Khizhniy I V, Frankowski M and Bondybey V E 2007 *Low Temp. Phys.* **33** 532–537
[11] Savchenko E, Khyzhniy I, Uyutnov S, Bludov M, Gumenchuk G and Bondybey V 2016 *Radiation Measurements* **90** 1–5
[12] Savchenko E V, Khyzhniy I V, Uyutnov S A, Barabashov A P, Gumenchuk G B, Ponomarov A N and Bondybey V E 2015 *Physica Status Solidi C* **12** 49–54
[13] Savchenko E V, Khyzhniy I V, Uyutnov S A, Barabashov A P, Gumenchuk G B, Beyer M K, Ponomaryov A N and Bondybey V E 2015 *J. Phys. Chem. A* **119** 2475–2482
[14] Savchenko E V, Khyzhniy I V, Uyutnov S A, Bludov M A, Barabashov A P, Gumenchuk G B and Bondybey V E 2017 *J. Low Temp. Phys.* **187** 62–70
[15] Pelmenev A A, Krushinskaya I N, Bykhalo I B and Boltnev R E 2016 *Low Temperature Physics* **42** 224–229
[16] Stenum B, Schou J, Sørensen H and Gürtler P 1993 *J. Chem. Phys.* **98** 126–134
[17] Meyer D M and Roth K C 1991 *The Astrophysical Journal* **376** L49
[18] Gordon E B, Mezhov-Deglin L P and Pugachev O F 1974 *JETP Lett.* **19** 63
[19] Gordon E B, Mezhov-Deglin L P, Pugachev O F and Khmelenko V V 1976 *Cryogenics* **16(9)** 555–557
[20] Kiryukhin V, Keimer B, Boltnev R E, Khmelenko V V and Gordon E B 1997 *Phys. Rev. Lett.* **79** 1774–1777
[21] Kiselev S I, Khmelenko V V, Lee D M, Kiryukhin V, Boltnev R E, Gordon E B and Keimer B 2002 *Phys. Rev. B* **65** 024517
[22] Kiryukhin V, Bernard E B, Khmelenko V V, Boltnev R E, Krainyukova N V and Lee D M 2007 *Phys. Rev. Lett.* **98** 195506–4
[23] Bernard E P, Khmelenko V V and Lee D M 2008 *J. Low Temp. Phys.* **150** 516–524
[24] Meraki A, Mao S, McColgan P T, Boltnev R E, Lee D M and Khmelenko V V 2016 *J. Low Temp. Phys.* **185** 269–286
[25] Boltnev R E, Bykhalo I B, Krushinskaya I N, Pelmenev A A, Mao S, Meraki A, McColgan P T, Lee D M and Khmelenko V V 2016 *Phys. Chem. Chem. Phys.* **18** 16013
[26] Fugol I Y, Poltoratski Y B and Rybalko Y I 1978 *Low Temp. Phys.* **4** 1048–1052
[27] Loveland R J, Comber P G L and Spear W E 1972 *Phys. Rev. B* **6**(8) 3121–3127
[28] Hiraoka K and Nakajima G 1988 *The Journal of Chemical Physics* **88** 7709–7714
[29] David D E, Magnera T F, Tian R, Stulik D and Michl J 1986 *Nuclear Instruments and Methods in Physics Research Section B: Beam Interactions with Materials and Atoms* **14** 378 – 391
[30] Bondybey V E and Brus L E 1975 *The Journal of Chemical Physics* **63** 794–804

# Low-temperature hydrogen absorption into V and Nb metals from liquid hydrogen

**H Takata[1], K Ienaga[2], M Shiga[1], Md S Islam[1], Y Inagaki[1], H Tsujii[3], K Hashizume[4] and T Kawae[1]**

[1] Department of Applied Quantum Physics, Kyushu University, Fukuoka 819-0395, Japan
[2] Department of physics, Tokyo Institute of Technology School of Science, Tokyo 152-8551, Japan
[3] Department of Education, Kanazawa University, Kanazawa 920-1192, Japan
[4] Department of Engineering Science, Kyushu University, Fukuoka 816-8580, Japan

E-mail: takata.h.919@s.kyushu-u.ac.jp

**Abstract.** We report experimental study on low-temperature hydrogen (H) absorption in vanadium (V) and niobium (Nb) nanocontacts below $T = 20$ K using a point-contact spectroscopy (PCS) technique. When a small bias voltage is applied between both sides of nanocontacts immersed in liquid $H_2$, the differential conductance ($dI/dV$) and the second derivative ($d^2I/dV^2$) are changed from those for pure V and Nb nanocontacts. Further, the spectra approach to those for a high concentrated phase of H with increasing the bias voltage. The results indicate that in-situ investigation of H absorption process from liquid $H_2$ is possible through $dI/dV$ and $d^2I/dV^2$ measurements using the PCS technique.

## 1. Introduction

Hydrogen (H) atoms absorbed in metals occupy interstitial sites of their host lattice, which shows strong quantum characters because of its light mass [1]. A most famous example is H diffusion via quantum tunnelling (QT). H diffusion inside metals is governed by thermal activation process at high temperature, while the diffusion due to QT becomes dominant below liquid nitrogen temperature [1]. However, it is difficult to detect QT of H experimentally. Thus most studies on QT of H are discussed in terms of the temperature dependence of the diffusion rate or the relaxation time [1,4,5]. The temperature dependence of the diffusion rate obeys the Arrhenius equation at high temperatures, while it deviates from the equation at lower temperatures due to the change of their diffusion process. These suggest that understanding of H diffusion process is not complete yet because the experimental evaluation of the diffusion rate depends on the sample quality.

We focus on point-contact spectroscopy (PCS) to investigate low-temperature properties of metallic hydrides [6]. In a nanoscale contact that size is smaller than the mean free path of conduction electrons, the electrons accelerated by a bias voltage are transported through the contact ballistically, and then lose their energy by inelastic scattering with excitations such as phonons. This leads to a nonlinear relationship between current ($I$) and voltage ($V$) character [7,8]. The existence of H atom affects the electron transport, enabling to explore the properties of hydrides by tracing $dI/dV$ and $d^2I/dV^2$ spectra using differential conductance measurements [6,9].

In addition, we found that PCS experiments advance H absorption in palladium (Pd) nanocontacts even below $T = 20$ K. When a small bias voltage is applied between the nanocontacts immersed in liquid

$H_2$, the d$I$/d$V$ and d$^2I$/d$V^2$ spectra show the time evolution from the pure Pd to the Pd hydride, indicating that a large amount of H atoms go into Pd [6]. The results suggest the following absorption process. First, $H_2$ molecules dissociate into atoms at the surface of Pd due to steering effect [10]. Note that the dissociation occurs without the activation energy in Pd according to the theoretical calculation [11]. Then, H atoms are absorbed in Pd via QT by applying the bias voltage because the thermal hopping process of H is prohibited at $T = 20$ K [6].

In this study, to investigate the low-temperature H absorption in different metals other than Pd, we have performed PCS experiments in vanadium (V) and niobium (Nb) nanocontacts, which are known as H absorption metals, immersed in liquid $H_2$. H absorption are confirmed for $T < 20$ K as in the case of Pd nanocontacts. The results suggest that applying a bias voltage between nanocontacts advances H absorption due to QT. Further, this method would be a powerful tool for exploring QT of H.

## 2. Experimental

The present experiments are carried out with a home-made cryostat attaching a MCBJ apparatus. Nanocontacts are made in vacuum and low-temperature environment to prevent surface contamination and oxidation. V and Nb polycrystalline wire (99.8 % and 99.9 % respectively, 0.25 mm diameter) were used for samples. Hydride samples, $VH_{0.43}$ and $NbH_{0.43}$, are prepared by loading H atoms into these wires at high temperatures with the gas absorption method [12]. The H contents are determined in accordance with each pressure-temperature phase diagram of V-H and Nb-H system. In MCBJ experiments, sample wires on a phosphor-bronze substrate is stretched by bending the substrate with mechanical force, which is applied by a screw and a piezo element as shown in figure 1. The size of nanocontacts is changed continuously from atomic scale to micrometres and is kept over several hours by avoiding thermal input to the sample space [13, 14]. The MCBJ apparatus is mounted in an inner cell with flexible bellows inside the vacuum chamber to immerse the nanocontacts in liquid $H_2$ [15]. It is worthy to note that the electrical conductivity measurements are performed in liquid $H_2$, preventing the temperature increase due to Joule heating in the nanocontacts.

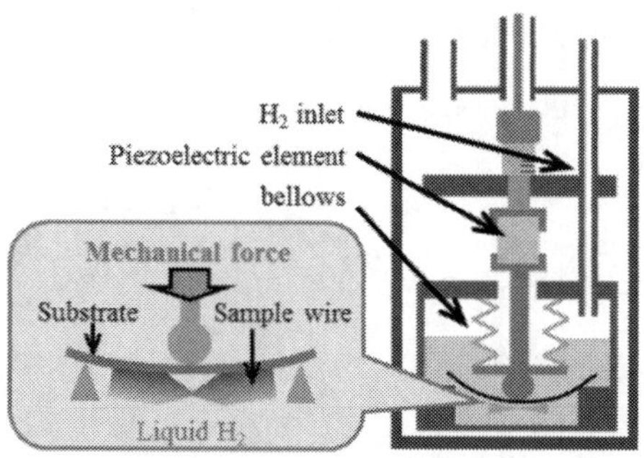


Figure 1. (Color online) Schematic illustration of MCBJ experimental cell. H absorption and diffusion experiments can be carried out in liquid $H_2$, avoiding the temperature increase due to Joule heating in the nanocontacts during the measurements.

The differential conductance d$I$/d$V$ spectra and the second derivative d$^2I$/d$V^2$ are measured as a function of bias voltage using a lock-in technique with 1 kHz modulation [7]. d$I$/d$V$ and d$^2I$/d$V^2$ are proportional to the first and second harmonics of the modulation frequency, respectively, so that the two spectra are measured simultaneously using two lock-in amplifiers.

## 3. Results

We show d$I$/d$V$ and d$^2I$/d$V^2$ spectra for a pure V nanocontact measured in vacuum at $T \sim 15$ K in figures 2(a) and 2(b), respectively. Here, the vertical axis is plotted in the universal conductance unit $G_0=2e^2/h$. In pure V nanocontacts, the intensity of d$I$/d$V$ spectra is lowered rapidly with increasing the bias voltage, which is caused by backscattering between the conduction electrons and phonon excitations [7]. d$^2I$/d$V^2$ spectra corresponding to the density of states for phonon excitation exhibit a sharp peak at $|V| \sim 20$ mV, which is consistent with the previous experiments [16,17].

The bias voltage dependence of the spectra is significantly modified by the existence of H atoms as shown in figures 2(a) and 2(b). The width of the d$I$/d$V$ spectrum is broadened. Besides, the peak at $|V| \sim 20$ mV in the d$^2I$/d$V^2$ spectrum is strongly suppressed. These features can be understood by the change of the transport character in the nanocontacts from ballistic to diffusive regime by hydrogenation [18,

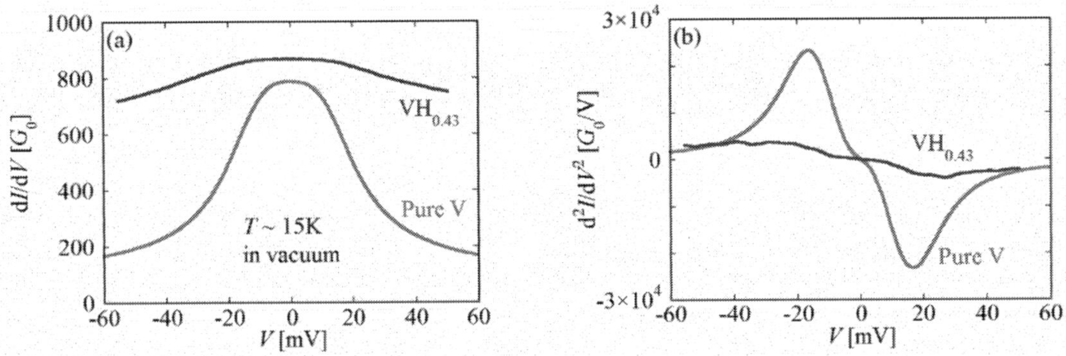


**Figure 2.** (Color online) (a) d$I$/d$V$ and (b) d$^2I$/d$V^2$ spectra for pure V and $VH_{0.43}$ nanocontacts in vacuum at 15K, where the vertical axis is the universal conductance unit $G_0$=$2e^2/h$.

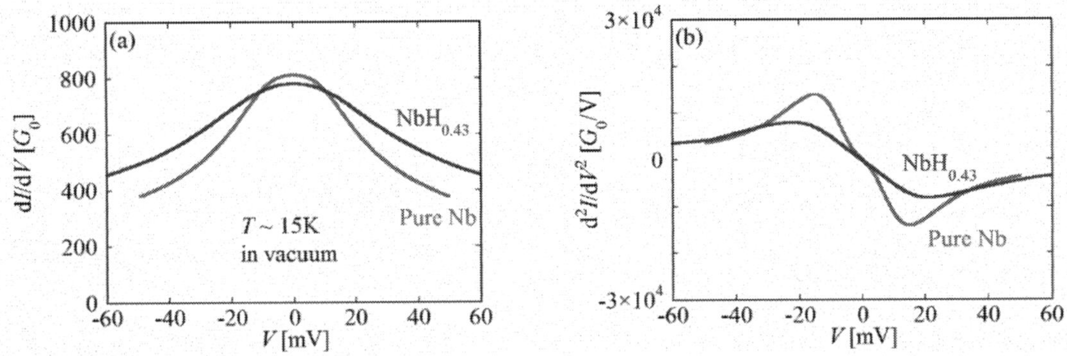


**Figure 3.** (Color online) (a) d$I$/d$V$ and (b) d$^2I$/d$V^2$ spectra for pure Nb and $NbH_{0.43}$ nanocontacts in vacuum at 15K.

19]. In hydride sample H atoms occupy interstitial lattice sites, causing the strain fields at around H atoms. This reduces the mean free path of conduction electrons.

The similar features are observed in pure Nb and $NbH_{0.43}$ nanocontacts as shown in figures 3(a) and 3(b). The width of the d$I$/d$V$ spectrum is broadened by hydrogenation. A sharp peak at $|V|$ ~ 15 mV in the d$^2I$/d$V^2$ spectrum of pure Nb is strongly suppressed. These results demonstrate that hydrogenation causes substantial change for d$I$/d$V$ and d$^2I$/d$V^2$ spectra because of scatterings between H atoms and conduction electrons, which enables to trace H absorption and diffusion process using the PCS technique as in the case of Pd.

Next we describe the results on low-temperature H absorption experiments in V and Nb nanocontacts. The procedure is done as follows. V and Nb nanocontacts are prepared at $T$ ~ 15 K in vacuum by MCBJ method, and then liquid $H_2$ is introduced into the inner cell of the cryostat. After immersing the nanocontacts in liquid $H_2$, we start the measurements. d$I$/d$V$ and d$^2I$/d$V^2$ spectra are measured at the bias voltage of $V_m$ = 20 mV, 30 mV, 40, mV, 60 mV and 80 mV, where $V_m$ is increased monotonously. During the measurements, the zero bias conductance $G$(0), which is obtained from d$I$/d$V$ at 0 mV, are kept to be 800 $G_0$ with feedback adjustment.

Figures 4(a) and 4(b) plot the bias voltage dependence for d$I$/d$V$ and d$^2I$/d$V^2$ spectra in V nanocontacts, respectively. After starting the measurements in liquid $H_2$ by applying $V_m$ = 20 mV, d$I$/d$V$ and d$^2I$/d$V^2$ spectra have been already changed from those for the pure nanocontact. On the other hand, the spectra show no change as long as the bias voltage is kept at $V_m$ = 20 mV. This indicates that H absorption is finished immediately after starting the measurements. When $V_m$ is increased to 30 mV, d$I$/d$V$ spectra are further broadened and the intensity of d$^2I$/d$V^2$ is decreased. The broadening of d$I$/d$V$ spectra is progressed with increasing $V_m$. It is easily seen that d$I$/d$V$ spectra approach to those for $VH_{0.43}$

in figures 2(a), indicating that H absorption is developed by applying the bias voltage. It takes about 30 min for the complete change after increasing $V_m$, which is much faster than that of Pd nanocontacts with more than 10 hours [6]. In Nb nanocontacts, qualitatively the same features are observed as shown in figures 5(a) and 5(b).

From these results, we conclude that H absorption is developed in V and Nb nanocontacts even below $T = 20$ K, and propose the following scenario for the low-temperature H absorption. First, $H_2$ molecules dissociate into atoms at the surfaces of V and Nb nanocontacts. It is well known that the activation energy for the dissociation is lowered not only by the high reactivity at the surface of transition metals, but also by the lattice distortions and defects, which are inevitable for the nanocontacts fabricated by MCBJ method. As a result, the kinetic energy of some $H_2$ molecules, which have a large energy distribution due to quantum effect [20], surpasses the activation energy, leading to the dissociation even at $T = 15$ K. Then, H atoms are absorbed when a small bias voltage is applied between the nanocontacts.

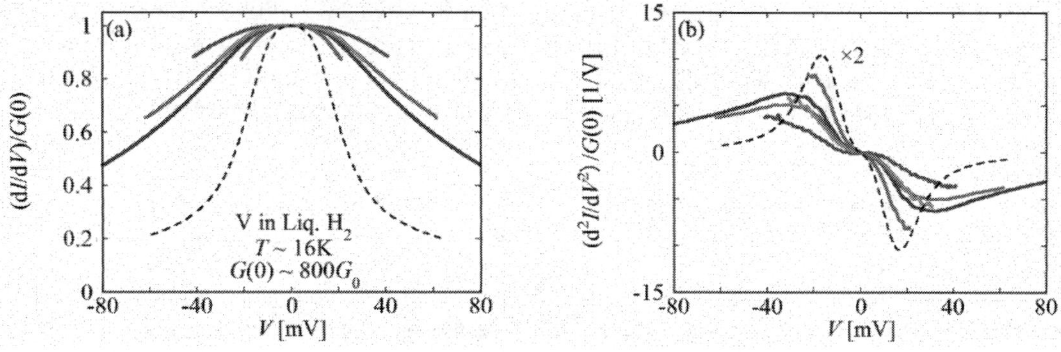


**Figure 4.** (Color online) The bias voltage dependence of (a) $dI/dV$ and (b) $d^2I/dV^2$ spectra in V nanocontacts immersed in liquid $H_2$. The spectra in (a) and (b) are measured for the bias amplitude at $V_m$ = 20, 30, 40, 60, 80 mV, after finishing the spectra changes by H absorption into the nanocontacts. For comparison, the spectra for pure nanocontact (dashed line) are plotted in the figures. The $d^2I/dV^2$ curve is scaled down one-half.

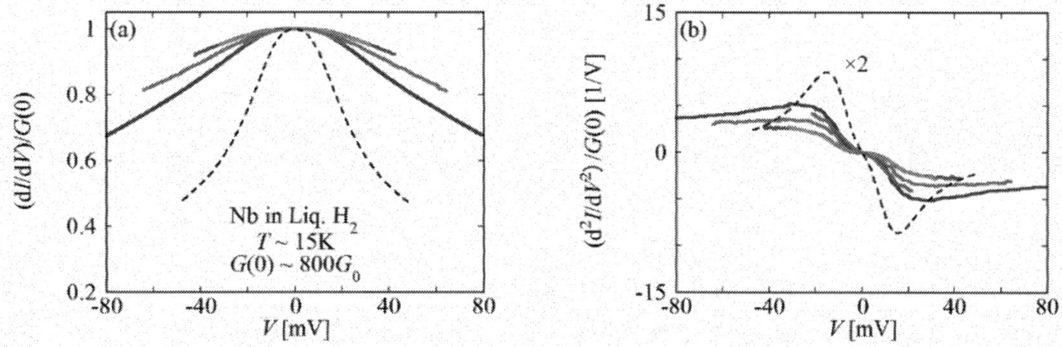


**Figure 5.** (Color online) The bias voltage dependence of (a) $dI/dV$ and (b) $d^2I/dV^2$ spectra in Nb nanocontacts immersed in liquid $H_2$. The spectra in (a) and (b) are measured for the bias amplitude at $V_m$ = 20, 30, 40, 60, 80 mV, after finishing the spectra changes by H absorption into the nanocontacts. For comparison, the spectra for pure nanocontact (dashed line) are plotted in the figures. The $d^2I/dV^2$ curve is scaled down one-half.

Additionally, it is important that the evolution can be traced using the PCS technique. From the fact that the thermal hopping process is frozen at $T = 20$ K, it is reasonable to consider that H absorption is caused by the phonon assisted tunnelling process which is excited by an in elastic scattering with the accelerated electrons [1]. Moreover, time for the complete change of the spectra in V and Nb experiments is much faster than that in the Pd nanocontacts. This difference is likely caused by the difference of crystal structure in the base metal. The wave function of H atom in bcc metals such as V and Nb is more extended than that in fcc metal of Pd [21], which increases the probability of H diffusion. At $V_m = 80$ mV, a broad peak is seen at around $V = 30$ mV, which is apparently higher than $V = 15 \sim 20$ mV in pure V and Nb nanocontacts. The increase of Debye temperature due to hydrogenation can be responsible for this shift. It was revealed that the Debye temperatures of $VH_x$ and $NbH_x$ are raised with the H contents [22]. This implies that the mean value of the phonon density of states should move to higher energy side by H absorption, resulting in the shift of the peaks in $d^2I/dV^2$ spectra.

Surprisingly, the intensity of $d^2I/dV^2$ spectra grows with increasing $V_m$ from 40 mV to 80 mV in the both nanocontacts. Generally, H absorption shortens the electron mean free path, reducing the intensity of $d^2I/dV^2$ spectra as mentioned above. Therefore, the growth suggests the increase of electron mean free path. This would be caused by the structural transition of H arrangement from $\alpha$ phase with a random occupation of the interstitial sites to $\beta$ phase with the periodic arrangement with increasing H contents. Actually, a resistivity drop is observed at the boundary between $\alpha$ phase to $\beta$ phase in bulk V [23, 24], implying the increase of the electron mean free path in $\beta$ phase. Recalling that $\beta$ phase appears H concentration higher than 0.5, we could perform in-situ investigation of the H absorption process via QT for a wide range of H concentration through $dI/dV$ and $d^2I/dV^2$ measurements using PCS.

## 4. Conclusion

We have performed H absorption experiments in vanadium (V) and niobium (Nb) nanocontacts using the PCS technique. The $dI/dV$ and $d^2I/dV^2$ spectra in hydride nanocontacts prepared by the gas absorption method are clearly different from those in the pure nanocontacts, indicating that PCS is used for tracing the H absorption and diffusion process in metals. When a small bias voltage is applied between the nanocontacts immersed in liquid $H_2$, the spectra shape is changed with increasing the bias voltage, demonstrating that H absorption into the nanocontacts is developed even below $T = 20$ K. Since the thermal hopping of H is frozen at $T = 20$ K, this should be explained by QT. Time for the complete change of the spectra in V and Nb experiments is much faster than that in the Pd nanocontacts, which is probably understood by the difference of crystal structure in the base metal. The bias voltage dependence of $d^2I/dV^2$ spectra implies that a high concentrated phase of H is realized by applying a high bias voltage between the nanocontacts. These results suggest that in-situ investigation of H absorption process via QT is possible through $dI/dV$ and $d^2I/dV^2$ measurements using PCS.

**Acknowledgments**
We would like to express our decent gratitude to the Japan Society for the Promotion of Science for providing the financial support by a Grant-in-Aid for Scientific Research (Grant Nos. 25220605, 25287076 and 26600102).

References

[1] Fukai Y 2005 *The Metal-Hydrogen System*, 2nd Edition (Springer, Berlin)
[2] Nelin G and Scöld K 1975 *J. Phys. Chem. Solids* **36** 1175
[3] Arons R R, Bohn H G and Lütgemeier H 1974 *Solid State Commun.* **14** 1203
[4] Minissale M, Congiu E, Baouche S, Chaabouni H, Moudens A, Dulieu F, Accolla M, Cazaux S, Manico G and Pirronello V 2013 *Phys. Rev. Lett.* **111** 053201
[5] Shannon R J, Blitz M A, Goddard A and Heard D E 2013 *Nature Chem.* **5** 745
[6] Ienaga K, Takata H, Onishi Y, Inagaki Y, Tsujii H, Kimura T and Kawae T 2015 *Appl. Phys. Lett.* **106** 021605
[7] Naidyuk Yu G and Yanson I K 2004 *Point-Contact Spectroscopy* (Springer, Berlin)
[8] Jansen A G M, Mueller F M and Wyder P 1977 *Phys. Rev. B* **16** 1325
[9] Csonka Sz, Halbritter A, Mihály G, Shklyarevskii O I., Speller S and van Kempen H 2004 *Phys. Rev. Lett.* **93** 016802
[10] Gross A, Wilke S and Scheffler M 1995 *Phys. Rev. Lett.* **75** 2718
[11] Wilke S and Scheffler M 1996 *Phys. Rev. B* **53** 4926

[12] Veleckis E and Edwards R K 1969 *J. Phys. Chem* 73 683
[13] Ienaga K, Nakashima N, Inagaki Y, Tsujii H, Honda S, Kimura T and Kawae T 2012 *Phys. Rev.* B **86** 064404 (2012).
[14] Ienaga K, Nakashima N, Inagaki Y, Tsujii H, Kimura T and Kawae T 2012 *Appl. Phys. Lett.* **101** 123114
[15] Ienaga K, Yokota T, Nakashima N, Inagaki Y, Tsujii H and Kawae T 2012 *J. Phys. : Conf. Ser.* **400** 042019
[16] Rybaltchenko L F, Yanson I K and Fisun V V 1980 *Sov. Phys.-Solid State* **22** 1182
[17] Yanson I K, Bobrov N L, Rybaltchenko L F and Fisun V V 1983 *Sov. J. Low Temp. Phys.* **9** 596
[18] Yanson I K, Kulik I O and Batrak A G 1981 *J. Low Temp. Phys.* **42** 527
[19] Lysykh A A, Yanson I K, Shklyarevski O I and Naydyuk Yu G 1980 *Solid State Commun.* **35** 987
[20] Boninsegni M 2009 *Phys. Rev. B* **79** 174203
[21] Sugimoto H and Fukai Y 1980 *Phys. Rev. B* **22** 670
[22] Ohlendorf D and Wicke E 1979 *Phys. Chem. Solids* **40** 721
[23] V Westlake D G, Ockers S T and Gray W R 1970 *Metall. Trans.* **1** 1361
[24] V Smith J F and Peterson D T 1982 *Bulletin of Alloy Phase Diagrams* **3** 55

# Onsager vortex formation in two-component Bose–Einstein condensates in two-dimensional traps

**Junsik Han and Makoto Tsubota**

Department of Physics, Osaka City University, Sumiyoshi-ku, Osaka 558-8585, Japan

E-mail: han@sci.osaka-cu.ac.jp

**Abstract.** We study numerically the dynamics of quantized vortices in two-dimensional one-component and two-component Bose–Einstein condensates (BECs) trapped by a harmonic and box potentials. In two-component miscible BECs, we confirmed the tendency of the formation of Onsager vortices in both traps. The vortices in one component separate spatially from those in the other component, which comes from their intercomponent-coupling. We also discuss the decay of the number of vortices.

## 1. Introduction

Turbulence shows strong dependence on the spatial dimension [1]. The three-dimensional (3D) classical turbulence sustains the direct cascade of energy transfered from large to small scales. Then, its energy spectrum obeys the Kolmogorov's -5/3 power law. On the other hand, Onsager predicted that the spontaneous formation of large-scale, long-lived vortices in two-dimensional (2D) turbulence [2, 3]. Kraichnan predicted that 2D classical turbulence shows the inverse energy cascade from small to large scales [4].

These statistics are also studied in quantum turbulence. In Bose–Einstein condensate (BEC), the system can be described by the macroscopic wave function $\psi$. As a result, a quantized vortex appear as a well-defined topological defect with the quantized circulation, being different from vortices in classical hydrodynamics. Various properties of quantum turbulence are attributed to dynamics of quantized vortices. In experiments of cold atoms, BECs are usually trapped by some potential like a harmonic one. Recently the box potential is realized and gives another important stage of quantum hydrodynamics [5, 6]. Controlling the potential enables us to realize 3D and 2D condensates. Three-dimensional quantum turbulence shows the same energy cascade and statistical Kolmogorov's -5/3 power law as 3D classical turbulence [7]. If the confinement along one direction is much stronger than other directions, the condensate is highly oblate and becomes two-dimensional. Two-dimensional quantum turbulence is expected to show the same inverse energy cascade as 2D classical turbulence and the clustering of like-sign vortices into large-scale Onsager vortex structure.

In the experimental [8] and numerical [9] studies, highly oblate BECs trapped by harmonic potentials did not show the tendency to form Onsager vortices. On the other hand, BECs in the box trap show strong tendency to form Onsager vortices [10]. Groszek *et al.* studied numerically the dynamics of quantized vortices in BECs trapped by both harmonic and box potentials [11]. They concluded that the vortices can form Onsager vortices in the box trap but their formation is suppressed in the harmonic trap.

We study whether Onsager vortices form or not in 2C BECs trapped by the harmonic and the box potentials. We investigate the decay of the vortex number, the development of the dipole moment and the distribution of vortices in 2C BECs. The main interest of this work is to know the effect of intercomponent-coupling on the dynamics of vortices. We also do the same calculation in 1C BEC for the comparison with 2C BECs.

## 2. Model

The macroscopic wave function $\psi$ of 2D 1C BEC satisfies the Gross–Pitaevskii equation (GPE)

$$(\imath - \gamma)\hbar\frac{\partial}{\partial t}\psi(\boldsymbol{r},t) = \left\{\frac{-\hbar^2}{2m}\nabla^2 + V_{\rm trap}(\boldsymbol{r}) + g|\psi(\boldsymbol{r},t)|^2\right\}\psi(\boldsymbol{r},t), \tag{1}$$

where $\gamma$ is the dimensionless damping parameter [12], $m$ is the mass of an atom, $V_{\rm trap}$ is the trapping potential and $g$ is the strength of interaction. We consider the harmonic potential

$$V_{\rm trap}(\mathbf{r}) = \frac{1}{2}m\omega_r^2|\boldsymbol{r}|^2, \tag{2}$$

with the radial harmonic trapping frequency $\omega_r$, and the box potential

$$V_{\rm trap}(\boldsymbol{r}) = \begin{cases} V_0 & (|\boldsymbol{r}| > R_0) \\ 0 & (|\boldsymbol{r}| < R_0) \end{cases}, \tag{3}$$

with an effective system radius $R_0$ and the potential height $V_0$. Two-dimensional BECs are described by the coupled GPEs for two macroscopic wave function $\psi_i$

$$\imath\hbar\frac{\partial}{\partial t}\psi_i(\boldsymbol{r},t) = \left\{\frac{-\hbar^2}{2m_i}\nabla^2 + V_{\rm trap}(\boldsymbol{r}) + \sum_{j=1,2} g_{ij}|\psi_j(\boldsymbol{r},t)|^2\right\}\psi_i(\boldsymbol{r},t), \quad (i=1,2) \tag{4}$$

where $m_i$ is the mass of the $i$th component's atom. Here, $g_{11}$ and $g_{22}$ are intracomponent-coupling constants, and $g_{12}$ is the intercomponent-coupling constant. In this study, we choose $m_1 = m_2 = m$, $g_{11} = g_{22} = g$ and $g_{12} = 0.9g$, where the two components are miscible.

We prepare the initial configuration of the vortices by the following method. For 1C BEC, we first imprint $N_v = 120$ vortices in the condensate by multiplying the ground state wave function by a phase factor $\Pi_i^{N_v}\exp(\imath\phi_i)$, with $\phi_i(x,y) = s_i\arctan\{(y-y_i)/(x-x_i)\}$. Here, the coordinate $(x_i, y_i)$ refers to the position of the $i$th vortex in the initial state and $s_i$ is its sign. We choose $x_i$ and $y_i$ randomly and imprint equal numbers of vortices ($s_i = 1$) and antivortices ($s_i = -1$). After the imprinting, the wave function is evolved in imaginary time to establish the structure of the vortex cores, and we solve the GPE in real time from this initial configuration. For 2C BECs system, we make vortices using the same method as 1C BEC system in each component. Then, we solve the GPE or the coupled GPEs by a fourth-order Fourier method on a $1024\times1024$ spatial grid.

The vortices are identified by measuring the phase singularities in the wave function. The vortex number is counted in a region $|\boldsymbol{r}| < 0.9R_{\rm TF}$ with $R_{\rm TF}$ is Thomas–Fermi radius in the harmonic trap, and $|\boldsymbol{r}| < 0.9R_0$ in the box trap in order to avoid counting ghost vortices[12]. We also calculate the dipole moment of the vortex distribution, which is defined as $d = |\boldsymbol{d}| = |\Sigma_i q_i\boldsymbol{r}_i|$, where $\boldsymbol{r}_i$ is the position of the $\imath$th vortex and $q_i = s_i\kappa = s_i h/m$ is its charge with $s_i = \pm1$. If the vortices are distributed uniformly, $d$ seldom grows. If like-sign vortices form an Onsager vortex, $d$ develops into some finite value.

In the previous studies [8, 9, 10, 11] and this work, the vortex number $N_v$ is found to decrease. Kwon *et al.* studied the decay of $N_v$ experimentlly and suggested a phenomenological rate equation for $N_v$ [8]. Groszek *et al.* also studied numerically the decay of $N_v$ by solving the GPE [11]. They found that the decay of $N_v$ was independent the geometry of the trapping potential. They extended the phenomenological rate equation introduced by Kwon *et al.* to

$$\frac{dN_v}{dt} = -\Gamma_1 N_v - \Gamma_2 N_v^2 - \Gamma_3 N_v^3 - \Gamma_4 N_v^4 - \cdots \quad , \tag{5}$$

in both systems of $\gamma = 0$ and $\gamma = 10^{-3}$. Here, the first term with $\Gamma_1$ means the drifting-out process, which means the vortices drift out of the condensate. The second term with $\Gamma_2$ refers to the vortex-antivortex pair annihilation [8]. The third and forth terms with $\Gamma_3$ and $\Gamma_4$ mean three-body and four-body loss, respectively. The three-body loss comes from the pair annihilation with a catalyst vortex and the four-body loss comes from the pair annihilation with two catalyst vortices. Groszek *et al.* fit their numerical results with Eq. (5) to study the effect of the thermal cloud [11]. From these simulation, they obtained $\Gamma_1 = 0.14\ s^{-1}, \Gamma_2 = 0.044\ s^{-1}, \Gamma_3 = \Gamma_4 = 0$ for the $\gamma = 10^{-3}$ case and found that the drifting-out and the pair annihilation were dominant. For the $\gamma = 0$ case, however, they obtained $\Gamma_1 = \Gamma_2 = 0, \Gamma_3 = 1.2 \times 10^{-4}\ s^{-1}, \Gamma_4 = 8.1 \times 10^{-7}\ s^{-1}$, and showed that the three-body and four-body losses were dominant.

## 3. Results

In this chapter, we first calculate $N_v$ and $d$. Then, we compare these between the harmonic and the box traps. We first address 1C BEC and then 2C BECs. We confine ourselves to the case of $\gamma = 0$.

The results in 1C BEC are shown by Figs. 1 and 2. Figure 1 (a) shows that the vortex number decays in both traps. We fit Eq. (5) with the two curves of $N_v$. In the harmonic trap, we obtain $\Gamma_1 = \Gamma_2 = 0, \Gamma_3 = 1.9 \times 10^{-5}\ s^{-1}, \Gamma_4 = 1.3 \times 10^{-7}\ s^{-1}$ (Fig. 1 (a) yellow (light) full thick line). In the box trap, we have $\Gamma_1 = \Gamma_2 = 0, \Gamma_3 = 3.2 \times 10^{-5}\ s^{-1}, \Gamma_4 = 2.2 \times 10^{-7}\ s^{-1}$ (Fig. 1 (a) purple (dark) full thick line). They are same order as the previously study [11]. The difference can come from the different sizes of the system. If the systems of us and Groszek *et al.* are normalized by each coherence length $\xi$, our condensate is larger than that of Groszek *et al.*, while the initial number of imprinted vortices are same. Then, the density of vortices in our system is smaller than that of Groszek *et al.* This may make our values of $\Gamma_i$ smaller than those of Groszek *et al.*

Figure 1 (b) and (c) show the time development of the amplitude $d$ of the dipole moment and its $x$ component $d_x$. Figure 2 shows how the distribution of vortices changes as time goes by. In the harmonic trap, $d$ hardly increases, and $d_x$ fluctuates around zero (Fig. 1 (b),(c) black (dark) dotted line). Then, the vortices are distributed uniformly at $t = 0$ (FIg. 2 (a)), and they are still uniform at $t = 1335$ (Fig. 2 (b)) even after the decay. On the other hand, in the box trap, $d$ and $d_x$ increase into some finite values (Fig. 1 (b),(c) light blue (light) dashed line). The initially uniform vortices (FIg. 2 (c)) lead to two Onsager vortices, one consisting of vortices ($s_i = 1$) and the other consisting of antivortices ($s_i = -1$) at $t = 1335$ (Fig. 2 (d)). These results are consistent with the previous simulation [9, 10, 11] and experiment [8]. In Fig. 1 (c), $d_x$ in the box trap is almost constant and positive from $t \simeq 400$ until $t \simeq 1000$, and then $d_x$ changes the sign with keeping $d$, which means that the two Onsager vortices revolve inside the condensate.

In 2C BECs, both the decay of $N_v$ and the development of $d$ are different from those in 1C BEC. Figure 3 shows $N_v$ and $d$ in the harmonic and the box traps.

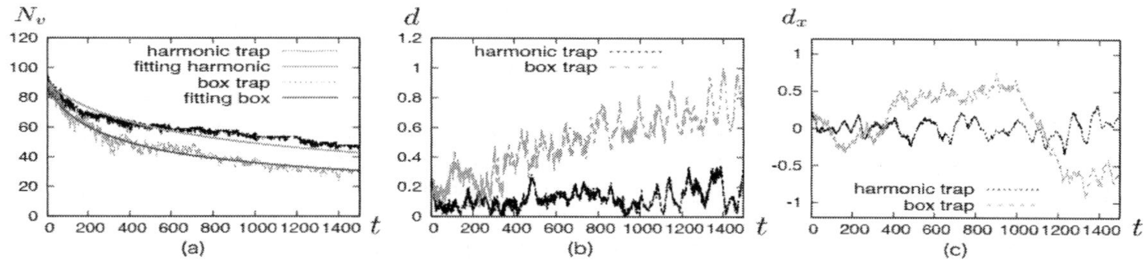


**Figure 1.** Time development of vortex number $N_v$ (a), the dipole moment $d$ (b) and its $x$ component $d_x$ (c) in 1C BEC. In (a), yellow (light) full thick line is fitted with decay in the harmonic trap and purple (dark) full thick line is fitted with decay in the box trap. Dipole moment is normalized by the effective system radius ($R_0$), the vortex charge ($\kappa$) and the vortex pair number ($N_v/2$).

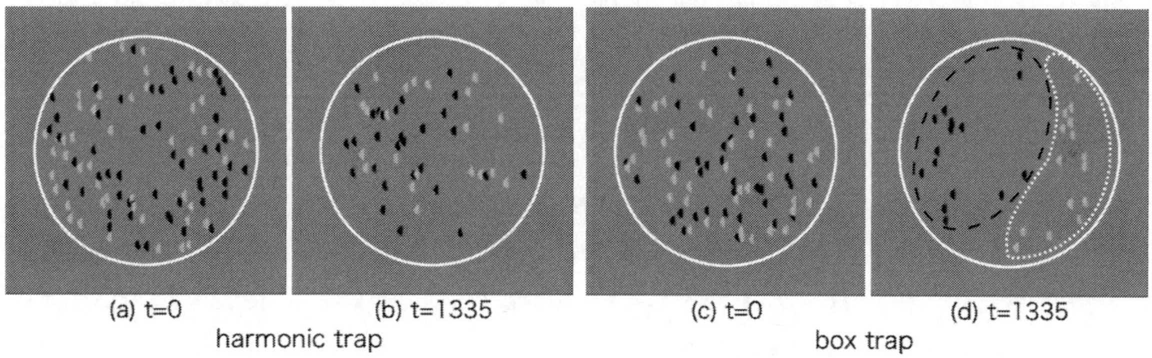


**Figure 2.** The distribution of vortices in 1C BEC in the harmonic trap at $t = 0$ (a) and at $t = 1335$ (b), and in the box trap at $t = 0$ (c) and at $t = 1335$ (d). Vortices and antivortices are denoted by black (dark) points and white (light) points, respectively. White (light) circles represent the boundary of condensate. In (d), vortices cluster in the area surrounded by dashed black (dark) line and antivortices cluster in the area surrounded by dotted white (light) line.

Decay of $N_v$ in 2C BECs (Fig. 3 (a), (c)) is much faster than in 1C BEC. This difference of the decays can be attributed to the forces $F^{\rm intra}$ between the vortices in the same component and $F^{\rm inter}$ between the vortices in different components [13]. In 1C BEC and 2C BECs, $F_{ij}^{\rm intra}$ is given by

$$F_{ij}^{\rm intra} = \frac{2\pi s_i s_j \hbar^2 n}{m}\frac{1}{R_{ij}}, \tag{6}$$

where $2R_{ij}$ is the distance between the $i$th and the $j$th vortices [13]. Then, $F_{ij}^{\rm intra}$ is repulsive between the vortices with the same sign and attractive between the vortices with different signs.

In 2C BECs, Eto *et al.* found that the $F_{ij'}^{\rm inter}$ can be written by [13]

$$F_{ij'}^{\rm inter} = \frac{\pi\hbar^4 g_{12}}{4m_1m_2(g_{11}g_{22}-g_{12}^2)}\frac{1}{R_{ij'}^3}\left(\ln\frac{R_{ij'}}{\xi}-\frac{1}{2}\right), \tag{7}$$

where $2R_{ij'}$ is the distance between the $i$th and $j'$th vortices belonging to different components. In this simulation, $g_{12} > 0$ and hence $F_{ij'}^{\rm inter} > 0$ for $R_{ij'} > \xi\sqrt{e}$, which means that $F_{ij'}^{\rm inter}$ is repulsive independently of the sign of vortices. When $R \gg \xi$, $F_{ij}^{\rm intra}$ is superior to $F_{ij'}^{\rm inter}$, while when $R \ll \xi$, $F_{ij'}^{\rm inter}$ is superior to $F_{ij}^{\rm intra}$. When $R_{ij'}$ is close to $\xi$, these two forces are comparable.

In 1C BEC, the total force between the $i$th vortex and all other vortices is written by

$$\boldsymbol{F}_i^{1\mathrm{C}} = \Sigma_{j \neq i} \boldsymbol{F}_{ij}^{\mathrm{intra}}. \tag{8}$$

On the other hand, in 2C BECs, the total force between the $i$th vortex and all other vortices is written by

$$\boldsymbol{F}_i^{2\mathrm{C}} = \Sigma_{j \neq i} \boldsymbol{F}_{ij}^{\mathrm{intra}} + \Sigma_{j'} \boldsymbol{F}_{ij'}^{\mathrm{inter}}, \tag{9}$$

where the first term is same as $\boldsymbol{F}_i^{1\mathrm{C}}$, and the second term could make $|\boldsymbol{F}_i^{1\mathrm{C}}| < |\boldsymbol{F}_i^{2\mathrm{C}}|$. In this condition, the average velocity of a vortex in 2C BECs becomes bigger than in 1C BEC. This makes the decay of $N_v$ in 2C BECs faster than in 1C BEC.

Although two components are equivalent in this study, the values of $d$ of two components differ after $t \simeq 100$ in the harmonic and box traps (Fig. 3 (b), (d)). Figure 4 shows the distribution of vortices after $t = 100$. In Fig. 4 (a), vortices cluster in the area surrounded by the dotted black ellipse in component1 and in the areas surrounded by dashed black ellipses in component2. We can see the similar tendency in Fig. 4 (b), (c), (d), which means that vortices in one component separate from those in the other component because of the repulsive force $F_{ij'}^{inter}$. We expect that this phenomenon can be treated as a kind of the phase separation about the distribution of vortices.

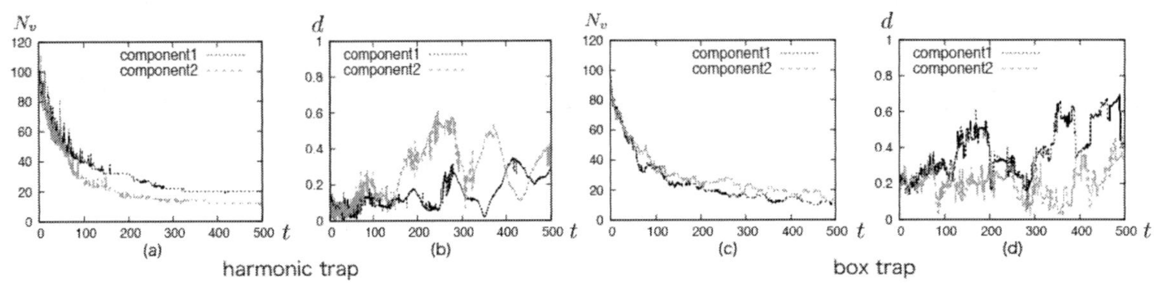


**Figure 3.** Time development of $N_v$ and $d$ in 2C BECs. Decay of $N_v$ of two components in harmonic trap and in box trap are shown in (a) and (c), and development of $d$ in (b) and (d), respectively..

## 4. Conclusions and discussions

We study the decay of $N_v$, the development of $d$ and the distribution of vortices in 1C BEC and 2C BECs. In 1C BEC, we obtain almost same results as the previously studies.

In 2C BECs, we can show the difference from in 1C BEC. In 2C BECs, the decay of $N_v$ is faster than in 1C BEC and vortices in one component are distributed separately from vortices in the other component, which comes from repulsive force $F_{ij'}^{inter}$ due to the positive $g_{12}$. Thus, $g_{12}$ affects greatly the dynamics of vortices and the decay of $N_v$. Thus, $d$ of one component increases, while the structure of Onsager vortex dose not appear obviously as in 1C BEC in the box trap. In this study, $g_{12}$ is relatively large, namely $g_{12} = 0.9g$, which suppresses the clear formation of Onsager vortices. If $g_{12}$ is smaller, the Onsager vortices should appear in the box trap.

This work is still just preliminary and it is necessary to study the dependence of the dynamics of vortices and the decay of $N_v$ on $g_{12}$. Further studies are also needed in order to investigate the phenomenological rate equation of the decay of $N_v$ in 2C BECs.

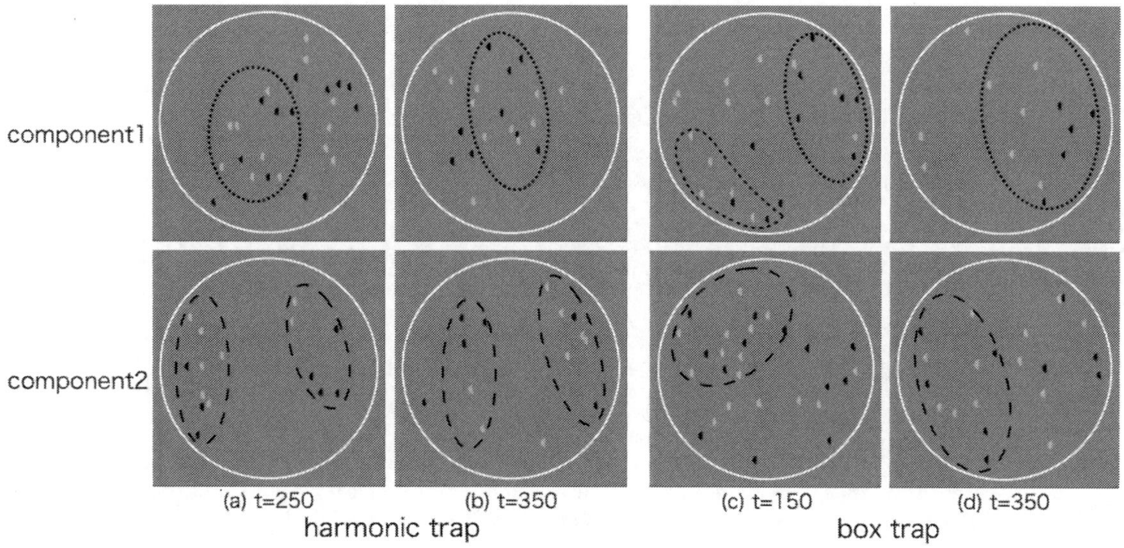


**Figure 4.** Top row shows the distribution of vortices in the component1 and the bottom row shows that in the component2. Vortices and antivortices are denoted by black (dark) points and white (light) points, respectively. White (light) circles represent the boundary of condensates. The distributions in the harmonic trap are shown in (a) ($t = 250$) and (b) ($t = 350$), in the box trap in (c) ($t = 150$) and (d) ($t = 350$).

## Acknowledgments

M.T. was supported by JSPS KAKENHI Grant No. 26400366 and MEXT KAKENHI "Fluctuation & Structure," Grant No. 16H00807. The authors thank S. Yui and T. Kanai for useful discussions.

## Reference

[1] Davidson P A 2015 *Turbulence an Introduction for Scientists and Engineers* (New York: Oxford University Press)
[2] Onsager L 1949 *Nuovo Cimento* **6** 279
[3] Eyink G L and Sreenivasan K R 2006 *Rev. Mod. Phys.* **78** 87
[4] Kraichnan R H 1967 *Phys. Fluids* **10** 1417
[5] Gaunt A L, Schmidutz T F, Gotlibovych I, Smith R P and Hadzibabic Z 2013 *Phys. Rev. Lett.* **110** 200406
[6] Navon N, Gaunt A L, Smith R P and Hadzibabic Z 2016 *Nature* **539** pp 72-79
[7] Kobayash M and Tsubota M 2007 *Pays. Rev.* A **76** 045603
[8] Kwon W J, Moon G, Choi J Y, Seo S W and Shin Y I 2014 *Phys. Rev.* A **90** 063627
[9] Stagg G W, Allen A J, Parker N G and Barenghi C F 2015 *Phys. Rev.* A **91** 013612
[10] Simula T, Davis M J and Helmerson K 2014 *Phys. Rev. Lett.* **113** 165302
[11] Groszek A J, Simula T P, Paganin D M and Helmerson K 2016 *Phys. Rev.* A **93** 043614
[12] Tsubota M, Kasamatsu K and Ueda M 2002 *Phys. Rev.* A **65** 023603
[13] Eto M, Kasamatsu K, Nitta M, Takeuchi H and Tsubota M 2011 *Phys. Rev.* A **83** 063603

# Two-parameter boundary condition applied to transverse acoustic impedance of a Fermi liquid

**J. A. Kuorelahti and E. V. Thuneberg**

Nano and molecular systems research unit, University of Oulu, Finland

**Abstract.** We discuss surface scattering in terms of bidirectional reflectance distribution function. We formulate a boundary condition that in addition to diffuse and specular reflection has tilted diffuse reflection. We apply the boundary condition to calculate the transverse acoustic impedance of a normal-state Fermi liquid. Some additional results on the impedance in a slab geometry are presented.

## 1. Introduction

In this article we are interested in the reflection of quasiparticles of a Fermi liquid from surfaces. Very often, only diffuse or specular reflection, or their combination, has been considered [1, 2, 3, 4, 5, 6]. The purpose is to test a boundary condition that goes beyond these. The reflectance of electromagnetic waves from a surface is a topic of long history [7]. The reflectance is commonly specified by *bidirectional reflectance distribution function* (BRDF) [8]. Especially in the connection of computer graphics, there is a need for a simple but physically realistic rendering, and several different reflectance models have been proposed, see e.g. Refs. [9, 10, 11]. We formulate a general boundary using BRDF and propose a simple generalization of the diffuse-specular boundary condition that has a tilt in the diffusive part. We apply the boundary condition to a normal-state Fermi liquid film on a transversely oscillation plane [12]. Besides presenting results of the new boundary condition, some additional results are shown.

## 2. Bidirectional reflectance distribution function

Let us consider elastic scattering from a surface. We study an approximately planar piece of the surface, which has area $A$ and surface normal $\hat{\boldsymbol{n}}$. The amount of both incoming and reflected monochromatic radiation in a given direction can be described by radiance. Radiance $L$ is the flux $\Phi$ of the radiation arriving to a surface area $A$ from a small solid angle $\Omega$ around the incoming direction $\hat{\boldsymbol{p}}$ per solid angle and per area projected in the direction of incidence. As a formula, $L = \Phi/\Omega A\,|\cos\theta|$, where $\cos\theta = \hat{\boldsymbol{n}}\cdot\hat{\boldsymbol{p}}$. The same formula describes also the radiance reflected from the surface element to a solid angle around the outgoing direction. The reflected radiance $L_r(\hat{\boldsymbol{p}})$ is given in terms of the incoming radiance $L_i(\hat{\boldsymbol{p}})$ as

$$L_r(\hat{\boldsymbol{p}}) = \int_{\text{in}} d\Omega' \left|\cos\theta'\right| f(\hat{\boldsymbol{p}},\hat{\boldsymbol{p}}')L_i(\hat{\boldsymbol{p}}'), \tag{1}$$

where $f(\hat{\boldsymbol{p}},\hat{\boldsymbol{p}}')$ is the *bidirectional reflectance distribution function* (BRDF). Here $\int_{\text{in}} d\Omega$ is the integral over the solid angle of incoming momenta, $\hat{\boldsymbol{n}}\cdot\hat{\boldsymbol{p}} = \cos\theta < 0$. The factor $|\cos\theta|$ is

IOP Conf. Series: Journal of Physics: Conf. Series **969** (2018) 012010 doi:10.1088/1742-6596/969/1/012010

included in (1) to compensate the same factor appearing in the definition of $L_i$. For a perfectly reflecting surface the flux conservation then sets

$$\int_{\text{in}} d\Omega' \left|\cos\theta'\right| f(\hat{\boldsymbol{p}}, \hat{\boldsymbol{p}}') = 1. \tag{2}$$

Apparently $f$ has to be non-negative and time reversal invariance requires

$$f(\hat{\boldsymbol{p}}, \hat{\boldsymbol{p}}') = f(-\hat{\boldsymbol{p}}', -\hat{\boldsymbol{p}}). \tag{3}$$

It is useful to notice that $\int_{\text{in}} d\Omega \left|\cos\theta\right|$ can be interpreted as an integral over *projected solid angle*. Thus it is equivalent to integration of over a unit disk. Writing in spherical coordinates $\hat{\boldsymbol{p}} = (\hat{\boldsymbol{x}}\cos\phi + \hat{\boldsymbol{y}}\sin\phi)\sin\theta + \hat{\boldsymbol{z}}\cos\theta$, where $\hat{\boldsymbol{z}} = \hat{\boldsymbol{n}}$, and defining $\rho = \sin\theta$, the integral is

$$\int_{\text{in}} d\Omega \left|\cos\theta\right| \ldots = \int_0^{2\pi} d\phi \int_0^1 d\rho\, \rho \ldots, \tag{4}$$

i.e. integration over the unit disk. It is also equivalent to an integral (or sum) over "transverse modes". This language is often used to describe electrical conductance in a ballistic channel, where each occupied mode contributes equally to the flux [13]. In a quantum treatment, BRDF can be presented as a square of a unitary scattering matrix.

The most common model forms of BRDF are *diffuse reflection*

$$f(\rho, \phi, \rho', \phi') = \frac{1}{\pi} \tag{5}$$

and *specular reflection*

$$f(\rho, \phi, \rho', \phi') = \frac{1}{\rho'}\delta(\rho - \rho')\delta(\phi - \phi'). \tag{6}$$

Diffuse reflection means that the scattered distribution is independent of the incoming direction and is equally probable in any direction with the same projected solid angle. Specular reflection means that the incoming flux in direction $\hat{\boldsymbol{p}}'$ is all scattered in the direction $\hat{\boldsymbol{p}} = \hat{\boldsymbol{p}}' - 2\hat{\boldsymbol{n}}(\hat{\boldsymbol{n}} \cdot \hat{\boldsymbol{p}}')$. A commonly used boundary condition is a combination of the diffuse and specular scattering,

$$f(\rho, \phi, \rho', \phi') = \frac{1 - s}{\pi} + \frac{s}{\rho'}\delta(\rho - \rho')\delta(\phi - \phi') \tag{7}$$

with parameter $s$ describing specularity, $0 \le s \le 1$.

Our plan is to look for a more general boundary condition than the combination of diffuse and specular scattering (7). One possibility is to spread the specular scattering $\delta$ functions to have a finite width. This is especially used in computer graphics [9, 10, 11]. However, we did not find a simple model that would satisfy the conditions (2) and (3) precisely. Therefore we turned our attention to the generalization of the diffuse limit. A general $f$ could be presented in terms of Zernike polynomials. They are an orthogonal set of functions defined on the unit disc (4). Instead of formulating such expansion in general, we limit here to the leading order generalization of (5), and we write our two-parameter boundary condition

$$f(\rho, \phi, \rho', \phi') = \frac{1 - s}{\pi} + a\rho\rho'\cos(\phi - \phi') + \frac{s}{\rho'}\delta(\rho - \rho')\delta(\phi - \phi'). \tag{8}$$

In order to ensure non-negativity, we require $|a| \le (1 - s)/\pi$. The new middle term describes tilt of the reflected distribution in the direction of in-plane component of the incoming momentum.

IOP Conf. Series: Journal of Physics: Conf. Series **969** (2018) 012010 doi:10.1088/1742-6596/969/1/012010

## 3. Transverse acoustic impedance of a Fermi liquid

We consider a degenerate Fermi system as described by Landau's theory [14]. As all quasiparticles effectively propagate at the Fermi velocity $v_F$, we can apply the general boundary condition (1) directly to the quasiparticle distribution function. We use the energy-integrated distribution $\psi_{\hat{\boldsymbol{p}}}$, which gives the local energy shift of the Fermi surface relative to the equilibrium Fermi level [5, 12]. Taking into account that the surface is moving with velocity $\boldsymbol{u}$, we have to replace $\psi_{\hat{\boldsymbol{p}}} \to \psi_{\hat{\boldsymbol{p}}} - p_F \hat{\boldsymbol{p}} \cdot \boldsymbol{u}$ [5]. The boundary condition (1) takes the form

$$\psi_{\hat{\boldsymbol{p}}} = \int_{\text{in}} d\Omega' \left|\cos\theta'\right| f(\hat{\boldsymbol{p}}, \hat{\boldsymbol{p}}')\psi_{\hat{\boldsymbol{p}}'} + p_F \boldsymbol{u} \cdot \left( \hat{\boldsymbol{p}} - \int_{\text{in}} d\Omega' \left|\cos\theta'\right| f(\hat{\boldsymbol{p}}, \hat{\boldsymbol{p}}')\hat{\boldsymbol{p}}' \right) \tag{9}$$

for $\hat{\boldsymbol{p}} \cdot \hat{\boldsymbol{n}} > 0$. For transverse impedance with $\boldsymbol{u} = u\hat{\boldsymbol{x}}$, the distribution takes the form $\psi_{\hat{\boldsymbol{p}}} = \cos\phi \sin\theta\, \psi(\cos\theta, \zeta)$, where $\zeta$ is the dimensionless vertical coordinate in the liquid [12]. The BRDF (8) then leads to the boundary condition

$$\psi(\mu > 0, 0) = \pi a \int_{-1}^{0} d\mu' \, |\mu'|(1 - \mu'^2)\psi(\mu', 0) + s\psi(-\mu, 0) + \left(1 - \tfrac{\pi}{4}a - s\right) p_F u. \tag{10}$$

Here the last term appears as the driving term in the calculation of the transverse acoustic impedance. We can define *momentum transfer efficiency* $q = 1 - \frac{\pi}{4}a - s$. We can expect that the main effect of changing $a$ and $s$ is through the change of $q$. Then we can study the remaining effect of $s$ at a constant $q$.

The calculation of the transverse acoustic impedance is described in Ref. [12]. The reflection on the lower, oscillating plane is described by parameters $q$ and $s$. The upper surface can be free ($q_2 = 0$, $s_2 = 1$) or a diffusely reflecting, stationary surface ($q_2 = 1$, $s_2 = 0$). In addition, the calculation depends on the following dimensionless parameters: the ratio of the mean free path $l$ to the thickness of the film $d$, $\Omega = \omega d / v_F(1 + \frac{1}{3}F_1)$ containing the angular frequency $\omega$ of the oscillation, the Fermi liquid parameters $F_1$ and $F_2$, and parameter $\xi_2$ describing the anisotropy of the quasiparticle relaxation rate. In the figures we study the effect of $q$ and $s$ as well as other parameters on the acoustic impedance $Z = Z' + iZ''$. $Z$ is expressed in units of $p_F n$ or $p_F n\Omega$, where $p_F$ is the Fermi momentum and $n$ the fermion number density.

Figures 1 and 2 display the case of small $\Omega$, meaning a low frequency or a thin film. We see that the new boundary condition has an effect only in the ballistic region $l/d > 1$, and even there the difference is small. Besides the hydrodynamic limit curve and ballistic limit points, Fig. 1 displays a half circle of radius 0.5 centered at $-0.5i$ (dash-dotted line). This curve is obtained if the liquid would be considered as a single rigid body that is dissipatively coupled to the substrate. We see that parts of the curves closely follow this "single body model" in the regime $l/d \lesssim 1$. We have checked that the flow profile in these regions indeed is like motion as a rigid body that slips from the motion of the substrate.

A model similar to the single-body model was used in Ref. [15] to fit experimental data on the decoupling of liquid $^3$He from a torsional oscillator. Both models give the same parametric curve, but the temperature dependences are different. The experimental finding in Ref. [15] was an increasing $Z''$ with decreasing temperature. In our model the temperature dependence is visible in Fig. 2 as $l/d$ increases with decreasing temperature. We see mainly decreasing $Z''$ and only a slight increase for $l/d \gg 1$. We see that the new boundary condition (8) gives similar result as the earlier one (7). The reasons for the difference between the experiment and our Fermi-liquid model remain open.

Figure 3 gives results for $\Omega = 1$, which could be relevant for experiments at higher frequencies. We see that there is a small effect of $s$ at a given $q$ in the ballistic region. The same figure also compares the effect of the relaxation anisotropy parameter $\xi_2$. It has effect particularly on $Z''$ but the effect vanishes in the ballistic limit, as it should be.

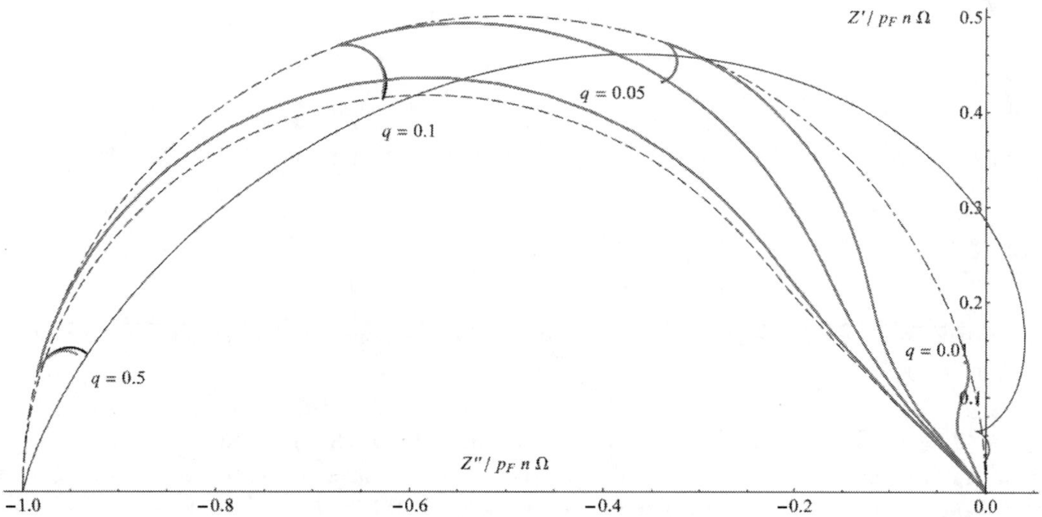


**Figure 1.** The acoustic impedance $Z$ of a Fermi liquid film with a free upper surface: the real and imaginary parts as a parametric plot with parameter $l/d$ [12]. The red/black curves are labelled by the momentum transfer efficiency $q = 0.5$, 0.1, 0.05 and 0.01. There is very little difference between curves with $s = 0$ (red) and $s = 1 - q$ (black, visible only if different from $s = 0$ case). Other parameters are $\Omega = 0.01$, $F_1 = 6$, $F_2 = 0$, and $\xi_2 = 0.35$. For comparison, the figure shows the single-body limit (dash-dotted), the hydrodynamic limit (dashed), and the ballistic gas limit (thin solid line).

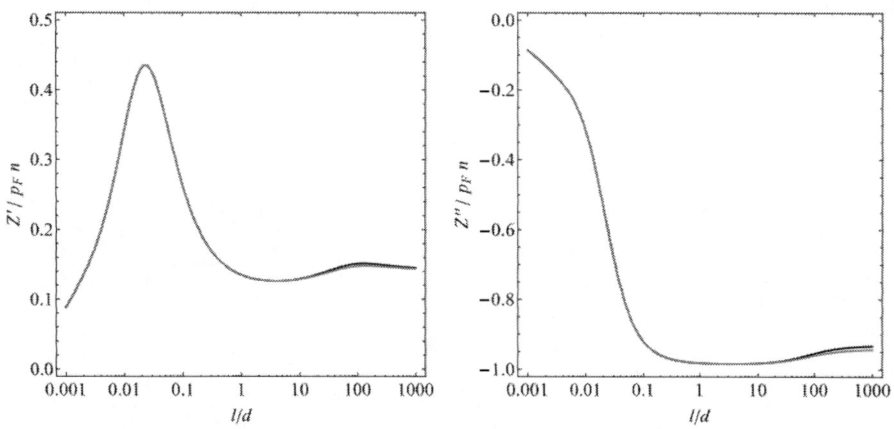


**Figure 2.** The real and imaginary parts of $Z$ of a Fermi liquid film with a free upper surface as a function of $l/d$. The curves and parameters are the same as for $q = 0.5$ in Fig. 1.

Figure 4 studies the case of a Fermi liquid in the space between a stationary and an oscillating surface, which both are diffusely reflecting. The shown results complement the special cases studied in Ref. [12].

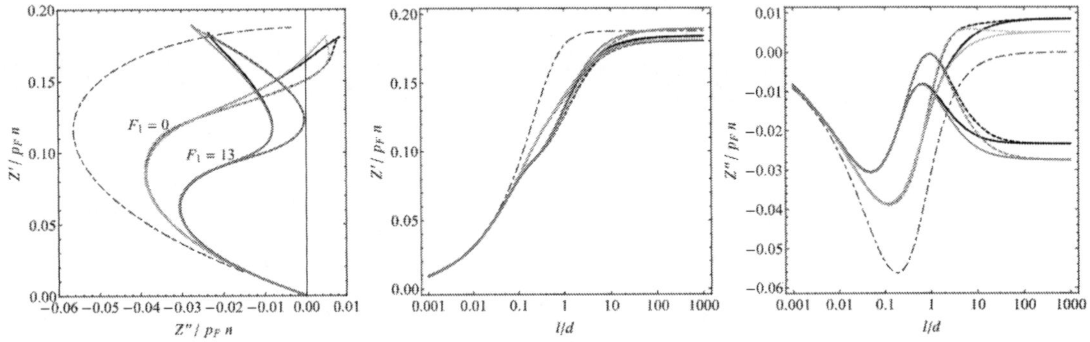


**Figure 3.** The acoustic impedance $Z$ of a Fermi liquid film with a free upper surface. Displayed are, from left to right, the real and imaginary parts as a parametric plot, the real part and the imaginary part as functions of $l/d$. The curves are for $s = 0$ (red and blue) and $s = 0.25$ (black), $\xi_2 = 0.35$ (dashed) and $\xi_2 = 1$ (solid), $F_1 = 0$ (blue and black) and $F_1 = 13$ (red and black). Other parameters are $q = 0.75$, $\Omega = 1$, and $F_2 = 0$. For comparison, the thick-film limit at $q = 1$, $F_1 = 0$, $\xi_2 = 1$ is shown by dash-dotted line.

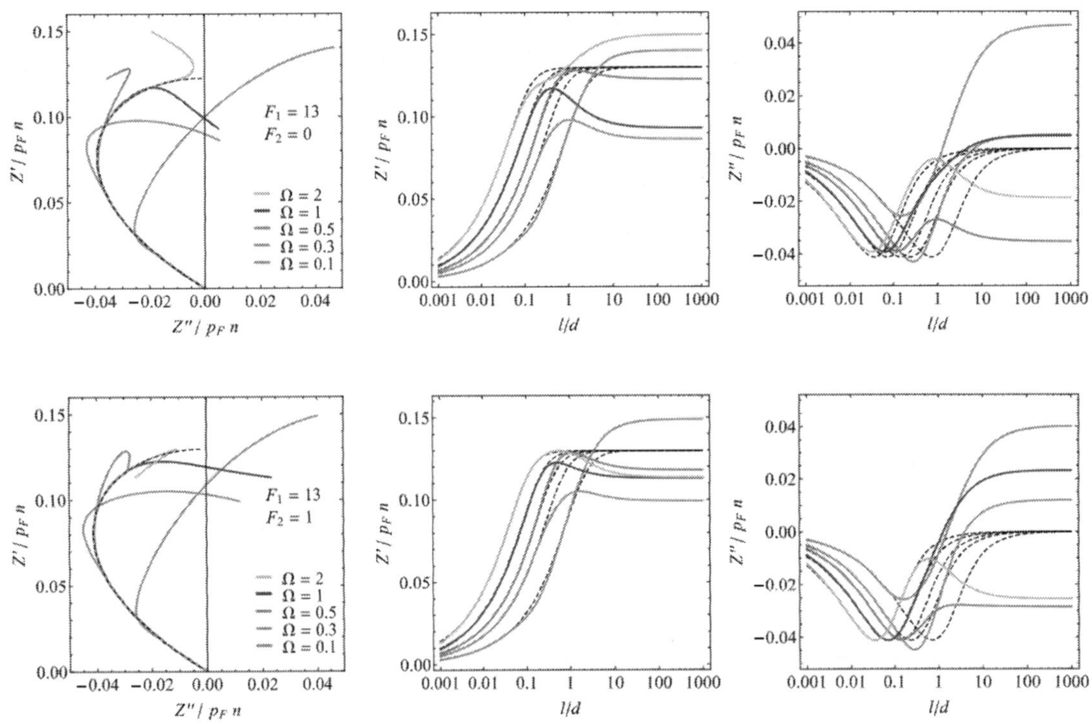


**Figure 4.** The acoustic impedance $Z$ of a Fermi liquid film with a stationary, diffuse upper surface. The thick-film limit is shown by a dashed line. The parameters are $q = 1$, $s = 0$, $\xi_2 = 0.35$.

IOP Conf. Series: Journal of Physics: Conf. Series **969** (2018) 012010 doi:10.1088/1742-6596/969/1/012010

## 4. Conclusions

We have suggested a boundary condition that goes beyond the often used combination of specular and diffuse reflection. In application to transverse acoustic impedance of a Fermi liquid, we see that at least in the cases studied, the boundary condition has only minor effect at constant momentum transfer efficiency of the reflection.

## Acknowledgments

We thank Jenny and Antti Wihuri foundation, Oskar Öflunds Stiftelse sr, the Academy of Finland and Tauno Tönning foundation for financial support.

## References

[1] Buchholtz L J and Zwicknagl G 1981 *Phys. Rev.* B **23** 5788
[2] Zhang W, Kurkijärvi J and Thuneberg E V 1987 *Phys. Rev.* B **36** 1987
[3] Vorontsov A B and Sauls J A 2007 *Phys. Rev. Lett.* **98** 045301
[4] Nagai K, Nagato Y, Yamamoto M and Higashitani S 2008 *J. Phys. Soc. Japan* **77** 111003
[5] Thuneberg E V and Virtanen T H 2011 *Phys. Rev.* B **83** 245137
[6] Mizushima T, Tsutsumi Y, Kawakami T, Sato M, Ichioka M and Machida K 2016 *J. Phys. Soc. Japan* **85** 022001
[7] Beckmann P and Spizzichino A 1963 *The Scattering of Electromagnetic Waves from Rough Surfaces* (New York, Pergamon Press)
[8] Nicodemus F E, Richmond J C, Hsia J J, Ginsberg I W and Limperis T 1977 *Geometrical considerations and nomenclature for reflectance* (US, National Bureau of Standards)
[9] Ward G 1992 *Computer Graphics* **26** 265
[10] Neumann L, Neumann A and Szirmay-Kalos L 1999 *Computer Graphics Forum* **18** 161
[11] Geisler-Moroder D and Dür A 2010 *Computer Graphics Forum* **29** 1391
[12] Kuorelahti J A, Tuorila J A and Thuneberg E V 2016 *Phys. Rev.* B **94** 184103
[13] Datta S 1995 *Electronic transport in mesoscopic systems* (Cambridge UK, Cambridge)
[14] Landau L D 1957 *Sov. Phys. JETP* **3** 920
[15] Casey A, Parpia J, Schanen R, Cowan B and Saunders J 2004 *Phys. Rev. Lett.* **92**, 255301

# Decoupling of Solid $^4$He Layers under the Superfluid Overlayer

**Kenji Ishibashi, Jo Hiraide, Junko Taniguchi, and Masaru Suzuki**
Department of Engineering Science, The University of Electro-Communications,
1-5-1 Chofugaoka, Chofu, Tokyo 182-8585, Japan
E-mail: i_kenji@phys.uec.ac.jp

**Abstract.**
It has been reported that in a large oscillation amplitude, the mass decoupling of multilayer $^4$He films adsorbed on graphite results from the depinning of the second solid atomic layer. This decoupling suddenly vanishes below a certain low temperature $T_D$ due to the cancellation of mass decoupling by the superfluid counterflow of the the overylayer. We studied the relaxation of the depinned state at various temperatures, after reduction of oscillation amplitude below $T_D$. It was found that above the superfluid transition temperature the mass decoupling revives with a relaxation time of several 100 s. It strongly supports that the depinned state of the second solid atomic layer remains underneath the superfluid overlayer.

## 1. Introduction

An atomically thin films adsorbed on a solid substrate exhibit various unique properties reflecting the geometry of the substrate, and provide a number of attractive topics in physics for many years. Recently, the nanofriction is added to this study. Krim and co-workers performed quartz crystal microbalance (QCM) measurements for several physisorbed films on metal substrates and found that the films undergo partial decoupling from the oscillating substrate.[1] Mistura and co-workers also observed this decoupling on metal substrates, and reported the pinning-depinning transition against the driving force of oscillating substrates.[2] They found that the decoupling state remains down to a lower driving force than the depinning transition. It indicates that the decoupling state is metastable below the depinning amplitude.

$^4$He film on graphite ($^4$He/Gr) is an ideal system for a study on decoupling mechanism, since the interaction between the adatom and the substrate is so small that the decoupling is expected to occur at a small driving force. In addition, $^4$He/Gr shows a layer-by-layer growth up to the five-atomic layers. We have studied multi-layer $^4$He films on graphite using the QCM technique. In $^4$He/Gr, we observed that the two- and three-atom-thick films start to decouple below a certain temperature $T_S$ at an oscillation amplitude larger than $A_S$.[3, 4] Furthermore, we confirmed that the decoupling state is metastable; after keeping an oscillation amplitude larger than $A_S$, we switched a sufficiently lower oscillation amplitude than $A_S$, and found that the decoupling state remains with a long relaxation time, which reaches several thousands seconds at low temperatures.

Figures 1(a) and (b) show our naive picture about the observations. In $^4$He/Gr, the areal density of the second solid atomic layer is smaller than that of the first solid atomic layer, which

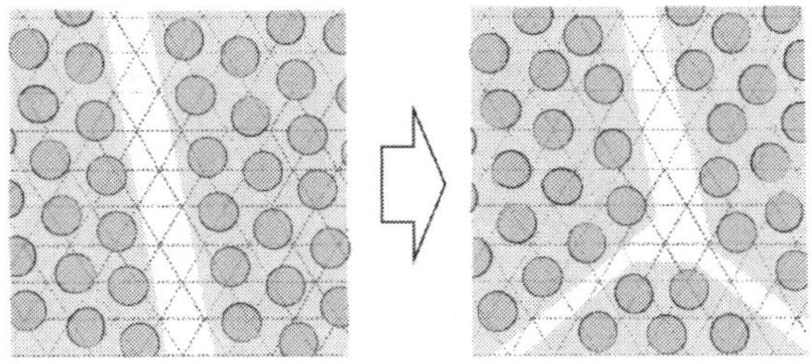


**Figure 1.** Picture of decoupling in the two- and three-atom-thick films on graphite. The bars show the lattice of the first solid atomic layer. The circles and the colored areas show $^4$He atoms and domains in the second solid atomic layer, respectively. (a) Stable structure where $^4$He atoms in the second solid atomic layer are locked to stable positions at the oscillation amplitude below $A_S$. (b) Disordered structure where $^4$He atoms are depinned from the stable positions at the oscillation amplitude larger than $A_S$. After the oscillation amplitude is switched to that below $A_S$, this remains with a long relaxation time at low temperatures.

causes edge dislocations between the layers. At an oscillation amplitude below $A_S$, $^4$He atoms in the second atomic layer cannot overcome the potential barrier owing to the first solid atomic layer. Thus, $^4$He atoms in the second atomic layer are locked to stable positions, as shown in Fig. 1(a). When the oscillation amplitude is larger than $A_S$, *i.e.* when the driving force is large enough to overcome the barrier, the $^4$He atoms in the second atomic layer are depinned from the stable positions, and forms a disordered (domain) structure, as shown in Fig. 1(b), resulting in the decoupling of the second solid atomic layer from the oscillation of the first solid atomic layer. After the oscillation amplitude is switched to that below $A_S$, the depinned $^4$He atoms in the second atomic layer relaxes to the stable positions with a certain relaxation time.

We have also performed QCM measurements on four-atom-thick films, where the third- and the four-atomic layers become a superfluid below the transition temperature $T_C$. We found that the decoupling suddenly vanishes below a certain temperature $T_D$, which is lower than $T_C$. This vanishment is explained by the cancellation of mass decoupling of the second atomic layer by the superfluid counterflow of the overlayer.[5] This explanation indicates that the $^4$He atoms in the second solid atomic layer is still in the depinned state, even though the entire mass decoupling vanishes below $T_D$.

It is of interest to clarify whether or not, the depinned state of the second atomic layer remains underneath the superfluid overlayer. Thus motivated, we studied the relaxation of the depinned state of the second solid atomic layer by developing a new procedure. In this paper, we report how we detect the relaxation, and then discuss the relaxation of the depinned state.

## 2. Experiment

We used the QCM technique to measure the mass decoupling. In the QCM technique, the coupled mass to the oscillating substrate is obtained from the change in the resonance frequency $\Delta f$ as

$$\frac{\Delta f}{f} = -\frac{m}{M}, \tag{1}$$

where $m$ is the coupled mass density of film, $M$ is the areal density of the crystal, and $f$ is the resonance frequency. When the film is decoupled from the oscillation, the coupled mass

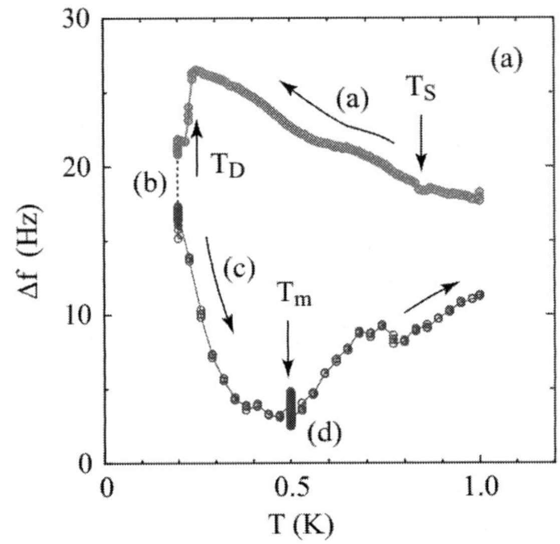


**Figure 2.** Experimental procedure. (a) First, in a large amplitude of 0.25 nm, the film was slowly cooled down from 1.0 K to 0.20 K (red). (b) At 0.20 K, the amplitude was switched to a small amplitude of 0.025 nm. (c) Then, the film was quickly warmed up to $T_m$ and (d) was kept at this temperature to measure the change in resonance frequency (blue).

decreases and the resonance frequency increases.

In the present experiments, the resonator is a 5.0 MHz AT-cut quartz crystal. The crystal was commercially available, and no special treatment was applied to the Ag electrode. At first, Grafoil (exfoliated graphite) was baked in a vacuum at 900°C for 3 h, and a 30-nm-thick film of Ag was deposited onto it. The crystal and Ag-plated Grafoil were pressed together and were heated in a vacuum at 350°C for 2 h. Then, Grafoil was bonded on both sides of the Ag electrode. After bonding, the excess amount of Grafoil was removed. To keep good thermal contact, the crystal was fixed to the metal holder with electrically conductive adhesive. After these processes, the $Q$ value of the crystal remained better than $10^4$, and the areal density of Grafoil was 7.30 g/m$^2$. After being heated in $2 \times 10^{-6}$ Pa at 130°C for 5 h, the crystal was mounted in the sample cell.

The resonance frequency was measured using a transmission circuit. In the circuit, the quartz crystal was placed in series with a coaxial line connecting a 50 Ω cw signal generator and a RF lock-in amplifier. The frequency of the signal generator was then controlled in order to keep the inphase output zero, and was locked to the resonance frequency. The quadrature output at this frequency is the resonance amplitude.

We prepared the four-atom-thick film of 32.0 atoms/nm$^2$ with $T_C$ of $\sim$ 0.5 K, and carried out relaxation experiments with the following procedure as shown in Fig. 2:
(a) The film was slowly cooled down from 1.0 K to 0.20 K in a large amplitude of 0.25 nm. The resonance frequency increases below $T_S$ of 0.85 K, *i.e.* the mass decoupling of solid layers occurs. Then, this frequency is suddenly dropped at $T_D$ of 0.23 K, *i.e.* the entire mass decoupling vanishes.
(b) At 0.20 K (below $T_D$), the amplitude was switched to a small amplitude of 0.025 nm.
(c) The film was quickly warmed up to the setting temperature of $T_m$ at the warming rate of 0.03 K/min.[6]
(d) At $T_m$, the change in resonance frequency was measured for 2 h.

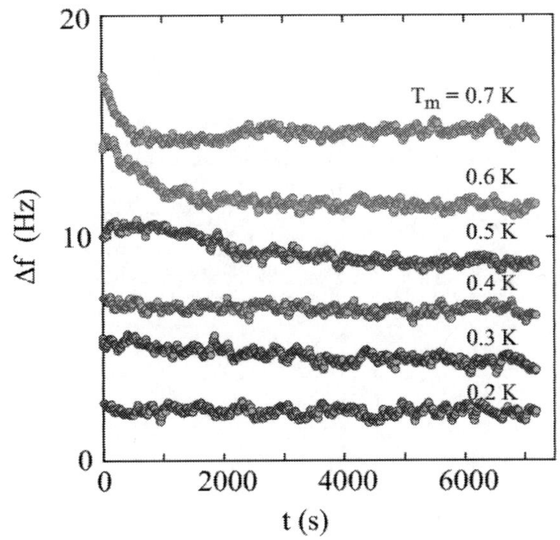


**Figure 3.** Change in the resonance frequency against waiting time at $T_m$. The resonance frequency decreases slowly at the beginning above 0.5 K, while it does not change greatly below 0.4 K.

Then, the film is warmed up to 1.0 K, and the procedure of (a)-(d) was repeated for different $T_m$. If the depinned state of the second solid atomic layer remains underneath the superfluid overlayer, the decoupling is expected to revive at $T_m > T_C$ on the condition that the warming time is sufficiently shorter than the relaxation time of the depinned state.

## 3. Results and Discussion

Figure 3 shows the change in the resonance frequency for several $T_m$. Below 0.4 K($< T_C$), the resonance frequency does not change greatly. In contrast, above $T_C \sim 0.5$ K the resonance frequency decreases at the beginning of waiting time, with a relaxation time of about several 100 s.[7] The observations clearly show that the decoupling revives above $T_C$, due to the lack of the cancellation by the superfluid counterflow. It means that the depinned state of the second solid atomic layer remains underneath the superfluid overlayer.

Next we focus on the relaxation time of the depinned state after the temperature becomes constant. In Fig. 3, the relaxation time for $T_m = 0.7$ K is shorter than that for 0.6 K. This tendency is consistent with that observed in the three-atom-thick films.[8]In the three-atom-thick films, the relaxation time obeys the Arrhenius law, indicating that the existence of energy barrier between the stable and the disordered structure in the second solid atomic layer. It remains a future issue to examine the energy barrier for the four-atom-thick films and clarify how the energy barrier varies by increasing the superfluid overlayer.

## 4. Summary

We measured the mass decoupling after the reduction in amplitude below $T_D$, by developing a new procedure. We found that the mass decoupling revives above $T_C$. It strongly supports that the depinned state in the second layer remains below $T_D$ underneath the superlfuid overlayer, even though the entire mass decoupling vanishes by the superlfuid counterflow.

## References

[1] J. Krim, D. H. Solina, and R. Chiarello, *Phys. Rev. Lett.* **66** 181 (1991).
[2] L. Bruschi, A. Carlin, and G. Mistura, *Phys. Rev. Lett.* **88** 036105 (2002).
[3] N. Hosomi, A. Tanabe, M. Suzuki, and M. Hieda, *Phys. Rev. B* **75** 064513 (2007).
[4] N. Hosomi and M. Suzuki, *Phys. Rev. B* **77** 024501 (2008).
[5] N. Hosomi, J. Taniguchi, M. Suzuki, and T. Minoguchi, *Phys. Rev. B* **79** 172503 (2009).
[6] The frequency drop with increasing temperature from 0.2 to 0.5 K at the oscillation amplitude of 0.025 nm is not from the change in decoupling amount of $^4$He film, but from the change in background peculiar to this amplitude.
[7] The decoupling amount at the beginning of relaxation measurements ($\sim$ 3 Hz) is about 60% of that at the large oscillation amplitude just above $T_D$. This reduction is attributable to the relaxation during the warming time (800 s for $T_m = 0.6$ K).
[8] K. Noda, K. Okamura, J. Taniguchi, M. Suzuki, and M. Hieda, *J. Low Temp. Phys.* **171** 638 (2013).

IOP Conf. Series: Journal of Physics: Conf. Series **969** (2018) 012012 doi:10.1088/1742-6596/969/1/012012

# Pairing fluctuations in a strongly interacting two-dimensional Fermi gas

**Morio Matsumoto[1], Ryo Hanai[2], Daisuke Inotani[1], and Yoji Ohashi[1]**

[1]Faculty of Science and Technology, Keio University, Yokohama, Kanagawa, Japan
[2]Department of Physics, Osaka University, Toyonaka, Osaka, Japan

E-mail: moriom@rk.phys.keio.ac.jp

**Abstract.** We theoretically investigate strong-coupling properties of a two-dimensional Fermi gas in the normal phase. Including pairing fluctuations within the framework of a self-consistent $T$-matrix approximation, we calculate the single-particle density of states (DOS) near the observed Berezinskii-Kosterlitz-Thouless phase transition temperature $T_{\rm BKT}$. We show that the pseudogap phenomenon associated with strong pairing fluctuations appears as an almost fully gapped single-particle density of states in the strong-coupling regime, indicating that amplitude fluctuations of the order parameter are almost absent and the system is dominated by phase fluctuations of the order parameter, as expected in a two-dimensional system near $T_{\rm BKT}$. On the other hand, although pairing fluctuations are enhanced by the low-dimensionality of the system, they are found to be still not strong enough to open such a full gap in the intermediate coupling (crossover) regime. In this regime, a dip structure is only seen in DOS around $\omega = 0$, indicating that amplitude of the superfluid order parameter is also fluctuating, in addition to phase fluctuations. In theoretically determining $T_{\rm BKT}$ in a two-dimensional Fermi gas, it is frequently assumed that, while phase fluctuations exist, the amplitude of the superfluid order parameter is fixed. However, our results indicate the necessity of including both amplitude and phase fluctuations of the order parameter in evaluating $T_{\rm BKT}$ in the crossover region. Since the BKT phase transition has recently been reported in a two-dimensional $^6$Li Fermi gas, our results would be useful for the theoretical study of this observed novel Fermi superfluid and effects of strong low-dimensional pairing fluctuations.

## 1. Introduction

Experimental techniques in an ultracold Fermi gas enable us to tune various physical parameters [1, 2]. In particular, a tunable pairing interaction associated with a Feshbach resonance has been used for the study of strong-coupling effects from the weak-coupling BCS (Bardeen-Cooper-Schrieffer) regime to the strong-coupling BEC (Bose-Einstein condensation) regime [2, 3, 4]. In the BCS-BEC crossover region, the so-called pseudogap phenomenon has been observed near the superfluid phase transition temperature [5, 6], which indicates the presence of strong pairing fluctuations.

The high tunability of an ultracold Fermi gas has also realized a two-dimensional system, by using an optical lattice technique. Since pairing fluctuations are enhanced by the low dimensionality of the system, this system is used for the study of, not only low-dimensional physics, strong-coupling physics. As a typical interesting many-body phenomenon, the possibility of Berezinskii-Kosterlitz-Thouless (BKT) phase [7, 8] has recently been discussed in this artificial low-dimensional Fermi system [9, 10, 11, 12, 13, 14, 15, 16].

In considering the BKT transition in a two-dimensional Fermi gas, a simple approach is that one takes into account phase fluctuations of the superfluid order parameter $\Delta$ with a fixed value of $|\Delta|$.

IOP Conf. Series: Journal of Physics: Conf. Series **969** (2018) 012012 doi:10.1088/1742-6596/969/1/012012

This is analogues to the two-dimensional XY-model, where the magnitude and the direction of spins correspond to $|\Delta|$ and the phase of the order parameter, respectively. Using this approach, it has recently been evaluated $T_{\rm BKT}$ from the weak- to strong-coupling regime of a two-dimensional Fermi gas [17], where phase fluctuations are included within the framework of the strong-coupling theory developed by Nozières and Schmitt-Rink[18]. In this paper, we assess the importance of amplitude fluctuations of the order parameter near $T_{\rm BKT}$, from the viewpoint of pseudogap phenomenon associated with pairing fluctuations. Since the magnitude of the order parameter $|\Delta|$ is deeply related to the binding energy of a Cooper pair, while the appearance of a full gap structure in the single particle density of states $\rho(\omega)$ around $\omega = 0$ indicates the establishment of a fixed value of $|\Delta|$, the so-called pseudogap structure (which is characterized by a dip structure with a non-zero DOS around $\omega = 0$) in $\rho(\omega)$ implying the fluctuating $|\Delta|$.

We note that effects of amplitude fluctuations have recently been discussed within the Gaussian fluctuations theory (NSR theory) [19, 20]. However, in addition to the fact that NSR theory cannot explain the recent experiments on thermodynamic quantities, it is also known to unphysically give *negative* DOS around the Fermi level in the BCS-BEC crossover region. This problem is solved by extending the NSR theory to the so-called non-selfconsistent $T$-matrix approximation (TMA), which has extensively been used to examine various pseudogap phenomena in a three-dimensional ultracold Fermi gas [21, 22]. However, in the present two-dimensional case, it has been shown[23] that TMA unphysically gives a clear gap structure in DOS with the gap size being comparable to the Fermi energy at low temperatures, even in the weak-coupling normal phase. Thus, to overcome this, in this paper, we employ the self-consistent $T$-matrix approximation(SCTMA)[24, 25], to include two-dimensional pairing fluctuations in a consistent manner. Although SCTMA cannot deal with the BKT phase transition, we expect that it can correctly describe strong-coupling effects associated with pairing fluctuations in the low temperature region, where the BKT transition has recently been reported in a $^6$Li Fermi gas[14, 15]. We briefly note SCTMA has been applied to evaluate DOS in the weak-coupling regime of a two-dimensional Fermi gas[25, 26]. In this paper, we investigate the pseudogap phenomenon over the entire interaction regime, to clarify how the pseudogap develops in DOS, with increasing the interaction strength.

Throughout this paper, we take $\hbar = k_{\rm B} = 1$ and the two-dimensional system area is taken to be unity, for simplicity.

## 2. Formulation

We consider a uniform two-dimensional Fermi gas, described by the model Hamiltonian

$$H = \sum_{\boldsymbol{p},\sigma} \xi_{\boldsymbol{p}} c^\dagger_{\boldsymbol{p},\sigma} c_{\boldsymbol{p},\sigma} - U \sum_{\boldsymbol{p},\boldsymbol{p}',\boldsymbol{q}} c^\dagger_{\boldsymbol{p}+\boldsymbol{q}/2,\uparrow} c^\dagger_{-\boldsymbol{p}+\boldsymbol{q}/2,\downarrow} c_{-\boldsymbol{p}'+\boldsymbol{q}/2,\downarrow} c_{\boldsymbol{p}'+\boldsymbol{q}/2,\uparrow}. \tag{1}$$

Here, $\boldsymbol{p} = (p_x, p_y)$ is the two-dimensional momentum, and $\xi_{\boldsymbol{p}} = p^2/(2m) - \mu$ is the kinetic energy measured from the Fermi chemical potential $\mu$, where $m$ is an atomic mass. $c^\dagger_{\boldsymbol{p},\sigma}$ is the creation operator of a Fermi atom, where pseudospin $\sigma = \uparrow, \downarrow$ describes two atomic hyper-fine states. $-U(<0)$ is a pairing interaction, which is related to the two-dimensional s-wave scattering length $a_{\rm 2D}$ as [27],

$$\frac{1}{U} = \frac{m}{2\pi} \ln(k_{\rm F} a_{\rm 2D}) + \sum_{p \geq k_{\rm F}} \frac{m}{p^2}. \tag{2}$$

Here, $k_{\rm F} = \sqrt{2\pi N}$ is the Fermi momentum, and $N$ is the total number of Fermi atoms. The strong-coupling regime corresponds to $\ln(k_{\rm F} a_{\rm 2D}) \lesssim -1$, and the weak-coupling regime corresponds to $\ln(k_{\rm F} a_{\rm 2D}) \gtrsim 1$. The region between the two, $-1 \lesssim \ln(k_{\rm F} a_{\rm 2D}) \lesssim 1$, is sometimes called the crossover region in the literature.

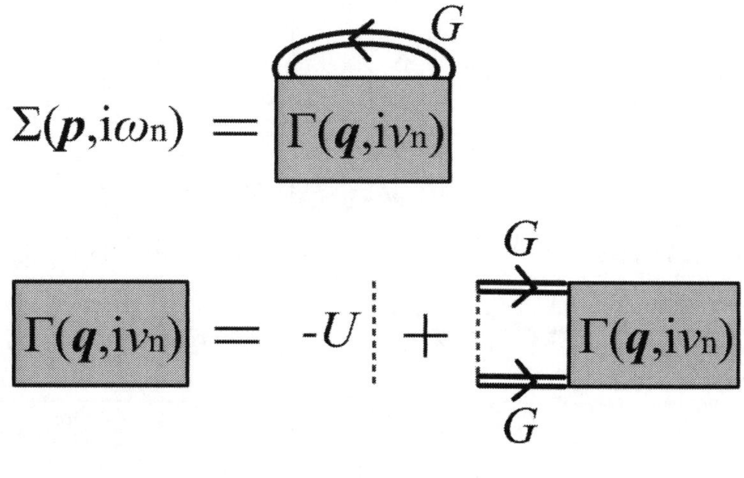


**Figure 1.** Self-energy $\Sigma(\boldsymbol{p}, i\omega_n)$ and the particle-particle scattering matrix $\Gamma$ in the self-consistent $T$-matrix approximation (SCTMA). The double-solid line is the SCTMA dressed Green's function $G$ and the dotted line represents the pairing interaction $-U(<0)$.

In considering single-particle properties of the system, the single-particle thermal Green's function,

$$G(\boldsymbol{p}, i\omega_n) = \frac{1}{i\omega_n - \xi_{\boldsymbol{p}} - \Sigma(\boldsymbol{p}, i\omega_n)}, \tag{3}$$

is useful. Here, $\omega_n$ is the fermion Matsubara frequency. Within the framework of the self-consistent $T$-matrix approximation (SCTMA), the self-energy $\Sigma(\boldsymbol{p}, i\omega_n)$ is diagrammatically described as Fig. 1, which gives [24, 25, 26],

$$\Sigma(\boldsymbol{p}, i\omega_n) = T \sum_{\boldsymbol{q}, i\nu_n} \Gamma(\boldsymbol{q}, i\nu_n) G(\boldsymbol{q} - \boldsymbol{p}, i\nu_n - i\omega_n), \tag{4}$$

$$\Gamma(\boldsymbol{q}, i\nu_n) = -\frac{U}{1 - U\Pi(\boldsymbol{q}, i\nu_n)}. \tag{5}$$

Here, $\Gamma(\boldsymbol{q}, i\nu_n)$ is the particle-particle scattering matrix, where $\nu_n$ is the boson Matsubara frequency, and

$$\Pi(\boldsymbol{q}, i\nu_n) = T \sum_{\boldsymbol{p}, i\omega_n} G\left(\boldsymbol{p} + \frac{\boldsymbol{q}}{2}, i\nu_n + i\omega_n\right) G\left(-\boldsymbol{p} + \frac{\boldsymbol{q}}{2}, -i\omega_n\right) \tag{6}$$

is a pair-correlation function, describing fluctuations in the Cooper channel.

As usual, the Fermi chemical potential $\mu$ is determined from the equation of the total number $N$ of Fermi atoms,

$$N = 2T \sum_{\boldsymbol{p}, i\omega_n} G(\boldsymbol{p}, i\omega_n). \tag{7}$$

Using the calculated $\mu$ from the number equation (7), we evaluate the single-particle density of states $\rho(\omega)$,

$$\rho(\omega) = -\frac{1}{\pi} \sum_{\boldsymbol{p}} \mathrm{Im} G(\boldsymbol{p}, i\omega_n \to \omega_+ = \omega + i\delta), \tag{8}$$

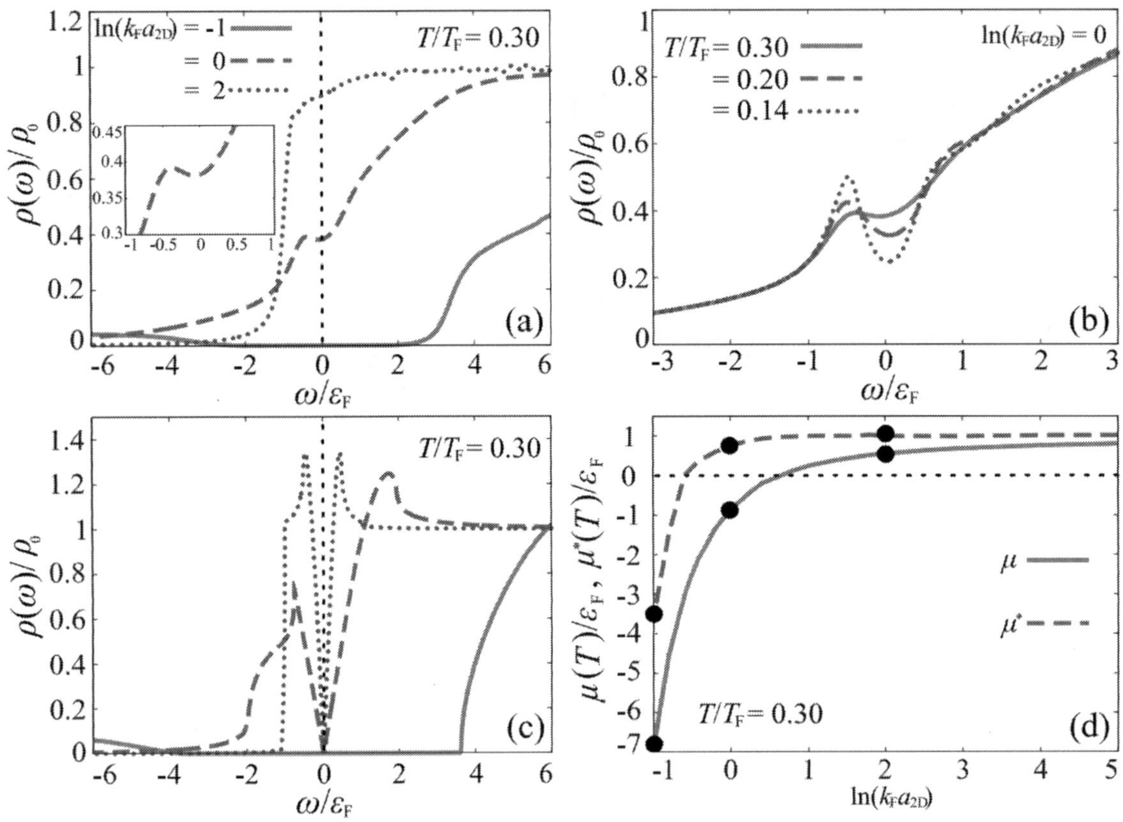


**Figure 2.** (a) Calculated single-particle density-of-states $\rho(\omega)$ in SCTMA when $T/T_F = 0.3$. $\rho_0 = m/(2\pi)$ is the density of states in a two-dimensional free Fermi gas. $T_F(\varepsilon_F)$ is the Fermi temperature (energy). The inset shows DOS around $\omega = 0$ when $\ln(k_F a_{2D}) = 0$, to clearly show the pseudogap structure in this case. (b) Density of states $\rho(\omega)$ at various temperatures. We take $\ln(k_F a_{2D}) = 0$ (crossover region). The lowest temperature ($T/T_F = 0.14$) is close to the BKT phase transition temperature ($T_{BKT}/T_F = 0.12$), predicted by a theory that only include phase fluctuations of the order parameter [17]. (c) Calculated single-particle density-of-states $\rho(\omega)$ by using the approximate Green's function in Eq. (10), when $T/T_F = 0.3$. (d) Chemical potential $\mu$, as well as the effective chemical potential $\mu^*$ determined from Eq. (11) as a function of the interaction strength $\ln(k_F a_{2D})$, when $T/T_F = 0.3$. The interaction strengths considered in panels (a) and (c) are shown as the filled circles. (Color figure online)

to examine the pseudogap phenomenon caused by strong two-dimensional pairing fluctuations.

We briefly note that SCTMA cannot describe the BKT phase transition temperature $T_{BKT}$. Thus, this paper only deals with the normal phase above $T_{BKT}$. $T_{BKT}$ and strong coupling effects have recently been discussed, from the weak-coupling to strong-coupling regime, by only including phase fluctuations of the superfluid order parameter (with a fixed value of the magnitude of the order parameter). In this regard, although the present SCTMA approach cannot treat $T_{BKT}$, both phase and amplitude fluctuations in the Cooper channel are taken into account on an equal footing.

IOP Conf. Series: Journal of Physics: Conf. Series **969** (2018) 012012 doi:10.1088/1742-6596/969/1/012012

## 3. Results

Figure 2(a) shows the SCTMA single-particle density of states (DOS) $\rho(\omega)$ in a two-dimensional Fermi gas ($T/T_{\rm F} = 0.3$). In the strong-coupling regime ($\ln(k_{\rm F}a_{\rm 2D}) = -1$), one sees a large full gap structure in $\rho(\omega)$ around $\omega = 0$, indicating the existence of large binding energy of a bound pair. Indeed, within a two-particle problem, the binding energy $E_{\rm bind} = 1/(ma_{2D}^2)$ is very large as $E_{\rm bind}/\varepsilon_{\rm F} \gg 1$. In this case, since the molecular dissociation is difficult to occur when the temperature is much lower than the gap energy seen in Fig. 2. In the language of the superfluid order parameter $\Delta$, the amplitude is established and the phase is only fluctuating, to destroy the superfluid (BKT) order.

In contrast, such a full-gap is not obtained in the intermediate coupling case ($\ln(k_{\rm F}a_{\rm 2D}) = 0$), as well as in the weak-coupling regime ($\ln(k_{\rm F}a_{\rm 2D}) = +2$), as shown in Fig. 2 (a). We only see the so-called pseudogap (dip) in $\rho(\omega)$ around $\omega = 0$ in the former (although $\rho(\omega \simeq 0)$ is also found to be modified by pairing fluctuations when $\ln(k_{\rm F}a_{\rm 2D}) = +2$). Even when one decreases the temperature down to be close to the predicted $T_{\rm BKT} = 0.12T_{\rm F}$ [17] when $\ln(k_{\rm F}a_{\rm 2D}) = 0$, Fig. 2 (b) shows that the non-zero value of $\rho(\omega = 0)$ continues to remain. These results imply that the assumption of a static magnitude of the superfluid order parameter in evaluating $T_{\rm BKT}$ should be improved in considering the intermediate and weak-coupling regime ($\ln(k_{\rm F}a_{\rm 2D}) \gtrsim 0$).

To relate the presented results in Fig. 2 (a) and (b) to the picture of "static amplitude $|\Delta|$" in a simple manner, it is convenient to employ the static approximation to pairing fluctuations [21, 28, 29], given by

$$\Sigma(\boldsymbol{p},\omega_+) = -\Delta_{\rm PG}^2 G(-\boldsymbol{p},-\omega_+) + F(\boldsymbol{p},\omega_+). \tag{9}$$

When $\Delta_{\rm PG} = \sqrt{-T\sum_{\boldsymbol{q},i\nu_n}\Gamma(\boldsymbol{q},i\nu_n)}$ is frequently referred to as the pseudogap parameter in the literature, which corresponds to $|\Delta|$ in the above discussions. We briefly note that $\Delta_{\rm PG}$ is also related to the Tan's contact $C$ as $C = \Delta_{\rm PG}^2$ [30]. For the remaining part $F(\boldsymbol{p},\omega_+)$ in Eq. (9), when we treat it with the mean-field level by approximately taking $F(\boldsymbol{p},\omega_+)$ to be a constant ($\equiv \alpha$), the analytic-continual SCTMA Green's function in Eq. (3) is approximated to

$$G(\boldsymbol{p},\omega_+) = \frac{\omega_+ + \xi'_{\boldsymbol{p}}}{\omega_+^2 - \xi'^2_{\boldsymbol{p}} - 2\Delta_{\rm PG}^2\left[1+\sqrt{1-\frac{4\Delta_{\rm PG}^2}{\omega_+^2-\xi'^2_{\boldsymbol{p}}}}\right]^{-1}}, \tag{10}$$

Here, $\xi'_{\boldsymbol{p}} = \xi_{\boldsymbol{p}} + \alpha$. In using Eqs. (9) and (10), " the mean-field shift $\alpha$" is determined so that " the effective chemical potential $\mu^*$ " shown in Fig. 2 (d) which is determined from the equation,

$$-\mu^* = -\mu + {\rm Re}[\Sigma(\boldsymbol{p}=\boldsymbol{0},\omega_+ = -\mu^* + i\delta)] \tag{11}$$

can be reproduced when the approximate self-energy in Eq. (9) is used for SCTMA $\Sigma(\boldsymbol{p},\omega_+)$ in Eq. (11). We briefly note that when we ignore $\Delta_{\rm PG}^2$ in $[\cdots]^{-1}$ in the denominator of Eq. (10), the Green's function in Eq. (10) has the same form as the diagonal component of the BCS Green's function when the pseudogap parameter $\Delta_{\rm PG}$ is replaced by the BCS superfluid order parameter $\Delta$.

The Green's function in the static approximation in Eq. (10) can explain the large gap structure in DOS in the strong coupling region ($\ln(k_{\rm F}a_{\rm 2D}) = -1$), as shown in Fig. 2 (c). We also find from this figure that the static approximation overestimates the pseudogap phenomenon when $\ln(k_{\rm F}a_{\rm 2D}) = 0$ and 2. (Note that $\rho(\omega = 0) = 0$ is always obtained in the present static approximation). Thus, as discussed previously, we need to treat both amplitude and phase fluctuating in this regime in considering the BKT transition. In this regard, we briefly note that, when we employ the Gaussian fluctuation theory for this purpose, $\rho(\omega)$ exhibits a full-gap structure near the predicted $T_{\rm BKT}$ even in the weak- and intermediate- coupling regime[31]. Thus, to examine amplitude fluctuations, self-consistent treatment of pairing fluctuations as we have done in this paper is important.

## 4. Summary

To summarize, we have discussed strong-coupling corrections to single-particle density of states $\rho(\omega)$ in a two-dimensional Fermi gas, from the strong-coupling regime to the weak-coupling regime. Including pairing fluctuations within the framework of a self-consistent $T$-matrix approximation (SCTMA), we showed that, while BCS-type gap develops in the strong-coupling region, in the crossover and weak-coupling region, the density of states $\rho(\omega)$ at $\omega = 0$ is still non-zero, exhibiting the pseudogap structure, even in the low temperature region where $T_{\rm BKT}$ is predicted. So far, it has frequently been assumed a fixed magnitude of the superfluid order parameter in theoretically determining $T_{\rm BKT}$ in this system. However, our results indicate the importance of amplitude fluctuations of the order parameter which lead to, not a fully gapped, but a pseudogapped DOS, in addition to phase fluctuations, in considering the BKT transition in the intermediate coupling regime. For this purpose, extension of the SCTMA to the theory which can deal with BKT transition would be a critical future problem. Since BKT transition is of great interest in cold Fermi gas physics, our results would contribute to the future developments of this field.

## 5. Acknowledgments

We thank H. Tajima and P. van Wyk for discussions. This work was supported by the KiPAS project in Keio university. M. M. was supported by KLL PhD Program Research Grant. D. I. was supported by Grant-in-Aid for Young Scientists (B) (No. JP16K17773) from JSPS. R. H. was supported by Grant-in-Aid for JSPS fellows. Y. O. was supported by Grand-in-Aid for Scientific Research from MEXT and JSPS in Japan (No. JP15K00178, No. JP15H00840, No. JP16K05503).

## 6. References

[1] V. Gurarie, and L. Radzihovsky, Ann. Phys. **332**, 2 (2007).
[2] I. Bloch, J. Dalibard, and W. Zwerger, Rev. Mod. Phys. **80**, 885 (2008).
[3] S. Giorgini, L. P. Pitaevskii, and S. Stringari, Rev. Mod. Phys. **80**, 1215 (2008).
[4] *The BCS-BEC Crossover and the Unitary Fermi Gas*, edited by W. Zwerger (Springer-Verlag, Berlin, 2012).
[5] J.T. Stewart, J.P. Gaebler, D.S. Jin, Nature **454**, 744 (2008).
[6] J.P. Gaebler, *et al.*, Nat. Phys. **6**, 569 (2010).
[7] V. L. Berezinskii, Sov. Phys. JETP **32**, 493 (1971).
[8] J. M. Kosterlitz, and D. J. Thouless, J. Phys. C **6**, 1181 (1973).
[9] K. Martiyanov, V. Makhalov, and A. Turlapov, Phys. Rev. Lett. **105**, 030404 (2010).
[10] M. Feld, *et al.*, Nature **480**, 75 (2011).
[11] B. Fröhlich, *et al.*, Phys. Rev. Lett. **106**, 105301 (2011).
[12] A. T. Sommer, *et al.*, Phys. Rev. Lett. **108**, 045302 (2012).
[13] V. Makhalov, K. Martiyanov, and A. Turlapov, Phys. Rev. Lett. **112**, 045301 (2014).
[14] M. G. Ries, *et. al.*, Phys. Rev. Lett. **114**, 230401 (2015).
[15] P. A. Murthy, *et al.*, Phys. Rev. Lett. **115**, 010401 (2015).
[16] K. Fenech, *et al.*, Phys. Rev. Lett. **116**, 045302 (2016).
[17] J. Tempere, S. N. Klimin, and J. T. Devreese, Phys. Rev. A **79**, 053637 (2009).
[18] P. Nozières and S. Schmitt-Rink, J. Low. Temp. Phys. **59**, 195 (1985).
[19] L. Salasnich, and G. Bighin, *J. Supercond. Nov. Magn*, (2016).
[20] B. C. Mulkerin, *et al.*, arXiv e-prints, 1702.07091 (2017).
[21] S. Tsuchiya, R. Watanabe, and Y. Ohashi, Phys. Rev. A **80**, 033613 (2009).
[22] A. Perali, P. Pieri, G. C. Strinati, and C. Castellani, Phys. Rev. B **66**, 024510 (2002).
[23] M. Matsumoto, R. Hanai, D. Inotani, and Y. Ohashi, J. Low Temp. Phys. pp. 1 (2017).
[24] R. Haussmann, Z. Phys. B: Condens. Matter **91**, 291 (1993).
[25] M. Bauer, M. M. Parish, T. Enss, Phys. Rev. Lett. **112**, 135302 (2014).
[26] B. C. Mulkerin, *et al.*, Phys. Rev. A **92**, 063636 (2015).
[27] S. A. Morgan, M. D. Lee, and K. Burnett, Phys. Rev. A **65**, 022706 (2002).
[28] R. Hanai and Y. Ohashi, Phys. Rev. A **90**, 043622 (2014).
[29] Q. J. Chen, and K. Levin, Phys. Rev. Lett. **102**, 190402 (2009).
[30] S. Tan, Ann. Phys. **323**, 2971 (2008).
[31] M. Matsumoto, and Y. Ohashi, J. Phys.: Conf. Ser. **568**, 012012 (2014).

# Specific heat and effects of strong pairing fluctuations in a superfluid Fermi atom gas in the BCS-BEC crossover region

**Pieter van Wyk, Daisuke Inotani, Yoji Ohashi**
Department of Physics, Keio University, 3-14-1 Hiyoshi, Kohoku-ku, Yokohama 223-8522, Japan

**Abstract.** We theoretically investigate the specific heat at constant volume $C_V$ in the BCS(Bardeen-Cooper-Schrieffer)-BEC(Bose-Einstein-condensation)-crossover regime of an ultracold Fermi gas, below the superfluid phase transition temperature $T_c$. Within the strong-coupling framework developed by Nozières and Schmitt-Rink, we show that the temperature dependence of $C_V$ drastically changes as one passes through the crossover region, and is sensitive to strong fluctuations in the Cooper channel near the unitarity limit. We also compare our results to a recent experiment on a $^6$Li unitary Fermi gas. Since fluctuation effects are a crucial key in the BCS-BEC-crossover phenomenon, our results would be helpful in considering how the fermionic BCS superfluid changes into BEC with increasing the interaction strength, from the viewpoint of specific heat.

## 1. Introduction

Superluid Fermi gases have recently attracted much theoretical and experimental attention as a highly controllable many-body quantum system[1, 2, 3, 4, 5, 6, 7, 8]. Using a tunable pairing interaction associated with a Feshbach resonance[6], we have realised the so-called BCS (Bardeen-Cooper-Schrieffer)-BEC (Bose Einstein condensation) crossover phenomenon, where we can study a Fermi superfluid and a Bose superfluid in a unified manner. In the BCS-BEC crossover region, the system is dominated by strong superfluid fluctuations, so that fluctuation corrections to physical properties in this regime have extensively been discussed[9, 10, 11]. Recently various thermodynamic quantities have become accessible in cold Fermi gas physics[12, 13, 14, 15], allowing us to directly compare strong-coupling theoretical results with experimental data observed in the BCS-BEC crossover region[10, 16, 17, 18].

The specific heat at constant volume $C_V$, which has recently become experimentally accessible in cold Fermi gas physics[18], was shown to be sensitive to pairing fluctuations above the phase transition temperature $T_c$ in the cases of an isotropic s-wave pairing interaction[19], as well as an anisotropic p-wave [20, 21, 22] one. Using this strong point, Refs.[19] and [20] succeded in identifying the region where pairing fluctuations dominate over normal-state properties, in the phase diagram of a Fermi gas in terms of the temperature and the s-wave and p-wave interaction strength, respectively. Thus, it is an interesting topic to see whether or not this thermodynamic quantity is also useful to detect pairing (superfluid) fluctuations below $T_c$. In this regard, we briefly note that a similar study has been done for the spin susceptibility in the BCS-BEC crossover regime of a superfluid Fermi gas[23].

In this paper, we theoretically investigate the specific heat at constant volume $C_V$ in the superfluid Fermi gas. We show that inclusion of superfluid fluctuations within the strong-coupling theory developed by Nozières and Schmitt-Rink (NSR)[24] extended to the superfluid phase bellow $T_c$ [25] naturally gives

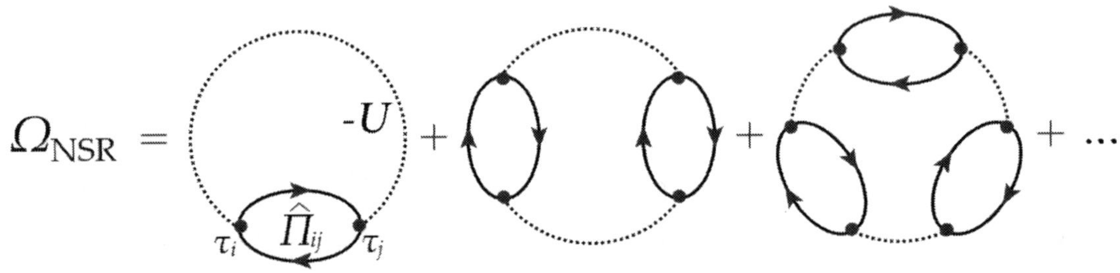


**Figure 1.** Feynman diagrams describing superfluid corrections $\Omega_{\rm NSR}$ to the thermodynamic potential $\Omega$. $-U<0$ is an attractive interaction. $\hat{\Pi}_{ij}$ is the $ij$ component of the particle-particle correlation function in Eq.(5).

a $T^3$ behaviour of $C_V$ at low temperatures in the unitarity limit, coming from collective excitations of a Goldstone mode with gapless linear dispersion. We point out that this temperature dependence is qualitatively different from the well-known $T^{\frac{3}{2}}$ behaviour in an ideal Bose gas expected in the strong coupling BEC limit, as well as the exponential suppression in the mean-field BCS result (where effects of single-particle excitations are only taken into account), so that $C_V$ may be used to identify the region where phase fluctuations of the superfluid order parameter dominate over low-energy properties of a superfluid Fermi gas in the BCS-BEC crossover region. Throughout this paper we take $\hbar = k_{\rm B} = 1$, and the system volume $V$ is taken to be unity, for simplicity.

## 2. Formalism

We consider a two component uniform superfluid Fermi gas, described by the ordinary BCS Hamiltonian given in the Nambu representation

$$H = \sum_p \Psi_p^\dagger [\xi_p \tau_3 - \Delta\tau_1] \Psi_p - U \sum_q \rho_+(\mathbf{q}) \rho_-(-\mathbf{q}), \tag{1}$$

where

$$\Psi_p = \begin{bmatrix} c_{p,\uparrow} \\ c^\dagger_{-p,\downarrow} \end{bmatrix}, \tag{2}$$

is the Nambu field acting on particle hole space.

$c_{p,\sigma}$ is the annihilation operator of a Fermi atom with pseudospin $\sigma = \uparrow, \downarrow$, describing two atomic hyperfine states contributing to Cooper-pairs. $\xi_p = \varepsilon_p - \mu = p^2/2m - \mu$ is the kinetic energy, measured from the Fermi chemical potential $\mu$ (where $m$ is an atomic mass), $\Delta$ is the superfluid order parameter, and $-U<0$ is an attractive contact type s-wave interaction. $\tau_j\,(j=1,2,3)$ are the Pauli matrices. In Eq.(1) $\rho_\pm(\mathbf{q}) = [\rho_1(\mathbf{q}) \pm i\rho_2(\mathbf{q})]/2$, where $\rho_j(\mathbf{q}) = \sum_p \Psi^\dagger_{p+q/2} \tau_j \Psi_{p-q/2}$ are generalised density operators describing amplitude ($j=1$) and phase ($j=2$) fluctuations of the superfluid order parameter $\Delta$[25, 26].

The interaction strength is measured in terms of the inverse $s$-wave scattering length $a_s^{-1}$, which is related to $-U$ through the renormalization prescription,

$$-\frac{1}{U} = -\frac{m}{4\pi a_s} + \sum_p \frac{1}{2\varepsilon_p}. \tag{3}$$

In the NSR theory, fluctuation corrections ($\equiv \Omega_{\rm NSR}$) to the thermodynamic potential $\Omega = \Omega_0 + \Omega_{\rm NSR}$ are diagrammatically given as Fig.1, where

$$\Omega_0 = \sum_p \left[ \xi_p - E_p + \frac{\Delta^2}{2\varepsilon_p} + 2T\ln[1 - n_{\rm F}(E_p)] \right] - \frac{m\Delta^2}{4\pi a_s} \tag{4}$$

is the thermodynamic potential in the mean-field BCS theory. In Eq.(4) $E_p = \sqrt{\xi_p^2 + \Delta^2}$, and $n_F(E_p) = \left[e^{E_p/T} + 1\right]^{-1}$ is the Fermi distribution function. Summing up the diagrams in Fig.1, we have

$$\Omega_{NSR} = \frac{T}{2} \sum_{q,i\nu_n} \mathrm{tr} \left[ \ln \left[ \frac{m}{4\pi a_s} - \hat{\Pi}(q, i\nu_n) + \sum_p \frac{1}{2\varepsilon_p} \right] \right], \tag{5}$$

where

$$\hat{\Pi}(q, i\nu_n) = \begin{bmatrix} \Pi^{-+}(q, i\nu_n) & \Pi^{--}(q, i\nu_n) \\ \Pi^{++}(q, i\nu_n) & \Pi^{+-}(q, i\nu_n) \end{bmatrix}, \tag{6}$$

with

$$\begin{aligned} \Pi^{-+}(q, i\nu_n) &= -\frac{1}{4} \sum_{p,s=\pm1} \frac{\tanh\frac{E_+}{2T} + s\tanh\frac{E_-}{2T}}{(E_+ + sE_-)^2 + \nu_n^2} \left[ (E_+ + sE_-)\left(1 + s\frac{\xi_+\xi_-}{E_+E_-}\right) + i\nu_n\left(\frac{\xi_+}{E_+} + s\frac{\xi_+}{E_+}\right) \right], \\ \Pi^{--}(q, i\nu_n) &= \frac{1}{4} \sum_{p,s=\pm1} \frac{\Delta^2}{E_+E_-} \frac{s(E_+ + sE_-)}{(E_+ + sE_-)^2 + \nu_n^2} \left[ \tanh\frac{E_+}{2T} + s\tanh\frac{E_-}{2T} \right], \end{aligned} \tag{7}$$

being the pair correlation function describing superfluid fluctuations (where $\xi_\pm = \xi_{p\pm p/2}$ and $E_\pm = E_{p\pm p/2}$).

The superfluid order parameter $\Delta$ and the Fermi chemical potential $\mu$ for a given temperature $T$ are self consistently determined from the coupled particle number equation $N = N_0 + N_{NSR}$ (where $N_0 = -\left(\frac{\partial\Omega_0}{\partial\mu}\right)_T$ and $N_{NSR} = -\left(\frac{\partial\Omega_{NSR}}{\partial\mu}\right)_{T,\Delta}$) with the so-called Thouless criterion[27]

$$\det \left[ \frac{m}{4\pi a_s} - \hat{\Pi}(q = \mathbf{0}, i\nu_n = 0) + \sum_p \frac{1}{2\varepsilon_p} \right] = 0. \tag{8}$$

The specific heat at constant volume $C_V$ is then calculated from the relation

$$C_V = \left(\frac{\partial E}{\partial T}\right)_N, \tag{9}$$

where the internal energy $E$ of the system is conveniently obtained from the Legendre transformation

$$E = \Omega - T\left(\frac{\partial\Omega}{\partial T}\right)_\mu - \mu\left(\frac{\partial\Omega}{\partial\mu}\right)_T. \tag{10}$$

## 3. Results

Figure.2 (a) shows the specific heat at constant volume $C_V$ as a function of temperature in a unitary superfluid Fermi gas. In this figure, we find that $C_V$ behaves as $C_V \sim T^3$ in the low temperature region ($T/T_c \lesssim 0.6$). Since the $T^3$-behaviour of $C_V$ is characteristic of a system with linear dispersion, this result indicates that the strong-coupling corrections in the NSR theory enables us to include effects of a collective Goldstone mode (having a phonon-like linear dispersion) associated with the broken $U(1)$ gauge symmetry in the superfluid phase. When the NSR term $\Omega_{NSR}$ in Eq.(5) is ignored, the resulting ordinary BCS theory gives the internal energy

$$\begin{aligned} E &\simeq \Omega_0 - T\left(\frac{\partial\Omega_0}{\partial T}\right)_\mu - \mu\left(\frac{\partial\Omega_0}{\partial\mu}\right)_T \\ &= \sum_p \left[ \xi_p - E_p + \frac{\Delta^2}{2\varepsilon_p} + 2E_p n_F(E_p) \right] - \frac{m\Delta^2}{4\pi a_s} + \mu N_0. \end{aligned} \tag{11}$$

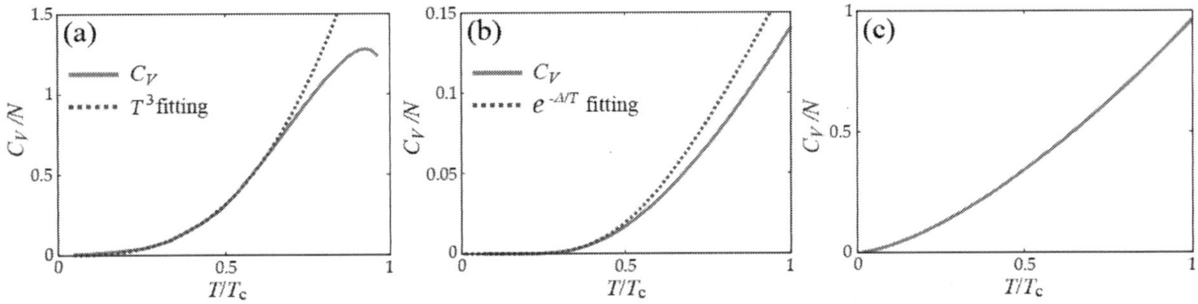


**Figure 2.** (a) Specific heat at constant volume $C_V$ in a unitary Fermi gas bellow $T_c$. (b) $C_V$ in the mean-field BCS theory, where collective excitations of a gapless Goldstone mode are ignored. (c) $C_V$ in an ideal gas of $N/2$ molecular bosons with a molecular mass $M = 2m$.

Then, as well-known in the BCS-type metallic superconductivity, the low-temperature specific heat exhibits exponential temperature dependence as $C_V = \sum_p 2E_p \frac{\partial n_F(E_p)}{\partial T} \simeq \sum_p \frac{2E_p^2}{T^2} e^{-E_p/T}$. Since the main contribution to this summation is from around $\xi_p = 0$, $C_V$ must be proportional to $e^{-\Delta/T}$. This case is shown in panel (b).

On the other hand, in the strong coupling BEC regime, our NSR formalism reduces to the Bogoliubov theory for a weakly repulsive Bose gas [26, 28], consisting of $N/2$ molecular bosons with mass $M = 2m$. In the extreme BEC limit $(k_F a_s)^{-1} \to \infty$, the molecular Bose system can be treated as a non-interacting gas, so that the internal energy in this limit is given by

$$E \simeq \sum_q \varepsilon_q^B n_B\left(\varepsilon_q^B\right) - \frac{E_{bi}}{2} N, \tag{12}$$

where $\varepsilon_q^B = q^2/2(2m)$ is the kinetic energy, $n_B\left(\varepsilon_q^B\right) = \left[e^{\varepsilon_q^B/T} - 1\right]^{-1}$ is the Bose distribution function, and $E_{bi} = 1/ma_s^2$ is the binding energy of a molecular boson. The specific heat calculated from Eq.(12) is simply given by $C_V = \frac{5}{\pi^2}(mT)^{3/2} \int_0^\infty dx \frac{x^{3/2}}{e^x - 1}$. This clearly shows that the specific heat in this regime is proportional to $T^{3/2}$, which we show in Fig.2 (c).

While one cannot see the $T^{3/2}$-behaviour of $C_V$ in Fig.2 (a) $C_V$ should continuously change into the BEC result shown in Fig.2 (c). Thus, when we examine how the temperature region where $C_V$ exhibits the $T^3$-behaviour gradually disappears as one enters the strong-coupling regime, it would give the region where phase fluctuations of the order parameter dominate over the superfluid properties of the system. On the other hand, since the velocity of the Goldstone mode approaches the value $v_F/\sqrt{3}$ of the Anderson-Bogoliubov mode in the weak-coupling BCS regime (where $v_F$ is the Fermi velocity), it is unclear whether the $T^3$-behaviour of $C_V$ seen in the unitarity limit is gradually replaced by the exponential temperature dependence shown in Fig.2 (b), as one decreases the interaction strength, which remains as our future problem.

Figure 3 compares our result with the recent experiment on a $^6$Li unitary Fermi gas [18]. We see that the calculated $C_V$ in the superfluid phase agrees with the observed data, supporting the validity of the present strong-coupling approach to this thermodynamic quantity.

## 4. Conclusion

In this paper, we have discussed the specific heat at constant volume $C_V$ in a superfluid Fermi gas within the NSR theory. We showed that $C_V$ at unitarity exhibits a $T^3$-behaviour in a wide temperature region below $T_c$, indicating that excitations are dominated by a collective Goldstone mode with gapless linear

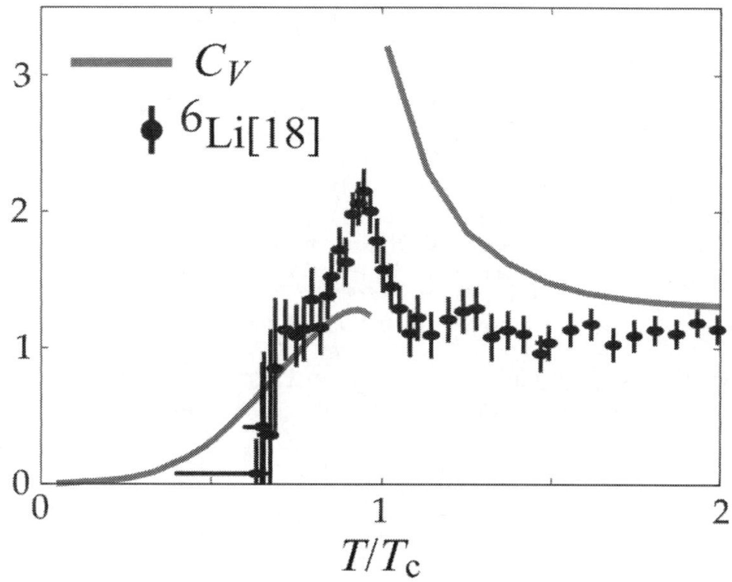


**Figure 3.** Comparison of the calculated specific heat $C_V$ in NSR theory with the recent experiment on a $^6$Li Fermi gas.

dispersion. Since this temperature dependence is not obtained in the BEC limit, as well as the mean-field BCS theory, we expect that the $T^3$-behaviour of this thermodynamic quantity would be useful for identifying the region where phase fluctuations of the superfluid order parameter are crucial for system properties in the phase diagram of an ultracold Fermi gas in terms of the temperature and interaction strength.

**Acknowledgments**
We would like to thank M. Matsumoto, D. Kagamihara, and D. Kharga for useful discussions. This work was supported by KiPAS project in Keio University. YO was also supported by Grant-in-Aid for Scientific research from MEXT and JSPS in Japan (No.15K00178, No.15H00840, No.16K05503).

## 5. References

[1] C. A. Regal, M. Greiner and D. S. Jin, Phys. Rev. Lett. **92**, 040403 (2004).
[2] J. Kinast, S. L. Hemmer, M. E. Gehm, A. Turlapov and J. E. Thomas, Phys. Rev. Lett. **92**, 150402 (2004).
[3] M. Bartenstein, A. Altmeyer, S. Riedl, S. Jochim, C. Chin, J. H. Denschlang and R. Grimm, Phys. Rev. Lett. **92**, 203201 (2004).
[4] M. W. Zwierlein, J. R. Abo-Shaeer, A. Schirotzek, C. H. Schunck and W. Ketterle, Nature **435**, 1047 (2005).
[5] W. Ketterle and M. W. Zwierlein, arXiv:cond-mat/0801.2500 (2008).
[6] C. Chin, R. Grimm, P. Julienne and E. Tiesinga, Rev. Mod. Phys. **82**, 1225 (2010).
[7] W. Zwerger, *The BCS-BEC Crossover and the Unitary Fermi Gas* (Springer - Verlag, Berlin, 2012).
[8] H. Hu, X. J. Lui and P. D. Drummond, Phys. Rev. A **73**, 023617 (2006).
[9] S.Tsuchiya, R. Watanabe, Y. Ohashi, Phys. Rev. A **80**, 033613 (2009).
[10] H. Hu, X. Lui and P. Drummond, Phys. Rev. A **77**, 061605 (2008).
[11] A. Perali, F. Palestini, P. Pieri, G. C. Strinati, J. T. Stewart, J. P. Gaebler, T. E. Drake and D. S. Jin, Phys. Rev. Lett. **106**, 060402 (2011).
[12] J. Kinast, A. Turpalov, J. E. Thomas, Q. Chen, J. Stajic and K. Levin, Science, **307**, 1296 (2005).
[13] L. Luo and J. E. Thomas, J. Low. Temp. Phys **154**, 1 (2009).
[14] M. Horikoshi, S. Nakajima, M. Ueda and T. Mukaiyama, Science **327**, 442 (2010).
[15] S. Nascimbene, N. Navon, K. J. Jiang, F. Chevy and C. Salomon, Nature **463**, 1057 (2010).

[16] R. Haussmann, W. Rantner, S. Cerrito and W. Zwerger, Phys. Rev. A **75**, 023610 (2007).
[17] A. Perali, F. Palestini, P. Pieri, G. C. Strinati, J. T. Stewart, J. P. Gaebler, T. E. Drake and D. S. Jin, Phys. Rev. Lett. **106**, 060402 (2011).
[18] M. J. H. Ku, A. T. Sommer, L. W. Cheuk, M. W. Zwierlein, Science **335**, 563 (2012).
[19] P. van Wyk, H. Tajima, R. Hanai and Y. Ohashi, Phys. Rev. A **93**, 013621 (2016).
[20] D. Inotani, P. van Wyk and Y. Ohashi, JPSJ **85**, 123301 (2016).
[21] D. Inotani, P. van Wyk and Y. Ohashi, JPSJ **86**, 024302 (2017).
[22] D. Inotani, P. van Wyk and Y. Ohashi, JPSJ **86**, 044301 (2017).
[23] H. Tajima, R. Hanai, and Y. Ohashi, Phys. Rev. A **93**, 013610 (2016).
[24] P. Nozières, and S. Schmitt-Rink, J. Low Temp. Phys. **59**, 195 (1985).
[25] Y. Ohashi and A. Griffin, Phys. Rev. A **67**, 063612 (2003).
[26] N. Fukushima, Y. Ohashi, E. Taylor and A. Griffin, Phys. Rev. A **75**, 033609 (2007).
[27] D. J. Thouless, Ann. Phys. **10**, 553 (1960).
[28] J. R. Engelbrecht, M. Randeria and C. A. R. Sáde Melo, Phys. Rev. B **55**, 15153 (1997).
[29] R. Haussmann, W. Rantner, S. Cerrito, and W. Zwerger, Phys. Rev. A **75**, 023610 (2007).

IOP Conf. Series: Journal of Physics: Conf. Series **969** (2018) 012014 doi:10.1088/1742-6596/969/1/012014

# Analysis of Bose system in spin-orbit coupled Bose-Fermi mixture to induce a spin current of fermions

**R Sakamoto, Y Ono, R Hatsuda, K Shiina, E Arahata, and H Mori**
Department of physics, Tokyo Metropolitan University, 1-1 Minamiosawa, Hachioji-shi, Tokyo 192-0397, Japan

E-mail: sakamoto-ryohei@ed.tmu.ac.jp

**Abstract.** We found that a spin current of fermions could be induced in spin-orbit coupled Bose-Fermi mixture at zero temperature. Since spatial change of the spin structure of the bosons is necessary to induce the spin current of the fermions, we analyzed the ground state of the bosons in the mixture system, using a variational method. The obtained phase diagram indicated the presence of a bosonic phase that allowed the fermions to have a spin current.

## 1. Introduction

Spin current is the counterflow of particles with opposite spins and a number of methods to induce local and global spin currents have been suggested[1, 2, 3, 4, 5]. Techniques to control electrons using a spin current have a potential to develop a novel device, called spintronics device. On the other hand, a neutral cold atom system with a synthetic spin-orbit (SO) coupling was engineered between two atomic spin states with a pair of Raman lasers[6, 7]. Since SO coupling is known to induce various phenomena involving a spin current, a number of studies of SO coupled cold atom systems have been reported[8, 9, 10, 11, 12, 13, 14, 15, 16, 17, 18]. In particular, it is known that a spiral spin structure can be induced in the SO coupled Bose-Einstein condensate (BEC)[9, 10, 11, 12]. One can expect from these studies that a spin current of fermions could be induced in a mixture of SO coupled BEC and spin-1/2 fermions.

The purpose of this paper is to reveal a possibility of inducing the spin current of fermions in the SO coupled Bose-Fermi mixture. First, we explain that a certain-type spatial spin structure of the bosons is necessary to induce the spin current of the fermions. Next, we show the ground state phase diagram of the bosons in the SO coupled Bose-Fermi mixture by using a variational method and the presence of a bosonic phase that allowed the fermions to have a spin current. Finally, we discuss the density dependence on the phase diagram.

IOP Conf. Series: Journal of Physics: Conf. Series **969** (2018) 012014 doi:10.1088/1742-6596/969/1/012014

## 2. Model and spin current

Let us consider a mixture of SO coupled pseudospin-1/2 bosons and spin-1/2 fermions. The total Hamiltonian $H$ is given by

$$H = H^{(b)} + H^{(f)} + H_0^{(bf)} + H_S^{(bf)}, \tag{1}$$

$$H^{(b)} = \sum_{j=1}^{N_b} h_{SO}^{(b)}(j) + \frac{1}{2}\sum_{\alpha,\,\beta}\int d\boldsymbol{r}\; g_{\alpha\beta} n_\alpha^{(b)} n_\beta^{(b)}, \tag{2}$$

$$H^{(f)} = \sum_{\sigma}\sum_{j=1}^{N_f} h_\sigma^{(f)}(j), \tag{3}$$

$$H_0^{(bf)} = \frac{u_0}{2}\int d\boldsymbol{r}\; n^{(b)} n^{(f)}, \tag{4}$$

$$H_S^{(bf)} = \frac{u_s}{2}\int d\boldsymbol{r}\; \boldsymbol{S}^{(b)}\cdot\boldsymbol{S}^{(f)}. \tag{5}$$

In the Bose Hamiltonian $H^{(b)}$, $\alpha$ and $\beta$ ($=\uparrow,\downarrow$) are the spin indices, $g_{\alpha\beta}$ are the spin-dependent strengths of the Bose interaction, $n_\alpha^{(b)}(\boldsymbol{r})$ is the number density of the bosons with spin $\alpha$, and

$$h_{SO}^{(b)} = \frac{1}{2m_b}(p_x + \hbar k_0\sigma_z)^2 + \frac{p_y^2 + p_z^2}{2m_b} + \frac{\Omega}{2}\sigma_x + \frac{\delta}{2}\sigma_z, \tag{6}$$

is the single-particle Hamiltonian of the bosons. Systems described by $h_{SO}^{(b)}$ have been already realized experimentally[6, 7]. In Eq. (6), $k_0$ is the strength of the SO coupling, $\Omega$ and $\delta$ are the strengths of the effective magnetic field, and $\sigma_i$, with $i = x, y, z$, is the $2\times 2$ Pauli matrix. Equation (3) expresses the Fermi Hamiltonian where $h_\alpha^{(f)} = \boldsymbol{p}_\alpha^2/(2m_f)$ is the kinetic energy of the fermions with spin $\alpha$. $H_0^{(bf)}$ and $H_S^{(bf)}$ are the Bose-Fermi particle and spin interaction Hamiltonians, respectively, where $u_0$ and $u_s$ are the spin-independent strengths of the Bose-Fermi particle and spin interaction, respectively, $n^{(f)}$ is the number density of the fermions, and $\boldsymbol{S}^{(b)}$ and $\boldsymbol{S}^{(f)}$ are the spin densities of the bosons and fermions, respectively.

The single-particle energy of the fermions including the Bose-Fermi interaction is given by

$$\varepsilon^{(f+bf)} = \int d\boldsymbol{r}\,\left[\psi^{(f)}\right]^*\left[-\frac{\hbar^2\nabla^2}{2m_f} + \frac{u_0}{2}n^{(b)} + \frac{u_s}{2}\boldsymbol{S}^{(b)}\cdot\boldsymbol{\sigma}\right]\psi^{(f)}, \tag{7}$$

where $\psi^{(f)}$ is the single-particle wave function of the fermions. This energy is just the one of the spin-1/2 Fermi system in the potential $u_0 n^{(b)}/2$ and the magnetic field $u_s\boldsymbol{S}^{(b)}/2$. The first point that we should discuss is the condition to induce the spin current in this Fermi system. To investigate the condition, we simply analyze the spin current by using a single-fermion spinor wave function[4]

$$\psi_1^{(f)} = \sqrt{n_1^{(f)}}\begin{pmatrix}\cos\frac{\gamma}{2}e^{i\phi_\uparrow}\\ \sin\frac{\gamma}{2}e^{i\phi_\downarrow}\end{pmatrix}, \tag{8}$$

where $n_1^{(f)}$ is the Fermi density normalized to unity: $\int d\boldsymbol{r}\, n_1^{(f)}(\boldsymbol{r}) = 1$. $\gamma$ is a parameter to determine the ratio between the two spinor components and $\phi_\uparrow$ and $\phi_\downarrow$ are the phases of the components. Substituting Eq. (8) into Eq. (7), we obtain the free energy of the Fermi system

$$F = \int d\boldsymbol{r} \left\{ \frac{n_1^{(f)}\hbar^2}{2m_f} \left[ \left( \frac{\nabla\gamma}{2} \right)^2 - \frac{\nabla^2 \sqrt{n_1^{(f)}}}{\sqrt{n_1^{(f)}}} + (\nabla\phi)^2 + (\nabla\phi_z)^2 - 2(\nabla\phi)(\nabla\phi_z)\cos\gamma \right] + \frac{n_1^{(f)} u_0}{2} n^{(b)} + \frac{n_1^{(f)} u_s}{2} \boldsymbol{S}^{(b)} \cdot \boldsymbol{S}_1^{(f)} \right\}, \tag{9}$$

where $\phi = \phi_\uparrow + \phi_\downarrow$, $\phi_z = \phi_\downarrow - \phi_\uparrow$, and

$$\boldsymbol{S}_1^{(f)} = \frac{\left[\psi_1^{(f)}\right]^* \boldsymbol{\sigma} \psi_1^{(f)}}{n_1^{(f)}} = (\sin\gamma\cos\phi_z, \sin\gamma\sin\phi_z, \cos\gamma). \tag{10}$$

Note that the spin quantization axis is along the $z$ axis. The spin current density is given by

$$\boldsymbol{j}_s^z = -\frac{1}{\hbar}\frac{\partial F}{\partial(\nabla\phi_z)} = \frac{n_1^{(f)}\hbar}{m_f}\left\{(\nabla\phi)\cos\gamma - \nabla\phi_z\right\}. \tag{11}$$

This equation shows that the spatial dependence of the phase $\phi_\uparrow$ or $\phi_\downarrow$ is necessary to induce the spin current. The spatial change of the phase of the fermions can be generated by a spatially-changing spin structure in the background. Therefore, for the realization of the spin current of the fermions, spatial dependence of the bosonic spins is at least necessary.

## 3. Phase diagram

Here, we analyze the ground state of the bosons in the SO coupled Bose-Fermi mixture by using a variational method and look for a state where the spin current could be induced. For simplicity, we assume that the bosons and fermions have the same mass, $m_b = m_f = m$, and that $\delta = 0$ and $g_{\uparrow\uparrow} = g_{\downarrow\downarrow} > g_{\uparrow\downarrow}$. In addition, we use a variational wave function of the bosons in the form[11, 12]

$$\psi^{(b)} = \begin{pmatrix} \psi_\uparrow^{(b)} \\ \psi_\downarrow^{(b)} \end{pmatrix} = \sqrt{n_b}\left[ C_+ \begin{pmatrix} \cos\theta \\ -\sin\theta \end{pmatrix} e^{ik_1x} + C_- \begin{pmatrix} \sin\theta \\ -\cos\theta \end{pmatrix} e^{-ik_1x} \right], \tag{12}$$

where $n_b$ is the mean number density of the bosons and $C_\pm, k_1$, and $\theta$ are variational parameters satisfying $|C_+|^2 + |C_-|^2 = 1$ and $0 \leq \theta \leq \pi/4$. For this variational function, we have the boson density

$$n_1^{(b)} = \frac{n^{(b)}}{n_b} = [1 + 2\beta\sin(2\theta)\cos(2k_1x + \eta)], \tag{13}$$

and the spin density

$$\boldsymbol{s}^{(b)} = \big(-\sin(2\theta) - 2\beta\cos(2k_1x+\eta), 2\beta\cos(2\theta)\sin(2k_1x+\eta), \alpha\cos(2\theta)\big), \tag{14}$$

where $\alpha = |C_+|^2 - |C_-|^2$, $\beta = |C_+||C_-|$, and $\eta$ is the phase difference between $C_+$ and $C_-$. Substituting Eq. (12) into Eq. (2), we can obtain the energy of the bosons as a function of the variational parameters. Without the fermions, the ground state phase diagram of the SO coupled Bose system was already calculated as in Fig. 1[11, 12]. Figure 1 shows three different phases, Phase I to III. In Phase I ($k_1 \neq 0, \alpha = 0, \beta = 1/2, 0 \leq \theta < \pi/4$), the bosons

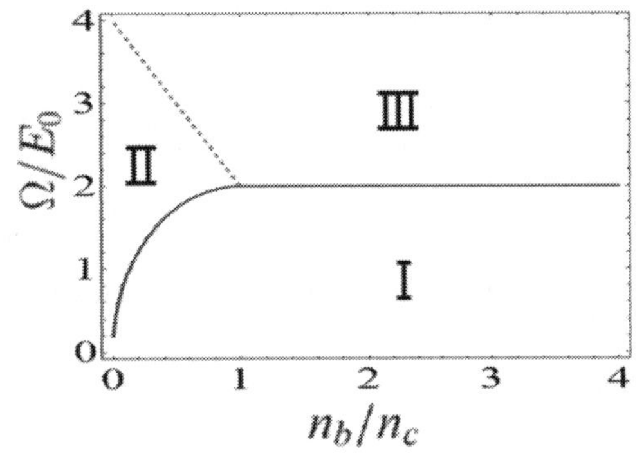


**Figure 1.** Phase diagram of the SO coupled BEC at zero temperature. $\Omega$ and $n_b$ are normalized with $E_0 = \hbar^2 k_0^2/2m$ and $n_c = E_0(g_{\uparrow\uparrow} + g_{\uparrow\downarrow})/[g_{\uparrow\uparrow}(g_{\uparrow\uparrow} - g_{\uparrow\downarrow})]$, respectively. The solid (blue) and dashed (red) lines indicate the first- and second-order transition lines, respectively. The parameters are the same as those in Ref. [18].

equally condensate in two finite-momentum states and have a spiral spin structure. In Phase II ($k_1 \neq 0, \alpha = \pm 1, \beta = 0, 0 \leq \theta < \pi/4$), all the bosons condensate in a single finite-momentum state. In Phase III ($k_1 = 0, \alpha = \pm 1, \beta = 0, \theta = \pi/4$), all the bosons condensate in a zero-momentum state. Since the spin structure does not change spatially in Phase II and III but do in Phase I, we can expect that, if fermions are introduced, the bosons in Phase I would induce a fermionic spin current. Therefore, for the Bose-Fermi mixture system to sustain a fermionic spin current, Phase I of the bosons has to remain even when the fermions are introduced to the system.

To find the ground state of the bosons in the SO coupled Bose-Fermi mixture, we calculate the energy of the Fermi system including the Bose-Fermi interaction by using the variational wave function of the bosons (12). Substituting Eqs. (13) and (14) into Eq. (7), Schrödinger equations of the fermions can be written as

$$i\hbar\frac{\partial}{\partial t}\psi^{(f)} = \begin{pmatrix} -\frac{\hbar^2\nabla^2}{2m} + U_0 n_1^{(b)} + U_s s_z^{(b)} & U_s\left(s_x^{(b)} - i s_y^{(b)}\right) \\ U_s\left(s_x^{(b)} + i s_y^{(b)}\right) & -\frac{\hbar^2\nabla^2}{2m} + U_0 n_1^{(b)} - U_s s_z^{(b)} \end{pmatrix}\psi^{(f)}, \tag{15}$$

where $U_0 = n_b u_0/2$ and $U_s = n_b u_s/2$.

### 3.1. $\Omega \sim 0$ *limit*

To calculate the energy from Eq. (15) analytically, we consider $\Omega \sim 0$ limit. In this limit, two spin components of the bosons are not coupled with each other and the variational wave function of the bosons can be written as

$$\psi^{(b)} = \begin{pmatrix} \psi_\uparrow^{(b)} \\ \psi_\downarrow^{(b)} \end{pmatrix} = \sqrt{n_b}\begin{pmatrix} C_+ e^{ik_1 x} \\ -C_- e^{-ik_1 x} \end{pmatrix}. \tag{16}$$

This function corresponds to Eq. (12) with $\theta = 0$. Therefore, Eq. (15) becomes

$$\begin{aligned} \hbar\omega\psi_{\uparrow,\boldsymbol{q}}^{(f)}(\omega) &= [\epsilon_{\boldsymbol{q}} + U_0 + U_s\alpha]\,\psi_{\uparrow,\boldsymbol{q}}^{(f)}(\omega) - 2U_s\beta e^{i\eta}\psi_{\downarrow,\boldsymbol{q}-2\boldsymbol{k}_1}(\omega), \\ \hbar\omega\psi_{\downarrow,\boldsymbol{q}}^{(f)}(\omega) &= -2U_s\beta e^{-i\eta}\psi_{\uparrow,\boldsymbol{q}+2\boldsymbol{k}_1}(\omega) + [\epsilon_{\boldsymbol{q}} + U_0 - U_s\alpha]\,\psi_{\downarrow,\boldsymbol{q}}^{(f)}(\omega), \end{aligned} \tag{17}$$

where $\psi_\sigma^{(f)}(\boldsymbol{r}, t) = \frac{1}{2\pi}\int d\omega \sum_{\boldsymbol{q}} \psi_{\sigma,\boldsymbol{q}}^{(f)}(\omega) e^{-i(\omega t + \boldsymbol{r}\cdot\boldsymbol{q})}$, $\epsilon_{\boldsymbol{q}} = \hbar^2\boldsymbol{q}^2/(2m)$, and $\boldsymbol{k}_1 = (k_1, 0, 0)$. Note that the fermions do not affect the phase transition of the bosons between Phases II and III because Eq. (17) for Phases II and III have the same form. From Eq. (17), we can obtain the

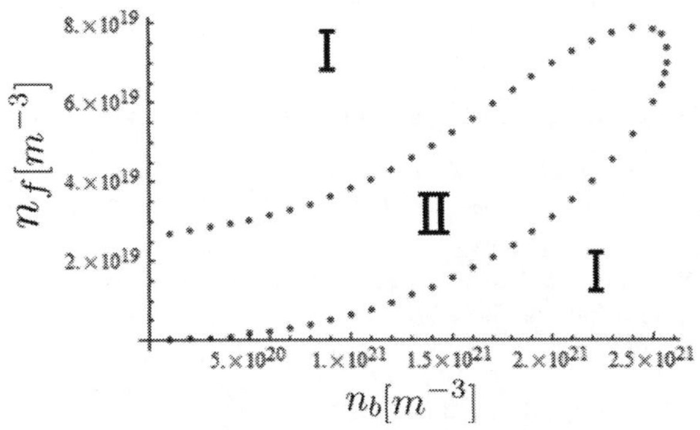


**Figure 2.** Phase diagram of the bosons in the SO coupled Bose-Fermi mixture for $\Omega \sim 0$ at zero temperature, where $u_0 = u_s = g_{\uparrow\uparrow}$. The phase transition is the first-order transition. The parameters are the same as those in Fig. 1.

single-fermion energy including Bose-Fermi interaction

$$\varepsilon_{\pm}^{(f+bf)}(\boldsymbol{q}; k_1, \alpha, \beta) = \frac{1}{2}\left[\epsilon_{\boldsymbol{q}+\boldsymbol{k}_1} + \epsilon_{\boldsymbol{q}-\boldsymbol{k}_1} + 2U_0 \pm \sqrt{(\epsilon_{\boldsymbol{q}+\boldsymbol{k}_1} - \epsilon_{\boldsymbol{q}-\boldsymbol{k}_1} - 2U_s\alpha)^2 + 16\beta^2 U_s^2}\right]. \quad (18)$$

The total energy of the Fermi system is given by $E^{(f+bf)} = \sum_{\sigma=\pm}\sum_{\boldsymbol{q}} \varepsilon_{\sigma}^{(f+bf)}\theta\left[\epsilon_F - \varepsilon_{\sigma}^{(f+bf)}\right]$, where $\theta(\epsilon)$ is the step function and $\epsilon_F$ is Fermi energy. Minimizing the total energy of the Bose-Fermi system, we can find the ground state of the bosons and obtain the ground state phase diagram of the bosons in the SO coupled Bose-Fermi mixture (Fig. 2).

Figure 2 shows that Phase I where the spin current could be induced can persist even in the presence of the fermions and Phase II partially appears. Figure 1 shows that the bosons always prefer Phase I when $\Omega \sim 0$. On the other hand, the fermions always prefer Phase II since the bosonic spins apply the spiral magnetic field in Phase I and a uniform magnetic field in Phase II to the fermions. Therefore, Phase II appears when a small amount of the fermions are first introduced to the bosons. However, as the number of the fermions increases, the energy profit that the fermions have from Phase II under the uniform magnetic field generated by the bosonic spins would decrease. This is because two Zeeman-split fermionic bands in the uniform magnetic field begin to be both filled if the fermions are added more and more. So Phase II turns into Phase I with the fermions added more to the bosons in order to let the bosons have the lower energy.

## 4. Conclusion

We investigated a possibility to induce the spin current of fermions in an SO coupled Bose-Fermi mixture at zero temperature. We focused on a zero magnetic field limit ($\Omega \sim 0$) and showed the ground state phase diagram of the bosons in the SO coupled Bose-Fermi mixture. As a result, it was found from the phase diagram that Phase I, where the bosons had a spiral spin structure and a fermionic spin current was expected, persisted even when the number of the spin-1/2 fermions increased. In addition, we discussed the density dependence on the ground state of the bosons. The present result showed that the spin current could be induced in the SO coupled Bose-Fermi mixture. The magnitude and controllability of the spin current will be studied elsewhere.

**Acknowledgments**
This work was partly supported by JSPS KAKENHI Grant Number JP 17J00717.

IOP Conf. Series: Journal of Physics: Conf. Series **969** (2018) 012014 doi:10.1088/1742-6596/969/1/012014

## References

[1] Murakami S, Nagaosa N and Zhang S C 2003 *Science* **301** 1348
[2] Bernevig B A, Orenstein J and Zhang S C 2006 *Phys. Rev. Lett.* **97** 236601
[3] Koralek J D, Weber C P, Orenstein J, Bernevig B A, Zhang S C, Mack S and Awschalom D D 2009 *Nature* **458** 610
[4] Sonin E B 2010 *Adv. Phys.* **59** 181
[5] Maekawa S, Adachi H, Uchida K, Ieda J and Saitoh E 2013 *J. Phys. Soc. Jpn.* **82** 102002
[6] Lin Y J, Jiménez-García K and Spielman I B 2011 *Nature* **471** 83
[7] Wang P, Yu Z Q, Fu Z, Miao J, Huang L, Chai S, Zhai H and Zhang J 2012 *Phys. Rev. Lett.* **109** 095301
[8] Stanescu T D, Anderson B and Galitski V 2008 *Phys. Rev.* A **78** 023616
[9] Wang C, Gao C, Jian C M and Zhai H 2010 *Phys. Rev. Lett.* **105** 160403
[10] Ho T L and Zhang S 2011 *Phys. Rev. Lett.* **107** 150403
[11] Li Y, Pitaevskii L P and Stringari S 2012 *Phys. Rev. Lett.* **108** 225301
[12] Martone G I, Li Y, Pitaevskii L P and Stringari S 2012 *Phys. Rev.* A **86** 063621
[13] Wu C J, Mondragon-Shem I and Zhou X F 2011 *Chinese Phys. Lett.* **28** 097102
[14] Beeler M C, Williams R A, Jiménez-García K, LeBlanc L J, Perry A R and Spielman I B 2013 *Nature* **498** 201
[15] Cheuk L W, Sommer A T, Hadzibabic Z, Yefsah T, Bakr W S and Zwierlein M W 2012 *Phys. Rev. Lett.* **109** 095302
[16] Gong M, Tewari S and Zhang C 2011 *Phys. Rev. Lett.* **107** 195303
[17] Zhai H 2015 *Rep. Prog. Phys.* **78** 026001
[18] Sakamoto R, Ono Y, Arahata E and Mori H 2016 *J. Phys. Soc. Jpn.* **85** 064401

# Thermophysical and Heat transfer Characteristics of R134a-$TiO_2$ Nanorefrigerant: A Numerical Investigation.

S S Sanukrishna , Nabeel Ajmal and M Jose Prakash.

University of Kerala, Department of Mechanical Engineering, TKM College of Engineering, Kerala, India.

Email: sanukrishna@sctce.ac.in

**Abstract**. Nanorefrigerants are new class of nanofluids in which nanoparticles are suspended in the refrigerant host fluid. They have superior thermophysical and heat transfer characteristics than conventional refrigerants. Present study investigates the thermophysical, heat transfer and pressure drop characteristics of $TiO_2$-R134a nanorefrigerants flow boiling in a mini tube. Studies have been conducted at various heat fluxes, vapour qualities and particle concentrations. From the studies it is observed that the presence of nanoparticles in refrigerant can enhance the thermophysical and heat transfer characteristics. Studies show that with $TiO_2$-R134a nanorefrigerant, the heat transfer coefficient increases by 30.2% when particle concentration is 1.5 %.

## 1. Introduction

Nanorefrigerants are a new research frontier described by the application of nanoparticles as additives into various refrigerants. The studies of nanorefrigerants are still in the embryonic stage. Recently scientists used nanoparticles in refrigeration and air conditioning systems because of its remarkable improvement in thermo-physical, and heat transfer capabilities. M. Akbari et al. [1] conducted a comparative study on heat transfer of nanofluids. CFD predictions of laminar mixed convection of $Al_2O_3$-water nanofluid in a horizontal tube with uniform heat flux by, single-phase and different two-phase models were compared. The numerical predictions of the convective heat transfer coefficient are compared with experimental data. Based on the study the authors reported that, the predictions by the three two-phase models are essentially the same and the less expensive model was VOF. Jacqueline et al. [2] conducted experimental studies on flow boiling heat transfer and pressure drop of pure R134a in a mini tube. The results show that in the low quality region, there is a significant influence of heat flux on heat transfer coefficient.The COP and freezing capacity of vapour compression system enhances by using refrigerant-nanolubricant mixtures [3]. Experimental studies on flow condensation heat transfer studies show an increased heat transfer coefficient for nanorefrigerant [4]. However, research works in the area of nanorefrigerants is scarce due to complexity of the phenomenon of flow boiling heat transfer. This necessitates more studies regarding thermophysical and heat transfer characteristics of nanorefrigerants through mini tubes. Present study includes Modelling of thermal conductivity of R134a based nanorefrigerants and numerical investigations by simulation of flow boiling of R134a-TiO2 nanorefrigerant through a mini channel using ANSYS FLUENT. Parametric studies also been performed to evaluate the effect of addition of nanoparticles to the refrigerants.

IOP Conf. Series: Journal of Physics: Conf. Series **969** (2018) 012015 doi:10.1088/1742-6596/969/1/012015

## 2. Modelling of Thermal conductivity of nanorefrigerant

Various theoretical models were developed by researchers to explain the unusual behaviour in thermophysical properties of nanofluids. In order to investigate the effect of nanoparticle on the thermal conductivity of TiO2-R134a nanorefrigerant, a Modelling study has been conducted with available classical models.

| **Thermal conductivity models** | | |
|---|---|---|
| Model | Correlation | Description |
| Maxwell | $\frac{k_{eff}}{k_{bf}} = \left[\frac{k_p + 2k_{bf} - 2\phi(k_{bf} - k_p)}{k_p + 2k_{bf} - \phi(k_{bf} - k_{pbf})}\right]$ | $k_{eff}$ and $k_{bf}$ -thermal conductivities of nanofluids and base fluid. $k_f$ and $k_p$ thermal conductivity of the liquid and the particle. $\phi$ is the particle volume fraction. |
| Hamilton and crosser | $\frac{k_{eff}}{k_{bf}} = \left[\frac{k_p + (n-1)k_{bf} - (n-1)\phi(k_{bf} - k_p)}{k_p + (n-1)k_{bf} - \phi(k_{bf} - k_{bf})}\right]$ | Modified Maxwell model. Introduced the shape factor (n) |
| Yu and Choi | $\frac{k_{eff}}{k_{bf}} = \left[\frac{k_p + 2k_{bf} + 2(k_p - k_{bf})\phi(1+\beta)^3}{k_p + 2k_{bf} - \phi(k_p - k_{bf})(1+\beta)^3}\right]$ | Introduced the effect of interfacial layering. |
| Timofeeva | $\frac{k_{eff}}{k_{bf}} = 1 + 3\phi$ | Suggested the effective medium theory |
| Koo and Kleinstreuer | $5X\,10^4\beta\phi\,\rho_{bf}C_p\sqrt{\frac{k_{bf}\,T(-6.04\phi + 0.4705)T + (1722.3\phi - 134.63)}{d_p\rho_p}}$ | Considered Brownian motion enhanced heat transfer |

## 3. Modelling and simulation of flow boiling.

The computational fluid dynamic code FLUENT is employed for the present study. For nanofluids, the number of particles in the computational domain, even for a very small particle volume fraction, is extremely large due to the very small size of the particles. This makes Lagrangian-Eulerian approach difficult to solve the nanofluid flow problems. Therefore, in the present study Eulerian-Eulerian VOF model has been selected. All the physical properties are calculated by taking weighted average of different phases based on their volume fraction throughout each control volume. The single set of momentum equations is solved to find the velocity components which are shared by all the phases. In the same manner, a shared temperature is calculated from a single energy equation. Tracking of the interfaces between the phases is accomplished by the solution of a continuity equation for the volume fraction of one (or more) of the phases. The k-ε turbulence model is used to simulate turbulent flow. This model provides general description of turbulence which allows for the effects of transport of turbulent properties by the mean flow and diffusion for the production and destruction of turbulence. Turbulent kinetic energy (K) and its dissipation rate (ε) are coupled to the governing equations via the turbulent viscosity relation. Second order upwind scheme is applied for solving the convective and diffusive terms. PISCO discretization was adopted for pressure velocity factor.

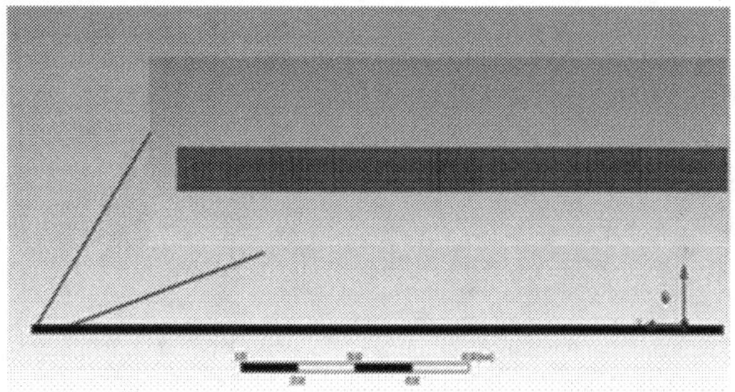

Fig. 1 Geometry of computational domain with mesh

A 2- D axisymmetric geometry is considered. Geometry of the computational domain is shown in Fig 1. Flow through the mini tube is assumed to be a steady and incompressible. Constant heat flux boundary condition is applied at the outer surface of the tube. The mesh used is of rectangular type. In order to solve the governing equations, appropriate boundary conditions were applied. Operating pressure is set as 6.0178 bar and accordingly the saturation temperature of the refrigerant was 295 K. The vapour quality is varied from 0.1 to 0.9 with an interval of 0.1. Studies have been conducted for a mass flux of 440 $kg/m^2s$. Constant heat flux wall boundary conditions are applied at the outer surface varying from 5000 to 47000 $W/m^2$. No slip boundary conditions are imposed at the wall.

## 4. Results and discussion

### *4.1 Thermal conductivity*

Modelling of thermal conductivity of $TiO_2$/R134a nanorefrigerant with theoretical correlations proposed by various researchers has been conducted and which is presented here. Broad range of particle concentrations from dilute level has been considered for the study.

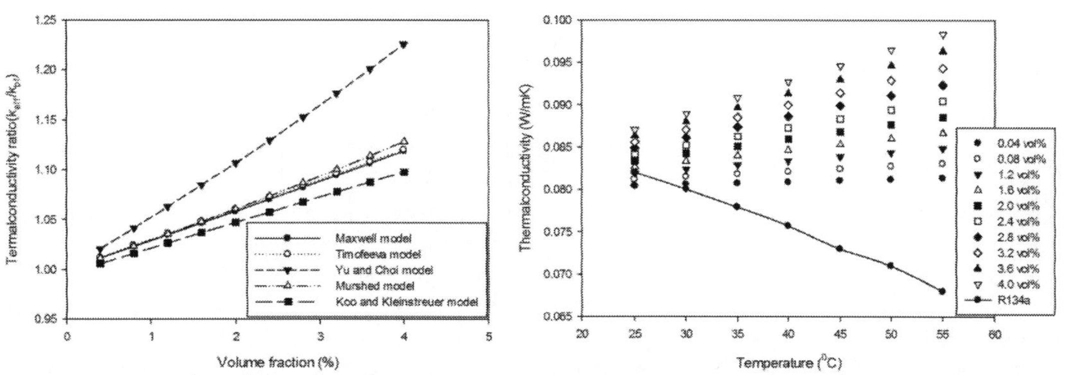


Fig. 2 Variation of thermal conductivity and thermal conductivity ratio of nanorefrigerant as a function of particle volume fraction and temperature.

From Fig. 2 it is evident that the thermal conductivity of $TiO_2$-R134a nanorefrigerant increases linearly with nanoparticle volume concentration of (0.4-4 vol. %). It can be concluded that the thermal conductivity of nanorefrigerant increases with the increase of particle volume fractions. Fig. 3 shows the variation of thermal conductivity of pure refrigerant and nanorefrigerant with temperature ranging from $25^0$C to $55^0$C. It can be seen that, the thermal conductivity of $TiO_2$/ R-134a nanorefrigerant increases linearly with temperature, while for pure refrigerant, thermal conductivity moderately decreases with increase in temperature. Besides, with the rise of temperature, the Brownian motion of nanoparticles will intensify and the contribution of micro convection in heat transport will increase, which results in the augmentation of thermal conductivity. For pure refrigerant, thermal conductivity was decreased with increasing temperature. This is due to the fact that when temperature increases, the liquid is evaporated, which causes the atoms to be positively charged and vibrate with greater amplitude. Results show that, the thermal conductivity enhancement is about 44% at a temperature of $55^0$C with higher nanoparticle concentration (4 vol. %) compared to pure R134a refrigerant. At lower particle concentration and temperature (0.4 vol. % and $25^0$C), the obtained result shows lowest increase in thermal conductivity value i.e. 2%.

### *4.2 Flow boiling heat transfer coefficient*

A grid independency test is conducted to guarantee the accuracy of the numerical results. Twelve different sets of uniform grids have been used to check for grid independency. Tests were carried out with pure refrigerant at different heat flux conditions for each grid. It can be seen from Fig. 3 that, the value of the heat transfer coefficient converges as the number of grid cells increases. Based on this test optimum grid was selected as 24360.

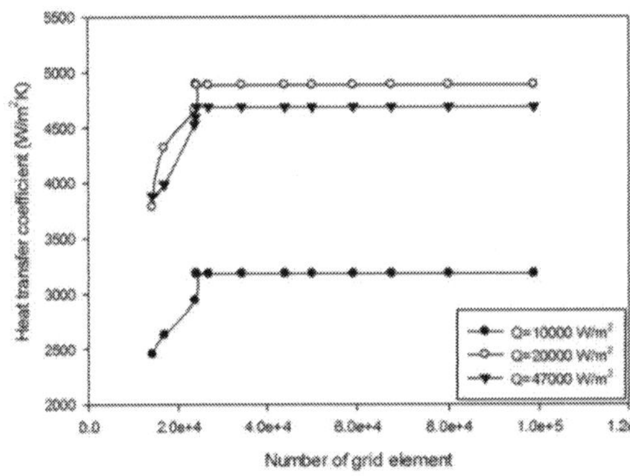


Fig. 3 Grid independent study.

Before conducting the numerical studies with nanorefrigerants, the validation of the numerical scheme is carried out by comparing the result with experimental results reported in literature (Jacqueline et al., 2011). Results show good agreement with experimental data. Subsequently, studies were extended with $TiO_2$-R134a nanorefrigerants in order to find out effect of nanoparticles on heat transfer coefficient and pressure drop characteristics of R134a flow boiling through horizontal mini channel. Fig. 4 shows the comparison of boiling heat transfer coefficient of pure R134a obtained from experimental results (Jaqciline.et al., 2011) and numerical predictions for a mass flux of440kg/$m^2$s at heat fluxes in the range of 10000W/m2 to 47000 W/$m^2$.The obtained heat transfer coefficient from the simulation shows good agreement with experimental results. From figures, it is clear that heat transfer coefficient increases with an increase in heat flux from 10000 W/$m^2$ to 47000W/m2. Both experimental and modelling results exhibit the similar trends.

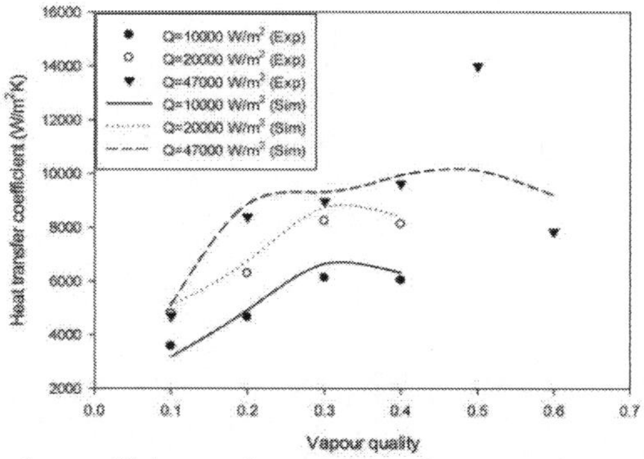


Fig. 4 Comparison of hear transfer coefficients of pure R134a from experiment and simulation at a mass flux of 440kg/m$^2$s .

Fig. 5(a) shows the comparison of heat transfer coefficient of pure R134a and $TiO_2$-R134a nanorefrigerant respectively at a particle concentration of 0.5 % at a mass flux of 440kg/m$^2$ and three different heat fluxes. From figure it is found that heat transfer coefficient increases with increase in inlet vapour quality up to about 0.4 and thereafter decreases with increase in vapour quality. The similar trend is observed for all heat flux conditions. There is an increase in heat transfer coefficient by 19% for TiO2 –R134a nanorefrigerant (0.5% volume fraction) compared to pure refrigerant at a heat flux 47000 W/m2 and mass flux of 440kg/m2s. At higher vapour quality region nanorefrigerant and pure refrigerant show similar trends, that is decrease in the heat transfer coefficient. This is due to the formation of the vapour barrier between the bulk fluid and the wall of the tube

.

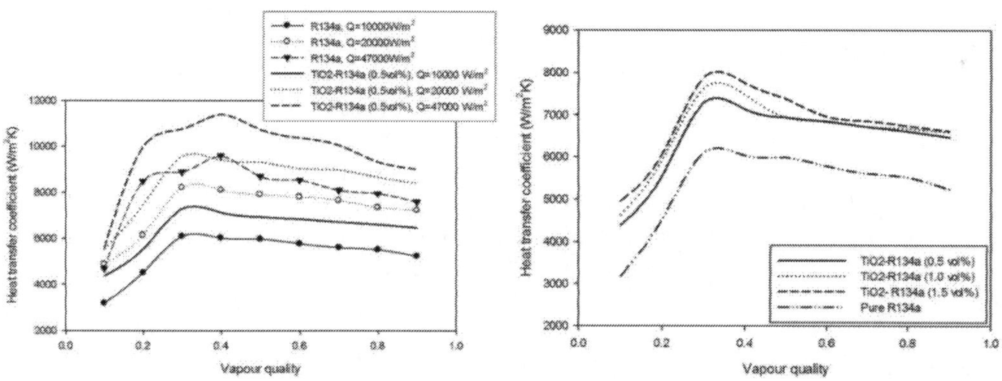


Fig. 5 Variation of heat transfer coefficient with (a) Vapour quality (b) particle concentrations.

Fig. 5(b) shows the variation of heat transfer coefficient at 3 particle concentrations of $TiO_2$ nanoparticles. While increasing the volume concentration of the nanoparticle the heat transfer coefficient increases in a noteworthy manner. In the case of $TiO_2$-R134a nanorefrigerant, heat transfer coefficient increases from 19% to 30.2% when particle concentration is increased from 0.5 to 1.5 vol. %. This anomalous enhancement in heat transfer coefficient is due to the improved thermophysical and boiling heat transfer properties of nanorefrigerant.

### *4.3 Pressure drop characteristics:*

The local pressure drop dp/dz, calculated by assuming pressure as the linear function of tube length. Fig. 6 shows the effect of particle concentrations on pressure drop. It is evident that, for nanorefrigerant, irrespective of the heat flux, the pressure drop was increased with increase in particle concentration. As the volume fraction increase from 0 .5 % to 1.5%, the pressure drop increases by 21.4 % 27.5% compared to pure refrigerant.

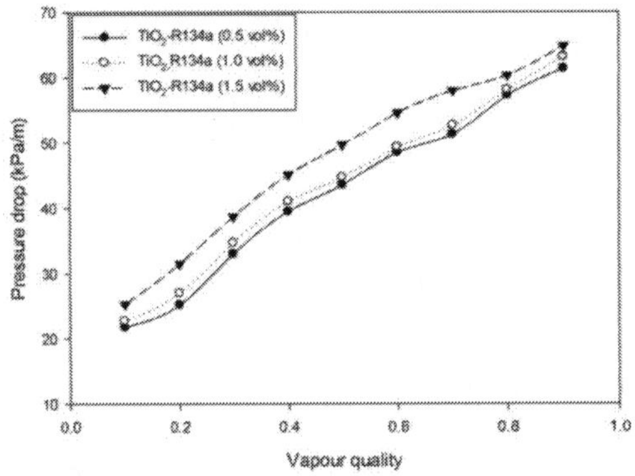


Fig 6 pressure drop of nanorefrigerants at different particle concentrations at a heat flux of 47000w/m$^2$.

## 5. Conclusions

Numerical investigations on heat transfer characteristics of nanorefrigerants flowing in a mini channel have been carried out. An increase in thermal conductivity by 44% for nanorefrigerant compared to pure refrigerant was manifested. The heat transfer coefficient increases with increase in heat flux and particle concentration. The heat transfer coefficient increases with increase in inlet vapour quality up to about 0.4 and thereafter decreases. The maximum increase in heat transfer coefficient of $TiO_2$–R134a nanorefrigerant compared to pure refrigerant was by 30.2% ,at a volume fraction 1.5%.. The maximum increase in pressure drop was observed as 27.5% at a volume fraction of 1.5 % .TiO2 based nanorefrigerant shows superior heat transfer performance than pure refrigerant.

## References

[1] M.Akbari, N.Galanis and A.Behzadmehr.,Comparative analysis of single and two- phase models for CFD studies of nanofluid heat transfer, Int .J thermal science, 50, 1343-1354.

[2] Jacqueline B.Copetti, Fiavia Zinaniand Nicole L.F.Kunsler., Flow boiling heat transfer and pressure drop of R-134a in a mini tube: An experimental investigation, Int. J. Thermal and fluid science, 35, 636-644.

[3] Subramani, and N., Prakash, M.J., (2011). Experimental studies on a vapour compression system using nanorefrigerants, Int. J. Eng. Science and technology, 3, 95–102

[4] M.A. Akhavan-Behabadi, M.K. Sadoughi , Milad Darzi and M. Fakoor-Pakdaman (2015) Experimental study on heat transfer characteristics of R600a/POE/CuO nanorefrigerant flow condensation, J. Thermal and fluid Science, 66,46-52

# Dynamics of vortex lattice formation in rotating two-component Bose–Einstein condensates

**Toshiaki Kanai**[1], **Makoto Tsubota**[1,2]

[1]Department of Physics, Osaka City University, 3-3-138 Sugimoto, Sumiyoshi-Ku, Osaka 558-8585, Japan

[2]The OCU Advanced Research Institute for Natural Science and Technology (OCARINA), Osaka City University, 3-3-138 Sugimoto, Sumiyoshi-Ku, Osaka 558-8585, Japan

**Abstract.** We study the dynamics of rotating two-component Bose–Einstein condensates(BECs) toward vortex lattice formation by numerically solving the two-dimensional Gross–Pitaevskii equations. The dynamics are similar to that of single-component BEC, and the whole changes of physical quantities in two cases are qualitatively similar. However, there are some differences between two cases. First, the characteristic time toward vortex entering depends on the inter-component interaction. The time toward vortex entering becomes shorter with increasing the inter-component interaction. Second, the angular momenta exchange between two components with keeping the total angular momentum.

## 1. Introduction

Quantized vortices appear in diverse fields of quantum condensate systems. They have been studied in superfluid $^4$He for a long time [1]. The theoretical treatments of $^4$He are complicated because of high density and strong interaction. However, the theoretical treatments of cold atom BEC are easy because of low density and weak interaction [2]. Moreover, this system has several experimental advantages compared to the other system. The first advantage is the visualization of the density distribution. The vortices in cold atom BEC can be visualized. The second advantage is the high controllability by optical technology. It is possible to control the potential and the strength of interaction. The third advantage is the realization of multicomponent BECs. Therefore, it is important to study the physics of quantized vortices in cold atoms.

Vortex lattice formation has been studied experimentally, numerically, and theoretically in single-component BEC [3]. By rotating an asymmetric potential, the condensate is distorted to an elliptic shape and oscillates. Then, the surface of the condensate becomes unstable, leading to the excitation of the surface wave. From the surface wave, the quantized vortices enter the condensate and form a triangular lattice [4, 5].

Multicomponent superfluids have been studied in superfluid $^3$He [6]. Recently, they are studied in cold atom BECs because of the experimental and theotitical advantages. The equilibrium states of two-component BECs are classified under miscible and phase-separated states depending on the strength of inter-component interaction. Rotating two-component BECs are known to have many kinds of equilibrium states of vortices [7]. Moreover, it is known that the inter-vortex forces of the inter-component BECs differ from that of the same component BEC [8]. For these reasons, it is expected that the inter-component interaction makes the dynamics of two-component BECs different from that of single-component BEC. Therefore, we

IOP Conf. Series: Journal of Physics: Conf. Series **969** (2018) 012016 doi:10.1088/1742-6596/969/1/012016

study the dynamics toward the equilibrium states numerically. We confine ourselves to the case of miscible condensates.

## 2. Formulation and Equilibrium State

In this section, we describe the formulation of two-dimensional two-component BECs. Two-component BECs are described by two macroscopic wavefunctions $\psi_j(\boldsymbol{r},t) = \sqrt{n_j(\boldsymbol{r},t)}e^{i\phi_j(\boldsymbol{r},t)}$ in the mean-field approximation at zero temperature, where the index $j$ refers to each component ($j = 1, 2$). Here $n_j$ and $\phi_j$ are the density and the phase of the $j$ th component. Wavefunctions are normalized by the total particle number $N_j = \int dxdy|\psi_j|^2$. Wavefunctions in a rotating frame satisfy the coupled Gross–Pitaevskii equations(GPEs) [4]:

$$(i-\gamma)\hbar\frac{\partial}{\partial t}\psi_j = \left(-\frac{\hbar^2}{2m_j}(\partial_x{}^2+\partial_y{}^2) + V_j(\boldsymbol{r}) + \sum_{k=1,2} g_{jk}|\psi_k|^2 - \mu - \Omega\hat{L}_z\right)\psi_j. \tag{1}$$

Here $\hat{L}_z = -i\hbar(x\partial_y - y\partial_x)$ is the angular momentum operator, $\Omega$ is the rotation frequency around z-axis, $m_j$ and $V_j(\boldsymbol{r})$ are the mass and the trapping potential of $j$ th component, and the coefficients $g_{jk}$ represent the strength of interactions between the $j$ th and $k$ th components. In this paper, we assume the interactions are repulsive ($g_{jk} > 0$). The term with $\gamma$ introduces the dissipation [10]. In this paper, we use $\gamma = 0.03$ unless the value is mentioned explicitly, following the previous studies in single-component BEC [5, 10]. The potentials are the asymmetric harmonic potentials :

$$V_j(x,y) = \frac{1}{2}m_j\omega^2\{(1+\epsilon)x^2 + y^2\}, \tag{2}$$

with $\epsilon = 0.05$. In this paper, the number of parameters is reduced for simplicity by assuming that the components are symmetric namely $g_{11} = g_{22} \equiv g$, $m_1 = m_2 \equiv m$, and $N_1 = N_2 \equiv N$, and we use the parameters of miscible condensates ($g_{12} < g$). In a rotating frame, the energy $F$ of condensates is given in terms of the energy $E$ in the non-rotating frame by

$$F = E - \Omega\sum_{j=1,2} < \psi_j|\hat{L}_z|\psi_j > . \tag{3}$$

Under rotation, the equilibrium states of single component BEC are the triangular lattices [5], and those of miscible two-component BECs are triangular or square lattices depending on the strength of inter-component interaction and the rotation frequency [7]. The two vortex lattices are interlocked in such a manner that a peak in the density of one component is located at the density hole of the other component.

## 3. The dynamics toward vortex lattice formation

We perform numerical simulations of miscible two-component BECs. The time, the length, and the wavefunctions are normalized :

$$\begin{aligned}\tilde{t} &= \omega t,\\ \tilde{x} &= \frac{x}{a},\\ \tilde{\psi}_j &= a\psi_j.\end{aligned}$$

Here $a = \sqrt{\hbar/m\omega}$ (the typical value of frequency is $\omega/2\pi = 219$Hz [9]). Figure 1 shows the time development of the density and the phase. We first prepare an equilibrium state of two-component BECs in a stationary potential (FIg.1 (a)).

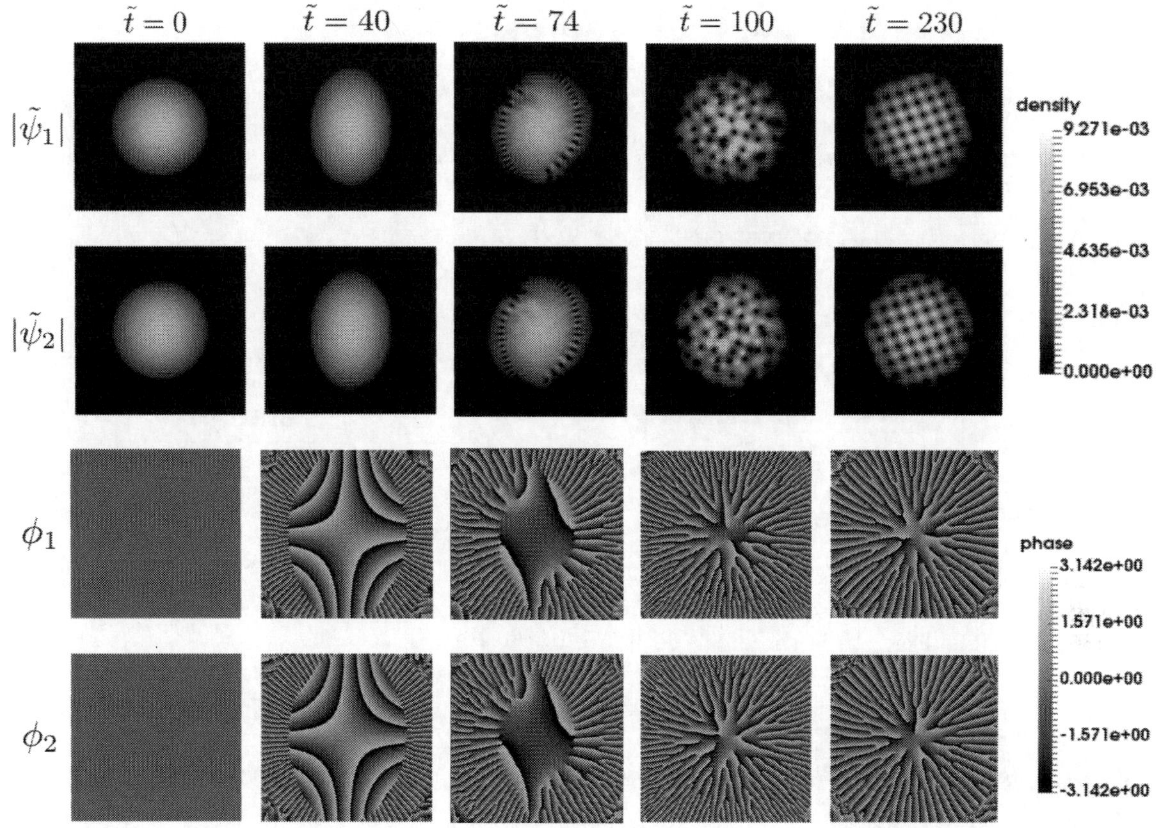


**Figure 1.** Time development of the condensates density $|\tilde{\psi}_1|^2$ and $|\tilde{\psi}_2|^2$, and phase $\theta_1$ and $\theta_2$ after the trapping potential begins to rotate suddenly with $\Omega/\omega_\perp = 0.7$, $g_{12}/g = 0.9$, and $g = 2000$. The time is $\tilde{t} =$ (a) 0, (b) 40, (c) 74, (d) 100, (e) 230. The equilibrium state is a square lattice with these parameters.

By rotating an asymmetric potential, two-component BECs are distorted to an elliptic shape and oscillate together (Fig.1 (b)), and their angular momenta temporarily increase together (Fig.2). Then, the surface of the condensate becomes unstable, leading to the excitation of the surface wave. Surfaces of each component wave alternately (Fig.1 (c)). From the surface wave, the vortices enter the condensate, and the angular momenta increase rapidly (Fig.2 (a)). The vortices in the condensates move randomly (Fig.1 (d)). Then, they form a square lattice in both components with arranging themselves alternately in order to reduce the energy of inter-component interaction (Fig.1 (e)).

Two-component BECs exchange the angular momenta after the vortices enter the condensates (Fig.4 (b)). The vortices on the inside of condensates come to rest early on forming a lattice, but they near the surface continue to fluctuate. In order to reveal the mechanism of the exchange of angular momenta between two-component BECs, we take a boundary in the radial direction and separate the angular momenta of the inner and outer parts (Fig.3). Here, we take the Thomas–Fermi radii without rotation for the boundary. Figure 4 shows that the oscillation of the angular momentum comes from the whole condensates before $\tilde{t} \sim 140$ but from the outer part of the condensates after about $\tilde{t} \sim 140$ and come to rest gradually because of the dissipation. The vortices near the surface go in and out the condensates fluently with keeping total angular

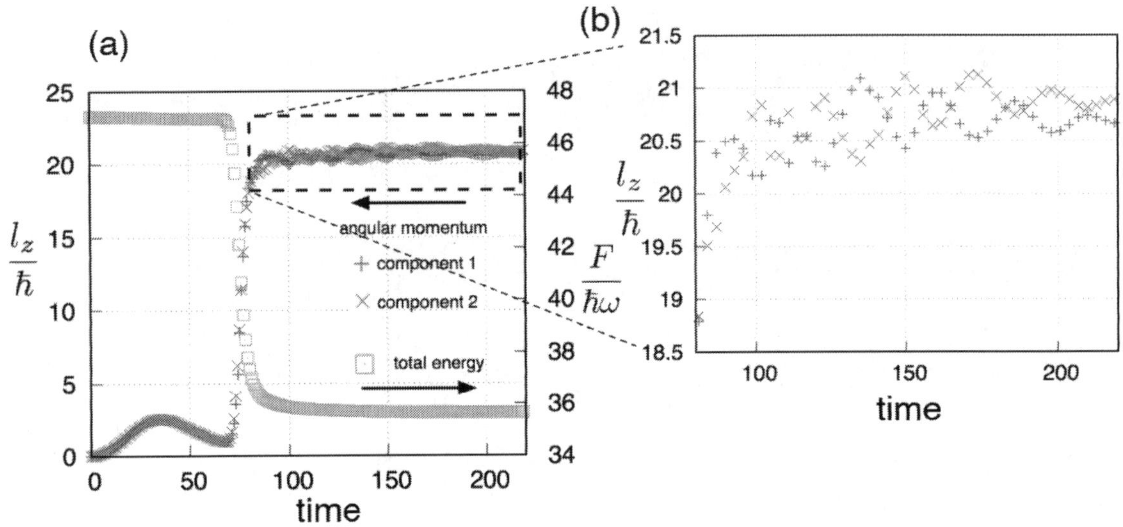


**Figure 2.** Time evolution of the energy and angular momenta per atom of component 1 and component 2. The Fig. (b) is the enlarged figure of the Fig. (a).

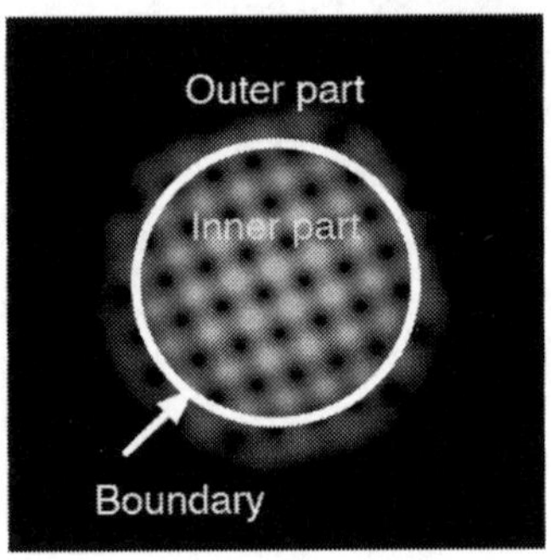


**Figure 3.** The boundary in the radial direction. We take the Thomas–Fermi radii without rotation for the boundary.

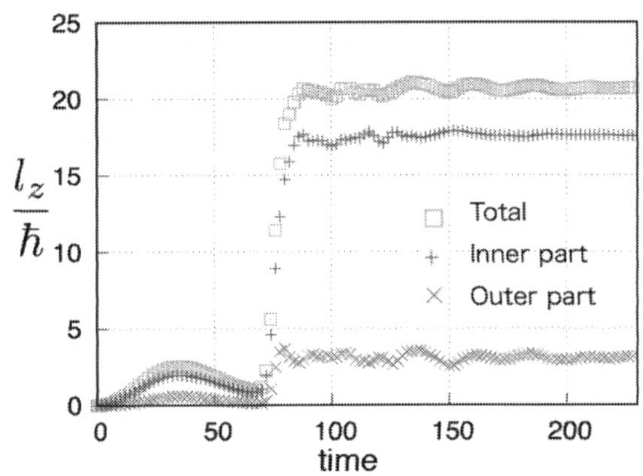


**Figure 4.** Time evolution of angular momentum per atom of component 1. The total angular momentum is separated into the contributions of the inner and outer parts.

momentum and cause the oscillation of each angular momenta. This oscillation of the angular momenta is a unique phenomenon in two-component BECs.

As the ratio $g_{12}/g$ increases with keeping $g = 2000$, the angular momentum $l_z$ of equilibrium state increases (Fig.5). Under the Thomas–Fermi approximation with rotation, the density profile of the sum of the condensates is

$$n_{TF} = \frac{2}{g+g_{12}} \left\{ \sqrt{\frac{N(g+g_{12})}{\pi} m(\omega^2 - \Omega^2)} - \frac{1}{2} m(\omega^2 - \Omega^2) r^2 \right\}. \tag{4}$$

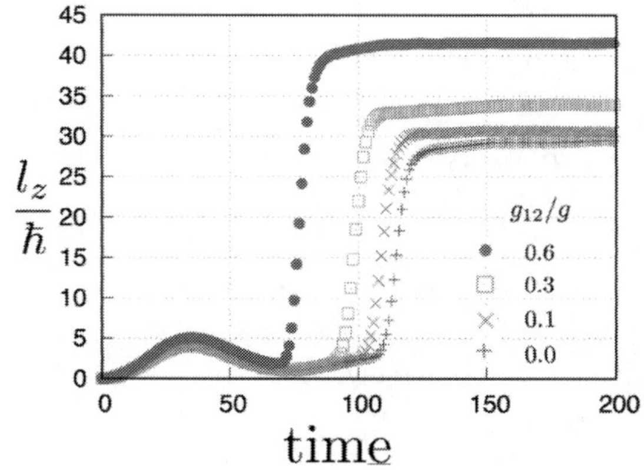


**Figure 5.** Time evolution of the sum of angular momentums per atom of component 1 and component 2. We change $g_{12}/g$ with keeping $g = 2000$.

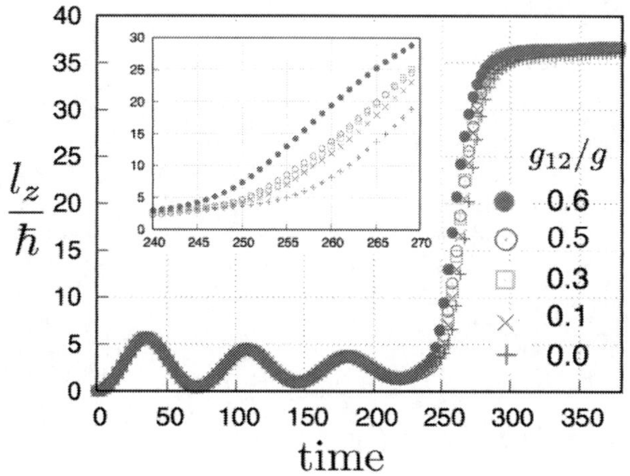


**Figure 6.** Time evolution of the sum of angular momentums per atom of component 1 and component 2 with $\gamma = 0.01$. We change $g_{12}/g$ with keeping $g + g_{12} = 3000$.

Thus the angular momentum of equilibrium state per atom is

$$l_z = \frac{m\Omega}{2N}\int dr\, r^3 n_{TF} \propto \sqrt{g + g_{12}}. \tag{5}$$

When the sum $g + g_{12}$ increases, the density distribution spreads widely, so that $l_z$ becomes large.

We found that the characteristic time toward the surface wave excitation and the vortices entering the condensate depends on the inter-component interaction (Figs.5 and 6). As the ratio $g_{12}/g$ increases with keeping $g = 2000$, the characteristic time becomes short (Fig.5). As the sum $g + g_{12}$ increases, the radii of the condensates becomes large, and the velocity near the surface becomes fast. Thus the vortices enter the condensates easier. As the ratio $g_{12}/g$ increases with keeping $g + g_{12} = 3000$, the characteristic time becomes short not as much as the previous case with keeping $g = 2000$ (Fig.5). The surface waves of each component are alternated and amplify that of the other component by the repulsive inter-component interaction. The larger the ratio $g_{12}/g$ is, the more remarkable this effect is. In these cases, the characteristic time depends on the sum $g + g_{12}$ strongly, and the ratio $g_{12}/g$ weakly.

## 4. Conclusion

The inter-component interaction makes the dynamics of two-component BECs different from that of single-component BEC. We found that the characteristic time and the behaviors are changed by the inter-component interaction. These behaviors reflect the differences of inter-vortex interaction between intra-component and inter-component. In this paper, we study the dynamics of two-component BECs under the symmetric parameters. Under the unsymmetric parameters, there are other equilibrium states [11]. Therefore, it is expected that there are unique behaviors under the unsymmetric parameters.

## Acknowledgement

M. Tsubota was supported by JSPS KAKENHI Grant No. 26400366 and MEXT KAKENHI " Fluctuation & Structure ", Grant No. 16H00807.

IOP Conf. Series: Journal of Physics: Conf. Series **969** (2018) 012016 doi:10.1088/1742-6596/969/1/012016

## References


[1] R. J. Donnelly, *Quantized Vortices in Helium II*, (Cambridge University Press, Cambridge, 1995).
[2] C. J. Pethick, and H. Smith, *Bose-Einstein Condensation in Dilute Gases*, 2nd ed. (Cambridge University Press, Cambridge, 2008).
[3] K. Kasamatsu and M. Tsubota, *Prog. Low Temp. Phys.*, ed. W. P. Halperin and M. Tsubota(Elsevier, Amsterdam, 2008) Vol. 16, pp. 349 - 401.
[4] K. Kasamatsu, M. Tsubota, and M. Ueda, *Int. J. Mod. Phys. B*, **19**, pp. 1835 - 1904 (2005).
[5] M. Tsubota, K. Kasamatsu, and M. Ueda, *Phys. Rev. A* **65**, 023603 (2002).
[6] D. Vollhardt, and P. Wölfle, *The Superfluid Phases Of Helium 3*, (Taylor & Francis, London, 1990).
[7] K. Kasamatsu, M. Tsubota, and M. Ueda, *Phys. Rev. Lett.* **91**, 150406 (2003).
[8] M. Eto, K. Kasamatsu, M. Nitta, H. Takeuchi, and M. Tsubota, *Phys. Rev. A* **83**, 063603 (2002).
[9] K. W. Madison, F. Chevy, W. Wohlleben, and J. Dalibard, *Phys. Rev. Lett.***84**, 806 (1999)
[10] S. Choi, S. A. Morgan, and K. Burnett, *Phys. Rev. A.* **57**, 4057 (1998).
[11] P. Mason, and A. Aftalion, *Phys. Rev. A* **84**, 033611 (2011).

IOP Conf. Series: Journal of Physics: Conf. Series **969** (2018) 012017 doi:10.1088/1742-6596/969/1/012017

# Closed-channel contribution in the BCS-BEC crossover regime of an ultracold Fermi gas with an orbital Feshbach resonance

**S Mondal[1], D Inotani[1] and Y Ohashi[1]**

Department of Physics, Faculty of Science and Technology, Keio University, 3-14-1 Hiyoshi, Kohoku-ku, Yokohama 223-8522, Japan

E-mail: soumita_phy@keio.jp

**Abstract.** We theoretically investigate strong-coupling properties of an ultracold Fermi gas with an orbital Feshbach resonance (OFR). Including tunable pairing interaction associated with an OFR within the framework of the strong-coupling theory developed by Nozières and Schmitt-Rink (NSR), we examine the occupation of the closed channel. We show that, although the importance of the closed channel is characteristic of the system with an OFR, the occupation number of the closed channel is found to actually be very small at the superfluid phase transition temperature $T_c$, in the whole BCS (Bardeen-Cooper-Schrieffer)-BEC (Bose-Einstein condensation) crossover region, when we use the scattering parameters for an ultracold $^{173}$Yb Fermi gas. The occupation of the closed channel increases with increasing the temperature above $T_c$, which is more remarkable for a stronger pairing interaction. We also present a prescription to remove effects of an experimentally inaccessible deep bound state from the NSR formalism, which we meet when we theoretically deal with a $^{173}$Yb Fermi gas with an OFR.

## 1. Introduction

A magnetic Feshbach resonance (MFR) has proved to be a prerequisite for the study of an ultracold Fermi gas of alkali-metal atoms over a wide range of the strength of a pairing interaction, because it enables us to experimentally tune the interaction strength[1]. One typical success by using this unique property is the experimental realization of the BCS (Bardeen-Cooper-Schrieffer)-BEC (Bose-Einstein condensation) crossover phenomenon[2, 3, 4] in $^{40}$K[5] and $^{6}$Li[6] Fermi gases, where the character of a Fermi superfluid continuously changes from the weak-coupling BCS-type to the BEC of tightly-bound molecules, with increasing the interaction strength.

However, this MFR-technique is not applicable to the cases of alkali-earth atoms like Sr, as well as alkali-earth-like atoms like Yb, because of the vanishing total electronic spin due to the fully occupied outer-most shell. To overcome this problem, the possibility of using an optical Feshbach resonance (OFR) has been explored[7]. However, this approach has serious problems, such as strong particle loss, as well as heating, so that it seems difficult to achieve the superfluid phase transition of a non-alkali metal Fermi gas by using this technique.

Latest measurements showing the scattering properties of $^1S_0$ (electronic spin-singlet) and $^3P_0$ (spin-triplet) atomic pairs of $^{173}$Yb have predicted the existence of a shallow molecular bound state[8, 9]. Very recently, a novel idea using the scattering between $^1S_0$ and $^3P_0$ in a $^{173}$Yb Fermi gas has been proposed[10]. This proposal uses different nuclear spin states of the atoms to tune the interaction strength, unlike a MFR that uses different electronic spin states. This so-called orbital Feshbach resonance (OFR) is expected to provide a tunable pairing interaction for an alkali-earth (like) atomic Fermi gas, where MFR cannot be deployed. This OFR has recently experimentally been confirmed[11, 12]. Measurements of anisotropic expansion of a gas cloud, as well as cross-thermalization rate, support the presence of OFR in an ultracold $^{173}$Yb Fermi gas at the field strengths $B = 41$ G and 55(8) G. A relatively long lifetime of the system against collapse has also been reported[11, 12]. Thus, the superfluid phase transition in an ultracold $^{173}$Yb Fermi gas with an OFR is promising.

A crucial issue in theoretically considering an OFR is that the energy difference between the open and closed channels is comparable to the Fermi energy $\varepsilon_{\rm F}$ of the system or even smaller than $\varepsilon_{\rm F}$, so that we need to explicitly deal with the both channels[10]. This situation is quite different from the case of a broad (magnetic) Feshbach resonance in $^6$Li and $^{40}$K Fermi gases[1, 5, 6], where the energy difference is so large that the close channel is only virtually occupied in the intermediate state of a Feshbach resonance. As a result, the latter can be treated by the simple single-channel BCS model, taking into account atoms only in the open channel. Effects of the closed channel only remains as a Feshbach-induced effective interaction between atoms in the open channel.

In this paper, we theoretically investigate to what extent the closed channel contributes to the BCS-BEC crossover physics in an ultracold Fermi gas with an OFR. For this purpose, we evaluate the number of atoms in the closed channel within the framework of the strong-couping theory developed by Nozières and Schmitt-Rink (NSR)[2]. We briefly note that the BCS-BEC crossover behavior of the superfluid phase transition temperature $T_{\rm c}$ in an ultracold Fermi gas with an OFR has been discussed[13] within the NSR theory[2], where higher superfluid transition temperature than the previous MFR case has been obtained in the weak-coupling regime.

## 2. Formulation

We consider a four-component Fermi gas with an OFR, described by the model Hamiltonian[10, 13],

$$H = K_{\rm o} + K_{\rm c} + \frac{1}{2}\sum_{\boldsymbol{q}}\Big[U_+ A_+^\dagger(\boldsymbol{q})A_+(-\boldsymbol{q}) + U_- A_-^\dagger(\boldsymbol{q})A_-(-\boldsymbol{q})\Big]. \tag{1}$$

Here, $K_{\rm o} = \sum_{\boldsymbol{p}} \xi^{\rm o}_{\boldsymbol{p}}[c^\dagger_{{\rm g}\downarrow,\boldsymbol{p}}c_{{\rm g}\downarrow,\boldsymbol{p}} + c^\dagger_{{\rm e}\uparrow,\boldsymbol{p}}c_{{\rm e}\uparrow,\boldsymbol{p}}]$ and $K_{\rm c} = \sum_{\boldsymbol{p}} \xi^{\rm c}_{\boldsymbol{p}}[c^\dagger_{{\rm g}\uparrow,\boldsymbol{p}}c_{{\rm g}\uparrow,\boldsymbol{p}} + c^\dagger_{{\rm e}\downarrow,\boldsymbol{p}}c_{{\rm e}\downarrow,\boldsymbol{p}}]$ are kinetic terms in the open channel ($|{\rm g}\downarrow\rangle$, $|{\rm e}\uparrow\rangle$) and closed channel ($|{\rm g}\uparrow\rangle$, $|{\rm e}\downarrow\rangle$), respectively. $\xi^{\rm o}_{\boldsymbol{p}} = \boldsymbol{p}^2/(2m) - \mu$ is the energy in the open channel, and $\xi^{\rm c}_{\boldsymbol{p}} = \boldsymbol{p}^2/(2m) + \delta/2 - \mu$ is that in the closed channel, with $m$ being the atomic mass. In the above, "g" and "e" denote two different orbital states, and pseudospins $\sigma = \uparrow, \downarrow$ describe two nuclear spin states. $\delta/2$ is the energy difference between the open and closed channel. The last two terms in Eq. (1) describe interactions between Fermi atoms, where

$$A_\pm(\boldsymbol{q}) = \sum_{\boldsymbol{p}}\big[c_{{\rm g}\uparrow,\boldsymbol{p}+\boldsymbol{q}/2}c_{{\rm e}\downarrow,-\boldsymbol{p}+\boldsymbol{q}/2} \mp c_{{\rm g}\downarrow,\boldsymbol{p}+\boldsymbol{q}/2}c_{{\rm e}\uparrow,-\boldsymbol{p}+\boldsymbol{q}/2}\big]. \tag{2}$$

The bare coupling strengths $U_\pm$ ($<0$) are related to the observable $s$-wave scattering length $a_{s,\pm}$ as,

$$\frac{4\pi a_{s,\pm}}{m} = \frac{U_\pm}{1 + U_\pm \sum_{\boldsymbol{p}} m/\boldsymbol{p}^2}. \tag{3}$$

Unlike a broad MFR (where the energy difference $\delta/2$ between the open and closed channels is much larger than the Fermi energy $\varepsilon_{\rm F}$), $\delta/2$ is comparable or even smaller than $\varepsilon_{\rm F}$ in the case of OFR[10]. Thus, we need to take into account scattering effects on both the open and the closed channels in the latter. In this paper, we include pairing fluctuations induced by the interactions $U_\pm$, within the framework of the strong-coupling (Gaussian fluctuation) theory developed by Nozières and Schmitt-Rink (NSR)[2]. For this purpose, it is convenient to rewrite the interaction terms ($\equiv H_{\rm int}$) in Eq. (1) to the sum of intra-channel

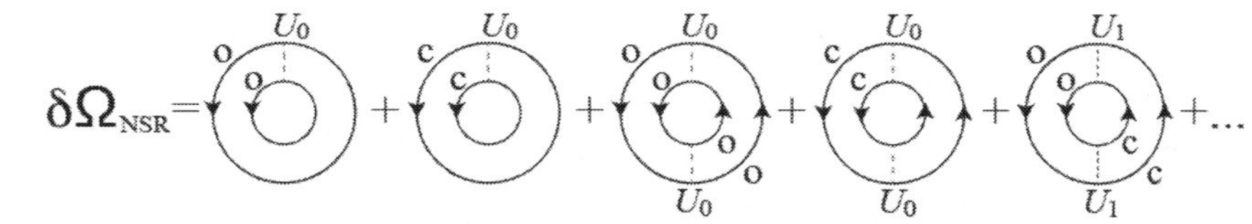


**Figure 1.** Strong-coupling corrections ($\delta\Omega_{\rm NSR}$) to the thermodynamic potential in NSR. The solid line with label "o" and "c" represent the bare single-particle thermal Green's function $G^{-1}_{\rm s=o,c}(\boldsymbol{p}, i\omega_n) = i\omega_n - \xi^{\rm s}_{\boldsymbol{p}}$ in the open and closed channel, respectively, where $\omega_n$ is the fermion Matsubara frequency. The red and blue dashed line denote the intra- and inter-channel interactions, respectively. (Color figure online)

and inter-channel interactions, as

$$\begin{aligned} H_{\rm int} &= U^{\rm open}_{\rm intra} \sum_{\boldsymbol{p},\boldsymbol{p}',\boldsymbol{q}} c^\dagger_{{\rm e}\uparrow,\boldsymbol{p}+\boldsymbol{q}/2} c^\dagger_{{\rm g}\downarrow,-\boldsymbol{p}+\boldsymbol{q}/2} c_{{\rm g}\downarrow,-\boldsymbol{p}'+\boldsymbol{q}/2} c_{{\rm e}\uparrow,\boldsymbol{p}'+\boldsymbol{q}/2} \\ &+ U^{\rm closed}_{\rm intra} \sum_{\boldsymbol{p},\boldsymbol{p}',\boldsymbol{q}} c^\dagger_{{\rm g}\uparrow,\boldsymbol{p}+\boldsymbol{q}/2} c^\dagger_{{\rm e}\downarrow,-\boldsymbol{p}+\boldsymbol{q}/2} c_{{\rm e}\downarrow,-\boldsymbol{p}'+\boldsymbol{q}/2} c_{{\rm g}\uparrow,\boldsymbol{p}'+\boldsymbol{q}/2} \\ &+ U_{\rm inter} \sum_{\boldsymbol{p},\boldsymbol{p}',\boldsymbol{q}} \left[ c^\dagger_{{\rm e}\uparrow,\boldsymbol{p}+\boldsymbol{q}/2} c^\dagger_{{\rm g}\downarrow,-\boldsymbol{p}+\boldsymbol{q}/2} c_{{\rm e}\downarrow,-\boldsymbol{p}'+\boldsymbol{q}/2} c_{{\rm g}\uparrow,\boldsymbol{p}'+\boldsymbol{q}/2} + {\rm h.c.} \right]. \end{aligned} \tag{4}$$

Here, $U^{\rm open}_{\rm intra}$, $U^{\rm closed}_{\rm intra}$, $U_{\rm inter}$ represent, respectively, the interaction in the open channel, that in the closed channel, and the inter-channel interaction. These are related to $U_\pm$ in Eq. (1) as $U^{\rm open}_{\rm intra} = U^{\rm closed}_{\rm intra} = [U_- + U_+]/2$ ($\equiv U_0$) and $U_{\rm inter} = [U_- - U_+]/2$ ($\equiv U_1$). Summing up the NSR diagrams in Fig. 1 describing fluctuation corrections ($\delta\Omega_{\rm NSR}$) to the thermodynamic potential $\Omega = \Omega_0 + \delta\Omega_{\rm NSR}$, we obtain

$$\Omega = \Omega_0 + T\sum_{\boldsymbol{q},\nu_n} \ln \left[ \det \left[ \begin{array}{cc} 1 - U_0\Pi_{\rm o}(\boldsymbol{q}, i\nu_n) & -U_1\Pi_{\rm o}(\boldsymbol{q}, i\nu_n) \\ -U_1\Pi_{\rm c}(\boldsymbol{q}, i\nu_n) & 1 - U_0\Pi_{\rm c}(\boldsymbol{q}, i\nu_n) \end{array} \right] \right], \tag{5}$$

where $\Omega_0 = -T\sum_{\boldsymbol{p},{\rm s=o,c}} \ln[1 + e^{-\xi^{\rm s}_{\boldsymbol{p}}/T}]$ is the thermodynamic potential in the non-interacting case, and $\nu_n$ is the boson Matsubara frequency. In Eq. (5),

$$\Pi_{\rm s=o,c}(\boldsymbol{q}, i\nu_n) = \sum_{\boldsymbol{p}} \frac{1 - f(\xi^{\rm s}_{\boldsymbol{p}+\boldsymbol{q}/2}) - f(\xi^{\rm s}_{-\boldsymbol{p}+\boldsymbol{q}/2})}{i\nu_n - \xi^{\rm s}_{\boldsymbol{p}+\boldsymbol{q}/2} - \xi^{\rm s}_{-\boldsymbol{p}+\boldsymbol{q}/2}} \tag{6}$$

is the lowest-order pair-correlation function, where $f(x)$ is the Fermi distribution function.

In the NSR scheme[2], the superfluid phase transition temperature $T_{\rm c}$ is determined from the Thouless criterion[14], stating that the particle-particle scattering matrix has a pole in the low-energy and low-momentum limit. When we naively use the NSR thermodynamic potential in Eq. (5), this condition is achieved when $\det[\cdots]$ in Eq. (5) vanishes at $\boldsymbol{q} = \nu_n = 0$, which gives,

$$1 - U_0[\Pi_{\rm o}(0,0) + \Pi_{\rm c}(0,0)] + (U_0^2 - U_1^2)\Pi_{\rm o}(0,0)\Pi_{\rm c}(0,0) = 0. \tag{7}$$

We then solve the $T_{\rm c}$-equation (7), together with the equation for the total number $N$ of Fermi atoms,

$$N = -\frac{\partial\Omega}{\partial\mu} = \sum_{\rm s=o,c} N_{\rm s} - T\sum_{\boldsymbol{q},\nu_n} \frac{\partial}{\partial\mu} \ln \left[ \det \left[ \begin{array}{cc} 1 - U_0\Pi_{\rm o}(\boldsymbol{q}, i\nu_n) & -U_1\Pi_{\rm o}(\boldsymbol{q}, i\nu_n) \\ -U_1\Pi_{\rm c}(\boldsymbol{q}, i\nu_n) & 1 - U_0\Pi_{\rm c}(\boldsymbol{q}, i\nu_n) \end{array} \right] \right] \tag{8}$$

(where $N_{\rm s} = 2\sum_{\boldsymbol{p}} f(\xi^{\rm s}_{\boldsymbol{p}})$), to determine $T_{\rm c}$ and $\mu(T_{\rm c})$ self-consistently.

However, when we apply this formalism to an ultracold $^{173}$Yb Fermi gas with an OFR ($a_{s,+}$=1900$a_0$ and $a_{s,-}$=200$a_0$, where $a_0 = 0.529$ A is the Bohr's radius[9, 12]), in addition to the experimentally accessible bound state with shallow binding energy $E_+^{\rm bind}$= $-1/ma_{s,+}^2$, one obtains another bound state which is experimentally inaccessible because of the very deep energy level $E_+^{\rm bind} = -1/ma_{s,-}^2 \ll E_-$[15]. Although the $T_{\rm c}$ equation (7) naively gives the Bose-Einstein condensation associated with the latter molecular state as the *highest* $T_{\rm c}$, we need to remove this "experimentally inaccessible part" from the theory, in order to focus on the experimentally accessible shallow bound state.

This attempt is achieved by diagonalizing the $2 \times 2$ matrix part in Eq. (5) as

$$\Omega = \Omega_0 + T \sum_{\boldsymbol{q}, \nu_n} \ln \left[ \det \left[ \begin{array}{cc} \lambda_+(\boldsymbol{q}, i\nu_n) & 0 \\ 0 & \lambda_-(\boldsymbol{q}, i\nu_n) \end{array} \right] \right], \tag{9}$$

where

$$\lambda_\pm(\boldsymbol{q}, i\nu_n) = 1 - \frac{U_0}{2}[\Pi_{\rm o} + \Pi_{\rm c}] \pm \frac{1}{2}\sqrt{U_0^2[\Pi_{\rm o}^2 + \Pi_{\rm c}^2] - 2[U_0^2 - 2U_1^2]\Pi_{\rm o}\Pi_{\rm c}}. \tag{10}$$

Between $\lambda_\pm$, one finds that $\lambda_+$ corresponds to the accessible shallow state. Only retaining this contribution, we obtain the number equation, as well as the $T_{\rm c}$-equation as, respectively,

$$N = \sum_{\rm s=o,c} N_{\rm s} - T \sum_{\boldsymbol{q}, \nu_n} \frac{\partial}{\partial \mu} \ln[\lambda_+(\boldsymbol{q}, i\nu_n)], \tag{11}$$

$$1 = \frac{U_0}{2}[\Pi_o(0,0) + \Pi_{\rm c}(0,0)] + \frac{1}{2}\sqrt{U_0^2[\Pi_{\rm o}^2(0,0) + \Pi_{\rm c}^2(0,0)] - 2[U_0^2 - 2U_1^2]\Pi_{\rm o}(0,0)\Pi_{\rm c}(0,0)}. \tag{12}$$

We briefly note that we only deal with the number equation (11), when we evaluate $\mu(T > T_{\rm c})$.

In contrast to the case of a broad magnetic Feshbach resonance, where the interaction effects are simply tuned by the scaled inverse $s$-wave scattering length $(k_{\rm F}a_s)^{-1}$ (where $k_{\rm F}$ is the Fermi momentum), the present OFR case is known to also depend on other parameters, such as the number density $n$[13]. In this paper, we set the number density as $n = k_{\rm F}^3/(3\pi^2) = 5 \times 10^{13}{\rm cm}^{-3}$[10, 13]. We also take the energy difference $\delta/2$ between the open and closed channel as $\delta/2 = (2\pi\hbar \times 56\Delta_m)B$ Hz[10, 13] (where $\Delta_m$=5 is the difference between the real nuclear-spin quantum numbers described by pseudospin $\sigma = \uparrow, \downarrow$, and $B$ [G] is an external magnetic field). As in the MFR case, we measure the interaction strength in terms of the scaled inverse $s$-wave scattering length $(k_{\rm F}a_s)^{-1}$ in the open channel, where $a_s$ is given by[10]

$$a_s = \frac{-a_{s0} + \sqrt{m\delta/\hbar^2}[a_{s0}^2 - a_{s1}^2]}{a_{s0}\sqrt{m\delta/\hbar^2} - 1} \quad (\delta > 0), \tag{13}$$

where $a_{s0} = [a_{s,+} + a_{s,-}]/2$ and $a_{s1} = [a_{s,-} - a_{s,+}]/2$.

## 3. Results

Figure 2(a) shows the superfluid phase transition temperature $T_{\rm c}$, obtained from the coupled equations (11) with (12). ($\mu(T_{\rm c})$ is shown as the dotted line in Fig. 2(b).) Our result agrees well with the previous work[13], confirming that our prescription for removing inaccessible deep bound state really works. As pointed out in Ref.[13], OFR gives higher $T_{\rm c}$ in the weak-coupling side, compared to the case of a broad magnetic Feshbach resonance (MFR).

To see to what extent atoms in the closed channel contribute to the superfluid instability, we plot the number $N_{\rm c}$ of atoms at $T_{\rm c}$ in Fig. 3(a). Then, we find that, although the necessity of explicit treatment of the closed channel is characteristic of OFR, $N_{\rm c}(T = T_{\rm c})$ is actually very small in the entire BCS-BEC crossover region. $N_{\rm c}$ reaches the number $N_{\rm o}$ of atoms in the open channel, when the energy difference $\delta/2$ between the open and closed channels vanishes at $(k_{\rm F}a_s)^{-1} = 1.58$ (see the inset in Fig. 3(a)). However, this strong-coupling regime is already dominated by tightly bound molecules described by the "NSR" term $N_{\rm NSR} = N - N_{\rm o} - N_{\rm c}$, as shown in Fig. 3(a). Thus, in the case of a $^{173}$Yb Fermi gas, the

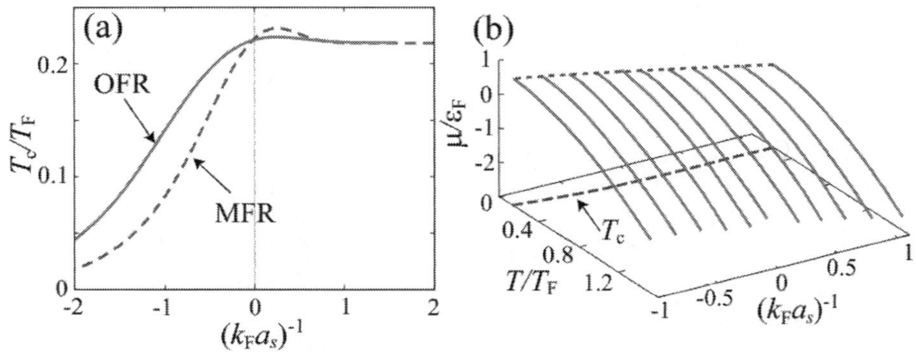


**Figure 2.** (a) Calculated $T_c$ in an ultracold Fermi gas with an OFR, as a function of the interaction strength $(k_F a_s)^{-1}$ . The dashed line shows the result in the MFR case. (b) Fermi chemical potential $\mu(T \geq T_c)$ in the BCS-BEC crossover region. The dashed line shows $\mu(T_c)$. (Color figure online)

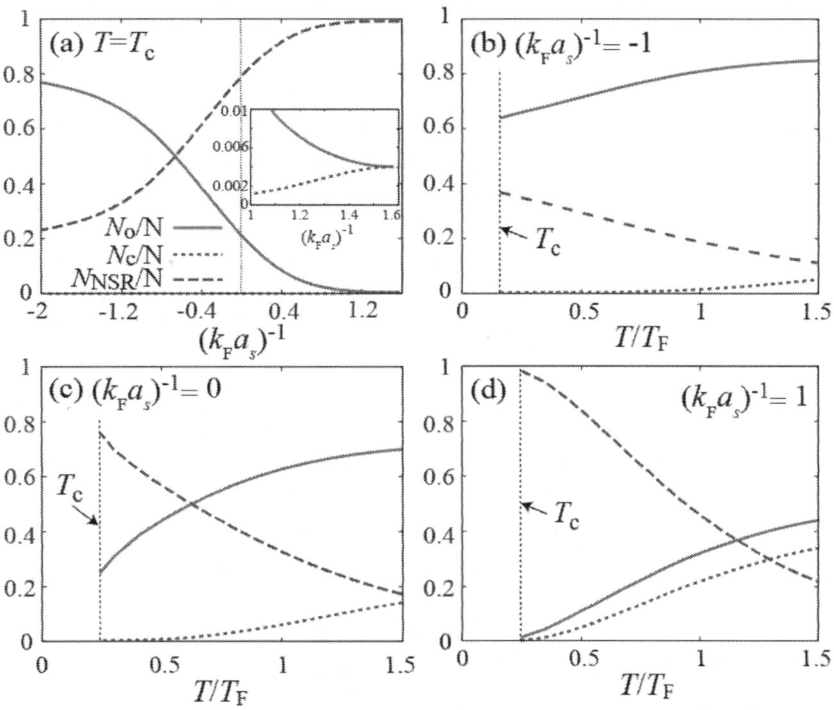


**Figure 3.** (a) The number of atoms in the open channel $N_o = \sum_p f(\xi_p^o)$, as well as that in the closed channel $N_c = \sum_p f(\xi_p^c)$ in the BCS-BEC crossover region at $T_c$. $N_{NSR} = N - N_o - N_c$ describes NSR fluctuation contribution. For clarity, we magnify the result in the strong-coupling regime in the inset. (b)-(d) Temperature dependence of $N_o$, $N_c$, and $N_{NSR}$: (b) weak-coupling regime. (c) unitarity limit. (d) strong-coupling regime. (Color figure online)

system properties at $T_c$ are dominated by the open channel ($N_o$) and fluctuation contribution ($N_{NSR}$), as in the case of a broad magnetic Feshbach resonance used in $^6$Li and $^{40}$K Fermi gases.

The occupation of the closed channel gradually increases with increasing the temperature above $T_c$, as shown in Figs. 3(b)-(d). Here, these results are obtained by using the Fermi chemical potential $\mu(T \geq T_c)$ shown in Fig. 2(b), calculated from the number equation (11). Since the energy difference $\delta/2$ between

IOP Conf. Series: Journal of Physics: Conf. Series **969** (2018) 012017 doi:10.1088/1742-6596/969/1/012017

the open and closed channels is smaller for a stronger pairing interaction (see Eq. (13)), the growth of $N_c$ is more remarkable in the strong-coupling side (Fig. 3(d)) than in the weak-coupling case (Fig. 3(b)). We briefly note that, in all the cases shown in Figs. 3(b)-(d), both $N_o$ and $N_c$ monotonically increase and the fluctuations contribution $N_{\rm NSR}$ only decreases, as one increases the temperature. This means that the atoms in the closed channel ($N_c$) are supplied by the decrease of atoms in the scattering states, as well as by the dissociation of tightly bound molecules (that are described by $N_{\rm NSR}$), rather than thermal excitations from the open channel to the closed channel. We briefly note that the closed-channel band $\xi^{\rm c}_{\boldsymbol{p}} = \boldsymbol{p}^2/(2m) + \delta/2 - \mu$ is higher than the open-channel band $\xi^{\rm o}_{\boldsymbol{p}} = \boldsymbol{p}^2/(2m) - \mu$ by $\delta/2 > 0$. Thus, when the temperature is much higher than this energy difference ($T \gg \delta/2$), the number $N_c$ of atoms in the closed would approach that in the open channel as $N_o \simeq N_c \simeq N/2$.

## 4. Conclusion

To summarize, we have discussed an ultracold Fermi gas with an orbital Feshbach resonance (OFR). Including both the open and closed channels, as well as pairing fluctuations, within the framework of the strong-coupling theory developed by Nozières and Schmitt-Rink (NSR), we have calculated the numbers of atoms in the open channel ($N_o$) and closed channel ($N_c$), as well as the NSR fluctuation contribution ($N_{\rm NSR} = N - N_o - N_c$), in the BCS-BEC crossover region above $T_c$. In this procedure, we have also presented an idea to remove effects of the experimentally inaccessible bound state which appears when a $^{173}$Yb Fermi gas is considered.

We showed that, when we take parameters for a $^{173}$Yb Fermi gas, the occupation of the closed channel is actually very small over the entire BCS-BEC crossover region at $T_c$. On the other hand, the occupation number of the closed channel gradually becomes remarkable, as one raises the temperature above $T_c$, especially in the strong-coupling regime. This result implies that atoms in the closed channel (which is characteristic of the OFR case) affect the temperature dependence of physical quantities. Since the superfluid phase transition in a $^{173}$Yb Fermi gas with an OFR is promising, our results would contribute to the research toward the realization of a Fermi superfluid of non-alkali metal atomic gas.

## Acknowledgements

We thank P. van Wyk, D. Kharga and R. Hanai for useful discussions. This work was supported by the KiPAS project in Keio university. DI was supported by Grant-in-Aid for Young Scientists (B) (No. JP16K17773) from JSPS. YO was supported by Grand-in-Aid for Scientific Research from MEXT and JSPS in Japan (No. JP15K00178, No. JP15H00840, No. JP16K05503).

## References

[1] Chin C, Grimm R, Julienne P and Tiesinga E 2010 *Rev. Mod. Phys.* **82** 1225
[2] Nozières P and Schmitt-Rink S 1985 *J. Low Temp. Phys.* **59**, 195
[3] Sa de Melo C, Randeria M and Engelbrecht J 1993 *Phys. Rev. Lett.* **71** 3202
[4] Ohashi Y and Griffin A 2002 *Phys. Rev. Lett.* **89** 130402.
[5] Regal C, Greiner M and Jin D 2004 *Phys. Rev. Lett.* **92** 040403
[6] Zwierlein M, Stan C, Schunck C, Raupach S, Kerman A and Ketterle W 2004 *Phys. Rev. Lett.* **92** 120403
[7] Enomoto K, Kasa K, Kitagawa M and Takahashi Y 2008 *Phys. Rev. Lett.* **101** 203201
[8] Scazza F, Hofrichter C, Höfer M, De Groot P C, Bloch I and Fïling S 2014 *Nature Physics.* **10** 779.
[9] Cappellini G, Mancini M, Pagano G, Lombardi P, Livi L, Siciliani de Cumis M, Cancio P, Pizzocaro M, Calonico D, Levi F, Sias C, Catani J, Inguscio M and Fallani L 2014 *Phys. Rev. Lett.* **113** 120402
[10] Zhang R, Cheng Y, Zhai H and Zhang P 2015 *Phys. Rev. Lett.* **115** 135301
[11] Pagano G, Mancini M, Cappellini G, Livi L, Sias C, Catani J, Inguscio M and Fallani L 2015 *Phys. Rev. Lett.* **115** 265302
[12] Höfer M, Riegger L, Scazza F, Hofrichter C, Fernandes D R, Parish M M, Levinsen J, Bloch I and Fölling S 2015 *Phys. Rev. Lett.* **115** 265302.
[13] Xu J, Zhang R, Cheng Y, Zhang P, Qi R and Zhai H 2016 *Phys. Rev. A.* **94** 033609
[14] Thouless D 1960 *Ann. Phys.* **10** 553
[15] He L, Wang J, Peng S G, Liu X J and Hu H 2016 *Phys. Rev. A.* **94** 043624

IOP Conf. Series: Journal of Physics: Conf. Series **969** (2018) 012018 doi:10.1088/1742-6596/969/1/012018

# Coalescence of Multielectron Bubbles in Liquid Helium

**Vaisakh Vadakkumbatt**[1] **and Ambarish Ghosh**[1,2,3]

[1] Department of Physics, Indian Institute of Science, Bangalore, India, 560012

[2] Department of Electrical Communication Engineering, Indian Institute of Science, Bangalore, India, 560012

[3] Centre for Nano Science and Engineering, Indian Institute of Science, Bangalore, India, 560012

E-mail: ambarish@ece.iisc.ernet.in

**Abstract.** In 1977, Volodin *et al.* observed that the electrohydrodynamic instability of charged helium surfaces can lead to the loss of the electrons from the surface in the form of bubbles. These are Multielectron Bubbles (MEBs) which contain electrons pinned on their inner surfaces. The MEBs form a model system for studying electrons on curved surfaces, which are predicted to have many interesting properties. Recent experiments showed that above the Lambda point, MEBs could be trapped using a Paul-trap and their sizes are mainly determined by the amount of vapour present inside. Here, we report the experimental observation of the coalescence of two MEBs which were moving upward in bulk liquid helium-4. The charge and radii of the MEBs were determined before and after the coalescence. The merging of two similar charged bubbles was possible because of the presence of vapour inside the merging MEBs.

## 1. Introduction

An electron present above a liquid helium surface are attracted to the liquid, which originates due to the image charges induced due to the finite polarizability of helium. However, to enter inside the bulk liquid, the electron has to overcome a potential barrier of 1 eV which arises due to Pauli exclusion principle. Hence all the low energy electrons are confined on the helium surface, and form a model two dimensional electron system. Compared to two dimensional electron systems on other semiconductor systems, 2DES on the helium surface are unique as they are governed by classical statistics. If the electron density on the helium surface is taken beyond a critical value of $2 \times 10^9$ electrons/cm$^2$, an electrohydrodynamic (EHD) instability develops[1], which results in charge loss from the surface in the form of bubbles. These bubbles contain a layer of electrons pressed on their inner surfaces and are known as multielectron bubbles (MEBs), which were first observed by Volodin [2] in 1977. Unlike the flat helium surface, the electron density of MEBs can be varied over a wide range, which could necessitate their behavior to be modeled by quantum statistics at high densities. This and many interesting electronic phases possible with MEBs have been predicted[3]. Of particular interest is the question whether these objects are stable against shape fluctuations, for which there are conflicting theoretical predictions[4, 5, 6, 7].

After the first observation of MEBs, there were very few experiments on MEBs[8, 9], where the properties such as charge, velocity, radius and drag of the MEBs were measured in a destructive measurement. There have also been reports of their generation[10, 11, 12, 13] using alternate methods. A scheme to observe and study MEBs non-destructively has been developed by our group[14, 15] using a Paul trap above the lambda point. Using the Paul trap, we were able to trap the MEBs for a longer period of time (a few hundred milliseconds), and the charge and radius of the MEBs were measured by analyzing the trajectories. The results showed that the MEBs were filled with vapour, and the size of the MEB was mainly determined by the amount of vapour inside. As we observe in our experiments, the MEBs collapse as the vapour condenses with time. This phenomenon has been modeled, and the time scale was found to match[16] with experimentally observed values of the collapse time.

In the Paul trap experiments, the charge of the MEBs were typically low, around $2 \times 10^3$ - $4 \times 10^4$ electrons, implying as the vapour condense they become too small to be resolved using our imaging setup. In our recent experiments[17], it was shown that the splitting of MEBs using high electric fields could result in the generation of MEBs with higher charge density. Here we describe coalescence, which is the opposite of splitting phenomena, to obtain MEBs with lower charge density.

## 2. Experimental Setup

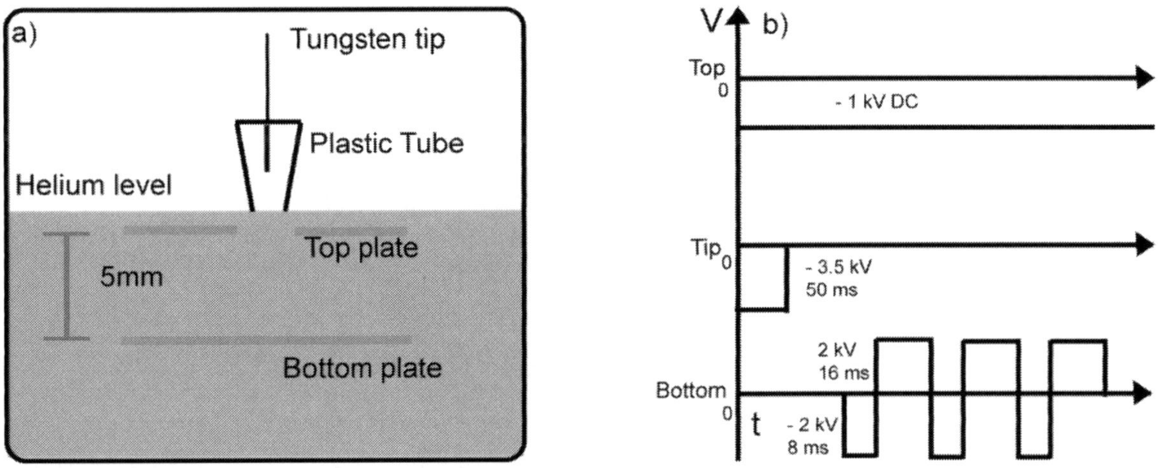


**Figure 1.** Schematic of the experimental set up inside the cell (a). Liquid helium was filled inside till the level touched the bottom of the plastic tube. Top and bottom plates were kept at a separation of 5 mm. b) Different wave forms applied to the tip, the bottom plate, and the top plate for the generation and the manipulation of MEBs.

Experiments were carried out at 2.5 K inside an optical cryostat. The experimental chamber was made of stainless steel with optical access from four sides. The schematic of the setup inside the chamber is shown in figure 1, which was similar to our previous experiments[17]. Electrons were generated on the helium surface by field emission from the tungsten tip. The emitted electrons were confined using a plastic tube of inner diameter 3 mm around the tungsten tip to achieve the electron densities required for the MEB generation. Two metal plates, top and bottom, were kept parallel to each other for generating the electric field required for controlling the motion of MEBs.

To create field emission, high voltage pulses in the order of -2 to -3.5 kV were applied to tip for duration varying between 30 to 100 milliseconds. When the charge density on the helium surface exceeded beyond critical value, an EHD instability developed on the surface. At the instability point, charged helium surface broke, and gave off MEBs. Negative voltage present on the tip pushed MEBs into the bulk of the liquid. Charge of the MEB was measured by applying appropriate voltage pulses to top and bottom metal plates, using a method described later. The motion of MEBs between the plates occurred within few milliseconds, hence we used a high speed camera with suitable light source for capturing the entire sequence of events. For the experiments mentioned here, the high speed camera was operated at a frame rate of 3600 fps. To obtain the trajectories of the MEBs, videos were processed using Mathlab image processing tools. The number of electrons inside the MEB was estimated by analyzing the trajectories.

## 3. Coalescence of MEBs

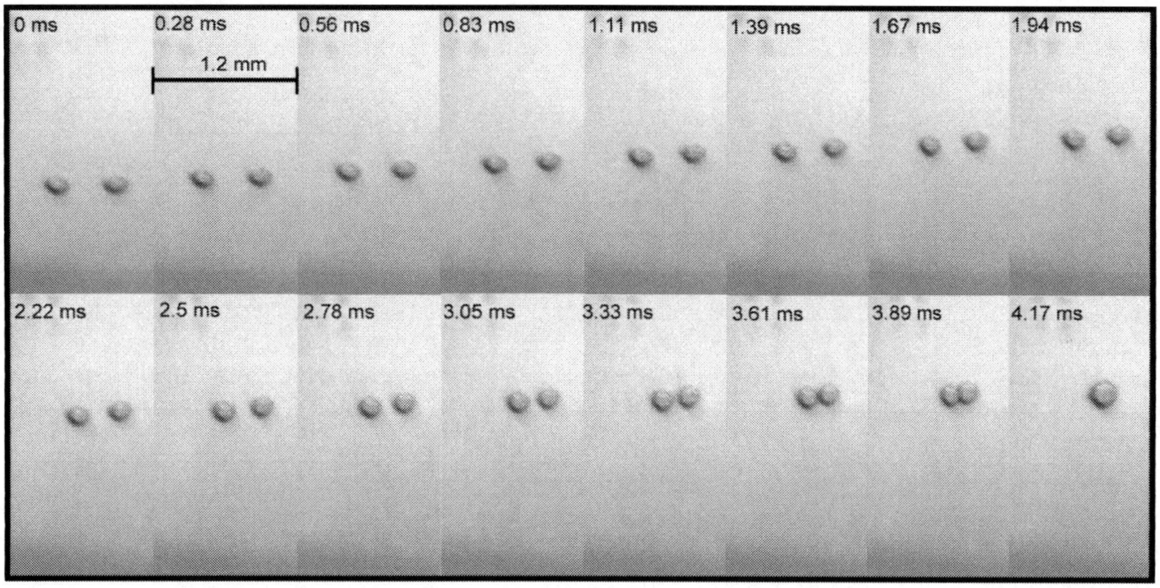


**Figure 2.** Series of images showing the coalescence of MEBs. Initially the two MEBs were separated by a distance of 0.53 mm. As they move up during the negative pulse, they came nearby and fused with each other. Each frame corresponds to a time scale of 0.277 milliseconds, total frame width is 1.2 mm and height is 2.4 mm.

After the generation, we applied a series of high voltage pulses to the bottom plate as shown in figure 1b for measuring the charge and radius of MEBs. During the experimental trials, we observed the coalescence two MEBs even though they were separated by a few millimetres initially. These MEBs were moving up during negative pulse, and when they were sufficiently close to each other, they merged. It appears their proximity during their upward motion is a random event made possible by a somewhat chaotic flow field inside the bulk. However, we can not rule out the fact that the MEBs may interact with each other through hydrodynamic interactions and this may bring them together. The sequence of images of the two MEBs fusing with each other is shown in figure 2. Just before coalescence, the MEBs were having radii 90 $\mu m$ and 95 $\mu m$ moving with velocities 16 cm/s and 18 cm/s respectively. The resultant MEB had a radius of 120 $\mu m$. After the coalescence, the resultant MEB showed a particular pattern of oscillation. This is shown in figure 3. Amplitude of the oscillations decayed in a few milliseconds.

For a detailed analysis of coalescence, charges of the MEBs before and after coalescence event were estimated. We used a methodology, where all the forces on the MEB are equated at the point where the MEB makes a turn (zero velocity). Total force acting on the MEB at any instant can be related to its acceleration, and is given by

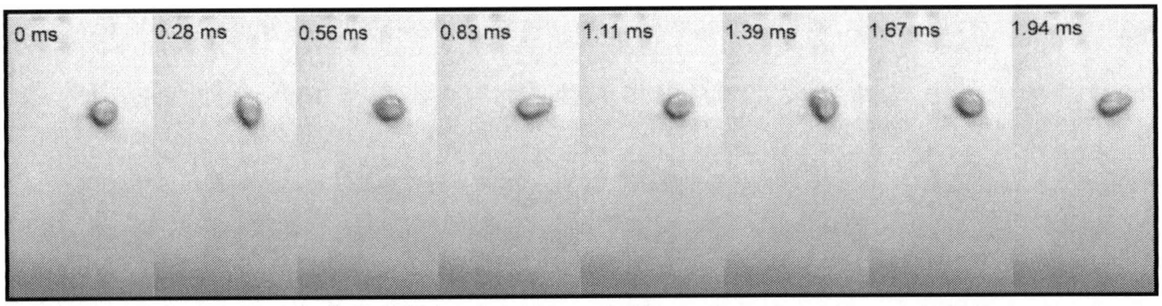


**Figure 3.** Series of pictures showing the two cycles of oscillation of the resultant MEB after the coalescence.

$$F_b + F_E - F_{drag} = m_{MEB}a \tag{1}$$

Here, $F_b$ is the buoyant force acting on the bubble and given by $\frac{4}{3}\pi\rho R^3 g$, $m_{MEB}$ is the effective hydrodynamic mass of the MEB moving through the liquid with acceleration $a$, and given by $\frac{2}{3}\pi\rho R^3$, where $R$ is the effective radius of the MEB, $\rho$ is the density of liquid helium. We compared the forces at the point where the MEB turned, implying it was a point of zero velocity and therefore $F_{drag} = 0$. The only unknown in this equation is the charge of the MEB, which can be determined by knowing the electric field at zero velocity point. We used a Comsol model for obtaining the electric field value at zero velocity point for the applied electrical conditions.

We followed this procedure and found that MEBs were having charges $1.38 \times 10^5$ electrons and $3 \times 10^5$ electrons before coalescence. After the coalescence, the resultant MEB contained approximately $5 \times 10^5$electrons, which proved the consistency of the method. The equilibrium radius of an MEB containing $Z$ electrons with no vapour is $R_{eq} = (\frac{Z^2e^2}{64\pi^2\epsilon\sigma})^{1/3}$, where $\sigma$ is the surface tension of helium at 2.5 K. For a given $Z$ of $5 \times 10^5$, $R_{eq}$ is 16.1 $\mu m$. This value is much smaller than the experimentally observed value of 120 $\mu m$, which indicates that the MEB was filled with vapour.

Since MEBs were having similar charges, their union was somewhat surprising. This can be explained by looking at different energy scales involved in the process. The MEB with a radius of 90 $\mu m$ was moving with a velocity 16 cm/s, and contained $1.38 \times 10^5$ number of electrons. For this MEB, the electrostatic and kinetic energies were 0.02 pJ and 2.8 pJ respectively. Similarly for the other MEB which moved at 18 cm/s and contained $3 \times 10^5$ number of electrons, these values were 0.1 pJ and 4.2 pJ respectively. For coalescence to occur, they needed to overcome the electrostatic repulsion, which resulted in an energy barrier of around 0.05 pJ. This was estimated by assuming the electrons in the individual MEBs to be effectively point charges and separated by a distance which was the sum of their individual radii. This was a justifiable assumption since the size of the MEBs were effectively governed by the vapor inside, which allowed reorganization of the electrons inside the bubbles with minimal energy cost. As the kinetic energy of the individual MEBs were sufficiently larger than the repulsive barrier of 0.05 pJ, coalescence was possible. The oscillation modes induced in the coalesced MEB, as shown in figure 3, was a consequence of the excess energy getting dissipated into the liquid.

## 4. Conclusion

We present experimental evidence of coalescence of two MEBs. The radii and charges of the MEBs were determined before and after the coalescence, which suggested significant amount of helium vapor inside the bubbles. The estimate of different energies of the MEBs suggest that coalescence of these like-charged systems was indeed energetically allowed.

## 5. Acknowledgements
We thank Emil M. Joseph and Humphrey J. Maris for helpful discussions. This work was supported by Nanomission, Science and Engineering Research Board, India. The usage of the facilities in Micro and Nano Characterization Facility (MNCF, CeNSE) at IISc is gratefully acknowledged, and the work is partially supported by the Ministry of Communication and Information Technology under a grant for the Centre of Excellence in Nanoelectronics, Phase II.

## References
[1] Gor'kov L and Chernikova D 1973 *Soviet Journal of Experimental and Theoretical Physics Letters* **18** 68
[2] Volodin A, Khaikin M and Édelman V 1977 *JETP Lett* **26** 11
[3] Tempere J, Silvera I and Devreese J T 2007 *Surface Science Reports* **62** 159–217
[4] Salomaa M M and Williams G A 1981 *Physical Review Letters* **47** 1730
[5] Tempere J, Silvera I F and Devreese J 2001 *Physical review letters* **87** 275301
[6] Tempere J, Silvera I and Devreese J 2003 *Physical Review B* **67** 035402
[7] Guo W, Jin D and Maris H J 2008 *Physical Review B* **78** 014511
[8] Albrecht U and Leiderer P 1987 *EPL (Europhysics Letters)* **3** 705
[9] Albrecht U and Leiderer P 1992 *Journal of low temperature physics* **86** 131–138
[10] Fang J, Dementyev A E, Tempere J and Silvera I F 2009 *Review of Scientific Instruments* **80** 043901
[11] Fang J, Dementyev A E, Tempere J and Silvera I F 2011 *Review of Scientific Instruments* **82** 033904
[12] Vadakkumbatt V, Joseph E, Pal A and Ghosh A 2013 *Journal of Low Temperature Physics* **171** 239–244
[13] Joseph E, Vadakkumbatt V, Pal A and Ghosh A 2014 *Journal of Low Temperature Physics* **175** 78–84
[14] Vadakkumbatt V, Joseph E, Pal A and Ghosh A 2014 *Nature communications* **5**
[15] Joseph E, Vadakkumbatt V, Pal A and Ghosh A 2017 *Journal of Low Temperature Physics* **187** 580–587
[16] Pal A, Joseph E, Vadakkumbatt V, Yadav N, Srinivasan V, Maris H J and Ghosh A 2017 *Journal of Low Temperature Physics* 1–11
[17] Vadakkumbatt V and Ghosh A 2017 *Journal of Low Temperature Physics* **187** 369–375

IOP Conf. Series: Journal of Physics: Conf. Series **969** (2018) 012019 doi:10.1088/1742-6596/969/1/012019

# Superfluidity of $^4$He in dense aerogel studied using quartz tuning fork

**K Matsumoto[1], R Okamoto[1], A Nakajima[1], S Abe[1]**

[1] Department of Physics, Kanazawa University, Kakuma-machi, Kanazawa 920-1192, Japan

E-mail: `k.matsu@staff.kanazawa-u.ac.jp`

**Abstract.** Superfluid $^4$He in aerogel is of interest because it has a normal component coupling to gel strand due to viscosity and a superfluid component with zero viscosity. Superfluid helium in aerogel has two sound modes, a slow critical mode and a fast one. In this study, quartz tuning fork was used in order to study acoustic properties of liquid $^4$He in aerogel with 90% porosity. Two pieces of aerogel were glued on both prongs of quartz tuning fork that had a resonance frequency of 33 kHz. The tuning fork was immersed in liquid $^4$He from 2 to 20 bar. The resonance frequency increased in the superfluid phase due to decrease in loaded mass. Temperature variation of resonance frequency was explained by that of superfluid density. Superfluid transition in aerogel was 2 mK lower than that without gel. Additional dissipation was observed in the temperature range between 1 K and transition temperature.

## 1. Introduction

Quartz tuning forks are piezoelectric mechanical oscillators consisted of two prongs and used as frequency standards for many devices with resonant frequencies typically in the kilohertz range. In recent years, quartz tuning forks have been used for the studies of quantum fluids such as viscosity[1, 2], turbulence[3, 4, 5], cavitation[6], coupling to acoustic modes[7, 8], and many others[9, 10]. The advantages of quartz tuning forks are highly sensitive, compact, and easy to install and operate. The $Q$ factors of quartz tuning forks are typically of order $10^5$ in vacuum at room temperature, and larger values can be attained at very low temperatures.

Acoustic property of superfluid helium is of interest because it has a normal component playing the role of viscous fluid and simultaneously a superfluid component with zero viscosity. The normal fluid in high-porous media like aerogel that is locked to the substrate by viscosity drags the substrate along. So, there are two sound modes, a slow critical mode and a fast one[11]. We have reported acoustic properties of liquid $^4$He in aerogel using longitudinal ultrasound and the different behavior from rigid porous material[12, 13]. An anomalous attenuation was observed in superfluid phase with 97% open aerogel at the frequency of 10 MHz[13]. Possibility of sound mode conversion was discussed in helium and porous material system[14]. In order to elucidate the acoustic properties in porous materials, it is useful to make measurements by changing the vibration frequency because coupling between aerogel and $^4$He could be changed by varying the viscous penetration depth.

We report our preliminary experiments on superfluidity of $^4$He in aerogel with 90% porosity using quartz tuning fork. In this study, quartz tuning fork that had a resonance frequency of 33 kHz in vacuum was used. The onset of superfluid transition was observed as an abrupt

IOP Conf. Series: Journal of Physics: Conf. Series **969** (2018) 012019 doi:10.1088/1742-6596/969/1/012019

increase in resonance frequency. Transition temperature in aerogel was 2 mK lower than that without gel. This suppression of transition temperature was consistent with other aerogels. Temperature variation of resonance frequency was explained by that of superfluid density. Additional dissipation was observed in the temperature range between 1 K and transition temperature.

## 2. Experimental procedures

Experiments were made using commercially available tuning fork (DT-38, Daishinku cooperation) which had a resonance frequency of 32,768 Hz at room temperature in vacuum. The arm of the tuning forks had length 3.03 mm and a rectangular cross-section of thickness 0.336 mm and width 0.474 mm shown in Fig. 1. The metal can was removed.

Two pieces of aerogel with 90% porosity were glued on the front surfaces of both prongs with Stycast 2850FT as shown in Fig. 1. Each piece had similar size but the shape wasn't identical because of difficulty in cutting the aerogel. The total volume of aerogel was about $8 \times 10^{-12}$ m$^3$.

The tuning fork was located in a copper cell that contacted with the mixing chamber of a dilution refrigerator. Temperature was measured with a $RuO_2$ resistance thermometer. Pressure of the cell was measured and regulated using a pressure sensor located at room temperature. In our experiment we have measured the resonance frequency and amplitude as well as the resonance line width of the quartz tuning forks using a standard Lock-in amplifier circuit. Resonance was kept track during temperature variation by adjusting oscillator frequency automatically with a computer algorithm of phase locked loop. Wide frequency sweep across the resonance peak was made at some fixed temperatures. The excitation of oscillation has been done with small amplitude of the signal. So, the response of the tuning fork was in linear region.

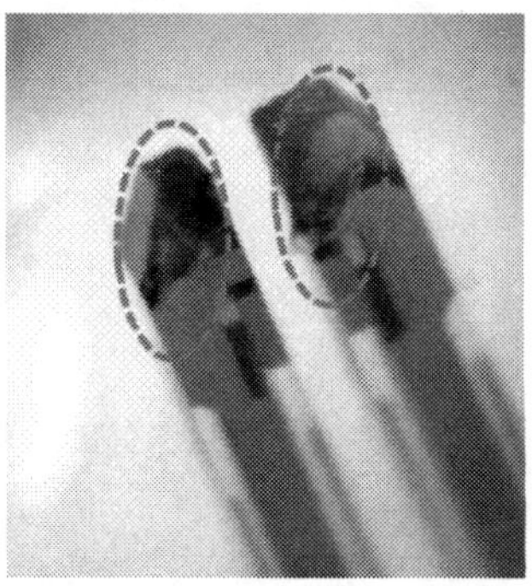

**Figure 1.** Photograph of quartz tuning fork with aerogel. The aerogels were shown in ellipses.

## 3. Results and discussion

### *3.1. Resonance frequency*

The resonance frequency with aerogel decreased to 32.1635 kHz in vacuum at room temperature due to the mass of aerogel and Stycast. Resonance frequency of tuning fork with aerogel $f_{\mathrm{w/}}$ was measured as functions of liquid helium pressure at 0.14, 1.3, and 3.0 K. At all temperatures, resonance frequency decreased linearly with helium density. That without aerogel $f_{\mathrm{w/o}}$ also linearly depended on helium density.

Resonance frequencies in vacuum for both without aerogel $f_{\mathrm{w/o}}^{\mathrm{vac}}$ and with aerogel $f_{\mathrm{w/}}^{\mathrm{vac}}$ showed negligible temperature dependence compared with those in liquid helium. Upper panel of Fig. 2 shows the resonance frequency change of the tuning fork with aerogel from that at 2 K in vacuum. In liquid helium at 20 bar, $f_{\mathrm{w/}}$ and $f_{\mathrm{w/o}}$ had a minimum $f_{\mathrm{w/}}^{\mathrm{min}} = 31.0105$ kHz and $f_{\mathrm{w/o}}^{\mathrm{min}} = 31.6772$ kHz at superfluid transition temperature $T_\lambda$, respectively. Lower panel of Fig. 2 shows $f_{\mathrm{w/}}$ and $f_{\mathrm{w/o}}$ as functions of temperature. In this panel, vertical axis is plotted as the resonance

IOP Conf. Series: Journal of Physics: Conf. Series **969** (2018) 012019 doi:10.1088/1742-6596/969/1/012019

frequency difference $\delta f$ from $f^{\mathrm{min}}$ for with and without aerogel such as $\delta f_{\mathrm{w/}} = f_{\mathrm{w/}} - f^{\mathrm{min}}_{\mathrm{w/}}$ and $\delta f_{\mathrm{w/o}} = f_{\mathrm{w/o}} - f^{\mathrm{min}}_{\mathrm{w/o}}$. Frequency change $\delta f_{\mathrm{w/}}$ and $\delta f_{\mathrm{w/o}}$ agree with each other in normal phase. But, those significantly differ in superfluid phase. In superfluid phase, increase in resonance frequency with aerogel is larger than that without aerogel because large amount of helium decoupled from the aerogel strand. Figure 3 shows the ratio of frequency change $\left|\frac{f-f^{\mathrm{vac}}}{f^{\mathrm{min}}-f^{\mathrm{vac}}}\right|$ with aerogel $f_{\mathrm{w/}}$ and without aerogel $f_{\mathrm{w/o}}$ at 20 bar. $|f^{\mathrm{min}} - f^{\mathrm{vac}}|$ corresponds to frequency change by filling helium at transition temperature. $\left|\frac{f_{\mathrm{w/o}}-f^{\mathrm{vac}}_{\mathrm{w/o}}}{f^{\mathrm{min}}_{\mathrm{w/o}}-f^{\mathrm{vac}}_{\mathrm{w/o}}}\right|$ and $\left|\frac{f_{\mathrm{w/}}-f^{\mathrm{vac}}_{\mathrm{w/}}}{f^{\mathrm{min}}_{\mathrm{w/}}-f^{\mathrm{vac}}_{\mathrm{w/}}}\right|$ at lowest temperature decrease about 6 and 8.5% from transition temperature, respectively. Resonace frequencies at lowest temperature didn't decrease to $f^{\mathrm{vac}}$ for both with and without aerogel case.

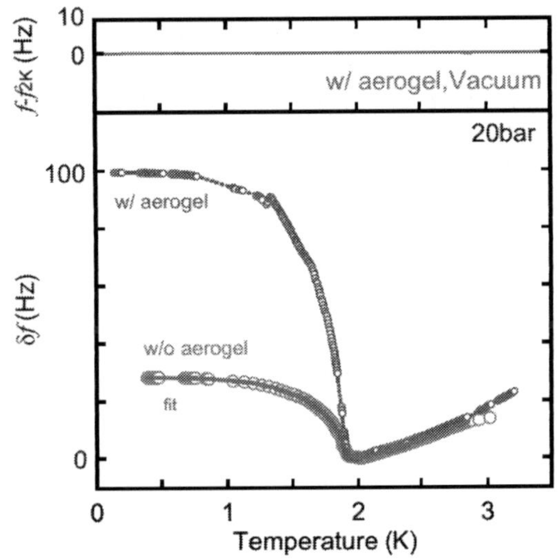


**Figure 2.** Resonance frequency of quartz tuning fork with and without aerogel at 20 bar. That with aerogel in vacuum is also represented in upper panel. The solid line is the fitting to the without aerogel data.

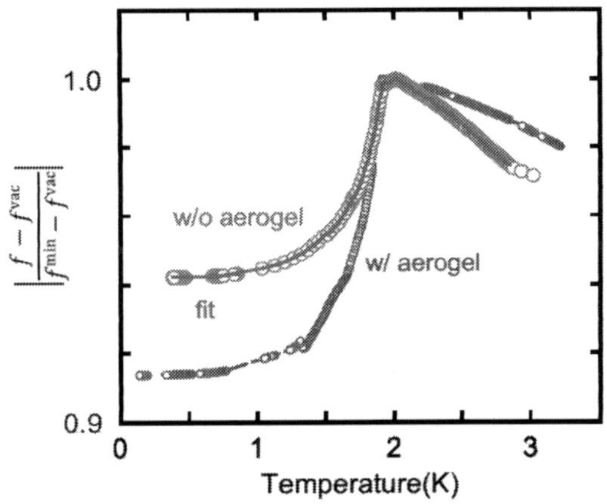


**Figure 3.** Temperature dependence of the ratio of frequency change $\left|\frac{f-f^{\mathrm{vac}}}{f^{\mathrm{min}}-f^{\mathrm{vac}}}\right|$ for with and without aerogel. The solid line is the fitting to the without aerogel data.

In order to analyze resonance frequency, we use the theoretical relation by R. Blaauwgeers[1]. The resonance frequency $f$ is represented as follows,

$$T > T_\lambda : \quad \left(\frac{f^{\mathrm{vac}}}{f}\right)^2 = 1 + \frac{\rho}{\rho_{\mathrm{q}}}\beta + B\frac{S}{V\rho_{\mathrm{q}}}\sqrt{\frac{\eta\rho}{\pi f}}, \tag{1}$$

$$T < T_\lambda : \quad \left(\frac{f^{\mathrm{vac}}}{f}\right)^2 = 1 + \frac{\rho}{\rho_{\mathrm{q}}}\beta + B\frac{S}{V\rho_{\mathrm{q}}}\sqrt{\frac{\eta\rho_{\mathrm{n}}}{\pi f}}. \tag{2}$$

Here $f$ and $f^{\mathrm{vac}}$ are resonance frequencies in liquid helium and vacuum. $\eta$, $\rho$ and $\rho_{\mathrm{n}}$ are viscosity coefficient, density and normal density of helium. $\rho_{\mathrm{q}}$ is density of quartz. $S$ and $V$ are surface area and volume of a prong of tuning fork. $\beta, B$ are fitting parameters.

$f_{\mathrm{w/o}}$ was well fitted with $\beta = 0.971$ and $B = 1.0$ in both normal and superfluid phase as shown in solid lines in Fig. 2 and 3. From this fitting, the second term $(\frac{\rho}{\rho_{\mathrm{q}}}\beta)$ in Eq. 2 due to potential flow is dominant in comparison with the third term $(B\frac{S}{V\rho_{\mathrm{q}}}\sqrt{\frac{\eta\rho_{\mathrm{n}}}{\pi f}})$ due to viscous drag.

When viscous drag effect (third term) dominantly contributed, the frequency ratio $\left|\frac{f_{w/o}-f_{w/o}^{vac}}{f_{w/o}^{min}-f_{w/o}^{vac}}\right|$ will decrease to nearly zero because $\rho_n$ becomes zero at lowest temperature.

$f_{w/}$ was well fitted in normal phase with the parameters close to the case without aerogel ($\beta = 1.108$ and $B = 1.0$). However, it was hard to fit $f_{w/}$ in superfluid phase using the same parameters. Large increase in $f_{w/}$ in superfluid phase was attributed to the mass decoupling of superfluid $^4$He in aerogel. Normal fluid in aerogel is considered to coupled to aerogel strand, comparing strand separation with viscous penetration depth. However, quantitative analysis of superfluid component is difficult because of the low accuracy of gel volume and dominant potential flow contribution. Then, we estimated temperature dependence of superfluid component in aerogel as follows. Resonance frequency had linear dependence on liquid density. The resonance frequency changed about 0.3% at the lowest temperature from that at transition temperature. So, it is considered that the frequency difference between with and without aerogel ($\delta f_{w/} - \delta f_{w/o}$) is proportional to the superfluid component in aerogel. Temperature dependence of $\delta f_{w/} - \delta f_{w/o}$ agreed with that of superfluid fraction of bulk helium as shown in Fig. 4. This suggests that superfluid fraction in aerogel of this study had the same temperature dependence as that of bulk.

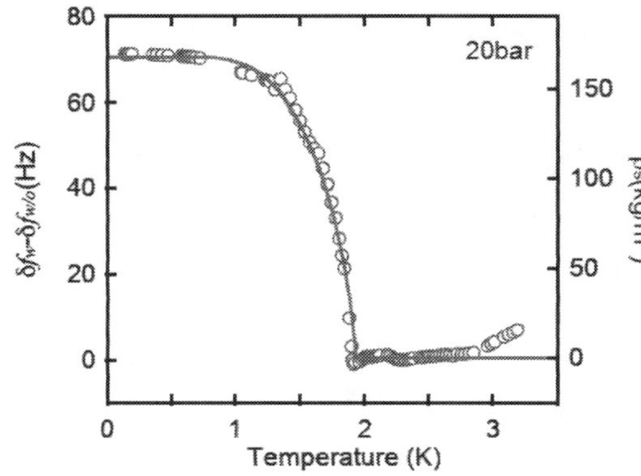


**Figure 4.** Resonance frequency difference $\delta f_{w/} - \delta f_{w/o}$ as a function of temperature compared with superfluid density.

Figure 5 shows resonance frequency in 20 bar expanded around superfluid transition. Superfluid transition was recognized as onset of increase in resonance frequency. Transition temperature in our aerogel decreased 2 mK which was in reasonable agreement with suppression of superfluid transition in other gels.

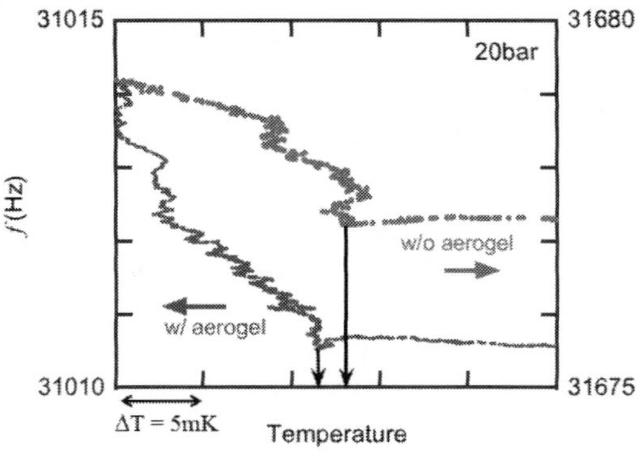


**Figure 5.** Resonance frequency of quartz tuning fork with and without aerogel at 20 bar around superfluid transition.

### *3.2. Resonance line width*

The resonance width of tuning fork $\Delta f$ with aerogel increased to 4.44 Hz in vacuum at room temperature due to additional mass and dissipation of Stycast and aerogel. $\Delta f$ decreased to 1.29 Hz at 1.3 K in vacuum. In resonance tracking measurements, we evaluated $\Delta f$ from resonance frequency $f$ and amplitude of the signal $A$ as $\frac{(2\pi)^2 F}{fA}$, the driving force $F$ was evaluated from excitation voltage and properties of our tuning fork. Figure 6 shows resonance width without aerogel $\Delta f_{\mathrm{w/o}}$ at 20 bar. $\Delta f_{\mathrm{w/o}}$ is almost constant in normal phase and begins to decrease rapidly at superfluid transition. Figure 7 shows resonance width with aerogel $\Delta f_{\mathrm{w/}}$ at 20 bar. $\Delta f_{\mathrm{w/}}$ also shows small temperature dependence in normal phase and rapid decrease in superfluid phase. $\Delta f_{\mathrm{w/}}$ (Fig. 7) was larger than $\Delta f_{\mathrm{w/o}}$ (Fig. 6) in both normal and superfluid phase. Large scatter of $\Delta f_{\mathrm{w/}}$ was observed in superfluid phase. Acoustic modes such as second sound is possible to couple to the fork resonance. For example, large jump in $f_{\mathrm{w/}}$ and $\Delta f_{\mathrm{w/}}$ were observed at 1.3 K. For this reason, we made frequency sweeps at some fixed temperatures. $\Delta f_{\mathrm{w/}}$ obtained by frequency sweep was represented as squares in Fig. 7 and coincide reasonably with that by tracking measurement.

We used the theoretical relation by R. Blaauwgeers[1] in order to analyze resonance width. The resonance width $\Delta f$ is represented as follows,

$$T > T_\lambda: \qquad \Delta f = \frac{1}{2}\sqrt{\frac{\rho\eta f}{\pi}} CS \frac{(f/f^{\mathrm{vac}})^2}{m_{\mathrm{vac}}}, \tag{3}$$

$$T < T_\lambda: \qquad \Delta f = \frac{1}{2}\sqrt{\frac{\rho_{\mathrm{n}}\eta f}{\pi}} CS \frac{(f/f^{\mathrm{vac}})^2}{m_{\mathrm{vac}}}, \tag{4}$$

where $m_{\mathrm{vac}}$ is effective mass of the prong, $C$ is a fitting parameter. We neglected resonance width in vacuum $\Delta f^{\mathrm{vac}}$ because this was small compared with $\Delta f$ in liquid helium.

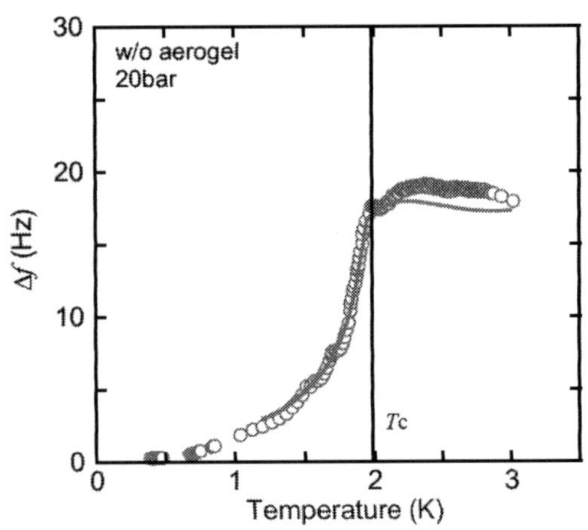


**Figure 6.** Resonance width of quartz tuning fork without aerogel $\Delta f_{\mathrm{w/o}}$ at 20 bar. The solid line is the fitting.

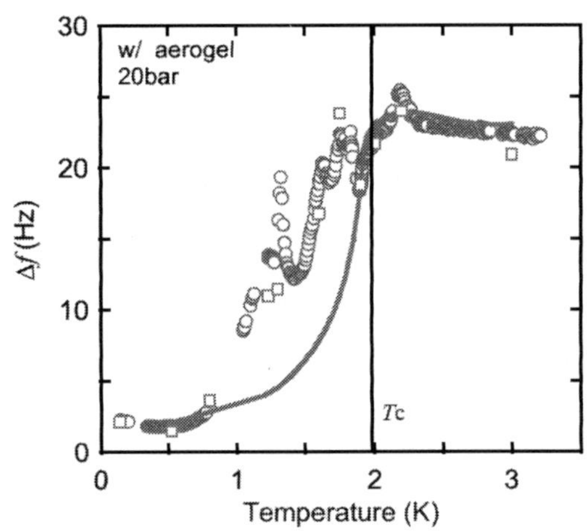


**Figure 7.** Resonance width of quartz tuning fork with aerogel $\Delta f_{\mathrm{w/}}$ at 20 bar. Those obtained frequency sweep are shown as squares. The solid line is the fitting.

$\Delta f_{\mathrm{w/o}}$ at 20 bar was well fitted in both normal and superfluid phase by the above equations and fitting parameter $C = 0.70$ as shown the line in Fig. 6. $\Delta f_{\mathrm{w/}}$ in normal phase was well fitted with $C = 0.96$ as shown the line in Fig. 7. However, $\Delta f_{\mathrm{w/}}$ doesn't agree with that from Eq. 4

in superfluid phase. Temperature variation of $\Delta f$ is expected to result from $\sqrt{\rho_n \eta}$. This implies that additional dissipation occurred between 1 K and $T_c$ in superfluid phase. We have observed anomalous sound attenuation in similar temperature region using several MHz ultrasound[13]. For further study on the acoustic properties of liquid $^4$He in aerogel, it is necessary to make experiments at various frequencies and with other aerogels.

## 4. Summary
We reported our preliminary experiments on superfluidity of helium in aerogel with 90% porosity using quartz tuning fork with the frequency of 33 kHz. Superfluid transition temperature in our aerogel decreased 2 mK, which was in reasonable agreement with suppression of superfluid transition in other gels. Temperature variation of resonance frequency was explained by that of superfluid density. Additional dissipation was observed in the temperature range between 1 K and transition temperature.

## Acknowledgments
This work was supported by JSPS KAKENHI. We thank Dr. K. Kanamori and Dr. K. Nakanishi at Kyoto University for providing aerogel. We thank K. Nunomura for technical assistance.

## References
[1] Blaauwgeers R, Blazkova M, Clovecko M, Eltsov V B, de Graaf R, Hosio J, Krusius M, Schmoranzer D, Schoepe W, Skrbek L, Skyba P, Solntsev R E and Zmeev D E 2007 *J. Low Temp. Phys.* **146** 537-62
[2] Bradley D I, Fear M J, Fisher S N, Guènault A M, Haley R P, Lawson C R, McClintock P V E, Pickett G R, Schanen R, Tsepelin V and Wheatland L A 2009 *J. Low Temp. Phys.* **156** 116
[3] Bradley D I, Fear M J, Fisher S N, Guènault A M, Haley R P, Lawson C R, Pickett G R, Schanen R, Tsepelin V and Wheatland L A 2014 *Phys. Rev.* B **89** 214503
[4] Ahlstrom S L, Bradley D I, Clovecko M, Fisher S N, Guènault A M, Guise E A, Haley R P, Kolosov O, McClintock P V E, Pickett G R, Poole M, Tsepelin V and Woods A J 2014 *Phys. Rev.* B **89** 014515
[5] Jackson M J, Kolosov O, Schmoranzer D, Skrbek L, Tsepelin V and Woods A J 2016 *J. Low Temp. Phys.* **183** 208
[6] Blažkova M, Chagovets T V, Rotter M, Schmoranzer D and Skrbek L 2008 *J. Low Temp. Phys.* **150** 194
[7] Rysti J and Tuoriniemi J 2014 *J. Low Temp. Phys.* **177** 133
[8] Schmoranzer D, LaMantia M, Sheshin G, Gritsenko I, Zadorozhko A, Rotter M and Skrbek L 2011 *J. Low Temp. Phys.* **163** 317
[9] Bradley D I, Clovecko M, Fisher S N, Garg D, Guènault A M, Guise E A, Haley R P, Pickett G R, Poole M and Tsepelin V 2013 *J. Low Temp. Phys.* **171** 750
[10] Todoshchenko I, Kaikkonen J P, Blaauwgeers R, Hakonen P J and Savin A 2014 *Rev. Sci. Instrum.* **85** 085106
[11] McKenna M J, Slawecki T and Maynard J D 1991 *Phys. Rev. Lett.* **66** 1878
[12] Nishikawa M, Yoshino K, Abe S, Suzuki H, Matsumoto K, Tayurskii D and Tajiri K 2005 *J. Phys. Chem. Solids* **66** 1506
[13] Matsumoto K, Tsuboya H, Yoshino K, Abe S, Suzuki H and Tayurskii D 2007 *J. Low Temp. Phys.* **148** 615
[14] Brusov P, Parpia J M, Brusov P and Lawes G 2001 *Phys. Rev.* B **63** 140507

# Correlation of Hall and Shubnikov-de Haas Oscillations and Impurity States in *Sn*- and *I*- Doped Single Crystals p-$Bi_2Te_3$

M Z Tahar[1], D I Popov[2] and S A Nemov[3]
[1]Department of Physics, State University of New York at Brockport, Brockport, NY 14420, USA

[2]ZAO NevInPat, per. Pirogova 7, St.-Petersburg 190000 Russia
[3]State Politechnical University, Politechnicheskaya ul.29, St.-Petersburg 195251 Russia

E-mail: mtahar@brockport.edu

**Abstract**. Oscillations of the Hall coefficient and Shubnikov-de Haas (SdH) were observed in *p*-$Bi_2Te_3$ crystals doped with *Sn* (acceptor) and with *I* (donor) in magnetic fields up to 9 T parallel to the C3 trigonal axis at low temperatures (2 K < T < 20K), which is an evidence of the spatial homogeneity of carriers in complex solid solutions. This supports the existence of a narrow band of *Sn* states (partially filled) against the background of the valence band acting as a reservoir with high density of states partially filled with electrons. Previously, in these systems in which the Fermi level was in the light-hole valence band, both large Hall and SdH oscillations were observed, with ~π phase shift between them, whereas when the Fermi level was in the heavy-hole valence band (larger acceptor content), no quantum oscillations were observed. It was concluded that the observed low amplitude quantum oscillations may be attributed to the shifting of the reservoir from the light-hole band to the heavy-hole, and the observed phase shift in the range 0 - π/2 between Hall and SdH oscillations may be attributed to filling factor of the reservoir with electrons, which varies with *I* content. Experimental results along with theoretical explanation of these correlations are presented.

## 1. Introduction

Bismuth telluride ($Bi_2Te_3$) is a narrow gap semiconductor with Eg ~ 0.1 eV. It is of practical importance since it is a foundation compound for developing high efficiency thermoelectric materials used for manufacturing cooling devices operating at room temperatures. $Bi_2Te_3$ belongs to the class of layered $R_{3mm}$ structures. The anisotropy of physical properties and peculiarities in its phase diagram result in the fact that even the most perfect crystals are characterized by a large quantity of various point defects and inhomogeneities ($10^{19}cm^{-3}$). This leads to considerable fluctuations of the electro-physical properties of crystals and their technical characteristics.

Usually impurities in semiconductors do not create impurity states in the allowed zone, but in the forbidden zone, which leads to statistical distribution of free carriers, contributing to spatial inhomogeneity and low-amplitude Shubnikov-de Haas (SdH) oscillations at high-field. However, a number of impurities in narrow gap semiconductors IV-VI ($A^4B^6$) and V-VI ($A^5B^6$) create quasi-local states on the background of the allowed zone of the conduction electrons or holes [1]. The same phenomenon has been observed by Kulbachinskii *et al* [2] in $Bi_2Te_3$-$Sb_2Te_3$ solid solutions, the most commonly used room temperature thermoelectric materials. For the first time quantum oscillations of SdH effect in a bulk sample (for these systems) were observed, with the tendency of magnetoresistance saturation but not to full saturation. Further, the SdH oscillations started only at 10 Tesla.

Doping of $Bi_2Te_3$ - $Sb_2Te_3$ with *Sn* leads to formation of a band of the quasi-local states on the background of the valence zone of $Bi_2Te_3$ - $Sb_2Te_3$. Also the impurity states of *Sn* stabilize the hole concentration in the valence zone and pin down the Fermi level. This results in considerable improvement of the spatial homogeneity of electro-physical properties of $Bi_2Te_3$:*Sn* [3], which allows for the observation of high-amplitude SdH oscillations in lower magnetic field (3 Tesla), see Laiho *et al.* [4]. This is an indication of the existence of the *Sn* impurity states in the valence band of $Bi_2Te_3$. The position of the Fermi level was estimated to be at the top of the second valence band (heavy holes). From the Hall Effect measurements the Fermi level was estimated at 0.015-0.025 eV, due to its pinning effect by the *Sn* impurities; this can be also considered as the average energy of the *Sn* states.

In order to study the behavior of the *Sn* impurity states, an additional iodine donor-like impurity was introduced to shift the Fermi level in order to understand what model can be used for the *Sn* states. In view of this, the present work is on samples of $Bi_2Te_3$ doped with 0.5% *Sn,* which were additionally doped with iodine. The iodine was introduced in the above crystals in the form of $SbI_3$ in concentrations of 0.05% (#97) and 0.1% (#98).

*Sn* is considered to be an acceptor impurity, while iodine is definitely a donor impurity, however, the picture is more complicated. This complication is due to the presence of many different point defects - vacancies (Schottky defects), pair defects (Frenkel) and anti-sites. $Bi_2Te_3$ has a layered structure in the form (*Te*(1)-*Bi*-*Te(2)*-*Bi*-*Te*(1)). So, based on the two *Te* different positions, an introduction of impurities shifts the equilibrium of defects, with hard to predict results since they depend on prevailing conditions (temperature, pressure, etc.) during doping. However, the effect of stabilization (pinning) of the Fermi level may be observed only if the impurity (*Sn*, here) band is partially filled with electrons. Then the band can donate and accept electrons, thus compensating for the influence of the various defects.

## 2. Experimental Details

The magnetoresistance (MR) and Hall Effect measurements were carried out in magnetic field (0 < H < 9 T) on thin samples (~ 80μm), which were cleaved off of a thicker slab using adhesive tape, which was subsequently removed using acetone. Four electrical contacts were made using a two component silver loaded epoxy which was cured at ambient temperature for 24 hrs. Special attention was made for the current contacts across the edge of the sample and the voltage contacts were made in the middle third of the sample to minimize the anisotropy effects and maximize the sensed voltage. Since the materials have thermoelectric properties that may depend on the kind of leads, and the temperature and its gradient across the leads, we chose AC excitation current and used a floating current source [5] and lock-in-detectors at relatively low frequency. Because we measured two samples at a time we chose 7.00 Hz and 13.0 Hz as two prime number frequencies which are neither multiples nor submultiples of the line frequency and cannot interfere with each other constructively.

To correlate the magnetoresistance and Hall Effect measurements we used six point contacts on a single sample and used two lock-in-detectors one for each signal. Additionally, to enhance the Hall Effect signal over the MR contribution, the ratio of the width of the sample to the distance between the longitudinal (MR) contacts was set at 5/6 – 6/7 and Hall contacts were aligned facing each other to better than 0.25 mm or ~1/20 the width of the sample. To avoid sample self-heating the excitation currents were so that the total power dissipation was under 5μW. As for the quantum oscillations, the samples were inserted in a cryostat with a superconducting magnet capable of generating a 9 Tesla field. The field ramping up and down was at a maximum rate of 0.6 Tesla/minute, while the temperature was kept constant to better than 0.05K.

## 3. Results and Discussion

First, from the Hall Effect measurements in low magnetic field and resistivity measurements (magnetoresistance at zero field) the concentration of holes was deduced (see Fig.1 and Fig.2 for #97 and #98, respectively). This is quite different from $Bi_2Te_3$ samples with no impurity states (no *Sn* in our case). The observed constant concentration of holes for a certain range of temperatures (especially well seen for #98) is a characteristic of impurity states pinning the Fermi level, so that scattering of charge carriers on impurities becomes dominant up to higher temperatures. When more additional donor impurity (*I* in our case) is introduced in a sample, intrinsic conductivity becomes less significant and the Fermi level stabilizes up to higher temperatures, which can be seen by comparison of Fig. 1 and Fig. 2.

Second, we present the results of magnetoresistance measurements in an applied magnetic field 0 < *H* < *9T*, for the two compositions at different temperatures (2 – 280K). The data is for field ramping up and down and as can be seen (Fig. 3 and Fig. 4), both curves show little or no hysteresis and a positive magnetoresistance with a quadratic field dependence, with a slight deviation at low fields. The non-oscillating part of magnetoresistance $\rho_{xx}(H_z)$ tends to saturate, and can be described by [4] the relation: $\frac{R(H)}{R_0} = \frac{\rho(H)}{\rho_0} = \frac{1+(R_0\sigma_0 H)^2/f_{||}}{1+(R_0\sigma_0 H)^2}$, which leads to $\frac{H^2}{(\rho(H)-\rho_0)/\rho_0} = a + bH^2$, where $a = \frac{1}{(\frac{1}{f_{||}}-1)(R_0\sigma_0)^2}$ and $b = (\frac{1}{f_{||}} - 1)$. Thus, from the above equations, the coefficients a and b can be evaluated, then using their value the coefficients of the Hall carrier mobility, $R_0\sigma_0$, and the Hall factor $f_{||}$ can be obtained.

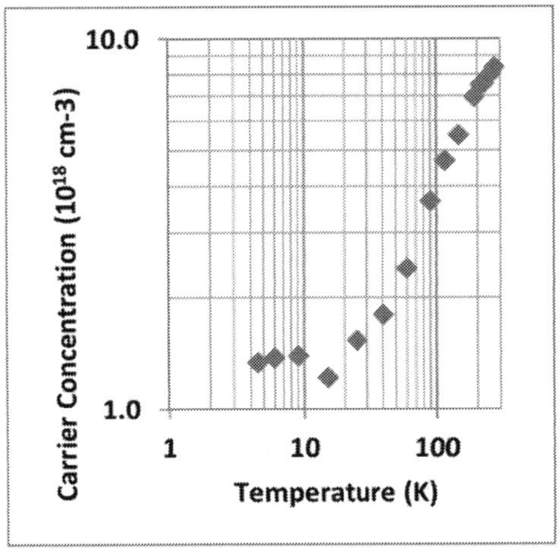


**Figure 1.** Carrier concentration of $Be_2Te_3$ doped with 0.5% *Sn* and 0.05% $SbI_3$ (#97) from 4.2K up to 280 K.

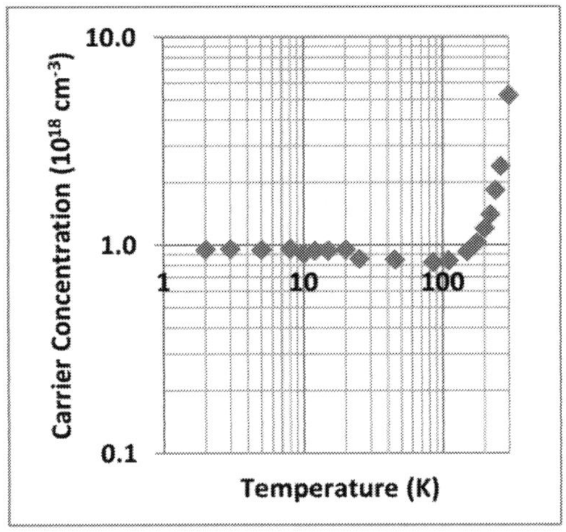


**Figure 2.** Carrier concentration of $Be_2Te_3$ doped with 0.5% *Sn* and 0.1% $SbI_3$ (#98) from 4.2K up to 280 K.

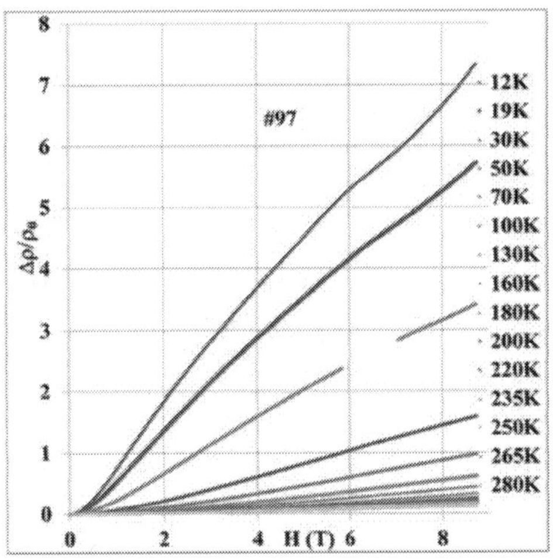


**Figure 3.** Magnetoresistance, $\Delta\rho(H)/\rho_0$ vs. H of $Be_2Te_3$ doped with 0.5% *Sn* and 0.05% $SbI_3$ (#97) at various temperatures.

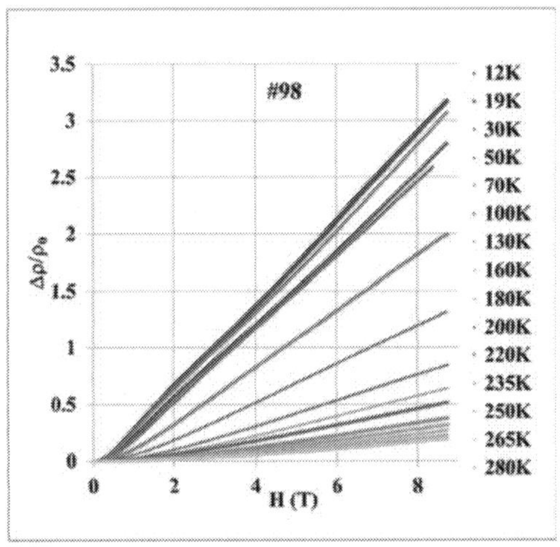


**Figure 4.** Magnetoresistance $\Delta\rho(H)/\rho_0$ vs. H of $Be_2Te_3$ doped with 0.5% *Sn* and 0.1% $SbI_3$ (#98) at various temperatures.

We found that the Hall factor has an average value of 0.35 (0.41 for #97 and 0.29 for #98), and is highly influenced by the anisotropy of effective mass and the orientation of the energy ellipsoids in the six-ellipsoid model of Drabble and Wolf [6] with respect to crystallographic axis. The difference in the Hall factor between the two compositions can also be explained by the deviation from classical square root dependence of the density of states on energy. From the above equations the mobility, $R_0\sigma_0$, was evaluated and plotted together with the hole mobility obtained from the Hall effect in low magnetic field (see Fig.5 and Fig.6).

The higher values of mobility at low temperature, in comparison with undoped samples [7] are an indication of less variety of defects and smaller fluctuation of volumetric concentration. The difference in mobility obtained from the Hall Effect measurements and from magnetoresistance measurements at high fields may be attributed, first of all, to temperature dependence of the Hall factor, however, different non-linear effects should not be overlooked.

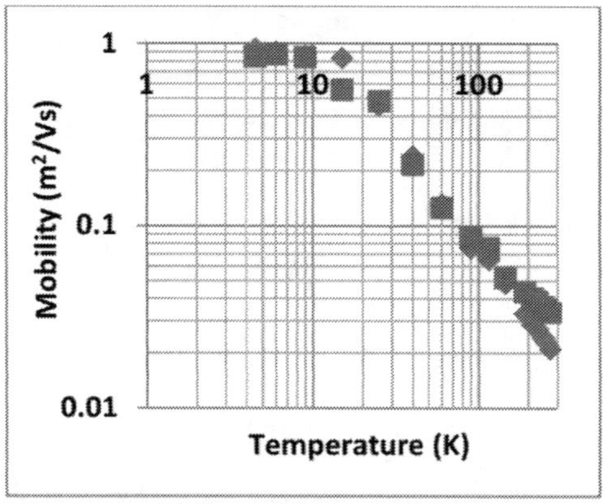


**Figure 5.** Hall mobility $\mu_H$ (diamonds) and magnetoresistance mobility $\mu_{MR}$ (squares) of $Be_2Te_3$ doped with 0.5% *Sn* and 0.05% $SbI_3$ (#97) from 4.2K up to 280 K.

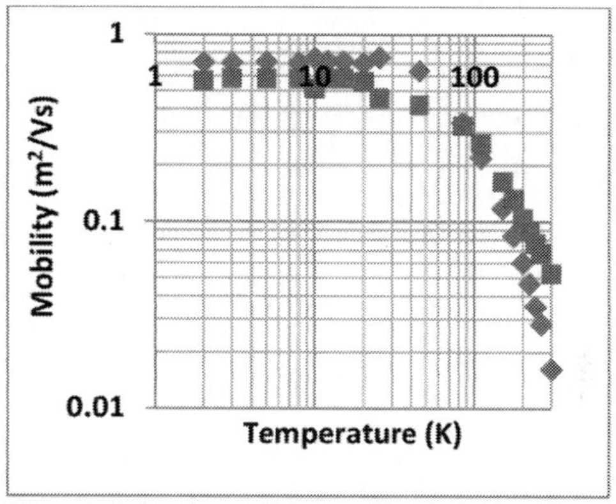


**Figure 6.** Hall mobility $\mu_H$ (diamonds) and magnetoresistance mobility $\mu_{MR}$ (squares) of $Be_2Te_3$ doped with 0.5% *Sn* and 0.1% $SbI_3$ (#98) from 4.2K up to 280 K

Third, we observed SdH and Hall oscillations in magnetic fields $0 < H < 9$ T for the two compositions below 4.2 K. The magnetoresistance for both samples starts exhibiting quantum oscillations for H ~ 2.5 T. The oscillating part of magnetoresistance for both samples (Fig.7) shows only one period of oscillations, which indicates that only one type of holes (light holes) contributes effectively to all transport properties in this system.

From the oscillation part of SdH and Hall Effect it was found that the period of oscillations in the inverse magnetic field P(1/H) is 0.09 $T^{-1}$ for the sample #97 and 0.1 $T^{-1}$ for the sample #98. As known [8], the period of quantum oscillations P(1/H) in SdH effect and Fermi surface cross section $S_F$ are related as $P\left(\frac{1}{H}\right) = \frac{2\pi e}{c\hbar S_F}$. Both materials (#97 and #98) exhibit relatively similar values of the Fermi surface cross-sections (11.1 and 10.2 Tesla, respectively), differing by 10%, which is similar to the difference in Hall concentrations for these samples, for T< 10K.

Therefore, the hole effective masses stay close for the two donor doping concentrations, which shows that iodine doping does not change the zone structure of $Bi_2Te_3$ much, if at all. It has been shown in [9] that for 6-ellipsoid model of Drabble and Wolfe [6] $S_F = \frac{\sqrt{\alpha_{11}\alpha_{22}}}{\pi\left(3\pi^2\frac{p}{K_e}\right)^{2/3}\sqrt[3]{\alpha_{11}(\alpha_{22}\alpha_{33}-\alpha_{23}^2)}}$,

where $K_e$ is the number of ellipsoids (6 in the Drabble-Wolfe model), $\alpha 11$, $\alpha 22$, $\alpha 33$, and $\alpha 23$ are the components of effective mass tensor, while p is the full concentration of holes. From the above equation it is possible to find the full concentration of holes p, assuming that the *Sn* impurity does not change the shape of the ellipsoids. It was found that p = 1.37*10$^{18}$ cm$^{-3}$ for the sample #97 and p = 1.15*10$^{18}$ cm$^{-3}$

IOP Conf. Series: Journal of Physics: Conf. Series **969** (2018) 012020 doi:10.1088/1742-6596/969/1/012020

for the sample #98. This is in good agreement with the concentration of holes at same low temperature obtained from the Hall Effect measurements (see Figs.1 and 2), where p = 1.34*10[18] cm[-3] for sample #97 and p = 0.94*10[18] cm[-3] for sample #98.

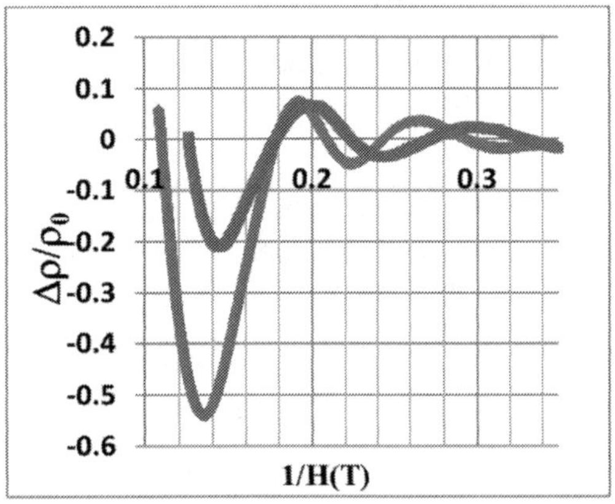


**Figure 7.** Oscillating part of magnetoresistance of $Be_2Te_3$ doped with 0.5% *Sn* and 0.05% $SbI_3$ (#97) (red) and 0.1% $SbI_3$ (#98) (blue) at 4 K

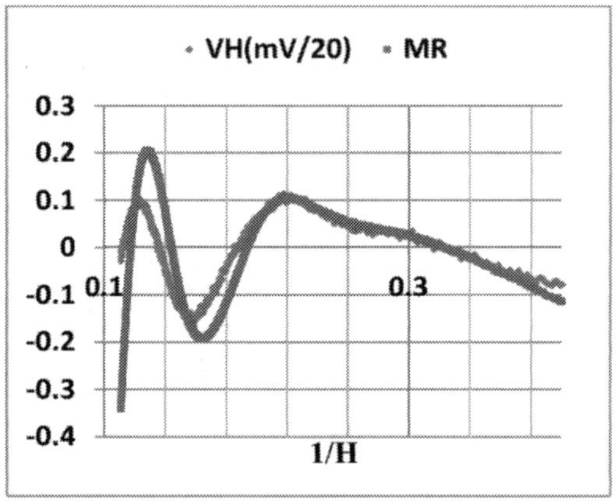


**Figure 8.** Oscillating part of magnetoresistance (red) and Hall Effect (blue) of $Be_2Te_3$ doped with 0.5% *Sn* and 0.05% $SbI_3$ (#97) at 4 K.

In connection with quantum oscillations an interesting fact should be mentioned. Quantum oscillations for sample #97 were observed in the range of temperatures 2K – 7K. The amplitude of these oscillations exponentially decreased with an increase in temperature (40% difference between 2K and 7K), whereas the oscillation frequency remained essentially the same (within 2%), which is another evidence of the fact that the Fermi level is pinned by the *Sn* impurity states and does not change with temperature, making the system electrically well-stabilized.

Whereas in [4] the phase shift between SdH and Hall effect oscillations was about $\pi$, in the present measurements the phase shifts of $\pi/8$ and $\pi/6$ were observed for sample #97 (see Fig.8) and #98, respectively, which is in correlation with the filling factor of Sn impurity states by electrons / holes.

Besides well-known conditions that should be met in order for quantum oscillations to be observed, i.e. the spacing between Landau levels should be higher that thermal broadening ($\hbar\omega_C \gg kT$); the broadening due to scattering is less than the spacing between Landau levels ($\omega_C\tau \gg 1$); and the Fermi level is higher than at least some of the Landau levels ($\varepsilon_F > \hbar\omega_C$), for heavily doped systems the following condition must be met [7]: $\sqrt{\overline{(\delta\varepsilon_F)^2}} \ll \hbar\omega_C$. This means that spatial fluctuations of chemical potential are much less than the spacing between Landau levels, i.e. the samples are electrically homogeneous.

## 4. Conclusion

The observation of quantum oscillations is a demonstration of excellent spatial homogeneity of electrical properties of quasi-local states of *Sn,* which act as a reservoir of charge carriers pinning the Fermi level and minimizing spatial fluctuations of chemical potential.

The results indicate that the most probable model for one-electron states in $Bi_2Te_3$ solid solutions doped with *Sn* is the presence of two impurity bands - one filled and one empty of carriers with the Fermi level pinned in between - is the one that can be described by disproportionate reaction $2Sn(1e) \rightarrow Sn(0) + Sn(2e)$ with respect to *Bi* sub-lattice.

**Acknowledgments**
One of the authors (MZT) wishes to acknowledge support and encouragement by the Department of Physics, The College at Brockport, State University of New York.

**References**

[1] S A Nemov and Y I Ravich 1998 Sov.Phys.Uspekhi **41**(8), 735
[2] Kulbachinskii *et al.* 2006 Physics of the Solid State **48**(5), 833
[3] Zhitinskaya, S A Nemov and Svechnikova 1997 Semiconductors **31**, 375
[4] R Laiho, S A Nemov, *et al.* 2007 Semiconductors **41**(5), 546
[5] P P Burin, C Pfleiderer 1996 Rev. Sci. Inst. **67**, 4024
[6] Drabble and Wolfe 1956 Pro. Phys. Soc., **69**, 1101.
[7] B.M.Goltzman, V.A.Kudiniv, I.A.Smirnov, Semiconductor Thermoelectric Materials based on *Bi2Te3*, Nauka, 1972.
[8] I.M.Lifshitz, A.M.Kosevich, 1956, Sov.Phys.JETP, **2**, 646
[9] N.A.Abdullaev *et al.* 2009 Semiconductors, **43**(2), 145

IOP Conf. Series: Journal of Physics: Conf. Series **969** (2018) 012021 doi:10.1088/1742-6596/969/1/012021

# Electron spectra of graphene with local and extended defects

**V.V. Eremenko[1], V.A. Sirenko[1], I.A. Gospodarev[1], E.S. Syrkin[1], S.B. Feodosyev[1], I.S. Bondar[1], K.A. Minakova[2], A. Feher[3]**

[1] B. Verkin Institute for Low Temperature Physics and Engineering of the National Academy of Sciences of Ukraine 47 Nauki Ave., Kharkov 61103, Ukraine
[2] National Technical University "Kharkiv Polytechnic Institute", 21 Bagaliya Str., Kharkov 61002, Ukraine
[3] Centre of Low Temperature Physics Faculty of Science P.J. Šafarik University &Institute of Experimental Physics SAS Park Angelinum 9, 041 54 Košice, Slovakia

E-mail: sirenko@ilt.kharkov.ua

**Abstract** We show theoretically a possibility to increase the superconducting transition temperature $T_c$ of graphene by introducing defects. Eventually, the peak of local electron densities shifts to the Fermi level with $T_c$ above ambient temperature in line with recent observations.

## 1. Introduction

Pronounced hybridization of atomic orbitals in a monolayer of graphene enhances its Debye temperature up to 2500 K, giving rise to the "low-temperature" features in vibrational thermodynamic properties of graphene and its derivatives, nano-tubes and nano-ribbons in a particular, at temperatures above ambient. In the BCS consideration, strongly evidenced for a broad variety of superconductors, the unconventional ones, in particular (see, e.g., [1] and references therein), high temperature of superconducting transition $T_c$ should be expected then in line with a mean frequency of phonons [2]. Within this approach the absence of superconductivity in conventional graphene materials is explained by a low density of charge carriers in the vicinity of Fermi energy ($\varepsilon_F$), together with a lack of phonons for basic contribution to the electron-phonon coupling constant. To enhance the latter, it is necessary to increase the number of conducting electrons and saturate the fraction of phonon spectrum responsible for Cooper pairing. In fact, graphite with intercalated metal layers has manifested a superconducting transition at temperatures increased by a growing number of quasi-flexural phonon modes with frequencies close to the K-point of the first Brillouin zone [3 - 5] in addition to increment of charge carriers number. In the works [6,7] it was demonstrated, that a similar increase of a charge carriers and corresponding phonons number takes place in thin graphene nano-film, i.e. bi- and tri-graphene with a "step-edge" boundary. In present contribution, evolution of the electron and phonon spectra of graphene on insulating substrate gains analytical and numerical analysis when the graphene layer is cut along a zig-zag line (Fig. 1*a*). It is shown, that such imperfection significantly increases both the population of electrons at Fermi level and the number of quasi-flexural phonons with frequencies about that, corresponding to K-point of first Brillouin zone.

## 2. Electron spectrum of graphene, cut along zig-zag line

In zero magnetic field, the electron spectrum of graphene is reasonably described within tight-binding approximation (see, e.g.[8 - 10]) and corresponding Hamiltonian, read as:

$$\mathscr{H} = \sum_i \varepsilon_i |i\rangle\langle i| - \sum_{i,j} J_{ij} |i\rangle\langle j| . \quad (1)$$

In the case (see, *e.g*,. [11]) of in-plane electron hopping restricted to the nearest neighbors $\forall J_{ij} = J \approx 2.8$ eV, and $\forall \varepsilon_i = \varepsilon_F = 3J$ (indices $i$ and $j$ label sites), the Hamiltonian (1) yields the following dispersion law:

$$\varepsilon_0(\mathbf{k}) = \pm J\sqrt{1+4\cos\left(\mathbf{k}\cdot\frac{\mathbf{a}_1-\mathbf{a}_2}{2}\right)\left[\cos\left(\mathbf{k}\cdot\frac{\mathbf{a}_1+\mathbf{a}_2}{2}\right)+\cos\left(\mathbf{k}\cdot\frac{\mathbf{a}_1-\mathbf{a}_2}{2}\right)\right]}. \quad (2)$$

For high-symmetry directions in $\mathbf{k}$ - space it is illustrated by Fig. 1*b* with Fermi energy as the reference one.

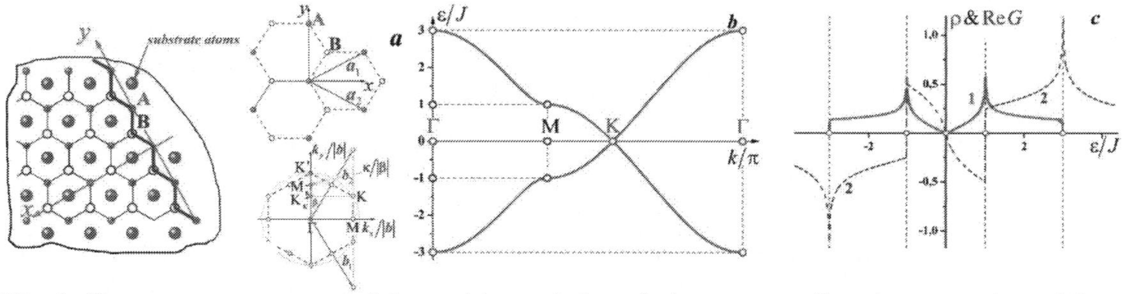

**Fig. 1:** Fragment ***a***: geometry of the problem, choice of elementary cell and construction of first Brillouin zone of graphene and 1D Brillouin zone of graphene with a zig-zag boundary. Fragment ***b***: dispersion curves of graphene along high-symmetry directions. Fragment ***c***: DOS of graphene and real part of its Green function (curves 1 and 2, respectively).

Due to specific 2D- crystal structure of graphene, the spectral branches coincide in a single point of reciprocal space (K-point of 2D first Brillouin zone). The other directions, outgoing from Γ point, are characterized by a finite gap between spectral branches, yielding a linear dispersion of electrons at such point, as $\varepsilon(\mathrm{K}) = \varepsilon_\mathrm{F}$ in graphene, and V-singularity of DOS at $\varepsilon_\mathrm{F}$ (curve 1 in Fig. 1*c*).The behavior of a real part of Green function $\mathrm{Re}\,G(\varepsilon)$ near the Fermi level, i.e. $\varepsilon(\mathrm{K}) = 0$ (curve 2 in Fig. 1*c*) demonstrates, that various sorts of defects should localize elementary excitations near this level [5].

It should be noted, that in perfect structure of graphene, the atoms of different sublattices (A (•) and B (○) - Fig. 1*a*) are physically equivalent, which means that their local Green functions, and eventually, local densities of states (LDOS) are coincident. This equivalency should be obviously broken by introduction of defects in one of sublattices. Then electron spectra for the atoms, pertained to different sublattices should differ. The imperfection, built by atoms of particular sublattice, is represented by a zig-zag -boundary. In this case, as is clearly demonstrated by Fig. 1*a*, similar to vacancy, the atomic bonds are broken in one sublattice only, say the sublattice A, Fig. 1*a*). In contrast to vacancy, which is the point defect and is breaking translational symmetry along all directions in the crystal [5 – 7], it is conserved along extended defect of zig-zag type. In this direction the one-dimensional vector of reciprocal lattice $\boldsymbol{\beta} = (0, \beta)$, quasi-wave vector $\boldsymbol{\kappa} = (0, \kappa)$ can be introduced, and Hamiltonian (1) in a form of Jacobi matrix [11,12] is written as:

$$\mathscr{H}_{ik} = a_i(\kappa)\delta_{ik} - b_i(\kappa)\left(\delta_{i,k+1} + \delta_{i+1,k}\right); \quad i,k = 0,1,2,\ldots,\infty, \quad (3)$$

where all diagonal matrix elements $a_i(\kappa)$ with a probable exception of the first one, are zero, while non-diagonal elements $b_i(\kappa) = \alpha(\kappa) = 2J\cos\frac{\kappa a}{2}$ for even values of $i$ and $b_i(\kappa) = \beta = J$ - for odd ones. Construction of 2D first Brillouin zone for perfect graphene and 1D first Brillouin zone for considered problem are presented in Fig. 1*a*. The non-zero value of a diagonal matrix element (3)

$a_0(\kappa)$ can result from relaxation of interactions at the boundary. Assuming these processes to occur on the atoms with dangling bonds only, it is written as

$$\Lambda(\kappa) = 2\tilde{J}\cos\kappa a \tag{4}$$

The local Green function $G^{(0)}(\varepsilon,\kappa,\Lambda)$, corresponding to the row of atoms of sublattice $\mathbb{A}$, which lie on zig-zag – boundary (taken as the reference row - $n=0$), can be written in a form of infinite continuous fraction [13, 14], which can be transformed as follows:

$$G^{(0)}(\varepsilon,\kappa,\Lambda) = \frac{1}{2}\cdot\frac{\varepsilon^2 + 2\Lambda(\kappa)\varepsilon + \beta^2 - \alpha^2(\kappa) + Z(\varepsilon)\cdot\sqrt{\left|\left(\varepsilon^2-\varepsilon_0^2\right)\left(\varepsilon^2-\varepsilon'^2\right)\right|}}{\Lambda(\kappa)\varepsilon^2 + \left[\Lambda^2(\kappa)+\beta^2\right]\varepsilon + \Lambda(\kappa)\left[\beta^2-\alpha^2(\kappa)\right]} \tag{5}$$

with the introduced variables: $\varepsilon_0^2 \equiv \left[\alpha(\kappa)+\beta\right]$ and $\varepsilon'^2 \equiv \left[\alpha(\kappa)-\beta\right]$. The function

$$\begin{aligned} Z(\varepsilon) \equiv\ & \Theta(-\varepsilon_0-\varepsilon) + i\cdot\Theta(\varepsilon+\varepsilon_0)\,\Theta(-|\varepsilon'|-\varepsilon) + \Theta(\varepsilon+|\varepsilon'|)\,\Theta(|\varepsilon'|-\varepsilon) + \\ & + i\cdot\Theta(\varepsilon-|\varepsilon'|)\,\Theta(\varepsilon_0-\varepsilon) - \Theta(\varepsilon-\varepsilon_0) \end{aligned} \tag{6}$$

is determined by a form of Green function both inside and out of double-connected band of quasi-continuous spectrum $\mathscr{D} = \left[-\varepsilon_0(\kappa), -|\varepsilon'(\kappa)|\right] \cup \left[|\varepsilon'(\kappa)|, \varepsilon_0(\kappa)\right]$. For $\varepsilon \in \mathscr{D}$ function $G^{(0)}(\varepsilon,\kappa,\Lambda)$ includes imaginary part, which determines an energy distribution inside the band of quasi-continuous spectrum of the atoms of row $n=0$.

In the absence of relaxation on the boundary, namely at $\Lambda(\kappa)=0$, denominator (5) equals $-2\beta^2(\kappa)\varepsilon$ and function (4) gains pole at $\varepsilon=0$, i.e. on the Fermi level, located in the gap $\left[-|\varepsilon'(\kappa)|, |\varepsilon'(\kappa)|\right]$. Among two roots of denominator (4) at $\Lambda(\kappa)\neq 0$ we should select then the one vanishing at $\Lambda \to 0$

$$\varepsilon_g(\kappa,\Lambda) = \frac{\sqrt{\left[\beta^2-\Lambda^2(\kappa)\right]^2 + 4\alpha^2(\kappa)\Lambda^2(\kappa)} - \beta^2 - \Lambda^2(\kappa)}{2\Lambda(\kappa)} \underset{\Lambda\to 0}{\to} 0 \tag{7}$$

determines a dispersion law of the "border" wave, split of the band of quasi-continuous spectrum.

Residue of function (5) in this pole, determines intensity of corresponding discrete level, and, eventually, the portion of quasi-particles moved to the split-off wave. It equals to

$$\mu_g^{(0)}(\kappa,\Lambda) = -\frac{\beta^2-\Lambda^2(\kappa)}{\Lambda(\kappa)}\cdot\frac{\varepsilon_g(\kappa,\Lambda)}{\sqrt{\left[\beta^2-\Lambda^2(\kappa)\right]^2+4\alpha^2(\kappa)\Lambda^2(\kappa)}}\cdot\Theta\left(-\frac{\beta^2-\Lambda^2(\kappa)}{\Lambda(\kappa)}\cdot\varepsilon_g(\kappa,\Lambda)\right). \tag{8}$$

Condition of existence of the wave is that of positive meaning of the argument of Heaviside Theta function $\frac{\beta^2-\Lambda^2(\kappa)}{\Lambda(\kappa)}\cdot\varepsilon_g(\kappa,\Lambda)<0$ and coincides with a condition of existence of non-dispersive gap level for the case $\Lambda=0$, i.e. $\kappa\in\left[-\mathrm{M}_\kappa,-\mathrm{K}_\kappa\right]\cup\left[\mathrm{K}_\kappa,\mathrm{M}_\kappa\right]$ (see Fig. 1*a*). A dispersion law near Fermi level is linear (relativistic) $\varepsilon_g(\mathrm{K}_\kappa+x,\Lambda)\approx\hbar v_F\cdot x$, where Fermi velocity $v_F$ equals to group velocity in the point $\mathrm{K}_\kappa$. The group velocity, corresponding to dispersion (7) as a function of quasi-wave vector $\kappa$, can be written in the following form:

$$v_{gr}(\kappa) = \frac{\partial}{\partial\kappa}\frac{\varepsilon(\kappa)}{\hbar} = -\frac{\partial\Lambda(\kappa)}{\partial\kappa}\cdot\frac{\mu_g^{(0)}(\kappa,\Lambda)}{\hbar} - \frac{aJ^2\sin a\kappa}{\hbar}\cdot\frac{\Lambda(\kappa)}{\sqrt{\left[\Lambda^2(\kappa)-J^2\right]^2+16J^2\Lambda^2(\kappa)\cos^2\frac{\kappa a}{2}}}. \tag{9}$$

Taking into account (4), $v_F^{(g)} = \frac{a}{\hbar}\frac{\sqrt{3}}{2}\frac{J^2\tilde{J}}{J^2+\tilde{J}^2} = \frac{v_F^{(0)}}{\sqrt{3}}\frac{J\tilde{J}}{J^2+\tilde{J}^2}$, where $v_F^{(0)} = 3aJ/2\hbar$ is the Fermi level of perfect graphene with dispersion (2). When the relation (4) is valid, the dispersion law (7) assumes the states density $g_g(\varepsilon) = \frac{a}{\pi\hbar}\cdot v_{gr}^{-1}(\varepsilon)$, where $v_{gr}(\varepsilon)$ is the group velocity, represented as energy function. Contribution of gap functions into the LDOS of atoms from different rows (with $n \geq 0$) is (see. [15]) $\rho_g^{(n)}(\varepsilon) = g_g(\varepsilon)\cdot\mu_g^{(n)}(\varepsilon)$, where

$$\mu_g^{(2m+1)} = \mu_g^{(0)}\cdot\left(-\frac{\beta}{\Lambda}\right)^{2m}\left[\frac{\sqrt{\left(\beta^2-\Lambda^2\right)^2+4\alpha^2\Lambda^2}-\left(\beta^2-\Lambda^2\right)}{2\Lambda\alpha}\right]^{2(m+1)} \underset{\Lambda\to 0}{\to} 0;$$

$$\mu_g^{(2m)} = \mu_g^{(0)}\cdot\left[-\frac{\beta}{\Lambda}\cdot\frac{\sqrt{\left(\beta^2-\Lambda^2\right)^2+4\alpha^2\Lambda^2}-\left(\beta^2-\Lambda^2\right)}{2\Lambda\alpha}\right]^{2m} \underset{\Lambda\to 0}{\to}\left(-\frac{\alpha}{\beta}\right)^{2m}. \tag{10}$$

In Fig. 2 the evolution of LDOS with increasing $n$ is presented for $\tilde{J} = 0.1J$.

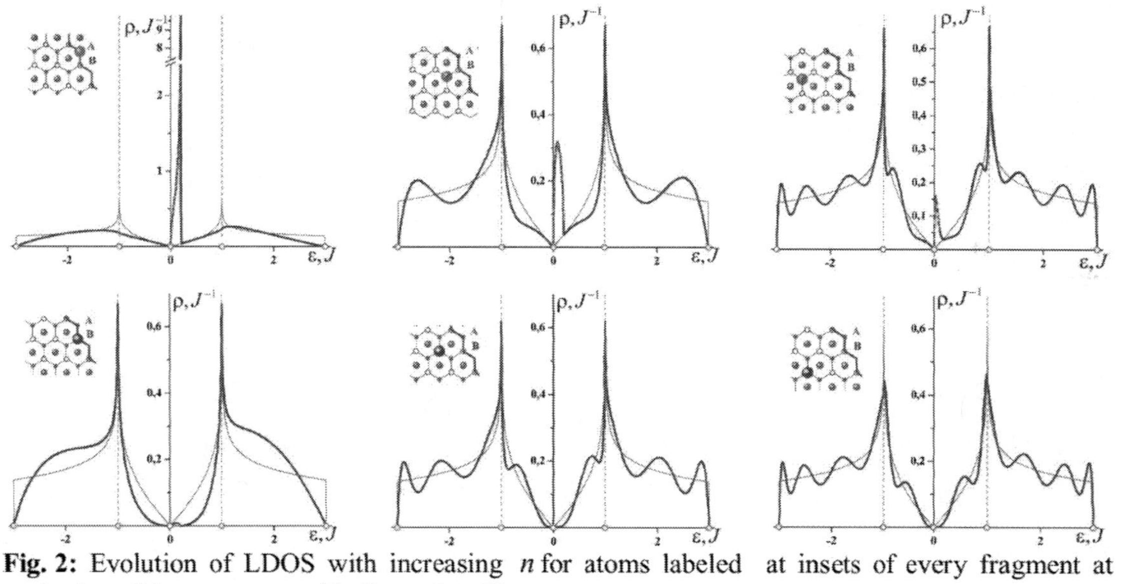

**Fig. 2:** Evolution of LDOS with increasing $n$ for atoms labeled at insets of every fragment at excitation of the gap wave with dispersion (7).

In a conduction band near $\varepsilon_F$, peaks of width $2\tilde{J}$ appear in LDOS of atoms, pertained to sublattice A. They coincide with the spectral density $\rho_g^{(n)}(\varepsilon)$ calculations (10) for even $n$. These functions are presented in fragments of Fig. 2 by dashed lines, which perfectly superimpose on LDOS dependences. It should be noted, that at the same time LDOS keep their relativistic (Dirac) behavior, corresponding to the Fermi velocity of perfect graphene $v_F^{(0)}$ at $\varepsilon\to -0$, while at $\varepsilon\to +0$ it corresponds to the Fermi velocity $v_F^{(g)}$ in one-dimensional system. It means, that angular coefficient of linear dependence $\rho_g^{(2m)}(\varepsilon\to +0)$ is related to the similar one $\rho_g^{(2m)}(\varepsilon\to -0)$ as $\sim\left[v_F^{(0)}/v_F^{(g)}\right]^2$.

In sublattice **B** only LDOS of the boundary atom ($n=1$) reveals a small peak of the width $2\tilde{J}$. It coincides completely with spectral density $\rho_g^{(1)}(\varepsilon)$. With a distance from the boundary the peaks become vanishingly small. Near Fermi level, the behavior of LDOS of the atoms from sublattice **B** is non-relativistic and correspondent to the quasi-particles of finite mass.

## 3. Electron spectrum of graphene with vacancy

Earlier [5 – 7], we have shown that the presence of isolated vacancy in graphene provides the sharp resonance in its electron spectrum at the Fermi level. This peak, similar to the case, considered above, takes place only in the LDOS of atoms from sub-lattice with dangling bonds on the atom, i.e. taking into account only the nearest-neighbor interactions in sublattice without vacancy. Atoms of another sublattice do not manifest such peak at Fermi level. The effect of near-to-nearest neighbor interactions over the whole structure of graphene was studied and demonstrated appearance of asymmetry in the DOS of graphene relative to Fermi level [7].

In Fig. 3, LDOS of the atoms, neighbor to vacancy, are presented for different distances from the vacancy. Relaxation of interactions between the nearest neighbors of vacancy is taken into account here. In consideration of interactions between the rest of the atoms we restricted ourselves to interactions of the nearest neighbors, but took into account interactions between vacancy's nearest neighbors with a single dangling bond. These atoms are the second neighbors of each other. Their coupling constant is assumed to be $0.1J$. All of the LDOS are calculated using Jacobi matrix technique, with Hamiltonian $\mathscr{H}$ (1) presented in the basis, obtained by orthonormalization of sequence $\left\{\mathscr{H}^p\Psi_0\right\}_{p=0}^{\infty}$. For generating function $\Psi_0$ we have chosen a single excitation of atoms, labeled in inset of each fragment.

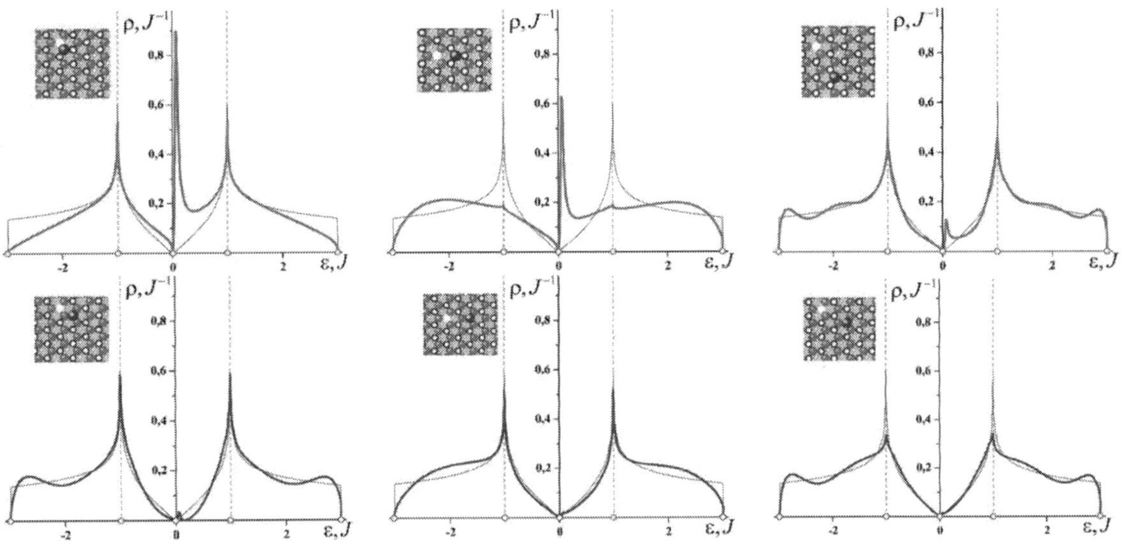


**Fig. 3:** Evolution of electron LDOS of the neighbors of vacancy with increasing distance from the latter.

It is clearly seen that LDOS of atoms from vacancy-free sublattice have peaks, similar to those in Fig. 2. At $\varepsilon \to -0$ these functions manifest a pronounced relativistic behavior. In another sublattice, LDOS of near-to-nearest neighbors of vacancy have small peaks, which rapidly decay with the distance from vacancy. This decay is accompanied by "restoration" of relativistic behavior of the considered dependences.

## 4. Conclusion

It is shown here, that in electron spectra of graphene with zig-zag boundary, there appear the waves, split of the bands of quasi-continuous spectrum, which propagate along the boundary and decay with a distance from it. They, moreover, propagate only via the atoms of sublattice, which contains atoms with dangling bond, appeared with a formation of boundary. Dispersion of these waves is determined by character of relaxation processes during formation of the boundary. Dispersion in electron spectrum is relativistic, but corresponds to the significantly less values of group velocity, compared to infinite monolayer of graphene. The similar behavior is demonstrated by electron spectra of graphene with isolated vacancy. The split gap waves lead to a formation on local densities of state of the sharp resonances, which enrich significanly electron spectrum near Fermi level, as well as phonon spectrum near the point of intersection of acoustic and optical branches, polarized normally to the plane of graphene monolayer [5,6]. These phonons within a considered frequency range do not practically interact with differently polarized phonons, and also possess high group velocities, which there dominant contribution to electron-phonon coupling. The presented results demonstrate a possibility to facilitate superconductivity in a graphene matter by a controlled creation of defects like vacancy or zig-zag boundary, which distort the atomic bonding in a particular sublattice of graphene monolayer.

The grant support by FFI NAS of Ukraine (grant 4/17 –H) is acknowledged.

**References:**

[1] Maksimov E G 2008 *Phys. Usp.* **51** 167.
[2] Èliashberg G M. 1960 *Sov. Phys. JETP* **11** 966; **12**1437
[3] Weller T E, Ellerby M, Saxena S S, Smith R P and Skipper N T 2005 *Nat. Phys.* **1** 39
[4] Dean M P M., Howard C A, Saxena S S and Ellerby M 2010 *Phys. Rev. B* **81** 045405.
[5] Feher A, Gospodarev I A , Grishaev V I , Kravchenko K V, Manzhelii E V, Syrkin E S and Feodosyev S.B. 2009 *Low. Temp. Phys.* **35** 679.
[6] Eremenko V V, Sirenko V A, Gospodarev I A, Syrkin E S, Feodosyev S B, Bondar I S, Saxena S S , Feher A and Minakova K A 2016 *Low Temp. Phys.* **42** 99.
[7] Eremenko V V, Sirenko V A, Gospodarev I A, Syrkin E S, Feodosyev S B, Bondar I S, and Minakova K A 2016 *Journal of Science: Advanced Materials and Devices* **1** 167
[8] Novoselov K S 2011 *Phys. Usp.* **54** 1299
[9] Oliveira M H, Lopes J M J, Schumann T, Galves L A, Ramsteiner M, Berlin K, Trampert A , Riechert H 2015 *Nat. Com.* DOI: 10.1038/ncomms8632 (9 Jul 2015)
[10] Shangluan Wu, Lei Jing, e.a. 2007 *arXiv: 0711.1018[cond. Mat.mes-hall]*, 7 Nov (2007).
[11] Peresada V I and Syrkin E.S 1977 *Sov. J. Low. Temp. Phys* **3** 113.
[12] Maradudin A A 1980 Surface waves In: *Mod. Probl. of Surf. Phys.* ISCMP pp. 11-400.
[13] Peresada V I, Afanas'ev V N and Borovikov V.S 1975 *Sov. J. Low. Temp. Phys.* **1** 227.
[14] Haydock R 1980 in: *Solid State Physics* **35**, Acad. Press, N- p. 129.
[15] Kotlyar O V and Feodosyev S B 2006 *Low. Temp. Phys.* **32** 256.

# Impurity states in mesoscopic SNS junctions with a point defect

**A. A. Bepalov and A. S. Mel'nikov**

Institute for Physics of Microstructures RAS – branch of Federal State Budgetary Scientific Institution "Federal Research Center The Institute of Applied Physics of the Russian Academy of Sciences" (IAP RAS), GSP-105, Nizhny Novgorod, 603950, Russia

E-mail: `bespalovaa@gmail.com`

**Abstract.** We study theoretically localized states in a clean three-dimensional SNS junction, containing one point defect (impurity) located in the normal layer. We find that the defect induces two quasibound states in a short junction, corresponding to different spin projections. The energies of these states coincide for a nonmagnetic defect and are generally different for a magnetic defect. The spatial structures of the impurity states resemble the structures of Shiba states in a bulk superconductor. We also consider the case when the impurity is located close to a flat edge of the junction. We find mesoscopic fluctuations of the energies of the quasibound states as a function of the distance between the impurity and the edge.

## 1. Introduction

The phenomenon of sample-to-sample fluctuations of transport characteristics is one of the key problems in modern mesoscopic physics and nanotechnology. A number of fascinating results in this field are related, for example, to universal conductance fluctuations in normal metal systems (see [1] for a review) and critical current fluctuations in short SNS junctions [2] and SFS junctions [3, 4]. Later on, similar problems have been analyzed within the nonlinear sigma model in longer SNS junctions: the fluctuating density of states [5] and Josephson current [6] have been determined. Most of these studies concentrate on the calculation of the root-mean-square deviation of the current from its average value. Such calculations can be compared with experiment by averaging over a large number of samples. For a given sample, it might be important to understand how a given realization of disorder, e.g. several impurities affect its characteristics. Individual impurities can be nicely observed by the STM technique [7] (see [8] for a recent study of magnetic impurities on a two-dimensional superconductor). It is the goal of the present study to analyze the electronic structure of a single impurity atom in a short Josephson junction with a large number of transport modes. The formation of localized impurity states in a bulk superconductor has been previously stated for a number of systems, including magnetic impurities [9, 10, 11] or nonmagnetic defects inside an unconventional superconductor [12]. In this work we use a more general approach which allows to take into account the inhomogeneity of the superconducting order parameter to find the localized impurity states as well as the whole density of states in the presence of a nonmagnetic or magnetic point defect. Our approach is applicable for arbitrary impurity positions and strengths, providing the basis for further studies of quasiparticle interference patterns in random impurity arrays.

## 2. Infinite SNS junction with a point defect

We will start by considering an infinite SNS junction with a nonmagnetic point defect located in the normal layer, as depicted in Fig. 1a. We assume a step-like order parmeter profile in the junction: the gap $\Delta(\mathbf{r})$ is equal to $|\Delta|\, e^{-i\varphi/2}$ in the left superconducting bank and $|\Delta|\, e^{i\varphi/2}$ in the right superconducting bank, where $|\Delta| = \mathrm{const}$ and $\varphi = \mathrm{const}$. Such a model is justified, for example, for junctions of the constriction type [13], where $\Delta(\mathbf{r})$ may vary strongly on a scale much smaller than the superconducting coherence length $\xi$. In the normal layer we put $\Delta = 0$.

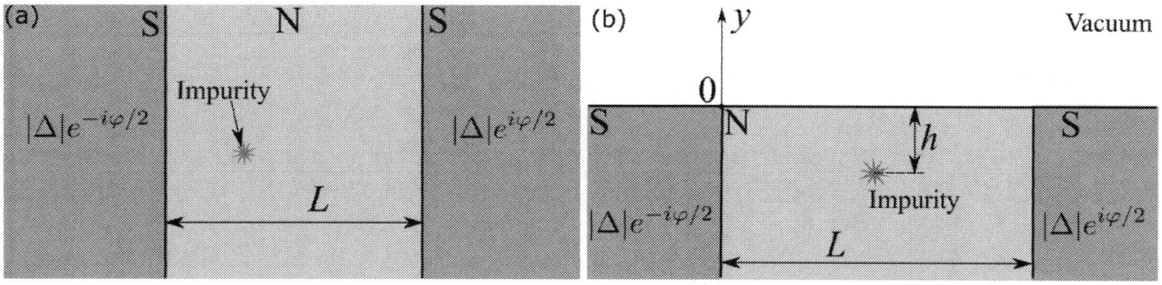


**Figure 1.** Infinite (a) and semi-infinite (b) SNS junction with a point defect.

We will analyze the spectral characteristics of subgap states in the junction. These characteristics can be extracted from the $2 \times 2$ matrix retarded Green function $\check{G}_E(\mathbf{r}, \mathbf{r}')$, which satisfies the Gor'kov equation [14]

$$\left\{ \hat{\tau}_0[H_0(\mathbf{r}) + U(\mathbf{r} - \mathbf{r}_1)] - \hat{\tau}_z(E + i\epsilon^+) + \begin{pmatrix} 0 & -\Delta(\mathbf{r}) \\ \Delta^*(\mathbf{r}) & 0 \end{pmatrix} \right\} \hat{G}_E(\mathbf{r}, \mathbf{r}') = \hat{\tau}_0 \delta(\mathbf{r} - \mathbf{r}'). \quad (1)$$

Here, $\hat{\tau}_0$ and $\hat{\tau}_z$ are a unit matrix and Pauli matrix in Nambu space, respectively, $H_0(\mathbf{r})$ is the normal-state Hamiltonian of a pure system, $U(\mathbf{r})$ is the electric potential of the impurity, the vector $\mathbf{r}_1$ indicates the position of the impurity, $E$ is the energy and $\epsilon^+$ is an infinitely small positive quantity. For our infinite system, we will use the simple normal-state Hamiltonian

$$H_0(\mathbf{r}) = -\frac{\hbar^2 \nabla^2}{2m} - \mu, \quad (2)$$

where $m$ is the electron mass, $\mu = \hbar^2 k_F^2/(2m)$ is the chemical potential, and $k_F$ is the Fermi wave number. Now we will use the fact that the defect is small: specifically, we mean that the range of the potential $U(\mathbf{r})$ is much smaller than $k_F^{-1}$. Then the impurity is essentially an isotropic point scatterer, which is completely characterized by an energy dependent scattering phase $\alpha(E)$. This means that the wave function of an electron will have the following form in the vicinity of the defect [15]:

$$\psi(\mathbf{r}) = \psi_{\mathrm{in}}(\mathbf{r}) + \frac{\psi_{\mathrm{in}}(\mathbf{r}_1) e^{ik|\mathbf{r}-\mathbf{r}_1| + i\alpha}}{k\,|\mathbf{r} - \mathbf{r}_1|} \sin\alpha, \quad (3)$$

where $\psi_{\mathrm{in}}(\mathbf{r})$ is a regular function, and $k$ is the energy dependent wave number. Equation (3) means that a point defect in the Schrödinger or Gor'kov equations can be incorporated as an effective boundary condition at the point $\mathbf{r} = \mathbf{r}_1$ [16]. This observation allows to solve Eq. (1) in terms of the Green function $\hat{G}_E^{(0)}(\mathbf{r}, \mathbf{r}')$ of the system without the defect:

$$\hat{G}_E^{(0)}(\mathbf{r}, \mathbf{r}') = \begin{pmatrix} G_E^{(0)}(\mathbf{r}, \mathbf{r}') & F_E^{(0)}(\mathbf{r}, \mathbf{r}') \\ -F_E^{\dagger(0)}(\mathbf{r}, \mathbf{r}') & \bar{G}_E^{(0)}(\mathbf{r}, \mathbf{r}') \end{pmatrix}. \quad (4)$$

IOP Conf. Series: Journal of Physics: Conf. Series **969** (2018) 012022 doi:10.1088/1742-6596/969/1/012022

Taking $E \lesssim |\Delta|$ and assuming that $\alpha \approx$ const for such energies, we obtain the following solution of Eq. (1):

$$\hat{G}_E(\mathbf{r},\mathbf{r}') = \hat{G}_E^{(0)}(\mathbf{r},\mathbf{r}') + \frac{\hat{G}_E^{(1)}(\mathbf{r},\mathbf{r}',\mathbf{r})}{\mathcal{D}(E,\mathbf{r}_1)}, \tag{5}$$

where

$$\mathcal{D}(E,\mathbf{r}_1) = \left[\frac{mk_F\cot\alpha}{2\pi\hbar^2} - G_{ER}^{(0)}(\mathbf{r}_1,\mathbf{r}_1)\right]\left[\frac{mk_F\cot\alpha}{2\pi\hbar^2} - G_{-ER}^{(0)*}(\mathbf{r}_1,\mathbf{r}_1)\right] + F_E^{\dagger(0)}(\mathbf{r}_1,\mathbf{r}_1)F_{-E}^{\dagger(0)*}(\mathbf{r}_1,\mathbf{r}_1), \tag{6}$$

$$G_{ER}^{(0)}(\mathbf{r},\mathbf{r}') = G_E^{(0)}(\mathbf{r},\mathbf{r}') - \frac{m}{2\pi\hbar^2|\mathbf{r}-\mathbf{r}'|}, \tag{7}$$

and $\hat{G}_E^{(1)}(\mathbf{r},\mathbf{r}',\mathbf{r})$ is a combination of the components of $\hat{G}_E^{(0)}(\mathbf{r},\mathbf{r}')$ which has only poles coinciding with those of $\hat{G}_E^{(0)}(\mathbf{r},\mathbf{r}')$. Additional poles of the Green function appear at energies where the denominator $\mathcal{D}(E,\mathbf{r}_1)$ vanishes. These are the energies of impurity-induced bound states. We will now analyze such states in our SNS junction. Generally speaking, the matrix $\hat{G}_E^{(0)}(\mathbf{r},\mathbf{r}')$ can be determined for a junction with arbitrary length $L$, however, the most interesting situation occurs in short junctions with $L \ll \xi$. We have then

$$G_{ER}^{(0)}(\mathbf{r}_1,\mathbf{r}_1) \approx \tfrac{mk_F}{4\pi\hbar^2}\left[\cot\left(\gamma(E)+\tfrac{\varphi}{2}\right) + \cot\left(\gamma(E)-\tfrac{\varphi}{2}\right)\right], \tag{8}$$

$$F_E^{\dagger(0)}(\mathbf{r}_1,\mathbf{r}_1) \approx \tfrac{mk_F}{4\pi\hbar^2}\left[\sin^{-1}\left(\gamma(E)+\tfrac{\varphi}{2}\right) + \sin^{-1}\left(\gamma(E)-\tfrac{\varphi}{2}\right)\right], \tag{9}$$

where

$$\gamma(E) = \arccos\left(\frac{E}{\Delta}\right). \tag{10}$$

One finds then that $\mathcal{D}(E,\mathbf{r}_1)$ vanishes at an energy

$$E_A = |\Delta|\sqrt{1-\sin^2\frac{\varphi}{2}\cos^2\alpha} - iE'_A(\alpha,\varphi,\mathbf{r}_1), \tag{11}$$

where $E'_A \sim |\Delta|\, L/\xi$. This corresponds to two spin-degenerate localized states. The presence of an imaginary part of the energy means that the impurity states are in fact resonances with a finite lifetime. This is easy to understand: a quasiparticle localized at the impurity will "leak" to infinity along the normal layer (note that in [2] in a short disordered SNS junction no quasibound states have been found, because a finite system has been considered). The quasibound states in the SNS junction are somewhat similar to Yu-Shiba-Rusinov states [9, 10, 11] in bulk superconductors: their spatial structure is roughly a bubble with a radius of the order of $\xi(1-E^2/|\Delta|^2)^{-1/2}$.

Equation (11) has an interesting interpretation in view of the theory of Beenakker [2] applicable to short SNS junction. It means that the point impurity blocks one transmission channel (per spin projection) in the normal layer, reducing its transparency from unity to $\tau = \cos^2\alpha$. All other channels remain completely transparent.

The theory developed above can be generalized for a magnetic impurity. In this case, a spin quantization axis can be chosen such that electrons with "spin up" and "spin down" will be scattered by the defect with phases $\alpha_\uparrow$ and $\alpha_\downarrow$, respectively. The real parts of the energies of impurity states with spin projection $\sigma$ are then given by

$$\frac{E_\sigma}{|\Delta|} \approx \sigma\sin\beta\sqrt{\cos\alpha_\uparrow\cos\alpha_\downarrow\sin^2\frac{\varphi}{2} + \sin^2\beta} \pm \cos\beta\sqrt{\cos^2\beta - \cos\alpha_\uparrow\cos\alpha_\downarrow\sin^2\frac{\varphi}{2}}, \tag{12}$$

where $\sigma = \pm 1$, and $\beta = (\alpha_\uparrow - \alpha_\downarrow)/2$. Generally, Eq. (12) yields two negative and two positive values of $E_\sigma$, the latter two corresponding to the true energies of the impurity states. Depending on relations between $\alpha_\uparrow$, $\alpha_\downarrow$ and $\varphi$, one may have two quasibound states with either the same spin or opposite spins. Some $E_\sigma$ vs. $\varphi$ graphs are shown in Fig. 2.

IOP Conf. Series: Journal of Physics: Conf. Series **969** (2018) 012022 doi:10.1088/1742-6596/969/1/012022

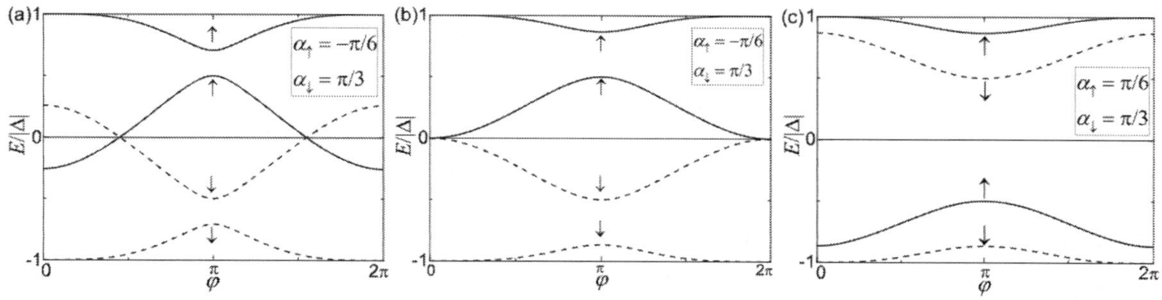


**Figure 2.** Energies of "spin-up" (↑) and "spin-down" (↓) quasilocalized states at a magnetic impurity.

## 3. Semi-infinite SNS junction with a point defect

Real Josephson junction have a finite size, thus it is important to understand how boundaries of the system influence the impurity states. In this section we consider a semi-infinite SNS junction, as shown in Fig. 1b. Here, the half-space $y < 0$ is occupied by the junction, and the half-space $y > 0$ is occupied by vacuum. On the edge of the junction we impose the boundary condition

$$\hat{G}_E(\mathbf{r}, \mathbf{r}')\Big|_{y=0} = 0. \tag{13}$$

Then, the Green function $G_E^{(0)}(\mathbf{r}, \mathbf{r}')$ can be obtained using the image method. We find then that the energies of the impurity states are still given by Eqs. (11) and (12) with the scattering phases replaced by their effective values $\alpha_{\text{eff}}$ defined as

$$\cot \alpha_{\text{eff}} = \frac{\cot \alpha + \frac{\cos(2k_F h)}{2k_F h}}{1 - \frac{\sin(2k_F h)}{2k_F h}}, \tag{14}$$

where $h$ is the distance between the impurity and the edge of the junction. Remarkably, the phase $\alpha_{\text{eff}}$ is an oscillating function of $h$, and hence the energies of the impurity states also oscillate when $h$ is changed. This is a simple illustration of mesoscopic fluctuations in a system with low disorder. Note that the oscillations decay on a scale $h \sim k_F$, so that even if the distance $h$ is much smaller than the spatial extent of the impurity states ($k_F^{-1} \ll h \ll \xi$), their energies may be very weakly affected by the surface of the junction. Some typical graphs of the impurity state energies vs. $h$ dependencies are shown in Fig. 3.

## 4. Conclusion

We have analyzed the localized states induced by a point impurity in a short SNS junction. The impurity generally supports two quasibound states, which may have the same spin or opposite spins (in the case of a magnetic impurity). If the defect is located close to the edge of the junction, the energies of the localized states become sample-specific: they exhibit mesoscopic fluctuations when the distance between the impurity and the sample edge is changed.

There are several observable effects related to the impurity states. First, since the defect blocks a transport channel in the normal layer, it reduces the critical current of the junction. For example, at low temperatures a unitary impurity with $\alpha \approx \pi/2$ should reduce the critical current by a value of the order of $e|\Delta|/\hbar$ (see [2]), where $e$ is the elementary charge. Another way to observe the impurity states is by using the STM technique, which allows to measure both the energies and the spatial structures of the localized states.

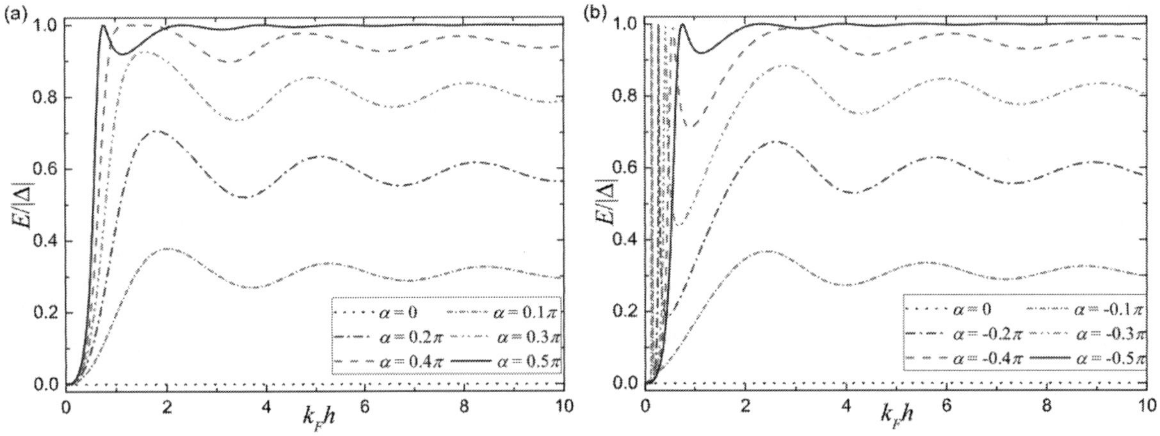


**Figure 3.** Dependencies of the impurity state energy $E$ on the distance $h$ between a nonmagnetic impurity and the surface of the SNS junction for $\varphi = \pi$. The scattering phases $\alpha$ are positive in graph (a) and negative in graph (b).

**Acknowledgments**
The work has been supported by RSF grant №17-12-01383.

## References

[1] Akkermans E and Montambaux G 2007 *Mesoscopic Physics of Electrons and Photons* (New York: Cambridge University Press)
[2] Beenakker C W J 1991 *Phys. Rev. Lett.* **67**(27) 3836–3839
[3] Zyuzin A Y, Spivak B and Hruška M 2003 *EPL* **62** 97
[4] Mel'nikov A S and Buzdin A I 2016 *Phys. Rev. Lett.* **117**(7) 077001
[5] Ostrovsky P M, Skvortsov M A and Feigel'man M V 2001 *Phys. Rev. Lett.* **87**(2) 027002
[6] Houzet M and Skvortsov M A 2008 *Phys. Rev. B* **77**(2) 024525
[7] Fischer Ø, Kugler M, Maggio-Aprile I, Berthod C and Renner C 2007 *Rev. Mod. Phys.* **79**(1) 353–419
[8] Menard G C, Guissart S, Brun C, Pons S, Stolyarov V S, Debontridder F, Leclerc M V, Janod E, Cario L, Roditchev D, Simon P and Cren T 2015 *Nat. Phys.* **11** 1013–1016
[9] Yu L 1965 *Acta Phys. Sin.* **21** 75 (pages 16)
[10] Shiba H 1968 *Prog. Theor. Phys.* **40** 435–451
[11] Rusinov A I 1969 *Sov. Phys. JETP* **29**(6) 1101–1106
[12] Balatsky A V, Vekhter I and Zhu J X 2006 *Rev. Mod. Phys.* **78**(2) 373–433
[13] Likharev K K 1979 *Rev. Mod. Phys.* **51**(1) 101–159
[14] Kopnin N 2001 *Theory of Nonequilibrium Superconductivity* International Series of Monographs on Physics (Clarendon Press) ISBN 9780198507888
[15] Landau L and Lifshits E 1977 *Quantum Mechanics: Non-relativistic Theory* Butterworth Heinemann (Oxford: Butterworth-Heinemann)
[16] Albeverio A, Gesztesy F, Høegh-Krohn R and Holden H 2005 *Solvable models in quantum mechanics* (Providence, Rhode Island: AMS Chelsea Publishing)

# Coexistence of topological charge density waves and superconductivity in a two-dimensional topological superconductor

**K. Tanaka**[1,†], **Yuki Nagai**[2,3], **S. L. Goertzen**[1], **Evan D. B. Smith**[1]

[1] Department of Physics and Engineering Physics, University of Saskatchewan, 116 Science Place, Saskatoon, SK, S7N 5E2 Canada
[2] CCSE, Japan Atomic Energy Agency, 178-4-4, Wakashiba, Kashiwa, Chiba, 277-0871, Japan
[3] Department of Physics, Massachusetts Institute of Technology, Cambridge, MA 02139, USA

E-mail: †kat221@campus.usask.ca

**Abstract.** We perform microscopic mean-field studies of topological order in a two-dimensional s-wave topological superconductor with Rashba spin-orbit coupling and Zeeman field by solving the Bogoliubov-de Gennes equations. By solving for the spin-dependent Hartree potential self-consistently along with the superconducting order parameter, we show that topological charge density waves (TCDW) can coexist with topological superconductivity (TSC) at half filling just as in a conventional s-wave superconductor. Furthermore, we examine the effects of nonmagnetic impurities – which tend to create spin-polarised midgap excitation and pin the phase of charge density modulations – on possible interplay of TCDW and TSC.

## 1. Introduction

Whether charge density waves (CDW) compete or cooperate with superconductivity has been debated intensely for the last several decades, especially for unconventional superconductors such as high-$T_c$ cuprates in recent years. It is a topic of great interest to pursue in the new and rapidly growing field of topological superconductivity [1, 2]. In particular, the two-dimensional (2D) $s$-wave TSC model with Rashba spin-orbit (SO) coupling and Zeeman field [3, 4, 5] is one of the most promising models for platforms that can lead to realisation of fault-tolerant topological quantum computation [1, 5, 6]. There has been recent progress on the material side towards realising 2D $s$-wave TSC [7, 8, 9].

## 2. Model

We use the tight-binding model of Sato, Takahashi, and Fujimoto for 2D $s$-wave TSC with Rashba SO coupling and Zeeman field [3, 4]. Our mean-field Hamiltonian is given by [10]

$$\begin{aligned}\mathcal{H} &= \sum_{\langle ij\rangle}\sum_{\sigma=\uparrow,\downarrow} t_{ij}c^\dagger_{i\sigma}c_{j\sigma} + \sum_i\sum_{\sigma=\uparrow,\downarrow}(\epsilon_i - \tilde{\mu} + \tilde{h}_\sigma + V^{(H)}_{i\bar{\sigma}} - \bar{V}^{(H)}_{\bar{\sigma}})c^\dagger_{i\sigma}c_{i\sigma} \\ &+ \frac{\alpha}{2}\left[\sum_i (c^\dagger_{i-\hat{x}\downarrow}c_{i\uparrow} - c^\dagger_{i+\hat{x}\downarrow}c_{i\uparrow}) + i(c^\dagger_{i-\hat{y}\downarrow}c_{i\uparrow} - c^\dagger_{i+\hat{y}\downarrow}c_{i\uparrow}) + \text{H.c.}\right]\end{aligned}$$

IOP Conf. Series: Journal of Physics: Conf. Series **969** (2018) 012023 doi:10.1088/1742-6596/969/1/012023

$$+ \sum_i (\Delta_i c^\dagger_{i\uparrow} c^\dagger_{i\downarrow} + \text{H.c.}) \,. \tag{1}$$

Here $c^\dagger_{i\sigma}$ and $c_{i\sigma}$ creates and annihilates, respectively, the electron at lattice site $i$ with spin $\sigma$, $\tilde{h}_\sigma = -\tilde{h}$ $(+\tilde{h})$ for $\sigma = \uparrow$ $(\sigma = \downarrow)$, and we have defined

$$\tilde{\mu} = \mu - \frac{\bar{V}^{(H)}_\uparrow + \bar{V}^{(H)}_\downarrow}{2}\,, \qquad \tilde{h} = h + \frac{\bar{V}^{(H)}_\uparrow - \bar{V}^{(H)}_\downarrow}{2}\,, \tag{2}$$

with the bare chemical potential $\mu$, the Zeeman energy $-h$ $(+h)$ for $\sigma = \uparrow$ $(\sigma = \downarrow)$, and the average Hartree potential for each spin component:

$$\bar{V}^{(H)}_\sigma = \frac{1}{N} \sum_i V^{(H)}_{i\sigma}\,, \tag{3}$$

where $N$ is the total number of lattice sites. We assume an intrinsic $s$-wave pairing interaction $U_i \equiv U < 0$ that is uniform and results in the superconducting order parameter $\Delta_i = U\langle c_{i\downarrow} c_{i\uparrow}\rangle$ and the Hartree potential $V^{(H)}_{i\sigma} = U\langle c^\dagger_{i\sigma} c_{i\sigma}\rangle$, which is the average potential created by the electrons with spin $\sigma$ and felt by those with the opposite spin $\bar{\sigma}$ at site $i$. In the Hamiltonian (1), we consider hopping among nearest-neighbour sites $\langle ij\rangle$ only with the probability amplitude $t_{ij} \equiv -t$, $\epsilon_i$ is the single-particle potential due to a nonmagnetic impurity at site $i$, $\alpha > 0$ is the Rashba SO coupling strength, $\hat{x}$ and $\hat{y}$ are the unit vectors in the $x$ and $y$ directions, and H.c. stands for the Hermitian conjugate. Length and energy are measured in units of the lattice constant and $t$, respectively.

An effective coupling constant $U$ that is the same for the Hartree potential and the superconducting order parameter can be generated, for example, via the $s$-wave Feshbach resonance in a superfluid Fermi gas [11]. For $h > 0$, typically there are more spin-up electrons than spin-down electrons and the Hartree potential effectively reduces the Zeeman field. Intuitively, the average energy gain by the pairing interaction with spin-up electrons makes the electron to have its spin down less costly in terms of the Zeeman energy.

When the system has translational symmetry, $V^{(H)}_{i\sigma} = \bar{V}^{(H)}_\sigma; \forall i$, and the Hamiltonian (1) can be expressed in momentum space as $\mathcal{H} = \frac{1}{2}\sum_{\boldsymbol{k}} \Psi^\dagger_{\boldsymbol{k}} \mathcal{H}(\boldsymbol{k}) \Psi_{\boldsymbol{k}}$, where $\Psi_{\boldsymbol{k}} = (c_{\boldsymbol{k}\uparrow}\; c_{\boldsymbol{k}\downarrow}\; c^\dagger_{-\boldsymbol{k}\uparrow}\; c^\dagger_{-\boldsymbol{k}\downarrow})^T$ and

$$\mathcal{H}(\boldsymbol{k}) = \begin{pmatrix} \epsilon(\boldsymbol{k}) - \tilde{h}\sigma_z + \alpha\mathcal{L}(\boldsymbol{k})\cdot\boldsymbol{\sigma} & i\Delta(\boldsymbol{k})\sigma_y \\ -i\Delta(\boldsymbol{k})^*\sigma_y & -\epsilon(\boldsymbol{k}) + \tilde{h}\sigma_z + \alpha\mathcal{L}(\boldsymbol{k})\cdot\boldsymbol{\sigma}^* \end{pmatrix}, \tag{4}$$

with $\epsilon(\boldsymbol{k}) = -2t(\cos k_x + \cos k_y) - \tilde{\mu}$ and $\mathcal{L}(\boldsymbol{k}) \equiv (\mathcal{L}_x, \mathcal{L}_y) = (\sin k_y, -\sin k_x)$. $c^\dagger_{\boldsymbol{k}\sigma}$ and $c_{\boldsymbol{k}\sigma}$ are the creation and annihilation operators of the electron with momentum $\boldsymbol{k} = (k_x, k_y)$ and spin $\sigma$, and $\boldsymbol{\sigma} \equiv (\sigma_x, \sigma_y)$ and $\sigma_z$ are the Pauli matrices. $\mathcal{H}(\boldsymbol{k})$ above reduces to the momentum-space Hamiltonian given in Ref. [4] when the Hartree potential is neglected, with $\tilde{\mu} \equiv \mu$ and $\tilde{h} \equiv h$. Thus, various topological phases as classified in Ref. [4] according to the first Chern number or the Thouless-Kohmoto-Nightingale-Nijs (TKNN) number [12] can be achieved by replacing the chemical potential and Zeeman field by $\tilde{\mu}$ and $\tilde{h}$, respectively, when the Hartree potential is taken into account. Depending on the values of $\tilde{\mu}$, $\tilde{h}$, $\alpha$, and the bulk order parameter, this model allows non-Abelian and Abelian phases with the TKNN number $\nu = \pm 1$ and $\nu = -2$, respectively, in addition to trivial phase with $\nu = 0$. Furthermore, topological density wave states such as spin density waves (TSDW) and charge density waves (TCDW) are also possible in Abelian phase [4].

In this 2D $s$-wave TSC model, there are two underlying chiralities, $\eta_\pm \equiv -(\mathcal{L}_x \pm i\mathcal{L}_y)/\sqrt{\mathcal{L}_x^2 + \mathcal{L}_y^2} = \pm i(\sin k_x \pm i \sin k_y)/\sqrt{\sin^2 k_x + \sin^2 k_y}$ [13, 14], which are analogous to the two chiralities $p_x \pm ip_y$ present in the nontrivial (non-Abelian) phase of the continuum model [15, 16, 17]. While in non-Abelian phase with a single Fermi surface, one chirality can be dominant over the other [4, 18, 19], the two chiralities $\eta_\pm$ associated with the two Fermi surfaces are always mixed in Abelian phase.

We solve the Bogoliubov-de Gennes (BdG) equations [20] on the effective Hamiltonian in Eq. (1) self-consistently for not only the superconducting order parameter, but also each spin component of the Hartree potential. We utilise numerically efficient algorithms for solving for the mean fields [21, 22] and for obtaining excitation spectrum and quasiparticle wavefunctions [23, 24], in both cases circumventing the high numerical demand of diagonalizing the BdG matrix. Self-consistent iterations are performed up to the $l$-th iteration step, where, e.g., the order parameter as a complex vector $\vec{\Delta}$ of length $N$ satisfies

$$\frac{\|\vec{\Delta}^{(l)} - \vec{\Delta}^{(l-1)}\|}{\|\vec{\Delta}^{(l-1)}\|} < 10^{-6} \tag{5}$$

and similarly for each spin component of the Hartree potential that can be regarded as a real vector of length $N$. All the calculation presented below has been performed for zero temperature.

**Table 1.** Converged order parameters and the ground-state energy per electron for each of the three degenerate TSC, TCDW, and TSC+TCDW states in a homogeneous 64×64-site lattice with $\tilde{\mu} = 0$, $U = -4t$, $h = 1.5t$ and $\alpha = t$ with PBC. The Zeeman field is effectively reduced to $\tilde{h} = 0.9613t$.

| State | $\Delta_0$ [$t$] | $\Delta_C$ [$t$] | $E/N_e$ [$t$] |
|---|---|---|---|
| SC | 0.675 | 0.000 | -1.100 |
| CDW | 0.000 | 0.675 | -1.100 |
| SC+CDW | 0.422 | 0.527 | -1.100 |

## 3. Results

It has been shown by solving for the superconducting order parameter and each spin component of the Hartree potential self-consistently [10] that $\tilde{\mu} = 0$ corresponds to exactly half filling (i.e., one electron per site on average) and that uniform TSC and TCDW states – the latter with wave vector $(\pm\pi, \pm\pi)$ – are degenerate ground states at half filling, just as in the attractive Hubbard model for conventional $s$-wave superconductivity for uniform superconducting and CDW states [25]. Therefore, any linear superposition of the TSC and TCDW states is also a ground state and thus the two kinds of order can coexist for $\tilde{\mu} = 0$, where the system is in Abelian phase. In Table 1 we summarise an example of degenerate TSC, TCDW, and a mixed TSC+TCDW states for $\tilde{\mu} = 0$, $U = -4t$, $h = 1.5t$, and $\alpha = t$ on a square lattice with 64×64 sites with the periodic boundary condition (PBC) and no impurity. Not only the ground-state energy, but also the excitation spectrum is identical among degenerate states, and in case of pure TSC and CDW states, the order parameter (superconducting or CDW) is also the same. When there are surface edges, there are two zero-energy bound states on each surface – as guaranteed by the bulk-edge correspondence to the topological invariant $\nu = -2$, but they are not Majorana fermions whenever there is nonzero CDW order parameter [4, 10].

It is known for conventional $s$-wave superconductivity that the presence of a nonmagnetic impurity, which tends to pin the phase of CDW, lifts the degeneracy in favour of CDW [25].

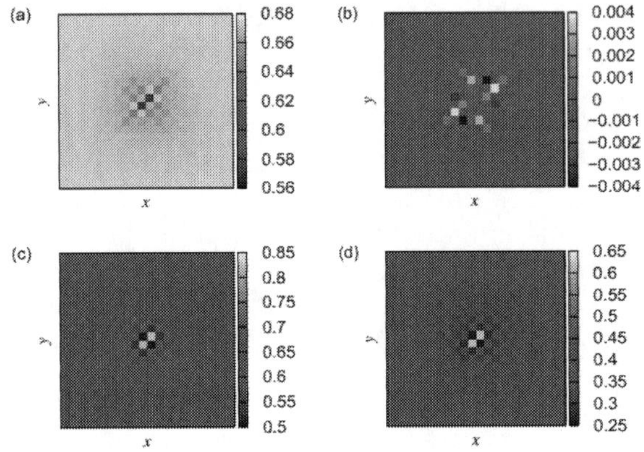


**Figure 1.** Real (a) and imaginary (b) parts of the superconducting order parameter and the average number of electrons with spin up (c) and spin down (d) in the TSC state for $\tilde{\mu} = 0$, $U = -4t$, $h = 1.5t$ and $\alpha = t$, with two nonmagnetic impurities with potential $V_{\rm imp} = -t$ placed diagonally at the centre of a 64×64-site lattice, for $22 \leq x, y \leq 44$.

Due to the two underlying chiralities $\eta_\pm$ present in the system, it is an intriguing question as to what happens to degenerate TSC and TCDW states if a nonmagnetic impurity is deposited. While a single nonmagnetic impurity does not result in any quasiparticle bound state in a conventional $s$-wave superconductor [26], a nonmagnetic impurity can create spin-polarised, midgap quasiparticle excitation and can cause the spin of the midgap state to flip [10]. In this 2D $s$-wave TSC system, a nonmagnetic impurity can even bind a quasiparticle, leading to a phase transition of the ground state [10], and hence act exactly like a magnetic impurity (classical spin) in a conventional $s$-wave superconductor [27, 28, 29].

In Fig. 1 we show the real (a) and imaginary (b) parts of the superconducting order parameter $\Delta_i \equiv \Delta(x,y)$ in units of $t$ and the average number of electrons with spin up (c) and spin down (d) in the TSC state for $\tilde{\mu} = 0$, $U = -4t$, $h = 1.5t$, and $\alpha = t$ in a 64×64-site lattice with PBC, where a nonmagnetic impurity with potential $\epsilon_i \equiv V_{\rm imp} = -t$ is placed diagonally at two sites, $(x,y) = (32,32)$ and $(33,33)$. The data are shown for $22 \leq x, y \leq 44$. Although the self-consistent calculation is started with uniform superconducting order parameter and Hartree potential, we can see in Fig. 1(c) and (d) that CDW with wave vector $(\pm\pi, \pm\pi)$ are induced by the two nonmagnetic impurities at the centre of the lattice. CDW extends over the entire lattice, albeit with nonuniform CDW order parameter: charge density modulations are most enhanced relatively close to the impurities. The superconducting order parameter is suppressed at the impurity sites and oscillates around them in its real part. Moreover, a small, oscillatory imaginary part is induced around the impurities, indicating the occurrence of spontaneous supercurrent: this is reminiscent of the influence of nonmagnetic impurities in a spin-triplet chiral $p$-wave superconductor [30].

Figure 2 illustrates the effects of four nonmagnetic impurities with potential $\epsilon_i \equiv V_{\rm imp} = -0.5t$ placed at the centre of the lattice, $(x,y) = (32,32)$, $(33,32)$, $(32,33)$, and $(33,33)$ in a 64×64-site system with PBC, for $\tilde{\mu} = 0$, $U = -4t$, $h = 1.5t$, and $\alpha = 2.5t$. The average number of electrons is shown for spin up (a) and down (b) in the TSC state, and for spin up (c) and down (d) in the pure TCDW state. Although the electron density is peaked at the impurity sites for both spin up and down, there is no impurity bound state in the TSC state for this potential strength.

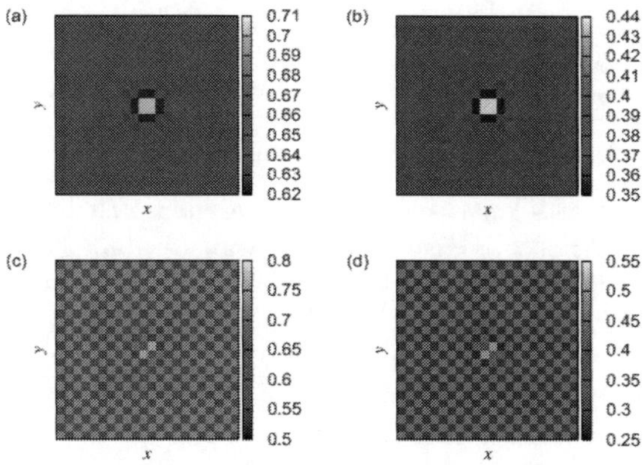


**Figure 2.** Average number of electrons with spin up [(a) and (c)] and spin down [(b) and (d)] in the TSC [(a) and (b)] and pure TCDW [(c) and (d)] states for $\tilde{\mu} = 0$, $U = -4t$, $h = 1.5t$ and $\alpha = 2.5t$, with four nonmagnetic impurities with potential $V_{\text{imp}} = -0.5t$ placed at the centre of a 64×64-site lattice, for $22 \leq x, y \leq 44$.

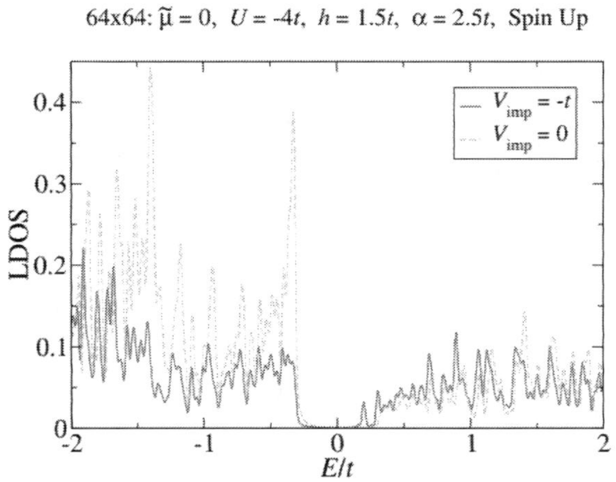


**Figure 3.** LDOS for spin up at one of the impurity site $(x, y) = (33, 33)$ with $V_{\text{imp}} = -t$ in the pure TCDW state for $\tilde{\mu} = 0$, $U = -4t$, $h = 1.5t$, and $\alpha = 2.5t$ in a 64×64-site lattice.

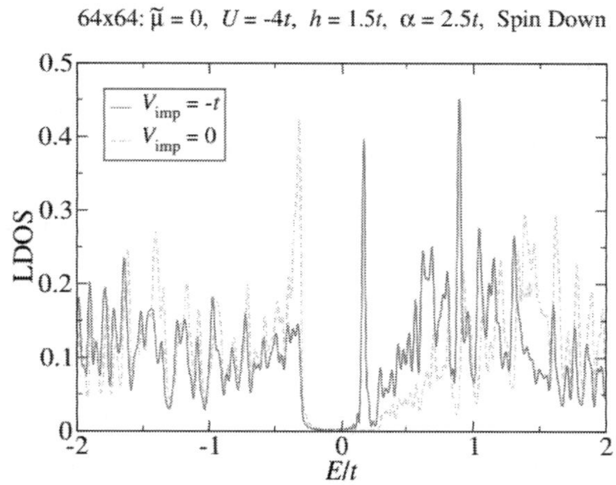


**Figure 4.** LDOS for spin down at one of the impurity site $(x, y) = (33, 33)$ with $V_{\text{imp}} = -t$ in the pure TCDW state for $\tilde{\mu} = 0$, $U = -4t$, $h = 1.5t$, and $\alpha = 2.5t$ in a 64×64-site lattice.

In contrast, there are two particle and two hole bound states close to the gap edge in the pure TCDW state. In inhomogeneous states with impurities, $\tilde{\mu} = 0$ does not necessarily correspond to the exact half filling. For the states presented in Fig. 2, the TCDW state corresponds to half filling within the accuracy of the mean fields, while the TSC state deviates from half filling by one order of magnitude larger than the convergence of the mean fields. The energy per electron is $E/N_e = -1.9687t$ and $-1.9685t$ for the TCDW and TSC states, respectively.

The presence of impurity bound states is illustrated in terms of the local density of states (LDOS) at site $(x, y) = (33, 33)$ in Figs. 3 and 4 for spin up and down, respectively, for the

TCDW state in the same system as shown in Fig. 2 – except that $V_{\rm imp} = -t$, which yields two particle bound states with energy $0.1702t$ and $0.1997t$. The LDOS in Figs. 3 and 4 prominently shows the bound state at energy $0.1702t$ and the fact that its spin is mostly polarised to be down. Impurity bound states can also occur in TSC states for stronger potential strength.

## 4. Conclusions

By solving the BdG equations self-consistently, we have demonstrated the coexistence of superconductivity and CDW in Abelian phase of a 2D $s$-wave topological superconductor, and have found that CDW can be induced in TSC states by nonmagnetic impurities. Further studies as to how nonmagnetic impurities lift the degeneracy of TSC and TCDW states and can lead to possible interplay of the two states will be presented in a future publication.

## Acknowledgements

K.T. is grateful to CCSE at the Japan Atomic Energy Agency for hospitality, where part of the research was performed. This work was enabled in part by support provided by ICT at the University of Saskatchewan, WestGrid (www.westgrid.ca) and Compute Canada Calcul Canada (www.computecanada.ca). Some preliminary calculation was also performed on the supercomputing system SGI ICE X at the Japan Atomic Energy Agency. The research was supported by the Natural Sciences and Engineering Research Council of Canada, and partially by JSPS KAKENHI Grant No. 26800197 and the "Topological Materials Science" (No. 16H00995) KAKENHI on Innovative Areas from JSPS of Japan.

## References

[1] J. Alicea, Rep. Prog. Phys. **75**, 076501 (2012).
[2] M. Sato and Y. Ando, Rep. Prog. Phys. **80**, 076501 (2017).
[3] M. Sato, Y. Takahashi, and S. Fujimoto, Phys. Rev. Lett. **103**, 020401 (2009).
[4] M. Sato, Y. Takahashi, and S. Fujimoto, Phys. Rev. B **82**, 134521 (2010).
[5] J. D. Sau, R. M. Lutchyn, S. Tewari, and S. Das Sarma, Phys. Rev. Lett. **104**, 040502 (2010).
[6] S. Tewari, J. D. Sau, S. Das Sarma, Ann. Phys. **325**, 219 (2010).
[7] A. V. Matetskiy *et al.*, Phys. Rev. Lett. **115**, 147003 (2015).
[8] L. J. Li *et al.*, Nature (London) **529**, 185 (2016).
[9] Y. Nagai, S. Hoshino, and Y. Ota, Phys. Rev. B **93**, 220505(R) (2016).
[10] S. L. Goertzen, K. Tanaka, Y. Nagai, Phys. Rev. B **95**, 064509 (2017).
[11] Y. Ohashi, Phys. Rev. A **83**, 063611 (2011).
[12] D. J. Thouless, M. Kohmoto, M. P. Nightingale, and M. den Nijs, Phys. Rev. Lett. **49**, 405 (1982).
[13] S. Fujimoto, Phys. Rev. B **77**, 220501(R) (2008).
[14] M. Sato and S. Fujimoto, Phys. Rev. B **79**, 094504 (2009).
[15] C. Zhang, S. Tewari, R. M. Lutchyn, and S. Das Sarma, Phys. Rev. Lett. **101**, 160401 (2008).
[16] J. Alicea, Phys. Rev. B **81**, 125318 (2010).
[17] A. Shitade and Y. Nagai, Phys. Rev. B **92**, 024502 (2015).
[18] L.-H. Wu, Q.-F. Liang, Z. Wang and X. Hu, J. Phys.: Conf. Ser. **393**, 012018 (2012).
[19] E. D. B. Smith, K. Tanaka, Y. Nagai, Phys. Rev. B **94**, 064515 (2016).
[20] P. G. de Gennes, *Superconductivity of Metals and Alloys* (Westview Press, Boulder, 1999).
[21] L. Covaci, F. M. Peeters, and M. Berciu, Phys. Rev. Lett. **105**, 167006 (2010).
[22] Y. Nagai, Y. Ota, and M. Machida, J. Phys. Soc. Jpn. **81**, 024710 (2012).
[23] T. Sakurai and H. Sugiura, J. Comput. Appl. Math. **159**, 119 (2003).
[24] Y. Nagai, Y. Shinohara, Y. Futamura, Y. Ota, and T. Sakurai, J. Phys. Soc. Jpn. **82**, 094701 (2013).
[25] K. Tanaka and F. Marsiglio, Phys. Rev. B **62**, 5345 (2000).
[26] A. V. Balatsky, I. Vekhter, J.-X. Zhu, Rev. Mod. Phys. **78**, 373 (2006).
[27] A. Sakurai, Prog. Theor. Phys. **44**, 1472 (1970).
[28] M. I. Salkola, A. V. Balatsky, J. R. Schrieffer, Phys. Rev. B **55**, 12648 (1997).
[29] M. E. Flatté and J. M. Byers, Phys. Rev. Lett. **78**, 3761 (1997); Phys. Rev. B **56**, 11213 (1997).
[30] Y. Okuno, J. Phys. Soc. Jpn. **69**, 858 (2000); see also references therein.

IOP Conf. Series: Journal of Physics: Conf. Series **969** (2018) 012024 doi:10.1088/1742-6596/969/1/012024

# Fe-impurity-induced magnetic excitations in heavily over-doped $La_{1.7}Sr_{0.3}Cu_{0.95}Fe_{0.05}O_4$

**K. Ikeuchi[1], T. Kikuchi[2], K. Nakajima[2], R. Kajimoto[2], S. Wakimoto[3], and M. Fujita[4]**

[1]Neutron Science and Technology Center, Comprehensive Research Organization for Science and Society (CROSS), Tokai, Ibaraki 319-1106, Japan

[2]Materials and Life Science Division (MLF), J-PARC Center, Tokai, Ibaraki 319-1195, Japan

[3]Materials Sciences Research Center, Japan Atomic Energy Agency (JAEA), Tokai, Ibaraki 319-1195, Japan

[4]Institute for Materials Research, Tohoku University, Katahira, Sendai 980-8577, Japan

E-mail: k_ikeuchi@cross.or.jp

**Abstract.** Inelastic neutron scattering experiments were performed to investigate the effect of Fe substitution on the magnetic excitation spectra in a heavily hole-doped cuprate system $La_{1.7}Sr_{0.3}Cu_{0.95}Fe_{0.05}O_4$, with an effective concentration of holes of $x_{eff} = 0.25$. At $T = 7$ K, well-defined low-energy magnetic excitations, associated with a static magnetic order, were observed at the incommensurate positions $\boldsymbol{Q}_{tet} = (0.5\pm\delta, 0.5)/(0.5, 0.5\pm\delta)$. Although the incommensurate structure resembles that observed in $La_{2-x}Sr_xCuO_4$ (LSCO) systems, the incommensurability ($\delta$) of 0.144 (r.l.u) was larger than that of ~0.125 (r.l.u) in LSCO with $x = 0.25$, suggesting an increase in $\delta$ by Fe substitution. In addition to the obvious effect on IC excitations, we show the emergence of ring-shaped magnetic excitations centered at $\boldsymbol{Q}_{tet} = (0.5, 0.5)$ in a constant energy spectrum by Fe-substitution. These ring-shaped excitations were observed for energies below ~8 meV, yielding a cylindrical continuum excitation in the energy and momentum spaces. Similar continuum excitations were recently observed in pristine LSCO with $x = 0.25$ at high temperatures, in our neutron experiments [1]. These results, which show the enhancement of $\delta$ (higher than $\delta = 1/8$) and the stabilization of continuum excitations by Fe substitution, are incompatible with a simple pinning scenario of the stripe order, which is characterized by local spin correlations. The metallic aspect of the low-energy part of the magnetic excitation, which underlies high-temperature superconductivity in over-doped cuprate systems, would be enhanced/induced by Fe substitution.

## 1. Introduction

To elucidate the mechanism of high-temperature superconductivity (high-$T_c$) in cuprate systems, it is important to understand the nature of spin correlations that couple with doped carriers in Mott insulators; whether localized and itinerant spin nature is necessary for realizing the ground state at a particular hole concentration. One of the key aspects is that, in the archetypal cuprate superconductor $La_{2-x}Sr_xCuO_4$ (LSCO) system, antiferromagnetic low-energy signal splits into spots with wave vectors of $(0.5\pm\delta, 0.5)/(0.5, 0.5\pm\delta)$ in the tetragonal notation in the superconducting phase, where the incommensurability $\delta$ is the half-distance between the pair of incommensurate (IC) magnetic peaks. The incommensurability increases linearly and almost saturates for the doping rates under and over $x$ = 0.12, respectively [2, 3], implying that magnetic correlations have different origins in the under- and over-doped regions.

Previous neutron scattering experiments reported that dilute doping of Fe into LSCO induces a static magnetic order in a wide superconducting phase [4, 5]. The IC wave vectors of the induced order resemble that observed in the static and/or low-energy channels in pristine systems. Angle-resolved photoemission spectroscopy (ARPES) measurements on over-hole-doped samples suggest that the induced magnetic peaks arise from the nesting of the Fermi surface, indicating that the magnetically ordered ground state is characterized by the spins of itinerant doped holes [5]. This stimulated the discussion on the long-studying issue in the high-$T_c$ superconductivity: which of the two, localized or itinerant spin pictures, are more adequate for describing the magnetism in the superconducting phase?

To elucidate the nature of magnetic correlations related to impurity-induced magnetism, we performed inelastic neutron scattering (INS) experiments on Fe-doped LSCO samples, using a chopper spectrometer with wide-area detectors and time-of-flight (TOF) analysis. Because the TOF method yields the intensity maps of scattered neutrons in a wide space of momentum transfer ($\boldsymbol{Q}$) and energy transfer ($\omega$), the overall dynamical susceptibility of continuums excitations, which is often discussed in studies of metallic magnetism, and of the collective excitation associated with localized spins can be effectively measured.

## 2. Experiments

One single crystalline rod of Fe-substituted and heavily hole-doped sample $La_{1.7}Sr_{0.3}Cu_{0.95}Fe_{0.05}O_4$ was grown using the traveling-solvent floating-zone method. Because substituting $Fe^{3+}$ for $Cu^{2+}$ reduces the number of holes that are doped by the $Sr^{2+}$ substitution for $La^{3+}$ [6], we can estimate the concentration of holes $x_{eff}$ in $La_{1.7}Sr_{0.3}Cu_{0.95}Fe_{0.05}O_4$ as $x_{eff} = 0.30-0.05 = 0.25$. The weight of the crystal was 22 g. To minimize the background, we used an Al sample-can free from the exchange gas. The lattice constants of $La_{1.7}Sr_{0.3}Cu_{0.95}Fe_{0.05}O_4$ are $a = b = 3.97$ Å and $c = 13.2$ Å at room temperature, in the tetragonal notation. To describe the experimental results, we employed the orthorhombic notation, because by using this natation, the integration range can be easily described. In this notation, the magnetic Γ point $\boldsymbol{Q}_{tet} = (0.5, 0.5, L)$ for the parent antiferromagnet in the tetragonal notation corresponds to $\boldsymbol{Q}_{ort} = (1, 0, L)$, and the IC positions can be written as $(1+\delta, \pm\delta)/(1-\delta, \pm\delta)$.

The INS experiments were performed using BL14 AMATERAS[7], a cold-neutron chopper spectrometer installed at the Materials and Life Science Experimental Facility (MLF) of J-PARC. The Utsusemi software suite was used to analyze the spectral data obtained using the pulsed-neutron scattering technique [8]. The sample was mounted on a 4-K-temperature cryostat, and $H_{ort}$ and $L$ axes in $\boldsymbol{Q}_{ort} = (H_{ort}, K_{ort}, L)$ were aligned in the horizontal plane of the spectrometer. The $L$ axis was at 30° to the direction of the incident neutron beam. This configuration allowed us to obtain the magnetic excitation effectively by projecting it onto the $H_{ort}$-$K_{ort}$ plane, owing to its 2-dimensional nature. The parameters for monochromatizing choppers were tuned to gain neutrons with the incident energy of $E_i$ = 10.52 meV and to realize an energy resolution of $\Delta\omega = 0.7$ meV at the elastic position. In these experimental setups, we measured magnetic excitations in the $\boldsymbol{Q}_{ort}$-$\omega$ space and their temperature dependence.

## 3. Results and Discussion

### *3.1. Incommensurate magnetic excitations*

Figure 1 (a) shows INS intensity maps of the magnetic excitation spectra taken at $T$ = 7 K, which is below the magnetic ordering temperature of 45 K. Excitations are clearly seen at IC positions around $K_{ort}$ = 0 r.l.u., similar to the results for pristine LSCO with $x$ = 0.25 [9]. At 300 K, the intensity below ~5meV is much weaker than that at 7 K (Fig. 1(b)), suggesting that low-energy excitations below ~5 meV strengthen at low temperatures. In the present system, excitations arose from the elastic position, while there was no magnetic Bragg peak in the over-doped pristine LSCO. The incommensurability of $La_{1.7}Sr_{0.3}Cu_{0.95}Fe_{0.05}O_4$ is $\delta$ = 0.144 (r.l.u) at $T$ = 7 K, higher than $\delta$ = 0.125 (r.l.u) for LSCO with $x$ = 0.25 measured in an inelastic region [2]. As mentioned above, in the pristine LSCO system, $\delta$ exhibits a linear dependence in the under-doped region and saturates with $\delta$ = 0.125 (r.l.u) for $x \geq 0.12$. Therefore, $\delta$ is increased by Fe substitution in the doping region while $\delta$ is almost constant in the pristine system. Therefore, variations in the concentration of holes do not explain $\delta$-enhancement larger than 0.125, and the stabilization of the stripe fluctuation in the pristine LSCO is not directly related to the appearance of magnetic order. We speculate that the modification of the Fermi surface topology by Fe substitution induces magnetic response at different positions in the reciprocal space. The appearance of the IC magnetic order for quite high $\delta$ ~ 0.2 has been also reported in $Bi_{1.75}Pb_{0.35}Sr_{1.90}Cu_{0.91}Fe_{0.09}O_6$, for the hole concentration $p$ = 0.25 [10]. Fe impurities probably have the same effect on the correlations between spins in Cu sites, and stabilize static magnetic order in the high-density-carrier phase of cuprate superconductors.

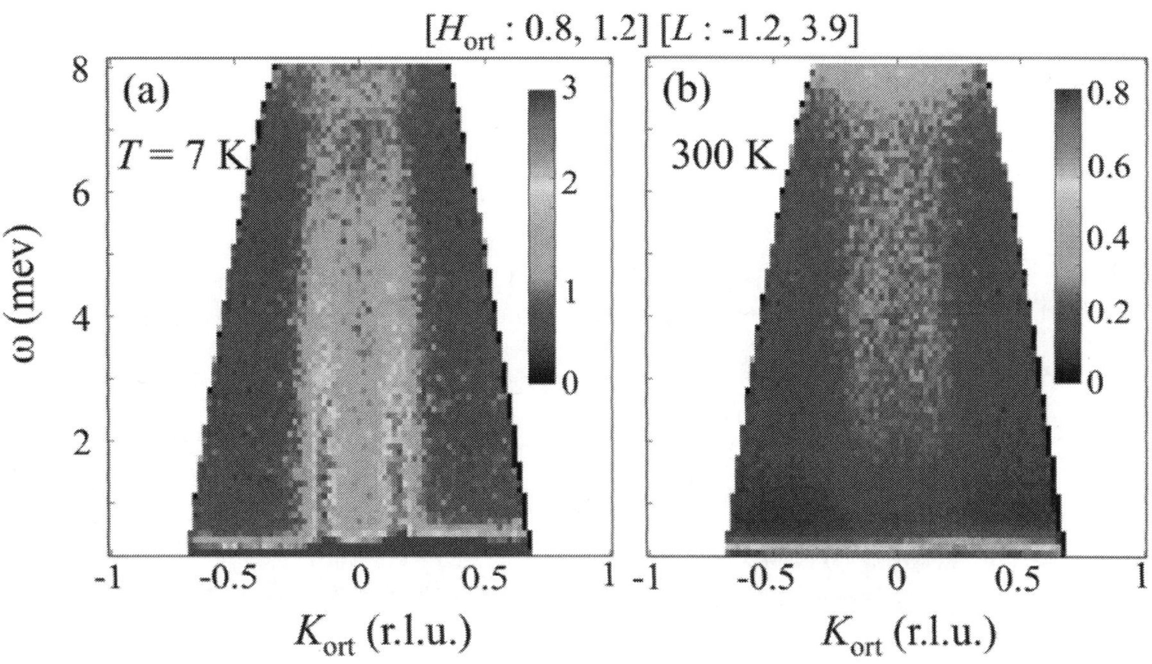


Fig. 1. INS intensity maps in the $K_{ort}$-ω plane, taken at $T$ = 7 K and 300 K. Integration ranges over $H_{ort}$ and $L$ were from 0.8 to 1.2 r.l.u. and from -1.2 to 3.9 r.l.u., respectively. These intensities were divided by the Bose population factor at each temperature.

### *3.2. Appearance of magnetic continuum excitations*

Figure 2 shows the INS intensity maps in the $H_{ort}$-$K_{ort}$ plane, taken at $T$ = 7 K and 300 K. As seen in this figure, at low temperatures there is an additional spectral weight between the IC positions with the ring-like intensity-distribution around $Q_{ort}$ = (1, 0). The radius of the ring-like signal was estimated as $r$ = 0.126 (r.l.u) in the tetragonal notation, comparable to δ for pristine LSCO with $x$ = 0.25. This ring-like signal would be the response of spins of itinerant electrons, because the distribution of the spectral weight cannot be simply reproduced by collective excitations from an ordered state of localized spins.

As shown in Fig. 1 and Fig. 2, the IC excitations disappeared at the high temperature $T$ = 300 K, whereas the ring-like signal remained, although the intensity distribution became blurred. In the case of the $YBa_2Cu_3O_{6.95}$ system, a metallic nature has been discussed to understand the low energy component of magnetic excitations; excitations in the IC structure arose from the Fermi surface nesting and became continuum excitations confined at around the magnetic Γ point, owing to the breakdown of the nesting condition in the high-temperature regime [11]. This consideration suggests that the magnetic excitation switches with temperature between the well-defined IC structure and continuum. In our case of Fe-doped LSCO, the difference in the incommensurability between the IC and ring-shaped signals and in their temperature dependence directly suggests that the low-energy magnetic excitations are composed of IC excitations and continua at the fixed energy transfer and at low temperatures. In the case of Fe 1% substituted LSCO with $x$ = 0.25 and with $x$ = 0.29, INS and ARPES experiments suggest the appearance of a spin density wave owing to the Fermi surface nesting [5]. This supports the notion of the metallic origin of the appearance of the magnetic order and related low-energy excitations. Although the evidence for continuum excitation has not been pointed out theoretically [5], it is interesting to determine whether continuum components can be reproduced from the viewpoint of Fermi surface topology.

Finally, we briefly report a comparison with the results for pristine LSCO with $x$ = 0.25. Quite recently, we found similar ring-like continuum excitations in LSCO with $x$ = 0.25 at high temperatures [1], whereas, as discussed previously [9], such excitation was not confirmed for low temperatures. Therefore, continuum excitations in the high-temperature region for pristine LSCO with $x$ = 0.25 would be stabilized by the Fe substitution even at low temperatures. These results suggest that elucidating the roles of Fe substitution can provide further understanding of inherent magnetic correlations underlying the superconductivity of the LSCO system, taking a different approach in high-$T_c$ studies on the cuprates.

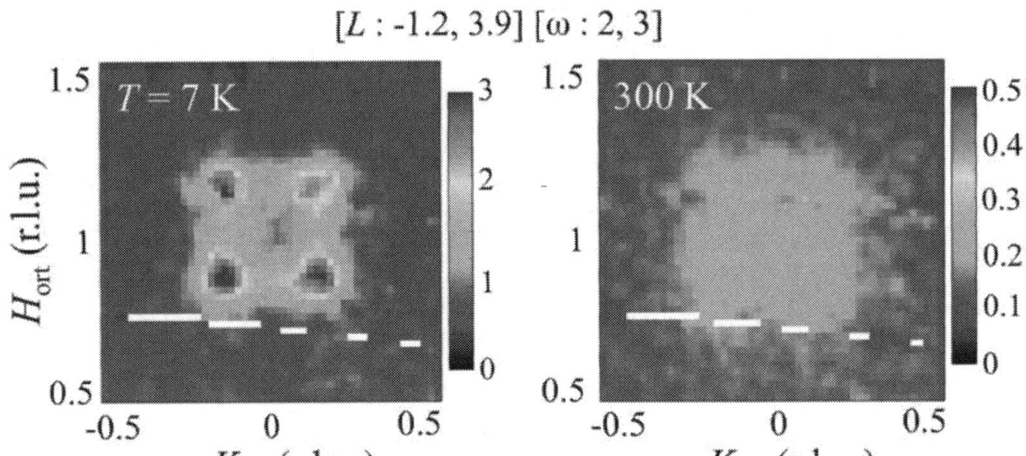


Fig. 2. INS intensity maps in the $H_{ort}$-$K_{ort}$ plane, taken at $T$ = 7 K and 300 K. Integration ranges over $L$ and ω were from -1.2 to 3.9 r.l.u. and from 2 to 3 meV, respectively. These intensities were divided by the Bose population factor at each temperature.

## 4. Summary

Using INS, magnetic excitation spectra were measured for a single crystal of Fe-substituted and heavily hole-doped $La_{1.7}Sr_{0.3}Cu_{0.95}Fe_{0.05}O_4$. On the $K_{ort}$-ω and $H_{ort}$-$K_{ort}$ maps experimentally obtained at $T$ = 7 K, we found that the excitation spectra are composed of incommensurate excitations and ring-like continua. The incommensurability of the induced magnetic peaks (δ = 0.144 (r.l.u)) was different from the radius of ring-shaped signal $r$ = 0.126 (r.l.u), and even larger than δ = 0.125 (r.l.u), for pristine LSCO with the same concentration of holes. At a high temperature of $T$ = 300 K, the incommensurate excitations disappeared, whereas the continuum excitations remained. Both excitations were intimately related to the inherent metallic nature of magnetic correlations in the heavily hole-doped LSCO system. Elucidating the effect of Fe substitution on spin correlations is an effective approach for obtaining key information on the coupling of electrons with strong spin correlation in high-$T_c$ cuprate superconductors.

## 5. Acknowledgment

The neutron-scattering experiment at AMATERAS was carried out under a research project organized by J-PARC and CROSS (Proposal No. 2016P0601). This work was performed under the Inter-university Cooperative Research Program of the Institute for Materials Research, Tohoku University (Proposal No. 15K0108, 16K0036 and 17K0088). K. I. and M. F. and are supported by Grant-in-Aid for Young Scientists (B) (17K14348) and for Scientific Research (A) (16H02125), respectively.

## References

[1] Ikeuchi K *et al*: Prepared as a proceedings paper of SCES 2017 (the International Conference on Strongly Correlated Electron Systems in Prague (17. - 21. 7. 2017)).
[2] Yamada K, Lee C H, Kurahashi K, Wada J, Wakimoto S, Ueki S, Kimura H, Endoh Y, Hosoya S, Shirane G, Birgeneau R J, Greven M, Kastner M A and Kim Y J 1998 *Phys. Rev.* B **57** 6165
[3] Fujita M, Yamada K, Hiraka H, Gehring P M, Lee S H, Wakimoto S and Shirane G 2002 *Phys. Rev.* B **65** 064505
[4] Fujita M, Enoki M and Yamada M 2008 *J. of Phys. and Chem. of Solids* **69** 3167– 3170
[5] He R-H, Fujita M, Enoki M, Hashimoto M, Iikubo S, Mo S-K, Yao H, Adachi T, Koike Y, Hussain Z, Shen Z-X and Yamada K 2011 *Phys. Rev. Lett.* **107** 127002
[6] Lu L, Pi L, Tan S and Zhang Y 2010 *Physica* C **470** 1920-1927
[7] Nakajima K, Ohira-Kawamura S, Kikuchi T, Nakamura M, Kajimoto R, Inamura Y, Takahashi N, Aizawa K, Suzuya K, Shibata K, Nakatani T, Soyama K, Maruyama R, Tanaka H, Kambara W, Iwahashi T, Itoh Y, Osakabe T, Wakimoto S, Kakurai K, Maekawa F, Harada M , Oikawa K, Lechner R E, Mezei F and Arai M 2011 *J. Phys. Soc. Jpn.* **80** SB028
[8] Inamura Y, Nakatani T, Suzuki J and Otomo T 2013 *J. Phys. Soc. Jpn.* **82** SA031
[9] Wakimoto S, Zhang H, Yamada K, Swainson I, Kim H and Birgeneau R J 2004 *Phys. Rev. Lett.* **92** 217004
[10] Hiraka H, Hayashi Y, Wakimoto S, Takeda M, Kakurai K, Adachi T, Koike Y, Yamada I, Miyazaki M, Hiraishi M, Takeshita S, Kohda A, Kadono R, Tranquada J M and Yamada K 2010 *Phys. Rev.* B **81** 144501
[11] Reznik D, Ismer J-P, Eremin I, Pintschovius L, Wolf T, Arai M, Endoh Y, Masui T and Tajima S 2008 *Phys. Rev.* B **78** 132503

# Electrical transport properties of epitaxial titanium nitride nanowire

**K Makise[1] and B Shinozaki[2]**

[1]National Institute of Advanced Industrial Science and Technology (AIST), Tsukuba 305-8568, Japan

[2] Department of Physics Kyushu University, Motooka, 744, Fukuoka 819-0395, Japan

kf-makise@aist.go.jp

**Abstract**. We have measured the transport properties of epitaxial titanium nitride (TiN) nanowires. Epitaxial TiN layer, deposited by dc magnetron sputtering on MgO(100) substrates at growth temperature $T = 1073$ K. Samples of nanowire were fabricated by e-beam lithography and reactive ion etching. Although TiN films with 100 nm-thickness have superconducting transition temperature $T_C \sim 5$ K, nanowires does not appear resistive transition until 0.15 K. The magnetoresistance (MR) are always negative. Furthermore for MR experimental results, we attempt to fit the data using one-dimensional weak localization theory. In addition we observed oscillations of magnetoresistance below 5 K.

kf-makise@aist.go.jp

## 1. Introduction

Recently, the application of mesoscopic superconductors in quantum devices are one of the major focuses in nano-science researches. Especially, nanowires and nano-tubes are important materials for the application to devises, because of nano-size function. In low dimensional disordered system, transport properties are strongly affected by quantum interference, which relate to the phase coherence of wave functions of electrons. Such effects of transport properties have often carried out by investigating the temperature and/or magnetic field dependence of resistance (or conductance). In mesoscopic system, magnetoresistance (MR) measurements have observed interesting quantum interference effect such as the Aharonov-Bohm(AB) and Al'tshuler-Aronov-Spivak(AAS) oscillations[1,2]. In this study, we present our experimental results of MR oscillation of epitaxial titanium nitride (TiN) nanowire. Previous our studies have shown that the superconducting critical temperature ($T_C$) of TiN films on MgO substrates is around 5 K and that of lattice constant and is about 0.424 nm, which is close to the lattice constant of MgO of 0.421 nm. As a results of MR measurements, we find that the MR oscillation of TIN nanowires are related to the AB and AAS effect by the fast Fourier transform (FFT) analysis

## 2. Experimental detail

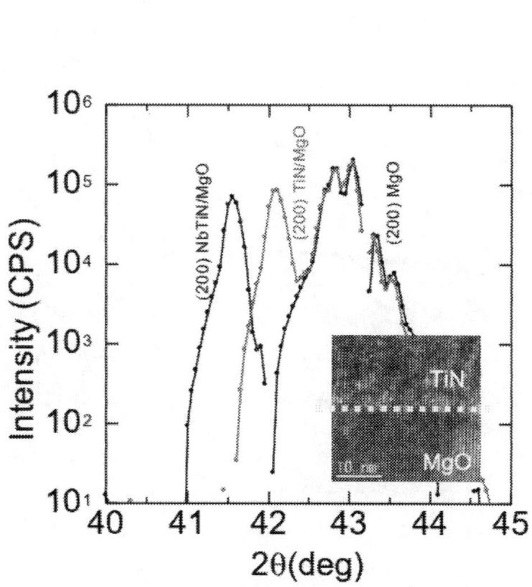


**Figure 1.** XRD patterns of TiN and NbTiN films. Inset: Cross-section TEM image

Firstly, TiN films were deposited at room temperature on 100 single crystal MgO substrates by DC reactive magnetron sputtering method. The background pressure of the chamber was maintained below $8 \times 10^{-8}$ Pa. The relative amounts of argon and nitrogen introduced during sputtering were controlled by mass flow controllers. The total pressure was maintained at $1.3 \times 10^{-1}$ Pa and the substrate was heated during deposition. The TiN nanowires were fabricated from 2D films with thickness $d \sim 2.5$ nm by conventional e-beam lithography method and reactive ion etching method with $CF_4$ plasma. The nanowire of length $L$ is 500 μm and width $w$ of nanowires are 100, 200 and 300 nm. To eliminate the influence of the contact resistance, measurements of transport properties carried out by four-probe method.

## 3. Results and discussion

XRD measurements of TiN and NbN on MgO, shown in Fig.1. The indexed XRD patterns show those for the NaCl structure of TiN. The lattice constants $a_0$ were determined using the XRD peaks at 200 for TiN on MgO substrates. The lattice constant of the TiN film has 4.24 Å [3]. On the other hand, the lattice mismatch of the NbN films is larger than that of TiN films. Superconducting transitions temperature of the films with 100 nm thickness showed a $T_C$ of 5.3 K and a resistivity of 3.5 μΩcm at 10 K. The detail of film preparation is described elsewhere [4] .

Figures 2 show logarithmic temperature dependence of sheet conductance, $\sigma_{sq}(T)$ with different width. And we found that the $R_{sq}(T)$ shows a monotonic decrease with decreasing temperature down to the lowest temperature in our measurements. We think that the nanowire do not have superconducting resistive transition due to thickness-tuned superconductor-insulator transition. All samples indicate metallic like temperature dependence of resistance, except for low temperatures, as shown in inset of Fig. 2(a). For 2D disordered system, the conductance is expect to $\sigma_{sq} = \sigma_0 + pe^2/2\pi^2\hbar \ln T$ by weak localization, WL and electron-electron interaction theory in the low temperature region. Here, pre-factor $p$ is the exponent of the temperature term in the inelastic scattering rate $1/\tau_{in} \propto T^p$. Therefore, we can estimate $p$ in this relation. In generally, the factor of $p$ has around 1 - 2, depending on material and the ratio of inelastic scattering time and spin-orbit scattering time. Although the $p$ became about 2-3, which

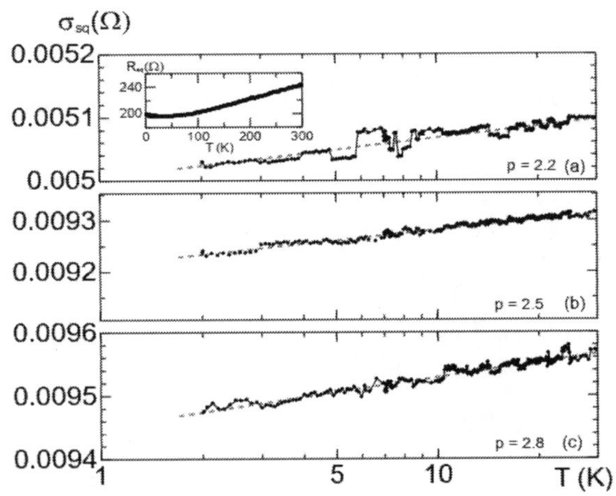


**Figure 2.** The logarithmic $T$ dependence of resistance sheet conductance $\sigma_{sq}$ for nanowires: (a) w = 100 nm, (b) 200 nm and (c) 300 nm. Dotted lines were calculated by least-squares method using the 2D localization theory and the numerals denote the corresponding pre-factor $p$ (see text.). Inset of Fig. 1(a) is $T$ dependence of sheet resistance for samples with $w$ = 100 nm.

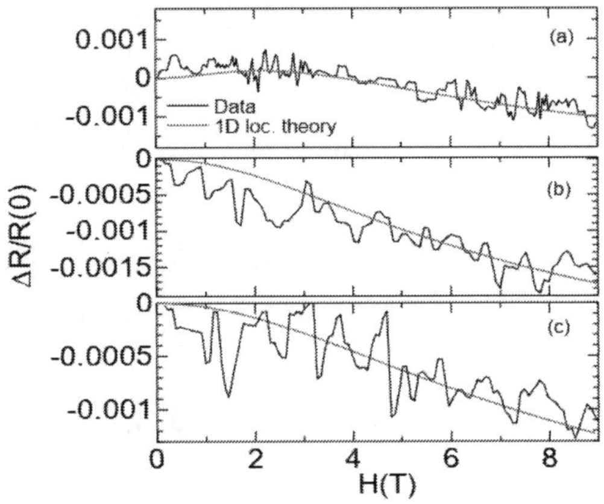


**Figure 3.** Normalized MR versus H of nanowire sample with $w$ = 100 nm at various temperature: (a) 2.0 K, (b) 2.4 K and (c) 3.6 K. The Solid line is a fit to the data using Eq. (1).

is almost consistent with the presence of 2D localization, the $p$ slightly increases as the width of nanowire increases.

In Fig. 3, we plot $H$ dependence of normalized magnetoresistance (MR), $\Delta R(H)$ =[$R$(H)-$R$(0)]/$R$(0) at several temperature at (a) 2.0 K , (b) 2.4 K and (c) 3.6 K, with w = 100 nm of sample. The samples have negative magnetoresistance, which means weak spin-orbit scattering effect. The MR in the WL has been well reported effect In addition we observed that the MR oscillations appear. Firstly the data were fitted to the one-dimensional weak-localization theoretical prediction [5]

$$\frac{\Delta R(B)}{R(0)} = \frac{e^2}{\pi\hbar}\frac{R}{L}\left\{\frac{3}{2}\left[\left(\frac{1}{L_\varphi^2}+\frac{4}{3L_{so}^2}+\frac{W^2}{3L_B^4}\right)^{-1/2} - \left(\frac{1}{L_\varphi^2}+\frac{4}{3L_{so}^2}\right)^{-1/2}\right] - \frac{1}{2}\left[\left(\frac{1}{L_\varphi^2}+\frac{W^2}{3L_B^4}\right)^{-1/2} - L_\varphi\right]\right\}, \quad (1)$$

whrere $R$, $W$ and $L$ are a nanowire of resistance, width and length, respectively. Also, $L_B$, $L_\varphi$ and $L_{so}$ are magnetic length, $(\hbar/eB)^{1/2}$, inelastic scattering length, $(D\tau_\varphi)^{1/2}$, and spin scattering length $(D\tau_{so})^{1/2}$. Here $D$ is diffusion constant. As a results, we found that the data can be described by 1D weak localization theory below 5 K. Next, we used fast-Fourier-transform (FFT) method for analysis of MR oscillations. In Fig. 4(a) the FFT amplitude spectrum was observed the period of about 0.9 T. A prominent period of about 0.6 T is shown for nanowire with w =300 nm(Fig. 4(c)). We see an increase in period with increase in width of the nanowire. The AAS effect with period h/2e [1]. The deviation and other period can be attribute to a contribution of the AB oscillations and aperiodic universal conductance fluctuation. Actually, such an oscillations are observed for silicon nanowires [6]. For 200 nm-width nanowire (Fig. 4 (b)), the 3.70 T peaks ($B_p$) corresponds to an AB ring of area ($S = \pi\hbar/eB_p$) of 0.55 $\times 10^{-15}$ m$^2$. This area is almost equals to cross section of area of nano-wire, that is 2.5 nm × 200 nm = 0.5 $\times 10^{-15}$ m$^2$. Similarly, we also obtain that the $S$ for 300 nm width ($B_p$=2.56 T) and cross section of area are 0.8 $\times 10^{-15}$ m$^2$ and 0.7 $\times 10^{-15}$ m$^2$, respectiviy.

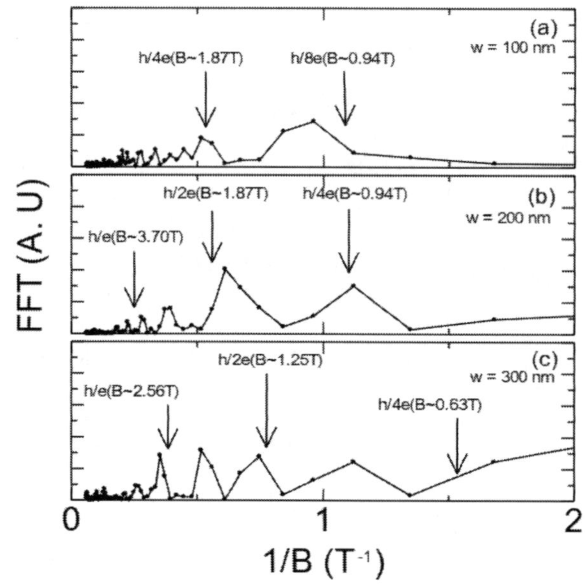


**Figure 4.** FFT of the MR with different width (a) w = 100 nm, (b) 200 nm and 300 nm at $T = 2$ K.

## 4. Summary

In summary, we have studied electrical transport properties using TiN nanowire with various width. From *R-T* measurements at low temperatures, we find that logarithmic temperature dependence, which means weak localization effect obey the data for our samples. Furthermore the MR experimental results can be fitted using prediction of one dimensional WL theory. As for MR oscillation, it seems that the data can be explained by AB of AAS effect by FFT analysis.

## References

[1] Altshuler B L, Aronov A G, and Spivak B Z 1983 *JETP Lett.* **33** 94
[2] Washiburn S and Webb R A1986 *Adv. Phys.* **35** 375422
[3] Krockenberger Y, Karimoto S, Yamamoto H and Semba K 2012 *J. Appl. Phys.* **112** 083920
[4] Makise K, Sun R, Terai H and Wang Z 2015 *IEEE Tran. Appl. Supercon.* **25** 1100204
[5] Pierre F, Gougam B, Anthore A, Pothier H, Esteve D and Birge N O 2003 *Phys. Rev B* **68** 085413
[6] Mtsuko D, Aslan T, Ncube S, Coleman C, Wamwangi D and Bhattachayya S 2016 *EPL* **113** *47002*

IOP Conf. Series: Journal of Physics: Conf. Series **969** (2018) 012026 doi:10.1088/1742-6596/969/1/012026

# Superconductor-Insulator transition in sputtered amorphous MoRu and MoRuN thin films

**K Makise[1], B Shinozaki[2], and F Ichikawa[3]**

[1]National Institute of Advanced Industrial Science and Technology (AIST), Tsukuba 305-8568, Japan

[2] Department of Physics Kyushu University, Motooka, 744, Fukuoka 819-0395, Japan

[3]Department of Physics Kumamoto University, Kurokami 2-39-1, Kumamoto 860-8555, Japan

kf-makise@aist.go.jp

**Abstract**. This work shows the experimental results of the superconductor-insulator (S-I) transition for amorphous molybdenum ruthenium (MoRu) and molybdenum ruthenium nitride (MoRuN) films. These amorphous films onto c-plane sapphire substrates have been interpreted to be homogeneous by XRD and AFM measurements. Electrical and superconducting properties measurements were carried out on MoRu and MoRuN thin films deposited by reactive sputtering technique. We have analysed the data on $R_{sq}$ $(T)$ based on excess conductivity of superconducting films by the AL and MT term and weak localization and electron-electron interaction for the conductance. MoRu films which offer the most homogeneous film morphology, showed a critical sheet resistance of transition, $Rc$, of ~ 2 kΩ. This values is smaller than those previously our reported for quench-condensed MoRu films on SiO underlayer held at liquid He temperature.

...

## 1. Introduction

In field of condensed matter physics, the origin of two dimensional superconducting-insulator transitions (2D-SIT) is still unresolved [1]. For example, is there a universal critical sheet resistance?, that is quantum resistance, $R_Q = h/4e^2$. If system is completely self-dual, the critical sheet resistance is a universal value, i. e. $R_Q$, independent of the kind of material and disorder. However, despite the amount of experimental work on SIT based on several 2D system over the past forty years, such a universal results had not found. Crauste et al. showed the relationship between effect of annealing and a changing of film thickness for SIT using amorphous NbSi films [2]. And they also reported the chemical composition is not changed and the film structure is still amorphous by annealing. However, by the influence of annealing, superconductivity of films is weaken due to the rearrangement of atoms. The effect appears though the sheet resistance, $R_{sq}$ by the variation of mean free path (mfp) $\ell$ of electron. Because $R_{sq}$ is defined by $R_{sq} = \rho/d$ and $\ell$ is obtain by $(\hbar/e^2\rho)\times(3\pi^2/n^2)^{1/2}$. Here $\rho$, $d$ and $n$ are resistivity, films thickness and carrier density, respectively. In fermionic picture of SIT, the theoretical prediction implicitly depend on mfp [3]. On the other hand, in bosonic picture, it is independent of kind of disorder for homogeneous 2D superconductor [4]. In this work, we compare results from two techniques on ex-situ method, using sputtering deposition, and in situ method, using an evaporator-cryostat composite, a series of amorphous MoRu films with various thickness.

## 2. Experimental detail

MoRu and MoRuN films were deposited at room temperature on c-plane sapphire substrates by RF reactive magnetron sputtering method. Collver and Hammond reported the relation between amorphous transition-metal alloys and transition temperature [5]. According to the results, when the electron-to-atom ratio (e/a) is close to 6.4, the $T_C$ has the maximum value of 9.4 K [6]. Therefore, we chosen the sputtering target of Mo75-Ru25 wt% alloys. The background pressure of the chamber was maintained at $5 \times 10^{-5}$ Pa. The relative amounts of argon and nitrogen introduced during sputtering were controlled by mass flow controllers. The total pressure was maintained at 0.6 Pa and the substrate was not heated intentionally during deposition. For MoRuN films, the flow rate of Ar gas was fixed at 50 sccm, while that of $N_2$ gas was of 25 sccm. The microscopic structure was investigated by X-ray diffraction measurement and atomic force microscopy (AFM). The XRD patterns show a broad peak which is the characteristic of amorphous films. And any granular structures are not observed with the surface roughness below 1 nm from the AFM measurements. The results indicate amorphous structure and smooth surface. Resistance and Hall effect measurements were carried out by four probe and Hall bar configuration. Samples were patterned using photolithography techniques and $CF_4$ reactive ion etching. Quench-condensed MoRu films was prepared using an evaporator-cryostat composite system. The results of details are described elsewhere [7,8].

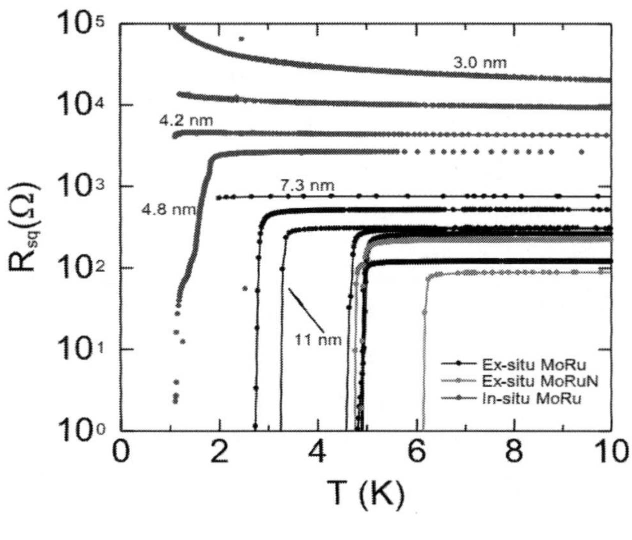


**Figure 1.** Temperature dependence of sheet resistance $R_{sq}$. with various film thickness. Numerals denote the film thickness.

## 3. Results and discussion

Figure 1 shows temperature dependence of sheet resistance, $R_{sq}$ $(T)$ with different thickness. We found that the $R_{sq}$ $(T)$ shows a monotonic decrease with decreasing temperature down to the lowest temperature in our measurements. For in-situ films, a slight increase in film thickness changes the transport properties from insulating to superconducting behavior. $R_{sq}(T)$ also did not have reentrant behavior as seen frequently for granular 2D films. We consider the relation between $T_C$ and $R^N{}_{sq}$. In Fig. 2, we plot the $T_C$ as a function of the $R_{sq}$ for in-situ [8] and ex-situ MoRu and MoRuN films. We used the $R$ at 10 K as $R^N{}_{sq}$ since $R$ was less dependent on temperature around 10K for ultrathin MoRu films with low $T_C$. This figure shows that $T_C$ decreases as $R_{sq}$ for all samples. For MoRu films both series, we fitted the experimental data using expression which applied to homogeneous two-dimensional disordered system with uniform thickness. According to Finkel'stein [3], lowering of $T_C$ from a bulk transition temperature $T_{C0}$ follows the expression:

$$\frac{T_C}{T_{C0}} = \exp\left(-\frac{1}{\gamma}\right)\left[\frac{\left(1+\dfrac{(r/2)^{1/2}}{\gamma - r/4}\right)}{\left(1-\dfrac{(r/2)^{1/2}}{\gamma - r/4}\right)}\right]^{1/\sqrt{2r}}, \quad (1)$$

where $\gamma = 1/\ln\,(k_B T_{c0}\tau_0/\hbar)$ which is characterized by $T_{C0}$ and the transport relaxation time (elastic scattering time) $\tau_0$. $\gamma$ is used as a fitting parameter. And the parameter $r$ is $(e^2/2\pi\hbar)R_{sq}$. We also used this theory on different system: amorphous Bi [7], Mo-alloy[8,9] and NbN [10]. The red and black theoretical curves for $\gamma = -0.17$ and $\gamma = -0.11$ are fitted to initial gradient of experimental data for in-situ and ex-situ MoRu films respectively. The mfp is obtain from $l = v_F\tau_0$, using the above value of $\tau_0$, where the Fermi velocity $v_F$ is estimated from the free electron model. From the value of $\gamma$ the transport relaxation time is evaluated to be the $1.1 \sim 2.2\times10^{-16}$ s which, using a typical Fermi velocity, gives a mfp of about few nano meter.

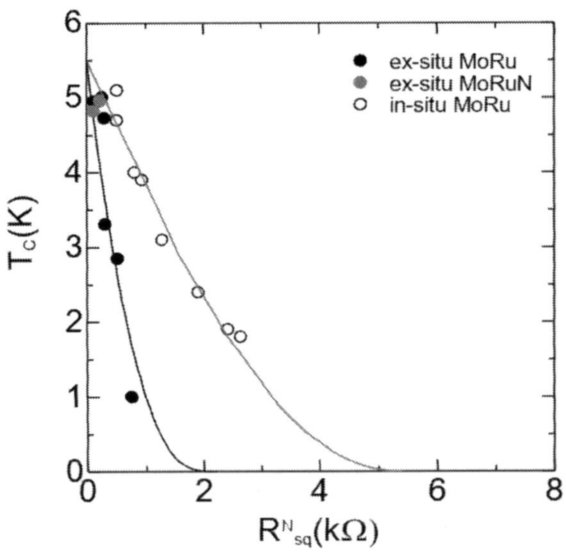


**Figure 2.** The $T_C$ as a function of the $R_{sq}$ for the ex-situ films (closed circle) and in-circle (empty circle) [8], and the red and black solid lines are enhanced Coulomb interaction model [3].

Comparing the reduction of $T_C$ for the two series, we found that change in ex-situ MoRu films are sharper than in-situ MoRu films. By theoretical curve of ref. 3, the value of $R_C$, which means $T_C(R_{sq}^N) = 0$ can be estimated to be nearly 2 kΩ for ex-situ films. On the other hand, it becomes about 5 kΩ for in-

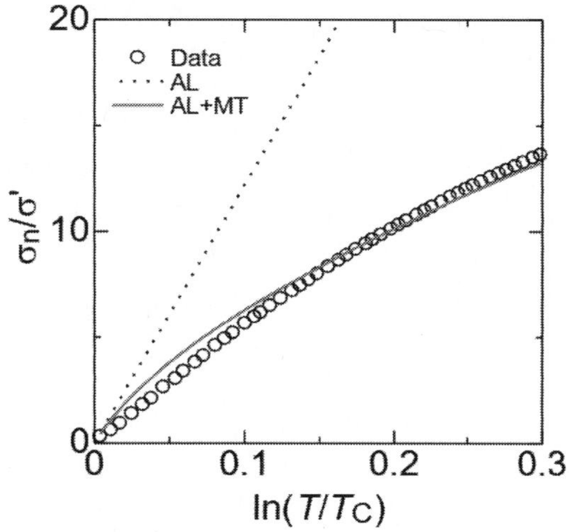


**Figure 3.** Reduced temperature $\ln(T/T_C)$ dependence fluctuation conductivity. The empty circle are experimental data. The dotted line means the contribution from only the AL theory. The Solid line is a fit to the data using Eq. (2).

situ films. This difference may reflects film structure, that is a homogeneous films for in-situ case and granular film for ex-situ case. The R($T$) of ultrathin granular films of conventional superconductors exhibited quasi-reentrant behaviour [8]. In ex-situ films, we can check the film morphology using AFM. On the other hand, it is difficult to observe the film surface of in-situ films using conventional AFM, because the amorphous structure of films change to poly-crystalline one around 180 K. To clarify the difference, we believe that it is necessary to observe the surface while keeping the film at low temperature and in vacuum.

Next, we have carried out the fluctuation analysis to evaluate the quantities. In our present results, the mechanism on suppression of superconductivity might be related to Coulomb interactions using Eq.(1) of Finkel'stein's theory. Therefore, we need to know these localization and interactions effect for $R_{sq}(T)$ of the region above the $T_C$, i.e., para-conductivity regime, where there is enhanced conductivity due to presence of superconducting fluctuation. The superconducting fluctuation theory is denoted as the form of the excess sheet conductance $1/R_{sq}(T)$ - $1/R_{sq}^N(T)$ for a uniform superconducting film in terms of the following fluctuation conductance

$$\frac{1}{R_{sq}(T)} = \frac{1}{R_{sq}^N} + \frac{e^2}{16\hbar}\frac{1}{\ln\left(\frac{T}{T_c}\right)} + \frac{e^2}{8\hbar\ln\left(\frac{T}{T_c}\right) - \delta}\ln\left(\frac{\ln(T/T_c)}{\delta}\right), \tag{2}$$

where, δ is a pair-breaking parameter, and $R_{sq}^N$ is the normal state sheet resistance. The second term can be described by the Aslamazov-Larkin (AL) model of fluctuating Cooperpair[11]. The third term is due to Maki-Thompson contributions by a pair-breaking interaction[12,13]. To analyze present data within this work, we use three parameters to describe each curve: $R_{sq}^N$, $T_C$ and δ. We determined $R_{sq}^N$ from the intercept of the straight line in the plots of $1/R_{sq}(T)$ vs $1/H$ around $T_C$. Figure 3 shows normalized inverse fluctuation conductivity $\sigma_N/\sigma'(= R_{sq}(T)/[R_{sq}^N - R_{sq}(T)])$ as a function of normalized reduced temperature $\ln(T/T_C)$ at zero magnetic field. The dotted and solid lines are determined from the only AL term and AL + MT terms, respectively. Experimental data cannot be fitted with only AL term, represented by the dotted line. Use of AL + MT term, the data can be explained well by the fluctuation theory. We think that the agreement means the homogeneity of films. According to weak localization theory, the $\delta$ is represent by $\delta = \pi\hbar/(8k_B\tau_{in})$ using the inelastic scattering rate $\tau_{in}$ [14]. In addition, the rate of $1/\tau_{in}$ can be expressed as

$$\frac{1}{\tau_{in}} = \frac{1}{\tau_{fluc}} + \frac{1}{\tau_{e-e}} + \frac{1}{\tau_{e-ph}} = \frac{e^2}{2\pi\hbar} R_{sq} k_B T \ln\left(\frac{2\ln 2}{\eta + \beta}\right) + \frac{e^2}{2\pi\hbar} R_{sq} k_B T \ln\left(\frac{\pi\hbar}{e^2 R_{sq}}\right) + \frac{14\pi\varsigma(2)\lambda k_B}{\hbar} \cdot \frac{T^3}{\Theta_D^2}, \quad (3)$$

where $\eta = \ln(T/T_C)$, $\varsigma$ is the zeta function, $\lambda$ is the electron-phonon interaction strength and $\Theta_D$ is the Debye temperature[15,16]. The first, second and third terms on the right side can be expressed as the $\delta_{fluc}$ due to the superconducting fluctuation effect, $\delta_{e-e}$ due to the electron-electron interaction, and $\delta_{e-ph}$ due to the electron-phonon interaction, respectively. The first and second term are given by $R_{sq}^N$ at a fixed value of $\eta$. The third term is the origin of $\delta_0$ in the above expression $\delta = \delta_0 + \alpha(R_{sq}^N)^\beta$. For details of the estimation procedure on of $\delta$, see our previous work [9]. As a results, we obtain that the red solid line in Fig. 4 is calculated from eq. (3) at $\eta = 0.1$ under the assumptions $\Theta_D = 750$ K which is the same as that of NbN [17] and MoN [9] and $\lambda = 1$. It is similar that the $R_{sq}^N$ dependence of $\delta$ show the effect of weak localization and Coulomb interaction in spite of there is deference of $R_C$ between in-situ films and ex-situ film. Though we do not know yet the $R_{sq}^N$ dependence. To obtain the qualitative explanation, it is need to study the magneto-conductance for clarification of the $R_{sq}^N$ dependence of $\tau_{in}$. We will discuss an analysis of magneto resistance for $\tau_{in}$ in the near future.

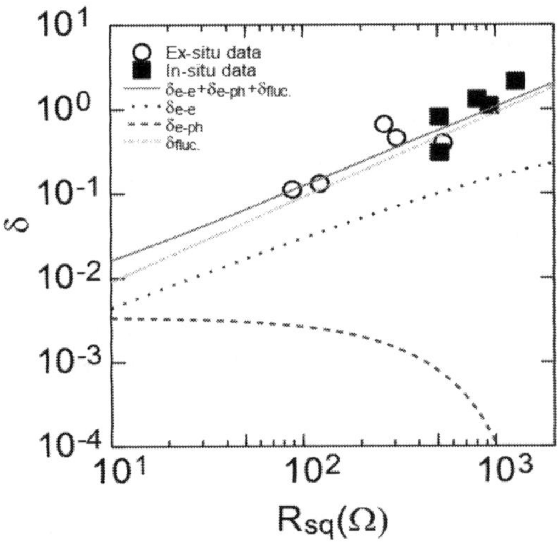


**Figure 4.** $R_{sq}$ dependence of the $\delta$ as a function of the $R_{sq}$ for ex-situ and in-situ MoRu films.

## 4. Summary

We have demonstrated S-I transition using MoRu and MoRuN films with various thickness deposited by reactive DC magnetron sputtering method. In addition, we compared the experimental results between in-situ and ex-situ films. As for $R_{sq}$ dependence of $T_C$, it seems that the data can be explained by enhanced Coulomb interaction theory. It is important to note that this result for the thickness-tuned SI transition has been obtained for MoRu films . Since there is no report, as far as we know, on the electron tunneling measurement of MoRu films near the thickness-tuned S–I transition, it would give us further information as to the effects due to the difference of films deposition method. Although we believe that this difference come from film structure, a full understanding of the thickness-tuned SI transition in a thin MoRu film would need further investigation from the point of view of the electrical transport properties combined film structure analysis.

**References**

[1] Goldman A M 2010 *Low Temp. Phys.* **36** 884
[2] Crauste O, Gentlis A, Couëdo F, Dolgorousky Y, Bergé, Collin S, Marrache-Kikuchi C A and Dumoulin 2013 *Phys. Rev. B* **87** 144514
[3] Finkel'stein A M 1987 *JETP Lett.* **45** 46
[4] Fisher M P A 1990 Phys. Rev. Lett. **65** 923
[5] Collver M M and Hammond R H 1973 *Phys. Rev. Lett.* **30** 92
[6] Rathnakaya K D D, Waston P W, Valys D T, Wang P S and Naugle D G 1994 Physica B **194-196** 1953
[7] Makise K, Kawaguti T and Shinozaki B 2007 *Physica E* **39** 30
[8] Hirakawa A, Makise K, Kawaguti T and Shinozaki B 2008 *J. Phys.: Condens. Matter* **20** 485206
[9] Tsuneoka T, Makise K, Maeda S, Shinozaki B and Ichikawa F 2017 *J. Phys.: Condens. Matter* **29** 015701
[10] Ezaki S, Makise K, Shinozaki B, Odo T, Asano T, Terai H, Yamashita T, Miki S and Wang Z 2012 *J. phys.:Condense. Matter.* **24** 475702
[11] Aslamazov L G and Larkin A I 1968 *Phys. Lett. A* **26** 238
[12] Maki K 1968 *Prog. Theor. Phys.* **39** 897
[13] Thompson R S 1970 *Phys. Rev. B* **1** 327
[14] Ebisawa H, Maekawa S and Fukuyama H 1983 *Solid State Commun.* **45** 75
[15] Altshuler B L, Arnonov A G and Khmelnitsky 1982 *J. Phys* C **15** 7367
[16] Brenig W, Chang M C and Abrahams 1985 *Phys. Rev. B* **31** 7001
[17] Zou Y, Wang X, Chen T, Li X, Qi X, Welch D, Zhu P, Liu B, Cui T and Li B 2015 *Sci. Rep.* **5** 10811

IOP Conf. Series: Journal of Physics: Conf. Series **969** (2018) 012027 doi:10.1088/1742-6596/969/1/012027

# Superconducting state in $(Eu_{1-x}Ca_x)RbFe_4As_4$ with 1144-type Structure

**K. Kawashima[1,2], S. Ishida[2], K. Oka[2], H. Kito[2], N. Takeshita[2], H. Fujihisa[2], Y. Gotoh[2], K. Kihou[2], H. Eisaki[2], Y. Yoshida[2] , and A. Iyo[2]**

[1] IMRA Material R&D Co., Ltd., Kariya, Aichi 448-0032, Japan
[2] National Institute of Advanced Industrial Science and technology (AIST), Tsukukba, Ibaraki 305-8568, Japan
E-mail: kenji.kawashima@aisin.co.jp

**Abstract.** We report the Ca substitution effect of the Fe-based superconductor $EuRbFe_4As_4$ with 1144-type structure (crystal system: tetragonal, space group: *P*4/mmm (No. 123)), in order to elucidate the relationship between the superconductivity and the magnetic order. The lattice constant systematically changed with the Ca concentration $x$. By Ca-substitution, the magnetic order is suppressed, while $T_c$ is almost unchanged. The results indicate that the superconductivity of $EuRbFe_4As_4$ is not sensitive to the magnetic order.

## 1. Introduction

Since the superconductivity in La(O,F)FeAs has discovered[1)], variety of Fe-based superconductors have been reported. . Recently discovered $AeAFe_4As_4$ ($Ae$ = Ca, Sr, Ba, Eu, $A$ = K, Rb, Cs), which is so-called 1144-type compounds, is one of the Fe-based superconductors[2-5)]. $AeAFe_4As_4$ has a tetragonal structure (space group: *P*4/mmm (No. 123)) that consists of alternate stacking of two inequivalent $ThCr_2Si_2$ structures, namely, $AeFe_2As_2$ and $AFe_2As_2$ (inset of Fig. 1(a)) and shows the superconductivity with superconducting transition temperature: $T_c$ = 31 ~ 36 K. The $T_c$ values of 1144 compounds are closed to that of 122-type superconductors, $Ae_{1-x}A_xFe_2As_2$ ($Ae$ = Ca, Sr, Ba, Eu, $A$ = Na, K)[7-9)]. However, the $Ae$ : $A$ ratio in the 1144-type crystal structure is fixed to be 1 : 1 , because the $Ae$ and the $A$ ions do not occupy crystallographically equivalent sites in the 1144-type crystal structure. In the case of $EuAFe_4As_4$ ($A$ = Rb, Cs), besides the bulk superconductivity with $T_c$ = ~36 K, the magnetic transition takes place at $T_m$ = ~15 K, indicating the coexistence of the superconductivity and the magnetic order. The magnetic order comes from the localized spin on $Eu^{2+}$ [3-5)]. The magnetic transition was only confirmed in the magnetic susceptibility data. In the case of the related 122-type compound, $EuFe_2As_2$, it is known that the antiferromagnetic order shows up below 20 K arising from the $Eu^{2+}$ ions[10)], and this magnetic order influences the superconductivity. By applying high pressure, zero resistivity was observed in a narrow pressure range of 2.6 GPa < $P$ < 2.7 GPa[11)]. By substituting non-magnetic elements $Ca^{2+}$ for $Eu^{2+}$, the magnetic order is suppressed and zero resistivity is observed in a wide pressure range of 1.2 GPa < $P$ < 2.4 GPa compared to pure $EuFe_2As_2$[12)]. In $EuFe_2As_2$, the antiferromagnetic order likely competes with the superconductivity. In $EuAFe_4As_4$, the relationship between the magnetic order and the superconductivity remains has not been investigated so far.

In this paper, we investigate the physical properties of the Ca-substituted samples of $(Eu_{1-x}Ca_x)RbFe_4As_4$ to clarify the competition between the superconductivity and the magnetic order in the 1144 system.

## 2. Experimental details

Polycrystalline samples of $Eu_{1-x}Ca_xRbFe_4As_4$ ($x$ = 0.0, 0.25, 0.5, 0.75, 1.0) were synthesized by a conventional solid state reaction using the stainless steel (SUS) pipe and cap method as described in Ref. 7. Details of the synthesis condition are given in Ref. 2. A powder x-ray diffraction (PXRD) pattern was measured at room temperature using CuKα radiation to evaluate the composition dependence of the lattice parameters. Intensity data were collected over a $2\theta$ range from 5 to 80° at 0.01° step width. Magnetic susceptibility measurements were performed under a magnetic field $H$ of 10 Oe using a magnetic-property measurement system (MPMS) (Quantum Design, MPMS-XL7). Data were collected during warming after zero-field cooling (ZFC) and then during field cooling (FC).

## 3. Results and discussions

Figure 1(a) shows the PXRD pattern of the $Eu_{0.5}Ca_{0.5}RbFe_4As_4$ sample. The main peaks can be indexed by employing a primitive tetragonal structure with space group $P4/mmm$ (No. 123), indicating that $Eu_{0.5}Ca_{0.5}RbFe_4As_4$ possesses the 1144-type crystal structure. There are extra reflections due to 122-type $RbFe_2As_2$ and $(Eu,Ca)Fe_2As_2$ impurity phase[14,15]. Similar characteristic diffraction patterns were also observed for other synthesized samples, ensuring that the samples have the 1144-type crystal structure. The calculated lattice constants $a$ and $c$ are plotted in Fig. 1(b), which decrease linearly with increasing $x$, as expected from the Vegard's low. No structural phase transition is recognized by substituting Ca for Eu.

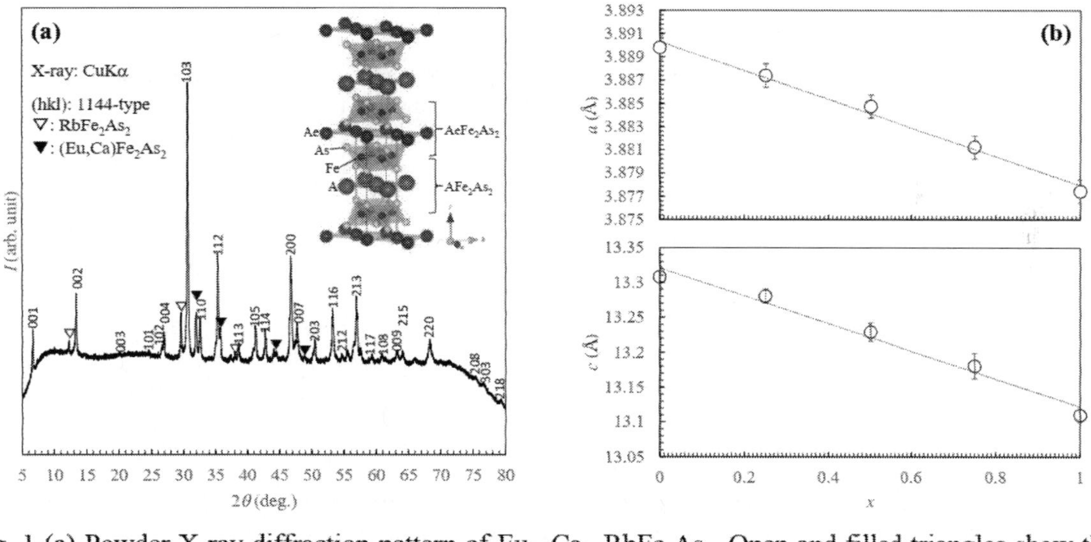


Fig. 1 (a) Powder X-ray diffraction pattern of $Eu_{0.5}Ca_{0.5}RbFe_4As_4$. Open and filled triangles show the impurity phases of $RbFe_2As_2$ and $(Eu,Ca)Fe_2As_2$, respectively. Inset shows the crystal structure of $AeAFe_4As_4$ (Program VESTA was used[13]). (b) Ca concentration $x$ dependence of the $a$- and $c$-axis lattice constants of $Eu_{1-x}Ca_xRbFe_4As_4$.

Figure 2(a) shows the temperature dependence of the magnetic susceptibility $\chi$ of $Eu_{1-x}Ca_xRbFe_4As_4$. The magnetic susceptibility data of all samples exhibits a large Meissner diamagnetic signal at approximately 36 K in both ZFC and ZC data, indicating the occurrence of bulk superconductivity. Because $T_c$ of the $RbFe_2As_2$ and $(Eu,Ca)Fe_2As_2$ impurity phase is below 2.6 K under ambient pressure[14,15], these impurity phases do not contribute for $T_c$ as high as 36 K. The kink behaviour in the magnetic susceptibility is observed below $T_c$ at $T$=15K for $x$ = 0.0, $T$=12K for $x$ =0.25, and $T$=7K for $x$

=0.5 , respectively. These features are the signatures of the magnetic phase transition and the transition temperature, $T_m$, decreases with increasing Ca concentration $x$ (Fig. 2(b)). For the $x$ = 0.25 sample, the magnetic susceptibility slightly increases near the lowest temperature. Observed $T_m$ values of $Eu_{1-x}Ca_xRbFe_4As_4$ are different from those of $Eu_{1-x}Ca_xFe_2As_2$ ($x \leqq 0.5$)[12,15], indicating that the behaviour is not due to the $Eu_{1-x}Ca_xFe_2As_2$ impurity phase.

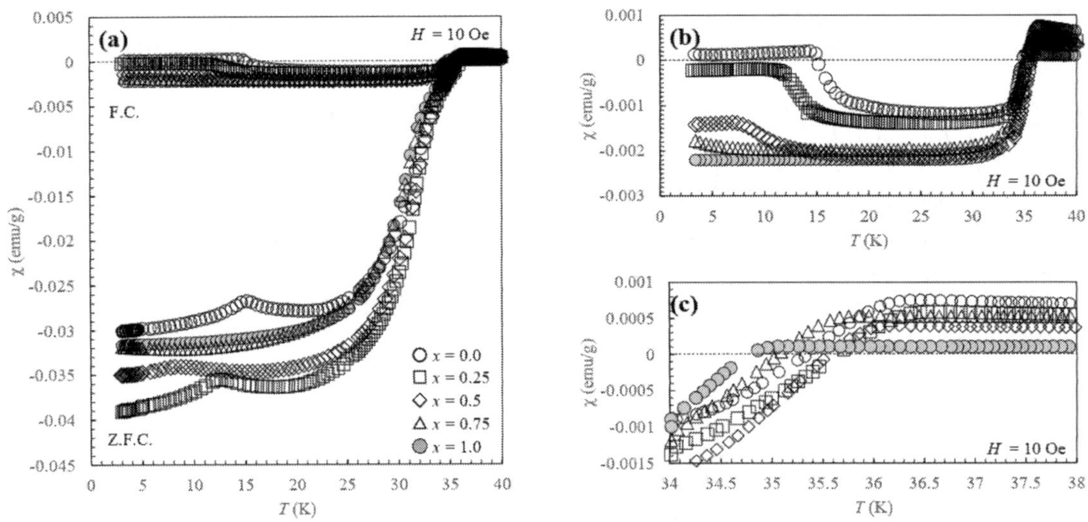


Fig. 2 (a) Magnetic susceptibility $\chi$ of $Eu_{1-x}Ca_xRbFe_4As_4$ as a function of temperature $T$. Enlarged view of the F.C. data and near $T_c$ are shown in Fig. 2(b) and Fig. 2(c), respectively.

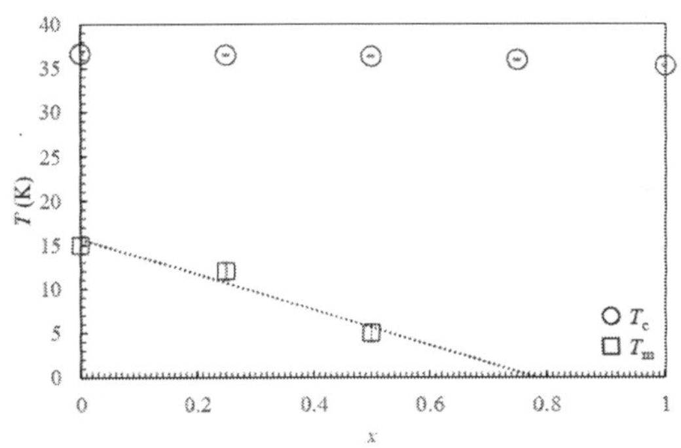


Fig. 3 $T_c$ and $T_m$ of $Eu_{1-x}Ca_xRbFe_4As_4$ as a function of Ca concentration $x$. $T_c$ and $T_m$ were defined from the $\chi$ -$T$ data of Fig. 2(a).

Figure 3 shows $T_c$ and $T_m$ as a function of $x$. Although $T_m$ monotonously decreases with increasing $x$, $T_c$ is almost unchanged. This behaviour is in contrast to the case of $Eu_{1-x}Ca_xFe_2As_2$, in which pressure-induced superconductivity shows up at lower pressures in the Ca-substituted samples. It is noted that the $c$-axis length of $EuFe_2As_2$ sublattice in $EuRbFe_4As_4$ is larger compared to $EuFe_2As_2$[4]. The longer distance between the Eu layer and the FeAs-layer in $EuRbFe_4As_4$,would weaken the influence of the $Eu^{2+}$ magnetic order to the superconductivity compared to $EuFe_2As_2$.

## 4. Summary

We synthesized the polycrystalline sample of $Eu_{1-x}Ca_xRbFe_4As_4$ and characterized their properties. The lattice constants show the linear Ca concentration $x$ dependence. Superconductivity was confirmed at around 36 K in all synthesized samples. The magnetic order shows up below 15K for $EuRbFe_4As_4$, which decreases with increasing $x$. On the other hand, $T_c$ does not change with $x$. These behaviors suggest that the magnetic order in the Eu-layer have little influence on the superconductivity in the 1144system.

## ACKNOWLEDGEMENT

This work was partially supported by JSPS KAKENHI Grant Number JP16H06439.

## References

[1] Y. Kamihara, T. Watanabe, M. Hirano, and H. Hosono, "Iron-Based Layered Superconductor La[$O_{1-x}F_x$]FeAs ($x$ = 0.05 - 0.12) with $T_c$ = 26 K," J. Am. Chem. Soc. **130** 3296 (2008).

[2] A. Iyo, K. Kawashima, T. Kinjo, T. Nishio, S. Ishida, H. Fujihisa, Y. Gotoh, K. Kihou, H. Eisaki, and Y. Yoshida, "New-Structure-Type Fe-Based Superconductors: $CaAFe_4As_4$ (A = K, Rb, Cs) and $SrAFe_4As_4$ (A = Rb, Cs)," J. Am. Chem. Soc. **138** 3410 (2016).

[3] K. Kawashima, T. Kinjo, T. Nishio, S. Ishida, H. Fujihisa, Y. Gotoh, K. Kihou, H. Eisaki, Y. Yoshida, and A. Iyo, "Superconductivity in Fe-Based Compound $EuAFe_4As_4$ (A = Rb and Cs)" J. Phys. Soc. Jpn. **85** 064710 (2016).

[4] Y. Liu, Y.-B. Liu, Z.-T. Tang, H. Jiang, Z.-C. Wang, A. Ablimit, W.-H. Jiao, Q. Tao, C.-M. Feng, Z.-A. Xu, and G.-H. Cao, "Superconductivity and ferromagnetism in hole-doped $RbEuFe_4As_4$" Phys. Rev. B **93** 214503 (2016).

[5] Y. Liu, Y.-B. Liu, Q. Chen, Z.-T. Tang, W.-H. Jiao, Q. Tao, Z.-A. Xu, G.-H. Cao, "A new ferromagnetic superconductor: $CsEuFe_4As_4$," Sci. Bull. **61**(15) 1213 (2016).

[6] K. Zhao, Q. Q. Liu, X. C. Wang, Z. Deng, Y, X. Lv, J. L. Zhu, F. Y. Li and C. Q. Jin, "Superconductivity above 33 K in $(Ca_{1-x}Na_x)Fe_2As_2$," J. Phys.: Condens. Matter **22** 222203 (2010).

[7] N. Shinohara, K. Tokiwa, H. Fujihisa, Y. Gotoh, S. Ishida, K. Kihou, C. H. Lee, H. Eisaki, Y. Yoshida, and A. Iyo, "Synthesis, structure, and phase diagram of $(Sr_{1-x}Na_x)Fe_2As_2$ superconductors," Supercond. Sci. Technol. **28** 062001 (2015).

[8] S. Avci, O. Chmaissem, D. Y. Chung, S. Rosenkranz, E. A. Goremychkin, J. P. Castellan, I. S. Todorov, J. A. Schlueter, H. Claus, A. Daoud-Aladine, D. D. Khalyavin, M. G. Kanatzidis, and R. Osborn, "Phase diagram of $Ba_{1-x}K_xFe_2As_2$," Phys. Rev. B **85** 184507 (2012).

[9] Y. Qi, L. Wang, Z. Gao, X. Zhang, D. Wang, C. Yao, C. Wang, C. Wang, and Y. Ma, "Superconductivity and upper fields in Na-doped iron arsenides $Eu_{1-x}Na_xFe_2As_2$," New J. Phys. **14** 033011 (2012).

[10] J. Herrero-Martín, V. Scagnoli, C. Mazzoli, Y. Su, R. Mittal, Y. Xiao, T. Brueckel, N. Kumar, S. K. Dhar, A. Thamizhavel, and L. Paolasini, "Magnetic structure of $EuFe_2As_2$ as determined by resonant x-ray scattering," Phys. Rev. B **80** 134411 (2009).

[11] T. Terashima, M. Kimata, H. Satsukawa, A. Harada, K. Hazama, S. UJI, H. S. Ssuzuki, T. Matsumoto, and K. Murata, "$EuFe_2As_2$ under High Pressure: An Antiferromagnetic Bulk Superconductor," J. Phys. Soc. Jpn. **78**(8) 083701 (2009).

[12] A. Mitsuda, T. Matoba, F. Ishikawa, Y. Yamada, and H. Wada, "Pressure-Induced Superconductivity in $Eu_{0.5}Ca_{0.5}Fe_2As_2$: Wide Zero-Resistivity Region Due to Suppression of Eu Magnetic Order and Chemical Pressure," J. Phys. Soc. Jpn. **79**(7) 073704 (2010).

[13] K. Momma and F. Izumi, "VESTA 3 for three-dimensional visualization of crystal, volumetric and morphology data," J. Appl. Cryst. **44** 1272 (2011).

[14] Z. Bukowski, S. Weyeneth, R. Puzniak, J. Karpinski, and B. Batlogg, "Bulk superconductivity at 2.6 K in undoped $RbFe_2As_2$," Physica C **470** S328 (2010).
[15] A. Mitsuda, S. Seike, T. Matoba, H. Wada, F. Ishikawa and Y. Yamada, "Competition between Fe-based superconductivity and antiferromagnetism of $Eu^{2+}$ in $Eu_{1-x}Ca_xFe_2As_2$," J. Phys.: Confer. Seri. **273** 012100 (2011).

# Possibility of Cooper-pair formation controlled by multi-terminal spin injection

**K Ohnishi[1,2], M Sakamoto[1], M Ishitaki[1], and T Kimura[1,2]**

[1] Department of Physics, Kyushu University, Fukuoka 819-0395, Japan
[2] Research Center for Quantum Nano-Spin Sciences, Kyushu University, Fukuoka 819-0395, Japan

E-mail: kohnishi@phys.kyushu-u.ac.jp

**Abstract.** A multi-terminal lateral spin valve consisting of three ferromagnetic nanopillars on a Cu/Nb bilayer has been fabricated. We investigated the influence of the spin injection on the superconducting properties at the Cu/Nb interface. The non-local spin valve signal exhibits a clear spin insulation signature due to the superconducting gap of the Nb. The magnitude of the spin signal is found to show the probe configuration dependence. From the careful analysis of the bias current dependence, we found the suppression of the superconductivity due to the exchange interaction between the Cooper pair and accumulated spin plays an important role in the multi-terminal spin injections. We also discuss about the possibility of the Cooper-pair formation due to the spin injection from the two injectors with the anti-parallel alignment.

## 1. Introduction

Ferromagnet/superconductor hybrid structure is a fantastic and attractive system because the combination between the spin and super-conducting transports is expected to produce various intriguing phenomena[1–7]. However, superconducting properties are, in general, easily destroyed by the direct contact with the ferromagnetic material[8–10]. This is because the Cooper pairs are deformed by the strong exchange interaction from the ferromagnet and/or other extrinsic effects such as the interface mixing and the diffusion of the magnetic impurities. These serious obstacles can be solved by introducing a normal metal spacer between the ferromagnetic metal and superconductor. Indeed, a conventional lateral spin valve combined with a superconductor enable us to inject the pure spin current into the superconductor[11, 12]. The advantage of the electrical spin injection is its precise tunability of the magnitude of the spin accumulation. However the relatively large electrical resistivity of the ferromagnetic injector produces non-negligible Joule heating, resulting in the suppression of the superconductivity at the interface[11]. To reduce the Joule heating from the ferromagnet, we have developed a lateral spin valve consisting of a nano-pillar spin injector/detector and the Cu/Nb bilayer channel. By using this novel device structure, we showed that the superconducting Nb acts as an insulator for a pure spin current and that the superconducting proximity effect is induced also in the Cu layer with the non-equilibrium spin accumulation[13]. Thus, a nano-pillar-based lateral spin valve with a normal metal/superconductor bilayer provides an ideal platform for investigating the interplay between superconductivity and ferromagnetism. In this paper, by extending the nano-pillar spin injector to the multi-terminal spin injectors as shown in Fig. 1, we control the

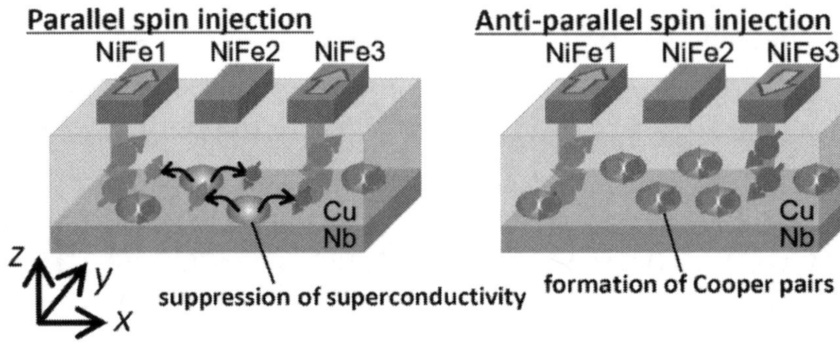


**Figure 1.** Schematic illustration of multi-terminal spin injection into a Cu/Nb bilayer. (a) Suppression of the superconductivity under parallel spin injection and (b) formation of the Cooper pairs under anti-parallel spin injection.

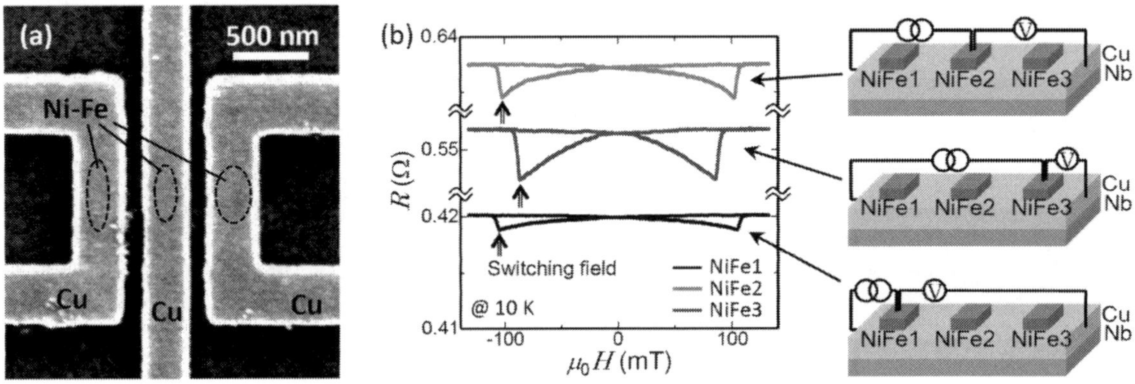


**Figure 2.** (a) Scanning electron microscope image of the measured sample. (b) Anisotropic magnetoresistance of each injectors. The recovery of the resistance to the high resistance state indicates that the flip of the magnetization.

spin current in a Cu/Nb bilayer channel by the magnetic alignment of the injectors and seek a novel control method of the superconducting properties using the spin injection.

## 2. Experimental

We have fabricated a nanopillar based lateral spin valve consisting of three Ni-Fe nanopillars and Cu/Nb bilayer channel. First, a Ni-Fe/Cu/Nb trilayer film as a ferromagnet/normal metal/superconductor by using a high vacuum thin-film growth system consisting of two magnetron sputterings and a Joule evaporator with the base pressure less than $5 \times 10^{-5}$ Pa. Here, the Ni-Fe and Nb are grown by the magnetron sputterings while the Cu is deposited by a Joule evaporation. The thicknesses for the Ni-Fe, Cu and Nb are, respectively, 20 nm, 100 nm and 45 nm. Then, the trilayer film was patterned into the nanopillar structure by using electron-beam lithography and Ar ion milling. Figure 2(a) is the scanning electron microscope image of the measured sample. The lateral dimensions for two ferromagnetic nanopillars are designed as $320 \times 60\,\mathrm{nm}^2$, $210 \times 50\,\mathrm{nm}^2$ and $210 \times 80\,\mathrm{nm}^2$, and the center-center distance between the neighboring ferromagnetic nanopillars is 500 nm. Here, the superconducting transition temperature of the fabricated Nb was 6.8 K.

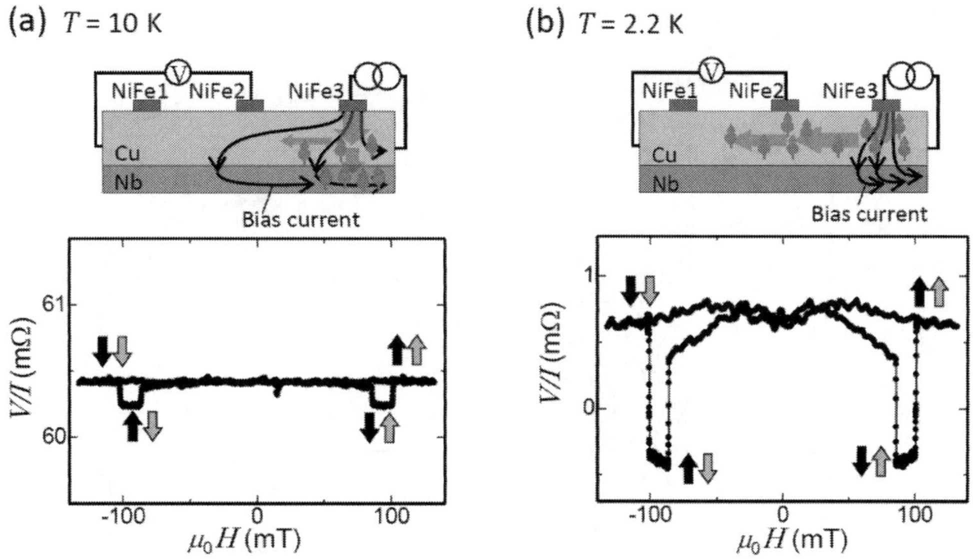


**Figure 3.** Schematic illustration of the spin accumulation in the Cu/Nb bilayer and magnetic field dependence of the non-local resistance at (a) 10 K and (b) 2.2 K. The black and gray arrows indicate the magnetization of the detector and the injector, respectively. The spin signal at 2.2 K is much larger than that at 10 K, which indicates the spin current absorption is effectively protected by the superconducting gap.

We investigate the spin transport properties of the Cu/Nb bilayer by measuring the non-local spin valve signals[14–16]. Here, the left- and right-hand-side Ni-Fe nanopillars NiFe1 and NiFe3 were used as the spin injectors and the middle Ni-Fe nanopillar NiFe2 was used as the spin detector. There are several advantages in this device structure. First, the interfaces have ideal conditions because the trilayer was prepared by in-situ multi-layered film grown system. As second advantage, a large spin accumulation can be generated by flowing a large current because the quasi-two-dimensional Cu heat sink strongly suppresses the temperature increase due to the spin injection. Third advantage is the variety of the magnetic configuration owing to the combination of triple ferromagnetic nanopillars with different switching fields.

## 3. Results and Discussion

First, we confirmed that three Ni-Fe nanopillars have different switching fields by measuring the anisotropic magnetoresistance (AMR) of each nanopillar with the probe configurations shown in the inset of Fig. 2(b). Figure 2(b) shows the AMR curves for three Ni-Fe nanopillars measured at 10 K under the magnetic field along the $y$ axis. Since the magnetization is laying in $xy$ plane, the magneto-resistance measured by the current along the $z$ axis should show no field dependence of the resistance. However, the inhomogeneous current distribution produces a small in-plane component of the current, resulting in a small field dependence of the resistance. As can be seen in Fig. 2(b), the magnetizations of three Ni-Fe nanopillars showed the different switching fields depending on their geometrical shape. Thus, the magnetization configuration can be controlled by the external magnetic field.

Next, we evaluate the spin transport properties at the Cu/Nb interface by measuring the non-local spin valve signals with the different temperature. Figures 3(a) and 3(b) show the non-local spin signals measured at 10 K and 2.2 K, respectively. Although the spin signal at 10 K

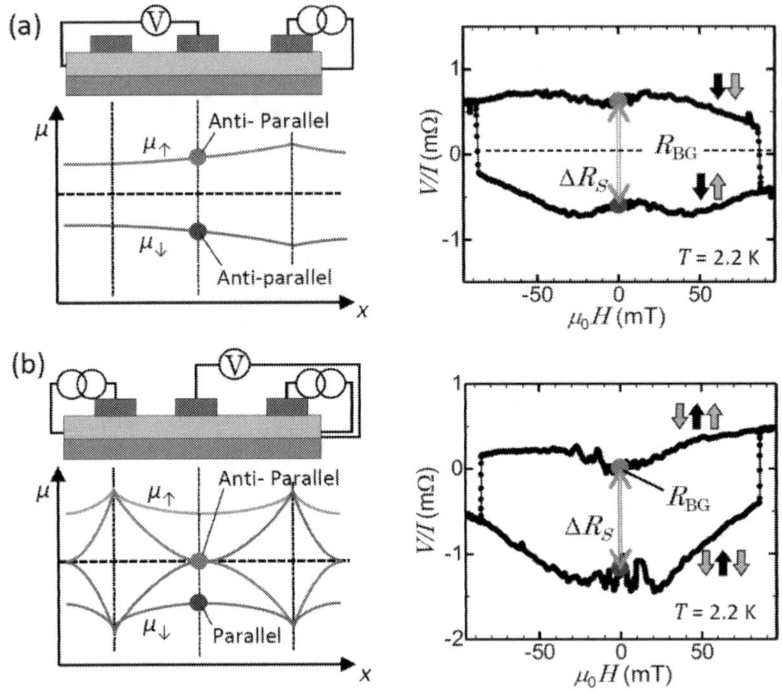


**Figure 4.** Electrochemical potentials in the Cu layer and the measured non-local resistance under (a) single- and (b) two-terminal spin injection. The black and gray arrows indicate the magnetization of the detectors and the injectors, respectively. The spin signal $\Delta R_S$ is defined by the difference of the resistance at the zero magnetic field. $R_{\mathrm{BG}}$ is the background resistance, which indicates the non-local resistance without spin accumulation.

was as small as 0.18 m$\Omega$ because of the strong spin absorption effect of the Nb metal, the spin signal at 2.2 K becomes 6 times larger than that at 10 K. This significant increase of the spin signal at lower temperature clearly indicates that the spin current absorption was effectively protected by the superconducting gap of the Nb at the interface. Here, the non-local spin signal at 2.2 K shows unconventional broad field dependence, where the resistance change between the parallel and anti-parallel states decreases with increasing the magnitude of the magnetic field. This can be understood by the reduction of the spin signal because the superconducting gap at the Nb/Cu interface decreases by the application of the magnetic field, leading to the less protection of the spin absorption. We also point out the relatively large background resistance in Fig. 3(a) because of the spreading of the injecting current in the two-dimensional Cu film. This background resistance is significantly suppressed by the superconductivity, as shown in Fig. 3(b).

To investigate the influence of the spin accumulation on the superconducting properties, we perform two kinds of the spin injection. One is a conventional single spin injection using a NiFe3 injector and NiFe2 detector, as schematically shown in Fig. 4(a). The other one is a dual spin injection (two-terminal injection) by using NiFe1 and NiFe3 injectors with NiFe2 detector, as schematically shown in Fig. 4(b). Figures 4(a) and 4(b) show the representative spin signals measured at 2.2 K with the low bias current of 0.26 mA. To evaluate the magnitude of the spin accumulation precisely, we measured the minor loops. For the two-terminal measurement, the vertical axis is the voltage divided by the current for each electrode. Therefore, the overall

IOP Conf. Series: Journal of Physics: Conf. Series **969** (2018) 012028 doi:10.1088/1742-6596/969/1/012028

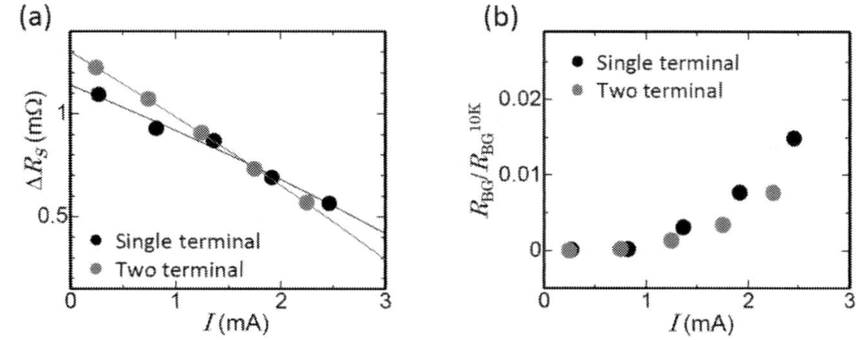


**Figure 5.** Bias current dependence of (a) spin signal $\Delta R_S$ and (b) background resistance $R_{\rm BG}$. The black and red circles indicate the signals with single- and two-terminal spin injections, respectively, and $R_{\rm BG}$ is normalized by the value at 10 K. The solid lines are the fitting curves.

resistance change for the two-terminal spin injection is approximately twice of the single-terminal one.

We then investigate the bias current dependence of the spin transport properties. Since the heating effect under the electrical spin injection is mainly generated from the Ni-Fe nanopillar, we assume that the influence of the Joule heating can be characterized by the local current of each electrode[17]. Therefore, we expect that the influence of the Joule heating under the two-terminal spin injection is almost same as that for the single-terminal one when the current flowing in each nanopillar is the same value. On the other hand, the influence of the spin accumulation for the two-terminal spin injection should be two times larger than that for the single-terminal one. So, these two effects should reduce the spin signal because both the temperature increase and the magnetic field suppress the superconductivity. However, the current dependence should show the different signature. Namely, the influence of the Joule heating is proportional to $I^2$ while that of the spin accumulation should be proportional to $I$. Therefore, by carefully analyzing the bias current dependence of the spin signals both for the single- and two-terminal spin injections, we can clarify the influence of the spin accumulation on the superconducting property.

Figure 5(a) shows the bias current dependences of the spin signals for the single- and two-terminal spin injections. Here, we plot the spin signal at the zero magnetic field $\Delta R_S$, in order to remove the extrinsic effect of the spin accumulation. As can be seen in the plot, the reduction of the spin signal for the two-terminal spin injection was found to be larger than that for the single-terminal spin injection. This is consistent with our expectation that the spin accumulation destroys the superconductivity. To analyze this more quantitatively, the bias current dependence was fitted with the following equation.

$$\Delta R_S = \Delta R_{S0} - aI^2 - bI, \qquad (1)$$

where $\Delta R_{S0}$ is the spin signal at zero bias, $a$ and $b$ are the coefficients for the Joule heating and spin accumulation, respectively. From the fittings, for the single case, $a$ and $b$ are $1.01 \times 10^{-2}\,\mathrm{m\Omega/mA^2}$ and $0.21\,\mathrm{m\Omega/mA}$, respectively. For the two-terminal spin injection case, we obtained that $a$ and $b$ are $1.06 \times 10^{-2}\,\mathrm{m\Omega/mA^2}$ and $0.30\,\mathrm{m\Omega/mA}$, respectively. Thus, the influence of the Joule heatings for both configurations were almost same. On the other hand, the influence of the spin accumulation in two-terminal spin injection is larger than that of the single spin injection, by a factor of 1.4. This is a clear evidence that the spin accumulation effectively suppresses superconductivity. We believe that the exchange interaction between the spin-polarized electron and the Cooper pair is a microscopic origin of this reduction, as schematically shown in Fig. 1(a).

Finally, we investigate the possibility of the Cooper-pair formation, by injecting the electrons with anti-parallel spins, as shown in Fig. 1(b). Since the two-terminal spin injection with the anti-parallel configuration does not induce the detectable spin signal, the spin signal does not provide any evidence of the Cooper-pair formation due to the anti-parallel spin injection. Therefore, we discuss about the bias current dependence of the background resistance $R_{BG}$, in order to pick up the signature of the Cooper-pair formation. As shown in Fig. 5(b), $R_{BG}$ gradually increases with increasing the bias current both for the single and two-terminal injections. However, for the two-terminal injection, in spite of the larger total current, the increase of $R_{BG}$ was smaller than that for the single terminal injection. This indicates that the suppression of the superconductivity was weaker in the two-terminal injection than that in the single-terminal injection. Since the effective conversion from the electrons to the Cooper pair tends to protect the spin absorption, the less increase of $R_{BG}$ suggests the effective Cooper-pair formation under the two-terminal spin injection with anti-parallel orientation. Additionally, as shown in Fig. 4(b), the broad field dependence of the non-local signal for the anti-parallel configuration seems to be smaller than that for the parallel configuration. Since the Cooper-pair formation effectively occurs under the anti-parallel configuration, the observed difference between two configurations can be explained by the difference in the formation rate. These results imply that the two-terminal spin injection with the anti-parallel orientation assists the Cooper-pair formation.

## 4. Summary

We have investigated the influence of the electrical spin injection from the multi-terminal Ni-Fe nanopillars on the superconducting properties in the Nb/Cu bilayer film. By comparing the non-local spin valve signal of the single-terminal spin injection with that of the two-terminal spin injection, we are able to analyze the influence of the spin injection and the Joule heating on the superconductivity separately. The non-equilibrium spin accumulation is found to suppress the superconductivity of the Nb. However, its magnitude is not simply proportional to the number of the injectors. This implies that the Cooper-pair formation process can be controlled by injecting the spin-polarized electrons.

## Acknowledgments

This work was partially supported by Grant-in-Aid for Scientific Research on Innovative Area, "Nano Spin Conversion Science" (26103002), that for Scientific Research (S) (25220605), and that for Young Scientists (B) (17K14109).

## References

[1] Cadden-Zimansky P and Chandrasekhar V 2006 *Phys. Rev. Lett.* **97** 237003
[2] Giazotto F, Taddei F, Beltram F and Fazio R 2006 *Phys. Rev. Lett.* **97** 087001
[3] Giazotto F and Taddei F 2008 *Phys. Rev. B* **77** 132501
[4] Miao G X, Ramos A V and Moodera J S 2008 *Phys. Rev. Lett.* **101** 137001
[5] Cadden-Zimansky P, Wei J and Chandrasekhar V 2009 *Nat. Phys.* **5** 393–397
[6] Eschrig M 2011 *Phys. Today* **64** 43
[7] Linder J and Robinson J W A 2015 *Nat. Phys.* **11** 307–315
[8] Izyumov Y A, Proshin Y N and Khusainov M G 2002 *Physics-Uspekhi* **45** 109–148
[9] Buzdin A I 2005 *Rev. Mod. Phys.* **77** 935–976
[10] Bergeret F S, Volkov A F and Efetov K B 2005 *Rev. Mod. Phys.* **77** 1321–1373
[11] Ohnishi K, Kimura T and Otani Y 2010 *Appl. Phys. Lett.* **96** 192509

[12] Wakamura T, Hasegawa N, Ohnishi K, Niimi Y and Otani Y 2014 *Phys. Rev. Lett.* **112** 036602
[13] Ohnishi K, Ono Y, Nomura T and Kimura T 2014 *Sci. Rep.* **4** 6260
[14] Jedema F J, Filip A T and van Wees B J 2001 *Nature* **410** 345–348
[15] Kimura T, Hamrle J and Otani Y 2005 *Phys. Rev. B* **72** 014461
[16] Takahashi S and Maekawa S 2003 *Phys. Rev. B* **67** 052409
[17] Nonoguchi S, Nomura T and Kimura T 2012 *Appl. Phys. Lett.* **100** 132401

# The electrical resistance of gold-capped chromium thin films

**Masashi Ohashi[1,2], Masaki Sawabu[2], Kohei Ohashi[2], Masahiro Miyagawa[2], Kae Maeta[2], Takahide Kubota[3] and Koki Takanashi[3]**
[1]Institute of Science and Engineering, Kanazawa University, Kakuma-machi, Kanazawa, 920-1192, Japan
[2]Graduate School of Natural Science and Technology, Kanazawa University, Kakuma-machi, Kanazawa, 920-1192, Japan
[3]Institute for Materials Research, Tohoku University, 2-1-1 Katahira, Aoba-ku, Sendai, 980-8577, Japan

E-mail: ohashi@se.kanazawa-u.ac.jp

**Abstract.** We studied the electrical resistance of polycrystalline chromium films capped by a gold layer. No anomaly was detected by resistance measurements of 10 nm thick film around room temperature, indicating that the antiferromagnetic interaction may be suppressed as decreasing the thickness of the chromium film. The sheet resistance $R_s(T)$ curves differ from polycrystalline chromium films in previous studies because of the electrical current flows through a gold capping layer. On the other hand, the resistance drop is observed at $T_C = 1.15 \pm 0.05$ K as that of polycrystalline chromium films in the previous report. It means that such resistance drop is not related to the chromium oxide layer on a polycrystalline chromium films. However, it is difficult to conclude that superconducting transition occurs because of the large residual resistance below the temperature where the resistance drop is observed.

## 1. Introduction

Superconducting thin films have attracted from both pure and applied research perspectives. However, in almost all of the existing studies of thin film superconductors, the superconductivity disappeared when film thickness is decreased[1, 2, 3, 4]. Superconducting electron pairs will likely form when the film becomes thinner than the mean free path of the electrons.

On the other hand, Schmidt et al. reported that thin films of chromium (Cr) metal suppress the antiferromagnetic ordering and become superconductive at $T_C \sim 1.5$ K[5, 6]. Considering that the bulk Cr is antiferromagnetic below $T_N = 311$ K[7] and show no superconductivity, the relationship between film thickness and superconductivity in Cr opposes that in other superconducting thin films.

Assuming that the magnetic correlation interaction is suppressed by controlling the film thickness, tuning the thickness of Cr film might reveal a magnetic order−disorder transition, and unconventional superconductivity may occur at the quantum critical point of $T \to 0$, where quantum fluctuation produces an electron pair. This idea is partially supported by superconductivity in some heavy fermion systems after suppression of magnetism to some extent under pressure[8, 9, 10, 11, 12].

Generally, three-dimensional electronic systems such as bulk compounds and two-dimensional ones such as thin films are described by completely different Hamiltonians. That is, film thickness is an effective parameter that directly controls the dimensionality of the system. Whether Cr is the first case of a new quantum phase transition tuned by film thickness is an interesting conjecture. Moreover, bulk superconductivity in Cr thin film would be remarkable finding, because it has not been reported in strongly correlated $3d$ transition-metal compounds such as Cr-based superconducting compounds, except for CrAs[13, 14].

However, experimental evidences of superconductivity in Cr thin film, such as resistivity drop and the Meissner effect, are lucking. So, we have studied the electrical resistance of Cr films[15]. Although we observed sudden decrease in the resistance around 1.5 K, it is difficult to conclude whether a superconducting transition occurs because the electrical resistivity is not zero in all films. It may come from the fact that some chromium oxide is produced on a film surface, and that partially show a superconducting transition. In the present study, we perform precise electrical resistance measurements of Cr thin films capped by a gold layer to clarify the electronic state in a wide temperature range.

## 2. Experimental

Several polycrystalline Cr films were deposited on silicon substrate using ion beam sputtering with a base pressure of about $8 \times 10^{-6}$ Pa. The working deposition gas was argon and a pressure was controlled between $1.15 \times 10^{-2}$ and $1.17 \times 10^{-2}$ Pa. A gold capping layer of 5 nm thick was deposited on the surface of the Cr films to avoid the oxidation. X-ray measurements were performed using a Rigaku SmartLab diffractometer with CuK$\alpha_1$ radiation ($\lambda$ = 1.54059 Å). The electrical resistance was measured by a four-point collinear four-probe dc method with the current direction on the film plane. Aluminum wires were bonded on the film plane by wire bonding in the same way of the previous report[15]. The temperature dependence of the electrical resistivity was measured using the Quantum-Design PPMS between 0.5 and 350 K in the low-temperature laboratory, Kanazawa University. The direction of the applied field was perpendicular to the film plane and the electrical current.

## 3. Results and Discussion

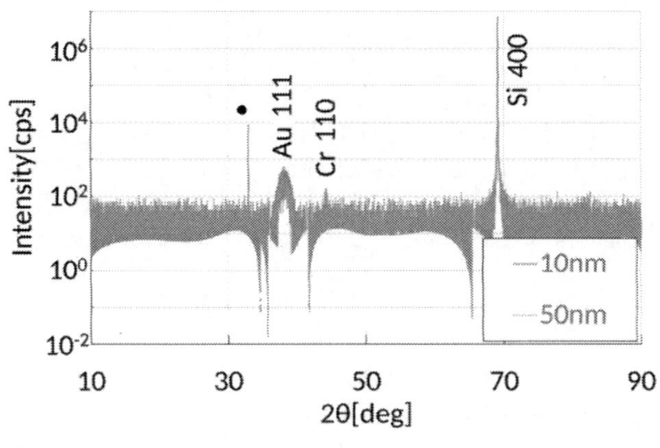


**Figure 1.** The X-ray diffraction patterns of Cr film of 10 nm and 50 thick capped by a gold layer.

Figure 1 shows the X-ray diffraction pattern of chromium films capped by a gold layer. We can observe the diffractions of Si (400), Au (111) and Cr (110) planes. For 50 nm thick film, the lattice constants of chromium are obtained to be 2.89 Å, which is comparable with that of bulk

IOP Conf. Series: Journal of Physics: Conf. Series **969** (2018) 012029 doi:10.1088/1742-6596/969/1/012029

chromium 2.88 Å[16] within the experimental error. An unknown peak is also observed around 32.96° which may correspond to Si (200) reflection.

Figure 2 shows the sheet resistance $R_s$ of 10 nm thick Cr film capped by a gold layer as a function of temperature between 0.5 and 350 K. The value $R_s = 46.4\ \Omega/\Box$ at 300 K is much smaller than that of a 10 nm thick polycrystalline film reported in the previous study[15]. Moreover, $R_s$ decreases with decreasing temperature while that of a 10 nm thick film in the previous study show a semiconducting behavior. It suggests that the electrical current of the gold-capped Cr film of 10 nm thick flows through both a Cr film and a gold capping layer in parallel. No hump is observed around 300 K related to the magnetic phase transition in the $R_s(T)$ curve, which is consistent with previous study.

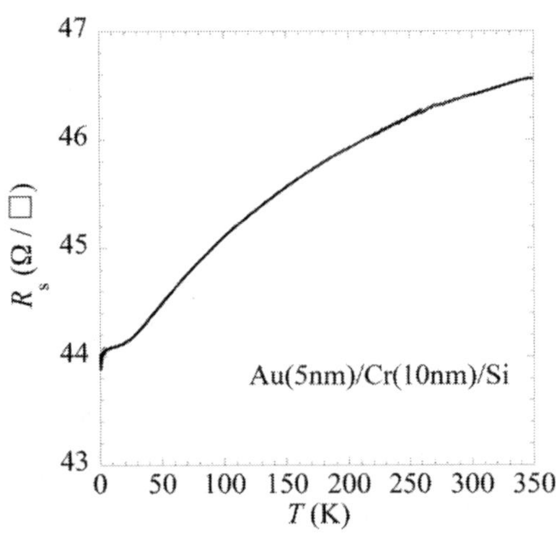


**Figure 2.** Sheet resistance $R_s$ of Cr film of 10 nm thick capped by a gold layer as a function of temperature.

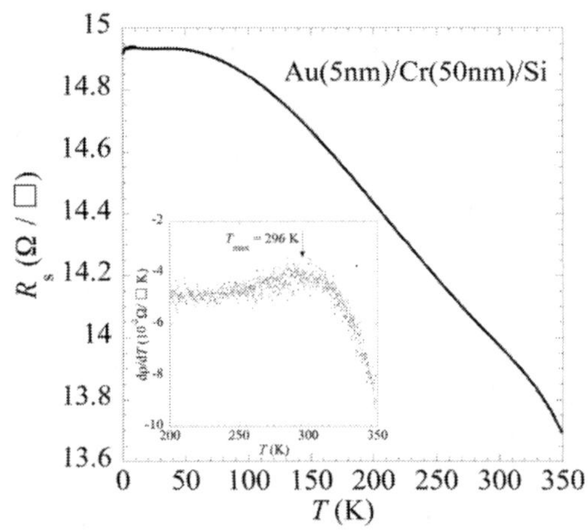


**Figure 3.** Sheet resistance $R_s$ of Cr film of 50 nm thick capped by a gold layer as a function of temperature. Inset shows the temperature derivative of $R_s$

Figure 3 shows the sheet resistance $R_s$ of 50 nm thick Cr film capped by a gold layer as a function of temperature between 0.5 and 350 K. The value $R_s = 14.0\ \Omega/\Box$ at 300 K is close to that of a 50 nm thick film reported previously[15]. $R_s$ increases with decreasing temperature between 10 and 350 K. Taking account that the fact the 50 nm thick Cr film is 10 times thicker than that of a gold-capping layer, the semiconducting conductivity of the 50 nm thick Cr film reported previously[15] may contribute to the electrical resistance compared with the metallic behavior of a gold-capping layer. Below 10 K, on the other hand, $R_s$ decreases with decreasing temperature.

A hump is observed around room temperature in the $R_s(T)$ curve. The inset of figure 3 shows the temperature differential of the electrical resistivity $dR_s/dT$ as a function of temperature. $dR_s/dT(T)$ curve shows a maximum at around 296 K. It may be related to the antiferromagnetic phase transition, which is observed in bulk Cr with $T_N = 311$ K[7, 17, 18].

Figure 4 and 5 shows the sheet resistance $R_s$ of 10 and 50 nm thick Cr film capped by a gold layer at low temperature, respectively. We found that the resistance drops are observed at $T_C = 1.15 \pm 0.05$ K for both films. However, $R_s$ is not zero in both films below $T_C$, and the magnitude of the resistivity drop ratio is very small, i.e., 0.16 and 0.05 % for the films of 10 and 50 nm thick, respectively. Similar behavior is also observed in polycrystalline Cr films although $T_C = 1.5$ K[15] differs from that of this work. It means that such resistance drop is not related to the presence or absence of chromium oxide layer on polycrystalline Cr films. On the other hand, taking account that aluminum becomes superconductive at $T_C = 1.1$ K[16], aluminum wires bonded on a sample may contribute to the electrical resistance. Although Schmidt *et al.* also reported that thin films of Cr metal suppress the antiferromagnetic ordering and become superconductive at $T_C \sim 1.5$ K[5, 6], it is difficult to conclude whether superconducting transition occurs.

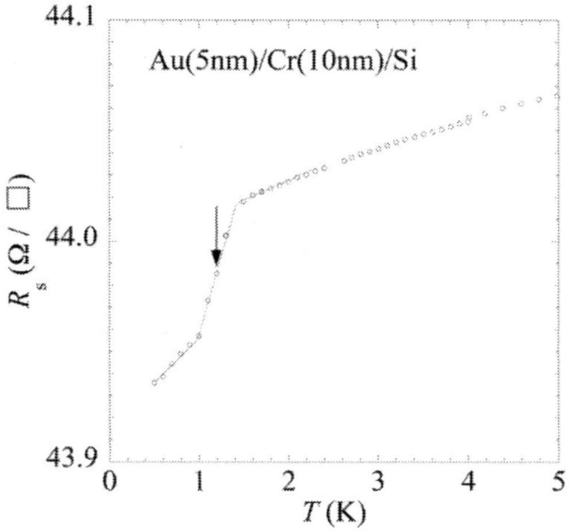


**Figure 4.** Sheet resistance $R_s$ of Cr film of 10 nm thick capped by a gold layer at low temperature. Arrow shows the temperature where $R_s$ shows a sudden decrease.

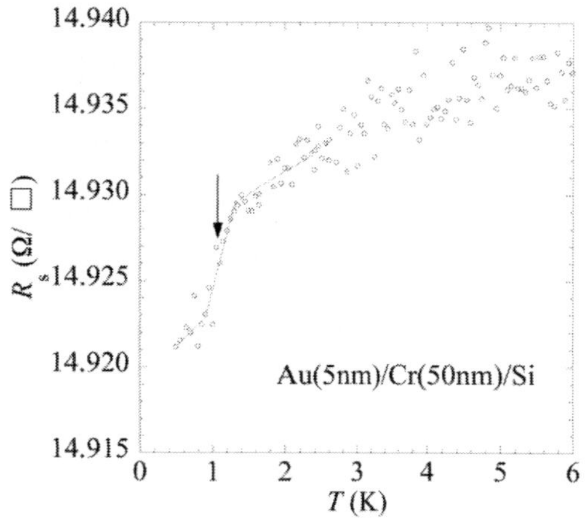


**Figure 5.** Sheet resistance $R_s$ of Cr film of 50 nm thick capped by a gold layer at low temperature. Arrow shows the temperature where $R_s$ shows a sudden decrease.

Figure 6 and 7 shows the sheet resistance $R_s$ of 10 and 50 nm thick Cr film capped by a gold layer as a function of magnetic field at 0.5 K, respectively. We found that the slope $d\mathrm{MR}/dH$ is positive and the MR tends to saturate above 2 kOe. At the 3 kOe field, the MR obtained is 0.16 and 0.05 % for samples 10 nm and 50 nm thick, respectively, which is identical to the magnitude of the electrical resistivity drop, as mentioned in Figs. 4 and 5. If we assume that the upper critical field $H_{C2} \sim 2$ kOe where the MR saturates for 10 nm thick Cr film, Ginzburg-Landau coherence length is obtained as $\xi = 162$ Å according to this relationship: $\mu_0 H_{C2}(0) = \Phi_0/2\pi\xi^2$, where $\Phi_0 = 2.067 \times 10^{-15}$ Wb is the magnetic flux quantum.

## 4. Summary

In this study, we performed the electrical resistance measurements on polycrystalline chromium films capped by a gold layer. The sheet resistance $R_s(T)$ curves differ from polycrystalline chromium films in previous studies because of the electrical current flows through a gold capping layer. On the other hand, the resistance drop is observed at low temperature. It means that the resistance drop is not related to the chromium oxide layer on a polycrystalline chromium films. However, it is difficult to conclude that superconducting transition occurs because of the large residual resistance below the temperature where the resistance drop is observed. To make it clear, more precise experiments are planed to clarify the transport properties of single crystal chromium films at low temperature.

**Acknowledgments**
This work was performed under the inter-university cooperative research program of the Cooperative Research and Development Center for Advanced Materials, Institute for Materials

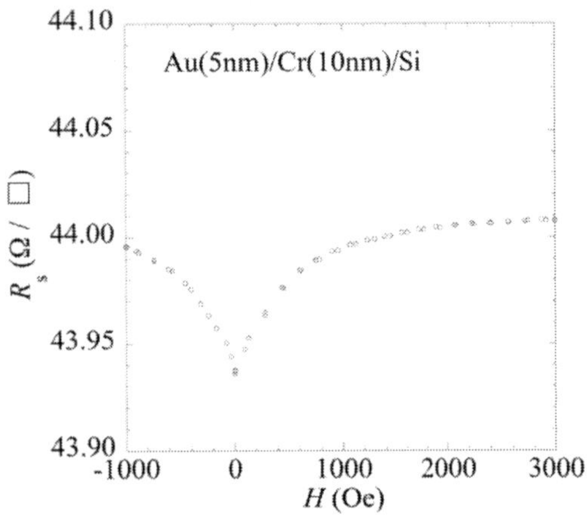


**Figure 6.** Sheet resistance $R_s$ of Cr film of 10 nm thick capped by a gold layer as a function of magnetic field at 0.5 K.

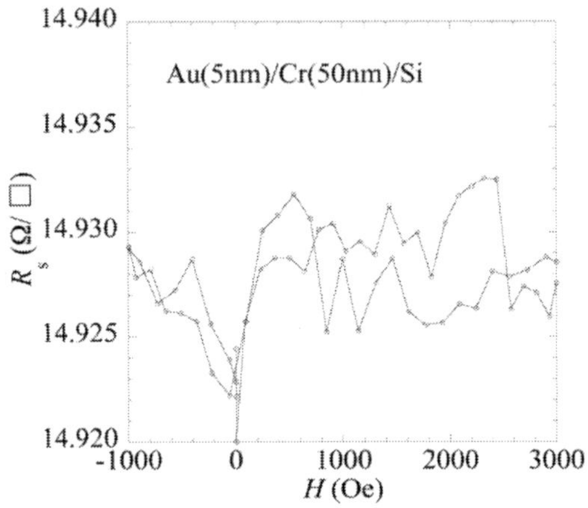


**Figure 7.** Sheet resistance $R_s$ of Cr film of 50 nm thick capped by a gold layer as a function of magnetic field at 0.5 K.

Research, Tohoku University (Proposal No. 16G0039). This work was supported in part by Futaba Electronics Memorial Foundation.

## References

[1] Chakravarty S, Ingold G L, Kivelson S and Luther A 1986 *Phys. Rev. Lett.* **56** 2303
[2] Orr B G, Jaeger H M and Goldman A M 1985 *Phys. Rev. B* **32** 7586(R)
[3] Jaeger H M, Haviland D B, Orr B G and Goldman A M 1989 *Phys. Rev. B* **40** 182
[4] Haviland D B, Liu Y, Nease B and Goldman A M 1990 *Phisica B* **165-166** 1457
[5] Schmidt P H, Castellano R N, Barz H and Matthias B T 1972 *Physics Letters* **41A** 367
[6] Schmidt P H, Castellano R N, Barz H, Cooper A S and Spencer E G 1973 *J. Appl. Phys.* **44** 1833
[7] Koehler W C, Moon R M, Trego A L and Machintosh A R 1966 *Phys. Rev.* **151** 405
[8] Uwatoko Y, Umehara I, Ohashi M, Nakano T and Oomi G 2012 *Handbook on the Physics and Chemistry of Rare Earths* vol 2 (Amsterdam: North-Holland) chap 252, p 1
[9] Nakano T, Ohashi M, Oomi G, Matsubayashi K and Uwatoko Y 2009 *Phys. Rev. B* **79** 172507
[10] Miyagawa H, Oomi G, Ohashi M, Satoh I, Komatsubara T, Hedo M and Uwatoko Y 2008 *Phys. Rev. B* **78** 064403
[11] Ohashi M, Oomi G, Koiwai S, Hedo M and Uwatoko Y 2003 *Phys. Rev. B* **68** 144428
[12] Ohashi M, Oomi G and Satoh I 2007 *J. Phys. Soc. Jpn.* **76** 114712
[13] Wu W, Cheng J, Matsubayashi K, Kong P, Lin F, Jin C, Wang N, Uwatoko U and Luo J 2014 *Nature Communications* **5** 5508
[14] Kotegawa H, Nakahara S, Tou H and Sugawara H 2014 *J. Phys. Soc. Jpn.* **83** 093702
[15] Ohashi M, Ohashi K, Sawabu M, Miyagawa M, Kubota T and Takanashi K 2016 *Physics Letters A* **380** 3133
[16] Kittel C 1996 *Introduction to Solid State Physics* 7th ed (John Wiley & Sons, Inc.)
[17] McWhan D B and Rice T M 1967 *Phys. Rev. Lett.* **19** 846
[18] Ohashi M and Oomi G 2009 *Jpn. J. Appl. Phys.* **48** 070221

# Electrical Transport between Superconducting Niobium and a Zinc-Oxide-based Electric Double-layer Transistor

**T Narita[1], H Takayanagi[2] and R Ishiguro[1,*]**

[1]Japan Women's University, Department of Mathematical and Physical Sciences, Tokyo, Japan

[2]Tokyo University of Science, Research institute of science and technology, Tokyo, Japan

[*]E-mail: ishiguror@fc.jwu.ac.jp

**Abstract.** We report electrical transport measurements between a metallic state induced on the surface of an insulating zinc oxide film by an electric double-layer transistor and superconducting niobium electrode. A superconductor/electric double-layer transistor channel/superconductor was created that can be a superconductor/metallic state/superconductor junction and is expected to function as a Josephson junction. The metallic state on the zinc oxide surface and a low contact resistance between niobium and zinc oxide were realized by high carrier doping arising from the electric field effect of the electric double-layer transistor. However, there was no supercurrent through the metallic state at the junction down to a temperature of 0.3 K. Nevertheless, temperature variations of the differential resistance of the junction below the superconducting transition temperature suggest the existence of a superconducting proximity effect between the superconducting Nb and the channel of the zinc oxide-based electric double-layer transistor.

## 1. Introduction

Recently, electric double-layer transistors (EDLTs) have attracted considerable attention as novel devices to control solid-state properties [1–10]. An EDLT is a type of electric field-effect transistor (FET) that uses ionic liquids, which are salts in liquid state without any additional solvents, as a gate insulating layer of conventional FETs. The upper limit of the sheet carrier density in conventional FETs is restricted by the dielectric breakdown field. By contrast, the upper limit of the sheet carrier density in EDLTs can be more than a factor of ten larger than that in FETs and can have values as large as $10^{14}$–$10^{15}$ cm$^{-2}$ [2,9]. Previous studies have reported that as a high-density carrier accumulation is induced at the surface by EDLTs, the insulating channel of EDLTs becomes a metal and moreover a superconductor [1,5].

Although the junction between the EDLT channel and the metal superconductor is a new interface where novel functions can arise, no studies have analyzed such junctions. Moreover, investigating the bonding between the new substance and the metal electrode should be very important for industrial applications.

In this study, the fabrication of a junction between zinc oxide (ZnO) and Niobium (Nb) is presented. Zinc oxide is a wide-gap semiconductor with a band gap of $E_g = 3.4$ eV at 300 K that has already demonstrated metallization at the surface due to high-density carrier accumulation induced by an

EDLT [2]. Niobium is a type II superconductor with a superconducting transition temperature $T_c = 9.2$ K. We present a study of the electrical transport properties of the junction in the temperature range of 0.3–300 K.

## 2. Fabrication

A schematic of our device is shown in figure 1. An epitaxially grown 500-nm-thick ZnO film was used on a single-crystal sapphire substrate, which is commercially available from MIT Co.

The fabrication of the ZnO-based EDLT involved the following three processes. In the first process, micropatterns of ZnO were made. Firstly, a 50-nm-thick $SiO_2$ film was deposited on the ZnO film to protect ZnO from over-etching. Micropatterns of ZnO were formed by etching both the ZnO and $SiO_2$ films in the outer region by using maskless laser lithography and a reactive ion etching (RIE) technique. The $SiO_2$ film on the micropattern was subsequently removed by using RIE.

In the second process, superconducting electrodes on the ZnO micropattern separated by a 100-nm-wide gap were made by using an electron beam lithography lift-off technique with niobium (Nb) sputter deposition, as shown in figure 2. These electrodes are expected to work as a superconductor-normal-superconductor (SNS) Josephson junction. Immediately before deposition of Nb, the contact surface was treated by argon sputter cleaning. The 100-nm-thick Nb electrodes were sputtered at a backing pressure of $\sim 10^{-6}$ Pa. Subsequently, an Au covering layer with a thickness of 5 nm was deposited on the Nb electrodes by electron beam physical evaporation.

Then, large structures such as bonding pads, electrodes and the gate electrode were formed by using maskless laser lithography and electron beam physical evaporation. The deposition of 10-nm-thick Ti and 150-nm-thick Au films was performed under a pressure of a few $10^{-5}$ Pa. Finally, immediately before performing measurements, ionic liquids were dropped as gate dielectrics on the device to cover the Nb/ZnO/Nb structure and a part of the gate electrode. In this work, *N,N*-Diethyl-*N*-methyl-*N*-(2-methoxyethyl)ammonium tetrafluoroborate(DEME-$BF_4$) was used as an ionic liquid.

Two samples, A and B, with the mostly same structure were fabricated to perform different experiments.

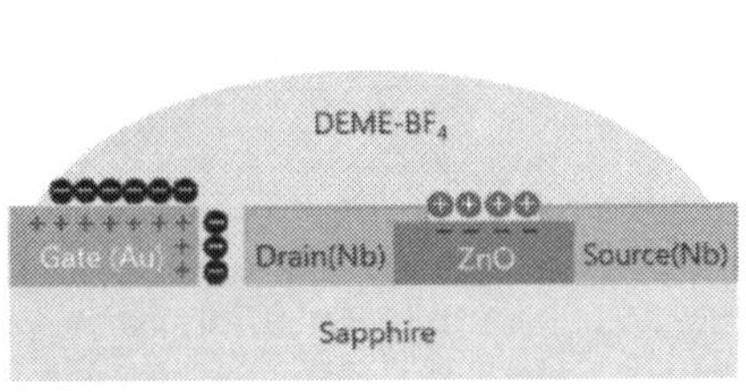


**Figure 1.** Schematic of a ZnO-based EDLT with an Au gate electrode and superconducting Nb electrodes.

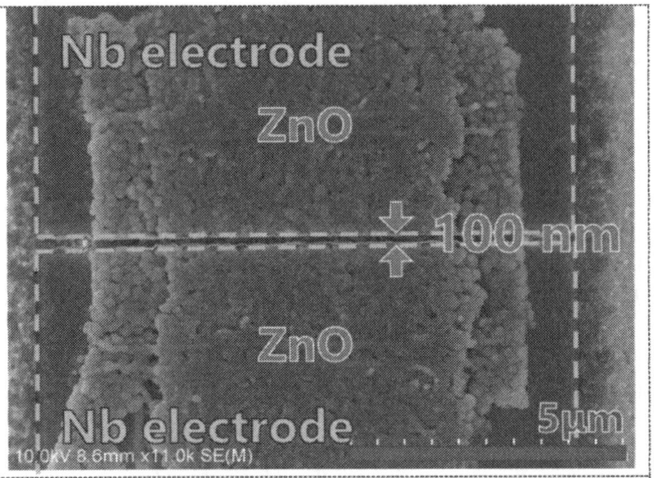


**Figure 2.** Scanning electron microscopy image of the Nb/ZnO/Nb junction structure. The gap length $L$ between the superconducting electrodes is 100 nm and the ZnO channel width $W$ is about 8 μm.

## 3. Results and discussion

All transport measurements were performed in a helium-3 evaporative refrigerator with a base temperature of ~ 300 mK, which was well below the superconducting transition temperature $T_c$ ~ 9 K of our samples. Several filtering systems were used, consisting of *LC* filters at room temperature and *RC* filters at the base temperature.

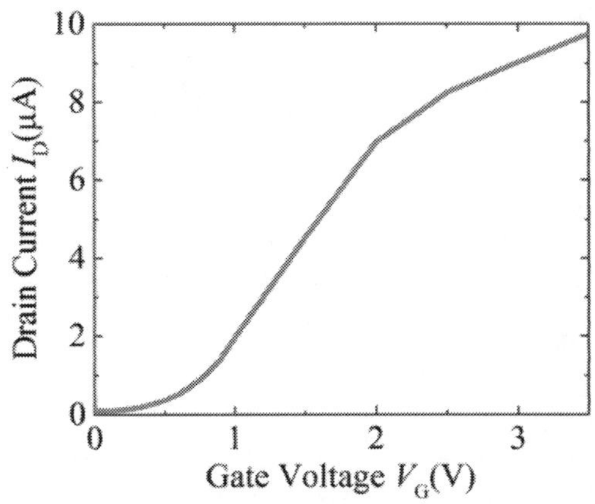


**Figure 3.** Transfer curve for the ZnO-based EDLT measured at room temperature and a source-drain voltage $V_{SD}$ = 10 mV.

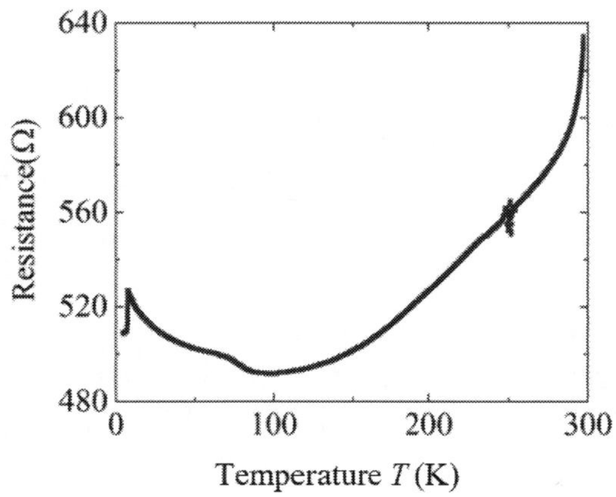


**Figure 4.** Temperature dependence of the resistance at a gate voltage $V_G$ = 3 V.

Figure 3 shows a transfer curve for the ZnO-based EDLT of sample A measured at room temperature and for a source-drain voltage $V_{SD}$ of 10 mV. The drain current $I_D$ is equal to 0.08 μA at $V_G$ = 0 V and increases monotonically for increasing $V_G$ up to a value of 9.7 μA at $V_G$ = 3.5 V. The on/off ratio of the device is about 120, indicating that DEME-$BF_4$ could penetrate into the 100-nm-wide gap between the Nb electrodes and the EDLT works properly in such a narrow gap.

The temperature dependence of the resistance at $V_G$ = 3 V for the ZnO-based EDLT of sample B containing a Nb/ZnO-based EDLT/Nb structure is plotted in figure 4. The resistance decreases with decreasing temperature for temperatures between 100 K and 300 K, indicating that a metallic state in the ZnO film was realized. Moreover, it can be inferred from the data that quantum tunneling of electrons between Nb to ZnO is more dominant than thermionic transfer and that the potential barrier between ZnO and Nb at $V_G$ = 3 V is small.

Figure 5 shows the differential resistance d$V$/d$I$ against the bias voltage of the Nb/ZnO-based EDLT channel/Nb structure of sample A at a temperature of 350 mK. The figure exhibits three sharp peaks and two valleys for absolute bias voltages |$V$| smaller than ~ 1 mV, which is about the same as the voltage corresponding to the superconducting gap. However, since there were voltage drops at the ZnO-based EDLT channel with a resistance of about 440 Ω, the voltage actually applied to the junction was sufficiently lower than the voltage corresponding to the superconducting gap of Nb in this measurement.

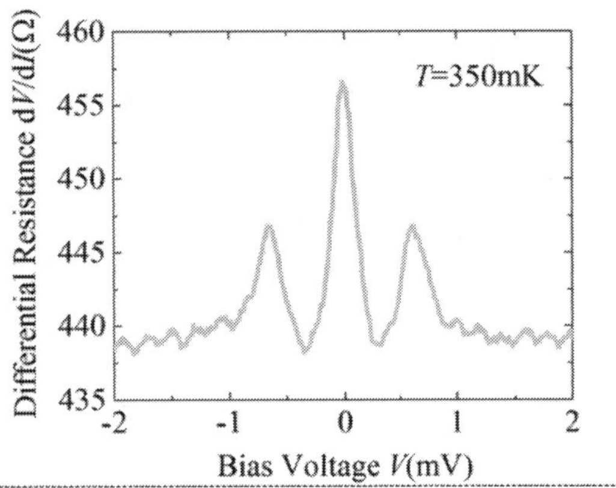


**Figure 5.** Differential resistance $dV/dI$ against bias voltage $V$ for the Nb/metallized ZnO100nm/Nb structure at 350 mK.

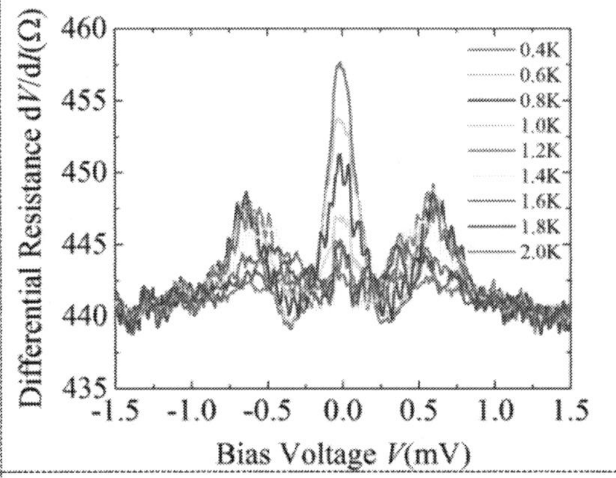


**Figure 6.** Temperature variation of the differential resistance $dV/dI$ against bias voltage $V$ for the Nb/metallized ZnO100nm/Nb structure at temperatures ranging from 0.4 K to 2 K.

As the maximum value of differential resistance at the zero bias was 455 Ω, it was impossible for the two superconducting electrodes to be coherently connected and for the superconducting current to flow.

Figure 6 shows the temperature variations of the differential resistance $dV/dI$ against the bias voltage $V$ for the Nb/metalized ZnO100nm/Nb structure at temperatures ranging from 0.4 K to 2 K. The differential resistances for voltages below the voltage corresponding to the superconducting gap vary significantly with temperature at these low temperatures. This means that there was a superconducting proximity effect in the SN junction between the Nb superconducting electrode and the carrier-doped metallic channel with high carrier concentration induced by the EDLT.

## 4. Conclusions

We fabricated a superconducting Nb/ZnO-based EDLT channel/superconducting junction and examined its transport characteristics. The ZnO-based EDLT channel with a length of 100 nm and a width of 5 μm was metalized under a gate voltage of 3 V.

Electrical transport measurements between the metallic state induced on the surface of the ZnO by the EDLT and the superconducting Nb electrodes showed that there is no supercurrent between the junctions down to a temperature of 0.3 K. However, measured data suggest that a superconducting proximity effect in the SN junction between the normal state in the EDLT and the superconducting electrode can be observed.

In order to realize coherent transport between the superconducting electrodes, improvement of the contact resistance is necessary, which could be accomplished by increasing the gate voltage or/and inserting a bonding layer such as a Ti layer between the Nb electrodes and the ZnO based EDLT channel.

## Acknowledgments

We would like to thank S. Nomura from the Univ. of Tsukuba, S. Kashiwaya and K. Tsumura from AIST, Y. Nago from Keio Univ. and D. Sakuma from the Tokyo Univ. of Science for stimulating discussions. This work was supported by JSPS KAKENHI C (Grant No. 17K05551). Device fabrication was supported by the NIMS Nanofabrication platform of the "Nanotechnology Platform Project", sponsored by MEXT.

## References

[1] Ueno K, Nakamura S, Shimotani H, Ohtomo A, Kimura N, Nojima T, Aoki H, Iwasa Y, and Kawasaki M 2008 *Nat. Mater.* **7** 855.
[2] Yuan H, Shimotani H, Tsukazaki A, Ohtomo A, Kawasaki M and Iwasa Y 2009 *Adv. Fanct. Mater.* **19** 1046.
[3] Ye J T, Inoue S, Kobayashi K, Kasahara Y, Yuan H T, Shimotani H and Iwasa Y 2010 *Physica C* **470** S682.
[4] Ye J T, Inoue S, Kobayashi K, Kasahara Y, Yuan H T, Shimotani H and Iwasa Y 2010 *Nat. Mater.* **9** 125.
[5] Ueno K, Nakamura S, Shimotani H, Yuan H T, Kimura N, Nojima T, Aoki H, Iwasa Y and Kawasaki M 2011 *Nat. Nanotech.* **6** 408.
[6] Yamada Y, Ueno K, Fukumura T, Yuan H T, Shimotani H, Iwasa Y, Gu L, Tsukimoto S, Ikuhara Y and Kawasaki M 2011 *Science* **332** 1065.
[7] Ye J, Zhang Y, Akashi R, Bahramy M, Arita R and Iwasa Y 2012 *Science* **338** 1193.
[8] Katase T, Hiramatsu H, Kamiya T and Hosono H 2014 *Proc. Nat. Acad. Sci. USA* **111** 3979.
[9] Du H, Lin X, Xu Z and Chu D 2015 *J. Mater. Sci.* **50** 5641.
[10] Saito Y, Nakamura Y, Bahramy M S, Kohama Y, Ye J, Kasahara Y, Nakagawa Y, Onga M, Tokunaga M, Nojima T, Yanase Y and Iwasa Y 2015 *Nat. Phys.* **12** 144.

IOP Conf. Series: Journal of Physics: Conf. Series **969** (2018) 012031 doi:10.1088/1742-6596/969/1/012031

# Different electronic states at crystallographically inequivalent $CuO_2$ planes on four-layered cuprates $HgBa_2Ca_3Cu_4O_{10+\delta}$

**R Sekine[1,2], S Nakayama[1], A Tsukada[2], S J Denholme[2], N Miyakawa[2], and K Tokiwa[1]**

[1] Department of Applied Electronics, Tokyo University of Science, 6-3-1 Niijuku, Katsushika-ku, Tokyo 125-8585, Japan

[2] Department of Applied Physics, Tokyo University of Science, 6-3-1 Niijuku, Katsushika-ku, Tokyo 125-8585, Japan

E-mail: rsekine@alumni.tus.ac.jp

**Abstract.** We report tunneling conductances due to two kinds of crystallographically inequivalent $CuO_2$ planes on multi-layered cuprates $Hg_{0.95}Ba_2Ca_3Cu_4O_{10.05}$ (Hg1234) measured by a point-contact tunneling technique. One is an outer $CuO_2$ plane (OP) which has a pyramidal five-oxygen coordination, another is an inner $CuO_2$ plane (IP) which has a square four oxygen coordination. These tunneling conductances exhibit two kinds of superconducting gaps $\Delta$ with clearly different sizes. That is, $\Delta$ on Hg1234 was $36 \pm 2$ meV and $55 \pm 2$ meV for OP and IP, respectively. Moreover, we report the correlation between the mode energy $\Omega$ and $\Delta$. The $\Omega/\Delta$ exhibits the common feature with other cuprates and does not exceed 2. This behaviour implies that the collective spin excitation is a candidate for the mediator on pair formation.

## 1. Introduction

High-$T_C$ cuprate superconductor has been over 30 years since its discovery, however mechanism for pair formation has not been elucidated yet, and therefore it is one of the challenging subject. Cuprates having three or more $CuO_2$ planes in a unit cell are called a multilayer cuprates (MLCs) and have two kinds of $CuO_2$ planes that are crystallographically inequivalent. In particular, the sufficient investigation for MLCs have not been collected since the complexity of crystal structure. Thus, it is one of the reasons why high-$T_C$ superconducting mechanisms remain unexplained. On the other hand, investigation for mono- or bilayered cuprates have been actively conducted. In particular, many common properties of cuprate superconductors have been revealed by spectroscopic studies such as point-contact/break-junction tunneling (PCT/BJT) [1, 2], scanning tunneling microscopy/spectroscopy (STM/STS) [3] and angle-resolved photoemission spectroscopy (ARPES) [4], since the electronic state can be observed directly. Specifically, for example on hole-doped cuprates, there is common understanding such as *d*-wave superconductor, bell-shaped superconducting phase diagram, and the decreasing of magnitude of the superconducting gap $\Delta$ with doping increasing.

As shown in Fig. 1, MLC such as $Hg_{0.95}Ba_2Ca_3Cu_4O_{10.05}$ (Hg1234) has two kinds of crystallographically inequivalent $CuO_2$ plane: an outer $CuO_2$ plane (OP) which has a pyramidal five-oxygen coordination, another is an inner $CuO_2$ plane (IP) which has a square four-oxygen coordination. Investigation for MLCs has been actively done in a few spectroscopic studies [5], and has been performed energetically in NMR [6, 7]. According to NMR results, the local carrier concentration at the OP is higher than that at the IP [6]. By combining NMR results with the doping dependence of the $\Delta$, it can be expected that $\Delta$s of different magnitudes will be observed in MLCs. However, in a spectroscopic experiment using a vacuum as a tunnel barrier, the information of the OP closest to the cleavage surface becomes the main and it is difficult to investigate the superconducting characteristics on the IP. On the other hand, the two kinds of superconducting gap originating at the OP and IP have been successfully observed by the PCT method by forming a direct junction to the OP and IP, as shown in the Fig. 1(b, c). Since the MLCs have $CuO_2$ planes (IP) distant from the charge supplying layer, there is a possibility that the nature of the essential superconducting $CuO_2$ surface can be clarified. Therefore, it is principal to investigate two kinds of $CuO_2$ planes of multi-layered cuprates. Hg-based cuprates are well known for having a flat $CuO_2$ plane and the higher $T_C$ among cuprate superconductors. Therefore, Hg1234 is one of the ideal material for the PCT method.

Despite numerous experimental studies, there is still no unified understanding for the mechanism of pair formation on cuprate superconductor [8]. As a probe to explore the mechanism of the pair formation, the dip structure observed outside the coherence peak of the tunneling conductance has been drawing attention. As one of the interpretation, the dip structure has been interpreted as a result of the contribution of bosonic excitations and has been reproduced by strong coupling analysis reflecting pairing interaction [1, 9, 10]. These result are consistent with the theory that the peak due to the mode energy $\Omega$ is located inside $2\Delta$ in the pairing function for the collective spin excitation model [11]. Furthermore, it has been found that the doping dependence of the $\Omega$ estimated from the dip structure coincides with the magnetic resonance mode energy observed in inelastic neutron scattering (INS) [1, 12]. Thus, the magnetic interaction has been interpreted one of the candidate playing a role of glue for the Cooper pair.

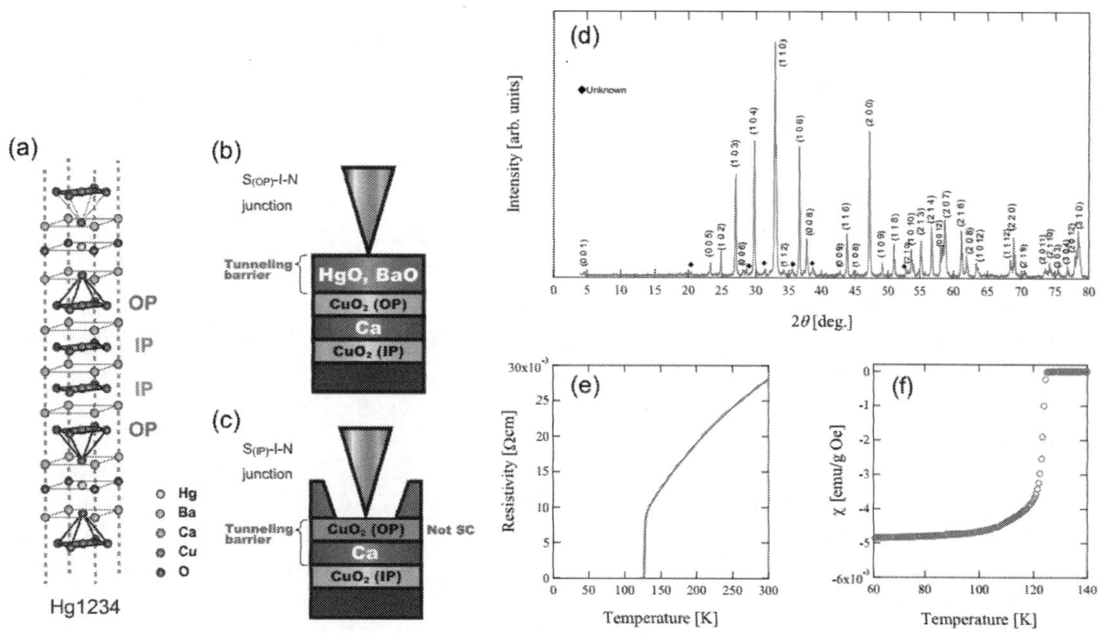


**Figure 1.** (a) Crystal structure for Hg1234. Hg1223 exhibits two kinds of $CuO_2$ plane [the inner plane (IP) with four-fold oxygen coordination and the outer plane (OP) with five-fold oxygen coordination]. (b, c) Schematic structure of SIN point contact tunneling junction at the OP (b) and IP (c) for Hg1234. (d-f) Characterization of Hg1234 sample. (d) X-ray diffraction measurement. Diamonds indicate unknown phase that is not cuprates with other layered. The lattice parameters of $a$ and $c$-axis are 3.854 Å and 19.043 Å, respectively. (e) Temperature dependence of DC resistivity and (f) magnetic susceptibility measurements. The $T_C$ for Hg1234 was determined as 127 K from both measurements.

However, other interpretations different from the bosonic excitation have also been mentioned, such as an energy-dependent gap function model [13], gap inhomogeneities [14], charge-density-wave ordering [15], and bilayer band splitting [16]. In addition, tunneling conductance can be reproduced in a wide range by a model considering the pair-pair interaction by Sacks et al. [17, 18]

## 2. Experimental details

The polycrystalline Hg1234 sample was synthesized under high pressure of 4.5 GPa at 850 °C for 1 h. The nominal composition of the sample corresponds to $Hg_{0.95}Ba_2Ca_3Cu_4O_{10.05}$ (Hg1234). HgO, BaO, $BaO_2$ and $Ca_2CuO_3$ was used as the source material. BaO was prepared by removing $CO_2$ from $BaCO_3$ with condition at 1025 °C for 24 h in vacuum of $1.2 \times 10^{-4}$ Torr. $BaO_2$ was prepared from the BaO at 600 °C for 24 h in $O_2$ atmosphere. $Ca_2CuO_3$ was synthesized from CuO and $CaCO_3$ whose mixture was heated at 950 °C for 20–40 h five times in air. It was confirmed that all precursors do not contain impurities by XRD measurement. The oxygen content of Hg1234 was adjusted by the ratio between BaO and $BaO_2$. MLCs such as four-layered Hg1234 are known to be easily mixed with superconducting impurities having different number of $CuO_2$ planes in a unit cell. However, as shown in Fig. 1(d), we succeeded to synthesize the Hg1234 which does not have superconducting impurities such as three-layered and five-layered Hg-based cuprates. The lattice parameter of Hg1234 corresponds to $a \sim 3.854$ Å and $c \sim 19.043$ Å, respectively. The $T_C$ for Hg1234 was determined as 127 K by resistivity measurement [Fig. 1(e)] and magnetic susceptibility measurement [Fig. 1(f)]. The average doping level between IP and OP was estimated as $p \sim 0.13$ (under-doped region) by iodometry method.

Tunneling conductance by a superconductor-insulator-normal metal (SIN) junction have been measured by the PCT method using a Au tip. The insulating layer corresponds to blocking layers such as HgO, BaO and Ca. $CuO_2$ planes are regarded as superconducting layer. Thus, the Au tip was gradually brought close to the sample surface, and the SIN junction was formed by using the block layer as the insulating layer. Figure 1(b, c) are the schematic illustrations for the two type of SIN junction. In the PCT measurement, we have reported that the electronic state at the OP and IP can be observed by forming a junction by contacting a tip on the sample surface. That is, in the case for Hg1234, as shown in Fig. 1(b), it is considered that the $S_{(OP)}$-I-N junction is formed by adopting the HgO and BaO layers as the insulating layer. On the other hand, as shown in Fig. 1(c), when the HgO layer supplying charge to $CuO_2$ planes are scraped off by the Au-tip, it is considered that the OP cannot become the superconducting state due to the carrier deficiency. In such the situation, it is considered that the $S_{(IP)}$-I-N junction is formed by adopting the non-superconducting OP and Ca layer as the insulating layer. $dI/dV$s were measured by an ac lock-in technique at 4.2 K. The negative (positive) bias in the tunneling conductance corresponds to the occupied (unoccupied) state of superconductivity.

## 3. Results and discussion

Two kinds of the superconducting gap originating from the OP and IP for MLCs have been reported by our PCT method [19–24]. As shown in Fig. 2, two kinds of spectra with different gap magnitudes have also been observed non-selectively for Hg1234. Figure 2(a) shows the typical tunneling conductance on Hg1234. In the PCT measurement, the electronic state at the OP and IP can be observed by forming a junction by contacting a tip on the sample surface. The tunneling conductance for Hg1234 exhibits similar features to the monolayer cuprates $Bi_2Sr_2CuO_y$ (Bi2201) [25], $Tl_2Ba_2CuO_y$ (Tl2201) [26] and the bilayer cuprates $Bi_2Sr_2CaCu_2O_y$ (Bi2212) [27], $TlBa_2CaCu_2O_y$ (Tl1212) [28]. That is the tunneling conductances exhibits not only the peak-dip-hump structure [as indicated by allows in Fig. 2(a)] but also the $d$-wave like shape sub-gap region. In Fig. 2(a), the upper curve corresponds to the tunneling conductance due to the electronic state at the OP and the lower curve corresponds to that at the IP, respectively. As shown in Fig. 2(a), the tunneling conductance at the OP exhibits sharp and high coherence peaks, and the dip structure (indicated by allows) was observed on both positive and negative bias sides. We note that the dip structure is clearly observed also on the positive bias side. On the other hand, the tunneling conductance at the IP is more asymmetric and exhibits the coherence peak broader

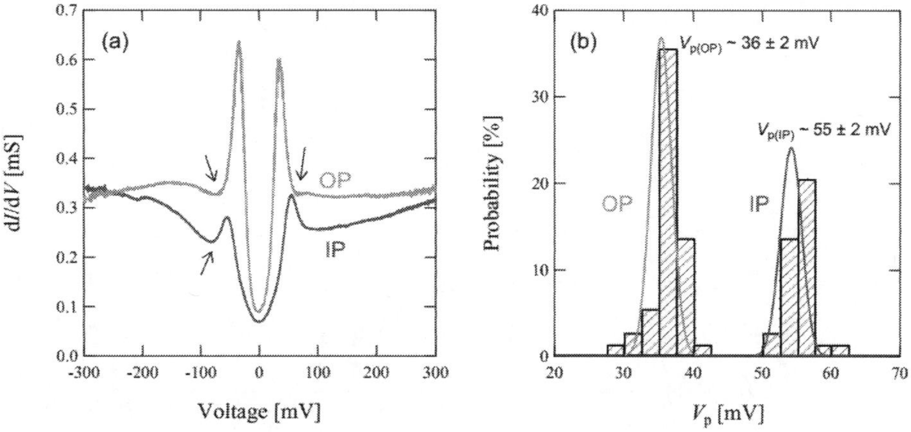


**Figure 2.** (a) Two kinds of the typical tunneling conductances originating from the OP (top) and IP (bottom) on Hg1234 at 4.2 K, respectively. Allows indicate the dip structure. (b) Histogram of the superconducting gap magnitudes at 4.2 K on Hg1234. The superconducting gap Δ was estimated as $\Delta = eV_p$, where $V_p$ is a differential voltage at the coherence peak of the tunneling conductance. Hg1234 exhibits two distinct distributions originating the OP and IP. The Δ on Hg1234 was 36 ± 2 meV at the OP and 55 ± 2 meV at the IP, respectively. The green and blue curves are fitted with a Gaussian distribution.

and lower than that at the OP. Furthermore, the dip structure is observed more strongly at the negative bias side than at the positive bias side. Such features have been discussed in strong coupling analysis considering Van Hove singularity at the *M* point in the Brillouin zone [10, 29]. That is, the small coupling constant reproduces the coherence peak which becomes higher on the negative bias side. On the other hand, the large coupling constant reproduces that the coherence peak on the positive bias side becomes higher than that on the negative bias side.

As shown in Fig. 2(b), the histogram of the gap magnitudes on Hg1234 exhibits two distinct distributions. That is, it indicates that there are two kinds of gap due to different electronic states at the OP and IP, respectively. Based on the statistical distribution, the gap magnitudes of Hg1234 are determined 36 ± 2 meV and 55 ± 2 meV for the OP and the IP, respectively. The existence of two kinds of superconducting gaps has also been observed by the PCT method [19–24]. According to the NMR results, MLCs exhibit different local carrier concentration at the OP and IP, and $p_{(OP)}$ is larger than $p_{(IP)}$ [7]. On the other hand, as investigated in spectroscopic studies, it is well known that the gap magnitude decreases with hole concentration increasing [2–4]. These two experimental facts lead to a conclusion; the gap magnitude at the OP is smaller than that at the IP. Therefore, the tunneling conductance which exhibits the smaller (larger) gap magnitude reflects to the electric state at the OP (IP).

As another of the characteristic common features of tunneling spectra of cuprate superconductors, a dip structure has been observed outside the coherence peak of tunneling conductance. This structure has been interpreted as an effect by collective excitation mode by strong coupling theory [1, 10]. This feature has also been observed for MLCs [24]. As shown in Fig. 2(a), Hg1234 also exhibits the dip structure outside the coherence peak. According to the collective spin excitation model, the resonance peak at the mode energy in the paring function corresponds to the position of the dip minimum [10, 11]. Thus, we can estimate the mode energy from the dip structure as $\Omega = eV_p - eV_{dip}$. The $V_p$ and $V_{dip}$ were estimated from the point that the second derivative of the tunneling current becomes zero at the peak and dip position, respectively.

We now discuss the correlation between mode energy and superconducting gap. Figure 3 exhibits the $\Omega/\Delta$ estimated from the dip structure are plotted as function of Δ. As shown in Fig. 3, we compare $\Omega/\Delta$ vs Δ on Hg1234 with that on other materials such as Bi2212 [1], $Bi_2Sr_2Ca_2Cu_3O_y$ (Bi2223) [10],

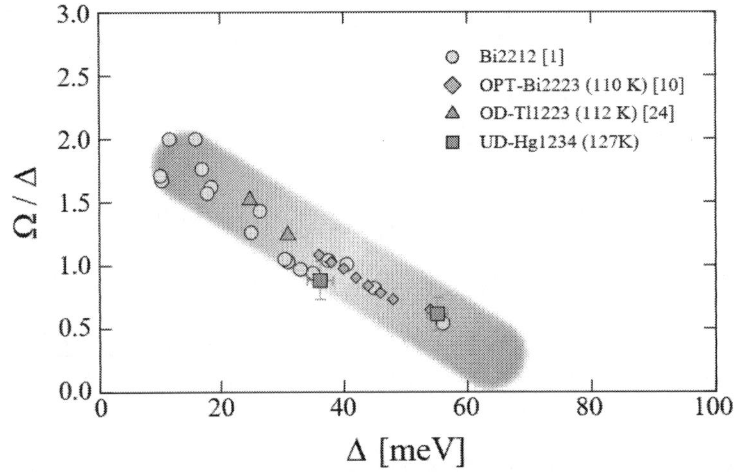


**Figure 3.** Comparison of the $\Omega/\Delta$ vs $\Delta$ between Bi2212 [1], Bi2223 [10], Tl1223 [24] and Hg1234. Squares are for Hg1234. Circles, diamonds and triangles represent results forBi2212 [1], Bi2223 [10] and Tl1223 [24], respectively. The shaded region is a guide to the eye.

$TlBa_2Ca_2Cu_3O_y$ (Tl1223) [24]. Squares correspond to the average value of $\Omega/\Delta$ on Hg1234 containing error bars. The dataset on Tl1223 includes values estimated at the OP and IP from the PCT results. As shown in Fig. 3, the $\Omega/\Delta$ vs $\Delta$ is in good agreement between the these cuprates despite the difference in the number of $CuO_2$ planes and the charge supplying layer. That is, the $\Omega$ correlates with the $\Delta$, indicating that $\Omega/\Delta$ increases with $\Delta$ decreasing, and its value is less than 2 for both IP and OP. Thus, this result is not inconsistent with the model that the excitation exhibits excitonic character. According to the result on Bi2212, it is concluded that this behavior is based on the spin resonance mode from the coincidence of the mode energy estimated from BJT [1] and that observed in INS [12]. Furthermore, these results are also consistent with the results on Bi2223 [10] and Tl1223 [24]. Although, there has not been report of the spin resonance mode on Hg1234 in INS measurement, it will be consistent with the mode energy on Hg1234 reported by our PCT.

## 4. Conclusion

We have succeeded in observing the two kinds of the supeconducting gap originating from the different electronic states at the OP and IP on Hg1234. The tunneling conductance on Hg1234 exhibits the common features with mono- and bilayered cuprates such as *d*-wave like gap shape and the dip structure. The $\Delta$ on Hg1234 was $36 \pm 2$ meV at the OP and $55 \pm 2$ meV at the IP, respectively. In the analysis of the mode energy estimated from the dip structure, the $\Omega/\Delta$ increases with $\Delta$ decreasing but does not exceed 2. The $\Omega/\Delta$ vs $\Delta$ for Hg1234 is in good agreement with that for other cuprates such as Bi2212, Bi2223, and Tl1223. This behavior is consistent with the model that the excitation exhibits excitonic character. Thus, this common feature implies that the collective spin excitation is a candidate for the mediator on pair formation for cuprate superconductors.

**Acknowledgment**
We would like to thank Prof. William Sacks for useful discussion.

**References**

[1] Zasadzinski J F, Ozyuzer L, Miyakawa N, Gray K E, Hinks D G and Kendziora C 2001 *Phys. Rev. Lett.* **87**(6) 067005

[2] Miyakawa N, Zasadzinski J F, Ozyuzer L, Kaneko T and Gray K E 2002 *Trends Appl. Spectrosc.* **4** 47

[3] Fischer Ø, Kugler M, Maggio-Aprile I, Berthod C and Renner C 2007 *Rev. Mod. Phys.* **79**(1) 353

[4] Damascelli A, Hussain Z and Shen Z X 2003 *Rev. Mod. Phys.* **75**(2) 473

[5] Ideta S, Takashima K, Hashimoto M, Yoshida T, Fujimori A, Anzai H, Fujita T, Nakashima Y, Ino A, Arita M, Namatame H, Taniguchi M, Ono K, Kubota M, Lu D H, Shen Z X, Kojima K M and Uchida S 2010 *Phys. Rev. Lett.* **104**(22) 227001

[6] Kotegawa H, Tokunaga Y, Ishida K, Zheng G q, Kitaoka Y, Kito H, Iyo A, Tokiwa K, Watanabe T and Ihara H 2001 *Phys. Rev.* B **64**(6) 064515

[7] Mukuda H, Shimizu S, Iyo A and Kitaoka Y 2012 *J. Phys. Soc. Jpn.* **81** 011008

[8] Carbotte J P, Timusk T and Hwang J 2011 *Rep. Prog. Phys.* **74** 066501

[9] Ahmadi O, Coffey L, Zasadzinski J F, Miyakawa N and Ozyuzer L 2011 *Phys. Rev. Lett.* **106**(16) 167005

[10] Berthod C, Fasano Y, Maggio-Aprile I, Piriou A, Giannini E, Levy de Castro G and Fischer Ø 2013 *Phys. Rev.* B **88**(1) 014528

[11] Abanov A and Chubukov A V 1999 *Phys. Rev. Lett.* **83**(8) 1652

[12] Capogna L, Fauqué B, Sidis Y, Ulrich C, Bourges P, Pailhès S, Ivanov A, Tallon J L, Liang B, Lin C T, Rykov A I and Keimer B 2007 *Phys. Rev.* B **75**(6) 060502

[13] Cren T, Roditchev D, Sacks W and Klein J 2000 *Europhys. Lett.* **52** 203

[14] Fang A C, Capriotti L, Scalapino D J, Kivelson S A, Kaneko N, Greven M and Kapitulnik A 2006 *Phys. Rev. Lett.* **96**(1) 017007

[15] Gabovich A M and Voitenko A I 2007 *Phys. Rev.* B **75**(6) 064516

[16] Feng D L, Armitage N P, Lu D H, Damascelli A, Hu J P, Bogdanov P, Lanzara A, Ronning F, Shen K M, Eisaki H, Kim C, Shen Z X, Shimoyama J i and Kishio K 2001 *Phys. Rev. Lett.* **86**(24) 5550

[17] Sacks W, Mauger A and Noat Y 2016 *Eur. Phys. J.* B **89** 183

[18] Sacks W, Mauger A and Noat Y 2017 *Solid State Commun.* **257** 1

[19] Miyakawa N, Hoya H, Iyo A, Tanaka Y, Tokiwa K, Watanabe T and Kaneko T 2007 *Int. J. Mod. Phys.* B **21** 3233

[20] Miyakawa N, Tokiwa K, Mikusu S, Zasadzinski J F, Ozyuzer L, Ishihara T, Kaneko T, Watanabe T and Gray K E 2003 *Int. J. Mod. Phys.* B **17** 3612

[21] Miyakawa N, Tokiwa K, Watanabe T, Iyo A and Tanaka Y 2006 *AIP Conf. Proc.* **850** 397

[22] Sekine R, Kawashima S, Minematsu M, Inose T, Mikusu S, Tokiwa K, Watanabe T, Tsukada A and Miyakawa N 2015 *Phys. Procedia* **65** 45

[23] Sekine R, Ogata K, Tsukada A and Miyakawa N 2014 *Phys. Procedia* **58** 82

[24] Sekine R, Denholme S J, Tsukada A, Kawashima S, Minematsu M, Inose T, Mikusu S, Tokiwa K, Watanabe T and Miyakawa N 2016 *J. Phys. Soc. Jpn.* **85** 024702

[25] Miyakawa N, Zasadzinski J, Oonuki S, Asano M, Henmi D, Kaneko T, Ozyuzer L and Gray K 2001 *Physica* C **364** 475

[26] Ozyuzer L, Yusof Z, Zasadzinski J F, Li T W, Hinks D G and Gray K 1999 *Physica* C **320** 9

[27] Miyakawa N, Zasadzinski J F, Ozyuzer L, Guptasarma P, Hinks D G, Kendziora C and Gray K E 1999 *Phys. Rev. Lett.* **83**(5) 1018

[28] Kawashima S, Inose T, Mikusu S, Tokiwa K, Watanabe T and Miyakawa N 2010 *Physica* C **470** S178

[29] Eschrig M and Norman M R 2000 *Phys. Rev. Lett.* **85**(15) 3261

IOP Conf. Series: Journal of Physics: Conf. Series **969** (2018) 012032 doi:10.1088/1742-6596/969/1/012032

# Transport Properties of the Layered Transition Metal Oxypnictide $Sr_2ScCo_{1-x}Fe_xPO_3$ with Fe-doped $Co_{1-x}Fe_xP$ layers

**S Okada**[1], **Y Kamihara**[2], **N Ohkubo**[1], **S Ban**[1] **and M Matoba**[2]

[1] Laboratory of Physics, College of Science and Technology, Nihon University, Funabashi, Japan

[2] Department of Applied Physics and Physico-Informatics, Faculty of Science and Technology, Keio University, Yokohama, Japan

E-mail: okada@phys.ge.cst.nihon-u.ac.jp

**Abstract.** Polycrystalline samples of novel oxypnictides, $Sr_2ScCo_{1-x}Fe_xPO_3$, with Fe-doped $Co_{1-x}Fe_xP$ layer were synthesized, and their resistivities and Seebeck coefficients were measured. These materials, $Sr_2ScCo_{1-x}Fe_xPO_3$, crystallize in a layered structure which represent an alternative stack of $ThCr_2Si_2$ type $Co_{1-x}Fe_x$ layers and $K_2NiF_4$ type $Sr_2ScO_3$ layers. The resistivities ($\rho$) decrease with decreasing temperature like a metal. The Seebeck coefficients ($S$) show negative values at room temperature, while the $S$ show positive values for $x$ = 0.04–0.7 at low temperature.

## 1. Introduction

Since the discovery of high $T_c$ superconductivity in F-doped La$FePnO$ ($Pn$ = P and As), many scientists forcus on the transport properties of oxypnictides and oxychalcogenides[1, 2]. These oxypnictides have a quasi-two-dimensional structure, in which an anti-fluorite transition metal ($M$ = Fe, Co, Mn, Ni, $\cdots$) pnictide layer is separated by a carrier blocking layer[3]. These materials are multi anion systems, containing $Pn$ in transition metal layer and O in a carrier blocking layer. The Fe$Pn$ series, in particular, have been intensively studied because the quasi-two-dimensional structure and the Fe$Pn$ layer plays an important role in their superconductivity.

In past research, we have focused on perovskite-type oxypnictides and oxychalcogenides that have a large blocking layer[4, 5], with particular interest in $Sr_2ScFePO_3$ with a $Sr_2ScO_3$ layer that was reported by Ogino *et al.*[6]. $Sr_2ScFePO_3$ crystallizes in an unusual intergrowth structure consisting of an anti-fluorite FeP layer and a perovskite related $K_2NiF_4$-type $Sr_2ScO_3$ layer . It was determined that $Sr_2ScFePO_3$ exhibited superconductivity below 17 K. In 2012, we have successfully synthesized and evaluated the transport properties of $Sr_2ScMPO_3$($M$ = Co[7], Mn, Ni and $Co_{0.5}Fe_{0.5}$[8]). The resistivity ($\rho$) of $Sr_2ScCoPO_3$ is 4.5 m$\Omega$cm at room temperature and decreases with decreasing temperature ($T$). The Seebeck coefficient ($S$) of $Sr_2ScCoPO_3$ is $-12\mu$V/K at room temperature, decreases with decreasing temperature, and rapidly increases with decreasing temperature below 50 K[7]. The $\rho$ of $Sr_2ScMnPO_3$ at room temperature is found to be a large value like that of an insulator[8]. For $Sr_2ScNiPO_3$ and $Sr_2ScCo_{0.5}Fe_{0.5}PO_3$, the values of $\rho$ at room temperature are 1.7 m$\Omega$cm ($M$ = Ni) and 6.0 m$\Omega$cm ($M$ = $Co_{0.5}Fe_{0.5}$), and the $\rho$ decrease with decreasing temperature like that of $Sr_2ScFePO_3$ although both compounds do

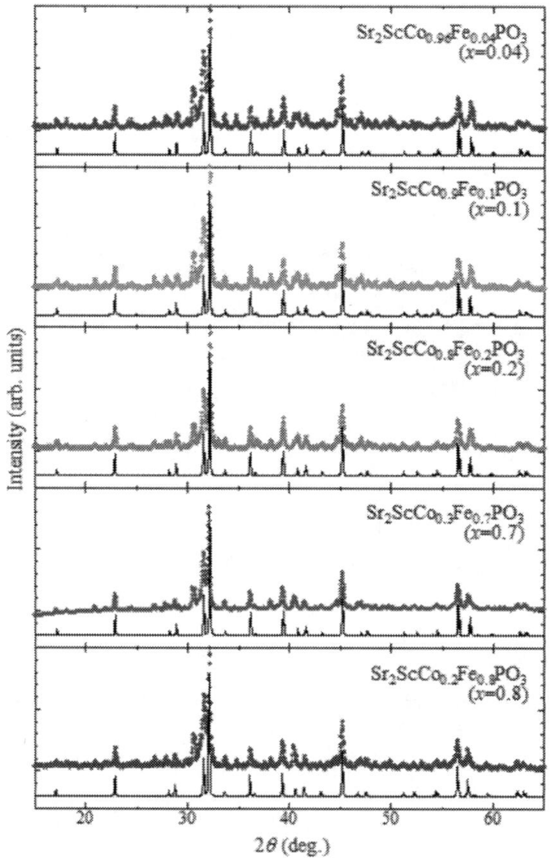


**Figure 1.** Powder XRD patterns of $Sr_2ScCo_{1-x}Fe_xPO_3$.

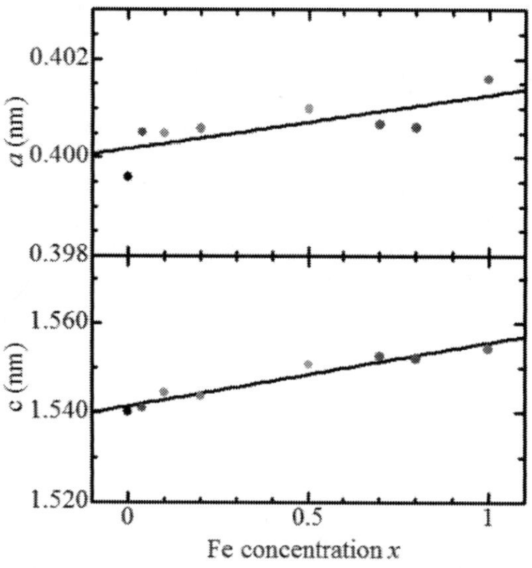


**Figure 2.** Fe concentration dependense of the lattice constants on $Sr_2ScCo_{1-x}Fe_xPO_3$. The solid lines are a guide to the eyes.

not show a superconducting phase. The $S$ of $Sr_2ScNiPO_3$ is $-1.2\mu V/K$ at room temperature, slightly decreases with decreasing temperature down to 50 K, and increases with decreasing temperature below 50 K. Although the similar $S-T$ curve is observed for $Sr_2ScCoPO_3$, the absolute value of $S$ is smaller than that of $Sr_2ScCoPO_3$. For $Sr_2ScCo_{0.5}Fe_{0.5}PO_3$, the values of $S$ at room temperature is $-0.6\mu V/K$, slightly increases with decreasing temperature down to 50 K, and rapidly increases with decreasing temperature below 50 K. The sign inversion of $S$ is obtained at about 270 K, and $S$ retains a positive value below 270 K[8]. This behavior is clearly different from that of $Sr_2ScCoPO_3$. In this study, we have successfully synthesized a Fe-doped oxypnictide, $Sr_2ScCo_{1-x}Fe_xPO_3$, and measured their $\rho$ and $S$ to evaluate electrical transport properties of Fe-doped samples.

## 2. Experiment

Polycrystalline samples of $Sr_2ScCo_{1-x}Fe_xPO_3$ ($x = 0.04$, 0.1, 0.2, 0.7 and 0.8) were synthesized by a solid state reaction in sealed quartz tubes. Precursors of the reaction are CoP, $Fe_2P$, P, SrO, Sr and $Sc_2O_3$. A mixture of starting materials with a molar ratio of CoP:$Fe_2P$:P:SrO:Sr:$Sc_2O_3$ = 1-$x$:$x$:$x$:3:1:1. The mixture of precursors was pressed into pellets. Weighing and pressing were done in Ar-filled glove box. CoP was prepared by a solid state reaction of Co and P with a molar ratio of 1:1 at 700 °C for 40 h. These pellets were calcined at 1200 °C for 30 h in a quartz tube filled with Ar gas under atmospheric pressure of 0.1 at room temperature.

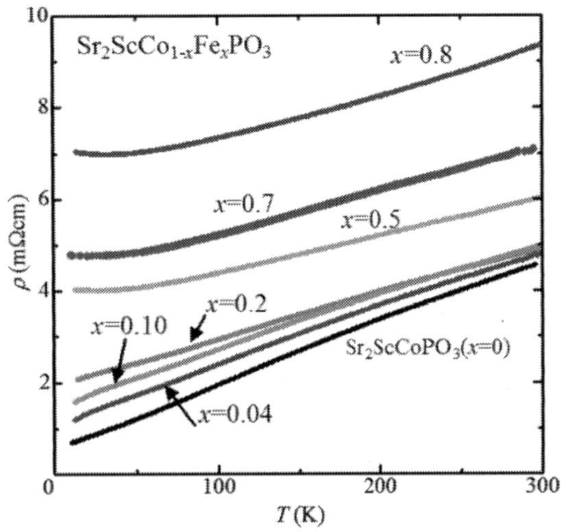


**Figure 3.** Temperature dependence of resistivity ($\rho$) on $Sr_2ScCo_{1-x}Fe_xPO_3$.

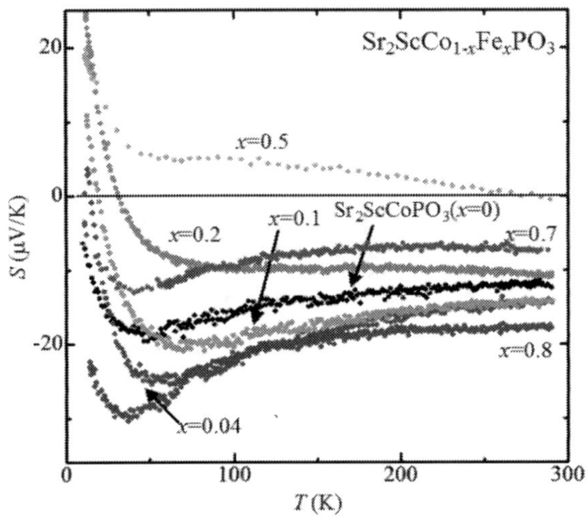


**Figure 4.** Temperature dependence of Seebeck coefficient ($S$) on $Sr_2ScCo_{1-x}Fe_xPO_3$.

Crystallographic phases were examined by powder X–ray diffraction (XRD, Rigaku, Rint2500 with Cu K$\alpha$ radiation). Lattice parameters were roughly obtained by a least–squares fitting method using the diffraction angles collected in the $2\theta$ range from 10 to 70°.

Resistivity measurements were performed by a four–probe method at temperatures from 20 to 300 K. Seebeck coefficients were measured using a steady–state technique at temperatures from 20 to 300 K. A temperature gradient of 0.3 or 0.1 K/mm was generated by a small resistive heater placed on one edge of the sample, and was monitored with a copper–constantan differential thermocouple. The thermopower of the voltage leads was carefully subtracted.

## 3. Result and Discussion

Figure 1 shows the powder XRD patterns of $Sr_2ScCo_{1-x}Fe_xPO_3$ ($x$ = 0.04, 0.1, 0.2, 0.7 and 0.8) with the accompanying Bragg positions calculated by the Rietveld refinement program RIETAN-FP[9]. These patterns are similar to the pattern of $Sr_2ScFePO_3$[6], which indicates that these compounds are isomorphic to $Sr_2ScFePO_3$. The powder XRD patterns were successfully indexed except for some peaks assigned to secondary phases. Figure 2 shows Fe concentration ($x$) dependense of the lattice constants on $Sr_2ScCo_{1-x}Fe_xPO_3$ with $Sr_2ScCoPO_3$ ($x$=0)[7], $Sr_2ScCo_{0.5}Fe_{0.5}PO_3$ ($x$=0.5)[8] and $Sr_2ScFePO_3$ ($x$=1.0)[6]. The lattice constants continuously increases with increasing $x$ and no structural anomaly is observed.

Figure 3 shows the temperature dependence of $\rho$ for $Sr_2ScCo_{1-x}Fe_xPO_3$ with that of $Sr_2ScCoPO_3$ ($x$=0) and $Sr_2ScCo_{0.5}Fe_{0.5}PO_3$ ($x$=0.5) reproduced from refs [7, 8]. In all samples, $\rho$ decreases with decreasing temperature. This metallic behavior is similar to that of $Sr_2ScFePO_3$[6]. For the samples of $x$=0.5, 0.7 and 0.8 below 50 K, the slopes of these $\rho - T$ curves decrease with decreasing temperature; i.e. residual resistivities increase for these samples. Such a enhancement of the residual resistivities is probably due to crystallographic and/or site specific disorder in the two dimensional metal layer. These samples do not show any superconducting phase at $T$ = 20–300 K. The value of $\rho$ at room temperature continuously increases with increasing Fe concentration $x$. This behavior is observed in other 1111-type Fe-based superconductors[10, 11].

Figure 4 shows the temperature dependence of $S$ for $Sr_2ScCo_{1-x}Fe_xPO_3$ with that of $Sr_2ScCoPO_3$ ($x$=0) and $Sr_2ScCo_{0.5}Fe_{0.5}PO_3$ ($x$=0.5) previously reported[7, 8]. The $S$ of all samples at room temperature are negative. These absolute values of $S$ at room temperature is observed to decrease with increasing $x$ up to 0.5, and decrease with increasing over 0.5. $S$ of each sample rapidly increase with decreasing temperature below 50 K. Although the $S$ of $x = 0.2$ and 0.7 samples slightly increase with decreasing temperature near room temperature, like that of $x = 0.5$, the $S$ of other $x$ samples slightly decrease with decreasing temperature like that of $x = 0$. In addition, the sign inversions for $S$ appear for $x = 0.04$ –0.7 below 50 K except for $x = 0.5$ at 270 K. This inversion of the polarity for $S$ is also observed for $Sr_2ScFeAsO_3$ at 100 K[12]. When it was even possible to measure the lower temperature, it is expected that the range of $x$ which is observed the sign inversion becomes wider. The observed differences of the $S-T$ curves reflect the influence of multiband electronic structures for the layered oxypnictides. According to the *ab initio* calculations for $Sr_2ScFePO_3$[13, 14], the electron structure of $Sr_2ScFePO_3$ has hole and electron bands near Fermi Energy ($E_F$). The crystal structures of $Sr_2ScCo_{1-x}Fe_xPO_3$ are isomorphic to $Sr_2ScFePO_3$, so we presume that the electron structures of $Sr_2ScCo_{1-x}Fe_xPO_3$ have also hole and electron bands near $E_F$. The polarity of $S$ indicates a majority carrier-type in conductors. Majority carriers are determined by electronic band structures near $E_F$ and the Fermi-Dirac distribution function. Therefore, the $S-T$ curves are understood as results due to the multi-band electronic structure. Although the lattice constants and the $\rho$ are continuously changed with $x$ and no anomaly is observed, the $S-T$ curves for $x = 0.5$ are different from those of other samples. Therefore, this behavior suggests that there is difference among those electronic structures.

## 4. Conclusion

In summary, we have successfully synthesized polycrystalline samples of novel oxypnictides, $Sr_2ScCo_{1-x}Fe_xPO_3$ ($x = 0.04$, 0.1, 0.2, 0.7 and 0.8). The transport properties of these samples are demonstrated. The value of $\rho$ at room temperature continuously increases with $x$. The $\rho$ of all samples decrease with decreasing temperature like a metal. The $S$ of all samples at room temperature are negative. Although the $S$ of $x = 0.2$ and 0.7 samples slightly increase with decreasing temperature down to 50 K like that of $x = 0.5$, the $S$ of other $x$ samples slightly decrease with decreasing temperature like that of $x = 0$. Inversions of the polarity for $S$ appear for $x = 0.04$ –0.7 below 50 K except for $x = 0.5$ at 270 K. In further study, local structural refinement and theoretical band structure calculations are required to quantitatively analyze the transport and magnetic properties of the materials. Higher accuracy and precisely Rietveld refinement and application of density functional theory[15, 16, 17, 18] are therefore chosen for future work.

## References

[1] Kamihara Y, Hiramatsu H, Hirano M, Kawamura R, Yanagi H, Kamiya T and Hosono H 2006 *J. Am. Chem. Soc.* **128** 10012.
[2] Kamihara Y, Watanabe T, Hirano M, and Hosono H 2008 *J. Am. Chem. Soc.* **130** 3296.
[3] Hosono H 2009 *Physica* **C 469** 314.
[4] Okada S, Matoba M, Fukumoto S, Soyano S, Kamihara Y, Takeuchi T, Yoshida H, Ohoyama K and Yamaguchi 2002 *J. Appl. Phys.* **91** 8861.
[5] Kamihara Y, Matoba M, Kyomen T and Itoh M 2002 *J. Appl. Phys.* **91** 8864.
[6] Ogino H, Matsumura Y, Katsura Y, Ushiyama K, Horii S,Kishio K and Shimoyama J 2009 *Supercond. Sci. Technol.* **22** 075008.
[7] Okada S, Kamihara Y, Ohkubo N, Ban S and Matoba M 2012 *J. Phys.: Conf. Ser.* **400** 022091.
[8] Okada S, Kamihara Y, Ohkubo N, Ban S and Matoba M 2014 *J. Phys.: Conf. Ser.* **568** 022034.
[9] Izumi F and Momma K 2007 *Solid State Phenom.* **130** 15.
[10] Nomra T, Inoue Y, Matsuishi S, Hirano M, Kim J E, Kato K, Tanaka M and Hosono H 2009 *Supercond. Sci. Technol.* **22** 055016.
[11] Cheng P, Xiang Z J, Ye G J, Lu X F, Lei B, Wang A F, Chen F and Luo X G 2014 *Supercond. Sci. Technol.* **27** 065012.

[12] Xie Y L, Liu R H, Wu T, Wu G, Song Y A, Tan D, Wang X F, Chen H, Ying J J, Yan Y J, Li Q J and Chen X H 2009 *Europhys. Lett.* **86** 57007.
[13] Shein I R and Ivanovskii A L 2009 *Phys. Rev.* **B 79** 245115.
[14] Nakamura H and Machida M 2010 *J. Phys. Soc. Jpn.* **79** 013705.
[15] Blochl E 1994 *Phys. Rev.* **B 50** 17953.
[16] Kresse G and Furthmuller J 1996 *Comput. Mater. Sci.* **6** 15.
[17] Kresse G and Furthmuller J 1996 *Phys. Rev.* **B 54** 11169.
[18] Kresse G and Joubert D 1999 *Phys. Rev.* **B 59** 1758.

IOP Conf. Series: Journal of Physics: Conf. Series **969** (2018) 012033 doi:10.1088/1742-6596/969/1/012033

# Ruderman-Kittel interaction between Si in $URu_2Si_2$

**S. Kambe, T. Hattori, Y. Tokunaga, H. Sakai, T.D. Matsuda,[1] Y. Haga and R.E. Walstedt[2]**

Advanced Science Research Center, Japan Atomic Energy Agency, Tokai-mura, Ibaraki 319-1195, Japan

[2]Physics Department, The University of Michigan, Ann Arbor, MI 48109, USA

E-mail: kambe.shinsaku@jaea.go.jp

**Abstract.** $^{29}$Si nuclear magnetic resonance (NMR) has been studied in a $^{29}$Si–enriched single crystal sample of $URu_2Si_2$. The spin-echo decay for applied field $H \parallel [110]$ and [001] directions has been measured at 50 K. A clear spin-echo decay oscillation is observed for both cases, reflecting the Ruderman-Kittel interaction between Si nuclei. Since the observed oscillation frequency depends on the direction of applied magnetic field, the Ruderman-Kittel and anisotropic pseudo-dipolar interactions may not be negligible in this compound. The origin of spin-echo decay oscillations is discussed.

## 1. Introduction

The heavy fermion compound $URu_2Si_2$ undergoes a phase transition [1, 2, 3] at $T_0 \sim 17.5$ K. Since the order parameter of the transition has not ever been clearly identified, such order has been termed "hidden order"[4] and has been studied intensively. In addition, the superconducting transition appears only in the hidden ordered state, indicating that the superconducting pairing glue can be related to fluctuations that occur within the hidden order. Previously, the hidden order and superconducting states have been studied by means of NMR Knight shift and spin-lattice relaxation measurements [5, 6, 7, 8]. Recently, we recognize that NMR spin-echo decay measurements are quite useful for investigating the electronic state of strongly correlated electron systems. It is well know that the spin-echo decay process oscillates due to the Ruderman-Kittel (RK) interaction [9, 10] between nuclei mediated by the conduction electrons. Thus, the modification of conduction bands can be measured via the spin-echo decay. Indeed, a strong change of spin-echo oscillation frequency is observed at low temperatures in $YbRh_2Si_2$ [11].

In the present study, we report data for the $^{29}$Si nuclear spin-echo decay in $URu_2Si_2$. The observed spin-echo decay curves show a clear oscillation, which we attribute to the RK interaction. The anisotropic oscillation frequency indicates that the contributions of the pseudo-dipolar and dipolar (PDD) interactions are not negligible.

## 2. Experimental

Experimental data have been presented and discussed in a previous report [6]. Single crystals were grown using the Czochralski method in a tetra-arc furnace under an argon gas atmosphere,

[1] Present address: Division of Physics, Tokyo Metropolitan University, Hachioji-shi, Tokyo 192-0397, Japan

as previously described. A single crystal sample with a nearly perfect cylindrical shape (1 mm $\theta \perp$ [001]×3 mm‖ [001]), with narrow facets on the (110) planes, has been used.

In the present study, high sample purity was confirmed by a residual resistivity of $\sim 5\ \mu\Omega$cm (RRR $\sim$ 70). The resistivity was measured in a piece cut ($2 \times 0.5 \times 0.5$ mm$^3$) from the same single crystal used for NMR measurements.

The natural abundance of the NMR isotope $^{29}$Si ($I = 1/2$, gyromagnetic ratio $^{29}\gamma/2\pi$=0.84577 kHz/Oe) is only 4.7%. This has prevented highly accurate Si NMR measurements in $URu_2Si_2$ in the past. For the present study, a single crystal sample with a 53% enriched $^{29}$Si isotope has been prepared, improving the NMR sensitivity by a factor $\sim$ 11. As there are no quadrupolar interactions for $I = 1/2$ nuclei, $^{29}$Si NMR spectra reflect only magnetic shift and broadening effects.

A single crystal specimen was mounted in a $^4$He cryostat with an NMR pickup coil. Using a standard $\pi/2 - \pi$ pulse sequence, the spin-echo intensity $m(2\tau)$ so generated was measured as a function of the time $\tau$ between excitation pulses (see Fig. 1). Here, a typical $\pi/2$ pulse width is $4 \sim 5\ \mu$sec. Since the resonance linewidth is quite narrow (*e.g.* $\sim$ 1 kHz at 2.5 T along the [110] axis in the paramagnetic state), all nuclear spins in the spectrum were quite uniformly excited by the radio-frequency pulses used. The pulse repetition time $t_{rep}$ was taken to be much longer than the previously determined $T_1$ [12].

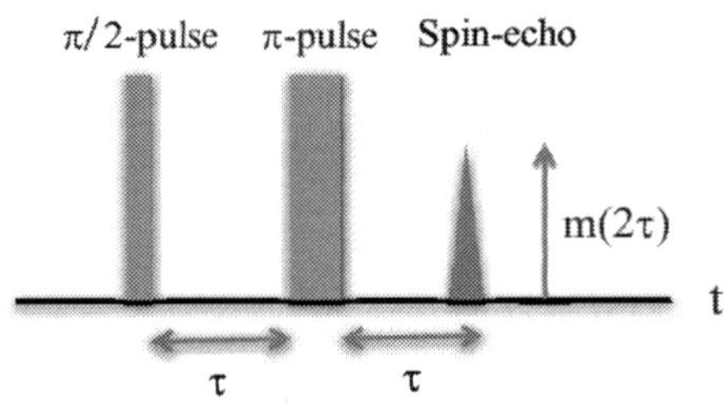


**Figure 1.** Schematic description of NMR spin-echo formation. A $\pi/2$ pulse rotates the nuclear magnetic moment by 90°, a $\pi$ pulse by 180°.

## 3. NMR Spin-echo decay

The spin-echo intensity $m$ decays with increasing time period $\tau$ between the first ($\pi/2$) and second ($\pi$) pluses due to several interactions between nuclei spins. Generally, the $\tau$ dependence of $m$ may be expressed;

$$\frac{m(2\tau)}{m(0)} = \exp\{-\frac{2\tau}{T_{1L}} - \frac{1}{2}\left(\frac{2\tau}{T_{2G}}\right)^2\}\Omega(2\tau), \qquad (1)$$

where $T_{1L}$ is the Lorentzian type relaxation. For the present case $T_{1L}$ can be replaced by the spin-lattice relaxation time $T_1$ determined previously [12]; $T_{2G}$ is the Gaussian spin-spin relaxation time. $\Omega(2\tau)$ is the modulation factor which is under investigation.

### *3.1. Spin-echo decay modulation due to RK and PDD interactions*

It is well known that the RK and PDD interactions can induce oscillations in the spin-echo decay curve [13, 14, 15]. In metals the indirect RK interaction between nuclear spins $\mathbf{I}_i, \mathbf{I}_j$ occurs through second–order scattering of conduction electrons [9], taking the scalar form

$$J_{ij}\mathbf{I}_i \cdot \mathbf{I}_j. \tag{2}$$

In contrast to the scalar RK interaction (Eq. 2), the PDD interaction is a tensor quantity [10],

$$B_{ij}(\mathbf{I}_i \cdot \mathbf{I}_j - 3I_j^z I_k^z), \tag{3}$$
$$B_{ij} = \frac{1}{2}b_{ij}(3\cos^2\delta_{ij} - 1), \tag{4}$$

where $\vec{r_{ij}}$ is the radius vector between the $i$th and the $j$th nuclei, $\delta_{ij}$ is the angle between $\vec{r_{ij}}$ and the applied magnetic field, and $b_{ij}$ is the effective coefficient for a dipolar–type interaction $B_{ij}$ between the $i$th and $j$th nuclei. Thus, the PDD interaction depends on the direction of applied field.

A complete formula for the spin-echo decay modulation factor $\Omega(2\tau)$ due to RK and PDD interactions is given by Alloul and Froidevaux [15]. As they have used polycrystalline samples, angle-integrated equations are given. Here, using their formulas, equations for the $I = 1/2$ case for a single crystal are presented. Considering the nearest neighbor (nn) sites among ions considered (i.e. Si for the present case),

$$\Omega(2\tau) = \sum_{r=0}^{N} P_r \sum_{k=1}^{n_r} \rho_{k,r} \sum_{\alpha=1}^{r} \cos^r(2\pi\tau G_{\alpha,k}), \tag{5}$$
$$G_{\alpha,k} \equiv J + B - 3B\cos^2\Theta_{\alpha,k}, \tag{6}$$
$$\sum_{k=1}^{n_r} \rho_{k,r} = 1, \tag{7}$$

where $J$ and $B$ are the RK and PDD interactions between nn nuclear spins, respectively; $N$ is the total number of nn sites; $r$ corresponds to the number of occupied nn sites (i.e. with $^{29}$Si nuclear spins); $P_r =_N C_r c^r(1-c)^{N-r}$ is the binomial coefficient; $c = 0.53$ is the concentration of $^{29}$Si isotope; $n_r$ is number of possible configurations of occupied nn sites for a given $r$, the configurations being indexed by $k$, $\rho_{k,r}$ is a distribution of situation $k$ of $r$ case; $\Theta_{\alpha,k}$ is the angle between the applied magnetic field $H$ and the direction to the $\alpha$th nn for the situation $k$. It should be noted that contributions of unlike spins are negligible in the present case.

### *3.2. $URu_2Si_2$ case*

For the case of $^{29}$Si in $URu_2Si_2$, there is only one nn site, namely along the $c$ axis, i.e $N = 1$ (Fig. 2), as is also the case for $YbRh_2Si_2$. The RK and PDD interactions (i.e. $J$ and $B$) are proportional to $R^{-3}$ ($R$ is distance between nn sites). Since the distance to four second nn sites (3.75Å) is comparable with that of the nn site (2.35Å) in $URu_2Si_2$, both cases are included here. The definition of angles $\phi, \theta$ and $\psi$ is given in Fig. 2.

*a) $\Omega_{1st}(2\tau)$ for the first nn case*

The nn case is simple since there is only one nn site ($N = 1$, $n_r = 1$), which was already discussed previously for $YbRh_2Si_2$ [11] (note: definition of $\theta$ in Ref. [11] corresponds to present $\phi$). There is a unique oscillation frequency $G$ for $\Omega_{1st}(2\tau)$. Thus,

$$\Omega_{1st}(2\tau) = (1-c) + c\cos(2\pi\tau G), \tag{8}$$
$$G \equiv \pm(J_1 + B_1 - 3B_1\cos^2\phi), \tag{9}$$

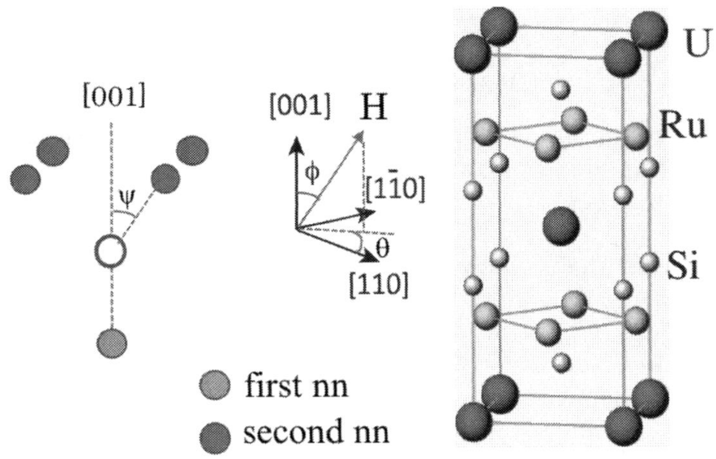


**Figure 2.** Left: The one nn and four second–neighbor Si sites to the addressed (open circle) Si site. Middle: The definition of angles $\phi$ and $\theta$ between applied magnetic field $H$ and the crystal axes. Right: Crystal structure of $URu_2Si_2$. As shown in the Left figure, the first nn Si site is found along the [001] axis.

where $J_1$ and $B_1$ are the RK and PDD interactions between the first nn sites, respectively. The double signs are coordinated. The signs in Eq. 9 cannot be determined by the present method; fortunately, they are not required in a fit to the data. For the nn site, $\Omega(2\tau)$ is independent of $\theta$ since the nn sites are located along the $c$ ([001]) axis.

*b) $\Omega_{2nd}(2\tau)$ for the second four nn case*
The second four nn sites ($N = 4$) case is more complicated. All possible configurations $n_r$ for $r = 0 - 4$ cases should be considered. Based on Eq. 5, $\Omega_{2nd}(2\tau)$ may be expressed;

$$\begin{aligned}\Omega_{2nd}(2\tau) =&(1-c)^4+\\ &2c(1-c)^3\{\cos(2\pi\tau G_I)+\cos(2\pi\tau G_{II})\}+\\ &c^2(1-c)^2\{\cos^2(2\pi\tau G_I)+\cos^2(2\pi\tau G_{II})\}+\\ &4c^2(1-c)^2\{\cos(2\pi\tau G_I)\cos(2\pi\tau G_{II})\}+\\ &2c^3(1-c)\{\cos^2(2\pi\tau G_I)\cos(2\pi\tau G_{II})+\cos(2\pi\tau G_I)\cos^2(2\pi\tau G_{II})\}+\\ &c^4\{\cos^2(2\pi\tau G_I)\cos^2(2\pi\tau G_{II})\}.\end{aligned} \tag{10}$$

$G_{I,II}$ are defined as;

$$\begin{aligned}G_I &\equiv \pm\{J_2+B_2-3B_2(\sin\psi\sin\phi\cos\theta+\cos\psi\cos\phi)^2\},\\ G_{II} &\equiv \pm\{J_2+B_2-3B_2(\sin\psi\sin\phi\sin\theta+\cos\psi\cos\phi)^2\},\end{aligned} \tag{11}$$

IOP Conf. Series: Journal of Physics: Conf. Series **969** (2018) 012033 doi:10.1088/1742-6596/969/1/012033

**Table 1.** Values of $1/T_{2G}$ and $G$ obtained from fits based on Eqs. 1 and 8, along with previously determined $1/T_1$ values [12], which are adopted as $1/T_{1L}$.

| | $1/T_{1L}$ (sec$^{-1}$) | $1/T_{2G}$ (sec$^{-1}$) | $G$(kHz) |
|---|---|---|---|
| $H \parallel [001]$ | 2.3 | $452 \pm 15$ | $1.026 \pm 0.02$ |
| $H \parallel [110]$ | 17.1 | $572 \pm 11$ | $0.583 \pm 0.01$ |

where $J_2$ and $B_2$ are the RK and PDD interactions with the second nn sites, respectively. The double signs are coordinated. The signs in Eq. 11 cannot be determined by the present method; fortunately, they are not required for a fit to the data. In $URu_2Si_2$, $\cos\psi = 0.626$. Compared to the first nn case, the second nn case contains many different oscillation frequencies. Note that Eq. 10 has 4-fold $\theta$ symmetry as expected.

Generally, the contribution from the nn site is dominant. If the second nn contributions are not negligible, $\Omega(2\tau)$ may be expressed [15];

$$\Omega(2\tau) = \Omega_{1st}(2\tau)\Omega_{2nd}(2\tau), \tag{12}$$

which is often a complicated function of $2\tau$.

## 4. Results and discussion

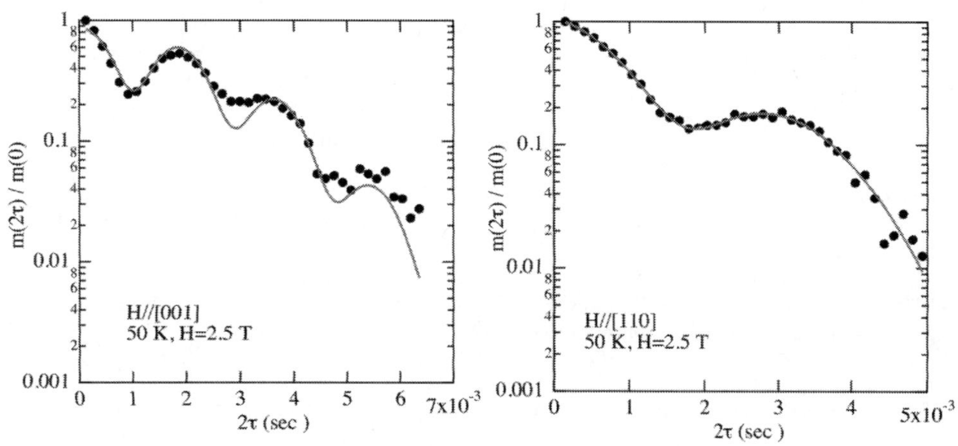


**Figure 3.** Spin-echo decay curves at 50 K for $H \parallel [001]$ ($\phi = 0$) and $H \parallel [110]$ ($\phi = 90°$, $\theta = 0$) in $URu_2Si_2$. Solid lines are fitting curves based on Eqs. 1 and 8. Parameters obtained are presented in Table 1.

Figure 3 shows spin-echo decay data for $H = 2.5\ T \parallel [001]$ and $[110]$ directions at 50 K. The data obtained can be well fitted using Eqs. 1 and 8, indicating that the contributions from the second nn sites are almost negligible. It should be noted that no better fitting can be obtained using Eq. 12, indicating that the first nn contribution dominates the spin-echo decay at the Si site in $URu_2Si_2$. As the $G$ values obtained depend on the direction of field, certainly the PDD interaction should be considered in the present case. Based on Eq. 9 and $G$ values from Table 1, equations $\pm(J_1 - 2B_1) = 1026$; $\pm(J_1 + B_1) = 583$ are deduced. From these equations, $|J_1| = 731$

Hz and $|B_1| = 148$ Hz are obtained. If we consider the nuclear dipole-dipole contribution of 324 Hz in this enriched $URu_2Si_2$, the purely pseudo-dipolar interaction is estimated $\pm 148 + 324 = 472$ or 176 Hz (see [11]).

## 5. Summary

In the present study, spin-echo oscillations have been observed in the paramagnetic state of $URu_2Si_2$. The results are well accounted for by the Ruderman-Kittel interaction. As the Ruderman-Kittel interaction is due to conduction electrons, the spin-echo oscillation frequency is considered to be sensitive to modifications of the Fermi surface in $URu_2Si_2$. This is a possible topic for future investigation.

## Acknowledgments

We are grateful for stimulating discussions with H. Ikeda, H. Harima, D. Aoki, H. Yamagami and J. Flouquet This work was supported by JSPS KAKENHI Grant Number 15H05884 (J-Physics) and No.15K05152 and the REIMEI Research Program of JAEA.

## References

[1] Palstra T T M, Menovsky AA, van den Berg J, Dirkmaat A J, Kes P H, Nieuwenhuys G J, and Mydosh J A 1985 *Phys. Rev. Lett.* **55** 2727.
[2] Schlabitz W, Baumann J, Pollit B, Rauchschwalbe U, Mayer H M, Ahlheim U and Bredl C D 1986 *Z. Phys. B Condensed Matter* **62** 171.
[3] Maple M B, Chen J W, Dalichaouch Y, Kohara T, Rossel C, Torikachvili M S, McElfresh M W and Thompson J D 1986 *Phys. Rev. Lett.* **56** 185.
[4] Tripathi V, Chandra P and Coleman P 2007 *Nature Phys.* **3** 78.
[5] Kambe S, Tokunaga Y, Sakai H, Matsuda T D, Haga Y, Fisk Z and Walstedt R E 2013 *Phys. Rev. Lett.* **110** 246406 and references therein.
[6] Kambe S, Tokunaga Y, Sakai H and Walstedt R E 2015 *Phys. Rev. B* **91** 035111.
[7] Walstedt R E, Kambe S, Tokunaga Y and Sakai H 2016 *Phys. Rev. B* **93** 045122.
[8] Hattori T, Sakai H, Tokunaga Y, Kambe S, Matsuda T D and Haga Y 2016 *J. Phys. Soc. Jpn.* **85** 073711.
[9] Ruderman M A and Kittel C 1954 *Phys. Rev.* **96** 99.
[10] Bloembergen N and Rowland T J 1955 *Phys. Rev.* **97** 1679.
[11] Kambe S, Sakai H, Tokunaga Y, Hattori T, Lapertot G, Matsuda T D, Knebel G, Flouquet J and Walstedt R E 2017 *Phys. Rev. B* **95** 195121.
[12] Kohori Y, Matsuda K and Kohara T 1996 *J. Phys. Soc. Jpn.* **65** 1083.
[13] Walstedt R E, Dowley M W, Hahn E L and Froidevaux C 1962 *Phys. Rev. Lett.* **8** 406.
[14] Froidevaux C and Weger M 1964 *Phys. Rev. Lett.* **12** 123.
[15] Alloul H and Froidevaux C 1967 *Phys. Rev.* **163** 324.

IOP Conf. Series: Journal of Physics: Conf. Series **969** (2018) 012034 doi:10.1088/1742-6596/969/1/012034

# Vortex states in a submicron Bi2212 crystal probed by intrinsic Josephson junctions

**S. Ooi, M. Tachiki, T. Mochiku, H. B. Wang, K. Komori, K. Hirata, and S. Arisawa**

National Institute for Materials Science, 1-2-1 Sengen, Tsukuba 305-0047, Japan

E-mail: OOI.Shuuichi@nims.go.jp

**Abstract.** To study the pancake-vortex states confined in a submicron $Bi_2Sr_2CaCu_2O_{8+y}$ (Bi2212) crystal, we have measured the $c$-axis resistance and $I$-$V$ characteristics of a stack of intrinsic Josephson junctions with a lateral dimension less than 1 $\mu$m. Although the stack was accidentally shunted by a parallel resistance of 7.5 k$\Omega$, the $I$-$V$ characteristics show homogeneous multiple branches after the subtraction of the component. The penetrations of single vortices into the submicron stack were clearly observed in the resistance measurements. A vortex phase diagram was constructed by mapping the $c$-axis resistance on an $H$-$T$ plane. Temperature dependence of the first-vortex penetration field is consistent with the theoretical estimation on the formation of a pancake-vortex stack in the center of a superconducting strip.

## 1. Introduction

With reducing the size of a small object, the increase of the surface-to-volume ratio and/or a smaller number of elements constituting the object influence their physical properties including thermodynamic ones [1]. This can be also expected in vortex matter confined in sufficiently smaller high-$T_c$ superconductors. While past extensive studies revealed very rich pancake-vortex states and phase transitions in bulk high-$T_c$ superconductors [2, 3], it has attracted interest how the vortex states and dynamical properties, such as penetrations and exits of vortices, are modified by the reduction of sample sizes. Actually, this subject was studied especially in $Bi_2Sr_2CaCu_2O_{8+y}$ (Bi2212) in a field perpendicular to the superconducting planes [4–8]. However, these observations were limited in small Bi2212 crystals down to 10 - 20 $\mu$m, probably because of the field sensitivity and/or the spatial resolution of the measurements using magnetic detection techniques, e.g., micro-Hall-sensor, magneto-optical imaging or Bitter decoration.

On the other hand, $c$-axis electrical transport measurements using a stack of intrinsic Josephson junctions (IJJs) give us an alternative mean to detect the vortex states [9–12]. In samples with several tens micron, a melting transition of vortex lattice has been observed as a resistive jump of the $c$-axis resistance $R_c$ and as an anomaly in the $c$-axis critical current $I_c$ in stacks of IJJs [10]. Moreover, the penetrations of individual vortices were found by $R_c$ measurement in smaller stacks with an in-plane area of 1.2 to 100 $\mu m^2$ [11, 13]. To check applicability of the vortex-detection method using IJJs for much smaller Bi2212 and to explore vortex states in such samples, we have investingated a submicron-sized stack of IJJs fabricated by a focused ion beam(FIB) in present study. In this proceeding, we report successful detection

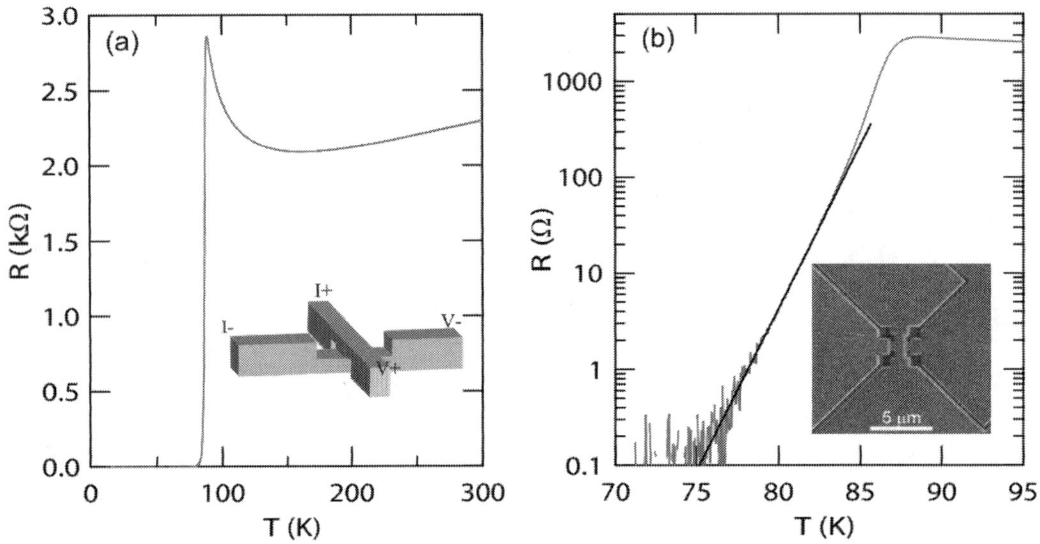


**Figure 1.** (a) Temperature dependence of $c$-axis resistance in a submicron stack ($0.75 \times 0.75$ $\mu m^2$) of IJJs with cross-type Bi2212 electrodes. The inset schematically shows a configuration of the measured stack and the electrodes. The central part between the crossed electrodes, indicated by red, is the stack of IJJs. (b) The same data as (a) on a semi-logarithmic scale, in which a tail of resistance is manifested below $T_c$. The solid straight line is a fitting line (see text). The inset is a SIM image of the submicron stack (top view).

of the penetrations and exits of individual vortices by $R_c$ measurements and the temperature dependence of the first-vortex penetration field $H_{p1}$ in the submicron stack of Bi2212.

## 2. Experimental

High-quality single crystals of Bi2212 were grown by the traveling-solvent floating-zone method [14]. In present study, we employed a cross-type structure of Bi2212 containing a stack of IJJs in the center (see the inset of Fig. 1(a)). In this structure, both current and voltage electrodes are directly attached on the stack so as to only measure the $c$-axis resistance of the stack without any series in-plane resistance. Our fabrication process is similar to the method used to fabricate whiskers by Latyshev, *et al.* [15], in which narrow grooves were fabricated on the top and bottom surfaces by a flipping of whiskers using FIB. Only the difference from their process is that we used a thin flake of single crystal instead of the whiskers. A scanning-ion-microscope (SIM) image of a fabricated sample is shown in the inset of Fig. 1(b). The lateral dimension of the square stack is $\sim 0.75\mu$m and the thickness is $\sim$ 30nm, estimated from the number of branches in $I$-$V$ characteristics. Transport measurements were performed using the four-probes technique with a current source (Keithley 6430) and a nanovoltmeter (Keithley 2182) in a chamber with a cryocooler. Magnetic fields were applied parallel to the $c$ axis of Bi2212. Data of the $c$-axis resistance were acquired in field-scans from negative to positive at various fixed temperatures.

## 3. Results and discussion

Temperature dependence of $R_c$ for the fabricated submicron stack is shown in Fig. 1(a) and (b). Upturn of $R_c$, a typical feature on the $c$-axis resistance of Bi2212, is observed above the superconducting transition temperature $T_c$. Although, even in this tiny stack, the superconducting transition accompanies rather sharp drop of the resistance as shown in Fig. 1(a),

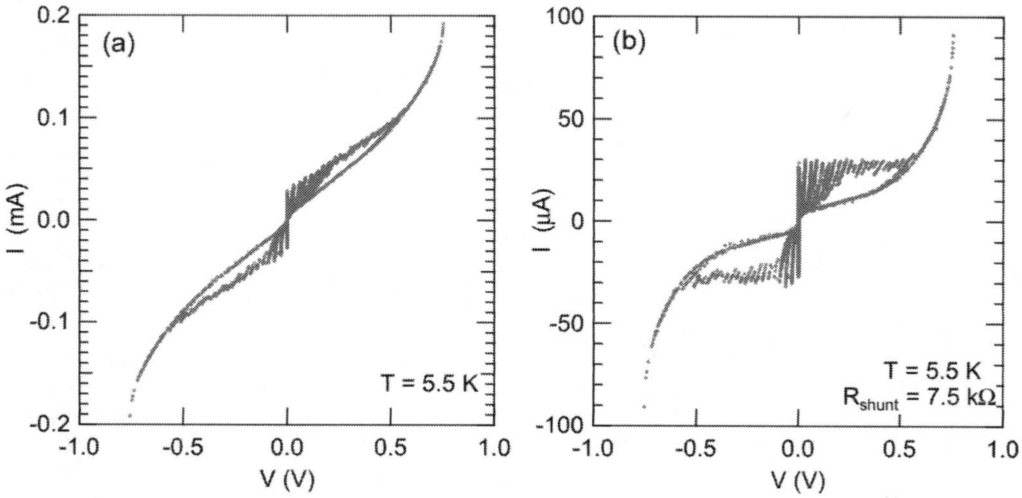


**Figure 2.** (a) $I$-$V$ characteristics of the submicron stack at 5.5 K in a zero magnetic field. (b) Corrected $I$-$V$ characteristics by subtraction of the contribution from a parallel resistance shunting the stack. The estimated value of the shunt resistance is 7.5 kΩ.

there remains a broad resistive tail below $T_c$ which is clarified in Fig. 1(b) of a semi-logarithmic plot, where the tail is well-fitted by a line of $R_c \propto \exp(\alpha T)$. The resistance on the tail does not depend on magnetic fields unless there are no vortices in the stack. Probably, the resistive tail is related to a thermally-activated phase fluctuation, because the Josephson coupling energy $I_c\Phi_0/2\pi$ ($\Phi_0$ is the flux quantum) is comparable with the thermal energy $k_BT$ in this temperature range for this sample.

Figure 2(a) shows $I$-$V$ characteristics of the submicron stack at 5.5 K in a zero field. Although it seems that $I_c$ of IJJs is inhomogeneous, this is because of a shunt resistance $R_{shunt}$ which may be accidentally introduced during the fabrication. After the correction on the current axis to subtract the contribution of $R_{shunt}$, i.e., $I - V/R_{shunt}$, we obtain $I$-$V$ characteristics with quite uniform multiple branches as shown in Fig. 2(b). The number of the branches is 21 and the branches have uniform value of $I_c \sim 30$ $\mu$A, corresponding to a critical current density $J_c \sim 4700$ A/cm$^2$, indicating this stack is in the overdoped regime. This high value of $J_c$ prevents the suppression of $J_c$ itself from the thermal fluctuation even in the tiny submicron stack of IJJs. More increase of the doping level of Bi2212, i.e., increase of $I_c$, is a route to realize much smaller stacks of IJJs without the suppression of $J_c$, which would be useful to study vortex state confined in a Bi2212 crystal of 100-nm or less size.

Vortex states in the submicron Bi2212 are studied by magnetic field dependence of $R_c$ at a fixed temperature. A representative example of $R$-$H$ curve is shown in Fig. 3(a). Around the zero field there is a small offset resistance which is equivalent to the resistance tail observed in $R$-$T$. The periodic peaks appeared above 100 Oe indicate the vorticity change, i.e., the penetration or exit of an individual vortex [12, 13]. For instance, the field interval of the neighboring third and fourth peaks from the zero field is 43 Oe, corresponding to the magnetic field to add $\Phi_0$ in an area of 0.69 $\mu$m square. Although this is a little smaller value than the fabricated sample dimension, estimated sizes of samples are usually smaller than actual ones, probably because of the existence of a screening current and a thin region around the edges damaged by FIB [13]. To unveil the whole features of the vortex phase diagram, we map log $R_c$ on a $H$-$T$ plane as shown in Fig. 3(b). The regions with a different vorticity are well separated, so that it is easy to distinguish

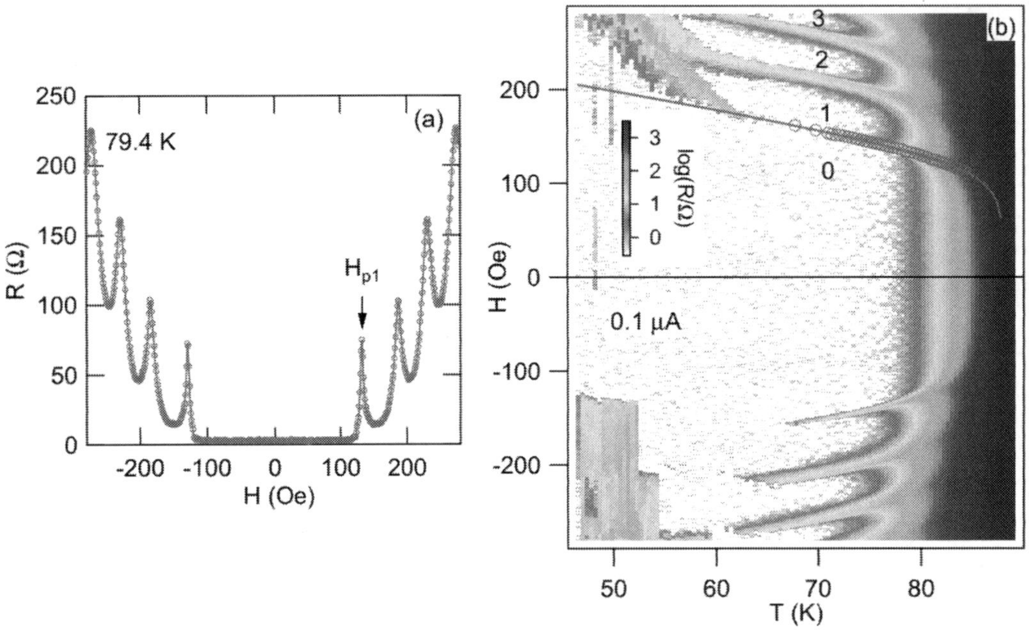


**Figure 3.** (a) $c$-axis resistance of the submicron stack as a function of magnetic field at 79.4 K with an applied current of 100 nA. Sharp resistance peaks emerge when an individual vortex penetrates or exits from the stack. (b) Mapping of log $R_c$ on a $H$-$T$ plane, indicating a vortex phase diagram in the submicron squared Bi2212. The fields of the resistance peak corresponding to the first-vortex penetration fields $H_{p1}$ are marked by the red circles. The red curve shows a fitting curve of Eq. (1) for the field where fast pancake-vortex stack formation starts in the center of a superconducting strip.

the number of vortex at a certain point on the map, especially at higher temperatures than 60 K. Below ~60 K, appearance of the log $R_c$ map is asymmetric with respect to the $H = 0$ line, which means hysteretic and irreversible behavior regarding the vortex penetrations and exits at the low temperatures. This is caused by the Bean-Livingston surface barrier and/or the pinning by quenched disorder whose influence becomes stronger at lower temperatures [2].

In the reversible region, the first-vortex penetration fields $H_{p1}$ are well-defined in the map, which are marked by red circles. $H_{p1}$ rapidlly drops in the vicinity of $T_c$ with increasing temperature. This temperature dependence is consistent with the theoretical analysis for the stability of pancake vortices. The field where a stack of pancake vortices is formed in the center of a strip-shaped layered superconductor with a width $w$ has been evaluated as

$$H_{s1} \approx \frac{\Phi_0}{4\pi\lambda^2}\left[\ln\left(\frac{w}{\pi\xi}\right)+0.5\right]\frac{\cosh(w/2\lambda)}{\cosh(w/2\lambda)-1}\,, \tag{1}$$

here $\lambda$ and $\xi$ are the penetration depth and the superconducting coherence length, respectively [16]. As shown in Fig. 3(b), the temperature dependence of $H_{p1}$ is nicely fitted by Eq. (1) with fixed parameters of $T_c = 88.0$ K and $\lambda/\xi = 100$, and a temperature dependence $(1 - T/T_c)^{-1/2}$ for $\lambda$ and $\xi$. Obtained fitting parameters are $w = 0.65$ $\mu$m and $\lambda = 0.23$ $\mu$m, both of which are quite reasonable values in spite of the difference of sample shapes (square and strip).

## 4. Summary

We have studied the pancake-vortex states confined in a submicron-sized Bi2212 single crystal by the $c$-axis electrical transport measurements using a square stack of IJJs with a lateral dimension of 0.75 $\mu$m. After the correction of the contribution from the parallel resistance, which was accidentally introduced during the fabrication, the $I$-$V$ characteristics shows homogeneous multiple branches. The penetrations(exits) of individual vortices into(from) the submicron stack were clearly observed in $R_c$-$H$ measurements. The vortex phase diagram constructed by a mapping $\log R_c$ on an $H$-$T$ plane shows a reversible feature for the penetrations of vortices above 60-70 K. The temperature dependence of the first-vortex penetration field is consistent with the theoretical estimation on the formation of a pancake-vortex stack in the center of a superconducting strip.

## Acknowledgements

Present works are partially supported by JSPS KAKENHI Grant Number JP16H04022, JP24560021.

## References

[1] Mei Q S and Lu K 2007 *Prog. Mater Sci.* **52** 1175
[2] Blatter G, Geshkenbein V B, Larkin A I and Vinokur V M 1994 *Rev. Mod. Phys.* **66** 1125
[3] Beidenkopf H, Avraham N, Myasoedov Y, Shtrikman H, Zeldov E, Rosenstein B, Brandt E H and Tamegai T 2005 *Phys. Rev. Lett.* **95** 257004
[4] Wang Y, Fuhrer M, Zettl A, Ooi S and Tamegai T 2001 *Phys. Rev. Lett.* **86** 3626
[5] Connolly M R, Milošević M V, Bending S J, Clem J R and Tamegai T 2009 *Europhys. Lett.* **85** 17008
[6] Dolz M I, Fasano Y, Cejas Bolecek N R, Pastoriza H, Konczykowski M and van der Beek C J 2014 *J. Phys: Conf. Ser.* **568** 022010
[7] Cejas Bolecek N R, Dolz M I, Kolton A, Pastoriza H, van der Beek C J, Konczykowski M, Menghini M, Nieva G and Fasano Y 2015 *J. Low Temp. Phys.* **179** 35
[8] Dolz M , Fasano Y, Cejas Bolecek N , Pastoriza H, Mosser V, Li M and Konczykowski M 2015 *Phys. Rev. Lett.* **115** 137003
[9] Yurgens A, Winkler D, Claeson T, Yang G, Parker I F G and Gough C E 1999 *Phys. Rev. B* **59** 7196
[10] Ooi S, Mochiku T and Hirata K 2007 *Phys. Rev. B* **76** 224527
[11] Kakeya I, Fukui K, Kawamata K, Yamamoto T and Kadowaki K 2008 *Physica C* **468** 669
[12] Ooi S, Mochiku T, Tachiki M and Hirata K 2015 *Phys. Rev. Lett.* **114** 087001
[13] Ooi S, Mochiku T, Tachiki M and Hirata K 2015 *Physics Procedia* **65** 109
[14] Mochiku T, Hirata K and Kadowaki K 1997 *Physica C* **282-287** 475
[15] Latyshev Y I and et al 1999 *Pis'ma Zh. Eksp. Teor. Fiz.* **69** 75
[16] Koshelev A E, Buzdin A I, Kakeya I, Yamamoto T and Kadowaki K 2011 *Phys. Rev. B* **83** 224515

# Two-photon absorption and amplitude mode in conventional superconductors with paramagnetic impurities

**Takanobu Jujo**

Graduate School of Materials Science, Nara Institute of Science and Technology, 8916-5, Takayama-cho, Ikoma, Nara 630-0101, Japan

E-mail: jujo@ms.aist-nara.ac.jp

**Abstract.** The two-photon absorption (TPA) spectrum of BCS superconductors is theoretically investigated. We discuss a possibility of observing the amplitude mode (one of collective modes in superconductors) in this spectrum. An absorption edge by this mode appears at the same frequency as that by quasiparticle excitations in the TPA spectrum of superconductors with time-reversal symmetry, and it will be difficult to distinguish the amplitude mode from quasiparticle excitations. This property originates from the fact that the magnitude of superconducting order parameter is one half of the energy gap in the linear absorption spectrum. In contrast to this one half of the energy gap is smaller than the magnitude of order parameter in superconductors with paramagnetic impurities. Though the amplitude mode is not well-defined owing to the broken time-reversal symmetry, a numerical calculation shows that the peak by this mode persists in the TPA spectrum. This peak appears at the frequency different from that of the absorption edge by quasiparticle excitations, and it is possible to specify the amplitude mode in the TPA spectrum.

## 1. Introduction

Microwave absorption spectrum reflects various properties in the superconducting state because the value of its frequency ($\omega$) is close to that of the superconducting gap ($\Delta_0$). The linear absorption spectrum shows that there is a finite absorption for $\omega > 2\Delta_0$, which is caused by quasiparticle excitations [1]. In the superconducting state collective modes (phase and amplitude modes of the order parameter) exist owing to the spontaneous symmetry breaking. The phase mode does not appear in the microwave absorption spectrum because it couples to the fluctuation of electron density and becomes a plasmon [2]. The amplitude mode is not effective in the linear absorption spectrum when superconductors have inversion symmetry. Then the amplitude mode [3, 4] has been studied in the superconductivity (SC) which coexists with the charge density wave (CDW) state, and was observed in the phonon spectrum through the coupling between SC and CDW [5, 6].

Recently the amplitude mode has been observed in a superconductor without the CDW state with use of the pump-probe spectroscopy [7]. In this experiment the amplitude mode is observed as an oscillation of the optical conductivity in the transient state. In contrast to this, a calculation of two-photon absorption (TPA) shows that there is a method to observe the amplitude mode in the stationary state [8]. By this calculation it is found that there is a peak around $\omega \simeq \Delta_0$ in the TPA spectrum owing to the amplitude mode. It is also shown that the quasiparticle excitation by TPA gives a finite absorption edge at $\omega = \Delta_0$ in contrast to the vanishing absorption edge in the case of the linear absorption. Then there will be a possibility that an obscurity arises as for the origin of the absorption edge at $\omega = \Delta_0$ in the TPA spectrum (the amplitude mode or the quasiparticle excitation).

In this paper we show that a peak by the amplitude mode appears at the frequency different from an absorption edge by quasiparticle excitations in superconductors with paramagnetic impurities. Though there is no well-defined amplitude mode [9] due to the broken time-reversal symmetry, it is shown that the peak by this mode remains in the TPA spectrum.

## 2. Formulation

A nonlinear current is written as follows [10].

$$J_\omega^{(3)\mu} = e \int_{\rm FS} v_k^\mu \frac{mk_F}{2\pi} \int \frac{d\epsilon}{4\pi i} \sum_{\nu=x,y,z} {\rm Tr}[v_k^\nu \hat{g}_\nu^{K(3)}(\epsilon+\omega,\epsilon)]. \tag{1}$$

Here, $^{(3)}$ in the superscript indicates the third order of external field. We assume that the system is isotropic, and omit indices indicating directions of velocity ($v_k$) and external field hereafter. $\hat{g}^{K(3)}(\epsilon+\omega,\epsilon) = \frac{1}{\pi}\int d\xi \hat{G}_k^{K(3)}(\epsilon+\omega,\epsilon)$ is the quasiclassical Keldysh Green's function. ($\hat{}$ and Tr indicate the Nambu representation and the summation of its diagonal elements, respectively. The summation over wave number is replaced by the integration as $\frac{1}{N^3}\sum_k = \frac{mk_F}{2\pi^2}\int_{FS}\int d\xi$ with $m$ the mass of electrons, $k_F$ the Fermi wave number, $N^3$ the number of sites and $\int_{FS}$ the integration over the Fermi surface. We put $\hbar = c = 1$ in this paper with $c$ the velocity of light.)

In the dirty limit $\hat{g}^{K(3)}(\epsilon+\omega,\epsilon)$ is derived from kinetic equations [11], and the result is written with use of quasiclassical Green functions including the second order of external field as follows.

$$\begin{aligned} J_\omega^{(3)} = -e^2 D_\alpha \tfrac{mk_F}{2\pi} \int \tfrac{d\epsilon}{4\pi i} \Big\{ & A_\omega {\rm Tr}\left[\hat{g}^+_{\epsilon+\omega}\hat{g}^K_{\epsilon,\epsilon} + \hat{g}^K_{\epsilon+\omega}\hat{g}^-_{\epsilon,\epsilon} + \hat{g}^+_{\epsilon,\epsilon}\hat{g}^K_{\epsilon-\omega} + \hat{g}^K_{\epsilon,\epsilon}\hat{g}^-_{\epsilon-\omega}\right] \\ & + A_{-\omega}{\rm Tr}\left[\hat{g}^+_\epsilon \hat{g}^K_{\epsilon+\omega,\epsilon-\omega} + \hat{g}^K_\epsilon \hat{g}^-_{\epsilon+\omega,\epsilon-\omega} + \hat{g}^+_{\epsilon+\omega,\epsilon-\omega}\hat{g}^K_\epsilon + \hat{g}^K_{\epsilon+\omega,\epsilon-\omega}\hat{g}^-_\epsilon\right]\Big\}. \end{aligned} \tag{2}$$

Here, we omitted terms which are ineffective in the absorption spectrum. $D_\alpha$ is the diffusion constant, $A_{\pm\omega}$ is the external vector potential and $\hat{\tau}_3 = \begin{pmatrix} 1 & 0 \\ 0 & -1 \end{pmatrix}$. $^+$ and $^-$ in the superscript indicate the retarded and advanced Green function, respectively. $\hat{g}$ with one variable is a quasiclassical Green function in the equilibrium state, and $\hat{g}$ with two variables is a quadratic function of external fields. In the dirty limit kinetic equations for the latter functions are written as

$$B^{+,+}_{\epsilon,\epsilon} = e^2 D_\alpha A_\omega A_{-\omega} \textstyle\sum_{\omega'=\pm\omega} \left(\hat{\tau}_3 \hat{g}^+_{\epsilon+\omega'}\hat{g}^+_\epsilon - \hat{g}^+_\epsilon \hat{g}^+_{\epsilon+\omega'}\hat{\tau}_3\right), \tag{3}$$

$$B^{+,+}_{\epsilon+\omega,\epsilon-\omega} = e^2 D_\alpha A_\omega A_\omega \left(\hat{\tau}_3 \hat{g}^+_\epsilon \hat{g}^+_{\epsilon-\omega} - \hat{g}^+_{\epsilon+\omega}\hat{g}^+_\epsilon \hat{\tau}_3\right), \tag{4}$$

$$B^{-,(a)}_{\epsilon,\epsilon} = 2ie^2 D_\alpha A_\omega A_{-\omega} \textstyle\sum_{\omega'=\pm\omega}(t^h_{\epsilon+\omega'} - t^h_\epsilon)\left[\hat{\tau}_3 {\rm Im}(\hat{g}^+_{\epsilon+\omega'})\hat{g}^-_\epsilon - \hat{g}^+_\epsilon {\rm Im}(\hat{g}^+_{\epsilon+\omega'})\hat{\tau}_3\right], \tag{5}$$

$$B^{-,(a)}_{\epsilon+\omega,\epsilon-\omega} = e^2 D_\alpha A_\omega A_\omega \textstyle\sum_{s=\pm}(-s)(t^h_{\epsilon-s\omega} - t^h_\epsilon)\left(\hat{\tau}_3 \hat{g}^s_{\epsilon,\epsilon}\hat{g}^-_{\epsilon-\omega} - \hat{g}^+_{\epsilon+\omega}\hat{g}^s_{\epsilon,\epsilon}\hat{\tau}_3\right). \tag{6}$$

Here, $B^{s,t}_{\epsilon_1,\epsilon_2} := \hat{\tau}_3\epsilon_1\hat{g}^t_{\epsilon_1,\epsilon_2} - \hat{g}^t_{\epsilon_1,\epsilon_2}\epsilon_2\hat{\tau}_3 - \hat{\tau}_3\hat{\Sigma}^+_{\epsilon_1}\hat{g}^t_{\epsilon_1,\epsilon_2} + \hat{g}^t_{\epsilon_1,\epsilon_2}\hat{\Sigma}^s_{\epsilon_2}\hat{\tau}_3 - \hat{\tau}_3\hat{\Sigma}^t_{\epsilon_1,\epsilon_2}\hat{g}^s_{\epsilon_2} + \hat{g}^+_{\epsilon_1}\hat{\Sigma}^t_{\epsilon_1,\epsilon_2}\hat{\tau}_3$. (The equations for advanced functions are the same as those for retarded ones with $^+$ replaced by $^-$. Im indicates the imaginary part.) $\hat{g}^{(a)}_{\epsilon,\epsilon'} = \hat{g}^K_{\epsilon,\epsilon'} - t^h_{\epsilon'}\hat{g}^+_{\epsilon,\epsilon'} + t^h_\epsilon \hat{g}^-_{\epsilon,\epsilon'}$ is the anomalous Green function, which vanishes in the equilibrium state. $t^h_\epsilon := \tanh\left(\frac{\epsilon}{2T}\right)$ ($T$ is the temperature). The self-energy is written as $\hat{\Sigma}_{\epsilon,\epsilon'} = \hat{\Sigma}^{(ep)}_{\epsilon,\epsilon'} + \hat{\Sigma}^{(ni)}_{\epsilon,\epsilon'} + \hat{\Sigma}^{(pi)}_{\epsilon,\epsilon'}$. We use Born approximation for

IOP Conf. Series: Journal of Physics: Conf. Series **969** (2018) 012035 doi:10.1088/1742-6596/969/1/012035

the impurity scattering [12, 13]. $\hat{\Sigma}^{(ni)}_{\epsilon,\epsilon'} = \alpha\hat{\tau}_3\hat{g}_{\epsilon,\epsilon'}\hat{\tau}_3$ and $\hat{\Sigma}^{(pi)}_{\epsilon,\epsilon'} = \alpha_p\hat{g}_{\epsilon,\epsilon'}$ ($\alpha = n_i u^2 mk_F/2\pi$ and $\alpha_p = n'_i u'^2 mk_F/2\pi$). $n_i$ ($n'_i$) and $u$ ($u'$) are the concentration and the magnitude of the potential of nonmagnetic (paramagnetic) impurities, respectively. $D_\alpha = v_F^2/6\alpha = v_F^2\tau/3$ with $\tau$ the relaxation time. We use the weak coupling approximation for the interaction between electrons and phonons. $\hat{\Sigma}^{(ep)}_{\epsilon,\epsilon'} = g_0\int\frac{d\epsilon_1}{2\pi i}\hat{\tau}_3\hat{g}^K_{\epsilon_1+(\epsilon-\epsilon')/2,\epsilon_1-(\epsilon-\epsilon')/2}\hat{\tau}_3$ with $g_0 = (g_{ph}^2/\omega_D)(mk_F/2\pi)$. $g_{ph}$ and $\omega_D$ are the coupling constant and Debye frequency, respectively. In this approximation $\Sigma^{(a)(ep)}_{\epsilon,\epsilon'} = (t^h_\epsilon - t^h_{\epsilon'})\Sigma^{(ep)}_{\epsilon,\epsilon'}$.

We introduce $\bar{g}$ and $\bar{f}$ as follows. $\hat{g}_{\epsilon,\epsilon} = e^2D_\alpha A_\omega A_{-\omega}\left(\bar{g}_{\epsilon(0)}\hat{\tau}_0 + \bar{f}_{\epsilon(0)}\hat{\tau}_1\right)$ and $\hat{g}_{\epsilon+\omega,\epsilon-\omega} = e^2D_\alpha A_\omega A_\omega\left(\bar{g}_{\epsilon(2\omega)}\hat{\tau}_0 + \bar{f}_{\epsilon(2\omega)}\hat{\tau}_1\right)$ with $\hat{\tau}_0$ the unit matrix and $\hat{\tau}_1 = \begin{pmatrix} 0 & 1 \\ 1 & 0 \end{pmatrix}$. By solving the kinetic equations with use of $\Sigma^{(ep)}_\epsilon = \Delta\hat{\tau}_1$ ($\Delta$ is the superconducting gap with paramagnetic impurities included), we obtain the following results.

$$\begin{pmatrix} \bar{g}^+_{\epsilon(0)} \\ \bar{f}^+_{\epsilon(0)} \end{pmatrix} = \hat{M}^{++}_{\epsilon,\epsilon}\begin{pmatrix} g^+_{\epsilon_1} + g^+_{\epsilon_2} \\ f^+_{\epsilon_1} + f^+_{\epsilon_2} + \bar{\Sigma}^{(ep)}_0 \end{pmatrix}, \quad \begin{pmatrix} \bar{g}^{(a)}_{\epsilon(0)} \\ \bar{f}^{(a)}_{\epsilon(0)} \end{pmatrix} = \sum_{\omega'=\pm\omega} 2i(t^h_{\epsilon+\omega'} - t^h_\epsilon)\hat{M}^{+-}_{\epsilon,\epsilon}\begin{pmatrix} \mathrm{Im}g^+_{\epsilon+\omega'} \\ \mathrm{Im}f^+_{\epsilon+\omega'} \end{pmatrix},$$

$$\begin{pmatrix} \bar{g}^+_{\epsilon(2\omega)} \\ \bar{f}^+_{\epsilon(2\omega)} \end{pmatrix} = \hat{M}^{++}_{\epsilon_1,\epsilon_2}\begin{pmatrix} g^+_\epsilon \\ f^+_\epsilon + \bar{\Sigma}^{(ep)}_{2\omega} \end{pmatrix}, \quad \begin{pmatrix} \bar{g}^{(a)}_{\epsilon(2\omega)} \\ \bar{f}^{(a)}_{\epsilon(2\omega)} \end{pmatrix} = \hat{M}^{+-}_{\epsilon_1,\epsilon_2}\sum_{s=\pm} s(t^h_\epsilon - t^h_{\epsilon-s\omega})\begin{pmatrix} g^s_\epsilon \\ f^s_\epsilon + \bar{\Sigma}^{(ep)}_{2\omega} \end{pmatrix}.$$

Here, $\epsilon_{1,2} = \epsilon \pm \omega$, $\hat{M}^{ab}_{\epsilon,\epsilon'} = Z^{ab}_{\epsilon,\epsilon'}\hat{\tau}_3 - X^{ab}_{\epsilon,\epsilon'}\hat{\tau}_0 - Y^{ab}_{\epsilon,\epsilon'}\hat{\tau}_1$ with $Z^{ab}_{\epsilon,\epsilon'} = i(z^a_\epsilon + z^b_{\epsilon'} + 2i\alpha_p X^{ab}_{\epsilon,\epsilon'}/Z^{ab}_{\epsilon,\epsilon'})^{-1}$, $X^{ab}_{\epsilon,\epsilon'} = Z^{ab}_{\epsilon,\epsilon'}(\epsilon^a_p\epsilon'^b_p + \Delta^a_\epsilon\Delta^b_{\epsilon'})/(z^a_\epsilon z^b_{\epsilon'})$ and $Y^{ab}_{\epsilon,\epsilon'} = Z^{ab}_{\epsilon,\epsilon'}(\epsilon^a_p\Delta^b_{\epsilon'} + \Delta^a_\epsilon\epsilon'^b_p)/(z^a_\epsilon z^b_{\epsilon'})$ ($\epsilon^\pm_p = \epsilon - \alpha_p g^\pm_\epsilon$, $\Delta^\pm_\epsilon = \Delta + \alpha_p f^\pm_\epsilon$, and $z^\pm_\epsilon = \sqrt{(\epsilon^\pm_p)^2 - (\Delta^\pm_\epsilon)^2}$). $g^\pm_\epsilon = -i\epsilon^\pm_p/z^\pm_\epsilon$ and $f^\pm_\epsilon = -i\Delta^\pm_\epsilon/z^\pm_\epsilon$. We obtain $\bar{\Sigma}^{(ep)}_0$ and $\bar{\Sigma}^{(ep)}_{2\omega}$ by solving the equations: $\bar{\Sigma}^{(ep)}_0 = -g_0\int\frac{d\epsilon}{2\pi i}\left(\bar{f}^{(a)}_{\epsilon(0)} + t^h_\epsilon\bar{f}^+_{\epsilon(0)} - t^h_\epsilon\bar{f}^-_{\epsilon(0)}\right)$ and $\bar{\Sigma}^{(ep)}_{2\omega} = -g_0\int\frac{d\epsilon}{2\pi i}\left(\bar{f}^{(a)}_{\epsilon(2\omega)} + t^h_{\epsilon-\omega}\bar{f}^+_{\epsilon(2\omega)} - t^h_{\epsilon+\omega}\bar{f}^-_{\epsilon(2\omega)}\right)$. The results are written as follows.

$$\begin{aligned}\bar{\Sigma}^{(ep)}_0 &= \tfrac{g_0}{\pi D_0}\int d\epsilon\sum\nolimits_{\omega'=\pm\omega}\Big\{(t^h_{\epsilon+\omega'} - t^h_\epsilon)\Big[Y^{+-}_{\epsilon,\epsilon}\mathrm{Im}g^+_{\epsilon+\omega'} + (Z^{+-}_{\epsilon,\epsilon} + X^{+-}_{\epsilon,\epsilon})\mathrm{Im}f^+_{\epsilon+\omega'}\Big] \\ &\quad + t^h_\epsilon\mathrm{Im}\Big[Y^{++}_{\epsilon,\epsilon}g^+_{\epsilon+\omega'} + (Z^{++}_{\epsilon,\epsilon} + X^{++}_{\epsilon,\epsilon})f^+_{\epsilon+\omega'}\Big]\Big\}, \quad (7)\\ \bar{\Sigma}^{(ep)}_{2\omega} &= \tfrac{g_0}{D_{2\omega}}\int\tfrac{d\epsilon}{2\pi i}\sum\nolimits_{s=\pm} s\Big\{t^h_{\epsilon-s\omega}\Big[Y^{ss}_{\epsilon+\omega,\epsilon-\omega}g^s_\epsilon + (Z^{ss}_{\epsilon+\omega,\epsilon-\omega} + X^{ss}_{\epsilon+\omega,\epsilon-\omega})f^s_\epsilon\Big] \\ &\quad + (t^h_\epsilon - t^h_{\epsilon-s\omega})\Big[Y^{+-}_{\epsilon+\omega,\epsilon-\omega}g^s_\epsilon + (Z^{+-}_{\epsilon+\omega,\epsilon-\omega} + X^{+-}_{\epsilon+\omega,\epsilon-\omega})f^s_\epsilon\Big]\Big\}. \quad (8)\end{aligned}$$

Here, $D_0 = 1 - \frac{g_0}{\pi}\int d\epsilon t^h_\epsilon\mathrm{Im}(Z^{++}_{\epsilon,\epsilon} + X^{++}_{\epsilon,\epsilon})$ and $D_{2\omega} = 1 - g_0\int\frac{d\epsilon}{2\pi i}\sum_{s=\pm}s[t^h_{\epsilon+\omega}(Z^{s-}_{\epsilon+\omega,\epsilon-\omega} + X^{s-}_{\epsilon+\omega,\epsilon-\omega}) + t^h_{\epsilon-\omega}(Z^{+s}_{\epsilon+\omega,\epsilon-\omega} + X^{+s}_{\epsilon+\omega,\epsilon-\omega})]$.

The nonlinear current is written as $J^{(3)}_\omega = -K^{(3)}_\omega A_\omega$, and then the two-photon absorption spectrum is given by $\mathrm{Re}\sigma^{(3)}_\omega = -\mathrm{Im}K^{(3)}_\omega/\omega$ with $-\mathrm{Im}K^{(3)}_\omega/\omega = e^4D^2_\alpha|E_\omega|^2\frac{mk_F}{8\pi^2}\bar{\sigma}_\omega$ and $\bar{\sigma}_\omega = \bar{\sigma}^{(qp)}_\omega + \bar{\sigma}^{(am)}_\omega$. ($E_\omega$ is the electric field, and Re indicates the real part.) Here, $\bar{\sigma}^{(qp)}_\omega = \mathrm{Re}\int\frac{d\epsilon}{\omega^3}\mathrm{Tr}\Big\{t^h_{\epsilon-\omega}2i\mathrm{Im}(\hat{g}^+_{\epsilon-\omega})\hat{\bar{g}}^+_{\epsilon(0)} + t^h_{\epsilon+\omega}2i\mathrm{Im}(\hat{g}^+_{\epsilon+\omega})\hat{\bar{g}}^-_{\epsilon(0)} + (\hat{g}^+_{\epsilon+\omega} + \hat{g}^-_{\epsilon-\omega})\Big[\hat{\bar{g}}^{(a)}_{\epsilon(0)} + t^h_\epsilon(\hat{\bar{g}}^+_{\epsilon(0)} - \hat{\bar{g}}^-_{\epsilon(0)})\Big]\Big\}$ and $\bar{\sigma}^{(am)}_\omega = 2\mathrm{Re}\int\frac{d\epsilon}{\omega^3}\mathrm{Tr}\Big[t^h_\epsilon i\mathrm{Im}(\hat{g}^+_\epsilon)\left(\hat{\bar{g}}^+_{\epsilon(2\omega)} + \hat{\bar{g}}^-_{\epsilon(2\omega)}\right) + \mathrm{Re}(\hat{g}^+_\epsilon)\left(\hat{\bar{g}}^{(a)}_{\epsilon(2\omega)} + t^h_{\epsilon-\omega}\hat{\bar{g}}^+_{\epsilon(2\omega)} - t^h_{\epsilon+\omega}\hat{\bar{g}}^-_{\epsilon(2\omega)}\right)\Big]$.

## 3. Numerical calculations

We perform numerical calculations at $T = 0$. There is no direct excitation of quasiparticles for $\omega < 2E_g$. ($2E_g$ is the energy gap for quasiparticle excitations [13].) The quantities introduced in the previous section do not depend on $\alpha$ (nonmagnetic impurities) except for $D_\alpha$. We do not specify the value of $\alpha$ in this section.

The nondiagonal part of $\hat{\Sigma}^{(ep)}_{\epsilon,\epsilon'}$ represents the superconducting gap and its correction under external fields. $\bar{\Sigma}^{(ep)}_0$ gives a static correction to the gap, and $\bar{\Sigma}^{(ep)}_{2\omega}$ includes a correction by the amplitude mode represented by $D_{2\omega}$. The calculated results for real and imaginary parts of $D_{2\omega}$ are shown in figures 1 and 2. $\Delta_0$ is the superconducting gap in the absence of paramagnetic

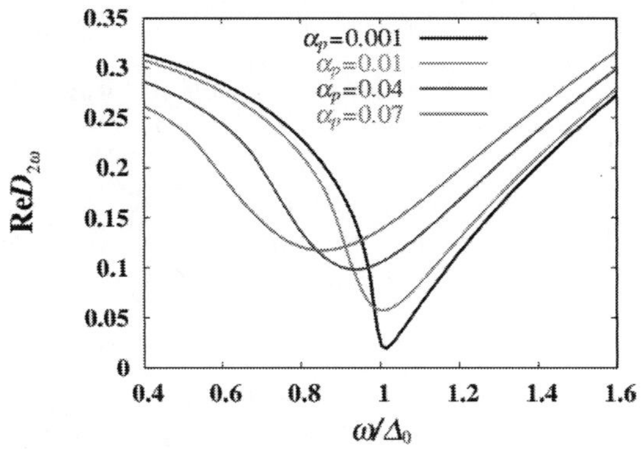


**Figure 1.** The dependences of Re$D_{2\omega}$ (The real part of $D_{2\omega}$ which is in the vertex correction under external field and indicates existence of the amplitude mode of the order parameter) on $\omega$ for various values of $\alpha_p$.

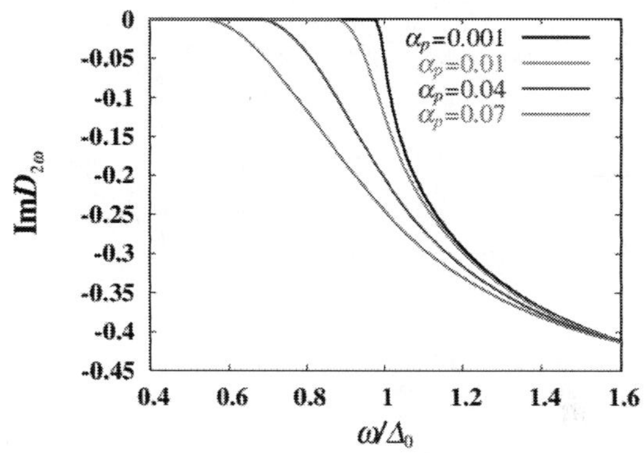


**Figure 2.** The dependences of Im$D_{2\omega}$ (The imaginary part of $D_{2\omega}$) on $\omega$ for various values of $\alpha_p$.

impurities, and we take $\Delta_0$ as the unit of energy ($\Delta_0 = 1$). In the presence of finite values of $\alpha_p$ (paramagnetic impurities) Re$D_{2\omega} \neq 0$ at $\omega = \Delta$, and there is no well-defined amplitude mode. (On the other hand Re$D_{2\omega} = 0$ at $\omega = \Delta_0$ for $\alpha_p = 0$.) The mininum value of Re$D_{2\omega}$ increases with increasing $\alpha_p$. $-$Im$D_{2\omega}$, which indicates the damping of the amplitude mode and originates from quasiparticle excitations, takes a finite value at smaller $\omega$ with increasing $\alpha_p$ because of decreasing $E_g$.

The nonlinear correction to the absorption spectrum is written as

$$\frac{\mathrm{Re}\sigma^{(3)}_\omega}{\sigma_0} = \frac{\Delta_0}{3\alpha}\left(\frac{\pi e\xi_0|E_\omega|}{4\Delta_0}\right)^2 \Delta_0^3\bar{\sigma}_\omega. \tag{9}$$

Here, $\xi_0 = v_F/\pi\Delta_0$ is the coherence length, and $\sigma_0 = e^2 n_e\tau/m$ is the conductivity in the normal state ($n_e = k_F^3/3\pi^2$ is electron density). $Z^{+-}_{\epsilon,\epsilon}$, which is a kind of diffusion propagator [14], diverges for $|\epsilon| > E_g$ because paramagnetic impurities also bring about elastic scattering as nonmagnetic impurities do [15]. The diffusion propagator is ineffective in the case that there is no quasiparticle excitation (its condition is $T = 0$ and $\omega < 2E_g$).

The calculated result of $\bar{\sigma}_\omega$ is shown in figure 3. (The calculation is restricted to $\omega < 2E_g$ as noted above. $\bar{\sigma}^{(am)}_\omega$ is not the case because it does not include $Z^{+-}_{\epsilon,\epsilon}$.) The TPA spectrum shows a peak at $\omega = \Delta_0$ and no absorption for $\omega < \Delta_0$ in the case of $\alpha_p \simeq 0$, which is the same result as in [8]. This indicates that the absorption edge by the amplitude mode is the same as that by quasiparticle excitations.

The peak in the TPA spectrum ($\omega_p$ is its frequency) shifts to lower frequency with increasing $\alpha_p$. The absorption edge also shifts to lower frequency further. This edge originates from quasiparticle excitations, and its frequency is given by $\omega = E_g$. The difference between $\omega_p$ and

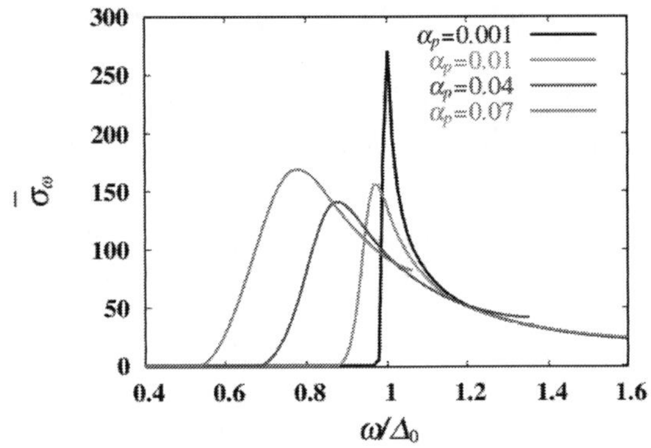


**Figure 3.** The dependences of $\bar{\sigma}_\omega$ (the TPA spectrum without a factor indicating the pump field) on $\omega$ for various values of $\alpha_p$.

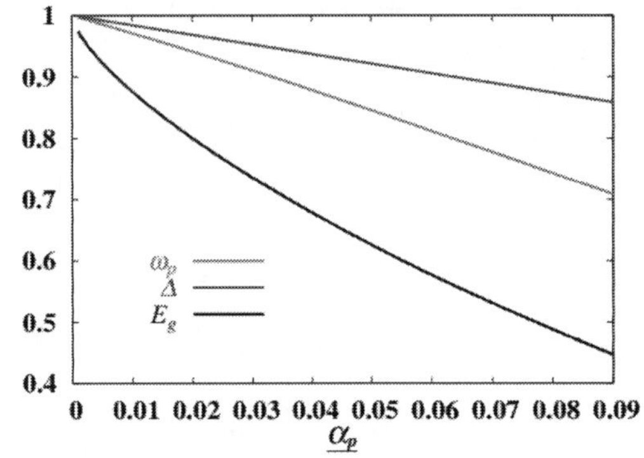


**Figure 4.** The dependences of $\omega_p$ (the frequency at which the TPA spectrum shows a peak by the amplitude mode) on $\alpha_p$.

$E_g$ becomes large as $\alpha_p$ increases as shown in figure 4. The peak in the TPA spectrum indicates the existence of the amplitude mode as shown below. This implies that the amplitude mode is separable from quasiparticle excitations by introducing paramagnetic impurities in contrast to the case of $\alpha_p = 0$.

Figure 4 shows that the frequency of the peak is different from the superconducting gap $\Delta$ for $\alpha_p \neq 0$. Though $\omega_p \neq \Delta$, the peak in the TPA spectrum indicates the existence of the amplitude mode as is seen by comparing figures 5 and 6. Figure 5 shows calculated results for $\bar{\sigma}_\omega^{(am)}$ which includes the amplitude mode $D_{2\omega}$ in $\mathrm{Re}\sigma_\omega^{(3)}$. $\bar{\sigma}_\omega^{(na)}$ shown in figure 6 is the term in which $D_{2\omega}$ in $\bar{\sigma}_\omega^{(am)}$ is replaced by 1. These figures show that the existence of $D_{2\omega}$ in $\bar{\sigma}_\omega$ is

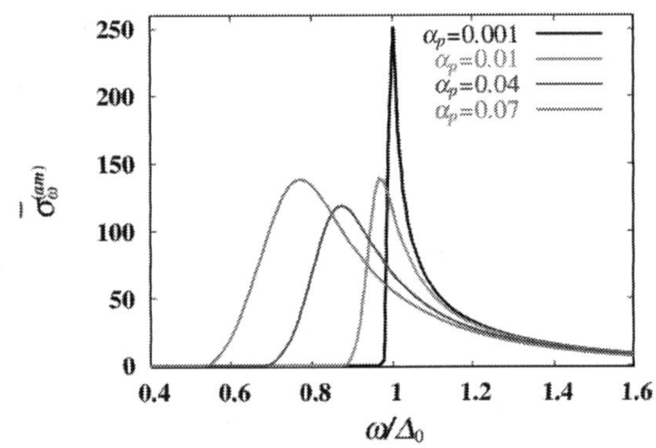


**Figure 5.** The dependences of $\bar{\sigma}_\omega^{(am)}$ (there is an amplitude mode in this term) on $\omega$ for various values of $\alpha_p$.

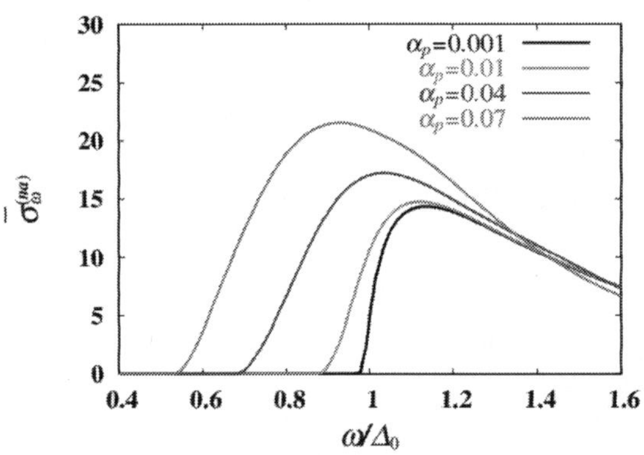


**Figure 6.** The dependences of $\bar{\sigma}_\omega^{(na)}$ (there is no amplitude mode in this term) on $\omega$ for various values of $\alpha_p$.

essential to the appearance of the peak in the TPA spectrum. The difference between $\omega_p$ and $\Delta$ arises from the fact that there is no well-defined amplitude mode for $\alpha_p \neq 0$. Our calculations show that there is a peak in the TPA spectrum even in the case of $\mathrm{Re}D_{2\omega} \neq 0$. The suppression of the peak by increasing $\mathrm{Re}D_{2\omega}$ for large values of $\alpha_p$ is compensated by the decreasing factor $\omega^3$ in the denominator of $\bar{\sigma}_\omega$.

$\bar{\sigma}_\omega^{(qp)} \left(= \bar{\sigma}_\omega - \bar{\sigma}_\omega^{(am)}\right)$ does not include the amplitude mode $D_{2\omega}$. This quantity is small as is seen by making a comparison between figures 3 and 5. $\bar{\sigma}_\omega^{(qp)}$ includes a diffusion propagator ($Z^{+-}_{\epsilon,\epsilon}$), and makes a large contribution to the TPA spectrum in the case of $\omega > 2E_g$ (there is a direct excitation of quasiparticles).

## 4. Discussion

When we put $\xi_0 = 5$ nm, $|E_\omega| = x$ kV/cm, and $\Delta_0 = 5$ meV to estimate $\mathrm{Re}\sigma_\omega^{(3)}/\sigma_0$ quantitatively, the dimensionless factor is calculated as $(\pi e \xi_0 |E_\omega|/4\Delta_0)^2 \Delta_0/(3\alpha) \simeq [2x^2/(\alpha/\Delta_0)] \times 10^{-3}$. Numerical calculations in the previous section show that $\Delta_0^3 \bar{\sigma}_\omega > 100$. Then $\mathrm{Re}\sigma_\omega^{(3)}/\sigma_0 \simeq 0.1$ for $|E_\omega| > 2$ kV/cm even in the dirty limit $\alpha = 10\Delta_0$. This result indicates that a nonlinear correction to the absorption spectrum will be observable in experiments.

In this paper we used a monochromatic external field to discuss the absorption spectrum in the steady state. The direct excitation of quasiparticles should be avoided to obtain meaningful results. This is the reason why we suppose $T = 0$ and $\omega < 2E_g$. In the presence of quasiparticle excitations we have to take account of mechanisms for the dissipation of energy such as inelastic scattering and nonlocality (nonconservation of momentum). In the case of $\omega > 2E_g$ the direct excitation of quasiparticles above the energy gap narrows the range of $|E_\omega|$ in which the perturbation expansion by external fields is valid. In contrast to this, the perturbation expansion is considered to be valid at $T \neq 0$ for $\omega < 2E_g$ (there are only thermally excited quasiparticles in this case) when there is a dissipation mechanism. The difference between $T = 0$ and $T \neq 0$ is that there is a negative nonlinear correction to the absorption spectrum for $\omega < E_g$ in the latter case [15]. Therefore, our calculation on the peak by the amplitude mode at $T = 0$ is considered to be valid at finite but low temperatures.

## Acknowledgments

The numerical computation in this work was carried out at the Yukawa Institute Computer Facility.

## References

[1] Mattis D C and Bardeen J 1958 *Phys. Rev.* **111** 412
[2] Anderson P W 1958 *Phys. Rev.* **112** 1900
[3] Schmid A 1968 *Phys. Kondens. Matter.* **8** 129
[4] Volkov A F and Kogan S M 1974 *Sov. Phys. JETP* **38** 1018
[5] Sooryakumar R and Klein M V 1980 *Phys. Rev. Lett.* **45** 660
[6] Littlewood P B and Varma C M 1982 *Phys. Rev.* B **26** 4883
[7] Matsunaga R, Hamada Y I, Makise K, Uzawa Y, Terai H, Wang Z and Shimano R 2013 *Phy. Rev. Lett.* **111** 057002
[8] Jujo T 2015 *J. Phys. Soc. Jpn.* **84** 114711
[9] Ivlev B I 1972 *JETP Lett.* **15** 313
[10] Éliashberg G M 1972 *Sov. Phys. JETP* **34** 668
[11] Usadel K D 1970 *Phys. Rev. Lett.* **25** 507
[12] Abrikosov A A and Gor'kov L P 1961 *Sov. Phys. JETP* **12** 1243
[13] Skalski S, Betbeder-Matibet O and Weiss P R 1964 *Phy. Rev.* **136** A1500
[14] Vollhardt D and Wölfle P 1980 *Phys. Rev.* B **22** 4666
[15] Jujo T 2017 *J. Phys. Soc. Jpn.* **86** 024709

IOP Conf. Series: Journal of Physics: Conf. Series **969** (2018) 012036 doi:10.1088/1742-6596/969/1/012036

# Proximity-Induced Superconductivity in a Ferromagnetic Semiconductor (In,Fe)As

**Taketomo Nakamura[1], Le Duc Anh[2], Yoshiaki Hashimoto[1], Yu Iwasaki[1], Shinobu Ohya[2,3], Masaaki Tanaka[2,3], and Shingo Katsumoto[1,3]**

[1] Institute for Solid State Physics, The University of Tokyo, 5-1-5 Kashiwanoha, Kashiwa, Chiba 277-8581, Japan

[2] Department of Electrical Engineering and Information Systems, The University of Tokyo 7-3-1 Hongo, Bunkyo-ku, Tokyo 113-8656, Japan

[3] Center for Spintronics Research Network, The University of Tokyo, 7-3-1 Hongo, Bunkyo-ku, Tokyo 113-8656, Japan

E-mail: taketomo@issp.u-tokyo.ac.jp

**Abstract.** We observed supercurrent through an $n$-type ferromagnetic semiconductor (In,Fe)As. For the purpose we prepared S/F/S (S: superconductor, F: ferromagnet) junctions, which exhibited zero resistance up to 1 $\mu$m of junction width with critical current over 1 $\mu$A below 1 K. The long reach of the superconducting proximity effect suggests that the induced superconductivity is brought about by spin-triplet paring in (In,Fe)As.

## 1. Introduction

The superconductor(S)/ferromagnet(F)/superconductor(S) structure is attracting attentions as a new platform of unconventional superconductivities. In a simple thought, the exchange interaction between spins destroys the conventional spin-singlet Cooper pairs in ferromagnetic materials[1]. In some limited cases, however, the proximity effect in the SFS devices enables superconductivity to go into ferromagnets far away from the interfaces[2]. The theoretical and experimental works suggest that the superconductivities deep inside the ferromagnets are driven by a spin-triplet type electron pairing, which is created in the vicinity of interfaces under certain conditions[2, 3, 4]. Such spin-triplet pairs particularly in so called half-metallic ferromagnets, i.e., materials with perfect spin polarization, can have the penetration depth as long as that of conventional spin-singlet pairs in normal metals[5, 6, 7].

One of the big findings of ferromagnetic materials is that of III-V based ferromagnetic semiconductors (FMSs), which have comparatively high transition temperatures and good connectivity to highly elaborated quantum structures. And above all, the magnetism can be controlled through the carrier concentration. An interesting idea here is to form an S/F/S junction with an FMS and superconducting metals. Many of the FMS have been reported to be half-metallic and the ferromagnetic interaction is mediated by electric carriers[8, 9]. For good electric contacts with metals, narrow-gap semiconductor like InAs is appropriate for the base material. InAs has natural surface two-dimensional electrons and, the carrier type should be $n$. Recently found (In,Fe)As just fits to these good-contact conditions. Here we report the transport properties of Nb/(In,Fe)As/Nb junctions with low interface barriers. They showed finite superconducting critical currents with long proximity length, which suggests spin-triplet type pairing.

## 2. Experimental

(In,Fe)As with Fe concentration of 6% was grown by low temperature molecular beam epitaxy on a (001) GaAs substrate with buffer layers[10]. The mobility and the concentration of carriers were estimated at 3.5 K to be 100 $cm^2/Vs$ and $8 \times 10^{18}$ cm, respectively. The Curie temperature was estimated to be about 120 K from the temperature dependence of the magnetic susceptibility. The sample structure is illustrated in Fig.1(a). Nb/Ti electrodes with the width of 30 $\mu$m patterned by electron-beam lithography were deposited immediately after surface cleaning by Ar plasma, on top of the (In,Fe)As film with ion-beam sputtering. An optical micrograph image of a junction is shown in Fig.1(b). We fabricated two junctions with the separation between the electrodes (gap) L of 1 $\mu$m (junctions A, B) and the one with 1.5 $\mu$m (junction C). The current directions were along [110] for junction A and $[\bar{1}10]$ for junctions B and C.

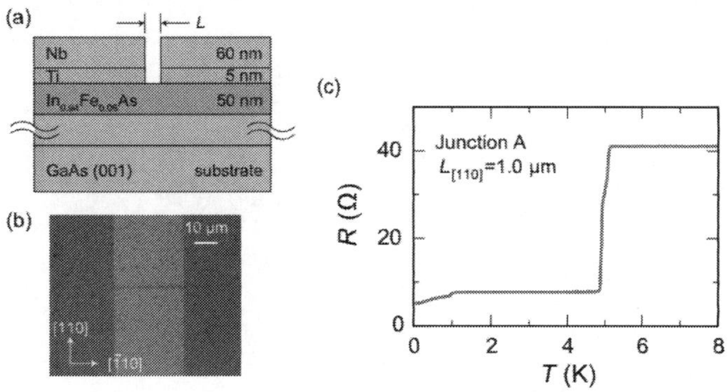


**Figure 1.** (a) Schematic cross sectional view of the samples. (b) Optical microscope image of junction A. (c) Temperature dependence of the zero-bias resistance of junction A. The AC current modulation is 100 nA-rms at the frequency of 19 Hz.

The junctions were cooled down in a dilution fridge down to 0.1 K. Each wire passes through a 2 k$\Omega$ resistor as a low-pass filter anchored at the mixing chamber. The differential resistance $R = dV/dI$ of the junctions were measured by a lock-in technique with AC current modulation. All data presented in this paper were taken under zero magnetic fields.

## 3. Results and discussion

The Nb/Ti electrodes undergo superconducting transition at around 5 K, to which zero-temperature superconducting gap $\Delta_0$ of 0.9 meV corresponds. Figure 1(c) displays the temperature dependence of the zero-bias resistance of junction A. The resistance is almost constant below the transition temperature and above 1.6 K, below which it decreases again with the temperature. This temperature dependence manifests that the superconducting coherence penetrates into the F-regions of the junctions via the proximity effect. At the lowest temperature around 0.1 K, all the junctions show non-linear $I$-$V$ characteristics, that is a dip structure in $dV/dI$ at zero-bias current. As shown in Fig.2, zero-bias resistance of each junction at the lowest temperature is less than half of its normal resistance, exceeding the conductance enhancement by the Andreev reflection. Especially, junction B exhibits clear zero resistance plateau at around zero bias.

IOP Conf. Series: Journal of Physics: Conf. Series **969** (2018) 012036 doi:10.1088/1742-6596/969/1/012036

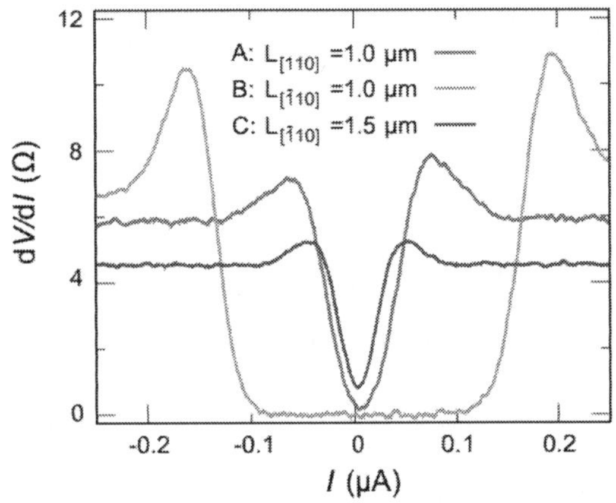


**Figure 2.** Differential resistance of junctions A, B and C as a function of the bias current. All the data were taken around 0.1 K without external magnetic field. The AC current modulation is 10 nA-rms at the frequency of 19 Hz.

The proximity length of 1 $\mu$m is even longer than that reported for ballistic two-dimensional electrons in a pure InAs quantum well[11, 12]. The supercurrent over such a long distance in ferromagnetic (In,Fe)As strongly suggests spin-triplet pairing because it is well known that spin-singlet pairs survive in ferromagnets only for a few tens of nanometers[13, 14, 15]. The spatial dependence of the spin-singlet order parameters in ferromagnets is calculated from the Usadel equations[16] or the Eilenberger equations[17]. The decay length $\xi_{\rm d}$ in dirty limit and $\xi_{\rm c}$ in the clean limit are written as follows:

$$\xi_{\rm d} = \sqrt{\frac{D}{\sqrt{(\pi T)^2 + E_{\rm ex}^2} \pm \pi T}}, \tag{1}$$

$$\text{and } \xi_{\rm c} = \frac{\xi_0 l}{\xi_0 + l}, \quad \xi_0 = \frac{v_{\rm F}}{2\pi T} \tag{2}$$

where $D$, $E_{\rm ex}$, $l$ and $v_{\rm F}$ are the diffusion coefficient, exchange field, mean free path, and the Fermi velocity in ferromagnets, respectively. In (In,Fe)As, from the parameters $D = 1.4 \times 10^{-3}$ m$^2$/s, $E_{\rm ex} = 31.7$-$50$ meV, $l = 3.7$ nm, and $v_{\rm F} = 1.1 \times 10^6$ m/s, the characteristics length $\xi_{\rm d}$ and $\xi_{\rm c}$ are calculated as 4.2-5.3 nm and 3.7 nm at 0.1 K, respectively[9].

The applicability of the above discussion can further be tested from the temperature dependence of the critical current $I_{\rm c}$. Because the rise of the resistance is rather soft as shown in Fig.2, we here define $I_{\rm c}$ as the bias current where the resistance is equal to 1 $\Omega$. The data in Fig.3 were obtained in the following way. After a magnetization process of (In,Fe)As, in which the magnetic field along [110] (in-plane) was cycled between 0 and 3 kOe with ending up at 0 Oe at the lowest temperature, the specimen was warmed up over 7 K and cooled down to the lowest temperature again at zero field in order to eliminate the magnetic flux pinning in superconducting electrodes. After this process, the $I_{\rm c}$ increased to 1 $\mu$A below 0.1 K. Then with increasing temperature, $I_{\rm c}$ was measured.

The $I_{\rm c}$ decreases monotonically with increasing temperature, in accordance with the previous works on S/ (half metallic) F/S junctions[6, 18]. The asymmetry of the $I_{\rm c}$ against the current direction reflects the broken time reversal symmetry due to the ferromagnetic exchange in (In,Fe)As. The curves of the $I_{\rm c}$ are slightly concave upward, indicating that the present junction is not in the tunnelling regime but in the weak-links regime, and the theories for weak-links are applicable[19]. This temperature dependence assures that we can apply the estimation of monotonic decay of the order parameter even though the (In,Fe)As is ferromagnetic, as predicted for the triplet pairing in ferromagnet[20]. Hence the results in Fig.3 are consistent with the emergence of spin-triplet superconductivity in the ferromagnetic (In,Fe)As layers.

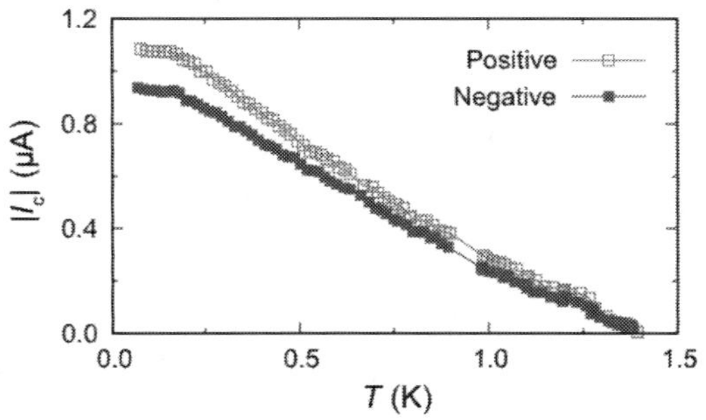


**Figure 3.** Temperature dependence of the critical current $I_c$ in junction B without external magnetic field. The AC current modulation for the measurement of $dI/dV$ vs $I$ characteristics to obtain the $I_c$ is 10 nA-rms at the frequency of 870 Hz.

For the spin-triplet superconductivity to emerge in the proximity effect, inhomogeneous magnetization around the interfaces is required as a spin mixing matrix, which interconnects the singlet pairs and the triplet pairs[3]. In this theory of proximity induced spin-triplet superconductivity, S/F'/F/F'/S structures are proposed to realize such inhomogeneous magnetization. However in the present case, such inhomogeneity is expected to be formed naturally at the surface of FMS due to the strain in the crystals generated from the layered structure. The crystal direction dependence of the critical current shown in Fig.2 probably comes from the directional dependence of the strain that gives difference in the amplitude of spin-mixing matrix elements.

## 4. Conclusion

We fabricated S/F/S junctions from an $n$-type III-V based FMS (In,Fe)As and observed finite supercurrents through them. This is the first realization of proximity driven superconductivity in FMSs. The supercurrents are presumably carried by spin-triplet Cooper pairs induced at the interfaces just like those observed in superconductor/half-metal junctions.

This work was supported by Grant-in-Aid for Scientific Research on Innovative Area, "Nano Spin Conversion Science" (Grant No.26103003), also by JSPS KAKENHI Grant Numbers JP25247051 and JP17K05492. Iwasaki was also supported by Japan Society for the Promotion of Science through Program for Leading Graduate Schools (MERIT).

## References

[1] Buzdin A I 2005 *Rev. Mod. Phys.* **77** 935
[2] Bergeret F S, Volkov A F and Efetov K B 2005 *Rev. Mod. Phys.* **77** 1321
[3] Eschrig M and Löfwander T 2008 *Nat. Phys.* **4** 138
[4] Robinson J W A, Witt J D S and Blamire M G 2010 *Science* **329** 59
[5] Pena V, Sefrioui Z, Arias D, Leon C, Santamaria J, Varela M, Pennycook S J and Martinez J L 2004 *Phys. Rev. B* **69** 224502
[6] Keizer R S, Goennenwein S T B, Klapwijk T M, Miao G, Xiao G and Gupta A 2006 *Nature* **439** 825
[7] Asano Y, Tanaka Y and Golubov A A 2007 *Phys. Rev. Lett.* **98** 107002
[8] Sato K, Bergqvist L, Kudrnovský J, Dederichs P H, Eriksson O, Turek I, Sanyal B, Bouzerar G, Katayama-Yoshida H, Dinh V A, Fukushima T, Kizaki H and Zeller R 2010 *Rev. Mod. Phys.* **82** 1633
[9] Anh L D, Hai P N and Tanaka M 2016 *Nat. Commun.* **7** 13810
[10] Hai P N, Anh L D, Mohan S, Tamegai T, Kodzuka M, Ohkubo T, Hono K and Tanaka M 2012 *Appl. Phys. Lett.* **101** 182403
[11] Schäpers T 2001 *Superconductor/Semiconductor Junctions* (Berlin: Springer) p 145
[12] Irie H, Harada Y, Sugiyama H and Akazaki T 2014 *Phys. Rev. B* **89** 165415
[13] Kontos T, Aprili M, Lesueur J, Genêt F, Stephanidis B and Boursier R 2002 *Phys. Rev. Lett* **89** 137007
[14] Weides M, Kemmler M, Goldobin E, Koelle D, Kleiner R, Kohlstedt H and Buzdin A 2006 *Appl. Phys. Lett.* **89** 122511
[15] Born F, Siegel M, Hollmann E K, Braak H, Golubov A A, Gusakova D Y and Kupriyanov M Y 2006 *Phys. Rev. B* **74** 140501
[16] Buzdin A I, Bujicic B and Kupriyanov M Y 1992 *Sov. Phys. JETP* **74** 124

[17] Eilenberger G 1968 *Z. Phys.* **214** 196
[18] Anwar M S, Czeschka F, Hesselberth M, Porcu M and Aarts J 2010 *Phys. Rev. B* **82** 100501
[19] Furusaki A and Tsukada M 1991 *Phys. Rev. B* **43** 10164
[20] Volkov A F, Bergeret F S and Efetov K B 2003 *Phys. Rev. Lett.* **90** 117006

IOP Conf. Series: Journal of Physics: Conf. Series **969** (2018) 012037 doi:10.1088/1742-6596/969/1/012037

# Spontaneous generation of fractional vortex-antivortex pairs at single edges of high-Tc superconductors

**P Holmvall, T Löfwander and M Fogelström**

Department of Microtechnology and Nanoscience – MC2, Chalmers University of Technology, SE-41296 Göteborg, Sweden

E-mail: holmvall@chalmers.se, tomas.lofwander@chalmers.se, mikael.fogelstrom@chalmers.se

**Abstract.** Unconventional $d$-wave superconductors with pair-breaking edges are predicted to have ground states with spontaneously broken time-reversal and translational symmetries. We use the quasiclassical theory of superconductivity to demonstrate that such phases can exist at any single pair-breaking facet. This implies that a greater variety of systems, not necessarily mesoscopic in size, should be unstable to such symmetry breaking. The density of states averaged over the facet displays a broad peak centered at zero energy, which is consistent with experimental findings of a broad zero-bias conductance peak with a temperature-independent width at low temperatures.

## 1. Introduction

It was established already in the 1990s that a number of high-temperature superconductors have an order parameter with $d_{x^2-y^2}$ symmetry [1]. In such materials, quasiparticle scattering at surfaces or off defects, where the sign of the $d$-wave order parameter changes for incoming and outgoing scattering trajectories, leads to the formation of Andreev bound states at zero energy [2–4]. For an ideal specular surface with [110]-orientation, all scattering trajectories include the sign change, and the spectral weight of these zero-energy Andreev bound states is very large: they form a flat band at zero energy as function of momentum parallel to the interface, $k_{\|}$. Shifting these mid-gap states to finite energies can lead to lowering of the free energy. Any mechanism providing such a shift can then lead to a phase transition into a new ground state with an associated broken symmetry [5, 6]. Several mechanisms have been proposed, all leading to spontaneous time-reversal symmetry breaking: development of a subdominant superconducting component of the order parameter in a time-reversal symmetry breaking combination with the dominant, e.g. $d_{x^2-y^2} + is$, [7–9]; magnetic ordering [10]; and, finally, spontaneous supercurrents [11–16]. The first two scenarios require an additional coupling constant leading to an associated mean-field order parameter, while the last does not. Which scenario that would be realized experimentally depends on material parameters, for instance the strength of the coupling constants. It was shown in Ref. [15] that the transition temperature within the third scenario is very large, of the order of 20% of the superconducting transition temperature $T_c$. As a consequence, the other scenarios can compete only if their corresponding coupling constants are very large, or if the phase with spontaneous supercurrents is suppressed for one reason or another.

So far, there are several transport experiments supporting spontaneous time-reversal symmetry breaking [17–23]. But direct measurements of the associated supercurrents and magnetic fields remain controversial [24, 25]. In our previous studies [15], we showed that this controversy could be related to the manner in which these currents and magnetic fields appear. We found a translational and time-reversal symmetry-breaking phase, in which a staggered pattern of fractional vortex-antivortex pairs forms like a necklace along the pair-breaking surface. The symmetric proportion of vortices to antivortices effectively eliminates any net current and magnetic flux, and the small size of the vortices of a few coherence lengths makes direct observation challenging.

Vorontsov found that a phase gradient can be generated through spontaneous time-reversal symmetry breaking in thin films [14, 26–29], caused by finite-size effects in the form of a proximity of two pair-breaking interfaces. In our previous work [15] we studied mesoscopic grains with only pair-breaking edges and found that the vortex-antivortex phase is more energetically favorable than the thin-film phase predicted by Vorontsov. In this study, we show that the vortex-antivortex phase can occur without finite-size effects. This is done by considering a system with a single pair-breaking edge.

## 2. Model and methods

We study a mesoscopic superconducting grain in vacuum and equilibrium, with a $d$-wave pairing symmetry. The sides of the system are perfectly aligned with the crystal $ab$-axes, except one facet which is misaligned by a 45° rotation (see Fig. 1). The facet gives rise to mid-gap states associated with surface pair-breaking, and has a side-length given in units of the superconducting coherence length $\xi_0 \equiv \hbar v_F/2\pi k_B T_c$. Furthermore, a clean superconductor and a cylindrical Fermi surface is assumed.

To study this system, the quasiclassical theory of superconductivity [30, 31] is used. In this formulation, the superconducting $d$-wave order parameter $\Delta_d(\boldsymbol{R})$ depends on the anomalous Green's function (pair propagator) $f(\boldsymbol{p}_F, \boldsymbol{R}; \epsilon_n)$ through the gap equation

$$\Delta_d(\boldsymbol{R}) = V_d k_B T \int \frac{d\theta_{p_F}}{2\pi} \eta_d^*(\theta_{p_F}) \sum_{|\epsilon_n| \leq \Omega_c} f(\boldsymbol{p}_F, \boldsymbol{R}; \epsilon_n), \tag{1}$$

at spatial coordinate $\boldsymbol{R}$, quasiparticle momentum $\boldsymbol{p}_F$ and Matsubara energy $\epsilon_n$ (these parameters will from now on be dropped for a compact notation). Here, $\theta_{p_F}$ is the angle between the Fermi momentum and the crystal $ab$-axes, $\eta_d(\theta_{p_F}) = \sqrt{2}\cos(2\theta_{p_F})$ the $d$-wave order parameter basis function, $V_d = -N_F \lambda_d$ the pair-potential, $N_F$ the normal-state density of states at the Fermi surface, $\lambda_d$ the pairing interaction, and $\Omega_c$ the cutoff energy. The anomalous Green's function is the off-diagonal component of the Matsubara Green's function

$$\hat{g} = \begin{pmatrix} g & f \\ -\tilde{f} & \tilde{g} \end{pmatrix}, \tag{2}$$

where *hat* denotes Nambu (electron-hole) space. The tilde operation denotes particle-hole conjugation, $\tilde{g}(\boldsymbol{p}_F, \boldsymbol{R}; \epsilon_n) = g^*(-\boldsymbol{p}_F, \boldsymbol{R}, \epsilon_n)$ [and the same for $\tilde{f}$]. The Green's function is obtained by solving the Eilenberger equation with the associated normalization condition

$$i\hbar \boldsymbol{v}_F \cdot \boldsymbol{\nabla}_R \hat{g} + \left[\hat{\tau}_3 z - \hat{\Delta}, \hat{g}\right] = \hat{0}, \tag{3}$$

$$\hat{g}^2 = -\pi^2 \hat{1}, \tag{4}$$

where $\hat{\Delta} = i(\hat{\tau}_2\Re\Delta_d + \hat{\tau}_1\Im\Delta_d)\eta_d(\theta_{p_F})$, and $\hat{\tau}_i$ $(i = 1, 2, 3)$ are the three Pauli matrices in Nambu space. The Eilenberger equation and the gap equation are solved self-consistently by the so-called Riccati technique (see for instance Ref. [32]). After self-consistency has been achieved, we compute observables such as the current density

$$\boldsymbol{j}(\boldsymbol{R}) = 4\pi e N_F k_B T \int \frac{d\theta_{p_F}}{2\pi} \boldsymbol{v}_F(\boldsymbol{p}_F) \sum_{\epsilon_n} g(\boldsymbol{p}_F, \boldsymbol{R}; \epsilon_n). \tag{5}$$

The magnetic flux density induced by the current density is calculated through Maxwells equations and Ampère's circuit law.

## 3. Results and discussion

Figure 1 shows the induced magnetic flux density for two different superconducting grains that both have a single pair-breaking facet. The flux is generated by the fractional vortex-antivortex phase, and the pair-breaking facet is formed by cutting away either a triangular corner or a triangular section in the middle of a square grain, as shown in Figs. 1 (a) and (b), respectively. Thus, in the latter case, the pair-breaking facet is completely surrounded by bulk superconductivity. The fact that the phase persists in these two systems clearly illustrates a contrast to the Vorontsov phase [14], which relies on the proximity of two pair-breaking edges. Figure 1 (c) shows a magnification of the pair-breaking facet in Figure 1 (a). As shown, there

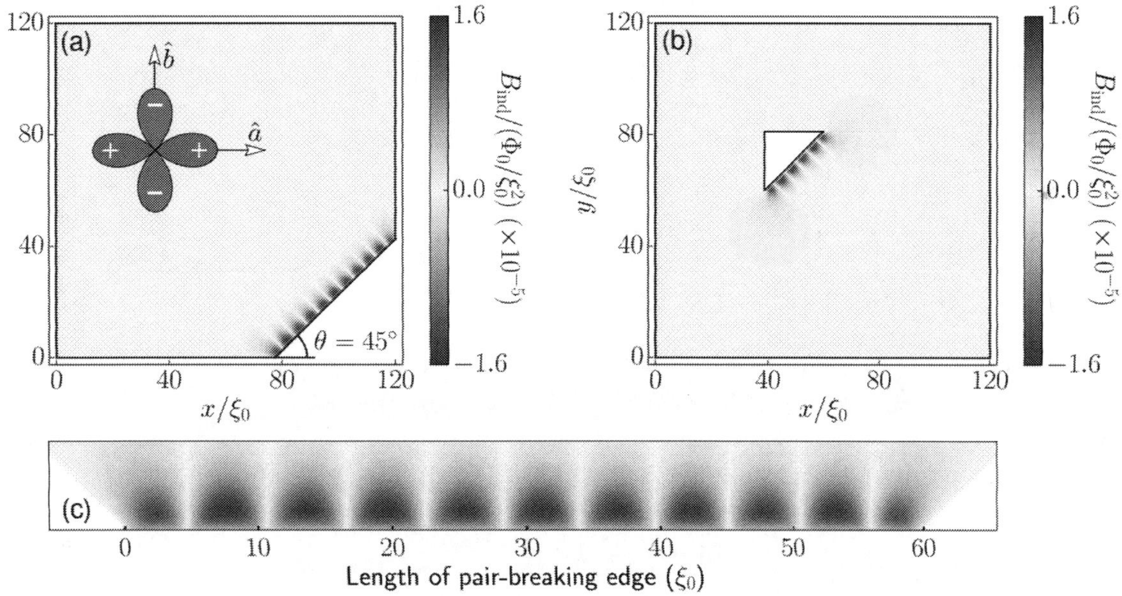


**Figure 1.** (Color online) (a) A *d*-wave superconducting grain at temperature $T = 0.1T_c$ with a spontaneously induced magnetic flux density, due to spontaneous generation of fractional vortices and antivortices. The latter breaks time-reversal and translational (along the facet) symmetries, and is linked to an energetically favorable Doppler shift of mid-gap states to finite energies. These mid-gap states are formed through pair-breaking along the diagonal grain facet, which is rotated 45° relative to the crystal *ab*-axes. All other grain edges are perfectly aligned with the crystal axes, as indicated by the graphics. In panel (b), a triangular portion of a square superconductor has been cut away, such that the pair-breaking facet is surrounded by bulk superconductivity. Panel (c) is a magnification of the pair-breaking facet in panel (a).

might be an unequal number of vortices and antivortices for certain sizes, although the flux density sums to zero. This is illustrated further in Fig. 2, where we vary the length of the pair-breaking facet in corner-cut systems. Each panel shows the induced flux along the pair-breaking facet in a square grain of side-length $120\xi_0$. The length of the facet varies from $2.5\xi_0$ in panel (a), to $66\xi_0$ in panel (f). There are two relevant regimes; one when the length of the facet is comparable to the fractional vortex size ($\sim 5\xi_0$), and another when it is much larger. In the latter case, the fractional vortices have a fairly constant diameter of $5\xi_0$, except the corner, or (outermost, vortices which are generally smaller. Lengthening the facet increases the size of the corner vortices, until they reach the same size as the central vortices, and new corner vortices are formed. Therefore, there might be an unequal number of vortices and antivortices for certain sizes. The flux density sums to zero, however, thanks to the corner vortices being much smaller. This again illustrates the fractionality of the vortices. The most striking feature, however, is that the phase survives even as the facet becomes smaller than $5\xi_0$, yielding a system with a single fractional vortex and a clear net flux. The system obviously finds it more favorable to shift the mid-gap states at the expense of having a net flux. Thus, the system with a single pair-breaking facet seems to lack a critical minimum size, in contrast to both the thin-film geometry [14], and the mesoscopic grain where all sides are pair-breaking [15].

Finally, Figs. 3 (a)–(b) show the density of states (DOS) along the facet for the systems in Figs. 2 (a)–(b), respectively. All other systems have an identical DOS as in panel (b). The solid

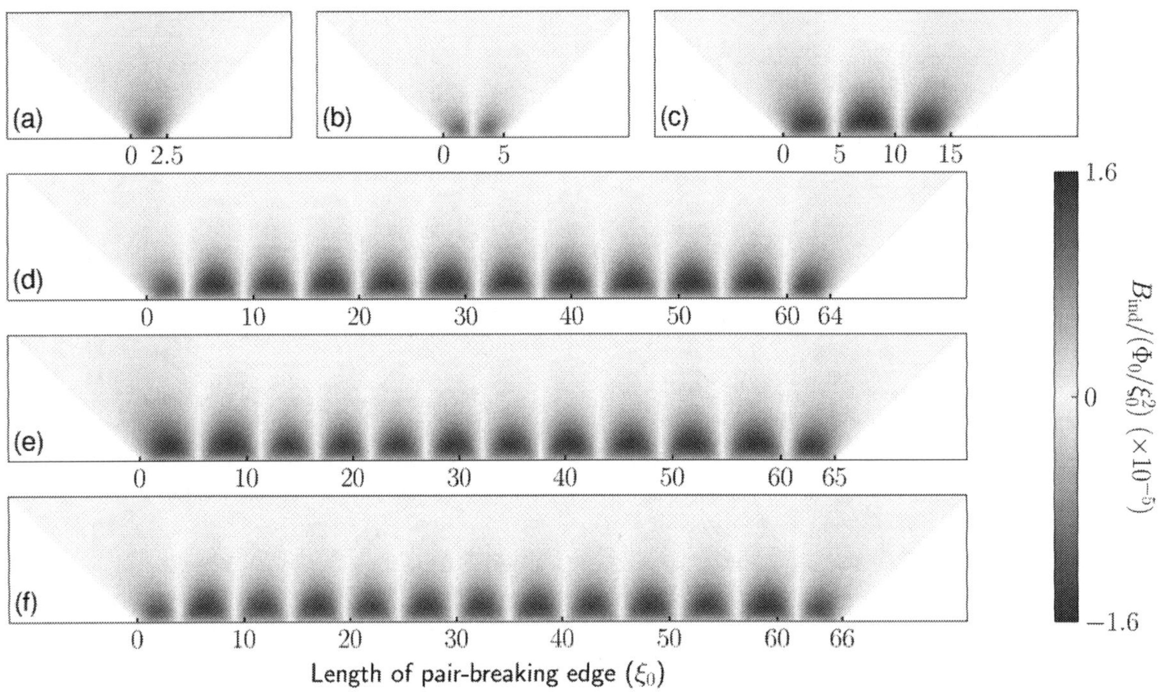


**Figure 2.** (Color online) Magnetic flux density due to spontaneous fractional vortices along a pair-breaking facet, where the length of the facet varies from panel (a) to (f). In each panel, the system is a square grain with a side-length of $120\xi_0$, with one of the corners cut off at a 45° angle to generate the pair-breaking facet, as illustrated in Fig. 1 (a). Due to finite-size effects, there might be an unequal amount of vortices and antivortices, although the total flux still adds to zero. The only exception is when the facet is smaller than the typical vortex size ($\sim 5\xi_0$) as in panel (a), at which point there is a single vortex and a net flux. The temperature is $T = 0.1T_c$.

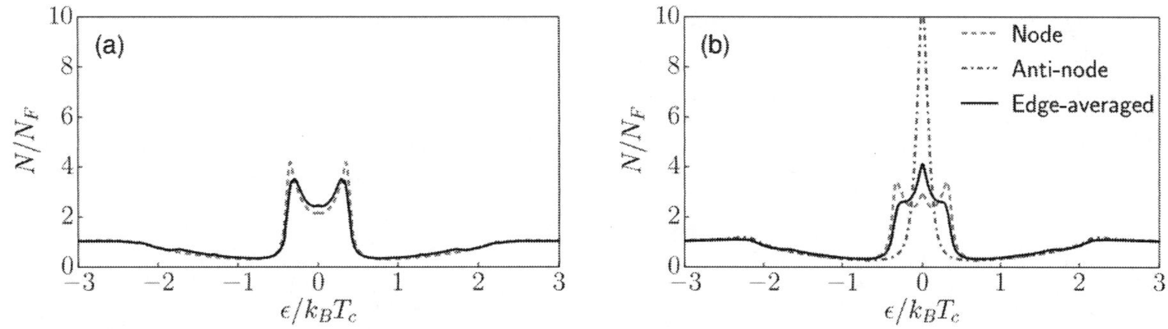


**Figure 3.** (Color online) Density of states as a function of energy at the pair-breaking facet, evaluated in the middle of a vortex current (dashed line) and between vortices (dot-dashed line). The solid line is the facet-averaged density of states. Panels (a)–(b) correspond to the systems in Figs. 2 (a)–(b), respectively. The rest of the systems have an identical DOS as in panel (b).

lines represent the facet-averaged DOS, the dashed lines the local DOS at a node (vortex) and the dot-dashed lines the local DOS at an anti-node (between vortices). System (a) has a single vortex, resulting in a fully split peak. All other systems show a wide peak in the facet-averaged DOS. This result would be observable in a tunneling experiment as a conductance peak centered at zero energy with a rather large width, that at low temperatures is temperature independent. Only for system (a), or with a very local probe (point contact with a diameter smaller than the superconducting coherence length) would a split conductance peak be observable.

## 4. Conclusions

We have used the quasiclassical theory of superconductivity to study a phase that spontaneously breaks translational and time-reversal symmetries at pair-breaking edges, in unconventional *d*-wave superconductors. Similar phases have been suggested by theory for quite some time, but up until now, have relied on finite-size effects and the proximity of two such pair-breaking edges. We have shown that such finite-size effects are not necessary for such a phase to exist, and that there is no clear critical size below which the phase disappears. This implies that any system with pair-breaking edges should be unstable to the formation of fractional vortices. Therefore, the phase should be present at a greater variety of systems than previously proposed, and lead to a broadening of zero-bias conductance peaks.

## Acknowledgments

We thank the Swedish Research Council (VR) for financial support. It is a pleasure to thank Mikael Håkansson, Niclas Wennerdal and Anton Vorontsov for valuable discussions.

## References

[1] Tsuei C C and Kirtley J R 2000 *Rev. Mod. Phys.* **72** 969
[2] Hu C R 1994 *Phys. Rev. Lett.* **72** 1526
[3] Kashiwaya S and Tanaka Y 2000 *Rep. Prog. Phys.* **63** 1641
[4] Löfwander T, Shumeiko V S and Wendin G 2001 *Supercond. Sci. Technol.* **14** R53
[5] Volovik G and Gor'kov L 1985 *Sov. Phys. JETP* **61** 843
[6] Sigrist M and Ueda K 1991 *Rev. Mod. Phys.* **63** 239
[7] Matsumoto M and Shiba H 1995 *J. Phys. Soc. Jpn.* **64** 3384
[8] Matsumoto M and Shiba H 1995 *J. Phys. Soc. Jpn.* **64** 4867

[9] Fogelström M, Rainer D and Sauls J A 1997 *Phys. Rev. Lett.* **79** 281
[10] Potter A C and Lee P A 2014 *Phys. Rev. Lett.* **112** 117002
[11] Higashitani S 1997 *J. Soc. Phys. Jpn.* **66** 2556
[12] Barash Y S, Kalenkov M S and Kurkijärvi J 2000 *Phys. Rev. B* **62** 6665
[13] Löfwander T, Shumeiko V S and Wendin G 2000 *Phys. Rev. B* **62** R14653
[14] Vorontsov A B 2009 *Phys. Rev. Lett.* **102** 177001
[15] Håkansson M, Löfwander T and Fogelström M 2015 *Nat. Phys.* **11** 755
[16] Nagai Y, Ota Y and Tanaka K 2017 *Phys. Rev. B* **96** 060503
[17] Covington M, Aprili M, Paraoanu E, Greene L H, Xu F, Zhu J and Mirkin C A 1997 *Phys. Rev. Lett.* **79** 277
[18] Krishana K, Ong N P, Li Q, Gu G D and Koshizuka N 1997 *Science* **277** 83
[19] Dagan Y and Deutscher G 2001 *Phys. Rev. Lett.* **87** 177004
[20] Gonnelli R, Calzolari A, Daghero D, Natale L, Ummarino G, Stepanov V and Ferretti M 2001 *Eur. Phys. J. B* **22** 411
[21] Elhalel G, Beck R, Leibovitch G and Deutscher G 2007 *Phys. Rev. Lett.* **98** 137002
[22] Gustafsson D, Golubev D, Fogelström M, Claeson T, Kubatkin S, Bauch T and Lombardi F 2013 *Nat. Nanotechnol.* **8** 25
[23] Watashige T, Tsutsumi Y, Hanaguri T, Kohsaka Y, Kasahara S, Furusaki A, Sigrist M, Meingast C, Wolf T, Löhneysen H v, Shibauchi T and Matsuda Y 2015 *Phys. Rev. X* **5** 031022
[24] Neils W K and Van Harlingen D J 2002 *Phys. Rev. Lett.* **88** 047001
[25] Kirtley J, Tsuei C, Ariando A, Verwijs C, Harkema S and Hilgenkamp H 2006 *Nat Phys* **2** 190
[26] Hachiya M, Aoyama K and Ikeda R 2013 *Phys. Rev. B* **88** 064519
[27] Higashitani S and Miyawaki N 2015 *J. Soc. Phys. Jpn.* **84** 033708
[28] Miyawaki N and Higashitani S 2015 *Phys. Proc.* **65** 25
[29] Miyawaki N and Higashitani S 2015 *Phys. Rev. B* **91** 094511
[30] Eilenberger G 1968 *Z. Phys.* **214** 195
[31] Larkin A I and Ovchinnikov Y N 1969 *Sov. Phys. JETP* **28** 1200
[32] Eschrig M 2009 *Phys. Rev. B* **80** 134511

IOP Conf. Series: Journal of Physics: Conf. Series **969** (2018) 012038 doi:10.1088/1742-6596/969/1/012038

# Irreversibility and critical current density of $FeSr_2ErCu_2O_{6+y}$

**Y. Hata[1], I. Iida[1], T. Mochiku[2] and H. Yasuoka[1]**

[1]Department of Applied Physics, National Defense Academy, Kanagawa, 239-8686, Japan
[2]Research Center for Advanced Measurement and Characterizaition, National Institute for Materials Science, Ibaraki, 305-0047, Japan

E-mail: hata@nda.ac.jp

**Abstract.** $FeSr_2ErCu_2O_{6+y}$ (ErFe1212) and non-superconducting $FeSr_2ErCu_{1.9}Zn_{0.1}O_{6+y}$ were synthesized to study the property of the superconductivity and the irreversibility of ErFe1212. A large irreversibility in the temperature dependence of magnetization and a hysteresis in the magnetization curve were observed in ErFe1212. By comparison with non-superconducting $FeSr_2ErCu_{1.9}Zn_{0.1}O_{6+y}$, it was found that the most part of the hysteresis at high magnetic field originates from the magnetism of Fe ion and some part of the hysteresis at low magnetic field originates from the superconductivity. Using the magnetization curve of ErFe1212 and $FeSr_2ErCu_{1.9}Zn_{0.1}O_{6+y}$, the $J_c$ of ErFe1212 in individual grains at 10 K under 0.1 T was estimated by the Bean model and $J_c^{\rm intra}$ was $2.6 \times 10^9$ A/m$^2$. The critical current density across inter-grain boundaries at 10 K estimated by $V - I$ measurement was $J_c^{\rm inter} = 5.7 \times 10^4$ A/m$^2$. A large difference between $J_c^{\rm inter}$ and $J_c^{\rm intra}$ was observed in ErFe1212. $J_c^{\rm intra}$ and $J_c^{\rm intra}$ of ErFe1212 are 2.2 and 5.2 times larger than these of YFe1212, respectively.

## 1. Introduction

The crystal structure of $FeSr_2YCu_2O_{6+y}$ (YFe1212) is related to that of $YBa_2Cu_3O_{7-\delta}$, in which Cu ion is partially replaced by Fe ion and Ba ion is fully replaced by Sr ion. In YFe1212, superconductivity has been found to be associated with $CuO_2$ plane and magnetism also has been associated with $FeO_\delta$ layers. YFe1212 shows superconductivity only after three separate annealing process [1]. The first annealing is performed in $N_2$ flow, and this relates to suppression of Fe substitution at Cu(2) sites. The second and third annealing are in $O_2$ flow and in a high-pressure $O_2$ atmosphere, respectively, to supply carriers to the $CuO_2$ planes[2, 3]. The optimized annealing condition is annealing in $N_2$ flow at 790 °C, annealing in $O_2$ flow and under high $O_2$ pressure at 270 °C[4]. Superconductivity of YFe1212 was much affected by the annealing temperature in $N_2$ flow rather than the annealing temperature in $O_2$ flow. After this annealing process, YFe1212 shows zero resistivity below 40 K.

Two stage transition was observed in resistivity measurement under the magnetic field. The critical current densities of YFe1212 in individual grains, $J_c^{\rm intra}$, and across inter-grain boundaries, $J_c^{\rm inter}$, were studied[5]. $J_c^{\rm inter}$ was about six orders of magnitude lower than $J_c^{\rm intra}$ at 2 K under 0.1 T and this seems to be responsible for the two stage transition.

The superconducting properties of $YBa_2Cu_3O_{7-\delta}$ are affected by the substitution of rare earth (RE) ions for Y ion, which are sandwiched between $CuO_2$ plane. Most type of RE ions substituted for Y ion and the effects of the substituting ions on the superconducting transition temperature[6] and magnetic irreversibility field[7] have been reported. The Y sites in YFe1212 is also occupied by RE ions. The crystal structure of $FeSr_2ErCu_2O_{6+y}$ (ErFe1212) and $FeSr_2NdCu_2O_{6+y}$ was studied[8] and the superconducting properties and Fe ion distribution of ErFe1212, $FeSr_2EuCu_2O_{6+y}$, $FeSr_2GdCu_2O_{6+y}$ and $FeSr_2NdCu_2O_{6+y}$ were studied[9]. A large irreversibility was observed in the temperature dependence of magnetization for ErFe1212 and the origin of this irreversibility was unclear. On the assumption that this large irreversibility originates from the superconductivity, it implies that ErFe1212 has large critical current density.

In this study, the proper annealing condition was determined and the magnetic and electric property of ErFe1212 were studied to reveal the origin of this irreversibility and the superconducting property.

## 2. Experimental

Polycrystalline samples of ErFe1212 were prepared by a solid state reaction using a stoichiometric mixture of $Er_2O_3$, $SrCO_3$, CuO and $Fe_2O_3$ powders. The mixture was calcined at 850 °C for 24 h in air, ground and then pressed into pellets. The pellets were sintered at 950 °C for 24 h in air. Then, they were subsequently annealed at $T_{N_2} = 730 \sim 850$ °C for 24 h in $N_2$ flow, at 270 °C for 24 h in $O_2$ flow, and finally at 270 °C for 24 h in $O_2$ atmosphere under a high pressure of 17.6 MPa. The non-superconducting $FeSr_2YCu_{1.9}Zn_{0.1}O_{6+y}$ were synthesized and annealed by the same process.

The crystal structure and phase purity of the samples were characterized using powder x-ray diffraction. The magnetization curves and the temperature dependence of magnetization were measured using a SQUID magnetometer (MPMS-XL5, Quantum Design). The resistivity was measured by a four-probe dc technique. The critical current density in individual grains, $J_c^{intra}$ was estimated by the Bean model[10] and that across inter-grain boundaries, $J_c^{inter}$ was estimated from voltage-current measurement.

## 3. Result and discussion

For as-synthesized sample, all of the peaks in x-ray diffraction pattern can be assigned to single-phase ErFe1212 with a tetragonal crystal structure. After the annealing in $N_2$ flow, the crystal structure changes to an orthorhombic structure and after the annealing in $O_2$ flow, it reverts to tetragonal structure.

The temperature dependence of the resistance for $FeSr_2ErCu_2O_{6+y}$ after various annealing temperature in $N_2$ flow is shown in Fig. 1. The samples were annealed in $O_2$ flow and under high $O_2$ pressure at 270 °C after the annealing in $N_2$ flow. The resistance began to decrease below around 60 K in the sample annealed below 810 °C in $N_2$ flow. Zero resistivity was observed below $T_c^{zero} = 27 \sim 44$ K for the sample annealed at $T_{N_2} = 750 \sim 810$ °C. The annealing temperature in $N_2$ flow dependence of the transition temperatures, $T_c^{onset}$ and $T_c^{zero}$, were shown in Fig. 2. A peak was observed in $T_{N_2}$ dependence of $T_c^{zero}$ and the maximum $T_c^{zero} = 44$ K was observed at $T_{N_2} = 790$ °C. Since superconductivity of YFe1212 was much affected by the annealing temperature in $N_2$ flow rather than that in $O_2$ flow [4], we concluded that the proper annealing temperature of $N_2$ flow and $O_2$ atmosphere is 790 °C and 270 °C, respectively. The sample annealed at this proper annealing temperature was used in following study.

Figure 3 shows the temperature dependence of the magnetization under an applied magnetic field of 1.0 mT after zero field cooling and field cooling. The magnetization of ErFe1212 begins to decrease at $T_c = 56.6$ K and diamagnetism was observed. The temperature dependence of magnetization consists of diamagnetism of superconductivity and the magnetism of Fe ion and paramagnetism of Er ion. A large irreversibility was observed between zero field cooled and field cooled curves below 26.6 K. On the other hand, no diamagnetism was observed and similar irreversibility was observed below 21.6 K in non-superconducting $FeSr_2YCu_{1.9}Zn_{0.1}O_{6+y}$.

IOP Conf. Series: Journal of Physics: Conf. Series **969** (2018) 012038 doi:10.1088/1742-6596/969/1/012038

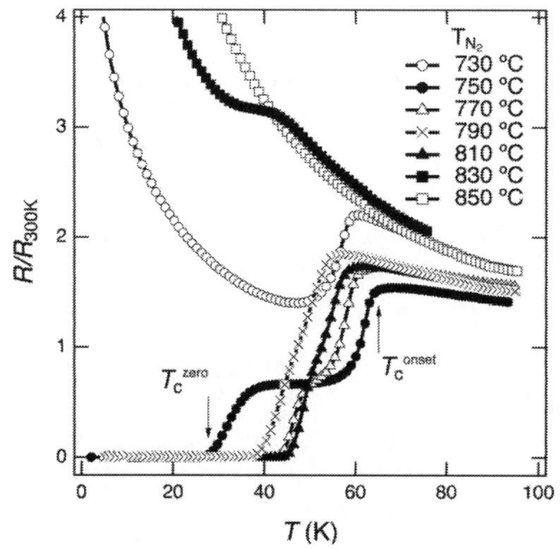


**Figure 1.** Temperature dependence of the resistances for $FeSr_2ErCu_2O_{6+y}$. The values of the resistivities are normalized with the resistances at 300 K.

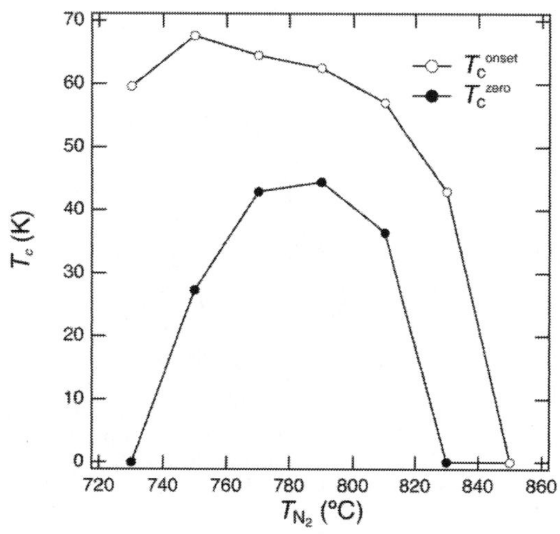


**Figure 2.** Relation between the transition temperature, $T_c^{onset}$ and $T_c^{zero}$, for $FeSr_2ErCu_2O_{6+y}$ and the annealing temperature in $N_2$ flow, $T_{N_2}$.

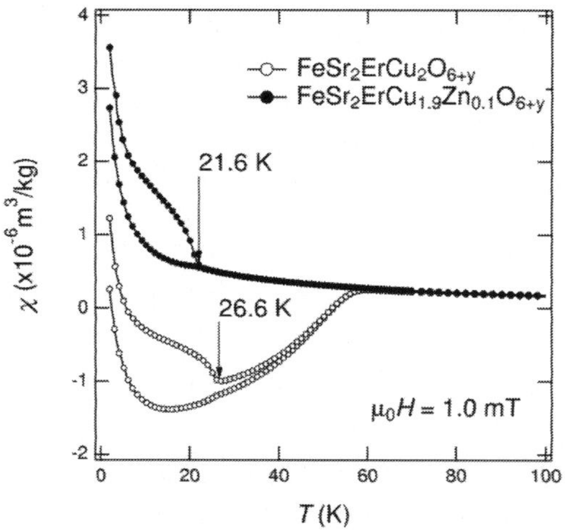


**Figure 3.** Temperature dependence of magnetization under 1.0 mT.

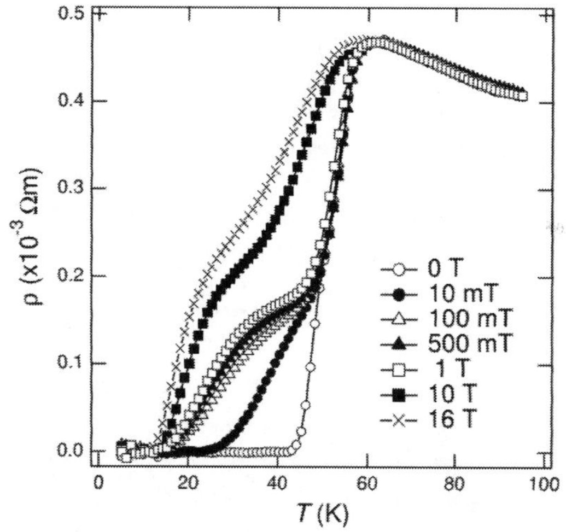


**Figure 4.** Temperature dependence of resistivity for $FeSr_2ErCu_2O_{6+y}$.

The temperature dependence of resistivity was shown in Fig. 4. In zero magnetic field, the resistivity was decreased due to the onset of the superconductivity at 60.5 K. Zero resistivity was observed below $T_c^{zero} = 43.3$ K. Under the magnetic field, two stage transition in the resistivity curves were observed and $T_c^{zero}$ was decreased with increase of the magnetic field.

The magnetization curves of ErFe1212 were shown in Fig. 5. The magnetization curve consists of the type-II superconductivity and the magnetism of Fe ion and paramagnetism of Er ion. Small hysteresis was observed and it becomes small with increase of temperature. The critical current density in individual grains, $J_c^{intra}$, was estimated by the width of magnetization in the hysteresis curve and the Bean model[10].

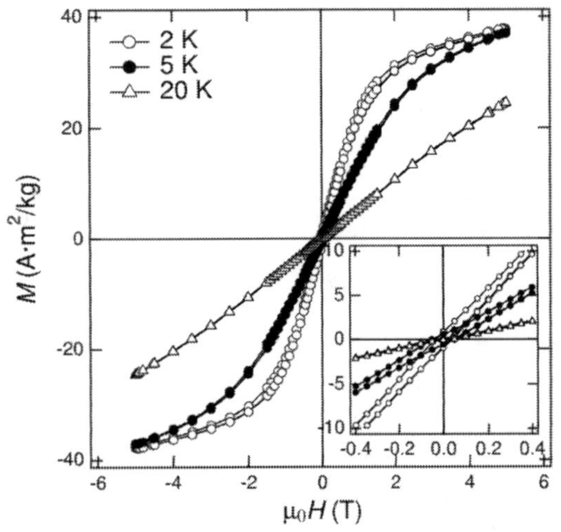


**Figure 5.** Magnetization curves for $FeSr_2ErCu_2O_{6+y}$.

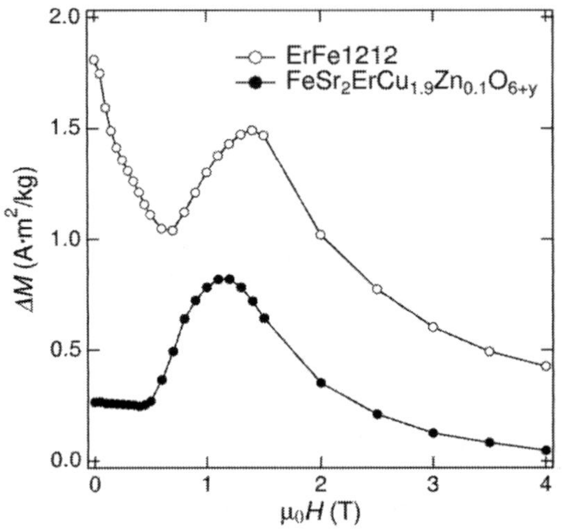


**Figure 6.** Magnetic field dependence of $\Delta M$ for ErFe1212 and $FeSr_2YCu_{1.9}Zn_{0.1}O_{6+y}$ at 2 K.

$$J_c^{\mathrm{intra}} = \frac{3}{2} \times \frac{\Delta M}{r} \tag{1}$$

$\Delta M$ is the difference of the magnetization per unit volume between ascending and descending field process and $r$ is the radius of the grain. The magnetic field dependence of $\Delta M$ is shown in Fig. 6. A peak was observed at around the magnetic field of 1.3 T in ErFe1212 and $FeSr_2YCu_{1.9}Zn_{0.1}O_{6+y}$. Since the magnetism of Er ion is paramagnetism down to 2 K[9], this peak may originate from the magnetism of Fe ion. The irreversibility of ErFe1212 under low magnetic field may mainly originate from the superconductivity and to avoid the effect of the irreversibility of magnetism, $J_c^{\mathrm{intra}}$ of ErFe1212 was deduced from

$$J_c^{\mathrm{intra}} = \frac{3}{2} \times \frac{\Delta M - \Delta M'}{r} \tag{2}$$

,where $\Delta M'$ is $\Delta M$ of $FeSr_2YCu_{1.9}Zn_{0.1}O_{6+y}$. $J_c^{\mathrm{intra}}$ of ErFe1212 at 10 K under 0.1 T is estimated as $2.6 \times 10^9 \mathrm{A/m^2}$ and this value is 2.2 times larger than that of YFe1212[5].

The voltage-current curve for ErFe1212 is shown in Fig. 7. The values of the voltage and current are normalized by the sample dimension. The critical current density across the inter-grain boundaries, $J_c^{\mathrm{inter}}$, was determined usign a 100 $\mu$V/m offset criterion. The temperature dependence of $J_c^{\mathrm{inter}}$ is shown in Fig. 8. $J_c^{\mathrm{inter}}$ at 10 K was $5.7 \times 10^4$ A/m$^2$ and was decreased linearly with increase of temperature. $J_c^{\mathrm{inter}}$ of ErFe1212 was 5.2 times larger than that of YFe1212[5]. $J_c^{\mathrm{inter}}$ was about five orders of magnitude lower than $J_c^{\mathrm{intra}}$ under 0.1 T. The broadening in resistivity measurement was probably due to this large difference bewteen $J_c^{\mathrm{inter}}$ and $J_c^{\mathrm{intra}}$[5].

## 4. Summary

ErFe1212 and non-superconducting $FeSr_2ErCu_{1.9}Zn_{0.1}O_{6+y}$ were prepared to study the property of the superconductivity and the irreversibility of ErFe1212. ErFe1212 exhibits superconductivity only when it is annealed in a $N_2$ atmoshpere and subsequently in an $O_2$ atmosphere. The proper annealing condition of ErFe1212 was determined in advance.

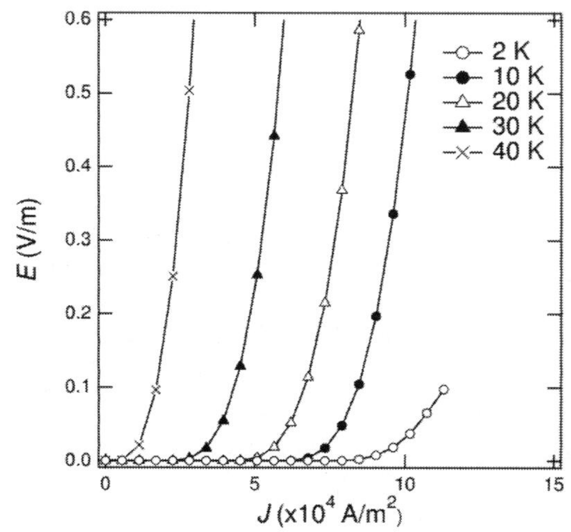


**Figure 7.** Voltage - current curves for $FeSr_2ErCu_2O_{6+y}$. The values of the voltage and current are normalized by the sample dimension.

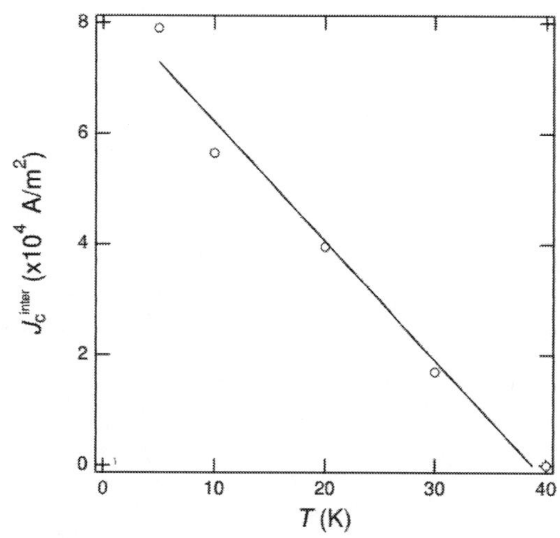


**Figure 8.** Temperature dependence of critical current density across inter-grain boundaries, $J_c^{inter}$ of $FeSr_2ErCu_2O_{6+y}$.

A large irreversibility was observed in the temperature dependence of magnetization below 26.6 K. The temperature dependence of magnetization consists of diamagnetism of superconductivity and the magnetism of Fe ion and Er ion. The magnetism of Er ion kept Curie-Weiss type paramagnetism down to 2 K.

The magnetization curve consists of the type-II superconductivity and the magnetism of Fe ion and paramagnetism of Er ion. A peak of $\Delta M$, that is the difference of the magnetization per unit volume between ascending and descending field process, was observed at around 1.3 T and large $\Delta M$ was observed at zero magnetic field. Since a similar irreversibility was observed for non-superconducting $FeSr_2ErCu_{1.9}Zn_{0.1}O_{6+y}$, the irreversibility at around 1.3 T may originate from the magnetism of the Fe ion.

Some part of the irreversibility under low magnetic field may originate from the superconductivity. Using the magnetization curve of ErFe1212 and $FeSr_2ErCu_{1.9}Zn_{0.1}O_{6+y}$, $J_c^{intra}$ of ErFe1212 at 10 K under 0.1 T was $2.6 \times 10^9$ A/m$^2$. $J_c^{inter}$ at 10 K was $5.7 \times 10^4$ A/m$^2$ estimated by $V - I$ measurement. A large difference between $J_c^{inter}$ and $J_c^{intra}$ was observed in ErFe1212. The broadening in resistivity measurement was probably due to this large difference. $J_c^{intra}$ and $J_c^{intra}$ of ErFe1212 are 2.2 and 5.2 times larger than these of YFe1212, repspectively.

## References

[1] J. Shimoyama, K.Otzschi, T. Hinouchi and K. Kishio 2000 *Physica* C **341-348** 563.
[2] T. Mochiku, Y. Mihara, Y. Hata, S. Kamisawa, M. Furuyama, J. Suzuki, K. Kadowaki, N. Metoki, H. Fujii and K. Hirata 2002 *J. Phys. Soc. Jpn.* **71** 790.
[3] Y. Hata, Y. Mihara, T. Mochiku, J. Suzuki, K. Kadowaki, E. Kita and H. Yasuoka 2004 *Physica* C **417** 17.
[4] K. Yamaguchi, Y. Hata, T. Mochiku and H. Yasuoka 2012 *Phys. Procedia* **27** 24.
[5] K. Yamaguchi, Y. Hata, T. Mochiku and H. Yasuoka 2013 *Physica* C **487** 56.
[6] J. G. Lin, C. Y. Huang, Y. Y. Xue, C. W. Chu, X. W. Cao, J. C. Ho 1995 *Phys. Rev.* **B 51** 12900.
[7] Y. Yasukawa, T. Nakane, M. Karppinen, H. Yamaguchi 2001 *Physica* C **357-360** 230.
[8] T. Mochiku, Y. Nakano, K. Oikawa, T. Kamiyama, H. Fujii, Y. Hata, J. Suzuki, I. Kakeya, K. Kadowaki and K. Hirata 2003 *Physica* C **400** 43.
[9] Y. Hata, K. Yamaguchi, K. Kinoshita, E. Kita, T. Mochiku and H. Yasuoka 2014 *Physica* C **507** 85.
[10] C. P. Bean 1962 *Phys. Rev. Lett.* **8** 250.

IOP Conf. Series: Journal of Physics: Conf. Series **969** (2018) 012039 doi:10.1088/1742-6596/969/1/012039

# Time-reversal symmetry breaking phase and gapped surface states in *d*-wave nanoscale superconductors

**Yuki Nagai[1,2], Yukihiro Ota[3], and K. Tanaka[4]**
[1]Japan Atomic Energy Agency, CCSE, Kashiwa, Japan
[2]Massachusetts Institute of Technology, Department of Physics, Cambridge, MA, USA
[3]Research Organization for Information Science and Technology (RIST), Kobe, Japan
[4]University of Saskatchewan, Department of Physics and Engineering Physics, Saskatoon, SK, Canada
E-mail: nagai.yuki@jaea.go.jp

**Abstract.** We solve the Bogoliubov-de Gennes equations self-consistently for *d*-wave superconductors with [110] surfaces. We find spontaneous breaking of time-reversal symmetry (TRS) caused by the spontaneous occurrence of a complex order parameter with an extended *s*-wave symmetry along the [110] surfaces. In the TRS-breaking phase, the *d*-wave order parameter itself becomes complex. We also show that vortex-antivortex pairs can appear along the [110] surfaces of nanoribbons. These pairs are detectable with surface sensitive probes.

## 1. Introduction

Time-reversal symmetry (TRS) and the corresponding topological phenomena are hot topics in condensed matter physics. The authors of Ref. [1] have shown that the bulk-edge correspondence with respect to a topological invariant protected by TRS underlies the zero-energy Andreev bound states on [110] surfaces of high-$T_c$ cuprates. While the Andreev bound states are usually understood as arising from the sign-changing property of the $d_{x^2-y^2}$-wave order parameter [2, 3], such zero-energy surface states are robust as the bulk-edge correspondence ensures the existence of gapless surface states in a topologically nontrivial system [4, 5].

The gapless surface states linked to TRS can become gapped by breaking of TRS [6, 7]. Spontaneous TRS breaking in *d*-wave superconductors has been discussed phenomenologically by introducing a secondary order parameter with relative phase $\pi/2$ to the predominant component that is stabilized near the [110] surfaces [8, 9]. This has been a controversial topic in the past two decades owing to contradicting experimental results on high-$T_c$ cuprates [10, 11]. Recent experiment, however, has clearly detected a full gap in the excitation spectrum that is consistent with spontaneously broken TRS in a nanoscale $YBa_2Cu_3O_{7-\delta}$ island [10].

In our previous paper [14], we have shown, by solving the Bogoliubov-de Gennes (BdG) equations self-consistently for the *d*-wave order parameter in diamond-shaped nanoislands with [110] surfaces, that spontaneous TRS breaking occurs at low temperatures via the spontaneous emergence of a new complex order parameter. The spontaneous disappearance of the topological protection is accompanied by this order parameter, induced on the surfaces below a critical temperature lower than the superconducting transition temperature, $T_c$. The additional order parameter has extended *s*-wave symmetry and characterizes the energy splitting of the Andreev

IOP Conf. Series: Journal of Physics: Conf. Series **969** (2018) 012039 doi:10.1088/1742-6596/969/1/012039

bound states on the surfaces. We have also shown that integer vortex-antivortex pairs are formed in the extended $s$-wave order parameter along the surfaces if the side length of a nanoisland is relatively large [14].

In this paper, we consider nanoribbons made of a $d$-wave superconductor by self-consistently solving the BdG equations along with the $d$-wave gap equation. The periodic boundary condition allows us to model nanoribbons of infinite length. We fix the temperature such that a TRS-breaking phase occurs. We find a spatial distribution of vortex-antivortex pairs along the [110] surfaces, which is similar to that in a large nanoisland [14].

## 2. Model and method

### 2.1. Model

We consider a tight-binding Hamiltonian $\mathcal{H} = \mathcal{H}_{\rm BCS} + \mathcal{H}_{\rm fab}$ on a two-dimensional square lattice. Here the generalized Bardeen-Cooper-Schrieffer (BCS) Hamiltonian for $d$-wave superconductivity is $\mathcal{H}_{\rm BCS} = \sum_{ij,\sigma}(-t_{ij} - \mu)c^{\dagger}_{i\sigma}c_{j\sigma} + \sum_{ij}\left[\Delta_{ij}c^{\dagger}_{i\uparrow}c^{\dagger}_{j\downarrow} + \text{H.c.}\right]$, where $c^{\dagger}_{i\sigma}$ creates the electron with spin $\sigma$ at site $i$ and $\mu$ denotes the chemical potential. For simplicity, we only consider the nearest-neighbor hopping. We use the unit system with $\hbar = k_{\rm B} = 1$. The Hamiltonian for the fabrication potential $\mathcal{H}_{\rm fab}$ can be written as $\mathcal{H}_{\rm fab} = \sum_{i,\sigma} V_0(\boldsymbol{r}_i)c^{\dagger}_{i\sigma}c_{i\sigma}$, where $\boldsymbol{r}_i$ is a position vector with index $i$ whose origin is a center of the system: $\boldsymbol{r}_i = (i_x - L_x/2 - 1/2, i_y - L_y/2 - 1/2)$. The fabrication potential term $\mathcal{H}_{\rm fab}$ enables us to create an arbitrary shape for the nanoscale superconductor. One diagonalizes $\mathcal{H}$ to solve the BdG equations expressed as

$$\sum_j \hat{H}_{ij}\begin{pmatrix} u_\gamma(\boldsymbol{r}_j) \\ v_\gamma(\boldsymbol{r}_j)\end{pmatrix} = E_\gamma \begin{pmatrix} u_\gamma(\boldsymbol{r}_i) \\ v_\gamma(\boldsymbol{r}_i)\end{pmatrix}, \tag{1}$$

with

$$\hat{H}_{ij} = \begin{pmatrix} \left[\hat{H}^{\rm N}\right]_{ij} & [\hat{\Delta}]_{ij} \\ \left[\hat{\Delta}^{\dagger}\right]_{ij} & -\left[\hat{H}^{\rm N*}\right]_{ij}\end{pmatrix}. \tag{2}$$

Here $\left[\hat{H}^{\rm N}\right]_{ij} = -t_{ij} - (\mu - V_0(\boldsymbol{r}_i))\delta_{ij}$ and $[\hat{\Delta}]_{ij} = V_{ij}\sum_{\gamma=1}^{2N} u_\gamma(\boldsymbol{r}_j)v^*_\gamma(\boldsymbol{r}_i)f(E_\gamma)$, where $N$ is the number of lattice sites, $V_{ij}$ denotes the pairing interaction, and $f(x)$ is the Fermi-Dirac distribution function. The $d$-wave order parameter at site $i$ is defined in terms of the mean fields as

$$\Delta_{d,i} = (\Delta_{\hat{x},i} + \Delta_{-\hat{x},i} - \Delta_{\hat{y},i} - \Delta_{-\hat{y},i})/4, \tag{3}$$

with $\Delta_{\pm\hat{e},i} = \Delta(\boldsymbol{r}_i, \boldsymbol{r}_i \pm \hat{e})$, where $\hat{x}$ and $\hat{y}$ denote the unit vectors in the square lattice. Considering two-dimensional systems, we assume that the field penetration depth $\lambda$ is infinity. We use complex mean fields $\{\Delta_{ij}\}$ with random phases as an initial guess for our calculation. For the results presented below, the chemical potential is fixed to be $\mu = -1.5t$ and the pairing interaction is nonzero only between nearest-neighbor sites, $V_{ij} \equiv U = -2t$, so that the resulting order parameter is purely of the $d_{x^2-y^2}$-wave symmetry in the TRS-preserved phase.

In a diamond-shaped square nanoisland of a $d$-wave superconductor, we find at a certain temperature that the system goes through a second-order phase transition into a ground state with the property that the extended $s$-wave order parameter defined by [12, 13]

$$\Delta_{s,i} = (\Delta_{\hat{x},i} + \Delta_{-\hat{x},i} + \Delta_{\hat{y},i} + \Delta_{-\hat{y},i})/4, \tag{4}$$

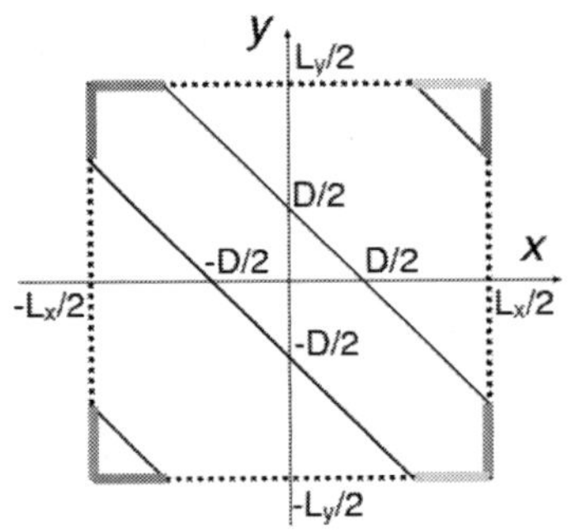


**Figure 1.** (Color online) Schematic figure of a nanoribbon superconductor. The periodic boundary condition is used.

becomes nonzero along the [110] surfaces [14]. We stress that no on-site attractive interaction is involved in our calculation: a $d + is$ state typically appears in the presence of such an on-site interaction [3, 15]. The fabrication potential of a nanoribbon is

$$V_0(\boldsymbol{r}_i) = \begin{cases} 0 & -x_i - \frac{D}{2} < y_i < -x_i + \frac{D}{2}, \\ 0 & -x_i - \left(\frac{L_x+L_y-D}{2}\right) < y_i < -x_i + \left(\frac{L_x+L_y-D}{2}\right), \\ 500t & \text{else}, \end{cases} \tag{5}$$

where the nanoribbon width is $D/\sqrt{2}$. Our previous study [14] indicates the presence of a suitable spatial scale for the formation of vortex-antivortex pairs; the width of a nanoribbon being roughly eight times the coherence length. Figure 1 shows our setup of a $d$-wave superconducting nanoribbon, with the periodic boundary condition.

### *2.2. Method*

To solve the BdG equations and the gap equation self-consistently, we calculate the mean fields, $[\hat{\Delta}]_{ij} = \langle c_i c_j \rangle$. With the use of analytic continuation, the mean fields can be rewritten as

$$[\hat{\Delta}]_{ij} = T \sum_{n=-\infty}^{\infty} [\check{G}(i\omega_n)]_{ji+N}, \tag{6}$$

with the fermion Matsubara frequency $\omega_n = \pi T(2n+1)$. Here, $\check{G}(i\omega_n)$ is a $2N \times 2N$ matrix form of the temperature-dependent Green's function,

$$\check{G}(i\omega_n) = \left[i\omega_n \check{1} - \check{H}\right]^{-1}. \tag{7}$$

Since the element is calculated as $\left[\check{G}(i\omega_n)\right]_{ij} = \boldsymbol{e}^i \check{G}(i\omega_n) \boldsymbol{e}^j$, we obtain

$$\left[\check{G}(i\omega_n)\right]_{ij} = \boldsymbol{e}^i \boldsymbol{x}^j(n), \tag{8}$$

with the solutions of the linear equations,

$$\left[i\omega_n \check{1} - \check{H}\right] \boldsymbol{x}^j(n) = \boldsymbol{e}^j, \tag{9}$$

where $[\boldsymbol{e}^i]_k = \delta_{ik}$. These linear equations for different frequencies can be solved simultaneously by the reduced-shifted-conjugate-gradient (RSCG) method [16]. The cutoff for the Matsubara

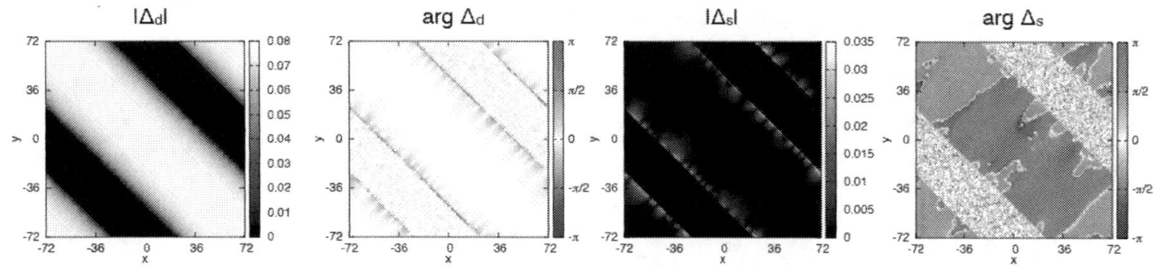


**Figure 2.** (Color online) $d$-wave and induced extended $s$-wave order parameters in the TRS-broken phase. The system size is $L_x \times L_y = 144 \times 144$ and the width of the nanoribbon is $D/\sqrt{2} = (2L_x/3)/\sqrt{2}$. The temperature is $T = 0.01t$.

frequency is $\omega_c = 60\pi = \pi T(2n_c + 1)$. The convergence criterion for the RSCG method is that the residual is smaller than 0.1, leading to a sufficient accuracy in the calculation of the Green's function [16].

To find the true ground state, we calculate the thermodynamic potential given by [14],

$$\Omega_s = -T \sum_{\gamma=1}^{2N} \ln \left[1 + \exp\left(\frac{E_\gamma}{T}\right)\right] - \sum_{ij} \frac{|[\hat{\Delta}]_{ij}|^2}{U}. \tag{10}$$

Once the mean fields are converged in the RSCG method, all the eigenvalues, $E_\gamma$, are obtained by the Sakurai-Sugiura (SS) method [17, 18]. The SS method allows us to extract the eigenvalues (and corresponding eigenvectors) of a generic matrix from a given domain on the complex plane [17]. We take the adequate number of small circular domains for calculating all of the eigenvalues.

### *2.3. Topological invariant*

A topological invariant characterizing $d$-wave superconductors with TRS is the one-dimensional (1D) topological invariant [1], defined by the accumulation of the phase of the quasiparticle wavefunction along a closed loop in the momentum space (i.e., winding number),

$$w(k_y^0) = \frac{1}{2}\mathrm{sgn}\left[\frac{\partial\epsilon(-k_x^0, k_y^0)}{\partial k_x}\right]\left(\mathrm{sgn}\left[\Delta(-k_x^0, k_y^0)\right] - \mathrm{sgn}\left[\Delta(k_x^0, k_y^0)\right]\right), \tag{11}$$

where $\epsilon(\boldsymbol{k})$ is the normal-state band dispersion and $(\pm k_x^0, k_y^0)$ denote the intersection points between the integration path with fixed $k_y^0$ and the Fermi surface. Thus, if the gap function $\Delta(\boldsymbol{k})$ satisfies

$$\Delta(-k_x^0, k_y^0)\Delta(k_x^0, k_y^0) < 0, \tag{12}$$

then the 1D topological invariant has a nonzero value and the zero-energy bound states exist because of the bulk-edge correspondence. Varying $k_y^0$ in the range with no gap closing, we can obtain the surface dispersion relation for the zero-energy bound states [1]. Since the time-reversal operation includes the complex conjugation operator, the gap function $\Delta$ should be real in the TRS-preserved phase. In contrast, if the gap function is complex (TRS-breaking phase), the above winding number becomes ill-defined [14].

IOP Conf. Series: Journal of Physics: Conf. Series **969** (2018) 012039 doi:10.1088/1742-6596/969/1/012039

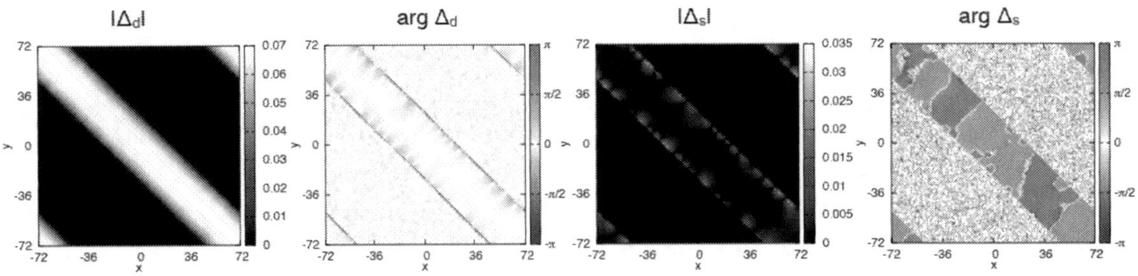


**Figure 3.** (Color online) $d$-wave and induced extended $s$-wave order parameters in the TRS-broken phase. The system size is $L_x \times L_y = 144 \times 144$ and the width of the nanoribbon is $D/\sqrt{2} = (L_x/3)/\sqrt{2}$. The temperature is $T = 0.01t$.

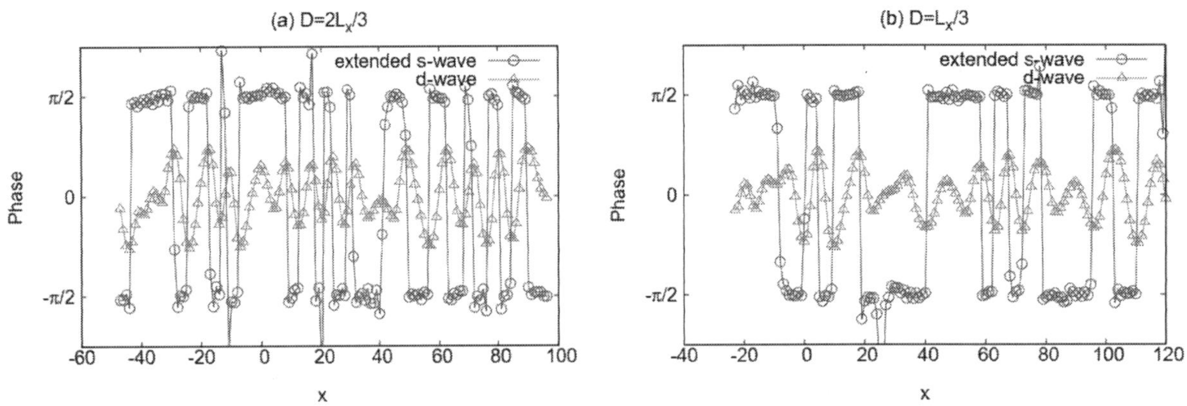


**Figure 4.** (Color online) The spatial distributions of the phases $\arg(\Delta_s)$ and $\arg(\Delta_d)$ on the surface edge ($y = -x + D/2$).

## 3. Results

We consider two kinds of $d$-wave superconducting nanoribbons, with width $D = 2L_x/3$ and $D = L_x/3$. Both of these systems have $L_x \times L_y = 144 \times 144$ lattice sites. We set the temperature $T = 0.01t$, where the TRS-breaking phase appears in a diamond-shape nanoisland [14]. Figures 2 ($D = 2L_x/3$) and 3 ($D = L_x/3$) show the amplitudes and phases of the order parameters. Focusing on the spatial distribution of $\arg(\Delta_d)$ and $\arg(\Delta_s)$, we find that in the extended $s$-wave order parameter the phase varies from $\pi/2$ to $-\pi/2$, and vice versa, with an alternating manner. Thus, the $d$-wave and extended $s$-wave order parameters are both complex and TRS is broken at this temperature. The ground state in this phase is fundamentally different from a $d + is$ state, which was discussed previously [8, 9, 15]. We also confirm that there is no zero-energy surface state in this phase.

Let us now focus on the spatial distribution of the order parameter phases in Figs. 2 and 3. To this end, we examine the phase distribution on the surface, $y = -x + D/2$, as seen in Fig. 4. It can be seen more clearly that $\arg(\Delta_s)$ spatially varies with $\pi$ shifts along the surfaces. In contrast, arg ($\Delta_d$) has no such phase shift and changes very little, although it is important to note that the phase is nonzero and hence $\Delta_d$ itself is complex. The amplitude of the extended $s$-wave order parameter gives us insight into this phenomenon: it vanishes along the surface as seen in Figs. 2 and 3, at positions where the $\pm\pi$ jump occurs in the phase. Thus, integer vortex-antivortex pairs appear in the extended $s$-wave order parameter along the surfaces. This result is similar to that for a larger-size $d$-wave superconducting nanoisland [14]. We note that the positions of the vortex-antivortex pairs depend on the initial condition of the randomly-distributed phase of the $d$-wave order parameter. Such vortex-antivortex pairs can be detected by surface-sensitive probes.

Vorontsov has proposed that films of a $d$-wave superconductor at low temperatures can exhibit unusual superconducting phases due to transverse confinement, associated with the spontaneous breaking of either TRS or continuous translational symmetry [19]. In the current work, the ground state still has the translational symmetry in the bulk region of the nanoribbons. According to Vorontsov's proposal, which is based on the quasiclassical Eilenberger approach, nanoribbons much narrower than the present ones can have a translational-symmetry breaking phase. The coherence length in our calculations would be too small to study such extremely narrow nanoribbons. An interesting future work is to examine such breaking of translational symmetry within the BdG framework by decreasing the width of nanoribbons.

In conclusion, we have shown by solving the BdG equations self-consistently for $d$-wave nanoribbons that spontaneous emergence of a complex, extended $s$-wave order parameter along the [110] surfaces breaks time-reversal symmetry. Such TRS-breaking phases can also sustain vortex-antivortex pairs along the surfaces.

## Acknowledgments

The calculations were performed on the supercomputing system SGI ICE X at the Japan Atomic Energy Agency. This study was partially supported by the "Topological Materials Science" (No. JP16H00995) KAKENHI on Innovative Areas from JSPS of Japan, and the Natural Sciences and Engineering Research Council of Canada.

## References

[1] Sato M, Tanaka Y, Yada K and Yokoyama T 2011, *Phys. Rev. B* **83** 224511
[2] Tsuei C C and Kirtley J R 2000 *Rev. Mod. Phys.* **72** 969
[3] Kashiwaya S and Tanaka Y 2000 *Rep. Prog. Phys.* **63** 1641
[4] Hasan M Z and Kane C L 2010 *Rev. Mod. Phys.* **82** 3045
[5] Alicea J, 2012 *Rep. Prog. Phys.* **75** 076501
[6] Chen Y L, Chu J-H, J. Analytis G, Liu Z K, Igarashi K, Kuo H-H, Qi X L, Mo S K, Moore R G, Lu D H, Hashimoto M, Sasagawa T, Zhang S C, Fisher I R, Hussain Z, and Shen Z X 2010 *Science* **329** 659
[7] Yu R, Zhang W, Zhang H-J, Zhang S-C, Dai X and Fang Z 2010 *Science* **329** 61
[8] Matsumoto M and Shiba H 1995 *J. Phys. Soc. Jpn.* **64** 3384.
[9] Fogelström M, Rainer D and Sauls J A 1997 *Phys. Rev. Lett.* **79** 281; see also references therein.
[10] Gustafsson D, Golubev D, Fogelström M, Claeson T, Kubatkin S, Bauch T and Lombardi F 2013 *Nat. Nanotech.* **8** 25
[11] Håkansson M, Löfwander T, Fogelström M 2015 *Nat. Phys.* **11** 755
[12] Kosztin I, Kos S, Stone M and Leggett A J 1998 *Phys. Rev. B* **58** 9365
[13] Hosseini M V 2015*Euro. Phys. Lett.* **110** 47010.
[14] Nagai Y, Ota Y and Tanaka K 2017 *Phys. Rev. B* **96** 060503(R)
[15] Black-Schaffer A M, Golubev D S, Bauch T, Lombardi F, and Fogelström M 2013 *Phys. Rev. Lett.* **110** 197001
[16] Nagai Y, Shinohara Y, Futamura Y, and Sakurai T 2017 *J. Phys. Soc. Jpn.* **86** 014708
[17] Sakurai T and Sugiura H 2003 *J. Comput. Appl. Math.* **159** 119
[18] Nagai Y, Shinohara Y, Futamura F, Ota Y and Sakurai T 2013 *J. Phys. Soc. Jpn.* **82** 094701
[19] Vorontsov A B 2009 *Phys. Rev. Lett.* **102** 177001

# Magnetization measurements of $Sr_2RuO_4$-Ru eutectic microplates using dc-SQUIDs

**Y. Nago[1,2,4], D. Sakuma[2,4], R. Ishiguro[3,4], S. Kashiwaya[5], S. Nomura [6], K. Kono [4], Y. Maeno[7] and H. Takayanagi[2]**

1 Department of Physics, Faculty of Science and Technology, Keio University, Yokohama, Japan
2 Department of Applied Physics, Faculty of Science, Tokyo University of Science, Tokyo, Japan
3 Department of Mathematical and Physical Science, Faculty of Science, Japan Women's University, Tokyo, Japan
4 Center for Emergent Matter Science, RIKEN, Wako, Japan
5 National Institute of Advanced Industrial Science and Technology (AIST), Tsukuba, Japan
6 Division of Physics, Faculty of Pure and Applied Sciences, University of Tsukuba, Tsukuba, Japan
7 Department of Physics, Graduate School of Science, Kyoto University, Kyoto, Japan

E-mail: usk.nago@gmail.com

**Abstract.** We report magnetization measurements of $Sr_2RuO_4$-Ru eutectic microplates using micro-dc-SQUIDs. $Sr_2RuO_4$ is considered as a chiral p-wave superconductor and hence $Sr_2RuO_4$-Ru eutectic becomes in an unstable state with a superconducting phase frustration between a chiral p-wave state of $Sr_2RuO_4$ and a s-wave state of Ru. To compensate the frustration, a single quantum vortex is spontaneously formed at the center of the Ru inclusion at sufficiently low temperatures. However, such a spontaneous vortex state has not been experimentally observed yet. In this study, we prepared a micro-dc-SQUID and a $Sr_2RuO_4$-Ru eutectic microplate containing a single Ru-inclusion at the center of the microplate. We performed magnetization measurements down below the superconducting transition temperature of the Ru inclusion to investigate the spontaneous Ru-center vortex state.

## 1. Introduction

Unconventional superconductivity is one of attractive subjects because it exhibits orbital- and spin-symmetry breaking [1]. Among unconventional superconductors, $Sr_2RuO_4$(SRO) is the strong candidate of a spin-triplet chiral p-wave pairing with broken time reversal symmetry in the orbital state [2, 3, 4]. This property exhibits a continuous phase winding from 0 to $2\pi$ following the momentum vector around $p_z$ axis. SRO-Ru eutectic crystal [5] is useful for research for chiral superconductivity of SRO because the Ru inclusion is surrounded by the SRO: the superconducting SRO shows a phase winding along the SRO/Ru interface due to its chiral property while the superconducting Ru has a constant phase, leading to a topological phase frustration between a s-wave state of the Ru and a chiral p-wave state of the SRO at the interface [6, 7, 8, 9]. The frustration causes a spontaneous supercurrent flowing through the interface to

IOP Conf. Series: Journal of Physics: Conf. Series **969** (2018) 012040 doi:10.1088/1742-6596/969/1/012040

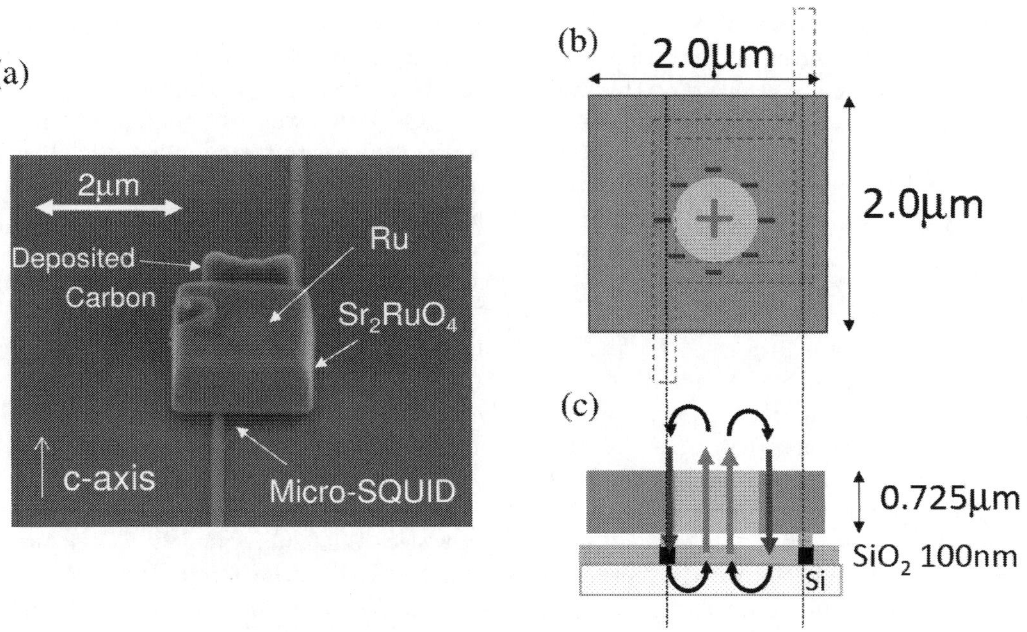


**Figure 1.** (a) Electron micrograph of the magnetization device. (b)(c) Top and side views of the device. The plus and minus signs in (b) indicates a flux direction in the Ru-center vortex state, corresponding to the arrows in (c).

minimize the interface free energy [10, 11]. With decreasing temperature, the spontaneous-current profile gradually changes according to growth of the frustrating coupling of both states, finally forming a single vortex state at the center of a Ru-inclusion [10]. The Ru-center vortex accompanies a quantized flux distributed along the SRO/Ru interface to compensate the phase frustration (see Fig.1(c) in Ref.[10]). Therefore detection of the spontaneous Ru-center vortex state can be one of evidences of chiral superconductivity of SRO. However, it has not been experimentally observed yet.

We have so far fabricated dc-SQUIDs with Nb/Ru/SRO junctions using SRO-Ru eutectic crystal and investigated the effects due to a topological mismatch of s- and chiral p-wave phases [9, 12]. However, an evidence of the Ru-center vortex state has not been obtained. Investigation of the vortes state requires a simpler high-sensitivity device to detect a magnetic flux directly. Recently, we have succesfully developed magnetization measurement device composed of a micro-dc-SQUID and a SRO microplate [13]. This device is useful to detect the magnetic flux induced on the SRO-Ru eutectic microplate. In this study, We prepared a micro-dc-SQUID and a SRO-Ru eutectic microplate with one Ru inclusion and performed magnetization measurements below the superconducting transition temperature of the Ru inclusion.

## 2. Experimental

Figure 1 exhibits the magnetization measurement device composed of the SRO-Ru eutectic microplate and the micro-dc-SQUID. A SRO-Ru eutectic crystal was fabricated into a micron-sized square plate with a focused Ga-ion beam (FIB) technique, based on our previous studies [13, 14]. The microplate was milled so that a cylindrical-shaped Ru inclusion was located nearly in the center of the plate as seen in Fig. 1 (a). Diameter of the Ru inclusion was less than $1\mu m$. We checked that a depth of proximal Ru-cylinders in our eutectic crystal is larger than 725 nm, corresponding to a height of our present microplate, and hence we sure that the Ru-inclusion cylinder in the microplate reaches the bottom.

The micro-dc-SQUID consists of $Al/AlO_x/Al$ tunnel-type Josephson junctions. The design of the SQUID was typically the same as those used in a previous study [13], wherein the Josephson junctions were located at each corner of the SQUID loop to prevent any undesirable magnetic flux from intruding into the SQUID electrodes. The electrode width was 100 nm and the inner square loop length was $1\mu m \times 1\mu m$. An $SiO_2$ mask 100 nm thick was sputtered onto the SQUIDs to prevent the Ga ions from reaching the SQUIDs during the FIB process.

The Ru-center vortex state consists of a magnetic flux quantum $\Phi_0$ at the Ru center and a quantized flux distributed along the SRO/Ru interface [10]. The direction of the vortex flux at the center of the Ru is opposite to that of the flux at the interface, satisfying flux conservation. Therefore a flux circulates in a very small scale in the vicinity of the plate surface as shown by the arrows and the signs in Figs 1 (b) and (c). This makes it hard to detect a flux induced by the vortex. In this study, the milled eutectic microplate was picked up using a Tungsten nano-probe of a nano-manipulator system and directly mounted onto the $SiO_2$-masked SQUID so that the center of the microplate was slightly displaced from the center of the SQUID loop, as in Figs 1 (b) and (c). This condition is expected to enable detection of the vortex state.

The device was cooled down below 1 K with a dilution cryostat holding an applied magnetic field $H$ at zero. We measured current-voltage characteristics and subsequently a Josephson critical current $I_c$ as a function of the field applied along c-axis of the microplate. We also measured a SQUID without a eutectic as a reference, separated by about $40\mu m$ from the SQUID with the eutectic microplate.

## 3. Results and discussion

Figure 2 shows the Josephson critical current $I_c$ of the SQUID with the eutectic microplate as a function of a external magnetic field $\mu_0 H$ at each temperature. The depicted field range is lower than the field of 3.3 mT where a quantum vortex intrusion into the microplate was observed. The modulation period for the reference SQUID was measured to be 1.71mT at 0.04 K, while the period of the SQUID with the eutectic microplate was estimated to be about 2 mT at all temperature range as seen in Fig. 2. This difference is attributed to be the exclusion of the magnetic flux from the loop of the SQUID owing to the Meissner effect of the superconducting SRO-Ru. The period $\Delta B$ and the offset $B_{\rm off}$ of the modulation patterns were estimated from fitting the data in Fig 2 by $I_c \propto \cos[\pi(\Phi - \Phi_{\rm off})/\Phi_0] = \cos[\pi(\mu_0 H - B_{\rm off})/\Delta B]$, where a flux density $B$ is obtained from a flux threading the SQUID $\Phi$ divided by the effective SQUID-loop area. Fig. 3 exhibits a temperature dependence of $\Delta B$ and $B_{\rm off}$. $\Delta B$ monotonically increases with decreasing temperature, which means a partial exclusion of a flux out of the SQUID loop. An external flux penetrates the eutectic microplate by the penetration depth $\lambda_{\rm L}^{\rm SRO}$ which is temperature dependent [9, 15]. The decrease of the depth with decreasing temperature leads to the exclusion of the flux, qualitatively in good agreement with the behavior in the lower panel of Fig. 3. However, one can see a kink at about 0.425 K. This indicates that the Ru-inclusion undergoes the superconducting transition at $T_{\rm c}^{\rm Ru} = 0.425$ K and subsequently the flux was pushed out further. In other words, a temperature dependence of the penetration depth of the Ru $\lambda_{\rm L}^{\rm Ru}$ reflects the excess increase of $\Delta B$ below $T_{\rm c}^{\rm Ru}$.

In contrast, the upper panel in Fig. 3 shows no remarkable change of $B_{\rm off}$ below $T_{\rm c}^{\rm Ru}$. The offset $B_{\rm off}$ includes contribution of spontaneous magnetization. Therefore, if the Ru-center vortex state exists below $T_{\rm c}^{\rm Ru}$, the SQUID should detect the excess flux induced by the vortex, resulting in a shift on the temperature dependence of $B_{\rm off}$. According to our numerical simulation of magnetic flux distribution solving Maxwell and London equations using the commercial package COMSOL (the detail will be described elsewhere [16]), the total flux by the Ru-center vortex detectable for the SQUID was estimated to be $0.09\Phi_0$ in our present device geometry.

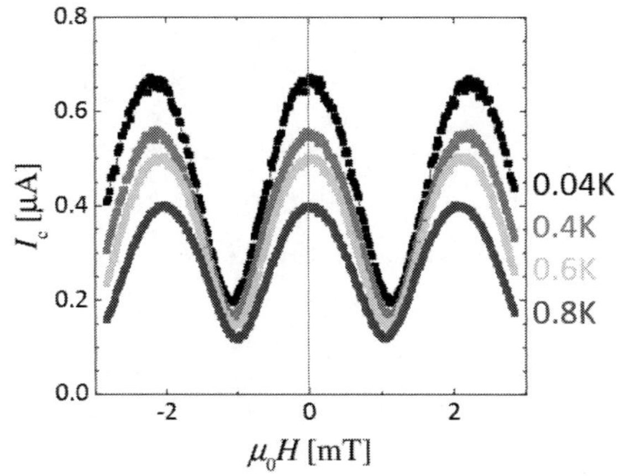


**Figure 2.** Josephson critical current $I_c$ as a function of external magnetic field at 0.8 K - 0.04 K.

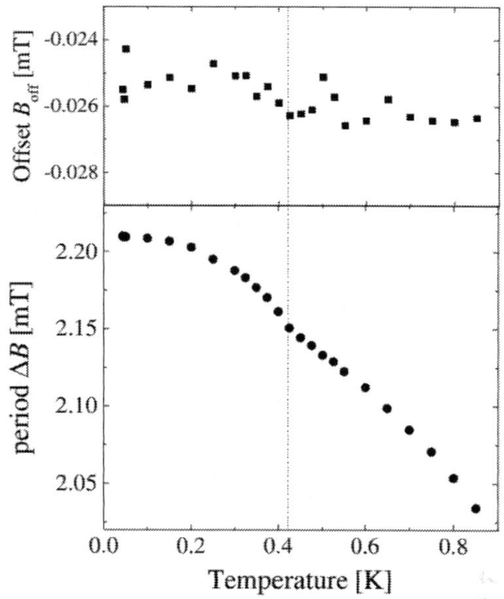


**Figure 3.** Temperature dependence of the period $\Delta B$ (lower panel) and the offset $B_{off}$ (upper panel) of the $I_c$ modulation. The broken line indicates the temperature corresponding to a kink in the lower panel.

However, our present result in Fig. 3 shows almost constant $B_{\text{off}}$ and the overall shift of $B_{\text{off}}$ at the measured temperature range corresponds to be less than $0.001\Phi_0$. Therefore we conclude that the spontaneous Ru-center vortex state does not exsist in our present eutectic microplate.

Our result showing absence of the Ru-center vortex state implies a weak coupling of proximity-induced superconductivity between SRO and Ru. A phase frustration between SRO and Ru superconducting states causes a spontaneous supercurrent at the SRO/Ru interface [10, 11]. If the superconducting coupling at the SRO/Ru interface is weak, the interface free energy in the spontaneous current state becomes smaller than that in the Ru-center vortex state and hence the Ru-center vortex state cannot be stably formed [10]. This feature at the SRO/Ru interface is consistent with the previous tunneling measurement on an SRO/Ru/Au junction which revealed suppression of the superconducting gap of the SRO in the Ru-inclusion [6].

We note that possibilities of non-chiral superconductivity of SRO or influence of a chiral domain state cannot be ruled out yet. Clarification of a magnetic flux distribution in the eutectic microplate and a phase frustration of both superconductivity requires further theoretical and experimental approaches.

## 4. Conclusion

We measured magnetization of $Sr_2RuO_4$-Ru eutectic microplates using micro-dc-SQUIDs at temperatures down below the superconducting transition temperature $T_c^{\text{Ru}}$ of the Ru inclusion. With decreasing temperature, the magnetic flux is gradually excluded from the microplate according to decrease of the penetration depth of superconducting $Sr_2RuO_4$ and pushed out further below $T_c^{\text{Ru}} = 0.425$ K owing to the Meissner diamagnetism of the superconducting Ru. Our present results show that spontaneous magnetization such as a Ru-center vortex state is absent in the present microplate, implying weak proximity-induced superconductivity at the $Sr_2RuO_4$/Ru interface.

**Acknowledgments**
This work was supported by Japan Society for the Promotion of Science (JSPS) through a Grants-in-Aid for Scientific Research KAKENHI (Nos. 15H05852, 15H05853, and 15K17708). The technical part of device fabrication was supported by the NIMS Nanofabrication Platform of the "Nanotechnology Platform Project" and the RIKEN Center for Emergent Matter Science, sponsored by the Ministry of Education, Culture, Sports, Science and Technology of Japan (MEXT).

**References**
[1] Sigrist M and Ueda K 1991 *Rev. Mod. Phys.* **63** 239
[2] Maeno Y, Hashimoto H, Yoshida K, Nishizaki S, Fujita T, Bednorz J G and Lichtenberg F 1994 *Nature* **372** 532
[3] Mackenzie A P and Maeno Y 2003 *Rev. Mod. Phys.* **372** 657
[4] Maeno Y, Kittaka S, Nomura T, Yonezawa S and Ishida K 2012 *J. Phys. Soc. Jpn.* **81** 011009
[5] Maeno Y, Ando T, Mori Y, Ohmichi E, Ikeda S, Nishizaki S and Nakatsuji S 1998 *Phys. Rev. Lett.* **81** 3765
[6] Ying Y A, Xin Y, Clouser B W, Hao E, Staley N E, Myers R J, Allard L F, Fobes D, Liu T, Mao Z Q and Liu Y 2009 *Phys. Rev. Lett.* **103** 247004
[7] Nakamura T, Nakagawa R, Yamagishi T, Terashima T, Yonezawa S, Sigrist M and Maeno Y 2011 *Phys. Rev.* B **84** 060512
[8] Anwar M S, Nakamura T, Yonezawa S, Yakabe M, Ishiguro R, Takayanagi H and Maeno Y 2013 *Sci. Rep.* **3** 2480
[9] Nago Y, Ishiguro R, Sakurai T, Yakabe M, Nakamura T, Yonezawa S, Kashiwaya S, Takayanagi H and Maeno Y 2016 *Phys. Rev.* B **94** 054501
[10] Kaneyasu H and Sigrist M 2010 *J. Phys. Soc. Jpn.* **79** 053706
[11] Etter S B, Kaneyasu H, Ossadnik M and Sigrist M 2014 *Phys. Rev.* B **90** 024515
[12] Nago Y, Ishiguro R, Sakurai T, Yakabe M, Nakamura T, Yonezawa S, Kashiwaya S, Takayanagi H and Maeno Y 2014 *J. Phys.: Conf. Ser.* **568** 022031
[13] Nago Y, Shinozaki T, Tsuchiya S, Ishiguro R, Kashiwaya H, Kashiwaya S, Nomura S, Kono K, Takayanagi H and Maeno Y 2016 *J. Low Temp. Phys.* **183** 292
[14] Tsuchiya S, Matsuno M, Ishiguro R, Kashiwaya H, Kashiwaya S, Nomura S, Kono K, Takayanagi H and Maeno Y 2014 *J. Phys. Soc. Jpn.* **83** 094715
[15] Bonalde I, Yanoff B D, Salamon M B, Van Harlingen D J, Chia E M E, Mao Z Q and Maeno Y 2000 *Phys. Rev. Lett.* **85** 4775
[16] Sakuma D, Nago Y, Ishiguro R, Kashiwaya S, Nomura S, Kono K, Maeno Y and Takayanagi H submitted

IOP Conf. Series: Journal of Physics: Conf. Series **969** (2018) 012041 doi:10.1088/1742-6596/969/1/012041

# Electronic states of domain structure in $1T$-$TaS_{2-x}Se_x$ observed by STM/STS

**D Fujii, T Iwasaki, K Akiyama, Y Fujisawa, S Demura, H Sakata**

Department of Physics, Tokyo University of Science, Shinjuku-ku, Tokyo 162-8601, Japan

1217637@ed.tus.ac.jp

**Abstract**. We report on a systematic scanning tunneling microscopy and spectroscopy (STM/STS) study on $1T$–$TaS_{2-x}Se_x$ (x = 0, 0.3, 1.0) at 4.2 K. While the compounds with $x = 0$ and 0.3, which undergoes the Mott transition, showed the commensurate charge density wave (CDW) with the period of $\sqrt{13}a_0\times\sqrt{13}a_0$ ($a_0$ is in-plane lattice constant), the compound with $x$ =1, which shows superconductivity at 3.5 K, exhibits anomalous domain structure: The domain structure consists of regions with regular array of David-stars divided by bright contrasted walls at positive bias voltage. We found the domain wall showed the different electronic state from that of the domain.

## 1. Introduction

Layered transition metal dichalcogenide (TMD) has been studied for a long time because of its plentiful physical properties such as superconductivity or charge density wave (CDW). Among many TMDs, $1T$-$TaS_2$ has attracted much attention due to various CDW states. In the commensurate CDW (CCDW) state below 180 K, $\sqrt{13}a_0\times\sqrt{13}a_0$ supermodulation ($a_0$ is in-plane lattice constant) is formed and 13 Ta atoms form a cluster called "David star" as shown in Fig.1(a) schematically. Furthermore, in this state, the Mott insulating state emerges due to the strong correlation between $d$-electrons in Ta at each David star. In the nearly CCDW (NCCDW) state above 180 K, the commensurate regions are separated into periodic hexagonal domains by domain walls. Simultaneously the Mott insulating state melts into the metallic state. Thus, the melting of the Mott state is closely related to the change in CDW state in this material.

It has been reported that several kinds of perturbations such as transition metal substitution [1] or applying hydrostatic pressure [2] melt the Mott state and induce superconductivity. Thus, the melting of the Mott state in this material is an interesting problem. Recently substitution of S with Se was found to melt the Mott state and induce superconductivity at 3.5K [1]. To elucidate the change in the CDW state with Se substitution, we have performed scanning tunnelling microscopy and spectroscopy (STM/STS) measurements in $1T$–$TaS_{2-x}Se_x$ ($0 \leq x \leq 1$) at 4.2 K.

## 2. Experimental details

Single crystals of $1T$-$TaS_{2-x}Se_x$ ($x$ = 0, 0.3, 1.0) were grown by chemical vapor transport method in a quartz tube with $I_2$ as transport gas. To obtain $1T$ type crystals, quenching into ice water was conducted after heating in the horizontal furnace. Electrical resistivity was measured to verify the ground state of each compound. Figure 1(b) shows the results. For the compound with $x = 0$ and 0.3, the sudden change in the resistivity by the Mott transition is observed, indicating the existence of the CCDW state. On the other hand, for the compound with $x = 1.0$, the Mott transition is completely suppressed and the drop of the resistivity due to superconductivity appeared at about 3.5 K. Figure 1(c) shows the result of DC magnetization measurements at $x = 1.0$ from 6 K down to 2 K. The diamagnetism due to superconductivity is clearly observed. The Superconducting transition temperature and the shielding volume fraction are estimated to be 3.5 K and 20 % at 2 K, which are consistent with the previous study [1].

STM/STS measurements were performed by using a laboratory-build scanning tunneling microscope. An electrochemically polished Au wire was used as an STM tip. STM/STS measurements were performed at 4.2K. A Sample surface was prepared by cleavage at 4.2 K *in situ*. STM images were obtained with constant current mode. $dI/dV$ spectra were obtained by numerical differentiation of the $I$-$V$ characteristics.

## 3. Results and Discussion

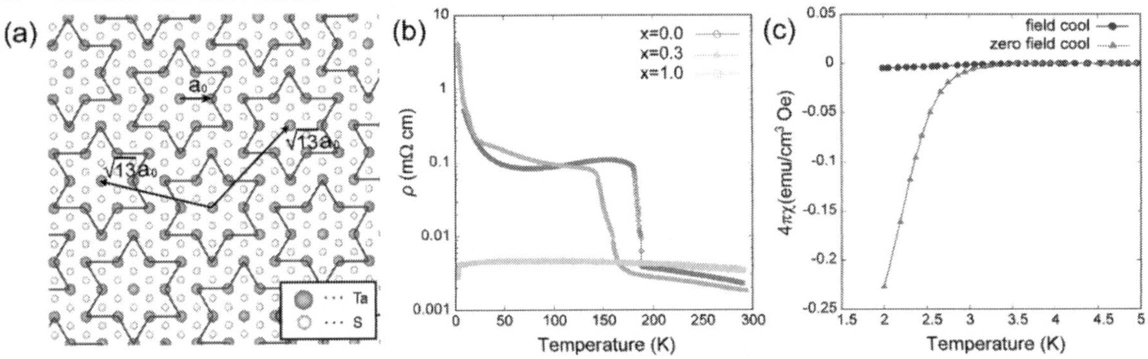


Figure 1: (a) Schematic CCDW structure in $1T$-$TaS_2$. (b) Temperature dependence of the resistivity in $1T$-$TaS_{2-x}Se_x$ measured on cooling process (c) Magnetic susceptibility in $1T$-TaSSe (x=1.0) at 10 Oe.

Figures 2(a), (b), and (c) show STM images of the compounds with $x = 0$, 0.3, and 1.0, respectively. The white dots seen in each figure correspond to David-stars. The compounds with $x = 0$ and 0.3 show an almost perfectly ordered triangular lattice of David-stars. This indicates that the samples are in the CCDW state at this temperature, which is consistent with the resistivity measurements. In the compound with $x = 0.3$, there are some defects of David-star or vague black contrasted region spread to a few nm. Because the number of defects or black contrast regions is almost same as that in the compound with $x = 0$, these structures seem to have no relation with the existence of Se atoms. Endo *et. al* reported STM images in $1T$-$TaS_{2-x}Se_x$ ($x = 0$, 0.3 and 0.5) at room temperature [3]. They found irregular domain structure at $x = 0.3$. They attributed the domain structure to a Moire pattern originating from the local variation of CDW stacking along the $c$-axis due to the lattice distortion. Because we did not observe the domain structure in the sample with $x = 0.3$ at 4.2 K, the observed domain structure at room temperature is not thought to be due to the lattice distortion.

In contrast, the compound with $x = 1.0$, which is in the NCCDW state at 4.2 K, shows domain structure. As shown in Fig. 2(c), the STM image of the compound with $x = 1.0$ consists of the regions with regular array of David-stars divided by bright contrasted walls at positive bias voltage. Since the configuration of the walls is irregular, this domain structure has different nature from that observed in the NCCDW state in pristine $1T$-$TaS_2$, which has periodic hexagonal domains, but rather resembles that observed in the sample with $x = 0.3$ at room temperature [3].

IOP Conf. Series: Journal of Physics: Conf. Series **969** (2018) 012041 doi:10.1088/1742-6596/969/1/012041

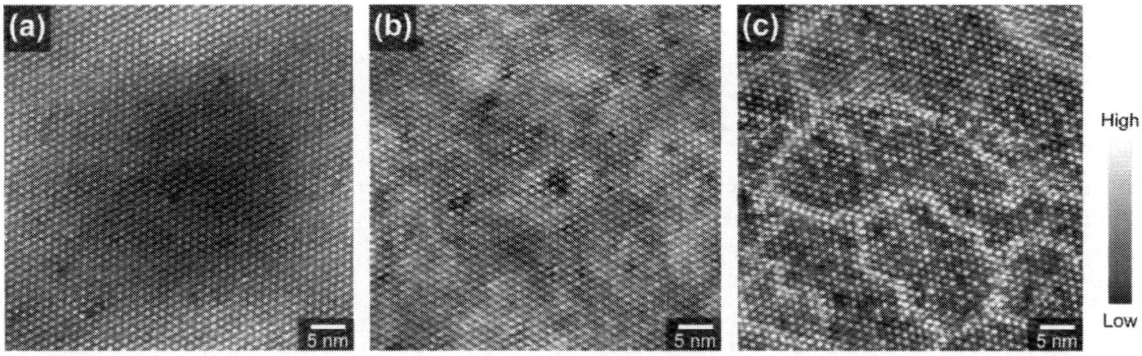

Figure 2: STM images of 1*T*-$TaS_{2-x}Se_x$ (a) $x = 0.0$ (b) $x = 0.3$ (c) $x = 1.0$ respectively. The bias voltage and the tunneling current are (a) +0.5V, 0.2nA (b) +0.5V 1.0nA (c) +0.5V 1.0nA, respectively.

Figure 3(a) shows a magnified STM image of a domain wall. This STM image suggests that the domain wall consists of a sequence of two pairs of David-stars close to each other, and behaves as a phase defect of CDW.

Figure 3(b) shows the STM image at bias voltage of -0.5 V at the same field of view. The bright contrast observed in the positive bias voltage can hardly be seen. Thus, the observed bright contrast reflects the electronic states at the wall. It is noted that this bias dependence is consistent with the previous STM study on the compound with $x = 0.3$ at room temperature [3]. This structure is similar to the metastable mosaic state in 1*T*-$TaS_2$ realized by the application of voltage pluses [4] or that observed in Fe doped 1*T*-$TaS_2$ [5]. Thus, the domain structure divided by the wall seems to be a common feature for the melting of the Mott state in this material. Very recent report shows the existence of such domain structure even in the CCDW state near NCCDW state [6].

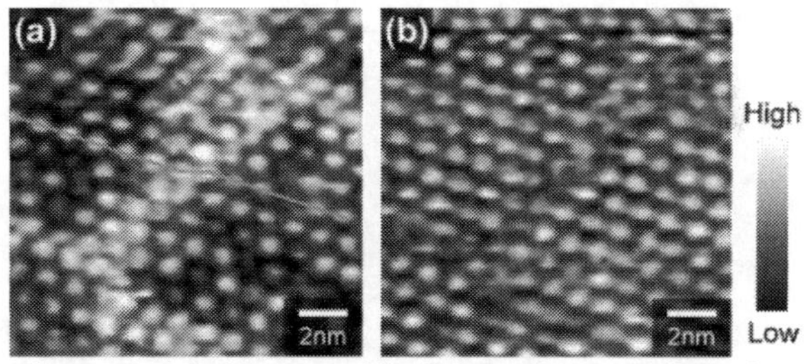

Figure3: (a),(b) Magnified STM images focusing on the domain wall. The sample bias voltage is (a) +0.5V (b) -0.5V respectively.

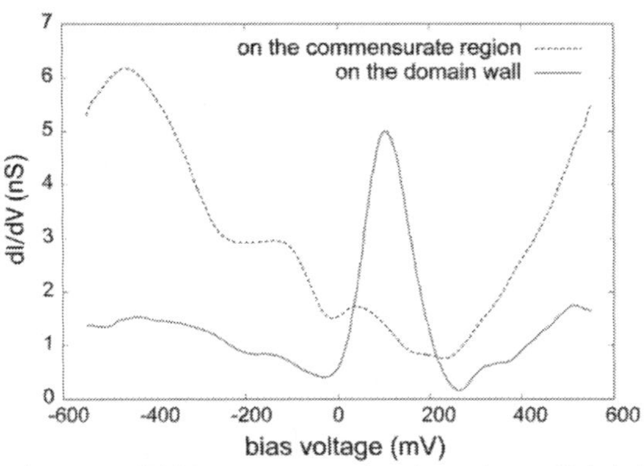

Figure4: *dI*/*dV* spectra on the domain wall (pink) and the commensurate region (green)

Figure 4 shows a spectrum averaged in a region surrounded by domain walls. A spectrum observed on a domain wall was also shown. We found different nature in spectra. The former has broad peaks at about -150 mV and +100 mV, which are considered to be remnants of the LDOS derived from the lower Hubbard band (LHB) and the upper Hubbard band (UHB) observed in the CCDW state in pristine 1*T*-$TaS_2$. On the contrary, the spectrum on the domain wall has a sharp peak at about +120 mV and no peak corresponding to the LHB can be seen. These spectra seem to indicate that the Mott nature survives in the region surrounded by the domain walls, while the Mott nature is suppressed at the domain wall. It is noted that these spectral features are similar to those observed in the metallic mosaic state in $TaS_2$ [4] or Fe-doped $TaS_2$ [5].

## 4. Summary

We have investigated the change of the CDW state with Se substitution in 1*T*-$TaS_{2-x}Se_x$ ($x$ = 0.0, 0.3 and 1.0) by using STM/STS. From STM measurements, the novel domain structure has been found at $x$ = 1.0, where the regions with regular array of David-stars are divided by bright contrasted walls at positive bias voltage. STS measurements revealed that there is a spatial evolution of spectra across the domain wall: The Mott nature survives in the region surrounded by the domain walls, while the Mott nature is suppressed at the domain wall.

## References

[1] Liu Y, Ang R, Lu W J, H. Song, Li L J and Sun Y P, *Appl. Phys. Lett.* **102**,192602 (2013)
[2] Sipos B, Kusmartseva A F, Akrap A, Berger H, Forrö L and Tutiš E , *Nature Materials* **7**, 960 - 965 (2008)
[3] Endo T, Yamaguchi W, Shiino O, Hasegawa T and Kitazawa K, *Surface Science* **453** (2000) 1-8
[4] Ma L, Ye C, Yu Y, Lu X F, Niu X, Kim S, Feng D, Tománek D, Son Y W, Chen X H and Zhang Y, *Nature Communications* **7**, 10956 (2016)
[5] Fujisawa Y, Shimabukuro T, Machida T, Demura S and Sakata H, to be submitted
[6] Qiao S, Li X, Wang N, Ruan W, Ye C, Cai P, Hao Z, Yao H, Chen X, Wu J, Wang Y and Liu Z, *arXiv*:1611.08957v2 [cond-mat.str-el]

IOP Conf. Series: Journal of Physics: Conf. Series **969** (2018) 012042 doi:10.1088/1742-6596/969/1/012042

# Thickness dependence of dendritic flux avalanches in $YBa_2Cu_3O_{7-x}$ films

**E. Baruch-El[1], M. Baziljevich[2], T.H. Johansen[2,3], A. Shaulov[1] and Y. Yeshurun[1]**

[1] Department of Physics, Institute of Superconductivity and Institute of Nanotechnology and Advanced Materials, Bar-Ilan University, Ramat-Gan 52900, Israel
[2] Department of Physics, University of Oslo, P.O. Box 1048, Blindern, N-0316 Oslo, Norway
[3] Institute for Superconducting and Electronic Materials, University of Wollongong, NSW 2522, Australia

**Abstract.** By implementing a unique magneto-optical system with ultrafast magnetic-field ramping-rate capability (up to 3 kT/s), we have been able to routinely generate and image dendritic flux instabilities in $YBa_2Cu_3O_{7-x}$ films. In the present work we study the effect of the film thickness on the dendritic instability. Dendritic avalanches in 50 - 600 nm thick films were magneto-optically imaged at 7 K, after ramping the magnetic field from zero to 60 mT at different rates. The data reveal a remarkable change in flux morphologies between the thin and the thicker films. While the former (50-250 nm) display well-developed dendritic patterns, the latter (350-600 nm) exhibit few avalanches with favored branch directions parallel to the film's edges. Several possible explanations for this behavior are discussed.

## 1. Introduction

Dendritic flux instabilities have been observed in superconducting films such as $MgB_2$ [1-3], Nb [4], NbN [5], and a-MoSi [6], after exposing them to an external perpendicular magnetic field. The phenomenon, explained in several theoretical works [7-9], reflects a thermomagnetic runaway. It occurs when thermal fluctuations locally release some vortices out of their pining sites, enabling the vortices to move and heat the superconductor, which in turn further reduces the pinning and increases vortex motion. If the heat generated in the film is not dissipated quickly enough into the substrate, a thermomagnetic runaway occurs, resulting in a dendritic flux avalanche pattern in the film.

Experiments have demonstrated that high-temperature superconductors, such as $YBa_2Cu_3O_{7-\delta}$ (YBCO), are very stable, preventing the formation of dendritic avalanches. In fact, it was shown that local laser-heating of the films was necessary to generate such avalanches [10, 11]. Nevertheless, we recently found that YBCO films become thermomagnetically unstable when applying fast ramping-rate of the external field [12-14]. Exploiting this technique, we investigated different parameters that control the dendritic penetration. In a recent publication we studied the effect of the magnetic field ramp rate, identifying the threshold rate for generating avalanches at different temperatures [13]. In the present work we focus on how the film thickness influences the avalanche behavior. Our results display dendritic avalanches in all the studied YBCO films. However, the thin (50-250 nm) films and the thicker (350-600 nm) ones show significantly different dendritic patterns. In contrast, the threshold ramp rate values do not show any clear trend as the film thickness varies.

## 2. Experimental

YBCO films (4x4 $mm^2$) with thickness in the range 50-600 nm were epitaxially grown by thermal reactive coevaporation [15] on yttrium stabilized zirconia (YSZ) substrate, resulting in films with the *c* axis perpendicular to the surface. The low thermal conductivity of YSZ facilitates the generation of dendritic avalanches [14]. Magnetic measurements, using a commercial Quantum Design magnetometer, reveal a transition temperature, $T_c \approx 87$ K. All the magneto-optical imaging (MOI) measurements described in the next sections were done using our custom-made MOI system that enables real-time imaging at rates up to 70,000 frames per second. Furthermore, the magnetic field in this MOI system can be ramped up to a field of 60 mT at rates up to 3 kT/s [16].

## 3. Results

MOI experiments using slow field ramping yield images with a smooth flux penetration front, consistent with the Bean model [17], see Figure 1. As the number of pinning sites increases with film thickness, the effective critical current also increases with thickness. Consequently, thinner films exhibit deeper flux penetration.

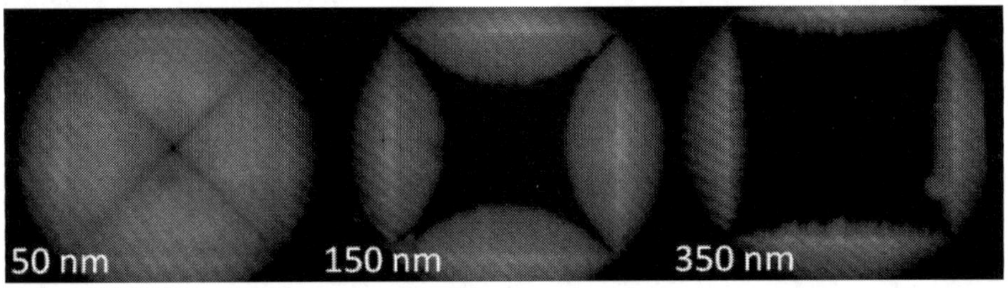


**Figure 1**. MO images of the YBCO films. The films were zero-field-cooled to 7 K and then exposed to a 60 mT magnetic field increased at a rate of several mT/s.

Repeating the same experiments with a fast field ramp rate (3 kT/s) produced dendritic patterns, as shown in Figure 2 and 3 for the thin (50 – 250 nm) and thicker (350 – 600 nm) YBCO films, respectively.

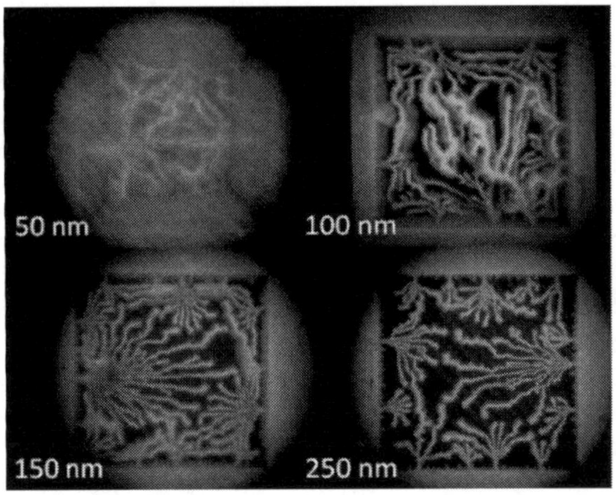


**Figure 2**. Dendritic avalanches in the thin YBCO films. The films were zero-field cooled to 7 K and then exposed to magnetic field ramped from zero to 60 mT at 3 kT/s.

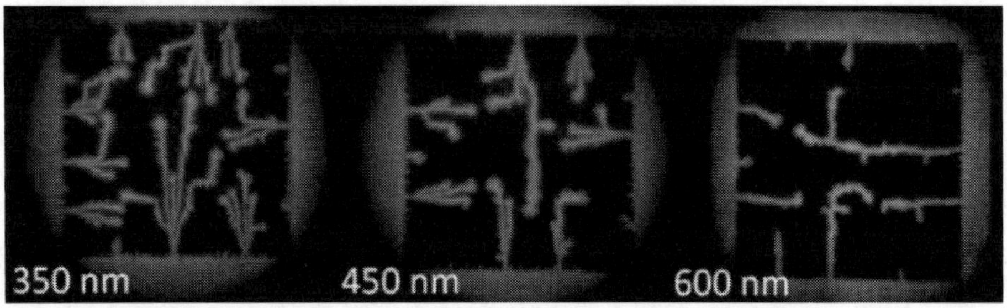


**Figure 3**. Dendritic avalanches in the thicker YBCO films. The films were zero-field cooled to 7 K and then exposed to magnetic field ramped from zero to 60 mT at 3 kT/s.

In the thinnest film (50 nm) the avalanches were superimposed on a Bean-like penetration profile, where most of the film was penetrated by magnetic flux. The Bean-like profile covers less of the area as the film thickness increases, virtually disappearing altogether at 150 nm. The 150 and 250 nm films exhibit similar avalanche morphology though it seems that the number of branches of each trunk reached a maximum at the 150 nm film. As discussed above, thicker films have more pinning sites and, therefore provide a stronger resistance against flux penetration; this may explain the slightly less developed trees in 250 nm film.

As shown in Figure 3, the thicker films (350-600 nm) have much different avalanche patterns. In those films, the penetrated flux appears to follow some preferred directions, parallel and perpendicular to the sample edges. Moreover, the avalanches do not here develop into tree-like structures.

We also made an effort to identify experimentally the effect of the film thickness on the minimum ramp rate $\dot{B}_{th}$ required for the appearance of the first avalanche [13]. In these experiments, the film was first zero-field-cooled (ZFC) to 15 K and then a perpendicular field, $B_a=60$ mT, was applied. The measured $\dot{B}_{th}$ values were fluctuating between 0.2 and 0.8 kT/s with no clear trend for the film thickness dependence. Additional efforts with different fields at different temperatures were made, giving essentially the same result.

## 4. Discussion

In the theoretical analysis of the thermomagnetic runaway, a 'stability parameter', $\tau=t_m/t_h$ is defined [7-9]. This parameter signifies the ratio between the magnetic ($t_m$) and thermal ($t_h$) diffusion times. A film with a smaller $\tau$ has higher tendency for the occurrence of an avalanche. The time $t_m$ is predicted to increase with the film thickness, $d$, consistent with the fact that the number of pinning sites is larger in thicker films. An indirect evidence for the increase of $t_m$ with the film thickness was found by dendritic velocity measurements, showing that the flux front velocity of the dendrite decreases as the film thickness increases [11]. The direct dependence of $t_m$ on $d$ implies an increase of $\tau$ with $d$. Thus, thicker films are expected to show better resistance to dendritic flux avalanches.

Our MO images, Figures 2 and 3, are consistent with the above prediction, as the avalanches in the thicker films are less developed and a larger part of these films remain superconducting. The above arguments lead also to an expectation for an increase of $\dot{B}_{th}$, the threshold field ramp rate, as the film thickness increases. However, as indicated above, we could not detect a clear trend for $\dot{B}_{th}$ as the film thickness varied. We believe that the absence of a clear thickness dependence results from local damages created during the generation of the dendritic avalanches [12]. Such damage often forms permanent gaps in the film, which in turn amplifies the local field, and has a significant impact on the following instabilities, perturbing the effect of the film thickness on $\dot{B}_{th}$.

IOP Conf. Series: Journal of Physics: Conf. Series **969** (2018) 012042 doi:10.1088/1742-6596/969/1/012042

We turn now to discuss the origin of the preferred directions for the dendritic branches in the thicker films (Figure 3). A possible origin may be traced to oriented cracks or similar defects in these films. Such defects, however, should also affect smooth flux entry. To test this possibility we performed extensive measurements at temperatures above the dendritic threshold temperature, where there is no dendritic instability [13]. In these experiments, the thick films were ZFC to 55 K, then exposed to an external magnetic field increased at various rates. In all these experiments, the flux penetration was smooth and homogeneous, and no indication for guiding defects was found. Additionally, those films were checked with scanning electron microscope (SEM) and no sign of such defects was found. It is important to note that the YBCO crystalline a-b-axes are aligned at 45 degrees with respect to the film edges, and although thicker films have a higher tendency to crack (as there is more elastic energy in the films), those cracks would also be oriented 45 degrees with respect to the edge of the film [18]. In light of the above, we conclude that defects cannot be responsible for the observed preferred direction of dendritic branches.

Another possible origin for the flux direction preference may be related to the square geometry of our films. Screening currents along the film edges apply on the penetrating flux a Lorentz force directed perpendicular to these edges (see, e.g., Figure 4 of Ref. [1]). To test this possibility, we etched (with Argon plasma) few of the 600 nm thick YBCO films, preparing the different geometries (trapezoid and circle) displayed in Figure 4. The films were field cooled (FC) with -60 mT to 7 K, then exposed to a 60 mT field ramped up at a rate of 3 kT/s. The FC procedure facilitates the triggering of more avalanches, due to extra heating from vortex-antivortex annihilation. As is apparent from the MO images of Figure 4, even though the sample geometries have been changed, the preferred directions of the avalanche branches still remain as they were in the case of the square sample (see Figure 3), i.e. parallel and perpendicular to the original square film edges. We thus conclude that also the film geometry cannot be causing the directed flux entry. Thus, the physical mechanism behind the different avalanches morphologies remains unresolved.

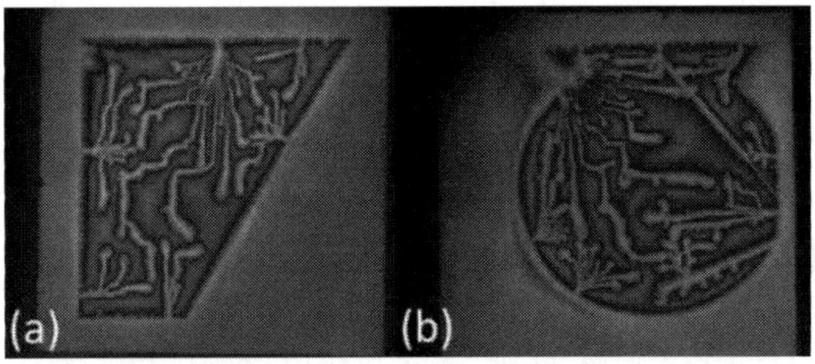


**Figure 4.** Flux avalanches in 600 nm YBCO films etched to form different geometries: (a) trapezoid, and (b) circle. The films were cooled to 7 K in the presence of -60 mT and the MO images were taken while ramping up the field to 60 mT at a rate of 3 kT/s. (The two long diagonal lines in the right edge of the circle sample are mechanical scratches in the YBCO).

Finally, we present preliminary results of fractal analysis of the different dendritic morphologies. Using the 'box-counting' method [19], the images were covered with a grid, and the number of boxes occupied with magnetic flux was counted. A graph of the number of occupied boxes versus the box size was then plotted. The slope of this graph represents the fractal dimension of the studied image. The above procedure was done for the 150 nm – 600 nm YBCO samples. The 50 nm and 100 nm films were not included in this study because of the significant Bean-like penetration that adds complexity to the fractal dimension analysis. The results of this analysis showed a monotonic decrease of the fractal dimension from 1.85 for the 150 nm film to 1.36 for the 600 nm film. The calculated fractal dimension may be used as a quantifying parameter of the films stability against dendritic avalanches. More experimental work combined with fractal analysis is needed to substantiate such a conclusion.

## 5. Summary

MO images of YBCO films exposed to an ultra-fast magnetic field ramp, revealed different dendritic morphology for the thin and thicker films. Specifically, we observed a change from dense, well-developed tree-like avalanches in the thin films, to avalanches with few branches penetrating along favored directions in the thicker ones. The origin of this change in the morphology is still unclear. The fractal dimension of the dendritic avalanches is suggested as a quantifying parameter of the sample stability against the flux avalanches. Quantifying the parameters governing the dendritic avalanches is also important for application of high-temperature superconducting films, as such thermomagnetic avalanches limit their potential use and may even cause permanent damage [12].

## Acknowledgments

The authors thank Dr. Robert Semerad for preparation of the films and for helpful discussions. We also acknowledge a financial support from the Israel Science Foundation (ISF-164/12) and the Norwegian Research Council.

## References

[1] Johansen T H, Baziljevich M, Shantsev D V, Goa P E, Kang W N, Kim H J, Choi E M, Kim M-S and Lee S I 2002 Dendritic magnetic instability in superconducting $MgB_2$ films *EPL (Europhysics Letters)* **59** 599

[2] Baziljevich M, Bobyl A V, Shantsev D V, Altshuler E, Johansen T H and Lee S I 2002 Origin of dendritic flux patterns in $MgB_2$ films *Physica C: Superconductivity* **369** 93

[3] Albrecht J, Matveev A, Djupmyr M, Schütz G, Stuhlhofer B and Habermeier H-U 2005 Bending of magnetic avalanches in $MgB_2$ thin films *Applied Physics Letters* **87** 182501

[4] Duran C, Gammel P, Miller R and Bishop D 1995 Observation of magnetic-field penetration via dendritic growth in superconducting niobium films *Physical Review B* **52** 75

[5] Yurchenko V V, Shantsev D V, Johansen T H, Nevala M R, Maasilta I J, Senapati K and Budhani R C 2007 Reentrant stability of superconducting films and the vanishing of dendritic flux instability *Physical Review B* **76** 092504

[6] Colauto F, Motta M, Palau A, Blamire M G, Johansen T H and Ortiz W A 2015 First observation of flux avalanches in a-MoSi superconducting thin films *IEEE Transactions on Applied Superconductivity* **25** 1

[7] Aranson I S, Gurevich A, Welling M S, Wijngaarden R J, Vlasko-Vlasov V K, Vinokur V M and Welp U 2005 Dendritic flux avalanches and nonlocal electrodynamics in thin superconducting films *Physical review letters* **94** 037002

[8] Denisov D V, Rakhmanov A L, Shantsev D V, Galperin Y M and Johansen T H 2006 Dendritic and uniform flux jumps in superconducting films *Physical Review B* **73** 014512

[9] Dvash E, Shapiro I and Shapiro B Y 2009 Dendritic instability of the magnetic flux in thermally anisotropic type-II superconductors *Physical Review B* **80** 134522

[10] Leiderer P, Boneberg J, Brüll P, Bujok V and Herminghaus S 1993 Nucleation and growth of a flux instability in superconducting $YBa_2Cu_3O_{7-x}$ films *Physical review letters* **71** 2646

[11] Bolz U, Biehler B, Schmidt D, Runge B-U and Leiderer P 2003 Dynamics of the dendritic flux instability in $YBa_2Cu_3O_{7-\delta}$ films *EPL (Europhysics Letters)* **64** 517

[12] Baziljevich M, Baruch-El E, Johansen T H and Yeshurun Y 2014 Dendritic instability in YBCO films triggered by transient magnetic fields *Applied Physics Letters* **105** 012602

[13] Baruch-El E, Baziljevich M, Shapiro B Y, Johansen T H, Shaulov A and Yeshurun Y 2016 Dendritic flux instabilities in $YBa_2Cu_3O_{7-x}$ films: Effects of temperature and magnetic field ramp rate *Physical Review B* **94** 054509

[14] Baruch-El E, Baziljevich M, Johansen T and Yeshurun Y 2015 Substrate Influence on Dendritic Flux Instability in YBCO Thin Films *Journal of Superconductivity and Novel Magnetism* **28** 379

[15] Utz B, Semerad R, Bauer M, Prusseit W, Berberich P and Kinder H 1997 Deposition of YBCO and NBCO films on areas of 9 inches in diameter *Applied Superconductivity, IEEE Transactions on* **7** 1272

[16] Baziljevich M, Barness D, Sinvani M, Perel E, Shaulov A and Yeshurun Y 2012 Magneto-optical system for high speed real time imaging *Review of Scientific Instruments* **83** 083707

[17] Schuster T, Kuhn H, Brandt E H, Indenbom M V, Kläser M, Müller-Vogt G, Habermeier H U, Kronmüller H and Forkl A 1995 Current and field pattern in rectangular and inhomogeneous superconductors *Physical Review B* **52** 10375

[18] Semerad R private communication

[19] Feder J 2013 *Fractals* (Springer Science & Business Media)

IOP Conf. Series: Journal of Physics: Conf. Series **969** (2018) 012043 doi:10.1088/1742-6596/969/1/012043

# Theoretical analysis for spin-polarized local density of states in the vortex state of helical *p*-wave superconductors

**Kenta K Tanaka**[1], **Masanori Ichioka**[1,2] **and Seiichiro Onari**[1,2]

[1] Department of Physics, Okayama University, Okayama 700-8530, Japan
[2] Research Institute for Interdisciplinary Science (RIIS), Okayama University, Okayama 700-8530, Japan

E-mail: ktanaka@mp.okayama-u.ac.jp

**Abstract.** Based on the quasi-classical Eilenberger theory, we calculate the site-dependence of order-parameter and spin-polarized local density of states (LDOS) in the vortex lattice state of helical *p*-wave superconductors. The spin-polarized LDOS is induced around the vortex core by the vorticity coupling to the chirality of up-spin pair or down-spin pair. In order to distinguish the *d*-vector symmetry, we investigate the magnetic field effect on the spatial structure of the spin-polarized LDOS.

## 1. Introduction

The spin-triplet superconductors have attracted great interest, since Majorana state is induced around the surface and vortex core. The ruthenate superconductor $Sr_2RuO_4$ is a leading candidate material for the spin-triplet superconductor. However, the detailed structure of *d*-vector in this material remains unclear. A lot of studies support that $Sr_2RuO_4$ is a spin-triplet chiral *p*-wave superconductor [1, 2]. On the other hand, the helical *p*-wave state also has been studied as another scenario [3, 4, 5, 6]. In addition, from the theoretical study employing the multi-orbital model of $Sr_2RuO_4$ [7], it is found that the condensation energy of chiral and helical *p*-wave states are very close. Therefore, we investigate behaviors of physical quantities depending on the difference between chiral and helical superconducting states, to distinguish the *d*-vector symmetry of $Sr_2RuO_4$ and other candidate materials for spin-triplet superconductors.

In the bulk state of chiral *p*-wave superconductor, time-reversal symmetry breaking state is realized by chirality of Cooper pair, i.e., orbital angular momentum $L_z = \pm 1$. According to the theoretical studies for the vortex state of chiral *p*-wave superconductors, the local density of states (LDOS) and the local NMR relaxation rate around the vortex core show the chirality dependent behaviors [8, 9, 10, 11, 12]. These chirality dependences of physical quantities are caused by the vorticity coupling to the chirality, depending on whether the chirality is parallel ($L_z = +1$) or anti-parallel ($L_z = -1$) to the vorticity ($W = +1$) [13, 14]. On the other hand, in the helical *p*-wave superconductor, a time-reversal-invariant state has a full pairing gap in the bulk. This is because the chirality $L_z = \pm 1$ are quenched with the degeneracy between up-spin and down-spin pairs. The order-parameter of up-spin (down-spin) pair $\Delta_{\uparrow\uparrow}(\Delta_{\downarrow\downarrow})$ has chirality $L_z = -1(+1)$ so that $L_z + S_z = 0$ [3]. $S_z = +1(-1)$ is the spin angular momentum of up-spin (down-spin) pair. Therefore, the spin components of LDOS in the vortex state of helical $p$-wave superconductor show unique behaviors, reflecting the vorticity coupling to the chirality of $\Delta_{\uparrow\uparrow}(L_z=-1)$ or $\Delta_{\downarrow\downarrow}(L_z=+1)$ [15].

In the various superconductors, the scanning tunneling microscopy and spectroscopy (STM/STS) measurements can detect the site- and energy-dependence of LDOS in the vortex state [16, 17]. Recently, the STM/STS measurement in the vortex state of topological insulator-superconductor $Bi_2Te_3/NbSe_2$ heterostructure has also performed [18]. From the theoretical studies for the measurement, the existence of Majorana state is suggested at the inside of the vortex core [19, 20]. The spin-polarized Majorana state in the vortex core has been also examined by the spin-polarized STM/STS measurement [21]. The spin-polarized STM/STS measurement can selectively observe the spin-dependent tunnelling conductance. In addition, by solving the Bogoliubov-de Gennes equation [22], the spin-polarized Majorana state in the vortex state of $Cu_xBi_2Si_3$ has been studied. Recently, we revealed that the spin-polarized LDOS appears in the vortex state of helical $p$-wave superconductors, and focused on the magnetic field $H$- and energy $E$-dependence of the spin-polarized LDOS [15]. We also confirmed the instability of the helical $p$-wave state at high fields.

In this paper, we focus on the spatial structure of the spin-polarized LDOS in the vortex lattice state of helical $p$-wave superconductor, and investigate the magnetic field effects on the spatial structure of the spin-polarized LDOS. We expect that these calculation results help to investigate the vortex state of helical $p$-wave superconductors and the spin-polarized Majorana state by the spin-polarized STM measurement.

This paper is organized as follows. After the introduction, we explain formulation of the Eilenberger equation in the vortex lattice state and the spin-resolved LDOS in section 2. In section 3, we study the site-dependence of order-parameter and spin-polarized LDOS. In section 4, we show magnetic field effect on the spatial structure of the spin-polarized LDOS around the vortex core. The last section is devoted to the summary.

## 2. Formulation of Eilenberger theory

We calculate the spatial structure of the vortex lattice state by the quasi-classical Eilenberger theory. The quasi-classical theory is valid when $k_F\xi$ is sufficiently larger than 1. $k_F$ is the Fermi wave number and $\xi$ is the coherence length. The quasi-classical condition is well satisfied for $Sr_2RuO_4$ [1, 2] and most superconductors.

We assume the helical $p$-wave pairing symmetry on the cylindrical Fermi surface, $\mathbf{k}=(k_x,k_y)=k_{\rm F}(\cos\theta_k,\sin\theta_k)$, and the Fermi velocity $\mathbf{v}_{\rm F}=v_{\rm F0}\mathbf{k}/k_{\rm F}$. The hat $\hat{}$ indicates the 2×2 matrix in spin space and the check $\check{}$ is for the 4×4 matrix in particle-hole and spin spaces.

We obtain quasi-classical Green's functions $\check{g}(\mathrm{i}\omega_n,\mathbf{r},\mathbf{k})$ in the vortex lattice state by solving the Riccati equation derived from Eilenberger equation [23, 24, 15]

$$-\mathrm{i}\mathbf{v}\cdot\nabla\check{g}(\mathrm{i}\omega_n,\mathbf{r},\mathbf{k})=\frac{1}{2}[(\mathrm{i}\omega_n-\mathbf{v}\cdot\mathbf{A})\check{\sigma}_z-\check{\Delta}(\mathbf{r},\mathbf{k}),\check{g}(\mathrm{i}\omega_n,\mathbf{r},\mathbf{k})] \tag{1}$$

in the clean limit, where $\mathbf{v}=\mathbf{v}_{\rm F}/v_{\rm F0}$, $\mathbf{r}$ is the center-of-mass coordinate of the Cooper pair, $\check{\sigma}_z$ is the Pauli matrix, and $\omega_n$ is Matsubara frequency. The quasi-classical Green's function and order parameter are defined by

$$\check{g}(\mathrm{i}\omega_n,\mathbf{r},\mathbf{k})=-\mathrm{i}\pi\begin{pmatrix}\hat{g}(\mathrm{i}\omega_n,\mathbf{r},\mathbf{k}) & \mathrm{i}\hat{f}(\mathrm{i}\omega_n,\mathbf{r},\mathbf{k})\\ -\mathrm{i}\underline{\hat{f}}(\mathrm{i}\omega_n,\mathbf{r},\mathbf{k}) & -\underline{\hat{g}}(\mathrm{i}\omega_n,\mathbf{r},\mathbf{k})\end{pmatrix}, \tag{2}$$

$$\check{\Delta}(\mathbf{r},\mathbf{k})=\begin{pmatrix}0 & \hat{\Delta}(\mathbf{r},\mathbf{k})\\ -\hat{\Delta}^\dagger(\mathbf{r},\mathbf{k}) & 0\end{pmatrix} \tag{3}$$

where $\check{g}^2 = -\pi^2\check{1}$. The matrix elements of $\hat{g}$ and $\hat{\Delta}$ are defined by $g_{\sigma\sigma'}(\mathrm{i}\omega_n, \mathbf{r}, \mathbf{k}) = [g_0(\mathrm{i}\omega_n, \mathbf{r}, \mathbf{k})\hat{1} + \sum_{\mu=x,y,z} g_\mu(\mathrm{i}\omega_n, \mathbf{r}, \mathbf{k})\hat{\sigma}_\mu]_{\sigma\sigma'}$ and $\Delta_{\sigma\sigma'}(\mathbf{r}, \mathbf{k}) = [\mathrm{i}\sum_{\mu=x,y,z}(d_\mu(\mathbf{r}, \mathbf{k})\cdot\hat{\sigma}_\mu)\hat{\sigma}_y]_{\sigma\sigma'}$ where $\sigma, \sigma' = \uparrow$(up-spin) or $\downarrow$(down-spin), and $d_\mu$ is $\mu$-component of $d$-vector. On the other hand, to consider the order-parameter $\Delta_{\pm,\sigma\sigma'}(\mathbf{r})$ about pairing function $\phi_{p\pm}(\mathbf{k}) = k_x \pm ik_y$ for $p_\pm$-state, the matrix elements of order-parameter are also defined by

$$\Delta_{\sigma\sigma'}(\mathbf{r}, \mathbf{k}) = \Delta_{+,\sigma\sigma'}(\mathbf{r})\phi_{p+}(\mathbf{k}) + \Delta_{-,\sigma\sigma'}(\mathbf{r})\phi_{p-}(\mathbf{k}). \tag{4}$$

Length, temperature, and magnetic field are, respectively, measured in unit of $\xi_0$, $T_c$, and $B_0$. Here, $\xi_0 = \hbar v_{F0}/2\pi k_B T_c$, $B_0 = \phi_0/2\pi\xi_0^2$ with the flux quantum $\phi_0$. $T_c$ is superconducting transition temperature at a zero field. The energy $E$, pair potential $\Delta$ and $\omega_n$ are normalized by $\pi k_B T_c$. In the following, we set $\hbar = k_B = 1$. In this work, all calculations are performed at a $T/T_c = 0.5$.

As magnetic fields are applied to the $\hat{\mathbf{z}}$ direction, the vector potential is set by $\mathbf{A}(\mathbf{r}) = \frac{1}{2}\mathbf{H}\times\mathbf{r} + \mathbf{a}(\mathbf{r})$ in the symmetric gauge, where $\mathbf{H} = (0, 0, H)$ is a average flux density, and $\mathbf{a}(\mathbf{r})$ is related to the internal magnetic field $\mathbf{B}(\mathbf{r}) = (0, 0, B(\mathbf{r})) = \mathbf{H} + \nabla\times\mathbf{a}(\mathbf{r})$. The unit cell is arranged to be square vortex lattice [1].

In order to obtain the spatial structure of the quasi-classical Green's functions and the order-parameter $\hat{\Delta}(\mathbf{r})$ selfconsistently, we calculate $\hat{\Delta}_\pm(\mathbf{r})$ by the gap equation

$$\hat{\Delta}_\pm(\mathbf{r}) = gN_0T \sum_{|\omega_n|\le\omega_{\mathrm{cut}}} \left\langle \phi^*_{p\pm}(\mathbf{k})\hat{f}(i\omega_n, \mathbf{r}, \mathbf{k})\right\rangle_{\mathbf{k}}, \tag{5}$$

where $(gN_0)^{-1} = \ln T + 2T\sum_{0<\omega_n\le\omega_{\mathrm{cut}}}\omega_n^{-1}$, $\langle\ldots\rangle_{\mathbf{k}}$ indicates Fermi surface average, and we set $\omega_{\mathrm{cut}} = 20k_B T_c$. In equation (5), $p$-wave pairing interaction is isotropic in spin space. We use the current equation $\nabla\times(\nabla\times\mathbf{A}) = -\frac{2T}{\kappa^2}\sum_{0<\omega_n}\langle\mathbf{v}\mathrm{Im}\{g_0\}\rangle_{\mathbf{k}}$ with the Ginzburg-Landau parameter $\kappa = B_0/\pi k_B T_c\sqrt{8\pi N_0}$ for the selfconsistent calculation of the vector potential. In the calculations, we set $\kappa = 2.7$ appropriate to $Sr_2RuO_4$ as a candidate superconductor for the chiral or helical state. Equations (1)-(5) for $\omega_n$ are calculated selfconsistently until the order-parameter converges. From this calculation, we obtain the results of $\mathbf{A}(\mathbf{r})$, $\hat{\Delta}(\mathbf{r})$ and the quasi-classical Green's functions of $\omega_n$ in the vortex lattice state.

$d$-vector in the helical $p$-wave superconductors is defined by $\mathbf{d}(\mathbf{k})\propto k_x\hat{x} + k_y\hat{y} = \phi_{p+}(\mathbf{k})\mathbf{d}_- + \phi_{p-}(\mathbf{k})\mathbf{d}_+$ at a zero field $H = 0$, with $\mathbf{d}_\pm(\mathbf{k}) = \frac{1}{2}(1, \pm\mathrm{i}, 0)$. Thus, when we iterate calculations of equations (1)-(5), as an initial value we set to be $\mathbf{d}(\mathbf{r}, \mathbf{k}) = d(\mathbf{r})(k_x\hat{x} + k_y\hat{y})$ where $d(\mathbf{r})$ is Abrikosov vortex lattice solution.

Next, we calculate $\check{g}(E \pm \mathrm{i}\eta, \mathbf{r}, \mathbf{k})$ for real energy $E$ by solving Eilenberger equation (1) with $\mathrm{i}\omega_n \to E \pm \mathrm{i}\eta$, using the selfconsistently obtained spatial structure of $\mathbf{A}(\mathbf{r})$ and $\Delta(\mathbf{r})$. $\eta$ is a small parameter as a smearing factor. We typically use $\eta = 0.01$. The spin-resolved LDOS $N_\sigma(E, \mathbf{r})$ is given by

$$N_\sigma(E, \mathbf{r}) = \langle\mathrm{Re}\{[\hat{g}(E + \mathrm{i}\eta, \mathbf{r}, \mathbf{k})]_{\sigma\sigma}\}\rangle_{\mathbf{k}}. \tag{6}$$

We define the LDOS as $N(E, \mathbf{r}) = N_\downarrow(E, \mathbf{r}) + N_\uparrow(E, \mathbf{r})$, and spin-polarized LDOS as $M(E, \mathbf{r}) = N_\downarrow(E, \mathbf{r}) - N_\uparrow(E, \mathbf{r})$.

## 3. Site-dependence of order-parameter and spin-polarized LDOS

As shown in figure 1(a), we present the $r$-dependence of the order-parameter amplitudes at low $H$ on a line between NNN vortices within $0 < r < 0.5a_x$. $a_x$ is distance between NNN vortices. Vortex center is at $r = 0$, and farthest site $r = 0.5a_x$ from the vortex center is the midpoint between NNN vortex. We present $|\Delta_{\pm,\sigma\sigma'}(\mathbf{r})|$ given by equation (4). In the vortex state of helical $p$-wave superconductor, up-spin pair is described by $\Delta_{\uparrow\uparrow}(\mathbf{r},\mathbf{k}) = \Delta_{-,\uparrow\uparrow}(\mathbf{r})\phi_{p-}(\mathbf{k}) + \Delta_{+,\uparrow\uparrow}(\mathbf{r})\phi_{p+}(\mathbf{k})$ with small component $\Delta_{+,\uparrow\uparrow}(\mathbf{r})$ induced around the vortex core as shown in figure 1(a). The chirality of main component $\Delta_{-,\uparrow\uparrow}(\mathbf{r})$ is $L_z = -1$. This is anti-parallel to vorticity $W = +1$ as $L_z + W = 0$. From the previous studies for the chiral $p$-wave superconductor [13, 14], the anti-parallel vortex state ($L_z + W = 0$) has larger amplitude of the order-parameter, compared with the parallel vortex state ($L_z + W = +2$). On the other hand, down-spin pair is described by $\Delta_{\downarrow\downarrow}(\mathbf{r},\mathbf{k}) = \Delta_{+,\downarrow\downarrow}(\mathbf{r})\phi_{p+}(\mathbf{k}) + \Delta_{-,\downarrow\downarrow}(\mathbf{r})\phi_{p-}(\mathbf{k})$ with small induced $\Delta_{-,\downarrow\downarrow}(\mathbf{r})$. This corresponds to the anti-parallel vortex state $L_z + W = +2$, since the chirality of the main component $\Delta_{+,\uparrow\uparrow}$ is $L_z = +1$. Therefore, $|\Delta_{-,\uparrow\uparrow}(\mathbf{r})|$ is larger than $|\Delta_{+,\downarrow\downarrow}(\mathbf{r})|$ at every position, reflecting the vorticity coupling to the chirality of $\Delta_{\uparrow\uparrow}(L_z = -1)$ or $\Delta_{\downarrow\downarrow}(L_z = +1)$.

In addition, we show the $r$-dependence of the zero-energy spin-polarized LDOS $M(E = 0, r)$ around the vortex core within $0 < r < 0.1a_x$ in figure 1(b). The spin-polarization of LDOS is induced by the vorticity coupling to the chirality of each spin-component of order-parameter, and $M(E = 0, \mathbf{r})$ has large amplitude near vortex center. Since the zero energy states localized inside the vortex core in the chiral and helical superconductors correspond to Majorana zero energy mode, figure 1(b) shows that the spin-polarized Majorana state is induced in the helical $p$-wave superconductors. This state is another type of spin-polarized Majorana state than that supposed in the vortex state of $Bi_2Te_3/NbSe_2$ [20] or $Cu_xBi_2Se_3$ [22]. In the theoretical studies based on the Bogoliubov de-Gennes equation [20, 22], vortex states were investigated in the topological superconductivity induced by Rashba spin-orbit coupling.

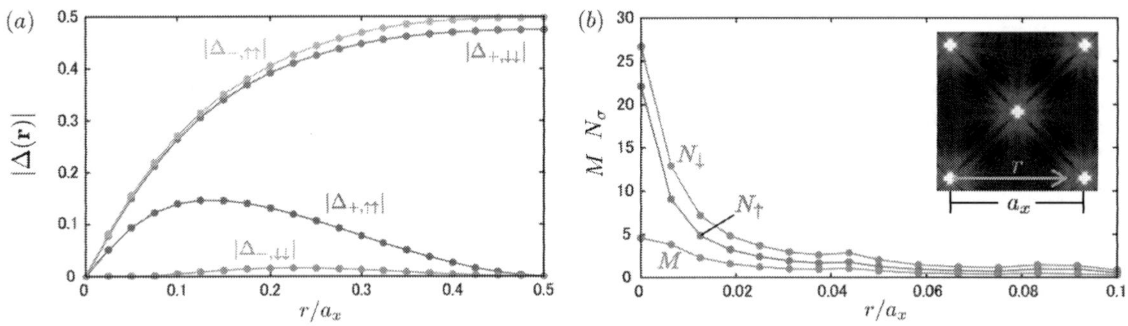


**Figure 1.** (a) $r$-dependence of the order-parameter amplitudes $|\Delta_{-,\downarrow\downarrow}(r)|$, $|\Delta_{+,\downarrow\downarrow}(r)|$, $|\Delta_{-,\uparrow\uparrow}(r)|$ and $|\Delta_{+,\uparrow\uparrow}(r)|$. $H/H_{c2}{\simeq}0.18$. (b) $r$-dependence of the spin-resolved LDOS $N_\sigma(E = 0, r)$ and spin-polarized LDOS $M(E = 0, r)$. $H/H_{c2}{\simeq}0.18$. $r$ axis is along the next-nearest-neighbor (NNN) direction from the vortex center. $a_x$ is NNN intervortex distance. The inset shows distribution of $M(E = 0, \mathbf{r})$. Brighter region has larger $M(E = 0, \mathbf{r})$.

## 4. Magnetic field effect on the spatial structure of spin-polarized LDOS

In figures 2, we present the spatial structure of the zero-energy spin-resolved LDOS $N_\sigma(E = 0, \mathbf{r})$, LDOS $N(E = 0, \mathbf{r})$ and spin-polarized LDOS $M(E = 0, \mathbf{r})$ along a line between NNN vortices within $0 < r < 0.1a_x$ at some magnetic fields. Here we study lower field range to avoid instability of helical state at high fields [15]. At a very low field $H/H_{c2} = 0.02$, $N_\sigma(E = 0, r)$ shows sharp peak at the vortex center $r = 0$. $N_\uparrow(E = 0, \mathbf{r})$ and $N_\downarrow(E = 0, \mathbf{r})$ show different behavior in the range $0.12 < H/H_{c2} < 0.24$. There, with increasing $H$, the peak at the vortex center decreases in $N_\downarrow(E = 0, \mathbf{r})$ in figure 2(a), but the magnetic field effect on $N_\downarrow(E = 0, \mathbf{r})$ is small in figure 2(b). Therefore, the spatial structure of LDOS $N(E = 0, \mathbf{r})$ in figure 2(c) are similar to $N_\downarrow(E = 0, \mathbf{r})$. Since $N_\uparrow(E = 0, \mathbf{r})$ is smaller than $N_\downarrow(E = 0, \mathbf{r})$ at every position in the field range considered in figure 2, $M(E=0,\mathbf{r})$ is positive value in figure 2(d). We note that the magnetic field effect on the spatial structure of $M(E=0,\mathbf{r})$ shows a unique behavior. At $H/H_{c2}\simeq 0.02$ and 0.12, $M(E=0,\mathbf{r})$ monotonically decreasing as a function of $r$. On the other hand, at $H/H_{c2}\simeq 0.18$ and 0.24, $M(E=0,\mathbf{r})$ shows convex behavior as a function of $r$. These unique behaviors of the helical state are expected to be examined by spin-polarized STM measurement.

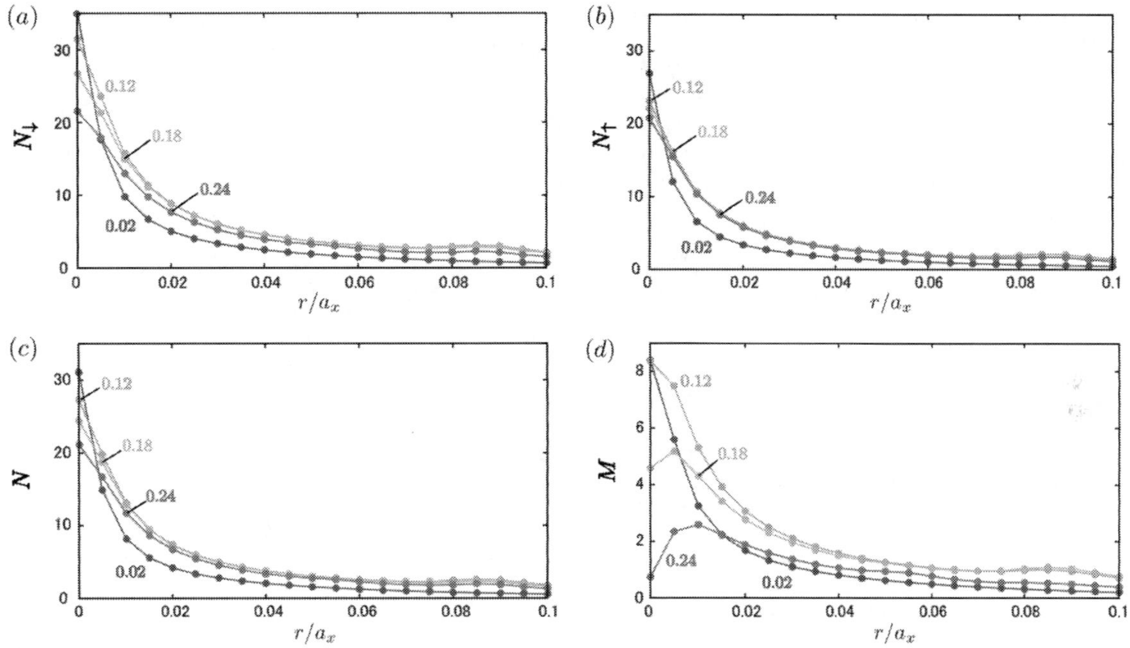


**Figure 2.** $r$-dependence of spin-resolved LDOS (a) $N_\downarrow(E=0,r)$, (b) $N_\uparrow(E=0,r)$, (c) LDOS $N(E=0,r)$, and (d) spin-polarized LDOS $M(E=0,r)$ at magnetic fields $H/H_{c2}\simeq$ 0.02, 0.12, 0.18, and 0.24 along the NNN direction within $0 < a_x < 0.1$.

## 5. Summary

We studied the spatial structure of the helical $p$-wave superconductors in the vortex lattice state by solving the quasi-classical Eilenberger equation. In the helical state, up-spin pairing order-parameter $\Delta_{\uparrow\uparrow}$ has larger amplitude compared to that of down-spin pairing $\Delta_{\downarrow\downarrow}$ by the contribution of coupling between vorticity and chirality. Therefore, spin-resolved zero-energy LDOS shows the differences between $N_\uparrow(E=0,\mathbf{r})$ and $N_\downarrow(E=0,\mathbf{r})$, inducing finite spin-polarized LDOS $M(E=0,\mathbf{r})$ around a vortex. The zero-energy state in the $p$-wave pairing is spin-polarized Majorana state localizing in a vortex core. We study how the spatial structure of $M(E=0,\mathbf{r})$ is affected by the magnitudes of magnetic fields in the vortex states. At very low field, $M(E=0,\mathbf{r})$ has a sharp peak at a vortex center. With increasing magnetic fields, the peak height decrease, and $M(E=0,\mathbf{r}=\mathbf{0})$ becomes local minimum at the vortex center. The spin-polarized Majorana state $M(E=0,\mathbf{r})$ may be a key features to identify the helical $p$-wave state. We hope that the spin-polarized STM measurements will be performed to observe $M(E=0,\mathbf{r})$ around vortices in candidate superconductors of helical $p$-wave state.

**Acknowledgments**
This work was supported by JSPS KAKENHI Grant Number JP16J05824.

## References

[1] Mackenzie A P and Maeno Y 2003 *Rev. Mod. Phys.* **75** 657
[2] Maeno Y, Kittaka S, Nomura T, Yonezawa S and Ishida K 2012 *J. Phys. Soc. Jpn.* **81** 011009
[3] Rice T M and Sigrist M 1995 *J. Phys.: Condens. Matter* **7** L643
[4] Takamatsu S and Yanase Y 2013 *J. Phys. Soc. Jpn.* **82** 063706
[5] Scaffidi T, Romers J C and Simon S H 2014 *Phys. Rev.* B **89** 220510(R)
[6] Zhang J, Lörscher C, Gu Q and Klemn R A 2014 *J. Phys.: Condens. Matter* **26** 252201
[7] Tsuchiizu M, Yamakawa Y, Onari S, Ohno Y and Kontani H 2015 *Phys. Rev.* B **91** 155103
[8] Kato Y and Hayashi N 2003 *Physica* C **388** 519
[9] Tanuma Y, Hayashi N, Tanaka Y and Golubov A A 2009 *Phys. Rev. Lett.* **102** 117003
[10] Kurosawa N, Hayashi N and Kato Y 2015 *J. Phys. Soc. Jpn.* **84** 114710
[11] Hayashi N and Kato Y 2003 *Physica* C **388** 513
[12] Tanaka K K, Ichioka M and Onari S 2016 *Phys. Rev.* B **93** 094507
[13] Ichioka M and Machida K 2002 *Phys. Rev.* B **65** 224517
[14] Heeb R and Agterberg D F 1999 *Phys. Rev.* B **59** 7076
[15] Tanaka K K, Ichioka M and Onari S 2017 *Phys. Rev.* B **95** 134502
[16] Hess H F, Robinson R B, Dynes R C, Valles J M, Jr. and Waszczak J V 1989 *Phys. Rev. Lett.* **62** 214
[17] Fischer Ø, Kugler M, Maggio-Aprile I, Berthod C and Renner C 2007 *Rev. Mod. Phys.* **79** 353
[18] Xu J -P, Wang M -X, Liu Z L, Ge J -F, Yang X, Liu C, Xu Z A, Guan D, Gao C L, Qian D, Liu Y, Wang Q -H, Zhang F -C, Xue Q -K and Jia J -F 2015 *Phys. Rev. Lett.* **114** 017001
[19] Li Z -Z, Zhang F -C and Wang Q -H 2014 *Sci. Rep.* **4**, 6363
[20] Kawakami T and Hu X 2015 *Phys. Rev. Lett.* **115** 177001
[21] Sun H -H, Zhang K -W, Hu L -H, Li C, Wang G -Y, Ma H -Y, Xu Z -A, Gao C -L, Guan D -D, Li Y -Y, Liu C, Qian D, Zhou Y, Fu L, Li S -C, Zhang F -C and Jia J -F 2016 *Phys. Rev. Lett.* **116** 257003
[22] Nagai Y, Nakamura H and Machida M 2014 *J. Phys. Soc. Jpn.* **83** 064703
[23] Eilenberger G 1968 *Z. Phys.* **214** 195
[24] Tsutsumi Y, Kawakami T, Shiozaki K, Sato M and Machida K 2015 *Phys. Rev.* B **91** 144504

# Superconductivity in the underdoped region of the T'-structure cuprates based on an effective two-band model

**Kunito Yamazaki**[1], **Hiroki Tsuchiura**[1,2], **Takuya Yoshioka**[1], **Masao Ogata**[3]

[1]Department of Applied Physics, Tohoku University, Sendai 980-8579, Japan
[2]Center for Spintronics Research Network, Tohoku University, Sendai 980-8577, Japan
[3]Department of Physics, University of Tokyo, Bunkyo, Tokyo 113-0033, Japan

E-mail: yamazaki@olive.apph.tohoku.ac.jp

**Abstract.** Electronic states in the underdoped (including non-doped) region of the so-called T'-structure cuprates are studied based on the *d-p* model and a newly proposed effective two-band model for those compounds. We find *d*-wave superconducting transition for both models when the charge transfer gap is small enough which is consistent with the recent experimental results for the properly reduced NCO films etc. We also show that the two-band model gives almost similar result for the superconducting transition temperature and the static spin susceptibility with those obtained in the *d-p* model, which implies the two-band model can give a good description of the T'-cuprates.

## 1. Introduction

Recently, growing attention has been paid to the reconstruction of the phase diagram of the T'-structure cuprates, that is, the so-called electron-doped curates. Several experimental studies have shown that the T'-type cuprates exhibit superconductivity with very low or without chemical carrier doping, only when excess oxygen which inevitably exists in as-grown samples are properly removed[1-8].

Theoretically, we have studied the superconducting transition around the underdoped region of these compounds based on an effective two-band model[9], which can be derived from the *d-p* model essentially along with the idea of Zhang and Rice[10], but explicitly consists of Cu *d*-orbitals and symmetrically hybridized O 2*p*-orbitals. Thus, the two-band model can be regarded as a simplified version of the *d-p* model. By applying the fluctuation-exchange approximation (FLEX)[11-12], we have found that the d-wave superconductivity appears around the underdoped region, and even at the non-doped point with one hole per Cu site, when the charge transfer gap $\Delta = U_d - \Delta_p$ is small enough, where $U_d$ is the on-site Coulomb repulsion on Cu site, and $\Delta_p$ is the energy difference between the O 2*p* and Cu 3*d* levels.

Although our theoretical results are qualitatively in agreement with the experimental observations for the T'-cuprates, the validity of the two-band model has not been assessed in detail. Thus the next question we must consider is whether the non-doped dSC is found even in the *d-p* model or not. In this paper, we study the superconducting transition at the non-doped point with one-hole per Cu site based on the *d-p* model, and compare the results such as dispersion relations, $T_c$'s, and spin susceptibilities with those obtained using the two-band model.

## 2. model and formalism

The Hamiltonian of the *d-p* model is written as

$$\mathcal{H}_{dp} = -t_{dp} \sum_{\langle i,j \rangle \sigma} \left( d^\dagger_{i\sigma} p_{j\sigma} + h.c. \right) - t_p \sum_{\langle j,j' \rangle \sigma} \left( p^\dagger_{j\sigma} p_{j'\sigma} + h.c. \right) + \Delta \sum_{j\sigma} n_{p,j\sigma} + U_d \sum_i n_{d,i\uparrow} n_{d,i\downarrow} \quad (1)$$

in the standard notation, where $d_{i\sigma}$ and $p_{j\sigma}$ are annihilation operators of holes with spin $\sigma$ on the Cu $3d_{x^2-y^2}$-orbital at site $i$ and on the O $2p_\sigma$-orbital at site $j$, respectively. $t_{dp}$ and $t_p$ are amplitudes of the nearest-neighbor Cu-O and O-O hopping integral, repectively, $U_d$ is the Coulomb repulsion on Cu site i, and $\Delta = U_d - \Delta_p$ is the lowest excitation energy in the localized limit, where $\Delta_p$ is the energy difference between the O $2p$ and Cu $3d$ levels.

An effective two-band model for the T ' -structure cuprates derived by following the idea of Zhang-Rice[10]. We keep the symmetrically hybridized O $2p$-band and omit the anti-symmetric hybridized one. The two-band model is written as,

$$\mathcal{H}^{\text{eff}} = \sum_{\boldsymbol{k}\sigma} \varepsilon_p(\boldsymbol{k}) p^{(s)\dagger}_{\boldsymbol{k}\sigma} p^{(s)}_{\boldsymbol{k}\sigma} + \sum_{\boldsymbol{k}\sigma} t_{dp} t_{\boldsymbol{k}} (d^\dagger_{\boldsymbol{k}\sigma} p^{(s)}_{\boldsymbol{k}\sigma} + h.c.) + \frac{U_d}{N_L} \sum_{\boldsymbol{k},\boldsymbol{k}',\boldsymbol{q}} d^\dagger_{\boldsymbol{k}+\boldsymbol{q},\uparrow} d^\dagger_{\boldsymbol{k}'-\boldsymbol{q},\downarrow} d_{\boldsymbol{k}',\downarrow} d_{\boldsymbol{k},\uparrow} \quad (2)$$

where

$$\varepsilon_p(\boldsymbol{k}) = \left\{ \Delta - 32 t_p \frac{1}{t_{\boldsymbol{k}^2}} \sin^2(\frac{k_x}{2}) \sin^2(\frac{k_y}{2}) \right\},$$

$$t_{\boldsymbol{k}} = 2 \left\{ \sin^2(\frac{k_x}{2}) + \sin^2(\frac{k_y}{2}) \right\}^{1/2}, \quad (3)$$

$d_{\boldsymbol{k}\sigma}$ and $p^{(s)}_{\boldsymbol{k}\sigma}$ are the Fourier transforms of the annihilation operators for the holes with spin $\sigma$ on the $d$- and symmetrized p-orbitals, respectively, and $N_L$ is the number of atoms in the crystal.

Here we show the noninteracting dispersion relations of the energy band in which the Fermi energy lies for both models in Fig. 1. It should be noted here that these dispersion relations are depicted in the hole picture. We can see that these two models have very similar dispersion relations all over the Brillouin zone.

We note here that the Coulomb interaction is considered only within a $d$-orbital for brevity. We use the parameters of NCO[13]($t_{dp} = 1.18\text{eV}, t_p = 0.621\text{eV}$), but $\Delta$ is changed($\Delta = 1.65\text{eV} - 1.80\text{eV}$) to take into account the effect of the excess oxygen in the T'-structure cuprates on both models. In order to obtain the renormalized Green's function, we carry out the FLEX calculation to the $d$-$p$ model and the two-band model. The transition temperature is determined as the temperature which the linearized Eliashberg equation using the Green's function,

$$\phi(\boldsymbol{k}, i\epsilon_n) = -\frac{T}{N_L} \sum_{\boldsymbol{k}',m} V^{(2)}(\boldsymbol{k} - \boldsymbol{k}', i\epsilon_n - i\epsilon_m) G_d(\boldsymbol{k}', i\epsilon_m) G_d(-\boldsymbol{k}', -i\epsilon_m) \phi(\boldsymbol{k}' i\epsilon_m) \quad (4)$$

has a non-trivial solution, and we assume that the d-wave superconductivity is realized in the T'-structure cuprates. Here, the effective interaction for singlet pairing is obtained as $V^{(2)}(k - k') = U_d + \frac{3}{2} U_d^2 \chi_d^{(s)}(k - k') - \frac{1}{2} U_d^2 \chi_d^{(c)}(k - k')$, where the spin and charge susceptibilities for the d-orbital $\chi_d^{(s)}$ and $\chi_d^{(c)}$. In this paper, we employ $128 \times 128$ $\boldsymbol{k}$-point and 16384 Matsubara frequencies.

IOP Conf. Series: Journal of Physics: Conf. Series **969** (2018) 012044 doi:10.1088/1742-6596/969/1/012044

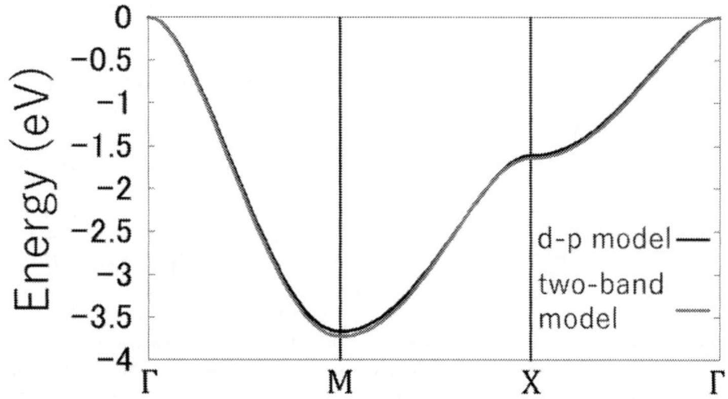


**Figure 1.** The energy dispersion around the Fermi energy of the *d-p* model and the two-band model for $t_{dp} = 1.18$eV, $t_p = 0.621$eV, and $\Delta = 1.75$eV.

## 3. Results and discussion

Let us look at first the superconducting transition temperatures $T_c$ for the *d-p* and the two-band model as functions of the charge transfer gap $\Delta$ at the non-doped point with one hole per Cu site. We can clearly see that the both models describe the superconducting transition even at the non-doped point. The $\Delta$ dependences are qualitatively the same, but the $T_c$'s for the *d-p* model is somewhat lower than those for the two-band model, except for one case ($\Delta = 1.8$eV). In any case, the superconducting transitions at the non-doped point is confirmed even in the *d-p* model.

In the present calculation, when $\Delta \geq 1.85$eV, we have confirmed that the spin susceptibility diverges at higher temperatures than $T_c$'s shown in Fig. 2, implying the antiferromagnetic(AFM) phase transition. But the coexisting state of AFM and d-wave superconductivity may still be possible as a candidate for a lower-temperature phase or the ground state of the present system. Although it is not possible to figure out if this is the case within the present theoretical framework, we note here that it is interesting to study the stability of the coexisting state by using variational Monte Carlo simulations.

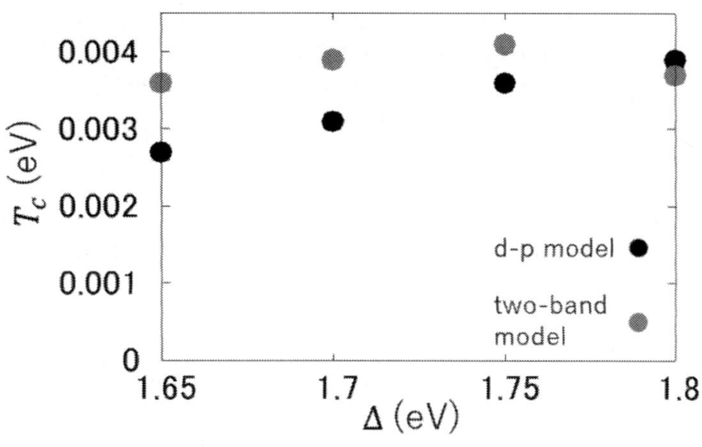


**Figure 2.** $T_c$ of the *d-p* model and the two-band model for $t_{dp} = 1.18$eV, $t_p = 0.621$eV, $U_d = 5.0$eV, and $\Delta = 1.65 - 1.80$eV.

Next, we compare the $\boldsymbol{q}$-dependences of the static spin susceptibility $\chi(\boldsymbol{q}, 0)$ for both models for several $\Delta$'s at the same temperature $T = 0.005$eV in Fig. 3. The antiferromagneteic peaks around the $X$ point are found for both cases, and the height of these increases as $\Delta$ increases. The peaks for the $d$-$p$ model are, however, smaller than those for the two-band model, which is consistent with the results for $T_c$'s shown in Fig. 2.

Since the discovery of the cuprate superconductors, the $t$-$J$ model and the single band Hubbard model have been used to analyze the low-energy physics in the $CuO_2$ planes. Because of the extensive theoretical studies, it is known that these single-band models provide a reasonable description of the phase diagram of cuprate superconductors, in particular of the hole-doped cuprates with T-structure. For the electron-doped cuprates with T ' -structure, however, the O $2p$ degrees of freedom plays an important role as implied by the present results. Thus the three-band $d$-$p$ model, or the two-band $d$-$p$ model should be used to analyze such materials. In particular, because of the less degrees of freedom, we believe that the two-band model is more preferable when we further examine the phase diagram of the electron-doped cuprates by using laborious computational techniques for many-body problems[14].

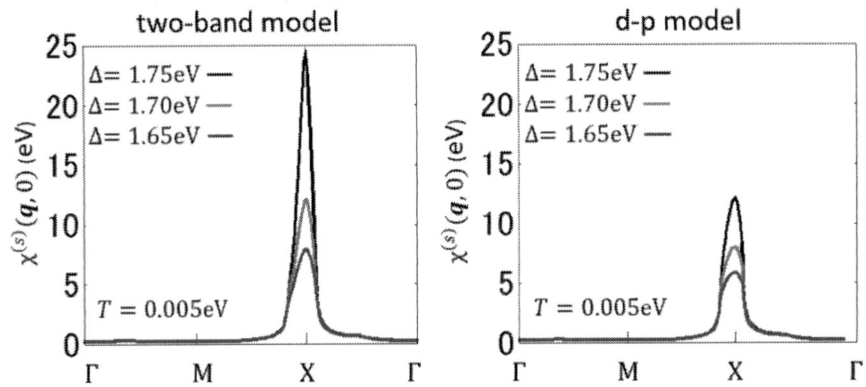


**Figure 3.** The statice spin susceptibility $\chi^s(\boldsymbol{q}, 0)$ of the $d$-$p$ model and the two-band model for $t_{dp} = 1.18$eV, $t_p = 0.621$eV, $U_d = 5.0$eV, and $\Delta = 1.65 - 1.75$eV.

## 4. Summary

We have studied the superconducting transition of the T'-cuprates based on the $d$-$p$ model and the effective two-band model, and found that these two models describe the $d$-wave superconductivity at the non-doped point when the charge transfer gap is small enough. The two band model is found to give $T_c$'s, spin susceptibilities which are in satisfactory agreement with those for the $d$-$p$ model.

## 5. Acknowledgments

This work was supported by JSPS KAKENHI Grant Number JP17K05528. The numerical computations were carried out at the Cyber-science Center, Tohoku University.

## References

[1] Tsukada A, Krockenberger Y, Noda M, Yamamoto H, Manske D, Alff L, and Naito M 2005 Solid State Commun. **133** 427
[2] Matsumoto O, Utsuki A, Tsukada A, Yamamoto H, Manabe T, and Naito M 2009 Phys. Rev. B **79** 100508(R)
[3] Asai S, Ueda S, and Naito M 2011 Physica C **471** 682
[4] Yamamoto H, Matsumoto O, Krockenberger Y, Yamagami K, and Naito M 2011 Solid State Commun. **151** 771
[5] Krockenberger Y, Yamamoto H, Tsukada A, Mitsuhashi M, and M. Naito 2012 Phys. Rev. B **85** 184502

[6] Takamatsu T, Kato M, Noji T, and Koike Y 2012 Appl. Phys. Express **5** 073101
[7] Brinkman M, Rex T, Bach H, and Westerholt K 1995 Phys. Rev. Lett. **74** 4927
[8] Adachi T, Mori Y, Takahashi A, Kato M, Nishizaki T, Sasaki T, Kobayashi N, and Koike Y 2013 J. Phys. Soc. Jpn. **82** 063713
[9] Yamazaki K, Yoshioka T, Tsuchiura H, and Ogata M 2017 J. Phys. Conf. Ser. **871** 012009
[10] Zhang F C and Rice T M 1988 Phys. Rev. B **37** 3759
[11] Koikegami S, Fujimoto S, and Yamada K 1997 J. Phys. Soc. Jpn. **66** 1438
[12] Takimoto T and Moriya T 1997 J. Phys. Soc. Jpn. **66** 2459
[13] Ogura D and Kuroki K 2015 Phys. Rev. B **92** 144511
[14] Ogata M and Fukuyama H 2008 Rep. Prog. Phys. **71** 036501

# Effect of carbon coating on the superconducting properties of Pb and Sn nano-spheres

L. Shani [1], V. B. Kumar [2], A. Gedanken [2], I.Shapiro[1], B.Y.Shapiro[1], A. Shaulov[1], and Y. Yeshurun [1]

[1] Department of Physics and Institute of Nanotechnology and Advanced Materials, Bar-Ilan University, 52900 Ramat-Gan, Israel
[2] Department of Chemistry and Institute of Nanotechnology and Advanced Materials, Bar-Ilan University, 52900 Ramat-Gan, Israel

**Abstract.** Carbon coated Sn and Pb spheres of radius smaller than the superconducting coherence length of the bulk materials were fabricated using a sonochemical technique. Both samples reveal Type-I superconducting behavior characterized by supercritical fields. Analysis of the data reveal the expected enhancement of the critical field, $H_c$, in the Pb but not in the Sn nano-spheres. The different behavior of the Sn spheres is attributed to carbon doping which results in a decrease of their effective coherence length to below the spheres size.

## 1. Introduction

Superconductors of size smaller than the coherence length exhibit significant different properties than the bulk materials[1, 2]. A notable example is the dramatically enhanced critical field, $H_c$, in nano-size Type-I superconductors [3-5]. A pitfall in characterizing nano-particles may arise from agglomeration and/or oxidation of the nano-particles. This is commonly prevented by coating the particles with various materials. In this paper we describe magnetic measurements in carbon coated Pb and Sn nano-spheres, indicating that the coating process may cause doping of the material, thus reducing the effective coherence length to below the particles size leading to elimination of size effects. Such a phenomenon occurs in Sn but absent in Pb for a reason to be explained below.

## 2. Experimental

Granules of pure Sn and Pb metals were melted in silicone oil in a separate quartz test tubes. The mixture was irradiated for 5 minutes with 20 kHz ultrasonic energy, forming a suspension of particles in the Si-oil. After precipitation, the particles were separated using a centrifuge, washed several times with n-hexane and dried in vacuum. X-ray diffraction of both elements showed the existence of only Pb and Sn confirming the purity of the samples. The size distribution of the nano-particles was measured using Dynamic Light Scattering (DLS). These measurements showed an average particle size of ~120 and ~70 nm with a full width at half maximum of ~20 nm for Sn and Pb, respectively. The magnetic measurements described in the next section were performed using a Superconducting Quantum Interference Device (SQUID - Quantum Design MPMS-5XL) magnetometer.

## 3. Results and Discussion

Figure 1 shows magnetization curves for the Pb (upper panel) and Sn samples measured at the indicated temperatures, for increasing and decreasing fields. Both samples exhibit characteristics of type-I superconductors, namely magnetic hysteresis with so called "superheating", $H_{sh}$, and "supercooling", $H_{sc}$, fields – terms borrowed from first order phase transition driven by temperature (these supercritical field are marked in Fig.1 by arrows for the lowest temperature of each sample) [6]. The temperature dependence of $H_{sh}$ and $H_{sc}$ for both samples is shown in Figure 2. From these data one can deduce $H_c$ and the Ginzburg-Landau parameter $\kappa$, based on the equations derived by Feder *et al.* for spherical type-I superconductors [7]:

$$H_{sh} = \frac{2}{3}2^{-1/4}\kappa^{-1/2}H_c \quad (1)$$

$$H_{sc} = 1.69\sqrt{2}\kappa H_c. \quad (2)$$

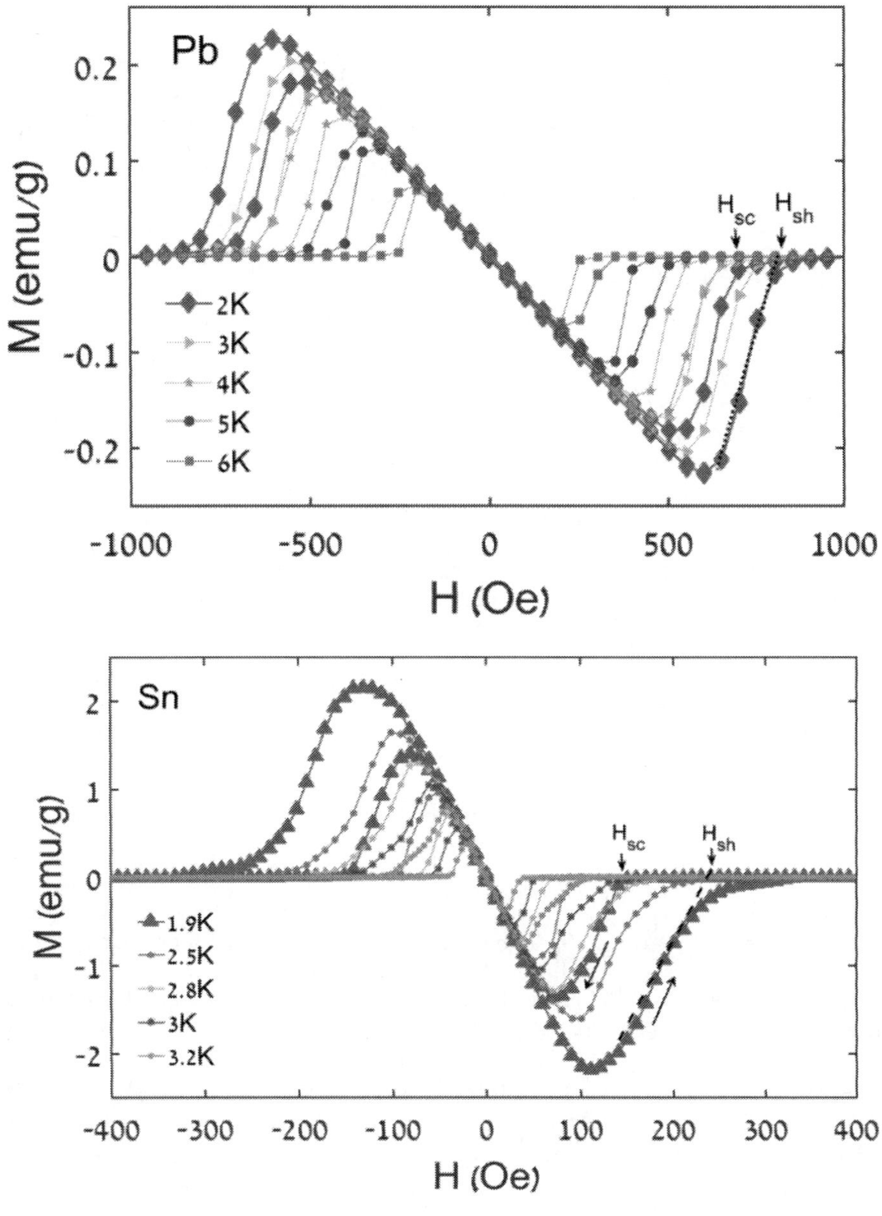


**Figure 1.** Magnetization curves at different temperatures for Pb (upper panel) and Sn nanospheres. Arrows indicate increasing and decreasing fields for the lowest temperatures.

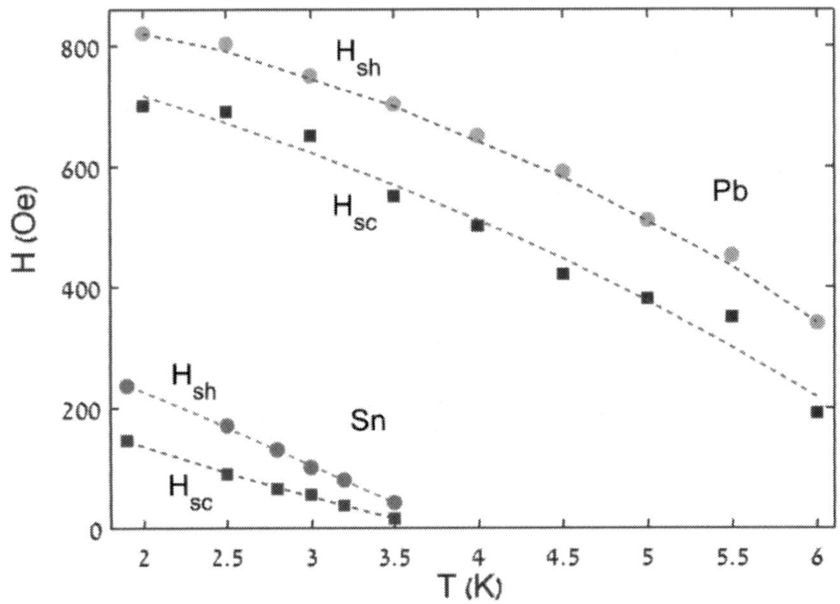


**Figure 2.** Temperature dependence of the superheating fields (circles) and the supercooling fields (squares) for the Pb and Sn samples. Dashed lines are guide to the eye.

These two equations yield:

$$H_c = 1.1(H_{sc}H_{sh}^2)^{1/3}, \tag{3}$$

and

$$\kappa = 0.38(H_{sc}/\ H_{sh}\ )^{2/3}. \tag{4}$$

Figure 3 shows the values of $H_c$ for both samples as a function of temperature, derived from Eq. (3). The solid lines are fits to the empirical approximation $H_c = H_c(0)(1 - (T/T_c)^2)$, yielding the values 305 and 950 Oe for $H_c(0)$ in the Sn and Pb nano-spheres, respectively. We note that while the $H_c(0)$ value for the Sn nano-spheres are close to that for bulk (300 Oe), $H_c(0)$ for the Pb nano-spheres is significantly higher than the bulk value (800 Oe) [8]. The increase of $H_c(0)$ for the Pb nano-spheres is consistent with previous reports of enhanced critical field in type-I nano-particles (see, e.g. [8, 9]). The result for Sn, however, is inconsistent with these reports (see, e.g.[3]).

We argue that the size effect in Sn is eliminated by a significant reduction of the coherence length to a value smaller than the size of the sample, due to carbon doping of the spheres. This is indicated by a significant increase in $\kappa(0)$ relative to a clean Sn bulk, estimated as follows. We approximate the value of $\kappa(T_c)$ by inserting in Eq. (4) the values of $H_{sh}$ and $H_{sc}$ at the highest measured temperature (3.2 K in Sn), yielding $\kappa(T_c) \approx 0.18$ . Using the empirical approximation $\kappa(T) = \kappa(0)/(1 + (T/T_c)^2)$ [6] yields $\kappa(0) \approx 0.36$, which is significantly higher than the reported value of 0.15 for clean bulk Sn [8]. To estimate the value of the effective coherence length, $\xi_{eff}$ for this sample we use the equation [10]:

$$\xi_{eff} = \xi_0 l/(\xi_0 + l).$$
(5)

where the value of the electron mean free path, $l$, is estimated using the dirty limit formula [6]

$$\kappa(T_c) = \frac{0.72\lambda_L(0)}{l}.$$
(6)

Insertion of the values $\kappa(T_c) \approx 0.18$ and $\lambda_L(0) = 34\ nm$ in Eq. (6), yields $l = 129\ nm$. With $\xi_0 \approx 230\ nm$, Eq. (5) yields $\xi_{eff} = 83$ nm. This value of $\xi_{eff}$ is smaller than the size of the sample (~120 nm), thus no size-effect is expected for the Sn spheres.

Similar calculation for the Pb sample yields $\kappa(T_c) \approx 0.23$, implying $\kappa(0) \approx 0.46$, which is close to that of a clean bulk sample (0.44 in [8]). This result suggest that the Pb nano-spheres are unaffected by the carbon coating and, therefore, exhibit the expected size effect in $H_c$ described above. The fact that the Pb spheres are 'clean' whereas the Sn spheres are 'dirty', although they went through the same carbon coating process, is explained by the fact that Pb, unlike Sn, does not dissolve carbon and does not interact with it to form chemical bonds[11].

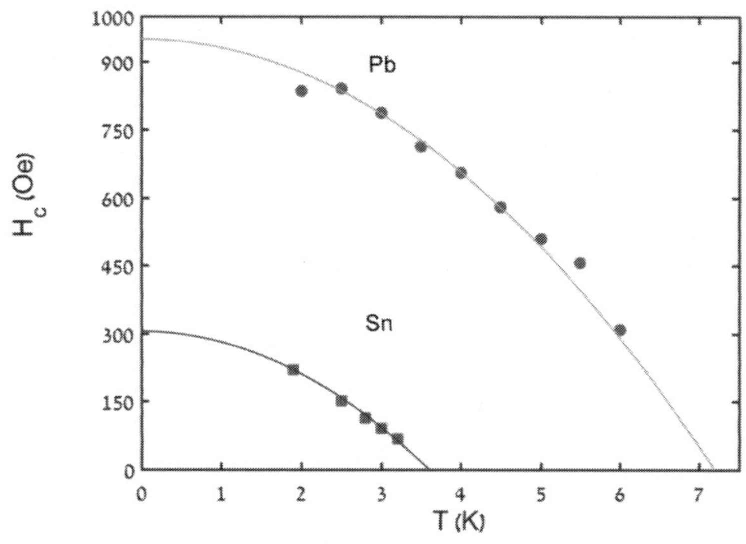


**Figure 3**. Calculated temperature dependence of the critical field, $H_c$ , for samples Pb and Sn, as deduced from Eq. (3), see text. The solid lines through the data points are fits to the empirical approximation $H_c = H_{c0}(1 - (T/T_c)^2)$ with $H_{c0} = 950$ and 305 Oe for Pb and Sn, respectively.

## 4. Summary and Conclusions

Magnetic measurements in carbon-coated Sn and Pb nano-spheres, with size smaller than the coherence length, $\xi_0$, of the clean material, show a significant increase of $H_c$ in the Pb but not in the Sn spheres. The absence of an increase in $H_c$ in the Sn spheres is attributed to carbon doping of the Sn during the coating process. The doping results in a decrease in the effective coherence length to below the spheres size, thus eliminating the size effect. Doping of the Pb nano-spheres by carbon is less effective because Pb does not dissolve carbon and does not interact with it to form chemical bonds.

**Acknowledgements**. This work was partially supported by the Israel Science Foundation (ISF) and the German-Israeli Foundation (GIF).

**References**

[1] Li W-H, Yang C, Tsao F, Wu S, Huang P, Chung M and Yao Y 2005 Enhancement of superconductivity by the small size effect in In nanoparticles *Physical Review B* **72** 214516

[2] Hagel J, Kelemen M, Fischer G, Pilawa B, Wosnitza J, Dormann E, Löhneysen H v, Schnepf A, Schnöckel H and Neisel U 2002 Superconductivity of a Crystalline Ga84-Cluster Compound *Journal of low temperature physics* **129** 133-42

[3] Giaever I and Zeller H 1968 Superconductivity of small tin particles measured by tunneling *Physical Review Letters* **20** 1504

[4] Zhang Y, Wong C H, Shen J, Sze S T, Zhang B, Zhang H, Dong Y, Xu H, Yan Z and Li Y 2016 Dramatic enhancement of superconductivity in single-crystalline nanowire arrays of Sn *Scientific Reports* **6**

[5] Wang X-L, Feygenson M, Aronson M C and Han W-Q 2010 Sn/SnO x Core– Shell Nanospheres: Synthesis, Anode Performance in Li Ion Batteries, and Superconductivity *The Journal of Physical Chemistry C* **114** 14697-703

[6] Tinkham M 2004 *Introduction to Superconductivity*: Dover Books on Physics)

[7] Feder J and McLachlan D S 1969 Superheating and supercooling in single spheres of tin, indium, and gold-plated indium *Physical Review* **177** 763

[8] Bose S and Ayyub P 2014 A review of finite size effects in quasi-zero dimensional superconductors *Reports on Progress in Physics* **77** 116503

[9] Reich S, Leitus G, Popovitz-Biro R and Schechter M 2003 Magnetization of small lead particles *Physical Review Letters* **91** 147001

[10] Huebener R 2001 *Magnetic flux structures in superconductors: extended reprint of a classic text*: Springer Science & Business Media)

[11] Banhart F, Hernandez E and Terrones M 2003 Extreme superheating and supercooling of encapsulated metals in fullerenelike shells *Physical review letters* **90** 185502

# Impurity effects of hydrogen and deuterium in vanadium nanoconstrictions

**M S Islam[1,2], Y Ueno[1], H Takata[1], Y Inagaki[1], K Hashizume[3] and T Kawae[1]**

[1] Department of Applied Quantum Physics, Kyushu University, Fukuoka 819-0395, Japan
[2] Department of Physics, Rajshahi University, Rajshahi 6205, Bangladesh
[3] Department of Advanced Energy Science and Engineering, Kyushu University, Fukuoka 816-8580, Japan

**Abstract.** To study non-magnetic and magnetic impurity effects at the superconductor-normal metal interfaces, we have measured the differential conductance d*I*/d*V* for Josephson contacts made by vanadium (V) nanoconstrictions with a different amount of hydrogen (H) and deuterium (D) impurities, which are prepared by a mechanically controllable break junction (MCBJ) technique at low temperature. Below the superconducting transition temperature $T_C$, we have found distinct peaks within and outside the gap, known as the sub-gap structure and an over-the-gap structure respectively, due to 5% concentration of atomic H and D on V nanocontact. Moreover, the temperature dependence of d*I*/d*V* spectra represents that both structures are survived until the critical temperature $T_C$, which is consistent with the prediction of BCS energy gap. On the other hand, a very high concentrating phase (30atom%) behaves as a normal metal.

Keywords: MCBJ, impurity effects, sub-gap structure and over-the-gap structure

## 1. Introduction

Hydrogen (H) is an ideal replacement for synthetic fuel because of its lightest weight, high abundance, high energy density and environmental cleanness of the oxidation product. Hence, understanding the role of H atoms in metals is a matter of much technological importance as well as scientific interest. Many hydrides drastically differ from their parent metals in terms of crystal structure, electronic structure and mechanical properties. Recently, superconductivity with the transition temperature $T_C$ over 200 K has been discovered in $H_2S$ [1], which would pave the way for a room temperature superconductor. To study the interaction between electrons and phonon vibrations assisted by H atoms from microscopic view points, we have investigated impurity effects of H and deuterium (D) in vanadium (V) nanoconstrictions and measured the differential conductance d*I*/d*V* spectra for Josephson contacts.

Group-V metals show superconductivity, which is other interesting aspect of this series of metals. The study of impurity doping on the superconductor is one of the most significant tools for understanding the nature of electronic properties of host metals. The main purpose of this article is to examine the influence of H and deuterium (D) on the electronic properties of vanadium to obtain information about the effects of interstitial H and D on the host metal at low temperature. Since the mass difference between H and D atoms provides the highest possible mass ratios for all elements, the

[1,2] Md. Saiful Islam (Email: sislamru@gmail.com)

superconducting properties could be affected by a significant modification of electron-phonon interaction. In addition, according to the net spin of the atom, D as an isotope of H, shows half-integer spin like fermion due to its constituent particles of neutron ($I = 1/2$). H atoms are generally absorbed to be almost neutral condition in metals. Therefore, the difference of neutron spin between H and D would lead to significant changes for the pair-breaking on scattering between conduction electrons and their impurities, implying the possibility to obtain a different qualitative behavior by the effect of neutron spin on superconductivity.

In the present work, we investigate the microscopic properties of metallic vanadium (V) containing several concentration of atomic H and D with the reduced dimensionality. When the material is being constricted, the electronic interactions slightly differ from those in the bulk. We measure electrical conductance of pure V constrictions down to ~ 2 K and what behaves actually does after doping a different amount of H and D on it. The differential conductance spectra show distinct peaks not only inside the superconducting gap, but also outside the gap below the superconducting transition temperature. The temperature dependence of $dI/dV$ spectra suggests that the outside peaks are related to the superconducting transition although the peaks appear at higher energies than the superconducting gap.

## 2. Sample Preparation

H atoms are loaded in the polycrystalline sample wire of V, which has a typical diameter of 0.25 mm with a purity of 99.8% consisting 200 ppm of Fe and 10 ppm of Mn impurities and other nonmagnetic impurities such as 1000 ppm of Cu and 500 ppm of Si, by the following procedure. By dissociating $H_2$ molecules into atoms while getting dissolved in metals, the hydrogenated and deuterated samples are prepared by a gas absorption method [2]. The H content was quantified in accordance with the pressure–temperature phase diagram of V-H and V-D.

Then, the hydrogenated and deuterated sample wires were intimately fixed on the top of the flexible phosphor-bronze substrate after making notch at the center by a commercial blade under a microscope and mounted in a triple point bending configuration in a vacuum chamber. After cooling at liquid helium temperature, the sample wire was stretched by bending the substrate mechanically. The nanocontact was made by employing a homemade MCBJ apparatus [3,4]. Due to the application of bias voltage, a piezo element was then used to control the constriction size (a few 100 nm ~ atomic-sized) precisely. Thereafter, we have measured the spectra of differential conductance $dI/dV$ as a function of the bias voltage using a lock-in technique with a frequency of 1 kHz and performed the whole experiment at cryogenic environment to keep a stable contact long time. Moreover, the cryogenic environment diminishes the thermal fluctuation and also prevents any contaminants at the contact.

## 3. Results and Analysis

First, we have measured the temperature dependence of resistance in pure V constrictions by using LR700 AC resistance bridge (Linear Research Inc) [5]. The resistance starts to decrease at around 5.7 K, which is consistent with the superconducting transition temperature of V, and shows almost the zero resistivity at around 3 K as illustrated in figure 1. The essence of this measurement is to observe the nature of superconductivity after the material becomes nano-sized. The differential conductance for pure V nanoconstrictions recorded as a function of bias voltage at 2 K as depicted in figure 2. A huge conductance peak at zero bias is caused by the Josephson current flowing through the identical V electrodes, as confirmed by the corresponding $I$-$V$ characteristics shown in the inset. In figure 2, the conductance enhancements indicated by the black arrows in $dI/dV$ spectrum, referred as sub-gap structure (SGS), appear at voltages $2\Delta/e$ ( ~ 1.7 mV), $\Delta/e$ ( ~ 1.1 mV) and $2\Delta/3e$ ( ~ 0.7 mV). These SGS peaks can be explained in terms of the first, second and third order multiple Andreev reflection

(MAR) at SN interfaces of this SNS junction that was successfully discussed in OBTK model by M Octavio *et al.* [6]. However, the Δ gap value is not exactly half of the 2Δ gap. In this case, the slight increase of Δ gap may be due to the effect of geometrical instability of the contact giving rise to a mismatch of the quasiparticle wavefunction at the SN interface, because the contact is prepared by MCBJ technique. Similar inconsistencies found in some superconductor-normal metal (SN), superconductor-semiconductor interfaces [7-10].

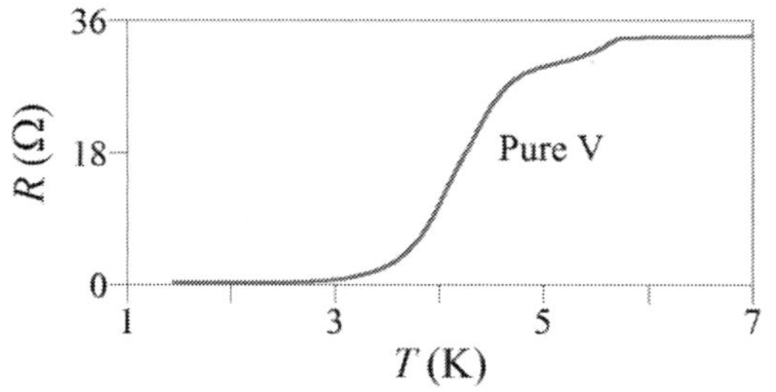


**Figure 1.** (Color online) Illustration of the zero resistivity as a function of temperature for pure V after being constricted in nano scale.

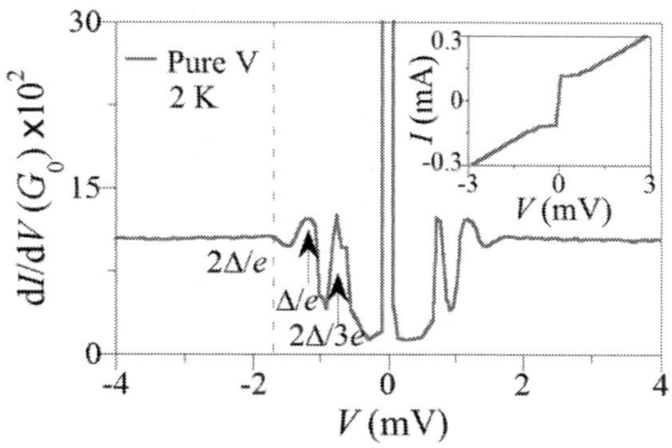


**Figure 2.** (Color online) The obtained d*I*/d*V* spectrum for a pure V constriction as a function of bias voltage. The arrows indicate the energies of the SGS's inside the gap. Inset shows the corresponding *I-V* characteristic.

After introducing a small proportion of atomic H and D on V nanoconstrictions, we measure the differential conductance to study their impurity effects. We show the temperature dependence of resistance for $VH_{0.05}$ and $VD_{0.05}$ in figure 3. In comparison with pure vanadium, the isotope effects in the resistivity are not so prominent for starting transition in *R-T* curves but the difference of the decrease trend of resistance shows the effect of atomic concentration of H and D in vanadium. Similar findings have been mentioned by Maier *et al.* [11]. In case of $VH_{0.05}$ and $VD_{0.05}$ constrictions with a contact size of about 9 nm, we present the spectra of d*I*/d*V* in figure 4. The conductance anomaly inside the superconducting energy gap 2Δ/*e* indicated by the dashed line, as can be obtained from the $R \sim T$ curves, is largely modified in both of these systems. Additionally, the corresponding Josephson currents in these nanosystems shown in the inset are affected especially at finite voltages by that amount of H and D impurities.

It is worthy to note that a series of peaks indicated by the black arrows appear outside the superconducting energy gap of 2Δ/*e*, referred as an over-the-gap structure (OGS), by introducing H and D impurities as in figure 4. On the other hand, the peak-shape anomalies due to SGS inside the gap are blurred, suggesting that the MAR is suppressed by H and D impurities. Although the anomalies due to OGS are clearly seen in the both systems, the bias voltage at the OGS in $VH_{0.05}$ is larger than that in $VD_{0.05}$. Similar anomalies mentioned experimentally in both the symmetric and asymmetric niobium (Nb) dimer contacts [12]. Moreover, the conductance spectra exhibits sharp anomalies at energies larger than the superconducting energy gap in various combinations between normal metals and low- and high-$T_C$ superconductors by a point contact Andreev reflection process [13-15]. Although a detailed satisfactory explanation of the origin of these dips is still lacking but they proposed the possible origin of such anomalies are the proximity effect [16] and the intergrain Josephson tunneling [17] for a contact made with a conventional *s*-wave superconductor.

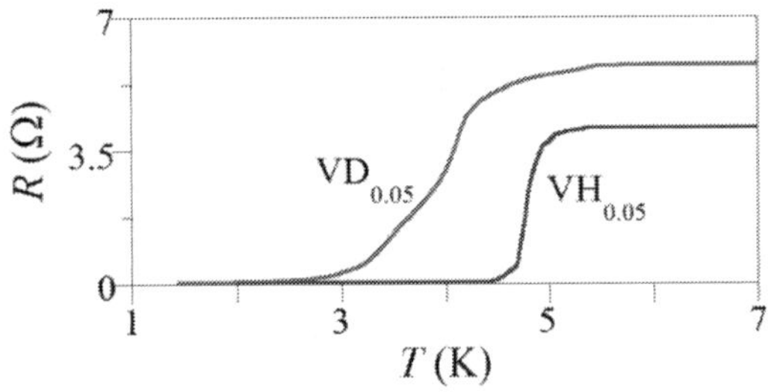


**Figure 3.** (Color online) The temperature dependence of the resistance for vanadium nanocontact with a five percent of atomic concentration of hydrogen and deuterium.

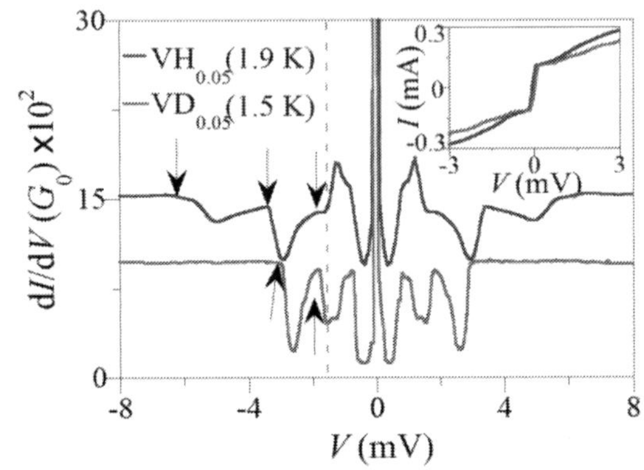


**Figure 4.** (Color online) The spectra of d$I$/d$V$ measured for 5% of atomic H and D in V constrictions. Both spectra show the SGS and OGS features and the clear observation of OGS's indicated by the arrows. Inset: *I-V* characteristics of the corresponding contacts.

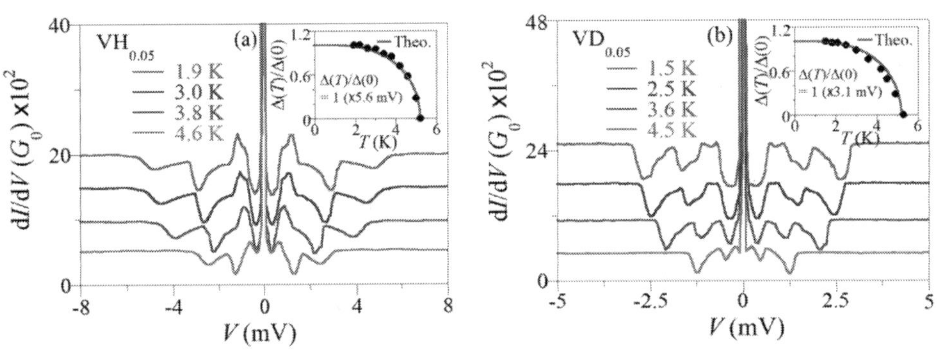


**Figure 5.** (Color online) Representative spectra of d$I$/d$V$ as a function of bias voltage for (a) $VH_{0.05}$ and (b) $VD_{0.05}$ depending on the temperature below $T_C$. Insets: the temperature dependence of the corresponding energy gap is fitted by the BCS gap function indicated by the solid red line.

To identify the origin of OGS's, the most authentic way is to measure their temperature dependence. The temperature dependence of d$I$/d$V$ spectra for $VH_{0.05}$ and $VD_{0.05}$ have been presented in figures 5 (a) and (b), respectively. Obviously, the positions at all anomalies due to OGS shift to lower bias voltages with increasing temperature in the both systems. Finally, the structures disappear when the temperature is increased above $T_C$. The corresponding normalized gap energies for the last OGS peak, shown in the insets, exactly follow the phenomenological approximate closed form of the expected BCS gap function near $T$ = 0 and = $T_C$ as indicated by the solid red lines [18]. According to Marchenkov *et al.*, the position of OGS peaks in Nb dimer nanocontact is independent of temperature but only the peak amplitudes reduce with temperature until $T_C$. They also proposed that such feature can be explained by the coupling resonance between the oscillating current and the dimer vibrational modes in both the symmetric and asymmetric contacts because of considering ac Josephson effect [12]. In our measurement, however, we only observed an over gap structure in $VH_{0.05}$ & $VD_{0.05}$ but not in pure vanadium, where the normal resistance of those contacts was at around 13 Ω. When the contact resistance is increased about 5 times (~ 65 Ω), the outside anomalies are only 2 times shifted at higher voltages. Moreover, the observed anomalies in $VH_{0.05}$ & $VD_{0.05}$ cannot be explained by the multiple Andreev reflection because those peaks are clearly inconsistent with the ratio $2\Delta/en$, where $n = 1, 2, 3$. In this context, we conclude that the OGS features for VH and VD nanosystems are due to the modification of quasiparticle excitation at finite voltages with a small proportion of H and D atoms. Although the observed gap energies are $2\Delta/e^{VH}(T = 1.9\text{ K}) = 5.6$ mV for VH and $2\Delta/e^{VD}(T = 1.5\text{ K}) = 3.1$ mV for VD nanosystems but the superconducting transition at 5.4 K, is almost similar for the both systems. The difference of modified gap energies for VH and VD system may be related to the effect of mass difference and spin interaction in vanadium metal.

When 30 percent of H and D are introduced in V nanoconstrictions, the d*I*/d*V* spectra have been reported in figure 6. The *I*-*V* character plotted in the inset shows that the both of the materials ($VH_{0.3}$ and $VD_{0.3}$) behave as a normal metal. Namely, such concentration completely destroys the superconductivity. From the X-ray diffraction study in VD system by Hardcastle *et al.* [19], a body-centered cubic (bcc) phase exists until 10% of atomic concentration of H and D dissolved into vanadium metal lattice but when the concentration increases from 10%, the structure is completely changed.

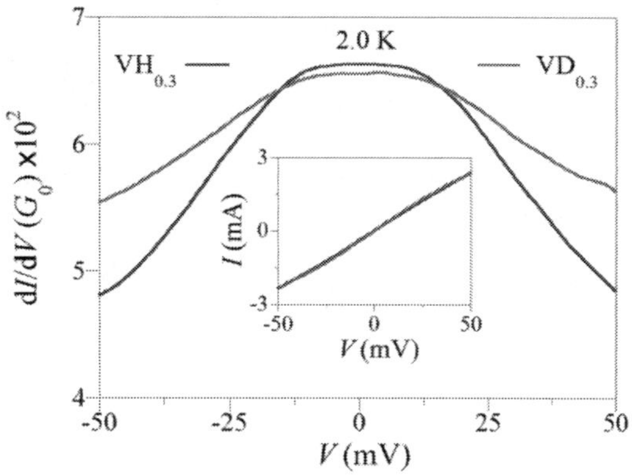


**Figure 6.** (Color online) The observed d*I*/d*V* spectra for 30 percent of atomic concentration of H and D in V nanoconstrictions exhibit normal metallic behavior even at 2 K.

## 4. Conclusion

We have measured the differential conductance d*I*/d*V* for Josephson contacts made by V nanoconstrictions with a different amount of H and D impurities. Below the superconducting transition temperature $T_C$, distinct peaks are observed within and outside the gap due to a small amount of H and D doping on V nanocontact. Moreover, the temperature dependence of d*I*/d*V* spectra represents that both structures are survived until the critical temperature $T_C$, which is consistent with BCS energy gap. On the other hand, a very high concentrating phase (30 atom%) behaves as a normal metallic state.

**Acknowledgments**

We would like to express our decent gratitude to the Japan Society for the Promotion of Science for providing the financial support by a Grant-in-Aid for Scientific Research (Grant Nos. 25220605, 25287076 and 26600102).

## References


[1] Drozdov A P, Eremets M I, Troyan I A, Ksenofontov V and Shylin S I 2015 *Nature* **525** 73
[2] Veleckis E and Edwards R K 1969 *J. Phys. Chem.* **73** 683
[3] Agraït N, Yeyati A L and van Ruitenbeek J M 2003 *Phys. Rep.* **377** 81
[4] Ienaga K, Nakashima N, Inagaki Y, Tsujii H, Honda S, Kimura T and Kawae T 2012 *Phys. Rev.* B **86** 064404
[5] Ienaga K, Nakashima N, Inagaki Y, Tsujii H, Kimura T and Kawae T 2012 *Appl. Phys. Lett.* **101** 123114
[6] Octavio M, Tinkham M, Blonder G E and Klapwijk T M 1983 *Phys. Rev.* B **27** 6739
[7] van Son P C, van Kempen H and Wyder P 1988 *Phys. Rev.* B **37** 5015
[8] Riedel R A and Bagwell P F 1993 *Phys. Rev.* B **48** 15198
[9] Günel H Y, Borgwardt N, Batov I E, Hardtdegen H, Sladek K, Panaitov G, Grützmacher D and Schäpers Th 2014 *Nano Lett.* **14** 4977
[10] Lachenmann S G, Forster A, Friedrich I, Uhlisch D, Schapers Th, Kastalsky A and Golubov A A 1998 *Appl. Supercond.* **6** 681
[11] Maier W and Wipf H 1977 *Scr. Mater.* **11** 695
[12] Marchenkov A, Dai Z, Donehoo B, Barnett R N and Landman U 2007 *Nature Nanotech.* **2** 481
[13] Srikanth H and Raychaudhuri A K 1992 *Phys. Rev.* B **46** 14713
[14] Soulen Jr. R J *et al.* 1998 *Science* **282** 85
[15] Li Z-Z, Tao H-J, Xuan Y, Ren Z-A, Che G-C and Zhao B-R 2002 *Phys. Rev.* B **66** 064513
[16] Strijkers G J, Ji Y, Yang F Y, Chien C L and Byers J M 2001 *Phys. Rev.* B **63** 104510
[17] Shan L, Tao H J, Gao H, Li Z Z, Ren Z A, Che G C and Wen H H 2003 *Phys. Rev.* B **68** 144510
[18] Romeo F, Giubileo F, Citro R, Bartolomeo A D, Attanasio C, Cirillo C, Polcari A and Romano P 2015 *Sci. Rep.* **5** 17544
[19] Hardcastle K I and Gibb Jr. T R P 1972 *J. Phys. Chem.* **76** 927

# Little-Parks oscillations in superconducting ring with Josephson junctions

**Omri J. Sharon[1], Amos Sharoni[1], Jorge Berger[2], Avner Shaulov[1], and Yosi Yeshurun[1]**

[1] Department of Physics and Institute of Nano Technology, Bar-Ilan University, 5290002 Ramat-Gan, Israel

[2] Department of Physics, Ort Braude College, 21982 Karmiel, Israel

omri_s@yahoo.com

**Abstract**. Nb nano-rings connected serially by Nb wires exhibit, at low bias currents, the typical parabolic Little-Parks magnetoresistance oscillations. As the bias current increases, these oscillations become sinusoidal. This result is ascribed to the generation of Josephson junctions caused by the combined effect of current-induced phase slips and the non-uniformity of the order parameter along each ring due to the Nb wires attached to it. This interpretation is validated by further increasing the bias current, which results in magnetoresistance oscillations typical of a SQUID.

## 1. Introduction

The Little-Parks magnetoresistance oscillations in superconducting rings have been extensively studied both theoretically and experimentally, see e.g. [1-13]. However, quite frequently the observed oscillations' waveform deviates from the predicted parabolic oscillations, exhibiting sinusoidal-like oscillations, see e.g. [2-4]. Such deviations can be related to a distribution of the ratio $\xi/R$ in a wide ring [2] ($\xi$ is the coherence length and R is the radius of the ring), or to a size distribution of rings in a network [3]. In this article we propose an alternative explanation associated with the existence of Josephson junctions (JJ) in a ring. The existence of such junctions is highly probable in superconducting nano-rings with superconducting leads ('arms') attached to them. Such superconducting structures comprising a ring with two arms are common in nano-fabrication in which the arms serve as leads to the ring. As shown by de-Gennes [14] and Alexander [15] the arms cause a non-uniform order parameter along the ring with two minima at equal distances from the connection points of the arms to the ring. In the presence of large enough bias-current, enhanced phase slips at these minima can generate Josephson junctions in the ring [16]. In this paper we show that in such a ring with Josephson junctions, the Little-Parks (LP) oscillations should become sinusoidal. We demonstrate this effect in Nb nano-rings by showing that the parabolic Little-Parks oscillations at low bias currents are switched into sinusoidal oscillations by increasing the bias current.

## 2. Experimental

E-beam lithography was used to fabricate Nb square loops (340x340 $nm^2$) connected serially by 66 nm wide Nb wires, see right panel of Figure 1. The ring rim (~57 nm) is of the order of the zero temperature coherence length in Nb, $\xi_0$ = 40 nm. For details of the fabrication process see Ref. [16]. Magnetoresistance measurements were performed using a commercial system (PPMS, Quantum-Design).

## 3. Results

Current-induced switching of the classical LP parabolic oscillations into sinusoidal ones is demonstrated in the left panel of Figure 1 which shows typical magnetoresistance oscillations measured at T = 7.1 K. At low currents (1 μA and below) parabolic LP oscillations are obtained [1], exhibiting upward cusps at odd multiples of $\Phi_0/2$, and a field-period of ~ 170 Oe, corresponding to the area of a single ring (~$1.2\cdot10^{-9}$ cm$^2$). For higher currents, (2 - 3 μA), the cusps disappear and the oscillations become sinusoidal. As we argue below, this change results from generation of Josephson junctions in the rings. A clear manifestation of the existence of these junctions is obtained by further increasing the current to 4 μA, yielding oscillations with downward cusps at multiples of $\Phi_0$, typical of the magnetoresistance response of a SQUID biased with a current that is equal to its maximum critical current [17].

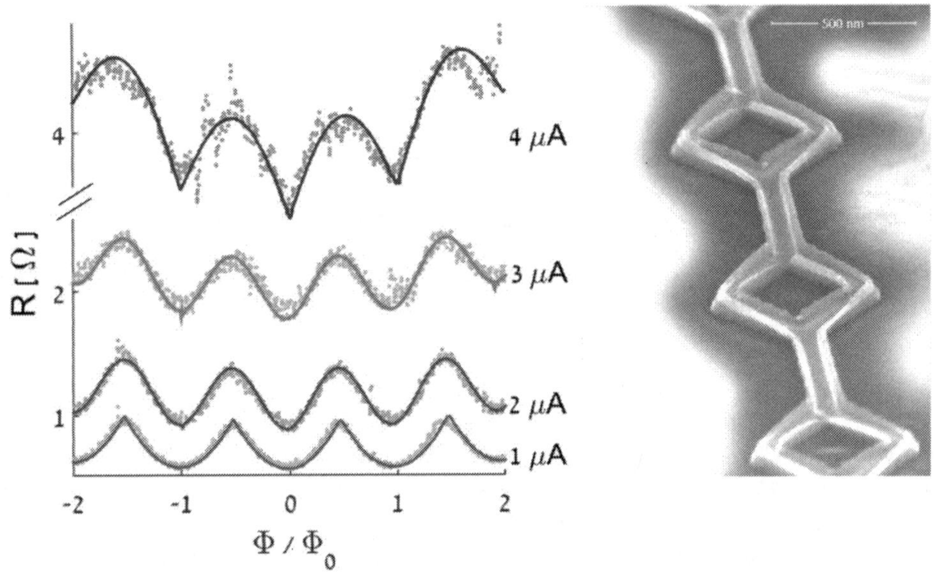


Figure 1. Left panel: Magnetoresistance of a single Nb ring measured at T = 7.1 K with currents between 1 and 4 μA. The magnetic flux, Φ, is normalized to the quantum flux, $\Phi_0$, and calculated for a ring area of $1.2\cdot10^{-9}$ cm$^2$. The solid curves through the data points are guide to the eye. Right panel: A scanning electron microscope image of the Nb rings.

## 4. Discussion

We begin by considering the effect of a single arm on a ring. As the arm is not affected by the magnetic flux, the order parameter along the ring has a maximum at the connection point and a minimum at the antipodal point. This minimum drops to zero when the flux becomes equal to a half flux quantum, $\Phi = \Phi_0/2$ [18, 19]. When two symmetrical arms are connected to a ring, the order parameter is maximum at the connection points and minimum at equal distances from these points [18, 19]. Current-induced phase slips at these minima further reduce the order parameter down to a level required for the creation of effective Josephson junctions.

We recall that screening current in a ring without Josephson junctions is linear with the magnetic flux, with discontinuities at odd multiplications of $\Phi_0/2$:

$$(2)\ I_s^{LP-JJ} = I_c \sin(\pi\Phi/\Phi_0)\cos(\pi n)\,;\ \left(n-\tfrac{1}{2}\right)\Phi_0 \le \Phi \le \left(n+\tfrac{1}{2}\right)\Phi_0, n = 0, \pm 1, \pm 2, \ldots$$

as shown by the dashed line in Figure 2. Consequently, the magnetoresistance, which is proportional to the kinetic energy $I_s^2$ [20-22] is parabolic (see the dashed line in Figure 3). However, for a ring with Josephson junctions of critical current $I_c$, the dependence of the screening current $I_s^{(LP\text{-}JJ)}$ on $\Phi/\Phi_0$ is [23]:

$$(2)\ I_s^{LP-JJ} = I_c \sin(\pi\Phi/\Phi_0)\cos(\pi n)\,;\ \left(n-\tfrac{1}{2}\right)\Phi_0 \le \Phi \le \left(n+\tfrac{1}{2}\right)\Phi_0, n = 0, \pm 1, \pm 2, \ldots$$

as described by the solid line in Figure 2. As a result, the magnetoresistance is sinusoidal, as shown by the solid line in Figure 3. The switching of the magnetoresistance waveform from parabolic to sinusoidal in our data can, therefore, be ascribed to the generation of Josephson junctions in the ring due to the combined effects of current induced phase slips and non-uniform order parameter along the ring caused by the superconducting arms. The existence of the Josephson junctions in the ring is clearly manifested by the SQUID-like magnetoresistance oscillations, with cusps down, obtained when the current is increased to 4 μA (see Fig. 1). At this current the magnetoresistance oscillations result from the flux dependence of the critical current of the SQUID rather than by oscillation of the critical temperature due to oscillations of the screening current.

Note that in a conventional SQUID it is assumed that the rim width is larger than the superconducting penetration depth, λ, and, therefore, the Little-Parks effect is unobservable. However, in our case, as is usually the case in most nano-rings, the rim width is smaller than λ. Thus, magnetoresistance oscillations due to Little-Parks effect, Eq. (2), are expected in such SQUIDs near Tc.

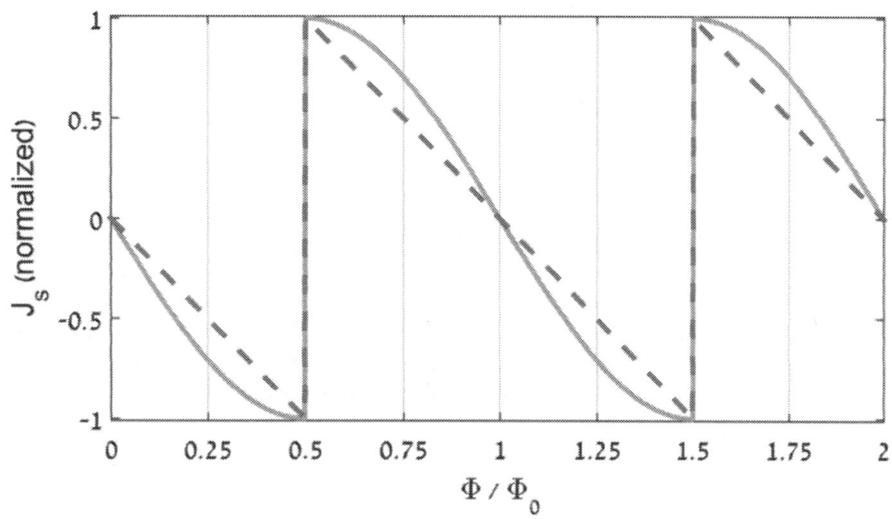


Figure 2. Flux dependence of the screening current in homogeneous ring and in a ring with Josephson junctions (dashed and solid lines, respectively).

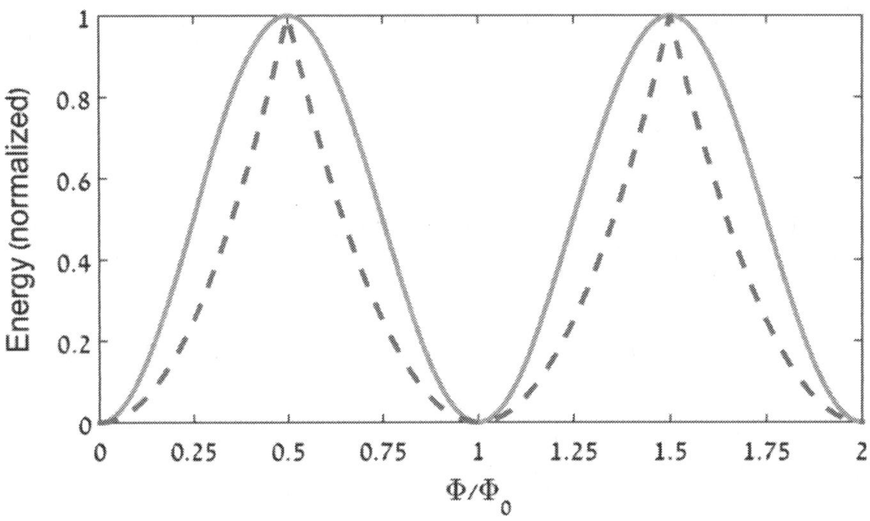


Figure 3. Flux dependence of the superconducting kinetic energy in homogeneous ring and in a ring with Josephson junctions (dashed and solid lines, respectively).

## 5. Summary and Conclusion

Little-Parks oscillations can transform from parabolic to sinusoidal when a Josephson junction is generated in the superconducting loop. We demonstrated that such an effect can be induced by external current in Nb nano-rings with two arms. Moreover, we demonstrated that such a ring exhibits SQUID-like magnetoresistance oscillations when large enough bias current is applied.

## Acknowledgement

This research was supported by a Grant from the GIF, the German-Israeli Foundation for Scientific Research and Development.

## References

[1] Little W and Parks R 1962 Observation of quantum periodicity in the transition temperature of a superconducting cylinder *Physical Review Letters* **9** 9

[2] Carillo F, Papari G, Stornaiuolo D, Born D, Montemurro D, Pingue P, Beltram F and Tafuri F 2010 Little-Parks effect in single nanoscale $YBa_2Cu_3O_{6+x}$ rings *Physical Review B* **81** 054505

[3] Sochnikov I, Shaulov A, Yeshurun Y, Logvenov G and Božović I 2010 Large oscillations of the magnetoresistance in nanopatterned high-temperature superconducting films *Nature Nanotechnology* **5** 516-9

[4] Gammel P, Polakos P, Rice C, Harriott L and Bishop D 1990 Little-Parks oscillations of $T_c$ in patterned microstructures of the oxide superconductor $YBa_2Cu_3O_7$: Experimental limits on fractional-statistics-particle theories *Physical Review B* **41** 2593

[5] Israeloff N, Yu F, Goldman A and Bojko R 1994 Nonlocal conductance of small superconducting loops in the paraconductive regime *Physica B: Condensed Matter* **203** 454-9

[6] Vodolazov D Y, Golubović D, Peeters F and Moshchalkov V 2007 Enhancement and decrease of critical current due to suppression of superconductivity by a magnetic field *Physical Review B* **76** 134505

IOP Conf. Series: Journal of Physics: Conf. Series **969** (2018) 012047 doi:10.1088/1742-6596/969/1/012047

[7] Snyder S, Dunn T, Erickson M, Kinney J, Lee Y, Nelson J and Goldman A 2013 High-resistance state of phase coherent nanoscale thin-film Al superconducting nanorings in magnetic fields *Physical Review B* **87** 144503

[8] Burlakov A, Gurtovoĭ V, Dubonos S, Nikulov A V and Tulin V 2007 Little-parks effect in a system of asymmetric superconducting rings *JETP Letters* **86** 517-21

[9] Gurtovoi V, Ilin A, Nikulov A and Tulin V 2010 Weak dissipation does not result in the disappearance of the persistent current *Low Temperature Physics* **36** 974-81

[10] Michotte S, Lucot D and Mailly D 2010 Fluxoid quantization in the critical current of a niobium superconducting loop far below the critical temperature *Physical Review B* **81** 100503

[11] Berdiyorov G, Milošević M and Peeters F 2010 Vortices induced in a superconducting loop by asymmetric kinetic inductance and their detection in transport measurements *Physical Review B* **81** 144511

[12] Burlakov A, Gurtovoi V, Ilin A, Nikulov A and Tulin V 2012 Possibility of persistent voltage observation in a system of asymmetric superconducting rings *Physics Letters A* **376** 2325-9

[13] Moshchalkov V, Gielen L, Dhallé M, Van Haesendonck C and Bruynseraede Y 1993 Quantum interference in a mesoscopic superconducting loop *Nature* **361** 617-20

[14] de Gennes P-G 1981 and C. R. Acad. Sci. Ser. B 292, 9 (1981) Suspensions colloïdales dans un mélange binaire critique *C. R. Acad. Sci. B* **292** 279

[15] Alexander S 1983 Superconductivity of networks. A percolation approach to the effects of disorder *Physical Review B* **27** 1541

[16] Sharon O J, Shaulov A, Berger J, Sharoni A and Yeshurun Y 2016 Current-induced SQUID behavior of superconducting Nb nano-rings *Scientific Reports* **6**

[17] Clarke J and Braginski A I, (Editors) 2006 *The SQUID Handbook, Volume I: Fundamentals and Technology of SQUIDs and SQUID Systems*: Wiley VCH Verlag GmbH & Co. KGaA, Weinheim)

[18] Kato M and Sato O 2013 Magnetic flux structures of finite superconducting networks *Superconductor Science and Technology* **26** 33001-26

[19] Rammal R, Lubensky T and Toulouse G 1983 Superconducting networks in a magnetic field *Physical Review B* **27** 2820

[20] Behrooz A, Burns M, Levine D, Whitehead B and Chaikin P 1987 Superconducting phase boundary of aperiodic networks in a magnetic field *Physical Review B* **35** 8396

[21] Chaikin P, Behrooz A, Itzler M, Wilks C, Whitehead B, Grest G and Levine D 1988 Studies of quasicrystalline superconducting networks *Physica B: Condensed Matter* **152** 113-24

[22] Itzler M, Behrooz A, Wilks C, Bojko R and Chaikin P 1990 Commensurate states in disordered networks *Physical Review B* **42** 8319

[23] Rose-Innes A. C. and Rhoderick E. H. 1978 *Introduction to Superconductivity, 2nd edition*: Pergamon Press New York

# Fluxoids behavior in superconducting ladders

**Omri J. Sharon, Noam Haham, Avner Shaulov and Yosef Yeshurun**

Department of Physics and Institute of Nano Technology, Bar-Ilan University, 5290002 Ramat-Gan, Israel

omri_s@yahoo.com

**Abstract**. The nature of the interaction between fluxoids and between them and the external magnetic field is studied in one-dimensional superconducting networks. An Ising like expression is derived for the energy of a network revealing that fluxoids behave as repulsively interacting objects driven towards the network center by the effective applied field. Competition between these two interactions determines the equilibrium arrangement of fluxoids in the network as a function of the applied field. It is demonstrated that the fluxoids configurations are not always commensurate to the network symmetry. Incommensurate, degenerated configurations may be formed even in networks with an odd number of loops.

## 1. Introduction

The macroscopic quantum nature of superconductivity is manifested in loops, and generally in multiply-connected superconductors, in quantization of the 'fluxoid' defined as: $(4\pi\lambda^2/c)\oint j \cdot dl + \Phi$, where j is the shielding current in a closed loop, λ is the penetration depth, and Φ is the magnetic flux threading the loop [1]. In the early days of superconductivity, it was predicted by Fritz London [2], and later confirmed experimentally by Little and Parks [3], that the fluxoid must be an integer multiple of the flux quantum $\phi_0 = hc/2e$. Fluxoid quantization effects have been studied extensively, both theoretically and experimentally, in a variety of superconducting networks [4-12]. However, most of these studies adopt the mean field approach, providing no intuitive understanding of the interaction between fluxoids and the mechanism governing their arrangement in superconducting networks. The purpose of the present work is to elucidate the nature of the interaction between fluxoids and to clarify the physics behind their arrangements in networks. Analyzing the simplest case of a superconducting one-dimensional network ('ladder'), we show that fluxoids act as repulsively interacting objects dragged towards the ladder center by their interaction with the externally applied field. A competition between these two interactions determines the equilibrium positions of the fluxoids in the network as a function of the applied field. To demonstrate this concept we present calculated results of fluxoid arrangements in several examples of finite 1D and 2D networks.

## 2. Analysis

Our analysis is based on the "current squared" model (known as the "$J^2$ model") [6], In which the kinetic energy of the network is calculated as the sum of the squared currents over all the network wires. The number and arrangement of the fluxoids is determined by the requirement of minimum energy.

Consider a superconducting ladder of finite length, consisting of N square loops of unit side, as shown in Figure 1.

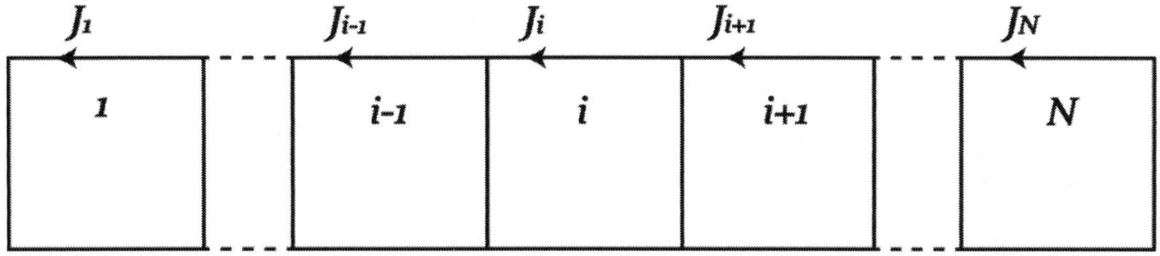


Figure 1: Finite ladder consisting of N square loops.

The fluxoid quantization equation for the loop $i$ reads:

$$4J_i - J_{i-1} - J_{i+1} = n_i - \phi/\phi_0, \tag{1}$$

where $n_i$ is the vorticity of the loop $i$, $\phi$ is the flux threading this loop, and by definition $J_i = 0$ for $0 > i$ or $i > N$. For simplicity, the coefficient $4\pi\lambda^2/c$ is taken as 1. According to Eq. (1), the set of the fluxoid quantization equations for all the loops can be written as a matrix equation:

$$\hat{A} \cdot \vec{J} = \vec{n} - \frac{\vec{\phi}}{\phi_0}, \tag{2}$$

where the elements of the matrix $\hat{A}$: $A_{ij} = 4\delta_{i,j} - \delta_{i,j-1} - \delta_{i,j+1}$, and $\delta_{i,j}$ is the Kronecker $\delta$. The current vector $\vec{J}$ can be calculated from Eq. (2) by inversion:

$$\vec{J} = \hat{A}^{-1}(\vec{n} - \frac{\vec{\phi}}{\phi_0}). \tag{3}$$

Denoting the matrix $\hat{A}^{-1}$ as $\hat{B}$, Eq. (3) can be written as a set of equations:

$$J_i = \sum_{j=1}^{N} B_{ij}\left(n_j - \frac{\phi}{\phi_0}\right), \quad i = 1..N. \tag{4}$$

Using the $J^2$ model, knowledge of $J_i$ allows calculation of the energy $E_i$ of the loop $i$:

$$\begin{aligned} E_i &= 2J_i^2 + \frac{1}{2}(J_i - J_{i-1})^2 + \frac{1}{2}(J_i - J_{i+1})^2 + \frac{1}{2}J_1^2\delta_{1,i} + \frac{1}{2}J_N^2\delta_{N,i} \\ &= J_i[3J_i - J_{i-1} - J_{i+1}] + \frac{1}{2}J_{i-1}^2 + \frac{1}{2}J_{i+1}^2 + \frac{1}{2}J_1^2\delta_{1,i} + \frac{1}{2}J_N^2\delta_{N,i}, \end{aligned}$$

and the total energy $E$ of the network:

$$E = \sum_{i=1}^{N} E_i = \sum_{i=1}^{N}\left\{J_i(3J_i - J_{i-1} - J_{i+1}) + \frac{1}{2}J_{i-1}^2 + \frac{1}{2}J_{i+1}^2\right\} + \frac{1}{2}J_1^2 + \frac{1}{2}J_N^2. \tag{5}$$

Using Eq. (1) and realizing that $\sum_{i=1}^{N}\left\{-J_i^2 + \frac{1}{2}J_{i-1}^2 + \frac{1}{2}J_{i+1}^2\right\} + \frac{1}{2}J_1^2 + \frac{1}{2}J_N^2 = 0$, Eq. (5) becomes

$$E = \sum_{i=1}^{N} J_i\left(n_i - \frac{\phi}{\phi_0}\right). \tag{6}$$

Inserting $J_i$ from Eq. (4) yields

$$E = \sum_{i=1}^{N}\sum_{j=1}^{N} B_{ij}\left(n_j - \frac{\phi}{\phi_0}\right)\left(n_i - \frac{\phi}{\phi_0}\right) = \sum_{ij} B_{ij}\left(n_i n_j - 2\frac{\phi}{\phi_0}n_i + \left(\frac{\phi}{\phi_0}\right)^2\right). \tag{7}$$

The above expression for the total energy, $E$, is reminiscent of the Ising model for the energy of a spin configuration, having the form $\sum_{ij} J_{ij} S_i S_j$ - $\mu \sum_j h_j S_j$ [13]; $n_j$, and $B_{ij}$ playing the role of the Ising spin $S_j$ and the exchange interaction term $J_{ij}$, respectively. The first term on the right hand side of Eq. (7), $\left(\sum_{ij} B_{ij} n_i n_j\right)$, represents the interaction between fluxoids, including the self-interactions $\sum_i B_{ii} n_i^2$. The second term, $\left(-2\frac{\phi}{\phi_0}\sum_{ij} n_i B_{ij}\right)$, expresses the interaction between the fluxoids and the effective magnetic field. The third term, $(\phi/\phi_0)^2 \sum_{ij} B_{ij}$ , is constant, independent of the vorticities, and thus may be ignored.

For the matrix $\hat{A}$ given in Eq. (2), $\hat{B} = \hat{A}^{-1}$ is a symmetric matrix with elements [14]:

$$B_{ij} = C\gamma^{i-j-1}\left(1-\eta^{i}\right)\left(1-\eta^{N+1-j}\right), \text{ for } i \le j. \tag{8}$$

where $\gamma = 2+\sqrt{3}$ , $\eta = (2-\sqrt{3})/(2+\sqrt{3})$ and $C = 1/(1-\eta)(1-\eta^{N+1})$. Due to the symmetry of $\hat{B}$, $B_{ij}$ for $i > j$ can be calculated as $B_{ji}$ using Eq. (8). Since $\eta \ll 1$, C is approximately 1 and $B_{ij}$ can be approximated as $\gamma^{-(|i-j|+1)}$, for all $i$ and $j$. Thus Eq. (7) becomes

$$E = \sum_{ij} \gamma^{-(|i-j|+1)}\left(n_i n_j - 2\frac{\phi}{\phi_0}\, n_i + \left(\frac{\phi}{\phi_0}\right)^2\right) . \tag{9}$$

The above expression shows that fluxoids can be treated as repulsively interacting objects, with interaction energy decreasing exponentially with their separation. In order to minimize the total energy, the repulsive interaction between fluxoids tends to keep them away from each other. On the other hand, the interaction between the fluxoids and the effective magnetic field, represented by the second term in Eq. (9), $\left(-2\phi/\phi_0 \sum_i n_i \sum_j \gamma_1^{-(|i-j|+1)}\right)$ tends to drive the fluxoids away from the network edges towards the network's center. This is because the effective magnetic field $-2(\phi/\phi_0)\sum_j \gamma_1^{-(|i-j|+1)} \propto \cosh\left(\ln(\gamma_1)\left((N+1)/2 - i\right)\right)$ is minimum at the center of the ladder $(i = (N+1)/2)$. Thus, while the external magnetic favors assembling the fluxoids near the ladder center, the fluxoids repel each other tending to keep themselves apart. Competition between these two opposite interactions determines the equilibrium arrangement of fluxoids in the network as a function of the applied field.

Considering the first fluxoid which enters the ladder, it always appears at the center of the network (or next to it, in a ladder with an even number of loops) as it is affected only by the external field which drives it to the center. As the field increases, a second fluxoid appears, pushing the first one out of its central position and both fluxoids arrange themselves in an optimum configuration around the center, keeping apart from each other and away from the network edges. The same principle determines the arrangements of the next fluxoids entering the ladder as the field further increases.

Our analysis of superconducting ladders can be extended to two dimensional superconducting networks. The basic idea that the fluxoid arrangements are determined by a competition between the fluxoid repulsive interaction and their interaction with the applied field, remains the same. However, computations of these interactions, and the resulting fluxoid equilibrium configurations in two dimensional networks become more complicated. The occupation process in 1D and 2D superconducting networks are demonstrated in the following section.

## 3. Simulations

As an example, we present calculated results for ladders with 7, 8 and 9 loops. In each case, the energy of the ladder, as a function of the loops vorticities and the external field, was calculated using Eq. (7) and the exact expression for the elements $B_{ij}$ (Eq. 8). For each given field the fluxoid arrangement $(n_1, n_2, \ldots, n_N)$ which minimizes the energy was determined.

IOP Conf. Series: Journal of Physics: Conf. Series **969** (2018) 012048 doi:10.1088/1742-6596/969/1/012048

The solid, dashed, dashed-dotted curves in Figure 2 show the minimum energy as a function of the normalized flux $\phi/\phi_0$ in ladders with 7,8 and 9 loops, respectively. The crests in each curve indicate a change in the fluxoids configurations in the ladder. Thus, in the ladders with 7,8 and 9 elements the total number of configurations is 7,8 and 9, respectively. It is interesting to note that the number of configurations is not necessarily equal to the number of elements. For example, in ladders with rectangular loops attached along their long side, a change of the applied field can cause rearrangement of the same number of fluxoids, giving rise to an access number of configurations [15].

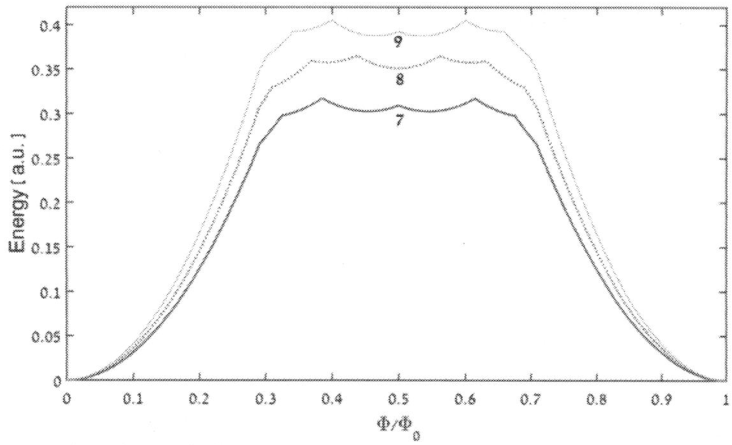


Figure 2: Energy as a function of the normalized magnetic flux in ladders with 7, 8 and 9 loops.

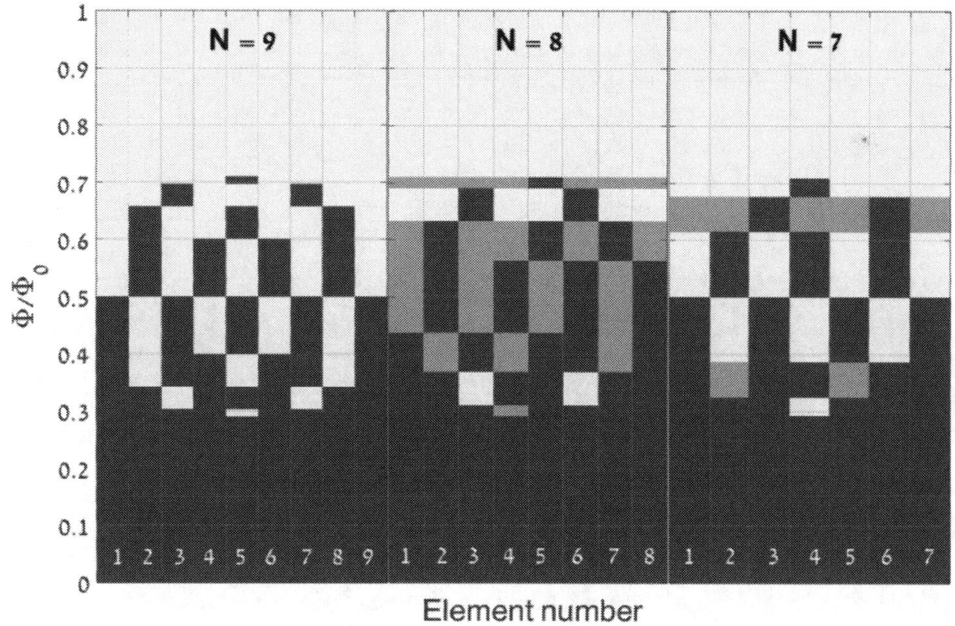


Figure 3: Fluxoid configuration as a function of magnetic flux in ladders with 7, 8 and 9 loops, in the first period. An empty loop is colored dark blue, and occupied loop is colored yellow or green. The green color indicates degenerated configurations which are incommensurate with the symmetry of the ladder.

The fluxoids arrangements as a function of field are illustrated in Figure 3. An empty loop is colored blue, and occupied loop is colored yellow or green. The green color indicates degenerated configurations which are incommensurate with the symmetry of the ladder. The ladders with 7 and 9 loops demonstrate the general rule that when the number of loops is odd, the first fluxoid always occupies the loop at the ladder's center. In ladders with an even number of loops, the first fluxoid occupies a degenerated state on either side of the center, as demonstrated by the ladder with 8 loops. As the field increases, a second fluxoid enters the ladder, pushing the first one out of its position and both fluxoids arrange themselves in an optimum configuration, keeping apart from each other and away from the network edges. In the ladder with 9 loops, this configuration conforms the symmetry of the ladder, however, this is not the case in the ladders with 7 and 8 elements. As more fluxoids enter the ladder with increasing field, rearrangement of fluxoids continues until the last fluxoid enters the ladder's center completing one period in which each loop is occupied with one fluxoid. Occupation of the loops in the following periods follows the same pattern.

Figure 4 illustrates calculated results for a 3x3 square network, based on the $J^2$ model. Note that the first fluxoid appears in the central loop of the network, as in a ladder. Also note that the number of different configurations (11) exceeds the number of loops in the network, due to rearrangement of the same number of fluxoids as the field increases. This situation occurs in configurations of 3 and 6 fluxoids. Among the 11 different configurations, there are 6 degenerated states that are incommensurate to the network symmetry (marked in green in Fig. 4). The degenerated configurations are obtained by applying the symmetry operations of the network. Thus, two degenerated configurations correspond to N=2,7, and four degenerated configurations correspond to each configuration with N=3,6.

We note that calculations based on the de-Gennes-Alexander equations for a network yield quite different results [16]. For example, following the appearance of the first fluxoid at the network central loop, the second fluxoid appears at the same loop creating a double fluxoid at the network center. This configuration has a higher energy than that the configuration of two separated fluxoids derived from the '$J^2$ model' (see Fig. 4). Involving the appearance of anti-fluxoids in the network, the calculation based on the de-Gennes-Alexander equations, predicts 9 configurations all of which are commensurate to the network symmetry. It should be noted, however, that by minimizing the Ginzburg–Landau free energy, asymmetric fluxoid patterns have been reported for a 10x10 network [17]. Finally we note that in our calculations based on the $J^2$ model, the fluxoid configurations are temperature independent, as the only temperature dependent factor in this model is $4\pi\lambda^2/c$ which scales the current density, and the square of this factor scales the energy.

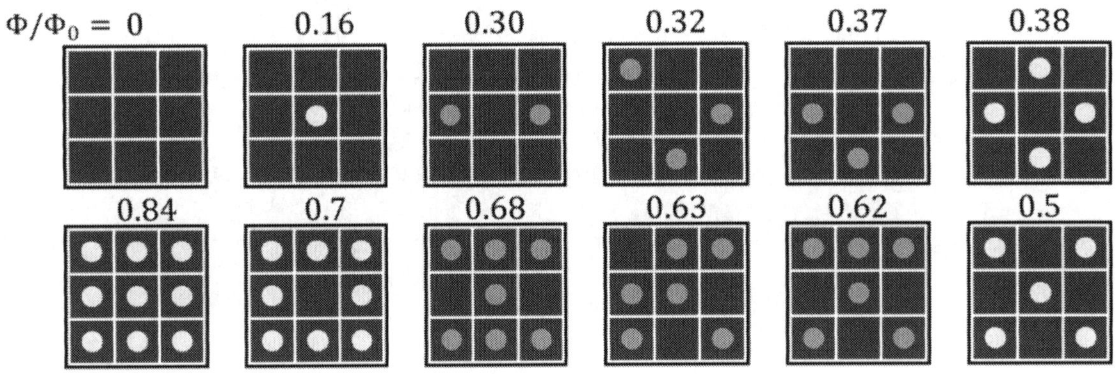


Figure 4: Fluxoid configurations in 3x3 square network calculated in the framework of the '$J^2$ model'. An empty loop is colored dark blue, and occupied loop is colored yellow or green. The green color indicates degenerated configurations which are incommensurate with the symmetry of the ladder.

IOP Conf. Series: Journal of Physics: Conf. Series **969** (2018) 012048 doi:10.1088/1742-6596/969/1/012048

## 4. Summary and conclusions

An Ising-like expression derived for the energy of fluxoids in a 1D superconducting network reveals that the fluxoids act as repulsively interacting objects with an interaction energy that decreases exponentially with their relative separation. In this expression, the effective magnetic field drives the fluxoids toward the network center. The competition between these two interactions determines the equilibrium configuration of the fluxoids in the ladders. These configurations may be incommensurate to the symmetry of the ladder, in ladders with even as well as odd number of loops. Fluxoids in 2D networks follow a similar pattern, i.e. they repel each other and are driven to the center by the applied field.

## Acknowledgement

This research was supported by the German-Israeli Foundation (GIF).

## References


[1] Tinkham M 1996 *Introduction to superconductivity* 2nd edition: McGraw-Hill, Inc.

[2] London F 1948 On the Problem of the Molecular Theory of Superconductivity *Physical Review* **74** 562-73

[3] Parks R and Little W 1964 Fluxoid quantization in a multiply-connected superconductor *Physical Review* **133** A97

[4] Pannetier B, Chaussy J, Rammal R and Villegier J 1984 Experimental fine tuning of frustration: Two-dimensional superconducting network in a magnetic field *Physical review letters* **53** 1845

[5] Behrooz A, Burns M, Levine D, Whitehead B and Chaikin P 1987 Superconducting phase boundary of aperiodic networks in a magnetic field *Physical Review B* **35** 8396

[6] Chaikin P, Behrooz A, Itzler M, Wilks C, Whitehead B, Grest G and Levine D 1988 Studies of quasicrystalline superconducting networks *Physica B: Condensed Matter* **152** 113-24

[7] Gordon J M, Goldman A M and Whitehead B 1987 Dimensionality crossover in superconducting wire networks *Physical review letters* **59** 2311

[8] Chi C, Santhanam P and Blöchl P 1992 Interacting loop-current model of superconducting networks *Journal of low temperature physics* **88** 163-95

[9] Bruyndoncx V, Strunk C, Moshchalkov V, Van Haesendonck C and Bruynseraede Y 1996 Fluxoid quantization effects in superconducting mesoscopic Al multiloop structures *EPL (Europhysics Letters)* **36** 449

[10] Reich D H, Silevitch D M, Chien C, Davidovic D and Field S 2000 Disorder and correlations in extended superconducting nanostructures *Journal of Alloys and Compounds* **303** 245-51

[11] Burlakov A, Gurtovoĭ V, Dubonos S, Nikulov A V and Tulin V 2007 Little-parks effect in a system of asymmetric superconducting rings *JETP Letters* **86** 517-21

[12] Sochnikov I, Shaulov A, Yeshurun Y, Logvenov G and Božović I 2010 Large oscillations of the magnetoresistance in nanopatterned high-temperature superconducting films *Nature Nanotechnology* **5** 516-9

[13] Giles R T and Kusmartsev F V 2000 Critical fields and devil's staircase in superconducting ladders *Physica B: Condensed Matter* **284** 1850-1

[14] Da Fonseca C and Petronilho J 2005 Explicit inverse of a tridiagonal k− Toeplitz matrix *Numerische Mathematik* **100** 457-82

[15] Sharon O J, Haham N, Shaulov A A and Yeshurun Y 2017 Fluxoids configurations in finite superconducting networks *Unpublished*

[16] Kato M and Sato O 2013 Magnetic flux structures of finite superconducting networks *Superconductor Science and Technology* **26** 33001-26

[17] Hayashi M, Ebisawa H and Kato M 2005 Vortex structures in mesoscopic superconducting network systems *Physica C: Superconductivity* **426** 136-40

# Design versions of HTS three-phase cables with the minimized value of AC losses

**V A Altov[1], N N Balashov[1], P N Degtyarenko[1, 2], S S Ivanov[1], S I Kopylov[1], DA Lipa[3], S V Samoilenkov[1, 2], V E Sytnikov[4, 5], V V Zheltov[1]**

[1]Joint Institute for High Temperatures, Russian Academy of Science, Izhorskaya str, 13, bd. 2, 125412 Moscow, Russia

[2]Joint Stock Company "SuperOx", 20-2 Nauchnyi proezd, 117246 Moscow, Russia

[3]FSBEI HE RGAZU, Entuziastov shosse, bld. 50, 143900 Balashicha, Moscow Region, Russia

[4]JSC R&D Center, FGC UES, Kashirskoe shosse bld. 2, 115201 Moscow, Russia

[5]National Research Center "Kurchatov institute", 1, Akademika Kurchatova pl., 123182 Moscow, Russia

E-mail: degtyarenkopn@gmail.com

**Abstract**. Design versions of HTS three-phase cables consisting of 2G HTS tapes have been investigated by the numerical simulation method with the aim of AC losses minimization. Two design versions of cables with the coaxial and extended rectangular cross-section shape are considered – the non-sectioned and sectioned one. In the latter each cable phase consists of sections connected in parallel. The optimal dimensions of sections and order of their alteration are chosen by appropriate calculations. The model used takes into account the current distribution between the sections and its non-uniformity within each single HTS tape as well. The following characteristics are varied: design version, dimension, positioning of extra copper layer in a cable, design of HTS tapes themselves and their mutual position. The dependence of AC losses on the latter two characteristics is considered in details, and the examples of cable designs optimized by the total set of characteristics for the medium class of voltages (10 – 60 kV) are given. At the critical current JC=5.1 кA per phase and current amplitudes lower than 0.85JC, the level of total AC losses does not exceed the natural cryostat heat losses.

## 1. Introduction

In [1], within the framework of the program of investigations aimed to minimization of AC losses in HTS-cables, there had been performed an optimization of coaxial three-phase cable designs by the variation of the cable geometry, mutual positions of HTS tapes layers and extra copper layers as well [2-7]. We also have investigated and used an opportunity to diminish AC losses by the minimization of gaps between the HTS tapes in each of their layers. It has been shown that, due to optimal choosing of these parameters, there is an opportunity to decrease AC losses in 2 - 3 times. We also considered the sectioning of cables allowing decreasing of losses by an order of magnitude and even more. In this paper, we investigate another opportunity of AC losses decreasing consisting in the lowering of distance between the superconducting layers positioned in HTS tapes themselves. As it was made before, the investigations were performed by the numerical simulation using the software described in [8, 9].

## 2. Cable design versions

The general view of the coaxial cable cross-section is shown in Fig. 1(a), and cable design versions described in details are given in Table 1. In Fig. 1(b) are shown the configurations of HTS tapes used as basic current-carrying elements of cables. All the cable designs considered here have one and the same mutual phases position, HTS tapes layers groups (2 or 3 layers per phase), insulation and extra copper layers. Extra copper (completing copper layers of HTS tapes themselves) is introduced to prevent the superconductor overheating in a short-circuit event (within a time interval of ca. 120 ms between the start of the event and the actuation of protective circuit-breakers). The extra copper cross-section area (ca. 122 mm$^2$ per phase) is chosen by averaging, but the exact value may be found only for specific characteristics of the given electric power network and its fault current limiting devices. The first three positions in Table 1 are occupied by cables Ph3R"x" whose designs were obtained by the optimization described in [1]. They differ by the rated insulation layers thickness ("x") – respectively, 2, 4 and 6 mm. In process of optimization, actually chosen insulation thicknesses were varied in order to ensure compact positioning of tapes in layers (certainly, by the increasing of the thickness with respect to the rated value). To the same end, the number of tapes in certain cable phases

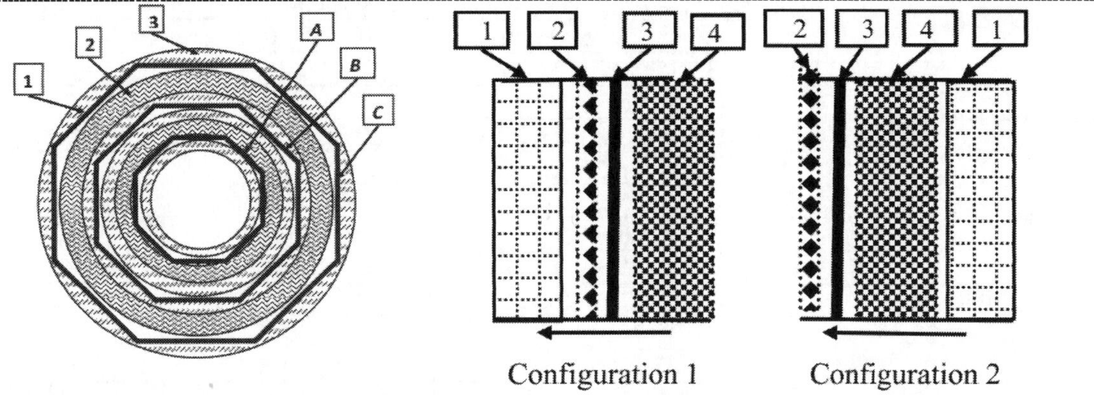


**Figure 1.** (a) The view of the cross-section shape of coaxial cables, given in Table 1: 1 - groups of HTS tapes layers, 2 - insulation layers, 3 – extra copper layers; (b) variants of the HTS material configuration in tapes: δ– is the HTS material thickness. The arrows show the orientation of tapes in cable phases. Configuration 1 – is one of the standard designs of SuperPower tapes [10, 11]. Configuration 2 – designs of tapes, considered here. 1- copper layer (for SuperPower tapes δ= 40 μm, for SuperOx tapes δ=60 μm); 2 – silver layer (δ= 2 μm); 3 – superconducting layer(for SuperPower tapes δ=1 μm, for SuperOx tapes δ=1,5 μm); 4 – hastelloy substrate for SuperPower tapes δ=50 μm, for SuperOx tapes δ=60 μm. The critical current of SuperPower tapes is 85 A, of SuperOx tapes - 127A, the width of both tapes is 4 mm.

was increased (resulting in the phase critical current relative to the established rated value of 5.1 кA). Since the cable critical current is determined by its weakest phase, this increasing must be interpreted as a compelled excessive consumption of the superconducting material allowed to diminish the AC losses value. In the first three cable designs of Table 1 is used standard HTS tapes configuration 1 (see

**Table 1.** Coaxial cables with the polygonal cross-section shape
(Designations of Table 1: d, D – are the internal and external cable diameters (D is given without the external cable insulation thickness) [mm], $d_S$ – is the superconducting layer diameter of HTS tapes (i.e., diameter of the circle inscribed into a polygon), $J_{CPh}$ – is the phase critical current [kA]. The general cable structure (from internal to external layers) is described as follows: $A_i$, $B_i$, $C_i$ – are the phases of a layer (i- is the layer number in a phase), Cu "xx" – is the extra copper layer thickness [mm], ins "xx" – is the insulation layer thickness [mm].)

| Cable type | Phase | $d_S$ of layers | Number of tapes in layers | $J_{CPh}$ |
|---|---|---|---|---|
| Ph3R2 d = 23 D = 41.16 | Cu1.5-$A_1$-$A_2$-$A_3$-ins, 2.3-Cu0.6-$B_1$-$B_2$-$B_3$-Cu0.6–ins 2.3-$C_1$-$C_2$-Cu1.0 | | | |
| | A | 26.08; 26.28; 26.48 | 20; 20; 20 | 5.1 |
| | B | 32.48; 32.68; 32.88 | 25; 25; 25 | 6.375 |
| | C | 38.88; 39.08 | 30; 30 | 5.1 |
| Ph3R4 d = 23 D = 49.76 | Cu1.5-$A_1$-$A_2$-$A_3$-ins 5.4-Cu0.5--$B_1$-$B_2$-Cu0.5–ins 4- $C_1$-$C_2$-Cu0.8 | | | |
| | A | 26.08; 26.28; 26.48 | 20; 20; 20 | 5.1 |
| | B | 38.48; 38.68 | 30; 30 | 5.1 |
| | C | 47.88; 48.08 | 37; 37 | 6.29 |
| Ph3R6 d = 23 D = 54.76 | Cu1.5- $A_1$-$A_2$-$A_3$-ins 6-Cu0.5- $B_1$-$B_2$-Cu0.5–ins 6- $C_1$-$C_2$-Cu0.7 | | | |
| | A | 26.08; 26.28; 26.48 | 20; 20; 20 | 5.1 |
| | B | 39.68; 39.88 | 30; 30 | 5.1 |
| | C | 53.08; 53.28 | 40; 40 | 6.8 |
| Ph3R2-V1 d = 23 D = 41.16 | Cu1.5-$A_1$-$A_2$-$A_3$-ins 2.3-Cu0.6-$B_1$-$B_2$-$B_3$-Cu0.6–ins 2.3-$C_1$-$C_2$-Cu1.0 | | | |
| | A | 26.16; 26.35; 26.36 | 20; 20; 20 | 5.1 |
| | B | 32.56; 32.73; 32.74 | 25; 25; 25 | 6.375 |
| | C | 38.96; 38.97 | 30; 30 | 5.1 |
| Ph3R4-V1 d = 23 D = 49.76 | Cu1.5-$A_1$-$A_2$-$A_3$-ins 5.4-Cu0.5-$B_1$-$B_2$-Cu0.5–ins 4- $C_1$-$C_2$-Cu0.8 | | | |
| | A | 26.16; 26.35; 26.36 | 20; 20; 20 | 5.1 |
| | B | 38.56; 38.57 | 30; 30 | 5.1 |
| | C | 48.06; 48.07 | 37; 37 | 6.29 |
| Ph3R6-V1 d=23 D=26.58 | Cu1.5- $A_1$-$A_2$-$A_3$-ins 6-Cu0.5- $B_1$-$B_2$-Cu0.5–ins 6- $C_1$-$C_2$-Cu0.7 | | | |
| | A | 26.16; 26.35; 26.36 | 20; 20; 20 | 5.1 |
| | B | 39.76; 39.77 | 30; 30 | 5.1 |
| | C | 53.17; 53.18 | 40; 40 | 6.8 |
| Ph3R2-V2 d = 23 D=41.15 | Cu1.5-$A_1$-$A_2$-ins 2.3-Cu0.6-$B_1$-$B_2$-Cu0.6–ins 2.3-$C_1$-$C_2$-Cu1.0 | | | |
| | A | 26.31; 26.32 | 20; 20 (SuperOx) | 5.1 |
| | B | 32.60; 32.61 | 25; 25 (SuperOx) | 6.375 |
| | C | 38.96; 38.97 | 30; 30 | 5.1 |
| Ph3R4-V2 d = 23 D=49.76 | Cu1.5--$A_1$-$A_2$-ins 5.4-Cu0.5--$B_1$-$B_2$-Cu0.5–ins 4- $C_1$-$C_2$-Cu0.8 | | | |
| | A | 26.31; 26.32 | 20; 20 (SuperOx) | 5.1 |
| | B | 38.56; 38.57 | 30; 30 | 5.1 |
| | C | 48.06; 48.07 | 37; 37 | 6.29 |
| Ph3R6-V2 d = 23 D = 54.76 | Cu1.5- $A_1$-$A_2$-ins 6-Cu0.5- $B_1$-$B_2$-Cu0.5–ins 6- $C_1$-$C_2$-Cu0.7 | | | |
| | A | 26.31; 26.32 | 20; 20 (SuperOx) | 5.1 |
| | B | 39.76; 39.77 | 30; 30 | 5.1 |
| | C | 53.17; 53.18 | 40; 40 | 6.8 |

Fig. 1b). In this paper were calculated the characteristics of cables denoted as Ph3R"x"-Vi. These cables differ from their counterparts only by the replacement of tapes with configuration 1 by ones with 2. That's why comparison of the results obtained here with those of [1] allows estimation of

effects, exclusively caused by decreasing the distance between the superconducting layers of HTS tapes.

Analytical expressions of [12, 13] show, that in a continuous superconductor with the ring cross-section shape the hysteresis losses drop with the decreasing of the ring thickness. The same is likely to expect in a cable, consisting of HTS tapes, when decreasing the distance between the HTS adjacent layers. In tapes of configuration 1, the superconducting layer is actually positioned on the tape axis of symmetry, and due to this, the distance between HTS layers weakly depends on the mutual orientation of tapes in adjacent layers. In tapes of configuration 2 the copper coating is repositioned on the substrate side. As a result, positioning the tapes of adjacent layers with the oppositely oriented silver coating reduces the distance between the HTS layers themselves to 4 μm. The transition of the copper coating part to the substrate side has been already made in some design versions of SuperPower tapes (having 20 μm of copper per each side). This design modification does not actually affect both the tapes stability against magnetic flux jumps and their mechanical characteristics. Some of our preliminary experiments showed the tape carrying AC current to remain stable and to have actually the same critical current even without copper coating. Therefore, the hypothetical design of configuration 2 may be realized in principle. However, the peculiarities of the cable manufacturing technology may require having a thin copper coating on the silver for additional mechanical and chemical protection of the superconductor's surface. In this case the efficiency of the technique proposed is lower, and the results obtained here should be considered as ultimately possible ones.

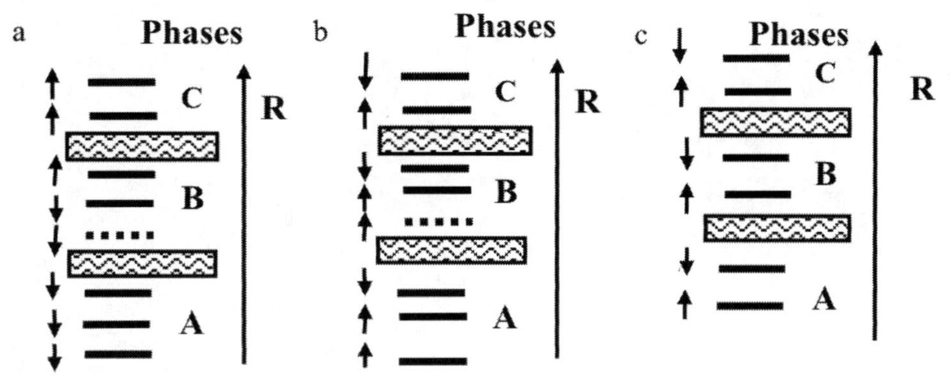


**Figure 2.** Schematic diagrams of HTS tapes layers positioning in cable phases. The arrows show the tapes orientation according to Fig. 1(b). The dashed line is the third layer of tapes in phases B of cables Ph3R2 and Ph3R2–V1. (a) The cables Ph3Ri, SuperPower tapes, configuration 1; (b) the cables Ph3Ri-V1, SuperPower tapes, configuration 2; (c) the cables Ph3Ri-V2, all the tapes are of configuration 2, in phases *A* of all cables and in phase *B* of cable Ph3R2-V2 are the SuperOx tapes, in other phases are the SuperPower tapes.

The schematic diagrams of positioning layers of tapes in various cables are shown in Fig. 2. In phases *A* of cable design versions V1 there are three layers of HTS tapes (for design Ph3R2-V1 – three layers of tapes are in phase B as well). That's why the effect of distance reduction between the layers is not totally used. In cable design V2, for these phases are used SuperOx tapes [14, 15], whose characteristics ensure the rated phase critical current when only two layers of tapes are present. This causes an additional reduction of hysteresis losses in these phases, and the general result is also improved.

## 3. Results of calculations.

AC losses in superconducting power transmission lines are the sum of hysteresis losses in superconductors and eddy current losses in other non-superconducting elements. Since in coaxial three-phase cables the external magnetic stray field is actually absent, eddy current losses consist only of losses in non-superconducting components of HTS tapes and ones in extra copper layers. The latter, having essentially greater thickness, predominate and are ca. 95% of the total eddy current losses.

In Fig. 3a are given the results of calculation of the AC losses power $P_S$ per 1 m of cable length at 50 Hz. The dimensionless current plotted on the x-axis is determined as $I = J_0/J_C$, where $J_0$ – is the current amplitude, $J_C$ – is the critical current of the weakest cable phase. Fig. 3b shows the contribution of eddy current losses in the total AC losses.

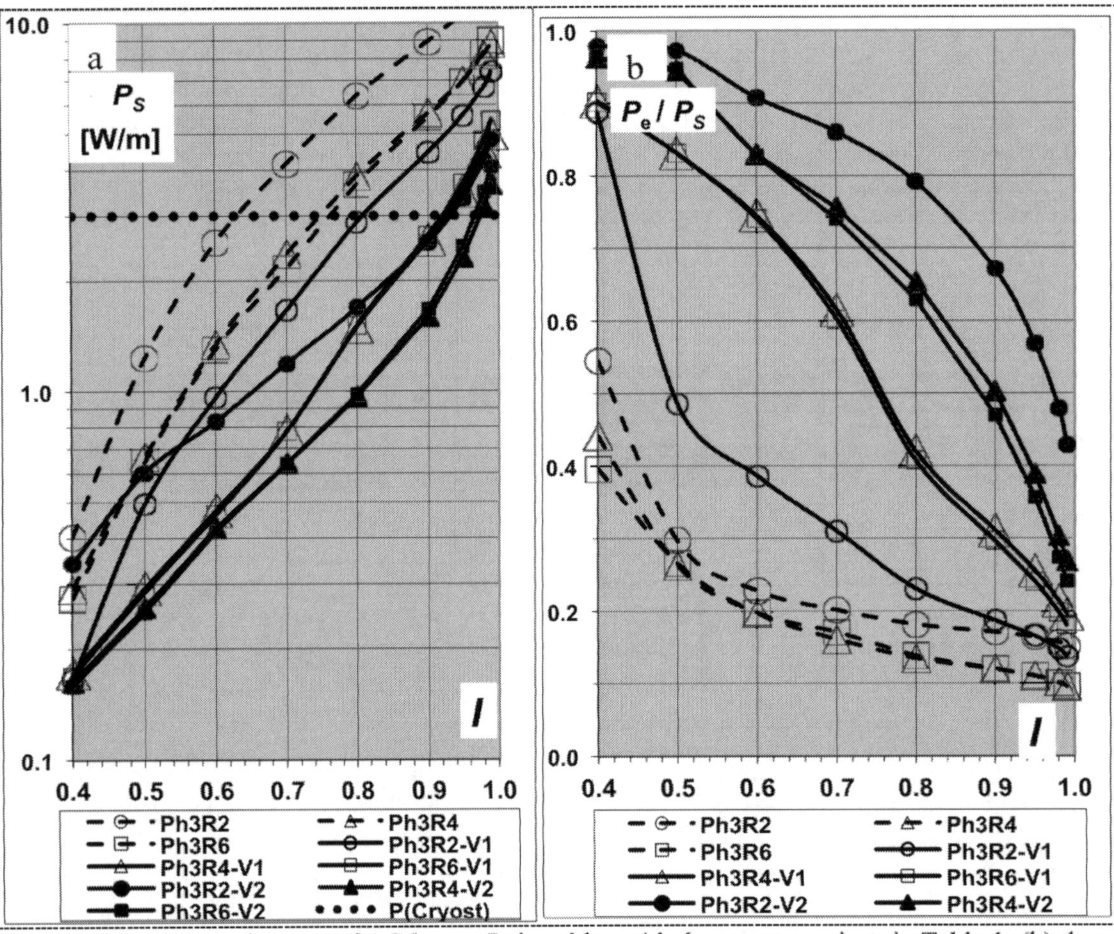


**Figure 3.** (a) The total power of AC losses $P_S$ in cables with the structure given in Table 1; (b) the ratio of the eddy current losses $P_e$ power to the total power of AC losses $P_S$.

AC losses in a cable do not affect essentially the economic parameters of the power system (including choosing between AC and DC cables), if they are much lower than the natural cryostat heat losses (denoted as $P_{(Cryost)}$ in Fig.3a). As it can be seen from Fig. 3a, within the actually significant range of $I \leq 0.8$, all the cables of design V2 satisfy this requirement. From Fig. 3b it follows, that in designs Ph3-V the contribution of eddy current losses is greater than in non-modified ones Ph3R. But their value does not change significantly, and the cause of it consists in an essential lowering of hysteresis losses. Hence, there is an additional opportunity of further $P_S$ decreasing by lowering the losses in extra copper layers. Since the latter have already been positioned optimally, this task can be solved either by decreasing of their thickness according to the specific power network characteristics or by the replacement of copper by a material having the lower electric conductivity, but, respectively, the greater specific heat capacity.

## 4. Conclusions

There had been proposed a technique to diminish the losses in AC HTS-cables by using HTS tapes with an asymmetrical position of the superconducting layer. It is shown that this modification enables the reduction of losses in 2 – 3.7 times, as compared with already optimized cables having the tapes with an ordinary structure. It should be especially noted that the maximum effect (decreasing in 3.7 times) takes place within the most important range of current amplitudes $0.7 \leq I \leq 0.8$. The effect obtained is lower than one reached by cable sectioning. However, unlike sectioning, the modification proposed does not increase the overall cable dimensions and does not deteriorate the cooling conditions of phases. It is simpler with respect to the cable manufacturing technology. Moreover, for cables, having high critical current, and, hence, many layers of tapes in each phase, there is an opportunity to combine sectioning with the tape structure modification in each section. In this case, the losses reduction coefficient is determined both by sectioning and the technique proposed here and may be greater than an order of magnitude.

## 5. Acknowledgement.

The work was supported by RFBR, Project Nr. 16-08-00304a.

## 6. References

[1] V.A. Altov et al., IEEE Trans. Appl. Supercond. V. 27(4), June 2017, 2658196.
[2] F. Grilli et al., arXiv:1306.6251v1 [cond-mat.supr-con].
[3] D.E. Daney et al., Advances in Cryogenic Engineering Materials, v. 44, 1998, pp 791-798.
[4] E. Shabagin et al., Cryogenics, V. 81, January 2017, Pages 24-32.
[5] Thomas Heiko et al., IEEE Transactions on Applied Superconductivity V. 27, Issue: 4, June 2017, 5400405
[7] K Mohit, et al., International Journal of Control Theory and Applications 9 (41), 309-317, 2017
[8] V.V. Zheltov, Elektrichestvo, no. 7, pp.61-67, 2009 *(in Russian)*.
[9] V.V. Zheltov, Superconductivity: research & development, no. 14, pp. 39-46, 2009 *(in Russian)*.
[10] D. W. Hazelton, LTHFSWS2012 Napa, CA, pp. 1 – 20, November 6, 2012.
[11] D.W. Hazelton, et al., IEEE Trans. Appl. Supercond., vol. 19, no. 3, part 2, pp. 2218 – 2222, 2009.
[12] H. London, *Physics Letters*, vol. 6, pp. 162–165, 1963.
[13] V.V. Zheltov, *Electrotekhnika*, no. 3, pp. 51-59, 2008 *(in Russian)*.
[14] S. Samoilenkov, et al., Supercond. Sci Technol. 29 (2016) 024001 (10pp).
[15] V.V. Sychugov, et al., Journal of Physics: Conference Series, Volume 747, Number 1, 747 (2016) 012048.

IOP Conf. Series: Journal of Physics: Conf. Series **969** (2018) 012050 doi:10.1088/1742-6596/969/1/012050

# Correlation of $T_c$ and coefficient of $T^2$ resistivity term of Fe-based pnictide & chalcogenide superconductors

**P. B. Castro, J. L. Ferreira, M. B. Silva Neto and M. ElMassalami**

Instituto de Fisica, Universidade Federal do Rio de Janeiro, Caixa Postal 68528, 21945-970 Rio de Janeiro, Brazil

E-mail: pedrobcst@if.ufrj.br

**Abstract.** Normal-state of many Fe-based pnictides and chalcogenides superconductors exhibit a quadratic-in-temperature, $\rho_{tot} - \rho_o - \rho_{ph} = AT^2$, over wide ranges of temperature and pressure. Moreover, these systems exhibit a correlation between their $T_c$ and $A$, namely $\ln(T_c/\theta) \propto A^{-1/2}$($\theta$ is an energy scale parameter), even when a control parameter such as pressure is widely varied. This manifestation, as well as that of $\frac{1}{T_c}\left(-\frac{dH_{c2}}{dT}\right)_{T_c} \propto \frac{A}{n}$ [Phys. Rev. B 89, 220509 (2014), $n$ is charge density, $H_{c2}$ is the upper critical field] suggests a common Landau Fermi Liquid scenario for both superconductivity and quadratic-in-$T$ contribution.

## 1. Introduction

The normal-state of various conventional superconductors manifest a quadratic-in-$T$ resistivity

$$\Delta\rho_{qt} = \rho_{tot} - \rho_o - \rho_{ph} = AT^2, \tag{1}$$

within the range $T_c < T << \epsilon$ [1–4]. $\epsilon$ is a temperature scale which is a fraction of Debye $\theta_D$ or Fermi $T_F$ temperature, $T_c$ is superconducting transition point, $\rho_o$ is residual resistivity and $\rho_{ph}$ is a phonon contribution. This same Eq.1 is also manifested in some unconventional superconductors such as heavy Fermion [5] and Fe-based pnictides and chalcogenides [6–17] (the main interest of this study). The latter observation motivated us to investigate whether there is any correlation between their $A$ and $T_c$ and whether such a correlation can be understood within the Landau-Fermi Liquid (LFL) scenario [1–3; 18–20].

The relation between $T_c$ and $A$ is usually developed through an analysis in terms of the Landau parameters $A_l^\alpha = F_l^\alpha / [1 + F_l^\alpha / (2l+1)]$ ($\alpha = s, a$ is the parity of the scattering process while $l = 0, 1, ..$ is the angular momentum). This analysis is usually limited, within the $sp$ model, to $A_0^s, A_0^a, A_1^s$ and $A_1^a$ terms [21]. Within this context, $A$ can be written as [4; 18; 22–24]

$$A = \frac{4\pi^4 k_B^2}{\hbar\omega_p^2 \epsilon_F^*}\lambda_t^2 u, \tag{2}$$

where $\omega_p^2 = 4\pi n e^2/m$, $n$ is the carrier density, $\epsilon_F^*$ is the Fermi energy, $u$ is a fraction of momentum transfer due to any possible momentum-transferring-to-lattice process [18; 22; 23] such as umklapp scattering. $\lambda_t = f(A_0^s, A_0^a, A_1^s, A_1^a)$ is the transport effective coupling parameter.

IOP Conf. Series: Journal of Physics: Conf. Series **969** (2018) 012050 doi:10.1088/1742-6596/969/1/012050

$T_c$, on the other hand, can be expressed in terms of a different coupling constant $\lambda_s = g(A_0^s, A_0^a, A_1^s, A_1^a)$ as [18; 20]

$$T_{c,s} = 1.13\omega_c \exp\left(-1/\lambda_s\right), \tag{3}$$

where $\omega_c$ is a cutoff energy of the pairing interaction and $\lambda_s$ describes either a singlet ($\lambda_{s=0}$) or a triplet ($\lambda_{s=1}$) pairing channel [2; 3; 18; 19].

Ideally, one needs to evaluate the involved $A_l^{\alpha\prime}$ parameters (say, empirically as outlined in Ref. [18]) as well as $u$ and the symmetry of the pairing. Presently we are unable to meet all these demands. But, for our present objectives, it is reasonable to assume that the effective pairing is mainly $\lambda_{s=0}$, that $u$ is close to 1, and that $A_1^s$ is the only nonzero (or the largest) parameter [18]. Then $m^*/m = (9-2A_1^s)/(9-3A_1^s)$ and $\lambda_t = \sqrt{3.5/9}\lambda_{s=0} = \sqrt{7/32}(1-m/m^*)$ [18]. Substitution of $\lambda_t$ and $\lambda_s$ into Eqs.(2) and (3), one obtains this simple analytical expression [2; 3; 20]:

$$T_c/\Theta = \exp\left(-\mathcal{F}/\sqrt{A}\right) \quad \rightarrow \qquad \ln(T_c/\Theta) \propto A_{tot}^{-1/2} \tag{4}$$

$\Theta$ and $\mathcal{F}$ are normalization factors that are assumed to depend only on type of superconductor but not on control parameters such as pressure $p$ or doping $x$.

It is worth recalling that $A$ is also related to other electronic parameters (such as Sommerfeld specific-heat coefficient via Kadowaki–Woods ratio). In this work we are interested in the correlation of $A$ to $\rho_o$. Recalling that $A$ can be written as two distinct contributions:

$$A_{tot} = A_{ee}^f + A_{ee}^i = A_{ee}^f + \rho_o B \tag{5}$$

$$A_{tot} \propto \rho_o \qquad (\text{if } A_{ee}^i \gg A_{ee}^f) \tag{6}$$

where $A_{ee}^f$ appears in defect-free samples and is associated with Coulomb and/or phonon-exchange (umklapp) electron-electron interactions [4; 25]. $A_{ee}^i$, on the other hand, is assumed to be due to all impurity/defects-driven electron-electron interaction processes, which include Koshino-Taylor ($B \approx 0.1{-}1{\times}10^{-5}\mathrm{K}^{-2}$) [26], Reizer-Sergeev [27–31], or dislocations [32–34].

Lawrence and Wilkins [23] studied the influence of impurities (of concentration $x$) on a system in which both electron-electron and electron-impurity scatterings are present. $A_{LW}$ is given as:

$$A_{LW}(x) \propto u_{tot}(x) = u\left(1\text{-}\frac{u}{10.4 + u + \frac{10\tau_o}{\pi^2\tau_{imp}}}\right),$$

$$\rho(T,x) = \frac{2\pi^2 m_{opt}}{3ne^2\tau_0} u_{tot}(x), \tag{7}$$

here $m_{opt}$ is the optical mass, $\tau_{imp}(x)$ the relaxation time due to impurities and $\tau_o$ is impurity-free electron-electron scattering time. Considering that $\tau_o \propto T^{-2}$, $\tau_{imp} \rightarrow \infty$ gives the usual quadratic-in-$T$ character of the pure limit [23]. Interestingly, both Eqs.(5) and (7) predict that as $x$ increases, both $A(x)$ and $T_c(x)$ are enhanced, given that Eq.(4) is still valid. The latter prediction will be checked below (see also Supplementary Materials [35] SM).

Eq.(4) together with the similar ones (see Refs. [17; 36; 37]) were often employed for checking the presence of common LFL scenario in various superconducting systems [2; 3; 18–20] including the under-study Fe-based superconductors [6; 7; 20]. Pnictides and chalcogenides systems are well suited for such investigation since there is a vast literature on the evolution of their $T_c(p,x)$, $A(p,x)$ and $\rho_o(p,x)$[8–17; 38–42]. Among these, $KFe_2As_2$[17; 38–40], LiFeAs [8; 14–16], and LiFeP [16] are of particular interest to this study since they are stoichiometric, nonmagnetic, have simple crystal structures, manifest relatively high $A$ and low $\rho_o$. Most relevantly, the quadratic-in-$T$ behavior is manifested over wide ranges of $T$, $p$, and $x$. In addition to such pnictides, chalcogenides superconductors, e.g. archetype $K_xFe_{2-y}Se_2$, are just as interesting since they manifest LFL scenario over wide ranges of $T$, $p$, $x$, $y$ and annealing conditions [43].

We analyze below the evolution of the reported $\rho(T,p,x)$ curves of various Fe-based superconductors with the aim of determining $A_{tot}(p,x)$, $T_c(p,x)$, and $\rho_o(p,x)$ parameters. Care was exercised when evaluating these parameters because some Fe-based superconductors manifest other resistive contributions (e.g. localization, Kondo-like or magnetic) that mask the intrinsic values of $A_{tot}$, $T_c$, and $\rho_o$ (see [35]). The empirically determined correlations among these parameters are confronted with the predictions of Eqs.(4)-(7).

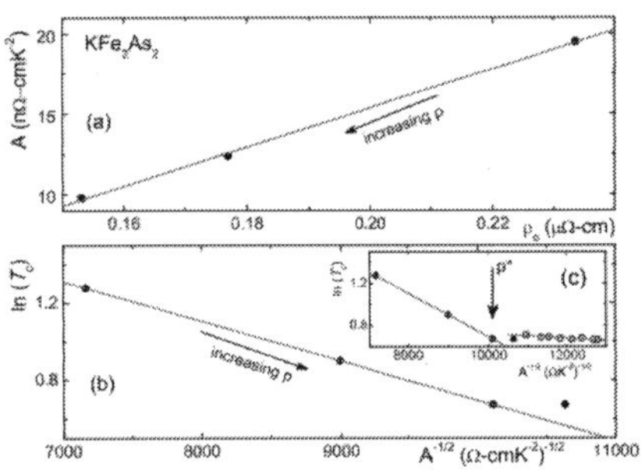


Fig. 1: Evolution of $T_c(p)$, $A(p)$ and $\rho_o(p)$ of $KFe_2As_2$. These parameters were obtained directly or after analysis of $\rho(T,p,H)$ curves of Taufour *et al.* [17] (a) $A(p \leq p^*)$ *versus* $\rho_o(p \leq p^*)$. The straight line represents a Koshino-Taylor-like linear dependence. (b) A plot of $\ln[T_c(p \leq p^*)]$ versus $[A(p \leq p^*)]^{-1/2}$. The straight line is a fit to Eq.4. (c) A plot of $\ln[T_c(p)]$ versus $[A(p)]^{-1/2}$.

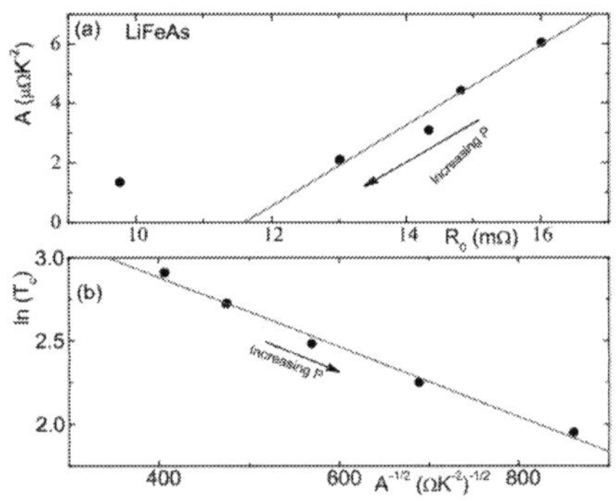


Fig. 2: Evolution of $T_c(p)$, $A(p)$ and residual resistance $R_o(p)$ of LiFeAs. These parameters were obtained after fitting Eq.1 to the pressure-dependent resistance curves of Zhang *et al.* [8]. (a) $A(p)$ plotted as a function of $R_o(p)$. The straight line represents a Koshino-Taylor linear dependence of $A(p)$ on $R_o(p)$. (b) A plot of $ln\,[T_c(p,x)]$ versus $[A(p,x)]^{-1/2}$. The straight line is a fit to Eq.4.

## 2. Analysis and Results

$KFe_2As_2$ is a stoichiometric and clean Fe-based superconductor [17; 38–40]. Taufour *et al.* [17] derived, based on a LFL scenario, the relation

$$\frac{1}{T_c}\left(-\frac{dH_{c2}}{dT}\right)_{T_c} \propto \frac{A_{tot}}{n} \tag{8}$$

wherein $n$ is charge density and $H_{c2}$ is the upper critical field. An alternative correlation between $T_c$ and $A_{tot}$ is provided by our analysis of the extensively studied $\rho(T,p,H)$ curves [17]: Fig.1(b) shows that $\ln[T_c(p \leq p^*)]$ is linearly related to $A_{tot}^{-1/2}$. This alternative confirmation of the common LFL origin [17] can be considered as a validation of our present analytical procedures. Interestingly, Fig.1(a) shows that $A(p \leq p^*)$ evolves linearly with $\rho_o(p \leq p^*)$ indicative of a strong Koshino-Taylor contribution, with a sizable $A^i_{ee}$.

LiFeAs is another clean and stoichiometric Fe-based superconductor with a well-established normal-state LFL character [8; 14–16]. Here, we establish the connection between this FL character and superconductivity based on our analysis of the extensive $\rho(T,p)$ curves of Zhang *et al.* [8] The obtained $A_{tot}(p)$ and $R_o(p)$ as well as $T_c(p)$ are shown in Fig.2 ($R_o \propto \rho_o$, see [35]). Within the low pressure range, $A(p)$ is related almost linearly to $R_o(p)$ suggesting a strong Koshino-Taylor contribution. Most importantly, Fig.2(b) shows that $\ln[T_c(p)]$ is related linearly to $A_{tot}^{-1/2}$. This agreement with Eq.(4) is indicative of a common LFL origin.

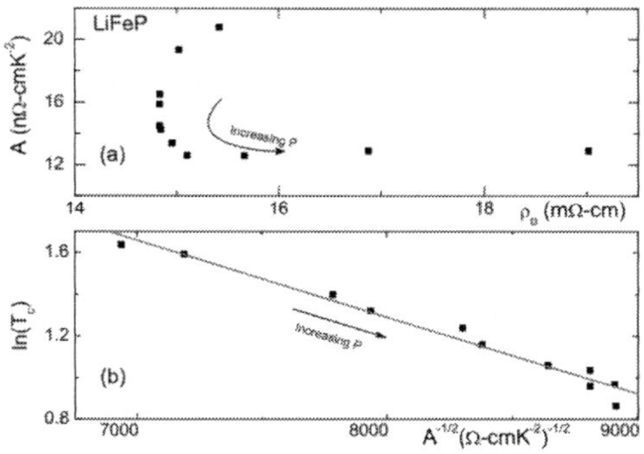


Fig. 3: The evolution of $T_c(p)$ and $A(p)$ and $\rho_o(p)$ of LiFeP as obtained from fitting Eq.1 to the data reported by Mydeen *et al.* [10]. (a) $A(p)$ is not monotonically related to $\rho_o$ suggesting that Koshino-Taylor relation is not obeyed. (b) A plot of $ln\,(T_c)$ *versus* $[A_{tot}]^{-1/2}$; the straight line fit emphasizes the applicability of Eq.4.

The analysis of $\rho(T,p)$ of the stoichiometric and clean LiFeP, an isomorph of LiFeAs [10], is shown Fig.3(a) which indicates an absence of any linear or any monotonic dependence of $A_{tot}$ on $\rho_o$. In spite of this unusual dependence of $A_{tot}$ on $\rho_o$, it is remarkable that Fig.3(b) reveals a linear plot of $\ln(T_c)$ *versus* $A_{tot}^{-1/2}$.

The analysis based on Eqs.(4)-(6) was extended to other pnictides, such as $BaFe_2As_2$ [42] and $Sr_2ScFePO_3$[44], as well as chalcogenides, such as $K_{0.8}Fe_{1.7}Se_2$ [13] and $Rb_{0.8}Fe_{2-y}Se_{2-x}Te_x$ [41]: Some additional examples are shown in [35].

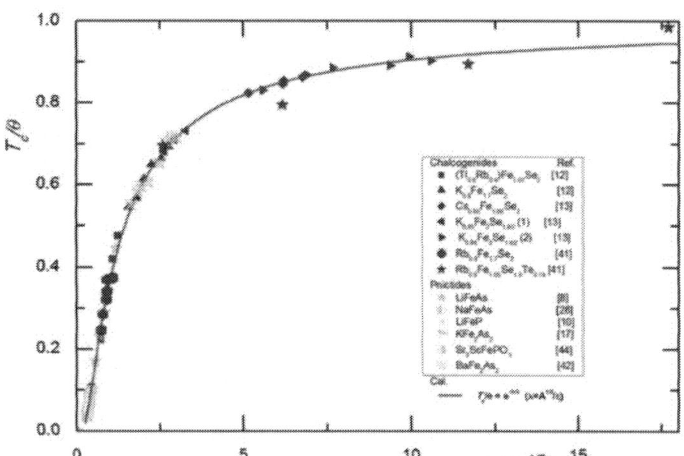


Fig. 4: Normalized $T_c(p)$ $/\theta$ plotted against $\sqrt{A_{tot}(p)}/\mathcal{F}$ for various Fe-based pnictides and chalcogenides superconductors. These were obtained after fitting $T_c(p)$ and $A_{tot}(p)$ to Eq.4, based on the resistivity curves reported in the references shown in inset table.

For all studied samples, we observed that a linearized $\ln(T_c)$ *versus* $A_{tot}^{-1/2}$ is manifested. Based on this linearity, it is no surprise that the normalized $T_c/\Theta$ *versus* effective coupling constant $\lambda_s = \frac{\sqrt{A_{tot}}}{\mathcal{F}}$ for all studied compounds do lie on the top of the very same curve of Eq.(4) (Fig.4). As mentioned before, there is no point in discussing the strength or evolution of $\Theta$, $\mathcal{F}$ and $\lambda_s$: such a discussion needs (i) a clarification of the pairing mechanism, which is outside the scope of this work, and that (ii) none of $\Theta$, $\omega_c$ or $\mathcal{F}$ has any strong dependence on p or x (in addition to the questions raised in the Supplementary Material [35]). The condition in (ii) is necessary for the validity of linearity of the ln $T_c$ *versus* $1/\sqrt{A}$ plot.

A closer look at the distribution of the normalized ($T_c$,$A_{tot}^{-\frac{1}{2}}$) along the curve of Fig.4 suggests that the pnictides are distributed within the region of lower $T_c/\Theta$ and $\lambda_s$ which is the same region occupied by conventional superconductors (see Fig.4 of Ref. [20]). It is surprising that $\lambda_s$ of chalcogenides reaches values which are much higher than 2.2, which is considered to be an upper limit for strong conventional superconductors [1] (see Fig.4 and the arguments about $\Theta$ and $\mathcal{F}$ in [35]).

Let us add some remarks regarding Eqs.(5) and (7): Figs. 1and 2 indicate that an increase of $p$ induces a correlated reduction of $\rho_o(p)$, $A(p)$ and $T_c(p)$. Fig.3 is an exception in that $\rho_o(p)$ is decreased and afterwards increased. During this study we did not observe a correlated enhancement of $A$ or $T_c$ (either pressure- or disorder-driven) in any of the studied systems.

## 3. Discussion and conclusion

As far as the reported Fe-based superconductors are concerned, it is remarkable that $T_c$ and $A_{tot}$ of more than a dozen superconductors do follow Eq.(4). Evidently, the list of Fig.4 does not include all Fe-based superconductors even though one expects, by generalization, that all should follow Eqs. (1) and (4). We suggest that the following reasons explains why some Fe-based superconductors (e.g. $SrFe_2As_2$) do not follow Eqs.(1) and (4). First the *quadratic-in-T* contributions may be masked due to the above-mentioned extrinsic non-quadratic-in-$T$ effects. Because of these extrinsic effects, $\lambda_t$ and $\lambda_s$ may be not related in a simple way and as such Eq.(4) is not obtainable. Secondly, some others exhibit doping/disorder-influenced $T_c(x)$ and $A(x)$ that do not follow Eq.(4): This is discussed in more details in [35]. It is recalled that similar non-inclusive list is encountered in conventional, strongly coupled, *phonon-driven* $A$-15 superconductors [1]: while a few of $A$-15 superconductors (such as $V_3Si$) do exhibit a $\Delta\rho_{qt}$ contribution (possibly $A$ is enhanced by the previously discussed Lawrence-Wilkins effect), the majority does not.

Finally, we discuss why in most of Fe-based superconductors the relation $\ln(T_c/\Theta) \propto A_{tot}^{-1/2}$ is manifested, but with no concomitant linear correlation between $A_{tot}$ and $\rho_o$. Eq.(6) indicates that $A_{tot} \propto \rho_o$ only if $A_{ee}^{f}$ ($\sim 10^{-11}$ $\Omega$-cmK$^{-2}$)$\ll A_{tot}$[1; 4] and that B is independent of $\rho_o$. Some experiments, e.g. Fig.3(a), indicated that relation between $A_{tot}$ and $\rho_o$ is more complicated than the simply linear Koshino-Taylor relation. In fact, not even Matthiessen rule is valid in such cases [23]. At any rate, the dominance of $A_{ee}^{i}$ in Eq.(6) should be reflected in Eq.(4) as $T_c/\Theta = \exp\left(-\mathcal{F}/\sqrt{A_{ee}^{i}}\right)$. Accordingly, if pressure and doping does not modify the functional dependence of $T_c$ and $A_{ee}^{i}$, then the validity of Eq.(4) will be maintained even if the Koshino-Taylor relation is violated.

In summary, we showed that most of the Fe-based pnictides and chalcogenides superconductors manifest ln ($T_c$) *verus* $A_{tot}^{-1/2}$ relation. This feature is taken to be an indicative of a common Landau Fermi-liquid scenario for their normal-state and superconducting properties. Extension of this analysis to unconventional heavy-fermion and conventional strong-coupled superconductors is under way.

**Acknowledgments**
We acknowledge partial financial support from CNPq.

**References**

[1] Gurvitch M 1986 *Phys. Rev. Lett.* **56** 647
[2] Patton B et. al 1975 *Physics Letters A* **55** 95 – 97

[3] Pethick C J et. al 1986 *Phys. Rev. Lett.* **57** 1955
[4] MacDonald A H et. al *Phys. Rev. B* **23**(6) 2718–2730
[5] Stewart G 1984 *Rev. Mod. Phys.* **56** 755
[6] Garbarino G et. al 2009 *Europhys. Lett.* **86** 27001
[7] Garbarino et. al 2008 *Phys. Rev. B* **78** 100507
[8] Zhang S J et. al 2009 *Phys. Rev. B* **80** 014506
[9] Wang A F et. al 2012 *New J. Phys.* **14** 113043
[10] Mydeen et. al 2010 *Phys. Rev. B* **82** 014514
[11] Okada H et. al H 2010 *Phys. Rev. B* **81** 054507
[12] Sun L et. al 2012 *Nature* **483** 67
[13] Ying J J et. al 2011 *N J Physics* **13** 033008
[14] Heyer O et. al 2011 *Phys. Rev. B* **84** 064512
[15] Rullier-Albenque F et. al 2012 *Phys. Rev. Lett.* **109** 187005
[16] Kasahara S et. al 2012 *Phys. Rev. B* **85** 060503
[17] Taufour V et. al 2014 *Phys. Rev. B* **89** 220509
[18] van der Marel D et. al 2011 *Phys. Rev. B* **84** 205111
[19] Klimin S N et. al 2012 *Phys. Rev. B* **86** 045113
[20] Nunez-Regueiro M et. al 2012 *J Phys: Conf. Series* **400** 022085
[21] Dy K S and Pethick C J 1969 *Phys. Rev.* **185** 373
[22] Lawrence W E 1976 *Phys. Rev. B* **13**(12) 5316
[23] Lawrence W E and Wilkins J W 1973 *Phys. Rev. B* **7** 2317
[24] MacDonald A H 1980 *Phys. Rev. Lett.* **44** 489
[25] Zheng X H and Walmsley D G 2014 *Physica Scripta* **89** 095803
[26] Mahan G D and Wang Z 1989 *Phys. Rev. B* **39** 4926
[27] Reizer M and Sergeev A 1987 *Zh. Eksp. Theor. Fiz. [Sov. Phys. JETP 65, 1291 (1987)].*
[28] Pautrat A and Kobayashi W 2012 *EPL (Europhysics Letters)* **97** 67003
[29] Ptitsina N G et. al 1997 *Phys. Rev. B* **56**(16) 10089–10096
[30] Fisher B et. al 2003 *Phys. Rev. B* **68** 014118
[31] Echternach P M, Gershenson M E and Bozler H M 1993 *Phys. Rev. B* **47** 13659
[32] Kaveh M and Wiser N 1984 *Adv. Phys.* **33** 257
[33] Kaveh M and Wiser N 1985 *J. Phys. F: Met. Phys.* **15** L195
[34] Levy B, Sinvani M and Greenfield A J 1979 *Phys. Rev. Lett.* **43**(24) 1822–1825
[35] See Supplemenatary Materials at xxx
[36] Miyake A, Aoki D and Flouquet J 2008 *J. Phys. Soc. Jpn.* **77** 094709
[37] Knebel G et. al 2010 *physica status solidi (b)* **247** 557
[38] Tafti F F et. al 2013 *Nat. Phys.* **9** 2013
[39] Tafti F F et. al 2015 *Phys. Rev. B* **91** 054511
[40] Terashima T et. al 2014 *Phys. Rev. B* **89**(13) 134520
[41] Gu D *et al.* 2015 *New Journal of Physics* **17** 073021
[42] Ishikawa F et al. 2009 *Physical Review B* **79** 172506
[43] Soares C. C *et al.* 2017 to be published
[44] Kotegawa H et. al 2009 *J. Phys. Soc. Jpn.* **78** 123707

IOP Conf. Series: Journal of Physics: Conf. Series **969** (2018) 012051 doi:10.1088/1742-6596/969/1/012051

# Reduction and oxidation annealing effects on Cu *K*-edge XAFS for electron-doped cuprate superconductors

**S Asano[1, 2], K M Suzuki[2], D Matsumura[3], K Ishii[4], T Ina[5] and M Fujita[2]**

[1]Department of Physics, Tohoku University, Sendai, Japan, [2]Institute for Materials Research, Tohoku University, Sendai, Japan, [3]Quantum Beam Science Center, Japan Atomic Energy Agency, Hyogo, Japan, [4]Synchrotron Radiation Research Center, National Institutes for Quantum and Radiological Science and Technology, Hyogo, Japan, [5]Japan Synchrotron Radiation Research Institute, Hyogo, Japan

E-mail: shun.asano@imr.tohoku.ac.jp

**Abstract.** Reduction and oxidation annealing effects on the electronic states around the copper sites for $Pr_{2-x}Ce_xCuO_4$ and $Nd_{2-x}Ce_xCuO_4$ with $x = 0$ and $x = 0.15$ were investigated by Cu *K*-edge x-ray absorption measurements. Cu *K* near-edge spectra were changed by the reduction annealing in a manner similar to the case of Ce substitution for both $x = 0$ and $x = 0.15$. This means an increase of electron density at the copper sites, indicating the aspect of electron doping in the reduction annealing. This reduction annealing effect on the near-edge spectra are reverted by the additional oxidation annealing. The amount of electron density around the copper sites is varied by the reduction and oxidation annealing, reversibly, corresponding to the reversible variation of the physical property from insulating to superconductivity.

## 1. Introduction

High temperature cuprate superconductivity emerges when electron or hole carriers are introduced into an antiferromagnetic Mott insulator. For the emergence of superconductivity in the electron-doped cuprate $RE_{2-x}Ce_xCuO_4$ (*RE* = Pr, Nd, Sm, Eu), adequate reduction annealing is necessary [1, 2]. As-sintered materials show antiferromagnetic orders and insulating behaviours despite the doping of sufficient number of electron carrier by Ce substitution. It is widely recognized that the reduction annealing induces structural modifications: removal of a small amount of oxygen [3, 4] and repair of Cu deficiency [5]. The superconductivity again disappears by oxidation annealing and turns into an insulator like as-sintered materials. Therefore, the physical property varies reversibly by the reduction and oxidation annealing [6]. However, the relationship between the structural change and the variation of the electric state has not been fully understood. Furthermore, the study of annealing effects was limited to the superconducting sample [3, 5, 7 − 9], and no systematic measurement of physical properties as well as the structure modified by the reduction and oxidation annealing was reported for parent and underdoped compounds. In the present work, we studied the reduction and oxidation annealing effects on the electronic states around the copper sites for $Pr_{2-x}Ce_xCuO_4$ and $Nd_{2-x}Ce_xCuO_4$ with $x = 0$ and 0.15 by Cu *K*-edge x-ray absorption measurements with transmission mode.

## 2. Experimental Details

Polycrystalline samples of $Pr_{2-x}Ce_xCuO_{4+\alpha-\delta}$, $Nd_{2-x}Ce_xCuO_{4+\alpha-\delta}$ with $x = 0$ and 0.15 were synthesized by a solid-state reaction method. The dried powder of $Pr_6O_{11}$, $Nd_2O_3$, $CeO_2$ and CuO were mixture with the proper molar ratio. The mixture was pressed into pellets and sintered at 1030 ℃ with a couple of intermediate grindings. The oxygen reduced samples were prepared by annealing in flowing Ar gas at temperatures between 750℃ and 940℃ for 12 hours. The amount of oxygen loss and the reduction annealing condition are summarized in Table 1. The $\alpha$ and $\delta$ are the amount of excess oxygen in the as-sintered samples and that of oxygen loss by the reduction annealing, respectively. The oxygenized samples were prepared by subsequent annealing for the oxygen reduced samples in flowing $O_2$ gas at 900 °C for 12 hours.

The x-ray absorption measurements were carried out at the BL01B1 in SPring-8. Cu *K*-edge absorption spectra were measured in a transmission mode using Si(1 1 1) double-crystal monochromator. All x-ray absorption measurements were performed at 300 K on pressurized pellets of a mixture of the powder samples and boron nitride.

Table 1. The amount of oxygen loss and the reduction annealing condition

| | Oxygen loss $\delta$ | Reduction annealing condition |
|---|---|---|
| $Pr_2CuO_{4+\alpha-\delta}$ | 0.035 | 920℃, 12 hours |
| $Pr_{1.85}Ce_{0.15}CuO_{4+\alpha-\delta}$ | 0.016 | 940℃, 12 hours |
| $Nd_2CuO_{4+\alpha-\delta}$ | 0.035 | 750℃, 18 hours |
| $Nd_{1.85}Ce_{0.15}CuO_{4+\alpha-\delta}$ | 0.011 | 940℃, 12 hours |

## 3. Results and Discussion

Fig. 1 (a) shows Cu *K* near-edge spectra for as-sintered (AS) $Pr_{2-x}Ce_xCuO_{4+\alpha-\delta}$ (PCCO) with $x = 0$ and 0.15. The intensity of spectra was normalized by the value at high-energy region, which is insensitive to the variation of electric state. In the figure, the near-edge spectrum for $Cu_2O$ is plotted as a reference of $Cu^+$. Following previous studies for $Pr_2CuO_4$ [8], $La_{2-x}Sr_xCuO_4$ and $Nd_{2-x}Ce_xCuO_4$ [10], the characteristic shoulder and peaks in the spectra were assigned; the shoulder at 8983 eV (labelled a) and 8991 eV (labelled b) corresponds to the dipole transition from 1*s* to 4*pπ*, and the peaks at 8994 eV (labelled c) and 9000 eV (labelled d) comes from the dipole transition from 1*s* to 4*pσ*. The final state for a and c is a well-screened state and that for b and d is a poorly-screened state. As reported previously for $Nd_{2-x}Ce_xCuO_4$ [8, 10, 11], the effect of Ce doping is observed around near-edge region; intensity of peak corresponding to 1*s*-4*pπ* transition increases, while that corresponding to 1*s*-4*pσ* transition decreases upon Ce doping. In Fig. 1 (d), the difference between the near-edge spectrum of AS PCCO $x = 0$ and $x = 0.15$, denoted as [AS PCCO with $x = 0.15$] − [AS PCCO with $x = 0$], is shown. This remnant spectra correspond to the signal induced by Ce-doping, namely, electron-doping. In the figure, the difference between the near-edge spectrum of $Cu_2O$ and that of AS PCCO with $x = 0$, denoted as [$Cu_2O$] − [AS PCCO with $x = 0$], is overdrawn. Assuming that the reduced (RE) parent sample is stoichiometry ($\alpha = \delta$), [$Cu_2O$] − [AS PCCO with $x = 0$] gives difference between the spectra for $Cu^+$ and $Cu^{2+}$. In the near-edge spectrum of $Cu_2O$, the peak at ~8981 eV corresponds to 1*s*-4*pπ* transition for $Cu^+$, and the peak structure at this energy becomes clear in the subtracted spectra shown in Fig. (d). This indicates that $Cu^+$ site is formed by the electron-doping and the intensity of peak A corresponds to the relative electron number ($Cu^+$ content) of the samples, as discussed previously [11].

Figs. 1 (b) and (c) show the Cu *K* near-edge spectra for AS, RE and oxygenized (OX) PCCO with $x = 0$ and $x = 0.15$. Comparing to the Ce substitution effect on the spectra, the degree of change in the spectra by the reduction annealing is smaller. However, as seen in the inset of Figs. 1 (a), (b) and (c), the near-edge spectra were certainly modified by the reduction and oxidation annealing. The spectral intensity around 8980 eV (labelled A) is increased by the reduction annealing in a similar manner to the Ce substitution and decreased by the oxidation annealing. In Figs. 1 (e) and (f), led (blue) lines represent the remnant spectra obtained by subtracting the data for RE PCCO from that for AS PCCO, denoted as [RE PCCO] − [AS PCCO] (by subtracting the data for OX PCCO from that for the AS PCCO, denoted as [OX PCCO] − [AS PCCO]). The resultant spectra are quantitatively similar to [AS PCCO with $x$ = 0.15] − [AS PCCO with $x$ = 0] in Fig. 1 (d), indicating that the reduction annealing effect on the near-edge spectra is similar to the Ce-substitution effect, that is, the electron-doping at the copper sites.

The difference between the spectrum for AS and OX PCCO is shown in Figs. 1 (e) and (f) as the subtracted spectrum (blue line). The overall spectral shape is rather flat compared with the difference between the data for RE and AS PCCO (red line), meaning that the near-edge spectra for RE PCCO are reverted to that for AS PCCO by additional oxidation annealing. Therefore, the

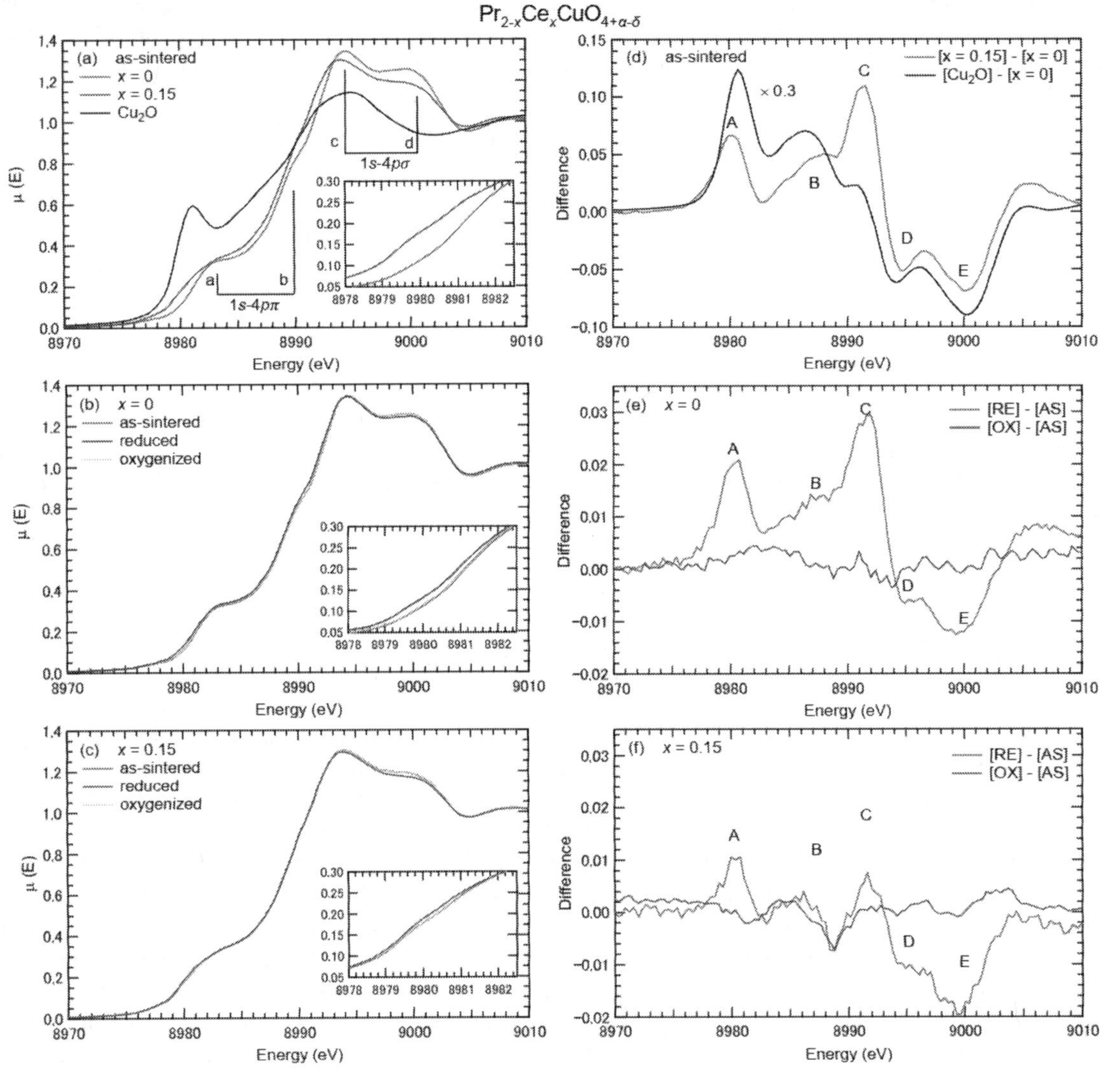


**Figure 1.** (a)−(c) Cu $K$ near-edge spectra for as-sintered (AS), reduced (RE) and oxygenized (OX) $Pr_{2-x}Ce_xCuO_{4+\alpha-\delta}$ (PCCO) with $x$ = 0 and $x$ = 0.15. (a) The spectra for AS PCCO with $x$ = 0 and 0.15, comparing with the spectrum for $Cu_2O$. (b) The spectra for AS, RD and OX PCCO with $x$ = 0. (c) The spectra for AS, RD and OX PCCO with $x$ = 0.15. Inset is the spectra for the energy between 8978 eV and 8982.5 eV. (d) The subtracted spectrum for AS PCCO with $x$ = 0 and 0.15, comparing with the subtracted near-edge spectrum for $Cu_2O$ and AS PCCO with $x$ = 0. The subtracted spectrum for AS PCCO and RE PCCO, comparing with the subtracted spectra for AS PCCO and OX PCCO, with (e) $x$ = 0 and (f) $x$ = 0.15.

oxidation annealing reduces the electron density, and the density is reversibly controlled through the reduction and oxidation annealing.

For the quantitative comparison of spectral change induced by Ce-substitution and annealing, we calculated the integrated spectral weight between 8976 and 8983 eV, which corresponds to the energy for peak A, and converted into the electron number ($n$). Fig. 2 (a) shows the evaluated $n$ as a function of Ce concentration ($x$) for all samples. Note that the carrier number in the as-sintered parent sample is assumed to be zero. In the AS PCCO and NCCO with $x = 0.15$, $n$ is almost comparable to $x$, supporting that the one Ce ion effectively dopes one electron. For the sample with fixed $x$, the $n$ is larger in the annealed sample. Therefore, the reduction annealing also works as the electron doping to $CuO_2$ plane even for the parent compounds. Fig. 2 (b) shows the $\delta$-dependence of increased electron number by Ar annealing ($\Delta_e$). The $\delta$ was calculaed from the weight loss of the sample after the annealing. The $\Delta_e$ is larger in the parent systems, in which $\delta$ is relatively large. The solid line in the figure represents a relation of $\Delta_e = 2\delta$, which is expected from the charge neutrality with an assumption that all carrier is doped onto $CuO_2$ plane. Except for NCCO with $x = 0.15$, $\Delta_e$ is slightly smaller than $2\delta$, meaning that the removed one oxygen ($O^{2-}$) introduce less than two electrons to $CuO_2$ plane. The possible reason for $\Delta_e < 2\delta$ is that some oxygen removes from the rare-earth layer with remaining electron density within the rare-earth layer.

## 4. Summary

The Reduction and oxidation annealing effects on the electronic states around the copper sites for the $Pr_{2-x}Ce_xCuO_4$ and $Nd_{2-x}Ce_xCuO_4$ with $x = 0$ and $x = 0.15$ were investigated by Cu $K$-edge x-ray absorption measurements. Cu $K$ near-edge spectra were changed by the reduction annealing in a manner similar to the case of Ce substitution for both $x = 0$ and $x = 0.15$ samples, that is, the increase of electron density at the copper sites. Therefore, the reduction annealing works as the electron doping. We furthermore found the decrease of the electron density by oxidation annealing. Qualitative analysis of the near-edge spectra clarified that one electron is effectively doped by one substituted Ce ion and a little less than two electrons by removed one oxygen ion.

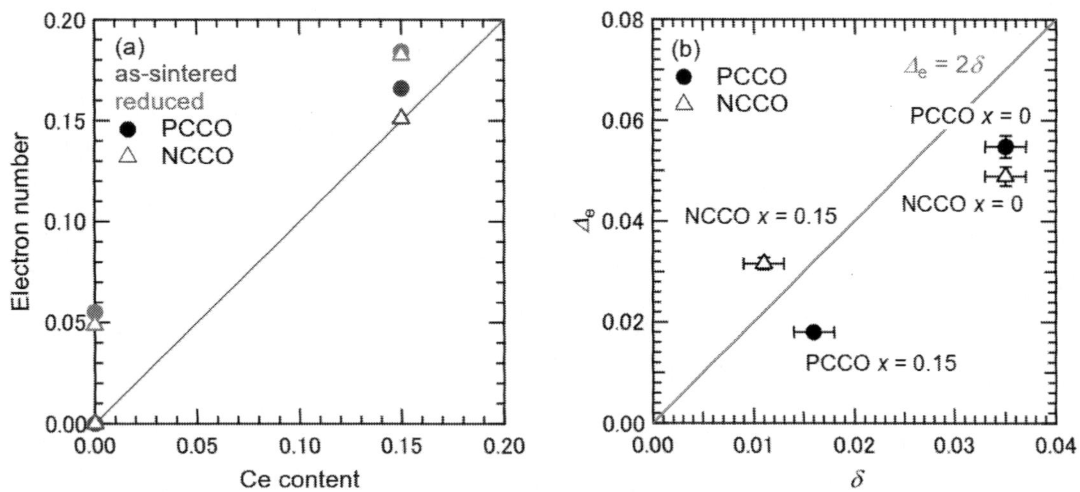


**Figure 2.** (a) Electron number ($n$) as a function of Ce concentration ($x$) for as-sintered (AS), reduced (RE) $Pr_{2-x}Ce_xCuO_{4+\alpha-\delta}$ (PCCO) and $Nd_{2-x}Ce_xCuO_{4+\alpha-\delta}$ (NCCO) with $x = 0$ and $x = 0.15$. The electron number was evaluated from the intensity of the near-edge spectra in the energy between 8976 eV and 8983 eV in Fig. 1 after normalizing by the near-edge spectrum for $Cu_2O$ standard. (b) The amount of increased electron number by the reduction annealing $\Delta_e$ as a function of the $\delta$. The $\delta$ is the amount of the oxygen loss by the reduction annealing.

**Acknowledgement**

We thank Y. Kimura for the fruitful discussion. The experiment at SPring-8 was carried out under Project No. 2016A1603. M.F. is supported by Grant-in-Aid for Scientific Research (A) (16H02125).

**References**

[1] Tokura Y, Takagi H, Uchida S 1989 *Nature* **337**, 345
[2] Takagi H, Uchida S, Tokura Y 1989 *Phys. Rev. Lett.* **62**, 1197
[3] Radaelli P G, Jorgensen J D, Schultz A J, Peng J L, and Greene R L 1994 *Phys. Rev.* B **49**, 15322
[4] Moran E, Nazzal A I, Huang T C, and Torrance J B 1989 *Physica* C **160**, 30
[5] Kang H J, Dai P, Campbell B J, Chupas P J, Rosenkranz S, Lee P L, Huang Q, Li S, Komiya S, and Ando Y 2007 *Nat. Mater.* **6**, 224
[6] Wang Y L, Huang Y, Shan L, Li S L, Dai P, Ren C and Wen H-H 2009 *Phys. Rev.* B **80**, 094513
[7] Song D, Park S R, Kim C, Kim Y, Leem C, Choi S, Jung W, Koh Y, Han G, Yoshida Y, Eisaki H, Lu D H, Shen Z-X and Kim C 2012 *Phys. Rev.* B **86**, 144520
[8] Oyanagi H, Yokoyama Y, Yamaguchi H, Kuwahara Y, Katayama T and Nishihara Y 1990 *Phys. Rev.* B **42**, 10136
[9] Wilson S D, Li S, Dai P, Bao W, Chung J-H, Kang H J, Lee S-H, Komiya S, Ando Y and Si Q 2006 *Phys. Rev.* B **74**, 144514
[10] Kosugi N, Tokura Y, Takagi H and Uchida S 1990 *Phys. Rev.* B **41**, 131
[11] Liang G, Guo Y, Badresingh D, Xu W, Tang Y, Croft M, Chen J, Sahiner A, Beom-hoan O and Markert J T 1995 *Phys. Rev.* B **51**, 1258

IOP Conf. Series: Journal of Physics: Conf. Series **969** (2018) 012052 doi:10.1088/1742-6596/969/1/012052

# Electronic phase diagram of the titanium oxypnictide superconductor $BaTi_2(Sb_{1-x}Bi_x)_2O$

**Wataru Ishii, Takeshi Yajima and Zenji Hiroi**
Institute for Solid State Physics, University of Tokyo, Kashiwa, Chiba 277–8581, Japan

E-mail: yajima@issp.u-tokyo.ac.jp

**Abstract**. The titanium oxypnictide superconductor is studied by resistivity, magnetization and heat capacity measurements down to 0.5 K on the powder samples of all proportional solid solution $BaTi_2(Sb_{1-x}Bi_x)_2O$. An unusual electronic phase diagram as a function of $x$ is obtained, in which a double superconducting dome with peaks at $T_c = 3.6$ and ~4 K appears across an intervening range with lower $T_c$s of ~1 K at $0.35 \leq x \leq 0.55$. Moreover, a sudden jump in $T_c$ is observed at $x_c$ between 0.55 and 0.60. These facts suggest unconventional mechanisms of the superconductivity of the titanium oxypnictides.

## 1. Introduction

Superconductivity is often observed in the proximity of other types of ordered states such as antiferromagnetic orders [1,2] and charge-density-wave (CDW) states [3]. When an ordered state is suppressed as a function of a control parameter such as carrier density or pressure, superconductivity can emerge or be enhanced with its $T_c$ increasing toward a maximum at around the vanishing point of the order. Then, the $T_c$ decreases with further increasing the control parameter. As a result, a superconducting phase with a "$T_c$ dome" occurs next to the ordered phase in the electronic phase diagram. In order to understand the mechanism of the superconductivity, it would be crucial to characterize the nature of the parent ordered phase and to elucidate its relevance to the superconducting phase. The superconducting phase is either competing with the ordered phase or is helped via fluctuations associated with the suppression of the relevant order.

The $T_c$ dome is not always a smooth single peak but can split into two peaks. In the copper oxide superconductor $La_{2-x}Ba_xCuO_4$, a dip in the $T_c$ dome is observed at $x \sim 0.125$, which may be due to the suppression of superconductivity by a charge-ordered phase [4]. In the iron oxypnictide superconductor $LaFeAs(O_{1-x}H_x)$, on the other hand, a distinct double dome appears as a function of electron doping by substituting $H^-$ for $O^{2-}$ ions [5]. The two peaks have been ascribed to either of different types of structural and magnetic orders existing at around $x = 0$ and 0.5 [6]. These observations reveal the presence of competing order parameters or reflect complex electronic structures arising from the multibands and may carry important information on the pairing mechanism.

The titanium oxypnictides with the general formula of $BaTi_2Pn_2O$ (Pn = As, Sb, Bi) include unique superconductors found in 2012 [7-9]. They crystallize in a layered structure comprising the $Ti_2Pn_2O$ layers separated by the Ba layers, as schematically depicted in Fig. 1. It is well established that the electronic properties are mostly governed by the Ti $3d$ states spreading in the $Ti_2O$ sheet [10]. Among the three members, the Sb and Bi compounds exhibit superconductivity at $T_c$ = 1.2 and 4.6 K, respectively, while the As compound remains normal down to 0.5 K. On the other hand, CDW-like phases are observed at $T_{CDW}$ = 200 and 50 K for the As and Sb compounds, respectively. According to the previous studies on the solid solution samples, the CDW order is gradually suppressed with increasing the Sb content starting from the As compound and disappears by a small substitution of Bi for Sb [11]. The nature of the order may not be a simple CDW type, because an accompanying superstructure has not been observed. Instead, a nematic charge order or an orbital order has been proposed [12,13].

An interesting finding in the previous study on the all proportional solid solution of $BaTi_2(Sb_{1-x}Bi_x)_2O$ is the presence of two peaks in the $x$ dependence of $T_c$ [11]: as $x$ increases, the $T_c$ increases from 1.2 K at $x = 0$ to 3.5 K at $x = 0.2$, rapidly decreases below 1.85 K (the lowest measured temperature) at $0.3 < x < 0.60$, and suddenly rises to 4.0 K at $x = 0.60$, followed by a gradual increase to 4.6 K at $x = 1$. The first peak at small $x$ values may be related to the suppression of the CDW order, while no such nearby order has been observed for the second peak. The origin of the two $T_c$ peaks has remained mystery.

In order to clarify the complex doping dependence of $T_c$, we have carried out a systematic study on the $BaTi_2(Sb_{1-x}Bi_x)_2O$ solid solution. Problems in the previous study are the limited number of samples with varying compositions and the limited measurement temperature range above 1.85 K. In addition, the difficulty in preparing samples and the easy degradation of the compounds in air have prevented one from obtaining reliable data thus far. We have carefully prepared a series of polycrystalline samples and extended the temperature range for measurements down to 0.5 K. As the result, it is revealed that the two $T_c$ peaks are not separated from each other but connected by an intervening range with low $T_c$s around 1.0 K. Moreover, the $T_c$ jumps up at $x_c$ between 0.55 and 0.60. It is suggested that a sudden change in the electronic structure and the superconductivity takes place at $x_c$.

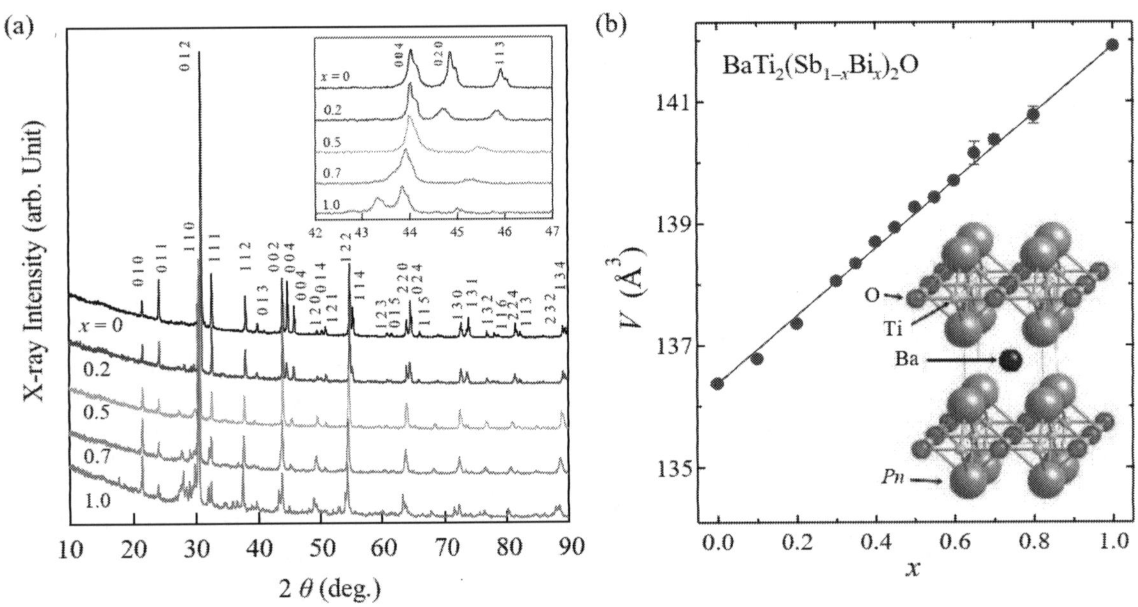


**Figure 1.** (a) Powder x-ray diffraction profiles and (b) the composition dependence of the unit-cell volume $V$ for the solid solution $BaTi_2(Sb_{1-x}Bi_x)_2O$. The crystal structure of the titanium oxypnictides $BaTi_2Pn_2O$ (Pn = As, Sb, Bi) is depicted in the inset of (b).

## 2. Experimental

A series of polycrystalline samples of $BaTi_2(Sb_{1-x}Bi_x)_2O$ ($x$ = 0, 0.10, 0.20, 0.30, 0.35, 0.40, 0.45, 0.50, 0.55, 0.60, 0.70, 0.80, 1) were synthesized by the conventional solid-state reaction method. A stoichiometric amount of BaO (99.99%, Aldrich), Ti (99.9%, Rare Metallic), Sb (99.99%, Rare Metallic), Bi (99.999%, Kojundo Chemical) was weighed, mixed, and pelletized in an Ar-filled glove box. The pellet was wrapped with a Ta foil, sealed in a quartz tube, and heated at 1000 °C for 50 hours. Then, it was slowly cooled to 300 °C at a rate of −3 °C/h before furnace-cooling to room temperature. The obtained samples were highly sensitive to air and moisture, more sensitive with increasing $x$. A special care was taken to minimize the degradation of the samples, as will be reported elsewhere. X-ray diffraction (XRD) experiments were performed at room temperature in a SmartLab (Rigaku) with Cu-K$\alpha$ radiation. The temperature dependence of electrical resistivity was measured by a standard four-probe method in a Physical Properties Measurement System (PPMS, Quantum Design). Heat capacity was measured in the temperature range from 0.5 to 10 K by the relaxation method in a PPMS equipped with a $^3$He refrigerator. In order to avoid the sample degradation, a pellet was wrapped by a 0.03 mm-thick Au foil in an Ar-filled glove box, and the contribution of the Au foil was subtracted from the data. Magnetic susceptibility was measured in the temperature range from 1.8 to 7 K under 5 Oe in a SQUID magnetometer (MPMS, Quantum Design) using samples wrapped by a Parafilm (Bemis).

## 3. Results and discussion

All the XRD patterns of the obtained samples are nearly monophasic, as shown in Fig. 1(a), and the lattice parameters vary almost linearly as a function of $x$ in the whole $x$ range, as typically shown for the unit-cell volume in Fig. 1(b). Thus, we have obtained a series of all proportional solid solution samples with varying Sb/Bi ratio. The electrical resistivities for $x = 0$, 0.10 and 0.20 show upturns due to CDW transitions, while the $x = 0.30$ sample shows a smooth temperature dependence (Fig. 2a). The transition temperature $T_{CDW}$ is determined from the temperature derivative of the resistivity curve, which decreases gradually with increasing $x$ (Fig. 2b).

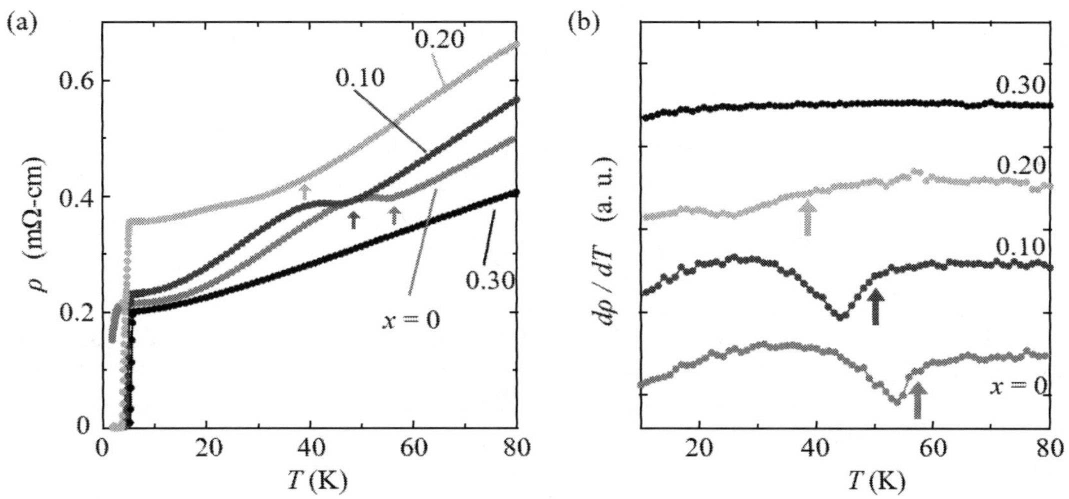


**Figure 2**. Evolution of (a) resistivity curves and (b) their temperature derivatives for the solid solution $BaTi_2(Sb_{1-x}Bi_x)_2O$ with $x = 0$, 0.10, 0.20 and 0.30.

At low temperatures, the resistivities show abrupt drops to zero at around 5.6 K except for $x = 0$, which may be due to the superconducting impurity phase of $BaBi_3$ with $T_c = 5.6$ K [14]: the amount of $BaBi_3$ in the samples must be small as it is not detected in the XRD patterns of Fig. 1(a), but a superconducting path may be generated in a filament. Superconducting transitions of the bulk nature are detected by magnetic susceptibility and heat capacity measurements. Figure 3 shows the temperature dependences of magnetic susceptibility down to 1.8 K for $0.10 \leq x \leq 0.30$ and $0.60 \leq x \leq 1$. Large diamagnetic signals due to superconducting transitions are observed in the zero-field cooling curves for all the samples. The shielding volume fractions exceed 40% at the lowest temperature, indicating of the bulk nature of the superconducting transitions. A small diamagnetic signal of 0.1% or less in volume fraction was also observed at 5.6 K from the impurity $BaBi_3$. The $T_c$s of $BaTi_2(Sb_{1-x}Bi_x)_2O$ are determined at the onsets; 2.4 K ($x$ = 0.10), 3.6 K (0.20), 2.6 K (0.30), 4.5 K (0.60), 4.5 K (0.70), 4.6 K (0.80) and 4.65 K (1). In contrast, no superconducting signals are observed above 1.8 K for $0.30 < x < 0.60$. The observed $x$ dependence of $T_c$ is consistent with that in the previous study [11].

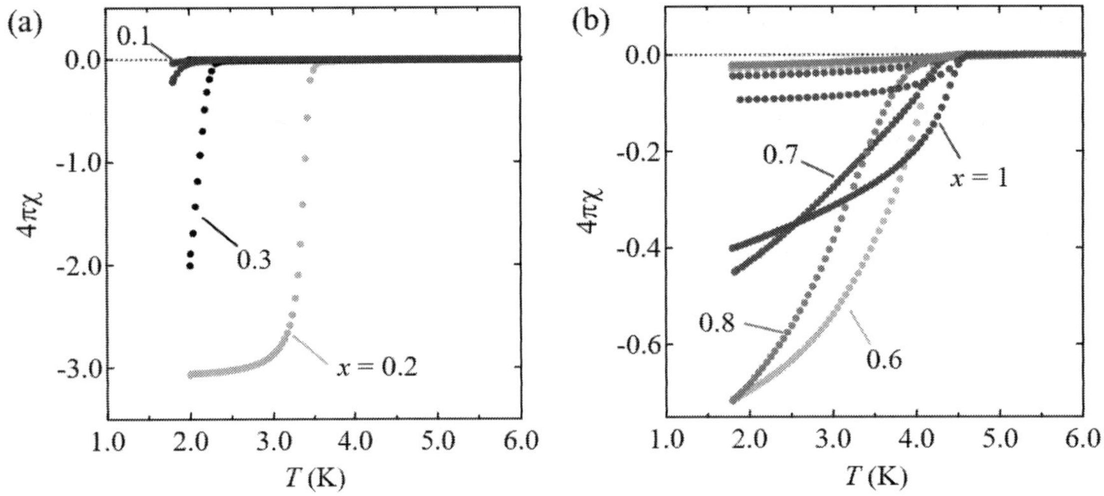


**Figure 3**. Magnetic susceptibility for (a) $x$ = 0.1, 0.2, 0.3 and (b) $x$ = 0.6, 0.7, 0.8, 1. Each measurement was carried in a magnetic field of 5 Oe upon heating after zero-field cooling and then upon cooling in the field.

In order to investigate the missing window of superconductivity at $0.30 < x < 0.60$, we have performed heat capacity measurements down to 0.5 K. Figure 4 shows the electronic heat capacity divided by temperature, $C_e/T$, for the samples with $x$ = 0.35, 0.40, 0.45, 0.50 and 0.55, which were obtained after the subtraction of the lattice contribution. For all the samples, distinct but broad peaks are observed, indicating superconducting transitions in bulk at low temperatures. The $T_c$s are estimated on the basis of the entropy conservation for the broad transitions, as typically shown for $x$ = 0.50 in Fig. 4(b): 1.45 K ($x$ = 0.35), 1.0 K (0.40), 0.95 K (0.45), 0.88K (0.50) and 0.88 K (0.55). The magnitude of the jump at $T_c$, $\Delta C_e(T_c)/\gamma T_c$, is 0.45 for $x$ = 0.50, which is significantly smaller than the weak-coupling BCS value of 1.43. The small jump and the broad transition may be partly due to degradation of the samples during the measurements.

Thus obtained $T_{CDW}$s and $T_c$s are plotted in the phase diagram of Fig. 5. The $T_{CDW}$ decreases gradually from 57 K at $x$ = 0 to 40 K at $x$ = 0.20 and seems to vanish at around $x$ = 0.30 in the resistivity data; it is not clear at which composition the $T_{CDW}$ really goes to zero. On the other hand, the $T_c$ shows a peak at $x$ = 0.20 and remains at ~1 K at $x$ = 0.4–0.55. Then, surprisingly, it shows a sudden jump to 4.5 K at $x$ = 0.60 and remains nearly constant to the Bi end. Note that the superconductivity is observed at $x$ = 0.4–0.55 in contrast to the previous result [11]. Moreover, it becomes clear that there is a discontinuous change in $T_c$ at $0.55 < x_c < 0.60$.

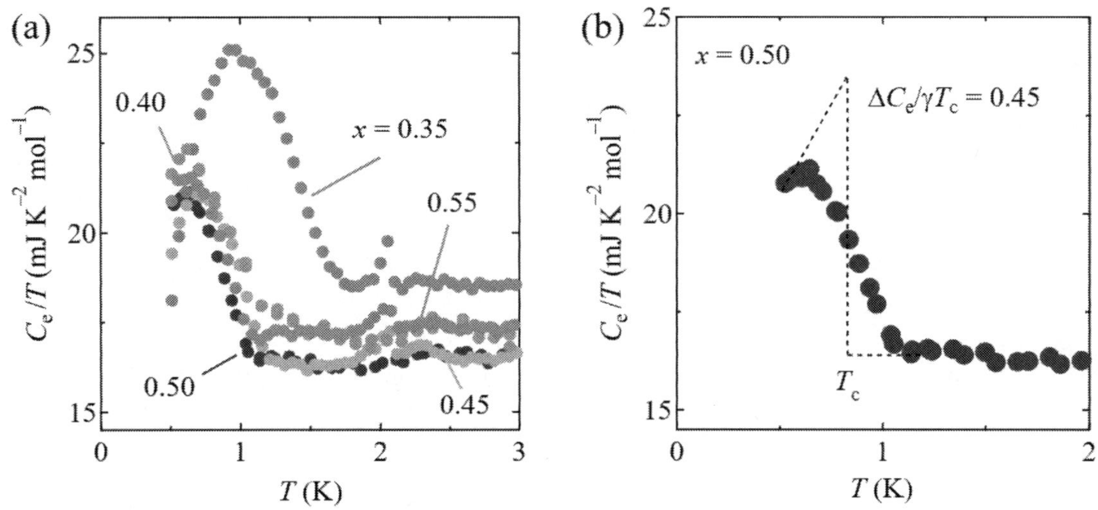


**Figure 4**. (a) Electronic heat capacity divided by temperature down to 0.5 K for $x = 0.35$, 0.40, 0.45, 0.50 and 0.55. (b) $x = 0.50$ data.

The double $T_c$ dome of $BaTi_2(Sb_{1-x}Bi_x)_2O$ is reminiscent of the similar double dome of $LaFeAs(O_{1-x}H_x)$ [6]. The first peak at smaller $x$ values is no doubt related to the adjacent CDW order: its suppression by the Bi substitution must cause the $T_c$ peak, although it is not clear whether the CDW phase competes with or helps the superconductivity. In contrast, the presence of the superconductivity at $x \geq 0.6$ seems not to have such a parent ordered phase. Alternatively, one might assume that there is a dip at $x = 0.4$–0.55 in the single $T_c$ dome spreading over the whole $x$ range. In this case, the dip could be introduced as a consequence of the stabilization of a certain competing order, as in the case of $La_{2-x}Ba_xCuO_4$ [4].

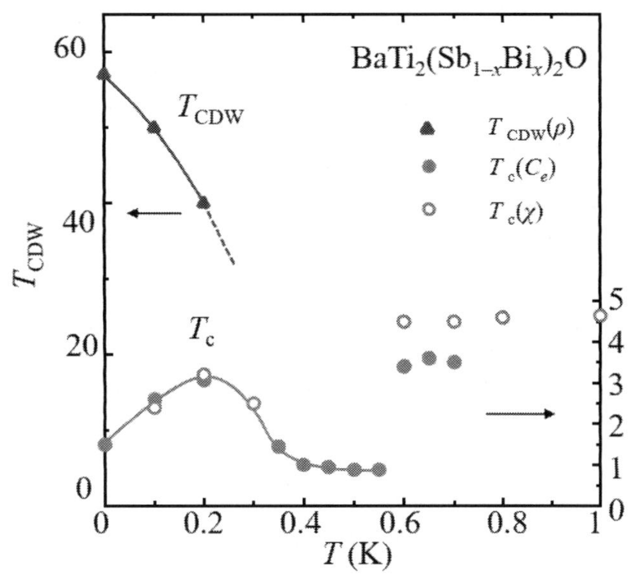


**Figure 5**. Phase diagram of $BaTi_2(Sb_{1-x}Bi_x)_2O$. The CDW transition temperatures $T_{CDW}$ determined from the resistivity data are plotted by the triangles, and the superconducting transition temperatures $T_c$ from the magnetic susceptibility and heat capacity data are shown by open and filled circles, respectively.

Finally, we would like to consider the origin of the sudden jump of $T_c$ at $x_c$. Note that the two end members of $BaTi_2(Sb_{1-x}Bi_x)_2O$ are structurally and electronically quite similar to each other. Their crystal structures at room temperature are basically the same, and no anomaly has been observed upon cooling in the solid solution. Thus, it is unlikely to assume a structural transition at $x_c$ to cause the $T_c$ jump, as in the cases of $Au_{1-x}Pd_xTe_2$ and $Ca(Fe_{1-x}Rh_x)_2As_2$ [15,16]. On the other hand, the electronic structures of the two end members also resemble each other, having similar Fermi surfaces [17], which make one believe the absence of a drastic change in the electronic structures at $x_c$ by the isovalent substitution. Further experiments are required for understanding this interesting phase diagram of $BaTi_2(Sb_{1-x}Bi_x)_2O$.

## 4. Summary

The electronic phase diagram of the titanium oxypnictide superconductor $BaTi_2(Sb_{1-x}Bi_x)_2O$ is studied. It is found that the $T_c$ exhibits a double dome with a dip at $0.3 < x < 0.6$ and a sudden jump at $0.55 < x_c < 0.60$.

**Acknowledgments**
We are grateful to D. Hirai for helpful discussion. This work was partially supported by JSPS KAKENHI Grant Number 15K17698 and the Core-to-Core Program for Advanced Research Networks given by the Japan Society for the Promotion of Science.

**References**

[1] Bednorz J G and Muller K A 1986 *Z. Phys. B* **64** 189
[2] Kamihara Y, Watanabe T, Hirano M and Hosono H 2008 *J. Am. Chem. Soc.* **130** 3296
[3] Sipos B, Kusmartseva F, Akrap A, Berger H, Forró L and Tutis E 2008 *Nat. Mater.* **7** 960
[4] Homes C C, Hucker M, Li Q, Xu Z J, Wen J S, Gu G D and Tranquada J M 2012 *Phys. Rev. B* **85** 134510
[5] Iimura S, Matsuishi S, Sato H, Hanna T, Muraba Y, Kim S W, Kim J E, Takata M and Hosono H 2012 *Nat. Commun.* **3** 943
[6] Hiraishi M, Iimura S, Kojima K M, Yamaura J, Hiraka H, Ikeda K, Miao P, Ishikawa Y, Torii S, Miyazaki M, Yamauchi I, Koda A, Ishii K, Yoshida M, Mizuki J, Kadono R, Kumai R, Kamiyama T, Otomo T, Murakami Y, Matsuishi S and Hosono H 2014 *Nat. Phys.* **10** 300
[7] Wang X F, Yan Y J, Ying J J, Li Q J, Zhang M, Xu N and Chen X H 2010 *J. Phys. Condens. Matter* **22** 075702
[8] Yajima T, Nakano K, Takairi F, Ono T, Hoshokoshi Y, Mathushita Y, Hester J, Kobayashi Y and Kageyama H 2012 *J. Phys. Soc. Jpn.* **81** 103706
[9] Yajima T, Nakano K, Takeiri F, Hester J, Yamamoto T, Kobayashi Y and Kageyama H 2013 *J. Phys. Soc. Jpn.* **82** 013703
[10] Singh D J 2012 *New. J. Phys.* **14** 123003
[11] Yajima T, Nakano K, Takeiri F, Nozaki Y, Kobayashi Y and Kageyama H 2013 *J. Phys. Soc. Jpn.* **82** 033705
[12] Frandsen Y, Bozin E S, Hu H, Zhu Y, Nozaki Y, Kageyama H, Uemura1 W J, Yin W and Billinge S L 2014 *Nat. Commun.* **5** 5761
[13] Nakaoka H, Yamakawa Y and Kontani H 2016 *Phys. Rev. B* **93** 245122
[14] Haldolaarachchige N, Kushwaha S K, Gibson Q and Cava R J 2014 *Supercond. Sci. Tech.* **27** 105001
[15] Kudo K, Ishii H and Nohara M 2016 *Phys. Rev. B* **93** 140505
[16] Danura M, Kudo K, Oshiro Y, Araki S, Kobayashi T C and Nohara M 2011 *J. Phys. Soc. Jpn.* **80** 103701
[17] Nakano K, Hongo K and Maezono R 2016 *Sci. Rep.* **6** 29661

# Superposition of $\sqrt{13} \times \sqrt{13}$ and $3 \times 3$ supermodulations in $TaS_2$ probed by scanning tunneling microscopy

**Y Fujisawa, T Iwasaki, D Fujii, S Ohta, J Iwashita, T Fujita, M Nakata, K Kishimoto, S Demura, and H Sakata**

Department of Physics, Tokyo University of Science, Shinjuku-ku, Tokyo 162-8601, Japan

1215702@ed.tus.ac.jp

**Abstract**. We report on a scanning tunnelling microscopy study of $TaS_2$ at 4.2 K. A surface prepared by cleavage showed a superimposed pattern of two types of charge density waves with $3a_0 \times 3a_0$ and $\sqrt{13}a_0 \times \sqrt{13}a_0$ periodicity, which had never been observed previously. We attribute the superposition to regular stacking of $4H_b$ polytypes or irregular stacking of $2H$ and $4H_b$ layers.

## 1. Introduction

Scanning tunneling microscopy (STM) is a technique for investigating the surface structure of materials with atomic resolution. STM can generally probe the top most layer. However, it occasionally probes not only the top most layer, but also the effect of the underlying layer. A Moiré pattern is an example of such effects and results from supermodulation due to two atomic layers that are stacked at an angle [1]. For example, a Moiré pattern is observed on the surface of highly ordered pyrolytic graphite, bilayer graphene, and graphene on transition metal substrates [2, 3]. On the other hand, there are few reports of such phenomena in other layered materials, such as transition metal dichalcogenides (TMDCs) [4].

One of the TMDCs, $TaS_2$, has several polytypes, such as $1T$, $2H$, and $4H_b$, and shows various charge density wave (CDW) states depending on the polytype. In $1T$-$TaS_2$, a layer is composed of edge sharing $TaS_6$ octahedra. It undergoes a phase transition from a nearly commensurate CDW state to the Mott insulating commensurate state with a period of $\sqrt{13}a_0 \times \sqrt{13}a_0$ at about 200 K [5, 6]. In $2H$-$TaS_2$, a layer is composed of edge shared $TaS_6$ prisms. It undergoes a phase transition to a commensurate CDW state with a period of $3a_0 \times 3a_0$ at 75 K. The electronic state is metallic and superconducting below 1 K [7, 8]. In $4H_b$-$TaS_2$, the structure consists of stacks of alternating $1T$ and $1H$ layers. Further, $4H_b$-$TaS_2$ undergoes CDW transitions and shows $\sqrt{13}a_0 \times \sqrt{13}a_0$ and $3a_0 \times 3a_0$ CDWs within each layer at 4.2 K. It also shows superconductivity at a transition temperature $T_c$ of 3.5 K, which is three times higher than that of $2H$-$TaS_2$ [9, 10]. Because $4H_b$-$TaS_2$ consists of alternating stacks of two different CDW layers, this material is expected to show superposition of the CDWs. However, previous reports presented STM images that showed both CDWs depending on the bias voltage but did not show superposition of the CDWs [11, 12]. In this study, we performed an STM study of a single crystal of $TaS_2$ and revealed a superimposed pattern composed of two types of CDWs with $\sqrt{13}a_0 \times \sqrt{13}a_0$ and $3a_0 \times 3a_0$ periodicity on the surface.

## 2. Experimental details

### *2.1. Crystal growth and characterization*

The single crystal used in this study is grown by the chemical vapor transport method with iodine as a transport agent. First, Ta and S powders are mixed at chemical stoichiometry, heated at 1000 °C for 4 days in an evacuated silica tube, and then quenched in water. Second, polycrystalline $TaS_2$ and iodine (5 mg/cm$^3$) are sealed in another quartz tube and heated for 7 days, where the source zone and the growth zone are fixed at 1000 and 850 °C, respectively. The obtained single crystal is evaluated by X-ray diffraction (XRD) using Cu $K_\alpha$ radiation and resistivity measurements. The XRD patterns are shown in Fig. 1(a) and (b). Figure 1(a) shows clear (00*l*) peaks without extra peaks. However, closer examination of the (008) peak [Fig. 1(b)] shows split peaks. Although the two peaks labeled as A′ and B′ are due to diffraction by Cu $K_{\alpha 2}$, peaks A and B are associated with two different lattice constants, which are estimated to be $6.05n$ and $5.97n$ Å, where $n$ is the number of stacks. Because these lattice constants coincide with those of $4H_b$ and $2H$, we conclude that our crystal consists of two polytypes; the major one is $4H_b$ and the minor one is $2H$. Figure 1(c) shows the temperature dependence of the resistivity of the single crystal. The resistivity decreases monotonically with decreasing temperature. This also indicates that our sample consists of metallic $2H$ and $4H_b$ polytypes. The inset shows the resistivity from 5 to 2 K. A rapid decrease due to superconductivity appears at about 3.5 K, which corresponds to the $T_c$ value of the $4H_b$ polytype.

### *2.2. Experimental setup for the STM/STS measurements*

STM/scanning tunnelling spectroscopy (STS) measurements are performed by a laboratory-made scanning tunneling microscope. All results are obtained at 4.2 K under a He gas environment at a few torr. Electrochemically polished Au wire is used as an STM tip. STM images are taken in the constant current mode. Tunneling spectra are obtained by numerical differentiation of the *I*-*V* characteristics. The sample surface is prepared by cleavage at 4.2 K. Because our sample consists of $2H$ and $4H_b$ polytypes, which have alternating stacks of $1T$ and $1H$ layers, the top most surface obtained by cleavage cannot be selected in a controlled manner.

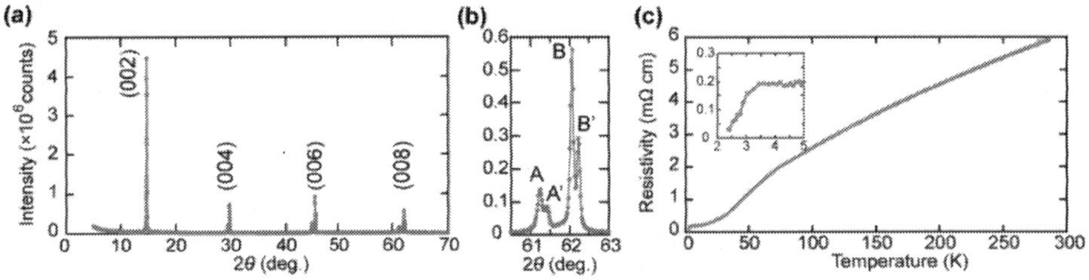


Figure 1(a) XRD profile of the single crystal used in this study. (b) Magnified view of XRD profile in (a). (c) Temperature dependence of resistivity of the single crystal used in this study.

## 3. Results and Discussion

Figure 2(a) shows an STM image obtained at a bias voltage of −50 mV. The observed pattern does not correspond to either $\sqrt{13}a_0 \times \sqrt{13}a_0$ or $3a_0 \times 3a_0$ CDWs; although $\sqrt{13}a_0 \times \sqrt{13}a_0$ and $3a_0 \times 3a_0$ CDWs have three fold symmetry, the observed image does not. To disentangle the complicated supermodulation, we performed fast Fourier transform (FFT) analysis. The inset of Fig. 2(b) shows the corresponding FFT image. One can see six Bragg peaks (black circles). This indicates that the top most S atoms form a regular triangular lattice. The main figure shows the magnified FFT image. This image captures two types of CDWs; the one marked by red circles corresponds to the $3a_0 \times 3a_0$ CDW, and the one marked by blue diamonds corresponds to the $\sqrt{13}a_0 \times \sqrt{13}a_0$ CDW. Whereas the wave vector of the $3a_0 \times 3a_0$ CDW is aligned with the S atoms, the wave vector of the $\sqrt{13}a_0 \times \sqrt{13}a_0$ CDW is rotated by 14° with respect to the S atoms. These results show that the STM image can be expressed as the superposition of the two types of CDWs with periods of $\sqrt{13}a_0 \times \sqrt{13}a_0$ and $3a_0 \times 3a_0$. Because each CDW comes from different layers, namely the *T* and *H* layers, respectively, the observed STM image is a result of the superposition of the effects of each layer. The lack of three fold symmetry in the obtained STM image is due to the difference in the intensity of the FFT spots of the $3a_0 \times 3a_0$ CDW; the intensity of spots in one direction is much stronger than that of spots in the other two directions. This is not the effect of the tip shape, because there is no difference between the intensity of the FFT spots of the other CDW or atoms. It is reported that, in $2H$-$NbSe_2$, owing to the strain of one layer perpendicular to the other layer, the $3q$ state was suppressed, and the $1q$ state was realized [13]. The non equivalence of the $3q$ state in this sample may result from strain caused by stacking of different types of layers.

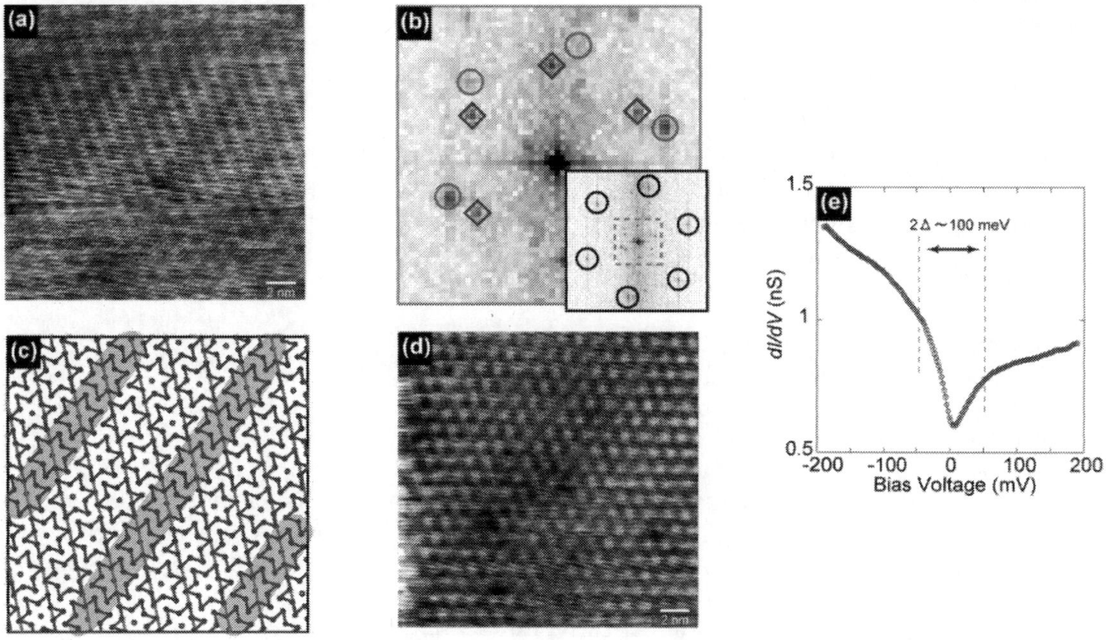


Figure 2(a) STM image at bias voltage of −50 mV and current of 200 pA. (b) Magnified FFT image of (a). Inset shows FFT image including Bragg peaks. (c) Possible model explaining the superposition pattern in (a). (d) STM image taken in almost the same area as (a). The set point is 500 mV and 200 pA. (e) Tunneling spectrum of the surface investigated by STM.

Figure 2(c) shows a schematic of the configuration of the two CDWs. Because one of the three components in the $3a_0 \times 3a_0$ CDW in the *H* layer is stronger, the superposition of the star of David pattern and one $3a_0$ component causes one-dimensional supermodulation with a period of about $8.1a_0$, which is consistent with the observed STM image. This indicates that there is no misorientation or slight relative displacement of the atomic lattices, but that STM image can be described as the superposition of two types of CDWs.

The relative strength of the two CDWs changed depending on the bias voltage. Figure 2(d) shows an STM image taken at a bias voltage of +500 mV. At this bias voltage, the $\sqrt{13}a_0 \times \sqrt{13}a_0$ CDW is greatly enhanced compared to its appearance in the STM image at −50 mV. The same tendency has been reported in a $4H_b$ sample, although the superposition of two CDWs has not been observed [11, 12]. The tunneling spectrum of this surface is shown in Fig. 2(e). The observed spectrum has a finite local density of states (LDOS) at the Fermi level ($E_F$) with a shallow gap of approximately 100 meV. This feature does not coincide with either the $1T$ surface of $4H_b$-$TaS_2$, where a clear energy gap opens, or the $1H$ surface, where the spectrum is metallic and flat near $E_F$, but rather is rather intermediate between the two. The shallow gap seems to come from the energy gap in the $1T$ layer, indicating a smaller LDOS contribution from the $1T$ layer at a low bias voltage. Thus, the low-bias STM would show the $3a_0 \times 3a_0$ CDW more strongly.

We found that the STM image shown in Fig. 2(a) is explained by the superposition of two CDWs. The superposition is due to stacking of $1T$ and $1H$ layers. One possible origin of this stacking is $4H_b$ structure, which consists of alternate stacks of $1T$ and $1H$ layers, although previous reports on $4H_b$ have not shown the superposition of two CDWs on both the $1T$ and $1H$ layers [11, 12]. Another possibility is irregular stacking. As shown in the XRD results in Fig. 1(b), our sample has a partial $2H$ polytype volume fraction, indicating possible irregular stacking of $2H$ and $4H_b$ layers, such as *H-H-H-T*. In this case, if cleavage revealed a $1T$ layer as the top most layer, the resultant structure below the top $1T$ layer is $2H$, which is different from $4H_b$ stacking. This stacking may produce superposition of two CDWs.

## 4. Summary

In conclusion, we report an STM study of $TaS_2$ that revealed the superposition of two types of CDWs with periods of $\sqrt{13}a_0 \times \sqrt{13}a_0$ and $3a_0 \times 3a_0$. We propose two possible explanations for the phenomenon. One is regular stacking of the $4H_b$ crystal structure. Another is some types of stacking fault such as *H-H-H-T* stacking.

## References

[1] Macdonald A and Bistritzer R 2011 *Nature* **474** 453
[2] Zhao Y R and Kuiper P 1993 *Phys. Rev. Lett.* **48** 17427
[3] N'Diaye A T, Bleikamp S, Feibelman P J and Michely T 2006 *Phys. Rev. Lett.* **97** 215501
[4] Rossnagel K 2011 *J. Phys.: Condens. Matter* **23** 213001
[5] Wilson J A, DiSalvo F J and Mahasan S 1975 *Adv. Phys.* **24** 117
[6] Kim J J, Yamaguchi W, Hasegawa T and Kitazawa K 1994 *Phys. Rev. Lett.* **73** 2103
[7] Li L, Deng X, Wang Z, Liu Y, Abeykoon M, Dooryhee E, Tomic A, Huang Y, Warren J B, Bozin E S, Billinge S J L, Sun Y, Zhu Y, Kotliar G and Petrovic C 2017 *npj Quantum Mater.* **2** 11
[8] Dai J, Calleja E, Alldredge J, Zhu X, Li X, Lu W, Sun Y, Wolf T, Berger H and McElroy K 2014 *Phys. Rev. B* **89** 165140
[9] DiSalvo F J, Bargley B G, Voorhoeve J M and Waszczak J V 1973 *J. Phys. Chem. Solids* **34** 1357
[10] Liu Y, Li L J, Lu W J, Ang R, Liu X Z and Sun Y 2014 *J. Appl. Phys.(Melville, NY, U. S.)* **115** 043915
[11] Han W, Hunt E R, Pankratov O, Frindt R F 1994 *Phys. Rev. B (R)* **50** 14746
[12] Ekvall I, Kim J J, Olin H 1997 *Phys. Rev. B* **55** 6758
[13] Soumyanarayanan A, Yee M M, He Y, Wezel J, Rahn D, Rossnagel K, Hudson E W, Norman M R and Hoffman J E 2013 *Proc. Natl. Acad. Sci. U. S. A* **110** 1623

28th International Conference on Low Temperature Physics (LT28) IOP Publishing
IOP Conf. Series: Journal of Physics: Conf. Series **969** (2018) 012054 doi:10.1088/1742-6596/969/1/012054

# Charged and uncharged vortices in quasiclassical theory

**Yusuke Masaki**[1] **and Yusuke Kato**[1,2]

[1]Department of Physics, The University of Tokyo, Bunkyo, Tokyo 113-0033, Japan
[2]Department of Basic Science, The University of Tokyo, Meguro, Tokyo 153-8902, Japan

E-mail: masaki@vortex.c.u-tokyo.ac.jp

**Abstract.** The charging effect of a superconducting vortex core is very important for transport properties of superconducting vortices. The chiral p-wave superconductor, known as a topological superconductor (SC), has a Majorana fermion in a vortex core and the charging effect has been studied using microscopic Bogoliubov–de Gennes (BdG) theory. According to calculations based on the BdG theory, one type of the vortex is charged as well as the vortex of the s-wave SC, while the other is uncharged. We reproduce this interesting charging effect using an augmented quasiclassical theory in chiral p-wave SCs, by which we can deal with particle-hole asymmetry in the quasiclassical approximation.

## 1. Introduction

Electronic bound states in a superconducting vortex core, which is called the Caroli–de Gennes–Matricon (CdGM) mode [1] or Andreev bound state [2, 3], cause a lot of fascinating phenomena. For example, the disorder-scattering bound states result in the viscous flow of the vortex, while the particle-hole asymmetry of the CdGM mode as well as other origins of the particle-hole asymmetry leads to the Hall effect of the flux flow states [4, 5, 6]. The charging effect of the vortex core is another consequence of the particle–hole asymmetry of the CdGM mode. Recently there is another interest in the zero energy bound state in a vortex core of a topological superconductor, called the Majorana state [7, 8]. Majorana bound states obey the non-Abelian statistics and enable a fault tolerant quantum computation [9], which will be done by exchanging the vortices [10]. Thus deeper understanding of the flux flow states of topological superconductors is highly required from both fundamental and applicational viewpoints.

As we mentioned above, the particle-hole asymmetry can be understood as a charging effect of the vortex core [11, 12], which has been studied using microscopic Bogoliubov–de Gennes (BdG) theory [13, 14] and very recently done using augmented quasiclassical theory [15, 16]. The charging effect may be regarded as the main mechanism leading to the Hall effect of the flux flow state [11]. It is numerically prohibitive to study the non-equilibrium phenomena with use of the BdG theory or Gor'kov theory. Because of this, the study of the flux-flow state has been done using the time dependent Ginzburg–Landau equation [17], the Usadel equation [18], and the Eilenberger equation [19, 20], the former two of which are valid only for dirty superconductors. The Eilenberger theory is a quasiclassical theory and a very powerful tool to study the flux flow state, but the conventional one is always particle-hole symmetric. Taking the next leading order of the quasiclassical approximation into account, we can deal with the charging effect and thereby the Hall state in the quasiclassical regime. Kopnin developed the convenient kinetic theory with respect to the gauge transformation within the Wigner representation [21], which was originally given in the normal state [22]. Subsequently Kita has improved the form of the phase transformation [23].

In this paper we apply this theory to the chiral p-wave SCs [24, 25], and study the charging effect. Here we show that the striking difference between two types of vortices in chiral p-wave SCs can be obtained also in this framework, which is known by BdG theory [14]. Our scheme can be extended to systems with disorder in non-equilibrium and thus we believe that our result in the present work becomes a first step toward consistent understanding of the charging effect and the Hall effect in chiral p-wave SCs.

In this paper, we use the unit that $k_{\mathrm{B}} = \hbar = 1$.

## 2. Formulation

Green's function is defined using the Nambu spinor $\vec{\Psi}(\boldsymbol{r},\tau) = {}^t[\psi_\uparrow(\boldsymbol{r},\tau), \psi_\downarrow^\dagger(\boldsymbol{r},\tau)]$ as

$$\widehat{G}^{\mathrm{M}}(\boldsymbol{r}_1,\boldsymbol{r}_2,\omega_n) = \begin{bmatrix} G^{\mathrm{M}}(\boldsymbol{r}_1,\boldsymbol{r}_2,\omega_n) & F^{\mathrm{M}}(\boldsymbol{r}_1,\boldsymbol{r}_2,\omega_n) \\ -\bar{F}^{\mathrm{M}}(\boldsymbol{r}_1,\boldsymbol{r}_2,\omega_n) & \bar{G}^{\mathrm{M}}(\boldsymbol{r}_1,\boldsymbol{r}_2,\omega_n) \end{bmatrix} = \int_0^\beta \mathrm{d}\tau e^{\mathrm{i}\omega_n\tau}\widehat{\tau}_3\left[\langle\vec{\Psi}(\boldsymbol{r}_1,\tau)\vec{\Psi}^\dagger(\boldsymbol{r}_2)\rangle\right]. \quad (1)$$

Here $\omega_n$ and $T$ are the fermion Matsubara frequency and temperature, respectively, and $\beta = 1/T$ is the inverse temperature. The symbol $\widehat{\cdot}$ denotes the 2 by 2 matrix, and $\widehat{\tau}_3$ is the third component of the 2 by 2 Pauli matrix. The Fourier transform with respect to the relative spatial and temporal coordinates leads to the Wigner representation. In this paper, we only consider the system in equilibrium, which means that Green's function depends only on the relative imaginary time. When we discuss the gauge transformation it is convenient to do the phase transformation introduced in Refs. [21, 23]. In this paper, we consider the phase transformation only in the space and vector potential, and consider the partially fixed gauge, where the electric field is described by the scalar potential only. The same transformation was done in Refs. [15, 16]. Within this scheme, we restrict ourselves to the equilibrium situation, namely, there are no time dependence even in the gauge field.

The Green's function in the space-momentum mixed representation and thereby the quasiclassical Green's function are introduced by the Fourier transform with the relative coordinate $\boldsymbol{r} = \boldsymbol{r}_1 - \boldsymbol{r}_2$ and the energy integration:

$$\begin{aligned} \widehat{g}^{\mathrm{M}}(\boldsymbol{k}_{\mathrm{F}},\boldsymbol{R};\mathrm{i}\omega_n) &= \oint_{C_{\mathrm{qc}}} \frac{\mathrm{d}\xi_{\boldsymbol{k}}}{\mathrm{i}\pi}\widehat{G}^{\mathrm{M}}(\boldsymbol{k},\boldsymbol{R};\mathrm{i}\omega_n) \\ &= \oint_{C_{\mathrm{qc}}} \frac{\mathrm{d}\xi_{\boldsymbol{k}}}{\mathrm{i}\pi}\int \mathrm{d}\boldsymbol{r} e^{\mathrm{i}I(\boldsymbol{R},\boldsymbol{r}_1)\widehat{\tau}_3}\widehat{G}^{\mathrm{M}}(\boldsymbol{r}_1,\boldsymbol{r}_2;\mathrm{i}\omega_n)e^{-\mathrm{i}I(\boldsymbol{R},\boldsymbol{r}_2)\widehat{\tau}_3}e^{-\mathrm{i}\boldsymbol{k}\cdot\boldsymbol{r}}, \end{aligned} \quad (2)$$

where $C_{\mathrm{qc}}$ is the contour used in Ref. [19] and the phase $I(\boldsymbol{R},\boldsymbol{r}_1)$ is defined as $I(\boldsymbol{R},\boldsymbol{r}_1) = -\frac{e}{c}\int_{\boldsymbol{r}_1}^{\boldsymbol{R}}\mathrm{d}\boldsymbol{s}\cdot\boldsymbol{A}(\boldsymbol{s})$ with vector potential $\boldsymbol{A}$, the center-of-mass coordinate $\boldsymbol{R} = (\boldsymbol{r}_1+\boldsymbol{r}_2)/2$, electron charge $e$ and the speed of light $c$. We describe the components of the quasiclassical Green's function as $\widehat{g}^{\mathrm{M}} = \begin{bmatrix} g & f \\ -\bar{f} & \bar{g} \end{bmatrix}$.

We obtain the Eilenberger equation by the following steps: (i) Expand the left- and right-sided Gor'kov equations for the above phase transformed Green's function up to the second order in the gradient expansion (ii) Subtract the right equation from the left equation (iii) Integrate the obtained equation along the above contour with the approximation that the momentum dependence of the self energy and pair potential terms can be treated by considering only the Fermi momentum $\boldsymbol{k}_{\mathrm{F}}$. The augmented Eilenberger equation is finally obtained as follows:

$$[\mathrm{i}\omega_n\widehat{\tau}_3 + \widehat{\sigma}^{\mathrm{M}}, \widehat{g}^{\mathrm{M}}]_\star + \mathrm{i}\boldsymbol{v}_{\boldsymbol{k}_{\mathrm{F}}}\cdot\partial_{\boldsymbol{R}}\widehat{g}^{\mathrm{M}} + \frac{e}{2}(\boldsymbol{v}_{\boldsymbol{k}_{\mathrm{F}}}\times\boldsymbol{B})\left(\mathrm{i}\frac{\partial}{\partial\boldsymbol{k}_{\mathrm{F},\parallel}}\right)\{\widehat{\tau}_3,\widehat{g}^{\mathrm{M}}\} = 0 \quad (3)$$

IOP Conf. Series: Journal of Physics: Conf. Series **969** (2018) 012054 doi:10.1088/1742-6596/969/1/012054

with the self energy consisting of the pair potential $\Delta$ and the impurity self-energy $\widehat{\sigma}^{\rm M}_{\rm imp}$

$$\widehat{\sigma}^{\rm M} = \begin{bmatrix} \sigma_{\rm d} & \sigma_{\rm od} \\ \bar{\sigma}_{\rm od} & \bar{\sigma}_{\rm d} \end{bmatrix} = \begin{bmatrix} 0 & \Delta(\boldsymbol{k}_{\rm F}, \boldsymbol{R}) \\ -\Delta^*(\boldsymbol{k}_{\rm F}, \boldsymbol{R}) & 0 \end{bmatrix} + \widehat{\sigma}^{\rm M}_{\rm imp} \tag{4}$$

and

$$\partial_{\boldsymbol{R}} = \begin{cases} \boldsymbol{\nabla} & \text{on } g, \bar{g}, \sigma_{\rm d}, \bar{\sigma}_{\rm d} \\ \boldsymbol{\nabla} - \dfrac{2{\rm i}e}{c}\boldsymbol{A}(\boldsymbol{R}) & \text{on } f, \sigma_{\rm od} \\ \boldsymbol{\nabla} + \dfrac{2{\rm i}e}{c}\boldsymbol{A}(\boldsymbol{R}) & \text{on } \bar{f}, \bar{\sigma}_{\rm od}. \end{cases} \tag{5}$$

The braces in the last term of Eq. (3) denote the anticommutator. The subscript $\|$ in the momentum derivative stands for the tangential component of the momentum to the Fermi surface. The subscript $\star$ of the first term in Eq. (3) means that the product of the commutation relation is replaced by the Moyal product in the spatial part ($\boldsymbol{r}-\boldsymbol{k}$ derivatives), whose definition is given by

$$[A, B]_\star \equiv A \star B - B \star A, \tag{6}$$

$$A \star B \equiv \lim_{\boldsymbol{R}' \to \boldsymbol{R}} \lim_{\boldsymbol{k}'_{\rm F} \to \boldsymbol{k}_{\rm F}} \exp\left[\frac{\rm i}{2}\left(\frac{\partial}{\partial \boldsymbol{k}'_{{\rm F},\|}} \cdot \partial_{\boldsymbol{R}} - \frac{\partial}{\partial \boldsymbol{k}_{{\rm F},\|}} \cdot \partial_{\boldsymbol{R}'}\right)\right] A(\boldsymbol{k}_{\rm F}, \boldsymbol{R}; {\rm i}\omega_n) B(\boldsymbol{k}'_{\rm F}, \boldsymbol{R}'; {\rm i}\omega_n). \tag{7}$$

In the above expression, we retain the terms up to the first order. We consider the case where the Fermi velocity $\boldsymbol{v}_{\rm F}$ is given by $\boldsymbol{k}_{\rm F}/m$.

We write down the augmented quasiclassical equation order by order with respect to the quasiclassical parameter $(k_{\rm F}\xi_0)^{-1}$ , where $\xi_0 = v_{\rm F}/\Delta_\infty$ and $\Delta_\infty = \lim_{R\to\infty} |\Delta(\boldsymbol{k}_{\rm F}, \boldsymbol{R})|$ are the coherence length and the magnitude of the gap in the bulk region, respectively. In the following, we neglect the magnetic field for simplicity. The Green's function as well as the self energy and the pair potential terms can be expanded as $\mathcal{O} = \mathcal{O}_0 + \mathcal{O}_1 + \cdots$, where $\mathcal{O}$ denotes any of $\widehat{g}^{\rm M}$, $\widehat{\sigma}^{\rm M}$, and $\Delta$. The lowest order equation is reduced to the conventional Eilenberger equation, which can be solved using the Riccati transformation supplemented with the gap equation and the explicit form of the self energy.

$$2{\rm i}\omega_n f_0 + {\rm i}\boldsymbol{v}_{\rm F} \cdot \partial_{\boldsymbol{R}} f_0 - 2\Delta_0 g_0 + 2{\rm i}\Gamma_{\rm n}[f_0 \langle g_0 \rangle_{\rm F} - \langle f_0 \rangle_{\rm F}\, g_0] = 0, \tag{8}$$

$$g_0 = {\rm sgn}(\omega_n)\left[1 - f_0 \bar{f}_0\right]^{1/2}, \tag{9}$$

$$\Delta_0(\boldsymbol{k}_{\rm F}, \boldsymbol{R}) = g\nu_0(\xi_{\rm F}){\rm i}\pi T \sum_n \left\langle V^{L_z}_{\boldsymbol{k}_{\rm F},\boldsymbol{k}'_{\rm F}} f^{\rm M}_0(\boldsymbol{k}'_{\rm F}, \boldsymbol{R}; {\rm i}\omega_n)\right\rangle_{\rm F'}, \tag{10}$$

$$\widehat{\sigma}^{\rm M}_{\rm imp,0}(\boldsymbol{k}_{\rm F}, \boldsymbol{R}; {\rm i}\omega_n) = {\rm i}\Gamma_{\rm n}\, \langle \widehat{g}^{\rm M}_0(\boldsymbol{k}'_{\rm F}, \boldsymbol{R}; {\rm i}\omega_n)\rangle_{\rm F'}, \tag{11}$$

where $V^{L_z}_{\boldsymbol{k}_{\rm F},\boldsymbol{k}'_{\rm F}} = 1$ for s-wave SC ($|L_z| = 1$) and $2\cos(\alpha - \alpha')$ for chiral p-wave SCs ($|L_z| = 2$ or 0). Note that $L_z$ describes the total angular momentum of the Cooper pair [26, 27, 28]. We have introduced $\alpha$ as a direction of the Fermi momentum in two dimensional (2D) space. The second equation describes the normalization condition for the zeroth order quasiclassical Green's function. We note that $\langle o(\boldsymbol{k}_{\rm F})\rangle_{\rm F}$ denotes the average over the 2D Fermi surface defined by $(2\pi)^{-1}\int_0^{2\pi} {\rm d}\alpha o(\boldsymbol{k}_{\rm F})$, and $\nu_0(\xi_{\rm F})$ and $g$ are the 2D density of states per spin at the Fermi level $\xi_{\rm F}$ and the coupling constant, respectively. The coupling constant is related to the transition temperature $T_{\rm c}$ as

$$\frac{1}{g\nu_0(\xi_{\rm F})} = \ln\frac{T}{T_{\rm c}} + 2\pi \sum_{0 \le n \le N_{\rm c}} \frac{1}{2n+1} \tag{12}$$

with a cut-off number $N_c$. The impurity scattering rate in the normal state is described by $\Gamma_n$.

The first order equation consists of two closed parts: One is an equation for $g_1 + \bar{g}_1$, and the other is a coupled equation for $g_1 - \bar{g}_1$, $f_1$, and $\bar{f}_1$. The charging effect, as we see next, is described by $g_1 + \bar{g}_1$, which obeys

$$\frac{1}{2} i v_F \frac{\partial(g_1 + \bar{g}_1)}{\partial s} = \frac{i\hat{e}_b}{2k_F} \cdot \left[ 2\left( \partial_{\boldsymbol{R}} g_0 \frac{\partial \sigma_{d,0}}{\partial \alpha} - \partial_{\boldsymbol{R}} \sigma_{d,0} \frac{\partial g_0}{\partial \alpha} \right) + \partial_{\boldsymbol{R}} \sigma_{od,0} \frac{\partial \bar{f}_0}{\partial \alpha} - \partial_{\boldsymbol{R}} \bar{\sigma}_{od,0} \frac{\partial f_0}{\partial \alpha} - \frac{\partial \sigma_{od,0}}{\partial \alpha} \partial_{\boldsymbol{R}} \bar{f}_0 + \frac{\partial \bar{\sigma}_{od,0}}{\partial \alpha} \partial_{\boldsymbol{R}} f_0 \right]. \quad (13)$$

Here $\hat{e}_b := \hat{e}_z \times \hat{e}_s$ with $\hat{e}_s = \boldsymbol{v}_F / v_F$ and $v_F = |\boldsymbol{v}_F|$, and $s$ is the coordinate along $\hat{e}_s$.

Next we see how the charging effect is obtained. The 2D charge density is given by [29, 30]

$$e(n(\boldsymbol{R}) - n_n) = -ei\pi T \sum_n \nu_0(\xi_F) \operatorname{Tr} \left[ \langle \widehat{g}^M \rangle_F - \langle \widehat{g}_n^M \rangle_F \right] - 2\nu_0(\xi_F) e^2 \Phi(\boldsymbol{R}), \quad (14)$$

where the scalar potential $\Phi$ on the right-hand side comes from the high-energy contribution. Note that if we do the phase transformation involved with the scalar potential, particularly in the Keldysh formalism, we do not have the scalar potential explicitly in the charge density, instead we have the electric field term in the Eilenberger equation [6, 31]. The left-hand side is described as the difference in the charge density between the superconducting state in question and the normal state without any gauge fields.

The charge expression, of course, includes the scalar potential, and hence we have to solve the complementary Poisson equation for the thickness $d$, given by

$$\nabla^2 \Phi(\boldsymbol{R}) = -\frac{4\pi e}{d} (n(\boldsymbol{R}) - n_n), \quad (15)$$

or

$$\left[ \nabla^2 - \lambda_{TF}^{-2} \right] \Phi(\boldsymbol{R}) = -\lambda_{TF}^{-2} \frac{2\pi T}{e} \sum_{\omega_n > 0} \left\langle \operatorname{Im} \left[ \frac{g_1 + \bar{g}_1}{2} \right] (\boldsymbol{k}_F, \boldsymbol{R}; i\omega_n) \right\rangle_F, \quad (16)$$

where we use Eq. (14) and the definition of the Thomas–Fermi length $\lambda_{TF} = [d/(8\pi\nu_0(\xi_F)e^2)]^{1/2}$. The second term in the left-hand side describes the screening effect. First we solve this Poisson equation through the perturbative Green's function, then we obtain the charge density from Eq. (14) or (15).

## 3. Numerical results

In this section we show the numerical results for two types of vortices of chiral p-wave SCs. Two types of vortices are characterized by the transformation properties; the pair potential for one type of the vortex acquires the additional phase $2\theta$ under the simultaneous $\theta$ rotation of real and momentum spaces, and we label it as $L_z = 2$. The pair potential for the other type of the vortex is invariant under the same transformation, and we label it as $L_z = 0$. The charge density is normalized by $2|e|T_c\nu_0(\xi_F)$ and use the notation $\delta\bar{\rho}(R) = e(n(R) - n_n)/(2|e|T_c\nu_0(\xi_F))$. Also we define the integral of the charge density from the origin to radius $R$ as

$$\bar{Q}(R)\xi_0^2 = \int_0^R dR' R' \delta\bar{\rho}(R'). \quad (17)$$

Here $\bar{Q}(R)$ can be interpreted as the actual charge within the radius $R$ and we have confirmed that the neutrality of the whole system holds ($\bar{Q}(R \to \infty) = 0$), although we do not show any

IOP Conf. Series: Journal of Physics: Conf. Series **969** (2018) 012054 doi:10.1088/1742-6596/969/1/012054

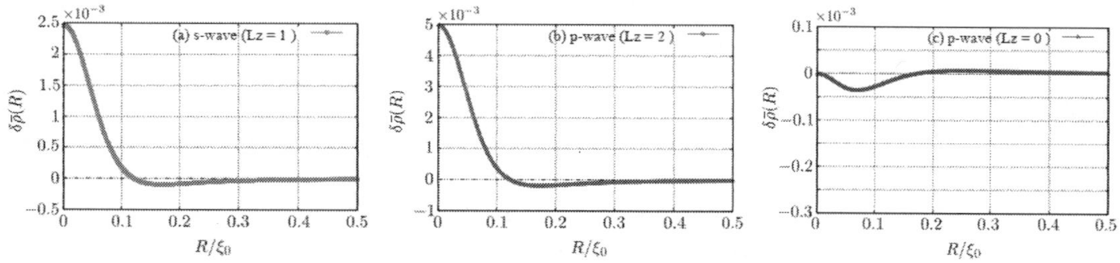


**Figure 1.** Charge density profiles for (a) the s-wave SC, (b) the chiral p-wave SC with $L_z = 2$, and (c) the chiral p-wave SC with $L_z = 0$.

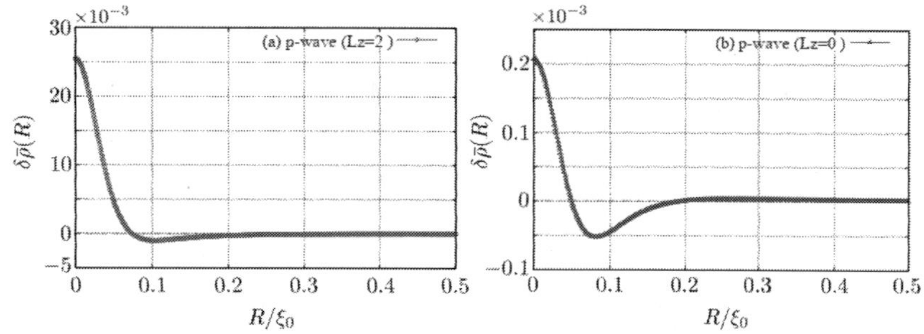


**Figure 2.** The charging profiles of chiral p-wave SCs with consideration of induced components of gap function. Note that the ranges of the vertical axes are different from those in Fig. 1.

data about this in the following. We set $T$ to $0.2T_c$, $(k_F\xi_0)^{-1}$ to 0.01, and $\Gamma_n$ to 0, in all results presented below.

We show the charge profile and the total charge as functions of the radial coordinate. We set the Thomas–Fermi length $\lambda_{TF}$ to $0.01\xi_0$. First we neglect the effects of the induced components in chiral p-wave SCs, which are discussed later. Figures 1(b) and 1(c) are for $L_z = 2$ and $L_z = 0$, respectively. Note that the s-wave SC is the case of $L_z = 1$, which we show in Fig. 1(a) as a reference. We see that the behavior of $L_z = 2$ is similar to that of $L_z = 1$, while there is conspicuous suppression in the case of $L_z = 0$. This can be understood as follows. The spatial derivative in the Moyal product includes radial and angular directions and the latter is proportional to the Cooper pair angular momentum $L_z$. When $L_z \neq 0$, the contribution from the angular component is dominant over that from the radial component. In the case of the zero angular momentum state $L_z = 0$, the contribution of the angular component vanishes, and this is the reason that the vortex charge of $L_z = 0$ is extremely small compared with those of $L_z = 1$ and 2. In addition, the charge of $L_z = 2$ is almost twice as large as that of $L_z = 1$. When we do not take account of the induced component, the gap functions as well as the Green's functions are quite similar except the difference of phase factors related to the angular momentum.

Next we discuss the effects of induced components. The profiles of the charge densities are shown in Figs. 2(a) and (b). Induced components do not affect the qualitative properties, namely, the vortex of $L_z = 2$ is charged while the vortex of $L_z = 0$ is hardly charged. There is, however, an interesting and quantitative difference in the case of $L_z = 2$. Figure 2 (a) shows that the charge density at the origin becomes five times as large as that of Fig. 1 (b) and thus the vortex charge of $L_z = 2$ is an order of magnitude greater than that of the s-wave SC. The dominant contribution is still from the angular derivative, but the changes of the gap function and Green's function cause this enhancement.

## 4. Conclusion

We have studied the vortex charging effect for chiral p-wave SCs using the augmented quasiclassical theory. There are two types of vortices in the chiral p-wave SCs, and in one case where $L_z = 0$, the charging effect is noticeably suppressed. Consistently to the calculation using the BdG theory, the dominant contribution is from the angular component of the spatial derivative, which vanishes for $L_z = 0$. The total angular momentum $L_z$ describes the cancellation of two phases in momentum and real spaces. The induced components affect the gap functions and the quasiclassical Green's functions, and thereby lead to the enhancement of the vortex charge in the case of $L_z = 2$. The effect of impurities should be discussed in subsequent work.

## Acknowledgments

We would like to thank Y. Tsutsumi for helpful discussions. This work was supported by a Grant-in-Aid for JSPS Fellows (Grant No. 16J03224) and JSPS KAKENHI Grant Number 15K05160. Y. M. was supported by the Program for Leading Graduate Schools, the Ministry of Education, Culture, Sports, Science and Technology, Japan.

## References

[1] Caroli C, de Gennes P G and Matricon J 1964 *Phys. Lett.* **9** 307–309
[2] Andreev A F 1965 *Sov. Phys. JETP* **20** 1490–1493
[3] Stone M 1996 *Phys. Rev. B* **54** 13222–13229
[4] Larkin A I and Ovchinnikov Y N 1986 *Nonequilibrium Supercond. (Modern Probl. Condens. Matter Sci.* ed Langenberg D N and Larkin A I (Amsterdam: Elsevier Science Ltd) chap 9, p 493
[5] Blatter G, Feigel'man M V, Geshkenbein V B, Larkin A I and Vinokur V M 1994 *Rev. Mod. Phys.* **66** 1125–1388
[6] Kopnin N B 2001 *Theory of Nonequilibrium Superconductivity (International Series of Monographs on Physics)* (NewYork: Oxford University Press)
[7] Hasan M Z and Kane C L 2010 *Rev. Mod. Phys.* **82** 3045–3067
[8] Qi X L and Zhang S C 2011 *Rev. Mod. Phys.* **83** 1057–1110
[9] Nayak C, Simon S H, Stern A, Freedman M and Das Sarma S 2008 *Rev. Mod. Phys.* **80** 1083–1159
[10] Ivanov B A, Stephanovich V A and Zhmudskii A A 1990 *J. Magn. Magn. Mater.* **88** 116–120
[11] Khomskii D I and Freimuth A 1995 *Phys. Rev. Lett.* **75** 1384–1386
[12] Blatter G, Feigel'man M, Geshkenbein V, Larkin A and van Otterlo A 1996 *Phys. Rev. Lett.* **77** 566–569
[13] Hayashi N, Ichioka M and Machida K 1998 *J. Phys. Soc. Jpn.* **67** 3368–3371
[14] Matsumoto M and Heeb R 2001 *Phys. Rev. B* **65** 014504
[15] Ueki H, Kohno W and Kita T 2016 *J. Phys. Soc. Jpn.* **85** 064702
[16] Ohuchi M, Ueki H and Kita T 2017 *J. Phys. Soc. Jpn.* **86** 073702
[17] Schmid A 1966 *Phys. Kondens. Mater.* **5** 302–317
[18] Usadel K D 1970 *Phys. Rev. Lett.* **25** 507–509
[19] Eilenberger G 1968 *Z. Phys.* **214** 195–213
[20] Larkin A I and Ovchinnikov Y N 1969 *Sov. Phys. JETP* **28** 1200–1205
[21] Kopnin N B 1994 *J. Low Temp. Phys.* **97** 157–179
[22] Levanda M and Fleurov V 1994 *J. Phys. Condens. Matter* **6** 7889–7908
[23] Kita T 2001 *Phys. Rev. B* **64** 054503
[24] Mackenzie A P and Maeno Y 2003 *Rev. Mod. Phys.* **75** 657–712
[25] Maeno Y, Kittaka S, Nomura T, Yonezawa S and Ishida K 2012 *J. Phys. Soc. Jpn.* **81** 11009
[26] Kato Y and Hayashi N 2001 *J. Phys. Soc. Jpn.* **70** 3368–3376
[27] Masaki Y and Kato Y 2015 *J. Phys. Soc. Jpn.* **84** 094701
[28] Masaki Y and Kato Y 2016 *J. Phys. Soc. Jpn.* **85** 014705
[29] Eliashberg G M 1972 *Sov. Phys. JETP* **34** 668–676
[30] Serene J and Rainer D 1983 *Phys. Rep.* **101** 221–311
[31] Arahata E and Kato Y 2014 *J. Low Temp. Phys.* **175** 346–352

# Effect of diffusion annealing regimes on the structure of $Nb_3Sn$ layers in ITER-type bronze-processed wires

**E G Valova-Zaharevskaya[1,*], E N Popova[1], I L Deryagina[1], I M Abdyukhanov[2], A S Tsapleva[2]**

[1]M.N. Miheev Institute of Metal Physics of Ural Branch of Russian Academy of Sciences, 18 S. Kovalevskaya Str., Ekaterinburg 620990, Russian Federation
[2]A.A. Bochvar High-Technology Research Institute of Inorganic Materials, 5a Rogov Str., Moscow 123098, Russian Federation

*E-mail: valova@imp.uran.ru

**Abstract**. The goal of the present study is to characterize the growth kinetics and structural parameters of the $Nb_3Sn$ layers formed under various regimes of the diffusion annealing of bronze-processed Nb/Cu-Sn composites. The structure of the superconducting layers is characterized by their thickness, average size of equiaxed grains and by the ratio of fractions of columnar and equiaxed grains. It was found that at higher diffusion annealing temperatures (above 650°C) thicker superconducting layers are obtained, but the average sizes of equiaxed $Nb_3Sn$ grains even under short exposures (10 h) are much larger than after the long low-temperature annealing. At the low-temperature (575 °C) annealing the relative fraction of columnar grains increases with increasing annealing time. Based on the data obtained, optimal regimes of the diffusion annealing can be chosen, which would on the one hand ensure complete transformation of Nb into $Nb_3Sn$ of close to the stoichiometric composition, and on the other hand prevent the formation of coarse and columnar grains.

## 1. Introduction

The $Nb_3Sn$-based superconducting composites under study were developed and manufactured by bronze technology at the Bochvar Institute of Inorganic Materials (VNIINM, Moscow, Russia) [1]. Superconductors of such design have high critical characteristics in magnetic fields up to 12 T, which allowed them to participate in the ITER project [2]. At present, they are the main candidates for use in the superconducting magnetic systems of future large accelerator projects at CERN and thermonuclear reactors. Particularly, $Nb_3Sn$ superconductors are planned to be used in the Future Circular Collider (FCC) project that considers the creation of magnetic fields up to 16 T [3]. To further increase the critical characteristics in high magnetic fields for similar applications attention should be paid to studying the growth of the superconducting layers in industrial conductors manufactured by up-to-date technologies.

The goal of the present study is to characterize the growth kinetics and structural parameters of the $Nb_3Sn$ layers formed under various regimes of the diffusion annealing.

The current-carrying capacity of superconductors in high magnetic fields depends on several basic parameters, such as the value of the upper critical field of the superconducting compound, and the amount and structure of the superconducting phase [4]. Other parameters of importance include

IOP Conf. Series: Journal of Physics: Conf. Series **969** (2018) 012055 doi:10.1088/1742-6596/969/1/012055

structural assembly features of a conductor, the average size of the initial niobium filaments before diffusion annealing, etc.

It is well known that in niobium-tin superconductors the highest values of the critical temperature $T_c$ and upper critical field $H_{c2}$ are achieved when the phase composition approaches a slight deviation from the stoichiometric composition of $Nb_3Sn$ in the region of 24.5 at.% [5]. The content of tin in the layer increases (this leads to an approach of its composition to the stoichiometric one) with an increase in the annealing temperature.

It is known that in filaments of composites produced according to the bronze technology, there arise concentric zones of $Nb_3Sn$ grains of different morphology [6]. A zone of equiaxed grains adjoined the bronze matrix has a certain gradient in the sizes of grains from the fine ones in the middle to the coarse ones near the bronze matrix; the zone of columnar grains is contiguous with the residual niobium. It has been well established that tin concentration is different in the $Nb_3Sn$ grains of different morphology, decreasing from 24.5–21 at.% in equiaxed grains to 21–18 at.% in columnar grains from the outside to inside of the $Nb_3Sn$ layers forming in Nb filaments under heat treatment [6–8]. Based on this, it can be concluded that the smaller fraction of columnar grains provides the higher average tin concentration in the superconducting layers.

The critical current density, in its turn, is determined by the pinning force. As has been well established, the main pinning centers in multifilamentary $Nb_3Sn$-based composites are grain boundaries [9, 10]. The $Nb_3Sn$ phase grain refinement results in larger area of grain boundaries, and, thus, in higher pinning force and $J_c$ [11]. Of not less importance than the grain sizes proper is the morphology of superconducting layers, because in equiaxed grains the pinning centers are distributed more uniformly than in columnar ones. Because of the variety of parameters that determine the properties of a superconductor, it becomes necessary to thoroughly investigate the microstructure of the $Nb_3Sn$ layers under various conditions.

## 2. Materials and experimental procedure

### 2.1. Samples

The samples studied are composite wires of overall diameter of 0.82 mm manufactured by bronze technology as mentioned above. They consist of 13,212 coupled niobium filaments artificially doped with 1.55 wt.% Ti and embedded in a bronze matrix (Cu–14%Sn). This assembled billet is surrounded with a niobium barrier with tantalum inserts and placed in a tube of stabilizing copper. Under the diffusion annealing superconducting $Nb_3Sn$ layers are formed starting from interfaces of niobium filaments and bronze matrix.

Different temperatures and times of annealing were applied to samples studied in this work. The first group of composite wires was annealed at 575 °C during 10, 50 and 100 hours (samples 1, 2 and 3, respectively). The second group (samples 4–6) was annealed at 650 °C for the same annealing times. The annealing temperature of the last group (samples 7 and 8) was increased up to 750 °C and the heat treatment was carried out for 10 and 50 hours.

Composites of this design with other regimes of the diffusion annealing were investigated in our previous studies [12–15].

### 2.2. Experimental methods

For scanning electron microscopy (SEM) polished transverse sections and fractures of strands were used to determine the thickness of the superconducting layers. SEM images were obtained in Inspect F (FEI) and Quanta–200 (FEI) microscopes in the Secondary Electrons and Back-scattered Electrons modes. Polished sections were produced by standard metallographic procedures. The morphology of grains in $Nb_3Sn$ layers was determined using SEM images of Nb filaments on transverse fractures of the wires prepared in liquid nitrogen. Quantitative image analysis was performed on SEM images of several Nb filaments from different (central and peripheral) areas of the strands.

IOP Conf. Series: Journal of Physics: Conf. Series **969** (2018) 012055 doi:10.1088/1742-6596/969/1/012055

The relative fractions of the zones of different morphology in the superconducting layers were determined by the program of image treatment ImageMagick (GNU). The areas of different morphology were painted in various colors, and the ratio was calculated from the relationships of pixel proportions of every color.

The $Nb_3Sn$ layers structure was also studied by transmission electron microscopy (TEM) of longitudinal foils in JEM-200CX electron microscope. The foils were prepared by a standard method of mechanical thinning of longitudinal sections with further chemical polishing in a mixture of $HNO_3$, $H_2SO_4$ and HF acids. TEM (longitudinal foils) and SEM (transverse fractures) images obtained were treated with the help of specialized program of statistical analysis of images SIAMS-600 (SIAMS) to calculate the average grain size of the $Nb_3Sn$ phase.

## 3. Results and discussion

An overall view of multifilamentary ITER-type composite is shown in figure 1a. As it is shown in figure 1b, after multiple cold drawing and diffusion annealing the coupled filaments retain their dumbbell shape, but the filament diameter varies along the wire axis and the thicker filaments have wider areas of the residual niobium.

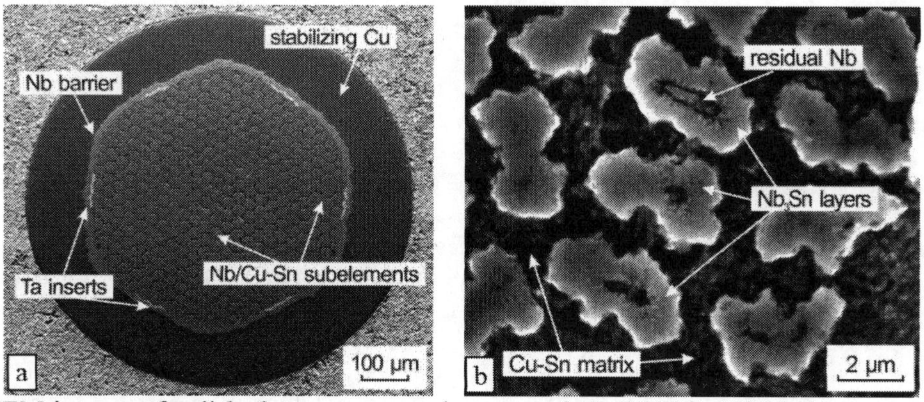


**Figure 1.** SEM images of polished transverse sections, sample 3 (575 °C, 100 h): (a) an overall view of multifilamentary ITER-type composite; (b) a group of filaments with different degree of Nb transformation into $Nb_3Sn$.

For each sample, fracture images were obtained, on which the boundary of the residual niobium is clearly visible, as well as the morphology of the grains of the superconducting phase. Thus, in the layers we can distinguish zones of different morphology and determine the ratio of columnar and equiaxed grains. An example of such an image is shown in figure 2. The proportion of grains of certain morphology was calculated as the area of this zone in the cross section of the filament reduced to the area of the entire filament. The values obtained were averaged over several filaments in order to reduce the error. In general, from the data obtained, it can be seen that the columnar grains account for slightly more than a third of the volume of the superconducting layer at all the temperatures studied, which is a common ratio for bronze-processed composites [14]. A slight increase in the fraction of columnar grains is observed with an increase in the annealing time at a temperature of 575 °C (figure 3). For more definite conclusions about the effect of temperature and time of single-stage annealing on the fraction of columnar grains, further investigations with a larger number of points, both over the temperatures and the annealing durations, will be required.

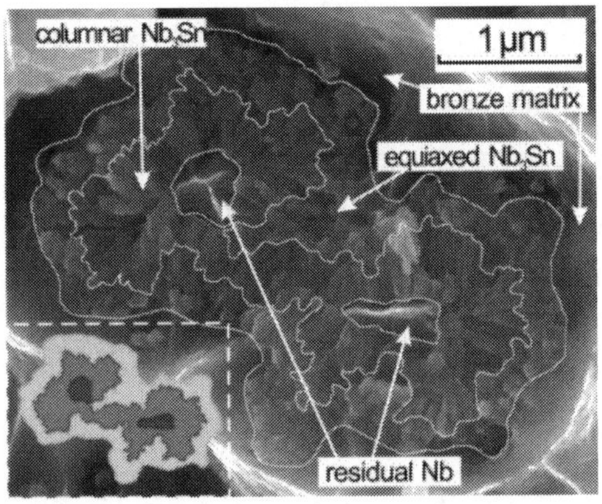


**Figure 2.** The fracture image of a filament with grain zones of different morphology.

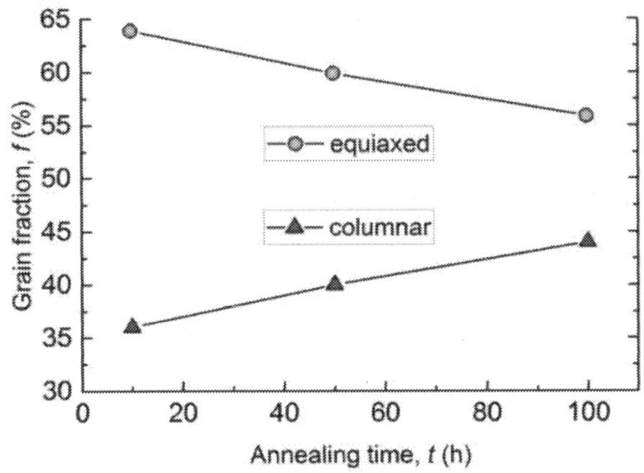


**Figure 3.** The fraction of equiaxed and columnar grains versus the annealing time at 575 °C.

The thickness of the superconducting layer is usually determined on the cross sections of the composite as the shortest distance between the phase boundaries "bronze matrix–$Nb_3Sn$" and "$Nb_3Sn$–residual Nb". Exact thickness measurements are complicated by irregularities of the Nb filaments surface. The error due to these irregularities is minimized by averaging a number of thickness measurements made on several cross sections of a composite wire. When the boundaries of the phase sections are strongly distorted and the thickness of the layer reaches the entire volume of the material, the task becomes much more complicated. Since two hexahedral rods are used as a billet for the creation of coupled filaments, the final dumbbell-shaped Nb filament can be approximated by two hexahedrons. SEM images of fractures of Nb filaments with $Nb_3Sn$ layers after the annealing at 575°C, 650 °C and 750 °C during 10 hours are shown in figure 4. The calculated values of the average layer thickness, $L$, with the standard deviation (SD) of this value, obtained during the processing of images, are presented as a graph in figure 5a.

It can be seen from figures 4 and 5a that at 750 °C the superconducting layer growth is much faster. After a short annealing (10 hours), slightly thicker layers are formed in sample 4 (650 °C) than in sample 1 (575 °C), but at the same time almost complete Nb transformation into $Nb_3Sn$ is achieved in sample 7 (750 °C) in the majority of filaments. At lower annealing temperatures, even after 100 hours exposure, a full transformation of niobium filaments is not achieved (although at 650 °C there are not many filaments with the residual niobium).

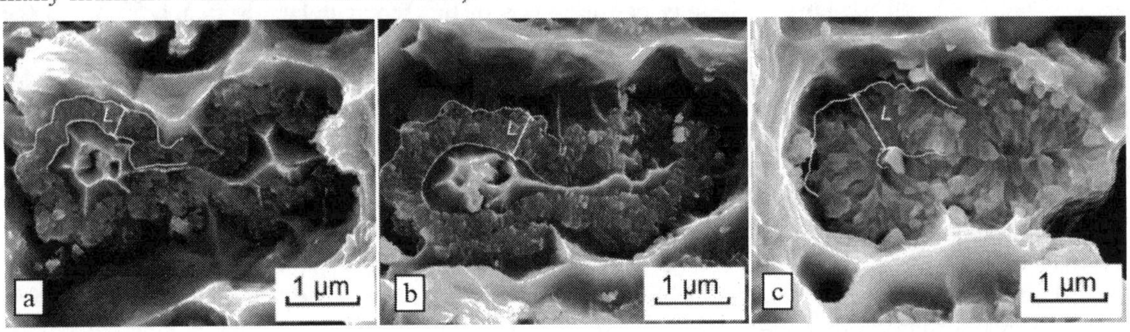


**Figure 4.** SEM images of fractures of Nb filaments with $Nb_3Sn$ layers after the annealing at 575 °C (a), 650 °C (b) and 750 °C (c) during 10 hours. L - segments denoting layer thickness.

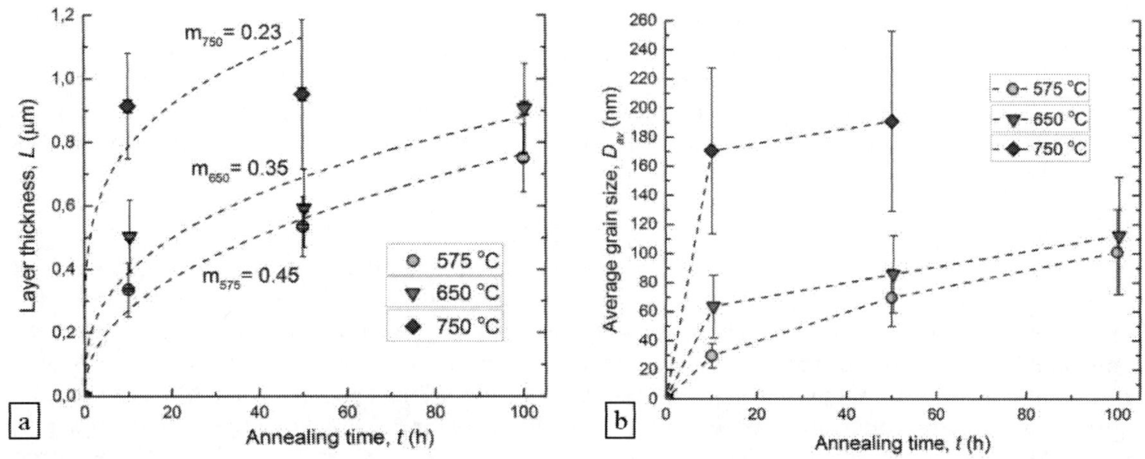


**Figure 5.** Parameters of $Nb_3Sn$ layers: (a) Layer thickness versus the annealing time. Wide bars show standard deviation of values obtained, narrow bars mark standard error of mean. Dashed lines demonstrate the allometric fittings $L(t)=bt^m$ for different temperatures. (b) Average size of equiaxed grains of the superconducting $Nb_3Sn$ phase as a function of the annealing time at different temperatures (according to SEM). Vertical bars indicate the standard deviation of the grain size distribution.

The common used fitting for time dependence of layer thickness $L(t)$ is $L=bt^m$, in which $m$ is a growth exponent, and $m = 1$ for a reaction-rate limited dependence and $m = 0.5$ for a diffusion-rate limited case. The SD of the data obtained in this system is very high, so it is very difficult to find the exact growth exponent. Nevertheless, an attempt was done, and the layer thickness data is shown in figure 5a with the fits of $bt^m$. Such a rough estimate shows that the layer growth at 575 °C corresponds most completely to a parabolic law ($m = 0.45$). At higher temperatures (650 °C and 750 °C) there is a significant deviation from this law ($m = 0.35$ and 0.23, respectively). It is known that a time-dependent increase in grain size can suppress the intermetallic layer growth below a parabolic rate when grain boundary diffusion is important [16]. This factor along with the depletion of bronze matrix can be a reason for the deviation of the layer growth from the parabolic law.

The results of grain size distribution were obtained according to the transmission and scanning microscopy. In the first case, we studied the grains from the images of longitudinal foils of composites. In the second case we used images of fractures of the cross sections. The calculated values of the average size of equiaxed grains (with the SD of this value) obtained during the processing of images are presented in figure 5b. The data from longitudinal foils reflect similar trends, except for the fact that the average grain size according to SEM data is slightly increased due to the fact that more large grains are counted in this calculation. As can be seen from the presented data, at a temperature of 750 °C, even with short annealing at 10 h, the average size of equiaxed grains considerably exceeds those after long low-temperature annealing.

The data obtained are of importance for the optimization of current-carrying capacity of the wires under consideration. As mentioned in the Introduction, columnar grains are undesirable for two reasons. Firstly, their composition is farther from stoichiometry, which results in lower upper critical field and, thus, in their poorer performance in high magnetic fields. Secondly, in grains of such shape the grain boundaries are distributed non-uniformly, which may cause current instabilities. The equiaxed grains are much more preferable as their composition is closer to stoichiometry and thus they have higher critical characteristics. Of not less importance than composition is grain size, for, as mentioned above, grain boundaries in such superconductors are the main pinning centers, and grain refinement is the crucial factor for the enhancement of the current-carrying capacity. As seen from the results obtained, some factors (higher annealing temperature) are beneficial for an improvement of composition and an increase of the total amount of superconducting phase. On the other hand, grain growth at high temperatures results in lower pinning force. At longer low-temperature annealing the relative fraction of columnar grains is increasing, which results in poorer performance of a composite for the reasons mentioned above. Thus, based on the data obtained one can find a reasonable compromise in annealing temperature and duration in order to achieve the best performance of the $Nb_3Sn$ strands.

## 4. Summary

Since at high annealing temperatures (750 °C) there is no increase in the fraction of columnar grains in the layer, we can expect that the average concentration of tin in the layer in these samples is not lower than in samples with a lower annealing temperature. This makes the high-temperature annealing regime preferable because of more complete transformation of the filaments. However, in the sample annealed at 650 °C, 100 h almost complete processing of niobium filaments is also observed, but the grain size of the superconducting phase is much smaller. This means that the pinning force is higher, which is also significant for an achievement of high critical characteristics. In high fields the current carrying capacity is determined mainly by the value of the upper critical field of the $Nb_3Sn$ intermetallic compound, whereas in intermediate fields the main role is played by pinning force, and thus the grain size is of great importance.

Therefore, when choosing the annealing regimes, different trends should be taken into account: on the one hand, the possibility of approaching stoichiometry and ensuring the maximum amount of the superconducting phase, and on the other hand, the growth of grains and the loss in the pinning force. The knowledge of these trends will allow finding optimal trade-offs to ensure the best current-carrying capacity of conductors, which makes the ongoing research of great importance.

## 5. Acknowledgements

This work has been done on the equipment of the Collaborative Access Center TC NTAM of the Institute of Metal Physics, Ural Branch of RAS, within the State assignment of FASO of Russia (topic “Pressure” No. AAAA-A18-118020190104-3) with partial support of RFBR (project No. 16-33-00430).

## References

[1] Pantsyrny V, Shikov A and Vorobieva A 2008 *Cryogenics* **48** 354
[2] Tronza V I et al. 2016 *IEEE Trans. Appl. Supercond.* **26** (4) 4200105
[3] Ballarino A and Bottura L 2015 *IEEE Trans. Appl. Supercond* **25** (3) 6000906
[4] Popova E N, Deryagina I L, Valova-Zakharevskaya E G and Patrakov E I 2016 *Phys. Met. and Metallogr.* **117** (10) 1028–37
[5] Flükiger R, Uglietti D, Senatore C and Buta F 2008 *Cryogenics* **48** 293–307
[6] Uglietti D, Abächerli V, Cantoni M and Flükiger R 2007 *IEEE Trans. Appl. Supercond* **17** (2) 2615
[7] Abächerli V, Uglietti D, Lezza P, Seeber B, Flükiger R, Cantoni M and Buffat P-A 2005 *IEEE Trans. Appl. Supercond.* **15** 3482
[8] Senatore C, Abächerli V, Cantoni M and Flükiger R 2007 *Supercond. Sci. Technol.* **20** S217
[9] Luhman T, Pande C S and Dew-Hughes D 1976 *J. Appl. Phys.* **47** (4) 1459
[10] Lee P J and Larbalestier D C 2008 *Cryogenics* **48** 283
[11] Larbalestier D C 1995 *Cryogenics* **35** S15–8
[12] Deryagina I L, Popova E N, Patrakov E I and Valova-Zaharevskaya E G 2017 *J. Magn. Magn. Mater.* **440** 119-122
[13] Deryagina I, Popova E, Patrakov E and Valova-Zaharevskaya E 2017 *J. Appl. Phys.* **121** 233901
[14] Popova E N, Deryagina I L and Valova-Zaharevskaya E G 2014 *Cryogenics* **63** 63
[15] Popova E N and Deryagina I L 2015 *Diff. Found.* **5** 199
[16] Farrell H H and Gilmer G H 1974 *J. Appl. Phys.* **45** 4025

IOP Conf. Series: Journal of Physics: Conf. Series **969** (2018) 012056 doi:10.1088/1742-6596/969/1/012056

# Influence of substitution, nonstoichiometry and annealing-conditions on superconductivity and normal conductivity of $Fe_{1+\delta}(Te_{1-x}X_x)$ ($X$=Se, S)

**M. S. L. Lima[1], M. ElMassalami[1], K. Deguchi[2], H. Takeya[2], Y. Takano[2]**

[1]Instituto de Fisica, Universidade Federal do Rio de Janeiro, Caixa Postal 68528, 21945-970 Rio de Janeiro, Brazil

[2] National Institute for Materials Science, 1-2-1, Sengen, Tsukuba, 305-0047, Japan

E-mail: marianalima@if.ufrj.br

**Abstract.** Thermal evolution of resistivity, $\rho(T,x)$, of as-prepared samples of $Fe_{1+\delta}Te_{1-x}S_x$ ($\delta \approx 0$, $x \leq 0.2$ = solubility limit) demonstrate a granular *log-in-T* character within $T_s < T <$300K, a Kondo-like resistive contribution within $T_c < T < T_s$ and granular superconductivity at low temperature ($T_s$= structural transition point of $Fe_{1+\delta}Te$, $T_c$=superconducting transition point). We attribute the *log-in-T* character as well as the nonbulk superconducting features of as-prepared samples to their granular superconductor nature. Annealing in oxygen removes Kondo-like contribution, annihilates pair-breaking centres and establishes bulk superconductivity but, in contrast, the high-temperature granular *log-in-T* character is hardly influenced. This analysis was successfully extended to the isomorphous $Fe_{1+\delta}Te_{1-x}Se_x$ as well as to other types of post-synthesis sample-treatment (e.g. annealing in different gas ambient or soaking in particular liquids).

## 1. Introduction

Extensive studies on the normal and superconducting phase diagrams of $Fe_{1+\delta}(Te_{1-x}X_x)$ ($X$=Se, S) [see e.g. Refs.[1, 2] and references therein] established that (i) $Fe_{1+\delta}Te$ is an antiferromagnetic nonsuperconducting metal, (ii) $Fe_{1+\delta}Se$ is a nonmagnetic superconductor and (iii) for $Fe_{1+\delta}(Te_{1-x}X_x)$ ($0.05 < x <$solubility limit), an increase in $x$ leads to a quench of magnetism, a surge of superconductivity and an enhancement of normal-state conductivity. Two additional factors are also influential in shaping the phase diagram of $Fe_{1+\delta}(Te_{1-x}X_x)$: sample nonstoichiometry [3, 4, 5, 6, 7] and sample treatment whether annealing in suitable gas or vapor (such as O, Te, Se, S, and I) or soaking in liquid (such as beverage or nitric acid) [1, 2].

Among these various treatment processes, annealing in oxygen is quite remarkable: While it has weak (negative, poisonous) influence on $Fe_{1+\delta}Te$ ($Fe_{1+\delta}Se$ [Ref.[3]]), it enhances tremendously both the superconductivity and low-$T$ normal-state conductivity of $Fe_{1+\delta}(Te_{1-x}X_x)$ $x > 0.05$ [8, 9, 10, 11]. The mechanism behind this O-annealing-induced influence is controversial but there is a general inclination towards attributing it to the following process: diffused oxygen "removes" interstitial Fe which manifests $\sim 2.5\mu_B$ [4, 6, 12, 13, 14, 15, 5, 7]. However, $^{57}Fe$ Mössbauer studies on $Fe_{1+\delta}(Te_{1-x}X_x)$ revealed only one asymmetric doublet indicative of nonmagnetic $Fe^{2+}$-like site and, moreover, their spectra before and after annealing are almost identical [16, 3, 17, 18, 19, 20]. Within the Mössbauer frequency window, this excludes all types of interstitial Fe states and, as well, any pair-breaking Fe moment; as such the influence of O-annealing can not be due to a "removal" of such inexistent Fe species. Nevertheless, experiments did reveal a magnetic pair-breaking agent which is removed by optimized annealing or soaking and, as a consequence, superconductivity is established, normal-state conductivity is enhanced and localization effects[4, 21, 22] are removed. It is worth adding that the process of loading and deloading oxygen and its influence on, say, the superconducting properties is reversible [8].

This work proposes that, for $Fe_{1+\delta}(Te_{1-x}X_x)_{1-y}$, $\delta \approx 0$, some compensating defect such as chalcogen vacancies (instead of additional interstitial Fe species) are created and that they are nonuniformly distributed within nano-sized stoichiometric metallic/superconducting granules; these granules are separated from each other by nonstoichiometric, less conducting and nonsuperconducting batches. This constitutes a granular metallic/superconducting system. According to this scenario, normal-state conductivity of $Fe_{1+\delta}(Te_{1-x}X_x)_{1-y}$ should exhibit a *granular log-in-T* character while its superconducting properties would be tuned by Josephson coupling of the superconducting granules [23, 24]. Furthermore, we attribute the observed magnetic moments to a trapping of odd electronic spins into the vacancies (reminiscent of paramagnetic color-centres); these paramagnetic scattering centers are considered to give rise to Kondo-like contribution within the normal-state phase and to Abrikosov-Gorkov pair-breaking effects within the superconducting phase. Optimized annealing (soaking) in favorable ambient causes a removal of these vacancies leading to a disappearance of Kondo-type localization effects, enhancement of normal-state conductivity, removal of magnetic pair-breaking centres and establishment of bulk superconductivity.

With the objective of demonstrating the validity of these arguments, we studied the normal and superconducting properties of various as-prepared and oxygen-annealed $Fe_{1+\delta}Te_{1-x}S_x$ samples. We show that while both as-prepared and O-annealed samples exhibit a similar characteristic *log-in-T* granular character at higher temperatures, their low-$T(< T_S)$ conductivities are distinctly different. We show that these same features are valid for isomorphous $Fe_{1+\delta}Te_{1-x}Se_x$ system.

## 2. Experimental

The procedures used for samples preparation, heat-treatment, as well as structural, resistivity and magnetization characterizations are given in the supplementary materials. Also given in the supplementary materials are the X-ray diffractograms of the studied as-prepared and O-annealed samples.

## 3. Results and analysis

Figure 1 shows $\rho(T,x)$ curves of as-prepared and O-annealed $Fe_{1+\delta}Te_{1-x}S_x$ ($x$=0, 0.05, 0.1, 0.15, 0.2). These are similar to the reported data of, e.g., Refs [8, 25, 26, 27, 28, 29]. Three main temperature ranges can be identified. Below, the evolution of $\rho(T,x)$ within each regime will be analyzed and discussed in terms of the scenario of granularity. We also discuss how granular normal conductivity and superconductivity are being influenced by O-annealing.

### *3.1. $T_s < T <$300 K regime*

Figure 1 indicates that, within $T_s < T <$300K, $\frac{\partial\rho}{\partial T} < 0$ for all $x$ as well as for both as-prepared and O-annealed samples. This evolution is not due to intrinsic semiconducting feature [30]; rather it is taken as a manifestation of granularity wherein metallic/superconducting granules are embedded within a less conducting nonsuperconducting matrix [23, 24]. As mentioned above the distribution of conductance is taken to be stemming from nonuniform distribution

IOP Conf. Series: Journal of Physics: Conf. Series **969** (2018) 012056 doi:10.1088/1742-6596/969/1/012056

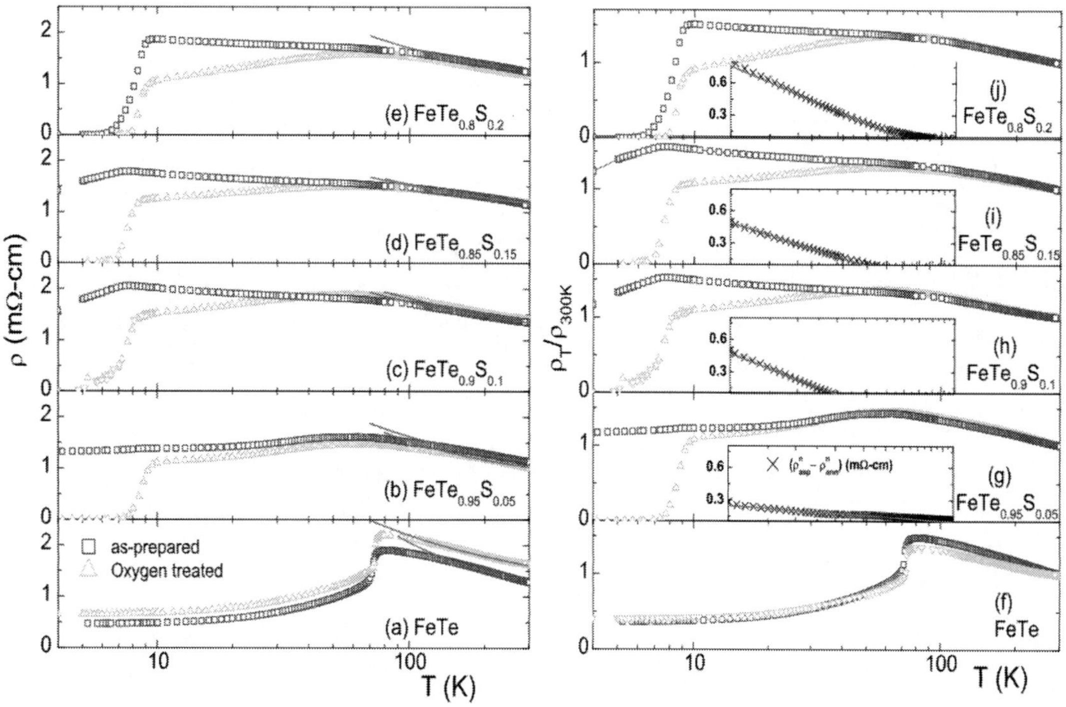


**Figure 1.** $\rho(T,x)$ curves of as-prepared (blue squares) and annealed (green triangles) $Fe_{1+\delta}Te_{1-x}S_x$ samples. (a-d) $\rho(T,x)$ curves. Solid red lines are fit to Eq.(1); (f-j) normalized $\rho(T,x)/\rho(300\text{K},x)$ curves showing the similarity of curves within the $T_s < T <$300K regime. The semilog normalized differences are shown as crosses within inset of (g-j): Solid red lines are fit to Eq.(2).

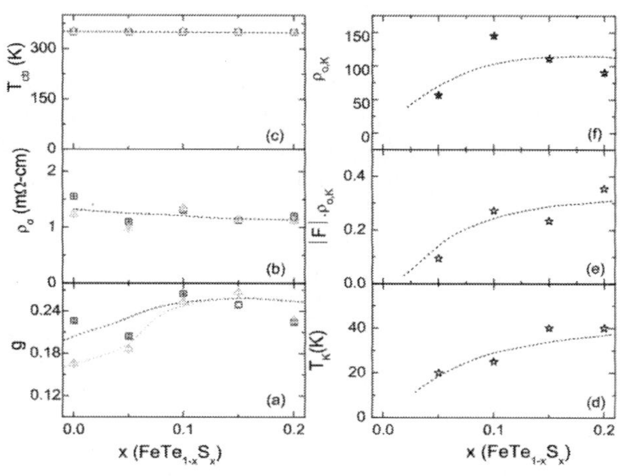


**Figure 2.** (a-c) Parameters obtained from the fit of Eq.(1) to $\rho(T,x)$ curves of as-prepared (blue squares) and O-annealed (green triangles) $Fe_{1+\delta}Te_{1-x}S_x$ samples: (a) $g(x)$; (b) $\rho_o(x)$; (c) $T_{cb}(x)$. (d-f) Parameters of the fit of Eq.(2) to normalized differences of the insets of Fig.1. (d) $T_K(x)$; (e) $|F|\,.\rho_{o,K}$ ; (f) $\rho_{o,K}(x)$ which, in contrast to that of panel (b), is the sum of all non-Kondo-type contributions that are different among the two normalized curves. Dashed lines are guides to the eye.

of nonstoichiometry [31] and defects. The thermal evolution of the resistivity within this range follows the characteristic granular *log-in-T* behavior [24]:

$$\rho(T) = \frac{\rho_o}{1 - \frac{1}{6\pi g}\ln(\frac{T_{cb}}{T})} \quad (1)$$

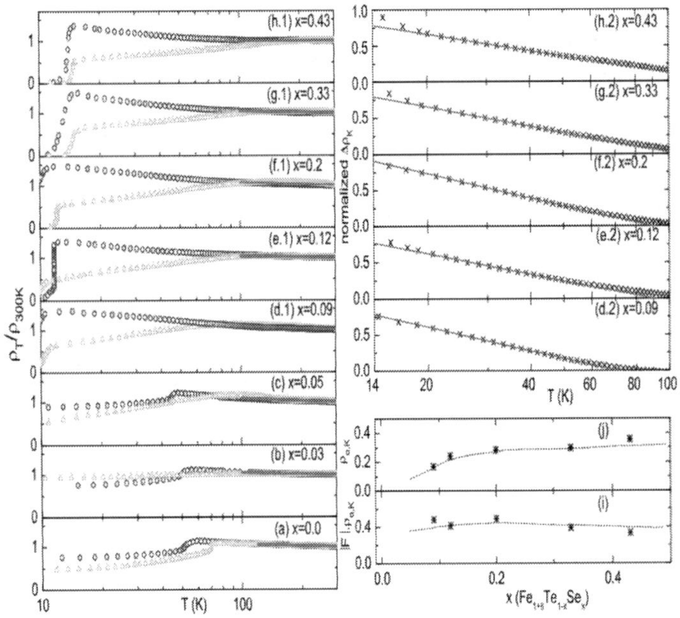

**Figure 3.** Normalized semilog plots of $\rho(T,x)/\rho(300\mathrm{K},x)$ curves of as-prepared (squares) and annealed (triangle) $Fe_{1+\delta}Te_{1-x}Se_x$ samples as reported in Ref. [6]. Solid red lines are fit to Eq.(1): the obtained parameters and their evolutions are similar to those of Fig.2(a-c). The normalized low-$T$ semi-log plots of differences are shown as crosses in Figs.(d.2 ..,h.2). Solid red lines are fit to Eq.(2): the parameters of the fits are shown in (i) $F.\rho_{o,\mathrm{K}}$ and (j) $\rho_{o,\mathrm{K}}$. $T_\mathrm{K}$ increases with $x$.

wherein $g$ is a tunneling conductance, $T_{cb}$ is a measure of the effective Coulomb blocking temperature which can be determined from the measured $\rho(T)$ curves and $\rho_o$ is the resistivity at $T_{cb}$. Applying this granular scenario to the high-$T$ resistivity of FeSe [32, 33], we identified its $T_{cb}$ to be 580 K while 350 K is a point of crossover from a *log-in-T* behavior into a homogeneously disordered granular metallic state. Since we were unsuccessful in experimentally determining $T_{cb}$ of $Fe_{1+\delta}Te_{1-x}S_x$ during our present work, then setting $T_{cb}(x)$=350K amounts to assigning $T_{cb}(x)$ to some possible value and as such ensures a minimum number of parameters when fitting Eq.1 to $\rho(T,x)$ curves of Fig.1. The obtained fit parameters are shown in Fig.2.

The evolution of both $g(x)$ and $\rho_o(x)$ reflects a weak dependence on $x$ as well as a similarity between the high-$T$ curves of as-prepared and annealed samples: exception is $Fe_{1+\delta}Te$ samples [compare the pair of curves in Fig.1(a) and Fig.1(f)] wherein there is a slight O-anneal-induced change in $g$ and $\rho_o$ but no drastic change in $T_s$ or magnetic contribution to resistivity. It is concluded that granular *log-in-T* behavior in $Fe_{1+\delta}Te_{1-x}S_x$ is is manifested for all $x$ as well as for both the as-prepared and O-annealed samples (below, similar conclusions are valid for $Fe_{1+\delta}Te_{1-x}Se_x$).

### *3.2. $T_c < T < T_s$ regime*

On approaching $T_s^+$, $\rho(T,x)$ of all samples start to deviate downwards away from the granular *log-in-T* behavior; the extent of deviation depends on $x$ and the annealing conditions. For both as-prepared and annealed $Fe_{1+\delta}Te$, $\rho(T,x{=}0)$ drops sharply at $T_s$ [1, 2]. Well below $T_s$, $\rho(T<T_s,x{=}0)$ is mostly governed by magnon scattering process which, at lower temperatures, is dominated by a *quadratic-in-T* behavior. In contrast, for sulfur doped samples, $\rho(T,x)$ evolves differently: First, there is no sharp drop in $\rho(T,x)$ that can be identified as an onset of structural transformation (most evident at $x$ <0.05) [31]. However, one observes a change of slope in the as-prepared $\rho(T,x)$ curves around 60∼100K (around $T_s$ of $Fe_{1+\delta}Te$). Most remarkably, a coherent peak is manifested in $\rho(T,x)$ curves of annealed samples such that below that peak a metallic, $\frac{\partial\rho}{\partial T}>0$, character is established. These features suggest an increase in conductance $g$ and this, in turn, is attributed to some short-ranged, on nano-sized level, structural modification (reminiscent of the case of $Fe_{1+\delta}Te$) which is hard to detect by conventional structural techniques.

The most important difference among $\rho(T_c < T < T_s, x)$ curves of as-prepared and annealed samples is the additional resistive contribution shown in insets of Figs.2(g-j): This contribution is evaluated as the difference $\Delta\rho_K = \frac{\rho_{asp}(T,x)}{\rho_{asp}(300K,x)} - \frac{\rho_{ann}(T,x)}{\rho_{ann}(300K,x)}$: this exclude any possible errors in estimating sample geometrical dimensions. Based on insets of Figs.2(g-j), $\Delta\rho_K$ is analyzed in terms of

$$\Delta\rho_K = \rho_{o,K}\left(1 + F\ln(\frac{T_K}{T})\right) \tag{2}$$

where $F$ and $T_K$ are parameters of the contribution which is assumed to be Kondo-type rather than quantum-correction-type (such as Anderson-type or Altshuler-Aronov-type) since the latter is expected to be manifested at much lower temperatures than this range (extending to an upper limit of $\sim$ 70K). As can be noticed from insets of Figs.1(g-j) and the fit parameters in Figs.2(d-f), both $F\rho_{o,K}$ and $T_K$ as well as the temperature range increase with $x$: further sulfur incorporation leads to an increase in the concentration of magnetic scatterer (the above-mentioned color-centre-type paramagnetic moments). Annealing in oxygen eliminates these centres and, as a consequence, removes $\Delta\rho_K$ contribution.

### *3.3. 2 K< $T < T_c$ regime*

Figure 1 indicates that, with the exception of $Fe_{1+\delta}Te$, all studied as-prepared samples exhibit an onset of superconductivity at $T_c^{onset} \simeq 8 \sim 9$ K indicative of granular superconductivity with no long-range coherence. This $T_c^{onset}$ event becomes more sharper and well enhanced when $x$ is increased: $x \to 0.2$, $T_{c,asp}^{zero} \to 6.8$ K and $T_{c,asp}^{onset} \to 9.1$ K. On O-annealing, long-range coherence, and as such bulk SC, is established with $T_c^{onset} \approx$9.1K, $T_c^{zero} \approx$8K. The similarity and difference among the onset and zero-resistivity points are typical of granular superconductivity.

## 4. Discussion and Summary

Let us recall our starting assumption that all as-prepared $Fe_{1+\delta}(Te_{1-x}S_x)$, $x > 0.05$, samples are granular superconductors wherein substitution of Te by S leads to a creation of additional nonuniformaly-distributed defects which strongly influence both the superconductivity and normal-state conductivity and, furthermore, that a reduction or a removal of these defects, by heat treatment or soaking in favorable ambient, enhances both superconductivity and normal-state conductivity. Based on this unified and consistent scenario, we explained various effects within both the normal and superconducting phases. Furthermore, this same scenario enables one to construct a generalized $x-T$ phase diagram of $Fe_{1+\delta}Te_{1-x}S_x$, not shown here, that clarifies how each state (namely antiferromagnetism, granularity with *log-in-T*, Kondo-like localization or superconductivity) evolves with concentration and annealing conditions. Indeed, such evolution can be seen in the various earlier reported diagrams [8, 10, 27, 26, 34] which indicate, just as mentioned above, that doping of sulfur destroys the magnetic and structural transition, introduces an additional low-$T$ Kondo-like contribution and induces granular superconductivity; similarly, O-annealing maintains high-$T$ granularity, establishes bulk SC and suppresses any low-$T$ anomalous resistive contribution.

It is interesting to extend this granular scenario to analyze the extensively studied normal and superconducting phase diagram of isomorphous $Fe_{1+\delta}Te_{1-x}Se_x$ which manifest similar structural, magnetic and superconducting properties [6, 11, 13, 35, 36, 37, 38]. Similarities of superconductivity and normal conductivities are best illustrated in Fig.3 which shows $\rho(T, x)$ curves of as-prepared and O-annealed $Fe_{1+\delta}Te_{1-x}Se_x$ samples as reported by Sun *et al.*[6] As evident, $\rho(T, x)$ curves of as-prepared samples reveal the *log-in-T* granularity, Kondo-type contribution and granular superconductivity. In contrast, $\rho(T, x)$ of O-annealed samples exhibit the induced enhancement of both the superconductivity and normal-state conductivity.

We verified that these granular features are manifested in most reported data on $Fe_{1+\delta}(Te_{1-x}X_x)$ [$X$= Se, S; see e.g. [4, 5, 6, 7, 12, 13, 14, 15]). We also verified that these features are manifested in polycrystalline[1, 2] as well as monocrystalline [see e.g. Refs.[4, 6, 12, 26, 39]] samples of these $Fe_{1+\delta}(Te_{1-x}X_x)$ systems.

In summary, we showed that the model of (defective) granular superconductor is an adequate unifying scenario for explaining the normal-state and superconducting properties of $Fe_{1+\delta}(Te_{1-x}X_x)$ systems. This is particularly valid for the followings: the *log-in-T* granular character of both as-prepared and treated samples, the Kondo-like behavior and weak superconductivity of as-prepared samples, as well as the metallic character below a coherent peak and bulk superconductivity of treated samples. It would be an interesting continuation of this work to investigate how applied pressure or intercalation of cations such as potassium would influence the granular character of both the superconducting and normal-state properties of these $Fe_{1+\delta}(Te_{1-x}X_x)$ ($X$= Te, Se) systems.

## Acknowledgments

We acknowledge partial financial support from CNPq and CAPES.

[1] Mizuguchi Y and Takano Y 2010 *J. Phys. Soc. Jpn.* **79** 102001
[2] Deguchi K, Takano Y and Mizuguchi Y 2012 *Sci. Technol. Adv. Mater.* **13** 054303
[3] McQueen T M *et al.* 2009 *Phys. Rev. B* **79** 014522
[4] Liu T J *et al.* 2009 *Phys. Rev. B* **80** 174509
[5] Zhang L *et al.* 2009 *Phys. Rev. B* **79** 012506
[6] Sun Y *et al.* 2016 *Scientific Reports* **6** 32290
[7] Hu J *et al.* 2012 *Supercond. Sci. Technol.* **25** 084011
[8] Mizuguchi Y *et al.* 2010 *Europhys. Lett.* **90** 57002
[9] Sun Y *et al.* 2014 *Sci. Rep.* **4** 4585
[10] Yamazaki T *et al.* 2016 *J. Phys. Soc. Jpn.* **85** 114712
[11] Kawasaki Y *et al.* 2012 *Solid State Commun.* **152** 1135
[12] Cao S *et al.* 2011 *J. Appl. Phys.* **110** 033914
[13] Dong C *et al.* 2011 *Phys. Rev. B* **84** 224506
[14] Ge J *et al.* J 2010 *Solid State Commun.* **150** 1641
[15] Bao W *et al.* 2009 *Phys. Rev. Lett.* **102** 247001
[16] Mizuguchi Y *et al.* 2010 *Physica C: Superconductivity* **470, Supplement 1** S338
[17] Medvedev *et al.* 2009 *Nat. Mater.* **8** 630
[18] McQueen T M *et al.* 2009 *Phys. Rev. Lett.* **103** 057002
[19] Blachowski A *et al.* 2010 *Journal of Alloys and Compounds* **494** 1
[20] Szymanski K *et al.* 2011 *Superconductor Science and Technology* **24** 105010
[21] Rossler S *et al.* 2010 *Phys. Rev. B* **82** 144523
[22] Chang H H *et al.* 2012 *Supercond. Sci. Technol.* **25** 035004
[23] Efetov K B and Tschersich A 2003 *Phys. Rev. B* **67** 174205
[24] Beloborodov I S *et al.* 2007 *Rev. Mod. Phys.* **79** 469
[25] Dong C *et al.* 2010 *Sc. Chin. ser. G* **53** 1216
[26] Dong C *et al.* 2013 *Journal of Physics: Condensed Matter* **25** 385701
[27] Sun Y *et al.* 2013 *Journal of the Physical Society of Japan* **82** 115002
[28] Mizuguchi Y *et al.*2009 *Appl. Phys. Lett.* **94** 012503
[29] Mizuguchi Y *et al.* 2010 *Phys. Rev. B* **81** 214510
[30] Finlayson D M *et al.* 1956 *Proceedings of the Physical Society. Section B* **69** 860
[31] Zajdel P *et al.* 2010 *J. Am. Chem. Soc.* **132** 13000
[32] Liu C J, Bhaskar A, Huang H J and Lin F H 2014 *Applied Physics Letters* **104** 252602
[33] Karlsson S *et al.* 2015 *Superconductor Science and Technology* **28** 105009
[34] ElMassalami M *et al.* 2014 *Journal of Physics: Conference Series* **568** 022012
[35] Hu J, Liu T J, Qian B and Mao Z Q 2013 *Phys. Rev. B* **88** 094505
[36] Liu T J *et al.* 2010 *Nat. Mater.* **9** 716
[37] Katayama N *et al.* 2010 *J. Phys. Soc. Jpn.* **79** 113702
[38] Martinelli A *et al.* 2010 *Phys. Rev. B* **81** 094115
[39] Noji T *et al.* Y 2010 *J. Phys. Soc. Jpn.* **79** 084711

IOP Conf. Series: Journal of Physics: Conf. Series **969** (2018) 012057 doi:10.1088/1742-6596/969/1/012057

# Electrical transport between a $MoS_2$-based electric double-layer transistor and normal and superconducting Al

**Y Aikawa[1], K Tsumura[2], T Narita[1], H Takayanagi[3], and R Ishiguro[1*]**

[1]Department of Physics, Japan Women's University, Bunkyo, Tokyo 112-8681, Japan
[2]Department of Applied Physics, Tokyo University of Science, Katsushika, Tokyo 125-8585, Japan
[3]Research Institute of Science & Tech., Tokyo University of Science, Katsushika, Tokyo 125-8585, Japan
E-mail: m1316001ay@gmail.com, ishiguror@fc.jwu.ac.jp*

**Abstract**. We present fabrication of a junction between an atomic layer molybdenum disulfide-based electric double layer transistor and an aluminum film, and electrical transport measurements between molybdenum disulfide ($MoS_2$) carriers induced by strong electric fields due to its electric double-layer transistor and a conventional superconducting aluminum though Schottky barrier at the interface. A superconducting gap structure was observed in the differential resistance measurement of the dilution refrigerator temperature. Since the gap structures disappeared at about 900mK, the temperature change of this structure is caused due to aluminum superconductivity. However, as for the magnetic field change of the gap structures, the possibility of the coexistence of two kinds of superconducting states at the junction would be considered.

## 1. Introduction

Since Geim and Novoselov succeeded in cleaving the monolayer of graphene by a Scotch tape method , studies on atomic layered materials have been intensively conducted [1-3]. Graphene has attracted attention as a two-dimensional semiconductor with a high mobility. However, it is not optimal as a semiconductor material because graphene has no band gap. Recently, transition metal dichalcogenides (TMDCs) as two-dimensional materials have attracted attention for diverse properties such as semiconductors, metals, semi-metals, superconductors and topological insulators [4-6]. Molybdenum disulphide ($MoS_2$), which is an abundant natural TMDC mineral and is known to easily realize large crystals and long been used as a lubricant in many applications, is an indirect bandgap semiconductor with $E_g$=1.2 eV at 300 K in bulk. Since the intermediate layer of TMDC is bonded by a weak van der Waals force, $MoS_2$ is also easy to be cleaved [7]. The band gap of a monolayer $MoS_2$ is 1.8 eV at 300K. $MoS_2$ is known to be chemically stable because of few dangling bonds on the cleavage surface.

In semiconductor electronics, a chemical carrier doping method has been commonly used for controlling carriers of semiconductor materials and forming ohmic contacts between the metal electrode and the semiconductor. However, since changes in crystal structure are inevitable in insertion of impurities into 2D material, it is not easy to apply chemical carrier doping method to atomic layer semiconductors to control the carrier. For this reason, electric field carrier doping is more suitable for carrier control and the contact resistance control of atomic layer semiconductor materials.

Field effect transistors (FETs) are best known as an electric field doping device, but in recent years, electric double layer transistors (EDLTs) have attracted attention as an electric field doping device capable of accumulating higher carriers. The EDLT is a kind of electric field-effect transistor (FET) using ionic liquid, which is a salt in liquid state, as gate insulating layer of ordinary FETs.

Electric field doping by EDLT is also effective for reducing contact resistance between metal and semiconductor. Normally, a Schottky barrier is formed at the interface between the semiconductor and the metal, resulting in an increase in contact resistance and rectification. The electric field effect by the EDLT can realize ohmic contact by thinning the Schottky barrier at the interface.

Modulating carrier density of $MoS_2$ and the potential barrier at the interface between $MoS_2$ and aluminum electrode by using EDLT, we aim at coherently connecting conventional superconductivity and the electric field induced state in $MoS_2$.

## 2. Experimental

As shown in Fig.1 (a), we fabricated a $MoS_2$–based EDLT with superconducting electrodes, which contain two junctions between $MoS_2$ and Al electrodes. Bulk natural $MoS_2$ was cleaved into double-layered flakes by using the Scotch tape method, and the flakes were transferred onto a Si substrate with a thermally grown $SiO_2$ layer of 290 nm. The double-layered flakes were identified by microscopic Raman measurements [8]. The flakes were shaped as rectangular ($2.4 \times 5.8\ \mu m^2$) by argon plasma etching and electron beam lithography, as shown in Figure 1(b). Then superconducting Al electrodes, which are composed a multilayered configuration of Ti/Al/Ti (100/1000/50 Å), were formed on the $MoS_2$ flakes by electron beam lithography and electron beam physical evaporation in a high-vacuum chamber with a base pressure of a few $10^{-6}$ Pa. The lower Ti layer was deposited for reducing the contact resistance between Al electrodes and the $MoS_2$. Gate electrode and bonding pads of Ti/Au (100/1000Å) were formed by laser lithography and electron beam physical evaporation. Immediately before the measurement, ionic liquids were dropped as a gate dielectric on the surface of the $MoS_2$ channel and a part of the gate electrode, as shown in Figure 1(a).N,N-diethyl-N-methyl-N-(2-methoxyethyl) ammonium bis (trifluoromethanesulfonyl) imide (DEME-TFSI) was used as ionic liquids.

A dilution refrigerator with a base temperature of 20 mK was used for cooling the sample to below the critical temperature of aluminum, which is $T_c = 1.2$ K. Chip ceramic capacitors for electrical noise filtering were mounted between the measurement line and electrical ground nearby the sample, and a copper powder filter was installed at the mixing chamber stage in the dilution refrigerator to minimize the electromagnetic noise. We had placed 4 superconducting electrodes on the $MoS_2$ flakes for performing four-terminal measurements method as shown in Fig.1 (b). Unfortunately, two of the four electrodes have broken out, so all measurements were done by two-terminal method. The differential resistance curves were numerically calculated from the *IV* curves.

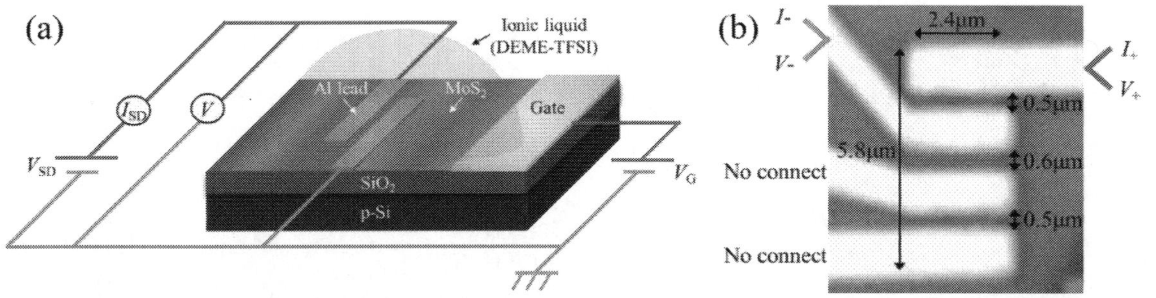


**Figure 1.** (a) Schematic illustration of the $MoS_2$-EDLT sample. A droplet of ionic liquid (DEME-TFSI) was applied to the surface of the channel and the gate electrode. (b) Optical micrograph of the double-layer $MoS_2$ with the Al electrodes as superconductor electrodes.

## 3. Results and discussion

Source-drain current $I_{SD}$ dependence of the gate voltage $V_G$ at $V_{SD} = 10$mV in the $MoS_2$ EDLT is shown in Figure 2(a). To avoid a chemical reaction between the ionic liquid and the $MoS_2$ channel and electrodes, $V_G$ of greater than 3 V was applied above 220 K, which was above the glass transition temperature of the ionic liquid. $I_{SD}$ was increased by applying $V_G = 6$ V; the on/off ratio was about 800 and the resistance was about 100kohm.

The gate voltage of 6 V would be sufficient to induce metallic and superconducting state in the $MoS_2$ [9]. However, compared with previous studies, this on / off ratio is somewhat worse, the resistance seems somewhat higher. This is because the applied effectively voltage is much lower than 6 V and the induced carrier density might be low. In fact, the resistance measured by the two-terminal method increased with decreasing temperature. This means that at least a part of the $MoS_2$ has semiconductor characteristics.

On the other hand, results of measurement of differential resistance at low temperature indicate that $MoS_2$ is a metallic state. Figure 2(b) shows the differential resistance–voltage curve ($dV/dI$–$V$ curve) at $T = 100$ mK to 9.2 K. There is a peak at zero-bias voltage and two valleys at about $|V| = 0.3$ mV in the differential resistances at $T < 1$K. It is thought that the cause of a slight shift of the differential resistance structures in the minus direction is the artifact of current-sweeping from plus to minus. These differential resistance structures would be resulted from the superconducting gap of Al: the structures disappear above the superconducting transition temperature of Al at 1.2K and the superconducting gap energy of Al at 0.1 meV. In other words, it is considered that the metallic state of $MoS_2$ exists near the interface, and a tunnel junction between the superconducting state of Al and the metallic state of $MoS_2$ are realized.

Furthermore, the structures of this differential resistance evolve with the magnetic fields. Figure 2(c) shows the $dV/dI$–$V$ curves when a magnetic field $B = 0$ to 24 mT was applied. Here, the peak at zero-bias voltage decreases with the applied magnetic field, and the two valleys at about $|V| = 0.3$ mV also decrease and the voltage positions of the valleys change in the direction to the zero-bias voltage with the applied magnetic field. Since the critical magnetic field of the Al superconductivity is about 10mT, these variations with the magnetic field seem consistent with suppression of the superconducting gap of Al by the magnetic field.

However, the voltage positions of the two valleys do not change from 7mT to 24mT, Figure 2(d) shows the variation of the voltage positions of the two valleys. Some structures of this differential resistance do not disappear even at $B = 24$ mT. There is small asymmetry with respect to $V = 0$ mV, which might be reflected the asymmetry of the normal resistance related to the Schottky barrier or measurement. The strong structure of the differential resistance against magnetic field might be related to superconducting state induced in a part of $MoS_2$ at the interface of the junction.

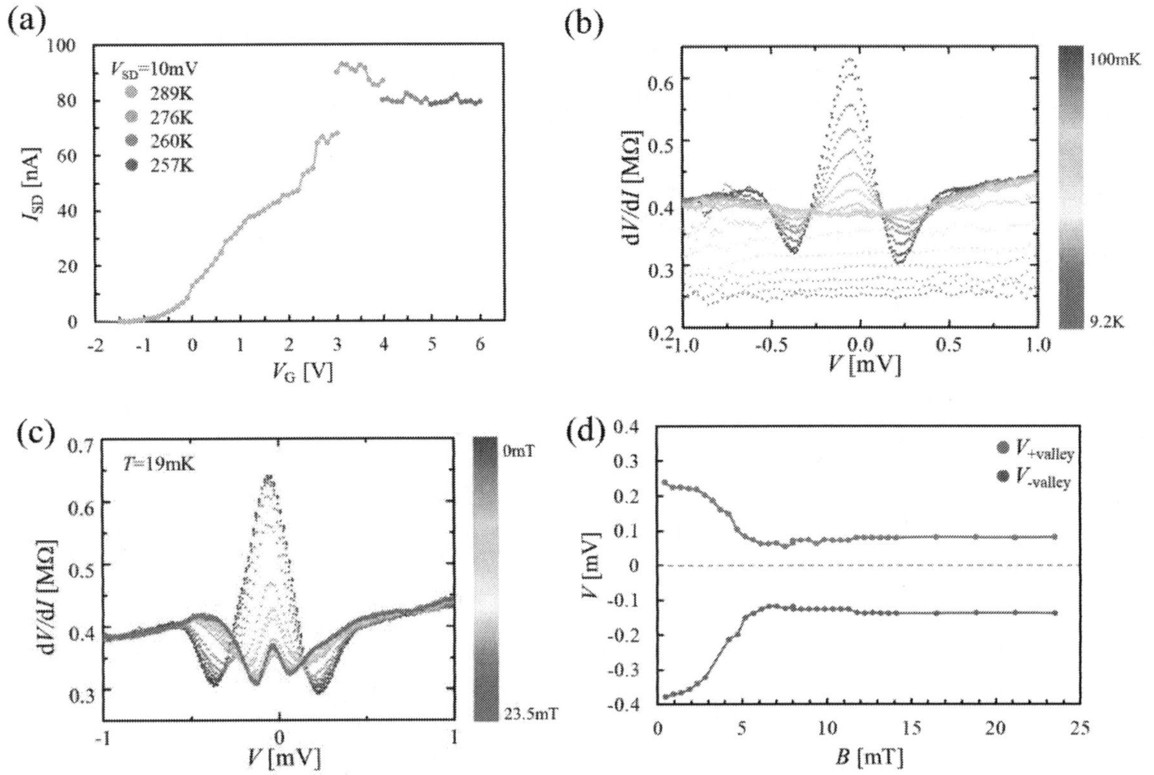


**Figure 2.** (a) Dependence of the source-drain current $I_{SD}$ on the gate voltage $V_G$ at $V_{SD}$ = 10mV in $MoS_2$ EDLT. (b) Differential resistance–voltage curve (d$V$/d$I$–$V$ curve) at $T$ = 100 mK to 9.2K. (c) d$V$/d$I$–$V$ curve at the applied magnetic field $B$ = 0 to 24 mT. (d) Voltage of +gap $V_{+valley}$ and –gap $V_{-valley}$ and dependence on the magnetic field $B$ = 0 to 24 mT.

## 4. Conclusions

We fabricated the $MoS_2$-based EDLT with a conventional superconductor Al electrodes. We measured the electrical transport of the $MoS_2$-based EDLT by two terminal method, which contain the junction between electrically induced state in the $MoS_2$ and a superconducting Al. The resistance of the EDLT at the gate voltage 6V is higher than the previous reports, and increased with decreasing temperature. This indicates that at least one part of the $MoS_2$ has semiconducting characteristics. On the other hand, the differential resistance indicates that metallic state and superconducting state are induced in the $MoS_2$. These results indicate that the channel of the $MoS_2$-EDLT is not uniform. In other words, semiconducting state, metallic state, and superconducting state coexist in the $MoS_2$. However, at least, the differential resistance structures related to the superconducting gap of aluminum is observed, which means that the tunnel barrier caused by the Schottky barrier between aluminum and $MoS_2$ can be controlled.

## Acknowledgments

We would like to thank S. Kashiwaya from AIST for the stimulating discussion. This work was supported by JSPS KAKENHI C (Grant No. 17K05551). Device fabrication was supported by the NIMS Nanofabrication platform of the "Nanotechnology Platform Project" sponsored by MEXT.

## References

[1] Novoselov K S, Geim A K, Morozov S V, Jiang D, Zhang Y, Dubonos S V, Griqorieva I V and Firsov A A 2004 Nature **306** 666
[2] Novoselov K S, Jiang D, Schedin F, Booth T J, Khotkevich V V, Morozov S V and Geim A K 2005 Proc Natl Acad Sci USA **102** 10451
[3] Novoselov K S 2011 Rev Mod Phys **83** 837
[4] Mak K F, Lee C, Hone J, Shan J and Heinz T F 2010 Phys Rev Lett **105** 136805
[5] Wang Q H, Zadeh K K, Kis A, Coleman J N and Strano M S 2012 Nat Nanotechnol **7** 699
[6] Allain A, Kang J, Banerjee K and Kis A 2015 Nat Mat **14** 1195
[7] Ataca C, Topsakal M, Aktürk E and Ciraci S 2011 J Phys chem C **115** 16354
[8] Lee C, Yan H, Brus L E, Heinz T F, Hone J and Ryu S 2010 ACS Nano **4** 2695.
[9] Ye J T, Zhang Y J, Akashi R, Bahramy M S, Arita R and Iwasa Y 2012 Science **338** 1193

# Supercurrent through $MoS_2$-based electric double-layer transistor sandwiched between conventional superconductors

**Y Aikawa[1], K Tsumura[2], T Narita[1], H Takayanagi[3] and R Ishiguro[1*]**

[1] Department of Physics. Japan Women's University, Bunkyo, Tokyo 112-8681, Japan
[2] Department of Applied Physics. Tokyo University of Science, University of Science, Tokyo 125-8585, Japan
[3]Research Institute of Science & Tech. Tokyo University of Science, Tokyo 125-8585, Japan

E-mail: m1316001ay@gmail.com, ishiguror@fc.jwu.ac.jp*

**Abstract**. Transition metal dichalcogenide molybdenum disulphide ($MoS_2$) is a two-dimensional layered crystal, which has a great potential as a flexible semiconductor device. Recently, manifestation of electrically induced superconductivity at the surface of the $MoS_2$-based electric double-layer transistor ($MoS_2$-EDLT) is reported. The superconducting properties, such as the ultra-high upper critical field, give a sign of unconventional spin-valley locking superconducting state. In this study, we aim to realize the coherent connection of the $MoS_2$-EDLT to conventional superconductors Aluminum (Al). Therefore, we first fabricated and evaluated the properties of Al/ $MoS_2$-EDLT configuration/Al junction. At $T = 18$ mK, it is considered that a superconducting current of about 30 pA flows into the Al/$MoS_2$-EDLT/Al junction. We discussed the magnetic field response of this supercurrent.

## 1. Introduction

Transition metal dichalcogenides (TMDCs) are two-dimensional material that are expected to be utilized in technologies such as field effect transistor (FET) and electric double-layer transistor (EDLT) due to its moderate energy gap [1-4]. Furthermore, TMDCs recently received a wide attention due to the peculiar properties in two-dimensional structure [5-7].

Recently, an FET configuration with atomic layer $MoS_2$ as a channel was fabricated. It was reported that the $MoS_2$-FET is an ultra-low power consumption FET with an on / off ratio of $1 \times 10^8$ at a low temperature [1]. Further, a high mobility of 700 $cm^2V^{-1}s^{-1}$ was obtained in a $MoS_2$-FET with $MoS_2$ channel of 10 nm in thickness using scandium as a contact [8].

A $MoS_2$-EDLT configuration using ionic liquid instead of a gate insulator of an FET was also fabricated. It was reported that an appearance of electrically induced superconductivity at the surface of the $MoS_2$-based EDLT by applying a gate voltage [9]. Furthermore, this superconducting state is robust against the magnetic field in the direction parallel to the layer, and it is maintained in an ultra-high magnetic field of 50 T or more. This is a unique case of superconductivity, in which the spin of the superconducting Cooper-pair is fixed in the direction of the plane due to the breaking of the inversion symmetry for the crystal structure in the single layer $MoS_2$ [6].

In general, it is known that superconductivity under a non-centrosymmetric crystal structure is often unconventional [10-12]. The method clearly demonstrates that non-centrosymmetric superconducting state is due to the mixing of spin singlet and triplet. It was proposed to fabricate a Josephson junction (S/S' junction) using a conventional superconductor (S) and non-centrosymmetric superconductor (S'), and measure the magnetic field change of a supercurrent through the junction [12, 13].

We choose $MoS_2$, which breaks the inversion symmetry of the crystal, and its unique superconducting state may be due to the mixing of spin singlet and triplet. A junction consisting of conventional superconductor and unconventional superconductor is often used for studying the anisotropic superconducting state.

In this research, we aim to realize a Josephson junction between $MoS_2$ is a unique superconductor and conventional superconductor, and elucidate specific superconducting state in $MoS_2$. To this end, it is necessary to reliably ascertain whether a junction is coherent, which the Cooper-pair in a conventional superconductor can sufficiently penetrate by the intermediary of the junction interface. Therefore, in this study, we aim to realize a coherent junction (Josephson junction) between conventional superconductors Aluminum (Al) and the MoS2-EDLT.

## 2. Experimental

We fabricated a conventional superconductor/$MoS_2$-EDLT configuration/conventional superconductor (S/$MoS_2$-EDLT/S) junction shown in Figure 1. Bulk $MoS_2$ was cleaved into thin flakes by using the Scotch tape method, and the flakes were transferred onto Si substrate with a thermally grown $SiO_2$ layer of 290 nm. Single- or triple-layered flakes were identified by microscopic Raman measurements [14]. The flakes were shaped to be rectangular ($2 \times 4.5\ \mu m^2$) by argon plasma etching as shown in Figure 1(a). Superconducting electrodes were composed of multilayered configuration of Ti/Al (100/1000Å), where Ti can reduce the contact resistance for $MoS_2$. The electrodes and the gate electrodes of Ti/Au (100/1000Å) were formed using electron beam lithography and laser lithography respectively.

A droplet of ionic liquid N,N-Diethyl-N-methyl-N-(2-methoxyethyl)ammonium bis(trifluoromethane sulfonyl)imide (abbreviated DEME-TFSI) was applied onto the surface of the channel and the gate electrodes as shown in Figure 1(c) and (d). Next, the sample was mounted on a dilution refrigerator and cooled down to 18 mK. For the transport measurement, copper power filters placed on the mixing chamber stage were employed to minimize the electromagnetic noise.

IOP Conf. Series: Journal of Physics: Conf. Series **969** (2018) 012058 doi:10.1088/1742-6596/969/1/012058

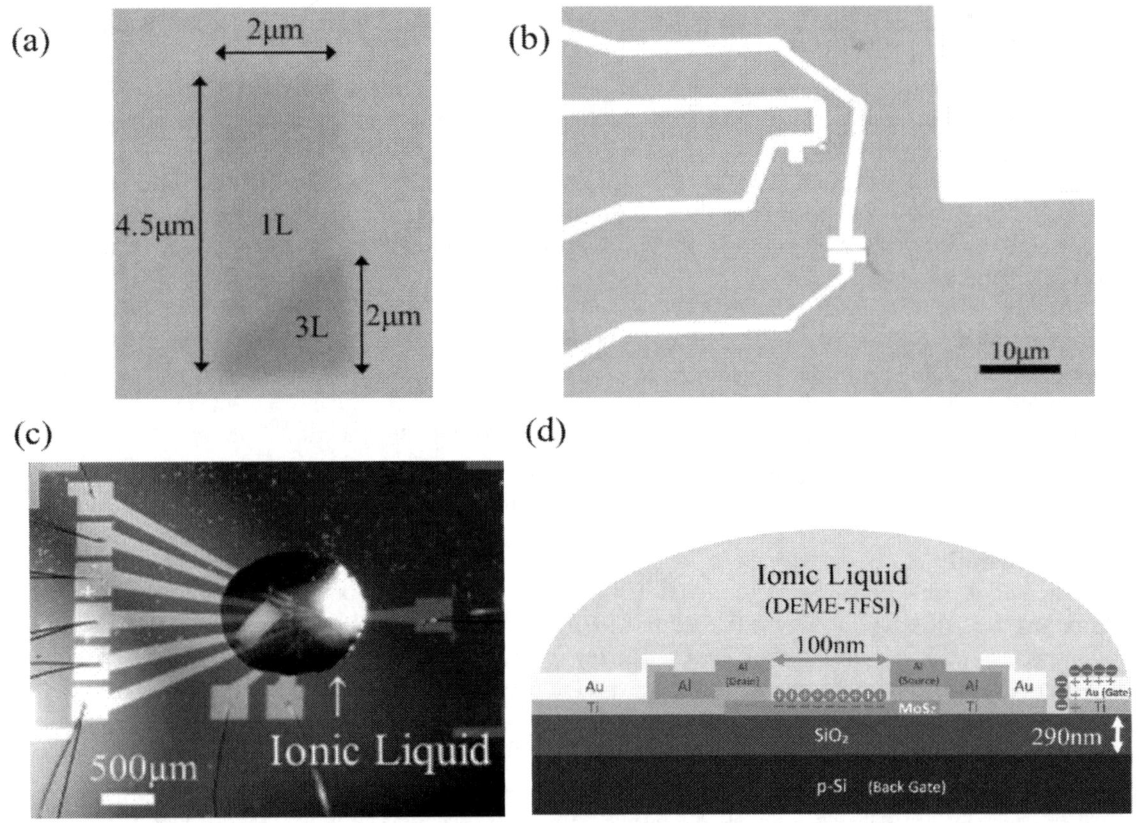


**Figure 1.** (a) After flakes of $MoS_2$ were shaped to be rectangular by argon plasma etching. The flake is mixed single- and triple-layer. (b) Optical micrograph of a $MoS_2$ thin flake patterned with superconductor electrodes. (c) Photograph of the $MoS_2$-EDLT sample in top view. (d) Schematic of the $MoS_2$-EDLT sample inside view.

## 3. Results and discussions

First, the gate voltage $V_G$ dependence of source-drain current $I_{SD}$ at $V_{SD} = 100$ mV was measured. To avoid a chemical reaction between ionic liquid and $MoS_2$ channel, and electrodes, $V_G$ higher than 3 V was applied at 220 K, which is above the glass transition temperature of the ionic liquid. The $I_{SD}$ increased by applying $V_G = 6$ V and the on/off ratio was about 400.

Figure 2(a) shows the current-voltage curve ($I$-$V$ curve) in magnetic field perpendicular to the $MoS_2$ plane $B = -14$ mT to 14 mT at $T = 18$ mK. It is clear from that a supercurrent with the critical current of $I_c \sim 30$ pA through S/$MoS_2$-EDLT/S junction. Moreover, $I_c$ changes by applying a magnetic field.

Figure 2(b) depicts the magnetic field $B$ dependence of $I_c$ at $T = 18$ mK. As can be seen from this Figure 2(b), $I_c$ changes with the application of a magnetic field. In addition, we can observe several jumps in $I_c$. The jumps might be due to the trapping of magnetic flux around the junction.

In addition, if the junction is a Josephson junction, a periodic pattern which is called Fraunhofer pattern can be observed by applying a magnetic field. It is clear from Figure 2(a), a period of magnetic field exhibit, however the next period_of magnetic field did not exhibit the expected Fraunhofer pattern. To discuss this issue, we focused on the shape and number of layers in $MoS_2$ sample used in this measurement. As shown in Figure 1(a), the sample has a single layer and three layers mixed in the $MoS_2$ of rectangular ($2 \times 4.5$ μ$m^2$). Here, we consider the magnetic field of a period in the Fraunhofer pattern in two cases where a current flows uniformly and only through a part of the triple layer, respectively.

As the junction junction length $L$ is 0.1 μm, the magnetic field penetration length of Al $\lambda$ is about 0.05 μm and the width $d$ is 4.5 μm, the effective area $S_1$ of the junction was calculated when flowing uniformly, $S_1 = (d + 2\lambda)\ and\ L = 0.9\ \mu m^2$. As the magnetic field $B$ for one period in the Josephson junction of the effective area $S$ is represented by $\varphi_0 / S_{eff}$, it follows that $B_1 = \frac{\varphi_0}{0.9} = \cong 2$ mT.
When it flows through only a part of the triple layer, the effective area $S_2$ of the junction was calculate to be~$0.2\ \mu\ m^2$, in a similar the case of uniform flow. We calculate the magnetic field for one period of the Fraunhofer pattern as, $B_2 = \frac{\varphi_0}{0.4} \cong 10$ mT using this.

From the above, the magnetic field of a period in the Fraunhofer pattern is 2 mT when a current flows uniformly in the $MoS_2$, and 10 mT when a current flows only through a part of the triple layer. It is considered that the magnetic field of one period in the Fraunhofer pattern is close to the case where the current flows only through a part of the triple layer in $MoS_2$.

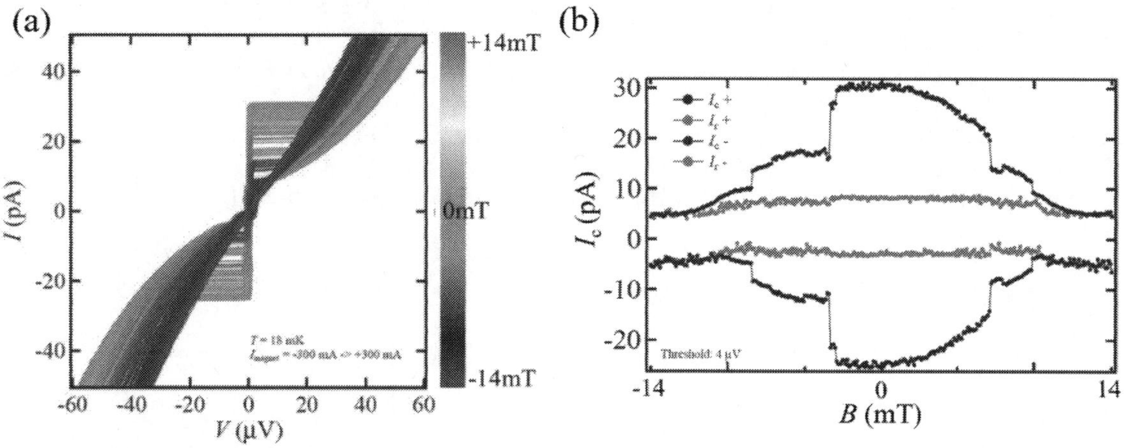


**Figure 2.** (a) Magnetic field dependence of *I-V* curve at 18 mK. The external magnetic field applied is parallel to the surface of the junction. (b) Magnetic field dependence of the critical current $I_c$ of the junction at 18 mK.

## 4. Conclusions

In conclusion, we demonstrated that a supercurrent through a $MoS_2$ sandwiched between conventional superconductors Al using an EDLT configuration. We verified that the device operates as the EDLT as the source-drain current is increased by applying the gate voltage at $V_{SD}$ =100 mV. At $T$ = 18 mK, it is considered that a superconducting current of about 30 pA flows into the Al/$MoS_2$-EDLT/Al junction. Furthermore, a vertical magnetic field between -14 mT to and +14 mT was applied to this junction, and the *I-V* characteristic was measured. As a result, the magnetic field response of the supercurrent was confirmed.

**Acknowledgements**
We would like to thank T. Sato, D. Sakuma from Tokyo Univ. of Science, S. Nomura from Univ. of Tsukuba and S. Kashiwaya from AIST for stimulating discussions. This work was supported by JSPS KAKENHI C (Grant No. 17K05551). Device fabrication was supported by the NIMS Nanofabrication platform of the "Nanotechnology Platform Project" sponsored by MEXT.

IOP Conf. Series: Journal of Physics: Conf. Series **969** (2018) 012058 doi:10.1088/1742-6596/969/1/012058

**References**

[1] Radisavljevic B, Radenovic A, Brivio J, Giacometti V, and Kis A 2011 Nat Nanotechnol **6** (3) 147.
[2] Perera M M, Lin M W, Chuang H J, Chamlagain B P, Wang C, Tan X, Cheng M M C, Tománek D, and Zhou Z 2013 ACS Nano **7** (5) 4449.
[3] Zhang Y, Ye J, Matsuhashi Y, and Iwasa Y 2012 Nano Lett **12** (3) 1136.
[4] Lee G H, Yu Y J, Cui X, Petrone N, Lee C H, Choi M S, Lee D Y, Lee C, Yoo W J, Watanabe K, Taniguchi T, Nuckolls C, Kim P, and Hone J 2013 ACS Nano **7** (9) 7931.
[5] Mak K F, Lee C, Hone J, Shan J, and Heinz T F 2010 Phys Rev Lett **105** (13) 136805.
[6] Saito Y, Nakamura Y, Bahramy M S, Kohama Y, Ye J, Kasahara Y, Nakagawa Y, Onga M, Tokunaga M, Nojima T, Yanase Y, and Iwasa Y 2015 Nature Phys **12** (2) 144.
[7] Saito Y, Nojima T, and Iwasa Y 2016 Nat Rev Mater **2** (1) 16094.
[8] Das S, Chen H Y, Penumatcha A V, and Appenzeller J 2013 Nano Lett **13** (1) 100.
[9] Ye J, Zhang Y, Akashi R, Bahramy M, Arita R, and Iwasa Y 2012 Science **338** (6111) 1193.
[10] Hsu Y T, Vaezi A, Fischer M H, and Kim E A 2017 Nat Commun **8** 14985.
[11] Qin F, Shi W, Ideue T, Yoshida M, Zak A, Tenne R, Kikitsu T, Inoue D, Hashizume D, and Iwasa Y 2017 Nat Commun **8** 14465.
[12] Kaur R P, Agterberg D F, and Sigrist M 2005 Phys Rev Lett **94** (13) 137002.
[13] Geshkenbein V, Larkin A, and Barone A 1987 Phys Rev B **36** (1) 235.
[14] Lee C, Yan H, Brus L E, Heinz T F, Hone J, and Ryu S 2010 ACS Nano **4** (5) 2695.

# Variations of superconducting transition temperature in $YbBa_2Cu_3O_{7-\delta}$ ceramics by Gd substitution

**K Matsushima, C Taka and A Nishida**

Department of Applied Physics, Fukuoka University, 8-19-1, Nanakuma, Jonan-ku, Fukuoka 814-0180, Japan

nishida@cis.fukuoka-u.ac.jp

**Abstract**. So-called RE-123 superconductors are said to generally exhibit superconductivity at the transition temperature $T_c$ around 90K with RE (rare-earth) ions from Nd to Lu. However there is some tendency that $T_c$ decreases with decrease in RE ionic radius. Especially for Yb-123, $T_c$ often becomes below 90K. We find that the smaller the ionic radius is, the lower the melting temperature becomes for RE-123 ceramics. In fact, Yb-123 ceramics often melt above 910°C, while Gd-123 ceramics can be sintered above 950°C without melting. From this tendency we suppose that such $T_c$ lowering is due to insufficient sintering temperature, resulting in insufficient solid state reaction and poor crystal growth. In this work, we try to raise $T_c$ in Yb-123 ceramics by increasing the sintering temperature through Gd substitution. We find that the optimum sintering temperature is 930°C in $Yb_{0.7}Gd_{0.3}Ba_2Cu_3O_{7-\delta}$ ceramics with the highest $T_c$ of 91.8K (zero resistivity).

## 1. Introduction

It is well known that the cuprate compound of $YBa_2Cu_3O_{7-\delta}$ is the first superconductor which has been discovered to exhibit superconductivity above liquid-nitrogen temperature [1]. Soon after this discovery, superconductivity in a series of $REBa_2Cu_3O_{7-\delta}$ (RE-123) compounds (RE = rare-earth element) has been investigated and the transition temperature has been reported to be nearly independent of the rare-earth ion [2]. Then, it seems that not much attention has been paid to the RE substitution effect in comparison to substitution effects of barium or copper ions [3, 4] with rather drastic phenomena.

However, when we look into detail, we find some tendency [5] that $T_c$ decreases with decrease in RE ionic radius as shown in figure 1 (the solid line is for eye-guide). Especially for Yb-123, $T_c$ often becomes below 90K. We find that the smaller the ionic radius is, the lower the melting temperature becomes for RE-123 ceramics. In fact, Yb-123 ceramics often melt above 910°C, while Gd-123 ceramics can be sintered above 950°C without melting. From this tendency we suppose that such $T_c$ lowering is due to insufficient sintering temperature, resulting in insufficient solid state reaction and poor crystal growth.

In this work, we try to raise $T_c$ in Yb-123 ceramics by increasing the sintering temperature through Gd substitution. We find that the optimum sintering temperature is 930°C in $Yb_{0.7}Gd_{0.3}Ba_2Cu_3O_{7-\delta}$ ceramics with the highest $T_c$ of 91.8K (zero resistivity).

## 2. Experimental

Pure and substituted samples were prepared by the solid state reaction method. The raw materials of $Yb_2O_3$, $Gd_2O_3$, $BaCO_3$ and CuO powders (all with 99.99% purity) were weighed to be in the ratio of Yb:Gd:Ba:Cu = 1-$x$:$x$:2:3, where $x$ = 0, 0.1, 0.2 and 0.3. The total of 8 g mixture with certain $x$ was ground carefully in the agate mortar, calcined at 880°C for 24 h in air and repeated this process twice. This makes a single batch of the calcined powder with a given $x$. Then, weighing about 0.8g out of the calcined powder, ground carefully, pelletized in a disk of 10mmϕ and sintered for 24 h in air at certain temperature between 885°C and 950°C. During the final cooling down, the intermediate annealing was applied at 500°C for 10 h in order to take enough oxygen. Thus, we were able to obtain up to 8 pellets of $Yb_{1-x}Gd_xBa_2Cu_3O_{7-\delta}$ ceramics with different sintering temperatures from a single batch of the calcined powder with a given $x$. Adherence to such single batch process can avoid possible complication due to undesirable variation in the calcination process.

Obtained pellets were cut into bar shape (typically 1x2x8 $mm^3$) for resistivity measurements with the standard four-probe method. Some other parts from the pellets were used for estimation of sample morphology with the scanning electron microscopy (SEM). Oxygen contents were also estimated by thermogravimetry (TG) measurements.

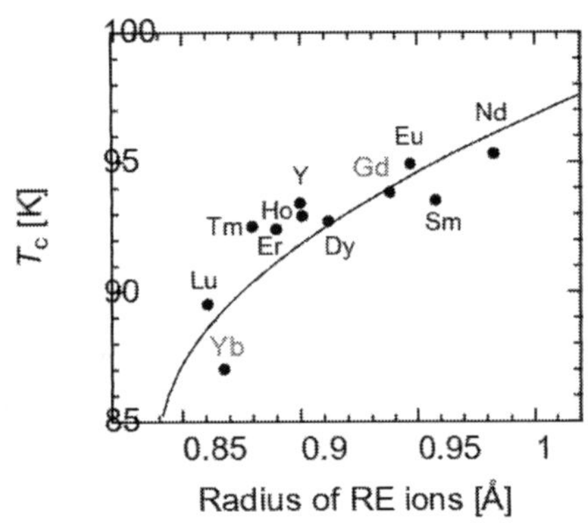


**Figure 1.** Variation of superconducting transition temperature $T_c$ as a function of rare-earth ionic radius.

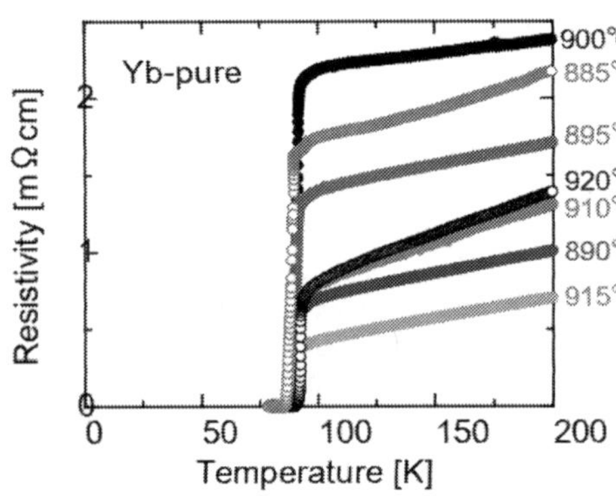


**Figure 2.** Electrical resistivity in the pure $YbBa_2Cu_3O_{7-\delta}$ (Yb-pure) as a function of temperature.

## 3. Results and discussion

### 3.1. *pure* $YbBa_2Cu_3O_{7-\delta}$

Temperature dependent resistivity in the pure $YbBa_2Cu_3O_{7-\delta}$ (Yb-pure) is shown in figure 2 for ceramics sintered between 885°C and 920°C with 5°C interval. From the results we first estimate the onset point of the superconducting transition by linearly extrapolating resistivity curves just around the onset region. In the sample sintered at 900°C, for example, we estimate the onset point as 92.2 K (onset $T_c$) and 2.1 mΩcm (onset resistivity, abbreviated as onset $R$). Next, we estimate the temperature where the resistivity shows just half of the onset R as midpoint $T_c$ (90.2 K for the sample sintered at 900°C). And finally, we estimate the offset $T_c$ where the resistivity becomes zero (87.5 K for the sample sintered at 900°C).

In figure 3 we present thus obtained transition temperature $T_c$'s in the pure $YbBa_2Cu_3O_{7-\delta}$ as a function of the sintering temperature $T_s$. Although the highest onset $T_c$ is obtained when the pellet was sintered at 920°C, the pellet was partially melted and the transition towards zero resistivity showed rather long tail. Thus we conclude that the optimum sintering temperature is 890°C for the pure $YbBa_2Cu_3O_{7-\delta}$, resulting in the highest offset (zero resistivity) $T_c$ of 90.4 K. Figure 4 indicates the onset $R$ as a function of $T_s$. The resistivity in the sample sintered at 890°C is fairly low, but the lowest resistivity was obtained when sintered at 905°C. Scattered values of the offset $T_c$ and the onset $R$

probably reflect difficulty in preparing the pure $YbBa_2Cu_3O_{7-\delta}$ which is easily melted at rather low sintering temperatures, resulting only in insufficient solid state reaction. Thus, we seek to raise the sintering temperature by the Gd substitution.

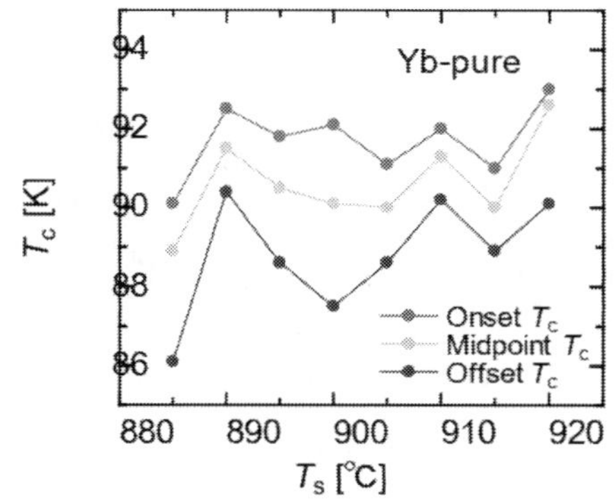


**Figure 3.** Variation of superconducting transition temperature $T_c$'s against the sintering temperature $T_s$ in Yb-pure sample.

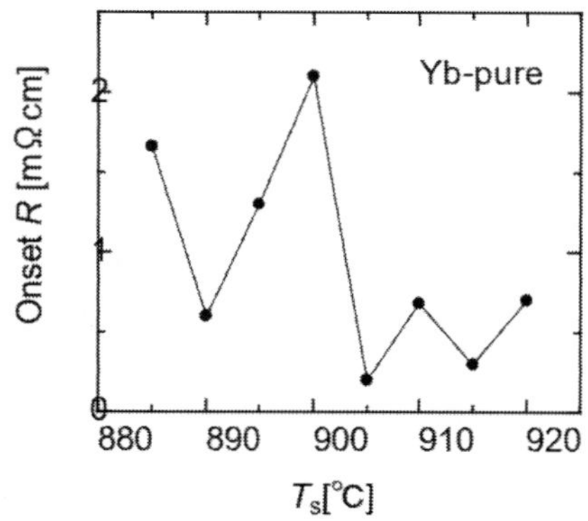


**Figure 4.** Variation of onset resistivity (Onset $R$) against the sintering temperature $T_s$ in Yb-pure sample.

*3.2. Yb-10%Gd*

In figure 5 we next show temperature dependent resistivity in 10%Gd-substituted ceramics; $Yb_{0.9}Gd_{0.1}Ba_2Cu_3O_{7-\delta}$ (Yb-10%Gd). Similarly to the previous subsection, we estimate the transition temperature $T_c$'s and onset $R$ as a function of the sintering temperature $T_s$, the results being given in figures 6 and 7, respectively. From the results we can conclude that the optimum sintering temperature is 900°C for the Yb-10% Gd sample with the highest zero resistivity $T_c$ of 91.1 K and very low onset $R$ of 0.35 mΩcm.

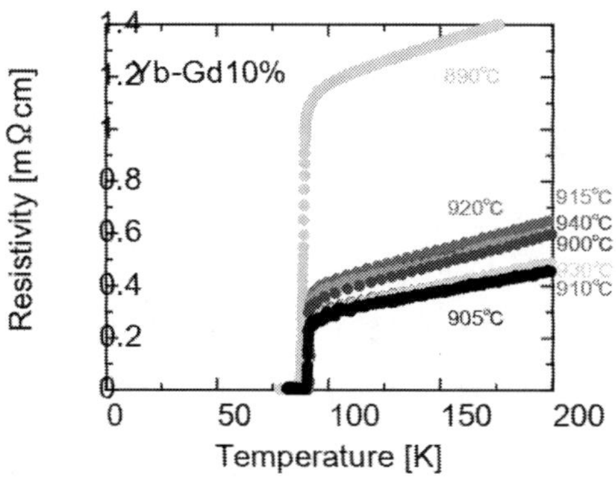


**Figure 5.** Electrical resistivity in 10%Gd-substituted ceramics; $Yb_{0.9}Gd_{0.1}Ba_2Cu_3O_{7-\delta}$ (Yb-10%Gd) as a function of temperature..

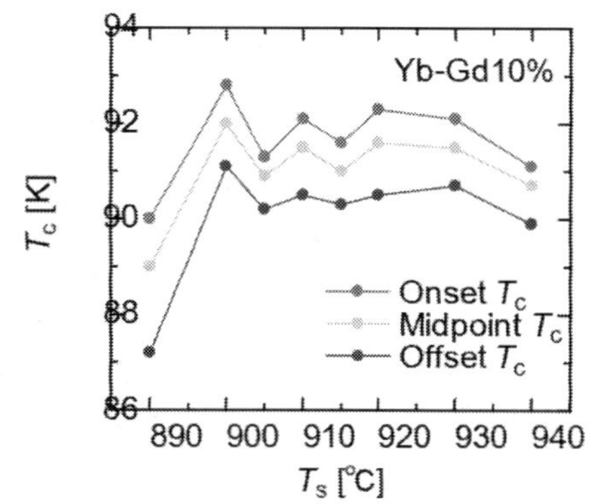


**Figure 6.** Variation of superconducting transition temperature $T_c$'s against the sintering temperature $T_s$ in Yb-10%Gd.

It should be noted that when compared with the pure sample the zero resistivity $T_c$ has increased by 0.7 K according to increased sintering temperature by the 10%Gd substitution. It is also noted that significant variations of $T_c$'s and onset $R$ in the pure sample have been drastically suppressed by the 10%Gd substitution. We infer the origin of these improvements as due to improved stability of the 123 structure by larger ion (Gd) substitution, which then enabled higher temperature sintering.

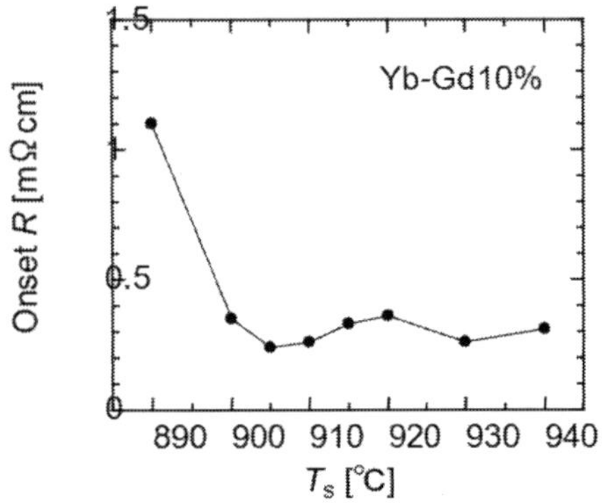


**Figure 7.** Variation of onset $R$ against the sintering temperature $T_s$ in Yb-10%Gd sample.

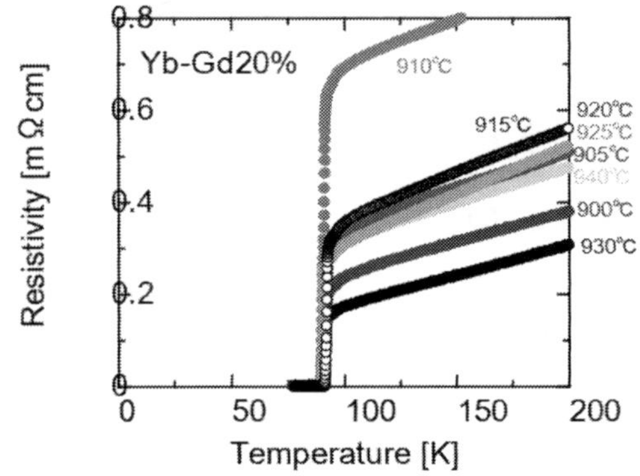


**Figure 8.** Electrical resistivity in Yb-20%Gd ceramics as a function of temperature.

### 3.3. *Yb-20%Gd*

In figure 8 we show temperature dependent resistivity in Yb-20%Gd ceramics. Estimated transition temperature $T_c$'s and onset $R$ as a function of the sintering temperature $T_s$ is given in figures 9 and 10, respectively. As can be seen from the figures, the optimum sintering temperature is 905°C for the Yb-20%Gd sample with the highest zero resistivity $T_c$ of 91.2 K and the very low onset $R$ of 0.28 mΩcm.

It is noted that the optimum sintering temperature has been even increased by the 20%Gd substitution, compared to 10%Gd substitution. Significant variations of $T_c$'s and onset $R$ in the pure sample have been again drastically suppressed by the 20%Gd substitution. In fact, onset $R$ for all 8 samples falls between 0.2 and 0.6 mΩcm, in spite of the wide variation of the sintering temperature from 900°C to 940°C.

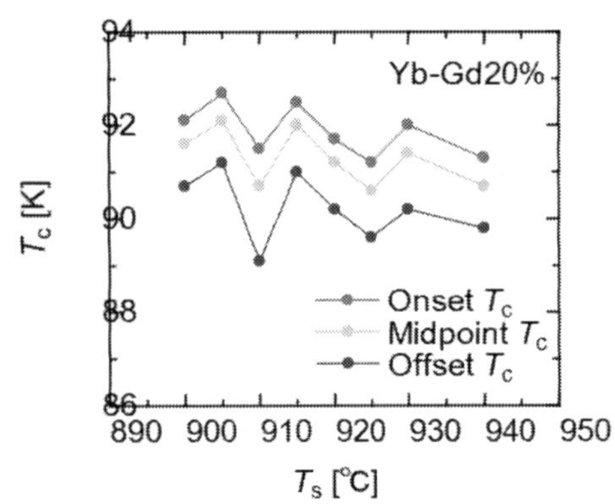


**Figure 9.** Variation of superconducting transition temperature $T_c$'s with the sintering temperature $T_s$ in Yb-20%Gd sample.

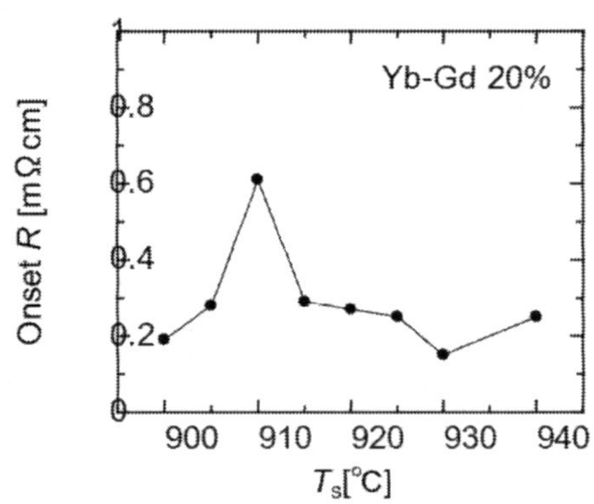


**Figure 10.** Variation of onset $R$ with the sintering temperature $T_s$ in Yb-20%Gd sample.

### *3.4. Yb-30%Gd*

Figure 11 shows temperature dependent resistivity in Yb-30%Gd ceramics. Estimated transition temperature $T_c$'s and onset $R$ as a function of the sintering temperature $T_s$ is given in figures 12 and 13, respectively. In this case, the optimum sintering temperature is 930°C for the Yb-30%Gd sample with the highest zero resistivity $T_c$ of 91.8 K and very low onset $R$ of 0.34 mΩcm. It should be remarkable that the zero resistivity $T_c$ has increased by 1.4 K when compared with the pure sample.

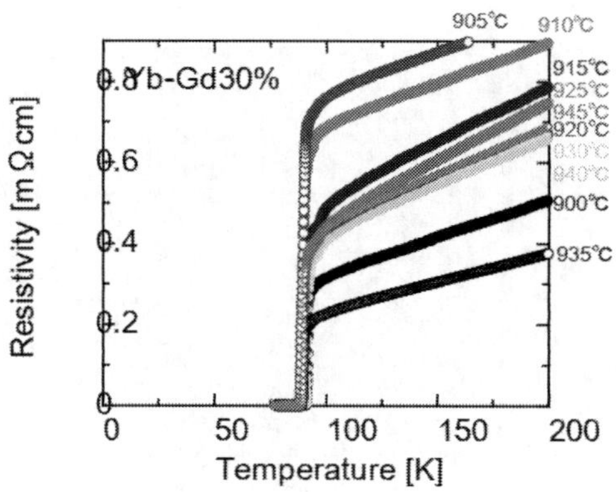


**Figure 11.** Electrical resistivity in Yb-30%Gd ceramics as a function of temperature.

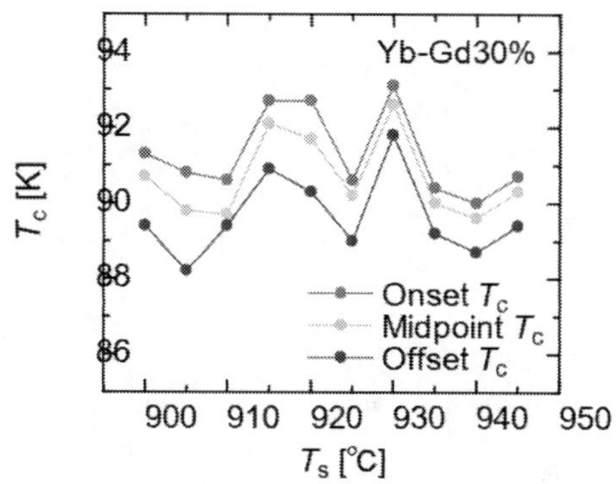


**Figure 12.** Variation of superconducting transition temperature $T_c$'s against the sintering temperature $T_s$ in Yb-30%Gd.

### *3.5. oxygen contents 7-δ*

In figure 14 we indicate estimation of oxygen contents 7-δ by the thermogravimetry (TG) measurements. Rather large variation of 7-δ in the pure sample again reflects difficulty in preparing the pure $YbBa_2Cu_3O_{7-\delta}$ which is easily melted at rather low sintering temperatures. On the other hand, 7-δ is well above 6.9 for Gd substituted samples, inferring stabilized 123 structure and pronounced crystal growth due to higher temperature sintering. It is also verified that the above estimation of the optimum sintering temperature based on the highest $T_c$ and the lowest onset-$R$ is not affected by undesirable variation of oxygen contents depending on different sintering temperatures.

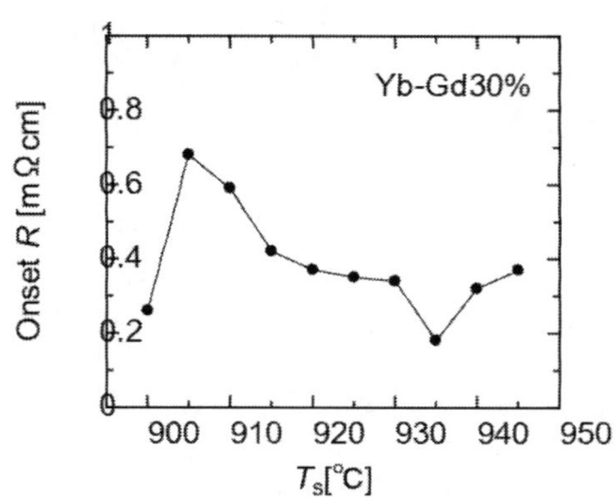


**Figure 13.** Variation of onset $R$ against the sintering temperature $T_s$ in Yb-30%Gd sample.

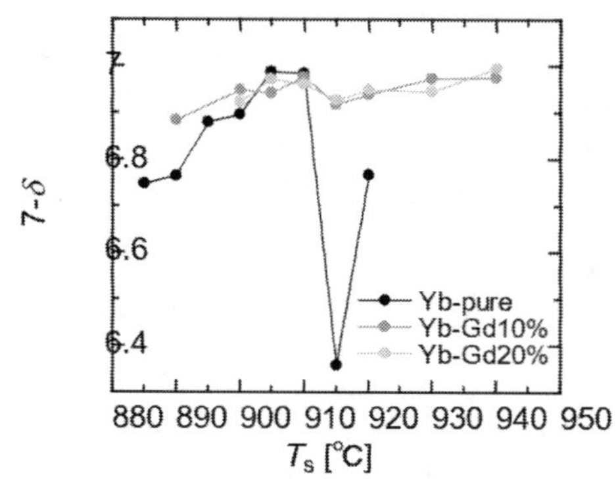


**Figure 14.** Oxygen contents 7-δ in Yb-pure, 10%Gd and 20%Gd ceramics as a function of sintering temperature $T_s$.

### *3.6. crystal morphology observed by SEM*

Such pronounced crystal growth is confirmed by the SEM images as compared in figures 15 and 16. When the Yb-10%Gd sample is sintered at 890°C, the size of microcrystals is small and the crystal facet is not clear as seen in figure 15. There are also rather many voids between crystals and the connectivity between grains is poor, too. On the other hand, when the Yb-10%Gd sample is sintered optimally at 900°C, the crystal growth is significant with the sharp facet, and the connectivity has also improved as shown in figure 16.

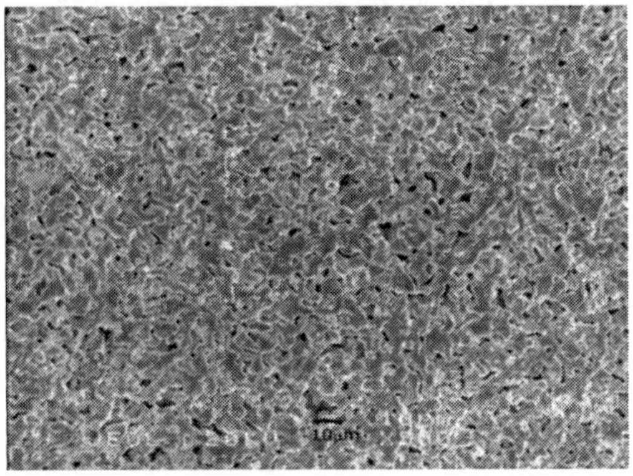

**Figure 15.** SEM image of Yb-10%Gd sample sintered at 890°C, showing poor crystal growth (magnification x500).

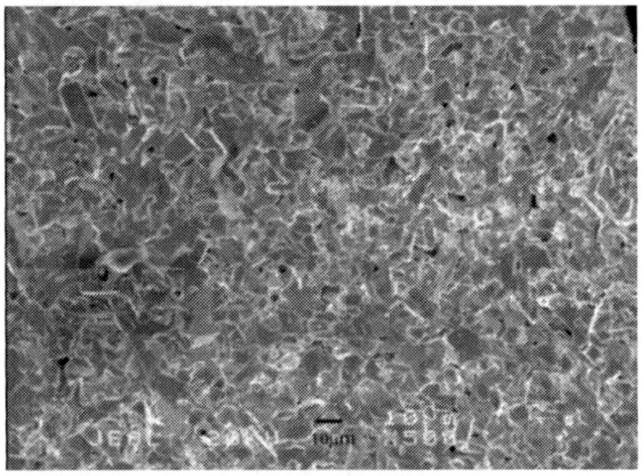

**Figure 16.** SEM image of Yb-10%Gd sample sintered at 900°C, indicating pronounced crystal growth (x500).

## 4. Conclusions

In summary, we prepared more than 30 samples of Gd substituted Yb-123 ceramics with various Gd concentration under various sintering temperatures, seeking to raise $T_c$ in Yb-123 ceramics by increasing the sintering temperature through Gd substitution. We, in fact, succeeded to obtain the pronounced crystal growth by higher temperature solid state reaction, resulting in the highest zero-resistivity superconducting transition at $T_c$ = 91.8K in $Yb_{0.7}Gd_{0.3}Ba_2Cu_3O_{7-\delta}$ ceramics, optimally sintered at 930°C. We infer that higher temperature sintering has been realized by improved stability of the 123 structure by the larger ion Gd substitution.

**Acknowledgments**

Authors wishing to acknowledge the Central Research Laboratory, Fukuoka University (Project #145004) for financial support.

**References**

[1] Wu M K, Ashburn J R, Torng C J, Hor P H, Meng R L, Gao L, Haung Z J, Wang Y Q and Chu C W 1987 *Phys. Rev. Lett.* **58** 908
[2] Shaheen S A, Jisrawi N, Lee Y H, Croft M, McLean W L, Zhen H, Rebelsky L and Horn S 1988 *J. Appl. Phys.* **63** 4187
[3] WadaT, Sakurai T, Suzuki N, Koriyama S, Yamauchi H and Tanaka S 1990 *Phys. Rev.* B **41** 11209
[4] Hussain M, Kuroda S and Takita K 1998 *Physica* C **297** 176
[5] Bourdillon A and Bourdillon N X Tan 1994 *High Temperature Superconductors: Processing and Science* (San Diego: Academic Press) p 46

IOP Conf. Series: Journal of Physics: Conf. Series **969** (2018) 012060 doi:10.1088/1742-6596/969/1/012060

# Superconducting critical temperature and upper magnetic field in Weyl semimetal.

**Dingping Li[1,2], B. Rosenstein[3], B. Ya. Shapiro[4] and I. Shapiro[4]**

[1]School of Physics, Peking University, Beijing 100871, China
[2]Collaborative Innovation Center of Quantum Matter, Beijing, China
[3]Department of Electrophysics, National Chiao Tung University, Hsinchu, Taiwan, R.O.C.
[4]Department of Physics, Institute of Superconductivity, Bar-Ilan University, Ramat-Gan 52900, Israel

**Abstract.** The influence of recently discovered topological transition between type I and type II Weyl semi-metals on superconductivity is considered. A set of Gorkov equations for weak superconductivity in Weyl semi-metal under topological phase transition is derived and solved. The critical temperature has spike in the point the transition point as function of the tilt parameter of the Dirac cone determined in turn by the material parameters like pressure. The second critical magnetic field is calculated for phase I. It shows narrow domains of reentrant superconductivity associated with Landau quantization relativistic 2D electron gas.

## 1. Introduction

Recently new class of 2D and 3D Weyl semi-metals were discovered. Most of them are characterized by linear dispersion relation near the Fermi surface (Dirac cone). The first material of this class, graphene[1], exhibits the highest symmetry leading to linear ultra - relativistic spectrum, however most of the other materials are anisotropic. Examples include 2D Weyl semi-metals (WSM) silicene, germanene and borophene [2] and 3D crystals[3] $Na_3Bi$ , and[4] $Cd_3As_2$ and numerous layered organic compounds[5, 6, 7]. Topological insulators (TI) like $Bi_2Se_3$ and other [8] generally have Dirac cones on their surfaces of topological insulators.

The pressure modifies the spin orbit coupling that in turn determines the topology of the Fermi surface of these novel materials [9]. Very recently the topological transition in Weyl semi - metal under pressure was observed [10].

Some Weyl materials are known to be superconducting. A detailed study of superconductivity in TI under hydrostatic pressure revealed a curious dependence of critical temperature of the superconducting transition on pressure [11] . In intercalated $Sr_{0.065}Bi_2Se_3$ single crystal [12] considered to be a pure material ambient weak superconductivity first is suppressed, but at high pressure of $6GPa$ reappears and reaches relatively high $T_c$ of $10K$ that persists till $80GPa$. The increase is not gradual, but rather abrupt in the region of $15GPa$ [12]. Superconductivity in similar TI compounds [13, 14],[15] $Bi_2Se_3$ , $Bi_2Te_3$ and 3D Weyl semi-metal[16] $HfTe_5$ were also studied experimentally. The critical temperature $T_c$ in some of these systems shows a sharp maximum as a function of pressure. Various mechanisms of superconductivity in Dirac semi – metals and topological insulators turned superconductors have been considered theoretically [17, 18, 19]. A theory predicted possibility of superconductivity in the type II Weyl semimetals was developed recently in the framework of Eliashberg model [20].

We calculated in this paper the critical temperature for arbitrary tilt and find its sharp increase at topological transition point from type I to type II Weyl semi-metal. The critical second magnetic field $H_{c2}(T)$ was calculated for type-I semi metals. It shows narrow domains of reentrant superconductivity associated with Landau quantization relativistic 2D electron gas.

## 2. Type I and type II Weyl semi - metal with local pairing interaction.

Weyl material typically possesses several sublattices. We exemplify the effect of the topological transition on superconductivity using the simplest possible model with just two sublattices denoted by $\alpha = 1, 2$. The band structure near the Fermi level of a 2D Weyl semi-metal is well captured by the following Hamiltonian,

$$K = \int_{\mathbf{r}} \psi_{\alpha}^{sL+}(\mathbf{r}) K_{\alpha\beta}^{L} \psi_{\beta}^{sL}(\mathbf{r}) + \psi_{\alpha}^{sR+}(\mathbf{r}) K_{\alpha\beta}^{R} \psi_{\beta}^{sR}(\mathbf{r}); \tag{1}$$
$$K_{\alpha\beta}^{L,R} = -i\hbar v \left(\nabla_x \sigma_{\alpha\beta}^{x} \mp \nabla_y \sigma_{\alpha\beta}^{y}\right) + (-i\hbar\mathbf{w}\cdot\nabla - \mu)\,\delta_{\gamma\delta}.$$

Here $v$ is Fermi velocity, $\mu$ - chemical potential, $\sigma$ are Pauli matrices in the sublattice space and $s$ is spin projection. We restrict ourself to the case of just one left handed $(L)$ and one right handed $(R)$ Dirac points, typically but not always separated in the Brillouin zone. The 2D velocity vector $\mathbf{w}$ defines the tilt of the cone. The effective electron-electron attraction due to either electron - phonon attraction and Coulomb repulsion (pseudopotential) or some unconventional pairing mechanism creates pairing. We assume that different valleys are paired independently and drop the valley indices (multiplying the density of states by $2N_f$). Further we assume the local singlet $s$-channel interaction Hamiltonian

$$V = \frac{g^2}{2} \int d\mathbf{r}\, \psi_{\alpha}^{+\uparrow}(\mathbf{r}) \psi_{\beta}^{\downarrow+}(\mathbf{r}) \psi_{\beta}^{\uparrow}(\mathbf{r}) \psi_{\alpha}^{\downarrow}(\mathbf{r}). \tag{2}$$

As usual the interaction has a cutoff frequency $\Omega$, so that it is active in an energy shell of width $2\hbar\Omega$ around the Fermi level [21]. For the phonon mechanism it is the Debye frequency.

## 3. Gor'kov Green functions.

Finite temperature properties of the superconducting condensate are described by the normal and the anomalous Matsubara Greens functions [21]:

$$G_{\alpha\beta}^{ts}(\mathbf{r}\tau, \mathbf{r}'\tau') = \delta^{ts} g_{\alpha\beta}(\mathbf{r}, \mathbf{r}', \tau - \tau');\ F_{\alpha\beta}^{ts}(\mathbf{r}\tau, \mathbf{r}'\tau') = -\varepsilon^{ts} f_{\alpha\beta}(\mathbf{r}, \mathbf{r}', \tau - \tau');\ F_{\alpha\beta}^{+ts}(\mathbf{r}\tau, \mathbf{r}'\tau') = \varepsilon^{ts} f_{\alpha\beta}^{+}(\mathbf{r}, \mathbf{r}', \tau - \tau')$$

Using the Fourier transform, $g_{\gamma\kappa}(\omega, \mathbf{r}) = T\sum_{\omega p} \exp[-i\omega\tau]\, g_{\gamma\kappa}(\omega, \mathbf{r})$, with Matsubara frequencies, $\omega_n = 2\pi T(n + 1/2)$, one obtains from equations of operator motion the set of Gor'kov equations [22]

$$i\omega g_{\gamma\kappa}(\mathbf{r}, \mathbf{r}', \omega) - iv\sigma_{\gamma\beta}^{i}\, \nabla_{\mathbf{r}}^{i} g_{\beta\kappa}(\mathbf{r}, \mathbf{r}', \omega) + \mu g_{\gamma\kappa}(\mathbf{r}, \mathbf{r}', \omega) + \Delta_{\alpha\gamma}(\mathbf{r}, 0)\, f_{\alpha\kappa}^{+}(\mathbf{r}, \mathbf{r}', \omega) = \delta^{\gamma\kappa}\delta(\mathbf{r} - \mathbf{r}') \tag{3}$$

$$-i\omega f_{\gamma\kappa}^{+}(\mathbf{r}, \mathbf{r}', \omega) - iv\sigma_{\alpha\gamma}^{i} \nabla_{\mathbf{r}}^{i} f_{\alpha\kappa}^{+}(\mathbf{r}, \mathbf{r}', \omega) + \mu f_{\gamma\kappa}^{+}(\mathbf{r}, \mathbf{r}', \omega) - \Delta_{\alpha\gamma}^{*}(\mathbf{r}, \tau = 0)\, g_{\alpha\kappa}(\mathbf{r}, \mathbf{r}', \omega) = 0 \tag{4}$$

Here $\Delta_{\alpha\gamma} \equiv \sigma_{\alpha\gamma}^{x}\Delta$, $\Delta = \frac{1}{2}\sigma_{\gamma\alpha}^{x}\Delta_{\alpha\gamma}$ is the gap Ansatz in s-wave channel, indexes $t, s$ and $\alpha, \beta$ relate to spin and pseudospin.

In matrix form (in sub-lattice space) we obtain the solution in uniform case [22]

$$\widehat{g}(\omega, \mathbf{p}) = \sigma^{x}\left(\mathbf{p}\cdot\sigma^{t} + (-i\omega + \mu - wp_x) I\right) M^{-1};\ \widehat{f}^{+}(\omega, \mathbf{p}) = M^{-1}\Delta^{*}, \tag{5}$$

where

$$M = (v\mathbf{p}\cdot\sigma + (i\omega + \mu - wp_x) I)\,\sigma^{x}\left(v\mathbf{p}\cdot\sigma^{t} + (-i\omega + \mu - wp_x) I\right) + |\Delta|^2\sigma^{x}, \tag{6}$$

and $I$ is the identity matrix.

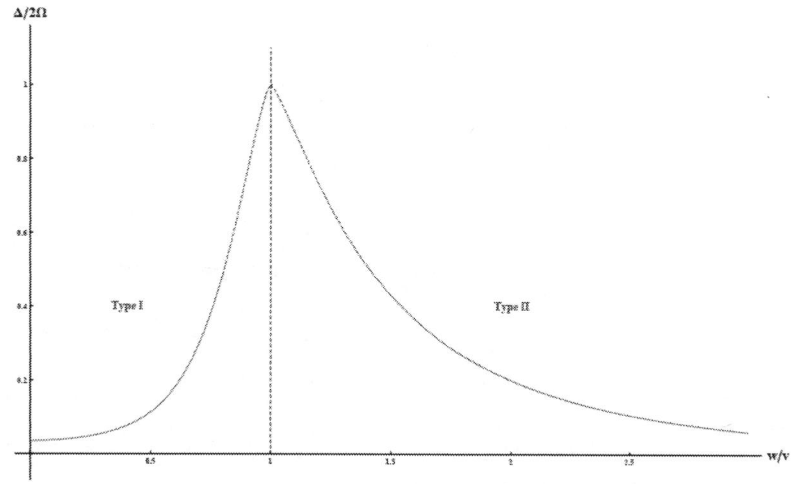


**Figure 1.** Superconducting gap at zero temperature versus cone tilted parameter $\kappa$. (Dash line marks region beyond assumption $\mu >> \Omega >> T_c$). Cut off parameter $\delta = 0.01, \mu = 3, \lambda = 0.15$.

## 4. Critical temperature.

The gap as function of the material parameters is determined by the gap equations

$$\Delta^* = \frac{\Delta^*}{2\left(2\pi\right)^3} T \sum_{\omega} \int d\mathbf{p} \ \mathrm{Tr}\left[\sigma_x M^{-1}\right], \tag{7}$$

The critical temperature is defined from the gap equation, when the gap $\Delta$ vanishes:

$$\frac{1}{g^2} = \frac{T}{\left(2\pi\right)^2} \int d^2\mathbf{p} \sum\nolimits_n \frac{v^2p^2 + \left(\mu - wp_x\right)^2 + \omega_n^2}{\left[\left(vp + wp_x - \mu\right)^2 + \omega_n^2\right]\left[\left(vp - wp_x + \mu\right)^2 + \omega_n^2\right]}. \tag{8}$$

The sum over Matsubara frequencies gives in the adiabatic approximation, $\mu >> \Omega$, after performing integration [22]

$$T_c = 1.14\Omega \exp\left[-\frac{1}{\lambda f\left(\kappa\right)}\right], \tag{9}$$

The function $f$ reads

$$f\left(\kappa\right) = \left\{ \begin{array}{cc} \frac{2}{\left(1-\kappa^2\right)^{3/2}} & \text{for } \kappa < 1 \\ \frac{2\kappa^2}{\pi\left(\kappa^2-1\right)^{3/2}}\left\{2\sqrt{1+\kappa} - 1 + \log\frac{2\left(\kappa^2-1\right)}{\kappa\left(1+\sqrt{1+\kappa}\right)^2\delta}\right\} & \text{for } \kappa > 1 \end{array} \right., \tag{10}$$

where $\kappa = w/v$ is the anisotropy parameter, $\delta = \pi a\Omega/v\hbar$ is the cutoff parameter, $a$ is the interatomic distance. (The effective density of states $f\left(\kappa\right)$ formally diverges at $\kappa = 1$. It's result of linear dispersion relation in the model Weyl Hamiltonian. Indeed, the linear approximation of the dispersion relation in the Weyl semimetal is valid only for small neighborhood of the Dirac point. At higher energies the band spectrum is nonlinear and cut off by the band width $1/a$. Fig.1 demonstrates that the critical temperature has a sharp spike at the transition point $w = v$. The ratio $\frac{2\Delta}{T_c}$ is still universal for any $w/v$. (Similar results were obtained in Ref [20] for small and large cone tilt).

## 5. Critical Magnetic Field

If the superconductor is subjected to external magnetic field $H$ then the gap equation 7 describing the critical curve $H_{c2}(T)$ has the form

$$\Delta(\mathbf{r}) = \frac{g^2}{2} T \sum_\omega \int_{r''} \Delta^*(\mathbf{r}'') \sigma^x_{\kappa\beta} G^2_{\beta\gamma}(\mathbf{r},\mathbf{r}'') \sigma^x_{\gamma\alpha} G^1_{\alpha\kappa}(\mathbf{r}'',\mathbf{r}) \tag{11}$$

here $G_{\alpha\kappa}(\mathbf{r}'',\mathbf{r})$ and $G^2_{\beta\gamma}(\mathbf{r},\mathbf{r}'')$ are the Green functions of the normal state obeyed the equations:

$$\left[v\left(-i\nabla - \frac{e}{c}\mathbf{A}\right)\cdot\sigma_{\gamma\beta} + (i\omega+\mu)\delta_{\gamma\beta}\right] G^1_{\beta\kappa}(\mathbf{r},\mathbf{r}') = \delta^{\gamma\kappa}\delta(\mathbf{r}-\mathbf{r}') \tag{12}$$

$$\left[v\left(-i\nabla - \frac{e}{c}\mathbf{A}\right)\cdot\sigma^t_{\gamma\beta} + (-i\omega+\mu)\delta_{\gamma\beta}\right] G^2_{\beta\kappa}(\mathbf{r}',\mathbf{r}) = \delta^{\gamma\kappa}\delta(\mathbf{r}-\mathbf{r}') \tag{13}$$

Here $\sigma^t$ is the transposed Pauli matrix. In the uniform magnetic field the Green's functions can be written in the Ansatz form:

$$G^{1,2}_{\beta\kappa}(\mathbf{R},\rho,\omega) = \exp\left(\pm\frac{e}{c} i\mathbf{A}(\mathbf{R})\rho\right) g^{1,2}_{\beta\kappa}(\rho,\omega) \tag{14}$$

where $\mathbf{R}=\frac{\mathbf{r}+\mathbf{r}'}{2}$, $\rho = \mathbf{r}-\mathbf{r}'$.The vector potential in the symmetric gauge reads $\mathbf{A}(\mathbf{r}) = \frac{1}{2}(-Hy, Hx, 0)$. Substituting this Ansatz into Eqs.(12),(13) one obtains:

$$\left\{(i\omega+\mu)\delta_{\gamma\beta} + v\sigma^x_{\gamma\beta}\Pi_x + v\sigma^y_{\gamma\beta}\Pi_y\right\} g^1_{\beta\kappa}(\rho,\omega) = \delta^{\gamma\kappa}\delta(\rho) \tag{15}$$

$$\left[(-i\omega+\mu)\delta_{\gamma\beta} + v\Pi_x\cdot\sigma^{tx}_{\gamma\beta} + v\Pi_y\cdot\sigma^{ty}_{\gamma\beta}\right] g^2_{\beta\kappa}(\rho,\omega) = \delta^{\gamma\kappa}\delta(\rho) \tag{16}$$

here

$$\Pi_x = \left(-i\frac{\partial}{\partial\rho_x} + \frac{e}{c}H\rho_y\right); \Pi_y = \left(-i\frac{\partial}{\partial\rho_y} - \frac{e}{c}H\rho_x\right) \tag{17}$$

are the Ladder operators related to creation and annihilation operators for bosonic field $a = \frac{l}{\sqrt{2}}(\Pi_x - i\Pi_y)$; $a^+ = \frac{l}{\sqrt{2}}(\Pi_x + i\Pi_y)$; $l = \sqrt{\frac{c}{eH}}$ where $l$ is the magnetic length.

Solving set of the equations (15) and (16) in polar coordinates and one performing summation on Matsubara frequencies one obtains [23] the equation for $H_{c2}$

$$\frac{1}{\lambda} = \frac{\overline{\omega}_c^2}{4\overline{\mu}}\left\{\sum_{n,m=1}\frac{(m+n)!\,f[n]\,f[m]}{2^{m+n+1}\,m!n!}s + \sum_{n=1}\frac{f[n]\,f[0]}{2^n}s_0 + \frac{f[0]^2}{2}s_{00}\right\} \tag{18}$$

where

$$f(n) = \frac{\overline{\Omega}_D^2}{\overline{\Omega}_D^2 + \left(\overline{\omega}_c\sqrt{n} - \overline{\mu}\right)^2} \tag{19}$$

here $\overline{\omega_c} \to \omega_c/T, \overline{\mu} \to \mu/T, \overline{\Omega}_D \to \Omega_D/T$ , $\omega_c = v\sqrt{2}/l$ are the reduced cyclotron frequency, chemical potential and Debye frequency correspondingly. Here $\lambda = \frac{g^2\mu}{4\pi\, v^2}$ is the dimensional electron-electron constant for 2D electron gas (per sublattice).

$$s = \left(A\left[\overline{\omega}_c^2(n+1), \overline{\omega}_c^2(m+1)\right] + A\left[\overline{\omega}_c^2 n, \overline{\omega}_c^2 m\right]\right) + \left(\begin{array}{c}\mu^2 B\left[\overline{\omega}_c^2(n+1), \overline{\omega}_c^2(m+1)\right] + \mu^2 B\left[\overline{\omega}_c^2 n, \overline{\omega}_c^2 m\right] \\ + n\overline{\omega}_c^2 B\left[\overline{\omega}_c^2(n+1), \overline{\omega}_c^2 m\right] + m\overline{\omega}_c^2 B\left[\overline{\omega}_c^2 n, \overline{\omega}_c^2(m+1)\right]\end{array}\right) \tag{20}$$

$$s_0 = A\left[\overline{\omega}_c^2(n+1), \overline{\omega}_c^2\right] + A\left[\overline{\omega}_c^2 n, 0\right] + \mu^2 B\left[\overline{\omega}_c^2(n+1), \overline{\omega}_c^2\right] + \mu^2 B\left[\overline{\omega}_c^2 n, 0\right] \tag{21}$$

$$s_{00} = A\left[\overline{\omega}_c^2, \overline{\omega}_c^2\right] + A[0,0] + \mu^2 B\left[\overline{\omega}_c^2, \overline{\omega}_c^2\right] + \mu^2 B[0,0]$$

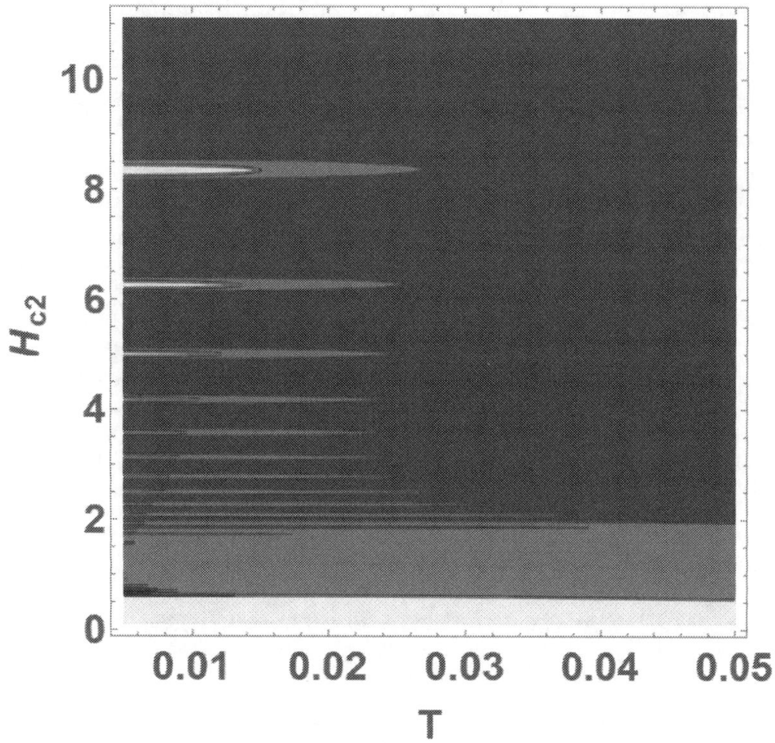


**Figure 2.** The upper critical field versus temperature. The figure is plotted for $\mu/\Omega_D = 5$. Domains of reentering superconductivity is presented for two different electron-electron strength $\lambda = 1$ (short light) and $\lambda = 0.66$ (long).

where

$$A[a,b] = \frac{(\sqrt{a}-\overline{\mu})^2 \tanh\left(\frac{\sqrt{a}-\overline{\mu}}{2}\right)}{4\sqrt{a}\left(-b+(\sqrt{a}-2\overline{\mu})^2\right)} + \frac{\left(\sqrt{b}-\overline{\mu}\right)^2 \tanh\left(\frac{\sqrt{b}-\overline{\mu}}{2}\right)}{4\sqrt{b}\left(-a+\left(\sqrt{b}-2\overline{\mu}\right)^2\right)} + \frac{(\sqrt{a}+\overline{\mu})^2 \tanh\left(\frac{\sqrt{a}+\overline{\mu}}{2}\right)}{4\sqrt{a}\left(-b+(\sqrt{a}+2\overline{\mu})^2\right)} + \frac{\left(\sqrt{b}+\overline{\mu}\right)^2 \tanh\left(\frac{\sqrt{b}+\overline{\mu}}{2}\right)}{4\sqrt{b}\left(-a+\left(\sqrt{b}+2\overline{\mu}\right)^2\right)} \quad (22)$$

$$B[a,b] = -\frac{\tanh\left(\frac{\sqrt{a}-\overline{\mu}}{2}\right)}{4\sqrt{a}\left(-b+\left(\sqrt{a}-2\overline{\mu}\right)^2\right)} - \frac{\tanh\left(\frac{\sqrt{b}-\overline{\mu}}{2}\right)}{4\sqrt{b}\left(-a+\left(\sqrt{b}-2\overline{\mu}\right)^2\right)} - \frac{\tanh\left(\frac{\sqrt{a}+\overline{\mu}}{2}\right)}{4\sqrt{a}\left(-b+\left(\sqrt{a}+2\overline{\mu}\right)^2\right)} - \frac{\tanh\left(\frac{\sqrt{b}+\overline{\mu}}{2}\right)}{4\sqrt{b}\left(-a+\left(\sqrt{b}+2\overline{\mu}\right)^2\right)} \tag{23}$$

Result is presented in Fig.2 where narrow domains of superconducting phase are presented. The figure is plotted for $\mu = 5\Omega_D$, temperature is in the units $\Omega_D$ while $H_{c2}$ is in the units of $\hbar c\Omega_D^2/2v^2e$.

Our result might provide an explanation of a spike in dependence of $T_c$ on pressure in single crystals observed in many superconducting semi-metals [13, 14]. Due to small number of carriers in this semimetal, the quasiclassical Werthamer-Helfand approximation breaks down and Landau quantization has to be taken into account. We found that the upper critical magnetic field $H_{c2}$ resonantly enhanced at the magnetic fields corresponding to full occupancy of the Landau levels in the shallow band. This enhancement is especially pronounced for the lowest Landau level. As a consequence, the reentrant superconducting regions in the temperature- field phase diagram emerge at low temperatures near the magnetic fields at which the chemical potential matches the Landau levels.

*Acknowledgements.*

We are grateful to T. Maniv, T. W. Luo, M. Lewkowicz, C. C. Hou, E. Farber for valuable discussions.

[1] Novoselov K.,Geim A. K., Morozov S. , Jiang D. , Katsnelson M.,Grigorieva I. , Dubonos S. , and Firsov A. , Nature 2005 **438**, 197.
[2] Feng B. et al. Nature Chemistry 2016 **8**, 563;
[3] Neupane M. ,Xu S.-Y. , Sankar R. , et al., Nat. Commun. 2014 **5** 3786;
[4] Wang Z. , Zhou X.-F. , Zhang X. , Zhu Q. , Dong H. , Zhao M. , and Oganov A. R. , Nano Lett. 2015 **15**, 6182; He L., Jia Y., Zhang S. ,Hong X. , Jin C. & Li S. , Quantum Materials 2016 **1** 16014.
[5] Singleton J. and Mielke C. , Cont. Phys. 2002 **43**, 63;Tajima N. ,Sugawara S. ,Kato R. ,Nishio Y. , and Kajita K. , Phys. Rev. Lett. 2009 **102**, 176403.
[6] Goerbig M. O. , Fuchs J. -N. , Montambaux G.,Piéchon F. , 2008 *Phys. Rev. B* **78**, 045415;
[7] Monteverde M. ,Goerbig M. O. ,Auban-Senzier P. ,Navarin F. ,Henck H. ,Pasquier C. R. ,Mèziére C. , and Batail P. , 2013 *Phys. Rev B* **87** 245110.
[8] Zhang H. ,Liu C-X ,Qi X.-L. ,Dai X. ,Fang Z. &Zhang S.-C. , 2009 *Nature Physics* **5** 438;
[9] Ruan J. ,Jian S.-K. ,Yao H. ,Zhang H. ,Zhang S.-C. &Xing D. , 2016 *Nature Com.* **7** 11136.
[10] Zhou Y. et al, 2016 *Phys. Rev. Let.*,**117** , 146402 .
[11] Qi Y. ,*et al* 2016 *Nature Comm.* **7**, 11038 .
[12] Zhou Y. et al. 2016 *Phys. Rev. B* **93** 144514 .
[13] Hamlin J. J. et al. 2012 *J. Phys. Cond. Mat.*, **24**, 035602 ;Kirshenbaum K. et al. 2013 *Phys. Rev. Let.* **111**, 087001 ;
[14] Kong P. P. , et al, 2013 *J. Phys. Condens. Matter* **25,** 362204 .
[15] Stillwell R. L. ,Jenei Z. ,Weir S. T. ,Vohra Y. K. , and Jeffries J. R. , 2016 *Phys. Rev. B* **93**, 094511
[16] Qi Y. et al. 2016 *Phys. Rev. B* **94**, 054517 .
[17] Das Sarma S. and Li Q. , 2013 *Phys. Rev. B* **88**, 081404(R) ;Brydon P.M.R. ,Das Sarma S. ,Hui H.-Y. , andSau J. D. , 2014 *Phys. Rev. B* **90**, 184512 ;
[18] Fu L. and Berg E. , 2010 *Phys. Rev. Lett.* **105**, 097001 .
[19] J.-L. Zhang et al. 2012 Front. Phys., **7**, 193 .
[20] M. Alidoust, K. Halterman, and A. A. Zyuzin, 2017 *Phys. Rev. B* 95 155124.
[21] Abrikosov A. A. ,Gor'kov L. P. ,Dzyaloshinskii I. E. , 1965 "Quantum field theoretical methods in statistical physics", Pergamon Press, New York .
[22] Li D.P. , Rosenstein B. ,Shapiro B. Ya. , and Shapiro I. , 2014 *Phys. Rev. B* **90**, 054517 .
[23] Maniv T., Zhuravlev V., Vagner I., Wyder P., 2001 Rev. Mod. Phys., **73,** 868.

IOP Conf. Series: Journal of Physics: Conf. Series **969** (2018) 012061 doi:10.1088/1742-6596/969/1/012061

# Superconducting Fluctuations above $T_c$ and pair breaking parameters of two dimensional Niobium Nitride Films

**B. Shinozaki [1], S. Ezaki [2], T. Odou[1], K. Makise [3], T. Asano[4]**

[1] Department of Physics Kyushu University, 744, Motooka, Fukuoka 819-0395, Japan

[2] National Astronomical Observatory of Japan National Astronomical Observatory of Japan, 2-21-1 Osawa, Mitaka, Tokyo 181-8588, Japan

[3] National Institute of Advanced Industrial Science and Technology, Tsukuba 305-8560 Japan

[4] Department of Applied Physics, University of Fukui, Fukui 910-8507, Japan

shinozaki@phys.kyushu-u.ac.jp

Transport properties have been investigated for the epitaxial superconducting NbN thin films. We analysed the excess conductance $\sigma' \equiv \sigma(T) - \sigma^N$ by the sum of the Aslamazov-Larkin (AL) and Maki-Thompson (MT) terms for thermal fluctuations above $T_c$, where the $\sigma^N \equiv 1/R_{sq}^N$ is the normal state sheet conductance. We have found that the theoretical expression $\sigma'_{theo}(T) = \sigma'_{AL}(T) + \sigma'_{MT}(T,\delta)$ can be well fitted to $\sigma'_{exp}(T)$ with use of the suitable value of the pair breaking parameter $\delta$ in the MT term relating to the inelastic scattering rate $1/\tau_{in}(T)$ as $\delta = \pi\hbar/8k_BT\tau_{in}$. The rate $1/\tau_{in}(T)$ given by the sum of $1/\tau_{fluc}(T)$, $1/\tau_{e\text{-}e}(T)$ and $1/\tau_{e\text{-}ph}(T)$ is determined from the analysis of the magneto-conductance $\Delta\sigma = \sigma(H) - \sigma(0)$ by the sum of AL, MT and the localization terms, where the first, second and third terms correspond to the rate due to the superconducting fluctuation effect, electron-electron and electron-phonon interactions, respectively. The $R_{sq}^N$ dependence of $\delta$ is expressed by $\delta = \delta_0 + \alpha R_{sq}^N$, where the first term $\delta_0$ due to $1/\tau_{e\text{-}ph}(T)$ and the second term due to the sum of $1/\tau_{fluc}(T)$ and $1/\tau_{e\text{-}e}(T)$. Although we obtained a reasonable value of Debye temperature $\Theta_D \approx 630$ K from the $\delta_0$, the magnitude of the $\alpha$ is about 5 times larger than the theoretical value.

## 1. Introduction

Transport properties of low-dimensional superconductors (LDSs) as ultra-thin films and nanowires are strongly affected by the thermal and quantum fluctuations. Then LDSs have been widely investigated from the viewpoints of not only device physics [1] but also fundamental physical interests as $R$-$T$ characteristics due to the excess conductances above superconducting transition temperature $T_c$, thermal [2] and quantum Phase slips. As for superconducting-insulator transition in two dimensional

(2D) weakly disordered NbN systems, we have already reported the transport properties of NbN films with different normal state resistance $R_{sq}^{N}$ prepared by tuning the film thickness $d$ of epitaxial film on MgO substrates.[3] We analyzed the suppression of $T_c$ by Finkel'stein formula[4] from the localization theory called as fermionic scenario. This theory shows that the $T_c$ depression is determined by the bulk $T_{c0}$ and elastic scattering time $\tau$, depending on materials.

Decades since 1970, superconducting thermal fluctuations at temperatures above $T_c$ have been intensively investigated. Conventionally, the temperature dependence of excess conductance $\sigma'(T) \equiv \sigma(T) - \sigma^N$ for 2D superconductors has been mainly analyzed by the sum of the two dominant Aslamazov-Larkin (AL) [5] and the Maki-Thompson (MT) [6] terms, where the $\sigma^N$ is the normal state conductance. One of the purposes for investigations of the $\sigma'(T)$ due to the superconducting fluctuations is the determination of the some inherent characteristics because the theoretical formula for $\sigma'(T)$ to fit the data contains several microscopic quantities as parameters. The MT term includes the pair breaking parameter $\delta$ introduced for removing the divergence from the Maki diagram.[7] The progress in the electron localization theory[8] has revealed that $\delta$ is essentially the same as the inverse of the inelastic scattering time $1/\tau_{in}(T)$, that is,

$$\delta = \frac{\pi\hbar}{8k_B T \tau_{in}(T)}. \qquad (1)$$

From fitting the theory to the experimental data of $\sigma'(T)$, we can determine $\delta(T)$, that is, $\tau_{in}(T)$. On the other hand, we can obtain $\tau_{in}(T)$ from the analysis of the magneto-conductance $\Delta\sigma$, defined as $\Delta\sigma(T,H) = \sigma(T,H) - \sigma(T,0)$. Following the previous investigations of $R_{sq}^{N}$ dependence of $T_c$ for the homogeneous 2D NbN/MgO films [3], in the present study we report the detailed result of the investigation of $\sigma'(T)$ at $H = 0$ and $\Delta\sigma(T,H)$.

## 2. Sample Preparation and Experimental Procedures

NbN specimen films were prepared by deposition on (100) single crystal MgO substrate with use of the DC-magnetron sputtering method. Film thickness $d$ is in the range from 2.0 nm to 25.8 nm. The $d$ was estimated from the sputtering time using the calibrated relation between the sputtering time and exact thickness by the atomic force micrometer in thick region. We measured the resistance $R(T,H)$ using a standard dc four-probe technique. We applied a magnetic field perpendicular to the film surface up to ± 5 T. The details of the preparation procedures and microcrystal structures have been shown in previous report.

## 3. Experimental result and discussion

Firstly, we discuss the temperature dependence of the excess conductance $\sigma'(T) = \sigma(T) - \sigma^N$ in the absence of the external magnetic field. To clarify the characteristic of $\sigma'(T)$ near $T_c$, Fig.1 shows the $T$ dependence of $\sigma^N/\sigma'(T) = R_{sq}(T)/[R_{sq}^{N} - R_{sq}(T)]$ for the typical film with $d = 5.2$nm and $R_{sq}^{N} = 257.7\ \Omega$, where the normal sheet resistance $R_{sq}^{N} \equiv 1/\sigma^N$ is defined as a resistance determined from the extrapolation of the nearly linear relation between $1/H$ and $1/R_{sq}(T)$ to $1/H = 0$ at temperatures near $T_c$. The excess conductance $\sigma'$ above $T_c$ for 2D superconductors is given by the sum of AL and MT terms as follows,

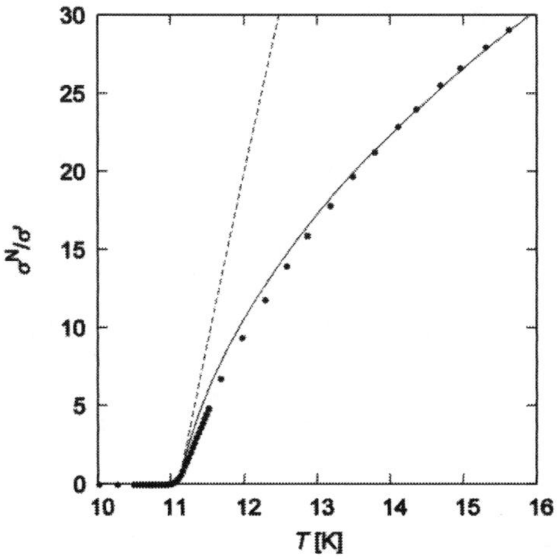

**Figure 1.** The ratio $\sigma^N/\sigma'$ vs. $T$ at $H$=0. Dotted and solid lines are theory from AL term and the sum of AL and MT terms with $\delta^F$=0.33, respectively.

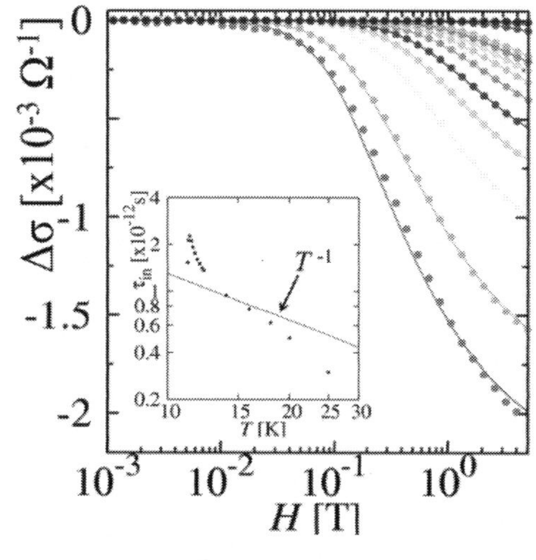

**Figure 2.** $\Delta\sigma(H)=\sigma(H)-\sigma(0)$ at various temperatures for the same film as in Fig. 1. Inset shows the temperature dependence of $\tau_{in}(T)$. Solid line shows $\tau_{in}(T) \propto T^{-1}$.

$$\sigma' = \sigma_{AL}' + \sigma_{MT}' = \left(\frac{e^2}{16\hbar}\right) \times \left[\frac{1}{\eta} + \frac{2}{\eta - \delta}\ln\left(\frac{\eta}{\delta}\right)\right], \qquad (2)$$

where $\eta = \ln(T/T_c)$ is the reduced temperature. Solid line is calculated from eq. (2) to fit the data for relatively wide temperature region with suitable values $T_c$ = 11.11 K and the constant $\delta$ = 0.33. Although the slight disagreements between the theory and the experimental data near $T_c$, the good fit of eq.(2) to data means the existence of the MT term, that is, the pair breaking of the cooper pairs in the present NbN films. The slight disagreements suggest the temperature dependence of $\delta$. To investigate the $\delta(T)$ in detail at temperature near $T_c$, the value of $\delta$ was determined at each temperature by fitting $\sigma_{MT}'$ in eq.(2) to data on the excess conductance. Before discussion of $\delta(T)$, we analyze $\Delta\sigma(T,H)$ for determination of $\tau_{in}(T)$ relating to the pair breaking parameter through eq.(1) in order to compare the $\delta^F(T)$ estimated from eq.(2) with $\delta^M(T)$ obtained from the combination of eq.(1) and the following eq.(3) including the $\tau_{in}(T)$. The $\Delta\sigma$ for 2D superconductors is given by the sum of the three terms as follows,

$$\Delta\sigma = \Delta\sigma_L(D\tau_{in}H, D\tau_{so}H) + \Delta\sigma_{MT}(D\tau_{in}H, DH, \eta, T) + \Delta\sigma_{AL}(DH, \eta, T), \qquad (3)$$

where each term corresponds to localization, Maki-Thomson and Aslamazov-Larkin terms, and $\tau_{so}$ and $D$ are the spin-orbit scattering time and diffusion constant, respectively. The exact expressions are given in the previous paper.[9] The quantity $\tau_{in}(T)$ can be determined from the fitting procedure of eq.(2) to the experimental data with $\tau_{in}(T)$ and $T$ independent $D$ regarding as fitting parameters.

Figure 2 shows the $H$ dependence of $\Delta\sigma$ for the same film as that shown in Fig. 1 at various temperatures from two typical temperatures $T$ =11.21 K and 11.35 K at bottoms to two typical temperatures 20.0 K and 25.0 K at uppers. From the experimental results of $\Delta\sigma < 0$ even at high temperatures and that the present system has a strong spin-orbit interaction, namely, $\tau_{zo}/\tau_{in}(T) \ll 1$ due to relatively heavy atoms of Nb and N, we assume that the magnitude of $\tau_{zo}$ is negligibly small compared with that of $\tau_{in}(T)$. The solid lines are calculated from eq. (3) with a suitable value $\tau_{in}(T)$ at each temperature and $D = 8.5 \times 10^{-5}\text{m}^2/\text{s}$. This value is almost the same as the value $D = 9.0 \times 10^{-5}\text{m}^2/\text{s}$ obtained from the magnitude of $dH_{c2}(T)/dT$ at temperatures near $T_c$ using the formula $D = (4k_B/\pi e) \times |dH_{c2}(T)/dT|^{-1}$. The inset shows the $T$ dependence of $\tau_{in}$ obtained from fitting procedures. The inelastic scattering rate $1/\tau_{in}$ is given by the sum of rates due to superconducting fluctuation, electron-electron (e-e) scattering [9] and electron-phonon (el-ph) scattering [10] mechanism as follows;

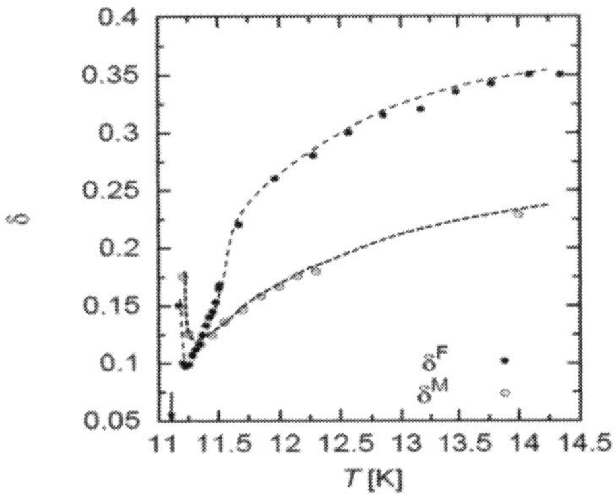


**Figure 3.** $T$ dependence of $\delta^F$ and $\delta^M$ determined from the analysis of $\sigma'(H = 0)$ and $\Delta\sigma$, respectively. Lines are guides for eyes.

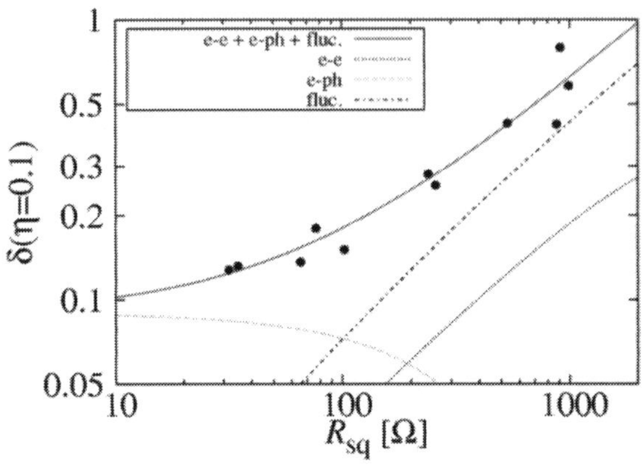


**Figure 4.** $R_{sq}$ dependence of $\delta^F$. Solid line is calculated from eq.(5) to fit the data with use of pre-factor $\alpha = 4.8$.

$$\begin{aligned} 1/\tau_{in} &= 1/\tau_{fluc} + 1/\tau_{el-el} + 1/\tau_{el-ph} \\ &= (k_B e^2 R_{sq}{}^N / 2\pi\hbar^2)[2\ln 2/(\eta+\gamma) + \ln(4\pi\hbar/e^2 R_{sq}{}^N)]T + [14\pi^2\varsigma(2) k_B \lambda / \hbar\Theta_D{}^2]T^3, \end{aligned} \quad (4)$$

where $\gamma = 4\ln 2/[(\varepsilon^2+128\hbar/e^2R_{sq}{}^N) - \varepsilon]$, $\varepsilon = \ln(\pi\hbar/e^2R_{sq}{}^N)$, $\lambda$ is the electron-phonon coupling constant and $\Theta_D$ is the Debye temperature. The solid line shows the relation $\tau_{in} \propto 1/T$ expected from the el-el scattering, the second term in eq. (4). The characteristic of $\tau_{in} \propto 1/T^n$ with $n > 1$ at high temperatures above $T \geq 17$ K comes from the combination of the el-ph and el-el scattering mechanisms. At the lowest temperatures, $\tau_{in}(T)$ shows drastic changes with decreasing temperature. The steep decrease may come from the first term, in eq.(4) that is, the superconducting fluctuation. However, the rapid increase below ~15K cannot be understood by eq.(4). Such anomalous behaviors of $\tau_{in}(T)$ are frequently observed for other films with different $R_{sq}{}^N$. At the present stage, we have not exact idea to explain the $\tau_{in}(T)$ data at low temperatures.

Figure 3 summarizes the temperature dependence of two pair breaking parameters $\delta^F(T)$ and $\delta^M(T)$. The $\delta^F(T)$ is obtained from the analysis of $\sigma^N/\sigma'(T)$ at $H = 0$ in Fig. 1 with eq. (2) and the $\delta^M(T)$ is obtained from the eq.(1) with use of $\tau_{in}$ $(T)$ in Fig. 2 determined from the analysis of $\Delta\sigma(H,T)$ by eq.(3). As temperature approaching $T_c$, both magnitudes of the $\delta^F(T)$ and $\delta^M(T)$ decrease and then seem to strongly increase. Although the ratio $\delta^F(T)/\delta^M(T)$ maximally has about 1.4 at high temperatures, two series of $\delta$ show qualitatively almost the same temperature dependence. The characteristic of $\delta$ depending on temperature shows that the inelastic scattering rate $1/\tau_{in}$ $(T)$ is given by the sum of different scattering mechanisms.

Figure 4 shows the $R_{sq}{}^N$ dependence of $\delta^F$ determined at $\eta = 0.1$ for all films in the present work. The magnitude of $\delta^F$ shows the gradual change at the low $R_{sq}{}^N$ region and seem to approach a certain value $\delta_0$. From the combination of eq. (1) and eq. (4), when $R_{sq}{}^N = 0$ only the el-ph inelastic scattering mechanism contributes to the theoretical value $\delta(0) = [7\pi^3\lambda\zeta(2)/4](T/\Theta_D)^2$. This means that the Debye temperature $\Theta_D$ can be estimated form the experimentally extrapolated value $\delta_{exp}(0)$. By assuming $\delta_{exp}(0) \approx 0.09$ and typical value $\lambda = 1.5$ for superconducting materials, we obtain $\Theta_D \approx 630$ K with use of $T = 1.1 \times T_{c0}( = 14.85$ K) in the above relation. Here, the factor 1.1 comes from the experimental condition of $\eta = \ln(T/T_c) = 0.1$ and $T_{c0}$ is the bulk superconducting transition temperature determined from the $R_{sq}{}^N$ dependence of $T_c$ in the previous report for the same NbN series. [3] Although the present value of $\Theta_D \approx 630$ K is lower than the bulk value $\Theta_{D,bulk} \approx 750$ K [11], it is quite reasonable that the $\Theta_{D,film}$ for thin films is generally low because of the specimen size effect. [12] With use of the value $\Theta_D \approx 630$ K obtained from the above estimation, the full expression for $\delta(R_{sq}{}^N)$ is given by

$$\delta(R_{sq}{}^{N}) = \alpha(e^2 R_{sq}{}^{N}/16\hbar)\times[\ln(\pi\hbar/e^2 R_{sq}{}^{N}) + 2\ln 2/(\eta+\gamma)] + (10.5\pi^3\zeta(2)/4)\times[1.1\times T_c(R_{sq}{}^{N})/630]^2, \quad (5)$$

where the factor $\alpha$ is the suitable parameters to fit the theory to data and $T_c(R_{sq}{}^{N})$ is given in the previous report. [3] The solid line in Fig.4 is fit of eq.(5) to data with use of the pre-factor $\alpha \approx 5$ and the other lines show each contribution due to the el-el, el-ph and fluc. inelastic scattering rates. Although the obtained value of $\Theta_D \approx 630$ K for NbN film series is reasonable, at the present stage we have no explanation for this difference of the coefficient $\alpha$ between the theory and experiment.

## 4. Conclusion

We prepared homogeneous NbN thin films with various $R_{sq}{}^{N}$ to investigate the superconducting transport properties above $T_c$. We measured the excess conductance σ' and magnetoconductance Δσ to estimate the pair breaking parameter δ in the MT term and inelastic scattering time $\tau_{in}(T)$. We analyze the $R_{sq}{}^{N}$ dependence of δ determined from σ' at $H = 0$ with the sum of three contributions due to the electron-electron, electron-phonon and superconducting fluctuation inelastic scattering rates, simply expressed as $\delta = \delta_0 + \alpha R_{sq}{}^{N}$. From the value of $\delta_0$ estimated form extrapolation of $R_{sq}{}^{N}$ to $R_{sq}{}^{N} = 0$ in the δ - $R_{sq}{}^{N}$ data, we obtained the Debye temperature $\Theta_D \approx 630$ K for the present NbN films. However, the pre-factor $\alpha$ is about 5 times larger than the theoretical value.

**Acknowledgment**

This work was supported by JSPS KAKENHI Grant Number 16K04877.

## References

[1] Mooji J E and Narozov Y V (2006) *Nat Phys.* **2** 169
[2] Skocpol W L and Tinkham (1975) *Rep. Prog. Phys.* **38** 1049
[3] Ezaki S, Makise K, Shinozaki B, Odou T, Asano Y, Terai H, Yamashita Y, Miki S and Wang Z, (2012) *J. Phys. Condens. Matter* **24** 475702
[4] Finkel'stein A M (1987) *JETP Lett.***45** 46
[5] Aslamazov L G and Larkin A I (1968) *Sov. Phys. Solid State* **10** 875
[6] Thompson R S (1970) *Phys. Rev.* **B1** 327
[7] Maki K (1968) *Theor. Phys.* (Kyoto) **40** 193
[8] Ebisawa H, Maekawa S and Fukuyama H (1983) *Solid State Commun.***45** 75
[9] Shinozaki B and Rinderer L (1988) *J Low Temp. Phys.* **73** 267
[10] Raffy H, Dumoulin L, Nedellec P and Burger J P (1985) *J. Phys. F* **15** L37
[11] Zou Y, Wang X, Chen T, Li X, Qi , Welch D, Zhu P, Liu B, Cui T and B Li *Scientific Reports* **5**:10811DOI:10.1038/srep10811
[12] Lang X Y, Jiang Q, *Solid State commun.* (2005) **134** 797

IOP Conf. Series: Journal of Physics: Conf. Series **969** (2018) 012062 doi:10.1088/1742-6596/969/1/012062

# Quantum phase slip and enhancement of superconductivity by magnetic field in NbN nanowires

**Bunju Shinozaki**[1], **Kazumasa Makise**[2], **Takayuki Asano**[3]

[1]Department of Physics, Kyushu University, 744 Motooka Fukuoka 819-0395, Japan

[2] National Institute of Advanced Industrial Science and Technology, Thukuba 305-8560, Japan

[3]Department of Applied Physics, University of Fukui, Fukui 910-8507, Japan

kf-makise@aist.go.jp

**Abstract**. Transport properties were investigated for quasi-one dimensional superconducting NbN nanowires with width $w$ = 20nm and thickness $d$=20nm on 3C-SiC substrate. It has been found that 1) the resistance $R$ shows the broad superconducting transition and the $R(T)$ characteristic at temperatures bellow $T_c$~13.6K can be well explained by the sum of the thermal activated phase slip and the quantum phase slip(QPS) of the superconducting order parameter. 2) The $R$(T) due to the QPS contribution is suppressed to enhance the superconductivity by the external magnetic field $H$. With decreasing temperature, the $R(T)$ of short specimen with length 600nm shows nearly zero resistance in the restricted temperature region depending on the magnitude of $H$ and recovers the QPS resistive state. 3) The critical magnetic field $H_{c2}(T)$ is fit to the theoretical expression $H_{c2,\mathrm{theo}}(T) = \phi_0 /[2\xi_{GL}(T) \times w] \propto (1-T/T_c)^{1/2}$, where the $\xi_{GL}(T)$ is the Ginzburg-Landau superconducting coherence length. Although the relation $H_{c2,\mathrm{exp}}(T) \propto (1-T/T_c)^{1/2}$ is satisfied, the magnitude of $H_{c2,\mathrm{exp}}(T)$ is about 5 times larger than that of the theoretical one.

## 1. Introduction

Recently, one-dimensional (1D) superconducting nanowires (SNWs) are being considered to develop superconducting computing devices.[1] SNW's can be strongly affected by thermally activated phase slip (TAPS) and/or quantum phase slips(QPS)[2,3] which play an important role in the transport properties of SNW's. The effect of TAPS to resistance $R(\Omega)$ sharply decays as a result of the temperature drop below $T_c$. On the other hand, the QPS represents the residual resistance to suppress the superconductivity at low temperatures even at $T \approx 0$. Although numerous studies have been conducted on SNWs, some investigations show no evidence of QPS behavior at very low temperatures.[4] On the other hand, there still are fundamental problems in the effect of external magnetic field on the temperature dependence of resistance $R(T)$ due to the TAPS and QPS on SNW specimens. Many investigation have reported the interesting field dependence of $R(T,H)$. [5-7] The typical behavior is the negative magneto-resistance (NMR) observed at relatively small magnetic fields. Further, recent work reports the magnetic field-enhanced superconductivity due to large NMR.[6] One of reasons for NMR is the presence of the magnetic impurity. Another mechanism is based on the suppression of the non-equilibrium charge imbalance process at normal and superconducting boundaries, namely, at the boundaries between phase-slipped and superconducting regions in SNWs. Although the exact mechanism is not sufficiently understood yet, it is reasonable to consider that the one of the reason for NMR concerns with the existence of phase slip centers originating the behaviors of TAPS and/or QPS.

In order to clarify the effect of the phase slips to the MR in SNWs, specimens are required to be homogeneous and satisfy the condition $d, w \leq \xi$ to avoid residual resistance due to inhomogeneities at low temperatures, where $d, w$ and $\xi$ are thickness, width of the specimen and superconducting coherence length, respectively. In the present work, we investigated mainly the effect of the magnetic field on the SNW specimens prepared from the NbN thin films on the 3C-SiC substrate. We observed large NMR to show the field-induced superconductivity.

## 2. Experimental Procedure

Superconducting NbN films were firstly prepared by deposition at ambient temperatures on (200)-3C-SiC substrates having small lattice mismatches comparing with MgO substrate. Details of the deposition procedures are previously reported. [3] The NbN SNWs were fabricated from 2D films with $d = 20$ nm by a conventional e-beam lithography method and a reactive ion etching method with $CF_4$ plasma. We prepared two specimens with the same width $w$= 20 nm but different length $L$ between voltage terminals. We measure transport properties by four-probe method with applying a constant current 10nA for all measurements.

## 3. Experimental results and discussion

Prior to showing the present data of NbN SNWs on 3C-SiC substrate, Fig.1 (a) shows the $R(T)$ of NbTiN SNW which has similar characteristics to those of NbN on the conventional MgO. The shape of this NbTiN-NW has $L$=500 nm and $w$=20nm which are almost the same sizes of the present NbN NWs. Although the anomalous negative magneto resistance (NMR), that is, $\Delta R=R(H)-R(0)<0$ typically appears in the low resistive region, $R(T)$ characteristic at high resistive region, for instance, at $R=R^N/2$ shown by arrow, monotonically shifts to low temperatures with increasing of the magnetic field except for low fields. Here the $R^N$ is the normal state resistance determined at higher temperatures where the resistance is independent of magnetic field. When the transition temperature $T_c(H)$ is estimated as a temperature where $R(T_c)=R^N/2$ is satisfied, the upper critical magnetic field $H_{c2}$ changes about 4T in a temperature range from 7K to 3.5K. On the other hand, the $R(T)$ of NbN on3C-SiC shows quite strange characteristics.

Figures 1 (b) and (c) show $R(T)$ characteristics of NbN NW on the 3C-SiC substrate with (b) $T_c$=12.7 K, $R^N$~15kΩand $L$=900nm and (c)$T_c$=13.6 K, $R^N$~10 kΩ and $L$ = 600nm, respectively. The arrows corresponding to $R=R^N/2$ are also shown as well as in Fig. 2(a). Comparing the $R(T)$ of (b) and (c) to that of (a) at high resistance regions, it is clear that both SNW have a strong superconducting characteristic against the magnetic field. The large enhancement of $\Delta H_{c2}$ = 9 T for SNW with $L$=900nm even in a small temperature range between ~12.2 K and ~12.7 K. As for another difference between SNW' on 3C-SiC and MgO substrates, it is found that the SNW's on 3C-SiC in the magnetic field show the V-shape $R(T)$ at the low resistive region, namely, the large NMR appears and then disappears with decrease of the temperature. In NbN NW with $L$=900nm, the V-shape NMR is clearly found and the value of the non-zero minimum resistance increases with increase of the magnetic field. For shorter SNW with $L$=600nm, as shown in Fig.1(c), the minimum resistance due NMR seems to be nearly zero in the error range of the voltage measurements.

IOP Conf. Series: Journal of Physics: Conf. Series **969** (2018) 012062 doi:10.1088/1742-6596/969/1/012062

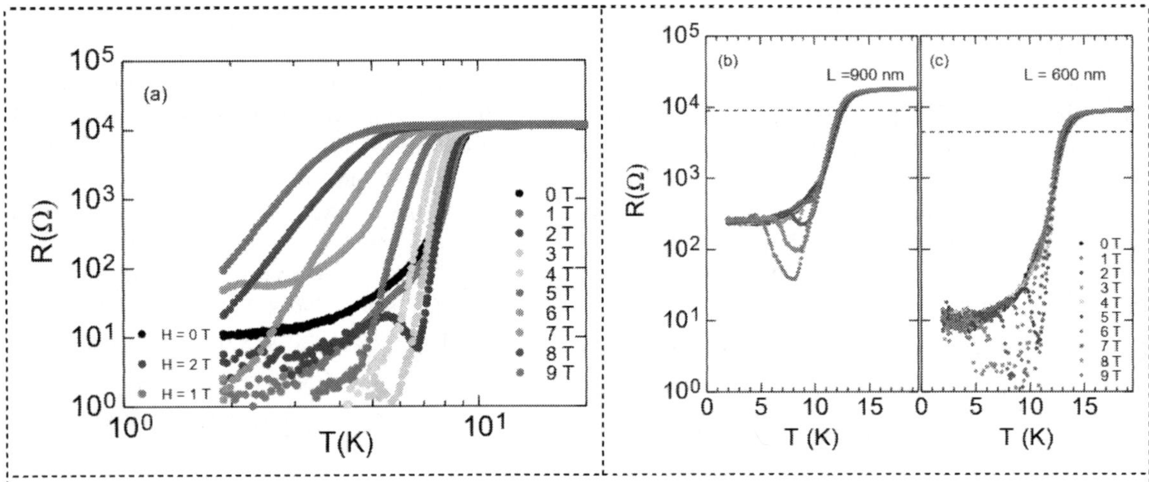


**Figure 1.** R(T) characteristics under the various magnetic fields for SNW's w=20nm (a)NiTiN with L=500nm on MgO substrate and (b) NbN with L=900nm and (c)NbN with L=600nm on 3C-SiC substrates.

Figure 2 shows the details of $R(T)$ characteristics in the low resistive region below 50 Ω for the SNW with $L$=600nm at various magnetic fields. The solid lines corresponding to each magnetic field above $H$=5 T show the results smoothly connected to data points to clear the characteristics. The resistances $R(T)$ clearly take zero values in the restricted temperature region depending on the magnitude of the external magnetic field. Here it is emphasized that the zero resistive state under any magnetic field $H$<9 T in this measurement recovers to the non-zero resistive state denoted by the QPS at $H$=0 with decreasing the temperature. This re-entrance characteristic suggests the field-induced superconductivity.

In order to clear the effect of the magnetic field on the TAPS and QPS controlling the $R(T)$ characteristics below $T_c$, Figure 3 shows $R(T)$ at $H$=0 (●) and at $H$= 6 T (□) as a typical high field. Theoretically, $R_{TAPS}$ $(T)$ is given as follows, [2]

$$R_{TAPS} = \phi_0 \Omega / (k_B T / \phi_0) \exp[-\Delta F / k_B T], \qquad (1)$$

where the magnetic barrier $\Delta F$, associated attempt frequency $\Omega$ and GL relaxation time $\tau_{GL}$ are given by $\Delta F = \sqrt{2}\sigma H_c^{\,2} \xi_{GL}(T)/3\pi$, $\Omega = (L/2\pi^2 \xi_{GL}(T)\tau_{GL})(3\pi\Delta F/k_B T)^{1/2}$ and $\tau_{GL}=h/16k_B(T_c-T)$, respectively. The GL coherence length $\xi_{GL}(T)$ is given by $\xi_{GL}(T) = (0)(1-T/T_c)^{-1/2}$. On the other hand, the $R_{QPS}(T)$ at $H$=0 is given by

$$R_{QPS} = (h/4e^2)(\hbar\Omega/(\hbar/\alpha\tau_{GL})\exp[-\Delta F/(\hbar/\alpha\tau_{GL})], \qquad (2)$$

The dotted and solid lines are calculated from eq.(1) and eq.(2), respectively. The coherence length $\xi_{GL}(0)$=6.3nm in eq.(1) was determined from the $H_{c2}(T)$ characteristic of the 2D film almost the same thickness and $T_c$ as those of this SNW L=600nm. The parameter $\alpha$ in eq.(2) was determined to be $\alpha$ =0.001 from well fitting of eq.(2) experimental data $R(T)$ at $H$=0T. This value of $\alpha$ is reasonable to compared that in previous work. The $R(T)$ at high magnetic field seems to be expressed only by the TAPS model. Present analyses suggest that the fields suppress the resistance due to QPS but not due to the TAPS. Even if the resistance due to TAPS does not be suppressed, almost the zero resistive state shown in Fig.3 can be understood because the resistance due to drastically decreases with decreasing of temperature. However, at the present stage, we have no explanation for the reason of the re-entrance to the QPS resistive state at low temperatures.

For 1D superconductors at temperatures near $T_c$ with both magnitudes of width $w$ and thickness $d$ smaller than $\xi_{GL}$, temperature dependence of $H_{c2}(T) \propto (1-T/T_c)^{1/2}$ is obtained from GL equation. In the

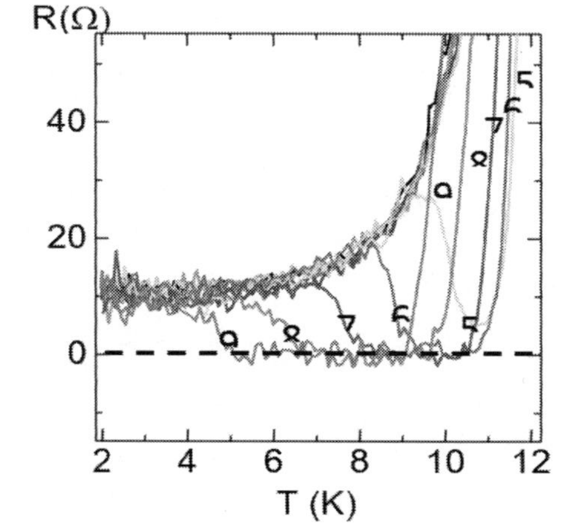


**Figure 2.** R(T) characteristics under the various magnetic fields for NbN SNW with L=600nm on 3C-SiC substrate. In a temperature interval depending on the strength of the magnetic field, the resistance shows nearly zero.

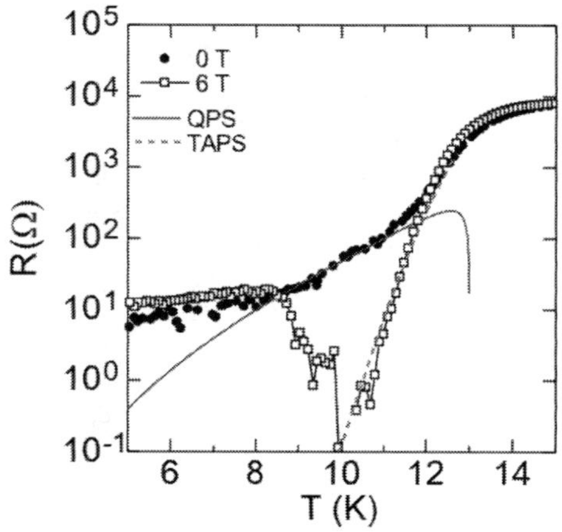


**Figure 3.** *R*(T) characteristics of NbN SNW with L=600nm on 3C-SiC at (●) H=0 and at(□) H=6 T. Solid and dotted lines are calculated from the theory of QPS model and TAPS model, respectively.

present thin film, we obtained a steep relation of $H_{c2}(T)$~$47(1-T/T_c)^{1/2}$[T] as recognized in Figs. 1(b) and 1(c). Although this relation $H_{c2}(T)$ obtained experimentally is not satisfied until absolute zero, there is a possibility that it exceeds the Pauli limit $H_p(0)=1.86\times T_c$[T]~25T.[8] On the other hand, we can get a relation $H_{c2}(T)=\varphi_0/[2\pi\xi_{GL}(T)w]\sim 8.2(1-T/T_c)^{1/2}$ [T] with use $\xi_{GL}(T)=6.3(1-T/T_c)^{-1/2}$ above mentioned. Experimental data are about 5 times as large as theoretical ones estimated from above relation. Taking account for the above relation $\xi_{GL}(T)$ for the present SNW, in the experimental temperature range of $H_{c2}(T)$, namely, 13 K<$T$<$T_c$~13.6K, the ratio $\xi/w$ is expected to exceeds ~2 and reach to ~200. Further understanding the reason for the extremely large $H_{c2}$, it is necessary to investigate the flux distribution in SNW near $T_c$.

## 4. Conclusions

For the NbN SNW's prepared on the 3C-SiC substrates, we observed the broad superconducting transition. The *R*(*T*) characteristics in the absence of the magnetic field *H* can be well explained by the quantum phase slip model. Under the large magnetic field, however, *R*(*T*) drastically decreases with decreasing the temperature. This *R*(*T*) characteristic suppressed to nearly zero resistance can be well explained by the thermal activated phase slip model. This behavior indicates the occurrence of the field induced superconductivity in the present SNW. As for the upper critical magnetic $H_{c2}(T)$, Experimental data experimental results are about 5 times as large as that estimated from GL relations with use of the suitable value of the diffusion constant D. This fact suggests that there is a possibility of exceeding the Pauli limit $H_p(0)$.

## References

[1] Mooji J E and Narozov Y V (2006) Nat Phys. 2 169
[2] Lau C N, Markovic N, Bockrath M, Bezryadin A and Tinkham M (2001) *Phys. Rev. Lett.* 19 217003
[3] Makise K, Terai H, Tominari Y, Tanaka S and Shinozaki B (2016) Scientific Report 6:27001

DOI 10.1638/srep27001
[4] Kim H, Jamari S and Rogachev A (2012) Phys. Rev. Lett. 109 027002
[5] Herzov A V,Xiong P and Dynes R C (1981) Phys. Rev. B58 14199
[6] Synder S D, Dunn T, Erikson M J, Kinney J, Leee Y, Nelson J and Goldman A M (2013)Phys. Rev. B87 144503
[7] Cordoba R et al. (2013) Nature comm.4.1437]
[8] Saint James D, Thomas E J and Sarma G (1969) Type II superconductivity Pergamon Press Ltd

IOP Conf. Series: Journal of Physics: Conf. Series **969** (2018) 012063 doi:10.1088/1742-6596/969/1/012063

# Flux-periodicity crossover from *h/2e* to *h/e* in aluminium nano-loops

**C. Espy[1], O. J. Sharon[2], J. Braun[1], R. Garreis[1], F. Strigl[1], A. Shaulov[2], P. Leiderer[1], E. Scheer[1,*], Y. Yeshurun[2]**

[1]Department of Physics, University of Konstanz, 78457 Konstanz, Germany
[2]Department of Physics and Institute of Nano Technology and Advanced Materials, Bar-Ilan University, 5290002 Ramat-Gan, Israel
*Elke.scheer@uni-konstanz.de

**Abstract.** We study the magnetoresistance of aluminium 'double-networks' formed by connecting the vertexes of nano-loops with relatively long wires, creating two interlaced subnetworks of small and large loops (SL and LL, respectively). Far below the critical temperature, Aharonov-Bohm like quantum interference effects are observed for both the LL and the SL subnetworks. When approaching $T_c$, both exhibit the usual Little-Parks oscillations, with periodicity of the superconducting flux quantum $\Phi_0=h/2e$. For one sample, with a relatively large coherence length, $\xi$, at temperatures very close to $T_c$, the $\Phi_0$ periodicity of the SL disappears, and the waveform of the first period is consistent with that predicted recently for loops with a size $a < \xi$, indicating a crossover to $2\Phi_0$ periodicity.

## 1. Introduction

In a multiply connected superconductor, the fluxoid, defined as the sum of the magnetic flux and the line integral of the screening current, is quantized in units of $\Phi_0 = h/2e$, where the *2e* is a hallmark of electron pairing in the superconductor. As a direct consequence of this fluxoid quantization, periodic oscillations of the critical temperature $T_c$ as a function of magnetic field, known as the Little-Parks effect manifest themselves as oscillations of the magnetoresistance (MR) close to $T_c$. The amplitudes of the critical temperature, $\Delta T_c$, and the magnetoresistance oscillations, $\Delta R$, are related by the slope of the resistance *vs.* temperature, $R(T)$: $\Delta R \approx (\mathrm{d}R/\mathrm{d}T)\Delta T_c$ [1].

Theoretical studies [2] have predicted that in superconducting nano-loops with a length-scale $a < \xi$ the dominant periodicity is *h/e* rather than *h/2e*. The same theories predict that for high-$T_c$ superconductors (HTS) with d-wave symmetry, the *h/e* periodicity is also expected for $a \gtrsim \xi$. Recent experiments ([3, 4]) failed to identify the *h/e* component in HTS, probably because $\xi_0$ in these materials is only ~2 nm and, therefore, $a >> \xi$ in most of the temperature range.

In the present study we focus on aluminium, a low-$T_c$ superconductor with a relatively large bulk coherence length ($\xi_0$ = 1.6 μm). In nanostructures made of diffusive thin films the coherence length is reduced due to the finite mean free path, and simultaneously the penetration depth $\lambda_L$ is enhanced. Typical values of $\xi$ in these aluminium nanostructures lie in the range of 100 to 200 nm. Close to $T_c$, the coherence length $\xi(T)$ diverges, allowing in principle to meet the criterion $a < \xi$ in nanostructures with circumferences in the order of several hundred nanometres. On the other hand, the critical field of

IOP Conf. Series: Journal of Physics: Conf. Series **969** (2018) 012063 doi:10.1088/1742-6596/969/1/012063

bulk Al amounts to only 10 mT, giving a strong limitation for the number of Little-Parks oscillations (LPO) that can be observed. In reduced dimension, *i.e.* when the lateral dimensions are in the order of the penetration depth $\lambda_L$, the critical field $B_c$ may increase to a few hundred mT. Taking these considerations together, we fabricated 'double-networks' [3], see figure 1, with small loops of order of 400 x 400 nm$^2$, connected by wires of ~ 1600 nm length. An applied field of ~ 10 mT corresponds to a flux of $h/2e$ through the small loops.

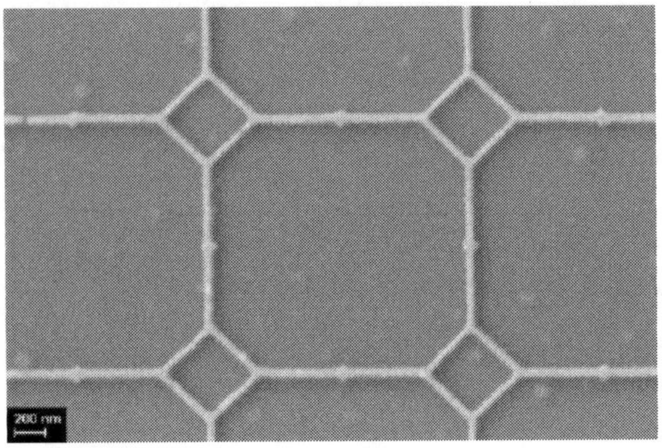


**Figure 1.** Scanning electron micrograph of an Al double-network (sample 1) with large (small) loop side length 1.71 µm (426 nm), line width $w = 50 \pm 5$ nm, thickness $d = 30$ nm.

## 2. Experimental

### 2.1. *Sample fabrication*

We use a lift-off electron beam lithography process in which Al is electron-beam evaporated onto a cooled, pre-patterned, oxidized Si substrate. The lithographic mask is then removed in warm acetone. The width of the lines is around 60 nm, the film thickness amounts to 30 nm. The arrays consist of roughly 10 x 10 loops. The samples feature normal state resistance of $R_n = 20 - 40\ \Omega$.

### 2.2. *Transport measurements*

Due to the pronounced temperature dependence of the LPO, particular care is taken to stabilize the cryostat temperature during the magnetic field sweeps. We use a combination of a carefully calibrated resistive thermometer to determine the absolute temperature and a capacitive sensor for keeping it constant within ± 1 mK around the set-point temperature. The set temperature spacing is adapted to the steepness of the $R(T)$ curves and amounts to a few mK around $T_c$. Before starting the magnetoresistance sweeps the temperature is allowed to stabilize for several minutes.

The cryostat is equipped with home-made high-frequency filtered cables to record the four-point differential resistance $dV/dI$ by biasing a DC bias current that is kept smaller than 500 nA superimposed with a small AC current in lock-in technique. We simultaneously measured the resistance $R = V/I$ which shows qualitatively the same but more noisy data than the $dV/dI$. The low temperature critical current of the samples amounts to 50 - 250 µA. Close to $T_c$, where the LPO are maximal, the zero-field critical current still amounts to more than 2 µA.

The magnetic field is applied perpendicular to the sample plane using a superconducting solenoid. For each temperature, measurements are performed at a sweep rate of ~ 5 mT/min. For each temperature we record a sweep with increasing and decreasing field, as a control for constant temperature throughout a sweep, and in order to be able to correct for small field offsets. The field range is adapted to the critical field at the set temperature.

**Table 1.** Dimensions and transport characteristics of the two samples presented in this article. LL: large loops, SL small loops. $w$: line width from electron micrographs, $w_{fit}$: line width from fit to eq. (1), $\xi$: coherence length from fit to eq. (1), $\xi_{cal}$: coherence length from fit to eq. (2), $T_c$: critical temperature, $I_c$: critical current at 300 mK, $B_c$: critical field extrapolated to $T = 0$ from the d$V$/d$I$($B$) measurements, $R_n$: normal resistance measured above $B_c$. For sample 2 the fits have been performed with the low-bias data shown in figure 3.

| Sample No. | SL size (nm) | LL size (nm) | $w$ (nm) | $w_{fit}$ (nm) | $\xi_{cal}$ (nm) | $\xi$ (nm) | $T_c$ (mK) | $I_c$ (μA) | $B_c$ (mT) | $R_n$ (Ω) |
|---|---|---|---|---|---|---|---|---|---|---|
| 1 | 426±10 | 1710±30 | 50±10 | 50±5 | 118±24 | 102±10 | 1408±5 | 55±5 | 194±10 | 33.2±1 |
| 2 | 324±10 | 1540±30 | 66±10 | 61±5 | 114±19 | 137±15 | 1457±5 | 240±10 | 152±10 | 23.3±1 |

## 3. Results

### 3.1. Critical temperature, critical field and coherence length

Figure 2 a) shows differential MR, d$V$/d$I$($B$), curves in the temperature range 380 - 1600 mK, recorded on sample 1 with 426 nm side length of the small loops and 1.71 μm of the large loops (see table 1 for sample dimensions). From the set of d$V$/d$I$($B$) curves we construct the envelope of the $B_c(T)$ phase boundary (without the oscillatory part) from which we deduce the coherence length, $\xi$, and $T_c$ using the relation [7]

$$T_c(B) = T_c\left[1 - \frac{\pi^2}{3}\left(\frac{w_{fit}\xi B}{\Phi_0}\right)^2\right], \qquad (1)$$

where $w_{fit}$ is the width of the wire. For comparison we also estimate the coherence length from the low temperature critical field, $B_c(0)$ assuming a thin slab in magnetic field resulting in the relation [10]:

$$\xi_{cal} = \sqrt{3}\Phi_0/(\pi w B_c(0)) \qquad (2)$$

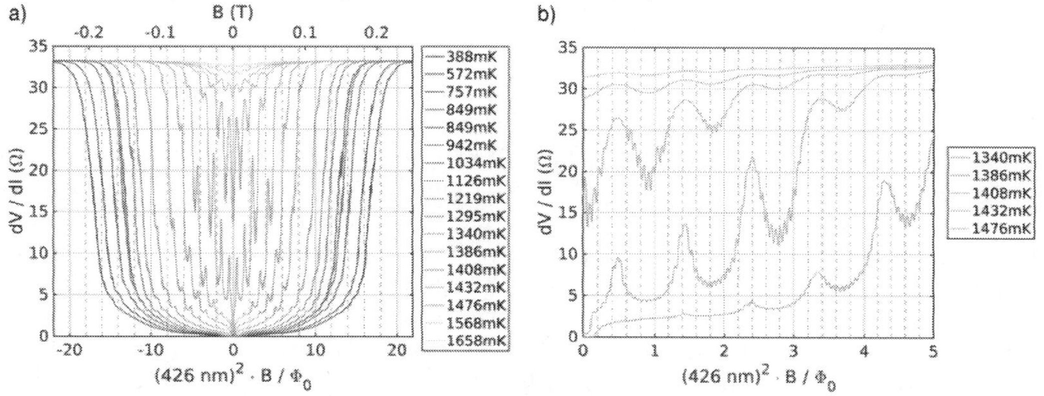


**Figure 2.** a) d$V$/d$I$($B$) curves of sample 1 covering the whole field range and the whole temperature range from 388 mK (dark blue) to 1658 mK (yellow), b) Zoom into the positive field direction of selected data from panel a) close to $T_c$.

The loop sizes have been determined by inspection of the electron micrographs and are additionally deduced from the observed flux periodicity assuming square-shaped loops. The numerical values for the various parameters are given in table 1. The values for $T_c$ correspond well with the position of the $R(T) = R_n/2$ in the temperature curves as well as with the d$V$/d$I$($B$) curve at which the largest oscillation amplitude is observed. Also, both estimates for the coherence length agree with each other within ~ 20% for both samples, supporting the suitability of the simplified models. Taking the average value of these two estimates, we find that for sample 1 (2) the ratio between the coherence length and the SL edge length $a$ amounts to $\xi/a \sim 0.26$ (0.39). Thus, we expect the transition to the $h/e$ periodicity being more likely to be observable in sample 2.

### 3.2. *Little Parks effect*

Figure 2 b) zooms into the positive field direction close to $T_c$. Figure 3 shows $dV/dI(B)$ data for sample 2 with SL size of 324 nm and LL size of 1.54 μm. For both samples and at all temperatures we observe symmetric periodic oscillations of $dV/dI(B)$ corresponding to $h/2e$ oscillations of the small loops (SL) and of the large loops (LL).

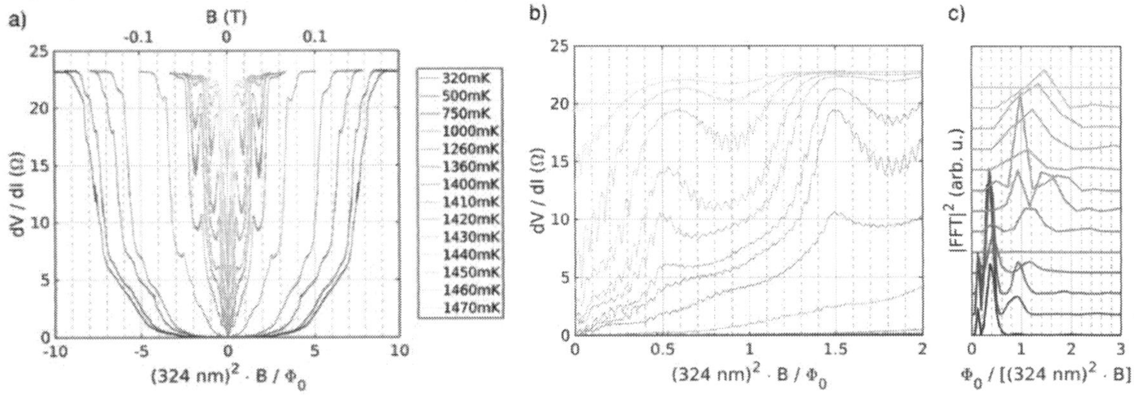


**Figure 3.** a) Selected $dV/dI(B)$ traces of sample 2. b) Same curves as in a) for a reduced range in positive field direction. c) FFT of the $dV/dI(B)$ data shown in a).

Closer inspection reveals for both samples a gradual change of the LL oscillations from a cusp-up form (typical of LPO) at low temperature, through a sinusoidal behaviour at intermediate temperature, into a cusp-down (SQUID-like) shape at higher temperatures. A similar crossover, induced by the bias current rather than by temperature, was reported by Sharon *et al.* for Nb chains of rings [5]. The effect is most pronounced at low fields when the flux through the SL is still smaller than $\Phi_0/2$. The amplitude of the LL oscillations vanishes at somewhat smaller temperatures than the SL oscillations. Both observations suggest that the coherence in the LLs is not fully developed and that weak links exist that act as Josephson junctions. Furthermore the amplitude of the LL oscillations is locally suppressed at odd multiples of $\Phi_0/2$ of the SLs. At these flux values the ring current in the SL is maximal. Hence, the suppression of the LL amplitude can be understood as an interruption of the ring current in the LLs. Similar observations were made in Al nanostructures consisting of several loops sharing the same strands [6, 7]. This observation is related to the Little-Parks-de Gennes effect [8] which describes the destruction and restauration of superconductivity in multiply connected superconductors around odd multiples of $\Phi_0/2$ [9].

### 3.3. *Little-Parks effect of small loops and transition to h/e periodicity*

We now turn to discuss the SL oscillations. These also show the typical cusp-up behaviour well below $T_c$ and a sinusoidal shape close to $T_c$. This shape transition is most pronounced in sample 2 and is marked by a broadening and a shift of the resistance maximum at $\Phi_0/2$ to larger fields. The Fast Fourier Transform (FFT) of the data (see figure 3 c)) reveals that besides the usual $h/2e$ component also an $h/3e$ component is present which starts to dominate close to $T_c$. The $h/3e$ component indicates a modulation of the amplitude of the regular $h/2e$ LPO. We interpret these findings as an onset of the transition to $h/e$ which is expected for very small loop sizes. Due to the limited field range that covers only slightly more than one period of the $h/e$, the fundamental $h/e$ component cannot be seen in the FFT, but rather a higher harmonic of it, *i.e.* the $h/3e$ component.

To investigate this observation further we performed $dV/dI(B)$ measurements on sample 2 under relatively high bias current of 2.1 μA. With this bias the apparent $T_c$ is somewhat smaller than in the low bias data. Two examples of differential resistance curves, one with low and one with high bias, are plotted in figure 4 a). In the high-bias data the oscillations appear more pronounced and $dV/dI(B)$ exceeds $R_n$ around $\Phi = \Phi_0/2$ and close to $B_c$(T). In these field ranges also irregular oscillations are visible although the LPOs are washed out in the low bias data. Both observations are consequences of the non-linear current-voltage characteristics in this range and the differential measurement scheme. We checked that the absolute resistance remains everywhere smaller than $R_n$.

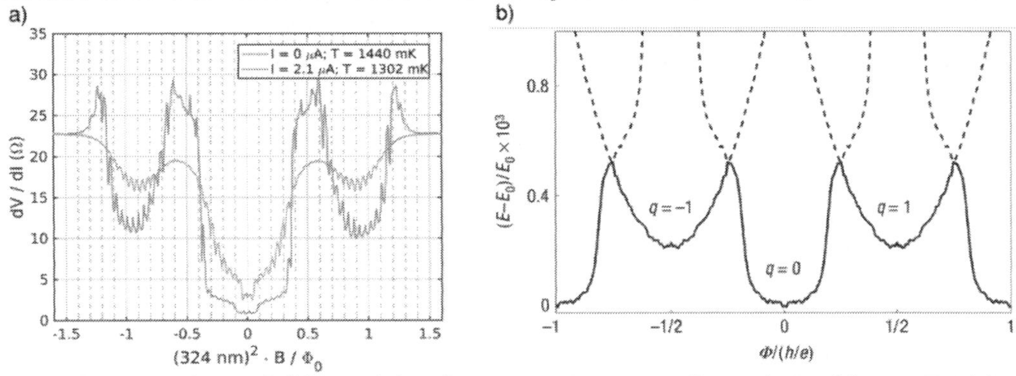


**Figure 4.** a) Comparison of differential resistance *vs.* $B$ curves of sample 2 with zero DC bias current and with finite DC bias current (2.1 μA) close to $T_c$. b) Energy vs. flux for a superconducting loop with $a < \xi$, showing the modulation of the odd LPOs. Reprinted with permission from MacMillan Publishers Ltd: Nature Physics [2], copyright 2008. In this right figure $\Phi$ is described in units of $\Phi_0 = h/e$.

For both curves, the shape of the central dip (at $\Phi = 0$) is much different from that of the side dips (at $\Phi = \Phi_0$), implying that the Little-Parks $h/2e$ periodicity is broken. On the other hand, these data show a striking similarity to the theoretical calculations of Loder *et al.* [2] predicting $h/e$ periodicity, as depicted in figure 4 b). Note that in this figure the flux is plotted in units of $h/e$. Unfortunately, we observe only one period as the field required to observe more periods exceeds the critical field, $B_c(0)$, of this sample. Nevertheless, the nearly parabolic shape of the side minima and the distorted shape of the parabola centred at $\Phi = 0$ may serve as a fingerprint of the predicted $h/e$ periodicity. Note that the theoretical prediction (right panel) describes the oscillation in the energy while the experimental results (left) describe the differential resistance. The two, however, are related since both are directly related to $\Delta T_c$ [10]. The similarity with the theoretical curve is most apparent for the high-bias $dV/dI$ curve. This suggests that under these measurement conditions, *i.e.* when the system is driven close to the transition to the normal state, the relation between differential resistance and energy is most direct.

### 3.4. *Aharonov-Bohm effect in small loops*

Finally we note that the temperature dependence of the amplitude of the SL oscillations is non-monotonous. As figure 3 c) shows, at low temperature, $T < \sim 1000$ mK, we observe $h/e$ and $h/2e$ oscillations of the small loops that almost disappear for intermediate temperatures, $\sim 1000 \text{ mK} < T < \sim 1200$ mK. Above 1200 mK dominantly the $h/2e$ and $h/3e$ components corresponding to the Little-Parks effect appear. We attribute the low-temperature oscillations to quantum interference effects, *i.e.* the Aharonov–Bohm effect, indicating that also the phase coherence length of the quasiparticles, $L_\Phi$, in this range is in the order of the perimeter of the SLs or larger. When increasing the temperature, inelastic scattering sets in that reduces $L_\Phi$ [11]. For completeness we mention that we do not observe such non-monotonous behaviour for the LL oscillations, indicating that $L_\Phi$ is always smaller than the perimeter of the LLs.

## 4. Conclusions

Summarizing, we have presented magnetoresistance data in interlaced networks comprising small and large loops made of aluminium. We observe the interplay between the magnetoresistance oscillations in these two subsets of loops. In particular, the $\Phi_0 = h/2e$ periodic Little-Parks oscillations of the large loops are modulated by the flux conditions in the small loops. For a sample with high ratio between coherence length and loop size we observe an onset of the transition from the conventional $h/2e$ to an $h/e$ periodicity, as predicted for long coherence length. A full experimental verification of the theory requires measurements of more than one period. However, the relatively low critical field of aluminium, and the large period of small loops, presently impedes achieving this condition. Further improvements of the sample design and material quality are currently underway to overcome this limitation.

## Acknowledgements

This research was supported by a Grant from the GIF, the German-Israeli Foundation for Scientific Research and Development. We thank Arno Kampf, Thilo Kopp, Jochen Mannhart, and Christoph Strunk for insightful discussion and Simon Haus for his help in sample preparation.

## References

[1] Little W, Parks R 1962 *Phys. Rev. Lett.* **9** 9.
[2] Loder F, Kampf A P, Kopp T, Mannhart J, Schneider, C W, Barash, Y S 2008 *Nature Phys* **4** 112.
[3] Sochnikov I, Shaulov A, Yeshurun Y, Logvenov G, Božović I 2010 *Nature Nanotechnol.* **5** 516.
[4] Carillo F, Papari G, Stornaiuolo D, Born D, Montemurro D, Pingue P, Beltram F, Tafuri F 2010 *Phys. Rev. B* **81** 054505.
[5] Sharon O J, Shaulov A, Berger J, Sharoni A, Yeshurun Y 2016 *Sci. Rep.* **6** 28320.
[6] Bruyndoncx V, Strunk C, Moshchalkov V V, van Haesendonck C, Bruynseraede Y 1996 *Europhys. Lett.* **36** 449.
[7] Strunk C, Bruyndoncx V, Moshchalkov V V, Van Haesendonck C, Bruynseraede Y Jonckheere R 1996 *Phys. Rev. B* **54** R12701.
[8] de Gennes P G, 1981 *C. R. Acad. Sci. Ser. B* **292** 9, and *ibid.* **292** 279.
[9] Staley N E, Liu Y 2012 *Proc. Nat. Acad. Sci* **109** 14819.
[10] Tinkham M 1996 *Introduction to Superconductivity* McGraw-Hill New York.
[11] Washburn S, Webb R A 1986, *Adv. Phys.* **35** 375.

IOP Conf. Series: Journal of Physics: Conf. Series **969** (2018) 012064 doi:10.1088/1742-6596/969/1/012064

# $T_c$ depression and superconductor-insulator transition in molybdenum nitride thin films

**F Ichikawa[1], K Makise[2], T Tsuneoka[1], S Maeda[1] and B Shinozaki[3]**

[1] Department of Physics, Kumamoto University, 2-39-1 Kurokami Chuo-ku, Kumamoto 860-8555, Japan

[2] National Institute of Advanced Industrial Science and Technology (AIST), Tsukuba 305-8568, Japan

[3] Department of Physics Kyushu University, 744, Motooka, Nishi-ku, Fukuoka, 819-0395, Japan

E-mail: ichikawa@kumamoto-u.ac.jp

**Abstract.** We have studied that the $T_c$ depression and the superconductor-insulator transition (SIT) in molybdenum nitride (MoN) thin films. Thin films were fabricated by reactive DC magnetron sputtering method onto (100) MgO substrates in the mixture of Ar and $N_2$ gases. Several dozen MoN thin films were prepared in the range of 3 nm $<$ thickness $d <$ 60 nm. The resistance was measured by a DC four-probe technique. It is found that $T_c$ decreases from 6.6 K for thick films with increase of the normal state sheet resistance $R_{\rm sq}^{\rm N}$ and experimental data were fitted to the Finkel'stein formula using the bulk superconducting transition temperature $T_{c0} = 6.45$ K and the elastic scattering time of electron $\tau = 1.6 \times 10^{-16}$ s. From this analysis the critical sheet resistance $R_c$ is found about 2 k$\Omega$, which is smaller than the quantum sheet resistance $R_{\rm Q}$. This value of $R_c$ is almost the same as those for 2D NbN films. The value of $\tau$ for MoN films is also the similar value for NbN films $1.0 \times 10^{-16}$ s, while $T_{c0}$ is different from that for NbN films 14.85 K. It is indicated that the mechanism of SIT for MoN films is similar to that of NbN films, while the mean free path $\ell$ for MoN films is larger than that for NbN films.

## 1. Introduction

To study low-dimensional superconducting devices like superconducting single-photon detectors [1, 2] and superconducting transition edge detector [3, 4], it is necessary to investigate the effects of disorder in low-dimensional superconductors on the superconductor-insulator transition (SIT). There are mainly two scenario to understand the SIT mechanism. One is the dirty boson scenario. According to this scenario the Cooper pairs exist even in the insulator side. They are localized at a sheet resistance close to the quantum sheet resistance $R_Q$ ($= h/4e^2 \sim 6.45$k$\Omega$) as observed in Bi [5, 6], Pb [5] and InO [7]. The other is the fermionic scenario. This scenario is based on the idea that the disorder-enhanced Coulomb repulsion prevents Cooper pairing. Both the energy gap $\Delta$ and $T_c$ decrease with the same rate to zero at SIT as the disorder, for instance, the sheet resistance $R_{\rm sq}$ increases. Finkel'stein [8] shows that the $T_c$ depression is determined from the values of bulk $T_{c0}$ and elastic scattering time $\tau$. This indicates that the $R_c$ at the SIT is not universal but depends on the material. $T_c$ depression of NbN [9, 10] and MoGe [11] induced by disorder has been explained by this scenario.

IOP Conf. Series: Journal of Physics: Conf. Series **969** (2018) 012064 doi:10.1088/1742-6596/969/1/012064

The thin film with lower $T_c$ is preferable for low-dimensional superconducting devices because the material with a smaller superconducting gap can show a higher sensitive response. However, the thin film with high resistivity $\rho$ have a disadvantage of slow response speeds because of the large kinetic inductance. Therefore it is necessary to search materials with lower $T_c$ and $\rho$ for the sensitive superconducting detectors. Molybdenum nitride (MoN) has attracted attentions because of the possibility for a new device-material to satisfy these demands. In this paper, we report the $T_c$ depression and the SIT in MoN thin films. The results of the suppression of $T_c$ are analyzed by the Finkel'stein formula from localization theory [8]. The parameters obtained by the analyze are compared with those for NbN films.

## 2. Experimental detail

MoN films were deposited on (100) MgO substrates by reactive DC magnetron sputtering method. Namely Mo was sputtered in the presence of a mixture of Ar and $N_2$ gases while keeping the total gas pressure at 0.27 Pa [12]. The gas ratio of $N_2$, $p(N_2)$, was controlled by the fixed flow rates of Ar $f(\mathrm{Ar})$ and $N_2$ $f(N_2)$ gases. The gas ratio $p(N_2)$ is defined as $p(N_2) = f(N_2)/(f(N_2) + f(\mathrm{Ar}))$. Table 1 shows the $p(N_2)$ dependences of chemical composition, resistivity at 300 K $\rho(300\mathrm{K})$ and $T_c$ in thin films [13]. The chemical composition in thin films was measured by wavelength dispersive X-ray spectroscopy (WDS). $N_2$ composition and $\rho(300\ \mathrm{K})$ monotonically increase as increasing $p(N_2)$. However, $T_c$ takes the maximum value at $p(N_2) = 9.1\%$. In this report we fabricated MoN films keeping the optimum value of $p(N_2) = 9.1\%$. The crystal structures of films were investigated by X-ray diffraction analysis. The films have clear (200) and (400) peaks with the lattice constant $d_0 = 0.4168$ nm. According to the previous research [14] for lattice constant $d_0$ and $T_c$, it was reported that MoN has $d_0 = 0.4213$ nm and $T_c \sim 12$ K and $\gamma$-$Mo_2N$ cubic has $d_0 = 0.4163$ nm and $T_c \sim 5$ K. Therefore, we suppose that present films take $\gamma$-$Mo_2N$ cubic structure.

**Table 1.** $N_2$ concentration, $\rho(300\ \mathrm{K})$ and $T_c$ as a function of the gas ratio for thick films.

| flow rate (SCCM) $f(\mathrm{Ar})$ | $f(N_2)$ | Gas Ratio $p(N_2)$ (%) | $N_2$ composition (%) | $\rho(300\ \mathrm{K})$ ($\mu\Omega$m) | $T_c$ (K) |
|---|---|---|---|---|---|
| 50 | 0 | 0 | 0 | 0.045 | 0.96 |
| 50 | 5 | 9.1 | 23 | 1.3 | 6.60 |
| 50 | 10 | 17 | 27 | 1.4 | 6.56 |
| 50 | 21.7 | 30 | 35 | 2.0 | 5.16 |
| 50 | 30 | 38 | 40 | 2.6 | 3.56 |

Various films with different thicknesses $d$ were prepared by changing the sputtering time because of the linear relation between the thickness and the sputtering time. Several dozen films were fabricated in the range of $3 < d < 60$ nm. The films were patterned in a thin strip using a photolithography and etching procedure, where the width $w$ and the length $l$ between the voltage terminals of the strip were 0.15 mm and 2 mm, respectively. The resistance $R$ was measured by a standard dc four-probe technique in a temperature range $2 \le T \le 300$ K. The sheet resistance $R_{\mathrm{sq}}$ is calculated through the equation $R_{\mathrm{sq}} = Rw/l$.

The sample pattern is also suitable for measuring the Hall voltage. We applied a magnetic field of up to $\pm 5$ T perpendicular to the film surface in order to obtain the carrier density from the Hall coefficient $R_{\mathrm{H}}$ based on the free electron model $R_{\mathrm{H}} = -1/ne$.

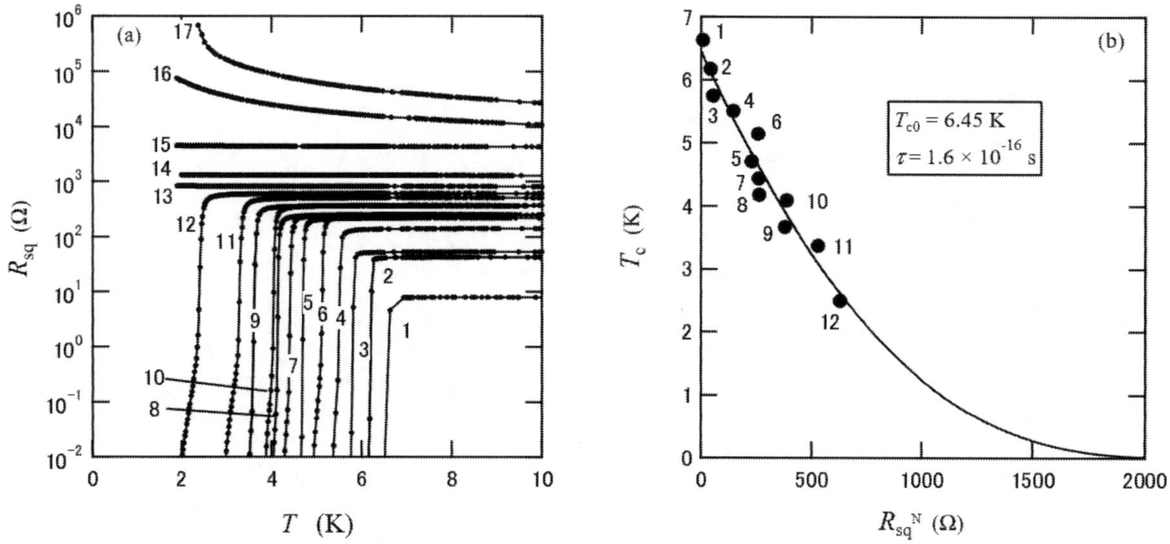


**Figure 1.** (a) $T$ dependence of $R_{sq}$ in MoN thin films. (b) $R_{sq}^{N}$ dependence of $T_c$ for MoN thin films. The solide line is calculated from Eq. (1), using $T_{c0}$ and $\tau$ as fitting parameters. Each figure corresponds to the specimen number.

## 3. Results and discussion

Figure 1(a) shows $R_{sq}$–$T$ curves for MoN films with various thicknesses [13]. Even for thin films with $d < 10$ nm (No. 9 ∼ 12), the $R_{sq}$–$T$ shows a sharp superconducting transition without quasi-reentrant behavior. This behavior indicates that all superconducting films are homogeneous. The determination of the critical sheet resistance $R_c$ from Fig. 1(a) is difficult because of the limitation of measurement temperature to over 2 K. However, the $R_c$ can be estimated from the expression for $T_c(R_{sq}^{N})$ derived by Finkel'stein in the following paragraph.

Figure 1(b) shows $R_{sq}^{N}$–$T_c$ curves. Here, the normal sheet resistance $R_{sq}^{N}$ is estimated by following procedures. The values of $R_{sq}^{N}$ for several films were obtained from the extrapolation of $1/R_{sq}(H)$ versus $1/H$ to $1/H = 0$. It is found that the empirical power law relation $R_{sq}^{N} = 0.967 R_{sq}(10\mathrm{K})^{1.01}$ is hold. $R_{sq}^{N}$ for all films in Fig. 1(b) were determined from $R_{sq}(10\mathrm{K})$ [13]. We can fit the theoretical model to experimental data in Fig. 1(b) to confirm the fermionic scenario and estimate the $R_c$ of the present MoN series. In the early stage of the investigation for the effect of disorder on $T_c$, Maekawa and Fukuyama proposed a theory for the suppression of $T_c$ due to enhanced Coulomb interactions in the presence of localization effects in a 2D disordered system [15]. Furthermore, Finkel'stein obtained the following expression from the renormalization group equation [8]

$$T_c = T_{c0}\exp(-1/\gamma)\left\{\left[1+\frac{\sqrt{r/2}}{\gamma-r/4}\right]\Big/\left[1-\frac{\sqrt{r/2}}{\gamma-r/4}\right]\right\}^{1/\sqrt{2r}}, \quad (1)$$

where $T_{c0}$ is the bulk superconducting transition temperature, $\gamma = 1/\ln(k_B T_{c0}\tau/\hbar)$, $r = R_{sq}^{N}/(2\pi^2\hbar/e^2)$ and $\tau$ is the elastic scattering time of electron. The solid line is calculated from Eq. (1) using the two values $T_{c0} = 6.45$ K and $\tau = 1.6\times10^{-16}$ s ($\gamma = 0.1128$) as fitting parameters. Although the $T_c$ data is restricted to the films with $T_c > 2$ K due to the measurement range of $T$, the value of $R_c$ can be estimated to be nearly ∼ 2 kΩ from the well extrapolation of Eq. (1) for $T_c(R_{sq}^{N})$ to zero.

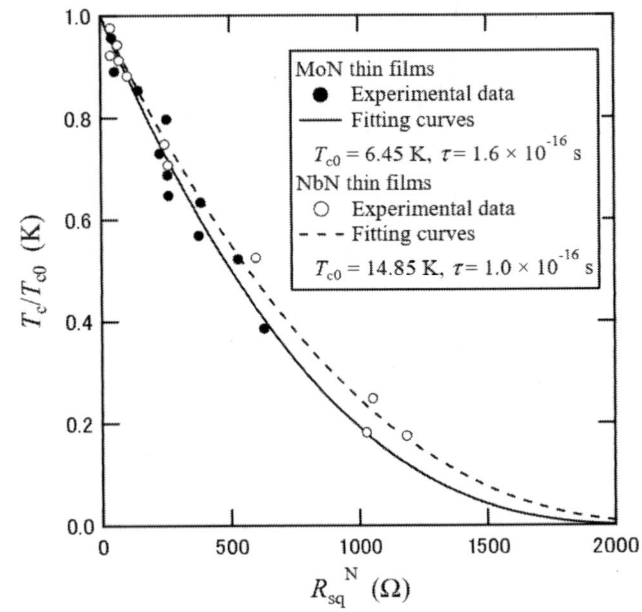


**Figure 2.** Experimental $R_{sq}^{N}$–$T_c/T_{c0}$ data for MoN thin films (○) and NbN thin films (●). The solid line is a curve fitted by Eq. (1) for data of MoN thin films and the dashed line is that for NbN thin films. The values of fitting parameters are shown in the figure.

In order to compare the $R_{sq}^{N}$ dependence of $T_c$ of MoN thin films with those of NbN thin films, Fig. 2 shows the normalized $T_c$ (i.e. $T_c/T_{c0}$) as a function of $R_{sq}^{N}$ for both MoN and NbN thin films [9, 13]. Though $T_{c0} = 6.45$ K for MoN thin films is different from that for NbN films 14.85 K, the $\tau = 1.6 \times 10^{-16}$ s for MoN films is also the similar value for NbN films $1.0 \times 10^{-16}$ s. The value of $R_c$ is also the similar to that for NbN [16]. The mechanism of SIT for MoN films should be similar to that of NbN films.

We also discuss about the difference of the mean free path $\ell$ between MoN and NbN thin films. We can estimate $\ell$ by three different methods. Two of them use the relation $\ell = v_F\tau$, where the Fermi velocity $v_F$ can be calculated assuming the free electron model $v_F = (\hbar/m)[(3\pi^2)n]^{1/3}$. The elastic scattering time $\tau$ can be estimated from the Drude model (the free electron model) $\tau = m/(ne^2\rho)$ or the fitting parameter of Finkel'stein formula. Therefore $n$ and the resistivity $\rho$ are used in order to obtain the mean free path by the first method $\ell_\rho = (\hbar/e^2\rho)[(3\pi^2)/n^2]^{1/3}$, while $n$ and $\tau$ are necessary to calculate it by the second method $\ell_\tau = (\hbar/m)[(3\pi^2)n]^{1/3}\tau$. The third method is used the relation $\ell_D = 3D/v_F = 3D(m/\hbar)[(3\pi^2)n]^{-1/3}$, where the diffusion constant $D$ can be determined from the slope $H_{c2}$–$T$ curves near $T_c$, given by $D = (4k_B/\pi e)|dH_{c2}/dT|^{-1}$.

The upper critical magnetic field $H_{c2}$ were measured for three typical superconducting films (No, 5, 7 and 8, 4 K $< T_c <$ 5 K) in order to calculated the diffusion constant $D$. Figure 3 shows $T$ dependences of $H_{c2}$ and the fitting lines to obtain the slope. The value of the slope is $|dH_{c2}/dT| = 2.6 \sim 3.0$ T/K, and $D$ is estimated to be $0.37 \sim 0.43$ cm$^2$/s. Because the sample dependence of $D$ is not so large, hereafter we use the data for the No. 8 film for simplicity.

The carrier density $n$ for the No. 8 film at $T = 20$ K is about $2.3 \times 10^{30}$ m$^{-3}$, this value is larger than reported values for NbN films ($3.0 \times 10^{28}$ m$^{-3}$ $< n <$ $1.4 \times 10^{29}$ m$^{-3}$) [16]. This discrepancy between MoN and NbN films cannot be understood in the present stage. It was estimated that $\rho$ for MoN films is about 3 $\mu\Omega$m from the relation between $R_{sq}$ and $1/d$ [13]. We can also use the elastic scattering time $\tau = 1.6 \times 10^{-16}$ s from the Finkel'stine fitting parameter and $D = 0.4$ cm$^2$/s for MoN thin films. The calculated mean free paths ($\ell_\rho$, $\ell_\tau$ and $\ell_D$) are shown in Table 2. There is some differences of $\ell$ for MoN films determined by the different

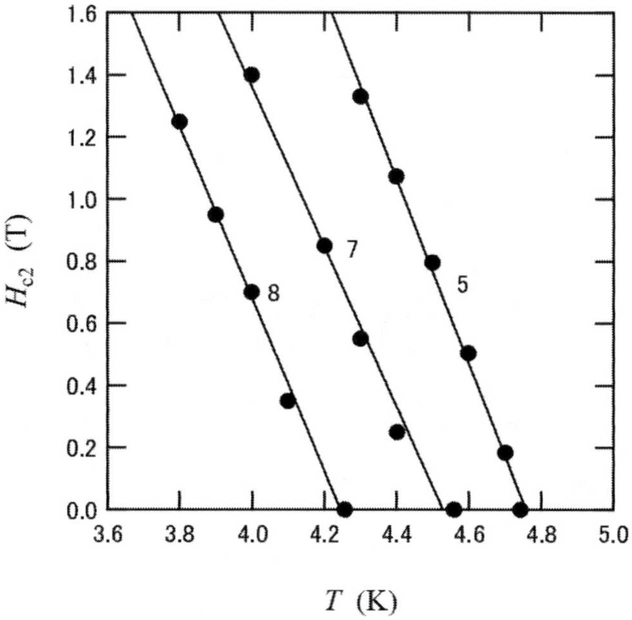


**Figure 3.** $H_{c2}$–$T$ data for MoN thin films (No. 5, 7 and 8). $|dH_{c2}/dT|$ can be obtained from the slope of solid lines obtained by fitting, 2.8, 2.6 and 3.0 T/K for No. 5, 7 and 8 films, respectively. The values of $D$ for No. 5, 7 and 8 films are calculated as follows: 0.37, 0.43 and 0.40 cm$^2$/s, respectively.

methods. It may be related to the large value of $n$. Because each formula has the different $n$ dependence as follows: $\ell_\rho \propto n^{-2/3}\rho(n)^{-1}$, $\ell_\tau \propto n^{1/3}\tau(n)$ and $\ell_D \propto n^{-1/3}D(n)$.

For NbN thin films it was reported that 0.1 nm $< \ell_\tau <$ 0.2 nm and 0.1 nm $< \ell_D <$ 0.3 nm [16]. Since $\rho$ for NbN films is about 1 $\mu\Omega$m [16], it was calculated that $\ell_\rho$ is in the range of 0.5 $\sim$ 1.2 nm. For NbN films the two values ($\ell_\tau$ and $\ell_D$) are nearly equal. However the values of $\ell_\tau$ and $\ell_D$ for MoN films determined in this work are larger than those for NbN films. It is possible that MoN films are more suitable for superconducting devices rather than NbN films from a viewpoint of the mean free path. While $\ell_\rho$ for MoN films is smaller than that for NbN films. These discrepancy should be related with the values of $\tau$ and $\rho$ since $\tau$ calculated from $\rho$ and $n$ for No.8 film is $4.9 \times 10^{-18}$ s and is very smaller than $1.6 \times 10^{-16}$ s. Further investigation needs because the thickness $d$ and other parameters dependence of $\ell$ are not understood.

**Table 2.** The mean free path obtained by three methods $\ell_\rho$, $\ell_\tau$ and $\ell_D$ for MoN and NbN films [13].

| materila | $\ell_\rho$ (nm) | $\ell_\tau$ (nm) | $\ell_D$ (nm) |
|---|---|---|---|
| MoN (No. 8) | 0.023 | 0.76 | 2.5 |
| NbN | 0.5 $\sim$ 1.2 | 0.1 $\sim$ 0.2 | 0.1 $\sim$ 0.3 |

## 4. Summary

We have investigated that the $T_c$ depression and the SIT in MoN thin films. Thin films in the range of $3 < d < 60$ nm were fabricated by DC reactive magnetron sputtering method. It is found that $T_c$ vs. $R_{\rm sq}^{\rm N}$ data were fitted to the Finkel'stein formula using $T_{c0} = 6.45$ K and $\tau = 1.6 \times 10^{-16}$ s. The critical sheet resistance $R_c$ is also found about 2 k$\Omega$, which is smaller than $R_Q$. The mechanism of SIT for MoN films may be similar to that of NbN films. The mean free path $\ell$ for MoN films is $0.6 \sim 3$ nm, which is larger than that for NbN films. Further studies are needed about the thickness dependence of $\ell$ for MoN films.

**Acknowledgments**
This work was partly supported by a Grant-in-Aid for Scientific Research from the Ministry of Education, Culture, Sports, Science, and Technology of Japan (grant No. 17K06390).

**References**

[1] Rosfjord K M, Yang J K W, Dauler E A, Kerman A J, Anant V, Boronov B M, Gol'tsman G N and Berggren K K 2006 *Opt. Express* **14** 527
[2] Zhang L, Zhao Q, Zhong Y, Chen J, Cao C, Xu W, Kang L, Wu P and Shi W 2009 *Appl. Phys.* B **97** 187
[3] Cabrera B, Clarke R M, Colling P, Miller A J, Nam S and Romani R W 1998 *Appl. Phys. Lett.* **73** 735
[4] Irwin K D, Hilton G C, Wollman D A and Martinis J M 1996 *Appl. Phys. Lett.* **69** 1945
[5] Haviland D B, Lui Y and Goldman A M 1989 *Phys. Rev. Lett.* **62** 2180
[6] Makise K, Kawaguti T and Shinozaki B 2007 *Physica* E **39** 30
[7] Hebard A F and Paalanen M A 1984 *Phys. Rev.* B **30** 4063
[8] Finkel'stein A M 1987 *JETP Lett.* **45** 46
[9] Ezaki S, Makise K, Shinozaki B, Odo T, Asano T, Terai H, Yamashita T, Miki S and Wang Z 2012 *J. Phys.:Condens. Matter.* **24** 475702
[10] Noat Y, Chekez V, Brun C, Cren T, Cabillet C, Debontridder F, Ilin K, Siegel M, Semenov A, H'ubers H-M and Roditchev D 2013 *Phys. Rev.* B **88** 014503
[11] Graybeal J M and Beasley M R 1984 *Phys. Rev.* B **29** 4167
[12] Wang Z, Kawakami A, Uzawa Y and Komiyama B 1996 *J. Appl. Phys.* **79** 7837
[13] Tsuneoka T, Makise K, Maeda S, Shinozaki B and Ichikawa F 2017 *J. Phys.: Condens. Matter* **29** 015701
[14] Inumaru K, Baba K and Yamanaka S 2006 *Phys. Rev.* B **73** 052504
[15] Maekawa S and Fukuyama H 1982 *J. Phys. Soc. Jpn.* **51** 1380
[16] Ezaki S, Makise K, Shinozaki B, Odo T, Asano T, Terai H, Yamashita T, Miki S and Wang Z 2012 *J. Phys.:Condens. Matter* **24** 475702

IOP Conf. Series: Journal of Physics: Conf. Series **969** (2018) 012065 doi:10.1088/1742-6596/969/1/012065

# Enhancement of macroscopic quantum tunneling in the higher-order phase switches of Bi2212 intrinsic Josephson junctions

**Haruhisa Kitano, Ayami Yamaguchi, Yusaku Takahashi, Shunpei Umegai, Yuji Watabe, Haruka Ohnuma, Kazutaka Hosaka, and Daiki Kakehi**

Department of Physics and Mathematics, Aoyama Gakuin University, Fuchinobe 5-10-1, Chuo-ku, Sagamihara, Kanagawa 252-5258, Japan

hkitano@phys.aoyama.ac.jp

**Abstract**. The macroscopic quantum tunneling (MQT) in the current-biased intrinsic Josephson junctions (IJJs) of high-$T_c$ cuprates has attracted much attention for decades. Although the MQT for the phase switches from the zero to the first voltage state (1st SW) in the multiple-branched *I-V* curves is well explained by the conventional theory, the occurrence of MQT for the higher order switches such as the switch from the 1st to 2nd voltage state (2nd SW) has been still debated. Here, we present an experimental study on the phase switches of small IJJs fabricated from underdoped $Bi_2Sr_2(Ca,Y)Cu_2O_y$. We observed the single photon transition between quantized energy levels in the 3rd phase switches at 59.15 GHz and 2 K. The comparison with the previous studies on the nearly optimal-doped $Bi_2Sr_2CaCu_2O_y$ clearly suggests a possibility that the MQT rate for the higher-order phase switches is commonly enhanced by the effective suppression of the energy barrier for the higher-order phase escape due to the phase-running state after the 1st SW, in spite of the large difference in a critical current density and $T_c$.

## 1. Introduction

Macroscopic quantum tunneling (MQT) in a superconducting Josephson junction (JJ) is one of interesting phenomena to study the quantum-mechanical behavior of a macroscopic degree of freedom [1]. MQT occurs in the stochastic switch from the zero voltage state to the finite voltage state in the current-biased JJ below a crossover temperature $T_{cr}$ [2] and is closely related with the energy level quantization (ELQ) in a potential well to describe the dynamics of the superconducting phase difference across the junction [3]. In early studies on MQT and ELQ [2, 3] providing the fundamental of superconducting qubits in quantum information technology of the moment, the single tunnel junctions, which were artificially fabricated on conventional superconducting films, have often been investigated.

On the other hand, the intrinsic Josephson junctions (IJJs) of high-$T_c$ cuprate superconductors provide another type of JJs that are naturally stacked in the layered crystal structure and mutually connected in an atomic scale [4]. Compared with the artificial JJs fabricated with integrated circuit technology, IJJs have great advantage such as a high critical current density $j_c$, an almost perfectly smooth tunnel junction, and the negligible influence of dissipative quasiparticles lying in the node of *d*-wave superconducting gap to the MQT rate [5]. These remarkable properties benefit an increase of

$T_{cr}$ to the quantum phase escape, as was successfully demonstrated in a small stack of IJJs made of $Bi_2Sr_2CaCu_2O_y$ (Bi2212) [6].

However, in contrast to the conventional JJs, the dynamics of the phase difference across each junction in IJJs is expected to be more complex, as suggested by the fact that the thickness of the superconducting layer and the distance between junctions in IJJs are much smaller than those of artificially-fabricated JJs. Actually, this fact gives a good explanation about the multiple-branched current-voltage (*I*-*V*) characteristics observed for stacked IJJs [7]. It seems that the small thickness of superconducting layers enhances the fluctuation of phase differences and that the small distance between JJs induces the strong correlation between the phase differences in IJJs. Thus, it is quite interesting to study MQT and ELQ in IJJs in the sense that the phase-running state in a phase-switched junction can affect the phase switching dynamics in another junction near by the junction. Unfortunately, in the previous studies on MQT in IJJs [6,8], only the quantum phase escape from the zero to the first voltage state (1st SW) was well established by the conventional theory, while the occurrence of MQT in the higher-order phase switches, such as the switch from the first to the second voltage state (2nd SW) or that from the second to the third state (3rd SW), has still been unresolved [9-12]. Note that the conventional theory [1] was based on single JJs and has not considered the strong correlation between junctions. The only exception is the recent theories discussing about the strong electromagnetic coupling between JJs [13,14], which is considered to play an important role in the uniform switches where several junctions in IJJs are simultaneously switched from the zero voltage state [15].

Thus, we focus on the dynamics of the higher-order phase switches in IJJs. Recently, we reported a study on the microwave-induced resonance in the switching current distribution $P(I)$ for the 2nd SW, surviving up to roughly ~10 K [16]. This strongly suggests that the macroscopic phase difference across the 2nd switched junction shows a quantum-mechanical behavior below ~10 K. This temperature is also close to a temperature $T^*$, below which the MQT-like behavior for the 2nd SW appears, although it is much higher than both values of $T_{cr}$ reported for the 1st SW [6,8] and predicted for the 2nd SW using the conventional theory. If a rapidly-oscillating supercurrent generating in the phase-running state after the 1st SW practically influences the phase dynamics in the 2nd switched junction through the strong correlation between junctions [17], the presence of the oscillating current is reminiscent of the strong irradiation of microwaves, which can enhance MQT rate through an effective suppression of the potential barrier for the phase escape [18]. This scenario seems to explain consistently our experimental results.

In this paper, we present a study on the higher-order phase switches in another stack of IJJs, fabricated from the underdoped $Bi_2Sr_2Ca_{0.7}Y_{0.3}Cu_2O_y$ (Y-doped Bi2212) with $T_c$ ~ 47 K, and compare it with the previous studies on the nearly optimal-doped Bi2212 with $T_c$ ~ 83-85 K [12,16]. We confirm that the MQT-like behavior and the microwave-induced resonance in $P(I)$ for the higher-order phase switches are commonly observed, independent of the carrier doping concentration. We also succeeded in detecting the single photon transition between quantized energy levels for the 3rd SW, which could be realized by the strong suppression of $j_c$ in Y-doped Bi2212. The comparison of $T^*$ for the 3rd to 5th SW in Y-doped Bi2212 and that for the 2nd to the 4th SW in the nearly optimal-doped Bi2212 clearly suggests that $T^*$ is not determined by $j_c$, in contrast to the conventional theory, rather it is roughly scaled with $T_c$. These results prove that the observed MQT-like behavior for the higher-order phase switches is not influenced by noise in our experimental setup and that it should be explained by a novel mechanism that has not been considered in the conventional theory.

## 2. Experiments and analyses

Y-doped Bi2212 single crystals were grown by the floating zone method and annealed in the same condition as the nearly optimal-doped Bi2212. We fabricated a small bridge-type IJJ device with a lateral size of 2.3×1.9 μm$^2$ by using the focused ion beam (FIB) milling techniques [12], as shown in the inset of Fig. 1(a). The switching current distribution $P(I)$ for the 3rd to the 5th SW was measured from 2 K to 24 K by applying a bias current with a constant rate $dI/dt$=5 mA/s under 5,000 or 10,000 switching events. The measured $P(I)$ and the switching rate $\Gamma(I)$ were analyzed by considering both the thermally-activated (TA) phase escape and the subsequent phase retrapping (PR) processes [17]. Other details on our experimental setup and analyses are described in our previous papers [10,12,16,17,19].

### 3. Results and discussion

Figure 1(a) shows the typical *I-V* characteristics of the bridge-type IJJs measured at 3 K, suggesting that the switching currents $I_{SW}$ for the 1st and the 2nd SWs are strongly suppressed, compared to those for other SWs higher than the 3rd SW. This is presumably due to the damage given by FIB milling or the disorder derived by Y substitution for Ca. The recent transmission electron microscope (TEM) study reports that the thickness of the FIB damage in stacked IJJs made of Bi2212 can be reduced to about 30 nm by using $Ga^+$ ion beam emitted at 50 pA and 30 kV [20]. Considering the fact that the bridge-type IJJs made of Bi2212 where the Ca site is not substituted by Y show much higher $I_{SW}$ for the 1st and 2nd SWs [6,10,12], the observed strong suppression of $I_{SW}$ for the 1st and 2nd SWs in Y-doped Bi2212 is probably attributed to strong disorders in the superconducting $CuO_2$ bilayers where the $Ca^{2+}$ ions sandwiched by two $CuO_2$ planes is partly replaced by $Y^{3+}$ ions.

On the other hand, the resistive branches higher than the 2nd branch are almost equally spaced and have relatively large switching currents, suggesting that the intrinsic junctions other than the weakest and next weakest junctions among IJJs are not degraded by Y substitution. Thus, $P(I)$ and $\Gamma(I)$ for the 3rd to the 5th SW were measured at several temperatures up to 24 K. Figures 1(b) and 1(c) show $P(I)$ and $\Gamma(I)$ for the 4th SW, respectively. They were successfully fitted to numerical results by a single junction model considering both the TA escape and the multiple PR process, as shown by solid lines in Figs. 1(b) and 1(c). We confirmed that the influence of PR events after a TA escape, which produced a negative curvature in logarithmic plots of $\Gamma(I)$, was more prominent with increasing the order of phase

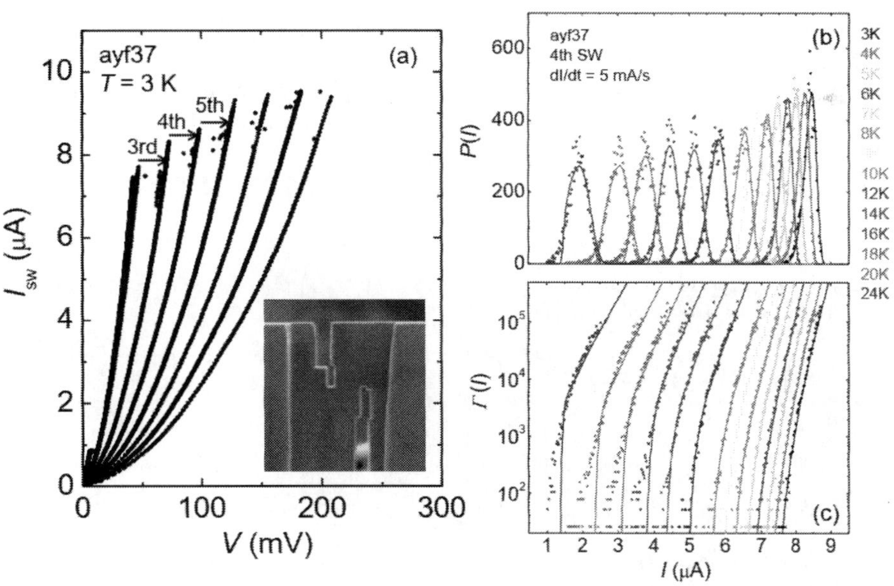


**Figure 1.** (a) *I-V* characteristic of the bridge-type Y-doped Bi2212-IJJs measured at 3 K. The inset shows an image of scanning ion microscope of the IJJ device. (b) (Color online) $P(I)$ for the 4th SW as a function of bias current. Solid circles and solid lines are experimental results and numerical calculations, respectively. (c) (Color online) $\Gamma(I)$ for the 4th SW as a function of bias current. Solid circles and solid lines are the same as (b).

switches from the 3rd SW to 5th SW. This trend is also confirmed by our previous study on the higher order phase switches for Bi2212-IJJs [17].

The microwave-induced resonance in *P*(*I*) for the 3rd SW was observed at 59.15 GHz and 2 K, as shown in Fig. 2(a). Note that there is almost no change in *P*(*I*) located at near 7 μA, corresponding to the primary peak of *P*(*I*), with increasing the irradiation power of microwaves up to a power at which the resonant double-peak structure appears in *P*(*I*). This behavior shows a sharp contrast to the recent results on the resonant phase escape due to the three-photon absorption for the 2nd SW in Bi2212-IJJs [16]. We estimated the energy-level separation between the first excited state and the ground state in a potential well corresponding to the 3rd SW, by using fitting parameters to *P*(*I*) and Γ(*I*) under the microwaves and a calculation approach proposed by Kopietz and Chakravarty [21]. The result supported that the observed resonance agreed with the single-photon absorption. For IJJs made of the nearly optimal-doped Bi2212, it has been quite difficult to observe the single-photon transition between quantized energy levels, because of the Josephson plasma frequency $\omega_p/2\pi$ higher than 100 GHz [10,15,16]. However, for the IJJs made of Y-doped Bi2212 with $T_c \sim 47$ K, we obtained the strongly suppressed $j_c$ (~ 200 A/cm$^2$), which was about an order of magnitude smaller than that for the nearly optimal-doped Bi2212. This leads to the first observation of single-photon transition in the higher-order phase switches.

Figure 2(b) shows the microwave power dependence of $I_{SW}$ giving maxima of *P*(*I*) at the same temperature and microwave frequency as those in Fig. 2(a). A solid line represents a numerical fit to the quantum-mechanical model given by the following equation [16],

$$\frac{I_{SW}}{I_{c0}} = 1 - \frac{4e^4}{C^2\hbar^2}\frac{(P_{mw}/k)}{\omega_p{}^2}\frac{1}{(\omega_p-\omega)^2+(\omega/2Q)^2} \qquad (1)$$

Here, $I_{c0}$ is a fluctuation-free critical current, $C$ is a capacitance for single-junction in IJJs, $P_{mw}$ is the applied power of microwaves, $k$ is a coupling coefficient which combines $P_{mw}$ with the ac current induced by microwaves, $Q$ is a quality factor of the resonance. We found that the double-peak structure in *P*(*I*) is successfully explained by Eq. (1) with $Q \sim 12$, as shown in Fig. 2(b). Note that this value of $Q$ is about twice larger than that for the three-photon absorption for the 2nd SW in the previous study [16], suggesting that the dissipation in the single-photon absorption is smaller than that in multiphoton absorption.

We also made sure that the double peak structure in *P*(*I*) for the 3rd SW was clearly observed up to 6 K and diminished at a temperature between 10 K and 16 K.

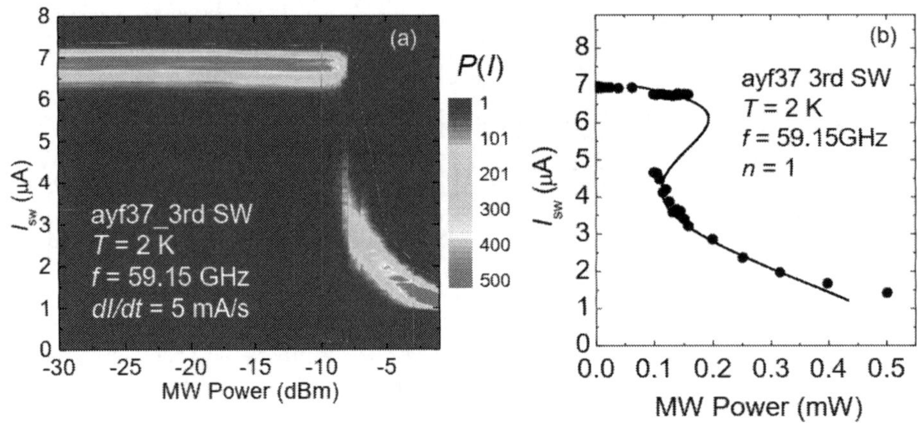


**Figure 2.** (a) (Color online) Density plots of the switching current versus the applied microwave power for the 3rd SW measured at 59.15 GHz and 2 K. (b) Microwave power dependence of the switching current giving maxima of *P*(*I*).

Finally, we estimated the effective escape temperature $T_{esc}$ for the 3rd to the 5th SW in the IJJs of Y-doped Bi2212 ($T_c$ ~ 47 K) at measured temperatures, by using the fitting results of $P(I)$ and $\Gamma(I)$ to numerical calculations based on both of TA escape and multiple PR process. We compared the plots of $T_{esc}$ versus the bath temperature with those reported in the previous study [12] on other IJJs made of Bi2212 ($T_c$ ~ 83 K), as shown in Fig. 3. Despite the fact that both samples have quite different $T_c$ and $j_c$, both plots shown in Fig. 3 indicate a common feature that the MQT-like behavior appears below $T^*$, corresponding to 0.1-0.2 times $T_c$. This proves that the observed MQT-like behavior is not limited by extrinsic noise in our experimental environment. Furthermore, the plots of $T_{esc}$ for the IJJs of Y-doped Bi2212 more clearly indicate a crossover from the classical to quantum phase escape. Note that the value of $T^*$(~ 6 K) for the 3rd to 5th SWs in Y-doped Bi2212 are very close to a temperature where the microwave-induced resonance in $P(I)$ disappears. Thus, compared to our previous studies [12,16], the present results on the IJJs made of Y-doped Bi2212 more strongly suggest that MQT and ELQ for the higher-order phase switches in IJJs can occur and survive up to $T^*$, which is much higher than $T_{cr}$ estimated by the conventional theory. Interestingly, the experimental values of $T^*$ seem to be roughly scaled with $T_c$, although the conventional theory predicts that $T_{cr}$ is determined by the current-dependent Josephson plasma frequency proportional to $\sqrt{j_c}$. On the other hand, the energy level spacing in ELQ is successfully explained by the conventional theory, suggesting that the parabolic approximation of the potential well is still useful to describe the transition between quantized levels. Therefore, our results suggest that the bottom region in the potential well is not essentially changed even for the higher-order phase switches while the potential barrier for the phase escape is strongly suppressed by some novel mechanism which becomes important for the higher-order phase switches.

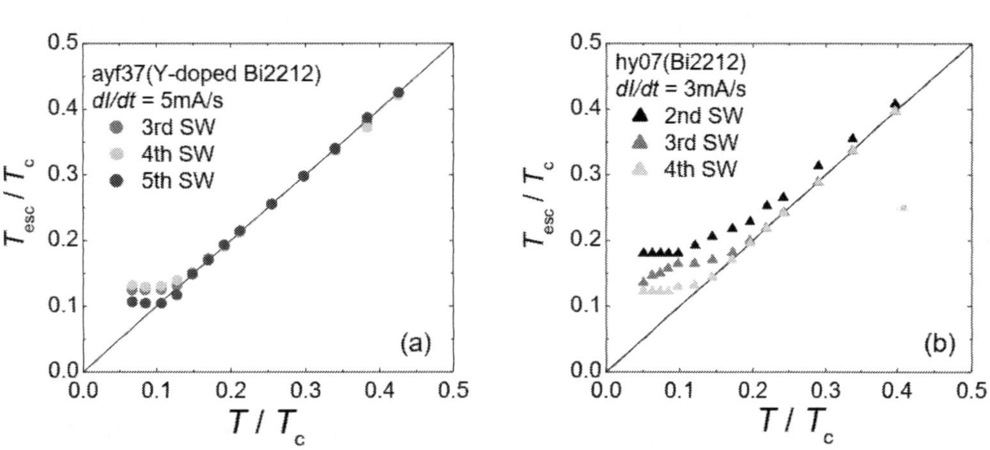


**Figure 3.** (a) (Color online) Plots of the effective escape temperature versus the bath temperature for the 3rd to 5th SW in IJJs made of Y-doped Bi2212. Both temperatures are normalized by $T_c$. A solid straight line represents that the effective escape temperature is equal to the bath temperature. (b) (Color online) Similar plots for the 2nd to 4th SW in IJJs made of nearly optimal-doped Bi2212.

## 4. Conclusion

We studied $P(I)$ and $\Gamma(I)$ for the higher-order phase switches in the IJJs made of underdoped $Bi_2Sr_2Ca_{0.7}Y_{0.3}Cu_2O_y$ and compared the experimental results with those measured previously in IJJs made of nearly optimal-doped Bi2212 [12,16,17,19]. We succeeded in observing the double-peak structure in $P(I)$ for the 3rd SW under the microwave irradiation, which was attributed to the single-photon transition between quantized energy levels. In spite of large differences in $T_c$ and $j_c$ between both the underdoped and optimal-doped IJJs, the distinctive properties suggesting the occurrence of MQT and ELQ in the higher-order phase switches were commonly observed below $T^*$. This confirms that our experimental results are not limited by extrinsic noise. Our results clearly show that the observed ELQ associated with MQT is successfully explained by the conventional theory while $T^*$ is much higher than $T_{cr}$ estimated by the conventional theory. This strongly suggests that the reason for disagreement between $T^*$ and $T_{cr}$ probably originates in the large enhancement of MQT rate due to the strong suppression of the energy barrier for the phase escape. Since many of the previous studies concluded that MQT and ELQ for the 1st SW were consistently explained by the conventional theory [6,8,9,10,15], the phase-running state after the 1st SW is considered to influence the suppression of the energy barrier. This situation is reminiscent of the quantum phase escape enhanced by the strong microwave irradiation proposed by Fistul *et al* [18], suggesting a possibility that a rapidly-oscillating current generating in the phase-running state suppresses the energy barrier for the higher-order phase switches.

**Acknowledgements**

We thank I. Kakeya, Y. Ota, T. Koyama, and M. Machida for fruitful discussion and valuable comments. This work was partly supported by Grants-in-Aid for Scientific Research from JSPS and MEXT of Japan (Grant Nos. 21560020, 24560017) and MEXT-Supported Program for the Strategic Research Foundation at Private Universities (2013-2017).

**References**

[1] Caldeira A O and Leggett A J 1981 *Phys. Rev. Lett.* **46** 211
[2] Voss R F and Webb R A 1981 *Phys. Rev. Lett.* **47** 265
[3] Martinis J, Devoret M and Clarke J 1985 *Phys. Rev. Lett.* **55** 1543
[4] Kleiner R, Müller P, Kohlstedt H, Pedersen N F, Sakai S, 1994 *Phys. Rev.* B **50** 3942
[5] Kawabata S, Kashiwaya S, Asano Y and Tanaka Y 2004 *Phys. Rev.* B **70** 132505
[6] Inomata K, Sato S, Nakajima K, Tanaka A, Takano Y, Wang H B, Nagao M, Hatano H and Kawabata S 2005 *Phys. Rev. Lett.* **95** 107005
[7] Machida M, Koyama T and Tachiki M 1999 *Phys. Rev. Lett.* **83** 4618
[8] Li S-X, Qiu W, Han S, Wei Y F, Zhu X B, Gu C Z, Zhao S P, Wang H B 2007 *Phys. Rev. Lett.* **99** 037002
[9] Kashiwaya H, Matsumoto T, Shibata H, Kashiwaya S, Eisaki H, Yoshida Y, Kawabata S and Tanaka Y, 2008 *J. Phys. Soc. Jpn.* **77** 104708
[10] Ota K, Hamada K, Takemura R, Ohmaki M, Machi T, Tanabe K, Suzuki M, Maeda A and Kitano H 2009 *Phys. Rev.* B **79** 134505
[11] Nomura Y, Mizuno T, Kambara H, Nakagawa Y and Kakeya I 2015 *J. Phys. Soc. Jpn.* **84** 013704
[12] Kakehi D, Takahashi Y, Yamaguchi H, Koizumi S, Ayukawa S and Kitano H 2016 *IEEE Trans. Appl. Supercond.* **26** (3) 1800204
[13] Sevel'ev S, Rakhmanov A L and Nori F 2007 *Phys. Rev. Lett.* **98** 077002; 2007 *Erratum*: **98** 269901
[14] Koyama T and Machida M 2008 *Physica C* **468** 695
[15] Jin X Y, Koval Y, Lukashenko A, Ustinov A V, and Müller P 2006 *Phys. Rev. Lett.* **96** 177003
[16] Takahashi Y, Kakehi D, Takekoshi S, Ishikawa K, Ayukawa S and Kitano H 2016 *J. Phys. Soc. Jpn.* **85** 073702
[17] Kitano H, Takahashi Y, Kakehi D, Yamaguchi H, Koizumi S and Ayukawa S 2016 *J. Phys. Soc. Jpn.* **85** 054703
[18] Fistul M V, Wallraff A and Ustinov A V 2003 *Phys. Rev.* B **68** 060504
[19] Kitano H, Yamaguchi A, Takahashi Y, Kakehi D and Ayukawa S *J. Phys: Conference Series (inpress)*
[20] Kakizaki Y, Koyama J, Yamaguchi A, Umegai S, Ayukawa S and Kitano H 2017 *Jpn. J. Appl. Phys.* **56** 043101
[21] Kopietz P and Chakravarty S 1988 *Phys. Rev.* B **38** 97

# Anomalous dispersion relations in the staggered flux state

**Hiroki Tsuchiura[1,2], Masao Ogata[3], Kunito Yamazaki[1], and Rui Asaoka[1,2]**

[1]Department of Applied Physics, Tohoku University, 6-6-05 Aoba, Sendai 980-8579, Japan
[2]Center for Spintronics Research Network, Tohoku University, Sendai 980-8577, Japan
[3]Department of Physics, University of Tokyo, Bunkyo, Tokyo 113-0033, Japan

E-mail: tsuchi@solid.apph.tohoku.ac.jp

**Abstract.** We study the quasiparticle properties in the d-wave superconductivity, antiferromagnetism, and the staggered flux state within a renormalized mean-field theory based on the two-dimensional $t$-$J$ model. In particular, we focus on the anomalous quasiparticle dispersion relations around the antinodal region of the Brillouin zone argued in Hashimoto *et al.*, Nature Phys. **6**, 414 (2010). We obtain the qualitatively consistent results with their observations when we take account of the SF state. The present analysis shows that the SF order can be a possible candidate of symmetry-breaking pseudogap states coexisting with the dSC.

## 1. Introduction

Recent experimental studies have indicated that the pseudogap phase in underdoped high-Tc cuprates is accompanied by some phase transitions with the breaking of time-reversal [1, 2], rotational [3], or translational symmetry. These symmetry breakings imply that there can be nematic order, charge order, or other instabilities in the pseudogap phase[4, 5, 6, 7, 8]. It has also been suggested that the pseudogap order and d-wave superconductivity (dSC) can coexist within the superconducting dome of the cuprates at $p < 0.19$, where p is the hole concentration.

One of the intriguing experimental evidences of the coexisting pseudogap phase with dSC is a gap-structure in the AREPES data opening with broken particle-hole symmetry found in Bi-2201 [9, 10]. They examined the quasi-particle dispersions along the Brillouin zone boundary, and claimed that the gap opens below the $T_c$ is distinct from dSC based on a phenomenological model calculation, and also that the gap can be understood by assuming the coexisting antiferromagnetic (AF) order with dSC. Soon after the that, Greco and Bejas [11] have carried out theoretical calculations of the quasiparticle properties in the pseudogap state incorporating the effects of $d$ charge-density-wave fluctuation, or the so-called staggered flux fluctuation) based on the $t$-$J$ model. They successfully described the anomalous dispersion relations along the zone-boundary without taking account of AF order or fluctuation.

Motivated by these backgrounds, we here study the quasiparticle properties in dSC, AF, and also the staggered flux (SF) states within the same theoretical framework. The SF state is a normal, or non-superconducting state with broken symmetries, and has been studied by many groups since the early stage of research on cuprate superconductors [13, 14, 15, 16, 17, 18]. It was considered firstly as a candidate of the ground state for the $CuO_2$ plain, then has been studied as a candidate for a non-superconducting state that causes the pseudogap found in underdoped cuprates. In this paper, we study the quasiparticle dispersions in the SF and dSC states by using the $t$-$J$ model within the Bogoliubov-de Gennes theory based on an extended version of the Gutzwiller approximation. We will see the SF state exhibits the qualitatively consistent dispersion relations with those observed in the ARPES experiments [9, 10].

## 2. model

In this study, we employ an advanced version of the renormalized mean-field theory for the $t$-$J$ model, the Hamiltonian of which is given as

$$\mathcal{H} = -\sum_{\langle i,j\rangle,\sigma} P_G(t_{ij}c_{i\sigma}^{\dagger}c_{j\sigma} + \text{h.c.})P_G + J\sum_{\langle i,j\rangle} \boldsymbol{S_i}\cdot\boldsymbol{S_j} - \mu\sum_{i,\sigma} c_{i\sigma}^{\dagger}c_{i\sigma} \tag{1}$$

in the standard notation where $\langle i,j\rangle$ means the summation over nearest-neighbor pairs. The Gutzwiller's projection operator $P_G$ is defined as $P_G = \Pi_i(1-n_{i\uparrow}n_{i\downarrow})$. The Bogoliubov-de Gennes (BdG) equation for a renormalized mean-field Hamiltonian based on an extended Gutzwiller approximation [19, 20, 21] is given as

$$\begin{pmatrix} H_{ij}^{\uparrow} & F_{ij} \\ F_{ji}^{*} & -H_{ji}^{\downarrow} \end{pmatrix} \begin{pmatrix} u_j^{\alpha} \\ v_j^{\alpha} \end{pmatrix} = E^{\alpha} \begin{pmatrix} u_i^{\alpha} \\ v_i^{\alpha} \end{pmatrix}, \tag{2}$$

with

$$\begin{aligned} H_{ij}^{\sigma} &= -\sum_{\tau}\left(t_{ij}^{\text{eff}} + J_{ij}^{\text{eff}}\tilde{\chi}_{ji}\right)\delta_{j,i+\tau} - \sum_{\nu} t_{ij}^{\prime\text{eff}}\delta_{j,i+\nu} + \sigma\delta_{ij}\sum_{\tau} h_{i,i+\tau}^{\text{eff}} - \mu\delta_{ij} \\ F_{ij}^{*} &= -\sum_{\tau} J_{ij}^{\text{eff}}\tilde{\Delta}_{ij}\delta_{j,i+\tau}, \end{aligned} \tag{3}$$

where $\sigma = \pm 1$, $i+\tau$ represents the nearest neighbor sites of the site $i$, while $i+\nu$ the 2nd neighbor sites.

The renormalized parameters $t_{ij}^{\text{eff}}$, $t_{ij}^{\prime\text{eff}}$, $J_{ij}^{\text{eff}}$ and $h_{ij}^{\text{eff}}$ have somewhat complicated forms depending on the local expectation values $\Delta_{ij} = \frac{1}{2}\langle c_{i\uparrow}^{\dagger}c_{j\downarrow}^{\dagger} - c_{i\downarrow}^{\dagger}c_{j\uparrow}^{\dagger}\rangle$, $\chi_{ij} = \langle c_{i\sigma}^{\dagger}c_{j\sigma}\rangle$, and $m_i = \frac{1}{2}\langle n_{i\uparrow} - n_{i\downarrow}\rangle$ [19, 20, 21]. These parameters are determined from cluster calculations to reproduce the variational Monte Carlo results [19]. Thus the present renormalized mean-field theory can give more reliable results than theories based on the conventional slave boson or Gutzwiller approximation [22]. In particular, the overestimation of AF commonly seen in the conventional mean-field theories is improved in the present theoretical treatment. Since the detailed derivation of these parameters are given in ref. [19], here we only show in Table 1 the self-consistently obtained numerical values of them for the $t$-$J$ model with $J/t = 0.3$, $t'/t = -0.3$, and the hole doping $\delta = 0.09$. In the present calculation, we regard the 2×2 square lattice as a unit cell to solve the BdG equation numerically.

## 3. Results and discussion

Let us look at firstly the quasiparticle states of dSC and SF states. Figure 1 shows the density of states (DOS) per each site for dSC and SF states. Here the DOS is given as

$$N_j(E) = \frac{1}{N_c}\sum_{\boldsymbol{k},\alpha}\left[\left|u_j^{\alpha}(\boldsymbol{k})\right|^2\delta(E^{\alpha}(\boldsymbol{k}) - E) + \left|v_j^{\alpha}(\boldsymbol{k})\right|^2\delta(E^{\alpha}(\boldsymbol{k}) + E)\right] \tag{4}$$

**Table 1.** Renormalized model parameters for the $t$-$J$ model with $J/t = 0.3$, $t'/t = -0.3$, and $\delta = 0.09$. Here we omit the site indices $i$ and $j$ because we are studying homogeneous electronic states.

| state | $t^{\rm eff}$ | $t'_{\rm eff}$ | $J^{\rm eff}\tilde{\Delta}$ | $J^{\rm eff}\tilde{\chi}$ | $h^{\rm eff}$ | $\mu$ |
|---|---|---|---|---|---|---|
| normal state | 0.1445 | -0.0495 | 0 | 0.0805 | 0 | -0.161 |
| dSC | 0.1403 | -0.0494 | 0.032 | 0.0642 | 0 | -0.143 |
| SF | 0.1416 | -0.0497 | 0 | 0.067±0.0332$i$ | 0 | -0.165 |
| dSC+AF | 0.1384 | -0.0491 | 0.030 | 0.062 | 0.0285 | -0.144 |

where $j$ represents a site where the local density of states is calculated, $\alpha$ is the eigenstate index, and $\boldsymbol{k}$ is the Bloch wave number of the present unit cells. The blue line in Fig. 1 represents the DOS for dSC, and the red one for SF states. The form of the FS implies that the DOS at zero-energy is nonzero, in contrast to the results for dSC, as previously found by Morr [16]. In particular, we can see that the DOS for the SF state is shifted downwards in energy, in comparison to the case of dSC. Thus we can expect the quasiparticle excitation energy for the SF is larger than that of dSC around the antinodal region of the Brillouin zone.

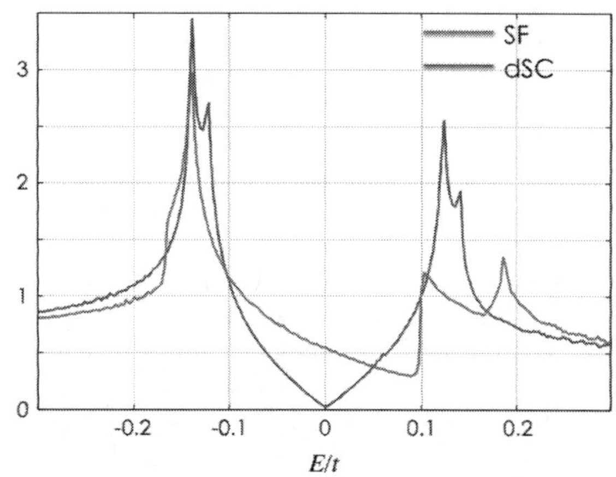


**Figure 1.** The density of states (DOS) per each site for dSC (blue) and SF (red) states.

Next, we show the antinodal quasiparticle dispersions $E_{\rm dSC}(\boldsymbol{k})$ for dSC (blue) and $E_{\rm SF}(\boldsymbol{k})$ for SF (red) [23] along the Brillouin zone boundary $k_x = \pi$ in Fig. 2. We also plot the normal state dispersion relation (green) for comparison. As expected, the dSC state shows the superconducting energy gap $\Delta E_{\rm dSC} \sim 0.12$ around $k_y/\pi \sim \pm 0.08$ as indicated by blue arrows. This momenta with the gap minima is slightly different from the Fermi wave number in the normal state due to the chemical potential shift accompanying the superconducting phase transition, as shown in Table 1. The SF state exhibits the larger excitation energy gap $\Delta E_{\rm SF} \sim 0.14$ around $k_y/\pi \sim \pm 1.1$, which is markedly away from those for the dSC state, as indicated by red arrows in Fig. 2. Since the chemical potential in the SF state is almost the same as the normal state (Table 1), the marked difference of the momenta with the gap minima from the Fermi wave number is due to the broken particle-hole symmetry in the SF state. These results are consistent with the recent ARPES observations [9, 10] in which an anomalous dispersion relation has been found at $T < T_c$ in an optimally doped Bi-2201 cuprate superconductor. Thus the SF state can be a possible candidate of symmetry-breaking pseudogap states coexisting with the dSC. We note here that, in the present calculation, the SF state does not have a lower energy than the dSC, and also that the coexisting state of these two (dSC+SF) has not been found. We can expect, however, that the SF or the SF+dSC state may be stabilized when we consider finite temperature effects, or intersite Coulomb repulsion [17].

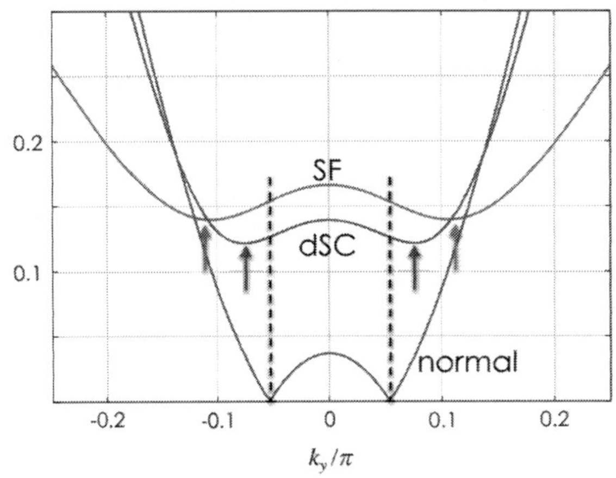


**Figure 2.** Quasiparticle dispersion relation for dSC (blue), SF (red), and the normal state (green) along the Brillouin zone boundary $k_x = \pi$.

Finally, let us mention the effects of antiferromagnetic (AF) order on the anomalous dispersion relation found in the zone boundary shown in Fig. 2. If the AF order is taken into account in our calculation, we find a small expectation value of $\langle S_j^z \rangle \sim \pm 0.05$ coexisting with the dSC state for the hole doping $\delta = 0.09$. Thus the antiferomagnetic gap is also very small, and we confirm that the DOS for this coexisting state is almost similar with that of the dSC, as shown in Fig. 3 (a). A small gap-like structure can be found around $E/t \sim 0.1$. In order to clarify the effect of this small gap structure in the DOS, we calculate the momentum distribution function $n(\boldsymbol{k})$ for the coexisting state (dSC+AF) and the dSC state. Figure 3 (b) shows the difference of the $n(\boldsymbol{k})$ between these two states in the full Brillouin zone. We can clearly see that the effects of the coexisting AF order is concentrated around the AF hot spots, and does not have a significant effect around the antinodal region of the Brillouin zone. Thus we speculate that the anomalous dispersion relation found in the ARPES experiments [9, 10] is not due to AF.

## 4. Summary

We have studied the quasiparticle dispersion relation around the antinodal region of the full Brillouin zone in the dSC and SF states within the renormalized mean-field theory based on the $t$-$J$ model. The present analysis has shown that the SF order can be a possible candidate of symmetry-breaking pseudogap states coexisting with the dSC.

## 5. Acknowledgments

This work was supported by JSPS KAKENHI Grant Number JP17K05528. The numerical computations were carried out at the Cyber-science Center, Tohoku University.

## References

[1] A. Kaminski, S. Rosenkranz, H. M. Fretwell, J. C. Campuzano, Z. Li, H. Raffy, W. G. Cullen, H. You, C. G. Olsonk, C. M. Varma, and H. Höchst, Nature **416**, 610 (2002).

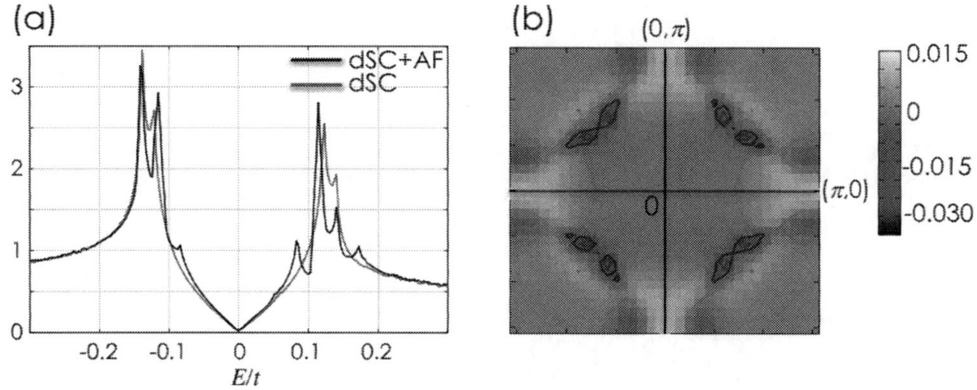


**Figure 3.** (a) The DOS for the dSC state coexisting with AF order. (b) The difference of the momentum distribution functions between the dSC and the coexisting state.

[2] Y. Li, G. Yu, M. K. Chan, V. Balédent, Y. Li, N. Barišić, X. Zhao, K. Kradil, R. A. Mole, Y. Sidis, P. Steffens, P. Bourges, and M. Greven, Nature Phys. **8**, 404 (2012).
[3] M. Fujita, H. Hiraka, M. Matsuda, M. Matsuura, J. M. Tranquada, S. Wakimoto, G. Xu, and K. Yamada, J. Phys. Soc. Jpn. bf 81, 011007 (2012).
[4] G. Ghiringhelli, M. Le Tacon, M. Minola, S. Blanco-Canosa, C. Mazzoli, N. B. Brookes, B. M. De Luca, A. Frano, D. G. Hawthorn, F. He, T. Loew, M. M. Sala, D. C. Peets, M. Salluzzo, E. Schierle, R. Sutarto, G. A. Sawatzky, E. Weschke, B. Keimer, and L. Braicovich, Science **337**, 821 (2012).
[5] K. Nakayama, T. Sato, T. Dobashi, K. Terashima, S. Souma, H. Matsui, T. Takahashi, J. C. Campuzano, K. Kudo, T. Sasaki, N. Kobayashi, T. Kondo, T. Takeuchi, K. Kadowaki, M. Kofu, and K. Hirota, Phys. Rev. B **74**, 054505 (2006).
[6] S. E. Sebastian, N. Harrison, E. Palm, T. P. Murphy, C. H. Mielke, R. Liang, D. A. Bonn, W. N. Hardy, and G. G. Lonzarich, Nature **454**, 200 (2008).
[7] E. Razzoli, Y. Sassa, G. Drachuck, M. Mansson, A. Keren, M. Shay, M. H. Berntsen, O. Tjernberg, M. Radovic, J. Chang, S. Pailhès, N. Momono, M. Oda, M. Ido, O. J. Lipscombe, S. M. Hayden, L. Patthey, J. Mesot, and M. Shi, New J. Phys. **12**, 125003 (2010).
[8] R.-H. He, X. J. Zhou, M. Hashimoto, T. Yoshida, K. Tanaka, S.-K. Mo, T. Sasagawa, N. Mannella, W. Meevasana, H. Yao, M. Fujita, T. Adachi, S. Komiya, S. Uchida, Y. Ando, F. Zhou, Z. X. Zhao, A. Fujimori, Y. Koike, K. Yamada, Z. Hussain, and Z.-X. Shen, New J. Phys. **13**, 013031 (2011).
[9] M. Hashimoto, R.-H. He, K. Tanaka, J.-P. Testaud, W. Meevasana, R. G. Moore, D. Lu, H. Yao, Y. Yoshida, H. Eisaki, T. P. Devereaux, Z. Hussain, and Z.-X. Shen, Nature Phys. **6**, 414 (2010).
[10] M. Hashimoto, I. M. Vishik, R.-H. He, T. P. Devereaux, and Z.-X. Shen, Nature Phys. **10**, 483 (2014).
[11] A. Greco and M. Bejas, Phys. Rev. B **83**, 212503 (2011).
[12] K. Fujita, A. R. Schmidt, E.-A. Kim, M. J. Lawler, D. H. Lee, J. C. Davis, H. Eisaki, and S. Uchida, J. Phys. Soc. Jpn. **81**, 011005 (2012).
[13] I. Affleck and J. B. Marston, Phys. Rev. B **37**, 3774 (1988).
[14] P. A. Lee and X. -G. Wen, Phys. Rev. B **63**, 224517 (2001).
[15] S. Chakravarty, R. B. Laughlin, D. K. Morr, and C. Nayak, Phys. Rev. B **63**, 094503 (2001).
[16] D. K. Morr, Phys. Rev. Lett. **89**, 106401 (2002).
[17] S. Zhou and Z. Wang, Phys. Rev. B **70**, 020501(R) (2004).
[18] H. Yokoyama, S. Tamura, and M. Ogata, J. Phys. Soc. Jpn. **85**, 124707 (2016).
[19] M. Ogata and A. Himeda, J. Phys. Soc. Jpn. **72**, 374 (2003).
[20] H. Tsuchiura, Y. Tanaka, M. Ogata, and S. Kashiwaya, Phys. Rev. B **64**, 140501(R) (2001).
[21] H. Tsuchiura, M. Ogata, Y. Tanaka, and S. Kashiwaya, Phys. Rev. B **68**, 012509 (2003).
[22] F. C. Zhang, C. Gros, T. M. Rice, and H. Shiba, Supercond. Sci. Technol. **1**, 36 (1988).
[23] A detailed derivation of the dispersion relation $E_{\rm SF}(\boldsymbol{k})$ is given, for example, in the Appendix A of [18].

# Composition and annealing effects on superconductivity in sintered and arc-melted $Fe_{1+\varepsilon}Te_{0.5}Se_{0.5}$

**M M Foreman, G Ponti, S Mozaffari and J T Markert**

*Department of Physics, The University of Texas at Austin, Austin, TX 78712 USA*

**Abstract.** We present the results of x-ray diffraction, electrical resistivity, and ac magnetic susceptibility measurements on specimens of the "11"-structure superconductor $Fe_{1+\varepsilon}Te_{0.50}Se_{0.50}$ ($0 \leq \varepsilon \leq 0.15$). Samples were initially either sintered in sealed quartz tubes or melted in a zirconium-gettered arc furnace. Sintered samples were fired two to three times at temperatures of 425°C, 600°C, or 675°C, while arc-melted samples were studied both as-melted and after annealing at 650°C. X-ray diffraction data show a predominant PbO-type tetragonal phase, with a secondary hexagonal NiAs-type phase; for sintered specimens annealed at 600°C, the secondary phase decreases as $\varepsilon$ increases over the range $0 \leq \varepsilon \leq 0.10$, with the composition $Fe_{1.10}Te_{0.5}Se_{0.5}$ exhibiting x-ray phase purity. A higher annealing temperature of 675°C provided such tetragonal phase purity at the composition $Fe_{1.05}Te_{0.5}Se_{0.5}$. The resistive superconducting transition temperature $T_c$ was nearly independent of the iron concentration $1+\varepsilon$, suggesting a single superconducting phase, while the magnetic screening fraction varied greatly with concentration and conditions, peaking at $\varepsilon = 0.07$, indicating that the amount of superconducting phase is strongly dependent on conditions. We propose that the behaviour can also be viewed in terms of an electron-doped, chalcogen-deficient stoichiometry.

## 1. Introduction

After the discovery of superconductivity in the simplest iron-based superconducting "11" structure, FeSe with $T_c \approx 8$ K [1], intense research on this and similar materials revealed a wealth of discoveries: the increase of $T_c$ in single-layer films to 65 K [2]; extremely large increases of $T_c$ with pressure [3]; and, recently, the substitution of other chalcogens for Se, namely S and Te [4,5], with an increase of the bulk $T_c$ of from 8 K in FeSe to 14 K in $FeTe_{0.5}Se_{0.5}$. Bulk FeTe is antiferromagnetic; the isoelectronic (no anion doping) Te-for-Se substitution results is the spectrum of behaviour from antiferromagnet to superconductor, presumably by shifts in electronic structure; a thorough review of electronic structure, transport, tunnelling, pressure dependence, and nuclear magnetic resonance recently appeared [6]. Detailed chalcogen substitutions in the Fe(Te,Se,S) system have revealed, for example, that an optimal $T_c$ occurs for Fe(Te,Se) near the midway composition $FeTe_{0.5}Se_{0.5}$.

Early studies of antiferromagnetic FeTe [7] indicate that the "FeTe" structure is actually two structures, with a PbO-type tetragonal phase stable for iron-rich stoichiometries near $Fe_{1.1}Te$ while an NiAs hexagonal structure is more stable near or just below $Fe_{1.0}Te$; annealing-temperature and quenching conditions also played a role in the composition dependence of the phases. More recent studies of the Fe(Te,Se) system [4,5] show a similar complex behaviour: some studies have concluded that such "excess" Fe can promote the purity of the tetragonal phase, while hindering bulk superconductivity [8]; others that oxygen annealing may help [9]. The recent studies suggest that phase evolution is most sensitive to the temperature of anneals and to other details affecting the diffusion of ions, but all suggest a complex interplay of phase purity, stability, and superconductivity. In many superconducting systems, a nearby instability plays an important role.

In this study, we examined the dependence of the structural and superconducting properties of the "11"-structure materials as a function of iron composition, for a series of specimens of composition $Fe_{1+\varepsilon}Te_{0.50}Se_{0.50}$ ($0 \le \varepsilon \le 0.15$). We also examine the route by which this achieved, including sintering in sealed quartz ampoules, annealing at higher temperatures, arc-melting, and arc-melting with post-annealing. We find that while the resistive $T_c \approx 14$ K varies only weakly across these compositions, indicating no appreciable electronic or doping changes in the superconducting phase, we do find strong variations in the magnetic superconducting screening fraction; thus, the *amount* of superconductivity varies noticeably, with volume fractions greater than 50% observed only for very particular conditions.

## 2. Experimental

For sintered samples, high-purity Fe (99.9%), Te (99.9999%), and FeSe (99.9%) (Alfa-Aesar) in stoichiometric ratios were weighed and thoroughly ground in an argon-atmosphere glove box using an agate mortar and pestle, and then pressed into pellets. First, a "survey" series of $Fe_{1+\varepsilon}Te_{0.50}Se_{0.50}$ with $\varepsilon = 0.00$, 0.05, 0.10, and 0.15 were sealed in quartz ampoules and fired at 425°C (below the melting point of Te) for 24 hours to react the tellurium, then fired again at 600°C for 24 hours. The samples were reground, pressed, and fired again at 600°C for another 24 hours. A portion of these specimens (labelled "A" below) were then reground, pressed, and annealed again, at 675°C. After structural and superconducting analyses, it was determined that the best structural and superconducting properties occurs in the range $\varepsilon = 0.05$–0.10; thus, a second detailed series of specimens was synthesized for $\varepsilon = 0.06$, 0.07, 0.08, and 0.09. The final specimens were dense polycrystalline specimens, approximately 90% of their x-ray density.

Arc-melted samples were synthesized from the same high-purity starting materials. For this technique, often a small percentage of the volatile Te is lost, evaporated onto the arc chamber interior. By weighing the sample before and after the melt, the amount lost, attributed to the most volatile component, tellurium (or, in previous studies, sulphur), could be determined. With extensive experience synthesizing FeTe and Fe(Se,S) compounds this way, a reproducible technique was developed: first, the starting materials were layered onto the copper hearth in the same order as their melting points: Te power on the bottom, FeSe powder next, and Fe powder on top. Then the arc flame was only brought near (~ 2 mm) but not touching, melting the Te gently and dissolving the FeSe and Fe in it. The ingot was turned and melted indirectly again, and then turned twice more, heated directly each time. The final ingots were shiny metallic with very high density, essentially 100% of their x-ray density. The tellurium loss was always about 2.0±0.5% using these techniques. Thus adjusted starting compositions of $Fe_{1+\varepsilon}Te_{0.54}Se_{0.50}$ (2% excess Te) were used, resulting in nearly $Fe_{1+\varepsilon}Te_{0.50}Se_{0.50}$ final compositions. Specimens with $\varepsilon = 0.01$, 0.08, and 0.10 were prepared by arc-melting and studied.

X-ray measurements were performed on a Philips X'PERT diffractometer using Cu $K_\alpha$ radiation. Four-wire resistivity data and ac susceptibility (17 Hz, milligauss) data were obtained on ~1x1x4 mm$^3$ specimens using either a Stanford Research Systems SIM921 ac resistance bridge or a Linear Research LR-400 ac resistance/mutual inductance bridge, calibrated with an identically-sized lead (Pb) standard.

## 3. Results

Typical x-ray diffraction data for the sintered samples are shown in Fig. 1. All specimens show the dominant tetragonal Pb-O structure phase with a nearly constant tetragonal $c$-axis lattice constant $c$ = 0.6010±0.0008 nm, and an $a$-axis lattice constant that increases from $a$ = 0.3785±0.0005 nm to $a$ = 0.3802±0.0005 nm as the Fe concentration in excess of unity, $\varepsilon$, increases from 0.00 to 0.10, in agreement with previous studies of $\varepsilon = 0.02$ [4] and 0.10 [5] samples. The peak just below $2\theta = 32°$ is the dominant (101) reflection of the hexagonal NiAs impurity phase. As is evident, for samples with $\varepsilon$ = 0.00–0.10 and final firings at 600°C, the impurity phase is largest at $\varepsilon = 0.00$, and tends to decrease with increasing Fe concentration; the $\varepsilon = 0.10$ samples appears to be phase-pure. The $\varepsilon = 0.05$ sample that was re-annealed at 675°C shows that the NiAs impurity can also be eliminated with annealing.

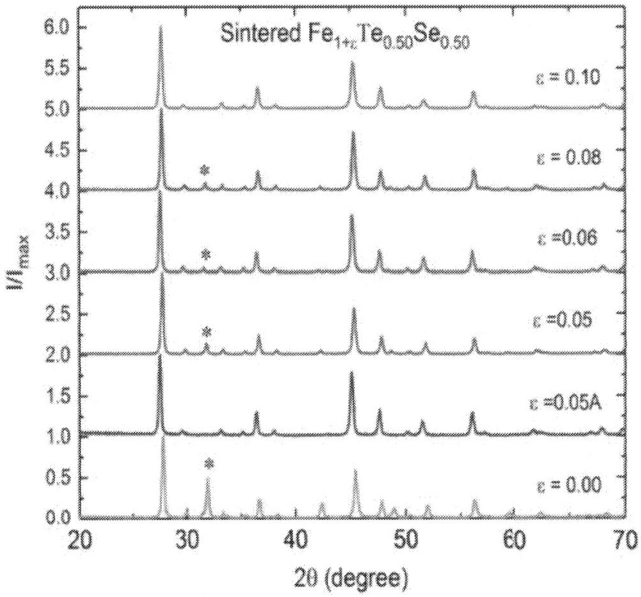


**Figure 1.** X-ray intensity as a function of angle 2θ in a θ-2θ scan for sintered $Fe_{1+\varepsilon}Te_{0.50}Se_{0.50}$ powders.

Figure 2 shows resistivity data as a function of temperature for the sintered samples. In the inset, arc-melted data shows the behaviour typical at high temperatures for all samples: as observed previously [4], beginning at room temperature, the resistivity of $Fe_{1+\varepsilon}Te_{0.5}Se_{0.5}$ slightly increases with decreasing temperature down to about 125–150 K, then decreases until the transition onset $T_c \approx 14$ K. Perhaps coincidentally, the antiferromagnetic transition in the parent alloy FeTe occurs at 130 K. For each curve, the temperatures at which the resistance reaches 90%, 50%, and 10% of the extrapolated normal state values were determined; the 10%–90% range defines the transition width $\Delta T_c$. These values are shown in Fig. 3. The most striking observation from the resistivity data is that $T_c$ does *not* vary appreciably. This is most likely an indication that the iron concentration is not directly changing the carrier density or other gross electronic properties of the superconducting phase; whatever that phase is, and however much of it is present, the resistive superconductivity measurements short the sample at essentially the same temperature for all ε. For the 675°C-annealed specimens (labelled "A"), even though their tetragonal phase purity was greater, it is obvious that annealing has broadened the superconducting transition, possibly due to non-optimal annealed chalcogen site occupancy.

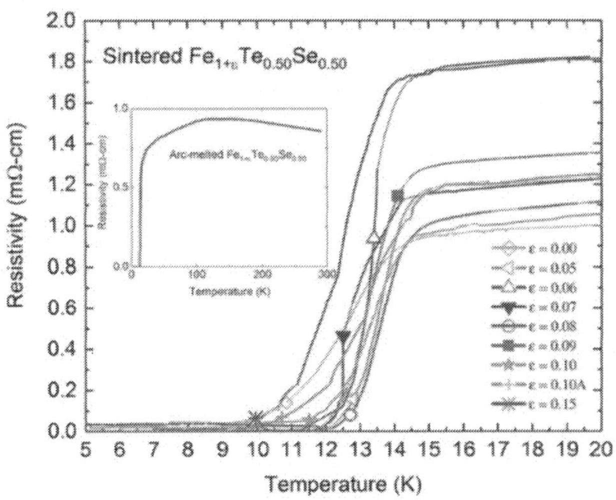


**Figure 2.** Electrical resistivity $R$ as a function of temperature $T$ over the range 5–20 K for $Fe_{1+\varepsilon}Te_{0.50}Se_{0.50}$ (ε=0.00–0.15) sintered samples. $T_c$ remains relatively constant. Inset: $R$ vs. $T$ for $T$ = 0–300 K for an arc-melted sample; such behaviour was typical of almost all sintered and arc-melted samples.

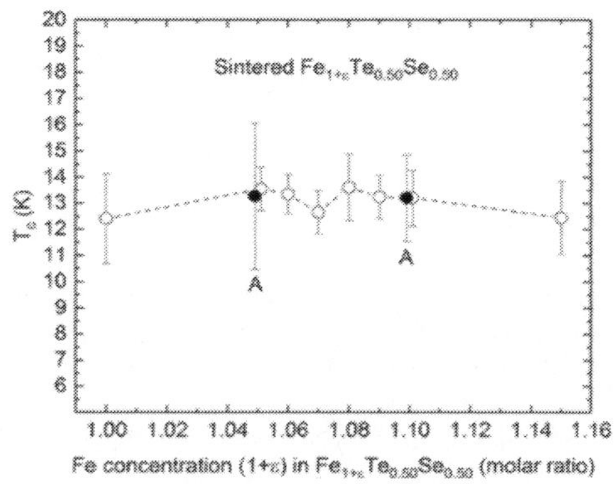


**Figure 3**. Temperatures at 90%, 50%, and 10% of the resistive superconducting transitions for sintered specimens of $Fe_{1+\varepsilon}Te_{0.50}Se_{0.50}$ annealed at 600°C (open symbols) or 675°C (solid symbols, A).

Quite different insight into the story of the "11" materials is provided by the ac magnetic susceptibility data shown in Fig. 4, as a function of temperature near the transition. While the resistivity showed little variation, large changes in the magnetic screening signal are evident. Perhaps similar to the resistivity, the magnetic onset temperature predominantly remains near 14 K; however, the screening fraction varies with Fe concentration and annealing temperature. The screening fraction at $T = 4.2$ K is shown in Fig. 5, for the first survey series (every $\Delta\varepsilon = 0.05$) in the left panel, and the second series, focusing on the central region with $\Delta\varepsilon = 0.01$, in the right. Evidently, for these synthesis conditions, a very large volume fraction is obtained for $\varepsilon = 0.07$.

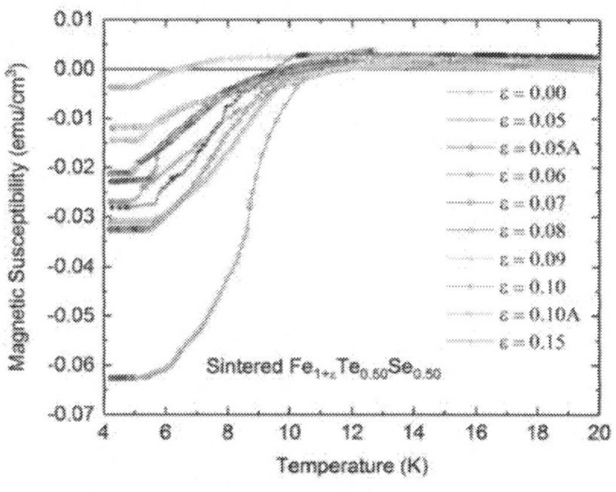


**Figure 4.** Magnetic susceptibility data vs. temperature for sintered $Fe_{1+\varepsilon}Te_{0.50}Se_{0.50}$ ($0 \leq \varepsilon \leq 0.15$).

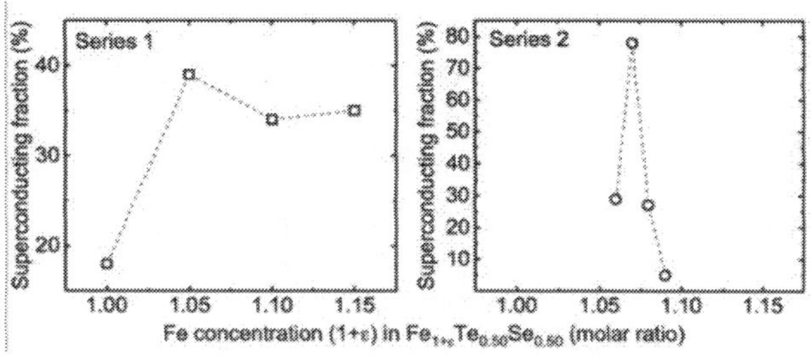


**Figure 5.** Superconducting screening fraction for a first ($\Delta\varepsilon = 0.05$) and a second ($\Delta\varepsilon = 0.01$) series of sintered $Fe_{1+\varepsilon}Te_{0.50}Se_{0.50}$ samples.

For the arc-melted samples, x-ray data are shown in Fig. 6, resistivity data in Fig. 7, and ac magnetic susceptibility data in Fig. 8. Compared to the sintered data, the rapid-quench conditions shift the apparent Fe ε-dependence significantly. However, the sharpest, large magnetic superconducting transition to date was observed with this method (Fig. 8); the diamagnetic drop represents an ~65% screening volume fraction. The nominal ε ~ 0.01 sample also exhibits a large, positive normal-state susceptibility, attributable to a small ferromagnetic (Fe) impurity. Note that evaporation in arc-melted samples results in somewhat uncertain concentrations (as discussed above, the quoted compositions assume all loss was Te). The lattice constants for the nominal ε ~ 0.01 arc-melted sample are $a$ = 0.3800±0.0005 nm and $c$ = 0.602±0.001 nm, similar to the ε = 0.07 sintered sample. We point out that that the rapid quench, in addition to promoting some impurity phases, results in some disorder, typically chalcogen vacancies, similar to behaviour in rapid-cooled oxygen-deficient copper oxides.

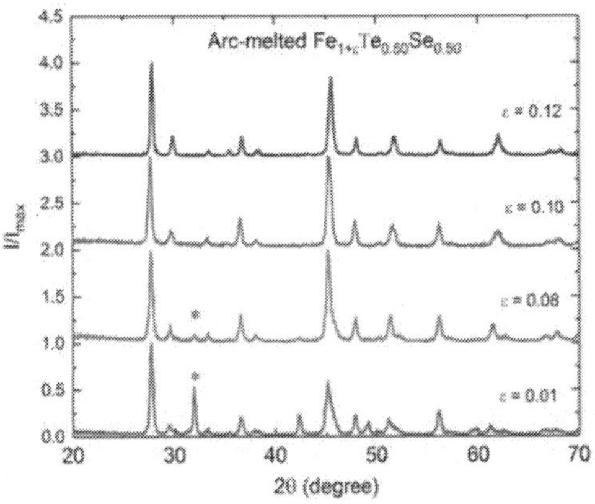


**Figure 6.** X-ray intensity as a function of angle 2θ in θ-2θ scans for various arc-melted $Fe_{1+\varepsilon}Te_{0.50}Se_{0.50}$.

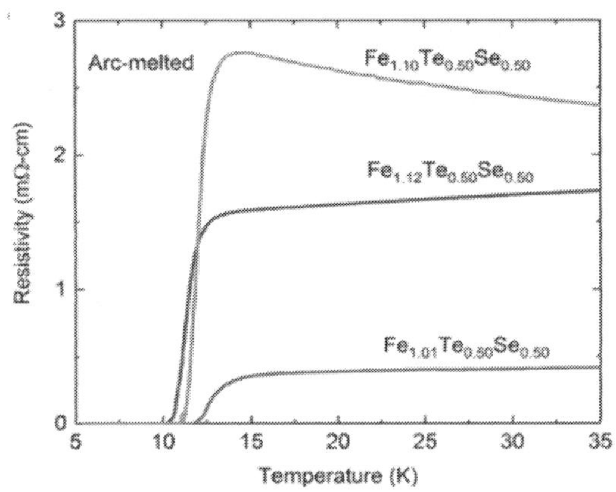


**Figure 7.** Electrical resistivity as a function of temperature $T$ over the range 5–35 K for arc-melted $Fe_{1+\varepsilon}Te_{0.50}Se_{0.50}$.

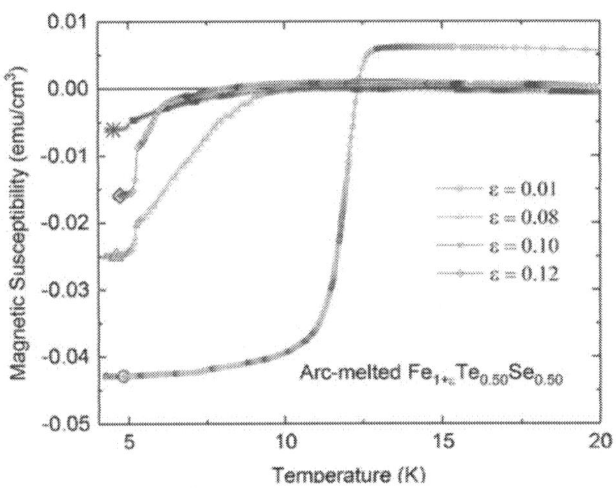


**Figure 8.** Magnetic susceptibility data as a function of temperature for arc-melted $Fe_{1+\varepsilon}Te_{0.50}Se_{0.50}$.

IOP Conf. Series: Journal of Physics: Conf. Series **969** (2018) 012067 doi:10.1088/1742-6596/969/1/012067

## 4. Discussion and Summary

The data for sintered samples indicate that the best superconducting volume fractions occur for samples near $\varepsilon = 0.07$, and that the *a*-axis lattice constant increases with increasing Fe concentration. We find that for our conditions, stabilizing the purest, tetragonal PbO-structure phase does *not* coincide with the best superconducting fraction; instead, superconductivity may best occur with some defect, or other instability, nearby. We have, following the literature, written these stoichiometries as $Fe_{1+\varepsilon}Te_{0.50}Se_{0.50}$; indeed, even the earliest literature [7] suggests the similar FeTe "11" structure actually has stoichiometry ~$Fe_{1.1}Te$, also previously written $Fe_9Te_8$, suggesting a line compound. We suggest that all these results are consistent with a picture of chalcogen vacancies, and could be written $Fe_{1.00}(Te_{0.50}Se_{0.50})_{1-\varepsilon}$, where we have used $(1+\varepsilon)^{-1} \approx 1-\varepsilon$. A chalcogen vacancy is, effectively, an electron donor (each anion would have accommodated two extra electrons; each vacancy thus dopes two electrons). Similar oxygen vacancy doping is common in the electron-doped copper oxides [10], where interstitial oxygen also sometimes plays a role. The increase in the *a*-axis lattice constant is in accord with this view, as each doped electron expands the planes, relieving bond tension; this is the primary difference between *T*-phase (hole) and *T'*-phase (electron) 2-1-4 copper oxides. Such doping behaviour contributes to the bond corrugation effects in Fe(Te,Se) reviewed previously [6]. The superconducting properties in the arc-melted samples, at shifted $\varepsilon$ compared to the sintered samples, likely indicate that the anion vacancy concentration increases with rapid quenching, or perhaps that quenching promotes both iron and chalcogen vacancies.

In summary, we have synthesized nominal $Fe_{1+\varepsilon}Te_{0.50}Se_{0.50}$ ($0.00 \leq \varepsilon$ 0.15) samples by sintering and by arc-melting, and have found that low normal-state resistivity (< 1 mΩ-cm) and high superconducting volume fraction specimens with $T_c \approx 14$ K can be obtained with both methods. Optimal behaviour was found around $\varepsilon \approx 0.07$ for the sintered samples, as well as for the arc-melted sample with lattice constants most similar to those of the $\varepsilon \approx 0.07$ sintered sample. Viewed instead as a possible chalcogen-deficient structure, our data suggests that the larger superconducting volume fractions are associated with larger *a*-axis lattice-constant structures, which could occur when a chalcogen vacancy acts as an electron donor. We observe no supercell reflections in our x-ray data; further work may determine whether ordered vacancy structures occur or can be synthesized, and neutron diffraction experiments are encouraged that could determine chalcogen site occupancy.


## Acknowledgments

We acknowledge support from the Robert A. Welch Foundation Grant No. F-1191 and from the University of Texas at Austin, College of Natural Sciences Freshman Research Initiative.


## 5. References


[1] Hsu F C, Luo J Y, Yeh K W *et al.* 2008 *Proc. Natl. Acad. Sci.* **105** 14262–14264.
[2] He S L, He J F, Zhang W H *et al.* 2013 *Nat. Mater.* **12** 605–610.
[3] Medvedev M, Troyan A, Palasyuk T, Eremets M I and Cava R J 2009 *Nat. Mater.* **8** 3.
[4] Abouhaswa A S, Merentsov A I and Baranov N V 2016 *Physica C* **527** 21–27.
[5] Jiu J, Li C, Zhang S, Feng J, Zhang P and Zhou L 2016 *Physica C* **527** 68–73.
[6] Mizuguchi Y and Takano Y 2010 *J. Phys. Soc. Jpn.* **74** 102001-1-18.
[7] Shunk F A 1974 *Constituion of Binary Alloys* (McGraw Hill, NY), pp. 345 and 351.
[8] Friederichs G M, Worsching M P B and Johrendt D 2015 *Supercond. Sci. Technol.* **28** 095005.
[9] Sun Y, Tsuchiya Y, Taen T *et al.* 2014 *Sci. Rep.* **4** 4585-1-7.
[10] Markert J T, Mochizuki M, and Elliott A V, 1996 *J. Low Temp. Phys.* **105**, 1367–1372.

# NMR study of the layered cobalt oxyphosphide $Sr_2Sc(Co_{1-x}Fe_x)PO_3$

**K Magishi[1], R Watanabe[1], A Hisada[1], T Saito[1], K Koyama[1], S Okada[2], Y Kamihara[3], N Ohkubo[2], S Ban[2] and M Matoba[3]**

[1] Graduate School of Technology, Industrial and Social Sciences, Tokushima University, Tokushima 770-8506, Japan

[2] Laboratory of Physics, College of Science and Technology, Nihon University, Funabashi 274-8501, Japan

[3] Department of Applied Physics and Physico-Informatics, Faculty of Science and Technology, Keio University, Yokohama 223-8522, Japan

E-mail: `magishi@tokushima-u.ac.jp`

**Abstract.** We report the results of $^{31}$P-nuclear magnetic resonance (NMR) measurements on the layered cobalt oxyphosphide $Sr_2Sc(Co_{1-x}Fe_x)PO_3$ in order to investigate the magnetic properties at low temperatures from a microscopic view point. The $^{31}$P-Knight shifts measured at the resonance peak maximum of $Sr_2Sc(Co_{1-x}Fe_x)PO_3$ have positive values and are $T$-independent in an entire temperature range, and the absolute value decreases with increasing Fe content. Also, the nuclear spin-lattice relaxation rate $1/T_1$ is almost proportional to the temperature at low temperatures. The magnitude of $1/T_1T$ decreases with increasing the Fe content, which suggests the decrease of the density of states around the Fermi level.

## 1. Introduction

Since the discovery of iron-based superconductivity in F-doped LaFe(P,As)O[1, 2], many researchers focus on physical properties on oxypnictides with quasi-two dimensional structure separated by a carrier blocking layer. In the case of Co-substitution in place of Fe, LaCoPO and LaCoAsO show weak itinerant ferromagnets with Curie temperatures of about 43 K and 66 K, respectively[3]. In these compounds, it has been proposed that spin fluctuations play an important role in the magnetic behavior[3, 4, 5, 6, 7].

Recently, we have successfully synthesized and evaluated the transport properties of a novel cobalt oxypnictide $Sr_2ScCoPO_3$[8], which crystallizes in an unusual intergrowth structure consisting of a CoP layer and a perovskite related $K_2NiF_4$-type $Sr_4Sc_2O_6$ layer. This structure is isomorphic to $Sr_2ScFePO_3$ which exhibits superconductivity below 17 K[9]. The electrical resistivity $\rho$ of $Sr_2ScCoPO_3$ decreases upon cooling like a metal. However, the superconducting transition is not observed on that of $Sr_2ScCoPO_3$. Also, we investigate the effect of changing the number of $3d$ electrons and show the detailed physical properties[10]. The electrical resistivity $\rho$'s for $Sr_2ScMPO_3$ ($M$ = Ni and $Co_{0.5}Fe_{0.5}$) show the metallic behavior similar to that of $Sr_2ScCoPO_3$. The value of $\rho$ for $Sr_2ScNiPO_3$, where the substitution of Ni at an Co site effectively dopes electrons, is smaller than that for $Sr_2ScCoPO_3$. In contrast, the value of $\rho$ for $Sr_2Sc(Co_{0.5}Fe_{0.5})PO_3$, where partially substitution of Fe at an Co site effectively reduces electrons, is larger than that for $Sr_2ScCoPO_3$. Quite recently, the transport

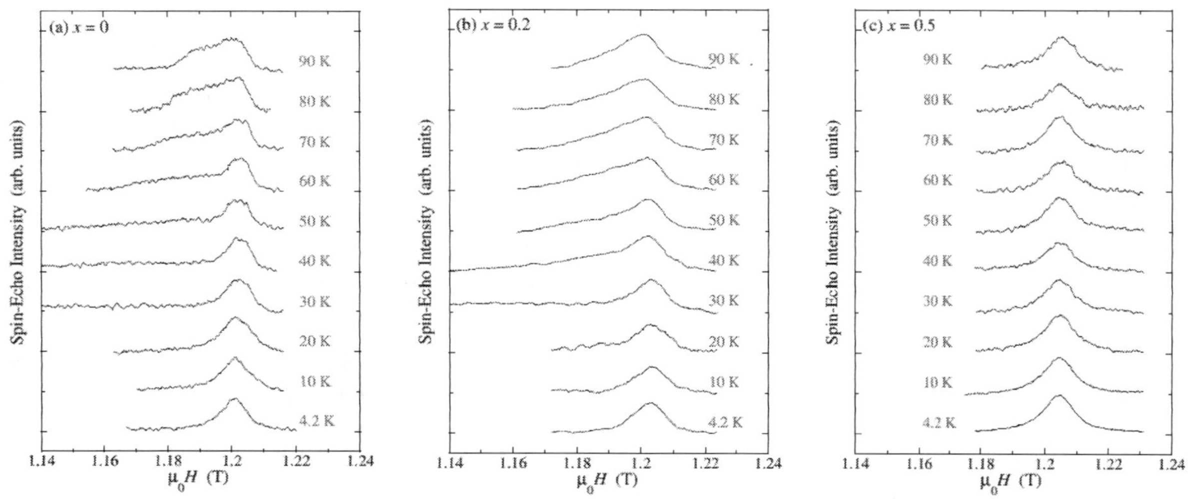


**Figure 1.** Temperature evolution of $^{31}$P-NMR spectra of $Sr_2Sc(Co_{1-x}Fe_x)PO_3$ for (a) $x = 0$, (b) $x = 0.2$, and (c) $x = 0.5$ in a temperature range of 4.2 K – 90 K at $f = 20.760$ MHz.

properties of $Sr_2Sc(Co_{1-x}Fe_x)PO_3$ ($0.04 \leq x \leq 0.8$) were reported[11]. The electrical resistivity $\rho$'s for $Sr_2Sc(Co_{1-x}Fe_x)PO_3$ decrease with decreasing temperature like a metal, but the superconducting transition is not observed. The value of $\rho$ increases with increasing the Fe content $x$.

In the paper, we will report on the $^{31}$P-nuclear magnetic resonance (NMR) results of $Sr_2ScCoPO_3$ and partially Fe-substituted $Sr_2Sc(Co_{1-x}Fe_x)PO_3$, and discuss the microscopic electronic states and the variation of the magnetic properties by changing the number of $3d$ electrons. It is quite helpful to make use of the NMR method for a study of local electronic properties.

## 2. Experimental

The polycrystalline sample of $Sr_2ScCoPO_3$ was synthesized by a standard solid state reaction method in a quartz tube from of CoP, SrO, Sr and $Sc_2O_3$. In case of $Sr_2Sc(Co_{1-x}Fe_x)PO_3$, the starting materails were $Co_{1-x}Fe_xP$, SrO, Sr and $Sc_2O_3$. The precursor $Co_{1-x}Fe_xP$ was prepared by a solid state reaction of Co, $Fe_2P$ and P. The detailed synthesis condition was described elsewhere[8, 10]. Samples were characterized by powder X-ray diffraction (XRD) at room temperature. The powder XRD patterns of all samples are similar to that of $Sr_2ScFePO_3$[9], which indicates that all samples are isomorphic to $Sr_2ScFePO_3$[10, 11]. For the NMR measurements, samples were crushed into powder. $^{31}$P-NMR ($I = 1/2$) measurements were performed in the temperature range of 1.5 K – 200 K in an external magnetic field of 1.2 T using a conventional phase-coherent pulsed spectrometer. NMR spectrum was obtained by sweeping magnetic field. The nuclear spin-lattice relaxation rate $1/T_1$ was measured by the saturation recovery method.

## 3. Results and discussion

### *3.1. NMR spectra*

Figure 1 shows the temperature evolution of $^{31}$P-NMR spectra of $Sr_2Sc(Co_{1-x}Fe_x)PO_3$ for (a) $x = 0$, (b) $x = 0.2$, and (c) $x = 0.5$ in a temperature range of 4.2 K – 90 K at $f = 20.760$ MHz. For the samples of $x = 0$ and 0.2, the $^{31}$P-NMR spectrum at high temperatures shows a typical powder pattern due to anisotropic Knight shift for a spin $I = 1/2$ nucleus experiencing

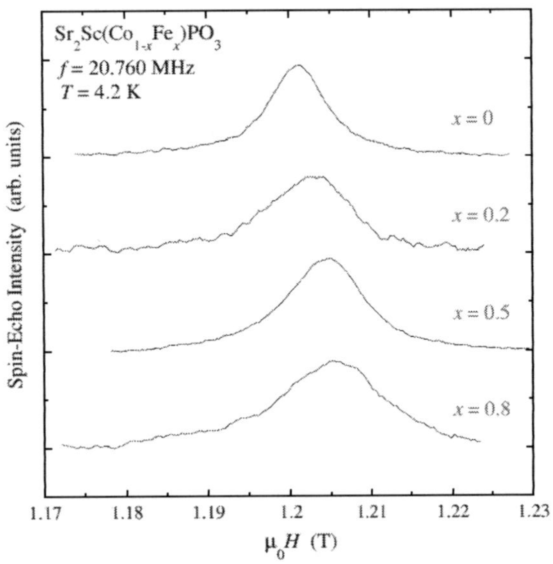


**Figure 2.** $^{31}$P-NMR spectra of $Sr_2Sc(Co_{1-x}Fe_x)PO_3$ for the compounds of $x = 0$, 0.2, 0.5 and 0.8 at $T = 4.2$ K and $f = 20.760$ MHz.

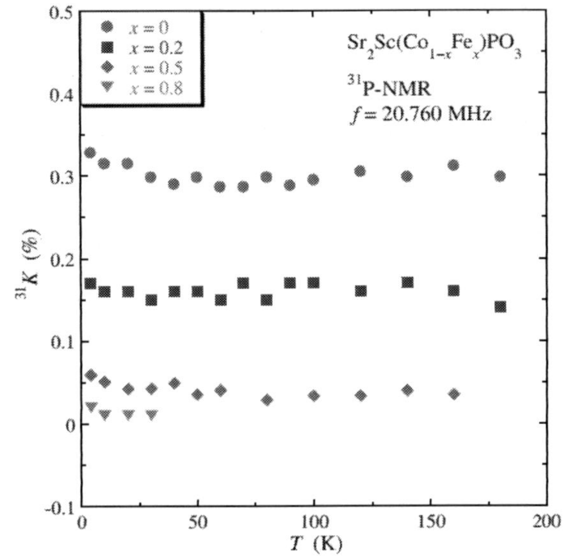


**Figure 3.** Temperature dependences of the $^{31}$P-Knight shift of $Sr_2Sc(Co_{1-x}Fe_x)PO_3$ for the compounds of $x = 0$, 0.2, 0.5 and 0.8 measured at $f = 20.760$ MHz.

an axially symmetric local magnetic field. The Knight shift components perpendicular $K_\perp$ and parallel $K_\parallel$ to the principal axis are derived from the peak maximum and the shoulder peak in the spectra, respectively. The component $K_\parallel$ shifts markedly upon cooling, and the spectrum becomes broader at low temperatures. In contrast, for the sample of $x = 0.5$, the spectrum is symmetric and almost independent of temperature. It suggests that the electronic states are changed by reducing the number of $3d$ electrons.

Figure 2 shows the $^{31}$P-NMR spectra of $Sr_2Sc(Co_{1-x}Fe_x)PO_3$ for the compounds of $x = 0$, 0.2, 0.5 and 0.8 at $T = 4.2$ K and a frequency of $f = 20.760$ MHz. At low temperature, the spectrum is symmetric, and the resonance magnetic field $H_{\rm res}$ shifts higher field side with increasing the substituted Fe content, where indicates the Knight shift decreases with increasing Fe content. The Knight shift is obtained from the relation $K = (2\pi f/\gamma_{\rm N} - H_{\rm res})/H_{\rm res}$, where $\gamma_{\rm N}$ is the gyromagnetic ratio of $^{31}$P ($^{31}\gamma_{\rm N}/2\pi = 17.237$ MHz/T). In general, the Knight shift is proportional to the spin susceptibility, $K = \frac{A_{hf}}{N_{\rm A}\mu_{\rm B}}\chi_s$, where $A_{hf}$ is the hyperfine coupling constant, $N_{\rm A}$ is the Avogadro number and $\mu_{\rm B}$ is the Bohr magneton.

Figure 3 indicates the temperature dependence of the Knight shift of the $^{31}$P nuclei, $^{31}K$, of $Sr_2Sc(Co_{1-x}Fe_x)PO_3$. The Knight shift was determined at the resonance peak maximum, which corresponds to the component $K_\perp$. The Knight shift is almost $T$-independent, which is mainly arised from conduction electron spin susceptibility. The spin susceptibility $\chi_s$ is proportional to the density of states at the Fermi level $\chi_s = \mu_{\rm B}^2 D(E_{\rm F})$. The values of the Knight shift are positive and small in an entire temperature range, which suggests the small hyperfine coupling constant. Also, the magnitude decreases with increasing Fe content, which indicates the decrease of the density of states around the Fermi level.

### *3.2. Nuclear relaxation rate* $1/T_1$

Figure 4 presents the recovery curves of the nuclear magnetization $M(t)$ of $^{31}$P in $Sr_2Sc(Co_{1-x}Fe_x)PO_3$ for (a) $x = 0$ and (b) $x = 0.8$ at $T = 4.2$ K. The nuclear spin-lattice

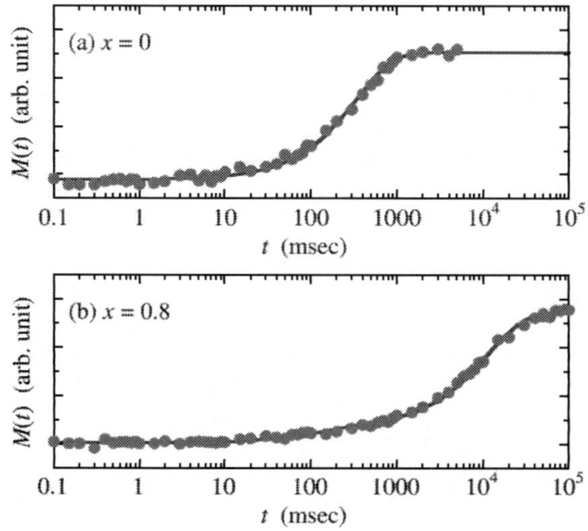


**Figure 4.** Recovery curves of the nuclear magnetization of $^{31}$P in $Sr_2Sc(Co_{1-x}Fe_x)PO_3$ for (a) $x = 0$ and (b) $x = 0.8$ at $T = 4.2$ K. The solid lines are the best fit using the formula described in the text.

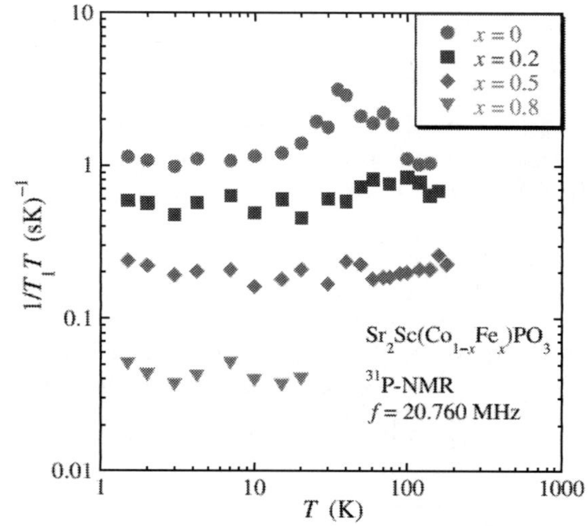


**Figure 5.** Temperature dependences of the nuclear spin-lattice relaxation rates devided by temperature $1/T_1T$ of $Sr_2Sc(Co_{1-x}Fe_x)PO_3$ in a temperature range of 1.5 K–200 K measured at $f$=20.760 MHz.

relaxation rate $1/T_1$ was measured at the resonance peak maximum. The recovery of the nuclear magnetization $M(t)$ is almost single component for the spin $I = 1/2$ of $^{31}$P nuclei as $M(t) = M_0[1 - \exp(-t/T_1)]$, where $M_0$ and $M(t)$ are the nuclear magnetization in the thermal equilibrium and at a time $t$ after the saturating pulse, respectively. As seen in the figure, the relaxation time $T_1$ for the sample of $x = 0.8$ is long more than one order compared to that for the sample of $x = 0$

Figure 5 presents the temperature dependences the nuclear spin-lattice relaxation rates devided by temperature $1/T_1T$ for $^{31}$P nuclei in $Sr_2Sc(Co_{1-x}Fe_x)PO_3$ in a temperature range of 1.5 K – 200 K. For the sample of $x = 0$, $1/T_1T$ shows a weak temperature dependence at high temperatures, which suggests the existence of weak spin fluctuations. As further decreasing temperature, $1/T_1T$ decreases with a small peak around $\sim$ 40 K and is almost independent of temperature at low temperatures. In contrast, for other samples, $1/T_1T$'s are roughly constant over the entire temperature range, as expected for simple metallic systems. Also, the magnitude of $1/T_1T$ decreases with increasing the Fe content.

Generally, $1/T_1T$ is expressed as $1/T_1T \propto \lim_{\omega\to 0}\sum_q |A_q|^2\frac{\chi"(q,\omega)}{\omega}$, where $\chi"(q,\omega)$ is the imaginary part of the dynamical susceptibility, $\omega$ is the resonance frequency, and $A_q$ is the Fourier $q$ component of the hyperfine coupling constant. Thus, $1/T_1T$ probes the $q$-sum of spin fluctuations $\sum_q \chi(q, \omega \to 0)$. If the $q$ dependence of spin fluctuations is small, $1/T_1T$ is given by $1/T_1T \propto \int N^2(E)f(E)[1 - f(E)]dE$, where $f(E)$ is the Fermi distribution function and $N(E)$ is the effective density of states around the Fermi level $E_F$. Thus, the decrease of $1/T_1T$ with increasing the Fe content indicates the decrease of the density of states around the Fermi level, which is consistent with the results of the Knight shift.

## 4. Summary

We have carried out the $^{31}$P-NMR experiments for $Sr_2Sc(Co_{1-x}Fe_x)PO_3$ in order to understand the magnetic properties at low temperatures from the microscopic viewpoints. The NMR spectrum of $Sr_2ScCoPO_3$ shows the uniaxially symmetric powder pattern due to the anisotropic

IOP Conf. Series: Journal of Physics: Conf. Series **969** (2018) 012068 doi:10.1088/1742-6596/969/1/012068

Knight shift at high temperatures. By substitution of Fe at an Co site, the spectrum changes to be symmetric. The $^{31}$P-Knight shift measured at the resonance peak maximum of $Sr_2Sc(Co_{1-x}Fe_x)PO_3$ is $T$-independent and has a positive value in an entire temperature range, and the absolute value decreases with increasing Fe content. Also, $1/T_1$ is almost proportional to the temperature at low temperatures., and the magnitude of $1/T_1T$ decreases with increasing the Fe content, which suggests the decrease of the density of states around the Fermi level.

## References


[1] Kamihara Y, Hiramatsu H, Hirano M, Kawamura R, Yanagi H, Kamiya T and Hosono H 2006 *J. Am. Chem. Soc.* **128** 10012

[2] Kamihara Y, Watanabe T, Hirano M and Hosono H 2008 *J. Am. Chem. Soc.* **130** 3296

[3] Yanagi H, Kawamura R, Kamiya T, Kamihara Y, Hirano M, Nakamura T, Osawa H and Hosono H 2008 *Phys. Rev. B* **77** 224431

[4] Sugawara H, Ishida K, Nakai Y, Yanagi H, Kamiya T, Kamihara Y, Hirano M and Hosono H 2009 *J. Phys. Soc Jpn.* **78** 113705

[5] Majumder M, Ghoshray K, Ghoshray A, Bandyopadhyay B, Pahari B and Banerjee S 2009 *Phys. Rev. B* **80** 212402

[6] Ohta H and Yoshimura K 2009 *Phys. Rev. B* **79** 184407

[7] Ohta H and Yoshimura K 2009 *Phys. Rev. B* **80** 184409

[8] Okada S, Kamihara Y, Ohkubo N, Ban S, Matoba M and Atou T 2012 *J. Phys.: Conf. Ser.* **400** 022091

[9] Ogino H, Matsumura Y, Katsura Y, Ushiyama K, Horii S, Kishio K and Shimoyama J 2009 *Supercond. Sci. Technol.* **22** 075008

[10] Okada S, Kamihara Y, Ohkubo N, Ban S and Matoba M 2014 *J. Phys.: Conf. Ser.* **568** 022034

[11] Okada S, Kamihara Y, Ohkubo N, Ban S and Matoba M 2017 *Presented at the 28th Int. Conf. on Low Temperature Physics* (Gothenburg)

IOP Conf. Series: Journal of Physics: Conf. Series **969** (2018) 012069 doi:10.1088/1742-6596/969/1/012069

# Strong Electron Self-Cooling in the Cold-Electron Bolometers Designed for CMB Measurements

**L.S. Kuzmin*[,1,2], A.L. Pankratov[2], A.V. Gordeeva[2], V.O. Zbrozhek[2], L.S. Revin[2], V.A. Shamporov[1,2], S. Masi[3] and P. de Bernardis[3]**

[1]Chalmers University of Technology, Gothenburg, Sweden,
[2]Nizhny Novgorod State Technical University, Nizhnij Novgorod, Russia,
[3]Sapienza University of Rome, Italy

*kuzmin@chalmers.se

**Abstract**. We have realized cold-electron bolometers (CEB) with direct electron self-cooling of the nanoabsorber by SIN (Superconductor-Insulator-Normal metal) tunnel junctions. This electron self-cooling acts as a strong negative electrothermal feedback, improving noise and dynamic properties. Due to this cooling the photon-noise-limited operation of CEBs was realized in array of bolometers developed for the 345 GHz channel of the OLIMPO Balloon Telescope in the power range from 10 pW to 20 pW at phonon temperature $T_{ph}$=310 mK. The negative electrothermal feedback in CEB is analogous to TES but instead of artificial heating we use cooling of the absorber. The high efficiency of the electron self-cooling to $T_e$ =100 mK without power load and to $Te$ =160 mK under power load is achieved by:
- a very small volume of the nanoabsorber (0.02 $\mu m^3$) and a large area of the SIN tunnel junctions,
- effective removal of hot quasiparticles by arranging double stock at both sides of the junctions and close position of the normal metal traps,
- self-protection of the 2D array of CEBs against interferences by dividing them between N series CEBs (for voltage interferences) and M parallel CEBs (for current interferences),
- suppression of Andreev reflection by a thin layer of Fe in the AlFe absorber.

As a result even under high power load the CEBs are working at electron temperature $T_e$ less than $T_{ph}$. To our knowledge, there is no analogue in the bolometers technology in the world for bolometers working at electron temperature colder than phonon temperature.

## 1. Introduction

Cosmic microwave background (CMB) measurements were ranked second by the journal Science among the top 10 Achievements of the Decade [1]. Starting with the BOOMERanG experiment [2], increasing precision maps of the CMB have been obtained, with improved sky and frequency coverage. These measurements culminated with the all-sky multi-frequency maps of the WMAP and Planck satellites, and the high resolution maps of the SPT and ACT telescopes. The current trend of CMB measurements focuses on polarization measurements (see e.g. the BICEP/Keck measurements) and on the measurement of the Sunyaev-Zeldovich effect (carried out with ground based telescopes like SPT and ACT, and multifrequency space missions like the Planck satellite). The OLIMPO experiment [3] is a mm-wave balloon-borne telescope, optimized for spectral measurements of the Sunyaev-Zeldovich effect (the distortion of the spectrum of the CMB in the direction of rich galaxy clusters) within four frequency bands: 150, 210, 345 and 480 GHz. The power of spectral measurements of the SZ effect is to solve the degeneracies between different sources of signals along the line of sight, as described in [4].

IOP Conf. Series: Journal of Physics: Conf. Series **969** (2018) 012069 doi:10.1088/1742-6596/969/1/012069

We have realized the photon-noise-limited 2D array of cold electron bolometers (CEB) [5-7] for the 345 GHz band of the OLIMPO, based on the strong direct electron self-cooling of the nanoabsorbers by SIN tunnel junctions of the CEBs (Fig. 1). The electron cooling is based on decoupling of electron and phonon systems at low temperatures [8]. This effect was observed for the first time in a SIN tunnel junction by Nahum et al. [9]. Then the cooling effect in a single SIN junction was demonstrated in several laboratories from 300 to 100 mK [10,11].

After that, the efforts were mainly directed towards the development of micro-refrigerators based on cooling platform [12-14] with the potential of cooling a bolometer in thermal contact with the platform at lower temperature down to 100 mK. However, this strategy of indirect cooling met severe problems related with creation of suspended membrane and cooling through two very weak electron-phonon (e-ph) and phonon-electron (ph-e) coupling. Practically speaking, creation of bolometers on cooling platform failed after 20 years of development [15]. Cooling of platform itself was realized only from 300 to 200 mK and no one bolometer was placed so far on the platform for testing [15].

On the contrary, we have avoided all these problems by developing a competing concept of CEB with direct electron self-cooling of the absorber by SIN junctions [5,6]. Coupling of external radiation is realized through capacitance of the tunnel junctions [6,16]. Electron cooling is serving as strong electrothermal feedback as in TES [6,17] but with principle difference: artificial heating of TES is replaced by strong electron cooling.

The high efficiency of electron cooling ($P_{cool}/IV = 40\%$) is achieved by the combination of the factors:

- a very small volume of the absorber (0.02 μm$^3$) and a large relative area of tunnel junctions (80% of an area of the absorber $S$=1.6 μm$^2$),
- effective removing of hot quasiparticles by double stock of them to superconducting external electrodes at both ends of the tunnel junctions,
- close position of the normal metal traps to SIN junctions at 1.6 μm to catch hot quasiparticles and remove them from tunnel junction to avoid recombination of them and back heating of the absorber.
- self-protection of the 2D array of CEBs against interferences by dividing them between N*M bolometers (N - series and M - parallel connections). For presented array N=48 and M=4 (Fig. 2),
- avoiding additional heating of the absorber by suppression of Andreev reflection by thin layer of Fe in AlFe absorber.

The structure is designed for the following requirements: bandwidth $\Delta\nu$= (330-360) GHz, sensitive to dual polarizations, power load from 20 to 90 pW (33 pW in photometer mode, 66 pW in spectrometer mode, 1.5 coefficient for margins), absorption efficiency not less than 50%, bolometer noise equivalent power $NEP_{CEB}$ is less than photon noise equivalent power $NEP_{phot}$, where $NEP^2_{phot} = 2h\nu P_0 + P_0^2/\Delta\nu$.

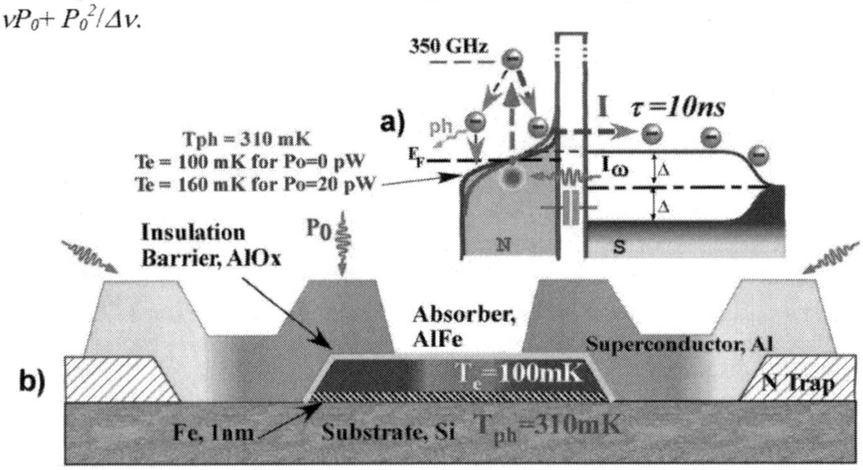


Fig. 1. a) Energy diagram of a CEB including normal absorber, isolation barrier and superconducting electrode, b) Cross-section of the CEB comprising two SIN tunnel junctions, an absorber and normal metal traps. The CEB demonstrates strong electron self-cooling from 310 mK to 100 mK providing perfect noise performance.

This paper is devoted to practical development of 2D arrays of CEBs for realization of photon-noise limited operation with bandwidth and efficiency required for OLIMPO Balloon Telescope. The comprehensive theoretical description can be found in [16], [18] and [19].

## 2. Array of CEBs in Current-Biased Mode.

Theoretical analysis has shown that 10% bandwidth can be achieved in the 2D array of CEBs [18] realized as meander-type structure [19] shown in Fig. 2. The geometry of the unit cell is created from the dense rectangular grid in a way that makes more uniform sparse distribution of dipoles over the unit cell. All CEBs are connected in series and parallel at DC for optimal matching with a JFET amplifier AD745 [20]. A pair of $\lambda/2$ dipoles with wide electrodes sensitive to the both orthogonal linear polarizations is connected to another pair by high-inductive 1 μm wide lines. The impedance of each bolometer was determined by 50 Ω resistance in series with 20 fF capacitance. The layout dimensions are optimized for the frequency band of 330-360 GHz, determined by band-pass filters.

The design of the pixel has started from the constructing of a numerical model and calculation of the absorption efficiency. The model consists of a horn, waveguide port, Si substrate, CEBs array with gold antennas and metallic backshort (Fig. 2). The thickness of the substrate and distance to the backshort are optimized to get maximal absorption at 345 GHz and to avoid excitation of substrate modes. The total number of bolometers and serial-parallel DC-connections had been determined by numerical simulations of the heat balance equations and in order to cover the most of the Airy disc increasing in this way the total absorbed power.

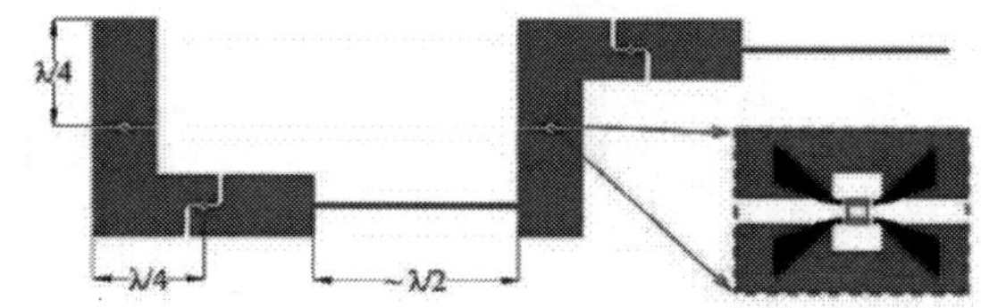


a)

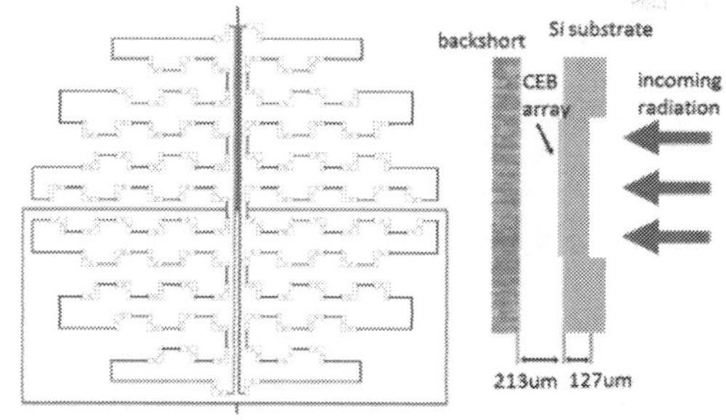


b)

Fig. 2. a) Meander structure of 2D array of $\lambda/2$ dipoles with CEBs inside; $\lambda$=260 μm; b) Schematic view of the array of CEBs, optimized for OLIMPO requirements. The array is placed on a silicon substrate of 127 um thick. The CEB array consists of 96 dipole double-polarization antennas, each with two cold-electron bolometers; so the total number of CEBs is 192. Schematics of the experiment.

The total absorbed power is calculated as a sum of powers absorbed by each dipole antenna individually in the required frequency band around 345 GHz. We calculated absorbed power for different horn modes and obtained reasonably good results (43 ±3%) for the absorption efficiency.

## 3. Results

The electron temperature as well as accepted power are calculated by fitting of the experimental IV-curves. The fitting procedure is described in [21], that allows to calculate the accepted power with the precision of order 0.2pW. The black body source (that can be heated up to 46K, keeping plate temperature at 0.3K) is placed inside the cryostat as shown in [21]. The measured CEB array demonstrates very effective electron cooling from 310 mK to 100 mK without optical power load, $P_0$, and from 330 mK to 160 mK with $P_0$= 20 pW at 345 GHz (Fig. 3). That means that even under high power load the CEBs are working at electron temperature less than the phonon temperature.

Dependences of cooling power, $P_{cool}$, electron-phonon power, $P_{e\text{-}ph}$, and absorbed power, $P_0$, obtained from two heat-balance equations [20] are shown in Fig. 4.

Horizontal lines show the phonon temperature, which is higher in case of incoming power due to microwave heating of the substrate. For $P_0$=0 (Fig. 4a), when we increase voltage $V$ and decrease $T_e$, $P_{e\text{-}ph}$ from phonons to electrons is increased and proportionally $P_{cool}$ is increased to consume this heat flow: $P_{e\text{-}ph} = \Sigma N (T_e^5 - T_{ph}^5)$, $\Sigma$ is the material constant, $N$ is the absorber volume. For $P_0$=20 pW (Fig. 4b) at zero V, the total signal power $P_0$ goes to phonons and $P_{e\text{-}ph}$ has negative sign. Then, when we start to increase $V$ and decrease $T_e$, $P_{e\text{-}ph}$ is increased and crossed zero at critical point around $V$=3.8 mV where cooling makes $T_e$ lower than $T_{ph}$. For voltages higher than 3.8 mV $P_{e\text{-}ph}$ is increased and proportionally $P_{cool}$ is increased keeping difference between them equal to $P_0$.

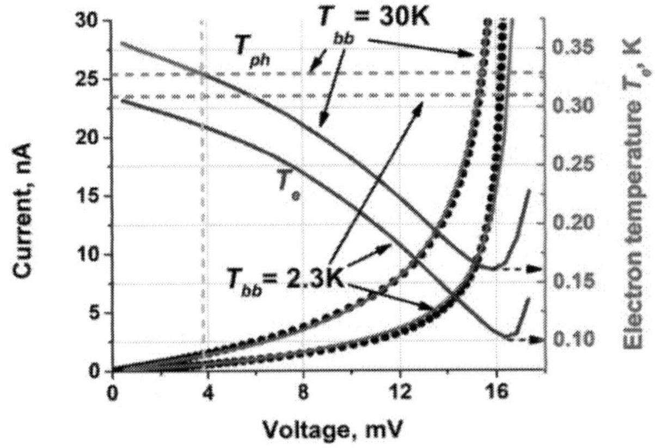


Fig. 3. IV-curves (black dots and solid red line) and dependences of electron temperature $T_e$ (blue curves) on voltage for low power load $P_0$ = 0 pW ($T_{bb}$ = 2.3 K) and high power load $P_0$ = 20 pW ($T_{bb}$ = 30 K). The horizontal dashed lines correspond to phonon temperature.

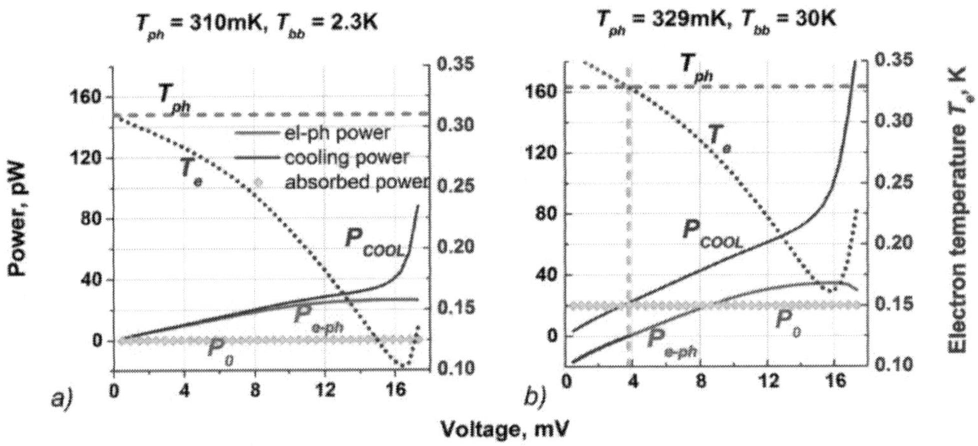


Fig. 4. a) Dependences of cooling power and e-ph power on voltage without optical power load, $P_0$=0; electron temperature $T_e$ is shown on the right; b) Dependences of cooling power and e-ph power on voltage with optical power load of $T_{bb}$=30K, $P_0$=20pW.

The electron-phonon thermal conductance, $G_{e\text{-}ph}$, and the cooling thermal conductance, $G_{cool}$, versus voltage for two different values of the incoming power are shown in Fig. 5a. Electron temperature is shown in Fig. 5b. Cooling conductance, $G_{cool}$, is increased with voltage and considerably overcomes $G_{e\text{-}ph}$. Both conductances are strongly increased for power load $P_0$=20 pW. The strength of negative electrothermal feedback, the loop gain $L$, (green lines)

$$L = G_{cool}/G_{e\text{-}ph} \tag{1}$$

is increased for larger voltage and achieves a value of 10 without power load $P_0$, and a value of 100 for power load $P_0$=20 pW.

Time response of the CEB ($\tau$) versus voltage for two values of incoming power is shown in Fig. 5b.

$$\tau_{tot} = C_\Lambda / G_{tot} = \tau_{e\text{-}ph}(L+1), \quad \text{with } G_{tot} = G_{cool} + G_{e\text{-}ph}, \tag{2}$$

Here $C_\Lambda = \Lambda\, C_v$ - is the heat capacity of the normal metal; $C_v = \gamma T_e$ is the specific heat capacity; $\Lambda$ - the volume of the absorber and $\tau_{e\text{-}ph}$ – the electron-phonon time constant.

IOP Conf. Series: Journal of Physics: Conf. Series **969** (2018) 012069 doi:10.1088/1742-6596/969/1/012069

As for TES [17], the time constant is strongly reduced by loop gain $L$ (1) of the negative electrothermal feedback. Total time constant is mainly determined by cooling conductance $G_{cool}$ dominated over $G_{e\text{-}ph}$. Maximum influence of electron cooling on time constant is for higher voltages where we have stronger cooling and lower $T_e$ than for lower voltages.

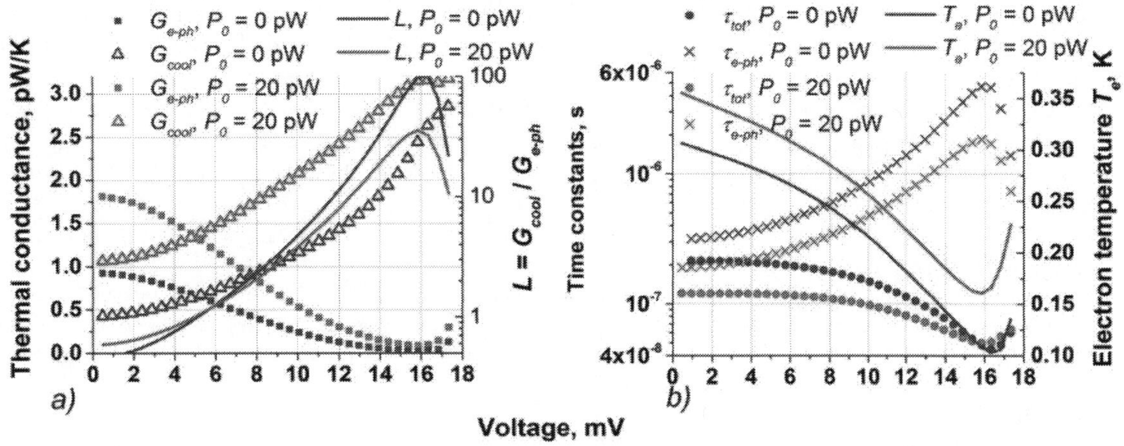


Fig. 5 a).Thermal conductance, $G_{cool}$ and $G_{e\text{-}ph}$, for $P_0$=0 (blue markers) and $P_0$ =20 pW (red markers) versus voltage of the CEB array. At the optimal operating point with minimum NEP, $V$=12 mV, $G_{cool}$ is considerably larger than $G_{e\text{-}ph}$ and the difference is increased when optical power load is applied. The strength of negative electrothermal feedback, the loop gain $L$ (solid cures), increases for larger $P_0$ and achieves value of 10 at the optimal point.

Fig. 5b.Time constants, $\tau_{tot}$ and $\tau_{e\text{-}ph}$, for $P_0$=0 (blue markers) and $P_0$ =20 pW (red markers) versus voltage of the CEB array. The $\tau$ is considerably shorter than $\tau_{e\text{-}ph}$ due to short $\tau_{cool}$.

Various components of theoretically estimated voltage noise versus CEB array voltage are shown in Fig. 6 by solid curves, while the measured total voltage noise is shown by symbols with error marks.

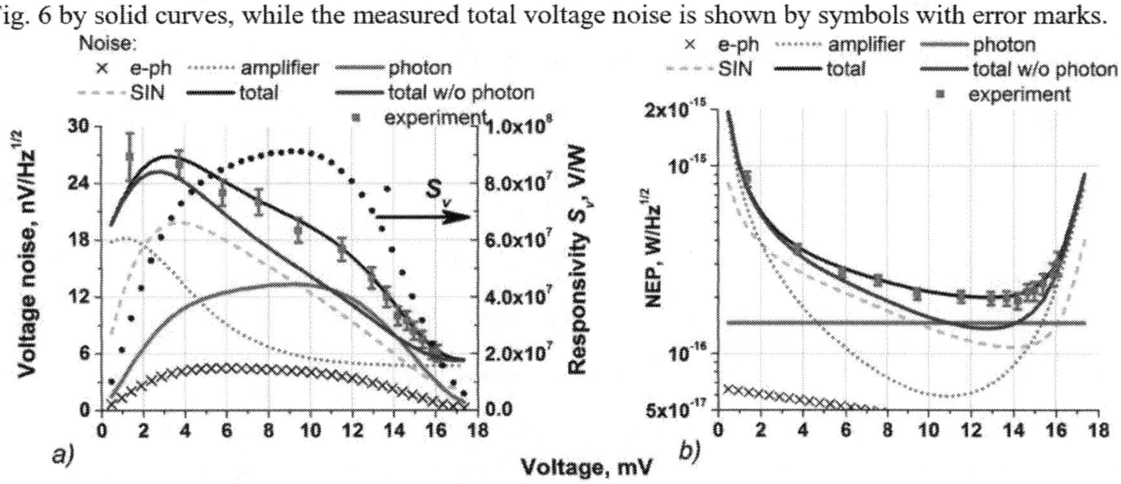


Fig. 6. Various components of voltage noise (a) and NEP (b) versus bolometer voltage. Curves - theory, red markers - experiment. Phonon temperature 330 mK, absorbed power 20 pW. Photon noise dominats over CEB noise from 11 to 14 mV.

Fig. 6 shows a good agreement between measured and theoretically estimated noise. In the voltage range, where photon noise exceeds the SIN and amplifier noise, one can see a small bump of the total noise at the voltages around (11-14) mV, which is well reproduced both in theory and experiment. We, therefore, can conclude that in this voltage range the photon noise dominates over the bolometer noise and is visible by the bolometer as in [20].

## 4. Conclusion

The CEB with strong direct electron self-cooling of the nanoabsorber by SIN tunnel junctions have been realized for the 345 GHz channel of the OLIMPO balloon experiment. The ultimate photon-noise-limited operation of 2D array of CEBs is achieved due to effective electron self-cooling of the absorber. This cooling is acting as the strong negative electrothermal feedback improving noise and dynamic characteristics. The absorption in the required frequency band is realized thanks to the 2D array of dipole antennas on a thinned Si substrate. This operation was demonstrated with $P_0$ in the range (10-20) pW at $T_{ph}$ from 310 to 330 mK. This means that even under relatively high power load the CEBs are working at electron temperature $T_e$=160 mK, i.e. much less than the phonon temperature. This is the first realization of the bolometer working at electron temperature colder than the phonon temperature under relatively high power load.

## 5. Acknowledgements

We would like to thank Mikhail Tarasov for help in fabrication of samples and Dmitry Golubev and Vyacheslav Vdovin for useful discussions. The work was supported by the Swedish National Space Board, SNSB, and Russian Science Foundation (Project No. 16-19-10468). Samples were fabricated at the Chalmers Nanotechnology Center.

## References

[1] Cho A: Insights of the decade: precision cosmology. Science 2010, 330:1615.
[2] Masi S, et al: Instrument, method, brightness and polarisation maps from the 2003 flight of BOOMERanG. Astron Astrophys 2006, 458:687–716. astro-ph/0507509.
[3] S. Masi, E. Battistelli, D. Brienza et al., "OLIMPO", Mem. S. It., vol. 79, p.887, 2008.
[4] P. de Bernardis, S. Colafrancesco, G. D'Alessandro, L. Lamagna, P. Marchegiani, S. Masi and A. Schillaci, Low-resolution spectroscopy of the Sunyaev-Zel'dovich effect and estimates of cluster parameters, Astronomy &Astrophysics, A86, 538 (2012).
[5] L. Kuzmin. "Optimization of the Hot-Electron Bolometer for Space Astronomy", SNED Proc. Vol. 1, pp. 145–154 (2002).
[6] L. Kuzmin, "Ultimate Cold-Electron Bolometer with Strong Electrothermal Feedback", in Proc. of the SPIE conference "Millimeters and Submillimeter Detectors", vol. 5498, pp. 349-36, 2004.
[7] M. Salatino, P. de Bernardis, L.S. Kuzmin, S. Mahashabde, S. Masi, "Sensitivity to cosmic rays of cold electron bolometers for space applications", J. Low Temp. Phys., vol. 176, p. 323, 2014.
[8] M. L. Roukes, M. R. Freeman, R. S. Germain, and M.B. Ketchen, Phys. Rev. Lett. 55, 422 (1985).
[9] M. Nahum, T. M. Eiles and J. M. Martinis, Appl. Phys. Lett. 65, 3123 (1994).
[10] M. M. Leivo, J. P. Pekola, and D. V. Averin, Appl. Phys. Lett. 68, 1996 (1996).
[11] L. Kuzmin, I. Agulo, M. Fominsky, A. Savin, and M. Tarasov. Optimization of the electron cooling by SIN tunnel junctions. Supercon. Sci. Technol., 17, S400 (2004).
[12] A. J. Manninen, M. M. Leivo, and J. P. Pekola Refrigeration of a dielectric membrane by SINIS tunneling. Appl. Phys. Lett. 70, 1885 (1997);
[13] N. A. Miller, G. C. O'Neil, J. A. Beall, G. C. Hilton, K. D.Irwin, D. R. Schmidt, L. R. Vale, and J. N. Ullom, Appl. Phys. Lett. 92, 163501 (2008).
[14] N. Vercruyssen, R. Barends, T. M. Klapwijk, J. T. Muhonen, M. Meschke, and J. P. Pekola, Appl. Phys. Lett. 99, 062509 (2011).
[15] H. Q. Nguyen,M. Meschke, and J. P. Pekola, A robust platform cooled by superconducting electronic refrigerators, Applied Physics Letters 106, 012601 (2015).
[16] L. Kuzmin and D. Golubev "On the concept of an optimal hot-electron bolometer with NIS tunnel junctions". Physica C 372-376, 378-382 (2002).
[17] K. Irwin. Applied Physics Letters 66, 1998 (1995).
[18] L. Kuzmin, "2D Array of Cold-Electron Nanobolometers with Double Polarization Cross-Dipole Antennae", Nanoscale Research Lett., 7:224 (2012).
[19] L. Kuzmin and A. Sobolev. "Wideband Double-Polarized Array of Cold-Electron Bolometers for OLIMPO Balloon Telescope", to be submitted to Nature Scientific Reports (December 2017).
[20] L. Kuzmin, "An array of cold-electron bolometers with SIN tunnel junctions and JFET readout for cosmology instruments," in Journal of Physics: Conference Series, vol. 97, 2008, p. 012310
[21] A.V. Gordeeva, et al., "Observation of photon noise by cold-electron bolometers", Appl. Phys. Lett., 110, 162603 (2017).

IOP Conf. Series: Journal of Physics: Conf. Series **969** (2018) 012070 doi:10.1088/1742-6596/969/1/012070

# Crystal Growth and Neutron Scattering Study of Spin Correlations of the T′-Structured $Pr_{2-x}Ca_xCuO_4$

**Masaki Fujita**
Institute for Materials Research, Tohoku University, Sendai, Miyagi 980-8577, Japan

**Kenji Tsutsumi, Tomohiro Miura**
Department of Physics, Tohoku University, Sendai, Miyagi 980-8578, Japan

**Sergey Danilkin**
Bragg Institute, ANSTO, Locked Bag 2001, Kirrawee DC NSW 2232, Australia

E-mail: fujita@imr.tohoku.ac.jp

**Abstract.** We studied Ca-doping effect on spin correlations in $T'$-structured cuprate oxide $RE_2CuO_4$ (RE: rear earth) with growing a sizable single crystal of $Pr_{2-x}Ca_xCuO_4$ ($x$=0.05 and 0.10) as well as synthesizing powder samples of $Pr_{1.90}Ca_{0.10}CuO_4$ and $Eu_{1.90}Ca_{0.10}CuO_4$. In the all as-prepared and annealed samples, no evidence of shielding signal associated with superconductivity was observed by magnetic susceptibility measurement. Elastic neutron scattering measurements on the as-grown $Pr_{1.90}Ca_{0.10}CuO_4$ clarified the existence of long-ranged magnetic order with commensurate spin correlation. The ordering temperature was determined to be ~290K, which is comparable to Néel temperature in the parent compound of $Pr_2CuO_4$. The existence of commensurate low-energy spin excitation was also confirmed by inelastic neutron scattering measurements for the annealed $Pr_{1.95}Ca_{0.05}CuO_4$. These results strongly suggest a negligible Ca-doping effect on the physical properties in $T'$-$RE_2CuO_4$, which is quite different from the drastic doping evolution of magnetism in $T$-$RE_2CuO_4$.

## 1. Introduction

High transition temperature (high-$T_c$) superconductivity in cuprate oxide emerges with doping of either types of carrier into an antiferromagnetically ordered Mott insulator. In the $RE_2CuO_4$ ($R$:rear earth) system, which has a simple crystal structure, the electron-hole symmetry of magnetism has been intensively studied in connection with a common mechanism of superconductivity. In $La_2CuO_4$ having distorted $K_2NiF_4$-type ($T$) crystal structure, long-ranged antiferromagnetic (AFM) order is destroyed by a small amount of hole-doping, while that in $Nd_2CuO_4$-type ($T'$) structured $RE_2CuO_4$ ($RE$=Pr, Nd, Pm, Sm, Gd) with no apical oxygen is quite robust against electron-doping [1]. In both systems, AFM spin fluctuations remain in the superconducting (SC) phase, indicating a common relation between the spin correlations and the superconductivity. However, a recent muon spin rotation/relaxation experiment on $T'$-$La_2CuO_4$ showed that Néel temperature ($T_N$) is much lower than that in $T$-$La_2CuO_4$ [2], implying that the crystal structure is an important factor to determine the physical properties. Furthermore, in the electron-doped $T'$-system, the as-grown sample does not show superconductivity and an adequate annealing procedure is required for the emergence of superconductivity, while the as-grown hole-doped $T$-system undergoes the superconducting (SC) transition. Thus, the study of physical properties in a system with identical crystal structure is required to clarify the genuine electron-hole symmetry.

Tsukada *et al.* first achieved Ce-doping into thin film of $T$-$La_2CuO_4$, which is never succeeded for bulk sample, and reported the resistivity in samples with different Ce concentrations [3]. They claimed a breaking down of electron-hole symmetry at zero doping from the enhancement of insulating nature by Ce doping. Furthermore, Segawa *et al.* reported a different stacking pattern of AFM layers in hole and electron sides of $Y_{1-z}La_z(Ba_{1-x}La_x)_2Cu_3O_y$ [4]. Recently, the abrupt drop of $T_N$ in the film of $T'$-$La_{2-x}Ce_xCuO_{4-\delta}$ similar to the sharp phase boundary of three- dimensional magnetic order in $T$-$La_{2-x}Sr_xCuO_4$ [5] and even the appearance of superconductivity in the $T'$-$La_{1.8-x}Eu_{0.2}Sr_xCuO_4$ [6] have been reported. Although these results are attractive for the study of electron-hole symmetry in the identical crystal structure, no systematic study of spin fluctuation on these compounds has been done, due to the difficulties in the availability of large sample and/or in the sufficient carrier doping for both electron and hole sides.

In order to make progress on this issue, we have explored the adequate system of which sizable single crystal can be available. In this paper, we report the result of crystal growth of hole-doped $T'$-system and the first neutron scattering measurement of $Pr_{2-x}Ca_xCuO_4$.

## 2. Experimental Details and Results

### *2.1. Sample preparation*

For the study of electron-hole symmetry of magnetism, we selected $T'$-$RE_2CuO_4$ (RE: rear earth) and attempted the hole-doping by substituting bivalent ion such as $Ca^{2+}$, $Sr^{2+}$ and $Ba^{2+}$ for $RE^{3+}$ ion. Although there are few reports which show the phase stability of $T'$-$(Pr,Nd)_{2-x}(Ca,Sr)_xCuO_4$ for small $x$ [7], no physical properties were so far obtained for single crystal. Therefore, we have made efforts to grow a sizable single crystal with the aim of performing the neutron scattering study of spin correlations. We first prepared the powder sample of $Pr_{2-x}Ca_xCuO_4$ and $Eu_{2-x}Ca_xCuO_4$ by a standard solid-state reaction. The appropriate ratio of dried $Pr_6O_{11}$, $Eu_2O_3$, $CaCO_3$, and CuO powders were mixed and sintered at 900 °C for 12 hour with a couple of intermediate re-grindings. Then, we applied a traveling-

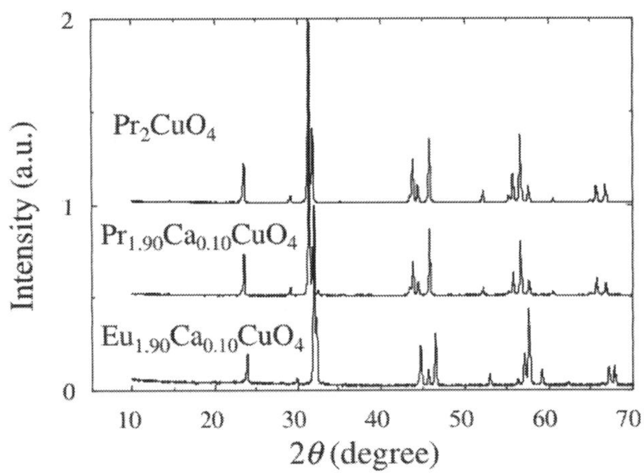


**Figure 1.** X-ray powder diffraction pattern of $T'$-structured $Pr_2CuO_4$, $Pr_{1.90}Ca_{0.10}CuO_4$ and $Eu_{1.90}Ca_{0.10}CuO_4$.

solvent floating-zone method to grow the single crystal of $Pr_{2-x}Ca_xCuO_4$ with $x = 0.05$ and 0.10. The growth condition is almost the same as that for $La_{2-x}Sr_xCuO_4$ [8]. The single crystal of $Pr_2CuO_4$ and and pressed powder of CuO were used as the seed rod and the solvent, respectively. The grown crystalline rod is columnar in the shape with length of ~100 mm and diameter of ~6 mm.

### *2.2. Sample characterization*

The actual concentrations of Pr, Ca and Cu in the single crystal were confirmed to be the same as nominal ones by inductively-coupled plasma. We checked the phase purity of samples by X-ray powder diffraction measurement. In Fig. 1, the diffraction pattern for powder sample of $Pr_{1.90}Ca_{0.10}CuO_4$ and $Eu_{1.90}Ca_{0.10}CuO_4$ is shown together with Ca-free $Pr_2CuO_4$. For the clear visualization, the result for $Pr_{1.90}Ca_{0.10}CuO_4$ and $Pr_2CuO_4$ is shifted along the vertical direction. Similar to the result of $Pr_2CuO_4$, all reflections in Ca-doped samples can be labeled in the I4/mmm symmetry, indicating that the single phased sample is successfully obtained. The evaluated in-plane (a) and out-plane (c) lattice constants are 3.949 Å and 12.20 Å for $Pr_{1.90}Ca_{0.10}CuO_4$, and 3.900 Å and 11.90 Å for $Eu_{1.90}Ca_{0.10}CuO_4$, respectively. The identical results were obtained for the powdered single crystal of $Pr_{1.90}Ca_{0.10}CuO_4$. The mosaic spread ($\theta_M$) of crystal was checked by neutron scattering. We confirmed that the width of Bragg peak is the resolution limited value, corresponding to $\theta_M$ less than 0.2° in angle unit. From these results, we conclude that a high-quality single crystal of hole-dope $T'$-$Pr_{2-x}Ca_xCuO_4$ was successfully obtained for the first time.

In order to examine whether the post-annealing affects superconductivity or induces superconductivity, single crystal of $Pr_{1.90}Ca_{0.10}CuO_4$ was annealed under oxidizing and deoxidizing conditions; (i) annealing under oxygen gas flow at 900 °C for 24 hours and (ii) annealing under argon gas flow at 900 °C for 10 hours. The variation of oxygen contents in molar ratio, which was evaluated from the weight change by annealing, was <0.01 (oxygen gain) and ~0.07 (oxygen loss) for conditions (i) and (ii), respectively. The magnetic susceptibility ($\chi$) was measured for all as-prepared and annealed samples by a SQUID magnetometer. Figure 2 shows

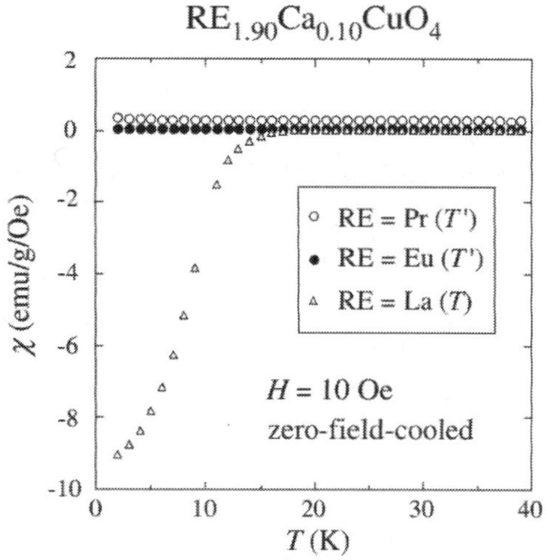


**Figure 2.** Magnetic susceptibility of $T'$-$Pr_{1.90}Ca_{0.10}CuO_4$ and $T'$-$Eu_{1.90}Ca_{0.10}CuO_4$ measured at 10 Oe after zero-field-cooling process. No evidence of superconducting transition was detected in these samples. For the comparison, our result for superconducting $T$-$La_{1.90}Ca_{0.10}CuO_4$ is shown.

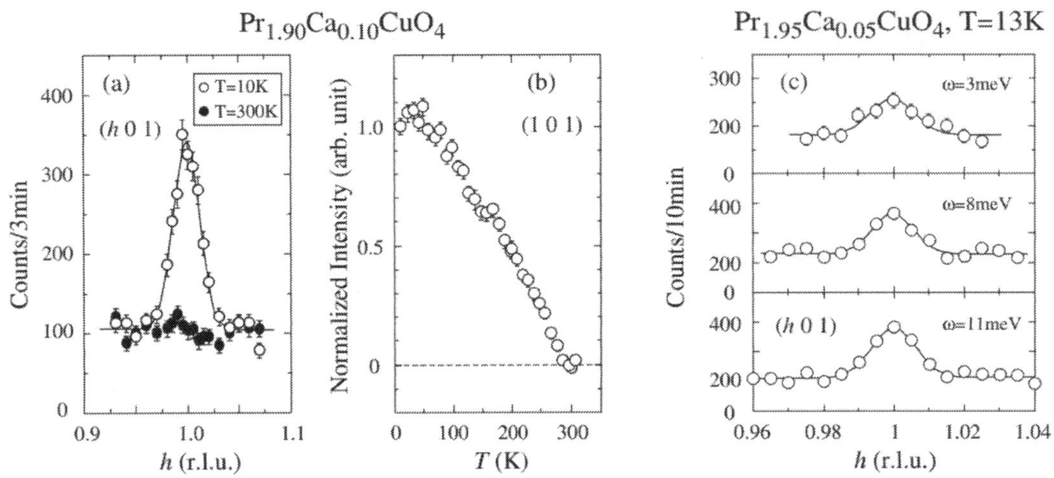


**Figure 3.** (a) Magnetic peak profile measured at 10K and 300K in the as-grown $T'$-$Pr_{1.90}Ca_{0.10}CuO_4$. (b) Temperature dependence of normalized peak-intensity at (1 0 1). (c) Inelastic neutron scattering spectra for annealed $Pr_{1.95}Ca_{0.05}CuO_4$ measured at 13 K and at 3, 8 and 11 meV.

the temperature dependence of $\chi$ of the as-grown samples measured under the magnetic field of 10 Oe after zero-field cooling process. As seen in the figure, no evidence of superconductivity was observed in the $T'$-$R_{1.90}Ca_{0.10}CuO_4$, while $\chi$ for the $T$-$La_{1.90}Ca_{0.10}CuO_4$ shows a shielding signal below ~15 K. Qualitatively the same result was obtained for both oxygen-annealed and argon-annealed $T'$-samples. Therefore, the emergence of superconductivity in the hole-doped system strongly depends on the crystal structure.

*2.3. Neutron scattering measurement*

Absence of superconductivity in the Ca substituted system implies the existence of competing magnetic order at low temperature. The nature of spin correlation, namely commensurate or incommensurate, is important to be clarified for understanding the relation between the magnetism and superconductivity. We, therefore, carried out neutron scattering, which is a momentum selective method, on the single crystal of $Pr_{2-x}Ca_xCuO_4$. The elastic and inelastic measurements were done with using triple-axis spectrometer ANAKE installed at the JRR-3 research reactor in Japan Atomic Energy Agency and TAIPAN at Bragg Institute in Australian Centre for Neutron Scattering, respectively. The incident (final) energy of neutron was fixed to be 19.6 meV (14.7 meV) at AKANE (TAIPAN) and the sample was mounted so that [1 0 0] and [0 0 1] orthorhombic crystalline directions are in the horizontal scattering plane. Note that in this paper we use the orthorhombic notation following the previous works for $T'$-$RE_{2-x}Ce_xCuO_4$ /citeKeimer92. In this notation, (1 0 1) corresponds to the magnetic Bragg positions in the pristine $Pr_2CuO_4$.

Figure 3(a) shows the profile of elastic neutron scattering from $Pr_{1.90}Ca_{0.10}CuO_4$ crystal. Scan was done along [1 0 0] direction through (1 0 1). At low temperature, clear magnetic Bragg peak with the resolution-limited width was observed, indicating the existence of long-ranged AFM order as observed in non-doped $Pr_2CuO_4$. The temperature dependence of normalized peak-intensity is plotted in Fig. 3(b).Upon cooling, the intensity starts to develop at ~290 K, which is the same as $T_N$ in $Pr_2CuO_4$.

To see the dynamical spin correlations, we measured the magnetic excitations in the annealed

$Pr_{1.95}Ca_{0.05}CuO_4$ up to 11 meV. The observed spectra for 3, 8 and 11 meV are shown in Fig. 3(c). In the measured energy region, a single peak center at (1 0 1) was detected, meaning that commensurate correlation is robust against the energy transfer. These results strongly suggest that the bulk magnetism is not affected by Ca-doping at least for $x \leq 0.1$. The negligible Ca-doping effect on magnetism is quite contrastive to the quick drop of $T_N$ by hole-doping in the $T$-structured system.

## 3. Discussion

Here, we discuss the origin of robust spin correlations against chemical doping and absence of superconductivity in the $T'$-$RE_2CuO_4$. In the electron-doped $T'$-$RE_{2-x}Ce_xCuO_4$, superconductivity emerges after the suppression of AFM order by reduction annealing. The removal of apical oxygen partially existing in the as-prepared sample is considered to be crucial for the appearance of superconductivity in $T'$-$RE_{2-x}Ce_xCuO_4$ [9]. In the present $T'$-system, both as-prepared and reduction annealed samples are non-superconductor. Thus, the existence of apical oxygen would not be the reason for emergence of commensurate spin correlations and absence of superconductivity. They would be intrinsic natures of $T'$-$RE_{2-x}Ce_xCuO_4$.

One possible explanation for the persistence of long-ranged magnetic order is that the mobile carrier is not effectively doped by Ca-substitution. In $YBa_2CuO_{7-\delta}$ (YBCO), Pr-substitution induces a remarkable suppression of superconductivity, and the hybridization of Pr $4f$ and O $2p$ orbitals is considered to be the origin of hole-localization [10]. Therefore, one may think that similar hole-trapping effect by Pr ions results into the robust AFM ordered phase. If only the Pr ion among the rear earth ions causes the hole-localization, we could expect the appearance of superconductivity in the other $4f$ ions-based $T'$-$RE_2CuO_4$ by Ca substitution. However, the present experimental fact showing that $Eu_{1.90}Ca_{0.10}CuO_4$ does not shows superconductivity is inconsistent with the above scenario. Thus, the orbital-hybridization is not the dominant factor for the persistence of AFM order against Ca(Sr)-doping. Another possible reason is that the hopping of holes within a $CuO_2$ plane is restricted due to the weak overlaps of Cu $3d$ and O $2p$ orbitals. The in-plane lattice constant in the present system is larger than that in $T$-$RE_2CuO_4$, resulting into the weaker overlaps of orbitals in $T'$-system. Since the mobility of holes is related with the reduction of $T_N$ [11], no increase of mobility leads the unchanged $T_N$. In this context, by applying chemical and/or physical pressures, the mobility of holes could be enhanced by the greater overlaps of orbitals. As a result, AFM order expected to degrade analogous to the case of $T$-LSCO.

We believe that further systematic studies of spin correlations for hole-doped $Pr_{2-x}Ca_xCuO_4$ combined with that for $La_{2-x}Sr_xCuO_4$ and electron-doped $Pr_{2-x}Ce_xCuO_4$ provides an important opportunity to clarify not only the common hole-doping effect on the spin correlations but also the genuine electron-hole symmetry of physical properties regardless of the crystal structure.

## Summary

We obtained the single phase of $T'$-structured $Pr_{1.90}Ca_{0.10}CuO_4$ and $Eu_{1.90}Ca_{0.10}CuO_4$ by a solid-state reaction and succeed in growing the single crystal of $Pr_{2-x}Ca_xCuO_4$ with $x$=0.05 and 0.10 for the first time. Magnetic susceptibility measurement clarified the absence of superconductivity in all as-grown and annealed samples. Elastic neutron scattering measurement showed the development of long-ranged AFM order in as-grown $Pr_{1.90}Ca_{0.10}CuO_4$ below ~290K. The inelastic magnetic signal was also confirmed at the commensurate (0.5, 0.5, 0) reciprocal position, as is the case for the parent $Pr_2CuO_4$. These results are quite different from the drastic doping evolution of superconductivity and magnetism in $T$-$La_{2-x}Sr_xCuO_4$, indicating a contrasting Ca doping effect on the physical properties in $T'$ and $T$ systems.

**Acknowledgments**

M.F. is supported by Grant-in-Aid for Scientific Research (A) (16H02125). Travel expenses for the inelastic scattering experiment performed using TAIPAN at ANSTO, Australia, were supported by General User Program for Neutron Scattering Experiments, Institute for Solid State Physics, The University of Tokyo (proposal no. 12539), at JRR-3, Japan Atomic Energy Agency, Tokai, Japan.

**References**

[1] Keimer B, Belk N, Birgeneau R J, Cassanho A, Chen C Y, Greven M, Kastner M A, Aharony A, Endoh Y, Erwin R W, Shirane G, 1992 Phys. Rev. B **46**, 14034
[2] Hord R, Luetkens H, Pascua G, Buckow A, Hofmann K, Krockenberger Y, Kurian J, Maeter H, Klauss H -H, Pomjakushin V, Suter A, Albert B, Alff, 2010 Phys. Rev. B **82**, 180508
[3] Tsukada A, Krockenberger Y, Noda M, Yamamoto H, Manske D, Alff L, Naito M, 2005 Solid State Comm. **133**, 427
[4] Segawa K, Kofu M, Lee S -H, Tsukada I, Hiraka H, Fujita M, Chang S, Yamada Y, Ando Y, 2010 Nature Physics **6**, 579
[5] Saadaoui H, Salman Z, Luetkens H, Prokscha T, Suter A, MacFarlane W A, Jiang Y, Jin K, Greene R L, Morenzoni E, Kiefl R F 2015 Nature Communications **6**, 6041
[6] Takamatsu T, Kato M, Noji T, Koike Y, 2012 Appl. Phys. Express **5**, 073101
[7] Hwang H Y, Cheong S -W, Cooper A S, Rupp Jr. L W, Batlogg B, 1992 Physica C **3-4**, 362, Rosseinky Matthew J, Prassides Kosmas, 1992 Physica B **180-181** 408
[8] Fujita M, Yamada K, Hiraka H, Gehring P M, Lee S H, Wakimoto S, Shirane G, 2002 Phys. Rev. B **65**, 064505
[9] Radaelli P G, Jorgensen J D, Schultz A J, Peng J L, Greene R, L, 1994 Phys. Rev. B **49**, 322, Schultz A, Jorgensen J, Peng J, Greene R, 1996 Phys. Rev. B **53**, 5157
[10] Fink J, Nücker N, Romberg H, Alexander M, Maple M B, Neumeier J J, Allen J W, 1990 Phys. Rev. B **42**, 4823
[11] Hücker M, Kataev M V, Pommer J, Harraß J, Hosni A, Pflitsch C, Gross R, Büchner B, 1999 Phys. Rev. B **59**, R725

IOP Conf. Series: Journal of Physics: Conf. Series **969** (2018) 012071 doi:10.1088/1742-6596/969/1/012071

# STM/STS study on electronic superstructures in the superconducting state of high-$T_c$ cuprate $Bi_2Sr_2CaCu_2O_{8+\delta}$

**S Mizuta[1], T Kurosawa[1], K Takeyama[2], N Momono[3], Y Ishii[1], H Yoshida[1], M Oda[1] and M Ido[1]**

[1] Department of Physics, Hokkaido University, Sapporo 060-0810, Japan
[2] Department of Physics, Asahikawa Medical University, Asahikawa 078-8510, Japan
[3] Department of Applied Sciences, Muroran Institute of Technology, Muroran 050-8585, Japan

E-mail: moda@sci.hokudai.ac.jp

**Abstract.** We report STM/STS measurements at 8 K in underdoped $Bi_2Sr_2CaCu_2O_{8+\delta}$ crystals ($T_c = 76$ K and hole-doping level $p \sim 0.12$) whose energy spectra around the Fermi level are characterized by a two-gap structure consisting of spatially inhomogeneous pseudogap (PG) and comparatively homogeneous superconducting gap (SCG). Two electronic superstructures, checkerboard modulation (CBM) and Cu-O-Cu bond-centered modulation (BCM), are observed with mapping spectral weights at low energies within the SCG and the ratio of spectral weights at $\pm\Delta_{PG}$ (PG energy), respectively. On the basis of the present findings, we suggest that the lower-energy scale CBM is an intrinsic property of Cu-O planes and can coexist with the BCM whose characteristic energy is $\sim\Delta_{PG}$ in identical regions in real space.

## 1. Introduction

In high-$T_c$ cuprates, it has been demonstrated by many spectroscopic techniques such as ARPES and STM/STS that the electronic states around the Fermi level exhibit distinct features in different momentum and/or energy regions [1–3]. The Fermi surface (FS) is roughly divided into two parts which are located around the nodal and antinodal points of a gap with $d_{x^2-y^2}$ symmetry, respectively (fig. 1(h)). In the antinodal region, consisting of incoherent electronic states, a pseudogap (PG) develops below around temperature $T^*$ higher than $T_c$, and another energy gap, which has been considered to be due to pairing, also develops successively below around temperature $T_{pair}$ [4, 5]. It has been revealed in STM/STS experiments on high-$T_c$ cuprates $Bi_2Sr_2CaCu_2O_{8+\delta}$ (Bi2212), $Ca_{2-x}Na_xCuO_2Cl_2$ (Na-CCOC) and so on that the antinodal PG is very sensitive to disorders and tends to be spatially inhomogeneous in nanometer scale, although the pairing gap, whose amplitude is hereafter referred to as $\Delta_0$, is comparatively homogeneous [3]. In particular, the PG inhomogeneity becomes marked in the underdoped (UD) region, in which the energy size $\Delta_{PG}$ seems to vary over a wide range from $\sim\Delta_0$ to a several times larger energy [3]. On the other hand, the nodal region, the so-called "Fermi arc," consisting of coherent electronic states, plays a role in the occurrence of superconductivity; a homogeneous $d$-wave superconducting (SC) gap (SCG) is formed in this region, which has been confirmed with the observation of energy-dependent Bogoliubov quasiparticle (QP) interference (QPI) patterns by STM/STS [6]. The nodal SCG smoothly connects with the antinodal pairing

IOP Conf. Series: Journal of Physics: Conf. Series **969** (2018) 012071 doi:10.1088/1742-6596/969/1/012071

gap. In the SC state, therefore, a *d*-wave gap seems to be formed apparently over the entire FS, in addition to the inhomogeneous PG developing in the antinodal region (fig. 1(h)).

It has been reported in UD Bi2212 and Na-CCOC crystals that besides the QPI patterns, other two electronic superstructures can also be observed by STM/STS. One of them is two-dimensional short-range checkerboard modulation (CBM) along the two Cu-O bond directions with a period of $4a \times 4a$ ($a$: the lattice constant) [3, 7–12]. The other is Cu-O-Cu bond-centered modulation (BCM) with $4a$-wide unidirectional nano-domains which are oriented randomly along the two Cu-O bond directions [6, 13]. The characteristic energies of these electronic superstructures are of great interest for the understanding of their origins and relationship with the superconductivity and the PG phenomenon. Kohsaka *et al.* have demonstrated that BCM is observed by mapping the ratio of STS spectral weights at $\pm\Delta_{\rm PG}$, or the ratio of conductance peak heights [6], suggesting that the characteristic energy of BCM is of the order of the PG size. On the other hand, it has yet to reach a consensus on the energy region in which the CBM is observed. Furthermore, it remains unclear whether these two electronic superstructures coexist in identical regions in real space.

In the present study, we performed low-temperature ($T$ = 8 K) STM/STS experiments in UD Bi2212 single crystals with $T_c$ = 76 K and $p \sim 0.12$, and clarified the above mentioned problems for the electronic superstructures.

## 2. Experimental

In the present STM/STS experiments, we used two methods to obtain conductance ($dI/dV$) images corresponding to images of the local density of states (LDOS) at specified energies. The first one is the conventional method for STS, in which the bias voltage ($V_s$) dependences of $dI/dV$, hereafter referred to as STS spectra, are first measured at many positions in an area and then, a $dI/dV$ image is constructed with the mapping of $dI/dV$ values at a specified $V_s$ in the measured STS spectra. In this method, the tip height is set every measurement position so as to give a constant current (CC) under a high set-up bias voltage. Thus, $dI/dV$ images in the conventional method are referred to as CC $dI/dV$ images. On the other hand, $dI/dV$ images in the other method are referred to as constant height (CH) $dI/dV$ images for the following reason: In this method, the bias voltage is modulated with a small amplitude while taking a current image in CH mode, and the output of a lock-in amplifier corresponding to $dI/dV$ is directly imaged [14]. The $dI/dV$ imaging in CH mode or the CH $dI/dV$ imaging will be substantially free from the so-called "set-point effect," which has been pointed out to appear on CC $dI/dV$ images obtained by the conventional method [13, 15].

## 3. Results and Discussion

Figure 1 shows results obtained with the conventional method in the UD Bi2212 crystal. STS spectra were measured at 128×128 points within a square region of ~13×13 nm$^2$. A representative STS spectrum, shown in fig. 1(f), is consistent with a two-gap structure; it exhibits broad PG peaks at $|V_s| \sim 65$ mV and a subgap peak or shoulder at $|V_s| \sim 25$ mV. The subgap size $\Delta_{\rm SG}$, which is comparatively homogeneous, 25 ~ 30 meV for the entire region examined, is smaller than the *d*-wave SCG amplitude or pairing gap size expected at the antinodal point $\Delta_0 \sim 40$ meV, which has been estimated in ARPES and STM/STS experiments [3, 16], but it is still close to $\Delta_0$. Such a subgap structure in STS spectra can be explained in terms of the superconductivity, because if the density of states near the Fermi level is completely suppressed in the vicinity of the antinodal point with the marked development of PG and no pairing gap opens up there, leading to a reduction of the effective pairing gap size.

Two-dimensional short-range CBM along the two Cu-O bond directions is observed in the CC $dI/dV$ image for $V_s$ = 25 mV (fig. 1(a)). In the Fourier transform (FT) map, where $q_x$ and $q_y$ axes are taken to be oriented towards the Bragg spots and reciprocal lattice units for both

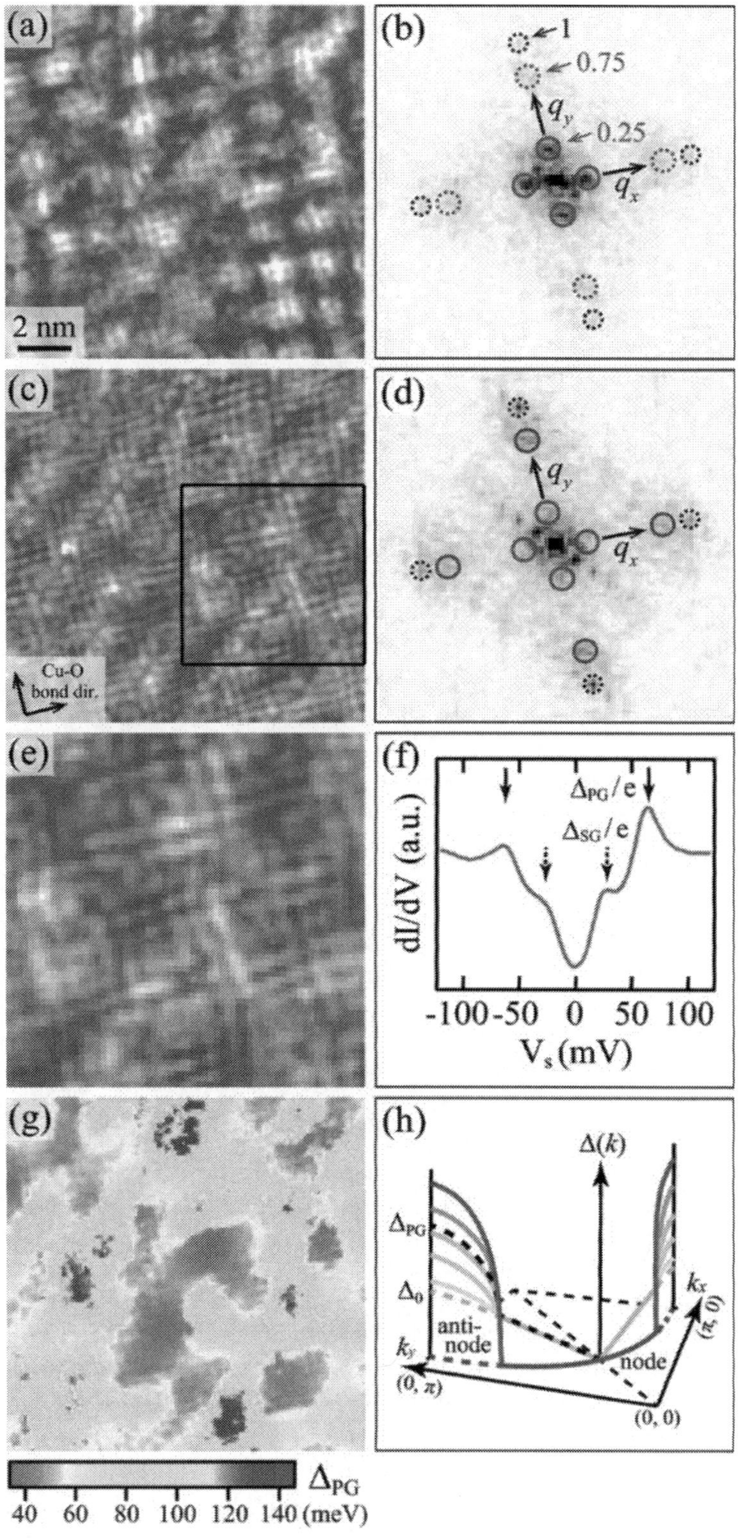


**Figure 1.** (a) CC $dI/dV$ image for $V_s$ = 25 mV and (b) the corresponding FT map. The black circles indicate the Bragg spots and, the red and blue ones denote Fourier spots at $|q| \sim 0.25$ and 0.75 on the $q_x$ and $q_y$ axes, respectively. (c) $Z(\boldsymbol{r}, e{=}1)$ map and (d) the corresponding FT map. (e) Magnified image of the boxed region in (c). (f) Representative example of STS spectra. (g) Map for PG size $\Delta_{PG}$. (h) Schematic illustration of energy gaps on the Fermi surface at $T \ll T_c$. The black broken line represents the spatial average of partly inhomogeneous gap dispersions.

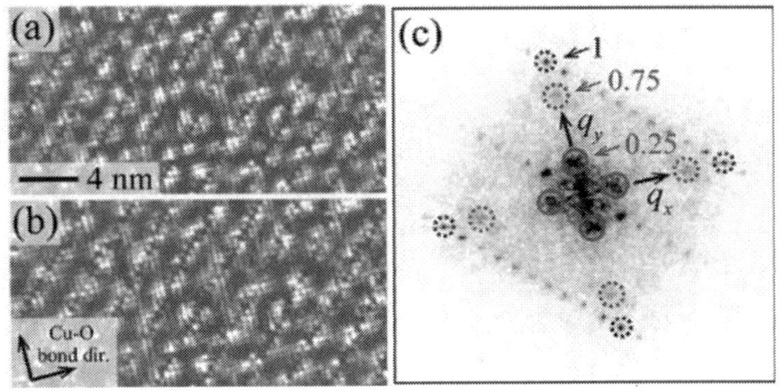


**Figure 2.** (a) CH $dI/dV$ image for $V_s = 20$ mV and (b) for $V_s = -20$ mV. (c) FT map of the image for $V_s = 20$ mV. The black circles indicate the Bragg spots and, the red and blue ones denote Fourier spots at $|q| \sim 0.25$ and 0.75 in the $q_x$ and $q_y$ axes, respectively.

axes are defined as $2\pi/a = 1$, strong spots corresponding to the CBM are located at $|q| \sim 0.25$ on the $q_x$ and $q_y$ axes (fig. 1(b)), and their positions are independent of $V_s$ (or energy); that is, the CBM is non-dispersive modulation. Thus, the period of CBM observed in the UD crystal is $\sim 4a \times 4a$. From analyses on the $V_s$ dependence of the spot intensity at $|q| \sim 0.25$, it was also found that the CBM tends to be marked around an energy of $\sim \Delta_{SG}$, which is comparable to the effective pairing gap size. Furthermore, fainter spots are located at $|q| \sim 0.75$ on both axes in the FT map (fig. 1(b)). This result might suggest the existence of an internal structure in the CBM as reported in previous STM/STS studies [9]. In our experiments, however, the $|q| \sim 0.75$ components included in $dI/dV$ images at $|V_s| \sim \Delta_{SG}/e$ were extremely weak, varying in their relative intensities with respect to the $|q| \sim 0.25$ ones (figs. 1(b) and 2(c)), and would be therefore extrinsic to the CBM. They could be rather explained in terms of the BCM whose characteristic energy is of the order of $\Delta_{PG}$, taking into account the following feature of PG: The PG is not a complete gap but the so-called "soft gap" and therefore, the single QP spectra with central energies of $\sim \Delta_{PG}$, which are responsible for the BCM, would be very broad and have small but finite weights even at low energies around $\Delta_{SG}$.

Previous STM/STS studies pointed out a possibility that the CBM in CC $dI/dV$ images is affected by the set-point effect [13, 15]; at low energies below $\sim \Delta_0$, it may result from such an effect in the conventional method for STS. However, the CBM can be demonstrated by the CH $dI/dV$ imaging which will be substantially free from the set-point effect. Figure 2 shows CH $dI/dV$ images measured in an identical area for two bias voltages with opposite polarities, $V_s = \pm 20$ mV, and the corresponding FT map for $V_s = 20$ mV; the area is within the same cleaved surface as in the CC $dI/dV$ image (fig. 1(a)). One can see in fig. 2 that CBM, which is characterized by very strong spots at $|q| \sim 0.25$ on the $q_x$ and $q_y$ axes in the FT map, is observed in CH $dI/dV$ images for $|V_s| \sim \Delta_{SG}/e$. It should be noted here that the amplitude and phase of CBM in the real-space images (figs. 2(a) and (b)) are almost independent of bias polarity, although Neto *et al.* have reported that such an electronic superstructure is much more intense for positive sample bias voltages where STS spectra reflect electron-like QP states [17]. In fact, the CBM disappeared with mapping the ratio of $dI/dV$ values at the two bias voltages with opposite polarities ($V_s = \pm 20$ mV). On the basis of our findings obtained by the CH $dI/dV$ imaging, it is suggested that the CBM will be an intrinsic property of both electron- and hole-like QP states around $|E| \sim \Delta_{SG}$, at least in such UD Bi2212 crystals as the PG is strongly inhomogeneous in nanometer scale (see fig. 1(g)), which will be discussed in the next paragraph.

Shown in fig. 1(g) is a map of the PG size $\Delta_{\rm PG}$ determined from the peak position in STS spectra of the UD Bi2212 crystal (fig. 1(f)), which are the same with those used to construct the CC $dI/dV$ image for $V_s = 25$ mV (fig. 1(a)). In contrast to the subgap $\Delta_{\rm SG}$, the PG $\Delta_{\rm PG}$ is strongly inhomogeneous, varying over a wide energy range from ~40 to ~140 meV. Figure 1(c) is a map of the ratio of spectral weights at energies $E = \pm\Delta_{\rm PG}$, $Z(\boldsymbol{r},\ e \equiv E/\Delta_{\rm PG} = 1) \equiv \frac{dI}{dV}|_{+\Delta_{\rm PG}} / \frac{dI}{dV}|_{-\Delta_{\rm PG}}$, that is, a $Z(\boldsymbol{r},\ e = 1)$ map [6]. In this map, one can see unidirectional nonometer-scale domains, partially developing in some regions as in the boxed one, which is magnified two times in fig. 1(e) to make it more visible. As shown in the FT map (fig. 1(d)), comparatively strong spots are located at $|q| \sim 0.75$ on the $q_x$ and $q_y$ axes; furthermore, $|q| \sim 0.25$ spots can be confirmed although they are very weak. These features for the electronic superstructure observed around $\Delta_{\rm PG}$ are consistent with Cu-O-Cu bond-centered modulation (BCM), which consists of randomly oriented $4a$-wide nano-domains including an internal structure with a period of $4a/3$ [6, 13]. Interestingly, it is found from a comparison of the two images for $E \sim \Delta_{\rm SG}$ and $\Delta_{\rm PG}$ (figs. 1(a) and (c)) that the CBM is clearly observed in such regions as the BCM develops. This suggests that the two electronic superstructures with different characteristic energies can coexist in real space.

As is well known, the CBM is similar to one of the seven independent components in the QPI pattern, represented by wavevector $\boldsymbol{q}_1$. The $\boldsymbol{q}_1$ modulation, which has been demonstrated in STM/STS experiments on optimally doped Bi2212 under magnetic fields to be much more intense, together with the BCM, around vortex cores in the mixed state than in the pure SC state, has a weak dispersion at low bias voltages below $\sim\Delta_0/e$ and may have a phase depending on bias polarity [18]. On the other hand, the CBM, which develops markedly in the pure SC state, is non-dispersive and has a phase independent of bias polarity. Thus, the CBM seems to be distinct in some properties from the $\boldsymbol{q}_1$ modulation of QPI.

## 4. Summary

We performed STM/STS measurements at 8 K in UD Bi2212 crystals with $T_c = 76$ K and $p \sim 0.12$ to examine the characteristic energy of CBM and its mutual relationship with the BCM whose characteristic energy is of the order of the spatially inhomogeneous PG. CBM, which has an energy-independent period of $\sim 4a \times 4a$, was observed at low energies around the subgap energy $\Delta_{\rm SG}$, comparable to the SCG size, not only by the conventional method (CC $dI/dV$ imaging) but also by the CH $dI/dV$ imaging that will be free from the set-point effect. This observation indicates that such a low-energy scale superstructure is an intrinsic property of Cu-O planes. Furthermore, it was demonstrated that the BCM is observed in some of the regions where the CBM appears clearly, suggesting that the two electronic superstructures with different characteristic energies can coexist in identical regions in real space.

## Acknowledgments

We thank Professor Y. Toda for useful discussions. This work was partly supported by Grant-in-Aid for Scientific Research from the Ministry of Education, Culture, Sports, Science and Technology of Japan.

## References

[1] Tanaka K *et al.* 2006 *Science* **314** 1910
[2] Kondo T *et al.* 2007 *Phys. Rev. Lett.* **98** 267004
[3] Kurosawa T *et al.* 2010 *Phys. Rev.* B **81** 094519
[4] Kondo T *et al.* 2011 *Nat. Phys.* **7** 21
[5] Kondo T *et al.* 2015 *Nat. Commun.* **6** 7699
[6] Kohsaka Y *et al.* 2008 *Nature* **454** 1072
[7] Howald C *et al.* 2003 *Phys. Rev.* B **67** 014533
[8] Vershinin M *et al.* 2004 *Science* **303** 1995

[9] Hanaguri T *et al.* 2004 *Nature* **430** 1001
[10] Momono N *et al.* 2005 *J. Phys. Soc. Jpn.* **74** 2400
[11] McElroy K *et al.* 2005 *Phys. Rev. Lett.* **94** 197005
[12] Liu Y H *et al.* 2007 *Phys. Rev.* B **75** 212507
[13] Kohsaka Y *et al.* 2007 *Science* **315** 1380
[14] Kurosawa T *et al.* 2016 *J. Phys. Soc. Jpn.* **85** 044709
[15] Hanaguri T *et al.* 2007 *Nat. Phys.* **3** 865
[16] Vishik I M *et al.* 2012 *Proc. Natl. Acad. Sci. USA* **109** 18332
[17] da Silva Neto E H *et al.* 2014 *Science* **343** 393
[18] Machida T, *et al.* 2016 *Nat. Commun.* **7** 11747

IOP Conf. Series: Journal of Physics: Conf. Series **969** (2018) 012072 doi:10.1088/1742-6596/969/1/012072

# Magnetic-field-induced 1st order transition to FFLO state at paramagnetic limit in 2D superconductors

**N A Fortune[1], C C Agosta[2], S T Hannahs[3] and J A Schleuter[4]**

[1] Department of Physics, Smith College, Northampton MA, USA
[2] Department of Physics, Clark University, Worcester MA, USA
[3] National High Magnetic Field Laboratory, DC Field Facility, Tallahassee FL, USA
[4] Argonne National Laboratory, Materials Science Division, Argonne IL, USA

E-mail: nfortune@smith.edu

**Abstract.** We have recently reported the first direct calorimetric observation of a magnetic-field-induced first-order phase transition into a high-field FFLO superconducting state at the Clogston-Chandrasekar 'Pauli' paramagnetic limit$H_p$ in a 2D superconductor $\kappa-(\text{BEDT-TTF})_2\text{Cu(NCS)}_2$. The high-field state is both higher entropy and strongly paramagnetic, as thermodynamically required for the FFLO state. Here we compare our results with theoretical predictions for the field dependence of the high-field FFLO state in the 2D limit, revealing tentative evidence for transitions between FFLO states of differing order parameter. We also present calorimetric evidence for a 1st order phase transition into the FFLO state for a second 2D organic superconductor: $\beta''-(\text{BEDT-TTF})_2\text{SF}_5(\text{CH})_2(\text{CF})_2(\text{SO})_3$.

## 1. Introduction

What is the highest possible applied magnetic field in which superconductivity can exist for an electronically 2D superconductor? One possible answer derives from the paramagnetic spin susceptibility of the conduction electrons. Because the electrons in BCS superconducting pairs have oppositely aligned spins with momenta $(\mathbf{k}\uparrow, -\mathbf{k}\downarrow)$, the reduction in Zeeman energy that arises from realignment of the electron spin in an applied magnetic field must eventually exceed the reduction in electronic energy available from the formation of superconducting Cooper pairs as the magnetic field increases in strength. The critical magnetic field at which this should occur is known variously as the Clogston-Chadrasakar or Pauli paramagnetic limit $H_p$.

In practice, orbital scattering from magnetic vortices typically limits superconductivity in 'clean' (impurity free) limit materials to magnetic fields much lower than $H_p$. For a 2D superconductor, however, this scattering can be suppressed by aligning the magnetic field exactly parallel to the 2D superconducting planes (thereby confining the vortices to the regions between the planes). This has been confirmed in angle-dependent studies of the Pauli-limited 2D organic superconductor $\alpha-(\text{BEDT-TTF})_2\text{NH}_4(\text{SCN})_4$ [1].

Perhaps surprisingly, then, it is possible for superconductivity to persist even above this expected paramagnetic limit. One way for this to happen is if Cooper pair formation above $H_p$ occurs between Zeeman-split parts of the Fermi surface with momenta $(\mathbf{k}\uparrow, -\mathbf{k}+\mathbf{q}\downarrow)$, as first proposed by Fulde and Ferrell [2] and independently by Larkin and Ovchinnikov [3].

This 'FFLO' phase is only energetically favored to exist for $H_p \leq H \leq H_{c2}$, where for a 2D superconductor, the zero temperature upper critical field $H_{c2}(0)$ has a predicted value of

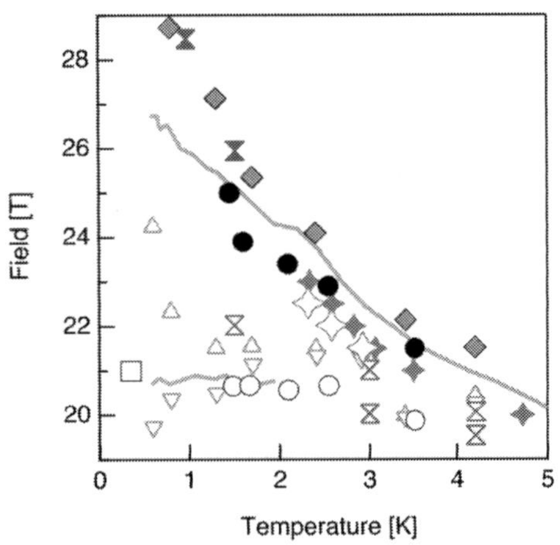


**Figure 1.** Previously reported transition points for the $H_{c2}$ (solid symbols) and FFLO (hollow symbols) phase boundaries for $\kappa-$(BEDT-TTF)$_2$Cu(NCS)$_2$ for a magnetic field applied parallel to the 2D superconducting layers. Our penetration depth measurements [4] are shown as green lines. Calorimetric measurements by Lortz [5] are shown as red stars. Magnetic-torque measurements by Bergk [6] and Tuschiya [7] are shown as blue diamonds/ triangles and purple hourglass symbols, respectively. NMR measurements by Wright [8] and Mayaffre [9] are shown as black squares and circles, respectively.

$H_{c2}^{2D}(0) \approx 1.42\ H_p$. [10–12]. In addition, the temperature cannot exceed a critical value $T^*$ where $T^* = 0.56\ T_c$ [11; 13]. For a 2D superconductor like $\kappa-(\mathrm{ET})_2\mathrm{Cu(NCS)}_2$ with a zero field superconducting critical temperature $T_c = 9.1 \pm 0.2$ K [4; 5] and a Pauli limit $H_p = 20.7 \pm 0.4$ T [4; 8], theory then predicts $T^* = 5.1 \pm 0.1$ K and $H_{c2}^{2D}(0) = 29.5 \pm 0.7$ T.

As noted by Matsuda and Shimahara [12], the $\mathbf{q}$ vector in the FFLO phase gives rise to spatial symmetry breaking, but because of the symmetry of the system, there is more than one equivalent $\mathbf{q}$ which gives the same upper critical field. In general, the superconducting order parameter $\Delta(\mathbf{r})$ at any particular magnetic field can be expressed as a linear combination of plane waves [3]. The 'FF' order parameter initially proposed by Fulde and Ferrell [2] corresponds to a single plane wave with $\Delta(\mathbf{r}) = \Delta_1 \exp^{i\mathbf{q}\cdot\mathbf{r}}$, which results in de-pairing of part of the Fermi surface at $H_p$ and an increase in entropy at the crossover from the BCS to the FF superconducting phase [2]. The 'LO' order parameter initially proposed by Larkin and Ovchinnikov [3] corresponds to the linear combination of two plane waves with $\Delta(\mathbf{r}) = 2\Delta_1 \cos(\mathbf{q}\cdot\mathbf{r})$, resulting in a 1D spatial oscillation of the order parameter with wavelength $2\pi/q$ and planes of paramagnetic spins at the nodes. Here too there is an increase in entropy upon crossing over to the LO phase [14; 15].

## 2. Experimental results for $\kappa-$(BEDT-TTF)$_2$Cu(NCS)$_2$

Previous calorimetric [5] and magnetic torque [6; 7] and measurements on this material have presented thermodynamic evidence for a change in curvature of the superconducting to normal state phase boundary $H_{c2}(T)$ at what was inferred to be the high temperature onset of an FFLO superconducting phase. They have, in addition, reported direct observations of an FFLO phase transition, but as shown in Fig. 1, the locations, curvatures, and temperature dependences of the reported phase boundaries are in contradiction with each other, with penetration depth measurements [4], and, at low temperature, with NMR measurements[8; 9]. The hollow red stars hugging the $H_{c2}(T)$ phase boundary represent a 1st order transition initially thought to represent the crossing of the FFLO phase boundary [5] but are now attributed [16] to a small out of plane component of the magnetic field and the corresponding emergence of a vortex phase near $H_{c2}$. The divergence in the magnetic torque measurements may reflect the difficulty in determining strongly magnetic-field-angle-dependent phase boundaries with a measurement that requires the rotation of a sample in magnetic field.

For comparison, our own recent calorimetric measurements of the magnetic-field-dependent specific heat for $\kappa-(\mathrm{ET})_2\mathrm{Cu(NCS)}_2$ for a magnetic field applied parallel to the 2D superconducting planes are shown in Figs. 2 and 3. Like the NMR measurements, our

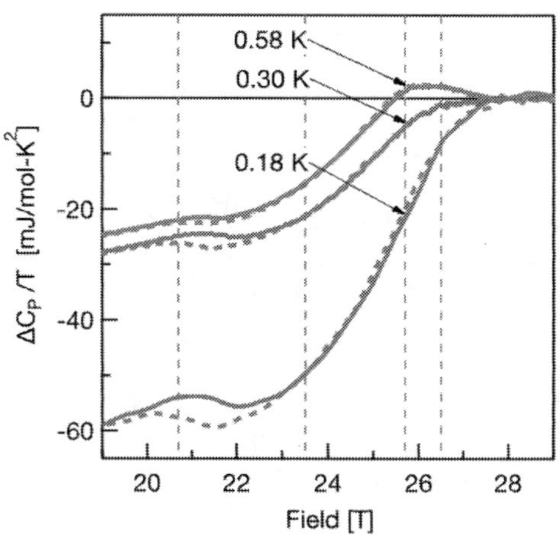


**Figure 2.** Magnetic-field induced change (solid for sweep up, dotted for sweep down) in the specific heat of $\kappa-(\mathrm{ET})_2\mathrm{Cu(NCS)}_2$ for fields applied parallel to the superconducting layers. $\Delta C/T$ is set equal to zero above $H_{c2}$. A hysteretic first order phase transition at $H_p = 20.7 \pm 0.4$ T into the FFLO superconducting phase followed by a rise in specific heat at the system approaches the normal state is visible at low temperature. The dashed vertical lines are guides to the eye marking $H_p$, the end of the hysteretic bubble in $C(H)$ above $H_p$, and breaks in the slope of $C(H)$ at fields corresponding to possible phase transitions within the FFLO region.

measurements have the advantage of being able to use the angle dependence of $H_{c2}(\theta)$ to orient the sample in field [17]. At low temperatures, as shown in Fig. 2, a hysteretic first order phase transition at $H_p = 20.7 \pm 0.4$ T into the FFLO superconducting phase followed by a rise in specific heat at the system approaches the normal state is visible, and as is shown the phase diagram presented in Fig. 4, the calorimetrically determined locations of both transitions are in good agreement with NMR measurements [8; 9]. Magnetocaloric measurements [18] indicate that the system above $H_p$ is paramagnetic, as expected for the FFLO state; the upward curvature of $C_p(H)$ is characteristic of strongly Pauli-paramagnetic superconductors at low T [19].

Looking closer within the FFLO superconducting region, we find additional fine structure in the form of (1) a hysteretic bubble in the specific heat $C_T(H)$ extending approximately 3 tesla above $H_p$ and (2) in the form of apparent changes in slope of $C_T(H)$ just below the upper critical field phase boundary $H_{c2}$ and the field-independent normal state. Interestingly, the collapse of the bubble in the 23 - 24 tesla region coincides with a pronounced drop in $|dM/dT|$ seen in the magnetocaloric effect. These phenomena are suggestive of an predicted change in the orientation of those spins at this field due to a change in the FFLO superconducting order parameter from 1D to 2D [13]. At higher temperatures, a broad peak emerges at high field, terminating at $H_{c2}(T)$. A strong temperature-dependent enhancement is seen in the NMR relaxation rate over the same field range at these temperatures [9], indicating that the Schottky-like peaks may arise from strong temperature-dependent spin flip scattering within the superconducting state.

### 3. Predictions for field-induced FFLO states in a 2D superconductor

The additional features seen in the specific heat and magnetocaloric effect above $H_p$ may indicate the existence of phase transitions within the FFLO region. Several theoretical proposals can be found in the literature [12] but we restrict ourselves here to a comparison of our experimental results with that expected for a 2D superconducting film in the FFLO state [13].

For a 2D superconducting film, Shimahara [13] predicts that the originally proposed 1D planar spatially modulated 'LO' order parameter [3] $\Delta(\mathbf{r}) = 2\Delta_1 \cos(\mathbf{q}\cdot\mathbf{r})$ remains the lowest energy configuration just below $T^*$ (where $T_c$ is the zero field superconducting critical temperature and $T^* = 0.56\ T_c$ is the FFLO superconducting critical temperature ) but as the temperature lowers still further, a sequence of transitions to other FFLO phases with larger numbers of nodes are predicted to occur. The argument is that increasing the number of plane waves increases the number of nodes, thereby increasing the area over which the superconducting order parameter is small, which in turn leads to a reduction in the Pauli paramagnetic energy of the quasiparticles

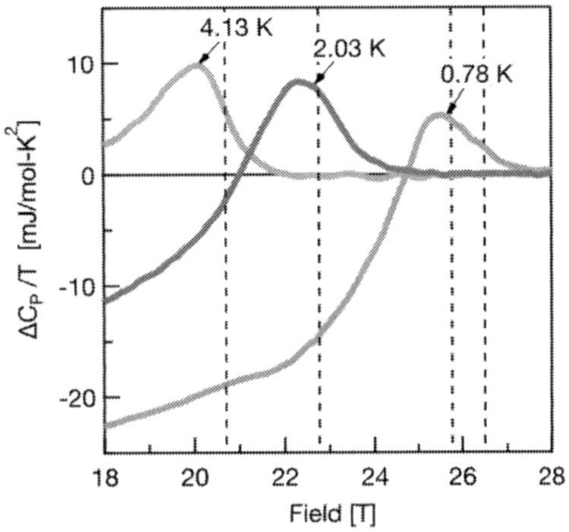


**Figure 3.** Relative change in the specific heat (scaled by temperature) of $\kappa-(\mathrm{BEDT\text{-}TTF})_2\mathrm{Cu(NCS)}_2$ with magnetic field for fields applied parallel to the superconducting layers. At higher temperatures, broad peak emerges at high field, terminating at $H_{c2}(T)$. A strong enhancement is seen in the NMR relaxation rate over the same field range at these temperatures [9]. The FFLO transition at $H_p$ is visible as a small bump at 0.78 K then as a slope change at higher T. Dashed lines are the same as in Fig. 2. Identified from changes in slope of $C(H)$ at lower T, they may correspond to the locations in field of phase transitions to FFLO states of different symmetries, as predicted for FFLO superconductors in the 2D limit [13].

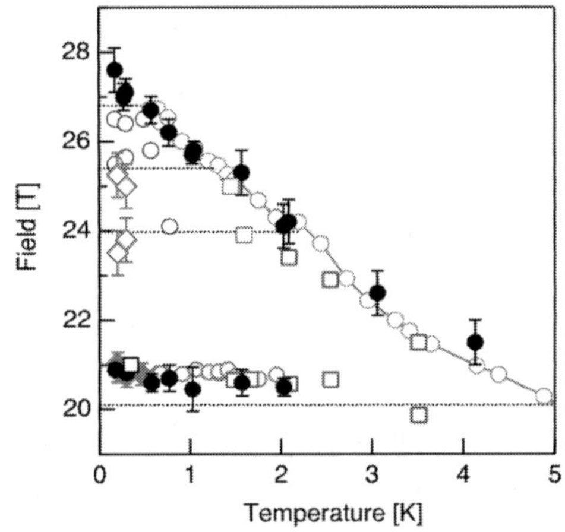


**Figure 4.** Phase diagram for $\kappa-(\mathrm{BEDT\text{-}TTF})_2\mathrm{Cu(NCS)}_2$, showing locations of the FFLO phase boundary at $H_p = 20.7$ T and $H_{c2}(T)$. Our calorimetric [18], magnetocaloric [18], and penetration depth [4] data are shown as solid black circles, solid red diamonds, and hollow green circles, respectively. NMR results are shown as hollow black [8] and blue [9] squares. Additional features in the calorimetric and magnetocaloric data are shown as hollow black circles and hollow red diamonds. Dashed lines represent predicted phase transitions [13] for a 2D superconductor with s-wave pairing to FFLO states of different symmetries. For d-wave pairing, only one transition within the FFLO state (at 26.5 T) should occur instead of the three shown here.

excited around the nodes and a consequent gain in spin-polarization energy [12; 13].

For a superconductor with s-wave pairing, four different FFLO phases are expected, beginning with the 1D planar LO phase for $0.24\ T_c \leq T \leq T^* = 0.56\ T_c$ followed by a state with triangular symmetry for $0.16\ T_c \leq T \leq 0.24\ T_c$, a state with square symmetry for $0.05\ T_c \leq T \leq 0.16\ T_c$, and finally a state with hexagonal symmetry for $T \leq 0.05\ T_c$. For a superconductor with d-wave pairing, the more restrictive symmetry of the superconducting pairing means that only two phases are expected: the LO phase for $0.06\ T_c \leq T < T^* = 0.56\ T_c$ and the square state for $T \leq 0.06\ T_c$. We can determine the fields at which field-induced transitions to the LO , triangle ($\triangle$), square ($\square$), and/or hexagonal ($\bigcirc$) states should occur by finding the value $H_{c2}(T)$ corresponding to each of these temperatures.

As the nature of the superconducting pairing in $\kappa-(\mathrm{BEDT\text{-}TTF})_2\mathrm{Cu(NCS)}_2$ is still a matter of debate [20], we consider each prediction in turn. For $\kappa-(\mathrm{BEDT\text{-}TTF})_2\mathrm{Cu(NCS)}_2$, Shimahara's theory for s-wave pairing predicts that field-induced transitions to various FFLO states should occur at $H_{LO} = 20.1$ T, $H_{\triangle} = 24.0$ T, $H_{\square} = 25.4$ T, and $H_{\bigcirc} = 26.8$ T. If we instead assume d-wave pairing, the transition to the 1D LO state again occurs at $H_{LO} = 20.1$ T followed by a single transition with the FFLO state to a 2D square state at $H_{\square} = 26.5$ T.

The critical fields for the predicted transitions within the FFLO state for s-wave pairing are shown as dashed lines in Fig. 4. As seen in Fig. 4, there is at least a plausible correspondence between theory and experiment, suggesting that there is indeed a cascade of field-induced phase transitions to FFLO states with differing 1D and 2D order parameters (provided that the features identified here in the specific heat and the magnetocaloric effect correspond to phase transitions and that the system can be properly modeled as a s-wave pairing superconductor in the 2D superconducting limit). Higher resolution measurements of specific heat and NMR as a function of field and temperature are needed to verify the tentative conclusion reached here. Of particular interest would be a comparison of measurements as a function of temperature for each of the proposed high field FFLO states.

## 4. Preliminary results for $\beta'' - (\mathrm{BEDT\text{-}TTF})_2\mathrm{SF}_5(\mathrm{CH})_2(\mathrm{CF})_2(\mathrm{SO})_3$

The calorimetric signature of the FFLO phase transition in $\kappa - (\mathrm{BEDT\text{-}TTF})_2\mathrm{Cu(NCS)}_2$ is strongly field-orientation dependent, disappearing for a rotation of less than 1° [18], as expected for a 2D superconductor where the exclusion of magnetic vortices from the superconducting layers needed to achieve the FFLO state relies on a parallel orientation of the applied field relative to the layers. If our interpretation is correct, however, then we should expect to see (1) corresponding first-order phase transitions at $H_p$ into FFLO states in related 2D organic superconductors in the clean impurity-scattering-free limit and (2) a strong field-angle-dependence of the FFLO states.

As a test, we have carried out preliminary results for angle-dependence of the heat capacity as a function of field $C(H)$ for the highly 2D superconductor $\beta'' - (\mathrm{BEDT\text{-}TTF})_2\mathrm{SF}_5(\mathrm{CH})_2(\mathrm{CF})_2(\mathrm{SO})_3$ with $T_c = 4.3$ K [16]) at 200 mK. As seen in Fig. 5, we observe for the field-parallel sample orientation a hysteric first order phase transition in a narrow region centered around $9.1 \pm 0.5$ tesla followed by transition to the normal state at approximately 13 tesla, in agreement with expectation. We also find that the calorimetric signature of the hysteretic 1st order FFLO phase transition disappears for a change in sample orientation $\Delta$ as small as 0.04° and that this change is also accompanied by a steep drop in the superconducting critical field. Although initial empirical determinations put $H_p$ and $H_{FFLO}$ for this material in the range of 9.7 T to 10.5 T [4; 16; 22], our lower value for $H_{FFLO}$ is consistent with both a recent NMR observation [21] of a phase transition from a uniform superconducting

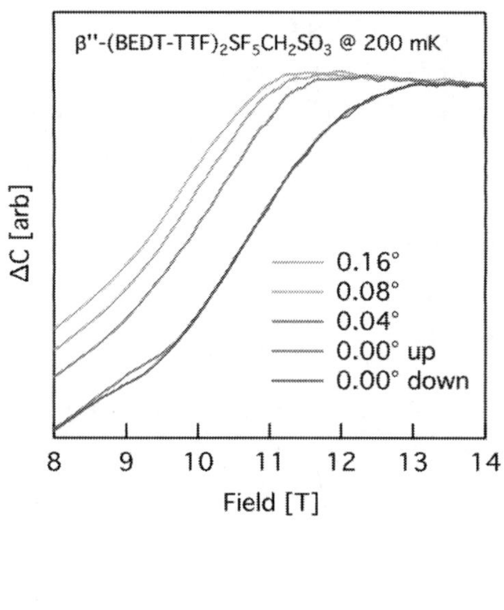


**Figure 5.** Dependence of the magnetic-field-dependent heat capacity of $\beta'' - (\mathrm{BEDT\text{-}TTF})_2\mathrm{SF}_5(\mathrm{CH})_2(\mathrm{CF})_2(\mathrm{SO})_3$ on magnetic field orientation at 200 mK. A first order transition to a field-induced FFLO superconducting state is observed for the 0° plane parallel orientation at a field consistent with a recent NMR determination for $H_{FFLO}$ of $9.3 \pm 0.1$ T [21]. The calorimetric signature of the hysteretic 1st order FFLO phase transition disappears for a change in sample orientation $\Delta\theta$ as small as 0.04°. This change is also accompanied by a steep drop in the critical field, as expected for a 2D superconductor.

to FFLO superconducting state at 9.3 ± 0.1 T and the theoretical expectation of an FFLO phase transition at a field $H_{FFLO} \leq H_{c2}(0.56\ T_c) = 9.2 \pm 0.2$ T. Here $H_{c2}(0.56\ T_c)$ has been empirically determined from the known $H_{c2}(T)$ phase boundary [22]. We note, however, that we are as of yet unable to resolve a phase transition at 10.5 T (corresponding to a $T^* = 0.24T_c$) reported in magnetic penetration depth measurements [22] and possibly also in NMR [21].

**Acknowledgments**
A portion of this work was performed at the National High Magnetic Field Laboratory (NHMFL), which is supported by National Science Foundation Cooperative Agreement No. DMR-11157490 and the State of Florida.

**References**

[1] Coffey T, Martin C, Agosta C C and Kinoshota T 2010 *Physical Review B* **82** 1278
[2] Fulde P and Ferrell R A 1964 *Physical Review* **135** A550 – A563
[3] Larkin A I and Ovchinnikov Y N 1965 *Soviet Physics JETP* **20** 762–770
[4] Agosta C C, Jin J, Coniglio W A, Smith B E, Cho K, Stroe I, Martin C, Tozer S W, Murphy T P, Palm E C, Schlueter J A and Kurmoo M 2012 *Physical Review B* **85** 762
[5] Lortz R, Wang Y, Demuer A, Böttger P, Bergk B, Zwicknagl G, Nakazawa Y and Wosnitza J 2007 *Physical Review Letters* **99** 187002
[6] Bergk B, Demuer A, Sheikin I, Wang Y, Wosnitza J, Nakazawa Y and Lortz R 2011 *Physical Review B* **83** 064506
[7] Tsuchiya S, Yamada J i, Sugii K, Graf D, Brooks J S, Terashima T and Uji S 2015 *Journal of The Physical Society of Japan* **84** 034703
[8] Wright J A, Green E, Kuhns P, Reyes A, Brooks J S, Schlueter J, Kato R, Yamamoto H, Kobayashi M and Brown S E 2011 *Physical Review Letters* **107** 087002
[9] Mayaffre H, Krämer S, Horvatic M, Berthier C, Miyagawa K, Kanoda K and Mitrovic V F 2014 *Nature Physics* **10** 928–932
[10] Bulaevskii L N 1974 *Sov. Phys. JETP* **37** 1133–1136 zh. Eksp. Teor. Fiz. 64, 2241-2247 (June 1973)
[11] Buzdin A I and Kachkachi H 1997 *Physics Letters A* **225** 341–348
[12] Matsuda Y and Shimahara H 2007 *Journal of The Physical Society of Japan* **76** 051005
[13] Shimahara H 1998 *Journal of The Physical Society of Japan* **67** 736–739
[14] Cai, Zi, Wang, Yupeng and Wu, Congjun 2011 *Physical Review A* **83**
[15] Tokiwa Y, Bauer E D and Gegenwart P 2012 *Physical Review Letters* **109** 116402
[16] Beyer R, Bergk B, Yasin S, Schlueter J A and Wosnitza J 2012 *Physical Review Letters* **109**
[17] Fortune, Nathanael A, Hannahs, S T and Hannahs, S T 2014 *Journal of Physics: Conference Series* **568** 032008
[18] Agosta C C, Fortune N A, Hannahs S T, Gu S, Liang L, Park J H and Schleuter J A 2017 *Physical Review Letters* **118** 267001
[19] Machida K and Ichioka M 2008 *Physical Review B* **77** 184515
[20] Wosnitza J 2012 *Crystals* **2** 248–265
[21] Koutroulakis G, Kühne H, Schlueter J A, Wosnitza J and Brown S E 2016 *Physical Review Letters* **116** 067003
[22] Cho K, Smith B E, Coniglio W A, Winter L E, Agosta C C and Schlueter J A 2009 *Physical Review B* **79** 220507

IOP Conf. Series: Journal of Physics: Conf. Series **969** (2018) 012073 doi:10.1088/1742-6596/969/1/012073

# Flux-flow instability in 2H-$NbSe_2$ superconducting thin crystals stamped on $SiO_2$/Si substrates

**J Yoshida, H Ando and N Kokubo**

Department of Engineering Science, University of Electro-Communications, Tokyo 182-8585, Japan

E-mail: kokubo@uec.ac.jp

**Abstract.** We have investigated the nonlinear flux flow in the mixed state of 2H-$NbSe_2$ superconducting thin crystals stamped on $SiO_2$/Si substrates. The nonlinearity in the flux flow behavior is marked with upward bending and subsequent voltage jump in the current-voltage characteristics. The results were analyzed within the framework of a modified Larkin-Ovchinnikov model, including the estimation of inelastic relaxation time of quasiparticles in the crystal and heat transfer coefficient between the crystal and $SiO_2$/Si substrate. The analysis provides an insight into the quality of acoustic match at the crystal-substrate interface.

## 1. Introduction

Nonlinear current-voltage ($I-V$) characteristics in the mixed state of type-II superconductors provides an unique opportunity to investigate the inelastic relaxation of quasiparticles in the flux-flow state. In an earlier theoretical model proposed by Larkin and Ovchinnikov (LO) [1], the nonlinearity is triggered by the shrinkage of a vortex core which leads to the reduction of a viscous damping coefficient for driven vortices and an increase in the vortex velocity (flux-flow voltage). The nonlinear flux flow has been investigated for decades in a variety of superconducting films like Sn [2], TiN [3], Nb [4], and shown discrepancies in critical parameters (like the critical vortex velocity) for the flux-flow instability. It turns out that the Joule heating of quasiparticles inherent in experiments becomes not negligible as vortex density increases. Bezuglyj and Shklovskij (BS) extended the LO model by taking into account the finite removal rate of the power dissipation through a substrate, and proposed two qualitatively different mechanisms of the instability depending on applied magnetic field $B$ [5]. One is the nonthermal mechanism of the instability proposed by LO ($B < B_T$) and another is dominated by the Joule heating ($B > B_T$). These are separated by a crossover field $B_T$, of which analytical result is given for the dirty limit as

$$B_T = 0.375\frac{e\rho_n h\tau_{qp}}{k_B d} \tag{1}$$

with the electronic charge $e$, (low-temperature) normal resistivity $\rho_n$, the Boltzmann constant $k_B$, the sample thickness $d$, the inelastic relaxation time of quasiparticles $\tau_{qp}$, and heat-transfer coefficient $h$ characterizing the heat flow from the sample to the substrate. Later, Peroz and Villard have made comparative studies on clean and dirty Nb superconducting films [6]. Although the dissipation process behind the flux-flow instability is different between in the two limits, they made qualitative determination of $\tau_{qp}$ for both films according to the BS model.

In this study, we report nonlinear flux flow in thin crystal films of a clean 2H-$NbSe_2$ superconductor. By employing techniques of mechanical exfoliation and optical identification developed by the success of the substrate-supported graphene research [7], we fabricated microbridges of $NbSe_2$ thin crystals stamped on $SiO_2$/Si substrates. We observed the nonlinear $I-V$ characteristics indicative of the flux-flow instability and analyzed them within the framework of the BS model. We estimated the inelastic relaxation time of quasiparticles and heat transfer coefficient and discussed the quality of acoustic transparency at the crystal-substrate interface.

**Table 1.** Properties of $NbSe_2$ thin crytal microbridges on $SiO_2$/Si substrates.

| Sample | $RRR$ | $\rho_n$(2K) ($\mu\Omega$m) | $d$ (nm) | $w$ ($\mu$m) | $p_0$ ($\mu$W$\mu$m$^{-3}$) | $B_T$ (mT) | $\tau_{qp}$ (ns) | $h$ (Wcm$^{-2}$K$^{-1}$) |
|---|---|---|---|---|---|---|---|---|
| A | 32 | 0.037 [a] | 91 [b] | 7.6 | 3.3 | 64 | 0.63 | 5.8 |
| B | 31 | 0.039 [a] | 43 [b] | 7.0 | 4.1 | 43 | 0.33 | 3.4 |

[a] $\rho_n$(2 K) was determined from the magnetic field dependence of resistance at 2 K.
[b] We estimated the thickness from room-temperature resistance $R$(295K) and resistivity $\rho_n$(295 K)= 1.2 $\mu\Omega$m [10].

## 2. Experimental

We used 2H-$NbSe_2$ crystals grown by iodide transport method [8]. Cleaved thin platelet flakes were gently rubbed onto a $SiO_2$ (290 nm) /Si substrate with an adhesive polydimethylsiloxane stamp [9]. Combining conventional lithographic and reactive dry etching techniques, we patterned thin flakes into microbridges with the dimension of 7-8 $\mu$m in width $w$ and $\sim$ 100 $\mu$m in length. Good electronic contacts (much less than 0.1 $\Omega$) were achieved by a subsequent deposition of Ag/NiCr films ($d_{Ag}/d_{NiCr} \approx 50$ nm/ 3 nm) on top of the microbridges with lift-off technique, as shown in the inset to Fig. 1. The distance $L$ between electrical voltage contacts was 31 $\mu$m. The flake thickness was calculated from the resistance at room temperature [10]. The values of the thickness are consistent with ones estimated from our optical identification color code for $NbSe_2$ flakes on $SiO_2$(290 nm)/Si substrates [11, 12]. The superconducting transition temperature $T_c$ was 7.1 K, the slope of the second critical field perpendicular to the planes $S(= -dB_{c2}(T)/dT|_{T\approx T_c})$ was 0.75 T/K, and the residual resistance rate $RRR$ was $\approx 30$ [13]. The Ginzburg-Landau coherence length parallel to the planes was determined from the second critical field $B_{c2}(0)$ extrapolated linearly to zero temperature as $\xi(0) \approx \sqrt{\Phi_0/2\pi B_{c2}(0)} \approx \sqrt{\Phi_0/2\pi T_c S} \approx 8$ nm, where $\Phi_0 = \pi\hbar/e$ is the magnetic flux quantum and $\hbar$ is the planck constant divided by $2\pi$. The mean free path $l$ was calculated as $l = (3\pi^2)^{1/3}\hbar/e^2 n^{2/3}(S/S_F)\rho_n \approx 70$ nm where we used the product of $n^{2/3}(S/S_F)$ with carrier density $n$ and a weighting factor $S/S_F$ for the Fermi surface with respect to the free-electron model reported in Ref. [8]. Some characteristic parameters of two samples (A and B) we focus in this study are summarized in Table 1.

We employed pulsed $I-V$ measurements to minimize the sample heating in the flux-flow state. We used an arbitrary wave function generator to generate stepwise current ramps with 40 $\mu$s width. The transient voltage signal from the sample was amplified with a low noise preamplifier (NF corporation LI 5307 and/or LI 75A) and traced, together with one from the standard resistor (500 $\Omega$), in a 2-channel digital oscilloscope (Yokogawa DL-5700). To eliminate an inductive response from the sample, we averaged the voltage signal over the middle of the steady state for each ramp, and obtained $I-V$ curves. To check any possible errors in the pulsed measurement, we also made conventional dc $I-V$ measurements for comparison. Good accordance between the two measurements were observed, except for high current regime where dc measurement was limited by the heating effect. For the good temperature stabilization, the sample space was immersed in superfluid of $^4$He.

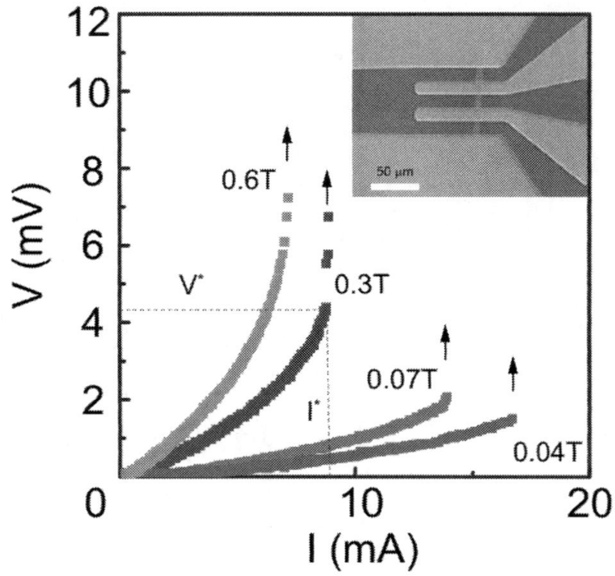


**Figure 1.** (Color online) Current($I$)-voltage($V$) characteristics of sample A taken in different magnetic fields at $T = 2$ K. Micrograph of the sample is given in the inset.

## 3. Results and discussions

Typical $I-V$ curves observed in the sample A at $T = 2.0$ K($\approx 0.3\ T_c$) in different magnetic fields are given in Fig. 1. At low currents (above the depinning current $I_c$), the $I-V$ curve, e.g. at 0.3 T, is almost linear, indicative of the conventional flux-flow state, while at high currents, an upward bending becomes remarkable and the nonlinearity ends with an abrupt voltage jump, indicative of the flux-flow instability. This behavior is a common feature observed in our samples studied.

Let us focus on the critical behavior at the voltage jump in nonlinear $I-V$ curves. As indicated, we define the critical current $I^*$ and the critical voltage $V^*$, and thus determined the critical power $P^*(= I^* \times V^*)$ at the anomaly. Shown in Fig. 2 is the results of the critical power density $p^*$ of the sample B plotted as function of magnetic field. In low fields, $p^*$ exhibits a monotonic increase with field, while in higher fields $p^*$ seems to be insensitive with field. The latter feature implies naively that the instability is set by the heat balance between the Joule heating by flux flow in the crystal and the heat removal through the substrate, as expected in the BS model, rather than the nonthermal instability dominated by the LO mechanism. This tempts us to analyze our results within the framework of the BS model. One can see a reasonable accordance between the experimental results and a theoretical curve (represented by a solid line) of the BS model, where we use the analytical result on the power density given as

$$p^*(B) = p_0\left(1 - \frac{1 + b + \sqrt{b^2 + 8b + 4}}{3(1 + 2b)}\right) \qquad (2)$$

with reduced magnetic field $b = B/B_T$. Two adjustable parameters of $B_{\rm T} = 43$ mT and $p_0 = 4.0\ \mu{\rm W}/\mu{\rm m}^3$ are obtained from the accordance.

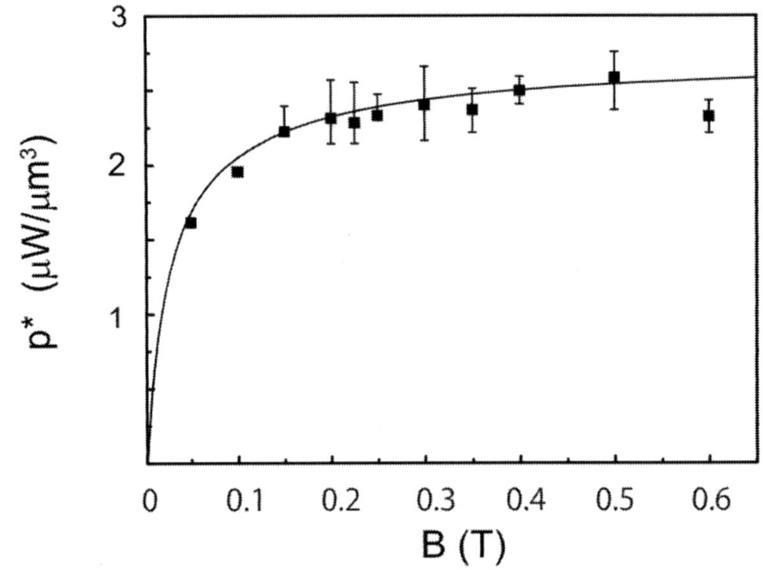


**Figure 2.** Critical power density vs magnetic field observed in sample B at $T = 2$ K.

The inelastic relaxation time of quasiparticles can be estimated from a relation given as

$$\tau_{\rm qp} = 2.67\frac{B_{\rm T}k_B(T_c - T)}{ep_0\rho_{\rm n}}. \tag{3}$$

This assumes that the quasiparticles dissipate within vortex cores because the mean free path $l$ is assumed to be shorter than the size of a vortex core (the coherence length $\xi$). Meanwhile, in a clean case ($l > \xi$), which corresponds to our case, quasiparticles escape from the vortex cores and diffuse around, resulting in nonequilibrium distribution of quasiparticles over a superconductor. Thus, $\tau_{\rm qp}$ becomes the nonequilibrium lifetime of quasiparticles [6]. Since a reasonable determination of $\tau_{\rm qp}$ is given in the clean Nb film [6], let us use Eq. (3) for the estimation of $\tau_{\rm qp}$ in our $NbSe_2$ thin crystals. Substituting $B_{\rm T}$ and $p_0$ in Eq. (3), we obtain $\tau_{\rm qp} = 0.63$ ns and 0.33 ns for samples A and B, respectively. These values are the same order of magnitude as $\tau_{\rm qp} = 3\Lambda^2/v_{\rm F}l \sim 0.9$ ns determined from the diffusion length $\Lambda \sim 4\ \mu$m of quasiparticles of $NbSe_2$ nanoribbons [14, 15] and Fermi velocity $v_{\rm F} = (\hbar/m)(3\pi^2 n\sqrt{S/S_F})^{1/3} \approx 8 \times 10^5$ m/s with electron rest mass $m$. Thus, at least for the order estimate of $\tau_{\rm qp}$ in $NbSe_2$ crystals, the BS model is useful.

Finally, we comment on the heat flow from the crystals to $SiO_2$/Si substrates due to their mutual phonon exchange. Substituting the values of $\tau_{\rm qp}$ and $B_{\rm T}$ in Eq. (1), we obtain the heat transfer coefficient $h = 5.8$ and 3.4 W/cm$^2$K for samples A and B, respectively. These values are decades larger than ones reported on $NbSe_2$ crystals adhered on sapphire substrates with GE varnish [16], but remain the same order of magnitude as ones found in many combinations of superconducting films on standard substrates which do not show good acoustic match [5]. As pointed out by BS, for low phonon transparency of the crystal-substrate interface the nonequilibrium phonons are reabsorbed in the film by quasiparticles and the temperature of the film (both electronic and phonon temperatures) becomes higher than the substrate temperature.

IOP Conf. Series: Journal of Physics: Conf. Series **969** (2018) 012073 doi:10.1088/1742-6596/969/1/012073

This is consistent with our observation that the flux-flow instability in higher fields is dominated by Joule heating.

In summary, we have presented the nonlinear flux flow of 2H-$NbSe_2$ superconducting thin crystals stamped on $SiO_2/Si$ substrates. Employing the modified LO model, we extracted the inelastic relaxation time of quasiparticles and the heat transfer coefficient from the critical parameters in the nonlinear $I-V$ characteristics. The results reveal the low phonon transparency of the crystal-substrate interface, which is likely responsible for the Joule heating mechanism governing the flux-flow instability in higher fields.

## Acknowledgments

N. K acknowledges K. Takita for providing us 2H-$NbSe_2$ crystals. This work was supported by JSPS KAKENHI Grant Numbers 26287075 and 17K05537, and NIMS Nanofabrication Platform in Nanotechnology Platform Project sponsored by the Ministry of Education, Culture, Sports, Science and Technology (MEXT), Japan.

## References

[1] Larkin A I and Ovchinnikov Y N 1975 *Sov. Phys. JETP* **41** 960
[2] Klein W, Huebener R P, Gauss S and Parisi J 1985 *J. Low Temp. Phys.* **61** 413
[3] Lefloch F, Hoffmann C, Demolliens O 1999 *Physica* C **319** 258
[4] Grimaldi G, Leo A, Nigro A, Pace S, Cirillo C and Attanasio C 2008 *Physica* C **468** 765
[5] Bezuglyj A I and Shklovskij V A 1992 *Physica* C **202** 234
[6] Peroz C and Villard C 2005 *Phys. Rev.* B **72** 014515
[7] Novoselov K S, Jiang D, Schedin F, Booth T J, Khotkevich V V, Morozov S V and Geim A K 2005 *Proc. Natl. Acad. Sci.* **102** 10451
[8] Takita K and Masuda K 1985 *J. Low Temp Phys.* **58** 127
[9] Tsen A W, Hunt B, Kim Y D, Yuan Z J, Jia S, Cava R J, Hone H, Kim P, Dean C R and Pasupathy A N 2016 *Nature Phys.* **12** 208
[10] Kokubo N, Kadowaki K and Takita K 2005 *Phys. Rev. Lett.* **95** 177005
[11] Kokubo N and Yoshida J, unpublished
[12] Benameur M M, Radisavljevic B, Héron J S, Sahoo S, Berger H and Kis A 2011 *Nanotechnology* **22** 125706
[13] Kokubo N, Asada T, Kadowaki K, Takita K, Sorop T G and Kes P H 2007 *Phys. Rev.* B **75** 184512
[14] Skocpol W J, Beasley M R and Tinkham M 1974 *J. Low. Temp. Phys.* **16** 145
[15] Falk A, Deshmukh M M, Prieto A L, Urban J J, Jonas A and Park H 2007 *Phys. Rev.* B **75** 020501(R)
[16] Xiao Z L, Andrei E Y, Shuk P and Greenblatt M 2001 *Phys. Rev.* B **64** 094511

IOP Conf. Series: Journal of Physics: Conf. Series **969** (2018) 012074 doi:10.1088/1742-6596/969/1/012074

# Direct observation of magnetic flux and interstitial vortices in perforated mesoscopic squares of superconducting films

**M Mitsuishi[1], N Kokubo[1], K Kitano[1], S Okayasu[2], T Nojima[3] and T Sasaki[3]**

[1] Department of Engineering Science, University of Electro-Communications, Tokyo 182-8585, Japan
[2] Advanced Science Research Center, Japan Atomic Research Institute, Ibaraki 319-1195, Japan
[3] Institute for Materials Research, Tohoku University, Sendai 980-8577, Japan

E-mail: kokubo@uec.ac.jp

**Abstract.** We report magnetic visualizations of fluxoid states in perforated mesoscopic squares of superconducting films with a scanning superconducting quantum interference device microscope. The observed magnetic images clearly reveal how the magnetic flux is distributed and trapped in a hole at different magnetic fields. The magnetization determined from the magnetic images exhibits a tilted sawtooth behavior with the magnetic field, indicating the multiple flux quanta trapped in the hole. Further increase of the magnetic field creates an interstitial vortex (vortices) between the hole and the sample edge. The penetration field $H_p$ of the interstitial vortex depends on the sample size and follows a simple relation $\mu_0 H_p \approx \Phi_0/w^2$ with the spacing $w$ between the hole and the sample edge.

## 1. Introduction

A mesoscopic superconducting ring provides an opportunity to investigate the fundamental property of fluxoid states in laterally modulated, perforated superconducting films. The fabrication of a regular array of antidots (holes) in superconducting films leads to unique anomalies in magnetic field dependences of superconducting transition temperature, critical current and magnetization at characteristic fields where the density of a magnetic flux lattice is harmonically related to the density of holes [1, 2, 3]. These are the hallmark features for the fluxoid quantization with magnetic flux quanta in holes and analogous to the Little-Parks effect in a superconducting cylinder (ring) [4]. The multiplicity of flux quanta is, however, limited by the radius $R$ of the hole with respect to the coherence length $\xi$ and characterized by the saturation number $N_s(\approx R/2\xi)$, above which vortices appear within interstices of the holes. In this case, additional anomalies occur due to the formation of supermatching flux-line lattices [5, 6, 7, 8].

When the array of holes is sparse, the situation is different. Irrespective of the saturation of magnetic flux quanta in holes, it is energetically more favorable to create (interstitial) vortices between holes than the multiple flux-quanta vortices in holes, provided that the density $n_c$ of holes is less than $\xi/R^3$ with the superconducting coherence length $\xi$ [9]. Moreover, as the superconducting film becomes thinner, the self-energy and mutual interaction of vortices depend on their positions with respect to the size and shape of the film [10, 11]. As Brandt argued [12], in a thin film ring with inner $a$ and outer radii $b$, a vortex can sit in the middle of the annular region, rather than to enter the hole when applied magnetic field exceeds a size-dependent penetration field $\mu_0 H_p \approx \Phi_0/w^2$ with the annular width $w = b - a$. More exact analytical results for the penetration field have been given to the limited case of a very narrow ring with $w \ll a$ [13, 14] and thin narrow slabs [15].

The understanding of the size-dependent penetration of interstitial vortices is important for practical applications like the reduction of 1/f noise in superconducting quantum interference devices (SQUIDs) [16, 17], while experimental evidences for the penetration field were limited to thin slabs [18, 19]. In this study, we report direct observation of fluxoid states in perforated superconducting *squares* films with a scanning SQUID microscope. We observed nonmonotonic evolution of magnetic flux accumulated in the hole with the applied magnetic field. Further increase of the magnetic field results in the penetration of the interstitial vortex (vortices) in the annular region. We found that the penetration field $H_\mathrm{p}$ of the vortex is size dependent and follows a simple relation $H_\mathrm{p} \approx \Phi_0/w^2$ with the spacing $w$ between the hole and the sample edge.

**Table 1.** Dimensions of perforated squares of amorphous $Mo_xGe_{1-x}$ thin films

| Sample | Thickness ($\mu$m) | Side length ($\mu$m) | $R$ ($\mu$m) |
|---|---|---|---|
| MoGe-A | 0.22 | 26 | 3.5 |
| MoGe-B | 0.42 | 55 | 4 |

## 2. Experimental

The samples we used mainly in this study were amorphous $Mo_xGe_{1-x}$ (MoGe) films with $x \approx 75$ %. The parameters for these films are as follows. The superconducting transition temperature $T_\mathrm{c}$ is $\approx 6.7$ K, normal resistivity $\rho_0$ (10 K) is $\approx 1.7$ $\mu\Omega$m, the slope $S$ of the second critical field $H_{\mathrm{c}2}$ near $T_\mathrm{c}$ is $\approx 2.2$ T/K. Using the dirty limit expressions, we estimated the zero-temperature magnetic penetration depth $\lambda(0) \approx 0.53$ $\mu$m and the zero-temperature coherence length $\xi(0) \approx 4.6$ nm [20]. We patterned the films into hundreds of small squares with different sizes and also fabricated holes in samples by ultraviolet lithographic and chemical etching techniques. Dimensions of two samples (A and B) we focus in this study are summarized in Table 1.

Magnetic images were taken with a scanning SQUID microscope (SQM-2000, SII Nanotechnology). The microscope had a dc SQUID sensor chip integrating Nb-based Josephson junctions and an inductively coupled, pick-up Nb coil with 10 $\mu$m in diameter. For the safe operation of the microscope, we tilted the sensor chip with respect to the sample surface by $\sim$ 10 degrees and kept the distance ($\sim$ 5 $\mu$m) between the pick coil and the sample surface during the measurements. The lateral resolution of the microscope was $\sim$ 4 $\mu$m, which allowed us to image a (Pearl) vortex ($\sim$ 10 $\mu$m in diameter) in MoGe thin superconducting films with reasonable lateral resolutions [21]. The samples and the sensor were surrounded by $\mu$-metal. The ambient magnetic field around the sample space was typically $\approx 2$ $\mu$T. All the magnetic images presented in this study were taken on flux states at $T \approx$ 3.7-3.9 K prepared by the field-cooled procedure. The details of the measurements are described in Refs. 21 and 22.

IOP Conf. Series: Journal of Physics: Conf. Series **969** (2018) 012074 doi:10.1088/1742-6596/969/1/012074

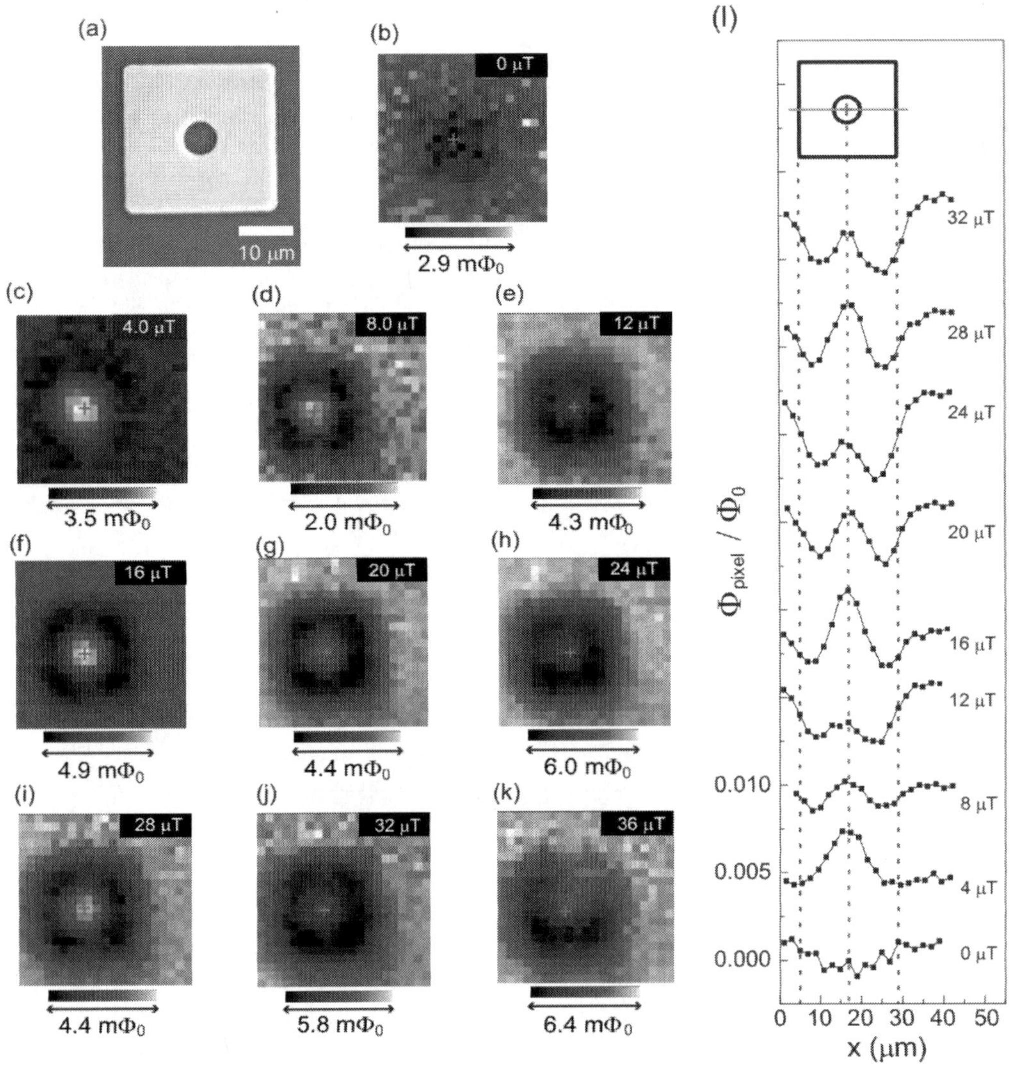


**Figure 1.** Magnetic images of fluxoid states observed in MoGe-A sample. (a) Micrograph of an identical sample to the one the magnetic images were taken. The sample size is 26 × 26 $\mu$m$^2$ with 0.22 $\mu$m in thickness. The radius of a centered hole is 3.5 $\mu$m. Images were taken at different applied magnetic fields of (b) 0, (c) 4, (d) 8, (e) 12, (f) 16, (g) 20, (h) 24, (i) 28, (j) 32, (k) 36 $\mu$T, respectively. All images are the same in size of 42 × 42 $\mu$m$^2$ (21 × 21 pixels) . The color bar represents the magnitude of magnetic flux $\Phi_{\rm pixel}$ per pixel (2×2 $\mu$m$^2$) divided by the flux quantum $\Phi_0$ (the dark color corresponds to the low value of $\Phi_{\rm pixel}$). For each image the hole center is marked with a red cross. (l) Profiles of magnetic images (a)-(j) across the hole center as illustrated in the inset. For clarity, the field profiles are shifted vertically (Some of them are also shifted horizontally). The ambient magnetic field around the sample stage is $\approx$ 2 $\mu$T.

## 3. Results and discussions

Figures 1b-1k show magnetic images of fluxoid states of the MoGe-A sample at different applied magnetic fields. The sample size is 26 × 26 $\mu$m$^2$ with a centered circular hole ($R$ = 3.5 $\mu$m). The color bar indicates the magnitude of magnetic flux $\Phi_{\rm pixel}$ over the area of 2 × 2 $\mu$m$^2$ divided by the flux quantum $\Phi_0$ (the dark color corresponds to the low value of $\Phi_{\rm pixel}$). For each image the hole center is marked with a red cross. One can see how the magnetic flux in the hole varies with the applied magnetic field. This behavior is also seen in a plot of magnetic profiles across the hole center shown in Fig. 1l. At 4 $\mu$T, the "dome" of magnetic flux suddenly appears in the hole. As the magnetic field is elevated, the flux dome shrinks with respect to the field outside the sample, but it again becomes large at 16 $\mu$T. This nonmonotonic sequence continues up to 32 $\mu$T, above which an interstitial vortex(vortices) appears between the hole and the sample edge (Fig. 1k). We note that the image size of a vortex in the thin film is $\sim$ 10 $\mu$m in diameter (which corresponds to $\sim$ 5 pixels of the image), while the narrowest (widest) spacing between the hole and the sample edge is 9.5 $\mu$m (15 $\mu$m). Thus, the penetration of the interstitial vortex between the hole and the sample edge can be recognized in magnetic images by the high intensity of magnetic flux filled in the annular region of the sample. Therefore, we can distinguish the penetration of interstitial vortices in the annular region and the flux accumulation in the centered hole.

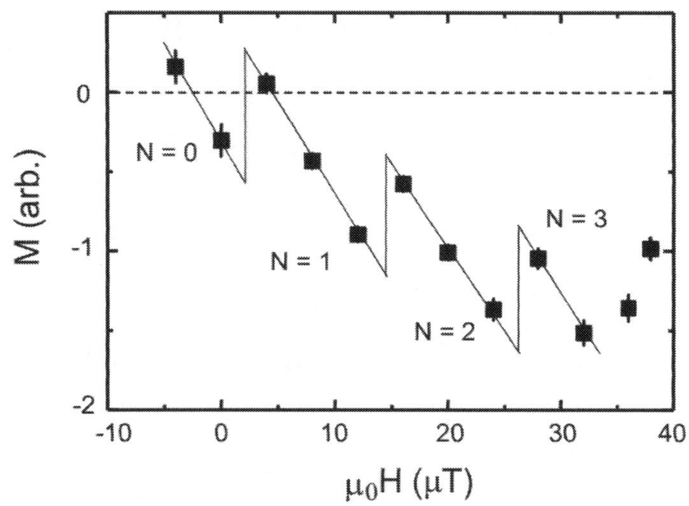


**Figure 2.** Magnetization curve of MoGe-A determined from magnetic images in Fig.1. A solid line is a guide to the eye. The ambient magnetic field perpendicular to the sample is $\approx 2$ $\mu$T as determined from the magnetization curve ($N = 0$) at $M = 0$.

Figure 2 shows the magnetization curve determined from magnetic images by following the procedure given in Ref. 22. As observed, the magnetization exhibits a tilted sawtooth behavior with the applied magnetic field up to 32 $\mu$T. Each magnetization jump indicates the entry of the quantized magnetic flux in the sample. Since the magnetic flux is accumulated in the hole without an interstitial vortex(vortices), we believe that the hole has the multiple flux quanta. The corresponding fluxoid number $N$ is indicated in Fig. 2.

With further increasing the applied magnetic field, the interstitial vortex (vortices) is created between the hole and the sample edge, while the fluxoid number in the hole remains constant. At this moment, we do not understand yet how the interstitial vortices should distribute in the annular region and also how the magnetization curve behaves. In the following, we focus on the first penetration of the interstitial vortex and determine the penetration field $H_p$ as the lowest magnetic field where the interstitial vortex appears in magnetic images. From the magnetic images taken in the two samples with different sizes we find that the penetration field $H_{\rm p}$ is size dependent, i.e., $\mu_0 H_{\rm p} = 36$ $\mu$T and 5 $\mu$T for the smaller (MoGe-A) and larger samples (MoGe-

B), respectively. The number of flux quanta, the fluxoid number $N = 3(2)$, accumulated in the hole for MoGe-A(-B) is much smaller than the corresponding saturation number $N_s \approx R/2\xi \approx 350\ (400)$. Thus, the argument of the saturation in the hole is not relevant to the appearance of the interstitial vortex observed in this study.

Let us interpret the present results according to the Brandt model [12]. In Fig. 3 the results of $H_\mathrm{p}$ are plotted against the spacing $w$ between the hole and the sample edge. Since there is no unique width for square loops, as illustrated in the inset, we take the widest spacing $w_1$ along the diagonal of the square (of which results are plotted by open symbols) and the narrowest one $w_2$ (which does with filled symbols). Here, the ambient magnetic field (perpendicular component determined from the magnetization curve in Fig. 2) is taken into account. Good accordance is seen between the results and theoretical curves representing relations $\mu_0 H_\mathrm{p} = C\Phi_0/w^2$ with $C = 3.9$ and 1.8 for $w_1$ and $w_2$, respectively. In the Brandt model the coefficient $C$ is of the order of unity. Thus, it is not possible to conclude which spacing should be relevant. A further experiment on superconducting rings with uniform annular width would clarify this issue.

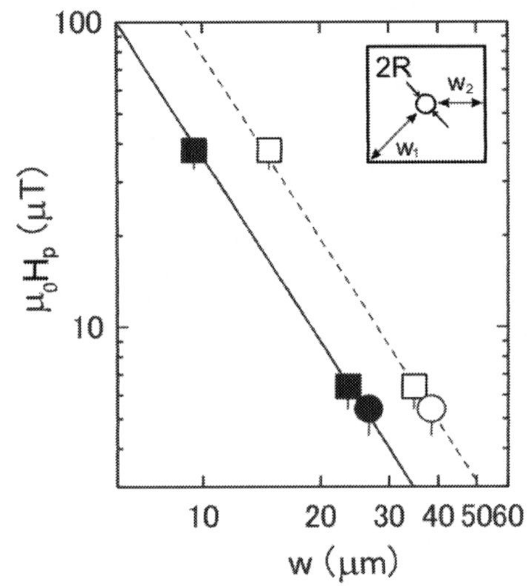


**Figure 3.** Size dependent penetration field. Inset illustrates how we define the spacing $w$ between the hole and the film edge. Open and filled symbols correspond to widest $w_1$ and narrowest spacings $w_2$, respectively. Square (circle) symbols indicate the results for amorphous MoGe square films (a polycrystalline Nb square film). Solid and dotted lines represent the Brandt relation with different coefficients [12].

Finally we comment on results made on a polycrystalline Nb square sample ($T_c \approx 9.2$ K) with a centered hole, of which size is nearly identical to the MoGe-B sample ($R \gg \xi$). We observed up to two flux quanta $N = 2(\ll N_\mathrm{s} \sim 100)$ accumulated in the hole and subsequent penetration of interstitial vortices in the annular region. As plotted in Fig. 3 with circle symbols, the results of the penetration field are in good accordance with the relation obtained in amorphous MoGe films. This agreement implies that the mechanism for the interstitial penetration is governed by the sample size, rather than the material parameters.

In summary, we have presented magnetic images of fluxoid states of perforated superconducting square films with the scanning SQUID microscope. We observed the accumulation of magnetic flux in the hole with increasing the applied magnetic field. The magnetization determined from the magnetic images exhibits the tilted sawtooth behavior with the applied magnetic field. These findings indicate the field evolution of multiple flux quanta through the hole. With further increasing field, we observed the interstitial vortex (vortices) between the hole and the sample edge. The corresponding penetration field $H_\mathrm{p}$ for the interstitial vortex is size dependent and follows the Brandt relation $\mu_0 H_\mathrm{p} \approx \Phi_0/w^2$ for both amorphous MoGe and polycrystalline Nb films.

## Acknowledgments

N. K acknowledges K. Makise for providing us Nb films. This work was supported by JSPS KAKENHI (Grant Numbers 26287075 and 17K05537), the Inter-university Cooperative Research Program of the Institute for Materials Research, Tohoku University (Proposal No. 16K0004 and 17K0051), and NIMS Nanofabrication Platform in Nanotechnology Platform Project sponsored by the Ministry of Education, Culture, Sports, Science and Technology (MEXT), Japan.

## References

[1] Bezryadin A, Ovchinnikov Y N and Pannetier B 1996 *Phys. Rev.* B **53** 8553
[2] Moshchalkov V V, Baert M, Metlushko V V, Rosseel E, Van Bael M J, Temst K, Bruynseraede Y and Jonckheere R 1998 *Phys. Rev.* B **57** 3615
[3] Baert M, Metlushko V V, Jonckheere R, Moshchalkov V V and Bruynseraede Y 1995 *Phys. Rev. Lett.* **74** 3269
[4] Little W A and Parks R D 1962 *Phys. Rev. lett.* **9** 9
[5] Metlushko V, Welp U, Crabtree G W, Osgood R, Bader S D, DeLong L E, Zhang Z, Brueck S R J, Ilic B, Chung K and Hesketh P J 1999 *Phys. Rev.* B **60** R12585
[6] Field S B, James S S, Barentine J, Metlushko V, Crabtree G, Shtrikman H, Ilic B and Brueck S R J 2002 *Phys. Rev. Lett.* **88** 067003
[7] Silhanek A V, Van Look L, Jonckheere R, Zhu B Y, Raedts S and Moshchalkov V V 2005 *Phys. Rev.* B **72** 014507
[8] Ooi S, Mochiku T and Hirata K 2011 *Physica C* **471** 804
[9] Buzdin A I 1993 *Phys. Rev.* B **47** 11416
[10] Kogan V G 1994 *Phys. Rev.* B **49** 15874
[11] Brandt E H 2005 *Phys. Rev.* B **72** 024529
[12] Brandt E H and Clem J R 2004 *Phys. Rev.* B **69** 184509
[13] Brojeny A A B and Clem J R 2003 *Phys. Rev.* B **68** 174514
[14] Kogan V G, Clem J R and Mints R G 2004 *Phys. Rev.* B **69** 064516
[15] Maksimova G M, 1998 *Phys. Solid State* **40** 1607
[16] Dantsker E, Tanaka S and Clarke J 1997 *Appl. Phys. Lett.* **70** 2037
[17] Wördenweber R and Selders P, 2002 *Physica* C **366** 135
[18] Stan G, Field S B and Martinis J M 2004 *Phys. Rev. Lett.* **92** 097003
[19] Kokanović I, Helzel A, Babić D, Sürgers C and Strunk C 2008 *Phys. Rev.* B **77** 172504
[20] Kes P H and Tsuei C C 1983 *Phys. Rev.* B **28** 5126.
[21] Kokubo N, Okayasu S, Kanda A and Shinozaki B 2010 *Phys. Rev.* B **82** 014501
[22] Kokubo N, Okayasu S and Kadowaki K 2017 *Multi-Vortex States in Mesoscopic Superconductors The Oxford Handbook of Small Superconductors* ed Narlikar A V (United Kindam, Oxford University Press) chapter 3 pp 81-107

IOP Conf. Series: Journal of Physics: Conf. Series **969** (2018) 012075 doi:10.1088/1742-6596/969/1/012075

# A doping dependent study of interplay between magnetic and superconducting properties in $BaFe_{2-x}Co_xAs_2$ single crystals

**Biplab Bag[1], Gorky Shaw[1,3], S S Banerjee[1,*], K Vinod[2], A Bharathi[2,4]**

[1]Department of Physics, Indian Institute of Technology, Kanpur-208016, India
[2]Condensed Matter Physics Division, Materials Science Group, Indira Gandhi Center for Atomic Research, Kalpakkam-603102, India
[3]Present address: Experimental Physics of Nanostructured Materials, Q-MAT, CESAM, Universitè de Liège,Sart Tilman, B-4000, Belgium
[4]Present address: UGC-DAE Consortium for Scientific Research, Kalpakkam-603104, India

**Abstract.** We show strong interplay between magnetic and superconducting order in three $BaFe_{2-x}Co_xAs_2$ single crystals with different $x$. Our study reveals the presence of magnetic fluctuations with superconducting order in our samples and the strength of the magnetic fluctuations as well as the pinning properties are found to be the strongest for the optimally doped sample and weakest for the overdoped sample. Using local magnetization measurements, we show that application of an external magnetic field in our samples suppresses the magnetic fluctuations and enhances the diamagnetic response. Further, we show presence of unusual superconducting fluctuations above $T_c$ in our samples which we find strongly depends on the strength of the magnetic fluctuations. We believe that our data suggest the possible role of magnetic fluctuations in mediating superconducting fluctuations above $T_c$ in our samples.

*satyajit@iitk.ac.in

## 1. Introduction

Iron (Fe) - arsenide (As) based 'Pnictide' superconductors in recent times have attracted intense interest due to some of their unusual and rich properties, like, significantly large superconducting transition temperatures ($T_c$), high critical current density ($J_c$) and critical field ($H_{c2}$), unconventional nature of superconducting symmetry, and a complex doping phase diagram which exhibit an interesting interplay between crystal structure, magnetism, and superconductivity [1,2,3]. The undoped parent compound is not superconducting but is a semimetal showing long range anti-ferromagnetic like ordering (viz., a spin density wave state) [4]. Doping (either electron or hole) the parent compound destroys long range magnetic order in the system and promotes the onset of superconductivity in these compounds [5]. An interesting aspect of iron pnictide compounds is presence of magnetic order along with superconductivity in the underdoped regime of these compounds. Experimental studies have shown complete suppression of magnetic order in the vicinity of optimum doping where $T_c$ is maximum [5]. However, recently we have reported presence of short range magnetic order (or magnetic fluctuations) in the optimally doped (and overdoped) $BaFe_{2-x}Co_xAs_2$ single crystals [6].

Here, by measuring the bulk and local magnetization ($M$) data, we show how the interplay between magnetic and superconducting order evolves with applied field ($H$) for three $BaFe_{2-x}Co_xAs_2$ single crystals with different $x$. The bulk magnetization data suggests that strength of the magnetic fluctuations as well as the pinning properties are strongest for the optimally doped sample (in which $T_c$ is maximum) and weakest for the overdoped sample. Our local $M(H)$ data suggests that application of $H$ in our samples suppresses the short range magnetic order and enhances the diamagnetic response. Further, we will show presence of unusual superconducting fluctuations above $T_c$ and how volume fraction of the regions with superconducting fluctuations ($v_{scf}$) changes with $H$ for our samples. The novelty of our data is we find that the extent of the superconducting fluctuations above $T_c$ strongly depends on the strength of the short range magnetic order.

## 2. Experimental details

For our study we investigate the bulk and local magnetization properties of the underdoped ($x = 0.10$), optimally doped ($x = 0.14$) and overdoped ($x = 0.20$) single crystals of $BaFe_{2-x}Co_xAs_2$. The bulk dc magnetization of the samples was measured using Quantum Design MPMS SQUID VSM magnetometer in a field ($H$) range of upto 7 T. In our measurements, $H$ is applied parallel to crystallographic $c$-axis of the single crystals. We also map the local magnetic field distribution ($B_z(x,y)$) using Faraday effect based high sensitivity magneto-optical imaging (MOI) technique [7]. MOI was performed in Advanced Research Systems closed cycle cryostat with magnetic field applied parallel to $c$-axis of the single crystal using a homemade copper coil solenoid magnet.

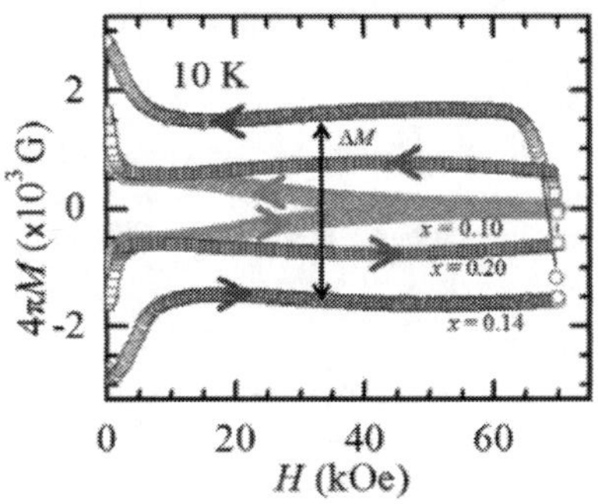


**Figure 1: Isothermal bulk $M(H)$ hysteresis loops for $x = 0.10$, 0.14 and 0.20 samples measured at 10 K. The arrows indicate the direction of $H$-sweep. Doubled sided black arrow shows the $\Delta M$ for $x = 0.14$ sample.**

## 3. Result

For chemical composition analysis of the samples, we have done careful Electron Probe Micro-Analysis (EPMA) studies across the sample. The EPMA data shows homogeneous and uniform chemical composition across the samples with Co-doping concentration determined as $x_{EPMA} = 0.1032 \pm 0.0003$ for $x = 0.10$, $x_{EPMA} = 0.1266 \pm 0.0002$ for $x = 0.14$ and $x_{EPMA} = 0.1986 \pm 0.0002$ for $x = 0.20$ samples. The critical temperature ($T_c$) of the sample is determined from four probe transport measurements with $H = 0$ Oe [6] which shows $T_c = 18.10 \pm 0.05$, $26.80 \pm 0.05$, and $25.99 \pm 0.05$ K for $x = 0.10$, 0.14, and 0.20 samples respectively. The bulk $M(T)$ and $M(H)$ data of the samples in the superconducting state exhibit bulk superconducting properties of the samples [6]. In figure 1 by plotting the bulk $M(H)$ response ($5^{th}$ and $6^{th}$ quadrant legs of a 6 quadrant $M(H)$ hysteresis loop) for all the samples together at 10 K, we have compared the pinning properties of the three samples. It is evident from figure 1 that all the three samples exhibit strong bulk pinning properties which is an important characteristic feature of a type II superconductor. By determining the maximum width, i.e., $\Delta M$ ($H$) of the $M(H)$ hysteresis loop (cf. figure 1) for the $x = 0.14$ sample, we have estimated the maximum critical current density,

$$J_c = 20\Delta M / a(1 - a/3b) \qquad \ldots(1)$$

(for a rectangular plate like crystal geometry, with in-plane crystal dimensions $a$, $b$ ($b > a$)) [8] which shows $J_c = 4\times10^5$ Amp/cm$^2$ at 10 K, which is comparable with similar values reported in $BaFe_{2-x}Co_xAs_2$ samples [9]. From figure 1, one observes that the $J_c (\propto \Delta M)$ shows a non-monotonic variation with doping concentration. Since doping introduces more disorder to the system, one should expect that $\Delta M$ should increase with doping. However, in figure 1 we see that at low $H$ (< 10 kOe) surprisingly $\Delta M$ is maximum for $x = 0.14$ sample (optimally doped), while $\Delta M$ are almost comparable for $x = 0.10$ (underdoped) and 0.20 (over doped) samples. The differences in $\Delta M$ could be a result of difference in effective thermal fluctuations in the three systems due to differences in $T/T_c$. The values of $T/T_c$ are 0.55, 0.37 and 0.38 for $x = 0.10$, 0.14 and 0.20 samples respectively. Therefore the differences in $T/T_c$ are not significant enough to explain the differences in $\Delta M$. Infact, although the $T/T_c$ values for $x = 0.14$ and 0.20 samples are almost similar, the $x = 0.20$ sample has weaker pinning (or smaller $\Delta M$) compared to the $x = 0.14$ sample, despite the fact that doping in $x = 0.20$ sample is much higher. This unusual non-monotonic behavior of pinning with respect to doping we will discuss subsequently.

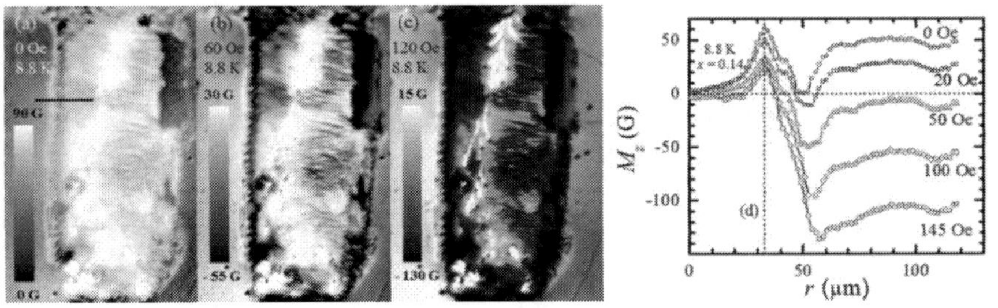


**Figure 2: Iso-thermal MO images for $x = 0.14$ at 8.8 K ($<T_c$) with (a) 0 Oe, (b) 60 Oe, and (c) 120 Oe with $H \parallel c$ measured on increasing $H$ (virgin forward) . The color bars in inset shows the mapping of contrast to local $M_z$ .(d) Variation of $M_z(r)$ profile at different $H$ at 8.8 K measured along the solid horizontal black line shown in (a). In (d) the edge of the sample is identified with the vertical dashed line.**

Figures 2(a)-(c) show the MOI images of the $x = 0.14$ sample captured at 8.8 K ($<<T_c$) with $H = 0$ Oe, 60 Oe, and 120 Oe respectively. Note that the intensity of Faraday rotated light in MOI is proportional to the local $B_z$. The dark contrast in a MO image corresponds to a region with $B_z = 0$ G and a white contrast corresponds to a region with finite $B_z$. For better understanding, we have converted the local $B_z$ values to local magnetization ($M_z$). From the $B_z(x,y)$ information, the local magnetization values are estimated using $M_z(x,y) = B_z(x,y) - H$. The intensity bars in figures 2(a)-(c) represent mapping of contrast to corresponding $M_z$. Generally, for a conventional type II superconductor, at low $H$ ($< H_{c1}$) and $T$($< T_c$) due to strong Meissner response, $B_z$ values should be zero (no Faraday rotation) inside and outside the sample and hence one should not observe any sample profile at low $H$. However, in figure 2(a) at $H = 0$ Oe and 8.8 K we observe a bright sample where the bright contrast indicates positive magnetization response emanating from the sample. With increase in $H$, while a conventional MOI response of a type II superconductor shows a transition of the intensity from blackish to greyish due to penetration of vortices, in figures 2(b),(c) we observe unusual darkening of the sample which is indicative of enhancement of the superconducting properties (or diamagnetic shielding response) with increasing $H$. To analyze the above features explicitly we have measured the $M_z(r)$ profiles (from the MO images captured at 8.8 K and different $H$) along the black line shown in figure 2(a) and have plotted in figure 2(d). In figure 2(d), the vertical dashed line identifies the sample edge. The $M_z(r)$ values close to the sample edge are significantly large due circulating Meissner shielding current at the sample edge. It shows that at $H = 0$ Oe, the $x = 0.14$ sample exhibits a positive magnetization response with $<M_z> \sim 45$ G. We have shown earlier that this positive $M_z$ response in the superconducting state of $x = 0.14$ sample appears due to presence of short range magnetic order (magnetic fluctuations) in the sample [6]. On application of a small field (at $H = 20$ Oe), $M_z$ decreases significantly though it remains positive. At $H = 50$ Oe, the local magnetization values decrease further and turn negative which is indicative of onset of a diamagnetic response with increasing $H$. On subsequent increase in $H$, $M_z$ becomes more negative signifying increase in diamagnetic response (superconducting response) with increasing $H$ as already indicated in figures 2(a)-(c). The above response suggests strong interplay between the magnetic and superconducting order in the optimally doped sample. The positive local magnetization deep in the superconducting state is feature of presence of magnetic fluctuations along with superconductivity [6]. Here, we must mention that the modulations in $M_z(r)$ profiles inside the sample in figure 2(d) are not due to any artifact, rather it is associated with a very inhomogeneous superconducting response in the $x$ = 0.14 sample due to presence of the short range magnetic order in the superconducting state [6].

Figure 3(a) shows bulk magnetization response with $T$ measured at $T \geq T_c$ with $H$ = 100 Oe for the zero field cooled (ZFC) state in $x$ = 0.14 sample. Unlike conventional superconducting response where the $M(T) \sim 0$ in the normal state, figure 3(a) shows an unusual positive magnetization with a peak at $T_p$ and a long tail at $T \geq T_c$ (in the normal state) in $x$ = 0.14 sample. The origin of this positive $4\pi M(T)$ response in the normal state of the $x$ = 0.14 sample is presence of short range magnetic order (magnetic fluctuations) in the normal state of the sample [6]. In figure 3(a), we identify $T_{mf}^{on} \sim 88$ K as the onset of the magnetic fluctuations and below $T_{mf}^{on}$ the $4\pi M(T)$ response increases with decreasing $T$. Here, we must mention that similar $4\pi M(T)$ response also have been found in $x$ = 0.10 and 0.20 samples [6]. Using the values of $4\pi M(T)$ from figure 3(a), in figure 3(b) we have plotted the $4\pi M$ response against $1/T$ for $T >> T_c$. Figure 3(b) shows that with decrease in $T$ (or increase in $1/T$) after the onset of positive magnetization response at $T_{mf}^{on} \sim 88$ K, the $4\pi M\left(\frac{1}{T}\right)$ response exhibits a linear

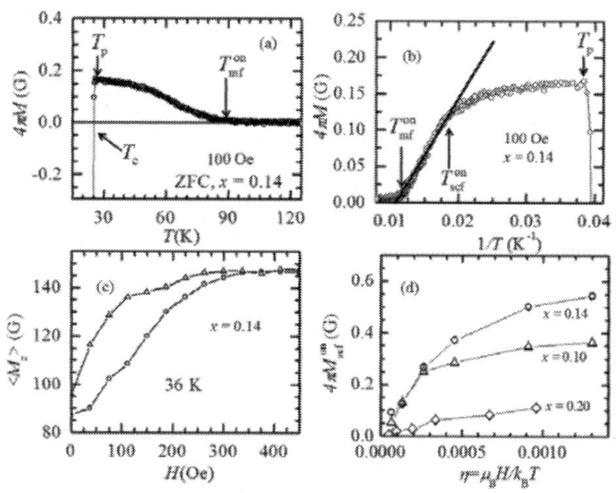


**Figure 3: (a) ZFC bulk $4\pi M(T)$ response at $T \geq T_c$ with $H$ = 100 Oe for $x$ = 0.14 sample. The locations of $T_{mf}^{on}$, $T_p$ and $T_c$ are marked with arrows. (b) Bulk $4\pi M(1/T)$ response at $T \geq T_c$ with $H$ = 100 Oe for $x$ = 0.14 sample. The characteristic temperatures are marked with arrows. The solid straight line represents paramagnetic like linear $4\pi M(1/T)$ response. (c) local $M_z(H)$ hysteresis loop at $T$ = 36 K for $x$ = 0.14 sample. (d) Variation of $4\pi M_{scf}^{on}$ for all samples with η.**

$4\pi M \propto \frac{1}{T}$ behavior (indicated by the thick black line in figure 3(b)). This linear $4\pi M\left(\frac{1}{T}\right)$ behavior indicates that the short range magnetic order in the normal state exhibits paramagnetic like behavior in the $x$ = 0.14 sample. However, the local $M_z(H)$ response (cf. figure 3(c), measured using MOI technique) for the virgin forward and reverse runs at $T$ = 36 K (> $T_c$) elucidates an unusual hysteresis in $M_z(H)$ and the $M_z(H)$ values saturate above 300 Oe. Following bulk and local magnetization responses in figures 3(b),(c) we argue that although the magnetic fluctuations exhibit paramagnetic like $4\pi M\left(\frac{1}{T}\right)$ behavior (cf. figure 3(b)), the hysteresis in local $M_z(H)$ (figure 3(c)) suggests it is unlike that of a conventional paramagnet.

Further, in figure 3(b) we observe that the $4\pi M\left(1/T\right)$ response deviates from the linear behavior below $T_{\rm scf}^{\rm on}$ ~ 55 K. Similar behavior also has been found in the $x$ = 0.10 and 0.20 samples [6]. This deviation from the linear $4\pi M\left(1/T\right)$ occurs due to appearance of superconducting fluctuations in the sample above $T_c$ [6] which compensates the effect of the magnetic fluctuations and in the $T$-region between $T_{\rm scf}^{\rm on}$ and $T_c$ superconducting fluctuations coexist with magnetic order. Hence at $T \geq T_{\rm scf}^{\rm on}$, the magnetization response is governed by short range magnetic order only [6]. Due to this above feature, we have estimated the $4\pi M\left(T = T_{\rm scf}^{\rm on}\right)$ values ($\equiv 4\pi M_{\rm scf}^{\rm on}$) at different $H$ for all the samples and the $4\pi M_{\rm scf}^{\rm on}$ values have been plotted as a function of $\eta = \mu_{\rm B}H / k_{\rm B}T$ in figure 3(d) where $\eta$ signifies the ratio of Zeeman energy and thermal energy. Figure 3(d) shows that in the low $H$-regime ($\mu_{\rm B}H << k_{\rm B}T$) the $4\pi M_{\rm scf}^{\rm on}$ values for all the samples are very small and comparable. However, with increase in $H$ (or $\eta$) we observe an increase in $4\pi M_{\rm scf}^{\rm on}(\eta)$ values for all the samples (cf. figure 3(d)) and it exhibits a saturation at $\eta > 0.001$. Hence, $4\pi M_{\rm scf}^{\rm on}(\eta)$ takes the form of Brillouin expression for magnetization. A noteworthy feature in figure 3(d) is that the $4\pi M_{\rm scf}^{\rm on}(\eta)$ values are maximum for the $x$ = 0.14 sample and minimum for the $x$ = 0.20 sample. Since the $4\pi M_{\rm scf}^{\rm on}(\eta)$ values are associated with the short range magnetic order, the above feature suggests that the effect of the short range magnetic order is maximum for the optimally doped sample and weakest for the overdoped sample.

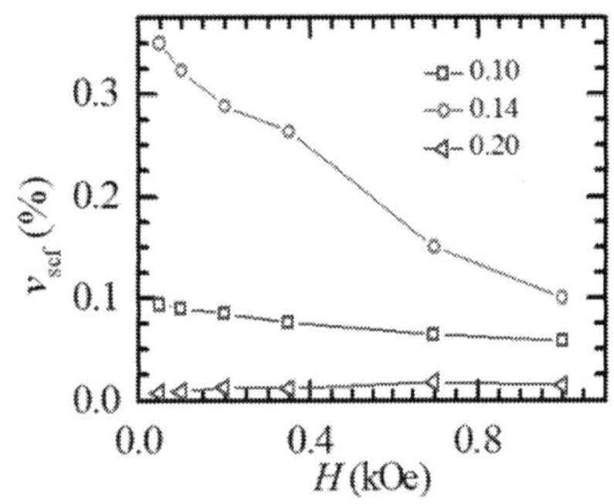


**Figure 4: $H$-dependence on $\nu_{\rm scf}$ for all the samples estimated from bulk $M(T)$ response.**

Note, in figure 3(b) we have shown a deviation from the linear $4\pi M\left(1/T\right)$ response due to onset of superconducting fluctuations at $T < T_{\rm scf}^{\rm on}$. From the deviation in $4\pi M\left(1/T\right)$, we have estimated the volume fraction of the regions with superconducting fluctuations ($\nu_{\rm scf}$) [6] above $T_c$ (at $T_p$, see figures 3(a),(b)) for different $H$ and the $\nu_{\rm scf}(H)$ response has been plotted in figure 4. It shows that at $H$ = 0 Oe $\nu_{\rm scf}$ values are small but non-zero and $\nu_{\rm scf}$ is maximum (0.346 %) for the $x$ = 0.14 sample (which also has the strongest effect of short range magnetic order) and minimum (0.008 %) for the $x$ = 0.20 sample (in which the short range magnetic order is the weakest). With increase in $H$, for $x$ = 0.10 and 0.14 samples there is a monotonous decrease in the $\nu_{\rm scf}(H)$ values which is indicative of the fact that the $H$ suppresses the weak superconducting fluctuations survive in the normal state of the samples. However, for $x$ = 0.20 sample, since the $\nu_{\rm scf}(H)$ values are extremely small, it is difficult to detect any variation in $\nu_{\rm scf}$ with $H$.

## 4. Discussion and conclusion

We have identified signatures of coexistence of short range magnetic order along with superconductivity in our $BaFe_{2-x}Co_xAs_2$ single crystals. The magnetic fluctuations which persist in the vicinity of $T_c$ sustain even deep inside the superconducting phase. We found the strength of the short range magnetic order is the strongest for the optimally doped sample and weakest for the overdoped sample. Further, we have also showed that the volume fraction of the superconducting fluctuations is maximum in the optimally doped sample and minimum in the overdoped sample. The above two information suggest that the short range magnetic order in our sample may mediate the superconducting fluctuations observed above $T_c$ in our samples. Deep in the superconducting state through our MOI measurements, we show the coexistence of effects related to superconductivity along with magnetic fluctuations. Note as argued earlier, the difference in pinning properties (see figure 1) cannot be attributed to a difference in $T/T_c$. Recall from figure 3(d) that the saturated value of $4\pi M_{scf}^{on}$ is the highest for $x = 0.14$ followed by $x = 0.10$ and is the least for $x = 0.20$ sample. This suggests that the presence of a larger contribution from magnetic fluctuations in the superconducting state of $x = 0.14$ sample. Since the presence of the magnetic fluctuations increases pinning of vortices in a superconductor and hence increases $\Delta M$, the larger magnetic fluctuations in $x = 0.14$ cause the width of the hysteresis loop ($\Delta M$) to be larger for $x = 0.14$ sample compared to $x = 0.20$ sample at 10 K. In iron pnictides, doping suppresses long range magnetic ordering. However, as doping concentration is increased from $x = 0.10$ to 0.14, we found that the effect of magnetic fluctuations increases and beyond $x = 0.14$ the effect magnetic fluctuations progressively diminishes. Concomitant with this behavior of magnetic fluctuations with doping we observe a similar corresponding effect not only on the superconducting $T_c$ but also on the superconducting volume fraction above $T_c$. We propose that short range magnetic fluctuations may play a crucial role in mediating superconductivity in these compounds. It may be noted that even in Co doped $CaFe_2As_2$ system we had shown the coexistence of magnetic fluctuations along with superconductivity [2].

## 5. Acknowledgement

SSB would like to acknowledge, the funding support from IIT Kanpur and DST (TSDP), help from SQUID – VSM facility in TIFR, Mumbai for a few measurements.

## 6. References

[1] Drew A J, *et al* 2009 *Nat. Mater.* **8**, 310; Shermadini Z, *et al* 2011 *Phys. Rev. Lett.* **106**, 117602
[2] Mandal P, *et al* 2012 *Europhys. Lett.* **100**, 47002
[3] Izyumov Y and Kurmaev E 2010 *High-Tc Superconductors Based on FeAs Compounds*, Springer Series in Material Science, Berlin, 143
[4] Cruz C R dela, *et al* 2008 *Nature* **453**, 899
[5] Ni N, *et al* 2008 *Phys. Rev. B* **78**, 214515; Nandi S, *et al* 2010 *Phys. Rev. Lett.* **104**, 057006; Chen H, *et al* 2009 *Europhys. Lett.* **85**, 17006
[6] Bag Biplab, Vinod K, Bharathi A and Banerjee S S 2016 *New. J. Phys.* **18**, 063025
[7] Johansen T H and Shantsev D V 2004 *Magneto-Optical Imaging*, Nato Sci. Ser. II, Vol. 142, Kluwer Academic Publisher, The Netherlands; Mandal P, *et al* 2012 *Rev. Sci. Instrum.* **83**, 123906
[8] Bean C P 1964 *Rev. Mod. Phys.* **36**, 31-39
[9] Nakajima Y, *et al* 2009 *Phys. Rev. B* **80**, 012510

IOP Conf. Series: Journal of Physics: Conf. Series **969** (2018) 012076 doi:10.1088/1742-6596/969/1/012076

# New ethylenediamine-intercalated superconductor $(C_2H_8N_2)_yTa_2PdSe_6$ with $T_c = 4.5$ K

**S. Nakamura, T. Noji, T. Hatakeda, K. Sato, T. Kawamata, M. Kato, and Y. Koike**

*Department of Applied Physics, Tohoku University, 6-6-05 Aoba, Aramaki, Aoba-ku, Sendai 980-8579, Japan*

noji@teion.apph.tohoku.ac.jp

**Abstract.** A new superconductor $(C_2H_8N_2)_yTa_2PdSe_6$ with $T_c = 4.5$ K has successfully been synthesized via the intercalation of ethylenediamine (EDA) into $Ta_2PdSe_6$. It has been found that the co-intercalation of lithium suppresses the superconductivity in $(C_2H_8N_2)_yTa_2PdSe_6$, indicating that the electron doping is harmful to the superconductivity. The origin of the appearance of superconductivity via the intercalation of only EDA into $Ta_2PdSe_6$ has been discussed.

## 1. Introduction

Superconductors including 4*d* and 5*d* transition metals such as $T_2Pd_xCh_5$ ($T$ = Nb, Ta ; $Ch$ = S, Se) have attracted attention owing to their extremely high upper critical fields $H_{c2}$'s exceeding their Pauli paramagnetic fields $H_P$'s, which are caused by the strong spin-orbit interaction [1-4]. The layered compound $Ta_2PdSe_6$ is not a superconductor but a degenerate semiconductor or a semimetal, though its crystal structure is analogous to that of $T_2Pd_xCh_5$, as shown in Fig. 1(a) [5]. Recently, it has also attracted great attention that the superconducting transition temperature $T_c$ of the layered superconductor FeSe dramatically increases from 8 K to ~ 45 K via the co-intercalation of alkali metal or alkali earth metal and ammonia or organic molecules [6-14]. Since $Ta_2PdSe_6$ has double Se layers weakly combined with the van der Waals force as well as FeSe, the appearance of superconductivity is expected in $Ta_2PdSe_6$ via such co-intercalation.

Here, we have tried to intercalate lithium and ethylenediamine (EDA) $C_2H_8N_2$ into $Ta_2PdSe_6$ to obtain a new intercalation superconductor.

## 2. Experimental

Polycrystalline samples of $Ta_2PdSe_6$ were prepared by the solid-state reaction method. Tantalum powder, palladium powder and selenium grains, which were weighed in a molar ratio of Ta : Pd : Se = 2 : 1 : 6.18, were mixed and pressed into pellets. The excess of Se was necessary to compensate for the loss of Se during the reaction due to its high vapor pressure. The pellets were sealed in an evacuated quartz tube and heated at 725°C for 300 h. The obtained pellets were thoroughly pulverized into powder, mixed and then pressed into pellets again and sealed and heated once more. Then, the obtained pellets of $Ta_2PdSe_6$ were pulverized into powder again to be used for the intercalation. The co-intercalation of lithium and EDA into the powdery $Ta_2PdSe_6$ was performed as follows. The powdery $Ta_2PdSe_6$ of 0.7g was placed in a beaker filled with a solution of pure lithium metal dissolved in 7.4 ml of EDA. The amount of Li was set at $0 \leq x \leq 1.5$ in a molar ratio of Li : $Ta_2PdSe_6 = x : 1$. The intercalation of only EDA without Li, namely, $x = 0$ was also tried. The intercalation reaction was carried out at 90°C for 10 days. The product was washed with fresh EDA. All the processes were performed in an argon-filled glove box. Both the host sample of $Ta_2PdSe_6$ and intercalated samples of $Li_x(C_2H_8N_2)_yTa_2PdSe_6$ were characterized by the powder x-ray diffraction using Cu K$\alpha$ radiation. For the intercalated samples, an airtight sample-holder was used. The diffraction patterns were analysed using RIETAN-FP [15]. The chemical composition was determined by the inductively coupled plasma optical emission spectrometry (ICP-OES). In order to detect the superconducting transition, the magnetic susceptibility $\chi$ was measured

IOP Conf. Series: Journal of Physics: Conf. Series **969** (2018) 012076 doi:10.1088/1742-6596/969/1/012076

using a superconducting quantum interference device (SQUID) magnetometer (Quantum Design MPMS). Measurements of the electrical resistivity $\rho$ were also carried out by the standard dc four-probe method. For the $\rho$ measurements, intercalated powdery samples were pressed into pellets at room temperature without heat treatment.

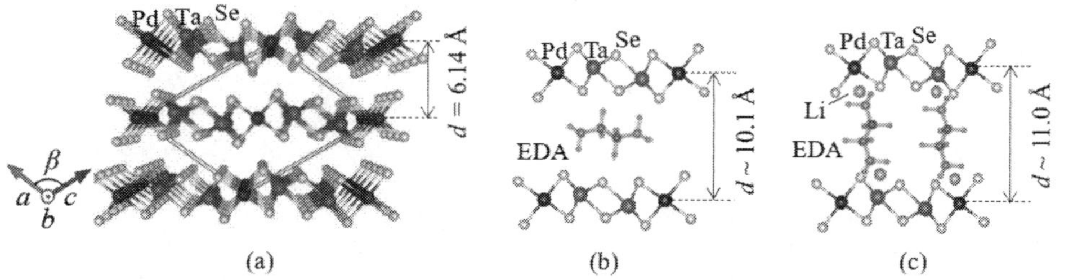


**Figure 1.** (a) Crystal structure of $Ta_2PdSe_6$ with the interlayer spacing between $Ta_2PdSe_6$ layers $d = 6.14$ Å. Solid lines represent the unit cell. Schematic views of crystal structures of (b) Phase I of $(C_2H_8N_2)_yTa_2PdSe_6$ with $d \sim 10.1$ Å and (c) Phase II of $Li_x(C_2H_8N_2)_yTa_2PdSe_6$ with $d \sim 11.0$ Å.

## 3. Results and discussion

A series of $Li_x(C_2H_8N_2)_yTa_2PdSe_6$ samples with nominal values of $x = 0$, 0.5, 1.2, and 1.5 have been confirmed to be synthesized by the powder x-ray diffraction. Figure 2 shows the powder x-ray diffraction patterns of the host sample of $Ta_2PdSe_6$ and the intercalated sample of $(C_2H_8N_2)_yTa_2PdSe_6$ with $x = 0$. The broad peak around $2\theta = 20°$ is due to the airtight sample holder. The obtained host sample of $Ta_2PdSe_6$ is found to be of the single phase, because all the Bragg peaks are able to be indexed on the basis of the monoclinic structure (space group: C2/m). For the intercalated sample of $(C_2H_8N_2)_yTa_2PdSe_6$, a sharp Bragg peak is observed at a very small value of $2\theta \sim 9°$, indicating a large expansion of the lattice constant via the intercalation of EDA. It is found that all the Bragg peaks are well indexed on the basis of the monoclinic structure (space group: C2/m), assuming that $Ta_2PdSe_6$ layers are much separated by the intercalation of EDA. The lattice constants of the intercalated sample of $(C_2H_8N_2)_yTa_2PdSe_6$ are calculated to be $a = 13.466(1)$ Å, $b = 3.3705(4)$ Å, $c = 14.183(5)$ Å, $\beta = 86.15(1)°$.

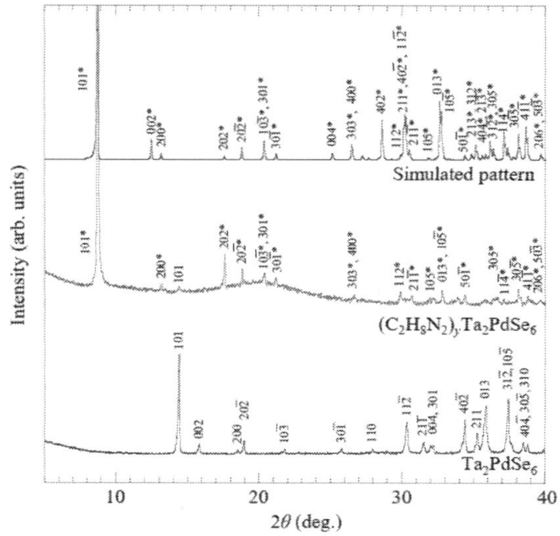


**Figure 2.** Powder x-ray diffraction patterns of the host sample of $Ta_2PdSe_6$ and the intercalated sample of $(C_2H_8N_2)_yTa_2PdSe_6$, using Cu K$\alpha$ radiation. For reference, the simulated powder x-ray diffraction pattern of the hypothetical compound of $Ta_2PdSe_6$ with $a = 13466(1)$ Å, $b = 3.3705(4)$ Å, $c = 14.183(5)$ Å, $\beta = 86.15(1)°$ is also shown. Indices without and with an asterisk are due to the host sample and the intercalated sample, respectively.

Accordingly, it is concluded that EDA has been intercalated between $Ta_2PdSe_6$ layers. The interlayer spacing between $Ta_2PdSe_6$ layers $d$ in $(C_2H_8N_2)_yTa_2PdSe_6$ is estimated from the 101 peak to be 10.1 Å and much larger than $d$ = 6.14 Å of the host sample of $Ta_2PdSe_6$. Since the enhancement of the $d$ value via the intercalation of EDA with the size of 3.7 Å in width and 5.4 Å in length is 4.0 Å, EDA is inferred to be intercalated in parallel between $Ta_2PdSe_6$ layers, as shown in Fig. 1(b).

Figure 3 shows the powder x-ray diffraction patterns of the host sample of $Ta_2PdSe_6$ and intercalated samples of $Li_x(C_2H_8N_2)_yTa_2PdSe_6$ with various nominal values of $x$. With increasing $x$, it is found that the 101 peak around 9° (due to Phase I) shifts to around 8° (due to Phase II). This indicates that Phase I with $d \sim 10.1$ Å of $x = 0$ changes to Phase II with $d \sim 11.0$ Å with increasing $x$. In $x = 0.5$, Phase I and Phase II are mixed. Phase I is only EDA-intercalated $(C_2H_8N_2)_yTa_2PdSe_6$, as mentioned above. Therefore, Phase II is regarded as both Li- and EDA-intercalated $Li_x(C_2H_8N_2)_yTa_2PdSe_6$. Since the enhancement of the $d$ value via the co-intercalation of Li and EDA in Phase II is ~ 4.9 Å and is close to the length of EDA, EDA is inferred to be intercalated perpendicularly to the $Ta_2PdSe_6$ layers as shown in Fig. 1(c), as in the case of both Li- and EDA-intercalated $Li_x(C_2H_8N_2)_yFe_{2-x}Se_2$ [10].

Figure 4 shows the temperature dependences of $\chi$ in a magnetic field of 10 Oe on zero-field cooling (ZFC) and on field cooling (FC) for intercalated samples of $Li_x(C_2H_8N_2)_yTa_2PdSe_6$ with various nominal values of $x$. A superconducting transition is observed around 4.5 K in $0 \leq x \leq 1.2$, while no superconducting transition is observed for $x = 1.5$. The superconducting volume fraction, simply estimated from the $\chi$ value at 2 K on ZFC, is as large as ~ 30 % in $x = 0$ and markedly decreases with increasing $x$. On the other hand, $T_c$ is almost constant in $0 \leq x \leq 1.2$. Accordingly, it is concluded that only Phase I of $(C_2H_8N_2)_yTa_2PdSe_6$ with $x = 0$ is a bulk superconductor with $T_c$ = 4.5 K, while Phase II of $Li_x(C_2H_8N_2)_yTa_2PdSe_6$ ($x \neq 0$) is not superconducting. That is, the increase of electron carriers by the co-intercalation of Li is harmful to the superconductivity in $(C_2H_8N_2)_yTa_2PdSe_6$.

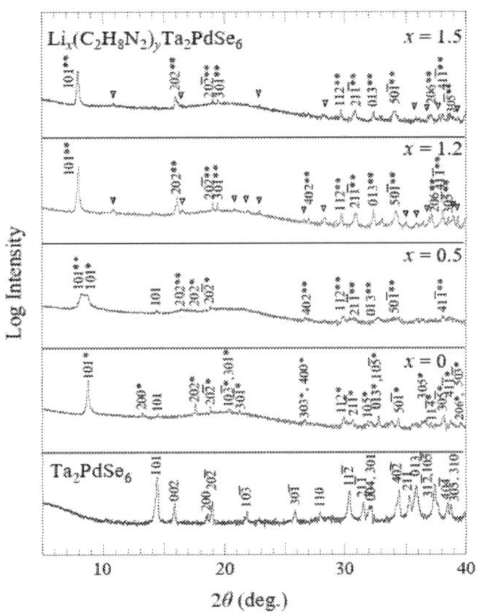


**Figure 3.** Powder x-ray diffraction patterns of the host sample of $Ta_2PdSe_6$ and intercalated samples of $Li_x(C_2H_8N_2)_yTa_2PdSe_6$ with various nominal values of $x$, using CuKα radiation. Indices without and with single and double asterisk marks are due to the host sample, Phase I and Phase II, respectively. Peaks marked by ▽ is due to an unknown compound.

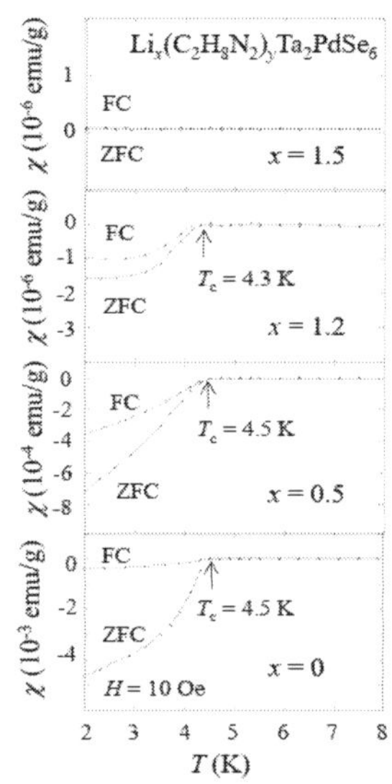


**Figure 4.** Temperature dependences of the magnetic susceptibility $\chi$ in a magnetic field of 10 Oe on zero-field cooling (ZFC) and on field cooling (FC) for intercalated samples of $Li_x(C_2H_8N_2)_yTa_2PdSe_6$ with various nominal values of $x$.

Figure 5 displays the temperature dependences of $\rho$ for the host sample of $Ta_2PdSe_6$ and intercalated samples of $Li_x(C_2H_8N_2)_yTa_2PdSe_6$ with typical nominal values of $x$. It is found that $\rho$ of $Ta_2PdSe_6$ shows a metallic temperature-dependence, while $\rho$ of intercalated samples shows a semiconducting temperature-dependence. A superconducting transition is observed at a low temperature below 4.5K for $x = 0$ and 0.5 including Phase I. The transitions are very broad and zero-resistivity is not observed above 2 K and moreover the absolute values of $\rho$ of intercalated samples are not so small. These are inferred to be due to grain-boundaries in the intercalated pellet samples and may be due to the degradation of the samples caused by the atmospheric exposure in the process of making four terminals on the sample surface for the $\rho$ measurement.

As for the origin of the appearance of superconductivity in $(C_2H_8N_2)_yTa_2PdSe_6$, several possibilities are supposed. One is the enhancement of the density of states at the Fermi level owing to the enhancement of the two-dimensionality of the electronic structure caused by the expansion of the interlayer spacing between $Ta_2PdSe_6$ layers. Another is the enhancement of the electron-phonon interaction owing to the isolation of each $Ta_2PdSe_6$ layer, as pointed out in single-layer FeSe [16]. The other is the contribution of the electronic polarization of EDA to the formation of Cooper pairs. Further investigation is necessary to clarity which the case is. On the other hand, the suppression of superconductivity by the co-intercalation of Li into $(C_2H_8N_2)_yTa_2PdSe_6$ may be due to the decrease in the density of states at the Fermi level caused by the charge transfer from intercalated Li to the $Ta_2PdSe_6$ layers.

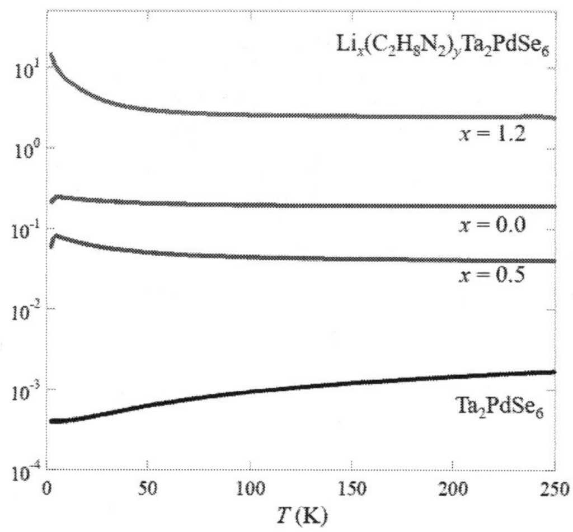


**Figure 5.** Temperature dependences of the electrical resistivity $\rho$ for the host sample of $Ta_2PdSe_6$ and intercalated samples of $Li_x(C_2H_8N_2)_yTa_2PdSe_6$ with typical nominal values of $x$.

## 4. Summary

We have succeeded in synthesizing new intercalation compounds of $Li_x(C_2H_8N_2)_yTa_2PdSe_6$ with $0 \leq x \leq 1.5$ via the intercalation of only EDA and the co- intercalation of Li and EDA into $Ta_2PdSe_6$. In only EDA-intercalated $(C_2H_8N_2)_yTa_2PdSe_6$, $d$ = 10.1 Å and EDA has been inferred to be intercalated in parallel between $Ta_2PdSe_6$ layers. In both Li- and EDA-intercalated $Li_x(C_2H_8N_2)_yTa_2PdSe_6$, with increasing $x$, it has been found that Phase I with $d \sim 10.1$ Å changes to Phase II with $d \sim 11.0$ Å, where EDA is inferred to be intercalated perpendicularly to $Ta_2PdSe_6$ layers. From $\chi$ and $\rho$ measurements, it has been concluded that only EDA-intercalated $(C_2H_8N_2)_yTa_2PdSe_6$, namely, Phase I is a bulk superconductor with $T_c = 4.5$ K and that both Li- and EDA-intercalated $Li_x(C_2H_8N_2)_yTa_2PdSe_6$, namely, Phase II is not superconducting. Therefore, the electron doping by the co-intercalation of Li is harmful to the superconductivity in $(C_2H_8N_2)_yTa_2PdSe_6$, which may be due to the decrease in the density of states at the Fermi level. The origin of the appearance of superconductivity in $(C_2H_8N_2)_yTa_2PdSe_6$ has been supposed as follows. One is the enhancement of the two-dimensionality of the electronic structure.

Another is the enhancement of the electron-phonon interaction. The other is the contribution of the electronic polarization of EDA to the formation of Cooper pairs.

**Acknowledgments**

We would like to thank M. Ishikuro of Institute for Materials Research, Tohoku University, for his aid in the ICP-OES analysis. This work was supported by JSPS KAKENHI (Grant Number 16K05429).

**References**

[1] Zhang Q, Li G, Rhodes D, Kiswandhi A, Besara T, Zeng B, Sun J, Siegrist T, Johannes M D and Balicas L 2013 *Sci. Rep.* **3** 1446
[2] Khim S, Lee B, Choi K Y, Jeon B G, Jang D H, Patil D, Patil S, Kim R, Choi E S, Lee S, Yu J and Kim K H 2013 *New J. Phys.* **15** 123031
[3] Lu Y, Takayama T, Bangura A F, Katsura Y, Hashizume D and Takagi H 2014 *J. Phys. Soc. Jpn.* **83** 023702
[4] Zhang J, Dong J K, Xu Y, Pan J, He L P, Zhang L J and Li S Y 2015 *Supercond. Sci. Technol.* **28** 115015
[5] Keszler D A, Squattrito P J, Brese N E, Ibers J A, Shang M and Lu J 1985 *Inorg. Chem.* **24** 3063
[6] Ying T P, Chen X L, Wang G, Jin S F, Zhou T T, Lai X F, Zhang H and Wang W Y 2012 *Sci. Rep.* **2** 426
[7] Krzton-Maziopa A, Pomjakushina E V, Pomjakushin V Y, Rohr F, Schilling A and Conder K 2012 *J. Phys. : Condens. Matter* **24** 382202
[8] Burrard-Lucas M, Free D G, Sedlmaier S J, Wright J D, Cassidy S J, Hara Y, Corkett A J, Lancaster T, Baker P J, Blundell S J and Clarke S J 2013 *Nat. Mater.* **12** 15
[9] Zheng L, Izumi M, Sakai Y, Eguchi R, Goto H, Takabayashi Y, Kambe T, Onji T, Araki S, Kobayashi T C, Kim J, Fujiwara A, and Kubozono Y 2013 *Phys. Rev. B* **88** 094521
[10] Hatakeda T, Noji T, Kawamata T, Kato M, and Koike Y 2013 *J. Phys. Soc. Jpn.* **82** 123705
[11] Noji T, Hatakeda T, Hosono S, Kawamata T, Kato M, and Koike Y *Physica C* 2014 **504** 8
[12] Hosono S, Noji T, Hatakeda T, Kawamata T, Kato M, and Koike Y 2014 *J. Phys. Soc. Jpn.* **83** 113704
[13] Hayashi F, Lei H, Guo J, and Hosono H 2015 *Inorg. Chem.* **54** 3346
[14] Hatakeda T, Noji T, Sato K, Kawamata T, Kato M and Koike Y 2016 *J. Phys. Soc. Jpn.* **85** 103702
[15] Izumi F and Momma K 2007 *Solid State Phenom.* **130** 15
[16] Peng R, Xu H C, Tan S Y, Cao Hy, Xia M, Shen X P, Huang Z C, Wen C H P, Song Q, Zhang T, Xie B P, Gong X G and Feng D L 2014 *Nat. Commun.* **5** 5044

IOP Conf. Series: Journal of Physics: Conf. Series **969** (2018) 012077 doi:10.1088/1742-6596/969/1/012077

# Superconductivity in Lithium- and Diamines-Intercalated $TiSe_2$ and $MoSe_2$

**K. Sato, T. Noji, T. Hatakeda, T. Kawamata, M. Kato, and Y. Koike**

Department of Applied Physics, Tohoku University, Sendai 980-8579, Japan

E-mail: noji@teion.apph.tohoku.ac.jp

**Abstract.** We have succeeded in synthesizing new superconductors $Li_x(C_2H_8N_2)_yMSe_2$ and $Li_x(C_6H_{16}N_2)_yMSe_2$ ($M$ = Ti, Mo) with $T_c$ = 2.4 – 6.0 K via the co-intercalation of lithium and diamines into 1T-$TiSe_2$ and 2H-$MoSe_2$. As for $M$ = Ti, it has been found that both the electron doping due to the Li intercalation and the expansion of the interlayer spacing between $TiSe_2$ layers due to the intercalation of diamines suppress the charge density wave transition, leading to the appearance of superconductivity. As for $M$ = Mo, $T_c$ has been found to be related not to the interlayer spacing but to the carrier density increased by the electron doping due to the Li intercalation.

## 1. Introduction

Several transition metal dichalcogenides $MX_2$ ($M$ = transition metal, $X$ = S, Se, or Te) with layered structures exhibit intriguing phenomena such as anisotropic superconductivity and charge density wave (CDW), owing to their quasi-two-dimensional electronic structures. It is also known that a variety of metal atoms and organic molecules are intercalated between weakly bound $MX_2$ layers, leading to drastic changes of the physical properties. Among $MX_2$, recently, 1T-$TiSe_2$ and 2H-$MoSe_2$ have attracted interest owing to the appearance of superconductivity via the intercalation or the application of pressure or electric field [1-6].

Recently, it has also attracted great interest that the superconducting transition temperature $T_c$ of the iron-based superconductor FeSe dramatically increases from 8 K to ~ 45 K through the co-intercalation of alkali or alkaline-earth metal and ammonia [7] or organic molecules such as pyridine [8], ethylenediamine (EDA) $C_2H_8N_2$ [9-10] and hexamethylenediamine (HMDA) $C_6H_{16}N_2$ [11-13] between FeSe layers. So far, however, there are few reports on such co-intercalation into $MX_2$ [6].

In this paper, we have tried to make co-intercalation of lithium and EDA or HMDA into 1T-$TiSe_2$ or 2H-$MoSe_2$ and succeeded in synthesizing new intercalation superconductors of $Li_x(C_2H_8N_2)_yMSe_2$ and $Li_x(C_6H_{16}N_2)_yMSe_2$ ($M$ = Ti, Mo) with $T_c$ = 2.4 – 6.0 K.

## 2. Experimental

Polycrystalline host samples of 1T-$TiSe_2$ and 2H-$MoSe_2$ were prepared by the solid-state reaction method. Starting materials were powders of Ti, Mo and grains of Se, which were weighted stoichiometrically, mixed thoroughly and pressed into pellets. The pellets were sealed in an evacuated quartz tube and heated at 800°C for 150 h. The obtained pellets of 1T-$TiSe_2$ and 2H-$MoSe_2$ were pulverized into powder to be used for the intercalation. Li- and EDA- or HMDA-intercalated samples of $Li_x(C_2H_8N_2)_yMSe_2$ or $Li_x(C_6H_{16}N_2)_yMSe_2$ ($M$ = Ti, Mo) were prepared as follows [9,11]. An appropriate amount of the powdery 1T-$TiSe_2$ or 2H-$MoSe_2$ was placed in a beaker filled with 0.01 – 0.2 M solution of pure Li metal dissolved in EDA or HMDA. The amount of 1T-$TiSe_2$ or 2H-$MoSe_2$ was calculated in a molar ratio of Li : 1T-$TiSe_2$ or 2H-$MoSe_2$ = 0 – 1 : 1. The reaction was carried out at 50°C for 2 weeks. The product in residual EDA was washed with fresh EDA. On the other hand, the separation of the product in residual HMDA was easily performed by the solidification of residual HMDA at the top cap of the beaker, keeping the temperature of the top cap of the beaker below the melting point of HMDA (42°C). All the processes were carried out in an argon-filled glove box. Only Li-intercalated samples of $Li_xTiSe_2$ were also prepared using 1.6 M solution of *n*-butyllithium in hexane. The powdery 1T-$TiSe_2$ was put into the solution in a molar ratio of Li : 1T-$TiSe_2$ = 1 : 1 and reacted at room temperature for 60 h in air. Here, it is noted that $x$ values in intercalated samples of $Li_xTiSe_2$, $Li_x(C_2H_8N_2)_yMSe_2$ and $Li_x(C_6H_{16}N_2)_yMSe_2$ ($M$ = Ti, Mo) are nominal ones based on the assumption that all $Li^+$ ions in the solution are intercalated into 1T-$TiSe_2$ or 2H-$MoSe_2$ [11].

Both host samples and intercalated samples were characterized by the powder x-ray diffraction using Cu K$\alpha$ radiation. For the intercalated samples, an airtight sample-holder was used. The magnetic susceptibility $\chi$ was measured using a superconducting quantum interference device (SQUID) magnetometer (Quantum Design, MPMS). The electrical resistivity $\rho$ was measured by the standard dc four-probe method. For the $\rho$ measurements, samples were pressed into pellets at room temperature without heat treatment.

## 3. Results and Discussion

First, we show the results of the intercalation into 1T-$TiSe_2$. Figure 1(a) shows the powder x-ray diffraction patterns of the host sample of 1T-$TiSe_2$ and intercalated samples of $Li_xTiSe_2$ ($x$ = 1), $(C_2H_8N_2)_yTiSe_2$, $Li_x(C_2H_8N_2)_yTiSe_2$ ($x$ = 0.2) and $Li_x(C_6H_{16}N_2)_yTiSe_2$ ($x$ = 0.2). The broad peak around $2\theta$ = 20° is due to the airtight sample-holder. The powder x-ray diffraction patterns of the host sample of 1T-$TiSe_2$ and all the intercalated samples are well indexed based on the 1T-type structure ($P\bar{3}m1$). The lattice parameters are estimated to be $a$ = 3.536(1) Å, $c$ = 6.021(5) Å for 1T-$TiSe_2$ and $a$ = 3.64(3) Å, $c$ = 6.45(7) Å for $Li_xTiSe_2$ ($x$ = 1). Here, we call the phase, in which only Li is intercalated, Phase I. For only EDA-intercalated $(C_2H_8N_2)_yTiSe_2$, new Bragg peaks are observed at $2\theta \sim$ 9° and 26°, though there still remain small Bragg peaks due to non-intercalated regions of 1T-$TiSe_2$. The $c$-axis length of $(C_2H_8N_2)_yTiSe_2$ is estimated to be 10.107(4) Å, so that the $c$-axis is expanded by ~ 4.1 Å through the intercalation of EDA. Since the width and length of EDA are approximately 3.7 Å and 5.1 Å, respectively, it seems that EDA is intercalated between $TiSe_2$ layers where the longitudinal direction of EDA is parallel to $TiSe_2$ layers. Here, we call this phase Phase II of $Li_x(C_2H_8N_2)_yTiSe_2$ ($x$ = 0). As for both Li- and EDA-intercalated $Li_x(C_2H_8N_2)_yTiSe_2$ ($x$ = 0.2), new Bragg peaks are observed at $2\theta \sim$ 8°, 9° and 27°. The new Bragg peaks are regarded as being due to Phase II and a different phase (we call this phase Phase III) with the $c$-axis lengths of 9.96(7) Å and 11.55(2) Å of $Li_x(C_2H_8N_2)_yTiSe_2$ ($x$ = 0.2), respectively. Phase II of this sample is the same as that observed in $(C_2H_8N_2)_yTiSe_2$, though $Li^+$ ions may be included dilutely. In the case of Phase III, on the other hand, the expansion of the $c$-axis is ~ 5.6 Å, so that EDA is intercalated between $TiSe_2$ layers where the longitudinal direction of EDA is perpendicular to $TiSe_2$ layers. As for both Li- and HMDA-intercalated $Li_x(C_6H_{16}N_2)_yTiSe_2$ ($x$ = 0.2), new Bragg peaks are observed at $2\theta \sim$ 6°, 12° and 25°. The $c$-axis length of $Li_x(C_6H_{16}N_2)_yTiSe_2$ ($x$ = 0.2) is estimated to be 14.08(9) Å. This $c$-axis length is much large than those of both Li- and EDA-intercalated $Li_x(C_2H_8N_2)_yTiSe_2$ ($x$ = 0.2), and the $c$-axis is expanded by ~ 8.1 Å through the co-intercalation of Li and HMDA. Since the width and length of HMDA are approximately 3.7 Å and 10.4 Å, respectively, HMDA is speculated to be intercalated between $TiSe_2$ layers where the longitudinal direction of HMDA is oblique to $TiSe_2$ layers.

Figure 1(b) shows the temperature dependences of $\chi$ in a magnetic field of 10 Oe on zero-field cooling (ZFC) and on field cooling (FC) for the host sample of 1T-$TiSe_2$ and intercalated samples of $Li_xTiSe_2$ ($x$ = 1), $(C_2H_8N_2)_yTiSe_2$, $Li_x(C_2H_8N_2)_yTiSe_2$ ($x$ = 0.2) and $Li_x(C_6H_{16}N_2)_yTiSe_2$ ($x$ = 0.2). Although the host sample of 1T-$TiSe_2$ shows no superconductivity, a superconducting transition is observed at 2.4 K and 2.8 K in only Li-intercalated $Li_xTiSe_2$ ($x$ = 1) and only EDA-intercalated

$(C_2H_8N_2)_yTiSe_2$, respectively. As for both Li- and EDA-intercalated $Li_x(C_2H_8N_2)_yTiSe_2$ ($x$ = 0.2), two steps of superconducting transition are observed: the first and second transitions are at 4.2 K and 2.4 K, respectively. Taking into account the powder x-ray diffraction results and $T_c$ = 2.8 K of $(C_2H_8N_2)_yTiSe_2$, the second at 2.4 K is regarded as being mainly due to Phase II of $Li_x(C_2H_8N_2)_yTiSe_2$, and therefore the first at 4.2 K is inferred to be due to Phase III of $Li_x(C_2H_8N_2)_yTiSe_2$. As for both Li- and HMDA-intercalated $Li_x(C_6H_{16}N_2)_yTiSe_2$ ($x$ = 0.2), a superconducting transition is observed at 4.2 K. The $T_c$ value is nearly the same as that of Phase III of both Li- and EDA-intercalated $Li_x(C_2H_8N_2)_yTiSe_2$ ($x$ = 0.2), though the interlayer distance is expanded more than that of Phase III of $Li_x(C_2H_8N_2)_yTiSe_2$ ($x$ = 0.2). This suggests that the electronic state is nearly the same between these samples. The result that the $T_c$ value of both Li- and EDA- or HMDA-intercalated $Li_x(C_2H_8N_2)_yTiSe_2$

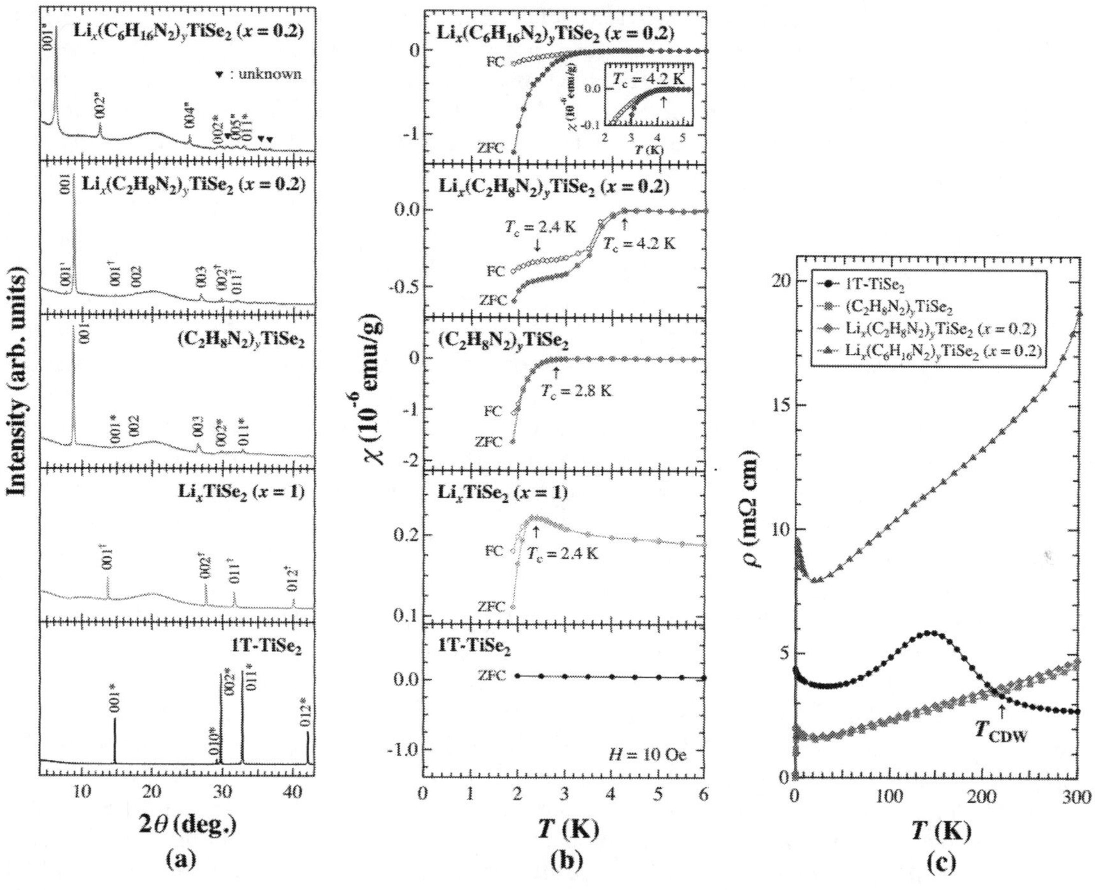


**Figure 1. (a)** Powder x-ray diffraction patterns of the host sample of 1T-$TiSe_2$ and intercalated samples of $Li_xTiSe_2$ ($x$ = 1), $Li_x(C_2H_8N_2)_yTiSe_2$ ($x$ = 0, 0.2) and $Li_x(C_6H_{16}N_2)_yTiSe_2$ ($x$ = 0.2), using Cu K$\alpha$ radiation. Indices with an asterisk are due to 1T-$TiSe_2$. Indices with a dagger, no mark and a quotation mark represent Phase I, Phase II and Phase III of $Li_x(C_2H_8N_2)_yTiSe_2$, respectively. Indices with a double quotation mark are due to $Li_x(C_6H_{16}N_2)_yTiSe_2$. Peaks marked by ▼ are due to an unknown compound. All indices are based on the 1T-type structure ($P\bar{3}m1$). **(b)** Temperature dependences of the magnetic susceptibility $\chi$ in a magnetic field of 10 Oe on zero-field cooling (ZFC) (closed circles) and on field cooling (FC) (open circles) for the host sample of 1T-$TiSe_2$ and intercalated samples of $Li_xTiSe_2$ ($x$ = 1), $Li_x(C_2H_8N_2)_yTiSe_2$ ($x$ = 0, 0.2) and $Li_x(C_6H_{16}N_2)_yTiSe_2$ ($x$ = 0.2). **(c)** Temperature dependences of the electrical resistivity $\rho$ for the host sample of 1T-$TiSe_2$ and intercalated samples of $Li_x(C_2H_8N_2)_yTiSe_2$ ($x$ = 0, 0.2) and $Li_x(C_6H_{16}N_2)_yTiSe_2$ ($x$ = 0.2), which were pelletized at room temperature without heat treatment.

or $Li_x(C_6H_{16}N_2)_yTiSe_2$ is a little higher than those of only Li-intercalated $Li_xTiSe_2$ ($x$ = 1) and only EDA-intercalated $(C_2H_8N_2)_yTiSe_2$ is speculated to be owing to the multiplier effect of (1) the enhancement of the two-dimensionality of the crystal structure and/or the electronic structure due to the expansion of the interlayer spacing between $TiSe_2$ layers and (2) the enlargement of the electronic density of states at the Fermi level due to the charge transfer from intercalated Li to $TiSe_2$ layers.

Figure 1(c) shows the temperature dependences of $\rho$ for the host sample of 1T-$TiSe_2$ and intercalated samples of $Li_x(C_2H_8N_2)_yTiSe_2$ ($x$ = 0, 0.2) and $Li_x(C_6H_{16}N_2)_yTiSe_2$ ($x$ = 0.2). It is found that 1T-$TiSe_2$ exhibits a behavior of $\rho$ well known in single-crystalline samples [14]. With decreasing temperature, a clear increase in $\rho$ is observed at low temperatures below the CDW transition temperature $T_{CDW}$ ~ 220 K, though an upturn behavior of $\rho$ due to insulating grain-boundaries in the pelletized powdery sample is observed at very low temperatures. As for intercalated samples of $Li_x(C_2H_8N_2)_yTiSe_2$ ($x$ = 0, 0.2) and $Li_x(C_6H_{16}N_2)_yTiSe_2$ ($x$ = 0.2), $\rho$ shows a metallic temperature-dependence except for the upturn of $\rho$ at very low temperatures due to insulating grain-boundaries. It is found that the CDW transition is completely suppressed and that a superconducting transition occurs at a low temperature. These results indicate that both the intercalation of EDA and the co-intercalation of Li and EDA or HMDA suppress the CDW transition and induce a metallic state, leading to the appearance of superconductivity.

Next, we show the results of the intercalation into 2H-$MoSe_2$. Figure 2(a) shows the powder x-ray diffraction patterns of the host sample of 2H-$MoSe_2$ and intercalated samples of $Li_x(C_2H_8N_2)_yMoSe_2$ ($x$ = 0.25, 1) and $Li_x(C_6H_{16}N_2)_yMoSe_2$ ($x$ = 0.25, 1). The powder x-ray diffraction patterns of the host sample of 2H-$MoSe_2$ and all the intercalated samples are well indexed based on the 2H-type structure ($P6_3/mmc$). The lattice parameters are estimated to be $a$ = 3.289(1) Å, $c$ = 12.92(3) Å for 2H-$MoSe_2$. As for both Li- and EDA-intercalated $Li_x(C_2H_8N_2)_yMoSe_2$ ($x$ = 0.25), new Bragg peaks are observed at $2\theta$ ~ 7.5° and 15°, though there still remain non-intercalated regions of 2H-$MoSe_2$. The $c$-axis length of $Li_x(C_2H_8N_2)_yMoSe_2$ ($x$ = 0.25) is estimated to be 23.7(2) Å, so that the interlayer distance is expanded by ~ 5.4 Å through the co-intercalation of Li and EDA. Taking into account the size of EDA, EDA is speculated to be intercalated between $MoSe_2$ layers where the longitudinal direction of EDA is perpendicular to $MoSe_2$ layers. Here, we call this phase Phase I of $Li_x(C_2H_8N_2)_yMoSe_2$. As for $Li_x(C_2H_8N_2)_yMoSe_2$ ($x$ = 1), new Bragg peaks are observed at $2\theta$ ~ 7.5°, 8.5° and 15°, 17°. The new Bragg peaks are regarded as being due to Phase I and a different phase (we call this phase Phase II) with the $c$-axis lengths of 23.6(8) Å and 20.56(3) Å of $Li_x(C_2H_8N_2)_yMoSe_2$, respectively. In the case of Phase II, on the other hand, the expansion of the interlayer distance is ~ 3.8 Å, so that EDA is speculated to be intercalated between $MoSe_2$ layers where the longitudinal direction of EDA is parallel to $MoSe_2$ layers. As for both Li- and HMDA-intercalated samples of $Li_x(C_6H_{16}N_2)_yMoSe_2$ ($x$ = 0.25, 1), new Bragg peaks are observed at $2\theta$ ~ 5.5°, 11° and 4.5°, 9°, respectively. The $c$-axis lengths of $Li_x(C_6H_{16}N_2)_yMoSe_2$ ($x$ = 0.25, 1) are estimated to be 30.8(1) Å and 37.5(4) Å, respectively. These $c$-axis lengths are much large than those of Li- and EDA-intercalated $Li_x(C_2H_8N_2)_yMoSe_2$, and the interlayer distances are expanded by ~ 8.9 Å and ~ 12.3 Å through the co-intercalation of Li and HMDA into 2H-$MoSe_2$. Taking into account the size of HMDA, the state of HMDA between $MoSe_2$ layers in $Li_x(C_6H_{16}N_2)_yMoSe_2$ ($x$ = 0.25, 1) is not determined definitely.

The temperature dependences of $\chi$ in a magnetic field of 10 Oe on ZFC and FC for the host sample of 2H-$MoSe_2$ and both Li- and EDA- or HMDA-intercalated samples of $Li_x(C_2H_8N_2)_yMoSe_2$ ($x$ = 0.25, 1) and $Li_x(C_6H_{16}N_2)_yMoSe_2$ ($x$ = 0.25, 1) are shown in Fig. 2(b). Although the host sample of 2H-$MoSe_2$ shows no superconductivity, superconducting transitions are observed at 3.2 – 6.0 K in Li- and EDA- or HMDA-intercalated samples. Both Li- and EDA-intercalated samples of $Li_x(C_2H_8N_2)_yMoSe_2$ ($x$ = 0.25, 1) show a superconducting transition at 4.2 K. On the other hand, both Li- and HMDA-intercalated samples of $Li_x(C_6H_{16}N_2)_yMoSe_2$ ($x$ = 0.25, 1) show a superconducting transition at 3.2 K and 6.0 K for $x$ = 0.25 and 1, respectively.

Figure 2(c) shows the temperature dependences of $\chi$ in a magnetic field of 1 T on ZFC for both for the host sample of 2H-$MoSe_2$ and intercalated samples of $Li_x(C_2H_8N_2)_yMoSe_2$ ($x$ = 0.25, 1) and $Li_x(C_6H_{16}N_2)_yMoSe_2$ ($x$ = 0.25, 1). It is found that $\chi$ tends to increase with increasing $x$. This suggests that intercalated Li supplies $MoSe_2$ layers with electron carriers, leading to the increase in the Pauli paramagnetism. Moreover, it has been found that the $T_c$ values are related not to the interlayer spacing between $MoSe_2$ layers but to the carrier density increased by the electron doping due to the Li intercalation, as shown in Fig. 2(d). Accordingly, it is concluded that first of all a sufficient amount of carrier density is necessary for the appearance of superconductivity rather than the two-dimensionality of the crystal structure and/or electronic structure, because 2H-$MoSe_2$ is a semiconductor.

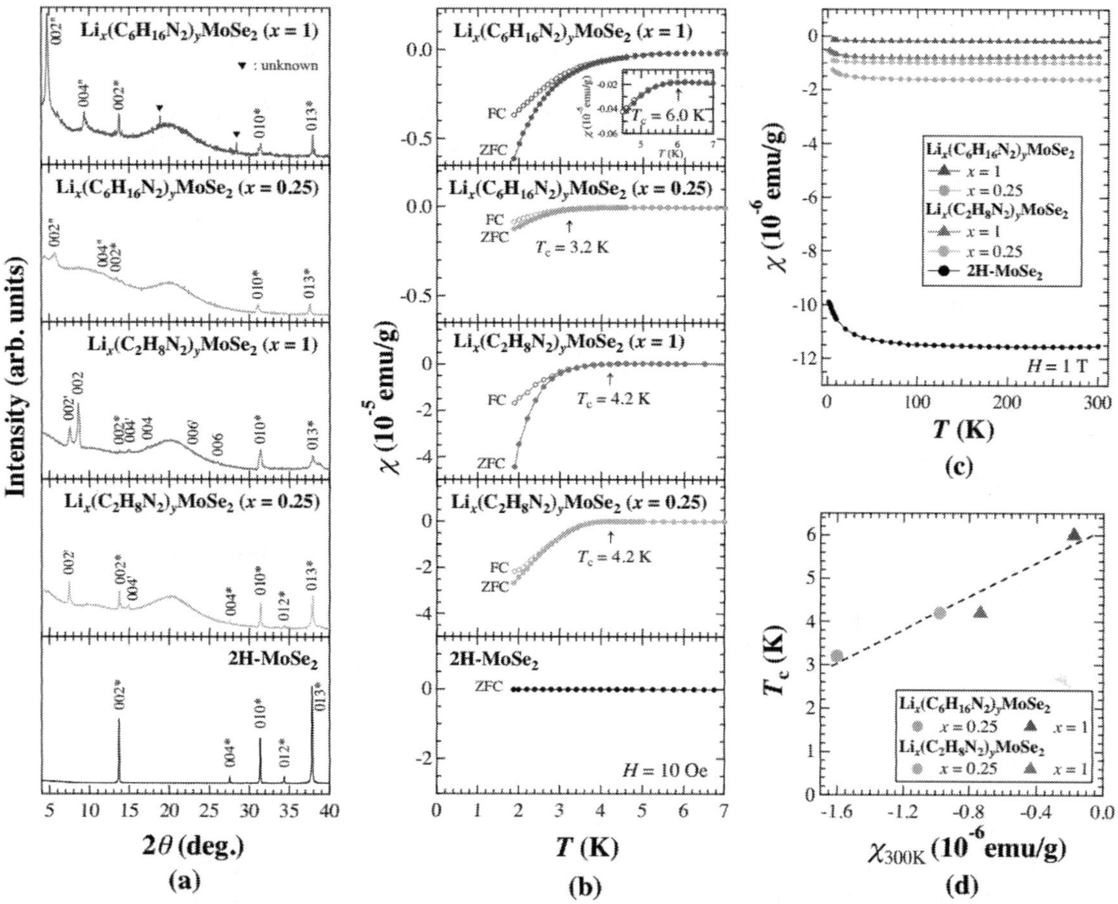


**Figure 2. (a)** Powder x-ray diffraction patterns of the host sample of 2H-$MoSe_2$ and intercalated samples of $Li_x(C_2H_8N_2)_yMoSe_2$ ($x$ = 0.25, 1) and $Li_x(C_6H_{16}N_2)_yMoSe_2$ ($x$ = 0.25, 1), using Cu K$\alpha$ radiation. Indices with an asterisk are due to 2H-$MoSe_2$. Indices with a quotation mark and no mark represent Phase I and Phase II of $Li_x(C_2H_8N_2)_yMoSe_2$, respectively. Indices with a double quotation mark are due to $Li_x(C_6H_{16}N_2)_yMoSe_2$. Peaks marked by ▼ are due to an unknown compound. All indices are based on the 2H-type structure ($P6_3/mmc$). **(b)** Temperature dependences of the magnetic susceptibility $\chi$ in a magnetic field of 10 Oe on zero-field cooling (ZFC) (closed circles) and on field cooling (FC) (open circles) for the host sample of 2H-$MoSe_2$ and intercalated samples of $Li_x(C_2H_8N_2)_yMoSe_2$ ($x$ = 0.25, 1) and $Li_x(C_6H_{16}N_2)_yMoSe_2$ ($x$ = 0.25, 1). **(c)** Temperature dependences of the magnetic susceptibility $\chi$ in a magnetic field of 1T on zero-field cooling for the host sample of 2H-$MoSe_2$ and intercalated samples of $Li_x(C_2H_8N_2)_yMoSe_2$ ($x$ = 0.25, 1) and $Li_x(C_6H_{16}N_2)_yMoSe_2$ ($x$ = 0.25, 1). **(d)** Relationship between $T_c$ and $\chi$ at 300 K, $\chi_{300K}$, for intercalated samples of $Li_x(C_2H_8N_2)_yMoSe_2$ ($x$ = 0.25, 1) and $Li_x(C_6H_{16}N_2)_yMoSe_2$ ($x$ = 0.25, 1). The dashed line is a guide to the eye.

## 4. Summary

We have succeeded in synthesizing new superconductors of $Li_x(C_2H_8N_2)_yTiSe_2$ and $Li_x(C_6H_{16}N_2)_yTiSe_2$ with $T_c$ = 4.2 K via the co-intercalation of Li and EDA or HMDA into semimetallic 1T-$TiSe_2$ exhibiting a CDW transition. Moreover, it has been found that $Li_xTiSe_2$ and $(C_2H_8N_2)_yTiSe_2$ also show superconductivity with $T_c$ = 2.4 K and 2.8 K, respectively. From the $\rho$ measurements, it has been found that both the intercalation of EDA and the co-intercalation of Li and EDA or HMDA completely suppress the CDW transition and induce a superconducting transition at a low temperature. These results indicate that both the electron doping due to the Li intercalation and the expansion of the interlayer spacing between $TiSe_2$ layers due to the intercalation of diamines suppress the CDW transition, leading to the appearance of superconductivity.

We have also succeeded in synthesizing new superconductors of $Li_x(C_2H_8N_2)_yMoSe_2$ and $Li_x(C_6H_{16}N_2)_yMoSe_2$ with $T_c$ = 4.2 K and 3.2 – 6.0 K, respectively, via the co-intercalation of Li and EDA or HMDA into semiconducting 2H-$MoSe_2$. It has been found that, the $T_c$ values are related not to the interlayer spacing but to the carrier density increased by the electron doping due to the Li intercalation. Accordingly, it has been concluded that first of all a sufficient amount of carrier density is necessary for the appearance of superconductivity in the 2H-$MoSe_2$ system rather than the two-dimensionality of the crystal structure and/or electronic structure.

## Acknowledgments

This work was supported by JSPS KAKENHI (Grant Numbers 15K13512 and 16K05429). One of the authors (K. S.) was supported by the Tohoku University Division for Interdisciplinary Advanced Research and Education.

## References

[1] E. Morosan, H. W. Zandbergen, B. S. Dennis, J. W. G. Bos, Y. Onose, T. Klimczuk, A. P. Ramirez, N. P. Ong, and R. J. Cava 2006 Nat. Phys. **2** 544
[2] E. Morosan, K. E. Wagner, L. L. Zhao, Y. Hor, A. J. Williams, J. Tao, Y. Zhu, and R. J. Cava 2010 Phys. Rev. B **81** 094524
[3] A. F. Kusmartseva, B. Sipos, H. Berger, L. Forró, and E. Tutiš 2009 Phys. Rev. Lett. **103** 236401
[4] L. J. Li, E. C. T. O'Farrell, K. P. Loh, G. Eda, B. Özyolmaz, and A. H. Castro Neto 2016 Nature **529** 185
[5] W. Shi, J. Ye, Y. Zhang, R. Suzuki, M. Yoshida, J. Miyazaki, N. Inoue, Y. Saito, and Y. Iwasa 2015 Sci. Rep. **5** 12534
[6] X. Miao, S. Nishiyama, L. Zheng, H. Goto, R. Eguchi, H. Ota, T. Kambe, K. Terashima, T. Yokoya, H. T. L. Nguyen, T. Kagayama, N. Hirao, Y. Ohishi, H. Ishii, Y.-F. Liao, and Y. Kubozono 2016 Sci. Rep. **6** 29292
[7] T. P. Ying, X. L. Chen, G. Wang, S. F. Jin, T. T. Zhou, X. F. Lai, H. Zhang, and W. Y. Wang 2012 Sci. Rep. **2** 426
[8] A. Krzton-Maziopa, E. V. Pomjakushima, V. Y. Pomjakushin, F. von Rohr, A. Schilling, and K. Conder 2012 J. Phys.: Condens. Matter **24** 382202
[9] T. Hatakeda, T. Noji, T. Kawamata, M. Kato, and Y. Koike 2013 J. Phys. Soc. Jpn. **82** 123705
[10] T. Noji, T. Hatakeda, S. Hosono, T. Kawamata, M. Kato, and Y. Koike 2014 Physica C **504** 8
[11] S. Hosono, T. Noji, T. Hatakeda, T. Kawamata, M. Kato, and Y. Koike 2014 J. Phys. Soc. Jpn. **83** 113704
[12] S. Hosono, T. Noji, T. Hatakeda, T. Kawamata, M. Kato, and Y. Koike 2016 J. Phys. Soc. Jpn. **85** 013702
[13] S. Hosono, T. Noji, T. Hatakeda, T. Kawamata, M. Kato, and Y. Koike 2016 J. Phys. Soc. Jpn. **85** 104701
[14] A. F. Kusmartseva, B. Sipos, H. Berger, L. Forró, and E. Tutiš 2009 Phys. Rev. Lett. **103** 236401

IOP Conf. Series: Journal of Physics: Conf. Series **969** (2018) 012078 doi:10.1088/1742-6596/969/1/012078

# Spatially dispersive surface impedance in the electrodynamics of conductors without dc dissipation

**N A Volchkov[1], A L Karuzskii[1], V A Kulbachinskii[2], Y A Mityagin[1], A V Perestoronin[1], A P Chernyaev[3]**

[1] P. N. Lebedev Physical Institute of RAS, Leninsky prospekt 53, Moscow 119991, Russia
[2] M.V. Lomonosov Moscow State University, 119991 GSP-1, Moscow, Russia
[3] Moscow Institute of Physics and Technology (State University), Institutsky Per. 9, Dolgoprudny, Moscow Region 141700, Russia

E-mail: karuz@sci.lebedev.ru

**Abstract.** We derive general frequency dependencies of the surface impedance modulus for conductors without the dc dissipation, i. e. for superconductors or perfect conductors. The frequency-dependent surface impedance was applied for the solutions corresponding to the spatially dispersive eigenvalues of the permittivity operator for conductors. We demonstrate that appropriately taken into account effects of the spatial dispersion can give the general frequency dependence of the surface impedance for the obtained solutions including that for superconductor. It is shown that an incorporation of the spatial dispersion leads to an appearance of the Meissner effect in perfect conductors in the same manner as in superconductors.

## 1. Introduction

Recently the frequency-dependent surface impedance was calculated for the spatially dispersive eigenvalues of the permittivity operator in conductors [1]. It is reasonable to give a brief description these formulations valid for both conductors and superconductors The electrodynamics of superconductors is supposed to be in principle reduced to those for conductors as the temperature approaches the critical temperature $T_c$ and beyond $T_c$. However, such reduction is not rather straightaway. Early the problem resulted in a supplement of the Maxwell equations by postulating additional London equations for an explanation the Meissner effect [2]. It discriminates the perfect conductor in non-zero-field cooling behavior as compared to the superconductor. Here we demonstrate that appropriately taken into account effects of the spatial dispersion can give general frequency dependencies of the surface impedance for the obtained in [1] solutions including those both for the normal conductor and for the superconductor or perfect conductor. It is shown that an incorporation of the spatial dispersion leads to an appearance of the Meissner effect in perfect conductors in the same manner as in superconductors.

## 2. Frequency dependencies

To evaluate the general frequency dependence of the eigenvalue modulus of absolute permittivity operator $|\tilde{\varepsilon}_a|$ and of the surface impedance modulus $|\widetilde{Z}|$, let us again [1] consider at first only the spatial effects in permittivity, assuming the problem is stationary $\omega = 0$ and taking into account only the spatial field inhomogenity in the form of a wave number $k''$. Preliminary results concerning the frequency dependencies were discussed in [3]. The designations of symbols in the equations below are generally accepted [1, 3–9] and can be found in [1, 3, 7, 8].

### *2.1. Normal conductivity*

The first of Maxwell equations can be written as

$$\text{rot}\,\widetilde{\mathbf{H}} = (\partial\widetilde{\mathbf{D}}/\partial t) + \widetilde{\mathbf{j}}. \tag{1}$$

The frequency $\omega_k'' = k''v_F$ can be associated with the spatial inhomogenity $k''$ [1], where $v_F$ is the Fermi velocity, i. e. the velocity of propagation of the external perturbation. Using the ambiguity of the representation of a right side of the previous expression [4–6] and integrating it over the time one can get from the constitutive equation [1] $\widetilde{\mathbf{D}}(\mathbf{r},t) = \widetilde{\varepsilon}_a\widetilde{\mathbf{E}}(\mathbf{r},t)$ an expression [7]

$$\widetilde{\varepsilon}_a = \widetilde{\varepsilon}_p\varepsilon_0 - i\widetilde{\sigma}/\omega_k'' = \widetilde{\varepsilon}_p\varepsilon_0 - i\widetilde{\sigma}/(k''v_F), \tag{2}$$

where $\widetilde{\sigma}$ is the static eigenvalue of conductivity operator, $\widetilde{\varepsilon}_p$ is the eigenvalue of relative permittivity operator of a lattice. The solutions found in [1] for conductors and superconductors have the fixed arguments of complex numbers $\widetilde{\varepsilon}_a$ and $\widetilde{Z} \sim 1/\widetilde{k}$ determined by conditions of the spatial force resonances. So for a finite frequency $\omega$ of an external field the spatial inhomogenity $k''$ is determined, according to [1, 7], by the frequency and by the eigenvalue of permittivity operator

$$k'' \sim \omega(\widetilde{\varepsilon}_a\mu_0)^{1/2} = \omega/\widetilde{v}_{ph}, \tag{3}$$

that may be assigned to a mean field approximation. The conductivity in the extremely anomalous limit is determined by the frequency $\widetilde{\sigma} \sim 1/\omega$ also [8]. The substitution of $k''$ and $\widetilde{\sigma}$ into the expression (2) for $\widetilde{\varepsilon}_a$ results in the proportionality $(\widetilde{\varepsilon}_a)^{3/2} \sim \omega^{-2}$ that corresponds to the frequency dependence of the impedance modulus $|\widetilde{Z}| \sim \omega^{2/3}$ for all solutions found for the conductor except that for the superconductor, as it was shown in [1].

### *2.2. Superconductors and perfect conductors*

To estimate the frequency dependencies of moduli $|\widetilde{Z}|$ and $|\widetilde{\varepsilon}_a|$ for the solution found for superconductors or perfect conductors having the phase of $\widetilde{\varepsilon}_a$ equal to $\beta_1 = \pi$ and the phase of $\widetilde{Z}$ equal to $\psi_1 = \pi/2$ (and additionally for the solution with $\alpha_0 = 0$ and $\varphi_0 = 0$) [1] one should note that for these solutions the real parts of eigenvalue $\widetilde{\sigma}$ of the conductivity operator in the equation (2) equal zero and the direct current does not dissipate with time. In this case one should take into account an influence of spatial inhomogenity of the field with the largest possible value of a wave number $k''$ or of a frequency $\omega_k'' = k''v_F$, regarding an absence of the dc dissipation, for all possible spatial scales, but not only for those corresponding to the frequency of an incident wave (cf. the Eq. (3)). The velocity of propagation of a perturbation in the quasi stationary environment without the dc dissipation could correspond to the spatially dispersive value $v$, which should not be necessarily associated with the Fermi velocity $v_F$, but may be equal to the speed of light $c$, for example, as in the Langmuir plasma frequency. The highest value of spatial inhomogenity of the field with the largest possible value of a wave number $k''$ or of an associated frequency $\omega_k'' = k''v$ in this case can approach the vertical-asymptote values equal to the $k_0''$ or to the frequency proportional to $\omega_{kp0}'' = k_0''v_F$, corresponding to the Langmuir plasma frequency

$$\omega_p = k_0''c = c/\lambda_0, \tag{4}$$

where

$$\lambda_0 = (k_0'')^{-1} = (m/\mu_0 ne^2)^{1/2} \tag{5}$$

is the London penetration depth. The absence of the dc dissipation should result in the feasibility of "acoustic" conditions ($v > 0$) for the spatially dispersive plasmon polariton frequency $\omega''_{kp} = k''v$.

This spatial inhomogenity will correspond to the highest value of the Abraham force (see Eq.(13) below and Ref. [1]). So it should result in the spatial structure dominating over the others possible structures [1, 7]. A value of this spatial inhomogenity will not depend on frequency values $\omega$ of any incident microwaves, while those could be considered as quasi stationary when $\omega < \omega''_{kp0} \ll \omega_p$. This can be illustrated by an expression of a current density after the end of an external perturbation obtained with the account of the spatial dispersion [8]. If one substitutes the spatial factor in Eq.(3) of Ref.[8] in the form of

$$j_0(\theta, r) = j_0(\theta)e^{i\kappa r}, \quad (6)$$

where $\theta$ is the reduced temperature $\theta = T/T_c$ [8, 9], it is reexpressed as the current density

$$j(\theta, r, t) = j_0(\theta)e^{-\omega'_k t}e^{i(\kappa r \pm \omega''_k t)}, \quad (7)$$

which remains after switching off an electric field $E = E_0 e^{i(kr+\omega t)}$. In a perfect conductor the direct current does not decay with time ($\omega'_k = 0$) and the stationary current solution (6) may be reconstructed from (7) by the proper time averaging of plasma oscillations with positive and negative signs of the plasmon polariton frequency $\omega''_k = \omega''_{kp0}$, by a finite time shift in their starting moments, for example.

The requirement of an independence of the spatial inhomogenity $k''$ on the frequency $\omega$ should result, according to the equation

$$k'' = \omega(|\widetilde{\varepsilon}_a|\mu_0)^{1/2} = \omega/|\widetilde{v}_{ph}|, \quad (8)$$

in the inverse ratio $\sqrt{|\widetilde{\varepsilon}_a|} \sim 1/\omega$ and in the dependence of the surface impedance (its modulus) linear on a frequency $\omega$

$$|\widetilde{Z}| = \sqrt{\frac{\mu_0}{|\widetilde{\varepsilon}_a|}} = \frac{\omega\mu_0}{k''}. \quad (9)$$

The frequency dependence of the eigenvalue $\widetilde{\varepsilon}_a$ in the right side of the Eq. (8) is determined by the requirement of the independent on $\omega$ wavenumber $k''$ in the left side, inversely to their correlations in the Eq. (3), where the wavenumber $k''$ is determined by the eigenvalue $\widetilde{\varepsilon}_a$ according to the mean-field approximation. The further temperature issue of how the electrodynamics of superconductors can be reduced to those for conductors as the temperature approaches and beyond the critical temperature may be clearly stated using, e. g., the generalized two-fluid principles [8, 9] or rigorous modeling [9, 10].

## 3. The spatial dispersion and Meissner effect in superconductors and perfect conductors

In the surface impedance approximation [4–7, 9, 10] let it will be the coordinate $z$ with the basis vector $\mathbf{k}$, which are directed into the conductor together with the unitary normal $\mathbf{n}$ to the boundary surface $z = 0$. A monochromatic transverse field at the normal incidence to the plane surface of a conductor will be considered. An electric field strength vector is directed tangentially along the $x$-axis ($\widetilde{\mathbf{E}}(z) = \widetilde{E}_x(z)\mathbf{i}$), and a magnetic induction vector is directed along the $y$-axis ($\widetilde{\mathbf{B}}(z) = \widetilde{B}_y(z)\mathbf{j}$). Here $\mathbf{i}$ and $\mathbf{j}$ are the basis vectors. The fields decrease into the depth of a conductor as $\widetilde{B}_y(z) = \widetilde{B}_y(0)e^{-z/\widetilde{\delta}}$ and $\widetilde{E}_x(z) = \widetilde{E}_x(0)e^{-z/\widetilde{\delta}}$, where $\widetilde{\delta}$ is the complex penetration depth, which defines the surface impedance by the relation $\widetilde{Z} = i\omega\mu_0\widetilde{\delta}$ [4–7, 9, 10], and $\mu_0$ is the permeability of vacuum. Since the surface impedance relates the tangential components of fields on the boundary surface [7, 9]

$$\widetilde{Z} = \frac{\widetilde{E}_x(0)}{\widetilde{H}_y(0)} = \frac{\widetilde{B}_y(0)}{\widetilde{D}_x(0)}, \tag{10}$$

the zero limit of the surface impedance $\widetilde{Z}|_{\omega\to 0} = 0$ at the zero frequency, deduced from the Eq. (9), means that the stationary tangential component of an electric field (or of a magnetic induction) should be equal zero $\widetilde{E}_x(0)|_{\omega\to 0} = 0$ (or $\widetilde{B}_y(0)|_{\omega\to 0} = 0$), but with a possible finite value of $\widetilde{H}_y(0)|_{\omega\to 0} \neq 0$ (or $\widetilde{D}_x(0)|_{\omega\to 0} \neq 0$).

Let us consider behaviour of the Abraham force in a perfect conductor (or in a superconductor) in the limit of zero frequency $\omega \to 0$. From the expression for the spatial density of the momentum flux [1, 7], which is the corresponding component of the Maxwellian field-stress tensor,

$$\widetilde{\Pi} = [\widetilde{\mathbf{D}} \times \widetilde{\mathbf{B}}] = (\mu_0^{1/2} \widetilde{\varepsilon}_a^{3/2}) \widetilde{E}_x^2(0)\mathbf{k}, \tag{11}$$

where $\mathbf{k}$ is the basis vector of the $z$-axis, the Abraham force can be derived [4, 6] neglecting the frequency dispersion:

$$\begin{aligned}
&\left(1 - \frac{1}{\widetilde{\varepsilon}}\right) \frac{\partial \widetilde{\Pi}}{\partial t} = \left(1 - \frac{1}{\widetilde{\varepsilon}}\right) \frac{\partial}{\partial t} [\widetilde{\mathbf{D}} \times \widetilde{\mathbf{B}}] = \\
&= \left(1 - \frac{1}{\widetilde{\varepsilon}}\right) \frac{\partial}{\partial t} \left[\widetilde{D}_x(0) e^{i\omega t}\mathbf{i} \times \widetilde{B}_y(0) e^{i\omega t}\mathbf{j}\right] = \\
&= 2i\omega \left(1 - \frac{1}{\widetilde{\varepsilon}}\right) e^{i2\omega t} \left[\widetilde{D}_x(0)\mathbf{i} \times \widetilde{B}_y(0)\mathbf{j}\right],
\end{aligned} \tag{12}$$

where $\widetilde{\varepsilon} = \widetilde{\varepsilon}_a/\varepsilon_0$ is the eigenvalue of relative permittivity operator. The surface impedance (10) defines the relation between electric and magnetic inductions $\widetilde{D}_x(0)$ and $\widetilde{B}_y(0)$ and may be substituted in the expression (12) for the Abraham force. Then using the Eq. (9) for the modulus of surface impedance and taking account of its phase $\psi_1 = \pi/2$ we obtain

$$\begin{aligned}
&2i\omega \left(1 - \frac{1}{\widetilde{\varepsilon}}\right) e^{i2\omega t} \left[\frac{\widetilde{B}_y(0)}{\widetilde{Z}}\mathbf{i} \times \widetilde{B}_y(0)\mathbf{j}\right] = \\
&= 2i\omega \left(1 - \frac{1}{\widetilde{\varepsilon}}\right) e^{i2\omega t} \left[\frac{-ik''_A}{\omega\mu_0} \widetilde{B}_y(0)\mathbf{i} \times \widetilde{B}_y(0)\mathbf{j}\right] = \\
&= 2\left(1 - \frac{1}{\widetilde{\varepsilon}}\right) e^{i2\omega t} \left[\frac{k''_A}{\mu_0} \widetilde{B}_y(0)\mathbf{i} \times \widetilde{B}_y(0)\mathbf{j}\right] = \\
&= 2\left(1 - \frac{1}{\widetilde{\varepsilon}}\right) e^{i2\omega t} \mu_0 \left[\mathrm{rot}\widetilde{\mathbf{H}}(0) \times \widetilde{\mathbf{H}}(0)\right].
\end{aligned} \tag{13}$$

The expression (13) shows that Abraham force at all frequencies $\omega \ll \omega_p$, as well as in the stationary limit $\omega \to 0$, achieves the highest magnitude when the spatial inhomogenity $k''_A$ gets the largest, frequency-independent value (see Sec. 2.2) corresponding to the London (or plasma) penetration depth $\lambda_0 = (k''_0)^{-1} = (m/\mu_0 n e^2)^{1/2}$. Such the spatial configuration of the fields related to the reconstruction of an electronic system, which corresponds to the highest magnitude of the Abraham force, is dominant due to the force correlations and the consequential decrease of the free energy [1, 7]. And respectively, the absolutely unstable spatial configuration, according to Eq. (13), corresponds to the geometry of the spatially homogeneous field and to the zero value of the spatial inhomogenity $k''_A = 0$ when the Abraham force is zero with no advantage in the free energy. So the conclusions obtained here in the surface impedance approximation, which is

IOP Conf. Series: Journal of Physics: Conf. Series **969** (2018) 012078 doi:10.1088/1742-6596/969/1/012078

quite adequate to the general description of Meissner effect [9], show that the spatial dispersion results in an appearance of the Meissner effect in perfect conductors in the same manner as in superconductors contrary to the preceding considerations [2], which do not incorporate the spatial dispersion effects. A driving force of its appearance is the Abraham force.

It should be noted that this conclusion may be derived more formally, without considerations of any frequency dependencies. Abraham force can be reexpressed as

$$\left(1-\frac{1}{\tilde{\varepsilon}}\right)\frac{\partial\tilde{\Pi}}{\partial t}=\left(1-\frac{1}{\tilde{\varepsilon}}\right)\frac{\partial}{\partial t}\left[\tilde{\mathbf{D}}\times\tilde{\mathbf{B}}\right]=$$
$$=\left(1-\frac{1}{\tilde{\varepsilon}}\right)\left[\frac{\partial\tilde{\mathbf{D}}}{\partial t}\times\tilde{\mathbf{B}}\right]+\left(1-\frac{1}{\tilde{\varepsilon}}\right)\left[\tilde{\mathbf{D}}\times\frac{\partial\tilde{\mathbf{B}}}{\partial t}\right] \quad (14)$$

For monochromatic fields the Maxwell equations in a homogeneous medium with the spatial dispersion may be written [1, 4–7] as

$$\mathrm{rot}\tilde{\mathbf{B}}=\mu_0\mathrm{rot}\tilde{\mathbf{H}}=\mu_0(\partial\tilde{\mathbf{D}}/\partial t);\quad \mathrm{rot}\tilde{\mathbf{E}}=-(\partial\tilde{\mathbf{B}}/\partial t); \quad (15)$$

$$\mathrm{div}\tilde{\mathbf{D}}=\mathrm{div}\tilde{\mathbf{E}}=0;\qquad \mathrm{div}\tilde{\mathbf{B}}=0. \quad (16)$$

Due to an absence of magnetic charges and an equality $\partial\tilde{\mathbf{B}}/\partial t=0$ in the second of Eqs (15) in the stationary approximation $\omega=0$ the second term in the sum (14) equals zero. Regarding the ambiguity of the representation of the first Maxwell equation (cf. the first of Eqs (15) and the Eq.(1) rot $\tilde{\mathbf{H}}=(\partial\tilde{\mathbf{D}}/\partial t)+\tilde{\mathbf{j}}$), one can replace the partial time derivative in the first term of the sum (14), which may be assigned formally to the "stationary displacement current" or simply to the current density $\tilde{\mathbf{j}}$, by the rot$\tilde{\mathbf{H}}$ from the first of Eqs (15), that results in

$$\left(1-\frac{1}{\tilde{\varepsilon}}\right)\left[\mathrm{rot}\tilde{\mathbf{H}}\times\tilde{\mathbf{B}}\right]=\left(1-\frac{1}{\tilde{\varepsilon}}\right)\left[\frac{k''_{ef}}{\mu_0}\tilde{B}_y(0)\mathbf{i}\times\tilde{B}_y(0)\mathbf{j}\right]. \quad (17)$$

This expression shows that Abraham force achieves the highest magnitude when the spatial inhomogenity $k''_{ef}$ gets the largest value. It corresponds to a half of the London penetration depth $\lambda_0/2=(2k''_0)^{-1}=(m/\mu_0 n4e^2)^{1/2}$ taking into account a factor of 2 that differentiates the Eqs (13) and (17) in the stationary approximation $\omega=0$ (cf. an introduction of the effective constitutive parameters in [3]).

## 4. London equations and Abraham force

The partial derivative of magnetic induction with respect to time in the second of Maxwell equations (15) may be represented as

$$(\partial\tilde{\mathbf{B}}/\partial t)=-\mathrm{rot}\ \tilde{\mathbf{E}}=-\mathrm{rot}\ (\tilde{\mathbf{j}}/\tilde{\sigma})=-(i\omega\mu_0/(k''_0)^2)\mathrm{rot}\ \tilde{\mathbf{j}}, \quad (18)$$

where the first London equation [2, 8, 9] $d\tilde{\mathbf{j}}(z,t)/dt=d(\tilde{\mathbf{j}}(z)e^{i\omega t})/dt=i\omega\tilde{\mathbf{j}}(z,t)=(1/\mu_0\lambda_0^2)\tilde{\mathbf{E}}=((k''_0)^2/\mu_0)\tilde{\mathbf{E}}$ has been used to derive the conductivity $\tilde{\sigma}$ from Ohm's law $\tilde{\mathbf{j}}=\tilde{\sigma}\tilde{\mathbf{E}}=(1/i\omega\mu_0\lambda_0^2)\tilde{\mathbf{E}}=((k''_0)^2/i\omega\mu_0)\tilde{\mathbf{E}}$. Substituting the Eq. (18) and second London equation [2, 9] rot $\tilde{\mathbf{j}}=-(1/\lambda_0^2)\tilde{\mathbf{H}}=-(k''_0)^2\tilde{\mathbf{H}}$ into the last term of Abraham force (14), one gets the reinforced expression instead of the Eq. (17)

$$\left(1-\frac{1}{\tilde{\varepsilon}}\right)\left[\frac{\partial\tilde{\mathbf{D}}}{\partial t}\times\tilde{\mathbf{B}}\right]+\left(1-\frac{1}{\tilde{\varepsilon}}\right)\left[\tilde{\mathbf{D}}\times\frac{\partial\tilde{\mathbf{B}}}{\partial t}\right]=$$

$$
\begin{aligned}
&= \left(1-\frac{1}{\tilde{\varepsilon}}\right)\left[\mathrm{rot}\,\tilde{\mathbf{H}}\times\tilde{\mathbf{B}}\right]-\left(1-\frac{1}{\tilde{\varepsilon}}\right)\left[\tilde{\mathbf{D}}\times\mathrm{rot}\,\tilde{\mathbf{E}}\right]= \\
&= \left(1-\frac{1}{\tilde{\varepsilon}}\right)\left[\mathrm{rot}\,\tilde{\mathbf{H}}\times\tilde{\mathbf{B}}\right]-\left(1-\frac{1}{\tilde{\varepsilon}}\right)\left[\tilde{\mathbf{D}}\times\frac{i\omega\mu_0}{(k_0'')^2}\,\mathrm{rot}\,\tilde{\mathbf{j}}\right]= \\
&= \left(1-\frac{1}{\tilde{\varepsilon}}\right)\left[\mathrm{rot}\,\tilde{\mathbf{H}}\times\tilde{\mathbf{B}}\right]-\left(1-\frac{1}{\tilde{\varepsilon}}\right)\left[\tilde{Z}\tilde{\mathbf{D}}\times\frac{1}{k_0''}\,\mathrm{rot}\,\tilde{\mathbf{j}}\right]= \\
&= \left(1-\frac{1}{\tilde{\varepsilon}}\right)\left[\mathrm{rot}\,\tilde{\mathbf{H}}\times\tilde{\mathbf{B}}\right]+\left(1-\frac{1}{\tilde{\varepsilon}}\right)\left[\tilde{B}_y(0)\mathbf{i}\times k_0''\tilde{\mathbf{H}}\right]= \qquad (19)\\
&= \left(1-\frac{1}{\tilde{\varepsilon}}\right)\mu_0\left[\mathrm{rot}\,\tilde{\mathbf{H}}\times\tilde{\mathbf{H}}\right]+\left(1-\frac{1}{\tilde{\varepsilon}}\right)\left[\frac{k_0''}{\mu_0}\tilde{B}_y(0)\mathbf{i}\times\tilde{B}_y(0)\mathbf{j}\right]= \\
&= 2\left(1-\frac{1}{\tilde{\varepsilon}}\right)\left[\frac{k_0''}{\mu_0}\tilde{B}_y(0)\mathbf{i}\times\tilde{B}_y(0)\mathbf{j}\right].
\end{aligned}
$$

Here the modulus (Eq. (9)) and argument $\psi_1 = \pi/2$ of the complex surface impedance $\tilde{Z}$ have been used similarly to the Eq. (13). Comparisons of the Eqs (13), (17) and (19) in the stationary approximation reveal the double ratios $k''_{ef} = 2k''_A = 2k''_0$ in the force expressions. These equalities demonstrate that finding the proper, empirically varified solution of the electrodynamic problem may require an introduction of the effective constitutive parameters (cf. [3]) or postulating the additional phenomenological equations (cf. Sec. 4) depending on the kind of representation of the original problem operator.

## 5. Conclusions

We derived the general frequency dependence of the frequency-dependent surface impedance for the solutions corresponding to the spatially dispersive eigenvalues of the permittivity operator for conductors for all solutions including that for superconductors. It is shown that an incorporation of the spatial dispersion leads to an appearance of the Meissner effect in perfect conductors in the same manner as in superconductors. This expanded conception is promising for applications in novel nanoelectronic devices exploiting the coherence, nonlocality of the superconducting-like state and for search of approaches to the problem of room temperature superconductivity.

## Acknowledgements

This research was supported by Grants of RFBR (15-02-09055, 14-02-00658), of Ministry of Education (16.513.11.3079), by Programmes of RAS (24, IV.12 and III.7).

## References

[1] Dresvyannikov M A, Chernyaev A P, Karuzskii A L, Mityagin Yu A, Perestoronin A V and Volchkov N A 2016 *J. Low Temp. Phys.* **185** 495
[2] London F 1961 *Superfluids, Vol. I: Macroscopic Theory of Superconductivity* (New York: Dover Publications)
[3] Dresvyannikov M A, Chernyaev A P, Karuzskii A L, Mityagin Yu A, Perestoronin A V, and Volchkov N A 2016 *Proc. of SPIE* **10224** 1022412
[4] Landau L D and Lifshiz E M 1992 *Theoretical Physics: Electrodynamics of Continuous Media* vol 8 (Moscow: Nauka) [in Russian]
[5] Lifshiz E M and Pitaevskii L P 1979 *Theoretical Physics: Physical Kinetics* vol 10 (Moscow: Nauka) [in Russian]
[6] Silin V P and Rukhadze A A 1961 *Electromagnetic Properties of Plasma and Plasma-Like Media* (Moscow: Gosatomizdat) [in Russian]; English translation 1965 (New York: Gordon and Breach).
[7] Karuzskii A L, Perestoronin A V, Volchkov N A, Zherikhina L N, and Chernyaev A P 2012 *J. Phys.: Conf. Ser.* **400** 022048
[8] Volchkov N A, Karuzskii A L, Perestoronin A V 2010 *Zh.Eksp.Teor.Fiz.* **138** 329 [in Russian]; *JETP* **111** 292
[9] Mende F F and Spitsyn A I 1985 *Surface Impedance of Superconductors* (Kiev: Naukova Dumka) [in Russian]
[10] Aude DPL 2010 *Modeling superconductors using surface impedance techniques* MIT Thesis (Mass.)

# Index theorem for the flat Andreev bound states at a dirty surface of a nodal superconductor

**Satoshi Ikegaya**[1] **and Yasuhiro Asano**[1,2,3]

[1] Department of Applied Physics, Hokkaido University, Sapporo, 060-8628, Japan
[2] Center of Topological Science and Technology, Hokkaido University, Sapporo 060-8628, Japan
[3] Moscow Institute of Physics and Technology, 141700 Dolgoprudny, Russia

E-mail: asano@eng.hokudai.ac.jp

**Abstract.** We discuss the stability of flat-band Andreev bound states appearing at a surface of a nodal unconventional superconductor. In the clean limit, the existence of the surface bound states is topologically characterized by a momentum-dependent topological invariant: one-dimensional winding number defined in the restricted Brillouin zone. Thus, such topological invariant is ill-defined in the presence of potential disorder which is inevitable in experiments. By paying attention to chiral symmetry of the Hamiltonian, we provide an alternative topological index $N_{\mathrm{ZES}}$ that predicts the number of Andreev bound states at a dirty surface of an unconventional superconductor. Moreover, we demonstrate that the zero-bias differential conductance in a normal metal/unconventional superconductor junction is quantized at $(4e^2/h)|N_{\mathrm{ZES}}|$ in the limit of strong impurity scattering in the normal metal.

## 1. Introduction

Exotic properties of unconventional superconductors have sparked a significant progress in physics of superconductivity. The essence of unconventional superconductivity is the internal phase of its pair potential. The sign changing of a pair potential on the Fermi surface causes the anomalous interference effect of a quasi-particle at the surface and results in the formation of the mid-gap Andreev bound states(ABSs) there [1, 2, 3, 4]. In a fully gapped unconventional superconductor, such as chiral and helical $p$-wave, the ABSs has the linear dispersion. On the other hand, a nodal unconventional superconductor, such as the $p_x$-wave, $d_{xy}$-wave and $d_{x^2-y^2}$-wave superconductor, hosts the dispersionless (flat) Andreev bound states. In the later case, the ABSs with the high degree of degeneracy at the Fermi level (zero energy) drastically affect the low-energy transport in the superconducting junctions. Actually, it has been shown that the flat ABSs causes the zero-bias anomaly in the differential conductance of a normal metal/superconductor junction [3, 4, 5, 6, 7] and the fractional Josephson effect in a superconductor/insulator/superconductor junction [8, 9, 10, 11, 12, 13].

Recently, the emergence of the flat ABSs has also been explained in terms of topology. Due to the presence of nodes in the superconducting gap, the well-known 10-fold topological classification [14] does not cover the nodal superconductors. Even though, a prescription called dimensional reduction enables us to topologically characterize the flat ABSs of a nodal superconductor [15, 16]. In a $d$-dimensional superconductor, it is possible to choose a one-dimensional Brillouin zone by fixing a $(d-1)$-dimensional wave number at a certain point (say $k$). When the momenta $k$ is away from the nodal points, it is possible to define the

IOP Conf. Series: Journal of Physics: Conf. Series **969** (2018) 012079 doi:10.1088/1742-6596/969/1/012079

one-dimensional winding number $w(k)$ for each restricted Brillouin zone [15]. From the bulk-boundary correspondence, the number of the ABSs should be $|w(k)|$ for each $k$. Therefore the flat ABSs are expected at a clear surface parallel to $k$.

In experiments, however, the potential disorder is inevitable in the vicinity of a surface or a junction interface of a superconductor. The one-dimensional Brillouin zone is ill-defined in the presence of disorder which breaks translational symmetry. Therefore, the winding number $w(k)$ no longer predict the number of ABSs at a dirty surface [17]. In other words, the potential disorder may lift the high degeneracy at the flat ABSs at a surface [18] and may wash out the unique transport properties to the flat ABSs. Such a situation requires a theoretical tool that measures the stability of flat ABSs in the presence of potential disorder. In this paper, we address this considerable issue.

We first study the stability of flat ABSs at the dirty surface of a nodal superconductor in Sec. 2. We consider three types of pairing symmetries such as $p_x$-wave, $d_{xy}$-wave, and $f$-wave symmetry. By taking into account chiral symmetry of the Bogoliubov-de Gennes Hamiltonian [19, 20, 21, 22, 23], we prove that the index $N_{\mathrm{ZES}} = \sum_k w(k)$ predicts the number of ABSs at a dirty surface of a nodal superconductor. The index $N_{\mathrm{ZES}}$ is a kind of Atiyah-Singer index, which relates a topological invariant and an invariant in terms of solutions of a differential equation. We also study the zero-bias differential conductance(ZBC) of a dirty normal-metal/superconductor (NS) junction in Sec. 3. As a consequence, we find that the ZBC decreases to the quantized value of $(4e^2/h)|N_{\mathrm{ZES}}|$ with the decrease of the normal conductance.

## 2. Stability of flat Andreev bound states

### 2.1. Flat Andreev bound states at the clean surface

We start our discussion with a brief summary of the topological property of a nodal superconductor in the clean limit. Let us consider the three types of unconventional superconductors with $p_x$-wave, $d_{xy}$-wave, and $f$-wave pairing symmetry. We describe the present superconductor by the $2 \times 2$ Bogoliubov-de Gennes Hamiltonian

$$\hat{H}_0(\boldsymbol{k}) = \begin{bmatrix} \xi(\boldsymbol{k}) & \Delta(\boldsymbol{k}) \\ \Delta(\boldsymbol{k}) & -\xi(\boldsymbol{k}) \end{bmatrix}, \quad \Delta(\boldsymbol{k}) = \begin{cases} \Delta_0 k_x/k_{\mathrm{F}} & p_x\text{-wave}, \\ \Delta_0 k_x k_y/k_{\mathrm{F}}^2 & d_{xy}\text{-wave}, \\ \Delta_0 k_x(k_{\mathrm{F}}^2 - 2k_y^2)/k_{\mathrm{F}}^3 & f\text{-wave}, \end{cases} \tag{1}$$

where $\xi(\boldsymbol{k}) = (\hbar^2\boldsymbol{k}^2/2m) - \mu_{\mathrm{F}}$, $m$ denotes the effective mass of an electron, $\mu_{\mathrm{F}}$ is the chemical potential, $\Delta_0$ is the amplitude of the pair potential at zero temperature, and $k_{\mathrm{F}} = \sqrt{2m\mu_{\mathrm{F}}}/\hbar$ represents the Fermi wave number. The Hamiltonian satisfies

$$\left\{\hat{\Gamma}, \hat{H}_0(\boldsymbol{k})\right\} = 0, \quad \hat{\Gamma} = \hat{\tau}_2, \tag{2}$$

which represents chiral symmetry of the Hamiltonian. Pauli matrices in Nambu space is denoted by $\tau_i$ for $i$ = 1-3. Since the pair potentials in Eq. (1) have nodes on the Fermi surface, it is impossible to topologically characterize such the superconductors in terms of wave functions of the whole Brillouin zone. Alternatively, we define a winding number in a restricted one-dimensional Brillouin zone as

$$w(\boldsymbol{k}_\parallel) = \frac{i}{4\pi}\int dk_\perp \mathrm{Tr}[\hat{\Gamma}\hat{H}_0^{-1}(\boldsymbol{k})\partial_{k_\perp}\hat{H}_0(\boldsymbol{k})], \tag{3}$$

where $\boldsymbol{k}_\parallel$ ($k_\perp$) is momenta parallel (perpendicular) to the surface we consider. The winding number $w(\boldsymbol{k}_\parallel)$ is ill-defined when the integration path along $k_\perp$ intersects the nodal points. Therefore, we have to choose $\boldsymbol{k}_\parallel$ so that $k_\perp$ can keep away from the nodal points. The winding number $w(\boldsymbol{k}_\parallel)$ can be nonzero in the finite region of $\boldsymbol{k}_\parallel$. In what follows, we consider the surface perpendicular to the $x$ direction. Namely, we chose $k_\perp = k_x$ and $\boldsymbol{k}_\parallel = k_y$, respectively. According to the bulk-boundary correspondence, the total number of topologically protected flat ABSs at the *clean* surface is calculated by

$$N_{\text{clean}} = {\sum_{k_y}}' |w(k_y)|, \tag{4}$$

where ${\sum_{k_y}}'$ denotes a summation over $k_y$ excluding the nodal points.

### *2.2. Flat Andreev bound states at the dirty surface*

Next, we study the degree of degeneracy in the flat ABSs at the *dirty* surface of an unconventional superconductor. The random impurity potentials in the bulk region strongly suppress the unconventional superconducting pair potential. Thus, we assume the potential disorder exists only near the surface. The BdG Hamiltonian in real space $\hat{H}_0(\boldsymbol{r})$ is obtained by replacing the momentum $\boldsymbol{k}$ by $-i\nabla_{\boldsymbol{r}}$ in Eq. (1). The non-magnetic random impurity potential in the vicinity of the surface is represented by $\hat{V}_{\text{imp}}(\boldsymbol{r}) = v(\boldsymbol{r})\hat{\tau}_3$, where the random potential $v(\boldsymbol{r})$ disappears rapidly with the decrease of distance from the surface. The total Hamiltonian is given by $\hat{H}(\boldsymbol{r}) = \hat{H}_0(\boldsymbol{r}) + \hat{V}_{\text{imp}}(\boldsymbol{r})$. The momentum $k_y$ is no longer a good quantum number because the impurity potential breaks translational symmetry. Therefore, it is impossible to define the one-dimensional winding number $w(k_y)$ in the presence of the potential disorder. However, the Hamiltonian $\hat{H}(\boldsymbol{r})$ preserves chiral symmetry in Eq. (2), which is the most important point in the argument below.

The central ingredient of our theory consists of the following two important properties of the eigenstates of a Hamiltonian $\hat{H}$ preserving chiral symmetry [15].

(i) First, the zero-energy states of $\hat{H}(\boldsymbol{r})$ are simultaneously the eigenstates of the chiral operator $\Gamma$. Since $\hat{\Gamma}^2 = 1$, the eigenvalue of $\hat{\Gamma}$ is either $\gamma = +1$ or $\gamma = -1$. Namely, a zero-energy states satisfying $\hat{H}(\boldsymbol{r})\varphi_\gamma(\boldsymbol{r}) = 0$ also satisfies $\hat{\Gamma}\varphi_\gamma(\boldsymbol{r}) = \gamma\varphi_\gamma(\boldsymbol{r})$. We refer to $\gamma$ as the chirality in the following.

(ii) Second, the nonzero energy states of $\hat{H}(\boldsymbol{r})$ are described by the linear combination of the two states: one has $\gamma = +1$ and the other has $\gamma = -1$. Namely, a nonzero energy state is described as $\varphi_{E\neq 0}(\boldsymbol{r}) = c_+\chi_+(\boldsymbol{r}) + c_-\chi_-(\boldsymbol{r})$, where $\Gamma\chi_\pm(\boldsymbol{r}) = \pm\chi_\pm(\boldsymbol{r})$. Moreover, the relation $|c_+| = |c_-|$ always holds.

Here, we define two integer numbers, $N_+$ and $N_-$, in the clean limit. According to property (i), we can immediately conclude that each ABSs at a clean surface belongs to either the positive chiral state or the negative chiral state. The integer $N_+$ ($N_-$) is the number of ABSs that have the positive (negative) chiral eigenvalue [See also Fig. 1(a)]. The total number of ABSs at the clean surface is represented by $N_+ + N_-$, which must be identical to $N_{\text{clean}}$ in Eq. (4).

The stability of the flat ABSs in the presence of impurities can be discussed by using property (ii). An ABS departs from the zero energy only when it can form a pair with its chiral partner. When $N_+ > N_-$, for example, $N_-$ negative chiral ABSs can couple to $N_-$ positive chiral ABSs under the potential disorder. As a result, they form nonzero energy states whose number is $2N_-$. However, $N_+ - N_-$ positive chiral states remain at the zero energy even in the presence of impurities because their chiral partner is absent. The integer number defined by $N_{\text{ZES}} = N_+ - N_-$, represents the number of ZESs that remain at a dirty surface. When $N_+ < N_-$, the number of ZESs at a dirty surface is given by $N_- - N_+$. Therefore, in general, $|N_{\text{ZES}}|$ is the degree of degeneracy at zero energy in the presence of potential disorder. The

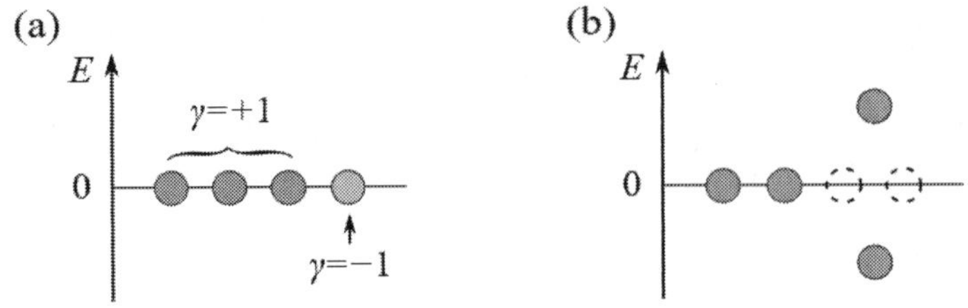


**Figure 1.** (Color online) (a) Four-fold degenerate zero-energy states (ZESs) in the absence of random potential. Three of them belong to the positive chiral eigenvalue, (i.e., $N_+ = 3$). One remaining ZES belongs to the negative chiral eigenvalue (i.e., $N_- = 1$). The four-fold degeneracy is protected by translational symmetry. (b) In the presence of random potential, a positive and a negative chiral ZES form a pair and departs from zero energy. However, two positive chiral ZESs remain at zero energy. The index $N_{\rm ZES} = N_+ - N_- = 2$ represents the number of ZESs remaining at zero energy in the presence of random potential.

essence of this argument is illustrated in Figs. 1 (a) and (b) with $N_+ = 3$ and $N_- = 1$. In Fig. 1 (a), we consider four ABSs at a clean surface. In Fig. 1 (b), we introduce the impurity potential at the surface.

At the end of this section, we discuss the topological aspect of $N_{\rm ZES}$. As examined in Ref. [15], the index theorem relates to the winding number $w(\boldsymbol{k}_\parallel)$ and the number of ABSs on a clean surface as follows

$$w(k_y) = \pm\left[n_+(k_y) - n_-(k_y)\right], \tag{5}$$

where $n_+(k_y)$ $[n_-(k_y)]$ denotes the number of positive (negative) chiral zero-energy states at $k_y$. There are two possible choices for the sign on the right-hand side of Eq. (5). When we consider the surface of a semi-infinite superconductor occupying $x_\perp \leq 0$, we should choose the positive sign. On the other hand, we should choose the negative sign at the surface of a semi-infinite superconductor occupying $x_\perp \geq 0$ [15]. However, this sign has no physical meaning because the number of ABSs is always given by $|n_+(k_y) - n_-(k_y)|$. As discussed above, the index $N_{\rm ZES}$ is represented by the difference between the total numbers of positive and negative chiral ABSs. Therefore, by taking Eq. (5) into account, we obtain an important relation

$$N_{\rm ZES} = {\sum_{k_y}}' w(k_y) = \pm(N_+ - N_-), \tag{6}$$

as a result of the index theorem. The index $N_{\rm ZES}$ is initially defined by using the solutions of the differential equation. From Eq. (6), we simultaneously find that the index $N_{\rm ZES}$ is a topological invariant. Mathematically, such an integer number is called the Atiyah-Singer index. Although the index $N_{\rm ZES}$ is defined in the presence of translational symmetry, it represents the degree of the degeneracy at the zero energy in the absence of translational symmetry. This is the first conclusion of this paper.

## 3. Quantization of conductance minimum

In this section, we numerically study the zero-bias differential conductance (ZBC) of a normal-metal/superconductor (NS) junction on the tight-binding model as shown in Fig. 2(a). The length of the normal metal segment is represented by $L_{\rm N}$. In the $y$ direction, the width of the junction is represented by $L_y$, and the periodic boundary condition is applied. In the normal segment, we introduce the random on-site potential $V_{\rm N}(\boldsymbol{r})\hat{\tau}_3$. ,where the amplitude of $\hat{V}_{\rm N}(\boldsymbol{r})$ is

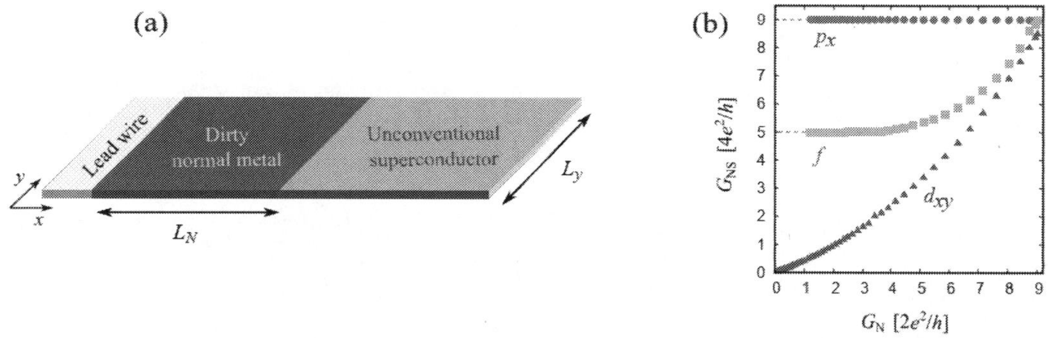


**Figure 2.** (Color online) (a)Schematic image of a NS junction. (b)Zero-bias conductance $G_{\rm NS}$ as a function of the normal conductance $G_{\rm N}$ is plotted. The index $N_{\rm ZES}$ = 9, 0, 5 for the $p_x$-wave, $d_{xy}$-wave, and $f$-wave superconductor, respectively.

given randomly in the range of $-V_{\rm X}/2 \leq V_{\rm N}(\boldsymbol{r}) \leq V_{\rm X}/2$. We calculate the zero-bias differential conductance $G_{\rm NS}$ of the NS junctions based on the Blonder-Tinkham-Klapwijk formula [24]. The normal and Andreev reflection coefficients are calculated by using the lattice Green's function method [25, 26].

In Fig. 2(b), we plot the zero-bias conductance $G_{\rm NS}$ versus the normal conductance $G_{\rm N}$, where $G_{\rm N}$ is calculated by setting $\Delta_0 = 0$. The parameters are chosen as $\mu_{\rm F}/t = 1.0$, $\Delta_0/t = 0.01$, $L_{\rm N}/a_0 = 30$, and $L_y/a_0 = 25$, where $t$ and $a_0$ denote the nearest neighbor hopping integral and the lattice constant, respectively. This parameter choice leads $N_{\rm clean} = 9$, 8, 9 and $N_{\rm ZES} = 9$, 0, 5 for the $p_x$-wave, $d_{xy}$-wave, and $f$-wave superconductor, respectively. The index $N_{\rm clean}$ ($N_{\rm ZES}$) is calculated by using Eq. (4) [ Eq. (6)]. In the $p_x$-wave case, the ZBC is quantized at $(4e^2/h)\times 9$ irrespective of potential disorder in the normal segment [5]. In the $d_{xy}$-wave case, the ZBC goes to zero with the decrease of $G_{\rm N}$ [10]. In the $f$-wave NS junction, we find that the ZBC decreases to the quantized value of $(4e^2/h)\times 5$ by decreasing $G_{\rm N}$. By summarizing these numerical results, we can conclude that

$$\lim_{G_{\rm N}\to 0} G_{\rm NS} \to \frac{4e^2}{h}|N_{\rm ZES}|. \tag{7}$$

The property of chiral symmetry enables us to explain such the characteristic transport property of Eq. (7). As shown in Ref. [7, 22], the zero-energy states in the normal segment become the eigenstates of $\Gamma$ due to the Andreev reflection. Namely, the Andreev reflection copies the chirality of the flat ABSs to the wave functions in the normal segment. Therefore, according to the chiral property (ii), the $|N_{\rm ZES}|$ zero-energy states can retain their degeneracy in the dirty normal segment and can form the resonant transmission channels there. The quantization of conductance minimum at $(4e^2/h)|N_{\rm ZES}|$ is the direct consequence from the existence of $|N_{\rm ZES}|$ resonant transmission channels protected by chiral symmetry. This is the second conclusion of this paper.

## 4. Conclusion

In conclusion, we have discussed the stability of flat Andreev bound states appearing at the dirty surface of a nodal superconductor. In the clean limit, the flat ABSs at the clean surface are characterized by the momentum dependent topological invariant $w(\boldsymbol{k}_\parallel)$. In a real superconductor, however, the potential disorder near the surface or junction interface is inevitable and may lift the degeneracy of flat ABSs at the zero energy. By taking into account chiral symmetry of the BdG Hamiltonian, we probe that the Atiyah-Singer index $N_{\rm ZES}$ predicts the number of the flat ABSs at the dirty surface. We have also discussed the zero-bias differential conductance $G_{\rm NS}$ in a normal metal/superconductor (NS) junction. As a consequence, we find that the minimum value of $G_{\rm NS}$ is quantized at $(4e^2/h)|N_{\rm ZES}|$.

**Acknowledgments**
The authors are grateful to M. Sato, Y. Tanaka, Ya. V. Fominov, and A. A. Golubov for useful discussions. This work was supported by “Topological Materials Science” (No. JP15H05852) and KAKENHI (No. JP15H03525) from the Ministry of Education, Culture, Sports, Science and Technology (MEXT) of Japan and by the Ministry of Education and Science of the Russian Federation (Grant No. 14Y.26.31.0007). S.I. are supported in part by a Grant-in-Aid for JSPS Fellows (Grants No. JP16J00956) by the Japan Society for the Promotion of Science (JSPS).

## References

[1] Buchholtz L J and Zwicknagl G 1981 *Phys. Rev.* B **23** 5788
[2] Hu C R 1994 *Phys. Rev. Lett* **72** 1526
[3] Tanaka Y and Kashiwaya S 1995 *Phys. Rev. Lett* **74** 3451
[4] Asano Y, Tanaka Y and Kashiwaya S 2004 *Phys. Rev.* B **69** 134501
[5] Tanaka Y and Kashiwaya S 2004 *Phys. Rev.* B **70** 012507
[6] Asano Y, Tanaka Y, Golubov A A and Kashiwaya S 2007 *Phys. Rev. Lett* **99** 067005
[7] Ikegaya S, Suzuki S-I, Tanaka Y and Asano Y 2016 *Phys. Rev.* B **94** 054512
[8] Tanaka Y and Kashiwaya S 1996 *Phys. Rev.* B **53** 11957(R)
[9] Barash Y S, Burkhardt H and Rainer D 1996 *Phys. Rev. Lett* **77** 4070
[10] Asano 2001 *Phys. Rev.* B **63** 052512
[11] Kwon H-J, Sengupta K and Yakovenko V M 2004 *Eur. Phys. J.* B **37** 349–361
[12] Asano Y, Tanaka Y and Kashiwaya S 2006 *Phys. Rev. Lett* **96** 097007
[13] Ikegaya S and Asano Y 2016 *J. Phys.: Condens. Matter* **28** 375702
[14] Schnyder A P, Ryu S, Furusaki A, Ludwig A W W 2008 *Phys. Rev.* B **78** 195125
[15] Sato M, Tanaka Y, Yada K and Yokoyama T 2011 *Phys. Rev.* B **83** 224511
[16] Schnyder A P and Brydon P M R 2015 *J. Phys.: Condens. Matter* **27** 243201
[17] Queiroz R and Schnyder A P 2014 *Phys. Rev.* B **89** 054501
[18] Black-Schaffer A M, Golubev D S, Bauch T, Lombardi F and Mikael Fogelström 2013 *Phys. Rev. Lett* **110** 197001
[19] Tewari S and Sau J D 2012 *Phys. Rev. Lett* **109** 150408
[20] Niu Y, Chung S-B, Hsu C-H, Mandal I, Raghu S and Chakravarty S 2012 *Phys. Rev.* B **85** 035110
[21] Diez M, Dahlhaus J P, Wimmer M and Beenakker C W J 2012 *Phys. Rev.* B **86** 094501
[22] Ikegaya S, Asano Y and Tanaka Y 2015 *Phys. Rev.* B **91** 174511
[23] Ikegaya S and Asano Y 2017 *Phys. Rev.* B **95** 214503
[24] Blonder G E, Tinkham M and Klapwijk T M 1982 *Phys. Rev.* B **25** 4515
[25] Lee P A and Fisher D S 1981 *Phys. Rev.* B **47** 882
[26] Ando T 1991 *Phys. Rev.* B **44** 8017

# $^4$He permeation and $H_2O$ uptake of cyanate ester resins — an alternative to commonly used epoxy resins at low temperature

**Sachiko Nakamura[1], Takenori Fujii[1], Shoji Matsukawa[2]*, Masayuki Katagiri[2] and Hiroshi Fukuyama[1,3]**

[1] Cryogenic Research Center, The University of Tokyo, 2-11-16, Yayoi, Bunkyo-ku, Tokyo 113-0032, Japan
[2] Mitsubishi Gas Chemical Company, Inc., Mitsubishi Building 2-5-2, Marunouchi, Chiyoda-ku, Tokyo 100-8324, Japan
[3] Department of Physics, The University of Tokyo, 7-3-1, Hongo, Bunkyo-ku, Tokyo 113-0033, Japan

E-mail: snakamura@crc.u-tokyo.ac.jp

**Abstract.** Cyanate ester (CE) thermoset is a polymer with a high glass-transition temperature of $\approx 300\,^\circ$C. CE is expected to be an alternative to Stycast 1266 as a sealing and casting glue for low temperature experiments, especially for adsorption experiments where baking of the substrate at $T > 100\,^\circ$C before cooling is required to eliminate surface contaminations. We experimentally confirmed that thermosets of CE monomers are non-porous and absorbs/desorbs water very little from measurements of (1) $^4$He permeation properties at temperatures from room temperature (RT) to 77 K and of (2) weight gains ($\delta W$) after storage for days in water and in air at RT. The $^4$He permeation is rather large at RT but negligibly small at $T \lessapprox 130$ K where the diffusion constant of $^4$He in CE is vanishingly small. $\delta W$ in water and air are 0.3–0.5% and 0.5–1.0%, respectively, which are much smaller than those of Stycast 1266. Therefore, cyanate ester is an excellent alternative to commonly used epoxy resins especially in surface-sensitive experiments at low temperature.

## 1. Introduction

To seal, glue, and cast parts of experimental apparatus used at low temperature, Stycast 1266 and 2850FT epoxy resins (EPs) [1] are the most commonly used materials. However, since their glass transition temperatures ($T_g$) are below 115 °C, the EPs are unusable for some applications which require baking at $T > 100$ °C to obtain higher vacuum before cooling down. Cyanate ester thermosets (CEs) [2] are a high temperature compatible and low outgassing polymer cured at 230–250 °C. Its $T_g$ ($\approx 300$ °C) is much higher than that of EPs. CE is also distinctive for small dielectric constant and small loss tangent among various thermosetting resin systems. In addition, with reinforcing carbon fibers, CE is focused on for its excellent cryogenic durability to microcracking [3]. For other properties such as mechanical properties, processability, and thermal expansion coefficients [4], CEs are almost similar to EPs. Thus CEs should be an excellent alternative to EPs in many applications at low temperature. However, current usage

* Present address: Mitsubishi Gas Chemical America, Inc., 655 Third Avenue, 24th Floor, New York, NY 10017

IOP Conf. Series: Journal of Physics: Conf. Series **969** (2018) 012080 doi:10.1088/1742-6596/969/1/012080

of CEs is almost limited to a matrix resin of FRPs (Fiber Reinforced Plastics) in electronic and aerospace industries, due to the lack of characterization and precedent. Recently, pure CE thermoset is also becoming available thanks to massive consumption as a matrix resin and 3D printing technologies.

Since we are planning to use CEs as a construction material and sealing glue in experimental apparatus for adsorbed helium [5], we measured various properties of thermosets of different species of CE monomers (TA: 2,2-bis(4-cyanatophenyl)propane, P201: 1,1-bis(4-cyanatophenyl)ethane, NCN: aralkyl type cyanate ester resin, and copolymer of TA and NCN [6]), along with Stycast 1266 and other polymers. Helium leak tightness is required for chamber walls at low temperature. Polymers sometimes have microcracks or pores which become more significant at low temperature. In addition, thermal stress can generate microcracks, which actually occured for our nylon cell after a few thermal cycles. Thus permeability of $^4$He through CEs was measured from room temperature to 77 K. Although finite solubility and diffusivity of $^4$He were observed at $T \gtrsim 130$ K, we did not find leakage due to microcracks or pores. Second, water/moisture uptake were measured because, if the material of the chamber walls uptakes water from dew or air, chamber walls can desorb water which contaminates the surface of the adsorption substrate. Other properties such as specific heat, thermal conductivity, and magnetic susceptibility at low temperature will be reported elsewhere [7]. Briefly, the thermal diffusivity is rather low among thermosets and magnetic susceptibility is as low as Stycast 1266.

## 2. $^4$He permeation properties

Permeation of $^4$He through polymer plates was measured using two experimental settings. One, hereafter RTcell, made of brass is for measurements at room temperature as shown in figure 1, and the other, VTcell, made of copper is for those at variable temperatures between 77–340 K. The temperature of VTcell was measured using a T-thermocouple and controlled using a ribbon heater with a temperature controller or using cold vapor from liquid nitrogen. In both settings, a polymer plate (20 mm-square or $\phi$20 mm-circular) is sandwiched by a pair of face-seal caps with O-rings. In RTcell, the O-rings are P-12 rubber rings while in VTcell, indium rings were used.

One side of the plate, upper side in figure 1, is exposed to $^4$He gas of a pressure $p_1$ (0.1–1 bar). The other side is evacuated to a pressure $p_2$ ($< 10^{-2}$ mbar). The $^4$He flow rate $Q$ through the plate was measured using a $^4$He leak detector. The measured $Q$ was between $10^{-9}$–$10^{-6}$ mbarL/s. Background $^4$He flow rate $Q_0$ through the seals instead of the plate was estimated using a metal plate sample through which $^4$He does not permeate. The $Q_0$ is less than the detection limit of the leak detector, $10^{-11}$ mbarL/s. Different samples of different thickness (0.3–1.6 mm) were measured for each CE monomer type. Beside them, Stycast 1266, Stycast 2850FT, two types of BT resins (copolymers of TA and bismaleimides), nylon 6, PTFE (polytetrafluoroethylene, Teflon), and PVDC (polyvinylidene chloride, Saran) were measured as shown in Table. 1.

Figure 2 shows an example of measurements of the flow rate $Q$ for a 1.12-mm-thick TA plate. The supplying pressure $p_1$ was abruptly changed at $t = 0$. After a certain time lag ($t_{\rm lag} \approx$100 s) due to the diffusion process through the sample, the flow rate $Q$ begins to increase. After about 1,800 seconds, $Q$ stops changing with the value of $Q_{\rm sat} \approx 1.5 \times 10^{-8}$ mbarL/s.

To derive permeation properties from such measurements, we develop a simple model for time ($t$) evolution of $^4$He concentration $\phi$ at a position $x$: $\phi(x,t)$. We take the origin of $x$ at the end of the plate of the detector side. At the sample surface of the supply side ($x = d$, where $d$ is the sample thickness), $^4$He atoms dissolve from the gas phase into the solid sample. $\phi(x = d)$ is conveniently expressed as $p_1 S$ in terms of the gas phase pressure $p_1$ and solubility $S$. According to the Fick's first law, diffusion flux in the $x$ direction $Q$ is described as

$$Q = -D\frac{\partial \phi}{\partial x} \tag{1}$$

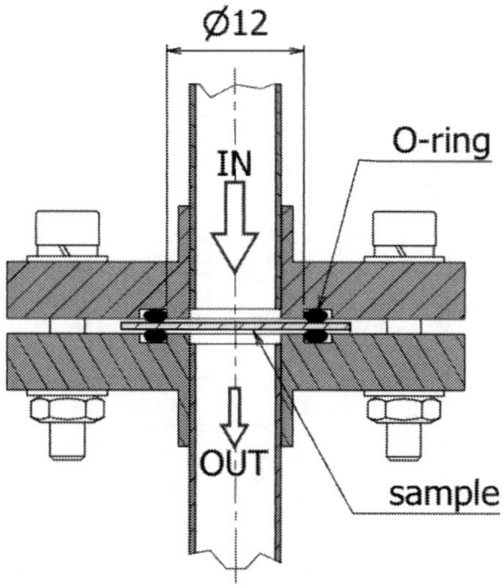


**Figure 1.** Schematic view of the experimental apparatus for measurements of $^4$He permeability at room temperature (RTcell).

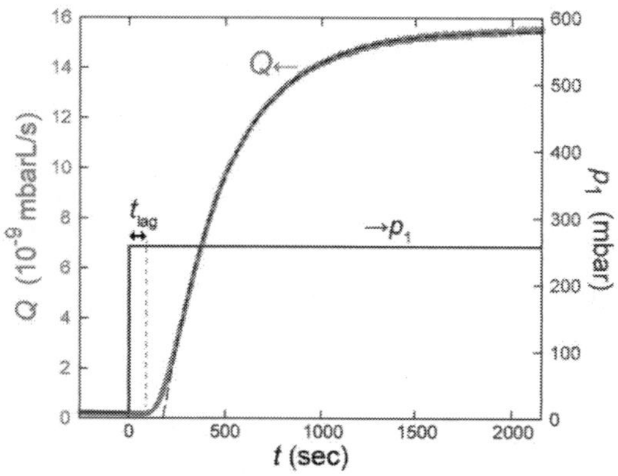


**Figure 2.** Time evolution of $^4$He flux $Q$ when the inlet pressure $p_1$ increased stepwisely at $t = 0$ for a 1.12-mm-thick sample of TA measured with RTcell.

**Table 1.** Permeability $P$, diffusivity $D$, solubility $S$ of $^4$He in various polymers, and number of samples $N$ measured at room temperature using RTcell.

| | | $P \times 10^{12}$ (m$^2$/s) | $D \times 10^9$ (m$^2$/s) | S | $N$ |
|---|---|---|---|---|---|
| CE | TA | 6.7(7) | 4.1(5) | 0.0017(2) | 7 |
| | P-201 | 3.6(8) | 2.8(2) | 0.0013(2) | 4 |
| | NCN | 3.1(8) | 3.1(2) | 0.0011(2) | 4 |
| | TA+NCN | 4.6(4) | 3.2(4) | 0.0015(2) | 4 |
| BT resin | TA+BMI70 | 6.4(5) | 4.1(1) | 0.0016(1) | 3 |
| | TA+BMI2300 [8] | 2.5(3) | 1.4(9) | 0.0020(11) | 3 |
| EP | Stycast 1266 | 0.5(1) | 1.20(6) | 0.00044(5) | 6 |
| | Stycast 2850FT | 0.4(1) | 0.11(2) | 0.0031(3) | 2 |
| Nylon | MC901 [9] | 0.1 | 0.6 | 0.0002 | 1 |
| PTFE | Teflon [10] | 21 | 9.4 | 0.0022 | 1 |
| PVDC | Saran wrap [11] | 0.3 | 0.004 | 0.067 | 1 |

using the diffusion constant $D$. Thus, in the steady state with constant $p_1$, the saturated flow rate $Q_{\rm sat}$ is described as

$$Q_{\rm sat} = DS\frac{A\Delta p}{d}, \tag{2}$$

where $A$ is the area inside the O-rings exposed to the inlet gas, and $\Delta p \equiv p_1 - p_2$ is the pressure difference ($\approx p_1$). For each sample, $Q_{\rm sat}$ values were measured for several different $p_1$, and the permeability $P$ (= $DS$) was determined by least squares fitting.

$D$ was determined independently by analyzing $t$-evolution of $Q$ based on the following diffusion equation with boundary conditions for sufficiently small $p_2$:

$$\frac{\partial \phi(x,t)}{\partial t} = D\frac{\partial^2 \phi(x,t)}{\partial x^2}, \qquad \phi(x,t) = \begin{cases} 0 & (t<0) \\ S\Delta p & (x=d, t>0) \\ 0 & (x=0). \end{cases} \tag{3}$$

The solution of the equation above is as follows:

$$\phi(x,t) = S\Delta p\left(\frac{x}{d} - 2\sum_{n=1}^{\infty}\frac{(-1)^{n-1}}{n\pi}\sin\frac{n\pi x}{d}\exp\left(-\frac{n^2\pi^2 D}{d^2}t\right)\right) \tag{4}$$

from which $^4$He flow rate $Q$ at the evacuated side is determined as

$$Q(t) = AD\left.\frac{\partial \phi}{\partial x}\right|_{x=0} = DS\frac{A\Delta p}{d}\left(1 - 2\sum_{n=1}^{\infty}(-1)^{n-1}\exp\left(-\frac{n^2\pi^2 D}{d^2}t\right)\right). \tag{5}$$

Here high-$n$ terms are significant only when $t$ is small, and a simple equation: $Q = Q_{\rm sat}(1 - \exp(-t/\tau))$ can represent the time evolution of $Q$ at $t \gtrsim 1.5\tau$ very well as indicated by the thin dashed line in figure 2. We determined $D$ from the relaxation time $\tau$ through

$$D = \frac{d^2}{\pi^2\tau} \tag{6}$$

and $S$ as well according to the relation $P = DS$. From numerical evaluation of equation 5, $t_{\rm lag}$ is $\approx 0.5\tau$ and proportional to $d^2$. It is confirmed by the actual measurements of CEs.

The values of $D$ and $S$ at room temperature determined by the above mentioned way for different species of CE and other polymers are summarized in figure 3 and table 1. According to the $P$-contours in the figure, nylon has the lowest $P$, PVDC and Stycast resins come next, then CEs, and finally PTFE has the highest. Both $D$ and $S$ are rather high in CEs compared to Stycast 1266 presumably because they have a larger free volume in the molecular structure ($\approx$ 2.2%) according to previous Fourier transform infrared spectrometry measurements [12]. Note that the sample of PVDC is very thin (0.011 mm thick) and was largely deformed by the pressure difference. Thus the values of $P$ and $D$ for this material should be considered as an upper limit.

Because Stycast 2850FT is an epoxy resin mixed with alumina powders, $^4$He permeation proceeds in two steps in this material characterized by $t_{\rm lag}$ much smaller than $0.5\tau$. $^4$He atoms first permeate through epoxy portion with larger $D$ and smaller $S$ and then alumina portion with smaller $D$ and larger $S$. Therefore, special care must be taken for leak testing on this epoxy: with $^4$He sprayed the permeability may look rather small but when packed in a bag filled with $^4$He gas, its long-time permeability is as large as that of Stycast 1266.

Next, using VTcell, we measured temperature dependencies of $P$, $S$, and $D$ for TA (0.9 mm thick), P201 (0.7 mm thick), and PTFE (0.5 mm thick) which is used in commercial demountable cryogenic seals. For PTFE, we had to retighten the screws of VTcell several times to compensate its large thermal contraction. The large dots in figures 4(a)–(c) are data taken at several discrete temperatures in the same way as at room temperature, while small dots in figure 4(a) are data taken on cooling by monitoring $Q$ with $p_1$ fixed. $Q$ responded immediately (within a second) to discrete changes of temperature through the whole temperature range investigated.

As shown in figure 4(a), $P$ diminishes exponentially with decreasing $T$ and becomes undetectable below 130 K. From figures 4(b) and (c), one can see that the exponential $T$-dependence of $P$ comes from that of $D$ not from $S$ nearly $T$-independent. It indicate that the permeation is associated with molecular scale diffusion not with gas flow through randomly-distributed microscopic pores where $D$ is expected to increase with decreasing $T$ due to decreasing viscosity [13] unlike the observation. Activation energy for the diffusion $E_{\rm d}$ is $2,000 \pm 200$ K for all the polymers investigated here.

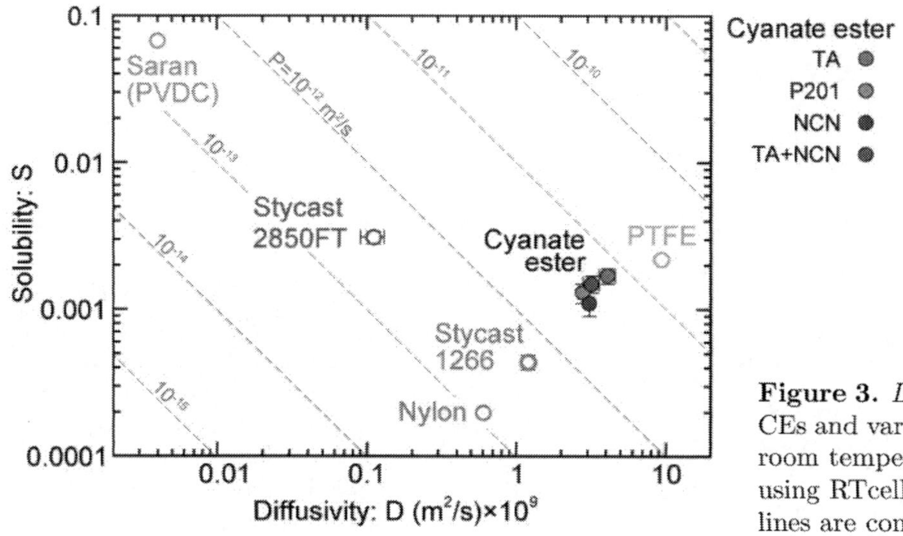


**Figure 3.** $D$ and $S$ of $^4He$ in CEs and various polymers at room temperature measured using RTcell. The thin solid lines are contours for $P$.

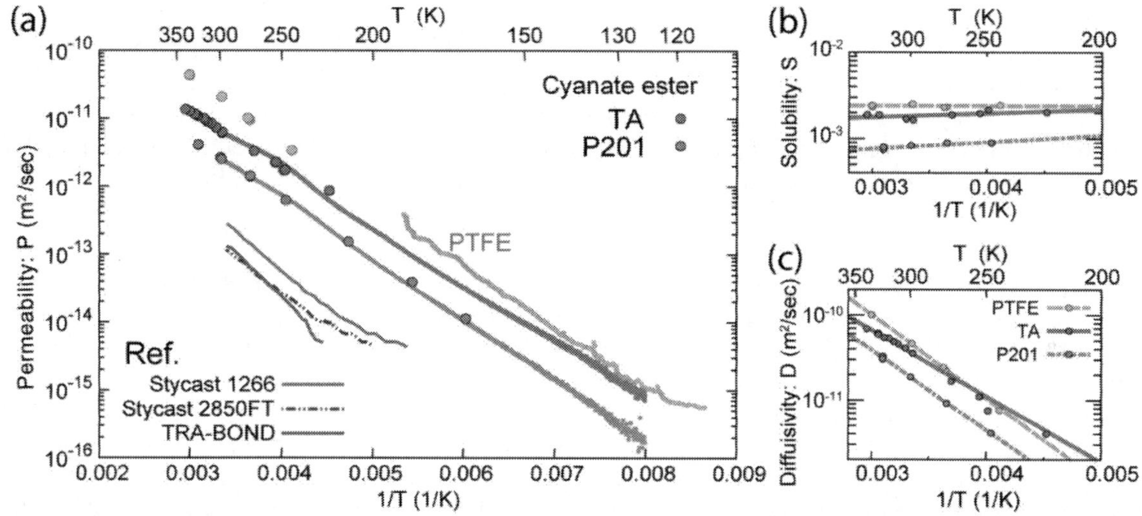


**Figure 4.** Temperature dependencies of (a) $P$, (b) $S$, and (c) $D$ of $^4He$ in CEs (TA, P201) and PTFE measured using VTcell. $P$ is normalized to the value measured using RTcell at room temperature because the exposed area $A$ is difficult to estimate with indium O-rings. Large dots are data obtained with temperature fixed, while small dots in (a) are those in cooling. Previous studies on $P$ for several species of epoxy glues are also plotted [14].

## 3. $H_2O$ uptake

$H_2O$ uptake of CE and other polymers was investigated by monitoring a weight change ($\delta W/W_0$) relative to the initial weight $W_0$ as shown in figure 5. In water at room temperature, $\delta W/W_0$ for nylon and Stycast 1266 are 4% and 5.5%, respectively, while those for CEs are only 0.5–1%. In air at room temperature (partial pressure of $H_2O$: 10–30 mbar), nylon still absorbs a large amount of $H_2O$ (3.5%) but Stycast 1266 does less (1.5%) and CEs do very little (0.3–0.5%).

Compared to helium-4, $H_2O$ dissolves very well in polymers. Even the lowest $\delta W/W_0$ (0.3%) in our measurements corresponds to a large $S$ ($\approx 10^2$). On the other hand, $D$ values of $H_2O$ are very small: $5.1\times10^{-13}$ (TA) and $3.6\times10^{-13}\,m^2/s$ (MC901) at room temperature, determined

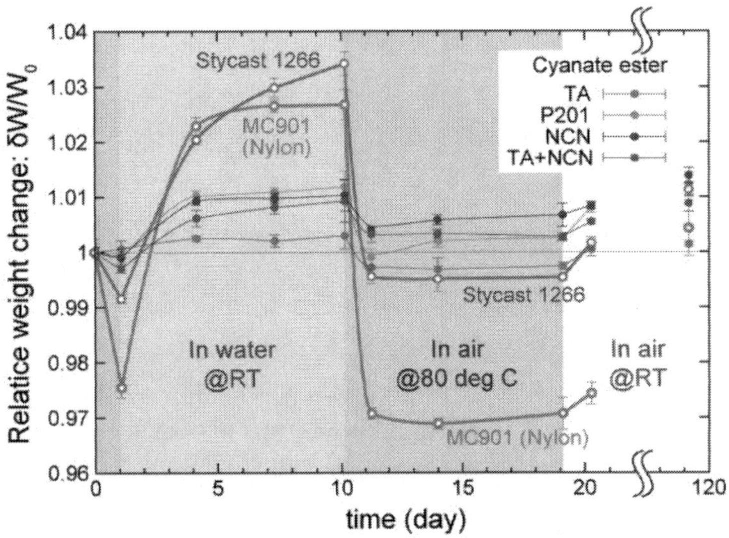


**Figure 5.** Relative weight changes of small samples (10–20 mg, 5 mm square, 0.7–0.9 mm thick) of CEs and other polymers. The samples were baked first at 80°C in air (red shaded), dropped into water (blue shaded), baked again (red shaded), and then kept at room temperature in air.

from similar measurements on thicker samples. For example, it takes a whole month for moisture in air to permeate in a 5-mm-thick chamber made of TA at room temperature. Thus, water release from polymer bulk at room temperature should be treated not as permeation but as desorption of dissolved water. The dissolved water can be removed efficiently by baking, for example, at $T \approx 160$ °C, where both $D^{-1}$ and $\delta W/W_0$ are expected to be reduced to one-tenth.

## 4. Conclusions

Permeation properties of $^4$He and $H_2O$ uptake for cyanate ester resins (CEs) were measured and compared with those of other polymers including Stycast epoxy resins. Although the permeability of $^4$He in CEs are rather high compared to the epoxies, it becomes negligibly low below $\approx$ 130 K. No vacuum leakage was detected for CEs even after 40 thermal cycles. It was found that CEs absorb $H_2O$ very little ($\lesssim$1 wt%) both in water and in air. Also, CEs are bakeable up to 250 °C, while the epoxy resins are tolerable only up to $\approx$100 °C. In conclusion, it is greatly advantageous to use cyanate ester resins as a sealing, gluing, and casting resin for surface-sensitive experiments such as adsorption experiments at low temperature in which one can reduce remnant moisture, a major contamination source, much more effectively than Stycast 1266 and 2850FT.

## References

[1] Emerson & Cuming, Inc. (present: Henkel AG & Co)
[2] Hamerton I (ed) 1994 *Chemistry and Technology of Cyanate Ester Resins* (Springer Netherlands)
[3] Brown T L 1998 *Microcracking of Materials for Space* Ph.D. thesis Virginia Polytechnic Inst. and State Univ.
[4] Walsh R P and Reed R P 1994 *Thermal Expansion Measurements of Resins (4 K–300 K)* (Boston, MA: Springer US) pp 1145–1151 ISBN 978-1-4757-9053-5
[5] Nakamura S, Matsui K, Matsui T and Fukuyama H 2012 *Journal of Physics: Conference Series*
[6] CYTESTER, manufactured by Mitsubishi Gas Chemical Company, Inc.
[7] Nakamura S, Fujii T, Matsukawa S, Katagiri M and Fukuyama H (to be published.)
[8] Daiwa Fine Chemicals Co., Ltd.
[9] QUADRANT POLYPENCO JAPAN LTD.
[10] E. I. du Pont de Nemours and Company.
[11] Asahi Kasei Corporation.
[12] Zeng M, Lu C, Wang B and Qi C 2010 *Radiation Physics and Chemistry* **79** 966–975 ISSN 0969-806X
[13] Johnston H L and Grilly E R 1942 *The Journal of Physical Chemistry* **46** 948–963
[14] Lovinger D J and Hallock R B 2012 *Journal of Physics: Conference Series* **400** 052006

# Experimental study of a SINIS detector response time at 350 GHz signal frequency

**S. Lemzyakov**[1,2,*], **M. Tarasov**[1,2,3,4], **S.Mahashabde**[4,5], **R. Yusupov**[2,3], **L. Kuzmin**[4], **V.Edelman**[1]

[1] Kapitza Institute for Physical Problems RAS, Kosygina 2, 119334 Moscow, Russia
[2] Moscow Institute of Physics and Technology (State University), Institutsky per. 9, 141700 Dolgoprudny, Russia
[3] Kotelnikov Institute of Radio Engineering and Electronics RAS, Mokhovaya 11/7, 125009 Moscow, Russia
[4] Chalmers University of Technology, Department of Microtechnology and Nanoscience (MC2), Kemivägen 9, Goteborg 41296, Sweden
[5] Department of Physics-Astrophysics, University of Oxford, Keble Road, OX1 3RH, Oxford, United Kingdom

*edelman@kapitza.ras.ru

**Abstract**. Response time constant of a SINIS bolometer integrated in an annular ring antenna was measured at a bath temperature of 100 mK. Samples comprising superconducting aluminium electrodes and normal-metal Al/Fe strip connected to electrodes via tunnel junctions were fabricated on oxidized Si substrate using shadow evaporation. The bolometer was illuminated by a fast black-body radiation source through a band-pass filter centered at 350 GHz with a passband of 7 GHz. Radiation source is a thin NiCr film on sapphire substrate. For rectangular 10÷100 μs current pulse the radiation front edge was rather sharp due to low thermal capacitance of NiCr film and low thermal conductivity of substrate at temperatures in the range 1-4 K. The rise time of the response was ~1-10 μs. This time presumably is limited by technical reasons: high dynamic resistance of series array of bolometers and capacitance of a long twisted pair wiring from SINIS bolometer to a room-temperature amplifier.

## 1. Introduction

Planar bolometers with Superconductor-Insulator-Normal metal-Insulator-Superconductor (SINIS) tunnel junctions due to tiny volume of normal metal absorber below 0.1 $\mu m^3$ and low thermal capacity about $10^{-17}$-$10^{-18}$ J/K at temperatures 0.1-0.2 K are expected to have a very short response time. Time constant for the input signal of about 10-100 ns is determined by thermalization of electron system due to electron-electron interactions [1, 2]. Potentially, the small time constant of the SINIS bolometers is advantageous compared to the well-established superconducting Transition Edge Sensors (TES) with time constant over millisecond [3]. In a previous study, the relaxation process in SINIS thermometer with absorber volume of 4.5 $\mu m^3$ for heating or cooling by a current pulse at a bath temperature of 270 mK was measured to be 1.2 μs [4]. A contradicting result for NIS thermometer measurements in the megahertz band with response time of 100 ms was obtained in [5]. There are no direct measurements of response time for SINIS bolometers illuminated with terahertz radiation. This is important for estimating the performance of RF readout and frequency multiplexing of multi-pixel systems.

## 2. Measurement setup

In our experiments, we use a thermal radiation source placed inside the cryostat close to the bolometer. This source comprises of a sapphire substrate and a NiCr resistive radiating film, and is mounted on thin wires providing bias current as well as heat sinking to the cold plate at 0.4 K. DC current can provide overheating by few Kelvin for applied power of the order of 10 μW. Temperature of the black body is monitored by a $RuO_2$ surface-mount resistor glued to the substrate. According to estimations in [6] the time constant of $RuO_2$ thermometer is about 100-200 μs. Radiation from this black body source is guided through an aperture 7 mm in diameter covered with a band-pass filter as in [1, 2], and focused by an extended hyper-hemisphere sapphire lens which is attached to the back side of the substrate with the bolometer.

In first series of experiments we study a series array of 25 annular ring antennas 307 μm in diameter each with two bolometers [7], see Fig. 1. Such antenna consists of two half-rings made of thick gold and connected with SINIS bolometers. Antennas are arranged in 5 rows with 5 elements each, covering an area of 3x3 $mm^2$, and connected in series for measurements of voltage response. Absorbers are formed of a Fe/Al bilayer strip with dimensions of 1x0.1x0.021 $\mu m^3$. The bilayer was formed by depositing 0.7 nm Fe followed by 14 nm of Al. This ensures that the Al is non-superconducting at experimental bath temperatures. The total volume of the absorber also includes the volume of the normal metal film under the two tunnel junctions with area of 0.4x1.8 $\mu m^2$ making the total volume of the normal metal electrode 0.03 $\mu m^3$ for one bolometer. The whole structure with the ring inductance and NIS junction capacitances is designed for central frequency of 350 GHz. For simple estimations of absorbed thermal radiation one can assume this structure as a single-mode receiver as in [1, 2]. Experimental results are presented for bath temperatures 90-100 mK.

In a second series of experiments, we studied SINIS bolometers with large area NIS junctions and a suspended Cu absorber integrated in a log-periodic antennas. In case of a suspended absorber, the heat link to the substrate is significantly reduced making thermalization of the electron system more effective.

## 3 Experimental results

For evaluation of the receiver performance, we first measured Current-Voltage characteristics (IV curves) at dc bias for different levels of irradiation from the source. For radiation source temperatures of 0.5 K, 5.8 K, 8.1 K, 11 K, and 14 K, we present IV curves in Fig. 2 (left panel), and voltage response in the right panel with corresponding numbers of the loaded curve. The "loaded" IV curves (curves 2-5) are subtracted from an unloaded IV curve (curve 1) to calculate the receiver response. The voltage response is the difference in response voltage as a function of bias current. We estimate the additional incoming power for one ring antenna as 1.5 pW, 3.5 pW, 6.5 pW, and 10 pW for curves 2, 3, 4, 5. It should be mentioned that an aperture of 7 mm reduces the actual incident power by a factor over 4. Taking into account the source temperature and geometry of experiment we can estimate voltage response on the absorbed power. Dynamic resistance of bolometer array at the response maximum is in the range of 1÷10 MΩ, a twisted pair wiring capacitance is 80-100 pF, the time constant of such RC circuit is 10-100 μs, that is too high. For pulse measurements, we choose the dc bias for which dynamic resistance of SINIS is in the range of 10÷100 kΩ, the response level is several times less compared to the maximum, and time constant of wiring is reduced down to 1 μs.

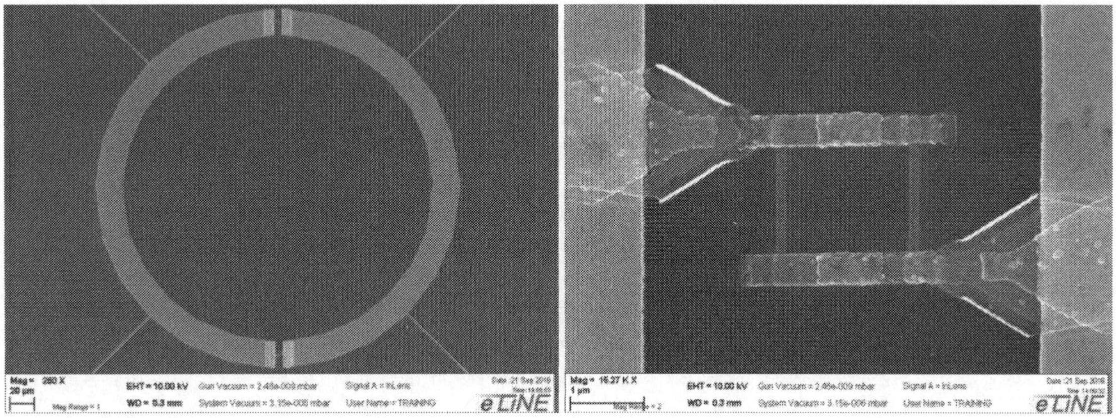

Figure 1. SEM view of single annular ring antenna (left panel) with two identical SINIS bolometers (right panel).

To study the thermodynamics of the bolometer, we irradiate the device with a heating pulse of relatively large power and very short rise time. In our case, the sapphire substrate with an area of 2 $cm^2$ and 0.3 mm thickness, the thermal capacity can be calculated as $29{*}T^3$ nJ/K. Thermal capacity of metal films less than 0.1 μm thick can be neglected. When applying an electric pulse with power of 1 W with duration of 1 μs, the source is heated up to 5 K and then preserves this temperature for a few hundred milliseconds. Source cooling back to bath temperature time can approach a second due to a very low thermal sink through bias and suspension wires.

Under experimental conditions, the black body source was heated by current pulse with an instant power of 0.4 W or less and corresponding energy below 4 μJ for 10 μs pulse duration. Current pulse was fed to black body through a symmetric line to reduce a crosstalk to amplifier input. The bolometer output voltage was preamplified and fed to an input of a 2-channel TEKTRONIX® TDS 1012B oscilloscope operated in a waiting mode. Synchronizing pulses were applied to the second input of the oscilloscope. For reducing pulse interferences that can be up to 1 μV measurements were done in the bolometer bias reversal mode +Vbias / -Vbias. When subtracting responses at positive and negative bias voltages we obtain response to radiation, and if summing up we get interference component and estimate of how much it can affect our measurements. For increasing the signal to noise ratio, we integrate many cycles keeping period in between that is enough to cool down the radiation source. Control over process was using a special program developed within LabView environment.

When NiCr film is heated by a current pulse, its temperature determines the instant radiation power. For maximum heating pulse power of 0.2 W/$cm^2$ the temperature gradient across the 15 nm thick film does not exceed 0.1 mK. Kapitza temperature drop between film and sapphire at T=4 K is below 0.1 K and decreases as $T^{-3}$. For this estimation we use Kapitza resistance for indium-sapphire interface from [8] that is higher compared to NiCr-sapphire case. Temperature variation in sapphire with its thermal conductivity about 100 W/(m*K) [9] is about 6 mK. So we assume that during current pulse, the temperature of radiating film is the same for the whole radiation source. Time constant of the source can be estimated from the following considerations. It operates at temperature up to 5 K, its thermal capacitance is $C=2.9{*}10^{-8}T^3$ J/K [9], thermal conductivity is proportional to $G{\sim}T^3$ [10] that makes at T=5 K the time constant $\tau=C/G=20\div30$ ns. This time is much less compared to bolometer together with readout time constant that is of the order of a microsecond.

Figure 3 shows the measurement results for heating pulse of 20 V during 10 μs to the radiation source with resistance of 900 Ω. Taking into account the response level at dc heating we can estimate the radiation source temperature as $T_{rad}=4\pm0.2$ K. Taking into account heat capacitance of sapphire and applied energy, estimation of black body source temperature brings $T_{rad}=4.5\pm0.3$ K.

For longer heating time of black body source about 100 μs the dependence is more complicated, see Fig. 4. In the initial part of dependence after heater switch-off the fitting corresponds to time constant of 16 μs, after that dependence approaching exponent with time constant of 120 μs. Such increase can be a manifestation that not only normal metal but also superconductor and silicon substrate are overheated.

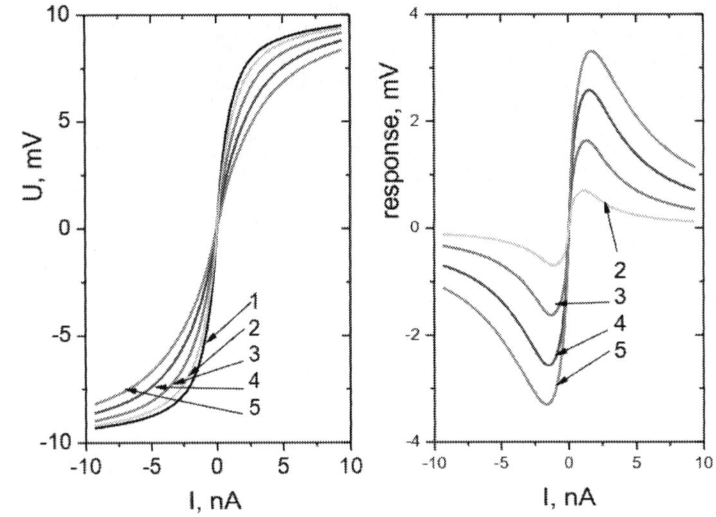


Figure 2. IV curves (left panel) and voltage response (right panel) for series array of 25 annular ring antennas with SINIS bolometers.

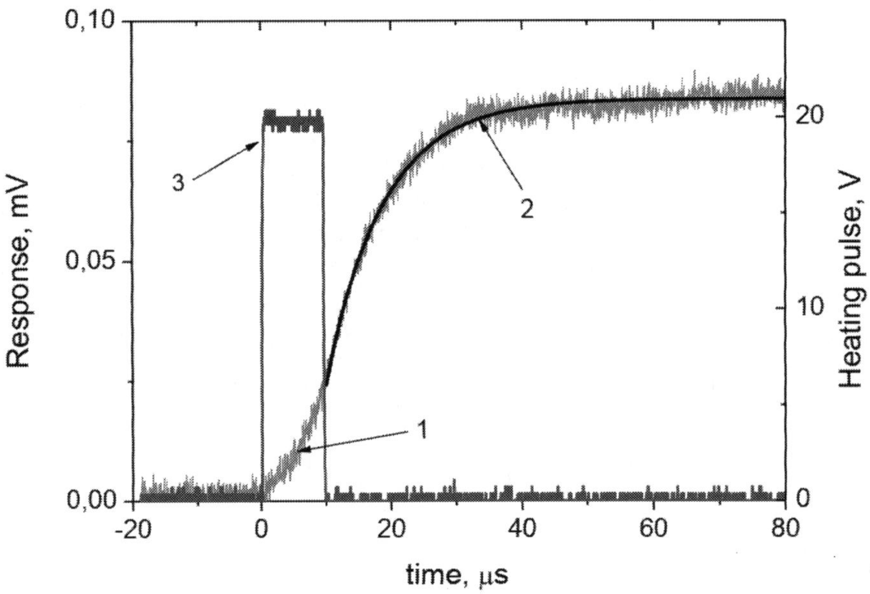


Figure 3. Voltage response of series array SINIS bolometer with dc bias ±7 nA for black body radiation source heated by rectangular shape pulse 20 V amplitude during 10 μs. Curve (1) is time dependence of response averaged for 128 cycles, (2) exponent that fitting dependence (1) after the end of warming pulse. Time constant of this exponent is 8.8 μs. Curve (3) initial heating pulse. Radiation power referred to one antenna with response of 85 μV is estimated to be about 0.5 pW.

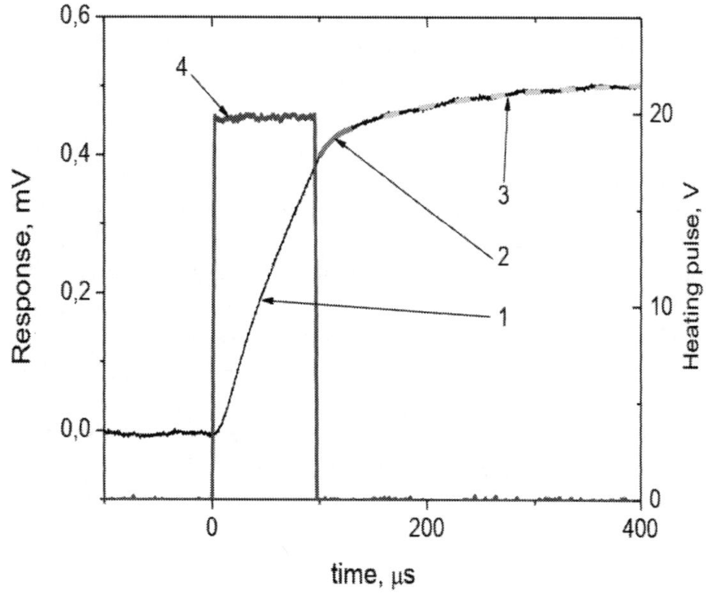


Fig. 4.. Voltage response of a SINIS array for pulse heating of radiation source with amplitude of 20 V and duration of 100 μs. Curve (1) is time dependence of bolometer response averaged for 128 pulses, (2) exponent fitting just after the end of heating pulse , with time constant of τ=16 μs, (3) exponent fitting after 50 μs from the end of current pulse, with time constant τ=120 μs, (4) the heating pulse.

We also measured the same dependence for SINIS bolometer with a suspended absorber and larger NIS junctions [11]. Bolometer is integrated in a log-periodic antenna. Level of signal was less because of just a single bolometer under test contrary to 25 bolometers in previous part, and dynamic resistance is 14 kΩ instead of 100 kΩ. Results are presented in Fig. 5. Time constant in this case decrease down to 2.3 μs.

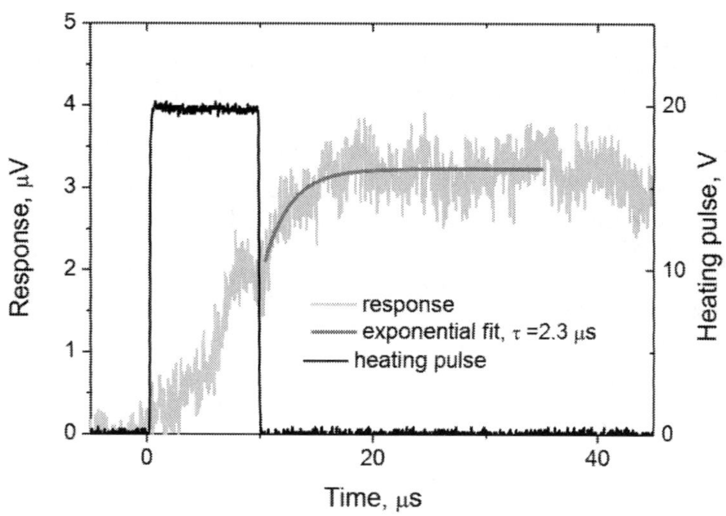


Figure 5 Response of bolometer with a suspended absorber to the radiation by heating pulse 20 V*10 μs. Response reaches exponentially the permanent value with τ = 2.3 μs.,

## 4. Conclusion

An important issue for SINIS bolometer applications is a proper estimation of its time constant for variations in the incoming Terahertz signal. According to our measurements for series array of SINIS bolometers at 100 mK bath temperature, the time constant is below 10 μs for short pulses and increase to 100 μs for a longer irradiation accompanied with heating of the wiring and substrate. For a low-resistance SINIS bolometer with suspended absorber, the measured time constant is below 1 μs. In our measurements resolution is limited by $\tau$=RC and can be substantially reduced by using parallel array of bolometers to reduce the resistance R and measurements with cold amplifier to avoid parasitic capacitance C of wiring. For higher bath temperatures, the time constant is reduced dramatically [4]. If, at 100 mK, the time constant it is over 10 μs, then at 300 mK it is 1 μs, and at 500 mK it is estimated to be 0.1 μs. In the case of the SINIS bolometer with strong electro-thermal feedback, known as Cold Electron Bolometer (CEB), the time constant can be reduced when bolometer is biased close to the energy gap and demonstrates electron cooling. According to [12] the intrinsic time constant of CEB can vary from 10 to 70 ns for signal power up to 1 pW. Such time constant is much shorter compared to Transition Edge Sensor (TES) for which it is in the range of milliseconds.

**Acknoledgements**
This work was done with support from Swedish National Space Board (CEB Proposal, dnr 196/15), European Space Agency (ESA, Project ITT 1-7393/13/NL/MH). . The clean-room processing was carried out in the Nanofabrication Laboratory on the equipment sponsored by the Knut and Alice Wallenberg Foundation.

**References**

[1] M.Tarasov, V.Edelman, S.Mahashabde, L.Kuzmin, JETPh,vol.119,no.1,pp.107, (2014).
[2] M. Tarasov,V Edelman,A Ermakov,S Mahashabde, L Kuzmin, IEEE Trans. Terahertz Sci.&Technol.,Vol.5, No.1, p. 44 (2015).
[3] W. S. Holland, D. Bintley, E. L. Chapin, et al., Mon. Not. R. Astron Soc, 430 (4): 2513 (2013).
[4] D. R. Schmidt, K. W. Lehnert, A. M. Clark, at al., Appl. Phys. Lett., 86, 053505 (2005).
[5] S. Gasparinetti, K. Viisanen, O.-P. Saira, T. Faivre, M. Arzeo, M. Meschke, J. Pekola, Phys. Rev. Appl., 3, 014007 (2015).
[6] Ya.E. Volokitin, R.C. Thiel, L.J. de Jongh, Cryogenics, Vol. 34, Issue 9, P. 771 (1994)
[7] Mahashabde, S., Sobolev, A., Bengtsson, et al.,, IEEE Trans. on Terahertz Science and Technol., 5(1), 145 (2015)
[8] D.A. Neeper, J.R. Dillinger, Phys. Rev. 135, A1028 (1964).
[9] T. Tomaru, T. Suzuki, T. Uchiyama, et al., Phys. Lett. A, vol. 301, Issue 3-4, p. 215 (2002).
[10] D.A.Ditmars, S.Ishihara, S.S. Chang, G.Bernstein, J.Res. of Nat. Bureau of Standards, vol. 87, No 2, p. 159 (1982)
[11] M. Tarasov, V. Edelman, S. Mahashabde, et al., Appl. Phys. Lett., 110, 242601 (2017).
[12] L. Kuzmin, Proc. of SPIE Astron. Telescopes and Instr. Conf., 5498, p. 349 (2004).

IOP Conf. Series: Journal of Physics: Conf. Series **969** (2018) 012082 doi:10.1088/1742-6596/969/1/012082

# Superconducting stripline resonators at frequencies up to 50 GHz for microwave spectroscopy applications

**Tobias Wollandt[1], Markus Thiemann[1], Martin Dressel[1], Marc Scheffler[1]**

[1] 1. Physikalisches Institut, Universität Stuttgart, Stuttgart, Germany

E-mail: `scheffl@pi1.physik.uni-stuttgart.de`

**Abstract.** Planar superconducting microwave resonators are commonly operated for various cryogenic applications, typically at frequencies between 2 and 15 GHz. But for certain applications, e.g. microwave spectroscopy to study the dynamics of charges and spins in solids, a wider spectral range is desired. Driving such microwave resonators at higher frequencies is challenging due to enhanced losses and reduced wavelength. Here we present superconducting stripline resonators operating at multiple frequencies as high as 50 GHz. These resonators are fabricated of Pb and can easily be adjusted to particular applications, including mK temperatures in a dilution refrigerator. Demonstrating such a resonator as a spectroscopic tool, we detect the superconducting transition of a Sn sample in a wide spectral range.

## 1. Introduction

Planar superconducting microwave resonators (based on stripline, microstrip, or coplanar transmission lines [1]) are employed in numerous cryogenic devices that play key roles in e.g. quantum information science [2, 3] or photon detection [4, 5, 6]. A different field of application is microwave spectroscopy on solids: several condensed matter systems that are at the focus of present research have characteristic energy or frequency scales that correspond to the microwave range (GHz frequencies), including conventional [7, 8, 9, 10] and unconventional superconductors [11, 12, 13, 14], strongly correlated metals [15, 16, 17], and magnetic [18, 19] as well as dielectric materials [20]. The energy scales of microwaves correspond to thermal energies of a few K or below ($21\,\mathrm{GHz} \cdot h \approx 1\,\mathrm{K} \cdot k_\mathrm{B}$, with Planck constant $h$ and Boltzmann constant $k_\mathrm{B}$), and thus microwave spectroscopy studies are particularly suited to probe low-temperature states in solids with similar characteristic energies. Depending on the particular requirements, several different techniques have been developed in recent years to perform microwave spectroscopy at low temperatures. While broadband spectrometers can continuously cover substantial frequency ranges [21, 22, 23, 24, 25], they require elaborate reference or calibration schemes [22, 23, 26, 27], and for many materials of interest they lack the sufficient sensitivity [23]. Cavity resonators on the other hand can be very sensitive [28], but hardly give any information on the frequency dependence. One option to combine high sensitivity with spectral resolution are resonators that can be operated at many different frequencies [29, 30], and one such approach is using stripline resonators [31, 32, 33, 34, 35, 36]. The main advantages of this technique are the compact size

IOP Conf. Series: Journal of Physics: Conf. Series **969** (2018) 012082 doi:10.1088/1742-6596/969/1/012082

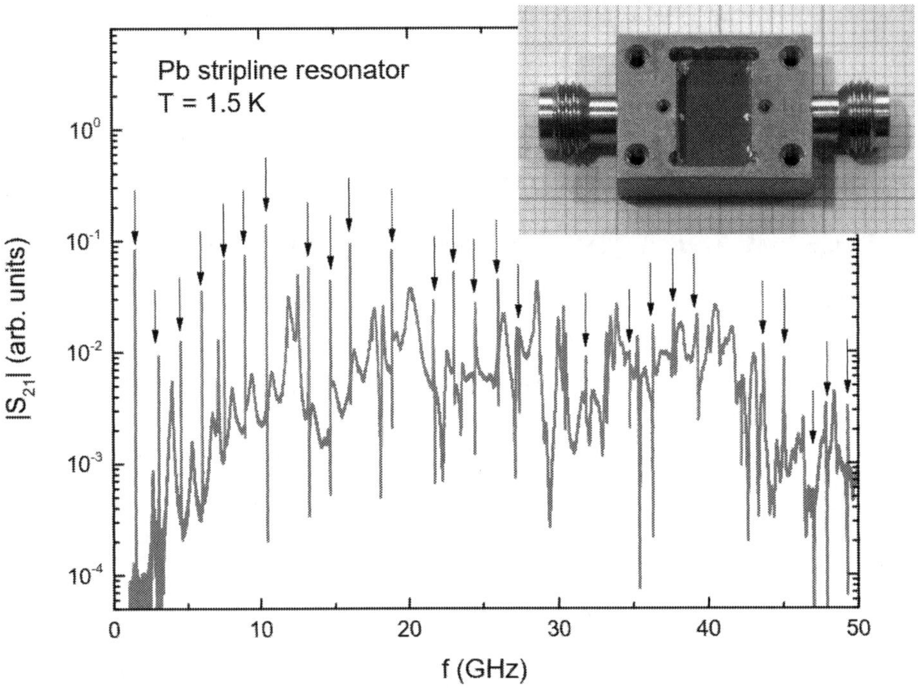


**Figure 1.** Measured transmission spectrum of a Pb superconducting stripline resonator at temperature 1.5 K. Numerous resonances are indicated by arrows. The photograph shows such a stripline resonator in its brass mounting.

which is easily compatible with measurements in a dilution refrigerator, the flexible fabrication in terms of resonator frequency, and the option to determine anisotropic material properties [36]. Such stripline resonators are particularly suited to study conductive materials such as metals and superconductors [31, 34, 35, 36, 37, 38], if the intrinsic resonator losses are very small. This requirement can be met for superconducting planar microwave resonators, which can have much higher quality factors than their metallic counterparts [39, 40], but at the expense of somewhat limited temperature and magnetic-field ranges for applications [41, 42, 43, 44].

Most low-temperature experiments with planar microwave devices are performed at frequencies between 2 and 15 GHz, and for certain applications this frequency range is fully sufficient. But in the field of microwave spectroscopy, the materials of interest might require a larger frequency base, and therefore there have been several developments to address a substantially wider spectral range [45, 46, 47], while many microwave spectroscopy setups remain limited to frequencies below roughly 20 GHz. Following the same motivation, in the present work we want to operate superconducting stripline resonators at much higher frequencies than before to make this technique available for microwave spectroscopy in an extended frequency range. Cryogenic microwave experiments quite generally become more challenging at higher frequencies due to the increasing losses in the microwave transmission lines and the larger role of standing waves due to the reduced wavelength. Furthermore, the electrodynamic losses in superconductors strongly increase with frequency [48], and thus the resonance linewidth of superconducting resonators becomes much broader, making the disturbance due to standing waves even more severe. Finally, appropriate cryogenic microwave equipment for higher frequencies is much less available and more costly than lower-frequency equivalents [49].

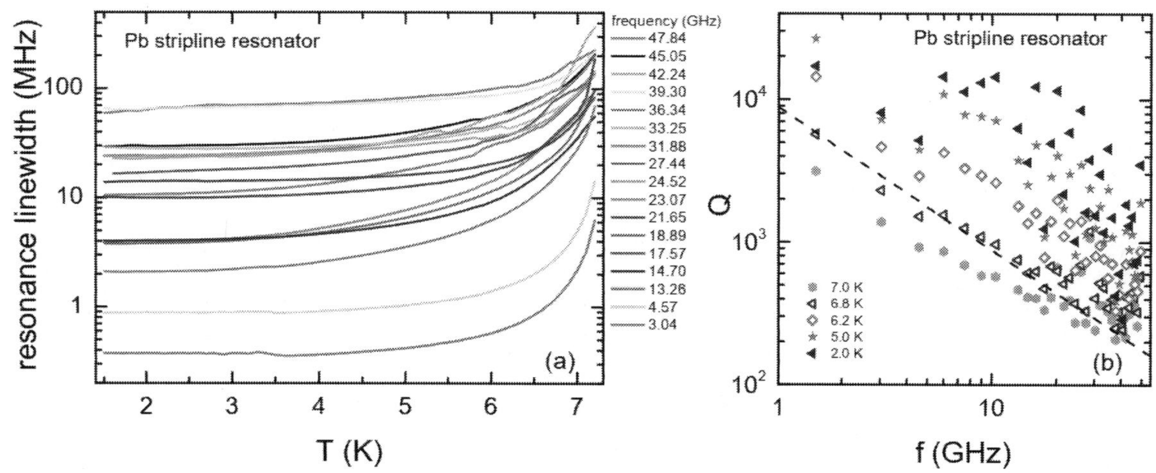


**Figure 2.** Microwave performance of a superconducting Pb stripline resonator. (a) The temperature dependence of resonance linewidth (for several modes) is almost flat at low temperature and increases strongly upon approaching $T_{c,Pb} = 7.2\,K$. (b) The quality factor $Q$ at higher temperatures exhibits an inverse dependence on frequency as expected for conventional superconductors, whereas at lower temperatures residual losses cause non-monotonic frequency dependence.

## 2. Experiment

A stripline is a microwave transmission line that consists of a center conductor sandwiched between two dielectric slabs, which in turn are surrounded by conductive ground planes [1]. In our case, the dielectrics are sapphire (127 $\mu$m thick), and all three conductive planes are fabricated of Pb (center conductor 1 $\mu$m thick evaporated film; ground planes bulk foils), which becomes superconducting below $T_{c,Pb} = 7.2\,K$ [35, 44]. The meander shape of the center conductor as well as the coupling gaps that define the resonator length (i.e. length of center conductor) and thus the resonator fundamental frequency are created using thermal evaporation through a shadow mask. The resonator components are stacked into purpose-designed brass boxes (see photograph in Fig. 1), which are smaller than those previously used [35]. The resonators are cooled with a $^4$He glass dewar (base temperature 1.5 K), and the microwave signal transmitted through the device is measured with a vector network analyzer.

## 3. Results

In Fig. 1 we show a typical (uncalibrated) transmission spectrum at base temperature. Clearly, a large number of roughly equidistant, pronounced resonances can be identified as indicated by the arrows. In this case the fundamental frequency is around 1.5 GHz, and higher harmonics throughout the complete spectral range up to 50 GHz can be studied. For such a superconducting resonator, dominant losses stem from thermally excited quasiparticles, and therefore a strong temperature dependence is expected. Such a behavior is indeed found, e.g. in the resonance linewidth, which is determined by fitting the complex transmission coefficient $S_{21}$ (assuming that the parasitic background can be modeled as a complex constant for frequencies near the resonance frequency [30]) and plotted in Fig. 2(a) for numerous resonator modes. As expected, the linewidth increases strongly with temperature when approaching $T_{c,Pb}$. Also, there is the general trend that the resonator linewidth increases with increasing frequency. This behavior is also expected, and its details play an important role concerning the applicability of such resonators as spectroscopic tools for large frequency ranges. To address the frequency

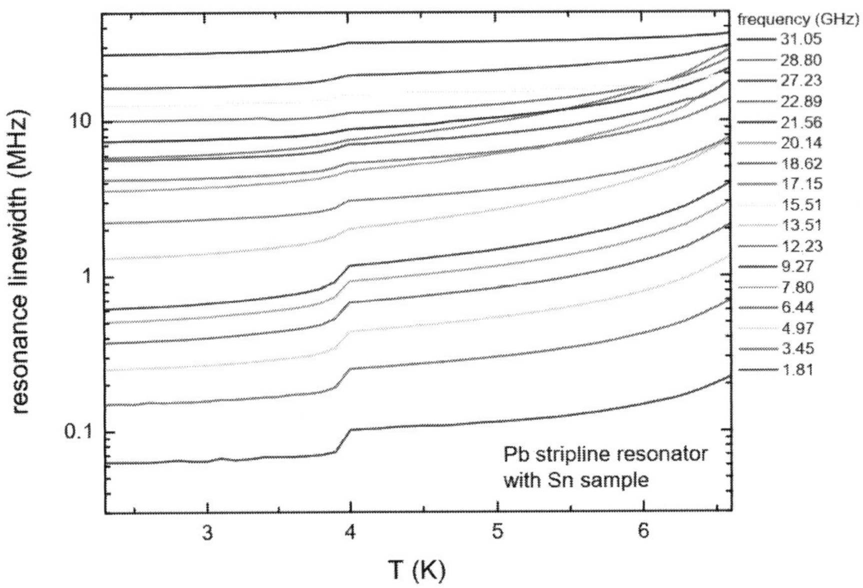


**Figure 3.** Temperature dependence of the resonance linewidth for several modes of a Pb stripline resonator with one ground plane replaced by a bulk Sn sample. The superconducting transition of the Sn below 4 K is visible in all modes.

dependence in more detail, in Fig. 2(b) we plot the resonator quality factor $Q = f_0/\Delta f$, with resonance frequency $f_0$ and resonance linewidth $\Delta f$ for each of the resonant modes, versus frequency (i.e. for all observed modes) for several temperatures. For an ideal superconducting resonator the microwave losses solely depend on the thermally excited quasiparticles. In this case, $Q \propto f/R_s \propto 1/f$ with frequency $f$ and surface resistance $R_s \propto f^2$ of a superconductor is expected [35]. Indeed, we find this frequency-dependent behavior for temperatures not much lower than $T_{c,\mathrm{Pb}} = 7.2\,\mathrm{K}$, as indicated by the dashed line in Fig. 2(b). But at temperatures much lower than $T_{c,\mathrm{Pb}}$, this dependence does not hold any more: here the conduction losses due to quasiparticles are so small that other loss mechanisms become dominant. In particular, residual low-temperature losses that are not well understood [50, 51] and can depend on details of material preparation, resonator geometry etc., and which are regularly found for such resonators [35, 52], have to be taken into account. But more relevant for our applications are the high $Q$ values that we find at the lowest temperatures throughout our complete spectral range, e.g. $Q \approx 3500$ at our highest resonator frequency of 49.32 GHz.

Such high $Q$ values make these stripline resonators appropriate probes for conductive bulk samples e.g. of unconventional (super-)conductors [34, 35, 36, 53]. To demonstrate such a possible application in spectroscopy, we have replaced one Pb ground plane of such a stripline resonator by a bulk sample of Sn. Again we have measured the resonator properties at cryogenic temperatures, and in Fig. 3 we show the temperature dependence of the linewidth of this Sn-loaded resonator. From Fig. 2 we know that the losses of the Pb are almost temperature independent below 4 K, and right in this regime we now find an abrupt decrease of the resonance linewidth upon cooling, i.e. a clear reduction of microwave losses. We assign this feature to the superconducting transition of the Sn sample [54]. These data demonstrate that the microwave response of a superconducting bulk sample is clearly discernible with a stripline resonator to much higher frequencies than shown previously [35, 53, 55]. Using such resonators for quantitative spectroscopy of samples with unknown microwave properties requires data analysis that can separate the loss contributions of Pb resonator and sample, respectively. In the simplest case, the total losses of the device (as quantified by $1/Q$ or by resonator linewidth) are governed completely by the sample of interest. This will apply if the sample is metallic and the temperatures of interest are well below $T_{c,\mathrm{Pb}}$ [35, 37, 56] or for superconducting samples with $T_c$ below 1 K [36], in particular if their Cooper pair density is rather low and thus their microwave losses are high (compared to other superconductors such as Pb) [53].

## 4. Conclusions

We have demonstrated superconducting stripline resonators made of Pb with operating frequencies up to 50 GHz. Even for these very high frequencies, quality factors of several thousand are achieved for temperature below 5 K. This substantially extends the frequency range that is accessible for cryogenic applications such as microwave spectroscopy, well beyond the well-established range up to 20 GHz, and we have shown an exemplary application by investigating a Sn bulk sample as a representative for a variety of conductive materials that can now be studied in this wide spectral range.

## Acknowledgements

We thank G. Untereiner for resonator fabrication and mounting. This work was supported by the Deutsche Forschungsgemeinschaft (DFG).

## References

[1] Pozar D M 1998 *Microwave Engineering* (New York: John Wiley & Sons)
[2] Frunzio L, Wallraff A, Schuster D, Majer J and Schoelkopf R 2005 *IEEE Tran. Appl. Supercond.* **15**, 860
[3] Göppl M, Fragner A, Baur M, Bianchetti R, Filipp S, Fink J M, Leek P J, Puebla G, Steffen L and Wallraff A 2008 *J. Appl. Phys.* **104** 113904
[4] Day P K, LeDuc H G, Mazin B A, Vayonakis A and Zmuidzinas J 2003 *Nature* **425** 817
[5] Hammer G, Wuensch S, Roesch M, Ilin K, Crocoll E and Siegel M 2007 *Supercond. Sci. Technol.* **20** S408
[6] Barends R, Baselmans J J A, Hovenier J N, Gao J R, Yates S J C, Klapwijk T M and Hoevers H F C 2007 *IEEE Trans. Appl. Supercond.* **17** 263
[7] Ohashi T, Kitano H, Maeda A, Akaike H and Fujimaki A 2006 *Phys. Rev. B* **73** 174522
[8] Crane R W, Armitage N P, Johansson A, Sambandamurthy G, Shahar D and Grüner 2007 *Phys. Rev. B* **75** 094506
[9] Steinberg K, Scheffler M and Dressel M 2008 *Phys. Rev. B* **77** 214517
[10] Pompeo N, Silva E, Sarti S, Attanasio C, Cirillo C 2010 *Physica C* **470** 901
[11] Wu D H, Booth J C, Anlage S M 1995 *Phys. Rev. Lett.* **75** 525
[12] Turner P J, Harris R, Kamal S, Hayden M E, Broun D M, Morgan D C, Hosseini A, Dosanjh P, Mullins G K, Preston J S, Liang R, Bonn D A and Hardy W N 2003 *Phys. Rev. Lett.* **90** 237005
[13] Ormeno R J, Hein M A, Barraclough T L, Sibley A, Gough C E, Mao Z Q, Nishizaki S, Maeno Y *Phys. Rev. B* **74** 092504
[14] Truncik C J S, Huttema W A, Turner P J, Özcan S, Murphy N C, Carrière P R, Thewalt E, Morse K J, Koenig A J, Sarrao J L and Broun D M 2013 *Nature Commun.* **4** 2477
[15] Scheffler M, Dressel M, Jourdan M and Adrian H 2005 *Nature* **438** 1135
[16] Scheffler M, Dressel M, Jourdan M and Adrian H 2006 *Physica B* **378-380** 993
[17] Basov D N, Averitt R D, van der Marel D, Dressel M and Haule K 2011 *Rev. Mod. Phys.* **83** 471
[18] Schwartz A, Scheffler M and Anlage S M 2000 *Phys. Rev. B* **61** R870
[19] Povarov K Y, Smirnov A I, Starykh O A, Petrov S V and Shapiro A Y 2011 *Phys. Rev. Lett.* **107** 037204
[20] Krupka J, 2006 *Meas. Sci. Technol.* **17** R55
[21] Booth J C, Wu D H and Anlage S M 1994 *Rev. Sci. Instrum.* **65** 2082
[22] Turner P J, Broun D M, Kamal S, Hayden M E, Bobowski J S, Harris R, Morgan D C, Preston J S, Bonn D A and Hardy W N 2004 *Rev. Sci. Instrum.* **75** 124
[23] Scheffler M and Dressel M 2005 *Rev. Sci. Instrum.* **76** 074702
[24] Steinberg K, Scheffler M and Dressel M 2012 *Rev. Sci. Instrum.* **83** 024704
[25] Liu W, Pan L and Armitage N P 2014 *Rev. Sci. Instrum.* **85** 093108
[26] Ranzani L, Spietz L, Popovic Z and Aumentado J 2013 *Rev. Sci. Instrum.* **84** 034704
[27] Yeh J-H and Anlage S M 2013 *Rev. Sci. Instrum.* **84** 034706
[28] Klein O, Donovan S, Dressel M and Grüner G 1993 *Int. J. Infrared Millim. Waves* **14** 2423

[29] Deri R J 1986 *Rev. Sci. Instrum.* **57** 82
[30] Huttema W A, Morgan B, Turner P J, Hardy W N, Zhou X, Bonn D A, Liang R and Broun D M 2006 *Rev. Sci. Instrum.* **77** 023901
[31] DiIorio M S, Anderson A C and Tsaur B-Y 1988 *Phys. Rev. B.* **38** 7019
[32] Revenaz S, Oates D E, Labbé-Lavigne D, Dresselhaus G and Dresselhaus M S 1994 *Phys. Rev. B.* **50** 1178
[33] Scheffler M, Fella C and Dressel M 2012 *J. Phys. Conf. Ser.* **400** 052031
[34] Scheffler M, Schlegel K, Clauss C, Hafner D, Fella C, Dressel M, Jourdan M, Sichelschmidt J, Krellner C, Geibel C and Steglich F 2013 *Phys. Status Solidi B* **250**, 439
[35] Hafner D, Dressel M and Scheffler M 2014 *Rev. Sci. Instrum.* **85** 014702
[36] Scheffler M, Felger M M, Thiemann M, Hafner D, Schlegel K, Dressel M, Ilin K S, Siegel M, Seiro S, Geibel C and Steglich F 2015 *Acta IMEKO* **4** 47
[37] Parkkinen K, Dressel M, Kliemt K, Krellner C, Geibel C, Steglich F and Scheffler M 2015 *Physics Procedia* **75** 340
[38] Beutel M H, Ebensperger N G, Thiemann M, Untereiner G, Fritz V, Javaheri M, Nägele J, Rösslhuber R, Dressel M and Scheffler M 2016 *Supercond. Sci. Technol.* **29** 085011
[39] Megrant A, Neill C, Barends R, Chiaro B, Chen Y, Feigl L, Kelly J, Lucero E, Mariantoni M, OMalley P J J, Sank D, Vainsencher A, Wenner J, White T C, Yin Y, Zhao J, Palmstrøm C J, Martinis J M and Cleland A N 2012 *Appl. Phys. Lett.* **100**, 113510
[40] Javaheri Rahim M, Lehleiter T, Bothner D, Krellner C, Koelle D, Kleiner R, Dressel M and Scheffler M 2016 *J. Phys. D: Appl. Phys.* **49** 395501
[41] Song C, DeFeo M P, Yu K and Plourde B L T, *Appl. Phys. Lett.* **95** 232501
[42] Bothner D, Clauss C, Koroknay E, Kemmler M, Gaber T, Jetter M, Scheffler M, Michler P, Dressel M, Koelle D and Kleiner R 2012 *Appl. Phys. Lett.* **100** 012601
[43] Ghirri A, Bonizzoni C, Gerace D, Sanna S, Cassinese A and Affronte M 2015 *Appl. Phys. Lett.* **106** 184101
[44] Ebensperger N G, Thiemann M, Dressel M and Scheffler M 2016 *Supercond. Sci. Technol.* **29** 115004
[45] Scheffler M, Kilic S and Dressel M 2007 *Rev. Sci. Instrum.* **78** 086106
[46] Clauss C, Bothner D, Koelle D, Kleiner R, Bogani L, Scheffler M and Dressel M 2013 *Appl. Phys. Lett.* **102** 162601
[47] Wiemann Y, Simmendinger J, Clauss C, Bogani L, Bothner D, Koelle D, Kleiner R, Dressel M and Scheffler M 2015 *Appl. Phys. Lett.* **106** 193505
[48] Pracht U S, Heintze E, Clauss C, Hafner D, Bek R, Werner D, Gelhorn S, Scheffler M, Dressel M, Sherman D, Gorshunov B, Ilin K S, Henrich D and Siegel M 2013 *IEEE Trans. THz Sci. Technol.* **3** 269
[49] The very popular SMA connectors are originally designed for operation up to 18 GHz, and they thus set an upper frequency limit for many established experimental setups.
[50] Halbritter J 1971 *J. Appl. Phys.* **42** 82
[51] Pierce J M 1973 *Journal of Applied Physics* **44** 1342
[52] Oates D E, Anderson A C, Chin C C, Derov J S, Dresselhaus G and Dresselhaus M S 1991 *Phys. Rev. B.* **43** 7655
[53] Thiemann M, Beutel M H, Dressel M, Lee-Hone N R, Broun D M, Fillis-Tsirakis E, Boschker H, Mannhart J and Scheffler M 2017 arXiv:1703.04716 [cond-mat.supr-con]
[54] The discrepancy between the temperature where we detect the superconducting transition in Sn (around 3.9 K) and the literature value of $T_{c,\mathrm{Sn}} = 3.72\,\mathrm{K}$ probably stems from the error in the temperature measurement.
[55] Thiemann M, Bothner D, Koelle D, Kleiner R, Dressel M and Scheffler M 2014 *J. Phys.: Conf. Ser.* **568** 022043
[56] Hafner D, Dressel M, Stockert O, Grube K, Löhneysen H v, Scheffler M 2014 *JPS Conf. Proc.* **3** 012016

# Cross-correlation limit of a SQUID-based noise thermometer of the pMFFT type

**A Kirste, J Engert**

Physikalisch-Technische Bundesanstalt (PTB), Abbestraße 2-12, 10587 Berlin, Germany

E-mail: alexander.kirste@ptb.de

**Abstract.** The primary magnetic field fluctuation thermometer (pMFFT) is a SQUID-based noise thermometer for temperatures below 1 K, which complies with metrological requirements. It combines two signal channels in order to apply the cross-correlation technique, but it requires statistically independent noise signals for proper operation. In order to check the limit of the cross-correlation readout, we have performed zero measurements in the millikelvin range in a setup that is identical to the pMFFT, except for the removed temperature sensor. We examined the influence of different parameters such as SQUID working point or flux-lock loop parameters on the minimum cross-correlation signal down to 24 mK and below 100 kHz. Depending on the configuration, typical minimum SQUID-referred cross-power spectral densities of $1.5 \times 10^{-15}$ $\Phi_0^2$/Hz or even smaller values were observed. For the pMFFT, considering its thermal noise spectrum, these flux densities correspond to a device noise temperature of $\lesssim$2.5 µK, thereby ensuring a negligible uncertainty contribution at the lower end of the PLTS-2000 (0.9 mK).

## 1. Introduction

The magnetic field fluctuation thermometer (MFFT) [1] is a noise thermometer, which derives temperature from the thermal motion of the electrons in a metallic body – the temperature sensor. For this purpose it measures the noise currents inside the temperature sensor by detecting the corresponding fluctuating magnetic field just above its surface using superconducting quantum interference devices (SQUIDs) [2]. The MFFT is based on the linear relationship between the power spectral density (PSD) $S_\Phi$ of the measured thermal magnetic flux noise $\Phi$ and thermodynamic temperature $T$, analogous to the well-known Nyquist relation for the Johnson noise in a resistor [3]. The pMFFT is an advanced version of the MFFT, which works in absolute primary mode, so that it does not require calibration at a known temperature [4].

If the thermal noise proportional to $T$ is interfered with any other added noise, this introduces an error (bias) to the measured temperature, leading to increased temperature values. This deviation gets worse for lower temperatures and finally sets a limit to the lowest measurable temperature. Often, that bias temperature is synonymous with the device noise temperature $T_\mathrm{N}$, which is reached if the averaged PSD from the thermal noise is equal to the averaged PSD from added non-thermal noise.

The cross-correlation technique (CCT) can provide a solution to this problem. It is applied in the pMFFT and uses two separate signal channels. Provided the two signal channels are completely independent, the CCT allows the measurement of the pure thermal noise identical in both detection coils. In the pMFFT, the background noise introduced in the signal channels is typically smaller than the thermal noise to be measured, though not completely negligible. The benefit of the CCT here is to achieve a lower measurement uncertainty by reducing the bias caused by any background noise. In real measurement setups, however, the two signal channels are not perfectly independent, and this will affect the ultimate performance (limit) of the CCT. Reasons for lifting the statistical independence of both channels are, for example, crosstalk or environmental fluctuations (due to external fields or temperature) acting simultaneously on both channels. The corresponding limit of the CCT caused by residual interactions is called hardware limit. Following first measurements of [5], we have characterized crosstalk and examined the hardware limit of the current pMFFT setup depending on different working conditions.

## 2. The cross-correlation technique

Here we introduce the basic concept of the cross-correlation method, the notation of quantities and selected properties used in the paper. An extensive treatment can be found in [6].

The CCT is based on a two-channel measurement of the common noise signal $c(t)$, where each channel adds its own background noise $a(t)$ and $b(t)$, respectively, cf. (1), (2). It is assumed that $c(t)$, $a(t)$ and $b(t)$ are statistically independent, $t$ is the time. The Fourier transform (FT) provides access to the PSD $S_x(f) \equiv S_{xx}(f)$ of a zero-mean finite-power process $\mathbf{x}(t)$. For the FT – inverse FT pair we use the notation $x(t) \leftrightarrow X(f)$ with an upper case for the FT and suppressing the frequency $f$ wherever possible.

$$\begin{aligned} \text{Time domain} &\leftrightarrow \text{Frequency domain} \\ x(t) = c(t) + a(t) &\leftrightarrow X = C + A \qquad (1) \\ y(t) = c(t) + b(t) &\leftrightarrow Y = C + B \qquad (2) \end{aligned}$$

In an actual experiment, the average $\langle \cdot \rangle_m$ on a suitable number of $m$ spectrum samples replaces the expectation operator $\mathbf{E}\{\cdot\}$. For the cross-power spectral density (CPSD) $S_{xy}(f)$ this reads

$$\langle S_{xy} \rangle_m = \tau^{-1} \langle XY^* \rangle_m, \qquad (3)$$

where an available truncated version $x_\tau(t) \leftrightarrow X_\tau(f)$ obtained in the finite measurement time $\tau$ is used for the entire process $\mathbf{x}(t)$ and the superscript '$*$' marks the complex conjugate. $S_{xy}$ is complex and can be decomposed into real and imaginary parts: $S_{xy} = S'_{xy} + \imath S''_{xy}$. Evaluating (3), the cross terms decrease proportional to $1/\sqrt{m}$ owing to the statistical independence:

$$\langle S_{xy} \rangle_m = \tau^{-1} \left[ \langle CC^* \rangle_m + \langle CB^* \rangle_m + \langle AC^* \rangle_m + \langle AB^* \rangle_m \right] = S_{cc} + O(\sqrt{1/m}). \qquad (4)$$

Hence, for a large number of averages $m$, $\langle S_{xy} \rangle_m$ converges to its limit $\mathbf{E}\{S_{xy}\} = S_{cc}$, while for small $m$ the cross terms can be the dominating part in (4) and cover small $S_{cc}$. The latter case is described by the statistical limit, assuming $c = 0$, $S_{cc} = 0$:

$$\langle S_{xy}^{\text{stat. lim.}} \rangle_m = \tau^{-1} \langle AB^* \rangle_m \approx \sqrt{\langle S_{xx} \rangle_m \langle S_{yy} \rangle_m / m}. \qquad (5)$$

In order to observe the hardware limit, the statistical limit must be made smaller by averaging over long enough times ($\Rightarrow$ large $m$). Assuming white Gaussian noise for $c(t)$, $a(t)$ and $b(t)$ in (1), (2), we can find expressions for the expectation values and variances (operator $\mathbf{V}\{\cdot\}$):

$$\mathbf{E}\{\langle S'_{xy} \rangle_m\} = \mathbf{E}\{S'_{xy}\} = S_{cc}, \qquad \mathbf{V}\{\langle S'_{xy} \rangle_m\} = \frac{1}{2m} \left( \mathbf{E}\{S'_{xy}\}^2 + \mathbf{E}\{S_{xx}\}\mathbf{E}\{S_{yy}\} \right), \qquad (6)$$

$$\mathbf{E}\{\langle S_{xx} \rangle_m\} = S_{cc} + S_{aa}, \qquad \mathbf{V}\{\langle S_{xx} \rangle_m\} = \frac{1}{m} \mathbf{E}\{S_{xx}\}^2. \qquad (7)$$

A subtle difference between (6) and (7) originates from the fact that $S_{xx}$ is always real and non-negative, whereas the real part $S'_{xy} = \Re\{S_{xy}\}$ can be both positive or negative. We use this estimator ($\Re\{\langle S_{xy} \rangle_m\}$) since it contains all the useful information but discards the imaginary

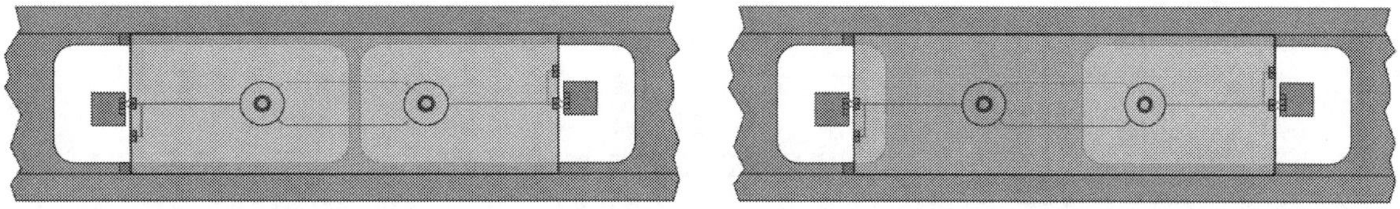

**Figure 1.** Setup for the zero measurements (left) in comparison to the pMFFT body (right). Except for the removed temperature sensor beneath the detection coils, both setups are identical. The chip carrying the detection coils is shown semi-transparently with interconnections drawn in different colours. The SQUID sensor chips next to it are mounted on PCBs (not shown).

part of the background noise. This improves the signal-to-noise ratio of the estimator and avoids a bias. Opposite to this, $\Im\{\langle S_{xy}\rangle_m\}$ is a good estimator for the background. Expressions (6) and (7) given here generalize those with a particular normalization chosen in [6].

A useful quantity describing the normalized measure of association is the coherency $C_{xy}(f)$:

$$C_{xy} = S_{xy}/\sqrt{S_{xx}S_{yy}}. \tag{8}$$

Its magnitude is referred to as coherence, and looking at its real part, $C'_{xy} = \Re\{C_{xy}\}$, it holds $-1 \leq C'_{xy} \leq +1$. Transforming (8) into $S_{xy} = C_{xy}\sqrt{S_{xx}S_{yy}}$ makes clear that for absolutely low values of $S_{xy}$ both a low residual correlation (coherence) and low background noise in both signal channels are required.

## 3. Experiment

### *3.1. Experimental setup*

The new zero measurements in the millikelvin range were performed in a dry dilution refrigerator, using the same setup and environment that is otherwise used for regular temperature measurements with the pMFFT. The only difference is a modified copper body of the pMFFT, in which the temperature sensor located directly beneath the detection coils is removed. Figure 1 depicts the setup for the zero measurement in comparison to that of the pMFFT.

Each signal channel consists of a detection coil, a SQUID sensor, wiring to the room-temperature SQUID electronics and an analogue-to-digital converter (ADC). The gradiometric detection coils have an estimated total inductance of ~400 nH. Details are described in [4]. SQUID current sensors with an input inductance of 400 nH (single-stage, double-transformer SQUIDs [7] of size "L") are used for the readout. Each input circuit is equipped with an additional RC damping ($R = 3.9\,\Omega$, $C = 47\,\text{nF}$). The flux-lock loop (FLL) based readout scheme is depicted in figure 2. In FLL mode, $S \equiv S_\Phi$ is measured.

A 3 m long, shielded cable (type Magnicon Cryocable CC-1/3 [8]) connects the SQUID sensors with the room temperature SQUID electronics (type Magnicon XXF-1/3 [8]). This cable contains three channels, each consisting of two twisted pairs and one twisted triple. All strands are combined in a common stainless steel braid. For the data acquisition we use a two-channel dynamic signal analyzer card (type National Instruments PXI-4461, 24 Bit, 204.8 kSa/s [8]). This enables us to measure spectra up to ~100 kHz, exceeding by far the typical ~5 kHz evaluated in the pMFFT for temperature determination.

### *3.2. Results and Discussion*

The spectra $S_1 \equiv \langle S_{11}\rangle_m$, $S_2 \equiv \langle S_{22}\rangle_m$ and $S'_{12} \equiv \langle S'_{12}\rangle_m$ were measured for various numbers of averages $m$ using the Welch periodogram [9]. If not otherwise stated, $T = 24\,\text{mK}$. In view of the scaling of its variance, it is useful to compare $S'_{12}$ with a reference spectrum that is normalized

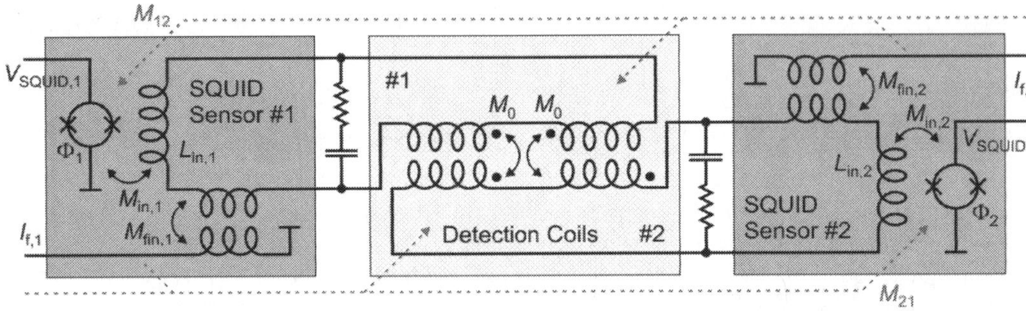


**Figure 2.** Schematic of the correlation-based pMFFT readout showing major couplings and parasitic interactions (dashed lines). Feedback ($I_\mathrm{f}$) is applied via $M_\mathrm{fin}$ into the input circuits.

with respect to $m$. We choose $S^\mathrm{ref} = \sqrt{\langle S_{11}\rangle_m \langle S_{22}\rangle_m/(2m)}$, which is equal to $\mathrm{dev}\{\langle S'_{12}\rangle_m\}$ in case of vanishing coherence (cf. (6) with $\mathbf{E}\{S'_{xy}\} = 0$ and $\mathrm{dev}\{\cdot\} = (\mathbf{V}\{\cdot\})^{1/2}$).

*3.2.1. Zero measurements* In order to distinguish between interferences from the pMFFT's SQUID readout and those from other origins (wiring to room temperature, SQUID electronics, ADCs), measurements of $S'_{12}$ were performed without SQUID operation. The SQUIDs were made insensitive to flux by a large bias current or by using an insensitive working point ($\mathrm{d}V_\mathrm{SQUID}/\mathrm{d}\Phi = 0$), and the SQUID electronics acted simply as a preamplifier (AMP mode, gain 2000). Results are identical, no matter which method is used. Both channels have a background white noise of $\approx 0.4\,\mathrm{nVHz}^{-1/2}$ with pronounced interference peaks at the mains frequency and odd harmonics up to 1050 Hz. For $S'_{12}$, it turns out that there are only few interference peaks above 1 kHz, apart from a pronounced peak centered at 59 kHz. Below 1 kHz, there is pronounced power only at the mains frequency and odd harmonics. The important outcome is that there is no obvious sign of a bias to $S'_{12}$ apart from the few discrete peaks.

*3.2.2. Crosstalk, mutual inductances* Crosstalk between both channels can, in principle, take place via different mechanisms and elements. If the cold-to-warm wiring of both channels is involved (each one consisting of a twisted triple and a twisted pair), several dual combinations exist. We confine our analysis to parasitic mutual induction between the feedback line of channel $i$ (current $I_{\mathrm{f},i}$) and a flux sensitive circuit of the other channel $j$ (causing the flux $\Phi_j$ in SQUID $j$), described by effective values $M_{12}$ and $M_{21}$, cf. figure 2. The feedback is described by $M_{\mathrm{f},1}$ and $M_{\mathrm{f},2}$.

$$\mathbf{\Phi} = \mathbf{M}\,\mathbf{I}_\mathrm{f}\,, \quad \mathbf{M} = \begin{pmatrix} M_{\mathrm{f},1} & M_{12} \\ M_{21} & M_{\mathrm{f},2} \end{pmatrix} = M_0 \begin{pmatrix} M_{\mathrm{f},1}/M_0 & M_{12}/M_0 \\ M_{21}/M_0 & M_{\mathrm{f},2}/M_0 \end{pmatrix} \tag{9}$$

The restriction to mutual inductances is consistent with the finding that the coupling is virtually independent of frequency when comparing results for 59 Hz, 1070 Hz and 4130 Hz. We can express our results relative to the scaling value $M_0^{-1} = 63.10\,\mu\mathrm{A}/\Phi_0$ ($\Phi_0 = h/2e$ is the flux quantum). For the parasitic coupling we obtained $M_{12}/M_0 = 7.4\times10^{-6}$ and $M_{21}/M_0 = 2.2\times10^{-5}$, which is almost 5 orders of magnitude smaller than the feedback, $M_{\mathrm{f},1}/M_0 = 0.9986$, $M_{\mathrm{f},2}/M_0 = 1.0016$.

*3.2.3. Influence of the SQUID working point* Although the SQUID working point does not affect the flux-to-voltage transfer coefficient of the FLL, it has an influence on noise properties and dynamic properties (bandwidth) of the FLL by means of the derivative of the SQUID characteristics ($V_\Phi \equiv \mathrm{d}V_\mathrm{SQUID}/\mathrm{d}\Phi$). Hence, it can be expected that the working point affects the limit of $S'_{12}$ and possibly also the coherence $|C'_{12}|$. We have checked this at the four working points (WP-$x$) given in table 1. The corresponding unity-gain frequencies $f_1$ of the FLL based on the given gain-bandwidth products (GBP) and on a feedback resistance of $R_f = 30\,\text{k}\Omega$ vary between 620 kHz and 1.9 MHz.

For further evaluation we define a frequency band $B$ in order to determine the statistical quantities of (6), (7) and (8). An averaging $\langle\cdot\rangle_f$ is done over all $n_{\text{bin}}$ bins within this band. By choosing $B = (1\ldots12)\,\text{kHz}$, we obtain the results summarized in table 2 with $n_{\text{bin}} = 451$ and $m(\text{WP-B, WP-D}) = 8788$ or $m(\text{WP-A, WP-C}) = 24413$. If the mean value $\mathbf{E}\{\langle S'_{12}\rangle_f\}$ is larger than the corresponding statistical limit ($= \sqrt{2}\,\langle S^{\text{ref}}\rangle_f$), it reflects the current hardware limit. This is clearly the case for WP-A, WP-C and WP-C/3. For the other working points (WP-B, WP-C/2, WP-D), longer measurement times (larger $m$) are necessary to reveal the hardware limit, and smaller absolute values ($|S'_{12}|$) can be expected. In comparison, WP-A, is constricted by the relatively large background noise, caused by the low $V_\Phi$. Although this results in a substantially increased $S'_{12}$ for equal $m$, the corresponding $C'_{12}$ as a normalized quantity is only moderately increased. With two exceptions (WP-C, WP-C/2), the coherence ($C'_{12}$) varies between ≈0.009 and ≈0.003, and a correlation with increasing $V_\Phi$ is apparent. The same correlation is apparent for $S'_{12}$. Consequently, working points with large $V_\Phi$ are favorable.

*3.2.4. Influence of FLL parameters ($R_f$, GBP)* Feedback resistance $R_f$ and GBP have a direct influence on the FLL bandwidth ($f_1 \propto \text{GBP}/R_f$) provided the cable delay is still negligible. It is known that the background noise can increase due to down mixing if the FLL bandwidth increases. A similar influence can be supposed to happen on $S'_{12}$. If down mixing is involved, the curvature of the SQUID characteristics ($\text{d}^2V_{\text{SQUID}}/\text{d}\Phi^2$) at the working point plays a role. For WP-C we tested two values for $R_f$: $30\,\text{k}\Omega$ (WP-C) and $100\,\text{k}\Omega$ (WP-C/2, $m = 48827$), keeping the GPB constant at 0.30 GHz. Results are given in table 2. For larger $R_f$, significantly reduced values for both $S'_{12}$ and $C'_{12}$ are found that have an even reversed sign. Although this finding fits into the simple picture outlined above, another, apparently contradicting measurement exists. Testing WP-C with an elevated GBP of 2.8 GHz (and thus increased FLL bandwidth) yields data very close to those of WP-C. For a comprehensive description of the influence of the FLL parameters on $S'_{12}$ and $C'_{12}$, further investigations are necessary.

**Table 1.** Characteristic properties of working points (WP-$x$) and FLL parameters.

| Parameter | WP-A | WP-B | WP-C | WP-D | Unit |
|---|---|---|---|---|---|
| $\text{d}V_{\text{SQUID}}/\text{d}\Phi$ (Ch. 1 \| Ch. 2) | 220 \| 190 | 840 \| 690 | 4300 \| 4200 | 12000 \| 7400 | $\mu\text{V}/\Phi_0$ |
| GBP (Ch. 1 & Ch. 2) | 6.2 | 2.8 | 0.30 | 0.30 | GHz |

**Table 2.** Cross-correlation results at different working points (WP-$x$) and other conditions evaluated for $(1\ldots12)\,\text{kHz}$. If no other values are given, $R_f = 30\,\text{k}\Omega$ and $T = 24\,\text{mK}$ apply.

| WP/Condition | WP-A | WP-B | WP-C | WP-C/2<br>$R_f$=100 kΩ | WP-C/3<br>$R_f$=100 kΩ<br>$T$=100 mK | WP-D | Unit |
|---|---|---|---|---|---|---|---|
| $\mathbf{E}\{\langle S'_{12}\rangle_f\}$ | 320 | 23 | 15 | -0.10 | 5.5 | 2.2 | $10^{-16}\Phi_0^2/\text{Hz}$ |
| $\langle S^{\text{ref}}\rangle_f$ | 160 | 26 | 3.8 | 2.7 | 3.5 | 5.4 | $10^{-16}\Phi_0^2/\text{Hz}$ |
| $\text{dev}\{\langle S'_{12}\rangle_f\}$ | 350 | 41 | 12 | 7.1 | 8.1 | 10 | $10^{-16}\Phi_0^2/\text{Hz}$ |
| $\mathbf{E}\{\langle C'_{12}\rangle_f\}$ | 9.1 | 6.6 | 18 | -0.14 | 5.2 | 3.1 | $10^{-3}$ |
| $\text{dev}\{\langle C'_{12}\rangle_f\}$ | 9.0 | 11 | 11 | 7.1 | 6.5 | 12 | $10^{-3}$ |

*3.2.5. Influence of temperature* In order to check if there is any influence of temperature, we performed two measurements under otherwise identical conditions. The two results for 24 mK (WP-C/2) and 100 mK (WP-C/3) are given in table 2. It is found that $S'_{12}$ and consequently $C'_{12}$ are increased at 100 mK in comparison to 24 mK. Between these two temperatures we can certainly assume that the SQUID working point does not change (because $T$ is much smaller than the critical temperature of the SQUID) and that the intrinsic SQUID noise does not change (because the actual SQUID temperature decouples already below ~300 mK). Since no direct influence of temperature on $C'_{12}$ is known, an obvious reason for this result is thermal noise sensed from the remaining copper body of the pMFFT. In this case we can assume a linear relationship for the background thermal noise: $\mathbf{E}\{\langle S'_{12,T}\rangle_f\} = S'_{12,0} + \alpha_{12}T$. Solving for $\alpha_{12}$ yields $\alpha_{12} = 7.1 \times 10^{-15}\,\Phi_0^2/(\mathrm{Hz\,K})$. The corresponding values of both signal channels (evaluated at 4.0 kHz only) are considerably larger: $\alpha_1 = 2.7 \times 10^{-13}\,\Phi_0^2/(\mathrm{Hz\,K})$ and $\alpha_2 = 3.4 \times 10^{-13}\,\Phi_0^2/(\mathrm{Hz\,K})$. These values must be compared to the sensitivity of the pMFFT, $\alpha_{\mathrm{pMFFT}}$. At 5 kHz, $\alpha_{\mathrm{pMFFT}} = 2.1 \times 10^{-10}\,\Phi_0^2/(\mathrm{Hz\,K})$. Thus, for pMFFT operation, the temperature effect is of the order $10^{-5}$.

*3.2.6. Device noise temperature predicted for the pMFFT* Now, in knowledge of the various effects on $S'_{12}$, we are able to estimate the device noise temperature $T_N$ for the pMFFT. We can neglect the temperature effect (subsection 3.2.5), but have to take into account influence from the working point and from FLL parameters (subsections 3.2.3, 3.2.4). Based on the mean value of WP-C, which can be considered as routine performance of the CCT, and on the spectral sensitivity of the pMFFT, we obtain $T_N = 2.5\,\mu\mathrm{K}$.

## 4. Conclusions

In conclusion, we have found that our cross-correlation setup, which is characterized by a rather small crosstalk between both signal channels, exhibits a typical coherence ranging between 0.3% and 0.9%. Although the absolute performance limit of the CCT depends on concrete working conditions, $40\,\mathrm{n}\Phi_0/\mathrm{Hz}^{1/2}$ can be considered as a typical average value. The best performance of the CCT is achieved for SQUID working points with large $V_\Phi$. When operating the pMFFT, the CCT performance can easily be optimized by monitoring the deviation of the imaginary parts (dev$\{\langle S''_{12}\rangle_m\}$ or dev$\{\langle C''_{12}\rangle_m\}$). The device noise temperature was estimated as $T_N = 2.5\,\mu\mathrm{K}$.

## Acknowledgments

The authors thank D. Heyer for technical assistance as well as PTB's Berlin Institute workshop for manufacturing the high-precision copper bodies. The research reported here has received funding from the EMPIR programme as an integrated part of Horizon 2020, the European Union's Framework Programme for Research and Innovation (grant number 15SIB02InK2).

## References

[1] Beyer J, Drung D, Kirste A, Engert J, Netsch A, Fleischmann A and Enss C 2007 *IEEE Trans. Appl. Supercond.* **17**, 760-763
[2] Netsch A, Hassinger E, Enss C and Fleischmann A 2006 *AIP Conf. Proc.* **850** 1593-1594
[3] Nyquist H 1928 *Phys. Rev.* **32** 110-113
[4] Kirste A and Engert J 2016 *Phil. Trans. R. Soc. A* **374** 20150050
[5] Kirste A, Regin M, Engert J, Drung D and Schurig Th 2014 *J. Phys.: Conf. Ser.* **568** 032012
[6] Rubiola E and Vernotte F 2010 http://arxiv.org/pdf/1003.0113v1.pdf
[7] Drung D, Aßmann C, Beyer J, Kirste A, Peters M, Ruede F, Schurig T. 2007 *IEEE Trans. Appl. Supercond.* **17** 699-704
[8] Identification of commercial equipment in this paper does not imply any recommendation or endorsement by the PTB.
[9] Welch PD 1967 *IEEE Transactions on Audio and Electroacoustics* **AU-15** 70-76

IOP Conf. Series: Journal of Physics: Conf. Series **969** (2018) 012084 doi:10.1088/1742-6596/969/1/012084

# Design of distributed JT (Joule-Thomson) effect heat exchanger for superfluid 2 K cooling device

**S Jeong, C Park, and K Kim**

Department of Mechanical Engineering (ME3039)
KAIST (Korea Advanced Institute of Science and Technology), Daejeon, Korea

skjeong@kaist.ac.kr

**Abstract.** Superfluid at 2 K or below is readily obtained from liquid helium at 4.2 K by reducing its vapour pressure. For better cooling performance, however, the cold energy of vaporized helium at 2 K chamber can be effectively utilized in a recuperator which is specially designed in this paper for accomplishing so-called the distributed Joule-Thomson (JT) expansion effect. This paper describes the design methodology of distributed JT effect heat exchanger for 2 K JT cooling device. The newly developed heat exchanger allows continuous significant pressure drop at high-pressure part of the recuperative heat exchanger by using a capillary tube. Being different from conventional recuperative heat exchangers, the efficient JT effect HX must consider the pressure drop effect as well as the heat transfer characteristic. The heat exchanger for the distributed JT effect actively utilizes continuous pressure loss at the hot stream of the heat exchanger by using an OD of 0.64 mm and an ID of 0.4 mm capillary tube. The analysis is performed by dividing the heat exchanger into the multiple sub-units of the heat exchange part and JT valve. For more accurate estimation of the pressure drop of spirally wound capillary tube, preliminary experiments are carried out to investigate the friction factor at high Reynolds number. By using the developed pressure drop correlation and the heat transfer correlation, the specification of the heat exchanger with distributed JT effect for 2 K JT refrigerator is determined.

## 1. Introduction

Cryogenic temperature of 2 K or below is utilized for low-temperature property measurement device, high magnetic field systems such as 45 Tesla hybrid magnet, LHC (Large hadron collider) to observe subatomic matter, NMR (Nuclear magnetic resonance) and various advanced particle accelerators [1-4]. A simple liquid helium system without recuperative heat exchanger can be employed with a vacuum pump to lower the vapour pressure of helium and produce superfluid to provide constant cooling effect. Liquid helium is continuously supplied to sub-atmospheric chamber while experiencing Joule-Thomson (JT) cooling effect. This JT expansion process can be effectively modified for a more efficient cooling device by incorporating a recuperative heat exchanger and an orifice to produce necessary JT cooling effect, which resultantly utilizes the cold energy of the pumped helium vapour. As pointed out by Ganni *et al.* [5], the sub-atmospheric JT cooling method can be further improved by combining multiple JT expansions and heat exchange processes in between. It is common that the large-size 2 K refrigerator requires very high electric power input in general such as 18 MW to run LHC (20 kW at 1.9 K as cooling capacity) and 5 MW to operate CEBAF (Continuous Electron Beam Accelerator Facility) accelerator (4.8 kW at 2 K as cooling capacity). Tons of liquid helium are also used to cooling such large-scale applications in conjunction with using JT expansions to create 2 K environments. Various studies, therefore, were seriously carried out to improve the efficiency of the 2 K JT refrigeration. Fig. 1 shows typical schematic diagrams of these refrigeration systems. In general, 2 K refrigerator is composed of a pre-cooler, a recuperative heat exchanger, a valve and an evaporator or 2 K chamber. The pre-cooler is required primarily to cool down helium from room temperature to approximately 4.5 K. After passing the pre-cooler, helium can be further cooled by a heat exchanger and an expansion valve. JLab (Jefferson laboratory) initially developed a recuperative heat exchanger to be utilized for the 2 K refrigerator in 1980s and SNS (Spallation neutron source) facility improved the efficiency of 2 K refrigeration system by switching the location of the heat exchanger in 2002. Specifically, as the heat in-leak from the environment is intercepted at higher temperature region than the previous system, the entropy generation of the system has been reduced. Knudsen *et al.* suggested another option using an additional valve and a heat exchanger in 2012, and, the experiment to verify their design was implemented in 2015. [5, 6] It is the most efficient 2 K JT refrigeration cycle up to now although multiple heat exchangers with expansion valves are expected to be more efficient.

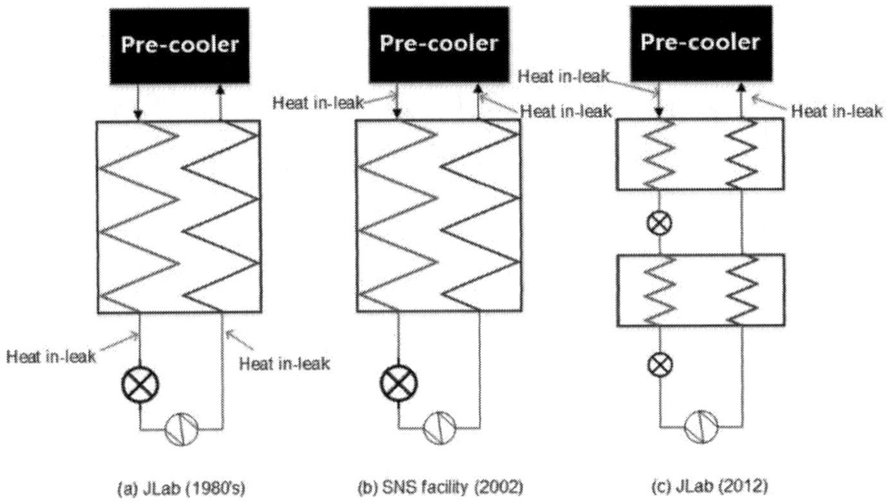


**Figure 1. Schematic diagrams of previous 2 K refrigeration systems.**

Various papers have dealt with recuperative heat exchangers with the distributed JT effect. Since the heat exchanger with the distributed JT effect actively utilizes the pressure loss effect in the high-pressure stream of the heat exchanger, a capillary tube is usually used to make the heat exchanger. Hwang *et al.* fabricated triplet heat exchanger by using capillary tubes and investigated the performance in 2009 [7]. Additionally, Hwang *et al.* referred the relation between the pressure losses in the heat exchanger and the effectiveness in 2010 [8]. Maytal *et al.* designed and verified the effect of the heat exchanger with the distributed JT effect by using mixed refrigerant in 2014 [9]. Hong *et al.* used effectiveness-NTU method to predict the effect of the heat exchanger utilizing a capillary tube as a hot stream in 2010 [10]. Damle *et al.* developed a numerical model to predict transient state of the heat exchanger with the distributed JT effect in 2015 [11]. However, the heat exchanger with the distributed JT effect for 2 K has not been made until now. Additionally, an existence of a JT valve in front of a heat exchanger with the distributed JT effect can regulate the mass flow rate. This paper proposes a 2 K cooling system which uses a JT valve and a distributed JT effect heat exchanger. The design methodology of the distributed JT effect heat exchanger is thoroughly addressed to be used for the efficient 2 K JT refrigerator.

## 2. Design methodology of recuperative heat exchanger

The distributed JT heat exchanger not only cools down the high-pressure stream passively by exchanging heat with the low-pressure stream but also allows the high-pressure stream to be cooled actively by its own JT expansion effect.

### *2.1. Efficiency improvement of 2 K cooling device*

The main purpose of 2 K cooling device is to create low temperature environment efficiently by using a high-pressure stream which is usually obtained from near 1 bar saturated liquid helium bath. Fig. 2 represents the overall refrigeration system suggested in this paper. Saturated liquid at 1.4 bar is supplied to the JT valve. After the JT expansion, two phase helium enters the recuperative heat exchanger with the distributed JT effect. At the end of the heat exchanger, the state of helium is to be 2 K of two phase state whose quality is 0.04. In the evaporator section, two phase helium is separated into liquid and vapour. The saturated vapour only returns to the cold stream of the heat exchanger after absorbing heat.

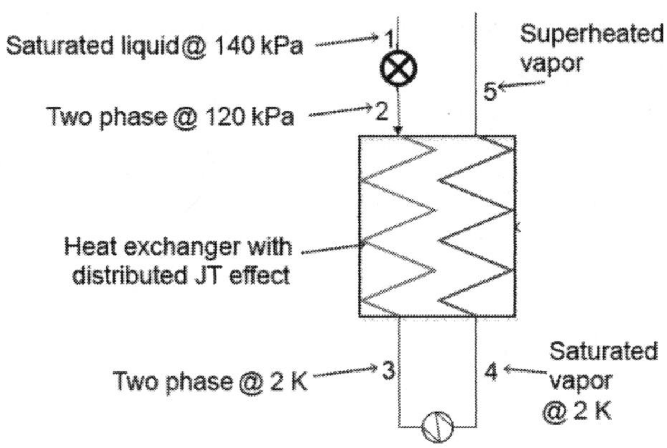


**Figure 2.** Schematic diagram of 2 K system using heat exchanger with distributed JT effect.

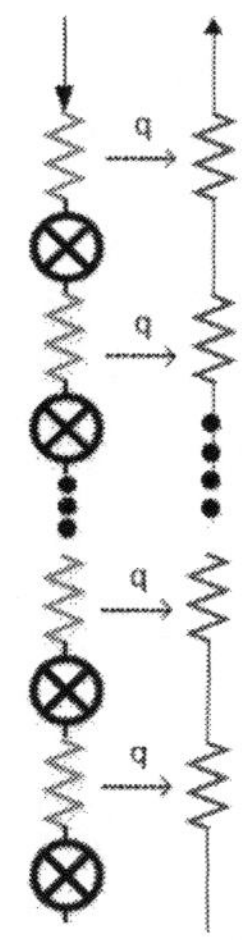


**Figure 3.** Design method of heat exchanger with distributed JT effect

Table 1 shows the state of each position. The system is ideally 2.4 % more efficient than the process suggested by Knudsen et al. The cooling efficiency is defined as (1).

$$Efficiency = \frac{refrigeration\ capacity}{required\ power} \tag{1}$$

The efficiency of the system is usually determined by only the potential refrigeration capacity which is inversely proportional to the enthalpy at the position 3. The continuous pressure drop process for this heat exchanger is simulated and carefully designed as illustrated in Fig. 3.

**Table 1.** Thermodynamic states of each positions indicated at Fig. 2.

| Position | Temperature (K) | Pressure (kPa) | Quality (-) |
|---|---|---|---|
| 1 | 4.6 | 140 | 0 (Saturated liquid) |
| 2 | 4.4 | 120 | 0.06 |
| 3 | 2 | 3.13 | 0.04 |
| 4 | 2 | 3.13 | 1 (Saturated vapor) |
| 5 | 3.8 | 3.13 | Superheated vapor |

Fig.4 shows the ideal pressure-enthalpy diagram of helium passing the high-pressure hot stream of the heat exchanger from 4.5 K to 2 K. The saw-toothed black line starts from 4.5 K and ends near 2 K by following the saturated liquid line closely. The estimated steady state temperature profiles in the heat exchanger are also plotted at Fig. 5. Fig. 5 (a) indicates the temperature profiles of the high-pressure (hot stream) and the low-pressure (cold stream) ones considering the pressure drop effect of the high-pressure stream (Fig. 2 and Fig. 3). Since the heat capacity rates of two streams are different, they are not parallel. Fig. 5 (b) is also depicted for comparison, which is the temperature profiles with no-pressure drop effect, but with two JT valves as shown in Fig. 1 (c). Fig. 5 (a) presents the smaller overall temperature difference between hot and cold streams than that of Fig. 5 (b), which may elucidate the better cooling efficiency of the distributed JT heat exchanger by producing less entropy due to heat transfer with finite temperature difference. NTU (Number of transfer unit) needed to this heat exchanger is precisely computed by applying general heat exchanger calculation algorithm for each sub-unit heat exchanger modelled as shown in Fig. 3. The total NTU is the summation of each NTU calculated for each sub-unit heat exchanger. The resultantly computed NTU value is 5.6. The design analysis demonstrates that the recuperative heat exchanger with the distributed JT effect should enable the 2 K refrigerator to maximize its cooling capacity at 2 K by creating near saturated liquid condition at the exit of the heat exchanger. The required pressure drop in the hot stream of the heat exchanger is 117 kPa from 4.5 to 2 K. It is also very crucial that the pressure drop less than 2 kPa is only permitted at the cold stream. The proper pressure drop and heat transfer coefficient correlations are, therefore, very important to design the heat exchanger with the distributed JT effect.

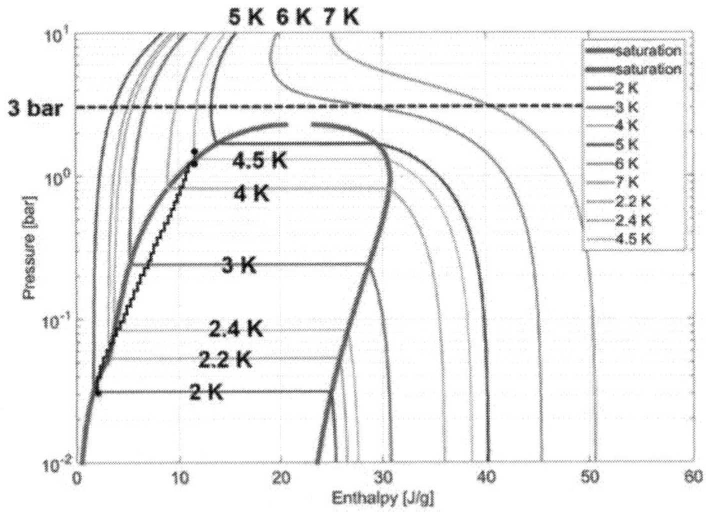


**Figure 4. Pressure-enthalpy diagram of helium passing the heat exchanger with continuous pressure loss.**

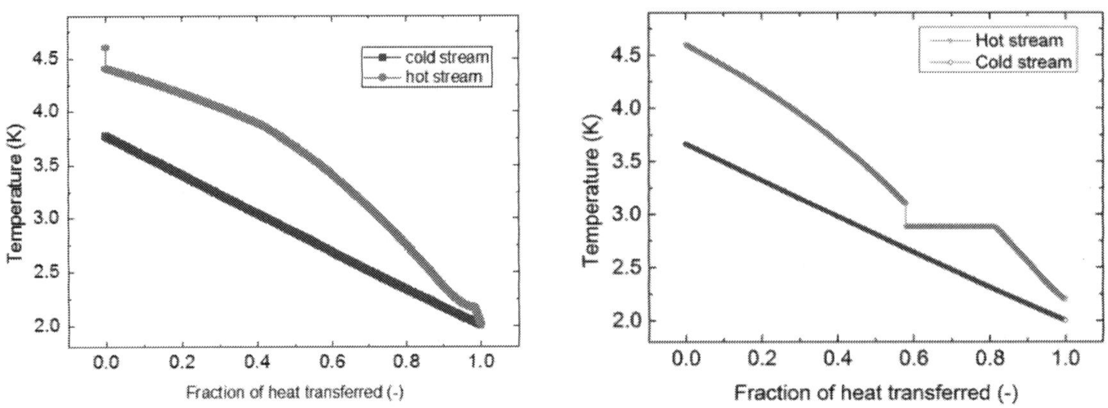


(a) Continuous pressure-drop type (Fig.2) (b) Conventional type with JT valve (Fig. 1c)

**Figure 5.** Steady state temperature profiles in the heat exchanger.

*2.2. Heat exchanger configuration*

It is clear that some extraordinary pressure drop in the high-pressure side of the heat exchanger can be deliberately induced so that the continuous J-T effect occurs during normal heat exchange process. The detailed heat exchanger analysis enables one to design a heat exchanger for producing near 2 K high-pressure stream at its cold end to create 2 K helium without further expansion. The physical configuration of the exemplary heat exchanger is similar to one found in a dilution refrigerator. The high-pressure helium flows through the spirally wound capillary tube to produce the desirable amount of pressure drop and accommodate sufficient heat transfer area. Fig. 6 (a) shows the test capillary tube with inner diameter of 0.4 mm before it is encapsulated by the larger spiral tube whose inner diameter is 2.4 mm. The larger tube holds the low-pressure helium stream for heat exchange. The fabricated heat exchanger is specifically for the mass flow rate of 0.07 g/s and also shown in Fig. 6 (b).

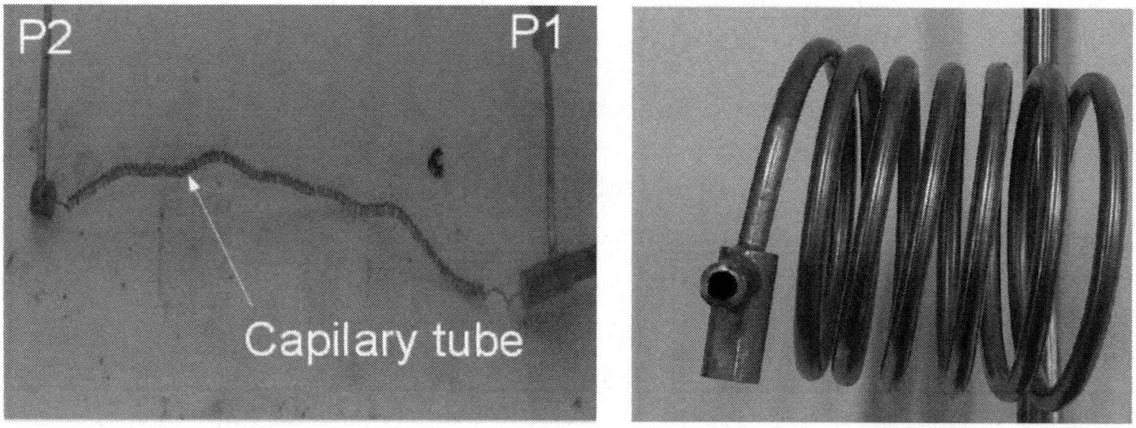


(a) Inner capillary tube of heat exchanger (b) Outer tube of heat exchanger

**Figure 6.** Fabricated spiral type recuperative heat exchanger.

*2.3. Pressure drop consideration*

The heat exchanger is specifically fabricated to produce precise continuous pressure drop effect as well as the appropriate heat transfer performance at the pre-determined mass flow rate of helium. In general, the empirical correlation to predict pressure drop of helical coil heat exchanger is suggested for capillary tube as written in (2) [12]. This correlation is utilized at various experimental conditions.

$$\frac{\Delta P}{\Delta L}=\frac{0.092G^2[1+3.5(D_e/D_h)]}{\mathrm{Re}^{0.2}D_e\rho} \tag{2}$$

Since the value of $\frac{D_h}{D_e}$ in this paper is relatively smaller than the usual cases, a preliminary experiment is carried out to verify and correct the correlation by using high pressure helium at liquid nitrogen temperature which is approximately 77 K. In the case of high Reynolds number range around 30,000 which is needed to design the heat exchanger with distributed JT effect for 2 K refrigerator, the correction factor of 1.46 is determined. When the two-phase flow occurs in the heat exchanger, the equation (3) is applied [13] to calculate the pressure drop.

$$\phi_l^2=\frac{(dp/dx)_{TP}}{(dp/dx)_l}=[1+x(\frac{\rho_l}{\rho_v}-1)][1+x(\frac{\mu_l}{\mu_v}-1)]^{-0.25} \tag{3}$$

*2.4. Heat transfer consideration*

The heat transfer requirement is less stringent than that of pressure drop. Using (5), the necessary UA value is computed at each small sub-unit heat exchanger. Convective heat transfer coefficient in the capillary tube is estimated by the following equation (6).

$$NTU=\frac{UA}{\dot{m}c_p} \tag{4}$$

$$h = 0.023\mathrm{Re}^{-0.2}\,\mathrm{Pr}^{-2/3}[1 + 3.5(D_e / D_h)] \tag{5}$$

If there is two-phase flow in the capillary tube, the homogeneous model is considered as suggested in the reference [13]. The convective heat transfer coefficient of low-pressure helium in the outer tube is calculated from the well-known Nusselt number relation [14] as shown in equation (11).

$$Nu = 0.023\mathrm{Re}^{0.8}\,\mathrm{Pr}^{0.4} \tag{6}$$

The required UA value at each sub-unit heat exchanger is calculated by the overall thermal resistances as in the following equation (12).

$$\frac{1}{UA} = \left(\frac{1}{hA}\right)_{HP} + \left(\frac{\ln\left(\frac{D_o}{D_i}\right)}{2\pi kL}\right) + \left(\frac{1}{hA}\right)_{LP} \tag{7}$$

## 3. Summary

The development methodology of distributed JT effect heat exchanger is described for 2 K cooling device. Different from conventional recuperative heat exchangers, the distributed JT effect heat exchanger actively utilizes the pressure loss of high-pressure stream as well as the heat transfer characteristic. For the lab scale application dealing with small mass flow rate such as 0.07 g/s, which can be used in low-temperature property measurement device, the heat exchanger can be accordingly designed and fabricated with two stainless steel 304 tubes and its dimension is precisely determined after performing pressure drop test. One for the high-pressure side is 1.7 m long spirally wound capillary tube of which the inner diameter is 0.4 mm and the outer diameter is 0.64 mm. The other for the low-pressure side is 1.1 m long and 0.8 mm thick tube with 4 mm outside diameter. The designed heat exchanger is expected to enable the 2 K refrigerator to maximize its potential cooling capacity at 2 K.

**Acknowledgments**
This work was supported by KBSI Grant D37614.

**References**

[1] Rode C H 1995 *Proc. 1995 Particle Accelerator Conf.* p 1994
[2] Dixon K, Ganni V, Knudsen P and Casagrande F 2015 *IOP Conf. Series: Materials Science and Engineering* **101** p 012071
[3] Miller J R 2003 *IEEE trans. on applied superconductivity* **13** p 1385
[4] Evans L 2007 *New Journal of Physics* **9** p 335
[5] Knudsen P and Ganni V 2012 *Adv. Cryo. Eng* **57** p 800
[6] Knudsen P and Ganni V 2015 *IOP Conf. Series: Materials Science and Engineering* **101** p 012105
[7] Hwang G and Jeong S 2009 *Progress in Superconductivity and Cryogenics* **11** p 60
[8] Hwang G and Jeong 2010 *Cryogenics* **50** p 13
[9] Maytal B 2014 *Cryogenics* **61** p 92
[10] Hong Y, Park S and Choi Y 2009 *Cryocoolers* **15** p 379
[11] Damle R and Atrey M 2015 *Cryogenics* **65** p 49
[12] Timmerhaus K D and Flynn T M 1989 *Cryogenic Process Engineering* (New York and London: Plenum press) p 194
[13] Van Sciver S W 2012 *Helium Cryogenics* (New York : Springer)
[14] DeWitt D P and Incropera F P 1996 *Fundamentals of heat and mass transfer*

IOP Conf. Series: Journal of Physics: Conf. Series **969** (2018) 012085 doi:10.1088/1742-6596/969/1/012085

# Photo-electron emission directly in superfluid helium

**V V Zavyalov**[1,2], **E B Pyurbeeva**[2,1] **and S I Khaldeev**[2,1]

[1] Kapitza Institute for Physical Problems, RAS, Moscow, Russian Federation
[2] Moscow Institute of Physics and Technology (State University), Department of General and Applied Physics, Moscow, Russian Federation

E-mail: zav@kapitza.ras.ru

**Abstract.** Despite the fact that electron transport in condensed helium has been studied for over half a century [1], observations of new intriguing effects still appear [2]. Alas, the traditional methods of injecting electrons into condensed helium (radioactive-sources, electrical discharge or field emission) lead to generation of helium ions, recombination of which is accompanied by emergence of a large number of excitations. As a result, interpretation of such experiments is not simple and sometimes may be questionable. In this respect, photoelectron emitters, which operate with energies substantially smaller than the ionization energy of helium, are preferable. However, immersion of the photocathode into condensed helium suppresses electron emission. Nevertheless, we managed to achieve electron currents (>20 fA) with the In photocathode immersed directly in liquid superfluid helium. The UV light ($\lambda$=254 nm) was guided to the photocathode through a two-meter long Al-covered quartz optical fiber.

## 1. Introduction

For many low-temperature experiments a source of electrons in liquid helium is required. Such a source should emit a controlled amount of electrons, whose energy should be sufficiently low not to create high energy excitations. In addition, the process of electron emission has to draw the minimum possible power into the low-temperature system.

There are several traditional methods to produce electrons with, namely a hot tungsten filament, a glow discharge, a radioactive source, or a field emission tip. Being placed in superfluid helium, a hot filament produces a significant flow of excitations. As to the other methods, they create high-energy electrons, causing ionization of helium. Resonant photons, arising in the ionization-recombination process, easily pass through retarding grids and carry sufficient energy to contribute to the creation of a large number of excimers and other metastable helium states [3, 4] in the entire volume of the experimental cell. Some of them are long lived excitations in liquid helium both above and below the $\lambda$-point [5, 6]. As a result, the interpretation of experiments is not simple and sometimes may be questionable. In this respect, the method of photoelectron emission, which operates with energies substantially smaller than the ionization energy of helium, seems preferable.

Unfortunately, when immersing the photocathode into liquid helium, the value of photo-injected electron current ($I_{\mathrm{liq}}$), in relation to the one measured under the same conditions in vacuum ($I_{\mathrm{vac}}$), becomes substantially smaller. This behavior can be characterized by the attenuation coefficient $J = I_{\mathrm{liq}}/I_{\mathrm{vac}}$.

In the simple and ingenious experiment[7], intended to measure the energy required to inject an electron into liquid helium, the cesium-antimony photocathode was immersed in liquid helium, cooled to a temperature of 1.1 K. In this case, the attenuation coefficient $J$ turned out to be equal to $10^{-6}$.

In the experimental work [8] the steady-state photoelectron current arising from an Au cathode, which was immersed in liquid He at 4.2 K and irradiated by a deuterium lamp, was studied. The attenuation $J$ was measured as a function of applied field strength, $E$. At $E$=$10^3$ V/cm the value of $J$ comprised $4 \cdot 10^{-6}$, and at $E$=$10^5$ V/cm the value of $J$ increased to $10^{-3}$. In this experiment, the UV radiation power falling on the photocathode was not explicitly specified, however, given the magnitude of the photocurrent in vacuum ($I_{\mathrm{vac}} = 3 \div 5$ nA) is already quite informative.

In [9] the cesium photocathode at $T$=1.33 K was used. Photoelectron currents through gaseous $^3$He or $^4$He and through thin liquid $^3$He films, were investigated. It was found that the growth of a $^3$He film on Cs can be studied up to a film thickness of 1.5 nm. Beyond that, the tunnel current is so low (<100 fA) that it is comparable with the noise in the electrometer used. As for $^4$He, the photocurrent vs film thickness dependency was not studied, because $^4$He does not wet Cs at $T$<2 K, even if the Cs is formed by quench condensation, which creates a relatively rough surface.

In the above-mentioned works the photocurrents in the bulk He were obtained at the very special conditions, namely, by using the UV radiation near the vacuum-ultraviolet edge [8] that makes it problematic to use optical fibers for delivery radiation to the photocathode, or using alkali metals that degrade easily and need to be *in situ* prepared [7, 9], which imposes strong limitations in their practical implementation.

The few practical realizations of cryogenic photoemissive electron sources has been described in [10, 11] but they were not used to work in the bulk He and so far have been mainly used above the helium surface to create a surface layer of electrons.

## 2. Experimental setup

A high-pressure mercury lamp was used as a source of 253 nm UV radiation. At a distance of ~ 60 mm from the lamp the Avantes$^{\mathrm{TM}}$COL collimating lens of 6 mm diameter and 8.7 mm confocal length was placed with an electromechanical shutter just before the lens. The UV-radiation was transmitted by the lens onto the end of Al-shielded 2 m long quartz fiber, whose diameter of core was 600 mkm and the numerical aperture was 0.22. According to the manufacturer information, absorption of the optical fiber at the working wavelength did not exceed 300 dB/km, which was quite acceptable. The optical fiber was passed into the cryostat through a silicone seal.

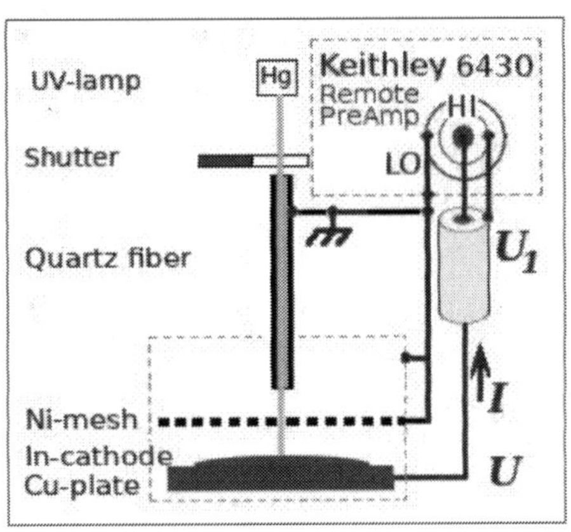


**Figure 1.** The measuring scheme. Radiation from the UV mercury lamp passes through the electromechanical TTL-driven shutter, enters the experimental cell through the quartz optical fiber, and generates photoelectrons at the cathode. The measuring scheme of a diode type with a grounded grid (mesh) was used. Sourcing $U$ potential and measuring $I$ current was provided by the Keithley$^{\mathrm{TM}}$6430 Sub-Femtoamp SourceMeter. The triaxial cable with a cable guard, supported at almost the same potential $U_1$ as the potential $U$, was used to minimize the problem of capacitance and leakage currents in measuring wiring. Shutter control was also performed by the SourceMeter with its digital I/O ports.

The lamp's radiation power passing into the experimental cell through the fiber was not exceed 20 uW. To measure this value, we used the platinum resistance thermometer which was installed in place of the photocathode and calibrated for power by measuring its current-voltage characteristics.

To study photoemission a diode scheme was used (see Figure 1). The UV light passed through the optical fiber, then through a nickel 130-mkm mesh (transparency of 70 percent) that served as a collector, and reached the photocathode. The distance between the mesh and the photocathode was 0.3 mm. The photocathode used was an indium drop, soldered on a copper substrate and mechanically cleaned before each low-temperature experiment session. Using the optical constants of In obtained in [12] it is possible to estimate that its reflectivity for 253 nm UV-light does not exceed 40%.

For measuring ultra-low photoemission currents the Keithley$^{TM}$Model 6430 Sub-Femtoamp SourceMeter with the remote preamplifier was used. It combines the voltage sourcing ($-200$ to $+200$ V) and high sensitivity measurement functions (0.4 fA noise).

To minimize the problem of capacitance and current leakage in measurement circuit, a shielded triaxial cable was used. The center conductor (HI) is surrounded by the inner shield, which acts as the cable guard. The source-meter keeps the shield at potential $U_1$ which practically the same as the potential $U$ at HI terminal ($U_1 - U < 1$ mV).

## 3. Experimental results

At first, the preliminary experiments measuring photoelectron current at $T$=300 K in vacuum were carried out. The necessary evacuation was produced with a standard backing pump. It turned out that, under continuous illumination of the cathode, the photocurrent decreased with time exponentially and went to saturation after 10–30 min (in relation with the pumping extent), thus relaxed from 0.6 to 0.16 nA, respectively. This effect could be explained by redistribution of charges on dielectric parts of the experimental cell. In order to bring the system to its initial state, it was sufficient to switch off the UV-lamp and to fill the cell with gaseous helium.

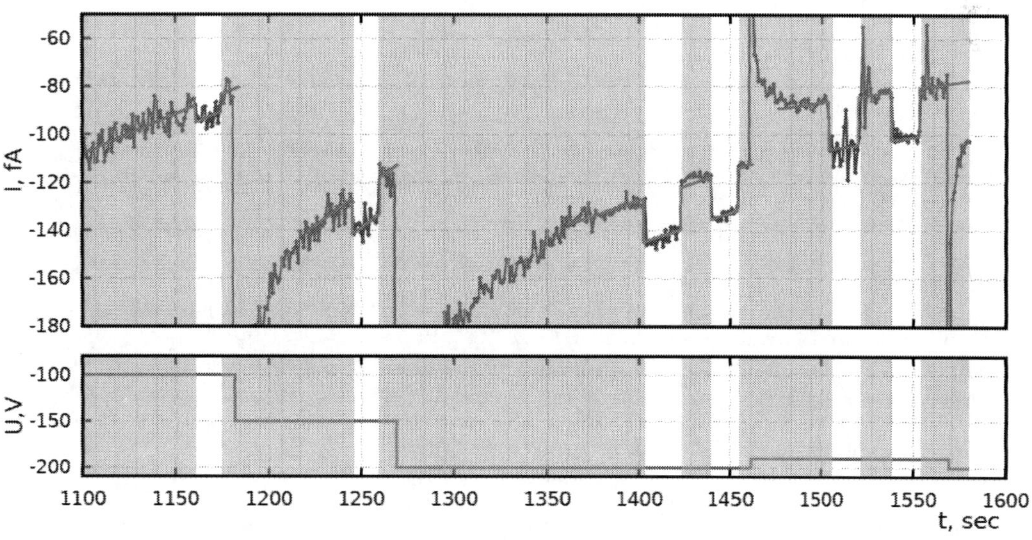


**Figure 2.** The time-chart fragment of the measured current, $I$ (blue) and its linear-piecewise approximation (red) clearly show the photocurrent steps. Dark and light areas correspond to the modulation of the UV beam with the electromechanical shutter. The cathode potential, $U$ is shown in the lower part of the diagram (brown).

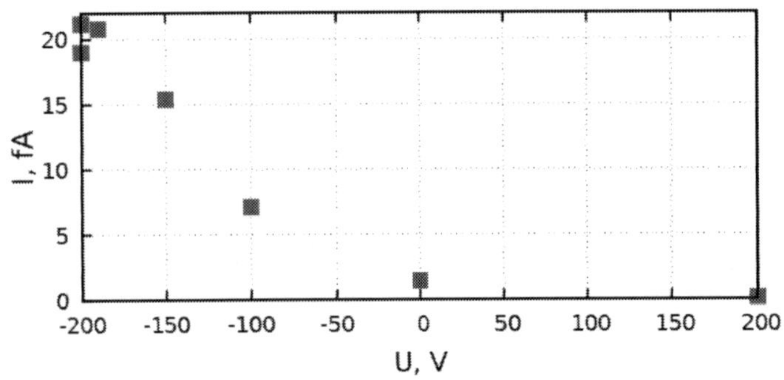


**Figure 3.** Photoelectron current vs. cathode potential. (The information obtained after the analysis of the measured charts.) The cathode was immersed in liquid helium at $T$=1.5 K .

Measurements in liquid helium at temperatures above the $\lambda$-point were not possible due to acoustic noise of boiling process, which led to electrical interference in the measuring circuit. That is why all the low temperature measurements were carried out at $T$=1.5 K.

Figure 2 shows a portion of a chart recording during the time interval, in which the potential of the cathode was changed four times. The graph demonstrates both the current relaxation processes caused by changes in the cathode potential, and the effect of UV-radiation modulation. This graph also shows the result of linear-piecewise approximation of the experimental data, which allowed us to reveal the contribution of photoelectrons with a sensitivity of $\approx$ 1 fA, that was close to a value corresponding to the highest sensitivity of the Keithley$^{\mathrm{TM}}$6430 electrometer.

Figure 3 presents the dependence of photoelectron current vs. cathode potential. The maximum value of 20 fA was observed at the cathode potential of $-200$ V, which corresponds to field strength of $E \approx 7000$ V/cm. Based on this data and considering that the measured value of the corresponding photoelectric current in vacuum was 0.16 nA, we can estimate the attenuation coefficient value, $J = 1.2 \cdot 10^{-4}$, which constitutes the coordinates of a point that surprising exactly falls on the LHe curve shown in Fig.2 in [8]. Note that the power of illumination of the photocathode is substantially less than in [8]. This can be judged by comparing the magnitudes of the corresponding photocurrents in vacuum. In [8] ($I_{\mathrm{vac}} = 3 \div 5$ nA) and in our case ($I_{\mathrm{vac}} = 0.6 \div 0.16$ nA).

We have previously mentioned the phenomenon of photoelectron current relaxation, which apparently can arise due to photoelectrons precipitating on dielectric parts of the experimental cell. It is hoped, that by optimizing the cell design, this parasitic effect can be reduced and, thus, we can count on a tenfold increase of the photoelectron currents with the same optical setup.

## 4. References

[1] Keshishev K O, Kovdrya Y Z, and Mezhdov-Deglin L P, and Shalnikov A I 1969 Mobility of charges in liquid helium up to solidification *Soviet J.Exp. Theor. Phys.* **29** 53.

[2] Wei W, Xie Z L, Yang Y, Seidel G M, and Maris H J 2014 Study of exotic ions in superfluid helium-4 using a carbon nanotube source *J. Phys.: Conf. Series* **568** 012011.

[3] Eltsov V B, Dzhosyuk S N, Parshin A Ya, and Todoshchenko I A 1998 Spectroscopic studies of the triplet-state $^4He_2^*$ molecules in superfluid helium up to the solidification pressure *J. Low Temp. Phys.* **110** 219-224.

[4] Tarchouna H, Bonifaci N, Aitken F, Shakhatov V, Atrazhev V, Eloranta J, von Haeften K, Sadeghi N, and Jomni F 2014 Molecular and atomic spectra emitted by normal liquid and supercritical 4He excited by corona discharge *IEEE Int. Conf.on Liquid Dielectrics (Bled, Slovenia, June 30 - July 3)*.

[5] Fischbach M R, Roberts H A, and Hereford F L 1969 Persistent metastable states and the inhibited scintillation of He II *Phys. Rev. Lett.* **23** 462-464.

[6] Keto J W, Soley F J, Stockton M, and Fitzsimmons W A 1974 Dynamic properties of neutral excitations produced in electron-bombarded superfluid helium. II. Afterglow fluorescence of excited helium molecules *Phis. Rev.* A **10** 887-896.

[7] Woolf M A and Rayfield G W 1965 Energy of negative ions in liquid helium by photoelectric injection *Phys. Rev. Lett.* **15** 235-237.
[8] Kojima H, Sakai Y, and Tagashira H 1987 Photo-injected electron current in cryogenic liquids: a comparison of liquid Ar, Ne, He, $N_2$, and $H_2$ near their boiling points *Chem. Phys. Lett.* **140** 631-635.
[9] Zech M, Bromberger H, Klier J, Leiderer P, and Wyatt A F G 2008 Electron tunneling through barriers of adjustable width: Role of the image potential and the wetting behavior of Cs by He *Phys. Rev.* B **78** 115113.
[10] Wilen L A and Giannetta R W 1985 Cryogenic photoemissive electron source *Rev. Sci. Instr.* **56** 2175-76.
[11] Shankar S, Sabouret G, and Lyon S 2010 A low power photoemission source for electrons on liquid helium *J. Low Temp. Phys.* **161** 410-416.
[12] McMahon J M, Schatz G C, and Gray S K 2013 Plasmonics in the ultraviolet with the poor metals Al, Ga, In, Sn, Tl, Pb, and Bi *Phys. Chem. Chem. Phys.* **15** 5415-23.

IOP Conf. Series: Journal of Physics: Conf. Series **969** (2018) 012086 doi:10.1088/1742-6596/969/1/012086

# Examination of cryogenic filters for multistage RF filtering in ultralow temperature experiments

**V V Zavyalov[1], S A Chernyaev[2], K V Shein[2], A G Shukaleva[2] and K Yu Arutyunov[1,2]**

[1] Kapitza Institute for Physical Problems, RAS, Moscow, Russian Federation
[2] National Research University Higher School of Economics, Moscow Institute of Electronics and Mathematics, 109028, Moscow, Russia

E-mail: zav@kapitza.ras.ru

**Abstract.** Cryo-filters are essential while studying electronic properties of nanoscale structures at very low temperatures. In this report we present the simple measuring methodology and experimental impedance characteristics of customized lumped filters cooled down to 4.2K in the 10 Hz-500 MHz frequency range. In particular, we tested the home-made permalloy-core RL filters, the Murata™Chip Ferrite Bead filter, and the Toshiba™Amobeads™cores. We use the high-frequency generalization of four-terminal sensing method to account for the wiring retardation effects, which are important when working with ultralow temperature systems.

## 1. Introduction

During electron transport measurements at very low temperatures ($T < 100$ mK) rather small external disturbances might lead to a noticeable overheating. The inevitable external electro-magnetic interference (EMI) noise, picked-up and transmitted through electric wires, results in a mismatch between the electron $T_e$ and the phonon $T_{ph}$ temperatures $P \sim W(T_e^5 - T_{ph}^5)$, where $P$ is the power dissipated at the sample with volume $W$[1]. Hence, for sufficiently small nanoelectronic systems the effect might be clearly pronounced. Multiple methods have been suggested to reduce the undesired electron heating [2]. Among them, the utilization of various RF filters to cut the impact of noisy EMI environment is the most common ones. So far this problem remains relevant [3, 4, 5, 6, 7].

To measure an RF filter's characteristics typically rather expensive broadband network analyzers should be used. The test sample needs to be directly connected to the analyzer via special certified cables, that is only possible for measurements conducted at room temperature.

To check the filters installed in the cryostat it is necessary to use additional cryo-wiring which leads to the mismatch of impedance between transmission line and load, that may distort the results, especially at high frequencies. Here we describe a simple high-frequency generalization of four-terminal sensing method to account for the wiring retardation effects. We also demonstrate the effect of cooling on the impedance-frequency characteristics of several home-made and customized filters.

## 2. The measuring scheme

The schematic diagram of the measurements is presented in Fig.1. For testing filters we use the top loading insert that operates in a standard liquid helium transport dewar with a 12 mm bore. The sample ($Z_x$) is mounted in a cylindrical copper shield at the bottom of the insert. Four thin coaxial cables ($Z$) of the same length of $1200 \pm 2$ mm pass from the oscilloscope's and generator's terminals to the sample. The two of them go to the sample directly and represent a current sensing circuit, the two others are connected in series with the resistors $R_1$ and represent a circuit for voltage measurements.

To measure the frequency response of samples we use the 4-channel oscilloscope Tektronix$^{TM}$TDS3054 with sampling rate of 5GS/s, 500 MHz bandwidth, and simultaneous recording of $10^4$ readings per channel with 9-bit resolution. This gives us the ability not only to find frequency, amplitudes and phases of the signals $U_1, U_2, U_3$ (Fig. 1), but also to control the presence of frequency harmonics in the case that the sample has nonlinear properties. For the oscilloscope's synchronization we use the signal $U_0$.

As RF signal source we use Agilent Technologies$^{TM}$PSG CW signal generator E8257D (for high frequencies) and Stanford Research Systems$^{TM}$15 MHz synthesized function generator DS340 (for lower frequencies). In our case, characteristics of the oscilloscope are significant. As for the RF source it does not impose any special requirements, and it can be replaced by any appropriate device.

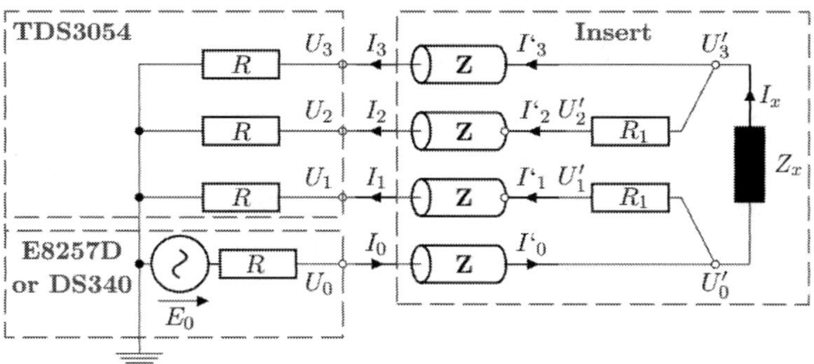


**Figure 1.** Schematic of the experiment. $R$=50 Ω corresponds to oscilloscope and oscillator impedances. $R_1$=250 Ω corresponds to low inductance metal resistors. **Z** stands for 1.2 meter long coaxial cables (PK50-0,6-23).

In this work we deal with lumped filter samples. To test them at liquid helium temperatures we need to use cables, and account for their retardation effects. We can circumvent this limitation by introducing the complex current/voltage transformation matrix **Z** :

$$\begin{bmatrix} U'_k \\ I'_k \end{bmatrix} = \begin{bmatrix} \cosh(\gamma l) & |Z|\sinh(\gamma l) \\ \dfrac{\sinh(\gamma l)}{|Z|} & \cosh(\gamma l) \end{bmatrix} \begin{bmatrix} U_k \\ I_k \end{bmatrix}, \qquad k = 0..3.$$

where $\gamma$ denotes the complex propagation constant, $l$ is the length, and $|Z|$ is the characteristic impedance of the cable. Using condition $I_k = U_k/R$, $k$=1..3 we get:

$$U'_k = \left(\cosh(\gamma l) + \frac{Z}{R}\sinh(\gamma l)\right) U_k \equiv a \cdot U_k, \qquad I'_k = \left(\frac{1}{Z}\sinh(\gamma l) + \frac{1}{R}\cosh(\gamma l)\right) U_k \equiv b \cdot U_k$$

Given that $U'_2 - U'_3 = I'_2 R_1$, and $U'_1 - U'_0 = I'_1 R_1$ we end up with a simple expression for the frequency-dependent impedance $Z_x$ as function of complex parameters $U_1, U_2, U_3$:

$$Z_x(f) = R_1 \frac{(U_1 - U_2)U_3}{(U_3 - U_2)(U_3 + U_2)} \tag{1}$$

IOP Conf. Series: Journal of Physics: Conf. Series **969** (2018) 012086 doi:10.1088/1742-6596/969/1/012086

Recall that the amplitudes and phases of $U_1, U_2, U_3$ as well as frequency $f$ are determined from the oscillograms.

The assumption that all four cables are identical allows to exclude their properties from the expression. Moreover, the result does not depend on amplitude of the testing signal, which allows the use of simple non-stabilized generators.

## 3. Frequency characteristics of various filters

Cryostat for ultralow temperature experiments should be equipped with a multistage filtration system which may consist of EMI filters at room, as well as, liquid He temperatures. One may also add some cryogenic low-pass filters thermally anchored to 1-K pot, to mixing chamber and experimental cell. Our task was to measure characteristics of various filters to evaluate the possibility of their usage in a multistage filtration system.

### *3.1. Permalloy-core inductor (RL filter).*

For fabrication of the magnetic core we used permalloy tape with thickness of 0.1 mm and width of about 11 mm, which was wound on a cylindrical template, forming the magnetic core of a toroidal shape with inner and outer diameters of 2 mm and 4.5 mm, respectively. The core was annealed in vacuum. On that core 20 turns of a NbTi/CuNi superconducting wire of 0.1 mm diameter were wound forming an inductance with room temperature DC resistance $R_{DC} = 23\ \Omega$.

One can notice that despite the expected pure reactive behavior of our filter, it is resistive in the range of $10 - 50\,\text{MHz}$, which can be attributed to eddy-currents in permalloy[8]. It is the positive observation for the filtering problem, though it is evident, that our filter design is far from optimal.

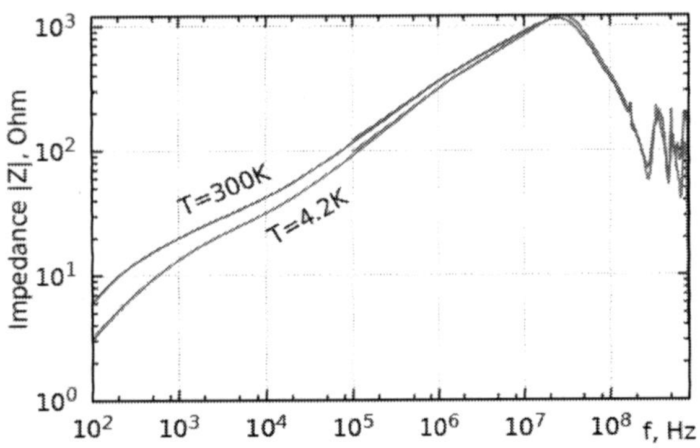


**Figure 2.** Measured frequency characteristics of the custom-made inductor with permalloy core: the amplitude values $|Z|$ of the impedance (left); real $Re(Z)$ and imaginary $Im(Z)$ parts of impedance (below). In all three graphs the upper (violet) lines represent the data obtained at room temperature and lower (red) lines represent the data obtained at 4.2K. Dashed lines indicate negative branches of the curves.

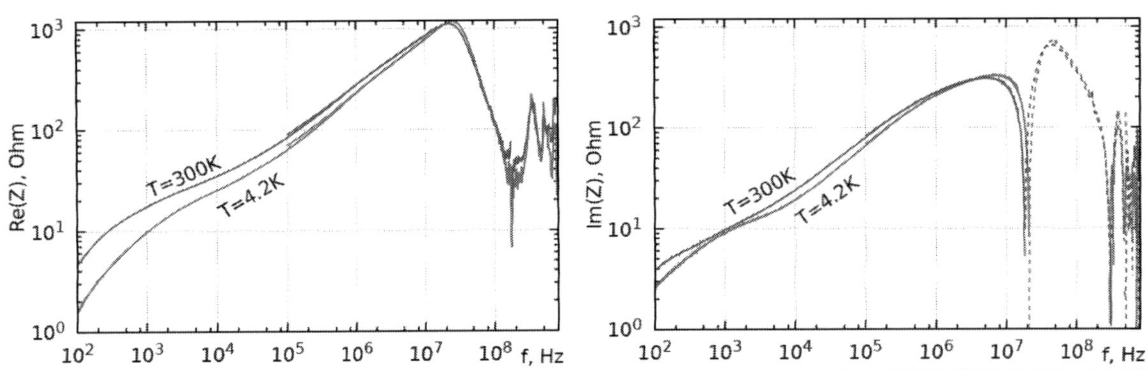

### *3.2. Murata$^{TM}$ Chip Ferrite Bead EMI suppression filter.*

EMI suppression filters from Murata$^{TM}$are widely used in modern electronics as a universal noise suppression component for frequencies ranging from a few MHz to a few GHz. They are easily available, cheap and well documented[9]. In the rated operational temperature range of -55C to +125C Chip Ferrite Bead acts like a micro inductance in the low frequency range. At high frequencies, however, the resistive component of the inductor constitutes the main part of the impedance.

For testing we choose the product BLM18HG102SN1 which is a EMI suppression filter of the Ferrite Beads type. It should be noted that experimentally measured characteristics of this filter (at room temperature) remarkably coincide with those obtained by modeling of its equivalent circuit [10].

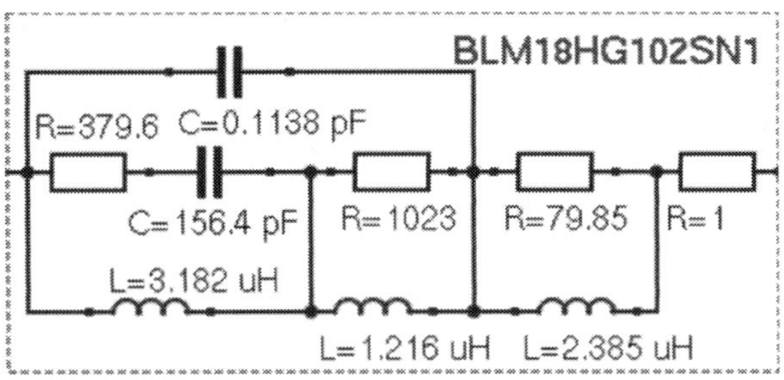


**Figure 3.** Equivalent circuit for BLM18HG102SN1 filter based on the parameters listed in [10] The manufacturer indicates the following information for the product:

- Shape: 0603 SMD.
- Operating temperature: (−55 to 125°C).
- DC resistance: 1.6 Ω.
- Impedances at 100 MHz and 1 GHz: 1 kΩ.

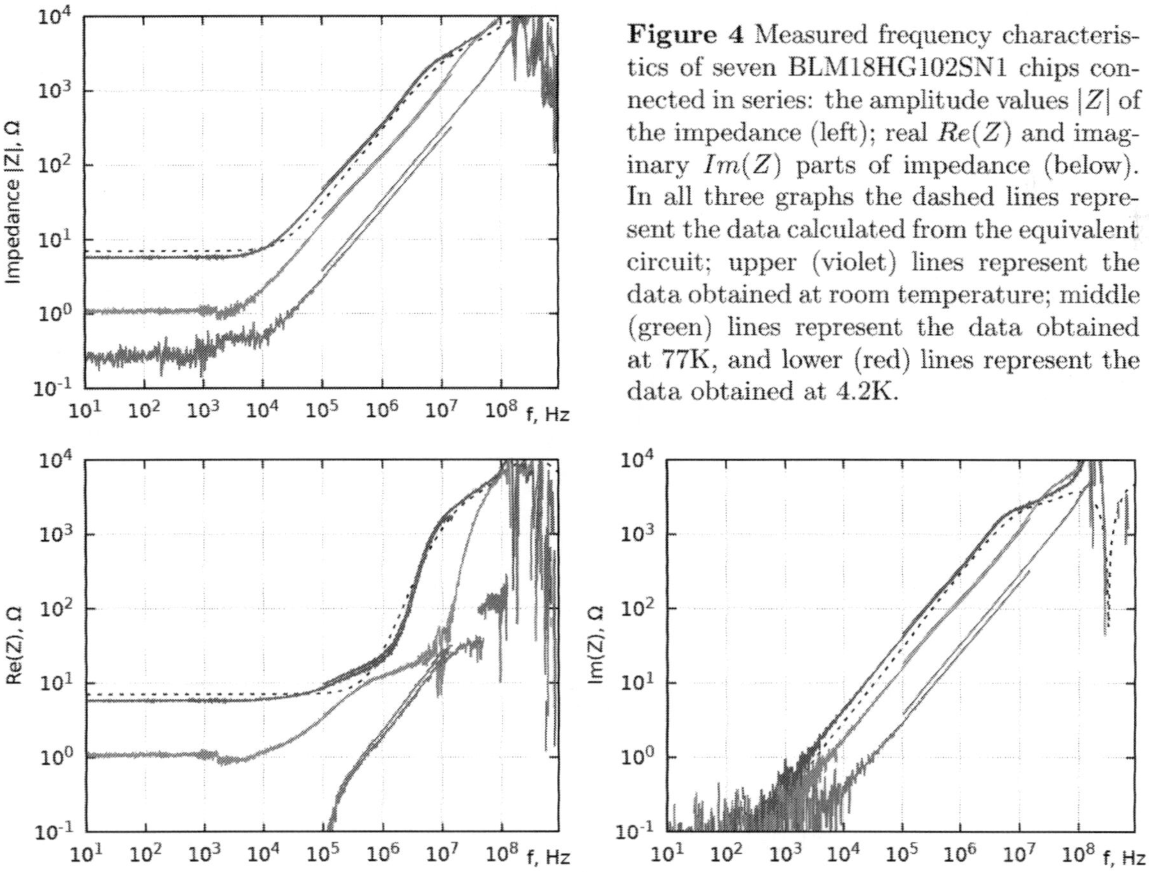


**Figure 4** Measured frequency characteristics of seven BLM18HG102SN1 chips connected in series: the amplitude values $|Z|$ of the impedance (left); real $Re(Z)$ and imaginary $Im(Z)$ parts of impedance (below). In all three graphs the dashed lines represent the data calculated from the equivalent circuit; upper (violet) lines represent the data obtained at room temperature; middle (green) lines represent the data obtained at 77K, and lower (red) lines represent the data obtained at 4.2K.

IOP Conf. Series: Journal of Physics: Conf. Series **969** (2018) 012086 doi:10.1088/1742-6596/969/1/012086

### *3.3. Toshiba™AMOBEADS™ferrite core AB3X2X6W.*

The amorphous noise suppression devices (AMOBEADS™) suppress rapid changes in current which could otherwise lead to electrical noise in various circuits. Because of their square-shaped magnetic hysteresis behavior, amorphous noise suppression devices provide very large inductance when the current crosses zero (i.e. changes sign). This large inductance effectively blocks any further current changes [11].

For testing we choose the ferrite core AB3X2X6W which is an amorphous noise suppression device. We were unable to find the manufacturer information about the minimal working temperature for this product but suspected that it can be low. The core is a hollow cylinder with inner and outer diameters of 1.5 mm and 4 mm, respectively; its length is 7.5 mm. For fabrication of the RL filter we wound 50 turns of a NbTi/CuNi superconducting wire of 0.1 mm diameter. The resistance of this coil comprises 36 Ω.

In the low frequency range, the specimen shows a nonlinear behavior, which is manifested in the form of a hump on the curves near 200 Hz. With the increase of the signal amplitude, the position of the hump moves towards higher frequencies, and its height increases. The graphs shown in Fig.5 was obtained for $U_0 = 0.2$ V.

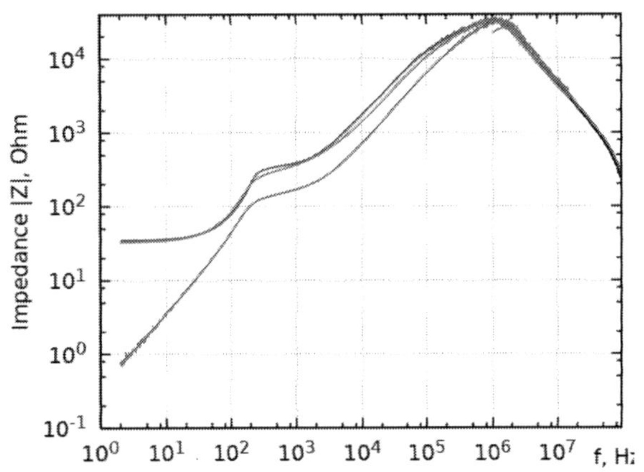


**Figure 5.** Measured frequency characteristics of inductor with a Toshiba™ Amobeads™ ferrite core AB3X2X6W: the amplitude values $|Z|$ of the impedance (left); real $Re(Z)$ and imaginary $Im(Z)$ parts of impedance (below). In all three graphs the upper (violet) lines represent the data obtained at room temperature; middle (green) lines represent the data obtained at 77K, and lower (red) lines represent the data obtained at 4.2K. Dashed lines indicate negative branches of the curves.

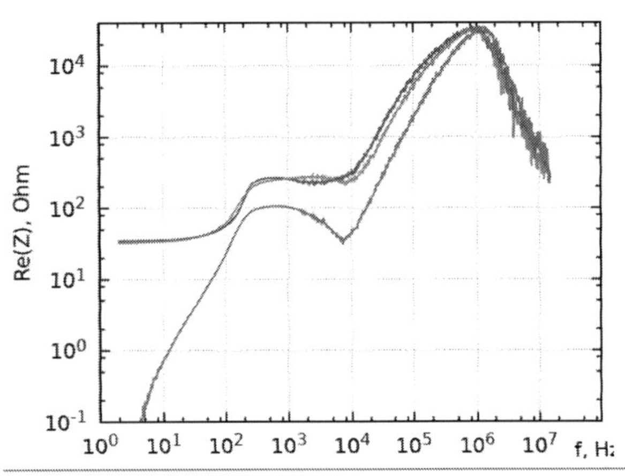


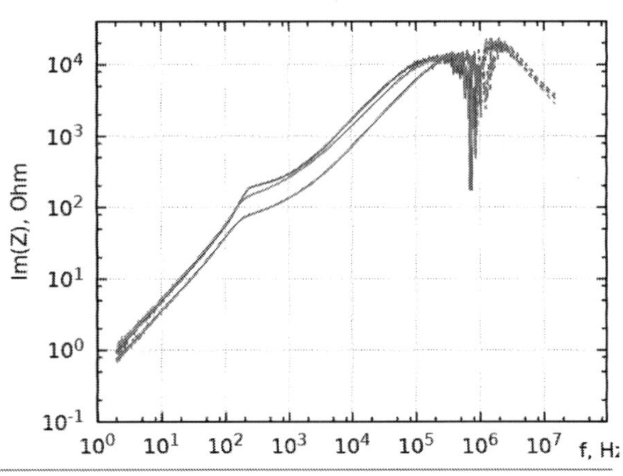


## 4. Conclusions

We have developed a simple method of measuring the frequency dependence of electrical impedances in a wide (10 Hz-500 MHz) frequency range without usage of expensive network analyzers. The method enables measurements to be made on samples cooled down to low temperatures.

As an example of using the method, Murata™Chip Ferrite Beads EMI filters and Toshiba™Amobeads™cores, whose low temperature characteristics were not rated previously, were measured. It was found, that in series connected Murata™BLM18HG102SN1 chips may serve as a high-efficiency EMI filter at room, as well as, liquid nitrogen temperatures. As for Toshiba™Amobeads™ferrite cores, they are useful to block rapid-transition currents (spikes) at room, as well as, at liquid He temperatures. We also tested a permalloy-based RL filter, and found, that it is promising, but needs to be optimized.

**Acknowledgments**
The experimental studies and theory assessment of the results were supported by Russian Science Foundation (project No.16-12-10521). Fabrication of the RL filters was made by S. A. Chernyaev, K. V. Shein and A. G. Shukaleva, who acknowledge the support of Higher School of Economics project No.16-05-0029 "Physics of low-dimensional quantum systems".

## 5. References

[1] Wellstood F C, Urbina C, and Clarke J 1994 Hot-electron effects in metals. *Phys. Rev.* B, **49** 5942-55.
[2] Bladh K, Gunnarsson D, Hurfeld E, Devi S, Kristoffersson C, Smlander B, Pehrson S, Claeson T, Delsing P, and Taslakov M 2003 Comparison of cryogenic filters for use in single electronics experiments. *Rev. Sci. Instr.* **74** 1323-27.
[3] Lee S H and Lee S-G 2016 Study on the fabrication of low-pass metal powder filters for use at cryogenic temperatures. *J. Korean Phys. Soc.* **69** 272-276.
[4] Evetts N, Dosanjh P, Zvyagintsev V, and Hardy W N 2015 Note: A cryogenic, ultra-high-vacuum, microwave filter which passes a narrow beam. *Rev. Sci. Instr.* **86** 126101.
[5] Kushino A and Kasai S 2015 Superconducting semi-rigid cable as low-pass filter for cryogenic readout. *J. Supercond. Novel Magn.* **28** 715-718.
[6] Shnyrkov V I, Korolev A M, Turutanov O G, Shulga V M, Lyakhno V Y, and Serebrovsky V V 2015 Isolation of a Josephson qubit from the electromagnetic environment. *Low Temp. Phys.* **41** 867-873.
[7] Tancredi G, Schmidlin S, and Meeson P J 2014 Note: Cryogenic coaxial microwave filters. *Rev. Sci. Instr.* **85** 026104.
[8] Kasagi T, Tsutaoka T, Hatakeyama K 1998 High Frequency Permeability of Permalloy and its Composite Materials. *J. Magn. Soc. Japan* **22** No. S1_ISFA_97 S1_295-297.
[9] Murata™product catalog 2014 "SMD/BLOCK Type EMI Suppression Filters EMIFIL®". CAT No. c31e-26
[10] Library for Cadence®PSpice®http://www.murata.com/tool/library/pspice.
[11] Toshiba: Noise Suppression - AMOBEADS®
https://toshiba.semicon-storage.com/us/product/amorphous/amobeads.html

# Pre-cooling of ton-scale particle detectors in low radioactivity environments

**L Cappelli[1,2], C E Pagliarone[1,3,4], C Bucci[1], D D'Aguanno[1,3,5], G Erme[1,3,4], P Gorla[1], S Kartal[6] and F Marignetti[1,3,5]**

[1] Gran Sasso National Laboratory, Via G. Acitelli, 22, 67100 Assergi (AQ), Italy
[2] Department of Physics, University of California, 366 LeConte Hall MC 7300, Berkeley, CA, 94720-7300, USA
[3] University of Cassino and Southern Lazio (UNICLAM), Via G. Di Biasio, 03043 Cassino (FR), Italy
[4] Department of Civil and Mechanical Engineering (DICeM), UNICLAM[2]
[5] Department of Electrical and Information Engineering (DIEI), UNICLAM[2]
[6] Department of Physics, High Energy and Plasma Division, Istanbul University, 34134 Vezneciler-İstanbul, Turkey

Email: luigi.cappelli@lngs.infn.it

**Abstract**. Low radioactivity sites are mandatory to perform searches for rare processes that cannot be studied with particle accelerators and requires low environmental backgrounds. Neutrino-less double β decay or Dark Matter searches must be performed in underground low radioactivity observatories. Large detectors are needed to increase the acceptances and proper cryogenic systems to run dedicated detectors. To reach the working temperatures, refrigerators as Pulse Tubes, Dilution Units are used inside complex cryostats. CUORE, Cryogenic Underground Observatory for Rare Events, is an experiment located at LNGS under the Gran Sasso mountain. So far, it's the coldest cubic meter and the largest cold mass ever realized. Its 998 $TeO_2$ bolometers need to be kept at temperatures T< 10 mK. Using only Pulse Tubes, CUORE needs several weeks to reach the baseline T. Then a Fast Cooling System has been designed and constructed for a faster precooling of the whole CUORE cold volume. The Fast Cooling System (FCS) consists of a cryostat with heat exchangers that use 3 Gifford-McMahon refrigerators, a $^4He$ compressor, a filtering module and several sensors that allow to monitor and control the system during CUORE cooldown. The present work describes the FCS and summarizes its performances during the first full CUORE cooldown.

## 1. Introduction

The search for neutrino-less double beta decay ($0\nu2\beta$) and Dark Matter (DM) requires the use of very sensitive low rate detectors. Currently crystal bolometers are the best detector candidates thanks to their high sensitivity and low background. A bolometer is a device for measuring the energy of incident radiation with a very low energy threshold. It consists of an absorptive element (absorber) connected to a heat sink (a body of constant temperature) through a thermal coupling, as shown Fig.1. Several dielectric and diamagnetic materials can act as a low energy particle absorber, and a wide range of such

IOP Conf. Series: Journal of Physics: Conf. Series **969** (2018) 012087 doi:10.1088/1742-6596/969/1/012087

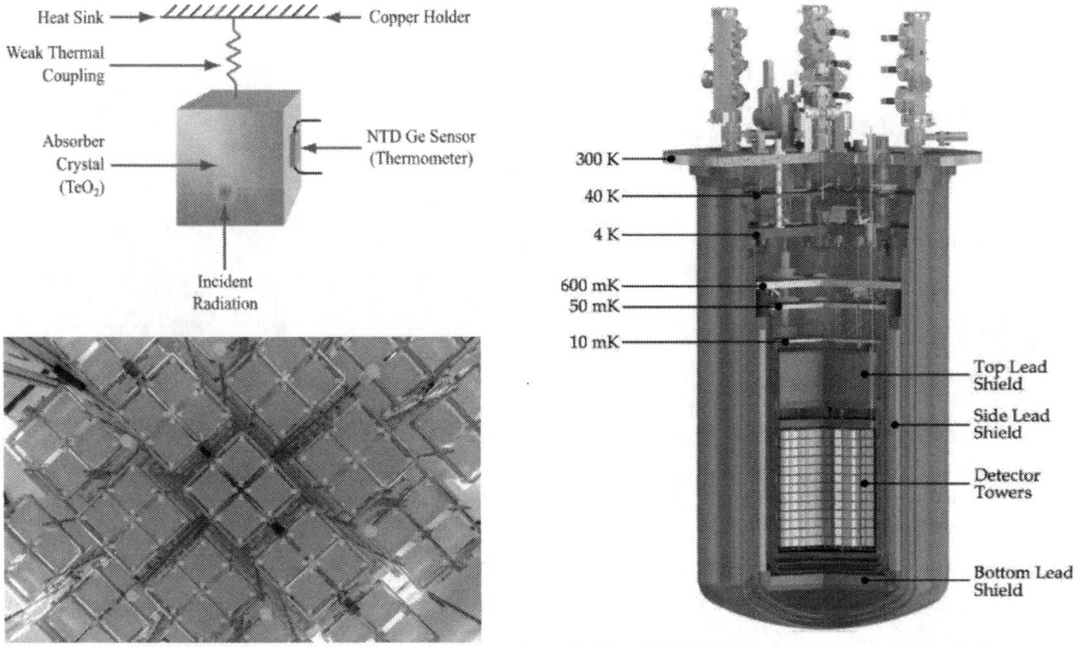


**Figure 1.** CUORE bolometer representation (top-left); CUORE Detector Assembled (bottom left); CUORE Cryostat (right).

materials are available as double beta decay emitters ($^{76}$Ge, $^{82}$Se, $^{100}$Mo, $^{116}$Cd, $^{130}$Te). Most of the bolometric detectors need to be operated at very low temperatures (~10 mK) while situated in a low radioactivity site. The acceptance for both $0\nu2\beta$ and DM events strongly depends on the detector mass (counts/keV·kg·yr). Increasing the detector mass also means a longer and more challenging cool down process to obtain a very low operating temperature. Thus, the cooling systems have a crucial role for this kind of research.

CUORE [1] is an experiment, currently taking data, at Laboratori Nazionali del Gran Sasso (LNGS) of Istituto Nazionale di Fisica Nucleare (INFN) in Italy. It is based on low temperature calorimeter technology, called bolometers [2]. At this time, it is the coldest cubic meter and the largest cold mass ever realized experimentally. Its purpose is to search for rare events such as $0\nu2\beta$ in $^{130}$Te. The CUORE detector consists of an array of 988 $TeO_2$ crystals arranged in a structure of 19 towers as shown in Fig.1 (left). The CUORE cryostat, see Fig.1 (right), is a custom apparatus consisting of six nested copper shields, related plates, a cryogen free cooling system with five Pulse Tubes (Cryomech PT415) and a very powerful Dilution Refrigerator (DR). The six shields are for the different temperature stages of the system: 300 K, 40 K, 4 K, 600 mK, 50 mK, and 10 mK. The PTs are thermally linked to 40 K and 4 K stages. The DR can operate once the temperature is below 4K. At base temperature, only four PTs are kept on; so one spare PT can be used in case of problems. The entire cryostat is suspended from a Main Support Plate (MSP), which bears the load of the system totaling close to 20 tons.

## 2. The Fast Cooling System

The Fast Cooling System (FCS) has been designed for a quicker pre-cooling of CUORE detector, from room temperature down to 40-50 K in about two weeks [3]. It consists of cryogen free coolers (the five PTs) and an external system that flows cold He inside the CUORE Inner Vacuum Chamber (IVC). This external system involves a Fast Cooling Unit (FCU), containing a primary and secondary heat exchanger, a gas compressor, double-walled flexible pipes, a filtering system and several other components and sensors. The FCU is cooled down by three Cryomech AL600 Gifford-McMahon

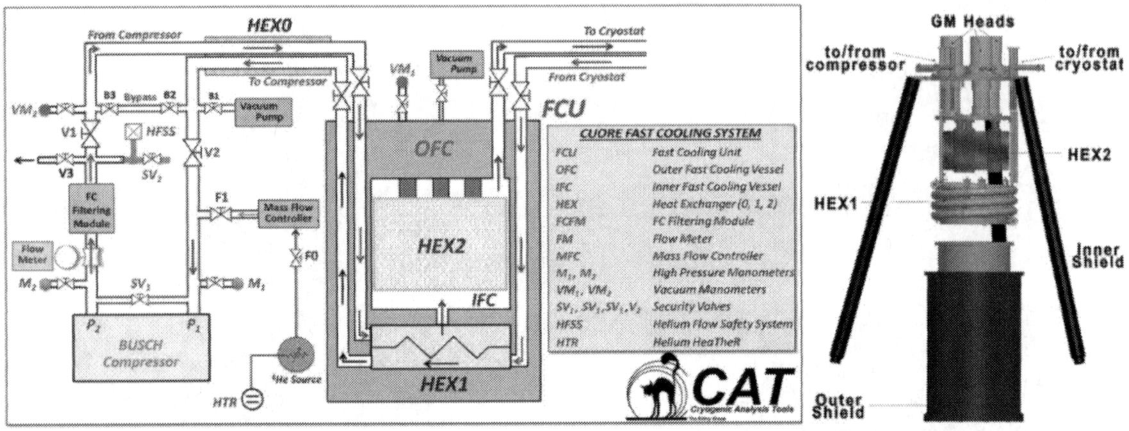


**Figure 2.** Schematic Overview of the Fast Cooling System (left); the Fast Cooling Unit (right).

cryocoolers (GM), each with a cooling power of about 600 W at 77 K. To avoid damages to the $TeO_2$ crystals, the temperature gradient ($T_{from\ cryostat} - T_{to\ cryostat}$) is kept below 40K, the pressure inside the entire system has to be 950 mbar < P < 1360 mbar and finally the He flux cannot exceed 5 g/s. The general schematic of the FCS is presented in Fig.2.

## 2.1. *Fast Cooling Unit*

The FCU mainly consists of two nested vessels containing two internal heat exchangers, cooled by the three GM heads. It has been constructed in The Netherlands by Leiden Cryogenics™ [4]. One internal heat exchanger (HEX) is made of copper and it is the coldest part of the FCU; the other is a counter-flow HEX; they correspond to HEX2 and HEX1 respectively as seen in Fig.2. The HEX1 is an 8 m long, counter-flow heat exchanger and has a spiral shape as shown in Fig.2 (right). It is made of metal double pipe bellows with an external diameter of 62 mm. The HEX2 is made of Oxygen-Free High thermal Conductivity (OFHC) copper and it consists of a stack of sixty-five disks, separated by 3 mm spacers, suspended by three copper rods; its total weight is about 300 kg. As shown in Fig.2, the HEX2 is connected to a steel flange that is thermally linked and mechanically supported by the three GM cryocoolers. To keep the inner part as cold as possible, HEX2 is then surrounded by an internal shield, called the Inner Fast Cooling Chamber (IFC) shown in Fig.2. In turn, the IFC is surrounded by the Outer Fast Cooling Chamber (OFC), which is kept under vacuum in order to prevent heat transfer by convection from room temperature to the IFC. Moreover, cryogenic multi-layer insulation is added over the entire internal surface to avoid heat transfer by radiation. The gas inlet and outlet valves, as well as all the sensor plugs, are located on the top flange of FCU.

## 2.2. *FCS Helium Circuit*

The Helium gas in the FCS is drawn from a dewar located at the ground floor of CUORE's hut (a 3 floor building). As shown in Fig.2 (left), the source of $^4He$ is connected to the FCS circuit by means of a Mass Flow Controller (MFC). This is used as a valve to inject He in the circuit both manually and automatically during the cool down phase. The He gas enters the compressor flowing through the compressor pre-filter at pressure $P_1$ and exits the compressor at pressure $P_2$. After the compressor, the gas flows through a Flow Meter (FM), which measure the g/s of He circulating in the FCS. It then enters the FC Filtering Module (FCFM) which consist of three different filters: Fine, Superfine and High-Efficiency Particulate Air (HEPA) filters. The MFC, the compressor, the FM, and the FCFM are located at the ground floor of the CUORE hut. The gas leaves the ground floor and reaches the second floor

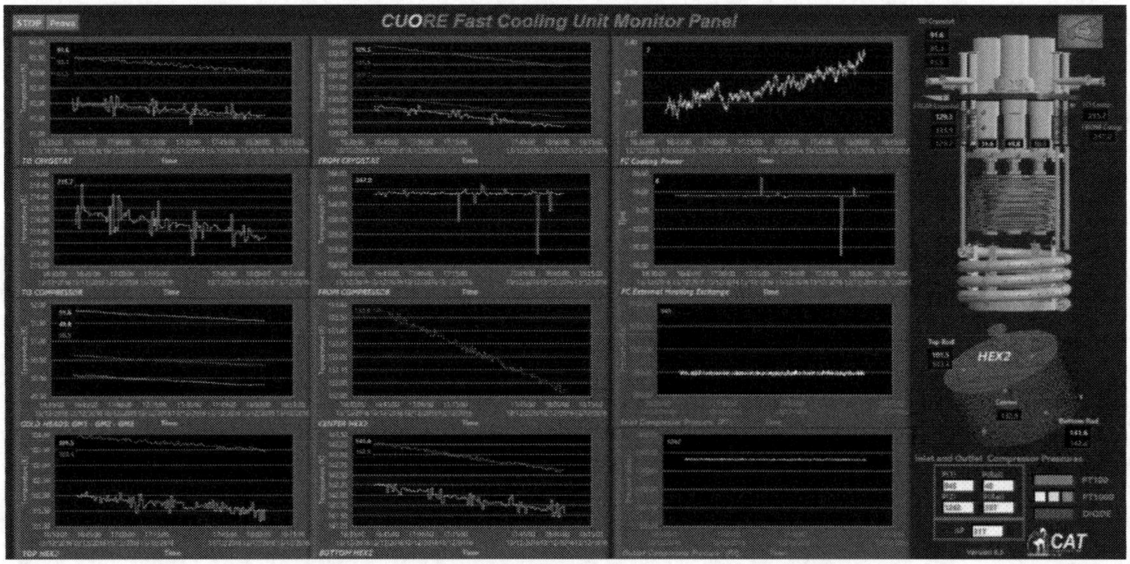


**Figure 3.** CUORE Fast Cooling Unit Monitor Panel, all critical parameters are redundantly monitored.

passing through a 12 m long bellows flex line. Then it enters a 4 m long counter-flow heat exchanger, HEX0, and finally reaches the FCU. The gas enters the FCU through a valve and reaches the second heat exchanger, HEX1, which is located in the bottom part of the FCU. The $^{4}He$ leaves HEX1 at a reduced temperature and enters the IFC where the coldest heat exchanger (HEX2) is located. After being cooled by HEX2, the gas leaves the IFC going to the CUORE cryostat. Since this is the coldest part of the circuit, a double-walled bellows flex line is used. After cooling the CUORE IVC, the gas goes back to the FCU where, passing through HEX1 and HEX0, it goes back to the compressor to start again the circulation. The gas is now at room temperature due to the heat exchanging with the inflowing gas on the way back.

### *2.3. FCS Data Acquisition System*
Temperature and pressure are the critical variables to be kept under control. The temperature of both the FCS and the CUORE cryostat is acquired via numerous thermometers all over the system. Three different types of thermometers are used: two Platinum (PT100, PT1000) and one Silicon Diode (DT470). In the most critical parts redundant thermometers are present for monitoring so that no excessive thermal stress is applied to the delicate crystal detector. The pressure of CUORE IVC during the cool down is strictly dependent on the frequency of the compressor and the pressure of the FCS. The lowest value of pressure needs to be always higher than atmospheric pressure in order to prevent air from entering the FCS and consequently the CUORE cryostat (contamination problems); however, the pressure in the FCS circuit cannot exceed a threshold set by the IVC o-ring seal to avoid damage. This threshold has been evaluated as 1.36 bar. Thus, the FCS is equipped with several pressure gauges in order to monitor the system behaviour and to have a feedback signal for self-regulation. Apart from the thermometry, which has a dedicated electronics setup, the remaining devices are monitored and controlled by means of three DAQ boards. All the instruments are then connected to a dedicated PC.

## 3. Fast Cooling Monitor and Control System
The whole FCS is monitored and controlled by a dedicated slow control system. The Fast Cooling Monitor and Control System (FCMCS) consists of several LabView based Virtual Instruments (VIs) organized in a more general slow control architecture. It not only acquires all the needed information from different instrumentation, with different protocols, but then supervises by using developed

IOP Conf. Series: Journal of Physics: Conf. Series **969** (2018) 012087 doi:10.1088/1742-6596/969/1/012087

**Figure 4.** CUORE Fast Cooling System Main Control Panel.

strategies and logics the FCS activities during the entire cool down process that is very critical and risky in several aspects. The temperature of the FCU, as well as the He entering the CUORE IVC, need to be continuously monitored and kept in a specific range. A CUORE Fast Cooling Unit Monitor Panel has been developed, as shown in Fig.3. The temperature values of all the thermometers related to the FCU, the high side and low side pressure in the He circuit, and their time evolution plots are displayed along with a map showing the position of each thermometer to help the operator be aware of the FCS status. The main control interface, called the Fast Cooling System Control Panel, is depicted in Fig.4. It contains all the crucial information to make the FCS work properly (i.e. compressor frequency, pressures, alarm thresholds, cryocoolers control). The operator can directly act on the system as well as activate the Automatic Control Algorithm (ACA). The ACA allows setting the min and max limits for the pressure resulting in an automatic refill of the He circuit by means of a Mass Flow Controller (i.e. the lower the temperature, the lower the pressure) and opens a safety valve in case of overpressure. Numerous safety alarms and actions have been developed which give the operator the ability to automatically adjust the compressor frequency in order to keep cooling the cryostat in the most efficient fashion. Several other interfaces allow extra features such monitoring the cryostat temperature, plotting the cooling speed for each single thermometer and a global temperature monitor with comparison plots.

## 4. CUORE Pre-cooling Performances

Several months of preparation and test runs have been performed between 2014 and 2015 in order to characterize the system. The final CUORE cryogenic cool down (without the ~1 ton crystal detector installed into it), took place in November 2015 and is referred to as Run 4 [5]. The cooldown started by means of the FCS alone (no PTs). After three days, the five PTs had been switched on. The FCS was then turned off at a temperature of approximately 60 K. The FCS brought the CUORE cryostat from 300 K to 60 K in about 10 days. A summary of the cooldown can be found in Fig.5 (top). During the summer of 2016 the complete detector was installed adding a mass of about 809 kg (742 kg of $TeO_2$ plus copper). In this configuration, the cooldown of the complete CUORE (Run I) started by means of the Fast Cooling System [6]. It was operated for 10 days, cooling the cryostat down to around 100 K as shown in Fig.5 (bottom). Three PTs, out of five, have been switched on along the cooldown process. At the beginning of Run I, the compressor frequency was progressively increased up to 18 Hz (corresponding to a He flow of about 2.8 g/s); it was kept around that frequency for several days. The cooling speed, where the bolometers are located, reached a peak value of (1.09±0.02) K/h, stabilizing

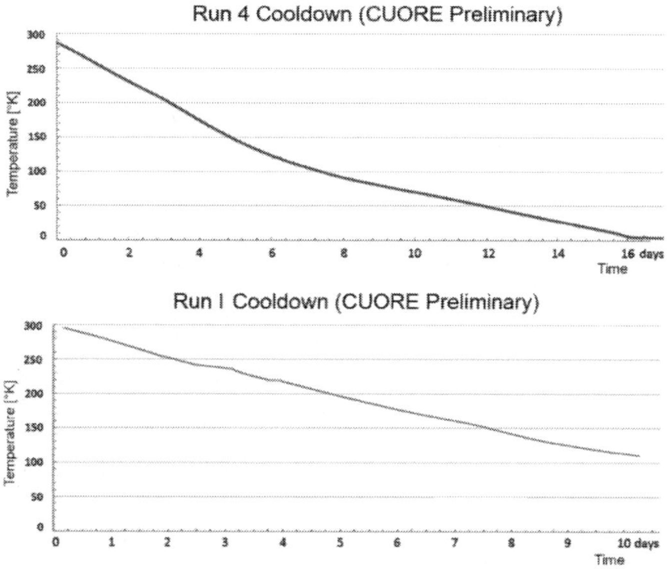


**Figure 5.** CUORE pre-cooldown performances using the FCS. IVC Run 4 pre-cooldown (top); temperature trend of one of CUORE crystal detector towers during Run I (bottom).

around (0.90±0.02) K/h for a majority of the time, and decreasing to (0.49±0.02) K/h in the last twelve hours of the pre-cooling. Taking into account the tremendous thermal inertia (~20 tons) of CUORE, the described performance can be considered good. In the last hours, the compressor frequency was progressively lowered to 13 Hz and then to 10 Hz before the final decision to switch off the system, at higher temperature than expected for testing of other apparatus. It has been proved that the design specifications have been matched [7].

## 5. Conclusions

A Fast Cooling System has been designed, constructed, tested and finally used for a faster pre-cooling of ton-scale bolometric experiments searching for Neutrino-less Double Beta Decay and Dark Matter. It has been operated on the CUORE cryostat (around 20 tons mass) at Laboratori Nazionali del Gran Sasso, Italy. The FCS brought the significantly large mass of CUORE's cryostat and crystal detectors below 100 K in 10 days successfully, with a mean cooling speed of approximately 0.71 K/h, proving the good performance of the system. The whole system has been equipped with numerous sensors in order to have precise control of the cool down process via the Fast Cooling Monitor and Control System software. It allows automatic and manual control of the entire FCS together with alarms and safety actions in case of dangerous anomalies. In future cooldowns, the possibility to simultaneously turn on FCU + 5 PTs at the beginning of the pre-cooling will guarantee better performance. Finally, this system is suitable for every experiment that needs a large cryostat (big mass) in a low radioactivity environment with the need to speed up the pre-cooling phase.

## References

[1] Artusa D R et al. (CUORE Collaboration) 2015 *Advances in High Energy Physics 2015* **1**
[2] Fiorini E, Niinikoski T, *Nucl.Instrum.Meth.*, vol. A224, 1984, **83**
[3] Pagliarone C E et al., The CUORE Fast Cooling System, *paper under submission*
[4] Leiden Cryogenics BV, 2311 VZ Leiden, The Netherlands http://www.leidencryogenics.com
[5] INFN LNGS, Laboratori Nazionali del Gran Sasso, *Annual Report*, 2015, **39-41**
[6] INFN LNGS, Laboratori Nazionali del Gran Sasso, *Annual Report*, 2016
[7] Pagliarone C E et al. 2017, *EPS Conference on High Energy Physics*

# SINIS bolometer with a suspended absorber

**M. Tarasov[1,2,3,*], V. Edelman[2], S. Mahashabde[4], M. Fominsky[1], S. Lemzyakov[2,5], A. Chekushkin[1,5], R. Yusupov[1,5], D. Winkler[3], and A. Yurgens[3]**

[1] Kotelnikov Institute of Radio Engineering and Electronics RAS, Mokhovaya 11/7, 125009 Moscow, Russia
[2] Kapitza Institute for Physical Problems RAS, Kosygina 2, 119334 Moscow, Russia
[3] Chalmers University of Technology, Microtechnology and Nanoscience – MC2, Kemivägen 9, Göteborg 41296, Sweden
[4] Department of Physics-Astrophysics, University of Oxford, Keble Road, OX1 3RH, Oxford, UK
[5] Moscow Institute of Physics and Technology (State University), Institutsky per. 9, 141700 Dolgoprudny, Russia

*tarasov@chalmers.se

**Abstract**. We have developed a Superconductor-Insulator-Normal Metal-Insulator-Superconductor (SINIS) bolometer with a suspended normal metal bridge. The suspended bridge acts as a bolometric absorber with reduced heat losses to the substrate. Such bolometers were characterized at 100-350 mK bath temperatures and electrical responsivity of over $10^9$ V/W was measured by dc heating the absorber through additional contacts. Suspended bolometers were also integrated in planar twin-slot and log-periodic antennas for operation in the submillimetre-band of radiation. The measured voltage response to radiation at 300 GHz and at 100 mK bath temperature is $3*10^8$ V/W and a current response is $1.1*10^4$ A/W which corresponds to a quantum efficiency of ~15 electrons per photon. An important feature of such suspended bolometers is the thermalization of electrons in the absorber heated by optical radiation, which in turn provides better quantum efficiency. This has been confirmed by comparison of bolometric response to dc and rf heating. We investigate the performance of direct SN traps and NIS traps with a tunnel barrier between the superconductor and normal metal trap. Increasing the volume of superconducting electrode helps to reduce overheating of superconductor. Influence of Andreev reflection and Kapitza resistance, as well as electron-phonon heat conductivity and thermal conductivity of N-wiring are estimated for such SINIS devices.

## 1. Introduction

Normal Metal-Insulator-Superconductor (NIS) tunnel junctions, as well as SIS junctions, are the main building blocks for superconducting electronics. Single NIS junctions and arrays are used in microwave bolometers, cryogenic thermometers, electron coolers, radiation detectors, etc. In such devices, junctions should fit different parameters for area, transparency, material properties, and thermal characteristics. This leads to various fabrication methods and technique for measurements. Estimations made for equilibrium measurement conditions can be controversial in the case of microwave bolometers, ultra-low temperature thermometers, electron coolers. In Terahertz bolometers, as well as in electron coolers, the energy distribution of electrons can differ from the Fermi-Dirac distribution. In bolometers illuminated with Terahertz radiation, the density of electrons will increase at higher energies, and in

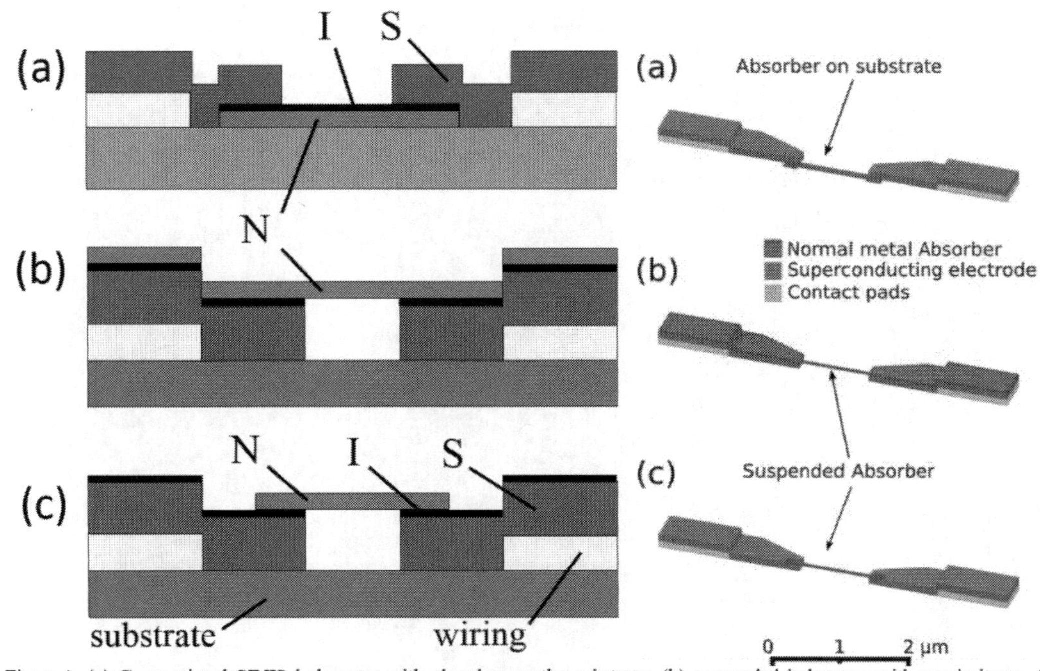


Figure 1. (a) Conventional SINIS bolometer with absorber on the substrate; (b) suspended bolometer with proximity to Au wiring, (c) suspended bolometer with reduced proximity (C). Samples are fabricated on Si substrates with superconducting Al electrodes (blue). The absorber (red) in (a) is oxidized to form a tunnel barrier to superconducting electrode, while in (b), (c) superconducting electrode is oxidized to create a tunnel barrier. Right panel: a 3-d view of the designs.

electron coolers, the distribution will have a reduced density at higher energies. Andreev reflection and proximity effects at the superconductor-normal metal interface induce changes in IV curve compared to simple SIN model and need to be accounted for in the heat-balance equations.

In superconducting bolometers, the electromagnetic radiation raises the temperature $T_e$ of the electron system in a normal-metal absorber. Superconducting leads connected to the absorber through insulating barriers form tunneling junctions. The tunneling current across such a junction is a strong function of $T_e$, which thereby offers a sensitive readout of the absorbed radiation power. At the low bath temperature $T$, the absorbed photon energy $hf >> k_BT$ is distributed among quasiparticles through their multiple interactions and is eventually lost into the cold bath through electrical contacts and through the substrate. Here $h$ is the Planck constant, $f$ is the frequency of incident radiation, $k_B$ is the Boltzmann constant. The balance between the heating radiation and cooling powers determines $T_e$. Decoupling the absorber from the heat sinks would increase $T_e$ and the detector response through the multiplication of excited electrons. It should be remembered that for the photon energy $hf >> k_BT_e$, the energy distribution of electrons can be substantially different from the Fermi distribution, depending on the quasiparticle interactions and tunneling rate of the excited electrons through the SIN junction. For estimations of bolometric sensitivity, it is usually assumed that the absorbed radiation is equivalent to a dc Joule heating of the same power.

## 2. Fabrication

The design of a bolometer with a suspended normal-metal absorber is described below (see Fig.1b and 1c). This fabrication process was inspired by [1]. The fabrication process is simpler compared to the conventional bolometer [2] (Fig. 1a) which uses the shadow evaporation technique. In the design presented here, thin films can be deposited by any physical vapor deposition method including

Figure 2. SEM image of fabricated sample with the test structure for dc calibration containing three suspended bridges.

sputtering, which is difficult to use for shadow evaporation. Patterning of the normal-metal- and superconducting layers can be done by the simple lift-off process. Fig. 1b shows a schematic view of the proposed layout. First, a trilayer of Ti (~10nm), Au (~50nm), and Pd (~10nm) is deposited on the substrate. This trilayer (in yellow) is patterned for wiring and contact pads followed by the deposition of the SIN structure. Superconducting Al (~100 nm) is evaporated and then oxidized in an atmosphere of pure $O_2$ at a pressure of 20 mbar and subsequently patterned into electrodes. Finally, the absorber normal-metal layer (20 – 30 nm of Pd or Cu) is deposited on the $AlO_x$ on top of the Al electrodes. To form the suspended absorber bridge, we selectively etch away the Al layer under the absorber in the region defined by a window in the resist. Aluminum under the bridge region is completely removed by etching in a weak base (Microposit™ MF CD-26 developer). The results of etching are clearly visible in the scanning electron microscope (see Fig. 2).

Initially, the suspended absorbers were made of Cu. It was observed that Cu thin films are soft and have a tendency to sag down to the substrate. Besides, Cu is a good conductor at cryogenic temperatures and forms relatively low resistivity bridges even at thicknesses approaching 20 nm. This restricts the impedance matching of the device with quasioptical antennas having impedances around 50 Ω. An absorber of Cu (~20nm) with a thin (~3nm) layer of Cr was instead used to create rigid suspended bridges. While Cr+Cu is a robust thin film, care needs to be taken during subsequent fabrication steps involving resist development and liftoff. Each of these steps need to be terminated using a critical point drying step in liquid $CO_2$ to prevent the collapse of already-formed bridges. Rinsing in acetone, ethanol, or methanol instead of water and blow-dry with nitrogen can be alternative to critical point dryer for hard and thick Pd and Hf suspended bridges.

To investigate the influence of proximity effect on superconductivity in the S electrode of NIS junction we added in fabrication one more step of photolithography and chemical etching of Cu in

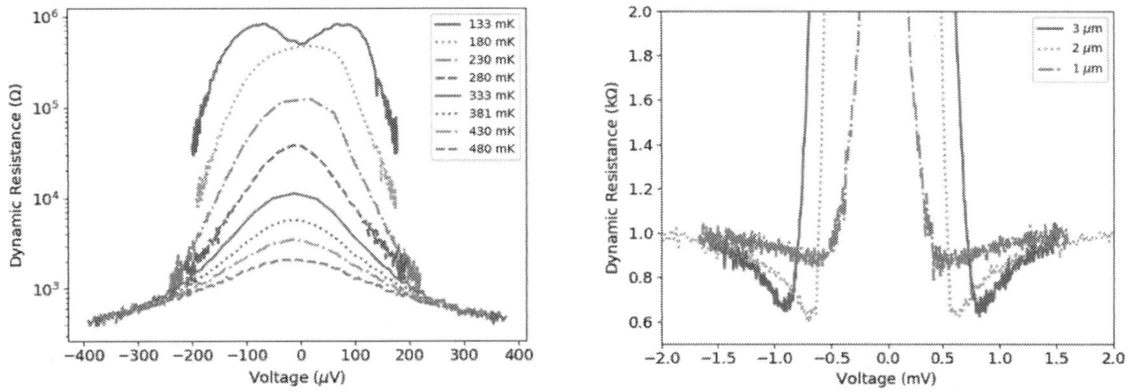


Figure 3. Dynamic resistance of the sample depicted in Fig. 1b measured in the temperature range 133-480 mK (left). Dynamic resistance of four NIS junctions connected in series with corresponding single energy gap of 0.1 mV, 0.155 mV, and 0.2 mV for samples with distance from normal-metal trap to superconducting electrode of 1, 2, and 3 μm measured at 300 mK.

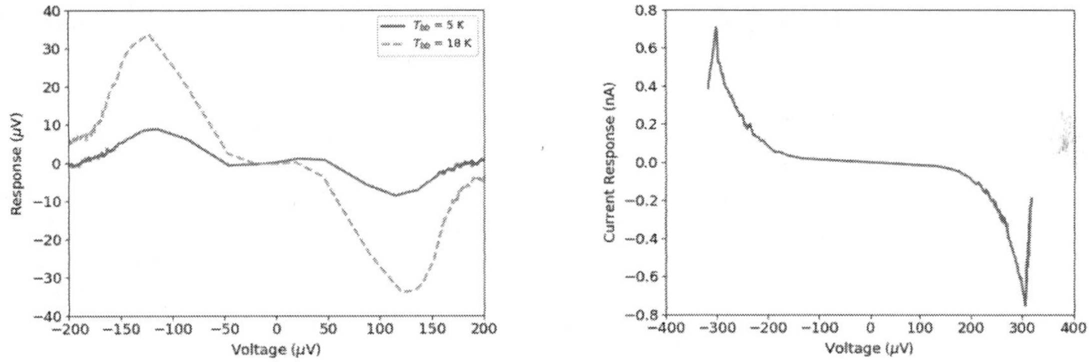


Figure 4. Voltage response of a SINIS bolometer with Hafnium suspended bridge approaching $3*10^8$ V/W (left). The corresponding blackbody temperatures are 5 K and 18 K. Maximum current response of 0.7 nA for the suspended bolometer, corresponding to a quantum efficiency of over 15 electrons per single photon at 350 GHz (right).

diluted 1:50 $HNO_3$. By this we were able to vary the distance from the golden wiring film (yellow) to the S electrode of the SIN junction (see Fig. 1c). SEM image of fabricated sample with suspended Cu bridge is presented in Fig 2. We also fabricated bolometers with Hf bridges using the above process. The higher resistivity of Hf improves the impedance matching to higher ohmic planar antennas.

## 3. Experimental procedure

With a 4-probe structure (Fig. 2), we were able to measure the electrical response of the bolometer by applying Joule heating across outer contacts and measuring IV curve and response for the middle contacts with single the suspended bridge. Results are presented in Fig. 3.

With Hf-based bolometers integrated in a twin-slot antenna, we measured optical response to radiation at 300 GHz. Radiation from a black body source, filtered by two band-pass filters and a neutral density filter, illuminated the sample with a twin-slot antenna which was placed on an extended hemisphere sapphire lens. Results are presented in Fig. 4 (left).

The current response can be a good indication of the quantum efficiency of the bolometer, i.e. the number of tunneling electrons $n_e$ induced by a single photon of radiation, $\eta=n_e/n_p$. For conventional SINIS bolometers, $\eta$=1.2-1.5 instead of the expected $\eta=hf/k_BT=50$ at f=350 GHz, as predicted by the theory for bolometric response [3]. In the case of the suspended bolometer, with current response as in

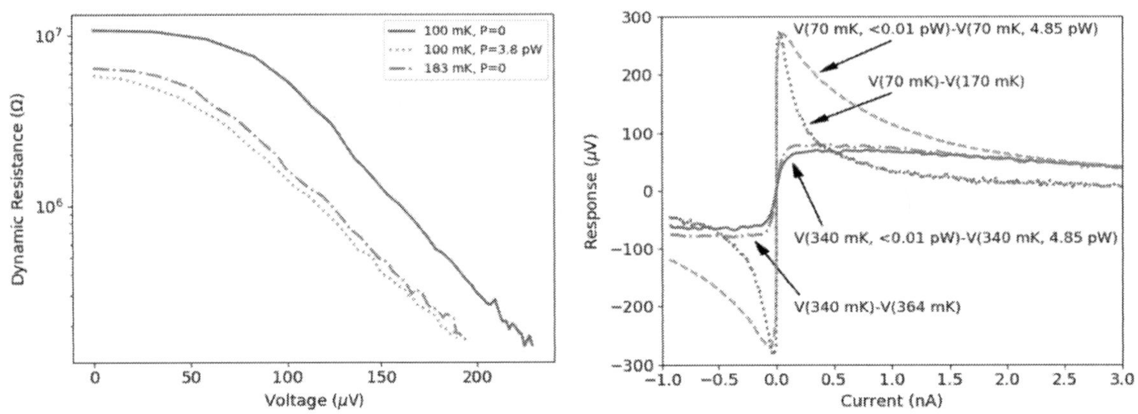


Figure 5. Dynamic resistance for phonon temperatures 100 mK and 183 mK without radiation, and for 100 mK with 3.8 pW optical power load for suspended absorber (left panel). Voltage response dependence on dc voltage for bath temperatures 70 mK and 340 mK for conventional bolometer with absorber on the substrate (right panel). Solid lines represent an estimated 4.85 pW irradiation, dashed lines represent response to equivalent increase in bath temperature without irradiation. A clear difference between thermal and optical response is visible due to non-Fermi distribution.

Fig. 4 of 0.7 nA at the incident power of 0.06 pW, the current responsivity is estimated to be $1.1*10^4$ A/W. The current of 0.7 nA corresponds to $4.3*10^9$ electrons per second and radiation power 0.06 pW at 350 GHz corresponds to $2.8*10^8$ photons per second. As a result, we have a quantum gain of over 15 electrons per photon, approaching the values predicted in [3]. This is direct proof of our initial design idea to disconnect the absorber from direct contact with the substrate for improving the thermal insulation and increasing the bolometer response. Another important feature of our suspended bolometer is the similarity in the shapes of dynamic resistance to thermal and optical heating (Fig. 5, left). This shows that in such bolometers, the electron system in the absorber heated by incident radiation is thermalized, which in turn provides better quantum efficiency. In conventional bolometers with absorber on substrate we observed nonthermal optical response [4], which is a reflection of a nonequilibrium energy distribution, and is different from the expected Fermi-Dirac distribution. By means of suspending the absorber and disconnecting the phonon system from the substrate, we have managed to better thermalize hot electrons, and as a result, obtained a higher quantum gain.

## 4. Discussion

An important issue of SINIS bolometer and electron cooler development is proper cooling of the superconducting electrode. The simplest way is to increase the thickness and area of S electrode which helps to spread injected hot quasiparticles in a larger volume, decreasing the local hot quasiparticle density. This is critical since quasiparticle lifetimes in a superconductor like Al can be of the order of μs. Another known method is the use of a normal metal trap. One pioneering experimental works [5] shows an improvement in the cooling of a thin 18 nm Al wire with a 28 nm Cu film deposited above the Al over its oxide barrier. An improvement was observed after reducing the distance from the tunnel junction to the Cu film from 8 μm to 0.2 μm. An alternative option used a direct contact of a normal metal trap [6] where the measured trapping efficiency (the ratio of absorbed to incident power), was about 10% near 100 mK. In our earlier paper [7], normal metal traps were made of 60 nm thick Au film, with a 65 nm thick Al superconducting electrode above it. In reality, contact between Au and Al is a kind of resistive intermetallic compound with reduced transparency. This contact can be viewed as a sort of a tunnel junction with relatively high transparency, similar to [5]. In the case of Figs. 1b, and 1c, Pd is present at the top of TiAuPd trilayer, and the top Pd makes perfect contact to Al. As a result, when Pd is very close to Al, as in Fig. 1b, we have a suppression of superconductivity and no strong evidence of electron cooling. When introducing a separation of 1-2 μm as in Fig. 1c, we observe the nominal value of a double energy gap of Al (~380 μeV).

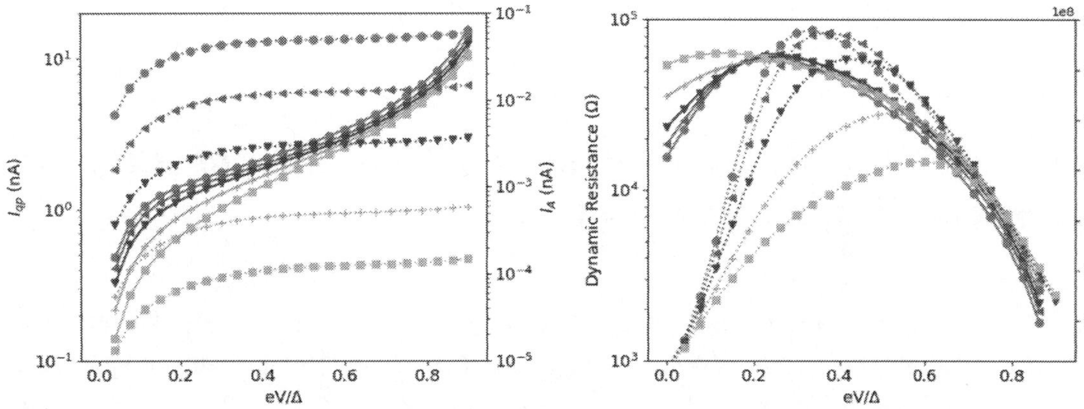


Figure 6. Relative contribution of quasiparticle ($I_{qp}$) and Andreev currents ($I_A$) to the IV characteristics of a typical SINIS bolometer with a low electron-phonon coupling absorber (left panel). Dynamic resistance and responsivity (right panel). Solid lines correspond to the left axis and dotted lines to the right axis. The various curves are simulated for normal state resistance of the NIS junction: 50 Ω (•), 100 Ω (◀), 200 Ω (▼), 500 Ω (+), and 1000 Ω (■). The electron-phonon coupling constant is assumed to be 0.5 nW/μm$^3$K$^5$, volume of the absorber 0.1 μm$^3$, incident power load 0.5 pW, and a bath temperature 200 mK.

In conventional SINIS bolometer model, the presence of Andreev currents is ignored [7]. Fig. 3 (left) shows the commonly observed zero bias conductance increase which can be explained by the presence of Andreev currents with non-negligible contribution to the IV characteristics at low temperatures. Using the model developed in [8], one can predict their magnitude and study their impact on the bolometer characteristics. Fig. 6 (left) shows simulations of the relative contributions of quasiparticle and Andreev currents to the total current running in the bolometer as a function of the bias voltage normalized to the superconducting energy gap. NIS junctions with high transparency show higher contribution of Andreev currents and better electron cooling. Electron cooling increases responsivity of the device but decreases the absolute value of voltage response. Fig. 6 (right) shows a representative simulation for devices with absorbers made of materials with low electron-phonon coupling (Hf, etc). The dynamic resistance is calculated as the slope of the IV characteristics at each bias point. More details are published in [9]. A detailed study of the normal state resistance of the tunnel junctions, of Andreev currents and their impact is necessary to decide the optimum operating point of the bolometer.

**Acknowledgements**
This work was done with support from The Swedish Foundation for International Cooperation in Research and Higher Education (STINT, grant IG2010-14) and the Swedish Research Council (VR, grant 621-2014-5468). The clean-room processing has been done in the Nanofabrication Laboratory on the equipment sponsored by the Knut and Alice Wallenberg Foundation.

**References**
[1] H. Nguyen, L. Pascal, Z. Peng, et al. , Appl. Phys. Lett., v.100, 252602 (2012).
[2] M. Tarasov, L. Kuzmin, N. Kaurova, Instr. and Experimental Techn., 52, No. 6 (2009).
[3] I. A. Devyatov, M. Yu. Kupriyanov, JETP Letters, **80**, No. 10, 646 (2004).
[4] M. Tarasov, V. Edelman, S. Mahashabde, L. Kuzmin, JETPh, **119**, No 1, 107 (2014).
[5] J. Pekola, D. Anghel, T. Suppula, et al., Appl. Phys.Lett., v.76, No 19, 2782, (2000)
[6] J. Ullom, P. Fisher, M. Nahum, Nucl. Instr.and Methods in Phys. Res., A 370, 98 (1996)
[7] L. Kuzmin, I. Agulo, M. Fominsky, et al., Superconductor Science and Technology, 17, S400-S405 (2004)
[8] F. J. W. Hekking, and Yu. V. Nazarov, Physical Review B, No. 49, 10 (1994).
[9] M. Tarasov, V. Edelman, S. Mahashabde, et al., Appl. Phys.Lett., v.100, 2426, (2017)

IOP Conf. Series: Journal of Physics: Conf. Series **969** (2018) 012089 doi:10.1088/1742-6596/969/1/012089

# Development of a pressure cell using a beta-titanium alloy for a Differential Scanning Calorimeter

**T Eto, K Akagi, E Takashima, A Noguchi, T Sakami, T Noda and F Nakamura**

Faculty of Engineering, Kurume Institute of Technology, Kurume, Fukuoka 830-0052, Japan
e-mail: teto@kurume-it.ac.jp

**Abstract**. We have developed downsizing a piston-cylinder clamp-cell to measure a specific heat under pressure using a commercial Differential Scanning Calorimeter (Shimadzu DSC-60 plus). The downsized cell with a diameter of 6.0 mm and a height of 5.0 mm has been achieved by a simplification of the cell design and using a high-yield-strength alloy of β-titanium (KOBELCO, KS15-5-3). The cell allows us to achieve pressures of 0.7 GPa at room temperature as our designed value. As a demonstration of the cell, we have measured the specific heat anomaly due to the first-order transition at around 357 K in $Ca_2RuO_4$. The pressure variation of the transition temperature indicates that our cell allows us to achieve pressure of ~0.3 GPa. Moreover, we have estimated that a heat capacity of ~0.6 mJ/K is fully possible to measure within β-Titanium cell as absolute accuracy.

## 1. Introduction

As well known, pressure is a key tool to tune the electronic states indirectly via the change of lattice parameters. In particular, a phase transition can be controlled by applying pressure. Measurements of electrical transport, magnetization and x-ray diffraction have been progressed by pressure techniques. In contrast, a specific heat, $C$, is the most suitable tool to understand the phase transitions thermodynamically. A specific heat at ambient pressure has generally been measured by using adiabatic [1], thermal relaxation [2], alternating current [3] and Differential Scanning Calorimetry; however, it is quite difficult to measure $C$ at around a first-order transition under pressure and in the wide temperature range.

Our attention has been paid to the metal-insulator (M-I) transition accompanied by structural change in the Mott insulator $Ca_2RuO_4$ (CRO) [4]. The M-I transition can be induced by heating above ~360 K, pressurising over than 0.5 GPa and applying electric field over than 40 V/cm. Moreover, CRO shows a variety of novel quantum phenomena, such as an antiferromagnetic (AFM) Mott insulator, a ferromagnetic quasi-two-dimensional metal and an unconventional superconductivity under pressure. In particular, there has recently been growing interest in the negative thermal expansion observed in the wide temperature range below ~360 K [5]. Thus, we have strong concern to understand the thermodynamics in CRO. However, there has been a lack of the thermodynamic data at around $T_{M-I}$. This is because the transition accompanied by a latent heat is difficult to detect by using a thermal relaxation calorimetry, which has been the most popular method for researchers in low temperature physics to measure a specific heat as the recent trend.

In contrast, thermodynamics information related to a transition accompanied by a latent heat can be obtained by using a Differential Scanning Calorimetry, which is popular method for chemists. However, the smallness of the sample space prevents us to equip a typical piston-cylinder clamp-cell. We report here a development of downsizing a piston-cylinder clamp-cell to measure a specific heat under pressure using a commercial Differential Scanning Calorimeter (DSC).

## 2. Design of the pressure cell

In order to measure $C(T)$ in the vicinity of the first-order transition in CRO, we have used a heat flux type DSC system (Shimadzu DSC-60 plus). The measurement was performed in the temperature range between 130 and 873 K. The sample space of our DSC system is $\phi$6.0 x 5.0 mm; however, it is too small to equip a typical size of a piston-cylinder clamp-cell (i.e., $\phi$22 x 40 mm for our PPMS). Thus, drastically downsizing the pressure cell is required. Moreover, our goal of pressurisation is above 0.5 GPa, and it might be achieved by using a high-yield-strength alloy and a simplification of the cell design. In the Differential Scanning Calorimetry, absolute value of $C(T)$ can precisely be obtained by a heat flux differential between a sample and a reference cells. For this, it is necessary to conform the cell's geometry and weight precisely.

The details of the sample chamber of our DSC-60plus instrument is illustrated by figure 1(a). Reference and sample cells are equipped with the left and the right sides of the chamber, respectively. Circle line indicates the maximum diameter of the reference cell is $\phi$6.0. In contrast, figure 1(b) shows a cross-sectional view of our piston-cylinder clamp-cell. The diameter and the height are $\phi$6.0 and 5.0 mm, respectively. A simplification of the cell design and using a high-yield-strength alloy is key of the downsizing. In concrete, a number of the parts was reduced from eleven to four. The downsized cell consists of an outer cylinder, a piston, a Teflon sheet and a Teflon tube. We took out a pressurisation system outside of the cell as mentioned next section, and the sample was pressurised by tightening an end screw used as the piston.

To establish both downsizing of the piston-cylinder clamp-cell and attain high sensitivity of the measurement, we must take into consideration both powerful tensile strength of materials and a small heat capacity. In table 1, some alloys CuBe, NiCrAl, WC, MP35N and β-Ti are compared as a candidate for the cell. We use the β-Titanium alloy (KOBELCO, KS15-5-3) [6] is the most suitable material for

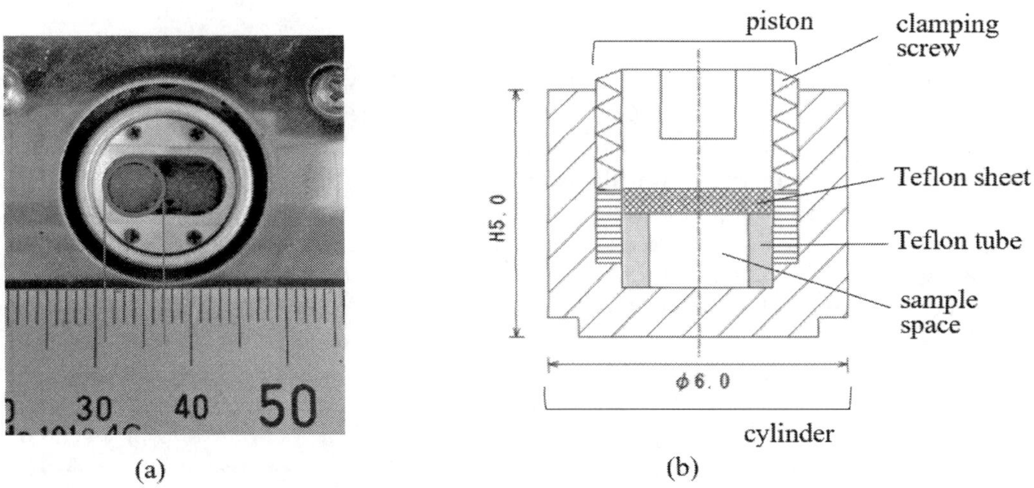


**Figure 1.** (a) Top view of the sample chamber of the DSC-60 plus. There are reference and furnaces of Circle indicates the maximum size of $\phi$6.0 diameter. (b) Cross-sectional view of our pressure cell.

**Table 1.** Mechanical and thermodynamic properties of the materials at room temperature (RT). The bulk densities and specific heats have been measured in the present study.

| | CuBe | NiCrAl | WC | MP35N | β-Titanium |
|---|---|---|---|---|---|
| Tensile Strength (GPa) | > 1.00 GPa[7] | 2.00 GPa[8] | 1.77 GPa[9] | 2.0 GPa[10] | 1.76 GPa[6] |
| Bulk density (g/cm$^3$) | 8.3 | 7.9 | 15.1 | 8.4 | 4.7 |
| Specific heat (J/K·g) | 0.36 | 0.32 | 0.17 | 0.39 | 0.45 |
| Heat capacity (J/K· cm$^3$) | 3.0 | 2.5 | 2.6 | 3.3 | 2.1 |

**Table 2.** Mass and thermodynamic properties of the materials used in the pressure cell at RT.

| | Daphne oil 7373 | Teflon tube & sheet | β-Titanium |
|---|---|---|---|
| Mass in a cell (mg) | ~5 | ~15 | ~400 |
| $C_p$ (J/K·g) | 2.0 | 0.78 | 0.45 |
| Heat capacity (mJ/K) | ~10 | ~12 | ~180 |

our cell because of good tensile strength and the smallest heat capacity per unit volume. By using this alloy, we can obtain the expected maximum pressure $P_{max}$ ~ 0.7 GPa from the following equation:

$$P_{max} = \frac{\sigma_y(b^2 - a^2)}{2b^2} \quad (1)$$

where we use the tensile strength $\sigma_y$=1760 MPa, the inner diameter $a$ = 4.0 mm and the outer diameter $b$ =6.0 mm.

## 3. Measurements

Two of pressure cells are equipped with each of the sample and the reference furnaces. In order to improve the accuracy of the measurement, differences in the weight of the each cells were balanced within 5 %. In figure 1(b), we illustrate a schematic view of our pressure cell. A Teflon tube with an inner space of ϕ3.0 diameter and 2 mm height was put into the cylindrical β-Titanium cell. The several pieces of crystal were put into the cell with Daphne oil 7373 (Idemitsu Kosan Co., Ltd.) as a pressure-transmitting medium [10]. Our pressurising procedures are shown in figure 2. The force from hydraulic press system was loaded to the screw cap via a hard metal sphere and a ratchet wrench. The sample was pressurised, and then it finally sealed by applying a torque to the Teflon-coated screw cap. Table 2 shows mass and thermodynamic properties of the materials used in a pressure cell at RT. The ratio of the oil and the Teflon components to the pressure cell was within ~10 %.

To demonstrate the performance of this cell, we used single crystals of CRO grown by a floating-zone method with $RuO_2$ self-flux. Several pieces of crystal with total masses of 3-5 mg were put into the pressure cell. Figure 3 shows specific heat $C$ in the vicinity of the M-I transition at several pressures as a function of temperature. The peak temperature is reduced with pressurising. Such reductions have been observed in the M-I transition as same as with temperature, electric field, and substitution of Sr for Ca[4,12,13]. The generated pressure $P$ can be estimated by determining the M-I transition temperature $T_{M\text{-}I}$ of CRO[14], the pressure dependence $dT_{M\text{-}I}/dP$ is well described as $dT_{M\text{-}I} / dP = -114$ K/GPa. In this measurement, hydrostaticity most likely be kept since there is little broadening of the peak in $C(T)$

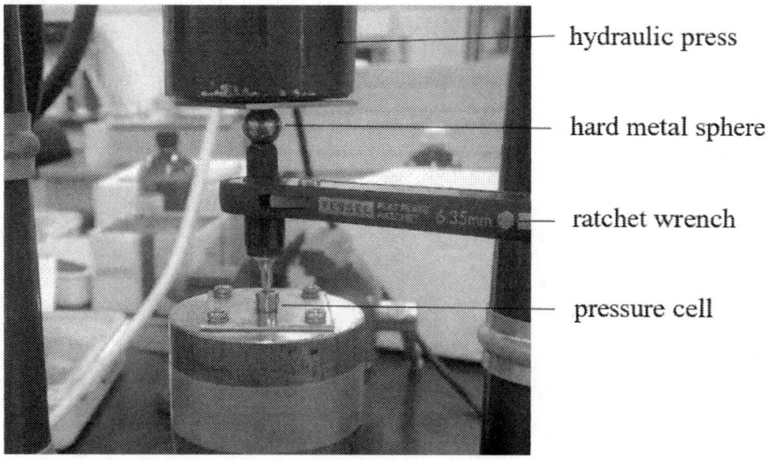


**Figure 2.** Our pressurising system of the pressure cell.

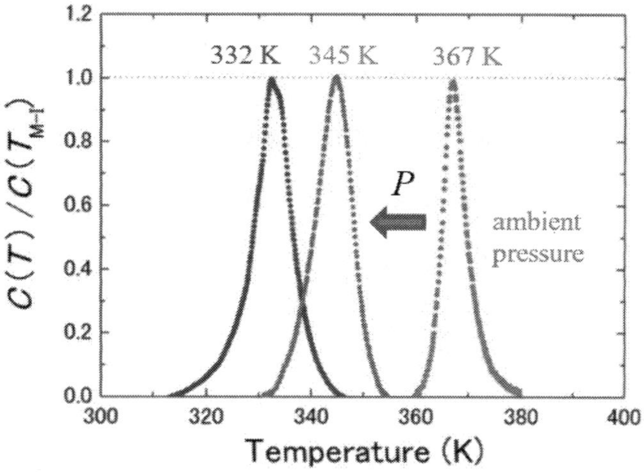


**Figure 3.** DSC profiles at various pressures. The curves normalized at $T_{M\text{-}I}$ correspond to $C(T)/C(T_{M\text{-}I})$. Red, Green and Blue colors indicate those of ambient pressure, pressurised and more pressurised, respectively. The figures on the profiles denote temperatures of peak positions.

curves. We can obtain the maximum pressure in this measurement to be 0.3 GPa from the shift of $\Delta T_{M\text{-}I}$ ~35 K, so far. The further high-pressure can fully expected from our designed value.

Let us next evaluate the accuracy of $C$ measurement by utilizing the β-Titanium cylinder. In principle, the DSC should not detect signals of the cell, the Teflon tube, the pressure medium and so forth by subtracting the heat flow of the reference cell from that of the sample one by a measurement. However, mechanical inhomogeneity of the cell and variations in the weights of loading materials in the cell, might influence on the background of $C$. In the DSC measurement under pressure, signals within ~0.2 J/g·K was occasionally obtained in background level. From the background we evaluate the minimum measurable value of the heat capacity within the β-Titanium cell to be ~0.6 mJ/K. That is, $T_{M\text{-}I}$ can be detected by using CRO samples over than ~3 mg, as absolute accuracy.

In conclusion, a pressure cell for a DSC has been developed by utilizing β-Titanium alloy; this cell allows us to pressurise up to 0.3 GPa. The minimum heat capacity measurable within the cell is ~0.6

mJ/K at ~RT. The maximum pressure that we can generate is now limited not by the cylinder but by the piston with poor toughness. Moreover, downsizing of pressure cell allows us to reduce the heat capacity of the cell and enables a wide variety of measurement at low temperature by using commercial instruments such as MPMS, PPMS (Quantum Design Inc.) and dilution refrigerator.

**Acknowledgment**
A part of this work was supported by JSPS KAKENHI Grant Numbers JP26247060 and JP17H06136.

**References**

[1] Morin F J and Maita J P 1963 *Phys. Rev.* **129** 1115
[2] Bachmann R, DiSalvo F J, Geballe T H, Greene R L, Howard R E, King C N, Kirsch H C, Lee K N, Schwall R E, Thomas H U and Zubeck R B 1972 *Rev. Sci. Instrum.* **43** 205
[3] Sullivan P F and Seidel G 1968 *Phys. Rev.* **173** 679
[4] Nakamura F, Sakaki M, Yamamoto Y, Tamaru S, Suzuki T and Maeno Y 2013 *Sci. Rep.* **3** 2536
[5] Braden M, Andre G, Nakatsuji S and Meno Y 1998 *Phys. Rev.* B **58** 848
[6] Kamishima K, Hagiwara M and Yoshida H. 2001 *Rev. Sci. Instrum.* **72** 1472
[7] Information from NGK Insulator, Ltd.
[8] Uwatoko Y, Todo S, Ueda K, Uchida A, Kosaka M, Mori N and Matsumoto T 2002 *J. Phys. :Condens. Matter.* **14** 11291
[9] Information from Fuji Die Co., Ltd
[10] Murata K, Yoshino H, Yadav H O, Honda Y, and Shirakawa N 1997 *Rev. Sci. Instrum.* **69**, 2490
[11] Walker I R 1999 *Rev. Sci. Instrum.* **70**, 3402
[12] Nakamura F, Nakai R, Takemoto T, Sakaki M, Suzuki T, Alireza P L, Nakatsuji S and Maeno Y 2009 *Phys. Rev.* B **80** 193103
[13] Nakatsuji S, Hall D, Balicas L, Fisk Z, Sugahara K, Yoshioka M and Maeno Y 2003 *Phys. Rev. Lett.* **90** 137202
[14] Nakamura F, Goko T, Ito M, Fujita T, Nakatsuji S, Fukazawa H, Yoshiteru M, Alireza P, Forsythe D and Julian S R 2002 *Phys. Rev.* B **65** 220402

IOP Conf. Series: Journal of Physics: Conf. Series **969** (2018) 012090 doi:10.1088/1742-6596/969/1/012090

# Normal metal - insulator - superconductor thermometers and coolers with titanium-gold bilayer as the normal metal

**I M W Räisänen, Z Geng, K M Kinnunen and I J Maasilta**

Nanoscience Center, Department of Physics, University of Jyväskylä, P.O. Box 35 (YFL), FI-40014 University of Jyväskylä, Finland

E-mail: `ilmo.raisanen@jyu.fi`

**Abstract.** We have fabricated superconductor - insulator - normal metal - insulator - superconductor (SINIS) tunnel junctions in which Al acts as the superconductor, AlOx is the insulator, and the normal metal consists of a thin Ti layer (5 nm) covered with a thicker Au layer (40 nm). We have characterized the junctions by measuring their current-voltage curves between 60 mK and 750 mK. For comparison, the same measurements have been performed for a SINIS junction pair whose normal metal is Cu. The Ti-Au bilayer decreases the SINIS tunneling resistance by an order of magnitude compared to junctions where Cu is used as normal metal, made with the same oxidation parameters. The Ti-Au devices are much more robust against chemical attacks, and their lower tunneling resistance makes them more robust against static charge. More significantly, they exhibit significantly stronger electron cooling than Cu devices with identical fabrication steps, when biased close to the energy gap of the superconducting Al. By using a self-consistent thermal model, we can fit the current-voltage characteristics well, and show an electron cooling from 200 mK to 110 mK, with a non-optimized device.

## 1. Introduction

Normal metal - insulator - superconductor (NIS) tunnel junctions are versatile and useful devices for thermometry, cooling, and metrological applications for the definition of the ampere [1, 2, 3]. In most applications, the normal metal lead needs to be highly conductive, so copper has been a common choice. However, Cu has its problems when the junctions have to go through additional fabrication steps, such as a second layer of resist coating, plasma etching etc. We have observed that the Cu electrode degrades strongly with heating, and in contact with many commonly used solvents in electron beam resists. The Cu lead can degrade even in the lift-off step, if acetone is used as the lift-off solvent and if it has absorbed some moisture, e.g., from air. With such conditions acetic acid is generated, which can etch some of the Cu [4]. This degradation effect is demonstrated in Figure 1a, where nanopores are visible in an NIS sample with a Cu normal metal electrode. Thus, it is desirable to find other alternatives for the normal metal.

The Cu lead can be replaced with Au. However, Au has a well known adhesion problem with the commonly used oxidized or nitridized Si substrates. In order to overcome this issue, a thin wetting layer of Cr or Ti underneath the Au layer is typically used. Here, we implement a thin Ti layer (5 nm) underneath a thicker Au layer (40 nm) in the normal metal fabrication, similar to Refs. [5, 6]. Even though Ti is a superconductor below 0.5 K [7], with such a large mismatch

IOP Conf. Series: Journal of Physics: Conf. Series **969** (2018) 012090 doi:10.1088/1742-6596/969/1/012090

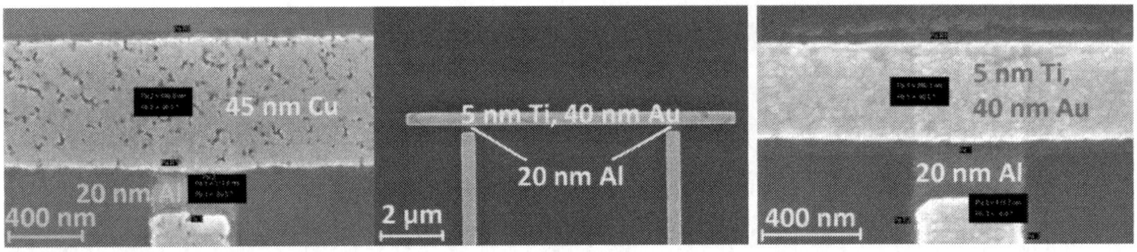


(a) (b) (c)

Figure 1: Scanning electron micrographs of (a) an NIS junction whose normal metal is Cu with clear degradation of Cu, (b) a SINIS junction pair whose normal metal is a Ti-Au bilayer, and (c) an NIS junction whose normal metal is Ti-Au.

of the layer thicknesses the direct contact with the Au layer causes the whole bilayer to be in normal state due to the inverse proximity effect [8].

In this article, we present the fabrication and measurement of SINIS tunnel junction pairs in which the superconductor is Al, the insulator is AlOx and the normal metal is a Ti-Au bilayer, see Figure 1b. The junctions are fabricated on a nitridized Si substrate using electron beam lithography, electron beam evaporation of the Al, Ti and Au films (thicknesses 20 nm, 5 nm and 40 nm, respectively), and in-situ oxidation of the Al in $\sim$ 40 mbar of pure oxygen for 4 min to form the tunnel barrier. As can be seen in Figure 1c, the Ti-Au island is smooth and uniform, and it never exhibited any obvious degradation after the lift-off in contrast to some Cu island samples (Figure 1a). We compare the current-voltage characteristics and cooling effects between 60 mK and 750 mK to otherwise identically fabricated SINIS junctions, but whose normal metal is Cu instead of Ti-Au. We found that the total tunneling resistances of the Ti-Au bilayer SINIS junctions were around 2.3 k$\Omega$ immediately after lift-off, and the resistances decreased by $\sim$ 10 % when the junctions were 6 weeks in normal atmosphere. This behavior is in contrast with Cu based SINIS junctions, which typically have an order of magnitude higher tunneling resistance with the same oxidation parameters, and a trend of increasing tunneling resistance with age.

## 2. Current-voltage characteristics

The current $I$ through a SINIS junction pair can be written in a symmetrized form as [1]

$$I(V) = \frac{1}{2eR_T} \int_{-\infty}^{\infty} N_S(E)[f_N(E-eV) - f_N(E+eV)]dE, \tag{1}$$

where $V$ is the total voltage, $R_T$ is the total tunneling resistance of the SINIS junction pair, and $f_N(E)$ denotes the Fermi-Dirac distribution in the normal metal lead at energy $E$. The quantity $N_S$ is the density of states (DOS) in the superconductor and it is typically given by

$$N_S(\epsilon, T_S) = \left| \mathrm{Re}\left( \frac{\epsilon + i\Gamma}{\sqrt{(\epsilon + i\Gamma)^2 - \Delta^2(T_S)}} \right) \right|, \tag{2}$$

where $\epsilon$ is the energy and $T_S$ is the temperature of the superconductor. The quantity $\Delta$ is the energy gap of the superconductor. The temperature dependence of $\Delta(T_S)$ can be calculated based on the BCS theory. Finally, $\Gamma$ denotes the Dynes parameter [9, 10], which is used to describe the broadening of the effective DOS, leading to sub-gap current through the junctions at voltages lower than $\Delta/e$. Below $T_S < 0.4\ T_C$ the gap has a very weak temperature dependence, in which case a SINIS device is sensitive to the normal metal electron temperature alone through $f_N(E)$.

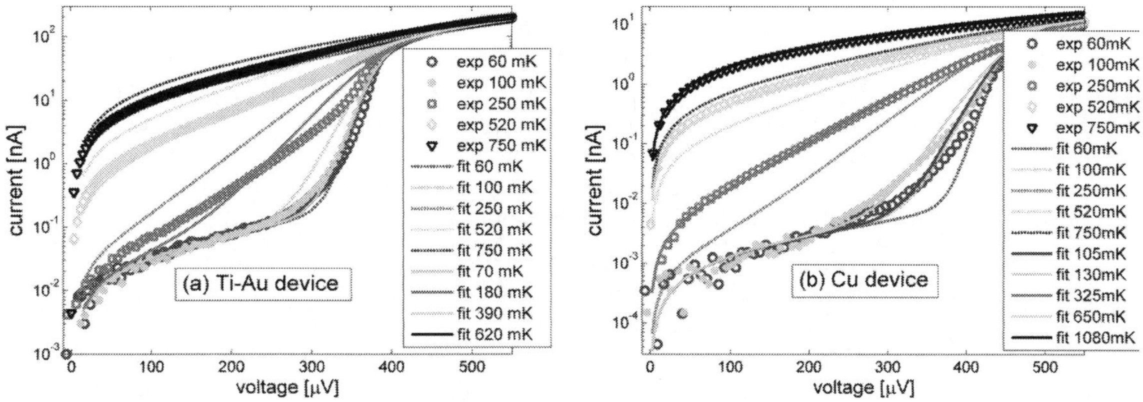


Figure 2: Measured current-voltage characteristics (open symbols). Theory fits of Equations (1) and (2) when assuming electron temperature to equal the cryostat temperature (dashed lines) and when electron temperature is used as a fit parameter (solid lines).

We measured the current-voltage characteristics of Ti-Au and Cu SINIS devices at five different temperatures between 60 mK and 750 mK. In Figure 2 we present the data in log-linear scale, together with equilibrium model ($T_S = T_N$) theory based on Equations (1) and (2). Figure 2a shows the data for the Ti-Au device, while Figure 2b presents the data for the Cu device. First, we make the theory fitting by assuming that the electron temperature in the normal metal equals the cryostat temperature, measured with a separate RuOx thermometer at the sample stage. These fits are shown by dashed lines. The quantities $\Gamma$ and $\Delta$, on the other hand, were used as fitting parameters. The fits give estimates $\Gamma = 6.5 \times 10^{-4} \times \Delta$ and $\Gamma = 3.5 \times 10^{-4} \times \Delta$ for the Ti-Au device and the Cu device, respectively, whereas for $\Delta$, we obtain $\Delta = 188$ $\mu$eV for the Ti-Au device and $\Delta = 220$ $\mu$eV for the Cu device. From Figure 2 we see that for higher bath temperatures, the fits shown by dashed lines are not good. Nevertheless, from the lowest 60 mK data, $\Gamma$ and $\Delta$ can be determined quite accurately. The differences in $\Gamma$ and $\Delta$ between the two devices are interesting. The variation of $\Gamma$ is perhaps not surprising, considering that the total tunneling resistance $R_T$, determined from the slope of the $I-V$ curves in a large bias regime (-7 mV ... +7 mV), varies by an order of magnitude between the two samples, 2.1 k$\Omega$ for the Ti-Au device and 28 k$\Omega$ for the Cu device. This can lead to different amount of absorbed noise power leading to variation of $\Gamma$ due to photon assisted tunneling [10]. The variation of $\Delta$, on the other hand, is an indication of differences in Al quality, a bit surprising result as the same evaporator and the same Al thickness were used.

The fits shown by dashed lines in Figure 2, where we assumed the electron temperature to equal the cryostat temperature, are not satisfactory, especially at temperatures above 100 mK. Therefore, we take the electron temperature as another fit parameter, see the solid lines in Figure 2. For the Ti-Au device, we see that the theoretical $I-V$ curves match the measurement data fairly well when the electron temperature is lowered from the measured bath temperature value. However, for the case of the Cu device, the electron temperature needs to be made higher than the bath temperature for the simulations to match the experimental data. Most notably, though, the fits are still not good at the lowest temperatures for both junctions types, most clearly seen for the Ti-Au device at bath temperature 250 mK.

## 3. Thermal model and cooling of electrons in the normal metal

To fully understand the measured $I-V$ curves, we must consider the cooling effect of an NIS junction. It is based on the fact that only electrons from the normal metal that have at least energy $\Delta$ can enter the superconductor. Since electrons of the normal metal obey the Fermi-Dirac distribution, there are hot electrons which can tunnel to the superconductor also at bias voltages $eV < \Delta$. Since only hot electrons escape from the normal metal, the temperature of remaining electrons in the normal metal lead decreases. The bias dependent heat flow from the normal metal to the superconductor in a single NIS junction is [1]

$$\dot{Q}_{\mathrm{cool}} = \frac{1}{e^2 R_T} \int_{-\infty}^{\infty} (E - eV) n_S(E) [f_N(E - eV) - f_S(E)] dE, \tag{3}$$

where $n_S(E)$ and $f_S(E)$ are the density of states and the Fermi-Dirac distribution in the superconductor at energy $E$, respectively. In dynamic equilibrium, cooling by the junctions in a SINIS device is balanced by the inflow of heat from the surroundings. This leads to the equation for a symmetric SINIS cooler [11]

$$2\dot{Q}_{\mathrm{cool}} = A(T_{\mathrm{bath}}^5 - T_N^5) + \beta(2\dot{Q}_{\mathrm{cool}} + IV) + I^2 R_N, \tag{4}$$

which describes three heating mechanisms: The first term on the right represents the heating from the substrate phonons that are at temperature $T_{\mathrm{bath}}$. The substrate phonons couple to the normal metal electrons that are at temperature $T_N$. The symbol $A$ denotes the coupling strength constant and it is given by $A = \Sigma\Omega$, where $\Sigma$ is the electron-phonon coupling constant and $\Omega$ is the volume of the normal metal lead. The second term on the right describes the flow of heat from the superconductor to the normal metal via phonons or back-tunneling, where $(0 \leq \beta \leq 1)$ is the fraction of total heat deposited in the superconductor that flows back to the normal metal. The third term on the right is the Joule heating of the normal metal whose resistance is $R_N$.

In Figure 3, we fit the $I-V$ curves with Equation (1), but including self-consistently the thermal model described by Equations (3) and (4) to the experimental data, using $A$, $T_S = T_{\mathrm{bath}}$, $\beta$ and $R_N$ as fitting parameters. For the Ti-Au device, the best fit is obtained for each bath temperature with $A = 5.78 \times 10^{-10}$ W/K$^5$, $\beta = 0.05$, and $R_N = 50\ \Omega$. On the other hand, for the Cu device, we get $A = 3.18 \times 10^{-10}$ W/K$^5$, $\beta = 0.05$, and $R_N = 40\ \Omega$. The fit quality improves significantly compared to those of the equilibrium model (Figure 2), specifically, the thermal model can account for the bias dependent curvature of the $I-V$ curves. The volume of the normal metal lead was measured with an atomic force microscope (AFM) and a scanning electron microscope (SEM), and it is 0.18 $\mu$m$^3$ for both junction types. We thus get estimates for the electron-phonon coupling constants, giving us for the Ti-Au bilayer $\Sigma_{\text{Ti-Au}} = (4.2 \pm 0.3) \times 10^9$ W/(K$^5$m$^3$), and for Cu $\Sigma_{\mathrm{Cu}} = (2.3 \pm 0.3) \times 10^9$ W/(K$^5$m$^3$). These values can be compared with literature values for Cu, Au and Ti thin films, with $\Sigma_{\mathrm{Cu}} = 2 \times 10^9$ W/(K$^5$m$^3$) [12, 13], $\Sigma_{\mathrm{Au}} = 2.4 \times 10^9$ W/(K$^5$m$^3$) [14] and $\Sigma_{\mathrm{Ti}} = 1.3 \times 10^9$ W/(K$^5$m$^3$) [15], respectively.

In Figure 4 we plot the bias dependent electron temperature $T_e$ of the normal metal obtained from the thermal model fits. We see that the electron temperature in the Ti-Au island (Figure 4a) decreases from 200 mK to 110 mK when the bias voltage is around 350 $\mu$V, with weaker effect at higher temperatures as expected. However, the Cu island clearly cools much less (Figure 4b). The main reason for the weaker cooling effect is the higher tunneling resistance (28 k$\Omega$) in the Cu device than in the Ti-Au device (2.1 k$\Omega$), implying a higher and/or wider tunneling barrier. However, cooling of the normal metal electrodes could be possibly made significantly stronger by increasing the junction area which is now only $(0.13 \pm 0.01)$ $\mu$m$^2$. The fact that the 60 mK and 100 mK bath temperature curves merge quickly after zero bias for the Ti-Au case in Figure 4a explains the similarity of the measured $I-V$ curves at 60 mK and 100 mK shown in Figure 3a.

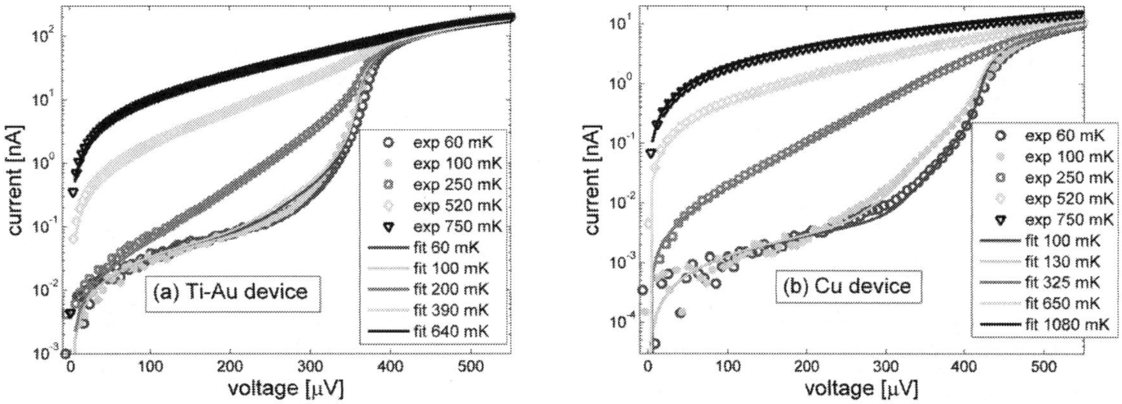


Figure 3: Measured current-voltage characteristics (open symbols) with thermal model fits using Equations (3) and (4) (solid lines).

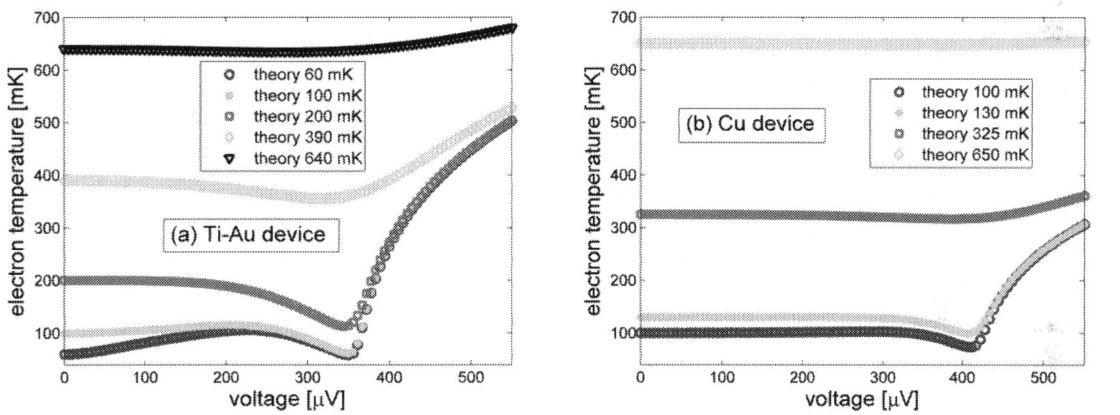


Figure 4: Electron temperature of the normal metal lead with respect to DC voltage bias obtained from the thermal model fits shown in Figure 3.

Figure 5 presents the cooling power $2\dot{Q}_{\mathrm{cool}}$ by the tunnel junctions and the corresponding coefficient of performance $\mathrm{COP} = 2\dot{Q}_{\mathrm{cool}}/(IV)$ of the Ti-Au and Cu cooler devices. The Ti-Au device has an order of magnitude higher cooling power than the Cu device as expected, and the cooling power is highest at highest temperatures for both devices. Around the optimal bias voltage for cooling, the cooling power and the COP match well with literature values $\dot{Q}_{\mathrm{cool}} = 0.06\Delta^2/(e^2 R_T)$ and $\mathrm{COP} = 0.25$ at $T = 0.25\Delta/k_B$ [1].

As the current-voltage characteristics of an NIS junction depend only on the temperature of the normal metal electrode when $T_S$ is well below $T_C$, which can be seen in Equation (1), a measurement of voltage $V$ at a constant current also gives the normal metal electron temperature in quasiequilibrium conditions [1]. We have also measured voltage vs. cryostat temperature curves to confirm that the Ti-Au junctions can also be operated in such a manner.

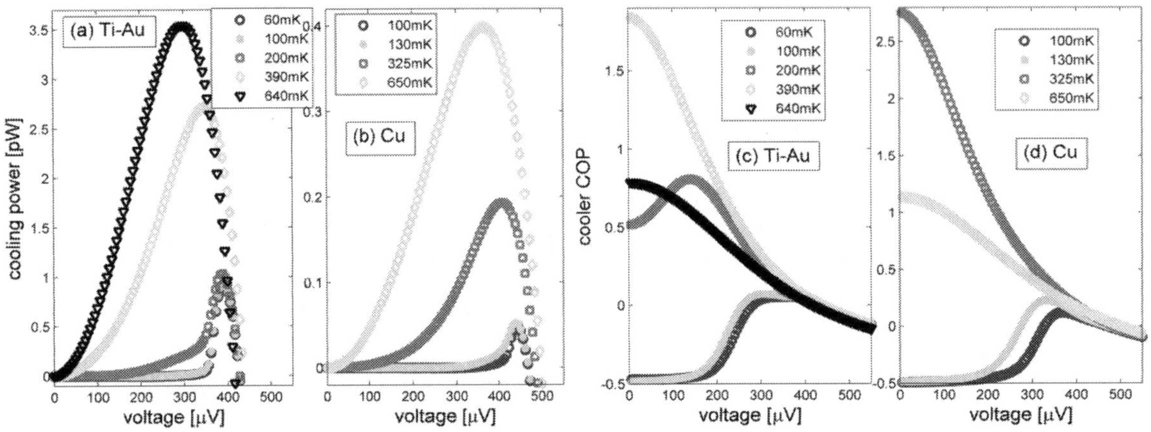


Figure 5: Cooling power and coefficient of performance of the Ti-Au and Cu cooler devices obtained from the thermal model fits shown in Figure 3.

## 4. Summary

This work shows that a Ti-Au bilayer is a promising material for the normal metal electrode in NIS junctions. The Ti-Au bilayer significantly improves the tolerance of the normal metal to solvents and thus further processing rounds, as compared to Cu. With the same processing parameters as with Cu devices, the tunneling resistance of Ti-Au devices is reduced by an order of magnitude, which, as we demonstrated, improves the electron cooling effect.

We managed to fit the experimental $I-V$ characteristics only with the help of a thermal model that took into account bias dependent cooling and heating of the NIS junctions, demonstrating a strong cooling effect even in a non-optimized structure. In the future, we plan to study the effects of the Ti-Au bilayer on the microscopic barrier properties in a more detailed manner, and optimize a cooler structure to demonstrate a much stronger cooling effect.

## Acknowledgments

This project is supported by Finnish Cultural Foundation and the Academy of Finland project number 298667.

## References

[1] Giazotto F, Heikkilä T T, Luukanen A, Savin A M and Pekola J P 2006 *Rev. Mod. Phys* **78** 217–74
[2] Pekola J P, Saira O-P, Maisi V F, Kemppinen A, Möttönen M, Pashkin Yu A and Averin D V 2013 *Rev. Mod. Phys.* **85** 1421–72
[3] Muhonen J T, Meschke M and Pekola J P 2012 *Rep. Prog. Phys.* **75** 046501
[4] Kagwade S V, Clayton C R, Chidambaram D and Halada G P 2001 *Electrochimica Acta* **46** 2337–42
[5] Koppinen P J, Isotalo T J and Maasilta I J 2009 *AIP Conf. Proc.* **1185** 318
[6] Faivre T, Golubev D S and Pekola J P 2015 *Appl. Phys. Lett.* **106** 182602
[7] Falge R L Jr 1963 *Phys. Rev. Lett.* **11** 248–50
[8] Fominov Ya V and Feigel'man M V 2001 *Phys. Rev.* B **63** 094518
[9] Dynes R C, Garno J P, Hertel G B and Orlando T P 1984 *Phys. Rev. Lett.* **53** 2437–40
[10] Pekola J P, Maisi V F, Kafanov S, Chekurov N, Kemppinen A, Pashkin Yu A, Saira O-P, Möttönen M and Tsai J S 2010 *Phys. Rev. Lett.* **105** 026803
[11] Chaudhuri S and Maasilta I J 2012 *Phys. Rev.* B **85** 014519
[12] Roukes M L, Freeman M R, Germain R S, Richardson R C and Ketchen M B 1985 *Phys. Rev. Lett.* **55** 422–5
[13] Karvonen J T, Taskinen L J and Maasilta I J 2007 *J. Low Temp. Phys.* **146** 213–26
[14] Echternach P M, Thoman M R, Gould C M and Bozler H M 1992 *Phys. Rev.* B **46** 10339–44
[15] Manninen A J, Suoknuuti J K, Leivo M M and Pekola J P 1999 *Appl. Phys. Lett.* **74** 3020–22

IOP Conf. Series: Journal of Physics: Conf. Series **969** (2018) 012091 doi:10.1088/1742-6596/969/1/012091

# A compact cryogen-free platform operating at 1 K or 50 mK

**A J Matthews, M Patton, T Marsh and H van der Vliet**

Oxford Instruments NanoScience, Tubney Woods, Abingdon, OX13 5QX, United Kingdom

E-mail: Anthony.Matthews@oxinst.com

**Abstract.** We report the design and performance characteristics of a compact cryogen-free platform. The system is based around a continuous 1 K pot which operates using a small (10 $m^3$ $h^{-1}$) room temperature circulation pump. The pot cools an experimental plate to $\approx$ 1.2 K, and has a cooling capacity of 100 mW at a temperature $\approx$ 1.9 K. Cooling the pot from room temperature to $<$ 2 K takes around 12 hours.

The temperature range of the platform can be lowered to $<$ 50 mK with the addition of a small dilution refrigerator, using the 1 K pot as a pre-cooling stage for the circulating $^3$He. The dilution stage has a typical (continuous) cooling capacity of 30 $\mu$W at 100 mK (300 $\mu$W at 250 mK) and is designed to operate with just 3 litres of (NTP) $^3$He.

## 1. Introduction

The cryogen free dilution refrigerator[1] has been an enabling technology for a range of experimental applications. The rapid uptake of these machines, particularly in the field of quantum information processing, has been driven by their ease of use, large experimental plates and the ability to integrate superconducting magnets[2]. The fact that these systems can be designed without an inner vacuum chamber means installing experimental services can be much more straightforward, and the desire to maximise the experimental payload has led to the design of ever larger and more powerful systems for these applications[3].

An alternative application space for milli-kelvin platforms, that has traditionally been satisfied with Adiabatic Demagnetization Refrigerators (ADRs)[4] or (complex) $^3$He refrigerators[5], is the development of photon detectors. For these applications the cooling capacity requirements at the milli-kelvin stage are much less demanding, so a small dilution refrigerator system would be suitable, likely with the added benefits of: continuous operation; lower base-temperatures; higher cooling-powers; and no requirement for a superconducting magnet (unless an experimental magnet is desired).

Dilution units with modest base temperature and cooling power requirements can be designed to operate at low $^3$He circulation rates. This low circulation rate constraint places a limit on the allowable power dissipation at the still. Increased cooling capacity at temperatures $\approx$ 1 K, which may be desirable for operating low-temperature amplifiers or microwave filters, can be provided by an additional $^4$He refrigeration stage (a '1 K pot')[6].

Additionally, a cryogen-free 1 K pot based system is useful in its own right: e.g. as a large sample-area platform to develop electronics that will subsequently be installed on the still of a

IOP Conf. Series: Journal of Physics: Conf. Series **969** (2018) 012091 doi:10.1088/1742-6596/969/1/012091

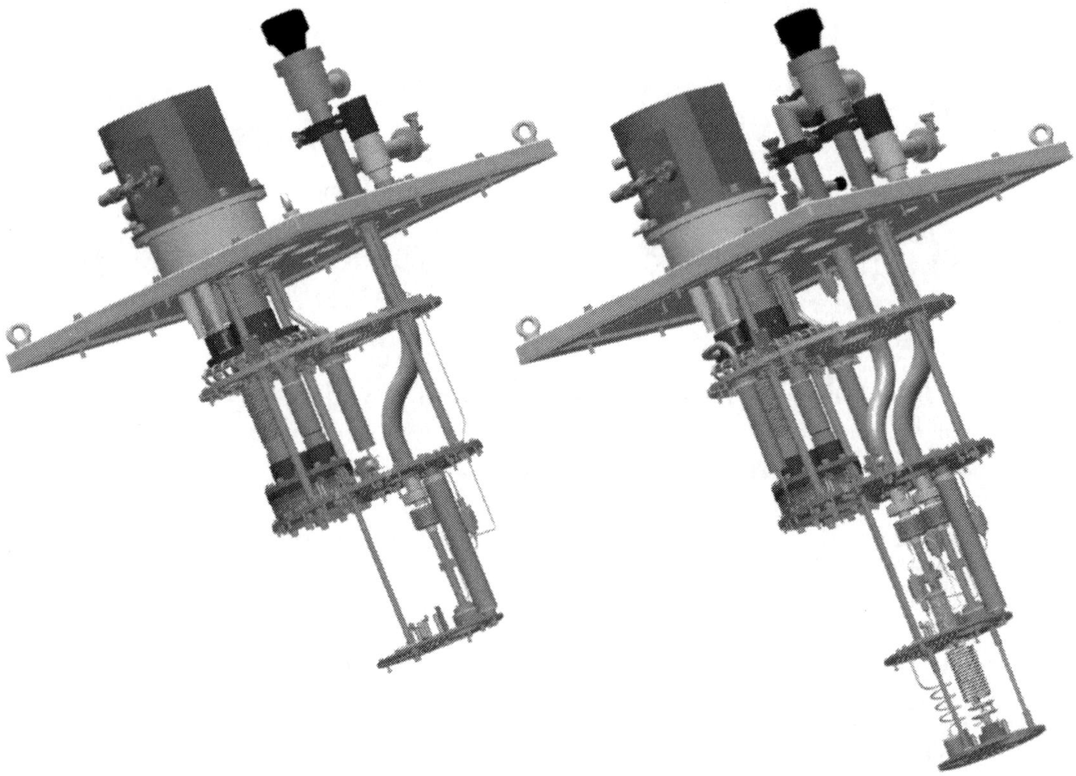

Figure 1: Models of both the 1 K platform (left) and the 50 mK platform (right) showing how the system can be configured.

dilution refrigerator; or as an alternative experimental platform to traditional (wet) Variable-Temperature Inserts (VTIs), where the available sample space is often limited.

## 2. System design

The top-level design concept for these systems is similar to that described previously[2], with a single vacuum space, however the size of the experimental plates has been reduced to keep the system compact. The cold environment is generated by a pulse tube cooler from Sumitomo[7] which has a remote rotary valve and a nominal cooling power at the second stage of 0.9 W at 4.2 K.

The system is configurable either as a '1 K' platform, or an additional dilution refrigerator stage can be added below the 1 K plate, figure 1. In the 50 mK configuration the still of the dilution refrigerator is in mechanical contact with, but thermally isolated from, the 1 K plate.

It is possible to integrate an experimental superconducting magnet, cooled by the pulse tube *via* a high thermal-conductivity link[2].

### *2.1. Pre-cooling*

To simplify the operation of the system (and the gas handling system, section 2.2) the cool down of the refrigerator is designed to be automatic (rather than automated). The pre-cooling from room temperature is affected by a heat switch, section 2.1.1, for the 1 K plate, and a heat pipe, section 2.1.2, for the mixing chamber plate (when the dilution stage is installed).

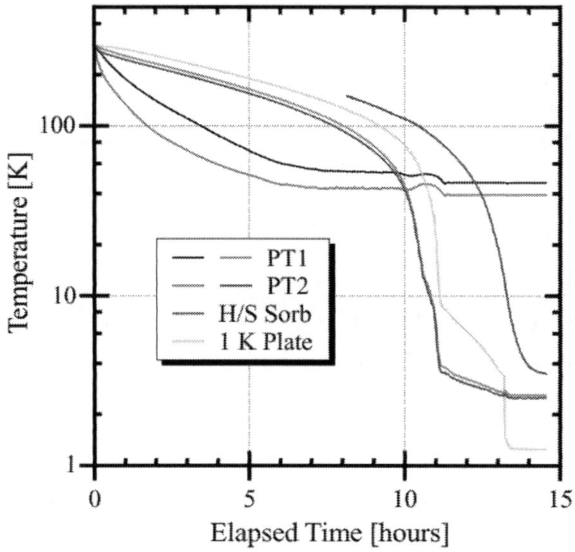


Figure 2: A plot showing the cool down of the 1 K stage (including the 1 K radiation shield) of the refrigerator from room temperature. The cooling of the heat switch (H/S) sorb was accomplished with a weak thermal link to the PT2 stage.

*2.1.1. Heat switch* Heat switches relying on the difference in thermal contraction between materials have been known for some time[8] and this principle has been used to manufacture gas-gap heat switches (GGHS) with small separations between conductors[9]. We have developed a robust manufacturing process for such thermal-contraction GGHSs which enables room temperature gaps between conductors to be zero[10]. A small adsorption pump (sorb), coupled to the cold end of the switch, is used to ensure that any residual gas is removed from the switch when the system is cold. During cool down, the conductance of the switch (and hence the heatload into the cold end) is sufficiently high that there is no need to actively temperature control the sorb, figure 2.

*2.1.2. Heat pipe* Heat pipes[11] can provide a very high thermal conductance over a selective temperature range, and have been shown to assist with the cool down of cryogen-free systems incorporating large magnets or experimental payloads[12]. We have deployed a helically-coiled-tube, gravity-driven heat pipe design (to minimise the residual conducted heat load to the mixing chamber plate) linking the 1 K plate and the mixing chamber plate, with an additional condensing heat exchanger at the PT2 stage. Using neon as the working fluid allows the mixing chamber plate to pre-cool to $\sim$ 25 K (in around six hours, once the PT2 stage has reached 25 K and the heat pipe is operational). Circulation of the mixture through the dilution circuit cools the system to temperatures below this.

Often heat pipes can be self-regulating, with the heat load at the condensing stage from the working fluid being sufficient to keep the system at its optimal operating temperature. Should the cooling power at the condensing stage (the PTR) be greater than the applied heat load, the heat pipe can be kept at its optimum temperature by applying additional heat. During the cool down of the system this situation can occur when the mixing chamber plate has cooled close to 25 K and the heat load on the condensing stage starts to fall; the additional heat can be applied automatically using the `alarm` and `relay` features of the temperature controller, a Lakeshore 372[13].

This technique has been verified with an additional mass of $\sim$ 3 kg attached to the mixing chamber plate to simulate a pressure cell for beam-line applications.

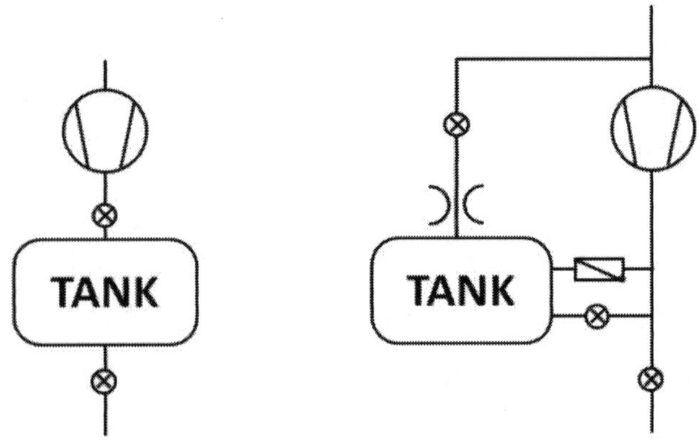


Figure 3: The layout of the two circulation circuits: $^4$He left; $^3$He right.

*2.2. Circulation circuits*
The automatic operation of the system extends to the two circulation circuits, the basic layout of which is shown in figure 3. The details of each circuit will be described in the following sections.

*2.2.1. $^4$He gas handling system* As $^4$He is relatively inexpensive, only a simple gas handling system (GHS) is required. A tank volume ($\sim$ 15 L) is connected directly to the exhaust of the circulation pump and charged at room temperature to a pressure $\sim$ 900 mbar. As the system cools and the gas is liquefied in the pot the tank, and hence condensing, pressure drops. An equilibrium state is achieved once the flow rate into the system (set by the value of the condensing impedance and condensing pressure) balances the rate at which $^4$He is boiled from the pot (determined by the total heat load into the 1 K stage).

*2.2.2. $^3$He/$^4$He gas handling system* Whilst a similar approach to the $^4$He GHS could, in principle, be followed for the $^3$He/$^4$He GHS this would be rather inefficient in the use of $^3$He. When operating at base temperature an ideal dilution refrigerator system will be circulating pure $^3$He. The condensing pressure will always be finite, and so buffering the condensing line with a tank volume could mean (at current prices) several thousand € of $^3$He 'sitting' in the tank.

To use the $^3$He more efficiently the storage tank should be empty in operation, to accomplish this in an automatic fashion the storage tank is connected to the front of the circulation pump *via* a flow restrictor (so that the condensing pressure can rise only slowly). The back of the pump is connected to the tank *via* a check valve (sensitive to the relative pressure difference between the condensing line and tank) which can discharge the condensing line to the tank quickly, should the condensing pressure rise above a threshold value. The size of the tank is then chosen (here a $\sim$ 50 L tank for $\sim$ 15 L of mix) such that the condensing pressure is automatically limited to the desired value, 700 mbar in this case. The system handles well the large changes in flow that occur as the main impedance cools, as well as the rapid liquefaction of the $^3$He/$^4$He mixture once the 1 K pot is cold, figure 4.

## 3. Cryogenic performance
*3.1. 1 K cooling loop*
The pot has a free volume of $\approx 8$ cm$^3$ and is coupled to a copper plate 180 mm in diameter. The pot pumping line incorporates a counter-flow heat exchanger to thermalise the returning $^4$He,

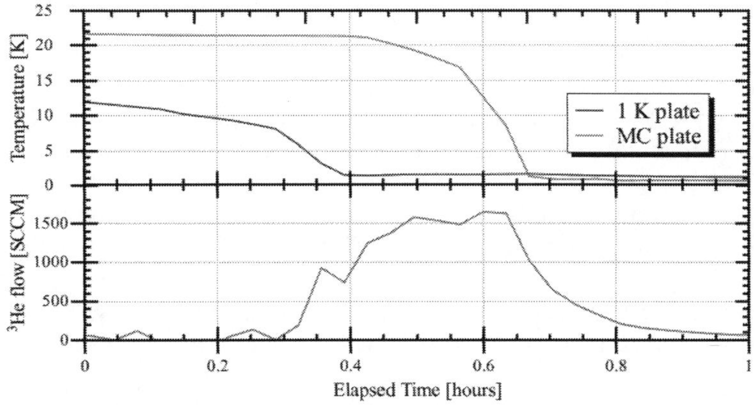


Figure 4: A plot showing how the automatic $^3$He condensing circuit allows large changes in flow as the 1 K pot cools.

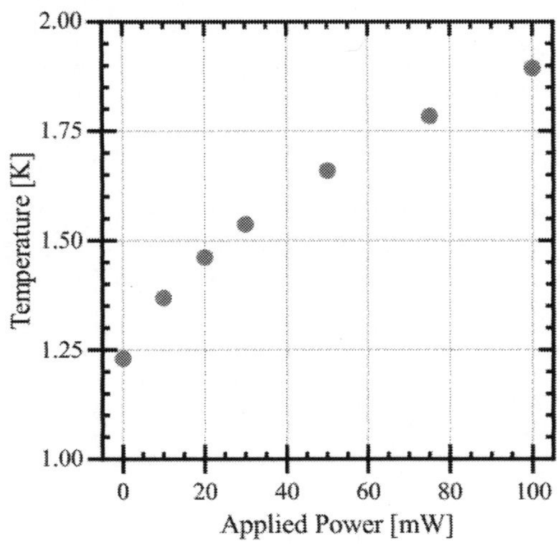


Figure 5: A plot showing the cooling capacity of the 1 K stage of the refrigerator.

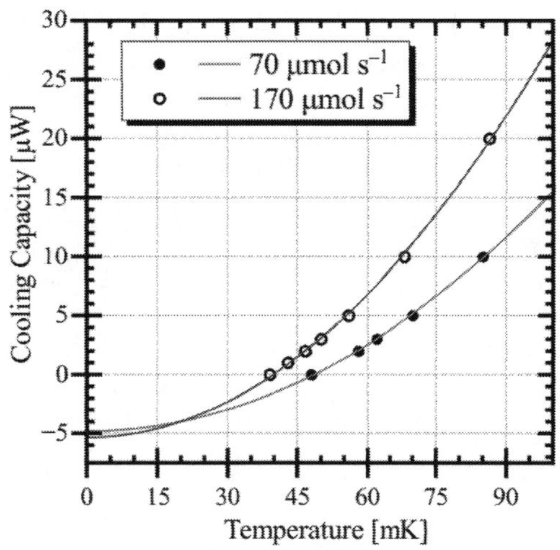


Figure 6: A plot showing cooling capacity of the dilution refrigerator stage as a function of temperature at different $^3$He circulation rates. The curves through the points are the usual $\dot{Q} = aT^2 - b$ trend expected, with $a$ and $b$ used as parameters in the fit.

and the cooling capacity available on the plate, when circulating with a 10 m$^3$ h$^{-1}$ dry pump, is as shown in figure 5.

For operation as a stand-alone '1 K' system, closed-loop temperature control of the plate is possible, at $\approx$ 1.3 K the measured stability was $1.30006 \pm 0.00031$ K.

*3.1.1. Operation with $^3$He* It is also possible to run the cooling loop with $^3$He, rather than $^4$He. Due to the higher vapour pressure of $^3$He, lower temperatures can be attained without altering the system configuration. Initial measurements have shown a base temperature of below 800 mK.

### *3.2. 50 mK stage*

The dilution unit has a diameter of $< 40$ mm and an overall length of $\approx 150$ mm. The mixing chamber is connected to an experimental plate 150 mm in diameter.

The circulating $^3$He is liquefied at the 1 K stage before heat-exchange with the liquid in the still, the continuous heat exchanger and two discrete (Ag sinter) heat exchangers prior to dilution in the mixing chamber.

The small size of the dilution unit means only $\sim$ 3 litres of (NTP) $^3$He are required to operate the refrigerator (with a total mixture volume $\sim$ 12-13 litres). The $^3$He is circulated using a 40 m$^3$ h$^{-1}$ dry pump.

The system attains base temperatures below 40 mK and typical cooling capacities at 100 mK of around 30 $\mu$W (rising to around 300 $\mu$W at 250 mK), as measured in vacuum on the experimental plate, figure 6.

The optimal $^3$He circulations rate is $\approx 200$ $\mu$mol s$^{-1}$ which equates to a required heat input to the still $\approx 5$ mW. Without the 1 K pot, this would be the upper limit on the allowed dissipation at temperatures around 1 K. The additional heat load on the 1 K pot from the circulating $^3$He is commensurate with the power applied to the still, whereas the cooling power available from the 1 K pot is much higher, figure 5, and so the addition of the $^4$He stage means an order of magnitude higher power dissipation at temperatures around 1 K is possible compared with using the still alone.

The relatively large background heat load into the mixing chamber, from the fits in figure 6 $\sim$ 4 $\mu$W, is due to the heat conducted through the support structure from the 1 K stage. The desire to make the support structure stiff and able to support heavy experimental payloads (10s of kg), coupled with the fact that these continuous cooling capacity figures (and the large cooling capacity at 1 K for thermalising experimental services) compare very favourably with the performance offered by ADR or $^3$He systems, means that no attempt has been made to further reduce the conducted load.

The (open loop) temperature stability of the mixing chamber stage is $< 20$ $\mu$K.

## Acknowledgments

The authors should like to thank D Gibbard for his contribution to the mechanical design of the systems described in this article, and A Czajkowski for his assistance with the build of the systems and subsequent acquisition of the experimental results presented.

## References

[1] Uhlig K 2002 *Cryogenics* **42** 73–77
[2] Batey G, Buehler M, Cuthbert M, Foster T, Matthews A J, Teleberg G and Twin A 2009 *Cryogenics* **49** 727 – 734
[3] Batey G, Matthews A J and Patton M 2014 *Journal of Physics: Conference Series* **568** 032014
[4] Hagmann C and Richards P 1995 *Cryogenics* **35** 303 – 309
[5] Klemencic G M, Ade P A R, Chase S, Sudiwala R and Woodcraft A L 2016 *Review of Scientific Instruments* **87** 045107
[6] Marx A, Hoess J and Uhlig K 2014 *Cryocoolers 18* ed S D Miller and R G Ross Jr (ICC Press) pp 481 – 486 ISBN 978-1-934-02104-0
[7] SHI Cryogenics Group URL `http://www.shicryogenics.com`
[8] Stuart R W and Hogan W H 1969 Thermal switch for cryogenic apparatus US Patent 3,430,455
[9] Franco J, Martins D, Catarino I and Bonfait G 2014 *Applied Thermal Engineering* **70** 115 – 121
[10] A full description of the design and operation of the switch will be the subject of a later publication.
[11] Grover G M, Cotter T P and Erickson G F 1964 *Journal of Applied Physics* **35** 1990 – 1991
[12] Teleberg G, Matthews A J, Batey G and Elford J 2012 *Journal of Physics: Conference Series* **400** 052036
[13] Lakeshore Cryotronics URL `http://www.lakeshore.com`

IOP Conf. Series: Journal of Physics: Conf. Series **969** (2018) 012092 doi:10.1088/1742-6596/969/1/012092

# Arcing time analysis of liquid nitrogen with respect to electrode materials

**Muhammad Junaid, Kun Yang, Hanming Ge, Jianhua Wang**

State Key Laboratory of Electrical Insulation and Power Equipment, School of Electrical Engineering, Xi'an Jiaotong University, Xi'an, P.R China

Email of the Corresponding Author: junaid@126.com

**Abstract.** Unlike sulphur hexafluoride ($SF_6$), liquid nitrogen ($LN_2$) is cost effective, environment friendly and cryogenic dielectric. It has astounding insulating properties with the potential to decrease power loss in switchgear applications due to its remarkably low temperatures. The basic research is however a necessity to observe the performance of $LN_2$ subjected to high luminance arcs. So far, there are no findings that refer to the arcing time inside the $LN_2$ environment. The objective of this work was to investigate the arcing times in $LN_2$ and compare the results with open air conditions using different electrode materials.

Experiments were conducted on different DC voltages and their arcing times were measured. Three different kinds of electrode materials, namely: pure copper (Cu), stainless used steel (SUS) and aluminium alloy (Al 6061) were tested under 1 atmospheric pressure.

The results revealed that $LN_2$ extinguishes arc in almost half the amount of time required by the open air insulation. With Al 6061 has the shortest arcing time, whilst Cu, the second best choice and SUS places last in the evaluation. It was encapsulated from the findings that $LN_2$ is a better choice than air insulation in terms of arc quenching and a better alternative to $SF_6$ when environment is the priority.

## 1. Introduction

Power system is the biggest man made marvel on earth. It is a major achievement in the history of science. Life would never have ameliorated to today's standards such that it has without it, as it saves time and human effort. Power system is subjected to severe environmental conditions so faults and annual maintenance is common practice. However, interruptions in any form are always undesirable. It results in blackouts, brownouts and instabilities in the power system [1]. In order to ensure continuous power, the switchgear protection apparatuses are employed at primary and secondary substations. These protection systems ensure the safety of the consumers and the expensive equipment installed.

Gas insulated switchgear (GIS) is conventionally used for high voltage power switchgear protection. It uses $SF_6$ gas as a quenching medium for all the components housed inside the metallic enclosure. Studies have shown that $SF_6$ gas has one of the best arc quenching properties. But despite its merits, there are some serious problems associated with it. The most important one is the issue of global warming, as it

IOP Conf. Series: Journal of Physics: Conf. Series **969** (2018) 012092 doi:10.1088/1742-6596/969/1/012092

is hazardous for the environment even though it is still exercised for the high voltage switchgear applications. In GIS, the $SF_6$ gas leakage routinely occurs during some of the below stated conditions;

- It has been observed that there is a continuous but slighter leakage of gas from the $SF_6$ bottle.
- In the course of maintenance, there is gas leakage.
- During fault conditions, especially after arc occurrence, other hazardous gases are produced which mixes up with $SF_6$ and it doesn't work properly afterwards. So as an end result, this mixture of gases has to be ejected. Subsequently, there will be an ejection of $SF_6$ gas into the air.

The Kyoto protocol listed $SF_6$ as greenhouse gas [2]. It has global warming potential of more than 23500 compared to carbon dioxide ($CO_2$) and has a lifetime in air of about 3200 years. So a quest for the alternative to $SF_6$ has so far been practiced. Since the switchgear system has very stringent rules, so the alternate insulation media should ensure high dielectric performance, rapid arc quenching ability, lower boiling point, high vapour pressure at low temperature, high heat dissipation rate, compatibility with the switchgear, design compatibility like GIS, and the list goes on and on. It also should meet the very strict health and safety prerequisites like non-flammability and lower toxicity. And most importantly the environmental conditions, ie, low global warming potential (GWP), no ozone depletion potential and minimal impact on the environment [3]. $LN_2$ is well thought out to be a promising alternative to the $SF_6$ gas. As it has outstanding dielectric properties and could also act as a coolant. Moreover, it can compensate the switchgear losses, i-e about 7~10% of electric power is lost in the grids [4, 5].

In June 2008, The editorial board of IEEE *Transactions on Dielectric and Electrical Insulation* (TDEI) selected Prof. T. Tanaka of Waseda University, Japan & Prof. I. Sauers of Oak Ridge National Laboratory, USA as guest editors to summarize the performance of cryogenic dielectrics. They concluded that $LN_2$ is a better and economical choice to Liquid Helium (LHe) [6]. Nitrogen even in gas state has far much better dielectric properties than helium gas and is easily available.

The above stated reasons clarify our commitment towards $LN_2$ over LHe. We are intended to come up with such a design that uses $LN_2$ as an insulating and cooling medium. We termed it as, Liquid Nitrogen Insulated metal enclosed Switchgear (LNIS) [7]. Figure 1 shows the conceptual design of the LNIS. Superconducting fault current limiter (SFCL) is connected in series with the fast acting vacuum interrupter to first limit and then interrupt the short circuit currents.

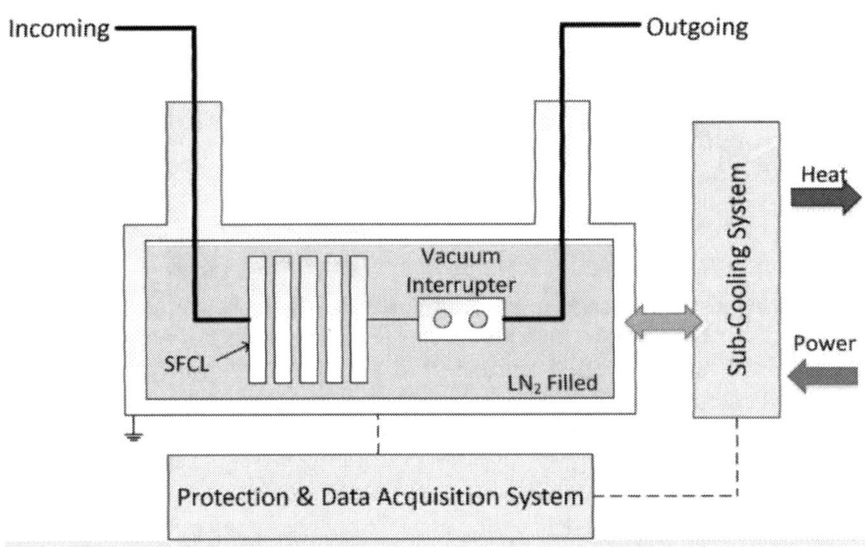


**Figure 1.** Liquid Nitrogen Insulated metal enclosed Switchgear (LNIS).

For the future design of LNIS, all of the GIS components should be tested with $LN_2$ as well. In case of fault, the short circuit current is interrupted by On-Load switches (such as circuit breakers), but in reality the Off-Load switches (ie isolators or disconnectors) also have to interrupt a comparatively small amount of current (roughly around 10 ~ 100 Amps). So investigation of the arcs in $LN_2$ environment is also very important. After going through the literature, we came up with the conclusion that no one has ever investigated high luminance arcs in $LN_2$. So the motivation of this very work was to analyse the arcing time of $LN_2$ and compare the results with the air insulation using different electrode materials.

## 2. Experimental setup

### *2.1. Schematic diagram*

Experiments were carried out on four different DC voltage levels (ie, 200, 300, 400 and 500 volts). An RLC current source was used to imitate DC situation. Hall effect sensors, high voltage probes and high performance oscilloscopes were used to measure the arc voltages and arc currents in the experimental setup. Optical observations were witnessed using a high speed camera. To ensure safety, all the tests were executed on 1 atmospheric pressure of $LN_2$ and air.

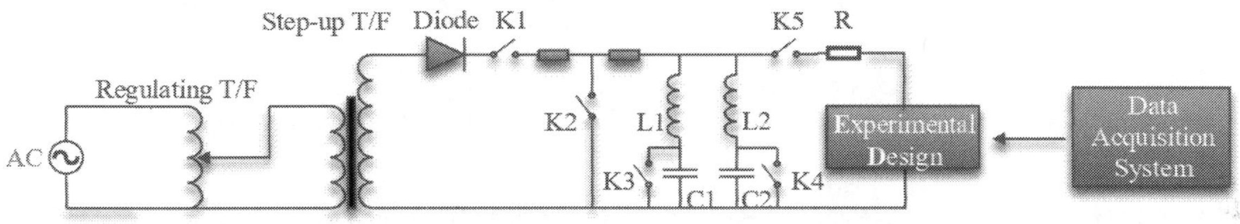


**Figure 2.** Schematic diagram of the experimental setup.

Figure 2 shows the schematic diagram of the experimental setup. Here C1 & C2 are connected in parallel giving an accumulated capacitance of 100mF. Similarly, the L1 & L2 are connected in parallel as well with the accumulated inductance of 0.101mH and R is 1 Ohm. K1, K2, K3 and K4 are vacuum contactors used for switching. Whereas, K5 is a 12 KV/47KA vacuum circuit breaker used as making switch (MS).

### *2.2. Experimental design*

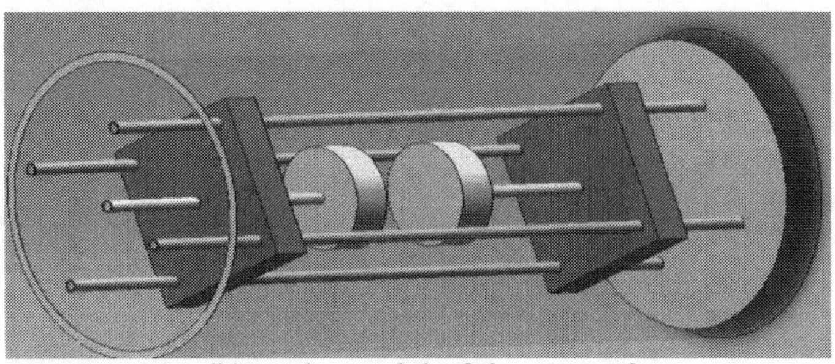

**Figure 3.** Solid-works model of the supporting structure.

Figure 3 shows the 3D model of the supporting structure with the electrodes. As it can be seen here, plane to plane (P-P) type electrodes geometry was employed for all the cases. The electrodes were symmetrical in nature. The diameter of the electrodes was 50mm and two 80mm long/10mm thick, square shaped nylon plates along with four 200mm long nylon thread rods with a diameter of 6mm each, were used as supporting insulators.

### *2.3. Methodology*

Figure 4 demonstrates the methodology of creating high luminance arcs. A 5 Amps fuse wire is placed between the two symmetrical P-P electrodes in such a way to initiate arc plasma. The experimental setup was planned very carefully so that no restrikes happens during the experiments. It was also made certain at first hand through calculations, MATLAB simulations and trail experiments that the RLC current source is not completely discharged. So that the RLC current source could be treated as a DC source.

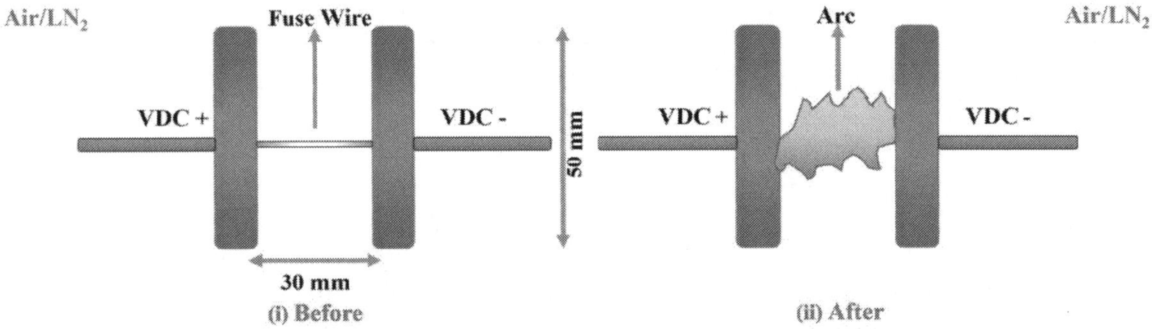


**Figure 4.** Arc plasma initiation technique.

## 3. Results

Below is the Table 1 for the different source voltages, interrupting currents, max. interrupting currents obtained and max. arcing time reduction rates obtained during the experiments;

**Table 1.** Statistical analysis of the findings.

| No. | Source Voltage (VDC) | Interrupting Current (Amps) | Max. Interrupting Current obtained (Amps) | Max. Arcing time Reduction rate by $LN_2$ (%) |
|---|---|---|---|---|
| 1 | 200 | 190 | 195 | ~79 |
| 2 | 300 | 280 | 277.5 | ~52 |
| 3 | 400 | 370 | 375 | ~46 |
| 4 | 500 | 460 | 474 | ~47 |

*Here ~ means approximate value.*

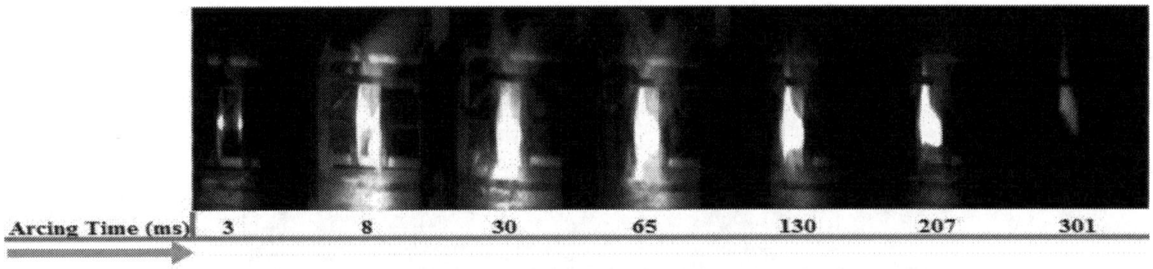


**Figure 5.** Transition of arc with time (Air insulation, 500 VDC).

Figure 5 shows the pictorial demonstration of the arcing time in air insulation. The applied voltage was 500 VDC, interrupting current 435 Amps, electrodes type P-P, electrodes material SUS, and the arc quenching time was 301 ms.

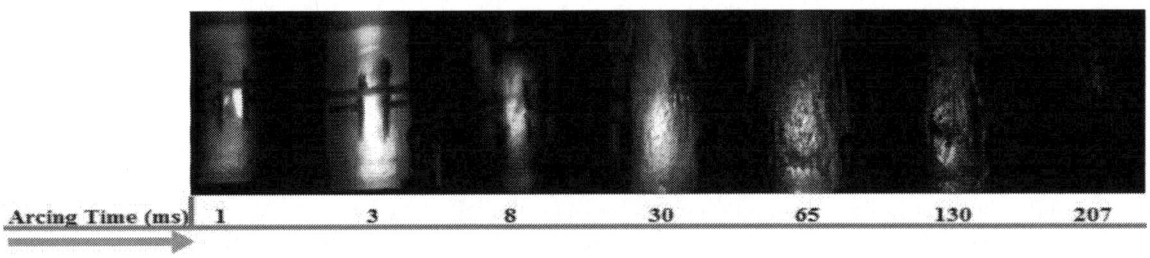


**Figure 6.** Transition of arc with time ($LN_2$ insulation, 500 VDC).

Figure 6 shows the pictorial demonstration of the arcing time in $LN_2$ insulation. The applied voltage was 500 VDC, interrupting current 453 Amps, electrodes type P-P, electrodes material SUS, and the arc quenching time obtained was 207.7 ms. It can be seen from the results of this case that the arcing time is reduced by 31% in $LN_2$.

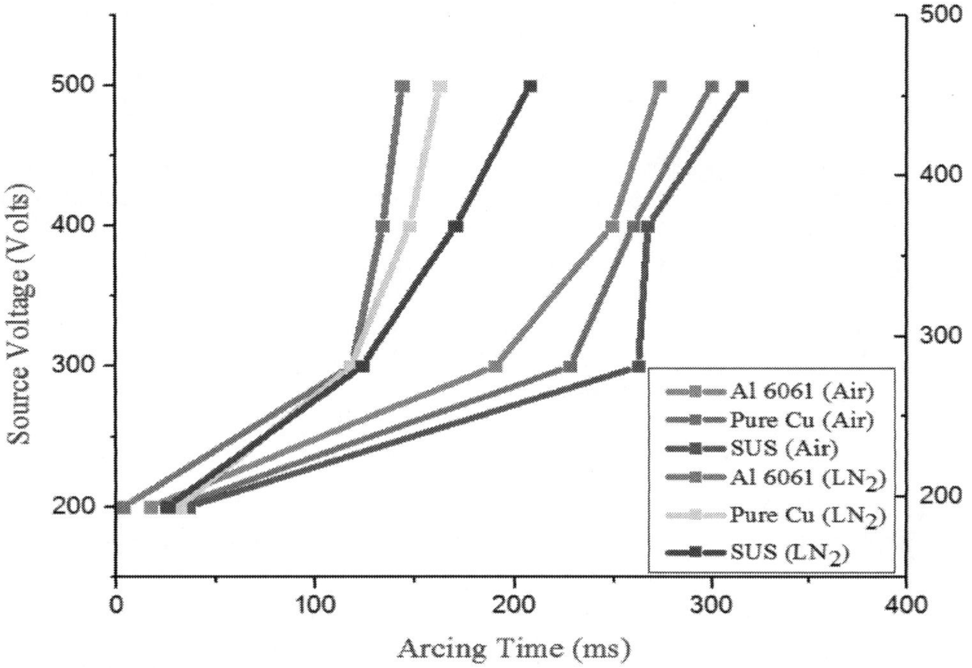


**Figure 7.** Source Voltage vs. Arcing time vs. Interrupting Current.

Figure 7 shows the relation between arcing time, applied voltage and interrupting current, it can clearly be seen that arcing times in $LN_2$ are quite shorter than in the case of air insulation. In terms of materials, Al 6061 have the shortest arcing time, whereas Cu is placed in second position, and SUS has the longest arcing time in comparison to the former ones.

## 4. Discussion

$LN_2$ increases the potential arc voltage which results in extinguishing the arcs quickly. So it may be considered an efficient approach to use it as arc quenching medium for future metal enclosed switchgear. Secondly, it can be observed from figures 5 and 6 that the arc luminance gets brighter very quickly in $LN_2$ compared to the case in air. It is due to different environmental conditions. Moreover, there is internal pressure build up in $LN_2$ due to high density arcs, and as an end result bubbles ($GN_2$) are generated which might show brighter luminance. The temperature of the arc is of interest here. And the temperature gradient may also have some influence on the arc behaviour. One thing worth noticing here is that air is the mixture of gases whereas $LN_2$ is the condensed form of nitrogen. So the arc behaviour will always be different, Hiroharu Kawasaki et Al., stated that the discharge characteristics is always the same regardless of the fact whether it's in LHe, $LN_2$ or even pure water [8]. Thus, the comparison of air insulation with $LN_2$ is justified.

Thirdly, all the three kinds of electrodes were perfectly smooth but SUS type was seemingly more polished as it is one of its property so it may influence the arcing time of this material. When there's arc between SUS electrodes, $CO_2$ gas is produced as a result of deterioration of the electrodes [9] and $CO_2$ has less arc quenching ability than $LN_2$ so these may be the reasons why it has the longest arcing time compared to the other two in the race. If we go through the literature of the contact materials for high voltage circuit breakers, we would realise that SUS is never used as electrode material because it produces $CO_2$ as by-products. Another reason that may have some influence on the arcing time is the *contact resistance* ($R_c$). The $R_c$ of SUS is much higher than that of Cu but according to [10], the $R_c$ of electrodes gets very lower in $LN_2$, so this argument is not applicable here.

## 5. Conclusions

131 tests were executed and interrupting currents up to 474 Amps were successfully interrupted in $LN_2$ environment. $LN_2$ extinguished arcs in almost half of the time required by open air insulation when P-P electrodes geometry is used. ie the arcing time is reduced by 40.829% (*Mean Value*) in $LN_2$. Materials wise, Al 6061 showed the shortest arcing time in both air and $LN_2$ while the Cu took a little longer than Al 6061 and SUS had the longest arcing time in comparison to the former ones.

It can be encapsulated from the findings that $LN_2$ has the potential to be used as insulating and coolant medium for future metal enclosed switchgear system. It is a better choice than air insulation in terms of arc quenching and a promising alternative to the $SF_6$ gas. The results established so far are expected to be considered as the base for future experiments.

## Acknowledgments

This work was solely supported by National Natural Science Foundation of the People's Republic of China under the Project # 51323012, entitled: "*The theory and technology of environment friendly electrical power equipment*". The Applied Superconductivity Lab used for experiments, is part of the State Key Laboratory of Electrical Insulation and Power Equipment.

## 6. References

[1] A. Castillo, "Risk analysis and management in power outage and restoration: A literature survey," *Electric Power Systems Research,* vol. 107, pp. 9-15, 2014.
[2] S. Oberthür and H. E. Ott, *The Kyoto Protocol: International Climate Policy for the 21st Century*, 1st ed. Germany: Springer-Verlag Berlin Heidelberg, 1999.
[3] Y. Kieffel, T. Irwin, P. Ponchon, and J. Owens, "Green Gas to Replace SF6 in Electrical Grids," *IEEE Power & Energy Magazine,* vol. 14, pp. 32-39, 2016.
[4] J. K. Sarrao, W-K Bozovic, I. Mazin, I. Seamus, J. C. Civale, L. Christen, D. Horwitz, J. Kellogg, G. Finnemore, D. Crabtree, G. Welp, U. Ashton, C. Herndon, B. Shapard, L. Nault, R. M., "Basic Research Needs for Superconductivity," *Office of Science, US DOE*, 2006.
[5] A. P. Malozemoff, "High Tc for the power grid," *Nature Materials,* vol. 6, pp. 617-619, 2007.
[6] T. Tanaka and I. Sauers, "Editorial - Cryogenic dielectrics," *IEEE Transactions on Dielectrics and Electrical Insulation,* vol. 15, pp. 619-619, 2008.
[7] J. Wang, Y. Tan, L. Zhang, Y. Geng, Z. Liu, and S. Wang, "Conceptual Design of a Liquid-Nitrogen-Insulated Metal-Enclosed Switchgear," *IEEE Transactions on Applied Superconductivity,* vol. 26, pp. 1-5, 2016.
[8] H. Kawasaki, T. Shigematsu, K. Imasaka, T. Ohshima, Y. Yagyu, and Y. Suda, "Discharge characteristics in liquid helium, liquid nitrogen and pure water preparatory to fabrication of carbon nanomaterials," *Journal of Physics: Conference Series,* vol. 400, p. 052010, 2012.
[9] J. X. Johnson, B. K. Reck, T. Wang, and T. E. Graedel, "The energy benefit of stainless steel recycling," *Energy Policy,* vol. 36, pp. 181-192, 2008.
[10] K. Golde and V. Hinrichsen, "Contact Resistance of Vacuum Interrupters for Electric Power Systems in Liquid Nitrogen at Direct and Alternating Current," in *ICEC 2014; The 27th International Conference on Electrical Contacts; Proceedings of*, 2014, pp. 1-5.

# Design and expected performance of a compact and continuous nuclear demagnetization refrigerator for sub-mK applications

**Ryo Toda**[1], **Satoshi Murakawa**[1], **and Hiroshi Fukuyama**[1,2]

[1]Cryogenic Research Center, The University of Tokyo, Tokyo 113-0032, Japan
[2]Department of Physics, The University of Tokyo, Tokyo 113-0033, Japan

E-mail: toda.ryo@mail.u-tokyo.ac.jp, hiroshi@phys.s.u-tokyo.ac.jp

**Abstract.** Sub-mK temperatures are achievable by a copper nuclear demagnetization refrigerator (NDR). Recently, research demands for such an ultra-low temperature environment are increasing not only in condensed matter physics but also in astrophysics. A standard NDR requires a specially designed room, a high-field superconducting magnet, and a high-power dilution refrigerator (DR). And it is a one-shot cooling apparatus. To reduce these requirements, we are developing a compact and continuous NDR with two $PrNi_5$ nuclear stages which occupies only a small space next to an appropriate pre-cooling stage such as DR. $PrNi_5$ has a large magnetic-field enhancement on $Pr^{3+}$ nuclei due to the strong hyperfine coupling. This enables us to enclose each stage in a miniature superconducting magnet and to locate two such sets in close proximity by surrounding them with high-permeability magnetic shields. The two stages are thermally connected in series to the pre-cooling stage by two Zn superconducting heat switches. A numerical analysis taking account of thermal resistances of all parts and an eddy current heating shows that the lowest sample temperature of 0.8 mK can be maintained continuously under a 10 nW ambient heat leak.

## 1. Introduction

In general, thermal fluctuations and specific heats of materials are substantially reduced at low temperatures near absolute zero. So one can improve the sensitivity of various particle detectors such as X-ray detector dramatically by cooling. It is also expected that unique ordered states and phase transitions of materials and new physical phenomena appear at lower temperatures. Nowadays, an automatically controllable dilution refrigerator (DR), which can produce a temperature as low as 10 mK continuously, is commercially available. Then many researchers, even who are not familiar with low temperature techniques, can make use of such very low temperatures easily. This contributes to the progress of science in wide areas. However, further low temperature experiments below 1 mK (sub-mK) are still highly demanding for non-specialists to use.

Usually, sub-mK temperatures are achievable with a copper nuclear demagnetization refrigerator (NDR) [1]. The principle of NDR is based on adiabatic demagnetization cooling of nuclear spins, a magnetocaloric effect. This method was established in the 1970s and has been contributed to many discoveries in the research field of quantum liquids and solids. The cooling procedure of NDR is as follows. First we apply a strong magnetic field $B_{\mathrm{max}}$ (typically

8 T) to copper nuclear spins ($I = 3/2$) and pre-cool them down to $T_\mathrm{i} = 10 \sim 15$ mK by a DR. In this process, the entropy of nuclear spin system is reduced by $\Delta S$ from its maximum value ($= Nk_\mathrm{B} \ln 4$). Here, $N$ is the number of spins. Then we thermally disconnect the spin system from DR by turning a superconducting heat switch (HSW) open and decrease $B$. During this adiabatic demagnetization process, the temperature of spin system decreases keeping $B/T$ constant in an ideal condition. Non-idealities come from, for example, heat loads to cool a sample and the copper electron system simultaneously and an eddy current heating of the copper stage. We can keep the lowest temperature, which is determined by the final field and the thermal loads, constant by reducing $B$ slowly so that an ambient heat leak is just absorbed as an entropy increase of spin system until $\Delta S$ becomes zero. Thus the cooling capacity of NDR is given by $\Delta S$, and this is a one-shot cooling method.

In order to obtain a large $\Delta S$, we need to reduce $T_\mathrm{i}$ and increase $B_\mathrm{max}$ and $N$ as much as possible. Typical sizes of the copper nuclear stage are more than 30 mm in diameter and 250 mm in length. We thus inevitably need to use a superconducting magnet of large bore diameter and of high magnetic field, and a powerful DR with a cooling power higher than 200 $\mu$W at 100 mK. At the same time, the sample space should be well isolated from fringing fields of the demagnetization magnet. As a result, the length of NDR becomes too long (more than 5 m when the DR insert is removed from a dewar) to fit in standard laboratory rooms demanding a specially designed room with a deep pit or a high ceiling. This is one of major barriers for application of the sub-mK environment to wider research fields.

Usage of recently developed cryogen-free DR (dry-DR) as a pre-cooler for NDR [2] can make the system size shorter since usually outer vacuum is separable in dry-DR. However, the large copper nuclear stage and superconducting magnet still limit the space conservation.

In this article, we describe design concept and results of feasibility study of the performance of a new compact demagnetization refrigerator with two nuclear stages of $PrNi_5$, a hyperfine enhanced nuclear magnet, connected in series to maintain the sub-mK environment continuously and easily.

## 2. Design concept of CNDR

If we have two independent magnetic refrigeration units, it is possible to keep a constant temperature below 1 mK continuously by using one of them as a heat absorber connected to the sample and the other as a shuttle stage to pump an entropy increase of that unit to DR. The two-unit setup is also good for minimizing the apparatus size because we don't have to maximize the cooling capacity of each unit. Such continuous refrigeration has previously been realized in an adiabatic demagnetization refrigerator for electronic spins (ADR) developed by Shirron et al. [3, 4]. By connecting one CPA and two GLF salt stages in series, they were successful to maintain the temperature of a soft X-ray calorimetric detector at $T = 50$ mK continuously on a satellite. In principle, it should be possible to extend this technique to the sub-mK region by replacing the electronic spins with nuclear spins.

The new continuous nuclear demagnetization refrigerator (CNDR) we propose here has two nuclear stages of $PrNi_5$. In this Van Vleck paramagnet, $Pr^{3+}$ nuclear spins ($I = 5/2$) feel an about 12 times higher effective magnetic field than the applied one due to the strong hyperfine coupling [1, 5]. This feature is essential to make the apparatus compact, since a sizable amount of $\Delta S$ can be obtained even in a relatively low field. For example, at $T_\mathrm{i} = 15$ mK, $\Delta S$ per unit volume of $PrNi_5$ in $B_\mathrm{max} = 1.2$ T is nearly the same as that of Cu in 8 T. The lowest achievable temperature of the $PrNi_5$ stage is limited to around 0.5 mK by its nuclear ferromagnetic ordering at 0.4 mK [6]. However, this would not be a serious problem for most realistic applications, and $PrNi_5$ has eventually been used as an alternative coolant to copper in previous single-stage NDRs [5, 7, 8, 9].

In our CNDR, one of the two nuclear stages will consist of three cylindrical $PrNi_5$ rods (total

IOP Conf. Series: Journal of Physics: Conf. Series **969** (2018) 012093 doi:10.1088/1742-6596/969/1/012093

volume $\approx$ 10 $cm^3$ or 80 g). The size of each rod is 6 mm in diameter and 120 mm in length. Since the size and $B_{max}$ of our demagnetization magnet are small (bore diameter = 22 mm, length = 140 mm, $B_{max}$ = 1.2 T), the fringing field can easily be suppressed by surrounding it with a high permeability magnetic shield of 40 mm in outer diameter. Such compactness enables us to install the two sets of magnetic units within a small space around the mixing chamber of DR. Since the demagnetization magnet is long enough compared to the $PrNi_5$ rods and the magnetic shielding further improves the field homogeneity, the difference between a root-mean-square field over the whole rods and the field at the magnet center is only 2%.

Although we are still in the process of fine tuning of geometry of the CNDR, the basic concept is to connect the two nuclear stages (NS1 and NS2) in series between the sample and mixing chamber through two HSWs (HSW1 and HSW2). Each HSW consists of 6 zinc foils of 0.25 mm thick contacting to 7 silver foils on both sides. After annealing separately, they will diffusively be bonded by heating to 230 °C. Zinc was chosen because of its low superconducting critical field ($B_c$ = 5 mT) and of the resultant low magnetic heat-release ($\propto B_c^2$) on turning open at the lowest temperature. The overall diameter of HSW including a superconducting solenoid and a magnetic shield will be 17 mm.

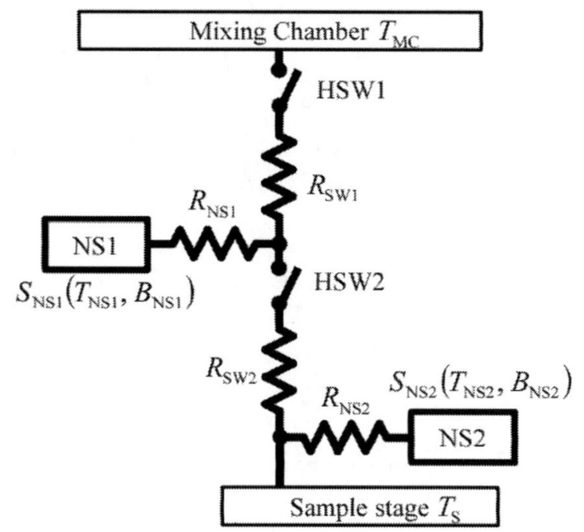


**Figure 1.** Block diagram of CNDR used for the thermodynamic simulations shown in figures 2 and 3.

## 3. Expected Performance of CNDR

All thermal links connecting each parts will be made of properly annealed pure silver. They will have tiny but non-negligible thermal resistances including contact resistances as shown in figure 1. For instance, $R_{NS1}$ consists of a resistance of a silver thermal link (15 silver wires of 1 mm in diameter for each $PrNi_5$ rod) and a contact resistance between the wires and rod by Zn soft soldering for NS1. $R_{SW1}$ is an overall resistance of the assembled HSW1. Based on the thermal block diagram shown in figure 1, we have evaluated the performance of CNDR numerically. In the evaluation, we assumed that $R_{SW1} = 100$ n$\Omega$ and $R_{NS1} = R_{NS2} = R_{SW2} = 50$ n$\Omega$ in the unit of electrical resistance, and thermal resistances were calculated from them through the Wiedemann-Franz law. These values are estimated from results of test annealing and measurements of short pieces [10]. Note that electrical resistance can be measured much more easily than thermal resistance. Various physical properties of $PrNi_5$ were taken from the book [1].

When the HSW1 or the HSW2 is kept open, a heat leak from the mixing chamber at 30 mK, the highest operation temperature, to sample stage will be less than 0.5 nW through it. On switching HSW2 to the open state, a latent heat of about 250 nJ will be released to the NS2 side. These are realistic estimations based on our previous experiences with a similar Zn HSW [11]. By employing standard thermal isolation techniques at sub-mK temperatures [1], a total heat leak to NS2 would be reduced to 1 nW in several weeks as was obtained in the previous single-stage $PrNi_5$ NDRs [7, 8]. A somewhat larger value ($\sim$ 5 nW) was reported if dry-DR is used as a pre-cooler presumably due to its larger mechanical vibrations [9].

Figure 2 shows a calculated entropy ($S$) vs. temperature ($T$) diagram of the two $PrNi_5$ stages in the continuous operation. Here we set a constant sample-stage temperature ($T_S$) at 0.8 mK assuming the ambient heat leak to the sample stage ($\dot{Q}_{leak}$) to be 10 nW, a much worse value than the ordinary cases. The red closed line in the figure is an entropy pumping cycle for NS1 which is thermally connected either to DR at high $T$ or the sample stage at low $T$. The temperature of NS1 ($T_{NS1}$) is pre-cooled down to $T_i$ = 17 mK in a magnetic field of 1.2 T by a DR which has a cooling power of 100 $\mu$W at $T$ = 100 mK and the lowest achievable temperature ($T_0$) of 15 mK. The cooling power of this DR is much lower than that for typical copper NDRs.

The blue upward and downward vertical arrows for NS2 in figure 2 correspond to absorbing $\dot{Q}_{leak}$ by slow demagnetizing and releasing an accumulated entropy to NS1 as a heat flow by slow magnetizing, respectively. Note that NS2 is directly connected to the sample stage. The temperature of NS2 ($T_{NS2}$) is slightly lower than $T_S$ during absorbing $\dot{Q}_{leak}$ and is slightly higher than $T_S$ during releasing the entropy to keep $T_S$ constant all the time. Appropriate temperature differences between $T_{NS2}$ and $T_S$ in the two processes depend on $\dot{Q}_{leak}$, $R_{NS1}$, $R_{NS2}$, and $R_{SW2}$. As the resistances increase, the heat (or entropy) transport rate from the sample stage to NS1 or NS2 decreases. This decreases the cooling power of CNDR, and finally we will not be able to keep $T_S$ = 0.8 mK in the presence of $\dot{Q}_{leak}$ = 10 nW. Thus it is essentially important to make the

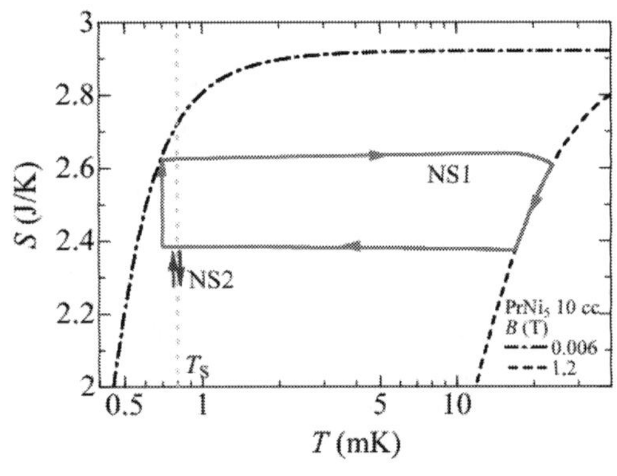


**Figure 2.** Simulation result for the entropy vs. temperature diagram for the CNDR designed in this work.

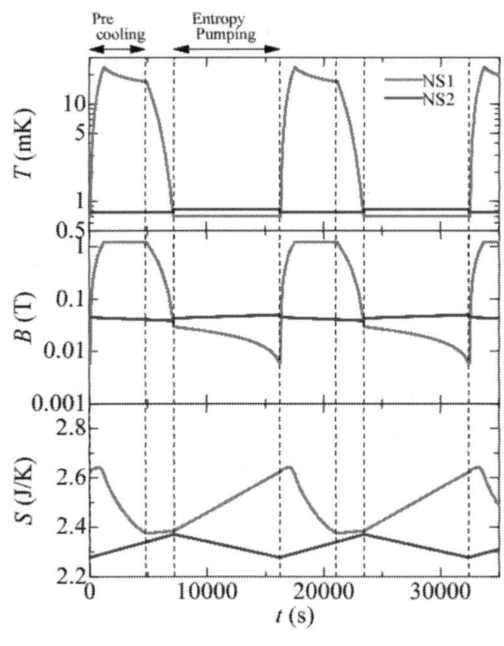


**Figure 3.** Simulation results for time evolutions of the temperatures, applied magnetic fields, and entropies of NS1 and NS2 for the CNDR designed in this work.

thermal resistances as low as possible for higher cooling power and lower operation temperature for the CNDR.

Figure 3 shows calculated time evolutions of the temperatures, applied magnetic fields and entropies of NS1 and NS2 in the continuous operation. In the CNDR, the cooling capacity divided by the cycle time (= 16210 s here) is more important than the cooling capacity itself of one cycle. Since the cooling power of DR is proportional to $T^2$ and thus the pre-cooling speed decreases near $T_0$, the cycle time becomes too long if we set $T_i$ too low. This degrades the performance of CNDR. On the other hand, if $T_i$ is set too high, we cannot make use of the cooling power of DR effectively, again degrading the performance. This is because the cycle time cannot be shorter than time necessary for demagnetization and magnetization. Therefore, there should be an optimum $T_i$ depending on $T_S$, $\dot{Q}_{leak}$, thermal resistances of each parts, and other factors. We shall discuss about the fine tuning of $T_i$ elsewhere [10].

From $t$ = 4830 to 7240 s in figure 3, NS1 is cooled from 17 to 0.7 mK by adiabatic demagnetization from $B_{NS1}$ = 1.2 T to 6 mT. During the demagnetization, $S_{NS1}$ increases by about 4% of $\Delta S$ due to the eddy current heating produced at the silver wires and Zn soldering layers around the $PrNi_5$ rods. An optimum sweep rate should be determined by competition between the eddy current loss and cycle time increase. At $t$ = 4830 and 16210 s when HSW1 and HSW2 are turned open, respectively, there must be small entropy losses in NS1 and NS2 due to the magnetic heat release. But they are invisible in the scale of figure 3.

Finally, let us briefly comment on the feasibility to use this CNDR on satellite. Cooling powers of current space-based DR [12] and ADR [4] are only 1–2 $\mu$W at $T$ = 50 mK. They are insufficient to pre-cool the $PrNi_5$ stages as is estimated in figure 2. So an additional platform kept around 10 mK consisting of a CMN salt stage [13] should be necessary in between them.

## 4. Summary

We designed a new compact and continuous nuclear demagnetization refrigerator (CNDR) using two $PrNi_5$ nuclear stages connected in series with two zinc superconducting heat switches. Because of the large hyperfine enhancement of applied magnetic field in $PrNi_5$, it was possible to design all the parts quite compact so that they can be installed in a relatively small space around the mixing chamber of a dilution refrigerator for pre-cooling. The size is substantially smaller than that of standard copper nuclear demagnetization refrigerators, and the cooling power of dilution refrigerator can be relatively low. We evaluated the expected performance of CNDR by numerical calculation based on a simple thermal model assuming realistic thermal resistances of each parts. The simulation results show that the CNDR can maintain an operation temperature of $T$ = 0.8 mK continuously with a cooling power of 10 nW by cyclic demagnetization and magnetization of the first nuclear stage between 1.2 T and 6 mT being pre-cooled by a dilution refrigerator with a cooling power of 100 $\mu$W at $T$ = 100 mK.

## 5. Acknowledgments

This work was financially supported by Grant-in-Aid for Challenging Exploratory Research (Grant No. 15K13398) from JSPS.

## References

[1] Pobell F 2007 *Matter and Methods at Low Temperatures 3rd. edition* (Belrin: Springer-Velrag)
[2] Todoshchenko I, Kaikkonen J P, Blaauwgeers R, Hakonen P J and Savin A 2014 *Rev. Sci. Instrum.* **85** 085106
[3] Shirron P J 2014 *Cryogenics* **62** 130
[4] Shirron P J, Kimball M O, James B L, Muench T, DiPirro M J, Letmate R V, Sampson M A, Bialas T G, Sneiderman G A, Porter F S and Kelley R L 2016 *Cryogenics* **74** 2
[5] Andres K and Darack S 1977 *Physica B+C* **86** 1071
[6] Kubota M, Folle H R, Buchal C, Mueller R M and Pobell F 1980 *Phys. Rev. Lett.* **45** 1812

[7] Parpia J M, Kirk W P, Kobiela P S, Rhodes T L, Olejniczak Z and Parker G N 1985 *Rev. Sci. Instrum.* **56** 437
[8] Greywall D S 1985 *Phys. Rev. B* **31** 2675
[9] Batey G, Casey A, Cuthbert M N, Matthews A J, Saunders J and Shibahara A 2013 *New J. Phys.* **15** 113034
[10] Toda R, Murakawa S, and Fukuyama H, to appear.
[11] Matsumoto Y, Murakawa S, Honkura K, Bäuerle C, Kambara H and Fukuyama H 2003 *Physica B: Condensed Matter* **329** 146
[12] Chaudhry G, Volpe A, Camus P, Triqueneaux S and Vermeulen G 2012 *Cryogenics* **52** 471
[13] Shirron P, Wegel D, DiPirro M and Sheldon S 2006 *Nuclear Instruments and Methods in Physics Research A* **559** 651

IOP Conf. Series: Journal of Physics: Conf. Series **969** (2018) 012094 doi:10.1088/1742-6596/969/1/012094

# Dielectric anomaly and magnetoelectric effect accompanying a charge ordering in the electron-doped manganite $Ca_{1-x}Sr_xMn_{0.85}Sb_{0.15}O_3$

**Haruka Taniguchi[1], Hidenori Takahashi[1], Akihiro Terui[1], Satoru Kobayashi[1], Michiaki Matsukawa[1] and Ramanathan Suryanarayanan[2,*]**

[1] Department of Physical Science and Materials Engineering, Iwate University, Morioka 020-8551, Japan
[2] ICMMO, SP2M, Université Paris-Sud, 91405 Orsay, France
[*] Retired, for correspondance sury39@gmail.com

E-mail: tanig@iwate-u.ac.jp

**Abstract.** We report on the dielectric and magnetic properties of the *A*-site substituted perovskite compounds $Ca_{1-x}Sr_xMn_{0.85}Sb_{0.15}O_3$ ($x$ = 0, 0.1, 0.2 and 0.3), to examine the magnetoelectric effect of electron-doped manganites. The dielectric constant of these compounds shows a broad peak around the magnetic transition temperature, which is probably associated with a charge ordering transition. Application of magnetic field suppresses the broad maximum of the dielectric constant, indicating the presence of the strong magnetoelectric effect. The observed dielectric anomaly and magnetoelectric effect have been discussed on the basis of the combined charge ordering model.

## 1. Introduction

Multiferroicity includes interesting physics and is expected to be applied to useful devices. One famous origin of the multiferroicity is inverse Dzyaloshinskii-Moriya (DM) interaction [1, 2], as widely observed in cycloid magnetic ordering systems, for example, $TbMnO_3$ [3], $Ni_3V_2O_8$ [4], $LiCu_2O_2$ [5], and $LiCuVO_4$ [6]. The multiferroicity of the inverse DM type is discovered also in $MnWO_4$ with an elliptical cycloid ordering [7, 8, 9] and $CoCr_2O_4$ with a transverse cone ordering [10]. The ferroelectricity in the low-temperature incommensurate ordering phase of $YMn_2O_5$ is categorized into the inverse DM type as well [11, 12]. Another origin of the multiferroicity is *d-p* hybridization [13], as reported in the magnets with a proper screw ordering such as $CuFeO_2$ [14] and $CuCrO_2$ [15]. Moreover, the multiferroicity of the *d-p* hybridization type is realized also in other magnetic orderings, for example, in $CuB_2O_4$ [16] and $Ba_2CoGe_2O_7$ [17]. The third origin of the multiferroicity is magnetic exchange striction, as observed in $YMnO_3$ [18], $HoMnO_3$ [18, 19], $GdFeO_3$ [20]. The ferroelectricity in the commensurate ordering phase of $YMn_2O_5$ is categorized into the magnetic exchange striction type as well [11, 12].

The effectiveness of some charge orderings (COs) is pointed out as the fourth origin of the multiferroicity. Importantly, the COs do not need atomic displacement in order to induce ferroelectricity. An example is the simultaneous presence of the site-centered CO and the bond-centered CO [21, 22, 23, 24, 25]. Here, the site-centered CO means the checkerboard pattern which is observed in manganites containing $Mn^{3+}$ and $Mn^{4+}$ with a ratio of 1:1. On the other hand, the bond-centered CO means Zener polaron which is proposed by A. Daoud-Aladine *et al.* [26]. The high-resolution transmission electron spectroscopy of $Pr_{1-x}Ca_xMnO_3$ revealed the existence of Zener polaron [27, 28]. Notably, the linear combination of the site-centered CO and the bond-centered CO breaks inversion symmetry and has a macroscopic electric polarization. A broad peak in the temperature dependence of the dielectric constant and the shift of the peak by a magnetic field are reported in $Pr_{1-x}Ca_xMnO_3$ ($x$ = 0.3, 0.33 and 0.4) [29, 30, 31, 32] and $Y_{1-x}Ca_xMnO_3$ ($x$ = 0.4, 0.45 and 0.5) [33]. These findings strongly indicate a spontaneous electric polarization and a magneto-electric coupling in hole-doped CO manganite systems. Moreover, the existence of ferroelectricity is directly proved by the remanent polarization measurement in $Pr_{1-x}Ca_xMnO_3$ [34].

In this study, in order to shed light on electron-doped systems as a new candidate for the multiferroics of the combined CO type, we have investigated the temperature dependence of the dielectric constant of polycrystalline $Ca_{1-x}Sr_xMn_{0.85}Sb_{0.15}O_3$ in magnetic field. On $Ca_{1-x}Sr_xMn_{1-y}Sb_yO_3$ system, previous studies exist for $y \leq 0.1$. From the X-ray photoelectron spectroscopy, it is confirmed that the valence of Sb is 5+ [35]. The substitution of $Sb^{5+}$ ion for $Mn^{4+}$ site causes one-electron doping with the chemical formula $Ca^{2+}Mn^{4+}_{1-2x}Mn^{3+}_xSb^{5+}_xO^{2-}_3$ accompanied by a monotonic increase of unit-cell volume as a function of $x$. The further doping of Sb ion breaks conduction paths along the Mn-O-Mn network, resulting in a carrier localization. In the magnetization and ac susceptibility measurements, a canted antiferromagnetic (AFM) ordering is detected below about 100 K [36, 37, 35, 38]. Moreover, by analogy with $Ca_{1-x}Ce_xMnO_3$ of the two electron doping [39, 40], CO is expected in a certain range of $y$. In this study, aiming for the combined CO, we deviate the $y$ value from 1/3 at which the site-centered CO is expected. We have observed a CO-like anomaly in the magnetization, and discovered a broad FE-like peak in the dielectric constant at almost the same temperature. Moreover, our investigations reveal a magnetoelectric effect on the dielectric anomaly. These results are consistent with the scenario that $Ca_{1-x}Sr_xMn_{0.85}Sb_{0.15}O_3$ is a multiferroics of the combined CO type.

## 2. Experimental

Polycrystalline samples of $Ca_{1-x}Sr_xMn_{0.85}Sb_{0.15}O_3$ ($x$ = 0, 0.1, 0.2 and 0.3) were prepared by a solid-state reaction method. The stoichiometric mixtures of $CaCO_3$, $SrCO_3$, $Mn_3O_4$ and $Sb_2O_3$ powders were calcined in air at 1000°C for 48 h. The products were ground and pressed into disk-like pellets. The pellets were sintered at 1350°C for 48 h. We performed X-ray diffraction measurements with an Ultima IV diffractometer (Rigaku) using Cu K$\alpha$ radiation.

As the main topic of this study, we measured the temperature dependence of the dielectric constant under DC magnetic field using the parallel mode of a LCR meter (Agilent, E4980A). Samples were cut into a parallel plate with $3.2 \times 6.0$ mm$^2$ area and 0.7 mm thickness, and Au wires for electric lead were connected by Ag paint (Dupont, 4929N). In order to improve electric conductivity, sample surfaces were polished to be flat using 9 $\mu$m diamond slurry, and Ag paint was heated at 110°C for 30 min. After field cooling, measurements were performed with AC voltage of 1 V and 10 kHz. In order to detect a CO, we also measured magnetization with a commercial SQUID (Quantum Design, MPMS) under 0.01 mT after field cooling.

## 3. Results and Discussion

Figure 1(a) presents X-ray diffraction spectrum of $Ca_{1-x}Sr_xMn_{0.85}Sb_{0.15}O_3$ ($x$ = 0, 0.1, 0.2 and 0.3). We have confirmed that all the samples are described as a *Pnma* orthorhombic structure and consist of a single phase. As shown in Fig. 1(b), (200) and (002) peaks are clearly split for $x$ = 0 and 0.1, whereas they form one combined peak for $x$ = 0.2 and 0.3. This substitution effect indicates that Sr ions with a larger radius than Ca ions reduce lattice distortion, and the value of the lattice parameter $c$ approaches that of $a$. Obtained lattice parameters and unit cell volume are plotted in Fig. 1(c). Quantitatively, the difference between $a$ and $c$ at $x$ = 0.3 is as about half as that at $x$ = 0. We have also observed that the values of $a$, $b$ and $c$ are enhanced by Sr substitution.

In Fig. 2, we show the temperature dependence of the dielectric constant $\epsilon_r'$ of $Ca_{1-x}Sr_xMn_{0.85}Sb_{0.15}O_3$ ($x$ = 0, 0.1, 0.2 and 0.3) under a magnetic field (0 T, 10 mT and 1 T). First of all, the $\epsilon_r'$ values of all the

IOP Conf. Series: Journal of Physics: Conf. Series **969** (2018) 012094 doi:10.1088/1742-6596/969/1/012094

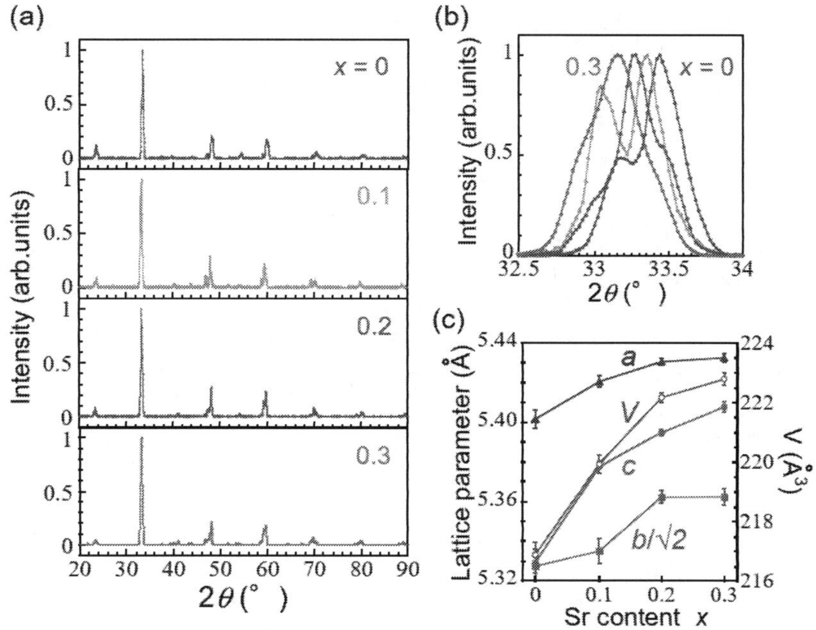


**Figure 1.** (a) X-ray diffraction spectrum of $Ca_{1-x}Sr_xMn_{0.85}Sb_{0.15}O_3$ ($x$ = 0, 0.1, 0.2 and 0.3). (b) Enlarged view of the (200) and (002) diffractions. (c) Lattice parameters and unit cell volume of $Ca_{1-x}Sr_xMn_{0.85}Sb_{0.15}O_3$ with error bars.

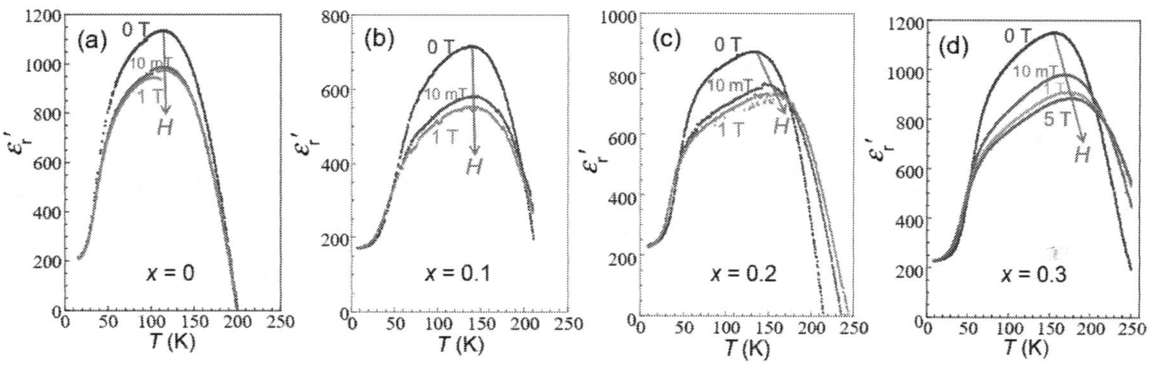


**Figure 2.** Temperature dependence of the dielectric constant of $Ca_{1-x}Sr_xMn_{0.85}Sb_{0.15}O_3$ ($x$ = 0, 0.1, 0.2 and 0.3) under AC voltage of 1 V and 10 kHz after field cooling. In every sample, a broad peak is suppressed by an applied magnetic field.

samples exhibit common broad peaks at intermediate temperatures and then rapidly decrease at low temperatures near 50 K. These trends are not changed under the application of magnetic fields. The peak structure which suggests that the response of dipole moments to an applied AC voltage becomes slow below the peak temperature $T_{\rm peak}$ can be interpreted as the result of a FE ordering, because an electric polarization is directly proved in $Pr_{1-x}Ca_xMnO_3$ [34] in which a similar dielectric peak is observed [29, 30, 32]. Moreover, we have revealed a magnetoelectric effect: the peak of $\epsilon_{\rm r}'(T)$ is shifted by magnetic field. The peak height $\epsilon_{\rm r}'(T_{\rm peak})$ is suppressed in all the samples, and $T_{\rm peak}$ is remarkably enhanced above $x$ of 0.2. The suppression of the peak height suggests that the direction of the electric polarization cannot completely follow the plus-minus switching of the AC voltage under magnetic field. The enhancement of $T_{\rm peak}$ indicates that magnetic field supports the FE ordering. Therefore, these results can be understood by the scenario that a spontaneous electric polarization along a certain direction is stabilized by magnetic field.

IOP Conf. Series: Journal of Physics: Conf. Series **969** (2018) 012094 doi:10.1088/1742-6596/969/1/012094

Focusing on the Sr substitution effect, we plotted $\epsilon_r'(T)$ curves of every sample together in Fig. 3(a). Here, we define ferroelectric temperature $T_{FE}$ as the temperature at which $\epsilon_r'(T)$ under 10 mT exhibits a peak. As indicated by arrows, we have clarified that $T_{FE}$ is enhanced by Sr substitution.

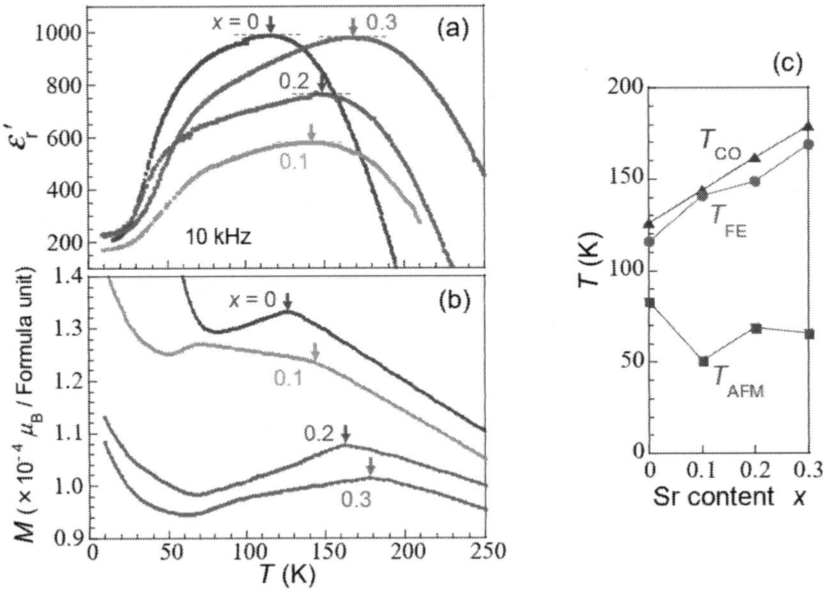


**Figure 3.** Comparison between dielectric anomalies and magnetic anomalies of $Ca_{1-x}Sr_xMn_{0.85}Sb_{0.15}O_3$ ($x$ = 0, 0.1, 0.2 and 0.3). (a) Dielectric constant measured under AC voltage of 1 V and 10 kHz after field cooling of 10 mT. Arrows indicate a ferroelectric-like peak for each sample. (b) Magnetization measured under 10 mT after field cooling. Arrows indicate a charge-ordering-like peak/kink for each sample. (c) Phase diagram of $Ca_{1-x}Sr_xMn_{0.85}Sb_{0.15}O_3$. Red circles present ferroelectric temperature $T_{FE}$ defined as the temperature at which dielectric constant exhibits a peak, blue triangles present charge ordering temperature $T_{CO}$ defined as the temperature at which magnetization exhibits a small peak/kink, and green rectangles present antiferromagnetic onset temperature $T_{AFM}$ defined as the temperature at which magnetization starts rapid increase toward lower temperature.

In order to clarify the origin of the dielectric anomaly and magnetoelectric effect in $Ca_{1-x}Sr_xMn_{0.85}Sb_{0.15}O_3$ ($x$ = 0, 0.1, 0.2 and 0.3), we investigated the magnetic properties. As shown in Fig. 3(b), the temperature dependence of the magnetization $M$ exhibits a remarkable increase below about 50 K, which is understood as an AFM transition by analogy with the AFM transition in $Ca_{1-x}Sr_xMn_{1-y}Sb_yO_3$ of $y \leq 0.1$ [36, 37, 35, 38]. Above 50 K, we observed a small peak or kink. The anomaly is expected to be caused by a CO, because similar anomalies are observed in the magnetization of $Ca_{1-x}Ce_xMnO_3$ [39] in which a CO is proved by the neutron diffraction [40]. The CO-like anomaly appears at two temperatures for $x$ = 0.1 and 0.3: at 69 and 144 K for $x$ = 0.1, and at 89 and 179 K for $x$ = 0.3. In this study, we focus on the anomaly at the higher temperature, because its temperature shown by arrows in Fig. 3(b) exhibits a systematic $x$ dependence. On the anomaly at the lower temperature, we consider specific heat measurement as a future work in order to obtain detailed information. Thus, we define CO temperature $T_{CO}$ as the temperature at which the the slope of $M(T)$ below 250 K changes. We also define AFM onset temperature $T_{AFM}$ as the temperature at which the negative $dM/dT$ from 0 K changes to positive one. The three characteristic temperatures $T_{FE}$, $T_{CO}$, and $T_{AFM}$ are summarized in Fig. 3(c).

Importantly, we have found that $T_{FE}$ and $T_{CO}$ exhibit a similar value and are enhanced together by Sr substitution. These results strongly indicate that the FE-like anomaly of $Ca_{1-x}Sr_xMn_{0.85}Sb_{0.15}O_3$ ($x$ = 0, 0.1, 0.2 and 0.3) is induced by the CO. Moreover, the AFM temperature roughly corresponds to the temperature at which $\epsilon_r'(T)$ exhibits the shoulder structure. This result suggests that a part of the CO region transfers to an AFM phase and the FE volume is suppressed. We note that the maximum value of $\epsilon_r'$ at $x = 0$ is larger than those at $x = 0.1$ and 0.2. These differences between the parent and lightly Sr substituted samples are related to the volume fraction of the original CO phase, which contributes to the formation of a ferroelectric state. The CO phase has a fragile nature against lattice disorder and is easily diminished in the Sr substituted samples. (For the $x = 0.3$ sample, the peak value is comparable to that of the parent sample. The lattice distortion is relaxed in the heavily Sr substituted sample because the value of the lattice parameter $c$ approaches that of $a$. We expect that the lattice relaxation prevents a decrease of the CO volume fraction, giving the comparable peak value with the parent sample.) As described in the introduction, the CO which causes the FE-like anomaly in $Ca_{1-x}Sr_xMn_{0.85}Sb_{0.15}O_3$ is expected to be the linearly combined state of the site-centered CO and the bond-centered CO. Since the inversion center of the site-centered CO is different from that of the bond-centered one, the combined CO lacks inversion symmetry and induces a spontaneous electric polarization [21, 22].

Moreover, a magnetoelectric effect is expected in the combined CO for the following reason. The site-centered CO and the bond-centered CO are accompanied by the CE magnetic ordering ($\Phi = 0$) and the $\perp$ magnetic ordering ($\Phi = \pi/2$), respectively [21, 22]. Since these two magnetic orderings are connected by a continuous rotation, the combined CO is expected to have a corresponding magnetic ordering ($0 < \Phi < \pi/2$). Therefore, a magneto-electric coupling is predicted under the combined CO. The observed suppression of the dielectric constant by magnetic field is consistent with this scenario. We would like to emphasize that this study has experimentally suggested the multiferroicity by the combined CO for the first time in electron-doped systems.

As a future issue, we are interested in the effect of Sb concentration $y$ of $Ca_{1-x}Sr_xMn_{1-y}Sb_yO_3$. Further discussion will be enabled by revealing the $y$ range in which the dielectric anomaly with the magnetoelectric effect is realized and the $y$ value at which $T_{FE}$ exhibits a maximum value. We are also trying a detailed investigation using single crystals in order to reveal anisotropy and perform quantitative discussion.

## 4. Summary

Our investigations reveal that the temperature dependence of the dielectric constant of $Ca_{1-x}Sr_xMn_{0.85}Sb_{0.15}O_3$ ($x$ = 0, 0.1, 0.2 and 0.3) exhibits a broad peak, which can be understood as a FE transition. Importantly, the peak temperature coincides well with the temperature of the CO-like anomaly in magnetization. Moreover, we have shown that the dielectric constant is suppressed by a magnetic field. These tendencies, which were commonly observed in $Ca_{1-x}Sr_xMn_{0.85}Sb_{0.15}O_3$ ($x$ = 0, 0.1, 0.2 and 0.3), are consistent with the scenario that the combined state of the site-centered CO and the bond-centered CO is realized in this system and induces a macroscopic electric polarization, which is stabilized by a magnetic field. This magnetoelectric effect suggests that the carrier doping which is deviated from the composition of the site-centered CO is a hopeful way to induce the multiferroicity via the combined CO state.

## Acknowledgments

This work is supported by Iwate University and JSPS KAKENHI Grant Number JP17K14101.

## References

[1] Katsura H, Nagaosa N and Balatsky A V 2005 *Phys. Rev. Lett.* **95** 057205
[2] Sergienko I A and Dagotto E 2006 *Phys. Rev. B* **73** 094434
[3] Kimura T, Goto T, Shintani H, Ishizaka K, Arima T and Tokura Y 2003 *Nature* **426** 55
[4] Lawes G, Kenzelmann M, Rogado N, Kim K H, Jorge G A, Cava R J, Aharony A, Entin-Wohlman O, Harris A B, Yildirim T, Huang Q Z, Park S, Broholm C and Ramirez A P 2004 *Phys. Rev. Lett.* **93** 247201

[5] Park S, Choi Y J, Zhang C L and Cheong S W 2007 *Phys. Rev. Lett.* **98** 057601
[6] Naito Y, Sato K, Yasui Y, Kobayashi Y, Kobayashi Y and Sato M 2007 *J. Phys. Soc. Jpn.* **76** 023708
[7] Taniguchi K, Abe N, Takenobu T, Iwasa Y and Arima T 2006 *Phys. Rev. Lett.* **97** 097203
[8] Heyer O, Hollmann N, Klassen I, Jodlauk S, Bohaty L, Becker P, Mydosh J A, Lorenz T and Khomskii D 2006 *J. Phys.: Condens. Matter* **18** L471
[9] Arkenbout A H, Palstra T T M, Siegrist T and Kimura T 2006 *Phys. Rev. B* **74** 184431
[10] Yamasaki Y, Miyasaka S, Kaneko Y, He J P, Arima T and Tokura Y 2006 *Phys. Rev. Lett.* **96** 207204
[11] Kobayashi S, Osawa T, Kimura H, Noda Y, Kagomiya I and Kohn K 2004 *J. Phys. Soc. Jpn.* **73** 1593
[12] Wakimoto S, Kimura H, Sakamoto Y, Fukunaga M, Noda Y, Takeda M and Kakurai K 2013 *Phys. Rev. B* **88** 140403(R)
[13] Arima T 2007 *J. Phys. Soc. Jpn.* **76** 073702
[14] Kimura T, Lashley J C and Ramirez A P 2006 *Phys. Rev. B* **73** 220401(R)
[15] Seki S, Onose Y and Tokura Y 2008 *Phys. Rev. Lett.* **101** 067204
[16] Saito M, Ishikawa K, Taniguchi K and Arima T 2008 *Phys. Rev. Lett.* **101** 117402
[17] Murakawa H, Onose Y, Miyahara S, Furukawa N and Tokura Y 2010 *Phys. Rev. Lett.* **105** 137202
[18] Lorentz B, Wang Y Q and Chu C W 2007 *Phys. Rev. B* **76** 104405
[19] Picozzi S, Yamauchi K, Sanyal B, Sergienko I A and Dagotto E 2007 *Phys. Rev. Lett.* **99** 227201
[20] Tokunaga Y, Furukawa N, Sakai H, Taguchi Y, Arima T and Tokura Y 2009 *Nature Materials* **8** 558
[21] Efremov D V, Brink J and Khomskii D I 2004 *Nature Materials* **3** 853
[22] Efremov D V, Brink J and Khomskii D I 2005 *Physica B* **359** 1433
[23] Khomskii D I 2006 *J. Magn. Magn. Mater.* **306** 1
[24] Brink J and Khomskii D I 2008 *J. Phys.: Condens. Matter* **20** 434217
[25] Giovannetti G, Kumar S, Brink J and Picozzi S 2009 *Phys. Rev. Lett.* **103** 037601
[26] Daoud-Aladine A, Rodríguez-Carvajal J, Pinsard-Gaudart L, Fernández-Díaz M T and Revcolevschi A 2002 *Phys. Rev. Lett.* **89** 097205
[27] Wu L, Klie R F, Zhu Y and Jooss C 2007 *Phys. Rev. B* **76** 174210
[28] Jooss C, Wu L, Beetz T, Klie R F, Beleggia M, Schofield M A, Schramm S, Hoffmann J and Zhu Y 2007 *Proc. Natl. Acad. Sci. U.S.A.* **104** 13597
[29] Jardón C, Rivadulla F, Hueso L E, Fondado A, López-Quintela M A, Rivas J, Zysler R, Causa M T and Sánchez R D 1999 *J. Magn. Magn. Mater.* **196** 475
[30] Mercone S, Wahl A, Pautrat A, Pollet M and Simon C 2004 *Phys. Rev. B* **69** 174433
[31] Freitas R S, Mitchell J F and Schiffer P 2005 *Phys. Rev. B* **72** 144429
[32] Serrao C R, Sundaresan A and Rao C N R 2007 *J. Phys.: Condens. Matter* **19** 496217
[33] Sahu J R, Serrao C R, Ghosh A, Sundaresan A and Rao C N R 2009 *Solid State Communications* **149** 49
[34] Shukla V K, Mukhopadhyay S, Das K, Sarma A and Das I 2014 *Phys. Rev. B* **90** 245126
[35] Fujiwara T, Matsukawa M, Ohuchi S, Kobayashi S, Nimori S and Suryanarayanan R 2013 *J. Kor. Phys. Soc.* **62** 1925
[36] Murano Y, Matsukawa M, Kobayashi S, Nimori S and Suryanarayanan R 2010 *J. Phys.: Conf. Ser.* **200** 012114
[37] Murano Y, Matsukawa M, Ohuchi S, Kobayashi S, Nimori S, Suryanarayanan R, Koyama K and Kobayashi N 2011 *Phys. Rev. B* **83** 054437
[38] Fujiwara T, Matsukawa M, Aoyagi T, Kobayashi S, Taniguchi H, Nimori S and Suryanarayanan R 2015 *J. Magn. Magn. Mater.* **378** 451
[39] Zeng Z, Greenblatt M and Croft M 2001 *Phys. Rev. B* **63** 224410
[40] Caspi E N, Avdeev M, Short S, Jorgensen J D, Lobanov M V, Zeng Z, Greenblatt M, Thiyagarajan P, Botez C E and Stephens P W 2004 *Phys. Rev. B* **69** 104402

IOP Conf. Series: Journal of Physics: Conf. Series **969** (2018) 012095 doi:10.1088/1742-6596/969/1/012095

# Phase diagram of the magnetic state obtained from a two-particle self-consistent analysis of the half-filled Hubbard model on a honeycomb lattice

**Hirohito Aizawa**

Institute of Physics, Kanagawa University, Yokohama, Kanagawa 221-8686, Japan

E-mail: aizawa@kanagawa-u.ac.jp

**Abstract.** Motivated by the fascinating magnetic properties in the half-filled Hubbard model on a honeycomb lattice, we calculate the spin susceptibility and estimate the magnetic transition by applying the two-particle self-consistent (TPSC) method. The diagonalized spin-susceptibility matrix has its maximum at $\boldsymbol{Q} = \mathbf{0}$, which indicates that an uniform magnetic order develops between the unit cells. The sign of the spin susceptibility between same sublattices is opposite to that between different sublattices. Therefore, an antiferromagnetic state develops in this system. We calculate the temperature dependence of the Stoner factor and estimate the critical temperature extrapolation. By applying the same analysis to several values of on-site interactions, we present the phase diagram of the magnetic state in the half-filled Hubbard model on a honeycomb lattice. Based on the phase diagram, the critical on-site interaction $U_c/t$ above which magnetism occurs at the ground state can be estimated as being approximately 3.81.

## 1. Introduction

Honeycomb lattices have interesting band structures such as a Dirac-like dispersion and the van Hove singularity. The physical properties of this type of lattice have been actively studied, and recently the Kane–Mele model [1] is attracted attentions. For the Hubbard model on a honeycomb lattice with half-filling electrons, the critical on-site interaction $U_c/t$ above which magnetism occurs at the ground state has been studied [2] and its value has been estimated to be in the range of 3.5–5.0 [3, 4, 5, 6, 7, 8, 9, 10, 11, 12]. The magnetic transition is thought to be caused by antiferromagnetism; however, recent studies suggest the possibility of a spin-liquid state [7]. Recent large-scale quantum Monte Carlo calculations have estimated the critical interaction $U_c/t$ to be $3.869 \pm 0.013$ [9] and $3.85 \pm 0.02$ [12]. The functional renormalization group indicates that $U_c/t$ is approximately 3.8 [5, 6]. The estimate of the crossover temperature obtained from the two-particle self-consistent (TPSC) method [13] is $U_c/t = 3.79 \pm 0.01$ [11].

We study the magnetic state in the half-filled Hubbard model on a honeycomb lattice by using the TPSC method. We confirm the wavenumber vector at which the spin susceptibility is maximized and extrapolate the temperature dependence of the Stoner factor that reaches unity at the critical temperature of the magnetic transition. Moreover, we estimate the critical temperature for several on-site interactions and obtain the phase diagram for the magnetic state in the interaction–temperature space. Based on the phase diagram, we estimate the critical on-site interaction $U_c/t$ at which the magnetic transition occurs at the ground state.

## 2. Method

We focus on the two-site Hubbard Hamiltonian $H$ given as follows:

$$H = \sum_{\langle i\alpha:j\beta\rangle,\sigma} \left\{ t_{i\alpha:j\beta} c^{\dagger}_{i\alpha\sigma} c_{j\beta\sigma} + \mathrm{H.c.} \right\} + \sum_{i\alpha} U n_{i\alpha\uparrow} n_{i\alpha\downarrow}, \tag{1}$$

where $i$ and $j$ are the unit cell indices, $\alpha$ and $\beta$ specify the sites in a unit cell, $c^{\dagger}_{i\alpha\sigma}$ ($c_{i\alpha\sigma}$) is the creation (annihilation) operator for spin $\sigma$ at site $\alpha$ in unit cell $i$, $t_{i\alpha:j\beta}$ is the electronic transfer energy between site $(i, \alpha)$ and site $(j, \beta)$, and $\langle i\alpha : j\beta\rangle$ represents the summation of the nearest-neighboring sites. The transfer energy is taken to be $t$ between the nearest-neighboring sites on the honeycomb lattice, as shown in Fig. 1(a). $U$ is the on-site interaction and $n_{i\alpha\sigma}$ is the number operator of electrons for spin $\sigma$ at site $\alpha$ in the unit cell $i$. By diagonalizing the first term in Eq. (1), we obtain the noninteracting electronic band structure $\xi(\boldsymbol{k})/t$, as shown in Fig. 1(b). The Dirac points are located at $\xi(\boldsymbol{k})/t = 0$ on the K point of the Brillouin zone. The density of states (DOS) is represented in Fig. 1(c). The DOS decreases continuously toward $\xi(\boldsymbol{k})/t = 0$ because the band structure is the linear dispersion around the Dirac points. Conversely, the van Hove singularities at the energy $\xi(\boldsymbol{k})/t = \pm 1$ on the M point bring about the divergence in the DOS. We set one electron per site, where the band structure is half-filled and the Fermi level coincides with the Dirac point.

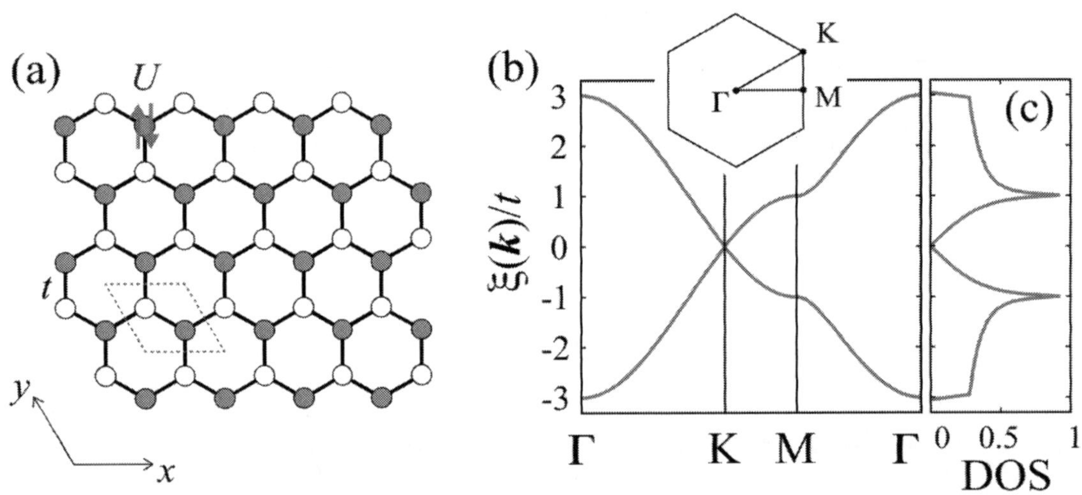


**Figure 1.** (a) Hubbard model on a honeycomb lattice, where the solid and open circles represent the A and B sublattices, the blue dashed rhombus indicates the unit cell, and the red up (down) arrow schematically indicates the up (down) spin of the electron. (b) The electronic band structure along the symmetric $k$-points, where the inset shows the Brillouin zone; and (c) the DOS in the honeycomb lattice.

To consider the effect of electron correlation, we obtained the bare susceptibility and the bare Green function in the site representation given by

$$\chi^0_{\alpha\beta}(q) = -\frac{T}{N_c} \sum_k G^0_{\alpha\beta}(k+q) G^0_{\beta\alpha}(k), \tag{2}$$

$$G^0_{\alpha\beta}(k) = \sum_{\gamma} d_{\alpha\gamma}(\boldsymbol{k}) d^{*}_{\beta\gamma}(\boldsymbol{k}) G^0_{\gamma}(k), \tag{3}$$

where $T$ and $N_c$ are the temperature and total number of unit cells, respectively, and $d_{\alpha\gamma}(\boldsymbol{k})$ is the unitary matrix. The bare Green function in the band representation, $G^0_\gamma(k)$, is given by

$$G^0_\gamma(k) = \frac{1}{\imath\varepsilon_n - \xi(\boldsymbol{k})}. \tag{4}$$

Here we introduce the abbreviations $k = (\boldsymbol{k}, \imath\varepsilon_n)$ and $q = (\boldsymbol{q}, \imath\omega_m)$ for the fermionic and bosonic Matsubara frequencies, respectively. The indices $\alpha\beta$ refer to an element $(\alpha\ \beta)$ of the matrices such as $\hat{\chi}^0(q)$.

The TPSC method has been applied to single-site systems [13, 14], multi-site systems [11, 15, 16, 17], and multi-orbital systems [18]. Within the TPSC scheme, using the bare susceptibility given by Eq. (2), the spin and charge susceptibilities are as follows:

$$\hat{\chi}^{\rm sp}(q) = \left[\hat{I} - \hat{\chi}^0(q)\hat{U}^{\rm sp}\right]^{-1}\hat{\chi}^0(q), \tag{5}$$

$$\hat{\chi}^{\rm ch}(q) = \left[\hat{I} + \hat{\chi}^0(q)\hat{U}^{\rm ch}\right]^{-1}\hat{\chi}^0(q), \tag{6}$$

where $\hat{U}^{\rm sp}$ ($\hat{U}^{\rm ch}$) is the local spin (charge) vertex, and $\hat{I}$ is the unit matrix. The local vertices are determined by satisfying two sum rules for the local moment:

$$\frac{2T}{N_c}\sum_q \chi^{\rm sp}_{\alpha\alpha}(q) = n_\alpha - 2\langle n_{\alpha\uparrow}n_{\alpha\downarrow}\rangle, \tag{7}$$

$$\frac{2T}{N_c}\sum_q \chi^{\rm ch}_{\alpha\alpha}(q) = n_\alpha + 2\langle n_{\alpha\uparrow}n_{\alpha\downarrow}\rangle - n_\alpha^2, \tag{8}$$

where $n_\alpha$ is the particle number at site $\alpha$. Here we used the relations $n_{\alpha\uparrow} = n_{\alpha\downarrow} = n/2$ and $n_{\alpha\sigma} = n^2_{\alpha\sigma}$ from the Pauli principle.

The local spin vertex $\hat{U}^{\rm sp}$ is related to the double occupancy $\langle n_{\alpha\uparrow}n_{\alpha\downarrow}\rangle$ by the ansatz

$$U^{\rm sp}_{\alpha\alpha} = \frac{\langle n_{\alpha\uparrow}n_{\alpha\downarrow}\rangle}{\langle n_{\alpha\uparrow}\rangle\langle n_{\alpha\downarrow}\rangle}U_{\alpha\alpha}, \tag{9}$$

where $U_{\alpha\alpha}$ is element $(\alpha\ \alpha)$ of the on-site interaction matrix $\hat{U}$. The double occupancy $D_\alpha = \langle n_{\alpha\uparrow}n_{\alpha\downarrow}\rangle$ is given by

$$D_\alpha = \frac{U^{\rm sp}_{\alpha\alpha}}{U_{\alpha\alpha}}\frac{n_\alpha^2}{4}. \tag{10}$$

Equations (5)–(10) give the set of self-consistent equations for the TPSC scheme.

In this study, we show the spin susceptibility obtained from the larger eigenvalue of the matrix to find the dominant magnetic order between the unit cells. In this calculation, we take the system size to be $96 \times 96$ $k$ meshes and 16384 Matsubara frequencies.

## 3. Results

Figure 2(a) [2(b)] shows the maximum (minimum) value of the diagonalized spin-susceptibility matrix for the on-site interaction $U/t = 3.82$ and temperature $T/t = 0.02$, respectively. We find that a uniform magnetic order between the unit cells develops because the spin susceptibility has its maximum at the wavenumber vector $\boldsymbol{Q} = \boldsymbol{0}$. The site-represented spin susceptibilities, which correspond to the elements of its matrix, are shown in Figs. 2(c) and 2(d). It is favorable

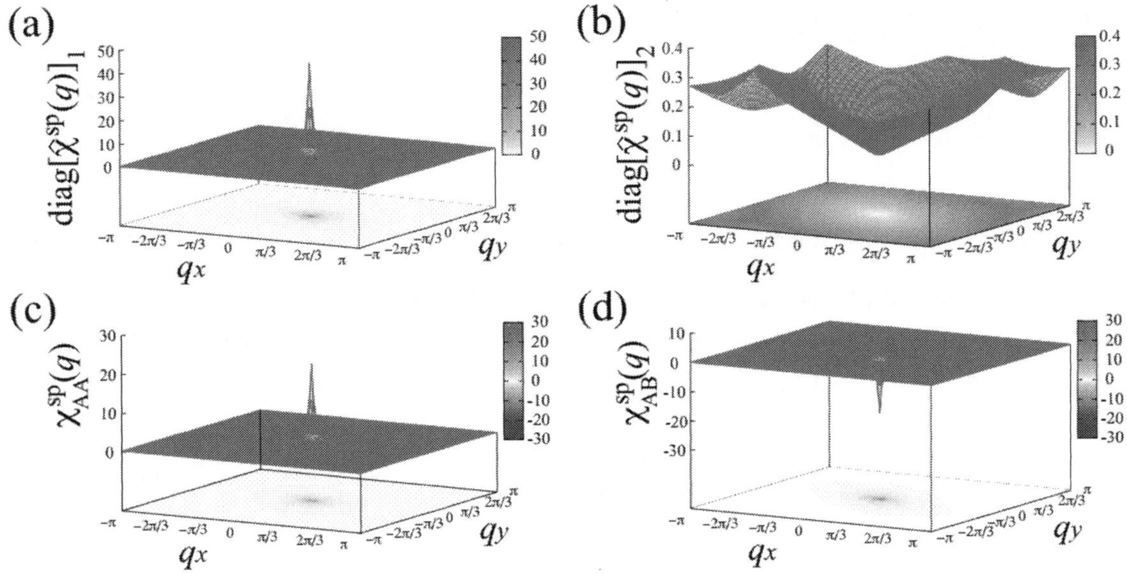


**Figure 2.** The (a) maximum and (b) minimum values of the diagonalized spin-susceptibility matrix. The site-represented spin susceptibility for (c) same sublattices and (d) different sublattices.

when the spin direction is parallel (opposite) between same (different) sublattices. Therefore, we confirmed that an antiferromagnetic order develops in this system.

In this paper, we regard the diagonalized $\hat{\chi}^0(Q)\hat{U}^{\rm sp}$ matrix as the Stoner factor and abbreviate it as $U_{\rm sp}\chi_0$, where the wavenumber $\boldsymbol{Q}$ gives the maximum value of the spin susceptibility and $\boldsymbol{Q} = \boldsymbol{0}$ from Fig. 2. The TPSC method satisfies the Mermin–Wagner theorem in two-dimensional system [13]; therefore, the true magnetic order does not occur in the presented model. In a previous study of the TPSC scheme in the Hubbard model on a honeycomb lattice, the critical temperature was estimated via the crossover temperature obtained from the magnetic correlation length [11]. In our study, we regard the temperature at which the line extrapolating the temperature dependence of the Stoner factor from the high temperature reaches unity as the critical magnetic temperature. In a previous study, we have shown that the estimate of the model reproduced by the first-principles band calculation for a $\beta$-type organic conductor is consistent with experimental results, i.e., the magnetic transition occurs at low temperature [16].

Figure 3(a) represents the Stoner factor $U_{\rm sp}\chi_0$ as a function of the temperature for several on-site interactions $U/t$ in the half-filled system. We can fit the temperature dependence of the Stoner factor with the extrapolating line shown in Fig. 3(a). The critical temperatures for each on-site interaction can be estimated. At the ground state, the extrapolating line below $U/t = 3.8$ cannot reach unity, that is, the magnetic transition appears to be absent below that interaction. The phase diagram in the $U/t$–$T/t$ space is shown in Fig. 3(b). Based on the interaction dependence of the critical temperature, we find that the estimate of the critical on-site interaction, $U_c/t$, is near 3.81, which is consistent with several previous studies [3, 4, 5, 6, 7, 8, 9, 10, 11, 12].

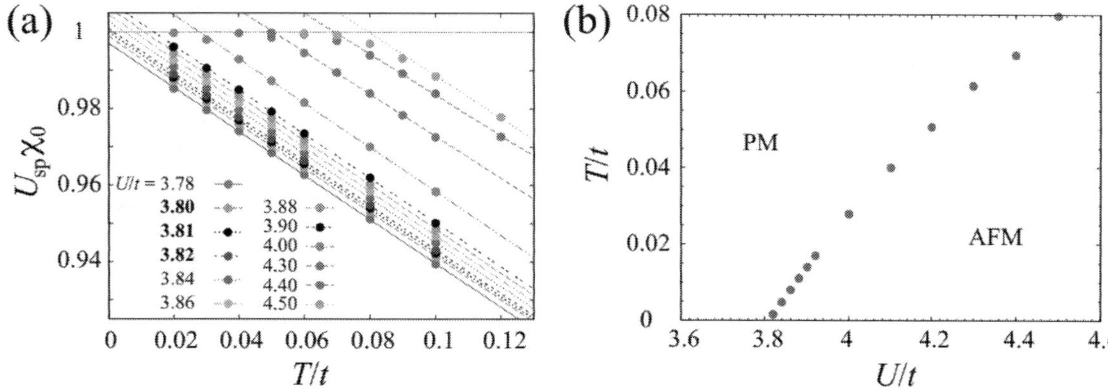


**Figure 3.** (a) Temperature dependence of the Stoner factor for several on-site interactions $U$ in the half-filled system. (b) The phase diagram in the $U/t$–$T/t$ space for the magnetic states, where PM represents the paramagnetic state and AFM represents the antiferromagnetic state.

## 4. Conclusions

We performed a TPSC analysis for the magnetic properties in the half-filled Hubbard model on a honeycomb lattice. The spin-susceptibility matrix indicates that an antiferromagnetic order develops in this system. We evaluated the temperature dependence of the Stoner factor and estimated the critical temperature by extrapolating this factor for the several on-site interactions. Plotting the critical temperature as a function of the on-site interaction, we presented the phase diagram of the magnetic state in the half-filled Hubbard model on a honeycomb lattice. From the phase diagram, we estimated the critical on-site interaction to be $U_c/t \approx 3.81$.

## Acknowledgments

This work was supported by JSPS KAKENHI Grant No. 16K17754.

## References

[1] Kane C L and Mele E J 2005 *Phys. Rev. Lett.* **95** 226801
[2] Sorella S and Tosatti E 1992 *Europhys. Lett.* **19** 699
[3] Furukawa N 2001 *J. Phys. Soc. Jpn.* **70** 1483
[4] Paiva T, Scalettar R T, Zheng W, Singh R R P and Oitmaa J 2005 *Phys. Rev. B* **72** 085123
[5] Honerkamp C 2008 *Phys. Rev. Lett.* **100** 146404
[6] Raghu S, Qi X L, Honerkamp C and Zhang S C 2008 *Phys. Rev. Lett.* **100** 156401
[7] Meng Z Y, Lang T C, Wessel S, Assaad F F and Muramatsu A 2010 *Nature* **464** 847
[8] Wu W, Chen Y H, Tao H S, Tong N H and Liu W M 2010 *Phys. Rev. B* **82** 245102
[9] Sorella S, Otsuka Y and Yunoki S 2012 *Sci. Rep.* **2** 992
[10] Assaad F F and Herbut I F 2013 *Phys. Rev. X* **3** 031010
[11] Arya S, Sriluckshmy P V, Hassan S R and Tremblay A M S 2015 *Phys. Rev. B* **92** 045111
[12] Otsuka Y, Yunoki S and Sorella S 2016 *Phys. Rev. X* **6** 011029
[13] Vilk Y M and Tremblay A M 1997 *J. Phys. I* **7** 1309
[14] Otsuki J 2012 *Phys. Rev. B* **85** 104513
[15] Ogura D and Kuroki K 2015 *Phys. Rev. B* **92** 144511
[16] Aizawa H, Kuroki K and Yamada J i 2015 *Phys. Rev. B* **92** 155108
[17] Aizawa H and Kuroki K 2016 *Phys. Procedia* **81** 21
[18] Miyahara H, Arita R and Ikeda H 2013 *Phys. Rev. B* **87** 045113

IOP Conf. Series: Journal of Physics: Conf. Series **969** (2018) 012096 doi:10.1088/1742-6596/969/1/012096

# Multipole interactions of $\Gamma_3$ non-Kramers doublet systems on cubic lattices

**Katsunori Kubo**[1] **and Takashi Hotta**[2]

[1]Advanced Science Research Center, Japan Atomic Energy Agency, Tokai, Ibaraki 319-1195, Japan

[2]Department of Physics, Tokyo Metropolitan University, Hachioji, Tokyo 192-0397, Japan

E-mail: kubo.katsunori@jaea.go.jp

**Abstract.** We investigate the multipole interactions between adjacent $f^2$ ions with the $\Gamma_3$ crystalline electric field (CEF) ground state. In this study, we consider only the $\Gamma_8$ states as the one-electron states. We construct the $\Gamma_3$ doublet state of the electrons in the $\Gamma_8$ states, and indeed, it can become the CEF ground state of a simple model Hamiltonian. To derive the multipole interactions, we apply the second-order perturbation theory with respect to the intersite hopping. We obtain no multipole interaction for a simple cubic lattice and octupole interactions for bcc and fcc lattices. We discuss the similarities and differences in the multipole interactions among the models with the local $\Gamma_3$ CEF states.

## 1. Introduction

In $f$-electron systems, multipole physics has been studied, in particular, on the $\Gamma_8$ systems such as $NpO_2$ [1–5] and $Ce_xLa_{1-x}B_6$ [6–12]. The $\Gamma_8$ quartet has sufficient degrees of freedom to possess higher-order multipoles such as octupoles and it is regarded as an ideal system for the multipole physics.

However, such large degeneracy is not a necessary condition to possess higher-order multipoles. If the crystalline electric field (CEF) ground state does not have the dipole but is not a singlet, this state inevitably has higher-order multipoles. Indeed, the $\Gamma_3$ non-Kramers doublet does not have the dipole but has quadrupole and octupole moments. In addition, the absence of the dipole moment is an advantage of the $\Gamma_3$ systems for multipole physics since we can focus only on the higher-order multipole moments.

Previously, we have investigated the multipole interactions between the $\Gamma_3$ ions by using a model in which the $f^2$-$\Gamma_3$ states are constructed as pseudospin singlets between the $\Gamma_7$ and $\Gamma_8$ orbitals of the total angular momentum $j = 5/2$ [13] [Fig.1(b)]. Then, we have found that the quadrupole interaction is dominant in the simple cubic (sc) lattice and the octupole interactions are dominant in the bcc and fcc lattices. These features depending on the lattice structure are similar to the multipole interactions in the $f^1$-$\Gamma_8$ system [4].

In this study, to clarify the influence of the choice of the model, we use another method for constructing the $f^2$-$\Gamma_3$ state. We construct the $f^2$-$\Gamma_3$ doublet state of the electrons only in the $\Gamma_8$ state [Fig.1(a)]. To derive the multipole interactions, we apply the second-order perturbation theory with respect to the intersite hopping. In general, the $f^2$-$\Gamma_3$ states composed of $j = 5/2$ electrons are superpositions of the $f^2$-$\Gamma_3$ states of the present study and of the previous study.

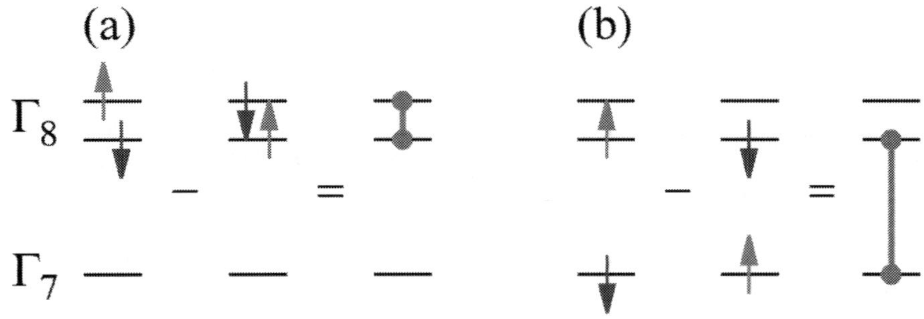


**Figure 1.** Electron configurations of the $f^2$-$\Gamma_3$ states (a) in the present study and (b) in the previous study [13]. The bold lines denote spin singlets.

Thus, if we obtain a common result between these studies, it will not depend on the details of the model.

## 2. Model and method

The $\Gamma_8$ states at site $\boldsymbol{r}$ are given by

$$c^\dagger_{\boldsymbol{r}\alpha\uparrow}|0\rangle \equiv \frac{1}{\sqrt{6}}\left(\sqrt{5}a^\dagger_{\boldsymbol{r}5/2} + a^\dagger_{\boldsymbol{r}-3/2}\right)|0\rangle, \tag{1a}$$

$$c^\dagger_{\boldsymbol{r}\alpha\downarrow}|0\rangle \equiv \frac{1}{\sqrt{6}}\left(\sqrt{5}a^\dagger_{\boldsymbol{r}-5/2} + a^\dagger_{\boldsymbol{r}3/2}\right)|0\rangle, \tag{1b}$$

$$c^\dagger_{\boldsymbol{r}\beta\uparrow}|0\rangle \equiv a^\dagger_{\boldsymbol{r}1/2}|0\rangle, \tag{1c}$$

$$c^\dagger_{\boldsymbol{r}\beta\downarrow}|0\rangle \equiv a^\dagger_{\boldsymbol{r}-1/2}|0\rangle, \tag{1d}$$

where $a^\dagger_{\boldsymbol{r}j_z}$ is the creation operator of the electron with the $z$-component $j_z$ of the total momentum $j = 5/2$ at $\boldsymbol{r}$ and $|0\rangle$ denotes the vacuum state. $\sigma = \uparrow$ or $\downarrow$ denotes the Kramers degeneracy of the one-electron states. While it is not a real spin, we call it spin in this paper.

The model Hamiltonian is

$$H = H_{\text{kin}} + H_{\text{loc}}, \tag{2}$$

where $H_{\text{kin}}$ is the kinetic energy term composed of the nearest-neighbor hopping of the $\Gamma_8$ electrons. $H_{\text{loc}}$ denotes the onsite interactions:

$$H_{\text{loc}} = U\sum_{\tau\boldsymbol{r}} n_{\boldsymbol{r}\tau\uparrow}n_{\boldsymbol{r}\tau\downarrow} + U'\sum_{\boldsymbol{r}} n_{\boldsymbol{r}\alpha}n_{\boldsymbol{r}\beta} + J\sum_{\boldsymbol{r}\sigma\sigma'} c^\dagger_{\boldsymbol{r}\alpha\sigma}c^\dagger_{\boldsymbol{r}\beta\sigma'}c_{\boldsymbol{r}\alpha\sigma'}c_{\boldsymbol{r}\beta\sigma} + J'\sum_{\tau\neq\tau'} c^\dagger_{\boldsymbol{r}\tau\uparrow}c^\dagger_{\boldsymbol{r}\tau\downarrow}c_{\boldsymbol{r}\tau'\downarrow}c_{\boldsymbol{r}\tau'\uparrow}, \tag{3}$$

where $n_{\boldsymbol{r}\tau\sigma} = c^\dagger_{\boldsymbol{r}\tau\sigma}c_{\boldsymbol{r}\tau\sigma}$ and $n_{\boldsymbol{r}\tau} = \sum_\sigma n_{\boldsymbol{r}\tau\sigma}$. When we consider only the Coulomb interaction between the electrons, $U$ is the intraorbital Coulomb interaction, $U'$ is the interorbital Coulomb interaction, $J$ is the exchange interaction, and $J'$ is the pair hopping interaction. In the present study, however, they should be regarded as effective interactions including the effects of the excited $j = 7/2$ states. Through the $j = 7/2$ states, the effects of the sixth-order terms of the CEF, which are indispensable to stabilize the $f^2$-$\Gamma_3$, can be taken into account [14].

Among the eigenstates of $H_{\text{loc}}$ of the $f^2$ states, we focus on the following singlets:

$$|z(\boldsymbol{r})\rangle \equiv \frac{1}{\sqrt{2}}(c^\dagger_{\boldsymbol{r}\alpha\uparrow}c^\dagger_{\boldsymbol{r}\alpha\downarrow} - c^\dagger_{\boldsymbol{r}\beta\uparrow}c^\dagger_{\boldsymbol{r}\beta\downarrow})|0\rangle, \tag{4}$$

IOP Conf. Series: Journal of Physics: Conf. Series **969** (2018) 012096 doi:10.1088/1742-6596/969/1/012096

with energy $U - J'$ and

$$|x(\boldsymbol{r})\rangle \equiv \frac{1}{\sqrt{2}}(c^\dagger_{\boldsymbol{r}\alpha\uparrow}c^\dagger_{\boldsymbol{r}\beta\downarrow} - c^\dagger_{\boldsymbol{r}\alpha\downarrow}c^\dagger_{\boldsymbol{r}\beta\uparrow})|0\rangle, \tag{5}$$

with energy $U' + J$. They are degenerate due to the relation $U = U' + J + J'$, which holds under the cubic symmetry. These two states constitute a basis of the $\Gamma_3$ representation. If the effective interactions satisfy $J < 0$ and $J' > 0$, the $f^2$ ground state is the $\Gamma_3$ state given above. We assume these conditions in the following, but the situation of $J < 0$ and $J' > 0$ is realized, if we appropriately include the effects of the sixth-order CEF terms into the effective interactions between the $\Gamma_8$ electrons.

For the formulation of the multipole interactions, it is convenient to rewrite the $\Gamma_3$ states as

$$|\mu(\boldsymbol{r})\rangle = \sum_{\tau\tau'\sigma\sigma'} B^\mu_{\tau\sigma;\tau'\sigma'}c^\dagger_{\boldsymbol{r}\tau\sigma}c^\dagger_{\boldsymbol{r}\tau'\sigma'}|0\rangle, \tag{6}$$

where

$$B^\mu_{\tau\sigma;\tau'\sigma'} = \frac{i}{2\sqrt{2}}\sigma^\mu_{\tau\tau'}\sigma^y_{\sigma\sigma'}. \tag{7}$$

$\sigma^\mu$ is the $\mu$-component of the Pauli matrices.

To derive the multipole interactions, we consider the intermediate $f^1$-$f^3$ configurations of the nearest-neighboring sites. The intermediate $f^1$ state is the $\Gamma_8$ state given by Eq. (1). The intermediate $f^3$ state is the one-hole state in the $\Gamma_8$ orbitals and obviously follows the $\Gamma_8$ symmetry. The energy of the $f^3$ state is $U + 2U' - J$. Note that, within the second-order perturbation theory with respect to the hopping, there is no other intermediate states in the present model.

The kinetic energy term of the Hamiltonian is given by

$$H_{\text{kin}} = \sum_{\boldsymbol{r},\boldsymbol{\xi},\tau,\sigma,\tau',\sigma'} c^\dagger_{\boldsymbol{r}+\boldsymbol{\xi}\tau\sigma}t^{\boldsymbol{\xi}}_{\tau\sigma;\tau'\sigma'}c_{\boldsymbol{r}\tau'\sigma'} = \sum_{\boldsymbol{r},\boldsymbol{\xi},\nu,\nu'} c^\dagger_{\boldsymbol{r}+\boldsymbol{\xi}\nu}t^{\boldsymbol{\xi}}_{\nu\nu'}c_{\boldsymbol{r}\nu'}, \tag{8}$$

where $\boldsymbol{\xi}$ denotes the vector connecting the nearest-neighboring sites. Here, we have introduced an abbreviation $\nu = (\tau, \sigma)$. Since $H_{\text{kin}}$ is Hermitian, $t^{\boldsymbol{\xi}*}_{\nu\nu'} = t^{-\boldsymbol{\xi}}_{\nu'\nu}$.

Concerning the hopping integrals, we consider only the $\sigma$ bonding $(ff\sigma)$ as in the previous study [13]. The hopping integrals were derived in Ref. [15] for the sc lattice and in Ref. [4] for the bcc and fcc lattices. To write out the hopping integral $t^{\boldsymbol{\xi}}$ for each lattice, it is convenient to introduce the following $4 \times 4$ matrices:

$$\tilde{1}_{\tau\sigma;\tau'\sigma'} \equiv \delta_{\tau\tau'}\delta_{\sigma\sigma'}, \tag{9a}$$
$$\tilde{\boldsymbol{\tau}}_{\tau\sigma;\tau'\sigma'} \equiv \boldsymbol{\sigma}_{\tau\tau'}\delta_{\sigma\sigma'}, \tag{9b}$$
$$\tilde{\boldsymbol{\sigma}}_{\tau\sigma;\tau'\sigma'} \equiv \delta_{\tau\tau'}\boldsymbol{\sigma}_{\sigma\sigma'}, \tag{9c}$$
$$\tilde{\eta}^\pm \equiv (\pm\sqrt{3}\tilde{\tau}^x - \tilde{\tau}^z)/2. \tag{9d}$$

Then, the hopping integrals for the sc lattice are written as

$$t^{(1,0,0)} = [\tilde{1} - \tilde{\eta}^+]t_1, \tag{10a}$$
$$t^{(0,1,0)} = [\tilde{1} - \tilde{\eta}^-]t_1, \tag{10b}$$
$$t^{(0,0,1)} = [\tilde{1} - \tilde{\tau}^z]t_1, \tag{10c}$$

where we have set the lattice constant as unity and $t_1$=3$(ff\sigma)/14$. For the bcc lattice,

$$t^{(1/2,1/2,1/2)} = [\tilde{1} + \tilde{\tau}^y(+\tilde{\sigma}^x + \tilde{\sigma}^y + \tilde{\sigma}^z)/\sqrt{3}]t_2, \tag{11a}$$
$$t^{(-1/2,1/2,1/2)} = [\tilde{1} + \tilde{\tau}^y(+\tilde{\sigma}^x - \tilde{\sigma}^y - \tilde{\sigma}^z)/\sqrt{3}]t_2, \tag{11b}$$
$$t^{(1/2,-1/2,1/2)} = [\tilde{1} + \tilde{\tau}^y(-\tilde{\sigma}^x + \tilde{\sigma}^y - \tilde{\sigma}^z)/\sqrt{3}]t_2, \tag{11c}$$
$$t^{(1/2,1/2,-1/2)} = [\tilde{1} + \tilde{\tau}^y(-\tilde{\sigma}^x - \tilde{\sigma}^y + \tilde{\sigma}^z)/\sqrt{3}]t_2, \tag{11d}$$

with $t_2$=2$(ff\sigma)/21$. For the fcc lattice,

$$t^{(0,1/2,1/2)} = [\tilde{1} + (\tilde{\eta}^+ - 4\sqrt{3}\tilde{\tau}^y\tilde{\sigma}^x)/7]t_3, \tag{12a}$$
$$t^{(1/2,0,1/2)} = [\tilde{1} + (\tilde{\eta}^- - 4\sqrt{3}\tilde{\tau}^y\tilde{\sigma}^y)/7]t_3, \tag{12b}$$
$$t^{(1/2,1/2,0)} = [\tilde{1} + (\tilde{\tau}^z - 4\sqrt{3}\tilde{\tau}^y\tilde{\sigma}^z)/7]t_3, \tag{12c}$$
$$t^{(0,1/2,-1/2)} = [\tilde{1} + (\tilde{\eta}^+ + 4\sqrt{3}\tilde{\tau}^y\tilde{\sigma}^x)/7]t_3, \tag{12d}$$
$$t^{(-1/2,0,1/2)} = [\tilde{1} + (\tilde{\eta}^- + 4\sqrt{3}\tilde{\tau}^y\tilde{\sigma}^y)/7]t_3, \tag{12e}$$
$$t^{(1/2,-1/2,0)} = [\tilde{1} + (\tilde{\tau}^z + 4\sqrt{3}\tilde{\tau}^y\tilde{\sigma}^z)/7]t_3, \tag{12f}$$

with $t_3$=$(ff\sigma)/8$. For the other directions such as $(-1,0,0)$, we can use the relation $t^{-\boldsymbol{\xi}} = t^{\boldsymbol{\xi}}$.

By employing the second-order perturbation theory with respect to $H_{\text{kin}}$, we derive the effective Hamiltonian. In general, it is given by

$$H^{(\text{eff})} = \sum_{a,b,u}\sum_{m\neq 0} |0,a\rangle\langle 0,a|H_{\text{kin}}\frac{|m,u\rangle\langle m,u|}{E_0 - E_m}H_{\text{kin}}|0,b\rangle\langle 0,b|. \tag{13}$$

Here, $|0,a\rangle$ is a ground state without $H_{\text{kin}}$ with the energy $E_0$ and $|m,u\rangle$ is an $m$-th excited state with the energy $E_m$. However, as noted above, we have only one excited level in the intermediate states, and the effective Hamiltonian can be rewritten as

$$H^{(\text{eff})} = -\frac{1}{\Delta E}\sum_{a,b}|0,a\rangle\langle 0,a|H_{\text{kin}}^2|0,b\rangle\langle 0,b|, \tag{14}$$

where

$$\Delta E = E_1 - E_0 = (U + 2U' - J) - 2(U' + J) = U - 3J. \tag{15}$$

If $\Delta E < 0$, the $f^2$ state is unstable. Thus, we assume a positive $\Delta E$. Then, the matrix elements of the effective interaction between ions at $\boldsymbol{r}_1$ and $\boldsymbol{r}_2$ is now evaluated by simply calculating the matrix elements of $H_{\text{kin}}^2$ between the ground states:

$$\begin{aligned} H^{(\text{eff})}_{\mu_1\mu_2;\mu_1'\mu_2'}(\boldsymbol{r}_1,\boldsymbol{r}_2) &= -\langle \mu_1(\boldsymbol{r}_1)\ \mu_2(\boldsymbol{r}_2)|H_{\text{kin}}^2|\mu_1'(\boldsymbol{r}_1)\ \mu_2'(\boldsymbol{r}_2)\rangle/\Delta E \\ &= -32\text{Re}\left[\text{Tr}\left(B^{\mu_2}t^{\boldsymbol{r}_2-\boldsymbol{r}_1}B^{\mu_1}B^{\mu_1'}t^{\boldsymbol{r}_1-\boldsymbol{r}_2}B^{\mu_2'}\right)\right]/\Delta E, \end{aligned} \tag{16}$$

where Re denotes the real part and Tr is the trace of the matrix.

The obtained effective Hamiltonian can be rewritten in terms of the multipole operators for the $f^2$-$\Gamma_3$ state defined by

$$O^0_{2\boldsymbol{r}} = \sum_{\mu\mu'}|\mu(\boldsymbol{r})\rangle\sigma^z_{\mu\mu'}\langle\mu'(\boldsymbol{r})|, \tag{17a}$$
$$O^2_{2\boldsymbol{r}} = \sum_{\mu\mu'}|\mu(\boldsymbol{r})\rangle\sigma^x_{\mu\mu'}\langle\mu'(\boldsymbol{r})|, \tag{17b}$$
$$T_{xyz\boldsymbol{r}} = \sum_{\mu\mu'}|\mu(\boldsymbol{r})\rangle\sigma^y_{\mu\mu'}\langle\mu'(\boldsymbol{r})|. \tag{17c}$$

$O^0_{2r}$ and $O^2_{2r}$ are the quadrupole moments with $\Gamma_{3g}$ symmetry and $T_{xyzr}$ is the octupole moment with $\Gamma_{2u}$ symmetry.

## 3. Results

For the sc lattice, we obtain no multipole interaction, that is, the second-order perturbation results in only an overall shift of energy. This result suggests that the present model is too simple to deal with the multipole interactions at least for the sc lattice, and the $\Gamma_7$ orbital should be included as in the previous study [13].

For the bcc lattice, we obtain only the following octupole interaction:

$$H^{(\mathrm{eff})} = 16 \sum_{\boldsymbol{q}} \cos(q_x/2) \cos(q_y/2) \cos(q_z/2) T_{xyz\boldsymbol{q}} T_{xyz-\boldsymbol{q}}, \tag{18}$$

in the unit of $t_2^2/\Delta E$. $T_{xyz\boldsymbol{q}}$ is the Fourier transform of $T_{xyz\boldsymbol{r}}$. We have not obtained the quadrupole interaction. Even in the previous study, we have obtained only the octupole interaction, and thus, we expect that the dominance of the octupole interaction on the bcc lattice does not depend on the details of the model.

For the fcc lattice, we again obtain only the octupole interaction:

$$H^{(\mathrm{eff})} = \frac{384}{49} \sum_{\boldsymbol{q}} \left[\cos(q_x/2) \cos(q_y/2) + \cos(q_y/2) \cos(q_z/2) + \cos(q_z/2) \cos(q_x/2)\right] T_{xyz\boldsymbol{q}} T_{xyz-\boldsymbol{q}}, \tag{19}$$

in the unit of $t_3^2/\Delta E$. In the previous study, we have obtained a weak but finite quadrupole interaction. Thus, the appearance of the quadrupole interaction, in addition to the octupole interaction, depends on the model to construct the $f^2$-$\Gamma_3$ state. However, even in the previous study, the quadrupole interaction is much weaker than the octupole interaction, and the dominance of the octupole interaction on the fcc lattice might not depend on the details of the model.

## 4. Summary and discussion

We have studied a simple model for the $f^2$-$\Gamma_3$ systems by considering only the $\Gamma_8$ orbitals as the one-electron states. We obtain no multipole interaction for the sc lattice and the octupole interactions for the bcc and fcc lattices. The dominance of the octupole interaction in the bcc and fcc lattices and the absence of the octupole interaction in the sc lattice are common between the present model and a model including the $\Gamma_7$ orbital in the one-electron states to construct the $f^2$-$\Gamma_3$ state [13]. Thus, these features will not be altered by changing the details of the model.

However, the absence of the multipole interactions in the sc lattice indicates that the present model including only the $\Gamma_8$ orbitals is too simple. In this regard, we should include at least the $\Gamma_7$ orbital to study the multipole physics of the $f^2$-$\Gamma_3$ systems from a microscopic viewpoint as in the previous study [13]. In a single ion model considering all the 14 $f$-orbitals with a realistic parameter set to obtain a $\Gamma_3$ ground state for $f^2$, we obtain the $\Gamma_6$ ground state when we accommodate three electrons [14]. Indeed, we obtain the $f^3$-$\Gamma_6$ CEF ground state in the previous model, but the $f^3$ states are $\Gamma_8$ in the present model. This problem on the $f^3$ states also indicates the importance of the $\Gamma_7$ orbital.

Note that we have implicitly assumed an insulating state to derive the multipole interactions. For a metallic state, it might be necessary to consider the conduction band explicitly and derive the multipole interactions as the Ruderman-Kittel-Kasuya-Yosida (RKKY) interactions. It is a future problem whether the multipole interactions as the RKKY interactions are much different or not.

Finally, from our studies, we notice that it is interesting to search the $\Gamma_3$ systems on the bcc lattices for the octupole ordering. The diamond structure is also a candidate to realize octupole ordering since the nearest-neighbor ions locate along [111] direction and so on as in the bcc lattice and we obtain only the octupole interaction in our models.

## Acknowledgments

This work was supported by JSPS KAKENHI Grant Numbers JP15K05191, JP16H04017, and JP16H01079 (J-Physics).

## References

[1] Paixão J A, Detlefs C, Longfield M J, Caciuffo R, Santini P, Bernhoeft N, Rebizant J and Lander G H 2002 *Phys. Rev. Lett.* **89** 187202
[2] Tokunaga Y, Homma Y, Kambe S, Aoki D, Sakai H, Yamamoto E, Nakamura A, Shiokawa Y, Walstedt R E and Yasuoka H 2005 *Phys. Rev. Lett.* **94** 137209
[3] Kubo K and Hotta T 2005 *Phys. Rev.* B **71** 140404(R)
[4] Kubo K and Hotta T 2005 *Phys. Rev.* B **72** 144401
[5] Kubo K and Hotta T 2005 *Phys. Rev.* B **72** 132411
[6] Akatsu M, Goto T, Nemoto Y, Suzuki O, Nakamura S and Kunii S 2003 *J. Phys. Soc. Jpn.* **72** 205
[7] Kubo K and Kuramoto Y 2003 *J. Phys. Soc. Jpn.* **72** 1859
[8] Kubo K and Kuramoto Y 2004 *J. Phys. Soc. Jpn.* **73** 216
[9] Morie T, Sakakibara T, Tayama T and Kunii S 2004 *J. Phys. Soc. Jpn.* **73** 2381
[10] Mannix D, Tanaka Y, Carbone D, Bernhoeft N and Kunii S 2005 *Phys. Rev. Lett.* **95** 117206
[11] Kuwahara K, Iwasa K, Kohgi M, Aso N, Sera M and Iga F 2007 *J. Phys. Soc. Jpn.* **76** 093702
[12] Inami T, Michimura S, Hayashi Y, Matsumura T, Sera M and Iga F 2014 *Phys. Rev.* B **90** 041108
[13] Kubo K and Hotta T 2017 *Phys. Rev.* B **95** 054425
[14] Hotta T and Harima H 2006 *J. Phys. Soc. Jpn.* **75** 124711
[15] Hotta T and Ueda K 2003 *Phys. Rev.* B **67** 104518

IOP Conf. Series: Journal of Physics: Conf. Series **969** (2018) 012097 doi:10.1088/1742-6596/969/1/012097

# Between metamagnetic transition and spin-flip behavior in Ce 122 system of (Ce-Gd)$Ru_2Si_2$

**K Yano[1], Y Amakai[2], Y Hara[3], K Sato[3], E Kita[3], H Takano[2], T Ohta[4] and S Murayama[2]**

[1] College of Science and Technology, Nihon University, Funabashi, Japan
[2] Materials Science, Muroran Insitute of Technology, Muroran, Japan
[3] National Institute of Technology, Ibaraki College, Hitachinaka, Japan
[4] Quantum Design Japan, Ikebukuro, Japan

E-mail: kyano@phys.ge.cst.nihon-u.ac.jp

**Abstract**. Aiming at getting some clues to the mechanism of meta-magnetic transition and surprisingly small magnetic moment of Ce along hard axis in $CeRu_2Si_2$, the (Ce-Gd)$Ru_2Si_2$ system where Ce was substituted by Gd were studied through magnetic properties mainly in Gd-rich regions. At Gd=0, i.e. in $CeRu_2Si_2$, the magnetic moment of Ce showed a symptom of saturation in *M-H* curve under *H*=90,000 Oe at 2 K and the Ce magnetic moment at 4.2 K can be nearly identical to that at 2 K employing 1/*H* plot. At Gd-rich content of 0.8, Ce magnetic moment coupled parallel to that of Gd, Ce ↑ Gd ↑ both in easy and hard axis and the extremely smallness of Ce magnetic moment in hard axis disappeared perfectly at x=0.8. Furthermore at Gd=1, $GdRu_2Si_2$, Gd magnetic moment caused 2-step like spin-flip in both easy and hard axis.

## 1. Introduction

Ce122 systems have been one of the most extensively studied heavy fermion materials and have revealed various properties such as superconductivity, ferromagnetism and so on depending on the electronic states of Ce [1,2]. Consequently, the Ce 122 systems have been attracting many researchers and remain to be one of the main topics at present. $CeRu_2Si_2$, which belongs to the Ce122 systems, is one of the most famous heavy fermion compounds and is known magnetically as paramagnetic [3]. On the other hand, $CeRu_2Ge_2$ where Si is replaced by Ge shows clear ferromagnetic behaviour with Curie temperature *Tc* of 7.9 K [2] and many substituted systems such as $CeRu_2$(Si-Ge)$_2$ [4] and Ce(Ru-Rh)$_2Si_2$ [5,6] have been investigated in detail and interesting results have been obtained. Furthermore these two compounds have been treated often as a good example of investigating a nature of quantum critical point (QCT) and the $CeRu_2Si_2$ have attracted many researchers [7,8] also from this point of view.

At the beginning of the study of $CeRu_2Si_2$, Haen *et al.* have carried out through investigations about this compound and pointed out 2 peculiar characteristics [9]. The first is the most famous "metamagnetic" transition, that is, the Ce magnetic moment increases sharply around the magnetic field of *H*=77,000 Oe applied in easy c axis under 4.2 K and the second is the unexpected smallness of Ce magnetic moment (< 0.1 $\mu_B$) along hard axis ($H \perp$ c-axis. Regarding the first feature of meta-magnetic transition in $CeRu_2Si_2$ along c-axis, some experiments and interpretation have been carried out and suggested [10,11]. One of the representatives was carried out using de Haas van Alphen (dHvA) effect and it was suggested that the 4*f* electron of Ce varied from itinerant to localized nature at meta-magnetic transition field $H_m$ with the increase of magnetic field [10]. On the other hand, the soft X-ray magnetic circular dichroism (XMCD) measurement seems to have suggested a little different picture [11]. At present, there have been no decisive pictures regarding the mechanism of metamagnetic transition and the electronic state of 4*f* electron in Ce before and after this transition.

In this study, we have aimed and tried to get some clues to the mechanism about metamagnetic transition and the unexpectedly small magnetic moment of Ce in $CeRu_2Si_2$ by using magnetic measurement for the substituted system of (Ce-Gd)$Ru_2Si_2$ where Gd was employed because of the simplest ($L$=0) and strongest molecular field in Rare Earth (RE) elements.

## 2. Experimental Procedure

Samples studied were all prepared by arch-melting the stoichiometrically weighted constituents 5 times and the obtained poly crystals were broken into pieces with a hammer. From these small pieces, some flakes with both sides glossy were picked up and checked by Laue photographs. The samples of single crystal like flakes obtained from these procedures were of the form of sheet and the magnetically easy axis of c-axis lied perpendicular to the sheets. These single-like crystals were magnetically measured. The purities of constituent elements used were 3N for Ce and Gd, 4N for Ru and 5N for Si, respectively.

Magnetic measurements were carried out for the above-mentioned samples from 2 K to 300 K under the magnetic field up to 90,000 Oe by the use of both SQUID and VSM system of Quantum Design Corporation.

## 3. Results and Discussion

The measurements of magnetic properties, here focusing on the *M-H* curves at various temperatures, were performed for mainly Gd-rich contents in both directions of easy axis and hard axis. Each result and discussion was described below, respectively.

### *3.1. In case of x=0 ( i.e. $CeRu_2Si_2$)*

At the beginning of the research of $CeRu_2Si_2$, Haen *et al.* have carried out detailed measurements regarding *M-H* curves in both directions of magnetically easy axis ($H \parallel$ c-axis) and hard axis ($H \perp$ c-axis). In their paper, they have pointed out two characteristic behaviors observed in this compound; the first is the metamagnetic transition around 77,000 Oe below 4.2 K in easy axis and this transition disappeared completely around 15 K and the second is the unexpected smallness of Ce magnetic moment in hard axis even in low temperatures.

Fig.1 (a), (b) show the present measured *M-H* curves for x=0 in easy axis ($H \parallel$ c-axis) and hard axis ($H \perp$ c-axis), respectively. From Fig.1 (a), the behaviors in *M-H* at 2 and 4.2 K are roughly identical to the detailed measurement for the single crystal by Sakakibara *et al.* [12] and the present result at 2 K shows a clearer symptom towards saturation in the magnetic field up to 90,000 Oe in comparison with their result below $H$=85,000 Oe. As described by many researchers, the present Ce magnetic moment increases linearly with the increase of magnetic field up to around $H$=50,000 Oe even in low temperatures. In this sense that the magnetic moment is nearly proportional to magnetic field $H$, Ce magnetic moment behaves like paramagnetism, however, the value of Ce magnetic moment is not small but rather large enough to be called magnetic to some degree by considering that Ni retains about 0.6 $\mu_B$ in Ni metal and 0.2 $\mu_B$ in Ni-RE (Rare Earth) compounds [13]. In Fig.1 (b), the Ce magnetic moment behavior at 3 K in hard axis is shown. The magnetic moment of Ce in low temperatures is considerably small and about 1/30 smaller than that in easy axis. The value and linear behavior in hard axis is nearly identical to the result of Haen *et al.* Fig.1 (c) and (d) show the *M-1/H* plot in easy axis at 2 K and 4.2 K, respectively. This *M-1/H* plot is not used frequently and used conveniently to assume and determine the saturation magnetization of the sample [14]. From both results of (c) and (d), Ce magnetic moment at 4.2 K can be nearly identical to that at 2 K and this result implies that the Ce 4*f* electronic state at 4.2 K is nearly identical to that at 2 K. This result is considered to be important to microscopical measurement since the lowest temperature obtained in the experiments can be often limited to 4.2 K because of the He cryostat.

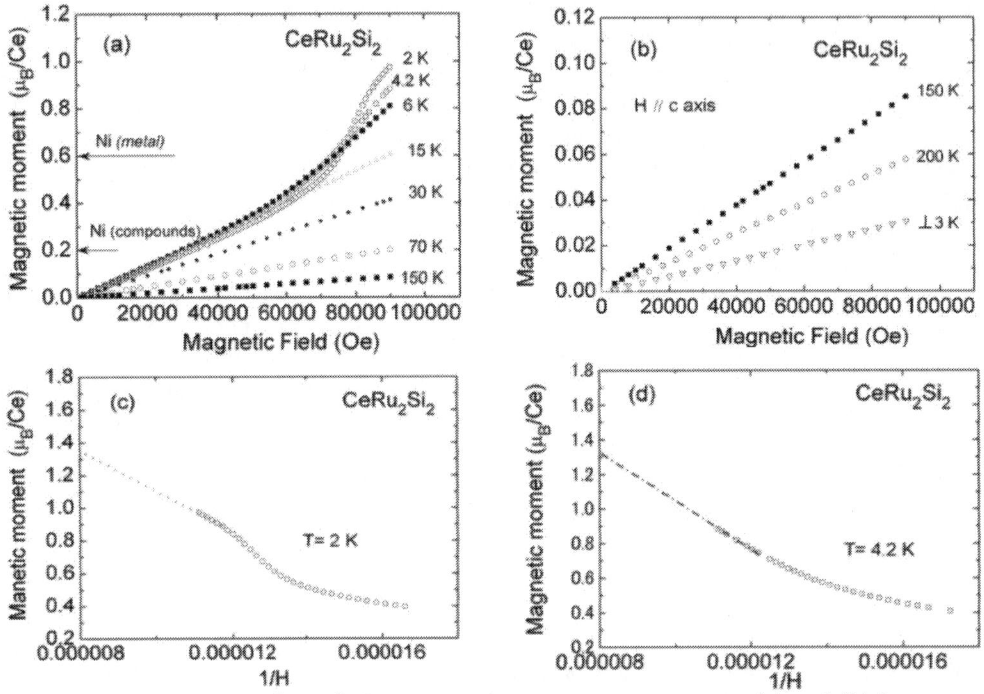


Figure 1. *M-H* curves for Gd=0 compound at some temperatures. (a) and (b) in easy axis. (b) includes also in hard axis at 3 K. (c) and (d) show *M-1/H* plot in easy axis.

### 3.2. *In case of x=0.8 ( i.e.* $(Ce_{0.2}Gd_{0.8})Ru_2Si_2$*)*

In case of Gd content of 0.8, the *M-H* curves at some temperatures between 2 K and 70 K up to the magnetic field of 70,000 Oe were measured in both directions of magnetically easy axis (*H* // c axis) and hard axis (*H* ⊥ c axis) and the results are shown in Fig.2 (a) and (b), respectively. In this content, total magnetization is dominated by that of Gd and both figures are scaled by magnetic moment of a Gd atom.

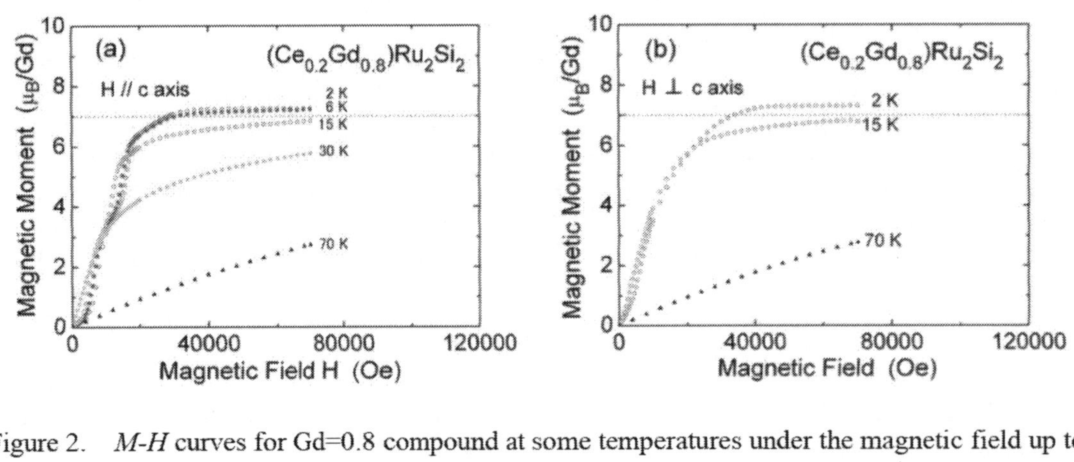


Figure 2. *M-H* curves for Gd=0.8 compound at some temperatures under the magnetic field up to 70,000 Oe. (a) in the easy axis (*H*// c axis) and (b) in the hard axis (*H* ⊥ c axis).

From Fig.2 (a) in easy direction ($H$ // c axis), the Gd magnetic moment, that is, the total magnetization, of this sample increases not linearly but 2-steps likely between 2 K and 15 K, which will be more clearly seen later in section 3.3 at x=1 as spin-flip behaviors. Over temperature of $T$=30 K, it can be found that the 2-steps like behaviors cannot be observed and disappear. It is remarkable that the Gd magnetic moment saturates well at 2 K and 6 K over the magnetic field of 30,000 Oe and from the value of saturation, the magnetic moment of other element can couple parallel with that of Gd under the conventional assumption that the Gd retains 7 $\mu_B$ since Gd magnetic moment is found to retain over 7 $\mu_B$ from the figure. The candidate of other magnetic element can be Ce and the Ce magnetic moment can be estimated to be 1.1 $\mu_B$ and the magnetic structure in this compound is found to be ferromagnetically coupled Gd↑ Ce↑. Taking into account that it is not popular that the Ce magnetic moment saturates completely in Ce122 systems, this estimated value of Ce magnetic moment is considerably valuable.

Fig.2 (b) shows the $M$-$H$ curves in hard axis ($H\perp$c axis) at three temperatures. Since Haen *et a*l. have pointed out the unexpected smallness of Ce magnetic moment in hard axis ($H\perp$c axis), this behavior seems to have attracted little attentions and there have been little reports piled up. From this figure, 2-steps like behavior in the $M$-$H$ curves tend to smear out a little by comparing with those in easy axis shown in (a) and the Gd magnetic moment can be clearly observed to saturate over 40,000 Oe at 2 K. Like in easy axis, the Ce is also found to magnetic and retain magnetic moment in hard axis and the magnetic moment of Ce can be estimated to be about 1.2 $\mu_B$ in the same way that in easy axis. In this figure, the total magnetization saturates well over $H$=40,000 Oe and the value derived out can be reliable. It follows that the Ce magnetic moment in hard axis is nearly identical to that in easy axis and this means that the unexpected smallness of Ce magnetic moment in hard axis does disappear perfectly in Gd content of 0.8. This finding can be the first time to confirm experimentally and this is expected to give an important clue to understand the electronic states of Ce in $CeRu_2Si_2$ and Ce122 systems.

*3.3. In case of x=1 ( i.e. $GdRu_2Si_2$)*

The $M$-$H$ curves in Gd content $x$=1 at some temperatures up to the magnetic field of 90,000 Oe were measured in both directions of easy axis ($H$//c axis) and hard axis ($H\perp$c axis) and the results are shown in Fig.3 (a) and (b), respectively.

From Fig.3 (a) in easy axis, 2-steps like spin-flip behaviors can be observed clearly between 2 K and 15 K and roughly 1-step like spin-flip behaviors can be confirmed between 20 K and 40 K. Over 50 K, spin-flip like behavior is found to disappear. The 2-steps like spin-flip behavior have been observed in $Nd_3Co$ [15]. Roughly speaking, Nd can be anisotropic since Nd atom retains a component of large angular momentum $L$=6 and Nd-TM (3$d$ transition metal) including Dy-TM and Ho-TM compounds and alloys are known to be anisotropic and retain generally large coercive force $Hc$ [16].

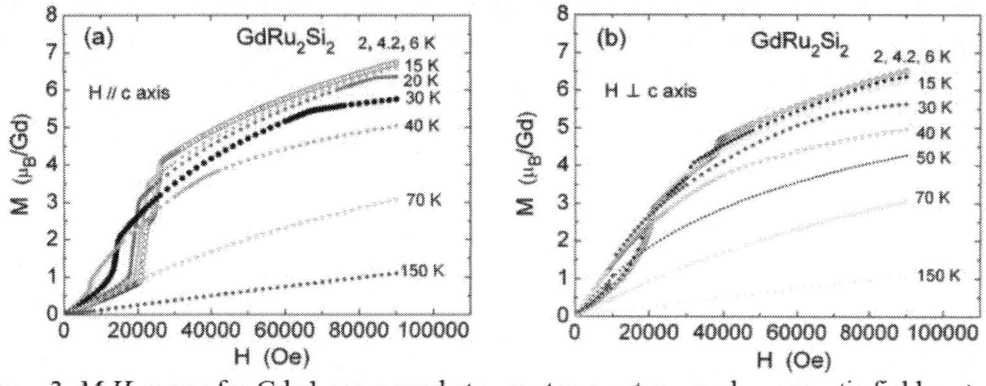


Figure 3. $M$-$H$ curves for Gd=1 compound at some temperatures under magnetic field up to 90,000 Oe. (a) in the easy axis ($H$// c axis) and (b) in the hard axis ($H\perp$ c axis).

On the contrary, Gd atom retains no angular momentum (ideally, $L$=0) and Gd-TM compounds and alloys retain little and negligible anisotropy and coercive force. That is, considering the above-mentioned, it is not self-evident that the $GdRu_2Si_2$ seems to be anisotropic and shows 2-steps spin-flip behaviors like $Nd_3Co$. It may be that Gd can retain angular momentum $L$ effectively in this compound and crystal structure [17]. After all, it is noteworthy that the $GdRu_2Si_2$ shows not a simple ferromagnetic behavior but a linear increase in $M$-$H$ curve at first and then 2-steps spin-flip behaviors in this compound. This crystal structure can tend to cause the spin-flip behavior in RE elements.

From Fig.3 (b) in hard axis, roughly speaking, the behaviors in $M$-$H$ curves in hard axis resemble to those in easy axis while the spin-flip behaviors in hard axis tend to smear out a little and not sharp in comparison with those in easy axis. Over $T$=50 K, the behaviors in $M$-$H$ curves in both easy and hard axis are considered to be nearly identical.

**Acknowledgement**
The authors would like to greatly appreciate Professor Sato for giving the data of 2-steps spin-flip behaviour in $Nd_3Co$.

**References**

[1] Steglich F, Aarta J, Bredle C D, Lieke W, Meschede D, Frants W and Schaefer H, 1979 Phys. Rev. Lett. **43** 1892
[2] Besnus M J, Kapler P, Lehmann P, Essaihi A, Hamdaoui N, Fischer G, Kappler J P, Meyer A, Pierre J, Haen P and Lejay P 1991 Physica B **171** 350
[3] Besnus M J, Kapler P, Lehmann P and Meyer A 1985 Solid State Commun. **55** 779
[4] Haen P, Bioud H and Fukuhara T 1999 Physica B **259-261** 85
[5] Sekine C, Sakakibara T, Amitsuka H, Miyako Y and Goto T 1992 J. Phys. Soc. Jpn. **61** 4536
[6] Murayama S, Sekine C, Yokoyanagi A, Hoshi K and Onuki Y 1997 Phys. Rev. B **56** 11092
[7] Flouquet J, Haen P, Raymond S, Aoki D and Knebel G 2002 Physica B **319** 251
[8] Lacerda A, Visser de, Haen P, Lejay P, and Flouquet J 1989 Phys. Rev. B **40** 8759
[9] Haen P, Flouquet J, Lapierre F, Lejay P and Remenyi G 1987 J Low Temp Phys. 391
[10] Aoki H, Uji S, Albessard A K and Onuki Y 1993 Phys.Rev.Lett. **71** 2110
[11] Okane T, Takeda Y, Yamagami H, Fujimori A, Matsumoto Y, Kimura N, Komatsubara T and Aoki H 2012 Phys.Rev.B **86** 125138-1
[12] Sakakibara T, Tayama T, Matsuhira K, Mitani H and Amitsuka H 1995 Phys.Rev.B **51** 12030
[13] Mizumaki M, Yano K, Umehara I, Ishikawa F, Sato K, Koizumi A, Sakai N and Muro T 2003 Phys. Rev. B **67** 132404-1
[14] Yano K, Ohta T and Sato K 2017 J Phys. Chem. of Solids **104** 13
[15] Umehara I, Lu Q, Endo M, Adachi Y and K Sato 1998 J. Phys. Soc. Jpn. **67** 4213
[16] Rhyne J J 1979 Amorphous magnetic rare earth alloys (ALLOYS AND INTERMETALLICS vol 2) ed Gschneidner and Eyring (Amsterdam: North-Holland) chapter 16 pp 259-294
[17] Yano K, Okane T, Takeda Y, Yamagami H, Fujjmori A, Nishimura K and Sato K 2017 Physica B **515** 118

IOP Conf. Series: Journal of Physics: Conf. Series **969** (2018) 012098 doi:10.1088/1742-6596/969/1/012098

# Magnetic properties of Heusler compound $Fe_{1.3}Mn_{1.7}Si$

**Masahiko Hiroi[1], Tomohito Nonoyama[1], Genta Adachi[1], Iduru Shigeta[1], Hirotaka Manaka[2], and Norio Terada[2]**

[1] Department of Physics and Astronomy, Graduate School of Science and Engineering, Kagoshima University, Kagoshima 890-0065, Japan

[2] Department of Electrical and Electronics Engineering, Graduate School of Science and Engineering, Kagoshima University, Kagoshima 890-0065, Japan

E-mail: hiroi@sci.kagoshima-u.ac.jp

**Abstract.** The magnetic properties of two Heusler compounds $Fe_{1.3}Mn_{1.7}Si$ and $Fe_{1.4}Mn_{1.6}Si$ are compared. $Fe_{1.4}Mn_{1.6}Si$ exhibits a ferromagnetic transition at $T_C \sim 117$ K and another transition to a phase with antiferromagnetic components at $T_A \sim 66$ K, which is the same as the behavior of $Fe_2MnSi$. Further, it is known that for $Fe_{3-x}Mn_xSi$, with increasing $x$, $T_C$ decreases but $T_A$ remains almost constant, and they appear to meet around $x \sim 1.7$. The careful measurements of magnetization at various magnetic fields reveal that $Fe_{1.3}Mn_{1.7}Si$ has three transitions at low fields; for instance, at 0.01 T, the transition temperatures are found to be 85 K, 67 K, and 60 K. The lowest transition at 60 K is characterized by a rapid decrease in magnetization with decreasing temperature and significant magnetic field dependence, in contrast to the transition at $T_A$, which is relatively independent of the magnetic field, for $x = 1$ and 1.6. The critical field at 0 K is $\sim$3 T. The transition at the lowest temperature was not found for $x \leq 1.6$, and it is considered a different transition from that at $T_A$. While the transition at the highest temperature 85 K is a ferromagnetic transition, the transition at 67 K has the characteristics of the transition at $T_A$, and it is found to be insensitive to magnetic fields. They are revealed to be the same transitions as those at $T_C$ and $T_A$ for $x \leq 1.6$.

## 1. Introduction

Half-metallic materials in Heusler compounds have attracted significant attention because of their potential applications in *spintronics* [1–3]. Half-metals are ferromagnetic metals in which at the Fermi energy the density of states exists only in one spin band and there is a gap in the other spin band; this implies that there are 100% spin-polarized conduction electrons. Half-metals are promising materials for magneto-resistive and spin-injection devices. The possibility of some Heusler compounds being half-metals was suggested some time ago. Heusler compounds have the formula $X_2YZ$, where $X$ and $Y$ are transition elements, $Z$ is an *sp* element, and the structure is $L2_1$ [1]. Among such compounds, $Co_2MnSi$ and $Fe_2MnSi$ were predicted to be half-metals based on first-principles band structure calculations [4, 5]. $Co_2MnSi$ has since been revealed to exhibit half-metallicity. Although $Fe_2MnSi$ shows a ferromagnetic transition at a temperature $T_C$ (Curie temperature) of $\sim$ 230 K, there is a transition to a phase with some antiferromagnetic components but with a large ferromagnetic component at a lower temperature of $\sim$ 60 K [6, 7]. In this paper, we refer to this transition temperature as $T_A$; the phase below $T_A$ as the AF phase; and the ferromagnetic phase between $T_A$ and $T_C$ as the F phase. These two transitions are observed not only in $Fe_2MnSi$ but also in compounds $Fe_{3-x}Mn_xSi$ for a wide $x$ range of $0.75 < x < 1.75$ [7]. Although $Fe_{3-x}Mn_xSi$ was also theoretically predicted to be a half-metal for some values of $x$ including $x = 1$ [8–10], the presence of an antiferromagnetic order implies that the ground state is not half-metallic ferromagnetism. A ferromagnetic transition exists in $Fe_3Si$ with $T_C \sim 800$ K. $T_C$ decreases as $x$ increases to 1.75. The AF transition does not exist for $x < 0.75$; it emerges abruptly around $x \sim 0.75$. $T_A$ does not change much with $x$ and is constant at around 60 K [11], although $T_C$ decreases with $x$ and seems to meet $T_A$ around $x = 1.7$. The transition and properties of the AF phase have been studied in magnetic and thermodynamic measurements privously [6–8, 11–14], and neutron scattering experiments have revealed a complex magnetic structure with the magnetic moment carried by Mn [11, 15]. According to theoretical calculations, some antiferromagnetic states are more energetically favorable than the ferromagnetic state [5]. However, the properties of the AF state have not been studied much, other than for $Fe_{3-x}Mn_xSi$ where $x = 1$.

For $x \sim 1.7$, when $T_C$ decreases and almost coincides with $T_A$, a meta-magnetic-like transition was observed in magnetization versus magnetic field [7, 16, 17]. The relation between the meta-magnetic transition fields and the temperatures of the observation was revealed to be similar as in an ordinary antiferromagnetic transition. From an extrapolation of the relation, the critical field at 0 K was found to be ~3 T, and the transition temperature at the zero field was found to be ~ 60 K and it almost agreed with the $T_A$ that was found for a lower $x$. However, this field dependence of the transition differs significantly from that of $T_A$ in $Fe_2MnSi$, where $T_A$ is not affected much by a magnetic field, and the critical field is estimated to be over ~ 70 T [18]. Previous studies appeared to assume this transition to be the AF transition implicitly and they did not address the relationship of this transition to the AF transition. Nevertheless, we believe that previous data suggests that this transition is not the same one as the AF transition. Although for $1 \leq x \leq 1.5$, the basic behavior was revealed to be the same as that for $x = 1$, that is, two transitions at $T_A$ and $T_C$ were observed, we believe that for $x \sim 1.7$, the AF phase may change to another phase.

In this study the magnetization of $Fe_{1.3}Mn_{1.7}Si$ is investigated in detail, where $T_A$ and $T_C$ are expected to almost agree, and it is compared with the results for $Fe_{1.4}Mn_{1.6}Si$, where $x$ is near 1.7 but the $T_C$ is expected to be significantly higher than $T_A$.

## 2. Experimental

Polycrystalline samples of $Fe_{1.3}Mn_{1.7}Si$ and $Fe_{1.4}Mn_{1.6}Si$ were prepared by arc-melting high-purity constituent elements in a high-purity argon atmosphere. Some parts of the samples were annealed at 800° C for 2 days. The crystal structures were examined using X-ray powder diffraction measurements with Cu K$\alpha$ radiation. X-ray diffraction measurements showed that the prepared samples were single-phase, and the crystal structure was $L2_1$. Magnetization was measured using SQUID magnetometers (MPMS: Quantum Design).

## 3. Results and Discussion

Figure 1 shows the temperature dependence of magnetization $M(T)$ of $Fe_{1.4}Mn_{1.6}Si$ for different values of magnetic fields (0 T$\leq B \leq$5 T). The behavior observed in $M(T)$ of $Fe_{1.4}Mn_{1.6}Si$ can be regarded as the same as that of $Fe_2MnSi$. The rapid increase in $M(T)$ with decreasing $T$ exhibits a ferromagnetic transition ( $T_C$ =117 K), which is lower than for $x$=0. The peak in $M(T)$ is judged to mark the transition to the AF phase at $T_A \sim 66$ K, which is not considerably dependent on $x$ in $Fe_{3-x}Mn_xSi$. At low fields hysteresis in $M(T)$ measured during a zero-field cooling process (ZFC) and a field cooling process (FC) was observed. At higher fields, hysteresis is almost negligible. These features agree with those that have been reported earlier [14].

Figure 2 (a) shows $M(T)$ for $Fe_{1.3}Mn_{1.7}Si$ for different values of magnetic fields (0 T$\leq B \leq$7 T). Figure 2 (b) shows $M(T)$ for $B \leq 1$ T in detail. In this figure, the $M(T)$ curves

IOP Conf. Series: Journal of Physics: Conf. Series **969** (2018) 012098 doi:10.1088/1742-6596/969/1/012098

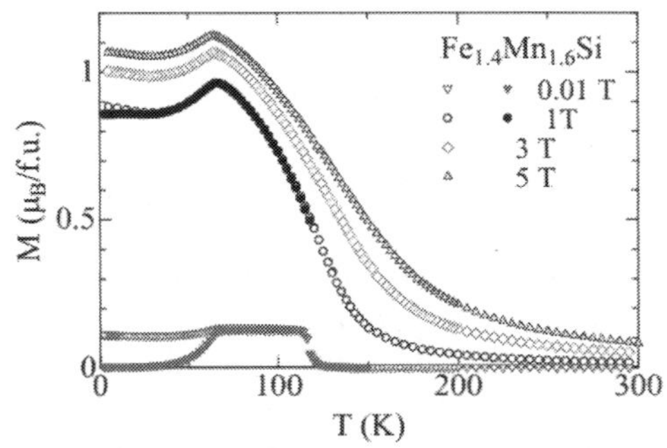


**Figure 1.** Temperature dependence of magnetization of $Fe_{1.4}Mn_{1.6}Si$ for different values of magnetic field. For $B = 0.01$ and 1 T data for zero-field cooling (solid symbols) and field cooling (open symbols) are shown.

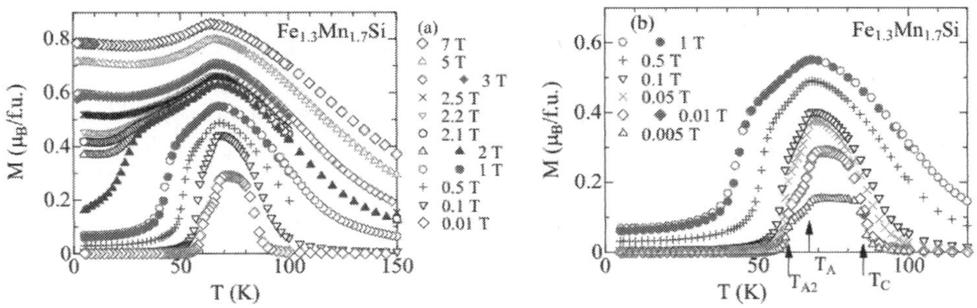


**Figure 2.** (a) Temperature dependence of magnetization of $Fe_{1.3}Mn_{1.7}Si$ $M(T)$ for different values of magnetic field. For some fields data for zero-field cooling (solid symbols) and field cooling (open symbols) are shown. (b) $M(T)$ of $Fe_{1.3}Mn_{1.7}Si$ for $B \leq 1\ T$ in detail. The arrows show three transitions at low fields. For $B = 0.01$ T the hysteresis is not observed because the data for ZFC are overlapped by those for FC.

for both ZFC and FC are shown for some values of magnetic fields. At $B >\sim 2.5$ T, the $M(T)$ curves show a peak around 60 K. These curves show the same behavior as those of $M(T)$ for $Fe_{1.4}Mn_{1.6}Si$ and also for lower $x$. The peaks appear to show the same AF transition as was found for $x = 1.0$ and $x = 1.6$. However, at lower fields, a different behavior was observed. For example, at 2 T, in addition to the peak in $M(T)$ at 67 K, which can be regarded as the transition at $T_A$, another kink was found at $\sim$31 K, below which $M$ decreases rather rapidly, and significant hysteresis in $M(T)$ between ZFC and FC was observed. It is noted that the temperatures at which the kink is observed for ZFC and FC are identical. At lower fields, for example, looking at the $M(T)$ curves at 0.01 T, a rapid increase of $M$ around 85 K was observed as temperatures decreased, which is regarded as the ferromagnetic transition at $T_C = 85$ K, as was observed for lower $x$. As the temperature decreased further, $M$ was revealed to stay almost constant, and to start to decrease at 68 K. Furthermore, it shows a rapid decrease below 60 K. This suggests that the phase below may be antiferromagnetic. The hysteresis becomes smaller at lower fields: at 1 T, the hysteresis becomes fairly small; at 0.01 T, the hysteresis is not observable, as is shown in Fig. 2 (b), where data for ZFC are overlapped by those for FC and hysteresis is almost negligible. Comparing this data with the data at other magnetic fields, the middle anomaly at 68 K is almost independent of $B$ and appears to connect to the peak observed at higher $B$, and thus, it is regarded as the AF transition. On the other hand, the temperature at the lowest anomaly, 60 K, changes with magnetic fields, and appears to connect to the anomaly at $\sim$31 K for the field of 2 T. No trace of this kind of anomaly was observed in $Fe_{1.4}Mn_{1.6}Si$ and for lower values of $x$. Hence, we regard this as another antiferromagnetic transition. We refer to this phase and the transition temperature as the AF2 phase and $T_{A2}$, respectively. In this measurement, $T_{A2}$ was observed below 2.2 T, but at the higher fields it was not observed. Below $T_{A2}$ significant hysteresis in $M(T)$ between ZFC and FC at $B \sim 2$T was observed, whereas at lower fields it becomes so small that it is not observable at very low fields. This is in contrast to the case for $T_A$ in $Fe_{3-x}Mn_xSi$ with lower $x$, where substantial hysteresis is observed only at low fields, as shown in Fig. 1.

Figure 3 shows the results for magnetization as a function of the magnetic field $M(B)$ measured at different temperatures. One significant feature is the appearance of meta-magnetic-like behavior at 2 K and 30 K. At 2 K, a rapid increase (decrease) in $M(B)$ was observed at ~2.5 T (1.2 T) with increasing (decreasing) $B$, which implies large hysteresis in $M(B)$. Although a rapid change in $M(B)$ was observed around 1.5 T at 30 K, the hysteresis becomes even smaller. At low magnetic fields below the meta-magnetic transition field, $M$ increases linearly with $B$. At 60 K the meta-magnetic-like behavior disappears; however, a rapid increase in $M(B)$ from 0 T was observed, which indicates the existence of spontaneous magnetization, and shows ferromagnetism. At 100 K spontaneous magnetization is not recognized. This can be easily observed by plotting this using the Arrott plot as $M^2$ versus $M/B$. These results are consistent with the $M(B)$ data previously reported for $Fe_{3-x}Mn_xSi$ ($x = 1.7$ and $x = 1.8$) [16, 17].

The transitions are plotted in a $B$-$T$ phase diagram, as shown in Fig. 4. The open circle shows the AF2 transition determined from the $M(T)$ measurements. As mentioned earlier, the transition points at low fields below 1 T smoothly connect to those around 2 T. It is noted that, nevertheless, significant hysteresis of $M(T)$ between ZFC and FC at $B \gtrsim 2$ T is observed, although the transition temperature for ZFC and FC itself is not different. The crosses, determined from the $M(B)$ measurement, indicate the points where the rapid increase in $M(B)$ with increasing $B$ ceases. These points coincide well with the open circles determined from the $M(T)$ measurements. The vertical lines show the width of the magnetic transition field between increasing and decreasing $B$ where the rapid change of $M(B)$ (meta-magnetic-like transition) occurs, as shown in Fig. 3. These indicate that the meta-magnetic-like transition found in $M(B)$ is associated with the transition that is observed at $T_{A2}$ in the $M(T)$ measurements. It seems that the magnetic field where the meta-magnetic-like transition occurs does not change at low

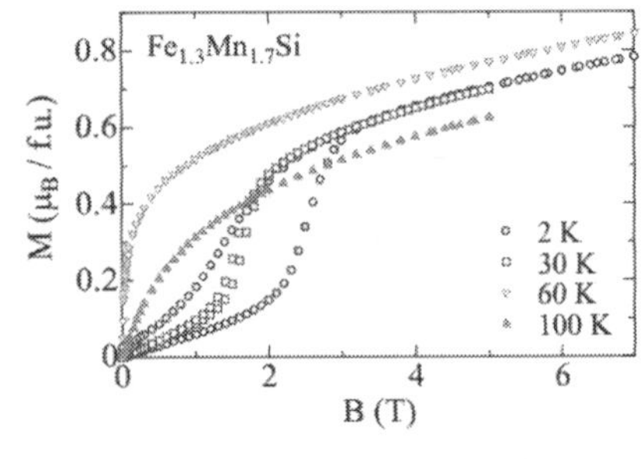


**Figure 3.** Magnetization of $Fe_{1.3}Mn_{1.7}Si$ at different temperatures. A meta-magnetic-like transition and hysteresis at the transition between increasing and decreasing magnetic fields is observed at 2 K (larger) and 30 K. At 60 K spontaneous magnetization is recognized but at 100 K it disappears.

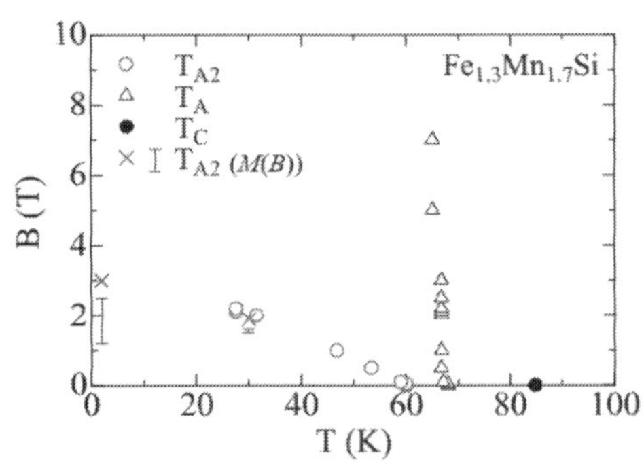


**Figure 4.** $B$-$T$ magnetic phase where transitions associated with $T_{A2}$, $T_A$, and $T_C$. The crosses and vertical lines are determined from $M(B)$ measurements and the others are from $M(T)$ measurements. See the text for details.

temperatures; only the degree of hysteresis increases.

We should note the effect of heat-treatment on the samples. The data we presented here are from heat-treated samples. In an as-cast sample, basically the same behavior was observed. In other words, anomalies at $T_{A2}$, $T_A$, and $T_C$ were observed. However, a slight ferromagnetic component, that is, finite magnetization at low fields, was observed, and the transition at $T_{A2}$ was less sharp, in as-cast samples.

Based on the above results, we can conclude that the observed three transition at different temperatures at $B \sim 0$ T are intrinsic. Furthermore, it is very likely that the AF2 phase below $T_{A2}$ at low magnetic fields has a typical antiferromagnetic character, because ferromagnetic components are not present and hysteresis behavior is not observed. On the other hand, in high magnetic fields, the AF2 phase becomes unusual because hysteresis behavior in $M(T)$ and $M(B)$ emerges. Although this might suggest that the properties of the AF2 phase may be changed significantly by the magnetic field, at present we have no hypothesis that explains these phenomena clearly. While the phase between $T_{A2}$ and $T_A$ has both ferromagnetic and antiferromagnetic components, this phase can be interpreted as similar to the AF phase found in $Fe_{3-x}Mn_xSi$ for smaller $x$. As clearly seen in Figs. 1 and 2, in $Fe_{3-x}Mn_xSi$, the AF2 phase appears rather abruptly between $x = 1.6$ and 1.7. To reveal the properties of the AF2 transition and the AF2 phase, thermodynamic and other measurements of physical properties are planned.

## 4. Conclusions

The magnetization of $Fe_{1.3}Mn_{1.7}Si$ was investigated in detail and compared with that of $Fe_{1.4}Mn_{1.6}Si$. It was concluded that at zero field three transitions, that is, the ferromagnetic transition at $T_C$, an antiferromagnetic transition to the AF phase at $T_A$, and another antiferromagnetic transition to the AF2 phase at $T_{A2}$ exist. They are rather close but clearly distinct. Although the transitions at $T_C$ and $T_A$ were observed in $Fe_{1.4}Mn_{1.6}Si$ and for other $Fe_{3-x}Mn_xSi$ with lower $x$, no sign of the transition at $T_{A2}$ was found in $Fe_{1.4}Mn_{1.6}Si$. This transition depends on the magnetic field and the critical field at 0 K is $\sim$3 T. The second antiferromagnetic phase, AF2, is quite a contrast to the AF phase which has large ferromagnetic components. The AF2 phase appears to be a typical antiferromagnetic phase at low magnetic fields, that is, it does not have any ferromagnetic components and does not show hysteresis behavior. However, at higher magnetic fields, it exhibits significant hysteresis at low temperatures.

## Acknowledgments

This study was carried out as a joint research in the Institute for Solid State Physics, the University of Tokyo (ISSP). We are grateful to T. Yamauchi at the Materials Design and Characterization Laboratory, ISSP, for advice and help. This work was supported by JSPS KAKENHI Grant Number JP17K06774.

## References

[1] Felser C, Fecher G and Balke B 2007 *Angewandte Chemie International Edition* **46** 668
[2] Galanakis I, Dederichs P H and Papanikolaou N 2002 *Phys. Rev. B* **66** 174429
[3] Katsnelson M I, Irkhin V Y, Chioncel L, Lichtenstein A I and de Groot R A 2008 *Rev. Mod. Phys.* **80** 315
[4] Ishida S, Fujii S, Kashiwagi S and Asano S 1995 *Journal of the Physical Society of Japan* **64** 2152
[5] Fujii S, Ishida S and Asano S 1995 *Journal of the Physical Society of Japan* **64** 185
[6] Ziebeck K R and Webster P J 1976 *Philosophical Magazine* **34** 973
[7] Yoon S and Booth J G 1977 *Journal of Physics F: Metal Physics* **7** 1079
[8] Ležaić M, Mavropoulos P, Blügel S and Ebert H 2011 *Phys. Rev. B* **83** 094434
[9] Hamad B, Khalifeh J, Abu Aljarayesh I, Demangeat C, Luo H B and Hu Q M 2010 *Journal of Applied Physics* **107** 093911
[10] Hamad B and Hu Q M 2011 *physica status solidi (b)* **248** 2893

[11] Yoon S and Booth J 1974 *Physics Letters A* **48** 381 – 382 ISSN 0375
[12] Smith T F, Bristow G K, Boughton C V and Hudson P R W 1980 *Journal of Physics F: Metal Physics* **10** 2867
[13] Miles J, Smith T and Finlayson T 1988 *Australian Journal of Physics* **41** 781
[14] Pal L, Suresh K G and Nigam A K 2013 *Journal of Applied Physics* **113** 093904
[15] Mohn P and Supanetz E 1998 *Philosophical Magazine Part B* **78** 629
[16] Al-Kanani H, Booth J G and Yoon S 1993 *Journal of Applied Physics* **73** 5671
[17] Al-Kanani H J, Booth J G, Cable J W and Fernandez-Baca J A 1994 *Journal of Applied Physics* **76** 6359
[18] Hiroi M, Tazoko T, Sano H, Shigeta I, Koyama K, Kondo A, Kindo K, Manaka H and Terada N 2017 *Phys. Rev. B* **95** 014410

IOP Conf. Series: Journal of Physics: Conf. Series **969** (2018) 012099 doi:10.1088/1742-6596/969/1/012099

# Magnetization and magnetic phase diagram of Heusler compounds $Fe_{3-y}(Mn_{1-x}V_x)_ySi$ ($y$ =1 and 1.5)

**Masahiko Hiroi[1], Syunya Ishikuma [1], Iduru Shigeta[1], Keiichi Koyama[1], Akihiro Kondo[2], Koichi Kindo[2], Hirotaka Manaka[3], and Norio Terada[3]**

[1] Department of Physics and Astronomy, Graduate School of Science and Engineering, Kagoshima University, Kagoshima 890-0065, Japan

[2] High Field Laboratory, Institute for Solid State Physics, the University of Tokyo, Kashiwa 277-8581, Japan

[3] Department of Electrical and Electronics Engineering, Graduate School of Science and Engineering, Kagoshima University, Kagoshima 890-0065, Japan

E-mail: hiroi@sci.kagoshima-u.ac.jp

**Abstract.** $Fe_2MnSi$ exhibits a ferromagnetic transition at $T_C$ and another transition to a phase with antiferromagnetic components at a low temperature of $T_A$. By substituting V for Mn, so as to obtain $Fe_2Mn_{1-x}V_xSi$, $T_A$ decreases with $x$ and then vanishes around $x \sim 0.2$. In this study, the magnetic transitions are investigated by magnetization measurements for $Fe_2Mn_{1-x}V_xSi$ ($y = 1$, $0 \leq x \leq 0.2$) in high magnetic fields up to ~70 T and for $Fe_{1.5}(Mn_{1-x}V_x)_{1.5}Si$ ($y = 1.5$, $0 \leq x \leq 0.1$) up to 5 T. For $y = 1.5$, with increasing $x$, $T_C$ increases and $T_A$ decreases as for $y = 1$, and the rate of the decrease in $T_A$ with $x$ is slightly smaller than for $y$=1. In the low magnetic field region, $T_A$ for $y = 1$ does not significantly change due to the magnetic field but a distinct decrease is observed in high fields. The critical field at 0 K of this transition in $Fe_2MnSi$ is found to be larger than 70 T, and with increasing $x$, the critical field decreases corresponding to the decrease of $T_A$ at zero field.

## 1. Introduction

Half-metallic materials in Heusler compounds have attracted consierable attention because of their potential applications in *spintronics* [1–3]. Half-metals are ferromagnetic metals wherein at the Fermi energy, the density of states (DOS) exists only in one spin band and there is a gap in the other spin band; this implies that there are 100% spin-polarized conduction electrons. Half-metals are thus promising materials for magneto-resistive and spin-injection devices. The possibility of some Heusler compounds being half-metals was suggested some time ago. $Co_2MnSi$ and $Fe_2MnSi$ were predicted to be half-metals based on first-principles band structure calculations [4–8]. $Co_2MnSi$ has since been revealed to exhibit half-metallicity. On the other hand, $Fe_2MnSi$ shows a ferromagnetic transition at a temperature $T_C$ (Curie temperature) of $\sim$ 230 K, and a transition occurs to a phase with antiferromagnetic components at a lower temperature of $\sim$ 60 K, defined as $T_A$; this phase, however, has a large ferromagnetic component. In this paper, we refer to this phase as the AF phase and to the ferromagnetic phase between $T_A$ and $T_C$ as the F phase. The transition and properties of the AF phase have been studied previously [6, 9–14], and neutron scattering experiments have revealed a complex magnetic structure with the magnetic moment carried by Mn [10, 15].

The presence of an antiferromagnetic order in $Fe_2MnSi$ shows that the ground state is not half-metallic ferromagnetism as has been predicted theoretically. Although the half-metallic ground state is not realized in $Fe_2MnSi$, the saturation magnetization of alloy systems containing $Fe_2MnSi$ was re-interpreted in terms of the Slater-Pauling (SP) rule in Heusler compounds, which holds when a substance is a half-metal [16–19]. For V-substituted $Fe_2Mn_{1-x}V_xSi$, as $x$ increases, $T_C$ increases and $T_A$ decreases, whereas $T_A$ vanishes around $x \sim 0.2$ [16, 20, 21]. Its saturation magnetization for $x > 0.2$ appears to obey the SP rule, but for $x \lesssim 0.2$ it deviates because of the appearance of the AF phase [22].

We investigated the magnetization of $Fe_2Mn_{1-x}V_xSi$ ($0 \leq x \leq 0.2$) at 4.2 K up to ~70 T with pulsed magnetic fields [23], and we observed that in the high field limit, where the F phase is supposed to be realized, the magnetization approaches the value indicated by the SP rule. This result suggests that the electronic structure of the high field F phase is intimately related to the half-metallic state, although the fact that magnetization does not agree completely with the SP rule implies that the half-metallic state is not completely realized in the high field F phase.

In this study the transition between the AF phase and the F phase is investigated by magnetization measurements for $Fe_2Mn_{1-x}V_xSi$ including high field regions with pulsed fields. Moreover, $Fe_{1.5}(Mn_{1-x}V_x)_{1.5}Si$ is studied in order to compare the effect of V. Although the $T_A$ of $Fe_2MnSi$ and $Fe_{1.5}Mn_{1.5}Si$ is almost the same, their $T_C$s are different. This difference may affect the properties of the AF phase. Because the increase of V content decreases $T_A$ at zero magnetic field and accordingly the critical magnetic field of the transition, this may allow us to study the entire phase boundary of the transition. Therefore, V substitution is expected to be a useful tool to study the AF transition and phase. Some of the results for $Fe_2Mn_{1-x}V_xSi$ have already been published [21, 23].

## 2. Experimental

Polycrystalline samples of $Fe_2Mn_{1-x}V_xSi$ ($0 \leq x \leq 0.2$) and $Fe_{1.5}(Mn_{1-x}V_x)_{1.5}Si$ ($0 \leq x \leq 0.1$) were prepared by arc-melting high-purity constituent elements in a high-purity argon atmosphere. Crystal structures were examined using X-ray powder diffraction measurements with Cu K$\alpha$ radiation. X-ray diffraction measurements showed that the prepared samples were single phase, and the crystal structure was $L2_1$. The profiles of the X-ray diffraction patterns were practically the same for all the samples. These results showed, at least in view of X-ray diffraction, the same sample quality of these samples. Magnetization was measured using SQUID magnetometers (MPMS: Quantum Design). Pulsed magnetic fields of up to 72 T were generated using non-destructive magnets with a duration of several milliseconds. Magnetization was measured by the induction method using a standard pick up coil in the magnetic fields. The absolute values of the data obtained by the pulsed magnets were calibrated by comparing the raw data obtained by the pulsed magnets with the absolute values of the magnetization below 7 T, which were measured by SQUID magnetometers.

## 3. Results

Figures 1 and 2 show the temperature dependence of magnetization $M(T)$ for $Fe_2Mn_{1-x}V_xSi$ ($x = 0$, 0.1, and 0.18) and $Fe_{1.5}(Mn_{1-x}V_x)_{1.5}Si$ ($x = 0$, 0.05, and 0.1) at a magnetic field $B$ of 1 T, respectively. These results exhibited the same behavior as that of $Fe_2MnSi$. With decreasing temperature $M$ increased rather rapidly. The increase shows the ferromagnetic transition to the F phase, and the transition temperature is $T_C$. And $M$ exhibits a peak or kink (the AF transition becomes less sharp and the peak collapses with increasing $x$). This anomaly shows the transition to the AF phase and we refer to this transition temperature as $T_A$.

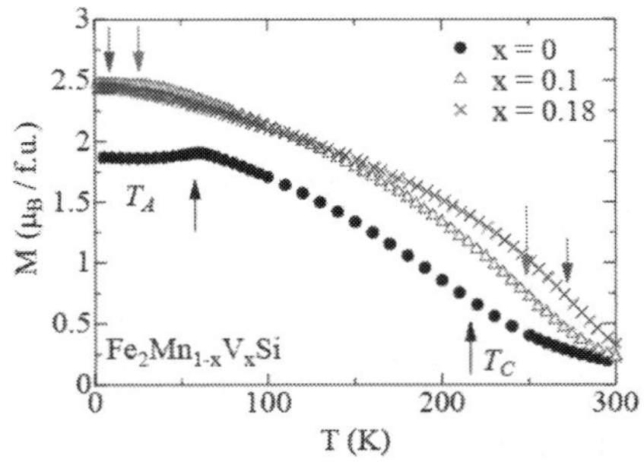


**Figure 1.** Temperature dependence of magnetization of $Fe_2Mn_{1-x}V_xSi$ ($x = 0$, 0.1, and 0.18) at $B = 1$ T.

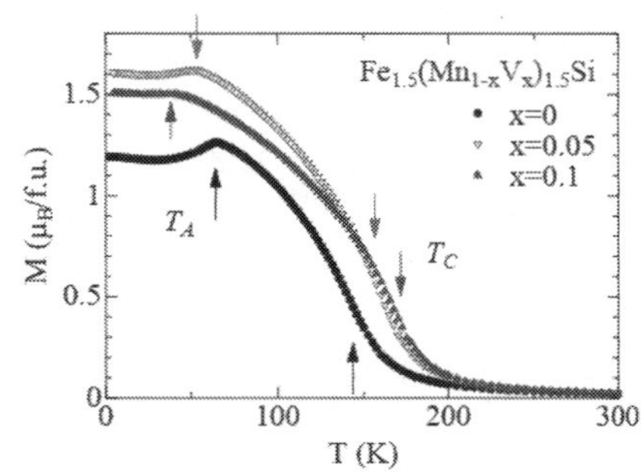


**Figure 2.** Temperature dependence of magnetization of $Fe_{1.5}(Mn_{1-x}V_x)_{1.5}Si$ ($x = 0$, 0.05, and 0.1) at $B = 1$ T.

In Fig. 3, reduced $T_C$ and $T_A$ as a function of $x$ are shown for $y = 1$ and 1.5. Reduced $T_C$ and $T_A$ are $T_A(x)/T_A(x = 0)$ and $T_C(x)/T_C(x = 0)$, respectively. $T_A(x = 0) = 65$ K for both $y$; $T_C(x = 0) = 220$ K for $y = 0$ and $T_C(x = 0) = 142$ K for $y = 1.5$. In addition, on the right axis, the scale of $T_A$ is also shown. With increasing V content, $x$, $T_C$ increases and $T_A$ decreases. $T_C$ for $y = 1$ is higher than for $y = 1.5$. The rate of the increase of reduced $T_C$ and that of the decrease of (reduced) $T_A$ against $x$ are not so largely different between the two series. Nevertheless, the rate of the decrease of $T_A$ for $y = 1.5$ appears slightly smaller than that for $y = 1$; $T_A$ for $y = 1$ vanishes around $x = 0.2$ [21].

Figure 4 (a) and (b) show $M(T)$ for $Fe_{1.5}Mn_{1.5}Si$ and $Fe_{1.5}(Mn_{0.9}V_{0.1})_{1.5}Si$ for magnetic fields $0 < B \leq 5$ T, respectively. In Fig.4 (a), at a low field, 0.01 T, hysteresis of $M(T)$ in a zero-field cooling and a field cooling process is observed. However, the hysteresis becomes much smaller when the magnetic field reaches 1 T. As seen in the figure, $T_A$ slightly changes with magnetic field. The sharpness of the anomaly at the transition at $T_A$ is also not considerably affected by the magnetic field, as exhibited remarkably in the derivatives $dM/dT$ shown in the insets. Compared with the case of $y = 0$, the AF transition in the case of $y = 1.5$ appears to be less affected by the magnetic field.

In order to determine the influence of the magnetic field on the transition, high field magnetization was measured for $Fe_2Mn_{1-x}V_xSi$ up to 70 T with pulsed magnetic fields. Figure 5 (a) shows, in an expanded scale, magnetization as a function of the magnetic field $M(B)$ at several temperatures between 4.2 K and 70 K for $Fe_2MnSi$ ($T_A^0 = 65$ K, where $T_A^0$ is the transition temperature at zero field.). In the inset of Fig. 5 (a), $M(B)$ is shown over the

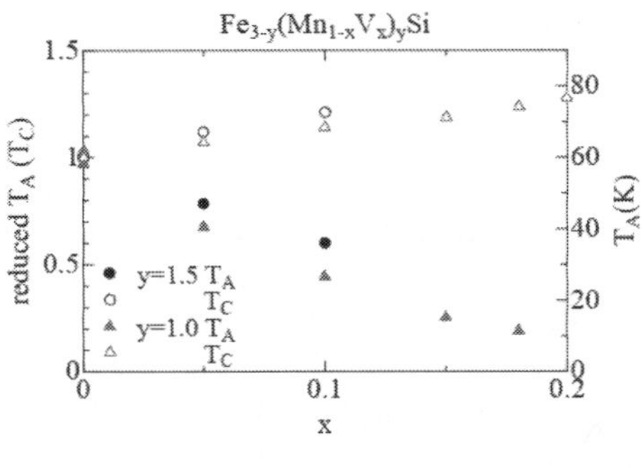


**Figure 3.** Reduced $T_A$ and $T_C$ as a function of $x$ ($T_A(x)/T_A(x{=}0)$ and $T_C(x)/T_C(x{=}0)$) are shown for $Fe_{3-y}(Mn_{1-x}V_x)_ySi$ ($y = 0$ and 1.5). The scale of $T_A$ is shown on the right axis because $T_A(x{=}0)$ is 65 K for both $y = 0$ and 1.5. $T_C(x{=}0) = 220$ K for $y = 0$ and $T_C(x{=}0) = 142$ K for $y = 1.5$.

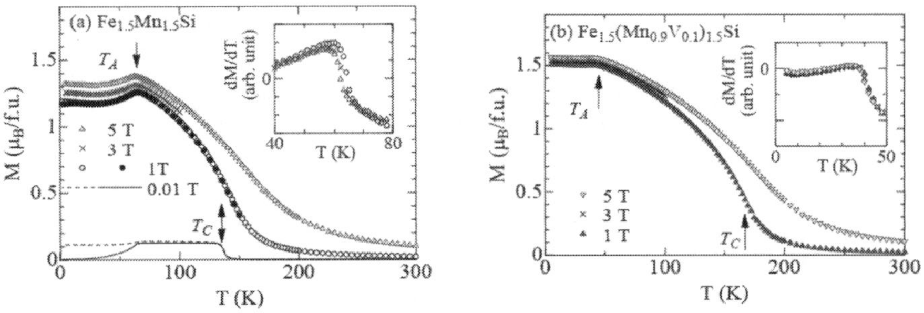


**Figure 4.** (a) Temperature dependence of magnetization $M(T)$ of $Fe_{1.5}Mn_{1.5}Si$ for different values of the magnetic field. For $B = 0.01$ T and 1 T, data for zero-field cooling (solid line and solid circles) and field cooling (broken line and open circles) are shown. The inset shows the derivative $dM/dT$. (b) Temperature dependence of magnetization $M(T)$ of $Fe_{1.5}(Mn_{0.9}V_{0.1})_{1.5}Si$ for different values of the magnetic field. The inset shows the derivative $dM/dT$.

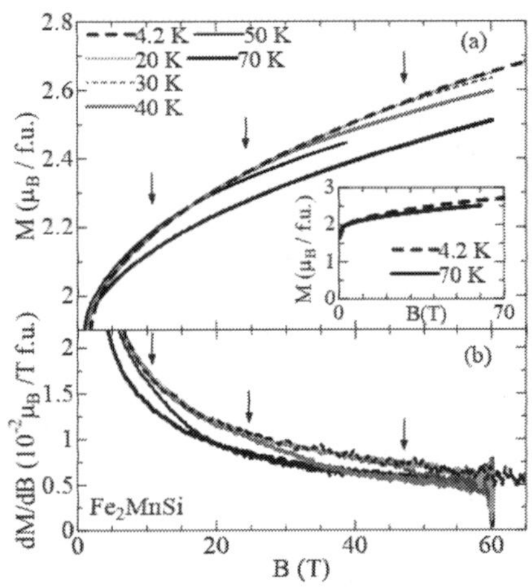


**Figure 5.** (a) Magnetization versus magnetic field $M(B)$ for $Fe_2MnSi$ ($T_A^0 = 65$ K) up to 70 T in an expanded scale. In the inset, $M(B)$ over the entire magnetic field range is shown. (b) Values of the derivative $dM/dB$. The arrows show the transitions.

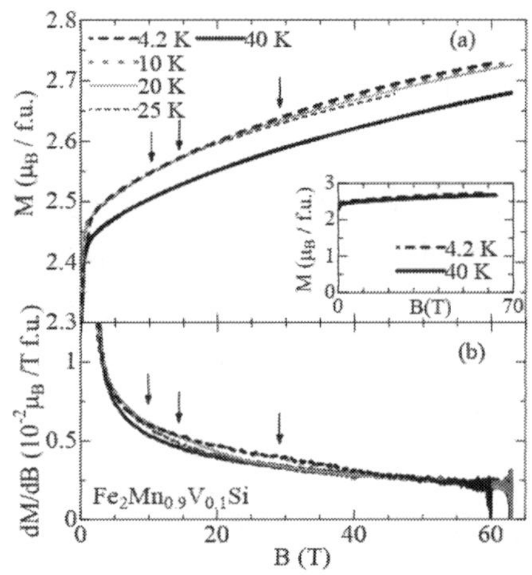


**Figure 6.** (a) Magnetization versus magnetic field $M(B)$ for $Fe_2Mn_{0.9}V_{0.1}Si$ ($T_A^0 = 30$ K) up to 63 T in an expanded scale. In the inset, $M(B)$ over the entire magnetic field range is shown. (b) Values of the derivative $dM/dB$. The arrows show the transitions.

entire field range. With increasing $B$, at low fields $M$ initially increases rapidly, reflecting its ferromagnetism, and slows down at higher fields. Although no anomaly appears to indicate a transition in the $M(B)$ curves at first glance, the derivative $dM/dB$, as shown in Fig. 5 (b), reveals that there is, in fact, a transition. The $dM/dB$ curves at $T < T_A^0$ similarly decrease at low fields, but at a field that depends on the temperature of measurement, they bend downward and then coincide with the $dM/dB$ curve at 70 K ($> T_A^0$), the point at which the F phase is considered to exist all over the field range and no transition occurs. From these results, we can

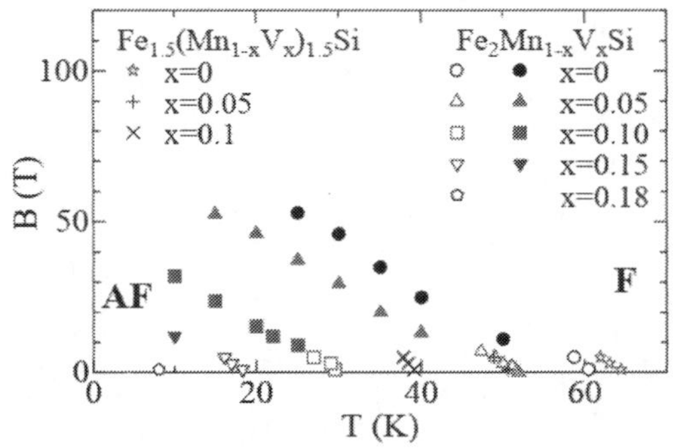


**Figure 7.** Magnetic phase diagram concerning the transitions between the AF and F phases up to $B \sim 70$ T for $Fe_{3-y}(Mn_{1-x}V_x)_ySi$ ($y$ =1 and 1.5). Open symbols show the transitions determined from the measurements using SQUID and solid symbols show those obtained using pulsed magnets for $y$ =1 ($0 \leq x \leq 0.18$). Stars, pluses, and crosses are for $y$ =1 ($0 \leq x \leq 0.1$) from the SQUID measurements.

recognize that this anomaly shows the transition from the AF to the F phase. In addition, the transition is actually discernible when the $M(B)$ curves are compared at different temperatures. As shown in Figure 5 (a), in the lower field range, the $M(B)$ curves at 40 K and 50 K, for example, almost coincide with that at 4.2 K, before starting to shift downward at around 25 T and 11 T, respectively. Although the transition around 47 T for 30 K can be identified, the transition becomes difficult to recognize at lower temperatures because the difference between $M(B)$ curves become even smaller and $M(B)$ and $dM/dB$ curves appear to almost coincide besides the inevitable enhancement of noise around the upper limit of the pulsed field.

Figure 6 shows $M(B)$ and $dM/dB$ for $Fe_2Mn_{0.9}V_{0.1}Si$ ($T_A^0 = 30$ K). As seen in the figures, the transition becomes smeared as $x$ increases and becomes more difficult to distinguish in $M(B)$. Nevertheless, as for $Fe_2Mn_2Si$ the $M(B)$ and $dM/dB$ curves allow us to spot the transition, as shown by the arrow in the figure. The magnetic field at the transition becomes smaller according to the decrease of $T_A^0$ with $x$.

Figure 7 shows the $B$-$T$ magnetic phase diagram of $Fe_{3-y}(Mn_{1-x}V_x)_ySi$. ($y$ =1 and 1,5, $0 \leq x \leq 0.18$). In the magnetic field range of up to $\sim 70$ T, the dependence of the AF transition on the magnetic field is clearly demonstrated. The phase boundaries between the AF and the F phases of $Fe_2Mn_{1-x}V_xSi$ almost look like straight lines up to the high field region, which is different from usual curved phase boundaries. However, at lower temperatures, the phase transition cannot be recognized mainly because the $M(B)$ curves are almost the same. This is because in the high field region, the magnetic states of the F phase and the AF phase are likely to be similar. Although a part of the data suggest that the straight phase boundaries bend down at lower temperatures toward 0 K, we could not draw reliable phase boundaries near 0 K due to the technical difficulty of measurements in the high field limits of pulsed magnetic fields. Nevertheless, we can estimate the order of the critical field of this transition at 0 K, $B_{CA}$, reliably. $B_{CA}$ is found to be $\sim k_B T_A^0/\mu_B$, which is expected of usual local moment antiferromagnets. The transition temperature $T_A$ in magnetic fields for $Fe_{1.5}(Mn_{1-x}V_x)_{1.5}Si$ exhibits similar behavior, whereas it appears to be less affected by a magnetic field below 5 T for the V-substituted samples. This may demonstrate more robustness to the magnetic field for $y = 1.5$ compared with the case for $y = 1$. However, data are limited to below 5 T for $Fe_{1.5}(Mn_{1-x}V_x)_{1.5}Si$. In this field range, notable magnetic field dependence was not recognized in $Fe_2Mn_{1-x}V_xSi$. In order to obtain definite conclusions, measurements in a higher field are desired.

## 4. Conclusions

The transition between the ferromagnetic phase (F phase) and the antiferromagnetic phase with large ferromagnetic component (AF phase) in $Fe_{1.5}(Mn_{1-x}V_x)_{1.5}Si$ ($y$ =1 and 1.5) was investigated in magnetic fields, including in a high field region of up to $\sim$70 T with pulsed fields, and a magnetic phase diagram is constructed. Although the phase boundaries of $Fe_2Mn_{1-x}V_xSi$ at low temperatures near the 0 K axis in the $B$-$T$ magnetic phase diagram were not clearly determined, the whole phase boundaries in the $B$-$T$ plain were obtained. Furthermore, magnetic transition in $Fe_{1.5}(Mn_{1-x}V_x)_{1.5}Si$ was investigated. The same type of transition as in $Fe_2Mn_{1-x}V_xSi$, i.e., the AF transition at $T_A$ and the ferromagnetic transition at $T_C$, were found. On the other hand, in $Fe_{1.5}(Mn_{1-x}V_x)_{1.5}Si$, although the measurements is limited to the magnetic field range below 5 T, $T_A$ appeared to be less affected by the magnetic field compared with the case for $y = 1$.

**Acknowledgments**
We would like to thank S. Fujii for useful discussion. This study was carried out as a joint research in the Institute for Solid State Physics, the University of Tokyo (ISSP). We are grateful to T. Yamauchi at the Materials Design and Characterization Laboratory, ISSP, for advice and help. This study was supported by JSPS KAKENHI Grant Number JP17K06774.

## References

[1] Felser C, Fecher G and Balke B 2007 *Angewandte Chemie International Edition* **46** 668
[2] Galanakis I, Dederichs P H and Papanikolaou N 2002 *Phys. Rev. B* **66** 174429
[3] Katsnelson M I, Irkhin V Y, Chioncel L, Lichtenstein A I and de Groot R A 2008 *Rev. Mod. Phys.* **80** 315
[4] Ishida S, Fujii S, Kashiwagi S and Asano S 1995 *Journal of the Physical Society of Japan* **64** 2152
[5] Fujii S, Ishida S and Asano S 1995 *Journal of the Physical Society of Japan* **64** 185
[6] Ležaić M, Mavropoulos P, Blügel S and Ebert H 2011 *Phys. Rev. B* **83** 094434
[7] Hamad B, Khalifeh J, Abu Aljarayesh I, Demangeat C, Luo H B and Hu Q M 2010 *Journal of Applied Physics* **107** 093911
[8] Hamad B and Hu Q M 2011 *physica status solidi (b)* **248** 2893
[9] Ziebeck K R and Webster P J 1976 *Philosophical Magazine* **34** 973
[10] Yoon S and Booth J 1974 *Physics Letters A* **48** 381
[11] Yoon S and Booth J G 1977 *Journal of Physics F: Metal Physics* **7** 1079
[12] Smith T F, Bristow G K, Boughton C V and Hudson P R W 1980 *Journal of Physics F: Metal Physics* **10** 2867
[13] Miles J, Smith T and Finlayson T 1988 *Australian Journal of Physics* **41** 781
[14] Pal L, Suresh K G and Nigam A K 2013 *Journal of Applied Physics* **113** 093904
[15] Mohn P and Supanetz E 1998 *Philosophical Magazine Part B* **78** 629
[16] Kondo Y, Yano I, Shigeta I, Ito M, Hiroi M, Manaka H and Terada N 2009 *Journal of Physics: Conference Series* **150** 042099
[17] Shigeta I, Urakawa S, Ito M and Hiroi M 2012 *Journal of Physics: Conference Series* **400** 032082
[18] Ishida S, Nagatomo D, Fujii S and Asano S 2008 *MATERIALS TRANSACTIONS* **49** 114
[19] Bainsla L, Raja M M, Nigam A K, Varaprasad B S D C S, Takahashi Y K, Suresh K G and Hono K 2015 *Journal of Physics D: Applied Physics* **48** 125002
[20] Kawakami M 1993 *Physica B: Condensed Matter* **186–188** 1037
[21] Hiroi M, Yano I, Sezaki K, Shigeta I, Ito M, Manaka H and Terada N 2012 *Journal of Physics: Conference Series* **400** 032021
[22] Sakon T, Koyama K, Kamiya O, Awaji S, Nakamura S, Nojima T, Nojiri H, Watanabe K and Hiroi M 2013 *Journal of the Physical Society of Japan* **82** 044802
[23] Hiroi M, Tazoko T, Sano H, Shigeta I, Koyama K, Kondo A, Kindo K, Manaka H and Terada N 2017 *Phys. Rev. B* **95** 014410

IOP Conf. Series: Journal of Physics: Conf. Series **969** (2018) 012100 doi:10.1088/1742-6596/969/1/012100

# Magnetic properties of the antiferromagnetic spin-$\frac{1}{2}$ tetramer compound $CuInVO_5$

**Masashi Hase[1], Masashige Matsumoto[2], Akira Matsuo[3] and Koichi Kindo[3]**

[1]Research Center for Advanced Measurement and Characterization, National Institute for Materials Science (NIMS), 1-2-1 Sengen, Tsukuba-shi, Ibaraki 305-0047, Japan
[2]Department of Physics, Shizuoka University, 836 Ohya, Suruga-ku, Shizuoka-shi, Shizuoka 422-8529, Japan
[3]The Institute for Solid State Physics (ISSP), The University of Tokyo, 5-1-5 Kashiwanoha, Kashiwa-shi, Chiba 277-8581, Japan

E-mail: HASE.Masashi@nims.go.jp

**Abstract.** We measured the temperature dependence of the magnetic susceptibility and specific heat and the magnetic-field dependence of the magnetization of $CuInVO_5$. An antiferromagnetically ordered state appears below $T_N = 2.7$ K. We observed a $\frac{1}{2}$ quantum magnetization plateau above 30 T at 1.3 K. The probable spin model for $CuInVO_5$ is an interacting spin-$\frac{1}{2}$ tetramer model. We evaluated the values of the intratetramer interactions as $J_1 = 240 \pm 20$ K (antiferromagnetic) and $J_2 = -142 \pm 10$ K (ferromagnetic) and the value of the effective intertetramer interaction as $J_{\rm eff} = 30 \pm 4$ K. The ground state of the isolated spin tetramer with the $J_1$ and $J_2$ values is spin singlet. The shrinkage of ordered magnetic moments by quantum fluctuation can be expected. Longitudinal-mode magnetic excitations may be observable in $CuInVO_5$.

## 1. Introduction

Two types of magnetic excitations exist in a magnetically ordered state. They are gapless transverse-mode (Nambu-Goldstone mode) [1] and gapped longitudinal-mode (amplitude Higgs mode) [2, 3, 4] excitations corresponding to fluctuations in directions perpendicular and parallel to ordered moments, respectively. The transverse-mode (T-mode) excitations are well known as spin wave excitations. Investigations of the longitudinal-mode (L-mode) are now in progress. The L-mode excitations were observed in a pressure-induced or magnetic-field-induced magnetically ordered state of interacting antiferromagnetic (AF) spin-$\frac{1}{2}$ dimer compounds $TlCuCl_3$ [5, 6, 7, 8, 9] and $KCuCl_3$ [10].

According to results of theoretical investigations, the L-mode excitations may be observed in an antiferromagnetically ordered state appearing on cooling at atmospheric pressure and zero magnetic field in interacting AF spin-cluster compounds [11]. A shrinkage of ordered magnetic moments by quantum fluctuations leads to a large intensity of the L-mode excitations. If the ground state (GS) of the corresponding isolated spin cluster is a spin-singlet state, the shrinkage of ordered moments can be expected in an ordered state generated by the introduction of intercluster interactions.

We focus on spin-$\frac{1}{2}$ tetramers. The Hamiltonian of the spin tetramer is expressed as

$$\mathcal{H} = J_1 S_2 \cdot S_3 + J_2(S_1 \cdot S_2 + S_3 \cdot S_4). \quad (1)$$

When $J_1 > 0$ and $J_2$ has negative or small positive values, the GS is the spin-singlet state and $\Delta/J_1$ can be sufficiently small [12]. Here $\Delta$ is the energy difference (spin gap) between the singlet GS and first-excited triplet states. An ordered state is possible by the introduction of weak intercluster interactions. An antiferromagnetically ordered state appears in spin-$\frac{1}{2}$ tetramer compounds $Cu_2CdB_2O_6$ [13, 14, 15] and $SeCuO_3$ [16].

Recently, we found an antiferromagnetically ordered state below $T_N = 2.7$ K in the spin-$\frac{1}{2}$ tetramer compound $CuInVO_5$. We report results of magnetization and specific heat.

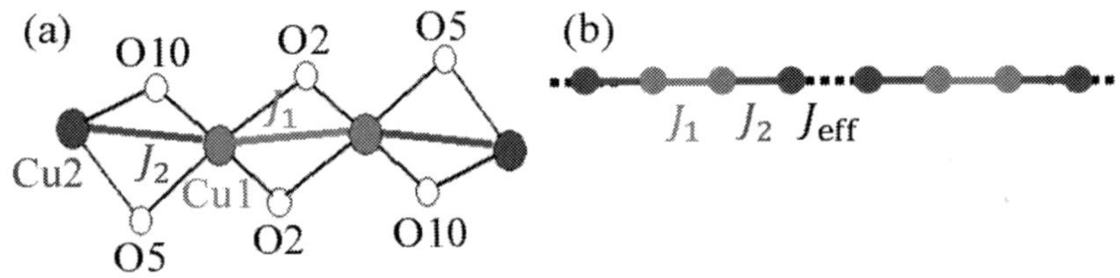


**Figure 1.** (a) Schematic drawing of positions of $Cu^{2+}$ ions having spin-$\frac{1}{2}$ and $O^{2-}$ ions connected to $Cu^{2+}$ ions in $CuInVO_5$ [17]. Red and blue bars represent the shortest and second-shortest Cu-Cu distances, respectively. We define $J_1$ and $J_2$ as the exchange interaction parameters for the Cu1-Cu1 and Cu1-Cu2 pairs, respectively. The $J_1$ and $J_2$ interactions form a spin-$\frac{1}{2}$ tetramer. (b) Interacting spin tetramer model used to calculate magnetization using a mean-field theory based on the tetramer unit (tetramer mean-field theory).

## 2. Experimental and Calculation Methods

Crystalline $CuInVO_5$ powder was synthesized by a solid-state reaction. Starting materials are CuO, $In_2O_3$, and $V_2O_5$ powder. Their purity is 99.99 %. A stoichiometric mixture of powder was sintered at 1,023 K in air for 100 h with intermediate grindings. We measured an x-ray powder diffraction pattern at room temperature using an x-ray diffractometer (RINT-TTR III; Rigaku). We confirmed that our sample was a nearly single phase of $CuInVO_5$. The space group is $P2_1/c$ (No. 14). The lattice constants are $a = 8.776(1)$, $b = 6.158(1)$, $c = 15.268(1)$ Å, and $\beta = 106.48(1)°$. These are almost the same as the values reported in the literature [17].

We measured the specific heat using a physical property measurement system (Quantum Design). We measured the magnetization in magnetic fields of up to 5 T using a superconducting quantum interference device magnetometer magnetic property measurement system (Quantum Design). High-field magnetization measurements were conducted using an induction method with a multilayer pulsed field magnet installed at the Institute for Solid State Physics (ISSP), the University of Tokyo.

We obtained the eigenenergies and eigenstates of isolated spin-$\frac{1}{2}$ tetramers using an exact diagonalization method [12]. We calculated the temperature $T$ dependence of the magnetic susceptibility and the specific heat and the magnetic-field $H$ dependence of the magnetization $M(H)$ using the eigenenergies. We calculated $M(H)$ for the model shown in Fig. 1(b) using a mean-field theory based on the tetramer unit (tetramer mean-field theory). Finite magnetic moments were initially assumed on the Cu sites in the tetramer. The mean-field Hamiltonian was then expressed by a $16 \times 16$ matrix form under consideration of the external magnetic field and the molecular field from the nearest-neighbor sites. The eigenstates of the mean-field Hamiltonian were used to calculate the expectation value of the ordered moments on the Cu sites. We continued this procedure until the values of the magnetic moments converged. We finally obtained a self-consistently determined solution for $M(H)$.

## 3. Results and discussion

The red circles in Figs. 2 and 3 show the $T$ dependence of the specific heat $C(T)$ of $CuInVO_5$ in zero magnetic field and the magnetic susceptibility $\chi(T)$ of $CuInVO_5$ in a magnetic field of $H = 0.01$ T, respectively. We can observe a peak in $C(T)$ at 2.7 K and a clear decrease in $\chi(T)$ below this temperature, indicating the occurrence of an AF long-range order. A broad maximum can be seen around 8 K in $C(T)$ and around 11 K in $\chi(T)$, indicating a low-dimensional AF spin system. As $T$ is increased, $\chi(T)$ decreases rapidly up to $T = 40$ K then decreases slowly at higher temperatures. We estimated the Weiss temperature to be 42 K from $\chi(T)$ above 200 K. Other phase transitions were not observed in $C(T)$ and $\chi(T)$ below 300 K.

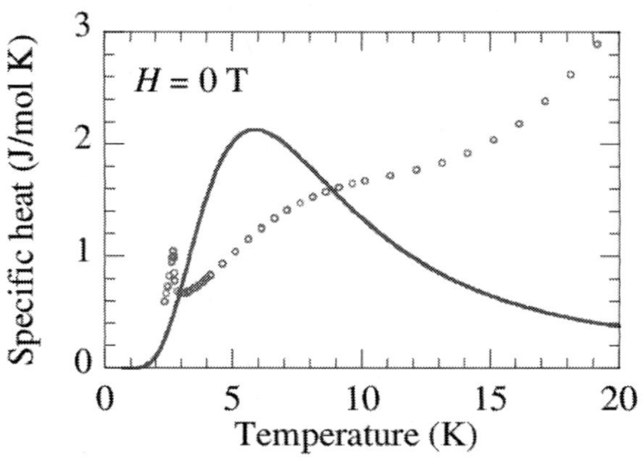


**Figure 2.** Temperature $T$ dependence of the specific heat $C(T)$ of $CuInVO_5$ (circles) in zero magnetic field. A green line indicates $C(T)$ calculated for an isolated spin-$\frac{1}{2}$ tetramer. The $J_1$ and $J_2$ values are listed in Table 1.

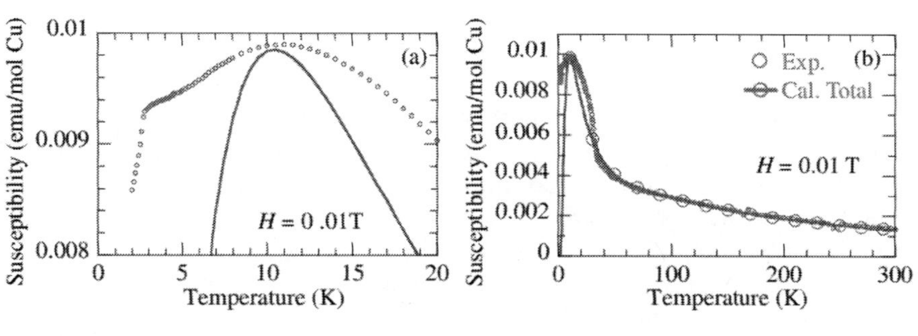


**Figure 3.** Temperature $T$ dependence of the magnetic susceptibility $\chi(T)$ of $CuInVO_5$ (circles) in a magnetic field of $H = 0.01$ T below 20 K (a) and 300 K (b). Green lines indicate $\chi(T)$ calculated for an isolated spin-$\frac{1}{2}$ tetramer. The $J_1$ and $J_2$ values are listed in Table 1.

The thick red line in Fig. 4 shows the $H$ dependence of the magnetization $M(H)$ of $CuInVO_5$ measured at 1.3 K. We can observe a $\frac{1}{2}$ quantum magnetization plateau above 30 T. The $g$ value was evaluated to be $2.09 \pm 0.02$ from the magnetization of the plateau.

We compare $\chi(T)$, $C(T)$, and $M(H)$ of $CuInVO_5$ with those calculated for isolated spin tetramers. The green line in Fig. 3 indicates $\chi(T)$ calculated for an isolated spin tetramer with

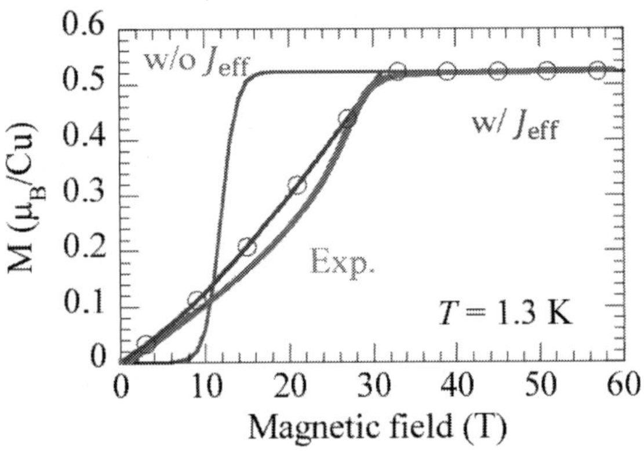


**Figure 4.** Magnetic-field dependence of the magnetization of $CuInVO_5$ (thick red line) at 1.3 K. Blue and green lines indicate the magnetization calculated for the interacting spin-$\frac{1}{2}$ tetramer model in Fig. 1(b) and for an isolated spin-$\frac{1}{2}$ tetramer. The labels w/$J_{\rm eff}$ and w/o$J_{\rm eff}$ mean the calculated $M(H)$ containing $J_{\rm eff}$ and non-containing $J_{\rm eff}$, respectively. The values of the exchange interactions are listed in Table 1.

**Table 1.** Values of exchange interaction parameters and $g$ value.

| $J_1$ (K) | $J_2$ (K) | $J_{\rm eff}$ (K) | g |
|---|---|---|---|
| $240 \pm 20$ | $-142 \pm 10$ | $30 \pm 4$ | $2.09 \pm 0.02$ |

$J_1 = 240$ and $J_2 = -142$ K. The $J_1$ and $J_2$ values are listed in Table 1. The agreement between the experimental and calculated $\chi(T)$ is nearly perfect above 30 K, whereas a discrepancy is seen below 30 K. The green line in Fig. 2 indicates $C(T)$ calculated for the same isolated spin tetramer. The positions of the broad maximum in the experimental and calculated $C(T)$ are close to each other. However, the specific heat around the broad maximum is larger in the calculated result. The green line in Fig. 4 indicates $M(H)$ calculated for the same isolated spin tetramer. The isolated spin tetramer model fails to reproduce the experimental $M(H)$ at 1.3 K. We could not find a set of exchange parameters that reproduced the experimental $M(H)$ at 1.3 K and $C(T)$ at low $T$ on the basis of an isolated tetramer.

We assumed intertetramer interactions between Cu2 spins as shown in Fig. 1(b) because the magnetization of Cu2 spins was expected to be large. The most dominant interaction is the $J_1$ interaction. The spin state of Cu1 spins is similar to the singlet state in AF dimers [18, 19, 20]. Therefore, the magnetization of Cu1 spins is small at low $T$. The two Cu2 spins in a tetramer are weakly and antiferromagnetically coupled to each other through a Cu1-Cu1 dimer in the same tetramer. Thus, the magnetization of Cu2 spins is large. Mean-field theories become less valid for calculation of susceptibility when the temperature approaches $T_{\rm N}$ owing to strong fluctuations. At low $T$, on the contrary, the ordered moment becomes substantial and the fluctuations are suppressed. Mean-field theories are reliable for $M(H)$. The blue line in Fig. 4 indicates $M(H)$ calculated for the interacting spin tetramer with $J_1 = 240$, $J_2 = -142$, and $J_{\rm eff} = 30$ K using the tetramer mean-field theory. The experimental and calculated magnetizations are in agreement with each other. Intertetramer interactions have a greater effect on the magnetization at lower $T$. Therefore, the discrepancy between the experimental results and those calculated for the isolated spin tetramer appears at low $T$. We evaluated $T_{\rm N} = 8.7$ K for the same model using the tetramer mean-field theory. The $T_{\rm N}$ value is larger than the experimental $T_{\rm N}$ (2.7 K). The difference is mainly caused by fluctuations that mean-field theories cannot treat properly. We will determine the magnetic structure. We will consider which intertetramer interactions are effective to stabilize the magnetic structure. We will calculate the magnetic susceptibility of a more realistic model using quantum Monte Carlo techniques.

We roughly estimated the errors of the $J_1$, $J_2$, $J_{\rm eff}$, and $g$ values and listed them in Table 1. A discrepancy between the experimental and calculated $\chi(T)$ appears around 80 K when $J_1$ deviates from 240 K. The experimental and calculated $\chi(T)$ are not in agreement with each other when $J_1 = 220$ or 260 K. A discrepancy between the experimental and calculated $\chi(T)$ appears around 11 K when $J_2$ deviates from -142 K. The peak heights of the experimental and calculated $\chi(T)$ are not in agreement with each other when $J_2 = -132$ or -152 K. The magnetic field at which the $\frac{1}{2}$ magnetization plateau appears increases with increasing $J_{\rm eff}$. We roughly estimated the error of $J_{\rm eff}$ to be $\pm 4$ K. The error of the magnetization of the plateau is less than 1 %. We estimated the error of $g$ to be 0.02.

We consider that the probable spin model for $\rm CuInVO_5$ is an interacting spin-$\frac{1}{2}$ tetramer model. This cuprate may be useful for studying the L mode under the atmospheric pressure. We can investigate the L mode by measuring a longitudinal or scalar susceptibility. It is difficult to observe the L mode in a longitudinal susceptibility that INS probes [21], because the longitudinal susceptibility exhibits an infrared singularity, which can obscure an amplitude peak at a finite energy. The L mode can be well-defined in terms of a scalar susceptibility, because the scalar susceptibility can display a sharp amplitude peak [21]. In magnetic systems, Raman spectra can measure the scalar susceptibility. One-magnon Raman scattering indicating L-mode magnetic excitations was actually observed in the pressure-induced ordered state of $\rm KCuCl_3$ [10, 9] and in the magnetic-field-induced ordered state of $\rm TlCuCl_3$ [5]. We intend to form single crystals of $\rm CuInVO_5$ and perform Raman scattering experiments on them.

## 4. Conclusion

We measured the temperature dependence of the magnetic susceptibility and specific heat and the magnetic-field dependence of the magnetization of $\rm CuInVO_5$. An antiferromagnetically ordered state appears below $T_{\rm N} = 2.7$ K. We observed a $\frac{1}{2}$ quantum magnetization plateau above 30 T at 1.3 K. An isolated antiferromagnetic spin-$\frac{1}{2}$ tetramer model with $J_1 = 240$ and $J_2 = -142$ K can account for the magnetic susceptibility above 30 K. The magnetization curves can be explained using the interacting spin tetramer model with the effective intertetramer interaction $J_{\rm eff} = 30$ K. The probable spin model for $\rm CuInVO_5$ is an interacting spin-$\frac{1}{2}$ tetramer model. Longitudinal-mode magnetic excitations may be observable in $\rm CuInVO_5$.

## References

[1] Goldstone J, Salam A and Weinberg S 1962 Broken symmetries *Phys. Rev.* **127** 965
[2] Sachdev S 2011 *Quantum Phase Transitions* 2nd ed. (Cambridge: Cambridge University Press)
[3] Sachdev S and Keimer B 2011 Quantum criticality *Physics Today* **64** 29
[4] Pekker D and Varma C M 2015 Amplitude/Higgs modes in condensed matter physics *Annu. Rev. Condens. Matter Phys.* **6** 269
[5] Kuroe H, Kusakabe K, Oosawa A, Sekine T, Yamada F, Tanaka H and MatsumotoM 2008 Magnetic field-induced one-magnon Raman scattering in the magnon Bose-Einstein condensation phase of $\rm TlCuCl_3$ *Phys. Rev.* B **77** 134420
[6] Rüegg Ch, Normand B, Matsumoto M, Furrer A, McMorrow D F, Krämer K W, Güdel H-U, Gvasaliya S N, Mutka H and Boehm M 2008 Quantum magnets under pressure: Controlling elementary excitations in $\rm TlCuCl_3$ *Phys. Rev. Lett.* **100** 205701
[7] Merchant P, Normand B, Krämer K W, Boehm M, McMorrow D F and Rüegg Ch 2014 Quantum and classical criticality in a dimerized quantum antiferromagnet *Nature Phys.* **10** 373.
[8] Matsumoto M, Normand B, Rice T M and Sigrist M 2004 Field- and pressure-induced magnetic quantum phase transitions in $\rm TlCuCl_3$ *Phys. Rev.* B **69** 054423
[9] Matsumoto M, Kuroe H, Oosawa A and Sekine T 2008 One-magnon Raman scattering as a probe of longitudinal excitation mode in spin dimer systems *J. Phys. Soc. Jpn.* **77** 033702
[10] Kuroe H, Takami N, Niwa N, Sekine T, Matsumoto M, Yamada F, Tanaka H and Takemura K 2012 Longitudinal magnetic excitation in $\rm KCuCl_3$ studied by Raman scattering under hydrostatic pressures *J. Phys.: Conf. Ser.* **400** 032042

[11] Matsumoto M, Kuroe H, Sekine T and Masuda T 2010 Transverse and longitudinal excitation modes in interacting multispin systems *J. Phys. Soc. Jpn.* **79** 084703
[12] Hase M, Etheredge K M S, Hwu S-J, Hirota K and Shirane G 1997 Spin-singlet ground state with energy gaps in $Cu_2PO_4$: Neutron-scattering, magnetic-susceptibility, and ESR measurements *Phys. Rev.* B **56** 3231 In this reference, the Hamiltonian is defined as $\mathcal{H} = \sum_{i,j} 2J_{ij}S_i \cdot S_j$ instead of $\mathcal{H} = \sum_{i,j} J_{ij}S_i \cdot S_j$ in the present paper.
[13] Hase M, Kohno M, Kitazawa H, Suzuki O, Ozawa K, Kido G, Imai M and Hu X 2005 Coexistence of a nearly spin-singlet state and antiferromagnetic long-range order in quantum spin system $Cu_2CdB_2O_6$ *Phys. Rev.* B **72** 172412
[14] Hase M, Dönni A, Pomjakushin V Yu, Keller L, Gozzo F, Cervellino A and Kohno M 2009 Magnetic structure of $Cu_2CdB_2O_6$ exhibiting a quantum-mechanical magnetization plateau and classical antiferromagnetic long-range order *Phys. Rev.* B **80** 104405
[15] Hase M, Nakajima K, Ohira-Kawamura S, Kawakita Y, Kikuchi T and Matsumoto M 2015 Magnetic excitations in the spin-$\frac{1}{2}$ tetramer substance $Cu_2^{114}Cd^{11}B_2O_6$ obtained by inelastic neutron scattering experiments *Phys. Rev.* B **92** 184412
[16] Živković I, Djokić D M, Herak M, Pajić D, Prša K, Pattison P, Dominko D, Micković Z, Cinčić D, Forró L, Berger H and Rønnow H M 2012 Site-selective quantum correlations revealed by magnetic anisotropy in the tetramer system $SeCuO_3$ *Phys. Rev.* B **86** 054405
[17] Moser P, Cirpus V and Jung W 1999 CuInOVO4 - Single Crystals of a Copper(II) Indium Oxide Vanadate by Oxidation of Cu/In/V Alloys *Z. Anorg. Allg. Chem.* **625** 714
[18] Hase M, Terasaki I and Uchinokura K 1993 Observation of the spin-Peierls transition in linear $Cu^{2+}$ (spin-$\frac{1}{2}$) chains in an inorganic compound $CuGeO_3$ *Phys. Rev. Lett.* **70** 3651
[19] Hase M, Terasaki I, Sasago Y, Uchinokura K and Obara H 1993 Effects of substitution of Zn for Cu in the spin-Peierls cuprate, $CuGeO_3$: The suppression of the spin-Peierls transition and the occurrence of a new spin-glass state *Phys. Rev. Lett.* **71** 4059
[20] Hase M, Terasaki I, Uchinokura K, Tokunaga M, Miura N and Obara H 1993 Magnetic phase diagram of the spin-Peierls cuprate $CuGeO_3$ *Phys. Rev.* B **48** 9616
[21] Podolsky D, Auerbach A and Arovas D. P 2011 Visibility of the amplitude (Higgs) mode in condensed matter *Phys. Rev.* B **84** 174522

IOP Conf. Series: Journal of Physics: Conf. Series **969** (2018) 012101 doi:10.1088/1742-6596/969/1/012101

# Magnetic properties of the $UNiGe_2$ at low temperature

**Kohei Ohashi[1], Masashi Ohashi[1,2], Masaki Sawabu[1], Masahiro Miyagawa[1], Kae Maeta[1] and Tomoo Yamamura[3]**

[1]Graduate School of Natural Science and Technology, Kanazawa University, Kakuma-machi, Kanazawa, 920-1192, Japan
[2]Institute of Science and Engineering, Kanazawa University, Kakuma-machi, Kanazawa, 920-1192, Japan
[3]Institute for Materials Research, Tohoku University, 2-1-1 Katahira, Aoba-ku, Sendai, 980-8577, Japan

E-mail: k-ohashi@stu.kanazawa-u.ac.jp

**Abstract.** We report on the magnetic characterization of a novel ternary uranium intermetallic $UNiGe_2$. When we assume that $UNiGe_2$ has the orthorhombic structure of $CeNiGe_2$-type which is same as that of $UNiSi_2$, the lattice constants were obtained to be $a$ = 3.97 Å, $b$ = 16.48 Å, and $c$ = 4.08 Å. The unit cell volume of $UNiGe_2$ is lager than that of $UNiSi_2$. It comes from the fact that the atomic radius of Ge is lager than that of Si. The temperature dependence of the magnetic susceptibility shows two peaks at $T_N$=45 K and $T_{N'}$= 65 K. Taking an account that $UNi_2Ge_2$ secondary phase exists in the compound, $UNiGe_2$ is an antiferromagnet below $T_N$ while $T_{N'}$ may come from the antiferomagnetic order of $UNi_2Ge_2$. At 5 K, the slope of the magnetization curve increases as increasing the magnetic field up to 5 T, indicating the presence of a metamagnetic transition. The residual magnetization remains on the magnetization curve at 5 K, which may come from a week ferromagnetism of $UNi_2Ge_2$ at low temperature.

## 1. Introduction

Intermetallic compounds including Ce or U atoms have been investigated extensively because these compounds give important information for studying the role of strong electron correlations in metallic systems[1, 2, 3, 4]. In these compounds, the ferromagnetic/antiferromagnetic interaction and Kondo effect compete each other.

The ternary compounds $CeTX_2$ (T = transition metal and X = Si, Ge, Sn) form a large family having the orthorhombic $CeNiSi_2$-type layered structure, which is constructed from deformed fragments of the $CeGa_2Al_2$ and $\alpha$-$ThSi_2$ structures[5]. The lattice parameter along $b$-axis is extremely large compared to those along $a$- and $c$- axes, and it is expected that highly anisotropic magnetic property exists. Indeed, these compounds have received considerable interest of a great variety of magnetic behaviors[6, 7, 8].

$UNiSi_2$ also crystallizes in orthorhombic $CeNiSi_2$-type layered structure, and is a ferromagnet at $T_C$= 95 K[9, 10]. Single crystals can be brown by Czochralski pulling method because $UNiSi_2$ melts congruently. The large anisotropic behavior is observed in the measurement of the magnetization of $UNiSi_2$ single crystal at low temperature ferromagnetic phase. The easy

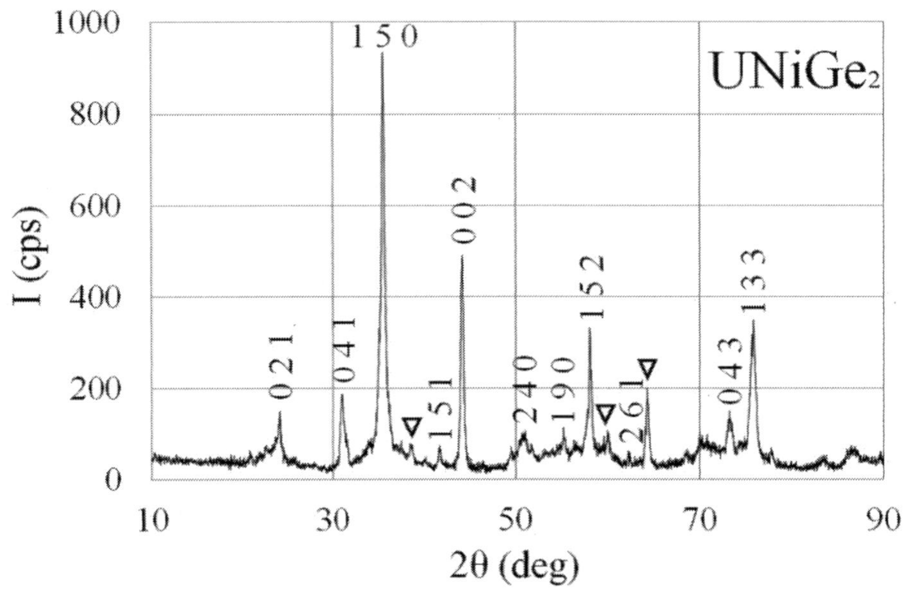


**Figure 1.** The X-ray powder diffraction pattern of $UNiGe_2$. Symbols ($\bigtriangledown$) show unknown peaks.

magnetization direction is in the *ac* plane. From the result of the magnetic susceptibility along the *ac* plane, the effective magnetic moment is obtained to be $\mu_{\text{eff}} \sim 2.47\mu_B$, which is smaller than the value expected for the free $U^{4+}$ or $U^{3+}$ ions. It may come from highly anisotropic magnetic property of $UNiSi_2$[11]. Recently a new compounds $UNiGe_2$ are also synthesized, but the crystal structure is unknown yet[12]. In the present work, we report on the synthesis of a ternary uranium compound $UNiGe_2$ and the magnetic property.

## 2. Experimental

Polycrystalline sample of $UNiGe_2$ was synthesized by arc melting with a stoichiometric composition in an Ar gas atmosphere. Weight losses were less than 0.3%. To improve homogeneity, the sample was turned over and re-melted several times. The sample was characterized by X-ray powder diffraction experiments using a Rigaku MiniFlex II diffractometer with Cu-$K_\alpha$ radiation. The dc magnetization was measured by using a Quantum Design MPMS-5 superconducting quantum interference device magnetometer.

## 3. Results and Discussion

### *3.1. X-ray diffraction*

Figure 1 shows the X-ray diffraction pattern of $UNiGe_2$ at room temperature. Here we assumed that $UNiGe_2$ has the same crystal structure as $UNiSi_2$ and indexed the Bragg peaks as the orthorhombic $CeNiSi_2$ type structure with the space group of *Cmcm*. The lattice constants were obtained to be $a = 3.97$ Å, $b = 16.48$ Å, and $c = 4.08$ Å for $UNiGe_2$. The unit cell volume is calculated to be 266 Å$^3$, which is lager than that of $UNiSi_2$, 258.83 Å$^3$[11]. It comes from the fact that the atomic radius of Ge is lager than that of Si. On the other hand, several unknown peaks are also observed in the X-ray diffraction pattern, and may correspond to the diffraction of $UNi_2Ge_2$ as a secondary phase[13]. Such behavior has been also observed in the previous report, in which it is suggested that a new dominant phase and $UNi_2Ge_2$ secondary phase exist in the compound[12].

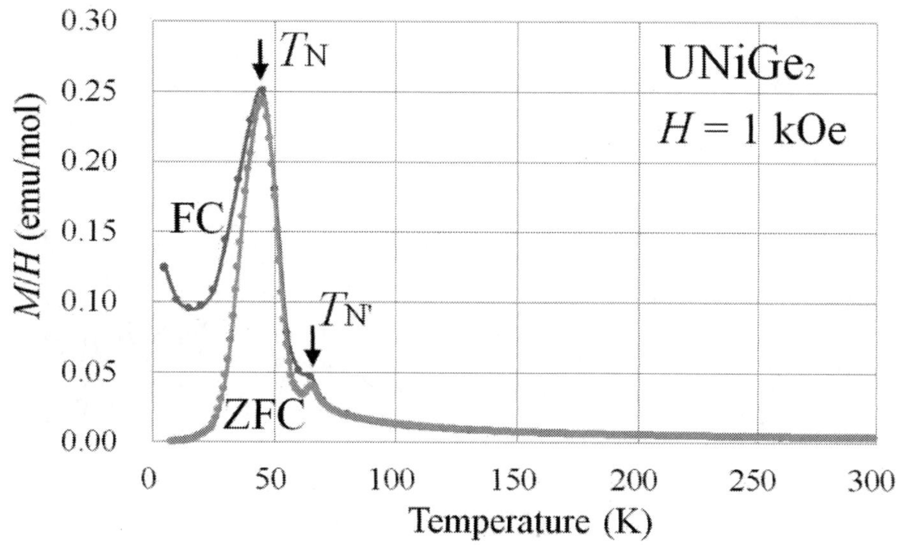


**Figure 2.** $M/H$ as a function of temperatture in a magnetic field of 1 kOe of $UNiGe_2$.

*3.2. Magnetic Properties*

Figure 2 shows the temperature dependence of $M/H$ of $UNiGe_2$ between 2 and 300 K in a magnetic field of 1 kOe. The ZFC $M/H$ curve is obtained by cooling in zero field from a high temperature while the FC one by cooling at small applied field. $M/H$ increases as temperature decreases and two peaks are visible on the $M/H(T)$ curve at $T_N = 45$ K and $T_{N'} = 65$ K. At low temperature below $T_N$, on the other hand, $M/H$ in the FC curve tends to increase as decreasing temperature while that in the ZFC one approaches to zero. It suggests that a dominant phase of $UNiGe_2$ is an antiferromagnet below $T_N$. Taking account that $UNi_2Ge_2$ is an antiferromagnet below 74 K and that a weak ferromagnet moment is also present at low temperature[13, 14], $T_{N'}$ and the residual magnetization in the FC curve may also come from a secondary phase in the sample.

Figure 3 displays the behavior of the magnetization $M$ of $UNiGe_2$ as a function of magnetic field $H$ up to 50 kOe at 5 K. $M$ increases as increasing magnetic field, and reaches 0.139 $\mu_B$, which is much smaller than the saturation magnetization $M_s = 1.12$ $\mu_B$ of $UNiSi_2$[9]. Since the slope of magnetization $\partial M/\partial H$ increases up to 50 kOe, a magnetic phase transition such as metamagnetism is expected at a high magnetic field below $T_N$. Moreover, hysteresis loop is also observed in the magnetization curve at 5 K, suggesting the the presence of some magnetic ordering phase such as weak ferromagnetism or canted magnetism. Taking account that $UNi_2Ge_2$ a weak ferromagnet moment is also present at low temperature[13], pure samples are needed to discuss the ground state of $UNiGe_2$.

## 4. Summary

In this study, we prepared a novel ternary uranium intermetallic $UNiGe_2$ during our investigation of $UNiX_2$ with the $CeNiSi_2$-type orthorhombic structure. The unit cell volume of $UNiGe_2$ is lager than that of $UNiSi_2$. Our preliminary result of the magnetization indicates that $UNiGe_2$ is an antiferromagnet of 45 K. Further experiments are in progress to synthesize $UNiSn_2$ and $UNiC_2$ to investigate magnetic properties of $UNiX_2$ (X = C, Si, Ge, Sn) series.

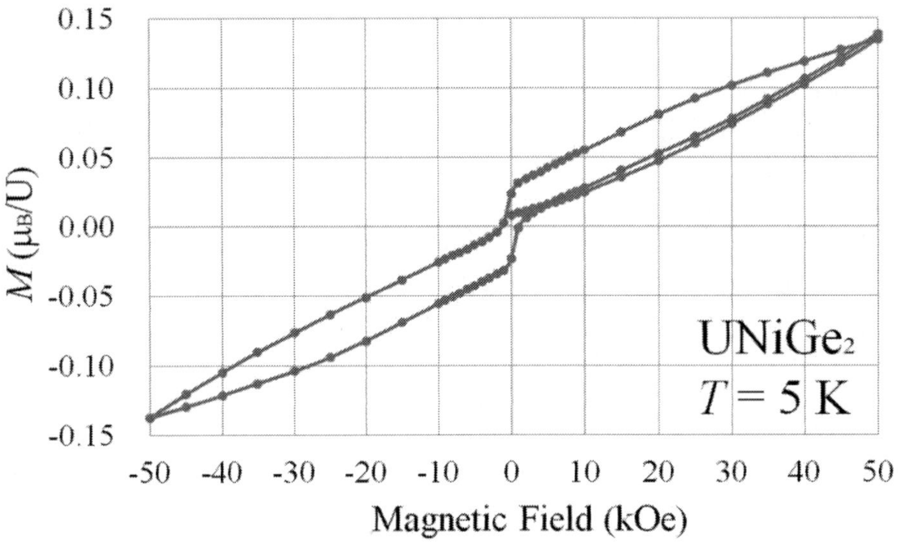


**Figure 3.** The field dependence of the magnetization of $UNiGe_2$ at 5.0 K.

**Acknowledgment**
This work was performed under the inter-university cooperative research program of the cooperative research program of the Institute for Materials Research, Tohoku University (Proposal No. 17K0057). This work was supported in part by Futaba Electronics Memorial Foundation and by Research Foundation for the Electrotechnology of Chubu.

## References

[1] Uwatoko Y, Umehara I, Ohashi M, Nakano T and Oomi G 2012 *Handbook on the Physics and Chemistry of Rare Earths* vol 2 (Amsterdam: North-Holland) chap 252, p 1
[2] Ohashi M, Oomi G, Koiwai S, Hedo M and Uwatoko Y 2003 *Phys. Rev. B* **68** 144428
[3] Miyagawa H, Oomi G, Ohashi M, Satoh I, Komatsubara T, Hedo M and Uwatoko Y 2008 *Phys. Rev. B* **78** 064403
[4] Ohashi M, Miyagawa H, Nakano T, Oomi G, Sechovsky V, Satoh I and Komatsubara T 2014 *J. Phys. Soc. Jpn.* **83** 024701
[5] Iandelli A and Palenzona A 1979 *Handbook on the Physics and Chemistry of Rare Earths* vol 2 (Amsterdam: North-Holland) chap 13, p 1
[6] Ohashi M, Oomi G, Ishida K, Satoh I, Komatsubara T, Kawae T and Takeda K 2006 *J. Alloys and Compounds* **408-412** 84
[7] Ohashi M, Oomi G and Satoh I 2007 *J. Phys. Soc. Jpn.* **76** 114712
[8] Nakano T, Ohashi M, Oomi G, Matsubayashi K and Uwatoko Y 2009 *Phys. Rev. B* **79** 172507
[9] Kaczorowski D 1996 *Solid State Commun.* **99** 949
[10] Taniguchi T, Morimoto H, Miyako Y and SRamakrishnan 1998 *J. Magn. Magn. Mater.* **177-181** 55
[11] Ohashi M, Oomi G, Ishida K and Satoh I 2006 *J. Phys. Soc. Jpn. Suppl.* **75** 124
[12] Molcanova Z, Jr M M, Reiffers M, Dzubinska A, Hurakova M, Kavecansky V, Paukov M and Havela L 2017 *Acta Physica Polonica A* **131** 994
[13] Ning Y B, Garret J D, Stager C V and Datars W R 1992 *Phys. Rev. B* **46** 8201
[14] Chelmichi L, Leciejewicz J and Zygmunt A 1985 *J. Phys. Chem. Solids* **46** 529

IOP Conf. Series: Journal of Physics: Conf. Series **969** (2018) 012102 doi:10.1088/1742-6596/969/1/012102

# Magnetoelectric properties of $Ca_2CoSi_2O_7$ studied by high-field electron spin resonance

**Mitsuru Akaki, Konatsu Kanaida, and Masayuki Hagiwara**

Center for Advanced High Magnetic Field Science, Graduate School of Science, Osaka University, Toyonaka, Osaka 560-0043, Japan

E-mail: akaki@ahmf.sci.osaka-u.ac.jp

**Abstract.** We have performed electron spin resonance (ESR) experiments of multiferroic $Ca_2CoSi_2O_7$ single crystals in high magnetic fields perpendicular to the easy plane. We observed many ESR absorption signals in the magnetization plateau phase below the saturation field which form branches with a linear magnetic field dependence. At zero magnetic field, large energy gaps with $\Delta = 1.7 \sim 1.8$ THz are identified by linear extrapolations of the resonance branches. We speculate that the high-energy ESR modes would be caused by electromagnons. With increasing magnetic field, some of these ESR modes decrease linearly with field even within the magnetization plateau phase and intersect the zero-frequency line at the magnetic field where the magnetization plateau terminates, suggesting a relationship between these energy gaps and the magnetization plateau.

## 1. Introduction

Magnetoelectric multiferroic materials have attracted both experimental and theoretical interest [1, 2, 3]. The mechanism of spin-dependent electric polarization is well explained by several microscopic models [4, 5, 6, 7]. A well-known model is the spin-current mechanism in which ferroelectricity originates from a cycloidal spin structure (e.g., $TbMnO_3$ [8]). On the other hand, up-up-down-down spin order induces electric polarization through exchange striction. This type of ferroelectricity is realized in highly-distorted orthorhombic $R$MnO$_3$ compounds ($R$=Y, Ho, ..., Lu) with $E$-type antiferromagnetism [9], $DyFeO_3$ [10], and $Ca_3(Co,Mn)_2O_6$ [11]. The third mechanism is the spin-dependent $p$-$d$ hybridization mechanism, where the hybridization between transition-metal $d$ and ligand $p$ orbitals is affected by the spin orientation relative to the bond direction between the transition metal and the ligand ions through the spin-orbit interaction. The variation of the hybridization generates a local electric dipole moment, and the sum of the local electric dipole moments is observed as spontaneous electric polarization. The remarkable feature of this mechanism is that the electric polarization is induced by one spin and its ligand ions without considering the spin-spin interactions. Therefore, the magnetic-field induced electric polarization caused by the $p$-$d$ hybridization mechanism has been observed not only in a magnetically long-range ordered state but also in a paramagnetic state [12]. In contrast, the other two mechanisms induce an electric polarization only in the magnetically ordered state because of the spin-spin interactions.

Recently, the Co åkermanite family, such as $Ca_2CoSi_2O_7$ [13, 14], $Sr_2CoSi_2O_7$ [12], $Ba_2CoGe_2O_7$ [15, 16], and $Sr_2CoGe_2O_7$ [17], have attracted attention as typical multiferroic examples of the $p$-$d$ hybridization mechanism. As seen from the schematic crystal structure

IOP Conf. Series: Journal of Physics: Conf. Series **969** (2018) 012102 doi:10.1088/1742-6596/969/1/012102

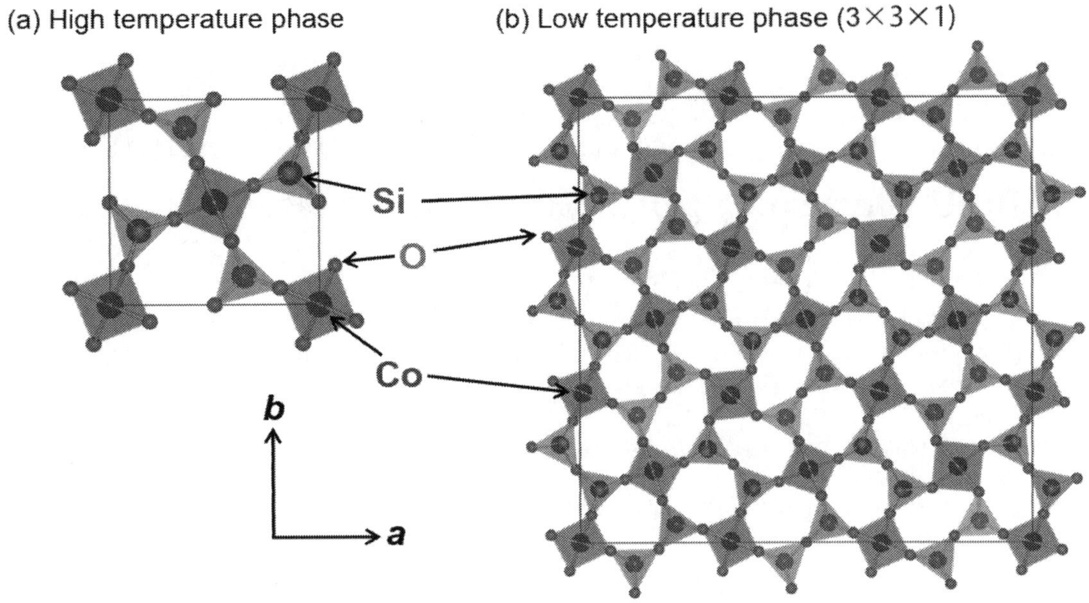


**Figure 1.** The crystal structure of $Ca_2CoSi_2O_7$ projected onto the *ab*-plane in (a) high temperature phase and (b) low temperature phase. Space group of high (low) temperature phase is $P\bar{4}2_1m$ ($P2_12_12$).

of åkermanite illustrated in Fig. 1(a), corner sharing $CoO_4$ and $SiO_4$ ($GeO_4$) tetrahedra form layers alternating with intervening layers of alkaline earth metal ions along the *c*-axis. The $Co^{2+}$ magnetic ions form diagonal square lattices. Among them, $Ca_2CoSi_2O_7$ undergoes a series of structural phase transitions as a function of temperature [18, 19, 20]. Above 500 K, it shows the normal åkermanite structure with tetragonal $P\bar{4}2_1m$ space group, as shown in Fig. 1(a). Between 150 and 500 K, the structure is incommensurate (IC). At 150 K, an IC-C (commensurate) transition occurs, and a $3 \times 3 \times 1$ supercell structure with orthorhombic $P2_12_12$ space group develops at lower temperatures, as shown in Fig. 1(b).

The magnetoelectric behavior is also complicated [13, 14]. An antiferromagnetic order takes place at the Neel temperature $T_N = 5.7$ K in $Ca_2CoSi_2O_7$. Magnetization processes for magnetic fields along the *a*- and *b*-axes show some anomalies. The magnetization along the [110] direction exhibits a steep increase at 11 T, accompanied by a reversal of electric polarization. As shown in Fig. 3(a), the magnetization along the *c*-axis increases linearly up to 18 T. Between 18 T and 50 T a magnetization plateau develops, the magnitude of which is slightly smaller (~85%) than the saturation magnetization, and a step-like increase to the saturation value at 60 T. To clarify the origin of this anomalous magnetization plateau state from a microscopic viewpoint, we have performed electron spin resonance (ESR) experiments of single crystals of $Ca_2CoSi_2O_7$ in high magnetic fields.

## 2. Experiment

We grew single crystalline samples of $Ca_2CoSi_2O_7$ using the floating zone method. X-ray diffraction measurements at room temperature assigned the crystal structure to the tetragonal $P\bar{4}2_1m$. All specimens used in the experiments were cut along the crystallographic principal

IOP Conf. Series: Journal of Physics: Conf. Series **969** (2018) 012102 doi:10.1088/1742-6596/969/1/012102

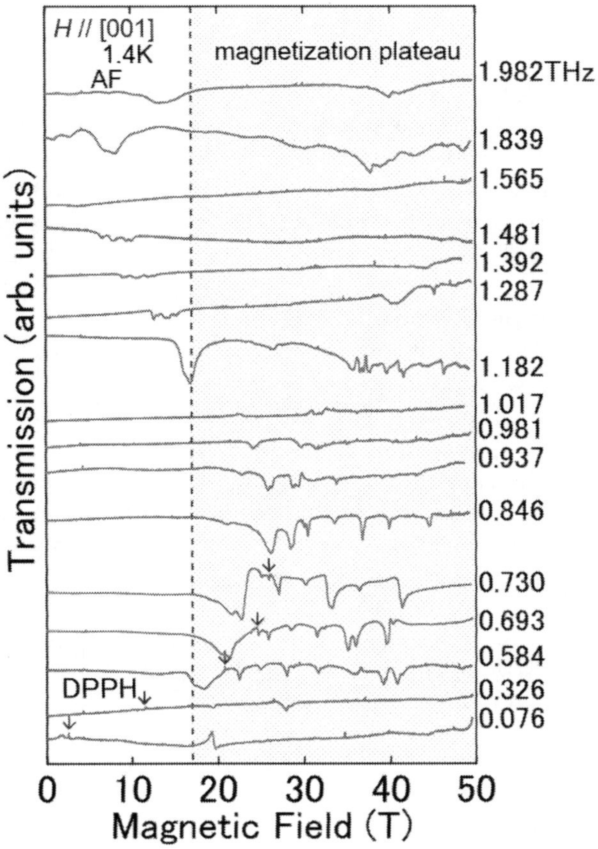


**Figure 2.** Frequency dependence of ESR absorption spectra of $Ca_2CoSi_2O_7$ for magnetic fields parallel to the $c$ direction. Vertical dashed line indicates the transition field into the magnetization plateau phase determined by magnetization measurements. Arrows mark the resonance fields of DPPH (2,2-diphenyl-1-picrylhydrazyl, ESR marker with $g = 2.0036$).

axes into platelike shapes by means of the X-ray back reflection Laue technique. High-field multi-frequency ESR measurements at 1.4 K in pulsed magnetic fields of up to 50 T were conducted by utilizing a far-infrared laser and Gunn oscillators with a frequency doubler to generate submillimeter and millimeter waves. We used a magnetically tuned InSb hot-electron bolometer as a detector. All the experiments were carried out using unpolarized light.

## 3. Results and Discussion

Figure 2 displays the ESR absorption spectra of $Ca_2CoSi_2O_7$ at 1.4 K for magnetic fields parallel to the $c$ direction. Several absorption signals were observed in the magnetization plateau phase (Fig. 3(a)). In contrast, below 17 T, absorption signals were detected mostly above 1 THz. The resonance fields are plotted on the frequency-magnetic field diagram as shown in Fig. 3(b). Resonance modes observed in the magnetization plateau phase are linear in magnetic fields ($H$) with $g = 2.2$. This $g$ value is consistent with the value of the saturation magnetization as shown in Fig. 3(a). The gap frequencies of these linear modes $\Delta_{pi}$ ($i = 0 - 5$) at the transition field into the magnetization plateau phase are 0, 0.14, 0.25, 0.34, 0.43, 0.53 THz. The origin of these split linear modes may be the different magnetic anisotropy at each site due to the distortion of the $CoO_4$ tetrahedra in the $3 \times 3 \times 1$ supercell.

Moreover, two-magnon excitation modes emerge in high-frequency region above 0.5 THz in the magnetization plateau phase and near zero field at about 1.7 THz. Similar excitation was observed in the high field saturated phase of $Sr_2CoGe_2O_7$ [17]. In $Sr_2CoGe_2O_7$, this mode is explained by a spin-quadrupolar excitation and could be observed due to electromagnetic

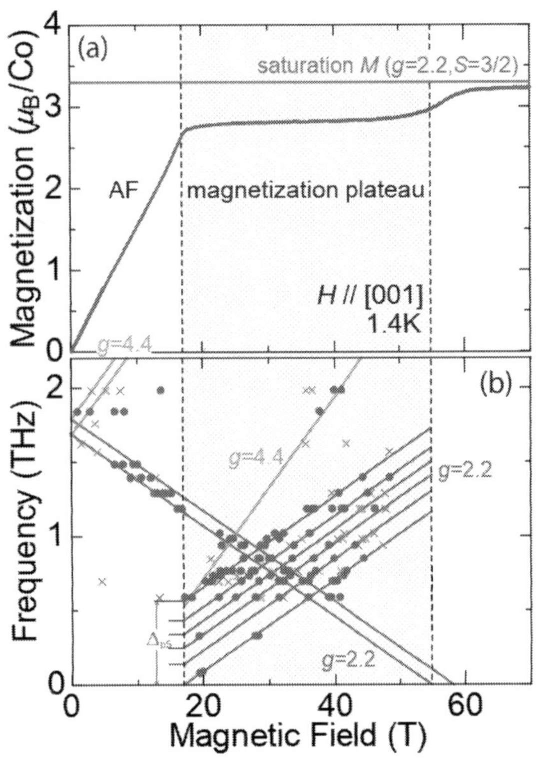


**Figure 3.** (a) Magnetization curve of $Ca_2CoSi_2O_7$ along the $c$-axis at 1.4 K [13]. Horizontal solid gray line indicates the saturation magnetization value given by $g = 2.2$ obtained from the ESR measurements mentioned below. (b) Frequency-magnetic field diagram of the ESR resonance fields of $Ca_2CoSi_2O_7$ for the magnetic fields parallel to the $c$ direction. Circles (crosses) show strong (weak) resonance signals. Vertical dashed lines at 17 T indicate the transition field into the magnetization plateau phase and at 55 T its termination field. The ESR points fall on straight lines (solid lines). From linear fits, we extracted the $g = 2.2$ value for the lowest 6 modes, and $g = 4.4$ for the highest mode. The descending straight lines with $g$=2.2 from the large energy gaps of about 1.7~1.8 THz are intersected near the termination field of the magnetization plateau, and the ascending straight lines from the same gaps seem to have the large $g$-value of 4.4.

coupling. Therefore, the two-magnon excitation modes in $Ca_2CoSi_2O_7$ are probably attributed to spin-quadrupolar excitations.

A weak signal with the zero field energy gap of 0.4 THz was observed in THz absorption spectroscopy [21]. However, this gapped mode could not be observed in our measurements, because the signal intensity must be too weak to be detected with our bolometer or this mode must be field independent along the [001] direction. In contrast, we observed other zero field energy gaps at $\Delta = 1.7 \sim 1.8$ THz by extrapolating the resonance modes to 0 T. In $Ba_2CoGe_2O_7$ ($\Delta = 0.5$ THz) [22] and $Sr_2CoGe_2O_7$ ($\Delta = 0.2$ THz) [17], the zero field energy gaps are explained by easy-plane type anisotropy. According to the spin-wave theory for the antiferomagnet with easy-plane anisotropy, the energy gap $\Delta$ is proportional to $\sqrt{2J\Lambda}$, where $J$ is the exchange constant between the nearest neighbor spins and $\Lambda$ is the single ion anisotropy constant. The antiferromagnetic transition temperatures $T_N$ are 5.7 K, 6.5 K and 6.7 K for $Ca_2CoSi_2O_7$ [14], $Sr_2CoGe_2O_7$ [23] and $Ba_2CoGe_2O_7$ [15], respectively. These similar values of $T_N$ suggest that the $J$ of $Ca_2CoSi_2O_7$ is not largely different from the $J$s in other Co åkermanite materials. On the other hand, the $\Lambda$ of $Ca_2CoSi_2O_7$ is probably small, because isotropic behavior was observed in its magnetization [13]. Therefore, the large zero-field energy gaps of $\Delta = 1.7 \sim 1.8$ THz are difficult to be understood as usual magnetic excitations, whereas the zero field gap of 0.4 THz is probably explained by the antiferromagnetic resonance mode with easy-plane type anisotropy. On applying magnetic fields, some ESR modes decrease linearly from the energy gaps of about $\Delta$. Assuming a linear field dependence with a slope $g$=2.2, these modes intersect the frequency-zero line at the magnetic fields between 54 and 58 T, which are in the vicinity of the magnetic field where the plateau terminates. Accordingly, this gapped resonance mode may be a clue to understand the emergence of magnetization plateau.

## 4. Conclusions

In conclusion, electromagnetic properties of $Ca_2CoSi_2O_7$ single crystals were investigated by high-field multi-frequency electron spin resonance (ESR) technique. Several ESR straight branches with $g = 2.2$ were observed in the magnetization plateau phase. We have found ESR modes with large zero-field energy gaps $\Delta = 1.7 \sim 1.8$ THz. These gapped ESR modes, which may be electromagnon excitations, suggest a strong relationship between this energy gap and the magnetization plateau. The studies of light polarization dependence of the ESR signals are necessary to clarify whether these high-energy gapped ESR modes are caused by electromagnons.

## Acknowledgment

The authors gratitude K. Penc for critical reading of the manuscript. This work was supported by JSPS KAKENHI Grants Numbers JP15K05145 and A25220803, and by JSPS Core-to-Core Program, A. Advanced Research Networks. M. A. would like to acknowledge the support from the Motizuki Fund of Yukawa Memorial Foundation.

## References

[1] M. Fiebig, J. Phys. D **38** R123 (2005).
[2] Y. Tokura, S. Seki and N. Nagaosa, Rep. Prog. Phys. **77** 076501 (2014).
[3] T. Arima, J. Phys. Soc. Jpn. **80** 052001 (2011).
[4] H. Katsura, N. Nagaosa, and A. V. Balatsky, Phys. Rev. Lett. **95**, 057205 (2005).
[5] I. A. Sergienko and E. Dagotto, Phys. Rev. B **73**, 094434 (2006).
[6] C. Jia, S. Onoda, N. Nagaosa, and J. H. Han, Phys. Rev. B **76**, 144424 (2007).
[7] T. Arima, J. Phys. Soc. Jpn. **76**, 073702 (2007).
[8] T. Kimura, T. Goto, H. Shintani, K. Ishizaka, T. Arima, and Y. Tokura, Nature (London) **426**, 55 (2003).
[9] S. Ishiwata, Y. Kaneko, Y. Tokunaga, Y. Taguchi, T. Arima, and Y. Tokura, Phys. Rev. B **81**, 100411(R) (2010).
[10] Y. Tokunaga, S. Iguchi, T. Arima, and Y. Tokura, Phys. Rev. Lett. **101**, 097205 (2008).
[11] Y. J. Choi, H. T. Yi, S. Lee, Q. Huang, V. Kiryukhin, and S.-W. Cheong, Phys. Rev. Lett. **100**, 047601 (2008).
[12] M. Akaki, H. Iwamoto, T. Kihara, M. Tokunaga, and H. Kuwahara, Phys. Rev. B **86**, 060413(R) (2012).
[13] M. Akaki, H. Kuwahara, A. Matsuo, K. Kindo, and M. Tokunaga, J. Phys. Soc. Jpn. **83**, 093704 (2014).
[14] M. Akaki, J. Tozawa, D. Akahishi, and H. Kuwahara, Appl. Phys. Lett. **94**, 212904 (2009).
[15] H. Y. Yi, Y. J. Choi, S. Lee, and S.-W. Cheong, Appl. Phys. Lett. **92**, 212904 (2008).
[16] H. Murakawa, Y. Onose, S. Miyahara, N. Furukawa, and Y. Tokura, Phys. Rev. Lett. **105**, 137202 (2010).
[17] M. Akaki, D. Yoshizawa, A. Okutani, T. Kida, J. Romhányi, K. Penc, and M. Hagiwara, arXiv:1702.04158.
[18] K. Hagiya, K. Kusaka, M. Ohmasa, and K. Iishi, Acta Crystallogr., Sect. B: Struct. Sci. **57**, 271 (2001).
[19] M. Riester and H. Böhm, Z. Kristallogr. **212**, 506 (1997).
[20] Z. H. Jia, A. K. Schaper, W. Massa, W. Treutmann, and H. Rager, Acta Crystallogr., Sect. B: Struct. Sci. **62**, 547 (2006).
[21] I. Kézsmárki, D. Szaller, S. Bordács, V. Kocsis, Y. Tokunaga, Y. Taguchi, H. Murakawa, Y. Tokura, H. Engelkamp, T. Rõõm, and U. Nagel, Nat. Commun. **5**, 3203 (2014).
[22] K. Penc, J. Romhányi, T. Rõõm, U. Nagel, Á. Antal, T. Fehér, A. Jánossy, H. Engelkamp, H. Murakawa, Y. Tokura, D. Szaller, S. Bordács, and I. Kézsmárki, Phys. Rev. Lett. **108**, 257203 (2012).
[23] T. Endo, Y. Doi, Y. Hinatsu, and K. Ohoyama, Inog. Chem. **51**, 3572-3578 (2012).

# Successive Phase Transitions and Magnetic Fluctuation in a Double-Perovskite $NdBaMn_2O_6$ Single Crystal

**S Yamada[1],H Sagayama[2,3],K Sugimoto[4] and T Arima[5]**

[1]Department of Materials System Science, Yokohama City University, Yokohama, 236-0027, Japan
[2]Institute of Materials Structure Science, High Energy Accelerator Research Organization, Tsukuba, Ibaraki 305-0801, Japan.
[3]Department of Materials Structure Science, The Graduate University for Advanced Studies, Tsukuba, Ibaraki 305-0801, Japan.
[4]SPring-8/JASRI, Kouto, Sayo, Hyogo, 679-5198, Japan
[5]Department of Advanced Materials Science, The University of Tokyo, Kashiwa, 277-8561, Japan

E-mail: shigeki@yokohama-cu.ac.jp

**Abstract.** We have succeeded in growing large high-quality single crystals of double-perovskite $NdBaMn_2O_6$ with $c$-axis aligned. Curie-Weiss paramagnetism and metallic conduction are observed above 290 K ($T_{MI}$). The magnetic susceptibility suddenly drops at $T_{MI}$ accompanied by a metal-insulator transition. Pervious studies using polycrystalline samples proposed that this material undergoes a ferromagnetic phase transition near 300K, and that the magnetic anomaly at $T_{MI}$ should be ascribed to layered antiferromagnetic phase transition. However, single-crystalline samples do not show any anomaly that indicates the ferromagnetic phase transition above $T_{MI}$. We assign the onset of magnetic anisotropy at 235 K as antiferromagnetic transition temperature $T_N$. Though the magnetization just above $T_{MI}$ shows the ferromagnetic-like magnetic-field dependence, the magnetization does not saturate under 70kOe at 300K. The magnetization behavior implies ferromagnetic fluctuation in the paramagnetic phase. The ferromagnetic fluctuation are also observed just below $T_{MI}$. Because a metamagnetic transition is observed at a higher magnetic field, the ferromagnetic fluctuation competes with antiferromagnetic fluctuation in this temperature range.

## 1. Introduction

The physical properties of a family of double-perovskite manganites $REBaMn_2O_6$ (where RE is divalent rare earth) were attracting much attention in terms of A-site-randomness free perovskite manganites.[1, 2, 3, 4, 5, 6, 7, 8, 9, 10, 11, 12] Because magnetic and electric ordering phases are stabilized by the A-site ordering, the phase diagram of $REBaMn_2O_6$ is completely different from that of A-site disordered perovskite manganites $RE_{0.5}Ba_{0.5}MnO_3$, where the magnetic ground state is paramagnetic or spin glass phase. When the ionic radius of RE is equal to or smaller than Sm, charge ordering phase is observed even above 380 K. The highest charge order is observed at 550 K in $YBaMn_2O_6$. This material has the potential for room temperature magnetoresistance device due to the melting of the charge ordering. Actually, fairly large magnetoresistance at room temperature was observed in A-site ordered $Sm_{1-x}La_{x+y}Ba_{1-y}Mn_2O_6$.[12] On the other hand, when the ionic radius of RE is equal to or larger than Nd, the charge ordering phase is not observed. The previous studies using the $NdBaMn_2O_6$ polycrystal suggested that a ferromagnetic phase transition was observed near 310 K, and A-type antiferromagnetic phase was stabilized below 290 K.[7, 11]

Recently, we succeeded in the single crystal growth of a double-perovskite manganite $SmBaMn_2O_6$, and revealed that the magnetic property was different from polycrystals.[13] Moreover, we found that the spatial inversion symmetry was broken in low-temperature charge ordering phase by means of the single crystal structure analysis using synchrotron x-ray diffraction.[14, 15] These results suggest the importance of the studies using single crystal for double perovskite manganites. We also succeeded in the single crystal growth of $NdBaMn_2O_6$.[16] The studies on the magnetic, electric and crystal structure using the single crystal provided new information about the phase transitions. In particular, the study on the magnetic anisotropy showed that a steep change of magnetic susceptibility near 290 K ($T_{MI}$) was not caused by the antiferromagnetic phase transition and the Néel temperature was near 235 K ($T_N$) which was more 50 K lower than $T_{MI}$. Additionally, we did not clearly observe the ferromagnetic transition near 300 K, which were proposed by previous papers. In general, two dimensional metallic conduction is observed in A-type antiferromagnetic phase[17]. Nevertheless, an insulating behavior was observed not only along the *c*-axis but also in the *ab*-plane below $T_{MI}$. The magnetic phases around $T_{MI}$ might be more complicated than reported by pervious studies. In this paper, we revise the magnetic phase diagram of $NdBaMn_2O_6$ through magnetization measurements using single crystals.

## 2. Experimental

Single crystals of $NdBaMn_2O_6$ were grown by floating zone (FZ) method. Powders of $Nd_2O_3$, $BaCO_3$, and $Mn_3O_4$ were mixed, ground, and calcined at 1290 °C for 48 hours in Ar (6N) atmosphere, and then pulverized. The resultant powder was shaped into a cylinder under a hydrostatic pressure of 30 MPa and sintered at 1290 °C for 12 hours in Ar (6N) atmosphere to form feeding and seeding rods. An FZ furnace equipped with two halogen incandescent lamps and hemielliptic focusing mirrors was used for the crystal growth. The molten zone was vertically scanned at a rate of 2 mm/h in Ar gas mixed with a tiny portion ($\leq 0.1$ %) of $H_2$. The melt grown bar was annealed in $O_2$ atmosphere at 500 °C for 48 hours. In order to measure anisotropic properties, we cut the crystal boules into some pieces, and the orientation of the *c*-axis was determined by x-ray diffraction. Laue photographs confirmed that every sample was a single crystal. Magnetization was measured using a commercial superconducting quantum interface device magnetometer (Quantum Design MPML-XL). The electrical resistivity was measured by a conventional four-probe method.

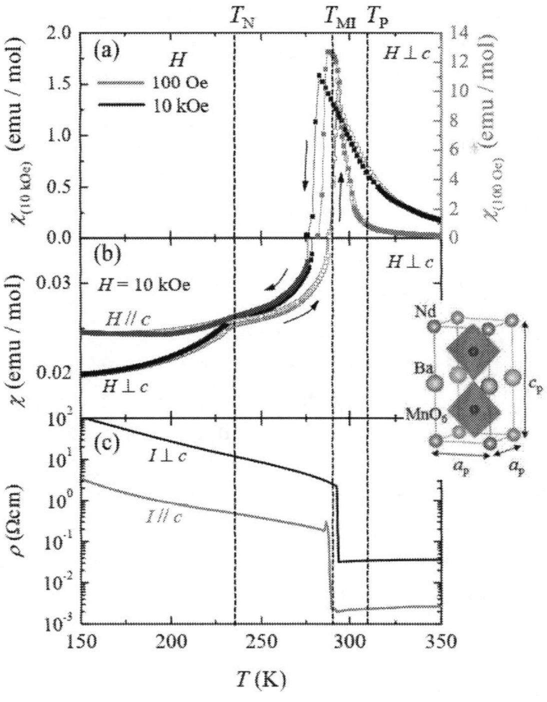


**Figure 1.** Temperature dependence of (a) magnetic susceptibility, (b) magnetic anisotropy, and (c) electrrical resistivity.[16] The solid and open squares in (a) and (b) denote the cooling and warming processes, respectively.

IOP Conf. Series: Journal of Physics: Conf. Series **969** (2018) 012103 doi:10.1088/1742-6596/969/1/012103

## 3. Results and Discussions

Figure 1 shows that the magnetic susceptibility steeply changes around 290 K ($T_{MI}$). This magnetic anomaly is accompanied by a metal-insulator transition. Though previous studies using poly-crystal samples suggested that $T_{MI}$ should correspond to A-type antiferromagnetic phase transition temperature, no clear magnetic anisotropy is observed down to 235 K. Therefore, we conclude that Néel temperature $T_N$ of $NdBaMn_2O_6$ is not identical to $T_{MI}$ but 235 K. When an applied magnetic field is 100 Oe, magnetic susceptibility steeply increases with a decrease of temperature around 310 K, as denoted by a large arrow in Fig. 1(a). On the other hand, when an applied magnetic field is 10 kOe. No anomaly is observed above $T_{MI}$. To reveal the magnetic phase just above $T_{MI}$, magnetic field dependence of magnetization was investigated in this temperature range. At 350 K which is sufficiently higher than $T_{MI}$, the magnetization linearly increases with an applied magnetic field, like a typical paramagnet. The magnetization curves become convex when approaching $T_{MI}$. Below 10 kOe, the magnetization steeply increases with the applied magnetic field. The increase becomes gradual above 20 kOe. The magnetization is smaller than the magnitude of full magnetic moments of $Mn^{3.5+}$ even when the applied magnetic field is 70 kOe. It suggests that ferromagnetic fluctuation in the paramagnetic phase is steeply enhanced down to $T_{MI}$. We conclude that the temperature at which the ferromagnetic fluctuation arises is 310 K ($T_P$) from Weiss temperature in our previous report.[16]

Next, we discuss the magnetic phase just below $T_{MI}$. Fig. 3 (a) shows that no magnetic anisotropy is observed below 70 kOe in this temperature range. The magnetization jumps near 20 kOe. The isotropic magnetic susceptibility below metamagnetic phase transition and the metamagnetic behavior imply that the long-range antiferromagnetic ordering is not present but the antiferromagnetically fluctuating domains exist in the magnetic ground state in this temperature range. The metamagnetic phase transition field depends on the measurement process. Because the phase transition at $T_{MI}$ is of the first-order, the higher magnetic field phase is essentially identical to that above $T_{MI}$. In low magnetic field range, the magnetization curve of sequence 3 is ferromagnetic-like and different from that of sequence 1. Fig. 3 (b) shows that, below 1 kOe, the magnetization curve after zero field cooling is different from the others. This magnetic field dependence of magnetization in a low magnetic field suggests that the application of a magnetic field induces a phase transition, and the initial state never revives by further field sweeping. At 270 K, which is a little lower than 286 K but still higher than $T_N$, the metamagnetic transition field shifts to higher one. Additionally, the ferromagnetic-like behavior in low magnetic field decreases. Hence the ferromagnetic fluctuation is suppressed with a decrease of temperature. As shown in Fig. 4, at 150 K which is below $T_N$, the magnetization linearly increases to an applied magnetic field, and the magnetic anisotropy is clearly observed. The ferromagnetic fluctuation is disappeared at $T_N$ and this phase below $T_N$ is antiferromagnetic single phase in which the magnetic moments of Mn ions lie in *ab* plane. In general, *ab* plane of nearly half doped perovskite manganite shows metallic conductivity in A-type antiferromagnetic phase.[17] However, as shown in Fig.1 (c), $NdBaMn_2O_6$ shows a metal-insulator transition at $T_{MI}$ not only along *c*-axis but also within *ab* plane. The complicated magnetic phase could be one of the origin of the three dimensional insulating conductivity below $T_{MI}$.

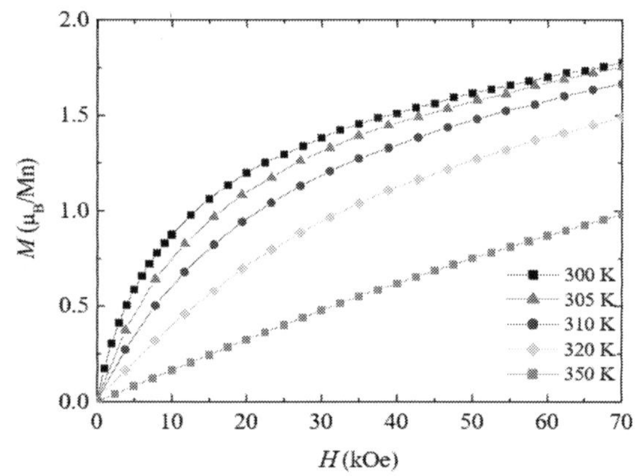


**Figure 2.** Magnetic field dependence of magnetization above $T_{MI}$.

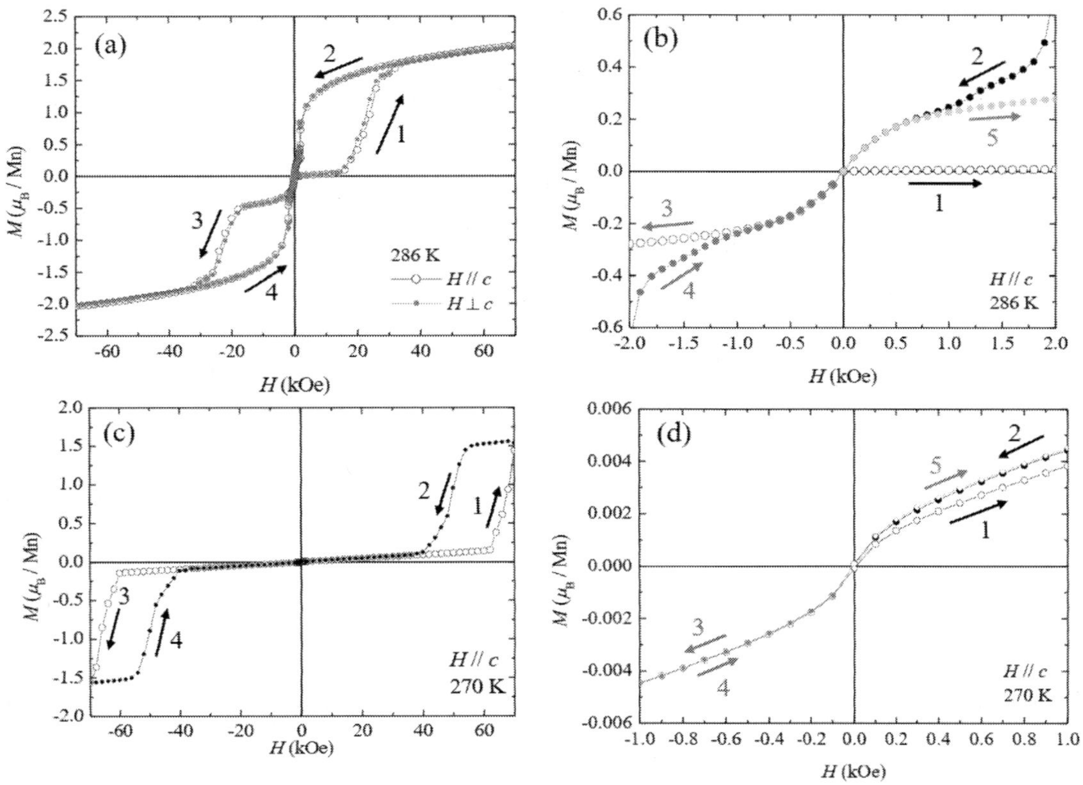


**Figure 3.** Magnetic field dependence of magnetization below $T_{MI}$. The numbers denote the measurement sequence.

Finally, we show magnetic phase diagram of $NdBaMn_2O_6$ which are concluded from the study. Though ferromagnetic fluctuation in paramagnetic phase steeply increase just below 310 K($T_P$), long-range ferromagnetic ordering is not realized above 290 K ($T_{MI}$). At $T_{MI}$, magnetic susceptibility suddenly drops due to antiferromagnetic fluctuation. However, just below $T_{MI}$, long-range antiferromagnetic ordering is not realized due in part to the remaining ferromagnetic fluctuation. The ferromagnetic fluctuation decreases with a decrease of temperature, and disappears at 235 K ($T_N$). Below $T_N$, A-type antiferromagnetic single phase is realized.

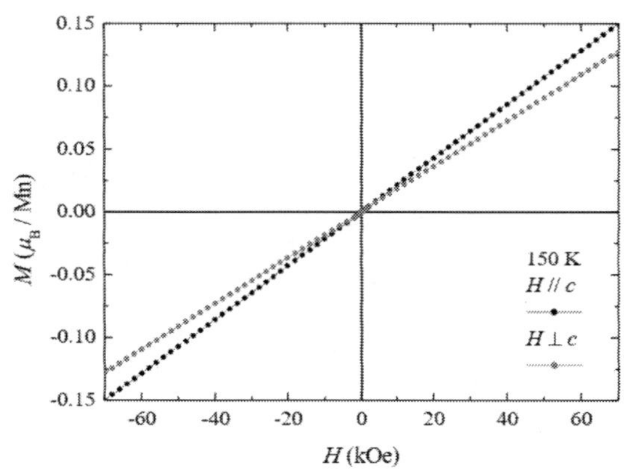


**Figure 4.** Magnetic field dependence of magnetization at 150 K ($< T_N$).

## 4. Conclusion

Through the measurements of the magnetic field dependence of magnetization around $T_{MI}$ in detail, the magnetic phases of $NdBaMn_2O_6$ are identified. Below 310 K, ferromagnetic fluctuation in paramagnetic phase grows as it gets closer to $T_{MI}$. On the other hand, the

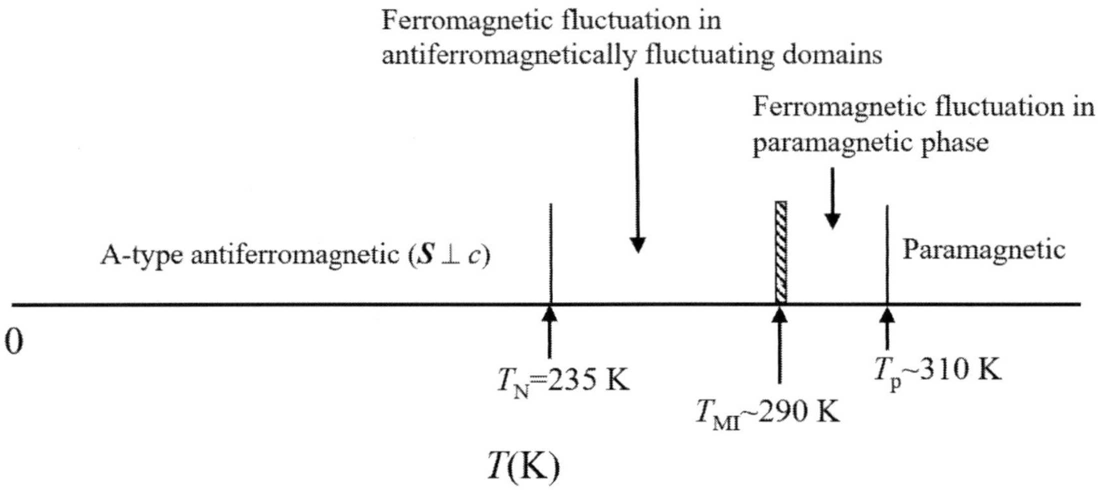


**Figure 5.** Magnetic phase diagram of $NdBaMn_2O_6$.$\boldsymbol{S}$ is Mn spin moments vector.

ferromagnetic fluctuation is observed in the temperature range between $T_{MI}$ and $T_N$. The metamagnetic phase transition which is caused by melting the antiferromagnetic domains is observed by relatively low magnetic field. Because these magnetic phases are strongly coupled with electric conduction properties, a large magnetoresistance effect is expected in this temperature range.

**Acknowledgments**
This study was partly supported by a Grant-in-Aid for Scientific Research (No. 24540380) from the Japan Society for the Promotion of Science and by the grant for Strategic Research Promotion(No.G2503) of Yokohama City University.

## References

[1] Millange F, Caignaert V, Domengès B, Raveau B and Suard E 1998 *Chem. Mater.* **10** 1974
[2] Caignaert V, Millange F, Domengès B, Raveau B and Suard E 1999 *Chem. Mater.* **11** 930
[3] Trukhanov S V, Troyanchuk I O, Hervieu M, H. Szymczak, and K. Bärner 2002 Phys. Rev. B **66** 184424
[4] Nakajima T, Kageyama H and Ueda Y 2002 *J. Phys. Chem. Solid.* **63,** 2843
[5] Williams A J, and Attfield J P 2002 *Phys. Rev.* B **66** 220405(R)
[6] Arima, T, Akahoshi, Oikawa D K, Kamiyama, Uchida T M, Matsui Y and Tokura Y, 2002 *Phys. Rev.* B **66** 140408(R)
[7] Nakajima T, Kageyama H, Yoshizawa H, Ohoyama K and Ueda Y 2003 *J. Phys. Soc. Jpn.* **72** 3237.
[8] Kageyama H, Nakajima, Ichihara T M, Ueda Y, Yoshizawa H and Ohoyama K 2003 *J. Phys. Soc. Jpn.* **72** 241
[9] Akahoshi D, Uchida M, Tomioka Y, Arima T, Matsui Y and Tokura Y 2003 Phys. *Rev. Lett.* **90** 177203
[10] Nakajima T, Yoshizawa H and Ueda Y 2004 *J. Phys. Soc. Jpn.* **73** 2283
[11] Akahoshi D, Okimoto, Kubota Y M, Kumai R, Arima T, Tomioka Y and Tokura Y 2004 *Phys. Rev.* B **70** 064418
[12] Nakajima T and Ueda Y 2005 *J. Appl. Phys.* **98** 046108
[13] Yamada S, Maeda Y and Arima T 2017 J. Phys. Soc. Jpn. B **81** 113711
[14] Morikawa D, Tsuda K , Maeda Y, Yamada S and Arima T 2012 J. Phys. Soc. Jpn. **81** 093602
[15] Sagayama H, Toyoda S, Sugimoto K, Maeda Y, Yamada S and Arima T 2017 Phys. Rev. B **90** 241113(R)
[16] Yamada S, Sagayama H, Higuchi K, Sasaki T, Sugimoto K and Arima T 2017 Phys. Rev. B **95** 035101
[17] Kuwahara H, Okuda T, Tomioka Y, Asamitsu A and Tokura Y 1999 *Phys. Rev. Lett.* **82** 4316

IOP Conf. Series: Journal of Physics: Conf. Series **969** (2018) 012104 doi:10.1088/1742-6596/969/1/012104

# Formation of the $S = 1$ paramagnetic centers in the bond-diluted spin-gap magnet $(C_4H_{12}N_2)(Cu_2Cl_6)$

**V N Glazkov[1,2], Yu V Krasnikova[1,2], D Hüvonen[3,4], A Zheludev[4]**

[1]P.Kapitza Institute for Physical Problems, Kosygin str. 2, 119334 Moscow, Russia
[2]International Laboratory for Condensed Matter Physics, Higher School of Economics, St.Basmannaya str. 21, 105066 Moscow, Russia
[3]National Institute of Chemical Physics and Biophysics, Akadeemia tee 23, 12618 Tallinn, Estonia
[4]Neutron Scattering and Magnetism, Laboratory for Solid State Physics, ETH Zürich, 8006 Zürich, Switzerland

E-mail: glazkov@kapitza.ras.ru

**Abstract.** Recently found quasi-two dimensional metalloorganic compound $(C_4H_{12}N_2)(Cu_2Cl_6)$ (abbreviated PHCC) is an example of a spin-gap magnet. Its ground state is a nonmagnetic singlet separated from the triplet excitations by an energy gap of approximately 1 meV. This compound allows partial substitution of chlorine ions by bromine, which results in the modulation of the affected exchange bonds. We have found by means of electron spin resonance spectroscopy that this doping results in the formation of the gapless $S = 1$ paramagnetic centers. These centers can be interpreted as triplet excitations trapped in a potential well created by doping.

## 1. Introduction

Spin-gap magnets are one of the focus topics of modern magnetism [1]. Due to a particular architecture of the exchange bonds, which frequently invokes dimer motives, these magnets remains in the disordered singlet ground state to the lowest temperatures despite strong antiferromagnetic interspin interaction. The stability of the singlet ground state versus conventional forms of (antiferro)magnetic order is granted by the energy gap of exchange origin separating singlet ground state from the excited triplet excitations. Quasi-two dimensional spin-gap magnet $(C_4H_{12}N_2)(Cu_2Cl_6)$ (also called PHCC, or piperazinium hexachlordicuprate) is one of the test compounds of this sort [2].

The magnetic subsystem of PHCC can be envisioned as layers of strongly coupled dimers, the inelastic neutron scattering experiments suggest that 6 to 8 relevant exchange bonds are present [2, 3]. All these exchange couplings are indirect superexchange pathes mediated by halogen ions. Thus, partial substitution of chlorine by bromine does not deplete magnetic subsystem of PHCC but, instead, modifies exchange bonds mediated by the substituted ion. The samples of $(C_4H_{12}N_2)(Cu_2Cl_{6(1-y)}Br_{6y})$ with nominal bromine content $x$ up to 12% can be grown. We will use nominal concentration for the sample identification as was used in previous works. Detailed structural analysis [4] shows that the actual bromine occupancy is linear with nominal one but slightly differs for different inequivalent positions, the bromine substituted PHCC has the same lattice symmetry for $x < 12\%$. Inelastic neutron scattering demonstrates [5] that excitation spectrum remains gapped and the energy gap increases with doping.

Electron spin resonance (ESR) is a sensitive technique to study low energy dynamics of magnetic systems. It allows to study fine details of the excitations spectrum with the routine resolution of less then 1 GHz (which corresponds to the energy of 5$\mu$eV) thus complementing powerful inelastic neutron scattering method. Since the ESR absorption spectra is recorded as a field scan, it becomes possible to separate contributions of different magnetic systems which is impossible to extract from the bulk magnetization measurements. Intensity of the ESR absorption is directly proportional to the static susceptibility and to the amount of spins. The shape of ESR absorption spectrum allows to definitively distinguish paramagnetic centers with different spin.

In the present report we will briefly discuss ESR of triplet excitations and impurity-induced multispin defects in pure and bromine-substituted PHCC. We will demonstrate that fine structure of the energy spectrum of collective excitations of the spin-gap magnet can be accessed by means of ESR and that bond-doping results in the formation of unusual multispin objects.

## 2. ESR of collective triplet excitations in the nominally pure sample

ESR on nominally pure PHCC was described in details in [6]. Representative absorption spectra are shown in figure 1. As expected for a spin-gap system, at temperatures below the energy scale of the gap in the excitation spectrum (in the case of PHCC the gap $\Delta = 1.02$meV or 12 K) the population of magnetic excitations decreases and the ESR absorption loses intensity.

On cooling below approximately 4K ESR absorption line splits in several subcomponents, which can be identified as a multicomponent signal from the triplet excitations (components "A", "B" and "C" in figure 1) and a signal from the residual paramagnetic defects (denoted "PM" in figure 1). Absorption from the triplet excitations continues to decrease on cooling and vanishes below 2K, while the signal from residual defects follows Curie law. The amount of residual paramagnetic centers was estimated from static magnetization measurements as less than 0.04% per copper ion.

The multicomponent absorption signal of triplet excitations is, in fact, a typical absorption spectrum of the $S = 1$ spin in an effective crystal field — it is due to the fine structure of the excitation spectrum because of anisotropic spin-spin interactions. The effective crystal field affecting triplet excitations can be described by anisotropic contribution to the triplet excitations' Hamiltonian:

$$\mathcal{H}_{anis} = D\hat{S}_z^2 + E(\hat{S}_x^2 - \hat{S}_y^2) \qquad (1)$$

here $x, y, z$ is a cartesian basis for anisotropic Hamiltonian, which is not directly related to the crystallographic frame of reference because of low (triclinic) symmetry of the crystal.

Joint analysis of the angular dependencies of split ESR absorption at T=3K [6] allowed to determine the magnitude of anisotropy constants $D = (-7.90 \pm 0.28)$GHz (-33$\mu$eV) and $E = (-1.34 \pm 0.19)$GHz (-5.6$\mu$eV) as well as orientation of anisotropy axes with respect to the crystallographic frame of reference. This means that the energies of the triplet sublevels at zero field are $\Delta$ and $\Delta + D \pm E$, here the gap $\Delta = 1.02$meV.

## 3. Low-temperature ESR of multispin paramagnetic centers in the bond-diluted PHCC

Investigation of the ESR in bond-diluted PHCC was reported in details in [7, 8]. Examples of ESR absorption spectra in the nominally 10% doped sample are shown in figure.1. The main difference from the pure sample is that the ESR absorption is observed at low temperatures down to 450 mK. This low-temperature ESR signal again can be separated into some sort of

IOP Conf. Series: Journal of Physics: Conf. Series **969** (2018) 012104 doi:10.1088/1742-6596/969/1/012104

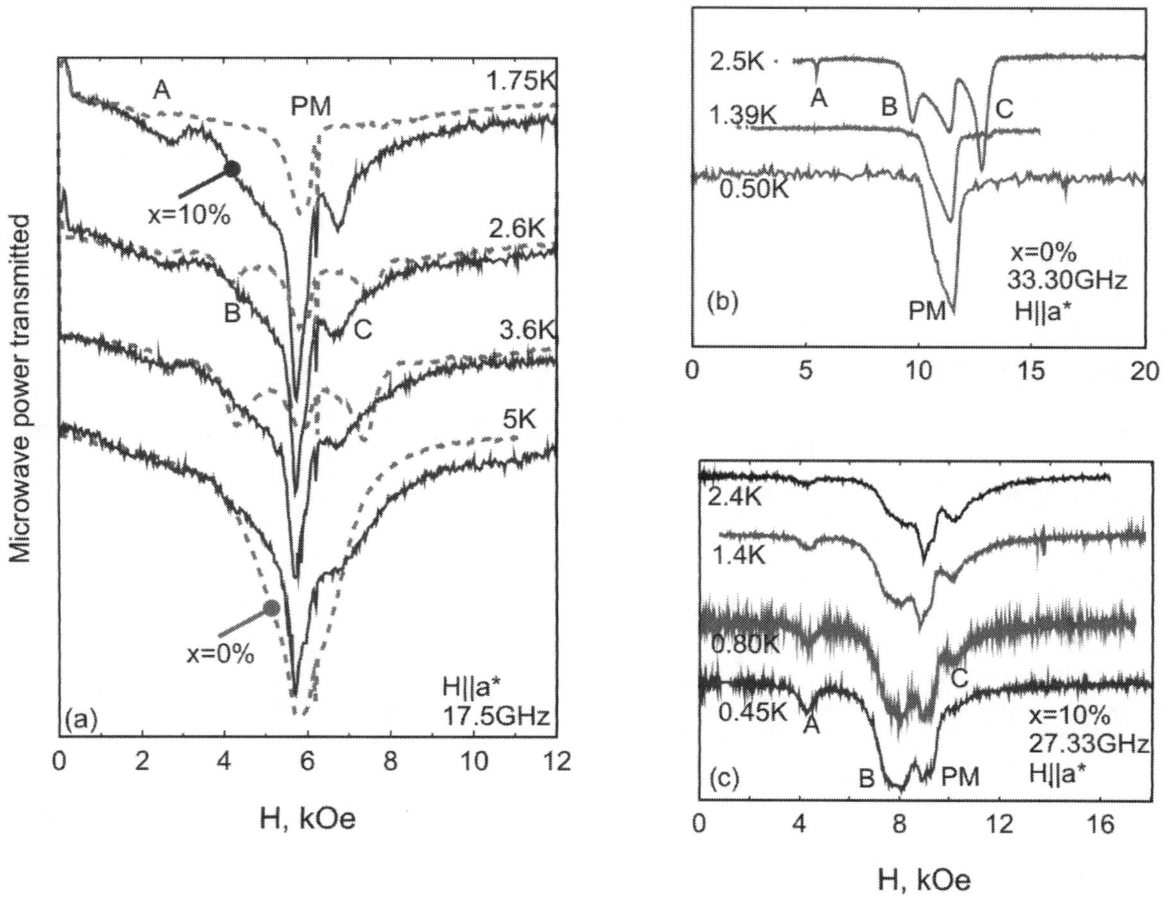


**Figure 1.** ESR absorption spectra in pure and nominally 10% bromine-diluted PHCC. Absorption components are marked as "A", "B", "C", signal from the residual defects and uncontrolled impurities is denoted as "PM". (a) Comparison of the absorption spectra at selected temperatures. Dashed lines — pure PHCC, solid lines - 10% Br-diluted sample. Data for x=0% and x=10% are scaled to yield the same integrated intensity at 5K. (b) and (c) Low temperature absorption spectra in pure and nominally 10% bromine-diluted samples.

irregularly shaped signal with $g \approx 2$ which is most likely due to some paramagnetic defects and three-subcomponents absorption which is typical for the $S = 1$ objects.

This low-temperature signal from $S = 1$ objects differs from that of triplet excitations in pure PHCC in certain ways. First, its temperature evolution indicates absence of the energy gap (or, at least, the gap is below 400 mK or 50$\mu$eV). Second, one can see that at low temperatures *left* component "B" dominates, while for the pure sample at 2.5—4 K *right* component "C" was more intense. Finally, splitting between "B" and "C" components is slightly different in comparison with pure PHCC. This means that the low-temperature absorption observed in doped samples is not due to delocalized triplet excitations. Note also that these $S = 1$ centers are formed under nonmagnetic bond-dilution in the system of strongly coupled $S = 1/2$ spins which has a non-magnetic (singlet) ground state.

Intensity of the ESR absorption can be scaled to static susceptibility measured on the same samples above 2K which allows to derive the amount of these new objects from the ESR data

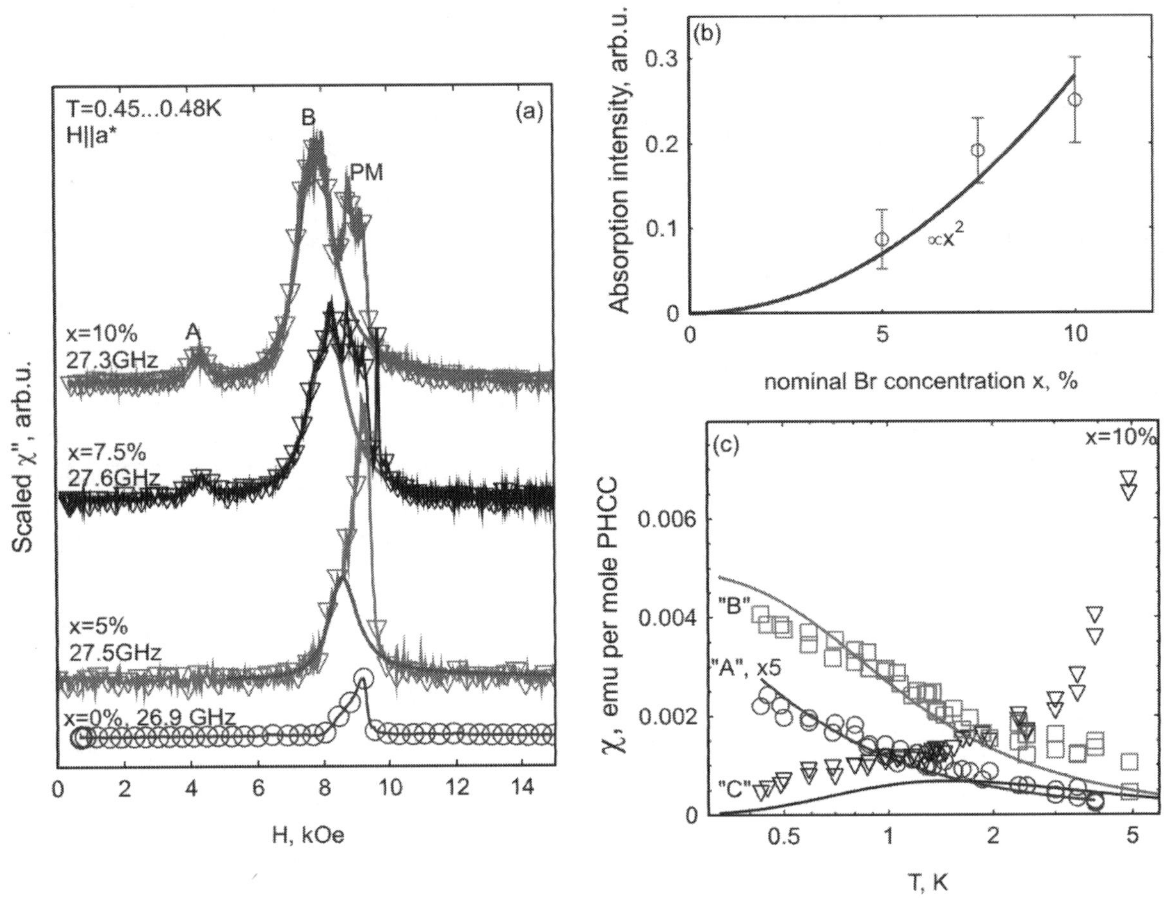


**Figure 2.** (a) Imaginary part of the high-frequency magnetic susceptibility $\chi''$ at 0.45K for the samples with different bromine content scaled to the same units. Absorption components due to the formed $S = 1$ objects are marked as "A" and "B", absorption component due to the uncontrolled defect is marked as "PM". Solid lines for $x \neq 0$ shows lorentzian fits of the "B" component used for intensity analysis on panel (b). (b) Concentration dependency of the "B" component intensity for the bromine-substituted PHCC at 0.45K. (c) Temperature dependency of the intensity of ESR absorption components for the 10% nominally bromine diluted sample. Solid lines — fits in the model of gapless $S = 1$ objects (see Ref.[7] for details).

(figure 2). We have found [7] that the concentration of $S = 1$ centers is $(0.4 \pm 0.1)\%$ per molecule of PHCC for the nominally 10% doped sample. As the building block of PHCC structure is a $Cu_2Cl_6$ dimer, we can conjecture that the $S = 1$ center is formed around the dimer with two bridging chlorines substituted by bromine. Structural refinement of Ref. [4] shows that actual bromine occupancy of interdimer halogen positions for the nominally 10% bromine-diluted sample is 7.5%, which yields probability of the double substitution about 0.56%.

To check this conjecture we have performed similar experiments for the samples with nominal bromine content of 5 and 7.5%: high-temperature (above 5K) absorption was scaled to the known static susceptibility which allowed to set the same scale to the absorption spectra measured on different samples (see figure.2). We observe increase of the $S = 1$ objects response with increasing concentration which is in agreement with $I \propto x^2$ dependence which follows from our "double substitution" conjecture. Note also that intensities of "PM" component in bromine-diluted samples do not strongly exceed that of pure sample, which is known to correspond to 0.04% of $S = 1/2$ defects per copper ion [6]. Thus, no additional paramagnetic centers in the amount comparable to the bromine concentration is formed.

The extreme scenario of formation of such a gapless $S = 1$ center is that the interdimer coupling turns ferromagnetic on double substitution. This is not utterly impossible as the bond angles there are close to 90° and as it is known that interdimer coupling in PHCC is the most susceptible to the hydrostatic pressure [9]. However, this effect can be re-interpreted in other way, which does not require extreme variation of coupling. Namely, bond doping changes local coupling constants, which is equivalent to formation of some potential relief for the propagating triplet excitations. If this potential happens to be attractive, then the potential well for the triplet excitation is formed. As the PHCC is two-dimensional magnet and 2D potential well always have a bound state, it can be possible that a low-lying bound state trapping one of the triplet excitations can be formed even without actual formation of ferromagnetic bond. The resulting $S = 1$ low-energy object is then a multispin defect spread in the certain vicinity of the attracting center.

**Acknowledgements**
Authors thank T.Yankova (ETH-Zurich and Moscow State University), G.Skoblin (Kapitza Institute) and J.Sichelschmidt (MPI CPFS, Dresden) for their involvement at the early stages of the project.

The work was supported by the Russian Foundation for Basic Research (Grant No.15-02-05918), Russian Scientific Foundation (Grant No. 17-12-01505) and by the Basic research Program of HSE. This work was partially supported by Swiss National Science Foundation, Division 2, the Estonian Ministry of Education and Research under Grant No. IUT23-03, the Estonian Research Council Grant No. PUT451, and the European Regional Development Fund project TK134.

## References

[1] Zapf V, Jaime M and Batista C D 2014 *Rev. Mod. Phys.* **86** 563
[2] Stone M B, Broholm C, Reich D H, Schiffer P, Tchernyshyov O, Vorderwisch P and Harrison N, New Journal of Physics **9**, 31 (2007)
[3] Stone M B, Zaliznyak I, Reich D H and Broholm C 2001 *Phys. Rev.* B **64** 144405
[4] Hüvonen D, Ballon G and Zheludev A 2013 *Phys. Rev.* B **88** 094402
[5] Hüvonen D, Zhao S, Månsson M, Yankova T, Ressouche E, Niedermayer C, Laver M, Gvasaliya S N and Zheludev A 2012 *Phys. Rev.* B **85** 100410(R)
[6] Glazkov V N, Yankova T S, Sichelschmidt J, Hüvonen D, Zheludev A 2012 *Phys. Rev.* B **85** 054415
[7] Glazkov V N, Skoblin G, Hüvonen D, Yankova T S, Zheludev A 2014 *J. Phys.: Condens. Matter* **26** 486002
[8] Glazkov V N, Krasnikova Yu V, Hüvonen D, Zheludev A. 2016 *J. Phys.: Condens. Matter* **28** 206003
[9] Perren G, Möller J S, Hüvonen D, Podlesnyak A A, and Zheludev A 2015 *Phys. Rev.* B **92** 054413

IOP Conf. Series: Journal of Physics: Conf. Series **969** (2018) 012105 doi:10.1088/1742-6596/969/1/012105

# Metal–insulator transition in a transition metal dichalcogenide: Dependence on metal contacts

**Y Shimazu[1], K Arai and T Iwabuchi**

Department of Physics, Yokohama National University
79-5 Tokiwadai, Hodogaya, Yokohama 240-8501, Japan

[1]E-mail: yshimazu@ynu.ac.jp

**Abstract**. Transition metal dichalcogenides are promising layered materials for realizing novel nanoelectronic and nano-optoelectronic devices. Molybdenum disulfide ($MoS_2$), a typical transition metal dichalcogenide, has been extensively investigated due to the presence of a sizable band gap, which enables the use of $MoS_2$ as a channel material in field-effect transistors (FET). The gate-voltage-tunable metal–insulator transition and superconductivity using $MoS_2$ have been demonstrated in previous studies. These interesting phenomena can be considered as quantum phase transitions in two-dimensional systems. In this study, we observed that the transport properties of thin $MoS_2$ flakes in FET geometry significantly depend on metal contacts. On comparing Ti/Au with Al contacts, it was found that the threshold voltages for FET switching and metal–insulator transition were considerably lower for the device with Al contacts. This result indicated the significant influence of the Al contacts on the properties of $MoS_2$ devices.

## 1. Introduction

Molybdenum disulfide ($MoS_2$) is a promising semiconducting material owing to its ultrathin structure that enables scaling of the $MoS_2$ devices to less than several nanometers [1-4]. From the perspective of fundamental physics, the superconductivity [5,6] and metal–insulator transition (MIT) [7-9] observed in $MoS_2$ have been given much attention, because these may be considered as quantum phase transitions in two-dimensional systems. In previous studies, Au [7,9] and Ti/Au [6,8] (Ti is the bottom layer) have been widely used as the contact metals for electrodes. The dependence of the transport properties on the contact metal has rarely been investigated. Particularly, the influence of the metal contacts on MIT has not been reported. We studied the transport properties of multilayer $MoS_2$ flakes with Au and Al contacts in a field-effect transistor (FET) geometry from room temperature (290 K) to 13 K. The reports on the transport properties of transition metal dichalcogenides that are contacted by Al have been very limited [10-12]. We found that the temperature dependences of the sheet resistance, relevant to the MIT, are different for the devices with the two types of contacts. The MIT at low temperatures was observed in back-gated FETs with Al contacts.

[1] To whom any correspondence should be addressed.

## 2. Experimental

$MoS_2$ flakes were exfoliated repeatedly from a bulk crystal (SPI Supplies) using adhesive tape [13]. Subsequently, thin flakes were transferred to a Si substrate with a 270-nm-thick $SiO_2$ layer via a gel sheet [14]. The highly n-doped Si substrate was used as a back gate. The source and drain electrodes were fabricated using photolithography and electron-beam deposition of metals. In the devices with Au contacts (referred as Au-contacted devices), Ti(12 nm)/Au(75 nm) were used as contact metals, where Ti was the bottom layer. In the devices with Al contacts (referred as Al-contacted devices), Al (90 nm) was used as a contact metal. For the Al-contacted devices, we deposited Ti (15 nm)/Au (50 nm) layers on the Al pads for connection of Au wires to the contact pads, where the Ti/Au layers were away from the $MoS_2$ flakes by more than 350 μm. The photographs of the Al-contacted device are shown in figure 1. We measured the drain–source current $I_d$ from room temperature to 13 K as a function of drain–source voltage $V_d$ and back-gate voltage $V_g$.

(a) (b)

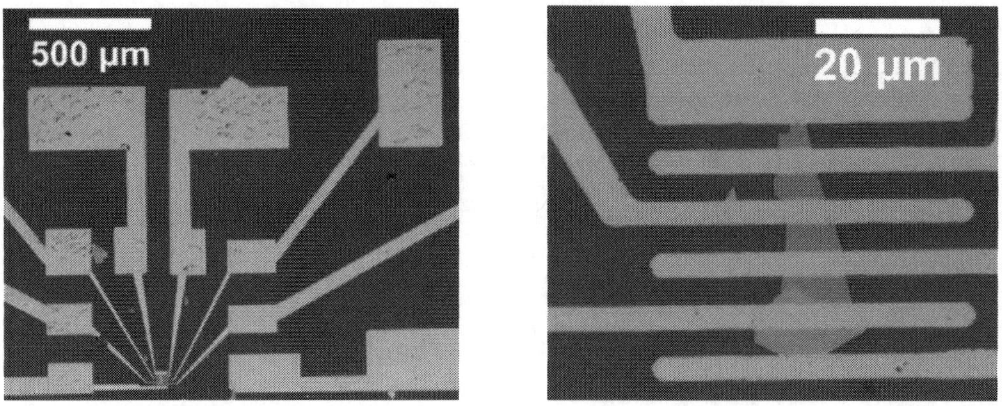


**Figure 1.** Photographs of the Al-contacted device with low magnification (a) and high magnification (b); in figure 1(a), Ti/Au layers (yellow) are deposited on Al layers (light yellow). The Al layers are in contact with an $MoS_2$ flake, which is located at the lower left of figure 1(a). The $MoS_2$ flake with the Al contacts is shown in figure 1(b). Four contacts were used to measure the sheet resistance in the middle of the flake (16 nm thick).

## 3. Results and discussion

The $I_d$–$V_d$ curves at room temperature for the Au-contacted and Al-contacted devices are shown in figures 2(a) and (b), respectively. The thicknesses of the $MoS_2$ flakes in these devices are 16 nm, as determined by an atomic force microscope. Both the devices exhibit n-channel FET characteristics; at low values of $V_g$, the off-state is realized. As shown in the figures, the threshold value of $V_g$ for the FET switching is lower in the Al-contacted device than that in the Au-contacted device. In the on-state, the $I_d$–$V_d$ curve is nearly linear for the Au-contacted device, while that for the Al-contacted device is nonlinear. The nonlinearity reflects the significance of Schottky barrier between the Al contacts and $MoS_2$ flakes. This result implies that the Schottky barrier associated with Al contacts is higher and/or thicker than that associated with Au contacts. This observation is inconsistent with that indicated by the work functions of Al (4.54 eV) and Au (5.40 eV) [15]; from the Schottky-Mott rule for the Schottky barrier height, the barrier height for Al should be lower than that for Au [10]. Existence of high-resistance Schottky barrier for Al contact has been reported in the literature [11] and is attributed to the lack of d-orbitals in Al, which leads to a small overlap of the electron orbitals with the d-orbitals in $MoS_2$ [11].

In contrast to the $I_d$–$V_d$ curve, the four-terminal current–voltage curves were nearly linear for both types of the devices from room temperature to 13 K. The sheet resistance is derived from the four-terminal measurements. The sheet resistances $R_s$ (at various values of $V_g$) as functions of temperature for the Au-contacted and Al-contacted devices are shown in figures 3(a) and (b), respectively. The temperature dependence of $R_s$ indicates a qualitative difference between the two types of devices. For the Au-contacted device, the temperature coefficient of $R_s$ at room temperature, $dR_s/dT$, is negative for $V_g < 27$ V and positive for $V_g > 27$ V. However, the insulating behavior ($dR_s/dT < 0$) is observed at lower temperatures for all the values of $V_g$. In contrast, for the Al-contacted device, the temperature dependence at lower temperatures (down to 13 K) shows a MIT with critical point at $V_g \sim -10$ V, while that near room temperature shows a metallic behavior ($dR_s/dT > 0$).

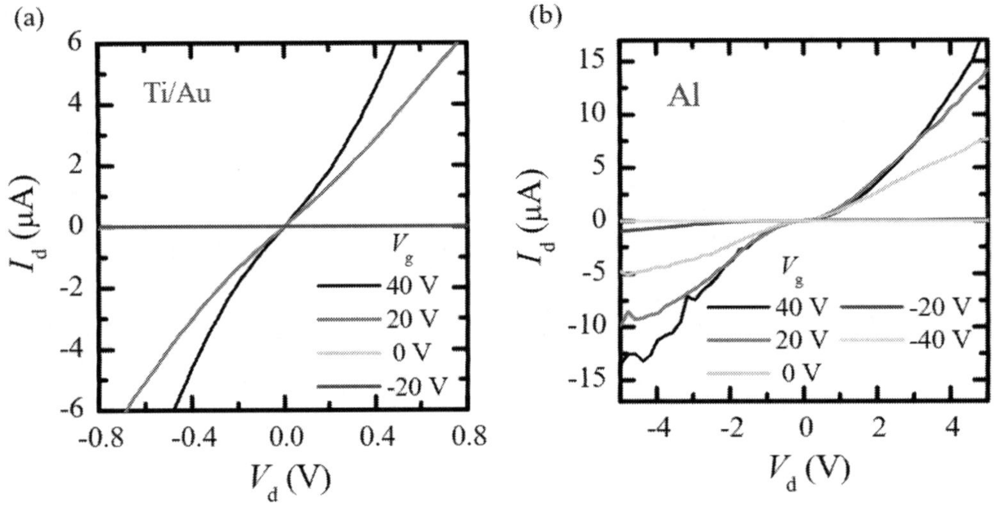


**Figure 2.** $I_d$–$V_g$ curves of $MoS_2$ FETs with Ti/Au contacts (a) and Al contacts (b) measured at room temperature for $V_g$ = 40, 20, 0, –20, and –40 V.

The difference, as shown in figures 3(a) and (b), can be explained by assuming that the carrier density in the Al-contacted device is higher than that in the Au-contacted device. This assumption is consistent with the difference in the threshold value of $V_g$ for FET switching in the devices with different contacts. The higher carrier density in the Al-contacted device may possibly be caused by electron transfer from Al to $MoS_2$, which is expected on the basis of the work functions of Al (4.54 eV) and Au (5.40 eV). For the Al-contacted devices, the carrier density in $MoS_2$ at equilibrium is influenced by the work function of Al, and not by the significant Schottky barrier at the Al contacts. We found that the presence of Ti/Au layer on the Al contact pads has a significant influence on the transport properties of the devices. Further study is required to elucidate the effects of Al contacts on the devices.

MITs in $MoS_2$ have been reported in the literature. In [7], MIT in monolayer $MoS_2$ is reported in a top-gated device using $HfO_2$ as a top-gate dielectric. Only insulating behavior is observed for back-gate voltages up to 40 V in this device. In [8,9], MITs in back-gated monolayer $MoS_2$ are reported, wherein the thicknesses of the $SiO_2$ dielectric are similar to that in our devices. For multilayer $MoS_2$ flakes, MIT and field-induced superconductivity are reported in [5,6], where significant electrostatic carrier doping is achieved using ionic liquids. In [5-9], the critical resistance of MIT is observed to be close to the quantum resistance $h/e^2$ (~25.8 kΩ) with an exception of [5] in which the critical sheet resistance is observed to be between 100 Ω and 1 kΩ. The quantum resistance is the most relevant unit of resistance in the analysis of MIT in two-dimensional electron systems [16]. In our results (shown in figures 3(a) and (b)), the critical resistance of MIT in the Al-contacted device is ~10 kΩ, and that of the MIT-like behavior near room temperature in the Au-contacted device is ~60 kΩ. Therefore, the critical resistance (~10 kΩ) of MIT observed in our study is near the quantum resistance, which is consistent with other reports. The critical resistance being close to the quantum resistance is expected from the Ioffe–Regel criterion for MIT, $k_F l_e \sim 1$, where $k_F$ and $l_e$ are the Fermi wave vector and mean free path of electrons, respectively. It has been revealed that this criterion is consistent with experimental data of MIT [7].

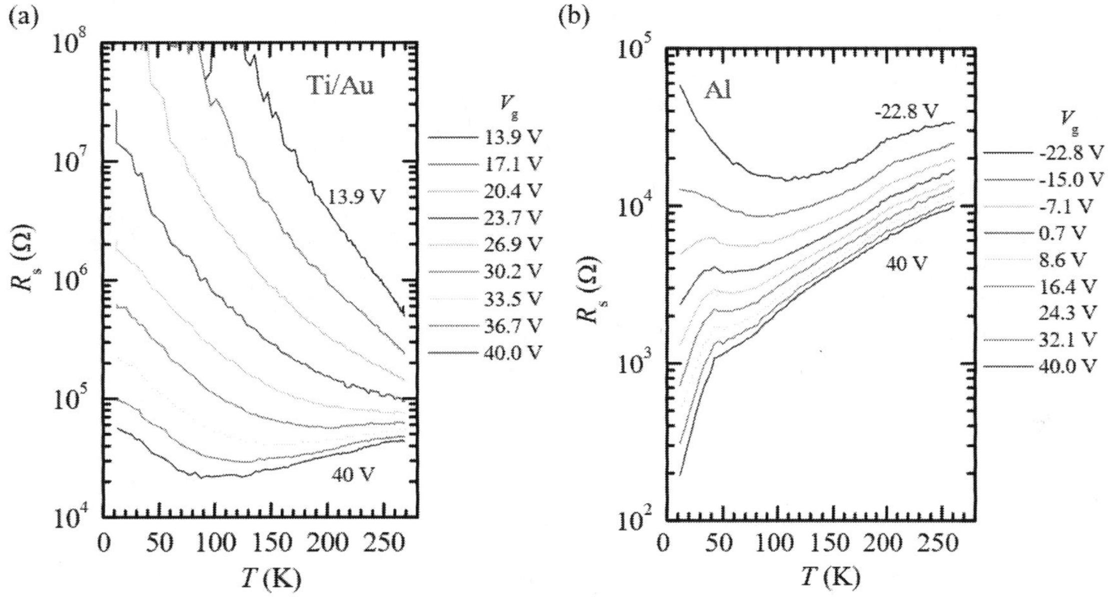


**Figure 3.** Sheet resistance of $MoS_2$ FETs with Ti/Au contacts (a) and Al contacts (b) as a function of temperature for various values of $V_g$; the resistance decreases with increasing $V_g$ at all the temperatures. MIT is observed at low temperatures for the Al-contacted device.

## 4. Conclusion

We examined the transport properties of multilayer $MoS_2$ flakes with Au and Al contacts in a back-gated FET geometry from room temperature to 13 K. For the Al-contacted devices, MIT was observed at lower temperatures (down to 13 K), with a critical resistance close to the quantum resistance, while the Au-contacted devices were insulating in the same range of temperature and for $V_g$ up to 40 V. This result implies that the MIT depends on the metal contacts and that the Al contacts induce higher carrier density in $MoS_2$ flakes.

**Acknowledgments**
This work was supported by JSPS KAKENHI Grant Number 15K13497.

**References**

[1] Wang Q H, Kalantar-Zadeh K, Kis A, Coleman J N and Strano M S 2012 *Nat. Nanotechnol.* **7** 699–712
[2] Ganatra R and Zhang Q 2014 *ACS Nano* **8** 4074–4099
[3] Tong X, Ashalley E, Lin F, Li H and Wang Z M 2015 *Nano-Micro Letters* **7** 203–218
[4] Wang H, Yuan H, Hong S S, Li Y and Cui Y 2015 *Chem. Soc. Rev.* **44** 2664–2680
[5] Taniguchi K, Matsumoto A, Shimotani H and Takagi H 2012 *Appl. Phys. Lett.* **101** 042603
[6] Ye J T, Zhang Y J, Akashi R, Bahramy M S, Arita R and Iwasa Y 2012 *Science* **338** 1193–1196
[7] Radisavljevic B and Kis A 2013 *Nat. Mater.* **12** 815–820
[8] Baugher B W H, Churchill H O H, Yang Y and Jarillo-Herrero P 2013 *Nano Lett.* **13** 4212–4216
[9] Schmidt H, Wang S, Chu L, M Toh, Kumar R, Zhao W, Castro Neto A H, Martin J, Adam S, Ozyilmaz B and Eda G 2014 *Nano Lett.* **14** 1909–1913
[10] Walia, S. *et al.* 2013 *Appl. Phys. Lett.* **103** 232105
[11] Liu W, Kang J, Sarkar D, Khatami Y, Jena D and Banerjee K 2013 *Nano Lett.* **13** 1983–1990
[12] Kwon J, Lee J Y, Yu Y J, Lee C H, Cui X, Hone J and Lee G H 2017 *Nanoscale* **9** 6151–6157
[13] Novoselov K S, Geim A K, Morozov S V, Jiang D, Zhang Y, Dubonos S V, Grigorieva I V and Firsov A A 2004 *Science* **306** 666–669
[14] Castellanos-Gomez A, Buscema M, Molenaar R, Singh V, Janssen L, van der Zant H S J and Steele G A 2014 *2D Materials* **1** 011002
[15] Skriver H L and Rosengaard N M 1992 *Phys. Rev.* B **46** 7157–7168
[16] Kravchenko S V and Sarachik M P 2004 *Rep. Prog. Phys.* **67** 1–44

IOP Conf. Series: Journal of Physics: Conf. Series **969** (2018) 012106 doi:10.1088/1742-6596/969/1/012106

# Tight-binding calculation of the magnetic moment of CrAs under pressure

**Carmine Autieri[1], Giuseppe Cuono[2], Filomena Forte[2,3] and Canio Noce[2,3]**

[1] Consiglio Nazionale delle Ricerche CNR-SPIN, UOS L'Aquila, Sede Temporanea di Chieti, I-66100 Chieti, Italy
[2]Dipartimento di Fisica "E. R. Caianiello", Università degli Studi di Salerno, I-84084 Fisciano (Salerno), Italy
[3] Consiglio Nazionale delle Ricerche CNR-SPIN, UOS Salerno, I-84084 Fisciano (Salerno), Italy

E-mail: gcuono@unisa.it

**Abstract.** We analyze the evolution of the local magnetic moment of the newly discovered pressure-induced superconductor CrAs, as a function of the applied pressure. Our theoretical method is based on a combination of the tight-binding approximation and the Löwdin down-folding procedure, which enables us to derive a low-energy effective Hamiltonian projected onto the Cr-subsector. We set up our calculations by considering several sets of *ab initio* derived hopping parameters, corresponding to different volumes of the unit cell, and use them to obtain the simulated pressure-dependence of the Cr magnetic moment, which is evaluated within a mean-field treatment of the Coulomb repulsion between the electrons at the Cr sites. Our calculations show good agreement with available experimental data, both for the normal phase measured 1.7 $\mu_B$ for Cr magnetic moment, and concerning the observed reduction of its amplitude for values that exceed the characteristic critical pressure.

## 1. Introduction

Recently, superconductivity was discovered in the Cr-based compound CrAs under pressure [1, 2]. Similarly to other superconducting materials, such as heavy-fermion compounds [3], high transition-temperature cuprate superconductors [4], strontium ruthenate superconductors [5] and iron-pnictide superconductors [6], the superconductivity in the CrAs appears near a magnetic quantum critical point. Wu *et al.* reported the appearance of superconductivity at $T_c \sim 2$ K for a critical pressure $P_c \sim 8$ Kbar, when the antiferromagnetic transition, that sits at $T_N \sim 265$ K under ambient pressure, is totally suppressed [1, 7]. According to Kotegawa *et al.*, the superconducting transition temperature has its maximum at about 1 GPa, and decreases at higher pressure, showing a well-known dome-shaped phase diagram [1, 2, 8, 9, 10, 11, 12, 13, 14, 15]. Furthermore, nuclear quadrupole resonance under pressure studies indicate a slow decrease, with a large value close to $P_c$, of the internal field in the magnetic state when the pressure increases. This indicates that the pressure-induced suppression of the magnetic order is a first order phase transition [8].

All these properties point towards a possible unconventional mechanism of formation of Cooper pairs, likely due to the interaction with magnetic fluctuations in the material near

IOP Conf. Series: Journal of Physics: Conf. Series **969** (2018) 012106 doi:10.1088/1742-6596/969/1/012106

the quantum critical point [1]. However, measurements of muon spin rotation on CrAs powder samples reveal that there is a coexistence region, in the intermediate pressure regime, where the magnetic and the superconducting phases are well separated, and this, together with the observation that the superfluid density $\rho_s$ scales with the critical temperature as $\sim T_c^{3.2}$, seems to bring towards a conventional mechanism of pairing in CrAs [16].

Neutron diffraction measurements [17, 18] establish that CrAs is a non collinear helimagnet, with a substantial Cr moment of 1.7 $\mu_B$ at ambient pressure that lies essentially within the *ab* plane, the propagation vector being directed along the *c* axis [18]. The magnetic structure is formed by four spirals along the *c* axis, one for every Cr-atom in the primitive cell, with a well-defined phase angle between the spirals. Measurements of the magnetic properties versus the pressure indicate that the average ordered magnetic moment decreases from 1.7 $\mu_B$ at ambient pressure to 0.4 $\mu_B$ at the critical pressure $P_c \sim 0.7$ GPa, where the magnetic order is totally suppressed [18]. Furthermore, Shen *et al.* showed that CrAs exhibits a spin reorientation in addition to an abrupt decrease of the magnetic propagation vector at the critical pressure. At this pressure the system moves from a configuration where the magnetic moments lie in the *ab* plane to one where they lie in the *ac* plane. This magnetic phase transition coincides with the emergence of bulk superconductivity. At higher pressure, CrAs approaches to the optimal superconductivity regime, where the magnetic moments tend to be aligned antiparallel between nearest neighbor ions [17].

Moreover, the helimagnetic order is correlated to structural and electronic degrees of freedom [17]. At $T_N \sim 265$ K, in addition to the magnetic transition, there are discontinuous changes of the lattice parameters [1], suggesting that electronic, magnetic and structural properties are correlated as in many transition metal compounds [19, 20, 21, 22].

As far as the theoretical results are concerned, *ab initio* calculations based on first-principles density functional theory strengthen the experimental trend that the magnetic and electronic properties strongly depend on the size of the unit cell [23]. Furthermore, the reasonable agreement for the magnetic moment obtained within this approach and the fact that the electric resistivity of CrAs doesn't seem to exceed the Mott-Ioffe-Regel limit [24] suggest that the CrAs could be considered as a weakly correlated material [23]. Indeed the Cr atoms in the CrAs compound can be considered closer to the metallic Cr than the strongly correlated ionic Cr compounds [25].

Besides, a method that combines the tight-binding approximation and the Löwdin down-folding procedure [26, 27] has been adopted to study some properties of CrAs [28]. Within this approach, the parameters entering the tight-binding matrix are the *ab initio* derived overlap integrals for orbitals, whereas the Löwdin technique allows us to integrate out all low-lying arsenic degrees of freedom and to get an Hamiltonian whose rank is definitely lower than that of the starting tight-binding matrix.

Taking advantage from this procedure, here we will focus on the magnetic properties of CrAs, investigating the behaviour of the Cr magnetic moment as a function of the variations of the volume of the primitive cell, as induced by applied external pressure. The paper is organized as follows: in the next section we will introduce the model Hamiltonian adopted and we will introduce the Löwdin down-folding procedure; Sec. III will be devoted to a discussion of the obtained results and to some remarks.

## 2. Model Hamiltonian and Löwdin down-folding procedure

CrAs exhibits either an hexagonal NiAs-type (B81) structure or an orthorhombic MnP-type (B31) structure. At ambient pressure, the system presents an orthorhombic MnP-type (B31) structure, and the unit-cell lattice parameters are $a$=5.649 Å, $b$=3.463 Å and $c$=6.2084 Å [1]. The primitive cell of the CrAs is reported in Fig. 1, and it consists of four Cr atoms and four As atoms. To study the magnetic properties of the material, and in particular the trend of the magnetization as a function of the external pressure, we start from the following tight-binding Hamiltonian:

$$H = \sum_{i,\sigma} \epsilon_i c^\dagger_{i\sigma} c_{i\sigma} - \sum_{i,j,\sigma} t_{ij} (c^\dagger_{i\sigma} c_{j\sigma} + h.c.) + U \sum_i n_{i\uparrow} n_{i\downarrow} \,. \tag{1}$$

This Hamiltonian is the sum of three contributions: the first one describes the on-site term, $\epsilon_i$ being the energy at each Cr or As lattice site; the second one takes into account the hopping processes of the electrons with spin $\sigma$ between the $i$ and $j$ sites of the Cr or As ions; finally the last term describes the Coulomb repulsion between electrons on the same Cr site having opposite spin.

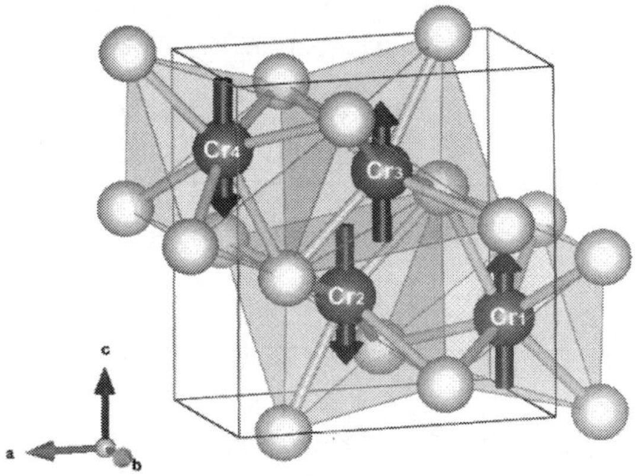


**Figure 1.** Primitive cell of the CrAs at ambient pressure. Cr and As atoms are shown as red and yellow spheres, respectively, while the arrows represent the spins of the Cr atoms in the magnetic ground state and in the collinear approximation.

Since the primitive cell contains four Cr atoms and four As atoms, and due to the fact that the orbitals entering the general Hamiltonian in Eq.(1) are the $d$ orbitals for Cr atoms and the $p$ orbitals for As atoms, the corresponding tight-binding matrix Hamiltonian is given by a $32 \times 32$ matrix. Nevertheless, since the As bands are located above and below 2 eV from the Fermi level [23], we use the Löwdin down-folding procedure to get the energy dispersion of the Cr $d$ bands only. Within this approach, the whole Hamiltonian is partitioned considering a $20 \times 20$ matrix that describes the $d$-$d$ hoppings among Cr ions, a $12 \times 12$ one describing the $p$-$p$ As ions hoppings, and two sub-matrices that correspond to the $d$-$p$ hoppings from Cr to As ions and vice-versa. The energy spectrum of the 3$d$ bands can be then obtained by integrating out all low-lying As 4$p$ degrees of freedom and diagonalizing the renormalized $20 \times 20$ Hamiltonian projected onto the Cr orbitals.

We point out that we have considered the nearest neighbour hoppings, the second nearest neighbour hoppings along the $x$-direction and the diagonal part of the second nearest neighbour hoppings along the $y$ and $z$ direction, confining the analysis to the parameters whose values range from 80 meV to 1 eV.

Since CrAs may be considered as a weakly correlated material, we can treat the Coulomb term in the mean-field approximation. Thus, the Hamiltonian of Eq.(1) becomes

$$H = -\sum_{i,j,\sigma} \tilde{t}_{ij} (\tilde{c}^\dagger_{i\sigma} \tilde{c}_{j\sigma} + h.c.) + U \sum_i (\tilde{n}_{i\uparrow} \langle \tilde{n}_{i\downarrow} \rangle + \tilde{n}_{i\downarrow} \langle \tilde{n}_{i\uparrow} \rangle - \langle \tilde{n}_{i\uparrow} \rangle \langle \tilde{n}_{i\downarrow} \rangle) \,, \tag{2}$$

where $\tilde{t}_{ij}$ are the renormalized hopping amplitudes obtained after the application of the Löwdin procedure, $\tilde{c}^\dagger_{i\sigma}$ and $\tilde{c}_{i\sigma}$ refer to the creation and annihilation operators of renormalized $d$ Cr

electrons, $\langle \tilde{n}_{i\uparrow} \rangle$ and $\langle \tilde{n}_{i\downarrow} \rangle$ are the average, in the ground state, of the electron number operator for spin up and down at the $i$ lattice site, respectively.
To obtain the magnetization in the G-type configuration emerging from *ab initio* calculations [23], we performed a self-consistent procedure according to the following scheme: we assign initial conditions for $\langle \tilde{n}_{i\uparrow} \rangle$ and $\langle \tilde{n}_{i\downarrow} \rangle$, and use them to evaluate improved expectation values, this procedure running until convergence with the requested accuracy is achieved.

## 3. Results and discussion

To simulate the effect of the external pressure acting on the system, we have performed our calculations at different values of the lattice constants, corresponding to different values of the volume of the primitive cell. This procedure is corroborated by the experimental data in Ref. [17] where it is reported the magnetic moment of the CrAs as a function of the pressure as well as of the lattice constants. From these data we have been able to plot in Fig. 2 the experimental pressure dependence of the volume of the primitive cell. We use these experimental results as input for our calculation.

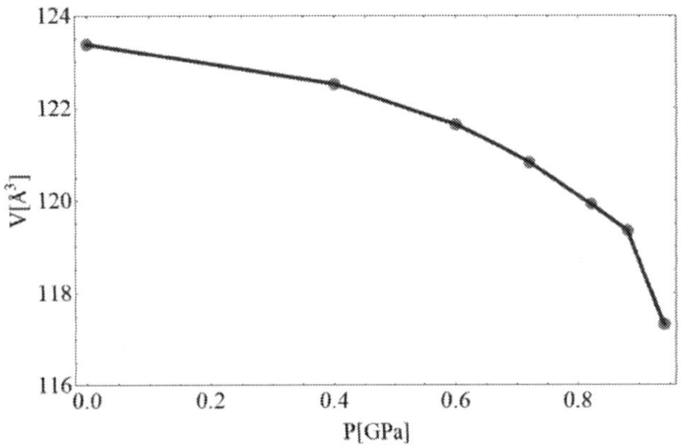


**Figure 2.** Volume of the primitive cell of the CrAs as a function of the external pressure acting on the system. The values are those reported in Ref. [17].

Within our scheme, to obtain the magnetic moment as a function of the external pressure, we choose the corresponding lattice parameters and perform the above mentioned self-consistent method. According to *ab initio* calculations, the value of the Coulomb interaction has been fixed at $U$=0.37 eV. The results of our simulations are reported in Fig. 3, together with the available experimental data [17] and the outcomes of the *ab initio* calculation [23].

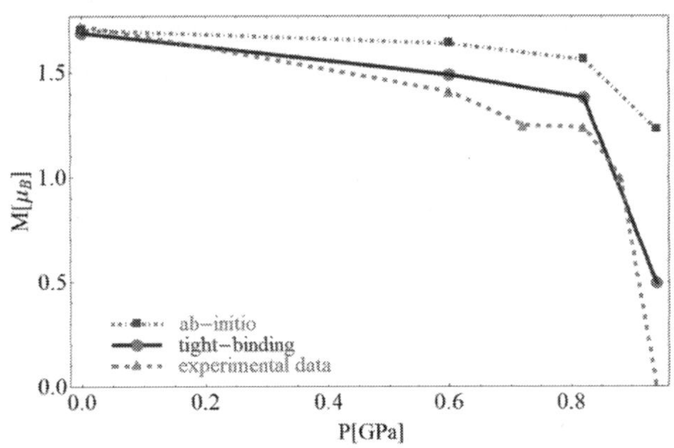


**Figure 3.** Magnetization of CrAs as a function of the external pressure. The blue line indicates the theoretical results, the brown dot-dashed line comes from the *ab initio* simulation [23], while the red dashed line is obtained from the experimental data in Ref. [17].

From this figure we notice that a general overall agreement with the experimental data is obtained as far as it concerns both the amplitude of the Cr magnetic moment and its reduction

IOP Conf. Series: Journal of Physics: Conf. Series **969** (2018) 012106 doi:10.1088/1742-6596/969/1/012106

upon applied pressure. According to the results of Refs. [23, 28], we infer that the magnetic moment reduction could be linked to the increase of the Cr-bandwidth. To clarify this point, we have plotted in Fig. 4 the Cr-bandwidth W as a function of the pressure for some values of the Coulomb repulsion, namely $U$=0.35, 0.37 and 0.39 eV. There are two trends that can be distinguished in Fig. 4. Below $P$=0.8 GPa the bandwidth smoothly decreases with increasing $P$, with a maximum percentage variation of 3% at 0.6 GPa. In correspondence, the evolution of the magnetization, as shown in Fig. 3, is almost constant. For external pressures that exceed $P$=0.8 GPa, the bandwidth displays a sudden increase corresponding to the dramatic reduction of the local magnetic moment, for all the values of $U$ investigated. We also notice that the minimum of the Cr-bandwidth at $P$=0.6 GPa coincides with the maximum strength of the antiferromagnetic order [23]. Nevertheless, we mention that other quantities may control the magnetic properties of CrAs, such as the bandwidth of the single bands crossing the Fermi level or the spin-orbit coupling. For completeness, we notice that in Refs. [7, 17] it is mentioned that the magnetic moments of the Cr atoms with shorter distance are antiparallel, indicating a strong antiferromagnetic coupling between these Cr atoms, in agreement with the results here presented. We finally point out that the local magnetic moment, evaluated via the mean-field treatment of the Coulomb repulsion within tight binding method, shows even a better agreement with the experiments, compared with the LDA calculations, as shown in Fig. 3 [23].

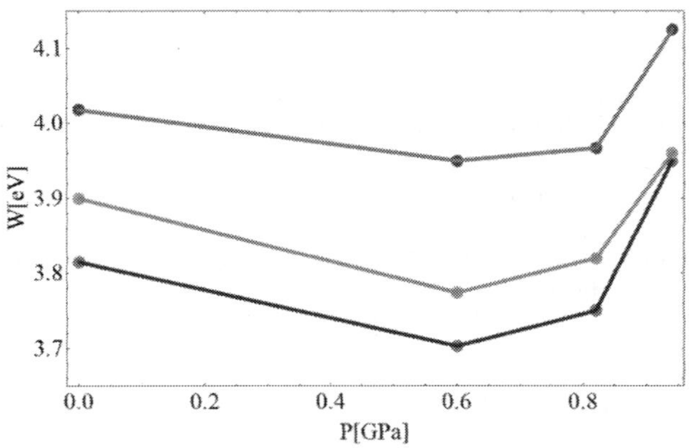


**Figure 4.** Bandwidth W of CrAs as function of the pressure for $U$=0.35 eV (blue line), 0.37 eV (green line) and 0.39 eV (red line).

It is worth stressing that even if *ab initio* calculations are available, they are not delivered in a form useful as the single-particle term of a model Hamiltonian that describes the low-energy excitations, whereas obtaining an effective Cr-bands only model is highly non-trivial [29]. Here, we have neglected the nontrivial details of the LDA band structure and have used the simplest possible tight-binding model with hopping integrals between only nearest and next-nearest neighbors. Hence, within the Löwdin procedure we have reduced the full tight-binding for As and Cr ions to Cr $d$ bands only, deriving manageable expressions for $d$-band dispersions near the Fermi level. This result turns out to be readily used in calculating physical quantities depending on the explicit form of the energy spectrum, as for instance the in-plane resistivity [28]. Furthermore, we are confident that this approach could be used as the single-particle term in more sophisticated Hamiltonians containing for instance Coulomb correlations, such as the Hund coupling, as well as to search for the superconducting instability within the standard broken-symmetry Hartree-Fock scheme [30, 31].

Finally, the overall agreement between the electric resistivity obtained using the procedure before specified and the experimental data as well as the Cr magnetic moment here discussed suggests that this material is a weakly correlated compound whose magnetic and electric properties are suitably described by a tight-binding approach. Nevertheless, to get the helimagnetic configuration exhibited by CrAs the As spin-orbit coupling has to be included.

IOP Conf. Series: Journal of Physics: Conf. Series **969** (2018) 012106 doi:10.1088/1742-6596/969/1/012106

Moreover, the introduction of the As spin-orbit coupling can generate an effective Dzyaloshinskii Moriya interaction between the Cr atoms [32]. Calculations in this direction are in progress and will be presented elsewhere.

## References

[1] Wu W, Cheng J, Matsubayashi K, Kong P, Lin F, Jin C, Wang N, Uwatoko Y and Luo J 2014 *Nat. Commun.* **5** 5508
[2] Kotegawa H, Nakahara S, Tou H and Sugawara H 2014 *J. Phys. Soc. Japan* **83** 093702
[3] Movshovich R, Jaime M, Thompson J D, Petrovic C, Fisk Z, Pagliuso P G and Sarrao J L 2001 *Phys. Rev. Lett.* **86** 5152
[4] Van Harlingen D J 1995 *Rev. Mod. Phys.* **67** 515
[5] Mackenzie A P and Maeno Y 2003 *Rev. Mod. Phys.* **75** 657
[6] Mazin I I, Singh D J, Johannes M D and Du M H 2008 *Phys. Rev. Lett.* **101** 057003
[7] Wu W, Zhang X, Yin Z, Zheng P, Wang N and Luo J 2010 *Science China Physics, Mechanics & Astronomy* **53** 1207
[8] Kotegawa H, Nakahara S, Akamatsu R, Tou H, Sugawara H and Harima H 2015 *Phys. Rev. Lett.* **114** 117002
[9] Varma C M 1999 *Phys. Rev. Lett.* **83** 3538
[10] Noce C, Busiello G and Cuoco M 2000 *Europhys. Lett.* **51** 195
[11] Noce C, Vecchione A, Cuoco M and Romano A (ed) 2002 *Ruthenate and Rutheno-Cuprate Materials Unconventional Superconductivity, Magnetism and Quantum Phase Transitions* (Berlin: Springer)
[12] Van der Marell D, Molegraaf H J A, Zaanen J, Nussinov Z, Carbone F, Damascelli A, Eisaki H, Greven M, Kes P H and Li M 2003 *Nature* **425** 271
[13] Jiang S, Xing H, Xuan G, Wang C, Ren Z, Feng C, Dai J, Xu Z and Cao G 2009 *J. Phys.: Condens. Matter* **21** 382203
[14] Shibauchi T, Carrington A and Matsuda Y 2014 *Annual Review of Condensed Matter Physics* **5** 113
[15] Seo S, Park E, Bauer E D, Ronning F, Kim J N, Shim J-H, Thompson J D and Park T 2015 *Nat. Commun.* **6** 6433
[16] Khasanov R, Guguchia Z, Eremin I, Luetkens H, Amato A, Biswas P K, Rüegg C, Susner M A, Sefat A S, Zhigadlo N D and Morenzoni E 2015 *Sci. Rep.* **5** 13788
[17] Shen Y, Wang Q, Hao Y, Pan B, Feng Y, Huang Q, Harriger L W, Leao J B, Zhao Y, Chisnell R M, Lynn J W, Cao H, Hu J and Zhao J 2016 *Phys. Rev.* B **93** 060503
[18] Keller L, White J S, Frontzek M, Babkevich P, Susner M A, Sims Z C, Sefat A S, Ronnow H M and Rüegg C 2015 *Phys. Rev.* B **91** 020409
[19] Cuoco M, Forte F and Noce C 2006 *Phys. Rev.* B **74** 195124
[20] Cuoco M, Forte F and Noce C 2006 *Phys. Rev.* B **73** 094428
[21] Cuoco M, Forte F and Noce C 2010 *Phys. Rev.* B **82** 155104
[22] Granata V, Capogna L, Forte F, Lepetit M-B, Fittipaldi R, Stunault A, Cuoco M and Vecchione A 2016 *Phys. Rev.* B **93** 115128
[23] Autieri C and Noce C 2017 *Phil. Mag.* DOI: 10.1080/14786435.2017.1375607
[24] Luo J 2017 private communication.
[25] Autieri C, Koch E and Pavarini E 2014 *Phys. Rev.* B **89** 155109
[26] Andersen O K, Liechtenstein A I, Jepsen O and Paulsen F 1995 *J. Phys. Chem. Solids* **56** 1573
[27] Noce C and Cuoco M 1999 *Phys. Rev.* B **59** 2659
[28] Autieri C, Cuono G, Forte F and Noce C 2017 *J. Phys.: Condens. Matter* **29** 224004
[29] Edelmann M, Sangiovanni G, Capone M and de' Medici L 2017 *Phys. Rev.* B **95** 205118
[30] Cuoco M, Romano A, Noce C and Gentile P 2008 *Phys. Rev.* B **78** 054503
[31] Autieri C, Cuoco M and Noce C 2012 *Phys. Rev.* B **85** 075126
[32] Fert A and Levy P M 1980 *Phys. Rev. Lett.* **44** 1538

IOP Conf. Series: Journal of Physics: Conf. Series **969** (2018) 012107 doi:10.1088/1742-6596/969/1/012107

# Surface topography and electrical properties in $Sr_2FeMoO_6$ films studied at cryogenic temperatures

**I. Angervo[1,2], M. Saloaro[1], J. Mäkelä[2,3], J. -P. Lehtiö[3], H. Huhtinen[1] and P. Paturi[1]**

[1] Wihuri Physical Laboratory, Department of Physics and Astronomy, FI-20014 University of Turku, Finland

[2] University of Turku Graduate School (UTUGS), University of Turku, FI-20014 Turku, Finland

[3] Materials Research Laboratory, Department of Physics and Astronomy, FI-20014 University of Turku, Finland

E-mail: ijange@utu.fi

**Abstract.** Pulsed laser deposited $Sr_2FeMoO_6$ thin films were investigated for the first time with scanning tunneling microscopy and spectroscopy. The results confirm atomic scale layer growth, with step-terrace structure corresponding to a single lattice cell scale. The spectroscopy research reveals a distribution of local electrical properties linked to structural deformation in the initial thin film layers at the film substrate interface. Significant hole structure giving rise to electrically distinctive regions in thinner film also seems to set a thickness limit for the thinnest films to be used in applications.

## 1. Introduction

Complex magnetic oxide material candidates, including magnetoresistive double perovskite $Sr_2FeMoO_6$ (SFMO), have risen to challenge the dominance of semiconductor based electronics, which is already close to its physical limits in size. In order to understand the physical phenomena in these substances enthusiastic research has taken place. Investigations of electrical properties have been previously conducted mainly by using techniques, which give a result over the entire sample, without local information. Scanning tunneling microscopy (STM) enables, not only examination of surface structure with outstanding sub-nanometer resolution, but also the electrical structure, a method known as scanning tunneling spectroscopy (STS) [1,2]. Ideally these methods provide a way to map the sample surface with atomic scale resolution and with information of the local density of states.

Magnetic and resistive properties in SFMO thin films have shown to be greatly affected for example by the substrate and film thickness, since the lattice mismatch between the film and the substrate distorts the initial layers of the film close to the film substrate interface [3–6]. Resistivity measurements have revealed a low temperature upturn, indicating semiconductive behaviour, which seems to be connected with the band structure [3, 4, 6–8]. Alterations in the band structure can be caused by lattice deformation like anti-site disorder (ASD), oxygen vacancies and strain [9–12]. In addition structural defects like dislocations could affect the transport properties [12]. In this study, we investigate the link between the surface microstructure and the local electronic structure of SFMO thin films.

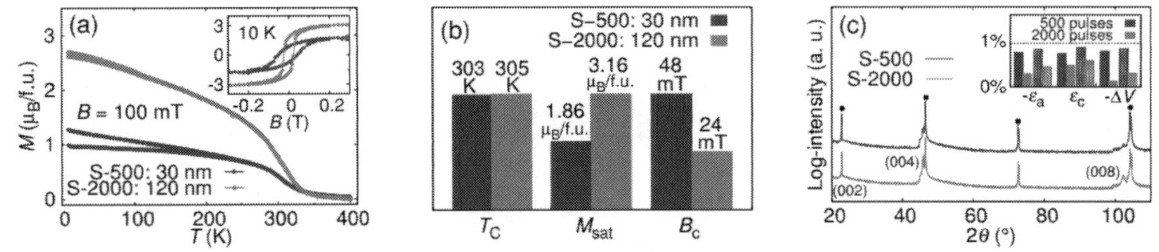


**Figure 1.** (a) ZFC/FC magnetization curves in 100 mT field for the 30 nm and 120 nm thick SFMO films. The inset shows the hysteresis loops at 10 K for the same samples. (b) The comparison of the main magnetic properties between the 30 nm and 120 nm thick SFMO films. (c) $\theta - 2\theta$-scans of the SFMO films with different thicknesses. The SFMO peaks are indexed and peaks arising from the substrate are marked with black dots. The inset shows our earlier results of in-plane strain, $\varepsilon_a$, out-of-plane strain, $\varepsilon_c$, and the change in the unit cell volume, $\Delta V$, for similar SFMO films deposited with 500 and 2000 pulses. The results from [15] are presented with solid bars and the results from [14] are presented with patterned bars.

## 2. Experimental details

A set of SFMO thin films was deposited on niobium doped $SrTiO_3$ (STO) single crystal substrates with pulsed laser deposition (PLD). The conductive niobium doped STO with small lattice mismatch was chosen to ensure the tunneling conductivity in the STM system along with expected good structural film quality. The deposition was carried out at 1050 °C temperature in 9 Pa Ar-atmosphere, the first film with 500 laser pulses and the second with 2000 pulses. The films fabricated with 500 and 2000 pulses are to be called S-500 and S-2000, respectively. Based on our recent analysis, 500 PLD laser pulses correspond to thin film thickness of 30 nm [13, 14]. Assuming linear growth rate, S-2000 has a thickness of 120 nm.

Phase purity and texture of the films was confirmed with $\theta - 2\theta$-scans X-ray diffraction (XRD) measurements between 20° and 110° as well as with the SFMO (204)-peak ($2\theta = 57.106°$), $SrMoO_4$ (112)-peak ($2\theta = 27.68°$) and Fe 110-peak ($2\theta = 44.98°$) $\phi - \psi$-scans. XRD measurements were performed with a Philips X'Pert Pro MPD diffractometer with a Schulz goniometer. Zero field cooled (ZFC) and field cooled (FC) magnetizations were measured as a function of temperature between 10 K and 400 K, with a Quantum Design MPMS SQUID magnetometer. The FC results were used to determine the Curie temperature ($T_C$). Magnetic hysteresis was measured between ±500 mT, at temperatures of 10 K and 400 K.

Thin film surface topography was investigated using OMICRON Fermi SPM ultra high vacuum scanning probe microscope. The equipment was also used for tunneling spectroscopy imaging. In order to enhance the thermal stability of the system and the resolution of the results, the system was cooled using liquid nitrogen. During the STM and STS measurements the sample temperature was 84 K.

## 3. Sample quality monitoring

The ZFC/FC magnetization curves and the hysteresis loops of both SFMO thin films are presented in figure 1 (a). Diamagnetic background arising from the substrate and the sample holder has been removed from the hysteresis loops. The results show clear ferro−paramagnetic phase transition with $T_C$ around 300 K. The Curie temperature was determined from the minimum of the first order derivative in FC measurement. Compared to our earlier results, the $T_C$ is decreased around 50 K [13, 14]. Even though the Curie temperature is decreased from

IOP Conf. Series: Journal of Physics: Conf. Series **969** (2018) 012107 doi:10.1088/1742-6596/969/1/012107

our earlier samples, due to the different Ar gas flow and temperature during the deposition, the saturation magnetization obtained from the hysteresis loop of S-2000 sample is rather high. Since our previous results have shown that ASD and oxygen vacancies affect the magnetic properties of SFMO films, the observed results suggest that we have been able to reduce the amount of structural defects by using a new sample holder [15].

In addition to the saturation magnetizations, $M_{sat}$, the coercivity fields, $B_c$, were also determined from the hysteresis loops shown in the inset of figure 1 (a). These values are presented together with the $T_C$ in figure 1 (b), which illustrates the differences in the magnetic properties between 30 nm and 120 nm thick SFMO films. Only a small difference is observed in the Curie temperature, which can be explained with the different level of substrate induced strain [15]. However, the saturation magnetization is clearly lower and the coercivity field clearly higher in the thinner S-500 film. The obvious difference in the low temperature part of the ZFC/FC curves between films shows larger magnetization deviation in S-500. These all suggest that there are more structural defects in the thinner S-500 film, which cause the domain wall formation and pinning. All in all, the values for $T_C$ and $M_{sat}$ presented here are comparable with our previous results, but still showing slightly higher $M_{sat}$ and lower $T_C$ than in general [3, 5, 6, 13].

The high quality of the films was confirmed with the X-ray diffraction measurements. The $\theta - 2\theta$-scans presented in figure 1 (c) together with the pole figures of SFMO 204-, Fe (110)- and $SrMoO_4$ (112) -peaks (not shown here) confirmed the SFMO films to be phase pure, fully textured and $c$-axis oriented. The inset of the figure 1 (c) is based on our earlier comprehensive XRD analysis of the similar 500 and 2000 pulses films as the S-500 and S-2000 samples in this work [14, 15]. These results are in good agreement with each other and they show that the in-plane strain, $\varepsilon_a$, out-of-plane strain, $\varepsilon_c$, and the change in the unit cell volume, $\Delta V$, are decreased with the increasing film thickness. Therefore, we can conclude that the thicker S-2000 film is less strained and the SFMO unit cell volume is closer to the literature value compared to the thinner S-500 film.

## 4. Topography and tunneling spectroscopy

STM measurements were conducted before any additional measurements. The images shown in figure 2 present the STM results for S-500, (a) and (b), and S-2000, (c). Multiple STM scans were taken across the film surface to achieve stable STM signal with the best possible resolution. However, all the images showed height scales very similar with the images presented here. It is evident, according to the STM results for S-500 in figure 2 (a), that SFMO seems to prefer island-like film growth combined with single layer atomic terraces. Island-like film growth seems to give rise to hole-like structures between separate islands. The hole depths are within the limits of image height scale, from approximately 5 nm to 15 nm. As already pointed by Borges *et.al.* [16], holes could oppose a serious obstacle for tunneling type spin valves, since atomically flat surfaces are preferred. Large structural defects at the multilayer interface can for example lead to diminished exchange coupling, which will affect the properties of multilayer spin valves [17]. Since similar holes were not observed in the thicker S-2000 sample (figure 2 (c)), this suggests that thickness is one of the limiting factors considering applications. The absence of hole structures in S-2000 indicates that the growth might be modified from initial island-like growth towards layer-by-layer growth [18].

In figure 2 (b) the results reveal a clear step-terrace structure presented for S-500 and the inset shows height profile from the line indicated in the figure. Steps are divided into two groups, with height of $\approx 0.4$ nm and $\approx 0.8$ nm. This corresponds well with either single perovskite step or double perovskite $c$-parameter lattice size [19]. Atomic scale step-terraces were identified also for S-2000 from figure 2 (c). However, steps seem to include more height scales exceeding the ones seen in figure 2 (c) for S-500. Similar results, showing step terrace structure with atomically flat surface, have been reported for epitaxially grown SFMO thin films [20]. More widely, the

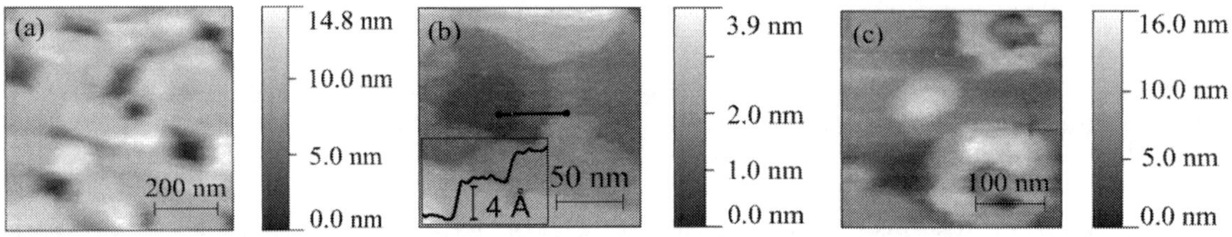


**Figure 2.** Surface microstructure of S-500, (a) and (b), and S-2000 SFMO thin films, (c), characterized by STM at 84 K temperature. The inset of (b) shows height profile from the region of the surface, which is indicated with black line.

terrace like structure without a clear evidence of atomically flat step size, has also been reported for SFMO thin films by various groups [14, 18, 21–23]. The atomically flat step-terrace structure provides our first evidence of atomic scale thin film growth in SFMO films and gives hope for the fabrication of SFMO spin-valves with atomically flat interface sections in multilayers.

Tunneling spectroscopy was used for local electrical characterization. STS measurements lack statistical signifigance, since STS-mapping was obtained from only single location with the best structural resolution. The SFMO film surface was scanned with STM, simultaneously mapping the surface area with scanning tunneling spectroscopy at 84 K. Figure 3 (a) presents the results obtained from STM topography, alike the results already presented in figure 2, along with STS mapping in figure 3 (b). The STS results present the $I(V)$-map corresponding to a threshold voltage of 1.0 V. Green areas, having lower current at 1.0 V, correspond well with the holes seen in STM topography. In order to prevent any of the influences from the tip instability possibly caused by too slow feedback, we analyze the results for two $(dI/dV)/\sqrt{(I/V)^2+b^2}$ curves shown in figure 3 (c) obtained from the areas indicated in the $I(V)$-map, to see differences in the local density of states (LDOS) which is proportional to normalized conductivity. The normalization method is similar as reported in [24]. The $b$ in $(dI/dV)/\sqrt{(I/V)^2+b^2}$ corresponds to a small constant, in the main panel of figure 3 (c) $b = 0.003$ nA/V, chosen to diminish noise around small current values. In the inset $b$ was set to 0 nA/V. The results reveal slight changes in normalized conductivity curves. Conductivity near the Fermi level is smaller when comparing green area in the $I(V)$-map to yellow area. The results for S-2000 did not reveal as clear connection with electrical properties and surface structure. It was also more difficult to obtain stable STS for S-2000, when for S-500 this was relatively straight forward.

According to our best knowledge, SFMO has not been studied with STS before, but few reports about STM and STS on manganites with similar properties, have been published [25–28]. Besides the atomically flat step terrace structure [25], intrinsic inhomogeneities have been identified along with alleged evidence of half-metallicity by STM and STS technique [25–27]. Wei *et.al.* concluded the peaks in the conductance data to arise from spin-polarized density of states [25]. Similar evidence of half-metallicity was not observed in our measurements, despite the measurements being conducted at 84 K, well below the Curie temperature (see figure 1 (a)), and despite the good structural data obtained with STM. Other reports show spatial distribution of insulating and conductive regions [26–28]. These results have been used to study the metal–insulator–transition, in the materials where it is observed. In addition lattice strain caused by the substrate material is known to influence the electric properties [28].

SFMO does not exhibit similar first order phase transition between metallic and insulating state as colossal magnetoresistive materials [29, 30]. However, the link between resistive and magnetic phenomena is close and studying local electric properties will reveal valuable information about SFMO thin films. Electric properties in SFMO are known to be influenced

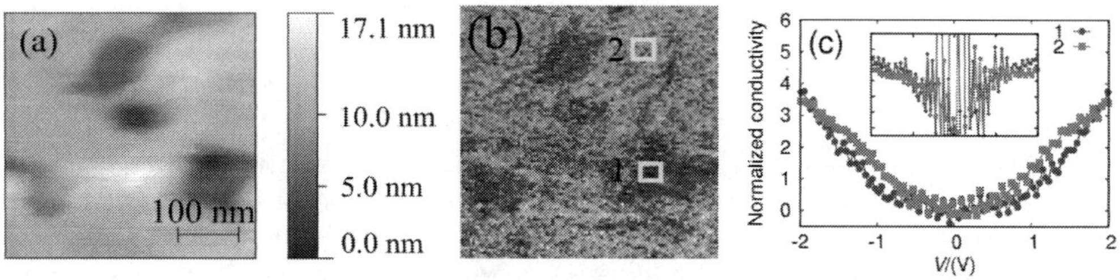


**Figure 3.** Surface of S-500 characterized by low temperature STM (a) along with STS $I(V)$-map (b). Aquamarine coloured squares illustrate the areas, which have been chosen to demonstrate the difference between yellow region and green region. The main panel of (c) shows an example of normalized $(dI/dV)/\sqrt{(I/V)^2+b^2}$, $b = 0.003$ nA/V (inset $b = 0$ nA/V), curves for the indicated map areas, red curve indicating the yellow region and blue the green region.

by structural deformation. This deformation is more dominant in thinner SFMO films [6, 15]. Our STM results have shown clear evidence of more significant structural deformity, in a form of hole structures. The results obtained in STS represent most likely evidence of locally diminished electrical properties, which give rise to more impaired properties measured from the entire sample. Lower conductivity around the Fermi level might suggest that energy gap in majority spin band is increased within the hole structures. This result would possibly contradict previous studies [3, 6, 11]. Therefore we believe that lower conductivity most likely arises, not from increased band gap in majority band, but from overall perishment of the density of states in both bands. Considering the height scale in STM images for S-500, the atomic layers within the hole structures are likely to be substitued with more deformed layers under more significant structural distortion and lattice strain, resulting with deviation in conductivity and in density of states. XRD measurements along with magnetic measurements support the conclusion of structural deformation in thinner film. The previous XRD results presented in the inset of figure 1 (c) for our other films, suggest that significant structural deformation is always present and the effect is more dominant in thinner films. Also dimished saturation magnetization suggests higher concentration of lattice defects, including ASD and oxygen vacancies.

Due to the samples' exposure to ambient air, the possibility of surface contamination cannot be totally excluded in our results. However, films' uniform surface quality suggests the samples having no major contamination. The measurements were repeated after storing the samples for approximately a week in ultra high vacuum at room temperature. An interesting result was that, compared to the initial results, the surface of both SFMO thin films appeared highly irregular and similar surface structures, like in the initial low temperature measurements, were not observed. This could mean that the long term exposure to the ultra high vacuum modifies the initial atomic layers, possibly by deoxidation. This could explain why stable STS for S-2000 was more difficult to obtain at the beginning of the measurements. Previous investigation has shown that vacuum annealing at 650 °C increases $T_C$ by reducing the oxygen content in the SFMO [15]. Sample contamination is insignificant considering the whole sample, based on sample quality monitoring, but could disturb the STM and STS measurements on the surface. Diminished tip quality could also play a role with worse results and possible sample contamination requires further investigation, but should be considered as a possibility [27].

Despite the sub-nanometer resolution in out-of-plane surface structure revealing atomic step-terrace structure, we have yet to demonstrate atomic-scale in-plane resolution. This proceeding would enable us to directly identify structural lattice defects, including strain, ASD and oxygen vacancies, combined with electrical analysis. Therefore, we will focus our following research, regarding the SFMO thin film investigation with scanning tunneling spectroscopy, to obtain greater stability in our system.

## 5. Conclusions

High quality SFMO films were fabricated using PLD and studied for the first time with low temperature scanning tunneling spectroscopy to obtain local information on electric properties accompanied by structural surface characters. The results reveal clear step-terrace structure on thin film surface indicating atomic level thin level film growth, which give hope for the fabrication atomically flat interface sections in multilayer structures. In addition, it was observed that thinner films had significant hole structure accompanied by the atomic steps. The holes could also be identified with STS from the rest of the film due to their electrical properties. The similar hole structure was not observed in thicker film, which indicates a limit for the minimum thickness for thin film, if used in spin-valves.

## 6. Acknowledgements

Jenny and Antti Wihuri foundation and University of Turku Graduate School are acknowledged for financial support.

## References

[1] Binnig G and Rohrer H 1983 *Surf. Sci* **126** 236
[2] Feenstra R M 1994 *Surf. Sci* **299** 965
[3] Saloaro M, Majumdar S, Huhtinen H and Paturi P 2012 *J. Phys. Cond. Mat.* **24** 366003
[4] Saloaro M, Majumdar S, Huhtinen H and Paturi P 2013 *EPJ Web of Conferences* **40** 15012
[5] Saloaro M, Deniz H, Huhtinen H, Palonen H, Majumdar S and Paturi P 2015 *J. Phys. Cond. Mat.* **27** 386001
[6] Angervo I, Saloaro M, Palonen H, Majumdar S, Huhtinen H and Paturi P 2015 *20th International Conference on Magnetism* **75** 1011
[7] Venimadhav A, Sher F, Attfield J P and Blamire M G 2004 *J. Magn. and Magn. Mater.* **269** 101
[8] Westerburg W, Reisinger D and Jakob G 2000 *Phys. Rev. B* **62** R767
[9] Stoeffler D and Colis S 2005 *J. Phys. Cond. Mat.* **17** 6415
[10] Stoeffler D and Colis S 2005 *J. Magn. and Magn. Mater.* **290-291** 400
[11] Hoffmann M, Antonov V N, Bekenov L V, Kokko K, Hergert W and Ernst A 2015 *arXiv:1504.02629*
[12] Fix T, Stoeffler D, Colis S, Ullhaq C, Versini G, Vola J P, Huber F and Dinia A 2005 *J. Appl. Phys.* **98** 023712
[13] Angervo I, Saloaro M, Tikkanen J, Huhtinen H and Paturi P 2017 *Appl. Surf. Sci.* **396** 754
[14] Angervo I, Saloaro M, , Huhtinen H and Paturi P 2017 *Appl. Surf. Sci.* **422** 682
[15] Saloaro M, Hoffmann M, Adeagbo W A, Granroth S, Deniz H, Palonen H, Huhtinen H, Majumdar S, Laukkanen P, Hergert W, Ernst A and Paturi P 2016 *ACS Appl. Mater. Interfaces* **8** 20440
[16] Borges R P, Lhostis S, Bari M A, Versluijs J J, Lunney J G, Coey J M D, Besse M and Contour J P 2003 *Thin Solid Films* **429** 5
[17] Coey J M D 2010 *Magnetism and magnetic materials* (Cambridge University Press)
[18] Jalili H, Heinig N F and Leung K T 2010 *J. Chem. Phys* **132** 204701
[19] Nakamura S and Oikawa K 2003 *J. Phys. Soc. Jpn.* **72** 3123
[20] Manako T, Izumi M, Konishi Y, Kobayashi K I, Kawasaki M and Tokura Y 1999 *Appl. Phys. Lett.* **74** 2215
[21] Song J H, Park J H and Jeong Y H 2005 *J. Appl. Phys.* **97** 46105–1
[22] Venimadhav A, Vickers M E and Blamire M G 2004 *Solid State Commun.* **130** 631
[23] Deniz H, Preziosi D, Alexe M, Hesse D, Eisenschmidt C, Schmidt G and Pintilie L 2015 *J. Mater. Sci.* **50** 3131
[24] Prietsch M, Samsavar A and Ludeke R 1991 *Phys. Rev. B* **43** 850
[25] Wei J Y T, Yeh N C and Vasquez R P 1997 *Phys. Rev. Lett.* **79** 5150
[26] Fäth M, Freisem S, Menovsky A A, Tomioka Y, Aarts J and Mydosh J A 1999 *Science* **285** 1540
[27] Becker T, Streng C, Luo Y, Moshnyaga V, Damaschke B, Shannon N and Samwer K 2002 *Phys. Rev. Lett.* **89** 23
[28] Okawa N, Tanaka H, Akiyama R, Matsumoto T and Kawai T 2000 *Solid.. State. Commun.* **114** 601
[29] Salamon M B and Jaime M 2001 *Rev. Mod. Phys* **73** 583
[30] Dagotto E, Hotta T and Moreo A 2001 *Phys. Rep.* **344** 1

# $^{133}$Cs-NMR Study on the Ground State of the Equilateral Triangular Spin Tube $CsCrF_4$

**K Matsui[1,4], T Goto[1], H Manaka[2] and Y Miura[3]**

[1] Physics Division, Sophia University, Tokyo 102-8644, Japan
[2] Graduate School of Science and Engineering, Kagoshima University, Kagoshima 890-0065, Japan
[3] Suzuka National College of Technology, Suzuka, Mie 510-0294, Japan
[4] Research Fellow of Japan Society for the Promotion of Science

E-mail: k703861@eagle.sophia.ac.jp

**Abstract.** We have investigated the hyperfine coupling between Cs and Cr on the $S = 3/2$ equilateral triangular spin tube $CsCrF_4$, utilizing $^{133}$Cs-NMR. At paramagnetic state above 80 K, we have obtained spectra containing a single peak, which reflects the single crystallographic Cs site. From the temperature dependence of the peak shift and peak width, we evaluated effective values of the isotropic and the anisotropic part of hyperfine coupling. The latter was compared with the calculated dipole contribution. Using obtained parameters with assumed spin structure, we tried to reproduce the broadened spectrum in the ordered state at 2.0 K. The preliminary analysis shows the 120-degree structure does not accord with the observed spectra at the ordered state.

## 1. Introduction

$S = 3/2$ spin tube compound $CsCrF_4$ consists of equilateral triangles of $Cr^{3+}$, stacked along $c$-axis without twisting. Tubes are separated by Cs atoms, forming a rhombic unit cell. The schematic structure (the space group $P\overline{6}2m$ [1]) is shown in Fig. 1.

From the theoretical view point, the ground state of the equilateral triangular spin tubes has been argued to have the gapped ground state. For the system with $S = 1/2$, the gap opens up by arranging the singlet dimers on the leg and rung direction [2, 3]. As for the $S = 3/2$ system, though the size is smaller than $S = 1/2$ system, the gap opens by forming singlet dimers in a similar way [4].

The synthesis of this compound was first reported in 1978 [1]. In 2009, Manaka *et al.* have reported its magnetic properties, such as uniform susceptibility, ESR and heat capacity [5, 6]. He argued that there is no magnetic order at low temperatures down to 1.5 K and that its ground state is the gapless spin-liquid. Quite recently, our group has reported the existence of a magnetic phase transition with $T_N \simeq 5.0$ K by the spectral broadening of NMR experiments [7]. Thus, $CsCrF_4$ is expected to be one of the simplest model compound of a playground of geometrical frustration and low-dimensionality.

Sofar, the spin exchange interactions were evaluated by the density functional theory calculation [8]. The dominant interactions are antiferromagnetic, that is, $J_{\text{rung}} = -19$ K for the triangular plane, $J_{\text{chain}} = -39$ K for the tube direction and $J_{\text{inter}} = -3.2 \times 10^{-3}$ K for the inter-tube exchange interaction.

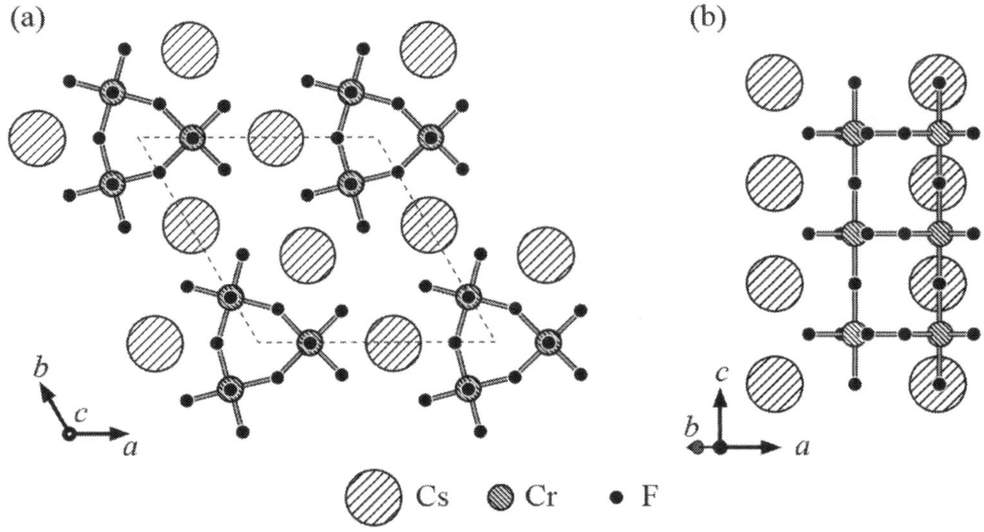


**Figure 1.** Schematic drawings of $CsCrF_4$ structure (a) on *ab*-plane and (b) along *c*-axis. Dashed lines show the unit cell. Cs atoms are located at interstitial positions between tubes.

Recently, Seki *et al.* suggested interesting approach with an inter-tube coupling. The spin tube system can be treated as Kagomé-triangular lattice, which has the same lattice topology in the *ab*-plane [9]. Depending on inter-tube coupling, various ordered phase are expected such as 120-degree structure, incommensurate order, ferromagnetic order and cuboc order, which has noncoplanar 120-degree spin structure.

Though there are many investigations theoretically and experimentally, the ground state of $CsCrF_4$ is still unknown. Particularly, the spin structure in the possible ordered state is not determined. This is mainly because no reports with neutron scattering has been given until now. Instead, we try to investigate the magnetic structure by NMR. For this purpose, it is essential to determine the hyperfine coupling tensor [10]. The preliminary result on the hyperfine coupling tensor for $^{19}$F sites has been reported by us recently [11]. In this paper, we investigate the hyperfine coupling tensor for $^{133}$Cs-site. By using the obtained hyperfine coupling tensor, we have performed the preliminary analysis in the ordered state.

## 2. Experimental

The powder sample is prepared by the conventional solid-state reaction method [5, 6]. $^{133}$Cs-NMR spectra were measured in wide temperature range between 2 and 300 K, under magnetic field around 5.67 T. The Cs site is located on interstitial position between tubes, and between $CrF_3$ layers. NMR spectra were obtained by recording the spin-echo amplitude against magnetic field, which was slowly varied within a narrow region around zero shift position [12, 13].

## 3. Results and Discussion

The typical profile of $^{133}$Cs-NMR spectra measured in various temperatures between 80 and 300 K are shown in Fig. 2. We have observed a single peak,which may reflect the single crystallographic site of $^{133}$Cs. The electrical quadrupole splitting $^{133}\nu_Q/^{133}\gamma$ is estimated to be smaller than 50 Oe. The spectral shape was slightly asymmetric as compared to Gaussian function. We treated this asymmetry in the terms of the uniaxial anisotropy to obtain the anisotropic part of Knight shift. This assumption is reasonable for the local symmetry around the

IOP Conf. Series: Journal of Physics: Conf. Series **969** (2018) 012108 doi:10.1088/1742-6596/969/1/012108

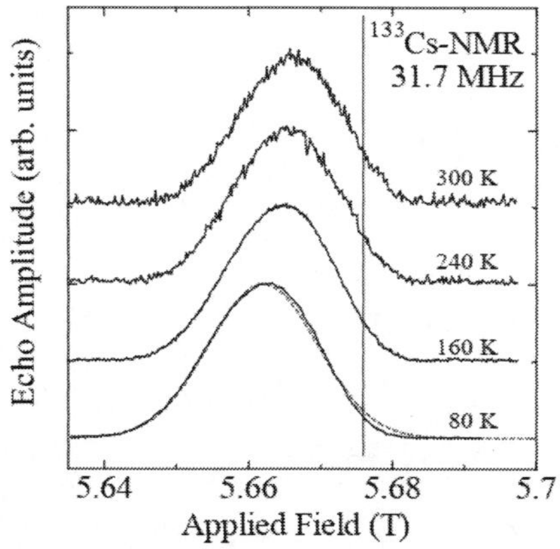


**Figure 2.** Typical profile of $^{133}$Cs-NMR spectra. The vertical line denotes zero shift. The dashed line denotes the Gaussian function.

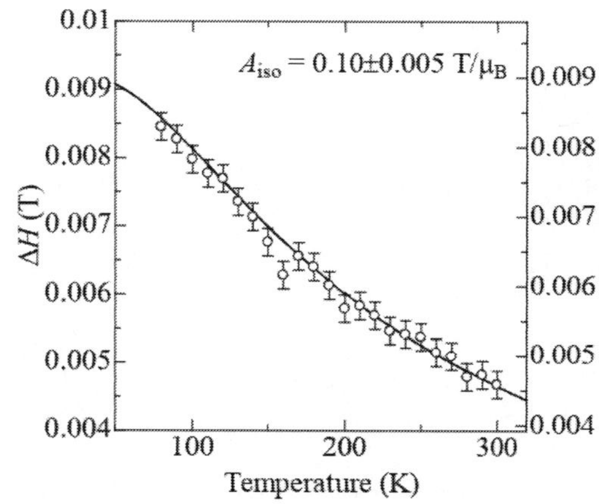


**Figure 3.** Temperature dependence of peak shift (open circle) and the uniform susceptibility (solid curve) [5].

Cs site (Fig. 1). We evaluated the peak shift from the peak position $H_{\rm peak}$ as ${}^{133}\nu_{\rm Q}/{}^{133}\gamma - H_{\rm peak}$. The peak shift decreases with increasing temperature. In Fig. 3 and Fig. 4, we show the temperature dependence of peak shift $\Delta H$ and full width at half maximum (FWHM) of the peak with that of the uniform susceptibility.

From these scaling plots, the coupling constants of the isotropic component of hyperfine coupling $A_{\rm iso}$ and the anisotropic component $A_{\rm an}$ were obtained to be $1000 \pm 50$ G/$\mu_{\rm B}$ and $60\pm40$ G/$\mu_{\rm B}$, respectively. Note that the latter includes both the contribution from the hyperfine coupling and the dipole-dipole coupling, and is too small compared with the contribution of dipole-dipole coupling, which was obtained by the calculation based on the classical dipole-dipole interaction as $A_{\rm dip} = 412$ G/$\mu_{\rm B}$. This means that the contribution of the superexchange interaction via F ion of $350 \pm 40$ G/$\mu_{\rm B}$ must exist, for the case that the principal axes of the two contributions are assumed to be identical.

Next, we try to reproduce the spectrum in the ordered state with these hyperfine coupling constants. The considerably large value $J_{\rm chain}$ and the three-fold symmetry invoke one, as the simplest assumption, the $q = 0$ 120-degree structure in $ab$-plane with antiferromagnetic structure along the tube direction. Before reproducing spectrum, we refer to contribution of hyperfine coupling in this ordered state. In this structure, the contribution of hyperfine coupling is geometrically canceled at Cs site due to antiferromagnetic two spins along the tube direction, so we can calculate the internal field by adding up the contribution of classical dipole-dipole interaction. We put at each Cr site an ordered moment with direction according with the 120-degree structure with chirality +1. The summation for the Cr site was taken for those with in the radius 400 Å. The simulated powder pattern is shown in Fig. 5. It is a single component of flat-topped shape, which reflects the fact that there is a unique magnetic $^{133}$Cs site in the assumed spin structure. This does not accord with the observed $^{133}$Cs-NMR spectrum at 2.0 K, irrespective of the assumed moment size. The latter consists of the two components, sharp center peak and shoulders at both sides, which indicates that there exist the two Cs sites in the unit cell.

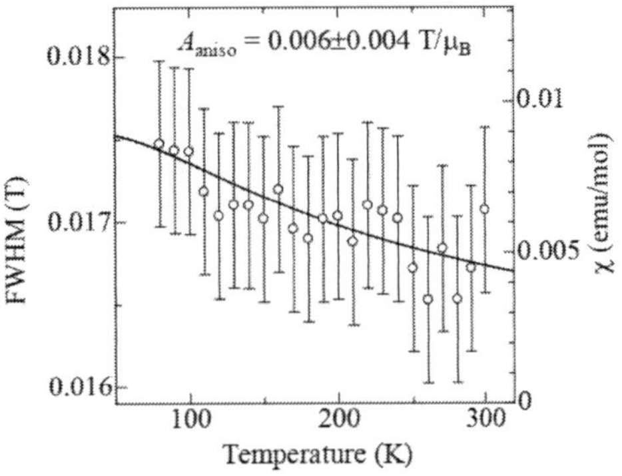


**Figure 4.** Temperature dependence of the peak width (open circle) and the uniform susceptibility (solid line) [5].

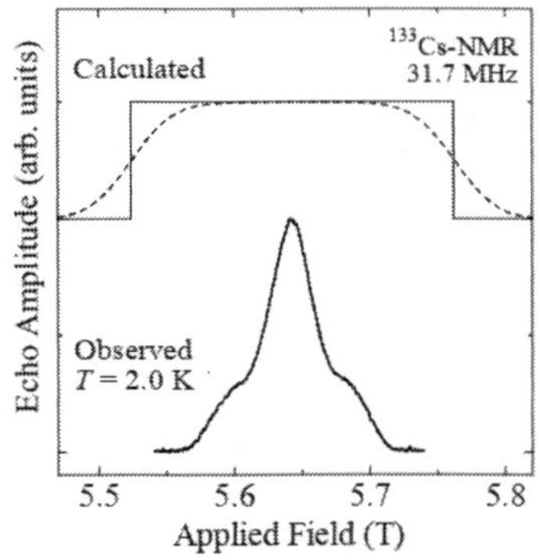


**Figure 5.** Calculated and observed $^{133}$Cs-NMR spectrum in an ordered state at 2.0 K. Calculation is based on 120-degree structure with the effective magnetic moment of $Cr^{3+}$, $3.84\mu_B$ [6]. Dashed curve denotes the calculated spectrum convoluted with Gaussian width of 0.05 T.

## 4. Summary

We have investigated the hyperfine coupling between Cs and Cr on the $S = 3/2$ equilateral triangular spin tube $CsCrF_4$, utilizing $^{133}$Cs-NMR. At paramagnetic state, we have obtained the isotropic and anisotropic parts of hyperfine coupling as $A_{\mathrm{iso}} = 1000 \pm 50$ G/$\mu_B$ and $A_{\mathrm{an}} = 60 \pm 40$ G/$\mu_B$, respectively. We also calculated the dipole coupling tensor contributing to the anisotropy in the hyperfine coupling, $A_{\mathrm{dip}} = 412$ G/$\mu_B$. By using these parameters, and by assuming the spin structure, we tried to reproduce the spectrum in the ordered state. The 120-degree structure was tested, and was found to disagree with the observation. In order to find a spin structure that matches the NMR observation, one has to assume the structure containing two inequivalent Cs sites, such as $q = 1/2$ within $ab$-plane or $\chi = -1$. Work is in progress to simulate NMR spectra for this structure.

## Acknowledgments

This work was partly supported by JSPS KAKENHI Grant Number 15J05303 and 15K05148.

## References

[1] Babel D and Knoke G 1978 *Zeitschrift fur anorganische und allgemeine Chemie* **442** 151–162 ISSN 1521-3749
[2] Nishimoto S and Arikawa M 2008 *Phys. Rev. B* **78**(5) 054421
[3] Sakai T, Sato M, Okunishi K, Otsuka Y, Okamoto K and Itoi C 2008 *Phys. Rev. B* **78**(18) 184415
[4] Nishimoto S, Fuji Y and Ohta Y 2011 *Phys. Rev. B* **83**(22) 224425
[5] Manaka H, Hirai Y, Hachigo Y, Mitsunaga M, Ito M and Terada N 2009 *J. Phys. Soc. Jpn.* **78** 093701 (*Preprint* `http://dx.doi.org/10.1143/JPSJ.78.093701`)
[6] Manaka H, Etoh T, Honda Y, Iwashita N, Ogata K, Terada N, Hisamatsu T, Ito M, Narumi Y, Kondo A, Kindo K and Miura Y 2011 *J. Phys. Soc. Jpn.* **80** 084714 (*Preprint* `http://dx.doi.org/10.1143/JPSJ.80.084714`)
[7] Matsui K, Goto T, Manaka H and Miura Y 2015 *Phys. Procedia* **75** 726 – 730 ISSN 1875-3892
[8] Koo H J 2012 *J. Magn. Magn. Mater.* **324** 2806 – 2808 ISSN 0304-8853
[9] Seki K and Okunishi K 2015 *Phys. Rev. B* **91**(22) 224403
[10] Kitagawa K, Katayama N, Ohgushi K, Yoshida M and Takigawa M 2008 *J. Phys. Soc. Jpn.* **77** 114709 (*Preprint* `http://dx.doi.org/10.1143/JPSJ.77.114709`)
[11] Matsui K, Goto T, Manaka H and Miura Y 2016 *Hyperfine Interactions* **237** 123 ISSN 1572-9540
[12] Matsui K, Fujisawa M, Hagiwara K, Hoshino Y, Goto T, Sasaki T, Tanaka H, Okubo S and Ohta H 2014

*Proc. Int. Conf. on Strongly Correl. Electron Syst. (SCES2013)* **3** 014011 (*Preprint* `http://journals.jps.jp/doi/pdf/10.7566/JPSCP.3.014011`)

[13] Kanada K, Saito T, Oosawa A, Goto T and Suzuki T 2007 *J. Phys. Soc. Jpn.* **76** 064706 (*Preprint* `http://dx.doi.org/10.1143/JPSJ.76.064706`)

# Thermal expansion and magnetostriction measurements of perovskite-type Co oxides $Sr_{1-x}Y_xCoO_{3-\delta}$ in high magnetic fields

S Imura[1], S Kimura[1], S Awaji[1], I Terasaki[2]

[1]Institute for Materials Research, Tohoku University, 2-1-1 Katahira, Aoba-ku Senndai 980-8577, Japan.
[2]Department of Physics, Nagoya University, Furocho, Chikusa, Nagoya 464-8602, Japan.

e-mail ; wp-tvi.imura@imr.tohoku.ac.jp

**Abstract.** We have measured the magnetization, linear thermal expansion and magnetostriction of polycrystalline samples of $Sr_{1-x}Y_xCoO_{3-\delta}$ ($x$ = 0.22, 0.25) in high magnetic fields, and found that the magnetization of $Sr_{1-x}Y_xCoO_{3-\delta}$ is closely related to the crystal lattice. This result agrees with our expectation that the large decrease of the magnetization in $Sr_{0.75}Y_{0.25}CoO_{3-\delta}$ with cooling in temperature region 150 – 250 K is caused by spin crossover.

## 1. Introduction

Magnetism of the perovskite-type Co oxides is characterized by $Co^{3+}$ ion with a $3d^6$ electronic configuration. The five-fold degeneracy of the $3d$ orbitals in $Co^{3+}$ ion under an octahedral crystal field is split into the two fold degenerate $e_g$ states and the three-fold degenerate $t_{2g}$ states. A competition between the crystal field and an electrostatic interaction between $d$ electrons occurs in the $3d^6$ ion. When the interaction between $d$ electrons is dominant, the $d$ electrons occupy both the $t_{2g}$ and $e_g$ levels in order to maximize the total spin number, resulting in the high-spin state (HS). On the other hand, when the crystal field is dominant, the $d$ electrons occupy the lower $t_{2g}$ levels as many as possible, resulting in the low-spin state (LS). The perovskite-type Co oxides show very rich physical behaviors associated with the spin crossover between the LS and HS or intermediate-spin state (IS).

Kobayashi *et al.* have found that polycrystalline samples of $Sr_{1-x}Y_xCoO_{3-\delta}$ show a ferromagnetic transition at 335 K in a narrow range of $0.2 \leq x \leq 0.25$, which is the highest Currie temperature among the perovskite-type Co oxides [1]. Magnetic properties of $Sr_{1-x}Y_xCoO_{3-\delta}$ are greatly changed by temperature, pressure or chemical composition of the sample. In particular, the $x = 0.25$ sample shows a curious magnetic behavior as follows. The magnetization of the $x = 0.25$ sample increases with cooling just below the Currie temperature as usual ferromagnetic materials. However, an abrupt decrease of magnetization is seen from 250 K as shown in Fig. 1(b). We consider that this decrease of the magnetization is due to the spin crossover from the HS to LS [2, 3]. On the other hand, Troyanchuk *et al.* argued that this abrupt decrease is caused by a transition from ferromagnet to antiferromagnet in the ordered state [4]. Thus, the origin of the decrease of the magnetization in the $x = 0.25$ sample is still under debate. Usually, a crystal lattice is largely changed by the spin crossover [5], because the Co-O bond length in the equilibrium state is different between the HS and LS. Therefore, in this study, we have measured the linear thermal expansion and magnetostriction in high magnetic fields in order to examine the relationship between the magnetism and the crystal lattice of $Sr_{1-x}Y_xCoO_{3-\delta}$ ($x = 0.25$, 0.22).

## 2. Experiment

Polycrystalline samples of $Sr_{1-x}Y_xCoO_{3-\delta}$ ($x = 0.22$, 0.25) were prepared by a solid state reaction method [1]. We used strain gages for linear thermal expansion and magnetostriction measurements. A 18 T superconducting magnet at Institute for Materials Research, Tohoku University was used for the measurements. The temperature dependences of magnetization were measured using a SQUID magnetometer (Quantum Design, MPMS).

## 3. Results and Discussion

Figures 1 (a) and (c) show a linear thermal expansion $\Delta L_T/L_0$ and a linear thermal expansion coefficient $\alpha$ at 0 T of $Sr_{1-x}Y_xCoO_{3-\delta}$ with the $x = 0.25$ and 0.22, respectively. The linear thermal expansion coefficient is calculated by differentiating the linear thermal expansion. The measurement starts from the highest temperature 380 K to cool down to 10 K, and then the sample is warmed up to 380 K. Temperature dependences of the magnetization at 0.1 T of the $x = 0.25$ and $x = 0.22$ are shown in Fig. 1 (c) and (d), respectively. As shown in Fig. 1(a), the crystal lattice of the $x = 0.25$ sample slightly increases just below $T_C \sim 350$ K, showing that the ferromagnetic transition causes lattice expansion, during the cooling process, and then shows a large contraction in a temperature region from 250 K to 100 K, where the magnetization decreases abruptly. In this temperature region, $\alpha$ shows a characteristic peak. Such a peak of $\alpha$ is also known to accompany the spin crossover in $LaCoO_3$, which is a representative of the

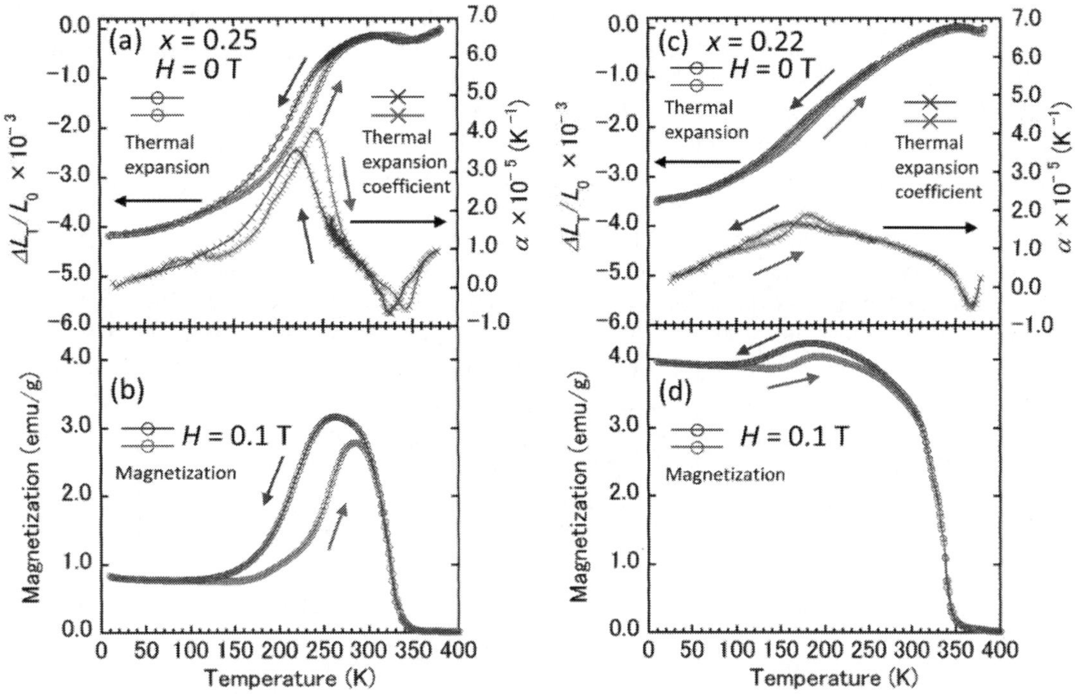


Figure 1. (a) Linear thermal expansion $\Delta L/L_0$ and linear thermal expansion coefficient $\alpha$ and (b) magnetization of $Sr_{1-x}Y_xCoO_{3-\delta}$ with the $x = 0.25$. (c) $\Delta L/L_0$ and $\alpha$ and (d) magnetization of $Sr_{1-x}Y_xCoO_{3-\delta}$ with the $x = 0.22$. The magnetization curves are measured at 0.1 T.

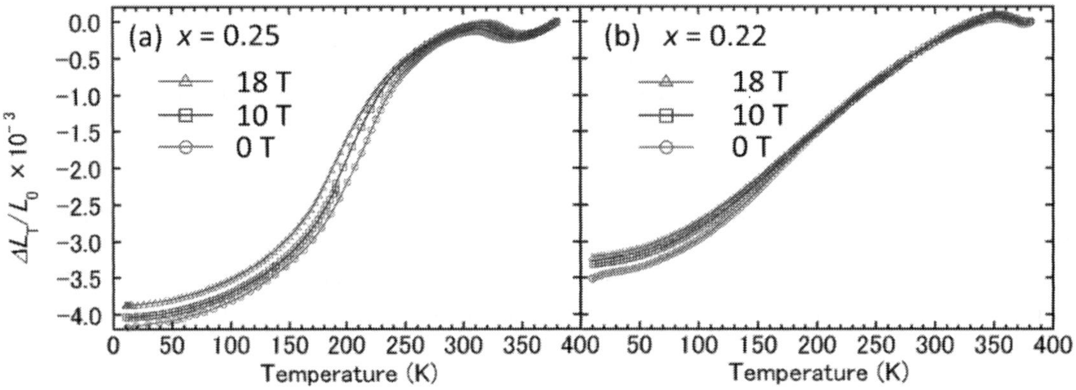


Figure 2. Linear thermal expansion in cooling process at 0 T, 10 T, 18 T of $Sr_{1-x}Y_xCoO_{3-\delta}$ with (a) the $x = 0.25$ and (b) $x = 0.22$.

perovskite-type Co spin crossover system [6]. We observe a large hysteresis of the thermal expansion at around 150 K – 300 K, where the magnetization curve also exhibits a hysteresis as shown in Fig. 1(b). These hysteresis behaviors indicate that a crossover from a high temperature metastable state with larger lattice volume and a larger magnetization to a low temperature stable state with smaller lattice volume and a smaller magnetization occurs in the cooling process at around 150 K – 300 K, and this crossover proceeds with a formation of domains of the stable state. Such hysteresis is a characteristic of spin crossover [5]. Temperature dependence of $\alpha$ shows a dip at $T_C$. The dip of $\alpha$ also indicates a hysteresis, reflecting the fact that the ferromagnetic transition of $Sr_{1-x}Y_xCoO_{3-\delta}$ is of first order with lowering of the crystal symmetry [7]. In the case of the $x = 0.22$ sample, which has the smaller drop of magnetization in the ferromagnetic state than that of the $x = 0.25$ sample, we observe a smaller thermal contraction of the crystal lattice with decreasing temperature. The height of the peak of $\alpha$ in the $x = 0.22$ sample is also lower than that in the $x = 0.25$ sample. Even though the decrease of magnetization in the $x = 0.22$ sample is small, however, a hysteresis is observed both in the thermal expansion and the magnetization curves below 250 K as well as the $x = 0.22$ sample. On the other hand, the dip of $\alpha$, observed at higher temperature, indicates higher $T_C$ of the $x = 0.22$ sample compared with the $x = 0.25$. Our results indicate a close relation between the magnetization and the crystal lattice in $Sr_{1-x}Y_xCoO_{3-\delta}$. The contraction of the lattice, accompanying the decrease of the magnetization, agrees with the spin crossover from the HS to LS, because the LS favors a smaller crystal lattice, which strengthens the crystal field.

Figure 2(a) and (b) shows the linear thermal expansion $\Delta L_T/L_0$ of the $x = 0.25$ and 0.22 samples, respectively, with cooling at 0 T, 10 T, 18 T. In both samples, the value of $\Delta L_T/L_0$ at 10 K increases with increasing the magnetic field. In addition, the thermal expansion curve of the $x = 0.25$ sample below

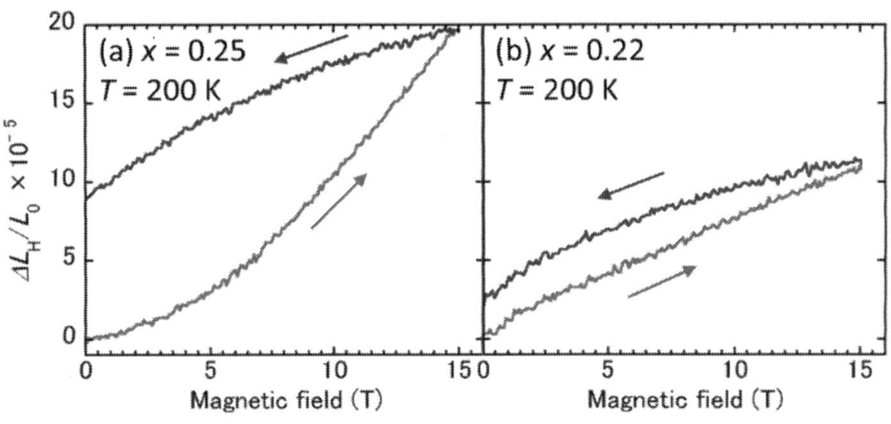


Figure 3. Magnetostriction of (a) $Sr_{0.75}Y_{0.25}CoO_{3-\delta}$ and (b) $Sr_{0.78}Y_{0.22}CoO_{3-\delta}$ at 200 K.

250 K shifts toward lower temperature, indicating that the high temperature state is stabilized by magnetic fields. This behavior is consistent with the crossover from the HS to LS by cooling, because the HS with larger magnetic moment is stabilized by increasing of the magnetic field. Such a shift of $\Delta L_T/L_0$ is less obvious in the $x = 0.22$ sample. The result that a minimum of $\Delta L_T/L_0$ around $T_C$ shifts toward higher temperature indicates that the ferromagnetic order in $Sr_{1-x}Y_xCoO_{3-\delta}$ is also stabilized by the magnetic field.

Figures 3 (a) and (b) shows the magnetostriction $\Delta L_H/L_0$ of the $x = 0.25$ and 0.22, respectively, at 200 K. We observe positive magnetostriction with a hysteresis. Magnetostriction curves of both samples in field decreasing process at 0 T show finite $\Delta L_H/L_0$. Therefore, the lattice deformation, caused by applied magnetic fields, remains even after removal of the external magnetic field because of long relaxation time. In addition, the magnetostriction of the $x = 0.25$ sample is larger than that of the $x = 0.22$ sample. Figure 4 shows temperature dependence of $\Delta L_H/L_0$ at 15 T observed in the $x = 0.25$ and 0.22 samples at 200 K. The magnetostriction is large in temperature region, where large thermal expansion coefficient $\alpha$ is observed, and almost disappears above $T_C$. We consider that the observed increase in the lattice volume is caused by the field-induced spin crossover from the LS to HS.

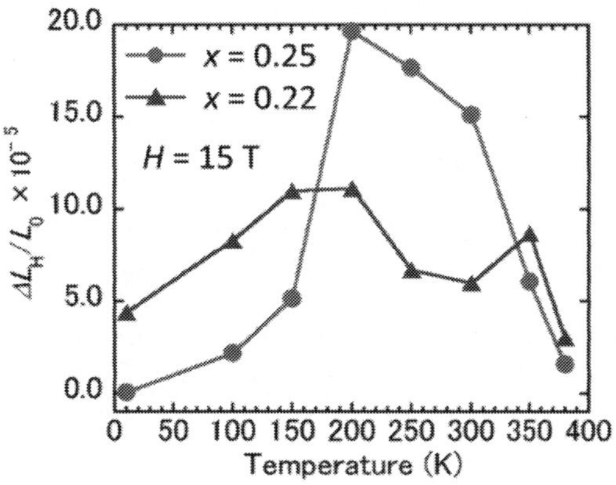


Figure 4. Magnetostriction $\Delta L_H/L_0$ at 15 T observed in the $x = 0.25$ and 0.22

## 4. Summary

We measured the magnetization, linear thermal expansion and magnetostriction to elucidate the origin of the curious behavior of the magnetization in the ferromagnetic phase in $Sr_{1-x}Y_xCoO_{3-\delta}$ with the $x = 0.25$, which shows abrupt decrease below 250 K with cooling. We found that magnetic properties of $Sr_{1-x}Y_xCoO_{3-\delta}$ are closely related to the crystal lattice. The experimental results agree with our expectation that the decrease in the magnetization of $Sr_{1-x}Y_xCoO_{3-\delta}$ with the $x = 0.25$ is caused by the spin crossover.

## References

[1] Kobayashi W, Ishiwata S, Terasaki I, Takano M, Grigoraviciute I, Yamauchi H and Karppinen M 2005 *Phys. Rev.* B **72** 104408.
[2] Kimura S, Maeda Y, Kashiwagi T, Yamaguchi H, Hagiwara M and Yoshida S 2008 *Phys. Rev.* B **78** 180403(R).
[3] Yoshida S, Kobayashi W, Nakano T and Terasaki I 2009 *J. Phys. Soc. Jpn.***78** 094711.
[4] Troyanchuk I O, Bushinsky M V, Dobryanskii V M and Pushkarev N V 2012 *Low Temp. Phys.* **38** 662.
[5] Gutlich P, Hauser A and Spiering H 1994 *Angrew. Chem. Int. Ed. Engl.* **33** 2024.
[6] Zobel C, Kriener M,Bruns D, Baier J, Grüinger M, Lorenz T, Reutler P and Revcolevschi A 2002 *Phys. Rev.* B **66** 020402(R).
[7] Ishiwata S, Kobayashi W, Terasaki I, Kato K and Takata M 2007 *Phys. Rev.* B **75** 220406(R).

# Magnetic and nonmagnetic impurity effects on $Cu_3Mo_2O_9$

**Y Ebukuro[1], M Noda[1], H Kuwahara[1], H Kuroe[1], M Hase[2], K Oka[3], T Ito[3] and H Eisaki[3]**

[1] Sophia University, Chiyoda-ku, Tokyo 102-8554, Japan
[2] National Institute for Materials Science (NIMS), Tsukuba, Ibaraki 305-0047, Japan
[3] National Institute of Advanced Industrial Science and Technology (AIST), Tsukuba, Ibaraki 305-8562, Japan

E-mail: y-ebukuro0116@eagle.sophia.ac.jp

**Abstract.** We compare the magnetic (Ni) substitution effects in $Cu_3Mo_2O_9$ to the nonmagnetic (Zn) ones. Changes in the lattice constants show similar behaviours for both impurity substituted samples. In specific heat measurements, we observe the anomalous specific heat at the phase transition temperature $T_N$. $T_N$ in the Ni substituted sample increases to 8.8 K; while $T_N$ in the Zn substituted sample decreases to 4.0 K. In magnetization measurement on the Ni substituted sample, we obtain the Van-Vleck type constant magnetization. We suggest the existence of the single-ion anisotropy of the substituted $Ni^{2+}$ ions, which is a possible candidate for the origin of the increment of $T_N$ in the Ni substituted sample.

## 1. Introduction

Impurity substitution effects have been extensively studied in the research field of magnetism. Controlling of magnetic properties by impurity substitution is an interesting issue not only in the viewpoint of science but also in the viewpoint of application. Sometimes impurity substitution induces a different long-range order from that in the parent material. This phenomenon has been studied in detail from the viewpoint of a new ordered state induced by disturbing the original ordered state in the parent material, i.e., "the order-by-disorder effect". As an example, we summarized the impurity substitution effect in an inorganic spin-Peierls material $CuGeO_3$ [1-3]. It has been reported that spin-Peierls transition is suppressed by nonmagnetic $Zn^{2+}$ ions substituted for $Cu^{2+}$ ions in this material; instead, another antiferromagnetic phase with the easy axis parallel to the *c*-axis appears at low temperatures.

Comparison between the magnetic and the nonmagnetic impurity effects is also interesting. In $CuGeO_3$, the effects of the magnetic impurity substitution have been reported [4-6]. In the spin-Peierls phase where $Cu^{2+}$ ions form nonmagnetic spin dimers, $S = 0$ $Zn^{2+}$ and $S = 1$ $Ni^{2+}$ ions introduce similar effects. i.e. the breaking of spin-singlet dimers. One difference occurs due to the single-ion anisotropy of $S = 1$ $Ni^{2+}$ ions: the change of the easy axis almost parallel to the *a*-axis in the low temperature antiferromagnetic phase in Ni substituted sample.

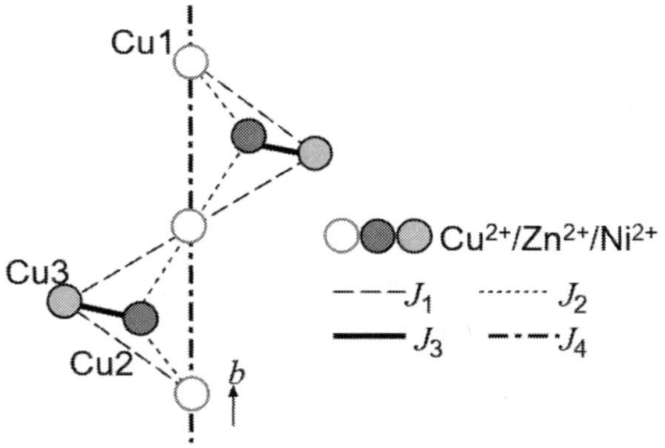


Figure 1: (color online) The schematics of the Cu sites forming the distorted tetrahedral spin chain of $Cu_3Mo_2O_9$. The colors of the circles and the numbers just after the symbols of chemical elements distinguish the crystallographically different sites from one another. The $Cu^{2+}$ ions are magnetically interacted through the $J_1$-$J_4$ super-exchange paths distinguished by different line types.

In this paper, we discuss the similarity and the difference between magnetic and nonmagnetic impurity substitution effects on $Cu_3Mo_2O_9$. Figure. 1 shows a distorted tetrahedral quasi-one-dimensional spin structure made from $Cu^{2+}$ ions in $Cu_3Mo_2O_9$. By substituting impurity ions ($Ni^{2+}$ and $Zn^{2+}$) for the $Cu^{2+}$ ions, it is possible to change the path of the super-exchange interaction directly and to affect the frustration effect of $Cu_3Mo_2O_9$ [7]. Because of the difference between the crystal radius of impurity ion and that of $Cu^{2+}$ one [8], we need to consider the chemical pressure effect.

## 2. Experiments

As a starting material, we synthesized polycrystalline powder of $(Cu,Zn,Ni)_3Mo_2O_9$ with various impurity concentrations from a stoichiometric mixture of raw materials, CuO, NiO, ZnO and $MoO_3$ by using a conventional solid-states reaction. We confirmed that the polycrystalline powder is single phase from the x-ray powder diffraction patterns. We obtain the lattice constants based on the orthorhombic crystal structure with the *Pnma* symmetry of the lattice. Recent result clearly shows that the monoclinic crystal structure with the $P2_1/m$ symmetry of the lattice is more suitable at room temperature [9]. Because of the limited accumulation time, we did not observe the peak in the diffraction pattern identifying the $P2_1/m$ lattice symmetry. The systematic change of the lattice constants at room temperature will be discussed later.

We prepared single crystals of $Cu_3Mo_2O_9$, $(Cu,Zn)_3Mo_2O_9$ and $(Cu,Ni)_3Mo_2O_9$ with 5.0% of Ni or Zn concentration. The single crystal of each sample was synthesized from the pressed powder with a rod shape by using the continuous solid-state crystallization method, which is based on the solid state reaction [10]. We assume that the impurity concentration in the single crystal is the same as that in the starting polycrystalline powder because the powder is not melt in this process. We determined the crystal axes by using Laue photograph and cut them into a rectangular parallelepiped of which surface vectors toward each crystal axis.

The temperature dependence of specific heat at 0 T was measured by using Quantum Design Physical Properties Measurement System (PPMS). The temperature dependences of magnetization under static magnetic field of 0.1 T were measured by using Quantum Design Magnetic Properties Measurement System (MPMS-XL).

## 3. Results

*3.1 Lattice constants at room temperature*

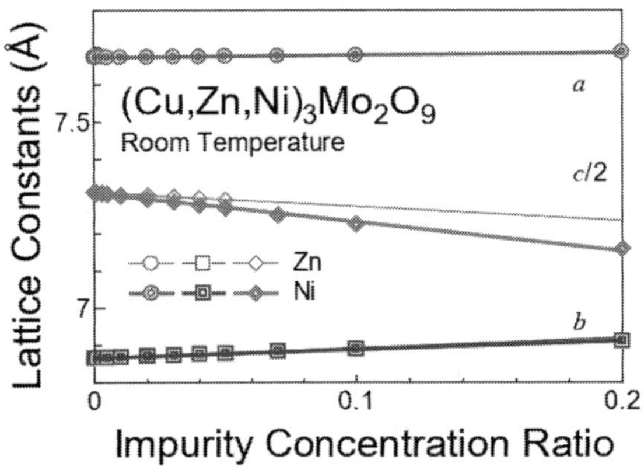


Figure 2: (color online) Lattice constants as functions of impurity concentration ratio.

**Table 1**: Spin moments, crystal radii referred from ref. [8] and the mass numbers of some ions. The crystal radii are determined by considering the number of oxygen ions around each ion, i.e., five (six) oxygen ions surround the Cu1 site (the Cu2 and Cu3 sites).

| Ions | *S* | Crystal Radius for Cu1 site (Å) | Crystal Radius for Cu2 and Cu3 sites (Å) | Mass Number |
|---|---|---|---|---|
| **$Ni^{2+}$** | 1 | 0.83 | 0.77 | 58.69 |
| **$Cu^{2+}$** | 1/2 | 0.87 | 0.79 | 63.55 |
| **$Zn^{2+}$** | 0 | 0.88 | 0.82 | 65.39 |

Figure 2 compares the lattice constants in Ni and Zn substituted samples as functions of impurity substitution ratio. The qualitative trends of the changes of the lattice constants in the Ni and the Zn substituted samples are similar with each other. The change of lattice constant along the *c*-axis in the Ni substituted samples is larger than that in the Zn substituted ones. The lattice constants along the *a*- and the *b*-axes have similar trends in both substituted samples. According to Table I, the crystal radius of $Zn^{2+}$ ion is larger than that of $Cu^{2+}$ ions but $Ni^{2+}$ one is smaller. It seems that this result was not expected from a standard chemical pressure model based on the packing of the rigid spheres with the average crystal radii. This is the unexpected similarity between magnetic and nonmagnetic impurity substitution effects.

### *3.2 Specific heat*

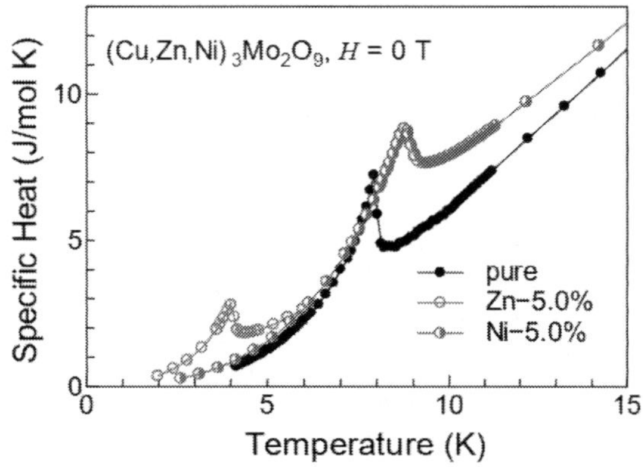


Figure 3: (color online) Temperature dependences of specific heats in $Cu_3Mo_2O_9$ and the Zn-5.0% and the Ni-5.0% substituted samples.

Figure 3 shows the temperature dependences of the specific heats (the $C$-$T$ curves) in $(Cu,Zn,Ni)_3Mo_2O_9$. We observed an anomalous specific heat, the jump of $C$ due to the second-order phase transition, in each $C$-$T$ curve. The peaks in the impurity substituted samples are rounded. We define $T_N$ as the temperature which gives the local maximum of the $C$-$T$ curve. We estimated $T_N$ in the Zn-5.0% and the Ni-5.0% substituted samples as 4.0 and 8.8 K, respectively, which are lower and higher than $T_N$ in the parent material. This is the unexpected difference between magnetic and nonmagnetic impurity substitution effects.

### *3.3 Magnetization*

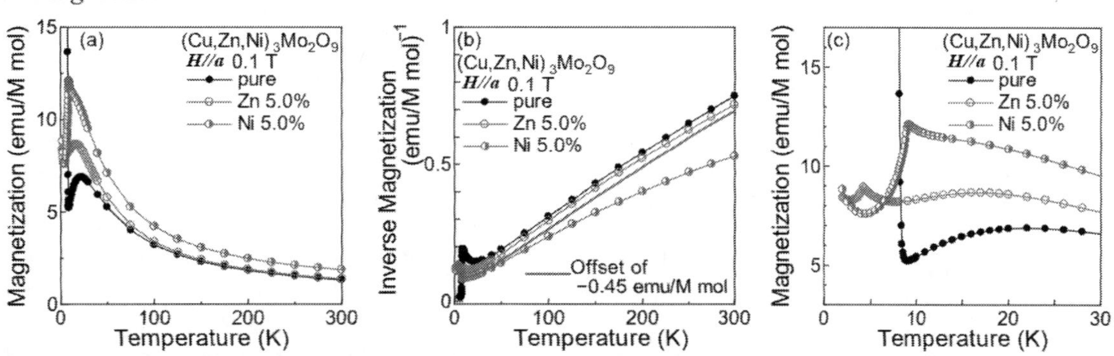


Figure 4: (color online) (a) Temperature dependences of magnetization at 0.1 T along the $a$-axis in $Cu_3Mo_2O_9$ and the Zn-5.0% and the Ni-5.0% substituted samples. (b) Inverse magnetization of (a). (c) Low temperature expansion of (a). M in the unit of the ordinate denotes magnetic ions ($Cu^{2+}$ and $Ni^{2+}$).

Figure 4(a) shows the temperature dependences of magnetization $M_a$ along the $a$-axis in $(Cu,Zn,Ni)_3Mo_2O_9$. In this work, we show $M_a$ for one mole of the magnetic ions. One can see in all panels in Fig. 4 that the Ni substitution increases $M_a$. The high temperature magnetization seems to be Curie-Weiss type. To estimate the gain and the offset parameters, we show $M_a^{-1}$ as functions of temperature in Fig. 4(b). One can see that the temperature dependence of $M_a^{-1}$ in the Ni substituted sample is very different from other two temperature dependences. When we subtract 0.45 emu/M mol from $M_a$ in the Ni substituted sample, as shown with the bold curve in Fig. 4(b), $(M_a - 0.45)^{-1}$ has a similar slope with $M_a^{-1}$ of $Cu_3Mo_2O_9$ and the Zn substituted sample in the wide temperature range. The small difference between $(M_a - 0.45)^{-1}$ in the Ni substituted sample and $M_a^{-1}$ in other two samples might be explained with the change of Weiss temperature. The origin of offset is Van-Vleck type temperature independent magnetization which is possible even in the isolated $S = 1$ $Ni^{2+}$ ions.

As shown in Fig. 4(c), the strong enhancement of $M_a$ below $T_N$ is observed in the parent material [7]. This is due to the order of weak ferromagnetic component of the spin moments under magnetic fields, which reflects the temperature dependence of the order parameter [11]. Instead of this increment, we observed the reduction of $M_a$ with decreasing temperature in the Zn and the Ni substituted samples. In our previous report [11], we explained this reduction in the Zn-5.0% substituted sample as cusp-shaped temperature dependence. In the Ni-5.0% substituted sample, we observed similar temperature dependence with very different $T_N$, i.e., $M_a$ rapidly decreases with decreasing temperatures just below $T_N$. We defined $T_N$ as the temperature at which $\partial M/\partial T$ had local maximum. $T_N$ in the Zn and the Ni substituted samples are 4.2 and 9.0 K, respectively. The values are slightly (~ 0.2 K) higher than $T_N$ estimated based on the results of the specific heat. This trend is consistent with our previous report [12]. Once the $M_a$ dropped just below $T_N$, a Curie-Weiss-type increment of $M_a$ with decreasing temperature was observed at the lowest temperature region in our measurements. This term in the Ni substituted sample is larger than that of Zn substituted one.

Above $T_N$, we observed broad peaks of $M_a$, which reflects the quasi-one dimensionality of $Cu_3Mo_2O_9$ [7]. In the impurity substituted samples, the temperature at which $M_a$ show local maximum is slightly lower than that in the parent material. This suggests that the quasi-one dimensionality of the magnetic system is suppressed or the amplitude of the effective magnetic interaction is lowered by impurity substitutions.

## 4. Discussion

Let us discuss the origin of the increment of $T_N$ upon the Ni substitution. When the easy axis or the easy plane induced by the single-ion anisotropy of $Ni^{2+}$ ions is consistent with (or not conflict with) the magnetic anisotropy in the antiferromagnetic phase of the parent material, Ni substitution has the potential to stabilize the magnetic order and to increase $T_N$. However, breaking of translational symmetry due to impurity substitution suppresses long-range order of the spin system, which decrease $T_N$. Due to the competitive balance of the above two effects, there is a possibility that the Ni and the Zn substituted samples show the different substitution-ratio dependences with each other. The difference of $T_N$ between the Ni and the Zn substituted $La_2CuO_4$ is a good example [13-15]. $La_2CuO_4$, which is known as the parent material of the high-temperature superconductor, has an antiferromagnetic $CuO_2$ plane and has the antiferromagnetic phase at which the $S = 1/2$ spins lying in the $CuO_2$ plane [16]. $La_2NiO_4$ has the similar crystal structure and has the antiferromagnetic phase at which the $S = 1$ spins lying in the $NiO_2$ plane [17]. Because of the single-ion anisotropy of the $Ni^{2+}$ ions, the spin direction in the antiferromagnetic phase is different from that in $La_2CuO_4$ [15]. When the $Ni^{2+}$ ions are introduced as a magnetic impurity into $La_2CuO_4$, they do not break the magnetic anisotropy in the $CuO_2$ plane of the parent material. Moreover, $Ni^{2+}$ ions interact antiferromagnetically with the surrounding $Cu^{2+}$ ions, which hardly breaks the antiferromagnetic long-range order. Even in the Ni-30% substituted sample, $T_N$ is almost unchanged. Zn substitution as a nonmagnetic impurity doping breaks the antiferromagnetic long-range order composed of the $Cu^{2+}$ ions, which decreases $T_N$ [13-16, 18].

We discuss the single-ion anisotropy in $Cu_3Mo_2O_9$. We observed Van-Vleck temperature-independent paramagnetic term in $M_a$ as shown in Figs. 4(a) - 4(c). According to the Pryce's Hamiltonian [19], single-ion anisotropy is also expected to appear in the Ni substituted sample because both of these come from the spin-orbit interaction. To discuss in more details, we should know the precise direction of the quantum axis of $Ni^{2+}$ ions, which is beyond the scope of this paper. Below $T_N$, spins of substituted $S = 1$ $Ni^{2+}$ ions point a fixed direction in the lattice, which generates an internal magnetic field. When this internal magnetic field encourages antiferromagnetic order, it makes $T_N$ higher. We conclude that this is the origin of the increment of $T_N$ upon the Ni substitution.

One may think that the systematic change of the crystal radius listed in Table 1 may cause a chemical pressure effect. Because the lattice constants in both of the Zn and the Ni substituted samples have the same impurity-concentration dependences as shown in Fig. 2. We consider it cannot be the origin of the change in $T_N$.

The Curie-Weiss type increment of $M_a$ at low temperatures is probably due to the almost free spin induced by the impurity substitution, which might be related to the magnetic structure of the parent material. According to the recent result of powder neutron diffraction [9], the spins at the Cu1 site is almost nonmagnetic. The magnitude of spin moments at the Cu2 and Cu3 sites are $(0.50 \sim 0.74)\mu_B$, where $\mu_B$ denotes the Bohr magneton. This strongly suggests that the magnetic ground state of this system includes the component of spin singlet state. In spin-Peierls system $CuGeO_3$ [1-3], as stated in the previous section, nonmagnetic impurities suppress the formation of spin singlet state and introduce spin moments. In case of the Ni substitution [4-6], the spin moment of $Ni^{2+}$ ion also breaks the spin-singlet state. The similarity between magnetic and nonmagnetic impurity substitution effects, the appearance of the Curie-Weiss term in both of the Ni and the Zn substituted samples, suggests the existence of the component of the spin singlet state in the magnetic ground state of $Cu_3Mo_2O_9$.

## 5. Conclusion

We compared the magnetic (Ni) substitution effects in $Cu_3Mo_2O_9$ to the nonmagnetic (Zn) ones. Changes in the lattice constant showed the similar behaviors qualitatively for both impurity substituted samples. In specific heat measurements, we estimated $T_N$ in the Zn-5.0% and the Ni-5.0% substituted samples as 4.0 and 8.8 K, respectively. In magnetization measurements on the Ni substituted sample, we obtained the Van-Vleck type temperature independent magnetization. We clarified the existence of the single-ion anisotropy of the substituted $Ni^{2+}$ ions, which is a possible candidate for the origin of the increment of $T_N$ in the Ni substituted sample.

## References

[1] Hase M *et al.* 1993 *Phys. Rev. Lett.* **70** 3651
[2] Hase M *et al.* 1993 *Phys. Rev. Lett.* **71** 4059
[3] Hase M *et al.* 1995 *Phsica* B **215** 164
[4] Koide N *et al.* 1998 arXiv:cond-mat/9805095 [cond-mat.str-el]
[5] Koide N *et al.* 1996 *Czech J. Phys.* **46** 1981
[6] Grenier B *et al.* 2002 *Phys. Rev.* B **65** 094425
[7] Hamasaki T *et al.* 2008 *Phys. Rev.* B **77** 134419
[8] Shannon R D 1976 *Acta. Cryst.* A **32** 751
[9] Hase M *et al.* 2015 *Phys. Rev.* B **92** 054425
[10] Oka K *et al.* 2011 J. *Cryst. Growth* **334** 108
[11] Kuroe H *et al.* 2014 *JPS Conf. Proc.* **2** 010206
[12] Hirata Y *et al.* 2015 *Phys. Procedia* **75** 134
[13] Ting S T *et al.* 1992 *Phys. Rev.* B **18** 11772
[14] Machi T *et al.* 2003 *Physica* C **233** 388
[15] Hiraka H *et al.* 2005 *J. Phys. Soc. Jpn.* **74** 2197
[16] Vaknin D *et al.* 1987 *Phys. Rev. Lett.* **58** 2802
[17] Aeppli G *et al.* 1988 *Phys. Rev. Lett.* **61** 203
[18] Uchinokura K *et al.* 1995 *Physica* B **205** 234
[19] Pryce M H L 1950 *Proc. Phys. Soc.* A **63** 25

# High-frequency electron-spin-resonance measurements on $Mn_xMg_{1-x}O$ ($x = 1.0 \times 10^{-4}$) and DPPH at very low temperatures

**Y Ishikawa[1], K Ohya[2], S Miura[1], Y Fujii[2], S Mitsudo[2], T Mizusaki[2], A Fukuda[3], A Matsubara[4], H Kikuchi[1], T Asano[1], H Yamamori[5], S Lee[6] and S Vasiliev[7]**

[1] Department of Applied Physics, University of Fukui, Fukui 910-8507, Japan
[2] Research Center for Development of Far-Infrared Region, University of Fukui, Fukui 910-8507, Japan
[3] Department of Physics, Hyogo College of Medicine, Nishinomiya 663-8501, Japan
[4] Department of Physics, Kyoto University, Kyoto 606-8501, Japan
[5] Technical Division, School of Engineering, University of Fukui, Fukui 910-8507, Japan
[6] Department of Physics, Korea Advanced Institute of Science and Technology, Daejeon 305-701, Korea
[7] Department of Physics and Astronomy, University of Turku, Turku 20014, Finland

E-mail: yuya-i@u-fukui.ac.jp

**Abstract.** We have developed a millimeter-wave electron-spin-resonance (ESR) system for very low temperatures ($T < 1$ K) that can be employed for nuclear-magnetic-resonance measurements by using dynamic nuclear polarization. The system uses a Fabry-Pérot resonator that works in the frequency range of 125 – 130 GHz and covers the temperature range of 0.09 – 6.5 K. We have performed ESR measurements in the frequency around 128 GHz by using $Mn_xMg_{1-x}O$ ($x = 1.0 \times 10^{-4}$) and free-radical samples of 1, 1-diphenyl-2-picrylhydrazyl (DPPH), because these samples have been proposed as field and sensitivity markers. Temperature dependence of the ESR signal intensity for $Mn_xMg_{1-x}O$ shows anomalies originating from magnetic order are found around 3.5 – 4 K. We estimate the sensitivity of the system for ESR detections to be $6 \times 10^{13}$ spins/G at 5.8 K. Because DPPH shows no observable shift in the magnetic field, we propose it as a useful standard marker for ESR measurements at very low temperatures.

## 1. Introduction

Magnetic-resonance techniques such as electron spin resonance (ESR) and nuclear magnetic resonance (NMR) are used in several research fields to provide useful information about the microscopic properties of various material samples. Dynamic nuclear polarization (DNP), in which a high electron-spin polarization is transferred to nuclear polarization through a hyperfine interaction, is a powerful technique for enhancing the sensitivity of NMR measurements. A typical example of DNP was suggested more than 50 years ago in silicon lightly doped with phosphorus (Si:P) [1]. However, it has been difficult to observe the $^{31}$P-NMR signal directly, because the critical concentration of P atoms in an insulator is too low. Moreover, recently it becomes intriguing to study spin dynamics, including DNP phenomena, for Si:P at mK temperatures because one of the most practical quantum computer designs proposed by Kane [2] uses Si:P at high magnetic-field strengths ($B$ > 3 T) and low temperatures ($T$ < 100 mK). Song *et al.* reported that $^{31}$P nuclear-spin polarization is enhanced by two or three orders of magnitude over the thermal-equilibrium value by using DNP induced by microwave irradiation above 80 GHz [3]. Fujii *et al.* have performed steady-state ESR experiments on the donor-electron spin in lightly doped Si:P (concentration $n = 6.5 \times 10^{16}$ cm$^{-3}$) at 3.6 T and 1.44 K. Their work indicates that the nuclear polarization can be changed by NMR pulse [4]. Recently, Järvinen *et al.* have performed extensive high-frequency ESR studies with a $^3$He/$^4$He dilution refrigerator (D/R) and have demonstrated $^{31}$P nuclear polarization of higher than 98% [5, 6]. We have been developing an ESR/NMR double-magnetic-resonance system enclosed in a D/R probe, which enables us to perform DNP-NMR and ENDOR measurements to investigate $^{31}$P spin dynamics. Developments for millimeter-wave ESR devices optimized for operation well below one Kelvin have been challenging and applied to, for example, studies on magnetic systems [7, 8]. In this paper, we report the current status of our developmental work and provide ESR measurements on $Mn_xMg_{1-x}O$ ($x = 1.0 \times 10^{-4}$) and DPPH. We have also determined the ESR sensitivity from measurements on $Mn_xMg_{1-x}O$ and have investigated the behavior of DPPH as a magnetic-field marker at very low temperatures.

## 2. Experiment

Figure 1 shows a schematic of our ESR apparatus in a $^3$He/$^4$He D/R (Kelvinox MX 400, Oxford Instruments). The D/R is installed in an 18-T superconducting magnet. A Gunn diode (PLO-6-128-01, Farran Technology) is used as an phase-locked oscillator with the output power of 10 mW around the

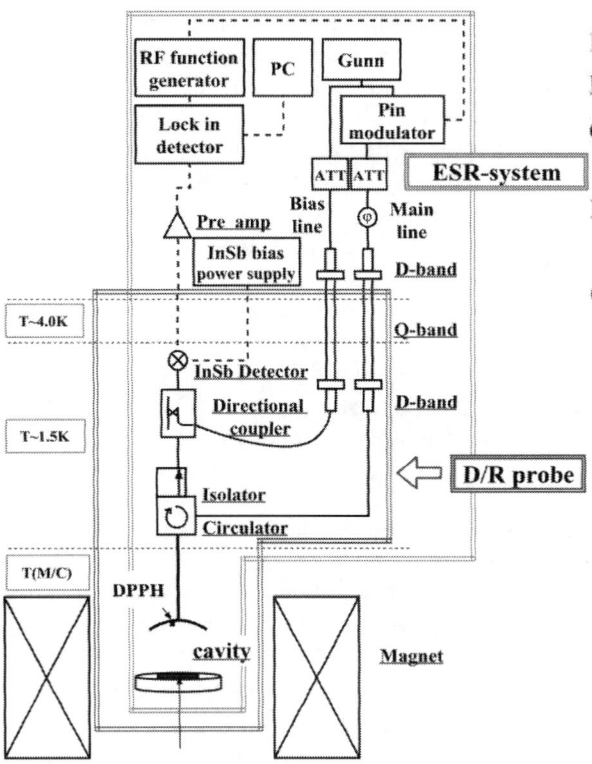


**Figure 1.** ESR system in the dilution refrigerator used for $Mn_xMg_{1-x}O$ and DPPH measurements.

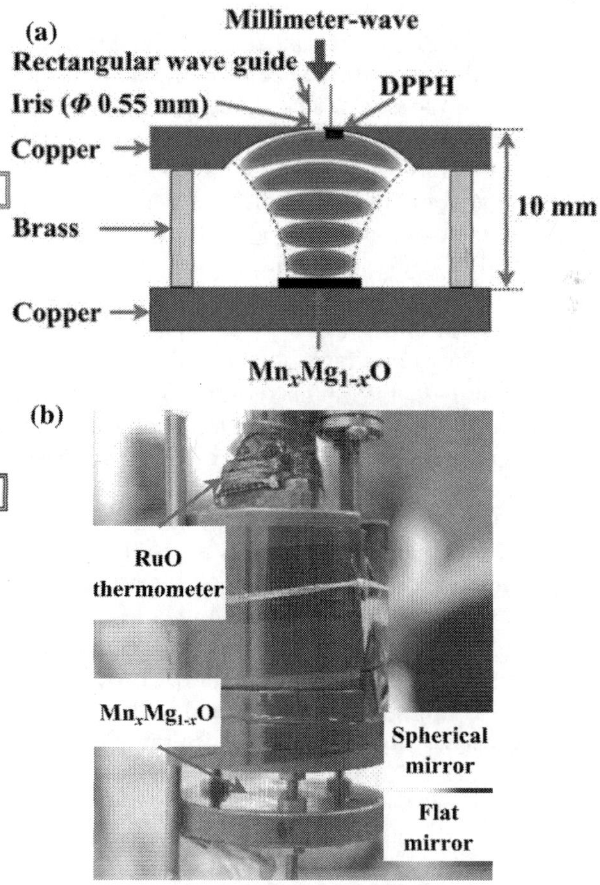


**Figure 2.** **(a)** Cross-section of the Fabry-Pérot resonator for our mm-wave ESR. **(b)** A photograph of the Fabry-Pérot resonator.

central frequency (~128 GHz). Millimeter waves from the oscillator are split by a directional coupler into main- and bias-lines in a homodyne detection system at cryogenic temperatures. ATT in Figure 1 stands attenuator. Temperature is measured using ruthenium oxide thermometers attached to a Fabry–Pérot resonator (FPR). The resonator frequency can be changed by changing the distance between the two mirrors [9]. DNP enhanced NMR or electron-nuclear double resonance experiments can be carried out with an FPR, though the double resonance experiments will be reported elsewhere. Figure 2 shows a schematic and a photograph of the FPR. The upper and lower parts of the FPR are a spherical and a flat mirror, respectively. The incident millimeter waves pass through the iris (coupling hole with 0.55 mm in diameter) and reflect between the mirrors, forming $TEM_{00q}$ resonant modes, where $q$ is an integer. The measurable temperature range extends down to 90 mK.

In order to evaluate the developed system, we employ two samples often used as standard samples of ESR [10]; $Mn_xMg_{1-x}O$ in which $Mn^{2+}$ is diluted in nonmagnetic MgO, and an organic radical crystal DPPH which has a radical of $S = 1/2$ (a powder sample, contains 10-20 % benzene, manufactured by Tokyo Chemical Industry). An isolated $Mn^{2+}$ ion (for which the outermost electron is in the $3d^5$state) has an ESR spectrum that is split into six components, which is known as its "hyperfine structure." This splitting is caused by the interaction of the isolated $Mn^{2+}$ ion with the $^{55}Mn$ nuclear spin ($I = 5/2$). For this reason, a sample containing $Mn^{2+}$ ions diluted in a nonmagnetic substance is useful as a magnetic-field marker [10-12], and $Mn_xMg_{1-x}O$ is thus used as such a marker in commercially available ESR apparatus. For $Mn_xMg_{1-x}O$ samples where $x$ is sufficiently small (so that the spin interaction is small), it is expected that magnetic correlations will not develop at extremely low temperatures. Therefore, the detection sensitivity can be determined from the intensity of the ESR signal at cryogenic temperatures. We prepare a sample of $Mn_xMg_{1-x}O$ ($x = 1.0 \times 10^{-4}$) by first weighing the MgO and MnO to achieve the desired molar ratio and then heating the appropriate mixture in an electric furnace at approximately 1000 °C for 72 hours. Using a powder X-ray diffraction apparatus, we confirm that the crystal structure of the sample is that of MgO.

DPPH is well known as a standard sample of ESR [13]. However, magnetic properties at low-temperature region strongly depend on its crystal solvents [14-18]. It is important to investigate commercially available DPPH for further analyses at very low temperatures. We acquire fine powder of DPPH by grinding tiny crystals in order to obtain powder average of anisotropic $g$-value of DPPH. The powder DPPH is mixed with Apiezon N grease (M&I Materials) and put on the surface of the spherical mirror as shown in Figure 2. A amount of heat generated around the sample is estimated by the following simple consideration: The attenuation of the electromagnetic wave from the oscillator output (~10 mW at maximum) to the entrance of the resonator is approximately 40 dB. Thus, in each mirror of the resonator, a caloric value of 0.5 μW will be generated at maximum. We consider that this value is sufficiently small to avoid significant temperature rise of the sample.

## 3. Results and discussions

The distance from the iris to the flat mirror have fixed at 10 mm, which corresponds to the $TEM_{007}$ mode for a Gaussian beam. As shown in Figure 2 (b), we affix a 10.8 mg sample of $Mn_xMg_{1-x}O$ to the center of the plane mirror. The $Mn_xMg_{1-x}O$ sample and the DPPH sample were separately positioned at the center of the superconducting magnet by raising and lowering the D/R probe and measuring the field strength. Figure 3 shows the ESR magnetic field as a function of D/R main body height for the two samples of $Mn_xMg_{1-x}O$ and DPPH. Here, data for the lowest field line '$Mn_{1st}$' among the six lines of $Mn_xMg_{1-x}O$ are plotted. Each sample is at the center of the magnet when the resonance field takes minimum. The magnetic field strength profile is approximately parabolic near the center. It is natural that the difference of the positions that give the minimum resonance fields for the two samples are 10 mm distant from each other reflecting on the vertical distance between the samples as shown in Figure 2 (a).

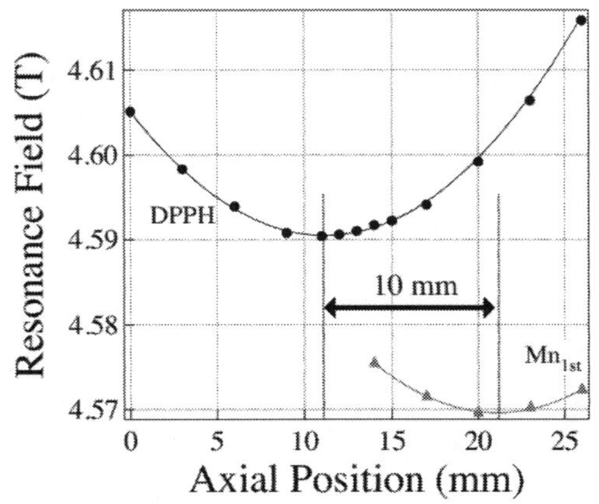


**Figure 3.** Magnetic-field measurements for DPPH and for the $Mn_{1st}$ signal (see definition below) in $Mn_xMg_{1-x}O$ as functions of the position of the D/R probe along the magnetic field axis. Observed resonance fields where the ESR line for DPPH and $Mn_xMg_{1-x}O$ appear as functions of axial position of the sample. Each sample is located at the center of the magnet when the field is a minimum. Each parabolic line shows the fitting result to experimental plot.

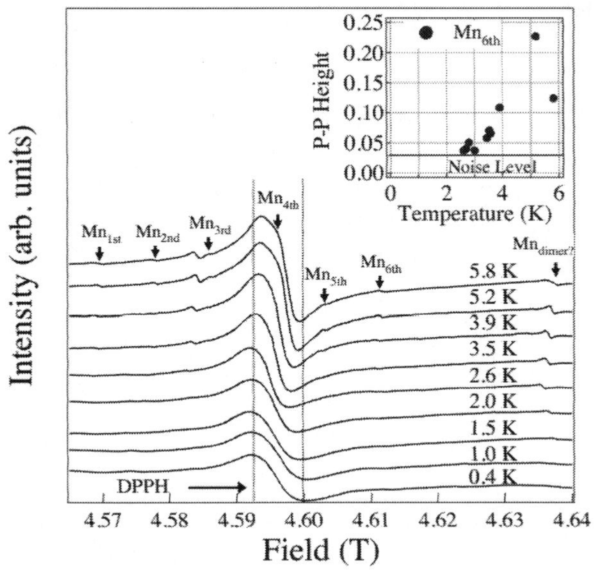


**Figure 4.** Temperature dependence of the dispersion signal of $Mn_xMg_{1-x}O$. The large central resonance line is from the DPPH radical in the FPR. The inset shows temperature dependence of the intensity (peak-to-peak height) of the $Mn_{6th}$ signal.

Figure 4 shows the temperature dependence of the ESR spectrum of $Mn_xMg_{1-x}O$. At 5.8 K, we can identify six dispersion curves due to $Mn^{2+}$. Moreover, we label them as $Mn_{1st}$ through $Mn_{6th}$ in the order of increasing magnetic–field strength. These signals are separated from each other by approximately 420 Oe; this interval is due to the hyperfine structure and is consistent with the previously reported value [11, 12]. Although these ESR signals are observed at 5.8 K, they gradually decrease in amplitude as the temperature is lowered and disappear completely at 0.4 K. Here, the peak-to-peak (P-P) width is defined as the width of peak to peak of dispersion signal. Figure 4 inset shows the temperature dependence of the signal intensity (peak-to-peak height) of the dispersion curve with focusing on the $Mn_{6th}$ signal. The signal intensity gradually decreases below 5.8 K and becomes invisible below 2.5 K. This is probably because the $Mn^{2+}$ ions are not isolated sufficiently in the sample; therefore, correlations between relatively close ions develop at low temperatures. We determin the ESR detection sensitivity, define as the number of spins observable per Oe, from the $Mn_{6th}$ signal at 5.8 K. ESR signal of unknown origin is detected around 4.365 T. This is likely to be a contribution from Mn – Mn dimer [19]. Here, the sensitivity of our device is estimated on the premise that all of the Mn of the sample are detected in the hyperfine split six ESR lines. The number of $Mn^{2+}$ spins in the 10.8 mg sample of $Mn_xMg_{1-x}O$ ($x = 1.0 \times 10^{-4}$) is $1.0 \times 10^{16}$. Because the $Mn^{2+}$ signal is divided into six spectra, the total number of spins must be divided by six. The P-P width of $Mn_{6th}$ is 11 Oe and the S/N ratio is 4, the detection sensitivity is $6 \times 10^{13}$ spins/G. We consider this value to be a

IOP Conf. Series: Journal of Physics: Conf. Series **969** (2018) 012111 doi:10.1088/1742-6596/969/1/012111

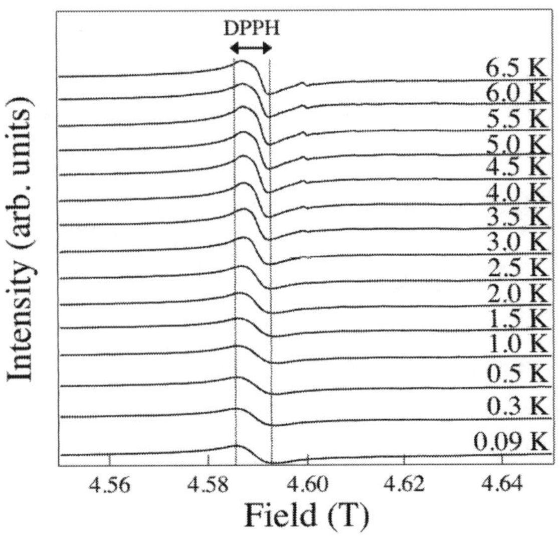


**Figure 5.** Dispersion of ESR signals from DPPH between 6.5 and 0.09 K. Signals from the $Mn_xMg_{1-x}O$ sample in the FPR are also observed.

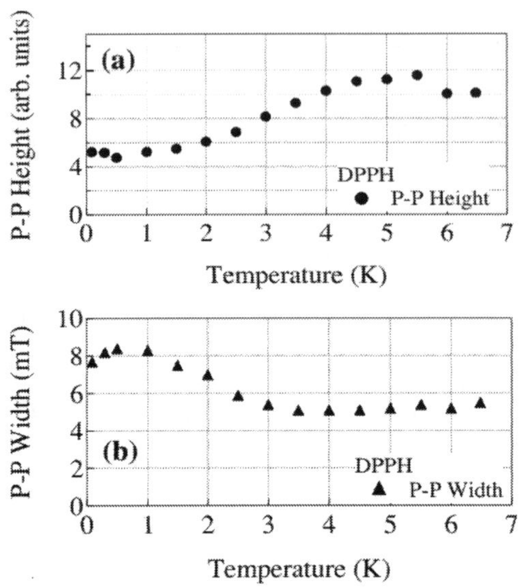


**Figure 6.** Temperature dependence of (a) the signal intensity (peak-to-peak height) and (b) peak to peak width of the ESR spectra of DPPH.

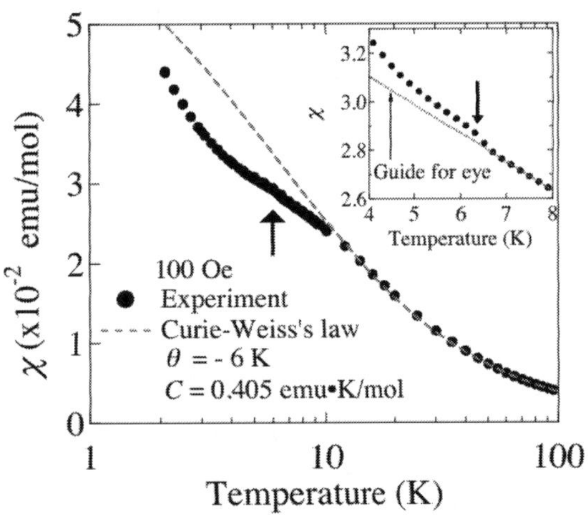


**Figure 7.** Temperature dependence of the magnetic susceptibility of DPPH. The arrow indicates an anomaly around 6 K. The inset shows the expansion below 8 K. The blue dot line shows Curie-Weiss's law $\chi = C/(T-\theta)$

lower limit to the actual sensitivity. Further experiments are necessary for more accurate quantitative discussion.

We conducted measurements on the DPPH sample over the temperature range of 6.5 – 0.09 K. As shown in Figure 5, the spectra clearly change over this temperature range. The changes in P-P width and signal intensity are complicated, as shown in Figure 6. We observe small anomalies both in the P-P width and the intensity near 6 K. As the temperature decreases below 3.5 K, the P-P width gradually increases, and the intensity decreases. Below approximately 0.5 K, the P-P width again decreases. We also measure the magnetic susceptibility $\chi$ of a 6.7 mg DPPH-powder sample in an applied magnetic field of 100 Oe by using the superconducting quantum interference device (SQUID) magnetometer (MPMS, Quantum Design) in the temperature range of 2 – 100 K. Results are shown in Figure 7. The susceptibility deviates from high-temperature Curie-Weiss type behavior below around 20 K, suggesting development of short-range antiferromagnetic ordering. We also have found a tiny anomaly in the susceptibility around 6 K. This anomaly possibly corresponds to small change of ESR intensity from DPPH at 6 K shown in Figure 6 (a), though its origin is unclear so far. Considering that low-temperature behavior of ESR spectrum related with the development of magnetic correlations in this material differs depending upon the type of solvent used [13-18], it is important to analyze crystal structures of our DPPH with benzene. Nevertheless, the observed ESR spectrum of our DPPH shows almost no shift in the resonance position over the temperature range of 6.5 – 0.09 K, although the P-P width increases by about a factor of 1.5. We therefore consider DPPH to function as a useful standard sample with a magnetic-field accuracy of a few tens of gauss over this temperature range.

## 4. Summary

We have performed ESR measurements on $Mn_xMg_{1-x}O$ ($x = 1.0 \times 10^{-4}$) and DPPH at approximately 128 GHz over the temperature range of 0.09 – 6.5 K by using a millimeter-wave ESR apparatus developed for DNP-NMR measurements. At 5.8 K, we can identify six dispersion curves due to $Mn^{2+}$. ESR signals from $Mn^{2+}$ gradually decrease in amplitude as the temperature is lowered and disappear completely at 0.4 K. This is because the $Mn^{2+}$ ions are not isolated sufficiently. This behavior is likely to be a contribution from Mn-Mn dimer. From the ESR signal due to $Mn^{2+}$ ions at 5.8 K, we also estimate the detection sensitivity of the ESR apparatus to be $6 \times 10^{13}$ spins/G. In ESR measurements on DPPH, we have observed small anomalies around 6 K. We also have found a tiny anomaly in the susceptibility around 6 K. This anomaly possibly corresponds to small change of ESR intensity from DPPH. As the temperature is lowered near 3.5 K, the ESR P-P width gradually increases, while the signal intensity decreases. The P-P width decreases again at 0.5 K, and the signal intensity becomes constant. Because no shift in the resonance position was found over this temperature range, we consider that DPPH can be used as an ESR magnetic-field marker over the temperature range covered by our D/R experiments.

## Acknowledgments

This work is partly supported by JSPS KAKENHI Grant Number 17K05514 and 26400331 and by the Cooperative Research Program of Research Center for Development of Far-Infrared Region, University of Fukui (No. H27FIRDM011E, H28FIRDM024A, H29FIRDM015B).

## References

[1] Abragam A 1961 *The Principles of Nuclear Magnetism* (Oxford: Clarendon Press) chapter 9-10
[2] Kane B E 1998 *Nature* **133** 393
[3] Song M *et al.* 2010 *J. Phys.: Condens. Matter* **22** 206001
[4] Fujii Y *et al.* 2015 *J. Phys.: Conf. Series* **568** 042005
[5] Järvinen J *et al.* 2014 *Phys. Rev.* B **90** 214401
[6] Järvinen J *et al.* 2015 *Phys. Rev.* B **92** 121202
[7] Sakon T *et al.* 2003 *J. Phys. Soc. Jpn* **72** 140
[8] Smirnov A I *et al.* 2012 *Phys. Rev. B* **85** 184423
[9] Kogelnik H and Li T 1966 *Proc. IEEE.* **54** 10
[10] Smith G M and Riedi P C 2000 Progress in high field EPR *Electron Paramagnetic Resonance* vol 17, ed Gilbert B C *et al.* (Cambridge: The Royal Society of Chemistry) chapter 6, p 178
[11] Miner G K *et al.* 1972 *Rev. Sci. Instr.* **43** 1297
[12] Ohya H and Yamauchi J 1989 *Electron Spin Resonance* (in Japanese) (Tokyo: Kodansha) p 43
[13] Krzystek J *et al.* 1997 *J. Magn. Reson.* **125** 207
[14] Žilić D *et al.* 2010 *J. Magn. Reson.* **207** 34
[15] Fujito T 1981 *Bull. Chem. Soc. Jpn.* **54** 3110
[16] Prokhorov A M and Fedorov V B 1963 *Soviet Physics JETP* **16** 1489
[17] Voesch W *et al.* 2015 *Physics Procedia* **75** 503
[18] Teitel'baum G B *et al.* 1981 *JETP Letters* **34** 555
[19] Coles B A *et al.* 1960 *Phys. Rev. Lett* **4** 3

IOP Conf. Series: Journal of Physics: Conf. Series **969** (2018) 012112 doi:10.1088/1742-6596/969/1/012112

# Numerical Study of the Kitaev-Heisenberg chain as a spin model of the K-intercalated $RuCl_3$

**Cliò Efthimia Agrapidis[1], Jeroen van den Brink[1,2], Satoshi Nishimoto[1,2]**

[1]Institute for Theoretical Solid State Physics, IFW Dresden, 01069 Dresden, Germany
[2]Department of Physics, Technical University Dresden, 01069 Dresden, Germany

E-mail: c.agrapidis@ifw-dresden.de

**Abstract.** We study the one-dimensional Kitaev-Heisenberg model as a possible realization of magnetic degrees of freedom of the K-intercalated honeycomb-lattice ruthenium trichloride $\alpha$-$RuCl_3$, denoted as $K_{0.5}RuCl_m$. First, we discuss the possible charge ordering patterns in $K_{0.5}RuCl_m$, where half of the $j = \frac{1}{2}$ spins are replaced by nonmagnetic ions in the honeycomb layer. The phase diagram is obtained as a function of nearest-, second-, and third-neighbor repulsions in the atomic limit. We find a strong competition between possible three charge ordering patterns when the repulsions are comparable. We further calculate the spin-spin correlation functions for the 1D Kitaev-Heisenberg model, which is one of the ordering patterns, and consider the effect of the Kitaev term by comparing the results to those for the antiferromagnetic SU(2) Heisenberg chain.

## 1. Introduction

Since the Kitaev model was proposed in 2006 [1], a quantum spin liquid (QSL) state induced by the Kitaev-type bond-dependent interactions is one of the most active fields in the condensed matter physics. In recent years, because of the emergence of candidate materials and the interest in the QSL realization, there has been a growing number of researches on the Kitaev model, at first, and on the Kitaev-Heisenberg (KH) after. Adding the Heisenberg-type interaction to the initial, exactly integrable, model is necessary and inevitable for any realistic description. However, the interplay between the Kitaev and Heisenberg interactions is always a nontrivial problem to solve. Nevertheless, the vast majority of these studies has been focused on two-dimensional lattices, while research on the KH chain is lacking.

The first candidates for the Kitaev materials are $5d^5$ honeycomb iridates $A_2IrO_3$ ($A$=Na or Li) [2, 3]. These materials are referred to as spin-orbit assisted Mott insulators and the magnetic degrees of freedom can be well described within an effective $j = \frac{1}{2}$ pseudospin model. Due to the existence of two Ir-O-Ir exchange paths with a $\sim 90°$ bond angle in the edge-sharing $IrO_6$ octahedra, a substantial anisotropic Kitaev-type coupling stems from the particular superexchange interactions between the $j = \frac{1}{2}$ pseudospins in addition to the isotropic Heisenberg coupling. It is known that the Kitaev QSL is most likely preempted by a certain level of the Heisenberg interactions. In fact, the above iridates exhibit long-range magnetic order at low temperatures [4, 5, 6, 7], possibly assisted by off-diagonal and/or long-range interactions.

At present, one of the most promising materials close to the Kitaev QSL is the $4d^5$ ruthenium trichloride $\alpha$-$RuCl_3$ in its honeycomb crystal phase. Neutron and Raman scattering studies gave evidence for fractionalized excitations typical for the Kitaev QSL [8, 9, 10] and both very recent theoretical [11] and experimental investigations [12, 13, 14, 15, 16] indicate in this material the presence of a transition into a QSL in presence of external magnetic field. For this compound the effective $j = \frac{1}{2}$ description, as a $Ru^{3+}$ $4d$ analogue to the iridates, seems to still give a good description of the magnetic properties, although some degree of admixture between $j = \frac{1}{2}$ and $j = \frac{3}{2}$ in the $t_{2g}^5$ states may arise from a weaker SOC for $4d$ electrons than in the Ir $5d$ orbitals.

Very recently, the temperature-dependent electron energy loss spectroscopy (EELS) measurement was performed in K-intercalated $RuCl_3$, denoted as $K_{0.5}RuCl_m$ for a fully K doped sample [17]. The intercalated $K^+$ ions may provide charge carriers, and however, a sharp peak (gap) was observed at $\sim$ 0.9 eV, indicating an insulating feature of $K_{0.5}RuCl_m$, instead of charge gap $E_g = 1.1{-}1.2$ eV for the undoped $\alpha$-$RuCl_3$ [18, 19, 20]. This has been interpreted as half of the $j = \frac{1}{2}$ pseudospins being replaced by nonmagnetic $d^6$ ions whereas we should see a pseudogap behavior in the EELS spectrum if the intercalated $K^+$ ions are randomly distributed. Therefore, this insulating state can be envisaged as a formation of superlattice by charge disproportionation (charge ordering). Different possible charge ordering patterns are shown in Fig. 1. Of the particular interest in the present context is the chain-type pattern.

In this paper, we consider the charge ordering patterns stabilized by the intersite repulsions and study the 1D KH model as a possible minimal spin model to describe the magnetic properties of the K-intercalated $RuCl_3$. Using the Lanczos exact diagonalization method, we calculate real-space spin-spin correlation functions. The effect of the Kitaev term is discussed by seeing a deviation from the isotropic antiferromagnetic Heisenberg chain. This paper is organized as follows. In Sec. 2, we discuss the possible charge ordering patterns in the classical limit and obtain the phase diagram. We also define the 1D KH chain as a spin model for the chain-type ordering. In Sec. 3 we show the results of spin-spin correlation functions and discuss the effect of the Kitaev term on the antiferromagnetic Heisenberg chain. Section 4 contains summary and conclusions.

## 2. Model

### 2.1. Charge ordering pattern in $K_{0.5}RuCl_m$

The insulating feature of $K_{0.5}RuCl_m$ with a sharp peak at $\sim$ 0.5 eV in the EELS spectrum may be explained by charge ordering associated with a superlattice formation rather than a random distribution of $K^+$ ions. The superlattice structures are simply constructed by replacing half of the $j_{\text{eff}} = 1/2$ pseudospins with nonmagnetic sites in the original hexagonal cluster. Practically, there are three possibilities of the charge ordering pattern; triangle-, dimer-, and chain-type charge orderings. So far, no crystal structural data of $K_{0.5}RuCl_m$ has been available and the detailed calculations are difficult. Thus, we simply consider the charge distribution to minimize the Coulomb repulsions, which are typically of larger order of magnitude than the magnetic interactions. The nearest- ($V_1$), second- ($V_2$), and third-neighbor ($V_3$) repulsions are taken into account (see Figure 1). In the atomic limit (hopping $t = 0$), the energy per particle is $E_t = 3V_2$, $E_d = \frac{1}{2}V_1 + V_2 + V_3$, and $E_c = V_1 + V_2$ for the triangle-, dimer-, and chain-type charge orderings, respectively. For a given set of $V$'s, the dominant configuration is determined as one having the lowest energy. In Figure 1, the obtained phase diagram is shown as a function of $v_1 = V_1/V_2$ and $v_3 = V_3/V_2$. We can see that the three ordering patterns strongly compete when the repulsions $V$'s are comparable. We note that the chain-type ordering always gives the lowest magnetic energy. To decide the detailed charge distribution, experimental observation and/or microscopic analysis including structural distortion by the K-intercalation are indeed required in future.

Since each charge has a $j_{\text{eff}} = 1/2$ pseudospin, the ordered charges are expected to interact magnetically. Recently, for exfoliated $\alpha$-$RuCl_3$, a first-order structural phase transition at $\sim$ 150 K between low-temperature $C2/\text{m}$ and high-temperature $P3_1$ structures was reported. [21] At

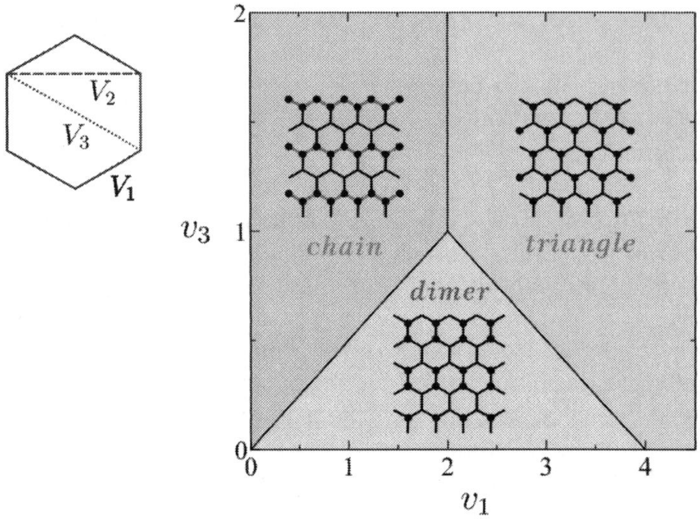


**Figure 1.** Classical phase diagram as a function of the nearest-neighbor ($v_1$) and third-neighbor ($v_3$) repulsions. The repulsions are normalized by the second-neighbor one.

the high temperature an emergence of charge ordering originating from anisotropy in the charge distribution along Ru-Cl-Ru hopping pathways was observed. It suggests a strong tendency to the bond-directional anisotropy along the Ru-Ru axes. In this sense a realization of the chain-type charge ordering may be most likely. Besides, if we assume a magnetic interaction between the nearest-neighbor sites, only the chain-type configuration derives a nontrivial spin model. The triangle-type ordering provides isolated spins and the dimer-type ordering isolated singlet/triplet pairs.

### *2.2. Hamiltonian*

It is generally believed that the magnetic properties of undoped $\alpha$-$RuCl_3$ is well described by the KH model on a honeycomb lattice. If we assume the chain-type charge ordering, the zigzag chains are well separated. The each chain is considered to be a 1D KH model, which is equivalent to a system obtained by removing the $z$-bonds from the honeycomb-lattice KH model. The Hamiltonian of 1D KH model is written as

$$\mathcal{H} = K\sum_{i=1}^{L/2}(S_{2i-1}^x S_{2i}^x + S_{2i}^y S_{2i+1}^y) + J\sum_{i=1}^{L}\vec{S}_i \cdot \vec{S}_{i+1} \tag{1}$$

for a system with $L$ sites, where $\vec{S}_i = (S_i^x, S_i^y, S_i^z)$ is a spin-1/2 operator at site $i$. The Kitaev and Heisenberg exchange couplings are denoted as $K$ and $J$, respectively. The system has two kinds of neighboring links and they appear alternately along the chain. Thus, the structural unit cell contains two lattice sites. Hereafter, we call the links $(2i-1,2i)$ and $(i,2i+1)$ as "$x$-link" and "$y$-link", respectively. By rewriting Eq. (1) to

$$\begin{aligned}\mathcal{H} &= \frac{2J+K}{4}\sum_{i=1}^{L}(S_i^+ S_{i+1}^- + S_i^- S_{i+1}^+) + J\sum_{i=1}^{L} S_i^z S_{i+1}^z + \frac{K}{4}\sum_{i}^{L}(-1)^i(S_i^+ S_{i+1}^+ + S_i^- S_{i+1}^-) \qquad (2)\\ &\equiv \mathcal{H}_{\rm ex} + \mathcal{H}_{\rm Ising} + \mathcal{H}_{\rm dsf},\end{aligned}$$

we can easily notice that a XXZ Heisenberg chain containing exchange ($\mathcal{H}_{\rm ex}$) and Ising ($\mathcal{H}_{\rm Ising}$) terms is disturbed by sign-alternating double-spin-flip ($\mathcal{H}_{\rm dsf}$) fluctuations.

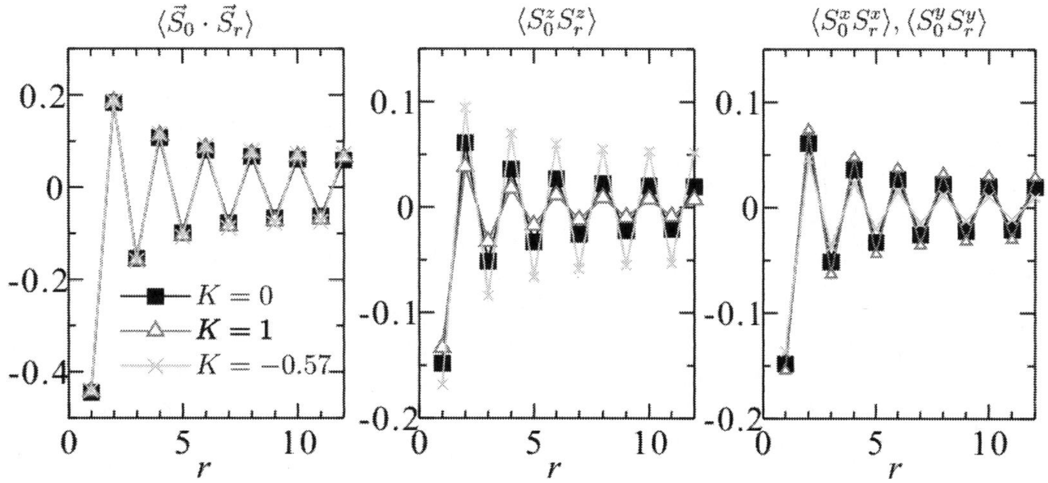


**Figure 2.** Spin-Spin correlation functions as a function of distance $r$. The Heisenberg term is fixed at $J = 1$.

## 3. Numerical results

To investigate the effect of the Kitaev term, we calculated the distance dependence of spin-spin correlation functions using the Lanczos exact diagonalization method for a periodic chain with 24 sites. The results for $K = 0$, 1, and $-0.57$ are plotted as a function of distance $r$ in Figure 2, where the Heisenberg interaction is fixed at $J = 1$. Therefore, we can see a deviation from the antiferomagnetic SU(2) Heisenberg chain ($K = 0$). Interestingly, the total spin-spin correlation function $\langle \vec{S}_0 \cdot \vec{S}_r \rangle$ is almost unchanged. The positive (negative) value of $K$ increases (decreases) quantum fluctuations on the z-axis and decreases (increases) along the xy-plane. As a result, their effects roughly cancell each other in $\langle \vec{S}_0 \cdot \vec{S}_r \rangle$. Whereas, the $z$-component $\langle S_0^z S_r^z \rangle$ is strongly affected by the Kitaev term. Especially, the staggered oscillation is enhanced in the case of negative $K$ value, although the double-spin-flip term is naively expected to weaken it. This means that the Ising term is still dominant in the region $-0.57 \leq K < 0$ with $J = 1$ due to a decrease of the exchange term, and the effect of double-spin-flip term seems not to emerge explicitly there. The other correlations $\langle S_0^y S_r^y \rangle$ and $\langle S_0^y S_r^y \rangle$ are oppositely-affected from $\langle S_0^z S_r^z \rangle$.

## 4. Summary

It is widely believed that the magnetic properties of $\alpha$-$RuCl_3$ are well described by the honeycomb-lattice $S = \frac{1}{2}$ KH model. First, we studied possible charge ordering patterns in the K-intercalated material $K_{0.5}RuCl_m$, where half of the $j = \frac{1}{2}$ spins are replaced by nonmagnetic ions in the honeycomb layer. The phase diagram was obtained as a function of nearest-, second-, and third-neighbor Coulomb repulsions in the atomic limit. We found a strong competition between possible three charge ordering patterns, namely, triangle-, dimer-, chain-type, when the repulsions $V$'s are comparable. Next, using the Lanczos exact diagonalization method we calculated the spin-spin correlation functions $\langle \vec{S}_0 \cdot \vec{S}_r \rangle$, $\langle S_0^x S_r^x \rangle$, $\langle S_0^y S_r^y \rangle$, $\langle S_0^z S_r^z \rangle$ for the 1D KH model, which is derived for the chain-type ordering pattern. Finally, we discussed the effect of the Kitaev term by comparing the correlations to those of the antiferromagnetic SU(2) Heisenberg chain.

## References

[1] Kitaev A 2006 Ann. Phys. **321** 2
[2] Jackeli G and Khaliullin G 2009 Phys. Rev. Lett. **102** 017205

[3] Chaloupka J, Jackeli G and Khaliullin G 2010 Phys. Rev. Lett. **105** 027204
[4] Singh Y and Gegenwart P 2010 Phys. Rev. B **82** 064412
[5] Ye F, Chi S, Cao H, Chakoumakos B C, Fernandez-Baca J A, Custelcean R, Qi T F, Korneta O B and Cao G 2012 Phys. Rev. B **85** 180403
[6] Choi S K, Coldea R, Kolmogorov A N, Lancaster T, Mazin I I, Blundell S J, Radaelli P G, Singh Y, Gegenwart P, Choi K R, Cheong S W, Baker P J, Stock C and Taylor J 2012 Phys. Rev. Lett. **108** 127204
[7] Takayama T, Kato A, Dinnebier R, Nuss J, Kono H, Veiga L, Fabbris G, Haskel D and Takagi H 2015 Phys. Rev. Lett. **114** 077202
[8] Banerjee A, Bridges C A, Yan J -Q, Aczel A A, Li L, Stone M B, Granroth G E, Lumsden M D, Yiu Y, Knolle J, Bhattacharjee S, Kovrizhin D L, Moessner R, Tennant D A, Mandrus D G and Nagler S E 2016 Nat. Mater. **15** 733
[9] Sandilands L J, Tian Y, Plumb K W, Kim Y -J and Burch K S 2015 Phys. Rev. Lett. **114** 147201
[10] Nasu J, Yoshitake J and Motome Y, ArXiv 2017 e-prints, arXiv:1703.10395
[11] Yadav R, Bogdanov N A, Katukuri V M, Nishimoto S, van den Brink J and Hozoi L 2016 Sci. Rep. **6** 37925
[12] Baek S -H, Do S -H, Choi K -Y, Kwon Y S, Wolter A U B, Nishimoto S, van den Brink J and Büchner B 2017 e-prints, arXiv:1702.01671
[13] Wolter A U B, Corredor L T, Janssen L, Nenkov K, Schönecker S, Do S -H, Choi K -Y, Albrecht R, Hunger J, Doert T, Vojta M and Büchner B 2017 e-prints, arXiv:1704.03475
[14] Hentrich R, Wolter A U B, Zotos X, Brenig W, Nowak D, Isaeva A, Doert T, Banerjee A, Lampen-Kelley P, Mandrus D G, Nagler S E, Sears J, Kim Y -J, Büchner B and Hess C 2017 e-prints, arXiv:1703.08623
[15] Zheng J, Ran K, Li T, Wang J, Wang P, Liu B, Liu Z, Normand B, Wen J and Yu W 2017 e-prints, arXiv:1703.08474
[16] Leahy I A, Pocs C A, Siegfried P E, Graf D, Do S -H, Choi K -Y, Normand B and Lee M 2016 e-prints, arXiv:1612.03881
[17] Knupfer M, private communication
[18] Koitzsch A, Habenicht C, Müller E, Knupfer M, Büchner B, Kandpal H C, van den Brink J, Nowak D, Isaeva A and Doert T 2016 Phys. Rev. Lett. **117** 126403
[19] Sandilands L J, Tian Y, Reijnders A A, Kim H -S, Plumb K W, Kim Y -J, Kee H -Y and Burch K S 2016 Phys. Rev. B **93** 075144
[20] Sandilands L J, Sohn C H, Park H J, Kim S Y, Kim K W, Sears J A, Kim Y -J and Noh T W 2016 Phys. Rev. B **94** 195156
[21] M. Ziatdinov, A. Banerjee, A. Maksov, T. Berlijn, W. Zhou, H.B. Cao, J.-Q. Yan, C.A. Bridges, D.G. Mandrus, S.E. Nagler, A.P. Baddorf, and S.V. Kalinin, Nat. Commun. **7**, 13774 (2016).

IOP Conf. Series: Journal of Physics: Conf. Series **969** (2018) 012113 doi:10.1088/1742-6596/969/1/012113

# Low temperature ESR in spin ladder $(C_7H_{10}N)_2Cu_{(1-x)}Zn_xBr_4$

**Yu V Krasnikova**[1,2], **V N Glazkov**[1,2], **M A Fayzullin**[3], **D Schmidiger**[4], **K Yu Povarov**[4], **S Galeski**[4], **A Zheludev**[4]

[1] P.L.Kapitza Institute for Physical Problems RAS, Kosygin str. 2, 119334 Moscow, Russia
[2] National Research University Higher School of Economics, 101000, Moscow, Russia
[3] Institute for Physics, Kazan Federal University, 420008, Kazan, Russia
[4] Neutron Scattering and Magnetism, Laboratory for Solid State Physics, ETH Zürich, 8006 Zürich, Switzerland

E-mail: krasnikova.mipt@gmail.com

**Abstract.** Spin-gap magnet $(C_7H_{10}N)_2Cu_{(1-x)}Zn_xBr_4$ (DIMPY) is an example of a strong-leg spin ladder. We report here the results of the ESR study of pure and diamagnetically diluted DIMPY. ESR study of the pure system (x=0) revealed that the spin dynamics of DIMPY is affected by uniform Dzyaloshinskii-Moriya (DM) interaction. We observe narrowing of the ESR absorption line in diamagnetically diluted DIMPY pointing to suppression of DM channel of spin relaxation by doping.

## 1. Introduction

Spin ladder is a model low-dimensional spin system. It consists of two spin chains (called 'legs') coupled by interchain exchange bonds (which are the 'rungs' of spin ladder). The ground state of such system is a singlet and exited states are triplets. Energy spectrum of spin chain has a gap for arbitrary ratio between in-chain and inter-chain antiferromagnetic exchange constants [1]. Recently discovered [2, 3] metallo-organic compound $(C_7H_{10}N)_2CuBr_4$ (called DIMPY for short) is a rare example of the strong leg spin ladder: in-chain exchange coupling $J = 1.42$ meV exceeds inter-chain coupling $j = 0.82$ meV [4]. Energy gap $\Delta$=0.33 meV was directly observed with inelastic neutron scattering [4] and in the specific heat and magnetization measurements [2, 3, 5]. Magnetic phase diagram of DIMPY features low temperature field-induced ordering below 300 mK in the fields above approx. 20 kOe [5].

The ESR method has high energy resolution (about 0.005 meV) and it could be used to detect weak anisotropic spin-spin interactions. Such anisotropic spin-spin interactions exist in our system: DIMPY crystal symmetry allows Dzyaloshinskii-Moriya (DM) interaction. Because of translation and inversion symmetry present, DM interaction is forbidden on the rungs of the ladder, it is uniform along the leg of the ladder and has a different sign on different legs (figure 1). This allows to build a compact microscopic model which includes two exchange coupling constants and three components of DM vector only:

$$\mathcal{H} = J\sum(\hat{\mathbf{S}}_{i,a}\hat{\mathbf{S}}_{i+1,a} + \hat{\mathbf{S}}_{i,b}\hat{\mathbf{S}}_{i+1,b}) + j\sum \hat{\mathbf{S}}_{i,a}\hat{\mathbf{S}}_{i,b} + \mathbf{D}\sum(\hat{\mathbf{S}}_{i,a}\times\hat{\mathbf{S}}_{i+1,a} - \hat{\mathbf{S}}_{i,b}\times\hat{\mathbf{S}}_{i+1,b}) \quad (1)$$

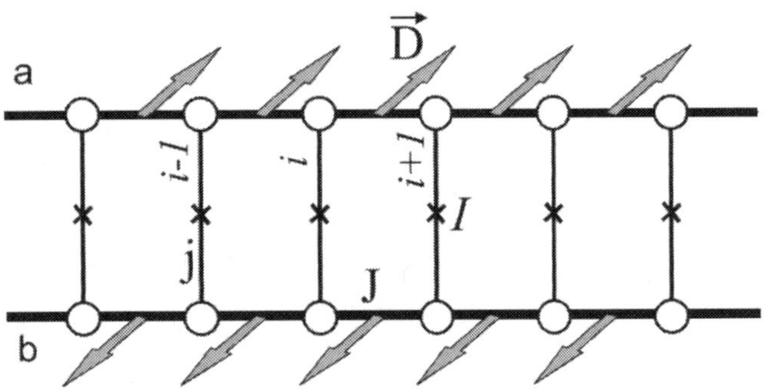


**Figure 1.** Model spin ladder with uniform Dzyaloshinckii-Moriya interaction along the legs of the ladder (Eqn.(1)). $J$, $j$ – in-chain and inter-chain exchange couplings, grey arrows schematically show directions of DM vectors $\vec{D}$; indices $a$ and $b$ enumerate legs of the ladder, $i$, $(i+1)$ etc. enumerate rungs of the ladder. Crosses on the rungs mark positions of inversion centers $I$.

here $i$ counts spins along the leg of the ladder and $a, b$ are legs' indices. Solving such a model remains a challenge, but supposedly it should yield both the fine structure of the excitations spectrum and relaxation of the spin excitations due to anisotropic spin-spin interactions.

Here we report on observations proving presence of DM interaction in pure DIMPY and on the effect of the diamagnetic dilution. We observe that diamagnetic dilution results in the narrowing of ESR absorption line, i. e. that introduction of structural disorder actually increases quasiparticles lifetime.

## 2. ESR in pure DIMPY

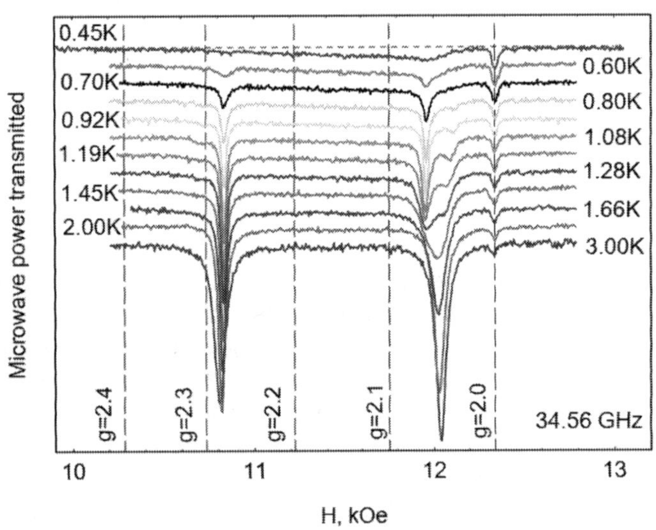


**Figure 2.** ESR absorption spectra of pure DIMPY at low temperatures. Magnetic field is applied at 45° from $a$ to $b$ axis, which corresponds to the maximum difference between absorption components. Vertical dashed lines show position of resonance absorption with certain $g$ - factors, narrow line at $g = 2.00$ is a DPPH marker.

ESR in pure DIMPY was described in details in [6]. We have measured ESR spectra for DIMPY at different temperatures (figure 2). The two main signals are because of two types of ladders in DIMPY, which are inequivalent for the chosen field direction. On cooling ESR absorption looses intensity as expected for the system with an energy gap. At the lowest temperature of 450 mK ESR signal vanishes. We have found additional splitting of ESR components at T≈1K (figure 2), which can be ascribed to the effect of zero field splitting of triplet sublevels. The estimate of zero field splitting is about 0.03 meV, as evaluated through maximum subcomponent splitting ΔH=150 Oe.

DM interaction could lead to such a zero field splitting. Additional proof of the presence of DM interaction can be obtained through the analysis of the high temperature ESR linewidth which is determined by anisotropic spin-spin interactions [7]. Analysis of linewidth angular dependences proved valuable contribution of this interaction, D≈0.03 meV [6]. Independently, high-frequency high-field ESR study [8] proved existence of DM interaction in DIMPY through observation of additional ESR mode which became allowed due to DM interaction.

Temperature dependence of the ESR linewidth was measured from 400 mK to 300 K [6]. Below 100 K ESR linewidth increases on cooling reaching its maximum value around 10 K. Below 10 K ESR line narrows with cooling as the gapped system enters to the regime of diluted quasiparticles gas (figure 3). Complete description of this temperature dependence still awaits theoretical effort, but we speculate that it can be explained in the compact model of Eqn.(1).

## 3. ESR in Zn-doped DIMPY

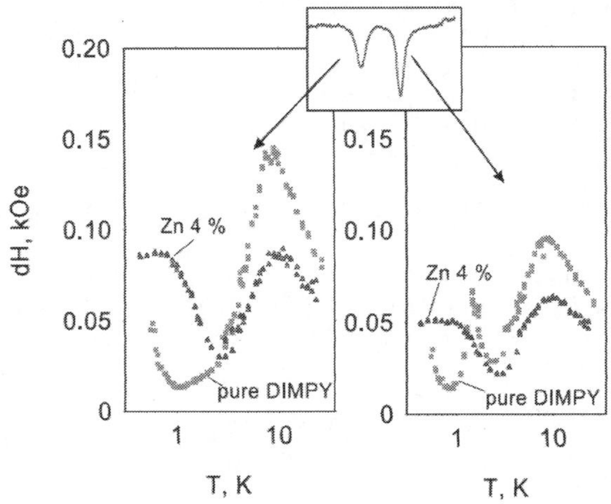


**Figure 3.** Temperature dependences of the ESR linewidth for the 4% Zn-doped sample (triangles) and for the pure DIMPY (squares). Upper panel shows typical ESR absorption spectrum with resolved absorption signals from the inequivalent ladders. Left and right panel show data for left and right absorption components, respectively.

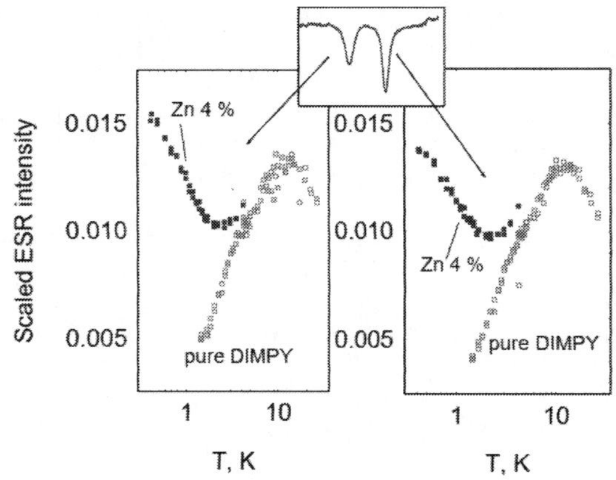


**Figure 4.** Temperature dependences of the ESR intensity for the 4% Zn-doped sample (closed squares) and for the pure DIMPY (open squares). Intensities are scaled to the known static susceptibility. Upper panel shows typical ESR absorption spectrum with resolved absorption signals from the inequivalent ladders. Left and right panel show data for left and right absorption components, respectively.

Recently, samples of DIMPY with part of the copper ions substituted by non-magnetic zinc were grown [9]. This diamagnetic dilution resulted in the formation of multispin defects around the impurity ion. These multispin defects determine the low temperature magnetic properties of the system. To check effect of this doping on spin dynamics of DIMPY we have measured ESR absorption for the sample containing 4% Zn in wide temperature range (0.5 K - 25 K).

ESR absorption spectra for the doped DIMPY have the same two components corresponding to different types of spin ladders. Intensity of the ESR absorption can be scaled to static magnetization thus allowing to compare absorption intensities for different samples (figure 4). We have found that above 4 K scaled intensities for pure and 4% Zn-diluted sample practically coincides. This proves that at these temperatures triplet excitations dominate. Below 4 K intensity of the ESR absorption for 4% Zn-diluted sample increases on cooling as paramagnetic multispin defects became dominant. ESR *g*-factors are the same in both regimes indicating that triplet excitations and multispin defects are formed from the same spins of $Cu^{2+}$ ions.

IOP Conf. Series: Journal of Physics: Conf. Series **969** (2018) 012113 doi:10.1088/1742-6596/969/1/012113

Surprisingly, ESR linewidth near maximum at $T \approx$10K decreases with doping (figure 3). It means that excitations lifetime increases with doping at this temperature, where the magnetic response arises from collective delocalized triplet excitations. Recalling that the ESR linewidth for pure DIMPY is, supposedly, due to DM interaction, we conjecture that disorder created by diamagnetic dilution of the spin ladder suppresses relaxation channel due to DM interaction.

## Acknowledgments

The work was supported by the Russian Foundation for Basic Research (Grant No.15-02-05981), Russian Scientific Foundation (Grant No. 17-12-01505) and by the Basic research Program of HSE. This work was partially supported by Swiss National Science Foundation Division 2.

## References

[1] Schmidiger D, Mühlbauer S, and Zheludev A 2013 *Phys. Rev. B* **88** 094411
[2] Shapiro A, Landee C P, Turnbul M Ml, Jornet J, Deumal M, Novoa J J, Robb M A, and Lewis W 2007 *J. Am. Chem. Soc.* **129** 952
[3] Hong T, Kim Y H, Hotta C, Takano Y, Tremelling G, Turnbull M M, Landee C P, Kang H J, Christensen N B, Lefmann K, Schmidt K P, Uhrig G S, and Broholm C 2010 *Phys. Rev. Lett.* **105** 137207
[4] Schmidiger D, Bouillot P, Guidi T, Bewley R, Kollath C, Giamarchi T, Zheludev A 2013 *Phys. Rev. Lett.* **111** 107202
[5] Schmidiger D, Bouillot P, Mühlbauer S, Gvasaliya S, Kollath C, Giamarchi T, and Zheludev A 2012 *Phys. Rev. Lett.* **108** 167201
[6] Glazkov V, Fayzullin M, Krasnikova Yu, Skoblin G, Schmidiger D, Zheludev A 2015 *Phys. Rev. B* **92** 184403
[7] Abragam A and Bleaney B 1970 *Electron Paramagnetic Resonance of Transition Ions* (Oxford: Clarendon Press)
[8] Ozerov M, Maksymenko M, Wosnitza J, Honecker A, Landee C P, Turnbull M M, Furuya c S, Giamarchi T, and Zvyagin S A 2015 *Phys. Rev. B* **92** 241113
[9] Schmidiger D, Povarov K Yu, Galeski S, Reynolds N, Bewley R, Guidi T, Ollivier J, Zheludev A 2016 *Phys. Rev. Lett.* **116** 257203

# Phase diagram of the frustrated $J_1 - J_2$ transverse field Ising model on the square lattice

**M. Sadrzadeh and A. Langari**

Department of Physics, Sharif University of Technology, P.O.Box 11155-9161, Tehran, Iran

E-mail: `langari@sharif.edu`

**Abstract.** We study the zero-temperature phase diagram of transverse field Ising model on the $J_1 - J_2$ square lattice. In zero magnetic field, the model has a classical Néel phase for $J_2/J_1 < 0.5$ and an antiferromagnetic collinear phase for $J_2/J_1 > 0.5$. We incorporate harmonic fluctuations by using linear spin wave theory (LSWT) with single spin flip excitations above a magnetic order background and obtain the phase diagram of the model in this approximation. We find that harmonic quantum fluctuations of LSWT fail to lift the large degeneracy at $J_2/J_1 = 0.5$ and exhibit some inconsistent regions on the phase diagram. However, we show that anharmonic fluctuations of cluster operator approach (COA) resolve the inconsistency of the LSWT, which reveals a string-valence bond solid ordered phase for the highly frustrated region.

## 1. Introduction

Searching for unusual phases in frustrated magnetic systems has received huge attention in recent years. Such exotic phases are completely different from the conventional magnetic orders known as ferromagnetic or antiferromagnetic Néel states. They can be spin liquid phases with no broken symmetry, or some resonating valence bond solid (VBS) phases with spatial symmetry breaking. One of the two-dimensional frustrated quantum magnets, which has been studied over decades, is spin-1/2 $J_1 - J_2$ antiferromagnetic (AFM) Heisenberg model on the square lattice. In the prior literature, the existence of a non-magnetic ground state (GS) has been established around the strongest frustration point $J_2/J_1 = 0.5$, between the Néel and striped AFM states, which occur for small and large $J_2/J_1$, respectively. However, the nature of such a non-magnetic phase is still under debate. Different candidate ground states have been proposed for the intermediate region such as columnar dimer VBS [1, 2], plaquette VBS [3, 4], gapless spin liquid [5, 6], or gapped $Z_2$ spin liquid phases [7, 8]. In all such VBS phases the internal SU(2) spin-rotation is preserved but spatial symmetries of the lattice are broken. However, by reduction of symmetry from SU(2) to Z2, the antiferromagnetic Ising model on the $J_1 - J_2$ square lattice can be studied as a prototype of frustrated systems, which gives interesting features as an interplay between frustration and quantum fluctuations. Specifically, the correct phase diagram and the true nature of ground state at the highly frustrated point $J_2/J_1 = 0.5$, are demanding issues to be discussed here. In this respect, we examine the spin-1/2 transverse field Ising (TFI) model on the $J_1 - J_2$ square lattice with the Hamiltonian

$$\mathcal{H} = J_1 \sum_{\langle i,j \rangle} S_i^z S_j^z + J_2 \sum_{\langle\langle i,j \rangle\rangle} S_i^z S_j^z - \Gamma \sum_i S_i^x, \tag{1}$$

where $J_1$ and $J_2$ are antiferromagnetic exchange couplings between nearest neighbor and next-to-nearest neighbor diagonal bonds, respectively. $\Gamma$ acts as a transverse magnetic field, which induces quantum fluctuations on the classical frustrated background. In the absence of transverse field $\Gamma$, the classical GS phase diagram of the model is well known[9, 10]. In fact, at $\Gamma = 0$ and $J_2 = 0$ a Néel phase forms the stable GS of the model, which persists up to the critical point $J_2/J_1 = 0.5$, where it shows a first order transition to a collinear AFM phase, called striped AFM order for $J_2/J_1 > 0.5$. However, at the highly frustrated point $J_2/J_1 = 0.5$, the classical system displays an exponential degeneracy for GS. Hence, we are interested to see the effect of quantum fluctuations on the such degenerate classical GS manifold. In fact, the ground state takes place on a constraint of two-up-two-down for spins on every crossed square of the lattice. Hence, by rotation of all spins along a vertical or horizontal direction of the lattice, the constraint two-up-two-down is remained and the system is again in the ground state. Hence, the ground state of the system at $J2/J1 = 0.5$ has $2^L$ exponential degeneracy, where $L$ is the linear size of the lattice. In other words, there are many zero-energy domain walls as excitations separating regions of Neel and striped states.

## 2. Linear Spin Wave Theory

We first obtain the quantum GS phase diagram of the model with linear spin-wave theory which is based on single-spin-flip harmonic excitations. In this respect, we begin with the classical limit ($S \to \infty$) of the Hamiltonian, Eq. 1. The corresponding GS configurations are used as elementary reference states of spin wave analysis for the quantum $S = 1/2$ case. Harmonic quantum fluctuations of LSWT around these classical reference states will reduce the magnitude of the classical order parameters and result in zero-point energy corrections. The classical spin vector components are given by $S_i^x = S\sin\theta$ and $S_i^z = \pm S\cos\theta$, where $\pm$ sign represents up and down spin in the $z$ direction. The angle $\theta$ increases with the strength of transverse field $\Gamma$ up to a maximum value of $\theta = \pi/2$ , which corresponds to the full polarization of the classical spins in the $+x$ direction for $\Gamma \geq \Gamma_c$. The total number of spins is N, hence, there are 2N bonds with coupling $J_1$ and 2N bonds of coupling $J_2$ on the whole square lattice. Thus, we can obtain the classical GS energy per spin for both the Néel and striped phases as

$$E_{cl}^{Néel} = 2S^2(J_2 - J_1)cos^2\theta - \Gamma S sin\theta, \quad E_{cl}^{striped} = -2S^2 J_2 cos^2\theta - \Gamma S sin\theta. \tag{2}$$

After minimizing the classical energy per spin with respect to $\theta$, we set $\theta = \pi/2$ to obtain the values of critical transverse field $\Gamma_c$. Accordingly, the classical phase diagram of the model is depicted in Fig.1-(a). It consists of three phases, canted Néel, canted striped and fully polarized ones. We now consider harmonic quantum fluctuations around these classical manifold of states according to a general LSWT analysis, similar to what has been done in Ref.[11]. Fig.1-(a) displays the schematic representations of the classical states. They all have long-range magnetic orders with a magnetic unit cell containing two spins in the Néel and striped cases and one spin for the polarized phase. We define $S_{l,p}$ as the $p$-th spin of the $l$-th cell, where $p = 1, 2$ for Néel and striped phases, while $p = 1$ for polarized phase. In the classical limit, an applied transverse field $\Gamma$ rotates all spins around the $y$ axis by an angle $\theta$. We now introduce a local rotation of spins, as $S_{l,p} \to \tilde{S}_{l,p}$, in such a way that all three classical states map to a simple ferromagnetic state in $z$ direction, i.e. $\tilde{S}_{l,p}^z = S$ everywhere. Accordingly, we define $\tilde{S}_{l,p} = \sigma_p R_y(\sigma_p\theta) S_{l,p}$, where $\sigma_p = \pm 1$ denotes the direction of $p$-th spin along the $z$ axis, and $R$ is the rotation matrix around $y$ axis by angle $\sigma_p\theta$. After rewriting the Hamiltonian in terms of new spin operators $\tilde{S}_{l,p}^z$ and $\tilde{S}_{l,p}^x$, we consider small quantum fluctuations around this general ferromagnetic classical reference state by the following linearized Holstein-Primakoff transformations,

$$\tilde{S}_{l,p}^z = S - a_{l,p}^\dagger a_{l,p}, \quad \tilde{S}_{l,p}^x \approx \sqrt{\frac{S}{2}} \left( a_{l,p}^\dagger + a_{l,p} \right), \tag{3}$$

IOP Conf. Series: Journal of Physics: Conf. Series **969** (2018) 012114 doi:10.1088/1742-6596/969/1/012114

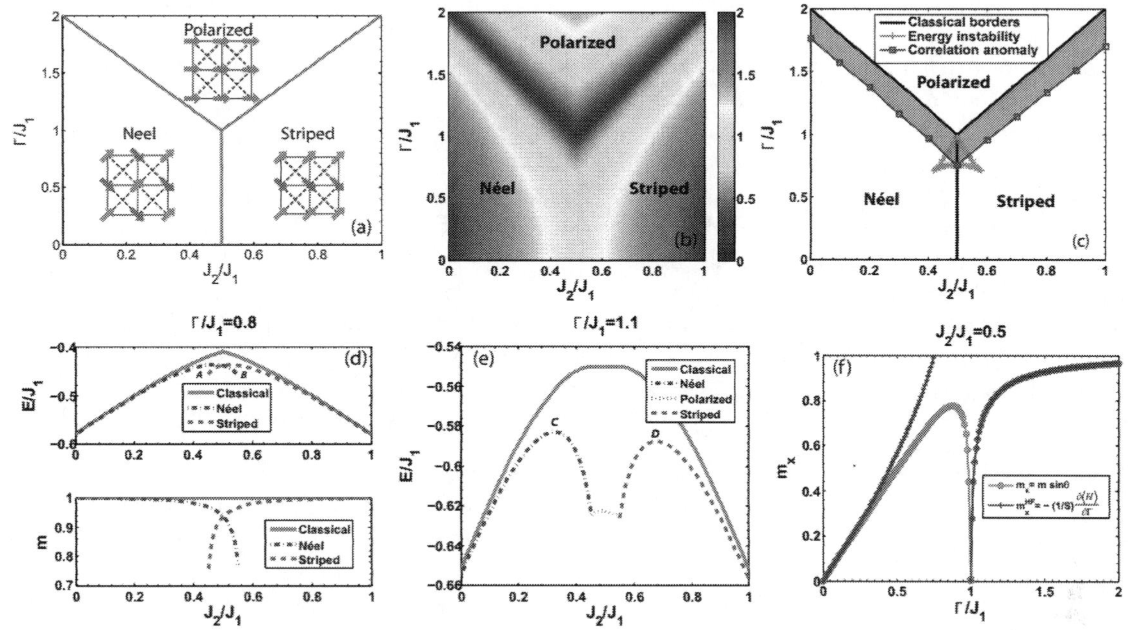


**Figure 1.** (color online) (a) Classical phase diagram of transverse field Ising model on the $J_1 - J_2$ square lattice. The classical spin configurations used as the background in LSWT are shown. Solid and dashed lines are $J_1$ and $J_2$ bonds, respectively. (b) Energy gap in a 2D-plot. (c) Quantum phase diagram obtained from LSWT. The hatched region indicates the failure of LSWT as explained in the main text. (d) Ground state energy and order parameter obtained from LSWT at $\Gamma = 0.8J_1$. (e) Ground state energy at $\Gamma = 1.1$, which shows a change of sign in the derivative of GS energy. (f) Transverse magnetization obtained from Hellmann-Feynman theorem, which is compared with the expectation value of magnetic order parameter.

where $a_{l,p}$ and $a^\dagger_{l,p}$ are bosonic operators with well-known commutation relations $[a_{l,p}, a^\dagger_{l,p}] = 1$. Finally, by expanding the Hamiltonian up to the quadratic order of bosonic operators followed by a transformation to momentum space, and performing an n-modes para-unitary Bogoliubov transformation [12], we obtain an effective diagonal quadratic Hamiltonian,

$$\mathcal{H}_{LSWT} = E_{cl} - \frac{N}{n}\sum_p \left(\frac{\tilde{h}_p}{2}\right) + \sum_{\mathbf{k},p} \omega_{\mathbf{k},p}\left(b^\dagger_{\mathbf{k},p} b_{\mathbf{k},p} + \frac{1}{2}\right), \tag{4}$$

where $n = 2$ for Néel and striped phases, while $n = 1$ for polarized phase, and

$$\tilde{h}_p \;=\; -\frac{1}{2}\sigma_p \cos^2(\theta) \sum_{\delta,\mathbf{p}'} \left[\tilde{J}(\delta)\right]_{pp'} \sigma_{p'} + \Gamma \sin\theta, \tag{5}$$

where $\tilde{J}(\delta)$ is a $n \times n$ matrix, containing the couplings between spins $p, p'$ of two unit cells $l, l'$ at a distance $\delta$. $\mathbf{k}$ sums over the first Brillouin zone of a lattice constructed from the magnetic unit cells of classical reference state and $\omega_{\mathbf{k},p}$ defines the spectrum of bosonic quasi-particles, created by $b^\dagger_{\mathbf{k},p}$, which is expressed in terms of $a_{l,p}$ and $a^\dagger_{l,p}$ via the Bogoliubov transformation.

The Minimum of the lowest band of excitation spectrum $\omega_{\mathbf{k},p}$, corresponding to the energy gap of the system is depicted in Fig.1-(b) for the whole phase diagram. There is no gap closing

IOP Conf. Series: Journal of Physics: Conf. Series **969** (2018) 012114 doi:10.1088/1742-6596/969/1/012114

between Néel and striped phases, while zero-gap borders from Néel and striped phases to the polarized phase completely conform to the classical borders. However, there are some regions of inconsistency around these phase boundaries, where LSWT fails to justify the phase diagram correctly, as shown in Fig.1-(c). As a first inconsistency of LSWT, having obtained the GS energy $E = \langle \mathcal{H}_{LSWT} \rangle$ and order parameter $m = \langle \tilde{S}^z \rangle / S$ for the whole phase diagram, we observe that for intermediate transverse fields $0.75 < \Gamma/J_1 < 1.0$ around $J_2/J_1 = 0.5$, when we are classically in the Néel (striped) phase, the GS energy of the striped (Néel) phase is lower than the Néel (striped) one. In other words, $E_{striped} < E_{N\acute{e}el}$ when $m_{N\acute{e}el} > m_{striped}$ and vise versa (see Fig.1-(d)). It means that quantum fluctuations beyond LSWT might cast the GS of system to an intermediate phase, which is neither a Néel nor a striped phase. This conjecture should be true as we do not observe any gap closing implying a direct phase transition, between the Néel and striped phases. On the other hand, according to Hellman-Feynman theorem, spin-spin correlation function between next-to-nearest neighbor spins is given by the energy derivative with respect to $J_2$, i.e. $C^{(2)} = \langle S_i^z S_j^z \rangle_{\langle\langle i,j \rangle\rangle} = \partial \langle \mathcal{H} \rangle / \partial J_2$. As we can see from the classical picture, this quantity has opposite signs for the Néel and striped phases. In fact, $C^{(2)}_{Neel} = S^2$, while $C^{(2)}_{striped} = -S^2$. However, energy calculations for $\Gamma/J_1 = 1.1$ indicate a change of sign of the energy derivative with respect to $J_2$, at an intermediate point inside a specified classical reference state (see Fig.1-(e)). The sign inversion of the next-to-nearest neighbor correlation, within a classical reference state, is again an indication of Néel and striped order destruction toward an intermediate phase. Fig.1-(c) displays the borders of such inconsistency, which anticipate a lower phase boundaries to the quantum paramagnet phase. The corrected borders should make a closer point to the well-known[13] critical point $\Gamma_c/J_1 \approx 1.52$ at $J_2 = 0$ limit. Consequently, the hatched regions of the quantum phase diagram obtained by LSWT imply that elementary Néel and striped reference states are not the true GS of model for these regions. Hence, it is surmised that anharmonic quantum fluctuations will render the system to the quantum paramagnet phase. In addition, we observe a violation of Hellmann-Feynman theorem at enough high fields before reaching the critical field $\Gamma/J_1 = 1.0$ at $J_2/J_1 = 0.5$. Indeed, Fig.1-(f) compares transverse magnetization obtained from Hellmann-Feynman theorem $\langle m_x \rangle = -\frac{1}{S} \partial \langle H \rangle / \partial \Gamma$ with the one obtained from the expectation value of magnetic order parameter, i.e. $\langle m_x \rangle = \langle m \rangle sin\theta$. It shows that by increasing transverse field $\Gamma$, before reaching the critical point $\Gamma_c/J_1 = 1.0$, both curves deviate from each other, signaling a violation of the Hellmann-Feynman theorem. All such inconsistencies may imply that quantum fluctuations near classical phase boundaries go beyond harmonic level of approximation considered in LSWT.

## 3. Highly frustrated regime $J_2/J_1 = 0.5$

Here, we discuss the highly frustrated regime $J_2/J_1 = 0.5$ at low fields. We found that both Néel and striped magnetic ordered backgrounds lead to the same zero-point eneregy corrections within LSWT. On the other hand, the lowest band of excitation spectrum obtained from LSWT above these classical reference states reveals a minimum at many different **k**-vectors besides the **k**-vectors corresponding to these classical reference states, which evidently demonstrates the large degeneracy of the system at point $J_2/J_1 = 0.5$. In other words, since both the state with maximal (Néel) and minimal (Striped) number of domain walls have the same zero point energy, the harmonic zero point energy cost of a domain wall is null[14]. Hence, there is the same zero-point eneregy correction for the entire family of states with different number of domain walls. Therefore, quantum corrections at harmonic level in LSWT can not distinguish between different ordered manifold of states, failing to lift the extensive degeneracy at $J_2/J_1 = 0.5$. Moreover, as Fig.1-(b) displays, there is no gap closing between the Néel and striped phases at $J_2/J_1 = 0.5$, which implies that there is no phase transition at this border. Taking into account anharmonic fluctuations we anticipate the emergence of a novel phase at low fields around $J_2/J_1 = 0.5$.

Therefore, we have applied a cluster operator approach (COA) [15, 16, 17, 18], which is based on multi-spin cluster-type of anharmonic quantum fluctuations above a non-magnetic cluster ordered background. Such multi-spin cluster type excitations represent anharmonic fluctuations of the original spin model, which resolve the incosistencies of LSWT. However, to recover the Hellmann-Feynman identities like $\langle m_x \rangle = -\frac{1}{S}\partial\langle H\rangle/\partial\Gamma$, one may consider non-linear corrections to the harmonic LSWT [19].

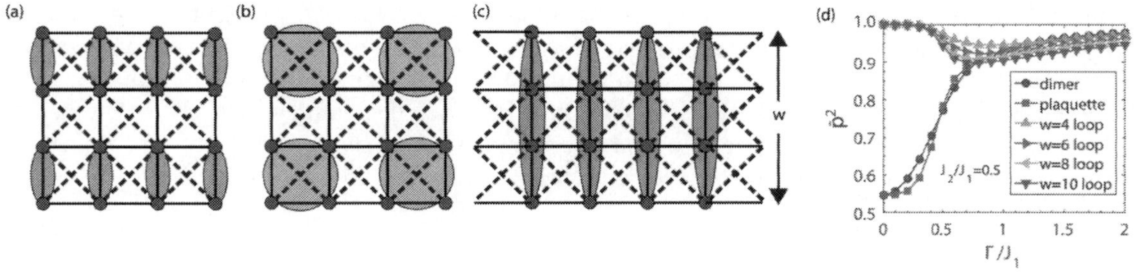


**Figure 2.** (color online) (a) Dimer valence-bond solid (VBS) state, (b) plaquette VBS state, covering the whole of a two-dimensional infinite lattice. (c) String-VBS state, which covers a two-dimensional finite width (w) lattice periodic in vertical (width) direction. The strutures in (a, b, c) form the background in cluster operator approach. Solid and dashed lines are $J_1$ and $J_2$ bonds, respectively. (d) The condensation probability $\bar{p}^2$ of different VBS orderings versus the transverse field ($\Gamma/J_1$) for $J_2/J_1 = 0.5$.

In COA as shown in Fig. 2-(a, b, c), the lattice is decomposed into independent finite-size clusters. Hamiltonian of each cluster is diagonalized exactly and a boson is associated to each eigenstate of the Hamiltonian [18]. In the absence of interaction between clusters, a direct product of GS of clusters represents the GS of the whole lattice, i.e. *a Bose-condensation of cluster ground states.* Let define the ground state of cluster-$I$ as $|\phi_0\rangle_I$ to be created by $c^\dagger_{I,0}$ boson, i.e. $|\phi_0\rangle_I = c^\dagger_{I,0}|0\rangle$, where $|0\rangle$ is the vacuum state. Hence, the Bose-condensation of cluster ordered background is measured in terms of

$$\langle c^\dagger_{I,0} c_{I,0}\rangle \equiv \bar{p}^2, \quad \forall I \tag{6}$$

where $\bar{p}$ is the condensation amplitude and $\bar{p}^2$ gives the probability of such condensation. Such a ground state does not necessarily show a magnetic long-range order, while it clearly represents a crystal structure formed by clusters, which could break rotational/translational symmetry of the original lattice. In the absence of inter-cluster interactions, $\bar{p}^2$ is equal to unity. The interactions between clusters do not commute with the cluster Hamiltonians, which hybridize the ground state with the excited states of clusters that leads to the creation of anharmonic excitations and consequently the reduction of $\bar{p}^2$ from unity. We obtain $\bar{p}^2$ from the minimization of ground-state energy in the presence of inter-cluster interactions within a self-consistent mean-field solution [18]. It has to be mentioned that this type of excitations comes from a cluster composed of several spins, which shows anharmonic fluctuations compared with the single-spin excitations of LSWT that are harmonic ones.

Our recent results [18] on different types of clusters, depicted in Fig. 2-(a, b, c) reveals that all finite-size clusters like dimers and plaquettes do not lead to a true Bose condensation of clusters, while a global periodic string (either horizontal or vertical) represents the Bose condensation in ground state. The self-consistent solution of $\bar{p}^2$ is shown in Fig. 2-(d) versus the transverse magnetic field ($\Gamma$) at the highly frustrated point $J_2/J_1 = 0.5$. The value of $\bar{p}^2$ is far from its maximum for dimers and plaquettes, while it is close to unity for any vertical global loop $w = 4, 6, 8, 10$. For the latter case, we considered a lattice of finite width w in vertical direction and infinite in the horizontal direction as shown in Fig. 2-(c). It shows that the string VBS state condensate at small magnetic field ($\Gamma < 0.5$) compared with dimer or plaquette VBS states, Fig. 2-(d). The increment of transverse field causes a transition from the string VBS phase to the quantum paramagnet one. In fact, the value of transverse magnetization is high enough (at high fields), imitating a quantum paramagnet (polarized) phase. Thus, we conclude that POA can reproduce the polarized phase of high fields in such a way that all types of clusters are in their polarized states, regardless of their cluster form. Finally, we conclude that the exponential degeneracy of the classical GS at $J_2/J_1 = 0.5$ is lifted toward a string VBS ordered phase which breaks rotational symmetry of the lattice with only two-fold degeneracy. On the other hand, the boundaries to the Néel and striped phases, corresponding to the vanishing of quasi-particle excitation gap should be determined more consistently via COA, in contrast to LSWT.

## 4. Concluding remarks

We have studied the zero-temperature phase diagram of the transverse field Ising model on the $J_1 - J_2$ square lattice, see Fig. 1. This model is a frustrated magnetic system, which has extensive degenerate classical GS at $J_2/J_1 = 0.5$. The LSWT analysis of the model fails to lift this classical degeneracy and the corresponding phase diagram shows inconsistencies near the classical phase boundaries. This implies that the harmonic fluctuations, coming from the single-spin-flip excitations of LSWT, does not incorporate the true quantum fluctuations of the system, close to the critical phase boundaries and around the highly frustrated region $J_2/J_1 = 0.5$. Therefore, we have applied a cluster operator approach, which is based on multi-spin cluster-type of anharmonic quantum fluctuations above a non-magnetic cluster ordered background. Our results confirm that the string VBS state condensate at small magnetic field ($\Gamma < 0.5$) compared with dimer or plaquette VBS states, Fig. 2-(d), where the exponential degeneracy of the classical GS at $J_2/J_1 = 0.5$ is lifted toward a string VBS ordered phase. Moreover, we anticipate to resolve the inconsistencies of LSWT within COA, in addition to obtain more accurate boundaries in the phase diagram.

We would like to thank S. S. Jahromi and R. Haghshenas for fruitful discussion.

## References

[1] Singh R R P, Weihong Z, Hamer C J and Oitmaa J 1999 *Phys. Rev. B* **60**(10) 7278–7283
[2] Morita S, Kaneko R and Imada M 2015 *Journal of the Physical Society of Japan* **84** 024720
[3] Doretto R 2014 *Physical Review B* **89** 104415
[4] Gong S S, Zhu W, Sheng D, Motrunich O I and Fisher M P 2014 *Physical review letters* **113** 027201
[5] Hu W J, Becca F, Parola A and Sorella S 2013 *Physical Review B* **88** 060402
[6] Wang L, Poilblanc D, Gu Z C, Wen X G and Verstraete F 2013 *Physical review letters* **111** 037202
[7] Jiang H C, Yao H and Balents L 2012 *Physical Review B* **86** 024424
[8] Mezzacapo F 2012 *Physical Review B* **86** 045115
[9] Morán-López J, Aguilera-Granja F and Sanchez J 1993 *Physical Review B* **48** 3519
[10] Kalz A, Honecker A, Fuchs S and Pruschke T 2009 *Journal of Physics: Conference Series* **145** 012051
[11] Henry L P, Holdsworth P C W, Mila F and Roscilde T 2012 *Phys. Rev. B* **85**(13) 134427
[12] Colpa J 1978 *Physica A: Statistical Mechanics and its Applications* **93** 327 – 353 ISSN 0378-4371
[13] Blöte H W J and Deng Y 2002 *Phys. Rev. E* **66**(6) 066110
[14] Coletta T, Picon J D, Korshunov S E and Mila F 2011 *Phys. Rev. B* **83**(5) 054402
[15] Sachdev S and Bhatt R N 1990 *Phys. Rev. B* **41**(13) 9323–9329
[16] Ganesh R, Nishimoto S and van den Brink J 2013 *Phys. Rev. B* **87**(5) 054413
[17] Sadrzadeh M and Langari A 2015 *The European Physical Journal B* **88** 1–12
[18] Sadrzadeh M, Haghshenas R, Jahromi S S and Langari A 2016 *Phys. Rev. B* **94**(21) 214419
[19] Coletta T, Laflorencie N and Mila F 2012 *Phys. Rev. B* **85**(10) 104421

IOP Conf. Series: Journal of Physics: Conf. Series **969** (2018) 012115 doi:10.1088/1742-6596/969/1/012115

# Competition between dynamic and structural disorder in a doped triangular antiferromagnet $RbFe(MoO_4)_2$

**A I Smirnov[1,2], T A Soldatov[1,3], O A Petrenko[4], A Takata[5], T Kida[5], M Hagiwara[5], M E Zhitomirsky[6] and A Ya Shapiro[7]**

[1] P.L. Kapitza Institute for Physical Problems RAS, Moscow 119334, Russia
[2] Higher School of Economics, 101000 Moscow, Russia
[3] Moscow Institute for Physics and Technology, Dolgoprudny 141700, Russia
[4] Department of Physics, University of Warwick, Coventry CV4 7AL, United Kingdom
[5] Center of Advanced High Magnetic Field Science (AHMF), Graduate School of Science, Osaka University, Osaka 560-0043, Japan
[6] Service de Physique Statistique, Magnetisme et Supraconductivite, UMR-E9001 CEA-INAC/UJF, 17 rue des Martyrs, Grenoble 38054, France
[7] A. V. Shubnikov Institute for Crystallography RAS, Moscow 119333, Russia

E-mail: smirnov@kapitza.ras.ru

**Abstract.** Magnetisation measurements and electron spin resonance (ESR) spectra of a doped quasi two dimensional (2D) antiferromagnet on a triangular lattice $Rb_{1-x}K_xFe(MoO_4)_2$ reveal a crucial change of the ground state spin configuration and a disappearance of a characteristic 1/3-magnetisation plateau at $x = 0.15$. According to theory for triangular antiferromagnets with a weak random modulation of the exchange bonds, this is a result of the competition between the structural and dynamic disorders. The dynamic zero-point or thermal fluctuations are known to lift the degeneracy of the mean field ground state of a triangular antiferromagnet and cause the spin configuration to be the most collinear, while the static disorder provides another selection of the ground state, with the least collinear structure. Low-level doping ($x \leq 0.15$) was found to decrease the Néel temperature and saturation field by only few percent, while the magnetisation plateau disappears completely and the spin configuration is drastically changed. ESR spectra confirm an impurity-induced change of the so-called Y-type structure to an inverted Y-structure for $x = 0.15$. For $x = 0.075$ the intermediate regime with the decrease of width and weakening of flattening of 1/3-plateau was found.

## 1. Introduction

Triangular lattice antiferromagnets (TLAF) provide challenges in contemporary magnetism due to the frustration of the exchange interaction and unusual magnetic phases, stabilised by fluctuations (for reviews, see, *e.g.*, Refs. [1–3]). The ground state of a TLAF in a magnetic field is strongly degenerate in a mean field approximation. The selection of the ground state is made through the so-called order-by-disorder mechanism, implying that thermal and quantum spin fluctuations lift the degeneracy of a classical ground state. A key feature of TLAF is the 1/3-magnetisation plateau with the collinear up-up-down (*uud*) spin structure stabilised by fluctuations [4–6]. This phase has been experimentally observed in several TLAFs [7–9].

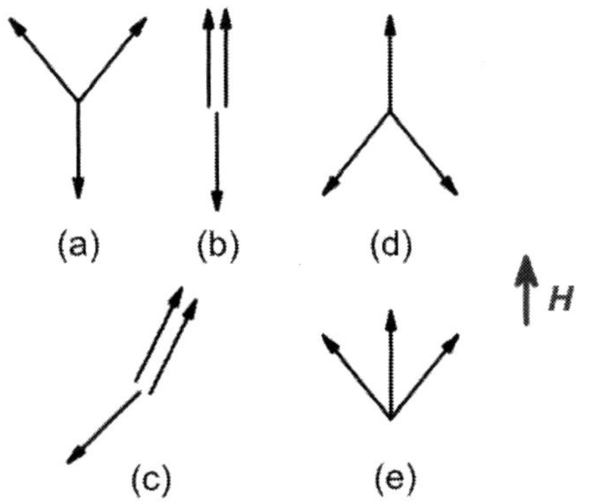


**Figure 1.** Spin structures stabilised in the pure TLAF in an applied field: (a) the Y-state, (b) the *uud*-state, (c) the 2:1 state. Spin structures of a planar TLAF with structural disorder: (d) inverted Y-state and (e) the fan state.

A recent theory [10, 11] predicted that the 1/3-plateau may be suppressed in the presence of a weak random modulation of the exchange bonds, which should lift the degeneracy. Apart from cancelling the plateau, a modification of the low- and high-field spin configuration is expected even at a very low doping: in low fields, the three-sublattice Y-structure should be changed for the inverted Y-state, while in high fields, the 2:1 structure should be replaced by the umbrella-type structure for a Heisenberg TLAF or by the fan-type structure for an XY TLAF, see the configuration sketches in Fig. 1. Formally, the impact of structural disorder on the degeneracy lifting in frustrated magnets may be represented by an effective biquadratic exchange with a positive sign [10–14]. The fluctuation contribution to the free energy is presented by a negative biquadratic term.

$RbFe(MoO_4)_2$ is a model TLAF formed by layers of magnetic $Fe^{3+}$ ions with a semiclassical $S = 5/2$ spin. At $T_N = 4.1$ K it exhibits a spin ordering within the layers with the three magnetic sublattices 120° apart, while the spin correlations in the adjacent layers correspond to an incommensurate structure, with the angle of 165° between the neighbouring spins [15]. The system has an easy-plane magnetic anisotropy with the plane parallel to layers and demonstrates a 1/3-magnetisation plateau for $H \perp c$ (the $c$ axis is perpendicular to the magnetic layers) [16,17]. The $H-T$ phase diagram corresponds well to the XY-TLAFM [16]. To perform a weak random modulation of exchange bonds we partially substitute Rb for K, disturbing the triangular layers by replacing nonmagnetic ions in the interlayer space.

In this paper we describe a study of K-doped samples $Rb_{1-x}K_xFe(MoO_4)_2$ with $x = 0, 0.075$ and 0.15 by means of electron spin resonance (ESR) spectroscopy and magnetisation measurements for the in-plane magnetic field. The mean-field parameters of the doped samples (the Néel temperature $T_N$, the saturation field $H_{sat}$ and the gap of antiferromagnetic resonance $\Delta$) were measured in a set of preliminary experiments, they showed a monotonic decrease by only a few percent for this range of impurity concentration, whereas the pulsed-field magnetisation curves demonstrated a complete disappearance of the plateau upon doping (the data are published elsewhere [18]). We report here the steady-field magnetisation curves, which have a higher signal to noise ratio compared to the pulsed field measurements and also correspond to a truly equilibrium regime. These data, as well as the ESR spectra, are compared for the pure sample, the intermediately doped sample ($x = 0.075$), and the sample with $x = 0.15$, the last is demonstrating a complete cancellation of the plateau and a change of the ground state.

## 2. Experimental methods

The crystals of $Rb_{1-x}K_xFe(MoO_4)_2$ were prepared as previously described [8]. The K content $x$ was determined by means of energy-dispersive X-ray spectroscopy.

The ESR signal is recorded as the field dependence of the microwave power transmitted through the resonator with the sample inside. A set of resonator-type spectrometers for the

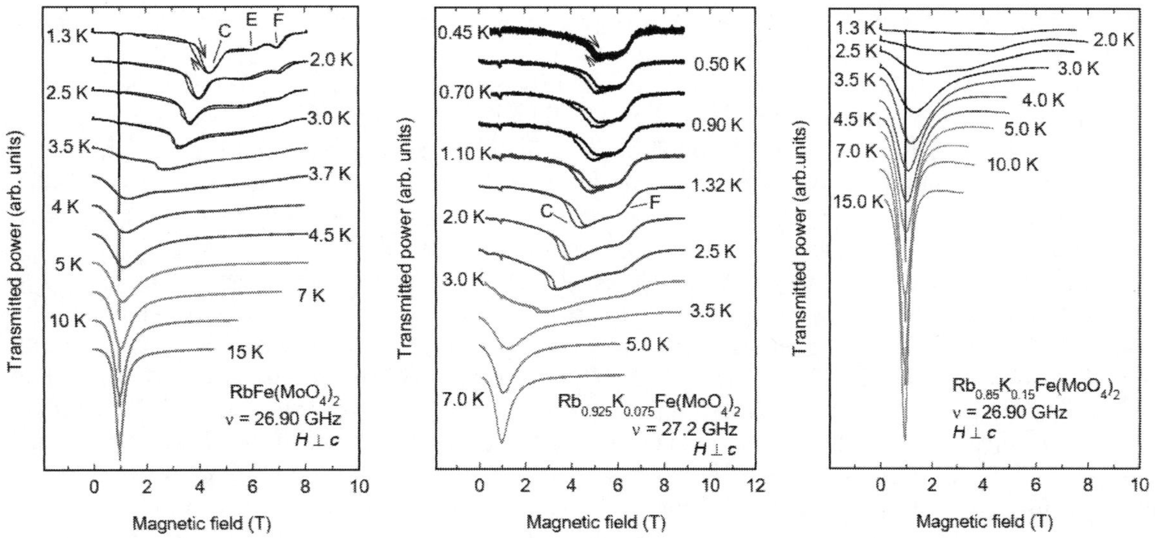


**Figure 2.** Temperature evolution of the ESR absorption curve for $x = 0$ (left panel), $x = 0.075$ (middle), and $x = 0.15$ (right panel).

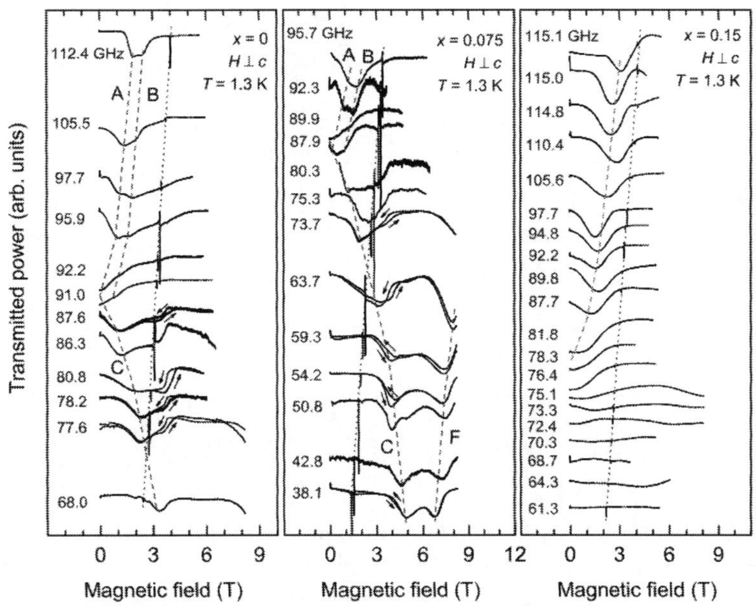


**Figure 3.** Normalised ESR absorption curves for the in-plane magnetic field for $Rb_{1-x}K_xFe(MoO_4)_2$ samples with $x = 0$, 0.075 and 0.15 for different frequencies. Dashed lines are a guide to the eye. The narrow ESR line is a DPPH label with $g = 2.00$. The gain is normalised to the intensity of the paramagnetic resonance signal at $T = 10$ K.

frequency range 25 - 120 GHz, combined with the $^4$He and $^3$He cryostats and superconducting solenoids was used. Frequency-field dependencies for the in-plane orientation of the magnetic field and temperature evolution of the ESR spectra were obtained. Magnetisation curves were measured using both a vibrating sample magnetometer and a pulsed field magnet.

## 3. Experimental results

As described in [8], on cooling the sample the ESR line broadens below 10 K. After passing through the Néel point, the ESR line shifts from the paramagnetic resonance position and again becomes narrow, exhibiting a splitting at some frequencies into a few antiferromagnetic resonance modes. A nonresonant change in the microwave susceptibility is also observed near a critical field with an abrupt change in the mutual orientation of spins in adjacent layers (4 T in the pure sample). This transition is accompanied by a hysteresis, marked by arrows denoting up and down field-sweep records in Figs. 2 and 3. The typical temperature evolution of an ESR line is presented in Fig. 2 for $x = 0, 0.075$ and 0.15. For different frequencies from the range 25-120 GHz, the field dependencies of the microwave transmission at $T = 1.3$ K for $x = 0, 0.075$ and 0.15 are shown in Fig. 3 for the in-plane magnetic field. The corresponding frequency-field dependencies of the spin-resonance modes are reconstructed in Fig. 4.

In correspondence with earlier observations [8], the ESR spectrum of a pure sample has an energy gap $\Delta_0 \simeq 90$ GHz and in low fields consists mainly of two branches. The frequency of the first branch rises with field, while the frequency of the second branch decreases. The ascending branch is split into two close branches (A) and (B) because of the interplane interaction [8]. The descending branch (C) goes to almost zero frequency near the lower boundary of the plateau. Another branch (F) at a frequency much lower than $\Delta_0$ arises at the upper boundary of the plateau. Thus the resonance fields of modes C and F observed at a low frequency of about 25 GHz mark the boundaries of the plateau. Another mode (D) appears near the onset of the plateau at a frequency higher than that of mode C. Mode E in the middle of the plateau

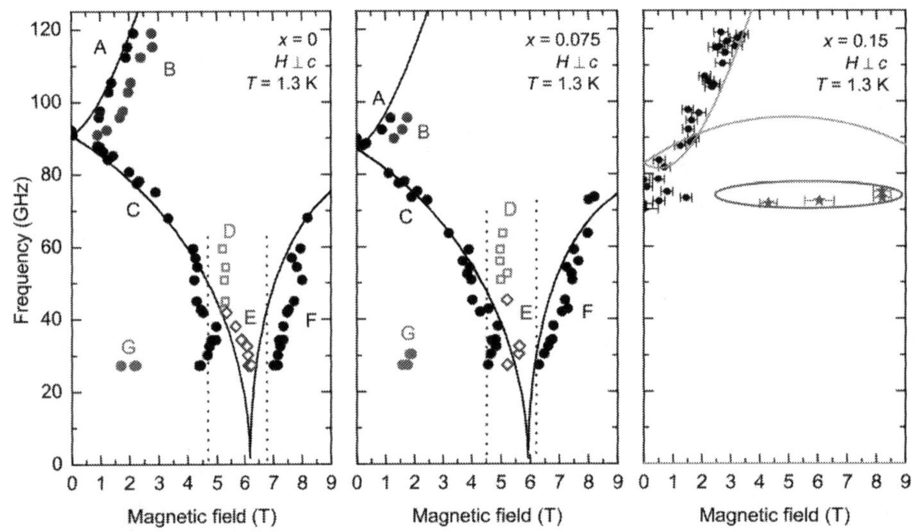


**Figure 4.** Frequency-field diagrams for the $Rb_{1-x}K_xFe(MoO_4)_2$ samples with different doping concentration $x$. The magnetic field is applied parallel to the magnetic layers. For $x = 0$ and 0.075, the solid lines represent theoretical calculations for the Y-type structure with parameters $J = 1.1J_0$ and $J = 1.03J_0$ respectively, while for the $x = 0.15$ sample the calculations are for the inverted Y state with $J = 0.9J_0$. Vertical dotted lines indicate 1/3-plateau boundaries evaluated from the $dM/dH$ curves (see Fig. 5). The ellipse covers a wide area of a weak absorption.

and the above mentioned modes A-D and F are all in accord with the theory [6] and previous experiments [8]. In addition, a weakly-absorbing mode (G) near 30 GHz exists due to a splitting of the zero-frequency mode [8].

The 27 GHz ESR records, which are sensitive to the plateau boundaries at the critical fields $H_{c1}$ and $H_{c2}$, are shown in Fig. 2. These data demonstrate that the features associated with the plateau in the ESR spectrum for the $x = 0.075$ sample are weaker, but still present, while for the $x = 0.15$ sample there is trace of the no plateau in the low temperature record. The intensities of

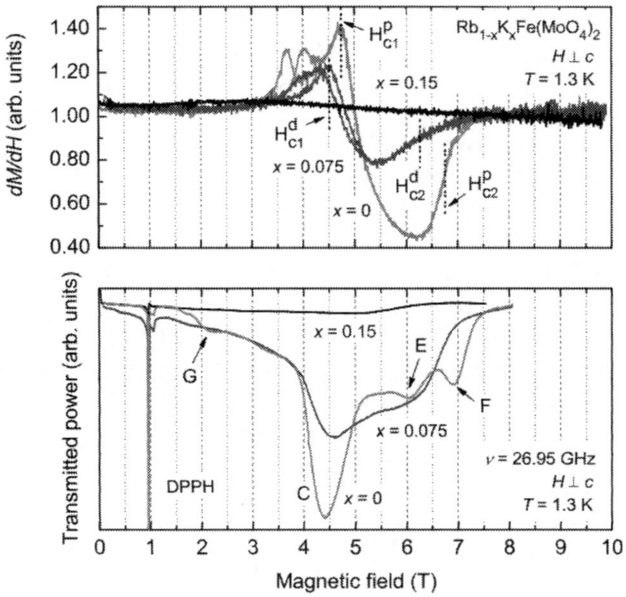


**Figure 5.** Upper panel: $dM/dH$ curves for $Rb_{1-x}K_xFe(MoO_4)_2$ samples with $x = 0$ and 0.075 measured in a static field by a vibrating sample magnetometer at $T = 1.3$ K. Lower panel: normalised ESR absorption curves for $x = 0$ and 0.075 samples. Resonance modes C, E, F, G are marked for the $x = 0$ sample in accordance with notation in Figs. 3 and 4.

resonance modes C and F are smaller in the intermediately doped sample ($x = 0.075$), and the width of the plateau, marked by these modes is reduced as compared to the pure sample. Figure 5 shows ESR response for all three concentrations at a low temperature $T = 1.3$ K, compared with $dM/dH$ curves of these samples. A long continuous drop of the $dM/dH$ curve marks the plateau range. We see that for $x = 0.075$ the drop of $dM/dH$ is much smaller, than for a pure sample, and for $x = 0.15$ the plateau disappears completely. At the same time the $dM/dH$ curve of the $x = 0.075$ sample marks distinctly the lower critical field $H_{c1}$, while the upper field $H_{c2}$ is smeared. We estimate the upper boundary of the plateau as the field where the change of $dM/dH$ reaches 1/2 of the total change during the return of $dM/dH$ to its nominal value. The estimated values of the critical fields both for a pure sample $H^{\rm p}_{c1,2}$ and for $x = 0.075$ sample $H^{\rm d}_{c1,2}$ correspond well to positions of ESR modes C and F. Both the drop of $dM/dH$ and ESR intensity of modes C and F demonstrate a vanishing in the $x = 0.075$ doped sample compared to a pure sample. This drop is much stronger than the diminishing of $T_{\rm N}$ or $H_{sat}$. The 27 GHz ESR response of the $x = 0.15$ sample is practically absent, as shown in Fig. 5, analogous to the absence of the drop of $dM/dH$. Thus, both the ESR signals and $M(H)$ curves of the $x = 0.15$ sample reveal the disappearance of the plateau, while the medium doped sample has a plateau reduced both in width and in flatness.

## 4. Disscussion

The ESR spectrum for the $x = 0.15$ doped sample demonstrates significant modifications in the entire range of applied magnetic fields. The descending ESR branch observed in the pure sample either completely disappears or transforms to a field-independent mode (see Figs. 3 and 4). This indicates a drastic change of the magnetic ground state with doping, and in particular, a full suppression of the ESR modes in the field range of the 1/3-magnetisation plateau and the complete absence of the descending branch C. At the same time, the frequency-field diagram for an intermediately doped sample indicates a narrowing of the plateau as well as a reduction of the energy gap by about of 5 percent, but remains broadly similar to that of a pure sample. We propose, that the crucial change of the ESR spectrum of the $x = 0.15$ sample is due to a change in the ground state spin configuration. To check this conjecture we have calculated the frequency-field dependencies using two models of the ground state, the Y-phase of a TLAF and the inverted Y-phase. The details of the theory of the ESR frequencies for these different spin structure models will be published elsewhere [18]. The results of the calculation are presented in Fig. 4 as solid lines. Calculations for the Y-type spin structure are shown in the left and middle panels in Fig. 4, and for the inverted Y-structure, in the right panel. The parameters used for the calculations for the pure compound are the main intraplane exchange integral, $J_0$ and the single-ion anisotropy constant, $D_0$. For the pure compound the parameters obtained from neutron scattering experiments [19] are: $J_0 = 0.086$ meV, $D_0 = 0.027$ meV. For fitting the theoretical frequency-field dependencies we used the parameters $J = 1.1J_0$ for $x = 0$, $J = 1.03J_0$ for $x = 0.075$ samples, and $J = 0.9J_0$ for $x = 0.15$ samples (the anisotropy term is kept equal to the $D_0$ of the pure sample). The decrease of the exchange interaction with doping agrees well with the observed decrease of $H_{sat}$ and $T_N$. From a comparison of the experimental data with the theoretical frequency-field dependencies we conclude, that the possible change of the ground state, induced by doping is from the Y- to inverted Y-phase, as predicted by [10, 11].

## 5. Conclusions

Measurements of the magnetisation versus field and the ESR spectra confirm the theoretical predictions that in a triangular antiferromagnet the 1/3 magnetisation plateau is dramatically suppressed by a moderate random modulation of the intraplane exchange interaction. The variation of magnetic resonance spectrum on doping, including the disappearance of the descending branch for $x = 0.15$, verifies the stabilisation of an inverted Y-state instead of Y structure.

## Acknowledgments

The work is supported by the Russian Foundation for Fundamental research, grant N 15-02-05918, Russian Science Foundation, grant N 17-1201505, by the Program of the Presidium of Russian Academy of Sciences, and by International Joint Research Promotion Program (short term) in Osaka University.

## References

[1] Collins M F and Petrenko O A 1997 *Can. J. Phys.* **75** 605
[2] Balents L 2010 *Nature* **464** 199
[3] Starykh O A 2015 *Rep. Prog. Phys.* **78** 052502
[4] Lee D H, Joannopoulos J D, Negele J W and Landau D P 1984 *Phys. Rev. Lett.* **52** 433
[5] Kawamura H and Miyashita S 1985 *J. Phys. Soc. Jpn.* **54** 4530
[6] Chubukov A V and Golosov D I 1991 *J. Phys.: Condens. Mat.* **3** 69
[7] Inami T, Ajiro Y and Goto T 1996 *J. Phys. Soc. Jpn.* **65** 2374
[8] Svistov L E, Smirnov A I, Prozorova L A, Petrenko O A, Demianets L N and Shapiro A Ya 2003 *Phys. Rev. B* **67** 094434
[9] Shirata Yu, Tanaka H, Matsuo A and Kindo K 2012 *Phys. Rev. Lett.* **108** 057205
[10] Maryasin V S and Zhitomirsky M E 2013 *Phys. Rev. Lett.* **111** 247201
[11] Maryasin V S and Zhitomirsky M E 2015 *J. Phys.: Confer. Ser.* **592** 012112
[12] Henley C L 1989 *Phys. Rev. Lett.* **62** 2056
[13] Long M W 1989 *J. Phys.: Condens. Mat.* **1** 2857
[14] Fyodorov Y V and Shender E F 1991 *J. Phys. Condens. Mat.* **3** 9123
[15] Kenzelman M, Lawes G, Harris A B, Gasparovic G, Broholm C, Ramirez A P, Jorge G A, Jaime M, Park S, Huang Q, Shapiro A Ya, Demianets L N 2007 *Phys. Rev. Lett.* **98**, 267205
[16] Svistov L E, Smirnov A I, Prozorova L A, Petrenko O A, Micheler A, Buttgen N, Shapiro A Ya and Demianets L N 2006 *Phys. Rev. B* **74** 024412
[17] Smirnov A I, Yashiro H, Kimura S, Hagiwara M, Narumi Y, Kindo K, Kikkawa A, Katsumata K, Shapiro A Ya and Demianets L N 2007 *Phys. Rev. B* **75** 134412
[18] Smirnov A I, Soldatov T A, Petrenko O A, Takata A, Kida T, Hagiwara M, Zhitomirsky M E, and Shapiro A Ya 2017 *Phys. Rev. Lett.* **119**, 047204

[19] White J S, Niedermayer Ch, Gasparovic G, Broholm C, Park J M S, Shapiro A Ya, Demianets L N and Kenzelmann M 2013 *Phys. Rev. B* **88** 060409

IOP Conf. Series: Journal of Physics: Conf. Series **969** (2018) 012116 doi:10.1088/1742-6596/969/1/012116

# Observation of spin superfluidity: YIG magnetic films and beyond

**Edouard Sonin**

Racah Institute of Physics, Hebrew University of Jerusalem, Givat Ram, Jerusalem 91904, Israel

E-mail: sonin@cc.huji.ac.il

**Abstract.** From topology of the order parameter of the magnon condensate observed in yttrium-iron-garnet (YIG) magnetic films one must not expect energetic barriers making spin supercurrents metastable. But we show that some barriers of dynamical origin are possible nevertheless until the gradient of the phase (angle of spin precession) does not exceed the critical value (analog of the Landau critical velocity in superfluids). On the other hand, recently published claims of experimental detection of spin superfluidity in YIG films and antiferromagnets are not justified, and spin superfluidity in magnetically ordered solids has not yet been experimentally confirmed.

## 1. Introduction

Spin superfluidity was suggested in the 1970s [1, 2] (see recent reviews [3, 4]). Manifestation of spin superfluidity is a stable spin supercurrent proportional to the gradient of the phase $\varphi$ (spin rotation angle in a plane) and not accompanied by dissipation, in contrast to a dissipative spin diffusion current proportional to the gradient of spin density.

In general the spin current proportional to the gradient of the phase $\varphi$ is ubiquitous and exists in any spin wave, although in these cases variation of the phase $\varphi$ is small [5]. Analogy with mass and charge persistent currents (supercurrents) arises when at long (macroscopical) spatial intervals along streamlines the phase variation is many times larger than $2\pi$. The supercurrent state is a helical spin structure, but in contrast to equilibrium helical structures is metastable.

An elementary process of relaxation of the supercurrent is phase slip. In this process a vortex with $2\pi$ phase variation around it crosses current streamlines decreasing the total phase variation along streamlines by $2\pi$. Phase slips are suppressed by energetic barriers for vortex creation, which disappear when phase gradients reach critical values determined by the Landau criterion. Emergence of superfluidity is conditioned by special topology of the order parameter space. This requires the easy-plane anisotropy for the spontaneous magnetization in ferromagnets or the antiferromagnetic vector in antiferromagnets [3, 4].

Experimentally evidence of spin superfluidity was obtained in the $B$-phase of superfluid $^3$He [6], in which spin degrees of freedom of the order parameter make it analogous to antiferromagnets. Recently Bozhko *et al.* [7] declared detection of spin supercurrent at room temperature in a coherent magnon condensate created in yttrium-iron-garnet (YIG) magnetic films by strong parametric pumping [8]. In YIG the equilibrium order parameter in the spin space is not confined to some easy plane as required for spin superfluidity. Nevertheless, Sun *et al.* [9]

argued that spin superfluidity is still possible. Indeed, one cannot rule out that metastability of supercurrent states is provided by barriers not connected with topology of the equilibrium order parameter, if the system is far from the equilibrium. In this work we confirm that it is possible indeed and determine critical values of possible supercurrents at which metastability is lost (the analogue of the Landau criterion).

In the end we overview experiments on experimental detection of spin superfluidity in various materials. We define the term *superfluidity* in its original meaning known from the times of Kamerlingh Ohnes and Kapitza: transport of some physical quantity (mass, charge, or spin) on macroscopical distances without essential dissipation. This definitely requires a large number of full $2\pi$ rotations along current streamlines as pointed out above. On the basis of this we conclude that recent claims of observation spin superfluidity in YIG films [7] and in antiferromagnets [10] were not justified, and further experiments are needed for observation of spin superfluidity in magnetically ordered solids.

## 2. Topology and superfluid spin currents

In a superfluid the order parameter is a complex wave function $\psi = \psi_0 e^{i\varphi}$, where the modulus $\psi_0$ of the wave function is a positive constant determined by minimization of the energy. The phase $\varphi$ is a degeneracy parameter since the energy does not depend on $\varphi$ because of gauge invariance. Any of degenerate ground states in a closed annular channel (torus) maps on some point at a circumference $|\psi| = \psi_0$ in the complex plane $\psi$, while a current state with the phase change $2\pi n$ around the torus maps onto a circumference (Fig. 1a) winding around the circumference $n$ times. It is evident that it is impossible to change $n$ keeping the path on the circumference $|\psi| = \psi_0$ all the time. In the language of topology states with different $n$ belong to different classes, and $n$ is a *topological charge.* Only a phase slip can change it when the path in the complex plane leaves the circumference. This should cost energy, which is spent on creation of a vortex crossing the cross-section of the torus channel and changing $n$ to $n-1$.

If we consider transport of spin parallel to the axis $z$ the analog of the phase of the superfluid order parameter is the rotation angle of the spin component in the plane $xy$, which we note also as $\varphi$. In isotropic ferromagnets the order parameter space is a sphere of radius equal to the absolute value of the magnetization vector $\boldsymbol{M}$ (Fig. 1b). All points on this sphere correspond to the same energy of the ground state. Suppose we created the spin current state with monotonously varying phase $\varphi$ in a torus. This state maps on the equatorial circumference on the order parameter sphere. Topology allows to continuously shift the circumference and to reduce it to the point of the northern pole. During this process shown in Fig. 1b the path remains on the sphere all the time and therefore no energetic barrier resists to the transformation. Thus metastability of the current state is not expected.

In a ferromagnet with easy-plane anisotropy the order parameter space contracts from the sphere to an equatorial circumference in the $xy$ plane. This makes the order parameter space topologically equivalent to that in superfluids (Fig. 1c). Now transformation of the equatorial circumference to the point shown in Fig. 1b costs anisotropy energy. This allows to expect metastable spin currents (supercurrents). They relax to the ground state via phase slips, in which magnetic vortices cross spin current streamlines. States with vortices maps on a hemisphere of radius $M$ either above or below the equator.

Up to now we considered states close to the equilibrium (ground) state. In a ferromagnet in a magnetic field the equilibrium magnetization is parallel to the field. However, by pumping magnons into the sample it is possible to tilt the magnetization with respect to the magnetic field. This creates a nonstationary state, in which the magnetization precesses around the magnetic field. Although the state is far from the true equilibrium, but it, nevertheless, is a state of minimal energy at fixed magnetization $M_z$. Because of inevitable spin relaxation the state of uniform precession requires permanent pumping of spin and energy. However, if

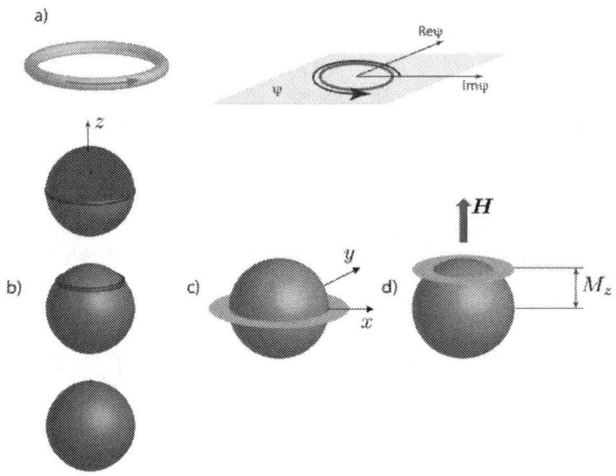


**Figure 1.** Mapping of current states on the order parameter space.
a) Mass currents in superfluids. The current state in torus maps on a circumference of radius $|\psi|$ on the complex plane $\psi$.
b) Spin currents in an isotropic ferromagnet. The current state in torus maps on an equatorial circumference on the sphere of radius $M$ (top). Continuous shift of the path on the sphere (middle) reduces it to a point (bottom), which is the ground state without currents.
c) Spin currents in an easy-plane ferromagnet. Easy-plane anisotropy contracts the order parameter space to an equatorial circumference in the $xy$ plane topologically equivalent to the order parameter space in superfluids.
d) Spin currents in an isotropic ferromagnet in a magnetic field $\boldsymbol{H}$ parallel to the axis $z$ with nonequilibrium magnetization $M_z$ supported by magnon pumping. Spin is confined in the plane normal to $\boldsymbol{H}$. This is an "easy plane" of dynamical origin.

processes violating the spin conservation law are weak, one can ignore them and treat the state as a quasi-equilibrium state. The state of uniform precession maps on a circumference parallel to the $xy$ plane, but in contrast to the easy-plane ferromagnet (Fig. 1c) the plane confining the precessing magnetization is much above the equator and not far the northern pole (Fig. 1d). One can consider also a current state, in which the phase (the rotation angle in the $xy$ plane) varies not only in time but also in space with a constant gradient. The current state will be metastable due to the same reason as in an easy-plane ferromagnet: in order to relax via phase slips the magnetization should go away from the circumference on which the state of uniform precession maps, and this increases the energy. Then the plane, in which the magnetization precesses, can be considered as an effective "easy plane" originating not from the equilibrium order parameter topology but created dynamically. The concept of dynamical easy plane is applicable for a magnon condensate in YIG films with some modifications. They take into account that the spin conservation law is not exact due to magnetostatic energy and the precession is not uniform since spin waves in YIG films have the energy minima at non-zero wave vectors.

In our discussion we assumed that phase gradients were small and ignored the gradient-dependent (kinetic) energy. At growing gradient we reach the critical gradient at which barriers making the supercurrent stable vanish. The critical gradient must be determined from the. criterion analogous to the famous Landau criterion [3, 11].

IOP Conf. Series: Journal of Physics: Conf. Series **969** (2018) 012116 doi:10.1088/1742-6596/969/1/012116

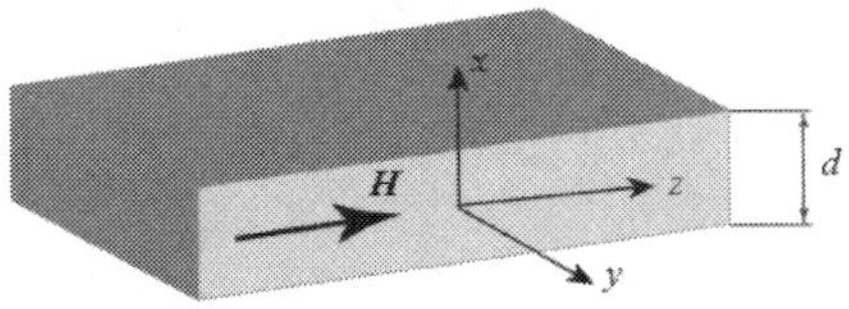


**Figure 2.** The YIG film of thickness $d$ in a magnetic field $\boldsymbol{H}$ parallel to the axis $z$.

## 3. Spin waves in YIG films

The coherent state of magnons is nothing else but a classical spin wave, and one can use the classical equations of the Landau–Lifshitz–Gilbert (LLG) theory. YIG is a ferrimagnet with complicated magnetic structure consisting of numerous sublattices. However at slow degrees of freedom relevant for our analysis one can treat it simply as an isotropic ferromagnet [12] with the spontaneous magnetization $\boldsymbol{M}$ described by the hamiltonian

$$\mathcal{H} = \int \left[-\boldsymbol{H}\cdot\boldsymbol{M} + D\frac{\nabla_i\boldsymbol{M}\cdot\nabla_i\boldsymbol{M}}{2}\right] d\boldsymbol{r} + \int \frac{\boldsymbol{\nabla}\cdot\boldsymbol{M}(\boldsymbol{r})\boldsymbol{\nabla}\cdot\boldsymbol{M}(\boldsymbol{r}_1)}{2|\boldsymbol{r}-\boldsymbol{r}_1|} d\boldsymbol{r}\, d\boldsymbol{r}_1. \tag{1}$$

Here the first term is the Zeeman energy in the magnetic field $\boldsymbol{H}$, the second term $\propto D$ is the inhomogeneous exchange energy, and the last one is the magnetostatic (dipolar) energy.

In a weak spin wave propagating in the plane $xz$ in a magnetic field $\boldsymbol{H}$ parallel to the axis $z$ $M_z \approx M - M_\perp^2/2M$ and $\boldsymbol{\nabla}\cdot\boldsymbol{M} \approx \nabla_x M_x$, where $M_\perp = \sqrt{M_x^2 + M_y^2}$. In the LLG theory the absolute value of the magnetization vector $\boldsymbol{M}$ does not vary in space and time, and the linearized LLG equations are reduced to two equations for only two independent magnetization components:

$$\dot{M}_x = -\gamma M \frac{\delta\mathcal{H}}{\delta M_y} = -\gamma H M_y + \gamma D M(\nabla_x^2 M_y + \nabla_z^2 M_y),$$

$$\dot{M}_y = \gamma M_z \frac{\delta\mathcal{H}}{\delta M_x} = \gamma H M_x - \gamma D M(\nabla_x^2 M_x + \nabla_z^2 M_x) - \gamma M \nabla_x \left(\int \frac{\nabla_x M_x(\boldsymbol{r}_1)}{|\boldsymbol{r}-\boldsymbol{r}_1|} d\boldsymbol{r}_1\right), \tag{2}$$

where $\gamma$ is the gyromagnetic ratio and $\delta\mathcal{H}/\delta M_x$ and $\delta\mathcal{H}/\delta M_y$ are functional derivatives. For the plane wave with the frequency $\omega$ and the wave vector $\boldsymbol{k}(k_x, 0, k_z)$ Eqs. (2) yield the spectrum

$$\omega(k) = \gamma\sqrt{(H + DMk^2)\left(H + DMk^2 + \frac{4\pi M k_x^2}{k^2}\right)}. \tag{3}$$

A spin wave propagating in the film of thickness $d$ parallel to the plane $yz$ (Fig. 2) must satisfy the boundary conditions at two film surfaces $x = \pm d/2$. In the situation when both the exchange and the dipole energies are of importance, the boundary problem is not trivial and was discussed in details in Ref. [11]. According to this analysis, the spectrum of the spin wave propagating along the magnetic field in the film is

$$\omega(k_z) \approx \gamma\left(H + DMk_z^2 + \frac{2\pi^3 M}{k_z^2 d^2}\right). \tag{4}$$

This follows from the spectrum Eq. (3) at $k_x = \pi/d$ and corresponds to a plane wave with weakly elliptical polarization propagating along the magnetic field (the axis $z$). The energy $E = \omega(M - \langle M_z\rangle)/\gamma$ and the frequency $\omega(k_z)$ given by Eq. (4) have two degenerate minima [12] at finite $k_z = \pm k_0$ where magnons can condense (Fig. 3). Here

$$k_0 = \left(\frac{2\pi^3}{Dd^2}\right)^{1/4} = \left(\frac{2\pi^2}{l_d^2 d^2}\right)^{1/4}, \tag{5}$$

where $l_d = \sqrt{D/\pi}$ is a small scale determined by the exchange energy.

## 4. Spin supercurrent in the YIG magnon condensate

In the linear theory the distribution of magnons between two condensates is arbitrary and does not affect the total energy at fixed magnetization $\langle M_z \rangle$. But non-linear corrections lift this degeneracy. The most important nonlinear term arises from the magnetostatic energy, which is maximal for the standing wave (two energy minima are equally populated by magnons):

$$M_x = 2\sqrt{2M(M-\langle M_z\rangle)}\cos\frac{\pi x}{d}\cos k_z z\cos\omega t,$$
$$M_y = 2\left(1+\frac{2\pi^3 M}{Hk_z^2d^2}\right)\sqrt{2M(M-\langle M_z\rangle)}\cos\frac{\pi x}{d}\cos k_z z\sin\omega t, \quad (6)$$

and is equal to

$$E_{ms} = \int\frac{\nabla_z M_z(\boldsymbol{r})\nabla_z M_z(\boldsymbol{r}_1))}{2|\boldsymbol{r}-\boldsymbol{r}_1|}d\boldsymbol{r}\,d\boldsymbol{r}_1 = \frac{3\pi(M-\langle M_z\rangle)^2}{2}. \quad (7)$$

The spin current appears if the wave numbers $k_z$ of two condensates differ from $\pm k_0$ (Fig. 3). The current is proportional to the phase gradient $\nabla_z\varphi = K = k_z - k_0 \ll k_0$. Keeping the magnetization $\langle M_z\rangle$ fixed as before and taking into account the nonlinear magnetostatic term (7) the energy in the spin-current state apart from some constant terms is

$$\Delta E = \frac{d^2\omega(k_z)}{dk_z^2}\frac{M-\langle M_z\rangle}{\gamma}\frac{(\nabla_z\varphi)^2}{2} + \frac{3\pi(M-\langle M_z\rangle)^2}{2}. \quad (8)$$

Stability of the spin-current state can be checked following the principal idea of the Landau criterion of superfluidity [3, 11]. Let us consider slowly varying in space weak perturbations $m_z = M_z - \langle M_z\rangle$ and $\nabla_z\varphi' = \nabla_z\varphi - K$. For stability of the supercurrent the quadratic form obtained from expansion of the energy (8) in perturbations $m_z$ and $\nabla_z\varphi'$ must be always positive. This takes place as far as $\nabla_z\varphi = K$ is less than the critical value

$$(\nabla_z\varphi)_{cr} = \sqrt{\frac{3\pi\gamma(M-\langle M_z\rangle)}{\frac{d^2\omega(k_0)}{dk_z^2}}} = \sqrt{\frac{3(M-\langle M_z\rangle)}{M}\frac{k_0^2 d}{4\pi}}. \quad (9)$$

Note that applying our course of derivation to superfluid hydrodynamics one obtains exactly the Landau critical velocity equal to the sound velocity (see Sec. 2.1 in Ref. [3]). One can find numerical estimation of the critical gradient and comparison with other theories in Ref. [11].

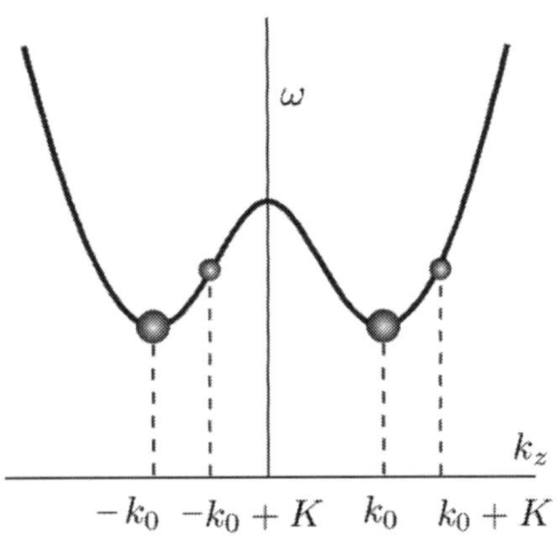


**Figure 3.** The spin-wave spectrum in a YIG film. In the ground state the magnon condensate occupies two minima in the $k$ space with $k_z = \pm k_0$ (large circles). In the current state two parts of the condensate are shifted to $k = \pm k_0 + K$ (small circles).

IOP Conf. Series: Journal of Physics: Conf. Series **969** (2018) 012116 doi:10.1088/1742-6596/969/1/012116

## 5. Experimental detection of spin superfluidity

A smoking gun of spin superfluidity in the $B$-phase of superfluid $^3$He was an experiment with a spin current through a long channel connecting two cells filled by the fluid with coherently precessing spins [13]. A small difference of the precession frequencies in two cells leads to a linear growth of difference of the precession phases in the cells and a phase gradient in the channel. When the gradient reaches the critical value phase slips must occur. Phase slips were detected experimentally at the total phase difference exceeding $2\pi$. A sharp $2\pi$ phase slip is reliable evidence of non-trivial spin supercurrents at phase gradients restricted by finite critical values.

Some time ago Bunkov *et al.* [10] declared that their observation of coherent spin precession in antiferromagnets demonstrated "high-$T_c$ spin superfluidity" since random inhomogeneity of their samples inevitably generates local spin currents. The very idea that random spin currents fluctuating at the disorder scale are the same as persistent currents, which are constant at macroscopic scales, is bizarre at least. Moreover, no shred of evidence, direct or indirect, of presence of any spin current, random or non-random, was presented. The spin current, or any quantity connected with it, is totally absent in any experimental data or formula. The only "evidence" was the statement that spin currents *must* be present in their experiment. Apparently Bunkov *et al.* do not see any difference between a theoretical prediction (*must* be present) and experimental confirmation of the prediction.

Recently Bozhko *et al.* [7] declared detection of spin superfluidity at high temperatures in a decaying magnon condensate in a YIG film. In their experiment the phase gradient emerged from spin precession difference produced by a temperature gradient. But the estimate made in Ref. [11] showed that the total phase difference across the magnon cloud in the experiment did not exceed about 1/3 of the full $2\pi$ rotation. Spin current relaxation via discrete $2\pi$ phase slips is impossible in this case independently from phase gradient. Therefore Bozhko *et al.* dealt with trivially stable spin currents present in any spin wave. Their existence does not require new confirmations after a half-a-century experiments with spin waves at *all* temperatures.

In summary, spin superfluidity, which was revealed in superfluid $^3$He-$B$, still waits its experimental confirmation in magnetically ordered solids.

## References

[1] Sonin E B 1978 *Zh. Eksp. Teor. Fiz.* **74** 2097–2111 [Sov. Phys.–JETP, **47**, 1091–1099 (1978)]
[2] Sonin E B 1982 *Usp. Fiz. Nauk* **137** 267 [Sov. Phys.–Usp., **25**, 409 (1982)]
[3] Sonin E B 2010 *Adv. Phys.* **59** 181–255
[4] Chen H and MacDonald A H 2017 Spin-superfluidity and spin-current mediated nonlocal transport, in: *Universal Themes of Bose-Einstein Condensation* ed Proukakis N P, Snoke D W and Littlewood P B (Cambridge University Press) pp 525–548; arXiv:1604.02429
[5] Halperin B I and Hohenberg P C 1969 *Phys. Rev.* **188** 898–918
[6] Bunkov Y 1995 *Progress of Low Temperature Physics* vol 14 ed Halperin W P (Elsevier) p 68
[7] Bozhko D A, Serga A A, Clausen P, Vasyuchka V I, Heussner F, Melkov G A, Pomyalov A, L'vov V S and Hillebrands B 2016 *Nat. Phys.* **12** 1057–1062
[8] Demokritov S O, Demidov V E, Dzyapko O, Melkov G A, Serga A A, Hillebrands B and Slavin A N 2006 *Nature* **443** 430–433 URL `http://dx.doi.org/10.1038/nature05117`
[9] Sun C, Nattermann T and Pokrovsky V L 2016 *Phys. Rev. Lett.* **116**(25) 257205
[10] Bunkov Y M, Alakshin E M, Gazizulin R R, Klochkov A V, Kuzmin V V, L'vov V S and Tagirov M S 2012 *Phys. Rev. Lett.* **108** 177002
[11] Sonin E B 2017 *Phys. Rev. B* 144432
[12] Gurevich A G and Melkov G A 1996 *Magnetization Oscillations and Waves* (CRC Press)
[13] Borovik-Romanov A S, Bunkov Y, Dmitriev V V, Mukharskii Y and Sergatskov D 1987 *Pis'ma Zh. Eksp. Teor. Fiz.* **45** 98 [JETP Lett. **45** 124 (1987)]

# Magnetic phase diagram of the frustrated $S$=1/2 triangular-lattice magnet $Cu_2(NO_3)(OH)_3$

**H. Kikuchi[1], N. Kasamatsu[1], Y. Ishikawa[1], Y. Koizumi[1], Y. Fujii[2], A. Matsuo[3], K. Kindo[3]**

[1]Department of Applied Physics, University of Fukui, Fukui 910-8507, Japan
[2]Research Center for Development of Far-Infrared Region, University of Fukui, Fukui 910-8507, Japan
[3]Institute for Solid State Physics, The University of Tokyo, 277-8581 Kashiwa, Japan

E-mail: kikuchi@u-fukui.ac.jp

**Abstract.** $Cu_2(NO_3)(OH)_3$ (rouaite) is a frustrated triangular lattice magnet. We measured the magnetic susceptibility, specific heat, $^1$H-NMR and high field magnetization of a powder sample of $Cu_2(NO_3)(OH)_3$ and found a clear evidence of the long range magnetic order at about 8 K. The transition temperature was found to decrease as the magnetic field $H$ increased. An $H$-$T$ phase diagram of $Cu_2(NO_3)(OH)_3$ was determined by specific heat, magnetization and $^1$H-NMR data. The high field magnetization curve was analyzed using a simple spin model.

## 1. Introduction

Geometrically frustrated magnets have attracted much interest recently, because some new physical states including spin liquids are expected to be realized in them [1]. Triangular lattice antiferromagnet is a representative example of the frustrated system and has been extensively studied [1]. $Cu_2(NO_3)(OH)_3$ (mineral name rouaite) is one of the triangular lattice antiferromagnets. The crystal structure of $Cu_2(NO_3)(OH)_3$ is monoclinic (space group : $P2_1$) and contains two crystallographically different sites of $Cu^{2+}$ ions [2]. Site one (Cu1) is surrounded by four OH groups and two $NO_3$ groups, and site two (Cu2) is surrounded by five OH groups and one $NO_3$ group.

The $Cu^{2+}$ ions are coupled via oxygen ions to form a distorted triangular lattice as shown in Fig. 1. Six different intralayer exchange couplings $J_i(i = 1 \sim 6)$ between $Cu^{2+}$ ions can be present. The values of these exchange couplings, which depend on the the bond length and bond angle, have been calculated using the Hückel model [3], density functional theory (DFT) and the quantum Monte Carlo method [4]. Detailed calculations of the electron density distribution of $Cu_2(NO_3)(OH)_3$ have also been reported [5]. The values obtained using the quantum Monte Carlo method were calculated as $J_1$(F)=14, $J_2$(AF)=-34.9, $J_3$(F)=6.9, $J_4$(F)=10.4, $J_5$(AF)=-4.9 and $J_6$(F) = 2.0 cm$^{-1}$ [4], where F and AF denotes ferromagnetic and antiferromagnetic interaction, respectively. These values suggest that the spin system in $Cu_2(NO_3)(OH)_3$ is composed of a dominant ferromagnetic ($J_1$) chain and an antiferromagnetic ($J_2$) chain. Both F and AF chains are coupled by other smaller frustrating exchange interactions ($J_3 \sim J_6$). Because the spin system of $Cu_2(NO_3)(OH)_3$ contains both properties of a one-dimensional quantum chain and a frustrated magnet, the spin system of this compound is very unique. The magnetic

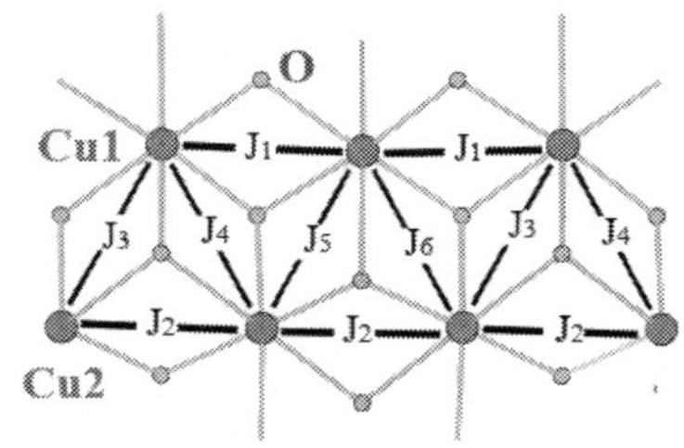


**Figure 1.** Crystal structure of $Cu_2(NO_3)(OH)_3$. Larger spheres represent Cu (red), and smaller ones represent O (light blue). $J_i$ ($i$=1∼6) (black line) represent different exchange constants between copper ions.

susceptibility $\chi(T)$ of $Cu_2(NO_3)(OH)_3$ was measured using either a powder sample [6] or a single crystal [3]. $\chi(T)$ of the powder sample has been reported to have a round maximum at 12 K, which has been discussed with respect to an occurrence of a long range order (LRO) although the possibility of a short range order (SRO) was not excluded because a neutron diffraction experiment on a deuterated sample did not detect the magnetic order [6]. $\chi(T)$ of the single crystal had a maximum at around 8 K and was analyzed using the antiferromagnetic chain model without a LRO. There is some controversy whether or not there is a magnetic LRO in $Cu_2(NO_3)(OH)_3$. The most convincing evidence of a thermodynamic phase transition is given by specific heat data. In this paper, we report measurements results of the specific heat of $Cu_2(NO_3)(OH)_3$ and present a clear evidence of the LRO. In addition, we measure magnetic susceptibility, high field magnetization and $^1$H-NMR to give the field-temperature ($H$-$T$) phase diagram of $Cu_2(NO_3)(OH)_3$.

## 2. Experiments

Powder samples of $Cu_2(NO_3)(OH)_3$ were synthesized by a hydrothermal method, according to the procedure reported in Ref. [7]. The quality of the specimen was investigated by powder X-ray diffraction. All the measurements were performed using the prepared powder sample. Magnetic susceptibility was measured using a superconducting quantum interference device (SQUID) magnetometer (MPMS, Quantum Design) in the temperature range of 2∼300 K. High-field magnetizations of up to about 60 T were measured using a pulsed high magnet at the Institute for Solid State Physics. Specific heat measurements were performed using a Quantum Design PPMS by the relaxation method under an applied field of up to 13 T. $^1$H-NMR spectra were measured by recording the spin-echo intensity while sweeping the magnetic field. Nuclear spin-lattice relaxation time $T_1$ was measured by the inversion recovery method.

## 3. Results and Discussion

Figure 2 shows the molar magnetic susceptibility $\chi(T) = M/H$, where $M$ and $H$ is magnetization and magnetic field, respectively, and the corresponding reciprocal susceptibility measured in a magnetic field of $H$ = 0.01 T for temperatures between 2 and 300 K. The higher temperature part of $\chi(T)$ follows the Curie-Weiss law. The Weiss temperature $\Theta$ was estimated to be about 6 K, which is slightly different from the reported value of - 8 K [3]. The Curie constant $C$ was evaluated to be 0.72 emu·K/mol. The effective magnetic moment of the $Cu^{2+}$ was estimated to be 1.70 $\mu_B$, which is comparable to the expected value of 1.73 $\mu_B$ for a $Cu^{2+}$ ion with $g = 2$. A broad peak is observed at around $T_{max}$= 11 K. This behavior agrees well with the reported result [6]. An inset of Fig. 2 shows the field dependence of $\chi(T)$ at low temperatures. As the magnetic field increases, the broad peak shifts toward a lower temperature.

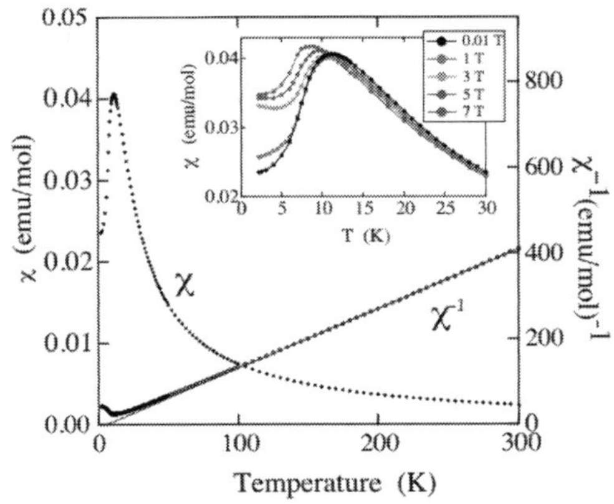


**Figure 2.** Temperature dependence of the magnetic susceptibility $\chi(T) = M/H$ and the corresponding reciprocal susceptibility measured at a magnetic field of $H = 0.01$T. Inset shows the low temperature part of $\chi(T)$ measured at various magnetic fields up to 7 T. Solid line is the Curie-Weiss fitting with Weiss temperature $\Theta = 6$ K and Curie constant $C = 0.72$ emu·K/mol .

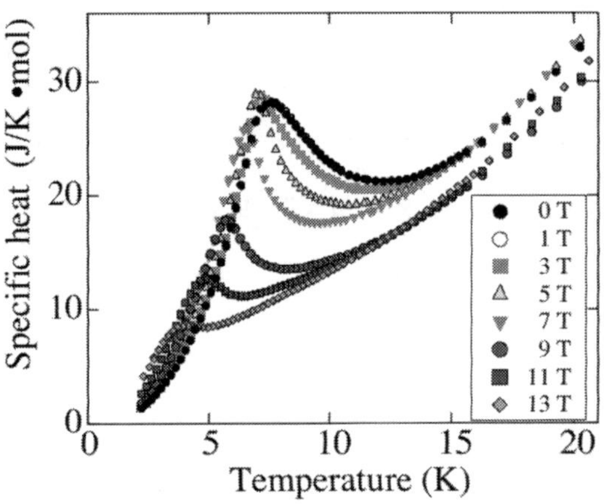


**Figure 3.** Specific heat data of $Cu_2(NO_3)(OH)_3$ measured under an applied magnetic field of up to 13 T.

Figure 3 shows the heat capacity measured at magnetic fields up to 13 T. Lattice contributions are not subtracted from the data because no isostructural material of $Cu_2(NO_3)(OH)_3$ without magnetic ions is known. A distinct $\lambda$-type anomaly was observed at $T_N = 8$ K in the zero field data. This finding is a clear evidence of the occurrence of magnetic LRO in this magnet.

Figure 4 shows the high field magnetization of $Cu_2(NO_3)(OH)_3$ measured at 4.2 and 1.3 K. The field derivative of the magnetization, $\mathrm{d}M/\mathrm{d}H$, is also shown. The saturation magnetization $M_s$ per chemical formula unit is given by $M_s = nSg\mu_B \approx 2\mu_B$ where $n = 2$ is the number of magnetic $Cu^{2+}$ ions in a formula unit, $g$ is Lande's g-factor, $\mu_B$ is the Bohr magneton and $S$ is the spin number, 1/2. $\mathrm{d}M/\mathrm{d}H$ shows very sharp peaks at around 2 T, and 14 T (1.3 K) or 11 T (4.2 K), which indicate the occurrence of field-induced magnetic transitions. The lower field transition at around 2 T is also observed in our magnetization measurement using the MPMS magnetometer and could be assigned to a spin-flip transition. The higher

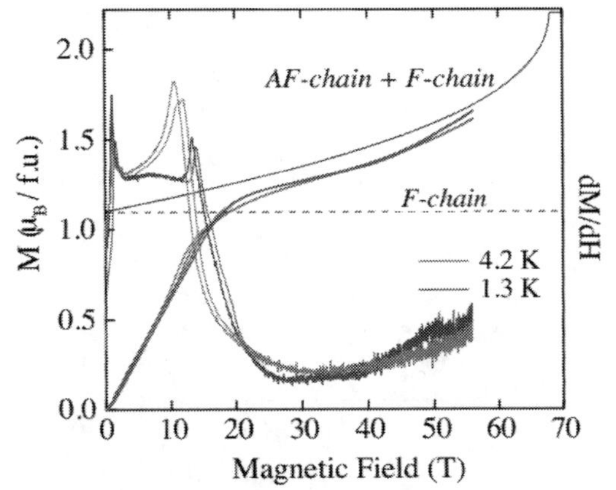


**Figure 4.** High field magnetization curve measured at 1.3 K (blue line) and 4.2 K (red line). The field derivative of the magnetization at 4.2 K is shown in the arbitral unit. The solid curve is the model calculation at $T$= 0 in which the spin system is assumed to be composed of mutually independent F- and AF- chains. The dotted line demonstrates the contribution from the fully polarized F-chain at $T = 0$ to the magnetization. The contribution from AF-chain is calculated with $J_2$ = -34.9 cm$^{-1}$ and g=2.23.

field transition corresponds to the LRO transition observed in the specific heat measurement. Besides these field-induced transitions, the magnetization curve has a unique property. It is easily seen that an increase rate of the magnetization becomes smaller above around 20 T where the magnetization reaches about one half of the saturation magnetization $M_s$. Because this field range is above the phase transition field, the magnet is in a paramagnetic phase, not in the classical magnetic ordered phase. We tried to explain this result using a simple model in which only the dominant $J_1$ and $J_2$ interactions are considered and other smaller interactions are neglected. In this simple model, the spin system is composed of the independent ferromagnetic and antiferromagnetic chains. The ferromagnetic chain is assumed to be fully polarized in this higher field range at low temperature and contributes to one half of the saturation magnetization. The magnetization curve originated from the antiferromagnetic chain is calculated using the theoretical result obtained by the density matrix renormalization group method [8] with $J_2$ =-34.9 cm$^{-1}$ and g=2.23 [4]. A saturation field $H_s$ is given by $H_s = 2|J_2|k_B/g\mu_B = 68$ T, where $k_B$ is the Boltzmann constant. The calculated curve based on the model is plotted in Fig. 4 as the solid line. Considering that our approximate model is very rough, the calculation qualitatively agrees with the observed curve in the higher field region. A point of note is that the magnetization above approximately 20 T follows the characteristic downward convex curve of the the 1D Heisenberg antiferromagnet. Remembering that such a feature represents the quantum nature of the 1D magnet, the quantum nature of the system is recovered in a state where the classical magnetic ordered phase is collapsed.

In order to study the magnetic order of $Cu_2(NO_3)(OH)_3$ from microscopic point view, we measured $^1$H-NMR spectra and $T_1$. Observed relaxation data against time $t$ are analyzed using the stretched exponential function, $\exp(-(t/T_1)^\beta)$ where $\beta$ is a stretch exponent. Figure 5 shows the temperature dependence of the spin-lattice relaxation rate $T_1^{-1}$ of $^1$H-NMR measured at operating radio frequencies of 42.3, 213 and 298 MHz. A very sharp peak is observed at around 7 K, reflecting the occurrence of the magnetic order. In the ordered phase below 7 K, $T_1^{-1}$ is found to follow the $T^3$ law, which is expected to be satisfied for the magnon relaxation process in the ordered antiferromagnet [9]. Value of $\beta$ is about one in the paramagnetic phase, and 0.6~0.7

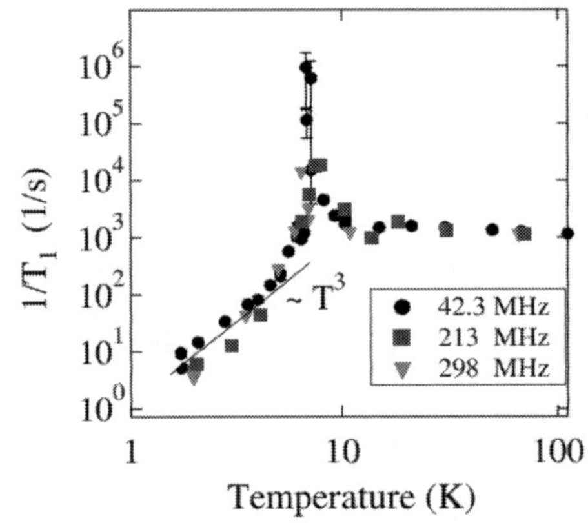


**Figure 5.** Temperature dependence of the spin-lattice relaxation rate of $^1$H-NMR of $Cu_2(NO_3)(OH)_3$ measured at various radio frequencies. The olid line indicates $T^3$ law.

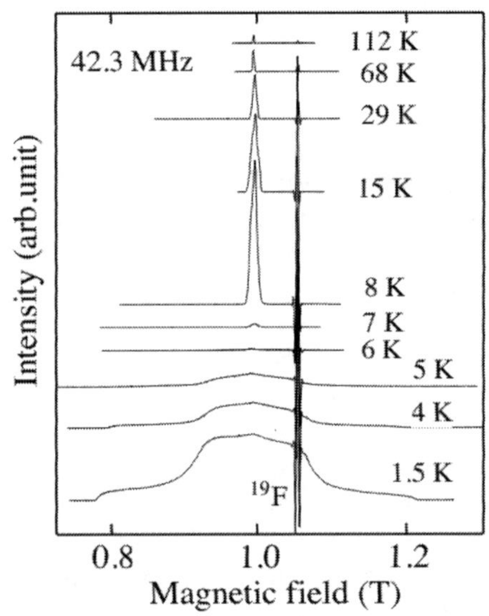


**Figure 6.** Temperature dependence of the $^1$H-NMR spectra measured at 42.3 MHz. Sharp signals at around 1.1 T originate from $^{19}$F signal contained in a sample holder.

in the ordered phase well below 7 K. Just at around $T_N$, $\beta$ takes small value down to about 0.2, suggesting that the relaxation times for protons in the sample are strongly inhomogeneous at $T \simeq T_N$ depending probably on the crystal orientation against the applied field.

Figure 6 shows $^1$H-NMR spectra measured at 42.3 MHz. As the temperature decreases, sharp peaks of $^1$H observed in the paramagnetic phase turn into very broadened pattern below $T_N$. The pattern at 1.5 K has characteristic rectangular pattern, which is expected for a powder sample with an antiferromagnetic ordered phase [10]. These NMR data indicate that the ordered phase of $Cu_2(NO_3)(OH)_3$ has a typical antiferromagnetic nature. We need further study to determine precise spin structure using a single crystal.

All collected data concerning the magnetic transition of $Cu_2(NO_3)(OH)_3$ is summarized in the

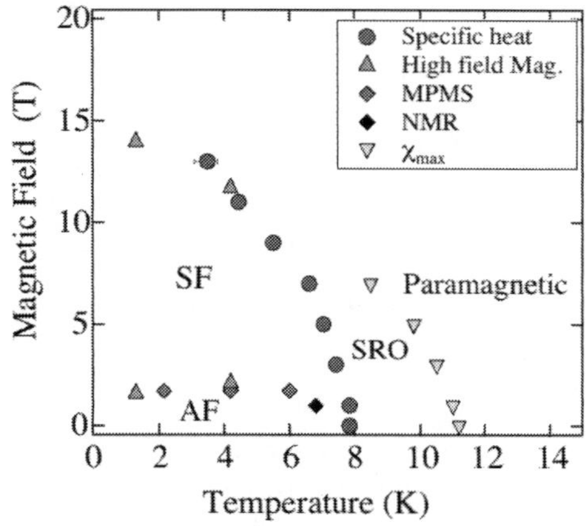


**Figure 7.** Field-temperature ($H$-$T$) phase diagram of $Cu_2(NO_3)(OH)_3$. Points are determined from heat capacity, magnetization and NMR measurements. SF and AF stand for spin-flip and antiferromagnetic phase, respectively.

$H$-$T$ phase diagram (Fig. 7). The field dependence of $T_{max}$ is also plotted in the phase diagram. The magnetic order temperature $T_N$ monotonically lowers as the magnetic field increases. Note should be paid that $T_N$ is lower than $T_{max}$. This result suggests that there is a crossover region in the phase diagram (the region described as 'SRO' in Fig. 7) where the LRO does not occur, probably due to the spin frustration, although the short range spin correlation is well developed. The same behavior has been observed in the frustrated triple-chain magnet $Cu_3(OH)_4SO_4$ (antlerite) [11], which has several exchange interactions. This behavior may be a universal feature of a frustrated system containing multiple magnetic interactions.

To conclude, we measured the magnetic susceptibility, specific heat, $^1$H-NMR spectra and high field magnetization of a powder sample of $Cu_2(NO_3)(OH)_3$ and found clear evidence of the long range magnetic order at about 8 K. The $H$-$T$ phase diagram of $Cu_2(NO_3)(OH)_3$ was determined by specific heat, magnetization and NMR data. The high field magnetization curve was analyzed using a simple spin model.

This research was supported by JSPS KAKENHI Grant Number 17K05515.

## References

[1] Diep H T ed. 2004 *Frustrated Spin Systems* (World Scientific)
[2] Effenberger H 1983 *Zeitschrift für Kristallographie* **165** 127–135
[3] Linder G.-G, Atanasov M and Pebler J 1995 *J. Solid State Chem.* **116** 1–7
[4] Ruiz E, Llunell M, Cano J, Rabu P, Drillon M and Massobrio C 2006 *J. Phys. Chem. B* **110** 115–118
[5] Pillet S, Soubassou M, Lecomte C, Rabu P, Drillon M, Massobrio C 2006 *Phys. Rev. B* **73** 115116
[6] Drillon M, Hornick C, Laget V, Rabu P, Romero F. M., Rouba S, Ulrich G and Ziessel R 1995 *Mol. Cryst. Liq. Cryst.* **273** 125–140
[7] Tanaka H and Terada S 1993 *J. Therm. Anal.* **39** 1011–1018
[8] Maeshima N and Okunishi K 2000 *Phys. Rev. B* **62** 934–939
[9] Nath R, Furukawa Y, Borsa F, Kauk E. E, Baenitz M, Geibel D and Johnston D. C 2009 *Phys. Rev. B* **80** 214430
[10] Yamada Y and Sakata A 1986 *J. Phys. Soc. Jpn.* **55** 1751–1758
[11] Vilminot S, Richard-Plouet M, André G, Swierczynski D, Guillot M, Bourée-Vigneron F, Drillon M 2003 *J. Solid State Chem.* **170** 255–264

IOP Conf. Series: Journal of Physics: Conf. Series **969** (2018) 012118 doi:10.1088/1742-6596/969/1/012118

# Resonant x-ray scattering study on electronic hybridization in unconventional ordered phase of $PrRu_4P_{12}$

**Hironori Nakao[1], Chihiro Tabata[1], Kazuaki Iwasa[2]**

[1] Condensed Matter Research Center and Photon Factory, Institute of Materials Structure Science, High Energy Accelerator Research Organization, Tsukuba 305-0801, Japan

[2] Frontier Research Center for Applied Atomic Sciences & Institute of Quantum Beam Science, Ibaraki University, Tokai, Naka, Ibaraki 319-1106, Japan

E-mail: hironori.nakao@kek.jp

**Abstract.** $PrRu_4P_{12}$, which undergoes a metal-insulator transition at $T_{MI} \sim 63$ K, has been investigated by resonant x-ray scattering technique at the Pr $L_3$-edge, the P $K$-edge, and the Ru $L_3$-edge, in which element and orbital selective electronic states can be clarified. Resonating energy spectra at the 111 reflection were found at the Pr $L_3$-edge and P $K$-edge, while it was not observed at the Rh $L_3$-edge. This is the first observation of quite large resonant signal at the P $K$-edge. This result is considered as a direct evidence for the strong Pr $4f$ and P $3p$ hybridization effect on the unconventional ordered phase accompanied by the insulator state.

## 1. Introduction

Filled skutterudites $RT_4X_{12}$ ($R$ = lanthanide and actinide elements, $T$ = transition metal, $X$ = pnictogen) have attracted great interests because of various physical properties, such as heavy-fermion superconductivity and multipole order [1, 2, 3, 4]. There the $R$ ion is surrounded by the $X$ ions forming an icosahedron, therefore it is expected that the mixing between $p$ state of $X$ ion and $f$ state of $R$ ion is responsible for the various physical properties.

$PrRu_4P_{12}$ exhibits a metal-insulator (MI) transition at $T_{MI} \sim 63$ K. X-ray and electron diffraction revealed a structural phase transition from the space group of $Im\bar{3}$ to that of $Pm\bar{3}$ at $T_{MI}$ [5]. Namley the reflections at the $hk\ell$ with $h + k + \ell$ = odd appear below $T_{MI}$. As an origin of the MI transition, the nesting of the Fermi surface was theoretically proposed [6]. However, $LaRu_4P_{12}$ does not show such the MI transition, even though it has a similar electronic structure except for the $4f$ electrons. It indicates that the $4f$ electron plays an important role in the MI transition. Then a possibility of antiferro-hexadecapole order of Pr $4f$ was theoretically proposed as an origin of the band-gap formation [7, 8].

Neutron scattering study on the crystal-field excitations elucidated the presence of strong orbital hybridization between Pr $4f$ and P $3p$ ($p - f$ hybridization), which causes the formation of charge-density-wave state associated with staggered $4f$-electron order below $T_{MI}$ [9]. Accordingly, the $p - f$ hybridization is expected to be essential for the MI transition. On the other hand, photoemission study reported that no notable anomaly was observed across the MI transition [10]. This result indicates that spatial structure of a particular electronic state

IOP Conf. Series: Journal of Physics: Conf. Series **969** (2018) 012118 doi:10.1088/1742-6596/969/1/012118

relevant to the MI transition should be investigated. Therefore, details of the $p-f$ hybridization effect on the ordered state in $PrRu_4P_{12}$ remains an important issue.

Resonant X-ray scattering (RXS) is a powerful tool for observing the spatial ordering of charge, orbital, and spin degrees of freedom [11]. $PrRu_4P_{12}$ has also been studied by the RXS to elucidate the unconventional ordered phase below $T_{MI}$ [12], although there was no clear evidence for the resonant features due to the electronic ordering. Furthermore, the RXS technique has a potential to elucidate the hybridized electronic state, since the resonant signal has element and orbital selective information depending on the absorption energy. In this study to investigate the hybridized electronic state below $T_{MI}$, the RXS measurement has been performed at the Pr $L_3$-edge, the P $K$-edge, and the Ru $L_3$-edge.

## 2. Experiments

Single crystal of $PrRu_4P_{12}$ was synthesized with Sn-flux method. The crystal structure is cubic with lattice constant $a$ = 8.042 Å at room temperature, which is consistent with previous report [13]. RXS and fluorescence measurements were carried out at BL-4C and BL-11B in Photon Factory, KEK. The incident beams were monochromatized by Si(111) double crystals. The experiments at the Pr $L_3$-edge were performed at BL-4C using a four-circle diffractometer with $\sigma$-polarization incident beam. The experiments at the P $K$-edge and at the Ru $L_3$-edge were performed at BL-11B using an in-vacuum two-circle diffractometer [14] with $\pi$-polarization incident beam. Polarization analysis of the scattered beam was not performed in these measurements.

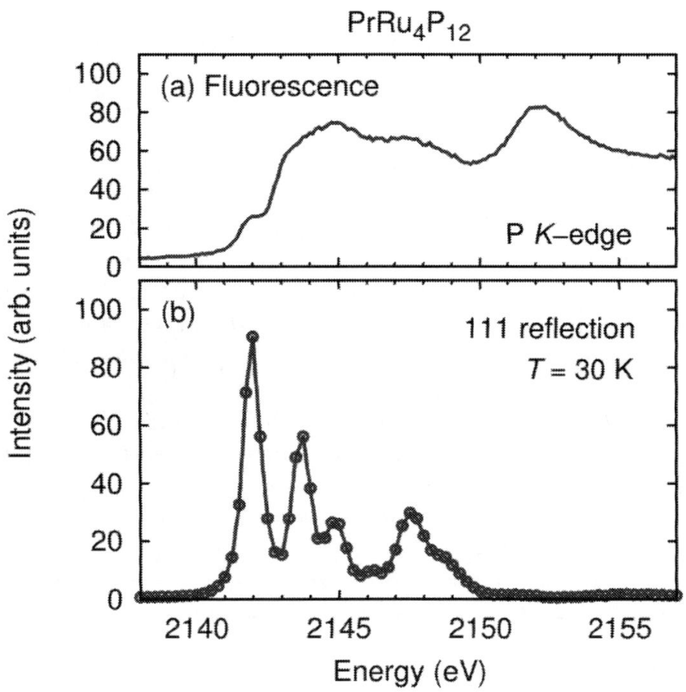


**Figure 1.** (a) Fluorescence spectrum at the P $K$-edge. (b) Energy dependence of the scattering intensity at the 111 reflection and at $T$ = 30 K.

IOP Conf. Series: Journal of Physics: Conf. Series **969** (2018) 012118 doi:10.1088/1742-6596/969/1/012118

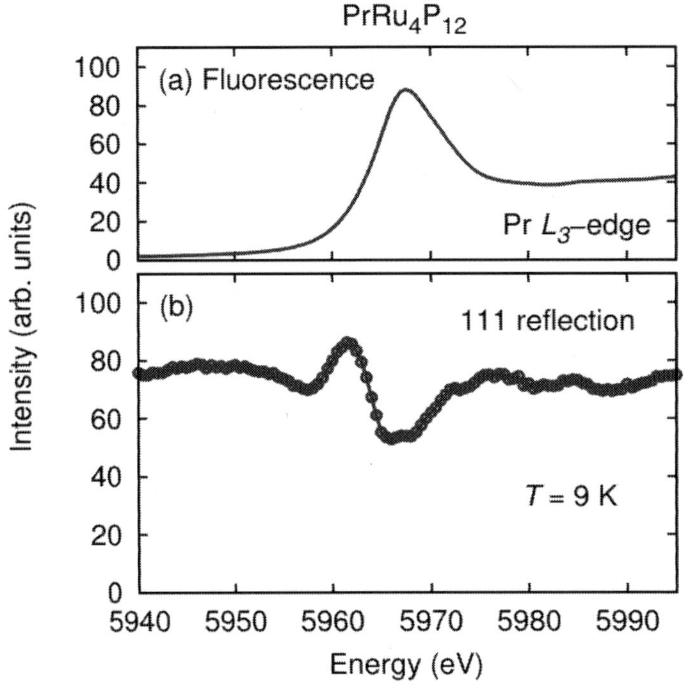


**Figure 2.** (a) Fluorescence spectrum at the Pr $L_3$-edge. (Data were taken from Ref. [12].) (b) Energy dependence of the scattering intensity at the 111 reflection and at $T = 9$ K.

## 3. Results and discussion

We have measured the RXS spectrum at the P $K$-edge to elucidate the P $3p$ electronic state in the unconventional ordered phase below $T_{MI}$. The fluorescence of $PrRu_4P_{12}$ was measured near the P $K$-edge ($1s \to 3p$ transition) as shown in Fig. 1(a). The spectrum reflects the unoccupied P $3p$ electronic state. Energy dependence of the scattering intensity was measured at the 111 reflection and at $T = 30$ K as shown in Fig. 1(b). The strongly resonated peaks were observed near the edge energy, while the scattering intensity exists even at 2135 eV (non-resonant energy). The latter is the Thomson scattering due to the structural distortion at the $T_{MI}$. On the other hand, the resonant signal at the P $K$-edge reflects the difference in the P $3p$ electronic states between two P icosahedrons due to the staggered $4f$-electron order. The strongest resonant peak exists at 2142 eV, which reflects the $3p$ state near the Fermi energy. The intensity is about $10^2$ times stronger than that at 2135 eV. This signal at the 111 reflection appears below $T_{MI}$, and no signal was observed above $T_{MI}$.

In order to compare with the resonant intensity at the Pr edge, we have measured the RXS signal at the Pr $L_3$-edge ($2p \to 5d$ transition). Energy dependence of the scattering intensity was measured at the 111 reflection as shown in Fig. 2(b). The spectrum has a small peak structure at 5961.6 eV, which reflects the staggered Pr $5d$-electronic state. The result is consistent with the previous report [12]. The component of the RXS signal is comparable to the Thomson scattering component at 5940 eV (non-resonant energy). It means that the change of the Pr $5d$ electronic states is much smaller than that of the P $3p$ state. It is noteworthy that the resonant signal at the Pr $L_3$-edge does not detect the $4f$ electronic state directly by the dipole transition ($E1$) process. The resonant signal at the quadrupole ($E2$) transition has not been reported so far [12], although a RXS signal associated with a hexadecapole order is theoretically proposed

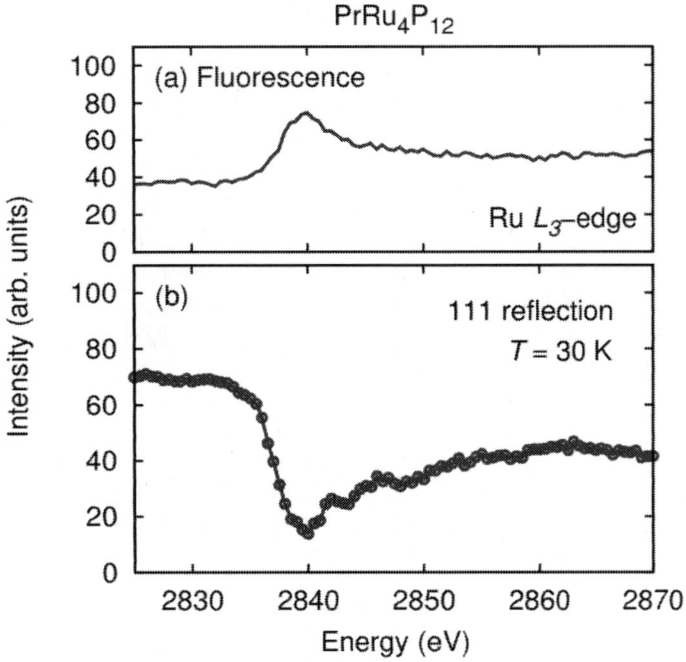


**Figure 3.** (a) Fluorescence spectrum at the Ru $L_3$-edge. (b) Energy dependence of the scattering intensity at the 111 reflection and at $T = 30$ K.

to appear at the $E2$ transition [15].

We have also measured the RXS signal at the Ru $L_3$-edge ($2p \rightarrow 4d$ transition). Since the atomic displacement of Ru on the structural change at $T_{MI}$ is comparable to the displacement of P atoms [16], we examined whether a RXS signal appears at the Ru $L_3$-edge or not. Energy spectrum at the 111 reflection is shown in Fig. 3(b). The intensities simply become weak near the absorption edge energy, and the spectrum is similar to that of the fluorescence (Fig. 3(a)). It indicates that any component of the RXS signal was not detected within the experimental error. Namely the change in the Ru $4d$ electronic state is quite small, which may indicate a less contribution of the Ru electrons to the unconventional ordered state below $T_{MI}$.

## 4. Conclution

We have performed the RXS measurements at the P $K$-edge, the Pr $L_3$-edge, and the Ru $L_3$-edge to elucidate the element and orbital selective electronic states in the unconventional ordered phase below $T_{MI}$. We newly found the strongly resonated peaks near the P $K$-edge, which directly evidence the alternative arrangement of P icosahedrons characterized by inequivalent $3p$ electronic states in the low-temperature $Pm\bar{3}$ phase. Moreover, the strongest resonant peak is considered to originate from the $3p$ state near the Fermi energy. The present result is in accordance with the gap formation accompanying the staggered $4f$ multipole order, which is driven by the strong $p - f$ hybridization. In contrast, because no resonant signal at the Ru $L_3$-edge was observed, the hybridization with the Pr $4f$ electrons is dominated by the phosphorus $p$ electrons. The resonant signal at the Pr $L_3$-edge is also quite weak. This result indicates that the Pr $5d$ state does not contribute to the ordered state. It is also consistent with the Pr $4f$ electrons play more important roles in the MI transition, because the $L_3$-edge measurement is

IOP Conf. Series: Journal of Physics: Conf. Series **969** (2018) 012118 doi:10.1088/1742-6596/969/1/012118

not a direct probe for the Pr $4f$ electrons. Finally, our result at the P $K$-edge is considered as a direct evidence for the strong Pr $4f$ and P $3p$ hybridization effect on the unconventional ordered phase below $T_{MI}$. In future work, the direct observation of the Pr $4f$ electronic state at the Pr $M$-edge becomes quite important to understand the $p - f$ hybridization effect.

## Acknowledgments
This research was partially supported by JSPS KAKENHI Grant (Nos. 21224008, 25286090, 15H05885 (J-Physics), and 17K05130), the Funding Program for World-Leading Innovative R&D on Science and Technology (FIRST), and the JST, Invitation for Applications for the Strategic Basic Research Programs (CREST). This study has been performed under the Photon Factory Program Advisory Committee (Proposal Nos. 2009S2-008, 2012S2-005, and 2017G553).

## References
[1] Sekine C, Uchiumi T, Shirotani I and Yagi T 1997 *Phys. Rev. Lett.* **79** 3218
[2] Sato H, Abe Y, Okada H, Matsuda T D, Abe K, Sugawara H and Aoki Y 2000 *Phys. Rev.* B **62** 15125
[3] Bauer E D, Frederik N A, Ho P-C Zapf V S and Maple M B 2002 *Phys. Rev.* B **65** 100506
[4] Kohgi M, Iwasa K, Nakajima M, Metoki N, Araki S, Bernhoeft N, Mignot J-M, Gukasov A, Sato H, Aoki Y and Sugawara H 2003 *J. Phys. Soc. Jpn.* **72** 1002
[5] Lee C H, Matsuhata H, Yamaguchi H, Sekine C, Kihou K, Suzuki T, Noro T and Shirotani I 2004 *Phys. Rev.* B **70** 153105
[6] Harima H and Takegahara K 2002 *Physica* B **312-313** 843
[7] Kuramoto Y, Otsuki J, Kiss A and Kusunose H 2005 *Prog. Theo. Phys. Suppl.* **166** 134.
[8] Takimoto T 2006 *J. Phys. Soc. Jpn.* **75** 034714
[9] Iwasa K, Hao L, Kuwahara K, Kohgi M, Saha S R, Sugawara H, Aoki Y Sato H, Tayama T and Sakakibara T 2005 *Phys. Rev.* B **72** 024414; Iwasa K, Hao L, Hasegawa T, Takagi T, Horiuchi K, Mori Y, Murakami Y, Kuwahara K, Kohgi M, Sugawara H, Saha S R, Aoki Y and Sato H 2005 *J. Phys. Soc. Jpn.* **74** 1930
[10] Yamasaki A, Imada S, Sekiyama A, Fujiwara H, Yano M, Yamaguchi J, Funahasi G, Sugawara H, Kikuchi D, Sato H, Muro T, Higashiya A, Tamasaku K, Yabashi M, Kobori H, Sugimura A, Ishikawa T, Harima H and Suga S 2008 *Phys. Rev.* B **77** 165125
[11] Matsumura T, Nakao H and Murakami Y 2013 *J. Phys. Soc. Jpn.* **82** 021007 and references therein
[12] Ishii K, Tsutsui S, Hao L, Hasegawa T, Iwasa K, Tsubota M, Inami T, Murakami Y, Saha S R, Sugawara H, Sato H 2007 *J. Magn. Magn. Mater.* **310** e178
[13] Jeitschiko W and Braun D 1997 *Acta Cryst.* B**33** 3401
[14] Nakao H, Yamasaki Y, Okamoto J, Sudayama T, Takahashi Y, Kobayashi K, Kumai R and Murakami Y 2014 *J. Phys.; Conf. Ser.* **502** 012015
[15] Kuramoto Y, Kusunose H, Kiss A 2006 *Physica* B **383** 5.
[16] Hao L, Iwasa K, Kuwahara K, Kohgi M, Saha S R, Sugawara H, Aoki Y, Sato H Sekine C, Lee C H and Harima H 2004 *J. Magn. Magn. Mater.* **272-276** e271

IOP Conf. Series: Journal of Physics: Conf. Series **969** (2018) 012119 doi:10.1088/1742-6596/969/1/012119

# Development of birefringence imaging techniques under high electric fields

**Hirotaka Manaka[1], Hirofumi Nozaki[1], and Yoko Miura[2]**

[1]Graduate School of Science and Engineering, Kagoshima University, Korimoto, Kagoshima 890-0065, Japan

[2]National Institute of Technology, Suzuka College, Shiroko-cho, Suzuka, Mie 510-0294, Japan

E-mail: `manaka@eee.kagoshima-u.ac.jp`

**Abstract.** We developed a suitable sample stage that is optimized for birefringence imaging measurements at high electric elds ($E \simeq 5$ kV/cm) and low temperatures ($T \simeq 20$ K) to directly observe electric- eld-induced ferroelectric domains in $SrTiO_3$. As a result, a huge retardance area corresponding to the ferroelectric domains appears at $E \geq 2$ kV/cm and $T \leq 60$ K even though the paraelectric domains partially remain. Furthermore, the fast-axis direction drastically rotates by 90° at the ferroelectric phase transition temperature because of an electrostrictive e ect in ferroelectrics. The phase diagram of the critical electric elds and temperatures agrees with previous reports obtained from dielectric and neutron scattering measurements.

A multi-critical phase transition in $SrTiO_3$ is well known to occur at $T = 105$ K[1-4]. From multiple investigations of the lattice dynamics, the soft-mode behavior, and the dielectric constant, $SrTiO_3$ below $T = 105$ K is known to belong to the incipient ferroelectricity class in which an electrically ordered state is suppressed by quantum uctuations. A further phase transition related to the coherent paraelectric state was revealed at $T \simeq 37$ K from electron paramagnetic resonance in a $Fe^{3+}$ doped sample[5, 6], Brillouin- and neutron-scattering measurements of the TA-phonon branches[6], and extended x-ray absorption ne structure (EXAFS) measurements in the Ti-O bond length uctuations[6]. Furthermore, the dielectric permittivity was found to show anomalies that indicate electric- eld-induced ferroelectricity above $E \simeq 2$ kV/cm and below $T \simeq 40$ K[7-10]. According to ref. 10, the electric- eld-induced ferroelectricity is still characterized by a large amount of disorder due to quantum uctuations. Therefore, it seems more appropriate to interpret the ordered state as an induced ferroelectric domain state. The phase transitions between the quantum paraelectric, coherent paraelectric and electric- eld-induced ferroelectric states may be governed by tetragonal domains with quantum uctuations[11]. It is necessary to observe the formation process of the ferroelectric domains at high electric elds and low temperatures to understand the quantum uctuation e ects at the electric- eld-induced ferroelectric phase transition. In this paper, we report observations of the electric- eld-induced ferroelectric domains in $SrTiO_3$ using birefringence imaging techniques.

The change in the birefringence ( $n$) under electric elds is known as the Pockels e ect in ferroelectricity, i.e., $n$ is proportional to the electric elds. As demonstrated in previous studies[12, 13], retardance ($\delta$) and fast-axis direction ($\psi$) images using birefringence imaging techniques are useful to investigate domain structures because $\delta$ levels are proportional to the degree of distortion and $\psi$ indicates not only optical anisotropy but also uniaxial stress in crystals. When the indices of refraction along the orthogonal optical principal axes are de ned as $n_1$ (for the slow axis) and $n_2$ (for the fast axis) with $n_1 \geq n_2 > 1$, the birefringence $n$ is expressed as

$$n \equiv n_1 - n_2. \tag{1}$$

In a material with $n_1 \neq n_2$, the transmitted light becomes elliptically polarized because the electric- eld component of the light along the fast axis travels faster than that along the slow axis, leading to a phase shift between the two components[14]. The e ect is described as a retardation between the electric- eld wave components along the fast and slow axes. The relationship between $n$ and $\delta$ is given by

$$\delta \equiv \quad n \cdot t, \tag{2}$$

where $t$ is the optical-path length. The value of $\delta$ corresponds to the circumferential length around the Poincare sphere from the polarization condition of the incident light to that of the transmitted light[14]. In this paper, we measured the $\delta$ and $\psi$ images at the same time under high electric elds and low temperatures. If a ferroelectric phase transition occurs, we expect $\delta$ to become very large and $\psi$ to give information about the polarization direction.

## 1. Materials and Methods

Birefringence images were obtained using a WPA-100 ellipsometer (Photonic Lattice, Inc.), which was employed to accurately characterize the polarization via the adoption of the Stokes parameters $(S_0, S_1, S_2, S_3)$. Furthermore, a polarimeter with a 6-W LED white light source exhibiting a circular polarization at $S_3/S_0 \simeq -1$ was used with an irradiation area equal to $100 \times 136$ mm$^2$. The wavelength was monochromated using three types of wavelength lters ($\lambda = 523$ nm, 543 nm, and 575 nm). Therefore, the light's energy density was very low, resulting in negligible increases in sample temperatures due to irradiation light. To simultaneously obtain $\delta$ and $\psi$ images at $384 \times 288$ pixels, the polarizer and wave plate generated by the autocloning techniques were arranged between the CCD camera and objective lens, with the Stokes parameters $(S_0, S_1, S_2, S_3)$ at the respective pixels given by a four-detector method[15]. As demonstrated in ref. 12, we need to correct the retardance and the fast-axis direction ($\delta_{\mathrm{C}}$ and $\psi_{\mathrm{C}}$) in $\delta \simeq 0$ compounds. However, if the retardance is beyond the inherent upper limit of birefringence ($\delta > \lambda/2$), the converted retardance and fast-axis direction ($\delta_{\mathrm{Cv}}$ and $\psi_{\mathrm{Cv}}$) are obtained using the multi-$\lambda$ mode as veri ed in ref. 13.

To observe the electric- eld-induced ferroelectric phase transition at low temperatures, the $SrTiO_3(110)$ substrate was prepared from a commercial substrate of superconducting thin lms (Shinkosha Co., ltd.). To obtain a large monodomain below the structural phase transition temperature, the substrate size was $7.0 \parallel [001] \times 2.0 \parallel [1\bar{1}0] \times 0.333 \parallel [110]$ mm$^3$ as mentioned by Muller[16].

## 2. Results and Discussion

Birefringence imaging measurements in a ferroelectric substrate under electric elds ($E$) up to $E \sim 5$ kV/cm at room temperature were reported by Jellison *et al.*[17]; however, there are no commercial products of suitable sample stages for applying high electric elds ($E > 2$ kV/cm) and decreasing temperatures ($T < 40$ K). Therefore, we assembled a sample stage that is optimized for the birefringence imaging measurements to observe the formation process of the electric- eld-induced ferroelectric phase transition in $SrTiO_3$.

From the birefringence measurements in the $SrTiO_3(110)$ substrate at $E = 0$ V/cm, $\psi_{\mathrm{C}}$ rotates at $T_{\mathrm{a}} = 103.8(5)$ K and a structural phase transition from cubic to tetragonal occurs at $T_c = 105.0(3)$ K[12]. To collect the birefringence images under high electric fields, external electric fields ($E = 1.3$ kV/cm, 2.5 kV/cm, 3.9 kV/cm, and 5.2 kV/cm) were applied at 125 K and then the temperature ($T$) was decreased to 20 K while maintaining $E$ (the electric poling process). Figures 1(a) and 1(b) show the corrected retardance ($\delta_{\mathrm{C}}^{575}$) images under $E = 0$ V/cm at 125 K and 20.4 K, respectively. These experimental data at $E = 0$ V/cm are highly reproducible and agree well with the previous results[12]. Even though there is no huge retardance that indicates the ferroelectricity, a homogeneous retardance that provides the tetragonal monodomain was observed. Figure 1(c) shows the converted retardance ($\delta_{\mathrm{Cv}}^{543}$) image under $E = 5.2$ kV/cm at 20.8 K. As can be seen in Fig. 1(c), three types of domains appear: a huge retardance area in the lower part, a $\delta_{\mathrm{Cv}}^{543} \sim 0$ area in the middle part, and a tetragonal domain area in the upper part. The boundary line between the upper and middle parts appears along the projection of [111] on (110). Figures 2(a) and 2(b) show the corrected fast-axis direction ($\psi_{\mathrm{C}}^{575}$) images under $E = 0$ V/cm at 125 K and 20.4 K, respectively. When $T$ decreases from 125 K to 20 K, $\psi_{\mathrm{C}}^{575}$ changes from $\boldsymbol{E} \parallel [001]$ to $\boldsymbol{E} \parallel [110]$[12]. Figure 2(c) shows the converted fast-axis direction ($\psi_{\mathrm{Cv}}^{543}$) image under $E = 5.2$ kV/cm at 20.8 K. In the upper

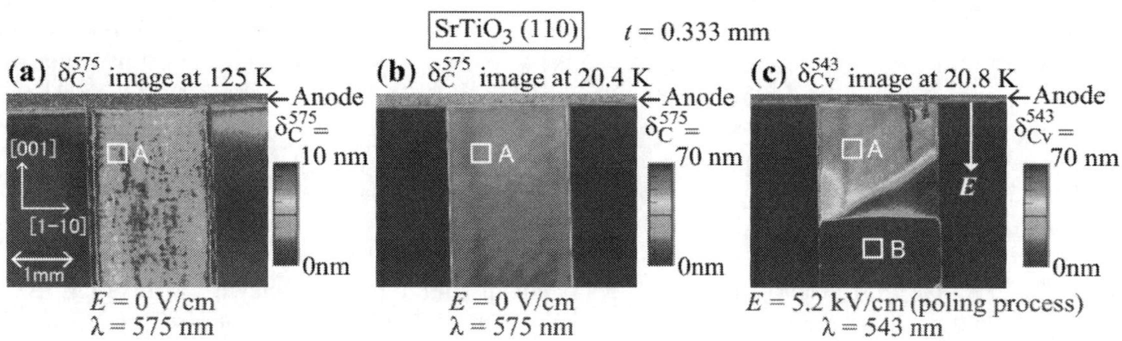


**Figure 1.** The corrected ($\delta_{\mathrm{C}}^{575}$) or converted ($\delta_{\mathrm{Cv}}^{543}$) retardance images at (a) $E = 0$ V/cm and $T = 125$ K, (b) 0 V/cm and 20.4 K, and (c) 5.2 kV/cm and 20.8 K for a $SrTiO_3(110)$ substrate. In panel (c), $E$ is applied at 125 K, and then $T$ decreases with the poling process. The squares indicate the sample analysis areas. The thickness of the substrate is $t = 0.333$ mm.

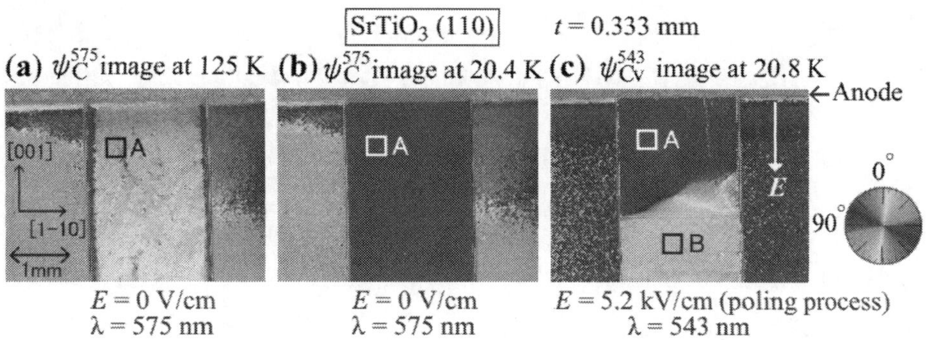


**Figure 2.** The corrected ($\psi_{\mathrm{C}}^{575}$) or converted ($\psi_{\mathrm{Cv}}^{543}$) fast-axis direction at (a) $E = 0$ V/cm and $T = 125$ K, (b) 0 V/cm and 20.4 K, and (c) 5.2 kV/cm and 20.8 K for a $SrTiO_3(110)$ substrate. In panel (c), $E$ is applied at 125 K, and then $T$ decreases with the poling process. The squares indicate the sample analysis areas. The thickness of the substrate is $t = 0.333$ mm.

part, because the $\delta_{\mathrm{Cv}}^{543}$ and $\psi_{\mathrm{Cv}}^{543}$ images at $E = 5.2$ kV/cm are the same as the $\delta_{\mathrm{C}}^{575}$ and $\psi_{\mathrm{C}}^{575}$ images at $E = 0$ V/cm as shown in Figs. 1(b) and 1(c), the tetragonal paraelectric state was realized. In the lower part, however, the direction of $\psi_{\mathrm{Cv}}^{543}$ at $E = 5.2$ kV/cm rotates by 90° and in the middle part that shows a gradual change. As a result, the large distortion that indicates the ferroelectricity may be generated in the lower part. In the middle part, it is likely that the tetragonal [100]-, [010]- and [001]-domains with paraelectricity overlap in (110) and then the retardance cancels out. After applying $E = 5.2$ kV/cm, the $\delta_{\mathrm{C}}^{575}$ and $\psi_{\mathrm{C}}^{575}$ images at $E = 0$ V/cm were measured between 20 K and 125 K again. Because these images agree well with those at $V_{\mathrm{E}} = 0$ V before applying $E$, there are no problems with sample damage by applying $E$.

Figure 3(a) shows the $T$ dependence of the corrected retardance ($|\delta_{\mathrm{C}}^{575}|$) and corrected fast-axis direction ($\psi_{\mathrm{C}}^{575}$) at $E = 0$ V/cm. Furthermore, the $|\delta_{\mathrm{C}}^{575}|(T)$ and $\psi_{\mathrm{C}}^{575}(T)$ curves as shown in Fig. 3(a) are the same as those at $E = 1.3$ kV/cm and 2.5 kV/cm. In a previous paper, the structural phase transition temperature ($T_{\mathrm{c}}$) was de ned as an in ection point of $\delta_{\mathrm{C}}^{575}(T)$, and $\psi_{\mathrm{C}}^{575}$ changes at $T_{\mathrm{a}}$ corresponding to the minimum value of $\delta_{\mathrm{C}}^{575}(T)$[12]. Therefore, the values of $T_{\mathrm{c}}$ and $T_{\mathrm{a}}$ were found to be constant below $E = 2.5$ kV/cm. Figure 3(b) shows the $T$ dependence of the converted retardance ($|\delta_{\mathrm{Cv}}^{543}|$) and converted fast-axis direction ($\psi_{\mathrm{Cv}}^{543}$) at $E = 5.2$ kV/cm in the area A (the upper part), as shown by the squares in Figs. 1(c) and 2(c). Even though the $|\delta_{\mathrm{Cv}}^{543}|(T)$ and the $\psi_{\mathrm{Cv}}^{543}(T)$ curves at $E = 5.2$ kV/cm show small undulations, the values of $T_{\mathrm{a}}$ and $T_{\mathrm{c}}$ agree well with those at $E = 0$ V/cm within our experimental errors. Therefore, the tetragonal paraelectric state was found to be realized in the upper part at $E = 5.2$ kV/cm. As can be seen in Fig. 3, there are no anomalies of the $|\delta_{\mathrm{Cv}}^{543}|(T)$ and $\psi_{\mathrm{Cv}}^{543}(T)$ curves at the phase transition between the quantum paraelectric and the coherent paraelectric states at $T \simeq 37$ K.

To examine the electric- eld-induced ferroelectric domains, we further analyzed the retardance and the fast-axis direction at $E = 5.2$ kV/cm in the lower part. Figure 4 shows the $|\delta_{\mathrm{Cv}}^{543}|(T)$ and $\psi_{\mathrm{Cv}}^{543}(T)$ curves at $E = 5.2$ kV/cm in the area B, as shown by the squares in Figs. 1(c) and 2(c). The transition temperatures $T_{\mathrm{c}} = 106.1(5)$ K and $T_{\mathrm{a}} = 99(2)$ K are obtained. Furthermore, the fast-axis direction changes again below $T \simeq 51$ K. When the minimum value of $\delta_{\mathrm{Cv}}^{543}(T)$ is de ned as $T_{\mathrm{F}}$, $T_{\mathrm{F}} = 51(1)$ K is obtained. Below $T_{\mathrm{F}}$, the value of $|\delta_{\mathrm{Cv}}^{543}|$ rapidly increases with decreasing $T$. According to refs. 18 and 19, $n(T)$ under $E = 0$ V/cm in ferroelectric

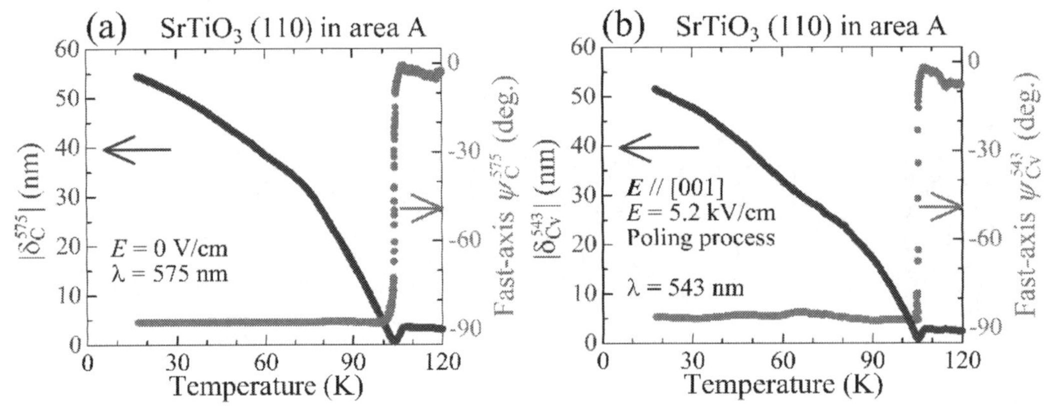


**Figure 3.** The corrected and converted retardance ($|\delta_{\mathrm{C}}^{575}|$ and $|\delta_{\mathrm{Cv}}^{543}|$) and fast-axis direction ($\psi_{\mathrm{C}}^{575}$ and $\psi_{\mathrm{Cv}}^{543}$) as a function of $T$ during cooling in a $\mathrm{SrTiO_3}(110)$ substrate in the area A shown by the squares in Figs. 1 and 2 at (a) $E = 0$ V/cm and (b) 5.2 kV/cm. The thickness of the substrate is $t = 0.333$ mm.

IOP Conf. Series: Journal of Physics: Conf. Series **969** (2018) 012119 doi:10.1088/1742-6596/969/1/012119

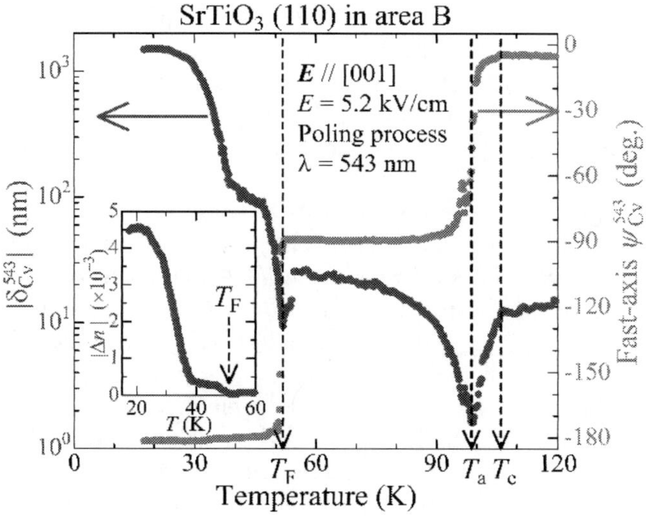


**Figure 4.** The converted retardance ($|\delta^{543}_{\mathrm{Cv}}|$) and the converted fast-axis direction ($\psi^{543}_{\mathrm{Cv}}$) as a function of $T$ during cooling in a $\mathrm{SrTiO_3(110)}$ substrate in the area B shown by the squares in Figs. 1(c) and 2(c) at $E = 5.2$ kV/cm. The transition temperatures are $T_c = 106.1(5)$ K, $T_a = 99(2)$ K, and $T_F = 51(1)$ K. The inset shows the $T$ dependence of the birefringence (| $n$|) below $T_F$. The thickness of the substrate is $t = 0.333$ mm.

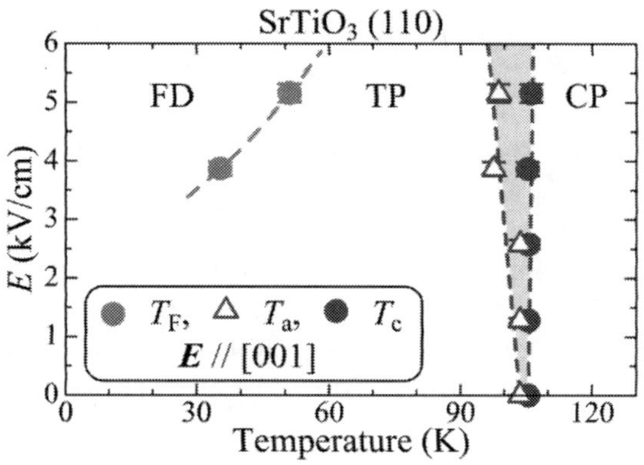


**Figure 5.** Phase diagram of $T$ versus $E$. Three phases appear: the cubic paraelectric (CP) phase, the tetragonal paraelectric (TP) phase, and the ferroelectric domain (FD) phase. The intermediate region between $T_a$ and $T_c$ indicates the tetragonal multi-domain state with paraelectricity. The dashed lines are visual guides.

$\mathrm{SrTi(^{16}O_{1-x}{}^{18}O_x)_3}$ substrates measured by the rotating analyzer method rapidly increases below 30 K and the maximum value of $n$ is approximately $1 \times 10^{-3}$ at $x = 0.99$[19]. Furthermore, when $E = 1.40$ kV/cm is applied parallel to [110] in the $x = 1$ substrate, the maximum value of $n$ is approximately $2.4 \times 10^{-3}$[20]. Because a change in the fast-axis direction cannot be detected by the rotating analyzer method, we do not know how the fast-axis direction changes at the ferroelectric phase transition temperature in $\mathrm{SrTi(^{16}O_{1-x}{}^{18}O_x)_3}$. As can be seen in the inset of Fig. 4, the appearance of the electric- eld-induced ferroelectric domains in $\mathrm{SrTiO_3(110)}$ was found to be successfully observed when using the birefringence imaging techniques. The $|\delta^{543}_{\mathrm{C}}|(T)$ curve at $E = 5.2$ kV/cm bends at $T \simeq 39$ K even though there is no anomaly in the $\psi^{543}_{\mathrm{Cv}}(T)$ curve below $T_F$. We believe that this anomaly below $T_F$ is extrinsic because the huge $n$ domains in the lower part are spread and moved with decreasing $T$ below $T_F$.

Figure 5 shows the phase diagram of the critical temperatures ($T_c$, $T_a$, and $T_F$). We found that the phase transition between the cubic paraelectric (CP) phase and the tetragonal paraelectric (TP) phase occurs at $T_c \simeq 105$ K independent of $E$ and the tetragonal multi-domain state with paraelectricity is realized in the intermediate region between $T_a$ and $T_c$. Furthermore, the electric- eld-induced ferroelectric domain (FD) phase appeared above $E \simeq 3$ kV/cm and below $T \simeq 60$ K. This phase diagram agrees with the data obtained by the dielectric measurements under high electric elds[7-10]. Even though the electric- eld-induced ferroelectric domains are successfully observed by the birefringence imaging techniques, we cannot distinguish between the quantum paraelectric state and the coherent paraelectric state in the TP phase. From the neutron-scattering measurements, the distribution of the tetragonal domains and its electric eld dependence were detected[21]. As a result, the orthorhombic ferroelectric phase de ned in ref. 21 corresponds to the FD phase. However, there are no anomalies in the $\delta$ and $\psi$ images in the intermediate phase between the tetragonal paraelectric phase and the orthorhombic ferroelectric phase. Consequently, we found that the huge $n$ region in the lower part shown in Fig. 1(c) indicated electric- eld-induced ferroelectric domains even though the tetragonal paraelectric domains remained in the upper part because the quantum uctuations were not completely suppressed by the external electric elds up to $E = 5.2$ kV/cm. When $T$ decreases below $T_F$, the direction of $\psi_{Cv}^{543}$ in the ferroelectric domains was found to rotate from $\boldsymbol{E} \perp$ the fast-axis to $\boldsymbol{E} \parallel$ the fast-axis; however, that in the paraelectric domains remained at $\boldsymbol{E} \perp$ the fast-axis. These results are explained by the electrostrictive e ect in ferroelectrics.

In conclusion, we successfully observed the electric- eld-induced ferroelectric domains in $SrTiO_3(110)$ at $E \geq 3$ kV/cm and $T \leq 60$ K even though the tetragonal paraelectric domains cannot be completely suppressed by $E$. To fully polarize the ferroelectric domains, it is necessary to generate much higher electric elds. Electric- eld-induced ferroelectric domains may be in uenced by the application of uniaxial stress because $SrTiO_3$ shows ferroelasticity. As in ref. 21, the electrostrictive deformation at 80 K and under $E = 4.6$ kV/cm was found to be reproduced by applying a uniaxial stress of $\sim 10$ MPa. We expect to observe the stress-assisted ferroelectric state with full polarization if both the electric eld and the uniaxial stress are applied along [001] simultaneously. In future studies, we will measure stress-induced birefringence.

## Acknowledgments

We are grateful to Y. Inoue, T. Sato, and T. Chiba in Photonic Lattice, Inc. for the developments of birefringence imaging system, and S. Kasai in Coax Co., ltd. for preparing the high voltage coaxial cable. We received valuable support from the Frontier Science Research Center in Kagoshima University for performing the X-ray di raction experiments. This research was partly supported by JSPS KAKENHI Grant Numbers JP25600133 and JP15K13400.

## References

[1] Unoki H and Sakudo T 1967 *J. Phys. Soc. Jpn.* **23** 546
[2] Cowley R A, Buyers W J L and Dolling G 1969 *Solid State Commun.* **8** 181
[3] Shirane G and Yamada Y 1969 *Phys. Rev.* **177** 858
[4] Müller K A and Berlinger W 1971 *Phys. Rev. Lett.* **26** 13
[5] Müller K A, Berlinger W and Tosatti E 1991 *Z. Phys. B* **48** 277
[6] Fischer M, Lahmar A, Maglione M, Miguel A S, Itie J P, Polian A and Bandelet F 1994 *Phys. Rev. B* **49** 12451
[7] Sawaguchi E, Kikuchi A and Kodera Y 1962 *J. Phys. Soc. Jpn.* **17** 1666
[8] Hegenbarth E 1964 *Phys. Stat. Sol.* **6** 333
[9] Hemberger J, Nicklas M, Viana R, Lunkenheimer P, Loidl A and Böohmer R 1996 *J. Phys.: Condens. Matter* **8** 4673
[10] Hemberger J, Lunkenheimer P, Viana R, Böhmer R and Loidl A 1995 *Phys. Rev. B* **52** 13159
[11] Martonak R and Tosatti E 1994 *Solid State Commun.* **92** 167
[12] Manaka H, Yagi G and Miura Y 2016 *Rev. Sci. Instrum.* **87** 073704
[13] Manaka H, Fukuda T and Miura Y 2016 *J. Phys. Soc. Jpn.* **85** 124701
[14] Azzam R M and Bashara N M 1977 *Ellipsometry and polarized light* (North Holland, Amsterdam)
[15] Sato T, Araki T, Sasaki Y, Tsuru T, Tadokoro T and Kawakami S 2007 *App. Opt.* **46** 4936
[16] Müller K A, Berlinger W, Capizzi M and Gränicher H 1970 *Solid State Commun.* **8** 549
[17] Jellison Jr G E, Griffiths C O and Holcomb D E 2002 *App. Phys. Lett.* **81** 1222.
[18] Yamanaka K, Wang R, Itoh M and Iio K, 2001 *J. Phys. Soc. Jpn.* **70** 3213
[19] Itoh M, Azuma T, KyôMen T, Iio K, Yamanaka K and Wang R 2004 *J. Phys. Soc. Jpn.* **73** 1377
[20] Dec J, Kleemann W, Boldyreva K and Itoh M 2005 *Ferroelectrics* **314** 7
[21] Sidoruk J, Leist J, Gibhardt H, Meven M, Hradil K and Eckold G 2010 *J. Phys.: Condens. Matter* **22** 235903

IOP Conf. Series: Journal of Physics: Conf. Series **969** (2018) 012120 doi:10.1088/1742-6596/969/1/012120

# Thermal expansion and magnetostriction of clathrate compound $Pr_3Pd_{20}Ge_6$

**K Matsumoto[1], Y Sekiguchi[1], O Iwakami[1], T Ono[1], S Abe[1], G Ano[2], M Akatsu[3], K Mitsumoto[2], Y Nemoto[2], T Goto[2], N Takeda[4], and H Kitazawa[5]**

[1] Department of Physics, Kanazawa University, Kakuma-machi, Kanazawa 920-1192, Japan
[2] Graduate School of Science and Technology, Niigata University, Japan
[3] Department of Physics, Niigata University, Japan
[4] Faculty of Engineering, Niigata University, Japan
[5] National Institute for Materials Science, Tsukuba, Japan

E-mail: k.matsu@staff.kanazawa-u.ac.jp

**Abstract.** In $Pr_3Pd_{20}Ge_6$, the Pr ions are located at two different crystallographic sites, 4a and 8c site. Antiferro-quadrupole ordering (AFQ) of the 8c site occurs at 250 mK. Ac susceptibility measurement indicated that antiferromagnetic ordering (AFM) of the 4a site and Hyperfine-enhanced Pr nuclear magnetic ordering of the 8c site occur at 77 and 9 mK, respectively. To clarify the magnetic and quadrupole properties of $Pr_3Pd_{20}Ge_6$, thermal expansion and magnetostriction measurements on single crystal sample were carried out along the [001] direction up to 8 T down to 500 $\mu$K using a capacitive dilatometer. In zero field, relative length change $\Delta L/L$ in [001] direction had a dip at AFQ and abrupt decrease at AFM ordering. From thermal expansion and isothermal magnetostriction measurements, magnetic phase diagram of $Pr_3Pd_{20}Ge_6$ along [001] direction was obtained.

## 1. Introduction

The clathrate compounds $R_3Pd_{20}X_6$ where $R$ = rare earth and $X$ = Si or Ge show either rattling motions of $R$ atoms, quadrupole order of $4f$ electrons, or both [1, 2]. Especially, $Pr_3Pd_{20}X_6$ have attracted attention because of hyperfine-coupled magnetic properties at very low temperatures. Hyperfine interaction through the $4f$-electrons and nuclei plays an important role in understanding magnetic ground stats of $Pr_3Pd_{20}X_6$ [3, 4].

$R_3Pd_{20}X_6$ have a $Cr_{23}C_6$-type cubic structure with a space group $Fm\bar{3}m$. The $R$ ions are located at two different crystallographic sites, where 4a and 8c site have a $O_h$ symmetry and a $T_d$ symmetry, respectively. In cubic Pr compounds, crystalline electric field (CEF) effects split the ninefold degenerate $J = 4$ multiplet of $Pr^{3+}$ ions into a $\Gamma_1$ singlet, non-Kramers $\Gamma_3$ doublet, $\Gamma_4$, and $\Gamma_5$ triplets. Interestingly, the ground states in $Pr_3Pd_{20}Ge_6$ are different for each sites [5, 6]. The ground state at the 4a site is a magnetic $\Gamma_5$ triplet which has three magnetic dipoles and five quadrupoles, while that at the 8c site is a non-magnetic $\Gamma_3$ doublet which has two quadrupoles and an octupole. The 4a and 8c site in the series of $R_3Pd_{20}X_6$ systems show quadrupole or magnetic ordering independently at different temperatures. Quadrupole and magnetic ordering in $Pr_3Pd_{20}Ge_6$ have been studied by ultrasound and ac susceptibility measurements [4, 7]. According to the analysis of the elastic constants, antiferro-quadrupole ordering of the 8c site occurs at $T_{\mathrm{AFQ}} = 250$ mK and ferro-quadrupole ordering occurs at 60 mK. It has been reported that quadrupole ordered phase closes in the critical field of $H_{\mathrm{AFQ}} = 2.5$ T along [001] direction. The ac susceptibility measurements indicated that magnetic ordering of the 4a site occurs at $T_{\mathrm{AFM}} = 77$ mK. Furthermore, the hyperfine-enhanced Pr nuclear magnetic ordering of the 8c site was observed at $T_{\mathrm{NMO}} = 9$ mK. However, the origin of the dissipation phenomenon observed in ac susceptibility measurements and the interplay between different sites are still unsolved.

Thermal expansion and magnetostriction are important to clarify the magnetic and quadrupole properties of $Pr_3Pd_{20}Ge_6$. In this paper, we report thermal expansion and magnetostriction measurements on single crystal of $Pr_3Pd_{20}Ge_6$ along [001] direction and discuss phase diagram.

## 2. Experimental procedures

Single crystal $Pr_3Pd_{20}Ge_6$ used in our study was obtained from the same batch of high quality samples, grown by a floating zone method, for ultrasound and ac susceptibility experiments. Thermal expansion and magnetostriction $\Delta L/L$ were measured along [001] direction of the cubic structure and magnetic field was applied along the same direction ($L \parallel H$). The sample length $L$ was 3.84 mm. Thermal expansion measurements on single crystal $La_3Pd_{20}Ge_6$ along [001] direction were also carried out to compare with $Pr_3Pd_{20}Ge_6$. The sample length was 3.5 mm along [001].

$\Delta L/L$ was measured by the capacitance method. The capacitive dilatometer was constructed in similar way to our previous studies [8, 9, 10]. We used a capacitance bridge based on ratio-transformer comparing the sample capacitance with a reference one. Thermal expansion and magnetostriction coefficients were obtained by differentiating the change rate of sample length $\Delta L/L$ with respect to temperature and magnetic field as $\alpha = \partial(\Delta L(T)/L)/\partial T$ and $\lambda = \partial(\Delta L(B)/L)/\partial B$, respectively.

Thermal expansion in zero magnetic field was measured using a copper nuclear demagnetization refrigerator and a $^3$He-$^4$He dilution refrigerator down to 500 $\mu$K. The temperature was measured by a Pt-NMR thermometer, a $^3$He melting curve thermometer, and a $RuO_2$ resistance thermometer. For measuring $\Delta L/L$ up to 8 T, the dilatometer was attached to an annealed copper thermal link which connected between the mixing chamber of $^3$He-$^4$He dilution refrigerator and the bore of a NbTi superconducting magnet. The sample was cooled down to 10 mK and the temperature of the sample was measured by a $^3$He melting curve thermometer and a carbon resistance thermometer.

## 3. Results and Discussion

### *3.1. Thermal expansion*

Relative length changes $\Delta L/L$ of $Pr_3Pd_{20}Ge_6$ and $La_3Pd_{20}Ge_6$ in zero magnetic field are shown in the lower panel of Fig. 1. Ac susceptibility of $Pr_3Pd_{20}Ge_6$ in zero static magnetic field is represented in the upper panel. It is shown that $La_3Pd_{20}Ge_6$ shows almost no change in $\Delta L/L$ below 1 K. Then, temperature dependent $\Delta L/L$ of $Pr_3Pd_{20}Ge_6$ is attributed to magnetic contribution of Pr atoms. At AFQ transition of 8c site ($T_{\mathrm{AFQ}}$= 250 mK ), no magnetic anomaly was observed. The dip of $\Delta L/L$ along [001] direction at AFQ transition had magnitude of $10^{-8}$ order so that lattice distortion is very small. These behaviors are expected from the nature of AFQ transition. In decreasing temperature, $\Delta L/L$ rapidly decreased near the antiferromagnetic (AFM) ordering of the 4a site and $\alpha$ had a sharp peak at 77 mK ($T_{\mathrm{AFM}}$), where susceptibility represented a peak. Below 15 mK, $\Delta L/L$ shows negative thermal expansion. However, there was no significant anomalous change at the Pr nuclear magnetic ordering of the 8c site ($T_{\mathrm{NMO}}$= 9 mK). Nuclear spin order possibly has only small effect on electron wave functions that cause lattice distortion.

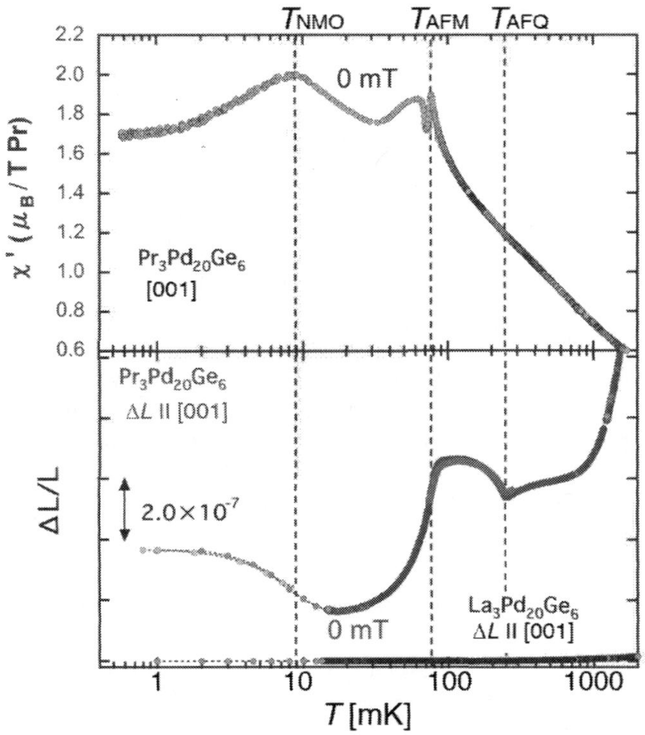


**Figure 1.** (Upper Panel) Ac susceptibility of single crystalline $Pr_3Pd_{20}Ge_6$ along the [001] axis in zero static magnetic field.
(Lower Panel) Thermal expansion $\Delta L/L$ of single crystalline $Pr_3Pd_{20}Ge_6$ and that of $La_3Pd_{20}Ge_6$ both along the [001] axis in zero field.

Figure 2 shows thermal expansion $\Delta L/L$ of $Pr_3Pd_{20}Ge_6$ along [001] direction when magnetic field was applied to the same direction. $\Delta L/L$ is plotted as relative change from 300 mK and shifted vertically for clarity. A dip in $\Delta L/L$ was observed at the AFQ transition in $B < 0.05$ T. In contrast to low field, AFQ transition was recognized as a peak in $\Delta L/L$ from 0.1 to 0.75 T and $T_{\rm AFQ}$ were represented as arrows. When $B \geq 1$ T, $\Delta L/L$ shows large temperature variation because of CEF effect. Then, AFQ transition was recognized as a peak in $\alpha$ as shown in the right inset of Fig. 2. AFQ transition was hardly recognized above 2 T in thermal expansion experiments because the phase boundary line became almost temperature independent. AFQ transition in low temperatures was identified by magnetostriction measurements explained in the next subsection.

At $T_{\rm AFM}$, $\alpha$ had a clear peak associated with magnetic ordering in magnetic field smaller than 0.2 T, but no peak could be observed above 0.2 T. $\Delta L/L$ showed negative thermal expansion below 50 mK in magnetic fields lower than 2.5 T and changed positive one above 2.5 T.

*3.2. magnetostriction*

Figure 3 shows magnetostriction $\Delta L/L$ of $Pr_3Pd_{20}Ge_6$ along [001] direction when magnetic field was applied to the same direction. Magnetostriction was small in high temperatures such as 10 K and increased with decreasing temperature. Eventually, $\Delta L/L$ had maximum at around 7 T in 2 K. The magnetic field at the maximum decreased with decreasing temperature and became 3.8 T at 283 mK where $Pr_3Pd_{20}Ge_6$ was in paramagnetic phase. The sign change in magnetostriction coefficient so called "reversed magnetostriction" was observed in other rare earth materials [11]. CEF and symmetric strain play an important role in reversed magnetostriction.

Below $T_{\rm AFQ}$, AFQ transition at $H_{\rm AFQ}$ was identified as inflection point of $\Delta L/L$ in Fig. 3. Below 200 mK, abrupt increase of $\Delta L/L$ was observed at about 1.4 T. This implies that structure of ordered quadrupole largely varied at this critical field $H^*$.

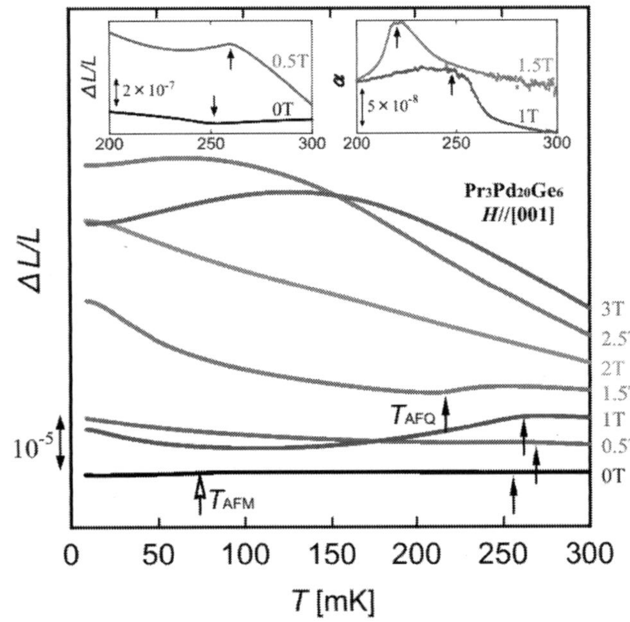


**Figure 2.** Thermal expansion $\Delta L/L$ of single crystalline $Pr_3Pd_{20}Ge_6$ along the [001] axis in 0, 0.5, 1, 1.5, 2, 2.5, and 3 T. $\Delta L/L$ is plotted as relative change from 300 mK and shifted for clarity. AFQ and AFM ordering of $4f$ electron are represented as solid arrows $T_{\rm AFQ}$ and open arrow $T_{\rm AFM}$, respectively. Right and left inset respectively show the expanded plot of $\alpha$ in 1, 1.5 T and $\Delta L/L$ in 0, 0.5 T around AFQ transition.

Figure 4 shows magnetostriction coefficient $\lambda$ in various temperatures. In paramagnetic phase, broad peak of $\lambda$ was observed due to CEF effect between 0 and 4 T. Critical magnetic field $H_{\rm AFQ}$ represented as arrows increased with decreasing temperature from 250 mK to about 150 mK and became almost constant with respect to temperature. Abrupt change in $\Delta L/L$ at $H^*$ gave rise to a huge peak of $\lambda$ that was $3.0 \times 10^{-4}$ $\rm T^{-1}$ at 38 mK. Critical field $H^*$ was shown to be 1.4 T and temperature independent in the quadrupole ordered phase. In ultrasound measurement [7], field dependence of elastic constant $C_{11}$ showed a sharp dip at $H_{\rm AFQ}$. However, no significant change in either $C_{11}$ or $C_{44}$ was observed at $H^*$.

### *3.3. Magnetic phase diagram*
Figure 5 shows the magnetic phase diagram of $Pr_3Pd_{20}Ge_6$ for magnetic fields applied along the [001] direction. Open circles are AFQ transition points observed by thermal expansion measurements in constant fields. Closed circles represent AFQ transition points obtained from the magnetostriction measurements at constant temperatures. The phase boundary of AFQ ordered phase obtained by this work is in reasonable agreement with that by ultrasound measurement[7]. $T_{\rm AFQ}$ increased with increasing field and then started decreasing above 0.75 T. This field dependence of $T_{\rm AFQ}$ is often observed in AFQ transitions for example $PrPb_3$, $CeB_6$, TmTe, and $DyB_2C_2$[12, 13, 14, 15]. The quadrupole ordered phase closes in critical field of about 2.4 T. There is an almost temperature independent phase boundary with critical field $H^*$ of 1.4 T as shown by squares. AFM transition of 4a site shown by solid triangles was observed by thermal expansion in magnetic field smaller than 0.2 T. This transition temperature appears to be almost field independent.

We observed hysteresis in thermal expansion measurements between 30 and 90 mK in low magnetic fields. In similar regions, dispersion was observed in magnetic susceptibility measurements[4] and hysteresis of elastic constant during magnetic field ramping was observed in ultrasonic experiments[7]. Interplay between magnetic dipole and electric quadrupole is considered as an open question and important issue.

It has been reported that magnetic phase diagram of $Pr_3Pd_{20}Ge_6$ has large anisotropy for field directions by ultrasound experiments[7]. Anisotropy for field directions has also been reported in

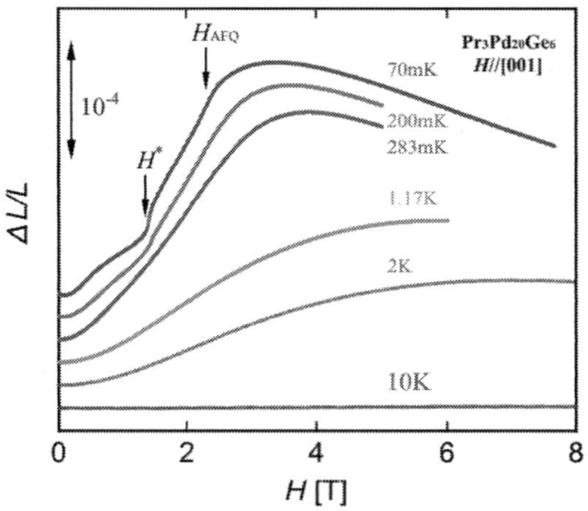


**Figure 3.** Magnetostriction $\Delta L/L$ of $Pr_3Pd_{20}Ge_6$ along [001] direction in paramagnetic and quadrupole ordered phase. $\Delta L/L$ is shifted vertically for clarity. Critical field of AFQ transition $H_{\rm AFQ}$ at 70 mK is shown as an arrow. $\Delta L/L$ shows abrupt change at $H^*$.

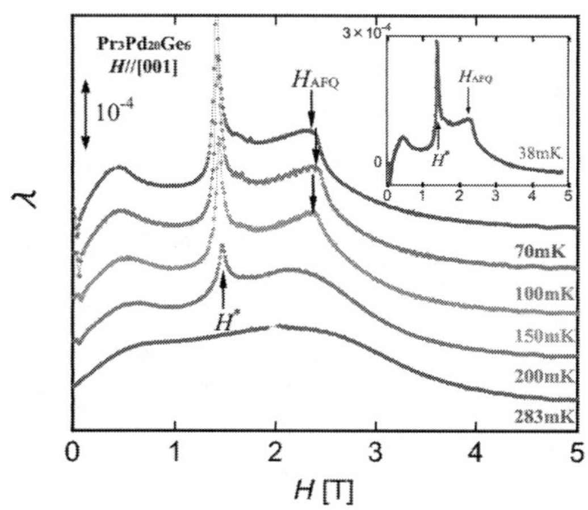


**Figure 4.** Magnetostriction coefficient $\lambda$ of $Pr_3Pd_{20}Ge_6$ along [001] direction in paramagnetic and quadrupole ordered phase. $\lambda$ is shifted vertically for clarity. Critical field of AFQ transition $H_{\rm AFQ}$ are shown as arrows. Inset shows overall view of $\lambda$ at 38 mK.

$CeB_6$[16], but the relative strength of critical fields along three directions are different from each other. Thermal expansion and magnetostriction study of $Pr_3Pd_{20}Ge_6$ along other directions are important for characterizing the order parameter of quadrupole ordered phase because relative length change $\Delta L/L$ along [001], [110], and [111] directions of the cubic symmetry are corresponding to symmetric strain $\epsilon_i$ as $\frac{\Delta L}{L}\mid_{[001]}=\frac{1}{3}\epsilon_{\rm B}+\frac{1}{\sqrt{3}}\epsilon_u$, $\frac{\Delta L}{L}\mid_{[110]}=\frac{1}{3}\epsilon_{\rm B}-\frac{1}{2\sqrt{3}}\epsilon_u+\epsilon_{xy}$, and $\frac{\Delta L}{L}\Big|_{[111]}=\frac{1}{3}\epsilon_{\rm B}+\frac{2}{3}(\epsilon_{xy}+\epsilon_{yz}+\epsilon_{zx})$, respectively. Here, $\epsilon_{\rm B}$ and $\epsilon_{\rm u}$ are the volume and tetragonal strain, respectively. $\epsilon_{xy}, \epsilon_{yz}, \epsilon_{zx}$ are the shear strains.

## 4. Summary

We studied thermal expansion and magnetostriction of $Pr_3Pd_{20}Ge_6$ using a capacitive dilatometer in order to clarify the magnetic and quadrupole properties. Thermal expansion and magnetostriction measurements on single crystal sample were carried out along the [001] in applied external fields up to 8 T and down to 500 $\mu$K. From these measurements, magnetic phase diagram was obtained. The quadrupole ordered phase closes in critical field of about 2.4 T that agreed with ultrasonic experiments. In magnetostriction measurements, an almost temperature independent phase boundary was observed and the critical field $H^*$ was 1.4 T.

## Acknowledgments

This work was supported by JSPS KAKENHI. We thank K. Nunomura for technical assistance.

## References

[1] Nemoto Y, Yamaguchi T, Horino T, Akatsu M, Yanagisawa T, Goto T, Suzuki O,Dönni A and Komatsubara T 2003 *Phys. Rev.* B **68** 184109
[2] Nemoto Y, Yanagisawa T, Yasumoto Y, Kobayashi H, Yamaguchi A, Tsuduku S, Goto T, Takeda N, Ochiai A, Sugawara H, Sato H and Kitazawa H 2008 *J. Phys. Soc. Jpn.* **77** 153-8

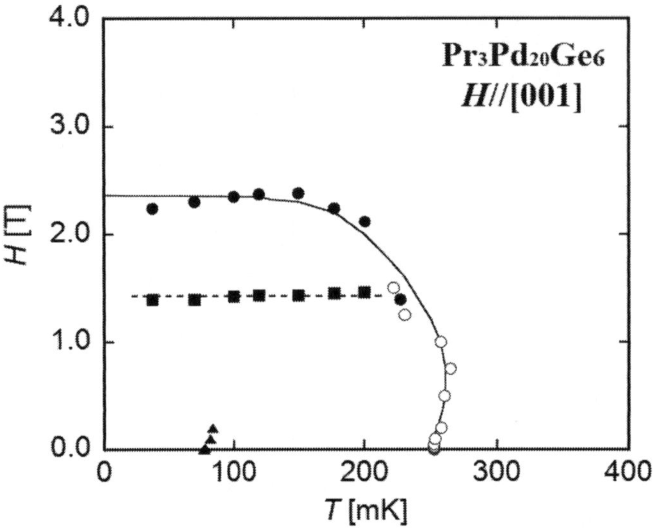


**Figure 5.** Magnetic Phase diagram of $Pr_3Pd_{20}Ge_6$ along [001] direction. AFQ transitions observed by thermal expansion and magnetostriction are plotted as open circles and closed circles. AFM transition of $4f$ electron in 4a site is shown as solid triangle. Critical points of abrupt change in $\Delta L/L$ observed by magnetostriction measurements are plotted as squares. Lines are guide to the eye.

[3] Steinke L, Mitsumoto K, Miclea C F, Weickert F, Dönni A, Akatsu M, Nemoto Y, Goto T, Kitazawa H, Thalmeier P and Brando M 2013 *Phys. Rev. Lett.* **111** 077202
[4] Iwakami O, Namisashi Y, Abe S, Matsumoto K, Ano G, Akatsu M, Mitsumoto K, Nemoto Y, Takeda N,Goto T and Kitazawa H 2014 *Phys. Rev.* B **90** 100402(R)
[5] Keller L, Dönni A, Zolliker M and Komatsubara T 1999 *Physica* B **259-61** 336-7
[6] Suzuki O, Kitazawa H, Kido G, Yamaguchi T, Nemoto Y and Goto T 2006 *J. Alloys Compounds* **408-12** 107-9
[7] Ano G, Akatsu M, Araki K, Matsuo K, Tachikawa Y, Mitsumoto K, Yamaguchi T, Nemoto Y, Goto T, Takeda N, Dönni A and Kitazawa H 2012 *J. Phys. Soc. Jpn.* **81** 034710
[8] Yoshida J, Abe S, Takahashi D, Segawa Y, Komai Y, Tsujii H, Matsumoto K, Suzuki H and Ōnuki Y 2008 *Phys. Rev. Lett.* **101** 256402
[9] Abe S, Sasaki F, Oonishi T, Inoue D, Yoshida J, Takahashi D, Tsujii H, Suzuki H and Matsumoto K 2012 *Cryogenics* **52** 452-6
[10] Iwakami O, Kawata N, Takeshita M, Yao Y, Abe S and Matsumoto K 2014 *J. Phys.: Conf. Series* **568** 032002
[11] Nieuwenhuys G J, Davidov D and Häfner H U 1982 *Phys. Rev. Lett.* **49** 1202-5
[12] Tayama T, Sakakibara T, Kitami K, Yokoyama M, Tenya K, Amitsuka H, Aoki D, Ōnuki Y and Kletowski Z 2001 *J. Phys. Soc. Jpn.* **70** 248-58
[13] Effantin J M, Rossat-Mignod J, Burlet P, Bartholin H, Kunii S and Kasuya T 1985 *J. Mag. Mag. Mater.* **47-8** 145-8
[14] Matsumura T , Nakamura S , Goto T, Shida H and Suzuki T 1996 *Physica* B **223-4** 385-8
[15] Yamauchi H, Onodera H, Ohoyama K, Onimaru T, Kosaka M, Ohashi M and Yamaguchi Y 1999 *J. Phys. Soc. Jpn.* **68** 2057-66
[16] Shiina R 2002 *J. Phys. Soc. Jpn.* **71** 2257-62

IOP Conf. Series: Journal of Physics: Conf. Series **969** (2018) 012121 doi:10.1088/1742-6596/969/1/012121

# Unusual Magnetic Response of an $S = 1$ Antiferromagetic Linear-Chain Material

**Jian-Sheng Xia[1], Andrzej Ozarowski[2], Peter M. Spurgeon[3], Adora G. Graham[3], Jamie L. Manson[3], Mark W. Meisel[1]**

[1]Department of Physics and the National High Magnetic Field Laboratory, University of Florida, Gainesville, FL 32611-8440, USA

[2]National High Magnetic Field Laboratory, Florida State University, Tallahassee, FL 32310, USA

[3]Department of Chemistry and Biochemistry, Eastern Washington University, Cheney, WA 99004, USA

E-mail: `meisel@phys.ufl.edu`

**Abstract.**

An $S = 1$ antiferromagnetic polymeric chain, $[Ni(HF_2)(3\text{-Clpy})_4]BF_4$ (py = pyridine), also referred to as NBCT, has previously been identified to have intrachain, nearest-neighbor antiferromagnetic interaction strength $J/k_B = 4.86$ K and single-ion anisotropy (zero-field splitting) $D/k_B = 4.3$ K, so the ratio $D/J = 0.88$ places this system close to the $D/J \approx 1$ gapless critical point between the topologically distinct Haldane and Large-$D$ phases. The magnetization was studied over a range of temperatures, 50 mK $\leq T \leq$ 1 K, and magnetic fields, $B \leq 10$ T, in an attempt to identify a critical field, $B_c$, associated with the closing of a gap, and the present work places an upper bound of $B_c \leq (35 \pm 10)$ mT. At higher fields, the observed magnetic response is qualitatively similar to the "excess" signal observed by other workers at 0.5 K and below 3 T. The high-field (up to 14.5 T), multi-frequency (nominally 200 GHz to 425 GHz) ESR spectra at 3 K reveal several features associated with the sample.

## 1. Introduction

After Haldane reported the fundamental differences between one-dimensional Heisenberg antiferromagnets with integer and half-integer spins [1, 2], theoretical [3–5] and experimental [6] activities focused on clarifying the magnetic phases and properties of a broad range of linear-chain magnets described by the Hamiltonian

$$\mathcal{H} = J \sum_i \{[S_i^x S_{i+1}^x + S_i^y S_{i+1}^y + \lambda\, S_i^z S_{i+1}^z] + D\,(S_i^z)^2 + E\,[(S_i^x)^2 - (S_i^y)^2]\} - \vec{B} \cdot \overleftrightarrow{g} \cdot \vec{S} \ , \quad (1)$$

where $J$ is the nearest-neighbor intrachain magnetic exchange parameter, $D$ and $E$ are the single-ion and rhombic anisotropies that are commonly referred to as zero-field splitting parameters arising from the crystal-field anisotropy, $g$ is the Lande $g$-factor tensor, $B$ is the applied magnetic field, $\lambda$ is a parameter distorting the exchange, and the inevitable interchain coupling $J'$ is not shown. For $S = 1$, the Haldane phase, which possesses an energy gap $\Delta = 0.41J$ when $D = 0$, and the Large-$D$ phase [3, 7] have been established by magnetization [8, 9], EPR [10–14], and inelastic neutron scattering studies [8, 15]. Recently, Brambleby *et al.* have demonstrated the power of employing a suite of experimental and numerical tools to untangle the single-ion anisotropy and exchange energies in the $S = 1$ quantum antiferromagnet $[Ni(HF_2)(pyz)_2]SbF_6$, where pyz = pyrazine, which was identified to have $D/J \approx 1.3$ and long-range magnetic order near 12 K [16]. To date, no real material has been identified as being close to the quantum critical point of $D/J \approx 1$ while also avoiding long-range order at $T \ll J/k_B$. This region of the phase diagram has attracted wide theoretical and numerical attention [17–27] and the ground-state properties have not been unambigously resolved. Nonetheless, there is converging consensus that the Haldane gap closes and the Large-$D$ gap opens at a gapless critical point, $(D/J)_c = 0.97$ when $\lambda = 1$, between the two topologically distinct gapped phases with different parity. In other words, this particular quantum critical region remains without any candidate systems whose properties may provide insight into the measurable thermodynamic and electromagnetic properties. Resolving this lacuna will provide new perspectives into quantum critical phenomena, and such experimental-theoretical advances are being realized in other systems [28, 29].

This paper reports experimental magnetization and EPR data on an intriguing $S = 1$ polymeric chain material, $[Ni(HF_2)(3\text{-}Clpy)_4]BF_4$ (py = pyridine) [30], Fig. 1, that appears to be located close to $(D/J)_c$. In 2012, $[Ni(HF_2)(3\text{-}Clpy)_4]BF_4$, now sometimes referred to as NBCT, was synthesized and characterized by a wide-range of spectroscopies [30]. Most notably, muon-spin relaxation spectra relax monotonically, and the lack of oscillations or a shift in the base line are interpreted as indicating the absence of long-range magnetic order. The low-field $B = 0.1$ T, high-temperature ($T \geq 1.8$ K) magnetic susceptibility possessed an antiferromagnetic component, which could be fit to $J/k_B = 4.86$ K with $g = 2.1$, and a paramagnetic Curie-like component that might be associated with uncoupled $Ni^{2+}$ spins and/or with $S = 1/2$ end-chain spins [5, 31–36]. In addition, analysis of the UV-vis spectra suggested $D/k_B \approx 4.3$ K, while heat capacity data were consistent with $D/J \lesssim 1$. Finally, magnetization data at $T = 0.5$ K and in $B < 3$ T revealed the presence of an "excess" amount of signal above the response measured at 1.6 K. The present studies were initiated to search for evidence of the critical field required to close a gap and to understand the origin of the magnetic signal at low temperatures.

## 2. Experimental Details

To experimentally explore the magnetization of $[Ni(HF_2)(3\text{-}Clpy)_4]BF_4$, nominally 4 ($\pm1$) mg of small cubic-like (with sides of a few mm), as-grown [30] polycrystalline pieces were randomly placed and rigidly held in one side of a mutual inductance cell, shown in Fig. 2, mounted to the dilution refrigerator operated in the Williamson Hall Annex of the High-B/T Facility of the National High Magnetic Field Laboratory (NHMFL) at the University of Florida. At a fixed temperature, the imbalance between the two secondary coils was monitored by a lock-in amplifier (Model NF5610B) at low frequency (52.93 Hz) and low ac field (0.1 mT) while sweeping a superconducting solenoid at a constant rate of 50 mT/min. The phase setting was adjusted to minimize one channel of the lock-in amplifier and remained constant for all runs. The data acquisition of the in-phase and out-of-phase voltages ($V_{in}$ and $V_{out}$) takes about 3.5 hours as the field is swept to or from 10 T. To within experimental resolution, no magnetic hysteresis was established at 50 mK, but the starting and final voltages differ by a few $\mu$V due to changes in the level of the liquid helium in the bath. With the absence of detectable hystersis at the lowest temperature, the runs at higher temperatures were performed by sweeping the field in only one direction. When the runs with the sample were complete, the sample was removed, and the measurements were repeated so "sample-in minus sample-out" analysis, commonly employed in pulsed field experiments, could be performed. It is noteworthy the sample was cooled by a liquid of pure $^3$He, which is known to provide excellent thermal coupling down to low temperatures and in high magnetic fields [39, 40], and a similar arrangement was used to cool another molecule-based magnetic system down to 40 mK [41]. In fact, the solvent associated with the present

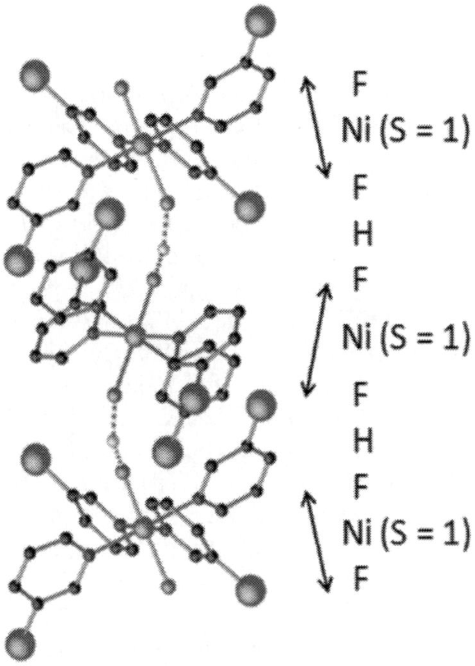


**Figure 1.** The crystal structure of $[Ni(HF_2)(3\text{-}Clpy)_4]BF_4$ (py = pyridine), also known as NBCT, space group $P2_1/c$, resembles a linear chain of octahedrally coordinated $Ni^{2+}$ ions linked by F–H–F bridges along the $c$-axis [30]. The 3-Clpyridine ($C_5H_4ClN$) ligands isolate the chains from neighboring ones, and the $BF_4$ anions required for charge neutralization are omitted for clarity. The local environment of the Ni ions switches axial orientation along the chain, as sketched by the arrows, and is an aspect reminescint of the staggered $g$-tensor in $Ni(C_2H_8N_2)_2NO_2(ClO_4)$ (commonly referred to as NENP) [37, 38].

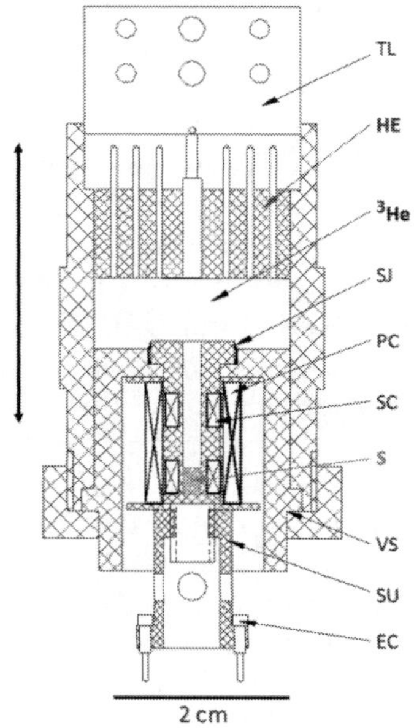


**Figure 2.** The cell used for magnetization studies is mostly made of plastics and is oriented along the direction of the magnetic fields represented by the left arrow. The main components are: a Ag thermal link (TL), with a Ag sintered powder heat exchanger (HE), a bath of pure $^3$He (fill line not shown) in direct contact with the sample (S), a Stycast 1266 joint (SJ) to the Kel-F (Neoflon PCTFE) housing, a primary coil (PC) consisting of 2100 turns of nominally 76 $\mu$m diameter NbTi multi-filamentary wire, a set of secondary coils (SC) each with 2500 turns of 51 $\mu$m diameter Cu wire, a lower plastic-to-plastic (reusable) vacuum seal (VS), and a supporting tower (SU) for the electrical connections (EC).

sample [30] eroded the thin Stycast 1266 wall of the previous cell, and a new one was constructed from an old piece of Kel-F by 3M (now available as Daikin Neoflon PCTFE).

## 3. Results and Discussion

The isothermal, low-field dependences of the magnitudes ($= \sqrt{V_{\text{in}}^2 + V_{\text{out}}^2}$), after background subtraction, are shown in Fig. 3. The overall experimental uncertainty, which is a combination of background subtraction and the bath level adjustment, is represented by the bar designating "+/−". The legend identifies the temperatures, and two runs at 50 mK were made. One 50 mK run was made at the start of the experiment (shown in blue), while the other 50 mK run (shown in black), which followed the measurement at 1000 mK, was made at the end of the study with the sample present. One striking aspect of the data is the apparent similarity of the data sets for $T \lesssim 400$ mK, and at first glance, this result seems to indicate the sample is not being cooled to the lowest temperatures. However, it is important to stress several points. Firstly, as previously mentioned, similar samples have been reproducibly cooled in other studies [41]. Secondly, difficulty in cooling a sample may indicate an increase of the specific heat near a quantum phase transition, where the accumulation of entropy is generally accepted as a generic consequence [43, 44]. In addition, the data at the lowest fields suggest the magnetic signal is absent or exceedingly small. Although the sensitivity of this cell is considered to be comparable to the one used in other work [41], the dynamic range of the detection coils was not quantitatively established, so a definitive value for the critical field necessary to close a gap, $B_c$, cannot be determined. Nevertheless, an upper bound of $B_c \leq (35 \pm 10)$ mT can be set as shown in Fig. 3. This result can be compared to the extrapolated Haldane gap values near $(D/J)_c = 0.968$, as given by Hu *et al.* [21], and the result is $B_c = 46$ mT, when using the aforementioned values for the parameters. In general, the "excess" signal, at $T = 0.5$ K and $B < 3$ T, reported by others [30] has been qualitatively reproduced in this work. Furthermore, the temperature-field dependences of the data sets shown in Fig. 3 cannot be modeled by non-interacting $S = 1/2$ end-chain spins. Quantitative insight about the magnetic response at low temperatures and in low magnetic fields will require data acquired with oriented crystals, where the impact of a staggered $g$-tensor, Fig. 1, may be revealed [37, 38].

High-field, high-frequency EPR spectra were acquired with a randomly-packed, purposely-powdered sample, with a total mass of approximately 50 mg, that was synthesized separately from the ones used in the magnetization studies. Using a constant flow cryostat operating at 3 K $\leq T \leq$ 20 K, the experiments were conducted with a home-built spectrometer at the electromagnetic resonance (EMR) facility of the NHMFL at Florida State University. The details

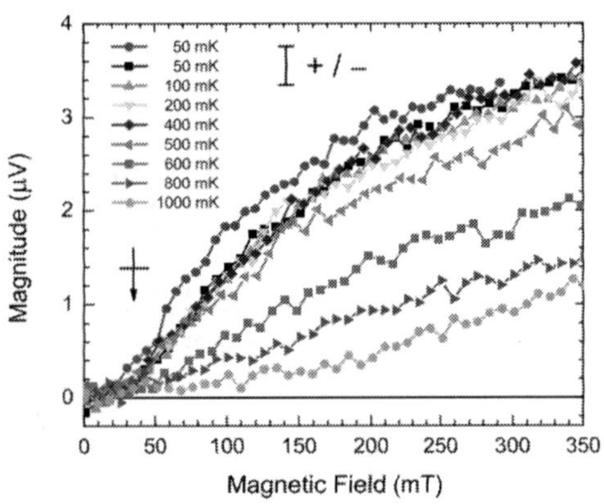


**Figure 3.** Magnitudes of the voltages, after background subtraction, are shown as a function of the magnetic field, and the uncertainty is given by the "+/−" bar. The two runs at 50 mK are discussed in the text. The degeneracy of the data sets for $T \lesssim$ 400 mK is conjectured to be evidence of the increase of entropy, see text. A critical field necessary to close a gap, if present, has an upper bound of $B_c \leq (35 \pm 10)$ mT.

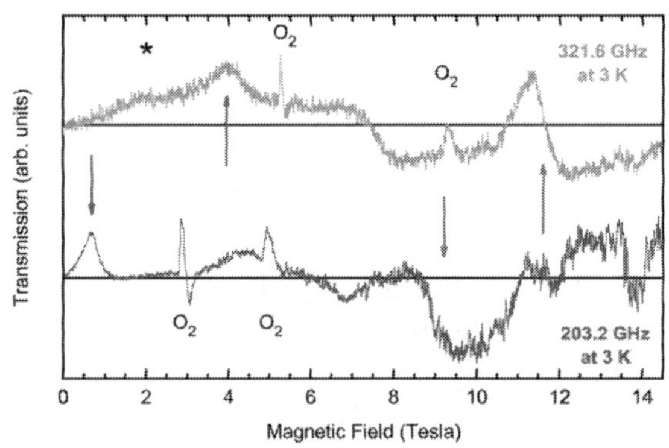


**Figure 4.** The EPR transmission signals at 321.6 GHz (top and organge trace) and 203.2 GHz (bottom and purple trace) are shown as a function of magnetic field. The arrows designate the features considered to arise from the Ni-chain, while fingerprints known to be indicative of trace amounts of oxygen in solid air are labeled as $O_2$ [42]. A weak feature, marked by an asterisk, may be a "forbidden" transition [14].

of the transmission-type instrument and spectrometer are described elsewhere [45], and for this work, spectra were acquired at several frequencies ranging from 203 GHz to 425 GHz. The spectra taken at $T = 3$ K and at 203.2 GHz and 321.6 GHz are shown in Fig. 4, where the broad features associated with the Ni-chains are identified. Since the thermal energy at $T = 3$ K is similar to the $J$ and $D$ energies, the full dynamics of the chain are not developed, especially when considering the gap is exceedingly small, $\Delta/k_B \lesssim 60$ mK, if it exists. The limited EPR data sets presented herein preclude any extended analysis that can resolve the magnetic parameters of the Hamiltonian, Eq. 1, as commonly done in these kinds of systems [11–14, 46–48]. However to date, no other EPR studies of NBCT have been reported, so typical spectra are included, Fig. 4, to serve as benchmarks for future investigations that will presumably employ oriented single crystal samples which can be studied over a wider range of frequency and temperature.

## 4. Conclusions

In summary, the magnetic properties of $[Ni(HF_2)(3\text{-}Clpy)_4]BF_4$ (py = pyridine) exhibit unusual behavior at low temperatures and in low fields. An upper bound for the critical field required to close a gap is established, $B_c \leq (35 \pm 10)$ mT, and this value is close to the one predicted for a Haldane system [21], 46 mT, when using the experimentally established values of $J$, $D$, and $g$. In low fields, the magnetic signal increases with decreasing temperature for 400 mK $\lesssim T \lesssim 800$ mK but, to within experimental resolution, is independent of temperature for 50 mK $\leq T \leq 400$ mK. This observation is consistent with a significant increase in the specific heat arising from the accumulation of entropy in the vicinity of the quantum critical transition near $D/J \approx 1$. These results provide new experimental access to this quantum critical regime, where the $D/J$ value might be tunable by pressure [49–53] and/or chemical modifications such as deuteration [54, 55]. Ultimately, future inelastic neutron scattering studies may clarify how the ground-state of $[Ni(HF_2)(3\text{-}Clpy)_4]BF_4$ compares to theoretical phase diagrams [26, 27].


## Acknowledgments

This work was supported by the NSF through DMR-1202033 and DMR-1708410 (MWM), DMR-1703003 (JLM), and DMR-1157490 (NHMFL). We acknowledge communications with M. Hagiwara, D. M. Pajerowski, M. Orendáč, A. Orendáčová, E. Čižmár, and V. N. Glazkov.


## References


[1] Haldane F 1983 *Physics Letters A* **93** 464–468
[2] Haldane F D M 1983 *Phys. Rev. Lett.* **50**(15) 1153–1156
[3] Schulz H J 1986 *Phys. Rev. B* **34**(9) 6372–6385
[4] A eck I, Kennedy T, Lieb E H and Tasaki H 1987 *Phys. Rev. Lett.* **59**(7) 799–802
[5] A eck I 1989 *Journal of Physics: Condensed Matter* **1** 3047
[6] Yamashita M, Ishii T and Matsuzaka H 2000 *Coordination Chemistry Reviews* **198** 347–366
[7] Botet R, Jullien R and Kolb M 1983 *Phys. Rev. B* **28**(7) 3914–3921
[8] Renard J P, Verdaguer M, Regnault L P, Erkelens W A C, Rossat-Mignod J and Stirling W G 1987 *EPL (Europhysics Letters)* **3** 945
[9] Orendáč M, Orendáčová A, Černák J, Feher A, Signore P J C, Meisel M W, Merah S and Verdaguer M 1995 *Phys. Rev. B* **52**(5) 3435–3440
[10] Orendáč M, Zvyagin S, Orendáčová A, Sieling M, Lüthi B, Feher A and Meisel M W 1999 *Phys. Rev. B* **60**(6) 4170–4175
[11] Sieling M, Löw U, Wolf B, Schmidt S, Zvyagin S and Lüthi B 2000 *Phys. Rev. B* **61**(1) 88–91
[12] Hagiwara M 2002 *Molecular Physics* **100** 1489–1495
[13] Čižmár E, Ozerov M, Ignatchik O, Papageorgiou T P, Wosnitza J, Zvyagin S A, Krzystek J, Zhou Z, Landee C P, Landry B R, Turnbull M M and Wikaira J L 2008 *New Journal of Physics* **10** 033008
[14] Smirnov A I, Glazkov V N, Kashiwagi T, Kimura S, Hagiwara M, Kindo K, Shapiro A Y and Demianets L N 2008 *Phys. Rev. B* **77**(10) 100401
[15] Zheludev A, Nagler S E, Shapiro S M, Chou L K, Talham D R and Meisel M W 1996 *Phys. Rev. B* **53**(22) 15004–15009

IOP Conf. Series: Journal of Physics: Conf. Series **969** (2018) 012121 doi:10.1088/1742-6596/969/1/012121

[16] Brambleby J, Manson J L, Goddard P A, Stone M B, Johnson R D, Manuel P, Villa J A, Brown C M, Lu H, Chikara S, Zapf V, Lapidus S H, Scatena R, Macchi P, Chen Y S, Wu L C and Singleton J 2017 *Phys. Rev. B* **95**(13) 134435
[17] Sakai T and Takahashi M 1990 *Phys. Rev. B* **42**(7) 4537–4543
[18] Tzeng Y C and Yang M F 2008 *Phys. Rev. A* **77**(1) 012311
[19] Albuquerque A F, Hamer C J and Oitmaa J 2009 *Phys. Rev. B* **79**(5) 054412
[20] Pollmann F, Turner A M, Berg E and Oshikawa M 2010 *Phys. Rev. B* **81**(6) 064439
[21] Hu S, Normand B, Wang X and Yu L 2011 *Phys. Rev. B* **84**(22) 220402
[22] Pollmann F, Berg E, Turner A M and Oshikawa M 2012 *Phys. Rev. B* **85**(7) 075125
[23] Li W, Weichselbaum A and von Delft J 2013 *Phys. Rev. B* **88**(24) 245121
[24] Langari A, Pollmann F and Siahatgar M 2013 *Journal of Physics: Condensed Matter* **25** 406002
[25] Wierschem K and Sengupta P 2014 *Phys. Rev. Lett.* **112**(24) 247203
[26] Wierschem K and Sengupta P 2014 *Modern Physics Letters B* **28** 1430017
[27] Tzeng Y C, Onishi H, Okubo T and Kao Y J 2017 *Phys. Rev. B* **96**(6) 060404
[28] Si Q and Steglich F 2010 *Science* **329** 1161–1166
[29] Kinross A W, Fu M, Munsie T J, Dabkowska H A, Luke G M, Sachdev S and Imai T 2014 *Phys. Rev. X* **4**(3) 031008
[30] Manson J L, Baldwin A G, Scott B L, Bendix J, Del Sesto R E, Goddard P A, Kohama Y, Tran H E, Ghannadzadeh S, Singleton J, Lancaster T, Möller J S, Blundell S J, Pratt F L, Zapf V S, Kang J, Lee C, Whangbo M H and Baines C 2012 *Inorganic Chemistry* **51** 7520–7528
[31] Hagiwara M, Katsumata K, A eck I, Halperin B I and Renard J P 1990 *Phys. Rev. Lett.* **65**(25) 3181–3184
[32] Glarum S H, Geschwind S, Lee K M, Kaplan M L and Michel J 1991 *Phys. Rev. Lett.* **67**(12) 1614–1617
[33] Avenel O, Xu J, Xia J S, Xu M F, Andraka B, Lang T, Moyland P L, Ni W, Signore P J C, van Woerkens C M C M, Adams E D, Ihas G G, Meisel M W, Nagler S E, Sullivan N S, Takano Y, Talham D R, Goto T and Fujiwara N 1992 *Phys. Rev. B* **46**(13) 8655–8658
[34] Avenel O, Xu J, Xia J, Xu M F, Andraka B, Lang T, Moyland P, Ni W, Signore P, Woerkens C, Adams E, Ihas G, Meisel M, Nagler S, Sullivan N, Takano Y, Talham D, Goto T and Fujiwara N 1992 *Journal of Low Temperature Physics* **89** 547–550
[35] Batista C D, Hallberg K and Aligia A A 1998 *Phys. Rev. B* **58**(14) 9248–9251
[36] Granroth G E, Maegawa S, Meisel M W, Krzystek J, Brunel L C, Bell N S, Adair J H, Ward B H, Fanucci G E, Chou L K and Talham D R 1998 *Phys. Rev. B* **58**(14) 9312–9315
[37] Chiba M, Ajiro Y, Kikuchi H, Kubo T and Morimoto T 1991 *Phys. Rev. B* **44**(6) 2838–2841
[38] Huang H and A eck I 2004 *Phys. Rev. B* **69**(18) 184414
[39] Hammel P C, Roukes M L, Hu Y, Gramila T J, Mamiya T and Richardson R C 1983 *Phys. Rev. Lett.* **51**(23) 2124–2127
[40] Schuhl A, Maegawa S, Meisel M W and Chapellier M 1987 *Phys. Rev. B* **36**(13) 6811–6836
[41] Orendáčová A, Čižmár E, Sedláková L, Hanko J, Kajňaková M, Orendáč M, Feher A, Xia J S, Yin L, Pajerowski D M, Meisel M W, Zeleňák V, Zvyagin S and Wosnitza J 2009 *Phys. Rev. B* **80**(14) 144418
[42] Pardi L A, Krzystek J, Telser J and Brunel L C 2000 *Journal of Magnetic Resonance* **146** 375–378
[43] Vidal G, Latorre J I, Rico E and Kitaev A 2003 *Phys. Rev. Lett.* **90**(22) 227902
[44] Zhu L, Garst M, Rosch A and Si Q 2003 *Phys. Rev. Lett.* **91**(6) 066404
[45] Hassan A, Pardi L, Krzystek J, Sienkiewicz A, Goy P, Rohrer M and Brunel L C 2000 *Journal of Magnetic Resonance* **142** 300–312
[46] Hagiwara M, Katsumata K, Yamada I and Suzuki H 1996 *Journal of Physics: Condensed Matter* **8** 7349
[47] Fanucci G E, Krzystek J, Meisel M W, Brunel L C and Talham D R 1998 *Journal of the American Chemical Society* **120** 5469–5479
[48] Furuya S C, Oshikawa M and A eck I 2011 *Phys. Rev. B* **83**(22) 224417
[49] Zaliznyak I A, Dender D C, Broholm C and Reich D H 1998 *Phys. Rev. B* **57**(9) 5200–5204
[50] Oosawa A, Kakurai K, Osakabe T, Nakamura M, Takeda M and Tanaka H 2004 *Journal of the Physical Society of Japan* **73** 1446–1449
[51] Rüegg C, Furrer A, Sheptyakov D, Strässle T, Krämer K W, Güdel H U and Mélési L 2004 *Phys. Rev. Lett.* **93**(25) 257201
[52] Hong T, Garlea V O, Zheludev A, Fernandez-Baca J A, Manaka H, Chang S, Leao J B and Poulton S J 2008 *Phys. Rev. B* **78**(22) 224409
[53] Thede M, Mannig A, Månsson M, Hüvonen D, Khasanov R, Morenzoni E and Zheludev A 2014 *Phys. Rev. Lett.* **112**(8) 087204
[54] Tsujii H, Honda Z, Andraka B, Katsumata K and Takano Y 2005 *Phys. Rev. B* **71**(1) 014426
[55] Goddard P A, Singleton J, Maitland C, Blundell S J, Lancaster T, Baker P J, McDonald R D, Cox S, Sengupta P, Manson J L, Funk K A and Schlueter J A 2008 *Phys. Rev. B* **78**(5) 052408

IOP Conf. Series: Journal of Physics: Conf. Series **969** (2018) 012122 doi:10.1088/1742-6596/969/1/012122

# Magnetic, thermodynamic and optical properties of Sb-substituted $Ba_2PrBiO_6$ double perovskite oxides

**K. Onodera, T. Kogawa, M. Matsukawa, H. Taniguchi, K. Nishidate**
Faculty of Science and Engineering, Iwate University, Morioka 020-8551, Japan
E-mail: `matsukawa@iwate-u.ac.jp`

**A. Matsushita, M. Shimoda**
National Institute for Materials Science, Tsukuba, Ibaraki 305-0047, Japan

**Abstract.** We demonstrated crystal structures, magnetic, thermodynamic and optical properties of the B-site substituted perovskite oxides $Ba_2Pr(Bi_{1-x},Sb_x)$ $O_6$ (x=0, 0.1 and 0.2). Polycrystalline samples of Sb-substituted $Ba_2PrBiO_6$ were prepared with the conventional solid-state reaction technique. The X-ray diffraction data revealed that the polycrystalline samples are an almost single phase with a monoclinic structure ($C2/m$). Substitution of smaller Sb ion at Bi site causes a monotonic decrease in both the lattice parameters and volume. Magnetization measurements at high temperatures above 200 K show that the effective magnetic moment is estimated to be around 3.15 $\mu_B$, which is close to that for $Pr^{3+}$ion. The X-ray photoemission spectroscopy analysis revealed that a prominent peak of $Pr^{3+}$ is dominant with a smaller shoulder structure of $Pr^{4+}$. A Schottky-like anomaly observed in the low-temperature specific heat measurement is explained by low-lying splitting of Pr ions under the crystal field effect. Optical spectra were measured using a diffuse-reflectance method. The band gaps were estimated from the optical data to be 0.977 eV and 1.073 eV, at $x$=0 and 0.2, respectively. The effect of band gap opening due to Sb substitution is examined by using the density functional theory.

## 1. Introduction

A large number of double perovskite oxides $A_2B'B''O_6$ have been widely studied due to their attractive physical properties and potential applications[1, 2]. Some of semiconducting $A_2B'B''O_6$ compounds exhibit photocatalytic properties such as hydrogen generation by water splitting and are taken as alternative materials for $TiO_2$ oxide [3]. In particular, $Ba_2PrBiO_6$ compound has been shown to possess high photocatalytic activity, which is probably related to the valence mixing[4, 5]. A previous study on the magnetic states of $Ba_2PrBiO_6$ compound suggests an anomalous valence situation for Pr ions.[6] The complicated ground states of rare earth ions such as Pr ion under the crystal field effect remain open question, not only in the physical properties of the double perovskite compound, but also in the view point of physics of $4f$ electron systems.

There are significant factors such as charge and size differences between B′ and B″ sites, to determine the B-site ordering of the double perovskite oxide[7]. Increase in lattice strain and/or increase in the electrostatic repulsion overcome the entropy contribution toward disordering, causing the alternate arrangement. For $A_2^{2+}B'^{3+}B''^{5+}O_6$ composition, the B sites tend to oder

IOP Conf. Series: Journal of Physics: Conf. Series **969** (2018) 012122 doi:10.1088/1742-6596/969/1/012122

with increasing the ion size difference $\Delta r_B = r_{B'} - r_{B''}$. If $\Delta r_B > 0.2$ Å, it is well known that the B sites of the compounds are almost ordered alternately in the all crystallographic axes[7]. In $Ba_2^{2+}Pr^{3+}Bi^{5+}O_6$ compound, the B-site ionic radius difference $\Delta r_B = 0.23$ Å meets the above condition, suggesting the B-site ordering. ($r_{B'}$ ($Pr^{3+}$)=0.99 Å and $r_{B''}$ ($Bi^{5+}$)=0.76 Å.[8] ) The density functional theory based calculation reveals the band structure and its associated orbital states of this compound, and predicts the stability of crystal structures of B site ordered states.

In this paper, we investigate crystal structures, magnetic, thermodynamic and optical properties of the $Ba_2Pr(Bi_{1-x},Sb_x)$ $O_6$ compounds ($x$=0, 0.1 and 0.2), to determine the physical properties of B-site substituted double perovskite oxide. For comparison, the physical properties of the B$'$site substituted $Ba_2LaBiO_6$ compound without magnetic ions are examined.

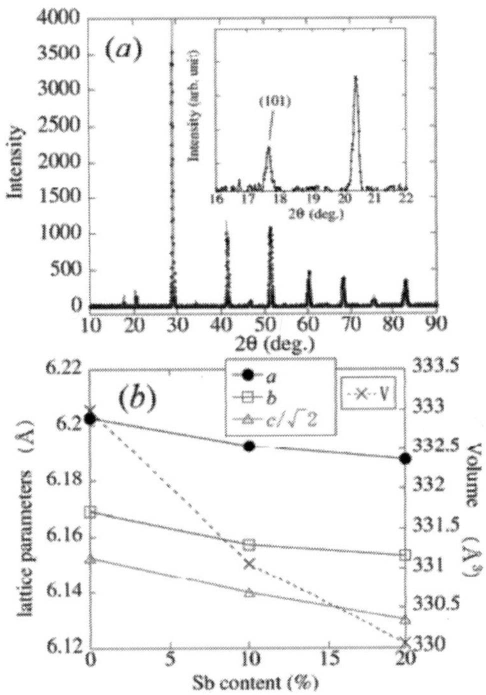


**Figure 1.** ($a$)X-ray diffraction patterns of the parent $Ba_2PrBiO_6$ compound. The (101) peak in the inset corresponds to one of typical Miller indexes of B-site ordered double-perovskite structure. The calculated curve was obtained by using the lattice parameters. ($b$)The lattice parameters and volume as a function of Sb content.

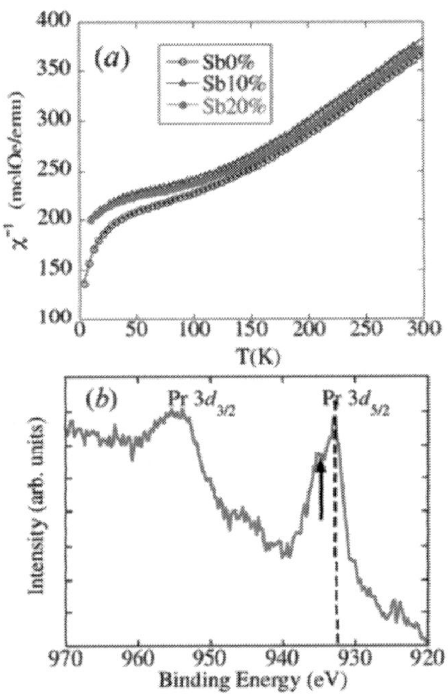


**Figure 2.** ($a$)Inverse magnetic susceptibilities ($\chi^{-1}$) for $Ba_2Pr(Bi_{1-x},Sb_x)O_6$ ($x$=0, 0.1 and 0.2). ($b$)the XPS spectrum of Pr $3d$ core level emission for the parent sample. The dashed line corresponds to a peak position in the $Pr^{3+}$ standard $Pr_2O_3$ spectrum. The arrow represents the $Pr^{4+}$ standard peak position.

## 2. Experiment

Polycrystalline samples of Sb-substituted $Ba_2PrBiO_6$ were prepared with the conventional solid-state reaction technique. Stoichiometric mixtures of $BaCO_3$, $Pr_6O_{11}$, $Bi_2O_3$, and $Sb_2O_3$ powders were ground, sintered in air at 800-900 °C for 2 days with intermediate grindings, and pressed into pellets. The pellet samples were annealed in air at 1000 °C for 4 days. The X-ray diffraction data revealed that the polycrystalline samples are an almost single phase with a monoclinic structure

($C2/m$). For the parent $Ba_2PrBiO_6$, the lattice parameters are $a$ = 6.2038Å, $b$ = 6.1689Å, $c$ = 8.7011Åand $\beta$ = 89.7303°, which are in fair agreement with those reported by a previous study[6]. The emergence of (101) reflection indicates B-cation ordering which is characteristic of B-site ordered double-perovskite structure. Substitution of smaller $Sb^{5+}$ (0.60 Å) ion at $Bi^{5+}$ (0.76 Å) site causes a monotonic decrease in both the lattice parameters and volume as shown in Fig.1 (b). The *dc* magnetization was conducted at a magnetic field of 1 T under the zero-field cooling process using a superconducting quantum interference device magnetometer (MPMS,Quantum Design). X-ray photoemission spectroscopy (XPS) measurements for Pr 3*d* core-level emission were performed in a VG ESCALAB MkII spectrometer by using MgK$\alpha$ X-ray irradiation (1254 eV). Specific heat measurements were conducted by a thermal relaxation method using a physical property measuring system (PPMS,Quantum Design). Optical spectra were measured by using a diffuse-reflectance method with a spectrophotometer (Hitachi U-3500) and $BaSO_4$ was used as the reference material. Optical band gaps for the powder samples were evaluated from reflectance spectral data using the conventional Kubelka-Munk functions[6, 9] . We calculated the band structure of $Ba_2PrBiO_6$ by using the density functional theory as implemented in the code VASP.

## 3. Results and Discussion

The inverse magnetic susceptibilities ($\chi^{-1}$)for $Ba_2Pr(Bi_{1-x},Sb_x)O_6$ ($x$=0, 0.1 and 0.2) are shown in Fig.2($a$) as a function of temperature at a magnetic field of 1 T. In the $Ba_2LaBiO_6$ compound, no magnetization signal was observed because of the absence of magnetic ions. From the magnetization data at high temperatures above 200 K, we estimate the effective magnetic moment according to the Curie-Weiss law. The value of effective magnetic moment $\mu_{eff}$ is evaluated by using the following formula,

$$C = N\mu_{eff}^2\mu_B^2/3k_B$$

where $C$, $N$, and $\mu_B$ denote the Curie constant, the number of magnetic atom per mol, and the Bohr magneton, respectively. Performing the calculation, we obtained that $\mu_{eff}$ = 3.15 $\mu_B$ and 3.12 $\mu_B$ for the parent and Sb20% samples, respectively. These values are close to the effective magnetic moment for the $Pr^{3+}$ free ion (3.57$\mu_B$). However, these findings do not exclude the presence of $Pr^{4+}$[5]. The XPS spectrum of Pr 3d core level for the parent sample is shown as a function of the binding energy in Fig 2 ($b$). The XPS profile strongly suggests that a prominent peak of $Pr^{3+}$ is dominant at the specified binding energy values around 933 eV, followed by a smaller shoulder structure of $Pr^{4+}$ around 935 eV. The dashed line corresponds to a peak position in the $Pr^{3+}$ standard $Pr_2O_3$ spectrum. The arrow points to the $Pr^{4+}$ standard $PrO_2$ spectrum.

In Fig. 3($a$), specific heats of the parent $Ba_2PrBiO_6$ and its related $Ba_2LaBiO_6$ compounds are shown as a function of temperature between 2 and 300 K. In addition, we plot in Fig. 3($b$) low-temperature specific heat data of both the samples, $C/T$(Pr) and $C/T$(La). Applying the Debye $T^3$ law to the low temperature data, we estimate the Debye temperature, $\theta$=112 K for $Ba_2LaBiO_6$. We assume that the lattice component in $C/T$ of the parent sample is almost equivalent to the values of $C/T$ of the nonmagnetic $Ba_2LaBiO_6$ compound since both samples share the same crystal structure. Thus, specific heat differences $dC/T$ between both compounds are taken as non lattice components of the parent sample. A Schottky-type specific heat with two energy levels [10] is given by

$$C_{\text{Sch}} = k_B(x/2)^2/\cosh^2(x/2) \text{ with } x = \Delta E/k_BT.$$

When the crystal field splitting between the ground and low-lying excited states of this compound is set to be $\Delta E$ =1.5 meV, we obtain a good agreement between the calculated curve and

IOP Conf. Series: Journal of Physics: Conf. Series **969** (2018) 012122 doi:10.1088/1742-6596/969/1/012122

experimental data. Accordingly, we attribute the observed broad peak in $C/T$ to the Schottky-type specific heat anomaly.

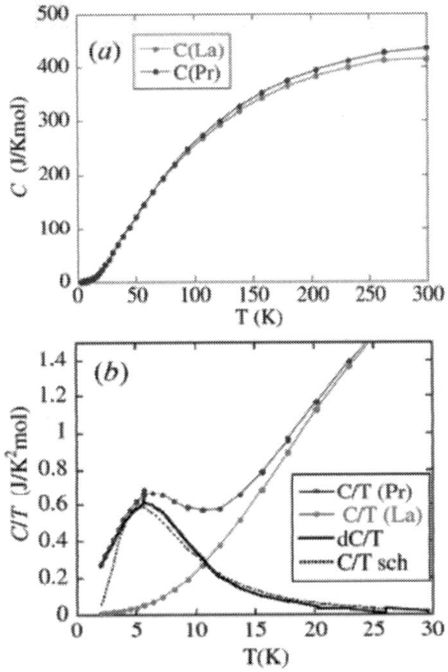


**Figure 3.** (*a*)Specific heats of the parent $Ba_2PrBiO_6$ and its related $Ba_2LaBiO_6$ compounds, $C$(Pr) and $C$(La). (*b*) Low temperature specific heats of both samples, $C/T$(Pr) and $C/T$(La). Specific heat differences $dC/T$ in both compounds are also plotted. Dashed curves represent the calculated Schottky heat anomaly with $\Delta E$ =1.5 meV.

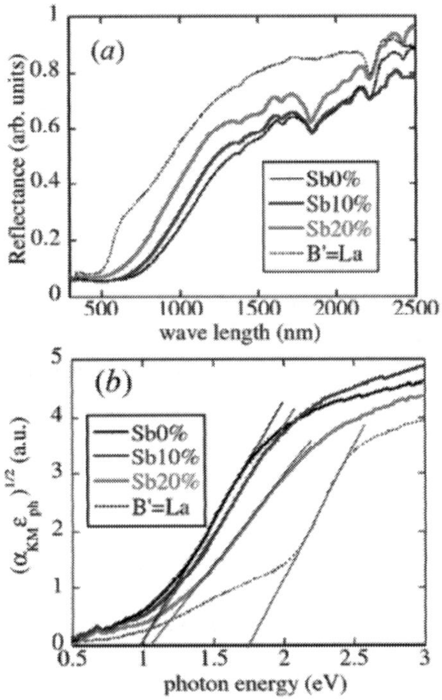


**Figure 4.** Optical measurements on $Ba_2Pr(Bi_{1-x},Sb_x)O_6$ ($x$=0, 0.1 and 0.2) powder samples. (*a*)Diffuse reflectance spectra as a function of wave length (*b*) Kubelka-Munk conversion of diffuse reflectance spectra as a function of photon energy. For the $x$=0 and $x$=0.2 samples, we evaluate $E_g = 0.977 \pm 0.023$ eV and $1.073 \pm 0.014$ eV, respectively. Straight lines denote the extrapolations to estimate the band gaps using the least squares method. For comparison, we exhibit the optical data of the $Ba_2LaBiO_6$ compound, giving $E_g = 1.767 \pm 0.028$ eV.

For $Ba_2Pr(Bi_{1-x},Sb_x)O_6$ ($x$=0, 0.1 and 0.2) powder samples, the diffuse reflectance spectra are shown in Fig.4 (*a*) as a function of wave length. First, the observed reflectance data for the powder samples are transformed to the absorption coefficient $\alpha_{KM}$ by using the conventional Kubelka-Munk function. Figure 4 (*b*) shows the square root of the absorption coefficient $(\alpha_{KM}\varepsilon_p)^{1/2}$ as a function of photon energy $\varepsilon_p$. For the Kubelka-Munk conversion data near the band edge, we extrapolate the tangent line to the $\varepsilon_p$ axis and evaluate the optical band gaps from the intersection according to the equation of

$$(\alpha_{KM}\varepsilon_p)^{1/2} \propto (\varepsilon_p - E_g)$$

IOP Conf. Series: Journal of Physics: Conf. Series **969** (2018) 012122 doi:10.1088/1742-6596/969/1/012122

, where $\alpha_{KM}$, $\varepsilon_p$, and $E_g$ are the absorption coefficient, photon energy, and band gap [4, 5]. We estimate the values of band gaps to be $E_g = 0.977 \pm 0.023$ eV at $x = 0$, $1.003 \pm 0.025$ eV at $x = 0.1$ and $1.073 \pm 0.014$ eV at $x = 0.2$. For comparison, we also exhibit the optical data of the $Ba_2LaBiO_6$ compound, resulting in $E_g = 1.767 \pm 0.028$ eV. All band gaps are evaluated within errors of plus or minus several percent.

The effect of the band gap opening by the atomic substitution is supported by our first-principles electric structure calculation. We have calculated the band structure of $Ba_2PrBiO_6$ by using the density functional theory as implemented in the code VASP. The estimated band gap energy of the parent $Ba_2PrBiO_6$ is around 0.5 eV as shown in Fig5 (*a*) , leading to an underestimation in comparison to the experimental value. However, the band gap of $Ba_2PrSbO_6$ becomes 4.1 eV when we substitute Sb for Bi (Fig5 (*b*)). In addition, we obtain $E_g$=2.5 eV for the $Ba_2LaBiO_6$ where B$'$ site is replaced by La (not shown here). These results probably predict the qualitative trends than the absolute values. The detailed analysis will be published elsewhere.

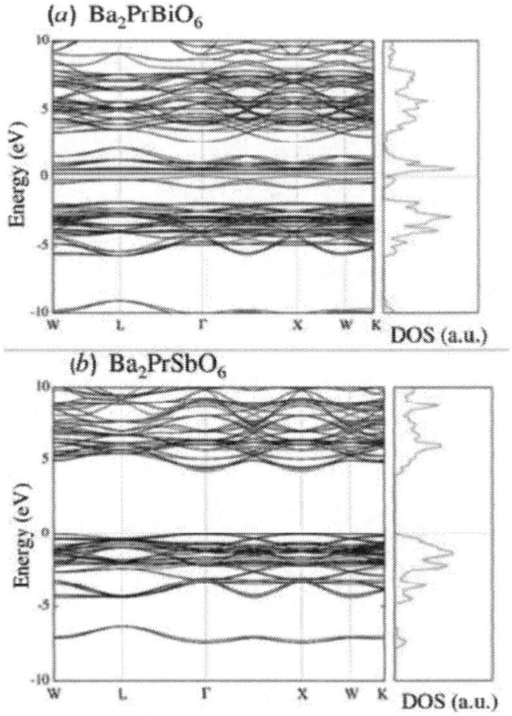


**Figure 5.** Energy band structures and total density of states (DOS) calculated by the density functional theory for (*a*) $Ba_2PrBiO_6$ and (*b*) $Ba_2PrSbO_6$ compounds. The band gaps are estimated to to be $\sim 0.5$ eV and 4.1 eV, for the parent $Ba_2PrBiO_6$ and Sb substituted $Ba_2PrSbO_6$ compounds, respectively.

## 4. Summary

We demonstrated crystal structures, magnetic, thermodynamic and optical properties of the B-site substituted perovskite oxides $Ba_2Pr(Bi_{1-x},Sb_x)\,O_6$ ($x$=0, 0.1 and 0.2). X-ray diffraction data point to the B-site ordered double-perovskite structure, which is consist with previous predictions based on the charge and ionic size differences. The magnetization and X-ray photoemission spectroscopy measurements considerably indicate the existence of $Pr^{3+}$, but do not exclude the presence of $Pr^{4+}$. The low temperature specific heat of the parent compound is well described by the Schottky heat anomaly due to crystal field splitting of Pr ion with $\triangle E$ =1.5 meV. The values of the band gap are evaluated to be 0.977±0.023 eV and 1.073±0.014 eV at the parent and Sb20% substituted samples, by using the conventional Kubelka-Munk function. The effect of band gap opening due to Sb substitution was examined by using the density functional theory.

## 5. Acknowledgment

The authors are grateful for Dr. K. Nonaka for his assistance in our experiments.

## 6. References

[1] Kobayashi K -I, Kimura T, Sawada H, Terakura K and Tokura Y, 1998 *Nature* **395** 677
[2] Ramesh R and Spaldin N A 2007 *Nature Materials* **6** 21
[3] Eng H W, Barnes P W, Auer B M and Woodward P M 2003 *J. Solid State Chem.* **175** 94
[4] Hatakeyama T, Takeda S, Ishikawa F, Ohmura A, Nakayama A, Yamada Y, Matsushita A and Yea J 2010 *J. Cer. Soc. Jpn.* **118** 91
[5] Matsushita A, Nakane T, Naka T, Isago H, Yamada Y and Yamada Yuh 2012 *Jpn. J. Appl. Phys.* **51** 121802
[6] Harrison W T A, Reis K P, Jacobson A J, Schneemeyer L F and Waszczak J V 1995 *Chem. Mater.* **7** 2161
[7] Vasala S and Karppinen M 2015 *Prog. Solid State Chem.* **43** 1
[8] Shannon R 1976 *Acta. Crystallogr.* **A32** 751
[9] Tang J W, Zou Z G and Ye J H 2007 *J. Phys. Chem.* **C111** 12779
[10] Kittel C 1976 *Introduction to Solid State Physics* sixth edition (New York: Wiley)

IOP Conf. Series: Journal of Physics: Conf. Series **969** (2018) 012123 doi:10.1088/1742-6596/969/1/012123

# Dynamical spin structure factors of $\alpha$-$RuCl_3$

**Takafumi Suzuki and Sei-ichiro Suga**
University of Hyogo, Himeji 671-2280, Japan
E-mail: suga@eng.u-hyogo.ac.jp

**Abstract.** Honeycomb-lattice magnet $\alpha$-$RuCl_3$ is considered to be a potential candidate of realizing Kitaev spin liquid, although this material undergoes a phase transition to the zigzag magnetically ordered state at $T_N \sim 7$ K. Quite recently, inelastic neutron-scattering experiments using single crystal $\alpha$-$RuCl_3$ have unveiled characteristic dynamical properties. We calculate dynamical spin structure factors of three *ab-initio* models for $\alpha$-$RuCl_3$ with an exact numerical diagonalization method. We also calculate temperature dependences of the specific heat by employing thermal pure quantum states. We compare our numerical results with the experiments and discuss characteristics obtained by using three *ab-initio* models.

## 1. Introduction

Since Kitaev presented an exactly solvable model for a spin-liquid ground state [1], Kitaev spin liquid (KSL) has attracted much attention because of its fascinating properties. Great efforts have been payed for realizing KSL in real materials. A layered honeycomb-lattice compound $\alpha$-$RuCl_3$ is considered to be a promising candidate for applying Kitaev physics [2]. In the specific heat measurements for $\alpha$-$RuCl_3$, a double-peak structure has been observed with decreasing temperature ($T$) [3, 4, 5]. The double-peak structure of the specific heat plays a key role in investigating the Kitaev physics [6, 7, 8], which has been first pointed out in the pure Kitaev model [6, 7]. When temperature is decreased, the spin fractionalization to the itinerant Majorana fermions and localized gauge fluxes is realized, leading to the double-peak structure. At each peak temperature, the entropy $\sim (R/2)\log 2$ is released, where $R$ is the gas constant [6, 7]. The double-peak structure is caused by the independent growth of the short-range and long-range spin correlation functions with decreasing temperature [7, 8]. Beyond the pure Kitaev model, it has been shown numerically that the specific heat shows a double-peak structure in the magnetically ordered state, when the system is located close to the KSL phase [8]. Thus, $\alpha$-$RuCl_3$ is probably located close to the KSL phase in spite of a phase transition to the zigzag magnetically ordered state at $T_N \sim 7$ K [9]. Several effective models for $\alpha$-$RuCl_3$ have been proposed so far. Although some effective models show qualitatively different features with each other, they have succeeded in explaining experimentally observed thermodynamic quantities and/or low-lying excitations. The adequate effective model is still controversial.

Inelastic neutron scattering (INS) experiments for single crystals provide us with ample information on excitation spectra via dynamical spin structure factors. Quite recently, INS experiments for single crystal $\alpha$-$RuCl_3$ were performed [5, 10]. The INS experiments have revealed following features, which offer clues to determine the adequate effective model.

- The magnetic Bragg peaks stand at the $M$ and $Y$ points (see Fig. 2 (d)) in the reciprocal space in $T < T_N$, reflecting the zigzag magnetic order [10]. This contributes to the

IOP Conf. Series: Journal of Physics: Conf. Series **969** (2018) 012123 doi:10.1088/1742-6596/969/1/012123

characteristic six-pointed star-shaped intensity centered at the Γ point in constant-energy cuts in low temperatures [5, 10].

- Around the Γ point, two peaks of the scattering intensity centered at $\omega \sim 3$ meV and $\sim 6$ meV appear in $T < T_{\rm N}$ [10]. The scattering intensity extends to $\omega \sim 15$ meV [10].
- In $T > T_{\rm N}$, the scattering intensity around the Γ point has a broad maximum centered at $\omega \sim 6$ meV [10]. The Γ-point intensity survives up to $T \sim 120$ K [10].

In this study, we calculate the dynamical spin structure factor (DSF) of three effective models [11, 12, 13] with an exact numerical diagonalization method. These effective models are obtained from *ab-initio* calculations. We call these three *ab-initio* models MODELs 1-3. We also calculate the temperature dependence of the specific heat by employing thermal pure quantum states [14, 15, 16]. By comparing the experimental results with our numerical results, we discuss characteristics of the three *ab-initio* models.

## 2. Model and methods

We consider a generalized Kitaev-Heisenberg model on a honeycomb lattice to describe MODELs 1-3 [11, 12, 13]. The Hamiltonian is written by

$$\mathcal{H} = \sum_{p} \sum_{\Gamma_p} \sum_{\langle ij \rangle \in \Gamma_p} \sum_{\mu,\nu=x,y,z} S_i^{\mu} \hat{\mathcal{J}}_{\Gamma_p}^{\mu\nu} S_j^{\nu}, \tag{1}$$

where $\hat{\mathcal{J}}_{\Gamma_p}^{\mu\nu}$ is a $3 \times 3$ matrix expressing the exchange coupling between $p$-th neighboring $i$ and $j$ sites on the bond $\Gamma_p$. For instance, the matrix elements of the nearest-neighbor pairs on a $Z$ bond (see Fig. 1 (a)) are written as

$$\begin{pmatrix} J_{\rm 1st}{}^{z} & \Gamma'_{\rm 1st}{}^{z} & \Gamma'_{\rm 1st}{}^{z} \\ \Gamma'_{\rm 1st}{}^{z} & J_{\rm 1st}{}^{z} & \Gamma_{\rm 1st}{}^{z} \\ \Gamma'_{\rm 1st}{}^{z} & \Gamma_{\rm 1st}{}^{z} & J_{\rm 1st}{}^{z} + K_{\rm 1st}{}^{z} \end{pmatrix},$$

where $J_p^{\mu}$ and $K_p^{\mu}$ denote coupling constants of the Heisenberg-type and the Kitaev-type interactions, respectively. We summarize the parameters of the three *ab-initio* models in table 1.

**Table 1.** Coupling constants of three *ab-initio* models. $J_p^{\mu}$ denotes coupling constants of Heisenberg-type interaction, while $K_p^{\mu}$ denotes those of Kitaev-type interaction. $\Gamma_p^{\mu}$ and $\Gamma_p^{\prime\mu}$ denote off-diagonal interactions. Positive (negative) values correspond to antiferromagnetic (ferromagnetic) interactions. Energy unit is meV.

| | $J_{\rm 1st}^{x/y}$ | $J_{\rm 1st}^{z}$ | $K_{\rm 1st}^{x/y}$ | $K_{\rm 1st}^{z}$ | $\Gamma_{\rm 1st}^{x/y}$ | $\Gamma_{\rm 1st}^{z}$ | $\Gamma_{\rm 1st}^{\prime x/y}$ | $\Gamma_{\rm 1st}^{\prime z}$ | $J_{\rm 2nd}$ | $J_{\rm 3rd}$ |
|---|---|---|---|---|---|---|---|---|---|---|
| MODEL 1 [11] | -1.55 | -1.49 | -6.47 | -6.71 | 5.24 | 5.28 | -1.08 | -0.69 | 0 | 0 |
| MODEL 2 [12] | -1.7 | -1.7 | -6.7 | -6.7 | 6.6 | 6.6 | -0.9 | -0.9 | 0 | 2.7 |
| MODEL 3 [13] | 1.2 | 1.2 | -5.6 | -5.6 | -1.2 | -1.2 | -0.7 | -0.7 | 0.25 | 0.25 |

In the three models we calculate the DSF, which is defined as

$$S^{\mu\nu}(\boldsymbol{Q},\omega;T) \equiv -\frac{1}{Z(T)} \sum_{n,m} \lim_{\epsilon\to+0} \frac{1}{\pi} {\rm Im} \frac{\langle \phi_n | \hat{S}_{\boldsymbol{Q}}^{\mu\dagger} | \phi_m \rangle \langle \phi_m | \hat{S}_{\boldsymbol{Q}}^{\nu} | \phi_n \rangle}{\omega + E_n + i\epsilon - E_m} e^{-(E_m - E_n)/k_B T}, \quad \mu,\nu = x,y,z, \tag{2}$$

where $Z(T)$ is a partition function, $\phi_n$ denotes an eigenstate of $\mathcal{H}$ with the eigenvalue $E_n$, and $\hat{S}_{\boldsymbol{Q}}^{\nu} = (1/N) \sum_{\boldsymbol{r}} S^{\nu} \exp(-i\boldsymbol{Q} \cdot \boldsymbol{r})$ with $N$ being the number of sites. Since we need all eigenstates and eigenvalues to calculate the DFSs at finite temperatures, the system size is limit to a small cluster. In this study, we calculate the DSF at finite temperatures for the $N$=8 clusters shown in Fig. 1 (b). At $T = 0$, the expression (2) is reduced to $S^{\mu\nu}(\boldsymbol{Q},\omega;0) \equiv -(1/\pi)\lim_{\epsilon\to+0}\mathrm{Im}\langle\phi_0|\left[\hat{S}^{\mu\dagger}_{\boldsymbol{Q}}\hat{S}^{\nu}_{\boldsymbol{Q}}/\left(\omega+E_0+i\epsilon-\mathcal{H}\right)\right]|\phi_0\rangle$, where $\phi_0$ is the ground state of $\mathcal{H}$ with the energy $E_0$. $\phi_0$ and $E_0$ are calculated by the Lanczos method, and then $S^{\mu\nu}(\boldsymbol{Q},\omega;0)$ is obtained by a continued fraction expansion [17]. We calculate the DSF at zero temperature for the $N$=24 clusters shown in Fig. 1 (c). In a similar manner to the INS experiments [5, 10], we evaluate the sum of the diagonal elements and set $S(\boldsymbol{Q},\omega;T) \equiv \sum_{\mu=x,y,z} S^{\mu\mu}(\boldsymbol{Q},\omega;T)$.

The specific heat is calculated by employing thermal pure quantum states [15, 16] for the $N = 24$ clusters. The construction of thermal pure quantum states is summarized in Ref. [8].

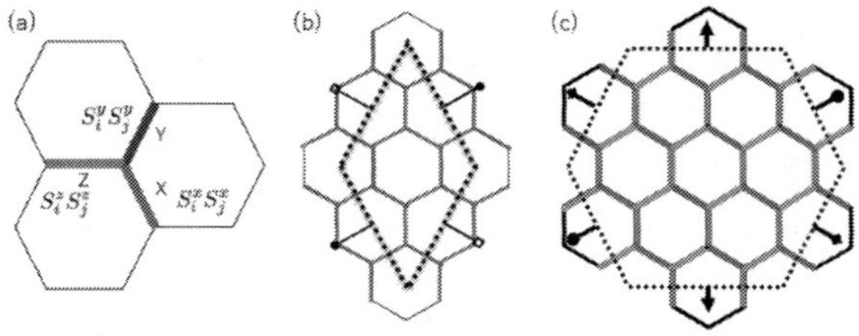


**Figure 1.** (a) Kitaev-type anisotropic interactions on a honeycomb lattice. (b) $N$=8 cluster and (c) $N$=24 cluster. Periodic boundary conditions are applied on dotted lines with same symbols.

## 3. Numerical results and discussions

### *3.1. Dynamical spin structure factors at $T = 0$*

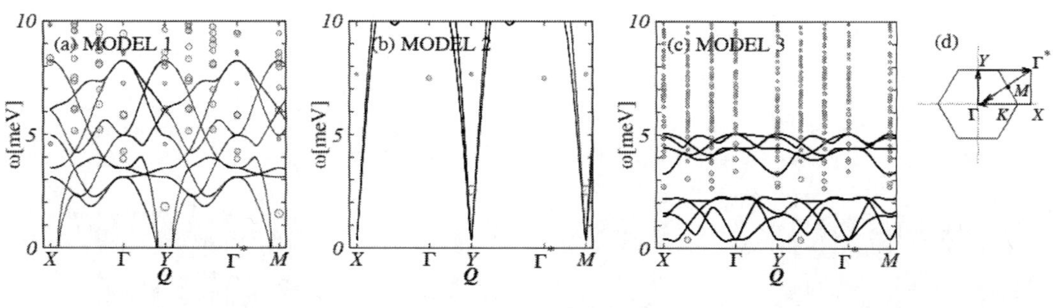


**Figure 2.** DSF $S(\boldsymbol{Q},\omega;0)$ for $N = 24$ cluster. Area of circle represents $\log S(\boldsymbol{Q},\omega;0)$. Solid curves are dispersion curves calculated from linearized spin-wave theory. In the calculations with linearized spin-wave theory, we start from the Ising-like zigzag state. Note that the ground state of the spherical model sometimes happens to yield a different ordered state from that obtained by the numerical exact diagonalization. Horizontal axis in (a)–(c) runs along arrows shown in Fig. 2 (d).

We first show the DSF at $T = 0$ in Fig. 2. The area of circle represents $\log S(\boldsymbol{Q},\omega;0)$. The solid curves are dispersion curves calculated from the linearized spin-wave theory by assuming

IOP Conf. Series: Journal of Physics: Conf. Series **969** (2018) 012123 doi:10.1088/1742-6596/969/1/012123

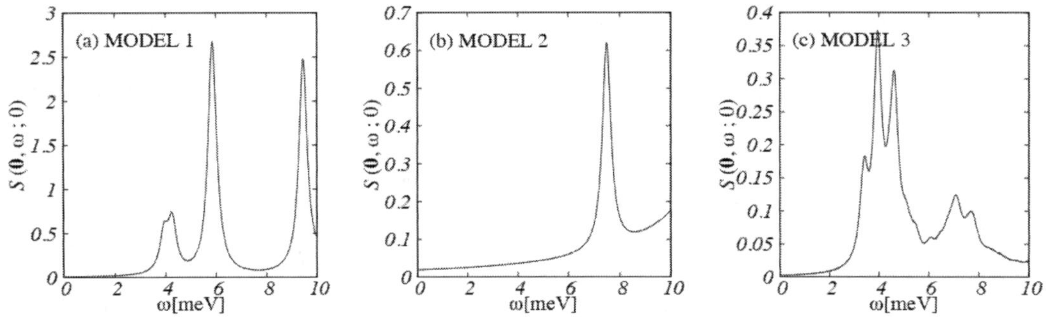


**Figure 3.** DSF $S(\mathbf{0},\omega;0)$ for $N = 24$ cluster. Half width of Lorentzian is fixed to be 0.2 meV.

the Ising-like zigzag state. We find that in MODELs 1-3 the linearlized spin-wave theory fails in explaining the DSFs. In MODELs 1 and 2, the largest intensities appear at the $M$ and $Y$ points at $\omega \sim 2$ meV, which stem from the zigzag magnetic order in agreement with the experiments [5, 10]. On the other hand, in MODEL 3, the largest intensities appear neither at the $M$ nor $Y$ points, but appear at the $K$ points at $\omega \sim 0.3$ meV. This result disagrees with the experiments. In MODEL 3, the Heisenberg-type long-range interactions $J_2$ and $J_3$ were introduced so as to reproduce the magnetization curves in the zigzag ordered state [13]. Another fitting of $J_2$ and $J_3$ that can induce strong intensities at the $M$ and $Y$ points may be effective.

In Fig. 3, we show the DSF at the Γ point, $S(\mathbf{0},\omega;0)$. In MODEL 1, lower two peaks appear at $\omega \sim 4$ meV and $\sim 6$ meV, which seem to describe the experimental results at the Γ point in $T < T_{\mathrm{N}}$. In MODEL 2, only a single peak appears at $\omega \sim 7.5$ meV in this energy scale, which deviates from the experiments. In MODEL 3, it is difficult to see the notable lower two peaks in $S(\mathbf{0},\omega;0)$. There are multipeaks in 3 meV $< \omega <$ 9 meV, which we consider to form an excitation continuum in the thermodynamic limit.

### *3.2. Temperature dependence of* $S(\mathbf{0},\omega;T)$

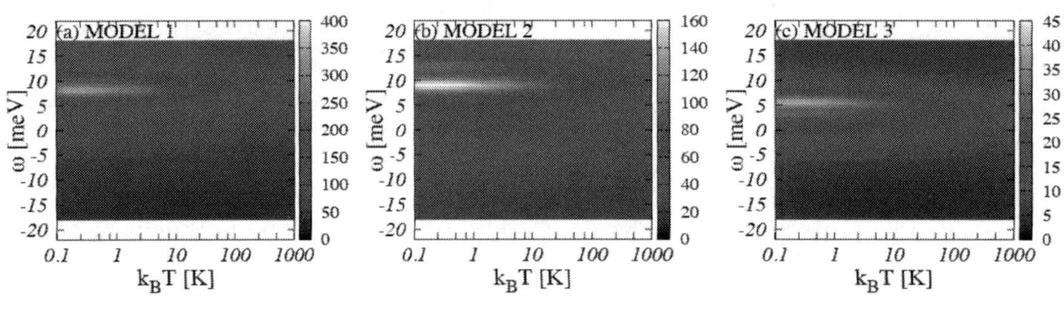


**Figure 4.** $S(\mathbf{0},\omega;T)$ at finite temperatures for the $N = 8$ cluster.

We show the temperature dependences of the DSFs at the Γ point for the $N = 8$ clusters in Fig. 4. In this calculation, it is difficult to get a good resolution with respect to $\omega$ because of the small-cluster calculations and the broad half width of used Lorenzians. Thus, in MODEL 1, the lower two peaks does not appear in $S(\mathbf{0},\omega;T)$, but a broad maximum appears centered at $\omega \sim 7$ meV. In MODEL 2, $S(\mathbf{0},\omega;T)$ has a broad maximum centered at $\omega \sim 8$ meV, which follows the peak in $S(\mathbf{0},\omega;0)$ shown in Fig. 3 (b). In MODEL 3, the peak intensity of $S(\mathbf{0},\omega;T)$ is smaller by approximately one order than that in MODEL 1. This is because, in MODEL 3, only the excitation continuum appears at the Γ point.

### *3.3. Temperature dependence of specific heat*

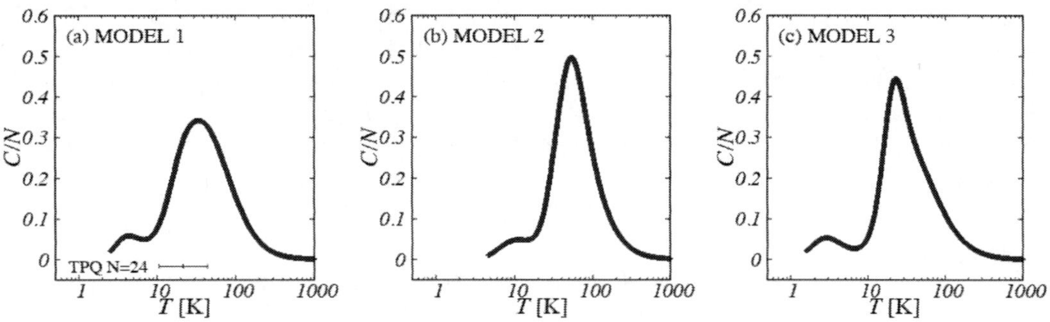


**Figure 5.** Temperature dependences of the specific heat $C(T)$ for the $N = 24$ cluster by employing thermal pure quantum states (TPQ) [15, 16]. All results are drawn with error bars that are smaller than symbol size.

We show the specific heat $C(T)$ for the $N = 24$ clusters obtained by using the thermal pure quantum states in Fig. 5. The double-peak structure appears in MODELs 1-3. We evaluate the temperature ratio of the low-temperature peak ($T_\mathrm{l}$) to the high-temperature one ($T_\mathrm{h}$). We obtain $T_\mathrm{l}/T_\mathrm{h} \sim 0.12$ in MODEL 1, $T_\mathrm{l}/T_\mathrm{h} \sim 0.20$ in MODEL 2, and $T_\mathrm{l}/T_\mathrm{h} \sim 0.12$ in MODEL 3. According to the criterion that has been elucidated in Ref. [8], MODELs 1 and 3 are located closer to the KSL phase than MODEL 2. Note that the ratios obtained here are larger than the experimentally observed ratios of $\alpha$-$RuCl_3$: $T_\mathrm{l}/T_\mathrm{h} \sim 0.09$ [3, 4] and $\sim 0.065$ [5].

In MODELs 1-3, $T_\mathrm{h}$'s are evaluated to be $T_\mathrm{h} \sim 30$ K, $\sim 50$ K, and $\sim 20$ K, respectively. We compare these values with the temperature regions where the large intensities of $S(\mathbf{0},\omega;T)$ appear in Fig. 4. We find that the large intensity of $S(\mathbf{0},\omega;T)$ appears in $T < T_\mathrm{h}$ in each MODEL. It has been shown that the nearest-neighbor spin correlation begins to develop at $T \sim T_\mathrm{h}$, as $T$ is lowered [7, 8]. The development of the nearest-neighbor spin correlation is a hallmark of the KSL [7, 8, 18]. Thus the large intensity of the DSF around the Γ point is associated with the Kitaev interaction, as has been pointed out in Refs. [5, 10].

The values of $T_\mathrm{h}$ in MODELs 1-3 shown above are lower than the experimental results: $T_\mathrm{h} \sim 85$ K [3, 4] and $\sim 100$ K [5]. From the numerical calculations, we find that $T_\mathrm{h}$ is mainly controlled by the dominant Kitaev interaction. These findings probably mean that the nearest-neighbor ferromagnetic Kitaev interaction is underestimated in these *ab-initio* calculations for explaining $T_\mathrm{h}$ observed in the experiments.

## 4. Summary

We have calculated the DSF and $C(T)$ in the three *ab-initio* models. In Refs. [11, 13] for MODELs 1 and 3, other parameter sets have been presented as well. We have also calculated the DSFs and $C(T)$ by using such parameter sets, and have employed the results that describe the experiments best in each model. According to our calculations, the INS experimental features are qualitatively described by MODEL 1, while the double-peak structure of $C(T)$ is qualitatively reproduced by MODELs 1-3. By comparing the numerical results for MODELs 1-3, we consider that the growths of the long-range spin correlation in MODELs 1 and 3 are suppressed as compared with that in MODEL 2 because of the absence of and the weak third neighbor Heisenberg interaction, respectively. This suppression makes $T_{\rm l}/T_{\rm h}$ of MODELs 1 and 3 small. In the DSF of MODEL 3, the excitation continuum appears in the low-energy region and it is difficult to expect the appearance of the localized mode observed in the experiments at $\omega \sim 3$ meV. The difference between MODEL 1 and MODEL 3 comes from the signs of the nearest-neighbor Heisenberg terms and the Γ terms. We consider that the signs of both terms affect the stability of the low-lying excitation mode. Thus, we conclude that MODEL 1 well describes the qualitative experimental features of $\alpha$-$RuCl_3$ that we have turned our attention to. We believe that it is effective to construct a phenomenological model for $\alpha$-$RuCl_3$ that can reproduce the experimental features of the DSF and $C(T)$ not only qualitatively but also quantitatively.

**Acknowledgments**
We thank Prof. Imada, Prof. Kawashima, and Prof. Yamaji. This work was supported by the Creation of new functional Devices and high-performance Materials to Support next-generation Industries (cdmsi), Challenge of Basic Science - Exploring Extremes through Multi-Scale Simulations (CBSM2), and KAKENHI (Grants No. 15K05232 and 16K17751) from MEXT Japan. TS thanks the computational resources of the K computer provided by the RIKEN Advanced Institute for Computational Science through the HPCI System Research project (hp160201, hp160262, hp170262, and hp170263). We also thank numerical resources in the ISSP Supercomputer Center at University of Tokyo and the Research Center for Nano-micro Structure Science and Engineering at University of Hyogo.

## References

[1] Kitaev A 2003 *Ann. Phys.* **303** 2
[2] Plumb K W, Clancy J P, Sandilands L J, Shankar V V, Hu Y F, Burch K S, Kee H-Y and Kim Y-J 2014 *Phys. Rev.* B **90** 041112(R)
[3] Kubota Y, Tanaka H, Ono T, Narumi Y and Kindo K 2015 *Phys. Rev.* B **91** 094422
[4] In Ref. [3], the double peak in the temperature dependence of the specific heat was not argued. However, we consider that the double peak appears in Fig. 9 in Ref. [3]. Using that result, we evaluate the temperature of the higher peak to be ~85K.
[5] Do S-H, Park S-Y, Yoshitake J, Nasu J, Motome Y, Kwon Y-S, Adroja D T, Voneshen D-J, Kim K, Jang T-H, Park J-H, Choi K-Y and Ji S *preprint* arXiv:1703. 01081
[6] Nasu J, Udagawa M and Motome Y 2014 *Phys. Rev. Lett.* **113** 197205
[7] Nasu J, Udagawa M and Motome Y 2015 *Phys. Rev.* B **92** 115122
[8] Yamaji Y, Suzuki T, Yamada T, Suga S, Kawashima N and Imada M 2016 *Phys. Rev.* B **93** 174425
[9] Sears J A, Songvilay M, Plumb K W, Clancy J P, Qiu Y, Zhao Y, Parshall D and Kim Y-J 2015 *Phys. Rev.* B **91** 144420
[10] Banerjee A, Yan J, Knolle J, Bridges C A, Stone M B, Lumsden M D, Mandrus D G, Tennant D A, Moessner R and Nagler S E *preprint* arXiv:1609.00103
[11] Kim H S and Kee H Y 2016 *Phys. Rev.* B **93** 155143
[12] Winter S M, Li Y, Jeschke H O and Valenti R 2016 *Phys. Rev.* B **93** 214431
[13] Yadav R, Bogdanov N A, Katukuri V M, Nishimoto S, van den Brink J and Hozoi L 2016 *Sci. Rep.* **6** 37925
[14] Hams A and Readt H D 2000 *Phys. Rev.* E **62** 4365
[15] Shimizu A and Sugiura S 2012 *Phys. Rev. Lett.* **108** 240401
[16] Shimizu A and Sugiura S 2013 *Phys. Rev. Lett.* **111** 010401
[17] Gagliano E R and Balseiro C A 1987 *Phys. Rev. Lett.* **59** 2999
[18] Baskaran G, Mandal S, and Shankar R 2007 *Phys. Rev. Lett.* **98** 247201

# Extremely anisotropic suppression of huge enhancement of electrical resistivity by magnetic field in $\alpha$-$R_2S_3$ ($R$ = Sm, Dy)

**S Ebisu, T Era, Q Guo and M Miyazaki**

Division of Applied Sciences, Muroran Inst. of Tech., 27-1 Mizumoto-cho, Muroran 050-8585, Hokkaido, Japan

ebisu@mmm.muroran-it.ac.jp

**Abstract**. The electrical resistivity $\rho$ and magnetization $M$ of $\alpha$-$R_2S_3$ ($R$ = Sm, Dy) single crystals have been measured in various magnetic fields lying in the *ac*-plane of an orthorhombic crystal structure. The measurements have been carried out by rotating the single crystals around the *b*-axis (being set parallel to a horizontal direction), corresponding to the longitudinal direction of needle-shaped single crystals, in vertical magnetic fields. The electrical resistivity under no magnetic field shows huge enhancement in a narrow temperature range around the successive magnetic transition temperatures, as reported previously. Such enhancement is suppressed rapidly by applying a magnetic field, which has been clearly seen as diminishing of the $\rho(T)$ peak. The magnetic field along the easy magnetization axis; $H_{easy}$, moves the $\rho(T)$ peak toward higher temperature with diminishing it, while the magnetic field along the hard magnetization axis; $H_{hard}$, moves it to lower temperature. The suppression effect is extremely anisotropic for the orientation of the magnetic field lying in the *ac*-plane. It has been concluded that $H_{easy}$ suppresses the $\rho$ enhancement most strongly. The suppression effect under $H_{easy}$ is more than 200 times larger at least in some cases than that under $H_{hard}$.

## 1. Introduction

Rare earth sesquisulfides $\alpha$-$R_2S_3$ ($R$ = Sm and Dy) have attracted much attention because of their novel physical properties brought about by successive magnetic transitions [1-6]. These compounds and other rare earth compounds $\alpha$-$R'_2S_3$ ($R'$ = La, Ce, Pr, Nd, Gd and Tb) have the $\alpha$-$Gd_2S_3$ type crystal structure with the *Pnma* symmetry shown in figure 1 [7-9]. There are two crystallographically inequivalent $R$ sites labeled with $R1$ and $R2$ in this structure. We have found fascinating physical properties in both series compounds $\alpha$-$R_2S_3$ and $\alpha$-$R'_2S_3$ (except $R'$ = La, Ce). Briefly speaking only about the magnetic properties in $\alpha$-$R'_2S_3$, they are as follows. The $\alpha$-$Gd_2S_3$ shows only one antiferromagnetic (AFM) transition at $T_N$ = 9.9 K [10-13], while the $\alpha$-$Tb_2S_3$ shows successive AFM transitions at $T_{N1}$ = 12.5 K and $T_{N2}$ = 3.5 K [3, 14, 15]. The $\alpha$-$Nd_2S_3$ also shows successive transitions at $T_N$ = 2.75 K and the temperature around 0.65 K suggesting magnetic ordering [16]. The $\alpha$-$Pr_2S_3$ shows an AFM transition at $T_N$ = 2.95 K, as well as Van Vleck paramagnetic contribution. The ground term of one Pr site atoms has been considered to be singlet and Pr atoms on the other site seem to demonstrate magnetic ordering partially at $T_N$. Remaining degrees of freedom have been inferred to disappear by ordering below 1 K from the specific heat data [16].

The physical properties of $\alpha$-$R_2S_3$ ($R$ = Sm and Dy) are outstandingly attractive. The $\alpha$-$Dy_2S_3$ also exhibits successive AFM transitions at $T_{N1}$ = 11.4 and $T_{N2}$ = 6.4 K [2]. As for $\alpha$-$Sm_2S_3$, weak ferromagnetic (WFM) properties were observed around 4 K in the temperature dependence of the magnetic susceptibility [4]. Recent our study on the specific heat of the $\alpha$-$Sm_2S_3$ has clarified that successive WFM transitions occur at $T_{C1}$ = 3.8 K and $T_{C2}$ = 1.9 K, which will be reported elsewhere. We have also found extremely broad hysteresis in magnetization process of $\alpha$-$Dy_2S_3$ after cooling in the magnetic field of 18 T [5]. Furthermore, easy magnetization axes in the *ac*-plane of both the $\alpha$-$Sm_2S_3$ and $\alpha$-$Dy_2S_3$ are surprisingly exchangeable between ***a*** and ***c*** after some processes of changing the magnetic field orientation and temperature, which will be reported by S Ebisu *et al.* elsewhere. The most fascinating property of $\alpha$-$R_2S_3$ is a huge enhancement and restoration of the electrical resistivity for the single crystals seen in a narrow temperature range of a few Kelvin just above $T_{N2}$ for $\alpha$-$Dy_2S_3$ and $T_{C1}$ for $\alpha$-$Sm_2S_3$ [1, 4]. The magnitude of the enhancement is more than 10 times for $\alpha$-$Dy_2S_3$ and 100 times for $\alpha$-$Sm_2S_3$, although the magnitude depends on single crystal sample (i.e. an amount of some lattice imperfections). This phenomenon forms a sharp peak in the temperature dependence of the electrical resistivity $\rho(T)$. Another small peak is also observed just below $T_{N2}$ or $T_{C1}$ in the $\rho(T)$ for each compound. Furthermore, the huge enhancement of the electrical resistivity measured along the *b*-axis corresponding longitudinal direction of the needle-shaped single crystal is suppressed rapidly by applying the magnetic field perpendicular to the *b*-axis. The needle-like shape of these single crystals suggests that these compounds have strong uniaxial anisotropies along the *b*-axis for physical properties. Moreover, considering the crystal structure of the $\alpha$-$R_2S_3$, anisotropies in the *ac*-plane do not seem to be negligible. In this paper, we will report anisotropic suppression effect of huge enhancement of the electrical resistivity in the $\alpha$-$Sm_2S_3$ and $\alpha$-$Dy_2S_3$ single crystals by the magnetic field lying in the *ac*-plane.

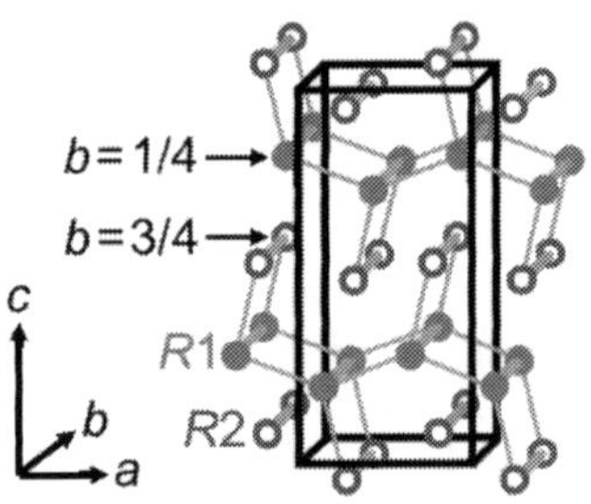


**Figure 1.** Crystal structure of $\alpha$-$R_2S_3$, in which only the $R$ atoms are shown for clarity. The $R$ atoms are located on two crystallographically inequivalent $R$ sites labelled with $R1$ (closed circles) and $R2$ (open circles). Thin lines are drawn between neighbor atoms on the same *ac*-plane ($b$ = 1/4 or 3/4). The orthorhombic unit cell is also shown as a box.

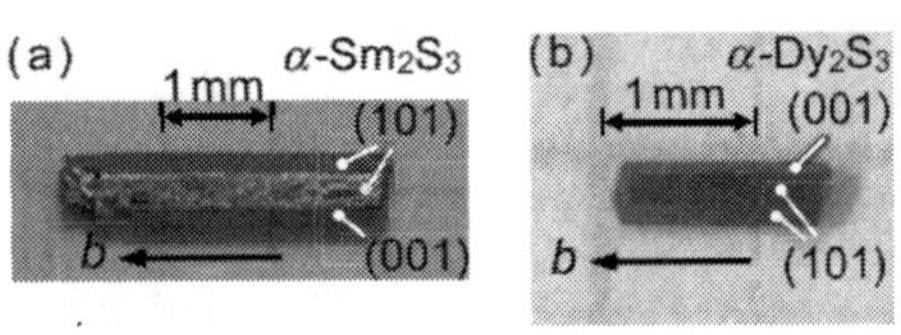


**Figure 2.** Single crystal of (a) $\alpha$-$Sm_2S_3$ and (b) $\alpha$-$Dy_2S_3$. The longitudinal direction of the needle-like shape corresponds to the *b*-axis of the orthorhombic crystal system. They have a hexagonal cross-section and one opposed pair of six side faces is the (001) crystal plane.

## 2. Experimental

The single crystals of $\alpha$-$Sm_2S_3$ and $\alpha$-$Dy_2S_3$ were grown by a chemical transport reaction method with iodine for carrier gas [10]. Identification of the powder sample and determination of the crystal orientation were performed by X-ray (Cu $K_\alpha$ radiation) diffraction measurements. The magnetic measurements were carried out using an *rf*-SQUID magnetometer with Magnetic Property Measurement System (MPMS, Quantum Design, USA). The needle-shaped single crystals having lengths of a few millimeters along the *b*-axis, distances of a few sub-millimeters across hexagonal cross section in the *ac*-plane and masses of several milligrams were mounted to the sample rotator with GE 7031 varnish. The sample rotator enables rotation of a sample around a horizontal axis in a vertical magnetic field. The electrical resistivity was measured by 2-probe DC method using Physical Properties Measurement System (PPMS, Quantum Design, USA). The very same single crystal sample that was used for magnetic measurements was used also for the electrical measurements. Also in the electrical measurements, the sample rotator was used to rotate the sample around a horizontal axis in a vertical magnetic field.

## 3. Results and discussion

Figure 3 shows the temperature and magnetic field dependence of the electrical resistivity $\rho$ for $\alpha$-$Sm_2S_3$ [(a) and (b)] and $\alpha$-$Dy_2S_3$ [(c) and (d)] in the low temperature range below 10 K. Here, the vertical axes denoting $\rho$ are in logarithmic scale and the horizontal axes denoting the temperature $T$ are in linear scale. The electrical resistivity was measured under various magnetic fields. The applied magnetic field to the $\alpha$-$Dy_2S_3$ single crystal is along to the $a$-axis for the case in figure 3(c) and along to the $c$-axis for the case in figure 3(d). As for $\alpha$-$Sm_2S_3$, the single crystal was mounted to the stage of the sample rotator in the state of inclining from the crystal axis about 10 degrees. Therefore, the direction of the magnetic field for the case in figure 3(a) is along $\boldsymbol{c}'$, which is inclined 10 degrees from the crystal axis $\boldsymbol{c}$. In the same way, the magnetic-field direction for the case in figure 3(b) is along $\boldsymbol{a}'$ being inclined 10 degrees from the crystal axis $\boldsymbol{a}$. In figure 3(a) and 3(b), the $\rho(T)$ curves under no magnetic field ($\mu_0 H$ = 0 T) are understandably the same. As seen in the $\rho(T)$ curve for 0 T, the value of $\rho$(5.7 K) is more than 500 times larger than the value of $\rho$(8.0 K). The value larger than $10^5$ $\Omega$ cm below $T$ = 5.7 K is out of the measureable range for normal PPMS, thus the experimental curve has

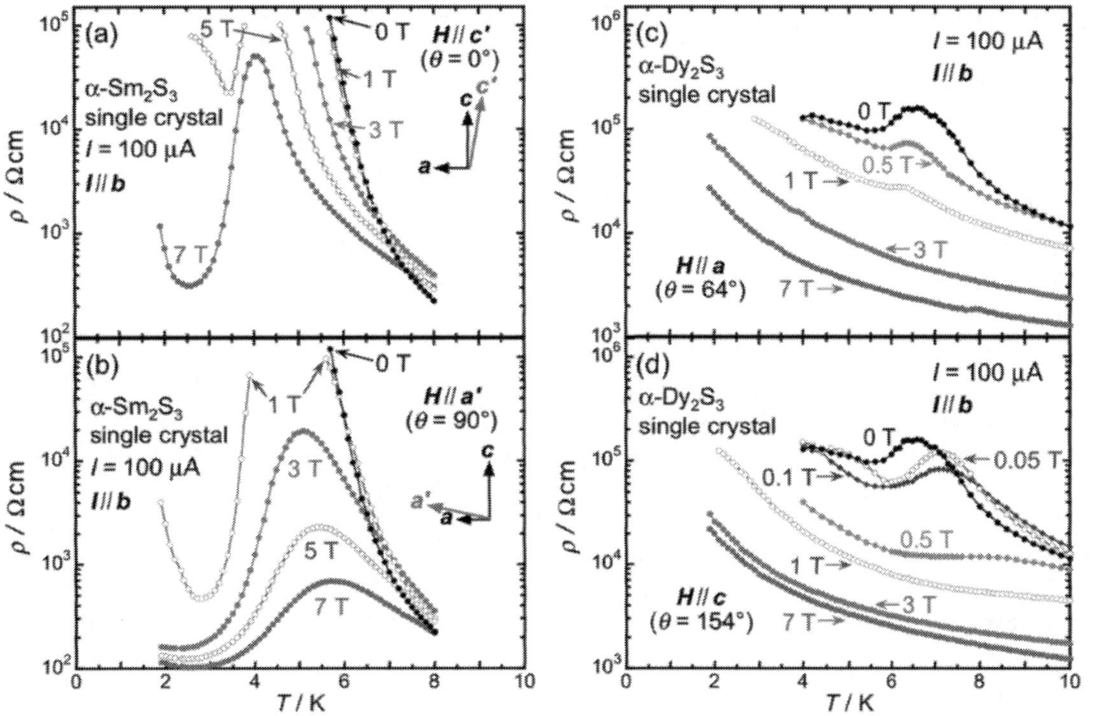


**Figure 3.** Temperature and magnetic field dependence of the electrical resistivity $\rho$ for $\alpha$-$Sm_2S_3$ [(a) and (b)] and $\alpha$-$Dy_2S_3$ [(c) and (d)] single crystals in the low temperature range below 10 K. The vertical axes are in logarithmic scale. In the upper figures (a) and (c), the magnetic field is applied; (c) along, or (a) in the near direction of, the hard magnetization axis. While, in the lower figures (b) and (d), the magnetic field is applied; (d) along, or (b) in the near direction of, the easy magnetization axis.

ended at this point. According to the previous data taken by another measurement system, this curve forms a sharp peak with a maximum at $T = 7.0$ K [4]. Similar behavior is seen in the curves measured under the magnetic field of 1 and 3 T along ***c'*** in figure 3(a). The curve for $\mu_0 H = 5$ T also resembles the curves mentioned above. However, in cooling process, the $\rho$ gets immeasurable high values once then it returns to measurable low values. We can see whole figure of the $\rho(T)$ curve for the case of $\mu_0 H = 7$ T. Thus, the electrical resistivity shows huge enhancement in a narrow temperature range and such enhancement is suppressed by a magnetic field. If we assume all the curves in figure 3(a) have peaks, the maximum point in each peak seems to shift to lower temperature with decreasing temperature. The easy magnetization axis in the *ac*-plane of used single crystal for $\alpha$-$Sm_2S_3$ is the *a*-axis and the hard magnetization axis is the *c*-axis, which will be shown later. Therefore, applying magnetic field in the near direction of the hard magnetization axis moves the enhanced $\rho(T)$ peak to lower temperature. Also in the case of applying magnetic field along ***a'***, the enhanced $\rho(T)$ peak shrinks with increasing magnetic field as shown in figure 3(b). Peaks of three curves for $\mu_0 H = 3$, 5 and 7 T are confirmed. Moreover, it is clearly seen that the peak shifts to higher temperature with increasing magnetic field in the near direction of the easy magnetization axis. Turning now to the case for $\alpha$-$Dy_2S_3$, although the $\rho(T)$ peak for the single crystal sample used for the present measurements is blunter than the previous result [1], the value of the ratio $\rho(6.8\ K)/\rho(10.0\ K)$ for $\mu_0 H = 0$ T is confirmed to be more than 10 in figure 3(c) and 3(d). The $\rho(T)$ peak is suppressed more rapidly by the magnetic field than the case of $\alpha$-$Sm_2S_3$, especially it is marked for ***H***//***c*** as shown in figure 3(d). The $\alpha$-$Dy_2S_3$ single crystal used has the easy magnetization axis along ***c*** and the hard axis along ***a*** in the *ac*-plane as shown in detail later. Therefore, also in the case of $\alpha$-$Dy_2S_3$, the magnetic field along the hard magnetization axis; $H_{hard}$, shifts the $\rho(T)$ peak to lower temperature, while the magnetic field along the easy axis; $H_{easy}$, shifts it to higher temperature. The higher magnetic field, the larger shift we observe.

Next, let us see the magnetic-field-direction effect to the suppression of the huge $\rho$ enhancement. We measured $\rho$ with rotating the sample stage in fixed vertical magnetic fields at some temperatures. The rotation angle $\theta$ was changed from 0 to 360°. It is adequate to show the results as an angle dependent electrical conductivity $\sigma(\theta)$ instead of $\rho(\theta)$ because the angle dependence of $\rho$ is not clear at the $\theta$ range in which the $\rho$ enhancement is strongly suppressed, i.e. the $\rho$ value in such $\theta$ range is too small relatively.

Figure 4 shows the sample-rotation-angle dependence of the electrical conductivity $\sigma$ (=$1/\rho$) and the magnetization $M$ for $\alpha$-$Sm_2S_3$ [(a) and (b)] and $\alpha$-$Dy_2S_3$ [(c) and (d)]. The $\sigma(\theta)$ has the same tendency regardless of the applied magnetic field strength, as well as the $M(\theta)$. Thus, we show the clearest figures for a certain magnetic field in each case. Figure 4(a), the $\sigma(\theta)$ of $\alpha$-$Sm_2S_3$ under the magnetic field of $\mu_0 H = 7$ T, demonstrates highly anisotropic features. The $\sigma(\theta)$ shows two humps feature having 2-fold symmetry. As for the 2-fold symmetry, it is natural because the space group *Pnma* of this crystal structure has 2-fold screw axis vertical to the *ac* plane. It has two maxima at $\theta$ = 100, 280° corresponding to ***H***//***a*** and two minima at $\theta$ = 10, 190° corresponding to ***H***//***c***. Although such feature is seen at the temperatures both below and above $T_{C1} = 3.8$ K, the anisotropy of $\sigma(\theta)$ below $T_{C1}$ is larger. Below $T_{C1}$, the value of $\sigma(10°)$ is not specified because the $\rho(10°)$ value is too large to measure. However, at least, the $\sigma(100°)$ value is more than 200 times larger than the $\sigma(10°)$ value at $T = 3.0$ K. Let us now reorganize as follows. The $\rho(T)$ shows a huge enhancement and restoration just above $T_{C1}$ and such enhancement is suppressed largely by a magnetic field as reported previously [4]. This suppression effect is highly anisotropic, and the suppression for the case of ***H***//***a*** is the strongest. Turning now to the $M(\theta)$ result for $\alpha$-$Sm_2S_3$ as shown for the case of $\mu_0 H = 5$ T in figure 4(b). We displace the $M(\theta)$ figure to the left side in order to arrange the magnetic field directions between the $M(\theta)$ and $\sigma(\theta)$. The $M(\theta)$ shows similar angle dependence to that of the $\sigma(\theta)$. In this temperature range, the easy magnetization axis in the *ac*-plane for this single crystal sample is

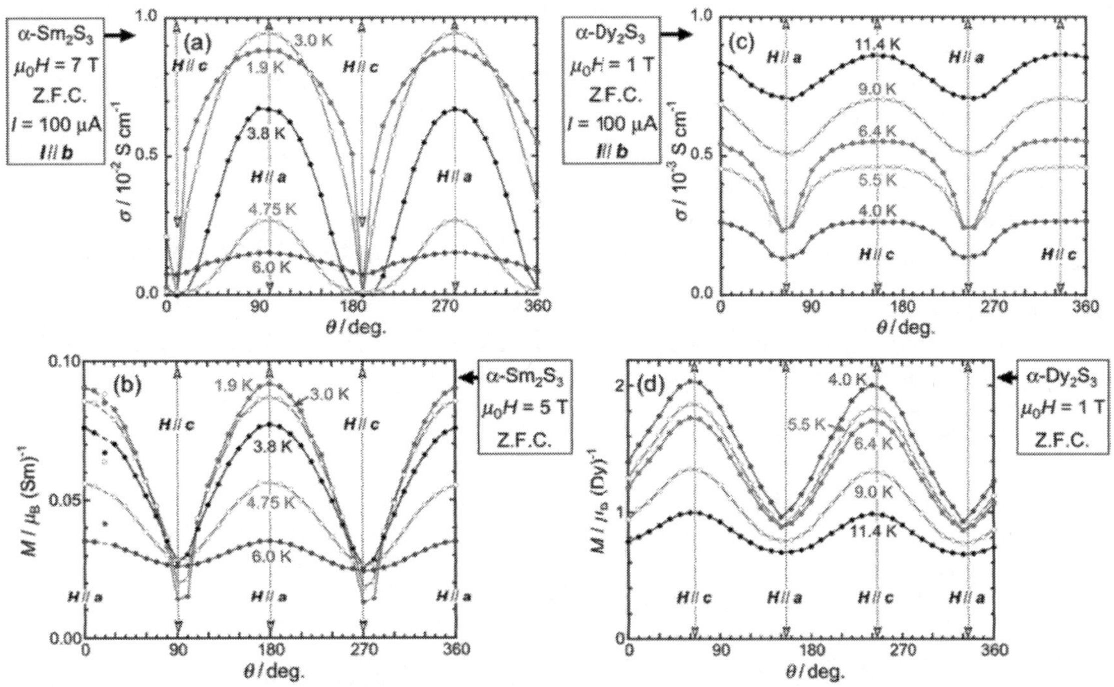


**Figure 4.** Sample-rotation-angle dependence of the electrical conductivity $\sigma$ and the magnetization $M$ for $\alpha$-$Sm_2S_3$ [(a) and (b)] and $\alpha$-$Dy_2S_3$ [(c) and (d)] single crystals at some temperatures around successive transition temperatures. The upper and lower figures are shifted relatively in order to arrange the magnetic field directions. The magnetic field directions are also shown in the figures.

the $a$-axis. Therefore, the WFM ordering of Sm moments affects the suppression of the electrical resistivity enhancement strongly. In other words, disordering state of Sm moments just before ordering is significantly related to the huge enhancement of the electrical resistivity.

Let us see the results for $\alpha$-$Dy_2S_3$. Since the wide surface parallel to the crystal plane (101) was agglutinated to each sample stage of the sample rotators, the angles of $64°+n{\cdot}90°$ ($n$ is an integer) in figure 4(c) and 4(d) correspond to the situation of the magnetic field parallel to the crystal axes. Figure 4(c) is the $\sigma(\theta)$ under $\mu_0H = 1$ T for $\alpha$-$Dy_2S_3$. Two humps feature is similar to that of $\alpha$-$Sm_2S_3$ qualitatively although the anisotropy is relatively small. Two maxima at $\theta$ = 154, 334° and two minima at $\theta$ = 64, 244° are observed. The former angles correspond to $\boldsymbol{H}//\boldsymbol{c}$ and the latter correspond to $\boldsymbol{H}//\boldsymbol{a}$. Common features are seen at the temperatures below $T_{N1}$ = 11.4 K. However, the anisotropy of $\sigma(\theta)$ below $T_{N2}$ = 6.4 K is relatively larger. The anisotropy for $T = T_{N2}$ is the largest among the measurements as $\rho(154°)/\rho(64°) = 2.3$. The suppression of the $\rho$ enhancement is the strongest when the magnetic field is applied along $\boldsymbol{c}$. The $\theta$ range, in which the strong suppression is observed, gets wider around 134° (and 314°) when the temperature goes below $T_{N2}$. Such behavior is confirmed also in the $\sigma(\theta)$ curve for $T = T_{C2}$ = 1.9 K of $\alpha$-$Sm_2S_3$ in figure 4(a). The $\boldsymbol{c}$ for $\alpha$-$Dy_2S_3$ single crystal used in the present measurement is the easy magnetization axis as shown in figure 4(d). Hence, the AFM ordering of Dy moments affects the suppression of the electrical resistivity enhancement strongly also in the case of $\alpha$-$Dy_2S_3$.

## 4. Summary

We have investigated the suppression effect of the anomalous electrical resistivity enhancement in the $\alpha$-$Sm_2S_3$ and $\alpha$-$Dy_2S_3$ single crystals by applying magnetic field. The enhancement of electrical resistivity is suppressed by magnetic fields regardless of their directions; however, the strength of the suppression significantly depends on the magnetic field direction in the *ac*-plane. It has been found that the magnetic field $H_{easy}$ affects the $\rho(T)$ peak to move toward higher temperature, while $H_{hard}$ moves the $\rho(T)$ peak toward lower temperature. We have examined the sample-rotation-angle dependent $\sigma(\theta)$ and $M(\theta)$ in fixed vertical magnetic fields in detail, and found highly anisotropic behavior. It has been clarified the enhanced $\rho$ is strongly suppressed by the magnetic field $H_{easy}$. We have concluded that the magnetic ordering, WFM ordering in $\alpha$-$Sm_2S_3$ and AFM ordering in $\alpha$-$Dy_2S_3$, of *R* moments suppresses the enhancement of $\rho$. In other words, disordering state of *R* moments just before ordering is significantly related to the huge enhancement of the electrical resistivity.

**Acknowledgments**
We thank Mr. S Nakano and Mr. T Otomo for their helpful assistance in measurements and Dr. Y Shibayama for useful discussions.

**References**

[1] Ebisu S, Narumi M and Nagata S 2006 *J. Phys. Soc. Jpn.* **75** 085002
[2] Ebisu S, Narumi M, Gorai M and Nagata S 2007 *J. Magn. Magn. Mater.* **310** 1741-3
[3] Ebisu S, Koyama K, Omote H and Nagata S 2009 *J. Phys.: Conf. Ser.* **150** 042027
[4] Ebisu S, Omote H and Nagata S 2010 *J. Phys.: Conf. Ser.* **200** 092005
[5] Ebisu S, Koyama K, Horikoshi T, Kokita M and Nagata S 2012 *J. Phys.: Conf. Ser.* **400** 032010
[6] Ebisu S, Ushiki Y and Takahashi S 2013 *J. Korean Phys. Soc.* **63** 571-4
[7] Prewitt C T and Sleight A W 1968 *Inorganic Chemistry* **7** 1090-3
[8] Picon M, Domange L, Flahaut J, Guittard M and Patrie M 1960 *Bull. Soc. Chim. France* **2** 221-8
[9] Sleight A W and Prewitt C T, 1968 *Inorg. Chem.* **7** 2282-8
[10] Ebisu S, Iijima Y, Iwasa T and Nagata S 2004 *J. Phys. Chem. Solids* **65** 1113-20
[11] Kikkawa A, Katsumata K, Ebisu S and Nagata S 2004 *J. Phys. Soc. Jpn.* **73** 2955-8
[12] Matsuda M, Kikkawa A, Katsumata K, Ebisu S and Nagata S 2005 *J. Phys. Soc. Jpn.* **74** 1412-5
[13] Katsumata K, Kikkawa A, Tanaka Y, Shimomura S, Ebisu S and Nagata S 2005 *J. Phys. Soc. Jpn.* **74** 1598-601
[14] Ebisu S, Gorai M, Maekawa K and Nagata S 2006 *AIP Conf. Proc.* **850** 1237-8
[15] Matsuda M, Kakurai K, Ebisu S and Nagata S 2006 *J. Phys. Soc. Jpn.* **75** 074710
[16] Ebisu S, Nagata T, Fuji K and Shibayama Y 2014 *J. Phys.: Conf. Ser.* **568** 042003

IOP Conf. Series: Journal of Physics: Conf. Series **969** (2018) 012125 doi:10.1088/1742-6596/969/1/012125

# $^{133}$Cs-NMR study on aligned powder of competing spin chain compound $Cs_2Cu_2Mo_3O_{12}$

**A Yagi[1] , K Matsui[1] T Goto[1], M Hase[2] and T Sasaki[3]**

[1] Sophia University, Physics Division, Tokyo, 102-8554, Japan
[2] National Institute for Materials Science (NIMS), Tsukuba, 305-0047, Japan
[3] Institute for Material Research (IMR), Tohoku University, Sendai, 980-8577, Japan

gotoo-t@sophia.ac.jp

**Abstract.**

$S$ = 1/2 competing spin chain compound $Cs_2Cu_2Mo_3O_{12}$ has two dominant exchange interactions of the nearest neighbouring ferromagnetic $J_1$= 93 K and the second nearest neighbouring antiferromagnetic $J_2$= +33 K, and is expected to show the nematic Tomonaga-Luttinger liquid (TLL) state under high magnetic field region. The recent theoretical study by Sato *et al.* has shown that in the nematic TLL state, the spin fluctuations are expected to be highly anisotropic, that is, its transverse component is suppressed. Our previous NMR study on the present system showed that the dominant contribution to nuclear spin relaxation comes from the longitudinal component. In order to conclude that the transverse component of spin fluctuations is suppressed, the knowledge of hyperfine coupling is indispensable. This article is solely devoted to investigate the hyperfine coupling of $^{133}$Cs-NMR site to prove that the anisotropic part of hyperfine coupling, which connects the nuclear spin relaxation with the transverse spin fluctuations is considerably large to be $A_{\mathrm{an}} = +770\ \mathrm{Oe}/\mu_{\mathrm{B}}$.

## 1. Introduction

The ground state of the competing spin chain, described by the $J_1$- $J_2$ model with two dominant exchange interactions of the nearest-neighboring ferromagnetic $J_1$ and the second nearest-neighboring antiferromagnetic $J_2$ has recently attracted much interest, because the existence of nematic Tomonaga-Luttinger state has been proposed theoretically[1-7]. The detection of this state, however requires to probe the four-spin correlation functions with high accuracy[1,2], so that it has been considered to be rather difficult. We address this problem with an assist of recent theoretical proposal by Sato *et al.*[2], who have proved that the nematic and ordinary TLL shows a distinct difference in the temperature dependence of NMR-$T_1$ in low temperatures. That is, while $T_1^{-1}$ for the latter is contributed both from the longitudinal and transverse spin correlations as $T^{2K-1}$ and $T^{1/2K-1}$, respectively, where $K$ is the Luttinger parameter[8-12], $T_1^{-1}$ for the former is contributed only from the longitudinal spin correlations. This clearly states that an experimental evidence for the nematic TLL state can be obtained, if only one observes that $T_1^{-1}$ decreases with lowering temperature under moderately high magnetic field, where $K$ becomes larger than 1/2 [8-12].

We have recently reported that $T_1^{-1}$ for the model compound $Cs_2Cu_2Mo_3O_{12}$ (Fig. 1) exhibits

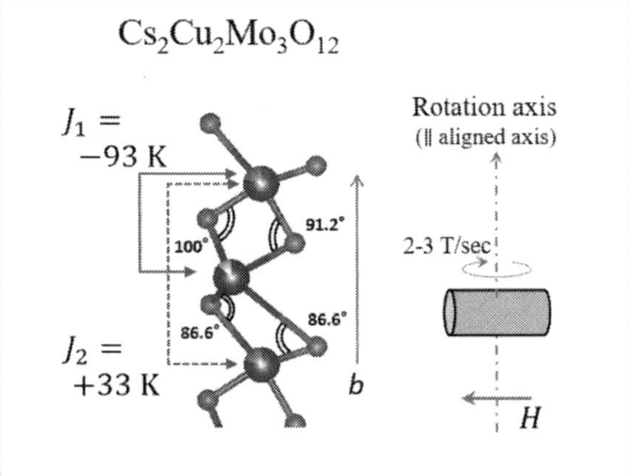


**Figure 1.** Schematic crystal structure of $Cs_2Cu_2Mo_3O_{12}$ (left) and the method of powder alignment for materials with easy plane anisotropy (right).

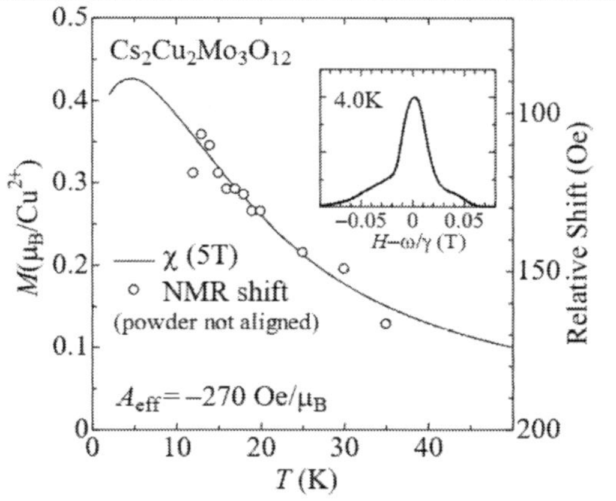


**Figure 2.** $K$-$\chi$ plot for the non-aligned powder sample[15]. The relative shift position is defined as the peak top of spectrum (inset).

nematic-TLL-like temperature dependence described above[13,14]. However, in order to conclude that the transverse spin fluctuations are suppressed, one must prove that there exists a finite hyperfine coupling. Generally, the nuclear spin relaxation is driven by the transverse component of field fluctuations at the nuclear site. The nuclear spin relaxation proceeds via both types of spin fluctuations, transverse and longitudinal, only when the hyperfine coupling bears a finite anisotropic part; in that case, the direction of fluctuating field changes from the electronic site to the nuclear site. In our previous NMR report[15,16], we have studied on the randomly oriented powder (Fig. 2) to obtain the hyperfine coupling constant $A_{\rm eff} = -270\,{\rm Oe}/\mu_{\rm B}$, which, however, is an averaged value for powder. For the discussion on spin nematic TLL state, it is indispensable to determine both the isotropic and anisotropic part of hyperfine coupling tensor. This article is solely devoted to its determination; we have utilized the uniaxially aligned powder sample and measured the paramagnetic NMR shift in the wide temperature region to determine separately the two components in the hyperfine coupling tensor of Cs-site.

Before describing the experimental method, we summarize the basic properties so far reported for the target compound $Cs_2Cu_2Mo_3O_{12}$ [13,14]. This quasi-one dimensional system is isomorphic to $Rb_2Cu_2Mo_3O_{12}$ [14], and is described by the $J_1$- $J_2$ model with $J_1 = -93$ K and $J_2 = +33$ K (Fig. 1) [13]. It is gapless under zero field, and has the saturation field $H_S = 9.15$ T [16]. Due to a possible tiny inter-chain interaction, there is a long range magnetic order at $T_N = 1.85$ K [15-19]. The nematic TLL state is expected to exist in a finite temperature region above $T_N$.

## 2. Experimental

$Cs_2Cu_2Mo_3O_{12}$ polycrystalline sample was synthesized by the conventional solid state reaction method [13,14]. In order to determine the anisotropic NMR shift, a uniaxially-aligned powder sample was prepared by the conventional method utilizing the paramagnetic anisotropy. The powder mixed with Stycast1266, the epoxy resin, was put into a Teflon tube, and was cured in the 6T horizontal magnetic field in the room temperature, while rotating slowly (2-3 turns/sec) the tube within a plane parallel with the field (Fig. 1). By this rotational method we have successfully aligned the paramagnetic powder with the easy-plane-type anisotropy. We note here that the powder did not align well when it was cured under field without rotation.

NMR measurements for $^{133}$Cs ($I = 7/2$) spectra were performed under field region 5 - 6 T and in the temperature range between 5 and 45 K. The uniform susceptibility was measured by SQUID magnetometer (Quantum Design Co. Ltd.) under the field of 5 T.

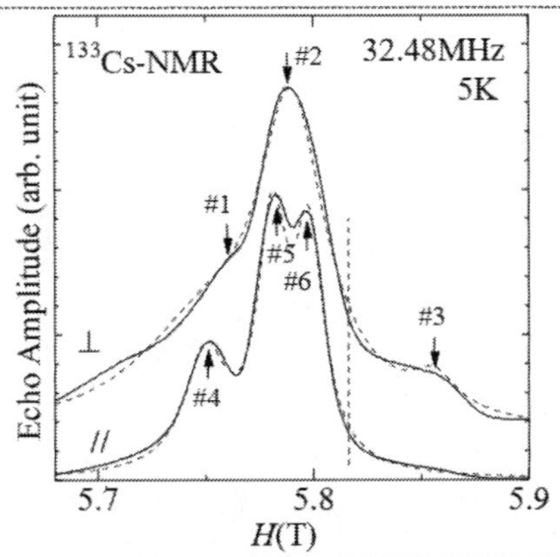


**Figure 3.** The comparison of two spectra for aligned powder sample with the field applied along or perpendicular to the aligned axis. Peak positions are indicated by arrows. The vertical line shows the zero shift position. Dashed curves show the sum of three Lorentzian peaks determined by curve fitting.

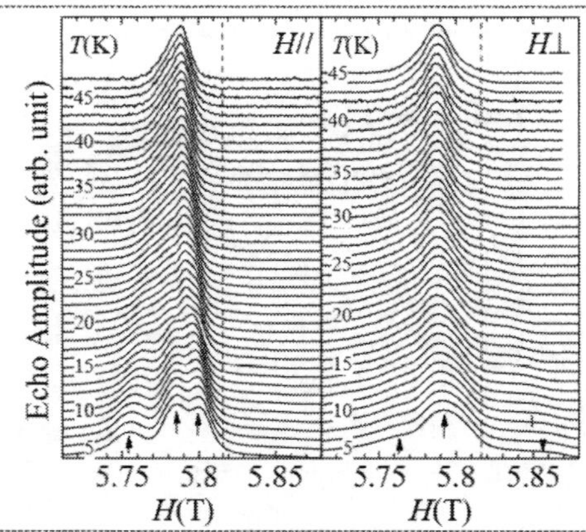


**Figure 4.** Temperature variation of the spectra of the aligned powder under field applied along parallel (*H//*) and perpendicular (*H*⊥) of the aligned axis. Arrows indicate the position of each peak.

## 3. Results and Discussion

Figure 3 shows the spectra measured under the two field directions, parallel ($H_{\parallel}$) or perpendicular ($H_{\perp}$) with the aligned axis. Note the distinct difference in the two profiles. Each two spectra in Fig. 3 are well described by a sum of the three Lorentzian peaks, denoted as #1 - #3 for $H_{\perp}$ and #4 - #6 for $H_{\parallel}$, respectively. The Lorentzian shape form comes simply from an inhomogeneity in the magnetic shift and also in the small quadrupolar splitting, which is negligibly small and hidden in each peak. These three peaks are considered to correspond to the three Cs sites in unit cell, *4d*, *4e* and *8f*. The symmetric shape of each peak assures that the powders are aligned well, and also that the hyperfine tensor of each Cs site possesses a uniaxial asymmetry. We determined the peak position by curve fitting and traced the temperature dependence of peak shift in the paramagnetic region up to 45 K as shown in Fig. 4 and 5. Their temperature dependences differ from one another, including the sign of gradient. That is, with decreasing temperatures, the shifts of #1 and #4 increase, while #3 and #6 decrease. This change in the sign clearly demonstrates the existence of anisotropic part within the hyperfine coupling. We proceed to the quantitative arguments in the following.

The NMR shift with the uniaxial-anisotropy is generally described as $K_{\parallel} = K_{\rm iso} + 2K_{\rm an}$ and $K_{\perp} = K_{\rm iso} - K_{\rm an}$, where $K_{\parallel}$ and $K_{\perp}$ are the magnetic shift under the field applied parallel or perpendicular to the principle axis of anisotropy, and $K_{\rm iso}$ and $K_{\rm an}$ are, as in usual denotation, the isotropic and anisotropic part of NMR shift tensor, respectively. So, for the case where $K_{\rm iso}$ is not very large compared to $K_{\rm an}$, $K_{\parallel}$ and $K_{\perp}$ are expected to have the temperature variation with ratio of nearly $-2$. This is what one observes for #3 and #4, indicating that the magnetic shifts of $\Delta H_3^{\perp}$ and $\Delta H_4^{\parallel}$ correspond to $K_{\parallel}$ and $K_{\perp}$, respectively. We also note here that the aligned axis is perpendicular to the principle axis of anisotropy. We have assumed that #3 and #4 belong to the identical Cs site and tentatively calculated the $K_{\rm iso}$ and $K_{\rm an}$, which are shown in Fig. 6. By scaling them with the uniform susceptibility, we further obtained the hyperfine coupling constants of isotropic and anisotropic parts as $A_{\rm iso} = -270\,{\rm Oe}/\mu_{\rm B}$ and $A_{\rm an} = -770\,{\rm Oe}/\mu_{\rm B}$. Note that the former just coincides with that obtained for the non-aligned powder sample. And also, the smallness of $A_{\rm iso}$ compared with $A_{\rm an}$ validates our assumption above.

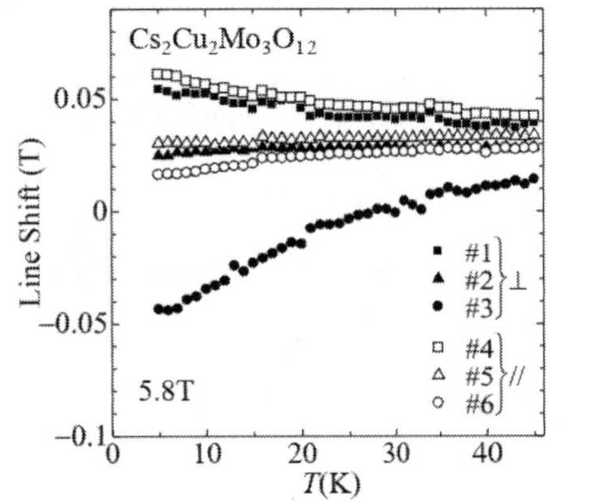


**Figure 5.** The temperature dependence of shift for each peak observed in Fig. 4.

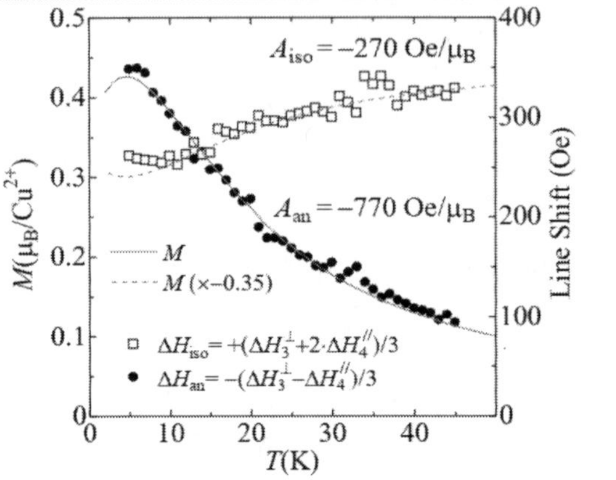


**Figure 6.** The temperature dependence of the isotropic and anisotropic shift deduced from the two-field configurations. Curves show uniform magnetization, scaled with the shifts.

Here, this considerably large $A_{an}$ indicate that the transverse component of spin fluctuations must be considerably suppressed, compared with the longitudinal one, to explain our previous observation of NMR-$T_1$, and hence, we can conclude the likeliness of the system being in the nematic TLL state[1,2,19].

Finally, we refer on the hyperfine coupling of the other two Cs-sites, and the origin of anisotropic hyperfine coupling. First, as for the other two sites, the value of hyperfine coupling constants must be much smaller than those for the Cs site assigned to #3 and #4, because they have smaller temperature dependence of the shift, as can be seen in Fig. 5. This will bring the considerable distribution in the relaxation rate among the three Cs sites. In fact, the nuclear spin relaxation for the non-aligned powder sample becomes quite inhomogeneous, and described by the stretched relaxation curve [15,17]. Next, as for the origin of $A_{an}$, a preliminary calculation has shown that the classical dipole-dipole interaction does not account for it, and that one must take into account the exchange interaction via $p$ orbitals of Cs ion. Detailed calculation of dipole tensor at Cs sites for the complete site assignment including a determination of principle axes of NMR shift tensor and also measurements of $T_1^{-1}$ for each site are now in the progress.

**Acknowledgement**
This work was partly supported by JSPS KAKENHI Grant Number 15K05148.

**References**

[1] T. Hikihara et al.: Phys. Rev. B78 (2008) 144404.
[2] M. Sato et al. Phys. Rev. B83 (2011) 064405, B79 (2009) 060406(R)
[3] S. F. Solodovnikov et al.: J. Struc. Chem. 38 (1997) 765.
[4] M. E. Zhitomirsky et al.: Euro Phys. Lett. 92 (2010) 37001.
[5] M. Hagiwara et al.: J. Phys. Conf. Ser. 320 (2011) 012049.
[6] H. Tsunetsugu et al.: J. Phys. Soc. Jap. 75 (2006) 083701.
[7] Lars Kecke et al.: Phys. Rev. B76, 060407R (2007).
[8] T. Sakai, J. Phys. Soc. Jpn. 64 (1995) 251.
[9] T. Suzuki *et al.*: Phys. Rev. B 74 (2006) 172410.
[10] T. Suzuki *et al.*: Phys. Rev. B70 (2004) 054419.

IOP Conf. Series: Journal of Physics: Conf. Series **969** (2018) 012125 doi:10.1088/1742-6596/969/1/012125

[11] M. Klanšjek et al.: Phys. Rev. Lett. 101 (2008) 13727.
[12] K. Ninios et al.: Phys. Rev. Lett. 108 (2012) 097201.
[13] M. Hase *et al.,* J. Appl. Phys. 97 (2005) 10B303.
[14] M. Hase *et al.,* Phys. Rev. B70 (2004) 104426.
[15] Y. Hoshino *et al.,* JPS Conf. Proc. 3 (2014) 014012.
[16] A. Fujimura *et al.,* IEEE Transactions on Magnetics 99 (2016) 1100503.
[17] T. Goto *et al.,* J. Phys. Conf. Ser. 828 (2017) 012017.
[18] A. Yagi *et al.,* J. Phys. Conf. Ser. 828 (2017) 012016.
[19] M. Sato et al.: Phys. Rev. Lett. 110, 077206 (2013).

# Spontaneous Magnetization of the Spin-1/2 Heisenberg Antiferromagnet on the Triangular Lattice with a Distortion

**Alisa Shimada[1], Tôru Sakai[2,3], Hiroki Nakano[2], Kazuyoshi Yoshimura[1]**

[1]Kyoto University, Graduate School of Science, Kyoto, Japan, [2]University of Hyogo, Graduate School of Material Science, Kamigori, Japan, [3]QST SPring-8, Sayo, Japan

E-mail: sakai@spring8.or.jp

**Abstract.** The spin-1/2 triangular-lattice Heisenberg antiferromagnet with a $\sqrt{3} \times \sqrt{3}$-type distortion is studied by the numerical-diagonalization method. We examine this model between the two cases, one is the undistorted triangular lattice and the other is the model on the honeycomb lattice with isolated spins. When the distortion is controlled, we find a nontrivial region where the ground states shows spontaneous magnetization; the magnitude increases gradually as the distortion is larger and is smaller than one third of the saturated magnetization.

## 1. Introduction

The triangular-lattice antiferromagnet is a typical system that has attracted much attention from its behavior owing to frustrations. In particular, such attention has increased explosively since Anderson pointed out that the quantum Heisenberg antiferromagnet on the triangular lattice is a candidate system of the spin liquid state[1]. Nowadays, it is widely believed that the so-called 120-degree structure is realized as a spin structure of its ground state. Extensive studies concerning this system have been carried out from various aspects[2, 3, 4, 5, 6, 7, 8, 9, 10, 11, 12, 13, 14, 15, 16, 17, 18, 19, 20, 21]. Some experiments reported that antiferromagnets have the triangular-lattice structure[22, 23, 24, 25]. Some of them, however, include a distortion in the triangular lattice[24, 25]; the type of this distortion is called the $\sqrt{3} \times \sqrt{3}$-type one. Unfortunately, theoretical understanding concerning effects due to the $\sqrt{3} \times \sqrt{3}$-type distortion in the quantum triangular-lattice antiferromagnet is still insufficient.

Under the above mentioned circumstances, a quite recent study[26] reported the spin-1/2 triangular-lattice antiferromagnet with the $\sqrt{3} \times \sqrt{3}$-type distortion shows an interesting intermediate spin state in the ground state between the undistorted case and the case of the dice lattice[27]. The dice-lattice antiferromagnet is a nonfrustrated system having a ferrimagnetic ground state of the up-up-down structure based on the so-called Lieb-Mattis theorem[28]. The intermediate state shows nonzero spontaneous magnetization but its magnitude is smaller than one third which is realized in the dice-lattice case. Unfortunately, the range of such the intermediate state is different from the experimental cases mentioned above[24, 25]. Therefore, it is an urgent task to clarify properties of the quantum system in the range of the distortion corresponding to the experimental cases.

The purpose of the present study is to clarify effects of the $\sqrt{3} \times \sqrt{3}$-type distortion in the spin-1/2 Heisenberg antiferromagnet on the triangular lattice illustrated in Fig. 1 by means of a numerical-diagonalization method of the Lanczos algorithm. In particular, we examine the model between the two cases, one is the undistorted triangular lattice and the other is the model on the honeycomb lattice with isolated spins. Our numerical results suggest that the ground states in the region close to the two limiting cases do not show a spontaneous magnetization; however, there appear nonzero spontaneous magnetizations in an intermediate region. The spontaneous magnetization is smaller than one third of the saturated magnetization and increases gradually as the distortion is larger.

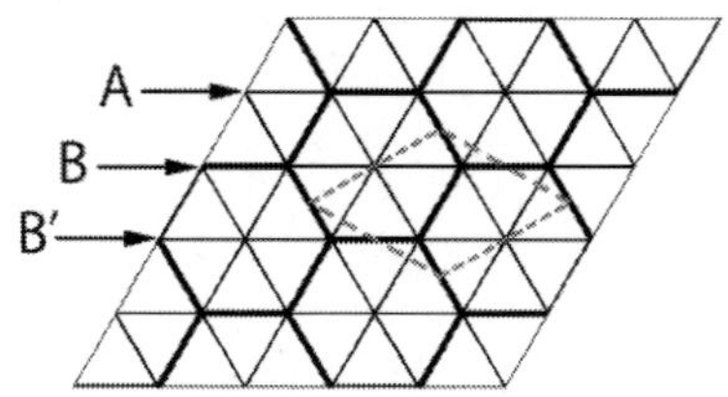


**Figure 1.** Lattice studied in the present paper. Its unit cell is shown by the red rhombus. The sublattices A, B, and B′ are also illustrated.

This paper is organized as follows. In the next section, the model Hamiltonian treated here and the methods used in this study are explained. The third section is devoted to the presentation and the discussion of our numerical results. In the final section, the summary is given.

## 2. Model Hamiltonian and methods

The Hamiltonian examined in this study is given by

$$\mathcal{H} = \sum_{i\in \mathrm{B}, j\in \mathrm{B}'} J_1 \boldsymbol{S}_i \cdot \boldsymbol{S}_j + \sum_{i\in \mathrm{A}, j\in \mathrm{B}} J_2 \boldsymbol{S}_i \cdot \boldsymbol{S}_j + \sum_{i\in \mathrm{A}, j\in \mathrm{B}'} J_2 \boldsymbol{S}_i \cdot \boldsymbol{S}_j. \quad (1)$$

Here, $\boldsymbol{S}_j$ represents the $S = 1/2$ spin operator at site $j$. Let us consider, in this study, the case of isotropic interaction in spin space. The site $j$ is assumed to be the vertices of the lattice depicted in Fig. 1. The number of spin sites is represented by $N_{\mathrm{s}}$. Note here that the vertices are divided into three sublattices A, B, and B′; each site $j_{\mathrm{A}}$ in the A sublattice is linked by six interaction bonds $J_2$ denoted by thin lines; each site $j_{\mathrm{B}}$ in the B or $j_{\mathrm{B}'}$ in the B′ sublattice is linked by three interaction bonds $J_2$ and three interaction bonds $J_1$, denoted by thick lines. The ratio of $J_2/J_1$ is denoted by $r$. All interactions are supposed to be antiferromagnetic, namely, $J_1 > 0$ and $J_2 > 0$. Energies are measured in units of $J_1$; hereafter, we set $J_1 = 1$. In particular, we examine the case of $J_2 \leq J_1$ in this study. Note that for $J_1 = J_2$, namely, $r = 1$, the present lattice is identical to the triangular lattice; the ground state of the system on this lattice is understood as the so-called 120-degree structure illustrated in Fig. 2(a). Therefore, this state clearly has no spontaneous magnetization. Just for $r = 0$, on the other hand, the network of the vertices becomes the honeycomb lattice accompanied by isolated spins. At zero temperature, the system shows degenerate ground states. One of the states is illustrated in Fig. 2(c). This state corresponds to the highest magnetization among the degenerate ground states. Note however that when infinitesimally small $J_2$ is switched on, it is a nontrivial issue whether or not the state illustrated in Fig. 2(c) becomes the lowest-energy state as a result of the perturbative effect of small $J_2$. Note also that, in the case of $J_2 \geq J_1$, a nontrivial intermediate phase was found between $r = 1$ for the triangular lattice and $r \to \infty$ for the dice lattice[26].

The finite-size clusters that we treat in the present study are depicted in Fig. 3(a)-(d). We examine the cases of $N_{\mathrm{s}} = 12$, 21, 27, and 36 under the periodic boundary condition. In all the cases, $N_{\mathrm{s}}/3$ is an integer; therefore, the number of spin sites in a sublattice is the same

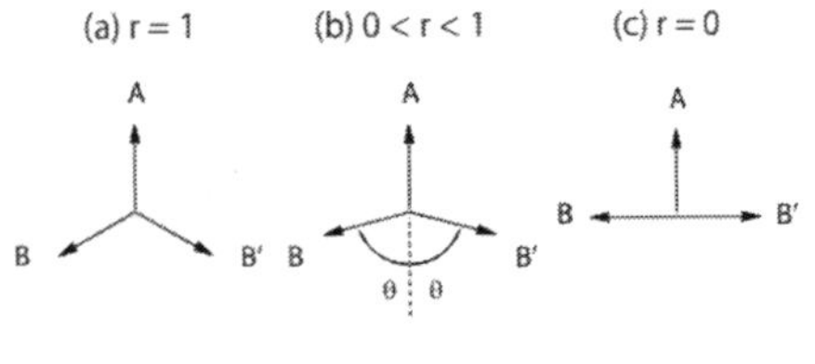


**Figure 2.** Possible spin states based on a classical picture for $r = 1$, $0 < r < 1$, and $r = 0$, respectively.

irrespective of sublattices. The clusters are rhombic and have an inner angle $\pi/3$; this shape allows us to capture two dimensionality well.

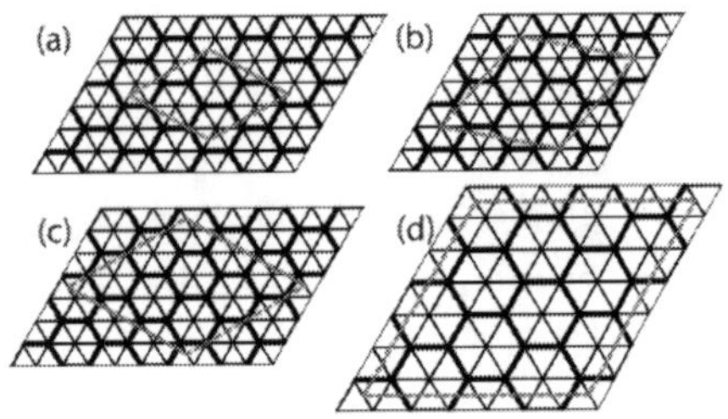


**Figure 3.** Finite-size clusters treated in the present study. Panels (a), (b), (c), and (d) denote cases for $N_s$ = 12, 21, 27, and 36, respectively.

We calculate the lowest energy of $\mathcal{H}$ in the subspace characterized by $\sum_j S_j^z = M$ by numerical diagonalizations based on the Lanczos algorithm and/or the Householder algorithm. The numerical-diagonalization calculations are unbiased against any approximations; thus we are able to obtain reliable information of the system. We denote the energy by $E(N_s, M)$, where $M$ takes an integer or a half odd integer up to the saturation value $M_{sat}$ ($= N_s/2$). We focus our attention on spontaneous magnetization $M_{spo}$; our calculations can determine $M_{spo}$ as the largest value of $M$ among the lowest-energy states. Note that for cases of odd $N_s$, the smallest $M_{spo}$ cannot rigorously vanish; the result of $M_{spo} = 1/2$ in the ground state indicates that the system shows no spontaneous magnetization. Part of the Lanczos diagonalizations were carried out using an MPI-parallelized code, which was originally developed in the study of Haldane gaps[29]. The usefulness of our program was confirmed in several large-scale parallelized calculations[30, 31, 32, 33].

## 3. Results and discussion

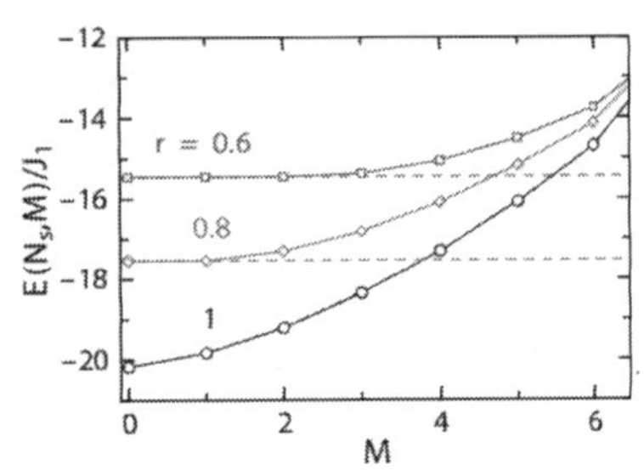


**Figure 4.** $M$-dependence of the ground-state energy for $N_s = 36$ when the distortion in the lattice is switched on. Black circles, red diamonds, and blue squares denote results for $r$ = 1, 0.8, and 0.6, respectively.

First, let us observe magnetization-dependence of the ground state energy which lets us recognize whether the spontaneous magnetization appears or not and which provides us with information about the magnitude of the spontaneous magnetization $M_{spo}$ if it appears; results for $N_s = 36$ are depicted in Fig. 4. For $r = 1$, the energy for $M = 0$ is lower than any other energies for $M > 0$, which indicates that the spontaneous magnetization does not appear in the ground state, namely, $M_{spo} = 0$. For $r = 0.8$ and 0.6, on the other hand, the ground states are degenerate; the degeneracy is illustrated by broken lines in Fig. 4. These degeneracies indicate that nonzero $M_{\mathrm{spo}}$ appears. The largest $M$ in the degenerated ground states corresponds to $M_{\mathrm{spo}}$; one finds $M_{\mathrm{spo}} = 1$ and $M_{\mathrm{spo}} = 2$ for $r = 0.8$ and $0.6$, respectively.

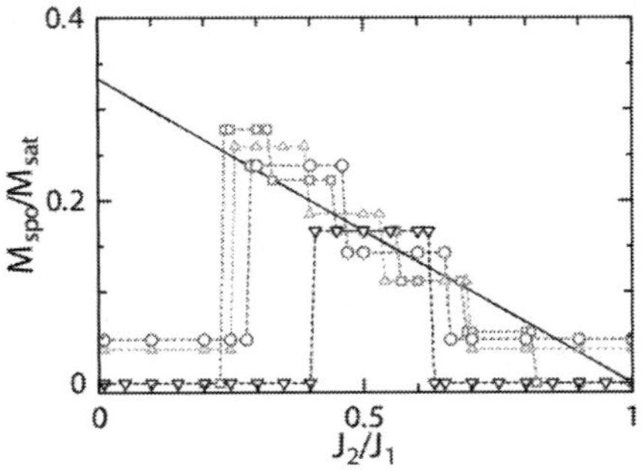


**Figure 5.** $r$-dependence of the spontaneous magnetization for various $N_{\mathrm{s}}$. Red squares, green triangles, blue circles, and black inversed triangles denote results for $N_{\mathrm{s}} = 36$, 27, 21, and 12, respectively. The line denotes the result from the classical picture illustrated in Fig. 2(b).

Next, we examine the change of the spontaneous magnetization when $r$ is controlled in $0 < r \leq 1$; results for various $N_{\mathrm{s}}$ are depicted in Fig. 5. For all the cases of $N_{\mathrm{s}}$=12, 21, 27, and 36, the region of nonzero $M_{\mathrm{spo}}$ certainly exists. Note that the nonzero $M_{\mathrm{spo}}$ gradually increases when $r$ is decreased. Note also that all possible values as $M_{\mathrm{spo}}$ are realized although the maximum of $M_{\mathrm{spo}}$ does not reach $(1/3)M_{\mathrm{sat}}$. The $r$-dependence of $M_{\mathrm{spo}}$ is compared with the result from the argument based on a classical picture illustrated in Fig. 2(b). In the region of $r$ with nonzero-$M_{\mathrm{spo}}$ up to $M_{\mathrm{spo}} = (1/3)M_{\mathrm{sat}} - 1$, brief dependence of the quantum case seems to agree with the classical line although step-like behavior owing to a finite-size effect exists. However, it is unclear within the present calculations whether the spin states in the quantum case are certainly understood well by the simple classical picture shown in Fig. 2(b) because such a example is actually known[26]. Intensive examination of the nonzero-$M_{\mathrm{spo}}$ spin states should be tackled in future studies. In the viewpoint of the appearance of states with continuously increasing $M_{\mathrm{spo}}$, similar behavior was observed in other various systems[34, 35, 36, 37, 38, 39, 40, 41]; these nontrivial states are considered to be so-called *non-Lieb-Mattis* ferrimagnetic ones. The relationship between the nonzero-$M_{\mathrm{spo}}$ states and the non-Lieb-Mattis ferrimagnetic states should be tackled in future studies. When $r$ is further decreased, one can noticeably observe that the spontaneous magnetization discontinuously disappears. A marked behavior is that the range of nonzero $M_{\mathrm{spo}}$ gradually widens as the system size is increased. With decreasing $r$, here, we define $r_{\mathrm{c2}}$ as the value of $r$ where $M_{\mathrm{spo}}$ changes from 0 or 1/2 to larger values and $r_{\mathrm{c1}}$ as the value of $r$ where $M_{\mathrm{spo}}$ decreases again to 0 or 1/2 from larger values. The system size dependences of $r_{\mathrm{c1}}$ and $r_{\mathrm{c2}}$ are depicted in Fig. 6. One finds that $r_{\mathrm{c2}}$ gradually increases with increasing $N_{\mathrm{s}}$ and that it seems to go to a value close to $r = 1$. On the other hand, $r_{\mathrm{c1}}$ gradually decreases with increasing $N_{\mathrm{s}}$, but it seems to converge to a nonzero value. It is unclear at present whether $r_{\mathrm{c1}}$ converges to zero or a nonzero value in the thermodynamic limit, which should be clarified in future studies. As a primary consequence of this study, the present calculations strongly suggest that the presence of the nonzero-$M_{\mathrm{spo}}$ region is evident.

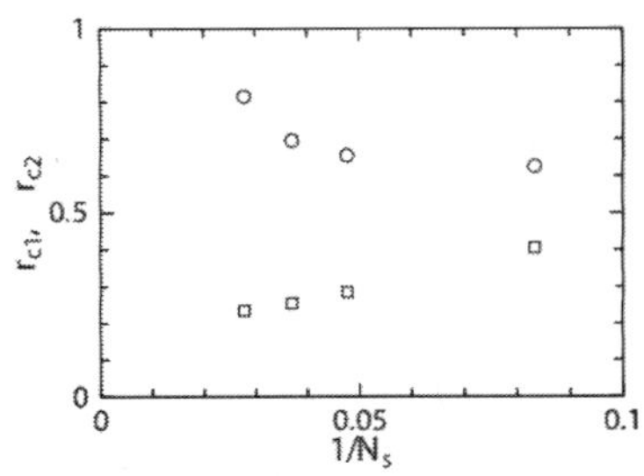


**Figure 6.** Size dependence of the boundaries of the nonzero-$M_{\mathrm{spo}}$ phase. Squares and circles denotes results for $r_{\mathrm{c1}}$ and $r_{\mathrm{c2}}$, respectively.

Next, let us discuss the behavior around $r = r_{c1}$ under the assumption that $r_{c1}$ converges to a nonzero value in the thermodynamic limit. We consider the $M$-dependence of the ground-state energy; our numerical results for $N_s = 36$ are depicted in Fig. 7. For $N_s = 36$, our calculations for $r = 0.24$ indicate $M_{spo} = 5$ and those for $r = 0.23$ indicate that $M_{spo}$ vanishes. For $r = 0.23$, however, the degeneracy from $M = 2$ to $M = 5$ still survives, which is illustrated by a broken line in Fig. 7. In our calculations for $r = 0.2$, on the other hand, the degeneracy appears from $M = 4$ to $M = 5$. One finds that the degeneracy gradually collapses from the side of smaller-$M$. The behavior around $r = r_{c1}$ is absolutely different from the behavior around $r = r_{c2}$ shown in Fig. 3. The spin state in the region of $0 < r < r_{c1}$ is unclear at present. To know the characteristics within numerical calculations, correlation functions for large systems are required. Another possible approach that is applicable to attack this issue may be a perturbation calculation from the case of $r = 0$. Such future studies would give additional information about this state.

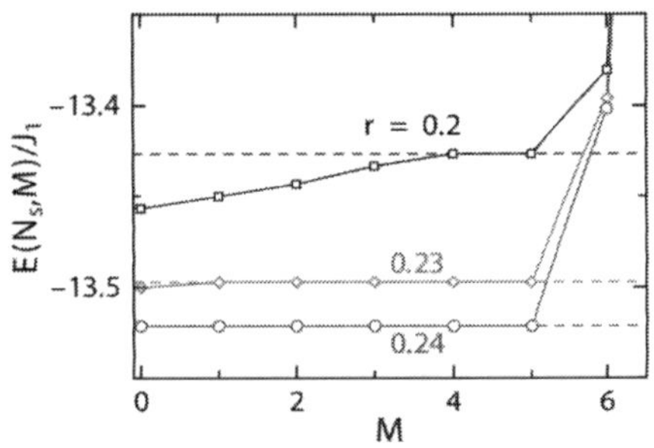


**Figure 7.** $M$-dependence of the ground-state energy for $N_s = 36$ in the region near $r = r_{c1}$ where the spontaneous magnetization is vanishing owing to a large distortion. Blue circles, red diamonds, and black squares denote results for $r = 0.24$, 0.23, and 0.2, respectively.

Finally, let us discuss our results based on the experimental results[24, 25]. Neutron scattering experiment in $RbVBr_3$ reported that the ratio of interactions is $r \sim 0.59$[24]. From our results in Fig. 5, $r \sim 0.59$ corresponds to the spontaneous magnetization of $M_{spo}/M_{sat} \sim 0.1$-0.2. Concerning $RbFeBr_3$, on the other hand, ref. [25] concluded $r < 1$ but did not give a specific value for $r$ unfortunately. Our numerical results will be helpful to analyze experimental results of these materials.

## 4. Summary

We studied the ground state of the spin-1/2 Heisenberg antiferromagnet on the triangular lattice with a $\sqrt{3} \times \sqrt{3}$-type distortion by the numerical-diagonalization method. We find a region of nontrivial states showing spontaneous magnetization between the case of undistorted triangular lattice and the case of the honeycomb-lattice system accompanied by isolated spins. Nontrivial spin states with spontaneous magnetization have been found in various systems of some modified cases from the kagome-lattice antiferromagnet, the square-kagome-lattice antiferromagnet, the Cairo-pentagon-lattice antiferromagnet, the Lieb-lattice antiferromagnet, and so on[39, 40, 42, 43, 44, 45, 46]. Comparison between our present finding and these previously known information contributes much for our understanding of frustration effects in magnetic materials.

**Acknowledgments**

We wish to thank Professors H. Sato and N. Todoroki for fruitful discussions. This work was partly supported by JSPS KAKENHI Grant Numbers 16K05418, 16K05419, and 16H01080(JPhysics). Nonhybrid thread-parallel calculations in numerical diagonalizations were based on TITPACK version 2 coded by H. Nishimori. This research used computational resources of the K computer provided by the RIKEN Advanced Institute for Computational Science through the HPCI System Research projects (Project ID: hp170017, hp170028, hp170070, and hp170207). Some of the computations were performed using facilities of the Department of Simulation Science, National Institute for Fusion Science; Institute for Solid State Physics, The University of Tokyo; and Supercomputing Division, Information Technology Center, The University of Tokyo. This work was partly supported by the Strategic Programs for Innovative Research; the Ministry of Education, Culture, Sports, Science and Technology of Japan; and the Computational Materials Science Initiative, Japan.

## References

[1] Anderson P W 1973 *Mater. Res. Bull.* **8** 153
[2] Huse D A and Elser V 1988 *Phys. Rev. Lett.* **60** 2531
[3] Jolicour Th and Le Guillou J C 1989 *Phys. Rev. B* **40** 2727
[4] Singh R R P and Huse D A 1992 *Phys. Rev. Lett.* **68** 1766
[5] Bernu B, Lhuillier C and Pierre L 1992 *Phys. Rev. Lett.* **69** 2590
[6] Bernu B, Lecheminant P, Lhuillier C and Pierre L 1994 *Phys. Rev. B* **50** 10048
[7] Leung P W and Runge K J 1993 *Phys. Rev. B* **47** 5861
[8] Weihong Z, McKenzie R H and Singh R R P 1999 *Phys. Rev. B* **59**, 14367
[9] Richter J, Schulenburg J and Honecker A 2004 *Lecture Notes in Physics* (Heidelberg: Springer) vol 645 p 85
[10] Yunoki S and Sollera S 2006 *Phys. Rev. B* **74** 014408
[11] Starykh O A and Balents L 2007 *Phys. Rev. Lett.* **98** 077205
[12] Heidarian D, Sollera S and Becca F 2009 *Phys. Rev. B* **80** 012404
[13] Sakai T and Nakano H 2011 *Phys. Rev. B* **83** 100405(R)
[14] Reuther J and Thomale R 2011 *Phys. Rev. B* **83** 024402
[15] Weichselbaum A and White S R 2011 *Phys. Rev. B* **84** 245130
[16] Ghamari S, Kallin C, Lee S S and Sorensen E S 2011 *Phys. Rev. B* **84** 174415
[17] Harada K 2012 *Phys. Rev. B* **86** 184421
[18] Nakano H, Todo S and Sakai T 2013 *J. Phys. Soc. Jpn.* **82** 043715
[19] Yamamoto D, Marmorini G and Danshita I 2014 *Phys. Rev. Lett.* **112** 127203
[20] Watanabe K, Kawamura H, Nakano H and Sakai T 2014 *J. Phys. Soc. Jpn.* **83** 034714
[21] Shimokawa T, Watanabe K and Kawamura H 2015 *Phys. Rev. B* **92** 134407
[22] Shirata Y, Tanaka H, Ono T, Matsuo A, Kindo K and Nakano H 2011 *J. Phys. Soc. Jpn.* **80** 093702
[23] Shirata Y, Tanaka H, Matsuo A and Kindo K 2012 *Phys. Rev. Lett.* **108** 057205
[24] Tanaka H and Kakurai K 1994 *J. Phys. Soc. Jpn.* **63** 3412
[25] Nishiwaki Y, Osawa A, Kakurai K, Kaneko K, Tokunaga M and Kato T 2011 *J. Phys. Soc. Jpn.* **80** 084711
[26] Nakano H and Sakai T 2017 *J. Phys. Soc. Jpn.* **86** 063702
[27] Jagannathan A and Szallas A 2013 *Eur. Phys. J. B* **86** 76
[28] Lieb E and Mattis D 1962 *J. Math. Phys. (N.Y.)* **3** 749
[29] Nakano H and Terai A 2009 *J. Phys. Soc. Jpn.* **78** 014003
[30] Nakano H and Sakai T 2011 *J. Phys. Soc. Jpn.* **80** 053704
[31] Nakano H and Sakai T 2014 *J. Phys. Soc. Jpn.* **83** 104710
[32] Nakano H and Sakai T 2015 *J. Phys. Soc. Jpn.* **84** 063705
[33] Nakano H, Hasegawa Y and Sakai T 2015 *J. Phys. Soc. Jpn.* **84** 114703
[34] Sachdev S and Senthil T 1996 *Ann. Phys.* **251** 76
[35] Tsukano M and Takahashi M 1996 *J. Phys. Soc. Jpn.* **66** 1153
[36] Ivanov N B and Richter J 2004 *Phys. Rev. B* **69** 214420
[37] Yoshikawa S and Miyashita S 2005 *J. Phys. Soc. Jpn. Suppl.* **74** 71
[38] Hida K 2007 *J. Phys.: Condens. Matter* **19** 145225
[39] Nakano H, Shimokawa T and Sakai T 2011 *J. Phys. Soc. Jpn.* **80** 033709
[40] Shimokawa T and Nakano H 2011 *J. Phys. Soc. Jpn.* **80** 043703
[41] Furuya S C and Giamarchi T 2014 *Phys. Rev. B* **89** 205131
[42] Nakano H and Sakai T 2013 *J. Phys. Soc. Jpn.* **82** 083709
[43] Nakano H, Sakai T and Hasegawa Y 2014 *J. Phys. Soc. Jpn.* **83** 084709
[44] Nakano H, Isoda M and Sakai T 2014 *J. Phys. Soc. Jpn.* **83** 053702
[45] Isoda M, Nakano H and Sakai T 2014 *J. Phys. Soc. Jpn.* **83** 084710
[46] Nakano H and Sakai T 2015 *Jpn. J. Appl. Phys.* **54** 030305

IOP Conf. Series: Journal of Physics: Conf. Series **969** (2018) 012127 doi:10.1088/1742-6596/969/1/012127

# Quantum Spin Fluid Behaviors of the Kagome- and Triangular-Lattice Antiferromagnets

**Tôru Sakai**[1,2] **and Hiroki Nakano**[1]

[1]Graduate School of Material Science, University of Hyogo, Hyogo 678-1297, Japan
[2]National Institute for Quantum and Radiological Science and Technology (QST), SPring-8, Hyogo 679-5148, Japan

E-mail: sakai@spring8.or.jp

**Abstract.** The S=1/2 kagome-lattice antiferromagnet is investigated using the numerical diagonalization of finite-size clusters. The analysis of the susceptibility at the zero magnetization indicates that the magnetic excitation of the system is gapless. It is consistent with our previous finite-scaling analysis of the spin gap. The application of the method for the triangular-lattice antiferromagnet confirms the validity of the analysis.

## 1. Introduction

The $S = 1/2$ kagome-lattice antiferromagnet has attracted a lot of interests in the field of the highly frustrated magnetism. Particularly since discoveries of several candidate materials; the herbertsmithite[1, 2], the volborthite[3, 4] and the vesignieite[5] for the kagome lattice, the study on the system has been accelerated. The quantum spin fluid behaviour of the system was predicted by many theoretical studies[6, 7, 8, 9, 10, 11, 12, 13, 14, 15, 16, 17, 18]. The $U(1)$ Dirac spin liquid theory[13] indicated a gapless spin excitation in the thermodynamic limit, which has been supported by the recent variational approach[19, 20]. Our recent numerical diagonalization study[21] also concluded that the system is gapless. On the other hand, the recent density matrix renormalization group (DMRG) analyses[22, 23, 24] suggested that the system has a finite spin gap even in the thermodynamic limit and supported the $Z_2$ topological spin liquid picture[6]. Thus whether the S=1/2 kagome-lattice antiferromagnet has a spin gap or not is still theoretically controversial, although the recent neutron scattering experiment of the single crystal of the herbertsmithite[25, 26] suggested that the system is gapless,

In this paper, using the recently developed susceptibility analysis based on the numerical diagonalization of finite-size clusters[27], we try to approach the spin-gap issue of the $S = 1/2$ kagome-lattice antiferromagnet. Particularly, we examine the gapless-feature result of the susceptibility analysis of the $S = 1/2$ kagome-lattice antiferromagnet by comparing the result with the result of the $S = 1/2$ triangular-lattice one, which is widely believed to be gapless. These two cases are typical among frustrated antiferromagnets. Although a randomness effect in these systems was additionally examined[28, 29], in the present study, properties of ideal systems without randomness are investigated as a fundamental issue.

IOP Conf. Series: Journal of Physics: Conf. Series **969** (2018) 012127 doi:10.1088/1742-6596/969/1/012127

## 2. Numerical diagonalization

Using the numerical exact diagonalization of finite-size clusters under periodic boundary condition, we investigate the $S = 1/2$ kagome-lattice antiferromagnet defined by the Hamiltonian

$$\mathcal{H} = \sum_{\langle i,j \rangle} \mathbf{S}_i \cdot \mathbf{S}_j, \tag{1}$$

where $\langle i, j \rangle$ means all the nearest neighbor pairs on the kagome lattice. For an $N$-site system, we consider subspaces characterized by $M = \sum_j S_j^z$; we obtain the lowest energy denoted by $E(N, M)$ of the Hamiltonian matrix in each subspace. We calculate all the values of $E(N, M)$ available for the clusters up to $N$ =42 by the numerical diagonalization.

## 3. Susceptibility analysis

In order to investigate the spin gap issue, we apply the susceptibility analysis which was developed in our previous work[27]. The argument of the analysis is briefly reviewed as follows: the effect of the applied external magnetic field $h$ is described by the Zeeman energy term

$$\mathcal{H}_Z = -h \sum_j S_j^z \tag{2}$$

The energy of $\mathcal{H}$ per site in the thermodynamic limit is defined as

$$\frac{E(N, M)}{N} \sim \epsilon(m) \qquad (N \to \infty) \tag{3}$$

where $m = M/(SN)$ is the magnetization normalized by the saturated magnetization $SN$. If we assume $\epsilon(m)$ is an analytic function of $m$, the spin excitation energy would become

$$E(N, M+1) - E(N, M) \sim \frac{1}{S}\left(\epsilon'(m) + \frac{1}{2}\epsilon''(m)\frac{1}{NS} + \cdots\right) \tag{4}$$

Thus, this equation gives the quantity corresponding to the width of the magnetization plateau at $m$ as follows,

$$(E(N, M+1) - E(N, M)) - (E(N, M) - E(N, M-1)) \sim \epsilon''(m)\frac{1}{NS^2} \tag{5}$$

Minimizing the energy of the total Hamiltonian $\mathcal{H} + \mathcal{H}_Z$, the ground state magnetization curve is derived by

$$h = \epsilon'(m)/S \tag{6}$$

The field derivative of the magnetization is defined as

$$\chi \equiv \frac{dm}{dh} = \frac{S}{\epsilon''(m)} \tag{7}$$

If we assume $\chi \neq 0$, namely $\epsilon''(m)$ is finite, the magnetization plateau at $m$ would vanish in the thermodynamic limit, because of (5). Thus a necessary condition for the existence of a magnetization plateau at $m$ is $\chi = 0$ in the thermodynamic limit. Now we apply this argument for the spin gap. We should examine the case of $m = 0$. In this case, the equation (5) can be rewritten as

$$2\Delta_N \sim \epsilon''(0)\frac{1}{NS^2} \tag{8}$$

where $\Delta_N = E(N, 1) - E(N, 0)$ is the spin gap for an $N$-spin cluster. Thus a necessary condition of the finite spin gap would be $\chi = 0$ at $m = 0$ in the thermodynamic limit.

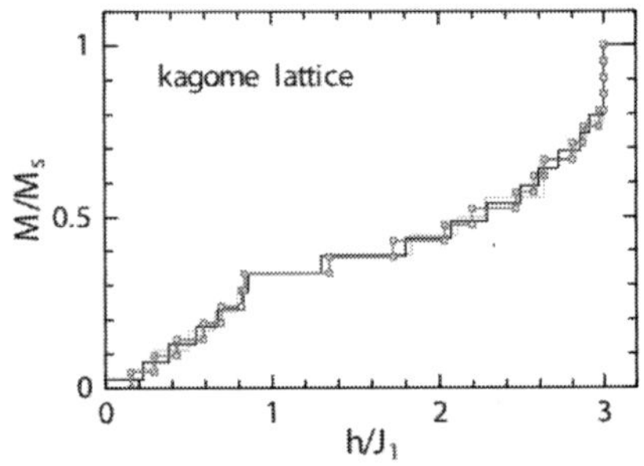


**Figure 1.** Magnetization curves of the kagome-lattice antiferromagnet for $N$ =36, 39 and 42. The results for $N$ =36 and 39 are denoted by green and black lines, respectively; that for $N$ =42 is denoted by red lines with squares.

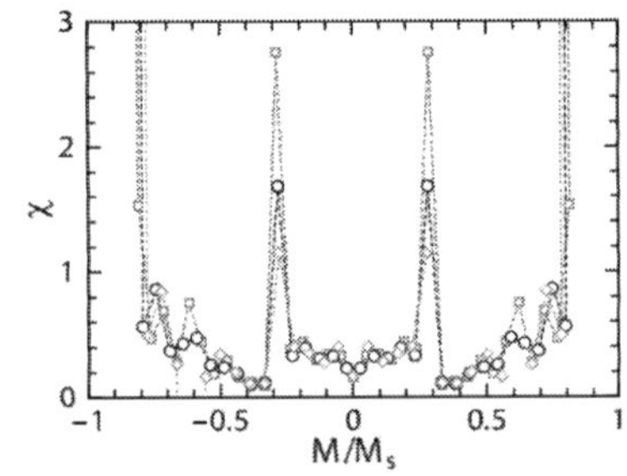


**Figure 2.** Magnetization dependence of $\chi$ of the kagome-lattice antiferromagnet for $N$ =36, 39 and 42. The results for $N$ =36, 39, and 42 are denoted by green diamonds, black circles, and red squares, respectively.

## 4. Result of the kagome-lattice antiferromagnet

We investigate the field derivative of the magnetization $\chi$ for the $S = 1/2$ kagome-lattice antiferromagnet. The ground state magnetization curves for $N$=36, 39 and 42 are shown in Figure 1; the results for $N$=39 and 42 were originally presented in [30] and [31], respectively. From the magnetization curves, it is difficult to determine whether the system is gapless or gapped. Next we show the magnetization dependence of the calculated $\chi$ for $N$=42, 39 and 36 in Figure 2. The behaviors around $m = 0$ are magnified in Figure 3. At least for $N$=36 and 42, the size dependence of $\chi$ at $m = 0$ is very small. $\chi$ at $m = 0$ is plotted versus $1/N$ and $1/\sqrt{N}$ for $N$=42, 36, 30, 24, 18, 12 in Figures 4(a) and (b), respectively. Although the system size dependence exhibits a slight oscillation, both plots clearly indicate that $\chi$ at $m = 0$ is still finite in the thermodynamic limit. Thus the system does not meet the condition for the finite spin gap. It should be one of strong evidences to justify that the $S = 1/2$ kagome-lattice antiferromagnet is gapless.

## 5. Triangular-lattice antiferromagnet

In order to test the validity of the present method, we apply it for the $S = 1/2$ triangular-lattice antiferromagnet, for which the triplet excitation is gapless. The ground-state magnetization curves calculated by the numerical exact diagonalization for $N$ =33, 36, and 39 are shown in Figure 5. The magnetization dependence of the susceptibility is shown in Figure 6. The behaviors around $m = 0$ are magnified in Figure 7. $\chi$ at $m = 0$ is plotted versus $1/N$ and $1/\sqrt{N}$ for $N$=36, 30, 24, 18, 12 in Figures 8(a) and (b), respectively. Both plots indicate that the extrapolated value in the thermodynamic limit seems to be finite ($\chi \neq 0$). It is consistent

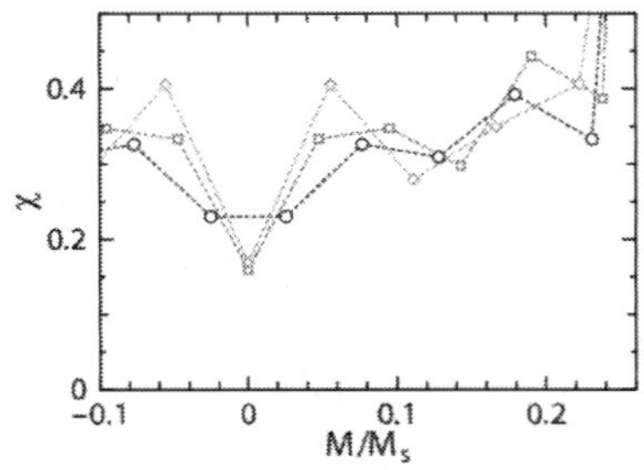


**Figure 3.** Magnetization dependence of $\chi$ of the kagome-lattice antiferromagnet for $N$ =36, 39 and 42 around $m = 0$. Symbols are the same as in Fig. 2.

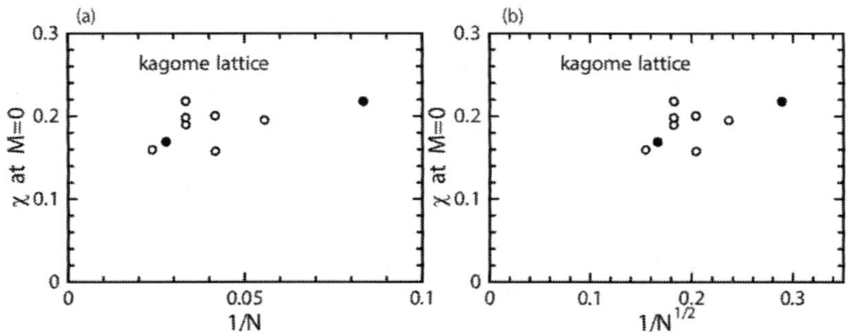


**Figure 4.** $\chi$ plotted versus $1/N$ (a) and $1/\sqrt{N}$ (b) for the kagome-lattice antiferromagnet. Closed symbols denote the results for the cases when the shape of the finite-size clusters are the rhombus with an interior angle $\pi/3$. When a finite-size cluster for a given $N$ cannot form the same rhombus even if the rhombus is tilted, on the other hand, the cluster forms a parallelogram; the results are given by open symbols. Both of (a) and (b) suggest that the extrapolated value in the thermodynamic limit seems to be finite ($\chi \neq 0$).

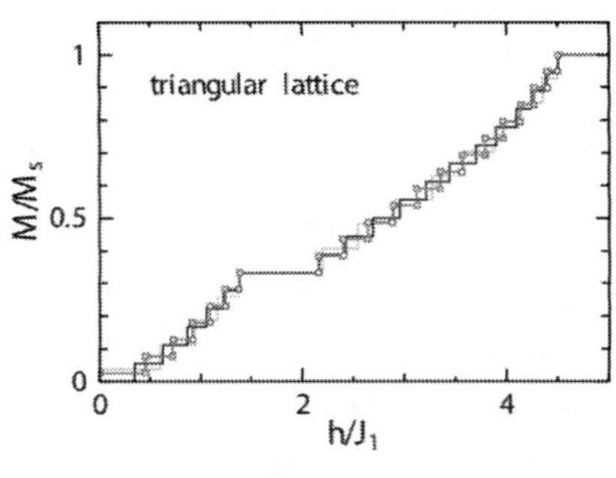


**Figure 5.** Magnetization curves of the triangular-lattice antiferromagnet for $N$ =33, 36 and 39. The results for $N$ =33 and 36 are denoted by green and black lines, respectively; that for $N$ =39 is denoted by red lines with squares.

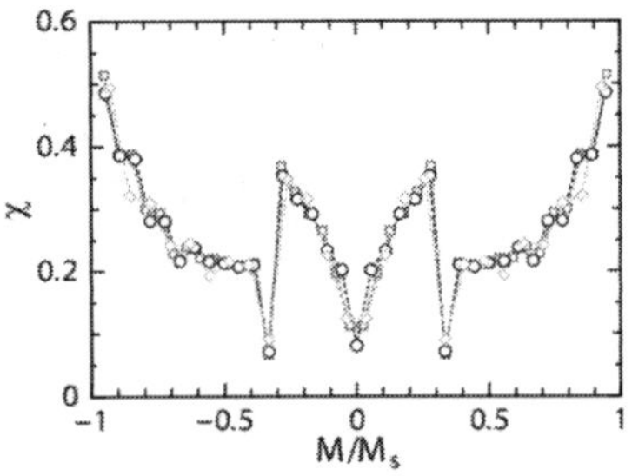


**Figure 6.** Magnetization dependence of $\chi$ of the triangular-lattice antiferromagnet for $N$ =33, 36 and 39. The results for $N$ =33, 36, and 39 are denoted by green diamonds, black circles, and red squares, respectively.

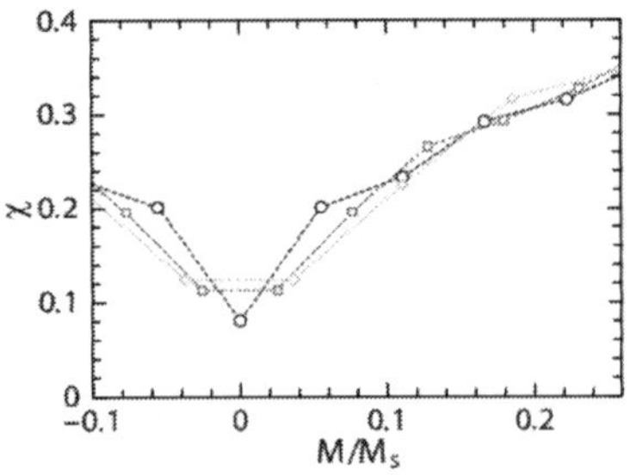


**Figure 7.** Magnetization dependence of $\chi$ of the kagome-lattice antiferromagnet for $N$ =33, 36 and 39 around $m = 0$. Symbols are the same as in Fig. 6.

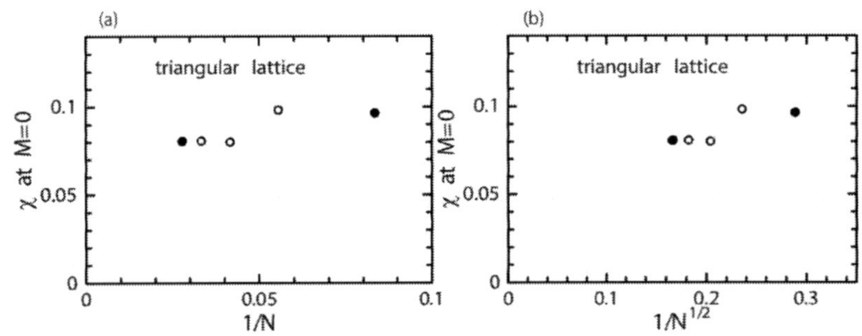


**Figure 8.** $\chi$ plotted versus $1/N$ (a) and $1/\sqrt{N}$ (b) for the triangular-lattice antiferromagnet. Both plots indicate that the extrapolated value in the thermodynamic limit seems to be finite ($\chi \neq 0$).

with the gapless feature of the triangular-lattice antiferromagnet. Thus it confirms the validity of the present method.

## 6. Summary

The spin gap issue of the kagome-lattice antiferromagnet is investigated using the numerical diagonalization up to $N = 42$. The analysis of the field derivative of the magnetization $\chi$ is successfully applied to the triangular-lattice antiferromagnet that is gapless in its spin excitation. The behavior of $\chi$ for the kagome-lattice antiferromagnet shares with that of triangular-lattice antiferromagnet. The agreement strongly suggests that the gapless spin excitation in the kagome-lattice antiferromagnet.

## Acknowledgments

This work has been partly supported by Grants-in-Aids for Scientific Research (Nos. 16H01080(JPhysics), 16K05418, and 16K05419) from the Ministry of Education, Culture, Sports, Science and Technology of Japan (MEXT), and Hyogo Science and Technology Association. This research used computational resources of the K computer provided by the RIKEN Advanced Institute for Computational Science through the HPCI System Research projects (Project IDs: hp130070, hp130098, hp150024, hp170017, hp170028, hp170070 and hp170207). We further thank the Supercomputer Center, Institute for Solid State Physics, University of Tokyo (UT); the Cyberscience Center, Tohoku University (TU); and the Computer Room, Yukawa Institute for Theoretical Physics, Kyoto University; the Department of Simulation Science, National Institute for Fusion Science; Center for Computational Materials Science, Institute for Materials Research, TU; Supercomputing Division, Information Technology Center, UT for computational facilities. This work was partly supported by the Strategic Programs for Innovative Research, MEXT, and the Computational Materials Science Initiative, Japan. The authors would like to express their sincere thanks to the staff members of the Center for Computational Materials Science of the Institute for Materials Research, TU for their continuous support of the SR16000 supercomputing facilities.

## References

[1] Shores M P, Nytko E A, Barlett B M and Nocera D G 2005 *J. Am. Chem. Soc.* **127** 13462
[2] Mendels P and Bert F 2010 *J. Phys. Soc. Jpn.* **79** 011001
[3] Yoshida H, Okamoto Y, Tayama T, Sakakibara T, Tokunaga M, Matsuo A, Narumi Y, Kindo K, Yoshida M, Takigawa M and Hiroi Z 2009 *J. Phys. Soc. Jpn.* **78** 043704
[4] Yoshida M, Takigawa M, Yoshida H, Okamoto Y and Hiroi Z 2009 *Phys. Phys. Lett.* **103** 077207
[5] Okamoto Y, Yoshida H and Hiroi Z 2009 *J. Phys. Soc. Jpn.* **78** 033701
[6] Sachdev S 1992 *Phys. Rev. B* **45** 12377
[7] Nakamura T and Miyashita S 1995 *Phys. Rev. B* **52** 9174
[8] Lecheminant P, Bernu B, Lhuillier C, Pierre L and Sindzingre P 1997 *Phys. Rev. B* **56** 2521
[9] Waldtmann Ch, Everts H -U, Bernu B, Lhuillier C, Sindzingre P, Lecheminant P and Pierre L 1998 *Eur. Phys. J. B* **2** 501
[10] Mila F 1998 *Phys. Rev. Lett.* **81** 2356
[11] Hermele M, Senthil T and Fisher M P A 2005 *Phys. Rev. B* **72** 104404
[12] Singh R R P and Huse D A 2007 *Phys. Rev. B* **76** 180407(R)
[13] Ran Y, Hermele M, Lee P A and Wen X -G 2007 *Phys. Rev. Lett.* **98** 117205
[14] Jiang H C, Weng Z Y and Sheng D N 2008 *Phys. Rev. Lett.* **98** 117203
[15] Cepas O, Fong C M, Leung P W and Lhuillier C 2008 *Phys. Rev. B* **78** 140405(R)
[16] Sindzingre P and Lhuillier C 2009 *Europhys. Lett.* **88** 27009
[17] Evenbly G and Vidal G 2010 *Phys. Rev. Lett.* **104** 187203
[18] Capponi S, Derzhko O, Honecker A, Läuchli A M and Richter J 2013 *Phys. Rev. B* **88** 144416
[19] Iqbal Y, Poilblanc D and Becca F 2014 *Phys. Rev. B* **89** 020407(R)
[20] Iqbal Y, Poilblanc D and Becca F 2015 *Phys. Rev. B* **91** 020402(R)
[21] Nakano H and Sakai T 2011 *J. Phys. Soc. Jpn.* **80** 053704
[22] Yan S, Huse D A and White S W 2011 *Science* **332** 1173
[23] Depenbrock S, McCulloch I P and Schollwöck U 2012 *Phys. Rev. Lett.* **109** 067201
[24] Nishimoto S, Shibata N and Hotta C 2013 *Nat. Commun.* **4** 2287
[25] Han T -H, Helton J S, Chu S, Nocera D G, Rodriguez-Rirera J A, Broholm C and Lee Y S 2012 *Nat.* **492** 406
[26] Han T -H, Chu S and Lee Y S 2012 *Phys. Rev. Lett.* **108** 157202
[27] Sakai T and Nakano H to appear in *J. Phys.: Conf. Ser.*
[28] Watanabe K, Kawamura H, Nakano H and Sakai T 2014 *J. Phys. Soc. Jpn.* **83** 034714
[29] Shimokawa T, Watanabe K and Kawamura H 2015 *Phys. Rev. B* **92** 134407
[30] Sakai T and Nakano H 2011 *Phys. Rev. B* **83** 100405(R)
[31] Nakano H and Sakai T 2014 *J. Phys. Soc. Jpn.* **83** 104710

IOP Conf. Series: Journal of Physics: Conf. Series **969** (2018) 012128 doi:10.1088/1742-6596/969/1/012128

# Low-temperature elastic properties of YbSbPt probed by ultrasound measurements

**Y Nakanishi[1], S. Takahashi[1], R. Ohyama[1], J. Hasegawa[1], M. Nakamura[1], H. Suzuki[2] and M. Yoshizawa[1]**

[1] Graduate School of Engineering, Iwate University, Morioka 020-8551, Japan
[2] Institute for Solid State Physics, The University of Tokyo, Kashiwa, Chiba 277-8581, Japan

E-mail: yoshiki@iwate-u.ac.jp

**Abstract.** The elastic properties of a single crystal of the half-Heusler compound YbSbPt have been investigated by means of the ultrasonic measurement. In particular, careful measurements of the temperature ($T$) dependent elastic constant $C_{11}(T)$ was performed in the vicinity of its phase transition point near $T_N$ of 0.5 K. A clear step-like anomaly accompanied by spin-density-wave type antiferromagnetic (AFM) phase transition was found in the $C_{11}(T)$ curve. The low-temperature magnetic phase diagram is proposed on the basis of the results. The phase diagram consists of, at least two main distinct phases: a low-field and high-field regime with a transition field of approximately 0.6 T at zero field. We discuss the low-temperature elastic property based on analysis of Landau-type free energy.

## 1. Introduction

The discovery of a massive electronic state in YbBiPt, and thereafter $YbCo_2Zn_{20}$ has stimulated experimental and theoretical activity aimed at understanding fundamental electronic state and magnetic properties of these and related Yb-based intermetallic compounds. [1-5] In particular, an intriguing aspect of YbBiPt is the existence of spin-density-wave type magnetic phase transition at 0.4 K determined by a small but rather sharp peak in its specific heat. Furthermore, a sudden steep jump was observed in the temperature ($T$) dependence of the electrical resistivity at 0.4 K, suggesting the formation of an energy gap at the Fermi level. [1-4] Recent neutron scattering measurement determined the averaged ordered moment of 0.8 $\mu_B$/Yb with the AFM propagation vector $q = (1/2, 1/2, 1/2)$ below 0.4 K. [6] Here, it should be noted that YbBiPt is a low-carrier-concentration semimetal, although most of systems with such a largely enhanced Sommerfeld coefficient $\gamma$ in the vicinity of the quantum critical point (QCP) favor a metallic state characterized by a renormalized effective mass. [7-10] The origin of the massive electronic state in the low-carrier-concentration system YbBiPt is of particular interest.From the structural analogy, a significant enhancement of the Sommerfeld coefficient $\gamma$ would be also expected in the isostructral system YbSbPt. The Sb substitution is regarded as a chemical pressure because of its smaller ionic radius than that of Bi, leading to the lattice contraction and its major effect on the system is to increase a hybridization between $4f$ states and valence electronic states. In fact, the massive electronic state and the similar magnetic phase transition have been clearly recognized in YbSbPt in the previous studies.[11] YbSbPt belongs to the series of cubic half-Heusler (space group F$43\bar{m}$)$R$PtX compounds ($R$ = rare-earth, X = Sb, Bi) with the magnetic Yb ions forming a face-centered-cubic (fcc) magnetic sublattice, and is one of the few stoichiometric Yb-based HF compounds. In fact, it was recently found that YbSbPt also exhibits an enormous low-temperature Sommerfeld coefficient, $\gamma\sim$ 9 J/mol $K^2$. Several hypotheses have been proposed to explain these observations. For example, the fcc structure produces frustrated Ruderman-Kittel-Kasuya-Yoshida interactions such that magnetic order cannot circumvent Kondo interactions. Very recently, several half-Heuslers are predicted to display band inversion at the $\Gamma$ point, leading to topologically nontrivial states.[12-14] These features could possibly lead to the formation of massive quasi-particles close to the Fermi level. Although there have been considerable experimental and theoretical efforts, the origin of the enormous $\gamma$ has not been yet clear at this stage.

An ultrasonic measurement is suited to study the $4f$ electronic state via quadrupolar response of the $4f$ electronic ground state split by the crystalline electric field effect if the 4f electrons are localized well. On the other hand, a narrow quasiparticle band formed by heavily renormalized quasiparticles causes elastic anomaly as well even if the $4f$ electrons are delocalized due to a deformation potential coupling to conduction bands.[15,16] In this paper, we report partial results of an investigation of the low temperature elastic properties of a single crystal of YbPtSb to deepen the understanding of the ground state properties in YbSbPt.

## 2. Experiment

The single crystal was grown from an excess Sb flux. The crystal structure was characterized by an X-ray diffraction measurement. The determined lattice parameters , $a$ = 6.4624 $\AA$ and the half-Heusler MgAgAs structure with space group F43$\bar{m}$ are in agreement with previous studies.[17] Specimen used for the present ultrasonic measurement was cut into a rectangular shape with two axes along the $\langle 100\rangle$- and $\langle 110\rangle$ -axis. The sound velocity, as the elastic constant was measured by an ultrasonic apparatus based on a phase comparison method. For low-temperature measurements, an Oxford helium dilution fridge adapted for ultrasonic measurements was employed with a 8 T superconducting magnet. Plates of $LiNbO_3$ was used for the piezoelectric transducer. The fundamental resonance frequency of $LiNbO_3$ transducer is 10 - 30 MHz. The transducer was glued on the parallel planes of the sample by an elastic polymer Thiokol. The absolute value of the sound velocity was obtained by measuring the delay time between the ultrasonic echo signals with an accuracy of a few percent. The elastic constant was calculated as $C = \rho v^2$ by using the sound velocity $v$ and the density $\rho$ of the crystal. The relative change of the $C_{11}$, $\Delta C_{11}$ of YbSbPt is used in this paper.

## 3. Experimental Results and Discussions

The relative variation of the elastic constant $C_{11}$ as a function of temperature is shown in Fig. 1 under selected fields along the $\langle 100\rangle$ axis. The data were normalized to their value at 1 K and the curves have been shifted vertically for clarity. The $C_{11}$ is one of the three principal independent elastic constants for the cubic crystal, and was measured by the longitudinal sound wave with frequencies of 10 - 30 MHz propagating along the $\langle 100\rangle$ axis. Although it is not shown here, the $C_{11}$ which denotes a pure compression along the $\langle 100\rangle$ axis smoothly increases at high temperature. [18] This temperature dependence appears normal in view of the expected stiffening following the progressive disappearance of phonon anharmonic effects upon lowering the temperature. However, the $C_{11}$ shows a noticeable softening below around 20 K, as already reported by our group. [19] Here, we put a focus on the elastic property involving the low-temperature phase. A step-like elastic anomaly at the phase transition is observed clearly at zero magnetic field. The transition temperature shifts slightly to higher temperatures and the shape of the anomaly becomes sharper with increasing the magnetic fields up to around 0.6 T. In contrary, the transition temperature shifts to lower temperatures with the further increase of the magnetic field. Surprisingly, the shape of the anomaly changes dramatically, namely a sudden

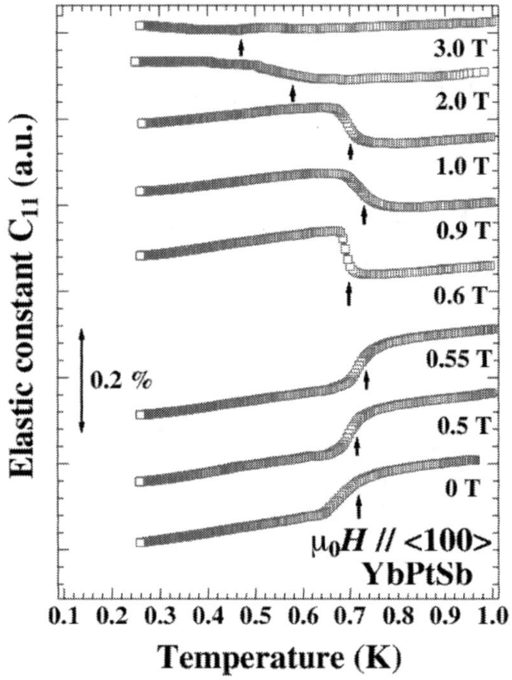


**Figure 1.** Temperature dependence of one of the principal elastic constants $C_{11}$ of YbSbPt, measured in different applied magnetic fields. The data were normalized to their value at 1 K and the curves have been shifted vertically for clarity. The arrows indicate the phase transition temperature point, derived from the minimum or maximum in $\mathrm{d}C_{11}(T)/\mathrm{d}T$.

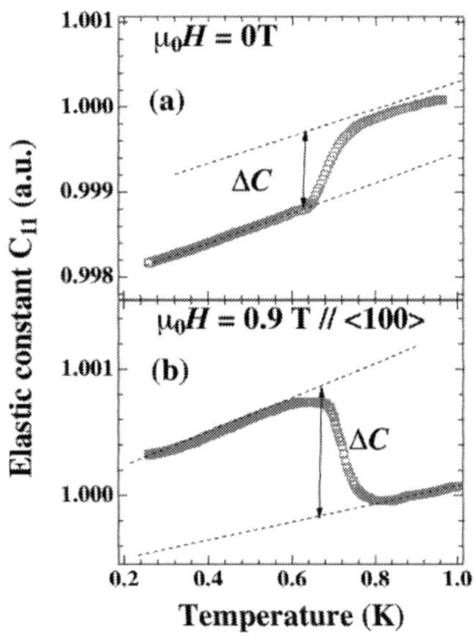


**Figure 2.** The relative change of the $C_{11}$, $\Delta C_{11}$ of YbSbPt in a zero field, (a) and in a field of 0.6 T, (b), being associated with the phase transition. $\Delta C_{11}$ is determined as shown the arrow.

increase (stiffening) appears at the transition temperature. The anomaly becomes suppressed with the further increase of the magnetic field, and invisible in the field above around 3 T. It should be noted that the precipitous increase of $C_{11}$ at the magnetic field of 0.6 T is characteristic of the low-temperature ordered phase. In order to discuss the magnetic field evolution of the temperature dependence of the elastic constant $C_{11}(T)$ in the vicinity of $T_{\mathrm{N}}$, a relative of change of $C_{11}$ : $\Delta C_{11}$ accompanied by the phase transition was estimated on the basis of the definition as shown in Figs 2(a) and (b). Figure 3 shows the obtained magnetic field dependence of the relative change of the elastic constant $\Delta C_{11}$ due to the phase transition.The dashed lines are

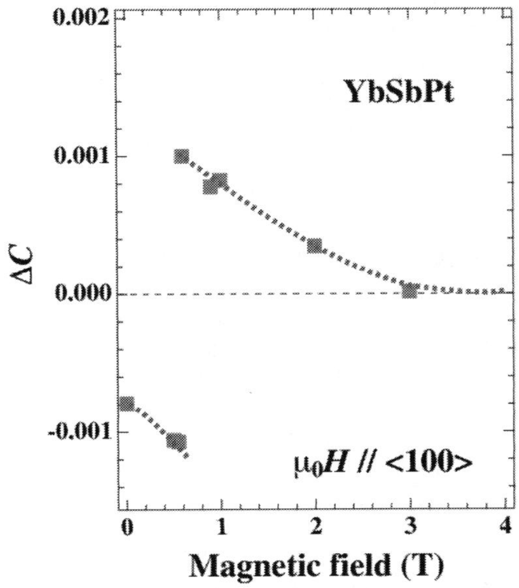


**Figure 3.** Magnetic field dependence of the relative change of the $C_{11}$, $\Delta C_{11}$ of YbSbPt. The dashed lines are a guide for the eye.

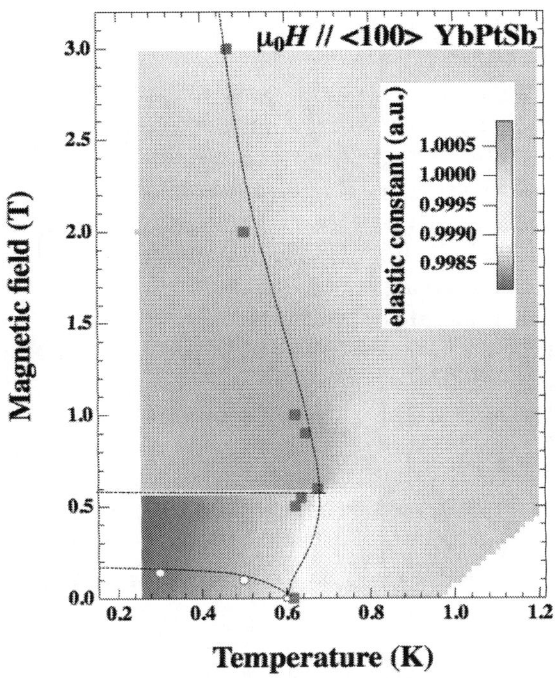


**Figure 4.** Magnetic phase diagram constructed and illustrated by data taken from the present measurements. The dashed lines are a guide for the eye. Orange regions indicate a low-field phase in which the $C_{11}(T)$ decreases below $T_N$ : AFM I. Dark blue regions, being the left side of the dashed line indicate a high-field phase in which the $C_{11}(T)$ increases below $T_N$ : AFM II.

given as guides to the eye. One can recognize that the critical magnetic field of $H_c \sim 0.6$ T exists and the system changes its elastic property when crossing $H_c$. Furthermore, it should be additionally noted here that a sharp increase of the resistivity observed around $T_N$ is completely suppressed by a magnetic field above$H_c$, suggesting disappearance of the energy gap.

Let us here discuss the obtained elastic anomalies from a coupling between elastic strain and oder parameters including magnetic moment, or enhanced quasi-particles based on analysis of Landau-type free energy. Recall the relative change of the $C_{11}$, $\Delta C_{11}$ of YbSbPt below and above $H_c$, due to the phase transition are shown in Fig 3. In general, the $\Delta C_{11}$ gives us microscopic information for the coupling between the elastic strain $\varepsilon_\Gamma$ and the oder parameter of the ordered phase $Q$ based on Landau phenomenological theory as follows,[20]

$$F = \frac{1}{2}aQ^2 + \frac{1}{4}bQ^4 - \zeta Q^2\varepsilon_\Gamma + \xi Q^2\varepsilon_\Gamma^2 + ..... \tag{1}$$

From the second derivative of the free energy $F$ with respect to the elastic strain, one can obtain the following formula

$$C_\Gamma(T) = \frac{\partial^2 F}{\partial \varepsilon_\Gamma^2} = C_\Gamma^{(0)}(T) - \frac{2\zeta^2}{b} - 2Q^2\xi \tag{2}$$

Here, ${C_\Gamma}^{(0)}(T)$ is the lattice part and originates mainly from anharmonic effect of the crystal. $a$ and $b$ denote the second- and fourth-order coefficient of free-energy. $\xi$ and $\zeta$ denote the fitting parameters related to the elastic softening and hardening, respectively. As mentioned above, the elastic softening of $C_{11}(T)$ below $H_c$ is primarily caused by $\xi$, while the elastic hardening above $H_c$ is primarily caused by $\zeta$ as shown in Figs. 2 (a) and (b), respectively. Furthermore, it is commonly known that the former corresponds to a possible change of transfer integrals of heavy quasi-particles formed near the Fermi energy, and the latter corresponds to a possible change of the coupling constant between the elastic strain and the order parameter $Q$, probably here being the magnetic moments ($M$) of the heavy quasi-particles. [21,22] If an energy gap appears i.e., the number of carriers decreases, it leads to an elastic hardening as seen in $PrRu_4P_{12}$, which is inconsistent with the behavior in the low-field regime below $H_c$.[23] This discrepancy remains unsettled at this stage. On the other hand, the significant decrease of the relative change of $C_{11}(T)$ above $H_c$ with increasing applied magnetic fields is probably due to the decrease of $\xi$, suggesting the suppression of spin-density-wave type magnetic phase with increasing the fields.

Viewed in this light, it is most likely that at least there are two different phases in the low-temperature ordered phase. The characteristics observed here are collected in the $H$ - $T$ phase diagram displayed in Fig. 4. The boundary was determined from the sharp peak position in $dC_{11}/dT$. The low-field phase is characterized by the elastic softening and semiconducting with an energy gap as labeled by AFM I. On the other hand, the high-field one is characterized by elastic stiffening and semi-metallic nature as labeled by AFM II. The lower-field phase is consistent with the previous results obtained from the magnetization, electric resistivity and specific heat measurements. The higher-field phase labeled by AF II, however does not fully agree with that determined by previous studies.[4] The AFM II boundary might be corresponding either to the crossover of $H^*$ on which the broad local maximum was observed in the $T$ dependence of the derivative $d\rho(T)/T$, or to $T_{FL}$ which was determined by the upper limit of the $T^2$ dependence of the electrical resistivity in the reference [4]. This is an open question.

Finally, we would like to comment on the possible surface state in YbSbPt due to topological nature of surface conducting channel, reported recently in half-Heusler systems such as HoPdBi, ErPdBi, LuPdBi and others.[24-29] Since ultrasound measurements are sensitive and reliable volume probes of ordering phenomena in the bulk of solids, discrepancies might occur in physical interpretations between results of the electrical resistivity and present elastic constant near the phase transition temperature. Namely, formation of the energy gap and a very weak induced magnetic moment below $T_N$ might be due to topological surface nature. Very sensitive and direct measurements at low temperatures, such as Scanning Tunneling Microscope and Angle-Resolved Photoemission Spectroscopy measurements are required in the future in order to make a sharp distinction between its surface- and bulk-nature.

## 4. Summary

In summary, we have investigated the low-temperature elastic properties of a single crystal of YbPtSb by means of the ultrasound measurement. A marked difference was found in the temperature dependence of the longitudinal elastic constant $C_{11}(T)$ between the low field- and the high field regime with $H_c = \sim 0.6$ T. The characteristic behavior would reflect the electronic state of strongly renormalized quasi particles including its energy gap and magnetic moment with the relatively lower carrier-concentration. The deeper investigation is necessary to understand its origin and to interpret properly the corresponding experimental data, including a possible scenario based on Dirac electrons formed in the vicinity of Fermi level in YbPtSb. For a future discussion, the same measurements for other kinds of elastic constants, $(C_{11}$ - $C_{12})/2$ and $C_{44}$ are currently in progress. We hope that these observation will stimulate theoretical studies to understand the remarkable elastic properties of YbPtSb.

## 5. Acknowledgement

The measurements have been performed in the Cryogenic Division of the Center for Instrumental Analysis, Iwate University. This work was supported by JSPS KAKENHI Grant Number JP15H05883 (J-Physics).

## 6. References

[1] Fisk Z *et al.* 1991 *Phys. Rev. Lett* **67** 3310.
[2] Robinson R. A. *et al.* 1994 *Phys. Rev* B **50** 9595.
[3] Hundley M. F. *et al.* 1997 *Phys. Rev* B **56** 8098.
[4] Mun E. D. *et al.* 2013 *Phys. Rev* B **87** 075120.
[5] For example, Torikachvili M S *et al.* 2007 *Proc. Natl. Acad. Sci. U. S. A.* **104** 9960
[6] Ueland B. G. *et al.* 2014 *Phys. Rev* B **89** 180403(R).
[7] Ueland B. G. *et al.* 2015 *Phys. Rev* B **92** 184111.
[8] Mun E. D. *et al.* 2015 *Phys. Rev* B **92** 085135.
[9] Eundeok M. *et al.* 2016 *Phys. Rev* B **93** 115134.
[10] Schilling M. B. *et al.* 2017 *Phys. Rev* B **95** 155201.
[11] Matsubayashi K. *et al.* 2012 *J. Phys.: Conf. Ser.* **391** 012040.
[12] Nakajima Y *et al.* 2015 *Sci. Adv.* **1** e1500242.
[13] Li Zhi *et al.* 2015 *Phys. Rev* B **91** 235128.
[14] Müller R. A. *et al.* 2015 *Phys. Rev* B **92** 184432.
[15] Lüthi B 1985 *J. Phys. Chem. Solids.* **52** 70.
[16] Thalmeier P 1988 *J. Magn. Magn. Mater.* **76 & 77** 299.
[17] Kasaya M. 1992 *J. Phys. Soc. Jpn.* **61** 4187.
[18] Nakanishi Y *et al.* (in preparation).
[19] Nakanishi Y *et al.* 2010 *Phys. Status Solidi B* **247** 754.
[20] For example, Landau L. D. and Lifshitz E. M. *Couse of Theoretical Physics* (Pergamon, Oxford, 1980) Vol.5, p.446.
[21] Imaduddin A. 2002 *J. Phys. Soc. Jpn.* **71** 1965.
[22] Nakanishi Y. *et al.* 2008 *J. Phys. Soc. Jpn. Suppl. A* **77** 216.
[23] Nakanishi Y. *et al.* 2006 *Phys. Rev* B **73** 165115.
[24] Nikitin A. M. *et al.* 2015 *J. Phys. Condens. Matter* **27** 275701.
[25] Pan Y. *et al.* 2013 *Europhys. Lett.* **104** 27001.
[26] Xu G. *et al.* 2014 *Sci. Rep.* **4** 5709.
[27] Pavlosiuk O. *et al.* 2015 *Sci. Rep.* **5** 9158.
[28] Müller R. *et al.* 2014 *Phys. Rev* B **90** 041109.
[29] Pavlosiuk O. *et al.* 2015 *Acta Phys. Pol. A* **127** 656.

IOP Conf. Series: Journal of Physics: Conf. Series **969** (2018) 012129 doi:10.1088/1742-6596/969/1/012129

# Frequency dependent ac transport of films of close-packed carbon nanotube arrays

**A Endo[1], S Katsumoto[1], K Matsuda[2,3], W Norimatsu[2] and M Kusunoki[3]**

[1] Institute for Solid State Physics, University of Tokyo, Kashiwa, Chiba 277-8581, Japan
[2] Graduate School of Engineering, Nagoya University, Nagoya, Aichi 464-8603, Japan
[3] Institute of Materials and Systems for Sustainability, Nagoya University, Nagoya, Aichi 464-8603, Japan

E-mail: akrendo@issp.u-tokyo.ac.jp

**Abstract.** We have measured low-temperature ac impedance of films of closely-packed, highly-aligned carbon nanotubes prepared by thermal decomposition of silicon carbide wafers. The measurement was performed on films with the thickness (the length of the nanotubes) ranging from 6.5 to 65 nm. We found that the impedance rapidly decreases with the frequency. This can be interpreted as resulting from the electric transport via capacitive coupling between adjacent nanotubes. We also found numbers of sharp spikes superposed on frequency vs. impedance curves, which presumably represent resonant frequencies seen in the calculated conductivity of random capacitance networks. Capacitive coupling between the nanotubes was reduced by the magnetic field perpendicular to the films at 8.2 mK, resulting in the transition from negative to positive magnetoresistance with an increase of the frequency.

## 1. Introduction

A film composed of a highly-aligned, close-packed array of nanotubes (CNT) can be grown on the surface of a SiC($000\bar{1}$) substrate by thermal decomposition; in the film, multi-walled CNTs having the diameter ranging from 2 to 5 nm are distributed randomly, with their tube axes aligned perpendicular to the substrate surface [1, 2, 3]. Such films provide interesting subjects for transport studies, owing to the intricate combination of inter- and intra-CNT transport in the network of randomly arranged CNTs. In the present paper, we report the measurements of ac impedance, focusing on the frequency, the temperature, and the magnetic-field dependence.

## 2. Experimental

A schematic drawing of a CNT film is shown in Fig. 1. Details of the growth of the CNT films are given in [1, 2, 3]. We measured four samples with different thickness (CNT length) $L$ tabulated in Fig. 1. Electrical contacts were made by balls of silver paste placed on the corners of a macroscopic ($\sim 5 \times 5$ mm$^2$) sample, as depicted in the insets of Figs. 2, 3, 4 (a). Measurements were performed by the standard ac lock-in technique. For measurements below 1 K, we used a dilution refrigerator (Oxford Kelvinox TLD) equipped with a superconducting magnet.

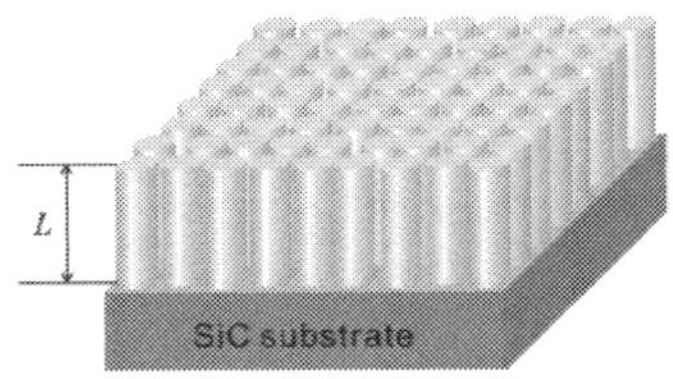


| Sample | $L$ (nm) |
|---|---|
| A | 6.5 |
| B | 22.6 |
| C | 25.0 |
| D | 65 |

**Figure 1.** Schematic drawing of a CNT-array film. Four samples with different thickness $L$ (the length of CNT), tabulated in the table, were examined.

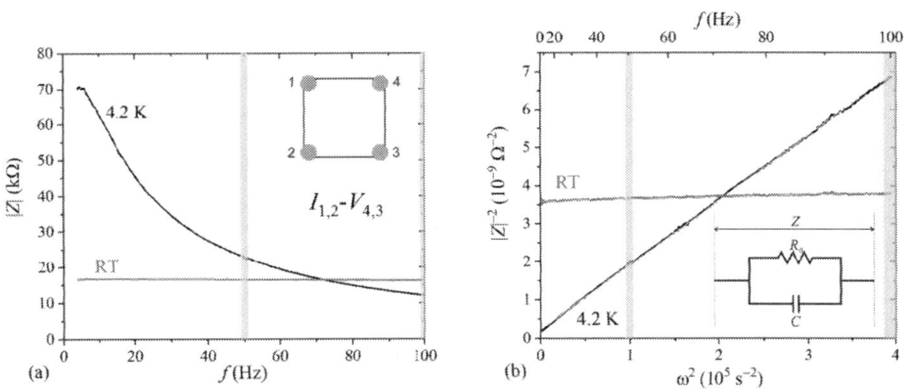


**Figure 2.** (a) Frequency dependence of the impedance $|Z|$ for sample A taken at RT and 4.2 K. Inset: schematic illustration of the contact geometry. $I_{j,k}$-$V_{l,m}$ designates the measurement configuration employing pairs of electrical contacts $j, k$ and $l, m$ as current and voltage probes, respectively. (b) Replot of the data in (a) as $|Z|^{-2}$ vs. $\omega^2$. The thin overlaid lines represent fitting to Eq. (2), using $R_0$ and $C$ as fitting parameters. Inset: equivalent circuit, represented by Eq. (1), modelling the conduction through the film. Shaded areas in both (a) and (b) indicate 50 Hz and its harmonics, where measurements are unreliable owing to the disturbance by the commercial power supply.

## 3. Results and discussion

### *3.1. Frequency dependence of the impedance at room temperature and 4.2 K*

Figure 2 (a) shows the ac impedance $|Z|$ of sample A ($L = 6.5$ nm), measured by varying the ac frequency $f$ from 4 to 100 Hz with the interval of 0.1 Hz per step. The traces taken both at room temperature (RT) and at 4.2 K are shown. At 4.2 K, $|Z|$ exhibits a remarkably rapid decrease with the increase of $f$ in this rather low frequency range, and becomes smaller than $|Z|$ for RT at $\sim$72 Hz, although the latter also slightly decreases with $f$. The rapid decrease of $|Z|$ with $f$ suggests that the capacitive coupling between adjacent CNTs plays a substantial role in the electric conduction. As the simplest possible model describing the conduction through the film, we consider an equivalent circuit containing the resistance $R_0$ and the capacitance $C$ in parallel, as depicted in the inset of Fig. 2 (b). In this model, the impedance $Z$ is given by

$$Z = \left(R_0^{-1} + i\omega C\right)^{-1}, \tag{1}$$

or, equivalently, by

$$|Z|^{-2} = R_0^{-2} + \omega^2 C^2, \tag{2}$$

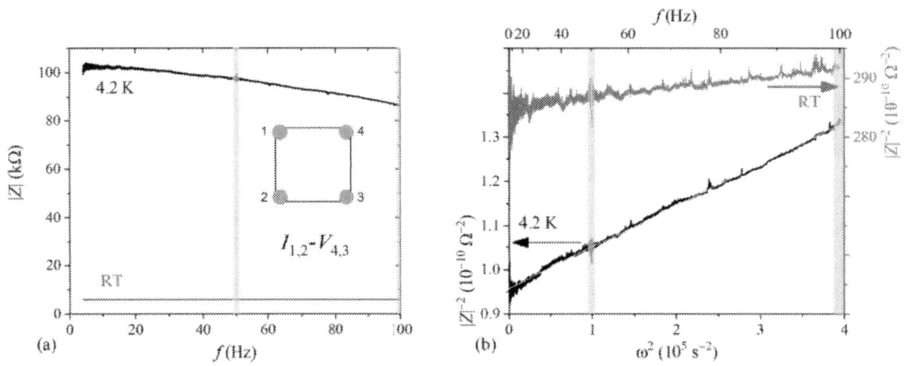


**Figure 3.** Similar to Fig. 2 for sample B. In (b), the right and left axes correspond to RT and 4.2 K traces, respectively.

with $\omega = 2\pi f$. The excellent linearity of the data replotted in the form of $|Z|^{-2}$ versus $\omega^2$ in Fig. 2 (b) demonstrates that the simple model actually describes the conduction of the film remarkably well. One can deduce the values of $R_0$ and $C$ from the intercept and the slope of the line, respectively. The values thus obtained are plotted in Fig. 5. In Figs. 3 and 4, we present the measurement results for samples B ($L = 22.6$ nm) and D ($L = 65$ nm), respectively. The decrease of $|Z|$ with $f$ becomes less apparent with the increase of the film thickness $L$. As can be seen in Figs. 3 (b) and 4 (b), the relation between $|Z|$ and $\omega$ can still basically be described by Eq. (2) for these samples (and also for sample C, not shown). The parameters $C$ and $R_0$ obtained by the fittings are plotted in Fig. 5 against $L$. At RT, $C$ increases and $R_0$ decreases monotonically with increasing $L$, as can be readily expected from the increase in the areas with which adjacent CNTs are facing. At 4.2 K, however, both $C$ and $R_0$ exhibit non-monotonic behavior, suggesting that electric conduction at low temperatures is more sensitive to details of the films, such as impurities or fluctuations in inter-CNT distances. By reducing the temperature from RT to 4.2 K, $R_0$ increases in all samples, consistent with thermally assisted inter-CNT hopping. In fact, Matsuda *et al.* observed temperature dependence consistent with variable range hopping (VRH) at 20 K $< T <$ 300 K by dc measurements [4]. The capacitance $C$ decreases except for sample A with the smallest $L$. The decrease is presumably caused by the increase in the inter-CNT

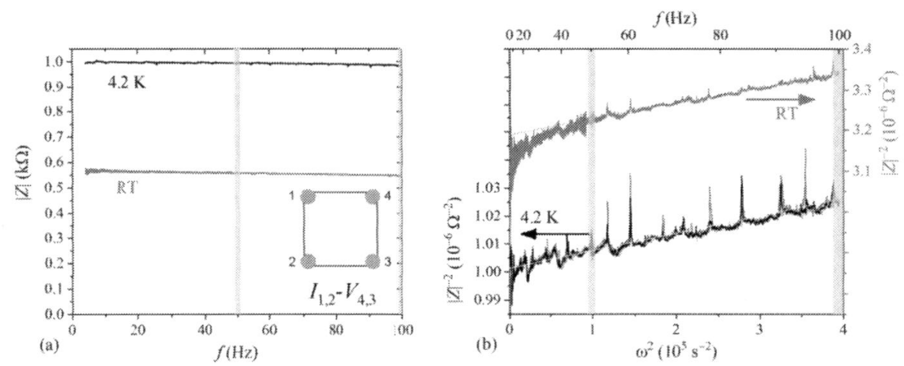


**Figure 4.** The same as Fig. 3 for sample D.

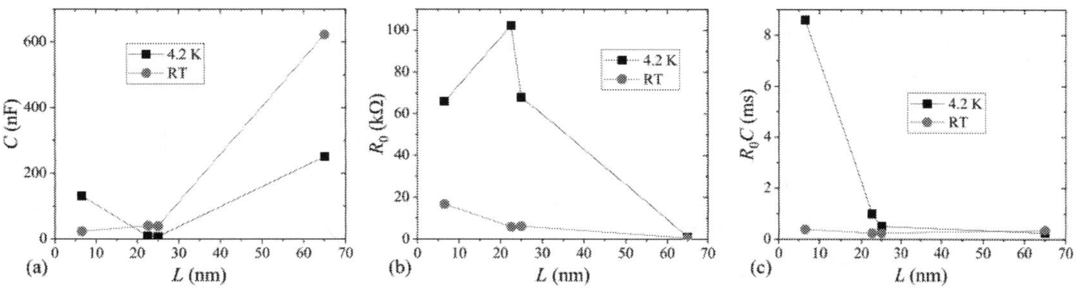


**Figure 5.** Parameters obtained by fitting to Eq. (2), plotted as a function $L$: $C$ (a), $R_0$ (b), and their product $R_0C$ (c).

distance owing to the radial contraction of CNTs on cooling. Deviation from this trend in the thinnest film may possibly be resulting from the strong influence of the SiC substrate on the CNT's elastic behavior. The product $R_0C$, which determines the rapidness of the decrease of $|Z|$ with $f$, monotonically decrease with increasing $L$.

In Fig. 4 (b), a number of sharp spikes are seen superposed on the linear background. Major peaks are observed at 42, 55, 60, 68, 73, 78, 84, 91, and 95 Hz in the trace taken at 4.2 K, and some of them (55, 60, 78 Hz) can be recognized even in the trace taken at RT. These spikes are reproducible. In fact, traces taken in three different $f$-sweeps are shown for 4.2 K in Fig. 4 (b) with slightly different grayscale tone, and all of them show the spikes at the same frequencies. Although much smaller in intensity, similar spikes are also observed in samples B (Fig. 3 (b)) and C. The spikes take place at the sample-specific random frequencies. The origin of these spikes is not clear at present, but is presumably related to resonant frequencies seen in the calculated conductivity of random capacitance networks [5, 6].

### *3.2. Frequency dependence of the impedance below 1 K*

In this and the following subsection, we focus on sample D. Figure 6 shows the temperature dependence of $|Z|$ taken below 1 K with $f = 4$ Hz. As can be seen, $|Z|$ steadily increases with decreasing temperature and starts to steeply rise below $T_r \sim 46$ mK. The frequency dependence of $|Z|$ at fixed temperatures is presented in Fig. 8 (a). For temperatures well above $T_r$, the

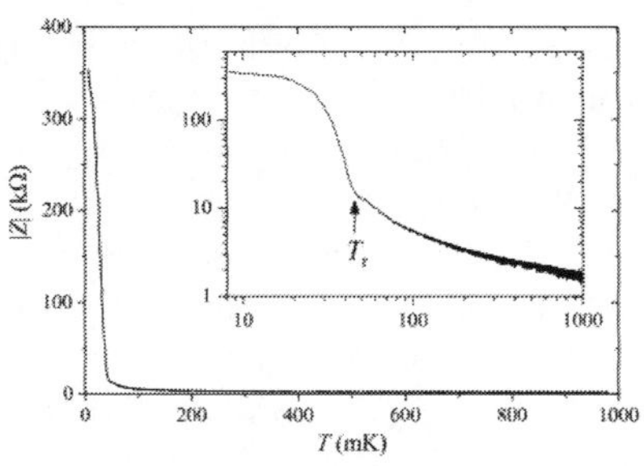


**Figure 6.** Temperature dependence of the impedance $|Z|$ taken at $f = 4$ Hz for sample D. Inset: log-log plot of the data shown in the main panel.

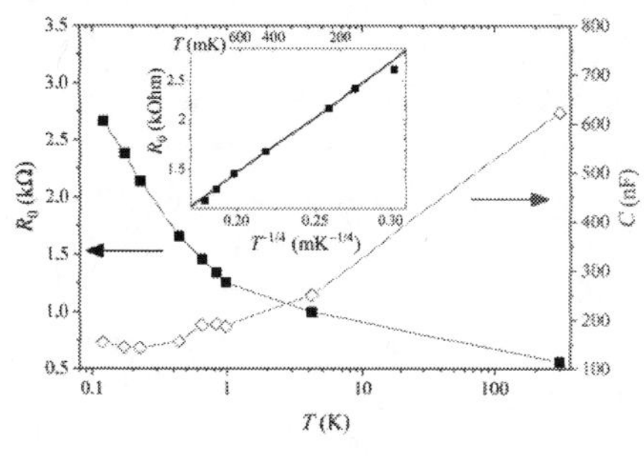


**Figure 7.** Temperature dependence of $R_0$ (left axis) and $C$ (right axis) obtained by fitting to Eq. (2) in sample D. Inset: $R_0$ in log scale vs. $T^{-1/4}$.

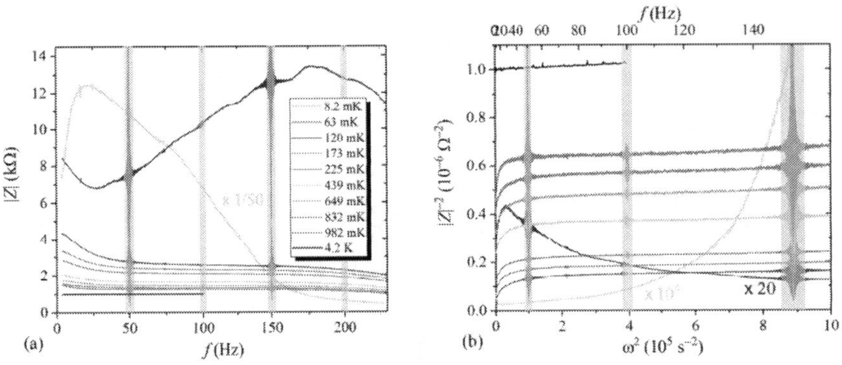


**Figure 8.** (a) Frequency dependence of the impedance $|Z|$ for sample D taken at temperatures ranging from 8.2 mK to 4.2 K. (b) Replot of the data in (a) as $|Z|^{-2}$ vs. $\omega^2$.

$|Z|^{-2}$ vs. $\omega^2$ plot shown in Fig. 8 (b) exhibits linearity above ~70 Hz, indicating that the simple equivalent circuit model, Eq. (1), is still valid in this temperature and frequency range. The parameters $C$ and $R_0$ deduced from the fittings to Eq. (2) are plotted in Fig. 7 along with those at 4.2 K and RT deduced above. $C$ and $R_0$ show monotonic increase and decrease with the temperature, respectively, revealing that the trend we saw in Fig. 5 continues down to lower temperatures. As shown in the inset of Fig. 7, $R_0$, corresponding to the resistance at $f \to 0$, is well described above ~200 mK by 3-dimensional VRH,

$$R_0 \propto \exp[(T_0/T)^{1/(d+1)}], \tag{3}$$

with $d = 3$, in agreement with the behavior observed at higher temperatures 20 K $< T <$ 300 K by dc measurements for films with $L = 65$ nm [4]. Linearity in $|Z|^{-2}$ vs. $\omega^2$ is lost around $T_r$ but restored for the lowest temperature (8.2 mK), albeit within a limited frequency range $25 \leq f \leq 80$ Hz.

### *3.3. The effect of magnetic field*

Finally, we examine the effect of the magnetic field applied perpendicular to the film. As can be seen in Fig. 9, negative magnetoresistance is observed, which becomes more prominent with decreasing temperature. At the lowest temperature (Fig. 9 (c)), $|Z|$ decreases more rapidly with $f$ around $B = 0$ T than at higher magnetic fields. Thus, the magnetoresistance turns from negative to positive at around $f = 103$ Hz. The parameters $R_0$ and $C$ deduced by the same procedure as in the previous subsections are plotted as a function of the magnetic field in Figs. 10 and 11 for $T = 225$ and 8.2 mK, respectively. For both temperatures, $R_0$ exhibits negative magnetoresistance resembling those observed by dc measurements on samples composed of bundles of randomly arranged CNTs [7, 8, 9], which is interpreted in terms of theories that take into account the quantum interference effect in the VRH conduction [10, 11].

The effective capacitance $C$ slightly increases with $B$ for the higher temperature, but decreases with $B$ for the lowest temperature. The decrease is responsible for the smaller $f$ dependence hence the transition from negative to positive magnetoresistance with $f$ observed at the lowest temperature, although the mechanism through which the magnetic field affects $C$ is not well understood at present.

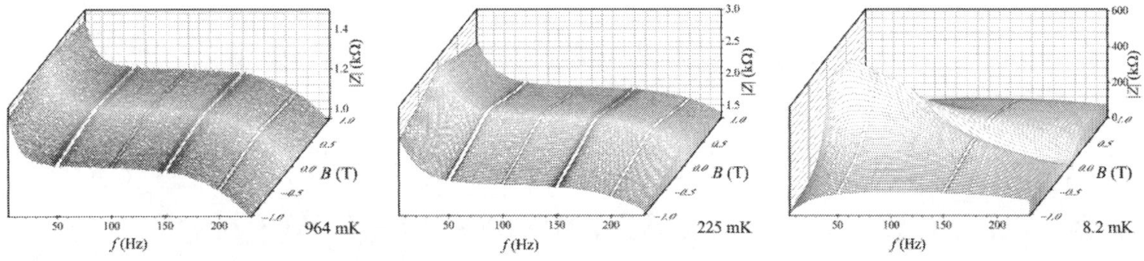


**Figure 9.** Frequency dependence of the impedance $|Z|$ for sample D taken at fixed magnetic fields ranging from −1 to 1 T with the step of 0.02 T. (a) 964 mK. (b) 225 mK. (c) 8.2 mK.

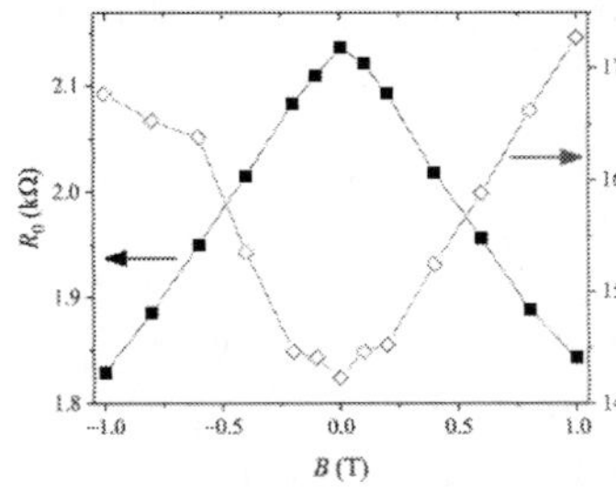


**Figure 10.** Magnetic-field dependence of $R_0$ (left axis) and $C$ (right axis) obtained by fitting to Eq. (2) at 225 mK in sample D.

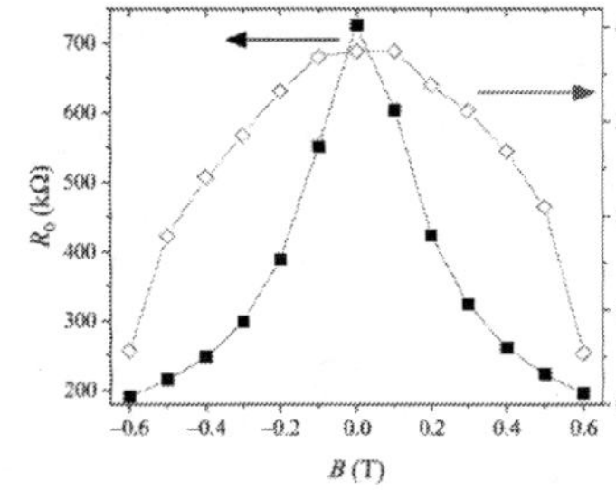


**Figure 11.** Magnetic-field dependence of $R_0$ (left axis) and $C$ (right axis) obtained by fitting to Eq. (2) at 8.2 mK in sample D.

## 4. Summary

We have measured the ac impedance $|Z|$ of films composed of closely-packed, highly-aligned carbon nanotubes. The observed decrease of $|Z|$ with the frequency $f$ was found to be described remarkably well by a simple equivalent circuit containing resistance $R_0$ and capacitance $C$ in parallel. The dependence of $R_0$ on the temperature and the magnetic field was consistent with those obtained by dc measurements. We also observed reproducible sharp spikes superposed on the $|Z|$ vs. $f$ traces.

## Acknowledgments

This work was carried out by the joint usage / research program of the Institute of Materials and Systems for Sustainability (IMaSS), Nagoya University. This work was supported by JSPS KAKENHI Grant Numbers JP25107003, JP26400311, and JP17K05491.

## References

[1] Kusunoki M, Shibata J, Rokkaku M and Hirayama T 1998 *Jpn. J. Appl. Phys.* **37** L605
[2] Kusunoki M, Suzuki T, Kaneko K and Ito M 1999 *Philos. Mag. Lett.* **79** 153–161
[3] Kusunoki M, Suzuki T, Hirayama T, Shibata N and Kaneko K 2000 *Appl. Phys. Lett.* **77** 531–533
[4] Matsuda K, Norimatsu W and Kusunoki M in preparation
[5] Olekhno N A, Beltukov Y M and Parshin D A 2015 *J. Phys. Conf. Ser.* **661** 012042
[6] Karki P and Loh Y L 2016 *J. Phys.: Condens. Matter* **28** 435701
[7] Yosida Y and Oguro I 1998 *J. Appl. Phys.* **83** 4985–4987
[8] Yosida Y and Oguro I 1999 *J. Appl. Phys.* **86** 999–1003
[9] Jaiswal M, Wang W, Fernando K A S, Sun Y P and Menon R 2007 *J. Phys.: Condens. Matter* **19** 446006
[10] Nguen V L, Spivak B Z and Shklovskii B I 1985 *Sov. Phys. JETP* **62** 1021
[11] Sivan U, Entin-Wohlman O and Imry Y 1988 *Phys. Rev. Lett.* **60** 1566–1569

IOP Conf. Series: Journal of Physics: Conf. Series **969** (2018) 012130 doi:10.1088/1742-6596/969/1/012130

# Magnetotransport of Monolayer Graphene with Inert Gas Adsorption in the Quantum Hall Regime

**A Fukuda[1], D Terasawa[1], A Fujimoto[2], Y Kanai[3], K Matsumoto[3]**

[1] Department of Physics, Hyogo College of Medicine, 1-1 Mukogawacho, Nishinomiya, Hyogo 663-8501, Japan

[2] Applied Physics, Faculty of Engineering, Osaka Institute of Technology, Osaka 535-8585, Japan

[3] The Institute of Scientific and Industrial Research, Osaka University, 8-1 Mihogaoka, Ibaraki, Osaka 567-0047, Japan

E-mail: `fuku@hyo-med.ac.jp`

**Abstract.** The surface of graphene is easily accessible from outside, and thus it is a suitable material to study the effects of molecular adsorption on the electric transport properties. We investigate the magnetotransport of inert-gas-adsorbed monolayer graphene at a temperature of 4.4 K under a magnetic field ranging from 0 to 7 T. We introduce $^4$He or Ar gas at low temperature to graphene kept inside a sample cell. The magnetoresistance change $\Delta R_{xx}$ and Hall resistance change $\Delta R_{xy}$ from the pristine graphene are measured as a function of gate voltage and magnetic field for one layer of adsorbates. $\Delta R_{xx}$ and $\Delta R_{xy}$ show oscillating patterns related to the constant filling factor lines in a Landau-fan diagram. Magnitudes of these quantities are relatively higher around a charge neutral point and may be mass-sensitive. These conditions could be optimized for development of a highly sensitive gas sensor.

## 1. Introduction

Graphene, a two-dimensional monolayer of carbon atoms, has been investigated both theoretically and experimentally, and is a topic of major interest in modern condensed-matter physics[1, 2, 3, 4]. A milestone in the study of graphene is the experimental evidence of an unusual quantum Hall effect (QHE) reported by two different groups[5, 6]. One of the major motivations for the enormous scientific interest in graphene is the technological potential of graphene such as an electric-field effect in graphene, i.e., the possibility of controlling the carrier density in a graphene sheet by application of a gate voltage. Apart from technological applications, another motivation to study graphene in fundamental research is the relativistic physics associated with the massless Dirac fermions. The relativistic QHE is understood in terms of Landau quantization for massless Dirac fermions. In contrast, in low magnetic fields, weak localization (WL) dominates the transport properties of graphene. A negative magnetoresistance as a signature of WL has been reported[7] and theoretically investigated[8]. Recent intensive investigation by our group reveals the clear relationship between universal conductance fluctuations (UCF) and the WL effect in monolayer graphene in terms of the ratio of the elastic intervalley scattering time to the inelastic dephasing time [9].

Another motivation of research on graphene is focused on its another remarkable feature - accessibility to the two-dimensional electron gas (2DEG) from outside. Direct access to the

IOP Conf. Series: Journal of Physics: Conf. Series **969** (2018) 012130 doi:10.1088/1742-6596/969/1/012130

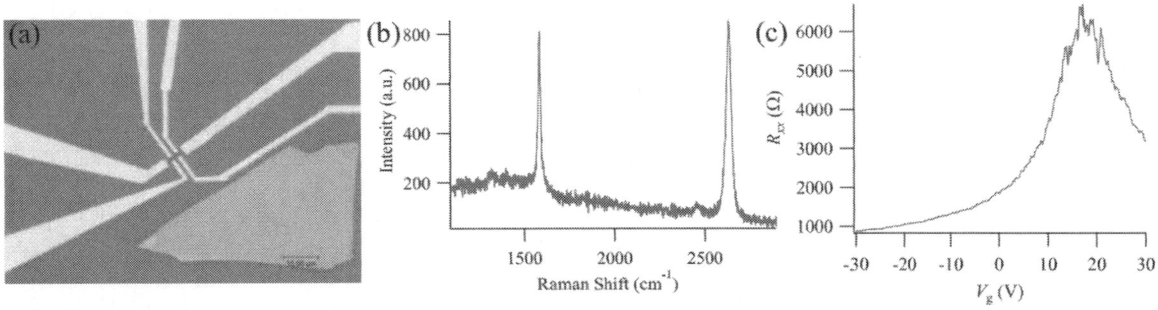


**Figure 1.** (a) Optical microscope image of our exfoliated monolayer graphene device. (b) Raman spectrum of monolayer graphene sample. (c) Zero-field resistance $R_{xx}$ as a function of gate voltage $V_g$ at 4.4 K.

2DEG using such as an optical or scanning tunnel microscope in conventional semiconductor heterostructures like GaAs/AlGaAs is extremely difficult due to the cap layers. Graphene is a suitable material to address this problem with earlier investigations on the effects of the adsorption or chemical adjunction of various kinds of molecules on its surface[10] and their applications[11]. Therefore, to study the transport properties with an adsorption or chemical modification by external molecules on graphene is intriguing. Shedin *et al.* show that sensors made from graphene can detect individual events when a gas molecule such as CO, $NH_3$, $H_2O$, or $NO_2$ attaches to or detaches from the surface of graphene[12]. Subsequently, the molecular adsorption has been investigated theoretically with a first-principles study[13]. Elias *et al.* found evidence of hydrogenated graphene called "graphane"[14], where the QHE is fully controllable by reversible hydrogenation. Their techniques are applied for such as an electrical detecting pH and protein adsorptions[15] or band-gap opening by water adsorption[16]. These works imply that graphene-based devices can be utilized as mass-sensitive or spicies-sensitive gas sensors with high sensitivity. Adsorption of an inert gas on graphene may result in extremely small changes in the transport properties, which is difficult to detect, and hence has not yet been thoroughly investigated. We previously reported magnetotransport measurements of $^4$He-adsorbed bilayer graphene at low temperatures under a magnetic field ranging from 0 to 4 T where even 10 % amount of one saturated layer of $^4$He was detectable associated with the magnetoresistance oscillation due to the WL and UCF in pristine graphene[17].

In this study, we observe the magnetotransport of monolayer graphene at low temperatures, and under a magnetic field $B$ ranging from 0 to 7 T. We measure the magnetoresistance $R_{xx}$ and Hall resistance $R_{xy}$, and the differences in the magnetoresistance $\Delta R_{xx}$ and Hall resistance $\Delta R_{xy}$ between pristine graphene and $^4$He or Ar-adsorbed graphene, where the layer number of adsorbed atoms is precisely controlled. We also discuss the possibility of utilizing the graphene device as a highly sensitive sensor for adsorbed gas.

## 2. Experiments

The monolayer graphene device is fabricated by the mechanical exfoliation method[18] from kish graphite onto the surface of $SiO_2$, 300 nm apart from the $n^+$ doped Si substrate. Optical microscope image of the exfoliated graphene device is shown in Fig. 1(a). We choose a large natural monolayer graphene flake that has an almost rectangular shape. We fabricate a Hall bar with the following dimension: 5 $\mu$m-long for the source - drain electrodes, 3 $\mu$m-long for the longitudinal voltage electrodes, and 2 $\mu$m-wide for the Hall voltage electrodes. The contact pattern is depicted using electron beam lithography. The ohmic contact materials (10-nm-thick

Pd followed by 100-nm-thick Au) are deposited by thermal evaporation followed by liftoff in warm 1-methyl-2-pyrrolidone. The number of layers is determined from the Raman shift spectra[19], which is displayed in Fig. 1(b). Using the 632.8 nm He-Ne laser, we observe two intense peaks - a G band peak at 1580 $cm^{-1}$ and a 2D band peak at 2640 $cm^{-1}$. Good Lorentzian fit and higher intensity of the 2D band peak ensures that the sample is a monolayer. The sample is first annealed at 700 K in a $H_2$ atmosphere for 30 minutes.

An overview of our cryostat for the transport measurements of graphene with gas adsorption is presented in our previous paper[17]. The sample is placed in a vacuum-tight cell with an external heater and a grafoil sheet having large effective areas of 1.2 $m^2$. The grafoil sheet is utilized to regulate and determine the precise number of layers of adsorbed molecules. The heater inside the sample cell enables *in situ* annealing even at low temperatures. The sample is annealed using the heater for several hours at 410 K. However, there is a concern that repeated thermal annealing may induce disorders and degrade the sample quality. Resistances are measured by a standard ac lock-in technique of 4-terminal measurements with a frequency of 37 Hz. Carrier densities are controlled by the back-gate voltage $V_g$ applied between the graphene sheet and the substrate. A helium-free Gifford-McMahone (GM) refrigerator is used to cool down the sample as well as the superconductor magnet with a maximum magnetic field of 8 T. A magnetic field up to 7 T is applied in this experiment.

Fig. 1(c) presents the $V_g$ dependence of the zero-field resistance $R_{xx}$ in the monolayer graphene device at 4.4 K. Initially, the charge neutral point (CNP) is located at approximately $V_{CNP} = 16.9$ V. The relationship between the carrier density $n$ and the gate voltage is given by the equation $n = (C/e) \cdot |V_g - V_{CNP}| = 7.2 \times 10^{10} \cdot |V_g - V_{CNP}|\,cm^{-2}$, where $C$ denotes the capacitance between the graphene sheet and the substrate and $e$ denotes the elementary charge. The carrier mobility differs slightly from the electron side $\mu_e$ ($V_g > V_{CNP}$) to the hole side $\mu_h$ ($V_g < V_{CNP}$), where $V_{CNP}$ is the gate voltage at the charge neutral point. Fitting of the 4-terminal resistance for $V_g < V_{CNP}$ and $V_g > V_{CNP}$ at room temperature and zero magnetic field gives $\mu_e = 0.25\,m^2V^{-1}s^{-1}$ and $\mu_h = 0.40\,m^2V^{-1}s^{-1}$, respectively. This behavior may be due to the invasive contacts with Pd [20]. It is also noteworthy that the resistance oscillation near the CNP emerges prominently due to UCF.

## 3. Results and discussion

Initially, we cool down the device of monolayer graphene to approximately 4.4 K in the vacuum-tight cell and investigate the magnetotransport properties of the pristine graphene. Figure 2(a) shows an image plot of the magnetoresistance $R_{xx} = R_{xx}^{\mathrm{pristine}}$ as a function of gate voltage $V_g$ and magnetic field $B$ up to 7 T. The data are taken for the sweep of $V_g$ at various constant values of $B$. The bright yellow area indicates the high-resistance state mainly emerging around the CNP. In high magnetic fields, oscillation in $R_{xx}$ manifests itself, though it is not perfectly quenched in the QHE due to the degradation of the sample by thermal annealing. The equi-resistance lines generally seem to be aligned along the equi-Landau filling factor $\nu$ lines, $B \propto |V_g - V_{CNP}|$, from $B = 0$ T to $B = 7$ T almost linearly, which forms the so-called Landau-fan diagram.

Next, we introduce $^4$He gas to the sample cell through a thin capillary of 0.6 mm diameter from a $^4$He tank with a constant volume at room temperature, monitoring the decrement of the gas pressure of the tank, which is translated to the amount of substance into the cell. The adsorption energy of a $^4$He atom on graphite is approximately 12.2 K[21]. Although the exact binding energy of a $^4$He atom on graphene has not yet been experimentally measured, theoretical calculation indicates that the adsorption energy on the monolayer graphene is approximately 13.4 K[22]. At our working temperature of $T = 4.4$ K, it is reasonable that almost all $^4$He atoms are homogeneously adsorbed on the grafoil together with the graphene sample, and we are able to estimate the layer number of adsorbed $^4$He atoms. Here we introduce the 1.0 layer $^4$He atoms on graphene. After equilibrium is reached, we measure $R_{xx}^{\mathrm{He}}$ as a function of $V_g$ and $B$ again.

IOP Conf. Series: Journal of Physics: Conf. Series **969** (2018) 012130 doi:10.1088/1742-6596/969/1/012130

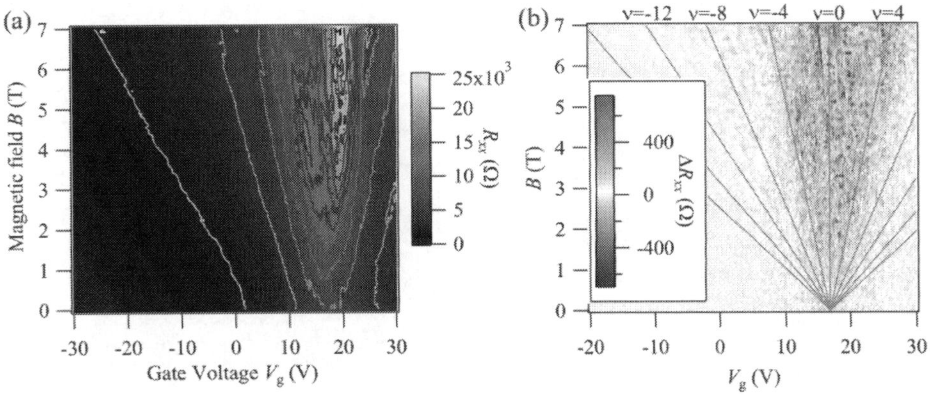


**Figure 2.** (a) Image plot of the magnetoresistance $R_{xx}$ as a function of gate voltage $V_\mathrm{g}$ and magnetic field $B$ with equi-resistance lines. (b) Image plot of the magnetoresistance change $\Delta R_{xx}$ of the monolayer graphene adsorbing one-layer $^4$He gas from the pristine one as a function of $V_\mathrm{g}$ and $B$. Green lines indicate the equi-Landau filling factor $\nu$ lines.

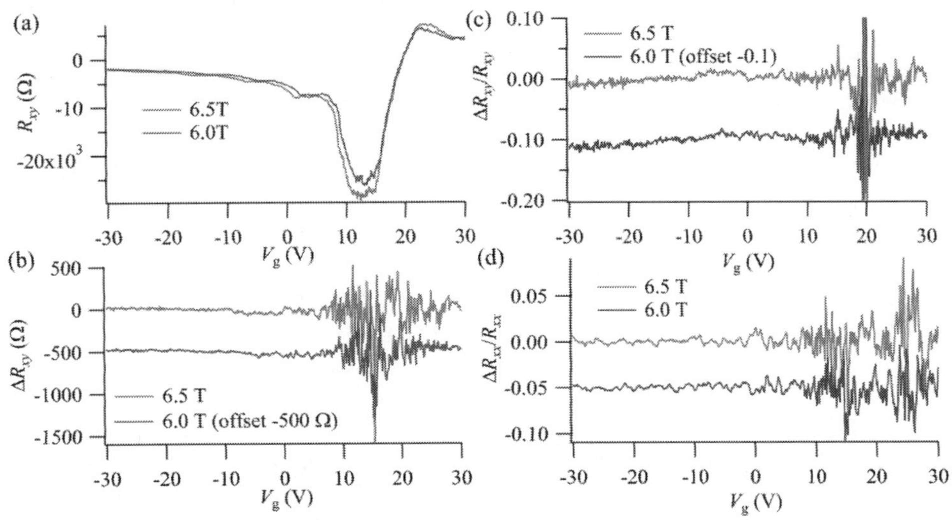


**Figure 3.** (a) Hall resistance $R_{xy}$ of the pristine graphene. (b) Hall resistance change $\Delta R_{xy}$ of the graphene adsorbing one-layer $^4$He gas from the pristine one. (c) Normalized Hall resistance change $\Delta R_{xy}/R_{xy}$. (d) Normalized magnetoresistance change $\Delta R_{xx}/R_{xx}$ as a function of $V_\mathrm{g}$ for $B = 6.0\,\mathrm{T}$ and $6.5\,\mathrm{T}$.

The resistance change $\Delta R_{xx} = R_{xx}^{\mathrm{He}} - R_{xx}^{\mathrm{pristine}}$ is demonstrated as an image plot for $B$ and $V_\mathrm{g}$ in Fig. 2(b). Blue and red colors represent the sign of $\Delta R_{xx}$. Maximum amplitude of the change ratio $|\Delta R_{xx}/R_{xx}|$ is approximately 4.0% at high magnetic field and near the CNP. Regions of each color emerge alternatively and their pattern also seems to be aligned along the equi-$\nu$ lines. It should be noted that this behavior is completely different from the bilayer case[17]. This may be because the effect of the adsorbate on the magnetotransport is larger in the monolayer adsorbent than in the bilayer one.

To investigate the effects of the $^4$He adsorption in detail, especially in the QHE region, we

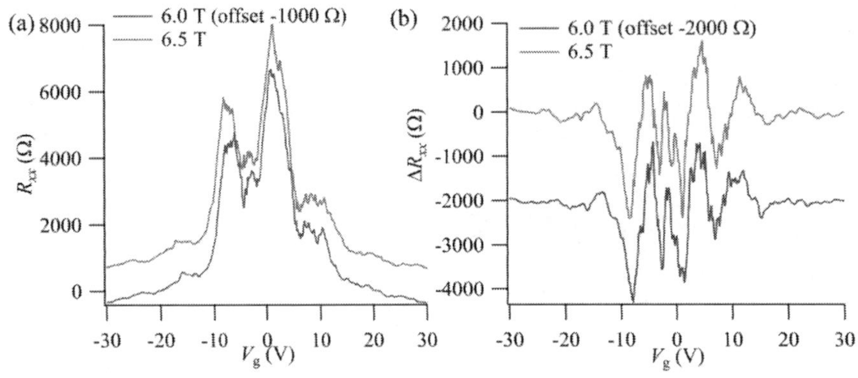


**Figure 4.** (a) Magnetoresistance $R_{xx}$ and (b) magnetoresistance change $\Delta R_{xx}$ of the graphene adsorbing one layer Ar gas from the pristine one.

study Hall resistance $R_{xy} = R_{xy}^{\text{pristine}}$. Fig. 3(a) presents the $V_g$ dependence of $R_{xy}$ for $B = 6.0\,\text{T}$ and 6.5 T. Non-perfect Hall plateaux are found, which may be affected by the disorders or impurities. Fig. 3(b) also indicates the Hall resistance change $\Delta R_{xy} = R_{xy}^{\text{He}} - R_{xy}^{\text{pristine}}$, where $R_{xy}^{\text{He}}$ denotes the Hall resistance of the graphene adsorbing one layer $^4$He gas. $\Delta R_{xy}$ has an oscillating structure around the CNP due to the UCF effects. The values of $V_g$ that yield the large amplitude of $\Delta R_{xy}$ are distributed around $V_{\text{CNP}}$. In contrast, $|\Delta R_{xy}|$ is relatively small in the quantum Hall regime; for instance, $2.1\,\text{V} < V_g < 7.0\,\text{V}$ for $B = 6.0\,\text{T}$, or $1.4\,\text{V} < V_g < 5.7\,\text{V}$ for $B = 6.5\,\text{T}$, which is reasonable due to the invariance of $R_{xy}$ in the quantum Hall plateau.

Next, we discuss the sensitivity of a monolayer graphene device for use as a gas sensor working at low temperatures and in high magnetic fields. To evaluate the resistance variation in more detail, we evaluate the normalized Hall resistance $r_{xy} = \Delta R_{xy}/R_{xy}$ as a function of $V_g$ in Fig. 3(c) together with the normalized magnetoresistance $r_{xx} = \Delta R_{xx}/R_{xx}$ in Fig. 3(d). In the case of $r_{xy}$, enormous enhancement of $|r_{xy}|$ appears at $V_g \approx 19.7\,\text{V}$, which is near the point where $R_{xy} = 0$, in contrast to the broad enhancement of $r_{xx}$ around $V_{\text{CNP}}$. Although the significance of the absolute resistance change or the normalized one is controversial, selectivity among $\Delta R_{xx}$, $\Delta R_{xy}$, $r_{xx}$ or $r_{xy}$ is beneficial for obtaining higher sensitivity as the situation demands for various parameters such as magnetic fields, temperatures, or gate voltages.

Finally, we present preliminary results regarding the mass effects on magnetotransport. Fig. 4(a) and (b) demonstrate the magnetoresistance $R_{xx}$ and its change $\Delta R_{xx}$ of the graphene adsorbing one layer Ar gas from the pristine one, respectively, as a function of $V_g$ for $B = 6.0\,\text{T}$ and 6.5 T. These data are taken after thermal annealing of the sample *in situ*; therefore, CNP moves to approximately 0 V. In Fig. 4(a), imperfect quantum oscillations originating in the QHE are also visible, as is the case with the initial state. Broader and larger-amplitude $\Delta R_{xx}$ oscillations are observed than in the case of $^4$He adsorption. The magnitude of $|\Delta R_{xx}/R_{xx}|$ is achieved up to several tenths of a percent. However, repeated thermal cycles of the sample may increase the disorder, and evacuations of the sample cell may introduce impurities on the graphene, which affect the transport properties. Consideration of disorder or impurities may be essential to the variation in $R_{xx}$ or $R_{xy}$. Though the data are still primitive, and further careful investigation into the effects of disorder or impurities on the magnetotransport using various grades of samples is required, the results imply that the monolayer graphene device is promising for the development of a highly sensitive gas sensor for heavier atoms.

In conclusion, we have carried out the magnetotransport measurements of $^4$He or Ar-adsorbed monolayer graphene with disorder or impurities at a temperature of 4.4 K under a magnetic field $B$ ranging from 0 to 7 T. The magnetoresistance difference $\Delta R_{xx}$ and the Hall resistance difference $\Delta R_{xy}$ from the pristine graphene oscillate with the gate voltage $V_g$ and $B$ for one layer of $^4$He or Ar-adsorbed graphene, and this behavior is related to the Landau filling factor. Magnitudes of both quantities strongly depend on $V_g$ and are generally larger in the region around the CNP while being smaller in the QHE regime. The results obtained in this study indicate that mass effects of inert atoms may facilitate development of highly sensitive gas sensors.

## Acknowledgments

We acknowledge A. Sawada and H. Yayama for the experimental setup at low temperature and high magnetic field with a helium-free cryostat. We thank T. Nakajima for machining the cryostat parts and the electronics of the experiments. We are thankful to T. Terashima and S. Sasaki for permitting us to use the clean room facilities in the LTM center of Kyoto University. This research was performed under the Cooperative Research Program of "Network Joint Research Center for Materials and Devices"(Grant No. 2011A10, No. 2013A18, No. 20161202, and No. 20161210). This work was supported by the "Topological Quantum Phenomena" (No. 25103722) Grant-in Aid for Scientific Research on Innovative Areas from the MEXT of Japan, JSPS KAKENHI Grant Number 24540331, No. 25870966 and No. 15K05135, Grants-in-Aid for Researchers from Hyogo College of Medicine (2010 for D. T. and 2011 for A. F.) and Grant for Basic Science Research Projects from the Sumitomo Foundation (Grant No. 100711).

## References

[1] A. H. Castro Neto, F. Guinea, N. M. R. Peres, K. S. Novoselov and A. K. Geim, *Rev. Mod. Phys.* **81**, 109 (2009).
[2] N. M. R. Peres, *Rev. Mod. Phys.* **82**, 2673 (2010).
[3] S. Das Sarma, S. Adam, E. H. Hwang and E. Rossi, *Rev. Mod. Phys.* **83**, 407 (2011).
[4] V. N. Kotov, B. Uchoa, V. M. Pereira, F. Guinea, and A. H. Castro Neto, *Rev. Mod. Phys.* **84**, 1067 (2012).
[5] Y. Zhang, Y. W. Tan, H. L. Stormer, and P. Kim, *Nature (London)* **438**, 201 (2005).
[6] K. S. Novoselov, A. K. Geim, S. V. Morozov, D. Jiang, M. I. Katsnelson, I. V. Grigorieva, S. V. Dubonos and A. A. Firsov, *Nature (London)* **438**, 197 (2005).
[7] S. V. Morozov, K. S. Novoselov, M. I. Katsnelson, F. Schedin, L. A. Ponomarenko, D. Jiang, and A. K. Geim, *Phys. Rev. Lett.* **97**, 016801 (2006).
[8] E. McCann, K. Kechedzhi, V. I. Fal ' ko, H. Suzuura, T. Ando, and B. L. Altshuler, *Phys. Rev. Lett.* **97**, 146805 (2006).
[9] D. Terasawa, A Fukuda, A. Fujimoto, Y. Ohno, Y. Kanai and K. Matsumoto, *Phys. Rev. B* **95**, 125427 (2017).
[10] L. Kong, A. Enders, T. S. Rahman and P. A. Dowben, *J. Phys: Condense. Matter* **26**, 443001 (2014).
[11] Y. Kim, Y. Kim, K. Novoselov and B. H. Hong, *2D Materials* **2**, 042001 (2015).
[12] F. Schedin, A. K. Geim, S. V. Morozov, E. W. Hill, P. Blake, M. I. Katsnelson and K. S. Novoselov, *Nat. Mat.* **6**, 652 (2007).
[13] O. Leenaerts, B. Partoens, and F. M. Peeters, *Phys. Rev. B* **77**, 125416 (2008).
[14] D. C. Elias, R. R. Nair, T. M. G. Mohiuddin, S. V. Morozov, P. Blake, M. P. Halsall, A. C. Ferrari, D. W. Boukhvalov, M. I. Katsnelson, A. K. Geim and K. S. Novoselov, *Science* **323**, 30 (2009).
[15] Y. Ohno, K. Maehashi, Y. Yamashiro, and K. Matsumoto, *Nano. Lett.* **9**, 3318 (2009).
[16] F. Yavari, C. Kritzinger, C. Gaire, L. Song, H. Gullapalli, T. Borca-Tasciuc, P. M. Ajayan, and N. Koratkar, *Small* **6**, 2535 (2010).
[17] A. Fukuda, D. Terasawa, Y. Ohno and K. Matsumoto, *J. Phys.: Conf. Ser.* **568**, 052009 (2014).
[18] K. S. Novoselov, A. K. Geim, S. V. Morozov, D. Jiang, Y. Zhang, S. V. Dubonos, I. V. Grigorieva, A. A. Firsov, *Science* **306**, 666 (2004).
[19] A. C. Ferrari, J. C. Meyer, V. Scardaci, C. Casiraghi, M. Lazzeri, F. Mauri, S. Piscanec, D. Jiang, K. S. Novoselov, S. Roth, and A. K. Geim, *Phys. Rev. Lett.* **97**, 187401 (2006).
[20] B. Huard, N. Stander, J. A. Sulpizio, and D. Goldhaber-Gordon, *Phys. Rev. B* **78**, 121402 (2008).
[21] M. W. Cole and D. R. Frankl, *Rev. Mod. Phys.* **53**, 199 (1981).
[22] M. C. Gordillo and J. Boronat, *Phys. Rev. Lett.* **102**, 085303 (2009).

IOP Conf. Series: Journal of Physics: Conf. Series **969** (2018) 012131 doi:10.1088/1742-6596/969/1/012131

# Noise and loss of superconducting aluminium resonators at single photon energies

**Jonathan Burnett, Andreas Bengtsson, David Niepce & Jonas Bylander**

Chalmers University of Technology, Department of Microtechnology and Nanoscience, Göteborg, Sweden

E-mail: burnett@chalmers.se

**Abstract.** The loss and noise mechanisms of superconducting resonators are useful tools for understanding decoherence in superconducting circuits. While the loss mechanisms have been heavily studied, noise in superconducting resonators has only recently been investigated. In particular, there is an absence of literature on noise in the single photon limit. Here, we measure the loss and noise of an aluminium on silicon quarter-wavelength ($\lambda/4$) resonator in the single photon regime.

## 1. Introduction

Over the past decade it has become clear that the main causes of decoherence in quantum circuits are bad microwave engineering, excess quasiparticles and parasitic two level systems (TLS). As the demands on quantum circuits increase, the tolerance to decoherence mechanisms decreases. Therefore, despite progress in tackling these issues, they remain the primary problem to overcome when attempting to manufacture large scale quantum systems.

Superconducting resonators have become a common tool in understanding the decoherence mechanisms of quantum circuits. When fabricated from the same materials, using the same processes, the resonator quality factor can provide a good estimate of the qubit relaxation rate [1]. This has motivated studies of the quality factor as a function of material [2], deposition [3], etching [4, 5], magnetic fields [6], microwave engineering [7] and quasiparticle trapping [8].

A prominent feature of these studies is related to the origin, nature and removal of TLS [9,10]. On the nature of TLS, recent results unambiguously demonstrated the the existence of a TLS-TLS interaction [11]. TLS-TLS interactions have been suggested within literature focusing on the low temperature properties of glasses. Specifically, such an interaction was proposed as the source of broadening in resonant absorption measurements in burosilicate glass [12] and as a potential explanation in dipole-gap measurements in thin film $SiO_x$ capacitors [13–15]. Within the context of superconducting devices, the interacting nature of TLS is revealed in the temperature dependence of $1/f$ noise [16–20]. These interactions result in TLS moving in and out of resonance with a superconducting device. This effect has been directly observed in superconducting qubits [21], where it leads to a time-varying qubit relaxation rate. Analogously, in superconducting resonators this would produce a time-varying single-photon $Q_\mathrm{i}$. While previous measurements have measured the TLS-TLS interaction, there is still no precise measurement of either the TLS-TLS interaction strength or the timescale for the interaction.

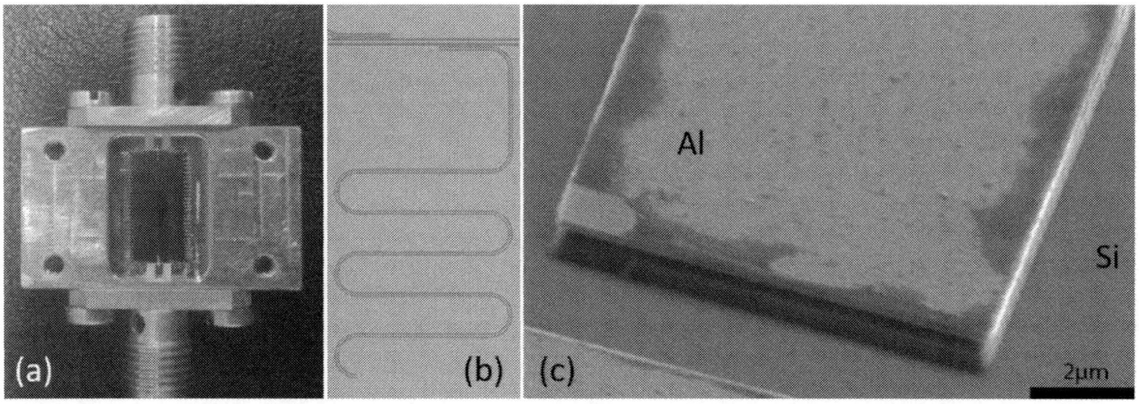


**Figure 1.** **(a)** An optical photograph of a copper sample enclosure, with sample in the centre.**(b)** An optical photograph of a $\lambda/4$ resonator, capacitively coupled to a microwave transmission line. The Si substrate is shown by the darker gray tone, while the Al superconductor is shown by the lighter gray tone. The ground plane contains holes for flux trapping. **(c)** A scanning electron micrograph of the open end of the central conductor of the resonator. The image is taken at an angle to demonstrate the trenching of the Si substrate.

In principle, these would be most directly found by performing noise measurements at single photon energies. However, to date noise studies have proved non trivial to perform for fewer than 10 photons in the resonator [16–18, 22].

Here, we study the loss and noise of a superconducting aluminium $\lambda/4$ resonator. We demonstrate a level of loss which is comparable to the literature. We then study the noise of this resonator at single photon energies. This opens up the possibility of directly measuring noise in superconducting qubits as well as further examining the nature of interacting TLS which are the limiting factor for many quantum circuits.

## 2. Sample details

Fabrication of the device begins with a solvent clean of a high resistivity silicon wafer. Following this, the wafer is submerged in a 2% hydrofluoric acid bath to remove the native surface oxide and passivate the surface with hydrogen. Within 3 minutes, the wafer is placed under vacuum inside the load lock of a Plassys MEB deposition system. The wafer is then heated to 300$^o$C while the vacuum chamber pumps. Once the wafer has cooled to room temperature and a base pressure of 1.1x10$^{-7}$ mbar is reached, 150 nm of Al is deposited at a rate of 0.5 nm/s. Next, the vacuum chamber is filled to 10 mbar of 99.99% pure molecular oxygen for 10 minutes, after this the chamber vented to atmosphere. A 1.2 $\mu$m thick layer of AZ1512HS photoresist is then patterned by direct-write laser lithography to realise the microwave circuitry. The photoresist is developed in AZ developer diluted with $H_2O$ 1:1, which minimises the parasitic etching of aluminium. This pattern is transferred into the Al film by a wet etch in a mixture of phosphoric, nitric, and acetic acids. Then, a reactive ion etch using an inductively coupled $NF_3$ plasma was used to isotropically etch the Si substrate, forming a 1 $\mu$m deep trench with a 400 nm undercut below the Al features. After dicing, the resulting chip is cleaned using hot solvents, then wirebonded within a light-tight connectorised copper sample enclosure (shown in Fig. 1a). This sample enclosure is then placed on a gold-plated copper cold finger at the 9.5 mK stage of a dilution refrigerator. A photograph of a typical microwave resonator, and a SEM image of an etched area, are shown in Fig. 1 b and c, respectively. The black residues close to the aluminium edge are indicative of burnt resist from the RIE process. This is supported by samples that only had

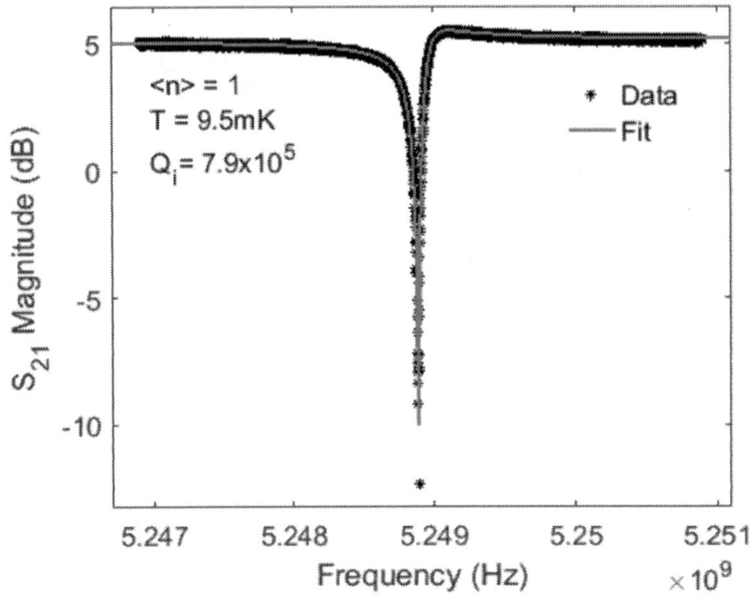


**Figure 2.** Plot of the $S_{21}$ magnitude response of the superconducting resonator. In red is a fit to the data which extracts the resonator parameters. The measurement was performed at 9.5 mK, with $P_{\rm app}$ = -146 dBm, which corresponds to $\langle n \rangle = 1$.

a wet etch not showing these residues.

## 3. Dielectric loss measurements

The $S_{21}$ transmission response of the superconducting resonator is measured at 9.5 mK while the microwave power is varied. A traceable fit routine [23] is used to extract the resonant frequency ($\omega_0$), internal quality factor ($Q_{\rm i}$) and coupling quality factor ($Q_{\rm c}$). Figure 2 shows the fitted $S_{21}$ magnitude response of the resonator at 9.5 mK. For an applied microwave power ($P_{\rm app}$) of -146 dBm, this reveals $\omega_0/2\pi$ = 5.24889 GHz, $Q_{\rm c} = 33\times10^3$ and $Q_{\rm i} = 7.9\times10^5$. As $P_{\rm app}$ is changed, $Q_{\rm i}$ is found to vary due to depolarisation of TLS [2–5, 16–18]. The average number of photons within the resonator ($\langle n \rangle$) varies with $P_{\rm app}$ as [5]

$$\langle n \rangle = \frac{\langle E_{\rm int} \rangle}{\hbar\omega_0} = \frac{2}{\hbar\omega_0^2}\frac{Z_0}{Z_{\rm r}}\frac{Q_{\rm l}^2}{Q_{\rm c}}P_{\rm app} \tag{1}$$

where $Z_0$ is the characteristic impedance ($Z_0 = 50\Omega$), $Z_r$ is the resonator impedance ($Z_r$ is chosen to be close to $50\Omega$), $\langle E_{\rm int} \rangle$ is the average energy stored within the resonator and $Q_{\rm l}$ is the loaded quality factor ($1/Q_{\rm l} = 1/Q_{\rm c} + 1/Q_{\rm i}$). Figure 3 shows the effect of TLS depolarisation in a measurement of $Q_{\rm i}$ as a function of $\langle n \rangle$. This TLS depolarisation can be fit to a TLS-based loss model [17]

$$\frac{1}{Q_{\rm i}} = F\delta^0_{\rm TLS}\frac{\tanh\left(\frac{\hbar\omega_0}{2k_BT}\right)}{\left(1+\left(\frac{\langle n \rangle}{n_c}\right)\right)^{\beta}} + \delta_{\rm other} \tag{2}$$

where $F$ is the filling factor describing the ratio of $E$-field in the TLS host volume to the total volume. $\delta^0_{\rm TLS} = 1/Q_{\rm TLS}$ is the TLS loss tangent, $n_c$ is the critical number of photons within

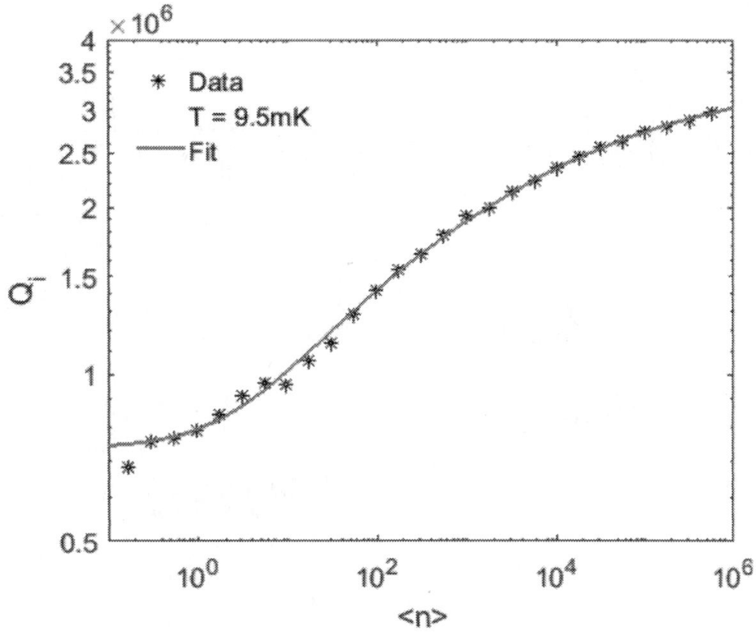


**Figure 3.** Plot of $Q_\mathrm{i}$ as a function of $\langle n \rangle$ for the resonator at 9.5 mK. In red is a fit to TLS losses described by Eq. (2).

the resonator to generate the $E$-field required to saturate one TLS. $\delta_\mathrm{other}$ is the contribution from non-TLS loss mechanisms, which are generally associated with loss at high microwave drives. A fit to Eq. (2) is shown in Fig. 3. This shows $\delta_\mathrm{other} \approx 1/(3.5\times10^6)$, while indicating that $F\delta^0_\mathrm{TLS} \approx 1/(8.7\times10^5)$. Both of these loss rates are comparable with those found for the best Al resonators [3]. Further improvements to $\delta_\mathrm{other}$ are possible by improving the infrared filtering [24] or the magnetic screening [6]. While further improvements to $\delta^0_\mathrm{TLS}$ are also possible with either increased trenching of the substrate [4,5] or by cleaning of resist residues [25] which are found in Fig. 1c.

## 4. Noise measurements

The loss induced by the TLS corresponds to the resonant absorption of microwave photons. Recent experiments have examined the role of these resonant TLS in contributing noise to the resonator [16,22]. The dependence of TLS-induced noise has been measured as a function of $P_\mathrm{app}$ and temperature. Within these studies, the temperature dependence has been thoroughly examined in the range of 50–700 mK. However, the span of $P_\mathrm{app}$ that was examined corresponds to $\langle n \rangle$ = 7–$10^4$. Consequently, the TLS noise in the limit of single-photon excitation has not been examined. Not only is this limit most relevant to dephasing in superconducting qubits, but it is also relevant to revealing properties of TLS in general.

A Pound setup is used to form a frequency locked loop which can continuously monitor $\omega_0(t)$. The Pound setup we use is identical to one previously used in the study of low frequency noise in superconducting resonators [16,26,27]. Here, the Pound setup is operated with a low bandwidth of 300 Hz. This improves the signal-to-noise, enabling the Pound setup to monitor a resonator at single photon energies. However, the low bandwidth means that this setup is only suitable for studying 'slow' fluctuations. Figure 4a shows a 500 s window of a measurement of the frequency jitter of the resonator, measured at $\langle n \rangle = 1$ and 9.5 mK. This frequency jitter can be better understood by examining the spectrum of frequency fluctuations ($S_y$). This is obtained using

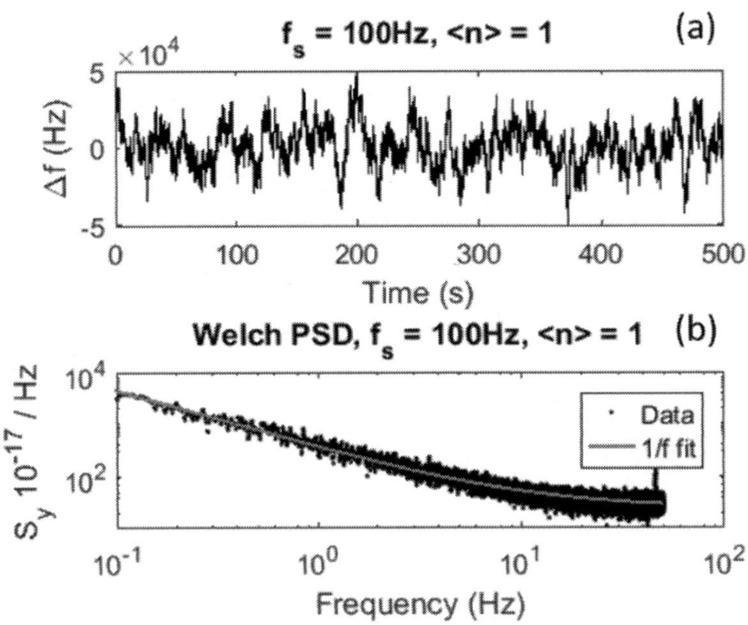


**Figure 4.** **(a)** A plot of the raw frequency jitter of the resonator measured at $\langle n \rangle = 1$ and at a temperature of 9.5 mK. **(b)** A plot of the spectrum of fractional frequency fluctuations ($S_Y$). In red is a fit to a $1/f$ noise model shown in Eq. (3).

the Welch spectral density estimate with a 50% overlap and a Hanning window. The resulting plot of $S_y$ is shown in Fig. 4b; this data is fit to a general noise model

$$S_y(f) = \frac{h_{-1}}{f^\alpha} + h_0 \qquad (3)$$

where $h_0$ is a white frequency noise level, $h_{-1}$ is a flicker frequency noise level and $\alpha$ is an exponent describing the strength of low frequency noise components. When $\alpha = 1$, the first term represents a true flicker noise process. From this fit we find that the white noise level is described by $h_0 = 2.5{\times}10^{-16}$, while the flicker noise level is described by $\alpha = 1.05$ and $h_{-1} = 3.5{\times}10^{-15}$. This level of noise is larger than that previously observed [16–18]. However, since we measure at both lower microwave powers and at lower temperatures, this is expected from the strong power and temperature dependence of dielectric noise [20].

## 5. Summary

In summary, we measured frequency noise of a high-Q superconducting resonator at single photon energies. This is an important step towards studies of the interacting nature of TLS, which are currently limiting the performance of many superconducting circuits. In showing that noise can be measured at single photon energies, the technique could be extended to the circuit-QED architecture. Here, a superconducting qubit in the dispersive regime shifts the resonator frequency with $\chi = -g^2/\Delta$, where $g$ is the qubit–resonator coupling and $\Delta$ is the frequency detuning between the qubit and the resonator. The frequency shift implies that any noise of the qubit frequency will get mapped to a frequency noise of the resonator. For normal values of $g$ and $\Delta$ the effective noise of the resonator will be between 0.01 and 0.001 times than that of the qubit. This gives a straightforward method to measure flux noise of qubits by measuring the frequency noise of the resonator, without needing to use advanced pulse sequences [28, 29]. It is important to point out that such measurement should be performed at sub single photon levels in the resonator, to avoid additional noise due to the AC-stark effect [30]. To have a stable locking of the Pound loop at such low energies, a parametric amplifier between the sample and the semiconductor amplifier, would be needed. We believe that this technique could lead to more efficient investigations of the origins of flux noise in superconducting circuits [31–33].

### Acknowledgments

We acknowledge fruitful discussions with Andrey Danilov and the mechanical work from Lars Jönsson. Financial support came from the Knut and Alice Wallenberg foundation, and the Swedish research council.

[1] A. Dunsworth *et al.*, arXiv preprint arXiv:1706.00879 (2017).
[2] J. M. Sage, V. Bolkhovsky, W. D. Oliver, B. Turek, and P. B. Welander, Journal of Applied Physics **109**, 063915 (2011).
[3] A. Megrant *et al.*, Applied Physics Letters **100**, 113510 (2012).
[4] M. Sandberg *et al.*, Applied Physics Letters **100**, 262605 (2012).
[5] A. Bruno *et al.*, Applied Physics Letters **106**, 182601 (2015).
[6] B. Chiaro *et al.*, Superconductor Science and Technology **29**, 104006 (2016).
[7] J. Wenner *et al.*, Superconductor Science and Technology **24**, 065001 (2011).
[8] I. Nsanzineza and B. L. T. Plourde, Phys. Rev. Lett. **113**, 117002 (2014).
[9] C. Müller, J. H. Cole, and J. Lisenfeld, arXiv preprint arXiv:1705.01108 (2017).
[10] A. Romanenko and D. I. Schuster, arXiv preprint arXiv:1705.05982 (2017).
[11] J. Lisenfeld *et al.*, Nature Communications **6**, 6182 (2015).
[12] W. Arnold and S. Hunklinger, Solid State Communications **17**, 883 (1975).
[13] D. J. Salvino, S. Rogge, B. Tigner, and D. D. Osheroff, Phys. Rev. Lett. **73**, 268 (1994).
[14] H. M. Carruzzo, E. R. Grannan, and C. C. Yu, Phys. Rev. B **50**, 6685 (1994).
[15] A. L. Burin, Journal of Low Temperature Physics **100**, 309 (1995).
[16] J. Burnett *et al.*, Nature Communications **5**, 4119 (2014).
[17] J. Burnett, L. Faoro, and T. Lindström, Superconductor Science and Technology **29**, 044008 (2016).
[18] A. Ramanayaka, B. Sarabi, and K. Osborn, arXiv preprint arXiv:1507.06043 (2015).
[19] L. Faoro and L. B. Ioffe, Physical Review Letters **109**, 157005 (2012).
[20] L. Faoro and L. B. Ioffe, Physical Review B **91**, 014201 (2015).
[21] C. Müller, J. Lisenfeld, A. Shnirman, and S. Poletto, Phys. Rev. B **92**, 035442 (2015).
[22] C. Neill *et al.*, Applied Physics Letters **103**, 072601 (2013).
[23] S. Probst, F. Song, P. Bushev, A. Ustinov, and M. Weides, Review of Scientific Instruments **86**, 024706 (2015).
[24] R. Barends *et al.*, Applied Physics Letters **99**, 113507 (2011).
[25] C. Quintana *et al.*, Applied Physics Letters **105**, 062601 (2014).
[26] T. Lindström, J. Burnett, M. Oxborrow, and A. Y. Tzalenchuk, Review of Scientific Instruments **82**, 104706 (2011).
[27] J. Burnett *et al.*, Physical Review B **87**, 140501 (2013).
[28] J. Bylander *et al.*, Nature Physics **7**, 565 (2011).
[29] F. Yan *et al.*, Phys. Rev. B **85**, 174521 (2012).
[30] D. I. Schuster *et al.*, Nature **445**, 515 (2007).
[31] P. Kumar *et al.*, Physical Review Applied **6**, 041001 (2016).
[32] S. de Graaf *et al.*, Physical Review Letters **118**, 057703 (2017).
[33] C. Quintana *et al.*, Physical Review Letters **118**, 057702 (2017).

IOP Conf. Series: Journal of Physics: Conf. Series **969** (2018) 012132 doi:10.1088/1742-6596/969/1/012132

# Effects of dynamic stress in magnetic superlattice of a monoaxial chiral magnet $Cr_{1/3}NbS_2$

**K Tsuruta[1,*], M Mito[1,2], Y Togawa[2,3], Y Kousaka[2,4], J Akimitsu[2,4], K Inoue[2,5,6]**

[1]Graduate School of Engineering, Kyushu Institute of Technology, Kitakyushu 804-8550, Japan
[2]Center for Chiral Science, Hiroshima University, Higashihiroshima, Hiroshima 739-8526, Japan
[3]Graduate School of Engineering, Osaka Prefecture University, Sakai, Osaka 599-8570, Japan
[4]Department of Physics, Faculty of Science, Okayama University, Okayama, 700-8530, Japan
[5]Department of Chemistry, Faculty of Science, Hiroshima University, Higashihiroshima, Hiroshima 739-8526, Japan
[6]Graduate School of Science, Hiroshima University, Higashihiroshima, Hiroshima 739-8526, Japan

*Current address: Research & Utilization Division, Japan Synchrotron Radiation Research Institute (JASRI), Sayo, Hyogo, 679-5198, Japan

Email: tsuruta@spring8.or.jp

**Abstract**. A monoaxial chiral magnet forms a kind of spin superlattice structure, termed chiral soliton lattice (CSL), by the application of a magnetic field $H$ perpendicular to a helical chiral axis. It has been reported that the CSL accompanies the magnetoresistance effect as well as a discrete change in magnetization and magnetoresistance. In order to verify the effect of the structural modification on the CSL state, we measured the magnetoresistance under the dynamic stress (DS) with a frequency of the order of MHz, which was applied by a piezoelectric ceramic oscillator. The steady application of DS while decreasing $H$ resulted in a suppression of the insertion of chiral soliton. On the other hand, the application of a pulse-like DS while $H$ decreased assisted the insertion of chiral soliton. These results demonstrate that DS modifies the spin structure of the monoaxial chiral magnet, and we can therefore change the activation energy for the insertion of chiral soliton while $H$ is decreased.

## 1. Introduction

Chirality refers to the fact that mirror images do not overlap like, as in the case of the relationship between right and left hands. In magnetic materials with spin chirality, the spin structure at zero magnetic field exhibits a helical structure with only a right- or left-handed helix owing to competition between symmetric interaction and antisymmetric Dzyaloshinskii–Moriya (DM) exchange interaction [1, 2]. A type B20 compound MnSi with a magnetic skyrmion state [3] and a monoaxial chiral magnetic material $Cr_{1/3}NbS_2$ with a chiral soliton lattice structure (CSL) under a magnetic field $H$ [4,

5] are well known as chiral magnets that present topological solitons. In millimeter-sized single crystals of $Cr_{1/3}NbS_2$, the characteristic shapes of the magnetic curves [6-8], magnetoresistance [9], and the discrete change in magnetization [10,11] due to the CSL formation have been reported. In micro-sized single crystals, successive discontinuity in magnetoresistance was observed, and in particular, for the condition of a decreasing $H$ at below critical magnetic field, there appeared to be a large discontinuity in terms of magnetoresistance [12]. While this phenomenon is recognized as a kind of supercooling with activation energy, details about its underlying mechanisms are not fully understood.

The magnetic properties of $Cr_{1/3}NbS_2$ are changed by structural modulations using hydrostatic pressure [13]. It is convincing that the spin texture is connected with the lattice system. We consider the resulting condition when the CSL stabilized under $H$ influences time-dependent stresses such as periodical and temporal stresses. We expect that they function as a perturbation to disturb the change in the topological soliton or as a stimulus to overcome the activation energy. It was reported that a dynamic stress (DS) with a frequency of the order of MHz affects physical properties in a semiconductor and a cupurate superconductor [14,15].

In this study, we focus on applying DS to the monoaxial chiral magnet $Cr_{1/3}NbS_2$ as a perturbation and/or stimulus to change the physical properties via the artificial manipulation of its magnetic structure. For instance, we perform two types of measurements for the electric resistance: (1) when DS is steadily applied and (2) when DS is applied in a pulse-like manner.

## 2. Experimental method

A single crystal of the monoaxial chiral magnet $Cr_{1/3}NbS_2$ was prepared for the experiments. This material belongs to space group $P6_322$, and the helical chiral axis is $c$-axis. Single crystals were prepared using the chemical transport method [16]. The single crystal was cut into sizes 10 μm ×5 μm ×0.5 μm using the focused-ion beam technique, and it was electrically connected to the gold pattern on the silicon substrate with tungsten paste. Figure 1(a) shows the SEM image of the micro-sized sample of the $Cr_{1/3}NbS_2$. The helical chiral axis, $c$-axis, was parallel to the current.

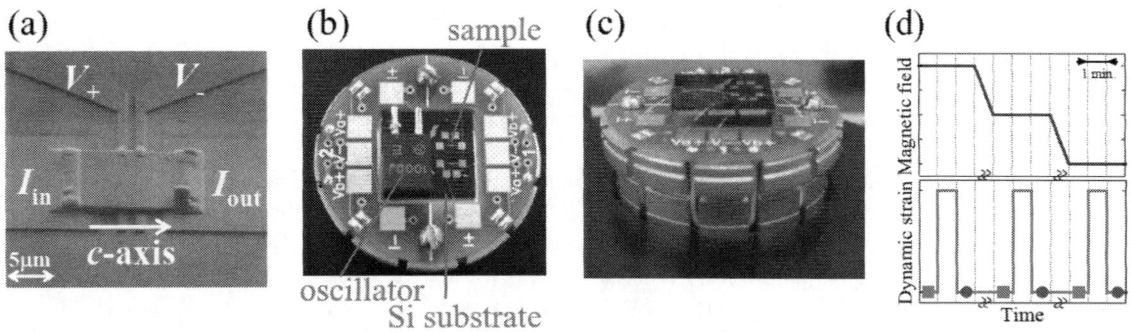


Figure 1. (color online) (a) SEM image of the micro-sized single crystal $Cr_{1/3}NbS_2$. (b)(c) Picture of the sample setting for experiments (1) and (2), respectively. (d) Relationship between timing of the applied DS and that of the changing magnetic field. Red squares and blue circles indicate the timing of measuring resistance.

A piezoelectric ceramic oscillator manufactured by Murata Manufacturing Co., Ltd. was used as the source of the DS. The resonance frequency of the oscillator was about 1 MHz. Because the resonance frequency changes with temperature, the frequency of the applied voltage to the oscillator was tuned at each measurement temperature. For the first experiment (1), the oscillator was placed beside the sample, as shown in Fig. 1(b), and for the second experiment (2), the oscillator was placed under the substrate, as shown in Fig. 1(c). In both experiments, the DS was propagated to the sample via the substrate. In (1), the $H$ dependence of the electric resistance $R$ was measured for both the increasing and decreasing $H$ under a constant application of DS. In (2), $R$ was measured at each $H$ for the decreasing $H$ immediately after applying the pulse DS for 1 min., as shown in Fig. 1(d). In both experiments, a current of 1 mA was applied to measure $R$.

## 3. Experimental results and discussion

### *3.1. Steady application of dynamic stress*

The magnetic transition temperature $T_c$ of the sample was approximately 124 K, and the critical field $H_c$ for the forced ferromagnetic state was about 2.5 kOe at a temperature ($T$) of 100 K. Figure 2(a) shows the $H$ dependence of $R$ for some magnitudes of the DS at 100 K. Here, "off" indicates the measurement without DS, and the voltage value ($V$) is the amplitude of the voltage applied to the oscillator. $V$ is treated as an equivalent to the magnitude of the DS. For zero DS, $R$ decreases as $H$ increases, while a finite hysteresis and large jump were observed as $H$ decreased. These behaviors of $R$ as a function of $H$ are consistent with results reported by Togawa et al. [12]. $H$ which large jump appeared tends to decrease with increasing $V$. Focusing on the area of hysteresis ($S$), it exhibits prominent increase for $V$ to more than 4 $V_{pp}$.

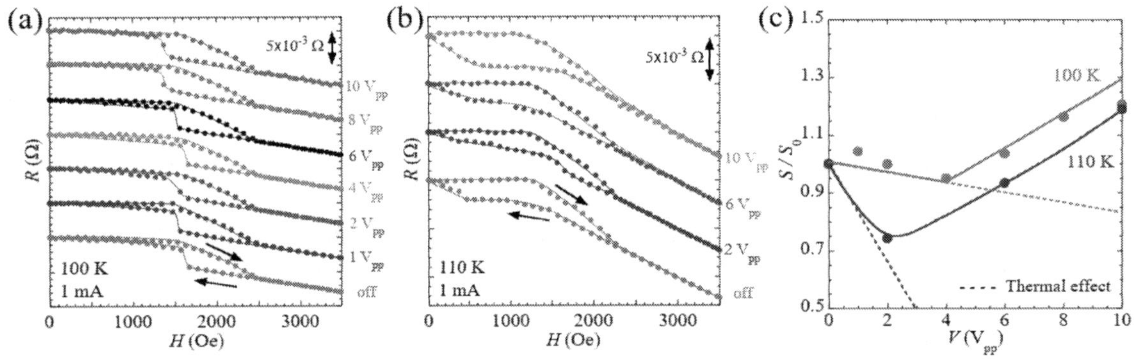


Figure 2. (color online) Magnetic-field $H$ dependence of electric resistance $R$ of $Cr_{1/3}NbS_2$ under a DS condition at $T = 100$ K (a) and 110 K (b). (c) DS dependences of the hysteresis area $S$ normalized with the hysteresis area without DS ($S_0$). The magnitude of DS is evaluated with the applied voltage $V$ to the oscillator. Broken lines are visual guides for the thermal effects caused by oscillations.

Figure 2(b) shows the $H$ dependence of $R$ for each magnitude of the DS at 110 K. By increasing $T$ from 100 K to 110 K, the shape changed and the overall hysteresis increased. When the magnitude of the DS was increased, $S$ decreased at a voltage less than 2 $V_{pp}$ whereas $S$ tended to increase at more than 2 $V_{pp}$. This behavior is qualitatively consistent with what was observed at 100 K.

Figure 2(c) shows the $V$ dependence of $S/S_0$. Indeed, when the $V$ was applied to the oscillator, the sample was slightly heated by oscillating. It was reported that the $S$ became smaller with increasing temperature to less than $T_c$ in the $T$ range, where the qualitative behavior of the $R$-$H$ curve did not change [12]. In the region where the $V$ applied to the oscillator was small, the hysteresis area became smaller with increasing $V$. This phenomenon is surely due to the heat generation caused by the oscillations, and its thermal effect is shown by visual guides in Fig. 2(c). However, when $V$ increases further, $S$ turns to increases, and this phenomenon cannot be explained by increasing the $T$ of the sample. This is considered to be the effect of stationary DS. Indeed, the ratio of the increase in $S/S_0$ as a function of $V$ is consistent with the data at 100 K and 110 K. According to the theoretical study, the elastic torsion can modify the DM vector [17]. In particular, when decreasing $H$ from the forced ferromagnetic state, the insertion of topological solitons may be restricted because the DM vector is spatially disturbed. If the above scenario is true, it is reasonable that the increase of $S$ results from the application of stationary DS.

IOP Conf. Series: Journal of Physics: Conf. Series **969** (2018) 012132 doi:10.1088/1742-6596/969/1/012132

### *3.2. Pulse-like application of dynamic stress*

We investigated the effect of a pulse-like DS at $T = 100$ K in the process of decreasing $H$, from 3500 Oe, as shown in Fig. 3. As seen in the red plots of Fig. 3(a), the DS was not applied until $H = 1645$ Oe. For reference, the green plots and broken line are the results obtained varying $H$ roughly in the situation without DS. Figure 3(b) shows an enlarged view for $H = 1640 \pm 10$ Oe. The pulse-like application of DS was carried out below 1645 Oe. The red plots show $R$ before applying the pulse-like DS, and the blue plots show the value after applying the pulse-like DS for each field. The latter measurement was conducted successively after the former measurement, and both measurements were carried out at the same $H$. The time interval of the pulse-like DS was less than 1 min., and so the heating can be neglected. From 1645 to 1636 Oe, as shown by black arrows in Fig. 3(b), $R$ did not change before or after applying the pulse-like DS. However, at $H = 1635$ Oe, as shown by the red arrow, the pulse-like application of DS changed the $R$ remarkably. This behavior is due to the magnetic transition from the forced ferromagnetic state to the CSL state. There, many topological solitons should be inserted collectively. In fact, after waiting for several tens of minutes and an increased temperature just before applying the pulse-like DS, there was no change in the electrical resistance. Thus, we confirmed that the above phenomenon does not originate in both the relaxation effect and heating one. A series of results implies that by applying pulse-like DS, we have succeeded in manipulating the activation energy for the transformation from the forced ferromagnetic state to the CSL state.

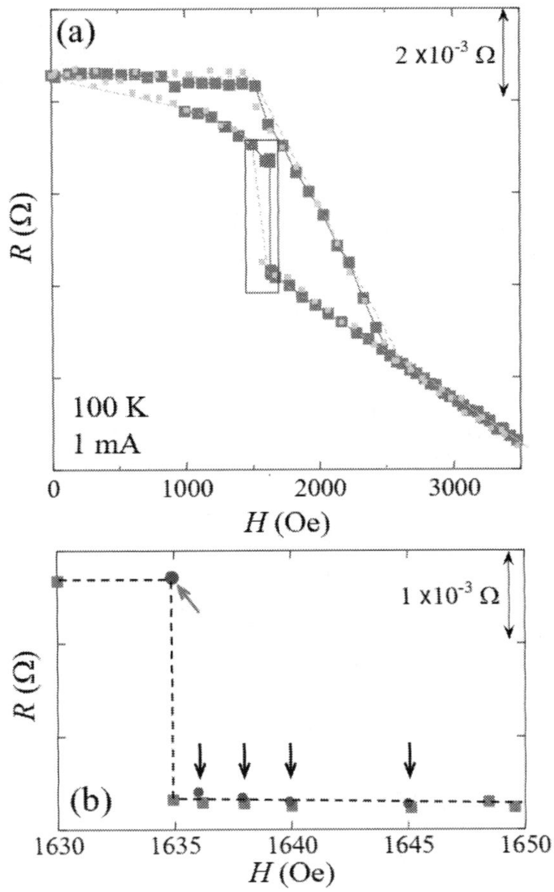


Figure 3. (color online) (a) Magnetoresistance $R(H)$ of $Cr_{1/3}NbS_2$ at $T = 100$ K in experiments involving the pulse-like application of DS. The pulse-like application was carried out below 1645 Oe. Based on preliminary experiments, the reference data shown by green plots and the broken line indicate a $R$-$H$ curve without DS. (b) $R$-$H$ curve in the process of decreasing $H$ for $H = 1640 \pm 10$ Oe, corresponding to the enlarged figure of the blue square in (a). Plots of red squares and blue circles indicate before and after the application of DS for each magnetic field, respectively. Some arrows show the field values for which the pulse-like application of DS was performed.

Dynamic stress, DS, modulates the crystal lattice. In $Cr_{1/3}NbS_2$, DM vectors could be modulated by DS. In the present experimental setting, the structural modulation has no specular directivity. Thus, we assume that there are two effects: One is heating effect that is equivalent to making a state instable thermodynamically, and it is inevitable effect in the case of stationary DS. Another is the effect of structural modulation, in which the phase transformation is restricted or motivated positively. In the present study, the stationary DS corresponds to the former, and it restricts the insertion of the topological solitons in the process of decreasing $H$. On the other hand, the pulse-like DS does to the latter, and it reduces any activation energy of the transition from the forced ferromagnetic state to the CSL state, resulting in finishing a kind of magnetic supercooling. We suppose that this activation energy of the transition from the forced ferromagnetic state to the CSL state is enhanced in thin film with clean surface. Indeed, the big jump in magnetoresistance has not been observed in the millimeter sized crystal. For any specimen for the DS experiment of $Cr_{1/3}NbS_2$, the thin film with clean surface is promising.

## 4. Conclusion

Magnetoresistance measurements were performed on a micro-sized single crystal of the monoaxial chiral magnet $Cr_{1/3}NbS_2$ in order to verify the dynamic stress (DS) effect. Based on the results, in the case of experiment (1), where the DS was steadily applied, the transition from the forced ferromagnetic state to the CSL state was suppressed during the process of decreasing magnetic field. We consider that the lattice would be modulated by the application of DS, resulting in the distortion of the DM vectors. We also succeeded in changing the state from the forced ferromagnetic state to the CSL state in experiment (2) when the DS was applied in pulses. There, we succeeded in the artificial transition from the forced ferromagnetic state to the CSL state by using dynamic stimulus. Thus, the spin-phase orders in $Cr_{1/3}NbS_2$ generated by the DM vectors are modulated by dynamic structural modulation.

**Acknowledgments**

This work was supported by a Grants-in-Aid for Scientific Research, Grant No. (S) 25220803 and Grant-in-Aid for Exploratory Research, Grant No. 15K13958 from the Ministry of Education, Culture, Sports, Science, and Technology (MEXT), Japan. This work was also supported by the Center for Chiral Science at Hiroshima University (the MEXT program for promoting the enhancement of research universities, Japan) and the JSPS Core-to-Core Program, A. Advanced Research Networks.

## References

[1] Dzyaloshinskii I E 1958 *J. Phys. Chem. Solids* **4**, 241
[2] Moriya T 1960 *Phys. Rev.* **120**, 91
[3] Mühlbauer S, Binz B, Jonietz F, Pfleiderer C, Rosch A, Neubauer A, Georgii R, and Böni P 2009 *Science* **323**, 915
[4] Kishine J, Inoue K, and Yoshida Y 2005 *Prog. Theor. Phys. Suppl.* **159** 82
[5] Togawa Y, Koyama T, Takayanagi K, Mori S, Kousaka Y, Akimitsu J, Nishihara S, Inoue K, Ovchinnikov A S, and Kishine J 2012 *Phys. Rev. Lett.* **108**, 107202
[6] Miyadai T, Kikuchi K, Kondo H, Sakka S, Arai M, and Ishikawa Y 1983 *J. Phys. Soc. Jpn.* **52**, 1394
[7] Ghimire N, McGuire M, Parker D, Sipos B, Tang S, Yan J -Q, Sales B, and Mandrus D 2013 *Phys. Rev. B* **87**, 104403
[8] Tsuruta K, Mito M, Deguchi H, Kishine J, Kousaka Y, Akimitsu J, and Inoue K 2016 *Phys. Rev. B* **93**, 104402
[9] Togawa Y, Kousaka Y, Nishihara S, Inoue K, Akimitsu J, Ovchinnikov A S, and Kishine J 2013 *Phys. Rev. Lett.* **111**, 197204
[10] Tsuruta K, Mito M, Kousaka Y, Akimitsu J, Kishine J, Togawa Y, Ohsumi H, and Inoue K 2016 *J. Phys. Soc. Jpn.* **85**, 013707

[11] Tsuruta K, Mito M, Kousaka Y, Akimitsu J, Kishine J, Togawa Y, and Inoue K 2016 *J. Appl. Phys.* **120**, 143901
[12] Togawa Y, Koyama T, Nishimori Y, Matsumoto Y, McVitie S, McGrouther D, Stamps R L, Kousaka Y, Akimitsu J, Nishihara S, Inoue K, Bostrem I G, Sinitsyn V E, Ovchinnikov A S, and Kishine J 2015 *Phys. Rev. B.* **92**, 220412
[13] Mito M, Tajiri T, Tsuruta K, Deguchi H, Kishine J, Inoue K, Kousaka Y, Nakao Y, Akimitsu, J 2015 *J. Appl. Phys.*, **117**(18), 183904
[14] Tsuruta K, Mito M, Nagano T, Katamune Y, and Yoshitake T 2014 *Jpn. J. Appl. Phys.* **53**, 07KC07
[15] Irie K, Mito M, Nagano T, Tsuruta K, and Nobukiyo S 2014 *Jpn. J. Appl. Phys.* **53**, 07KC05
[16] Kousaka Y, Nakao Y, Kishine J, Akita M, Inoue K, and Akimitsu J 2009 *Nucl. Instrum. Methods Phys. Res.*, Sect. A **600**, 250
[17] Fedorov V I, Gukasov A G, Kozlov V, Maleyev S V, Plakhty V P, and Zobkalo I A. 1997 Interaction between the spin chirality and the elastic torsion *Phys. Lett. A* **224** 372

IOP Conf. Series: Journal of Physics: Conf. Series **969** (2018) 012133 doi:10.1088/1742-6596/969/1/012133

# Bulk-edge correspondence in topological transport and pumping

**Ken-Ichiro Imura**[1], **Yukinori Yoshimura**[1,2], **Takahiro Fukui**[3], **Yasuhiro Hatsugai**[4]

[1]Department of Quantum Matter, AdSM, Hiroshima University, 739-8530, Japan
[2]MathAM-OIL, AIST & Tohoku University, Sendai 980-8577, Japan
[3]Department of Physics, Ibaraki University, Mito 310-8512, Japan
[4]Institute of Physics, University of Tsukuba, Tsukuba 305-8571, Japan

E-mail: imura@hiroshima-u.ac.jp

**Abstract.** The bulk-edge correspondence (BEC) refers to a one-to-one relation between the bulk and edge properties ubiquitous in topologically nontrivial systems. Depending on the setup, BEC manifests in different forms and govern the spectral and transport properties of topological insulators and semimetals. Although the topological pump is theoretically old, BEC in the pump has been established just recently [1] motivated by the state-of-the-art experiments using cold atoms [2, 3]. The center of mass (CM) of a system with boundaries shows a sequence of quantized jumps in the adiabatic limit associated with the edge states. Despite that the bulk is adiabatic, the edge is inevitably non-adiabatic in the experimental setup or in any numerical simulations. Still the pumped charge is quantized and carried by the bulk. Its quantization is guaranteed by a compensation between the bulk and edges. We show that in the presence of disorder the pumped charge continues to be quantized despite the appearance of non-quantized jumps.

## 1. Introduction

Recently, the role of topology is often highlighted in condensed-matter physics. This new trend in condensed-matter dates back to a few papers published three decades ago. One of them is the so-called TKNN paper [4], in which topological interpretation was given to quantum Hall effect (QHE), which had been discovered experimentally a few years earlier. The concept of topological pump is a variant of the idea of TKNN, applied to a temporal evolution of the system, instead of to the Brillouin zone. In Ref. [4], quantization of the Hall plateaus was given an interpretation as manifestation of an underlying topological order, which encodes a nontrivial phase property of the bulk wave function. Almost at the same time, Thouless, one of the authors of Ref. [4] had the idea of applying the same scenario to the system of adiabatic pump [5]. Since then, there have been both theoretical [6] and experimental [7, 8] efforts to realize and quantify quantum mechanical pumping in electronic systems. Yet, another experimental breakthrough was brought about recently in the field of cold atoms [2, 3]. In these papers two experimental groups reported experimental realization of the idea of topological pumping *a la* Thouless in the system of cold atoms. There have been some attempts to realize quantized pumping in other physical systems [9, 10, 11]. One of the issues yet to be investigated in the experimental studies is on the role of disorder in pumping, which we focus on in this paper.

IOP Conf. Series: Journal of Physics: Conf. Series **969** (2018) 012133 doi:10.1088/1742-6596/969/1/012133

## 2. Contribution of the bulk vs. the edge: the bulk-edge correspondence

A natural way to quantify pumping is to keep truck of a temporal evolution of the many body wave function. In case of topological pump this temporal evolution is adiabatic so that following the snapshot $\bar{x}(t)$ of the center of mass (CM) of the ground state

$$\bar{x}(t) = \sum_i x_i n_i \tag{1}$$

is enough to describe the pumping where $n_i$ is a many body particle number at $x_i$ [1]. Here taking a rescaling as

$$x_i = \frac{i - i_0}{L}, \quad i_0 = \frac{L}{2}, \; (i = 1, \cdots, L) \tag{2}$$

is essential ($L$ is the system size). To demonstrate this we consider a $1+1$-dimensional (i.e., 1 spatial + 1 temporal dimensions) model; a one-dimensional (1D) model with an (effective) time-dependent potential [see Eq. (7)]. For the practical numerical simulation we employ the pump version of the so-called Harper, or Aubry-Andre model. [1] Then, we impose the boundary condition such that the system is periodic in time, while it is open i.e., with boundaries [1] in the space direction. This means that the snapshots of the CM is well-defined and periodic in time:

$$\bar{x}(t_0 + T) - \bar{x}(t_0) = 0, \tag{3}$$

where $t_0$: initial time, $T$: pumping cycle. This might seem to imply that it is impossible to quantify pumping in this way. Let us recall, however, in the typical situation we consider in topological pump, the (one-body) spectrum of the system is characterized by the existence of edge modes traversing the bulk energy gap [see FIG. 1 (a)]. Therefore, the Fermi energy set typically in the gap intersects with such edge modes in the course of the time evolution. Then, if we consider an evolution of CM of the (many-body) ground state $\bar{x}(t)$, it has two distinct parts which can be readily separable; patches of continuous curves and discrete jumps [see FIG. 1 (b)]. The jumps are necesarilly associated with edge modes *in the clean limit*, and their magnitudes $\Delta\bar{x}_{\text{jump}}$ are always quantized to be *half integral*: $\Delta\bar{x}_{\text{jump}} = \pm 1/2$. [1]

In topological insulators and related systems a one-to-one relation can be established between the appearance of edge/surface modes and the topological non-triviality in the bulk. The bulk-edge correspondence (BEC) refers to this one-to-one relation. [12] Here, in our choice of the boundary condition (open in one and periodic in the other), which was also the case in the so-called Laughlin's argument, [13] the effects of the bulk and edges are *superposed* and *interconnected*. The evolution of the CM in continuous patches is due to the bulk, while the jumps are due to the edge modes. To quantify pumping and reveal the compensating roles of the bulk and the edge, we attempt to separate the effect of the bulk and that of the edges.

To concretize this BEC we reconnect the discrete patches of the CM curve by eliminating the discontinuities, and form a continuous CM curve over the cycle. Note that the resulting continuous curve is no longer periodic in time but it acquires a net gain (or loss) $\Delta\bar{x}_{\text{net}}$ per cycle. One can interpret this $\Delta\bar{x}_{\text{net}}$ as the net pumped charge, transported *through the bulk*. This is a polarization of the bulk. Since the net effect of bulk and edge contributions cancel after a cycle [see Eq. (3)],

$$\Delta\bar{x}_{\text{net}} = -\sum_{\{j_n\}} \Delta\bar{x}_{\text{jump}}(t_{j_n}), \tag{4}$$

where the summation is over the jumps, i.e., discontinuities of $\bar{x}(t)$ due to the "appearance" or "disappearance" of an edge mode in the ground state subspace. Here, we are in a "grand-canonical point of view," [1] in which all the states below $\epsilon_F$ is occupied (in the ground state).

[1] may sound paradoxical, but this seems to be the standard terminology in the field.

IOP Conf. Series: Journal of Physics: Conf. Series **969** (2018) 012133 doi:10.1088/1742-6596/969/1/012133

$\{j_n\} = \{j_1, j_2, \cdots\}$ represents a set of time slices where the jumps occur (see Sec. 3 for its precise definition). The number $N_e$ of the occupied states below $\epsilon_F$, i.e., the number of electrons changes by 1 at such jumps. Since after a complete cycle of time $T$ this number must get back to the original value, the number of times an edge state "appears" must be equal to the number of times an edge state "disappears" so that the total number $N_{\rm jump}$ of jumps is *even*. This immediately results in that the pumped charge $\Delta\bar{x}_{\rm net}$ per cycle is quantized to be an *integer*[1].

This integral quantization of $\Delta\bar{x}_{\rm net}$ has a profound mathematical meaning. In parallel with the case of QHE one can directly relate $\Delta\bar{x}_{\rm net}$ to a topological (Chern) number [5]. In QHE, quantization of the Hall conductance $\sigma_{xy}$ was attributed to the existence of an underlying topological number. [4, 14] Here, quantization of $\Delta\bar{x}_{\rm net}$ has the same mathematical origin. In case of FIG. 1 more than two: $n_F \geq 2$ bands are fully occupied below the Fermi energy $\epsilon_F$, which is set to be between the $n_F$-th and $n_F + 1$-th bands. Then, $\Delta\bar{x}_{\rm net}$ becomes the sum of all the Chern numbers associated with a filled band:

$$\Delta\bar{x}_{\rm net} = C(n_F), \quad C(n_F) = \sum_{n=1}^{n_F} C_n, \tag{5}$$

where $C_n$ is a Chern number associated with $n$th band. Or if one rather defines $I(n_F) = -\sum_{\{j_n\}} \Delta\bar{x}_{\rm jump}(t_{j_n})$, then $I(n_F)$ represents the number of (the pair of) edge modes with suitable sign that appear at $\epsilon_F$. The edge quantity $I(n_F)$ is connected to the bulk quantity $C(n_F)$ through Eq.(4). This is the BEC relation[1, 12, 15] in the topological pump:

$$I(n_F) = C(n_F), \quad I(n_F) - I(n_F - 1) = C_{n_F}. \tag{6}$$

## 3. Quantify pumping by the snapshots

To implement the features of topological pump, it is convenient to work on a simple theoretical model. Even experimentally, such an approach is proven to be useful, e.g., in the system of cold atoms, [2, 3] already referred to. In Refs. [2, 3] the so-called Rice-Mele model was presumed, and an effective situation in which a description by the Rice-Mele model has been realized experimentally in a optical lattice. On the other hand, Ref. [1] considers the case of Harper model, in which situations represented by a high ($\geq 2$) Chern number can be readily realized. In the practical implementation we also consider this model, but for the time being, we can still work on a general 1D model with a time-dependent potential:

$$H(k_y) = \sum_{i=1}^{L} \Big( |x_{i+1}\rangle t_x \langle x_i| + |x_i\rangle t_x^* \langle x_{i+1}| + |x_i\rangle (V(x_i, k_y) + W(x_i)) \langle x_i| \Big), \tag{7}$$

where the replacement: $k_y \to 2\pi t/T$ is presumed. [1, 5] In case of the Harper model the potential term becomes $V(x_i, k_y) = 2t_y \cos(k_y - 2\pi\phi i)$, where in this original 2D representation, $\phi$ represents the strength of a magnetic flux piercing a plaquette, while $\phi$ enters the 2D hopping Hamiltonian through Peierls substitution. $W(x)$ is a site random potential distributed uniformly in the range $W(x) \in [-W_0/2, W_0/2]$ [2]. Note that $W(x)$ is randomly distributed in space, while once this distribution is chosen, it stays *static*. $W_0$ specify the strength of disorder. Eq. (7) is a Fourier transform of that 2D-hopping Hamiltonian. In Eq. (7), $|x\rangle$ represents a Bloch state $|x, k_y\rangle = \sum_y e^{ik_y y}|x, y\rangle$, while we make the replacement: $k_y \to 2\pi t/T$, in order to make Eq. (7) a 1D pump Hamiltonian.

[2] This one dimensional randomness was discussed as a fictitious one in the original 2D problem[15]. It is now realized as it is in the 1D topological pump.

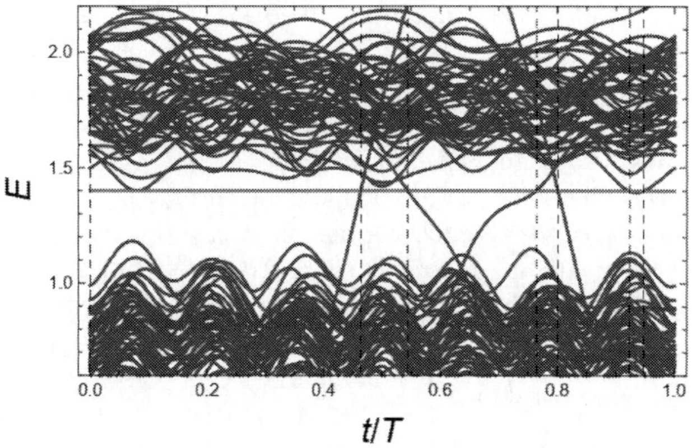


(a)

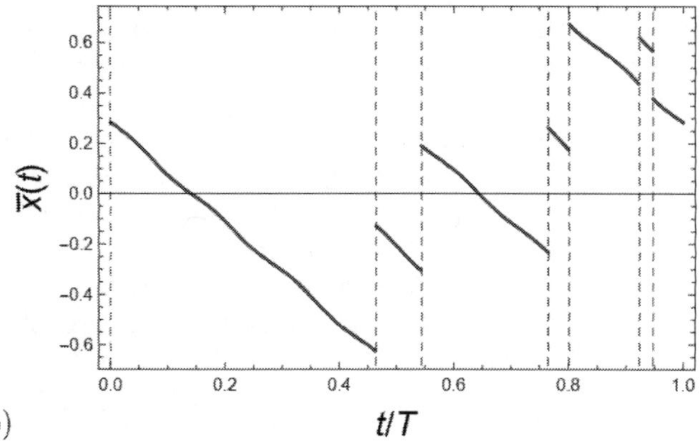


(b)

**Figure 1.** Time evolution of (a) the snapshot spectrum $\epsilon_\alpha(t)$, and (b) the center of mass $\bar{x}(t)$ of the ground state in case of the disordered Harper model. In panel (b) the magnitude of the 6 jumps are in the order of their appearance 0.497119, 0.494151, 0.494077, 0.496461, 0.182738, −0.188663. Model is given in Eq. (7). Parameters are specified toward the end of Sec. 3.

Let us consider snapshots of such a Hamiltonian as given in Eq. (7) with $k_y = 2\pi t/T$ at $t = j\Delta t$ ($j = 1, 2, 3, \cdots$). At each time slice $t_j = j\Delta t$ we diagonalize the Hamiltonian Eq. (7) and find eigenstates, which include both the bulk and the edge states. Then, we consider the ground state of the system in which all the states below $\epsilon_F$ is occupied; both at the edge and in the bulk. We typically consider the case in which the Fermi energy $\epsilon_F$ is in the gap, since the pumped charge is topologically quantized in this case. Since the present case is non interacting, the center of mass $\bar{x}(t_j)$ of the many body ground state is given as

$$\bar{x}(t_j) = \sum_{i=1}^{L} x_i n_i(t_j),\ n_i(t_j) = \sum_\alpha{}' |\psi_\alpha(x_i, t_j)|^2, \tag{8}$$

at different time slices $t_j$, where the summation $\sum_\alpha{}'$ is taken over all the occupied states $\alpha$ in the ground state. $\psi_\alpha(x_i, t_j)$ represents the eigenwavefunction corresponding to the eigenenergy $\epsilon_\alpha(t_j)$.

FIG. 1 shows the evolution of the spectrum $\epsilon_\alpha(t)$ [panel (a)]; here, the erratic behavior of the spectrum is due to the disorder potential, and that of the center of mass $\bar{x}(t)$ in the ground state [panel (b)]. In practice, we plot simply the spectrum $\epsilon_\alpha(t_j)$ and $\bar{x}(t_j)$ at different time slices to visualize their evolution. In panel (a) one can observe that four branches of edges modes appear and traverse the energy gap. In panel (b) the CM curve $\bar{x}(t)$ shows predominantly a continuous evolution except at a few [actually, six in the specific case of panel (b)] discontinuities (jumps), which occur typically when the equi-energy line at $\epsilon_F$ intersects with either of these edge branches. In panel (b) this is the case at the first four jumps, while the remaining jumps are due to impurities. Suppose that at $t = t_{j_1}$ a "right" edge mode localized in the vicinity of the right edge at $x = +1/2$. becomes *available* in the subspace of the ground state: $\epsilon \le \epsilon_F$.

Then, the number $N_e(t) = \sum_i n_i(t)$ of occupied states increases by one at this time slice: $N_e(t_{j_1}) - N_e(t_{j_1-1}) = 1$. Correspondingly, $\bar{x}(t)$ shows a quantized jump of $+1/2$; i.e., $\Delta\bar{x}_{j_1} = +1/2$. In panel (b) this seems to happen at the second and at the fourth jump. Generally, such a change of the occupied states $N_e(t)$ occurs at a set of a finite number of time slices: $t = t_{j_1}, t_{j_2}, \cdots$, and there, $\bar{x}(t)$ possibly shows discontinuities. $\{j_n\}$ in Eq. (4) specify the set of time slices $\{j_1, j_2, \cdots\}$ at which the jumps occur. In the clean limit and if $\epsilon_F$ is in the gap, all of such intersections are associated with an edge mode, and at each $t_{j_n}$ $(n = 1, 2, \cdots)$ $\bar{x}(t)$ shows a quantized jump of magnitude 1/2, since the localization length tends to zero after rescaling Eq.(2) at the extremity of the system. The sign of the jump depends on the location and the slope of the edge mode: [1]

$$\Delta\bar{x} = \frac{1}{2}\mathrm{sgn}(x_{\mathrm{edge}})[-\mathrm{sgn}(\mathrm{slope})], \tag{9}$$

where $x_{\mathrm{edge}} = \pm 1/2$ represents the location of the edge state, while when its slope is positive (negative) the state becomes empty (occupied) at the time slice in question. In other words, the factor $[-\mathrm{sgn}(\mathrm{slope})]$ is a measure of the appearance/disappearance of the state in question in/from the ground state.

In the case of panel (b) in FIG. 1 the contribution of the first four jumps (associated with an edge state) to the summation on the right hand side of Eq. (4) is close to +2, i.e., $\sum'_{\{j_n\}} \Delta\bar{x}_{\mathrm{jump}}(t_{j_n}) = +2$, where we used the notation $\sum'_{\{j_n\}}$ to make explicit that only the contribution from the edge states is considered (by just counting the discontinuities). Recall that these half-integral quantized jumps always appear *in pairs*. Accordingly, the pumped charge in the bulk becomes $\Delta\bar{x}_{\mathrm{net}} = -2$, which is indeed identical to the bulk topological (Chern) number. In the situation of panel (b) the flux $\phi$ and $\epsilon_F$ are chosen such that $\phi = 1/7$, $L = 351$ and $\epsilon_F = 1.4$ with $t_x = t_y = 1$. As a result, 5 of 7 bands are fully occupied; $\epsilon_F$ is in the gap between the 5th and 6th bands. The Chern number $C_n$ characterizing the occupied bands are $+1, +1, +1, -6, +1, +1, +1$ from the bottom to the valence band so that they sum up to $-2$.

## 4. Role of disorder: non-quantized jumps vs. quantized pumped charge

In FIG. 1(b) one can observe that in addition to the quantized jumps we have focused on so far, there are additional jumps which are not quantized and appear "trivially" in pairs. The fifth and sixth jumps in FIG. 1(b) fall on this category. These additional jumps are due to impurity states. The direction and magnitude of such jumps are such that

$$\Delta\bar{x} = x_{\mathrm{imp}}[-\mathrm{sgn}(\mathrm{slope})], \tag{10}$$

where $x_{\mathrm{imp}}$ [3] represents the location (CM) of the impurity state, while the factor $[-\mathrm{sgn}(\mathrm{slope})]$ indicates whether that appear or disappear in/from the ground state at the particular time slice. Another implication of this factor is that a given impurity state gives a pair of contributions to the summation on the right hand side of Eq. (4) with the same magnitude but with opposite signs, so that their contributions simply cancel each other. This is contrasting to the case of half-integral quantized jumps due to edge modes which also appear "in pairs" but in a different sense. In the case of quantized jumps the factor $\mathrm{sgn}(x_{\mathrm{edge}})$ allows them to evade this cancellation and can still give a non-vanishing (though integral quantized) contribution to $\Delta\bar{x}_{\mathrm{net}}$. Thanks to this cancellation, the calculated value of $\Delta\bar{x}_{\mathrm{net}}$ in the case of FIG. 1(b) is $\Delta\bar{x}_{\mathrm{net}} = -1.97588$, which is close to the ideal value $-2$ in the clean limit [in the case of two panels in FIG. 1 the strength of disorder $W_0$ is set as $W_0 = 1$]. This example shows that despite the appearance of non-quantized jumps the pumped charge can still be quantized in the presence of disorder.

[3] Due to the scaling Eq.(2), the position of the localized state is unambiguously specified as far as the localization length is finite.

## 5. Concluding remarks: comments on the experimental situations

Let us summarize what we have argued so far. For the actual time evolution to be strictly identical to the collection of snapshots, the system must be adiabatic. This is the case when the pumping cycle $T$ is long enough, satisfying the inequality $T \gg \hbar/\Delta\epsilon$, where $\Delta\epsilon$ is the characteristic energy scale. In the "bulk regime" in which $\bar{x}(t)$ shows a continuous evolution, $\Delta\epsilon = \epsilon_g$ (scale of the bulk energy gap), i.e., the adiabaticity is controlled by $\epsilon_g$. In the vicinity of the jumps, on contrary, or in the "edge regime" $\Delta\epsilon_{\rm edge} \to 0$, since the edge is gapless, so that the typical time $t_{\rm edge} = \hbar/\Delta\epsilon_{\rm edge}$ tends to be infinity. This means that for the adiabatic condition to be strictly satisfied at the edge the pumping cycle $T$ must be infinite. Fortunately, this condition will not be satisfied experimentally; the pumping cycle $T = T_{\rm exp}$ is well between the above two time scales: $t_{\rm bulk} \ll T_{\rm exp} \ll t_{\rm edge}$, i.e., the bulk is adiabatic, while the edges in experiments are non adiabatic, that is, described by the "sudden" approximation. Therefore, the jumps are not seen in the the experiments, even not in the numerical simulations of time evolution, while the "reconnection" is justified; i.e., one can safely skip the jumps. This is because in the regime of $T_{\rm exp} \ll t_{\rm edge}$, the sudden approximation is fully justified at the edge. [4] The system behaves before and after the jump as if the jump does not exist.

Finally, let us recall that the CM is only well-defined for an open system although the pumped charge is described by the polarization of the bulk, which is compensated by the discontinuities due to the edge states in the extreme adiabatic limit. Even though we may not see the jumps in the experiment, *this part must underlie* for the whole phenomenon to occur.

## Acknowledgments

The authors are supported by KAKENHI: 15K05131 (KI), 15H0370001 (KI), 17H06138 (KI, TF, YH) and 16K13845 (YH).

## References

[1] Hatsugai Y and Fukui T 2016 Phys. Rev. B **94**(4) 041102
[2] Nakajima S, Tomita T, Taie S, Ichinose T, Ozawa H, Wang L, Troyer M and Takahashi Y 2016 Nat Phys **12** 296–300
[3] Lohse M, Schweizer C, Zilberberg O, Aidelsburger M and Bloch I 2016 Nat Phys **12** 350–354
[4] Thouless D J, Kohmoto M, Nightingale M P and den Nijs M 1982 Phys. Rev. Lett. **49**(6) 405–408
[5] Thouless D J 1983 Phys. Rev. B **27**(10) 6083–6087
[6] Brouwer P W 1998 Phys. Rev. B **58**(16) R10135–R10138
[7] Switkes M, Marcus C M, Campman K and Gossard A C 1999 Science **283** 1905–1908 ISSN 0036-8075
[8] Giazotto F, Spathis P, Roddaro S, Biswas S, Taddei F, Governale M and Sorba L 2011 Nat Phys **7** 857–861
[9] Leek P J, Buitelaar M R, Talyanskii V I, Smith C G, Anderson D, Jones G A C, Wei J and Cobden D H 2005 Phys. Rev. Lett. **95**(25) 256802
[10] Kraus Y E, Lahini Y, Ringel Z, Verbin M and Zilberberg O 2012 Phys. Rev. Lett. **109**(10) 106402
[11] Hu W, Pillay J C, Wu K, Pasek M, Shum P P and Chong Y D 2015 Phys. Rev. X **5**(1) 011012
[12] Hatsugai Y 1993 Phys. Rev. Lett. **71**(22) 3697–3700
[13] Laughlin R B 1981 Phys. Rev. B **23**(10) 5632–5633
[14] Kohmoto M 1985 Annals of Physics **160** 343 – 354 ISSN 0003-4916
[15] Hatsugai Y 1993 Phys. Rev. B **48**(16) 11851–11862

[4] A. Messiah, "Quantum mechanics", chap. XVII, §7.

IOP Conf. Series: Journal of Physics: Conf. Series **969** (2018) 012134 doi:10.1088/1742-6596/969/1/012134

# Extremely large magnetoresistance in a high-quality $WTe_2$ grown by flux method

**K. Tsumura[1], R. Yano[1], H. Kashiwaya[1], M. Koyanagi[1], S. Masubuchi[2], T. Machida[2], H. Namiki[3,4], T. Sasagawa[3], and S. Kashiwaya[1]**

[1]National Institute of Advanced Industrial Science and Technology, 1-1-1 Umezono, Tsukuba, Ibaraki 305-8568, Japan

[2]Institute of Industrial Science, The University of Tokyo, 4-6-1 Komaba, Meguro-ku, Tokyo 153-8505, Japan

[3]Laboratory for Materials and Structures, Tokyo Institute of Technology, 4259 Nagatsuta, Midori-ku, Yokohama 226-8503, Japan

E-mail: kohei.tsumura@aist.go.jp

**Abstract.** We have grown single crystals of $WTe_2$ by a self-flux method and evaluated the quality of the crystals. A Hall bar-type device was fabricated from an as-exfoliated film on a Si substrate and longitudinal resistance $R_{xx}$ was measured. $R_{xx}$ increased with an applied perpendicular magnetic field without saturation and an extremely large magnetoresistance as high as 376,059 % was observed at 8.27 T and 1.7 K.

## 1. Introduction

Tungsten ditelluride $WTe_2$ is a layered transition metal dichalcogenide and has been the center of attention from various aspects. $WTe_2$ consists of tungsten chains placed between two tellurium sheets. Each layer is combined together by a weak van der Waals interaction between Te-Te, and neighboring layers are rotated by 180° due to the buckling of layers originating from a strong intermetallic bonding [1]. Monolayer is expected to be a quantum spin Hall insulator and much experimental effort has been made to reveal the topological nature [2, 3]. Bulk is considered as a type-II Weyl semimetal where the Lorentz invariance of Weyl fermion is broken [4]. One of the properties of Weyl semimetal, the chiral anomaly, has been observed as a negative magnetoresistance (MR) in a thin film [5]. On the other hand, an extremely large magnetoresistance (XMR) without saturation has been reported for a thick $WTe_2$ film [6]. A perfect electron-hole compensation is thought to be responsible for the XMR [7], however, there is a controversial argument which suggests alternative explanations [8].

Here we report on the magnetoresistance MR and the quality of $WTe_2$ grown by a flux method. We have measured a longitudinal resistance $R_{xx}$ as functions of temperature $T$ and a magnetic field $B$. Residual resistivity ratio ($RRR$) was 704 and the XMR up to 376,059 % was observed at $B$ = 8.27 T and $T$ = 1.7 K. Moreover, quantum oscillations on the $R_{xx}$ became visible from the magnetic field as small as 2.5 T. These results confirm the high quality of

[4] Present address: Tokyo Metropolitan Industrial Technology Research Institute, 2-4-10 Aomi, Koto-ku, Tokyo 135-0064, Japan

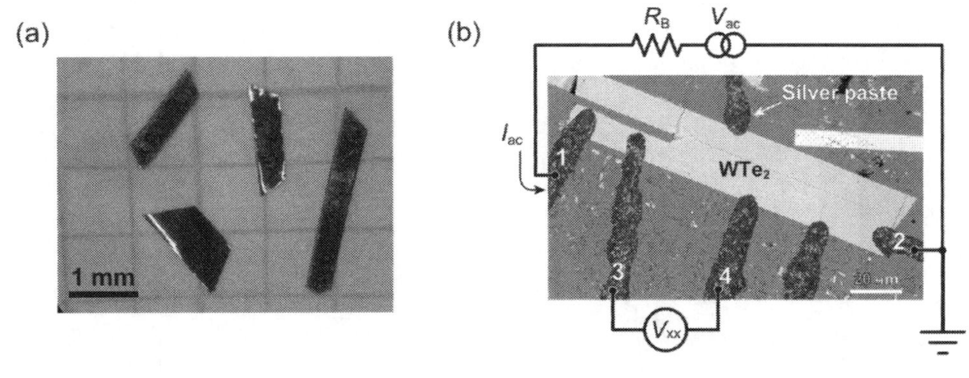


**Figure 1.** Sample structures. (a) A typical as-grown $WT_2$ by the self-flux method. (b) The sample fabricated on a $SiO_2$/Si substrate. Wirings used for transport measurements are also shown.

our crystal, which will serve as a prime physical system to study the topological properties of materials.

## 2. Experimental

Single crystals were grown by a self-flux method reported by several groups [8, 9, 10]. The mixture of polycrystalline $WTe_2$ (0.1 g) and Te (5 g) were loaded onto a silica tube and heated up to 835 °C. Then, the tube was cooled down to 535 °C at the rate of 1 °C/h. Excess Te was removed by centrifuging and the following annealing. As a result, sizable platelet crystals with a typical size of $1 \times 3$ mm$^2$ shown in Fig. 1(a) was obtained.

$WTe_2$ film was prepared by mechanical exfoliation on a Si substrate covered with a thermally oxidized $SiO_2$ layer. Au wires were directly connected to one of the exfoliated $WTe_2$ flakes using silver paste immediately after the exfoliation. Figure 1(b) shows the optical microscope image of the fabricated Hall bar-type device. The thickness of the flake is estimated to be around 100 nm judging from the brightness and the color of the flake though it has yet to be confirmed.

Transport measurements were carried out in a variable-temperature cryostat using the wirings shown in Fig. 1(b). $R_{xx} = R_{34}$ was measured by a standard lock-in technique as functions of temperature and applied magnetic fields. The magnetic field was applied both in-plane and perpendicular to the *ab*-plane up to $B = 8.27$ T.

## 3. Results and Discussion

Figure 2 shows the temperature dependence of $R_{xx}$ in a perpendicular magnetic field at $B = 0$ and 8.27 T. $R_{xx}$ decreases with temperature drop at $B = 0$ T. $RRR$ define as $R_{300\mathrm{K}}/R_{2\mathrm{K}}$ is 704. $RRR$ is the indicator of crystal quality, such as the amount of impurities and crystallographic defects. The obtained $RRR$ is not the best but fairly large compared with previous reports, which assures the high-quality of our $WTe_2$ grown by the flux method [6, 9, 11, 12]. When the field at $B = 8.27$ T is applied, the $R_{xx}$ deviates from the zero-field curve and starts to increase at $T^* = 50.5$ K, which is called the turn-on temperature [6]. We will mention this behavior later.

MR versus $B$ under an in-plane and a perpendicular magnetic fields at $T = 1.7$ K are shown in Fig. 3. Here, the MR is defined by the following equation; $\mathrm{MR} = [R_{xx}(B) - R_{xx}(0\ \mathrm{T})]/R_{xx}(0\ \mathrm{T})$. The MR in the in-plane field is no more than 6,600 %, on the other hand, the MR increases without saturation in the perpendicular field and the XMR of 376,059 % is observed at $B =$ 8.27 T. It has been shown that $WTe_2$ is a semimetal, so that both the conduction band and the

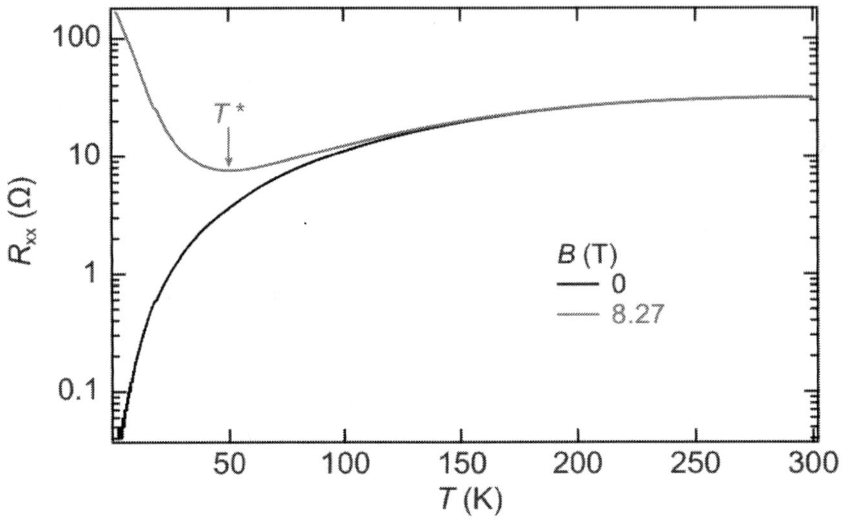


**Figure 2.** Temperature dependence of $R_{xx}$ in a perpendicular magnetic field at $B = 0$ and 8.27 T. The temperature $T^*$ where $R_{xx}$ becomes minimum is indicated by arrow.

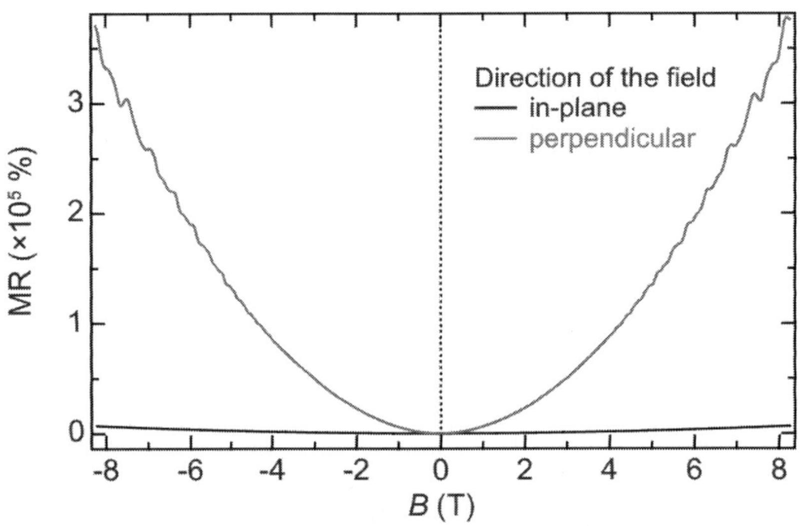


**Figure 3.** Magnetoresistance MR versus applied magnetic field $B$ at $T = 1.7$ K.

valence band cross the Fermi energy at the same time [6]. Moreover, it has been also identified that the electron density $n_e$ is perfectly compensated with the hole density $n_h$ in $WTe_2$. In the carrier compensated system, the longitudinal resistivity $\rho_{xx}$ can be described by the two-carrier model [13]:

$$\rho_{xx} = \frac{1}{e}\frac{(n_h\mu_h + n_e\mu_e) + (n_e\mu_e\mu_h^2 + n_h\mu_h\mu_e^2)B^2}{(n_h\mu_h + n_e\mu_e)^2 + \mu_h^2\mu_e^2B^2(n_h - n_e)^2}, \tag{1}$$

where $e$ is the elementary charge, and $\mu_e$ ($\mu_h$) is the mobility of electrons (holes). As observed in our experiment, the MR increases without saturation in the perfect $n$-$p$ compensated system

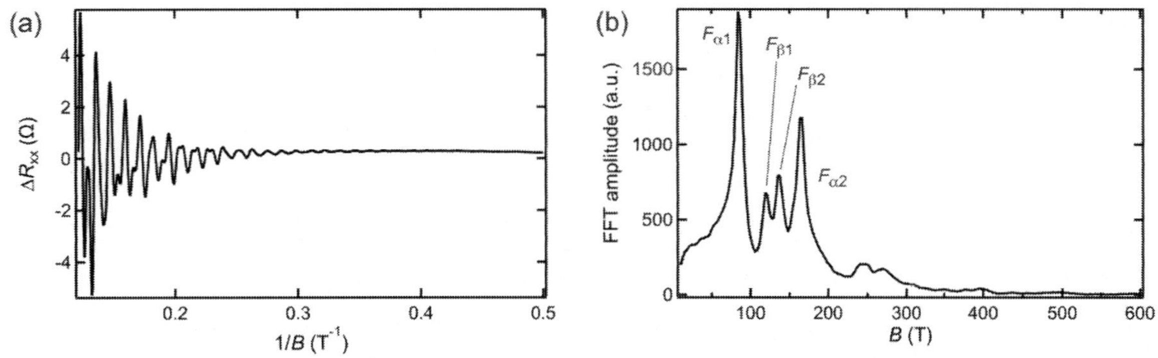


**Figure 4.** Qunatum oscillations on $R_{xx}$. (a) $\Delta R_{xx}$ versus $B^{-1}$. $\Delta R_{xx}$ was obtained after the subtraction of a smoothly increasing component from $R_{xx}$. (b) Spectrum obtained by fast Fourier transformation analysis on $\Delta R_{xx}$.

because $B^2$ term in the denominator of eq. (1) vanishes. When the temperature is increased, the balance between $n_e$ and $n_h$ is lost. As a result, the XMR effect is turned off and it is observed as the turn-on behavior of $R_{xx}$ [7]. However it has been also pointed out that the non-saturating XMR cannot be explained solely by the perfect carrier compensation because the compensation is no longer maintained in high magnetic fields [8, 14]. This issue has not been settled yet. Note that although the thickness of the sample was not uniform as seen in Fig. 1(b), the observed XMR and $RRR$ values are consistent with a previous study, where the XMR is proportional to the square of the $RRR$. [10]. This implies that the effect of inhomogeneous thickness is not significant in our experiment.

Finally, we examine quantum oscillations on $R_{xx}$ in the perpendicular field. Figure 4(a) shows $\Delta R_{xx}$ as a function of $B^{-1}$. Here, the $\Delta R_{xx}$ is extracted by subtracting a smoothly increasing component obtained by fit to eq. (1) from $R_{xx}$. The oscillations become visible from $B \sim 2.5$ T, which also implies the high mobility in our crystal. We have performed fast Fourier transformation (FFT) analysis on $\Delta R_{xx}$ and obtained an FFT spectrum shown in Fig. 4(b). Four pronounced peaks at $F_{\alpha 1} = 84$ T, $F_{\beta 1} = 120$ T, $F_{\beta 2} = 145$ T, and $F_{\alpha 2} = 160$ T are observed. On the Fermi surface of the semimetallic $WTe_2$, there are two electron and two hole pockets due to spin-orbit coupling, which has been revealed by band calculation [4]. $F_\alpha$ and $F_\beta$ in the spectrum are attributed to those hole and electron pockets, respectively. This result is consistent with previous reports [9, 14].

## 4. Conclusions

We have grown single crystals of $WTe_2$ by the self-flux method and fabricated a hall bar-type sample. The XMR as high as 376,059 % was observed at $B = 8.27$ T and the $RRR$ was 704. These values are fairy large, which guarantees the quality of our crystals. Such high-quality enabled the manifestation of quantum oscillations from low magnetic fields and revealed four Fermi pockets corresponding to the spin-orbit split electron and hole pockets.

## Acknowledgements

This study was supported by CREST, Japan Science and Technology Agency (Grant Nos. JPMJCR16F2 and JPMJCR15F3), a Grant-in-Aid for Scientific Research on Innovative Areas, Topological Material Science (Grant No. JP15H05853), and JSPS KAKENHI (Grant Nos. JP25107003 and JP16H00982).

IOP Conf. Series: Journal of Physics: Conf. Series **969** (2018) 012134 doi:10.1088/1742-6596/969/1/012134

## References

[1] Augustin J, Eyert V, Böker T, Frentrup W, Dwelk H, Janowitz C and Manzke R 2000 *Phys. Rev. B* **62** 10812
[2] Qian X, Liu J, Fu L and Li J 2014 *Science* **346** 1344
[3] Fei Z, Palomaki T, Wu S, Zhao W, Cai X, Sun B, Nguyen P, Finney J, Xu X and Cobden D H *Nat. Phys.* in press, arXiv:1610.07924
[4] Soluyanov A A, Gresch D, Wang Z, Wu Q, Troyer M, Dai X and Bernevig B A 2015 *Nature* **527** 495
[5] Wang Y, Liu E, Liu H, Pan Y, Zhang L, Zeng J, Fu Y, Wang M, Xu K, Huang Z, Wang Z, Lu H Z, Xing D, Wang B, Wan X and Miao F 2016 *Nat. Commun.* **7** 13142
[6] Ali M N, Xiong J, Flynn S, Tao J, Gibson Q D, Schoop L M, Liang T, Haldolaarachchige N, Hirschberger M, Ong N P and Cava R J 2014 *Nature* **514** 205
[7] Pletikosić I, Ali M N, Fedorov A V, Cava R J and Valla T 2014 *Phys. Rev. Lett.* **113** 216601
[8] Wang Y, Wang K, Reutt-Robey J, Paglione J and Fuhrer M S 2016 *Phys. Rev. B* **93** 121108
[9] Zhu Z, Lin X, Liu J, Fauqué B, Tao Q, Yang C, Shi Y and Behnia K 2015 *Phys. Rev. Lett.* **114** 176601
[10] Ali M N, Schoop L M, Xiong J, Flynn S, Gibson Q D, Hirschberger M, Ong N P and Cava R J 2015 *Europhys. Lett.* **110** 67002
[11] Cai P L, Hu J, He L P, Pan J, Hong X C, Zhang Z, Zhang J, Wei J, Mao Z Q and Li S Y 2015 *Phys. Rev. Lett.* **115** 057202
[12] Wu Y, Jo N H, Ochi M, Huang L, Mou D, Bud'ko S L, Canfield P C, Trivedi N, Arita R and Kaminski A 2015 *Phys. Rev. Lett.* **115** 166602
[13] Murzin S S, Dorozhkin S I, Landwehr G and Gossard A C 1998 *J. Exp. Theor. Phys. Lett.* **67** 113
[14] Rhodes D, Das S, Zhang Q R, Zeng B, Pradhan N R, Kikugawa N, Manousakis E and Balicas L 2015 *Phys. Rev. B* **92** 125152

# Controlling entanglement in the interferometry of driven coupled flux qubits

**Ana Laura Gramajo, Daniel Domínguez and María José Sánchez**

Centro Atómico Bariloche and Instituto Balseiro, 8400 San Carlos de Bariloche, Argentina.

**Abstract.** We study the manipulation of quantum entanglement by periodic external fields in the Landau-Zener-Stuckelberg interferometry of coupled flux qubits. As an entanglement measure we compute numerically the concurrence of two coupled flux qubits, both driven by a dc+ac magnetic flux. We show that it is possible to create or destroy entanglement in a controlled way by tuning the system at or near multiphoton resonances.

## 1. Introduction

Among the rich variety of qubits that have been explored as candidates for quantum computation, solid state superconducting devices based on Josephson junctions circuits are promising due to their microfabrication techniques and downscalability. [1, 2, 3, 4]

To implement a quantum algorithm, one must be able to entangle qubits by means of an interaction term in the Hamiltonian describing a two qubit system. For superconducting flux qubits [1, 5], the natural interaction is between the magnetic fluxes, providing a coupling through their mutual inductance.[1, 5, 6]

Although coupled superconducting qubits have been experimentally realized [6, 7, 8], the generation and control of entanglement can be quite complicated and demanding, even requiring sequences of single- and two-qubit operations. Along this line, pulse sequences have been implemented for several superconducting qubits with fixed interaction energies [9, 10]. However, entangling operations can be much more efficient if the interaction can be varied and, ideally, turned off during parts of the manipulation. Some tunable coupling schemes have been proposed, [11, 12, 13, 14] but these approaches failed in controlling the coupling entirely.

Alternatively, engineering selection rules of transitions among different energy levels is a possible strategy for coupling and decoupling superconducting qubits. [15] Following this idea, a method to create artificial selection rules- by suppressing and/or exciting specific transitions- has been developed for a pair of superconducting flux qubits [11], simultaneously driven by a single resonant frequency with different amplitudes and phases.[16] Also, entangling gates via proper microwave driving have been considered. [17]

Our main goal is to study the possibility of manipulating entanglement between two flux qubits, by external driving fields of variable amplitude and fixed frequency non resonant with any specific transition. The sensitivity of the energy levels of a flux qubit driven by an external (dc+ac) magnetic field, has been extensively studied in recent years.[18, 19, 20, 21, 22] The experimental implementation of Landau-Zener-Stückelberg interferometry [5] has become a tool to analyze quantum coherence under strong driving and to access the multilevel structure of flux qubits, which exhibits several avoided crossings as a function of the magnetic flux.[22] In most of these cases the Floquet formalism has been employed to solve the system dynamics in terms of quasienergies and Floquet states. [23, 24, 25]

In Ref.[26] it has been recently shown how to generate entangled Floquet states in weakly interacting two level systems by controlling the amplitude of the external driving field. However Floquet states are not accessible experimentally. Here we extend the study for arbitrary coupling beyond the weak interaction case, and for general pure states that can be prepared as initial states, like the ground state of the two qubit system or states of the computational basis.

The system of work consists in two coupled flux qubits driven by (the same) microwave field. As usual, each qubit is represented by a two level system [27, 28, 29] and we focus here on the case of negligible dissipation and/or interaction with the environment. Therefore, our results are valid for time scales smaller than the decoherence time of the qubits. As a measure of entanglement we choose the *concurrence* [30] which increases monotonically from 0 for non-entangled states to 1, for maximally entangled states. Experimentally there are some evidences of entanglement measurements in solid state physics, see for example [31], [32].

## 2. Concurrence for the two coupled flux qubits model

The dynamics of two inductively coupled flux qubits can be described by the Hamiltonian [33, 34]

$$\hat{H}_0 = -\frac{1}{2}\sum_{i=1}^{2}\left(\epsilon_i\sigma_z^{(i)} + \Delta_i\sigma_x^{(i)}\right) - \frac{J^z}{2}\sigma_z^{(1)}\otimes\sigma_z^{(2)}, \tag{1}$$

where $\epsilon_i$ is the detuning energy (which is proportional to the difference between the magnetic flux through the qubit and half the quantum of flux), $\Delta_i$ is the tunnel splitting energy and $\sigma_z^{(i)}, \sigma_x^{(i)}$ the Pauli matrices , with $i = 1, 2$ the index of each qubit, with $J^z$ the qubit-qubit coupling constant. The coupling corresponds to an inductive coupling [6, 13], since $J^z = \pm M I_1 I_2$ can be written in terms of the mutual inductance $M$ and the qubits currents $I_{1,2}$. For $J^z < 0$ ($J^z > 0$) the coupling is antiferromagnetic (ferromagnetic).

To study the dynamics in the presence of driving fields, we replace as usual $\epsilon_i \to \epsilon_i(t) = \epsilon_i + f(t)$ [26, 29, 33, 34] where $f(t) = A\cos(\omega t)$ is the microwave field (magnetic flux) of amplitude $A$ and frequency $\omega$ applied to each qubit.

The resulting Hamiltonian is thus periodic in time, $\hat{H}(t) = \hat{H}(t+T)$ with period $T = 2\pi/\omega$. According to the Floquet theorem [23, 27, 28], the solution of the Schrödinger equation can be spanned in the Floquet basis $\{|u_\alpha(t)\rangle\}$ as $|\Psi(t)\rangle = \sum_\alpha a_\alpha(t_0)e^{-i\gamma_\alpha t/\hbar}|u_\alpha(t)\rangle$, with $\gamma_\alpha$ the quasienergies and $\alpha$ the index labeling the eigenstates of the time independent problem. For an initial condition $|\Psi(t_0)\rangle$ at time $t_0$, we define the coefficients $a_\alpha(t_0) = \langle u_\alpha(t_0)|\Psi(t_0)\rangle$. The time-evolution for a Floquet state is given by $(H(t) - i\hbar\frac{\partial}{\partial t})|u_\alpha(t)\rangle = \gamma_\alpha|u_\alpha(t)\rangle$, and the satisfy $|u_\alpha(t+T)\rangle = |u_\alpha(t)\rangle$. Therefore after expanding the time periodic Floquet states in the Fourier basis, $|u_\alpha(t)\rangle = \sum_k e^{ik\omega t}|u_\alpha(k)\rangle$ the time-dependent problem is reduced to a time-independent eigenvalue problem.

An entanglement measure quantifies the degree of quantum correlations present in a given quantum state. In the case of pure states $|\Psi(t)\rangle$ a useful quantity is the concurrence [30],

$$C(t, t_0) = |\langle\Psi(t)|^*\sigma_y^{(1)}\otimes\sigma_y^{(2)}|\Psi(t)\rangle|, \tag{2}$$

that goes from 0 for non-entangled states, to 1 for maximally entangled states. Notice that Eq.(2) depends implicitly on the initial time $t_0$ through $|\Psi(t)\rangle$.

Using the extended Floquet basis in Fourier space $\{|u_\alpha(k)\rangle\}$ with $k \in \mathbb{Z}$, Eq.(2) can be written as

$$C(t, t_0) = |\sum_{\alpha\beta kk'qq'}\tilde{C}_{\alpha\beta}(k, k')f_{\alpha\beta}(q, q')e^{-i\varphi_{\alpha\beta}^{kk'qq'}(t,t_0)}|, \tag{3}$$

where $\varphi_{\alpha\beta}^{kk'qq'}(t,t_0) = (\gamma_\beta + \gamma_\alpha - (k'+k)\omega)t - (\gamma_\beta + \gamma_\alpha - (q'+q)\omega)t_0$, with $\gamma_{\alpha \atop \beta}$ the quasienergies, $\tilde{C}_{\alpha\beta}(k,k') = \langle u_\alpha(k)|^* \sigma_y^{(1)} \otimes \sigma_y^{(2)} |u_\beta(k')\rangle$ and $f_{\alpha\beta}(q,q') = a_\alpha(q) a_\beta(q')$, with $a_{\alpha \atop \beta}({q \atop q'}) = \langle u_{\alpha \atop \beta}({q \atop q'})|\Psi(t_0)\rangle$ Under general conditions, the initial time should be averaged out, thus we compute the time-averaged of Eq.(3) over $t_0$,

$$\overline{C}(t) = \frac{1}{T}\int_0^T dt_0\, C(t,t_0), \tag{4}$$

which still presents an oscillating behaviour with time t, typical for time dependent driven systems. However and unlike the occupation probability, the concurrence is not a periodic function of the driving period, as can be easily checked from its definition, Eq. (3). In order to obtain a representative value of concurrence, we eliminate the dependence on time t by performing an additional time average, obtaining

$$\overline{\overline{C}} = \lim_{t' \to \infty} \frac{1}{t'}\int_0^{t'} dt\, \overline{C}(t). \tag{5}$$

## 3. Results

After computing numerically the Floquet states and quasienergies, we calculate the concurrence using Eq.(3) and the respective averages over $t_0$ and $t$, given in Eqs. (4) and (5). Along this work, we fix $\epsilon_i = \epsilon_0$, $\forall i = 1, 2$, and choose without loss of generality, $\Delta_1/\omega = 0.1$ and $\Delta_2/\omega = 0.15$. We take $\hbar = 1$ and energy scales are normalized by $\omega$.

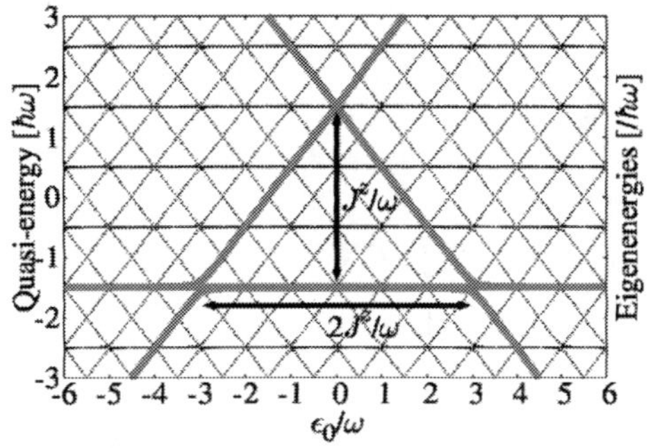


Figure 1: Eigenenergies in the absence of driving, $A/\omega = 0$ (red lines) and quasienergies for $A/\omega = 3.8$ (black lines) as a function of $\epsilon_0/\omega$, for the coupling strength $J^z/\omega = -3$. Parameters are $\Delta_1/\omega = 0.1$ and $\Delta_2/\omega = 0.15$.

Fig.1 shows as a function of the detuning $\epsilon_0/\omega$ and for the coupling strength $J^z/\omega = -3$, the eigenenergies $E_i$, $i = 0, ..., 3$ for the time independent Hamiltonian (red lines), and in black the quasienergies for the driven Hamiltonian for $A/\omega = 3.8$. We work with the (antiferromagnetic) coupling $J^z < 0$, which reduces the energy of states $|01\rangle, |10\rangle$ while it increases the energy of states $|00\rangle, |11\rangle$, with $\mathcal{E} = \{|00\rangle, |01\rangle, |10\rangle, |11\rangle\}$ the computational basis in the product space of the two qubits. The quasienergies display avoided crossings (quasidegeneracies), where the Floquet states are strongly mixed. These quasidegeneracies will play a central role in the structure of the concurrence as we will discuss in the following. In the limit of $\Delta_i/\omega \to 0$, the quasidegeneracies transform in exact crossings of quasienergies, giving the resonance condition $\gamma_\alpha - \gamma_\beta = n\omega$, $n \in \mathbb{Z}$. [27] The quasienergies of the two qubit system computed analytically for $\Delta_i/\omega \to 0$[27] are $\gamma_\alpha \sim \pm\epsilon_0 + J^z/2 + m\omega$ and the (quasi) degenerate pair $-J^z/2 + m\omega$ with $m \in \mathbb{Z}$. The resonance conditions are thus satisfied respectively for $2\epsilon_0 \sim n\omega$ and $\epsilon_0 \pm J^z \sim n\omega$, and correspond to multiphoton processes where the population probability is modulated by the zeros of the Bessel functions of order $n$, $J_n(A/\omega) = 0$.[34, 27] Notice that while the first condition gives half integer and integer values of $\epsilon_0/\omega$, the second one depends on $J^z/\omega$. For integer values of $J^z/\omega$ the quasidegeneracies are located at integer values of $\epsilon_0/\omega$, as can be clearly seen from Fig.1.

In Fig.2 we show $\overline{\overline{C}}$ as a function of $\epsilon_0/\omega$. The initial state is chosen as the ground state for the correspondent $\epsilon_0/\omega$. In the absence of driving, $A/\omega = 0$ (black line), the ground state is entangled for detuning energies satisfying $|\epsilon_0/\omega| \lesssim |J^z/\omega| = 3$, where the concurrence takes values close to 1. In particular, for $\epsilon_0 = 0$, the ground state is the Bell's state $(|01\rangle + |10\rangle)/\sqrt{2}$, which is known to be a maximally entangled state.[30] On the other hand, for values of detuning energy $|\epsilon_0/\omega| > |J^z/\omega| = 3$ the ground state is almost disentangled, i.e. for large values of $\epsilon_0$ the ground state is asymptotically a separable state of the computational basis, corresponding to $|00\rangle$ for $\epsilon_0 \gg 0$ and $|11\rangle$ for $\epsilon_0 \ll 0$, see Fig.1.

When the driving is turn on ($A/\omega = 3.8$), a pattern of resonances is clearly visible in $\overline{\overline{C}}$, where entanglement is either created or destroyed. For $|\epsilon_0/\omega| > |J^z/\omega| = 3$, where the initial condition corresponds to a separable state, we see that it is possible to generate entanglement in a controlled way around a given resonance. Otherwise entanglement is reduced.

The positions of the resonances in $\overline{\overline{C}}$ are determined from the already mentioned conditions: $2\epsilon_0/\omega \sim n$ and $\epsilon_0/\omega + J^z/\omega \sim n$, with $n \in \mathbb{Z}$. Therefore is around a (quasi) degeneracy where the Floquet states can be strongly mixed given rise to significant deviations in the behaviour of the concurrence compared to the undriven case.

Notice that in Fig.2, the resonances are at integer and half integer values of $\epsilon_0/\omega$, since $J^z/\omega = -3$. Furthermore, when studying C as a function of A, for a particular multiphoton resonance, we have found that the concurrence is modulated by the driving amplitude, where full (or partial) recovery of the initial entanglement is possible. [35]

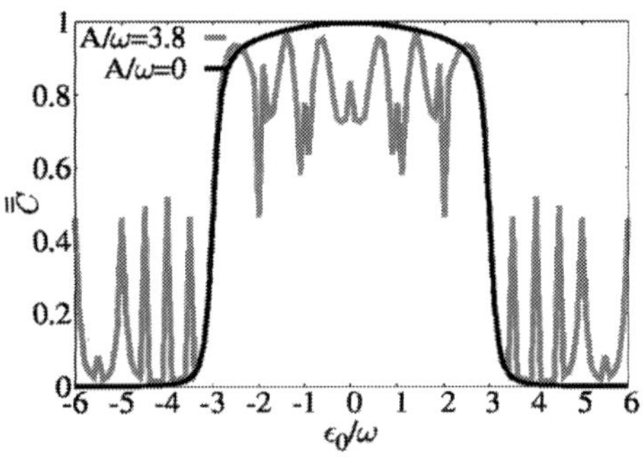


Figure 2: Plots of $\overline{\overline{C}}$ versus $\epsilon_0/\omega$ for $A/\omega = 0$ (black line) and $A/\omega = 3.8$ (red line). The initial condition corresponds to the ground state for the correspondent $\epsilon_0/\omega$. $J^z/\omega = -3$ is the coupling strength and other qubits parameters are the same as in Fig. 1.

So far, we have studied the entanglement for a fixed value of the coupling strength. However in practical implementations with flux qubits the intensity of the inductive coupling can be controlled, and it is interesting to analyse whether a different static coupling would induce qualitative changes in the above description, taking into account that the spectrum of quasienergies is sensitive to this change (see Fig. 1). In the absence of the microwave field two well separated behaviours are observed, corresponding to positive and negative values of $J^z$ respectively. [35] For $J^z < 0$ (antiferromagnetic coupling) the ground state is entangled for $|\epsilon_0| < |J^z|$ as we already described. On the other hand, for the ferromagnetic coupling $J^z > 0$, which increases the energy of states $|01\rangle, |10\rangle$ and decreases the energy of states $|00\rangle, |11\rangle$, the ground state is entangled only for values $\epsilon_0 \sim 0$, being approximately the Bell's state

$(|00\rangle + |11\rangle)/\sqrt{2}$. When the microwave field is on, $\overline{\overline{C}}$ exhibits the structure of resonances where the entanglement is created or destroyed in a well controlled way with resonances located at half integer or integer values of $\epsilon_0/\omega$ and others located at positions determined by the values of $J^z/\omega$, as we already mentioned. For integer values of $J^z/\omega$ the resonances are at integers $\epsilon_0/\omega$, while for arbitrary real values of $J^z/\omega$ they are respectively shifted to non integer values of $\epsilon_0/\omega$. These features are clearly observed in Fig.3, where $\overline{\overline{C}}$ versus $\epsilon_0/\omega$ is plotted for $J^z/\omega > 0$ and $A/\omega = 3.8$.

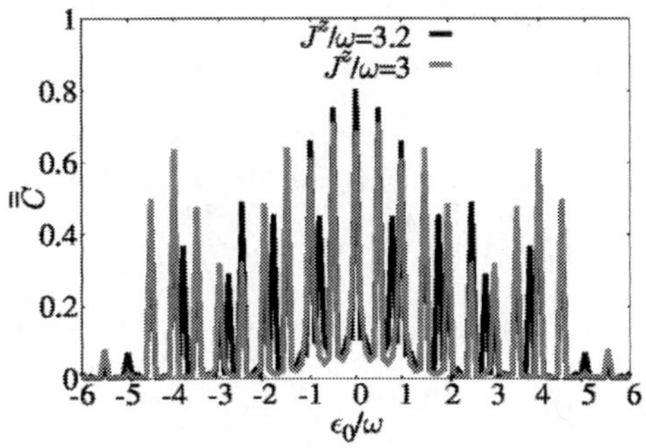


Figure 3: Plot of $\overline{\overline{C}}$ versus $\epsilon_0/\omega$ for $A/\omega = 3.8$ for ferromagnetic coupling strengths $J^z > 0$. See text for details.

## 4. Conclusions

In this work we have shown that entanglement can be manipulated by external periodic driving fields. In particular we presented numerical results for the concurrence of a system composed by two coupled flux qubits driven by an external ac magnetic flux.

The main result is that when the system is tuned at or near a multiphoton resonance full control of entanglement is possible: (a) when the initial state is disentangled one can drive it towards a highly entangled state, and (b) when the initial state is entangled one can strongly reduce entanglement with the driving. This will show, for example, as well-defined lines of "entanglement resonance " in the $\{\epsilon_0, J^z\}$ plane. One advantage of this case is that once the resonance is exactly tuned, the control of entanglement is almost complete.

## Acknowlegments

We acknowledge support from CNEA, CONICET, UNCuyo (P 06/C455) and ANPCyT (PICT2014-1382).

## References

[1] Orlando T P, Mooij J E, Tian L, van der Wal C H, Levitov L S, Lloyd S and Mazo J J, 1999 *Phys. Rev. B* **60**, 15398.
[2] Friedman J R, Patel V, Chen W, Tolpygo S K, and Lukens J E, 2000 *Nature* **406**, 43.
[3] van der Wal C H, ter Haar A C J, F K Wilhelm, Schouten R N, Harmans C J P M, Orlando T P, Lloyd S and Mooij J E, 2000 *Science* **290**, 773.
[4] Martinis J M, Nam S, Aumentado J, and Urbina C, 2002 *Phys. Rev. Lett.* **89**, 117901.
[5] Oliver W D, Yu Y, Lee J C, Berggren K K, Levitov L S and Orlando T P, 2005 *Science* **310**, 1653.
[6] Majer J B , Paauw F G, ter Haar A C J, Harmans C J P M and Mooij J E, 2005 *Phys. Rev. Lett.* **94**, 090501.
[7] Izmalkov A, Grajcar M, Ilichev E, Wagner T, Meyer H G , Smirnov A Y, Amin M H S , van den Brink A M and Zagoskin A M, 2004 *Phys. Rev. Lett.* **93**, 037003.
[8] Grajcar M, Izmalkov A, van der Ploeg S H W, Linzen S, Ilichev E, Wagner T, H bner U, Meyer H G, Maassen van den Brink A, Uchaikin S and Zagoskin A M, 2005 *Phys. Rev. B* **72**, 020503.

[9] Yamamoto T, Pashkin Y A, Astafiev O, Nakamura Y, and Tsai J S, 2003 *Nature* **425**, 941.
[10] Strauch F W, Johnson P R, Dragt A J, Lobb C J, Anderson J R and Wellstood F C, 2003 *Phys. Rev. Lett.* **91**, 167005.
[11] Mooij J E, Orlando T P, Levitov L, Tian L, van der Wal C H and Lloyd S, 1999 *Science* **285**, 1036.
[12] van der Ploeg S H W, Izmalkov A, van den Brink A M, Hübner U, Grajcar M, Ilichev E, Meyer H G and Zagoskin A M, 2007 *Phys. Rev. Lett.* **98**, 057004.
[13] Superconducting Quantum Circuits, Qubits and Computing, Wendin G and Shumeiko V, arXiv:cond-mat/0508729v1, 2005.
[14] Plourde B L T, Zhang J, Whaley K B, F K Wilhelm, Robertson T L, Hime T, Linzen S, Reichardt P A, Wu C E, and Clarke J, 2004 *Phys. Rev. B* **70**, 140501.
[15] Harrabi K, Yoshihara F, Niskanen A O, Nakamura Y, and Tsai J S, 2009 *Phys. Rev. B* **79**, 020507.
[16] de Groot P C, Lisenfeld J, Schouten R N, Ashhab S, Lupascu A, Harmans C J P M and Mooij J E, 2010 *Nat Phys* **6**, 763
[17] Poletto S, Gambetta J M, Merkel S T, Smolin J A, Chow J M, Córcoles A D, Keefe G A, Rothwell M B, Rozen J R, Abraham D W, Rigetti C and Steffen M, 2012 *Phys. Rev. Lett.* 109, 240505.
[18] Shevchenko S, Ashhab S and Nori F, 2010 *Physics Reports* **492**, 1.
[19] Berns D M, Oliver W D, Valenzuela S O, Shytov A V, Berggren K K, Levitov L S, and Orlando T P, 2006 *Phys. Rev. Lett.* **97**, 150502.
[20] M S Rudner, Shytov A V, Levitov L S, Berns D M, Oliver W D, Valenzuela S O and Orlando T P, 2008 *Phys. Rev. Lett.* **101**, 190502.
[21] Izmalkov A, van der Ploeg S H W, Shevchenko S N, Grajcar M, Ilâichev E, Hübner U, Omelyanchouk A N, and Meyer H G, 2008 *Phys. Rev. Lett.* **101**, 017003 .
[22] Ferrón A , Domínguez D and Sánchez M J, 2010 *Phys. Rev. B* **82**, 134522.
[23] Shirley J H, 1965 *Phys. Rev.* **138**, B979.
[24] Ferrón A, Domínguez D and Sánchez M J, 2012 *Phys. Rev. Lett.* **109**, 237005 .
[25] Ferrón A , Domínguez D and Sánchez M J, 2016 *Phys. Rev. B* **93**, 064521.
[26] Sauer S, Mintert F, Gneiting C and Buchleitner A, 2012 *Journal of Physics B: Atomic, Molecular and Optical Physics* **45**, 154011.
[27] Son S K, Han S and Chu S I, 2009 *Phys. Rev. A* **79**, 032301.
[28] Hausinger J and Grifoni M, 2010 *Phys. Rev. A* **81**, 022117.
[29] Ashhab S, Johansson J R, Zagoskin A M and Nori F, 2007 *Phys. Rev. A* **75**, 063414.
[30] Wootters W K, 1998 *Phys. Rev. Lett.* **80**, 2245.
[31] Chow J M, DiCarlo L, Gambetta J M, Nunnenkamp A, Bishop L S, Frunzio L, Devoret M H, Girvin S M, and Schoelkopf R J, 2010 *Phys. Rev. A* **81**, 062325 .
[32] Shulman M D, Dial O E, Harvey S P, Bluhm H, Umansky V and Yacoby A, 2012 *Science* **336**, 202.
[33] Macroscopic quantum resonance of coupled flux qubits: a quantum computation scheme, Akisato H, 2008.
[34] Satanin A M, Denisenko M V, Ashhab S and Nori F, 2012 *Phys. Rev. B* **85**, 184524.
[35] Gramajo A L, Domínguez D and Sánchez M J, to be published.

IOP Conf. Series: Journal of Physics: Conf. Series **969** (2018) 012136 doi:10.1088/1742-6596/969/1/012136

# Anomalous amplitude of the quantum oscillations in the longitudinal magneto-thermoelectric power

**N Satoh**

Department of Electronics and Computer Science, Iwaki Maisie University, 5-5-1 Iino chuoudai, Iwaki, Fukushima 970-8551, Japan

E-mail: nsatoh10015@cg8.so-net.ne.jp

**Abstract.** Longitudinal magneto-thermoelectric power $S_{yy}(y)$ of a pure bismuth single crystal was measured in magnetic fields up to 8T at several fixed temperatures between 1.4 and 15 K to investigate the magneto-phonon effect in the longitudinal magneto-thermoelectric power (MTP). The oscillation patterns of the longitudinal MTP was similar to that of the longitudinal Shubnikov-de Haas (SdH) effect, expectedly. However, the observed amplitude of oscillations showed a curious temperature dependence. That is, in the range of temperature $T > 4.2$ K, the amplitude has a maximum around 9K, which is well described by considering the inter-Landau level scattering of electrons. On the contrary, in the range of $T < 4.2$K, the observed amplitude is enhanced markedly although that of the longitudinal SdH oscillations becomes less pronounced by lowering temperature. This discrepancy may be attributed to the effect of the surface (wrapping) current and to the energy dependence of the electron relaxation time.

## 1. Introduction

It is well known that bismuth has frequently played an important pioneer-role in the history of solid state physics. Fundamental phenomena such as de Haas-Alphen effect, Shubnikov-de Haas (SdH) effect, giant magnetoresistance and oscillatory magnetostriction as well as Nernst effect were all first observed in bismuth [1]. Bismuth is the archetypical semimetal and is still attracting new attention owing to its unique properties, following the development of study on the Dirac Fermions in bismuth-based of topological insulators [2]. Nevertheless, there is no doubt that several fundamental problems concerning the transport phenomena in bulk bismuth remain still unsolved. One of them, for instance, is such that what brings about so large amplitude in the SdH oscillations. Also, we do not know exactly the reason why magnetic field can markedly enhance the resistivity by a factor of $10^6$ without saturation. We have been investigating the longitudinal SdH oscillations of semimetal, especially diffusion size effect, a small amount of charged (neutral) impurity effect [3-5] and surface recombination effect on the SdH oscillation amplitudes [6] to find a key to the unsolved problems.

We present here our study on the longitudinal magneto-thermoelectric power (MTP) $S_{yy}(y)$ in bismuth as an extension of our investigation. Although studies on the ordinary Seebeck effect of bismuth in strong magnetic fields were previously performed decades ago [7], but our experiment would serve as a good testing ground to investigate how the inter- and intra-Landau level scatterings, the so-called "magneto-phono effect", can play a role in the longitudinal MTP oscillations.

## 2. Experimental

Brief description of our sample preparation is as follows. A piece of commercially available pure bismuth ingot (6N grade) was purified further by zone-melting, and was finally single crystallized. The obtained single crystal was cut with an electric discharge cutting machine along the bisectrix ($y$) axis of bismuth into a rectangle shape ($3.0 \times 3.0 \times 18$ mm$^3$) after the crystal orientation was checked by X-ray diffraction analysis. The sample was annealed isothermally at 265°C for two weeks to remove strain. In our experimental configuration, both the temperature gradient $\nabla T$ and the magnetic field $B$ were parallel to the bisectrix axis of the sample. Thermoelectric voltages were detected in a conventional way using a thermocouple (0.07 at % Fe-Au verse Ag) that had extremely low magnetic-field-induced-error (less than 0.1% at 9T). Temperature was measured by a calibrated ceramic thin film resistance thermometer.

## 3. Result and discussions

Figure 1 shows our typical results of the oscillatory longitudinal MTP for bismuth single crystal along bisectrix axis as a function of inverse magnetic field at several fixed temperatures of 1.4K, 4.2K, 5.6K and 7.5K, respectively. Originally, the thermoelectric power $S_{yy}(y)$ is defined in terms of the thermal diffusion coefficient $A$ and the conductivity tensor $\sigma_{yy}$ as $S_{yy} = A/\sigma_{yy}$. Then, as would be expected, SdH-like oscillations certainly appeared in our measurements. In fact, it is clear by visual inspection that the oscillation patterns in figure 1 are the same as our previous work [3-5]. For instance, a series of pronounced oscillation peaks were observed at the same magnetic fields of 1.25, 0.625, 0.42 T.

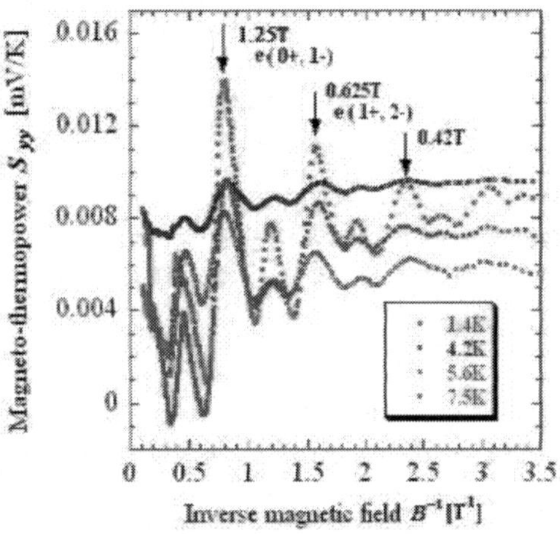


**Figure 1.** Typical magneto-thermopower $S_{yy}$ as a function of inverse magnetic field $B^{-1}$ at 1.4, 4.2, 5.6 and 7.5 K. Quantum number of the peaks indicated by arrows are electron levels assigned by Smith, *et.al.* [8]

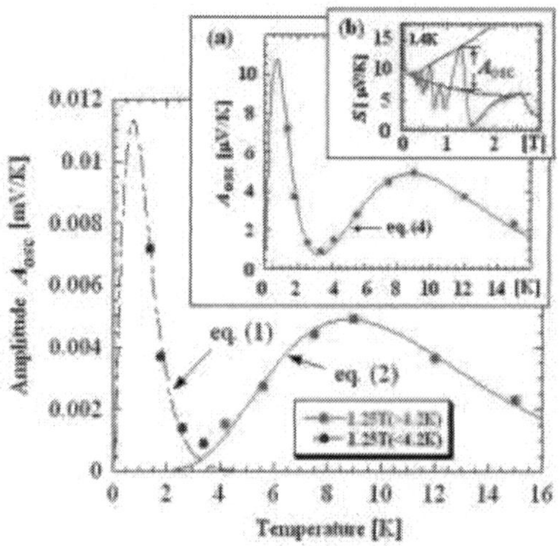


**Figure 2.** Temperature dependence of the $A_{osc}$ at $B$ =1.25T. Dotted and solid line are equations (1) and (2), respectively. Inset (a) shows the simulated curve, and (b) illustrates the definition of the amplitude $A_{osc}$.

Secondly, the temperature dependence of the oscillation amplitude (figure 1) seems to be unusual. To visualize this anomaly more clearly, we have picked up oscillation amplitudes $A_{osc}$ at a magnetic field of 1.25T and shown them in figure 2 as a function of temperature. Inset (b) of figure 2 illustrates the definition of the $A_{osc}$; this arises only from a piece of electron pockets located at point $L$ in the trigonal-bisectrix plane of the Brillouin zone. Our method of extracting $A_{osc}$ from those peaks is the same as that of Tanuma and Inada [9] and our previous works. Figure 2 certainly shows that the observed $A_{osc}$ becomes larger in magnitude and has a maximum around 9 K above 4.2 K. But on the other hand, below 4.2 K, the $A_{osc}$ enhances markedly although the amplitude of the longitudinal SdH oscillation become less pronounced by lowering temperature [3]. In the present paper, attention will be paid mainly to this anomalous temperature dependence of the amplitude.

Quantum oscillations in the MTP are treated differently in the literature. According to Young [10] and Fletcher [11], the oscillatory MTP depends on the derivative of the density of states at the Fermi level. Also, Pantsulaya and Varlamov [12] calculated the oscillatory longitudinal MTP considering the energy dependence of the electron relaxation time. The amplitude $A_{osc}$ of oscillation is given by the derivative of the Lifshiz-Kosevich formula as follows,

$$A_{osc}(T) \propto (X \coth X - 1)/\sinh X \ , \tag{1}$$

where
$$X = 2\pi^2 r k_B T/\hbar\omega_c \, .$$

In equation (1), $\omega_c$ (=e$B/m_c$) is the cyclotron frequency, $m_c$ the cyclotron effective mass, $r$ the number of harmonics. It is difficult to describe the whole temperature dependence (figure 2) only by equation (1). To understand this curious temperature dependence, we separate, for convenience, the observed data in figure 2 into two regions of temperature. One is upper branch ($T > 4.2$ K) and the other is lower branch ($T < 4.2$ K).

### 3.1. *Upper branch ($T > 4.2$ K)*

We shall, for the moment, confine ourselves to the amplitude $A_{osc}$ at a magnetic field of 1.25 T above 4.2 K. It is established that the longitudinal MTP oscillation is mostly explained by inter-Landau level scattering of electrons just as the longitudinal SdH oscillation. That is, the thermally excited electrons diffusing along bisectrix axis are scattered by excited phonons, which cause the large amplitude. Such inter-Landau level scattering of electrons is assisted by longitudinal acoustic phonons illustrated an arrow $\boldsymbol{q}$ in figure 3(a), because of the momentum conservation law. The magnitude of wave number $q$ of phonons is given by a distance between two Landau tubes, i.e., (0 -) and (0 +,1 -), which remain in a piece of electron pocket at point $L$. Those phonons increase in number with rising temperature. Therefore, as pointed out by Tanuma and Inada [9], we need to put a proportional factor on equation (1): the excitation $\omega_q^2 N_q$ of phonons only having the wave number $q$ where $N_q$ is Planck's function, $\omega_q$ the frequency of phonons. Also, the Dingle factors $e^{-\delta}$ is put to express Landau level broadening. Then, the amplitude $A_{osc}$ should be given as follows,

$$A_{osc}(T) = RB^{1/2}\{(X \coth X - 1)/\sinh X\}\omega_q^2 N_q e^{-\delta}, \tag{2}$$

where
$$N_q = \{\exp(\hbar\omega_q/k_B T) - 1\}^{-1},$$
$$\delta = \frac{2\pi^2 r k_B}{\hbar\omega_c}(T_D + T) \, .$$

In equation (2), $R$ is an adjustable constant and $T_D$ denotes the Dingle Temperature. The wave number $q$ (figure 3(a)), the corresponding values of the frequency $\omega_q$ and the energy $\hbar\omega_q$ of phonons are given to be $8.46\times10^8$ m$^{-1}$, $2.17\times10^{12}$ s$^{-1}$ and 1.43 meV, respectively [9]. By fitting equation (2) on the experimentally observed amplitudes ($T > 4.2$K in figure 2) using the Levenberg-Marquardt algorithm, the cyclotron frequency $\omega_c$ is obtained to be $1.21\times10^{13}$ s$^{-1}$. Then, the cyclotron effective mass $m_c$ is to be 0.0180 $m_0$ where $m_0$ is the electron rest mass. This value is almost 9.1% larger than the value of 0.0165 $m_0$, which is the cyclotron effective mass of heavy bisectrix and binary electrons [13]. This large mass can be understood as a measure of electron-phonon interaction (phonon drag) in the concerned rang of temperature. Dingle temperature $T_D$ obtained from the so-called "Dingle plot" is to be 1.2 K. Then, the relaxation time $\tau$ (= $\hbar/2\pi k_B T_D$) is to be $6.7\times10^{-11}$ s. The fact that the observed

IOP Conf. Series: Journal of Physics: Conf. Series **969** (2018) 012136 doi:10.1088/1742-6596/969/1/012136

amplitudes agree with equation (2) excellently is conclusive evidence that inter-level scatterings by excited phonons play a dominant role in the oscillatory longitudinal MTP above 4.2 K (figure 2).

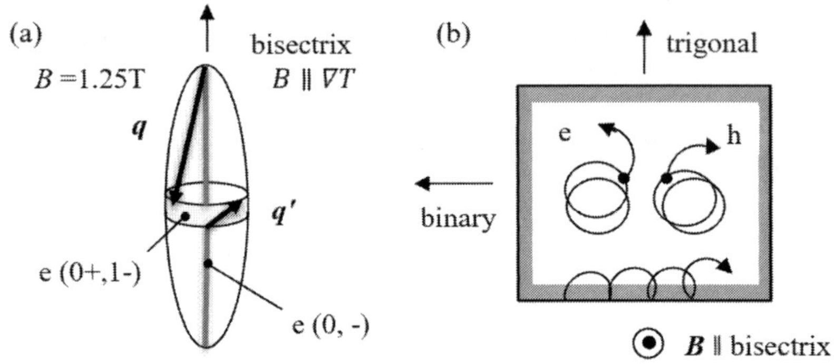


**Figure 3.** (a) Electron Fermi surface at 1.25 T being parallel to bisectrix axis. ***q*** and ***q'*** are wave number vector for inter- and intra-level transition, respectively. (b) Schematic drawing of the cyclotron orbits of carriers, of which centre migrate toward the sample boundary by means of scattering.

*3.2. Lower branch ($T < 4.2$ K)*

To the best of our knowledge, no reports have been made such enhanced data as our amplitude in the oscillatory longitudinal MTP below 4.2 K, especially in bismuth. To begin with, the MTP must vanish at $T = 0$ K, because of entropy flow by carriers. Then, a local maximum of the amplitude will surely appear at a certain temperature $T_{max}$ below our lowest data (1.4 K). Equation (1) will show a maximum for $X$=1.62 [10], i.e., at

$$T_{max} = 0.110\,B/rm_c\ . \tag{3}$$

We have fitted equation (1) on the observed amplitudes to determine the cyclotron effective mass for the lower branch ($T < 4.2$ K). The obtained mass $m_c$ is to be $0.187m_0$, which is an order of magnitude heavier than that of the upper branch. A heavy mass may suggest the metallic surface state ($\approx 0.2m_0$) of bismuth [14]. If we employ the following values: $B$ =1.25T, $r$ =1, and $m_c = 0.187m_0$ in equation (3), the value of $T_{max}$ is to be 0.74 K, which is a reasonable one as shown in figure 2.

To carry out numerical simulation for the $A_{osc}$ in full range of temperature, we have apparently obtained the following expression of $A_{osc}$ ($T$) from combing equation (1) with (2) as

$$A_{osc}(T) = RB^{1/2}\{(X\coth X - 1)/\sinh X\}\left[1 + \omega_q^2 N_q e^{-\delta}\right]. \tag{4}$$

The resultant simulated curve is drawn in the inset (a) of figure 2. In calculation, a weighted effective mass $pm_0$ ($p$: parameter) was assumed; that is, electron-phonon interaction changes only the mass.

Although the temperature dependencies of the experimentally observed amplitudes are in good agreement with equation (1) and (2) in each temperature range, but we do not fully understand yet what is essence. Some other process may play a role below 4.2 K. So, we will consider the motion of electrons semi-classically as follows.

Now, figure 3(b) shows a projection of the sample ($x$-$z$ plane) schematically. In a strong magnetic field, as is well known, the centre of cyclotron orbits of each carrier migrates toward sample boundary after colliding with scatterers such as phonons or impurities. Such a scattering is depicted by an arrow ***q'*** in figure 3(a). Consequently, there will be a density and a steady-state electric potential difference along the bisectrix axis and the other axes as well, because no current can flow. Moreover, near the boundaries the trajectories of electrons are skipping orbits by means of elastic reflection from the surface: there exists surface (wrapping) current. The increased spatial confinement at the boundary has to entail a strong enhancement of the energy of the Landan levels at the surface: metallic surface state.

In our case of extremely pure bismuth, the magnitude of wave number $q'$ (figure 3(a)) and the frequency $\omega_{q'}$ of phonons must satisfy the inequalities of $q' \ll 1.07\times 10^6$ cm$^{-1}$ and $\omega_{q'} \ll 2.9\times 10^{11}$ s$^{-1}$, respectively. Those values are obtained from the average radii of an electron pocket at point $L$ in the Brillouin zone [15] and the average sound velocities as $2.4\times 10^5$ cm/s. Therefore, a high temperature approximation $\hbar\omega_{q'} < k_B T$ holds at temperature below 4.2 K in our experiment. Then, according to the work of Roth and Argyres [16], a relaxation time $\tau(\varepsilon)$ for acoustic phonon scattering is expressed as:

$$\frac{1}{\tau(\varepsilon)} = (2\pi E^2 k_B T/\hbar\mu v^2)D(\varepsilon), \tag{5}$$

where $E$ is the deformation potential, $\mu$ the mass density, $v$ the sound velocity, $D(\varepsilon)$ the density of state in the presence of magnetic field. Since electrons exhibit long relaxation time by lowering temperature (equation (5)), there would arise an increase in diamagnetic surface (wrapping) current that gives a correction to the thermal diffusion coefficient $A$ in a quantizing magnetic field [17]. Also, the fact that the observed amplitude data are in good agreement with equation (1) provides strong evidence that the energy dependence of the electron relaxation time brings about giant oscillations of the longitudinal MTP below 4.2 K [12]. But, the energy dependence of $\tau(\varepsilon)$ is not well understood [18].

In conclusion, the observed curious temperature dependence of amplitude in the longitudinal MTP may be explained by assuming that inter-level scatterings by phonons play a dominant role above 4.2 K, and that both the diamagnetism based on surface current and the energy dependence of the electron relaxation time give rise to the pronounced oscillations of the longitudinal MTP below 4.2 K.

## References

[1] Edelman V S 1976 *Adv. Phys.* **25** 555
[2] Fu L and Kane C L 2007 *Phys. Rev.* B **76** 045302
[3] Satoh N 2003 *Physica* B **336** 290
[4] Satoh N, Kitamura Y and Takenaka H 2007 *Physica* B **391** 244
[5] Satoh N, Takenaka H 2008 *Physica* B **403** 3705
[6] Satoh N 2009 *J. Phys.: Conf. Ser.* **150** 022072
[7] Mangez J H, Issi J P and Heremans J 1976 *Phys. Rev.* B **14** 4381
[8] Smith G E, Baraff G A and Rowell J M 1964 *Phys. Rev.* **135** A1118
[9] Tanuma S and Inada R 1975 *J. Phys.: Con.Mat.* **19** 95
[10] Young R C 1973 *J. Phys. F: Met. Phys.* **3** 721
[11] Fletcher R 1981 *J. Phys. F: Met. Phys.* **11** 1093
[12] Pantsulaya A V and Varlamov A A 1989 *Phys. Lett.* A **136** 317
[13] Dinger R J and Lawson A W 1973 *Phys. Rev.* B **7** 5215
[14] Hofmann Ph 2006 *Prog. Sur. Sci.* **81** 191
[15] Brown R N, Hartman R and Koenig S H 1968 *Phys. Rev.* **172** 598
[16] Roth L M and Argyres P N 1966 *Semiconductors and Semimetals* vol 1, ed R K Willardson and A C Beer (San Diego: Academic Press) chapter 6
[17] Obraztsov Yu N 1965 *Sovi. Phys. Solid State* **7** 455
[18] Ashcroft N W and Mermin N D 1976 *Solid State Physics* (Philadelphia: Saunders College) Chap 13 p 258

## Acknowledgements

The author would like to thank Professor Y Ishizawa for valuable discussions and encouragement.

IOP Conf. Series: Journal of Physics: Conf. Series **969** (2018) 012137 doi:10.1088/1742-6596/969/1/012137

# Topology in single-wall carbon nanotube of zigzag and armchair type

**Rin Okuyama**[1], **Wataru Izumida**[2], **and Mikio Eto**[1]

[1] Faculty of Science and Technology, Keio University, 3-14-1 Hiyoshi, Kohoku-ku, Yokohama 223-8522, Japan

[2] Department of Physics, Tohoku University, 6-3 Aramaki-Aza-Aoba, Aoba-ku, Sendai 980-8578, Japan

E-mail: `rokuyama@rk.phys.keio.ac.jp`

**Abstract.** Single-wall carbon nanotubes can be one-dimensional (1D) topological insulators except for armchair nanotubes. The electronic states are characterized by a non-zero topological invariant, winding number, which is related to the number of 0D edge states via the bulk-edge correspondence. In the present study, we theoretically examine zigzag and armchair nanotubes to elucidate the emergence and absence of edge states. The effective 1D lattice model is employed in order to describe the fine structures due to the finite curvature of tube surface and spin-orbit interaction. We show that the lattice model for a zigzag nanotube is equivalent to the Su-Schrieffer-Heeger model, by which the formation of edge states is explained. An armchair nanotube is described by a ladder model, on the other hand, which does not host any edge states owing to the $\sigma_h$ symmetry.

## 1. Introduction

The single-wall carbon nanotube (SWNT) is a unique one-dimensional (1D) system made by rolling up a graphene sheet. It can be either metallic or semiconducting depending on its chirality [1]. Even for a metallic SWNT, a narrow energy gap opens between conduction and valence bands due to the mixing of $\sigma$ and $\pi$ orbitals by the finite curvature of tube surface [2, 3] and spin-orbit (SO) interaction [4].

Both metallic and semiconducting SWNTs can be regarded as 1D topological insulators owing to the sublattice symmetry for $A$ and $B$ lattice sites. The electronic states are characterized by a $\mathbb{Z}$ topological invariant, winding number [5], in both the absence (class BDI) and presence (AIII) of magnetic field [6]. The winding number is related to the number of 0D edge states in finite SWNTs with a suitable boundary condition via the bulk-edge correspondence [7]. The present authors revealed that SWNTs except for armchair nanotubes may be topologically non-trivial in general [8]. Besides, in metallic SWNTs, the small band gap is closed by applying a feasible magnetic field $B$ in the axial direction, which results in the topological phase transition. This phase transition should be observed as a change in the number of edge states, e.g., using scanning tunneling spectroscopy [9].

In this work, we elucidate the mechanism of emergence and absence of the edge states in SWNTs. For this purpose, we examine a 1D lattice model proposed in Ref. [8], which effectively describes the curvature effects and SO interaction. As an example of non-trivial topological

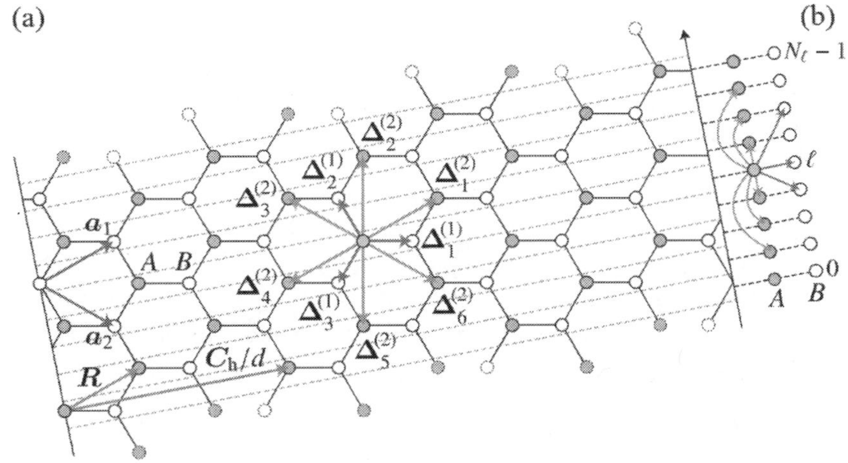

**Figure 1.** (a) Primitive lattice vectors of graphene, $\boldsymbol{a}_1$ and $\boldsymbol{a}_2$, and mapping of $(n,m)$-SWNT to the graphene sheet [8]. The chiral vector, $\boldsymbol{C}_{\rm h} = n\boldsymbol{a}_1 + m\boldsymbol{a}_2$, indicates the circumference of the tube. The three vectors $\boldsymbol{\Delta}_j^{(1)}$ $(j = 1,2,3)$ connect the nearest-neighbor atoms, whereas the six vectors $\boldsymbol{\Delta}_j^{(2)}$ $(j = 1,2,\cdots,6)$ connect the second nearest-neighbor atoms. The rotational symmetry $C_d$ around the tube axis (helical symmetry) corresponds to the translational symmetry of $\boldsymbol{C}_{\rm h}/d$ $(\boldsymbol{R} = p\boldsymbol{a}_1 + q\boldsymbol{a}_2)$, where $d = \gcd(n,m)$, and $p$ and $q$ are integers satisfying $mp - nq = d$. This figure shows the case of $(n,m) = (6,3)$ with $d = 3$, $p = 1$, and $q = 0$. (b) An effective 1D lattice model in which $A$ and $B$ atoms are aligned in the axial direction. The distant hoppings $\Delta\ell_j^{(i)}$ with phase factor $\Delta\nu_j^{(i)}$ of $H_{\mu,s}$ in Eq. (1) are determined from $\boldsymbol{\Delta}_j^{(i)} = \Delta\ell_j^{(i)}\boldsymbol{R} + \Delta\nu_j^{(i)}(\boldsymbol{C}_{\rm h}/d)$ [12].

insulators, a zigzag nanotube is studied. Its lattice model is shown to be equivalent to the Su-Schrieffer-Heeger (SSH) model, which is known as a fundamental model for 1D topological insulators [10]. The formation of edge states and topological phase transition are clearly explained in terms of the SSH model. For an armchair nanotube, on the other hand, the winding number is always zero and thus it is topologically trivial. Its lattice model is mapped to a ladder model, which is extended from that in Ref. [11] to take into account the SO interaction as well as axial symmetry. We show that it does not host any edge states due to the $\sigma_h$ symmetry [7].

## 2. Model

We consider a SWNT whose circumference is specified by the chiral vector, $\boldsymbol{C}_{\rm h} = n\boldsymbol{a}_1 + m\boldsymbol{a}_2 \equiv (n,m)$, with $\boldsymbol{a}_1$ and $\boldsymbol{a}_2$ being the primitive lattice vectors of graphene shown in Fig. 1 (a) [1]. For zigzag and armchair nanotubes, we use $(n,0)$- and $(n,n)$-SWNTs with $n > 0$, respectively. The diameter is given by $d_{\rm t} = |\boldsymbol{C}_{\rm h}|/\pi = a\sqrt{n^2 + nm + m^2}/\pi$ with the lattice constant $a = 0.246$ nm of graphene. The chiral angle $\theta$ is defined as the angle between $\boldsymbol{C}_{\rm h}$ and $\boldsymbol{a}_1$: $\theta = 0$ and $\pi/6$ for zigzag and armchair types, respectively.

$(n,m)$-SWNT has the $d$-fold symmetry around the tube axis, where $d = \gcd(n,m)$ is the greatest common divisor of $n$ and $m$. It also has the helical symmetry with translation along the tube axis with a rotation around it [13]. Due to these symmetries, the electronic states in $(n,m)$-SWNT with orbital angular momentum $\mu = 0,1,2,\ldots,d-1$ is described by a 1D lattice model shown in Fig. 1 (b). Starting from the $\boldsymbol{k}\cdot\boldsymbol{p}$ theory to consider the curvature effects and SO interaction [14], we construct a 1D lattice model to describe the electronic states around the $K$ and $K'$ points [8]. The Hamiltonian for electrons with angular momentum $\mu$ and spin $s = \pm 1$

IOP Conf. Series: Journal of Physics: Conf. Series **969** (2018) 012137 doi:10.1088/1742-6596/969/1/012137

is given by

$$H_{\mu,s} = \sum_{\sigma,\ell} \frac{1}{2} g_s \mu_B B s c^{\mu,s\dagger}_{\sigma,\ell} c^{\mu,s}_{\sigma,\ell} + \sum_{\ell} \sum_{j=1}^{3} \left( \gamma^{(1)}_{s,j} e^{i2\pi\mu\Delta\nu^{(1)}_j/d} c^{\mu,s\dagger}_{A,\ell} c^{\mu,s}_{B,\ell} + \text{h.c.} \right)$$
$$+ \sum_{\sigma,\ell} \sum_{j=1}^{3} \left( \gamma^{(2)}_{s,j} e^{i2\pi\mu\Delta\nu^{(2)}_j/d} c^{\mu,s\dagger}_{\sigma,\ell} c^{\mu,s}_{\sigma,\ell} + \text{h.c.} \right). \quad (1)$$

Here, $c^{\mu,s}_{\sigma,\ell}$ is the field operator of an electron with angular momentum $\mu$ and spin $s$ at atom $\sigma = A, B$ of site index $\ell$. The hopping to the first [second] nearest-neighbor atoms in Fig. 1 (a) gives rise to the hopping to the site separated by $\Delta\ell^{(1)}_j$ [$\Delta\ell^{(2)}_j$] with phase factor $\Delta\nu^{(1)}_j$ [$\Delta\nu^{(2)}_j$], as depicted in Fig. 1 (b). The hopping integrals $\gamma^{(1)}_{s,j}$ and $\gamma^{(2)}_{s,j}$ are given by

$$\gamma^{(1)}_{s,j} = \gamma\{1 + \Delta k_c a_{\rm CC} \sin\phi_j - [\Delta k_z + i(s\Delta k_{\rm so} + \Delta k_\phi)] a_{\rm CC} \cos\phi_j\}, \quad (2)$$

$$\gamma^{(2)}_{s,j} = i\frac{(-1)^{j+1}}{3\sqrt{3}} s\epsilon_{\rm so}, \quad (3)$$

respectively, where $\phi_j = \theta - (5\pi/6) + (2\pi/3)j$, $a_{\rm CC} = a/\sqrt{3}$ is the interatomic distance, and $\gamma = -2\hbar v_{\rm F}/(3a_{\rm CC})$ with $v_{\rm F} = 8.32 \times 10^5$ m/s being the Fermi velocity. The hybridization between $\sigma$ and $\pi$ orbitals due to the finite curvature is taking into account by

$$\Delta k_c = \beta' \frac{\cos 3\theta}{d_t^2}, \quad \Delta k_z = \zeta \frac{\sin 3\theta}{d_t^2}, \quad (4)$$

with $\beta' = 0.0436$ nm and $\zeta = -0.185$ nm. The SO interaction yields

$$\Delta k_{\rm so} = \alpha'_1 V_{\rm so} \frac{1}{d_t}, \quad \epsilon_{\rm so} = \alpha_2 V_{\rm so} \frac{\cos 3\theta}{d_t}, \quad (5)$$

with $\alpha'_1 = 8.8 \times 10^{-5}$ meV$^{-1}$, $\alpha_2 = -0.045$ nm, and $V_{\rm so} = 6$ meV being the SO interaction for 2p orbital in carbon atom. The Aharonov-Bohm (AB) phase by magnetic field $B$ appears as

$$\Delta k_\phi = -\frac{eB}{4\hbar} d_t. \quad (6)$$

For a metallic SWNT, the local gap in the vicinity of $K/K'$ points yields

$$E_g = 2\hbar v_{\rm F} |\pm \Delta k_c + s\Delta k_{\rm so} + \Delta k_\phi|, \quad (7)$$

for electrons with spin $s$.

## 3. Zigzag nanotube

We consider $(n, 0)$-SWNT as a zigzag nanotube with $n > 0$. It is semiconducting when $n \neq 0$ (mod 3) and metallic when $n = 0$ (mod 3). For a semiconducting nanotube, the band gap is given by the boundary condition in the circumference direction, and the fine structures are irrelevant. For a metallic nanotube, the band gap is dominantly determined by $\Delta k_c$, whereas the SO interaction brings about small corrections. We therefore neglect the SO interaction as well as Zeeman energy for a while. The effective 1D lattice model in Eq. (1) is reduced to [15],

$$H_{\mu,s} = \sum_\ell \left( \gamma^{\mu,s}_{\rm intra} c^{\mu,s}_{A,\ell} c^{\mu,s}_{B,\ell} + \gamma^{\mu,s}_{\rm inter} c^{\mu,s}_{A,\ell} c^{\mu,s}_{B,\ell+1} \right) + \text{h.c.}, \quad (8)$$

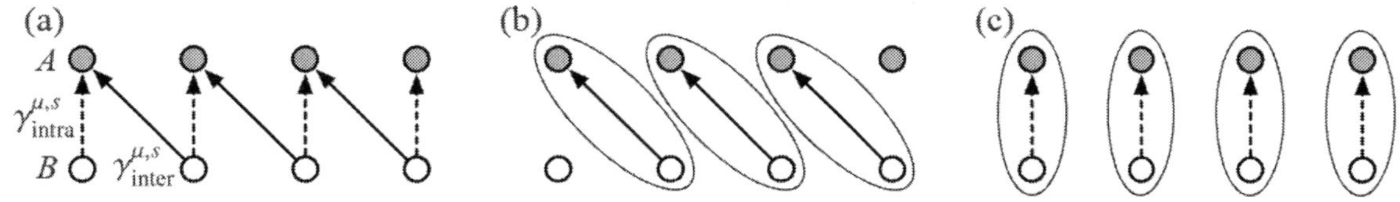


**Figure 2.** (a) The effective 1D lattice model for a zigzag nanotube. The SO interaction is neglected. This is equivalent to the SSH model with staggered hopping integrals, determined by $\mu$, $d_{\rm t}$, and $B$. (b) When the hopping amplitude within the unit cell is weaker than that between the adjacent cells, $|\gamma^{\mu,s}_{\rm inter}| < |\gamma^{\mu,s}_{\rm intra}|$, the model is topologically equivalent to the inter-cell dimerization with 0D edge states. (c) On the other hand, for $|\gamma^{\mu,s}_{\rm inter}| > |\gamma^{\mu,s}_{\rm intra}|$, the system hosts no edge states.

with

$$\gamma^{\mu,s}_{\rm intra} = 2\gamma {\rm e}^{-{\rm i}\frac{\pi\mu}{n} - \frac{1}{2}\Delta k_c a_{\rm CC}} \cos\left(\frac{\pi\mu}{n} - \frac{\sqrt{3}}{2}\Delta k_\phi a_{\rm CC}\right), \quad \gamma^{\mu s}_{\rm inter} = \gamma {\rm e}^{\Delta k_c a_{\rm CC}}. \tag{9}$$

This model, as shown in Fig. 2 (a), is equivalent to the SSH model for polyacetylene, which is a well known 1D topological insulator [10]. The topological nature of SSH model is classified by the ratio of the hopping amplitude within the unit cell to that between adjacent cells,

$$\alpha = \left|\frac{\gamma^{\mu,s}_{\rm intra}}{\gamma^{\mu,s}_{\rm inter}}\right| = 2{\rm e}^{-\frac{3}{2}\Delta k_c a_{\rm CC}} \left|\cos\left(\frac{\pi\mu}{n} - \frac{\sqrt{3}}{2}\Delta k_\phi a_{\rm CC}\right)\right|. \tag{10}$$

For angular momenta at the $K/K'$ points ($\mu_{K/K'} = \pm 2n/3$), we have

$$\alpha = {\rm e}^{-\frac{3}{2}(\Delta k_c \pm \Delta k_\phi) a_{\rm CC}}. \tag{11}$$

When $\alpha = 1$, the gap is closed and the winding number cannot be defined. For $\alpha < 1$, the system can be adiabatically deformed into that with $\gamma_{\rm intra} = 0$, as shown in Fig. 2 (b). The system has 0D edge states at zero energy. At the same time, the winding number becomes finite, and the system is topologically non-trivial. On the other hand, for $\alpha > 1$, the system can be deformed into that with $\gamma_{\rm inter} = 0$, as depicted in Fig. 2 (c), which has no edge states at zero energy. As a result, the winding number becomes zero, and the system is a trivial insulator.

By counting the angular momenta $\mu$ that satisfy $\alpha < 1$, we find that the number of edge states is

$$4[(n/3) + 1] \text{ for } n = 0 \pmod 3, \tag{12}$$
$$4[\lfloor n/3 \rfloor + {\rm mod}(n,3) - 1] \text{ for } n \neq 0 \pmod 3, \tag{13}$$

at $B = 0$ with $\lfloor x \rfloor$ being the greatest integer not exceeding $x$. Consequently, a zigzag nanotube is shown to be a topological insulator at $B = 0$.

Equation (11) clearly explains the topological phase transition in a metallic zigzag nanotube, where the local gap around the $K$ point disappears for $B = B^*$ at which $\Delta k_c + \Delta k_\phi = 0$ is satisfied. When $B < B^*$, $\alpha < 1$ for $\mu = \mu_K$, and the system hosts edge states from the $K$ valleys. When $B > B^*$, $\alpha > 1$ for $\mu = \mu_K$, and the edge states from the $K$ valley are delocalized. This is the topological phase transition discussed in Ref. [8]. Note that a similar phase transition takes place at $B = -B^*$ for the $K'$ valley.

We also consider the affects of the SO interaction and Zeeman energy. Our numerical calculation shows that edge states in typical zigzag nanotubes with $d_{\rm t} \sim 1$ nm can well be explained by the above-mentioned scenario.

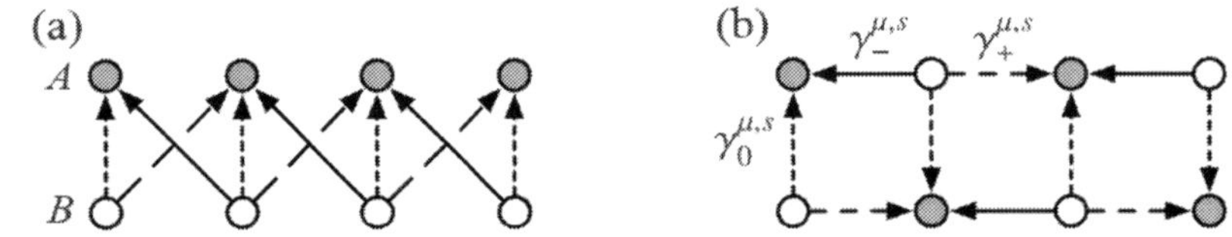


**Figure 3.** (a) The effective 1D lattice model for an armchair nanotube. Note that $|\gamma_+^{\mu,s}| = |\gamma_-^{\mu,s}|$ owing to the $\sigma_h$ symmetry. (b) The equivalent ladder model after swapping $A$ and $B$ atoms at even sites.

## 4. Armchair nanotube

Next, we consider $(n,n)$-SWNT as an armchair nanotube with $n > 0$. It is always metallic independent of $n$. Since $\Delta k_c = 0$, the small energy gap is determined by $\Delta k_{\rm so}$. Neglecting the small Zeeman splitting, the corresponding 1D lattice model yields [16],

$$H_{\mu,s} = \sum_\ell \left( \gamma_0^{\mu,s} c_{A,\ell}^{\mu,s\dagger} c_{B,\ell}^{\mu,s} + \gamma_+^{\mu,s} c_{A,\ell}^{\mu,s\dagger} c_{B,\ell+1}^{\mu,s} + \gamma_-^{\mu,s} c_{A,\ell}^{\mu,s\dagger} c_{B,\ell-1}^{\mu,s} \right) + \text{h.c.}, \quad (14)$$

with

$$\gamma_0^{\mu,s} = \gamma \mathrm{e}^{-[\mathrm{i}(\Delta k_\phi + s\Delta k_{\rm so}) + \Delta k_z] a_{\rm CC}}, \quad \mathrm{e}^{\mathrm{i}2\pi\mu/d} \gamma_+^{\mu,s} = \gamma_-^{\mu,s} = \gamma \mathrm{e}^{\frac{1}{2}[\mathrm{i}(\Delta k_\phi + s\Delta k_{\rm so}) + \Delta k_z] a_{\rm CC}}, \quad (15)$$

as shown in Fig. 3 (a). Note that $|\gamma_+^{\mu,s}| = |\gamma_-^{\mu,s}|$ due to the mirror reflection, $\sigma_h$, symmetry with a horizontal mirror plane perpendicular to the nanotube axis. This model is depicted in Fig. 3 (b), if $A$ and $B$ sublattices are exchanged in a similar manner to Ref. [11]. The gap disappears at a magnetic field which satisfies $\Delta k_\phi + s\Delta k_{\rm so} = 0$ with $s = +1$ or $-1$. Otherwise, the system becomes an insulator, on which we focus hereafter. We find that, unlike the case of a zigzag nanotube, there cannot be any edge state; the system can be adiabatically deformed into that with $|\gamma_+^{\mu,s}| = |\gamma_-^{\mu,s}| = 0$, which is the same situation as shown in Fig. 2 (c). This fact clearly explains that an armchair nanotube cannot host edge states for any $\mu$ and $s$ and thus it is a trivial insulator. Note that the Zeeman splitting is irrelevant also in the case of an armchair nanotube.

## 5. Conclusions

We have theoretically examined the properties of zigzag and armchair SWNTs as topological insulators, by observing edge states in an effective 1D lattice model to take into account the finite curvature and SO interaction. The lattice model for a zigzag nanotube is equivalent to the SSH model, by which the emergence of edge states as well as magnetic-field-induced topological phase transition have been clearly explained. The lattice model for an armchair nanotube is a ladder model, on the other hand, which does not host any edge states owing to the $\sigma_h$ symmetry.

## Acknowledgments

The authors acknowledge fruitful discussion with M. Grifoni. This work was partially supported by JSPS KAKENHI Grant Numbers JP16H01046, JP15H05870, JP15K05118, JP15KK0147, and JP26220711.

## References

[1] Saito R, Dresselhaus G and Dresselhaus M S 1998 *Physical Properties of Carbon Nanotubes* (London: Imperial College Press)
[2] Hamada N, Sawada S and Oshiyama A 1992 *Phys. Rev. Lett.* **68** 1579

[3] Saito R, Fujita M, Dresselhaus G and Dresselhaus M S 1992 *Phys. Rev. B* **46** 1804
[4] Ando T 2000 *J. Phys. Soc. Jpn* **69** 1757
[5] Wen X G and Zee A 1989 *Nucl. Phys. B* **316** 641
[6] Schnyder A P, Ryu S, Furusaki A and Ludwig A W W 2008 *Phys. Rev. B* **78** 195125
[7] Izumida W, Okuyama R, Yamakage A and Saito R 2016 *Phys. Rev. B* **93** 195442
[8] Okuyama R, Izumida W and Eto M 2017 *J. Phys. Soc. Jpn* **86** 2
[9] Kobayashi Y, Fukui K, Enoki T, Kusakabe K and Kaburagi Y 2005 *Phys. Rev. B* **71** 193406
[10] Asbóth J K, Oroszlány L and Pályi A 2015 *A Short Course on Topological Insulators: Band-structure topology and edge states in one and two dimensions* (Berlin: Springer)
[11] Balents L and Fisher M P 1997 *Phys. Rev. B* **55** R11973
[12] Explicitly, $\Delta\ell_j^{(1)} = 0$, $n/d$, $-m/d$ and $\Delta\nu_j^{(1)} = 0$, $-p$, $q$ for $j = 1, 2, 3$. $\Delta\ell_j^{(2)} = m/d$, $(n+m)/d$, $n/d$, $-m/d$, $-(n+m)/d$, $-n/d$ and $\Delta\nu_j^{(2)} = -q$, $-(p+q)$, $-p$, $q$, $(p+q)$, $p$ for $j = 1, 2, \cdots, 6$.
[13] White C T, Robertson D H and Mintmire J W 1993 *Phys. Rev. B* **47** 5485
[14] Izumida W, Sato K and Saito R 2009 *J. Phys. Soc. Jpn* **78** 074707
[15] Here, we have used $(p, q) = (0, -1)$, which minimizes the amplitude of helical vector, $|\boldsymbol{R}|$, in Fig. 1 (a).
[16] We choose $(p, q) = (1, 0)$ to minimize $|\boldsymbol{R}|$.

IOP Conf. Series: Journal of Physics: Conf. Series **969** (2018) 012138 doi:10.1088/1742-6596/969/1/012138

# Statistics of the quantized microwave electromagnetic field in mesoscopic elements at low temperature

**Stéphane Virally, Jean Olivier Simoneau, Christian Lupien and Bertrand Reulet**

Université de Sherbrooke, Physics Department,
2500, Boulevard de l'Université, Sherbrooke, QC J1K 2R1, Canada

E-mail: Stephane.Virally@USherbrooke.ca

**Abstract.** The quantum behavior of the electromagnetic field in mesoscopic elements is intimately linked to the quantization of the charge. In order to probe nonclassical aspects of the field in those elements, it is essential that thermal noise be reduced to the quantum level, i.e. to scales where $kT \lesssim h\nu$. This is easily achieved in dilution refrigerators for frequencies of a few GHz, i.e. in the microwave domain. Several recent experiments have highlighted the link between discrete charge transport and discrete photon emission in simple mesoscopic elements such as a tunnel junction. Photocount statistics are inferred from the measurement of continuous variables such as the quadratures of the field.

## 1. Introduction

Although charge currents in conductors are transported by fermions, they are always linked to bosonic excitations of the electromagnetic field. At the mesoscopic scale, currents exhibit properties that reflect the quantization of the electric charge (such as electron shot noise [1]), and it is reasonable to expect that the electromagnetic field also behaves in a quantum manner, i.e. that photon shot noise should be observed [2, 3, 4]. This is indeed the case and it can be observed in conditions where thermal noise is reduced to or below the level of the vacuum fluctuations of the field. Such conditions are met for microwave signals of a few GHz emitted by mesoscopic elements cooled down to a few milliKelvin, typically in dilution refrigerators [5].

Photon counters are difficult to come by in the microwave domain [6, 7], as there are no natural systems with the right energy gaps. On the other hand, continuous observables such as the voltage can be measured with precision [8]. This contrasts with the case of the electromagnetic field in the visible and near infrared spectra, where the field itself is difficult to measure directly. Consequently, mesoscopic systems at low tempreature constitute a very interesting field of study for continuous variable quantum optics. This paper details some of the techniques required for the study of statistical properties of the quantized electromagnetic field in the microwave domain. It summarizes a recent series of experiments on one of the simplest mesoscopic systems, a tunnel junction cooled down to ~20 mK and biased with both dc and ac signals [9, 10, 11, 12, 13, 14]. It also shows that questions surrounding wideband signals, that are hard to answer when studying quantum optics at room temperature with visible or infrared signals, can be studied in details in mesoscopic systems.

## 2. Narrowband signals

The main observable in electric circuits is the voltage

$$\hat{v}(t) = i\sqrt{\frac{2}{Z}} \int_0^{+\infty} d\nu \, \sqrt{h\nu} \left[\hat{a}^\dagger(\nu) \, e^{i2\pi\nu t} - \hat{a}(\nu) \, e^{-i2\pi\nu t}\right], \tag{1}$$

where $Z$ is the characteristic impedance of the transmission line in which the electromagnetic field propagates, and $\hat{a}^{(\dagger)}(\nu)$ are the usual bosonic annihilation (creation) operators. When the detection bandwidth is narrow, the voltage observable, viewed in the rotating frame (i.e. with the use of a mixer), reduces to a single quadrature of the field,

$$\hat{x}_\theta = \frac{\hat{a}^\dagger \, e^{i\theta} + \hat{a} \, e^{-i\theta}}{\sqrt{2}}, \tag{2}$$

with $\theta$ a phase angle that can be selected with a phase shifter on the local oscillator signal. It can also be continuously changed by using a local oscillator slightly detuned from the signal. In other cases, the phase varies randomly during detection. In any case where the phase $\theta$ can be averaged, expectations of all odd centered moments of the measured signal are zero, while expectations of the $2k^{\text{th}}$ centered moments are

$$\left\langle x_\theta^{2k} \right\rangle_\theta = \left(\frac{1}{2}\right)^k \left\langle \sum_{\text{c.s.}} \hat{a}^k \, \hat{a}^{\dagger k} \right\rangle, \tag{3}$$

with "c.s." standing for completely symmetric (i.e., the sum is taken on all possible permutations of the non-commuting operators). Here, $\langle \bullet \rangle$ represents averaging over the quantum ensemble, while $\langle \bullet \rangle_\theta$ represents averaging over both the quantum ensemble and the phase $\theta$. To simplify notations, we drop the $\theta$ indices everywhere below.

The completely symmetric sum of Eq. 3 can be explicitely evaluated and yields [15]

$$\sum_{\text{c.s.}} \hat{a}^k \, \hat{a}^{\dagger k} = \sum_{i=0}^{k} \left(\frac{1}{2}\right)^{k-i} \frac{(2k)!}{(i!)^2 (k-i)!} \prod_{j=0}^{i-1} (\hat{n} - j), \tag{4}$$

with $\hat{n} = \hat{a}^\dagger \, \hat{a}$ the usual number operator. Hence, the even moments $\langle x^{2k} \rangle$ of the continuous distribution are explicitely linked to the moments $\langle n^\ell \rangle$ of the discrete photocount distribution. As a measurement of $\langle x_\theta^k \rangle$ for all $k$ and all $\theta$ enables a full reconstruction (tomography) of the quantum state, the less stringent measurement of $\langle x^{2k} \rangle$ for all $k$ leads to a full reconstruction of the photocount distribution. This is expected, as we simply lose the phase information.

It is usually better to compute the cumulants of a distribution, rather than its moments, as cumulants are additive for independent distributions. Hence, noise cumulants can be readily subtracted from signal cumulants. The first moments of the photocount distribution are obtained from the first cumulants of the continuous voltage distribution as

$$\langle n \rangle = \langle\!\langle x^2 \rangle\!\rangle - \frac{1}{2}; \tag{5}$$

$$\langle \delta n^2 \rangle = \frac{2}{3} \langle\!\langle x^4 \rangle\!\rangle + \langle\!\langle x^2 \rangle\!\rangle^2 - \frac{1}{4}; \tag{6}$$

$$\langle \delta n^3 \rangle = \frac{2}{5} \langle\!\langle x^6 \rangle\!\rangle + 4 \langle\!\langle x^4 \rangle\!\rangle \langle\!\langle x^2 \rangle\!\rangle + 2 \langle\!\langle x^2 \rangle\!\rangle^3 - \frac{1}{2} \langle\!\langle x^2 \rangle\!\rangle, \tag{7}$$

where $\langle\!\langle \bullet \rangle\!\rangle$ are the cumulants, as opposed to the central moments ($\langle \bullet \rangle$) of the distribution.

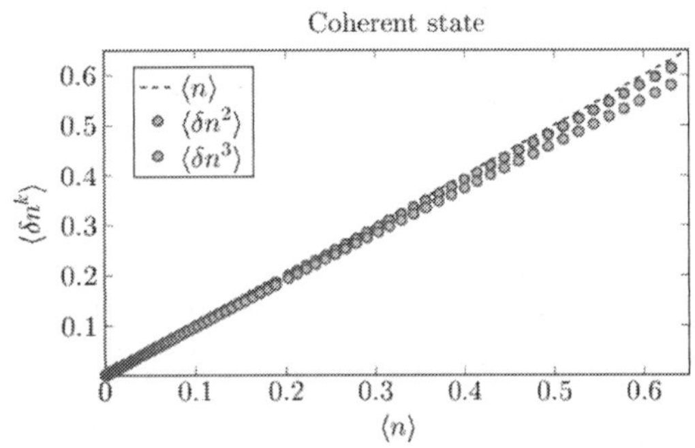


(a) Experimental determination of the first moments of the photocount distribution for a coherent state.

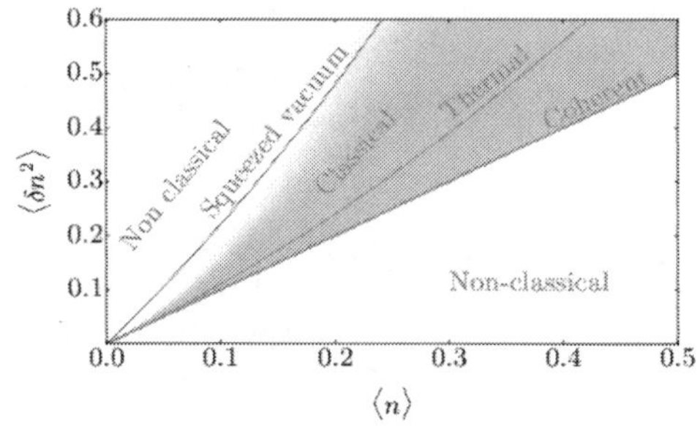


(b) Classical and nonclassical domains in the $(\langle n \rangle, \langle \delta n^2 \rangle)$ plane.

**Figure 1.** $\langle \delta n^k \rangle$ vs. $\langle n \rangle$.

As a proof-of-principle experiment, we measured the first three moments of the photocount distribution of a narrow-band coherent signal (i.e. a sine wave) in the limit $\langle n \rangle < 1$ [8]. In order to do so, we had to place an "emitter" (the last attenuator of a long transmission line) in a cryogenic environment to ensure $h\nu \gg k_B T$, with $\nu$ the frequency of the excitation ($\sim$ 6GHz) and $T$ the temperature ($\sim$ 10mK). The results are shown on Fig. 1a. For such a state, we expect $\langle \delta n^2 \rangle = \langle \delta n^3 \rangle = \langle n \rangle$.

Coherent states are only one variety of states that exhibit classical properties for the photocount distribution. Among those properties are

$$\langle \delta n^2 \rangle \geq \langle n \rangle ; \tag{8}$$

$$\langle \delta n^2 \rangle \to \langle n \rangle \text{ as } \langle n \rangle \to 0; \tag{9}$$

$$\langle \delta n^3 \rangle \geq \langle n \rangle + 3 \left( \langle \delta n^2 \rangle - \langle n \rangle \right) \left( 1 - \langle n \rangle \right). \tag{10}$$

In particular, it is possible to show domains of nonclassicality on the $(\langle n \rangle, \langle \delta n^2 \rangle)$ plane (see Fig. 1b).

The noise emitted by a tunnel junction biased in dc and ac exhibits some of the nonclassical characteristics defined above. In the narrow band case at half the frequency of the ac excitation, this noise exhibits a Fano factor $\mathcal{F} = \langle \delta n^2 \rangle / \langle n \rangle$ greater than unity, even in the limit of vanishing $\langle n \rangle$ (at least at $T = 0$), in contradiction with the second classical condition above. This was demonstrated in a recent series of experiments showing that the electromagnetic noise is the result of a squeezing of the vacuum modes of the electromagnetic field [9, 10, 11, 12, 13]. In particular, a Fano factor above unity was observed in Ref. [14]. Squeezing is produced when pairs of photons are generated simultaneously, which is the root of the greater than unity Fano factor (it would be 2 for a pure source of photon pairs, down to $\langle n \rangle = 0$ at $T = 0$). The signature of squeezing is an asymetry between the noise in two orthogonal quadratures ($\hat{x}_\theta \equiv \hat{x}$ and $\hat{x}_{\theta+\pi/2} \equiv \hat{p}$). Compared to the symetric vacuum noise, squeezed signals sub-vacuum noise in one quadrature, at the expense of added noise to the orthogonal quadrature. This is illustrated in Fig. 2.

## 3. Wideband signals

The voltage is directly proportional to the electric field. In regular quantum optics, the field itself is difficult to probe as its oscillations are much faster than the response time of the detectors. In

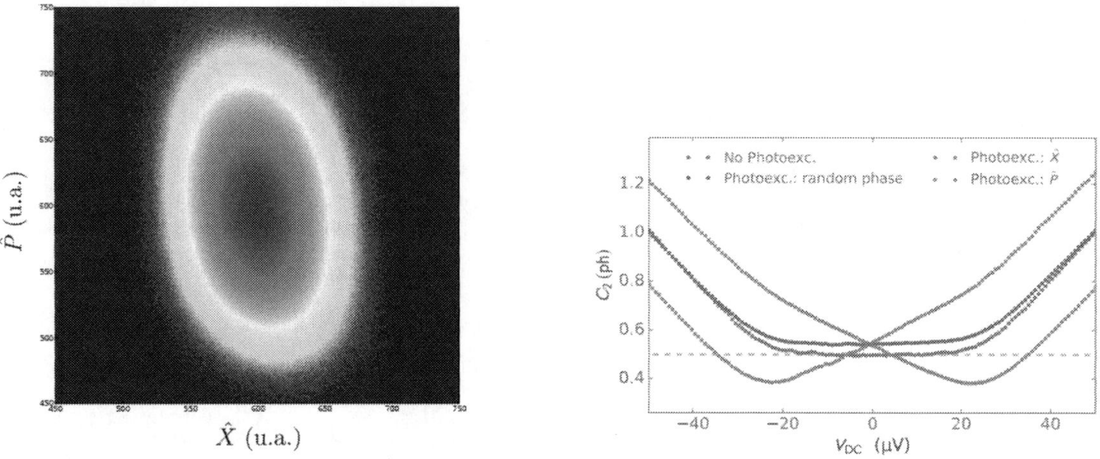


(a) Tomography of the squeezed vacuum

(b) Noise in both quadratures.

**Figure 2.** Squeezed vacuum state produced by an ac+dc bias of a tunnel junction.

contrast, mesoscopic systems at low temperature offer great opportunities to directly probe field oscillations in real time, as the GHz signals required for these experiments are directly accessible through fast sampling oscilloscopes. Interestingly, one can even probe a wideband signal with a setup allowing detection from dc to a few GHz. This is not possible in visible or near infrared quantum optics, where detection bands are always narrow compared to the value of the central frequency of the signal. Of course, probing of the quantum behavior of such a signal requires that $k_B T \ll h\Delta\nu$, where $\Delta\nu$ is the bandwidth of the signal.

For wideband signals, it is not immediately obvious what a photon is. A single excitation of the electromagnetic field is in a superposition of many wavelengths, and the question of how many photons are present in a single pulse is not trivial. However, one can imagine an experiment that determines such a number, and there must be a theory predicting the results. The thought experiment is as follows: imagine a gapless material (e.g. graphene) absorbing photons from a short, wideband pulse. All frequencies in the pulse can lead to an electron being ejected from the valence band and injected into the conduction band, before being amplified by a cascading process. Hence, it is theoretically possible to "count" the number of photons in such a pulse. If one were to carry such an experiment, all information on the energy of each photon would be lost, as dictated by Heisenberg's time/energy uncertainty principle. On the other end, it is possible to simply measure the energy content of the pulse, with a bolometer for instance. In this case, all timing information would be erased.

As in the narrow band case, it is in fact possible to obtain information on the continuous energy distribution and on the discrete photocount distribution by simply measuring the voltage on a sufficiently fast oscilloscope. It is found that the energy observable is simply

$$\frac{1}{Z}\int_{-\infty}^{+\infty} dt\ \hat{v}^2(t) = \int_0^{+\infty} d\nu\ h\nu\ \hat{a}^\dagger(\nu)\hat{a}(\nu), \tag{11}$$

while the observable for the number of photons is

$$\frac{2}{h\,Z}\int_{-\infty}^{+\infty} dt\ \hat{w}^2(t) = \int_0^{+\infty} d\nu\ \hat{a}^\dagger(\nu)\hat{a}(\nu), \tag{12}$$

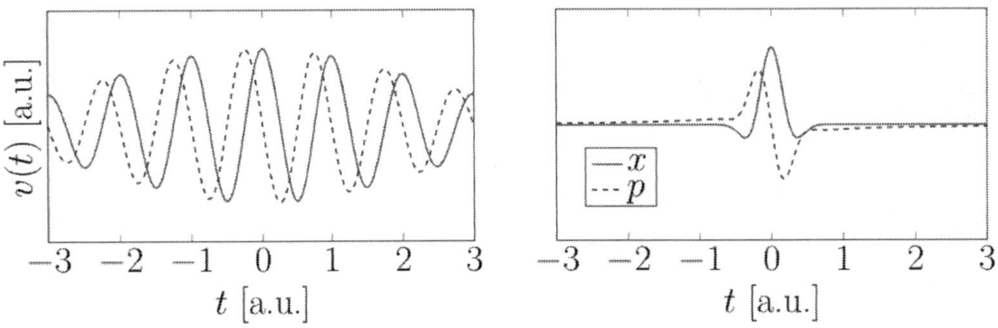


**Figure 3.** Quadratures of narrow band (left) and wideband (right) signals.

where we have defined the causal transform

$$\hat{w}(t) = \int_0^{+\infty} d\tau \, \frac{\hat{v}(t-\tau)}{\sqrt{\tau}}. \tag{13}$$

It turns out that the number of photons can always be written as a sum of the squares of two quadratures

$$\frac{1}{h\,Z} \int_{-\infty}^{+\infty} dt \left[\hat{x}^2(t) + \hat{p}^2(t)\right], \tag{14}$$

and that these quadratures are Hilbert transforms of one another ($\hat{w}$ is one of these quadratures). Fig. 3 illustrates the quadratures of signals in the narrow band and wideband limits. In the narrow band limit, we find, as expected, that the quadratures are essentially the sine and cosine at the central frequency.

## 4. Conclusion

Mesoscopic systems at low temperature provide interesting platforms for quantum optics. Time-resolved voltage measurements can fully characterize quantum states of the electromagnetic field, on the condition that the emitters be at low enough temperature that the condition $h\nu \gg k_B T$ (narrow band), or $h\Delta\nu \gg k_B T$ (wideband) be respected. Some mesoscopic conductors have already been shown to exhibit purely quantum behaviors. Such is the case for a tunnel junction appropriately biased in ac and dc. Interestingly, quantum optics in the microwave regime also offers interesting perspective for time domain quantum optics, as the fluctuations of the field can be fully resolved in time, contrary to visible or near-infrared optics. We have shown that the tools already exist for this exploration, and that exciting new physics can be probed in the near future in those systems.

**Acknowledgements**

This work was supported by the Canada Excellence Research Chairs program, Canada Foundation for Innovation, Canada NSERC, Québec MEIE, Québec FRQNT via INTRIQ, and Université de Sherbrooke via EPIQ.

## References

[1] Blanter Y and Büttiker M 2000 *Physics Reports* **336** 1–166
[2] Beenakker C W J and Schomerus H 2004 *Physical Review Letters* **93** 096801
[3] Girvin S M, Devoret M H and Schoelkopf R J 2009 *Physica Scripta* **T137** 014012

[4] Lebedev A V, Lesovik G B and Blatter G 2010 *Physical Review B* **81** 155421
[5] Grimsmo A L, Qassemi F, Reulet B and Blais A 2016 *Physical Review Letters* **116** 043602
[6] Romero G, Garcia-Ripoll J J and Solano E 2009 *Physical Review Letters* **102** 173602
[7] Govenius J, Lake R, Tan K and Möttönen M 2016 *Physical Review Letters* **117** 030802
[8] Virally S, Simoneau J O, Lupien C and Reulet B 2016 *Physical Review A* **93** 043813
[9] Gasse G, Spietz L, Lupien C and Reulet B 2013 *Physical Review B* **88** 241402
[10] Gasse G, Lupien C and Reulet B 2013 *Physical Review Letters* **111** 136601
[11] Forgues J C, Sane F B, Blanchard S, Spietz L, Lupien C and Reulet B 2013 *Scientific reports* **3** 2869
[12] Forgues J C, Lupien C and Reulet B 2014 *Physical Review Letters* **113** 043602
[13] Forgues J C, Lupien C and Reulet B 2015 *Physical Review Letters* **114** 130403
[14] Simoneau J O, Virally S, Lupien C and Reulet B 2017 *Physical Review B* **95** 060301
[15] Cahill K and Glauber R J 1969 *Physical Review* **177** 1857–1881

# Nonlinear heat transport in ferromagnetic-quantum dot-superconducting systems

**Sun-Yong Hwang**[1] **and David Sánchez**[2]

[1] Theoretische Physik, Universität Duisburg-Essen and CENIDE, D-47048 Duisburg, Germany
[2] Institute for Cross-Disciplinary Physics and Complex Systems IFISC (CSIC-UIB), E-07122 Palma de Mallorca, Spain

E-mail: sunnyong.hwang@gmail.com

**Abstract.** We analyze the heat current traversing a quantum dot sandwiched between a ferromagnetic and a superconducting electrode. The heat flow generated in response to a voltage bias presents rectification as a function of the gate potential applied to the quantum dot. Remarkably, in the thermally driven case the heat shows a strong diode effect with large asymmetry ratios that can be externally tuned with magnetic fields or spin-polarized tunneling. Our results thus demonstrate the importance of hybrid systems as promising candidates for thermal applications.

## 1. Introduction

Control of heat flow is a key goal in modern quantum electronics [1, 2]. Electrons carry energy in addition to charge and their transport can then be manipulated electrically [3, 4, 5, 6, 7, 8, 9, 10, 11, 12, 13] or thermally [14, 15, 16, 17, 18, 19]. It is thus highly desirable to possess a great variety of mesoscopic platforms where energy flow in response to various driving fields can be generated and detected [20, 21, 22, 23].

Here, we investigate a ferromagnetic-quantum dot-superconducting (F-D-S) junction and show that this system can work as an efficient heat diode both for charge and spin transport. These hybrid systems have recently received a good deal of attention due to their excellent thermoelectric properties [24, 25, 26, 27, 28]. Here, we show that they might also be attractive for thermal applications.

## 2. Theory of nonlinear heat transport

The total Hamiltonian describing the F-D-S system (left panel of Fig. 1) is given by [28]

$$\mathcal{H} = \mathcal{H}_F + \mathcal{H}_S + \mathcal{H}_D + \mathcal{H}_T\,, \tag{1}$$

where $\mathcal{H}_F = \sum_{k\sigma} \varepsilon_{Fk\sigma} c^\dagger_{Fk\sigma} c_{Fk\sigma}$ accounts for charge carriers in the ferromagnet with momentum $k$ and spin $\sigma = \uparrow, \downarrow$, and $\mathcal{H}_S = \sum_{k\sigma} \varepsilon_{Sk\sigma} c^\dagger_{Sk\sigma} c_{Sk\sigma} + \sum_k [\Delta c^\dagger_{S,-k\uparrow} c^\dagger_{Sk\downarrow} + \text{H.c.}]$ depicts a superconductor reservoir with the energy gap $\Delta$ as an order parameter. Importantly, in the dot Hamiltonian $\mathcal{H}_D = \sum_\sigma (\varepsilon_{d\sigma} - eU_\sigma) d^\dagger_\sigma d_\sigma$, not only the energy level can be Zeeman split with magnetic fields, viz. $\varepsilon_{d\sigma} = \varepsilon_d + \sigma\Delta_Z$, but it can also be renormalized by the spin-dependent interaction potential $U_\sigma$. We determine $U_\sigma$ within a self-consistent (Hartree) approach [11] to find the nonlinear heat transport in this setup. The tunneling between the dot and each lead is given by $\mathcal{H}_T = \sum_{k\sigma} t_{F\sigma} c^\dagger_{Fk\sigma} d_\sigma + \sum_{k\sigma} t_{S\sigma} c^\dagger_{Sk\sigma} d_\sigma$ + H.c., which leads to broadenings $\Gamma_{F\sigma} = 2\pi |t_{F\sigma}|^2 \sum_k \delta(\varepsilon - \varepsilon_{Fk\sigma})$ [parametrized as $\Gamma_{F\sigma} = \Gamma_F(1+\sigma p)$ with $p = (\nu_\uparrow - \nu_\downarrow)/(\nu_\uparrow + \nu_\downarrow)$ the $F$ polarization in terms of the density of states $\nu_\sigma = \sum_k \delta(\varepsilon - \varepsilon_{Fk\sigma})$ and $\Gamma_F = (\Gamma_{F\uparrow} + \Gamma_{F\downarrow})/2$] and $\Gamma_{S\sigma} = \Gamma_S = 2\pi |t_{S\sigma}|^2 \sum_p \delta(\varepsilon - \varepsilon_{Sp\sigma})$.

The spin-resolved heat current can be evaluated from the rate of energy flow per each spin ($\mathcal{H}_{F\sigma} = \sum_k \varepsilon_{Fk\sigma} c^\dagger_{Fk\sigma} c_{Fk\sigma}$) at the F side and the Joule heating in the presence of the voltage bias $V$

$$J_\sigma = -(i/\hbar)\langle[\mathcal{H}, \mathcal{H}_{F\sigma}]\rangle - I_\sigma V, \tag{2}$$

where the spin-resolved electric current is given by $I_\sigma = -(ie/\hbar)\langle[\mathcal{H}, N_{F\sigma}]\rangle$ with the charge number $N_{F\sigma} = \sum_k c^\dagger_{Fk\sigma} c_{Fk\sigma}$ for carriers with spin $\sigma$. Due to coupling to the superconductor, the spin-resolved heat transport ($J_\sigma = J^\sigma_A + J^\sigma_Q$) has two separate contributions (with $J_A \equiv J^\uparrow_A + J^\downarrow_A$, $J^s_A \equiv J^\uparrow_A - J^\downarrow_A$, $J_Q \equiv J^\uparrow_Q + J^\downarrow_Q$, $J^s_Q \equiv J^\uparrow_Q - J^\downarrow_Q$)

$$J^\sigma_A = \frac{-2eV}{h} \int d\varepsilon\; T^\sigma_A(\varepsilon)\big[f_F(\varepsilon - eV) - f_F(\varepsilon + eV)\big] = -2V I^\sigma_A, \tag{3}$$

$$J^\sigma_Q = \frac{1}{h} \int d\varepsilon\; (\varepsilon - eV)\; T^\sigma_Q(\varepsilon)\big[f_F(\varepsilon - eV) - f_S(\varepsilon)\big], \tag{4}$$

where $J^\sigma_A$ refers to the spin-resolved Andreev heat current dominant for the subgap transport $|\varepsilon| < \Delta$, while $J^\sigma_Q$ is that of quasiparticle contributions beyond the gap $|\varepsilon| > \Delta$. Here, $f_{\alpha=F,S}(\varepsilon \pm eV) = \{1 + \exp[(\varepsilon \pm eV - E_F)/k_B T_\alpha]\}^{-1}$ is the Fermi-Dirac distribution with the applied voltage to the ferromagnet $V = V_F$ and temperature $T_\alpha = T + \theta_\alpha$ ($T$: average temperature, $\theta_F = \theta$: thermal bias, $V_S = \theta_S = 0$). The respective transmission functions $T^\sigma_A(\varepsilon)$ and $T^\sigma_Q(\varepsilon)$ are evaluated using the Green's function approach (see, e.g., Refs. [28, 29, 30] for the explicit expressions).

In a previous work [28], we reported a large thermopower in the same setup but only linear response was considered. Thus, the potential shift $U_\sigma$ in $\mathcal{H}_D$ of Eq. (1) could be neglected. Further, the linear Andreev heat current was shown to be zero. Here, we consider the nonlinear case, where $U_\sigma$ is determined in a weakly nonequilibrium condition [30]. In Eq. (3) the Andreev electric current reads $I^\sigma_A = V(G^\sigma_{A0} + G^\sigma_{A1} V + M^\sigma_A \theta + \dots)$ (purely thermoelectric terms vanish due to the particle-hole symmetry [31]). Then, the Andreev heat flow becomes

$$J^\sigma_A = -2V^2(G^\sigma_{A0} + G^\sigma_{A1} V + M^\sigma_A \theta + \dots). \tag{5}$$

Note that the leading order nonvanishing Andreev thermal conductance can only be given by the cross coupling term $M^\sigma_A$ similarly to the subgap nonlinear electric current [31]. This implies that in the isoelectric case $V = 0$ the subgap thermal transport is entirely blocked and the thermal heat current will be activated by the quasiparticle contributions only. This effect has exactly the same origin as the recently proposed Seebeck diodes [32]. However, if we apply a finite voltage bias $V$ the subgap heat current compete with the quasiparticle contributions and hence Andreev-Peltier effects can become important. This will be discussed below for pure isothermal cases. Finally, if we apply high enough thermal gradient for a nonzero $V$, quasiparticles dominantly contribute to the heat transport after the competing regime is over where $J_Q \simeq -J_A$ and $J^s_Q \simeq -J^s_A$ (see Fig. 3). Beyond this competing regime, large heat and spin heat currents can be generated in our device from quasiparticle tunneling.

Having the aforementioned points in mind, we below discuss the total heat flux $J = J_\uparrow + J_\downarrow$ and the spin-polarized heat current $J_s = J_\uparrow - J_\downarrow$ where $J_\sigma = J^\sigma_A + J^\sigma_Q$.

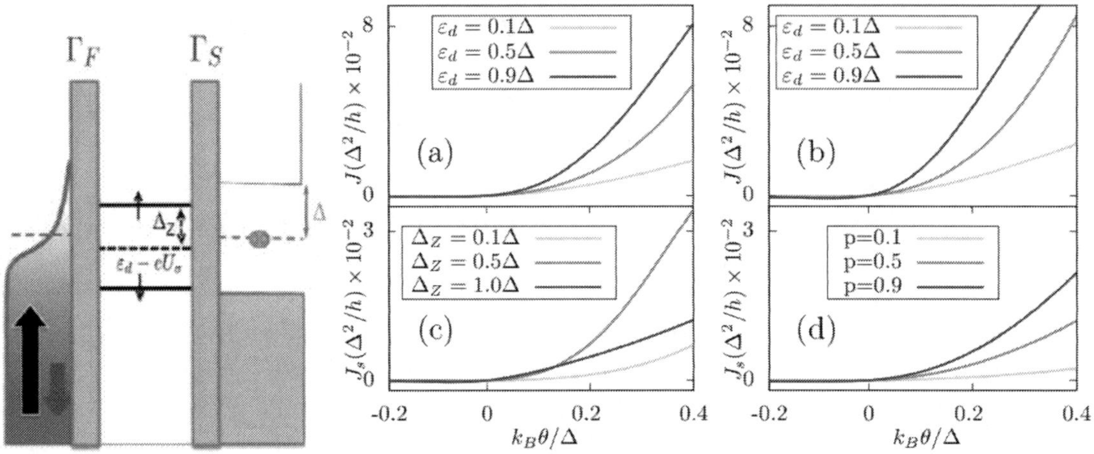


**Figure 1.** Left panel: sketch of our F-D-S device energy diagram. The superconducting reservoir shows an energy gap $\Delta$. The left ferromagnet has a spin polarization $p$ with different amounts of up and down spin component generating spin-dependent tunneling rates. $\Gamma_{F,S}$ is the spin-averaged tunnel broadening to the quantum dot from F or S contact. Thermal and electric biases are applied only to the ferromagnet. The dot level $\varepsilon_d$ depends on $\sigma = \uparrow, \downarrow$ due to either Zeeman splitting $\Delta_Z$ or Coulomb potential $U_\sigma$, which is a self-consistent function of voltage and temperature biases. Right panel: thermal [(a),(b)] and spin thermal [(c),(d)] diode effects of our device ($V = 0$). Charge heat flux for (a) $\Gamma_F = 0.1\Delta$, $\Gamma_S = 0.5\Delta$, (b) $\Gamma_F = 0.5\Delta$, $\Gamma_S = 0.1\Delta$, and spin heat current at $\varepsilon_d = 0.3\Delta$ with $\Gamma_F = 0.1\Delta$, $\Gamma_S = 0.5\Delta$ for (c) $p = 0$, (d) $\Delta_Z = 0$ are shown as a function of thermal bias $\theta$. Base temperature is $k_BT = 0.2\Delta$.

| | $k_B\theta_0 = 0.05\Delta$ | $k_B\theta_0 = 0.10\Delta$ | $k_B\theta_0 = 0.15\Delta$ |
|---|---|---|---|
| (a) $\varepsilon_d = 0.5\Delta$ | 3.25 | 9.41 | 21.6 |
| (b) $\varepsilon_d = 0.5\Delta$ | 3.02 | 8.34 | 18.7 |
| (c) $\Delta_Z = 0.5\Delta$ | 4.71 | 19.9 | 67.5 |
| (d) $p = 0.5$ | 2.89 | 7.46 | 15.3 |

**Table 1.** Asymmetry ratio $R = \frac{|J_{(s)}(\theta_0)|}{|J_{(s)}(-\theta_0)|}$ with denoted parameter values from Fig. 1.

## 3. Results and discussion

### *3.1. Thermal diode effects*

Figure 1 shows the thermal diode effects appearing in the isoelectric case ($V = 0$) where subgap heat transport is completely blocked. This device thus provides means to control the heat flow in a unidirectional way. This is akin to the diode effects of the thermoelectric currents [32]. Moreover, the spin polarized quasiparticle heat currents can be rectified if spin symmetry is broken by magnetic fields or coupling to the F lead [Figs. 1(c) and 1(d)]. In Figs. 1(a) and 1(b), the heat flux signal is slightly larger in F-dominant case [(b) $\Gamma_F > \Gamma_S$] but the rectification efficiency is higher in S-dominant case [(a) $\Gamma_F < \Gamma_S$], as shown in Table 1 with the asymmetry ratio $R$ at $\varepsilon_d = 0.5\Delta$. In Fig. 1, we use the average temperature $k_BT = 0.2\Delta$ but the efficiency quickly increases as we lower $T$. At $k_BT = 0.1\Delta$, for example, $R$ in Table 1 increases for (a) 3.25 $\rightarrow$ 49.7 and (b) 3.02 $\rightarrow$ 43 at $k_B\theta_0 = 0.05\Delta$. In Fig. 1(c), the spin heat current for $\theta > 0$ (say at $k_B\theta = 0.2\Delta$) displays a nonmonotonic dependence on the Zeeman splittings as it firstly increases with $\Delta_Z$ and can be maximized and then decreases again, because the dot level splitting by $\Delta_Z$ generates a strong energy dependence of the heat transport. Consequently, a strong spin heat

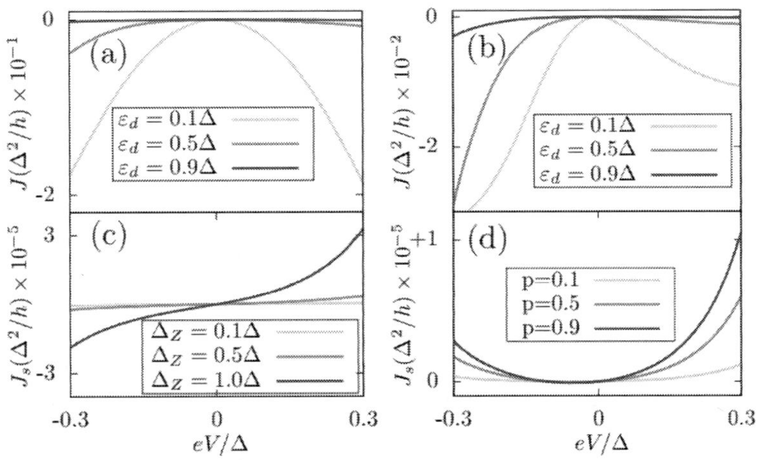


**Figure 2.** Nonlinear Peltier effects of our device in the isothermal case $\theta = 0$. Charge heat flux for (a) $\Gamma_F = 0.1\Delta$, $\Gamma_S = 0.5\Delta$, (b) $\Gamma_F = 0.5\Delta$, $\Gamma_S = 0.1\Delta$, and spin heat current at $\varepsilon_d = 0.3\Delta$ with $\Gamma_F = 0.1\Delta$, $\Gamma_S = 0.5\Delta$ for (c) $p = 0$, (d) $\Delta_Z = 0$ are shown as a function of the voltage $V$. Background temperature is $k_B T = 0.1\Delta$.

rectification is possible in this optimal Zeeman splitting as shown in Table 1 at $k_B\theta_0 = 0.15\Delta$ where $R \simeq 70$. This, however, is not the case for the spin asymmetry created solely from the F polarization $p$, which monotonously enhances the spin heat flow as $p$ increases [Fig. 1(d)].

*3.2. Nonlinear Peltier effect*

An applied voltage bias $V$ is detrimental to the diode effects described above as the subgap heat transport becomes appreciable beyond the linear response [Eq. (5)]. In Fig. 2, the charge heat [(a) $\Gamma_F < \Gamma_S$, (b) $\Gamma_F > \Gamma_S$] and the spin heat [either (c) $\Delta_Z \neq 0$ or (d) $p \neq 0$] fluxes are displayed in the isothermal case ($\theta = 0$). The S-dominant case in Fig. 2(a) has much larger heating currents (about one order of magnitude) than the opposite coupling limit shown in Fig. 2(b). At optimal gate potential, e.g., $\varepsilon_d = 0.5\Delta$, the heat can be rectified. Our device can thus act as Peltier diodes and the asymmetry ratio $R = \frac{|J(-V_0)|}{|J(V_0)|}$ for a given $V_0 > 0$ is much larger in the F-dominant case as displayed in (b). However, this diode effect tends to be fragile because the Andreev Joule heating is now manifested over the subgap energy range ($|\varepsilon| < \Delta$). As shown in Fig. 2(c), the sign of the spin polarized heat flux can be positive or negative depending on the bias voltage direction when the spin asymmetry is generated by $\Delta_Z$ in contrast to the asymmetry due to $p$ [Fig. 2(d)]. Notice that the voltage-driven spin heat flows in Figs. 2(c),(d) are several orders of magnitude smaller than the voltage-driven total heat flux [Figs. 2(a),(b)] or the (positive) temperature-driven spin heat flows [Figs. 1(c),(d); for direct comparisons we also generate same figures at $k_B T = 0.1\Delta$ (not shown) that give smaller values than those at $k_B T = 0.2\Delta$ but within the same order of magnitude]. Therefore, one needs to apply high enough thermal gradients even with finite voltages in order to activate spin-polarized quasiparticles. A quick understanding of this can be supplied by observing Figs. 3(b) and 3(c), where one can find $J_Q^s \simeq -J_A^s$ below $\theta \simeq T$ hence $J_s = J_A^s + J_Q^s \simeq 0$ albeit the applied voltage $eV = k_B T$. Now we will further discuss the results in Fig. 3.

*3.3. Competition between Andreev and quasiparticle contributions*

It should be again noted that there is no linear subgap Peltier effect in our system due to Onsager symmetry since there is no corresponding linear Andreev thermocurrent because of the particle-

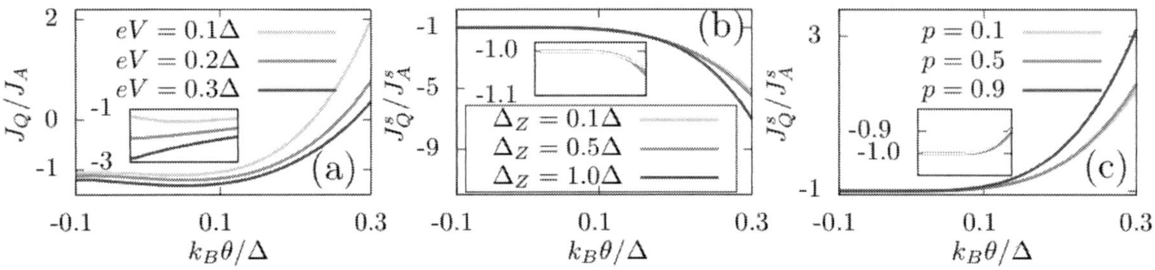


**Figure 3.** Competition with the subgap Joule heating in generic thermoelectric configurations. We use $k_BT = \Gamma_F = 0.1\Delta$, $\Gamma_S = \varepsilon_d = 0.5\Delta$, and $eV = 0.1\Delta$ in (b) $p = 0$ and (c) $\Delta_Z = 0$. The competing regime can be identified with $J_Q \simeq -J_A$ and $J_Q^s \simeq -J_A^s$ canceling the total heat currents. Each inset is explained in the main text.

hole symmetry [31]. Indeed, in Eq. (3) the energy carried by particles and holes exactly cancel each other and only the Joule heating remains [28]. The latter $J_A^{(s)}$ which appears only beyond the linear response can now compete with the normal heat flow $J_Q^{(s)}$ carried by quasiparticles. Hence, $J_A^{(s)}$ can create the tendencies of cooling with positive voltages while at the same time $J_Q^{(s)}$ tends to heat, or vice versa.

In Fig. 3(a), one finds $J_Q \simeq -J_A$ for a low voltage bias, e.g., $eV = 0.1\Delta$ below $k_B\theta \simeq 0.1\Delta = k_BT$ (the actual value is about $J_Q = -1.06J_A$). In this region, we have a very small total flux since $J = J_A + J_Q \simeq -0.06J_A \simeq 0$. But as $V$ is increased, quasiparticles slightly dominate and at $eV = 0.3\Delta$ we have $J_Q \simeq -1.2J_A$. At $\theta = 0$, this can be compared with Fig. 2(a) at $\varepsilon_d = 0.5\Delta$, in which one can notice a tendency of $J$ to increase (as $V > 0$ is applied) to negative values since $J \simeq -0.2J_A < 0$. A small detuning of the dot level is more beneficial for this purpose as shown in Fig. 2(a) with $\varepsilon_d = 0.1\Delta$. Inset of Fig. 3(a) shows the corresponding curves in the range $-0.1\Delta < k_B\theta < 0.1\Delta$ but with $\varepsilon_d = 0.1\Delta$. Indeed, one can notice an appreciable contribution from quasiparticles as the voltage is applied reaching $J_Q \simeq -2.4J_A$ at $eV = 0.3\Delta$ and $\theta = 0$, which explains the behavior in Fig. 2(a). As thermal gradients further increases, see Fig. 3(a), a compensating regime appears where $J_Q = 0$ due to the combined thermoelectric configurations (a voltage-driven quasiparticle heat compensates for the temperature-driven flow), after which $J_Q$ tends to collaborate with $J_A$ giving rise to a net heating. At $eV = 0.1\Delta$ and $k_B\theta = 0.3\Delta$, for instance, $J_Q \simeq 2J_A$ and hence the total flux is given by $J \simeq 3J_A$.

In Fig. 3(b), when the spin heat flow is generated by magnetic fields, there is no compensating or collaborating regime but the amplitude of $J_Q^s$ keeps increasing against $J_A^s$ reaching $J_Q^s \simeq -5J_A^s$ at $k_B\theta = 0.3\Delta$ and $\Delta_Z = eV = 0.1\Delta$. However, the polarization-driven spin heat currents in Fig. 3(c) exhibit the similar tendency with separate regimes as the curves in Fig. 3(a), i.e., crossing the zero and eventually becoming positive. Finally, the insets of Figs. 3(b) and 3(c) show the voltage dependence of the curves in the range $-0.1\Delta < k_B\theta < 0.1\Delta$. The red and yellow lines respectively refer to $eV = 0.2\Delta$ and $eV = 0.3\Delta$ at (b) $\Delta_Z = 0.5\Delta$, and (c) $p = 0.9$. In stark contrast to the voltage dependence of the total heat flux in Fig. 3(a), the competing effects are robust with respect to the applied $V$ in these spin heat cases, maintaining $J_Q^s \simeq -J_A^s$. This explains the vanishingly small amplitudes of $J_s$ in Figs. 2(c) and 2(d) as one specific example of the isothermal case $\theta = 0$. Indeed, in a broad range of $-0.1\Delta < k_B\theta < 0.06\Delta$, the voltage dependence is very small calling for a substantial amount of thermal gradients to observe the spin polarized heat in a general thermoelectric bias configuration.

It should be emphasized that the above discussions are meaningful only with the finite voltage $V \neq 0$, otherwise $J_A^{(s)} = 0$ [Eq. (5)] and the ratio $J_Q^{(s)}/J_A^{(s)}$ in Fig. 3 will diverge. Since the discussed effects are intrinsically nonlinear, the competition between $J_A^{(s)}$ and $J_Q^{(s)}$ appears even for small driving fields. This is completely different from the case for the electric currents, where one should have $I_A^{(s)} \gg I_Q^{(s)} \simeq 0$ for a low bias regime and the competition between the Andreev and quasiparticle parts in this case will be meaningless.

## 4. Conclusions

To summarize, we have examined the heat transport properties of a quantum dot attached to ferromagnetic and superconducting contacts. We have unveiled strong thermal diode effects that can be manipulated with external magnetic fields or magnetization of the ferromagnetic reservoir. In these systems, it is crucial to distinguish between Andreev and quasiparticle currents, which determine the specific transport mechanism. Our work thus represents an important step toward a full characterization of heat transport in hybrid systems.

## Acknowledgments

This work was supported by MINECO under Grant No. FIS2014-52564 and the Ministry of Innovation NRW.

## References

[1] Benenti G, Casati G, Saito K and Whitney R S 2017 *Phys. Rep.* **694** 1
[2] Sánchez D and López R 2016 *C. R. Physique* **17** 1060
[3] Kulik I O 1994 *J. Phys.: Condens. Matter* **6** 9737
[4] Bogachek E N, Scherbakov A G and Landman U 1999 *Phys. Rev. B* **60** 11678
[5] Çipiloğlu M A, Turgut S and Tomak M 2004 *Phys. Stat. Sol. (b)* **241** 2575
[6] Freericks J K and Zlatic V 2006 *Condensed Matter Physics* **9** 603
[7] Zeberjadi M, Esfarjani K and Shakouri A 2007 *Appl. Phys. Lett.* **91** 122104
[8] Leijnse M, Wegewijs M R and Flensberg K 2010 *Phys. Rev. B* **82** 045412
[9] López R and Sánchez D 2013 *Phys. Rev. B* **88** 045129
[10] Whitney R S 2013 *Phys. Rev. B* **88** 064302
[11] Hwang S-Y, López R, Lee M and Sánchez D 2014 *Phys. Rev. B* **90** 115301
[12] Jiang J H, Kulkarni M, Segal D and Imry Y 2015 *Phys. Rev. B* **92** 045309
[13] Zimbovskaya N 2016 *J. Phys.: Condens. Matter* **28** 183002
[14] Wang L and Li B 2007 *Phys. Rev. Lett.* **99** 177208
[15] Segal D 2008 *Phys. Rev. Lett.* **100** 105901
[16] Ruokola T, Ojanen T and Jauho A P 2009 *Phys. Rev. B* **79** 144306
[17] Hwang S-Y, Sánchez D, Lee M and López R 2013 *New J. Phys.* **15** 105012
[18] Sánchez R, Sothmann B and Jordan A N 2015 *New J. Phys.* **17** 075006
[19] Sierra M A and Sánchez D 2015 *Materials Today: Proceedings* **2** 483
[20] Molenkamp L W, Gravier Th, van Houten H, Buijk O J A, Mabesoone M A A and Foxon C T 1992 *Phys. Rev. Lett.* **68** 3765
[21] Chiatti O, Nicholls J T, Proskuryakov Y Y, Lumpkin N, Farrer I and Ritchie D A 2006 *Phys. Rev. Lett.* **97** 056601
[22] Meschke M, Guichard W and Pekola J P 2009 *Nature* **444** 187
[23] Jezouin S, Parmentier F D, Anthore A, Gennser U, Cavanna A, Jin Y and Pierre F 2013 *Science* **342** 601
[24] Ozaeta A, Virtanen P, Bergeret F S and Heikkila T T 2014 *Phys. Rev. Lett.* **112** 057001
[25] Kolenda S, Wolf M J and Beckmann D 2016 *Phys. Rev. Lett.* **116** 097001
[26] Kalenkov M S and Zaikin A D 2014 *Phys. Rev. B* **90** 134502
[27] Machon P, Eschrig M and Belzig W 2014 *New J. Phys.* **16** 073002
[28] Hwang S-Y, López R and Sánchez D 2016 *Phys. Rev. B* **94** 054506
[29] Cao X, Shi Y, Song X, Zhou S and Chen H 2004 *Phys. Rev. B* **70** 235341
[30] Hwang S-Y, Sánchez D and López R 2017 arXiv:1708.02451 (to appear in *Eur. Phys. J. B*)
[31] Hwang S-Y, López R and Sánchez D 2015 *Phys. Rev. B* **91** 104518
[32] Hwang S-Y, Sánchez D and López R 2016 *New J. Phys.* **18** 093024

IOP Conf. Series: Journal of Physics: Conf. Series **969** (2018) 012140 doi:10.1088/1742-6596/969/1/012140

# Asymmetries in the spectral density of an interaction-quenched Luttinger liquid

**A. Calzona**[1,2,3], **F. M. Gambetta**[1,2], **M. Carrega**[4], **F. Cavaliere**[1,2], **and M. Sassetti**[1,2]

[1]Dipartimento di Fisica, Università di Genova, Via Dodecaneso 33, 16146 Genova, Italy.
[2]SPIN-CNR, Via Dodecaneso 33, 16146, Genova, Italy.
[3]Physics and Materials Science Research Unit, University of Luxembourg, 162a avenue de la Faïencerie, L-1511 Luxembourg, Luxembourg.
[4]NEST, Istituto Nanoscienze-CNR and Scuola Normale Superiore, I-56127 Pisa, Italy.

E-mail: `calzona@fisica.unige.it`

**Abstract.** The spectral density of an interaction-quenched one-dimensional system is investigated. Both direct and inverse quench protocols are considered and it is found that the former leads to stronger effects on the spectral density with respect to the latter. Such asymmetry is directly reflected on transport properties of the system, namely the charge and energy current flowing to the system from a tunnel coupled biased probe. In particular, the injection of particles from the probe to the right-moving channel of the system is considered. The resulting fractionalization phenomena are strongly affected by the quench protocol and display asymmetries in the case of direct and inverse quench. Transport properties therefore emerge as natural probes for the observation of this quench-induced behavior.

## 1. Introduction

Isolated one-dimensional (1D) systems are an ideal playground where both non-equilibrium phenomena and interaction effects can be explored [1, 2, 3, 4]. Many of them fall in the class of integrable systems, in which the presence of an infinite number of local conserved quantities lead to peculiar relaxation properties after they have been driven far from equilibrium. Indeed, it has been shown that integrable systems reach a steady state that is locally described by a non-thermal density matrix, which can be obtained within the framework of the generalized Gibbs ensemble (GGE), retaining strong memory of initial conditions [5]. To drive the system out-of-equilibrium and study its dynamics, it is possible to perform a *quantum quench* [6]. Here, the system is initially prepared in the ground state of a certain Hamiltonian and then some of its parameters are changed in time. Such a procedure can be implemented i.e. in state-of-the-art setups of cold atoms, where the system parameters can be varied with high precision and in a time-dependent way [7, 8, 9].

It is well established that the equilibrium low energy sector of interacting 1D systems is described by the Luttinger Liquid (LL) model [10]. Already in equilibrium, LLs exhibit peculiar spectral properties, with characteristic power-law behavior, stemming from their non-Fermi liquid nature [11, 12, 13, 14, 15, 16, 17, 18]. In addition, interacting 1D systems exhibit charge and energy fractionalization [16, 19, 20, 21, 22], i.e. the split-up of a particle injected into an interacting 1D system in two counter-propagating excitations which carry a fraction of the original charge and energy of the particle. In the last decade, the LL model has also been successfully employed to study the dynamics of 1D system after a quench of the inter-particle interaction [23, 24]. In particular, many efforts have been devoted to investigate how the unusual LL features are modified in the presence of such an out-of-equilibrium situation [25, 26, 27, 28, 29, 30, 31, 32], usually focusing on the case of a "direct" quench. This consists in a quench protocol which brings the system towards a state with stronger interactions. In this paper we instead compare the direct quench scenario with the case of an "inverse" quench, which reduces the interaction strength. In particular, we analyze the spectral density [31] of a quenched 1D system and show that, given two interaction strengths, the direct quench always leads to a larger spectral density with respect to the inverse protocol. We argue that this strong asymmetry between the two cases is directly reflected also on transport properties. In particular, we consider a non-interacting biased 1D probe, whose left-moving particles are coupled to the right moving channel of the system through a non-local tunneling region, and we focus on the charge and energy differential conductance. The paper is organized as follows: in Sec. 2 the model of the system, the quenched spectral density and the charge and energy differential conductances are introduced. In Sec. 3 the effect of a direct and inverse quench on the spectral density of the system is explored and its influence on the transport properties is outlined. Section 4 contains our conclusions.

## 2. Model and methods

We consider a 1D system of interacting fermions with two counter-propagating channels, labeled by the index $r = R, L$. Its Hamiltonian is $H_s(t) = H_s^{(0)} + H_s^{(1)}(t)$, where ($\hbar = 1$)

$$H_s^{(0)} = v_F \sum_{r=R,L} \theta_r \int_{-\infty}^{\infty} \mathrm{d}x \; \psi_r^\dagger(x)(-i\partial_x)\psi_r(x) \,, \tag{1}$$

is the free kinetic part, with $v_F$ the Fermi velocity, $\psi_r(x)$ the Fermi field operator, and $\theta_{R/L} = \pm 1$. The interaction term is

$$H_s^{(1)}(t) = \frac{g_4(t)}{2} \sum_{r=R,L} \int_{-\infty}^{\infty} \mathrm{d}x \; [n_r(x)]^2 + g_2(t) \int_{-\infty}^{\infty} \mathrm{d}x \; n_R(x) n_L(x) \,, \tag{2}$$

with $n_r(x) =: \psi_r^\dagger(x)\psi_r(x) :$ the electron density of the $r$-th channel and $g_{2,4}(t)$ the intra/inter-channel interaction parameters [10]. Due to a sudden quench of the interactions [24, 29, 31], the latter parameters rapidly switch in time according to

$$g_{2,4}(t) = g_{2,4}^{(i)} \vartheta(-t) + g_{2,4}^{(f)} \vartheta(t), \tag{3}$$

with $\vartheta(x)$ the step function and the indices $i/f$ referring to situation before and after the quench. Standard bosonization procedure is applied [10], with the Fermi operator of the system expressed in terms of free bosonic fields $\phi_r(x)$ as

$$\psi_r^\dagger(x) = \frac{1}{\sqrt{2\pi a}} e^{i\sqrt{2\pi}\phi_r(x)} e^{-i\theta_r k_F x} \,, \tag{4}$$

with $a \propto k_F^{-1}$ a short length cutoff. The bosonized Hamiltonian density of the system reads

$$\mathcal{H}_s^{(\nu)}(x,t) = \frac{u_\nu}{2} \sum_{\eta=\pm} [\partial_x \phi_{\nu,\eta}(x - \eta u_\nu t)]^2 \tag{5}$$

with $\nu = i$ $(f)$ for $t < 0$ $(t > 0)$. Note that the fields $\phi_{\nu,\eta}(x)$ are chiral, with $\eta$ denoting their propagation direction and $u_\nu = (2\pi)^{-1}\sqrt{\left[2\pi v_F + g_4^{(\nu)}\right]^2 - \left[g_2^{(\nu)}\right]^2}$ their propagation velocity. For $t > 0$ one has the relation

$$\phi_r(x) = \sum_{\eta=\pm} A_{\eta\theta_r}\phi_{f,\eta}(x) \quad A_\pm = \frac{1}{2}\left[\frac{1}{\sqrt{K_f}} \pm \sqrt{K_f}\right], \tag{6}$$

with $K_f = \sqrt{\frac{2\pi v_F + g_4^{(f)} - g_2^{(f)}}{2\pi v_F + g_4^{(f)} + g_2^{(f)}}}$. The boson operators before quench $\phi_{i,\eta}(x)$ are different, due to the different interaction strength $K_i$, and connected to the post-quench operators via $\phi_{f,\eta}(x) = \sum_{\xi=\pm}\alpha_{\eta\xi}\phi_{i,\xi}(x)$, with

$$\alpha_\pm = \frac{1}{2}\left[\sqrt{\frac{K_i}{K_f}} \pm \sqrt{\frac{K_f}{K_i}}\right]. \tag{7}$$

We focus on the stationary regime reached long after the quench, whose properties are captured by the non-equilibrium quenched spectral function [27, 31] of the $R$ channel

$$A(k,\epsilon) = \int_{-\infty}^{\infty} \mathrm{d}x\mathrm{d}\tau\; e^{-i(kx-\epsilon\tau)} \lim_{t\to\infty} A(x,\tau,t), \tag{8}$$

with

$$A(x,\tau,t) = \langle\left\{\psi_R(x,t+\tau), \psi_R^\dagger(0,t)\right\}\rangle. \tag{9}$$

Here, $\langle\ldots\rangle$ denotes the average over the pre-quench ground state and the time evolution is performed in the Heisenberg picture. The spectral function evaluates to (for details, see [31])

$$A(k,\epsilon) = \pi\left(\frac{a}{u_f}\right)^{(A_-^2+A_+^2)(\alpha_-^2+\alpha_+^2)-1} \sum_{j=\pm}\prod_{\xi=\pm}\frac{I(A_\xi^2\alpha_+^2, A_\xi^2\alpha_-^2, j\Delta_\xi)}{\Gamma(A_\xi^2\alpha_+^2)\Gamma(A_\xi^2\alpha_-^2)}, \tag{10}$$

where $2\Delta_\pm = \epsilon \pm u_f k$ and

$$I(\mu_1,\mu_2,\Delta) = \int_0^\infty \mathrm{d}E\; E^{\mu_1-1}(E-\Delta)^{\mu_2-1}\vartheta(E-\Delta)e^{a(\Delta-2E)/u_f}. \tag{11}$$

To address transport properties we now consider a setup obtained by tunnel-coupling the above system to a 1D reservoir of non-interacting fermions (henceforth, the probe). The probe is modeled as a pair counter-propagating channels, labeled themselves by the index $r = R, L$, and is described by the Hamiltonian

$$H_p^{(0)} = v_F\sum_{r=R,L}\theta_r\int_{-\infty}^{\infty}\mathrm{d}x\;\chi_r^\dagger(x)(-i\partial_x)\chi_r(x), \tag{12}$$

with $\chi_r(x)$ the Fermi operator in the probe. The latter is subject to a bias voltage $V$ with respect to the system. Non-local tunneling between the system and the probe is switched on right after the quench. An interesting scenario is the one which couples only the $L$ (or $R$) movers of the probe to the $R$ (or $L$) channel of the system, allowing to study the charge and energy fractionalization induced by interactions. In solid state devices, this can be achieved via momentum-resolved tunneling, employing a magnetic field $B$ orthogonal to the plane of the two channels [33, 34]. The magnetic field also controls the momentum $k_0 \propto B$ of the tunneling

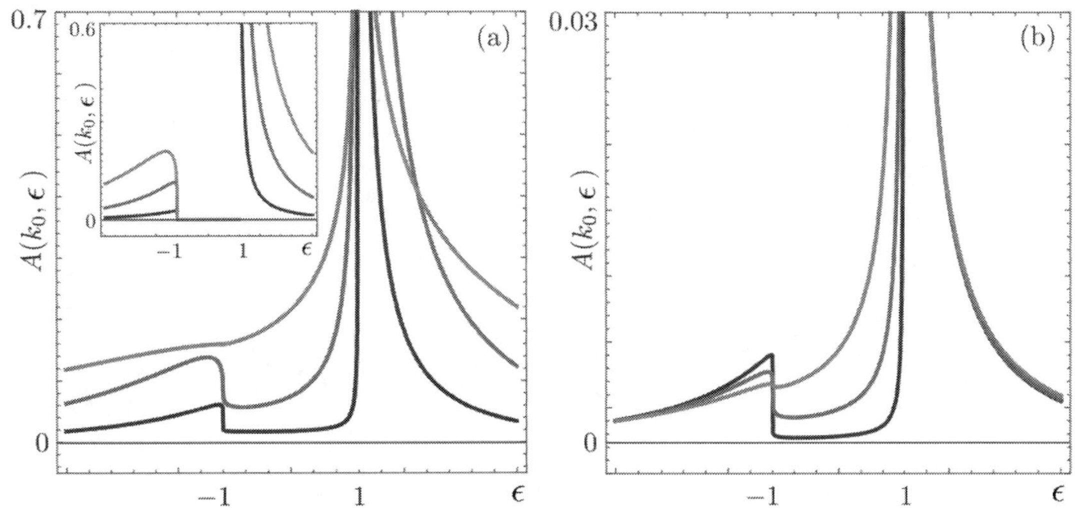


**Figure 1.** Spectral density $A(k_0, \epsilon)$ (units $au_f^{-1}$) as a function of $\epsilon$ (units $u_f k_0$) for the case of (a) direct quench with $K_i = 0.9$ and $K_f = 0.7$ (blue), $K_f = 0.5$ (red), $K_f = 0.3$ (green); (b) inverse quench with $K_f = 0.9$ and $K_i = 0.7$ (blue), $K_i = 0.5$ (red), $K_i = 0.3$ (green). In the inset, the spectral density for the non-quenched case ($K_i = K_f$) with $K_i = 0.9$ (blue), $K_i = 0.7$ (red), $K_i = 0.5$ (green). In all panels, $k_0 = 0.3a^{-1}$.

electrons (measured with respect to the Fermi momentum $k_F$). Considering a non-local tunneling along the channels and assuming the coupling of the $L$ probe states to the $R$ system channels only, the tunneling Hamiltonian is [19, 31, 35, 36]

$$H_t = \lambda\vartheta(t) \int_T \mathrm{d}x \left[\psi_R^\dagger(x)\chi_L(x) + \chi_L^\dagger(x)\psi_R(x)\right] , \tag{13}$$

with $\lambda$ the tunneling amplitude and $T$ the extended tunneling region with length $L_T$, which is greater than any other length scale of the system. In the stationary state, long after the quench and the switching on of the tunnel-coupling, from the knowledge of the spectral density of the system of Eq. (10) one can obtain the chiral charge [37] $G_\eta^Q(V) = \partial I_\eta^Q/\partial V$ and energy [19] $G_\eta^E(V) = \partial I_\eta^E/\partial V$ differential conductances,

$$G_\eta^Q(V) = \frac{e^2|\lambda|^2 L_T \sqrt{K_f} A_\eta}{4\pi^2 v_F} A(k_0, V) , \quad G_\eta^E(V) = \frac{e|\lambda|^2 L_T}{4\pi^2 v_F}\left(\frac{eV + \eta u_f k_0}{2}\right) A(k_0, V) , \tag{14}$$

where $I_\eta^{Q/E}(V)$ is the chiral charge/energy current [31].

## 3. Results and discussion

We start the discussion showing results for the quenched spectral density in the two cases of direct and inverse quench protocol. We focus on the stationary regime ($t \to \infty$) obtained long after the sudden quench. Figure 1 shows $A(k_0, \epsilon)$ in the case of a "direct" quench – i.e. for $K_f < K_i$ (panel (a)) – and for an "inverse" quench with $K_f > K_i$ (panel (b)). In both cases, the main difference with respect to the non-quenched case (shown in the inset of panel (a)) is that the spectral density is non-vanishing in the energy window $|\epsilon| < u_f k_0$. Indeed, a quench of the interactions effectively injects energy into the system, which is then no longer in the ground state

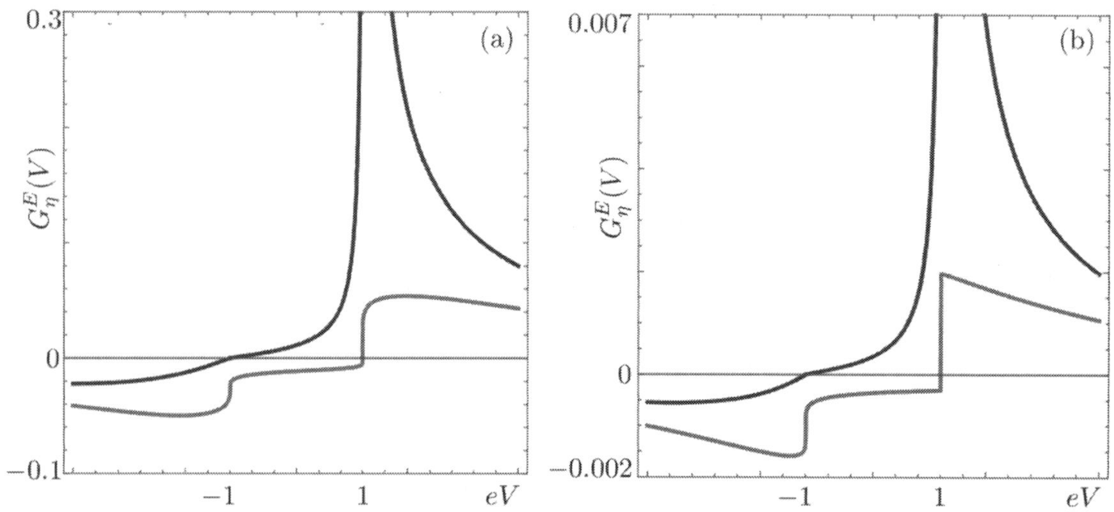


**Figure 2.** Chiral energy conductance $G_\eta^E(V)$ (units $e|\lambda|^2 L_T/(4\pi^2 v_F)$) as a function of $eV$ (units $u_f k_0$) for the case of (a) direct quench with $K_i = 0.9$ and $K_f = 0.5$ and (b) inverse quench with $K_i = 0.5$ and $K_f = 0.9$. In all panels, blue is for $\eta = +$, red for $\eta = -$ and $k_0 = 0.3a^{-1}$.

of the final Hamiltonian. The larger the difference between $K_i$ and $K_f$, the larger the injected energy [29], which allows for a larger spectral weight at low energy. The spectral density is clearly modified also in the "allowed" energy range $|\epsilon| > u_f k_0$, where crossings between the curves corresponding to different quench protocols can even be observed, in sharp contrast to the non-quenched case.

Interestingly, we observe the strong asymmetry between the direct and the inverse quench protocols. Indeed, as can be seen in panel (a) and (b), in the former case $A(k_0, \epsilon)$ is strongly enhanced with respect to the inverse case, resulting in stronger amplitude. This fact can be explained as follows: after an interaction quench, the steady state of the system can be characterized by a generalized Gibbs ensemble [5] which, roughly speaking, behaves as an excited Luttinger liquid with an effective interaction parameter $K_{\text{eff}}$ *intermediate* between $K_i$ and $K_f$ [29]. In particular, in the case of a direct quench the system settles onto a state which is effectively *more* interacting and thus with a larger spectral function (compare the inset of Figure 1(a)). On the other hand, for an inverse quench the effective interactions in the steady state of the system are smaller, resulting into a smaller spectral function. We also note that in the direct case the behavior of the spectral density for $|\epsilon| > u_F k_0$ can be better resolved in comparison to the inverse one.

The strong asymmetry between the direct and inverse case discussed above is also reflected in the behavior of transport properties of the system. The charge differential conductance needs no further discussion, being simply proportional to $A(k_0, eV)$ through a constant which only depends on $K_i$ and $K_f$ - see Eq.(14). The energy differential conductance shows even more marked differences between the quench protocols, as one can see in panels (a) and (b) of Fig. 2, which report $G_\eta^E(V)$ for the case of a direct and of an inverse quench, respectively. It is also worthwhile to notice that for small bias $|eV| < u_f k_0$, the sign of the two chiral energy conductances is always opposite regardless the quench protocol, in contrast to the large bias regime. This implies that the energy flow in the two counter-propagating channels of the system is always opposite in this regime.

IOP Conf. Series: Journal of Physics: Conf. Series **969** (2018) 012140 doi:10.1088/1742-6596/969/1/012140

## 4. Conclusions

In this paper we have analyzed the steady state of an interaction-quenched two channels Luttinger liquid, coupled to a non-interacting 1D probe via a extended tunnel region. Exploiting momentum-resolved tunneling, we have considered the injection of a left-mover from the probe into the right-moving channel of the system. We have studied the quenched spectral density of the system, showing that it exhibits different behavior depending on the quench protocols. In particular, fixed two interaction strengths, a direct quench (increasing interactions) results in a larger amplitude of the spectral density with respect to the opposite process. These features are directly observed in related transport properties such as charge and energy differential conductances. As a consequence, these properties are natural candidates to experimentally probe this asymmetry.

## References


[1] Polkovnikov A, Sengupta K, Silva A and Vengalattore M 2011 *Rev. Mod. Phys.* **83** 863
[2] Eisert J, Friesdorf M and Gogolin C 2015 *Nature Physics* **11** 124
[3] D'Alessio L, Kafri Y, Polkovnikov A and Rigol M 2016 *Adv. Phys.* **65** 239
[4] Essler F H L and Fagotti M 2016 *J. Stat. Mech.* 064002
[5] Rigol M, Dunjko V, Yurovsky V and Olshanii M 2007 *Phys. Rev. Lett.* **98** 050405; Rigol M, Dunjko V and Olshanii M 2008 *Nature* **452** 854; Vidmar L and Rigol M 2016 *J. Stat. Mech.* 064007
[6] Calabrese P and Cardy J 2006 *Phys. Rev. Lett.* **96**, 136801
[7] Kinoshita T, Wenger T and Weiss D S 2006 *Nature* **440** 900
[8] Bloch I, Dalibard J and Zwerger W 2009 *Rev. Mod. Phys.* **80**, 885; Bloch I 2008 *Science* **29** 1202; Bloch I, Dalibard J and Nascimbéne S 2012 *Nature Physics* **8** 267
[9] Langen T, Gasenzer T and Schmiedmayer J 2016 *J. Stat. Mech.* 064009
[10] Voit J 1995 *Rep. Prog. Phys.* **58** 977; von Delft J and Schoeller H 1998 *Ann. Phys.* **7** 225; Giamarchi T 2004 *Quantum Physics in One Dimension* (New York: Oxford University Press)
[11] Kane C L and Fisher M P A 1992 *Phys. Rev. Lett.* **68** 1220
[12] Matveev K A and Glazman L I 1993 *Phys. Rev. Lett.* **70** 990
[13] Voit J 1993 *J. Phys.: Condens. Matter* **5** 8305
[14] Furusaki A and Matveev K A 2002 *Phys. Rev. Lett.* **88** 226404
[15] Kleimann T, Cavaliere F, Sassetti M and Kramer B 2002 *Phys. Rev. B* **66** 165311
[16] Deshpande V V, Bockrath M, Glazman L I and Yacoby A 2010 *Nature* **464** 209
[17] Carrega M, Ferraro D, Braggio A, Magnoli N, and Sassetti M 2012 *New J. Phys.* **14** 023017
[18] Braggio A, Ferraro D, Carrega M, Magnoli N, and Sassetti M 2012 *New J. Phys.* **14** 093032
[19] Karzig T, Refael G, Glazman L I and Von Oppen F 2011 *Phys. Rev. Lett.* **107** 176403
[20] Calzona A, Carrega M, Dolcetto G and Sassetti M 2015 *Phys. Rev. B* **92** 195414; Calzona A, Carrega M, Dolcetto G and Sassetti M 2015 *Physica E* **74** 630
[21] Calzona A, Acciai M, Carrega M, Cavaliere F and Sassetti M 2016 *Phys. Rev. B* **94** 035404
[22] Acciai M, Calzona A, Dolcetto G, Schmidt T, and Sassetti M 2017 *Phys. Rev. B***96** 075144
[23] Cazalilla M A 2006 *Phys. Rev. Lett.* **97** 156403; Iucci A and Cazalilla M A 2009 *Phys. Rev. A* **80** 063619
[24] Cazalilla M A and Chung M-C 2016 *J. Stat. Mech.* 064004
[25] Perfetto E, Stefanucci G and Cini M 2010 *Phys. Rev. Lett.* **105** 156802
[26] Kennes D M and Meden V 2013 *Phys. Rev. B* **88** 165131
[27] Kennes D M, Klöckner C and Meden V 2014 *Phys. Rev. Lett.* **113** 116401
[28] Schiró M and Mitra A 2015 *Phys. Rev. B* **91** 235126
[29] Gambetta F M, Cavaliere F, Citro R and Sassetti M 2016 *Phys. Rev. B* **94** 045104
[30] Porta S, Gambetta F M, Cavaliere F, Traverso Ziani N and Sassetti M 2016 *Phys. Rev. B* **94** 085122
[31] Calzona A, Gambetta F M, Carrega M, Cavaliere F and Sassetti M 2017 *Phys. Rev. B* **95** 085101
[32] Calzona A, Gambetta F M, Cavaliere F, Carrega M and Sassetti M 2017 *Phys. Rev. B* **96** 085423
[33] Auslaender O M, Steinberg H, Yacoby A, Tserkovnyak Y, Halperin B I, Baldwin K W, Pfeiffer L N and West K W 2005 *Science* **308** 88
[34] Barak G, Steinberg H, Pfeiffer L N, West K W, Glazman L, Von Oppen F and Yacoby A 2010 *Nat. Phys.* **6** 489
[35] Dolcetto G, Barbarino S, Ferraro D, Magnoli N and Sassetti M 2012 *Phys.Rev.B* **85** 195138
[36] Vannucci L, Ronetti F, Dolcetto G, Carrega M and Sassetti M 2015 *Phys. Rev. B* **92** 075446
[37] Bruus H and Flensberg K 2004 *Many-Body Quantum Theory in Condensed Matter Physics* (New York: Oxford University Press)

IOP Conf. Series: Journal of Physics: Conf. Series **969** (2018) 012141 doi:10.1088/1742-6596/969/1/012141

# Cross-coupling effects in circuit-QED stimulated Raman adiabatic passage

**A. Vepsäläinen and G. S. Paraoanu**

Department of Applied Physics, Aalto University, Puumiehenkuja 2 B, 02150, Espoo, Finland

E-mail: antti.vepsalainen@aalto.fi

**Abstract.** Stimulated Raman adiabatic passage is a quantum protocol that can be used for robust state preparation in a three-level system. It has been commonly employed in quantum optics, but recently this technique has drawn attention also in circuit quantum electrodynamics. The protocol relies on two slowly varying drive pulses that couple the initial and the target state via an intermediate state, which remains unpopulated. Here we study the detrimental effect of the parasitic couplings of the drives into transitions other than those required by the protocol. The effect is most prominent in systems with almost harmonic energy level structure, such as the transmon. We show that under these conditions in the presence of decoherence there exists an optimal STIRAP amplitude for population transfer.

## 1. Introduction

Adiabatic pulse sequences can be used to transfer population between the states of a quantum system. As opposed to direct Rabi pulses, adiabatic protocols are not sensitive to the shape or frequency of the pulses and can therefore be used to realize robust quantum state preparation [1, 2]. The downside of the adiabatic methods is their inherent slowness: population transfer is based on the adiabatic theorem [3], according to which the system remains in its instantenous eigenstate as long as the Hamiltonian governing the evolution of the system changes slow enough.

Historically, adiabatic methods have been used in quantum optics [4], but recently they have started to gain traction also in other fields requiring accurate control of quantum state, such as circuit quantum electrodynamics (QED) where stimulated Raman adiabatic passage (STIRAP) has been studied theoretically [5] and experimentally [6]. In the experiments it has been shown that STIRAP can be used to prepare the system to the second excited state with high fidelity while being robust to the control parameters. However, when the process fidelity is further optimized an additional problem arises. Unlike atomic systems, in circuit QED the energy level structure of the devices is often relatively close to a harmonic oscillator. This is especially true for a transmon [7], which is a prominent candidate for being a building block of a quantum computer. Consequently, the drive pulses are coupled off-resonantly to all other transitions. If the drive pulses are strong, the parasitic couplings start to play a detrimental role in the efficiency of the population transfer. This factor limits the amplitude of all the driving pulses, but it is especially important in the context of STIRAP protocol.

### *1.1. STIRAP*

In a three-level system STIRAP can be used to transfer population from the ground state $|0\rangle$ to the second excited state $|2\rangle$ without having excitations in state $|1\rangle$ during the process. In this situation it is possible to realize $0-2$ population transfer even in systems where that would otherwise be a forbidden transition, as is the case of the transmon. The STIRAP protocol consists of two microwave drives that are coupled into the $0-1$ and $1-2$ transitions. In the adiabatic framework, the drives modify the system Hamiltonian such that its instantenous eigenstates slowly change. By properly designing the drives, it is possible to ensure that the initial state of the system evolves adiabatically to the desired target state. If the adiabatic condition of slow evolution is satisfied, the system does not get excited away from this state, and as a result it reaches the target state at the end of the protocol. We can see this by looking at the Hamiltonian of a three-level ladder system in a doubly-rotating frame, rotating with the two drives

$$H=\frac{\hbar}{2}\begin{bmatrix} 0 & \Omega_{0,1}(t) & 0 \\ \Omega_{0,1}(t) & 2\delta_{0,1} & \Omega_{1,2}(t) \\ 0 & \Omega_{1,2}(t) & 2(\delta_{0,1}+\delta_{1,2}) \end{bmatrix}. \tag{1}$$

The amplitudes of the drives are given by $\Omega_{0,1}(t)$ and $\Omega_{1,2}(t)$ and they are detuned from their corresponding transitions by $\delta_{i,j}=\omega_{i,j}^{(\mathrm{d})}-\omega_{i,j}$, where $\omega_{i,j}^{(\mathrm{d})}/(2\pi)$ are the drive frequencies and $\omega_{i,j}/(2\pi)$ are the system transition frequencies. The Hamiltonian can be diagonalized yielding the instantaneous eigenstates of the system

$$\begin{aligned} |+\rangle &= \sin\Phi|B\rangle+\cos\Phi|1\rangle, \\ |-\rangle &= \cos\Phi|B\rangle-\sin\Phi|1\rangle, \\ |D\rangle &= \cos\Theta|0\rangle-\sin\Theta|2\rangle, \end{aligned} \tag{2}$$

with eigenvalues

$$\begin{aligned} \hbar\omega_+ &= \hbar\left(\delta_{0,1}+\sqrt{\delta_{0,1}^2+\Omega_{0,1}(t)^2+\Omega_{1,2}(t)^2}\right)/2, \\ \hbar\omega_- &= \hbar\left(\delta_{0,1}-\sqrt{\delta_{0,1}^2+\Omega_{0,1}(t)^2+\Omega_{1,2}(t)^2}\right)/2, \\ \hbar\omega_D &= 0. \end{aligned} \tag{3}$$

The bright state is $|B\rangle=\sin\Theta|0\rangle+\cos\Theta|2\rangle$ with the mixing angle defined as

$$\tan\Theta=\Omega_{0,1}(t)/\Omega_{1,2}(t) \tag{4}$$

while

$$\tan\Phi=\frac{\sqrt{\Omega_{0,1}(t)^2+\Omega_{1,2}(t)^2}}{\sqrt{\Omega_{0,1}(t)^2+\Omega_{1,2}(t)^2+\delta_{0,1}^2}+\delta_{0,1}}. \tag{5}$$

Here we have assumed that $\delta_{0,1}+\delta_{1,2}=0$, realizing the two-photon resonance condition. Now we can see that as the ratio $\Omega_{0,1}(t)/\Omega_{1,2}(t)$ changes from 0 to $\infty$, the mixing angle $\Theta$ goes from 0 to $\pi/2$. As a result, the state $|\mathrm{D}\rangle$ changes from $|0\rangle$ to $|2\rangle$, realizing the required population inversion.

There are many possible choices for the shapes of the pulses $\Omega_{0,1}(t)$ and $\Omega_{1,2}(t)$ [8], but a common one is to take them as Gaussians

$$\begin{aligned} \Omega_{0,1}(t) &= \Omega_{0,1}\exp\left[-\frac{t^2}{2\sigma^2}\right], \\ \Omega_{1,2}(t) &= \Omega_{1,2}\exp\left[-\frac{(t-t_{\mathrm{s}})^2}{2\sigma^2}\right]. \end{aligned} \tag{6}$$

IOP Conf. Series: Journal of Physics: Conf. Series **969** (2018) 012141 doi:10.1088/1742-6596/969/1/012141

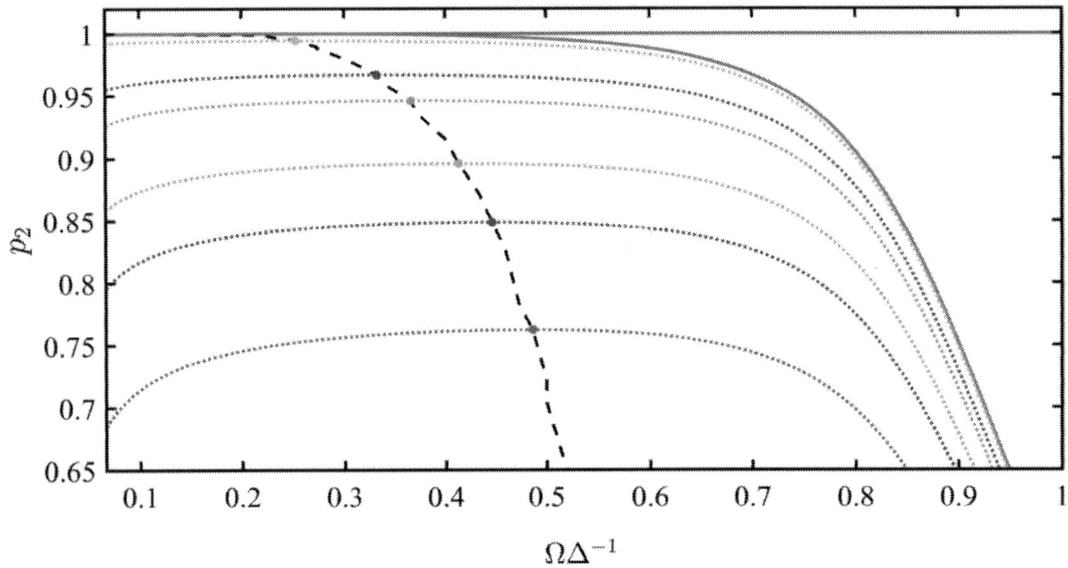


**Figure 1.** Transferred population $p_2$ as a function of $\Omega\Delta^{-1}$, where $\Omega = \Omega_{0,1} = \Omega_{1,2}$, with cross-couplings (red solid line) and without cross-couplings (blue solid line). The dotted lines show the transferred populations in the presence of decoherence. The lines correspond to dimensionless decoherence rates $2\pi\Gamma\Delta^{-1} * 10^{-4} = \{1, 10, 20, 30, 50, 100\}$, starting from the highest line. The dashed black line shows the optimum $\Omega_{0,1}\Delta^{-1}$ for different values of $\Gamma$.

While $t_s$ is negative the pulses are applied in an order where $\Omega_{1,2}$ pulse comes before $\Omega_{0,1}$ enabling the adiabatic population transfer. If the Hamiltonian (1) changes slowly enough, *i.e.*,

$$\sqrt{\Omega_{0,1}^2 + \Omega_{1,2}^2}\,\sigma \gg 1, \tag{7}$$

the system does not get excited away from state $|D\rangle$. This means that for a fixed $\sigma$ it is favourable for the process to have as high drive amplitudes as possible in order to minimize the diabatic losses.

## 2. Results

Until this point we have neglected the fact that both drive pulses couple off-resonantly also to the other transitions. In order to understand how this affects STIRAP, we start with the full Hamiltonian

$$H_{\text{full}} = \hbar \sum_{k=1}^{2} (k\omega_{0,1} - (k-1)\Delta)\,|k\rangle\langle k| + \frac{\hbar}{2} \sum_{k=0}^{1} \sum_{l=0}^{1} \left( \Omega_{k,k+1}(t) \mathrm{e}^{-i\omega_{k,k+1}^{(\mathrm{d})} t} |l+1\rangle\langle l| + \mathrm{H.c} \right), \tag{8}$$

where $\Delta$ is the energy anharmonicity of the system. Note that this reduces to Eq. (1) under the rotating wave approximation given that $\Delta \gg \Omega_{k,k+1}$ for all $k$. If that is not the case, it is informative to consider the Hamiltonian describing only the parasitic couplings. Under the rotating wave approximation in the frame moving with the parasitic drives, we get

$$H_{\rm p} = \frac{\hbar}{2}\begin{bmatrix} 0 & \Omega_{1,2}(t) & 0 \\ \Omega_{1,2}(t) & -2\Delta + 2\delta_{1,2} & \Omega_{0,1}(t) \\ 0 & \Omega_{0,1}(t) & 2(\delta_{0,1}+\delta_{1,2}) \end{bmatrix}. \tag{9}$$

If the two-photon resonance condition ($\delta_{0,1} = -\delta_{1,2}$) is satisfied, the above Hamiltonian describes a STIRAP process from state $|2\rangle$ to state $|0\rangle$ with the roles of the drives reversed. As a result, the parasitic couplings create a competing STIRAP process, though with a significant single photon detuning. However, this process is not negligible, because STIRAP is robust against the single-photon detuning. Ironically, one of the main advantages of STIRAP here acts against it. Note that as long as $\Delta - \delta_{1,2} \gg \Omega_{1,2}(t)$ and $\Delta + \delta_{0,1} \gg \Omega_{0,1}(t)$ there are no diabatic excitations to state $|1\rangle$ even though the $1-2$ drive is acting on the $0-1$ transition and vise versa.

In order to understand the combined effect of the Hamiltonians (1) and (9) we numerically solve the time-evolution of the system under the full Hamiltonian (8) from

$$\dot{\rho}(t) = -\frac{i}{\hbar}[\rho(t), H_{\rm full}], \tag{10}$$

with $\rho(-\infty) = |0\rangle\langle 0|$. For simplicity we assume that $\Omega_{0,1} = \Omega_{1,2} = \Omega$, $t_s = -1.5\sigma$ and $\delta_{0,1} = \delta_{1,2} = 0$. With these conventions Eq. (7) becomes $\Omega\sigma \gg 1$ and thus we can introduce a dimensionless parameter $a = \Omega\sigma$ which ideally determines the transfer fidelity of the STIRAP process. This is demonstrated in Fig. 1, where we show the population $p_2 = \langle 2|\rho(\infty)|2\rangle$ for $a = 3\pi$ when $\Omega$ and $\sigma$ are varied while keeping the anharmonicity $\Delta$ constant. The blue line describes the case without parasitic couplings. As expected, the fidelity of the process does not depend on the value of $\Omega$ because $\sigma$ is also changed accordingly. However, when the parasitic couplings are turned on (red line), the fidelity drops for larger values of $\Omega$ because the competing STIRAP process gets stronger as $\Omega$ approaches $\Delta$. Thus it would appear that it is optimal to make $\Omega$ as small as possible while allowing $\sigma$ to become longer. However, in the presence of decoherence the situation changes because longer $\sigma$ allows decoherence to affect more, suggesting that there exists an optimal value of $\sigma$ and $\Omega$ as a function of the decoherence rates. This is indeed the case and in Fig. 1 we show $p_2$ for several different values of the energy relaxation rate $\Gamma$, where $p_2$ is solved from the Lindblad equation

$$\dot{\rho}(t) = -\frac{i}{\hbar}[\rho(t), H_{\rm full}] + \mathcal{L}[\rho], \tag{11}$$

with $\mathcal{L}[\rho] = -\Gamma\rho_{22}|2\rangle\langle 2| - (\Gamma\rho_{11} - \Gamma\rho_{22})|1\rangle\langle 1| + \Gamma\rho_{11}|0\rangle\langle 0|$. The dashed black line shows the optimal value for the normalized $\Omega$. It is apparent that it is beneficial to use larger $\Omega$ for systems with higher decoherence rates. For example, in a typical transmon circuit we can take $\Delta/(2\pi) = 300$ MHz and $\Gamma = 0.5$ MHz, resulting in optimum of $\Omega/(2\pi) = 110$ MHz with $\sigma = 14$ ns.

## 3. Conclusions

We have shown that in systems with low anharmonicity the drives used in STIRAP protocol couple also to other transitions in a way that cannot be fully neglected. Even though in systems without decoherence the effect can be avoided by making the amplitudes of the drives very small and the duration of the pulses long, that is not optimal in general. We have demonstrated that the optimal value of the drive amplitude depends on the decoherence rate of the system, which we have numerically solved for several different relaxation rates. We believe that these results are useful when designing experiments exploiting STIRAP.

## Acknowledgements

We thank Väisälä Foundation, Academy of Finland CoE LTQ (project 250280), and Centre for Quantum Engineering at Aalto University for financial support.

IOP Conf. Series: Journal of Physics: Conf. Series **969** (2018) 012141 doi:10.1088/1742-6596/969/1/012141

## References

[1] Vitanov N V, Rangelov A A, Shore B W and Bergmann K 2017 *Rev. Mod. Phys.* **89**(1) 015006 URL https://link.aps.org/doi/10.1103/RevModPhys.89.015006
[2] Falci G, Di Stefano P G, Ridolfo A, D'Arrigo A, Paraoanu G S and Paladino E 2017 *Fortschritte der Physik* **65** 1600077–n/a ISSN 1521-3978 1600077 URL http://dx.doi.org/10.1002/prop.201600077
[3] Born M and Fock V 1928 *Zeitschrift für Physik* **51** 165–180 ISSN 0044-3328 URL http://dx.doi.org/10.1007/BF01343193
[4] Gaubatz U, Rudecki P, Schiemann S and Bergmann K 1990 *The Journal of Chemical Physics* **92** 5363–5376 URL http://scitation.aip.org/content/aip/journal/jcp/92/9/10.1063/1.458514
[5] Vepsäläinen A, Danilin S, Paladino E, Falci G and Paraoanu G S 2016 *Photonics* **3** ISSN 2304-6732 URL http://www.mdpi.com/2304-6732/3/4/62
[6] Kumar K S, Vepsäläinen A, Danilin S and Paraoanu G S 2016 *Nature Communications* **7** URL http://dx.doi.org/10.1038/ncomms10628
[7] Koch J, Yu T M, Gambetta J, Houck A A, Schuster D I, Majer J, Blais A, Devoret M H, Girvin S M and Schoelkopf R J 2007 *Phys. Rev. A* **76** 042319 URL http://journals.aps.org/pra/abstract/10.1103/PhysRevA.76.042319
[8] Vasilev G S, Kuhn A and Vitanov N V 2009 *Phys. Rev. A* **80**(1) 013417 URL http://link.aps.org/doi/10.1103/PhysRevA.80.013417

IOP Conf. Series: Journal of Physics: Conf. Series **969** (2018) 012142 doi:10.1088/1742-6596/969/1/012142

# Quantum magnetotransport and de Haas–van Alphen measurements in the three-dimensional Dirac semimetal $Pb_{0.83}Sn_{0.17}Se$

**F Orbanić[1], M Novak[1], S Pleslić[1] and I Kokanović[1]**

[1]Department of Physics, Faculty of Science, University of Zagreb, Bijenička cesta 32, Zagreb, Croatia.

[1]Department of Applied Physics, Faculty of Electrical Engineering and Computing, University of Zagreb, Unska 3, Zagreb, Croatia.

E-mail: forbanic@phy.hr

**Abstract.** Magnetoresistance (MR), Hall resistance and magnetization of the $Pb_{0.83}Sn_{0.17}Se$ crystals have been measured in magnetic field up to 15 T and 5 T, respectively, and temperatures from 1.7 K up to 300 K. A large linear and temperature dependent MR is observed in magnetic field up to 15 T. The de Haas–van Alphen (dHvA) and Shubnikov de Haas effects (SdH) of $Pb_{0.83}Sn_{0.17}Se$ crystals have been clearly seen in the temperature range from 30 K down to 1.7 K and magnetic field as low as 2 T. The dHvA and SdH oscillations reveal single frequency of around 8 T which confirms the existence of a single Fermi surface cross section. Influence of isothermal annealing of $Pb_{0.83}Sn_{0.17}Se$ crystals in Se vapours has been investigated. By increasing the annealing temperature from 433 °C up to 440 °C , transition from n to p-type conductivity has been observed. The dHvA and SdH effects clearly reflect the existence of a nontrivial Berry's phase owing to the linear band dispersion which is the signature of a three-dimensional Dirac fermion in the $Pb_{0.83}Sn_{0.17}Se$ crystals [1].

## 1. Introduction

Recent theoretical and experimental studies have suggested the existence of topological crystalline insulators (TCI), a novel class of topological insulators in which crystalline symmetry replaces the role of time-reversal symmetry in topological protection of their metallic surface states [2, 3]. These surface states are observed in the rocksalt IV–VI semiconductors (Pb,Sn)Te, (Pb,Sn)Se and SnTe using angle resolved photoemission spectroscopy (ARPES) [4, 5, 6]. Energy gap in the $Pb_{1-x}Sn_xSe$ semicoductor is temperature and Sn-content dependent [7]. The substitution of Sn for Pb strongly changes the spin-orbit effects and results in a compositional evolution of their band structures. An increasing Sn content in $Pb_{1-x}Sn_xSe$ semiconductor leads to a closing of the bulk bandgap at a critical value of $x = x_c \approx 0.17$ (for $T < 77K$) and the system undergoes gap inversion when x exceeds $x_c$ [7, 6]. In the $Pb_{1-x}Sn_xSe$ semiconductor, the chemical potential can be tuned during the crystal growth or annealing to yield n-type or p-type conductivity, which makes them more suitable for experimental investigations of the TCI state [6]. Very recently, we have found that $Pb_{0.83}Sn_{0.17}Se$ single crystal is a three-dimensional Dirac semimetal [1].

Here we report the low temperature magnetotransport and magnetization data measured from 300 K down to 1.8 K in high magnetic filed up to 15 T for semiconductor $Pb_{0.83}Sn_{0.17}Se$ single crystals which achieve the state of transition between the ordinary insulator and the TCI where the existence of a zero gap state is observed [10]. To investigate the tuning of the chemical potential, single crystals of $Pb_{0.83}Sn_{0.17}Se$ have been annealed in the Se vapours at temperatures between 433 and 440 °C . The $Pb_{0.83}Sn_{0.17}Se$ single crystals have peculiar magnetic and transport properties, such as a large negative unconventional magnetization, linear MR (LMR) in perpendicular magnetic fields and ultra high mobility. We have observed the linear magnetization and LMR accompanied by quantum oscillations at low temperature and in magnetic field up to 5 T and 15 T, respectively. The MR data reveal discernible Fermi surface change caused by the Zeeman splitting in the magnetic field range from 4 T up to quantum limit field of about 8 T. We have analysed the magnetization and magnetotransport experimental data using Lifshitz-Kosevich theory which predicts the magnitude of the observed dHvA and SdH effects as well as their temperature dependence and allow us to characterize the three-dimensional Fermi surface and its relevant parameters. Also, by studying the carrier density and mobility dependence of the observed LMR in perpendicular magnetic fields possible physical origins are discussed.

## 2. Experimental results and discussion

The $Pb_{0.83}Sn_{0.17}Se$ single crystals were grown by a modified Brigman method [1]. Evidence for the quality of the $Pb_{0.83}Sn_{0.17}Se$ crystals includes extremely sharp x-ray peaks observed at room temperature, and a substantial mean free path extracted from the quantum oscillations observed in the magnetization and magnetotransport data. Samples from several batches were annealed for 48 h in the Se vapours at temperatures from 433 to 440 °C (433, 435, 436, 438 and 440 °C ) and then slowly cooled down. The magnetization was measured using a superconducting quantum interference device magnetometer SQUID (Quantum Design). A long plastic straw was used as the sample holder [1]. The samples were first mounted and cooled in the SQUID magnetometer down to 5 K. The magnetization of $Pb_{0.83}Sn_{0.17}Se$ crystals has been measured at different temperatures from 300 K down to 5 K with B parallel to the [001] direction and shown in figure 1.

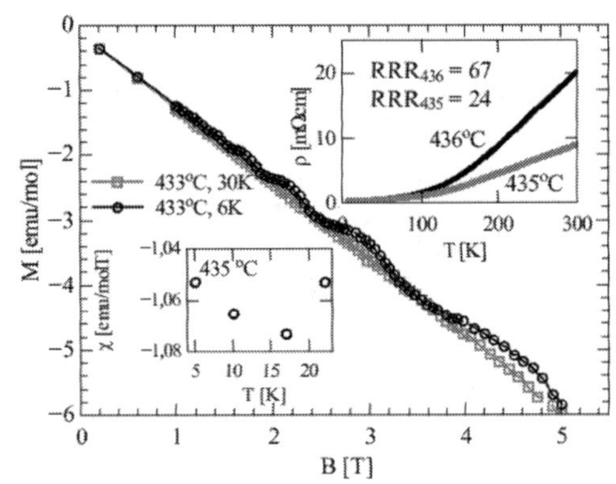


**Figure 1.** The magnetic field dependence of magnetization at two temperatures when the field was applied along the [001] direction for $Pb_{0.83}Sn_{0.17}Se$ crystal annealed at 433°C . The upper inset shows the temperature dependence of the resistivity of $Pb_{0.83}Sn_{0.17}Se$ crystals annealed at 436°C and 435 °C and the lower inset shows the magnetic susceptibility in the temperature range from 5-25 K of sample annealed at 435 °C .

It is found that the samples are diamagnetic. The linear behaviour of the magnetization vs. B was observed for all measured samples in the temperature range from 5 K up to 300 K. Any intercepts at B = 0 T were less than $1.6 \cdot 10^{-5}$ emu for the measured crystals, and have been subtracted from the data reported here. The magnetic susceptibility is weakly temperature dependent down to 5 K. A shallow minimum exists between 40 and 5 K, see the inset to figure

IOP Conf. Series: Journal of Physics: Conf. Series **969** (2018) 012142 doi:10.1088/1742-6596/969/1/012142

1. It can be seen from figure 1 that the magnetization displays a very clear dHvA effect. The oscillations are clearly seen down to 1 T and up to 30 K.

From Hall measurements of annealed samples at room temperature we found that samples annealed at 433 °C possess pure n-type conductivity while samples annealed at 440 °C show p-type conductivity. $Pb_{0.83}Sn_{0.17}Se$ p-type crystals show no quantum oscillations. So samples annealed at $435 \pm 1$ °C are close to the crossover from n to p-type conductivity and the chemical potential is then closest to the band touching point. Further transport measurements are done only on samples annealed at 435 and 436 °C .

Transport measurements were carried out in a $^4$He cryostat, using a constant bias current (dc, 10 mA) and a nanovoltmeter with a 10 G input resistance up to 12 V of the signal. For transport measurements, the electrical contacts in the six-point geometry were made by spot welding which resulted in contact resistance of the order of 0.5 Ω. The temperature dependence of longitudinal electrical resistivity of the $Pb_{0.83}Sn_{0.17}Se$ crystals is presented in the inset to figure 1. The longitudinal resistivity shows a typical metallic behaviour with the residual resistivity ratio $R(300K)/R(1.8K) = 67$ and 24 with residual resistivities $\sim 0.28$ mΩcm and $\sim 0.37$ mΩcm at 1.8 K, for the $Pb_{0.83}Sn_{0.17}Se$ crystals annealed at 436°C and 435°C , respectively. Hall signal mixed into measured longitudinal resistivity $\rho'_{xx}$ is removed by averaging over the positive and negative field directions. From the raw $\rho'_{xx}$ data measured at the indicated temperatures, we obtained the symmetrized (intrinsic) longitudinal MR, which are plotted as $MR = (\rho_{xx}(B) - \rho_{xx}(0))/\rho_{xx}(0) \cdot 100\%$ in figure 2 (where $\rho_{xx}$ is symmetrized longitudinal resistivity and $\rho_{xx}(0)$ denotes the longitudinal resistivity in the zero magnetic field). Here, current was along the [100] direction, and the applied magnetic field was perpendicular to current in [001] direction.

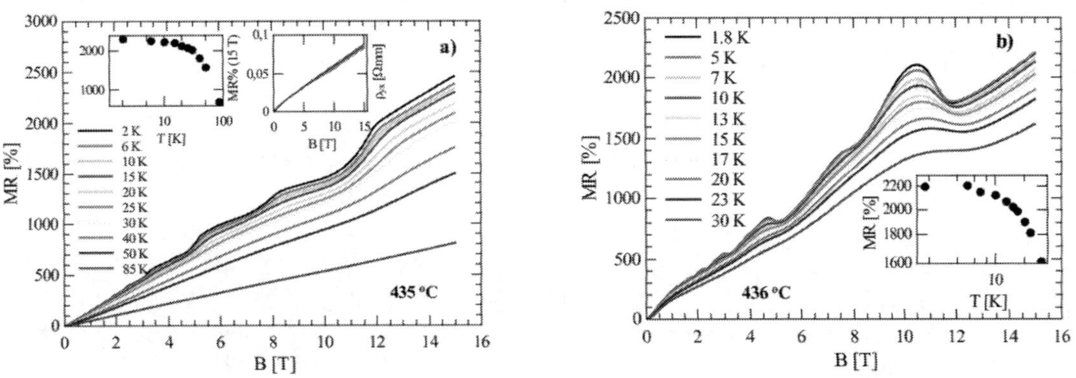


**Figure 2.** Magnetoresistivity (MR) of $Pb_{0.83}Sn_{0.17}Se$ single crystals annealed at 435 °C (a) and 436 °C (b), respectively, at the different temperatures in magnetic fields applied along the [001] direction. The right inset to a) displays the Hall resistivity vs. magnetic field data of the sample annealed at 435 °C . The insets to a) and b) display the MR versus temperature at 15 T for the $Pb_{0.83}Sn_{0.17}Se$ single crystals annealed at 435 °C and 436 °C , respectively.

The MR reveals a linear behaviour from very small magnetic fields for the both samples, figure 2. Even at temperature of 85 K and magnetic field of 15 T MR shows no sign of saturation, and its value is as high as 800%. Such a large LMR at 85 K is quite unusual and needs further theoretical study to understand the physical origin. To gain more physical insights into the observed linear MR we further study the temperature dependence of MR(B, T). In the inset of figure 2a) and 2b) we show the temperature dependence of MR (without subtracting the oscillatory part) at 15 T in log-log scale. With decreasing temperature, MR increases rapidly up

IOP Conf. Series: Journal of Physics: Conf. Series **969** (2018) 012142 doi:10.1088/1742-6596/969/1/012142

to 2500% at 10 K in magnetic field of 15 T. It is clear from the insets that, unlike conventional metal, MR shows a saturation-like behaviour in the low-temperature region below 20 K. Similar behaviour has been observed in $Cd_3As_2$ crystals [11]. We have also performed Hall resistivity measurements for the both samples annealed at 435 °C and 436 °C . Shown in the inset to figure 2a), the Hall resistivity of the sample annealed at 435 °C reveals nonlinear behaviour at low magnetic fields with crossover at $B_c = 2.5$ T and then changes behaviour in linear at high magnetic fields (there is obviously a small contribution of holes in transport). The Hall resistivity of the sample annealed at 436 °C has the linear magnetic dependence from 0 up to 15 T. From the slope of Hall resistivity data the electron density is extracted: $n_H = 11.6 \cdot 10^{17}$ cm$^{-3}$ for sample annealed at 435 °C , whereas for the sample annealed at 436 °C , $n_H = 5.9 \cdot 10^{17}$ cm$^{-3}$. Further, the Hall resistivity data for the both samples reveal very weak temperature behaviour in the temperature range from 300 K down to 1.8 K. Using the single carrier Drude band model and the above extracted electron density, we have obtained the electron mobility $\mu(1.8\text{K}) = 1/en\rho_{xx} = 3.74 \cdot 10^4$ cm$^2$/Vs for the sample annealed at 436 °C and $1.45 \cdot 10^4$ cm$^2$/Vs for 435 °C .

As it is seen from figures 1 and 2, the quantum oscillations are clearly observed in the magnetization and MR data of $Pb_{0.83}Sn_{0.17}Se$ crystals. Subtracting a smooth and linear background from the magnetization (figure 1) yields the dHvA oscillatory component, figure 3a). Using the MR and Hall resistivity data we have calculated magnetoconductivity $\sigma_{xx} = \rho_{xx}/(\rho_{xx}^2 + \rho_{xy}^2)$, its oscillatory part is shown in figure 3b). Figures 3a) and 3b) display the oscillatory components of magnetization and magnetoconductivity versus 1/B and reveal how the oscillation amplitude decreases with the annealed temperature of the samples and temperature, respectively. The inset to figure 3a) shows the oscillation frequencies obtained by the FFT analysis of the magnetization data. It is clearly seen that there is only one oscillation frequency which slightly changes in the range from 8.6 to 6.9 T for the annealed temperatures. In oscillatory part of magnetoconductivity the effect of Zeeman splitting in higher magnetic fields (above 4 T) can be clearly seen.

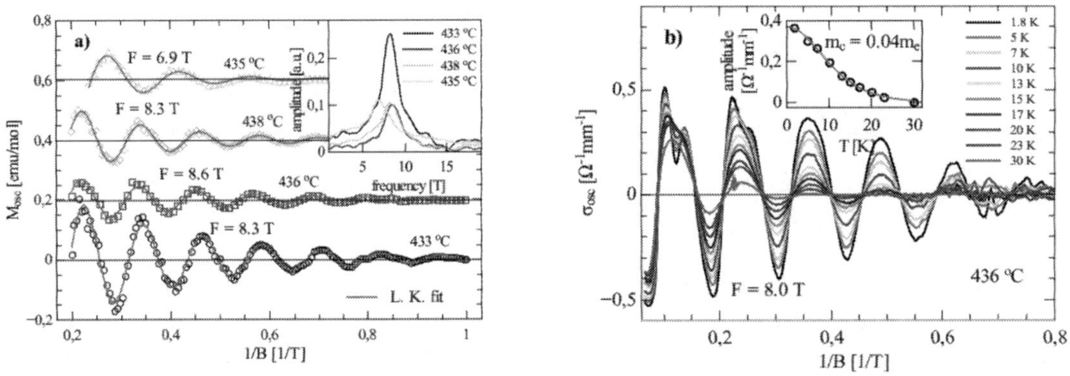


**Figure 3.** a) the oscillatory components of magnetization for $Pb_{0.83}Sn_{0.17}Se$ single crystals at temperature 5 K and b) the quantum oscillation contribution to magnetoconductivity for $Pb_{0.83}Sn_{0.17}Se$ single crystal versus 1/B. The inset to a) shows the oscillation frequency obtained by the FFT analysis of the magnetization curves of the samples. The inset to b) shows the temperature dependence of amplitude of oscillations from which the effective cyclotron mass of carriers can be obtained [1].

To explain the origin of LMR, two types of mechanisms are proposed: (i) a classical model which explains the LMR by modelling the material as a network due to the strong inhomogeneity in the carrier density hence disorder induced mobility fluctuation [12] and (ii) a quantum model due to the linear energy dispersion of carrier at the band touching point proposed by Abrikosov [13]. In the quantum model, the LMR also occurs when multiple Landau levels are filled, even when thermal smearing is severe, as long as the system has a positive g factor [14]. However, in our samples the LMR is observed from very small magnetic fields which are much lower than the onset of the quantum oscillations observed at 2 T. Thus, we can argue that the LMR in our samples arises because the local current density acquires spatial fluctuations in both magnitude and direction, as a result of the microstructure caused by inhomogeneous carrier density and mobility distribution.

From weak temperature dependence of Hall resistivity (inset to figure 2a)) one can conclude that the carrier density is also temperature independent. The decrease of MR at 15 T with temperature, shown in insets to figure 2, can than be explained with the change in mobility of carriers. Relative change of MR(15 T) and mobility $\mu$, calculated as $\mu(\mathrm{T}) = 1/\mathrm{en_H}\rho_{\mathrm{xx}}(\mathrm{T})$, from 85 K to 6 K for sample annealed at 435 °C is shown in figure 4. It can clearly be seen that temperature change of mobility follows change in MR at 15 T.

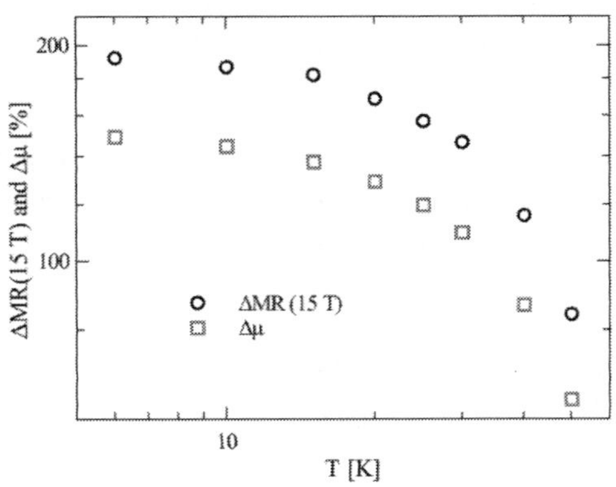


**Figure 4.** Relative change of MR(15 T) and mobility from 85 K to 6 K for $Pb_{0.83}Sn_{0.17}Se$ crystal annealed at 435 °C . $\Delta\mathrm{MR}(\mu) = (\mathrm{MR}(\mu)(\mathrm{T}) - \mathrm{MR}(\mu)(85\mathrm{K}))/\mathrm{MR}(\mu)(85\mathrm{K}) \cdot 100\%$

Quantum oscillation phenomenon is a powerful technique for measuring low temperature Fermi surface properties. Quantum oscillations can be understood quantitatively within the Lifshitz-Kosevich (LK) formalism [15]. Oscillation of different physical quantities with inverse magnetic field for three-dimensional electron gas can be described by the same functional form:

$$\Delta X(B,T) = A_0 A_T A_D A_S \left(\frac{B}{F}\right)^{1/2} \cos\left(2\pi\frac{\mathrm{F}}{\mathrm{B}} + \phi_{\mathrm{p}} + \phi_{\mathrm{M}}\right). \qquad (1)$$

$\Delta X$ stands for oscillatory part of magnetization, magnetic susceptibility or conductivity (resistivity). $F$ is the oscillation frequency and $\phi_p$ and $\phi_M$ are phase factors [1]. $A_T$, $A_D$ and $A_S$ are temperature, Dingle and spin dimensionless prefactors and $A_0$ is dimensional constant. $A_T = 2\pi^2(\mathrm{k_B T}/\hbar\omega_c)/\sinh[2\pi^2(\mathrm{k_B T}/\hbar\omega_c)]$, $A_D = \exp[2\pi^2(\mathrm{k_B T_D}/\hbar\omega_c)]$, where $T_D$ is the Dingle temperature related to quantum scattering time, $A_S = \cos[\pi \mathrm{gm_e}/(2\mathrm{m_c})]$. $g$ is the Lande factor, $m_c$ is the effective cyclotron mass, $m_e$ is the free electron mass and $\omega_c = eB/m_c$ is cyclotron frequency. In the inset to figure 3b) the temperature dependence of amplitude of oscillations in conductivity is shown. By fitting the temperature dependence of amplitude to functional form of $A_T$ the effective cyclotron mass of carriers can be obtained, $m_c = 0.04 m_e$. We have done a detailed analysis of dHvA and SdH oscillations in given $Pb_{0.83}Sn_{0.17}Se$ samples in [1]. Fermi surface parameters obtained from dHvA and SdH oscillations [1] for sample annealed at 436 °C are the following: $k_F = 0.016$ Å$^{-1}$, $n_{osc} = 1.4 \cdot 10^{17}$ cm$^{-3}$, $v_F = 4.6 \cdot 10^5$ m/s, $E_F = 48$ meV, $T_D = 3.6$ K (quantum scattering time $\tau_Q = 3.4 \cdot 10^{-13}$ s). Nearly the same values are obtained also for the sample annealed at 435 °C . We can see that for the sample annealed at 436°C charge density determined from quantum oscillations and Hall resistivity are in good agreement taking into account four electron pockets at the L point ([111] direction, B-field is in [001] direction) of the Brillouin zone [6]. Because of the small hole contribution (bending of Hall resistivity at small magnetic fields) in the sample annealed at 435 °C two charge densities are not in such a good agreement. By performing the Landau level (LL) diagram [1] a nontrivial Berry's phase, associated with the 3D Dirac semiconductor phase in $Pb_{0.83}Sn_{0.17}Se$ crystal is confirmed. Also, the magnetic field of quantum limit is found to be around 8 T.

## 3. Conclusion

In conclusion, we have studied the low temperature magnetotransport and magnetization data measured from 300 K down to 1.8 K in high magnetic filed up to 15 T for semimetal $Pb_{0.83}Sn_{0.17}Se$ single crystals. We have observed the linear magnetization and linear MR accompanied by quantum oscillations at low temperature. We have found that the annealing of the $Pb_{0.83}Sn_{0.17}Se$ samples in the Se vapours at temperatures from 433 to 440 °C results in the crossover from n to p-type conductivity. The MR data reveal discernible Fermi surface change caused by the Zeeman splitting in the magnetic field range from 4 T up to quantum limit field of about 8 T. We have extracted the Fermi surface parameters using the LK theory. It has been found that the LMR scales with electron mobility in the $Pb_{0.83}Sn_{0.17}Se$ crystals and originates from fluctuating mobility regions due to inhomogeneous current paths.

## Acknowledgments

This work has been fully supported by Croatian Science Foundation under the project No. 6216.

## References

[1] Orbanić F, Novak M, Baćani M and Kokanović I 2017 *Phys. Rev.* B **95** 035208
[2] Fu L 2011 *Phys. Rev. Lett.* **106** 106802
[3] Ando Y and Fu L 2015 *Annu. Rev. Condens. Matter Phys.* **6** 361
[4] Liang T, Gibson Q, Xiong J, Hirschberger M, Koduvayur S P, Cava R J and Ong N P 2013 *Nat. Commun.* **4** 2696
[5] Xu S Y et al *Nat. Commun.* 2012 **3** 1192
[6] Dziawa P et al 2012 *Nat. Mater.* **11** 1023
[7] Strauss A J 1967 *Phys. Rev.* **157** 608
[8] Tanaka Y, Ren Z, Sato T, Nakayama K, Souma S, Takahashi T, Segawa K and Ando Y 2012 *Nat. Phys.* **8** 800
[9] Ravich Y I, Efimova B A and Smirnov I A 1970 *Semiconducting Lead Chalcogenides* (New York: Plenum Press)
[10] Zeljkovic I et al 2015 *Nat. Mater.* **14** 318
[11] Pariari A, Dutta P and Mandal P 2015 *Phys. Rev.* B **91** 155139
[12] Parish M M and Littlewood P B 2003 *Nature* **426** 162
[13] Abrikosov A 1998 *Phys. Rev.* B **58** 2788
[14] Wang C M and Lei X L 2012 *Phys. Rev.* B **86** 035442
[15] Shoenberg D 1984 *Magnetic Oscillations in Metals* (Cambridge: Cambridge University Press)

IOP Conf. Series: Journal of Physics: Conf. Series **969** (2018) 012143 doi:10.1088/1742-6596/969/1/012143

# Photoassisted shot noise spectroscopy at fractional filling factor

**Luca Vannucci[1,2], Flavio Ronetti[1,2,3], Dario Ferraro[3], Jérôme Rech[3], Thibaut Jonckheere[3], Thierry Martin[3] and Maura Sassetti[1,2]**

[1]Dipartimento di Fisica, Università di Genova, Via Dodecaneso 33, 16146, Genova, Italy
[2]CNR-SPIN, Via Dodecaneso 33, 16146, Genova, Italy
[3]Aix Marseille Univ, Université de Toulon, CNRS, CPT, Marseille, France

E-mail: vannucci@fisica.unige.it

**Abstract.**
We study the photoassisted shot noise generated by a periodic voltage in the fractional quantum Hall regime. Fluctuations of the current are due to the presence of a quantum point contact operating in the weak backscattering regime. We show how to reconstruct the photoassisted absorption and emission probabilities by varying independently the dc and ac contributions to the voltage drive. This is made possible by the peculiar power-law behavior of the tunneling rates in the chiral Luttinger liquid theory, which allow to approximate the typical infinite sums of the photoassisted transport formalism in a simple and particularly convenient way.

## 1. Introduction

With the ultimate goal of controlling coherent few-particle excitations in quantum conductors, the condensed matter community has been paying an ever increasing attention to ac transport in mesoscopic devices over the last twenty years. This has lead to the development of full counting statistics [1, 2], photoassisted transport formalism [3, 4], Floquet scattering matrix approach [5] and challenging high-frequency experimental techniques [6, 7, 8, 9], culminated in the new paradigm of electron quantum optics (EQO) [10, 11].

However, less attention has been devoted to the role of electron-electron interactions, which often play a crucial role in low-dimensional systems. For instance, the fractional quantum Hall (FQH) effect emerges as a consequence of strong repulsive interactions that give rise to exotic quasi-particles carrying fractional charge and statistics [12]. At the same time, the presence of dissipationless topological edge modes makes the FQH phase a good candidate to study photoassisted quantum transport in interacting systems [13, 14, 15, 16].

In this paper we investigate the photoassisted shot noise (PASN) in the FQH regime. We consider a FQH system where periodic voltage pulses are injected from one of the terminals in presence of a quantum point contact (QPC). In this geometry, excitations incoming from the leads are partitioned at the QPC, in a protocol that is reminiscent of the Hanbury-Brown and Twiss optical experiment [10]. We evaluate the shot-noise at the first relevant order in the tunneling, considering the QPC in a weak backscattering regime. While the expression for the shot noise generally consists of an infinite superposition of dc contributions, each one weighted by the corresponding photoassisted probability, we show that the FQH physics allows to extract each single contribution to the PASN in a surprisingly simple fashion. We provide a recipe to reconstruct the typical absorption and emission probabilities of the photoassisted formalism by independently tuning the ac and dc components of the voltage drive. We also discuss the

IOP Conf. Series: Journal of Physics: Conf. Series **969** (2018) 012143 doi:10.1088/1742-6596/969/1/012143

experimental feasibility of this study, identifying a set of reasonable experimental parameters under which our protocol should be applicable.

## 2. Model

We consider the edge states of a quantum Hall system with filling factor in the Laughlin sequence $\nu = 1/(2n+1)$, $n \in \mathbb{N}$ [12]. The low energy excitations of such a quantum state are described by the chiral Luttinger liquid theory [17], with a pair of counter-propagating chiral modes (one for each edge of the sample) represented by bosonic fields $\phi_{R/L}(x)$. For an infinite Hall bar they satisfy $[\phi_{R/L}(x), \phi_{R/L}(y)] = \pm i\pi \text{sign}(x-y)$. They are linked to creation and annihilation of fractional quasi-particles with charge $e^* = \nu e$ through the bosonization identity

$$\psi_{R/L}(x) = \frac{U_{R/L}}{\sqrt{2\pi a}} e^{\pm i k_F x} e^{-i\sqrt{\nu}\phi_{R/L}(x)}, \tag{1}$$

where the parameter $a$ is a short-distance cutoff and $U_{R/L}$ are the Klein factors. The quasi-particle density operators read $\rho_{R/L}(x) = \mp\frac{\sqrt{\nu}}{2\pi}\partial_x \phi_{R/L}(x)$ and the effective low-energy Hamiltonian is given by ($\hbar = 1$)

$$H = \frac{\pi v}{\nu} \int_{-\infty}^{+\infty} dx \left[\rho_R^2(x) + \rho_L^2(x) + \Theta(-x-d)V(t)e\rho_R(x)\right], \tag{2}$$

with $v$ the propagation velocity of the free chiral fields. Here, we have included an additional coupling between the right-moving density and the time dependent voltage drive $V(t)$, which is used to drive the system out of equilibrium. The voltage $V(t)$ is assumed to be spatially homogeneous in the interval $(-\infty, -d)$, with $d > 0$. It is easy to verify that the bosonic field $\phi_R(x)$ in the presence of $V(t)$ acquires an additional term, namely

$$\phi_R(x,t) = \phi_R^{(0)}(x - vt, 0) + e\sqrt{\nu} \int_0^{t-\frac{x}{v}-\frac{d}{v}} dt' V(t'), \tag{3}$$

where $\phi_R^{(0)}(x,t)$ is the field at equilibrium [i.e. $V(t) = 0$]. To model tunneling between the upper and the lower edge we introduce a QPC at $x = 0$, represented by the bosonized tunneling Hamiltonian

$$H_{\text{tun}}(t) = \lambda \exp\left[ie^* \int_0^t dt'\, V(t')\right] e^{i\sqrt{\nu}\phi_R^{(0)}(0,t)} e^{-i\sqrt{\nu}\phi_L^{(0)}(0,t)} + \text{h.c.} \tag{4}$$

The constant $\lambda$ plays the role of the bare tunneling amplitude [in which we have also absorbed the factor $1/(2\pi a)$], while the time dependent phase is generated by the voltage drive as shown in Eq. (3). It is worth noticing that we have neglected the delay $d/v$ due to the finite distance between the contact and the QPC, which has no effect on the time-averaged quantities we are interested in. We have also omitted the unnecessary Klein factors [18].

We will consider two types of periodic voltage drive, with period $T = 2\pi/\omega$. The first one is the sinusoidal wave $V_{\text{sin}}(t) = V_{\text{dc}} - V_{\text{ac}}\cos(\omega t)$ while the second one is a Lorentzian drive given by

$$V_{\text{Lor}} = V_{\text{dc}} + V_{\text{ac}} \left[\frac{1}{\pi} \sum_{k=-\infty}^{+\infty} \frac{\eta}{\eta^2 + (t/T + k)^2} - 1\right]. \tag{5}$$

The parameter $\eta$ gives the half-width of each pulse, $W = \eta T$. The Lorentzian voltage represents a particularly interesting choice since it gives rise to minimal excitation states both in the integer and the FQH regime, called levitons [9, 15, 19]. Following a common procedure in the framework of the photoassisted transport [4, 20], we write the exponential in Eq. (4) as a Fourier series

$$\exp\left\{-ie^* \int_0^t dt'\, V(t')\right\} = e^{-iq\omega t} \sum_{l=-\infty}^{+\infty} p_l(\alpha) e^{-il\omega t}, \tag{6}$$

where $q = e^*V_{\rm dc}/\omega$ and $\alpha = e^*V_{\rm ac}/\omega$ are linked to the dc and ac amplitudes respectively. Focusing on the signals introduced above we get $p_l(\alpha) = J_{-l}(\alpha)$ [13] for the sinusoidal voltage [$J_n(z)$ is the Bessel function of the first kind] and

$$p_l(\alpha) = \alpha e^{-2\pi l\eta}\sum_{s=0}^{\infty}(-1)^s\frac{\Gamma(l+s+\alpha)}{\Gamma(1-s+\alpha)}\frac{e^{-4\pi s\eta}}{s!(s+l)!} \qquad (7)$$

for the Lorentzian drive [21, 22] .

## 3. Backscattering current and noise

In the weak backscattering regime, the QPC is almost fully transparent and only a small portion of the current is reflected back. Such a backscattering current can be calculated perturbatively at the first relevant order in the tunneling. We define the quasi-particle current operator as $I_B = e^*\dot{N}_L$, with the number of quasi-particle in the lower edge given by $N_L = \int dx\rho_L(x)$. This gives

$$I_B(t) = ie^*\lambda\exp\left[ie^*\int_0^t dt'\,V(t')\right]e^{i\sqrt{\nu}\phi_R^{(0)}(0,t)}e^{-i\sqrt{\nu}\phi_L^{(0)}(0,t)} + \text{h.c.} \qquad (8)$$

Fluctuations of the backscattering current are encoded in the zero-frequency shot noise defined as

$$\mathcal{S} = 2\int_0^T\frac{dt}{T}\int_{-\infty}^{+\infty}dt'\left[\langle I_B(t)I_B(t')\rangle - \langle I_B(t)\rangle\langle I_B(t')\rangle\right]. \qquad (9)$$

To calculate $\langle I_B\rangle$ and $\mathcal{S}$ one is asked to evaluate quantum averages involving quasi-particle fields. Invoking the bosonization identity Eq. (1), they can be related to the bosonic Green's function $\mathcal{G}(\tau)$, which is equal for left and right moving particles and is given by ($\kappa_{\rm B} = 1$) [23, 24]

$$\mathcal{G}(\tau) = \langle\phi(\tau)\phi(0)\rangle - \langle\phi^2(0)\rangle = \ln\left[\frac{\pi\theta\tau}{\sinh(\pi\theta\tau)(1+i\omega_c\tau)}\right]. \qquad (10)$$

Here we have introduced the high-frequency cutoff $\omega_c = v/a$ and the temperature $\theta$, and we consider the limit $\theta/\omega_c \ll 1$. Finally, we make use of the Fourier transform

$$\hat{P}_g(E) = \int_{-\infty}^{+\infty}dt\,e^{iEt}e^{g\mathcal{G}(t)} = \left(\frac{2\pi\theta}{\omega_c}\right)^{g-1}\frac{e^{E/(2\theta)}}{\Gamma(g)\omega_c}\left|\Gamma\left(\frac{g}{2} - i\frac{E}{2\pi\theta}\right)\right|^2 \qquad (11)$$

and the series in Eq. (6) to get

$$\mathcal{S} = 2(e^*)^2|\lambda|^2\sum_{l=-\infty}^{+\infty}|p_l(\alpha)|^2\left\{\hat{P}_{2\nu}\left[(q+l)\,\omega\right] + \hat{P}_{2\nu}\left[-(q+l)\,\omega\right]\right\}. \qquad (12)$$

This is the photoassisted expression for the shot noise. It can be viewed as a superposition of several dc contributions, whose effective bias is shifted by an amount $l\omega$ with respect to the dc value $q\omega$ and weighted by a probability $|p_l(\alpha)|^2$, which is nothing but the probability for a quasi-particle to absorb or emit $l$ energy quanta [21]. We note that the ac and dc amplitudes are well separated in Eq. (12). Indeed, the former emerges as the argument $\alpha$ of the coefficients $p_l$, while the latter appears in the functions $\hat{P}_{2\nu}(E)$ via the parameter $q$.

## 4. Results

In what follows we study the behavior of the PASN given in Eq. (12) when $q$ and $\alpha$ are varied independently. To begin, let us first discuss a set of reasonable values for $\theta$, $\omega$ and $\eta$. Experiments testing levitons in two-dimensional electron gases are usually performed at $\theta = 10 - 100$ mK [9, 25, 26], which also happens to be a range of temperature where well defined FQH states can be spotted [27]. The Lorentzian voltage drive usually operates at a frequency $f = \frac{\omega}{2\pi} = 5 - 6$ GHz, with dimensionless width of each pulse $\eta = 0.1 - 0.2$ [9, 25, 26]. Higher frequencies are also used

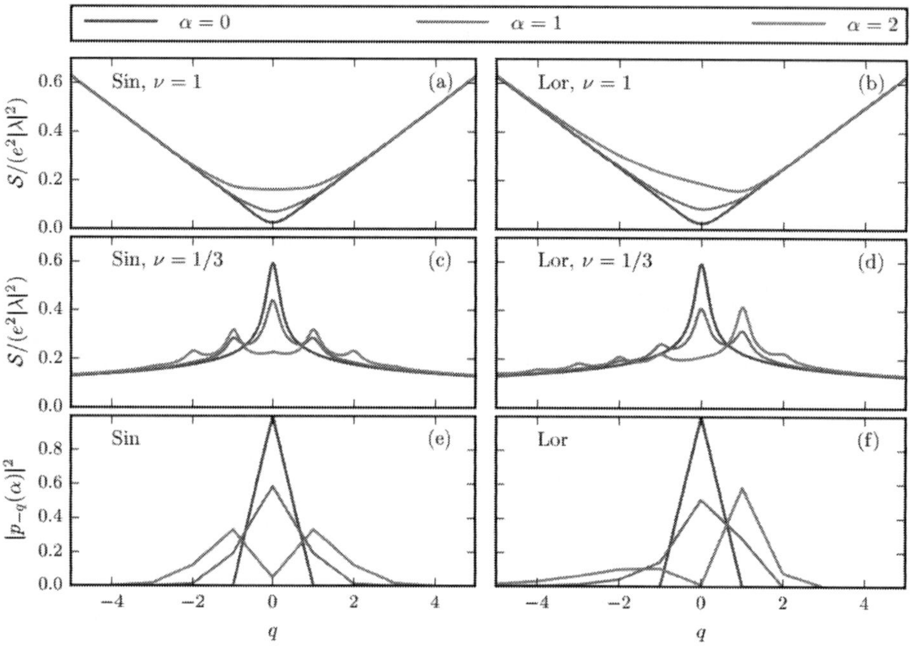


**Figure 1.** (a)-(d) PASN as a function of the dc amplitude $q$ for $\nu = 1$ and $\nu = 1/3$ and different values of the ac amplitude $\alpha$. (e)-(f) Coefficients $|p_{-q}(\alpha)|^2$ for different values of $\alpha$ as a function of $q$. In (e) and (f) the continuous lines are guides for the eyes, with the coefficients $|p_{-q}(\alpha)|^2$ only defined for integer values of $q$.

for the sinusoidal wave [9], and photoassisted transport in graphene nanoribbons illuminated with THz radiation was recently reported [28]. We initially set $\eta = 0.1$ and $\theta = 0.1\,\omega$, but lower values of the ratio $\theta/\omega$ can be reached in principle and will be discussed later.

Figure 1 shows the behavior of $\mathcal{S}$ as a function of $q$ for fixed values of $\alpha$. In panels (a) and (b) we report the case of sinusoidal and Lorentzian voltage pulses at integer filling factor $\nu = 1$. The blue curve represents the pure dc case where no ac component is present in the voltage drive ($\alpha = 0$). In this case $\mathcal{S}$ grows linearly with $q$ as expected, since $\mathcal{S} \propto q \coth(\frac{q\omega}{2\theta}) \approx |q|$ in the integer quantum Hall regime at sufficiently low temperature. Conversely, when a finite ac component is present ($\alpha = 1, 2$) the behavior at low $q$ is clearly non-linear and some excess noise due to the presence of the oscillating drive can be identified. Switching to the FQH regime [panels (c) and (d)], the linear (or almost linear) profile of the integer case is replaced by a strongly non-linear behavior, even for a dc voltage drive. This is a typical signature of the chiral Luttinger liquid theory. In particular the $\alpha = 0$ curve, which is proportional to $\hat{P}_{2\nu}(q\omega) + \hat{P}_{2\nu}(-q\omega)$, shows a sharp peak around $q = 0$. Such a structure is visible for $\alpha = 1$ and $\alpha = 2$ as well, with additional peaks arising for integer values of $q$. The different peaks in Fig. 1 (c) and (d) reproduce the features of the $\alpha = 0$ curve at shifted values $q + l$, since the photoassisted transport can be interpreted as an infinite superposition of shifted dc cases weighted by the probabilities $|p_l(\alpha)|^2$, as we mentioned before. Indeed, due to the sharply peaked structure of the tunneling rates in the FQH regime, the dominant contribution to the noise around integer $q$ is given by the photoassisted amplitudes with $l = -q$. In such a case the PASN is well approximated by

$$\mathcal{S} \approx 4(e^*)^2|\lambda|^2\hat{P}_{2\nu}(0)\,|p_{-q}(\alpha)|^2\,, \tag{13}$$

allowing to reconstruct the probabilities $|p_l(\alpha)|^2$ from the relative height of the different peaks. Thus, by fixing the ac component of the voltage drive and tuning the dc component we can

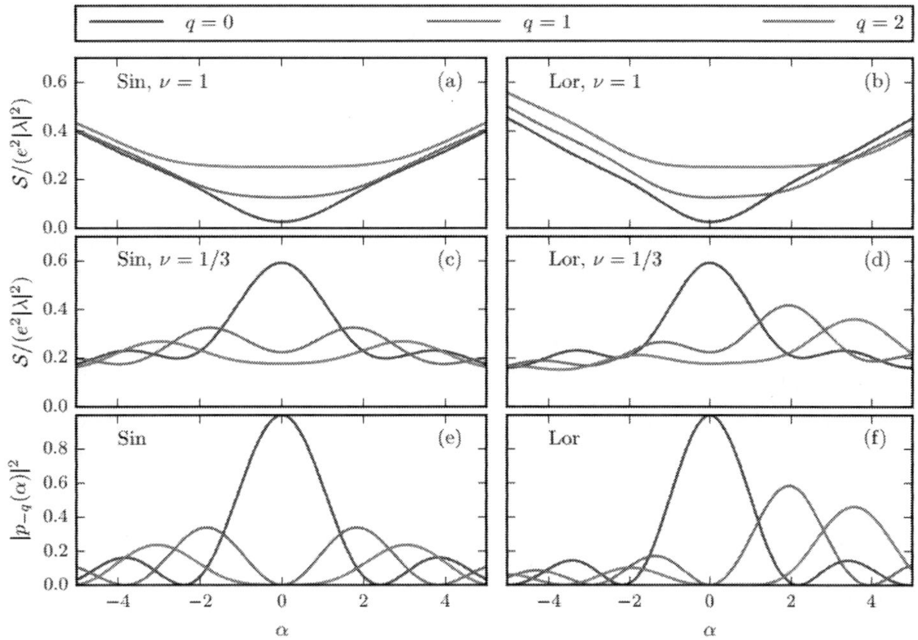


**Figure 2.** (a)-(d) PASN for $q = 0, 1, 2$ as a function of $\alpha$ at $\nu = 1$ and $\nu = 1/3$. (e)-(f) Coefficients $|p_{-q}(\alpha)|^2$ for $q = 0, 1, 2$ as a function of $\alpha$.

explore all the coefficients $|p_l(\alpha)|^2$ for $l = 0, \pm 1, \pm 2 \dots$. This is similar to the spectroscopic protocol developed by Dubois *et al.* in Refs. [9, 21] for the free Fermion case, although the fractional regime treated in the present work makes it much more effective and easy to visualize, due to the peculiar structure of the tunneling rate $\hat{P}_{2\nu}(E)$ at fractional filling (see also a similar analysis in the framework of finite frequency noise spectroscopy discussed in Refs. [29, 30]). For instance, Fig. 1 (d) suggests that all $p_{-q}(\alpha)$ with $q > \alpha$ vanish for a Lorentzian drive with integer $\alpha$, since we cannot see any further peak at $q > \alpha$. This is the striking property that allows the Lorentzian voltage pulse to generate a single electron above the Fermi level, with no disturbance below it [2, 19]. Conversely, no cancellation arises in the sinusoidal case, where Fourier coefficients manifestly satisfy $|p_q(\alpha)|^2 = |p_{-q}(\alpha)|^2$ [see Fig. 1 (c)]. To check the validity of our spectroscopic protocol we also show the coefficients $|p_{-q}(\alpha)|^2$ for $\alpha = 0, 1, 2$ in Fig. 1 (e) and (f). One can easily see that the relative heights of all the peaks in Fig. 1 (c) and (d) are very well reproduced by the coefficients $|p_{-q}(\alpha)|^2$. As an example, the absence (almost total) of peaks at $q = 0$ for both the sinusoidal and the Lorentzian drive with $\alpha = 2$ is linked to the fact that $|p_0(2)|^2 \ll 1$ for both signals. Moreover, the value of $|p_{-1}(2)|^2$ explains the high asymmetric peak at $q = 1$ for the Lorentzian drive with $\alpha = 2$.

We now turn to the opposite case, in which the dc component is fixed and we allow the parameter $\alpha$ to vary continuously. As shown in Fig. 2 (a) and (b), at $\nu = 1$ we get a linear behavior at high values of $|\alpha|$ both for the sinusoidal and the Lorentzian drive. In the vicinity of $\alpha = 0$, the curves deviate from the linear regime and the noise is more or less proportional to $q$. We note once again the sharp asymmetry for $q \neq 0$ of the Lorentzian voltage drive, as opposed to the symmetric profile of the sinusoidal wave. For fractional filling factor $\nu = 1/3$ [panels (c) and (d)] the curves are evidently non-linear and oscillate in a non-monotonous fashion as a function of $\alpha$. However, the behavior at $\nu = 1/3$ is much more interesting since we can link the value of $\mathcal{S}$ to the probability $|p_{-q}(\alpha)|^2$, following Eq. (13). In contrast with the case of fixed ac component, in this case we can explore the dependence of the $q$-th Fourier coefficients upon its

IOP Conf. Series: Journal of Physics: Conf. Series **969** (2018) 012143 doi:10.1088/1742-6596/969/1/012143

argument $\alpha$. Comparing with Fig. 2 (e) and (f), where we report the coefficients $|p_{-q}(\alpha)|^2$ as a function of $\alpha$, we observe that the approximation works well from a qualitative point of view, although an additional contribution due to finite temperature effects is present in all curves at $\nu = 1/3$, thus preventing us from getting a good quantitative match.

In order to improve the spectroscopic protocol from a quantitative point of view, let us remark that the peculiar peak of the function $\hat{P}_{2\nu}(E) + \hat{P}_{2\nu}(-E)$ around $E = 0$ becomes more and more pronounced as the ratio $\theta/\omega$ decreases. Indeed, it is well known that the Luttinger liquid theory can lead to a diverging power-law behavior in the limit $\theta \to 0$. Thus, the approximation Eq. (13) becomes more efficient at lower temperatures (or higher frequencies), since the relative weight of the term $l = -q$ in the PASN with respect to all other terms $l' \neq -q$ is given by $2\hat{P}_{2\nu}(0)/[\hat{P}_{2\nu}(l'\omega + q\omega) + \hat{P}_{2\nu}(-l'\omega - q\omega)]$. The recent exploration of PASN in the THz regime [28] suggests that our results could be tested in the near future in EQO experiments at fractional filling factor [31].

## 5. Conclusions

We have considered the PASN in a FQH system generated by periodic voltage pulses impinging on a QPC. Due to the typical non-linear behavior of the Green's function in the FQH regime, the photoassisted expression for the noise can be approximated in a remarkably simple way. This allows for a full spectroscopy of the photoassisted probabilities by varying both the dc and the ac amplitude of the voltage pulses. The spectroscopic technique presented in this work is within reach of current experimental technologies.

## References

[1] Levitov L S and Lesovik G B 1993 *JETP Lett.* **58** 230
[2] Levitov L S, Lee H and Lesovik G B 1996 *J. Math. Phys.* **37** 4845
[3] Pedersen M H and Büttiker M 1998 *Phys. Rev. B* **58** 12993
[4] Platero G and Aguado R 2004 *Phys. Rep.* **395** 1
[5] Moskalets M and Büttiker M 2002 *Phys. Rev. B* **66** 205320
[6] Reydellet L H, Roche P, Glattli D C, Etienne B and Jin Y 2003 *Phys. Rev. Lett.* **90** 176803
[7] Fève G, Mahé A, Berroir J M, Kontos T, Plaçais B, Glattli D C, Cavanna A, Etienne B and Jin Y 2007 *Science* **316** 1169
[8] Gabelli J and Reulet B 2013 *Phys. Rev. B* **87** 075403
[9] Dubois J, Jullien T, Portier F, Roche P, Cavanna A, Jin Y, Wegscheider W, Roulleau P and Glattli D C 2013 *Nature (London)* **502** 659
[10] Bocquillon E, Parmentier F D, Grenier C, Berroir J M, Degiovanni P, Glattli D C, Plaçais B, Cavanna A, Jin Y and Fève G 2012 *Phys. Rev. Lett.* **108** 196803
[11] Bocquillon E, Freulon V, Parmentier F D, Berroir J M, Plaçais B, Wahl C, Rech J, Jonckheere T, Martin T, Grenier C, Ferraro D, Degiovanni P and Fève G 2014 *Ann. Phys. (Berlin)* **526** 1
[12] Laughlin R B 1983 *Phys. Rev. Lett.* **50** 1395
[13] Crépieux A, Devillard P and Martin T 2004 *Phys. Rev. B* **69** 205302
[14] Dolcetto G, Vannucci L, Braggio A, Raimondi R and Sassetti M 2014 *Phys. Rev. B* **90** 165401
[15] Rech J, Ferraro D, Jonckheere T, Vannucci L, Sassetti M and Martin T 2017 *Phys. Rev. Lett.* **118** 076801
[16] Vannucci L, Ronetti F, Rech J, Ferraro D, Jonckheere T, Martin T and Sassetti M 2017 *Phys. Rev. B* **95** 245415
[17] Wen X G 1990 *Phys. Rev. B* **41** 12838
[18] Guyon R, Devillard P, Martin T and Safi I 2002 *Phys. Rev. B* **65** 153304
[19] Keeling J, Klich I and Levitov L S 2006 *Phys. Rev. Lett.* **97** 116403
[20] Tien P K and Gordon J P 1963 *Phys. Rev.* **129** 647
[21] Dubois J, Jullien T, Grenier C, Degiovanni P, Roulleau P and Glattli D C 2013 *Phys. Rev. B* **88** 085301
[22] Grenier C, Dubois J, Jullien T, Roulleau P, Glattli D C and Degiovanni P 2013 *Phys. Rev. B* **88** 085302
[23] Ferraro D, Braggio A, Magnoli N and Sassetti M 2010 *Phys. Rev. B* **82** 085323
[24] Ferraro D, Braggio A, Magnoli N and Sassetti M 2010 *New J. Phys.* **12** 013012
[25] Jullien T, Roulleau P, Roche B, Cavanna A, Jin Y and Glattli D C 2014 *Nature (London)* **514** 603
[26] Glattli D C and Roulleau P 2016 *Phys. E* **76** 216
[27] Tsui D C 1999 *Rev. Mod. Phys.* **71** 891
[28] Parmentier F D, Serkovic-Loli L N, Roulleau P and Glattli D C 2016 *Phys. Rev. Lett.* **116** 227401
[29] Carrega M, Ferraro D, Braggio A, Magnoli N and Sassetti M 2012 *New J. Phys.* **14** 023017
[30] Ferraro D, Carrega M, Braggio A and Sassetti M 2014 *New J. Phys.* **16** 043018
[31] Glattli D C and Roulleau P 2017 *Phys. Status Solidi B* **254** 1600650

IOP Conf. Series: Journal of Physics: Conf. Series **969** (2018) 012144 doi:10.1088/1742-6596/969/1/012144

# Heat current through an artificial Kondo impurity beyond linear response

**Miguel A. Sierra and David Sánchez**

IFISC (UIB-CSIC), Campus Universitat Illes Balears, E-07122 Palma de Mallorca, Spain

E-mail: msierra@ifisc.uib-csic.es

**Abstract.** We investigate the heat current of a strongly interacting quantum dot in the presence of a voltage bias in the Kondo regime. Using the slave-boson mean-field theory, we discuss the behavior of the energy flow and the Joule heating. We find that both contributions to the heat current display interesting symmetry properties under reversal of the applied dc bias. We show that the symmetries arise from the behavior of the dot transmission function. Importantly, the transmission probability is a function of both energy and voltage. This allows us to analyze the heat current in the nonlinear regime of transport. We observe that nonlinearities appear already for voltages smaller than the Kondo temperature. Finally, we suggest to use the contact and electric symmetry coefficients as a way to measure pure energy currents.

## 1. Introduction

There is a growing interest in controlling and manipulating heat currents flowing in nanoscale devices [1]. Efficient heat-to-work transformation, low-temperature thermometry, heat rectification and quantum cooling are but a few examples of possible beneficial applications foreseen within the realm of mesoscopic conductors [2]. An equally important motivation is the study of the fundamental mechanisms that govern energy transfer and dissipation beyond a purely (semi)classical description of particle transport.

In this work, we are concerned with heat fluxes driven by voltage biases. Current carriers (electrons) also carry energy and represent the main contribution to heat in quantum electronics at low temperatures. For very small shifts, the generated Peltier heat is reversible, its sign being dependent of the electric current direction. For larger shifts, heat is dominated by Joule power, which is always positive since it is connected to dissipation. The measured heat current then consists of two terms (energy flux and Joule heating) and it is thus natural to analyze their relative importance based on the transmission properties of the system. Here, we consider a single-level quantum dot setup, a prototypical mesoscopic system that allows us to analyze the role of strong electron-electron interactions in the generation of heat far from equilibrium.

The topic of voltage-driven quantum heat currents is interesting [3] and has been examined in a number of different systems [4, 5, 6, 7, 8, 9, 10, 11, 12]. Quantum dot setups are specially appealing due to their experimental tunability and theoretical simplicity. Nonlinear heat current in quantum dots were addressed in Ref. [13, 14]. It was argued that nonequilibrium screening effects lead to charge buildups that affect the heat current–voltage characteristic beyond linear response. The scattering model considered weak interactions and was therefore valid for large dots only. Coulomb blockade effects were then analyzed by us in Refs. [15, 16]. We found that Joule heating quickly surpasses the Peltier contribution as the applied voltage increases. The problem was also investigated recently [17, 18]. We note in passing that the study of heat currents is strongly related to thermoelectrics due to reciprocity. In linear response, the Seebeck and Peltier coefficients are connected through the Kelvin-Onsager relation. Discussions beyond linear response are available in the literature [16, 19, 20, 21, 22, 23]. The role of electron-electron interactions is crucial for the breakdown of reciprocal relations when the driving fields are large [20, 24].

Quantum dots with strong Coulomb interactions act as artificial quantum impurities. In the limit of very low temperatures and strong coupling, the many-body spin interaction between the localized electron in the quantum dot and the delocalized carriers in the attached reservoirs leads to the Kondo effect, which can be detected via electric conductance measurements [25, 26, 27]. It is worth mentioning that studies of the generated heat current in Kondo systems are scarce [28, 29] and focusing only on the linear response regime. Our aim here is to extend our work in Ref. [30] and investigate beyond linear response the dissipated power of a quantum dot in the Fermi liquid fixed point where charge fluctuations are quenched and the Kondo singlet is well formed. This case is relevant for very low temperatures. We find that the symmetry properties of the local density of states with respect to voltage determine the energy flux and the dissipated power. Our results are relevant for the evaluation of heat asymmetries that take place when voltage is reversed or when heat is measured in different contacts, thus providing a better understanding for the nonlinear electrothermal response of a Kondo-correlated system.

## 2. Model Hamiltonian

We consider a Kondo impurity in a quantum dot connected to two leads $\alpha = \{L, R\}$ characterized by their electrochemical potential $\mu_\alpha = \varepsilon_F + eV_\alpha$ and temperature $T_\alpha = T$. In the limit where interactions are infinite (charging energy $U \to \infty$), we consider the Anderson Hamiltonian in the slave-boson formalism [31]. The full Hamiltonian reads

$$\mathcal{H} = \sum_{\alpha k\sigma} \varepsilon_{\alpha k} C^\dagger_{\alpha k\sigma} C_{\alpha k\sigma} + \sum_{\sigma} \varepsilon_d f^\dagger_\sigma f_\sigma + \sum_{\alpha k\sigma} \mathcal{V}_{\alpha k} \left( C^\dagger_{\alpha k\sigma} b^\dagger f_\sigma + \text{H.c.} \right) + \lambda \left( b^\dagger b + \sum_\sigma f^\dagger_\sigma f_\sigma - 1 \right) . \quad (1)$$

Here, the first term in the right-hand side represents the fermionic reservoirs. The operator $C^\dagger_{\alpha k\sigma}$ ($C_{\alpha k\sigma}$) creates (destroys) electrons in the $\alpha$ reservoir with energy $\varepsilon_{\alpha k}$, where $k$ and $\sigma$ are the wavenumber and the spin, respectively. The next term is the dot Hamiltonian. In this case, $f^\dagger_\sigma$ ($f_\sigma$) creates (annihilates) a pseudofermion with energy $\varepsilon_d$ and spin $\sigma$. The tunneling between the dot and the reservoirs is described in the next term. $\mathcal{V}_{\alpha k}$ is the amplitude of electrons hopping on and off the reservoir $\alpha$. Additionally, we add the auxiliary bosonic field $b^\dagger$, which creates an empty state in the dot. Finally, in order to ensure that the quantum dot is occupied with only one electron (we recall that $U \to \infty$) we include a Lagrange multiplier $\lambda$ term in the Hamiltonian $\mathcal{H}$.

Let us consider the mean-field approach to the slave-boson Hamiltonian. This amounts to taking the leading order in a $1/N$ expansion ($N$ being the spin degeneracy). As a consequence, we replace the boson operator by its mean value $\langle b \rangle \equiv \tilde{b}$. This approximation neglects charge fluctuations since they are completely screened out. We remark that this approximation is valid in the Fermi liquid regime where temperatures are lower than the Kondo temperature $T_K$ and dot energy level lies below the Fermi energy $\varepsilon_F$.

Within the nonequilibrium Green's function framework, we derive the mean-field equations for the expectation value of $b$ and the Lagrange multiplier $\lambda$. First, we compute the evolution of the boson operator in the stationary limit, $\sum_{\alpha k\sigma} \tilde{\mathcal{V}}_{\alpha k} G^<_{f\sigma,\alpha k\sigma}(t,t) = -iN\lambda|\tilde{b}|^2/\hbar$, where $\tilde{\mathcal{V}}_{\alpha k} = b_\alpha \mathcal{V}_{\alpha k}$ is the normalized tunneling amplitude through the dot and $G^<_{f\sigma,\alpha k\sigma}(t',t) =$

$-(1/\hbar)\langle C^\dagger_{\alpha k\sigma}(t')f_\sigma(t)\rangle$ is the lesser Green's function for the tunneling process. The next expression corresponds to the single-occupied state condition in the Lagrange term. Its mean-field equation reads $\sum_\sigma G^<_{f\sigma,f\sigma}(t,t) = i(1-N|b|^2)/\hbar$, where $G^<_{f\sigma,f\sigma}(t',t) = -(i/\hbar)\langle f^\dagger_\sigma(t')f_\sigma(t)\rangle$ is the dot pseudofermion lesser Green's function. Both mean-field equations can be combined in a complex self-consistent equation in the Fourier space,

$$\frac{2}{\pi}\int_{-D}^{D} d\omega \frac{\mathcal{F}(\omega)}{\omega-\tilde{\varepsilon}_d+i\tilde{\Gamma}} = (\varepsilon_d-\tilde{\varepsilon}_d)\frac{N}{\Gamma} - i\left(1-N\frac{\tilde{\Gamma}}{\Gamma}\right). \tag{2}$$

The parameters $\tilde{\varepsilon}_d = \varepsilon_d + \lambda$ and $\tilde{\Gamma} = |\tilde{b}|^2\Gamma$ are the renormalized level position and width of the Kondo resonance at the dot. $\Gamma = \Gamma_L + \Gamma_R$ is the hybridization of the energy level due to tunneling, where $\Gamma_\alpha = \pi\rho_\alpha(\omega)|\mathcal{V}_\alpha(\omega)|^2$ depends on the tunneling amplitudes and the density of states $\rho_\alpha(\omega)$ of the lead $\alpha$. In the wide band limit, $\rho$ is constant and nonzero inside the bandwidth $\omega < |D|$. Equation (2) also depends on the nonequilibrium distribution function $\mathcal{F}(\omega) = \sum_\alpha(\Gamma_\alpha/\Gamma)f_\alpha(\omega)$ which is, in this case, a weighted sum of the Fermi-Dirac distribution functions $f_\alpha(\omega) = 1/\{1+\exp[(\omega-\mu_\alpha)/(k_BT_\alpha)]\}$ [32]. In the slave-boson mean-field approach, the Kondo resonance arises when the auxiliary boson condenses and, as result, $\tilde{\Gamma}$ becomes $k_BT_K$ and $\lambda$ shifts $\varepsilon_d$ up to the Fermi level [33].

Once we solve Eq. (2) for both $\lambda$ and $|\tilde{b}|$, the transport properties can be readily analyzed. We focus on the calculation of the heat current through the left reservoir $J \equiv J_L$, which is split into two different contributions: $J = J_E + J_I$. The first term is the energy current flowing through the lead, $J_E = d(\sum_{k\sigma}\varepsilon_{Lk}C^\dagger_{Lk\sigma}C_{Lk\sigma})/dt$. The second term is the Joule heating, $J_I = -I_LV_L$, where $V_L$ is the voltage applied to the left reservoir and $I_L = -ed(\sum_{k\sigma}C^\dagger_{Lk\sigma}C_{Lk\sigma})/dt$ is the charge current given by the evolution of the left lead occupation. Both charge and energy currents are conserved in the steady state whilst the heat fluxes satisfy the relation $\sum_\alpha(J_\alpha + I_\alpha V_\alpha) = 0$.

We find

$$J_E = \frac{2}{h}\int_{-D}^{D} d\omega\, \mathcal{T}(\omega)(\omega-\varepsilon_F)[f_L(\omega)-f_R(\omega)], \tag{3}$$

$$J_I = -\frac{2eV_L}{h}\int_{-D}^{D} d\omega\, \mathcal{T}(\omega)[f_L(\omega)-f_R(\omega)]. \tag{4}$$

Both currents depend on the Fermi functions of the leads and the transmission function, which takes a particularly simple form: $\mathcal{T}(\omega) = 4\tilde{\Gamma}_L\tilde{\Gamma}_R/[(\omega-\tilde{\varepsilon}_d)^2+\tilde{\Gamma}^2]$. It represents a Breit-Wigner lineshape with renormalized parameters ($\tilde{\varepsilon}_d$ and $\tilde{\Gamma}$). Unlike the noninteracting case, however, the parameters of this transmission function depends on the applied voltages and must be calculated from Eq. (2) for each dc bias.

## 3. Results

We investigate the energy and Joule currents given by Eqs. (3) and (4) in response to a symmetrically applied voltage bias ($\mu_L = eV/2$ and $\mu_R = -eV/2$ setting $\varepsilon_F = 0$ as the reference energy). Hereafter, we assume symmetric tunnel couplings $\Gamma_L = \Gamma_R = \Gamma/2$. In Fig 1, we depict $J_E$ and $J_I$ as a function of $V$ for a quantum dot energy level in the Kondo regime ($\varepsilon_d = -2.5\Gamma$). We focus on voltages smaller than the Kondo temperature $k_BT_K = D\exp(-|\varepsilon_d|\pi/\Gamma)$. For $V$ larger than $k_BT_K/e$ the boson field vanishes and the mean-field approach breaks down. For voltages smaller than $k_BT_K/e$ the boson is nonzero and the results are thus reliable.

The energy current in Fig. 1(a) is an increasing function of $V$ since for positive voltages the energy flows from the left lead and this energy increases for higher values of $V$. For $V < 0$ carriers predominantly impinge from the right lead and $J_E$ decreases as $V$ becomes more negative.

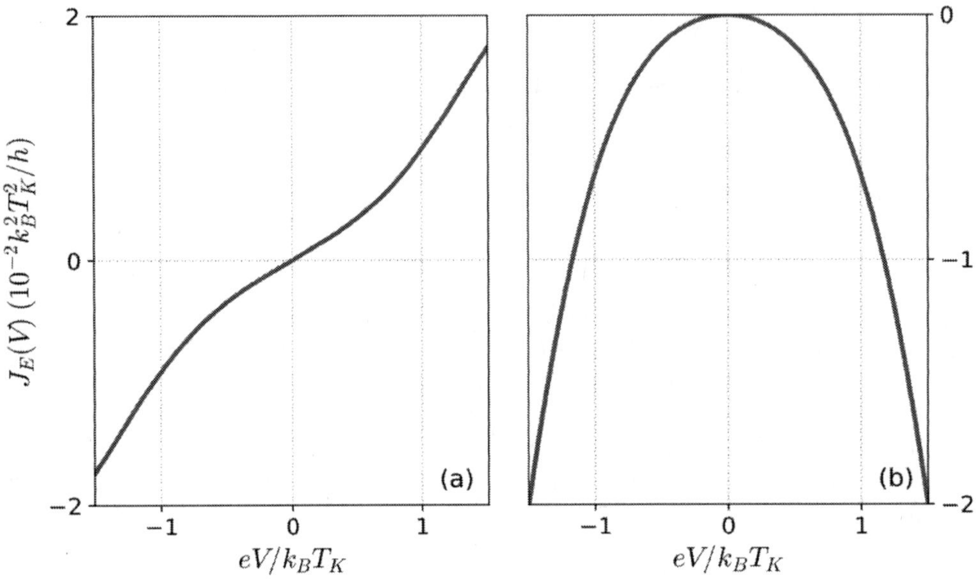


**Figure 1.** (a) Energy current $J_E$ and (b) Joule dissipation current $J_I$ as a function of the applied voltage bias $V$ for a given value of the energy level of the dot ($\varepsilon_d = -2.5\Gamma$). Parameters: temperature $k_BT = 0.01\Gamma$ and bandwidth $D = 100\Gamma$.

Clearly, the energy current shows an antisymmetric shape around $V = 0$. In contrast, the Joule term is always negative (with our sign convention, dissipation is negative) and symmetric under the replacement $V \to -V$. Let us examine in more detail the origin of these symmetry properties. First, we note that the difference of the Fermi functions in Eqs. (3) and (4), $f_L - f_R$, is an odd function of $V$ whereas the transmission function is even $\mathcal{T}(\omega, V) = \mathcal{T}(\omega, -V)$ because the mean-field parameters that characterize the Kondo resonance are also even functions of the applied voltage [see Figs. 2(a) and 2(b)]. These findings are expected since the renormalized width should not depend on the direction of the voltage bias. Further, the position of the Kondo peak lies near the Fermi level and is a function that weakly depends on $V$. As a result of the combination of an odd and an even function, we infer that the energy current is antisymmetric when $V \to -V$. On the other hand, the Joule term $J_I(V)$ is symmetric due to the fact that there is an additional $V$ factor in front of the integral in Eq. (4).

The total heat current is shown in Fig. 2(c). At very low voltages, $J_E(V)$ dominates the heat transport over $J_I(V)$, showing a linear dependence [dashed line in Fig. 2(c)], which is the hallmark of the Peltier effect. Then, we can approximate $J(V) \simeq M_0V$ with $M_0$ the electrothermal conductance [34], directly connected to the thermopower via the Kelvin-Onsager relation. Now, the thermopower is nonzero only for asymmetric density of states. In our case, we find nonzero electrothermal conductances because the Kondo peak is not exactly located at the Fermi energy $\tilde{\varepsilon}_d \simeq \varepsilon_F$ [see Fig. 2(a)] due to the potential scattering term in the quantum dot. This results in a nonsymmetric transmission function $\mathcal{T}(\omega, V) \neq \mathcal{T}(-\omega, V)$. Additionally, we find a positive heat current at positive voltages indicating $M_0 > 0$. Therefore, the Kondo dot at low $V > 0$ might serve as a cooler, although this property is quickly dominated by Joule heating at increasing voltages. As seen in Fig. 2(c), the Joule term overcomes the energy current inducing a negative heat flow. Importantly, $J(V)$ is not symmetric under the transformation $V \to -V$ because $J$ arises from the addition of symmetric and antisymmetric functions.

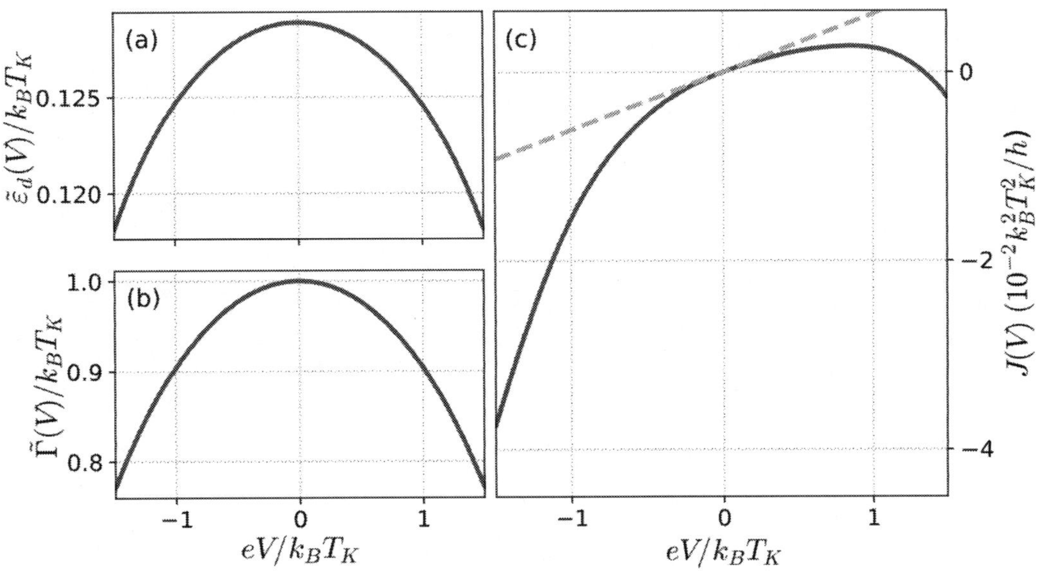


**Figure 2.** (a) Renormalized level position $\tilde{\varepsilon}_d$ and (b) renormalized width $\tilde{\Gamma}$ of the Kondo resonance as a function of the voltage bias $V$ for values lower than the Kondo temperature. (c) Heat current $J$ as a function of the voltage bias $V$. The dashed green line exhibits the Peltier contribution to the heat current. The parameters are the same as in Fig. 1.

## 4. Discussion

The asymmetries shown in the calculated heat current can be analyzed in terms of contact and electric asymmetries, as suggested in Ref. [35] for molecular tunnel junctions. The contact asymmetry $\Delta_C = J(V) - J(-V)$ measures the dissipated power under the reversal of the dc bias. The electric asymmetry $\Delta_E = J_L(V) - J_R(V)$ quantifies the different heat dissipation in both terminals as a function of $V$. Using the even and odd properties of $J_E(V)$ and $J_I(V)$ [Eqs. (3) and (4)], we straightforwardly arrive at the relation $\Delta_C(V) = 2J_E(V)$. This implies that the behavior of the contact asymmetry is in fact given by the results in Fig. 1a. Remarkably, we find that the electric asymmetry is also proportional to the energy current $\Delta_E(V) = 2J_E(V)$. This result is general [36] in the case of both a symmetric voltage bias $V_L = -V_R = V/2$ and a symmetric transmission function. Any quantum conductor that satisfies these conditions will show similar contact and electric asymmetries. Here, we find that these symmetry conditions are satisfied for an artifitial Kondo impurity due to the symmetric renormalization of the mean-field parameters. Therefore, the Kondo effect does not break these symmetries at least in the Fermi liquid regime. As a byproduct, our result would facilitate the experimental detection of the energy current from a measurement of either the heat current for different voltage directions or the heat current at different reservoirs for the same dc bias.

## 5. Conclusions

To sum up, we have examined the heat flow through an artificial Kondo impurity connected to two leads with a symmetric applied voltage. Within the mean-field slave-boson formalism, we have obtained the expression for the transmission and have computed both the energy current and the Joule dissipation current. We find an antisymmetric energy current and a symmetric Joule term as a function of the dc bias. This behavior can be understood from the symmetry properties of the transmission function which arises from the symmetry of the renormalized width and energy position of the Kondo resonance. Additionally, we show that at voltages smaller than the Kondo temperature the Peltier effect first dominates the heat current while Joule heating governs the heat current at higher biases. Then, the crossover from linear to quadratic behavior is controlled by the Kondo temperature. Our results could be tested with present techniques since there is a variety of available methods that measure heat currents in quantum conductors [35, 37, 38, 39, 40, 41].

We acknowledge support from MINECO under grant No. FIS2014-52564 and a PhD grant from CAIB.

## References

[1] Benenti G, Casati G, Saito K, and Whitney R S 2017 *Phys. Rep.* **694** 1
[2] Giazotto F, Heikkilä T T, Luukanen A , Savin A M and Pekola J P 2006 *Rev. Mod. Phys.* **78** 217
[3] Sánchez D and López R 2016 *C. R. Physique* **17** 1060
[4] Kulik I O 1994 *J. Phys.: Condens. Matter* **6** 9737
[5] Bogachek E N, Scherbakov A G, and Landman U 1999 *Phys. Rev. B* **60** 11678
[6] Çipiloğlu M A, Turgut S, and Tomak M 2004 *Phys. Stat. Sol. (b)* **241** 2575
[7] Freericks J K and Zlatic V, *Condensed Matter Physics* **9** 603
[8] Zeberjadi M, Esfarjani K, and Shakouri A 2007 *Appl. Phys. Lett.* **91** 122104
[9] Leijnse M, Wegewijs M R, and Flensberg K 2010 *Phys. Rev. B* **82** 045412
[10] Whitney R S 2013 *Phys. Rev. B* **88** 064302
[11] Jiang J H, Kulkarni M, Segal D, and Imry Y 2015 *Phys. Rev. B* **92** 045309
[12] Zimbovskaya N 2016 *J. Phys.: Condens. Matter* **28** 183002
[13] López R and Sánchez D 2013 *Phys. Rev. B* **88** 045129
[14] Meair J and Jacquod P 2013 *J. Phys.: Condens. Matter* **25** 082201
[15] Sierra M A and Sánchez D 2014 *Phys. Rev. B* **90** 115313
[16] Sierra M A and Sánchez D 2015 *Materials Today: Proceedings* **2** 483
[17] Gergs N M, Hörig C B M, Wegewijs M R, and Schuricht D 2015 *Phys. Rev. B* **91** 201107(R)
[18] Yamamoto K and Hatano N 2015 *Phys. Rev. E* **92** 042165
[19] Iyoda E, Utsumi Y, and Kato T 2010 *J. Phys. Soc. Jpn.* **79** 045003
[20] Hwang S Y, Sánchez D, Lee M, and López R 2013 *New J. Phys.* **15** 105012
[21] Bedkihal S, Bandyopadhyay M, and Segal D 2013 *Eur. Phys. J. B* **86** 506
[22] Cimmelli V A, Sellitto A, and Jou D 2014 *Proc. Royal Soc. A* **470** 0265
[23] Sellitto A 2014 *Physica D* **283** 56
[24] Matthews J, Battista F, Sánchez D, Samuelsson P, and Linke H 2014 *Phys. Rev. B* **90** 165428
[25] Goldhaber-Gordon D *et al.* 1998 *Nature* **391** 156
[26] Cronenwett S M, Oosterkamp T H, and Kouwenhoven L P 1998 *Science* **281** 540
[27] Schmid J, Weis J, Eberl K, and Klitzing K v 1998 *Physica B* **256** 182
[28] Boese D and Fazio R 2001 *Europhys. Lett.* **56** 576
[29] Saito K and Kato T 2013 *Phys. Rev. Lett.* **111** 214301
[30] Sierra M A, López R and Sánchez D 2017 *Phys. Rev. B* **96** 085416
[31] Hewson A C 1997 *The Kondo Problem to Heavy Fermions* (Cambridge University Press)
[32] Balseiro C A, Usaj G, and Sánchez M J 2010 *J. Phys.: Condens. Matter* **22** 425602
[33] Coleman P 1984 *Phys. Rev. B* **29** 3035
[34] Butcher P N 1990 *J. Phys.: Condens. Matter* **2** 4869
[35] Lee W, Kim K, Jeong W, Zotti L A, Pauly F, Cuevas J C, and Reddy P 2013 *Nature* **498** 209
[36] Argüello-Luengo J, Sánchez D, and López R 2015 *Phys. Rev. B* **91** 165431
[37] Molenkamp L W, Gravier T, van Houten H, Buijk O J A, Mabesoone M A A, and Foxon C T 1992 *Phys. Rev. Lett.* **68** 3765
[38] Chiatti O, Nicholls J T, Proskuryakov Y Y, Lumpkin N, Farrer I, and Ritchie D A 2006 *Phys. Rev. Lett.* **97** 056601
[39] Meschke M, Guichard W, and Pekola J P 2009 *Nature* **444** 187
[40] Jezouin S, Parmentier F D, Anthore A, Gennser U, Cavanna A, Jin Y, and Pierre F 2013 *Science* **342** 601
[41] Cui L, Jeong W, Hur S, Matt M, Klöckner J C, Pauly F, Nielaba P, Cuevas J C, Meyhofer E, and Reddy P 2017 *Science* **355** 1192

# Equilibrium finite-frequency noise of an interacting mesoscopic capacitor studied in time-dependent density functional theory

**Niklas Dittmann[1,2,3], Janine Splettstoesser[1] and Nicole Helbig[3]**

[1] Department of Microtechnology and Nanoscience (MC2), Chalmers University of Technology, SE-41258 Göteborg, Sweden

[2] Institute for Theory of Statistical Physics, RWTH Aachen, D-52056 Aachen, Germany

[3] Peter-Grünberg Institut and Institute for Advanced Simulation, Forschungszentrum Jülich, D-52425 Jülich, Germany

E-mail: dittmann@physik.rwth-aachen.de

**Abstract.** We calculate the frequency-dependent equilibrium noise of a mesoscopic capacitor in time-dependent density functional theory (TDDFT). The capacitor is modeled as a single-level quantum dot with on-site Coulomb interaction and tunnel coupling to a nearby reservoir. The noise spectra are derived from linear-response conductances via the fluctuation-dissipation theorem. Thereby, we analyze the performance of a recently derived exchange-correlation potential with time-nonlocal density dependence in the finite-frequency linear-response regime. We compare our TDDFT noise spectra with real-time perturbation theory and find excellent agreement for noise frequencies below the reservoir temperature.

## 1. Introduction

The presence of thermal and quantum fluctuations in mesoscopic conductors results in noise, even if the mean current through a sample vanishes [1, 2]. The noise spectrum gives insight into temporal correlations between fluctuation events. Measured in equilibrium[1], it reveals dissipative properties of a system, i. e. the real part of the finite-frequency linear-response conductance, by means of the fluctuation-dissipation theorem [5].

In this paper, we present the calculation of equilibrium finite-frequency noise spectra in time-dependent density functional theory (TDDFT) [6–8]. In TDDFT, the dynamics of an interacting quantum system is simulated by a non-interacting Kohn-Sham (KS) system [9], which leads to cost-efficient numerics even for complex systems. The KS potential, which consists of the external potential of the interacting system, modified by the Hartree (H) and the exchange-correlation (XC) contributions, is chosen such that the KS and the interacting systems share the same time-dependent electronic density. The H potential accounts for electrostatics, while all further many-body effects are included in the universal XC potential. The development and characterization of approximations for the latter is among the central challenges for TDDFT, and also a focus of this work.

[1] For systems which are out of equilibrium, the noise includes information which is not accessible by conductance measurements alone, see e. g. the experiments reported in Refs. [3, 4].

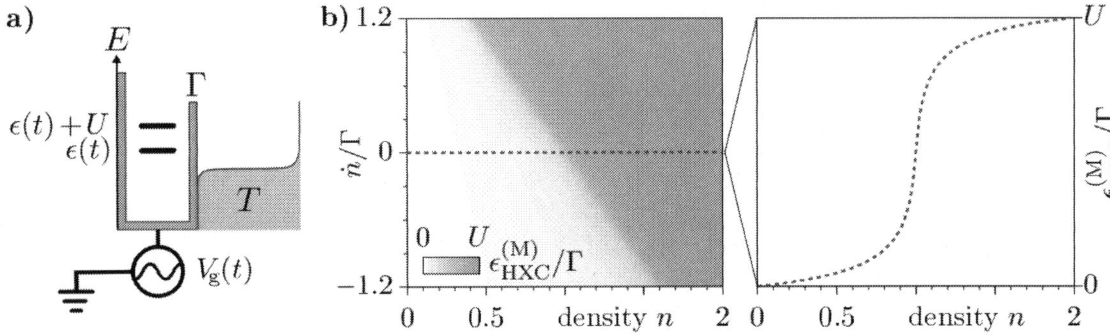


**Figure 1.** (a) Energy diagram of a single-level quantum dot with tunnel coupling, $\Gamma$, to a reservoir; (b) non-adiabatic HXC potential, $\epsilon_{\rm HXC}^{\rm (M)}$, see Ref. [14], for $U = 16\Gamma$, $T = 2\Gamma$; dotted line marks the line cut shown on the r. h. s.

As an instructive system with experimental relevance [10], we analyze an interacting mesoscopic capacitor. For this system, the finite-frequency noise has been studied theoretically, e. g. [11–13], and in experiment [3]. We model the interacting mesoscopic capacitor by a single-level quantum dot with on-site Coulomb interaction and tunnel coupling to an adjacent reservoir, see Fig. 1 (a). Based on this model, we recently developed a non-adiabatic, i. e. time nonlocal, XC potential for TDDFT simulations of single-electron tunneling devices [14]. The XC potential was derived by expressing the dynamics of both the system, and its KS counterpart, in terms of time-dependent Markovian[2] Master equations. It thereby includes information on the dynamics of electron tunneling. In Ref. [14], we applied the non-adiabatic XC potential to describe the operation mode of a mesoscopic capacitor as a single-electron source, i. e. its non-linear response to a time-periodic gate voltage. A key finding was that a related *adiabatic* XC potential leads to electron dynamics on time scales which correspond to non-interacting electrons, while our *non-adiabatic* XC potential describes electron dynamics on the respective time scales for interacting electrons. Remarkably, our non-adiabatic XC potential thereby only depends on the time-local density of the quantum dot and its first time derivative.

In the present paper, we add an analysis of finite-frequency linear-response physics to our previous discussion. For this purpose, we derive finite-frequency conductances of the system at hand with TDDFT. We then consider the real parts of the obtained conductances, which we express as equilibrium noise by means of the fluctuation-dissipation theorem. To test the performance of our non-adiabatic XC potential over a broad frequency range, we compare our results with noise spectra derived with perturbation theory.

The paper is structured as follows: We define our model for the interacting mesoscopic capacitor in Sec. 2 and its KS counterpart in Sec. 3, where we also review the non-adiabatic XC potential from Ref. [14]. The calculation of equilibrium noise in TDDFT is described in Sec. 4. In Sec. 5 we present TDDFT noise spectra, before we conclude our work in Sec. 6.

## 2. The interacting mesoscopic capacitor

We model the mesoscopic capacitor by a quantum dot with a single energy level, which is tunnel coupled to a nearby reservoir, see Fig. 1 (a). Therefore, the dot can be either empty, singly or doubly occupied. The energy level, $\epsilon(t) = -\alpha V_{\rm g}(t)$, is assumed to depend linearly on the gate voltage, $V_{\rm g}(t)$, with $\alpha > 0$. All energies are defined with respect to the Fermi energy and $e, \hbar$ and $k_{\rm B}$ are set to one. We consider strong Coulomb repulsion between electrons occupying the

[2] The Markov approximation neglects memory in the reservoir. The latter becomes relevant for dynamics on time scales smaller than the reservoir memory time, which is given by the inverse of the reservoir temperature $T$.

IOP Conf. Series: Journal of Physics: Conf. Series **969** (2018) 012145 doi:10.1088/1742-6596/969/1/012145

quantum dot, by taking into account a charging energy $U$ in the case of double occupation. In addition, we assume the electrons in the reservoir to be non-interacting, set the electrochemical potential to zero, and denote the reservoir temperature by $T$. The resulting Hamiltonian reads

$$H = \sum_{\sigma} \epsilon(t) d_\sigma^\dagger d_\sigma + U d_\uparrow^\dagger d_\uparrow d_\downarrow^\dagger d_\downarrow + \sum_{k,\sigma} \epsilon_k c_{k\sigma}^\dagger c_{k\sigma} + \left( \gamma \sum_{k,\sigma} c_{k\sigma} d_\sigma^\dagger + \text{h. c.} \right), \tag{1}$$

where $d_\sigma^{(\dagger)}$ denote the annihilation (creation) operators for quantum-dot states with spin index $\sigma = \uparrow, \downarrow$ and $c_{k\sigma}^{(\dagger)}$ denote the annihilation (creation) operators for reservoir states with spin $\sigma$ and momentum $k$. We assume spin-degeneracy and consider a single energy band, $\epsilon_k$, and an energy-independent tunnel coupling, $\gamma$. Besides that, we define $\Gamma = 2\pi|\gamma|^2\nu_0$ as the tunnel-coupling strength, where $\nu_0$ denotes the density-of-states (DOS) at the Fermi energy. The density on the quantum dot is given by $n(t) = \sum_\sigma \left\langle d_\sigma^\dagger d_\sigma \right\rangle (t)$, and the current into the quantum dot reads $I(t) = -\dot{n}(t)$ due to charge conservation.

## 3. Representation by a non-interacting system

We now describe the interacting mesoscopic capacitor defined in Eq. (1) by a non-interacting KS system, assuming non-interacting $v$-representability [6–8]. The H and XC potentials are taken into account via a combined shift of the quantum-dot's energy level by the amount $\epsilon_{\text{HXC}}[n](t)$, which we call the HXC potential of our model system. By furthermore setting the on-site interaction in Eq. (1) to zero, we obtain the KS Hamiltonian,

$$H_{\text{KS}} = \sum_{\sigma} \epsilon_{\text{KS}}[n](t) d_\sigma^\dagger d_\sigma + \sum_{k,\sigma} \epsilon_k c_{k\sigma}^\dagger c_{k\sigma} + \left( \gamma \sum_{k,\sigma} c_{k\sigma} d_\sigma^\dagger + \text{h. c.} \right), \tag{2}$$

with the KS energy level, $\epsilon_{\text{KS}}[n](t) = \epsilon(t) + \epsilon_{\text{HXC}}[n](t)$, and where a general functional dependence on the density is denoted by square brackets, $[n]$. For the HXC potential in Eq. (2), we analyze the non-adiabatic approximation derived in Ref. [14], $\epsilon_{\text{HXC}}^{(\text{M})}(n(t), \dot{n}(t))(t)$, and its adiabatic counterpart, $\epsilon_{\text{HXC}}^{(\text{A})}(n(t))(t)$, see also Ref. [15]. The two approximations read

$$\epsilon_{\text{HXC}}^{(\text{M})}(n(t), \dot{n}(t))(t) = \frac{1}{\beta} \log\left(C(n(t), \dot{n}(t))\right) \tag{3}$$

and $\epsilon_{\text{HXC}}^{(\text{A})}(n(t))(t) = \epsilon_{\text{HXC}}^{(\text{M})}(n(t), 0)(t)$, with the inverse temperature, $\beta = 1/T$ and

$$\frac{1}{C(n,\dot{n})} = \frac{\dot{n} + e^{U\beta}\left(\dot{n} + 2\Gamma(n-1)\right)}{2e^{U\beta}(\dot{n} + \Gamma(n-2))} \left( 1 - \left( 1 - \frac{4e^{U\beta}\left((\dot{n} + \Gamma n)^2 - 2\Gamma(\dot{n} + \Gamma n)\right)}{\left(\dot{n} + e^{U\beta}\left(\dot{n} + 2\Gamma(n-1)\right)\right)^2} \right)^{\frac{1}{2}} \right). \tag{4}$$

The adiabatic HXC potential is presented in the line cut in Fig. 1 (b), where we observe a smeared-out step in the potential at single occupation of the quantum dot [15, 16]. This step is known in TDDFT and DFT as the derivative discontinuity [17], which is, e. g., relevant for a proper description of Coulomb-blockade physics in a non-interacting Kohn-Sham system [16, 18]. The density plot in Fig. 1 (b), showing the non-adiabatic HXC potential, reveals that a non-zero time derivative of the density shifts this step to a different position. The physical interpretation of this shift is explained in detail in Ref. [14]: its magnitude is linked to the difference of relaxation times of the interacting and the non-interacting system. The accomplishment of this dynamical step is an improved description of the relaxation dynamics of the *interacting* mesoscopic capacitor by the *non-interacting* KS system. This is not achieved by considering the adiabatic HXC potential only, where relaxation occurs on a time scale which is characteristic for a non-interacting mesoscopic capacitor [14].

## 4. Equilibrium finite-frequency noise in TDDFT

Within TDDFT, we calculate the equilibrium finite-frequency noise, $S(\omega)$, of the interacting mesoscopic capacitor, which we define as the Fourier transform of the symmetric current-current fluctuations, $S(t-t') = \langle I(t)I(t') + I(t')I(t)\rangle$. In this work, we extract the noise spectrum from the linear-response conductance $G(\omega) = \frac{\partial I(\omega)}{\partial \epsilon(\omega)}\Big|_{\text{eq}}$, where the subscript denotes evaluation at the equilibrium density, $n_{\text{eq}}$. This is possible due to the fluctuation-dissipation theorem,

$$S(\omega) = 2\omega \coth\left(\frac{\beta\omega}{2}\right) \operatorname{Re} G(\omega), \tag{5}$$

which explicitly links the dissipative part of the conductance to the equilibrium noise [1]. To obtain $G(\omega)$ in TDDFT, we first calculate the conductance in the KS system, which is defined as $G_{\text{KS}}(\omega) = \frac{\partial I(\omega)}{\partial \epsilon_{\text{KS}}[n](\omega)}\Big|_{\text{eq}}$. The latter depends on the position of the KS energy level, $\epsilon_{\text{KS}}[n_{\text{eq}}]$, and is derived exactly, because the KS system is non-interacting, see e. g. the result in Ref. [19]. The density $n_{\text{eq}}$ is calculated self-consistently from the exact KS expression for the quantum-dot's equilibrium density, $n_{\text{eq}} = \frac{\Gamma}{\pi}\int dq \frac{d_q}{(\epsilon_{\text{KS}}[n_{\text{eq}}]-q)^2+\Gamma^2/4}$, with the Fermi function, $d_q = 1/(1+e^{\beta q})$. In TDDFT linear-response theory, the KS conductance is related to the conductance of the interacting system by a Dyson equation. This leads to

$$G(\omega) = \frac{\omega G_{\text{KS}}(\omega)}{\omega + i f_{\text{HXC}}(n_{\text{eq}}, \omega) G_{\text{KS}}(\omega)}, \tag{6}$$

with the frequency-dependent HXC kernel, $f_{\text{HXC}}(n_{\text{eq}}, \omega)$, defined as the Fourier transform of $f_{\text{HXC}}(n_{\text{eq}}, t-t') = \frac{\delta \epsilon_{\text{HXC}}[n](t)}{\delta n(t')}\Big|_{\text{eq}}$, see e. g. [6–8]. The HXC kernels related to our specific approximations in Eqs. (3) read

$$f^{(\text{M})}_{\text{HXC}}(n_{\text{eq}}, \omega) = f^{(0)}_{\text{HXC}}(n_{\text{eq}}) - i\omega\tau_{\text{ch}} f^{(1)}_{\text{HXC}}(n_{\text{eq}}) \tag{7}$$

and $f^{(\text{A})}_{\text{HXC}}(n_{\text{eq}}) = f^{(\text{M})}_{\text{HXC}}(n_{\text{eq}}, 0)$, with $f^{(0)}_{\text{HXC}}(n_{\text{eq}}) = \frac{\partial \epsilon^{(\text{M})}_{\text{HXC}}(n,\dot{n})}{\partial n}\Big|_{\text{eq}}$ and $f^{(1)}_{\text{HXC}}(n_{\text{eq}}) = \tau_{\text{ch}}^{-1} \frac{\partial \epsilon^{(\text{M})}_{\text{HXC}}(n,\dot{n})}{\partial \dot{n}}\Big|_{\text{eq}}$. Here, we introduce a characteristic time scale, $\tau_{\text{ch}}$, which for our approximations is convenient to define as $\tau_{\text{ch}} = \frac{1}{\Gamma} - \frac{1}{\Gamma_c}$, with the charge relaxation rate at the electron-hole symmetric point,

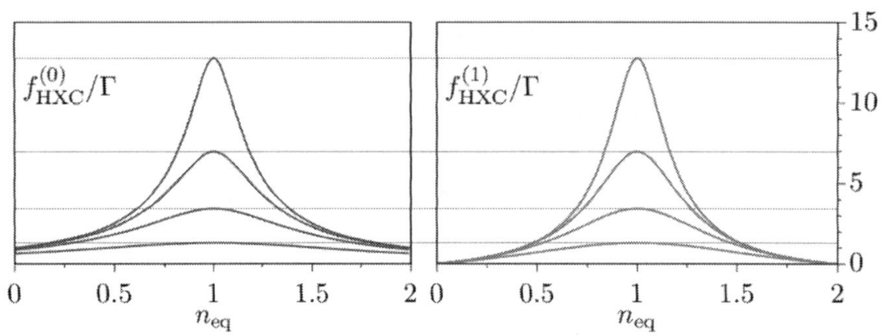


**Figure 2.** Coefficients of the HXC kernels in Eq. (7) for on-site interaction $U = 2\Gamma, 4\Gamma, 6\Gamma, 8\Gamma$ (lower to upper lines) and $T = 2\Gamma$. Thanks to the characteristic time scale $\tau_{\text{ch}}$ (defined in the main text), the peak heights are identical for both coefficients, while differences appear for $n_{\text{eq}} \to 0$ and $n_{\text{eq}} \to 2$.

$\Gamma_c = \Gamma \cdot \left(1 + d_{-U/2} - d_{U/2}\right)$. The scale $\tau_{\mathrm{ch}}$, which also quantifies the dynamical step visible in Fig. 1 (b) [14], is thus given by the difference between the charge relaxation times of the non-interacting and the interacting system. Fig. 2 shows that both real coefficients, which appear on the r. h. s. in Eq. (7), develop a peak at single occupation for increasing on-site interaction $U$. Therefore, the frequency dependence of $f^{(\mathrm{M})}_{\mathrm{HXC}}$ adds a linear and imaginary contribution to the peak of the frequency-independent HXC kernel, $f^{(\mathrm{A})}_{\mathrm{HXC}}$.

## 5. TDDFT noise spectra of the interacting mesoscopic capacitor

Equilibrium noise spectra of the interacting mesoscopic capacitor, using Eq. (5), are plotted in Fig. 3 for two different temperatures (left and right). As a reference, the black solid lines present the result of an analytic calculation based on perturbation theory, see e. g. Refs. [20, 21]. This method applies an expansion in the tunnel coupling. For our system, the noise derived with perturbation theory features smeared-out steps at noise frequencies which equal the absolute value of $\epsilon$ or $\epsilon + U$. Besides that, the noise tends to zero in the zero-frequency limit. The single step for the parameters in Fig. 3 is readily visible in the left plot by the black solid line, and strongly smeared-out in the right plot due to the increased temperature. In both plots in Fig. 3, three TDDFT curves are shown, which respectively correspond to the noise in the KS system, obtained from $G_{\mathrm{KS}}(\omega)$, and the noise calculated from the corrected conductances, see Eq. (6), using the two HXC kernels defined in Eq. (7). For both temperatures, the noise calculated in the KS system (dashed-dotted lines) deviates strongly from the reference noise, as it is expected for the unphysical KS system. For the chosen parameters, the KS energy level is positioned at the Fermi energy, and the smeared-out step in the noise spectrum is thus removed in the KS system. The dotted lines show noise spectra, which are derived from conductances, corrected with the HXC kernel of the *adiabatic* HXC potential. Here, the discrepancy with the perturbative result is reduced for very low frequencies. From Eq. (7) and Fig. 2 we estimate, that the frequency independent HXC kernel, $f^{(\mathrm{A})}_{\mathrm{HXC}}$, is reasonable for $\omega \ll 1/\tau_{\mathrm{ch}}$, i. e. where the frequency-dependent part of $f^{(\mathrm{M})}_{\mathrm{HXC}}$ is small. On the contrary, the kernel of the *non-adiabatic* HXC potential significantly improves the TDDFT result in the low and medium frequency regime (dashed lines). The agreement with the reference noise is excellent for frequencies $\omega \lesssim T$, i. e. for thermal noise, as it is visible in both plots in Fig. 3. We emphasize that already the linear frequency dependence of the associated HXC kernel, $f^{(\mathrm{M})}_{\mathrm{HXC}}$, is sufficient for the observed

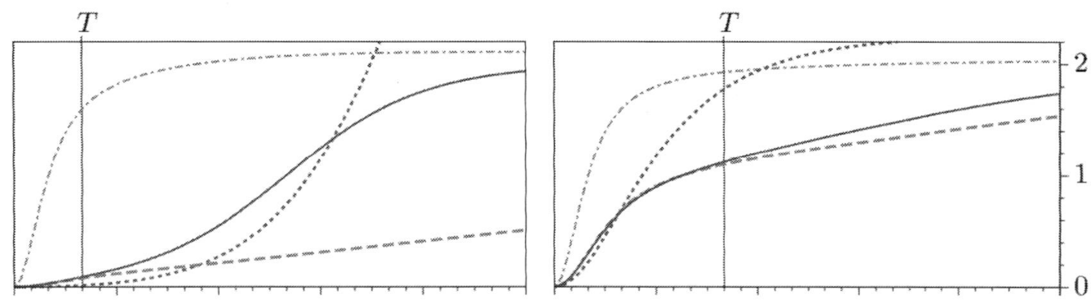


**Figure 3.** Equilibrium noise spectra for two temperatures, $T = 2\Gamma$ (left) and $T = 5\Gamma$ (right); Shown are the uncorrected noise spectra of the KS systems (purple dashed-dotted lines), and the corrected noise spectra (Eq. (6)), using the HXC kernels $f^{(\mathrm{A})}_{\mathrm{HXC}}$ (blue dotted lines) and $f^{(\mathrm{M})}_{\mathrm{HXC}}$ (red dashed lines). Further parameters are $\epsilon = -8\Gamma, U = 16\Gamma$. The black solid lines are shown for reference and present results obtained with perturbation theory [20, 21].

IOP Conf. Series: Journal of Physics: Conf. Series **969** (2018) 012145 doi:10.1088/1742-6596/969/1/012145

improvement. The failure for frequencies $\omega > T$ has its root in the derivation of the related HXC potential, which was based on a Markov approximation [14].

## 6. Conclusion

We calculated noise spectra of an interacting mesoscopic capacitor in TDDFT using a recently developed non-adiabatic approximation for the HXC potential [14]. Thereby, we extended the discussion of our previous work to linear-response physics and found an HXC kernel with linear frequency dependence. The derived TDDFT noise agrees well with reference data from perturbation theory for thermal noise frequencies, $\omega \lesssim T$.

## Acknowledgments

Support from the Deutsche Forschungsgemeinschaft via RTG1995 (ND, JS) and an Emmy-Noether grant (NH), as well as from the Knut and Alice Wallenberg foundation and the Swedish VR (JS), is gratefully acknowledged.

## References

[1] Blanter Y and Büttiker M 2000 *Phys. Rep.* **336** 1–166
[2] Clerk A A, Devoret M H, Girvin S M, Marquardt F and Schoelkopf R J 2010 *Rev. Mod. Phys.* **82**(2) 1155–1208
[3] Parmentier F D, Bocquillon E, Berroir J M, Glattli D C, Plaçais B, Fève G, Albert M, Flindt C and Büttiker M 2012 *Phys. Rev. B* **85**(16) 165438
[4] Ubbelohde N, Fricke C, Flindt C, Hohls F and Haug R J 2012 *Nat. Comm.* **3** 612
[5] Callen H B and Welton T A 1951 *Phys. Rev.* **83**(1) 34–40
[6] Miguel AL Marques, Neepa T Maitra, Fernando MS Nogueira, EKU Gross, Angel Rubio 2012 *Fundamentals of Time-Dependent Density Functional Theory* vol 827 (Springer Berlin Heidelberg)
[7] Carsten A Ullrich 2011 *Time-Dependent Density-Functional Theory: Concepts and Applications* (Oxford University Press)
[8] Maitra N T 2016 *J. Chem. Phys.* **144** 220901
[9] Runge E and Gross E K U 1984 *Phys. Rev. Lett.* **52**(12) 997–1000
[10] Fève G, Mahé A, Berroir J M, Kontos T, Plaçais B, Glattli D C, Cavanna A, Etienne B and Jin Y 2007 *Science* **316** 1169–1172
[11] Büttiker M 1992 *Phys. Rev. B* **46**(19) 12485–12507
[12] Rothstein E A, Entin-Wohlman O and Aharony A 2009 *Phys. Rev. B* **79**(7) 075307
[13] Gabdank N, Rothstein E A, Entin-Wohlman O and Aharony A 2011 *Phys. Rev. B* **84**(23) 235435
[14] Dittmann N, Splettstoesser J and Helbig N 2017 *Preprint* arxiv:1706.04547
[15] Stefanucci G and Kurth S 2011 *Phys. Rev. Lett.* **107**(21) 216401
[16] Evers F and Schmitteckert P 2011 *Phys. Chem. Chem. Phys.* **13**(32) 14417–14420
[17] Perdew J P, Parr R G, Levy M and Balduz J L 1982 *Phys. Rev. Lett.* **49**(23) 1691–1694
[18] Kurth S, Stefanucci G, Khosravi E, Verdozzi C and Gross E K U 2010 *Phys. Rev. Lett.* **104**(23) 236801
[19] Jauho A P, Wingreen N S and Meir Y 1994 *Phys. Rev. B* **50**(8) 5528–5544
[20] Braun M, König J and Martinek J 2006 *Phys. Rev. B* **74**(7) 075328
[21] Droste S, Splettstoesser J and Governale M 2015 *Phys. Rev. B* **91**(12) 125401

IOP Conf. Series: Journal of Physics: Conf. Series **969** (2018) 012146 doi:10.1088/1742-6596/969/1/012146

# Microwave photon generation in a doubly tunable superconducting resonator

**I-M. Svensson[1], M. Pierre, M. Simoen, W. Wustmann, P. Krantz, A. Bengtsson, G. Johansson, J. Bylander, V. Shumeiko and P. Delsing**

Microtechnology and Nanoscience, Chalmers University of Technology, Göteborg, Sweden

E-mail: [1]`ida-maria.svensson@chalmers.se`

**Abstract.** We have created a doubly tunable resonator, with the intention to simulate relativistic motion of the resonator boundaries in real space. Our device is a superconducting coplanar-waveguide microwave resonator, with fundamental resonant frequency $\omega_1/(2\pi) \sim 5\,\text{GHz}$. Both of its ends are terminated to ground via dc-SQUIDs, which serve as magnetic-flux-controlled inductances. Applying a flux to either SQUID allows the tuning of $\omega_1/(2\pi)$ by approximately 700 MHz. Using two separate on-chip magnetic-flux lines, we modulate the SQUIDs with two tones of equal frequency, close to $2\omega_1$. We observe photon generation, at $\omega_1$, above a certain pump amplitude threshold. By varying the relative phase of the two pumps we are able to control this threshold, in good agreement with a theoretical model. At the same time, some of our observations deviate from the theoretical predictions, which we attribute to parasitic couplings resulting in current driving of the SQUIDs.

## 1. Introduction

Vacuum is commonly considered to be empty space. However, in quantum theory, it contains vacuum fluctuations of the electromagnetic field. Due to these fluctuations, two perfectly conducting mirrors at rest, placed in close vicinity of each other, can exhibit radiation pressure forces, known as the Casimir effect [1]. Furthermore, if the mirrors are moved with a speed close to the speed of light, real photons can be generated as excitations of the vacuum fluctuations, a phenomenon called the dynamical Casimir effect (DCE) [2]. In fact, photon generation through the DCE requires only one rapidly moving mirror to produce photons [3, 4].

Using superconducting circuits, the physical conditions equivalent to a mirror moving at about 1/4 of the speed of light can be created [5]. This is done by placing a superconducting quantum interference device (SQUID) at the end of a transmission line. The SQUID acts as a tunable inductance, $L_J(\Phi_{ext}, I_s) = \Phi_0/\left(2\pi|\cos(\Phi_{ext}\pi/\Phi_0)|\sqrt{I_c^2 - I_s^2}\right)$, where $\Phi_0$ is the magnetic flux quantum, $\Phi_{ext} = \Phi_{dc} + \Phi_{ac}(t)$ is the applied external magnetic flux, $I_c$ is the SQUID's critical current, and $I_s$ the current through the SQUID. The SQUID inductance can be modulated either by flux pumping, through $\Phi_{ac}$, which is a direct modulation of the resonator boundary condition and the analogue of a moving mirror, or by ac driving the SQUID current $I_s$. The generation of DCE photons using a flux-pumped SQUID at the end of a transmission line was suggested in Ref. [6] and demonstrated in Ref. [7].

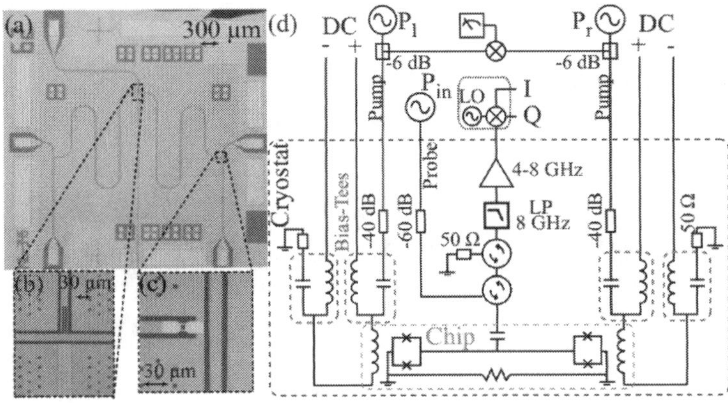


**Figure 1.** (a) Micrograph of the doubly tunable resonator chip. (b) Resonator middle with a gold-filled slot (below) and a coupling capacitor and probe (above). (c) Resonator end with the SQUID and on-chip flux line. (d) Schematic of the measurement setup. It is a reflection setup with circulators to probe the resonant frequency and also to measure the output when the resonator is pumped through the on-chip flux lines.

If a SQUID is included in a resonator and flux-modulated around twice the resonant frequency, the system is the equivalent of a parametric oscillator (PO) [8–10], *i.e.* a harmonic oscillator driven by the modulation of a system parameter, here the resonant frequency. The PO has a flux-pump amplitude threshold, determined by the system damping,above which self-sustained oscillations are generated [9]. Below threshold, the system can be operated as a parametric amplifier in which small input signals near its resonant frequency are amplified [11–14].

In this paper we use a superconducting coplanar waveguide $\lambda/2$ resonator, with each end grounded via a SQUID. If driven separately, both SQUIDs can generate photons individually through the DCE. When driven together at the same frequency, the resonator can be thought of as a vibrating resonator or a breathing resonator, depending on the phase difference between the two drive signals. When flux pumping both SQUIDs around $2\omega_1$, theory predicts constructive interference for the breathing mode, leading to a low threshold for photon generation, and destructive interference for the vibrating mode, *i.e.* no photon generation [15–19]. In addition to investigations of the DCE this device opens up doors for future interesting experiments, for example, measurements of the twin paradox [20], where a microwave signal could be sent on a "space trip" in a vibrating resonator, and the generation of cluster states [21].

## 2. Experimental setup

Our circuit is placed on a sapphire substrate (Fig. 1). The SQUIDs are made of aluminium and deposited by two-angle evaporation, while the rest of the circuit is etched in niobium. The resonator is meandered and grounded in both ends. To avoid a parasitic superconducting loop through resonator and ground plane, we made a slot in the ground plane. To keep good electrical contact we bridged the slot with normal metal (gold), see Fig. 1(b).

We use a reflection setup with circulators to allow for proper attenuation of an input probe signal and amplification of the resonator output signal (Fig. 1(d)). The flux-line setup enable both dc biasing and fast modulation (pumping) through separate lines that are combined in bias-Tees at the mixing chamber stage of the cryostat. The pump signals are generated in two sources, $P_{l/r}$, phase locked by a 10 MHz reference. To measure the phase difference between the sources, their output signals are divided in power splitters and compared using a mixer. Provided that the two pumps have the same frequency, the mixer output is a dc voltage with varying amplitude, depending on the phase difference between the pumps. The resonator output is down-converted and sampled in a digitizer, which records both the in- and out-of-phase quadratures.

## 3. Measurement results - Resonator characterization

We tune the resonant frequency by controlling the two dc-fluxes, $\Phi_{dc,l/r}$. The first resonator mode is probed by measuring the reflection coefficient of a microwave signal incident on the resonator

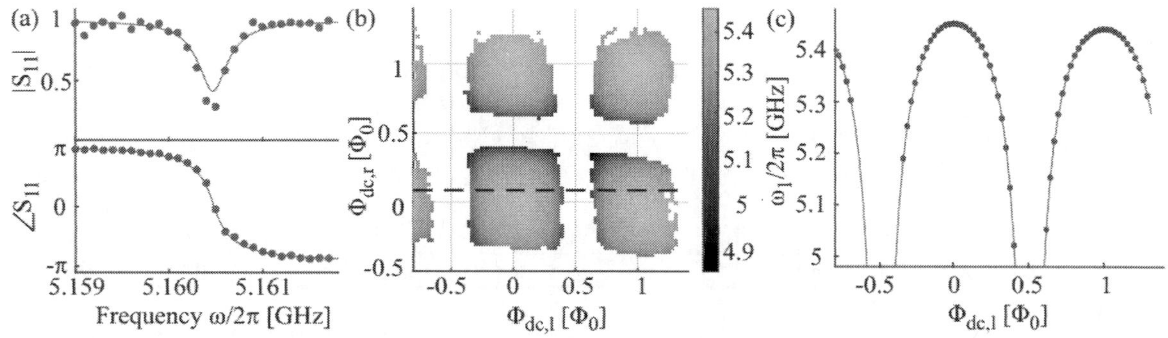


**Figure 2.** (a) Reflection measurement (blue dots) at $\Phi_{dc} = (0.3, 0.3)\,\Phi_0$ and a fit to the model $S_{11} = (1/Q_{ext} - 1/Q_{int} - 2i(\omega - \omega_1)/\omega_1)/(1/Q_{ext} + 1/Q_{int} + 2i(\omega - \omega_1)/\omega_1)$. For this bias point we can extract $\omega_1/2\pi = 5.1605\,\text{GHz}$, $Q_{ext} = 15.4 \cdot 10^3$ and $Q_{int} = 36.9 \cdot 10^3$. (b) dc tuning of the resonant frequency by both magnetic flux biases. (c) Linecut from (b) indicated by the black dashed line. The blue dots are data and the red line is a fit to the model in Eq. (1).

**Table 1.** Two-tone spectroscopy of the second resonator mode using parametric up-conversion. The first column indicates the flux bias point, $\omega_1$ and $\omega_2$ are the two lowest mode frequencies and the last column is the spectrum anharmonicity.

| $(\Phi_{dc,l}, \Phi_{dc,r})$ [$\Phi_0$] | $\omega_1/2\pi$ [GHz] | $\omega_2/2\pi$ [GHz] | $(2\omega_1 - \omega_2)/2\pi$ [MHz] |
|---|---|---|---|
| (0.01,0.01) | 5.459 | 10.867 | 47 |
| (0.21,-0.19) | 5.360 | 10.668 | 52 |
| (0.31,-0.29) | 5.184 | 10.323 | 45 |

(Fig. 2(a)); the extracted resonant frequencies are presented in Fig. 2(b). The pattern is slightly tilted due to a small inductive crosstalk.

The second resonator mode is outside the frequency band of our setup, but its resonant frequency can be found using parametric up-conversion [22, 23], a two-photon process which we implement by letting a weak current-probe signal resonantly excite the first mode while simultaneously flux-pumping one of the SQUIDs at a lower frequency. When this pump tone hits the difference frequency $\omega_2 - \omega_1$, photons are up-converted from $\omega_1$ to $\omega_2$, resulting in an observed level-avoided crossing in the reflected signal, from which $\omega_2$ can be determined. We list three measured points in Table 1. We conclude that the anharmonicity is much larger than the linewidth ($2\Gamma$) of the resonator. Importantly, this means that our pump tone, applied at frequency $2\omega_1$, should not excite the second harmonic.

By straightforward extension of the results in Ref. [10], the spectrum of the doubly tunable resonator is described by the equation

$$\frac{\omega_n}{v} d \tan\left(\frac{\omega_n}{v} d\right) \left[1 - \left(\frac{v}{\omega_n d}\right)^2 \left(\frac{1}{\gamma_l} - c\left(\frac{\omega_n}{v} d\right)^2\right) \left(\frac{1}{\gamma_r} - c\left(\frac{\omega_n}{v} d\right)^2\right)\right] = \frac{1}{\gamma_l} + \frac{1}{\gamma_r} - 2c\left(\frac{\omega_n}{v} d\right)^2 . \tag{1}$$

The subscripts $l/r$ correspond to the left and right SQUID, $\omega_n$ is the frequency of mode $n$, $d = 10.133\,\text{mm}$ is the resonator length and $v = 1/\sqrt{C_0 L_0}$ is the phase velocity. $\gamma_{l/r} = L_{J,l/r}/(L_0 d)$ is the inductive participation ratio, where the SQUID inductance is $L_{J,l/r} = \Phi_0/(2\pi I_c|\cos(\Phi_{dc,l/r}\pi/\Phi_0)|)$, assuming low signal levels, $I_s \ll I_c$, and $\Phi_{dc,l/r}$ is the static magnetic flux bias. The capacitive participation ratio is $c = C_J/C_0 d$. Here we have assumed that the two SQUIDs are nominally identical, $\gamma_0 = \gamma_{0,l} = \gamma_{0,r}$ and $C_J = C_{J,l} = C_{J,r}$.

The two-dimensional dc tuning, Fig. 2(b), together with the measurements of the second mode in Table 1, can be fitted using Eq. (1). A linecut of Fig. 2(b) with a fit is found in

**Table 2.** Extracted parameters for the resonator. The inductive participation ratio is $\gamma_0 = L_{J,0}/L_0 d$, $I_c$ the SQUID critical current, $C_J$ the SQUID capacitance and $\xi_{l/r}$ represents the dc-crosstalk. $C_0$ and $L_0$ are the capacitance and inductance per unit length of the coplanar waveguide. $\omega_1$ is the resonant frequency of the lowest mode, $\Gamma$ is the photon loss rate and $Q_{\text{int}}$ and $Q_{\text{ext}}$ are the quality factors of the resonator at $\Phi_{dc} = (0,0)\,\Phi_0$. The translation between the loss rate $\Gamma$ and the Q-values is $2\Gamma = \omega_1/Q_{\text{int}} + \omega_1/Q_{\text{ext}}$.

| $\gamma_0$ | $I_c$ | $C_J$ | $\xi_l$ | $\xi_r$ | $C_0$ | $L_0$ | $\omega_1(0)/(2\pi)$ | $2\Gamma(0)/(2\pi)$ | $Q_{\text{int}}(0)$ | $Q_{\text{ext}}(0)$ |
|---|---|---|---|---|---|---|---|---|---|---|
| [%] | [$\mu$A] | [fF] | [%] | [%] | [$\frac{\text{nF}}{\text{m}}$] | [$\frac{\mu\text{H}}{\text{m}}$] | [GHz] | [MHz] | [$10^3$] | [$10^3$] |
| 4.64 | 1.64 | 89 | 3.64 | 4.19 | 0.159 | 0.427 | 5.459 | 0.56 | 400 | 9.6 |

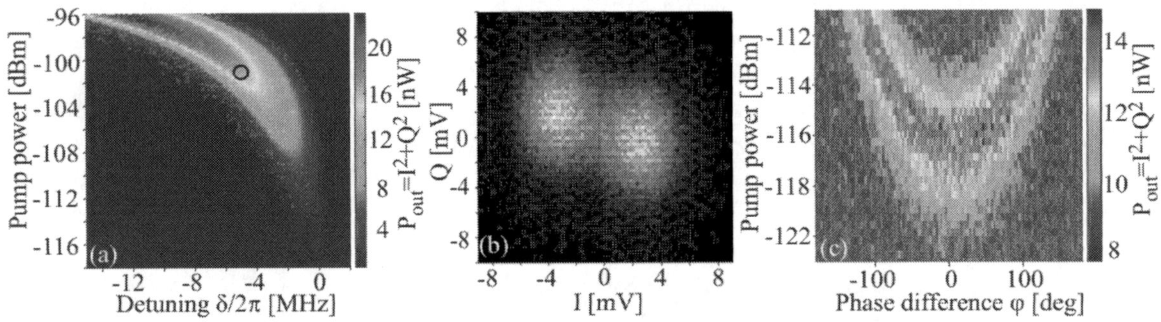


**Figure 3.** (a) Photon down-conversion, measured with a single pump applied to the left flux line at the bias point $(0.3, 0.3)\,\Phi_0$. (b) Histogram taken at the point marked with a black circle in (a). We measure two $\pi$-shifted states. (c) Double-pump measurement, where the phase difference $\varphi$ between the pump signals is varied. Here the SQUID bias is $(0.2, 0.2)\,\Phi_0$ and $\delta = -1\,$MHz.

Fig. 2(c). Extracted resonator and SQUID parameters are found in Table 2. $\xi_{l/r}$ are the dc crosstalks, *i.e.* how much each SQUID is affected by the opposite flux line (only a few percent of the coupling from the closest flux line). The resonant frequency can be tuned over a wide frequency range: the limiting factor is the photon loss rate, which increases as $\Phi_{dc,l/r}$ approaches $\Phi_0/2$. However, as seen in Fig. 2(b), resonant frequencies below 4.9 GHz are measurable.

## 4. Measurement results - Pumping

By applying a pump tone to one of the ac flux lines at a frequency close to $2\omega_1$, we expect parametric oscillations at $\omega_1$. We measure the quadrature components of the output signal, and calculate the total output power, $P_{out} = \langle I^2 \rangle + \langle Q^2 \rangle$. Fig. 3(a) shows photon down-conversion from $2\omega_1$ to $\omega_1$ in a range of detuning and pump power. The detuning is denoted $\delta = \omega_p/2 - \omega_1$, where $\omega_p$ is the pump frequency. Furthermore, we sample the individual quadratures, $\langle I(t) \rangle$ and $\langle Q(t) \rangle$, and histogram $1 \cdot 10^5$ samples, see Fig. 3(b). The histogram shows two stable $\pi$-shifted states with the same amplitude, characteristic of parametric oscillations [8, 9, 24].

We can also apply pump signals to both flux lines simultaneously. The amplitudes are adjusted such that the effective pump strengths of the two individual SQUIDs are equal. This was done by measuring single-pump thresholds, which for the bias point $(0.2, 0.2)\,\Phi_0$ should be equal. We find that, depending on the phase difference $\varphi = \varphi_r - \varphi_l$, the threshold for photon generation changes, see Fig. 3(c).

The theoretical parametric oscillation threshold is $\epsilon_{th} = \sqrt{\Gamma^2 + \delta^2}$ both for the single and double pump cases. The measured oscillation regions are asymmetric in $\delta$, due to a pump-induced frequency shift because of the resonator nonlinearity, shifting the resonant frequency

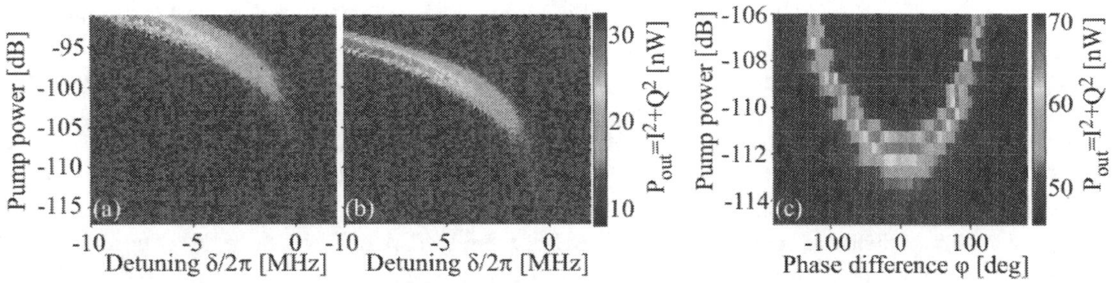


**Figure 4.** Measurement results at SQUID bias $(0, 0.2)\,\Phi_0$, in both cases using a single pump, coupled closest to the left (a), and the right (b) flux line respectively. (c) Double-pump measurement of a $\lambda/2$-resonator with only one SQUID, biased at $0.18\,\Phi_0$. Here $\delta = -6\,\text{MHz}$.

towards red detuning. The threshold is reached at an effective pump strength $\epsilon_{eff} = \epsilon_{th}$. Following the formalism [10] and extending the results to the double pump case, the effective pump strength is a superposition of complex amplitudes of flux modulation in the left and right SQUIDs, $\Phi_{ac,l/r} = |\Phi_{ac,l/r}|e^{i\varphi_{l/r}}$, so that $\epsilon_{eff} = A(\omega_1)(k_l\Phi_{ac,l} + k_r\Phi_{ac,r})$. The coefficients are, $k_{l/r} = |\tan(\Phi_{dc,l/r}\pi/\Phi_0)|/\gamma_{l/r}$. This gives an expected minimum threshold and therefore maximum photon generation in the breathing mode, $\varphi = 0°$, but cancellation and consequently no photons in the vibrating mode, $\varphi = \pm 180°$, in agreement with the measured data.

## 5. Discussion

We find qualitative agreement between Fig. 3 and the doubly flux-pumped resonator theory as well as some interesting deviations. In Fig. 4(a) and (b), we present regions of photon down-conversion, at the bias point $(0, 0.2)\,\Phi_0$. A single pump tone is applied to the left flux line in (a) and to the right in (b). Since for (a) the pumping is around zero flux and in (b) around $0.2\,\Phi_0$, different results are expected. However, the shapes of the oscillation regions in the two graphs are rather similar, although the thresholds differ by around 5 dB. Setup attenuation differences cannot explain this large number. The observation of parametric oscillations at zero flux bias is surprising, since this contradicts theoretical predictions [10]. We attribute this effect to a possible strong inductive ac crosstalk or a parasitic coupling. Even though the crosstalk at dc is negligible, it could be large at microwave frequencies, due to differences in signal distribution on the chip for dc and microwave signals. In Fig. 4(a), the pump power applied to the left pump line would actually also pump the right SQUID, with a pump power leakage of 5 dB. Assuming that the ac flux is proportional to the pump amplitude, *i.e.* $\Phi_{ac} \propto 10^{(P_{pump})/20}$, this corresponds to 56 % crosstalk.

A parasitic coupling from the flux pump to the SQUID current could occur, due to the presence of the low impedance loop through the resonator center conductor and the ground plane. This loop is $\sim 4000$ times larger than the SQUID loop, which corresponds to a significantly larger inductance. A coupling to this loop could cause circulating currents, and thereby directly drive the SQUID current. A possible solution, making the loop less parasitic, would be to increase its impedance by changing the geometry of the gold-bridge and slot.

Another issue is the threshold pump strength. In experiments with a $\lambda/4$-resonator with identical SQUID flux-line design and similar resonant frequency, the single pump threshold is at least 20 dB higher than what is measured here. The length difference of a $\lambda/2$ and $\lambda/4$-resonator could account for maximum a few dB of difference. Therefore the differing thresholds have to be explained, either by differing pumping mechanisms or significantly differing flux line to SQUID coupling. However, the latter can be ruled out since the couplings are designed to be identical.

To find an explanation of the mentioned discrepancies, we performed a control experiment to probe the ac crosstalk. A similar resonator was fabricated with only one SQUID, *i.e.*, the other end was shorted to ground. Both resonator ends were equipped with on-chip flux lines, to allow for double-pump experiments. Surprisingly, we observe the same qualitative behaviour, independently of whether the resonator has two (Fig. 3(c)) or one (Fig. 4(c)) SQUID. There are some differences in output power and oscillation region widths, but this is because the measurements were performed with different samples, and at different bias points and detunings. This suggests an additional mechanism of down-conversion, possibly related to the microwave field filling the cavity and producing a current-pumping effect [25] such as that used in many parametric amplifiers [11, 26]. The difference between flux and current pumping has been discussed in Ref. [24]. The phase dependence of the threshold in Fig. 4(c), could, for instance, be explained by direct interference of the two pump signals.

## 6. Conclusion

Using a $\lambda/2$ resonator with two magnetic-flux-tunable boundary conditions, we demonstrated photon generation by degenerate downconversion of a pump tone. When pumping with two signals at the same frequency, we observed a pump-phase dependence of the instability threshold for photon generation. This is in agreement with a theoretical model for modulation of the boundary conditions. We also observed non-ideal results attributable to ac crosstalk and parasitic couplings resulting in current driving of the SQUIDs.

## Acknowledgements

The authors acknowledge financial support from the European Research Council, the European project PROMISCE, the Swedish Research Council and the Wallenberg Foundation. J.B. acknowledges partial support by the EU under REA grant agreement no. CIG-618353.

## References

[1] Casimir H B G 1948 *P. Roy. Neth. Acad. Arts Sci.* 793
[2] Moore G T 1970 *J. Math. Phys.* **11** 2679
[3] DeWitt B S 1975 *Phys. Rep.* **19** 295
[4] Fulling S A and Davies P C W 1976 *P. Roy. Soc. Lond. A Mat.* **348**
[5] Sandberg M *et al.* 2008 *Appl. Phys. Lett.* **92** 203501
[6] Johansson J R, Johansson G, Wilson C M and Nori F 2010 *Phys. Rev. A* **82** 052509
[7] Wilson C M *et al.* 2011 *Nature* **479** 376
[8] Dykman M I, Maloney C M, Smelyanskiy V N and Silverstein M 1998 *Phys. Rev. E* **57** 5202
[9] Wilson C M, Duty T, Sandberg M, Persson F, Shumeiko V and Delsing P 2010 *Phys. Rev. Lett.* **105** 233907
[10] Wustmann W and Shumeiko V 2013 *Phys. Rev. B* **87**(18) 184501
[11] Castellanos-Beltran M A and Lehnert K W 2007 *Appl. Phys. Lett.* **91** 083509
[12] Tholén E A *et al.* 2007 *Appl. Phys. Lett.* **90** 253509
[13] Yamamoto T *et al.* 2008 *Appl. Phys. Lett.* **93** 042510
[14] Roch N *et al.* 2012 *Phys. Rev. Lett.* **108**(14) 147701
[15] Lambrecht A, Jaekel M T and Reynaud S 1996 *Phys. Rev. Lett.* **77**(4) 615–618
[16] Ji J Y, Jung H H and Soh K S 1998 *Phys. Rev. A* **57** 4952
[17] Dodonov V V 1998 *J. Phys. A-Math. Gen.* **31** 9835
[18] Lambrecht A, Jaekel M T and Reynaud S 1998 *Eur. Phys. J. D* **3**(1)
[19] Dalvit D A R and Mazzitelli F D 1999 *Phys. Rev. A* **59** 3049
[20] Lindkvist J *et al.* 2014 *Phys. Rev. A* **90** 052113
[21] Bruschi D E, Sabín C, Kok P, Johansson G, Delsing P and Fuentes I 2016 *Sci. Rep.* **6** 18349
[22] Zakka-Bajjani E, Nguyen F, Lee M, Vale L R, Simmonds R W and Aumentado J 2011 *Nat. Phys.* **7**(8) 599
[23] Wustmann W and Shumeiko V 2017 *arXiv:1704.05083*
[24] Krantz P *et al.* 2016 *Nat. Commun.* **7** 11417
[25] Yurke B *et al.* 1988 *Phys. Rev. Lett.* **60**(9) 764
[26] Siddiqi I *et al.* 2005 *Phys. Rev. Lett.* **94**(2) 027005

IOP Conf. Series: Journal of Physics: Conf. Series **969** (2018) 012147 doi:10.1088/1742-6596/969/1/012147

# Electron transport in a bilayer graphene/layered superconductor $NbSe_2$ junction: effect of work function difference

**Katsuhide Yarimizu,[1] Hikari Tomori,[1,2] Kenji Watanabe,[3] Takashi Taniguchi,[3] Akinobu Kanda[1]**

[1] Division of Physics and TIMS, Faculty of Pure and Applied Sciences, University of Tsukuba, Tennodai, Tsukuba, Ibaraki 305-8571, Japan

[2] PRESTO, Japan Science and Technology Agency, Kawaguchi 332-0012, Japan

[3] National Institute for Materials Science (NIMS), Tsukuba 305-0047, Japan

E-mail: kanda.akinobu.gm@u.tsukuba.ac.jp

**Abstract.** We have experimentally studied electron transport in a bilayer graphene (BLG)/layered superconductor $NbSe_2$ junction encapsulated with hexagonal boron nitride. The junction exhibits nonlinear current-voltage characteristics which strongly depend on the gate voltage around the charge neutrality point (CNP) of the BLG. Besides, we observe that the gate voltage dependence of electron transport in the BLG portion close to the junction interface is different from that of the BLG portion apart from the interface, indicating that the spatial variation of the Dirac point in the charge transfer region due to the difference in work function between superconductor and graphene needs to be considered in the analysis of the superconducting proximity effect.

## 1. Introduction

At a superconductor/normal metal interface, an incident electron ($-e$) from the normal metal enters the superconductor forming a Cooper pair ($-2e$) and a hole ($+e$) is reflected back to the normal metal due to the charge conservation. This process is called the Andreev reflection.[1] Usually, because both the incident electron and the reflected hole belong to the conduction band of the normal metal, the reflected hole goes back along the same path of the incident electron (intraband retro-reflection). On the other hand, when graphene is used as the normal metal, the reflected hole is in the valence band for the Fermi level close to the Dirac point (DP), leading the interband *specular* Andreev reflection. Theory[2] tells that when the Fermi level of graphene is swept across the DP, the transitions from retro- to specular and from specular to retro-reflections cause an anomalous behavior in the current-voltage ($I$-$V$) characteristics. This can be used as an experimental proof of the existence of the specular reflection.

In experiments, the specular Andreev reflection in (single layer) graphene has not been confirmed for a long time,[3, 4, 5, 6, 7, 8, 9] presumably due to the spatial fluctuation of the carrier density (electron/hole puddles) in graphene caused by the charged impurities[10], which allows the retro- and specular reflections to coexist. Actually, the magnitude of the DP fluctuation in graphene placed on $SiO_2$ is around 50 meV, which further exceeds the superconducting gap energy of conventional superconductors. Besides, due to the difference in work function between

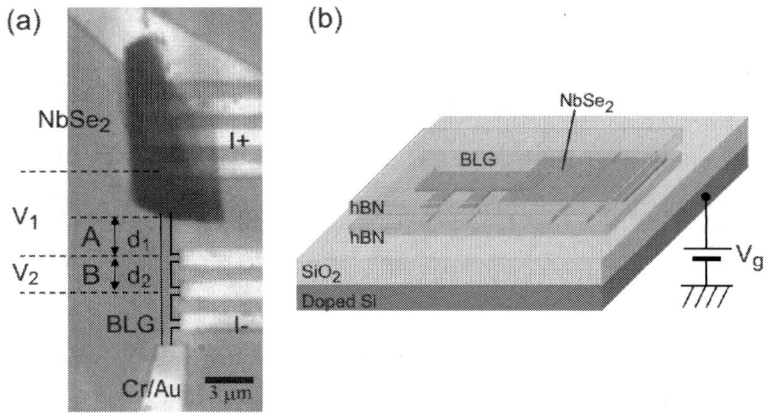


**Figure 1.** (a) Optical image of the sample. Due to poor contrast, the BLG is indicated by black lines. (b) Schematic of the device structure.

graphene and the superconductor, charge carriers (electrons or holes) are doped to graphene in the vicinity of the interface (charge transfer region).[11, 12, 13] The charge transfer region is reported to extend over $\sim$ 500 nm with the energy difference of 60 meV for a Ni/graphene interface.[13] The effect of the work function difference and the charge transfer region were not taken into account in the original theory of Andreev reflections in graphene.[2] Recently, Efetov *et al.*[14] reported observation of the anomalous $I$-$V$ characteristics due to the transition between specular and retro-reflections. In their device, they successfully decreased the number of charged impurities by encapsulating the junction with hexagonal boron nitride (hBN), and further reduced the spatial fluctuation of the DP by using bilayer graphene (BLG) instead of single layer graphene. Here, BLG is a zero-gap semiconductor with larger density of states around the DP due to the parabolic bands. It is noted that in their analysis, the experimental result is compared with the theoretical one which does not take into account the effect of the work function difference and the charge transfer region, and it is not clear how the work function difference affects the electron transport in the graphene/superconductor junctions.

In this study, to elucidate the effect of the work function difference between graphene and superconductor, we investigate the electron transport in a superconductor/BLG device which has a device structure similar to that reported in Ref. [14].

## 2. Experiment

We fabricated an $NbSe_2$/BLG junction encapsulated with hBN using the mechanical exfoliation of layered materials and van der Waals dry transfer technique[15] in a glove box filled with Ar to prevent the oxidization of $NbSe_2$. Here, $NbSe_2$ is a layered superconductor with the critical temperature of 7.0 K. The device structure is shown in Fig. 1. The BLG is rectangular-shaped with several side arms for the four terminal measurement. A flake of $NbSe_2$ with thickness of $\sim$ 35 nm is placed on Cr/Au electrodes and the BLG. We measured voltage $V_1$ across the junction (with the BLG portion "A" with length $d_1 = 2.51$ $\mu$m, see Fig. 1(a)), and voltage $V_2$ across the side arms in the BLG portion "B" (with length $d_2 = 2.20$ $\mu$m) simultaneously in the four terminal configuration, as a function of the bias voltage (applied between leads I+ and I− in Fig. 1) and the gate voltage $V_g$ which was applied to the highly doped Si substrate.

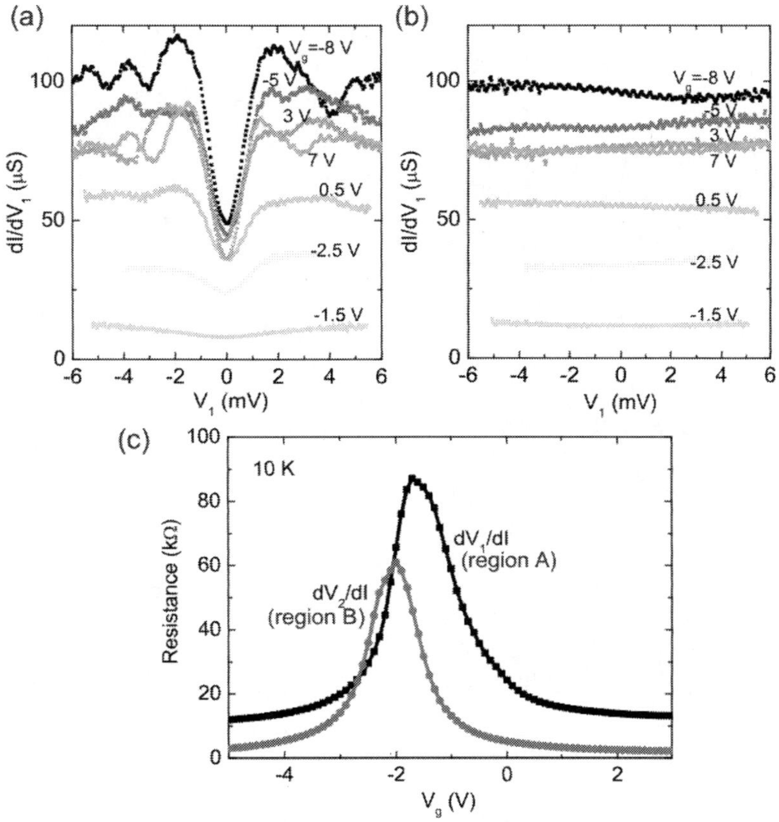


**Figure 2.** Voltage $V_1$ dependence of the differential conductance of the junction, $dI/dV_1$, for several gate voltages in the superconducting state at 4.2 K (a) and in the normal state at 10 K (b). (c) The differential resistances $dV_1/dI$ and $dV_2/dI$ at 10 K as a function of the gate voltage.

## 3. Results and discussion

Figure 2 shows the differential conductance of the junction, $dI/dV_1$, as a function of $V_1$ for several gate voltages with $NbSe_2$ in the superconducting state at 4.2 K (Fig. 2(a)) and in the normal state at 10 K (Fig. 2(b)). In Fig. 2(b), the conductance is almost independent of $V_1$, but strongly depends on the gate voltage due to the modulation of the carrier density of BLG. The gate voltage which gives the minimum of $dI/dV_1$ corresponds to the charge neutrality point (CNP) of the BLG. In the superconducting state (Fig. 2(a)), while the value of $dI/dV_1$ at $|V_1| \gtrsim 3$ mV is close to the value of $dI/dV_1$ at the same $V_g$ in Fig. 2(b), a dip appears in $dI/dV_1$ at $|V_1| \lesssim 1$ mV. This is due to the Andreev retro-reflection, corresponding to the superconducting energy gap of $NbSe_2$ ($\Delta \sim 1.1$ meV). Here, we note that the depth of the dip strongly depends on the gate voltage, *i.e.*, the dip becomes shallower as the gate voltage approaches the CNP. The origin of this gate voltage dependence is not clear at this moment. Figure 2(c) shows the differential resistances $dV_1/dI$ and $dV_2/dI$ at 10 K as a function of the gate voltage. The resistance peak appears at $V_g = -1.7$ V and $-2.0$ V for $dV_1/dI$ and $dV_2/dI$, respectively. The difference in the gate voltage dependence indicates the existence of the charge transfer region near the interface. Actually, since the work function of BLG ($\sim$ 4.5 eV[16]) is smaller than that of $NbSe_2$ (5.9 eV[17]), holes are doped to BLG, so that the gate voltage

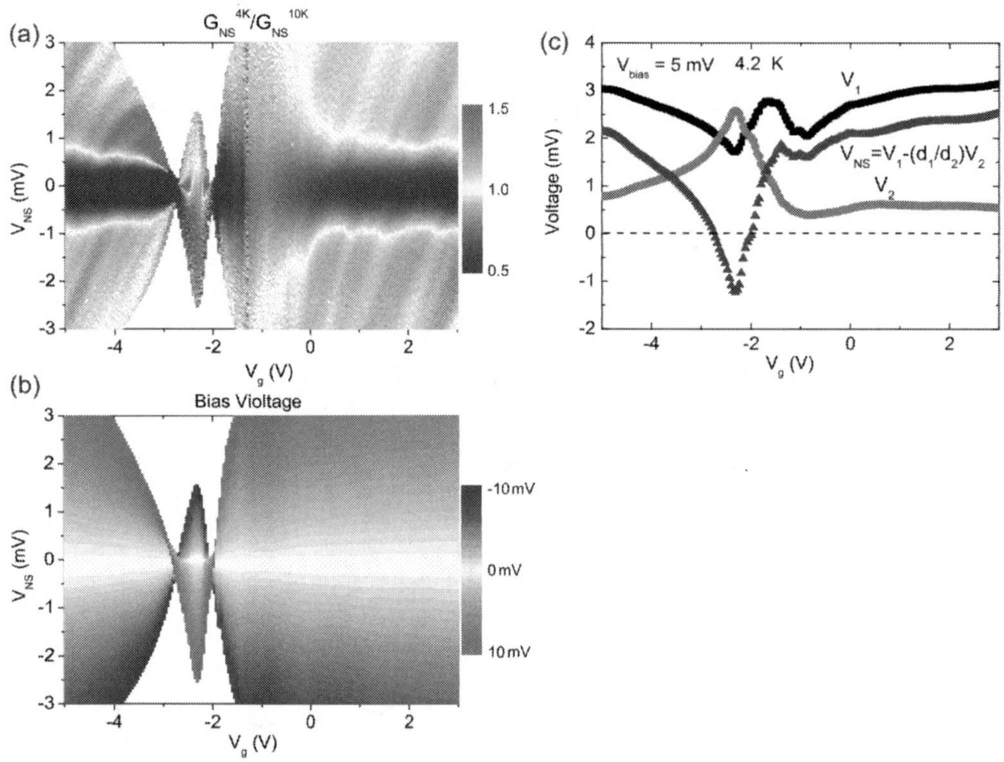


**Figure 3.** (a) The ratio of $G_{NS}$ in the superconducting state (4 K) to $G_{NS}$ in the normal state (10 K) plotted in the $V_{NS} - V_g$ plane. (b) The bias voltage plotted in the $V_{NS} - V_g$ plane. Red and blue correspond to positive and negative bias voltages. (c) The gate-voltage dependence of the measured voltages, $V_1$ and $V_2$ and the calculated interface voltage, $V_{NS}$, for bias voltage of +5 mV and temperature of 4.2 K.

corresponding to the resistance peak of region A near the interface is expected to be larger than that of region B apart from the interface, which agrees with the experimental finding in Fig. 2 (c).

Following the analysis in Ref. [14], we derive the voltage of the interface, $V_{NS}$, by subtracting the nominal voltage drop in the BLG portion A from $V_1$ as $V_{NS} = V_1 - (d_1/d_2)V_2$ and calculated the differential conductance of the junction interface, $G_{NS} = dI/dV_{NS}$.

Figure 3(a) shows the ratio of $G_{NS}$ in the superconducting state (4.2 K) to $G_{NS}$ in the normal state (10 K) plotted in the $V_{NS} - V_g$ plane. At $|V_{NS}| \lesssim 1$ mV and large $|V_g|$, suppression of $G^{4K}_{NS}/G^{10K}_{NS}$ is seen, which agrees with the behavior of the Andreev retro-reflection. Besides, anomalous behavior is seen around $V_g = -2.5$ V. Here, we point out that inside the anomaly region ($-2.7$ V $\leq V_g \leq -2.0$ V), the sign of the calculated interface voltage $V_{NS}$ is opposite to that of the bias voltage applied between leads I+ and I−, as shown in Figs. 3(b) and 3(c), which is physically incorrect. This occurs because the gate voltage dependence of the resistance of the BLG portion A is different from that of the BLG portion B due to the carrier doping from the superconductor, and thus, $V_{NS} = V_1 - (d_1/d_2)V_2$ does not hold. Since our device has a large $d_1/d_2$ ratio, this difference is clearly seen as the sign inversion of $V_{NS}$.

In recent theoretical works,[18, 19] the effect of the work junction difference on the differential conductance of BLG/superconductor interface have been discussed. In the model of Ref. [18], a part of BLG is covered with a superconductor, and the covered region exhibits a potential (corresponding to a DP shift) due to the work function difference. Here, a stepwise change of the potential is assumed, so that the gradual spatial variation of the DP near the interface within the charge transfer region is not taken into account. Another theoretical work by Takane *et al.* has adopted a similar model but taken into account the spatial variation of the DP in BLG close to the interface.[19] They predicted that even when the Andreev retro-reflection occurs at the interface, the incident electron and/or the reflected hole exhibits diffraction when passing over a naturally formed $p-n$ junction near the interface, leading to the seemingly specular Andreev reflection. The interpretation of our results based on Refs. [18] and [19] is in progress.

## 4. Conclusions

We have experimentally investigated electron transport in a BLG/superconductor junction. We observe that the gate voltage dependence of the resistance of the BLG portion near the junction interface is different from that of the BLG portion apart from the interface. This is presumably due to the work function difference between BLG and the superconductor and the existence of the charge transfer region. We point out that the analysis of the electron transport in BLG/superconductor junctions needs to take into account the work function difference and the charge transfer region.

## Acknowledgments

This work was supported by JSPS KAKENHI Grant Numbers JP16H00897 [Science of Atomic Layers (SATL)] and JP15H05867 [Hybrid Quantum Systems]. The authors gratefully acknowledge fruitful discussions with Y. Takane, M. Hayashi, H. Yoshioka and N. Nagamura.

## References

[1] Tinkham M 1996 *Introduction to Superconductivity* (New York: McGraw-Hill)
[2] Beenakker C W J 2006 *Phys. Rev. Lett.* **97** 067007
[3] Heersche H B, Jarillo-Herrero P, Oostinga J B, Vandersypen L M K and Morpurgo A F 2007 *Nature* **446** 56
[4] Miao F, Wijeratne S, Zhang Y, Coskun U C, Bao W and Lau C N 2007 *Science* **317** 1530
[5] Du X, Skachko I and Andrei E Y 2008 *Phys. Rev.* B **77** 184507
[6] Miao F, Bao W, Zhang H and Lau C N 2009 *Solid State Commun.* **149** 1046
[7] Kanda A, Sato T, Goto H, Tomori H, Tanaka S, Ootuka Y and Tsukagoshi K 2010 *Physica* C **470** 1477
[8] Calado V E, Goswami S, Nanda G, Diez M, Akhmerov A R, Watanabe K, Taniguchi T, Klapwijk T M and Vandersypen L M K 2015 *Nat. Nanotechnol.* **10** 761
[9] Shalom M B, Zhu M J , Fal ' ko V I, Mishchenko A, Kretinin A V, Novoselov K S, Woods C R, Watanabe K, Taniguchi T and Geim A K and Prance J R 2016 *Nat. Phys.* **12** 318
[10] Martin J, Akerman N, Ulbricht G, Lohmann T, Smet J H, von Klitzing K and Yacoby A 2008 *Nat. Phys.* **4** 144
[11] Mueller T, Xia F, Freitag M, Tsang J and Avouris Ph 2009 *Phys. Rev.* B **79** 245430
[12] Kanda A 2013 *Experimental approaches to graphene electron transport for device applications Physics and Chemistry of Graphene: Graphene to Nanographene* ed T Enoki and T Ando (Singapore: Pan Stanford) chapter 3 pp 89-205
[13] Nagamura N, Horiba K, Toyoda S, Kurosumi S, Shinohara T, Oshima M, Fukidome H, Suemitsu M, Nagashio K, and Toriumi A. 2013 *Appl. Phys. Lett.* **102** 241604
[14] Efetov D K, Wang L, Handschin C, Efetov K B, Shuang J, Cava R, Taniguchi T, Watanabe K, Hone J, Dean C R and Kim P 2016 *Nat. Phys.* **12** 328
[15] Wang L, Meric I, Huang P Y, Gao Q, Gao Y, Tran H, Taniguchi T, Watanabe K, Campos L M, Muller D A, Guo J, Kim P, Hone J, Shepard KL and Dean CR 2013 *Science* **342** 614
[16] Murata Y, Nie S, Ebnonnasir A, Starodub E, Kappes B B, McCarty K F, Ciobanu C V and Kodambaka S 2012 *Phys. Rev.* B **84** 205443
[17] Shimada T, Ohuchi F S and Parkinson B A 1994 *Jpn J. Appl. Phys.* **33** 2696
[18] Efetov D K and Efetov K B 2016 *Phys. Rev.* B **94** 075403
[19] Takane Y, Yarimizu K and Kanda A 2017 *J. Phys. Soc. Jpn* **86** 064707

IOP Conf. Series: Journal of Physics: Conf. Series **969** (2018) 012148 doi:10.1088/1742-6596/969/1/012148

# Magneto-Transport Measurements on Multilayered Graphene on SiC(000$\bar{1}$)

**Mohammed Z Tahar**
State University of New York, The College at Brockport, Brockport, NY 14420 USA
E-mail: mtahar@brockport.edu

**Abstract.** This report is on transport properties of a multilayer graphene on SiC(000$\bar{1}$) sample. Both the magnetoresistance (MR) and Hall Effect measurements were carried out at ten constant temperatures (2 K - 300 K) and magnetic field scans 0 to 9 T. The results exhibit a negative MR and a sign reversal at low field (< 0.4 T) and temperatures above 50 K and no sign reversal below that up to $H \sim 2$ T. Additionally, for low temperatures (< 20 K) and high field (> 2 T), the MR signal shows SdH oscillations of the same frequency, $f = 65$ T. The Hall Effect signal shows a linear increase with magnetic field ($H < 0.3$ T) and a leveling off at higher fields. Using the two signals and the metallic resistance with residual resistance ratio ($RRR = 1.7$), one deduces the carrier concentration, $3.3 - 1.0 \times 10^{13}$ cm$^{-2}$, and mobility $3,700 - 21,000$ cm$^2$/V·s at 300 K and below 20 K, respectively.

## 1. Introduction

Graphene is one of the many allotropes of the element carbon and is a single sheet of 'graphite' that is one atom thick. Graphene is probably the ultimate 2D material, is important both from pure physics and device application point of view, and can be fabricated using three main processes: exfoliation (repeated pealing) of graphite [1], chemical vapor deposition (CVD) on metal substrates [2], and epitaxial growth on SiC [3, 4].

Epitaxial growth lends itself to micro-fabrication for possible electronics applications much better then both exfoliation and CVD. However, in epitaxial growth both the environment and the substrate have an effect on the final product as well as the number of layers and growth on SiC can be on either the Si or C terminated face, SiC(0001) and SiC(000$\bar{1}$), respectively.

Graphene's exotic properties are due to its 2D hexagonal structure, which leads to a linear energy-momentum relationship, that gives rise to massless Dirac Fermions and valley and chiral symmetries. Experimentally, these properties are known to be dependent on the growth methods and substrates used, as well as the number of of layers and their stacking correlation. Magneto-transport studies can shed some light on the characterization of these intricate dependencies. Graphene physical properties are reviewed in the article by Castro Neto *et al* [5], where one can find an extensive theoretical treatment of single and multi-layer graphene of finite sheet size. For treatment of magneto-transport in graphene one is referred to the Physical Review Letter by McCann *et al.* [6].

This is a report on magneto-transport measurements on a sample of graphene of six layers and 2$\mu$m width grown on SiC(000$\bar{1}$) in the temperature range 2 K - 300 K and in magnetic fields up to 9 T. The results are analyzed in light of the above references and compared to others in the literature.

IOP Conf. Series: Journal of Physics: Conf. Series **969** (2018) 012148 doi:10.1088/1742-6596/969/1/012148

## 2. Experimental Details

Graphene was grown by low pressure sublimation in an inductively heated quartz tube furnace. The SiC(000$\bar{1}$) substrate was placed in a graphite box, and pumped overnight with a turbomolecular pump. For this particular sample, growth was performed in 50 mbar of Ar flowing at 5 standard liters per minute (SLPM). The temperature was ramped from room temperature to a growth temperature of 1525 °C, and held for 25 minutes. After growth was completed, the power supply controlling the heater was shut off, and the pressure was ramped to 900 mbar. The elevated Ar pressure impeded the growth of additional graphene as the reactor cooled. Full details of the growth process, the number of layers determination using XPS and optical measurements, and the absence of layer stacking correlation using low energy electron diffraction (LEED) have been described and discussed by Robinson *et al.* [4].

The transport measurements were carried out on the graphene sample using the multi-point contacts method. The contacts on the sample were made using lithography to larger pads onto which gold was wire-bonded, as shown in Fig. 1. The thin and short gold wires were connected to stronger copper wires, for soldering onto the sample holder pads, using a two-component silver loaded epoxy which was cured at ambient temperature for 24 hrs. Because these different metal-metal joints can give rise to thermoelectric voltages that depend on the metals, the temperature, and its gradient across the leads, an AC excitation from a floating current source [7] was used, along with lock-in-detectors at low frequency. At 7.000 Hz the frequency is low enough for the measurements to be considered DC measurements. To avoid the sample's self heating, the excitation current ($< 10\mu$A) was so that the total power dissipation was under $0.01\mu$W. The sample holder was inserted in a Quantum Design's Physical Properties Measurement System (PPMS) cryostat with a superconducting magnet capable of ramping the magnetic field at 0.6 T/min. up to 9 T and sample's temperature control for setting and holding with reproducibility $< 1\%$ or better, all in high vacuum. The applied magnetic field is perpendicular to the basal plane of the graphene sample and the probing current is parallel to it.

**Figure 1.** Optical image of contacts to the sample and the bond-pad layout. The current is right to left, the magnetic field is out of the plane of the image.

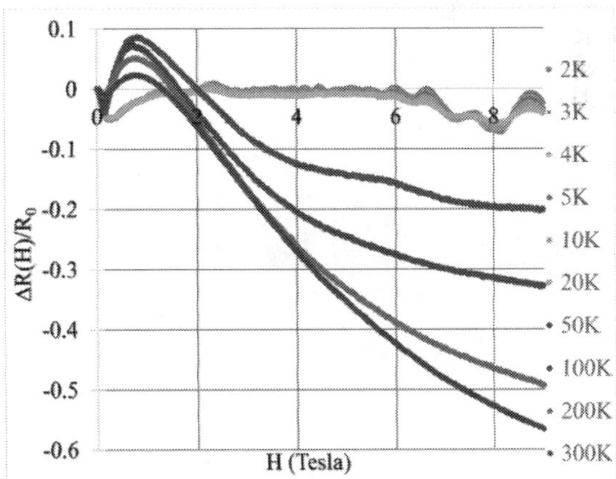


**Figure 2.** Magnetoresistance, $\frac{\Delta R(H)}{R(0)}$, with field (in Tesla) scans at different temperatures (2K - 300K) as indicated by the legend on the right of plot.

Referring to Fig. 1, the current was injected from the right middle pad and returned through the left pad, the longitudinal (magnetoresistance) voltage leads were connected to the top two outer pads, and the transverse (Hall Effect) voltage contacts were connected to the middle pads, with the top as positive and the bottom as negative. Notice the Hall contacts are not directly facing each other. The applied magnetic field is perpendicular and coming out of the plane of the image. The measurements were carried out as follows. The PPMS computer set and maintained a stable sample temperature and ramped the magnetic field up and down to the set field values, while another PC collected the different voltages off of two lock-in-detectors, one for the Hall voltage and the other for the resistance voltage.

## 3. Results

Figure 2 shows all the magnetoresistance, $\frac{\Delta\rho(H)}{\rho(0)} = \frac{\Delta R(H)}{R(0)}$, results for field scan-ups (0 to 9 T) at discrete temperatures of 300, 200, 100, 50, 20, 10, 5, 4, 3, and 2 K, as indicated by the legend on the right hand of the graph. One can distinguish two behaviors of the magnetoresistance with field at the different temperatures. First, a high temperature ($T > 50$ K) field dependence, where the magnetoresistance dips below zero for low fields, goes through a minimum at $H_{min}$, rises above zero to reach a maximum, and then slowly tapers off below zero for higher fields. This behavior is clearly seen in Fig. 3, which is a graph of the same MR data of $\frac{\Delta R(H)}{R(0)}$ vs. $H$, but only up to 2 T for more details.

The first zero crossings (at $H_1$), from negative to positive magnetoresistance, occur in the field range $0.2 < H < 0.3$ T and the second and reverse crossings (at $H_2$) occur in the range $1.3 < H < 2.1$ T. The magnetic fields at which the first crossings ($H_1$) occur seem to display a weak temperature dependence for $T > 50$ K and more than quadrupling for the temperatures below that. This weak temperature dependence is also mimicked by the field values at which the minima occur. This is shown in Fig. 4, as a plot of $H_{min}$, $H_1$, and $H_2$ as functions of temperature. The magnetoresistance also displays a broad maximum at $H \sim 0.8$ T, with similar but much weaker temperature dependence.

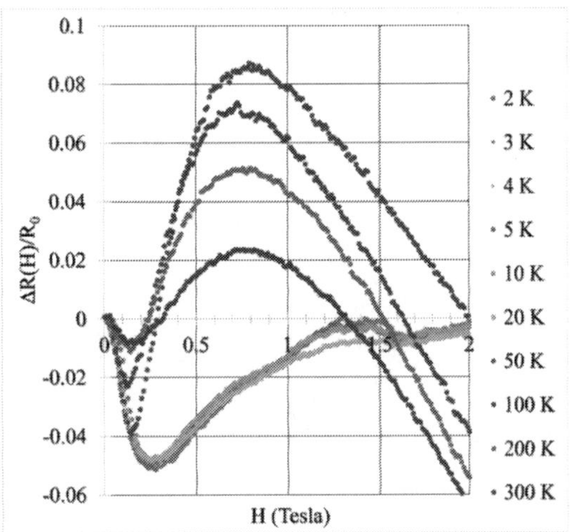


**Figure 3.** $\frac{\Delta R(H)}{R(0)}$ vs. $H$, the low field ($H < 2$ T) behavior to highlight the minima and the horizontal axis crossings.

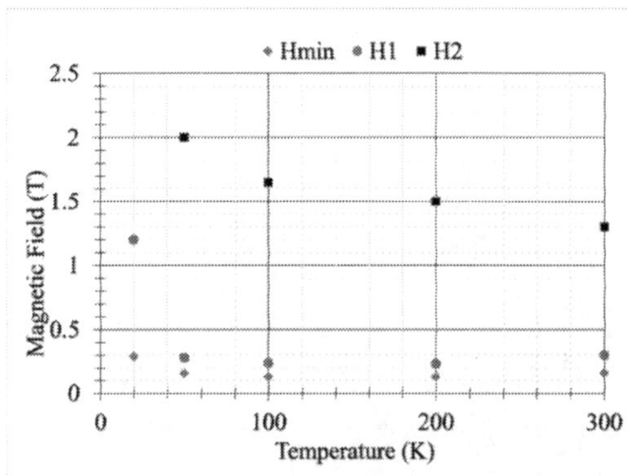


**Figure 4.** Characteristic Fields, $H_{min}(T)$, $H_1(T)$, and $H_2(T)$, for MR minimum, first crossing, and second crossing, respectively.

Second, at low temperatures ($T < 20$ K), there is a distinct field dependence, where the data dips below zero for low fields, goes through a minimum, and then rises but remains mainly below zero up to high fields, where Shubnikhov-de Haas (SdH) oscillations are evident.

Figure 5 shows all the results for the Hall Effect voltage as a function of applied magnetic field at different temperatures. Due to the misalignment of the Hall contacts (see Fig. 1) a magnetoresistance correction was applied. In Fig. 5, the Hall Effect voltage is always positive

IOP Conf. Series: Journal of Physics: Conf. Series **969** (2018) 012148 doi:10.1088/1742-6596/969/1/012148

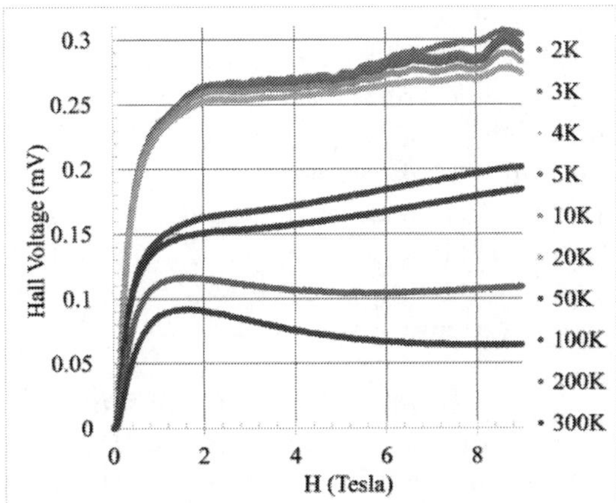


**Figure 5.** The corrected Hall Effect voltage with field scans at different temperatures (2K - 300K).

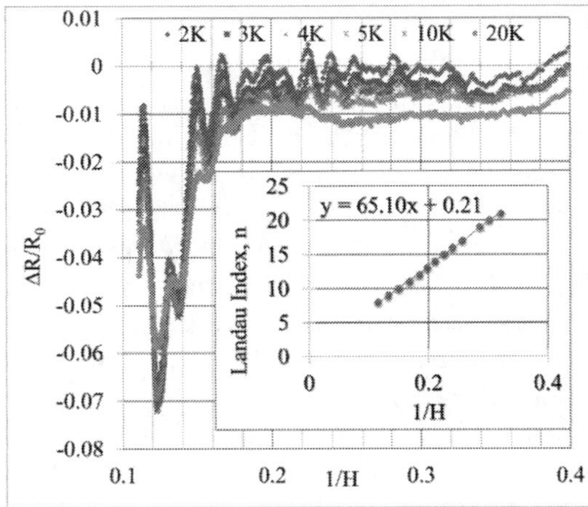


**Figure 6.** SdH oscillations of MR signal vs. $\frac{1}{H}$ ($T^{-1}$) for $T < 20$ K. The inset is a plot of the Landau Level index, n vs. $\frac{1}{H}$; its slope gives the frequency of oscillations.

for all temperatures, thus determining the carriers' positive polarity. One can distinguish the linear dependence of the Hall voltage on the applied field below 0.4 T for all temperatures, and subsequent different saturation/levelling off behavior for the high temperatures ($T > 50$ K). The Hall voltage linear region is used to obtain a slope, which in turn is used to infer the carrier concentration at that temperature. Further, the carrier concentration, the sample's measured resistance at that temperature and $H = 0$ T, the width, and the distance between the contacts are used to calculate the mobility of the carriers.

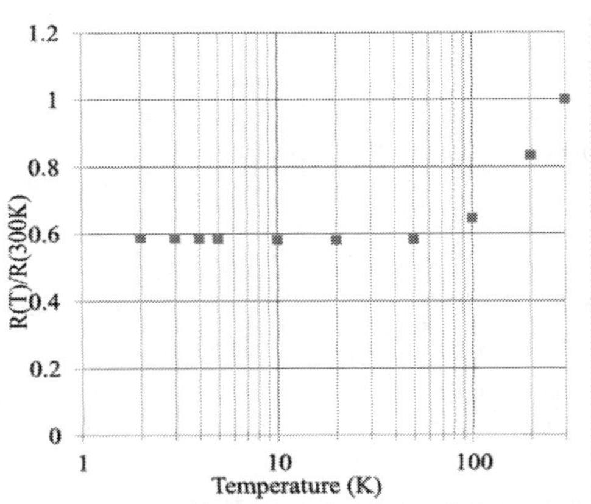


**Figure 7.** $R(T)/R(300K)$, with a room temperature resistance of 36Ω.

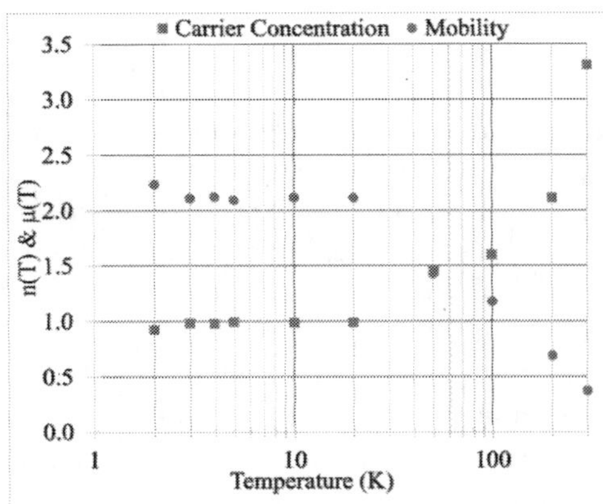


**Figure 8.** Carrier concentration, $n(T)$, in $10^{13}$ cm$^{-2}$ (blue) and mobility, $\mu(T)$, in $10^4$ cm$^2$/V·s (red).

Figure 6 shows the low temperature ($T < 20$ K) behavior of the magnetoresistance and its Shubnikov-de Haas oscillations, as $\frac{\Delta R(H)}{R(0)}$ vs. $\frac{1}{H}$. As can be seen, there are definite oscillations in $\frac{1}{H}$. However, their weakness, which may be due the micro width of the sample, prevents performing a Fourier transform on them to obtain the frequency(s). Instead one can label the maxima (peaks) as Landau levels and obtain the $\frac{1}{H}$ value at which they occur, as in Wu *et al.* [8]. This results in the inset of Fig. 6, as a plot of the Landau level index, n, as a function of $\frac{1}{H}$. A linear least squares fit to that data set yields a frequency of 65 T and an intercept of $0.21 \pm 0.34$, which may be indicative of the anomalous Berry's phase.

Figure 7 shows the resistivity temperature dependence in a semi-log plot of $R(T)/R(300K)$ vs. T. The sample has a room temperature resistance of 36$\Omega$ and a residual resistivity ratio (RRR) of 1.7, as can be deduced from the graph. This graph clearly shows the metal-like behavior of the resistivity of the sample. The resistivity of the sample is reasonably constant at low temperatures, up to 50 K, and increases with temperature above that, which is consistent with Matthiessen Rule.

Figure 8 shows a semi-log plot of the resulting carriers' concentration, $n$, in units of $10^{13}$ $cm^{-2}$ (blue) and their mobility, $\mu$, in units of $10^4$ $cm^2/V{\cdot}s$ (red) as functions of temperature, on the same graph. Both the concentration and the mobility of the carriers exhibit a metal-like behavior. They are both constant for temperatures below 50 K, and the carriers concentration subsequently increases with temperature due to thermal excitation above 50 K while the mobility decreases due to phonon scattering, as the phonons population increases with temperature.

## 4. Discussion/Conclusion

According to Hass *et al.* [9], multilayer graphene on SiC(000$\bar{1}$) behaves like a single sheet graphene. The behavior of the magnetoresistance results seem to qualitatively agree with theoretical prediction(s) by McCann *et al.* [6] as given in their Fig. 2, especially for field $H < 2$ T. The two predicted regimes correspond to the low temperature ($T < 20$ K) behavior before the onset of the SdH oscillations and the high temperature ($T > 50$ K) behavior. The characteristic magnetic fields ($H_{min}(T)$ and $H_1(T)$) and their temperature dependence in Fig. 4 may be linked to characteristic scattering times associated with phase-coherent graphene [6] by $H = \frac{hc}{8\pi De}\tau^{-1}$, with $D$ as the carrier diffusion constant.

The period of the SdH oscillations, ($\frac{1}{f} = \Delta(1/H)$), is related to the area of an extremal orbit [10] by:

$$A_e = \frac{4\pi^2 e}{h}/\Delta(1/H) = \frac{4\pi^2 e}{h} f \quad (1)$$

where $e$ is the electronic charge and $h$ is Plank's constant. Here, the orbit is a circle whose radius is $k_F$, the Fermi momentum, and whose area is $A_e = \pi k_F^2$. From the theoretical review of Castro Neto *et al.* [5], the Fermi momentum and the carrier concentration are related by $k_F^2 = \pi n$. $n$ is the carrier concentration, even when the carrier polarity is positive.

All this in turn yields the carrier concentration as $n = \frac{4e}{h} f$. Using the SdH frequency 65 T, the result is $n = 6.3 \times 10^{12}$ $cm^{-2}$, which compares favorably to $9.2 \times 10^{12}$ $cm^{-2}$, obtained from an average of the Hall Effect measurements at low temperatures. These carrier concentration values are comparable to the lower values obtained by Wu *et al.* [8] and consistent with the number of graphene layers of the sample. Additionally, the low temperature mobility values are lower but comparable to the results by Berger *et al.* [11] on epitaxial graphene, which may be due to the high carrier concentration deduced from the Hall Effect measurements.

Further, considering the residual scattering, one can infer the carrier mean free path and compare to the width of sample. Using the relation between the conductivity, mobility, and mean free path, $\sigma = en\mu = \frac{2e^2}{h}(k_F l)$, one obtains the mean free path, $l = \frac{h}{2e}\sqrt{\frac{n}{\pi}}\mu$. Use of the average numerical values for $n$ and $\mu$ at low temperature yields 1$\mu$m, a factor of two smaller than the width of the sample.

**Acknowledgment**
The author is grateful to Dr. Z. R. Robinson (SUNY Brockport Physics) and Dr. A. Nath (Navy Research Lab. Washington, DC) for providing the sample as well as fruitful discussions, and to the Department of Physics for local institutional support.

## References

[1] Kovoselov K S, Geim A K, Morozov S V, Jiang D, Zhang Y, Dubonos S V, Grigorieva I V, and Firsov A A, *Science* **306**, 666 (2004).
[2] Winterlin J and Bocquet M L, *Surf. Sci.* **603**, 1841 (2009).
[3] de Heer W A, *et al.*, *Solid State Commun.* **143**, 92 (2007).
[4] Robinson Z R, *et al.*, *Carbon* **81**, 73 (2015).
[5] Castro Neto A H, Guinea F, Peres N M R, and Novoselov K S, *Rev. Mod. Phys.* **81**, 109 (2009).
[6] McCann E, Kechedzhi K, Fal'ko V I, Suzuura H, Ando T, and Altshuler B L, *Phys Rev. Lett.* **97**, 146805 (2006).
[7] Burin P P and Pfleiderer C, *Rev. Sci. Inst.* **67**, 4024 (1996).
[8] Wu X, Li X, Song Z, Berger C, and de Heer W A, *Phys Rev. Lett.* **98**, 136801 (2007)
[9] Hass J, *et al.*, *Phys Rev. Lett.* **100**, 125504 (2008).
[10] Kittel C, *Intrduction to Solid State Physics*, 2005 (Wiley & Sons).
[11] Berger C, *et al.*, *Science* **312**, 1191 (2006)

# Magneto-photoconductivity of three dimensional topological insulator bismuth telluride

**Bingchen Cao[1], Mustafa Eginligil[2*], Ting Yu[1]**

[1] Division of Physics and Applied Physics, School of Physical and Mathematical Sciences, Nanyang Technological University, 637371, Singapore
[2] Key Laboratory of Flexible Electronics (KLOFE) & Institute of Advanced Materials (IAM), Jiangsu National Synergetic Innovation Center for Advanced Materials (SICAM), Nanjing Tech University (NanjingTech), 30 South Puzhu Road, Nanjing 211816, China

[*] iameginligil@njtech.edu.cn

**Abstract**. Magnetic field dependence of the photocurrent in a 3D topological insulator is studied. Among the 3D topological insulators bismuth telluride has unique hexagonal warping and spin texture which has been studied by photoemission, scanning tunnelling microscopy and transport. Here, we report on low temperature magneto-photoconductivity, up to 7 T, of two metallic bismuth telluride topological insulator samples with 68 and 110 nm thicknesses excited by 2.33 eV photon energy along the magnetic field perpendicular to the sample plane. At 4 K, both samples exhibit negative magneto-photoconductance below 4 T, which is as a result of weak-antilocalization of Dirac fermions similar to the previous observations in electrical transport. However the thinner sample shows positive magneto-photoconductance above 4 T. This can be attributed to the coupling of surface states. On the other hand, the thicker sample shows no positive magneto-photoconductance up to 7 T since there is only one surface state at play. By fitting the magneto-photoconductivity data of the thicker sample to the localization formula, we obtain weak antilocalization behaviour at 4, 10, and 20 K, as expected; however, weak localization behaviour at 30 K, which is a sign of surface states masked by bulk states. Also, from the temperature dependence of phase coherence length bulk carrier-carrier interaction is identified separately from the surface states. Therefore, it is possible to distinguish surface states by magneto-photoconductivity at low temperature, even in metallic samples.

## 1. Introduction

Topological insulator (TI) is an exotic phase of quantum matter with an insulating bulk state as well as time-reversal symmetry-protected Dirac-like surface states. The materials having TI phase belong to a large family of material systems ranging from 2D to 3D and from HgTe quantum wells to bismuth, selenium, tellurium, and antimony based alloys [1]. Well-known examples of bismuth based TI materials are bismuth selenide ($Bi_2Se_3$) and bismuth telluride ($Bi_2Te_3$) which are 3D TI for thickness above 5 nm (5 quintuples) [2]. One of the intriguing differences between these two similar tetradymite type structure 3D TI systems, is the unconventional hexagonal warping of the surface states in $Bi_2Te_3$ [3] which is the reason behind the observations of snowflake shape of the Fermi surface in angle resolved photoemission spectroscopy [4] and surface band oscillations in scanning tunnelling microscopy [5]. A rich spin texture is a natural expectation of the warping of the surface states. In order to make use of the spin properties of TI systems, as a first step, the conduction of surface states shouldn't be masked by bulk conduction. Although $Bi_2Te_3$ is usually rather metallic and the Fermi energy is way above the Dirac point, it is possible to make $Bi_2Te_3$ non-metallic and Shubnikov-de Haas (SdH) oscillations can be observed; which is a clear sign of the surface state conduction as shown at low temperature transport studies [6]. Another transport feature of TI materials is the weak-antilocalization (WAL) observed as a negative magneto-conductance in $Bi_2Te_3$ thin films [7] due to the absence of backscattering by nonmagnetic impurities [8]. At high magnetic fields, the WAL is suppressed. In magneto-transport of 3D $Bi_2Te_3$, a detailed and careful analysis is required to extract information about surface state conduction. In these two transport signatures, the SdH oscillations and the WAL, distinguishing bulk conduction from surface state conduction remains as a challenge.

On the other hand, opto-electronic response of the surface states can be obtained from helicity-dependent photocurrents by using circularly polarized light [9]. As demonstrated in 250 nm thick 3D TI $Bi_2Se_3$, the sign of spin-dependent photocurrent changes based on light helicity as a result of circular photogalvanic effect (CPGE) acting on the helical Dirac surface states [10]. This shows the importance of photocurrent studies in TI to identify the surface state conduction. The CPGE can also be observed at normal incidence if a magnetic field is applied [9]. Therefore it is possible to observe surface state conduction by photocurrent at non zero magnetic field. The photo-excited carriers in the surface states can couple with the magnetic field and there will be a net photocurrent which has magnetic field dependence. In this work, we present low temperature magneto-photoconductance (MPC) of two 3D $Bi_2Te_3$ samples, with different thickness, excited by polarized light with a normal incidence on to the sample plane and the magnetic field applied in the illumination direction. We show that even at zero degree angle of photon polarization it is possible to distinguish surface states from bulk states by MPC and provide a discussion in terms of localization effects. Then we further elaborate the temperature dependence of MPC in the thicker sample which shows unexpected localization behaviour. We interpret this observation in terms of separation of surface state conduction from bulk conduction.

## 2. Experimental

Exfoliated flakes of $Bi_2Te_3$ were fabricated into four terminal devices. Two samples were picked for MPC measurements, with thickness 68 nm (S1) and 110 nm (S2) as determined by AFM measurements. The current voltage (IV) characteristics and the laser power dependence of photocurrent of the two samples are similar and comparable to the previously reported ones [11]. Figure 1(a) and (b) shows the IV at room and 4 K, and photocurrent at various laser powers at 4 K, respectively, for S1.

Measurements are performed in a cryostat with a micrometer precision stage. After locating the devices, IV is performed by four probe technique. The measured linear IV indicates the contacts are ohmic. Then, a beam spot of 1 micron is focused at the center of the two electrodes separated by 2 microns. A linearly polarized solid state laser with wavelength 532 nm chopped by 137 Hz frequency and the magnitude and phase of photocurrent signal are collected by a low noise current pre-amplifier connected to a lock-in amplifier. A quarter waveplate is set initially to zero degree photon polarization. Photocurrent is measured for laser powers lying in the linear response regime. The magnetic field is applied perpendicular to the sample plane.

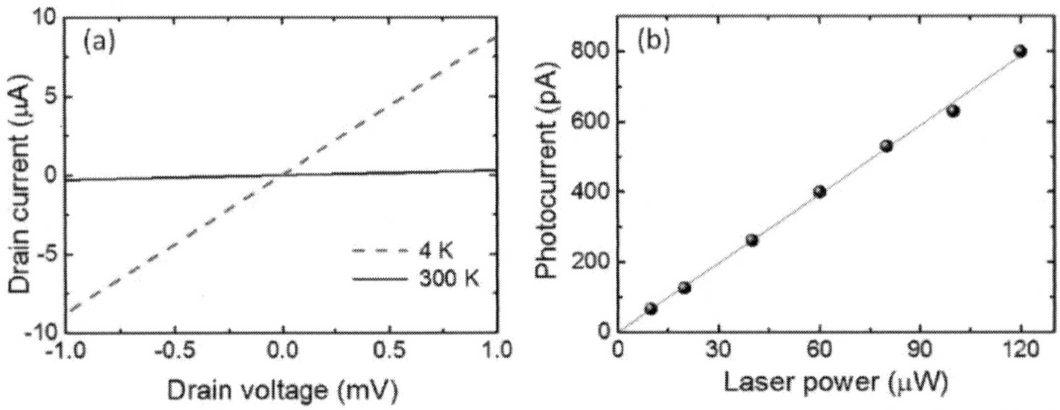


Figure 1 (a) IV characteristics of sample 1 (S1) and (b) photocurrent as a function of incident laser for S1. The photocurrent response is linear with laser power.

## 3. Results and discussion

### *3.1. Laser power and thickness dependence of the MPC*

In Fig. 2(a) the laser power dependence of the MPC of the thinner sample, S1 for 80 and 120 μW laser powers at 4 K is given., The MPC behaviors at two different laser powers are similar and based on the linear dependence of photocurrent of Fig. 1(b), it seems to be proportional at all fields. There is a negative MPC up to 4 T, which turns to a positive MPC at higher fields. The negative MPC is qualitatively similar to the previously observed WAL in transport [7]. As the field increases, the negative MPC is suppressed by spin-dependent transport term due to the surface states since there are no magnetic impurities in the system. But also, as the 3D quantum limit is exceeded at high fields [12] and two surface states can be coupled as the thickness get smaller [13], it is possible to observe a positive MPC. This turnover from negative MPC to positive MPC at 4 T, agrees well with the previous transport in low carrier density $Bi_2Se_3$ [12].

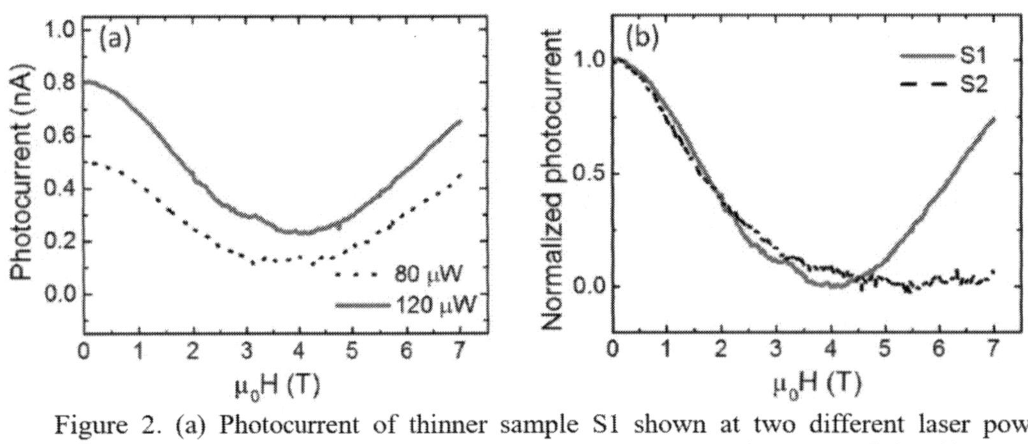


Figure 2. (a) Photocurrent of thinner sample S1 shown at two different laser power (b) normalized photocurrent at 120 μW of two samples are compared. The MPC of the thinner sample dramatically changes sign around 4 T, while thicker sample becomes constant.

While S1 exhibits a turnover as a function of field and several photocurrent contributions take place; the thicker sample, S2, exhibits a simple negative MPC behavior. In Fig. 2(b), the photocurrent values of S1 and S2 at 4 K and 120 μW laser power are compared. The photocurrent value of S1 at zero field (4 T) is taken as 1 (0) and the photocurrent value of S2 at zero field (5.5 T) is taken as 1 (0). The negative MPC in S2 seems to saturate around 5.5 T. But at higher fields, it looks like constant rather than a positive MPC. This could be due to the increase in the ratio of bulk state conduction to surface state conduction as the thickness of TI increases. Since there is no positive MPC at higher fields, one can think that there is only one surface state at play and the analysis of the thicker sample S2 can be simpler. Similar to magneto-transport, this effect can be taken as WAL effect and useful parameters can be extracted.

### *3.2. Temperature dependence of MPC*

The magnitude of negative MPC of S2 drops as temperature increases as seen in Fig.3. There are some fluctuations of MPC, which may remind the Aharanov-Bohm oscillations observed in topological insulator nanoribbons for in-plane magnetic fields [14]. On the other hand, S1 doesn't seem to have these fluctuations. As S2 is twice as thick, there could be contribution from the side surface states coupling with magnetic field. Although there is no clear trend in the fluctuations in S2, they are up to 10 pA. Given the fact that the change in photocurrent as a function of time at zero field within half an hour is not more than 1 pA, these fluctuations are not only noise. This need to be further studied in a different setup in which the applied field can be varied from perpendicular to the sample plane to in-plane.

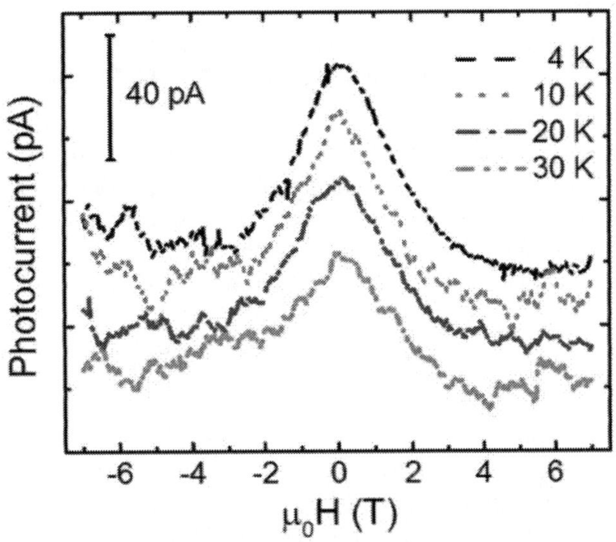


Figure 3. Temperature dependence of MPC at several temperatures shown.

The temperature dependent MPC data can be fit to localization formula [15] for photocurrent as follows,

$$\Delta PC = -\frac{\alpha e^2}{2\pi^2\hbar}\left[ln\frac{\hbar}{4Bel_\phi^2} - \psi\left(\frac{1}{2}+\frac{\hbar}{4Bel_\phi^2}\right)\right]$$

where e is the electronic charge, $\hbar$ is the Planck's constant, $l_\phi$ is the phase coherent length, and $\psi$ is the digamma function, $\alpha$ should be equal to 1 and -1/2 for the orthogonal and symplectic cases, respectively. The fitting in Fig. 4(a) yields, $\alpha$ = -0.65, -0.56, -0.32, and 0.26 at 4 K, 10 K, 20 K, and 30 K, respectively. If $\alpha$ is -1/2 there should be one surface state, if it is equal to -1 there should be two surface states involved in the conduction of this strong spin-orbit coupling system. At temperatures below 20 K, there seems to be one surface state contributing to the conduction, which also explains why there is no positive MPC at higher fields. As temperature increases $\alpha$ drops because the surface state conduction becomes smaller. Eventually, at 30 K, there is a sign change of $\alpha$ which indicates, bulk states completely mask the surface states conduction.

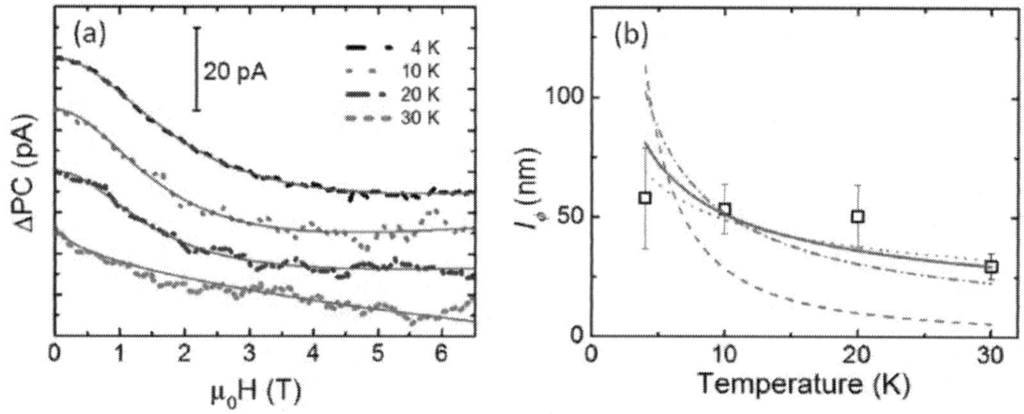


Figure 4. (a) The difference between the photocurrent PC at a field and the PC at zero field, $\Delta PC$, as a function of magnetic field at several temperatures is given, the curves are repositioned for a better presentation, the red solid lines are the localization fittings (b) phase coherence lengths plotted against temperature and fitted to $T^{-p/2}$ for p = 0.75 (dot), 1 (solid), 1.5 (dash dot), and 3 (dash).

In Figure 4(b) the phase coherence lengths $l_\phi$ that are extracted from the localization fitting are plotted. The temperature dependence of $l_\phi \sim T^{-\frac{p}{2}}$ can provide information about the carrier-carrier (p = 3/2) or carrier-phonon interactions (p=3) [16]. Several possible fittings are shown: p=3 (dash) seems to be worse; while p=0.75 (dot) and 1 (solid) are better ones (which have no physical meaning), p=1.5 (dash dot) is a better fit than p=3. This may indicate the electron-electron interaction, which is due to the bulk carriers may also play a role at low temperatures. This means in addition to weak-antilocalization due to the surface states, bulk state conduction can be identified by carrier-carrier interaction [17]. However, there is only one data point under 10 K and this data point deviates from the expectation of p=1.5 curve. Therefore there is a need for more data points less than 10 K to have a better understanding of this MPC feature.

## 4. Summary and conclusion

Briefly, this work presents the difference in magneto-photoconductivity behavior of two metallic 3D $Bi_2Te_3$ samples with different thickness for 2.33 eV linearly polarized excitation with zero photon polarization. A negative MPC for both samples at fields lower than 4 T can be understood as WAL similar to transport. At higher fields, 68 nm thick sample shows positive MPC due to the coupling of surface states. The 110 nm thick sample shows photocurrent due to one surface state and as shown by the localization fitting analysis of MPC data at several temperatures, bulk conduction can be identified separately. This demonstrates another way to identify surface states, in addition to the photocurrent at zero magnetic field and oblique incidence. Similarly, the control over surface states by MPC can be studied by right and left circularly polarized excitations, as well as wavelength dependence of MPC.

**Acknowledgements**

This work is supported in part by the National Synergetic Innovation Centre for Advanced Materials (SICAM), the start-up fund by Nanjing Tech University, and Jiangsu 100 Talent. T. Yu acknowledges the support from the Singapore Ministry of Education under MOE2015-T1-001-065 and MOE2015-T1-002-126.

## References

[1] Yan B, and Zhang S-C, *Rep. Prog. Phys.* **75**, 096501 (2012).
[2] Zhang Y *et al.*, *Nat. Phys.* **6** 584–588 (2010).
[3] Fu L *Phys.Rev. Lett.* **103**, 266801 (2009).
[4] Chen Y L *et al.*, *Science* **325**, 178 (2009).
[5] Alpichshev Z *et al.*, *Phys.Rev. Lett.* **104**, 016401 (2010).
[6] Qu D-X, Hor Y S, Xiong J, Cava R J, and Ong N P, *Science* **329**, 821 (2010).
[7] He H-T *et al.*, *Phys.Rev. Lett.* **106**, 166805 (2011).
[8] Zhang T *et al.*, *Phys.Rev. Lett.* **103**, 266803 (2009).
[9] Hosur P, *Phys.Rev. B* **83**, 035309 (2011).
[10] McIver J M, Hsieh D, Steinberg H, Jarillo-Herrero P, and Gedik N, *Nat. Nanotechnol.* **7**, 96 (2012).
[11] Yao J D, Shao J M, Li S W, Bao D H, and Yang G W, *Sci. Reports* **5**, 14184 (2015).
[12] Analytis J G, McDonald R D, Riggs S C, Chu J-H, Boebinger G S, and Fisher I R, *Nat. Phys.* **6**, 960-964 (2010).
[13] Akiyama R, Fujisawa K, Sakurai R, and Kuroda S, *Journal of Physics: Conference Series*, **568**, 052001 (2014).
[14] Peng H, *Nat. Mater.* **9**, 225 (2010).
[15] Hikami S, Larkin A I, and Nagaoka Y, *Prog. Theor. Phys.* **63**, 707 (1980).
[16] Lu H-Z, AND Shen S-Q, *Front. Phys.* **12**(3), 127201 (2017).
[17] Wang J *et al.*, Phys. Rev. B **83**, 245438 (2011)

IOP Conf. Series: Journal of Physics: Conf. Series **969** (2018) 012150 doi:10.1088/1742-6596/969/1/012150

# Magnetoresistance measurements of tetralayer graphene device with single gate electrode

**T Hirahara[1], R Ebisuoka[1], K Watanabe[2], T Taniguchi[2] and R Yagi[1]**

[1]Graduate School of Advanced Sciences of Matter, Hiroshima University, Higashi-Hiroshima, Japan
[2]National Institute for Material Science, Tsukuba, Japan

yagi@hiroshima-u.ac.jp

**Abstract.** We have studied low temperature magnetoresistance of tetralayer graphene. By analysing quantum magnetoresistance oscillation in detail, carrier density dependence of the fermi surfaces of orbits indicated that tetralayer graphene has two bilayer like bands whose energy of bottom of the bands were different. We estimated ratio of their band mass using carrier density dependence of magnetoresistance.

## 1. Introduction

Graphene has been studied intensively since the successful discovery of a method to create graphene flake with adhesive tape [1]. Electronic band structure of monolayer, bilayer and trilayer have been uncovered by experiments [1-6], showing unique properties for massless and massive Dirac fermions. Tetralayer, on the other hand, has not been fully understood. It was expected to have two bilayer like bands [7,8]. There are a few experimental works: insulating behavior near charge neutrality points studied by double-gate suspended graphene [9], and landau level structure studied by capacitance measurements [10]. In this paper, we measured Shubnikov-de Haas oscillation (SdHO) by a resistance measurement and studied electronic band structure of tetralayer graphene with a Bernal stacking (i.e., ABAB stacking) focusing on gate voltage dependence of carrier density of fermi surfaces.

## 2. Methods

We fabricated graphene by mechanical exfoliating high-quality Kish graphite with adhesive tape. Layer umber was determined by color intensity (RGB digits) of optical images of graphene. A large number of graphene samples were measured to obtain distribution of RGB digits of graphene color. The distribution exhibited quantization due to discrete change of graphene's color, which allowed us to determine layer number precisely. To improve quality, graphene flakes were transferred onto high-quality *h*-BN flakes using a washer method using PVA (poly vinyl alcohol) acrylic resin [6] as shown in Figure 1. Electric leads to measure resistivity were made by a standard electron beam lithography technique. An optical micrograph of a typical sample is displayed in Fig. 2. The graphene is indicated with lines for clarity. The Graphene flake consisted of tetralayer graphene. Fewer layer graphene was also present at the left regime. However we could measure resistance of tetralayer graphene with a four terminal measurement using Electrodes 3 and 1 for current injection and drain, and Electrodes 5 and 6 for voltage measurement, which allowed us to exclude picking up voltage drop due to fewer graphene layers.

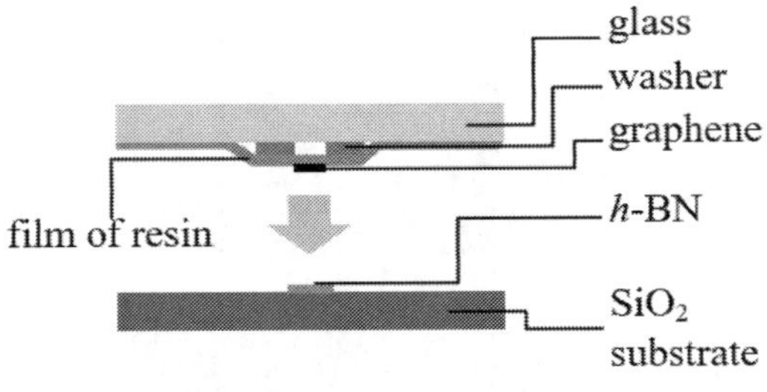


**Figure 1.** A schematic diagram of washer method to transfer graphene on *h*-BN.

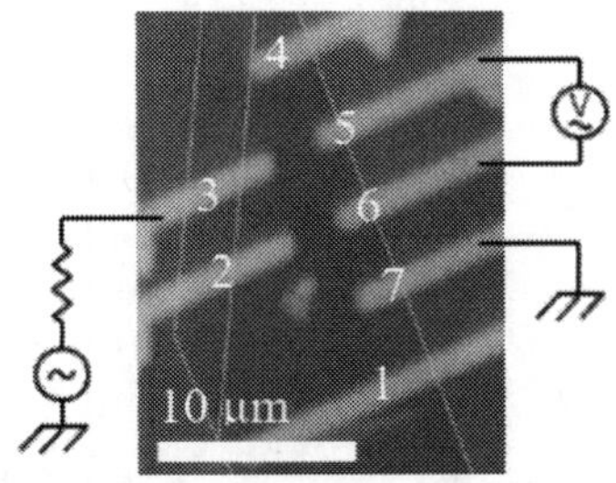


**Figure 2.** Optical micrograph of a tetralayer graphene sample. Graphene is indicated with lines.

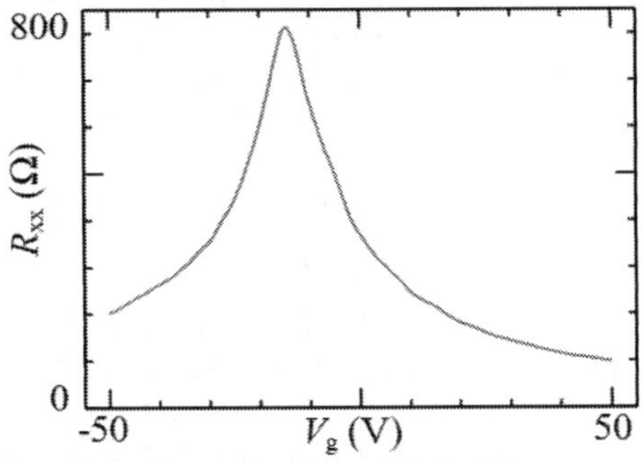


**Figure 3.** $V_g$-dependence of resistivity measured at $B = 0$ T and $T = 4.2$ K.

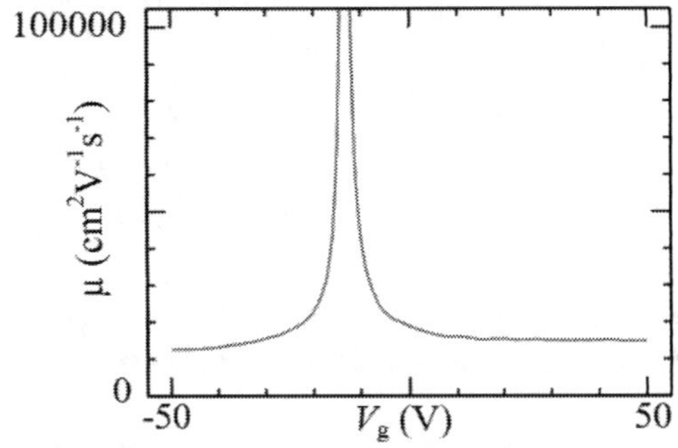


**Figure 4.** $V_g$-dependence of mobility.

Carrier density was tuned by a back gate voltage. A specific back gate capacitance was $C_g = 108$ aF$\mu$m$^{-2}$. Magnetic fields were applied perpendicular to graphene devices using a superconducting solenoid. Measurements were done at a liquid helium temperature, $T = 4.2$ K. Electric resistance was measured using a standard lock-in technique with excitation current of 1 $\mu$A.

## 3. Sample characterization

$V_g$-dependence of resistance at $B = 0$ T is shown in Fig. 3. Resistance took a peak (about 800 $\Omega$) at about $V_g = -15$ V, which apparently indicated a semiconducting or a semi-metallic behavior rather than a strong insulating one. Sample quality was estimated by a mobility, $\mu = 1/ne\rho$, as shown in Fig. 4. At high carrier density regime, mobility was about 10,000 cm$^2$V$^{-1}$s$^{-1}$, which was moderately higher than those that were expected for graphene samples fabricated on $SiO_2$ substrates.

## 4. Magnetotransport measurements

Left panel of Figure 5 shows traces of magnetoresistance for different carrier densities. Oscillations are due to Shubnikov-de Hass (SdH) effect arising from Landau quantization. Oscillations are periodic in $1/B$ as shown in a central panel of Fig. 5, showing a replot against reciprocal magnetic fields $1/B$. Oscillations looked complicated: they composed of more than one frequency. Fast Fourier Transform (FFT) was employed to determine frequency of the oscillations. Right panel shows Fourier spectra. Complicated peak structures in low frequency regime were possibly due to back ground of magnetoresistance. At higher frequencies, we could discern two salient peaks that varied with gate voltages. At electron regime, (*i.e.* $V_g > -13.3$ V) two peaks were observed. Apparently these were not higher order harmonics, but would originate from fermi surfaces of different subbands. Figure 6 shows carrier density estimated from the frequency of SdHO. Here we used formulas, $n_{osc} = gef_B/h$ and $f_B = 1/\Delta(B^{-1})$, where $\Delta(B^{-1})$ is a period, $g$ is degeneracy of landau level which we assumed to be 4 considering valley and spin. First, we focus our attention in electron regime where two frequency components were observed. These two would be light mass and heavy mass bilayer bands in tetralayer graphene. From the $V_g$-dependence of carrier density, red and blue points were identified to be light

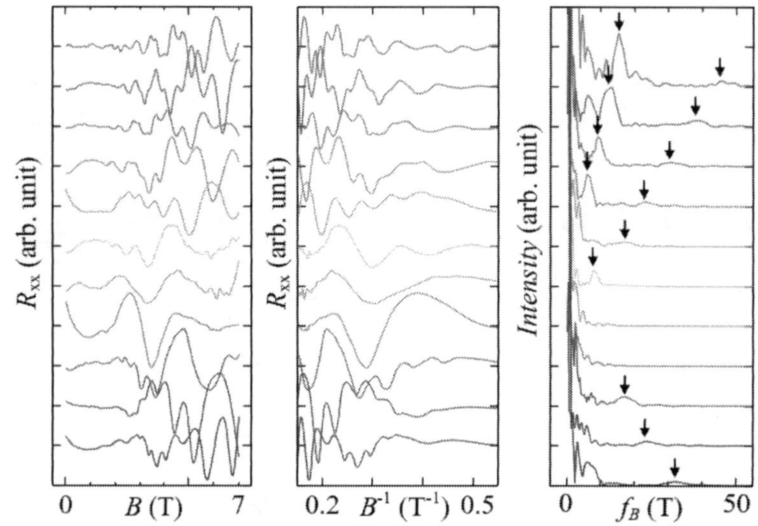


**Figure 5.** (Left) Magnetoresistance as a function of magnetic fields. Back ground was subtracted to obtain oscillatory components. (Center) Replot against $B^{-1}$. (Right) FFT results. Data were offset. From bottom to top $V_g$ was varied from -70 V (blue) to 70 V.

and heavy mass bilayer bands, respectively. Rate of variation of $n_{osc}$ for each orbit as a function of total carrier density $n_{tot}$ was dependent on the mass of the subbands since $\delta n_{osc} \propto m\delta E_f$ where $m$ is a band mass and $E_f$ is Fermi energy. Here we assumed that dispersion relation of bilayer bands can be approximated by $E - E_0 \approx \hbar k^2/2m$ where $E_0$ is the energy of the band bottom and $m$ is a band mass. From the slope of $V_g$-dependence of $n_{osc}$, ratio of the masses were estimated to be $m_{light}/m_{heavy} \sim 0.4$.

The sum of carrier density for light mass and heavy mass bilayer bands yields total carrier density. Black points in Fig. 6 is a sum of these carrier density determined from oscillation. The sum is approximately the same as $n_{tot}$ (large about factor of 1.025). This indicated that there are only two bands in this system.

Moreover, we could extract information about the dispersion relation from $n_{tot}$-dependence of $n_{osc}$. As mentioned earlier $n_{osc}$ for two bands both varied linearly with $n_{tot}$. This was a characteristic feature of subbands with dispersion, $E \propto k^2$. For other dispersion relation, such as $E - E_0 \propto k$, or $E - E_0 \propto k^3$, $n_{osc}$ varies as $(E_f - E_0)^2$ or $(E_f - E_0)^{2/3}$, resulting in non-linear $n_{tot}$ dependence of $n_{osc}$.

Energy of band bottom for light mass bilayer band was larger than that for heavy mass bilayer bands. This can be seen by the $n_{tot}$ value at $n_{osc}$, which was determined by extrapolating a relation between $n_{osc}$ and $n_{tot}$. A least square fitting for a liner line for light mass bilayer bands (red) intercepts the ordinate at a large value of $n_{tot}$ than that for heavy mass bilayer bands (blue).

Detailed landau level structure was revealed by the mapping magnetoresistance as a function of magnetic field and carrier density as shown in Fig. 7. We could discern characteristic landau level structure of tetralayer graphene. Stripes corresponding to landau levels with different indices arose from SdH. The shape of this fan diagram was different from monolayer, bilayer and trilayer graphene. The fan diagram consisted of two sets of fans which are superposed. This is a direct consequence of above mentioned linearity in $n_{osc}$ to $n_{tot}$.

Offset of conduction band bottoms were observed as an offset of zero-mode landau levels. We could easily identify a zero-mode landau level for a heavy mass bilayer band at $n_{tot} \sim 0$ cm$^{-2}$. White broken lines in the figure indicates filling factors $\nu = 4i$, where $i$ denotes integers. Salient stripes at high magnetic fields are mostly due to heavy mass bilayer bands. Near $n_{tot} = 1\times10^{12}$ cm$^{-2}$, $N = 0$ landau level of the light mass bilayer band gave rise to jumps in the filling factors of the landau levels for the heavy mass. Apparently the filling factors varied by 8, showing the $N = 0$ landau level had eight-fold degeneracy as expected for bilayer bands.

IOP Conf. Series: Journal of Physics: Conf. Series **969** (2018) 012150 doi:10.1088/1742-6596/969/1/012150

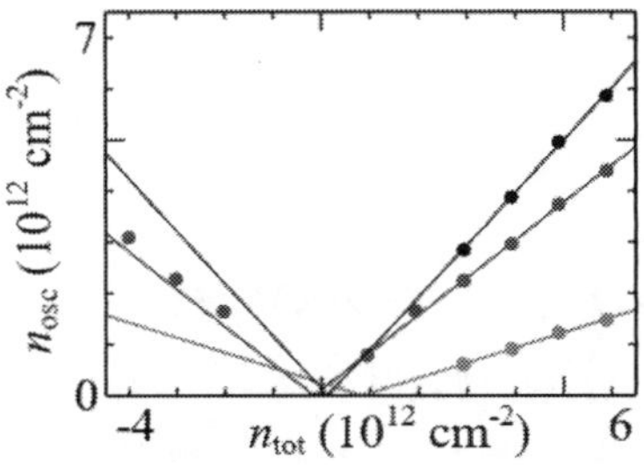


**Figure 6.** Carrier density calculated from frequency of SdHO. Horizontal axis is carrier density calculated form gate voltage. Red points are carrier density for light-mass bilayer band. Blue points are for heavy-mass bilayer band. Black points are sum of these.

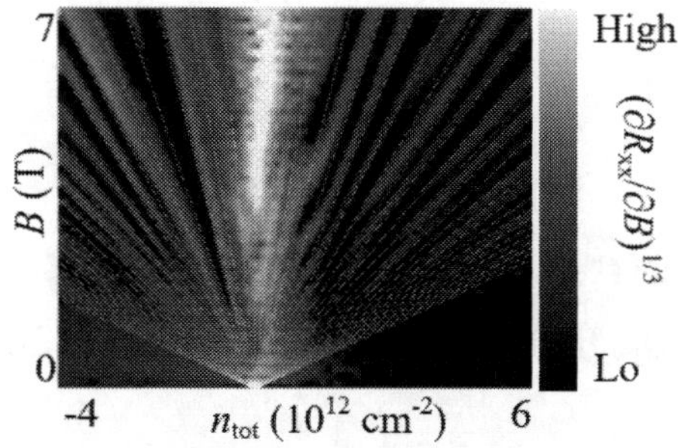


**Figure 7.** Mapping of magnetoresistance as a function of magnetic fields and carrier density. d$R_{xx}$/d$B$ was plotted. Broken lines indicate filling factor of landau levels.

From the above results, there are two sets of bilayer band. There are three different stacking structures, ABAB, ABCA, and ABCB in tetralayer graphene. These are expected to have different electronic structures, number of bands and their dispersions are different to one another. The ABAB stacking is the only stacking that has two-sets of bilayer bands.

## 5. Conclusion

We have succeeded in observing landau level structure of Bernal stacked tetralayer graphene by magnetotransport measurements. Tetralayer graphene has verified to have to sets of bilayer bands. The light mass bilayer band had larger energy for conduction band bottom than that for the heavy mass bilayer band. A ratio of band masses was estimated.

## References

[1] Novoselov K S, Geim A K, Morozov S V, Jiang D, Katsnelson M I, Grigorieva I V, Dubonos S V and Firsov A A 2005 *Nature* **438** 197-200

[2] Zhang Y, Tan Y-W, Stormer H L and Kim P 2005 *Nature* **438** 201-204

[3] Novoselov K S, McCann E, Morozov S V, Fal'ko V I, Katsnelson M I, Zeitler U, Jiang D, Schedin F and Geim A K 2006 *Nat. Phys.* **2** 177-180

[4] Zhang L M, Li Z Q, Basov D N, Fogler M M, Hao Z and Martin M C 2008 *Phys. Rev.* B **78** 235408

[5] Ohta T, Bostwick A, Seyller T, Horn K and Rotenberg E 2006 *Science* **313** 5789

[6] Taychatanapat T, Watanabe K, Taniguchi T and Jarillo-Herrero P 2011 *Nat. Phys.* **7** 621-625

[7] Koshino M and McCann E 2011 *Phys. Rev.* B **83** 165443

[8] Aoki M and Amawashi H 2007 *Soli State Commun.* **142** 3 123-127

[9] Grushina A L, Ki D K, Koshino M, Nicolet A A, Faugeras C, Mccann E, Potemski M and Morpurgo A F 2015 *Nat. Commun.* **6** 6419

[10] Wu Z F, Han Y, Lin J X Z, Zhu W, He M Q, Xu S G, Chen X L, Lu H H, Ye W G, Han T Y, Wu Y Y, Long G, Shen J Y, Huang R, Wang L, He Y H, Cai Y, Lortz R, Su D S and Wang N 2015 *Phys. Rev.* B **92** 075408

**Acknowledgements**
This work was supported by JSPS KAKENHI Grant Number JP25107003 from MEXT Japan.

# Annealing effect on effective mass of two-dimensional electrons in InGaAsN/GaAsSb type II quantum well

**Shuichi Kawamata[1,2], Sho Tanaka[1], Akira Hibino[1] and Yuichi Kawamura[1,2]**

[1] Graduate School of Engineering, Osaka Prefecture University, Sakai, 599-8531, Japan
[2] Organization for Research Promotion, Osaka Prefecture University, Sakai, 599-8570, Japan

s-kawamata@riast.osakafu-u.ac.jp

**Abstract**. The InP-based InGaAs/GaAsSb type II multiple quantum well is the system for developing optical devices for 2 – 3 μm wavelength regions. By doping nitrogen into InGaAs layers, the system becomes effective to fabricate the optical devices with longer wavelength. The epitaxial layers of InGaAsN/GaAsSb on InP substrates are grown by the molecular beam epitaxy. The electrical resistance has been measured as a function of the magnetic field up to 9 Tesla at several temperatures between 2 and 8 K. The effective mass is obtained from the temperature dependence of the amplitude of the Shubnikov-de Haas oscillations. We have reported the nitrogen concentration dependence of the effective mass on the InGaAsN/GaAsSb type II system. The effective mass increases as the nitrogen concentration increases from 0.0 to 1.5 %. In this report, the annealing effect on the effective mass is investigated. The effective mass decreases by the annealing. This result suggests that some amount of nitrogen atoms of the InGaAsN layers are considered to diffuse to the GaAsSb layers by the annealing.

## 1. Introduction

The mid-infrared laser technology by compound semiconductors grown on the InP substrate has been well developed [1-4]. Therefore, the InP-based compounds have been extensively studied in order to develop optical devices. Since mid-infrared light has absorption bands for several molecules such as $NO_x$, $SO_x$, $CO_2$, a lot of potential applications are expected for pollution monitoring, chemical gas analysis, and medical diagnostics. The InP-based InGaAs/GaAsSb type II multiple quantum well system has been investigated in order to develop the optical devices for 2 - 3 μm wavelength regions [5-7]. According to measurements of photoluminescence and photoreflectance for the InGaAsN mono layer on the InP substrate, the band gap energy becomes smaller by doping nitrogen into InGaAs layers [5]. Therefore, the InGaAsN/GaAsSb system becomes more effective in fabricating the optical devices for a longer wavelength by doping nitrogen into the InGaAs layers. In fact, the light emitting diodes have been developed by using the InGaAsN/GaAsSb type II quantum well on the InP substrate. The wave length corresponding to the peak energy of the electro-luminescence increases from 2.34 μm for N = 0.0 % to 2.86 μm for N = 1.0 % [8]. On the other hand, terahertz quantum cascade lasers based on type II InGaAs/GaAsSb/InP [9] have been developed in addition to the mid-infrared lasers [10].

IOP Conf. Series: Journal of Physics: Conf. Series **969** (2018) 012151 doi:10.1088/1742-6596/969/1/012151

Substituting nitrogen for arsenic in the InGaAs layers introduces carriers of electrons in the layers. We have reported the nitrogen concentration dependence of the effective mass on two-dimensional electrons of the InGaAsN/GaAsSb type II system. The effective masses of the epitaxial layers with several nitrogen concentrations N = 0.0, 0.6, 1.2 and 1.5 % are obtained from the temperature dependence of the amplitude of the Shubnikov-de Haas (SdH) oscillations. The effective mass increases as the nitrogen concentration increases from N = 0.0 to 1.2 % then saturates from N = 1.2 to 1.5 % [11]. On the other hand, the nitrogen concentration dependence of the bandgap energy, $E_g$ has been investigated by optical absorption measurements for the InGaAsN epitaxial layers with nitrogen concentrations N = 0.0, 0.5, 1.1, 1.6, 1.8 and 1.9 %. $E_g$ decreases as the nitrogen concentration increases from N = 0.0 to 1.1 % then saturates from N = 1.1 to 1.9 % [11]. These results are consistent with the band anticrossing model [12,13]. The mass enhancement corresponding to the reduction of the bandgap energy with increasing the nitrogen concentration is explained in the following way: The hybridization of the original conduction band and the narrow resonant band by nitrogen causes splitting into two subbands. As a result, the bottom energy of the conduction band decreases to reduce the bandgap energy. Simultaneously, flattening of the dispersion relation curve, $E(k)$ occurs to make the effective mass heavier.

In this report, the effective mass of two-dimensional electrons are reported on the annealed InGaAsN/GaAsSb type II system. The effective mass is obtained from the temperature dependence of the amplitude of the SdH oscillations observed in the magnetic field dependence of the electrical resistance. Annealing effect on the effective mass is discussed.

## 2. Experimental

The multiple quantum well (MQW) InGaAsN/GaAsSb epitaxial layers on semi-insulating Fe-doped (100) InP substrates were grown by molecular beam epitaxy (MBE) method at 480 degrees Celsius. The InAlAs buffer layer with the thickness of 0.10 μm was grown on the InP substrate first. Then, the InGaAsN and GaAsSb layers were grown with 50 periods. The thickness of each layer was 7 nm for both of the InGaAsN and GaAsSb layers. This thickness of 7 nm was determined in order to obtain the light emission with the relatively longer wave length. The growth rates of the InGaAsN layers were 1.45 μm/h and those of GaAsSb layers were 0.70 μm/h. Arsenic was supplied by needle-valve cracking cells, where the cracking temperature is set to be 600 degrees Celsius, so that tetramer $As_4$ was used. The tetramer $Sb_4$ was supplied by a conventional effusion cell. Indium and gallium were supplied by conventional effusion cells. Nitrogen was supplied by using an RF plasma cell. The composition ratio of the epitaxial layer is $In_{0.53}Ga_{0.47}As_{1-N}N_N/GaAs_{0.5}Sb_{0.5}$. The composition ratio of $In_{0.53}Ga_{0.47}$ was established in order to achieve lattice matching to the InP substrate. There is no compensation doping in the epitaxial layers.

The X-ray diffraction measurements show that lattice mismatching is less than 0.1 % by introducing nitrogen into the InGaAs layers. Post-growth thermal annealing was carried out for 30 sec in nitrogen cover-gas environments. The InGaAsN/GaAsSb MQW with nitrogen concentration N = 1.2 % were annealed at 550 degrees Celsius and 600 degrees Celsius, and that with N = 1.5 % were annealed at 600 degrees Celsius. No difference between before and after the annealing was observed in the X-ray diffraction measurements.

The longitudinal electrical resistance was measured by an AC four probe method with a current of 0.1 mA and frequency of 23 Hz. The measurements have been carried out as a function of the magnetic field, $B$ up to 9 Tesla at several temperatures, $T$ between 2 and 8 K. In order to remove the inference on the longitudinal resistance by the transverse component due to the Hall effect, the longitudinal electrical resistance at magnetic field $B$, $R(B)$ was obtained as the average of the resistance at $+B$ and that at $-B$, $(R(B)+R(-B))/2$.

## 3. Results and discussion

The magnetic field dependencies of the electrical resistance at 2 K for the InGaAsN/GaAsSb with N = 1.2 % are shown in figure 1. Those for the InGaAsN/GaAsSb with N = 1.5 % are shown in figure 2. The dotted, broken and solid lines indicate data for the epitaxial layers of as-grown at 480 degrees Celsius, annealed at 550 degrees Celsius and annealed at 600 degrees Celsius, respectively. The SdH oscillations are observed for all five data. The SdH oscillations also appear at all temperatures between 2 and 8 K for all epitaxial layers.

At each temperature, the amplitude of the SdH oscillation is obtained by the fast Fourier transform of the invers magnetic field dependence of the resistance, $R(1/B)$ in the magnetic fields between 2 and 9 Tesla. The temperature dependence of the amplitude of the SdH oscillation, $A$ is described by the Lifshitz-Kosevich formula [14]:

$$A \propto \frac{T \exp\left(-\lambda m^* T_D / B\right)}{\sinh\left(\lambda m^* T / B\right)}, \tag{1}$$

where $m^*$ is the effective mass, $T_D$ is the Dingle temperature and $\lambda = 2\pi^2 k_B / e\hbar$. The detail of fitting to the temperature dependence of the SdH oscillation amplitude by using equation (1) is described elsewhere [11]. Each data is well fitted and the value of $m^*$ is obtained as the fitting parameter. The annealing effect of $m^*$ normalized by the free electron mass, $m_0$ is summarized in table 1. The value of $m^*/m_0$ decreases by the annealing for the epitaxial layers with both nitrogen concentrations of N = 1.2 and 1.5 %. The sheet carrier densities, $n$ obtained from the period of the SdH oscillations are listed in table 2. They also decrease by the annealing for both nitrogen concentrations of N = 1.2 and 1.5 %. Since the effective mass increases with increasing the nitrogen concentration as reported previously [11], these results suggest that some amount of nitrogen atoms of the InGaAsN layers are considered to diffuse to the GaAsSb layers by the annealing. The results are not due to the compositional variations during the annealing because the X-ray diffraction measurements shows no difference between before and after the annealing.

**Table 1.** Effective mass $m^*$ normalized by the free electron mass $m_0$, $m^*/m_0$.

| N concentration (%) | As grown at 480℃ | Annealed at 550℃ | Annealed at 600℃ |
|---|---|---|---|
| 1.2 | $m^*/m_0 = 0.062$ | $m^*/m_0 = 0.056$ | $m^*/m_0 = 0.051$ |
| 1.5 | $m^*/m_0 = 0.062$ | | $m^*/m_0 = 0.046$ |

**Table 2.** Sheet carrier density, $n$ ($\times 10^{11}$ cm$^{-2}$).

| N concentration (%) | As grown at 480℃ | Annealed at 550℃ | Annealed at 600℃ |
|---|---|---|---|
| 1.2 | $n = 8.7$ | $n = 8.3$ | $n = 6.5$ |
| 1.5 | $n = 6.2$ | | $n = 6.1$ |

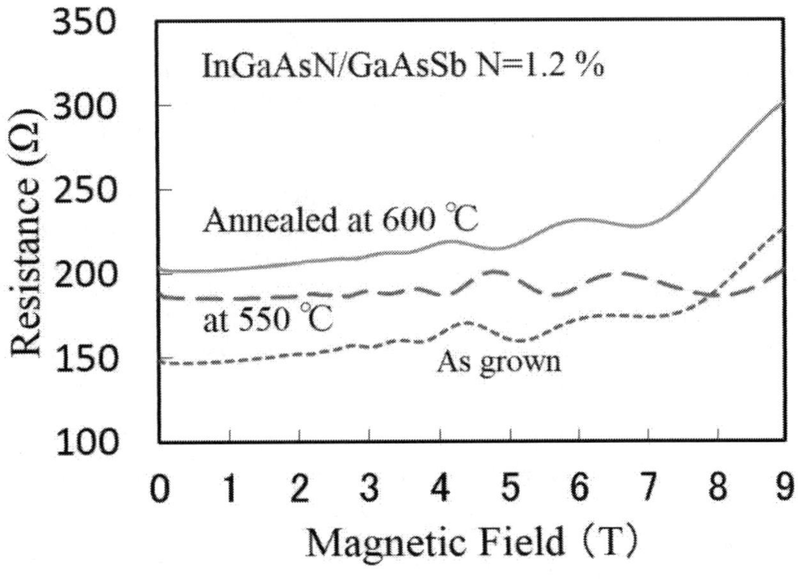


**Figure 1.** Magnetic field dependencies of electrical resistance at 2 K for InGaAsN/GaAsSb with N = 1.2 %. The dotted, broken and solid lines indicate data for the epitaxial layers of as-grown at 480 °C, annealed at 550 °C and annealed at 600 °C, respectively.

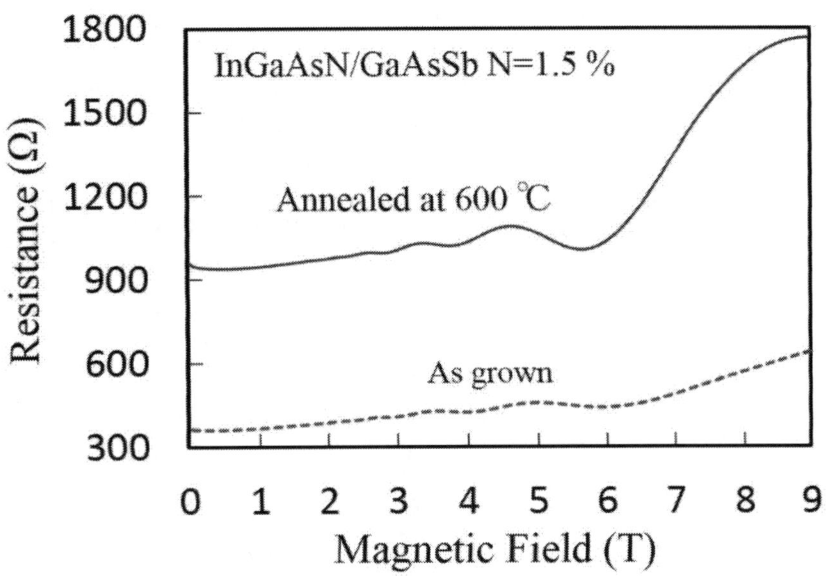


**Figure 2.** Magnetic field dependencies of electrical resistance at 2 K for InGaAsN/GaAsSb with N = 1.5 %. The dotted and solid lines indicate data for the epitaxial layers of as-grown at 480 °C and annealed at 600 °C, respectively.

## 4. Conclusion

In summary, the annealing effect has been investigated for the InGaAsN/GaAsSb epitaxial layers with the nitrogen concentrations of N = 1.2 and 1.5 %. The electrical resistance has been measured as a function of the magnetic field up to 9 Tesla at several temperatures between 2 and 8 K. The oscillation of the resistance due to the SdH effect is observed at each temperature. The effective mass is obtained from the temperature dependence of the amplitude of the SdH oscillations. The value of the effective mass decreases by the annealing. The sheet carrier densities obtained from the period of the SdH oscillations also decrease by the annealing. These results suggest that some amount of nitrogen atoms of the InGaAsN layers are considered to diffuse to the GaAsSb layers by the annealing.

**Acknowledgments**
Authors would like to thank Dr. K. Miura for preparations of the epitaxial layers.

## References

[1] Shih D - K, Lin H - H and Lin Y H 2001 *Electron. Lett.* **37** 1342
[2] Köhler K, Wagner J, Ganser P, Serries D, Geppert T, Maier M and Kirste L 2004 *J. Phys. Condens. Matter* **16** S2995
[3] Kawamura Y, Nakagawa T and Inoue N 2005 *Jpn. J. Appl. Phys.* **44** L1112
[4] Shono T, Mizuta S and Kawamura Y 2013 *J. Cryst. Growth* **378** 69
[5] Yoshikawa M, Miura K, Iguchi Y and Kawamura Y 2009 *J. Cryst. Growth* **311** 1745
[6] Yonezawa Y, Hiraike R, Miura K, Iguchi Y and Kawamura Y 2010 *Physica* E **42** 2781
[7] Miura K, Iguchi Y, Tsubokura M and Kawamura Y 2013 *J. Appl. Phys.* **113** 143506
[8] Kawamura Y and Sahashi T 2014 *Jpn. J. Appl. Phys.* **53** 028004
[9] Deutsch C, Benz A, Detz H, Klang P, Nobile M 2011 *Appl. Phys. Lett.* **97**, 261110
[10] Nobile M, Klang P, Mujagić E, Detz H, Andrews A M, Schrenk W, and Strasser G 2009 *Electron. Lett.* **45** 1031.
[11] Kawamata S, Hibino A, Tanaka S and Kawamura Y 2016 *J. Appl. Phys.* **120** 142109
[12] Shan W, Walukiewicz W, J. W. Ager III J W, Haller E E, Geisz J F, Friedman D J, Olson J M and Kurtz S R 1999 *Phys. Rev. Lett.* **82** 1221
[13] Skierbiszewski C, Perlin P, Wisniewski P, Knap W, Suski T, Walukiewicz W, Shan W, Yu K M, Ager J W, Haller E E, Geisz J F and Olson J M 2000 *Appl. Phys. Lett.* **76** 2409
[14] Shoenberg D 1984 *Magnetic Oscillation in Metals* (Cambridge: Cambridge University Press) chapter 2.3 pp 36-67

IOP Conf. Series: Journal of Physics: Conf. Series **969** (2018) 012152 doi:10.1088/1742-6596/969/1/012152

# Superconductivity, Magnetoresistance, Magnetic Anomaly and Crystal Structure of New Phases of Topological Insulators $Bi_2Se_3$ and $Sb_2Te_3$

**V.A. Kulbachinskii[1,2*], S.G. Buga[3,4], N.R. Serebryanaya[3,4], N.S. Perov[1], V.G. Kytin[1], S.A. Tarelkin[3], R.H. Bagramov[3], N.N. Eliseev[3,4] and V.D. Blank[3,4]**
[1]M.V. Lomonosov Moscow State University, Low Temperature Physics Department, 119991, Moscow, Russia
[2]Moscow Engineering Physics Institute (MEPhI)—National Nuclear Research University, 115409 Moscow, Russia
[3]Technological Institute for Superhard and Novel Carbon Materials, 142190 Troitsk, Moscow, Russia
[4]Moscow Institute of Physics and Technology, 141700 Dolgoprudny, Moscow Region, Russia

*E-mail: kulb@mig.phys.msu.ru

**Abstract.** We synthesized a new metastable phase of $Bi_2Se_3$ topological insulator by a rapid quenching after a high-pressure-high-temperature treatment at $P \approx 7.7$ GPa; $673 < T < 1400$ K. The structure of metastable phase is monoclinic $P2_1/m$ type. We observed the zero-field magnetic susceptibility cusp and linear positive magnetoresistance indicating the topological insulator state. The annealing at 673 K during 2 hours resulted in complete reversible transformation into the initial crystalline rhombohedral structure. Also we synthesized bulk polycrystalline samples of metastable at ambient conditions monoclinic (C2/m) phase of $Sb_2Te_3$ by rapid quenching after a high-pressure–high-temperature treatment at P=3.7–7.7 GPa; T=873 K and found superconductivity with $T_c$=2 K. A zero-field magnetic susceptibility cusp and linear positive magnetoresistance indicate a topological insulator state.

## 1. Introduction

Experimental studies of bismuth and antimony chalcogenides substantially intensified last years due to discovery of their topological insulator (TI) properties [1-13] predicted earlier theoretically. Among other unusual properties of these materials superconductivity attracts much attention as far as it may possess the unconventional p-wave type. The unconventional superconductivity features have been indeed observed in Cu doped $Bi_2Se_3$ [7,8] and specially sintered $Sb_2Te_3$ [13]. Mostly the superconductivity in TIs was observed "in situ" under high-pressure [8,9]. In particular, in $Bi_2Se_3$ superconductivity appeared under pressure of 11 GPa with the critical temperature rising from 0.5 to 7 K under pressure up to 30 GPa, and the upper critical magnetic field of transition increased from 0.3 up to 4 T in the pressure range of 13-30 GPa [8]. The reduced critical field $h*(T)=(H_{c2}(T)/T_c)/(dH_{c2}(T)/dT)|_{Tc}$ was calculated and compared to the models for orbitally limited s-wave and spin-triplet p-wave superconductors and found that the experimentally defined value of about $h*(0) = 0.9$ exceeded the upper limits of 0.7 for s-wave and 0.8 for the spin-triplet p-wave case. Thus the p-wave superconductivity type was concluded.

By applying high-pressure the superconducting phases have been also found in other TIs like tellurides of Bi [9], Sb [10], In [12] and selenides $Sb_2Se_3$ [11]. However, the transitions to superconducting phases were reversible as well as in $Bi_2Se_3$. After the pressure releases they transformed back to non-superconducting phases. Besides superconductivity the peculiar properties of TIs like linear positive magnetoresistance effect and the zero-field paramagnetic cusp were observed [13,14].

Here we present synthesis of new $Bi_2Se_3$ metastable phases with monoclinic crystal structures by quenching after high-pressure-high-temperature treatment at high pressure and different temperatures and metastable superconducting phase $Sb_2Te_3$, linear positive magnetoresistance effect at low temperatures and the zero-field paramagnetic cusp at T=100 K and 300 K indicating topological insulator state.

## 2. Experimental

We used the commercially available high-purity (99.999%) $Bi_2Se_3$ and $Sb_2Te_3$ alloys with rhombohedral structure [15]. We have synthesized metastable high-pressure phases by rapid quenching after the electrical current thermoresistive heating in the "anvil with cavity"-type high-pressure apparatus [14]. The high-pressure-high-temperature experiments were carried out up to 7.7 GPa pressure with the heating up to 1673 K. The sample cooling rate was $\approx$ 60°/min, and its pressure reduction rate in the reaction cell $\approx$ 1 GPa/min. The output samples were 2.5 mm thick and 4.5 mm in diameter. Metastable phases m-$Sb_2Te_3$ has monoclinic *C2/m* structure [16].

For the analysis of the crystal structure and physical properties of the samples we employed the powder X-ray diffraction, electrical, magnetic, and heat capacity measurements. PANalytical diffractometer with CuKα radiation source was used. We studied the temperature dependencies of the electrical resistivity of the samples via a conventional 4-probe method down to 1.8 K, current-voltage characteristics, effect of magnetic field up to 9 T and the heat capacity by differential scanning calorimetry in the temperature range of 1.8 - 400 K using the Quantum Design® physical properties measurement system (PPMS) 3.

## 3. Experimental results

### *3.1. Crystal structure of metastable phase of m-$Bi_2Se_3$.*

Fig. 1 shows the change in the diffraction patterns of $Bi_2Se_3$ samples quenched after treatment at a pressure of 7.7 GPa and heating at different temperatures. The structure of the reference sample is rhombohedral (Fig. 1, No 1), $R\bar{3}m$, $a = 4.143$ Å, $c = 28.636$ Å, the cell parameters are the same as that of ICDD database PDF-2 [17]. The peaks in the diffraction patterns of samples, obtained at temperatures below 1173 *K* (Fig 1, No 2-4) were broad, indicating disorder or amorphization of the structure. After treatment at higher temperatures 1373 - 1473 *K* (Fig 1, No 6-7) the crystal structure became more ordered, these peaks are split and narrowed.

Since the diffraction peaks of the metastable phase of m-$Bi_2Se_3$ did not match to the high-pressure phase peaks obtained in the high-pressure diamond-anvil cell [5] we carried out a crystal-chemical analysis of the structures of all known sesquioxide chalcogenides with large size cations. Almost all sulphides and some selenides of rare-earth elements (RE) crystallize in the orthorhombic structure with the space group *Pnma*. Metastable phases with the orthorhombic structure *Pnma* were found after high pressures and high temperatures treatment of RE sulphides like $Tm_2S$ with initial monoclinic $P2_1/m$ structure [18]. The *Pnma* structure was previously detected in the metastable phase of $Bi_2Se_3$ obtained after the treatment at lower pressures and temperatures [19] and in the present work at P = 4 GPa and T = 673 *K*. Here it should be mentioned, that after annealing at the temperature of 120°C we observed transformation of the metastable high-pressure monoclinic phase m-$Bi_2Se_3$i into an orthorhombic phase with a *Pnma* structure. Thus we assumed that the structure of the new metastable phase of m-$Bi_2Se_3$ would be similar to the monoclinic structure of thulium sulfide $P2_1/m$ and chose this structure as a prototype.

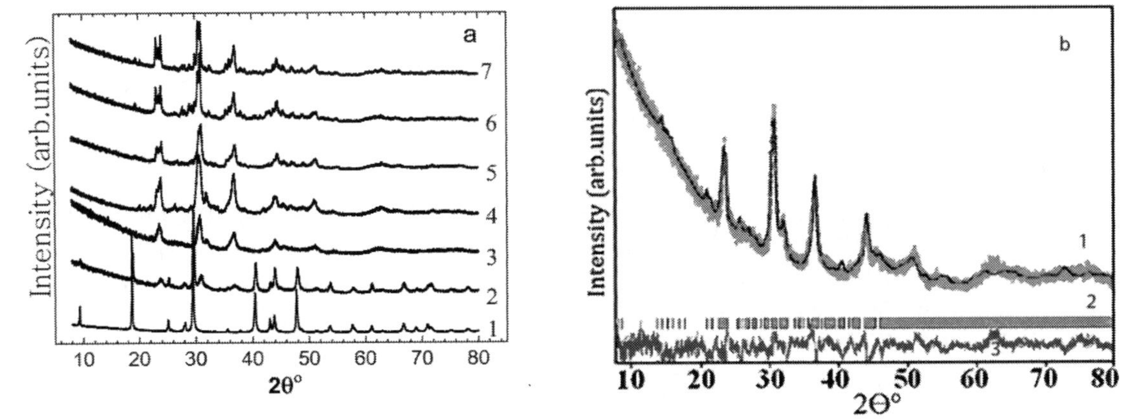


**Figure 1.** The evolution of $Bi_2Se_3$ X-ray data after HPHT treatment at a pressure of 7.7 GPa and temperatures: a) 1 - reference sample at ambient conditions; 2 – 473 K; 3 – 673 K; 4 – 873 K; 5 - 1173 K; 6 – 1373 K; 7 – 1473 K; (b) X-ray diffraction pattern of the sample No 3 (7.7 GPa and 673*K*) after Rietveld refinement. The grey line is the experimental data, black line is the calculated pattern (1), Ticks correspond to the 2 theta Bragg angles of m-$Bi_2Se_3$ (2). The bottom line is the difference pattern (3, exptl. – calcd.). The reliability parameter is equal $R_{Bragg}$=18%.

The parameters of the unit cell and the coordinates of the atoms in the structure of the new phase (Fig.1, No. 3) with the space group $P2_1/m$ were obtained by the method of full-profile analysis by the use of FULLIPROF program [21]. Here are its cell sizes: space group $P2_1/m$, $a$=12.25(6)Å, $b$ = 4.106(8) Å, c = 11.49(7) Å, β = 115.088°, Z = 4. The coordinates of 10 independent atoms are also defined. Thus, by quenching after HPHT treatment we obtained new metastable structures of m-$Bi_2Se_3$ ($P2_1/m$). The structure remains layered, but the interlayer distance decreased. The main difference is the decrease in the distance between metal atoms Bi-Bi 3.5-3.7Å.

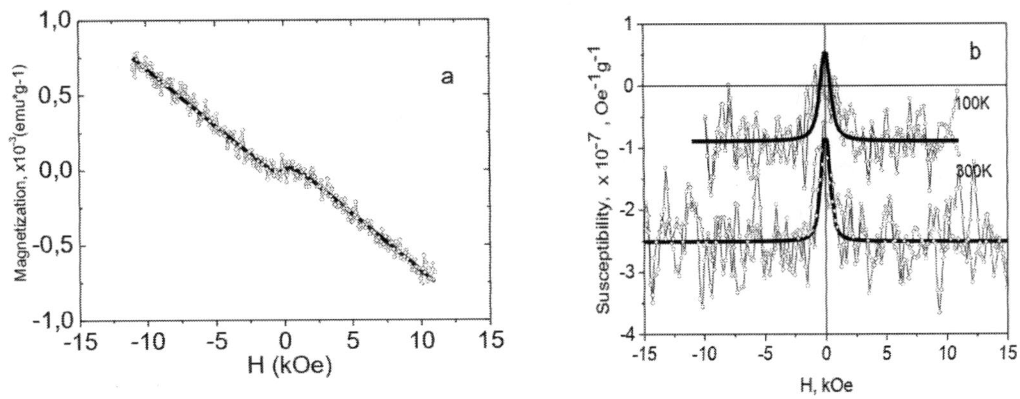


**Figure 2**. Magnetic moment of m-$Bi_2Se_3$ sample (Fig. 1, No. 7) vs magnetic field at room temperature (a) and susceptibility vs magnetic field at $T$ = 100$K$ and 300 $K$ (b). Solid lines show Langevin fits.

The coordination numbers are 6, 7, 8. Coordinate polyhedrons of metal atoms form columns of octahedral ($BiSe_6$) and mono- and double-capped trigonal prisms ($BiSe_7$ and $BiSe_8$) with the common ribs along the short axis "b". The interatomic distances of Bi-Bi in the $P2_1/m$ structure are much smaller than in the *Pnma* structure. After the annealing of metastable phase at T = 300 ° C during 1 hour structure of the sample returned back into initial rhombohedral structure.

### *3.2. Magnetization*

Magnetization versus magnetic field dependence of m-$Bi_2Se_3$ sample (Fig. 1a, No.7) shows general diamagnetic behavior with paramagnetic contribution (Fig. 2a). Fig. 2b displays the magnetic susceptibility with Langevin approximation. The approximation definitely shows the zero-field paramagnetic anomaly in the field less of about 2 *kOe*. Such an anomaly, a zero-field paramagnetic cusp, is typical for topological insulators [22].

The same cusp was observed in m-$Sb_2Te_3$ samples, as shown in Fig. 3a. In m-$Sb_2Te_3$ we observed superconductivity at T=2K which was suppressed by low magnetic field as shown in Fig. 3b. While electrical and magnetic measurements clearly indicate transition into superconducting state, the DSC study did not show any peak in heat capacity at critical temperature. That may evidence two-dimensional character of superconductivity.

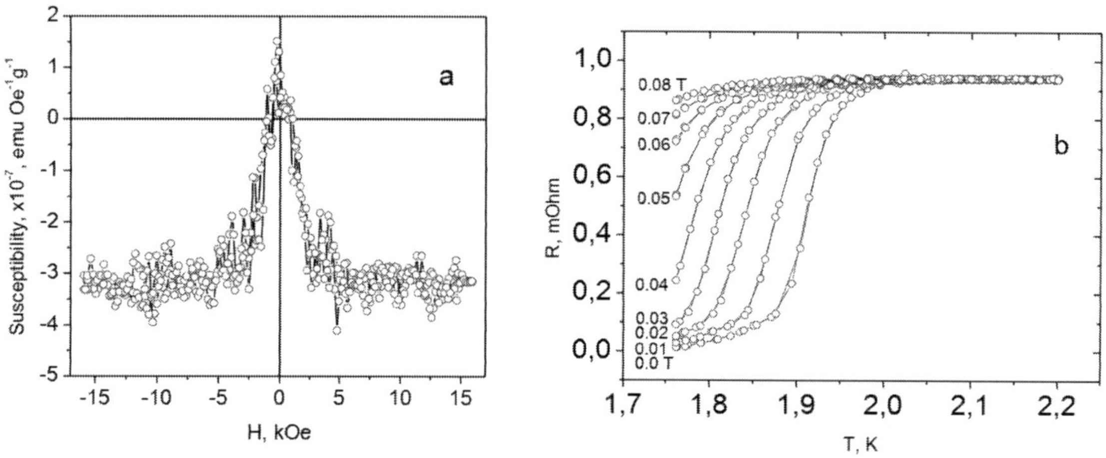


**Figure 3.** Magnetic susceptibility of m-$Sb_2Te_3$ phase vs magnetic field at T=300*K* (a) and magnetic field effect on superconductivity transition temperature in m-$Sb_2Te_3$ phase (b)

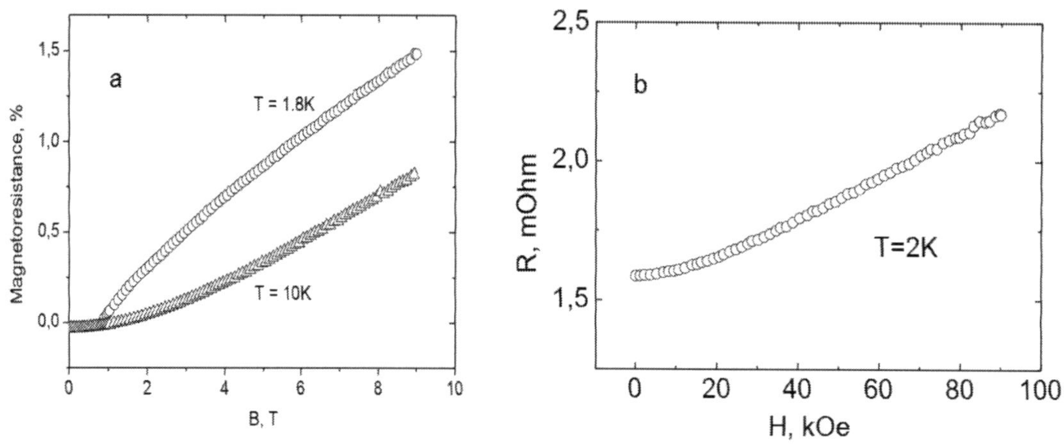


**Figure 4.** Linear positive transverse magnetoresistance in polycrystalline metastable m-$Bi_2Se_3$ (Fig. 1, No. 7) (a) and m-$Sb_2Te_3$ phases (b).

Applying strong magnetic field at fixed temperatures T=1.8 and T=10 *K* we observed linear positive transverse magnetoresistance in metastable phase m-$Bi_2Se_3$ in strong magnetic field (Fig. 4a). In m-$Sb_2Te_3$ transverse magnetoresistance is positive and increases linearly with the magnetic field at H>25 *kOe* as shown in Fig. 4b.

## 4. Discussion

We synthesized new metastable phases m-$Bi_2Se_3$ and m-$Sb_2Te_3$, and determined the crystal structure of m-$Bi_2Se_3$ phase. Earlier superconductivity in high-pressure phases was observed "*in situ*" [4,5,8,9], but after the pressure release the structures transformed back into non-superconductive. The quenched metastable phase m-$Sb_2Te_3$ possess superconductivity transition at normal pressure. It is totally superconductive below $T_c$ = 1.75 *K*.

It should be noted, that the *R(T)* dependence in our case display not simply metallic behavior, it looks more likely as degenerated semiconductor. The linear magnetoresistance (LMR) effect in high fields (see Fig. 4) and the zero-field paramagnetic cusp (see Fig. 2b and 3a) are typical in the topological insulators. Both m-$Bi_2Se_3$ and m-$Sb_2Te_3$ have positive linear magnetoresistance at H>25 *kOe*. It should be mentioned, that the disorder in narrow-gap semiconductors also causes linear magnetoresistance [23], but a relatively narrow temperature range of superconductivity transition can indicate good degree of crystallinity in our case. Thus we believe that this effect as well as the zero-field paramagnetic cusp manifests a topological insulator state.

## 5. Conclusion

We synthesized bulk polycrystalline samples of new metastable phase of $Bi_2Se_3$ topological insulator using a high-pressure and high-temperature treatment at P = 7.7 *GPa*; T = 673-1473 *K* with the subsequent quenching and investigated their electrical and magnetic properties. The monoclinic phase has space group $P2_1/m$, and the unit cell parameters $a$=12.25(6)Å, $b$ = 4.106 (8) Å, c = 11.49(7) Å, β = 115.088°, Z = 4. We observed the linear positive transverse magnetoresistance effect and the zero-field paramagnetic cusp which are proper to topological insulators. Also we synthesized bulk polycrystalline samples of metastable at ambient conditions monoclinic (C2/m) phase of $Sb_2Te_3$. We observed superconductivity at T=2K in m-$Sb_2Te_3$ The critical current value of about 3 *mA* and an absence of the heat capacity peak at superconductivity transition indicate low-dimensional character of superconductivity in the case of m-$Sb_2Te_3$ phase.

**Acknowledgement**
The work was financially supported by the Ministry of Education and Science of the Russian Federation, scientific project RFMEFI57714X0118 (Grant #14.577.21.0118).

## 6. References

[1] D. Hsieh,Y. Xia , D. Qian, et al., Nature Let., **460**, 1101 (2009).
[2] Y. Xia, D. Qian, D. Hsieh, et al., Nat. Phys., **5**, 398 (2009).
[3] H. Zhang, C.X. Liu, X.L. Qi, et al., Nat. Phys., **5**, 438 (2009).
[4] R. Vilaplana, D. Santamaría-Pérez, O. Gomis, et. al., Phys. Rev. B, **84**, 184110 (2011).
[5] J. Zhao, H. Liu, Lars Ehm, et. al., Journal Phys.: Cond. Mat., **25**, 125602 (2013).
[6] G. Liu, L. Zhu, Y. Ma, et. al., J. Phys. Chem. C, **117**, 10045 (2013).
[7]. T.V.Bay, T.Naka, Y.K.Huang, et al., PRL 108, 057001(2012)
[8] K. Kirshenbaum, P.S. Syers, A.P. Hope, et al., Phys. Rev. Lett., **111,** 087001 (2013).
[9] S. J. Zhang, J. L. Zhang, X. H. Yu, et al, J. of Appl. Phys. **111**,112630 (2012)
[10] J. Zhu, J. L. Zhang, P. P. Kong, et al., Sci. Rep. **3**, 2016 (2013).
[11] P.P .Kong, F. Sun, L.Y. Xing et al., Sci. Rep. **4**, 6679 (2014).
[12] S. Geller, A. Jayaraman, G.W. Hull Jr., J. Phys. Chem. Sol., **26**, 353 (1965).
[13] L. Zhao, H. Deng, I. Korzhovska, et. Al., Nature communications, 6:8279 (2015).

IOP Conf. Series: Journal of Physics: Conf. Series **969** (2018) 012152 doi:10.1088/1742-6596/969/1/012152

[14] ] S.G. Buga, V.A. Kulbachinskii, V.G. Kytin, et. al., Chem. Phys. Let., **631**, 97 (2015)
[15] R.W.G. Wyckoff. Crystal structures. Krieger, Melburne, FL, 1986, v.2.
[16] N. Serebryanaya, E. Tatyanin, S. Buga, I. Kruglov, N. Lvova,V. Blank, Phys. Stat. Sol. B, **252**, No. 2, 267–273 (2015).
[17] No 33-0214 of PDF-2 of International Centre for Diffraction Data.
[18] K.-J. Range und R. Leeв, Z.Naturforsch., **30b**, 889-895 (1975)
[19] E.Ya. Atabaeva, S.A.Mashkov, S.V. Popova, //Soviet Physics, Crystallography (Kristallografiya), **18**(1), 104-105 (1973).
[20] Klaus-Jürgen Range und Richard Leeв, Z. Naturforsch., **31b**, 311-314 (1976).
[21] Rodriguez-Carvajal J (2001) Recent Developments of the Program FULLPROF. Newsletter 26:12-19
[22] L. Zhao, H. Deng, I. Korzhovska, et.al., Nature Materials, **13**, 3962 (2014).
[23] M. Veldhorst, M. Snelder, M. Hoek, et al., Phys. Stat. Sol. 7, 26 (2013).

IOP Conf. Series: Journal of Physics: Conf. Series **969** (2018) 012153 doi:10.1088/1742-6596/969/1/012153

# Observation of ferroelastic domains in layered magnetic compounds using birefringence imaging

**Yoko Miura[1], Kazuya Okumura[2] and Hirotaka Manaka[2]**

[1]National Institute of Technology, Suzuka College, Shiroko-cho, Suzuka, Mie 510-0294, Japan
[2]Graduate School of Science and Engineering, Kagoshima University, Korimoto, Kagoshima 890-0065, Japan

E-mail: miura@genl.suzuka-ct.ac.jp

**Abstract.** The two-dimensional Heisenberg antiferromagnet $(C_2H_5NH_3)_2CuCl_4$ is a candidate compound for the coexistence of ferroelectricity and ferroelasticity; however, the microscopic observations of multiferroic domains may still be unclear. In-plane birefringence imaging measurements were performed to observe the manner in which the ferroelectric and the ferroelastic domains change during phase transitions between 15 K and 300 K. It was found that 90° ferroelastic domains appeared in the *ab*-plane at 300 K. As the temperature decreased toward 15 K, each domain inverted at a certain temperature ($T_a$) without structural or magnetic phase transitions. The value of $T_a$ was found to be significantly influenced by external stresses; therefore, birefringence imaging techniques are useful for investigating variations in ferroelastic domains with temperature. Furthermore, a structural phase transition from orthorhombic to monoclinic or triclinic occurred at 230 $\sim$ 240 K; however, no spontaneous polarization appeared in the *ab*-plane over the entire investigated range.

## 1. Introduction

The layered magnetic compound $(C_2H_5NH_3)_2CuCl_4$ (abbreviated as EACuCl) is known as a representative two-dimensional Heisenberg magnet[1]. This compound essentially comprises $CuCl_4$ layers that are well separated from one another by a bilayer of alkylammonium chains $(C_2H_5NH_3)$. The intralayer exchange interaction is ferromagnetic owing to the antiferrodistortive arrangement of the Jahn-Teller distorted $CuCl_6$ octahedra. The antiferromagnetic interlayer exchange interaction is very weak, and the antiferromagnetic phase transition occurs at $T_N = 10.3$ K[1, 2]. In contrast, the structural phase transition occurs at $T_4 = 234$ K wherein the crystal symmetry changes from orthorhombic (*Pbca*) to monoclinic with the *a*- or the *b*-axis as the unique axis or triclinic when the temperature decreases[3-5]. Even though the space group *Pbca* is known to be a non-polar point group, the ferroelectric phase transition has been reported to occur at $T_5 = 247$ K[6, 7, 8]. Below $T_5$, a large remnant polarization (37 $\mu$C/cm$^2$) appears when electric fields are applied parallel to *c*-axis. There are two types of domains above $T_5$; one of these vanishes below $T_5$, while all other domains disappear below $T_4$. Under such circumstances, the intermediate phase between $T_4$ and $T_5$ would be considered to be an improper ferroelectric state. It is likely to be important to observe the manner in which the ferroelastic and the ferroelectric domains in EACuCl change during the structural and ferroelectric phase transitions at $T_4$ and $T_5$ should be essentially observed.

IOP Conf. Series: Journal of Physics: Conf. Series **969** (2018) 012153 doi:10.1088/1742-6596/969/1/012153

Birefringence imaging microscopy is suited for comprehensively investigating various structural phase transitions. In ferroelastic systems, the temperature dependence of birefringence is directly proportional to the order parameter[9]. As demonstrated in our previous studies[10, 11], retardance ($\delta$) and fast-axis direction ($\psi$) images created using birefringence imaging techniques are useful for investigating domain structures because $\delta$ levels are proportional to the degree of distortion and $\psi$ indicates not only optical anisotropy but also uniaxial stress in crystals. In this study, birefringence imaging measurements were performed on EACuCl to observe the ferroelastic domains with ferroelectric properties.

When the indices of refraction along the orthogonal optical principal axes are defined as $n_1$ (for the slow axis) and $n_2$ (for the fast axis) with $n_1 \geq n_2 > 1$, the birefringence $\Delta n$ is expressed as

$$\Delta n \equiv n_1 - n_2. \tag{1}$$

In a material with $n_1 \neq n_2$, the transmitted light becomes elliptically polarized because the electric-field component of the light along the fast axis travels faster than that along the slow axis, leading to a phase shift between the two components[12]. The effect can be described as a retardation between the electric-field wave components along the fast and slow axes. The relation between $\Delta n$ and $\delta$ is given by following equation:

$$\delta \equiv \Delta n \cdot t, \tag{2}$$

where $t$ is the optical-path length. If a ferroelectric phase transition occurs, $\delta$ is expected to be very large and $\psi$ gives information about the polarization direction. In our previous study, a converted retardance method using different $\lambda$ values was developed to exceed the inherent upper limit of birefringence[11]. In this study, we discuss the birefringence imaging data using the converted retardance ($\delta^{543}_{\mathrm{Cv}}$) and the fast-axis direction ($\psi^{543}_{\mathrm{Cv}}$) at $\lambda = 543$ nm.

## 2. Materials and Methods

We used an ellipsometer (Photonic Lattice, Inc. WPA-100) to accurately characterize the polarization via the Stokes parameters ($S_0, S_1, S_2, S_3$). In addition, we used a polarimeter with a 6-W LED white-light source with an irradiation area of $100 \times 136$ mm$^2$ that exhibited circular polarization at $S_3/S_0 \simeq -1$. Thus, the energy density of the light was very low, resulting in negligible increases in the sample temperature due to irradiation. The wavelength was monochromated using three different wavelength filters ($\lambda = 523$, 543, and 575 nm). Sample temperatures were varied using a GM refrigerator with a range of 12 $\sim$ 315 K. The samples were mounted with grease on a copper plate with a thermometer to achieve good thermal contact but prevent stress-induced birefringence. A high-accuracy temperature controller with a silicon diode thermometer was used to ramp the temperature at a rate of $\pm 0.1$ K/min over the entire investigated range of 12 $\sim$ 315 K.

Single crystals of EACuCl were prepared via the slow evaporation of water solutions of $CuCl_2 \cdot 2H_2O$ and $C_2H_5NH_3 \cdot HCl$ at 30°C. The samples appeared yellow. To measure a "transmission-type" birefringence imaging system, thin plates had be prepared. The cleavage plane corresponding to (001) at room temperature was peeled off in the form of flakes because cutting, beating, or polishing causes internal damage of the sample and internal strain must be avoided due to ferroelastic compound. The incident light was perpendicular to (001) and, as in previous studies, we define the fast-axis direction in (001) as the $a$-axis at room temperature[3, 4].

## 3. Experimental Results and Discussion

In ferroelastic materials, 90° domains are expected to appear and can be controlled by applying external uniaxial stress. Thus, we used two different sample setup configurations with the sample laid (A) perpendicularly and (B) diagonally across the gap between opposite copper plates (see Fig. 1). In setup A [Fig. 1(a)], the $a$-axis (the fast axis at room temperature) is at 45° to the copper plates. Conversely, in setup B [Fig. 1(b)], the $a$-axis is rotated by 45° with respect to setup A. Figure 1(c) shows a side view of the sample stage. As the temperature decreases and the copper sample stage shrinks, it is likely that external stresses are generated. The resulting stress may be applied along both the $a$-axis and $b$-axis directions in setup A, but only along one or other of the two axes directions in setup B.

Figure 2 shows birefringence images of EACuCl for setup A. Although the $\delta^{543}_{\mathrm{Cv}}$ distributions are uniform over the entire temperature range, the 90° domains, which indicate a ferroelastic state, appear at 300 K[13]. The $\psi^{543}_{\mathrm{Cv}}$ direction in each domain is inverted between 300 K and 15.0 K. As can be seen in Figs. 2(c) and 2(f), the transmittance is inhomogeneous because the peeled sample contains many defects, cracks, and grooves[14]. To avoid complications due to crystal quality, the sample analysis area (10×10 pixels) was selected not only in the single $\psi^{543}_{\mathrm{Cv}}$ domain area but also in the highly homogeneous transmittance area, as shown by the small square in Fig. 2(b). Figure 3 shows the temperature dependence of $|\Delta n_{001}|(\equiv |n_b - n_a|)$ and $\psi^{543}_{\mathrm{Cv}}$, where $n_a$ and $n_b$ are refractive indices along the $a$- and $b$-axis, respectively. When the minimum value of $|\Delta n_{001}(T)|$ is defined as $T_{\mathrm{a}}$, switching between the fast- and slow-axis occurs at $T_{\mathrm{a}} = 196.0 \sim 198.9$ K. This result agrees well with previous results observed using the rotating analyzer method[3, 4]. Because the 90°-jump of $\psi^{543}_{\mathrm{Cv}}$ at $T_{\mathrm{a}}$ may become unclear regardless of the ferroelastic 90° domains, birefringence imaging techniques are very useful for the ferroelastic materials.

Furthermore, the birefringence experiments for setup B were performed using another sample. Figure 4 shows the temperature dependence of $|\Delta n_{001}|$ and $\psi^{543}_{\mathrm{Cv}}$ in setup B. The value of $|\Delta n_{001}|$ at 300 K is larger than that at 15 K in setup A, and the opposite is true for setup B. In a transmission birefringence imaging system, the value of $\Delta n_{001}$ is averaged along [001], i.e., the beam axis. When the 90° domains stack along the beam axis in our samples, the value of $\Delta n_{001}$ becomes small because the opposite components cancel out. The $|\Delta n_{001}(T)|$ curves during cooling and heating coincide with each other in setup A but show thermal hysteresis in setup B. The $T_{\mathrm{a}}$ values differ by over 30 K in setups A and B although they show small thermal hysteresis. Because of the ferroelastic state in EACuCl, its physical properties are strongly affected by the arrangement of the $(C_2H_5NH_3)$ chains, but the weak N-H- $\cdots$ -Cl hydrogen bonds are affected by sample quality and internal strain. Thus, the structural phase transition temperatures observed by different methods frequently differ by over 20 K[15]. To confirm the dependence of $T_{\mathrm{a}}$ on sample quality, all the samples were measured in both setups. The results shown in Fig. 5 indicate that the sample quality dependence is negligible, but the values of $T_{\mathrm{a}}$

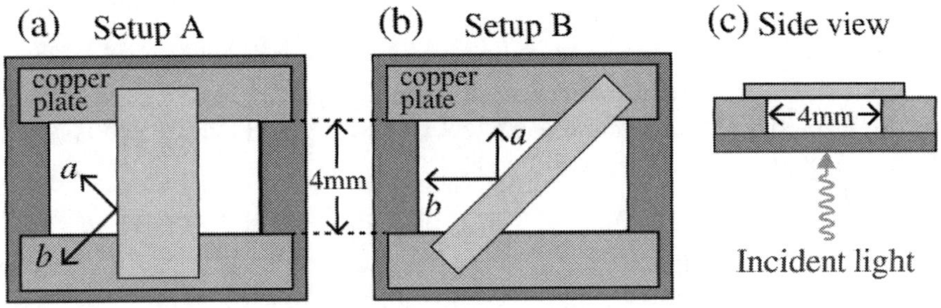


**Figure 1.** Schematic of the sample setup configurations. (a) The $a$-axis is a 45° to the copper plate (b) The $a$-axis is rotated by 45° with respect to the copper plate (c) Side view of the sample stage.

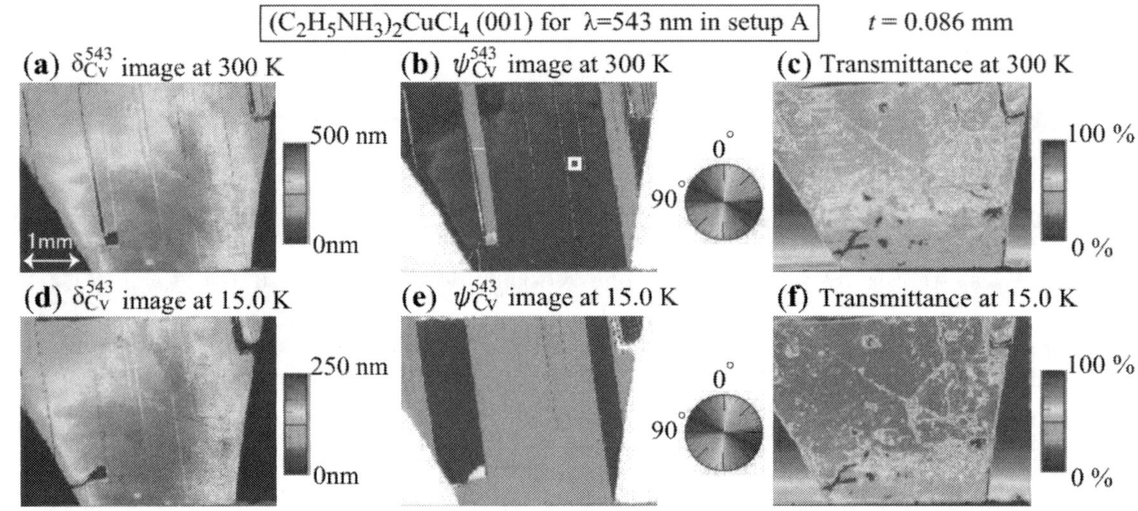


**Figure 2.** Converted ($\delta^{543}_{\mathrm{Cv}}$) retardance images at (a) 300 K and (d) 15.0 K for a $(C_2H_5NH_3)_2CuCl_4$(001) plate in setup A. Converted ($\psi^{543}_{\mathrm{Cv}}$) fast-axis images at (b) 300 K and (e) 15.0 K. Transmittance images at (c) 300 K and (f) 15.0 K. The square in (b) indicates the sample analysis area. The thickness $t$ is 0.086 mm.

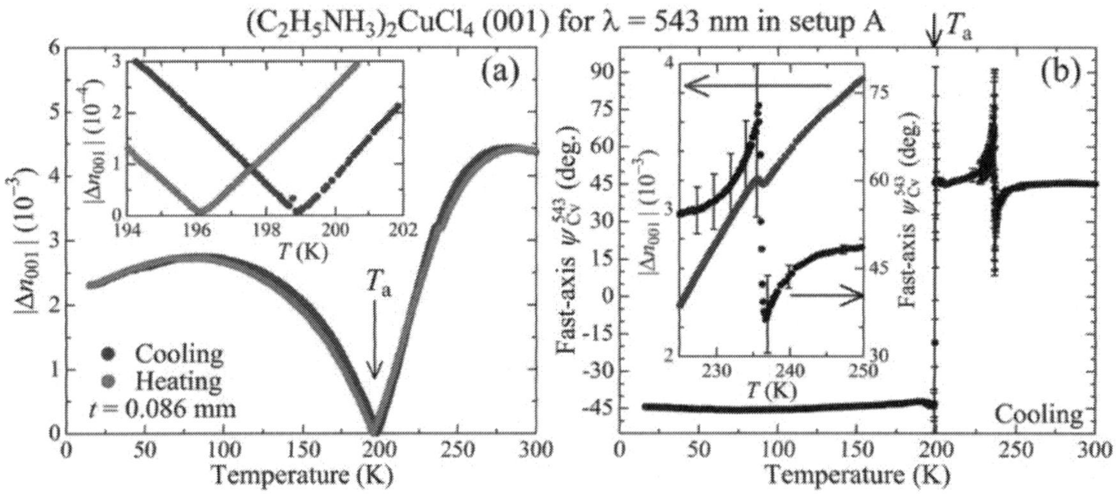


**Figure 3.** Temperature dependence of (a) $|\Delta n_{001}|$ and (b) $\psi^{543}_{\mathrm{Cv}}$ for a $(C_2H_5NH_3)_2CuCl_4$(001) plate in setup A. Inset to (a) is a zoomed-in view of the $|\Delta n_{001}(T)|$ curve around $T_a$. Inset to (b) is a zoomed-in view of the $|\Delta n_{001}(T)|$ and $\psi^{543}_{\mathrm{Cv}}(T)$ curves around $T_4$ and $T_5$.

can be classified by setup configuration. The average value of $T_a$ is 197(3) K in setup A but 226(3) K in setup B. The hysteresis of $T_a$ seems to be opposite in the two setups (i.e., inverted in setup A and normal in setup B). This result indicates that the value of $T_a$ is suppressed by the external stress because the 90°-jump in $\psi^{543}_{\mathrm{Cv}}$ is caused by the tilt of the $c$-axis in the $ab$-plane below $T_4$[3-5].

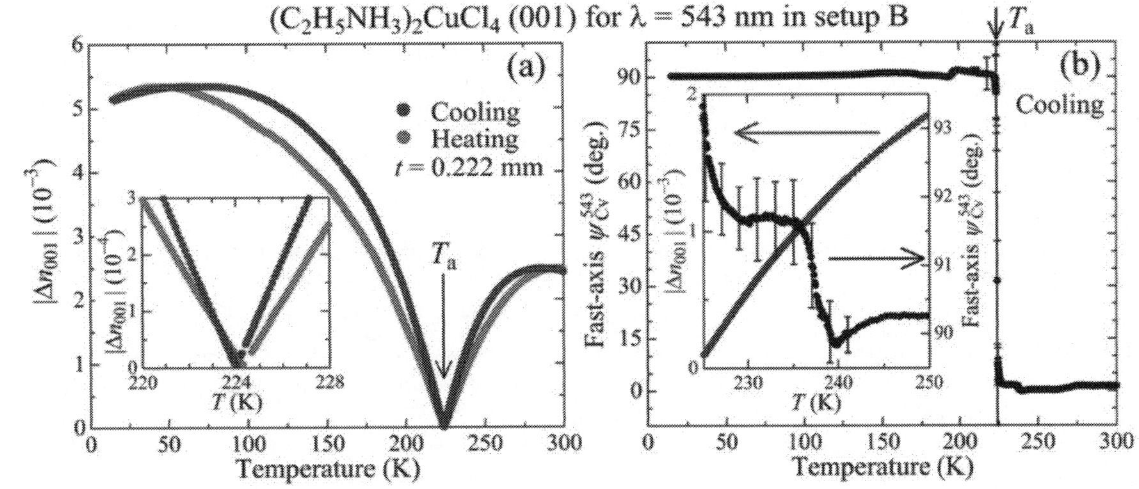


**Figure 4.** Temperature dependence of (a) $|\Delta n_{001}|$ and (b) $\psi_{\mathrm{Cv}}^{543}$ for a $(C_2H_5NH_3)_2CuCl_4(001)$ plate in setup B. Inset to (a) is a zoomed-in view of the $|\Delta n_{001}(T)|$ curve around $T_a$. Inset to (b) is a zoomed-in view of the $|\Delta n_{001}(T)|$ and $\psi_{\mathrm{Cv}}^{543}(T)$ curves around $T_4$ and $T_5$.

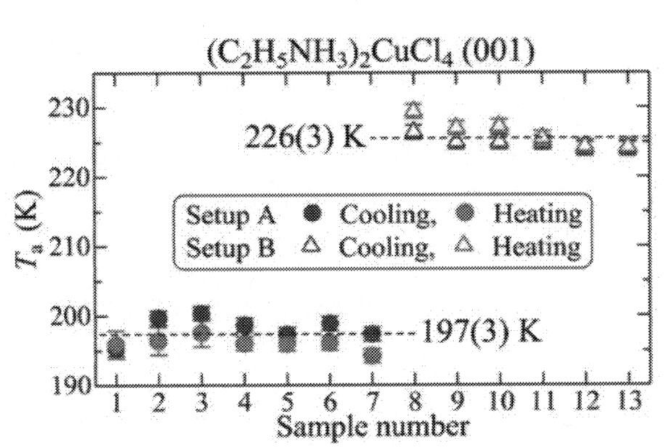


**Figure 5.** Variation of $T_a$ across all the $(C_2H_5NH_3)_2CuCl_4(001)$ plates in the two-setup configurations.

As can be seen in the inset of Fig. 3(b), a kink in the $|\Delta n_{001}(T)|$ curve and a large anomaly in the $\psi_{\mathrm{Cv}}^{543}$ direction appear at 230$\sim$240 K in setup A. In setup B, by contrast, it can be seen in the inset of Fig. 4(b) that no anomaly appears in the $|\Delta n_{001}(T)|$ curve and the change in $\psi_{\mathrm{Cv}}^{543}$ direction is less than 2°. From our previous studies on isomorphous compounds[11], the $\psi_{\mathrm{Cv}}^{543}$ directions are closely related to the arrangements of the alkylammonium chains ($C_nH_{2n+1}NH_3$). Thus, the fluctuations in $\psi_{\mathrm{Cv}}^{543}$ are probably due to rotation of the ($C_2H_5NH_3$) chains accompanied by the structural phase transition around $T_4$. As a result, the gradual change in the $|\Delta n_{001}(T)|$ curve indicates a second-order structural phase transition at 230$\sim$240 K corresponding to $T_4$. The crystal symmetry below $T_4$ can be characterized as monoclinic with the *a*- or the *b*-axis as the unique axis[3] or triclinic with $\gamma$ close to 90° because a large change in $\Delta n_{001}$ should occur at $T_4$ if $\gamma$ is far from 90°. In addition, the 90°-jump in $\psi_{\mathrm{Cv}}^{543}$ at $T_a$ may be due to $\gamma = 90°$. If spontaneous polarization appears along the *c*-axis below $T_5$, ferroelectric polarization should appear in the *ab*-plane because the *c*-axis tilts toward the *ab*-plane below $T_4$. However, no anomalies in the in-plane birefringence, which could indicate the existence of ferroelectric polarization, appeared between 15 K and 300 K. To directly observe spontaneous polarization along the *c*-axis, it would be necessary to measure the birefringence in the plane that included the *c*-axis. In ref.

IOP Conf. Series: Journal of Physics: Conf. Series **969** (2018) 012153 doi:10.1088/1742-6596/969/1/012153

3, the temperature dependence of birefringence in (111) was observed by the rotating analyzer method and the $\Delta n_{111}(T)$ curve was found to be almost the same as that of the $\Delta n_{001}(T)$ curve; however, no $\psi_{\mathrm{Cv}}^{543}(T)$ data were obtained. In future our study, an observation method using oblique incident light will be developed to obtain birefringence images in (111) with 90° ferroelastic domains in the *ab*-plane.

## 4. Conclusion

We successfully observed 90° ferroelastic domains in the *ab*-plane using the birefringence imaging techniques. From the temperature dependence of $|\Delta n_{001}|$ and $\psi_{\mathrm{Cv}}^{543}$, the 90°-jump in $\psi_{\mathrm{Cv}}^{543}$ appears at $T_{\mathrm{a}} = 197(3)$ K in setup A and $T_{\mathrm{a}} = 226(3)$ K in setup B without structural or magnetic phase transitions because the *c*-axis tilts toward the *ab*-plane below $T_4$. Furthermore, the structural phase transition from orthorhombic to monoclinic or triclinic was observed at $T_4$, but no ferroelectric phase transition was observed at $\sim T_5$. Because the value of $T_{\mathrm{a}}$ is significantly influenced by external stresses, we found that it was important to carefully observe the changes in the ferroelastic domains with temperature in stress-free conditions. Conversely, if large uniaxial stress is applied in the *ab*-plane, stress-induced ferroelectric phase transitions are expected to occur around $T_4$ and $T_5$. In future studies, we will endeavor to measure stress-induced birefringence.

## Acknowledgments

We received valuable support from the Frontier Science Research Center in Kagoshima University for performing the X-ray diffraction experiments. This research was partly supported by JSPS KAKENHI Grant Numbers JP25600133 and JP15K13400. One of the authors (Y.M.) is indebted to the Kurata Memorial Hitachi Science for paying for her travel expenses to Gothenburg, Sweden.

## References

[1] Jongh L J de 1990 *Magnetic Properties of Layered Transition Metal Compounds* (Klumer Academic Publishers, Dordrecht)
[2] Manaka H, Yamada I, Nishi M and Goto T 2001 *J. Phys. Soc. Jpn.* **70** 241
[3] Kleemann W, Schäfer F J, Käräjämäki E, Laiho R and Levola T 1983 *Physica B* **119** 269
[4] Jahn I R, Knorr K and Ihringer J 1989 *J. Phys.: Condens. Matter* **1** 6005
[5] Kapustianyk V, Rudyk V and Partyka M 2007 *Phys. Status Solidi B* **244** 2151
[6] Kundys B, Lappas A, Viret M, Kapustianyk V, Rudyk V, Semak S, Simon Ch and Bakaimi I 2010 *Phys. Rev. B* **81** 224434
[7] Hang T, Zhang W, Yea H -Y and Xiong R -G 2011 *Chem. Soc. Rev.* **40** 3577
[8] Kapustianyk V 2013 *J. Phys. Studies* **17** 1702
[9] Ferré J 1983 *J. Phys. C: Solid State Phys.* **16** 3971
[10] Manaka H, Yagi G and Miura Y 2016 *Rev. Sci. Instrum.* **87** 073704
[11] Manaka H, Fukuda T and Miura Y 2016 *J. Phys. Soc. Jpn.* **85** 124701
[12] Azzam R M and Bashara N M 1977 *Ellipsometry and Polarized Light* (North Holland, Amsterdam)
[13] Heygster G and Kleemann W 1977 *Physica B* **89** 165
[14] Mischgofsky F H 1978 *J. Crystal Growth* **43** 549
[15] Pabst I, Fuess H and Bats J W 1987 *Acta Crystallogr., Sect. C* **43** 413

# Magnetoresistance in single-electron transistors comprising a superconducting island with ferromagnetic leads

**Yoshinao Mizugaki, Masashi Takiguchi, Nobuyuki Tamura, and Hiroshi Shimada**

The University of Electro-Communications (UEC Tokyo), Chofu, Tokyo 182-8585, Japan

E-mail: y.mizugaki@uec.ac.jp

**Abstract.** We report electric and magnetic field responses of single-electron transistors (SETs) comprising a superconducting island with ferromagnetic (FM) leads. We fabricated two SETs, one of which had relatively high resistance and the other had relatively low resistance. The SETs had two states for the gate charge: SET-ON or SET-OFF. They also had two states for the FM lead magnetization: parallel (P) or anti-parallel (AP) configuration. Current-voltage characteristics of four SET states ("P & SET-ON," "P & SET-OFF," "AP & SET-ON," and "AP & SET-OFF") were measured at approximately 0.1 K in a compact dilution refrigerator. Magnetoresistance ratio ($MRR$) values were obtained for the SET-ON and SET-OFF states, respectively. The higher-resistance SET1 exhibited positive $MRR$ values for all measured bias voltages. The $MRR$ enhancement was confirmed for the SET-OFF state, which agreed well with the co-tunneling model. The lower-resistance SET2, on the other hand, exhibited negative and positive $MRR$ values for higher and lower bias voltage conditions, respectively. The bias voltage for the $MRR$ polarity reversal was changed by the gate voltage. It was also confirmed that the co-tunneling model was partially valid for negative $MRR$ values.

## 1. Introduction

A single-electron transistor (SET), which comprises two tiny tunnel junctions and a capacitive gate, is the most sensitive charge sensor, because its Coulomb blockade (CB) is periodically modulated by the gate charge with a period of the primary charge $e$ [1]. Since the first experimental demonstration [2], aluminum (Al) tunnel junctions have been widely used. Besides Al, ferromagnetic (FM) materials such as nickel (Ni) and cobalt (Co) are also used for electrodes of SETs that exhibit magnetoresistance effects. Such FM SETs respond not only to electric gate charge but also external magnetic field. In addition, it was reported that magnetoresistance effects were significantly enhanced in FM SETs under the CB state [3].

In this paper, we report SETs comprising a superconducting (SC) Al island with FM Co leads. FM-SC-FM SETs have been experimentally investigated since the 2000s [4, 5], in which several unique properties of SC islands are demonstrated.

- Since the spin lifetime in SC Al electrodes exceeds 0.1 ms [6], spin-dependent transport in a long ($\geq 1$ $\mu$m) SC island electrode is possible.

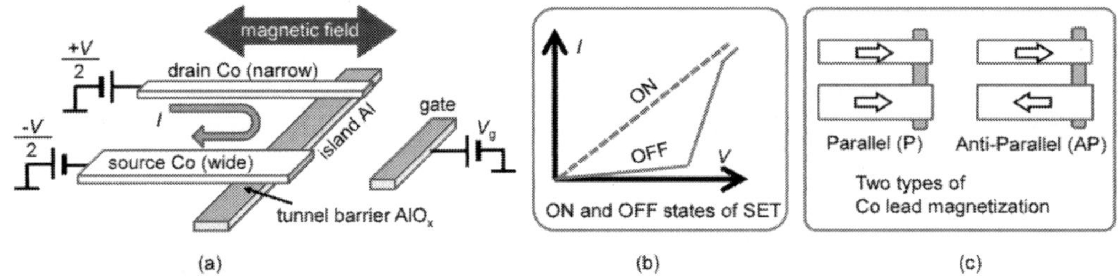


**Figure 1.** (a) Configuration of the FM-SC-FM SET with external bias sources. (b) Electric properties of an SET switched into either "ON" or "OFF" by tuning the gate charge. (c) Magnetic configurations of FM leads switched into either "Parallel (P)" or "Anti-Parallel (AP)" alignment by using magnetic hysteresis for two different coercive forces.

- Superconductivity in an SC island electrode is modified by spin injection from FM leads. One typical example is that the spin injection reduces superconducting energy gaps, which results in a negative magnetoresistance ratio ($MRR$) [5, 7].
- The polarity reversal of magnetoresistance can be modulated by the gate charge [8, 9].

Below, we first briefly explain the characteristics of an FM-SC-FM SET from the viewpoint of superconductivity with polarized spin injection. Next, experimental methods including sample fabrication and measurement are described. We present two FM-SC-FM SETs, one of which (SET1) has relatively high resistance and the other (SET2) has relatively low resistance. The polarity reversal as well as the $MRR$ enhancement are evaluated. SET1 exhibits positive $MRR$ values for all measured bias voltages. Those of SET2, on the other hand, are negative and positive respectively for high and low biasing conditions, where the bias voltage for the polarity reversal is changed by the gate voltage. Enhancement of negative $MRR$ is evaluated for the first time as functions of the bias voltages.

## 2. Magnetoresistance of FM-SC-FM SET

Figure 1(a) shows a schematic drawing of an FM-SC-FM SET that we fabricated. Two Co leads are contacted to an Al island through a thin $AlO_x$ tunnel barrier. A gate electrode is capacitively coupled to the Al island.

In measurements, two voltage sources are connected to the source and drain electrode. By applying voltages of $\pm V/2$, the FM-SC-FM SET is symmetrically biased. The gate charge switches the SET between ON and OFF state as shown in figure 1(b). Besides, external magnetic field is applied parallel to the FM leads. Different widths of the source and drain FM leads result in different coercive forces. Then, by using magnetic hysteresis, two types of magnetization, the parallel (P) and anti-parallel (AP) alignments, are realized as shown in figure 1(c). Combination of the electric and magnetic switchings brings about the four states of the FM-SC-FM SET: "P & SET-ON," "P & SET-OFF," "AP & SET-ON," and "AP & SET-OFF."

Magnetoresistance of the FM-SC-FM SET is described as follows. When spin-polarized electrons are transferred between the FM source and drain electrodes, the magnetization alignment (P or AP) changes the source-drain resistance, which is often evaluated using an $MRR$ value. In this paper, we define the $MRR$ value as

$$MRR = \frac{R_{\mathrm{AP}} - R_{\mathrm{P}}}{R_{\mathrm{P}}} = \frac{I_{\mathrm{P}} - I_{\mathrm{AP}}}{I_{\mathrm{AP}}}, \tag{1}$$

where $R_{\mathrm{AP}}$, $I_{\mathrm{AP}}$, $R_{\mathrm{P}}$, and $I_{\mathrm{P}}$ are the resistance and current values under a certain bias voltage for the AP and P alignment, respectively.

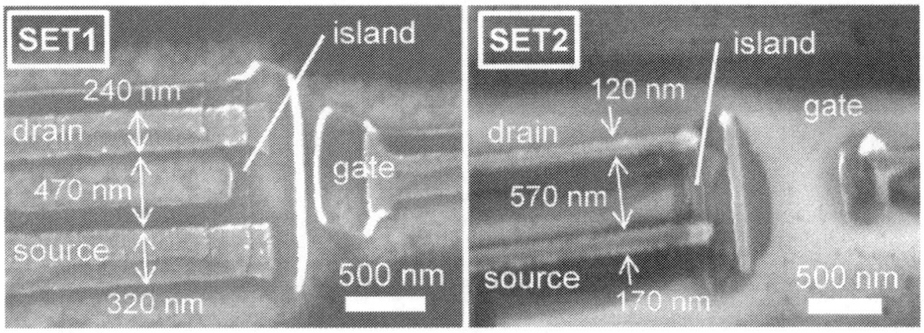


**Figure 2.** Scanning electron micrograph (SEM) images of two fabricated SETs. SET1 and SET2 have higher and lower tunnel resistance, respectively.

For an FM-FM tunnel junction, the relationship of $R_{\mathrm{AP}} > R_{\mathrm{P}}$ is usually observed because of the difference in the density of states for major and minor spins. The situation for an FM-SC-FM SET, however, is not so simple, because the spin imbalance in the SC island may reduce the superconducting energy gap. If the reduction of the superconducting energy gap is dominant, the relationship of $R_{\mathrm{P}} > R_{\mathrm{AP}}$ is observed, which results in a negative $MRR$ value [5, 6, 7].

In addition, enhancement of magnetoresistance is often observed in FM SETs [3]. This effect can be explained by a model employing cotunneling events in the CB state [10, 11, 12, 13, 14]. In this model, the 1st-order electron tunneling is dominant for the SET-ON state, where the current is inversely proportional to the tunnel resistance $R_{\mathrm{T}}$. On the other hand, the 2nd-order electron tunneling (co-tunneling) is dominant for the SET-OFF state, where the current is inversely proportional to $R_{\mathrm{T}}^2$. Then, the magnetoresistance of this model, $MRR_{\mathrm{cot}}$, is expressed as

$$MRR_{\mathrm{cot}} = \frac{I_{\mathrm{P,off}} - I_{\mathrm{AP,off}}}{I_{\mathrm{AP,off}}} = \frac{1/R_{\mathrm{P,on}}^2 - 1/R_{\mathrm{AP,on}}^2}{1/R_{\mathrm{AP,on}}^2} = \frac{R_{\mathrm{AP,on}}^2 - R_{\mathrm{P,on}}^2}{R_{\mathrm{P,on}}^2}, \tag{2}$$

where the suffixes "P," "AP," "on," and "off" mean the P alignment, AP alignment, ON state, and OFF state, respectively. That is, $MRR_{\mathrm{cot}}$ can be expressed as a function of the resistance values for the SET-ON state.

## 3. Experimental Method

FM-SC-FM SETs were fabricated on a thermally oxidized silicon substrate by electron beam lithography with a bilayer resist followed by the two-angle shadow evaporation technique [15]. Scanning electron micrograph (SEM) images of two fabricated SETs are presented in figure 2. The thicknesses of the Al islands were 25 and 30 nm for SET1 and SET2, respectively. Different oxidization conditions for tunnel barriers were used for SET1 and SET2: The tunnel barrier of SET1 was formed in 27 Pa of pure oxygen gas for 120 seconds, whereas that of SET2 was formed in 13 Pa of pure oxygen gas for 30 seconds. The thicknesses of the Co leads were respectively 45 and 40 nm for SET1 and SET2.

In measurements, samples were cooled down to 0.1 K in a compact dilution refrigerator [16], in which an external magnetic field $H$ was applied along the Co leads by an SC magnet.

## 4. Results and Discussion

### *4.1. Responses to gate voltage and magnetic field*

We first checked whether the samples worked as an SET. Figure 3 shows the $I$–$V_{\mathrm{g}}$ characteristics of SET1 and SET2. In measurements, relatively high magnetic field of 10–20 kOe was applied

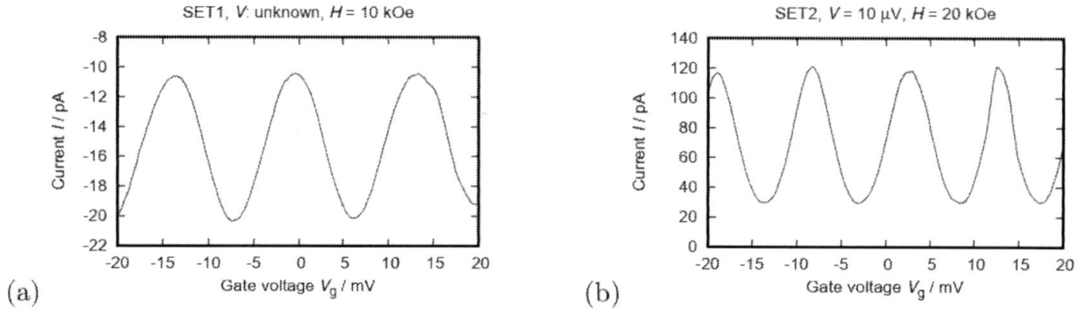


**Figure 3.** Coulomb oscillations ($I$–$V_g$ characteristics) of two SETs. The measurement temperatures were around 0.1 K.

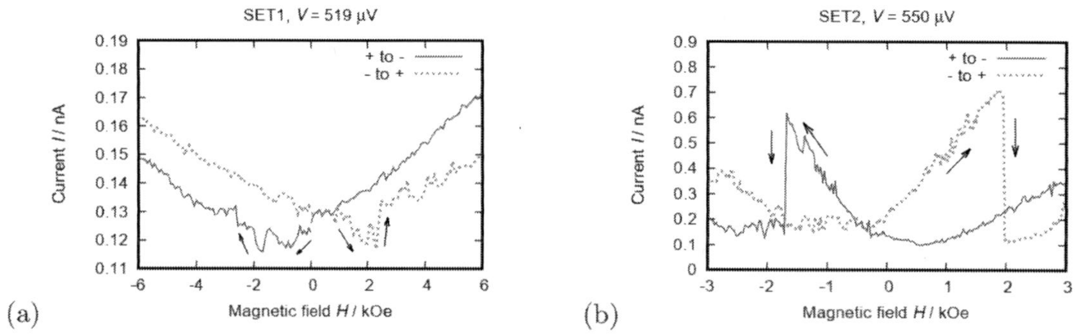


**Figure 4.** Magnetic field responses ($I$–$H$ characteristics) of two SETs. The measurement temperatures were around 0.1 K.

to break the superconductivity of the Al film. Periodic current oscillation (Coulomb oscillation) was confirmed for both SET1 and SET2.

Next, we measured $I$–$H$ characteristics, of which the results are shown in figure 4. In these measurements, we did not apply the gate voltage. Hystereses around 1–2 kOe suggest the difference between the P and AP alignment, although magnetization switching from P to AP is gradual. The gradual switching can be attributed to relatively wide Co (drain and source) electrodes (> 100 nm) and their insufficient extension beyond the Al island. (They could respectively result in multiple magnetic domains in the junction areas and suppression of superconductivity due to the fringing field [5, 17].) On the other hand, magnetoresistance observed in these SETs proves that the spin flip lengths in the Al islands were longer than a half micron, which is in agreement with our previous experimental results of 1.0–2.7 $\mu$m [18] obtained by means of a non-local measurement [19].

It should be noted that the current of SET1 is decreased at the AP alignment whereas that of SET2 is increased. The former and latter result in positive and negative $MRR$ values as described below.

### *4.2. Magnetoresistance ratio MRR*

$I$–$V$ characteristics for four states, "P & SET-ON," "P & SET-OFF," "AP & SET-ON," and "AP & SET-OFF," are shown in figure 5.

For both SET1 and SET2, the current values for "P & SET-ON" are always larger than those for "P & SET-OFF," which corresponds to our definition for "ON" and "OFF." The same

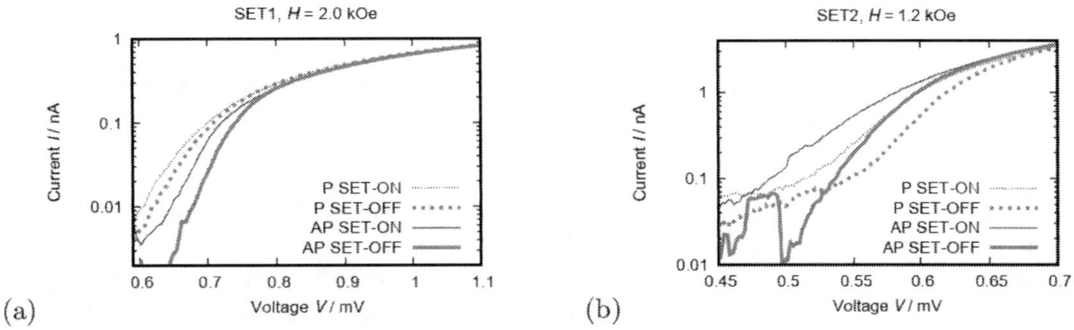


**Figure 5.** Four $I$–$V$ characteristics of two SETs for four states of "P & SET-ON," "P & SET-OFF," "AP & SET-ON," "AP & SET-OFF." The measurement temperatures were around 0.1 K.

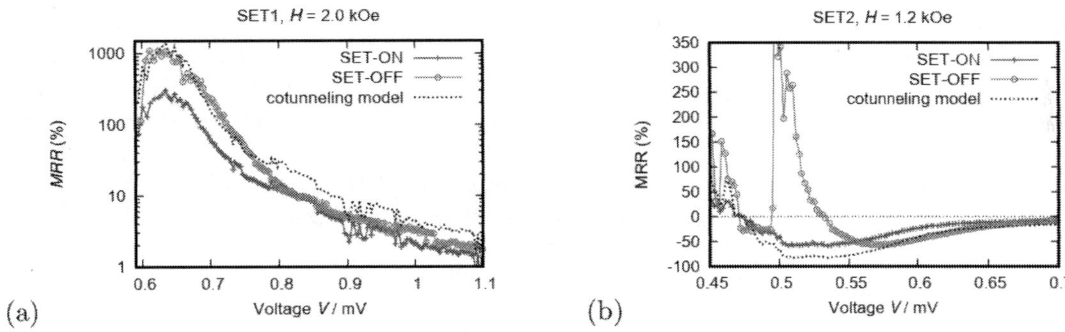


**Figure 6.** $MRR$–$V$ characteristics of two SETs for "SET-ON" and "SET-OFF" states. Dotted curves are obtained using equation (2) with the results for "SET-ON". It should be noted that the vertical axes are quite different for two SETs.

relationship is confirmed for "AP & SET-ON" and "AP & SET-OFF."

Next, to evaluate $MRR$, we compare $I$–$V$ characteristics of different alignments for the same "ON" and "OFF" state.

The $I$–$V$ characteristics of SET1 shown in figure 5(a) demonstrate that the current values for "AP & SET-ON" and "AP & SET-OFF' are always smaller than those of "P & SET-ON" and "P & SET-OFF," respectively. These are general magnetoresistance effects, for which the calculated $MRR$ values are plotted in figure 6(a). (Equation (1) was used for calculation.) For the voltage range of $V > 0.8$ mV, $MRR$ values for the SET-ON and SET-OFF state are comparable. They increase as $V$ decreases, although they are less than 10%. For the voltage range of $V < 0.8$ mV, $MRR$ values rapidly increases with decreasing $V$. Especially, $MRR$ values for the SET-OFF state are enhanced up to 1000%, whereas those for the SET-ON state are 300% at maximum. Such difference is explained using $MRR_{\rm cot}$ of equation (2), which is plotted in figure 6(a) as a dotted curve.

The reason why negative $MRR$ was not obtained for SET1 can be attributed to the large tunnel resistance that reduces the spin accumulation in an SC electrode [6, 9].

For SET2, on the other hand, the situations are not as simple as those of SET1. That is, the current values for the AP alignment are larger than those for the P alignment in the voltage ranges of $V > 0.47$ mV and $V > 0.53$ mV for the SET-ON and SET-OFF state, respectively. In these voltage ranges, $MRR$ values become negative as shown in figure 6(b). Anomalous current increase for "AP & SET-OFF" is also confirmed at around $V = 0.46$–$0.49$ mV, which might come from sudden shift of background charge.

In the SET-ON state, negative $MRR$ values are obtained for the voltage range of $V > 0.47$ mV. The dip value is $-59\%$ at $V = 0.53$ mV. The $MRR$ value increases as $V$ decreases below 0.50 mV, and it becomes positive at 0.47 mV. In the SET-OFF state, on the other hand, the voltage for the $MRR$ polarity reversal is 0.53 mV, which is larger than that in the SET-ON state. Besides, $MRR_{\mathrm{cot}}$ of equation (2) is in good agreement with experimental results for $0.58 < V < 0.65$ mV, while they are negative values.

Finally, it should be noted that the polarities of $MRR$ values for the voltage range of $0.50 < V < 0.53$ mV are negative and positive for the SET-ON and SET-OFF states, which suggests that not only the $MRR$ value but also its polarity can be controlled by the gate voltage.

## 5. Conclusion

We reported electric and magnetic responses of FM-SC-FM SETs. SET1 and SET2, which respectively had higher and lower tunnel resistance, were fabricated with an SC Al island and FM Co leads. $I$–$V$ characteristics for four states of "P & SET-ON," "P & SET-OFF," "AP & SET-ON," and "AP & SET-OFF" were measured at around 0.1 K. $MRR$ values were derived from $I$–$V$ characteristics and plotted as functions of $V$. SET1 exhibited positive $MRR$ values for all measured bias voltages, where the $MRR$ enhancement was also confirmed for the CB state. $MRR$ values of SET2, on the other hand, were negative and positive for higher and lower bias voltages. Furthermore, it was demonstrated that the voltage for the $MRR$ polarity reversal was changed by the gate voltage. It was also confirmed that $MRR_{\mathrm{cot}}$ of equation (2) was partially valid for negative $MRR$ values.

## Acknowledgments

This work was partly supported by Takayanagi Memorial Foundation for Electronics Science and Technology, The Asahi Glass Foundation, JSPS KAKENHI 18080004, 21360040, 24340067, 15K13999, 17K04979, and CREST-JST Grant Number JPMJCR14F3. The stable supply of liquid He from the Coordinated Center for UEC Research Facilities is also acknowledged.

## References

[1] Likharev K K 1999 *Proc. IEEE* **87** 606
[2] Fulton T A and Dolan D J 1987 *Phys. Rev. Lett.* **59** 109
[3] Ono K, Shimada H and Ootuka Y 1997 *J. Phys. Soc. Jpn.* **66** 1261
[4] Chen C D, Kuo W, Chung D S, Shyu J H and Wu C S 2002 *Phys. Rev. Lett.* **88** 047004; Johansson J, Urech M, Haviland D, and Korenivski V 2003 *Phys. Rev. Lett.* **91** 149701; Chen C D, Kuo W, Chung D S, Shyu J H and Wu C S 2003 *Phys. Rev. Lett.* **91** 149702
[5] Johansson J, Urech M, Haviland D and Korenivski V 2003 *J. Appl. Phys.* **93** 8650
[6] Yang H, Yang S, Takahashi S, Maekawa S and Parkin S S P 2010 *Nature Mat.* **9** 586
[7] Takahashi S, Imamura H and Maekawa S 1999 *Phys. Rev. Lett.* **82** 6237
[8] Takiguchi M, Sato T, Shimada H and Mizugaki Y 2014 *JPS Conf. Proc.* **1** 012004
[9] Takiguchi M, Shimada H and Mizugaki Y 2014 *Jpn. J. Appl. Phys.* **53** 043101
[10] Brückl H and Reiss G 1998 *Phys. Rev.* B **58** R8893
[11] So H -M, Kim J K and Kim J -J 2003 *J. Korean Phys. Soc.* **43** 578
[12] Takahashi S and Maekawa S 1998 *Phys. Rev. Lett.* **80** 1758
[13] Karlsson C and Wang X H 2000 *Appl. Phys. Lett.* **77** 3618
[14] Mizugaki Y, Tamura N and Shimada H 2013 *J. Appl. Phys.* **113** 156101
[15] Dolan G J 1977 *Appl. Phys. Lett.* **31** 337
[16] Ootsuka Y, Uchiyama T and Shimada H 1993 *Cryogenics* **33** 923
[17] Clinton T W and Johnson M 2000 *Appl. Phys. Lett.* **76** 2116
[18] Sato T, Takiguchi M, Shimada H and Mizugaki Y 2015 *Ext. Abstr. 62nd Jpn. Soc. Appl. Phys. Spring Meeting* (Hiratsuka: Japan) p 08-126 (in Japanese)
[19] Jedema F J, Filip A T and van Wees B J 2001 *Nature* **410** 345

IOP Conf. Series: Journal of Physics: Conf. Series **969** (2018) 012155 doi:10.1088/1742-6596/969/1/012155

# Andreev reflection in a proximity junction of graphene: Influence of a naturally formed *pn* junction

**Yositake Takane**[1] **and Akinobu Kanda**[2]

[1]Department of Quantum Matter, Graduate School of Advanced Sciences of Matter, Hiroshima University, Higashihiroshima, Hiroshima 739-8530, Japan

[2]Division of Physics and TIMS, Faculty of Pure and Applied Sciences, University of Tsukuba, Tsukuba 305-8571, Japan

E-mail: takane@hiroshima-u.ac.jp

**Abstract.** We theoretically study the electron transport in a proximity junction made by partially covering a graphene sheet with a bulk superconductor. Such a system serves as a normal conductor–superconductor (NS) junction and its electron transport properties are mainly governed by Andreev reflection (AR). In this NS junction, the charge neutrality point (CNP) of graphene must spatially vary as a result of the penetration of excess carriers induced from the superconductor. By taking this into account in the electron-doped case, we calculate the differential conductance $G_{\rm NS}$ of the graphene NS junction. Owing to the variation of the CNP in the uncovered region, a smooth *pn* junction is naturally formed at the point where the Fermi energy crosses the CNP. The resulting *pn* junction causes unusual behavior of $G_{\rm NS}$. For example, a resonant peak structure appears in $G_{\rm NS}$ reflecting the fact that quasi-bound states are created by the *pn* junction.

## 1. Introduction

Andreev reflection (AR) is a scattering process in which an electron (hole) is reflected back as a hole (electron) [1]. This typically takes place at the interface between a normal conductor (N) and a superconductor (S). An anomalous feature of the AR is that a reflected particle traces back the path of an incident particle. This is referred to as retroreflection. As the AR process accompanies the transfer of two electrons across the interface, it greatly affects the conductance of an NS junction. Since nearly a decade ago, experimentalists have intensively studied the NS junction made by partially covering a graphene sheet with a superconductor [2, 3, 4, 5, 6, 7, 8, 9]. It has been theoretically predicted that, in the graphene NS junction, an incident electron can be reflected back as a hole in a specular manner [10]. This is referred to as specular reflection.

In the graphene NS junction, the charge neutrality point (CNP) of graphene must spatially vary as a result of the penetration of excess carriers induced from the superconductor. Previous theoretical studies on the AR have assumed that the CNP uniformly shifts only in the covered region (i.e., S electrode) [10, 11, 12]. However, in actual situations, the CNP should vary even in the uncovered region (i.e., N electrode) over a screening length $\lambda$ from the NS interface [13]. By taking this into account in the electron-doped case, the present authors have studied the AR process in an NS junction of *bilayer graphene* [14]. It should be noted that, in the presence of the spatial variation of the CNP in the uncovered region, a smooth *pn* junction [15, 16] is naturally formed at the point where the Fermi energy $E_{\rm F}$ crosses the CNP [see Fig. 1(a)]. This *pn* junction acts as a weak barrier causing diffraction of electrons; thus, it significantly affects the differential conductance $G_{\rm NS}$ of a graphene NS junction. It is shown in Ref. [14] that $G_{\rm NS}$ is significantly reduced when the AR takes place in a specular manner. That is, $G_{\rm NS}$ in the specular-reflection regime is much smaller than that in the retroreflection regime. It is also shown that quasi-bound states created by the *pn* junction give rise to resonant peaks in $G_{\rm NS}$.

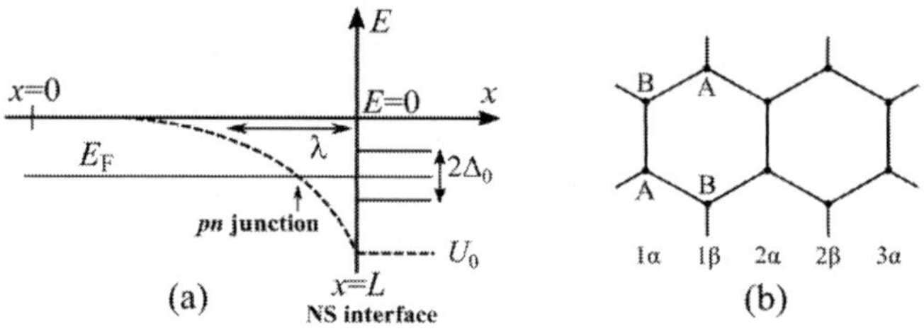


**Figure 1.** (a) Profile of the potential (the broken line) in a graphene NS junction. (b) Hexagonal lattice for a monolayer graphene sheet.

It is probable that these features of $G_{\rm NS}$ are observed in a monolayer graphene system as well because a smooth *pn* junction can also be formed. However, it is unclear how they appear in a monolayer system as the transport properties of a *pn* junction are qualitatively different between the bilayer and monolayer cases. In this paper, we calculate $G_{\rm NS}$ in the graphene NS junction and study how its features resemble or differ from between the bilayer and monolayer cases. We set $\hbar = 1$ hereafter.

## 2. Model

Let us introduce a graphene sheet of width $W$ infinitely long in the $x$-direction, where the region of $x > L$ is covered with a bulk superconductor. We assume that the potential $U(x)$ [see Fig. 1(a)], which determines the location of the CNP, is given by

$$U(x) = \begin{cases} -U_0\, e^{(x-L)/\lambda} & \text{if} \quad x \le L, \\ -U_0 & \text{if} \quad x > L, \end{cases} \tag{1}$$

where $U(x)$ vanishes in the asymptotic region of $x \le 0$ if $L \gg \lambda$. The pair potential induced in a graphene sheet is assumed as $\Delta(x) = 0$ if $x \le L$ and $\Delta(x) = \Delta_0$ if $x > L$. Quasiparticles are described by a tight-binding model [17, 18] on a hexagonal lattice [see Fig. 1(b)], in which the transfer integral between nearest neighbor sites is denoted by $\gamma_0$. By using the Bogoliubov–de Gennes equation in a tight-binding form [14], we consider the scattering problem for an electron with energy $E$ measured from $E_{\rm F}$ incident to the NS interface from the asymptotic region. Imposing the periodic boundary condition in the $y$-direction, we specify the $y$ dependence of a wave function with a transverse wave number $q$. For any $q$, the number of conducting channels in the asymptotic region is 2 as there are two energy valleys. Using a recursive Green's function technique, we can numerically determine the $2 \times 2$ Andreev reflection amplitude $r_{\rm A}(E_{\rm F} + E)$ and the $2 \times 2$ normal reflection amplitude $r_{\rm N}(E_{\rm F} + E)$. The differential conductance as a function of bias voltage $V$ is given by the Blonder–Tinkham–Klapwijk formula as [19, 20]

$$G_{\rm NS}(E_{\rm F} + eV) = \frac{2e^2}{\pi}\frac{W}{\pi a} g_{\rm NS}(E_{\rm F} + eV), \tag{2}$$

where $a$ is the lattice constant and $g_{\rm NS}$ is the dimensionless conductance per unit width:

$$g_{\rm NS}(E_{\rm F} + eV) = \frac{k_{\rm F}(E_{\rm F} + eV)a}{2} \int_0^{\frac{\pi}{2}} d\phi \cos\phi \;{\rm tr}\left\{1_{2\times 2} - r_{\rm N}^{\dagger} r_{\rm N} + r_{\rm A}^{\dagger} r_{\rm A}\right\}\Big|_{E_{\rm F}+eV}. \tag{3}$$

IOP Conf. Series: Journal of Physics: Conf. Series **969** (2018) 012155 doi:10.1088/1742-6596/969/1/012155

Here, $1_{2\times 2}$ is the $2\times 2$ unit matrix, $k_{\mathrm{F}}(E_{\mathrm{F}}+eV)$ is the Fermi wave number, and $\phi = \arccos(q/k_{\mathrm{F}})$ is the angle of incidence. Details of numerical methods are given in Ref. [14] for the bilayer case.

## 3. Numerical results

We present the numerical results of $g_{\mathrm{NS}}$ as a function of $eV$ for $E_{\mathrm{F}}/\Delta_0 = 0.25$ and $-5.0$. The other parameters are $\gamma_0 = 3.16$ eV, $\Delta_0 = 1.2$ meV, $U_0 = 1.0$ eV, and $\lambda = 100$ or 400 nm. In the bilayer case, the interlayer transfer integral $\gamma_1$ is set equal to 0.39 eV.

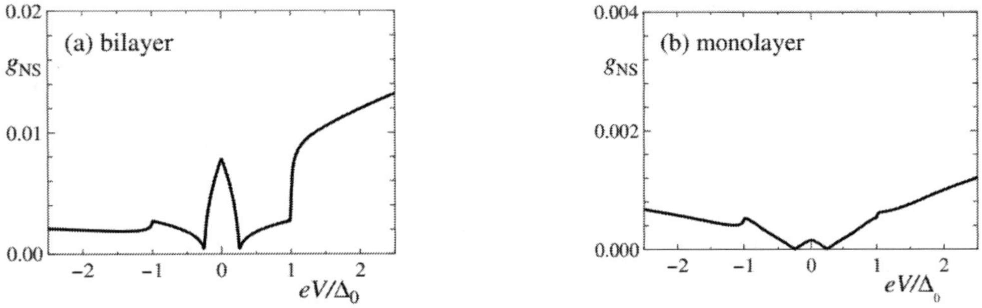


**Figure 2.** $eV$ dependence of $g_{\mathrm{NS}}(E_{\mathrm{F}} + eV)$ at $E_{\mathrm{F}}/\Delta_0 = 0.25$ with $\lambda = 100$ nm in the cases of (a) bilayer graphene and (b) monolayer graphene.

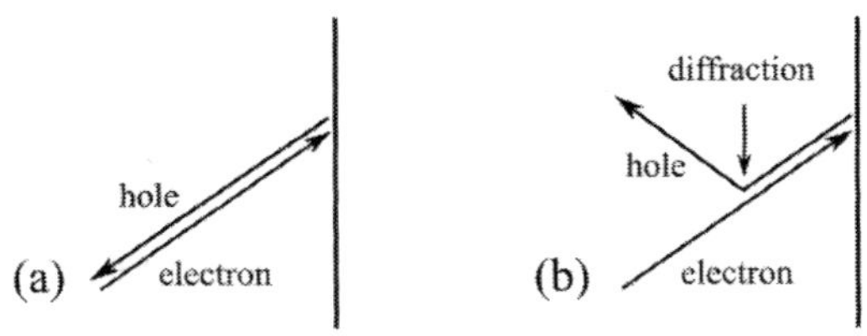


**Figure 3.** Trajectories of an incident electron and a reflected hole in the cases of (a) retroreflection ($E_{\mathrm{F}} > eV \geq 0$) and (b) specular-reflection ($\Delta_0 > eV > E_{\mathrm{F}}$). In the specular case, the reflected hole is diffracted by the *pn* junction.

Figure 2 shows $g_{\mathrm{NS}}(E_{\mathrm{F}} + eV)$ in the bilayer and monolayer cases for $E_{\mathrm{F}}/\Delta_0 = 0.25$ with $\lambda = 100$ nm. In Fig. 2(a) for the bilayer case, we observe that $g_{\mathrm{NS}}$ shows an unusual asymmetric behavior [i.e., $g_{\mathrm{NS}}(E_{\mathrm{F}} + e|V|) \gg g_{\mathrm{NS}}(E_{\mathrm{F}} - e|V|)$] in the above gap region of $e|V| > \Delta_0$ [14]. This is caused by a naturally formed *pn* junction, which reduces the transmission of an electron only when its energy (i.e., $E_{\mathrm{F}} + eV$) is less than zero. Though this asymmetry is also observed in Fig. 2(b) for the monolayer case, it is weaker than that in the bilayer case. The reason for this is that, roughly speaking, the transmission probability of an electron through a *pn* junction is larger in the monolayer case than in the bilayer case. Let us turn to the subgap region of $\Delta_0 > e|V|$, where retro AR takes place when $E_{\mathrm{F}} > e|V| \geq 0$, while specular AR is expected to appear when $\Delta_0 > e|V| > E_{\mathrm{F}}$ (see Fig. 3) [14]. In the bilayer case, $g_{\mathrm{NS}}$ is much larger in the retroreflection regime than in the specular-reflection regime. This is simply explained by the fact that neither incident nor reflected particle passes through the *pn* junction in the retroreflection regime, while one of them passes through it in the specular-reflection regime, resulting in the reduction of the AR probability. In Fig. 2(b) for the monolayer case, we do not observe clear difference of $g_{\mathrm{NS}}$ between the retroreflection and specular-reflection regimes. This should be attributed to the nature of a *pn* junction in the monolayer case that the transmission probability is not significantly reduced when the angle of incidence is small.

Figures 4 and 5 respectively show $g_{\mathrm{NS}}(E_{\mathrm{F}} + eV)$ in the bilayer and monolayer cases for $E_{\mathrm{F}}/\Delta_0 = -5.0$. We observe that resonant peaks appear in $g_{\mathrm{NS}}$ in both the bilayer and monolayer cases. These peaks directly reflect the appearance of quasi-bound states, which are created by

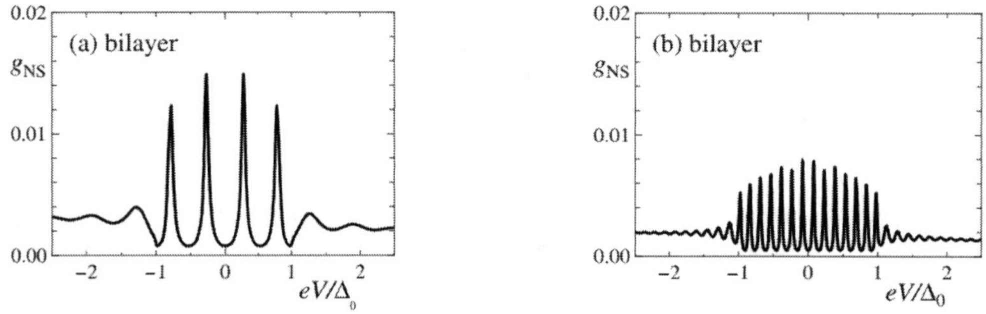


**Figure 4.** $eV$ dependence of $g_{NS}(E_F + eV)$ in the bilayer case for $E_F/\Delta_0 = -5.0$ with (a) $\lambda = 100$ nm and (b) 400 nm.

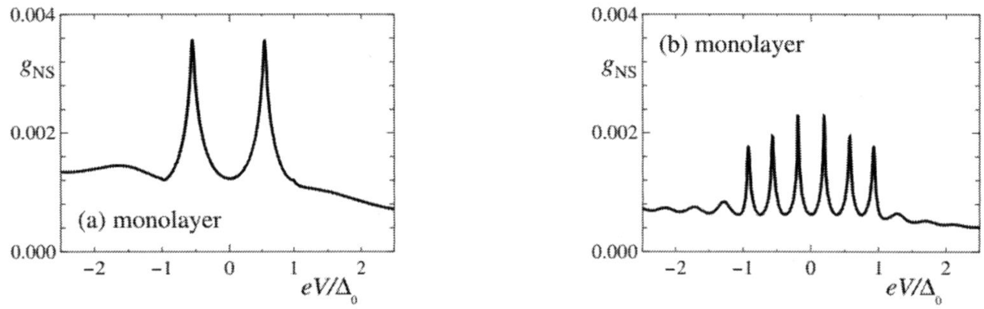


**Figure 5.** $eV$ dependence of $g_{NS}(E_F + eV)$ in the monolayer case for $E_F/\Delta_0 = -5.0$ with (a) $\lambda = 100$ nm and (b) 400 nm.

weak confinement potential due to the smooth $pn$ junction (see Fig. 6). As this potential depends on the details of junction parameters, the profile of resonant peaks significantly differs in each case.

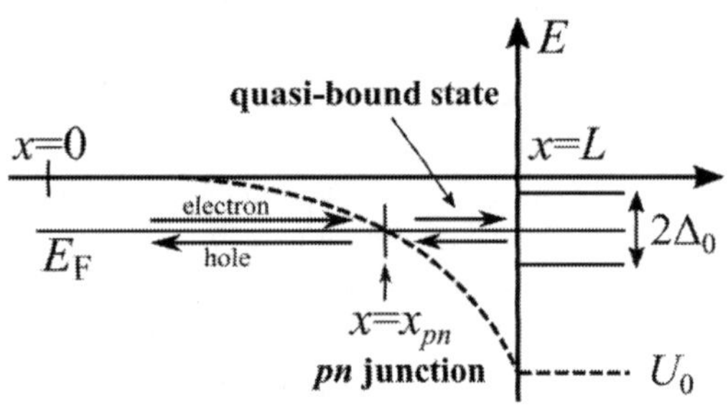


**Figure 6.** Quasi-bound states are created between the $pn$ junction at $x = x_{pn}$ and the NS interface at $x = L$.

## 4. Summary

We studied the behavior of $G_{NS}$ in a graphene NS junction in the bilayer and monolayer cases by taking account of a naturally formed $pn$ junction. It was shown that, in the bilayer case, $G_{NS}$ is significantly reduced in the specular-reflection regime while such behavior does not appear in the monolayer case. It was also shown that $G_{NS}$ shows a resonant peak structure in both the bilayer and monolayer cases reflecting the appearance of quasi-bound states created by the $pn$ junction.

**Acknowledgments**
This work was supported by JSPS KAKENHI Grant Numbers JP15K05130, and JP16H00897 [Science of Atomic Layers (SATL)].

## References

[1] Andreev A F, 1964 *Sov. Phys. JETP* **19** 1228.
[2] Heersche H B, Jarillo-Herrero P, Oostinga J B, Vandersypen L M K and Morpurgo A F, 2007 *Nature* **446** 56
[3] Miao F, Wijeratne S, Zhang Y, Coskun U C, Bao W and Lau C N, 2007 *Science* **317** 1530
[4] Sato T, Moriki T, Tanaka S, Kanda A, Goto H, Miyazaki H, Odaka S, Ootuka Y, Tsukagoshi K and Aoyagi Y, 2008 *Physica* E **40** 1495
[5] Du X, Skachko I and Andrei E Y, 2008 *Phys. Rev.* B **77** 184507
[6] Jeong D, Choi J H, Lee G H, Jo S, Doh Y J and Lee H J, 2011 *Phys. Rev.* B **83** 094503
[7] Shalom M B, Zhu M J, Fal'ko V I, Mishchenko A, Kretinin A V, Novoselov K S, Woods C R, Watanabe K, Taniguchi T, Geim A K and Prance J R, 2016 *Nat. Phys.* **12** 318
[8] D. K. Efetov, L. Wang, C. Handschin, K. B. Efetov, J. Shuang, R. Cava, T. Taniguchi, K. Watanabe, J. Hone, C. R. Dean, and P. Kim, 2016 *Nat. Phys.* **12** 328
[9] Borzenets I V, Amet F, Ke C T, Draelos A W, Wei M T, Seredinski A, Watanabe K, Taniguchi T, Bomze Y, Yamamoto M, Tarucha S and Finkelstein G, 2016 *Phys. Rev. Lett.* **117** 237002
[10] Beenakker C W J, 2006 *Phys. Rev. Lett.* **97** 067007
[11] Ludwig T, 2007 *Phys. Rev. B* **75** 195322
[12] Efetov D K and Efetov K B, 2016 *Phys. Rev. B* **94** 075403
[13] Mueller T, Xia F, Freitag M, Tsang J and Avouris Ph, 2009 *Phys. Rev.* B **79** 245430
[14] Takane Y, Yarimizu K and Kanda A, 2017 *J. Phys. Soc. Jpn.* **86** 064707
[15] Cheianov V V and Fal'ko V I, 2006 *Phys. Rev.* B **74** 041403
[16] Cheianov V V, Fal'ko V and Altshuler B L, 2007 *Science* **315** 1252
[17] Wallace P R, 1947 *Phys. Rev.* **71** 622
[18] Slonczewski J C and Weiss P R, 1958 *Phys. Rev.* **109** 272
[19] Blonder G E, Tinkham M and Klapwijk T M, 1982 *Phys. Rev.* B **25** 4515
[20] Takane Y and Ebisawa H, 1992 *J. Phys. Soc. Jpn.* **61** 1685

**Institute of Physics**
Dirac House, Temple Back
Bristol BS1 6BE UK

ISSN: 1742-6588
ISBN 978-1-5108-6236-4

9 781510 862364

CURRAN ASSOCIATES INC.
proceedings.com